VOLUME ONE HUNDRED AND THIRTEEN

ADVANCES IN PROTEIN CHEMISTRY AND STRUCTURAL BIOLOGY

Computational Molecular Modelling in Structural Biology

VOLUME ONE HUNDRED AND THIRTEEN

ADVANCES IN PROTEIN CHEMISTRY AND STRUCTURAL BIOLOGY

Computational Molecular Modelling in Structural Biology

Edited by

TATYANA G. KARABENCHEVA-CHRISTOVA
Department of Chemistry, Michigan Technological University, Houghton, MI, United States

CHRISTO Z. CHRISTOV
Department of Chemistry, Michigan Technological University, Houghton, MI, United States

ELSEVIER

ACADEMIC PRESS
An imprint of Elsevier

Academic Press is an imprint of Elsevier
50 Hampshire Street, 5th Floor, Cambridge, MA 02139, United States
525 B Street, Suite 1650, San Diego, CA 92101, United States
The Boulevard, Langford Lane, Kidlington, Oxford OX5 1GB, United Kingdom
125 London Wall, London, EC2Y 5AS, United Kingdom

First edition 2018

ISBN: 978-0-12-813916-5
ISSN: 1876-1623

For information on all Academic Press publications
visit our website at https://www.elsevier.com/books-and-journals

Publisher: Zoe Kruze
Acquisition Editor: Ashlie M. Jackman
Editorial Project Manager: Teresa Pons-Ferrer
Production Project Manager: James Selvam
Cover Designer: Greg Harris

Typeset by SPi Global, India

CONTENTS

CONTRIBUTORS

Jon Ainsley
Department of Applied Sciences, Faculty of Health and Life Sciences, Northumbria University, Newcastle upon Tyne, United Kingdom

Canan Atilgan
Faculty of Engineering and Natural Sciences, Sabancı University, Tuzla, Istanbul, Turkey

Joaquin Barroso-Flores
Instituto de Química, Universidad Nacional Autónoma de México, Mexico City, Mexico

Durbis J. Castillo-Pazos
Instituto de Química, Universidad Nacional Autónoma de México, Mexico City, Mexico

Christo Z. Christov
Department of Chemistry, Michigan Technological University, Houghton, MI, United States

Fernando Cortes-Guzman
Instituto de Química, Universidad Nacional Autónoma de México, Mexico City, Mexico

Eduardo M. Cortés-Ruiz
Facultad de Química, Universidad Nacional Autónoma de México, Mexico City, Mexico

Bertha Espinoza
Instituto de Investigaciones Biomédicas, Universidad Nacional Autónoma de México, Mexico City, Mexico

Gabriela Gómez-Jiménez
Instituto de Química, Universidad Nacional Autónoma de México, Mexico City, Mexico

Karla Gonzalez-Ponce
Instituto de Química, Universidad Nacional Autónoma de México, Mexico City, Mexico

Tatyana G. Karabencheva-Christova
Department of Chemistry, Michigan Technological University, Houghton, MI, United States

Alessio Lodola
Pharmacy Department, Università di Parma, Parma, Italy

Abraham Madariaga-Mazon
Instituto de Química, Universidad Nacional Autónoma de México, Mexico City, Mexico

Karina Martinez-Mayorga
Instituto de Química, Universidad Nacional Autónoma de México, Mexico City, Mexico

José L. Medina-Franco
Facultad de Química, Universidad Nacional Autónoma de México, Mexico City, Mexico

Adrian J. Mulholland
Centre for Computational Chemistry, School of Chemistry, University of Bristol, Bristol, United Kingdom

J. Jesús Naveja
Facultad de Química; PECEM, Facultad de Medicina, Universidad Nacional Autónoma de México, Mexico City, Mexico

C. Iluhí Oviedo-Osornio
Facultad de Química, Universidad Nacional Autónoma de México, Mexico City, Mexico

Oscar Palomino-Hernández
Facultad de Química, Universidad Nacional Autónoma de México, Mexico City, Mexico

Karla Daniela Rodríguez-Hernández
Instituto de Investigaciones Biomédicas, Universidad Nacional Autónoma de México, Mexico City, Mexico

PREFACE

Molecular modeling methods have received a fast growth over the last years due to the developments in CPU speed, the development of GPUs, the growth in high-performance computing capabilities, and new computational algorithms and theories. The applications of computational modeling methods in structural and chemical biology, and pharmaceutical chemistry also received a remarkable increase in integration with experimental, biochemical, structural, pharmaceutical, and spectroscopic studies. Nowadays, microsecond molecular dynamics simulations of large solvated protein–ligand complexes and membrane-based systems became available. The methods for dealing with the electronic structure of molecules and biomolecules became more accurate and applicable to larger systems. The combined quantum mechanics and molecular mechanics (QM/MM) methods received even broader applications to study enzymes and found novel implementations in drug design and chemical biology.

This volume of *Advances in Protein Chemistry and Structural Biology* is focused on some of the recent applications of computational modeling methods in the area of structural biology. In particular, one of the review chapters presents the state of the art of the QM/MM methods and their applications to study enzyme mechanisms and inhibition. Novel methods for efficient sampling of protein landscapes and disclosing allosteric regions are subject of another chapter. Further, the current state of the art of novel methods for design of inhibitors toward bromodomain BRD4 and QSAR models for knowledge discovery in databases is revealed. Finally, the modern computational methods for drug discovery of treatment of Chagas diseases are presented. Contributions from leading researchers in the field were selected to provide a good balance between the development of the computational methods and their implementations in the areas of structural biology, drug design, and enzymology.

Tatyana G. Karabencheva-Christova and Christo Z. Christov
Department of Chemistry,
Michigan Technological University,
Houghton, MI, United States

CHAPTER ONE

Combined Quantum Mechanics and Molecular Mechanics Studies of Enzymatic Reaction Mechanisms

Jon Ainsley*, Alessio Lodola†, Adrian J. Mulholland‡, Christo Z. Christov§,1, Tatyana G. Karabencheva-Christova§,1

*Department of Applied Sciences, Faculty of Health and Life Sciences, Northumbria University, Newcastle upon Tyne, United Kingdom
†Pharmacy Department, Università di Parma, Parma, Italy
‡Centre for Computational Chemistry, School of Chemistry, University of Bristol, Bristol, United Kingdom
§Department of Chemistry, Michigan Technological University, Houghton, MI, United States
[1]Corresponding authors: e-mail address: christov@mtu.edu; tatyanak@mtu.edu

Contents

Abstract

The combined quantum mechanics/molecular mechanics (QM/MM) methods have become a valuable tool in computational biochemistry and received versatile applications for studying the reaction mechanisms of enzymes. The approach combines the calculations of the electronic structure of the active site by QM, with modeling of

Advances in Protein Chemistry and Structural Biology, Volume 113
ISSN 1876-1623
https://doi.org/10.1016/bs.apcsb.2018.07.001

the protein environment using MM force field, which allows the long-range electrostatics and steric effects on the enzyme reactivity to be accounted for. In this review, we review some key theoretical and computational aspects of the method and we also present some applications to particular enzymatic reactions such as tryptophan-7-halogenase, cyclooxygenase-1, and the epidermal growth factor receptor.

1. INTRODUCTION

Enzymes are large and complex biological molecules which catalyze almost all biochemical metabolic reactions in the cellular environment crucial for life, they are able to greatly accelerate the rate of a reaction and show high specificity toward a substrate. In addition the enzymes as proteins are characterized with diverse conformational flexibility that relates their structure to functions (Karplus & McCammon, 2002). Enzymology was traditionally a purely experimental field; however, there are limits to the experimental approach when studying enzymatic reactions at the atomistic level (Carvalho et al., 2014). When the first three-dimensional atomistic structure of an enzyme was elucidated in 1965 (Blake et al., 1965) by X-ray crystallography, it was realized that the study of these static structures could not describe the complex atomistic, electronic structural, and conformational changes that occur during the enzymatic reaction. To fully understand an enzyme's reaction mechanisms and functions we need to model the reaction path that converts reactants to products and the relevant transition states and intermediates which is extremely hard or impossible to be done experimentally. Computational simulations can provide these details, opening up a new dimension to the study of the structure–function relationships underlying enzyme catalysis.

In 1972 Warshel and Karplus published a study describing a new method for calculating the potential energy surfaces for conjugated molecules in excited states (Warshel & Karplus, 1972); this was made possible by the differential treatment of the σ- and π-electrons in the system. The energy of the σ-electrons was estimated using an empirical potential function and the conjugated π-electron system was treated using a semiempirical level of theory. This provided an advantage over purely empirical calculations which could not accurately describe the delocalization effect of the excited state potential energy surfaces as an electronic level of calculation encompassing the whole molecule was computationally unfeasible at the time. This separation of a system into two portions to be treated with different levels of computational methods was first applied to an enzymatic system in 1976 to study the reaction of lysozyme by Warshel and Levitt (1976). In this paper, the groundwork was laid for a combined quantum mechanics/molecular mechanics (QM/MM) approach to the study of enzymatic reactions. The study of the electronic structure of the enzyme–substrate complex was achieved by treating the energy contributions of the substrate using a quantum mechanical (QM) level of calculation, the energetic influence of the protein and solvent were calculated using classical mechanics (MM). At this time it was known that the protein environment influenced the reaction with charged interactions and steric effects, however, the relative contribution of each of these factors to the overall reaction was unknown (Warshel, 1978). For lysozyme it was found that the charge stabilization of the transition state was the most significant factor that caused catalysis of the lysozyme reaction which caused a revision of the older mechanism proposed by Phillips (1966) which suggested that the enzyme mainly relied on steric effects to catalyze the reaction. This pioneering contributions laid the foundation for the eventual establishment of the QM/MM approach which together with other biomolecular simulation methods and earned Warshel, Levitt, and Karplus the 2013 Nobel prize in Chemistry for the development of multilevel modeling of complex chemical systems (Foundation, T.N, 2013).

The following years saw few other publications related to the application of the QM/MM technique, it was not until 1990 with the publication of a paper describing the combination of the semiempirical AM1 calculations with the empirical CHARMM force field (Field, Bash, & Karplus, 1990) that significant interest in QM/MM was generated. This paper showed a new method in which the QM/MM technique could be more easily integrated into geometry optimizations, molecular dynamics, and Monte Carlo simulations of enzymatic systems. A marked increase in the number of publications surrounding QM/MM techniques was seen in the years following 1991 (Fig. 1).

2. THE QM/MM METHOD

QM/MM simulations of enzymatic reactions have developed significantly as a technique since their first implementation nearly 40 years ago;

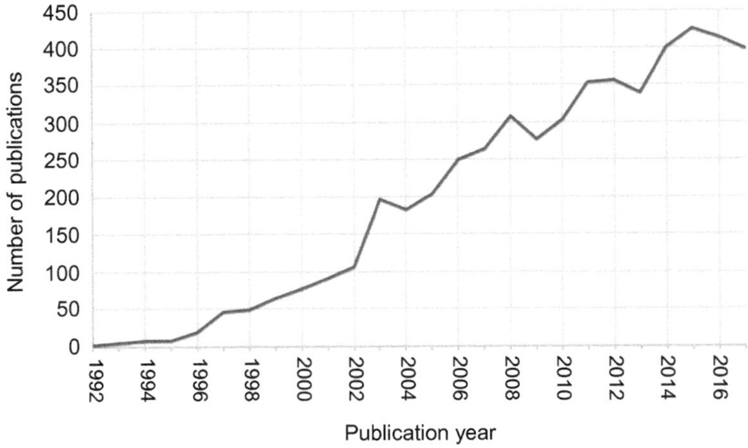

Fig. 1 The number of publications with QM/MM listed as a topic between the years 1992 and 2017 as created by the ISI web of science (Clarivate Analytics, 2018).

however, the overall concept has remained the same. A detailed electronic structure simulation of a relatively small number of atoms that make up the reacting portion of the system, this QM region often encompasses the substrate, cofactor as well as surrounding reacting amino acid residues. The QM region is embedded in a larger region that is simulated with an MM force field; this MM region consists of the remainder of the protein structure and the solvent environment (Fig. 2). The purpose of this approach is to treat the important reactive group of atoms in the QM system with the scrutiny of a high level of calculation theory such as ab initio or density functional theory (DFT), without incurring the computational cost associated with applying this calculation method to the larger number of atoms that represent the entirety of the system. In this way the bond formation/breaking that occurs in the QM region can be accurately described while also accounting for the interactions contributed by the rest of the protein and solvent environment which is simulated at the MM level (Meier et al., 2013). The combined QM/MM simulation of enzymes in this way can provide insights into the mechanistic details of the enzymatic reaction. QM/MM simulations can be used to model the structures along the reaction path (Mulholland, 2005), predict the structure of transition states (Ranaghan et al., 2004), and quantify the energetic changes associated with these transitions (Blumberger, 2008; Hu & Yang, 2008); these insights cannot be easily gained from experimental study alone

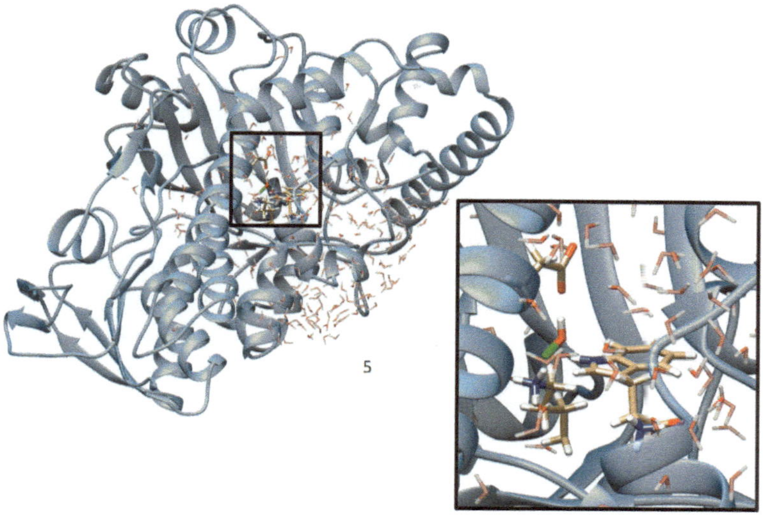

Fig. 2 Structure of the tryptophan 7-halogenase (PrnA) displayed to demonstrate the QM and MM regions. *Left*, the whole protein structure is rendered with *blue transparent ribbons* (the remainder of the enzyme atoms are not displayed for clarity purposes) along with the solvent displayed with transparent tubes this set of atoms constitutes the MM region. *Right*, a panel showing a zoomed image of the QM part of the system: the tryptophan substrate, hypochlorous acid, and the side chains of Glu346 and Lys79. Only solvent molecules within 15 Å of the QM region are displayed. Atoms are *colored* according to element: carbon—*beige*, hydrogen—*white*, nitrogen—*blue*, oxygen—*red*, and chlorine—*green*.

(Friesner & Guallar, 2005). This review aims to give an introduction of the theory underpinning QM/MM techniques and their applications, as well as giving a review of three recent publications that apply QM/MM techniques in order to study enzymatic reaction mechanisms. Other recommended examples of recent publications using QM/MM techniques are available (Bennie et al., 2016; Callegari et al., 2018; Dubey, Wang, & Shaik, 2016; Faponle, Seebeck, & de Visser, 2017; Lever et al., 2014; Liu et al., 2014; Quesne et al., 2014; Zhang et al., 2018).

3. CALCULATION OF QM/MM ENERGY

The total energy for the system cannot be summed up as a simple addition of the energies of the QM region and MM region, another term needs to be introduced to account for the interaction between the two regions. In the additive method for QM/MM calculations (Eq. 1) the total energy of the QM region (E_{QM}) and the MM region (E_{MM}) are calculated separately and then added together. This addition alone does not fully define the total energy of the whole system so a coupling term ($H_{QM/MM}$) is needed; this represents the bonding, electrostatic, and Van der Waals interactions between the QM and MM regions (Senn & Thiel, 2007). The additive method is the most popular from the point of view of biomolecular simulations because it does not require the energy of the QM region to be evaluated with an MM level of theory, which can be problematic when studying reactive systems for which no parameters exist (Sousa, Fernandes, & Ramos, 2012). A distinct disadvantage of the additive method is the need for the $H_{QM/MM}$ coupling term, the calculation of which can create its own unique difficulties.

$$E_{all} = E_{QM} + E_{MM} + E_{QM/MM} \tag{1}$$

Equation 1: The additive method for calculating the energy of a QM/MM simulation system.

The subtractive method (Eq. 2) starts with the energy of the whole system ($E_{MM\ \text{whole system}}$) being first calculated at the MM level and the energy contribution of the QM region calculated at the MM level ($E_{MM/QM\ \text{system}}$) is then subtracted from this. To this subtracted amount the energy contribution of the QM region, as calculated using QM methods ($E_{QM/QM\ \text{region}}$), is added; this gives the total energy of the whole system as calculated with two distinct levels of theory (Kang & Tateno, 2012).

$$E_{all} = \left(E_{MM\ \text{whole system}} - E_{MM/QM\ \text{system}}\right) + E_{QM/QM\ \text{region}} \tag{2}$$

Equation 2: The subtractive method for calculating energy of a QM/MM simulation system.

The subtractive energy calculation scheme has advantages over the additive one that it can more easily accommodate more than two levels of calculation and can in fact be extended to an "*n*" number of layers each represented by a different level of theory (Svensson et al., 1996). Unlike the additive method, the QM region of the subtractive method must also be calculated at the MM level. Problems can arise as to the treatment of nonparametrized atom types such as metal atoms or transition states, without a successful calculation of this region with an MM force field the total energy of system cannot be calculated using the subtractive method.

4. TREATMENT OF BONDS AT THE QM/MM BOUNDARY

In systems where the QM region is covalently bound to the MM region, for example, when a side chain of an important amino acid needs to be included in the QM region without also including the entire protein. Creating a QM/MM partition along a covalent bond to include one portion of a molecule in the QM and the other in the MM region requires the bond to be divided; this leaves the boundary atoms with unsatisfied valence. Not all bonds can be broken to form partitions in this way, conjugated polar or aromatic bonds should not be cut across as the uneven distribution of electrons in these will lead to erroneous results in the simulation (Lonsdale, Harvey, & Mulholland, 2012).

A simple solution originally suggested in a paper by Singh and Kollman (1986) proposed saturating the unoccupied valences with link atoms as this method is relatively easily implemented. The link atom added can be a hydrogen atom, a methyl group, or even pseudohalogen atoms to the partitioned atoms. However, the addition of these extra atoms can cause problems, changing the system by adding unnatural atoms into the QM region may distort the natural reaction intended to be simulated (Hu & Yang, 2009). These extra atoms can also negatively affect the thermodynamics and geometry of the QM region so should be considered with care (Lin & Truhlar, 2007).

Another alternative method for treating boundary atoms is the use of pseudoatoms instead of link atoms, new atoms with only a single valence can be added to satisfy the valence of the system and can possess properties to more accurately reflect the bond they are imitating (Zhang, Lee, & Yang, 1999). The pseudoatom bonded to the boundary atom of the QM region can be parametrized with the use of model systems, to have the same bond length, strength, and electronic properties as the bond that it is replacing. It is "visible" to both the QM and MM region, in the QM region it replicates the bond which has been severed and in the MM region it functions as a normal MM atom (Senn & Thiel, 2009). The pseduoatom method suffers from some of the same problems as the link atom method in that it introduces artificial atoms into the simulation; however, it allows more control over the properties of these introduced atoms. Problems can also arise with the parametrization process for the creation of the pseudoatom, if done incorrectly it could result in a QM system that is not an accurate representation of the real system.

The frozen local orbital approach is another method for treating the partition of bonds, where an orbital is introduced to the unoccupied valence of the boundary atom to close the QM system (Philipp & Friesner, 1999). The orbital is excluded from calculations and acts as lone pair pointing toward the severed MM atom. In this way it does not affect the other orbitals in the QM region and can be specifically designed and adjusted in such a way to minimize its effect on the charge distribution, position, or interactions surrounding the boundary atoms (Amara et al., 2000). Although the frozen orbital method is more difficult to implement it would provide a better description of the QM/MM boundary as it does not introduce artificial atoms. The generalized hybrid orbital approach is a modification of this method, the QM atom remains attached to the MM atom and the MM atom has multiple orbitals added to it. In this way, the bonds between the MM atom and other MM atoms are preserved as frozen orbitals, and the orbital between the QM and MM atom remains open. Because of this the electron density across the partition can readjust over the course QM/MM calculation rather than to remain fixed (Gao et al., 1998). The choice of the treatment of boundary regions is a crucial part of the QM/MM simulation process and is often a balance between simplicity and accurately representing the systems of interest and their properties.

The broadest application so far received the link atoms method.

5. QM/MM EMBEDDING TECHNIQUES

The representation of the interactions between the QM and MM region defines the embedding technique that is used. In the original and simpler technique of mechanical embedding, the charges from the MM region are evaluated and their influence added to the atoms of the QM region as point charges without the need for QM terms (Vreven et al., 2006). However, this can create problems as both QM and MM regions must be parameterized in an MM force field so that the partial charges of the two regions can be combined in this way, transition state atom types and metals are not normally included in MM force fields and therefore need parametrization and this can create problems (Lin & Truhlar, 2007).

The newer and more complex alternative for treating the QM/MM interactions is the electrostatic embedding. Instead of MM electrostatic interactions being added to the QM system as partial charges, the charge distribution of the MM region is directly allowed to influence the wavefunction in the calculation of the QM region (Sherwood et al., 2003). Although requiring more computational effort this embedding scheme allows to directly represent the polarization of the QM region by the MM atomic charges, by including them in the QM Hamiltonian. The electrostatic embedding has distinct advantages over mechanical embedding, it does not require the parametrization of the QM system in an MM force field, and additionally it allows the QM systems charge distribution to constantly adjust with any changes in the MM part (Vreven et al., 2003).

6. QM/MM MODELING OF REACTION MECHANISMS

6.1 Transition State Modeling

Transition states are in key importance to catalysis. In their structures the new bonds are partly formed and the old are partly broken and correspond to the highest energy point in the reaction path (Ranaghan et al., 2004). The catalytic power of the enzymes comes from their stabilizing effect on the transition state of the reaction by steric or electrostatic means which come from the influence of residues in the protein structure (Schramm, 2005). Owing to their unstable nature transition states only exist for incredibly short periods of time, typically in the femto- to pico-second range and for this reason they cannot be directly observed with experimental techniques (Wolfenden, 1972). Aside from offering vital information on the mechanistic details of the enzyme reaction, transition states make excellent drug targets (Amaro & Mulholland, 2018; Ruf et al., 1996). Naturally the transition state structure will bind more tightly than the natural substrate so designing a drug that is structurally similar to the transition state can produce useful compounds for effective enzyme inhibition (Copeland, 2013; Ruf et al., 1996). These types of noncovalent inhibitors are highly enzyme specific and can be some of the most potent inhibitors (Schramm, 1998).

Transition state structures can be found at the energy maxima along a reaction path, calculated along a particular reaction coordinate, however, additional method is needed to ascertain where exactly on the potential energy surface it is located (Baker, 1986). By making iterative changes to the molecular structure, optimizing these new structures, and then using an algorithm to direct the stepwise structural changes upward along the energy gradient toward the maxima the structure of the TS can be elucidated (Turner & Williams, 1999). Caution should be taken as other reactive intermediate structures will exist as peaks along the path but are not the actual transition state (Baker, 1986). A correct initial reaction coordinates are vital to finding the correct transition state and proper conformational sampling must be carried out beforehand to ensure this is found. Transition state modeling has been used to successfully model the transition states of a variety of enzymes, with lactate dehydrogenase (Moliner, Turner, & Williams, 1997; Turner & Williams, 1999) and chorismate mutase (Ranaghan et al., 2003, 2004) being two of the most intensively studied examples.

6.2 Reaction Path Modeling

Reaction path calculations include exploration of the full reaction path, deducing not only the TS but the whole range of intermediary structures involved in the reaction (Blumberger, 2008; Hu & Yang, 2008; Woodcock et al., 2007). It is important to start with an experimental foundation from structural, kinetic, and mutational studies (Kraut, Carroll, & Herschlag, 2003), these studies usually produce several hypothetical reaction mechanisms that can then be validated or rejected upon analysis of the reaction path modeled. From a detailed mechanistic simulation the relative structural and electronic contributions of specific active site residues on catalysis can be analyzed which might be difficult to be done using purely experimental techniques (van der Kamp & Mulholland, 2013; Warshel et al., 2006).

An important point impacting the enzyme efficiency is the role of the enzyme conformational dynamics. Enzymes are able to efficiently bind the substrate as well as to hold reactive intermediates and release the products; this is all made possible by dynamic motions and is evidenced by their excellent efficiency relative to rigid artificial catalysts (Glowacki, Harvey, & Mulholland, 2012). There is a discussion if the reaction path starts from a single structure with a distinct conformation binding the substrate or start from one of several low energy points that all reach a common product (Glowacki et al., 2012). In addition the choice of the initial structure is of vital importance, if an incorrect starting conformation is adopted by any subsequent modeling is made much more difficult and the results obtained can contain bias and be incorrect (Bolhuis, Dellago, & Chandler, 2000). For this reason a thorough sampling of multiple enzyme–substrate complex orientations should be undertaken, this can be undertaken with molecular dynamics, Monte Carlo simulations, or other enhanced conformational sampling techniques (McGeagh, Ranaghan, & Mulholland, 2011). Once a structure of the active conformation has been selected, a method which will seek to move from the starting conformation to the next lowest energy point via the lowest energy pathway possible is required. A variety of computational techniques have been developed for this purpose.

Broadly applied method for reaction path calculations is the coordinate driving (adiabatic mapping) method, in which a selected distance (or group of distances, angles, or torsions) in the starting structure (can be either reactant or product) coordinates is constrained and fixed using a force constant. The coordinate is then changed in a step-wise manner and at each step the structure is allowed to relax by means of an energy minimization. Steps are taken toward a saddle point by the path of least ascent, this is done using both the starting and product coordinates (Liu, Zhang, & Zhan, 2009). Although this method may seem the most straightforward it requires the correct initial reaction coordinates and these are not always available but could be deduced with thorough conformational sampling (Williams & Maggiora, 1982). However, conformational sampling of enzymes using the QM/MM methodology can be very computationally intense when treating the QM region, newer techniques to alleviate this issue are becoming more available that utilize highly accurate empirical potentials produced by QM/MM calculations (Duarte et al., 2015).

Another reaction path technique is the Nudged Elastic Band method which requires known starting and end point structures and linearly interpolating sets of transition structures between two known stable structures and then optimizing these toward the lowest energy pathway (Ayala & Schlegel, 1997). It is comparable to a stretched elastic band between two points being nudged to form a desired shape. The structure, however, may become trapped in a local minima, and therefore not reach a global minima which can be avoided by comprehensive sampling of the potential energy landscape of the reaction which can be computationally intensive (Klähn et al., 2005).

Some enzyme reactions, however, can be more challenging for reaction path simulations, for example, enzymatic reactions that involve the quantum tunneling of protons, electrons, or hydrogen atoms and would require special treatment to account for the free-energy barriers of the tunneling process (Masgrau et al., 2006).

6.3 QM/MM Free Energy Calculation Methods

Measuring the standard free energy of a system quantifies its stability, and the standard free energy change along reaction quantifies its driving force.

Nowadays these changes can be calculated using simulation methods (Beveridge & DiCapua, 1989). In a chemical transformation where an enzyme is involved, the catalytic effect provided by the enzyme is due to the formation of a transition state that is lower in energy, than that of the transition state for the same chemical process in solution. This is due to the specific interactions of the preorganized protein environment with the substrate and transition state. The free energy of activation is the difference between free energy of the transition state and the free energy of the reactants, and the free energy of the reaction is defined as the difference between the free energy of the reactants and the products (Atkins & De Paula, 2010). Calculating the free energy changes for a reaction is very important when establishing the correct mechanism from multiple suggested mechanisms (Sousa et al., 2012). The free energy is provided as combination from enthalpic and entropic contributions (Fig. 3). The separation of the contributions provided by entropic (ΔS) and enthalpic (ΔH) effects that describe the overall free energy change of the reaction is not possible through experimental observations alone, but can be achieved by simulation methods (Garcia-Viloca et al., 2004). This type of information can be vital to understanding which interactions contribute most to the catalysis of a reaction, especially in an active site environment and this information that is particularly useful when designing inhibitors.

$$\Delta G = \Delta H - T\Delta S \tag{3}$$

Equation 3: The Gibbs free energy equation used to calculate the change in free energy G for a reaction. H is the enthalpy, T is the temperature, and S is the entropy.

The accurate calculation of the free energy profile of a full reaction path using QM/MM methodology, with high enough levels of QM theory to allow the results to be reliably compared to those obtained from experiment,

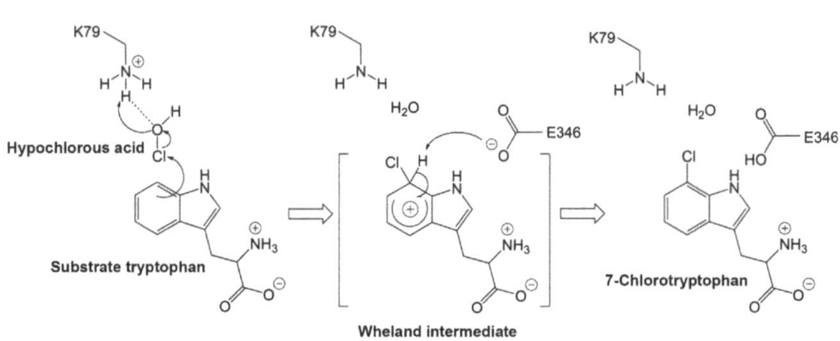

Fig. 3 The proposed reaction mechanism of the tryptophan 7-halogenase PrnA.

is a very computationally expensive task. Owing primarily to the need for thorough conformational sampling to estimate the entropic contribution (ΔS), instead sampling can be achieved using cheaper methods such as semiempirical QM or even fully empirical methods. Once the phase space has been adequately sampled with these computationally cheaper methods, higher levels of theory can be employed to calculate the enthalpic component (ΔH) of the reaction to produce a final calculated free energy that is accurate enough to be compared to experimentally generated values (Lonsdale et al., 2012).

7. QM/MM APPLICATIONS IN PROTEIN–LIGAND DOCKING

Molecular docking was primarily designed to predict the binding of small drug-like molecules to target proteins. Many diseases are caused by the malfunction of proteins and therapies are focused on the inhibition or activation of the target proteins (Taylor, Jewsbury, & Essex, 2002). Traditional lead generation methods for drug discovery normally entail assaying a large variety of interesting compounds against a specific protein known to be a disease target and hoping to observe a binding interaction (Drews, 2000). Docking can be used to virtually screen new compounds in a similar way to experimental high-throughput screening as well as offering atomistic level insight to facilitate structure-based drug design (Amaro & Mulholland, 2018; Gao et al., 1998), where the aim is find stable binding conformations between a ligand and a receptor (Morris & Lim-Wilby, 2008). In the same way that high-throughput screening attempts to test as many compounds as possible, molecular docking methods are also focused in this regard and therefore traditionally utilize only MM or knowledge-based scoring functions (Doman et al., 2002). However, when a more detailed descriptor of binding energy is required QM/MM methods can be used, this allows for a more accurate representation of the electronic structure of the docking site and ligand while still taking into account the electrostatic contributions of the rest of the protein environment at the MM level (Cho et al., 2005). The large number of compounds screened would normally make QM/MM techniques forbiddingly computationally expensive, however, this more computationally expensive method could be used for challenging docking problems and has been shown in recent studies that give greater agreement with experimental results (Khandelwal & Balaz, 2007; Khandelwal et al., 2005).

8. PRACTICAL QM/MM APPLICATIONS FOR ENZYME REACTIVITY

8.1 Reaction Mechanism of Tryptophan 7-Halogenase

Tryptophan 7-halogenase (PrnA) is an enzyme responsible for the highly regioselective chlorination of tryptophan to produce 7-chlorotryptophan in the biosynthesis of the antifungal natural product pyrrolnitrin. PrnA exclusively produces 7-chlorotryptophan using reduced FAD, dioxygen, and chloride ions (Sascha et al., 2000). Chlorination of tryptophan in solution does not spontaneously occur when chloride ions are used as the chlorinating species; when stronger chlorinating agents are used, a mixture of 1, 5, and 7 substituted chlorotryptophan products are produced (Somei, 2006). The high levels of regioselectivity and use of mild reagents are desirable properties for industrial processes to imitate, as they reduce costs and produce high yields. Many pharmaceutical compounds of interest such as the antibiotics, chlortetracycline and vancomycin, the antifungal compound pyrrolnitrin and chemotherapeutics—salinosporamide A and rebeccamycin all contain chlorine atoms (Ainsley et al., 2018). Current strategies for the synthetic production of these natural products rely on potentially expensive metal catalysts and the use of protecting groups which lower overall yields (Lang et al., 2011; Young & Baran, 2009). To mimic the kind of strategies used by PrnA in its highly regioselective reaction, a thorough understanding of the mechanism at an electronic level is needed. A study of the reaction mechanism of PrnA was accomplished by using both QM/MM potential energy scans (PES) and QM/MM free energy scans (FES) the reaction path (Karabencheva-Christova et al., 2017).

Several proposals on the raction mechanism have been done, speculated that chloride ions would react with FAD and oxygen to produce a chlorinated FAD intermediate which would react with tryptophan (Dong et al., 2005; Yeh, Garneau, & Walsh, 2005). When the first crystal structure of PrnA was published, it showed a ~10 Å separation between the proposed site of chloride ion reaction and the binding site of tryptophan. Although large scale conformational changes could occur to bring the two molecules into close enough proximity to react, this idea seemed unlikely based on observations on the X-ray crystalographic struycture oly. Recently it has been shown through 1 μs MD simulations that large conformational change that brings the FAD-binding site and TRP-binding site together is not observed (Ainsley et al., 2018). Mutational studies showed that the residues

Lys79 and Glu346 have key importance to the enzyme activity and their mutations greatly reduced or totally removed tryptophan halogenating activity. Some alternative mechanisms suggested that intermediate halogenation of Lys79 or Glu346 by the chloro-FAD intermediate could occur; however, MD simulations provided atomistic insights in the role of these mutations and showed this alternative to be not probable (Ainsley et al., 2018). To date, the most likely reaction mechanism suggests that a small mobile intermediate chlorianting agent is produced, hypochlorous acid is produced by the reaction of FAD, oxygen, and chloride (Dong et al., 2005). However, hypochlorous acid alone cannot chlorinate tryptophan, therefore it is suggested that the key catalytic residues Lys79 and Glu346 are involved in activating hypochlorous acid to chlorinate tryptophan. Another alternative mechanism consisted of hypochlorous acid reacting with Lys79 to form a chloramine (Wagner, El Omari, & König, 2009; Yeh et al., 2007), however, chloramines are less reactive toward trytophan than hypochlorous acid making this mechanism unlikely (Flecks et al., 2008). Based on all of these previous studies of the reaction mechanism of PrnA, the study was focused on the activation of hypochlorous acid by Lys79 and the stabilization and subsequent deprotonation of the Wheland intermdiate by means of both QM/MM PES and FES using DFT level of theory for the QM part (Fig. 3) (Karabencheva-Christova et al., 2017). Importantly the study involved not only DFT QM model for the PES studies but also FES was explored using novel QM/MM-MD method including again DFT level for the QM part.

A model of the reactant complex was constructed using the X-ray crystal structure (Dong et al., 2005) (PDBID: 2AR8) by modifying the product chlorotryptophan to produce the substrate tryptophan and hypochlorous acid. The system was partitioned with the substrate tryptophan, hypochlorous acid, and side chains (cut between the Cα and Cβ atoms) of Lys79 and Glu346 being included in the QM region, whereas the remainder of the system was designated as the MM region. The cut bonds of the Lys79 and Glu346 side chains had their valences satisfied via the linked pair approach. The potential energy reaction path was first modeled using adiabatic mapping of the two reaction coordinates RC1 and RC2 (Fig. 4) with a B3LYP/6-31+G(d) functional and basis set combined with the CHARMM27 force field (MacKerell et al., 1998) using the QoMMMA interface (Harvey, 2004) to combine the Jaguar software package (Schrödinger, 2005) for the QM calculation. The transition states were estimated to be the highest energy points of the calculated reaction path.

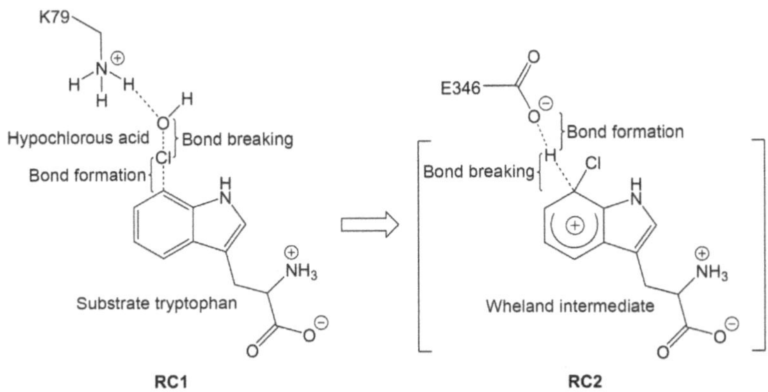

Fig. 4 The studied reaction coordinates RC1 and RC2. *From Karabencheva-Christova, T.G., et al. (2017). Mechanistic insights into the reaction of chlorination of tryptophan catalysed by tryptophan 7-halogenase. Scientific Reports, 7(1), p. 17395.*

A QM/MM-MD simulation of the same proposed reaction path was then performed using the multiple steered MD method, using AMBER–PUPIL–NWCHEM interface (Case et al., 2014; Torras et al., 2015; Valiev et al., 2010), in this way the free energy changes of the reaction path could also be calculated. This calculation, however, used a slightly different QM region, encompassing the whole of Glu346, link atoms were added to the neighboring residues Leu345 and Ser347. Additionally, to reduce the size of the QM region the substrate tryptophan was cut between the Cβ and Cγ atoms. A short classical MD simulation for 5 ns performed with AMBER (Case et al., 2014) and was used to sample different reactant complex structures, from this 15 different initial reactant complexes were extracted. All 15 of these structures then underwent relaxation using the QM/MM-MD approach for 0.5 ps prior to the reaction path coordinate pulling. A force constant of 5000 kcal mol^{-1} was applied to move the atoms along the two reaction coordinates RC1 and RC2 over 3000 steps with a step size of 0.5 fs for each reaction coordinate. For each of the 15 final structures a reverse reaction coordinate pulling was performed to pull the products back into their reactant complexes, resulting in the production of the free energy reaction path. The calculated PES and FES are shown in Fig. 5.

The potential energy-based adiabatic mapping and the free energy QM/MM-MD calculated reaction path agree that the rate limiting step of the reaction is the proton abstraction of the Wheland intermediate by Glu346 (simulated by RC2, Fig. 4). Good agreement was also found in

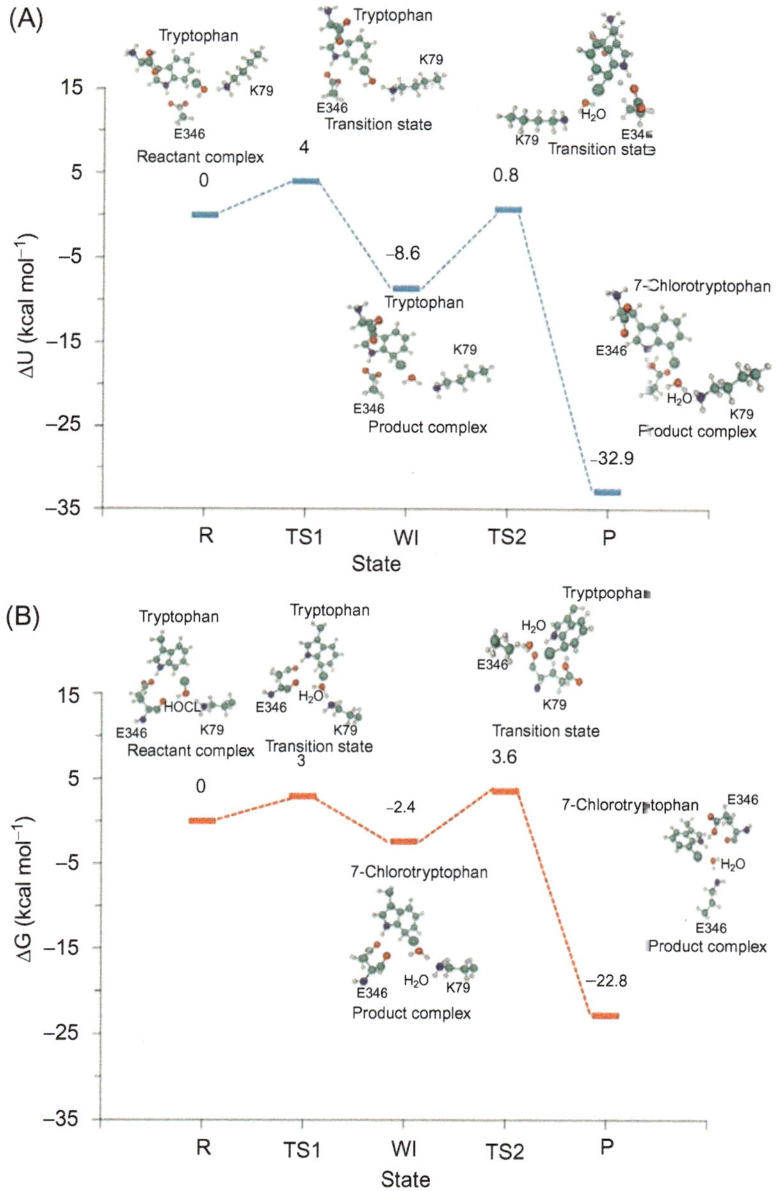

Fig. 5 (A) Potential energy and (B) free energy diagrams derived from potential energy and free energy paths. The stationary states/structures of RC, TS, and PCs are given for the two steps of the overall reaction. *Reproduced with permission from Karabencheva-Christova, T.G., et al. (2017). Mechanistic insights into the reaction of chlorination of tryptophan catalysed by tryptophan 7-halogenase. Scientific Reports, 7(1), p. 17395.*

the energy barrier for the formation of the Wheland intermediate (RC1, Fig. 4), however, the free energy reaction path was higher in energy for the Wheland intermediate structure. The stationary points along both reaction paths show similar geometries and also possess similar stabilizing hydrogen-bonding interactions. The study provides an atomistic insight into the experimentally suggested roles of Lys79 and Glu346 in the halogenating reaction of *PrnA*.

8.2 pro-S Hydrogen Abstraction Reaction in Cycoloxygenase-1

Cyclooxygenase (COX) enzymes are responsible for the conversion of arachidonic acid (ACD) to prostaglandin H2 (PGH2), an important precursor to the biosynthesis of prostaglandins and thromboxanes (Jakoby, 1998). These two lipid type molecules prostaglandins and thromboxanes fulfill a broad range of hormone-like roles related to the inflammatory response to injury: pain, inflammation, blood clotting as well as normal physiological roles such as platelet function, renal blood flow regulation, vasoconstriction, and the maintenance of the mucous membrane of the gastric system (Crofford, 1997). Targeting COX enzymes has long been a strategy for drugs hoping to reduce the inflammation and pain associated with injuries, common drugs such as Ibuprofen, Aspirin, and Paracetamol all target COX in some way (Hinz, Cheremina, & Brune, 2008). There are two main forms of COX found in humans: COX-1 is mainly responsible for the normal physiological functions and is expressed in most tissues in the body; COX-2 is more tightly regulated and should only become expressed on a local level in response to injury these two isoenzymes share around a 60% sequence identity (Kurumbail et al., 1996). Inhibition of COX-2 is usually the desired effect of a drug as inhibition of COX-1 can lead to side effects such as gastric ulceration and possible renal failure, due to this there is a desire to design drugs that selectively target COX-2.

The COX enzymes are membrane bound and actually contain two active sites, as the overall reaction is performed in two distinct stages. In the first stage of the reaction ACD reacts with dioxygen in a cyclooxygenase reaction to produce the intermediate compound Hydroperoxy Endoperoxide prostaglandin (PGG2). This is then reduced at a heme containing peroxidase site to produce PGH2 (Vane, Bakhle, & Botting, 1998). The reaction shown in Fig. 6 depicts the hydrogen atom abstraction from ACD at the 13-pro-S position by the Tyr-385 tyrosyl radical, this is thought to be the rate limiting

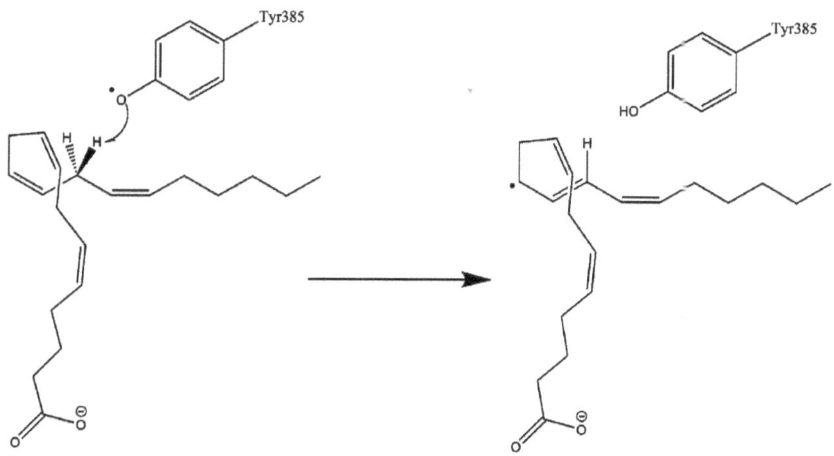

Fig. 6 The proposed mechanism for pro-S hydrogen abstraction by the Tyr385 tyrosyl radical. *Reproduced with permission from Christov, C. Z., et al. (2013). Conformational effects on the pro-S hydrogen abstraction reaction in cyclooxygenase-1: An integrated QM/MM and MD study. Biophysical Journal. 104(5), L5–L7.*

step of the reaction (Silva, Fernandes, & Ramos, 2003). The resulting pentadienyl radical then reacts with dioxygen creating a peroxide bridge forming PGG2.

The pro-S hydrogen atom abstraction step of the reaction (Fig. 6) was studied in a combined MD QM/MM study (Christov et al., 2013) with aim to improve upon previous QM studies by taking into account the influence of the enzymes conformational flexibility on the potential energy surface a factor not accounted for in the small cluster structures used previously (Blomberg et al., 2003; Silva et al., 2003). To sample the conformational flexibility of the protein, two MD simulations of COX-1 bound to a lipid bilayer were performed using the CHARMM version 27b2 code (Brooks et al., 1983), one simulation using stochastic boundary conditions for a total of 250 ps the other using periodic boundary conditions for 25 ns. Cluster analysis was then used to select representative structures from the MD trajectories in which the distance between the *pro*-S hydrogen atom and Tyr-385 tyrosyl radical oxygen atom ([O—H_s] distance) was below 3.0 Å, only structures 300 ps apart were considered to ensure a more diverse range of protein and ACD conformations.

Based on the selection criteria five structures were chosen to undergo QM/MM simulation, four were from the periodic boundary condition simulation (2180 ps, 3984 ps, 4960 ps, and 5344 ps) and one from the stochastic boundary conditions (SB-250 ps). The system was partitioned so that ACD and the side chain of Tyr385 tyrosyl radical were in the QM region (containing 68 atoms) was treated with a B3LYP/6-31G(d) level of theory; the rest of the system was modeled with the CHARMM27 force field. The link region between the Cα and Cβ of Tyr-385 was capped with a hydrogen link atom and the system was electrostatically embedded. The five structures taken from the MD simulations were minimized and a reaction path was modeled for each using the adiabatic mapping method. It was noted that all the reaction paths were similar position along the reaction coordinate (Fig. 7).

The calculated reaction barriers, shown in Fig. 7, range between 14.3 and 25.9 $kcal\,mol^{-1}$ with the Boltzmann-average value being 15.2 $kcal\,mol^{-1}$ which is in strong agreement with the experimentally calculated reaction barrier of 15.0 $kcal\,mol^{-1}$ (Kulmacz, Pendleton, & Lands, 1994). To assess the differences between the starting reactant complexes and the TS structures various geometrical descriptors (bond lengths and angles) were measured and a correlation with the reaction energy barrier height was found (Lodola et al., 2010). The measured geometric descriptors are recorded in Table 1.

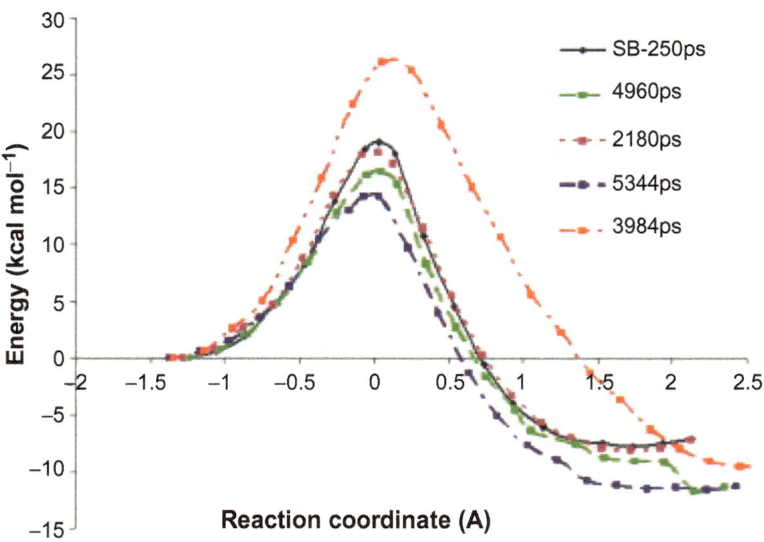

Fig. 7 The five different reaction paths calculated for the d1[C13—Hs] − d2[Hs—O] reaction coordinate. *Reproduced with permission from Christov, C. Z., et al. (2013). Conformational effects on the pro-S hydrogen abstraction reaction in cyclooxygenase-1: An integrated QM/MM and MD study. Biophysical Journal. 104(5), L5–L7.*

Table 1 A Table of the Different Geometric Descriptors Measured for Each of the Reaction Paths

Structure	[O,HS]RC (Å)	[O,HS]TS (Å)	α1RC (°)	α1TS (°)	Ea (kcal mol^{-1})
SB250 ps	2.37	1.37	109	110.2	18.4
2180 ps	2.38	1.3	95.2	108	18.2
3984 ps	2.47	1.41	111	114.7	25.9
4960 ps	2.36	1.29	89.3	108.4	16.4
5344 ps	2.48	1.37	83.4	106.7	14.3

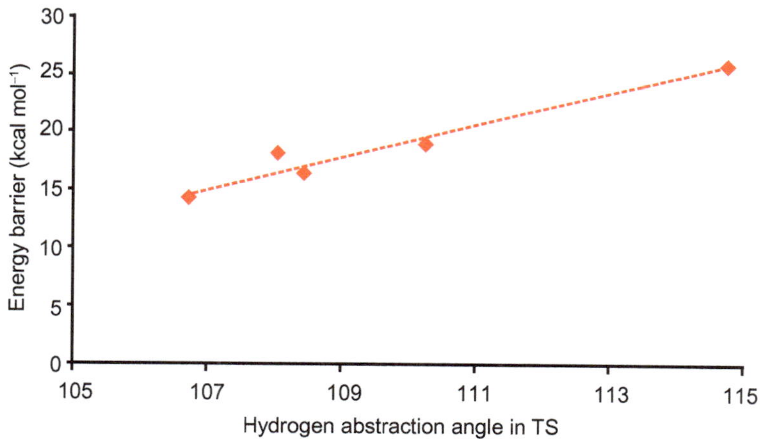

Fig. 8 A graph showing the correlation between the angle of hydrogen abstraction and energy barrier for abstraction of the pro-S hydrogen. *Reproduced with permission from Christov, C. Z., et al. (2013). Conformational effects on the pro-S hydrogen abstraction reaction in cyclooxygenase-1: An integrated QM/MM and MD study. Biophysical Journal.* 104*(5), L5–L7.*

It was found that there was a negative correlation between the energy barrier height and hydrogen abstraction angle α1 in the TS (Fig. 8) for the five analyzed structures. A similar yet weaker correlation was also observed in the starting complex structures. This suggests that the geometry of the transition state is directly related to the geometry of the starting complex, and the geometry of the transition state is directly related to the energy barrier height of the reaction. This observation reveals the need to account for conformational flexibility of the starting complex and the need to move toward free energy QM/MM simulation methods that would inherently account for these factors (Plotnikov, Kamerlin, & Warshel, 2011).

8.3 Mechanism of Covalent Addition Between EGFR-Cysteine 797 and *N*-(4-Anilinoquinazolin-6-yl) Acrylamide

The human irreversible epidermal growth factor receptor (EGFR) is a cell surface receptor that controls cell proliferation. When the gene encoding for EGFR mutates (i.e., the L858R mutation), or when this receptor is overexpressed, several types of carcinoma can arise (Olayioye et al., 2000). The first generation of drugs targeting the kinase domain of the EGFR receptor were the 4-anilinoquinazolinde derivatives, gefitinib (Barker et al., 2001) and erlotinib (Moyer et al., 1997). These compounds inhibit EGFR by displacing ATP from its binding site leading to the blockage of the cell signaling cascade responsible for the proliferation of carcinoma cells. In the T790M mutant form of the EGFR receptor the binding affinity for ATP is increased, while that of kinase drugs is decreased. This may explain the loss of potency of these first generation drugs on the EGFR L858R/T790M mutants as well as the resulting lack of efficacy in patients affected by nonsmall-cell lung carcinoma (NSCLC) (Yun et al., 2008). Newer generations of drugs targeting EGFR attempt to overcome the resistance due to T790M by using a different mechanism of inhibition. By utilizing an acrylamide Michael Acceptor (MA) group, they are able to covalently and thus irreversibly bind to a noncatalytic cysteine residue (Cys797), in a position that blocks the EGFR receptor site and prevents its activation (Carmi et al., 2012a). Examples of compounds working with this mechanism are afatinib (which unfortunately failed to be reach clinical use due to reported adverse side effects) and osimertinib, which has resulted in a safe and effective treatment of NSCLC patients affected by L858R/T790M mutation.

Although X-ray structures exist demonstrating the covalent adducts of EGFR- and MA-based drugs (Blair et al., 2007), the exact mechanism of the bonding reaction remained unknown. Kinetic studies had alluded to the reactivity of the acrylamide group being only one of several factors contributing to the inhibitory power of the EGFR acrylamide-based covalent inhibitors (Schwartz et al., 2014). Capoferri et al. sought to elucidate the exact mechanism of the Michael addition reaction between the cysteine residue of EGFR and the drug, to do this a variety of computational techniques: MD, umbrella sampling, QM/MM reaction path modeling, and free energy calculations were employed (Capoferri et al., 2015).

To simplify the modeling of the reaction a model compound (13-JAB) was used, it has the basic features of other acrylamide-based drugs without any unnecessary R groups to complicate the equation. In this way, the reaction between the acrylamide group and cysteine could be studied at its most basic level. Knowing the reaction between a thiol group and a MA passes through a significant free-energy barrier in solution (Paasche et al., 2010). Capoferri et al. looked to identify a residue in proximity to Cys797 that could facilitate the deprotonation of the thiol group. This deprotonation of the thiol and resultant formation of a thiolate would lead to a species that is much more reactive toward acrylamide groups.

Several X-ray structures were visually inspected to identify a residue acting as a general acid/base catalyst of the reaction; it was found that Asp800 consistently crystallized close to the thiol group of Cys797 so was found to be the most probable candidate. In addition to this crystallographic data, a mechanism not involving ketonization was favored due to the known large energy barrier of the ketonization process required by mechanism B to produce the known covalent adduct structure from the enolate containing intermediate (Paasche et al., 2010). It was hypothesized that the acrylamide-based inhibitors react through a direct addition mechanism (Fig. 9).

Molecular dynamics simulations of the solvated system of 13-JAB in complex with EGFR were carried out using the AMBER99SB force field (Hornak et al., 2006) for 100 ns, with the initial positioning of 13-JAB was based on previous docking simulations (Carmi et al., 2010). A similar simulation of the covalent adduct of 13-JAB was also performed under identical conditions for 100 ns and used as a comparison to the Michaelis-like complex. Reaction path calculations focused on recreating a mechanism where Asp800 deprotonated Cys797 as an initial stage followed by the nucleophilic attack of the β-carbon of acrylamide by the thiolate ion to produce the C—S bond of the covalent adduct. The final stage of the reaction would be the protonation of the α-carbanion by the side chain COOH of Asp800 thus regenerating it to its aspartate COO^- form. The Path Collective Variables

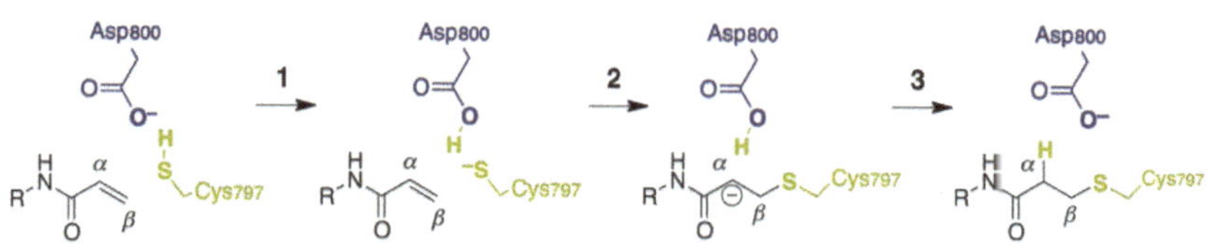

Fig. 9 The proposed mechanism of reaction for the deprotonation of Cys797 by Asp800 followed by the subsequent attack of the 13-JAB acrylamide group to form the covalently bound drug–protein adduct. *Reproduced with permission from Capoferri, L., et al. (2015). Quantum mechanics/molecular mechanics modelling of covalent addition between EGFR-cysteine 797 and N-(4-anilinoquinazolin-6-yl) acrylamide. Journal of Chemical Information and Modeling, 55(3), 589–599.*

(PCV) method was used to construct two collective variables S and V, and further details of how the two variables S and Z are calculated are available in the Branduardi et al. paper. The reaction path was modeled as a steered QM/MM-MD path using a SCC-DFTB/AMBER99SB potential (Seabra et al., 2007), the S and V PCVs were used to certify the path moved toward the product complex. Umbrella sampling was carried out at points along the converged reaction pathway as a method of estimating the free energy changes involved, this also served to validate the calculated path. It was found that it shared the same shape as the converged steered MD pathway.

The final reaction path showed five distinct phases (I–V) in Fig. 10 correspond to the chemical changes experienced by the system over the course of the reaction path, with the overall peak being 8.6 kcal mol^{-1} above the

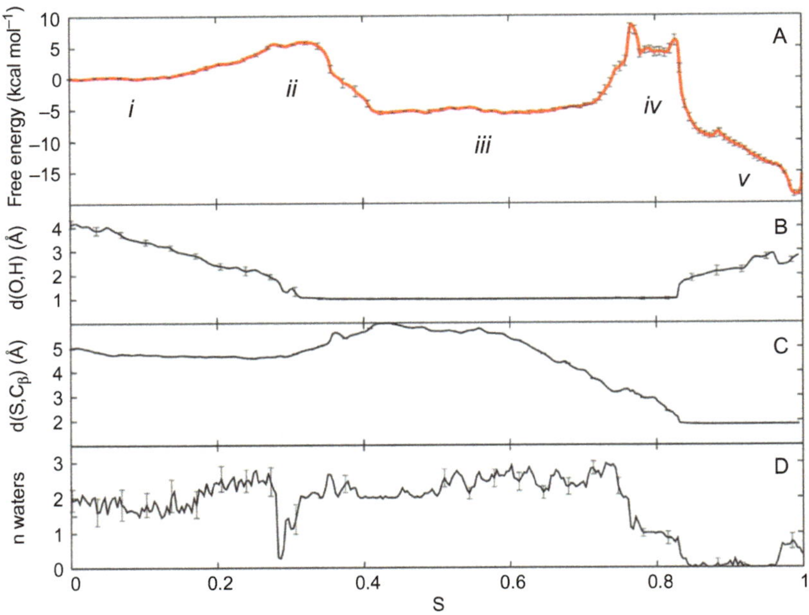

Fig. 10 (A) The calculated free energy profile of EGFR alkylation by 13-JAB. (B) Plot of the distance between the oxygen of Asp800 and the hydrogen of Cys797 against the collective variable S. (C) Plot of the distance between the sulfur of Cys797 and the Cβ of the acrylamide group of 13-JAB against the collective variable S. (D) The number of water molecules within 3.5 Å of the sulfur atom of Cys797. *Reproduced with permission from Capoferri, L., et al. (2015). Quantum mechanics/molecular mechanics modelling of covalent addition between EGFR-cysteine 797 and N-(4-anilinoquinazolin-6-yl) Acrylamide. Journal of Chemical Information and Modeling, 55(3), 589–599.*

energy of the reactants. Accounting for the energy of the Michaelis complex the overall activation barrier is +14.6 kcal mol^{-1}, this reasonably agrees with the experimentally calculated barrier value of ~20 kcal mol^{-1} (Schwartz et al., 2014) especially when accounting for the fact that SCC-DFTB methods are known from to underestimate energy barriers by around 25% (Mujika, Lopez, & Mulholland, 2012; Woodcock, Hodošček, & Brooks, 2007; Xu, Guo, & Cui, 2007). The direct addition of thiolate to the acrylamide β-carbon to form the C—S bond and the protonation of the α-carbon by aspartic acid happens in a concerted manner. This is consistent with the findings of a similar reaction path modeling study of cysteine reacting with an α,β-unsaturated compound (Grazioso et al., 2012). The stable covalent adduct, product complex, is 18 kcal mol^{-1} lower than the starting reactant complex and −12 kcal mol^{-1} lower than the most stable intermediate structure. The free energy differences indicate that the overall reaction is both spontaneous and irreversible, this is consistent with previous kinetic (Carmi et al., 2012b; Schwartz et al., 2014), structural (Blair et al., 2007), and pharmacological (Fry et al., 1998) studies of acrylamide-based drug interactions with EGFR. Similar results were also obtained for the reaction between EGFR and the lung cancer drug osimbertinib (Callegari et al., 2018). Based on the findings of the reaction path modeling the authors recommended that a terminal tertiary amine of some of the acrylamide-based drugs did not participate in the reaction mechanism and instead improved the rate of inactivation by improving the noncovalent binding of the drugs to form the Michaelis-like complex (Carmi et al., 2011).

9. CONCLUSIONS

The QM/MM methods received broad applications for modeling different aspects of reaction mechanisms of enzymes and became one of the most intensively applied computational chemistry methods nowadays. They are routinely applied to determine the structures of transition states and reaction intermediates, to discriminate between alternative reaction mechanisms, to understand the effects of mutations on the reaction mechanisms, and to simulate free energies of enzyme reaction mechanisms. The development and applications of the QM/MM methods depend on the development of more effective optimization algorithms, more accurate QM methods, and new force fields. In addition, the development of new computational chemistry codes and high-performance computers helps for the broader implementation of QM/MM in the area of chemical biology and pharmaceutical chemistry. Based on the increased number of applications and enhanced synergy with experimental methods in the near future, the QM/MM methods will become an even more accurate, predictive, and valuable tool in chemical and biological studies of enzymes and their applications in drug design.

REFERENCES

Ainsley, J., et al. (2018). Structural insights from molecular dynamics simulations of tryptophan 7-halogenase and tryptophan 5-halogenase. *ACS Omega*, *3*(5), 4847–4859.

Amara, P., et al. (2000). The generalized hybrid orbital method for combined quantum mechanical/molecular mechanical calculations: Formulation and tests of the analytical derivatives. *Theoretical Chemistry Accounts*, *104*(5), 336–343.

Amaro, R. E., & Mulholland, A. J. (2018). Multiscale methods in drug design bridge chemical and biological complexity in the search for cures. *Nature Reviews Chemistry*, *2*(4), 0148.

Atkins, P., & De Paula, J. (2010). *Physical chemistry for the life sciences*. Oxford University Press.

Ayala, P. Y., & Schlegel, H. B. (1997). A combined method for determining reaction paths, minima, and transition state geometries. *The Journal of Chemical Physics*, *107*(2), 375–384.

Baker, J. (1986). An algorithm for the location of transition states. *Journal of Computational Chemistry*, 7(4), 385–395.

Barker, A. J., et al. (2001). Studies leading to the identification of ZD1839 (Iressa™): An orally active, selective epidermal growth factor receptor tyrosine kinase inhibitor targeted to the treatment of cancer. *Bioorganic & Medicinal Chemistry Letters*, *11*(14), 1911–1914.

Bennie, S. J., et al. (2016). A projector-embedding approach for multiscale coupled-cluster calculations applied to citrate synthase. *Journal of Chemical Theory and Computation*, *12*(6), 2689–2697.

Beveridge, D. L., & DiCapua, F. (1989). Free energy via molecular simulation: Applications to chemical and biomolecular systems. *Annual Review of Biophysics and Biophysical Chemistry*, *18*(1), 431–492.

Blair, J. A., et al. (2007). Structure-guided development of affinity probes for tyrosine kinases using chemical genetics. *Nature Chemical Biology*, *3*(4), 229–238.

Blake, C., et al. (1965). Structure of hen egg-white lysozyme: A three-dimensional Fourier synthesis at 2 Å resolution. *Nature*, *206*(4986), 757–761.

Blomberg, L. M., et al. (2003). A quantum chemical study of the synthesis of prostaglandin G_2 by the cyclooxygenase active site in prostaglandin endoperoxide H synthase 1. *The Journal of Physical Chemistry B*, *107*(14), 3297–3308.

Blumberger, J. (2008). Free energies for biological electron transfer from QM/MM calculation: Method, application and critical assessment. *Physical Chemistry Chemical Physics*, *10*(37), 5651–5667.

Bolhuis, P. G., Dellago, C., & Chandler, D. (2000). Reaction coordinates of biomolecular isomerization. *Proceedings of the National Academy of Sciences of the United States of America*, *97*(11), 5877–5882.

Brooks, B. R., et al. (1983). CHARMM: A program for macromolecular energy, minimization, and dynamics calculations. *Journal of Computational Chemistry*, *4*(2), 187–217.

Callegari, D., et al. (2018). L718Q mutant EGFR escapes covalent inhibition by stabilizing a non-reactive conformation of the lung cancer drug osimertinib. *Chemical Science*, *9*(10), 2740–2749.

Capoferri, L., et al. (2015). Quantum mechanics/molecular mechanics modeling of covalent addition between EGFR–cysteine 797 and N-(4-anilinoquinazolin-6-yl) acrylamide. *Journal of Chemical Information and Modeling, 55*(3), 589–599.

Carmi, C., et al. (2010). Novel irreversible epidermal growth factor receptor inhibitors by chemical modulation of the cysteine-trap portion. *Journal of Medicinal Chemistry, 53*(5), 2038–2050.

Carmi, C., et al. (2011). Epidermal growth factor receptor irreversible inhibitors: Chemical exploration of the cysteine-trap portion. *Mini Reviews in Medicinal Chemistry, 11*(12), 1019–1030.

Carmi, C., et al. (2012a). Clinical perspectives for irreversible tyrosine kinase inhibitors in cancer. *Biochemical Pharmacology, 84*(11), 1388–1399.

Carmi, C., et al. (2012b). Irreversible inhibition of epidermal growth factor receptor activity by 3-aminopropanamides. *Journal of Medicinal Chemistry, 55*(5), 2251–2264.

Carvalho, A. T., et al. (2014). Challenges in computational studies of enzyme structure, function and dynamics. *Journal of Molecular Graphics and Modelling, 54*, 62–79.

Case, D. A., et al. (2014). *AMBER 14*. San Francisco: University of California.

Cho, A. E., et al. (2005). Importance of accurate charges in molecular docking: Quantum mechanical/molecular mechanical (QM/MM) approach. *Journal of Computational Chemistry, 26*(9), 915–931.

Christov, C. Z., et al. (2013). Conformational effects on the pro-S hydrogen abstraction reaction in cyclooxygenase-1: An integrated QM/MM and MD study. *Biophysical Journal, 104*(5), L5–L7.

Clarivate Analytics. (2018). *Web of science*. Clarivate Analytics. 01/06/2018; Available from: http://wok.mimas.ac.uk/.

Copeland, R. A. (2013). *Evaluation of enzyme inhibitors in drug discovery: a guide for medicinal chemists and pharmacologists*. John Wiley & Sons.

Crofford, L. J. (1997). COX-1 and COX-2 tissue expression: Implications and predictions. *The Journal of Rheumatology. Supplement, 49*, 15–19.

Doman, T. N., et al. (2002). Molecular docking and high-throughput screening for novel inhibitors of protein tyrosine phosphatase-1B. *Journal of Medicinal Chemistry, 45*(11), 2213–2221.

Dong, C., et al. (2005). Tryptophan 7-halogenase (PrnA) structure suggests a mechanism for regioselective chlorination. *Science, 309*(5744), 2216–2219.

Drews, J. (2000). Drug discovery: A historical perspective. *Science, 287*(5460), 1960–1964.

Duarte, F., et al. (2015). Recent advances in QM/MM free energy calculations using reference potentials. *Biochimica et Biophysica Acta (BBA) - General Subjects, 1850*(5), 954–965.

Dubey, K. D., Wang, B. J., & Shaik, S. (2016). Molecular dynamics and QM/MM calculations predict the substrate-induced gating of cytochrome P450 BM3 and the regio- and stereoselectivity of fatty acid hydroxylation. *Journal of the American Chemical Society, 138*(3), 837–845.

Faponle, A. S., Seebeck, F. P., & de Visser, S. P. (2017). Sulfoxide synthase versus cysteine dioxygenase reactivity in a nonheme iron enzyme. *Journal of the American Chemical Society, 139*(27), 9259–9270.

Field, M. J., Bash, P. A., & Karplus, M. (1990). A combined quantum mechanical and molecular mechanical potential for molecular dynamics simulations. *Journal of Computational Chemistry, 11*(6), 700–733.

Flecks, S., et al. (2008). New insights into the mechanism of enzymatic chlorination of tryptophan. *Angewandte Chemie (International Ed. in English), 47*(49), 9533–9536.

Foundation, T.N. (2013). *Nobel prize in chemsitry 2013*. 19/01/15; Available from: https://www.nobelprize.org/nobel_prizes/chemistry/laureates/2013/.

Friesner, R. A., & Guallar, V. (2005). Ab initio quantum chemical and mixed quantum mechanics/molecular mechanics (QM/MM) methods for studying enzymatic catalysis. *Annual Review of Physical Chemistry, 56*, 389–427.

Fry, D. W., et al. (1998). Specific, irreversible inactivation of the epidermal growth factor receptor and erbB2, by a new class of tyrosine kinase inhibitor. *Proceedings of the National Academy of Sciences of the United States of America, 95*(20), 12022–12027.
Gao, J., et al. (1998). A generalized hybrid orbital (GHO) method for the treatment of boundary atoms in combined QM/MM calculations. *The Journal of Physical Chemistry A, 102*(24), 4714–4721.
Garcia-Viloca, M., et al. (2004). How enzymes work: Analysis by modern rate theory and computer simulations. *Science, 303*(5655), 186–195.
Glowacki, D. R., Harvey, J. N., & Mulholland, A. J. (2012). Taking Ockham's razor to enzyme dynamics and catalysis. *Nature Chemistry, 4*(3), 169–176.
Grazioso, G., et al. (2012). Mechanism of falcipain-2 inhibition by α, β-unsaturated benzo [1, 4] diazepin-2-one methyl ester. *Journal of Computer-Aided Molecular Design, 26*(9), 1035–1043.
Harvey, J. N. (2004). Spin-forbidden CO ligand recombination in myoglobin. *Faraday Discussions, 127*, 165–177.
Hinz, B., Cheremina, O., & Brune, K. (2008). Acetaminophen (paracetamol) is a selective cyclooxygenase-2 inhibitor in man. *The FASEB Journal, 22*(2), 383–390.
Hornak, V., et al. (2006). Comparison of multiple Amber force fields and development of improved protein backbone parameters. *Proteins: Structure, Function, and Bioinformatics, 65*(3), 712–725.
Hu, H., & Yang, W. T. (2008). Free energies of chemical reactions in solution and in enzymes with ab initio quantum mechanics/molecular mechanics methods. *Annual Review of Physical Chemistry, 59*, 573–601.
Hu, H., & Yang, W. T. (2009). Development and application of ab initio QM/MM methods for mechanistic simulation of reactions in solution and in enzymes. *Journal of Molecular Structure Theochem, 898*(1–3), 17–30.
Jakoby, W. B. (1998). In A. D. Smith, S. P. Datta, G. H. Smith, P. N. Campbell, R. Bentley, & H. A. McKenzie (Eds.), *Oxford dictionary of biochemistry and molecular biology*. Academic Press.
Kang, J., & Tateno, M. (2012). Recent applications of hybrid Ab initio quantum mechanics–molecular mechanics simulations to biological macromolecules. In M. R. Pahlavani (Ed.), *Some applications of quantum mechanics*. InTech.
Karabencheva-Christova, T. G., et al. (2017). Mechanistic insights into the reaction of chlorination of tryptophan catalyzed by tryptophan 7-halogenase. *Scientific Reports*, 7(1), 17395.
Karplus, M., & McCammon, J. A. (2002). Molecular dynamics simulations of biomolecules. *Nature Structural & Molecular Biology, 9*(9), 646–652.
Khandelwal, A., & Balaz, S. (2007). QM/MM linear response method distinguishes ligand affinities for closely related metalloproteins. *Proteins: Structure, Function, and Bioinformatics, 69*(2), 326–339.
Khandelwal, A., et al. (2005). A combination of docking, QM/MM methods, and MD simulation for binding affinity estimation of metalloprotein ligands. *Journal of Medicinal Chemistry, 48*(17), 5437–5447.
Klähn, M., et al. (2005). On possible pitfalls in ab initio quantum mechanics/molecular mechanics minimization approaches for studies of enzymatic reactions. *The Journal of Physical Chemistry B, 109*(32), 15645–15650.
Kraut, D. A., Carroll, K. S., & Herschlag, D. (2003). Challenges in enzyme mechanism and energetics. *Annual Review of Biochemistry, 72*(1), 517–571.
Kulmacz, R. J., Pendleton, R. B., & Lands, W. (1994). Interaction between peroxidase and cyclooxygenase activities in prostaglandin-endoperoxide synthase. Interpretation of reaction kinetics. *Journal of Biological Chemistry, 269*(8), 5527–5536.

Kurumbail, R. G., et al. (1996). Structural basis for selective inhibition of cyclooxygenase-2 by anti-inflammatory agents. *Nature*, *384*(6610), 644–648.

Lang, A., et al. (2011). Changing the regioselectivity of the tryptophan 7-halogenase PrnA by site-directed mutagenesis. *Angewandte Chemie International Edition*, *50*(13), 2951–2953.

Lever, G., et al. (2014). Large-scale density functional theory transition state searching in enzymes. *The Journal of Physical Chemistry Letters*, *5*(21), 3614–3619.

Lin, H., & Truhlar, D. G. (2007). QM/MM: What have we learned, where are we, and where do we go from here? *Theoretical Chemistry Accounts*, *117*(2), 185–199.

Liu, J., Zhang, Y., & Zhan, C.-G. (2009). Reaction pathway and free-energy barrier for reactivation of dimethylphosphoryl-inhibited human acetylcholinesterase. *The Journal of Physical Chemistry B*, *113*(50), 16226–16236.

Liu, C. T., et al. (2014). Probing the electrostatics of active site microenvironments along the catalytic cycle for *Escherichia coli* dihydrofolate reductase. *Journal of the American Chemical Society*, *136*(29), 10349–10360.

Lodola, A., et al. (2010). Structural fluctuations in enzyme-catalyzed reactions: Determinants of reactivity in fatty acid amide hydrolase from multivariate statistical analysis of quantum mechanics/molecular mechanics paths. *Journal of Chemical Theory and Computation*, *6*(9), 2948–2960.

Lonsdale, R., Harvey, J. N., & Mulholland, A. J. (2012). A practical guide to modelling enzyme-catalysed reactions. *Chemical Society Reviews*, *41*(8), 3025–3038.

Lonsdale, R., et al. (2012). Determinants of reactivity and selectivity in soluble epoxide hydrolase from quantum mechanics/molecular mechanics modeling. *Biochemistry*, *51*(8), 1774–1786.

MacKerell, A. D., Jr., et al. (1998). All-atom empirical potential for molecular modeling and dynamics studies of proteins. *The Journal of Physical Chemistry B*, *102*(18), 3586–3616.

Masgrau, L., et al. (2006). Atomic description of an enzyme reaction dominated by proton tunneling. *Science*, *312*(5771), 237–241.

McGeagh, J. D., Ranaghan, K. E., & Mulholland, A. J. (2011). Protein dynamics and enzyme catalysis: Insights from simulations. *Biochimica et Biophysica Acta (BBA) - Proteins and Proteomics*, *1814*(8), 1077–1092.

Meier, K., et al. (2013). Multi-resolution simulation of biomolecular systems: A review of methodological issues. .*Angewandte Chemie (International ed. in English)*, *52*(10), 2820–2834.

Moliner, V., Turner, A. J., & Williams, I. H. (1997). Transition-state structural refinement with GRACE and CHARMM: Realistic modelling of lactate dehydrogenase using a combined quantum/classical method. *Chemical Communications*, *14*, 1271–1272.

Morris, G. M., & Lim-Wilby, M. (2008). Molecular docking. *Methods in Molecular Biology*, *443*, 365–382.

Moyer, J. D., et al. (1997). Induction of apoptosis and cell cycle arrest by CP-358,774, an inhibitor of epidermal growth factor receptor tyrosine kinase. *Cancer Research*, *57*(21), 4838–4848.

Mujika, J. I., Lopez, X., & Mulholland, A. J. (2012). Mechanism of C-terminal intein cleavage in protein splicing from QM/MM molecular dynamics simulations. *Organic & Biomolecular Chemistry*, *10*(6), 1207–1218.

Mulholland, A. J. (2005). Modelling enzyme reaction mechanisms, specificity and catalysis. *Drug Discovery Today*, *10*(20), 1393–1402.

Olayioye, M. A., et al. (2000). The ErbB signaling network: Receptor heterodimerization in development and cancer. *The EMBO Journal*, *19*(13), 3159–3167.

Paasche, A., et al. (2010). Mechanistic study of the reaction of thiol-containing enzymes with α, β-unsaturated carbonyl substrates by computation and chemoassays. *ChemMedChem*, *5*(6), 869–880.

Philipp, D. M., & Friesner, R. A. (1999). Mixed ab initio QM/MM modeling using frozen orbitals and tests with alanine dipeptide and tetrapeptide. *Journal of Computational Chemistry, 20*(14), 1468–1494.
Phillips, D. C. (1966). *The three-dimensional structure of an enzyme molecule.* WH Freeman and Company.
Plotnikov, N. V., Kamerlin, S. C., & Warshel, A. (2011). Paradynamics: An effective and reliable model for ab initio QM/MM free-energy calculations and related tasks. *The Journal of Physical Chemistry B, 115*(24), 7950–7962.
Quesne, M. G., et al. (2014). Quantum mechanics/molecular mechanics study on the oxygen binding and substrate hydroxylation step in AlkB repair enzymes. *Chemistry—A European Journal, 20*(2), 435–446.
Ranaghan, K. E., et al. (2003). Insights into enzyme catalysis from QM/MM modelling: Transition state stabilization in chorismate mutase. *Molecular Physics, 101*(17), 2695–2714.
Ranaghan, K. E., et al. (2004). Transition state stabilization and substrate strain in enzyme catalysis: Ab initio QM/MM modelling of the chorismate mutase reaction. *Organic & Biomolecular Chemistry, 2*(7), 968–980.
Ruf, M., et al. (1996). Modelling transition state analogues and enzyme-inhibitor complexes of zinc-containing class II aldolases and metalloproteases. *Inorganica Chimica Acta, 250*(1), 271–281.
Sascha, K., et al. (2000). Purification and partial characterization of tryptophan 7-halogenase (PrnA) from *Pseudomonas fluorescens. Angewandte Chemie International Edition, 39*(13), 2300–2302.
Schramm, V. L. (1998). Enzymatic transition states and transition state analog design. *Annual Review of Biochemistry, 67*(1), 693–720.
Schramm, V. L. (2005). Enzymatic transition states and transition state analogues. *Current Opinion in Structural Biology, 15*(6), 604–613.
Schrödinger, L. (2005). *Jaguar, 6.0; Schrödinger.* New York: LLC. There is no corresponding record for this reference.
Schwartz, P. A., et al. (2014). Covalent EGFR inhibitor analysis reveals importance of reversible interactions to potency and mechanisms of drug resistance. *Proceedings of the National Academy of Sciences of the United States of America, 111*(1), 173–178.
Seabra, G. d. M., et al. (2007). Implementation of the SCC-DFTB method for hybrid QM/MM simulations within the Amber molecular dynamics package. *.The Journal of Physical Chemistry. A, 111*(26), 5655–5664.
Senn, H. M., & Thiel, W. (2007). QM/MM studies of enzymes. *Current Opinion in Chemical Biology, 11*(2), 182–187.
Senn, H. M., & Thiel, W. (2009). QM/MM methods for biomolecular systems. *Angewandte Chemie International Edition, 48*(7), 1198–1229.
Sherwood, P., et al. (2003). QUASI: A general purpose implementation of the QM/MM approach and its application to problems in catalysis. *Journal of Molecular Structure: THEOCHEM, 632*(1–3), 1–28.
Silva, P. J., Fernandes, P. A., & Ramos, M. J. (2003). A theoretical study of radical-only and combined radical/carbocationic mechanisms of arachidonic acid cyclooxygenation by prostaglandin H synthase. *Theoretical Chemistry Accounts, 110*(5), 345–351.
Singh, U. C., & Kollman, P. A. (1986). A combined ab initio quantum mechanical and molecular mechanical method for carrying out simulations on complex molecular systems: Applications to the CH3Cl+ Cl− exchange reaction and gas phase protonation of polyethers. *Journal of Computational Chemistry,* 7(6), 718–730.
Somei, M. (2006). A Frontier in indole chemistry: 1-Hydroxyindoles, 1-hydroxytryptamines, and 1-hydroxytryptophans. In S. Eguchi (Ed.), *Bioactive heterocycles I* (pp. 77–111). Berlin, Heidelberg: Springer Berlin Heidelberg.

Sousa, S. F., Fernandes, P. A., & Ramos, M. J. (2012). Computational enzymatic catalysis—Clarifying enzymatic mechanisms with the help of computers. *Physical Chemistry Chemical Physics*, *14*(36), 12431–12441.

Svensson, M., et al. (1996). ONIOM: A multilayered integrated MO+ MM method for geometry optimizations and single point energy predictions. A test for Diels-Alder reactions and Pt (P (t-Bu) 3) 2+ H2 oxidative addition. *The Journal of Physical Chemistry*, *100*(50), 19357–19363.

Taylor, R. D., Jewsbury, P. J., & Essex, J. W. (2002). A review of protein-small molecule docking methods. *Journal of Computer-Aided Molecular Design*, *16*(3), 151–166.

Torras, J., et al. (2015). PUPIL: A software integration system for multi-scale QM/MM-MD simulations and its application to biomolecular systems. *Advances in Protein Chemistry and Structural Biology*, *100*, 1–31.

Turner, A., & Williams, I. (1999). Transition-state structural refinement with GRACE and CHARMM: Flexible QM/MM modelling for lactate dehydrogenase. *Physical Chemistry Chemical Physics*, *1*(6), 1323–1331.

Valiev, M., et al. (2010). NWChem: A comprehensive and scalable open-source solution for large scale molecular simulations. *Computer Physics Communications*, *181*(9), 1477–1489.

van der Kamp, M. W., & Mulholland, A. J. (2013). Combined quantum mechanics/molecular mechanics (QM/MM) methods in computational enzymology. *Biochemistry*, *52*(16), 2708–2728.

Vane, J. R., Bakhle, Y. S., & Botting, R. M. (1998). Cyclooxygenases 1 and 2. *Annual Review of Pharmacology and Toxicology*, *38*, 97–120.

Vreven, T., et al. (2003). Geometry optimization with QM/MM, ONIOM, and other combined methods. I. Microiterations and constraints. *Journal of Computational Chemistry*, *24*(6), 760–769.

Vreven, T., et al. (2006). Combining quantum mechanics methods with molecular mechanics methods in ONIOM. *Journal of Chemical Theory and Computation*, *2*(3), 815–826.

Wagner, C., El Omari, M., & König, G. M. (2009). Biohalogenation: Nature's way to synthesize halogenated metabolites. *Journal of Natural Products*, *72*(3), 540–553.

Warshel, A. (1978). Energetics of enzyme catalysis. *Proceedings of the National Academy of Sciences of the United States of America*, *75*(11), 5250–5254.

Warshel, A., & Karplus, M. (1972). Calculation of ground and excited state potential surfaces of conjugated molecules. I. Formulation and parametrization. *Journal of the American Chemical Society*, *94*(16), 5612–5625.

Warshel, A., & Levitt, M. (1976). Theoretical studies of enzymic reactions: Dielectric, electrostatic and steric stabilization of the carbonium ion in the reaction of lysozyme. *Journal of Molecular Biology*, *103*(2), 227–249.

Warshel, A., et al. (2006). Electrostatic basis for enzyme catalysis. *Chemical Reviews*, *106*(8), 3210–3235.

Williams, I. H., & Maggiora, G. M. (1982). Use and abuse of the distinguished-coordinate method for transition-state structure searching. *Journal of Molecular Structure: THEOCHEM*, *89*(3–4), 365–378.

Wolfenden, R. (1972). Analog approaches to the structure of the transition state in enzyme reactions. *Accounts of Chemical Research*, *5*(1), 10–18.

Woodcock, H. L., Hodoščček, M., & Brooks, B. R. (2007). Exploring SCC-DFTB paths for mapping QM/MM reaction mechanisms. *The Journal of Physical Chemistry A*, *111*(26), 5720–5728.

Woodcock, H. L., et al. (2007). Interfacing Q-Chem and CHARMM to perform QM/MM reaction path calculations. *Journal of Computational Chemistry*, *28*(9), 1485–1502.

Xu, D., Guo, H., & Cui, Q. (2007). Antibiotic deactivation by a dizinc β-lactamase: Mechanistic insights from QM/MM and DFT studies. *Journal of the American Chemical Society*, *129*(35), 10814–10822.

Yeh, E., Garneau, S., & Walsh, C. T. (2005). Robust in vitro activity of RebF and RebH, a two-component reductase/halogenase, generating 7-chlorotryptophan during rebeccamycin biosynthesis. *Proceedings of the National Academy of Sciences of the United States of America*, *102*(11), 3960–3965.

Yeh, E., et al. (2007). Chlorination by a long-lived intermediate in the mechanism of flavin-dependent halogenases. *Biochemistry*, *46*(5), 1284–1292.

Young, I. S., & Baran, P. S. (2009). Protecting-group-free synthesis as an opportunity for invention. *Nature Chemistry*, *1*(3), 193.

Yun, C.-H., et al. (2008). The T790M mutation in EGFR kinase causes drug resistance by increasing the affinity for ATP. *Proceedings of the National Academy of Sciences*, *105*(6), 2070–2075.

Zhang, Y., Lee, T.-S., & Yang, W. (1999). A pseudobond approach to combining quantum mechanical and molecular mechanical methods. *The Journal of Chemical Physics*, *110*(1), 46–54.

Zhang, X., et al. (2018). Multiscale analysis of enantioselectivity in enzyme-catalysed 'lethal synthesis' using projector-based embedding. *Royal Society Open Science*, *5*(2).

CHAPTER TWO

Computational Methods for Efficient Sampling of Protein Landscapes and Disclosing Allosteric Regions

Canan Atilgan[1]
Faculty of Engineering and Natural Sciences, Sabancı University, Tuzla, Istanbul, Turkey
[1]Corresponding author: e-mail address: canan@sabanciuniv.edu

Contents

Abstract

Methods developed toward computational exploration of protein landscapes have become standardized tools to assess biophysical experimental findings. They are also used on their own right to discover the workings of the protein as a molecular machine, potential sites of interest for protein functioning, allosteric regions in proteins, and communication pathways between different sites on a protein. With the development of reliable force fields that describe interactions in biomolecules, molecular dynamics

Advances in Protein Chemistry and Structural Biology, Volume 113
ISSN 1876-1623
https://doi.org/10.1016/bs.apcsb.2018.06.001

(MD) simulations have become the prime tool for this purpose. While it is now straightforward to carry out MD simulations up to microseconds with current computers readily available to researchers, many processes of biological interest occur on several of orders of magnitudes slower timescales. Thus, the latter problems are attackable through MD by a handful of researchers that have access to the most powerful computers. Alternatively, physics-based methods to interrogate the protein energy landscape are in continuous development to circumvent this problem. In addition to opening the routes for advancement to a large number of researchers that have access to modest computational resources, they have the advantage of providing an understanding of the mechanisms that govern protein dynamics. Here we discuss network-based approaches geared toward understanding protein dynamics. These include (i) construction of residue networks which view proteins as networks of nodes connected through local interactions and (ii) construction of proteins as elastic networks whose modes of motion may be manipulated to achieve allowed conformational changes. Limitations of the methods as well as opportunities for future exploitation are described.

1. WHY IS THERE NEED FOR DEVELOPING EFFICIENT COMPUTATIONAL METHODS FOR PROTEINS?

Molecular biologists are interested in the detailed functioning of biomolecules under the ever-changing conditions in their natural environments, particularly in the presence of many other biomolecules. This desire is usually not fulfilled, given the limitations of the experimental methods that may be used to access biomolecular structure such as X-ray crystallography, nuclear magnetic resonance (NMR), or, more recently, cryo-electron microscopy. The functions, on the other hand, are scrutinized by a large array of in vivo and in vitro biophysical methods. It has become increasingly clear over time that while proteins function in a small subset of the whole conformational space available to them, a knowledge of the static snapshots of their structures does not suffice to decipher their workings or how their function may be externally manipulated to fulfill designed purposes.

As mechanisms of protein function are associated with conformational changes that are inherently related to structural details, computational tools have been pivotal in relating protein structures to their function. One type of problem that has been scrutinized has been to monitor the dynamics of a protein, while it is within a single accessible state (Karplus & Kuriyan, 2005). For example, the bacterial ferric binding protein A (FbpA) is a transport protein that shuttles iron in the periplasmic space, from the cytoplasmic side to the inner membrane side (Abdizadeh, Atilgan, Atilgan, & Dedeoglu, 2017).

The iron capture is followed by a simple hinge bending motion that can be easily understood via many computational analyses. What is more puzzling to understand is how the iron is released at the inner membrane side, since iron has extremely high binding affinity to this protein with binding constants on the order of 10^{19} M^{-1} (Taboy, Vaughan, Mietzner, Aisen, & Crumbliss, 2001). Thus, understanding how conformational fluctuations are manipulated and binding affinities are tuned while the protein still sits in a single state is one example that may be scrutinized via computational means. Similarly, even in enzymes forming well-known complexes with their substrates, it turns out that the static picture provided by X-ray structures supplies part of the big picture, and the entropy–enthalpy compensation that leads to the final binding energies is a result of the dynamical processes at play. An example of the enzyme dihydrofolate reductase (DHFR) is displayed in Fig. 1 whereby the protein makes six hydrogen bonds with the substrate in the X-ray structure, whereas dynamically, three of these are lost and reestablished in the course of an MD trajectory. Only when these dynamics are taken into account are the relative binding free energies of the ligand vs the inhibitor in competitive inhibition of DHFR truly accounted for (Abdizadeh, Atilgan, et al., 2017; Abdizadeh, Tamer, et al., 2017).

Other problems that are of interest and may be attacked via computational means involve biomolecules displaying different functions at distinct conformations. In such problems, while the multiple conformations may be known experimentally, the (free) energy barriers between them may be very high to circumvent under ambient conditions. It is therefore of interest to scrutinize the environmental niches that enable the conformational transitions. One such example is provided by calmodulin (CaM) which is known to perform over 200 functions in the cell environment, and occupies a plethora of conformers; e.g., an NMR ensemble of CaM has 160 plausible structures (Gsponer et al., 2008), some of which coinciding with the known X-ray structures as well as some others occupying different parts of the conformer space (Aykut, Atilgan, & Atilgan, 2013). Two of these are displayed in Fig. 2, and it has been possible to interconvert between these structures on timescales much shorter than expected under physiological conditions by a mere modification of the ionic strength or a point mutation at a key location. Other examples studied include the apo ↔ holo (Fe-bound) transitions in FbpA (Atilgan & Atilgan, 2009) and human transferrin (Abdizadeh, Atilgan, & Atilgan, 2015), conversion to the RNA release intermediate in enteroviruses (Ross, Atilgan, Bishop, & Atilgan, 2018), and conformational transitions in the allosteric heat shock proteins Hsp70

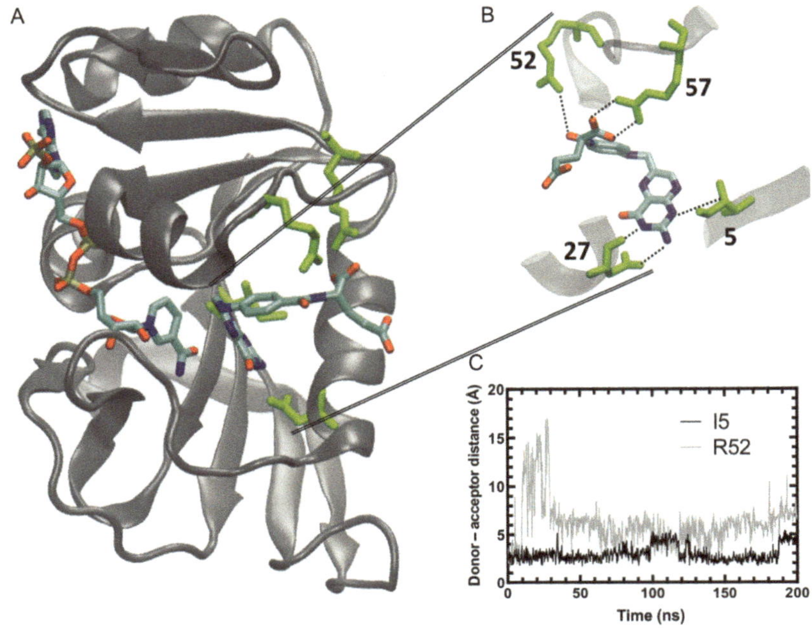

Fig. 1 Dynamics of DHFR affects hydrogen bonding patterns to the ligand. (A) The X-ray structure of *E. coli* DHFR (PDB code 1rx2); the ligand folate and the cofactor NADPH are displayed in *stick representation*. (B) The folate binding site is magnified. In the crystal structure, there are six fundamental hydrogen bonds between the enzyme and folate. The hydrogen bonding partners on the enzyme consist of the backbone oxygen of I5, the two side chain oxygens of D27, a sidechain nitrogen of R52, and the two-side chain nitrogens of R57; these are shown as *dashed lines*. (C) In a 200-ns long dynamical trajectory, following minimization and 10 ns equilibration (data not shown), the hydrogen bonds are indeed dynamical as exemplified for two selected cases (I5 and D52). *Details of MD simulations from Abdizadeh, H., Atilgan, A. R., Atilgan, C., & Dedeoglu, B. (2017). Computational approaches for deciphering the equilibrium and kinetic properties of iron transport proteins.* Metallomics, *9(11), 1513–1533; Abdizadeh, H., Tamer, Y. T., Acar, O., Toprak, E., Atilgan, A. R., & Atilgan, C. (2017). Increased substrate affinity in the Escherichia coli L28R dihydrofolate reductase mutant causes trimethoprim resistance.* Physical Chemistry Chemical Physics, 19*(18), 11416–11428.*

(Ozbaykal, Atilgan, & Atilgan, 2015; Penkler, Sensoy, Atilgan, & Bishop, 2017) and Hsp90 (Penkler, Atilgan, & Bishop, 2018).

In computational studies, molecular dynamics (MD) simulations that generate a trajectory of protein coordinates over time have been at the center stage. In practice, the conformational space that may be sampled is limited by days of computer time required by MD simulations of proteins, whereas the low-resolution models provide an immense amount of savings (Lopez-Blanco & Chacon, 2016). The main limitation of the crude

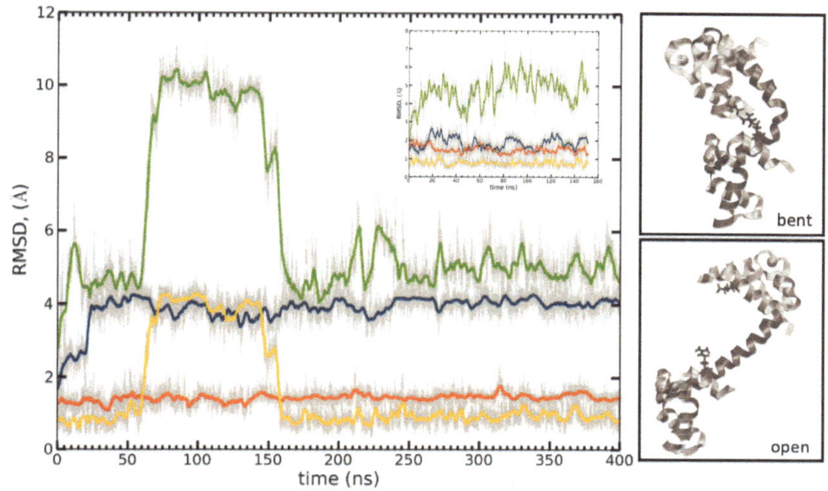

Fig. 2 CaM, while having flexible dynamics under physiological conditions, is stable around its initial structure (labeled open, *lower left*) in a usual MD trajectory (RMSD in a 160 ns trajectory is displayed in the *inset*; it reaches a value of ~5 Å). However, CaM is triggered to undergo a conformational change toward a bent structure by the E31A mutation (*upper left*). In the trajectories, moving average for the whole protein is displayed in *green*, the linker is in *yellow*, the N-lobe is in *blue*, and the C-lobe is in *red*; the detailed time points are shown by the *gray traces*. The structure stabilized by a salt bridge between E47 on the N-lobe and R86 on the linker whose side chains are displayed, remains bent for ~100 ns in the time interval 50–150 ns before switching back to the open structure. The initial PDB structure in the MD trajectories is coded 3cln.

modeling approaches has been viewed as their lack of ability to reproduce information whenever specificity is involved and methods to remedy such accuracy have been proposed over time (Na, Jernigan, & Song, 2015). In fact, even at the crudest level of coarse graining, studies have correlated function to structural fluctuations in proteins (Atilgan et al., 2001). This is because the packing density and coordination geometry of residues are the two dominating elements of structural arrangements which also accommodate global characteristics owing to long-range contacts (Atilgan, Akan, & Baysal, 2004; Baysal & Atilgan, 2001). This fact has been verified repeatedly on many proteins treated by low-resolution models that operate on simplistic harmonic potential energy functions accounting for packing density and geometry, while disregarding any information related to specificity. Using such approaches, in addition to reproducing the experimentally observed residue fluctuations from structural data, regions pivotal in orchestrating collective motions have been determined by associating them with low frequency modes obtained via modal decomposition (see Atilgan, Okan, & Atilgan, 2012 for a review).

It has been instructive to see that much of the protein dynamics around the equilibrium structure relies on packing density and order, by comparing results from crude models with those from detailed atomistic simulations. Thus, it is now possible to get information on very large systems, or to identify the distinguishing properties of proteins within a given family in reasonably short times. In this chapter, we will first discuss the possibilities opened up by MD simulations along with the limitations of the method in Section 2. We will then outline approaches whereby proteins are coarse grained at the level of the residue in Section 3, and we will discuss the functional information that may be deduced by these "residue network" approaches. In Section 4, we will discuss normal mode analyses of protein structures with emphasis on the coarse-grained elastic network models. Section 5 is devoted to the application of linear response theory (LRT) to protein structures having two end-point conformations, e.g., apo and holo forms; we then describe how perturbation-response scanning (PRS) based on LRT may be used to deduce sites that drive the conformational change between these end-points. Finally, in Section 6 we briefly discuss open problems and future directions in the study of protein structures along the guidelines provided by these approaches.

2. MARVELS AND LIMITATIONS OF MD SIMULATIONS

MD simulations have been instrumental in moving from the static view of proteins provided by structural data that has been mainly accessible through the protein data bank (PDB) for several decades (Berman et al., 2000), to a dynamical view where the motions of the protein have been observed and associated with function (Karplus & McCammon, 2002). These have at first been on the subnanosecond timescales in the early stages of MD methods development (Karplus & Petsko, 1990) but have expanded through to millisecond motions now accessible by the use of heavy duty computers configured solely for this purpose (Piana, Klepeis, & Shaw, 2014). Our confidence in the force fields used in the MD simulations has also grown over time (Lopes, Guvench, & MacKerell, 2015). Currently, it is possible to quantitatively predict ligand binding energies (Abel, Wang, Harder, Berne, & Friesner, 2017; Aldeghi, Heifetz, Bodkin, Knappcd, & Biggin, 2016) or protein folding kinetics using additive force fields (Lindorff-Larsen, Piana, Dror, & Shaw, 2011).

2.1 Brief Overview of MD Simulations

Briefly, MD simulations solve the Newton's equation of motion for all the atoms included in the simulation box that is treated by periodic boundary conditions, propagated by time steps on the order of femtoseconds. Thus, for a moderate sized protein that is simulated in an environment minimally mimicking the cell conditions, one not only includes the protein atoms, but also a layer of water that is typically padding the protein in all directions by at least 10 Å, harboring ions that represent ionic strength conditions in the cell, typically up to a couple of hundred ions for 150 mm. Thus, a typical MD simulation box includes $N \sim 10^5$–10^6 atoms whose coordinates are recalculated in time steps of $\sim 10^{-15}$ s by solving the equation $\mathbf{F} = m\mathbf{a}$ for each particle at every time step. Here, the force is calculated for all interacting atom pairs through the negative gradient of the potential energy that is described by the force field. Thus, the most time-consuming step of the simulation is to recalculate these interactions that are on the order of N^2. To circumvent some of these problems, various tricks have been developed whose main features are outlined in the classical book (Allen & Tildesley, 1989) and are implemented in standard MD simulation packages.

In sum, to reach the nanosecond timescale which begins to be relevant for protein function, these calculations need to be repeated a million times. MD simulations provide the coordinates and velocities of all particles in the system, completely defining its phase space. Thus, by using proper statistical mechanical relationships, it is in principle possible to calculate any property that is also observable by experiments, e.g., various relaxation times from NMR or FRET, dielectric properties, all measurables in different X-ray scattering experiments, to name a few. There also lies the main limitation of MD, since the computational cost is daunting if one needs to study observables on microsecond or longer timescales. The problem is worse, if one needs to do a comparative study, e.g., to study the effect of varying environmental conditions, or the dynamics in the presence of a variety of ligands.

2.2 Tour-de-force Calculations Employing MD

MD simulations are used to monitor protein stability and function by providing dynamical information on the conformations attainable under prevailing conditions, transitions between known conformations, interactions within the protein and with its environment. The method is also useful in aiding protein design efforts (Childers & Daggett, 2017).

An important goal has been to simulate large biomolecular assemblies to gain insight into the workings of molecular machines. As early as 2005, the entire ribosome, a system of 2.64 million atoms, was simulated to study the critical decoding step during protein synthesis, by utilizing 768 processors at the time (Sanbonmatsu, Joseph, & Tung, 2005). These 1.6 ns long calculations involved targeted MD (Schlitter, Engels, & Kruger, 1994) to enhance the sampling and to understand the movement of tRNA into the ribosome, as the relevant timescale for these motions is on the order of seconds. More recently, full-scale simulations of large virus capsids, with ~64 million atoms were conducted on a timescale of 100 ns (Zhao et al., 2013). These were run on a system of 128,000 cores with a speed of 5–9 ns/day of production, depending on usage of the cluster. A most impressive calculation involving all-atom MD simulations includes three atomistic cytoplasmic models explicitly containing proteins, nucleic acids, metabolites, ions, and water modeling roughly 1/10 of a total cell (Yu et al., 2016). The largest simulation therein had 103 million atoms, carried out for 140 ns, utilizing the K-computer at RIKEN. The simulations made possible the study of the effects of crowding on the dynamics of metabolites.

Instead of targeting sub-microsecond dynamics of extremely large systems, large clusters of computers have also been utilized to study the long-term dynamics of small to moderate sized proteins. An important outcome of such simulations has been to demonstrate the reliability of currently used force fields (Piana et al., 2014), as it had been debated that force fields developed for folded proteins may not be good enough to describe other states. The folding simulations came of age with the development of the ANTON computer specifically configured to efficiently run these simulations which led to the emergence of the first 1-ms long single trajectory of a protein (Shaw et al., 2010), so that sub-millisecond folding times could now be predicted using conventional MD. On the other hand, aggregate trajectories of total length surpassing milliseconds have been used to describe folding events (Voelz, Bowman, Beauchamp, & Pande, 2010) as well as the binding modes of drugs to their target binding site through observation of multiple binding events (Shan et al., 2011).

2.3 Limitations of MD

The simulations presented in the previous subsection may only be carried out by a handful of researchers who have access to the large computational power required to complete them. Even in those cases, studying multiple conditions for the same protein, such as the effect of changes in pH or ionic strength of the environment on the dynamics, or the binding of multiple drugs, poses the problem of requiring prohibitive resources. Moreover, analyzing the generated trajectories to distill scientifically meaningful information from them presents additional challenges (Lane, Shukla, Beauchamp, & Pande, 2013).

Thus, alternative methodologies to efficiently sample the conformational space available to the protein have been developed. Among these, Markov state models have provided a popular approach to stitching together a set of MD simulations to get information on the kinetics of conformational transitions and characterization of stationary processes (Chodera & Noe, 2014). Alternatively, in the reaction coordinate scheme, a single coordinate along which the conformational change occurs is selected and transition states as one progresses along that reaction coordinate are searched (Rohrdanz, Zheng, & Clementi, 2013).

Another direction has been taken toward developing force fields suitable for coarse-grained simulations (see, e.g., Das & Baker, 2008; de Jong et al., 2013). In these approaches, several atoms are grouped together to form a bead; the bead–bead interactions are then obtained, e.g., from potential of mean force calculations (Reith, Putz, & Muller-Plathe, 2003), and the system is propagated by employing a classical equation of motion or by Monte Carlo schemes. The more atoms are lumped into a bead, the larger a time step can be selected in a coarse-grained MD simulation. Along with the reduction in computer time gained through the much smaller number of interactions that require calculation, the conformational space may be sampled in a reasonable amount of computer time. For the particular problem of obtaining detailed interactions in the distinct conformations of the protein, reverse mapping schemes have to be used to map the atomistic detail back on the obtained conformation (Heath, Kavraki, & Clementi, 2007). This operation must then be followed by all-atom MD simulations started from various conformations. The field of coarse-grained MD simulations is beyond the scope of this chapter. The reader is referred to an excellent review that outlines the state-of-the-art, capabilities, limitations, and future development areas of these approaches (Kmiecik et al., 2016)

Last, but not least, as a standard, studies utilizing unbiased MD simulations are now expected to span microsecond timescales with multiple samplings to affirm interesting observations made based on the trajectories. Thus, while unbiased MD will continue to be the benchmark for studying protein dynamics, other approaches that efficiently deduce dynamical information based on structures localized in a given region of the conformational space have inevitably become popular. These physics-based approaches have not only enhanced our understanding of the biophysics of the phenomena, they have also allowed researchers with modest computational resources to make predictions that aid experimental work. The rest of the chapter is dedicated to such methods that have offered new ways of sampling the conformational space of proteins and have been pivotal in predicting sites that modulate their dynamics.

3. VIEWING PROTEINS AS NETWORKS OF INTERACTING RESIDUES

It is relatively straightforward to develop the idea that the three-dimensional structure of proteins depends on the primary sequence, particularly once the relationship between the structure of DNA and its nucleic acid sequence had been disclosed. However, how to handle the sequence in interpreting the structure–function relationships requires a much deeper understanding of the way proteins work. Earlier studies concentrated on distinguishing protein structural classes (α, β, α/β, or $\alpha+\beta$) through an analysis of their amino acid distributions. In particular, the classes were shown to display significantly different distributions in their nonbonded contacts (Bahar, Atilgan, Jernigan, & Erman, 1997). Moreover, efforts to predict protein folding through the use of contact maps were also quite popular (see, for example, Najmanovich & Domany, 1999). Methods that would enable direct prediction of contact maps have also shown a degree of success (Fariselli & Casadio, 1999).

In this chapter, our interest is not in predicting the contact map, though. Since the three-dimensional structure of the protein may be summarized into the contact map, it also holds the key to function. Thus, in residue network (RN) approaches, the starting point is in fact the contact map of a known protein structure, which is then subjected to mathematical operations to interrogate, e.g., key residues that affect binding, or regions that modulate large motions.

3.1 Construction of RNs

After the observation of real-life networks, such as the world-wide web (Huberman & Adamic, 1999), internet (Vazquez, Pastor-Satorras, & Vespignani, 2002), power-grids (Watts & Strogatz, 1998), and math coauthorship (Barabasi et al., 2002), having different degree distributions and different statistical measures than classical Erdös–Rényi random networks (Erdös & Rényi, 1961), the interest in complex networks heightened. With the proposition of new algorithms to produce networks that generate real-life-like networks (Barabasi & Albert, 1999; Watts & Strogatz, 1998), this area has been exploited extensively.

Construction of a network from a real system requires (i) selection of nodes and (ii) defining criteria by which the links directly connecting pairs of nodes are identified. For a system such as a power grid, these definitions are self-evident due to the presence of physical structures that are the nodes (e.g., power stations) and physical direct connections between the nodes (e.g., power lines). In a protein, while the atoms may seem to be the evident choice for nodes, there have been different approaches to node selection, such as representing each residue by a single node centered on its C_α or C_β atom, or on the geometric center of all its atoms. Other approaches include all heavy atoms of the protein as separate nodes in the network. The definition of a link gets even more intriguing, as interactions between atoms occur through distance-dependent continuous potentials, inevitably requiring discretization in the case of network construction. Thus, nodes within a selected cutoff distance of each other are assumed to be in contact. For nodes selected as heavy atoms, a distance that corresponds to the sum of the van der Waals radii might provide a suitable cutoff. For protein structures coarse-grained at the level of the residue, radial distribution functions (RDFs) between nodes (e.g., C_α or C_β) provide a crude approximation to the cutoff distance to be used. The RDF obtained by averaging over 595 nonhomologous proteins and taking the C_β atom locations of the residues (C_α for glycines) is displayed in Fig. 3A. The cutoff distance is selected as the range at which the first coordination shell ends, i.e., 6.7 Å in this case (Atilgan et al., 2004). A sample RN is displayed in Fig. 3B for the enzyme DHFR.

The topological properties of RN have been studied in detail and compared with those of other networks (Estrada, 2010; Turgut, Atilgan, & Atilgan, 2010). Mathematically, the RN is defined by the so-called adjacency matrix, **A**, whose elements are given by

$$A_{ij} = \begin{cases} H(r_c - r_{ij}) & i \neq j \\ 0 & i = j \end{cases} \qquad (1)$$

Here, r_{ij} is the distance between the ith and jth nodes, $H(x)$ is the Heaviside step function given by $H(x) = 1$ for $x > 0$ and $H(x) = 0$ for $x \leq 0$.

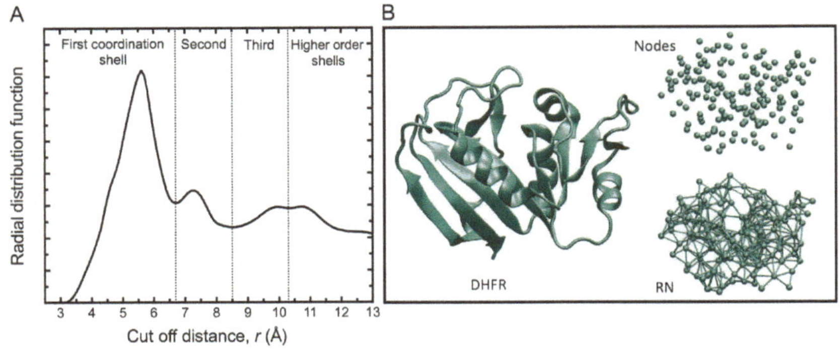

Fig. 3 (A) RDF between the C_β atoms of 595 nonhomologous proteins. The positions of the first, second, and third coordination shells are marked. Higher order shells cannot be distinguished. (B) The protein DHFR (*left*), its nodes centered on the C_β atoms obtained from the X-ray structure (PDB code 1rx2) are displayed on the *upper right*, while the full RN constructed by assigning links if two nodes are within 6.7 Å of each other is displayed on the *lower right*.

3.2 Network Parameters Useful in RN

The networks are quantified by local and global parameters, all of which can be derived from the adjacency matrix (Eq. 1). The degree k_i of residue i is the number of neighbors of that residue:

$$k_i = \frac{1}{N}\sum_{j=1}^{N} A_{ij} \tag{2}$$

The characteristic path length, L, of a network is a global measure of the average reachability of a node in the network. It is the average over the minimum number of connections that must be crossed to connect residue pair i and j. In computing the shortest path between a pair of nodes, we make use of the fact that the number of different paths connecting a pair of nodes i and j in n steps is given by $B_{ij} = (\mathbf{A}^n)_{ij}$. Thus, the shortest path between nodes i and j, L_{ij}, is given by the minimum power, m, of $\mathbf{A}$ for which $(\mathbf{A}^m)_{ij}$ is nonzero. In practice, L_{ij} is calculated via greedy algorithms. The characteristic path length of the network is the average,

$$L = \frac{2}{N(N-1)}\sum_{i=1}^{N-1}\sum_{j=i+1}^{N} L_{ij} \tag{3}$$

The clustering coefficient, C, on the other hand, is a measure of the local order reflecting the probability that the neighbors of a node are also neighbors of each other. For residue i this probability may be computed by

$$C_i = \frac{\frac{1}{2}\sum_{j=1}^{N}\sum_{k=1}^{N} A_{ij}A_{ik}A_{kj}}{{}_{ki}C_2} \tag{4}$$

Here ${}_{k_i}C_2$ is the combinatorial coefficient, and k_i is defined in Eq. 2. The clustering coefficient of the network is the average $C = \langle C_i \rangle$.

The final network parameter that has proven very useful for distinguishing residues with specific roles in RN is the betweenness centrality (BC) which indicates the ability of the node to control information transfer in a network. Under the assumption that information exchange occurs via shortest paths on a graph, the BC for an arbitrary node is calculated by,

$$\mathrm{BC}(\mathrm{Node}) = \sum \left(\frac{P(a,b|\,\mathrm{Node})}{P(a,b)} \right) \frac{2}{(n-1)(n-2)} \tag{5}$$

where $P(a,b\,|\,\mathrm{Node})$ is the number of the shortest paths which go through this node, $P(a,b)$ is the number of all shortest paths between nodes a and b, and n is number of the nodes in network. BC takes on values in the range [0,1], with a value of 1 if all the shortest paths between a and b pass through a node.

3.3 Applications of RN for Function Assignment to Residues

Several studies have shown that RNs have in fact small-world topology (Atilgan et al., 2004; Bagler & Sinha, 2005; Greene & Higman, 2003; Vendruscolo, Dokholyan, Paci, & Karplus, 2002). These networks are characterized by logarithmically scaling average path lengths and considerably high clustering coefficients. Further studies also utilized RN to predict hot spots, conserved sites, domain motions, functional residues, and protein–protein interactions (see, e.g., Amitai et al., 2004; Bahar, Atilgan, Demirel, & Erman, 1998; Bahar, Atilgan, & Erman, 1997; Bahar, Erman, Jernigan, Atilgan, & Covell, 1999; Baysal & Atilgan, 2001; Brinda & Vishveshwara, 2005; del Sol, Fujihashi, Amoros, & Nussinov, 2006; Demirel, Atilgan, Jernigan, Erman, & Bahar, 1998). In particular BC has emerged as a key parameter that distinguishes residues controlling communication within and between domains, with high BC nodes residing on the largest number of paths connecting pairs of residues in the protein

(Ozbaykal et al., 2015; Penkler et al., 2018). Thus, communication between functionally linked regions of a network of residues is controlled by those few frequently found along optimized paths, at least in some proteins.

While RN seems to be a static method in that it operates on a single conformation of a protein, in recent years the changes in RN properties have been used to assign function to regions in the three-dimensional structure. Some of these approaches monitor the change in L and BC as time progresses in a typical MD trajectory (Penkler et al., 2018; Stetz & Verkhivker, 2017). This analysis has recently been implemented in a software suite called MD-TASK (Brown, Penkler, et al., 2017). Other approaches are based on implementing systematic mutations in the protein structure, such as alanine scanning or SNP analyses, and monitoring how the local effects are sensed by the RN properties (Brown, Amamuddy, & Bishop, 2017; Ozbaykal et al., 2015). These studies offer RN as a promising efficient tool to pinpoint regions of interest in proteins that are not self-evident by simply monitoring the usual suspects such as the binding pockets.

Furthermore, it is possible to introduce inhomogeneity to the edges as weights via known residue-pair potentials such as those derived in Thomas and Dill (1996). Instead of treating all links as equivalent, a strong path, for example, selects the shortest path that minimizes the maximum weights over all possible paths between any selected pair of nodes (Atilgan, Turgut, & Atilgan, 2007). Using this approach, conserved residues regulating allostery in rhodopsin were distinguished (Atilgan & Atilgan, 2012; Atilgan et al., 2007), in agreement with experimental observations (Lockless & Ranganathan, 1999). However, the weighted RN approach has never been applied to dynamical changes in a protein or to the effects of point mutations in select regions discussed previously. Thus, it constitutes an interesting future direction to pursue in the applications of RN.

4. ELASTIC NETWORK MODELS OF PROTEINS

While the RN approach that utilizes network theory to develop a better understanding of folded protein structures has found widespread use, another independent network approach has been to utilize statistical mechanics by assigning harmonic potentials to the links. The groundbreaking work by Tirion showed that a single-parameter potential is sufficient to reproduce the large amplitude motions of proteins reduced to one-dimensional networks (Tirion, 1996) which was further scrutinized, with analogy to polymeric networks, as the Gaussian network model (Bahar, Atilgan, & Erman, 1997). This view was later adopted to three dimensions and has been very popular in studying protein motions under the name of anisotropic network model (ANM) (Atilgan et al., 2001). In this section, we focus on a description of ANM, which is based on the more general normal mode analysis (NMA).

4.1 Practical Implementations of ANM and Related Approaches

Here we will provide a general description of ANM and some practical aspects of the implementations of the method. The reader is referred to Bahar, Lezon, Yang, and Eyal (2010) for a detailed review. NMA which analyses the oscillations of a structure is useful for studying the large amplitude motions of proteins that samples a selected conformation. Thus, the motions are assumed to be harmonic and each normal mode, which is a concerted motion of many atoms, acts as a simple harmonic oscillator, independent of all the other normal modes. For a harmonic oscillator having a mass m supported on a spring with force constant k, the potential energy of the system is simply given by $V = kx^2/2$. Thus, an extension from the equilibrium position by an amount x leads to the restoring Hooke's Law force $\mathbf{F}$,

$$\mathbf{F} = -\frac{dV}{dx} = -kx \tag{6}$$

By substitution to Newton's Law, $\mathbf{F} = m\mathbf{a}$, one gets the differential equation and the solution, with ν being the frequency of the vibration,

$$-kx = m\frac{d^2x}{dt^2} = -4\pi^2\nu^2 mx \tag{7}$$

In three dimensions and for a set of N atoms, one has the corresponding generalized solution,

$$-\mathbf{HX} = -4\pi^2\nu^2\mathbf{X} \tag{8}$$

where $\mathbf{H}$ is the $3N \times 3N$ Hessian symmetric matrix of force constants, and $\mathbf{X}$ is the $3N \times 1$ vector of the positions of the atoms. The solutions of Eq. 8 may be cast in the form of an eigenvalue decomposition of $\mathbf{H}$ where the eigenvectors are the independent vibrational motions of the collection of atoms while the eigenvalues are the negative of the squared normal mode frequencies of each mode. The six zero eigenvalues of $\mathbf{H}$ represent the translations and rotations of the molecule in three-dimensional space.

Inasmuch as the motions around the folded protein structure are assumed to occur near the local potential energy minimum, **H** may be obtained by a simple harmonic potential such that residues are connected to their nearest neighbors via elastic springs. By further employing the assumption that the spring constants γ are equivalent in all pairs of interactions, one arrives at ANM (Atilgan et al., 2001).

In ANM, as in RN, the structure of the protein is coarse-grained at the level of a residue, represented either by the C_α or C_β coordinates of the protein atoms deposited in the PDB. Moreover, pairs of nodes are assumed to interact if they are within a preselected cutoff distance. **H** is thus arranged into the $N \times N$ matrix of superelements as,

$$\mathbf{H} = \begin{bmatrix} \mathbf{H_{11}} & \mathbf{H_{12}} & \cdots & \mathbf{H_{1N}} \\ \mathbf{H_{21}} & & & \mathbf{H_{2N}} \\ \vdots & & & \vdots \\ \mathbf{H_{N1}} & & & \mathbf{H_{NN}} \end{bmatrix} \tag{9}$$

where the orientation of nodes i,j is stored in each superelement $\mathbf{H_{ij}}$ as a 3×3 matrix,

$$\mathbf{H_{ij}} = \begin{bmatrix} \partial^2 V/\partial X_i \partial X_j & \partial^2 V/\partial X_i \partial Y_j & \partial^2 V/\partial X_i \partial Z_j \\ \partial^2 V/\partial Y_i \partial X_j & \partial^2 V/\partial Y_i \partial Y_j & \partial^2 V/\partial Y_i \partial Z_j \\ \partial^2 V/\partial Z_i \partial X_j & \partial^2 V/\partial Z_i \partial Y_j & \partial^2 V/\partial Z_i \partial Z_j \end{bmatrix} \tag{10}$$

With the distance between two nodes i and j denoted by S_{ij}, and Q, $P = X$, Y, or Z, the off-diagonal superelements $\mathbf{H_{ij}}$ are obtained from the coordinates as,

$$\partial^2 V/\partial Q_i \partial P_j = -\gamma \left(Q_j - Q_i\right)\left(P_j - P_i\right)/S_{ij}^{2.}$$

Similarly, the diagonal superelements $\mathbf{H_{ii}}$ are obtained through the summations,

$$\partial^2 V/\partial Q_i^2 = \gamma \sum_j \left(Q_j - Q_i\right)^2/S_{ij}^2$$

for the unmixed terms, and

$$\partial^2 V/\partial Q_i \partial P_j = \gamma \sum_j \left(Q_j - Q_i\right)\left(P_j - P_i\right)/S_{ij}^2$$

for the mixed terms.

We note that for large protein complexes, further coarse graining beyond the level of the residue might still describe the most collective modes of motion. However, care must be taken to ensure the coarse graining employs physics-based rules and exploits symmetries in the structure (Ross, Atilgan, et al., 2018).

The cutoff distance selection for ANM applications has been debated in the literature. Unlike the RN, connecting the residues that are within the first coordination shell of each other (Fig. 3A) does not suffice to get the correct modes of motion. In fact, distances in the 10–25 Å range have been employed in the literature, usually with arbitrary selection criteria. It seems that, so long as the six zero eigenvalues of the Hessian are satisfied, the collective modes of motion are relatively insensitive to the cutoff used. The origins of this insensitivity have been studied systematically where it was shown that selection of too large a value leads to results that are more consistent with the experimental B-factors (Atilgan, Okan, & Atilgan, 2010). Therein, it was argued that **H** may be expressed as the sum of an essential part that includes all the local information necessary for the correct representation of the modes, and a trivial part that does not contribute to the slow modes. In the latter, the effects from the long-range neighbors cancel, due to the symmetries in a protein originating from the orientational order of closely packed spheres (Atilgan et al., 2012). Thus, large cutoffs ensure all of the essential part is included in **H**. We note that Hessian construction free of cutoff distances, employing inverse power-dependent distances, has also been developed (Yang, Song, & Jernigan, 2009).

Alternatively, instead of using a network construction based on a single PDB structure, the inverse of the Hessian may be approximated by the variance–covariance matrix of the protein. This is justified via statistical mechanical treatments employing the harmonic assumption showing that the cross-correlations between residue displacements are expressed as,

$$\langle \Delta \mathbf{R}_i \cdot \Delta \mathbf{R}_j \rangle = 3k_B T \cdot \mathbf{H}^{-1} \qquad (11)$$

The left-hand side is the variance–covariance matrix and should be obtained from an MD simulation of suitable length. This length should ensure that the molecule completely samples only the potential energy minimum of interest, without spilling over into a neighboring well (Atilgan et al., 2012). For equilibrated MD trajectory lengths of 20–40 ns, we find this criterion that stems from the harmonic assumption is satisfied. The different implementations of the construction of the inverse Hessian are implemented in the freely available software suite MODE-TASK (Ross, Nizami, et al., 2018).

4.2 Extracting Function Related Information From Elastic Network Models

The original interest in elastic network models stemmed from the fact that they were able to predict the experimentally measured residue fluctuations, reported as the B-factors in X-ray structures, with high accuracy, noting that the self-terms of Eq. (11), $\langle \Delta \mathbf{R}_i \cdot \Delta \mathbf{R}_i \rangle = \langle \Delta R_i^2 \rangle$ is proportional to the B-factors. Once it was realized that such a simple model that connects neighboring nodes with springs in constructing the Hessian reproduces a fundamental experimental measurable, the predictions of cross-correlations between different regions of the proteins were investigated with high confidence. Furthermore, modal decomposition of the Hessian and selecting the contributions of different eigenvectors to the overall motions opened the way to discuss the fundamental functions of proteins by association to the most collective modes of motion. In Fig. 4, the first and third most collective mode CaM is displayed as examples. The former represents the bending of the two lobes toward each other, while the latter is a rocking motion of the two lobes around the linker region.

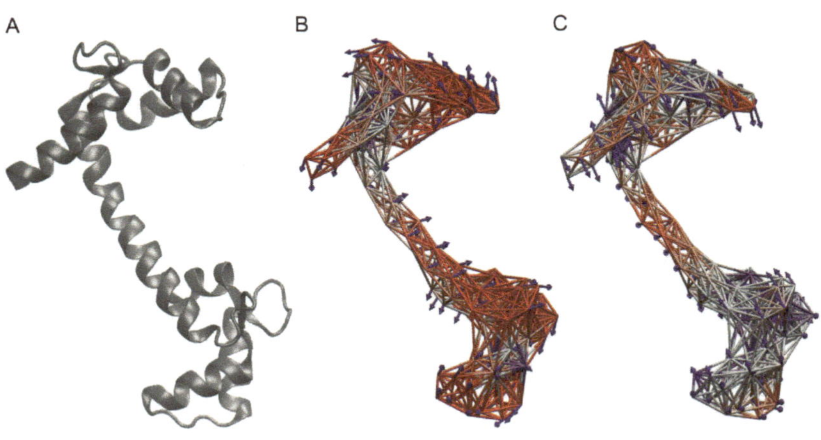

Fig. 4 (A) Structure of CaM (PDB code 3cln), also discussed in Fig. 2. (B and C) The motions of the first and third slowest modes of CaM, predicted by constructing an elastic network model by coarse graining the protein at the level of C_β atoms and using a cutoff distance of 10 Å. The more mobile regions of the network are displayed in *red*, the relative size and direction of motion of each node is shown by the *blue arrows*. In the slowest, most collective mode (B), the motion may be best described as the bending of the two lobes toward each other, while in the third most collective mode (C), a rocking motion of the two lobes along the axis of the linker is displayed. Movies of the motions may be generated on the web server provided at http://midst.sabanciuniv.edu/anm/.

In other applications of the method, MD simulations were biased along the collective modes to encourage efficient sampling of the phase space (Zhang, Shi, & Liu, 2003), and approaches for the efficient generation of pathways along the eigenvectors defining collective motions to accomplish conformational transitions were described (Kim, Jernigan, & Chirikjian, 2002). Collective modes were also used for refining electron density maps (Hinsen, Reuter, Navaza, Stokes, & Lacapere, 2005) and low resolution cryo-EM maps (Chacon, Tama, & Wriggers, 2003). The elastic network models are now well-exploited tools in structural studies of proteins; we therefore do not go into further details on current applications. The reader is referred to two recent reviews discussing the state-of-the-art in relating function to collective modes (Mahajan & Sanejouand, 2015), and protein–protein interactions and evolvability (Haliloglu & Bahar, 2015).

5. BEYOND ELASTIC NETWORKS: PRS

Elastic network models outlined in Section 4.2 are widely utilized. However, in many cases, proteins act by making use of several modes at once. In fact, it was elegantly shown that there is a spectrum of motions whereby some conformational changes require the operation of a few collective modes while others may only be described by including hundreds of them (Petrone & Pande, 2006). Thus, alternative methodologies that invoke the relevant modes of motion had to be developed.

By utilizing the techniques developed for adaptive structures, an analytical approach was developed to determine regions controlling conformational changes, particularly loop motions (Yilmaz & Atilgan, 2000). The motivation is based on realizing that the flexible regions in the functional protein which operate during binding must be controlled by rigid parts modulating the conformational change. Therein, it was demonstrated that, by inserting fluctuations in tertiary contacts supported by the flexible core, one can regulate the displacement of residues in loop regions, while the positional fluctuations of flexible parts are annihilated by the rest of the protein. Thus, the perturbation and the associated response are not reciprocal. By using an all atom force field, it is also possible to use molecular mechanics by inserting selected sizes of displacements to selected regions and recording the associated response (Baysal & Atilgan, 2001). Later, it was put forth that a linear relationship between the equilibrium fluctuations and the structural change upon ligand binding might be operative, and the proof of concept was demonstrated on three proteins of different sizes whereby the perturbations were inserted in regions selected heuristically (Ikeguchi, Ueno, Sato, & Kidera, 2005). The model is based on LRT and opens ways to explore the regions operative on the conformational change without assuming motions along essential collective modes. We have later expanded the idea to eliminate the heuristic step, by scanning the protein residue-by-residue and selecting the perturbations that display the best overlaps with the response (Atilgan & Atilgan, 2009). Since the procedure is repeated sequentially for all the residues in the protein, we name the methodology perturbation-response scanning. In this section, we briefly outline the details of the PRS technique followed by discussing application areas.

5.1 PRS Methodology

We first construct the protein as a network of N nodes that are centered on the C_α or C_β atoms. Any given pair of nodes are assumed to interact via a harmonic potential if they are within a cutoff distance of each other as described in Section 4.1. We assume that the network is elastic, described by Hooke's Law. Thus, we use the analogy applied in passing from Eqs. (7) and (8) for the force constant and the $3N \times 3N$ Hessian matrix, $\mathbf{H}$. As a result, Eq. (6) may be modified as,

$$\mathbf{F} = -\mathbf{H}\,\Delta\mathbf{R} \tag{12}$$

Here $\Delta\mathbf{R}$ represents the $3N \times 1$ matrix holding the displacements of residues from their equilibrium positions and the $3N \times 1$ matrix $\mathbf{F}$ is the associated force generation. One may pose the inverse problem and seek the displacements generated by forces inserted at a given location along the protein:

$$\Delta\mathbf{R} = -\mathbf{H}^{-1}\,\mathbf{F} \tag{13}$$

The $\mathbf{F}$ vector now contains the components of the externally applied fictitious force vectors on the selected residues. Since $\mathbf{H}^{-1}$ is the inverse Hessian also used in the ANM method described in Section 4.1, one may generate it by either constructing the elastic network (Eq. 10) or obtain it from the MD trajectory of suitable length by evaluating the variance–covariance matrix (Eq. 11).

The PRS method is based on the systematic application of Eq. (13). We insert a force on the C_α or C_β atom of each residue by forming the $\mathbf{F}$ vector in such a way that all the entries, except those corresponding to the residue being perturbed, are equal to zero. For a selected residue i, the force $\mathbf{F}^i$ is (F_x^i, F_y^i, F_z^i) so that the external force vector is constructed as,

$$\mathbf{F}^{\mathrm{T}} = \left\{0\,0\,0 \ldots F_x^i\, F_y^i\, F_z^i \ldots 0\,0\,0\right\} \tag{14}$$

Since one does know a priori the residues or the directions that will best mimic the targeted conformational change $\Delta\mathbf{R}^*$, we scan the protein residue-by-residue and insert random forces in random directions (typically several hundred) on each residue. Note that since we operate on the same Hessian that is only constructed once, based on the initial conformation of the molecule, the scanning procedure is computationally very fast, usually taking a few minutes of computer time on the largest proteins.

We thus obtain a large set of $\Delta\mathbf{R}$ vectors of the protein through Eq. (13), each of which is then compared to the targeted conformational change $\Delta\mathbf{R}^*$ to select the residues and force directions that best represent the fictitious forces expected to be introduced in the actual conformation change scenario. To assess the quality of the predicted displacements of all residues resulting from a force applied on selected residue i, we use the overlap of the predicted and experimental displacements:

$$O^i = \frac{(\Delta\mathbf{R})^i \cdot \Delta\mathbf{R}^*}{\left|(\Delta\mathbf{R} \cdot \Delta\mathbf{R}^{\mathrm{T}})^i (\Delta\mathbf{R}^* \cdot \Delta\mathbf{R}^{*\mathrm{T}})\right|^{1/2}} \tag{15}$$

The residues and force directions whose overlaps are above a predetermined threshold are then subjected to further analysis to assess their possible role in the functioning of the protein. The PRS method is implemented in the ProDy computational toolbox available as a plugin (Bakan et al., 2014) and as a standalone Python code in the MD-TASK software suite (Brown, Penkler, et al., 2017).

5.2 Selection of Allosteric Sites and Functional Regions via PRS

The PRS method has initially been compared with ANM to assess the additional information that may be obtained by the use of LRT (Atilgan, Gerek, Ozkan, & Atilgan, 2010). It was shown that for simple motions that operate through a single dominant mode, such as hinge bending, both approaches lead to similarly large overlaps. However, PRS provides the additional information on perturbation of which single residues are effective in propagating the motion. Furthermore, for proteins with complicated motions such as shearing, PRS outperforms ANM eigenvectors in associating motions with the produced candidate directions.

PRS has been notorious in predicting allosteric sites. In the study of PDZ domains, it was shown that residues identified via PRS coincide with those having experimentally determined allosteric roles (Gerek & Ozkan, 2011). It was also used along with crystallographic studies of glutamate receptors and the predicted dynamical motions were then confirmed via cross-linking experiments (Dutta et al., 2015). In cases where experimental data lack, PRS determined residues with functional implications were subjected to fully atomistic MD simulations to prove their role in manipulating conformational change. This is how residue E31 in CaM that was discussed in Fig. 2 has originally been targeted as a possible site of functional significance (Atilgan, Aykut, & Atilgan, 2011; Aykut et al., 2013; Negi, Aykut, Atilgan, & Atilgan, 2012). In Fig. 5, we display the two sets of

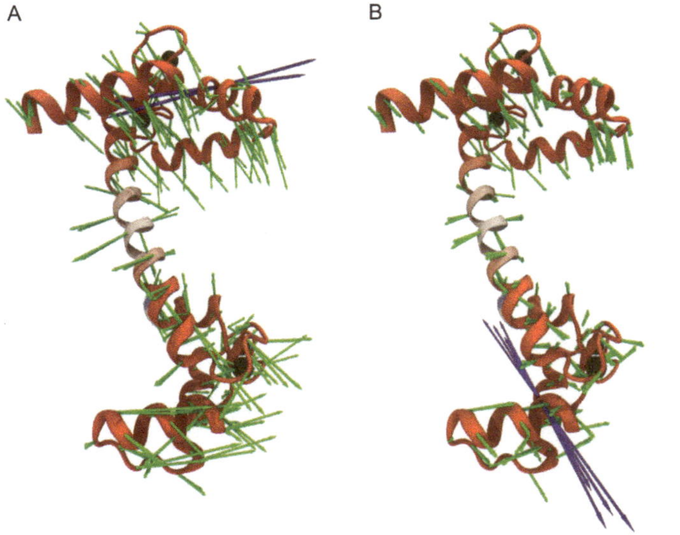

Fig. 5 Results of PRS analysis on CaM. In this application, the Hessian is using Eq. (11), obtained from a 40 ns MD trajectory with the initial PDB structure 3cln. The target structure is the closed form with PDB code 1prw. The protein is colored with shades of *red* according to the experimental B-factors. The highest overlapping perturbations with a value of ~0.7 on the selected residue are displayed in *blue*, and the responses are shown in *green*. (A) PRS results on D30. The best perturbations overlapping with the displacement vector between the initial and target structures (*blue*) and the resulting responses (*green*) are clearly aligned. The response causes a simultaneous twist around the linker as well as bending of the two lobes toward each other. (B) PRS results on D105. The response is similar to that observed in (A) but is less accentuated. Movies of the motions may be generated on the web server provided at http://midst.sabanciuniv.edu/prs/.

residues/forces that display good overlaps with the conformational change of CaM discussed in Fig. 2. This complicated motion requires simultaneous twisting and bending motions that are absent in any single mode (Fig. 4).

The approach has proven to be very useful in determining functional sites in large allosteric proteins such as Hsp70 and Hsp90. The first study on these proteins focused on the nucleotide binding domain of Hsp70 (General et al., 2014) whose findings were later confirmed in an investigation of the full-length protein (Penkler et al, 2017). The latter study also disclosed regions that operate in communicating the nucleotide binding domain to the substrate binding domain, supplementing findings obtained by the BC analysis in the RN approach (Ozbaykal et al., 2015). For Hsp90, several studies targeting the overall dynamics of the protein (Penkler et al, 2018), differentiating its kinase dependent clients (Stetz, Tse, & Verkhivker, 2017), and posttranslational modification sites (Stetz, Tse, & Verkhivker, 2018) were conducted, offering novel interpretations for the conformational cycle of the protein.

Finally, the problem of long range communication in large protein assemblies has been addressed by the PRS methodology. For example, the directional forces originating at the ATP hydrolysis step driving GroEL conformational change were disclosed using this approach (Liu, Sankar, Wang, Jia, & Jernigan, 2017). Still at the larger scale, the RNA release step of the enterovirus 71 uncoating was modeled via PRS which revealed that a protomer consisting of the assembly of four proteins cannot serve as a functional unit to explain deformations of the capsid (Ross, Atilgan et al., 2018). Instead, it revealed the pentamer (containing 20 proteins) as the minimum functional unit required to investigate changes within the capsid. It also enabled locating a hotspot region surrounding the fivefold axis of the virus whose range of effect extends to 140 Å. The perturbation of internal capsid residues in this region displayed greatest propensity to capsid expansion, thus indicating the significant role that the RNA genome may play in triggering uncoating.

PRS has also been used as an initial tool that enables other analysis methods to be developed. For example, it was used to determine a "dynamic flexibility index" relating structural dynamics to function and evolution at the proteomic scale (Gerek & Ozkan, 2011; Kim et al., 2015). In another application of the methodology, PRS was used as the input generation step for a flexible docking scheme to explore protein–ligand interactions based on the unbound conformations (Bolia, Gerek, & Ozkan, 2014; Bolia & Ozkan, 2016).

6. NEW DIRECTIONS FOR EFFICIENT SAMPLING OF CONFORMATIONAL LANDSCAPES

The combinations and extensions of the methods discussed in Sections 2–5 of this chapter hold great promise for the efficient sampling of conformational landscapes of proteins, as well as the detection of allosteric spots that may be targeted to promote select conformations. In the domain of RN (Section 3), the weighted network approach (Atilgan et al., 2007) has yet to be exploited to obtain a better understanding of communication routes in a protein or between interacting proteins. In particular, methods that translate residue–residue correlation information obtained from MD simulations, to weights on the links should be developed. Furthermore, even with the current weight assignment strategies (Thomas & Dill, 1996), weighted RNs have never been applied to dynamical changes occurring during an MD trajectory or to the effects of point mutations in select regions. Such problems point to additional future directions.

In the domain of elastic networks and beyond (Sections 4 and 5), the methods are already intertwined: Generation of the Hessian matrix through Eq. (11) by eliminating the cutoff distance selection problem requires relatively short lengths of MD trajectories. Thus, this is now becoming the method of choice over ANM, making MD simulations an integral part of elastic network models. Nevertheless, in some very large systems for which obtaining equilibrated MD trajectories is still computationally prohibitive, constructing the Hessian via Eq. (10) is a viable option for the analysis of the systems.

Furthermore, a thorough analysis of a given system may require a combination of the three approaches, i.e., MD simulations, elastic network models, and PRS. Case in point, the complete understanding of the iron binding and release cycle of human transferrin necessitates simulating the protein in different environments (Abdizadeh, Atilgan, et al., 2017; Abdizadeh, Tamer, et al., 2017), e.g., serum conditions at 7.4 pH vs endosomal conditions at acidic pH, or apo vs monoferric and diferric forms. Changes in its dynamics when bound to its receptor at the cell surface also bring up interesting questions. Finally, since this protein is a target by bacteria for iron sequestration, its complex with bacterial receptors also attracts attention. In fact, MD simulations have been utilized for distinguishing the dynamics of the protein loaded with none, 1 or 2 irons, as well as under different pH conditions (Abdizadeh, Atilgan, & Atilgan, 2015). On the other hand, based on the Hessians obtained from these MD trajectories, collective modes of motion operative for the same protein under the different environments were used to rationalize why the protein prefers to bind its receptor at the cell surface, why it is expected to readily release iron in the acidic endosome, and why the apo transferrin–receptor complex remains bound until they are recycled back to the cell surface (Abdizadeh & Atilgan, 2016). All these processes occur on timescales that are much longer than the hundreds of nanoseconds sampled by MD, but the proper analyses provide physics-based explanations for the observed behavior. Meanwhile, PRS on the apo transferrin has implicated a primary binding location for the synergistic phosphate ion, and the proposed closing motion has been confirmed using all atom MD simulations by docking phosphate at this location (Abdizadeh & Atilgan, 2016). Finally, PRS on the high resolution complex of transferrin and the bacterial receptor has been used to find allosteric locations operative in triggering iron dissociation, enabling a mechanistic explanation for the rapidly evolving distal residues (Abdizadeh & Atilgan, 2016).

While the methods have reached a level of maturity for applications to physiologically relevant problems, other developments are necessary to extend the simulations to mimic experimental conditions; e.g., in many equipment, frequency-dependent perturbations are introduced to systems which require modifying PRS to mimic this scenario. Very recently, a generalization of the technique to the analysis of the displacements of the residues under force excitations with varying frequency has been described (Varolgunes & Demir, 2018); the formulation reduces to PRS by setting the frequency of the excitation to zero. In other real-life scenarios, binding of a single ligand is expected to perturb several residues simultaneously which is already encoded in PRS formulation. In principle any number of combinations of forces may be inserted as the force in Eq. (14). However, the combinatorics may be overwhelming, and the compute times may be exhaustive. A recent development, called double force scanning, has been used to identify the second-site residues that make the protein structure most resilient to the effect of pathogenic mutations (Tiberti, Pandini, Fraternali, & Fornili, 2018), but this field needs further breakthroughs for the implementation of a comprehensive multiple point scanning.

It is also necessary to investigate the extent to which LRT applies to proteins. To this end, toy models that include symmetries and deliberately introduced asymmetries should be studied. Furthermore, PRS in its current form requires two end-point structures to be experimentally known to distill regions of functional interest. Once there is a deeper understanding on the limits of applicability of LRT, an extension of PRS would include starting from an initial structure only and clustering the response vectors so as to predict possible regions that may be accessed from that state.

These approaches will be continued to be developed in the future, not the least because they offer research routes to a large number of researchers that have access to modest computational resources, but also because they provide a physics-based understanding of the mechanisms that govern protein dynamics.

ACKNOWLEDGMENTS

This work was supported by the Scientific and Technological Research Council of Turkey (Grant Number 116F229). I thank Dr. Ali Rana Atilgan for the many discussions we have had on the topic over the years. Tandac Furkan Guclu and Farzaneh Jalalypour have provided help in preparing Figs. 3–5.

REFERENCES

Abdizadeh, H., & Atilgan, C. (2016). Predicting long term cooperativity and specific modulators of receptor interactions in human transferrin from dynamics within a single microstate. *Physical Chemistry Chemical Physics*, *18*(11), 7916–7926.

Abdizadeh, H., Atilgan, A. R., & Atilgan, C. (2015). Detailed molecular dynamics simulations of human transferrin provide insights into iron release dynamics at serum and endosomal pH. *Journal of Biological Inorganic Chemistry*, *20*(4), 705–718.

Abdizadeh, H., Atilgan, A. R., Atilgan, C., & Dedeoglu, B. (2017). Computational approaches for deciphering the equilibrium and kinetic properties of iron transport proteins. *Metallomics*, *9*(11), 1513–1533.

Abdizadeh, H., Tamer, Y. T., Acar, O., Toprak, E., Atilgan, A. R., & Atilgan, C. (2017). Increased substrate affinity in the Escherichia coli L28R dihydrofolate reductase mutant causes trimethoprim resistance. *Physical Chemistry Chemical Physics*, *19*(18), 11416–11428.

Abel, R., Wang, L., Harder, E. D., Berne, B. J., & Friesner, R. A. (2017). Advancing drug discovery through enhanced free energy calculations. *Accounts of Chemical Research*, *50*(7), 1625–1632.

Aldeghi, M., Heifetz, A., Bodkin, M. J., Knappcd, S., & Biggin, P. C. (2016). Accurate calculation of the absolute free energy of binding for drug molecules. *Chemical Science*, *7*(1), 207–218.

Allen, M. P., & Tildesley, D. J. (1989). *Computer simulation of liquids*. New York, NY: Clarendon Press.

Amitai, G., Shemesh, A., Sitbon, E., Shklar, M., Netanely, D., Venger, I., et al. (2004). Network analysis of protein structures identifies functional residues. *Journal of Molecular Biology*, *344*(4), 1135–1146.

Atilgan, A. R., Akan, P., & Baysal, C. (2004). Small-world communication of residues and significance for protein dynamics. *Biophysical Journal*, *86*(1), 85–91.

Atilgan, C., & Atilgan, A. R. (2009). Perturbation-response scanning reveals ligand entry-exit mechanisms of ferric binding protein. *PLoS Computational Biology*, *5*(10).

Atilgan, A. R., & Atilgan, C. (2012). Local motifs in proteins combine to generate global functional moves. *Briefings in Functional Genomics*, *11*(6), 479–488.

Atilgan, A. R., Aykut, A. O., & Atilgan, C. (2011). Subtle pH differences trigger single residue motions for moderating conformations of calmodulin. *Journal of Chemical Physics*, *135*(15), 155102.

Atilgan, A. R., Durell, S. R., Jernigan, R. L., Demirel, M. C., Keskin, O., & Bahar, I. (2001). Anisotropy of fluctuation dynamics of proteins with an elastic network model. *Biophysical Journal*, *80*(1), 505–515.

Atilgan, C., Gerek, Z. N., Ozkan, S. B., & Atilgan, A. R. (2010). Manipulation of conformational change in proteins by single-residue perturbations. *Biophysical Journal*, *99*(3), 933–943.

Atilgan, C., Okan, O. B., & Atilgan, A. R. (2010). How orientational order governs collectivity of folded proteins. *Proteins: Structure, Function, and Bioinformatics*, *78*(16), 3363–3375.

Atilgan, C., Okan, O. B., & Atilgan, A. R. (2012). Network-based models as tools hinting at nonevident protein functionality. In D. C. Rees (Ed.), *Annual review of biophysics: Vol. 41* (pp. 205–225).

Atilgan, A. R., Turgut, D., & Atilgan, C. (2007). Screened nonbonded interactions in native proteins manipulate optimal paths for robust residue communication. *Biophysical Journal*, *92*(9), 3052–3062.

Aykut, A. O., Atilgan, A. R., & Atilgan, C. (2013). Designing molecular dynamics simulations to shift populations of the conformational states of calmodulin. *PLoS Computational Biology*, *9*(12).

Bagler, G., & Sinha, S. (2005). Network properties of protein structures. *Physica A-Statistical Mechanics and Its Applications*, *346*(1–2), 27–33.

Bahar, I., Atilgan, A. R., Demirel, M. C., & Erman, B. (1998). Vibrational dynamics of folded proteins: Significance of slow and fast motions in relation to function and stability. *Physical Review Letters*, *80*(12), 2733–2736.

Bahar, I., Atilgan, A. R., & Erman, B. (1997). Direct evaluation of thermal fluctuations in proteins using a single-parameter harmonic potential. *Folding & Design*, *2*(3), 173–181.

Bahar, I., Atilgan, A. R., Jernigan, R. L., & Erman, B. (1997). Understanding the recognition of protein structural classes by amino acid composition. *Proteins: Structure, Function, and Genetics*, *29*(2), 172–185.

Bahar, I., Erman, B., Jernigan, R. L., Atilgan, A. R., & Covell, D. G. (1999). Collective motions in HIV-1 reverse transcriptase: Examination of flexibility and enzyme function. *Journal of Molecular Biology*, *285*(3), 1023–1037.

Bahar, I., Lezon, T. R., Yang, L.-W., & Eyal, E. (2010). Global dynamics of proteins: Bridging between structure and function. In D. C. Rees, K. A. Dill, & J. R. Williamson (Eds.), *Annual review of biophysics: Vol. 39.* (pp. 23–42). Annual Reviews.

Bakan, A., Dutta, A., Mao, W., Liu, Y., Chennubhotla, C., Lezon, T. R., et al. (2014). Evol and ProDy for bridging protein sequence evolution and structural dynamics. *Bioinformatics*, *30*(18), 2681–2683.

Barabasi, A. L., & Albert, R. (1999). Emergence of scaling in random networks. *Science*, *286*(5439), 509–512.

Barabasi, A. L., Jeong, H., Neda, Z., Ravasz, E., Schubert, A., & Vicsek, T. (2002). Evolution of the social network of scientific collaborations. *Physica A*, *311*(3–4), 590–614.

Baysal, C., & Atilgan, A. R. (2001). Coordination topology and stability for the native and binding conformers of chymotrypsin inhibitor 2. *Proteins: Structure, Function, and Genetics*, *45*(1), 62–70.

Berman, H. M., Westbrook, J., Feng, Z., Gilliland, G., Bhat, T. N., Weissig, H., et al. (2000). The protein data bank. *Nucleic Acids Research*, *28*(1), 235–242.

Bolia, A., Gerek, Z. N., & Ozkan, S. B. (2014). BP-dock: A flexible docking scheme for exploring protein ligand interactions based on unbound structures. *Journal of Chemical Information and Modeling*, *54*(3), 913–925.

Bolia, A., & Ozkan, S. B. (2016). Adaptive BP-dock: An induced fit docking approach for full receptor flexibility. *Journal of Chemical Information and Modeling*, *56*(4), 734–746.
Brinda, K. V., & Vishveshwara, S. (2005). Oligomeric protein structure networks: Insights into protein-protein interactions. *BMC Bioinformatics*, *6*, 296.
Brown, D. K., Amamuddy, O. S., & Bishop, O. T. (2017). Structure-based analysis of single nucleotide variants in the renin-angiotensinogen complex. *Global Heart*, *12*(2), 121–132.
Brown, D. K., Penkler, D. L., Amamuddy, O. S., Ross, C., Atilgan, A. R., Atilgan, C., et al. (2017). MD-TASK: A software suite for analyzing molecular dynamics trajectories. *Bioinformatics*, *33*(17), 2768–2771.
Chacon, P., Tama, F., & Wriggers, W. (2003). Mega-dalton biomolecular motion captured from electron microscopy reconstructions. *Journal of Molecular Biology*, *326*(2), 485–492.
Childers, M. C., & Daggett, V. (2017). Insights from molecular dynamics simulations for computational protein design. *Molecular Systems Design & Engineering*, *2*(1), 9–33.
Chodera, J. D., & Noe, F. (2014). Markov state models of biomolecular conformational dynamics. *Current Opinion in Structural Biology*, *25*, 135–144.
Das, R., & Baker, D. (2008). Macromolecular modeling with Rosetta. *Annual Review of Biochemistry*, *77*, 363–382.
de Jong, D. H., Singh, G., Bennett, W. F. D., Arnarez, C., Wassenaar, T. A., Schafer, L. V., et al. (2013). Improved parameters for the martini coarse-grained protein force field. *Journal of Chemical Theory and Computation*, *9*(1), 687–697.
del Sol, A., Fujihashi, H., Amoros, D., & Nussinov, R. (2006). Residue centrality, functionally important residues, and active site shape: Analysis of enzyme and non-enzyme families. *Protein Science*, *15*(9), 2120–2128.
Demirel, M. C., Atilgan, A. R., Jernigan, R. L., Erman, B., & Bahar, I. (1998). Identification of kinetically hot residues in proteins. *Protein Science*, 7(12), 2522–2532.
Dutta, A., Krieger, J., Lee, J. Y., Garcia-Nafria, J., Greger, I. H., & Bahar, I. (2015). Cooperative dynamics of intact AMPA and NMDA glutamate receptors: Similarities and subfamily-specific differences. *Structure*, *23*(9), 1692–1704.
Erdös, P., & Rényi, A. (1961). On the evolution of random graphs. *Bulletin of the International Statistical Institute*, *38*, 343–347.
Estrada, E. (2010). Universality in protein residue networks. *Biophysical Journal*, *98*(5), 890–900.
Fariselli, P., & Casadio, R. (1999). A neural network based predictor of residue contacts in proteins. *Protein Engineering*, *12*(1), 15–21.
General, I. J., Liu, Y., Blackburn, M. E., Mao, W., Gierasch, L. M., & Bahar, I. (2014). ATPase subdomain IA is a mediator of interdomain allostery in Hsp70 molecular chaperones. *PLoS Computational Biology*, *10*(5).
Gerek, Z. N., & Ozkan, S. B. (2011). Change in allosteric network affects binding affinities of PDZ domains: Analysis through perturbation response scanning. *PLoS Computational Biology*, 7(10).
Greene, L. H., & Higman, V. A. (2003). Uncovering network systems within protein structures. *Journal of Molecular Biology*, *334*(4), 781–791.
Gsponer, J., Christodoulou, J., Cavalli, A., Bui, J. M., Richter, B., Dobson, C. M., et al. (2008). A coupled equilibrium shift mechanism in calmodulin-mediated signal transduction. *Structure*, *16*(5), 736–746.
Haliloglu, T., & Bahar, I. (2015). Adaptability of protein structures to enable functional interactions and evolutionary implications. *Current Opinion in Structural Biology*, *35*, 17–23.
Heath, A. P., Kavraki, L. E., & Clementi, C. (2007). From coarse-grain to all-atom: Toward multiscale analysis of protein landscapes. *Proteins: Structure, Function, and Bioinformatics*, *68*(3), 646–661.
Hinsen, K., Reuter, N., Navaza, J., Stokes, D. L., & Lacapere, J. J. (2005). Normal mode-based fitting of atomic structure into electron density maps: Application to sarcoplasmic reticulum Ca-ATPase. *Biophysical Journal*, *88*(2), 818–827.

Huberman, B. A., & Adamic, L. A. (1999). Internet—Growth dynamics of the world-wide web. *Nature, 401*(6749), 131.

Ikeguchi, M., Ueno, J., Sato, M., & Kidera, A. (2005). Protein structural change upon ligand binding: Linear response theory. *Physical Review Letters, 94*(7), 078102.

Karplus, M., & Kuriyan, J. (2005). Molecular dynamics and protein function. *Proceedings of the National Academy of Sciences of the United States of America, 102*(19), 6579–6685.

Karplus, M., & McCammon, J. A. (2002). Molecular dynamics simulations of biomolecules. *Nature Structural Biology, 9*(9), 646–652.

Karplus, M., & Petsko, G. A. (1990). Molecular-dynamics simulations in biology. *Nature, 347*(6294), 631–639.

Kim, M. K., Jernigan, R. L., & Chirikjian, G. S. (2002). Efficient generation of feasible pathways for protein conformational transitions. *Biophysical Journal, 83*(3), 1620–1630.

Kim, H., Zou, T., Modi, C., Doerner, K., Grunkemeyer, T. J., Chen, L., et al. (2015). A hinge migration mechanism unlocks the evolution of green-to-red photoconversion in GFP-like proteins. *Structure, 23*(1), 34–43.

Kmiecik, S., Gront, D., Kolinski, M., Wieteska, L., Dawid, A. E., & Kolinski, A. (2016). Coarse-grained protein models and their applications. *Chemical Reviews, 116*(14), 7898–7936.

Lane, T. J., Shukla, D., Beauchamp, K. A., & Pande, V. S. (2013). To milliseconds and beyond: Challenges in the simulation of protein folding. *Current Opinion in Structural Biology, 23*(1), 58–65.

Lindorff-Larsen, K., Piana, S., Dror, R. O., & Shaw, D. E. (2011). How fast-folding proteins fold. *Science, 334*(6055), 517–520.

Liu, J., Sankar, K., Wang, Y., Jia, K., & Jernigan, R. L. (2017). Directional force originating from ATP hydrolysis drives the GroEL conformational change. *Biophysical Journal, 112*(8), 1561–1570.

Lockless, S. W., & Ranganathan, R. (1999). Evolutionarily conserved pathways of energetic connectivity in protein families. *Science, 286*(5438), 295–299.

Lopes, P. E. M., Guvench, O., & MacKerell, A. D. (2015). Current status of protein force fields for molecular dynamics simulations. In A. Kukol (Ed.), *Molecular modeling of proteins* (pp. 47–71). New York, NY: Springer New York.

Lopez-Blanco, J. R., & Chacon, P. (2016). New generation of elastic network models. *Current Opinion in Structural Biology, 37*, 46–53.

Mahajan, S., & Sanejouand, Y.-H. (2015). On the relationship between low-frequency normal modes and the large-scale conformational changes of proteins. *Archives of Biochemistry and Biophysics, 567*, 59–65.

Na, H., Jernigan, R. L., & Song, G. (2015). Bridging between NMA and elastic network models: Preserving all-atom accuracy in coarse-grained models. *PLoS Computational Biology, 11*(10).

Najmanovich, R., & Domany, E. (1999). Protein folding in contact map space. *Physical Review Letters, 82*(3), 659.

Negi, S., Aykut, A. O., Atilgan, A. R., & Atilgan, C. (2012). Calmodulin readily switches conformation upon protonating high pK(a) acidic residues. *Journal of Physical Chemistry B, 116*(24), 7145–7153.

Ozbaykal, G., Atilgan, A. R., & Atilgan, C. (2015). In silico mutational studies of Hsp70 disclose sites with distinct functional attributes. *Proteins: Structure, Function, and Bioinformatics, 83*(11), 2077–2090.

Penkler, D. L., Atilgan, C., & Bishop, O. T. (2018). Allosteric modulation of human Hsp90 alpha conformational dynamics. *Journal of Chemical Information and Modeling, 58*(2), 383–404.

Penkler, D., Sensoy, O., Atilgan, C., & Bishop, O. T. (2017). Perturbation-response scanning reveals key residues for allosteric control in Hsp70. *Journal of Chemical Information and Modeling, 57*(6), 1359–1374.

Petrone, P., & Pande, V. S. (2006). Can conformational change be described by only a few normal modes? *Biophysical Journal, 90*(5), 1583–1593.

Piana, S., Klepeis, J. L., & Shaw, D. E. (2014). Assessing the accuracy of physical models used in protein-folding simulations: Quantitative evidence from long molecular dynamics simulations. *Current Opinion in Structural Biology, 24*, 98–105.

Reith, D., Putz, M., & Muller-Plathe, F. (2003). Deriving effective mesoscale potentials from atomistic simulations. *Journal of Computational Chemistry, 24*(13), 1624–1636.

Rohrdanz, M. A., Zheng, W., & Clementi, C. (2013). Discovering mountain passes via torchlight: Methods for the definition of reaction coordinates and pathways in complex macromolecular reactions. In M. A. Johnson & T. J. Martinez (Eds.), *Annual review of physical chemistry: Vol. 64* (pp. 295–316).

Ross, C. J., Atilgan, A. R., Bishop, O. T., & Atilgan, C. (2018). Unraveling the motions behind enterovirus 71 uncoating. *Biophysical Journal, 114*(4), 822–838.

Ross, C., Nizami, B., Glenister, M., Amamuddy, O. S., Atilgan, A. R., Atilgan, C., et al. (2018). MODE-TASK: Large-scale protein motion tools. *Bioinformatics*. https://doi.org/10.1093/bioinformatics/bty427. (in press).

Sanbonmatsu, K. Y., Joseph, S., & Tung, C. S. (2005). Simulating movement of tRNA into the ribosome during decoding. *Proceedings of the National Academy of Sciences of the United States of America, 102*(44), 15854–15859.

Schlitter, J., Engels, M., & Kruger, P. (1994). Targeted molecular-dynamics—A new approach for searching pathways of conformational transitions. *Journal of Molecular Graphics, 12*(2), 84–89.

Shan, Y., Kim, E. T., Eastwood, M. P., Dror, R. O., Seeliger, M. A., & Shaw, D. E. (2011). How does a drug molecule find its target binding site? *Journal of the American Chemical Society, 133*(24), 9181–9183.

Shaw, D. E., Maragakis, P., Lindorff-Larsen, K., Piana, S., Dror, R. O., Eastwood, M. P., et al. (2010). Atomic-level characterization of the structural dynamics of proteins. *Science, 330*(6002), 341–346.

Stetz, G., Tse, A., & Verkhivker, G. M. (2017). Ensemble-based modeling and rigidity decomposition of allosteric interaction networks and communication pathways in cyclin-dependent kinases: Differentiating kinase clients of the Hsp90-Cdc37 chaperone. *PLoS One, 12*(11).

Stetz, G., Tse, A., & Verkhivker, G. M. (2018). Dissecting structure-encoded determinants of allosteric cross-talk between post-translational modification sites in the Hsp90 chaperones. *Scientific Reports, 8*.

Stetz, G., & Verkhivker, G. M. (2017). Computational analysis of residue interaction networks and coevolutionary relationships in the Hsp70 chaperones: A community-hopping model of allosteric regulation and communication. *PLoS Computational Biology, 13*(1).

Taboy, C. H., Vaughan, K. G., Mietzner, T. A., Aisen, P., & Crumbliss, A. L. (2001). Fe3+ coordination and redox properties of a bacterial transferrin. *The Journal of Biological Chemistry, 276*(4), 2719–2724.

Thomas, P. D., & Dill, K. A. (1996). An iterative method for extracting energy-like quantities from protein structures. *Proceedings of the National Academy of Sciences of the United States of America, 93*(21), 11628–11633.

Tiberti, M., Pandini, A., Fraternali, F., & Fornili, A. (2018). In silico identification of rescue sites by double force scanning. *Bioinformatics, 34*(2), 207–214.

Tirion, M. M. (1996). Large amplitude elastic motions in proteins from a single-parameter, atomic analysis. *Physical Review Letters*, 77(9), 1905–1908.

Turgut, D., Atilgan, A. R., & Atilgan, C. (2010). Assortative mixing in close-packed spatial networks. *PLoS One, 5*(12), e15551.

Varolgunes, Y. B., & Demir, A. (2018). ProteinAC: A frequency domain technique for analyzing protein dynamics. *Physical Biology*, *15*(2), 026009.

Vazquez, A., Pastor-Satorras, R., & Vespignani, A. (2002). Large-scale topological and dynamical properties of the internet. *Physical Review E*, *65*(6), 066130.

Vendruscolo, M., Dokholyan, N. V., Paci, E., & Karplus, M. (2002). Small-world view of the amino acids that play a key role in protein folding. *Physical Review E*, *65*(6), 061910.

Voelz, V. A., Bowman, G. R., Beauchamp, K., & Pande, V. S. (2010). Molecular simulation of ab initio protein folding for a millisecond folder NTL9(1-39). *Journal of the American Chemical Society*, *132*(5), 1526–1528.

Watts, D. J., & Strogatz, S. H. (1998). Collective dynamics of'small-world networks. *Nature*, *393*(6684), 440–442.

Yang, L., Song, G., & Jernigan, R. L. (2009). Protein elastic network models and the ranges of cooperativity. *Proceedings of the National Academy of Sciences of the United States of America*, *106*(30), 12347–12352.

Yilmaz, L. S., & Atilgan, A. R. (2000). Identifying the adaptive mechanism in globular proteins: Fluctuations in densely packed regions manipulate flexible parts. *Journal of Chemical Physics*, *113*(10), 4454–4464.

Yu, I., Mori, T., Ando, T., Harada, R., Jung, J., Sugita, Y., et al. (2016). Biomolecular interactions modulate macromolecular structure and dynamics in atomistic model of a bacterial cytoplasm. *eLife*, *5*, e19274.

Zhang, Z. Y., Shi, Y. Y., & Liu, H. Y. (2003). Molecular dynamics simulations of peptides and proteins with amplified collective motions. *Biophysical Journal*, *84*(6), 3583–3593.

Zhao, G., Perilla, J. R., Yufenyuy, E. L., Meng, X., Chen, B., Ning, J., et al. (2013). Mature HIV-1 capsid structure by cryo-electron microscopy and all-atom molecular dynamics. *Nature*, *497*(7451), 643–646.

CHAPTER THREE

Computational Methods for Epigenetic Drug Discovery: A Focus on Activity Landscape Modeling

J. Jesús Naveja*,†, C. Iluhí Oviedo-Osornio*, José L. Medina-Franco*,[1]

*Facultad de Química, Universidad Nacional Autónoma de México, Mexico City, Mexico
†PECEM, Facultad de Medicina, Universidad Nacional Autónoma de México, Mexico City, Mexico
[1]Corresponding author: e-mail addresses: medinajl@unam.mx; jose.medina.franco@gmail.com

Contents

Abstract

Epigenetic drug discovery is an emerging strategy against several chronic and complex diseases. The increased interest in epigenetics has boosted the development and maintenance of large information on structure–epigenetic activity relationships for several epigenetic targets. In turn, such large databases—many in the public domain—are a rich source of information to explore their structure–activity relationships (SARs). Herein, we conducted a large-scale analysis of the SAR of epigenetic targets using the concept of activity landscape modeling. A comprehensive quantitative analysis and a novel visual representation of the *epigenetic activity landscape* enabled the rapid identification of regions of targets with continuous and discontinuous SAR. This information led to the identification of epigenetic targets for which it is anticipated an easier or a more difficult drug-discovery program using conventional hit-to-lead approaches. The insights of this work also enabled the identification of specific structural changes associated with a

Advances in Protein Chemistry and Structural Biology, Volume 113
ISSN 1876-1623
https://doi.org/10.1016/bs.apcsb.2018.01.001

large shift in biological activity. To the best of our knowledge, this work represents the largest comprehensive SAR analysis of several epigenetic targets and contributes to the better understanding of the *epigenetic activity landscape*.

1. INTRODUCTION

Epigenetics, despite of all its intrinsic complexity, is quite promising in drug discovery, given the expected potential of modifying or even reversing complex gene expression patterns associated to chronic diseases (Dueñas-González, Naveja, & Medina-Franco, 2016). Also, the possibility of inhibiting epigenetic processes at writing, reading, and erasing times adds to the diversity of potential therapies (Chung, 2015). Briefly, epigenetic writers add marks to either DNA or histones, which are in turn transduced by readers, therefore leading to a cell response. In antagonism to writers, epigenetic erasers remove marks and thereby inhibit the subsequent signal (Allis & Jenuwein, 2016). Examples of writers are DNA methyltransferases (DNMTs) and histone acetyltransferases (HATs); examples of readers are bromodomains (BRDs), while histone deacetylases (HDACs) account for the most studied epigenetic erasers.

The increasing awareness of the role of epigenetic targets for the treatment of several diseases has boosted the development of compounds that can act as potential inhibitors. Indeed, the number of compounds tested and the number of targets under study have been increasing (Lundstrom, 2017). This far, the chemical space of small molecules targeting major epigenetic targets has been explored. Examples are inhibitors of HDACs and BRDs (Prieto-Martínez, Gortari, Méndez-Lucio, & Medina-Franco, 2016). These studies have contributed to the charting of the so-called Epigenetic Relevant Chemical Space (ERCS) (Gortari & Medina-Franco, 2015). More recently, the chemical space of 52 epigenetic targets has been characterized (Naveja & Medina-Franco, 2018). In that study, the recently developed concept of *database fingerprints* was used to map targets according to their pharmacological similarity (Fernández-de Gortari, García-Jacas, Martinez-Mayorga, & Medina-Franco, 2017). Further analysis of this type might confirm the feasibility of rational design of epi-polypharmacological compounds.

A major next step to develop effective compounds for epigenetic targets or epi-drugs is the analysis of the structure–activity relationships (SARs) associated with active molecules. One computational approach to systematically characterize the SAR of compound data sets is activity landscape modeling. Activity landscape modeling can be conceptualized as the association between the chemical space and biological activity. This far, the activity landscape of few, although major, epigenetic targets has been analyzed. For instance, Naveja et al. analyzed the activity landscape of DNMTs (Naveja & Medina-Franco, 2015a, 2015b). Saldívar-González et al. explored the landscape of HDACs (Saldívar-González, Naveja, Palomino-Hernández, & Medina-Franco, 2017), and García-Sánchez et al. the landscape of BRD inhibitors (García-Sánchez, Cruz-Monteagudo, & Medina-Franco, 2017). However, there is a need to analyze the activity landscape of many more relevant epigenetic targets associated with the ERCS.

The main goal of this study is to characterize the epigenetic activity landscape associated with ERCS currently known. The specific goals are: (a) to identify the epigenetic targets with more continuous and discontinuous SAR; (b) to quantify the proportion of activity cliffs for each target while identifying the most frequent activity cliffs, i.e., activity cliff generators (ACGs); and (c) to provide a structure-based rationale of the ACGs and epigenetic targets with more discontinuous SAR. To achieve these goals Structure–Activity Similarity (SAS) maps and Structure–Activity Landscape Index (SALI) values were employed. These are well-known computational approaches for activity landscape modeling (Guha & Van Drie, 2008; Méndez-Lucio, Pérez-Villanueva, Castillo, & Medina-Franco, 2012).

This chapter is organized into four major sections. After this introduction, Section 2 discusses the major epigenetic targets covered in this chapter. Section 3 presents the results of the activity landscape modeling of the epigenetic targets. This section is further divided into several subsections each discussing the major components of the epigenetic activity landscape. Section 4 presents conclusions and perspectives.

2. EPIGENETIC TARGETS

The activity landscape modeling presented in this work builds upon a recently published preliminary study on the chemical space of epigenetic targets (Naveja & Medina-Franco, 2018). In that work, the chemical libraries of major families of epigenetic targets, i.e., HDACs, DNMTs, HATs, BRDs, KMeR, PRMTs, HKMs, KDMs, and others (see summary in Table 1), revealed interesting similarity associations. For instance, inhibitors of HDACs tended to form well-defined clusters, while other families of targets, such as DNMTs, had little resemblance among each other (Naveja & Medina-Franco, 2018). Regarding SAR analysis, enrichment factors for

Table 1 Main Epigenetic Targets Considered in This Work

Family	Function	Targets
BRD	Histone acetylation reading	BAZ2B, BRD2, BRD3, BRD4, BRD9, BRPF1
DNMT	DNA methylation writing	DNMT1, DNMT3A, DNMT3B
HAT	Histone acetylation writing	CREBBP, EP300, KAT2A, KAT2B, NCOA1, NCOA3
HDAC	Histone acetylation erasing	HDAC: 1–11
HKM	Histone lysine methylation writing	DOT1L, EHMT1, EHMT2, KMT5A, SMYD2
KDM	Histone lysine methylation erasing	KDM: 1A, 2A, 3A, 4A, 4C, 4E, 5A, 5C
KMeR	Histone lysine methylation reading	L3MBTL: 1, 3, 4 CBX7, TP53BP1, WDR5
PRMT	Histone arginine methylation writing	PRMT: 1, 6, 8 CARM1
Others	Miscellaneous	MAP3K7 (kinase), MGEA5 (histone *O*-*N*-acetylglucosamine transferase), SMARCA2 (chromatin remodeler)

Bemis–Murcko scaffolds were calculated, finding distinct molecular motifs susceptible of further development. However, an exhaustive SAR analysis of all these epigenetic targets has not been performed until now. An overview of the epigenetic targets included in this work, aggregated by family, is presented in Table 1. Of note, epigenetic writers, readers, and erasers are represented in this list. For detailed reviews on epigenetic targets and their functions, please see Allis and Jenuwein (2016), Jenuwein and Allis (2001), Medvedeva et al. (2015), and citations therein.

3. ACTIVITY LANDSCAPE MODELING

Activity landscape modeling is a useful strategy in medicinal chemistry and drug discovery to explore and describe the SAR of chemical data sets (García-Sánchez et al., 2017). This computational strategy is valuable to guide lead-optimization efforts and to develop in silico models, such as

QSAR and similarity-based virtual screening (Medina-Franco, 2012; Peltason & Bajorath, 2007).

Activity landscape can be conceptualized as the association between the chemical space and the activity data (Wassermann, Wawer, & Bajorath, 2010). With this approach we can directly assess the similarity principle, i.e., whether compounds in the data set that are more similar structurally are also alike in activity. A continuous SAR is one where small changes in activity due to small changes in molecular structure can be found, while a discontinuous SAR has many cases where small changes in molecular structure can lead to a significant change in the activity (Pérez-Villanueva et al., 2010). Fig. 1 depicts the chemical space of epigenetic targets as it resulted in Naveja and Medina-Franco (2018). Data points are colored according to the proportion of activity cliffs in the data set using a continuous color scale from low (dark blue) to high (light blue) proportion of cliffs. Two examples of SAS maps (activity landscape depictions—see Sections 3.1 and 3.3) are illustrated: (a) DOT1L, which has a quite discontinuous SAR, and (b) HDAC9, in contrast, with a rather continuous SAR. In SAS maps, each point represents a paired comparison of structure and activity similarity, where points in the upper right quadrant are activity cliffs (Medina-Franco, 2012). Supplementary Fig. S1 in the online version at https://doi.org/10.1016/bs.apcsb.2018.01.001 shows a 3D representation of the SAR for these targets, along with relevant average physicochemical and complexity properties. Interestingly, regions in the chemical spaces enriched with discontinuous SAR targets seem to exist, and these same regions tend to have higher complexity (*n* chiral atoms and FCSP3) (Fig. 1 and Supplementary Fig. S1 in the online version at https://doi.org/10.1016/bs.apcsb.2018.01.001).

During the last few years, efforts have been made to develop quantitative and visual methods to model activity landscapes. Examples of these approaches are SAS maps, Dual-Activity Difference (DAD) maps, SALI, and Structure–Activity Relationship Index (SARI), to name a few (Méndez-Lucio, 2016). These approaches have been applied to a large number of compound data sets of relevance in medicinal chemistry (e.g., Medina-Franco, 2012; Méndez-Lucio et al., 2012; Naveja & Medina-Franco, 2015a).

3.1 SAS Maps

Shanmugasundaram and Maggiora introduced the SAS maps, a 2D activity landscape representation which compares structural similarity (ECFP, MACCS) and activity similarity (for example, pIC_{50} or pK_i) on the basis

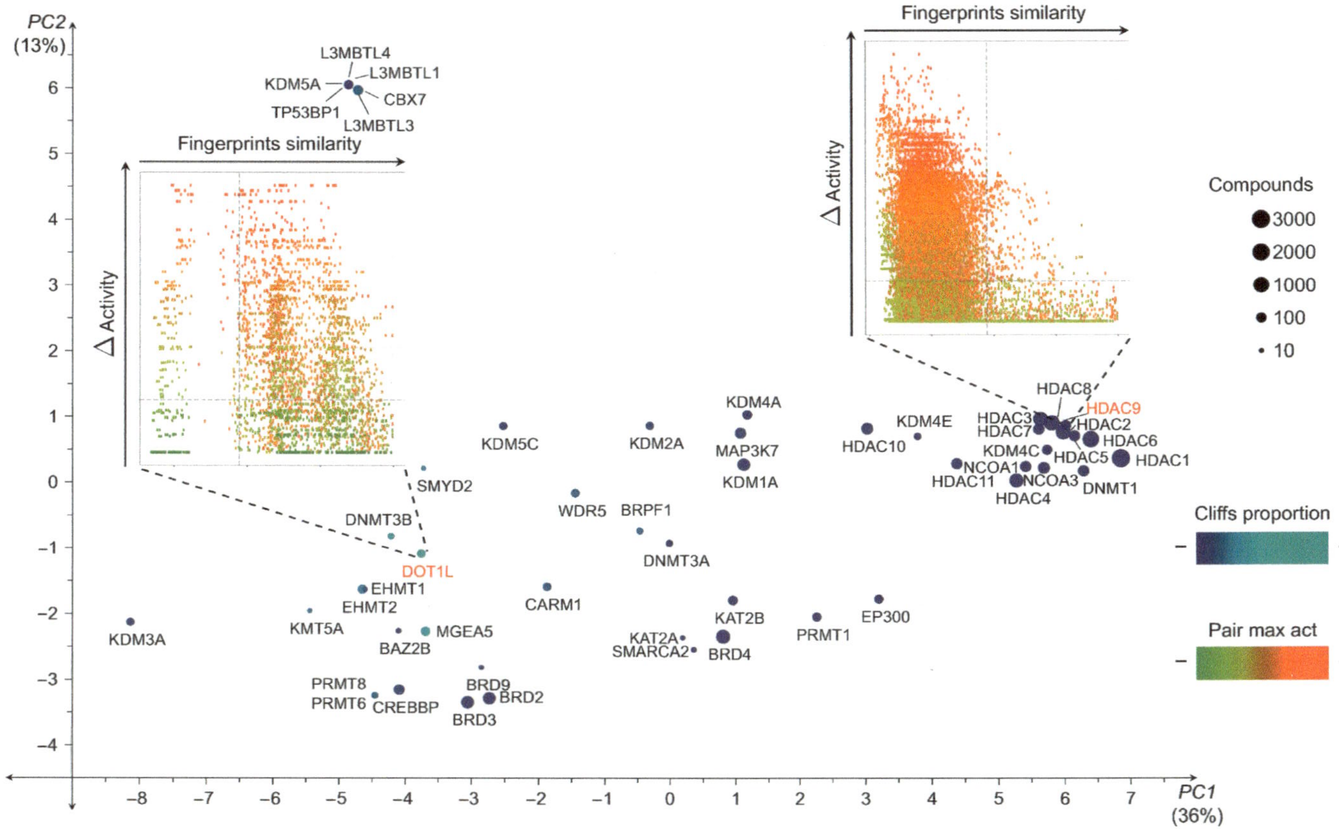

Fig. 1 Chemical space of epigenetic targets colored by the proportion of cliffs in each dataset. Two representative SAS maps (for DOT1L and HDAC9) are shown.

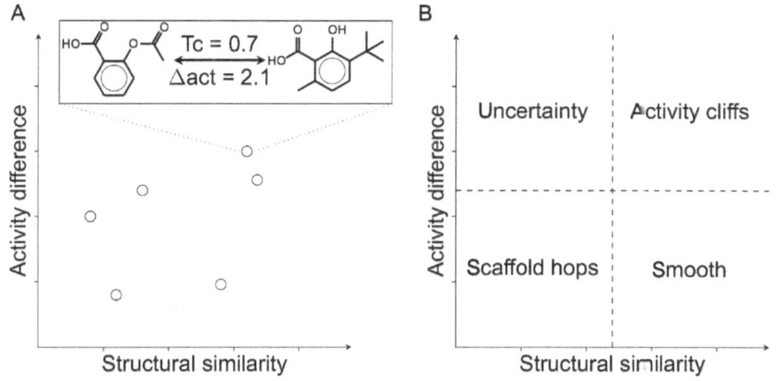

Fig. 2 (A) Schematic SAS map. Note that each point represents a pair of compounds. (B) Four major quadrants in a SAS map.

of systematic pairwise compound comparisons (Shanmugasundaram & Maggiora, 2001). Each point in a SAS map represents a pair of compounds and is colored according to the most active compound of the pair. The resultant plot (schematically illustrated in Fig. 2) can be roughly divided into four quadrants with thresholds defined a priori: (a) smooth (high structural similarity and low activity difference), (b) activity cliffs (high structural similarity but high activity difference), (c) scaffold hops (low structural similarity but low activity difference), and (d) uncertainty (low structural similarity and high activity difference) (Bajorath et al., 2009; Guha, 2012; Medina-Franco, 2012).

3.2 Structure–Activity Landscape Index

SALI was designed to identify activity cliffs and compounds that represent key inflection points on activity landscapes. This metric gives a score to a pair of compounds based on the comparison of the structural similarity and the difference between their potency. SALI is calculated as follows:

$$\mathrm{SALI}_{i,j} = \frac{|A_j - A_i|}{1 - \mathrm{sim}(i,j)}$$

where A_i and A_j are the potency values of the ith and jth molecule, respectively, and sim(i,j) is the similarity of the two molecules (Bajorath et al., 2009; Guha, 2012). Higher values of SALI correspond to more pronounced activity cliffs (Bajorath et al., 2009; Guha, 2012).

3.3 SAS Maps of Epigenetic Targets

As discussed in Section 3.1, exploring the SAR of compound data sets provides key information to determine the approach to further developing the compounds. Thus, if the SAR is continuous (e.g., similar compounds have similar activity), it is possible to implement methods based on the similarity principle. On the other hand, if the data set has a discontinuous SAR, insights from the activity landscape might provide specific information on the pharmacophore features that will be key for further compound development.

Herein, SAS maps for all the 52 epigenetic targets were generated and analyzed (Supplementary Fig. S2 in the online version at https://doi.org/10.1016/bs.apcsb.2018.01.001). It was found that the epigenetic targets DOT1L, DNMT3B, and MGEA5 have a significant percentage of cliffs (overall more than 30%) with respect to other epigenetic targets studied in this analysis (Fig. 3). These results suggest that, in general, small changes on the structure of the inhibitors of DOT1L, DNMT3B, and MGEA5 could generate large changes on their biological activity.

DOT1L is the epigenetic target that has the largest percentage of activity cliffs. In addition, similar to the inhibitors of EHMT2, inhibitors of DOT1L have a large percentage of "deep cliffs," i.e., a small modification in the structure will modify the activity in more than two logarithmic units. In comparison, DNMT3B and PRMT8 are the epigenetic targets with the largest percentage of shallow cliffs, i.e., activity differences between one and two logarithmic units. Such targets with many activity cliffs and therefore a discontinuous SAR are less suited for applying predictive methods that rely upon the similarity principle.

In contrast, other epigenetic targets have an overall continuous SAR. Remarkable examples are most of the HDACs. As shown in Fig. 3B, the majority of the pairs are in the smooth SAR region. This is no surprise, since HDACs are among the most well-understood epigenetic targets. A polarized example is provided by EHMT, which has both ~51% of compounds in the smooth SAR region and ~19% of activity cliffs. Interestingly, SMARCA2 and NCOA1 had a large proportion of scaffold hops, perhaps pointing toward multiple binding sites on the enzymes. Table 2 presents the targets with the highest percentages of points in each of the SAS maps quadrants, for each of the studied epigenetic families. Of note, although HDAC family is the one with the largest number of evaluated compounds, HKM family (and particularly DOT1L and EHMT2) has a larger proportion of

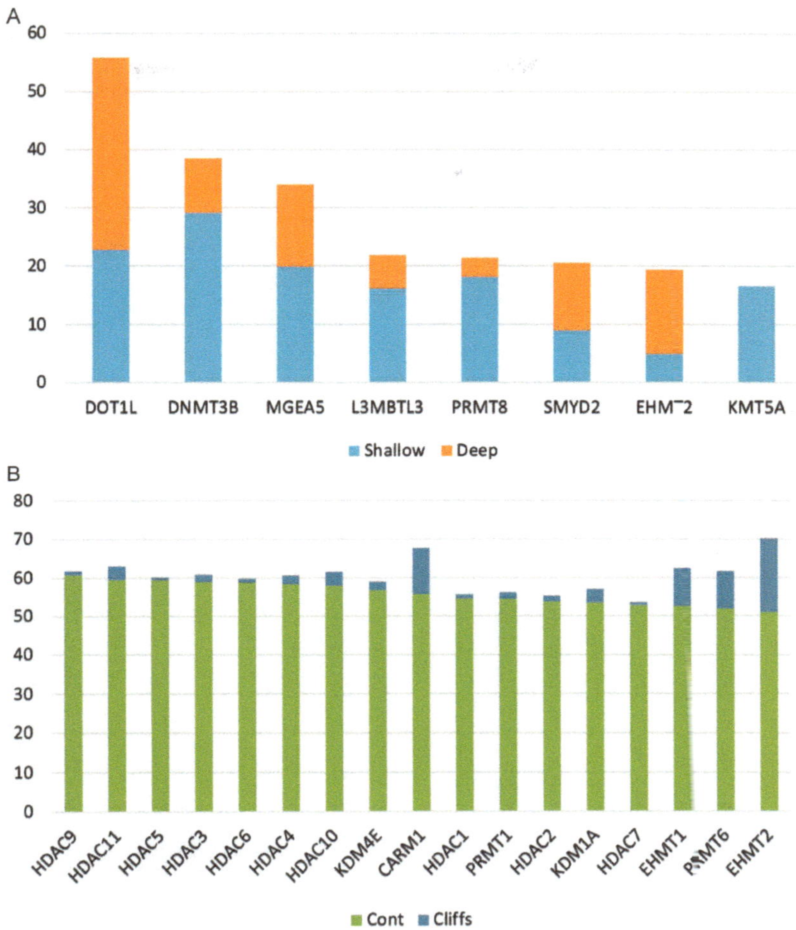

Fig. 3 Epigenetic targets with the largest proportion of pairs of compounds: (A) in the activity cliff region (shallow and deep cliffs are differentiated) and (B) in the continuous region of the SAS maps.

activity cliffs. Mean SALI values are not quite linearly correlated with the proportion of activity cliffs of each target ($r = 0.18$). However, WDR5 has the highest mean SALI, and it has a discontinuous SAR (Supplementary Figs. S1 and S2 in the online version at https://doi.org/10.1016/bs.apcsb.2018.01.001).

3.4 Activity Cliffs Generators

An ACG is defined as "a molecule with high probability to form activity cliffs with structurally similar molecules tested in the same biological assay"

Table 2 Percentage of Data Points in Different Regions of the SAS Maps and Mean SALI

		Activity Cliffs (%)					
Target Family	***n***	**Total**	**Deep**	**Shallow**	**Scaffold Hops (%)**	**Continuous (%)**	**SALI Mean**
BRD	200,028 BRD4	16.21 BRPF1	2.37 BRPF1	13.83 BRPF1	43.98 BAZ2B	48.79 BRD4	16.72 BRPF1
DNMT	29,646 DNMT1	38.72 DNMT3B	9.36 DNMT3B	29.10 DNMT3B	67.03 DNMT1	43.57 DNMT3A	4.70 DNMT3B
HAT	158,766 NCOA1NCOA3	7.94 CREBBP	0.08 EP300	7.80 CREBBP	84.49 NCOA1	39.44 NCOA3	1.75 CREBBP
HDAC	5,227,761 HDAC1	3.60 HDAC11	1.80 HDAC11	1.79 HDAC11	46.44 HDAC8	60.76 HDAC9	2.57 HDAC11
HKM	5995 EHMT2	56.11 DOT1L	33.06 DOT1L	22.78 DOT1L	22.13 EHMT1	52.57 EHMT1	17.38 KMT5A
KDM	102,378 KDM1A	7.54 KDM2A	1.10 KDM1A	6.23 KDM2A	66.17 KDM4A	56.80 KDM4E	6.45 KDM4A
KMeR	7750 L3MBTL1	21.83 L3MBTL3	5.70 L3MBTL3	16.13 L3MBTL3	38.08 L3MBTL1	34.67 WDR5	25.60 WDR5
PRMT	9870 PRMT1	21.74 PRMT8	4.19 CARM1	18.12 PRMT8	39.52 PRMT1	55.71 CARM1	5.70 CARM1
Others	24,090 SMARCA2	34.09 MGEA5	14.16 MGEA5	19.86 MGEA5	90.25 SMARCA2	37.79 MAP3K7	15.12 MGEA5

(Pérez-Villanueva, Méndez-Lucio, Soria-Arteche, & Medina-Franco, 2015). This kind of molecules provides relevant insights on the pharmacophoric regions. Below, examples of ACGs found in the database are described in detail and illustrated using the chemical neighborhood graphs devised by Namasivayam, Iyer, and Bajorath (2012).

3.4.1 Histone Deacetylases

HDACs are enzymes that regulate the gene transcription by catalyzing the deacetylation of the ε-amino group of lysine side chains of histone and non-histone proteins. Deacetylation promotes a stronger interaction between positively charged histones and negatively charged DNA, leading to a more condensed chromatin structure and gene transcription silencing (Ragno, 2016; Richon, 2006).

There are 18 human HDACs subdivided into four classes (Xu, Parmigiani, & Marks, 2007). Classes I (HDAC 1, 2, 3, and 8), IIa (HDAC 4, 5, 7, and 9), IIb (HDAC 6 and 10), and IV (HDAC 11) are metal dependent and share the catalytic system (Botta et al., 2011; Negmeldin, Padige, Bieliauskas, & Pflum, 2017). Some types of cancer have an altered gene expression of these enzymes. For example, there is an overexpression of HDAC1 in prostate, gastric, colon, and breast carcinomas; in turn, HDAC2 is overexpressed in colorectal, gastric, and cervical cancer (Kim & Bae, 2011; Kral et al., 2014). On the other hand, their inhibition causes histone hyperacetylation associated with cell cycle arrest or apoptosis in tumor cells (Xu et al., 2007). In addition, the acetylation of p53 by HATs increments its binding to DNA, leading to the expression of p53-regulated genes (Botta et al., 2011; Kral et al., 2014; Richon, 2006).

Some HDAC inhibitors have already been approved by the FDA as anticancer drugs: e.g., vorinostat, belinostat, panobinostat, and romidepsin. The former three contain a hydroxamic acid group and the latter is a cyclic peptide (Kim & Bae, 2011; Negmeldin et al., 2017). Moreover, HDAC inhibitors might be beneficial in different types of neurodegenerative and cardiovascular diseases, as well as in inflammatory disorders (Kim & Bae, 2011). Hydroxamic acid derivative inhibitors have a similar scaffold which can be divided into three parts: (1) a Zn^{2+} binding group (ZBG) that coordinates to the catalytic metal atom in the "tube-like" active site; (2) a linker, which helps to find the correct position and fits into the hydrophobic "tube"; and (3) a "cap" region, which interacts with the rim of the active pocket (Botta et al., 2011; Hou et al., 2012).

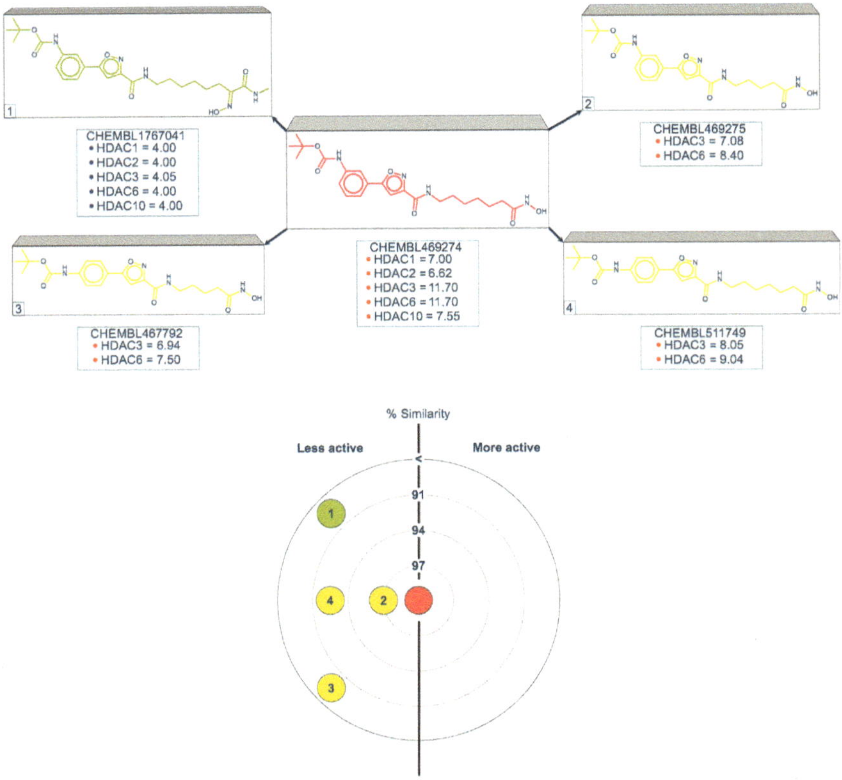

Fig. 4 Example of an activity cliff generator (CHEMBL469274) that is active against a variety of HDACs.

CHEMBL469274 is an ACG that is active against a variety of HDACs, but it is prone to forming activity cliffs (Fig. 4). The most dramatic drop in activity is shown by CHEMBL1767041, which lacks an efficient ZBG (Ragno, 2016). On the other hand, although the ZBG is conserved in CHEMBL469275, CHEMBL467792, and CHEMBL511749, the large decrease in the activity values seems to be a result of modifications in the linker length (CHEMBL469275 and CHEMBL467792) or regioisomerism (CHEMBL469275 and CHEMBL469274), which might prevent an adequate fitting in the binding pocket of the enzyme.

3.4.2 Histone Methyltransferases EHMT2 and DOT1L

Histone methyltransferases (HMTs) are currently under research, given the dysregulation of this metabolic pathway in cancer cells (Curry et al., 2015; Lu et al., 2013). Also, DOT1L has been described as a potentially selective therapy against some types of leukemia (Daigle et al., 2011; Sarkaria, Christopher, Klco, & Ley, 2014).

Some interesting ACGs were identified as inhibitors of HMTs and are presented in Figs. 5 and 6. The most pronounced activity cliffs from an ACG are those in Fig. 5. CHEMBL2441080 is a compound active against EHMT2, a HMT that has been found to be linked to carcinogenic processes (Lu et al., 2013). Fig. 5 shows four structurally similar compounds that

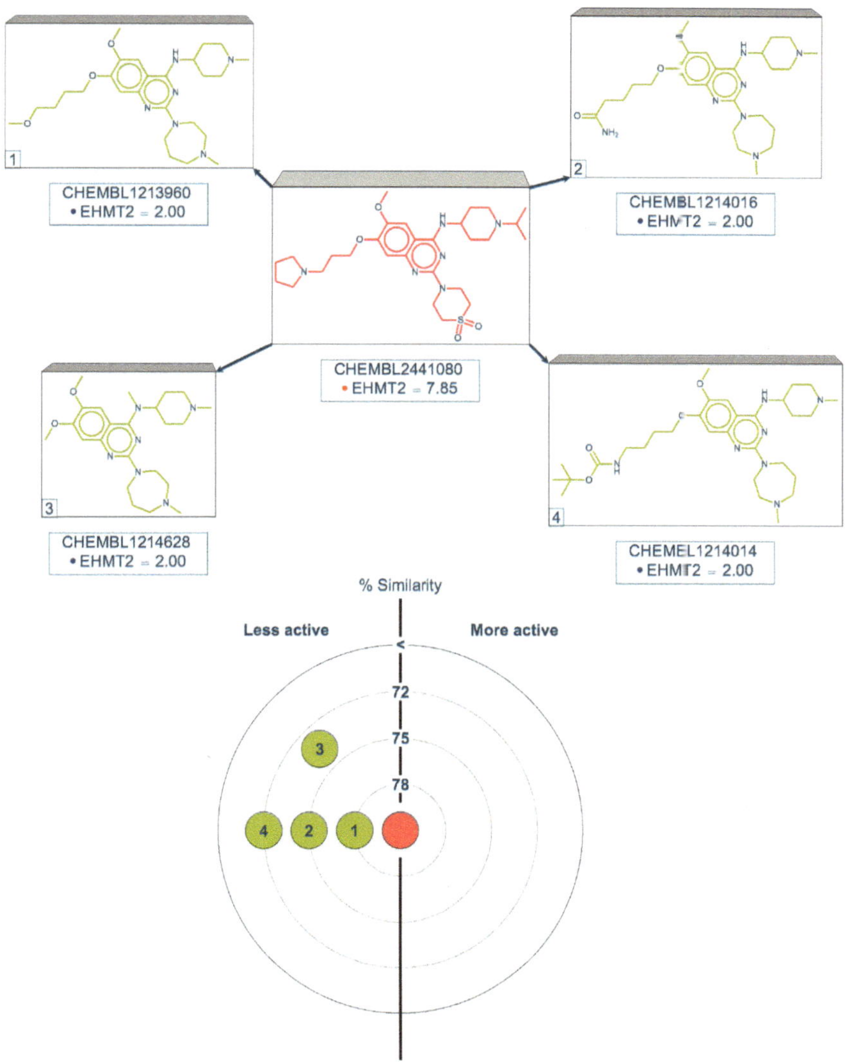

Fig. 5 Example of an activity cliff generator (CHEMBL2441080) that is active against EHMT2.

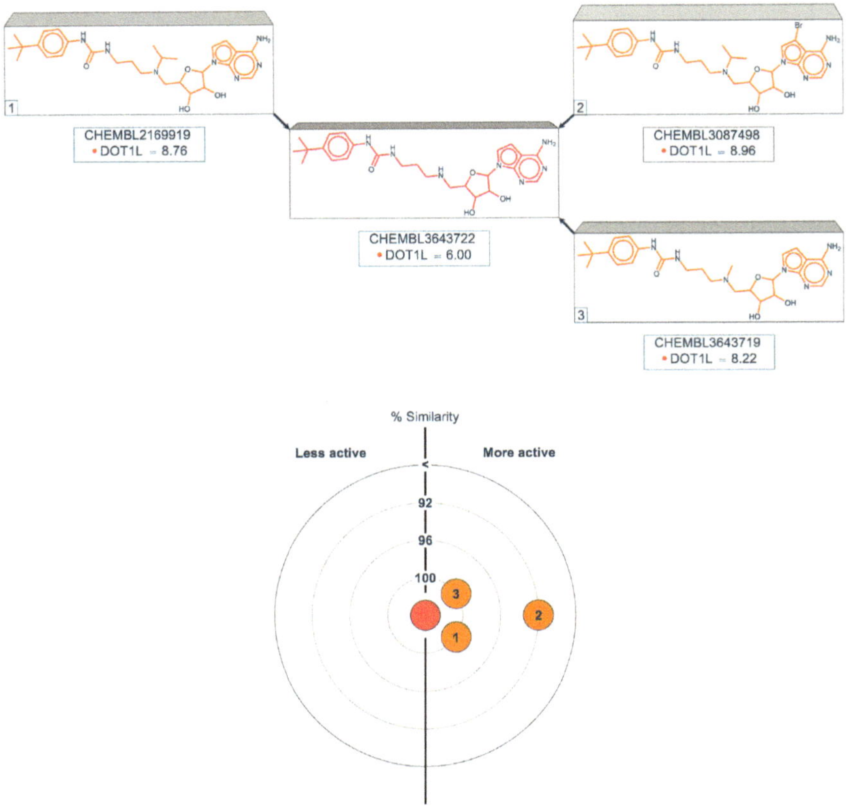

Fig. 6 Example of an activity cliff generator (CHEMBL3643722) that is active against DOT1L, although less active than its pairs.

have lost activity. They all share the elimination of a sulfone functional group, addition of a carbon to the ring where the sulfone is present in the ACG, a substitution of an isopropyl in a lateral chain by a methyl, and they differ in the substitutions done on the longest lateral chain. Despite the high chemical similarity identified through molecular fingerprints, the relatively high number of modifications make difficult to categorically telling which is the culprit of the loss of activity. A study on the SAR of 2,4-diamino-7-aminoalkoxy-quinazolines, such as the activity cliffs surrounding this ACG, showed that, in fact, there are compounds with this scaffold that are very active (Liu et al., 2010). Therefore, the long lateral chain mostly plays a predominant role on the differences observed in Fig. 5. Indeed, as per the crystallographic structure presented therein, the lack of the nitrogen four positions away from the oxygen on this same chain may disrupt electrostatic and cation–pi interactions with Leu1086 and Tyr1154, respectively (Liu et al., 2010).

A final illustrative example is the case of the ACG CHEMBL3643722 (Fig. 6). This is a compound with moderate activity against DOT1L, which is another HMT. In this case, turning the secondary amine into a tertiary amine enhances activity in more than two logarithmic units. Also, substituting with an isopropyl seems to provide higher activity than with a methyl group.

A SAR study on adenosine analogs as inhibitors of DOT1L found that CHEMBL2169919 (also known as EPZ004777) amine group can be replaced without significant loss of activity by —S—, similar to SAM, the enzyme's cofactor required for methylation (Anglin et al., 2012). However, the ACG identified in this study further suggests that the substitution of the amine is relevant in the activity.

3.5 Epigenetic Targets With Continuous SAR

According to Table 2, HDACs present the most continuous SAR in the database. The molecules in this region have high structural similarity and low potency differences (Fig. 2B), i.e., they follow the similarity principle. This property can be used to optimize the scaffold of active compounds in order to generate more potent and selective analogs. Overall, HDAC inhibitors are more predominant in this area (HDAC9 has the maximum, ~60% of pairs in the continuous region). Besides, they also have low presence of activity cliffs (HDAC11 has the maximum, ~4%). These facts, in principle, allow the efficient development of in silico predictive models, such as QSAR. Moreover, some HDAC inhibitors have a complex molecular structure, favoring the generation of a large number of analogs from the same synthetic route. However, the scaffold hop region of HDAC inhibitors is of considerable size (up to 46% in HDAC8), which could suggest that either these targets have multiple binding sites, or there are multiple chemotypes able to inhibit the same binding site.

3.6 Epigenetic Targets With Scaffold Hops

In clear contrast to the epigenetic targets with a large proportion of activity cliffs and data points in the continuous region of the SAR, SMARCA2 and HAT are the targets with the two largest proportions of data points in the region of the scaffold hops of the landscape (Table 2). This result suggests there is a large diversity of compounds with similar activity toward SMARC2 and HAT. If these compounds are active, there would be a significant opportunity to develop the most promising scaffolds with adequate drug-like properties. Also, having different chemical scaffolds with similar activity opens up more venues to synthesize analogs of the most tractable scaffolds and increases the chances to generate intellectual property (that is a sensitive point in particular to the pharmaceutical industry). From the point of view of the epigenetic targets, those targets with a large proportion of activity cliffs mean that they might be more promiscuous and more adaptable (flexible to accommodate ligands with different scaffolds). In addition, the large proportion of scaffold hops for SMARCA2 and HAT led to the hypothesis that active compounds toward these epigenetic targets may act through different mechanisms of action, e.g., bind in different binding sites. Despite the fact the structure-based interpretation of the scaffolds hops (and other regions of the landscape) is addressed in a separate study, the quantification of the activity landscape (e.g., Table 2 and SAS maps in Supplementary Fig. S2 in the online version at https://doi.org/10.1016/bs.apcsb.2018.01.001) points to specific targets that have a distinct landscape.

4. CONCLUSIONS AND PERSPECTIVES

An in-depth epigenetic activity landscape study for 52 epigenetic targets discussed in this work rapidly identified DOT1L and DNMT3B as the epigenetic targets with more discontinuous SAR, e.g., the largest proportion of activity cliffs. The study also found that HDACs are the targets with the most continuous SAR in epigenetics. The significance of this work is manifold: (a) it contributes to identify HDACs, in general, as those epigenetic targets are most suitable to perform a hit-to-lead optimization program (e.g., following the similarity principle), (b) it helped to identify or confirm small structural changes that have a large impact in the biological activity; and (c) it aided to uncover the epigenetic targets suitable to conduct traditional predictive computational approaches (such as QSAR) and those targets prone to scaffold hopping: SMARCA2 and HAT.

A major perspective of this work is to conduct structure–multiepigenetic activity relationships. A second major perspective is to rationalize activity cliffs using structure-based approaches such as molecular docking.

Supplementary data to this article can be found online at https://doi.org/10.1016/bs.apcsb.2018.01.001.

ACKNOWLEDGMENTS

This work was supported by the *Programa de Apoyo a Proyectos de Investigación e Innovación Tecnológica* (PAPIIT) grant IA203718, Universidad Nacional Autónoma de México (UNAM), *Consejo Nacional de Ciencia y Tecnología (CONACyT)* grant 282785, and the *Programa de Apoyo a la Investigación y el Posgrado* (PAIP) grant 5000-9163, Facultad de Química, UNAM. J.J.N. is thankful to *Consejo Nacional de Ciencia y Tecnología* (CONACyT) for the granted scholarship number 622969.

REFERENCES

Allis, C. D., & Jenuwein, T. (2016). The molecular hallmarks of epigenetic control. *Nature Reviews. Genetics*, *17*(8), 487–500. https://doi.org/10.1038/nrg.2016.59.

Anglin, J. L., Deng, L., Yao, Y., Cai, G., Liu, Z., Jiang, H., et al. (2012). Synthesis and structure-activity relationship investigation of adenosine-containing inhibitors of histone methyltransferase DOT1L. *Journal of Medicinal Chemistry*, *55*(18), 8066–8074. https://doi.org/10.1021/jm300917h.

Bajorath, J., Peltason, L., Wawer, M., Guha, R., Lajiness, M. S., & Van Drie, J. H. (2009). Navigating structure-activity landscapes. *Drug Discovery Today*, *14*(13–14), 698–705. https://doi.org/10.1016/j.drudis.2009.04.003.

Botta, C. B., Cabri, W., Cini, E., De Cesare, L., Fattorusso, C., Giannini, G., et al. (2011). Oxime amides as a novel zinc binding group in histone deacetylase inhibitors: Synthesis, biological activity, and computational evaluation. *Journal of Medicinal Chemistry*, *54*(7), 2165–2182. https://doi.org/10.1021/jm101373a.

Chung, C. (2015). Epigenetic drug discovery. In G. Scapin, D. Patel, & E. Arnold (Eds.), *Multifaceted roles of crystallography in modern drug discovery* (pp. 27–40). Dordrecht: Springer Netherlands. https://doi.org/10.1007/978-94-017-9719-1_3.

Curry, E., Green, I., Chapman-Rothe, N., Shamsaei, E., Kandil, S., Cherblanc, F. L., et al. (2015). Dual EZH2 and EHMT2 histone methyltransferase inhibition increases biological efficacy in breast cancer cells. *Clinical Epigenetics*, 7(84). https://doi.org/10.1186/s13148-015-0118-9.

Daigle, S. R., Olhava, E. J., Therkelsen, C. A., Majer, C. R., Sneeringer, C. J., Song, J., et al. (2011). Selective killing of mixed lineage leukemia cells by a potent small-molecule DOT1L inhibitor. *Cancer Cell*, *20*(1), 53–65. https://doi.org/10.1016/j.ccr.2011.06.009.

Dueñas-González, A., Naveja, J. J., & Medina-Franco, J. L. (2016). Introduction of epigenetic targets in drug discovery and current status of epi-drugs and epi-probes. In *Epi-Informatics* (pp. 1–20). Elsevier. https://doi.org/10.1016/B978-0-12-802808-7.00001-0.

Fernández-de Gortari, E., García-Jacas, C. R., Martinez-Mayorga, K., & Medina-Franco, J. L. (2017). Database fingerprint (DFP): An approach to represent molecular databases. *Journal of Cheminformatics*, *9*(1, 9). https://doi.org/10.1186/s13321-017-0195-1.

García-Sánchez, M. O., Cruz-Monteagudo, M., & Medina-Franco, J. L. (2017). Quantitative structure-epigenetic activity relationships. In K. Roy (Ed.), *Advances in QSAR modeling* (pp. 303–338). : *Vol. 24*. (pp. 303–338). Cham: Springer International Publishing. https://doi.org/10.1007/978-3-319-56850-8_8.

Gortari, E. F., & Medina-Franco, J. L. (2015). Epigenetic relevant chemical space: A chemoinformatic characterization of inhibitors of DNA methyltransferases. *RSC Advances*, *5*(106), 87465–87476. https://doi.org/10.1039/C5RA19611F.

Guha, R. (2012). Exploring structure-activity data using the landscape paradigm. *Wiley Interdisciplinary Reviews. Computational Molecular Science*, *2*(6). https://doi.org/10.1002/wcms.1087.

Guha, R., & Van Drie, J. H. (2008). Structure–activity landscape index: Identifying and quantifying activity cliffs. *Journal of Chemical Information and Modeling*, *48*(3), 646–658. https://doi.org/10.1021/ci7004093.

Hou, J., Li, Z., Fang, Q., Feng, C., Zhang, H., Guo, W., et al. (2012). Discovery and extensive in vitro evaluations of NK-HDAC-1: A chiral histone deacetylase inhibitor as a promising lead. *Journal of Medicinal Chemistry*, *55*(7), 3066–3075. https://doi.org/10.1021/jm201496g.

Jenuwein, T., & Allis, C. D. (2001). Translating the histone code. *Science*, *293*(5532), 1074–1080. https://doi.org/10.1126/science.1063127.

Kim, H.-J., & Bae, S.-C. (2011). Histone deacetylase inhibitors: Molecular mechanisms of action and clinical trials as anti-cancer drugs. *American Journal of Translational Research*, *3*(2), 166–179.

Kral, A. M., Ozerova, N., Close, J., Jung, J., Chenard, M., Fleming, J., et al. (2014). Divergent kinetics differentiate the mechanism of action of two HDAC inhibitors. *Biochemistry*, *53*(4), 725–734. https://doi.org/10.1021/bi400936h.

Liu, F., Chen, X., Allali-Hassani, A., Quinn, A. M., Wigle, T. J., Wasney, G. A., et al. (2010). Protein lysine methyltransferase G9a inhibitors: Design, synthesis, and structure activity relationships of 2,4-diamino-7-aminoalkoxy-quinazolines. *Journal of Medicinal Chemistry*, *53*(15), 5844–5857. https://doi.org/10.1021/jm100478y.

Lu, Z., Tian, Y., Salwen, H. R., Chlenski, A., Godley, L. A., Raj, J. U., et al. (2013). Histone-lysine methyltransferase EHMT2 is involved in proliferation, apoptosis, cell invasion, and DNA methylation of human neuroblastoma cells. *Anti-Cancer Drugs*, *24*(5), 484–493. https://doi.org/10.1097/CAD.0b013e32835ffdbb.

Lundstrom, K. (2017). Epigenetics: New possibilities for drug discovery. *Future Medicinal Chemistry*, *9*(5), 437–441. https://doi.org/10.4155/fmc-2017-0015.

Medina-Franco, J. L. (2012). Scanning structure-activity relationships with structure-activity similarity and related maps: From consensus activity cliffs to selectivity switches. *Journal of Chemical Information and Modeling*, *52*(10), 2485–2493. https://doi.org/10.1021/ci300362x.

Medvedeva, Y. A., Lennartsson, A., Ehsani, R., Kulakovskiy, I. V., Vorontsov, I. E., Panahandeh, P., et al. (2015). EpiFactors: A comprehensive database of human epigenetic factors and complexes. *Database: The Journal of Biological Databases and Curation*, *2015*, bav067. https://doi.org/10.1093/database/bav067.

Méndez-Lucio, O. (2016). Computational structure–activity relationship studies of epigenetic target inhibitors. In *Epi-Informatics* (pp. 359–384). Elsevier. https://doi.org/10.1016/B978-0-12-802808-7.00013-7.

Méndez-Lucio, O., Pérez-Villanueva, J., Castillo, R., & Medina-Franco, J. L. (2012). Identifying activity cliff generators of PPAR ligands using SAS maps. *Molecular Informatics*, *31*(11–12), 837–846. https://doi.org/10.1002/minf.201200078.

Namasivayam, V., Iyer, P., & Bajorath, J. (2012). Exploring SAR continuity in the vicinity of activity cliffs. *Chemical Biology & Drug Design*, *79*(1), 22–29. https://doi.org/10.1111/j.1747-0285.2011.01256.x.

Naveja, J. J., & Medina-Franco, J. L. (2015a). Activity landscape of DNA methyltransferase inhibitors bridges chemoinformatics with epigenetic drug discovery. *Expert Opinion on Drug Discovery*, *10*(10), 1059–1070. https://doi.org/10.1517/17460441.2015.1073257.

Naveja, J. J., & Medina-Franco, J. L. (2015b). Activity landscape sweeping: Insights into the mechanism of inhibition and optimization of DNMT1 inhibitors. *RSC Advances*, *5*(78), 63882–63895. https://doi.org/10.1039/C5RA12339A.

Naveja, J. J., & Medina-Franco, J. L. (2018). Insights from pharmacological similarity of epigenetic targets in epi-polypharmacology. *Drug Discovery Today*, *23*(1), 141–150. https://doi.org/10.1016/j.drudis.2017.10.006.

Negmeldin, A. T., Padige, G., Bieliauskas, A. V., & Pflum, M. K. H. (2017). Structural requirements of HDAC inhibitors: SAHA analogues modified at the C2 position display HDAC6/8 selectivity. *ACS Medicinal Chemistry Letters*, *8*(3), 281–286. https://doi.org/10.1021/acsmedchemlett.6b00124.

Peltason, L., & Bajorath, J. (2007). SAR index: Quantifying the nature of structure-activity relationships. *Journal of Medicinal Chemistry*, *50*(23), 5571–5578. https://doi.org/10.1021/jm0705713.

Pérez-Villanueva, J., Méndez-Lucio, O., Soria-Arteche, O., & Medina-Franco, J. L. (2015). Activity cliffs and activity cliff generators based on chemotype-related activity landscapes. *Molecular Diversity*, *19*(4), 1021–1035. https://doi.org/10.1007/s11030-015-9609-z.

Pérez-Villanueva, J., Santos, R., Hernández-Campos, A., Giulianotti, M. A., Castillo, R., & Medina-Franco, J. L. (2010). Towards a systematic characterization of the antiprotozoal activity landscape of benzimidazole derivatives. *Bioorganic & Medicinal Chemistry*, *18*(21), 7380–7391. https://doi.org/10.1016/j.bmc.2010.09.019.

Prieto-Martínez, F. D., Gortari, E. F., Méndez-Lucio, O., & Medina-Franco, J. L. (2016). A chemical space odyssey of inhibitors of histone deacetylases and bromodomains. *RSC Advances*, *6*(61), 56225–56239. https://doi.org/10.1039/C6RA07224K.

Ragno, R. (2016). Structure-based modeling of histone deacetylases inhibitors. In *Epi-Informatics* (pp. 155–212): Elsevier. https://doi.org/10.1016/B978-0-12-802808-7.00006-X.

Richon, V. M. (2006). Cancer biology: Mechanism of antitumour action of vorinostat (suberoylanilide hydroxamic acid), a novel histone deacetylase inhibitor. *British Journal of Cancer*, *95*, S2–S6. https://doi.org/10.1038/sj.bjc.6603463.

Saldívar-González, F. I., Naveja, J. J., Palomino-Hernández, O., & Medina-Franco, J. L. (2017). Getting SMARt in drug discovery: Chemoinformatics approaches for mining structure–multiple activity relationships. *RSC Advances*, 7(2), 632–641. https://doi.org/10.1039/C6RA26230A.

Sarkaria, S. M., Christopher, M. J., Klco, J. M., & Ley, T. J. (2014). Primary acute myeloid leukemia cells with IDH1 or IDH2 mutations respond to a DOT1L inhibitor in vitro. *Leukemia*, *28*(12), 2403–2406. https://doi.org/10.1038/leu.2014.235.

Shanmugasundaram, V., & Maggiora, G. M. (2001). In *Characterizing property and activity landscapes using an information-theoretic approach CINF-032. Presented at the 222nd ACS national meeting, Chicago, IL.*

Wassermann, A. M., Wawer, M., & Bajorath, J. (2010). Activity landscape representations for structure-activity relationship analysis. *Journal of Medicinal Chemistry*, *53*(23), 8209–8223. https://doi.org/10.1021/jm100933w.

Xu, W. S., Parmigiani, R. B., & Marks, P. A. (2007). Histone deacetylase inhibitors: Molecular mechanisms of action. *Oncogene*, *26*(37), 5541–5552. https://doi.org/10.1038/sj.onc.1210620.

CHAPTER FOUR

The OECD Principles for (Q)SAR Models in the Context of Knowledge Discovery in Databases (KDD)

Gabriela Gómez-Jiménez, Karla Gonzalez-Ponce, Durbis J. Castillo-Pazos, Abraham Madariaga-Mazon, Joaquin Barroso-Flores, Fernando Cortes-Guzman, Karina Martinez-Mayorga[1]

Instituto de Química, Universidad Nacional Autónoma de México, Mexico City, Mexico
[1]Corresponding author: e-mail address: kmtzm@unam.mx

Contents

Abstract

The steps followed in the knowledge discovery in databases (KDD) process are well documented and are widely used in different areas where exploration of data is used for decision making. In turn, while different workflows for developing quantitative structure–activity relationship (QSAR) models have been proposed, including combinatorial use of QSAR, there is now agreement on common requirements for building trustable predictive models. In this work, we analyze and confront the steps involved in KDD and QSAR and present how they comply with the OECD principles for the validation, for regulatory purposes, of QSAR models.

Advances in Protein Chemistry and Structural Biology, Volume 113
ISSN 1876-1623
https://doi.org/10.1016/bs.apcsb.2018.04.001

1. INTRODUCTION

The evolution of QSAR models (Tropsha & Golbraikh, 2007; Tropsha & Wang, 2007) led to its acceptance of validated models in the prediction of toxicity endpoints for regulatory purposes (Veith, 2004). A practical survey on common mistakes on the development of QSAR models and how they relate to the OECD principles is provided by Dearden, Cronin, and Kaiser (2009). A perspective of the field, provided by Cherkasov et al. (2014), has proven adequate 10 years after its publication, and a more recent opinion by Rusyn and Daston (2010) discusses how the expectations have been met, the challenges ahead, and the potentials for further development and applications on the field. The OECD guidelines for the testing of chemicals are internationally accepted as standard methods for safety testing and assessment of pesticides, personal care products, industrial chemicals, and even to aid on decision making in emergency responses (Demchuk, Ruiz, Chou, & Fowler, 2011; Ruiz, Begluitti, Tincher, Wheeler, & Mumtaz, 2012). These guidelines are regularly updated with the assistance of thousands of experts from OECD member countries. In November 2004, the OECD member countries agreed on the "OECD Principles for the Validation, for Regulation Purposes, of (Quantitative) Structure-Activity Relationship Models" (OECD, 2004). Some of the member countries and corresponding regulatory agencies are listed in Table 1. How and where the steps of KDD and QSAR methodologies fit in the OECD principles are summarized in Table 2. Initial steps in the KDD process include the incorporation, extraction, integration, and preparation of the data. This allows estimating the adequacy, quantity, quality, structure, and format of the data. Notably, details on the preparation of the data are not explicitly described on the OECD principles. However, this is indirectly evaluated, as it affects, for example, the applicability domain. Although Table 2 sets the steps on KDD and QSAR side by side to the different OECD principles, the recursive nature of these methodologies should be acknowledged.

2. DEFINITION OF THE GOALS

The first step in KDD is to define the goal of the study; this corresponds to the first principle of the OECD guidelines "A defined endpoint" and also the first step in QSAR models. For regulatory purposes, to define

Table 1 Regulatory Agencies and International Commissions and Regulations

OECD http://www.oecd.org/	The Organization for Economic Cooperation and Development (OECD) is an international cooperation body, made up of 35 states, whose objective is to coordinate their economic and social policies goals. The main objectives of the organization are: Contribute to a healthy economic expansion in the member countries, as well as nonmembers, in the process of economic development. Favor the expansion of world trade on a multilateral and nondiscriminatory basis in line with international obligations. Make the greatest possible expansion of the economy and employment and progress in the standard of living within the member countries, maintaining financial stability and thus contributing to the development of the world economy
ICH http://www.ich.org/products/guidelines/quality/article/quality-guidelines.html	The International Council for Harmonisation of Technical Requirements for Pharmaceuticals for Human Use (ICH) brings together the regulatory authorities and pharmaceutical industry to discuss scientific and technical aspects of drug registration. Since its inception in 1990, ICH has gradually evolved, to respond to the increasingly global face of drug development. ICH's mission is to achieve greater harmonization worldwide to ensure that safe, effective, and high-quality medicines are developed and registered in the most resource-efficient manner
Europe	
ECHA–REACH https://echa.europa.eu/	The European Chemicals Agency (ECHA) is an agency of the European Union, which manages the technical, scientific, and administrative aspects of the Registration, Evaluation, Authorisation, and Restriction of Chemicals (REACH) of the system
EMA http://www.ema.europa.eu/ema/	The European Medicines Agency (EMA) is a decentralized agency of the European Union (EU). It began operating in 1995. The Agency is responsible for the scientific evaluation, supervision, and safety monitoring of medicines in the EU. EMA protects public and animal health in 28 EU Member States, as well as the countries of the European Economic Area, by ensuring that all medicines available on the EU market are safe, effective, and of high quality

Continued

Table 1 Regulatory Agencies and International Commissions and Regulations—cont'd

Mexico	
COFEPRIS http://www.gob.mx/cofepris/	The Federal Commission for the Protection against Sanitary Risk (COFEPRIS) is a decentralized organ of the Mexican Department of Health with technical, administrative and operational autonomy, whose mission is to protect the population against sanitary risks, through sanitary regulation, control, and promotion
United States	
EPA https://www.epa.gov/	The EPA's mission is to protect human health and the environment. The purpose of the EPA is to ensure that: All Americans are protected from significant risks to human health and the environment where they live, learn, and work; National efforts to reduce environmental risk are based on the best available scientific information; Federal laws protecting human health and the environment are applied fairly and effectively
FDA https://www.fda.gov/	The Food and Drug Administration (FDA) is responsible for protecting and promoting public health through the control and supervision of food safety, tobacco products, dietary supplements, prescription and over-the-counter pharmaceutical drugs (medications), vaccines, biopharmaceuticals, blood transfusions, medical devices, electromagnetic radiation emitting devices (ERED), cosmetics, animal foods and feed, and veterinary products
MHLW (Japan) http://www.mhle.go.jp/	The Ministry of Health, Labour and Welfare is a governmental body responsible for pharmaceutical affairs. It provides regulations on maximum residue limits for agricultural chemicals in foods, basic food and regulations, standards for foods and food additives, among others

Table 2 Steps Involved in the KDD and QSAR Methodologies and the Related OECD Principles

KDD	QSAR	OECD Principles
Definition of the goals	Defined endpoint	Defined endpoint
Creation or selection of a data set	Data preparation	
Data cleaning and preprocessing		
Data reduction and projection		
Selection of data mining methods		
Exploratory analysis and model/ hypothesis selection	Data splitting/model generation/internal validation	A defined domain of applicability
Data mining	Model generation	Unambiguous algorithm
Evaluation	External validation/ experimental evaluation	Appropriate measures of goodness of fit, robustness, and predictivity
Interpretation/ utilization		Mechanistic interpretation, if possible

which endpoints can be predicted requires specific considerations. The endpoints that are candidates for predictive models, according to different regulatory agencies are listed in Table 3. Such regulation is admitted in member countries located in Europe, America (Canada, Mexico, and the United States), and Asia (Japan), among others. The current list of OECD member countries is publically accessible (OECD, 2008). To provide a validated prediction, endpoints should be toxicologically relevant. Also, there should be sufficient information in the structural and biological spaces, and preferentially, a wealth of biological information should be available to provide a mechanistic interpretation.

The complexity of an endpoint affects its transferability and predictability. By definition, complex systems cannot be derived (explained) from its parts. Molecules are complex systems and as such cannot be explained, for example, from one representation of its parts. To overcome this problem,

Table 3 Toxicological Endpoints Requested by Regulatory Agencies

OECD—Health Effects
Acute oral toxicity
Acute dermal toxicity
Acute inhalation toxicity
Acute dermal irritation/corrosion
Acute eye irritation/corrosion
Skin sensitization
Repeated dose 28-day oral toxicity study in rodents
Repeated dose 90-day oral toxicity study in rodents
Repeated dose 90-day oral toxicity study in nonrodents
Repeated dose dermal toxicity: 21/28-day study
Subchronic dermal toxicity: 90-day study
Subacute inhalation toxicity: 28-day study
Subchronic inhalation toxicity: 90-day study
Prenatal development toxicity study
One-generation reproduction toxicity study
Two-generation reproduction toxicity
Toxicokinetics
Delayed neurotoxicity of organophosphorus substances following acute exposure
Delayed neurotoxicity of organophosphorus substances: 28-day repeated dose study
Acute oral toxicity—fixed dose procedure
Reproduction/developmental toxicity screening test
Combined repeated dose toxicity study with the reproduction/developmental toxicity screening test
Acute oral toxicity—acute toxic class method
Neurotoxicity study in rodents
Acute oral toxicity: Up-and-down procedure
Developmental neurotoxicity study
Skin absorption: In vivo method

Table 3 Toxicological Endpoints Requested by Regulatory Agencies—cont'd

OECD—Health Effects
Skin absorption: In vitro method
In vitro skin corrosion: Transcutaneous electrical resistance test (TER)
In vitro skin corrosion: Human skin model test
In vitro 3T3 NRU phototoxicity test
In vitro membrane barrier test method for skin corrosion
Acute inhalation toxicity—acute toxic class method
Bovine corneal opacity and permeability test method for identifying (i) chemicals inducing serious eye damage and (ii) chemicals not requiring classification for eye irritation or serious eye damage
Isolated chicken eye test method for identifying ocular corrosives and severe irritants
In vitro skin irritation
Uterotrophic bioassay in rodents
Hershberger bioassay in rats
Extended one-generation reproductive toxicity study
Carcinogenicity studies
Chronic toxicity studies
Combined chronic toxicity/carcinogenicity studies
Performance-based test guideline for stably transfected transactivation in vitro assays to detect estrogen receptor agonists
H295R steroidogenesis assay
BG1Luc estrogen receptor transactivation test method for identifying estrogen receptor agonists and antagonists
Stably transfected human androgen receptor transcriptional activation assay for detection of androgenic agonist and antagonist activity of chemicals
Fluorescein leakage test method for identifying ocular corrosives and severe irritants
Bacterial reverse mutation test
In vitro mammalian chromosomal aberration test
Mammalian erythrocyte micronucleus test
Mammalian bone marrow chromosomal aberration test

Continued

Table 3 Toxicological Endpoints Requested by Regulatory Agencies—cont'd

OECD—Health Effects	
In vitro mammalian cell gene mutation test	
Genetic toxicology: Sex-linked recessive lethal test in *Drosophila melanogaster*	
Genetic toxicology: Rodent dominant lethal test	
Rodent dominant lethal test	
Genetic toxicology: In vitro sister chromatid exchange assay in mammalian cells	
Genetic toxicology: *Saccharomyces cerevisiae*, gene mutation assay	
Genetic toxicology: *Saccharomyces cerevisiae*, mitotic recombination assay	
Genetic toxicology: DNA Damage and repair, unscheduled DNA synthesis in mammalian cells in vitro	
Mammalian spermatogonial chromosomal aberration test	
Genetic toxicology: Mouse spot test	
Genetic toxicology, mouse heritable translocation assay	
Unscheduled DNA synthesis (UDS) test with mammalian liver cells in vivo	
In vitro mammalian cell micronucleus test	
Transgenic rodent somatic and germ cell gene mutation assays	
In vivo mammalian alkaline comet assay	
In vitro mammalian cell gene mutation tests using the thymidine kinase gene	
Short-time exposure in vitro test method for identifying (i) chemicals inducing serious eye damage and (ii) chemicals not Requiring classification for eye irritation or serious eye damage	
Reconstructed human cornea-like epithelium (RhCE) test method for identifying chemicals not requiring classification and labeling for eye irritation or serious eye damage	
Performance-based test guideline for human recombinant estrogen receptor (hrER) in vitro assays to detect chemicals with ER-binding affinity	
ECHA—REACH	
Toxicological information	
Skin irritation or skin corrosion	In vivo skin irritation
Eye irritation	In vivo eye irritation
Mutagenicity	In vitro gene mutation study in bacteria

Table 3 Toxicological Endpoints Requested by Regulatory Agencies—cont'd
ECHA—REACH

	In vitro cytogenicity study in mammalian cells or in vitro micronucleus study
	In vitro gene mutation study in mammalian cells
Acute toxicity	By oral route
	By inhalation
	By dermal route
Repeated dose toxicity	Short-term repeated dose toxicity study (28 days)
	Subchronic toxicity study (90 days)
	Long-term repeated toxicity study (≥12 months)
	Further repeated dose toxicity studies
Reproductive toxicity	Screening for reproductive/developmental toxicity
	Prenatal developmental toxicity study
	Developmental toxicity study
	Two-generation reproductive toxicity study
Toxicokinetics	Assessment of the toxicokinetic behavior of the substance to the extent that can be derived from the relevant available information
Carcinogenicity	Carcinogenicity study
Ecotoxicological information	
Aquatic toxicity	Short-term toxicity testing on invertebrates (preferred species *Daphnia*)
	Growth inhibition study aquatic plants (algae preferred)
	Short-term toxicity testing on fish
	Activated sludge respiration inhibition testing
	Long-term toxicity testing on invertebrates (preferred species *Daphnia*)
	Long-term toxicity testing on fish
	Fish early-life stage (FELS) toxicity test
	Fish short-term toxicity test on embryo and sac-fry stages
	Fish, juvenile growth test

Continued

Table 3 Toxicological Endpoints Requested by Regulatory Agencies—cont'd

ECHA—REACH	
Degradation	Biotic
	Ready biodegradability
	Simulation testing on ultimate degradation in surface water
	Soil simulation testing (for substances with a high potential for adsorption to soil)
	Sediment simulation testing (for substances with a high potential for adsorption to sediment)
	Abiotic
	Hydrolysis as a function of pH
	Identification of degradation products
Fate and behavior in the environment	Adsorption/desorption screening
	Bioaccumulation in aquatic species, preferably fish
	Further information on adsorption/desorption
	Further information on the environmental fate and behavior of the substance and/or degradation products
Effects on terrestrial organisms	Short-term toxicity to invertebrates
	Effects on soil microorganisms
	Short-term toxicity to plants
	Long-term toxicity testing on invertebrates
	Long-term toxicity testing on plants
Effects on sediment organisms	Long-term toxicity to sediment organisms
Toxicity to birds	Long-term or reproductive toxicity to birds
COFEPRIS	
For new drugs	Subacute toxicity (in vivo (7 days to several months))
	Acute (in vivo: unique dose, LD_{50})
	Chronic (in vivo (7 days to several months))
	Carcinogenicity (in vivo)
	Genotoxicity (in vitro)
	Teratogenicity (if carcinogenicity or genotoxicity are positive)

Table 3 Toxicological Endpoints Requested by Regulatory Agencies—cont'd

COFEPRIS	
For pesticides	Acute toxicity (oral LD50, dermal LD50, inhalation LC50, skin sensitization (for corrosive material), hypersensibility)
	Subchronic (oral 90 days in two species of animals, one should be rodent)
	Chronic toxicity
	Carcinogenicity (in vivo)
	Reproductive toxicity
	Teratogenicity (in two species of animals, one should be rodent)
	Neurotoxicity
EPA	
Skin irritation or skin corrosion	In vivo skin irritation
Eye irritation	In vivo eye irritation
Mutagenicity	In vitro gene mutation study in bacteria
	In vitro cytogenicity study in mammalian cells or in vitro micronucleus study
	In vitro gene mutation study in mammalian cells
Acute toxicity	By oral route
	By inhalation
	By dermal route
Repeated dose toxicity/chronic	Short-term repeated dose toxicity study (1–7 days)
	Intermediate-term toxicity study (7 days to several months)
	Long-term repeated toxicity study (several months to lifetime)
Reproductive toxicity	Screening for reproductive/developmental toxicity
	Prenatal developmental toxicity study
	Developmental toxicity study
	Two-generation reproductive toxicity study
Toxicokinetics	Assessment of the toxicokinetic behavior of the substance to the extent that can be derived from the relevant available information
Carcinogenicity	Carcinogenicity study

Continued

Table 3 Toxicological Endpoints Requested by Regulatory Agencies—cont'd
Japan

Ecotoxicity tests	Acute toxicity to fish
	Acute toxicity to aquatic invertebrates
	Toxicity to aquatic plants
Human health effects	Acute toxicity
	Repeated dose toxicity
	Genetic toxicity
	Reproductive toxicity

multiple descriptors are required, each one providing parts of the chemical information contained in the molecules. Similarly, biological endpoints are complex systems and require the development of multiple (and/or complex) models to account for different aspects of the biological response; this highlights the multidimensional nature of how biological systems work. Nonetheless, accumulation of reliable information has increased the acceptance of predictions to more complex endpoints. Predictions have evolved from physicochemical properties to acute oral toxicity in rodents, and more recently to mutagenicity (Amberg et al., 2016; Barber et al., 2015; Powley, 2015). Following this path, it is expected that more complex endpoints, such as developmental and reproductive toxicity, will be undertaken.

Progress on QSAR modeling has benefit from the availability of reliable data and the development of descriptors and methods. Moreover, development of statistical measures for the validation of external sets is expected to further improve the models, hence widening its acceptance and applications. Clearly, this highlights the need for harmonization and collective efforts, as is already taking place.

2.1 Creation or Selection of a Dataset

In KDD, the consideration of "the goal of the study" might have a variety of definitions, from insurance risk assessment to health care applications and to biological activity prediction, which is the topic of this work. In the prediction of biological properties, there are two important considerations to take into account, the first one is the nature of the dependent variable (i.e., a toxicological endpoint), which per se is typically complex and conveys different degrees of experimental uncertainty. The other important consideration is that the models are built upon chemical information, some of it might be obtained from experimental data and the other (the most commonly used) is derived from molecular descriptors (see later). Thus, besides the statistical considerations recommended by KDD for the creation of the database from which the models will be built, the chemical nature should be considered.

Molecular datasets may contain a specific chemical class. Such datasets typically contain 10–50 molecules obtained and evaluated under the same experimental conditions and are used for the generation of local models. All these characteristics make these datasets trustable and less prone to errors. Albeit the models from these datasets span a local area of chemical space, their predictive power tends to be high. Larger datasets are useful to produce general models, covering a broader area of chemical space. The scientific community has made available databases annotated with toxicity information that are useful for that purpose. Table 4 shows representative molecular databases used in toxicological studies.

If one decides to compile data from different sources, the accuracy, homogeneity and veracity of the data should be carefully revised. For example, in biological evaluations (such as toxicity assays) there might be differences in the measured activities, even from the same laboratory; this is due to variations in experimental conditions, such as cell reproducibility. In such cases, it is important to incorporate internal controls in the experimental design. Another source of variations might be the equipment used. All in all, it is important to assess if the measurements are comparable. These aspects might be a source of error and will negatively affect the performance of the predictive models. Selection of the set of molecules used for the development of models is crucial because it gives the foundation of the scope of the study and its limitations.

2.2 Data Cleaning and Preprocessing

Data cleaning is typically a tedious but important step; it regularly requires manual adjustments, for example, correction of typographical errors and removal of uncertain or blank values (in the prediction variable or toxicity endpoint). If there are missing values on a particular independent variable, but there is enough information from the dataset, it is possible to impute data (Nisbet, Elder, & Miner, 2009), and in all cases, the data should be homogenized to the same units.

Table 4 Molecular Databases Annotated With Toxicological Information

Name/Web Page	Description
ISSSTY http://www.iss.it/ampp/index.php?lang=1&id=233&tipo=7	Chemical toxicity: Structures and experimental data The ISSTOX databases are curated by the Institute Superiore di Sanità, and contain experimental results relative to various types of chemical toxicity. The available are: 1. long-term carcinogenicity bioassay on rodents (rat, mouse) (ISSCAN) 2. in vitro mutagenicity in *Salmonella typhimurium* (Ames test) (ISSSTY) 3. in vivo mutagenicity (micronucleus test) (ISSMIC) 4. cell transformation (ISSCTA) 5. long-term carcinogenicity bioassay on rodents and in vitro mutagenicity in *Salmonella typhimurium* (Ames test) of biocides and plant protection products (ISSBIOC)
T.E.S.T. https://www.epa.gov/chemical-research/toxicity-estimation-software-tool-test	Program developed by the EPA to estimate acute toxicity from the structure of chemical compounds. Provides the databases (over 5000 compounds) with the toxicological endpoints used in the development of predictions models
ToxNet—United States https://toxnet.nlm.nih.gov/	Toxicology Data Network (ToxNet) comprises a group of databases of chemicals and drugs, diseases and the environment, environmental health, occupational safety and health, poisoning risk assessment and regulations, and toxicology. Is useful to find information about specific toxic effects of chemicals in organisms

Leadscope http://www.leadscope.com/toxicity_database/	The Leadscope Toxicity Database contains information of over 400,000 toxicity study results for 180,000 chemical compounds. Toxicological data can be accessed by chemical structure (including exact match, substructure, and similarity), by type of study, toxic effect, species, sex, dosage, duration, or route of exposure
Lhasa Limited https://www.lhasalimited.org/	Is a nonprofit organization that specializes in the development of expert computer systems for toxicity and metabolism prediction. Facilitates collaborative data sharing projects in the pharmaceutical, cosmetics, and chemistry-related industries
PESTIMEP	Publically available database of pesticides annotated with multiple toxicological endpoints
NTP Natural products database http://www.hc-sc.gc.ca/dhp-mps/prodnatur/applications/licen-prod/lnhpd-bdpsnh-eng.php	The Licensed Natural Product Database contains information on natural health products that Health Canada has issued a product license. Licensed products have been evaluated by Health Canada and found to be safe, effective, and of high quality under the recommended conditions of use

Molecular structures are available in public databases with different levels of curation (Gozalbes, Ortiz, & López, 2014). It is always a good practice to further verify the accuracy of the chemical structures and the corresponding annotated information (Fourches, Muratov, & Tropsha, 2010; Young, Martin, Venkatapathy, & Harten, 2008). It has been estimated that as much as 10% of the chemical structures in databases have errors of different sorts (Waldman, Fraczkiewicz, & Clark, 2015). Molecular databases can also be built from scratch or from the compilation of datasets from different sources. Data curation is strongly recommended by the removal of inorganic compounds, salts, counterions, and duplicated structures. In the case of mixtures, one should select the molecule that is responsible of the biological activity; typically it consists of the one that is present in larger quantity. If the mixture consists of geometrical isomers and there is no indication on which one is responsible for the biological activity, one should either remove the molecule from the set or use descriptors that do not depend on the three-dimensional structure. Tautomers should be homogenized and protomers should be assigned at a specific pH value. Lastly, it is recommended to energy minimize the geometries and make a final visual inspection of the structures.

A descriptive statistical analysis is commonly done as a preprocessing step to explore the data. Initially, a univariate analysis is performed to identify trends and atypical data (also called outliers). This analysis includes the calculation of the median, standard deviation, minimum and maximum values, as well as frequency tables and histograms. Variables that deviate from a normal distribution are not good for prediction. In addition, to evaluate the utility of a variable it is convenient to calculate the Gini coefficient, which is a measure of statistical dispersion, in other words, measures the inequality among values of a frequency distribution. Gini coefficient takes values between 0 and 1. A value of 0 corresponds to perfect equality, while a value of 1 corresponds to total inequality. In the context of the distribution of the variables, as the Gini index approaches 1, the variable will be less predictive.

Many models in QSAR studies are based on multilinear regressions. Linear correlations are typically studied with parametric statistical modeling. The parametric model was developed by Fisher and is based on the following assumptions (Nisbet et al., 2009):

1. The data fits a known distribution: The assumption of a particular known distribution is necessary in order to draw upon the characteristics of the distribution function for making inferences.

2. Factor independency: The dependent variable (*Y*) is considered as a function of predictor variables (*X*'s) that are assumed to have independent effects on *Y*. That is, the effect of each *X*-variable on *Y* is not dependent on effects on *Y* of any other *X*-variable.
3. Linear additivity: Not only must the *X*-variables be independent, their effects on *Y* must be cumulative and linear. Assumption of linear additivity for nonlinear relationships may cause large errors in the predicted outputs.
4. Homoscedasticity: The variance throughout the range of each variable is assumed to be constant. If this is not satisfied, the error in the predicted value can be quite significant.
5. Variables must be numerical and continuous: Classical parametric statistical methods are not valid for use with discrete data; however, both scientists and business analysts use them.

Thus, when using the parametric model, one should analyze to which extent the case study falls into these assumptions. Alternatively, nonlinear relations can be transformed to linear relations and use the parametric statistics; else, nonlinear models can be developed. Nonparametric models may involve linear and nonlinear relations. Examples of nonparametric algorithms are CART and Bagging and Boosted trees (Nisbet et al., 2009).

Performing a descriptive statistical analysis informs on the uniformity of the data, which is important for assessing how balanced is the data and for the identification of outliers. Ideally, the data should be balanced, meaning that the distribution of toxic and nontoxic compounds should be even. Similarly, an even distribution in the chemical domain would provide a more general model. Else, a fit-to-purpose model would be developed, which is acceptable, as long as the predictive power is high and the boundaries are recognized. Admittedly, balance data are difficult to get (Chavez-Gomez et al., 2018). Identifying and removing outliers is an important step. Outliers can be rare cases (sampled less than 5% of the time) or can be wrong data. Outliers that correspond to rare cases might be relevant, for example, to inform of a different mechanism of action. How influential is an outlier depends on its leverage and how extreme it is in the *Y*-axis. If the outlier is influential, it may increase the noise in the data and reduce predictability and robustness of the model. If an outlier influences the conclusions, one could report the model and conclusions with and without the outlier(s); else one should limit the analysis to the data for which the relation is maintained.

Additionally, if there is enough information, data can be weighed to increase or decrease the relevance of particular datapoints. For example, this applies to datapoints that are measured with a particular accuracy; if it is known that the accuracy is twice as good, the weight used for those datapoints could be multiplied by 2. An advantage of machine learning methods is that they use random selection of weights to continuously optimize its values. This feature avoids ignoring underrepresented cases.

Weighing, balancing, and imputation of data should be performed with caution, only when it is necessary and when there is enough information to support the decisions.

2.3 Data Reduction and Projection

Once the molecules have been compiled into a database and annotated with the endpoint under study, the next step is to generate the descriptors associated to each molecule. Then, descriptive statistics is used to explore the data and to select the most relevant descriptors by data reduction methods. It should be noted that the descriptors should be previously normalized, in order to get all numerical values to a common range. There are different methods for data normalization reported in the literature (Nisbet et al., 2009). Then a bivariate analysis can be performed across variables, for example, pairwise relationships to identify highly related variables. These analyses allow identifying redundant variables. Reduction of dimensionality allows removing unnecessary variables. This can be performed by calculating the pairwise correlation coefficients across the variables. A correlation above 0.9 is considered collinear. Principal component analysis (PCA) and genetic algorithms (GAs) can also be used. However, for large number of variables GAs might be unnecessarily time consuming. An easy workaround can be performed as a first filter. The filter consists of the calculation of the correlation of each independent variable with the dependent variable. Then the variables with "best" correlations are used for developing models using GA. This methodology is used in the program QSARINS and has shown good performance (Gramatica, Chirico, Papa, Cassani, & Kovarich, 2013). Selecting the most relevant descriptors is crucial for the performance of the models. The use of unnecessary descriptors leads to overfitted models. Such models will fit the data but will also fit noise, decreasing the predictive power of the models.

3. SELECTION OF DATA MINING METHODS

The rule of thumb for the selection of the data mining methods to use is to start with simple models then, if required, move to more complex relations from a wide variety of options (Roy, Kar, & Das, 2015).

A number of software companies are available and a description of commonly used programs is described in the literature (Cronin et al., 2003). There are programs with Build-in models that allow the prediction of toxicity endpoints by incorporating the chemical structure. Such programs report the toxicity value, if the compound falls within the applicability domain and the statistics of the prediction are reported. Examples of those programs are the Toxicity Estimation Software Tool (T.E.S.T.) (US EPA, 2016), developed by USA EPA, and the lazy Structure–Activity Relationship (lazar), developed at Helma's group (Maunz et al., 2013).

Models can also be built from scratch. Programs particularly developed for this purpose are the Safety Evaluation Ultimately Replacing Animal Testing (SEURAT), currently developed under the project EU-ToxRisk (Schwarz & Gocht, 2011), and QSAR-INSUBRIA (QSARINS) developed at Gramatica's group (Gramatica et al., 2013). Commercial software is also available and a variety of toxicity endpoints can be predicted. Computational programs to cover these purposes are summarized in Table 5. Comparisons of programs commonly used for particular endpoints are reported in the literature, for example, for the estimation of rat oral acute toxicity (Gonella Diaza et al., 2015; Lo Piparo & Worth, 2010) and acute and chronic systemic toxicity (Lapenna, Fuart-Gatnik, & Worth, 2010).

Regarding the methods used for developing the models, the most commonly used are described in the QSAR DataBank (QSARDB): multiple linear regression, followed by *k*-nearest neighbors and decision trees. Examples of models deposited in QSARDB, under the classification of Human Health effects and Acute Inhalation Toxicity and Mutagenicity, are shown in Table 6. Notwithstanding although machine learning methods consistently provide highly predictive models, they are often not interpretable.

4. EXPLORATORY ANALYSIS AND MODEL/HYPOTHESIS SELECTION

In KDD as well as in QSAR, several models are developed. Thus, one should take into account factors such as the confidence of the models, the consensus among models, and the domain of applicability. A common definition of applicability domain is: the response and chemical structure spaces in which the model makes predictions with a given reliability (boundaries, likelihood, applicability, and reliability). Common methods for applicability domain look at molecular classes, and features, and at agreement in the chemical and response domains; a typical measure of the applicability

Table 5 Software for the Development of QSAR Models or for the Prediction of Toxicological and Other Biologically Relevant Endpoints From Precomputed Models

Name/Website	Short Description
Developed/Maintained by Governments	
TEST http://www.cpa.gov/nrmrl/std/qsar/qsar.html#TEST	Program developed by the EPA to estimate acute toxicity from the structure of chemical compounds. Predictions are made from precomputed models using different methodologies. *Endpoints*: 96-h fathead minnow 50% lethal concentration (LC_{50}), 48-h *Daphnia magna* 50% lethal concentration (LC_{50}), *Tetrahymena pyriformis* 50% growth inhibition concentration (IGC_{50}), oral rat 50% lethal dose (LD_{50}), bioconcentration factor (BCF), developmental toxicity (DevTox), and Ames mutagenicity (mutagenicity)
QSAR Toolbox http://www.qsartoolbox.org/	The QSAR Toolbox is a software intended to be used by governments, the chemical industry, and other stakeholders to fill gaps in (eco-) toxicity data needed for assessing the hazards of chemicals. The Toolbox incorporates information and tools from various sources into a logical workflow. Grouping chemicals into chemical categories is crucial to this workflow. *Endpoints:* carcinogenicity, repeated dose toxicity, reproductive/developmental toxicity, skin–eye irritation/corrosion, skin sensitization
CAESAR—VEGA http://www.caesar-project.eu/	CAESAR was an EC-funded project with the objective of developing QSAR models adapted to REACH legislation. Five predictive models were implemented for five highly relevant properties in REACH: Bioconcentration factor, skin sensitization, carcinogenicity, mutagenicity, and developmental toxicity

Build-in models free access	
Toxtree http://toxtree.sourceforge.net/	Toxtree is a full-featured and flexible user-friendly open-source application, which is able to estimate toxic hazard by applying a decision tree approach. *Endpoints:* Cramer rules and Cramer rules with extensions, skin irritation prediction, eye irritation prediction, mutagenicity, and carcinogenicity
Lazar http://lazar.in-silico.ch/predict	In Web interface through which different predictions of toxicity can be generated from a chemical structure. It includes a rational report on such predictions, applicability domain estimates, and validation results
VirtualToxLab http://www.chemic.unibas.ch/~vbc/molmod/virtualtox/index.html	Online tool for the prediction of toxicological potential in 16 biological proteins known for their ability to trigger adverse effects, using a combination of computational methods
Build-in models under commercial or academic license	
Lhasa https://www.lhasalimited.org/	Lhasa Limited is a not-for-profit organization and educational charity that facilitates collaborative data sharing projects in the pharmaceutical, cosmetics, and chemistry-related industries. Specializes in the development of expert computer systems for toxicity and metabolism prediction. *Endpoints*: Carcinogenicity, mutagenicity, genotoxicity, skin sensitization (EC3 potency prediction), teratogenicity, irritation, respiratory sensitization, and reproductive toxicity
Leadscope http://www.leadscope.com/	Leadscope is a chemo-computing platform. Allows integrated access to data mining based on chemistry, visualization, and predictive models, among computer techniques. *Endpoints:* Genetox Expert Alerts, Genetox Statistical Models, Carcinogenicity Statistical Models, Reproductive Toxicity Statistical models, Developmental Toxicity Statistical Models, Neurotoxicity Statistical Models

Continued

Table 5 Software for the Development of QSAR Models or for the Prediction of Toxicological and Other Biologically Relevant Endpoints From Precomputed Models—cont'd

Name/Website	Short Description
Simulations Plus http://www.simulations-plus.com/	Simulations Plus provides modeling and simulation software and consulting services from discovery through clinical development, with special emphasis on the QSAR-PBPK of GastroPlus™ & ADMET Predictor™. *Endpoints*: Absorption, distribution, metabolism, excretion, and toxicity
MultiCASE Inc. http://www.multicase.com/	MultiCASE Inc. develops computer programs for assessing toxicological and pharmacological potential of chemicals and pharmaceuticals. Allows in silico safety assessment and ICH M7, FDA, OECD submissions. Maintains strong academic ties in the areas of computational, medicinal, and environmental chemistry
Build models from scratch	
QSARINS (license upon request) http://www.qsar.it/	QSARINS is a new software for the development and validation of multiple linear regression models, quantitative structure–activity relation (QSAR) by ordinary least squares method and genetic algorithm for the selection of variables. This program focuses mainly on the external validation of the QSAR models
TIBCO Statistica/Spotfire	TIBCO Statistica is a statistical package useful for generating models through several analytic tools including: data mining, machine learning, clustering, correlations, fitting data, and forecasting, among others. Spotfire is a potent visualizing platform for data, with built-in analytic tools

Origin Pro	Software for data analysis and graphing. Enables analytic tools for fitting models (regression) and multivariate analysis (principal component analysis, *k*-means clustering, hierarchical cluster, discriminant analysis, and partial least square)
KNIME	KNIME is an open source for data analytics, reporting, and integration platform. Integrates various components for machine learning and data mining through its modular data pipeline concept. Allows assembly of nodes for data preprocessing, for modeling and data analysis and visualization
Seurat Asar (Free access) http://satijalab.org/seurat/	Seurat is an "R" package designed for QC, analysis, and exploration of single-cell RNA-seq data. It enables widely used analytical techniques, including the identification of highly variable genes, dimensionality reduction (PCA, ICA, t-SNE), standard unsupervised clustering algorithms (density clustering, hierarchical clustering, *k*-means), and the discovery of differentially expressed genes and markers
CORAL.SEA http://www.insilico.eu/coral/CORALSEA.html	Simple interface, possibility of generating regression models or binary classification
Weka www.cs.waikato.ac.nz/ml/weka/	Weka is an open-source software with a collection of machine learning algorithms for data mining tasks. Provides tools for data preprocessing, classification, regression, clustering, association rules, and visualization
Miscellaneous/specialized software	
RmSquare http://aptsoftware.co.in/rmsquare/	It is a web tool to estimate the predictive capacity of a model from a chemical structure. It includes a rational report on such predictions, applicability domain estimates, and validation results

Continued

Table 5 Software for the Development of QSAR Models or for the Prediction of Toxicological and Other Biologically Relevant Endpoints From Precomputed Models—cont'd

Name/Website	Short Description
CosmoLogic (Commercial) http://www.cosmologic.de/	COSMO simulates chemical processes based on quantum chemistry and thermodynamics. Predicts thermodynamical properties such as: solvent screening, partitioning, vapor pressures, and free energy of solvation
ChemAxon (Licensed free of charge for academic use) https://chemaxon.com/	Software specialized in chemoinformatics. It offers tools for visualization and drawing molecules, chemical database searching and management, and for drug discovery. Provides chemical calculations and predictions (logP, logD, pKa, NMR); chemical name conversion and data extraction from processed text on the web (www.chemicalize.com), through an online platform
Dragon (Commercial) http://www.talete.mi.it/products/dragon_description.htm	Software for the calculation of molecular descriptors. It calculates over 5000 descriptors, including simplest atom types, functional groups and fragment counts, topological and geometrical descriptors, and three-dimensional descriptors; it also calculates several properties estimation (such as logP) and drug-like and lead-like alerts (such as the Lipinski's alert). Dragon provides advanced tools to analyze the calculated descriptors (extended univariate statistics, pairwise correlation, principal component analysis)
MNAM (Commercial) www.mn-am.com	Molecular Networks GmbH and Altamira, LLC offer chemoinformatics software products, consulting, development, and research services in the chemical, pharmaceutical, and biotechnology areas. It focuses on applications for in silico profiling, evaluating, and assessing chemical compounds and finding new chemical entities.

Table 6 Examples of QSAR Models Deposited in QSARDB, Under the Classification Human Health Effects and Acute Inhalation Toxicity and Mutagenicity

Property	Compounds	Predictions	Model	Type	*n*	R^2	*s*	Ref
Acute inhalation toxicity								
pLC50mouse: Mice acute inhalation toxicity as pLC50 [−log(mmol/m^3)]	56	1	Regression model (regression)	Training	56	0.798	0.690	
pLC50rat: Rat acute inhalation toxicity as pLC50 [−log (mmol/m^3)]	52	1	Regression model (regression)	Training	52	0.781	0.796	
Property M12.pLC50: Acute inhalation toxicity in mice as log(1/LC50) [−log(mmol/m^3)]	56	1	Regression model (regression)	Training	56	0.786	0.711	
Property M13.pLC50: Acute inhalation toxicity in rat as log(1/LC50) [−log(mmol/m^3)]	52	1	Regression model (regression)	Training	52	0.787	0.785	
Property logLC50exp: Acute inhalation toxicity in rats as log(LC50) [log(mmol/m^3)]	41	1	Regression model (regression)	Training	36	0.908	0.191	
Human health effects 4.10. Mutagenicity								
Property mutagenicity: Mutagenic class	111	1	Regression model (classification)	Training	111	0.856		
Property M8.logTA100: The mutagenicity potency in TA100 (without the S9 activation system) as log(TA100)	48	1	Regression model (regression)	Training	48	0.828	0.719	
Property logTA100exp: Mutagenicity potency in TA100 without the S9 activation system [log(revertants/nmol)]	48	1	Regression model (regression)	Training	48	0.814	0.751	
Property mutagenicity: Mutagenicity on human h1A1v2 cells	70	2	*k*-Nearest neighbors (classification)	Training	70	0.814		
Property mutagenicity: Mutagenicity on human h1A1v2 cells	70	2	Decision tree (classification)	Training	70	0.829		

domain is based on distances within those domains. Exploration of the domain of applicability should answer if a model can be applied to a query compound. Thus, the domain is the model's boundaries. Having a notion of the applicability domain is good but is not enough. The problem is that applicability domain is not always well defined or it is not defined at all. Moreover, there is no a unique way for defining the applicability domain. If the molecule is out of domain, it would be necessary to consider expert knowledge to come to a decision (Sahigara et al., 2012). Recently, the lack of consensus of how to define the applicability domain has led to the proposal of harmonization (Hanser, Barber, Marchaland, & Werner, 2016). In that work, the authors suggest the separation of concepts to clarify and formalize the notion of applicability domain with the purpose of initiating a discussion among the QSAR community for building together a common understanding. Software companies and the QSAR community work on the harmonization and how to define and report the applicability domain to the community. As more data are available, models can (and should) be continuously enhanced.

The next step in the process is the validation. The internal validation is widely used, and the need for an external validation is been highlighted (Golbraikh, Shen, Xiao, Lee, & Tropsha, 2003; Tropsha, Gramatica, & Gombar, 2003). Accordingly, the data are divided into three groups: training set, internal test set, and external test set. Selecting sets is not a trivial task; suggested approaches for the selection of sets are available in the literature (Golbraikh et al., 2003). The internal validation involves the training set and the internal test set. Usually the leave-one-out and leave-many-out methods are used. For local models, structural similarity is the first important prerequisite for the selection of molecules for the training set, but is not sufficient. The training set should cover the wider possible chemical and toxicological spaces, such that the internal test set will fall into these two spaces. Training a model allows to adjust parameters and learn from the data. To obtain high-quality training sets, it is important to know which relevant properties determine similarity for endpoints, and to choose analogues that share features, chemical environment, and activity to the compounds of interest. Splitting can be performed multiple times so ensembles from splitting the data can be used.

5. DATA MINING

In the case of the development of models from scratch, once the different data mining methods have been explored, based on the nature of the data, the data mining process is performed. This includes the generation of series of models, and it is even possible to select an ensemble of models (Kuncheva, 2004). A description and classification of methods can be found in the literature (Kuncheva, 2004), and a survey of data mining algorithms has been previously revised (Helma, Gottmann, & Kramer, 2000).

6. EVALUATION

Statistical models are meant for finding correlations, but the researcher needs to decide if it is a coincidence or a cause. In other words, reliability and likelihood are not exchangeable concepts. Statistical models do not inform of why a correlation exists. So, the researcher needs to weight the results to arrive with a conclusion.

An internally validated model does not equal a valid prediction. The need for an external validation in QSAR models has been discussed in the literature (Alexander, Tropsha, & Winkler, 2015; Golbraikh et al., 2003). For the external validation, a group of molecules not used in the development of the model is evaluated. These molecules might correspond to the external validation set or an entirely new group of molecules, which have been experimentally evaluated in the toxicological endpoint under study. The group of molecules used for the external validation should be in the same applicability domain than the training set model and should be representative of the chemical space covered. There are different ways for selecting an external test set, for instance by random or stratified selection, which involves clustering followed by selecting cluster representatives, self-organizing maps, *k*-means, etc. To validate the performance of the model, one should look at reliability, quantity, and quality of the prediction and the prediction boundaries. The performance of the models is reported using statistical parameters, such as R^2 and Q^2. These parameters are also useful to identify overtrained models, which are characterized by having particularly good R^2 but low Q^2. Table 6 exemplifies and summarizes statistics of QSAR models collected in the QSAR Data Bank, QSARDB.

7. INTERPRETATION/UTILIZATION

In regulatory settings, the lack of transparency and documentation is unacceptable. The raw QSAR output should be provided as appendix information, along with any information that supports expert knowledge decisions. Adding interpretation and evidence helps supporting the decision made, framing the scope from which it is possible to make decisions, which is called the decision domain. If the available information, from statistical and knowledge can help to arrive to a clear choice, then an unequivocal decision can be made.

8. READ-ACROSS

In a broader sense, QSAR models can be used for different purposes, as stated by the OECD (Worth, 2007; Worth & Patlewics, 2007):

(a) To identify groups of similar chemicals and express this similarity in qualitative and/or quantitative terms, thereby supporting the formation of chemical categories and the application of read-across.

(b) To provide mechanistic information, thereby supporting the interpretation of experimental data.

(c) To fill data gaps, thereby replacing the need for (animal) testing.

(d) To supplement available test data, thereby supporting a weight-of-evidence (WoE) assessment.

(e) To identify chemicals of potential concern, in order to guide or prioritize testing.

The Registration, Evaluation, Authorisation and Restriction of Chemicals (REACH) establishes that information on intrinsic properties of substances shall be generated whenever possible by means other than vertebrate animals test, through the use of alternative methods (ECHA, 2009).

Combination of expertise, classical toxicological practices, and new rational approaches through computational chemistry has given birth to these alternative methods, namely QSAR, and read-across analysis.

Read-across is defined as (ECHA, 2009): "A technique for data-gap filling where endpoint information from one chemical is used to predict the same endpoint for another chemical which is considered to be similar in some important aspect relating to that endpoint, e.g. mode of action, toxicokinetics, metabolism etc. Read-across may be for a qualitative or quantitative result." Depending on the information available, there are different ways to fill data gaps, see for instance Table 7

Cronin et al. (2003) suggest that the vast majority of well-characterized chemicals with simple endpoints can be predicted from read-across. However, it is expected that many compounds might not be studied with this approach, leaving many gaps of information. Nonetheless, read-across studies are valuable for a rapid prediction of endpoints filling gaps of information.

Table 7 Example of a Matrix of Data Availability and Three Approaches That Could Be Used to Fill in Data Gaps, Including Read-Across

	Compound 1	Compound 2	Compound 3	Compound 4	Approach
Activity 1	2.0 →	?	6.7 →	?	SAR/read-across
Activity 2	1.5 →	?	?	← 11.0	Interpolation
Activity 3	?	← 3.1	7.2 →	?	Extrapolation

In read-across, there are two major approaches, one is based on analogues, and the other one is based on categories. A chemical category is defined as "a group of chemicals whose physicochemical and human health and/or environmental toxicological and/or environmental fate properties are likely to be similar or follow a regular pattern as a result of structural similarity (or other similarity characteristic)" (Worth, 2007; Worth & Patlewics, 2007).

Following the OECD guidance on grouping of chemicals, both analogue and category approaches might be based on the following characteristics (Worth, 2007; Worth & Patlewics, 2007):

1. Common functional group(s) (e.g., aldehyde, epoxide ester, specific metal ion).
2. A common mode or mechanism of action or adverse outcome pathway.
3. Common constituents or chemical classes, similar carbon range numbers. This is frequently the case with complex substances often known as "substances of unknown or variable composition, complex reaction products or biological material" (UVCB substances).
4. The likelihood of common precursors and/or breakdown products via physical or biological processes that result in structurally similar chemicals (e.g., the "metabolic pathway approach" of examining related chemicals such as acid/ester/salt); or
5. An incremental and constant change across the category (e.g., a chain-length category), often observed in physical chemical properties, e.g., boiling point range.

There are nine recommended steps to assign a chemical category to assess a new chemical (Puzyn, Leszcynski, & Cronin, 2010):

Step 0: Check whether the chemical is a member of an existing category.

Step 1: Develop category hypothesis and definition, and identify category members.

Step 2: Gather data for each category member.

Step 3: Evaluate available data for adequacy.

Step 4: Construct a matrix of data availability.

Step 5: Perform a preliminary evaluation of the category and fill data gaps.

Step 6: Perform and/or propose testing (if the information available cannot assess all members).

Step 7: Perform a further evaluation of the category.

Step 8: Document the finalized category and refine the category rationale.

Read-across can be performed qualitatively or quantitatively, using several approaches: One-to-one (one analogue use to estimate a single compound), many-to-one (several analogues used to estimate a single compound), one-to-many (one analogue to estimate several compounds), and many-to-many (two or more analogues to estimate two or more compounds). The difference between qualitative and quantitative analyses relies on the type of information that will be compared; while quantitative read-across uses known values of a property for the estimation, the qualitative read-across gives a binary answer based on the presence or absence of a certain quality or activity (Worth, 2007; Worth & Patlewics, 2007).

Here, we describe the steps involved in the development of QSAR models, and how they relate to the OECD principles for its validation. This analysis highlights the need of good practices should the predicted values be used for regulatory purposes.

ACKNOWLEDGMENTS

This work was supported by Instituto de Quimica-UNAM, CONACyT (project numbers 286854, 220392) and DGAPA-UNAM (PAPIIT IN210518). Authors thank Cielo Flores for assistance. K.M.-M. thanks Senosiain Laboratories for financial support. A.M.-M. thanks DGAPA for postdoctoral fellowship. Authors thank QSARINS, ChemAxon, StatSoft (Statistica and Spotfire), and KNIME for kindly providing academic licenses of their software.

REFERENCES

Alexander, D. L. J., Tropsha, A., & Winkler, D. A. (2015). Beware of R(2): Simple, unambiguous assessment of the prediction accuracy of QSAR and QSPR models. *Journal of Chemical Information and Modeling*, *55*(7), 1316–1322. https://doi.org/10.1021/acs.jcim.5b00206.

Amberg, A., Beilke, L., Bercu, J., Bower, D., Brigo, A., Cross, K. P., et al. (2016). Principles and procedures for implementation of ICH M7 recommended (Q)SAR analyses. *Regulatory Toxicology and Pharmacology*, 77(Suppl. C), 13–24. https://doi.org/10.1016/j.yrtph.2016.02.004.

Barber, C., Amberg, A., Custer, L., Dobo, K. L., Glowienke, S., Van Gompel, J., ... Wichard, J. (2015). Establishing best practise in the application of expert review of mutagenicity under ICH M7. Regulatory Toxicology and Pharmacology, 73(1), 367–377. https://doi.org/10.1016/j.yrtph.2015.07.018

Chavez-Gomez, A., Madariaga-Mazon, A., Garcia-Ramos, J. C., Castillo-Pazos, D. J., Cortes-Guzman, F., & Martinez-Mayorga, K. (2018). Interspecies toxicity relations of the multiple endpoint pesticide database, PESTIMEP. *Journal of the Mexican Chemical Society*(under review).

Cherkasov, A., Muratov, E. N., Fourches, D., Varnek, A., Baskin, I. I., Cronin, M., et al. (2014). QSAR modeling: Where have you been? Where are you going to? *Journal of Medicinal Chemistry*, *57*(12), 4977–5010. https://doi.org/10.1021/jm4004285.

Cronin, M. T. D., Jaworska, J. S., Walker, J. D., Comber, M. H. I., Watts, C. D., & Worth, A. P. (2003). Use of QSARs in international decision-making frameworks to predict health effects of chemical substances. *Environmental Health Perspectives*, *111*(10), 1391–1401.

Dearden, J. C., Cronin, M. T. D., & Kaiser, K. L. E. (2009). How not to develop a quantitative structure–activity or structure–property relationship (QSAR/QSPR). *SAR and QSAR in Environmental Research*, *20*(3–4), 241–266. https://doi.org/10.1080/10629360902949567.

Demchuk, E., Ruiz, P., Chou, S., & Fowler, B. A. (2011). SAR/QSAR methods in public health practice. *Toxicology and Applied Pharmacology*, *254*(2), 192–197. https://doi.org/10.1016/j.taap.2010.10.017.

ECHA. (2009). *How to report read-across and categories. In Practical guide 6, ECHA guidance.* Retrieved from http://echa.europa.eu/documents/10162/13655/pg_report_readacross_en.pdf.

Fourches, D., Muratov, E., & Tropsha, A. (2010). Trust, but verify: On the importance of chemical structure curation in cheminformatics and QSAR modeling research. *Journal of Chemical Information and Modeling*, *50*(7), 1189–1204. https://doi.org/10.1021/ci100176x.

Golbraikh, A., Shen, M., Xiao, Z., Lee, K., & Tropsha, A. (2003). Rational selection of training and test sets for the development of validated QSAR models. *Journal of Computer-Aided Molecular Design*, *17*, 241–253.

Gonella Diaza, R., Manganelli, S., Esposito, A., Roncaglioni, A., Manganaro, A., & Benfenati, E. (2015). Comparison of in silico tools for evaluating rat oral acute toxicity. *SAR and QSAR in Environmental Research*, *26*(1), 1–27. https://doi.org/10.1080/1062936x.2014.977819.

Gozalbes, R., Ortiz, J., & López, F. (2014). Computational methods in predictive toxicology: Application to the reduction of animal tests in the context of REACH community legislation. *Reviews in Toxicology*, *31*, 157–167.

Gramatica, P., Chirico, N., Papa, E., Cassani, S., & Kovarich, S. (2013). QSARINS: A new software for the development, analysis, and validation of QSAR MLR models. *Journal of Computational Chemistry*, *34*(24), 2121–2132. https://doi.org/10.1002/jcc.23361.

Hanser, T., Barber, C., Marchaland, J. F., & Werner, S. (2016). Applicability domain: Towards a more formal definition. *SAR and QSAR in Environmental Research*, *27*(11), 865–881. https://doi.org/10.1080/1062936x.2016.1250229.

Helma, C., Gottmann, E., & Kramer, S. (2000). Knowledge discovery and data mining in toxicology. *Statistical Methods in Medical Research*, *9*(4), 329–358. https://doi.org/10.1177/096228020000900403.

Kuncheva, L. I. (2004). Combining pattern classifiers. In *Methods and algorithms*. Wiley.

Lapenna, S., Fuart-Gatnik, M., & Worth, A. (2010). *Review of QSAR models and software tools for predicting acute and chronic systemic toxicity*. Retrieved from Luxemburg.

Lo Piparo, E., & Worth, A. (2010). *Review of QSAR models and software tools for predicting developmental and reproductive toxicology*. Retrieved from Luxemburg.

Maunz, A., Gütlein, M., Rautenberg, M., Vorgrimmler, D., Gebele, D., & Helma, C. (2013). Lazar: A modular predictive toxicology framework. *Frontiers in Pharmacology*, *4*, 38. https://doi.org/10.3389/fphar.2013.00038.

Nisbet, R., Elder, J., & Miner, G. (2009). *Handbook of statistical analysis and data mining applications*. Elsevier.

OECD. (2004). *OECD principles for the validation, for regulatory purposes, of (quantitative) structure-acivity relationship models*. Retrieved from http://www.oecd.org/chemicalsafety/risk-assessment/37849783.pdf.

OECD. (2008). *Using the OECD principles of corporate governance*. Retrieved from https://www.oecd.org/corporate/ca/corporategovernanceprinciples/40823806.pdf.

Powley, M. W. (2015). (Q)SAR assessments of potentially mutagenic impurities: A regulatory perspective on the utility of expert knowledge and data submission. *Regulatory Toxicology and Pharmacology*, *71*(2), 295–300. https://doi.org/10.1016/j.yrtph.2014.12.012.

Puzyn, T., Leszcynski, J., & Cronin, M. T. D. (2010). *Recent advances in QSAR studies*. Springer.

Roy, K., Kar, S., & Das, R. N. (2015). *A primer on QSAR/QSPR modeling fundamental concepts*. Springer.

Ruiz, P., Begluitti, G., Tincher, T., Wheeler, J., & Mumtaz, M. (2012). Prediction of acute mamalian toxicity using QSAr methods: A case study of sulfur mustard and its breakdown products. *Molecules*, *17*, 8982–9001.

Rusyn, I., & Daston, G. P. (2010). Computational toxicology: Realizing the promise of the toxicity testing in the 21st century. *Environmental Health Perspectives*, *118*(8), 1047–1050. https://doi.org/10.1289/ehp.1001925.

Sahigara, F., Mansouri, K., Ballabio, D., Mauri, A., Consonni, V., & Todeschini, R. (2012). Comparison of different approaches to define the applicability domain of QSAR models. *Molecules*, *17*(5), 4791.

Schwarz, M., & Gocht, T. (2011). Towards the replecement of in vivo repeated dose systemic toxicity testing. *In Annual report 2011 SEURAT-1: Vol. 1*. Retrieved from http://www.seurat-1.eu/pages/library/seurat-1-annual-report.php. Accessed 26 December 2018.

Tropsha, & Golbraikh. (2007). Predictive QSAR modeling workflow, model applicability domains, and virtual screening. *Current Pharmaceutical Design*, *13*(34), 3494–3504. https://doi.org/10.2174/138161207782794257.

Tropsha, Gramatica, P., & Gombar, V. K. (2003). The importance of being earnest: Validation is the absolute essential for successful application and interpretation of QSPR models. *QSAR & Combinatorial Science*, *22*(1), 69–77. https://doi.org/10.1002/qsar.200390007.

Tropsha, & Wang, S. X. (2007). QSAR modeling of GPCR ligands: Methodologies and examples of applications. In H. Bourne, R. Horuk, J. Kuhnke, & H. Michel (Eds.), *GPCRs: From deorphanization to lead structure identification, Vol. 2006/2* Berlin, Heidelberg: Springer.

US EPA. (2016). *User's guide for T.E.S.T. (version 4.2) (toxicity estimation software tool): A program to estimate toxicity from molecular structure*.

Veith, G. D. (2004). On the nature, evolution and future of quantitative structure-activity relationships (QSAR) in toxicology. *SAR and QSAR in Environmental Research*, *15*(5–6), 323–330. https://doi.org/10.1080/10629360412331297380.

Waldman, M., Fraczkiewicz, R., & Clark, R. D. (2015). Tales from the war on error: The art and science of curating QSAR data. *Journal of Computer-Aided Molecular Design*, *29*(9), 897–910. https://doi.org/10.1007/s10822-015-9865-0.

Worth, A. (2007). *The use of computational methods in the grouping and assessment of chemicals—Preliminary investigations*. Retrieved from Luxemburg, https://eurl-ecvam.jrc.ec.europa.eu/laboratories-research/predictive_toxicology/doc/EUR_22941_EN.pdf.

Worth, A., & Patlewics, G. (2007). *A compendium of case studies that helped to shape the REACH guidance on chemical categories and read across.* Retrieved from http://publications.jrc.ec.europa.eu/repository/bitstream/JRC37212/7212%20-%20Compendium%20of%20case%20studies_final_060707.pdf.

Young, D., Martin, T., Venkatapathy, R., & Harten, P. (2008). Are the chemical structures in your QSAR correct? *QSAR & Combinatorial Science*, *27*(11 – 12), 1337–1345. https://doi.org/10.1002/qsar.200810084.

CHAPTER FIVE

Computational Methods to Discover Compounds for the Treatment of Chagas Disease☆

Eduardo M. Cortés-Ruiz*, Oscar Palomino-Hernández*, Karla Daniela Rodríguez-Hernández†, Bertha Espinoza†, José L. Medina-Franco*,1

*Facultad de Química, Universidad Nacional Autónoma de México, Mexico City, Mexico
†Instituto de Investigaciones Biomédicas, Universidad Nacional Autónoma de México, Mexico City, Mexico
[1]Corresponding author: e-mail addresses: medinajl@unam.mx; jose.medina.franco@gmail.com

Contents

Abstract

Infectious diseases continue to be a major public health. Among these diseases, American trypanosomiasis or Chagas disease (CD) is a major cause of morbidity and death for millions of people in Latin America. The two drugs currently available for the treatment of CD have poor efficacy and major side effects. Thus, there is a pressing need to develop safe and effective drugs against this disease. Herein we review the diversity and coverage of chemical space of compounds tested as inhibitors of *Trypanosoma cruzi*, a parasite causing CD. We also review major molecular targets currently pursued to kill the parasite and recent computational approaches to identify inhibitors for such targets.

☆This work is dedicated to the loving memory of Nicolás Medina Sandoval.

Advances in Protein Chemistry and Structural Biology, Volume 113
ISSN 1876-1623
https://doi.org/10.1016/bs.apcsb.2018.03.005

1. INTRODUCTION

According to the World Health Organization (WHO), infectious diseases continue to be a major public health issue. Infectious diseases are caused by pathogenic microorganisms, such as bacteria, viruses, parasites, or fungi. The diseases can spread, directly or indirectly, from one person to another (World Health Organization, 2017).

Chagas disease (CD), or American trypanosomiasis (Bern, 2015), is an infection identified in 1909 by Chagas (1909) is caused by the parasite *Trypanosoma cruzi* (*T. cruzi*) and belongs to the group of neglected tropical diseases, and vector-borne diseases, a group of diseases that disproportionately affect marginalized populations in tropical and subtropical areas with lack effective treatments or vaccines (Hotez et al., 2007; World Health Organization, 2017). It is transmitted by many hematophagous triatomine species (Bern, 2015; Rassi, Rassi, & Marcondes de Rezende, 2012). It can also be transmitted by nonvector mechanisms as blood transfusion, organ and bone marrow transplantation, congenital transmission, and oral transmission with ingestion of contaminated food with vectors feces (Rassi et al., 2012; World Health Organization, 2017). Although CD is endemic of 21 countries in Latin America, this affection has spread to North America, Europe, and Australia due to large scale rural to urban migration over the past 50 years (Bern, 2015; Chatelain, 2017; Jackson, Pinto, & Pett, 2014).

T. cruzi is a protozoa parasite that belongs to the kinetoplastid group. It is a complex organism with an intricate life cycle that can be divided into two main principal stages: the vector stage, where we find epimastigotes as the main replication stage in the midgut of vector, and infective metacyclic trypomastigotes, which are excreted with the feces of the vector. In human, the metacyclic trypomastigotes stage invade many types of nucleated cells and are differentiated into intracellular amastigote form, which is the main replicative form in the vertebrate host (Bern, 2015; García-Torres & Pérez-Monfort, 2011). *T. cruzi* is a parasite that shows a high rate of genetic polymorphism and has been classified into six discrete typing units (DTUs) based on diverse molecular markers (Zingales et al. 2009). Other factor that difficult the treatment of CD is the differential DTUs susceptibilities to drugs, as has been demonstrated for several *T. cruzi* strains (Grosso et al. 2010, Martínez, Nogueda, Martínez, Espinoza, 2013).

CD occurs in two phases: acute and chronic; the clinical hallmarks for acute phase included detectable parasitemia, fever and a skin nodule (chagoma) or painless prolonged eyelid edema (Romaña's sign that indicates the site of inoculation). The chronic phase is characterized by cardiomyopathy, megaesophagus, megacolon, and alterations of peripheral nervous system that can cause the death (Bern, 2015; Rassi et al. 2012).

Despite these facts there has been a significant progress in the research in particular in the last 5 years, but there is a lot of missing information. There are still major questions that need to be answered related to the progression of the disease, its pathologies, and evaluation of the treatment efficacy. It is also needed the identification of markers of cure or efficacy of the treatment (Chatelain, 2017).

Only two old nitro-heterocyclic drugs are available for treatment of CD in acute phase, nifurtimox and benznidazole (Fig. 1). Both have a cure rate is 80% in the acute phase, but less than 20% in chronic stage (Maya et al. 2007). Nifurtimox is no longer prescribed in some regions of America included the United States of America (USA) owing to reports of gastrointestinal tract side effects, genotoxicity, low efficacy against particular *T. cruzi* strains, and the emergence of neurotoxicity.

For a long time, benznidazole has been used for the treatment of CD and is more tolerated than nifurtimox. Nevertheless, it is associated with serious side effects such as dermatological reactions, polyneuropathy, agranulocytosis, and mutagenesis (the latter found in some trials). The prolonged use of these drugs generates resistance on *T. cruzi* (García-Torres & Pérez-Monfort, 2011; Maya et al., 2007; Patterson and Wyllie, 2014). Since nifurtimox and benznidazole have several adverse side effects, new orally available and safe drugs are urgently required. Moreover, CD has a large impact and repercussions in socio-economic and human health (Beltran-Hortelano, Perez-Silanes, Galionao, 2017). There is also an urgent need to improve the mechanisms to make accessible the treatment to a major number of patients (Chatelain, 2017).

Several recent excellent reviews have been published addressing the current status of drug discovery for the treatment of CD (Beltran-Hortelano

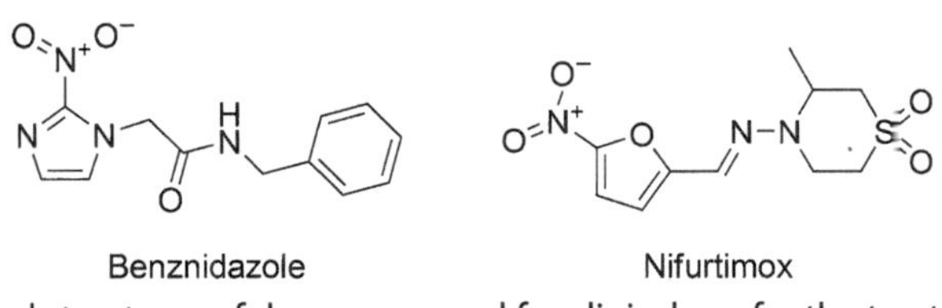

Fig. 1 Chemical structures of drugs approved for clinical use for the treatment of infections with *Trypanosoma cruzi*.

et al., 2017; Ferreira & Andricopulo, 2016; Katsuno et al., 2015; Leonardo, Marcelo, & Adriano, 2016; Moraes & Franco, 2016). The main goal of this chapter is to discuss the recent progress of computational approaches to advance the discovery and development of active compounds against *T. cruzi*. The chapter is organized in five major sections. After this introduction, we discuss the main and novel biomolecular targets that are being pursued. Section 3 presents a cheminformatics analysis of public information of chemical compounds that have been tested against *T. cruzi*. Section 4 discusses advances in computational approaches to identify bioactive compounds for the biomolecular targets. Section 5 presents concluding remarks and perspectives.

2. BIOLOGICAL RELEVANT SPACE

More than 30 molecular targets have been pursed to develop inhibitors of *T. cruzi* for the treatment of CD. Several of these targets are associated with the enzymes involved in *T. cruzi* metabolism. Therefore, inhibition of such enzymes is a relatively straightforward approach to kill the parasite. Of course, the compounds should be selective enough if there is a human-related enzyme that could lead to unwanted side effects. Beltrán-Hortelano et al. have recently collected these targets in a rich review (Beltran-Hortelano et al., 2017). Representative molecular targets are summarized in Table 1. Among these major targets, there are enzymes related to the defense system against oxidative damage: trypanothione reductase (TR) and the iron-containing superoxide dismutase (Fe-SOD); the proteases *trans*-sialidase (TS) and cruzipain or cruzain (Martinez-Mayorga, Byler, Ramirez-Hernandez, & Terrazas-Alvares, 2015). Other molecular targets are enzymes that interfere with the glycolysis (glycolytic pathway) such as triose phosphate isomerase (TPI), glyceraldehyde-3-phosphate dehydrogenase (GAPDH), and hexokinase (HK); enzymes that interfere with the sterol synthesis such as farnesyl pyrophosphate synthase (FFPS) and sterol C14-α-demethylase (CYP51); and enzymes that interfere with the purine salvage and nucleotide synthesis and metabolism such as dihydrofolate reductase (DHFR).

Two of the most promising targets are TR and Fe-SOD (Beltran-Hortelano et al., 2017). Additional widely pursued targets are cruzain, TPI, and DHFR. Commonly used targets for the computer-aided identification of hit compounds (discussed in Section 4) are *trans*-sialidase, trypanothione reductase, and cruzipain or cruzain. Other emerging targets are addressed in Section 5.

Table 1 Summary of Representative Biological Targets for Inhibition of *T. cruzi*

Sterol synthesis	Hydroxymethylglutaryl coenzyme A reductase (HMG CoA R)
	Farnesyl pyrophosphate synthase (FFPS)
	Squalene synthase (SQS)
	Squalene epoxidase (SQE)
	Oxidosqualene cyclase (OSC) or lanosterol synthase
	Sterol C14 demethylase (CYP51)
Proteases	Cruzipain or cruzain
	Trans-sialidase (TS)
	Prolyl-specific Tc80 proteinase
Glycolytic pathway (Glycolysis)	Glyceraldehyde-3-phosphate dehydrogenase (GAPDH)
	Hexokinase (HK)
	Hydroxyacid dehydrogenase (HADH)
	Triose phosphate isomerase (TPI)
	Malate dehydrogenase (MDH)
	Pyruvate kinase (PK)
Pentose phosphate pathway	Glucose-6-phosphate dehydrogenase (G6PDH)
	Ribose-5-phosphate isomerase (R5PI)
Purine salvage and nucleotide synthesis and metabolism	Hypoxanthine-guanine phosphoribosyltransferase (HGPRT)
	Dihydrofolate reductase (DHFR)
	Pteridine reductase (PTR)
	Thymidylate synthase (TMS)
Biosynthesis and metabolism of fatty acids	Synthesis of phosphatidylcholine
	Type II fatty acid synthase (FAS II)
DNA binders	DNA topoisomerases I and II
	Kinetoplastid DNA (kDNA)

Continued

Table 1 Summary of Representative Biological Targets for Inhibition of *T. cruzi*—cont'd

Defense system against oxidative damage	Trypanothione reductase (TR)
	Iron-containing superoxide dismutase (Fe-SOD)
Other dehydrogenases	Succinate dehydrogenase (SDH)
	Dihydrolipoamide dehydrogenase (DHLDH)
	Glutamate dehydrogenase (GDH)
	Dihydroorotate dehydrogenase (DHODH)
Other reductases	Fumarate reductase (FR)
Other kinase	Arginine kinase (ArgK)
Other decarboxylases	Arginine decarboxylase (ADC)

3. CHEMICAL SPACE

3.1 Chemical Libraries

A large number of inhibitors of the different target enzymes of *T. cruzi* have been reported. Perhaps, one of the most comprehensive sources of public information is the ChEMBL database (Bento et al., 2014). In order to review the available chemical space of inhibitors of major targets of *T. cruzi* reported to date, herein we assembled two datasets. One set of compounds with reported activity against a particular stain of the whole organism/parasite e.g., a specific DTUs of *T. cruzi*, and the second set of compounds with reported activity of a specific molecular target. All compounds were obtained from ChEMBL, version 23, Binding Database (Liu, Lin, Wen, Jorissen, & Gilson, 2007), and literature searching. The initial raw dataset was curated following stablished protocols in KNIME (Berthold et al., 2008) and MOE (MOE, 2014). The molecules were washed using a protocol implemented in MOE that involves removing salts, neutralizing the protonation state, and keeping the largest fragment. In a second step, the duplicates in each dataset were removed. The curated datasets contained 6312 compounds (activity vs whole organism) and 30,139 compounds (activity vs a specific protein). This large number of molecules in the public domain further emphasizes the global efforts (mostly in academia) to develop molecules against this parasite. Table 2 summarizes the

Table 2 Datasets to Explore the Chemical Space of Compounds With Reported Activity Against the Whole Parasite or Specific Targets

Dataset		Size	Activity (Range)	
			Max/Min	Mean/Median
In vitro activity vs whole organism	DTU			
	TcI	367	2.68/7.10	4.53/4.56
	TcII	1382	1.76/7.82	4.44/4.45
	TcVI	4563	1.60/9.82	5.45/5.07
Activity vs enzymes	Protein			
	Cruzipain	29,487	2.00/10.00	4.66/4.5
	Trans-sialidase (TS)	29	3.51/6.22	4.30/3.88
	Trypanothione reductase (TR)	325	2.41/6.52	4.53/4.38
	6-Phosphofrukto-1-kinase (6PF1K)	43	4.24/7.39	5.55/5.59
	Sterol C14-α-demethylase (CYP51)	60	4.29/7.90	6.21/6.27
	Bifunctional dihydrofolate reductase (BFDFR)	7	6.38/7.70	7.33/7.57
	Farnesyl pyrophosphate synthase (FFPS)	65	4.00/7.89	5.34/5.00
	Glyceraldehyde-3-phosphate dehydrogenase (G3PD)	42	3.16/7.00	3.95/3.69
	Hexokinase (HK)	29	4.34/6.09	5.80/6.09
	Triose phosphate isomerase (TPI)	9	3.00/5.46	4.75/4.89
	Phosphodiesterase	63	4.00/6.64	5.12/5.19

information of the compound datasets assembled. The table includes the distribution of compounds regarding the species (activity vs whole organisms) or the molecular targets (activity vs molecular targets). The table also shows the range of activity values of compounds with reported activity for a specific DTU or target protein. Of note, in the datasets collected, the most investigated DTUs are TcVI (4563 compounds) followed by TcII (1382). The protein with a significant large number of compounds is cruzipain (29,487 compounds) followed in a distant second place by TR (325).

3.2 Chemical Space: Diversity and Content

The dataset of 6312 compounds reported with in vitro activity vs different DTUs was analyzed in terms of physicochemical properties of pharmaceutical relevance, molecular fingerprints, and molecular scaffolds.

3.2.1 Physicochemical Properties and Visualization of Chemical Space

The following properties were computed with MOE: molecular weight (MW), log P, number of hydrogen bond donors (HBD) and acceptors (HBA), number of rotatable bonds (RB), and topological surface area (TPSA). These properties has a very importance in pharmaceutical studies because are associated with absorption, distribution, metabolism, and excretion (ADME) properties, the compounds that have the sufficiently acceptable properties of ADME and very low or acceptable toxicity are considered as "drug-like" compounds (Lipinski, 2000). MW, HBDs, HBAs, logP are related with solubility and permeability; TPSA, in addition to participating in the processes described earlier, is associated with the transport of chemical compounds. RB are associated with the molecule flexibility (Lipinski, 2000; Lipinski, Lombardo, Dominy, Feeney, 1997; Veber et al., 2002). These properties are commonly used to profile compound datasets with therapeutic relevance (González-Medina & Medina-Franco, 2017; López-Vallejo et al., 2011). Fig. 2 shows box plots with the distribution of the six properties for compounds tested with TcI (367 molecules), TcII (1382), and TcVI (4563). The figure indicates that compounds tested with the three DTUs have, in general, similar distribution of physicochemical properties, in particular logP and HBD. The distribution of RB indicates that compounds tested with TcI are more rigid.

Fig. 3 shows a visual representation of the chemical space of 600 compounds tested with TcI (200 molecules), TcII (200), and TcVI (200). The compounds were selected at random from the complete datasets. The visual representation was obtained by principal component analysis of the six

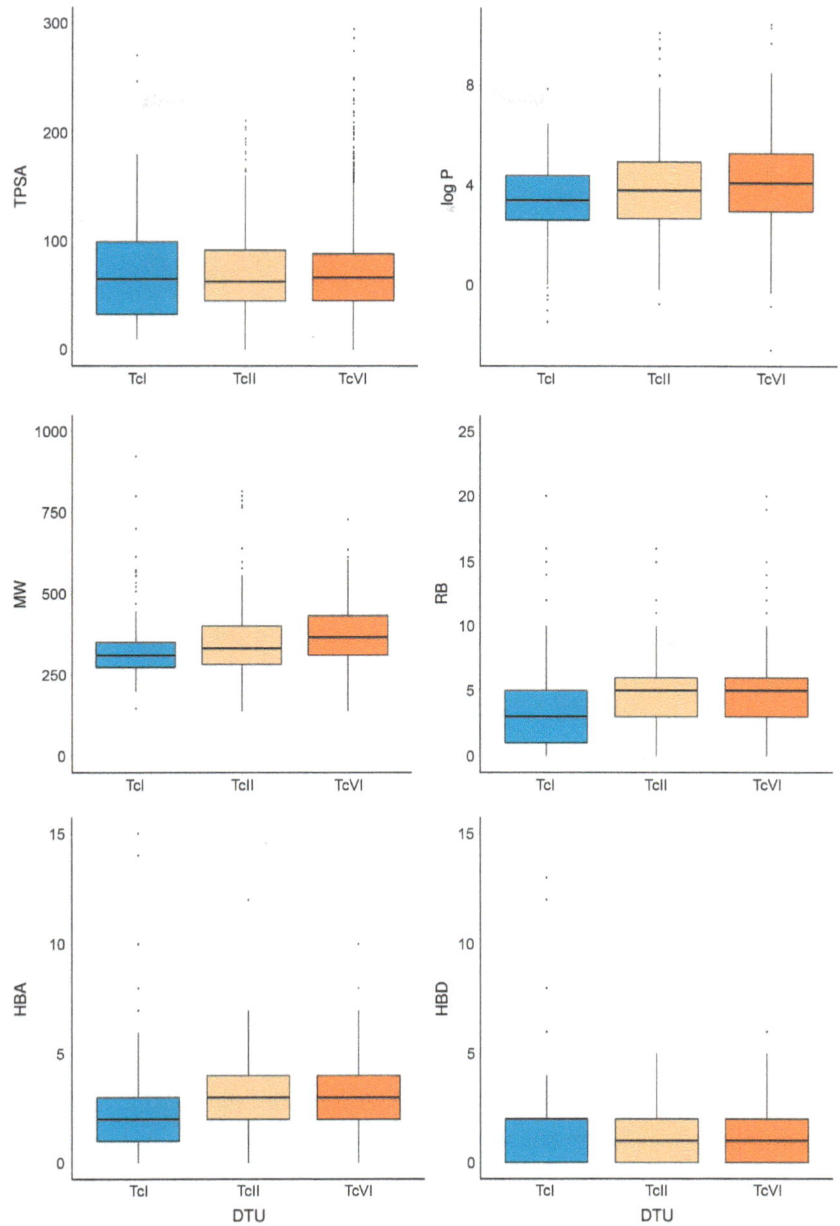

Fig. 2 *Box plots* of six physicochemical properties of pharmaceutical relevance of compounds tested with TcI, TcII, and TcVI. Topological surface area (TPSA), octanol–water partition coefficient (log P), molecular weight (MW), number of rotatable bonds (RB), hydrogen bond acceptors (HBA), and hydrogen bond donors (HBD).

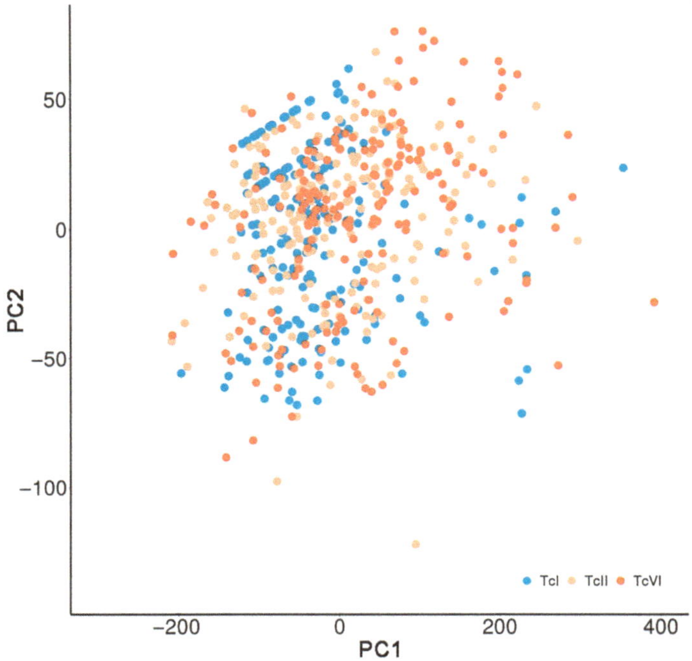

	a_acc	a_don	b_rotN	logP (o/w)	TPSA	Weight
1. Eigenvector	0.012	0.005	0.02	0.006	0.19	0.982
2. Eigenvector	−0.027	−0.023	−0.013	0.032	−0.981	0.19

Fig. 3 Visual representation of chemical space of three datasets of compounds tested with TcI (200 compounds), TcII (200 compounds), and TcVI (200 compounds). The visual representation is based on six properties of pharmaceutical relevance. The first two principal components capture 98% of the variance. The table summarizes the contribution of each of the six properties to the first two principal components.

autoscaled physicochemical properties. The first two components capture 98% of the variance. The property with the largest contribution to the first principal component (PC1) is MW. TPSA has the largest contribution to PC2. The figure indicates that compounds tested with TcI, TcII, and TcVI have, overall, similar distribution in the chemical space.

3.2.2 Molecular Fingerprints

Chemoinformatic analysis of the internal structural diversity of the compounds tested with TcI, TcII, and TcVI (e.g., intradataset similarity) (Fig. 4) showed that all three datasets have comparable structural diversity as measured with the extended connectivity fingerprint of radius 2 and the Tanimoto coefficient

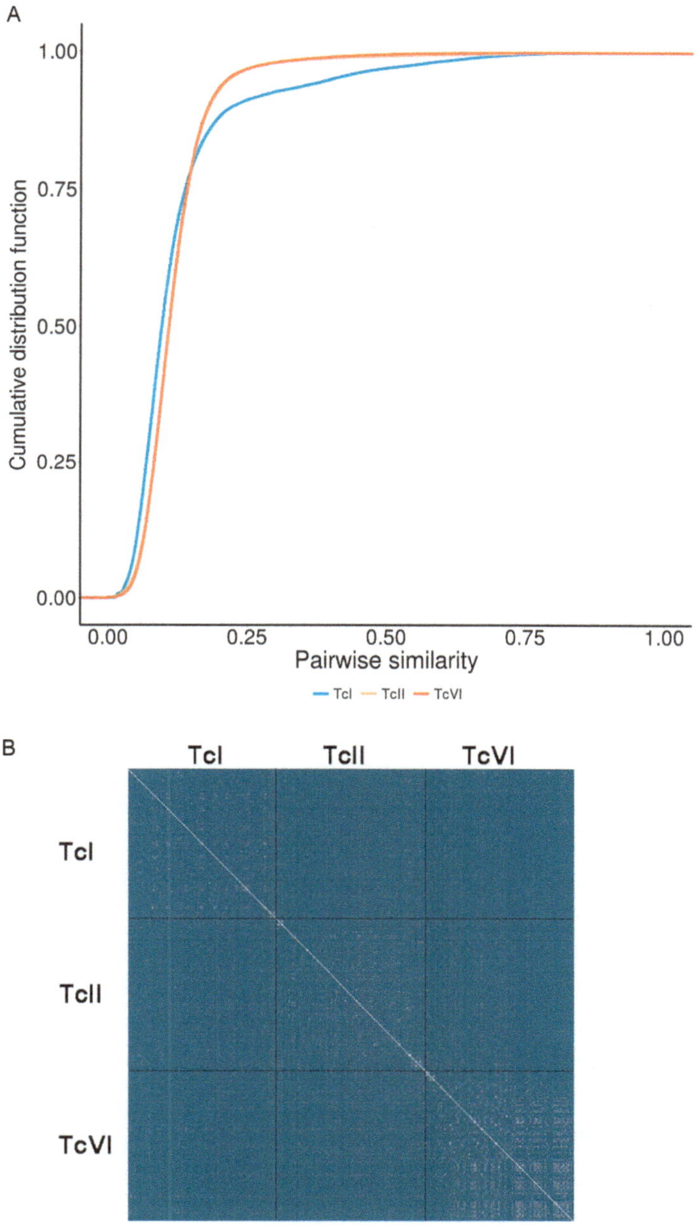

Fig. 4 Results of the fingerprint-based diversity analysis of compounds tested with TcI (367 compounds), TcII (1382 compounds), and TcVI (1505 compounds). Diversity analysis was computed with extended connectivity fingerprints, radius 2, and the Tanimoto coefficient. (A) Cumulative distribution function of the pairwise structure similarities. (B) Similarity matrix of a representative set of 250 compounds from each dataset: TcI, TcII, and TcVI. The similarity values are represented in a continuous scale from high values (*white*) to low values (*dark blue*).

(Fig. 4A). Compounds in TcVI are slightly less diverse than compounds in the other two sets. Comparison between compounds datasets (interdataset similarity) indicted that the compounds datasets have also have comparable diversity, e.g., compounds are similar to each other (Fig. 4B).

4. COMPUTATIONAL APPROACHES FOR LEAD IDENTIFICATION

There are several computational strategies that have been developed to identify hit compounds, optimize their activity and/or, in general, improve their pharmacokinetics and pharmacodynamics profile. Usually, the method to be used depend on the experimental information of the system available (Saldívar-González, Prieto-Martínez, & Medina-Franco, 2017). Computational approaches can be broadly divided into structure-based and ligand-based depending on the three-dimensional structure of the molecular target available. Whenever, possible it is desirable to use more than one approach to get to consensus conclusions. Furthermore, due to the increased awareness of polypharmacology (Medina-Franco, Giulianotti, Welmaker, & Houghten, 2013; Rosini, 2014), it is desirable to implement approaches focused on multitarget drug design (Méndez-Lucio, Naveja, Vite-Caritino, Prieto-Martínez, & Medina-Franco, 2016). In general, a common technique to identify hit compounds is virtual screening that can be based on the target, the ligand, or be a combination of both. It is also remarkable to mention that in recent years several online tools freely available have been developed to do virtual screening, and, in general, mining chemogenomic spaces (Gonzalez-Medina, Naveja, Sanchez-Cruz, & Medina-Franco, 2017).

Table 3 summarizes recent examples of computational strategies used to identify hit compounds for the treatment of CD. Representative examples are further discussed in the following sections. The chemical structures of selected compounds are shown in Fig. 5.

4.1 Docking-Based Virtual Screening

Regarding *T. cruzi*, computational studies combined both in vitro and in vivo evaluation of *trans*-sialidase and cruzain (Lara-Ramirez et al., 2017; Palos et al., 2017). For these works, authors performed a structure-based virtual screening of a database of 3180 drugs approved for clinical use. In vitro and in vivo assays were done for the computational hits against trypomastigotes from the INC-5 and NINOA strains of *T. cruzi*. For the *trans*-sialidase study, it was found that four drugs already approved with antihistaminic, antihypertensive,

Table 3 Summary of Recent Examples of Computational Strategies to Discover or Develop Candidate Compounds for the Treatment of Chagas Disease

Target	Computational Approach	Major Outcome	References
Trans-sialidase	Structure-based virtual screening of 3180 drugs	Sulfasalazine showed the best trypanocidal effects in in vivo experiments	Lara-Ramirez et al. (2017)
TcPAT12	Docking-based virtual screening of 3000 drugs	Isotretinoin inhibited the polyamine transport as well as other amino acid transporters from the same protein family	Reigada et al. (2017)
GAPDH of *T. cruzi*	Protein-based pharmacophore and structure-based virtual	Inhibitors with new chemotypes with activity in the micromolar range were identified	Maluf, Andricopulo, Oliva, and Guido et al. (2013)
Trans-sialidase of *T. cruzi*	*E*-pharmacophore-based virtual screening of more than 4 million compounds from the ZINC database	ZINC13359679 and ZINC02576132 were identified as the most promising inhibitors	Miller and Roitberg (2013)
Polyamine transport	Virtual screening of DrugBank and Sweetlead databases based on classification models	Triclabendazole, sertaconazole, and paroxetine displayed inhibitory effects on the proliferation of *T. cruzi* and the uptake of putrescine	Alberca et al. (2016)
In vivo mouse efficacy model of Chagas disease	Virtual screening of compound libraries based on Bayesian machine learning models	The antimalarial pyronaridine possessed 85.2% efficacy in the acute Chagas mouse model	Ekins et al. (2015)

Continued

Table 3 Summary of Recent Examples of Computational Strategies to Discover or Develop Candidate Compounds for the Treatment of Chagas Disease—cont'd

Target	Computational Approach	Major Outcome	References
TR	Screening of a highly diverse set followed by similarity searching of 200,000 compounds	82 novel TR inhibitors with activities in the nanomolar range	Beig et al. (2015)
TR	Virtual screening cascade protocol for identification of novel inhibitors	Identification of inhibitory effect of molecules with dibenzothiepine and other novel scaffolds	Perez-Pineiro et al. (2009)
GAPDH of *T. cruzi*	Target- and ligand-based virtual screening, isothermal titration calorimetry, and molecular dynamics	Nequimed125 was identified as a potent inhibitor of GAPDH	Prokopczyk et al. (2013)
Cruzain	Multistep ligand- and structure-based virtual screening of compounds in the ZINC database	Nequimed176 was identified as the most promising candidate compound for later optimization	Wiggers et al. (2013)

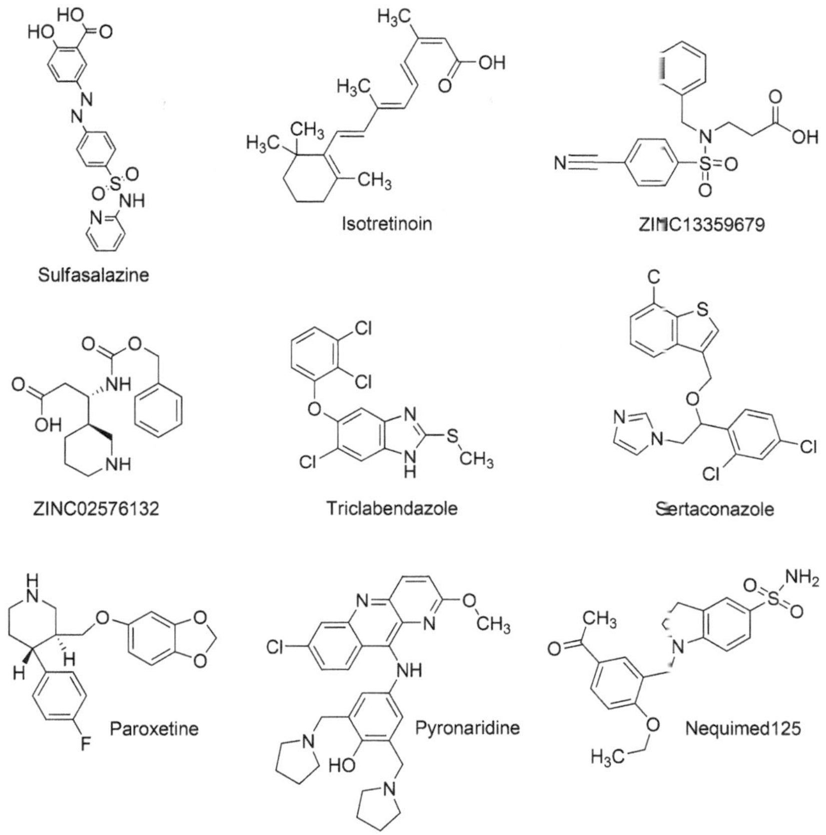

Fig. 5 Examples of active and other hit compounds identified with the aid of computational approaches for the potential treatment of Chagas disease. See Table 3.

and antibiotic properties had better trypanocidal effects than nifurtimox and benznidazole in both strains in the in vitro model. The antiinflammatory drug, sulfasalazine, had moderate inhibition (37.6%) of sialylation in a *trans*-sialidase enzyme inhibition reaction. Sulfasalazine also showed the best trypanocidal effects in short-term in vivo experiments on infected mice. For the cruzain study, the four selected compounds consisting of three beta-lactam antibiotics and one clofibrate derivate, showed better trypanocidal activity than the drugs of reference. In vivo testing indicated that three of the four compounds (flucloxacillin, cefoperazone, and etofylline clofibrate) maintained trypanocidal effects on mice infected with NINOA strains, indicating that these structures can be used as starting point for the design of new chemical entities.

Valera Vera et al. (2016) analyzed a set of structurally diverse polyphenols from ZINC as plausible agents against arginine kinase. From the computational hits, resveratrol was selected for further studies for its multiple beneficial properties. Inhibition assays with recombinant arginine kinase showed that the IC_{50} of resveratrol was 325 μM, while in vivo inhibition for Y strain epimastigotes was found to be $IC_{50} = 98$ μM. The results were in agreement with recent reports that resveratrol possesses antipromastigote and antiamastigote effects in *Leishmania amazonensis* (Ferreira et al., 2014).

A virtual screening by Paixão and Pita (2016) analyzed the compounds in the Natural Products Database of Bahia State (NatProDB) as possible inhibitors of TR. The results of this study showed novel triterpenic chemotypes which also interacted with key amino acids previously reported. Thus, these natural products can be used as starting points for modulators of the activity of this enzyme.

Yoshino et al. (2017) performed a virtual screening of 4.8 million compounds obtained from Namiki Shoji Co., Ltd. (Tokyo, Japan) and Astellas Pharma Inc. (Ibaraki, Japan) against spermidine synthase. A total of 176 compounds were selected as computational hits, which were further analyzed in terms of their inhibition activity toward this enzyme. Four different compounds were selected. Fragment molecular orbital analysis and molecular dynamics were performed to analyze relevant interactions between the protein and the newly identified inhibitors.

Also in a recent work, Reigada et al. (2017) described a combined approach of computational screening of 3000 drugs approved for clinical use with in vitro assays finding inhibitors of TcPAT12, the only polyamine transporter described so far in *T. cruzi*. In the docking-based virtual screening, the drug used for acne treatment isotretinoin showed the best interaction score with TcPAT12 and was selected for further in vitro studies. Isotretinoin inhibited the polyamine transport as well as other amino acid transporters from the same protein family. The drug also had a strong inhibitory activity of trypomastigote burst from infected cells but it was significantly less effective on the epimastigote stage.

4.2 Pharmacophore-Based VS

Maluf et al. (2013) reported a combined protein-based pharmacophore and structure-based virtual screening to identify three hits from three chemical classes with moderate inhibitory activity against GAPDH of *T. cruzi*. The inhibitors showed IC_{50} values in the high micromolar range. Authors concluded that the newly identified chemotypes are attractive molecules for later medicinal chemistry efforts aimed at developing new lead compounds for CD.

In an independent work, Miller and Roitberg (2013) reported a theoretical work focused on the generation of e-pharmacophore hypothesis that highlights potential important residues of *trans*-sialidase of *T. cruzi* for ligand binding. Using the e-pharmacophore hypothesis, authors conducted the virtual screening of the ZINC Clean Leads database with more than 4 million compounds. Authors selected 82 compounds as potential inhibitors of *trans*-sialidase. After performing molecular dynamics and free energy binding calculations, two molecules from the ZINC collection (ZINC13359679 and ZINC02576132) were identified as the most promising inhibitors.

Yoshino et al. (2015) constructed a pharmacophore model based on ororate analogs directed toward dihydroorotate dehydrogenase. In this work, 45 protein–ligand crystal structures were subjected to single point calculations and fragment molecular orbital analysis in order to analyze relevant amino acid interactions for these chemotypes. The results of this study uncovered that interactions with Lys43, Asn67, and Asn194 were relevant for molecular recognition.

In a more recent work, de V. C. Sinatti et al. (2017) designed five ribose 5-phosphate isomerase (RpiB) pharmacophore hypothesis based on an active subset of inhibitors and protein–ligand complexes. These models were validated and used as an input in the ZINCPharmer web server, which retrieved 18,038 hit compounds. The hit molecules were further ranked with the docking algorithm implemented in Glide. Three compounds with the best docking scores and no reported activity in RpiB yet were subjected to molecular dynamics simulations, which allowed observing the prevalence of the molecular interactions. This study suggested that ZINC36975961 and ZINC43763931 maintained key interactions and were considered as potential inhibitors.

4.3 Ligand-Based, Data Mining, Similarity Searching

Alberca et al. (2016) described the development of classification models based on a 268-compound database containing polyamine analogs with and without inhibitory effect on *T. cruzi*. The classification models were later applied to virtual screen drug approved for clinical use included in the DrugBank and Sweetlead databases. As a result, five compounds were selected as computational hits and further tested in cellular models from different pathogenic trypanosomatids (*T. cruzi* wt, *T. cruzi* PAT12, *T. brucei*, and *Leishmania infantum*), and in vitro models of aminoacid/polyamine transport assays and trypanothione synthetase inhibition assay. Three drugs triclabendazole, sertaconazole, and paroxetine displayed inhibitory effects on the proliferation of *T. cruzi* (epimastigotes) and the uptake of putrescine by the parasite.

Ekins et al. (2015) reported the development of an in silico method based on public results of whole-cell, phenotypic high-throughput screens that have been completed for *T. cruzi*. Part of the source database contained the results of more than 300,000 molecules tested as probes. Authors also compiled predicted metabolites of *T. cruzi* metabolic pathways. All source data were employed to generate a Pathway Genome Data Base for *T. cruzi*. Based on the compiled information, Ekins et al. constructed Bayesian machine learning models that were employed for the virtual screening of compound libraries. As a result of the virtual screening, 97 molecules were chosen for in vitro testing and 11 were finally identified to have $EC_{50} < 10\,\mu M$. Then, five compounds were moved forward to an in vivo mouse efficacy model of CD and validated that the machine-learning model could identify in vitro active compounds not in the training set. The antimalarial pyronaridine possessed 85.2% efficacy in the acute Chagas mouse model. Authors also proposed putative targets of the active molecule.

With the final goal of identifying inhibitors of TR, Beig et al. (2015) published a combination of computational studies followed by experimental evaluation. Authors started from a collection of 2816 molecules with high diversity leading to 21 novel TR inhibitors. The 21 hits were later used as reference molecules for a second round of computational studies. The second round was based on similarity searching of a database with 200,000 compounds available for screening. A total of 1204 compounds were selected and assayed in vitro leading to 61 active molecules. Overall, 82 novel TR inhibitors with activities in the nanomolar range were identified.

Guerra et al. (2017) used a previously reported neural network model (Angela et al., 2013) based on a three-layered 4-4-1 architecture to predict anti-*T. cruzi* activity against the epimastigote form of the Tulahuen strain for a series of imidazothiadiazines. Four selected derivatives showing promising predicted activities were tested against the whole cell and as inhibitors of cruzipain. Although the inhibitors showed variable degrees of cruzipain inhibition, most of the computational hits possess excellent trypanocidal activity with potency in some cases greater than nifurtimox.

4.4 Combined, Integrative Multistrategy

Prokopczyk et al. (2013) reported the discovery of inhibitors of GAPDH of *T. cruzi* based on target- and ligand-based virtual screening protocols using isothermal titration calorimetry and molecular dynamics. In that work, molecular dynamics simulations were employed to gain insights into the binding poses of newly identified inhibitors acting at the TcGAPDH substrate (G3P) site. Nequimed125 was identified as a potent inhibitor of GAPDH of *T. cruzi*. Moreover, a new class of dihydroindole compounds that act upon GAPDH of *T. cruzi* through a competitive mechanism of inhibition, also kills *T. cruzi*.

Wiggers et al. (2013) described an integrated multistep ligand- and structure-based virtual screening of compounds in the ZINC database. Twelve molecules were identified as noncovalent competitive inhibitors of cruzain having affinity values in the low micromolar range. The mechanism of action was confirmed tough the determination of the binding mode of the cruzain inhibitor Nequimed176 through X-ray crystallographic studies. Two compounds were found to have trypanocidal activity in an assay with *T. cruzi* infected cells. Using compound Nequimed42 as precursor, a SAR was established in which the 2-acetamidothiophene-3-carboxamide group was identified as essential for enzyme and parasite inhibition activities. The IC_{50} value for compound Nequimed42 acting against the trypomastigote form of the Tulahuen lacZ strain was found to be $10.6 \pm 0.1\,\mu M$, 10-fold lower than that for benznidazole. A smaller compound Nequimed176, which was derived from Nequimed42, was identified as the most promising candidate compound for later optimization.

In a separate work, Perez-Pineiro et al. (2009) proposed a virtual screening cascade protocol to identify potential inhibitors of the parasitic enzyme TR. First, the authors performed a ligand-similarity search of the ZINC database using reversible inhibitors of TR as reference. The second step was filtering the compounds according to ADME/Tox criteria. The resultant compound library was docked into the binding site of TR and the most promising compounds were considered for bioscreening. A scaffold hoping approach was followed to identify new chemotypes. This was the first work about the inhibitory effect of molecules-containing dibenzothiepine, dibenzooxathiepine, dibenzodithiepine, thiaazatetracyclo, and 2,4,6-trioxohexahydro pyrimidine on TR. These heterocyclic molecules have been used as inhibitors of tumor necrosis factor-α, DNA helicases, as well as neurotropic and psychotropic agents. ZINC04128838, a derivate of these heterocyclic compounds, was tested in vitro showing IC_{50} values of 11.5 and 14.6 μM for *T. cruzi* and *T. brucei* TR, respectively.

5. CONCLUSIONS

Several computational techniques have been employed toward the identification and optimization of small molecules for the treatment of CD. Such techniques are structure- or ligand based. Common techniques recently used are molecular docking, molecular dynamics, pharmacophore modeling, and chemoinformatic methods such as similarity searching. In particular, the large amount of structural information of a number of validated targets to kill *T. cruzi* has facilitated the application of structure-based approaches.

The computational approaches have been integrated in virtual screening campaigns, more frequently in combination with more than one computational technique (e.g., sequential screenings) followed by experimental testing of the computational hits. Several recent successful cases demonstrate that computational methods lead to active molecules. In several instances, the compounds screened are drugs already approved for clinical use leading to successful cases of in silico guided drug repurposing. Indeed, drug repurposing is a very attractive technique for the treatment of CD and other infectious diseases poorly addressed by other methods (Ferreira & Andricopulo, 2016). In addition, chemoinformatic approaches are playing a key role for the quantification of the chemical diversity of compound collections available in the public domain containing activity data.

6. PERSPECTIVES

From the computational perspective, one of the major goals on this field is the development of robust protocols to conduct either virtual screening or target fishing. Such protocols could be implemented online (Gonzalez-Medina et al., 2017). This would speed up the drug discovery process. It is anticipated that with such protocol, experimental-oriented scientists (working on, for instance, isolation and characterization of natural products or organic synthesis) test in silico their compounds to select the most promising for experimental validation.

In addition, it is expected the continued growth of public databases. Such resources will continue to facilitate the discovery process. One more perspective is the development of inhibitors of novel molecular targets such as alpha-carbonic anhydrases from *T. cruzi* (Supuran, 2016) and those related to epigenetics. Despite the fact these targets have been largely explored for cancer and other therapeutic indications (Naveja & Medina-Franco, 2018), their applications as targets for the treatment of CD are emerging. However, some targets are becoming promising (Klug, Gelb, & Pollastri, 2016; Sacconnay et al., 2014). Other promising approach is phenotypic screening (Alonso-Padilla & Rodríguez, 2014; Leonardo et al., 2016).

ACKNOWLEDGMENTS

This work was supported by the program *Nuevas Alternativas para el Tratamiento de Enfermedades Infecciosas* (NUATEI) and the *Programa de Apoyo a la Investigación y el Posgrado* (PAIP) Grant 5000-9163, Facultad de Química, UNAM. E.M.C.-R. acknowledges scholarship from *Programa de Apoyo a Proyectos de Investigación e Innovación Tecnológica* (PAPIIT) Grant IA203718. O.P.-H. is grateful to CONACyT for the fellowship granted No. 765897/606003. K.D.R.-H. is grateful with the PhD program in Ciencias Biomédicas of Universidad Nacional Autónoma de México (UNAM) and CONACyT for providing the scholarship number 663037.

REFERENCES

Alberca, L. N., Sbaraglini, M. L., Balcazar, D., Fraccaroli, L., Carrillo, C., Medeiros, A., et al. (2016). Discovery of novel polyamine analogs with anti-protozoal activity by computer guided drug repositioning. *Journal of Computer-Aided Molecular Design, 30*, 305–321.

Alonso-Padilla, J., & Rodríguez, A. (2014). High throughput screening for anti-trypanosoma cruzi drug discovery. *PLoS Neglected Tropical Diseases, 8*, e3259.

Angela, G., Pedro, G.-N., Nuria, E. C., Hugo, C., Mercedes, G., & Juan, A. P. (2013). Artificial neural networks based on CODES descriptors in pharmacology: Identification of novel trypanocidal drugs against chagas disease. *Current Computer-Aided Drug Design, 9*, 130–140.

Beig, M., Oellien, F., Garoff, L., Noack, S., Krauth-Siegel, R. L., & Selzer, P. M. (2015). Trypanothione reductase: A target protein for a combined in vitro and in silico screening approach. *PLoS Neglected Tropical Diseases, 9*, e0003773.

Beltran-Hortelano, I., Perez-Silanes, S., & Galiano, S. (2017). Trypanothione reductase and superoxide dismutase as current drug targets for trypanosoma cruzi: An overview of compounds with activity against Chagas disease. *Current Medicinal Chemistry, 24*, 1066–1138.

Bento, A. P., Gaulton, A., Hersey, A., Bellis, L. J., Chambers, J., Davies, M., et al. (2014). The ChEMBL bioactivity database: An update. *Nucleic Acids Research, 42*, D1083–D1090.

Bern, C. (2015). Chagas' disease. *New England Journal of Medicine, 373*, 456–466.

Berthold, M. R., Cebron, N., Dill, F., Gabriel, T. R., Kötter, T., Meinl, T., et al. (2008). KNIME: The Konstanz information miner. In C. Preisach, H. Burkhardt, L. Schmidt-Thieme, & R. Decker (Eds.), *Data analysis, machine learning and applications: Proceedings of the 31st annual conference of the Gesellschaft für Klassifikation e.V., Albert-Ludwigs-Universität Freiburg, March 7–9, 2007* (pp. 319–326). Berlin, Heidelberg: Springer Berlin Heidelberg.

Chagas, C. (1909). Nova tripanozomiaze humana: estudos sobre a morfolojia e o ciclo evolutivo do Schizotrypanum cruzi n. gen., n. sp., ajente etiolojico de nova entidade morbida do homem. *Memórias do Instituto Oswaldo Cruz, 1*, 159–218.

Chatelain, E. (2017). Chagas disease research and development: Is there light at the end of the tunnel? *Computational and Structural Biotechnology Journal, 15*, 98–103.

de V. C. Sinatti, V., R Baptista, L. P., Alves-Ferreira, M., Dardenne, L., Hermínio Martins da Silva, J., & Guimarães, A. C. (2017). In silico identification of inhibitors of ribose 5-phosphate isomerase from *Trypanosoma cruzi* using ligand and structure based approaches. *Journal of Molecular Graphics and Modelling, 77*, 168–180.

Ekins, S., Lage de Siqueira-Neto, J., McCall, L.-I., Sarker, M., Yadav, M., Ponder, E. L., et al. (2015). Machine learning models and pathway genome data base for *Trypanosoma cruzi* drug discovery. *PLoS Neglected Tropical Diseases, 9*, e0003878.

Ferreira, L. G., & Andricopulo, A. D. (2016). Drug repositioning approaches to parasitic diseases: A medicinal chemistry perspective. *Drug Discovery Today, 21*, 1699–1710.

Ferreira, C., Soares, D. C., Nascimento, M. T. C. D., Pinto-da-Silva, L. H., Sarzedas, C. G., Tinoco, L. W., et al. (2014). Resveratrol is active against *Leishmania amazonensis*: In vitro effect of its association with amphotericin B. *Antimicrobial Agents and Chemotherapy, 58*, 6197–6208.

García-Torres, I., & Pérez-Monfort, R. (2011). Avances en la identificación de blancos terapéuticos y el diseño racional de fármacos contra la enfermedad de chagas. *Revista de Educación Bioquímica, 30*, 68–81.

González-Medina, M., & Medina-Franco, J. L. (2017). Platform for unified molecular analysis: PUMA. *Journal of Chemical Information and Modeling, 57*, 1735–1740.

Gonzalez-Medina, M., Naveja, J. J., Sanchez-Cruz, N., & Medina-Franco, J. L. (2017). Open chemoinformatic resources to explore the structure, properties and chemical space of molecules. *RSC Advances, 7*, 54153–54163.

Grosso, N. L., Bua, J., Perrone, A. E., Gonzalez, M. N., Bustos, P. L., Postan, M., et al. (2010). *Trypanosoma cruzi*: Biological characterization of a isolate from an endemic area and its susceptibility to conventional drugs. *Experimental Parasitology, 126*, 239–244.

Guerra, A., Gonzalez-Naranjo, P., Campillo, N. E., Varela, J., Lavaggi, M. L., Merlino, A., et al. (2017). Novel Imidazo[4,5-c][1,2,6]thiadiazine 2,2-dioxides as antiproliferative *Trypanosoma cruzi* drugs: Computational screening from neural network, synthesis and in vivo biological properties. *European Journal of Medicinal Chemistry, 136*, 223–234.

Hotez, P. J., Molyneux, D. H., Fenwick, A., Kumaresan, J., Sachs, S. E., Sachs, J. D., et al. (2007). Control of neglected tropical diseases. *New England Journal of Medicine, 357*, 1018–1027.

Jackson, Y., Pinto, A., & Pett, S. (2014). Chagas disease in Australia and New Zealand: Risks and needs for public health interventions. *Tropical Medicine & International Health, 19*, 212–218.

Katsuno, K., Burrows, J. N., Duncan, K., van Huijsduijnen, R. H., Kaneko, T., Kita, K., et al. (2015). Hit and lead criteria in drug discovery for infectious diseases of the developing world. *Nature Reviews Drug Discovery, 14*, 751–758.

Klug, D. M., Gelb, M. H., & Pollastri, M. P. (2016). Repurposing strategies for tropical disease drug discovery. *Bioorganic & Medicinal Chemistry Letters, 26*, 2569–2576.

Lara-Ramirez, E. E., López-Cedillo, J. C., Nogueda-Torres, B., Kashif, M., Garcia-Perez, C., Bocanegra-Garcia, V., et al. (2017). An in vitro and in vivo evaluation of new potential trans-sialidase inhibitors of *Trypanosoma cruzi* predicted by a computational drug repositioning method. *European Journal of Medicinal Chemistry, 132*, 249–261.

Leonardo, G. F., Marcelo, T. d. O., & Adriano, D. A. (2016). Advances and progress in chagas disease drug discovery. *Current Topics in Medicinal Chemistry, 16*, 2290–2302.

Lipinski, C. A. (2000). Drug-like properties and the causes of poor solubility and poor permeability. *Journal of Pharmacological and Toxicological Methods, 44*, 235–249.

Lipinski, C. A., Lombardo, F., Dominy, B. W., & Feeney, P. J. (1997). Experimental and computational approaches to estimate solubility and permeability in drug discovery and development settings. *Advanced Drug Delivery Reviews, 23*, 3–25.

Liu, T. Q., Lin, Y. M., Wen, X., Jorissen, R. N., & Gilson, M. K. (2007). BindingDB: A web-accessible database of experimentally determined protein-ligand binding affinities. *Nucleic Acids Research, 35*, D198–D201.

López-Vallejo, F., Nefzi, A., Bender, A., Owen, J. R., Nabney, I. T., Houghten, R. A., et al. (2011). Increased diversity of libraries from libraries: Chemoinformatic analysis of bis-diazacyclic libraries. *Chemical Biology & Drug Design*, *77*, 328–342.

Maluf, F. V., Andricopulo, A. D., Oliva, G., & Guido, R. V. C. (2013). A pharmacophore-based virtual screening approach for the discovery of *Trypanosoma cruzi* GAPDH inhibitors. *Future Medicinal Chemistry*, *5*, 2019–2035.

Martínez, I., Nogueda, B., Martínez, F., & Espinoza, B. (2013). Microsatellite and mini-exon analysis of human DTU I *Trypanosoma cruzi* strains and their susceptibility to Nifurtimox and Benznidazole. *Vector Borne and Zoonotic Diseases*, *13*, 181–187.

Martinez-Mayorga, K., Byler, K. G., Ramirez-Hernandez, A. I., & Terrazas-Alvares, D. E. (2015). Cruzain inhibitors: Efforts made, current leads and a structural outlook of new hits. *Drug Discovery Today*, *20*, 890–898.

Maya, J. D., Cassels, B. K., Iturriaga-Vásquez, P., Ferreira, J., Faúndez, M., Galanti, N., et al. (2007). Mode of action of natural and synthetic drugs against Trypanosoma cruzi and their interaction with the mammalian host. *Comparative Biochemistry and Physiology*, *146*, 601–620.

Medina-Franco, J. L., Giulianotti, M. A., Welmaker, G. S., & Houghten, R. A. (2013). Shifting from the single to the multitarget paradigm in drug discovery. *Drug Discovery Today*, *18*, 495–501.

Méndez-Lucio, O., Naveja, J. J., Vite-Caritino, H., Prieto-Martínez, F. D., & Medina-Franco, J. L. (2016). One drug for multiple targets: A computational perspective. *Journal of the Mexican Chemical Society*, *60*, 168–181.

Miller, B. R., & Roitberg, A. E. (2013). Design of e-pharmacophore models using compound fragments for the trans-sialidase of *Trypanosoma cruzi*: Screening for novel inhibitor scaffolds. *Journal of Molecular Graphics and Modelling*, *45*, 84–97.

MOE. (2014). *Molecular Operating Environment version 2014.08*. Montreal, Quebec, Canada: Chemical Computing Group Inc. Available at http://www.chemcomp.com.

Moraes, C. B., & Franco, C. H. (2016). Novel drug discovery for Chagas disease. *Expert Opinion on Drug Discovery*, *11*, 447–455.

Naveja, J. J., & Medina-Franco, J. L. (2018). Insights from pharmacological similarity of epigenetic targets in epipolypharmacology. *Drug Discovery Today*, *23*, 141–150.

Paixão, V. G., & Pita, S. S. R. (2016). Triagem virtual aplicada na busca de inibidores da tripanotiona redutase de *Trypanosoma cruzi* utilizando a base de dados de produtos naturais do semiárido baiano (NatProDB). *Revista Virtual de Quimica*, *8*, 1289–1310.

Palos, I., Lara-Ramirez, E. E., Lopez-Cedillo, J. C., Garcia-Perez, C., Kashif, M., Bocanegra-Garcia, V., et al. (2017). Repositioning FDA drugs as potential cruzain inhibitors from trypanosoma cruzi: Virtual screening, in vitro and in vivo studies. *Molecules*, *22*, 1015.

Patterson, S., & Wyllie, S. (2014). Nitro drugs for the treatment of trypanosomatid diseases: Past, present, and future prospects. *Trends in Parasitology*, *30*, 289–298.

Perez-Pineiro, R., Burgos, A., Jones, D. C., Andrew, L. C., Rodriguez, H., Suares, M., et al. (2009). Development of a novel virtual screening cascade protocol to identify potential trypanothionereductase inhibitors. *Journal of Medicinal Chemistry*, *52*, 1670–1680.

Prokopczyk, I. M., Ribeiro, J. F. R., Sartori, G. R., Sesti-Costa, R., Silva, J. S., Freitas, R. F., et al. (2013). Integration of methods in cheminformatics and biocalorimetry for the design of trypanosomatid enzyme inhibitors. *Future Medicinal Chemistry*, *6*, 17–33.

Rassi, A., Rassi, A., & Marcondes de Rezende, J. (2012). American trypanosomiasis (Chagas disease). *Infectious Disease Clinics of North America*, *26*, 275–291.

Reigada, C., Valera-Vera, E. A., Sayé, M., Errasti, A. E., Avila, C. C , Miranda, M. R., et al. (2017). Trypanocidal effect of isotretinoin through the inhibition of polyamine and amino acid transporters in *Trypanosoma cruzi*. *PLoS Neglected Tropical Diseases*, *11*, e0005472.

Rosini, M. (2014). Polypharmacology: The rise of multitarget drugs over combination therapies. *Future Medicinal Chemistry*, *6*, 485–487.

Sacconnay, L., Angleviel, M., Randazzo, G. M., Marçal Ferreira Queiroz, M., Ferreira Queiroz, E., Wolfender, J.-L., et al. (2014). Computational studies on sirtuins from *Trypanosoma cruzi*: Structures, conformations and interactions with phytochemicals. *PLoS Neglected Tropical Diseases*, *8*, e2689.

Saldívar-González, F., Prieto-Martínez, F. D., & Medina-Franco, J. L. (2017). Descubrimiento y desarrollo de fármacos: un enfoque computacional. *Educación Química*, *28*, 51–58.

Supuran, C. T. (2016). Inhibition of carbonic anhydrase from *Trypanosoma cruzi* for the management of Chagas disease: An underexplored therapeutic opportunity. *Future Medicinal Chemistry*, *8*, 311–324.

Valera Vera, E. A., Sayé, M., Reigada, C., Damasceno, F. S., Silber, A. M., Miranda, M. R., et al. (2016). Resveratrol inhibits *Trypanosoma cruzi* arginine kinase and exerts a trypanocidal activity. *International Journal of Biological Macromolecules*, *87*, 498–503.

Veber, D. F., Johnson, S. R., Cheng, H. Y., Smith, B. R., Ward, K. W., & Kopple, K. D. (2002). Molecular properties that influence the oral bioavailability of drug candidates. *Journal of Medicinal Chemistry*, *45*, 2615–2623.

Wiggers, H. J., Rocha, J. R., Fernandes, W. B., Sesti-Costa, R., Carneiro, Z. A., Cheleski, J., et al. (2013). Non-peptidic cruzain inhibitors with trypanocidal activity discovered by virtual screening and in vitro assay. *PLoS Neglected Tropical Diseases*, 7, e2370.

World Health Organization. (2017). *WHO fact sheet no. 340*. Retrieved Accessed December 7, 2017, from http://www.who.int/mediacentre/factsheets/fs340/en/.

Yoshino, R., Yasuo, N., Hagiwara, Y., Ishida, T., Inaoka, D. K., Amano, Y., et al. (2017). In silico, in vitro, X-ray crystallography, and integrated strategies for discovering spermidine synthase inhibitors for Chagas disease. *Scientific Reports*, 7, 6666.

Yoshino, R., Yasuo, N., Inaoka, D. K., Hagiwara, Y., Ohno, K., Orita, M., et al. (2015). Pharmacophore modeling for anti-chagas drug design using the fragment molecular orbital method. *PLoS One*, *10*, e0125829.

Zingales, B., Andrade, S. G., Briones, M. R. S., Campbell, D. A., Chiari, E., Fernandes, O., et al. (2009). A new consensus for *Trypanosoma cruzi* intraspecific nomenclature: Second revision meeting recommends TcI to TcVI. *Memórias do Instituto Oswaldo Cruz*, *104*, 1051–1054.

FSC
www.fsc.org
MIX
Papier | Fördert
gute Waldnutzung
FSC® C083411

Kolibri 360 GmbH
Ferdinand-Jühlke-Straße 7
99095 Erfurt, Deutschland
produktsicherheit@kolibri360.de

Die Werkzeuge und Arbeitsverfahren der Pressen

Völlige Neubearbeitung des Buches
„Punches, dies and tools for manufacturing in presses"
von Joseph V. Woodworth

von

Privatdozent Dr. techn. Max Kurrein
Betriebsingenieur des Versuchsfeldes für Werkzeugmaschinen
an der Kgl. Technischen Hochschule zu Berlin

Mit 683 Textfiguren und einer Tafel

Springer-Verlag Berlin Heidelberg GmbH
1914

Additional material to this book can be downloaded from http://extras.springer.com.

ISBN 978-3-662-23911-7 ISBN 978-3-662-26023-4 (eBook)
DOI 10.1007/978-3-662-26023-4

Softcover reprint of the hardcover 1st edition 1914

Vorwort.

Das Gebiet der Pressenwerkzeuge und Arbeitsverfahren hat sich in den letzten Jahren derartig entwickelt, daß die Anzahl der gebauten Sonderwerkzeuge, selbst wenn man nur die Veröffentlichungen in den Fachzeitschriften berücksichtigt, fast unübersehbar geworden ist. Die in der Praxis entstandenen Einteilungen dieser Werkzeuge sind in den Veröffentlichungen über diesen Gegenstand durchweg übernommen worden, lassen aber den Zusammenhang untereinander und mit den grundlegenden technologischen Vorgängen nur in sehr wenigen Fällen erkennen. Ein Zusammenhang mit den nicht sehr zahlreichen vorhandenen Forschungen technologischer Natur, die immerhin aber bemerkenswerte Ergebnisse für eine spätere theoretische Behandlung dieses Gebietes gezeitigt haben, ist aber kaum irgendwo zu finden.

Deshalb versucht der Verfasser in diesem Buch, das aus einer ursprünglich geplanten Übersetzung des Buches „Punches, Dies and Tools" von Joseph V. Woodworth entstanden ist, die Werkzeuge und Arbeitsverfahren der Pressen auf neuer, systematischer Grundlage zu behandeln, die für die Haupteinteilung die technologischen Vorgänge im verarbeiteten Material und für die Unterteilungen die aus der Praxis entstandenen konstruktiven Merkmale der Werkzeuge verwendet.

Da das Buch auch eine Übersetzung des amerikanischen Buches ersetzen sollte, hat der Verfasser die in Woodworths Buch angeführten amerikanischen Konstruktionen fast vollständig übernommen, sie aber den entsprechenden deutschen Bauarten, soweit das Material zur Verfügung stand, gegenübergestellt und damit versucht, Gelegenheit zu einem Vergleich der in beiden Ländern herrschenden Anschauungen zu geben und den Stand der selbständigen Entwicklung Deutschlands auf diesem Gebiete darzutun.

Aus diesen beiden Hauptgesichtspunkten ergab sich eine vollständige Neubearbeitung des Stoffes, bei der der Verfasser die in dem Woodworthschen Buche vorausgesetzte Entwicklung der Werkzeuge durch Originalbeiträge der Firmen dieses Gebietes und Veröffentlichungen in Fachzeitschriften gebildet hat. Denn solange sich die Kenntnis dieses Gebietes aus einer Kenntnis einzelner Werkzeuge zusammensetzt, solange für die Arbeitsverfahren nicht die fehlenden technologischen Grundlagen gewonnen sind, hat man es mit Stückwerk zu tun, bei dem eine Unsumme geistiger Kraft und materieller Kosten an vielen Stellen gleichzeitig und der Allgemeinheit unerreichbar vergeudet werden. Erst wenn die Grundlagen der Arbeitsverfahren auf ihre Elemente zurückgeführt, diese

Arbeiten wieder mit den durch Rechnung und Versuch belegten Gesetzen der Technologie in Übereinstimmung gebracht werden können, wird das in diesem Buche niedergelegte System vollkommen beendet sein.

In noch höherem Maße als die Werkzeuge setzen sich die Arbeitsverfahren aus einer Reihe Einzelarbeiten zusammen, weshalb der Besprechung der Werkzeuge 33 Tabellen der Herstellungsgänge aus den verschiedensten Gebieten vorausgeschickt worden sind, die auf einen Blick die zur Fertigstellung des angeführten Teiles notwendigen Einzelarbeiten erkennen lassen. Der Verfasser glaubt, daß durch eine derartige Zusammenstellung der Leser am schnellsten ein Urteil über die Zweckmäßigkeit des gewählten Arbeitsverfahrens erhält und derjenige, der sich erst die Kenntnis dieses Arbeitsverfahrens verschaffen will, daraus die Grenzen des Gebietes am besten ersehen wird.

Das dritte Kapitel des Buches, „Besondere Herstellungsverfahren", ist mit unbedeutenden Zusätzen aus dem amerikanischen Buche entnommen worden. Dagegen erschien es dem Verfasser notwendig, den engen Zusammenhang zwischen Werkzeug, Maschine und Arbeitsverfahren bei dem heutigen Stand der Entwicklung dieses Gebietes durch eine Anführung einzelner Vertreter für jede Gruppe der Maschinen und Zuführungen im vierten Teil darzutun, wofür auch wieder Originalzeichnungen der einzelnen Werke verwendet wurden.

Es wäre noch etwas über die gewählten Bezeichnungen der Werkzeuge und ihrer Teile zu sagen. Es ergab sich ganz in Übereinstimmung mit der Entstehung dieses Fachgebietes eine derartige Fülle verschiedener Ausdrücke für denselben Teil und dieselbe Arbeit, daß versucht wurde, eine einheitliche Bezeichnung durchzuführen. Dabei wurde der Grundsatz befolgt, daß die in der Praxis üblichen Ausdrücke eindeutig einen Arbeitsvorgang bezeichnen sollten.

An dieser Stelle möchte der Verfasser noch der angenehmen Pflicht nachkommen, allen jenen, die seine Arbeit durch Überlassung von Material oder durch Ratschläge unterstützt haben, seinen verbindlichsten Dank zu sagen, vor allem Herrn Prof. Dr.-Ing. G. Schlesinger-Charlottenburg, der sowohl persönlich den Verfasser mit vielen wertvollen Ratschlägen unterstützte, als ihm auch in seiner Eigenschaft als Schriftleiter der Werkstattstechnik die Verwendung des zahlreichen, vorbildlichen Materiales dieser Zeitschrift gestattete. Ebenso dankt der Verfasser bestens den Firmen, die in der liebenswürdigsten Weise ihm entweder die Besichtigung ihrer Werke gestattet oder Material zur Verfügung gestellt haben: Sie sind an den betreffenden Stellen mit Namen angeführt.

Charlottenburg, im Juni 1914.

Kurrein.

Inhaltsverzeichnis.

Benutzte Zeitschriften:

Werkstattstechnik	abgekürzt:	=	W. T.
Zeitschrift des Vereins deutscher Ingenieure	,,	=	Z. d. V. d. I.
Stahl und Eisen	,,	=	St. u. E.
Illustrierte Zeitung für Blechindustrie	,,	=	Ill. Zeit. f. Blechindustrie
Zeitschrift für Werkzeugmaschinen	,,	=	Z. f. Werkzeugm.
Uhland, Technische Rundschau . .	,,	=	Uhland, Techn. Rundsch.
Dinglers Polytechnisches Journal . .	,,	=	Dingler

I. Tabellen der Arbeitsgänge.

Tabelle Nr. 1.

Teile für einen Hauswecker (C. Lorenz, Berlin).
I. Grundplatte.

Arbeitsgang	Gegenstand	Arbeitsverfahren	Werkzeug siehe Seite:	Bemerkung
1		Ausschneiden des Blanketts	einfaches Schnittwerkzeug S. 43	
2	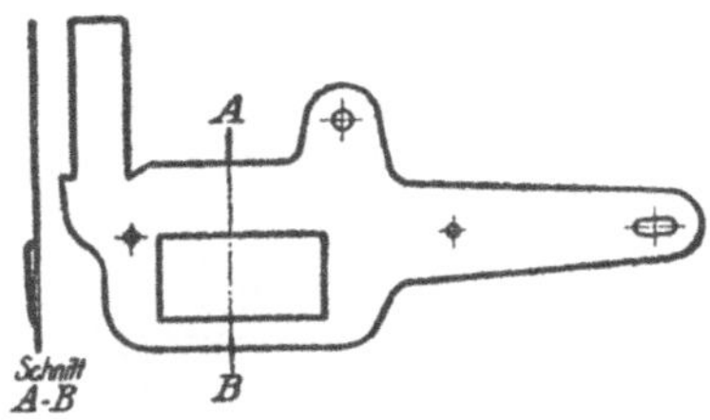	Lochen, Ausschneiden und Aufbiegen	Verbundwerkzeug S. 457	
3		Biegen des Endes	Einfaches Biegewerkzeug S. 352	
4	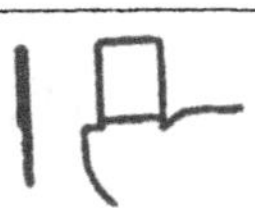	Zusammenbiegen desselben Endes	Einfaches Biegewerkzeug S. 352	
5	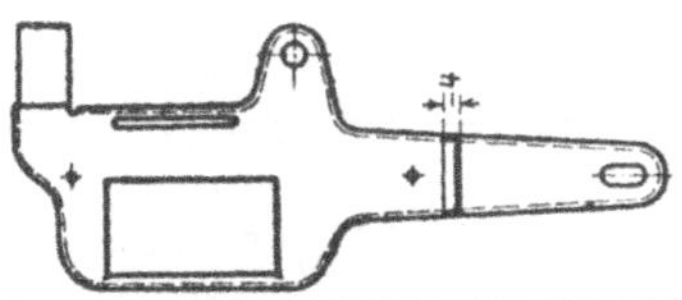	Ziehen des Randes und Prägen der Rippe	Verbund-, Zieh- und Prägewerkzeug S. 477	
6	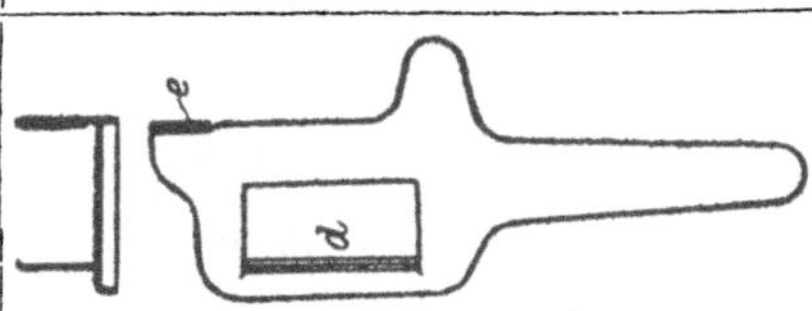	Aufbiegen der Teile e und d	Mehrfach-Biegewerkzeug S. 355	

Tabelle Nr. 1. (Fortsetzung.)

I. Grundplatte.

Arbeitsgang	Gegenstand	Arbeitsverfahren	Werkzeug siehe Seite:	Bemerkung
7	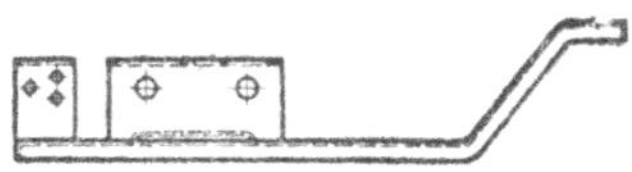	Biegen des Endes für die Glocke	Präge- u. Biegewerkzeug S. 479	
8	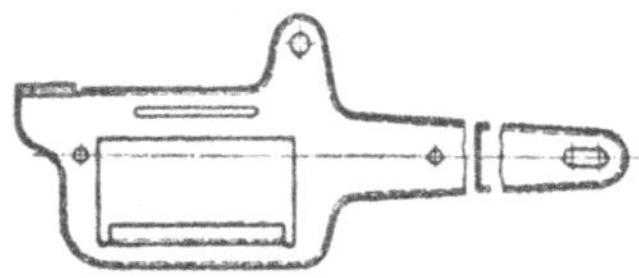	In schwarzen Lack tauchen		

II. Feder für Hammer.

Arbeitsgang	Gegenstand	Arbeitsverfahren	Werkzeug siehe Seite:	Bemerkung
1		Ausschneiden, Lochen	Folgewerkzeug S. 89	
2	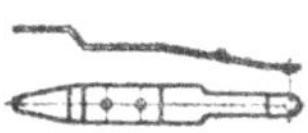	Biegen	einfaches Biegewerkzeug S. 365	

III. Hammer.

Arbeitsgang	Gegenstand	Arbeitsverfahren	Werkzeug siehe Seite:	Bemerkung
1		Ausschneiden, Lochen	Folgewerkzeug S. 88	
2	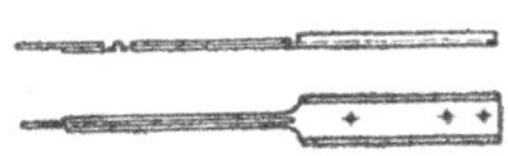	Biegen der Seiten, Prägen der Rippe	Verbund-, Biege- u. Prägewerkzeug S. 483	

IV. Ständer für Unterbrecherschraube.

Arbeitsgang	Gegenstand	Arbeitsverfahren	Werkzeug siehe Seite:	Bemerkung
1		Ausschneiden und Lochen	Folgewerkzeug S. 88	
2	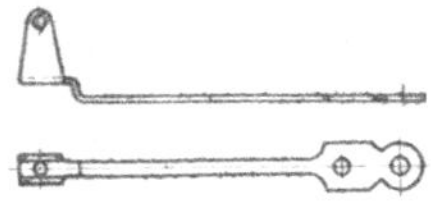	Mehrfachbiegen	Biegewerkzeug S. 369	

Tabelle Nr. 2.

Klinkenrohr. W. T. 1908, S. 405.

Arbeitsgang	Gegenstand	Arbeitsverfahren	Bemerkung
1	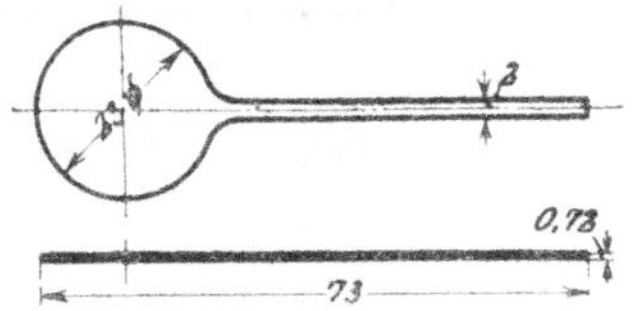	Ausschneiden	
2	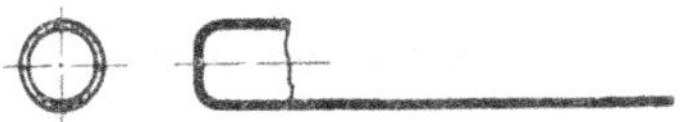	1. Ziehen	
3		Glühen	
4		2. Ziehen	
5		Glühen	
6		3. Ziehen	
7		Glühen	
8	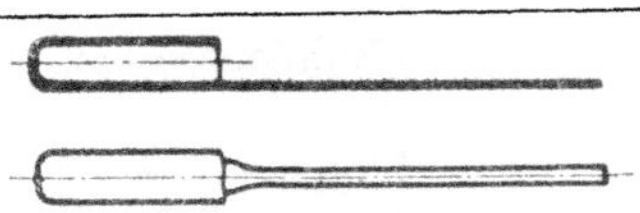	4. Ziehen	
9	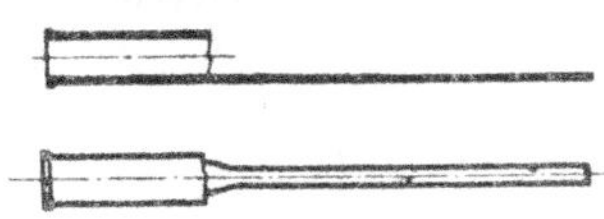	Abstechen und Bördeln	
10	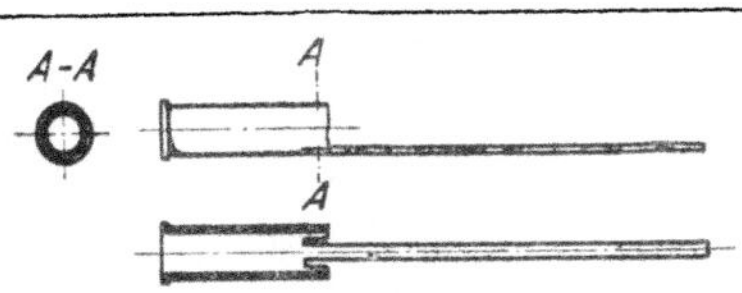	Einschneiden	
11	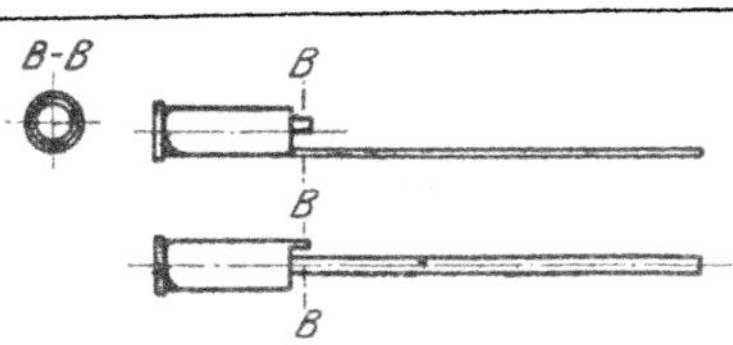	Beschneiden des Rohres	

Tabelle Nr. 2. (Fortsetzung.)

Klinkenrohr.

Arbeits-gang	Gegenstand	Arbeitsverfahren	Bemerkung
12		Schaft rundpressen	
13	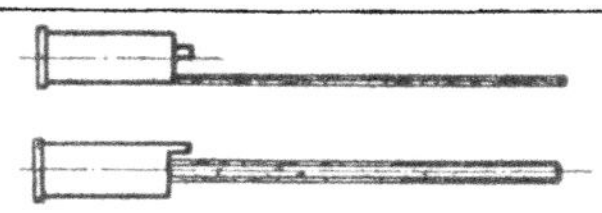	Schaft beschneiden	
14	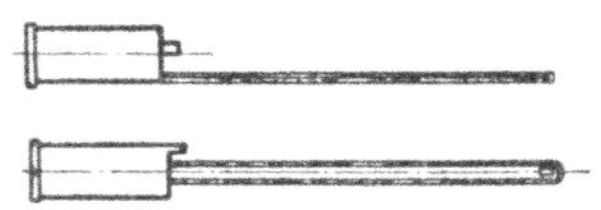	Schaft lochen	
15		Rohr nachdornen	
16		Schaft feilen	
17		Rohr feilen	
18		Rohrlappen biegen	
19		beizen	

Tabelle Nr. 3.

Induktorkurbel. W. T. 1908, S. 402.

Arbeits-Gang	Gegenstand	Arbeitsverfahren	Bemerkung
1	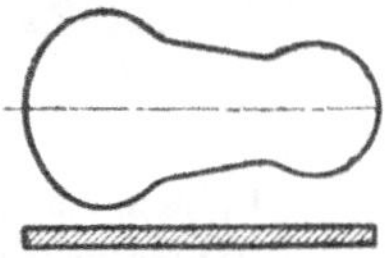	Streifen zuschneiden und ausschneiden	
2	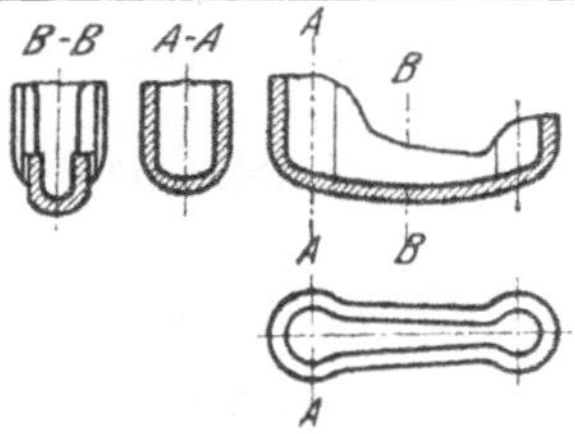	1. Ziehen	
3	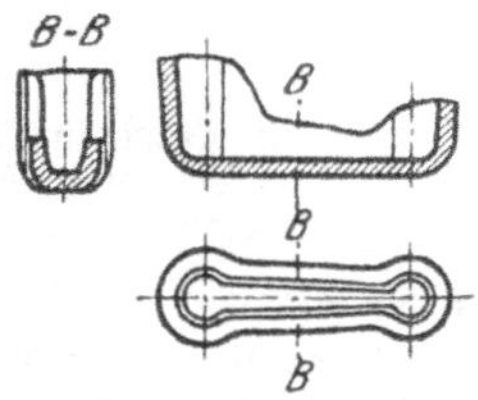	Fläche anpressen	
4	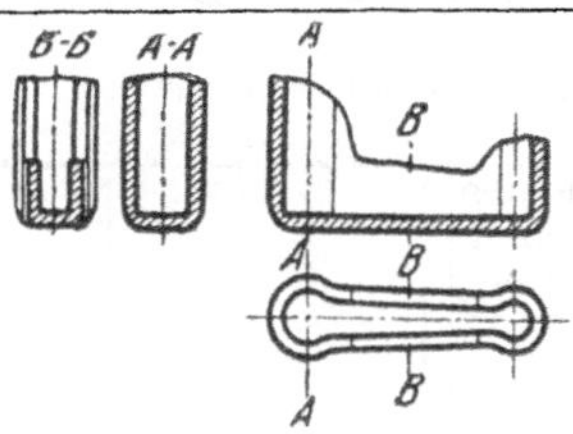	2. Ziehen	
5	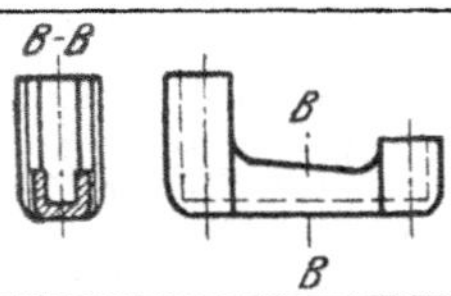	Profilfräsen	
6		Gratabnehmen und beizen	
7	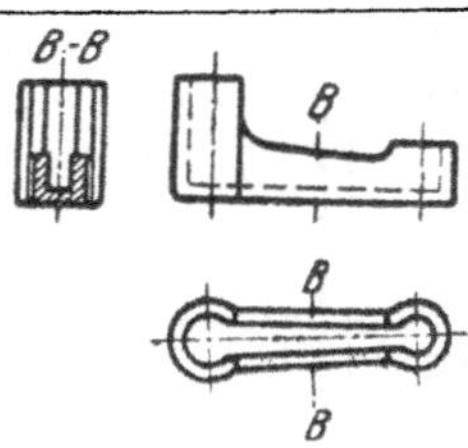	Fertig pressen	

Tabelle Nr. 3. (Fortsetzung).

Induktorkurbel.

Arbeits-gang	Gegenstand	Arbeitsverfahren	Bemerkung
8		Große Fläche schleifen	
9	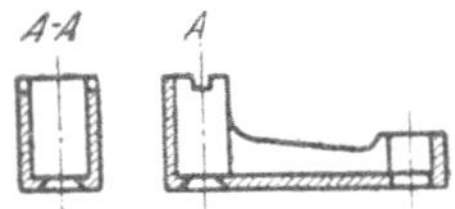	Bohren, aufreiben, abfräsen, senken	
10	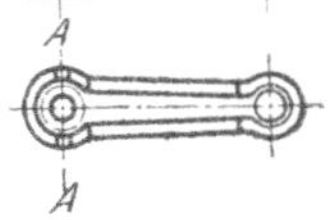	Schlitz fräsen	
11	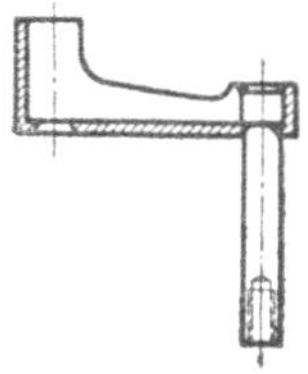	Vorschleifen 100 Kurbelstifte anfertigen	
12		desgl. in Kurbel einnieten, Kurbel richten, nachpressen	

Tabelle Nr. 4.

Herstellung einer Tülle mit gezacktem Rand in einem Werkzeug.

Arbeits-gang	Gegenstand	Arbeits-verfahren	Gesenk siehe Seite:	Bemerkung
1		Ausschneiden u. Prägen d. Vertiefung	Folgewerkzeug S. 399	
2		Lochen		
3		Ziehen d. Tülle		
4		Prägen d. Flansches		
5	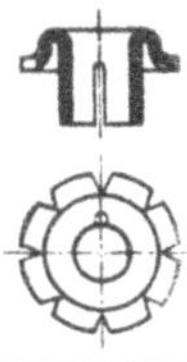	Ausschneiden d. Zwickel		
6	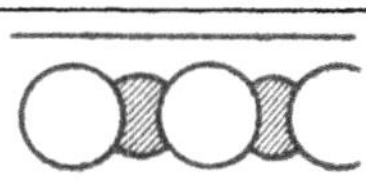	Ausschneiden des Steges im Abfall		

Tabelle Nr. 5.

Herstellung der Stahlfedern.

Arbeits-gang	Gegenstand	Arbeits-verfahren	Werkzeug siehe Seite:	Bemerkung
1	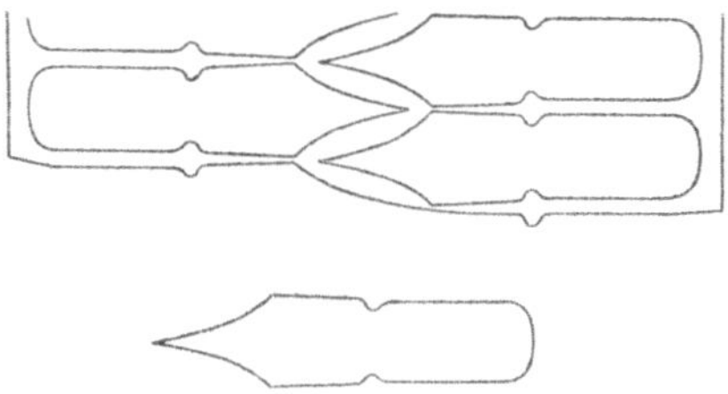	Ausschneiden	Mehrfach-schnitt, selbsttätige Zufuhr von der Rolle Seite 540	Vgl. Stanz-automat von Rob. Tümmler, Döbeln
2		Lochen 1. Ausstechen		
3		Seitl. Schlitze schneiden 2. Ausstechen		
4		Ausglühen		
5	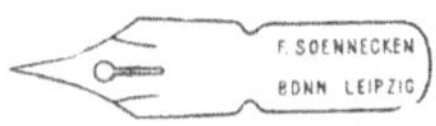	Beschriftung einprägen		„Stem-peln"
6		Hohl prägen		„Stamp-fen"
7		1. Härten		
8		2. Härten		
9		Blank scheuern		
10		Schleifen		
11	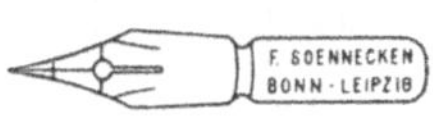	Spalten	Seite 493	Fallwerk
12		Färben		
13		Aussuchen		

Tabelle Nr. 6.

Schmieden einer Gabel von der Stange. (Schmiedemaschine Ajax-de Fries, Düsseldorf.)

Arbeitsgang	Gegenstand	Arbeitsverfahren	Werkzeug siehe Seite:	Bemerkung
1		Schlitzen	Warmsäge	
2	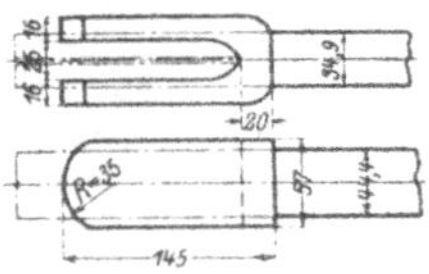	Vorschmieden der Gabel	1. Gesenk S. 175	Schmiedemaschine
3	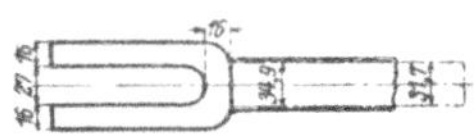	Fertigschmieden der Gabel	2. Gesenk S. 176	
4	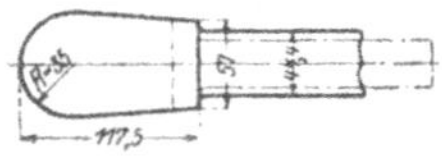	Ausrecken und Rundschmieden der Stange	Fallhammergesenk	

Tabelle Nr. 7.

Schmieden einer Büchse von der Stange. (Schmiedemaschine Hasenclever, Düsseldorf.)

Arbeitsgang	Gegenstand	Arbeitsverfahren	Werkzeug siehe Seite:	Bemerkung
1	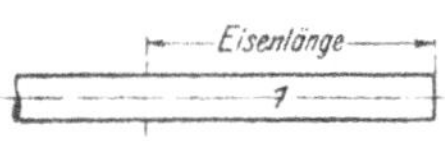	Einschieben der Stangenlänge		
2	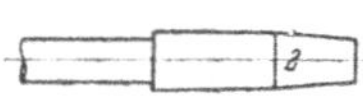	Anstauchen der Stange und des konischen Kopfes	1. Gesenk S. 176	Alle drei Gesenke sind gleichzeitig in der Schmiedemaschine eingespannt.
3	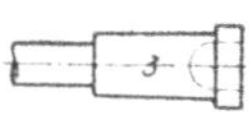	Verlochen und Stauchen des Flansches	2. Gesenk S. 176	
4	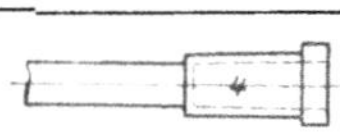	Zurücklochen der inneren Höhlung	3. Gesenk S. 176	
5	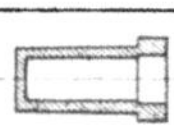	Abschneiden der fertigen Büchse	Warmsäge	

Tabelle Nr. 8.

Schmieden einer Ventilbrücke von der Stange. (Schmiedemaschine Hasenclever, Düsseldorf.)

Arbeitsgang	Gegenstand	Arbeitsverfahren	Werkzeug siehe Seite:	Bemerkung
1	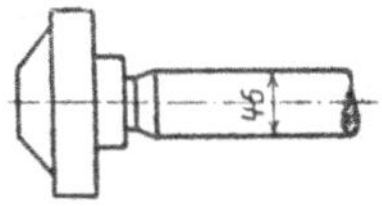	Anstauchen des Kopfes, Einkniffen		Schmiedemaschine
2	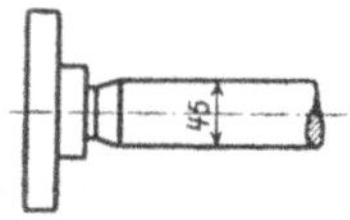	Fertigschmieden der Brücke		
3	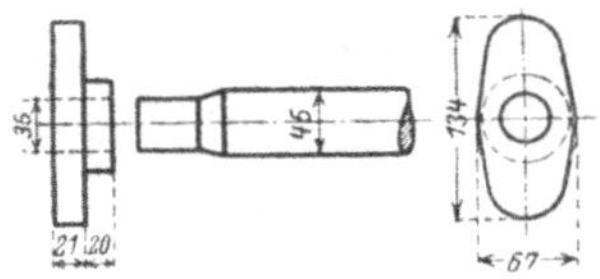	Zurücklochen und gleichzeitig Abtrennen		

Tabelle Nr. 9.

Warmpressen einer Flügelmutter aus Messing. (Kabelwerk Oberspree der A.E.G.)

Arbeitsgang	Gegenstand	Arbeitsverfahren	Werkzeug siehe Seite:	Bemerkung
1		Blankett von Stange abgeschnitten	Strangpresse Fig. 199 Warmsäge	
2	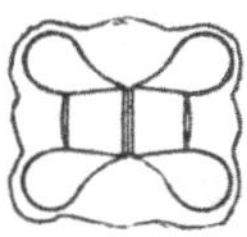	Pressen 2 Blanketts gleichzeitig	Warmpresse S. 203	
3	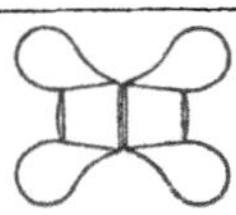	Abgraten	Abgratpresse	
4		Entzweisägen	Kaltsäge	
5	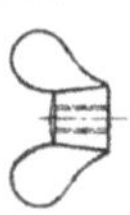	Bohren	Automat	
6		Gewindeschneiden		

Tabelle Nr. 10.

Warmpressen eines Heizungsgriffes aus Messing. (Kabelwerk Oberspree der A.E.G.)

Arbeitsgang	Gegenstand	Arbeitsverfahren	Werkzeug siehe Seite:	Bemerkung
1		Abschneiden		
2		Vorpressen des Griffes	S. 204	
3		Abgraten		
4	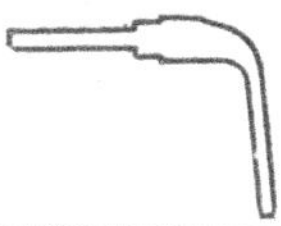	Abbiegen		
5	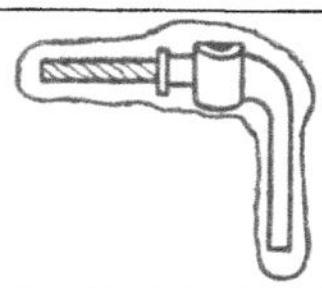	Fertigpressen	S. 204	
6	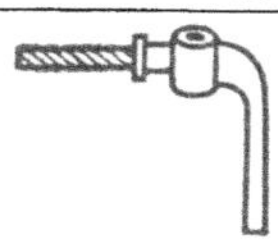	Abgraten		

Tabelle Nr. 11.

Warmpressen eines Schuhes. (Kabelwerk Oberspree der A.E.G.)

Arbeitsgang	Gegenstand	Arbeitsverfahren	Werkzeug siehe Seite:	Bemerkung
1		Abschneiden	Warmsäge	
2	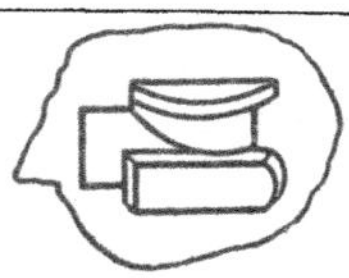	Vorpressen	S. 205	
3	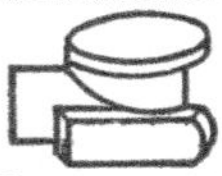	Abgraten		
4	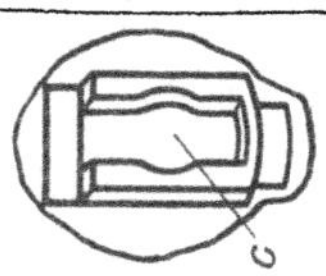	Fertigpressen	S. 206	
5		Abgraten		

Tabelle Nr. 12.

Herstellung eines Teelöffels.

Arbeitsgang	Gegenstand	Arbeitsverfahren	Werkzeug siehe Seite:	Bemerkung
1	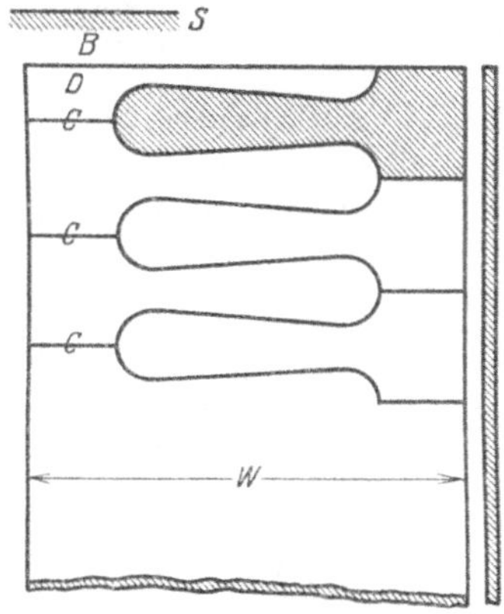	Abfallfreies Ausschneiden mit Schnitt- und Hilfsstempel (Form I)	S. 45	
2	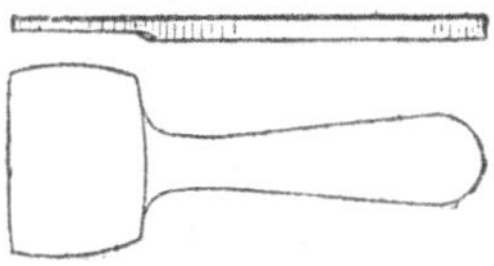	Querwalzen des Pfannenendes (Form II)		
3	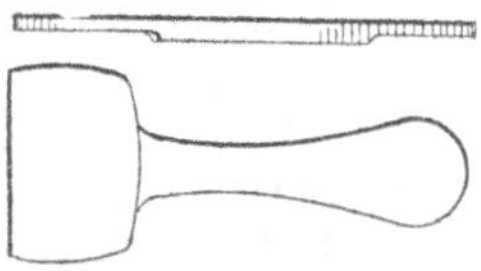	Querwalzen des Handgriffes (Form III)		
4	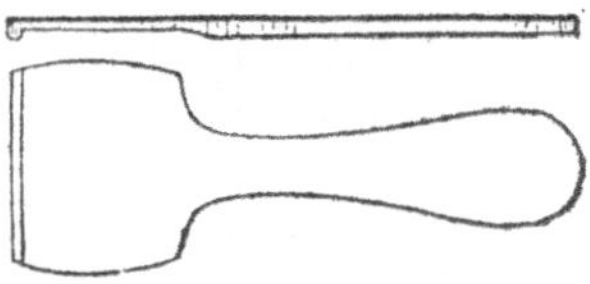	Längen auf den Streckwalzen (Form IV)	S. 233	
5 6 7		I. Ausglühen, Beizen, Reinigen		
8	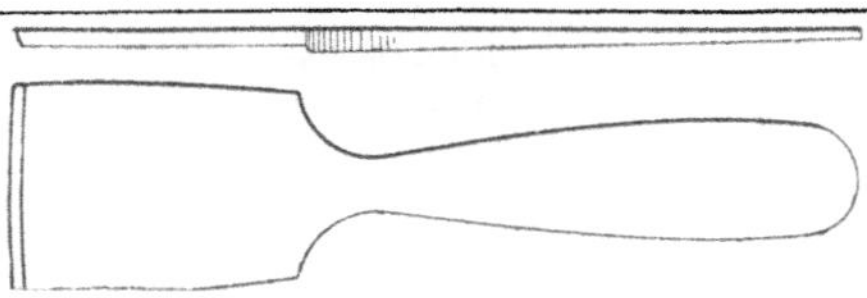	Letzter Gang durch die Streckwalzen (Form V)		
9		Ausglühen		

Tabelle Nr. 12. (Fortsetzung.)

Herstellung eines Teelöffels.

Arbeitsgang	Gegenstand	Arbeitsverfahren	Werkzeug siehe Seite:	Bemerkung
10	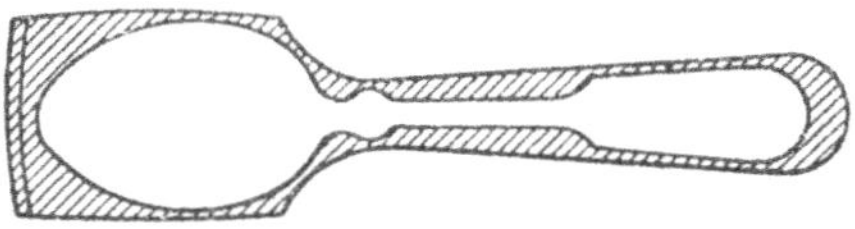	Ausschneiden (Form VI)		
11		Grat entfernen		
12	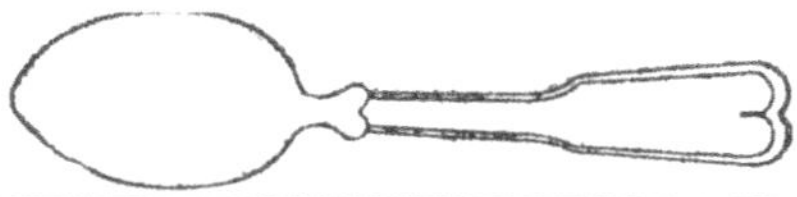	Prägen des Handgriffes (Form VII)	S. 234	
13		Prägegrat entfernen		
14		Aufbiegen des Handgriffes (Form VIII)		
15	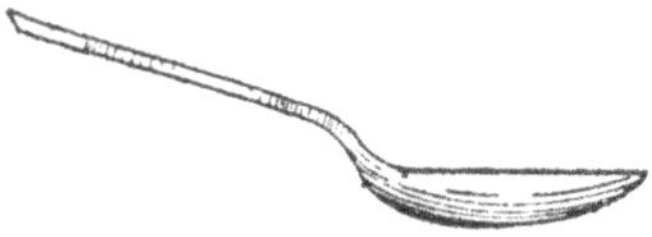	Prägen der Pfanne und des Stieles (Form IX)	S. 234	
16		Fertig polieren		
17		Fertig prägen i. Holzgesenk (Form X)		
18		Markenzeichen einschlagen		
19		Verschönerungsarbeiten usw.		

Tabelle 13.

Ziehen eines Gefäßes mit Rand aus weichem Messing.
(23,8 mm lang, 20,6 mm Durchmesser.)

Arbeits-gang	Gegenstand	Arbeitsverfahren	Werkzeug siehe Seite:	Bemerkung
1	Scheibe 47,6 Φ, 0,8 dick.	Ausschneiden	Mehrfach-schnitt	
2		Ziehen des Näpfchens		
3	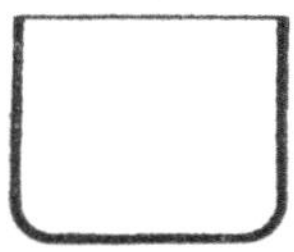	1. Ziehen	einfaches Zieh-werkzeug	
4		2. Ziehen	einfaches Zieh-werkzeug	
5		Rand abschneiden	Abstechbank	
6	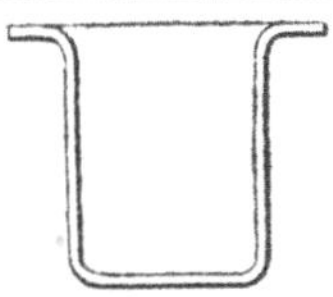	Umlegen des Flansches, Kalibrieren des Bodens	S. 333	

Tabelle Nr. 14.

Ziehen von Gewehrpatronen.

Arbeits-gang	Gegenstand	Arbeitsverfahren	Werkzeug siehe Seite:	Bemerkung
1	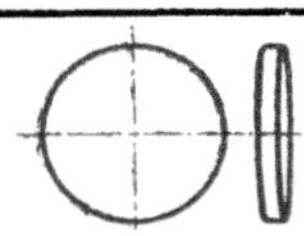	Ausschneiden der Scheibe	Verbund-, Schnitt- und Ziehwerkzeug	Doppelt-wirkende Presse; Ausglühen nach jedem Ziehen
2		Ziehen d. Näpfchens	S. 263 ff.	
3		Nachziehen d. Näpfchens		

Tabelle Nr. 14. (Fortsetzung.)

Ziehen von Gewehrpatronen.

Arbeitsgang	Gegenstand	Arbeitsverfahren	Werkzeug siehe Seite:	Bemerkung
4	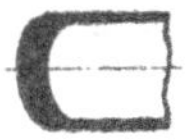	1. Ziehen	Ein Stempel mit zwei Ziehringen untereinander	
5		2. Ziehen		
6 7		3. Ziehen Abschneiden		
8		Boden anpressen		
9 10		4. Ziehen Abschneiden		
11		5. Ziehen		
12		Vertiefung f. d. Zünder		
13		Fertigpressen des Kopfes		
14		Lochen f. d. Zünder	S. 265	
15		Einziehen	S. 265	
16		Kalibrieren		
17 18		Abstechen Auswerfernut einfräsen	(hinterdrehter Fräser)	
19		Kalibrieren		
20		Einsenken d. Öffnung		

Tabelle Nr. 15.

Ziehen von Patronenhülsen (Fritz Werner, Berlin).

Arbeitsgang	Gegenstand	Arbeitsverfahren	Werkzeug siehe Seite:	Bemerkung
1	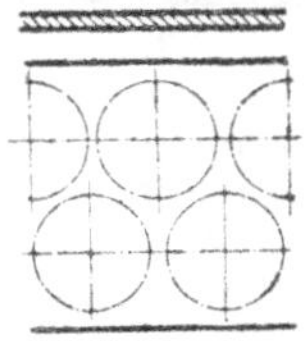	Ausschneiden der Scheiben	Mehrfachschnitt mit Bandzuführung von der Rolle	Ausglühen nach den einzelnen Zügen und Scheuern usw. nicht in Tabelle eingefügt
2		Ziehen des Näpfchens	Seite 269 ff.	
3		1. Ziehen		
4		2. Ziehen		
5		3. Ziehen		
6		Abschneiden	Kreisschere S. 279	
7		Vorpressen des Bodens		
8		4. Ziehen		
9		Abschneiden	Kreisschere S. 279	
10		Fertigpressen des Bodens		
11		Einziehen und Kalibrieren des Geschoßraumes		
12		Bohren der Zündlöcher	Schnellbohren mit Vorrichtung	
13		Einstechen der Ausziehnute, Geradedrehen des Bodens, Abfräsen auf Länge	Selbsttätige Sondermaschine	

Tabelle Nr. 16.

Herstellung der Cu-Ni-Mantelgeschosse **(Nordamerika).**

Arbeitsgang	Gegenstand	Arbeitsverfahren	Werkzeug siehe Seite:	Bemerkung
1		Ausschneiden der Scheibe	**Mehrfachschnitt, Zuführung von der Rolle**	
2	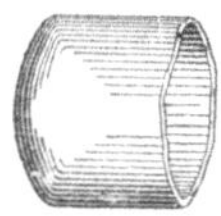	**Ziehen des Näpfchens**		
3	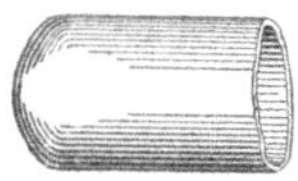	**1. Ziehen der Hülse**		
4		**2. Ziehen der Hülse**		**Ziehen der Mäntel S. 271**
5	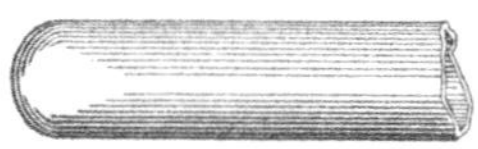	**3. Ziehen der Hülse**		
6	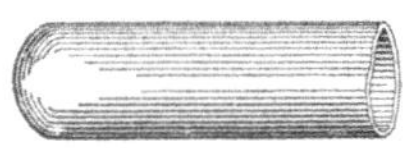	Abschneiden		
7	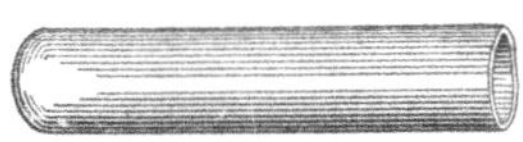	**4. Ziehen der Hülse**		
8	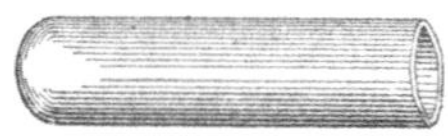	**Abschneiden auf genaue Länge**		
9		**Scheuern und Reinigen der Mäntel**		

Tabelle Nr. 16. (Fortsetzung.)
Herstellung der Cu-Ni-Mantelgeschosse (Nordamerika).

Arbeitsgang	Gegenstand	Arbeitsverfahren	Werkzeug siehe Seite:	Bemerkung
10	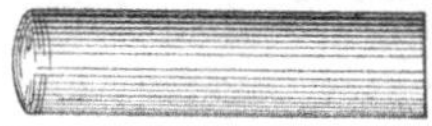	Vorpressen der Spitze		
11		Fertigpressen der Spitze		
12		Einsetzen des Bleikernes		Pressen und Füllen der Mäntel auf der sechsstempligen Revolverpresse
13	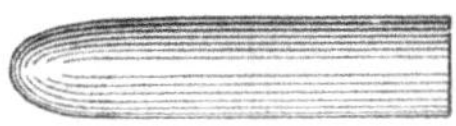	Anbörteln des offenen Endes		
14		Fertig-Umbörteln des offenen Endes		
15		Auswerfen des fertigen Geschosses		

Tabelle Nr. 17.
Ziehen von großen Pulverhülsen für 6″ Schnellfeuergeschütze (Spanien).

Arbeitsgang	Gegenstand	Arbeitsverfahren	Werkzeug siehe Seite:	Bemerkung
1	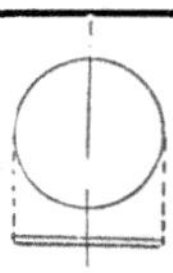	Anschneiden der Scheibe	S. 275 ff.	
2		Ziehen des Näpfchens		
3		Nachziehen des Näpfchens		Ausglühen, Beizen und Waschen nach allen Ziehgängen
4		1. Ziehen d. Hülse		stehende Presse
5		2. Ziehen d. Hülse		
6		3. Ziehen d. Hülse		
7		Flachpressen des Bodens	Werkzeug wie Nr. 8 mit unterer Druckplatte	

Tabelle Nr. 17. (Fortsetzung.)

Ziehen von großen Pulverhülsen für 6″ Schnellfeuergeschütze (Spanien).

Arbeitsgang	Gegenstand	Arbeitsverfahren	Werkzeug siehe Seite:	Bemerkung
8		4. Ziehen d. Hülse		stehende Presse
9		5. Ziehen d. Hülse		
10		Prägen d. Vertiefung für Zünder		
11		6. Ziehen d. Hülse		18″ liegende Presse
12		7. Ziehen d. Hülse		
13		8. Ziehen d. Hülse		
14		9. Ziehen d. Hülse		
15		10. Ziehen der Hülse		Nach dem 10. Zug folgt kein Ausglühen
16		Vorpressen des Bodens		2500 t Presse
17		Flachpressen der Bodenmitte		
18		Fertig- (Flach-) Pressen des Bodens		
19		Einziehen der Hülse		Liegende hydr. Presse
20		Abdrehen, Bohren d. Zündloches, Fräsen auf Länge, Fertigarbeiten		

Tabelle Nr. 18.

Ziehen eines konischen Bechers aus Aluminium (Erzgebirgische Schnittwerkzeug- und Maschinenfabrik).

Arbeitsgang	Gegenstand	Arbeitsverfahren	Werkzeug siehe Seite:	Bemerkung
1 2	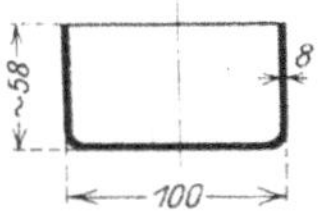	Ausschneiden Ziehen des Näpfchens	Verbund-Schnitt- und Ziehwerkzeug	
3	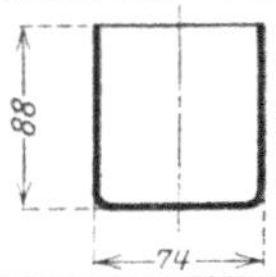	Ziehen der Hülse		
4	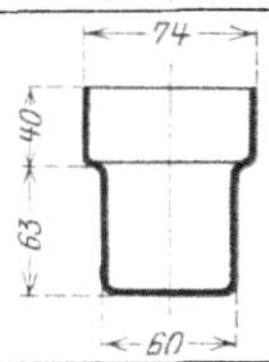	Absetzen durch Ziehen		
5		Abstechen		
6	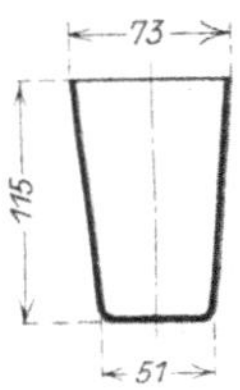	Konisch Fertigziehen	Zieh- und Kalibrier-Werkzeug	

Tabelle Nr. 19.

Ziehen eines konischen Messingnäpfchens mit Rippen.

1	Scheibe 58,8 Ⓓ 2,38 mm dick	Ausschneiden	Mehrfach-schnitt	
2	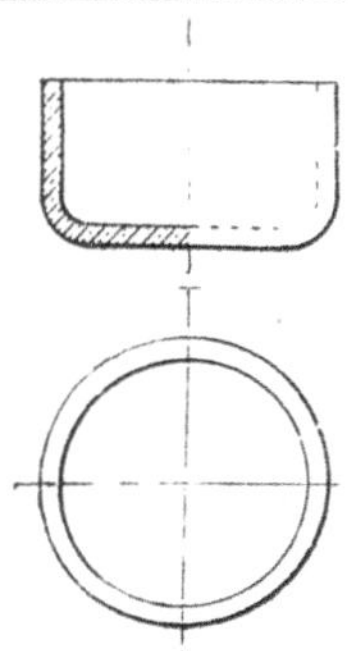	1. Zug*) Ziehen des zyl. Näpfchens	Normales Zieh-werkzeug	*) Ausglühen nach jedem Zug
3		Abschneiden		

Tabelle Nr. 19. (Fortsetzung.)

Ziehen eines konischen Messingnäpfchens mit Rippen.

Arbeits-gang	Gegenstand	Arbeitsverfahren	Werkzeug siehe Seite:	Bemerkung
3	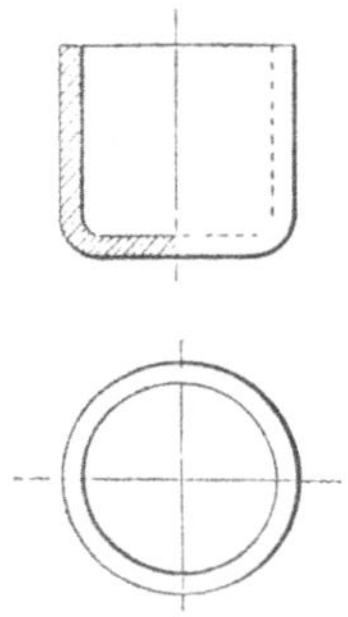	II. Zug Nachziehen des zylindrischen Näpfchens	Normales Zieh-werkzeug	
4		Abschneiden		
5	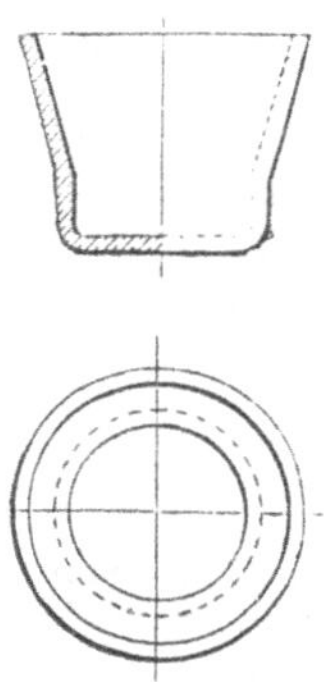	III. Zug Konisch ziehen	Wie Nr. 6, nur ohne Rippen	
6	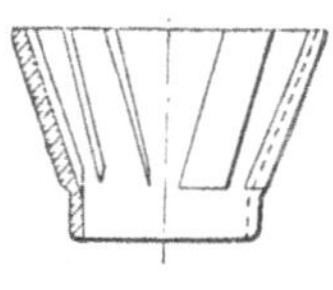	Rippen einprägen	S. 285	
7	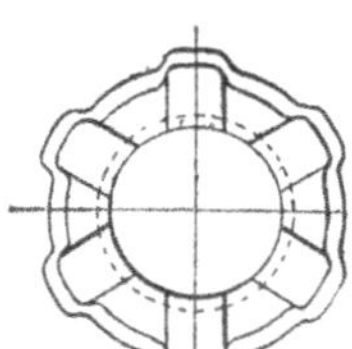	Boden ausschneiden	Einfaches Schnitt-werkzeug	

Tabelle Nr. 20.

Ziehen von Brennerteilen für Petroleumlampen (Ed. Sommerfeld, Berlin).

a) Oberteil.

Arbeitsgang	Gegenstand	Arbeitsverfahren	Werkzeug siehe Seite:	Bemerkung
1	82 Φ 0,35 dick	Ausschneiden der Scheibe	Mehrfachschnitt	Weißblech
2		Ziehen des Näpfchens	S. 289	Liegende Ziehpresse Reiß & Martin Berlin SW.
3		1. Nachziehen	S. 289	
4	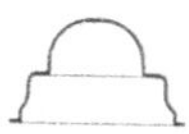	2. Nachziehen	S. 291	
5	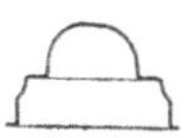	Fertigmachen	S. 291	
6		Schlitzherstellen	Lochwerkzeug	
7		Rand abschneiden, Kalibrieren	Rundschnitt S. 292	

b) Unterteil.

Arbeitsgang	Gegenstand	Arbeitsverfahren	Werkzeug siehe Seite:	Bemerkung
1	66 Φ 0,35 dick	Ausschneiden der Scheibe	Mehrfachschnitt	Weißblech
2		Ziehen des Näpfchens	S. 292	
3		Nachziehen	S. 293	
4	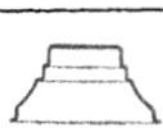	Kalibrieren	S. 293	
5		Rand abschneiden	Rundschnitt	

Tabelle Nr. 21.

Ziehen einer tiefen viereckigen Büchse aus Kupfer mit scharfen Ecken und Kanten.

Arbeitsgang	Gegenstand	Arbeitsverfahren	Werkzeug siehe Seite:	Bemerkung
1	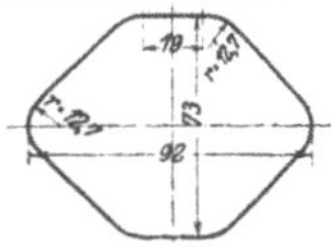	Ausschneiden	Verbund-Schnitt- und Ziehwerkzeug Seite 413	
1a	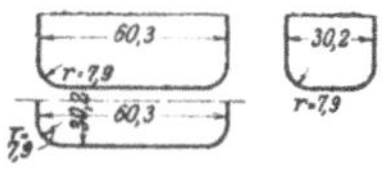	Ziehen des Näpfchens Ausglühen		
2		1. Ziehen Ausglühen	Einfaches Ziehwerkzeug gewöhnlicher Bauart	
3	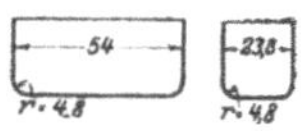	2. Ziehen Ausglühen		
4	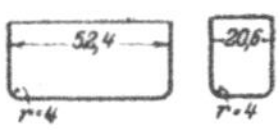	3. Ziehen Ausglühen		
5	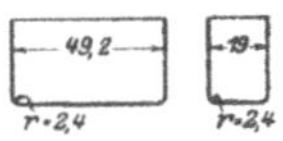	4. Ziehen Ausglühen		
6	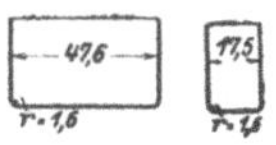	5. Ziehen Ausglühen		
7	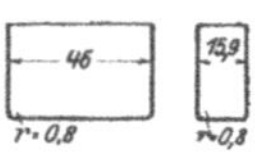	6. Ziehen		
8	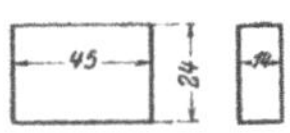	Kalibrieren		

Tabelle Nr. 22.

Herstellung der Edisongewinde (A. E. G.)

W. T. 1909, S. 13.

Arbeits-gang	Gegenstand	Arbeitsverfahren	Werkzeug siehe Seite:	Bemerkung
1	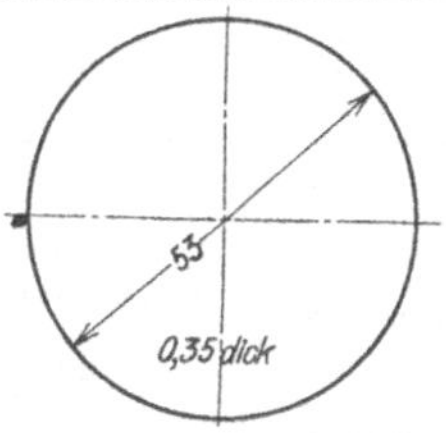	Ausschneiden der Scheiben	Mehrfach-schnitt vom Streifen oder Rollen-material mit selbsttätiger Zuführung	*) evtl. Verbund-zieh- u. Schneide-werkzeug für Gang 1/2 S. 406 Fig. 522
2	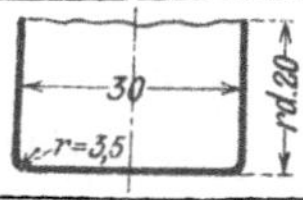	Ziehen des Näpfchens*)	einfaches Ziehwerkzeug	einstemplige Revolverpresse
3		Glühen, Beizen usw.		
4	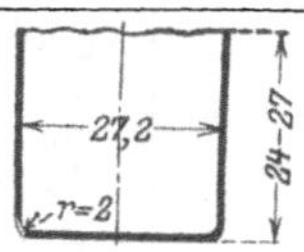	Nachziehen, Ausglühen	wie Nr. 2	
5	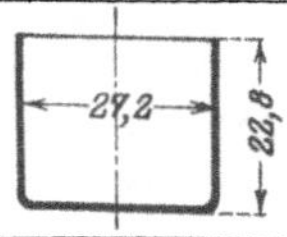	Abschneiden	selbsttätige Abstechbank	W. T. 1909, Fig. 9, S. 19
6	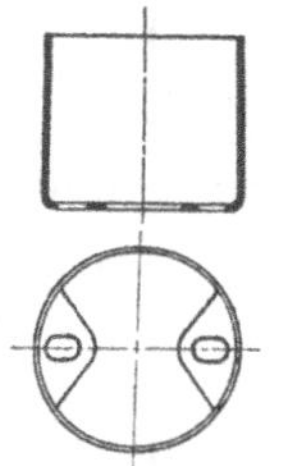	Ausschneiden des Bodens und Lochen	Verbund-schnitt- und Lochwerk-zeug	Exzenterpresse
7	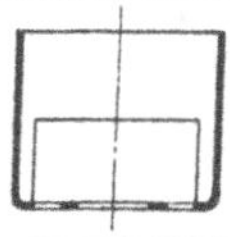	Ausschneiden der Seiten		
8	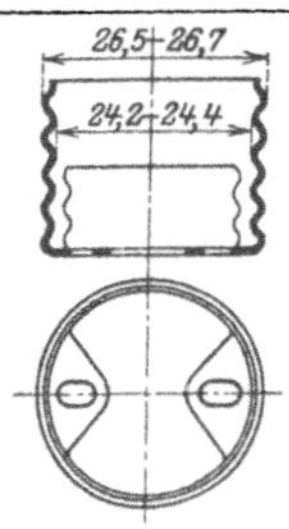	Gewinde-drücken	Drückrolle	Selbsttätige Gewinde-drückmaschine wie in Tabelle Nr. 23
9		Beizen, Galvanis.		

Tabelle Nr. 23.

Unterkasten für einen Kosmos-Petroleumbrenner.
(Reiß & Martin, Berlin.)

Arbeitsgang	Gegenstand	Arbeitsverfahren	Werkzeug siehe Seite:	Bemerkung
1	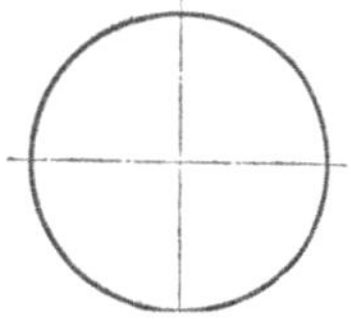	Ausschneiden der Scheibe	Mehrfachschnitt	Kurbelpresse mit selbsttätiger Zuführung des Streifens von der Rolle
2	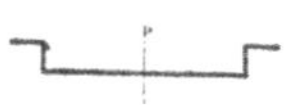	Ziehen		Schnellaufende schräge Kurbelpresse
3		Gewinde anrollen	Drückrolle S. 344	Selbsttätige Gewinde-Drückmaschine mit selbsttätiger Zuführung
4		Rippen einprägen	S. 430	
5	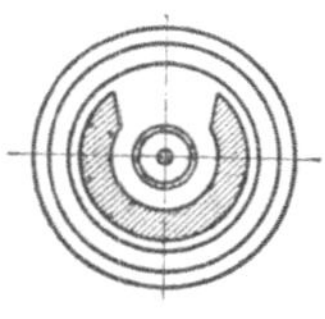	Innen ausschneiden	Verbundschnittwerkzeug S. 430	dreistemplige Revolverpresse von Reiß & Martin Berlin SW.
6	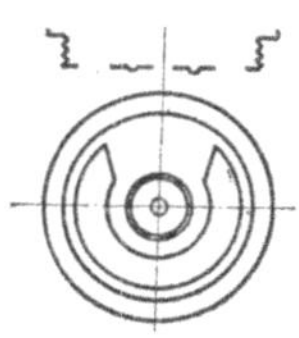	Rand abschneiden	Einfacher Schnitt und Gegenschnitt S. 430	

Tabelle 24.

Herstellung einer Schnauze für eine Blechkanne.

Arbeitsgang	Gegenstand	Arbeitsverfahren	Werkzeug siehe Seite:	Bemerkung
1	Scheibe	Ausschneiden d. Scheibe v. Bandmat. m. unmittelbar folgendem	Verbund-Schnitt- und Ziehwerkzeug S. 409	
2	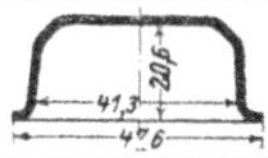	Ziehen des Näpfchens		
3	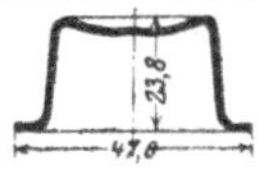	Nachziehen und Prägen d. Bodens	Verbund-Zieh-u. Prägewerkzeug S. 488	
4	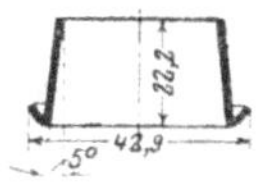	Rand abschn. 2. Nachziehen Ausschneiden des Bodens	Verbundwerkzeug S. 409	
5	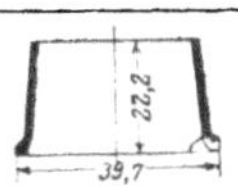	Einrollen und Niederprägen des unteren Randes	Verbund-Einroll- und Prägewerkzeug S. 489	

Tabelle Nr. 25.

Herstellung der Teile eines Uhrgehäuses.[1])

1.) Glasring.

Arbeitsgang	Gegenstand	Arbeitsverfahren	Werkzeug siehe Seite:	Bemerkung
1	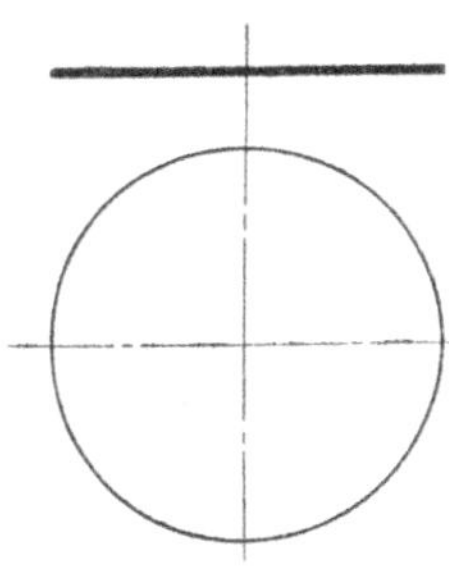	Ausschneiden	Verbund-Schnitt- und Ziehwerkzeug	
2		Ziehen		
3	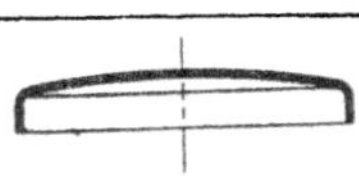	Abschneiden	Kreisschere	Vgl. Patronenziehen, Tabelle Nr. 15, G. 6

[1]) W. T. 1910, S. 405.

Tabelle Nr. 25. (Fortsetzung.)

Herstellung der Teile eines Uhrgehäuses.

1.) Glasring.

Arbeitsgang	Gegenstand	Arbeitsverfahren	Werkzeug siehe Seite:	Bemerkung
4	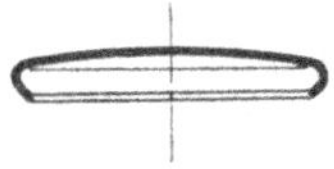	1. Einrollen		
5		2. Einrollen	S. 337	
6	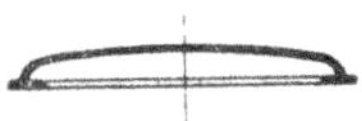	Drücken	Drückrolle	
7	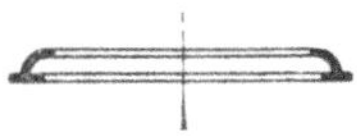	Ausschneiden	S. 337	Revolverpresse

Tabelle Nr. 26.

Herstellung der Teile eines Uhrgehäuses.

2.) Glasring.

Arbeitsgang	Gegenstand	Arbeitsverfahren	Werkzeug siehe Seite:	Bemerkung
1		Scheibe auschneiden	Verbund-Schnitt- und Prägewerkzeug	
2	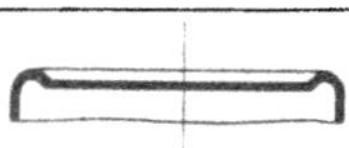	Prägen		
3		Abschneiden	Kreisschere	Vgl. Tab. Nr. 15 G. 6
4	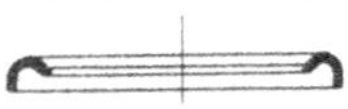	Ausschneiden		Revolverpresse
5	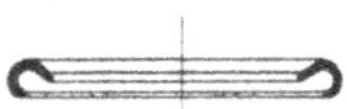	Einrollen		
6	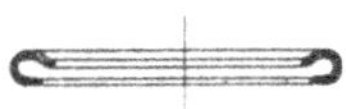	2. Einrollen		
7	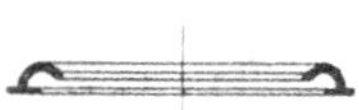	Drücken	Drückrolle	

Tabelle Nr. 27.

Warmbiegen einer Stahlfeder.

Arbeitsgang	Gegenstand	Arbeitsverfahren	Werkzeug siehe Seite:	Bemerkung
1	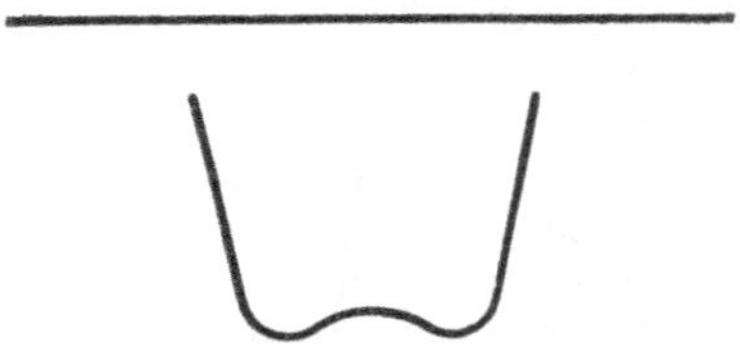	Abschneiden des Flachstahles 11 × 2,4 mm		
2		Anwärmen		
3		Vorbiegen	S. 358	
4		Anwärmen		
5		Fertigbiegen	S. 360	

Tabelle Nr. 28.

Biegen einer Blechrolle.

Arbeitsgang	Gegenstand	Arbeitsverfahren	Werkzeug siehe Seite:	Bemerkung
1	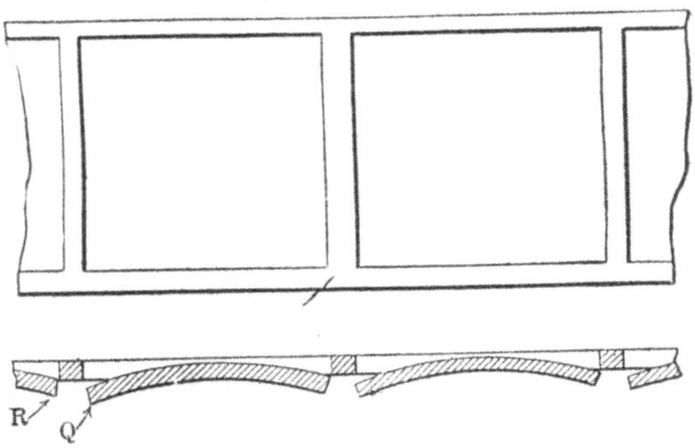	Ausschneiden und Anbiegen	Folgewerkzeug S. 443	
2	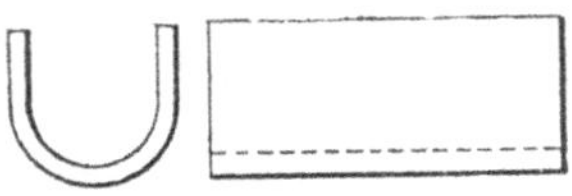	Biegen		
3	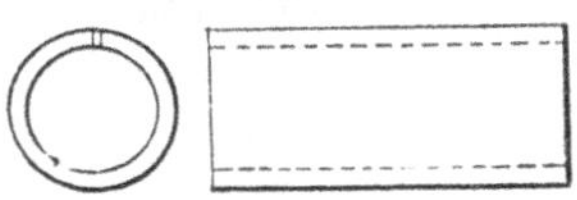	Schließen		Vgl. W. T. 1907 S. 193

Tabelle 29.

Kaltbiegen eines Blechfußes.

Arbeits-gang	Gegenstand	Arbeitsverfahren	Werkzeug siehe Seite:	Bemerkung
1	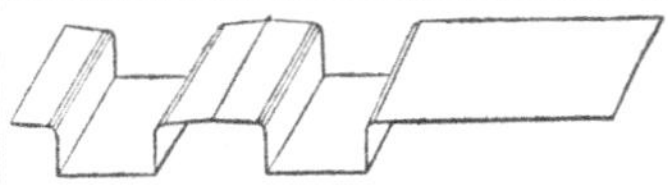	Vorbiegen des Streifens von der Rolle		
2	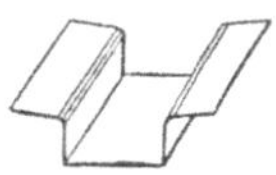	Abschneiden d. vorgebogenen Teiles	Verbund-Werkzeug mit Seitenstempeln S. 448	
3	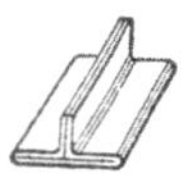	Fertigbiegen		

Tabelle Nr. 30.

Mehrfaches Biegen eines Blechteiles.

Arbeits gang	Gegenstand	Arbeitsverfahren	Werkzeug-siehe Seite:	Bemerkung
1		Abfallfreies Ausschneiden des Streifens v. d. Rolle		
2	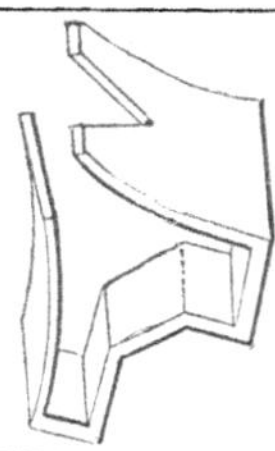	Anbiegen	Biege-werkzeug m. Seiten-stempeln S. 385	
3	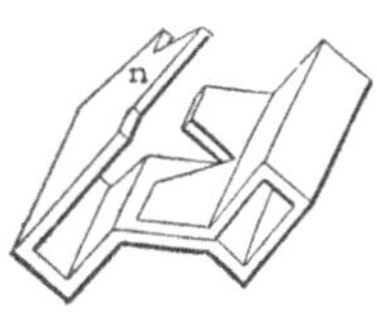	2. Biegen	Biege-werkzeug mit Dorn S. 386	
4	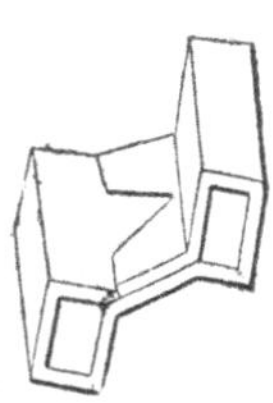	Fertigbiegen	S. 386	

Tabelle Nr. 31.

Herstellung eines Schließteiles.

Arbeitsgang	Gegenstand	Arbeitsverfahren	Werkzeug siehe Seite:	Bemerkung
1	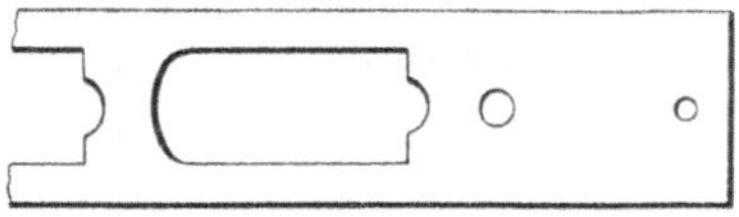	Abfallstreifen	S. 91	
	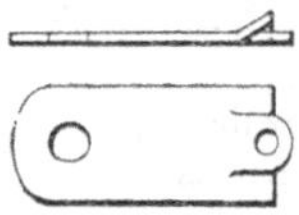	Ausschneiden u. Lochen, Aufbiegen der Enden	S. 91	
2	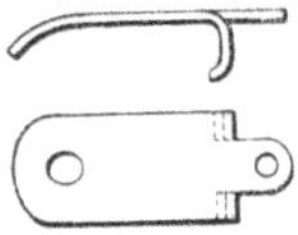	Biegen der Scharnierenden	S. 362	
3		Abschneiden u. Biegen des Drahtbügels	S. 441	
4	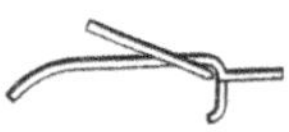	Einrollen der Scharnierenden f. Drahtbügel	S. 487	
5	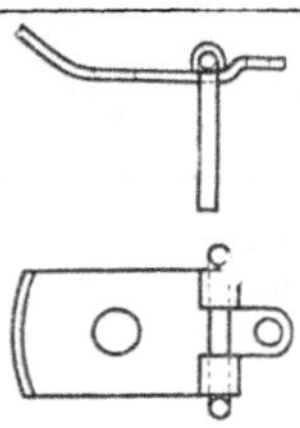	Der fertige Teil		

Tabelle Nr. 32.

Biegen eines Teiles aus blankgezogenem Stahlblech von 1 mm Dicke.

Arbeitsgang	Gegenstand	Arbeitsverfahren	Werkzeug siehe Seite:	Bemerkung
1		Ausschneiden		
2		Anbiegen des Auges	S. 356	
3		Einrollen des Auges	S. 332	
4		Biegen	S. 363	
5	 Schnitt A-B 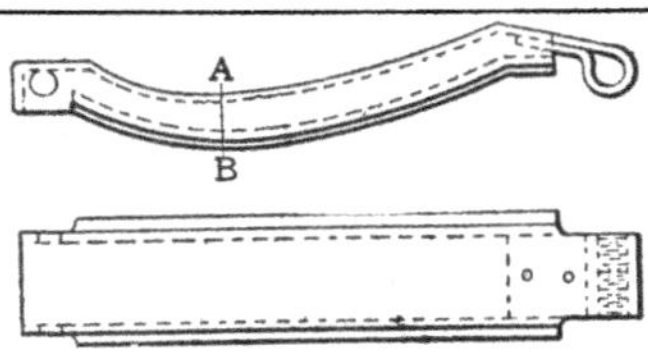	Ausprägen	S. 480	

Tabelle Nr. 33.

Herstellung der Knallkapseln.

Arbeitsgang	Gegenstand	Arbeitsverfahren	Werkzeug siehe Seite:	Bemerkung
1		Ausschneiden u. Prägen des Oberteils		
2		Ausschneiden u. Ziehen des Unterteils		
3		Abschneiden desselben		
4	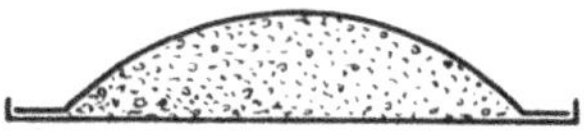	Ober- u. Unterteil ineinanderlegen u. Füllen m. Explosivstoff		
5		Schließen der gefüllten Kapsel	S. 334	

II. Werkzeuge.

A. Arbeiten mit Materialabtrennung.

Den gesamten Arbeiten, die in der Praxis als Schneiden, Lochen, Stanzen usw. bezeichnet werden, liegt der Schervorgang zugrunde. In der Materialprüfung und in der technologischen Mechanik, welche versuchen, die im täglichen Leben vorkommenden Bearbeitungen der Materialien auf möglichst einfache Einzelvorgänge zurückzuführen, wird unter Scheren jene Abtrennung bezweckende Arbeit verstanden, die eine Verschiebung der in zwei unmittelbar nebeneinander liegenden Flächen befindlichen Körperteilchen parallel zu diesen Flächen bewirkt. Der Widerstand, den der Körper dieser Verschiebung seiner Teilchen entgegensetzt, ist die Scherfestigkeit. Wäre es möglich, die beiden Scherblätter unendlich dünn zu machen und gleichzeitig unendlich nahe aneinander zu legen, so würde sich die Scherfestigkeit mit der Schubfestigkeit decken, es würde ein „reines" Scheren auftreten. Da aber die Scherbacken stets eine gewisse Dicke und meistens auch eine Schräge aufweisen, so fallen die beiden Kräfte nicht mehr in eine Ebene und erzeugen neben der Schubbeanspruchung noch ein Biegemoment, wie Fig. 1[1]) zeigt. Für die Art der Arbeit ist es gleichgültig, ob ein geradliniges offenes oder ein geschlossenes Scherblatt, ein zylindrischer Stempel, also ein Lochstempel, zur Anwendung kommt. Von besonderem Einfluß dagegen ist die Materialdicke und die Schnittgeschwindigkeit auf den Kraftverbrauch. Je dünner das zu lochende Blech im Verhältnis zum Durchmesser des Loch- bzw. Schnittstempels ist, um so mehr nähert sich die Scher- der Schubfestigkeit. Bei dickeren Blechen treten aber die Biegebeanspruchungen in bedeutend größerem Maße hervor.

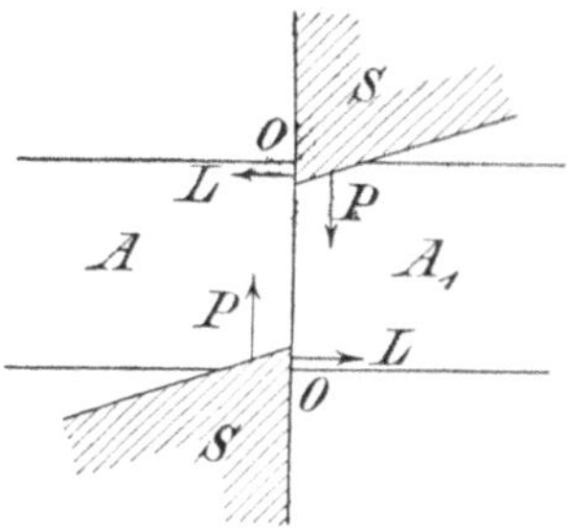

Fig. 1.

---

[1]) Martens, Materialienkunde, Fig. 155, S. 149.

so daß das Bild der Materialverschiebung sich nach Fig. 2[1]) ändert. Martens führt dies[2]) an derselben Stelle aus: „Solange die Kraft P nicht groß genug ist, um in dem punktiert angedeuteten Zylindermantel Abscherung zu bewirken, wird das Material unter dem Stempel zur Seite und nach oben hin ausweichen, und zwar am meisten dahin, wo es den geringsten Widerstand findet, wie in Fig. 2 punktiert angedeutet ist. Man wird bemerken, daß die Reibung an den Auflagerflächen und an den Stempelflächen zum Ausdruck kommt, und daß das Material neben dem Stempel mehr aufquillt als in einigem Abstand davon, weil die radial wirkenden Spannungen im Körper mit wachsender Entfernung von der Achse abnehmen und schließlich kein Fließen in radialer Richtung bewirken können. Ist der äußere Durchmesser des Körpers sehr groß gemacht, so wird nur Fließen in den Hohlraum der Matrize und Aufquellen nach oben stattfinden können. Das Eindringen kann in merkbarer Weise erst stattfinden, wenn die Quetschgrenze in merkbarer Weise überschritten worden ist, weil hier die freie Bewegung der Masse behindert ist. Ist die hier gültige Spannung an der Quetschgrenze k_s, so kann ein nennenswertes Fließen erst stattfinden, wenn die Kraft P den Wert $\frac{\pi d^2}{4} \cdot k_s$ übersteigt, und ein Abscheren erst eintreten, wenn die Höhe x so weit vermindert ist, daß die Kraft P größer als der Wert $\pi d \cdot x \cdot k_s$ wird.

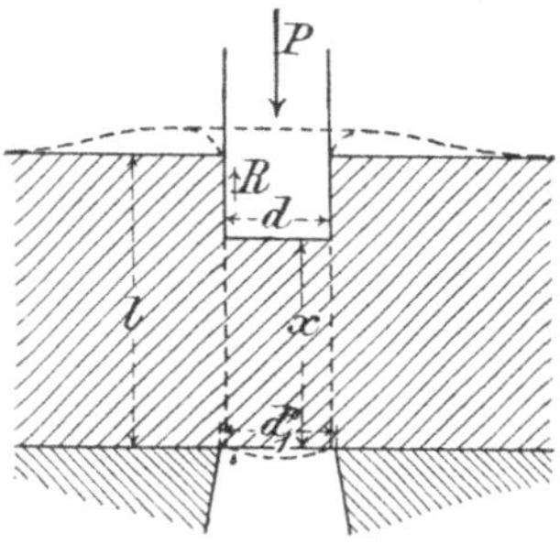

Fig. 2.

Daraus ergibt sich, daß die äußere Form des zu lochenden Stückes einen Einfluß auf den Kraftverbrauch haben kann, wenn das stehenbleibende Material nicht groß genug ist, um seitliches Ausfließen zu verhindern.“

Daß sich die Materialbewegung im Blech tatsächlich in der geschilderten Weise abspielt, zeigen die Ätzbilder Fig. 3—6[3]), welche die Formänderungen des Bleches und Putzens zu verschiedenen Zeiten des Lochvorganges darstellen. Man erkennt aus dem Verlauf der Schichtungen des Schweißeisens ganz deutlich den Einfluß der oben erwähnten Biegebeanspruchungen, das Umlegen der Fasern längs des Stempels und die gleichzeitige Arbeit der Matrize. Andrerseits ist aber daraus zu ersehen, daß der einfache „Schnitt und Gegenschnitt“[4]),

---

[1]) Martens, Materialienkunde, Fig. 163, S. 155.
[2]) Martens, Materialienkunde, S. 156.
[3]) Preuß, Die praktische Nutzanwendung usw., S. 29, Fig. 30a—d.
[4]) Werkstattstechnik.

wie ich mit der Quelle alle Werkzeuge nennen will, bei denen der eben erwähnte Schervorgang durch zwei einander entsprechende Scherwerkzeuge, Stempel und Schnittplatte, erzeugt wird, keine einfache Berechnung auf Grund der Festigkeitsformeln für die Schubfestigkeit zuläßt.

Dagegen läßt sich nach dem „Gesetz der proportionalen Widerstände" von Kick auf dem Versuchswege ein Resultat erreichen. Dieses Gesetz besagt:

„Zwei geometrisch ähnliche Körper aus demselben Material erfahren unter gleichen Umständen durch die gleichen Spannungen geometrisch ähnliche Formveränderungen."

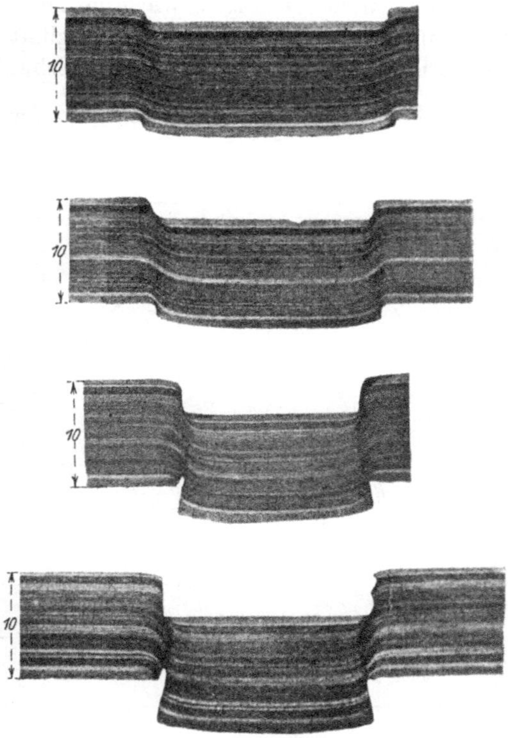

Fig. 3—6. Lochen von Schweißeisen.

Da es sich bei allen Arbeiten der Schnittwerkzeuge um Schnittflächen handelt, die abgewickelt Rechtecke ergeben, so werden zwei Schnitte geometrisch ähnlich sein, wenn bei ähnlichem Grundriß des Stempels das Verhältnis von Umfang zu Blechdicke gleich bleibt. Natürlich gilt dieser Satz nur für die Kräfte am Schnitt selbst und nicht für den Kraftaufwand der Presse, da sich die Verluste im Triebwerk in anderer Weise ändern. Auf Grund dieses Satzes lassen sich Versuchszahlen verallgemeinern.

Die Feststellung des reinen Kraftverbrauches am Werkzeug muß Gegenstand von eigenen Versuchen sein, die im allgemeinen, selbst unter richtigen Arbeitsbedingungen, nicht so schwierig sind, als es den Anschein hat[1]). Da sich der Kraftverbrauch beim Schnitt vom Beginn bis zum Ende dauernd ändert, so ist das Arbeitsdiagramm

[1]) Wormser, Über das Stanzen von Leder und Pappe. Dr.-Diss. 1913. Peiseler, W. T. 1913, S. 711.

Kraft-Weg-Diagramm, zur Beurteilung und Auswertung zu verwenden. Martens[1]) hat, wenn auch nicht unter den für Schnittwerkzeuge geltenden Arbeitsbedingungen, derartige Versuche gemacht, deren Arbeitsdiagramme in Fig. 7 wiedergegeben sind. Die Versuche erstreckten sich auf Feinkorneisen, geglühtes Kupferblech, Messingblech und gegossenes Blei. In dem Schaubild sind die spezifischen Spannungen $\tau_{\prime\prime} = \frac{P}{f}$ als Ordinaten und die Eindrucktiefen in Teilen

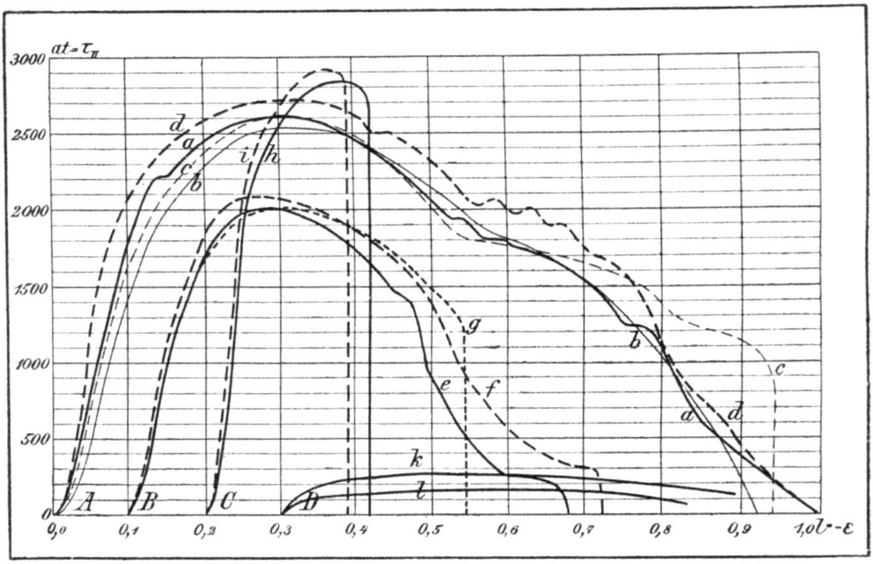

Fig. 7.

der Blechdicke als Abszissen aufgetragen. Die von den Linien a—l eingeschlossenen Flächen sind also proportional der beim Lochen aufgewendeten Arbeit am Werkzeug und werden ihr gleich, wenn die Kräfte selbst als Ordinaten und die Eindrucktiefen als Abszissen aufgetragen werden. Die Linien a b c d und h i zeigen deutlich den Einfluß der Arbeitsgeschwindigkeit, der bei den gebräuchlichen Arbeitsgeschwindigkeiten der Pressen noch stärker hervortreten wird. Auch der Einfluß des ausgehöhlten Stempels ist in den schwachen Linien zu ersehen. In nebenstehender Tabelle sind die näheren Angaben zu dem Schaubild eingetragen.

Auf diese Weise läßt sich für ein beliebiges Schnittwerkzeug, das in eine mit richtiger Versuchseinrichtung ausgestattete Presse eingebaut wird, der Kraftverbrauch am Werkzeug bestimmen, wobei der höchste im Diagramm auftretende Druck für die Festigkeitsbe-

[1]) Martens, Materialienkunde, Fig. 166, S. 158.

rechnung und die mittlere Leistung des Diagramms für den Arbeitsbedarf der Presse maßgebend ist.

Schaubilder von Lochversuchen.

Material	Linie	Blechdicke l cm	Durchmesser Stempel d cm	Durchmesser Matrize d_1 cm	$\frac{l}{d}$	$\frac{d'}{d}$	Zeit des Lochens Min.	Stempelfläche eben	Stempelfläche hohl
Feinkorneisen 5,20 · 1,35 cm	a	1,35	2,00	2,10	0,675	1,050	12–14	e	
	b	1,35	2,00	2,10	0,675	1,050	12–14		h
	c	1,35	2,00	2,10	0,675	1,050	weniger als 1 Min.		h
	d	1,35	2,00	2,10	0,675	1,050	„	e	
Kupferblech, geglüht	e	1,01	2,00	2,01	0,51	1,005	2–4	e	
	f	1,01	2,00	2,05	0,51	1,025	2–4	e	
	g	1,01	2,00	2,10	0,51	1,050	2–4	e	
Messingblech . .	h	1,00	2,50	2,60	0,40	1,040	2	e	
	i	1,00	2,50	2,55	0,40	1,020	13	e	
Blei, gegossen { Φ 7,18 cm	k	4,10	2,00	2,05	2,05	1,025	–	e	
Φ 10,03 cm	l	2,91	2,00	2,01	1,46	1,005	–	e	
Körper nicht ganz durchlocht									

Die Bestimmung dieser Ziffern mittels Crushers[1]), die im Materialprüfungswesen sehr verbreitet ist, ist für eine Auswertung der ganzen Presse[2]) mit genügender Berücksichtigung der Arbeitsgeschwindigkeit brauchbar, wird jedoch zur Bestimmung der Arbeitsleistung am Werkzeug versagen, da sich der Wirkungsgrad der Presse mit der Belastung ändert.

Ein gewisser Spielraum zwischen Stempel und Matrize ist nicht allein, wie oben angedeutet, für den Kraftbedarf am Werkzeug von Einfluß, sondern auch für die Erzielung eines glatten Schnittes, wie allgemein bekannt, von Wichtigkeit. In vorstehender Tabelle sind nach Woodworth ziffernmäßige Werte für das Spiel zwischen Stempel und Schnittplatte für genaue Schnittarbeit gegeben.

Die Angaben beziehen sich auf das Ausschneiden und Lochen von ebenem flachen Material in der Kraftpresse zur Herstellung von Teilen für Rechenmaschinen, Registrierkassen, Schreibmaschinen usw.

Bei dieser Arbeit verlangt man im allgemeinen von den verwendeten Werkzeugen zwei verschiedene Schnitte: Erstens soll die Schnittfläche der Scheibe beinahe glatt und gratfrei sein, scharfe Kanten und möglichst geringe Abrundungen an der Schnittseite haben, zweitens sollen die Löcher und Nuten, die in die Teile gestanzt werden, so glatt

[1]) Crushers sind kleine Stauchzylinder aus sehr gleichmäßigem Kupfer, deren Zusammendrückung unter einer bestimmten Belastung bekannt ist.

[2]) Kannegießer, Das Pressen, Stanzen und Prägen der Metalle, S. 373

und gerade wie möglich sein und die richtige Größe haben. Diese Tabelle ist das Ergebnis einiger Jahre Erfahrung in dieser Arbeit und in weiteren drei Jahren nach ihrer Zusammenstellung erprobt worden.

Spiel zwischen Stempel und Schnittplatte.

Metalldicke	in mm für		
mm	Messing, weicher Stahl	mittelhart gewalzter Stahl	hartgewalzter Stahl
0,254	0,013	0,015	0,018
0,508	0,025	0,031	0,036
0,762	0,038	0,046	0,053
1,016	0,051	0,061	0,071
1,270	0,064	0,076	0,089
1,524	0,076	0,091	0,107
1,778	0,089	0,107	0,124
2,032	0,102	0,122	0,142
2,286	0,114	0,137	0,160
2,54	0,127	0,152	0,178
2,794	0,140	0,168	0,196
3,048	0,152	0,183	0,213
3,302	0,165	0,198	0,231
3,556	0,178	0,213	0,249
3,810	0,191	0,229	0,267
4,064	0,203	0,244	0,285
4,318	0,216	0,259	0,302
4,572	0,229	0,274	0,320
4,826	0,241	0,290	0,338
5,08	0,254	0,305	0,356

Die Schnittplatte bestimmt immer die Größe des Arbeitsstückes, das durch sie durchgehen kann. Der Stempel bestimmt die Größe des Arbeitsstückes, das durchgehen soll. Beim Ausschneiden wird die Schnittplatte entsprechend der Größe des verlangten Stückes und der Stempel kleiner gemacht. Beim Lochen wird der Stempel nach der Größe des Arbeitsstückes und die Schnittplatte größer gemacht. Das Spiel zwischen Schnittplatte und Stempel bestimmt das Endergebnis. Bei der Zusammenstellung der Tabelle ist angestrebt worden, die Werte in • so einfache Form zu bringen, daß man danach ein Werkzeug mit gutem Erfolge herstellen kann. Die Fig. 8/11 zeigen die Anwendung der Tabelle beim Bestimmen des Spieles für das Ausschneiden oder Lochen von hartge-

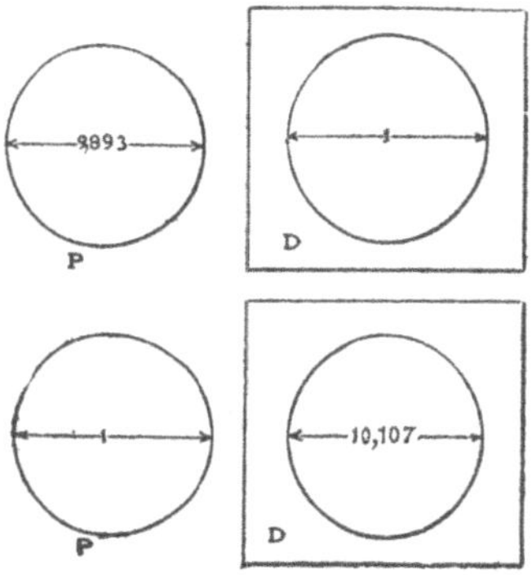

Fig. 8 bis 11.

walztem Stahl von 1,5 mm Dicke. Die Tabelle gibt das Spiel für Metalldicken von 0,254 mm bis 5,08 mm, und die Skizzen zeigen, daß zum Ausschneiden eines Stückes von dem Durchmesser 10 mm der Durchmesser des Stempels 9,893 werden muß, während zum Lochen des gleichen Durchmessers der Durchmesser der Schnittplatte um den gleichen Betrag vergrößert werden muß. Soll auf „Schiebesitz" gearbeitet werden, so macht man Stempel und Schnitt 0,006 mm bis 0,012 mm größer; für „Preßsitz" um 0,012 bis 0,035 mm kleiner für einen Stempel von 25 mm Durchmesser.

1. Schnitt und Gegenschnitt.

a) Der einfache Schnitt.

In der einfachsten Form besteht das Schnittwerkzeug aus dem Stempel und der Schnittplatte, zwischen denen das zu stanzende Blech eingelegt wird, worauf der Stempel geradlinig und senkrecht auf das Blech und die Schnittplatte geführt wird, bis er durch das Blech hindurchgetreten ist. Bei einem derartigen Werkzeuge wird also in einem Arbeitsgang nur ein Schnitt durch das Blech geführt. Denkt man sich den Querschnitt in Fig. 1 parallel zu sich selbst eine gewisse Strecke verschoben, so entsteht der gerade Schnitt, läßt man dagegen den Querschnitt um eine von der Linie OO um die Strecke r entfernte Achse rotieren, so entsteht der Rundschnitt, dessen Ergebnis eine Scheibe vom Durchmesser 2 r ist. Bewegt sich der Querschnitt längs irgendeiner geschlossenen oder offenen Leitlinie, so entsteht die einfachste Form eines allgemeinen Schnittwerkzeuges.

Ein derartiges Schnittwerkzeug heißt „offen". Wenn aber die Führung des Stempels durch den Pressenstößel für die Genauigkeit der Arbeit oder wegen Bruchgefahr nicht genügt, so daß der Stempel in einer mit der Schnittplatte fest verbundenen Führungsplatte geführt werden muß, so entsteht der „geschlossene" oder „Führungsschnitt".

Da es sich hier um den konstruktiven Aufbau der Schnittwerkzeuge handelt, ist die letzte Einteilung für die einfachen Schnittwerkzeuge gewählt worden.

α) Der offene Schnitt

tritt als gerader offener Schnitt selten als eigenes Werkzeug auf, sondern wird in den verschiedenen Scheren erzeugt, deren Besprechung hier übergangen werden soll. Weiter erscheint er häufig als Teil eines Folgewerkzeuges oder Verbundwerkzeuges, z. B. beim abfallosen Stanzen von Sechskantmuttern, S. 93, bei der Herstellung des Hebels für einen Momentausschalter, S. 464, in Fig. 21, S. 45, u. a. m. In Fig. 13 erscheint er als eigenes Werkzeug, welches der

Herstellung von Scharnieren zugehört, wenn man von der unvollkommenen Führung im letzten Teil des Hubes an den seitlichen Lappen der Schnittplatte absieht.

Zwei Arten Scharniere sind in Gebrauch und wahrscheinlich alle abgeänderten Formen von diesen beiden abgeleitet. In der einen Gruppe steht das Gelenk gleichmäßig nach beiden Seiten über beiden Blättern, in der anderen liegt es nur nach einer Seite. Um ein Scharnier der letzteren Art handelt es sich hier.

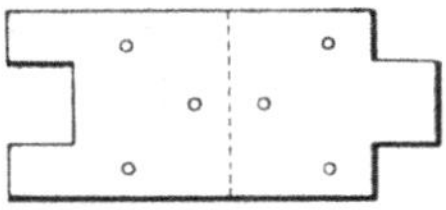

Fig. 12.

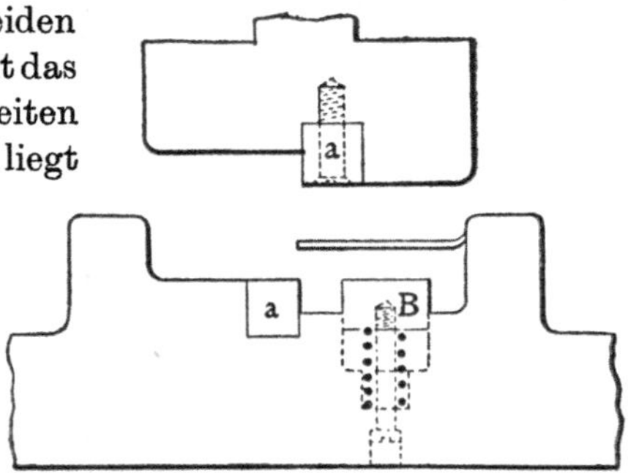

Fig. 13.

Die beiden Teile eines Scharnieres werden im ersten Arbeitsgang, um an Rohmaterial und Arbeitszeit zu sparen, in einem Stück ausgeschnitten und gelocht, Fig. 12. Im folgenden Arbeitsgang werden die beiden Scharnierhälften voneinander getrennt. Stempel und Schnitt für diesen Zweck sind in Fig. 13 dargestellt; sie sind aus Gußeisen gemacht, die Schneidbacken a a aus Werkzeugstahl sind eingesetzt. B ist der Federauswerfer, der die untere Scharnierhälfte nach dem Schnitt aus dem Werkzeug entfernt. Für die Herstellung von Scharnieren oder Angeln bringt die Verwendung eines solchen Werkzeuges keinerlei Verlust an Stoff mit sich. Als dritter Arbeitsgang folgt das Einrollen der Enden, das auf S. 331 beschrieben ist.

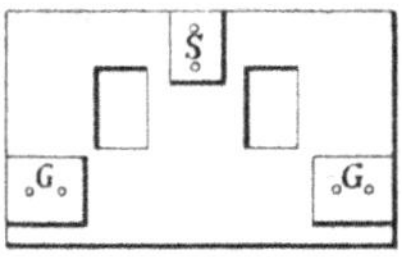

Fig. 14.

Der letzte Gang ist das Zuschneiden der einzelnen Scharnierteile, damit sie leichter zusammengebaut werden können. Zu dem Zweck kommt ein Schnittwerkzeug nach Fig. 14 mit einem passenden Stempel zur Anwendung. Die Zeichnung stellt die Schnittplatte mit den Seitenanschlägen und dem Endanschlag dar, zwischen die das zuzuschneidende Stück eingelegt wird. Streng genommen enthält das Werkzeug zwei gleiche, nebeneinander arbeitende Einzelschnitte.

Eine weitere Anwendung des geraden offenen Schnittes ist der *Ecken-Ausschneideapparat* der Erzgebirgischen Schnittwerkzeug- und

Maschinenfabrik in Schwarzenberg[1]). Er ist in Ansicht in Fig. 15 wiedergegeben und besteht in der Vereinigung zweier einstellbarer Schnitte. Sowohl Stempel wie Schnittplatten sind in den Schlitzführungen von 60° bis 120° verstellbar, wobei man noch Winkel von 1°

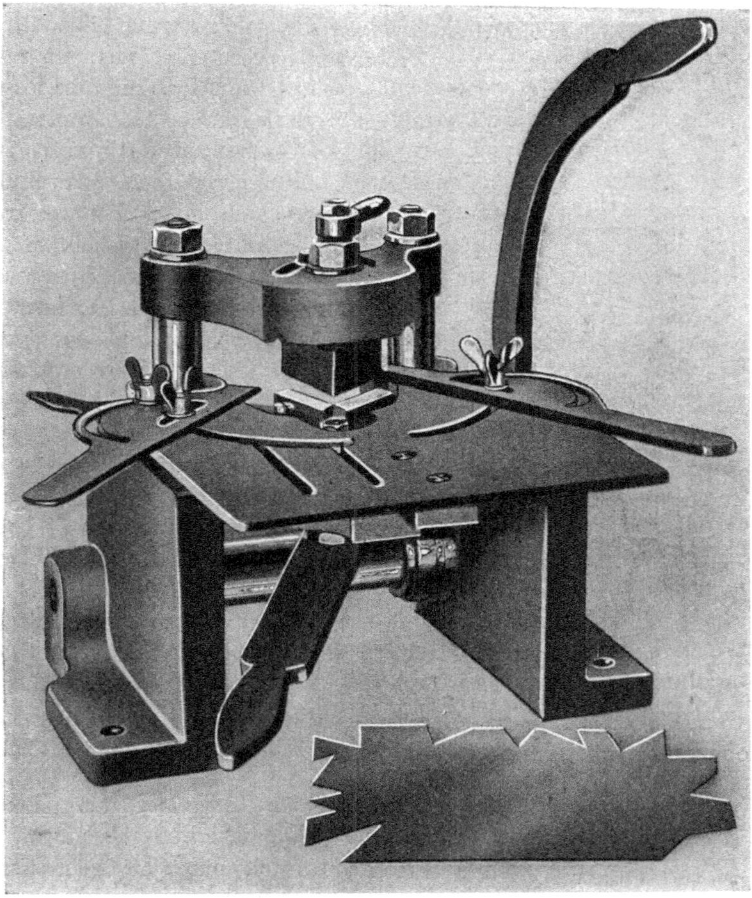

Fig. 15. Ecken-Ausschneideapparat.

einstellen kann. Durch den rechts ersichtlichen Handhebel mit Exzenterantrieb wird der Stempel herab gegen die Schnittplatte bewegt. Die schneidenden Teile sind leicht herausnehmbar befestigt, so daß man beim Nachschleifen immer die nötige Genauigkeit erhalten kann.

Der offene Rundschnitt ist wohl das am häufigsten angewendete Werkzeug der Blechwarenfabrikation, sowohl bei der Herstellung

---

[1]) Vgl. W. T. 1910, S. 499.

runder Scheiben und Ringe als Fertigprodukt als auch als Vorprodukt bzw. als erster Gang der verschiedensten Zieharbeiten. An ihm lassen sich fast noch besser als an dem zuerst angeführten geraden Schnitt die grundlegenden Teile eines Werkzeuges erkennen. Fig. 16[1]) stellt die einfachste Form eines Rundschnittes dar. a ist der Schnittstempel, der bei kleineren Werkzeugen aus gehärtetem Gußstahl, bei größeren aus Gußeisen mit aufgesetztem Schnittring aus gehärtetem Stahl hergestellt wird. Die Befestigung am Pressenstößel erfolgt mittels zylindrischen Zapfens und Druckschraube. Die Unterfläche des Stempels wird nach dem Härten hohl angeschliffen. Die Schnittplatte b ist entsprechend dem Stempeldurchmesser ausgebohrt, gehärtet und geschliffen und wird in einer gußeisernen Grundplatte, der Froschplatte, befestigt, die erst mit den Löchern und Aufspannschlitzen zur Befestigung des ganzen Werkzeugunterteils am Tisch der Presse versehen ist. Die Befestigung der Schnittplatte in der Froschplatte ist verschiedenartig, in Fig. 16 ist die erste abgeschrägt, in anderen Fällen wird sie gerade eingepaßt oder mit Druckschrauben gehalten.

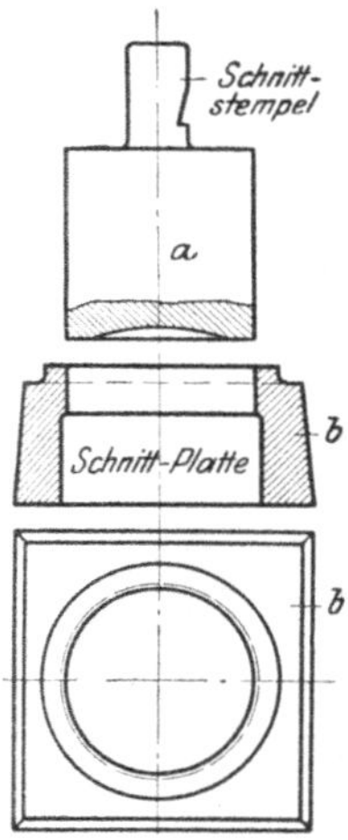

Fig. 16.

Bei der Herstellung der Schnittplatte ist Sorge zu tragen, daß die Auflagerfläche genau senkrecht zu dem oberen geschliffenen Teil der Bohrung steht. Dieser ist ungefähr 3—5 mm hoch und geht zum Zweck des leichteren Durchfallens der ausgeschnittenen Stücke in einen konischen oder erweitert zylindrischen Teil über. Gelegentlich wird bei der Lieferung von ausgeschnittenen Scheiben die Aufgabe gestellt, sie eben und frei von Buckeln usw. zu erhalten, vgl. dazu S. 238.

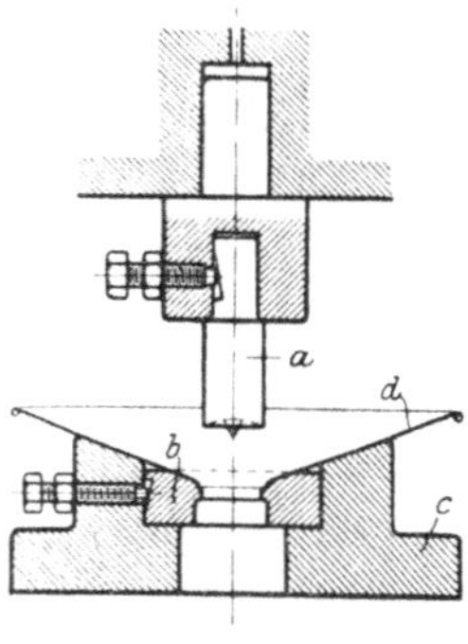

Fig. 17.

Lochwerkzeug für Lampenschirme. (Bonner Maschinenfabrik Mönkemöller G. m. b. H.) Zum Lochen des Bodens eines Lampenschirmes dient das Werkzeug nach Fig. 17. Hier ist der Lochstempel a — aus gehärtetem Werkzeugstahl — in einen Tragkörper, den Stempelkopf, eingesetzt, so daß einmal an Material gespart wird und außerdem derselbe Stempelkopf für Lochstempel verschiedener Durchmesser verwendet werden kann. Der zylindrische Zapfen des Stempels a ist in den Stempelkopf eingepaßt und durch eine Druckschraube mit Gegen-

[1]) W. T. 1909, S. 577, Fig. 1.

mutter gesichert. Die Zentrierspitze und der hohle Ausschliff des Stempels sind deutlich erkennbar. Die Schnittplatte b, ebenfalls aus gehärtetem Werkzeugstahl, ist in gleicher Weise in die gußeiserne Froschplatte c eingepaßt, doch sind hier drei Druckschrauben vorgesehen, so daß ein Verziehen der Schnittplatte möglichst vermieden wird. Das Lochen des fertig gezogenen und eingerollten Schirmes verlangt eine besondere Abschrägung als Auflage, die genau nach der Schablone des gezogenen Teiles hergestellt wird und teils auf der Froschplatte, teils auf der Schnittplatte selbst angearbeitet ist. In letzterer ist eine Ausdrehung von 34,9 mm Durchmesser oberhalb des Schnittringes von 34 mm Durchmesser als Zentrierung vorgesehen.

Den Übergang von dem offenen zu dem geschlossenen oder Führungsschnitt zeigt Fig. 18. Der Stempel a ist nicht mehr vollständig frei, sondern wird von der Platte g, dem Abstreifer, umgeben. Selbst wenn der Stempel in dieser Ausbildung in dem Ausschnitt der Platte g wirklich geführt gehen sollte, so ist doch die konstruktive Verbindung von g mit c und b zu schwach, als daß man dies als Führung bezeichnen könnte.

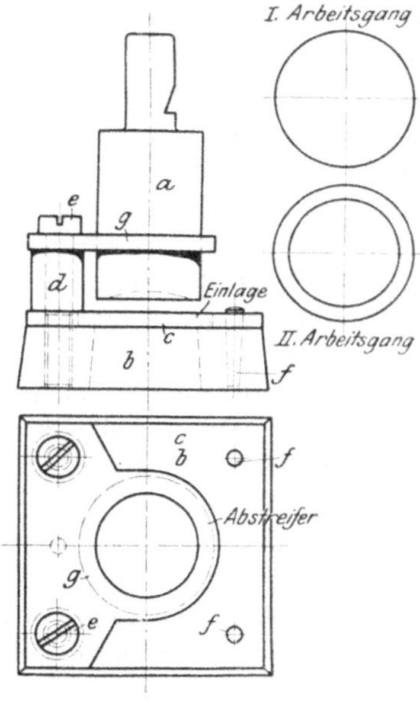

Fig. 18.

Der Rundschnitt mit Einlage[1]). Zur Herstellung eines Ringes nach Fig. 18 wird zuerst die Scheibe gestanzt und dann ebenfalls mittels eines Rundschnittes die innere Umgrenzung hergestellt. Damit der innere Schnitt konzentrisch mit dem äußeren werde, verwendet man ein Werkzeug nach Fig. 18. a ist wieder der Stempel, b die Schnittplatte, während die Teile c, die Einlage, und g, der Abstreifer, den neuen Arbeitsbedingungen entsprechend, dazukommen. Die Einlage c ist eine Führungsplatte, deren Bohrung genau zentrisch zu jener der Schnittplatte durch Prisonstifte gesichert ist, so daß die im ersten Arbeitsgang gestanzte Scheibe ihre genaue, richtige Stellung erhält, sobald sie in der Einlage liegt. Der andere Teil, der Abstreifer, ist eine den Stempel

---

[1]) W. T. 1909, S. 579, Fig. 1.

ziemlich genau umgreifende, feststehende Platte, die den fertigen Ring, der beim Aufgang des Stempels an diesem hängen bleibt, von ihm abstreift, so daß er leicht aus dem Werkzeug entfernt werden kann, wenn er nicht bereits durch eine schräge Lage des ganzen Werkzeuges herausfällt, sobald der Stempel durch den Abstreifer getreten ist.

β) Der geschlossene oder Führungsschnitt.

Sobald es sich aber um genaue Arbeit, schwieriges Material oder verwickelte Umrisse, schnellere Herstellung durch erhöhte Hubzahl der Pressen handelt, ebenso wie bei den später besprochenen Mehrfachwerkzeugen, muß der Stempel eine gesonderte Führung, die fest mit der Schnittplatte verbunden bleibt, erhalten, aus der er nie heraustritt. Dann wird die Gefahr des Durchscherens der Stempelkanten durch die Schnittplatte oder eines Brechens der Teile stark vermindert.

Diese Gruppe umfaßt die unendlich verschiedenen Formen, von denen kaum zwei einander gleich sind, die alle dadurch entstanden gedacht werden können, daß ein Schnitt durch die beiden arbeitenden Werkzeugteile sich längs einer Leitlinie, dem zu stanzenden Umriß, entlang und senkrecht auf dessen Ebene bewegt. Während die Herstellung der in den beiden ersten Gruppen besprochenen Werkzeuge der Hauptsache nach durch die beiden einfachsten Arbeiten, Drehen und Hobeln bzw. Eben- und Rundschleifen, erfolgen kann, kommen bei der Herstellung der allgemeinen Schnittwerkzeuge in erhöhtem Maße zu der Handarbeit oft sehr verwickelte Arbeiten, bei denen alle nur verfügbaren Arbeitsweisen der Werkstatt angewendet werden müssen.

Das Schnittwerkzeug liefert in den meisten Fällen kein Fertigprodukt, sondern nur das Blankett (Scheibe, Rondell usw.), aus dem in weiteren Arbeitsgängen das fertige Stück erhalten wird. Deshalb wird das Werkzeug so gebaut, daß es erstens der gebräuchlichen Form des angelieferten Rohmaterials möglichst angepaßt ist, zweitens aus diesem die Blanketts mit möglichst wenig Abfall geschnitten werden können, und drittens die Zuführung des Rohmaterials und die Entfernung der Blanketts sowie des Abfalles möglichst einfach bewerkstelligt werden kann.

Als Rohmaterial kommt in den weitaus meisten Fällen Bandmaterial in Streifen oder Rollen zur Verarbeitung, Platten erst in zweiter Linie. Um einen geringen Abfall zu erzielen, müssen die einzelnen Schnitte so nahe aneinandergesetzt werden können, als es sich mit der Festigkeit des übrigbleibenden Steges verträgt. Dies wird bei den von Hand bedienten Werkzeugen durch Anwendung von Anschlägen,

Seitenschneiden u. ä. m. erreicht, bei den selbsttätig wirkenden Pressen durch besondere Zuführungsvorrichtungen, von denen einige weiter unten angeführt sind[1]).

Werkzeug einfachster Form. Ein solches Werkzeug zur Verarbeitung von Bandmaterial zeigt Fig. 19[2]). Das auszuschneidende Stück ist ein gleichseitiges Dreieck, das gleichzeitig den Querschnitt des Schnittstempels a darstellt. Der Unterteil des Werkzeuges besteht aus der Schnittplatte b, auf der die Führungsplatte c verschraubt und durch Prisonstifte f gesichert ist. Auf der Unterseite der Führungsplatte ist eine Nut, entsprechend der Breite des verarbeiteten Streifens, gehobelt, während auf der Vorderseite bei d eine Aussparung vorgesehen ist, unterhalb welcher der Aufhängestift (Anschlagstift, Holperstift) in die Schnittplatte eingesetzt ist. Die Vereinigung aller dieser Teile heißt Schnittkasten.

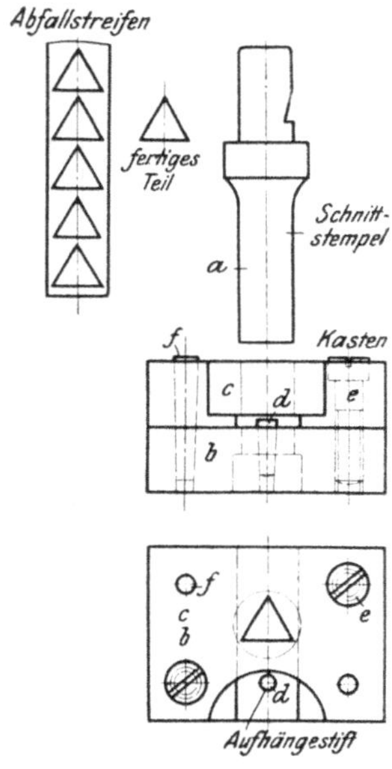

Fig. 19.

Der Aufhängestift dient dem Arbeiter, der nach jedem Pressenhub den Streifen von Hand vorzieht, als Wegbegrenzung, indem er das zuletzt ausgeschnittene Dreieck des Abfallstreifens über diesen Stift hängt und gegen ihn drückt. Dadurch wird nicht allein der Abfall auf ein verhältnismäßig genau bestimmbares Mindestmaß gebracht, sondern auch die Sicherheit des Arbeiters erhöht, da er mit seinen Fingern nichts unter den arbeitenden Teilen des Werkzeuges zu tun hat.

Schnittwerkzeug für die Grundplatte eines Hausweckers. (C. Lorenz, A.-G., Berlin SO 26.) Das Werkzeug, Fig. 20, wird beim ersten Herstellungsgang der Grundplatte eines elektrischen Hausweckers verwendet, vgl. Tabelle Nr. 1 - I. In den Ausschnitt der Schnittplatte ist ein Grundriß des Stempels eingezeichnet, der die Lage der Befestigungsschrauben, des Stempelzapfens und Kopfes in ihrer bezüglichen Lage zum Umriß des Blanketts andeutet. Die Zeichnung ist im Maßstab von ungefähr 1 : 3,4 wiedergegeben. Der eigentliche Schnittstempel und der Stempelkopf

[1]) Vgl. auch: Gugel, Materialzuführungsvorrichtungen. Springer, Berlin 1912
[2]) W. T. 1909, S. 578, Fig. 3.

werden hier getrennt hergestellt und beide Teile miteinander durch fünf Zylinderkopfschrauben verbunden, so daß die Bearbeitung des immerhin verwickelten Schnittstempels und seine Härtung erleichtert und Material erspart wird. Der Schnittkasten besteht aus der Schnittplatte aus ge-

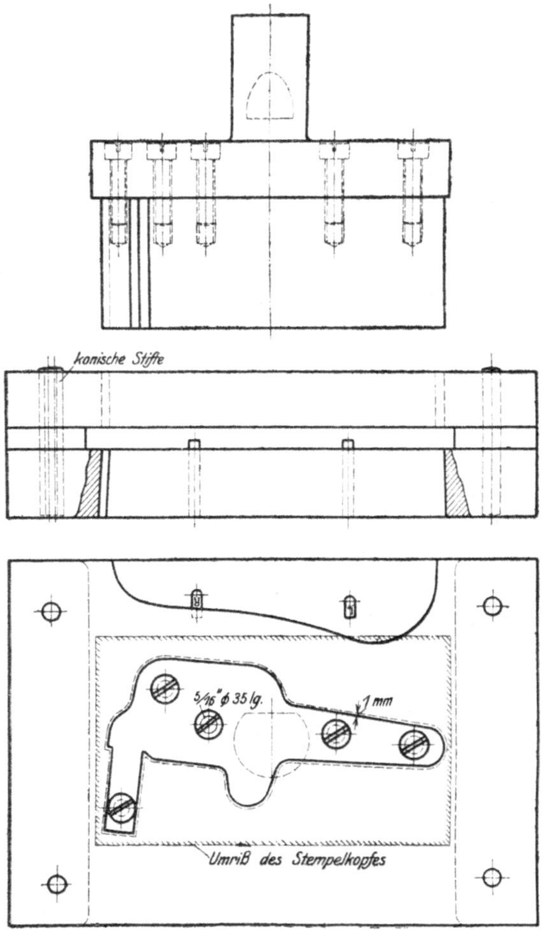

Fig. 20.

härtetem Werkzeugstahl, zwei seitlichen Führungsleisten, auf denen erst die Führungsplatte ruht, die alle durch vier konische Stifte miteinander zu einem festen Ganzen verbunden sind. Da es sich um eine ziemlich sperrige Form des Blanketts handelt, sind für den Abfallstreifen zwei Aufhängestifte vorgesehen und der Ausschnitt in der Führungsplatte dementsprechend gelegt worden.

Das Bestreben, den Abfall beim Ausschneiden der Blanketts möglichst gering zu halten, hat in vielen Fällen dazu geführt, dem Blankett eine derartige Form zu geben, daß je zwei aufeinander folgende Schnitte einander zur vollen Fläche des Streifens ergänzen. Die im folgenden beschriebenen

Schnittwerkzeuge zur Herstellung von Löffeln, Gabeln und Zierlöffeln gehören insoweit in diese Gruppe, als der eigentliche Schnittstempel ein vollständiges Blankett für sich ausschneidet.

In Tabelle Nr. 12 ist der Herstellungsgang der Löffel angegeben, wobei das Ausschneiden der Blanketts der erste oder einer der ersten Arbeitsgänge ist. Werkzeuge dafür sind im folgenden und auf S. 229—234 beschrieben.

Bei der Herstellung der Löffel aus Neusilber wird das Bandmaterial von den Walzwerken genau nach den Größen der herzustellenden Löffel bestellt. Breite, Dicke und Länge werden nach der Erfahrung festgelegt, so daß das Material gleich in verarbeitungsfähigem Zustand für den ersten Schnitt ankommt. Das Ausschneiden der Blanketts ohne jeden Abfall ist eine äußerst interessante Arbeit, da bei jedem Preßhub zwei Blanketts mit nur einem Schnittwerkzeug hergestellt werden. Dem Werkzeug ist ein Abschneider B beigegeben, der in der richtigen Entfernung von dem Schnittstempel A den Schnitt zur Herstellung des zweiten Blanketts führt. Fig. 21 zeigt die Verarbeitung des Materials beim Ausschneiden. A ist der Umriß des Schnittstempels am unteren Ende des Werkzeuges. Der schmale Streifen D ist der ganze Abfall am Anfang und Ende der langen Streifen. Die Linie C gibt die Schnittlinie des Abschneiders, nachdem der Streifen zum Ausschneiden der folgenden Blanketts vorgeschoben worden ist. Die Werkzeuge erhalten selbstverständlich die Abstreifer, Anschläge und alle Erfordernisse für schnelles Arbeiten.

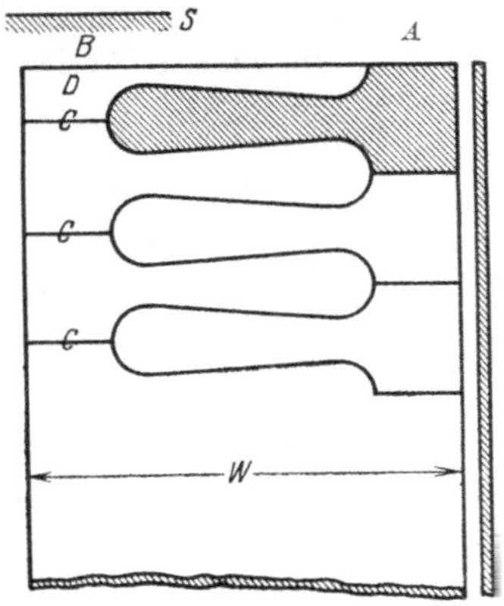

Fig. 21.
A = Schnittstempel.
B = Abschneider.
D = Abfall.
W = Breite des Bleches.

Dieser Arbeit folgt der auf S. 229 ausführlich beschriebene Herstellungsgang.

Bei der Herstellung von Neusilbergabeln bleiben die meisten Vorgänge dieselben, nur tritt an Stelle des Ausschneidens der Pfanne das Schneiden der Zinken, Fig. 22. Dies geschieht gewöhnlich in einer besonderen für diesen Zweck eingerichteten Presse, die jedesmal einen Zwischenraum zwischen den Zinken schneidet, und zwar den mittleren zuerst, und sich nach selbsttätigem Schneiden der beiden seitlichen Zwischenräume auch selbsttätig abstellt. Die Einrichtung hierfür ist gut durchdacht und arbeitet sehr genau und schnell. Die Enden der Zinken bleiben zusammen hängen, damit sie sich nicht während des Prägens spreizen. Es wird dadurch auch verhindert, daß die sonst scharfen Spitzen die Hand des Arbeiters verletzen oder die Polierriemen zerreißen. Dieses Ende wird erst knapp vor der letzten Polierarbeit, gewöhnlich in einer kleinen Fußpresse, weggenommen.

Fig. 22.

Bei der Herstellung der Zierlöffel (S. 232) tritt insoweit eine Abweichung ein, als Pfanne und Griff gesondert ausgeschnitten werden. Nachdem beide Enden des Blanketts keilförmig gestreckt worden sind, wird es in einem Werkzeug nach Fig. 23 ausgeschnitten, wodurch

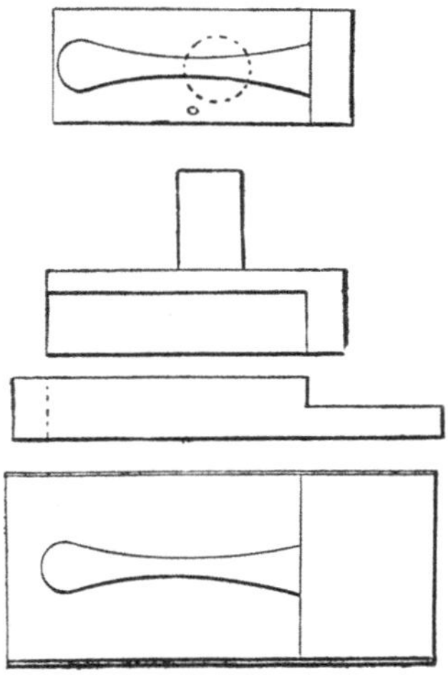

Fig. 23.

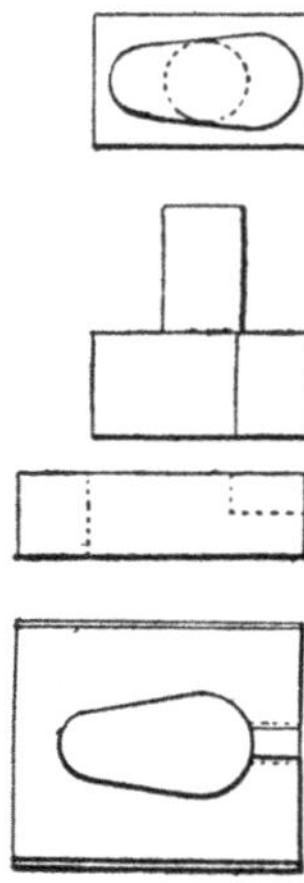

Fig. 24.

der erste Umriß des Handgriffes hergestellt, das Pfannenende jedoch so belassen wird, wie es aus den Streckwalzen gekommen ist. Der Grund für diesen Arbeitsgang liegt in folgendem: Im allgemeinen besitzt eine Fabrik, die derartige Löffel herstellt — man darf dies aber nicht mit der Herstellung der gewöhnlichen Handelsware vergleichen — eine ganze Reihe verschiedener Muster von Handgriffen, ebenso auch einen Satz Gesenke verschiedener Größe und Gestalt zur Herstellung der Pfanne. Wenn nun ein Auftrag hereinkommt, der verschiedene Muster, sei es im Handgriff, sei es in der Pfanne, enthält, so kann man bei der oben beschriebenen Arbeitstrennung jedes Griffmuster mit jedem Pfannenmuster vereinen. Nach dem Ausschneiden des Handgriffes wird demnach ein anderes Werkzeug zum Ausschneiden der Pfanne, Fig. 24, verwendet.

Die Herstellung von Glasklammern für Augengläser und deren Werkzeuge. Fig. 25 zeigt den Herstellungsgang der Glasklammern für Augengläser. Selbstverständlich gibt es viele Formen dieser Teile, die fast bei jeder Fassung verschieden sind; im allgemeinen dagegen sind sie, entsprechend Fig. 25, einander mehr oder weniger ähnlich. Die im folgenden beschriebene Form gilt als die beste dieser Art, weil das mit Gewinde versehene Auge verstärkt ist. Zur Herstellung der Klammer sind vier Arbeitsgänge nötig, die in dem Schnittwerkzeug, Fig. 26, dem Prägewerkzeug, Fig. 283, dem Lochwerkzeug, Fig. 27, und dem Aufbug, Fig. 479, ausgeführt werden. Die Hauptbedingungen sind, daß das Stück, welches aus Gold hergestellt wird, ein bestimmtes Gewicht nicht überschreiten soll, und daß die gesamte Oberfläche des Stückes von dem Prägewerkzeug bearbeitet werden muß, so daß sie glatt wird; die zugelassene Abweichung der Abmessungen darf 0,025 mm nicht überschreiten.

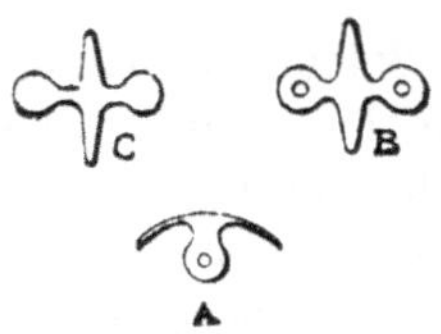

Fig. 25.

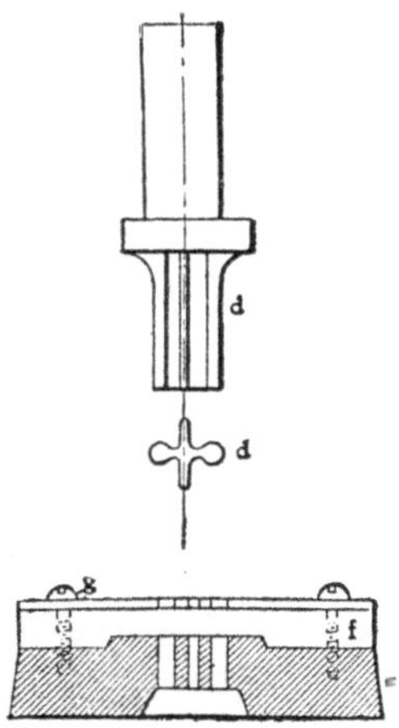

Fig. 26.

Bei Arbeiten dieser Art wird das Prägegesenk nach Fig. 283 zuerst fertiggestellt.

Zur Herstellung des Schnittwerkzeuges werden zwei Musterstücke ungefähr auf Maß gefeilt. Eines wird versuchsweise im Prägegesenk geschlagen. Man kann dabei irgendwelche Fehler in dem andern Musterstück verbessern und macht ein drittes Muster, entsprechend dem ausgebesserten Nr. 2, als „Original". Man versucht dies letztere wieder im Prägegesenk und bringt an Nr. 3 alle notwendigen Verbesserungen an. Diesen Weg führt man durch, bis man ein vollständig einwandfreies Musterstück erhalten hat. C in Fig. 25 zeigt den richtigen Gegenstand, der an die Stirnfläche des Stempels angelötet wird, nachdem die Schnittplatte e, Fig. 26, entsprechend hergestellt worden ist. Zur Herstellung der Glasklammern werden Gold und Neusilber verwendet, zwei Metalle, die ziemlich schwierig für diese feinkantigen Schnitte sind, so daß diese fast ganz gerade durchgefeilt werden. Die Oberfläche der Schnittplatte ist zurückgesetzt, damit man sie nicht ganz schleifen muß. Der Anschlag ist bei f und der Abstreifer aus Stahlblech bei g in Fig. 26 ersichtlich.

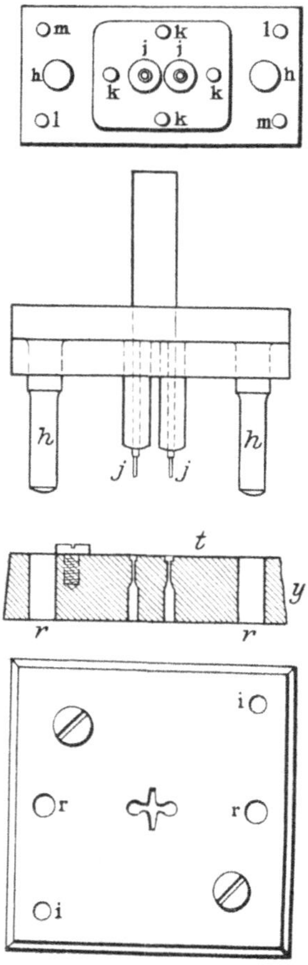

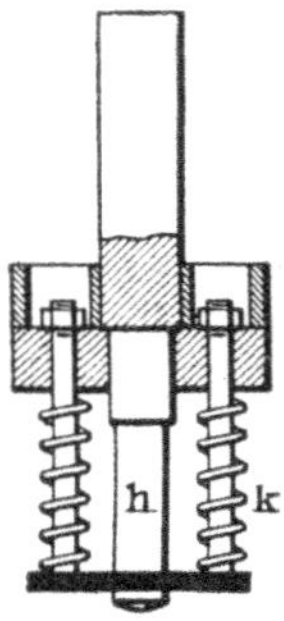

Fig. 27.

Fig. 27 stellt das Werkzeug zum Lochen der Schließe B, Fig. 25, dar. Ein richtiges Musterstück wird gebohrt und die Löcher auf der Schnittplatte y angerissen. Sie werden auch auf der Rückseite der Platte angezeichnet und mit einem größeren Bohrer ungefähr auf die halbe Plattendicke durchgebohrt, wie im Schnitt, Fig. 27, ersichtlich ist. Dann werden sie mit einem Bohrer, der ungefähr 0,075 mm kleiner im Durchmesser als das verlangte Loch ist, gebohrt und von rückwärts mit einer konischen Reibahle ausgerieben. Solche Reibahlen kann man aus Bohrerstahl herstellen, den man erst rund konisch und dann viereckig fertig feilt. Die Einlage t wird aus Stahlblech auf folgende Weise hergestellt: Nachdem das Stück viereckig gefeilt worden ist, wird es mit der Schnittplatte zusammengespannt und die Löcher für die Schrauben von der Rückseite aus gebohrt. Die Klammer wird nun abgenommen und die Löcher für die Prisonstifte i in der gleichen Weise gebohrt. Diese werden dann in ihrer Stellung vernietet, die Einlage auf der Schnittplatte verschraubt und die Löcher für die Lochstempel von rückwärts durch die Platte in die Einlage t gebohrt. Dann wird diese abgenommen und die Löcher auf die richtige Größe erweitert. Bevor man dies macht, wird jedoch die Umrißlinie der Schließe auf der Einlage angerissen, indem man das Muster mittels zweier Bohrer oder Stifte in den Löchern feststellt. Nun wird die Umrißlinie der Klammer gut ausgefeilt, so daß, wenn alle Arbeiten sorgfältig ausgeführt worden sind, die Löcher sich nach dem Lochen in der

Mitte der Klammer befinden werden. Wenn es sich als notwendig erweist, die Schnittplatte nachzuschleifen oder Späne und Schmutz unterhalb der Einlage zu entfernen, kann man sie abnehmen und wieder anschrauben, ohne fürchten zu müssen, an der Ausrichtung derselben etwas zu ändern. Fig. 27 zeigt drei Ansichten des Stempels. In der Vorderansicht fehlt der Abstreifer. Der Schnitt zeigt den Abstreifer in seiner Ruhestellung, während darüber eine Ansicht von unten gegeben ist. Zwei starke Führungsstifte h gleiten in der Schnittplatte bei r, wenn sich das Werkzeug in Arbeitsstellung befindet. Die Lochstempel passen in die Stempelhalter j, sind von rückwärts eingesetzt und vernietet. Die beiden Teile der Kopfplatte werden durch die Schrauben l und die Prisonstifte m zusammengehalten. Die Löcher für die Führungsstifte h und die Stempelhalter j werden von rückwärts durch die Schnittplatte angerissen, wobei die Unterfläche der Kopfplatte mit der Stirnfläche der Schnittplatte y zusammengespannt wird.

Das Lochwerkzeug arbeitet in folgender Weise: Ein geprägtes Blankett wird in die Öffnung der Einlage gelegt. Beim Abwärtsgang der Presse treten die Führungsstifte h in die Schnittplatte, bis der Abstreifer auf die Einlage stößt, worauf die vier Federn k zusammengedrückt werden und deren Führungsstangen durch die Löcher im oberen Teil der Kopfplatte treten. Während der Aufwärtsbewegung der Presse öffnen sich die Federn wieder und streifen das Blankett von den Stempeln ab.

Goldarbeiterstanzen und -stempel. Die Stempel nach Fig. 28 werden fast immer mit runden Zapfen von 9,5 mm bis ungefähr 16 mm Durchmesser gemacht und ungehärtet verwendet, während konische Zapfen fast nie in der Juwelierwerkstätte zu finden sind. Die Schnittplatten werden in der gewöhnlichen Weise hergestellt, nur sind sie aus sehr dünnem Stahlblech von 6,4 bis 12,7 mm Dicke geschnitten und erhalten viel Spiel, da Gold und Silber sehr leicht zu stanzen sind und auch das Material der Stanzen wenig abnutzen. Es wurde die Beobachtung gemacht, daß Werkzeuge über ein Jahr im Gebrauch gewesen sind und immer noch auf der Oberfläche die Anlaßfarben zeigten. Die Werkzeuge werden auf einer Fräsmaschine — jede Juwelierwerkstatt, ja vielleicht jede Werkstätte sollte eine solche besitzen — unter Benutzung eines

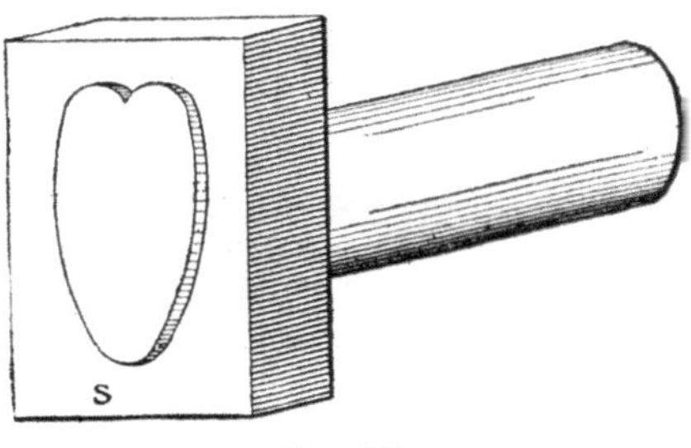

Fig. 28.

sehr kleinen Fräsers, der oft an der Spitze nur 1,6 mm im Durchmesser hat, ausgefräst, so daß sehr wenig Feilarbeit übrig bleibt.

Die gesamte Feilarbeit geschieht senkrecht[1]), eine Arbeitsweise, die man oft bei der Herstellung der Werkzeuge vorziehen würde, wenn man sie nur versuchen würde. Man nimmt dabei die Feile so, daß der Daumen ein gutes Stück oben am Feilenkörper anliegt, und läßt den linken Vorderfinger auf dem unteren Teil des Schraubstockes anliegen, so daß man eine Führung erhält und sicher ist, daß die Feile in einer senkrechten Linie bewegt wird. Man arbeitet an der weitest abgelegenen Seite und wird nach kurzer Übung finden, um wieviel diese Arbeitsweise der althergebrachten vorzuziehen ist.

Die Stempel werden oft nach Fig. 28 aufgebaut, wobei der eigentliche arbeitende Teil S aus Stahl von 4,7 bis 6,4 mm Dicke hergestellt und dann an den Zapfen hart angelötet wird. Solange diese Werkzeuge für Gold, Silber und vielleicht noch Messing verwendet werden, arbeiten sie gut. Sie kosten ungefähr die Hälfte eines aus dem vollen herausgearbeiteten Stempels und können, wenn die dünne Stahlauflage abgenützt ist, leicht erneuert werden. In dieser Weise sind schon äußerst verwickelte Werkzeuge in Juwelierwerkstätten hergestellt worden.

Schnitt- und Lochwerkzeug für Gold. Die Herstellung eines kleinen Winkels mit Loch aus Gold geschieht in drei Arbeitsgängen nach Fig. 29.

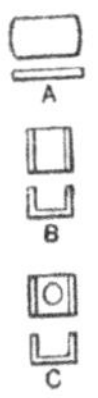

Fig. 29.

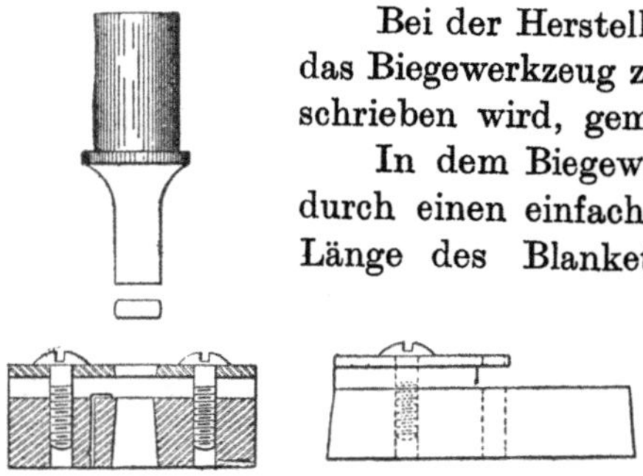

Fig. 30.

Bei der Herstellung der Werkzeuge wird das Biegewerkzeug zuerst, wie auf S. 364 beschrieben wird, gemacht.

In dem Biegewerkzeug kann man nun durch einen einfachen Versuch die richtige Länge des Blanketts feststellen, wonach Schnittstempel und Schnittplatte hergestellt werden. Der Stempel wird einfach abgedreht und auf die richtige Breite gefräst, und damit das Loch in der Schnittplatte geräumt. Immer, wenn der Stempel bearbeitet werden kann, ist es vorteilhaft, zuerst den Stempel und damit auf die oben beschriebene Weise die Schnittplatte herzustellen; man erspart jedenfalls damit viel Zeit. Stempel und Schnittplatte sind in Fig. 30 dargestellt und wohl ohne weiteres verständlich. Das Lochwerkzeug für den letzten Arbeitsgang ist in Fig. 31 abgebildet. Eine

[1]) Vgl. Feilmaschinen, S. 577/80; auch W. T. 1912, S. 343.

Abstreiferplatte K wird auf der Gesenkplatte J mittels Schrauben und Prisonstiften befestigt und hat einen überhängenden Fortsatz L, durch welchen ein Loch, entsprechend der Größe des Lochstempels, in Linie mit dem Loch in der gehärteten Lochplatte gebohrt ist. Die Teilscheibe ist aus Stahl von ungefähr 2,4 mm Dicke mit genau bearbeiteter Ober- und Unterfläche und am Umfang kordiert, damit sie leichter gedreht werden kann. Der Drehzapfen M ist an passender Stelle eingeschraubt und erhält eine Lage Papier unter dem Kopf, so daß eine genügende Oberflächenreibung zwischen Kopf und Scheibe erhalten wird. Das Werkzeug wird dann auf der Fräsmaschine eingespannt, und die Nut für den Hebel N in die Teilscheibe eingefräst, wobei die Nut in der gehärteten Gesenkplatte als Lehre benützt wird. Nun wird das Werkzeug um ein Viertel Umdrehung gedreht, und der nächste Schlitz in die Teilscheibe eingefräst usw., bis alle vier Schlitze fertig gestellt sind. Dabei verwendet man denselben Fräser wie zur Herstellung des Schlitzes in der Gesenkplatte.

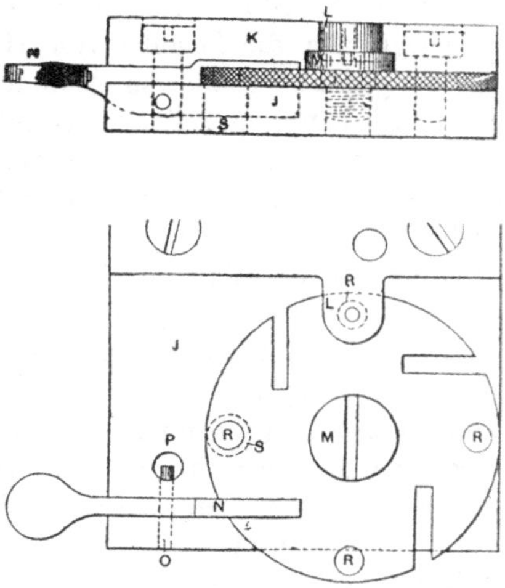

Fig. 31.

Der Schalthebel N ist aus einem Stück Werkzeugstahl geschmiedet und paßt gut in den Schlitz der Gesenkplatte, während der Teil desselben, der in die Teilscheibe paßt, ein wenig über deren Oberfläche hervorragt und leicht abgeschrägt ist, um jede mögliche Abnützung aufzunehmen. Der Hebel schwingt um den Bolzen O, der innerhalb des Loches P endigt, so daß man ihn, wenn nötig, leicht entfernen kann. Die Besonderheit dieses Werkzeuges war die Austeilung der vier Löcher R, die das Werkstück während des Lochens zu halten hatten. Man muß die Kanten der Schlitze sehr sorgfältig vom Grat befreien und den Schalthebel härten, bevor man diese Löcher bohrt. Nachdem der Hebel in einen der Schlitze eingedrückt worden ist, wird ein der Bohrung in der Schnittplatte entsprechendes Loch durch die Teilscheibe von der Unterseite her gebohrt. Die Scheibe wird nun bis zur nächsten Nut weitergedreht, und die übrigen drei Löcher in gleicher Weise gebohrt, wobei die Scheibe jedesmal in der richtigen Stellung in dem bezüglichen Schlitz gehalten wird. Nach dem Bohren werden die Löcher entweder durch Ausbohren oder Anwendung verschiedener

Bohrer vergrößert, so daß das geformte Arbeitsstück ohne besondere Anstrengung, aber auch ohne Spiel eingelegt werden kann. Die obere Fläche wird ein wenig versenkt, so daß sich die Stücke leicht einstellen können. Mit diesem Werkzeug kommt ein Stempel der gewöhnlichen Form zur Anwendung, so daß eine Abbildung desselben unterbleiben kann.

Beim Arbeiten wird ein gestanztes Stück in eines der Löcher R eingelegt und die Scheibe von Hand herumgedreht, bis das Stück gerade gegenüber und unter dem Loch in der Abstreiferplatte ist. Dann wird mit der linken Hand der Schalthebel N in den entsprechenden Schlitz der Teilscheibe eingesenkt und mit dem Fuß die Presse betätigt, so daß das Stück gelocht wird. Hierauf wird ein anderes Stück in das folgende Loch R eingelegt, der Schalthebel herausgezogen, die Scheibe um eine Viertelumdrehung weitergeschaltet, so daß das neue Stück unter den Lochstempel kommt. Dabei bleibt das bereits gelochte Stück in der Teilscheibe liegen, bis es über das Loch S in der Gesenkplatte kommt; hier stößt es ein Stift, der am Stempelkopf befestigt ist, beim Abwärtsgang der Presse durch, so daß es in einen Kasten unterhalb der Presse fällt. Man kann aus dieser Einrichtung ersehen, daß eine ganze Menge Zeit gespart wird, da es unnötig ist, die Stücke aus dem Werkzeug herauszunehmen, so daß der Arbeiter seine ganze Zeit darauf verwenden kann, die Stücke an der Vorderseite der Teilscheibe in diese einzulegen, während die auf der Rückseite befindlichen gelocht werden.

Werkzeug mit geteilter Schnittplatte. Das Ausschneiden von Teilen aus Werkzeugstahl, Fig. 32, von ungefähr 4 · 22 · 7,5 mm Abmessung, welche ganz genau sein müssen, verlangt Werkzeuge von äußerst genauer Konstruktion, die, falls etwas schadhaft wird, leicht und billig auszubessern sind Ein Satz solcher Werkzeuge ist in Fig. 33/34, das dazugehörige Blankett vor und nach dem Ausschneiden in Fig. 32 abgebildet.

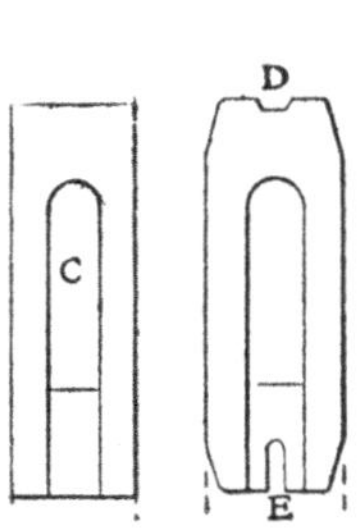

Fig. 32.

Das Blankett ist rechteckig mit scharfen Ecken, allseitig gefräst, hat eine auf beiden Seiten, A und B, eingefräste Nut und bei C ein gestanztes Loch. Das hier besprochene Werkzeug versieht das Stück mit einem Schlitz bei D und E und schneidet die Ecken, wie angedeutet, symmetrisch ab.

Mit Rücksicht auf die Qualität und Dicke des zu verarbeitenden Materials, die erforderlichen genauen Abmessungen des fertigen Stückes im Verein mit der Tatsache, daß wöchentlich Tausende dieser

Stücke gebraucht werden, und dem geringen Akkordlohn, der dafür gezahlt werden konnte — 17 Pf. für hundert —, ist es nötig, ein Werkzeug zu entwerfen, das leicht und billig ausgebessert werden kann, die Austauschbarkeit der Erzeugnisse gewährleistet und schließlich eine schnelle Herstellung gestattet.

F in Fig. 34 is teine Gesenkplatte aus Maschinenstahl, auf der alle Teile der Schnittplatte und sonstigen arbeitenden Teile befestigt sind. H, I, J und K sind die vier gehärteten, angelassenen und geschliffenen Teile der Schnittplatte aus Werkzeugstahl, die die Enden des Blanketts und die Schlitze ausschneiden. Diese Teile werden in genau ausgearbeitete Sitze in F eingelassen und mittels Zylinderkopfschrauben befestigt — wie im Grundriß zu ersehen ist — während deren Seiten und Enden sich gegen die Stellschrauben P, N, Q und O stemmen. LLLL sind enge Schlitze, die in der Gesenkplatte bis zur angezeigten Tiefe ausgefräst werden und deren

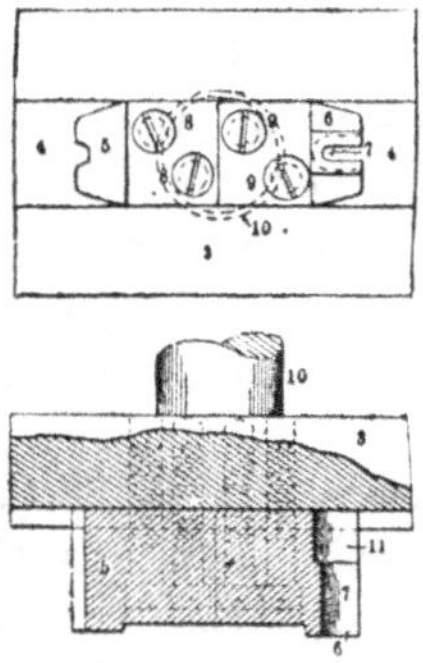

Fig. 33.

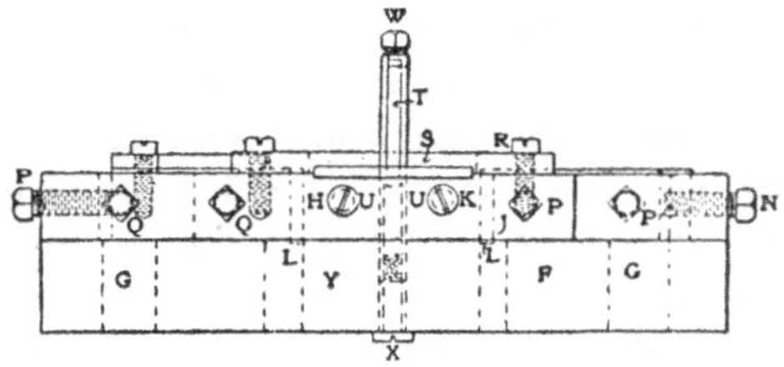

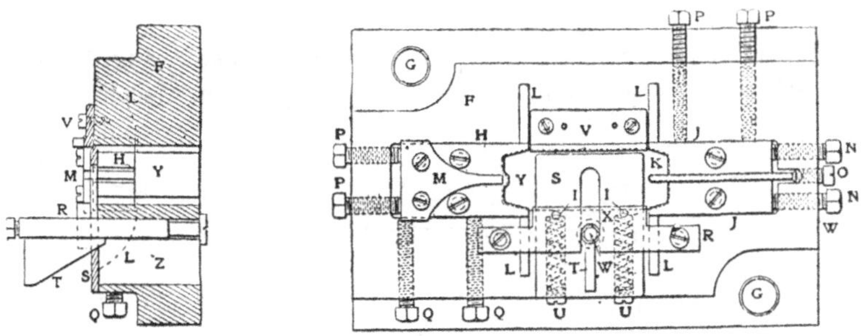

Fig. 34.

innere Seiten eine ganz genau bestimmte Entfernung voneinander haben, so daß sie als Anschläge zum Ausrichten der Enden der Schnittplattenteile H und J benutzt werden können. Die Mitte der Gesenkplatte erhält ein Loch, damit die fertigen Stücke frei durchfallen können.

Die übrigen Teile des Werzeuges, die erwähnt werden müssen, sind der hin- und hergehende Schlitten S, der die Auflage für das Stück vor dem Stanzen bildet, die Anschläge V und R, die das Stück seitwärts sichern, und die Einrichtung, die den Schlitten von dem Pressenstößel aus in Tätigkeit setzt.

Der Querschnitt durch den Werkzeugunterteil zeigt die Konstruktion der verschiedenen Teile und die Arbeitsweise des Schlittens, während der Grundriß die Federn bei I, das Keilstück T mit der Ein stellschraube W und die Schrauben ohne Köpfe U, die die Nachstellung der Federn bewirken, erkennen läßt.

Der obere Teil des Werkzeuges, der Stempel, ist in Fig. 33 dargestellt. Die Kopfplatte 3 aus Maschinenstahl mit dem Zapfen 10 erhält eine rechteckige Nut 4 auf ihrer ganzen Unterfläche eingefräst, in der die beiden Stempelteile 5 und 6 mit Zylinderkopfschrauben befestigt sind.

Das Werkzeug wird in die Presse eingebaut und die Stellschraube W in T so eingestellt, daß das Keilstück den Schlitten S berührt und zwingt, nach rückwärts dem Stempel aus dem Weg zu gehen, bevor dieser in die Schnittplatte eintritt. Nun wird vom Arbeiter ein Blankett auf den Schlitten S gelegt und zwischen den Anschlägen V und R sowie gegen den Endanschlag M eingestellt. Beim Einschalten der Presse stößt die Unterfläche des Stempels beim Abwärtsgang an die Stellschraube W und drückt den Keil T nach abwärts, so daß der Schlitten unter dem Werkstück weggezogen wird und den Platz frei läßt, damit der Stempel die Enden zuschneiden und das Stück in die Schnittplatte drücken kann. Beim Aufwärtsgang des Stößels bringen die Federn bei I den Schlitten S und den Keil T in ihre Anfangslage zurück, so daß ein neues Blankett eingelegt werden kann.

Zusammengesetztes Werkzeug für Ankersegmente. Fig. 35 ist das Blankett, das vollständig in einem Arbeitsgang in diesem Werkzeug

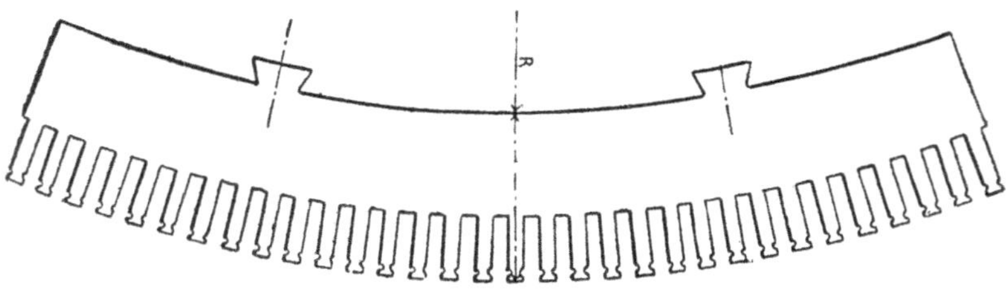

Fig. 35.

hergestellt wird. Acht solcher Segmente bilden dann den vollständigen Kreis.

Das in den Fig. 36/37 abgebildete Werkzeug wird in folgender Weise aufgebaut:

Nachdem die gußeiserne Gesenkplatte F, Fig. 37, oben und unten gehobelt worden ist, werden die Nuten für die Seitenstücke D eingearbeitet. Das Stück wird dann in ein Drehwerk eingespannt, um die Ringnut für die Stahlplatte, welche die kleinen Schnittstempel trägt, wie bei m und im Schnitt durch C-C ersichtlich ist, auszudrehen.

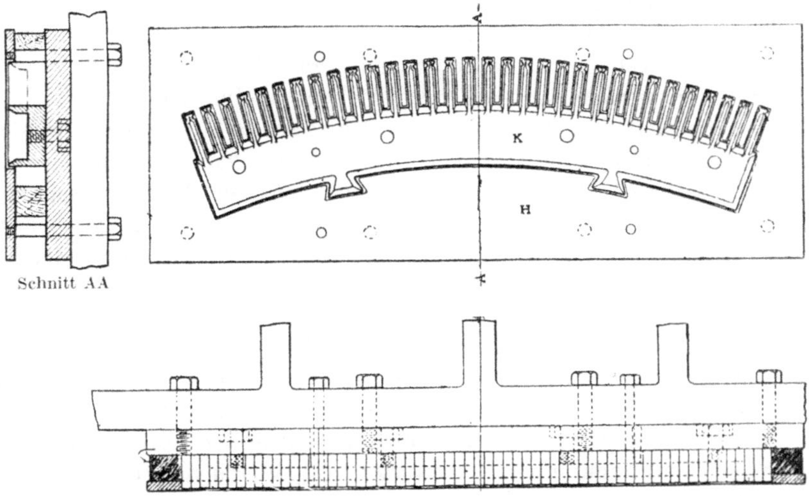

Fig. 36.

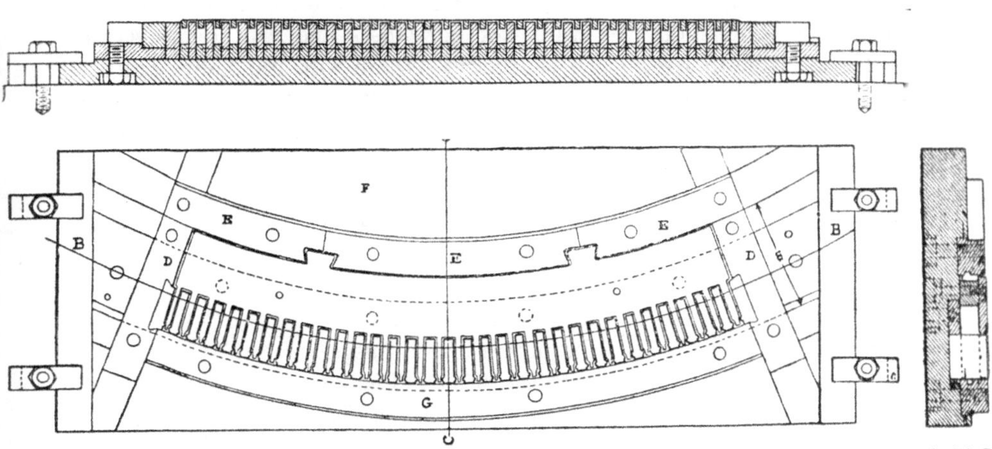

Schnitt CC

Fig. 37.

Die Stahlplatte m wird in die gußeiserne Grundplatte F eingepaßt und durch Prisonstifte und Schrauben in ihrer Stellung gesichert. Hierauf wird die Platte F wieder auf dem Dreh- und Bohrwerk eingespannt und die Nuten für die gekrümmten Schnittplattenteile aus Werkzeugstahl, E und G, eingearbeitet.

Diese Stücke werden zuerst erhitzt, ungefähr auf den verlangten Krümmungsradius gebogen, und deren Ober- und Unterseiten gehobelt, worauf die Seitenflächen nach dem Kreisbogen bearbeitet werden. Die Schwalbenschwänze in E werden gestoßen, und die Stücke in der Mittellinie der Schwalbenschwänze geteilt, wobei man eine Zugabe für das Schleifen nach dem Härten vorsieht. Hernach werden die Schraubenlöcher gebohrt und in E und G Gewinde eingeschnitten, worauf diese Teile gehärtet und angelassen werden. Die Ober- und Unterflächen werden dann auf der Flächenschleifmaschine geschliffen. Die Stücke sind nun für die radiale Schleifarbeit vorbereitet, bei der man sie zuerst mit der Oberseite nach unten einspannt und an den im Querschnitt CC, Fig. 37, mit Pfeilen bezeichneten Stellen anschleift. Die Schwalbenschwänze in E werden nach einer Lehre gearbeitet, nach welcher E und G in der Grundplatte F ausgerichtet und befestigt werden. Die Platte wird dann wieder auf das Drehwerk gebracht und der Werkzeugunterteil passend zur Lehre geschliffen. Die seitlichen Stücke D werden auf der Flächenschleifmaschine fertiggestellt. Wie die einzelnen Schnittstempel aus Werkzeugstahl in der Stahlplatte m gehalten werden, ist aus Fig. 39 ersichtlich. Die halben Stempel an den Enden werden mit Schwalbenschwanz in den Seitenstücken D gehalten. Unter dem inneren Auswerfer wird Gummi als Federung verwendet.

Der Stempel K, Fig. 36, ist ein Schmiedestück aus einem Stück Werkzeugstahl. Er wird auf der Ober- und Unterseite sowie auf dem äußeren Kreisbogen bearbeitet. Dann wird die Lehre auf die Oberfläche des Stempels gebracht und die Nuten, der Innenbogen und die Seiten sorgfältig angerissen. Hierauf wird der Stempel nach den Linien vorgestoßen, so daß, falls diese Arbeit sorgfältig durchgeführt worden ist, sehr wenig Nachfeilen nötig wird. Nun wird die innere Partie ausgearbeitet, wie aus Fig. 36 oder deutlicher aus Fig. 38 zu ersehen ist, so daß eine Wand von ungefähr 13 mm Dicke längs der Kante des Stempels verbleibt; diese Wand wird schräg fast zu einer scharfen Kante ausgefräst. Der volle Teil zwischen den Nuten wird ebenfalls schräg ausgefräst, so daß ringsherum an der Schnittkante eine schmale Fläche von ungefähr 0,8 mm Breite bleibt. Der Stempel wird nicht gehärtet; wenn er nach einigem Gebrauch Zeichen der Abnützung aufweist, wird er herausgenommen und die Schnittflächen

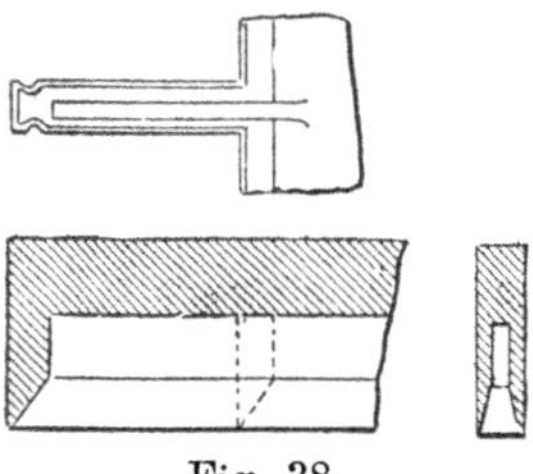

Fig. 38.

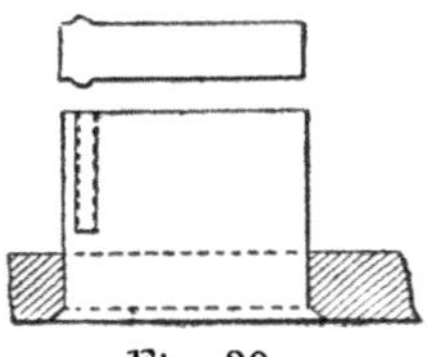

Fig. 39.

mit dem Hammer abgeflacht, dann wieder in seiner richtigen Stellung in die Presse gebracht und durch die Schnittplatte gedrückt, so daß die Schnittkanten genau auf Maß abgeschert werden. Für den Stempel ist ein äußerer Abstreifer vorgesehen, wobei wie oben Gummi als Abfederung verwendet wird.

Ein weiterer Punkt ist der Erwähnung wert. Man sieht, daß die Köpfe der Schrauben, welche die Schnittplattenstücke halten, bündig mit der Unterseite der Grundplatte sind, Fig. 36/37, ebenso die Kopfschrauben, welche den Stempel an der Kopfplatte befestigen. Dies ist gemacht worden, um jede Möglichkeit eines Sichlosarbeitens einzelner Teile während ihrer Arbeit zu verhindern. Werkzeuge dieser Art sind für Ankerdurchmesser von 1000 mm aufwärts gemacht worden.

Schnittplatte aus hohlgepreßtem Stahlblech. (Erzgebirgische Schnittwerkzeug- und Maschinenfabrik.) Hauptsächlich durch die Kosten der großen Schnittlager hervorgerufen, sind bereits verschiedene Vorschläge aufgetaucht, die schweren, vollen Schnittplatten aus Werkzeugstahl durch solche aus Stahlblech zu ersetzen. Es werden wohl noch verschiedene andere Gründe für und gegen den Ersatz jener durch solche aus gehärtetem Stahlblech ins Treffen geführt[1]). Von allen derartigen Vorschlägen hat aber nur die in Fig. 40 abgebildete Form der eingangs erwähnten Firma mehr Verbreitung gefunden. Der Hauptvorteil liegt in der Verwendung einer hohlgepreßten Stahlplatte, an Stelle des flachen Stahlbleches der früheren Schnittplatten. Dadurch erhält die Schnittplatte eine höhere Festigkeit und Steifigkeit, so daß sie allein ohne eigentliche Gesenkplatte verwendet werden kann. So unleugbar ihre Vorteile gegenüber der einfachen Schnittplatte aus Stahlblech sind, so ist doch eine Überlegenheit, besonders bei größeren Schnitten und verwickelten Formen, der festen, vollen Platte gegenüber nicht so ohne weiteres zu bejahen. Sie ist wohl leichter, kostet weniger an Material, läßt sich infolge ihres geringeren Gewichtes leichter handhaben und aufbewahren; ob sie aber nicht leichter als die volle Platte Beschädigungen aus-

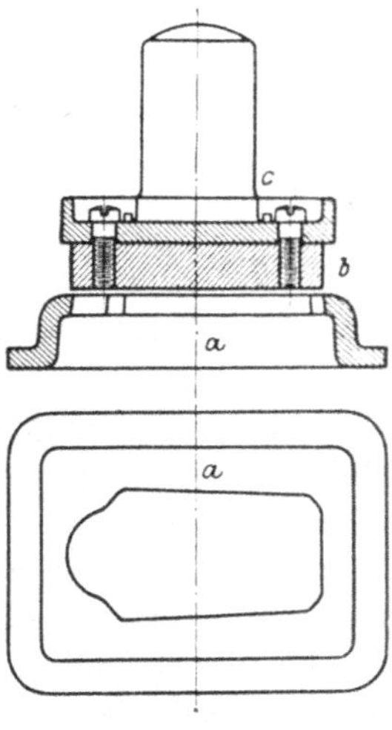

Fig. 40.

[1]) W. T. 1910, S. 497; Z. f. Werkzeugm. 1908, S. 95, 124, 321.

gesetzt ist, sich bei feineren Umrissen nicht leichter abnützt, ist ebenso fraglich wie die Erniedrigung der Herstellungskosten. An der angegebenen Stelle sind Druckversuche veröffentlicht, die durch Analogieschlüsse Belastungen bis zu 23 000 kg bei einem Schnitt von 240 mm Umfang auf Eisenblech von 3,05 mm Dicke feststellen. Wenn man nach der oben angeführten Quelle bei dem in Fig. 40 abgebildeten Werkzeug einen Umfang an der Schnittkante von 240 mm annimmt, so ergibt sich bei Zugrundelegung eines Maßstabes von 1 : 4 eine Fläche von ungefähr 35 cm^2 und somit beim Aufsetzen des Schnittstempels eine über die Fläche gleichmäßig verteilte Druckkraft von 650 kg/cm^2, was als ganz normale Materialbelastung bezeichnet werden muß.

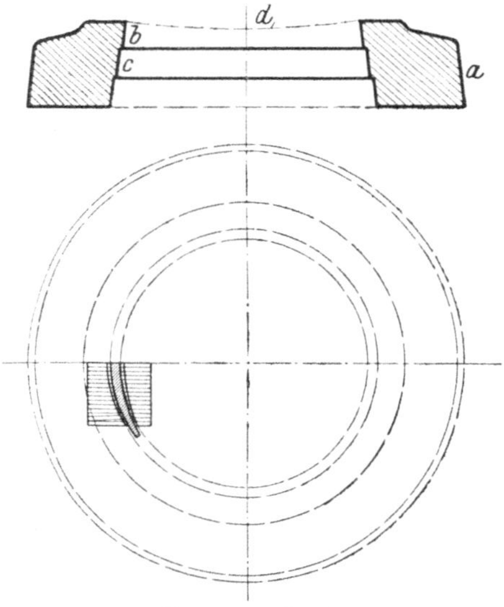

Fig. 41 und 42.

Die Herstellung von großen Rundschnitten[1]). Es handelte sich in diesem Fall um einen Schnitt für ein Kettenrad von 310 mm Durchmesser und 5 mm Dicke, für den als Schnittplatte ein Ring von 60—80 mm Höhe und 100 bis 120 mm Wandstärke in Betracht kam. Da das Schmieden eines solchen Ringes wegen des größeren Abfalles, der möglichen Risse und des Überhitzens der Schneidkante beim Schmieden, was sich erst am fertigen gehärteten Stück zeigt, als ungünstig abgelehnt wurde, griff man zum Ausstechen des Ringes aus einer vollen Scheibe von 500—600 mm Durchmesser. Der Ring, Fig. 41, wird gleich mit der Schräge ausgestochen, wobei man auf der unteren Fläche beginnt und diese gleichzeitig eben dreht. Hierauf wird die Außenfläche der Schnittplatte, bei a, 2^0 konisch angedreht. Wenn man bis auf die halbe Dicke durchgestochen hat, spannt man um, zentriert wieder und sticht mit derselben Schräge den Rest aus. Gleichzeitig wird der Schnitt oben eben gedreht. Es empfiehlt sich, die Schnitte mit 3—4^0 Anzug und die Fläche c möglichst genau auf Maß zu drehen, damit man nach dem Härten nicht zu viel wegzuschleifen hat. Der Oberteil ist bei derartig starkem Blech und großem Durchmesser um

---

[1]) **Illustr. Zeitung f. Blechindustrie 1906, S. 1326.**

0,15 mm schwächer zu halten, wird zylindrisch gedreht und gehärtet. Die Schnittplatte ist mit einem Pfeil von 8—10 mm konkav zu halten.

Zu Fig. 42 seien einige Angaben über die Herstellung des Kreisausstechstahles wiedergegeben. Der Stahl ist ungefähr 4 mm stark, genau der Krümmung entsprechend geformt und mit unterem und seitlichem Hinterschliff zu versehen, der am besten nur durch Schleifen und nicht durch Schmieden angearbeitet wird. Als Stahl wurde ein Schnelldrehstahl 30 · 10 verwendet und in eine Stange Vierkanteisen eingepaßt, vernietet, hart verlötet und gleich dabei gehärtet. Das Ausstechen der Schnittplatte von obigem Durchmesser und 75 mm Höhe dauerte 7 Stunden.

Stanzwerkzeuge der Kartonnagenindustrie. (Karl Krause, Leipzig.) Die Ausbildung der Schnittwerkzeuge der Kartonnagenindustrie, soweit Schnitte mit Ober- und Untermesser, usw. zur Verwendung kommen, ist in ihren Einzelheiten von den eben besprochenen Metallschnittwerkzeugen sehr verschieden. Der größte Gebrauch wird bei der Herstellung der verschiedenen Schachteln von diesen Werkzeugen gemacht, was wiederum aus Wirtschaftlichkeitsrücksichten verlangt, daß man innerhalb gewisser Grenzen die Werkzeuge verstellbar machen kann, so daß verschiedene Schachtelgrößen mit denselben Schnitten hergestellt werden können. Aus diesem Grunde werden die einzelnen Schnittstempel, „Messer", nicht in dem Werkzeug, sondern am Stößel vereinigt, so daß durch ihre verschiedene Befestigung am Stößel verschiedene Schnitte entstehen. In gleicher Weise werden die einzelnen Teile der Schnittplatte einzeln auf dem Pressentisch befestigt. Aus diesem Grunde sind diese Werkzeuge auch hier aufgeführt worden, da sie eine Anzahl einzelner nebeneinander arbeitender Werkzeuge mit eigenem Stempelkopf darstellen, deren jedes ohne Zweifel in diese Gruppe gehört.

Die betreffenden Maschinen werden auf S. 551/5 besprochen.

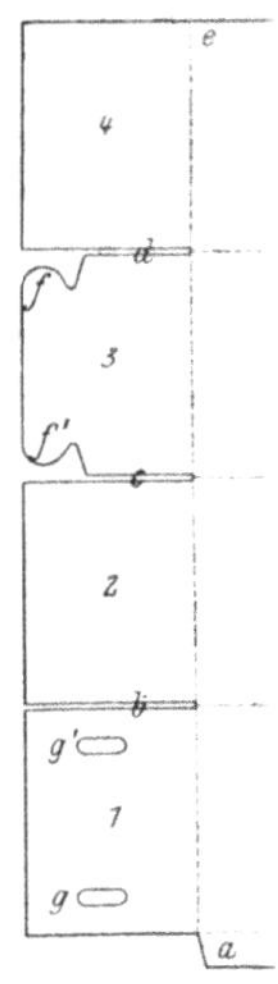

Fig. 43.

Faltschachtel-Apparate. In Fig. 43 ist der abgewickelte Schnitt einer halben Faltschachtel gegeben. Bei a ist der sogenannte Kleberand, der Streifen, mittels dessen die Verbindung der Kanten a und e später durch Kleben oder Heften erfolgt. Die Schnitte bei b, c und d trennen die vier einzelnen Deckelteile, deren Entfernung voneinander beliebig eingestellt werden kann, so daß man Schachteln von verschiedenem Querschnitt erhält. Während die

Deckelteile 2 und 4 durch diese Schnitte bereits fertiggestellt sind, werden an den Teilen 1 und 3 noch durch besondere Messer die Schlitze g und g′ sowie die entsprechenden Augen f und f′ ausgestanzt, die zum Schließen der Schachtel dienen. Da die Kartonstreifen bereits auf richtige Breite zugeschnitten worden sind, fällt der Längsschnitt sowie der Endschnitt bei e als besondere Werkzeuge weg.

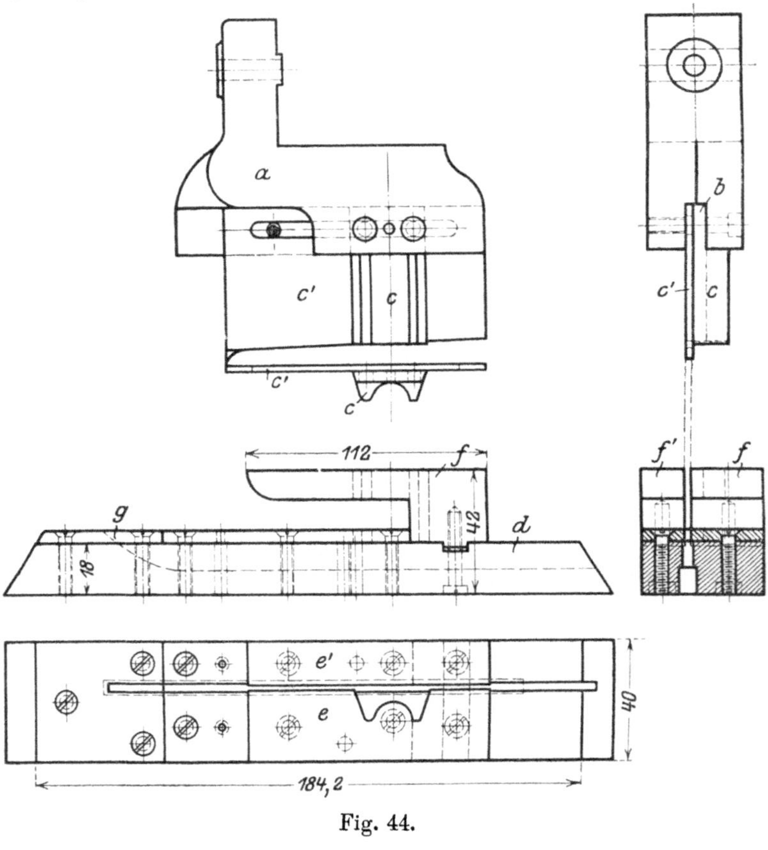

Fig. 44.

In Fig. **44** ist ein solches Teilwerkzeug für den Schlitz c und das Auge f′ gezeichnet. Der Stempelkopf, „Obermesserhalter“, a wird mittels Feder und Nut durch Schrauben an dem als Schiene ausgebildeten Stößel befestigt, gegen den er sich mit seiner Oberfläche anlegt. Demnach steht der Stempel in seiner Hauptausdehnung unter rechtem Winkel zum Stößel. An der Unterfläche des Stempelkopfes sind Schlitze b vorgesehen, in welche die betreffenden Stempel, „Obermesser“, c c′, eingesetzt werden, worauf sie durch Zylinderkopfschrauben in ihrer Stellung gesichert werden. Das Schlitzmesser c′ wird mittels Langschlitzes befestigt, so daß es den verschiedenen Faltschachtelabmessungen entsprechend eingestellt werden kann. Die Schnittstempel sind sämtlich an der Unterfläche abgeschrägt, das Maß der Schräge ergibt sich aus den Zeichnungen. Der Werkzeugunterteil

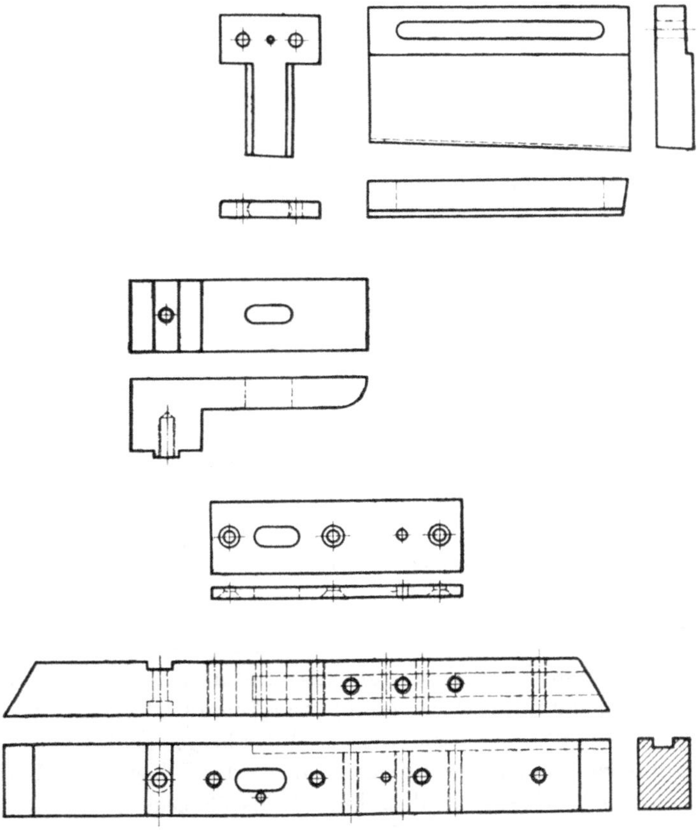

Fig. 45—49. Werkzeug für g.

ist ein vollständiger Schnittkasten, bestehend aus Abstreifer, Schnittplatte und Gesenkplatte. Die Gesenkplatte d wird aus gezogenem S.-M. Stahl hergestellt und die beiden Schrägen, die sich in der Schwalbenschwanznut des Tisches führen, nach der Schablone dort eingepaßt. Die Schlitze zum Entfernen des Abfalles werden bis zur Mitte der Plattendicke mit einem 3 mm breiten Scheibenfräser ausgefräst, worauf die Platte umgedreht wird und der unten erweiterte Schlitz mit einem 6 mm breiten Fräser ausgearbeitet wird. Auf der Gesenkplatte d werden die Schnittplatten e e′, eine für das Auge f′ und die eine Seite des Schlitzes c, die andere für dessen zweite Seite, mit Schrauben und Prisonstiften befestigt. Die Schnittplatten sind aus Werkzeugstahl, gehärtet, angelassen und geschliffen. Als Endanschlag dienen die Abstreifer f f′, die sich unmittelbar an die Schnittplatte anschließen und durch Feder

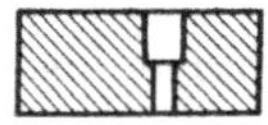

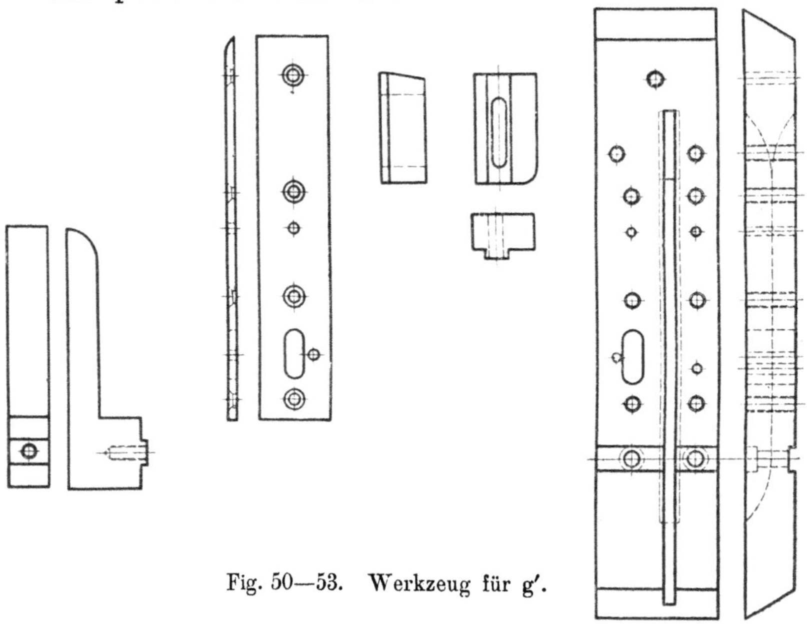

Fig. 50—53. Werkzeug für g′.

und Nut in ihrer Stellung gesichert erscheinen. Die Befestigung erfolgt durch Zylinderkopfschrauben von der Unterseite der Gesenkplatte aus. Am vorderen Ende der Gesenkplatte ist eine Deckplatte g befestigt, die ebenso wie der Abstreifer aus S.-M.-Stahl gemacht ist.

Die zur Herstellung der übrigen Schnitte des Faltschachtelblanketts erforderlichen Werkzeuge sind in Fig. 45 bis 53 gegeben, wobei noch zu bemerken ist, daß das in Fig. 44 abgebildete Werkzeug zweimal, einmal rechts und einmal links, zur Ausführung kommt, und außerdem die Teile c′, e′ f′ und g, Fig. 50/53, noch je einmal ausgeführt werden. Dazu kommen die verschiedenen Stempelköpfe, Obermesserhalter, entsprechend a in Fig. 44, alles in allem vier Stück aus Gußeisen.

b) Der Mehrfachschnitt.

Der Mehrfachschnitt entsteht dadurch, daß eine Anzahl gleicher Stempel in einer Kopfplatte, bzw. gleicher Schnitte in einer Schnittplatte vereinigt werden und so bei einem Pressenhub die entsprechende

Anzahl gleicher Stücke gleichzeitig ausschneiden. Bei der Konstruktion dieser Werkzeuge sind zwei Fälle möglich; entweder werden alle der Breite des Streifens entsprechenden Blanketts bei einem einmaligen Durchgang des Streifens ausgeschnitten, oder der Streifen muß zwei- oder mehrmals durch die Presse gehen, da es konstruktiv nicht möglich ist, so viele Stempel gleichzeitig nebeneinander in das Werkzeug einzubauen, als den Ausschnitten der vollen Blechbreite entspricht, oder dadurch die Kosten des Werkzeuges wesentlich herabgedrückt werden. Die Anordnung der Stempel in den Werkzeugen der ersten Gruppe kann auch so sein, daß die der vollen Breite entsprechenden Ausschnitte wohl in einem einmaligen Durchgang, aber erst nach je zwei oder mehr Stößelhüben erhalten werden. Schließlich werden, besonders beim Perforieren der Blechtafeln, die verschiedensten Verbindungen von Vorschub und einer senkrecht dazu verlaufenden Schaltbewegung ausgeführt. In diesem Fall kommen eigene Perforiermaschinen, Zickzackpressen u. ä., zur Anwendung. Weiter werden die Schnittplatten dieser Werkzeuge des öfteren der billigeren und sicheren Herstellung halber, und um im Falle eines Bruches nicht die ganze verwickelte Platte ersetzen zu müssen, aus mehreren Teilen zusammengesetzt, die meistens untereinander austauschbar gehalten werden. Um ein sicheres Arbeiten zu ermöglichen, sind außerdem am Oberteil eigene Führungssäulen angebracht oder die Stempel in einer eigenen Führungsplatte des Schnittkastens, wie die Vereinigung aller Teile des Unterteiles des Werkzeuges genannt wird, geführt.

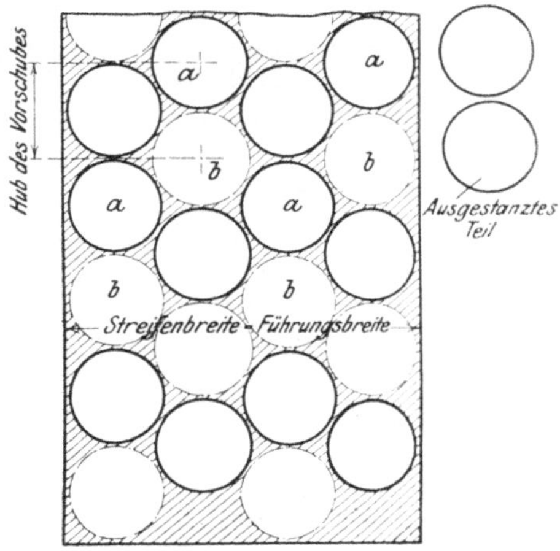

Fig. 54.

Bei den im folgenden beschriebenen Werkzeugen für Mehrfachschnitte tritt der Unterschied sehr deutlich auf, ob nämlich das ausgestanzte Blankett oder die übrigbleibende Blechtafel zur weiteren Verwendung kommen. Bezüglich der dabei auftretenden Unterschiede in den Abmessungen von Stempel und Schnittplattendurchmesser sei auf die Tabelle S. 36 verwiesen.

Als Schulbeispiel sei in Fig. 55 ein Mehrfachschnitt mit vier Stempeln[1]), in der Quelle „Massenschnitt" genannt, beschrieben,

[1]) W. T. 1909, S. 578.

der die konstruktive und grundsätzliche Ausführung derartiger Werkzeuge zeigt. Die volle Ausnutzung des Blechstreifens, aus dem die Scheiben, Fig. 54, mit dem abgebildeten Abfallstreifen ausgestanzt werden, erfolgt in einem Durchgang, doch wird erst nach zwei Schaltungen die volle Blechbreite verarbeitet. Die vier mit a bezeichneten Schnittstempel sind mittels Konus mit der Kopfplatte c

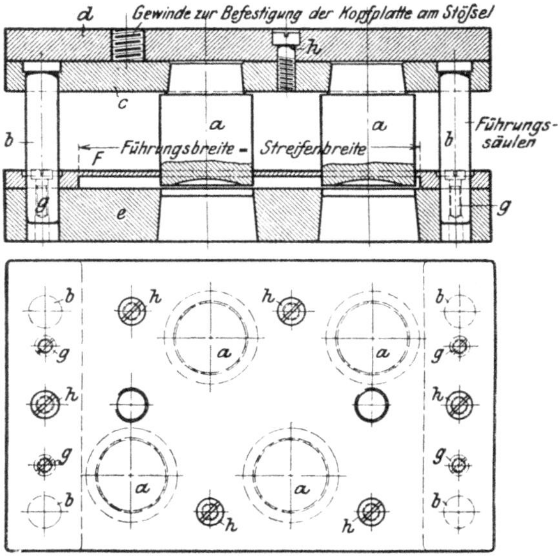

Fig. 55.

verbunden, die durch sechs Zylinderkopfschrauben h an dem Stempelkopf d befestigt ist. In der Kopfplatte c sind vier Führungssäulen b, die in entsprechende Bohrungen im Schnittkasten eintreten, befestigt. Der Abstreifer F ist auf der Schnittplatte verschraubt und dient gleichzeitig als Führung für den Blechstreifen. Diese Werkzeuge werden gewöhnlich mit einer Walzenzuführung[1]) verbunden, deren Vorschub in diesem Falle $^2/_3$ der Stempelentfernung, in der Vorschubrichtung gemessen, beträgt. Muß Tafelmaterial verarbeitet werden, so richtet man die Stempel so ein, daß die Tafel höchstens in 3, gewöhnlich aber in nur 2 Streifen zerschnitten wird. Bei kleinen Durchmessern verwendet man ausschließlich Bandmaterial.

Das Lochen von Blechen. Vielleicht die wichtigste Anwendung der Sperradschaltung an der Kraftpresse findet beim Lochen

---

[1]) Fig. 612/4.

von Blechen statt, für welche Arbeit Fig. 56 ein vollständiges Werkzeug darstellt. Der Stempelkopf T aus Maschinenstahl ist an dem Pressenstößel befestigt, trägt mittels Schrauben die Kopfplatte H und dient als Widerlager für die Lochstempel P. Die Kopfplatte H, auch aus Maschinenstahl, ist mit gebohrten Löchern für die Stempel P versehen, deren Köpfe auf der Oberseite der

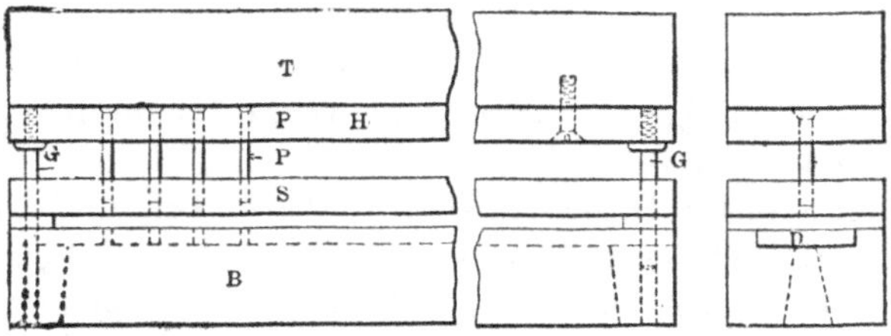

Fig. 56.

Kopfplatte versenkt sind. Nachdem die Stempel auf diese Weise in die Kopfplatte eingesetzt worden sind, wird diese sicher an dem Stempelkopf befestigt, womit der Stempelteil fertig ist. Die Stempel führen sich mit ihrem freien Ende im Abstreifer S, der nach der Kopfplatte gebohrt wird. Der Abstreifer wird auf eine der Presse angepaßte Weise unverrückbar befestigt, so daß er in einer gewissen Höhe über der Schnittplatte steht, damit das Blech gerade durchgehen kann.

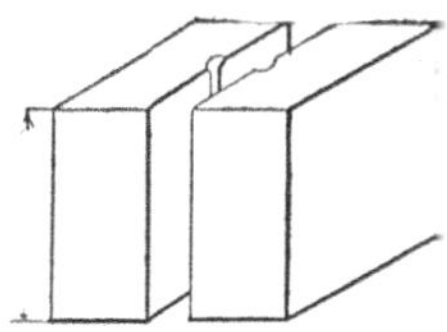

Fig. 57.

Die kleinen Lochstempel werden von Stangen aus Werkzeugstahl auf Länge geschnitten und die Köpfe in einem Nietgesenk gestaucht, Fig. 57. Dies kann von Hand oder in der Presse geschehen, doch ist das letzte vorzuziehen. Nachdem die Stempel mit den Köpfen versehen worden sind, werden sie gehärtet; manche ziehen vor, diese kleinen Werkzeuge im Salzbad zu härten, andere ziehen das Ölbad vor, wobei die Stempel gewöhnlich auf Strohgelb angelassen werden. Hier spricht jedoch die Erfahrung und eine genaue Kenntnis der von diesen Stempeln zu leistenden Arbeit stark mit. Nach dem Härten wird die Schnittfläche des Stempels geschliffen; sind sehr große Löcher zu stanzen, so werden die Stempel mit Vorteil hohl geschliffen, daß die Kanten zuerst schneiden, oder man macht zwei Gruppen der Stempel verschieden hoch, so daß zuerst die erste Hälfte der Stempel in das Blech eintritt und so der Anfangsdruck stark vermindert wird. Die Stempel P müssen ganz stramm in die Kopfplatte H passen, werden aber trotzdem durch den Abstreifer S geführt.

Die Schnittplatte D, deren Länge durch die Stärke der Presse bestimmt ist, mag in unserem Beispiel mit 762 mm Länge angenommen werden und ist aus Werkzeugstahl hergestellt. In manchen Fällen wird sie — nach Angabe der betreffenden Fabrikanten — aus naturhartem Stahl gemacht und in diesem Zustand verwendet. Sie wird durch Schleifen der Oberfläche, wie alle anderen flachen Schnittplatten, schnittfähig gemacht. Man kann die Scherwirkung dadurch unterstützen, daß man die Plattenoberfläche leicht krümmt, so daß sie in der Linie der Löcher am höchsten ist. Man kann dies jedoch nur bei großen Arbeitsstücken machen, und dann noch mit Vorsicht, da in manchen Fällen einzig und allein eine flache Schnittplatte ein Verbiegen des Materials verhindert.

Die Schnittplatte D, die ungefähr 3,2 mm dick und 51 mm breit ist, wird bündig mit der gußeisernen Gesenkplatte B gelegt und durch Schrauben, die in ungefähr 100 mm Teilung stehen, gehalten. Es ist nicht praktisch, irgend eine Regel für die Größe der Schrauben und anderen verwendeten Teile zu geben, da die Bedingungen für die verschiedenen Pressen wechseln. Der soweit fertiggestellte Schnitt kasten wird nun auf der Aufspannplatte der Presse mittels Schrauben, die in der Gesenkplatte sitzen, befestigt. Um ein Anbohren der Gesenkplatte zu sparen, wird sie mit einer eingegossenen Nut versehen, die durch Querrippen verstärkt ist.

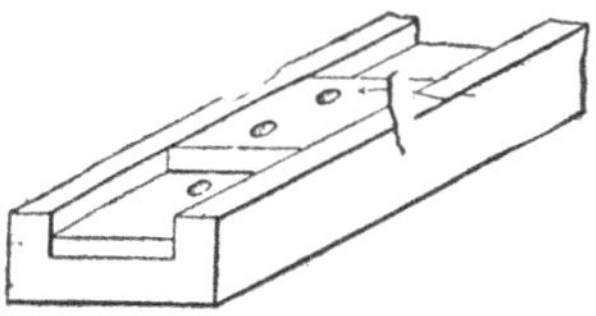

Fig. 58.

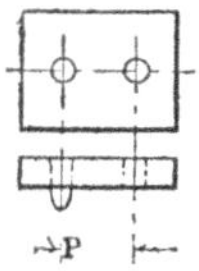

Fig. 59.

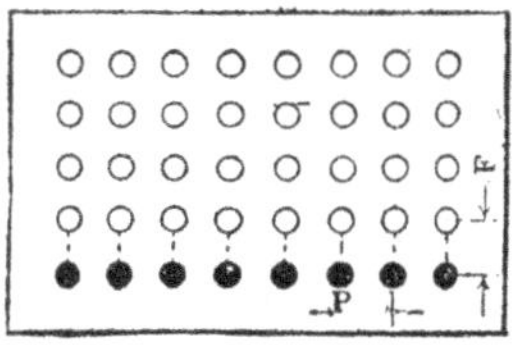

Fig. 60.

Beim Bohren dieser Teile wird der Abstreifer zuerst gebohrt, wobei oft eine Lehre nach Fig. 58 benutzt wird. Die beiden Löcher in dem Gleitschlitten, Fig. 58/59, sind genau nach der verlangten Teilung P gebohrt; es wird nach dem Bohren eines Loches im Abstreifer der Stift dort eingelegt, so daß die Lehre auf das nächste Loch eingestellt ist. Nach dem Abstreifer S wird die Schnittplatte D und die Kopfplatte H ausgeteilt.

Die Werkzeuge werden manchmal mit einer einzigen Reihe Löcher, oft aber auch mit einer doppelten Reihe versehen. Bei der „einfachsten" Teilung, einer einzigen Reihe Löcher und einem Vorschub F gleich der Teilung P, wird ein rechtwinkliges Muster, wie Fig. 60, hergestellt. Mit versetzten Löchern in der Schnittplatte und dem gezeichneten Vorschub F, Fig. 61, wird ein Diagonalmuster erzielt. Mit derselben versetzten Schnittplatte, doch unter Verwendung des halben Vorschubes F, erhält man dieselbe Anordnung der Löcher wie in Fig. 60 bezw. 62. Mit einer einzigen Lochreihe in der Schnittplatte wird ein Diagonalmuster dadurch erzielt, daß man den Vorschub F mit einer Querschaltung abwechseln läßt, Fig. 63. Manche Pressen sind von vornherein mit beiden Bewegungen versehen.

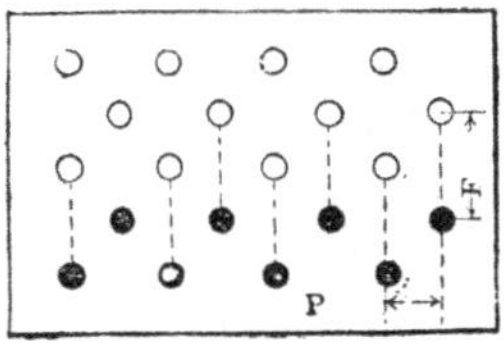

Fig. 61.

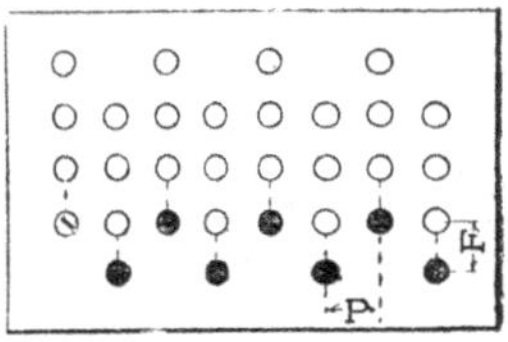

Fig. 62.

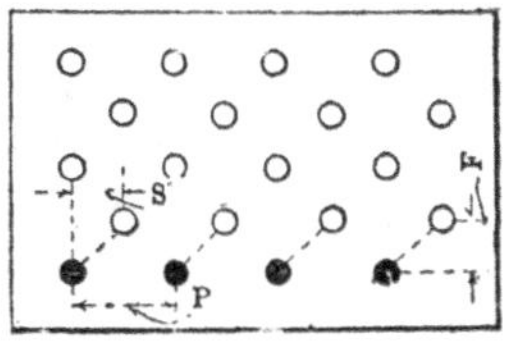

Fig. 63.

Werkzeug für Mehrfachschnitt. Einige der interessantesten Mehrfachschnitte finden sich in der Knopffabrikation, wie das im folgenden beschriebene Werkzeug zur Herstellung eines Teiles eines billigen Metallknopfes. Die Blanketts, die ausgeschnitten werden sollen, sind aus Stahlblech und dienen zum Ausfüllen des Raumes zwischen Kappe und Boden der Knöpfe, wenn diese in der Fußpresse zu einem Ganzen vereinigt werden. Fig. 64 gibt das Schema des betreffenden Werkzeuges.

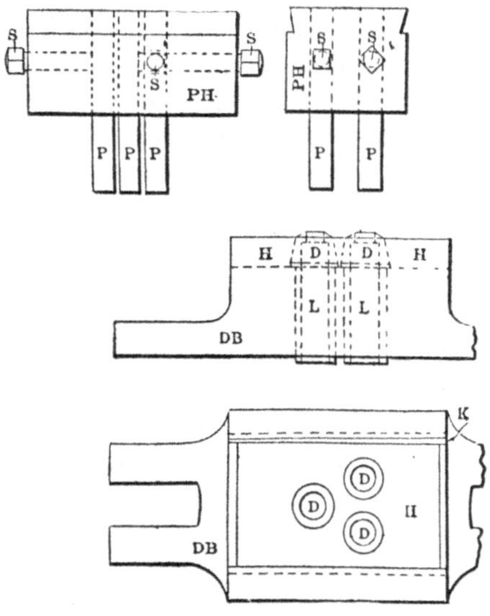

Fig. 64.

Der gußeiserne Stempelkopf PH trägt die drei Stempel P, die durch Kopfschrauben S gehalten werden. Die Ansichten der gußeisernen Froschplatte DB zeigen, wie die Schnitte selbst mittels einer Gesenkplatte H aus Maschinenstahl gehalten werden, während die Gesenkplatte H

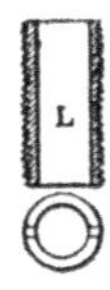

Fig. 65.

selbst im Bett durch einen Schrägkeil K festgespannt wird. L sind die Schrauben, die die Schnitte D von der Unterseite der Froschplatte aus in ihrer Stellung halten. Die gestanzten Blanketts fallen durch die Schnitte und die Halteschrauben L in einen Kasten unter der Presse. Diese Halteschrauben L sind überdies noch im Schnitt Fig. 65 gezeichnet. Sowohl die Schnitte wie die Stempel sind aus dem besten Werkzeugstahl zu machen, in einem Falle wurde steirischer Stahl verwendet. Die Stempel sind gehärtet und angelassen, die Schnitte gehärtet, angelassen und geschliffen.

Die Schnitte werden an der Oberseite konvex gemacht, so daß sie viel leichter geschliffen werden können, wenn sie stumpf werden. Sie sind so angeordnet, daß von einem gegebenen Metallstreifen bei größter Blankettzahl der geringste Abfall übrig bleibt. Die Unterseite, bzw. die Auflagerfläche der Schnitte muß nach dem Härten genau winklig geschliffen werden. Die Presse, in der das Werkzeug zur Anwendung kam, war mit einer selbsttätigen Zuführung mit mehrfachen Sperrklinken ausgestattet, so daß die Vorschublänge leicht um ein kleines Stück vermehrt oder vermindert werden konnte. Die verschiedenen Sperrklinken wurden durch einen Hebel gleichzeitig in und außer Eingriff gebracht.

Herstellung kleiner Löcher in Blanketts aus Werkzeugstahl. Zum genauen Nuten, Stanzen und Lochen kleiner Teile aus Blech sind Werkzeuge von sorgfältiger und zwangläufig wirkender Bauart erforderlich. Häufig verlangt die Konstruktion derselben die Vereinigung vieler, sorgfältig durchdachter Einrichtungen, um eine genaue Arbeit zu sichern und wirklich austauschbare Teile zu erhalten. Wenn die Nuten oder Löcher sehr klein sind, müssen die Stempel sehr sorgfältig befestigt und geführt werden, damit man das immer vorhandene Bestreben derselben, die Schnittplatten durchzuscheren, unter dem Druck abzubiegen oder selbst beim Durchgang durch das Arbeitsstück abzubrechen, verhindert. Das Werkzeug, das in Fig. 68 / 69 wiedergegeben ist, dient zur Herstellung von fünf Löchern von 1,25 mm Durchmesser in Werkzeugstahl von 1,58 mm Dicke. Man wird wohl zugeben, daß diese Arbeit ungewöhnliche Schwierigkeiten bietet, da der besonders kleine Durchmesser der Löcher im Vergleich zur Qualität und Dicke des verarbeiteten Materials den Gebrauch gewöhnlicher Werkzeuge einfach ausschließt. Fig. 66 und 67 stellen die zwei Teile dar, in welche mit diesen Werkzeugen, wie die schwarzen Punkte angeben, fünf Löcher zu stanzen sind. Der Teil in Fig. 67 ist ein kleines Stück Werkzeugstahl, schwalbenschwanzförmig zugeschnitten, das mit dem Teil in Fig. 66 zusammengenietet einen Schlitten bildet, der in einer allgemein ver-

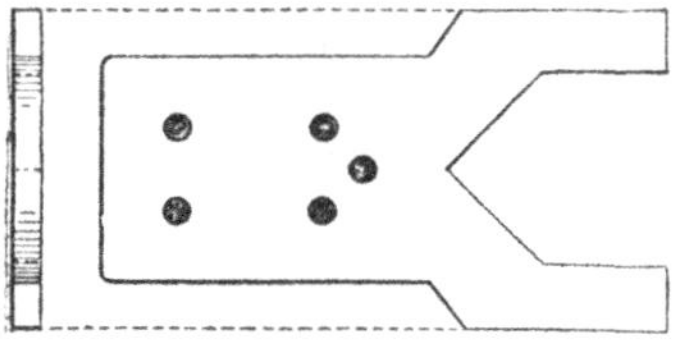

Fig. 66.

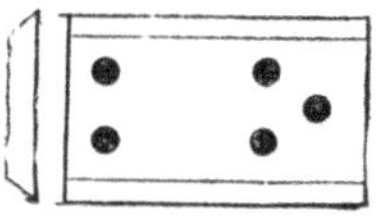

Fig. 67.

breiteten Druckereimaschine verwendet wird. Nachdem beide Teile miteinander vernietet worden sind, wird das Ganze gehärtet, angelassen und geschliffen, wobei der Schwalbenschwanz auf genaues Maß geschabt wird, da die zugelassene Abweichung an jeder Stelle nur 0,012 mm betragen darf.

Fig. 68 zeigt den Stempel, bei dessen Konstruktion hauptsächlich darauf Rücksicht genommen ist, das Bestreben solch kleiner Stempel, leicht abzubiegen, abzubrechen oder durch die Schnittplatte zu scheren, hinten anzuhalten, und zwar in folgender Weise:

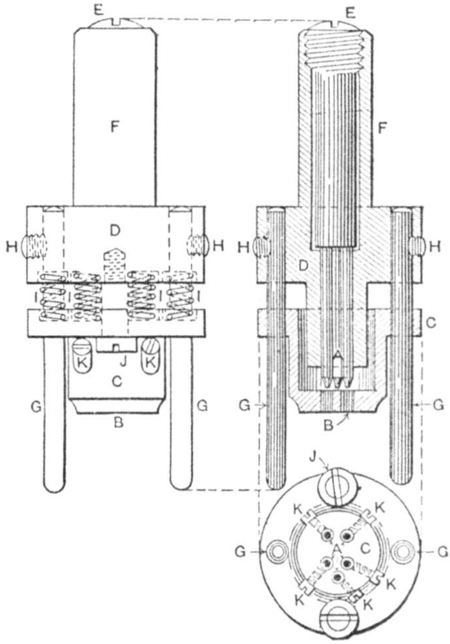

Fig. 68.

D ist der Stempelkopf aus Maschinenstahl, in dem die fünf Stempel A eingesetzt und befestigt sind. E die einstellbare Gegenschraube, gegen welche sich die Stempel stützen, und G zwei Führungsstifte. C ist eine Vereinigung von Führungsplatte und Abstreifer mit fünf Löchern in der Unterfläche B, in welche die Stempel gut passen. Fünf Zylinderkopfschrauben K halten die Stempel in ihren Sitzen. Die zwei Führungsbolzen G sind in genau ausgeriebenen Löchern durch zwei Stellschrauben befestigt. Acht kleine Wurstfedern, von denen nur vier in der Zeichnung erscheinen, sind in ausgebohrten Löchern des Stempelkopfes D und der Führungsplatte C gehalten und pressen letztere während des Lochens hart gegen das Blankett; sie dienen auch zum Abstreifen des Blanketts nach jedem Hub. Wie deutlich ersichtlich, ist die Konstruktion so gewählt worden, daß eine vollkommene Führung mit dem Unterteil und eine unnachgiebige, zwangläufig wirkende Unterstützung der Stempel bis zu dem Punkt erzielt wird, an dem sie in das zu lochende Material eintreten. Die Kopfschrauben J begrenzen die Abwärtsbewegung der Abstreif- und Führungsplatte C.

Der Unterteil, die Schnittplatte, ist in Fig. 69 im Grundriß und Vorderansicht abgebildet.

L ist eine Grundplatte aus Stahl, R die Schnittplatte mit den fünf Lochschnitten, die durch die fünf schwarzen Löcher angedeutet sind. Die Art und Weise, wie diese Schnittplatte in der Grundplatte befestigt und gesichert ist, so daß sie unverrückbar befestigt bleibt, zeigt der Grundriß deutlich. In die eine Seite der Schnittplatte R ist eine Keilnut eingearbeitet, in welche der Bolzen S mit seinem Keilende

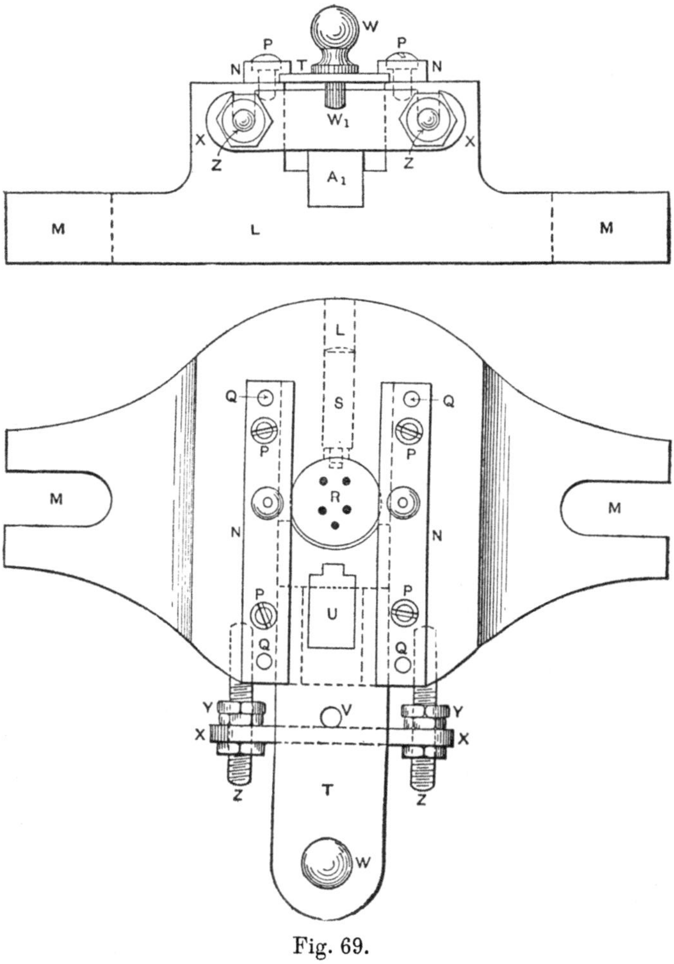

Fig. 69.

paßt. In der Zeichnung nicht ersichtlich ist ein Loch in der Grundplatte unterhalb der Schnittplatte, das auf dem größten Teil seiner Länge mit Gewinde für eine hohle Stellschraube versehen ist, welche das Auflager für R abgibt. Wenn es nötig wird, die Schnittplatte nachzuschleifen, wird sie mit dieser Schraube heraufgedrückt, bis sie wieder bündig mit der Grundplattenoberfläche steht. In der Fig. 69

sind zwei Führungsschienen aus Werkzeugstahl N ersichtlich, in denen der Schlitten T, der das Blankett auf der Werkzeugoberfläche einstellt, gleitet.

Diese Führungen besitzen ausgeriebene Löcher O für die Führungsstifte G und sind auf der Grundplatte mit je zwei Schrauben P und zwei Prisonstiften Q befestigt und gesichert. V, X, Y und Z sind die Bestandteile eines einstellbaren Anschlages, welcher die Bewegung des Schlittens T einstellt und dauernd sichert. W ist ein kleiner Knopf, den der Arbeiter benützt, um den Schlitten während des Lochens vor- und zurückzuschieben. V ist ein Anschlag, der eine zu große Auswärtsbewegung des Schlittens aus den Führungen hintenanhält, während W_1 eine zu große Einwärtsbewegung desselben über die richtige Stellung der Blanketts hinaus verhindert.

Wenn das Werkzeug arbeitet, wird der Stempel so eingestellt, daß die Führungsstifte G in den Löchern O der Führungen stehen. Das Werkzeug wird in seiner Stellung in der Presse befestigt und deren Hub so geregelt, daß die Führungsstifte die Führungen beim Aufgang des Pressenstößels nicht verlassen. Der Arbeiter läßt nun eines der kleinen Blanketts, Fig. 67, in den Sitz U des Schlittens gleiten, stößt den Schlitten hinein, bis der Stift W_1 gegen X stößt, und setzt die Presse in Tätigkeit. Beim Aufwärtsgang des Stößels wird das Stück von den Stempeln durch C abgestreift und bleibt auf U liegen. Wenn der Arbeiter nun den Schlitten zurückzieht, fällt das Blankett in die Ausfallrinne A_1, Fig. 69.

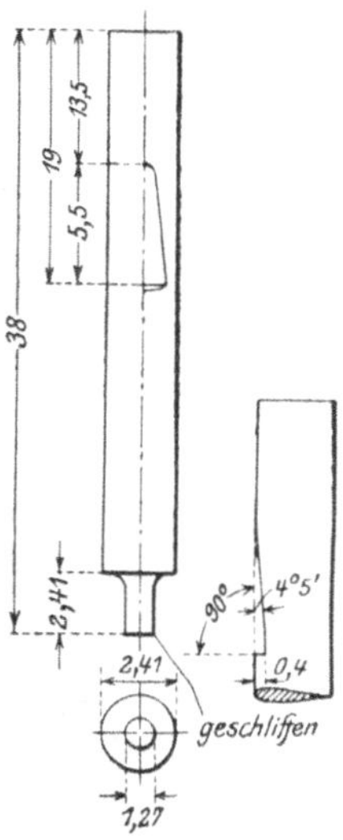

Fig. 70.

Fig. 70 zeigt die Einzelheiten des Stempels. Diese werden in Sätzen von 500 aus Bohrerstahl gemacht, der genau den Durchmesser des Stempelzapfens hat. Im Automaten wird dann das Ende des Stempels auf genaues Maß fertig gemacht.

In einem Werk, das diese Werkzeuge verwendet, werden in einem Jahr über 100 000 Blanketts beider Formen, Fig. 66 und 67, fertiggestellt. Um das größere Blankett Fig. 66 mit den fünf Löchern zu versehen, ist es nur notwendig, einen anderen Schlitten an Stelle des oben beschriebenen einzuführen.

Die an dem Stempel in Fig. 68 sichtbare äußere Parallelführung kann auch in anderer Form gelöst werden, welche die besonderen Schrauben J unnötig macht. In Fig. 71 ist ein Werkzeug mit einer

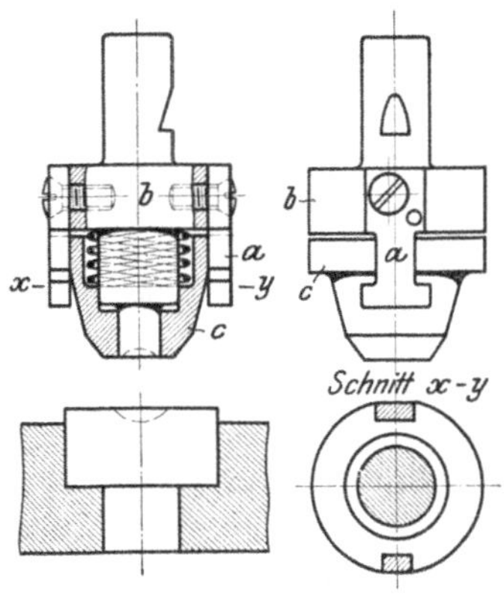

Fig. 71.

derartigen äußeren Parallelführung gegeben[1]). Die Führungsschienen a, welche hier die Stelle der zylindrischen Führungsstangen G in Fig. 68 übernehmen, werden am Stempel b mit Schrauben und Prisonstiften befestigt. Der vereinigte Niederhalter und Abstreifer c wird durch die innenliegende Spiralfeder nach abwärts gedrückt und durch die Fortsätze von a, die sich gegen den äußeren zylindrischen Kopf von c legen, in seiner Abwärtsbewegung begrenzt.

Eine ähnliche äußere Führung ist an dem Werkzeug zur Herstellung der Messer für einen Momenthebelschalter, Fig. 575, gezeichnet.

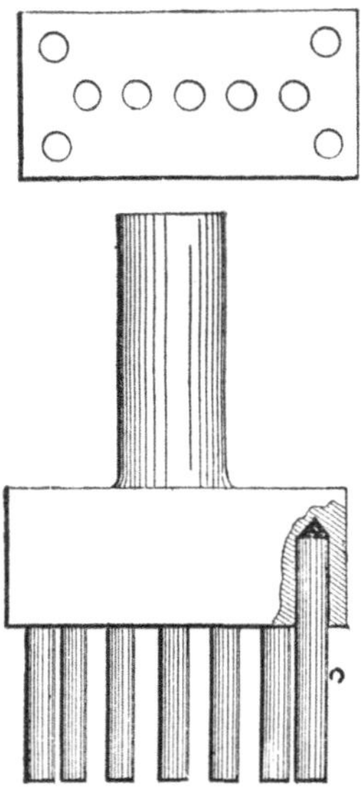

Fig. 72.

Die Konstruktion eines mehrfachen Stempels für dünne Metalle. Die im folgenden beschrie bene Methode ist von Wert, wenn es sich darum handelt, billig Mehrfachstempel, Fig. 72, zum Lochen dünner Metalle herzustellen.

Wenn sich die Notwendigkeit herausstellt, einen Stempel zur Herstellung einer Reihe kleiner Löcher, z. B. von ungefähr 4,7 mm Durchmesser, in derartigem Metall herzustellen, so verwendet man am billigsten Stücke von Stubs-Stahldraht, die in eine gußeiserne Kopfplatte eingesetzt werden.

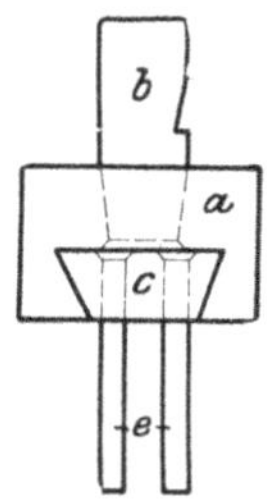

Fig. 73.

Die Art und Weise, wie diese Stempel eingesetzt werden, ist einigermaßen überraschend, doch arbeitet sie gut und zufriedenstellend, wenn es sich um Stempel für dünnes Material handelt. Die Löcher werden nach der Schnittplatte in die Kopfplatte gebohrt, und zwar ungefähr 0,4 mm größer als der zu den Stempeln verwendete Draht. Diese Löcher werden dann mit Zinnlot gefüllt und die Platte im Schraubstock geebnet. Dann werden die Stempel an den einzusetzenden Enden

[1]) W. T. 1909, S. 582, Fig. 10.

gut verzinnt und in die Schnittplatte gesteckt, wo sie genau passen. Nachdem man die Kopfplatte erwärmt hat, bis das Lot flüssig geworden ist, setzt man die Schnittplatte mit den Stempeln über dieselbe und schlägt die Stempel bis auf den Grund in die Löcher. Nach dem Erkalten ist das Werkzeug fertig. Nach bisherigen Erfahrungen sind derart befestigte Stempel noch nicht lose geworden.

Eine andere Befestigung für mehrere kleine Lochstempel ist aus Fig. 73[1]) zu ersehen, die gerade das Gegenteil der eben besprochenen Konstruktion ist. Die Stempel c werden in einer mit Schwalbenschwanz versehenen Kopfplatte vernietet, die erst wieder in dem Stempelkopf gehalten wird. Auch der Stempelzapfen ist gesondert in die Kopfplatte des Stempelkopfes eingesetzt und vernietet. Diese beiden Konstruktionen bilden Gegensätze in bezug auf billige Herstellung und billigen Ersatz der Einzelteile.

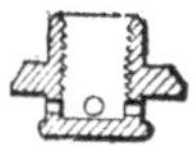

Fig. 74.

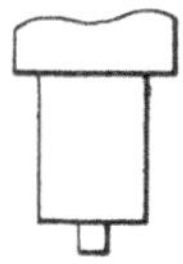

Fig. 75.

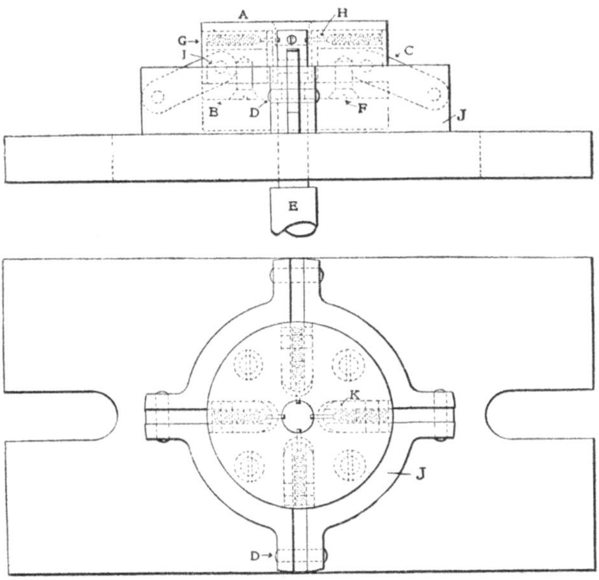

Fig. 76.

Mehrfachlochwerkzeug. Die Zeichnungen Fig. 75/76 zeigen ein Lochwerkzeug, mit dem man in ein Stück, das in Fig. 74 vergrößert gezeichnet

[1]) W. T. 1909, S. 580.

ist, vier Löcher unter rechtem Winkel zueinander stanzen kann. J ist die Gesenkplatte, die am äußeren Umfang ausgefräst ist, damit man die Löcher für die Stifte D leicht anzeichnen und bohren kann; an diesen Stiften hängen die Laschen C. Die Innenseite ist zur Aufnahme der Stücke A und B, Fig. 76/77, ausgedreht, und zwar tief genug, daß die Schlitten K genug Weg machen können, um die Stempel H durch das Blankett durchzutreiben. Das Stück A ist wiederum zur Aufnahme der Schlitten oder Stempelhalter K ausgefräst, die gut in A eingepaßt sind und durch die Platte B in ihrer Stellung gehalten werden. Die Schlitten erhalten am Grunde eine Ausfräsung für die Laschen C, die in ihnen durch die Stifte I angelenkt sind. Die Schlitten K sind auch, wie in der Schnittzeichnung, Fig. 77, ersichtlich ist, zur Aufnahme der Stempel H, welche durch die Schrauben G gehalten werden, ausgebohrt. Die Stempel H sind auch in den kleinen Löchern in A geführt und gehen am Rückweg aus diesen nie ganz heraus. Fig. 75 zeigt den Niederhalter, der an den Preßstößel angepaßt wird, mit einem abgedrehten Ende, welches leicht in das konische Loch des zu verarbeitenden Blanketts paßt und nur so lang ist, daß die kleinen Lochstempel, sobald sie durch das Material getreten sind, vorbeigehen können.

Das Blankett wird in das Loch auf der Oberfläche von A gelegt, das gerade tief genug ist, damit der Niederhalter, Fig. 75, so hart wie möglich auf das Stück A drücken kann. Die Platte B ist durch vier Schrauben F an A befestigt und enthält eingefräste Aussparungen, in welche die Laschen C bei der Aufwärtsbewegung von A eintreten. Wenn das Blankett richtig zum Lochen eingelegt ist, wird die Presse nach abwärts gesteuert, wodurch das Stück A niedergedrückt wird, so daß die Schlitten K vorwärtsgehen müssen, bis die Lochstempel das Blankett gelocht haben. Durch den Bolzen E, der durch einen nicht sichtbaren Hebel gesteuert wird, wird der Stempel A wieder in seine Anfangsstellung zurückgebracht.

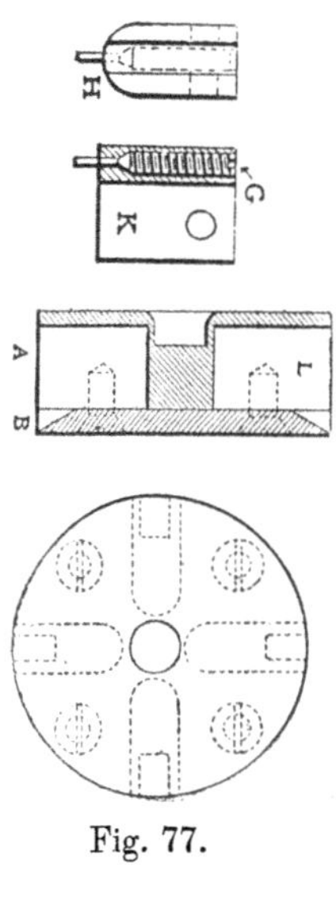

Fig. 77.

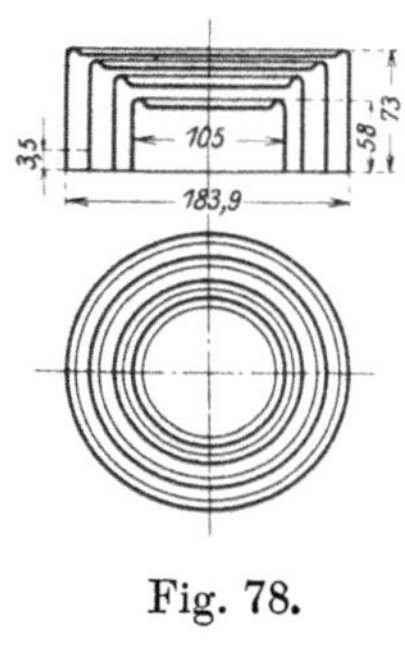

Fig. 78.

Vierfachlochwerkzeug für verschiedene Kappendurchmesser. Dieses Werkzeug, das aus der deutschen Praxis stammt, vollführt eine gleiche Arbeit wie das eben besprochene, jedoch mit ganz verschiedenen mechanischen Mitteln.

Fig. 78 zeigt die verschiedenen Kappengrößen, die auf diesem Werkzeug gelocht werden können, Fig. 79 die verschiedenen Ansichten des Werkzeuges und Fig. 80 eine Zeichnung des eigentlichen Lochwerkzeuges in größerem Maßstab.

Die Arbeitsweise ist folgende: Durch einen Schalthebel-Antrieb a wird die senkrechte Welle b in schwingende Bewegung versetzt, die sich auf die mit ihr verkeilte Nockenscheibe o überträgt. Diese hat auf ihrem Umfang vier Erhöhungen, gegen welche sich die Enden der Stempelführungen d, Fig. 80, durch die Federn e angedrückt, ständig anlegen, so daß die Stempel bei jedem Hube des Schaltwerkes einmal vor- und rückwärts gehen.

Die zu lochende Kappe m wird nur auf die — der Befestigung und Anordnung des Antriebes wegen — als Winkel ausgebildete Grundplatte zwischen die Teile k und i, wie die Zeichnung zeigt, eingelegt und die Presse eingeschaltet, wodurch die vier Löcher in der oben beschriebenen Weise hergestellt werden.

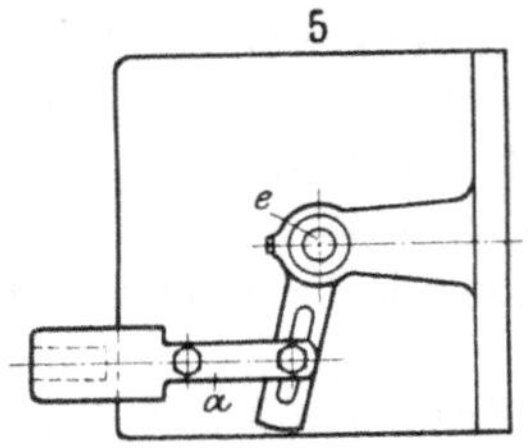

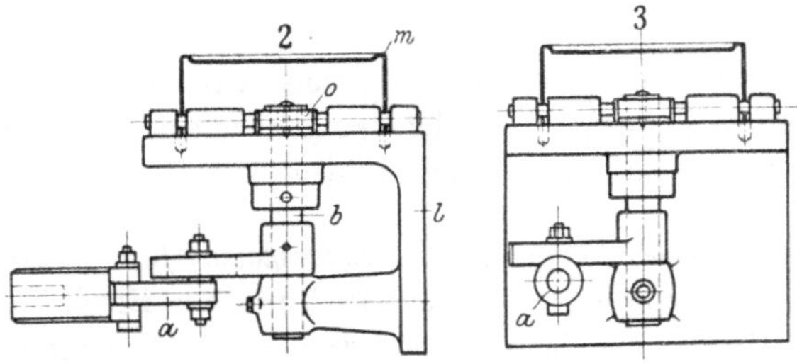

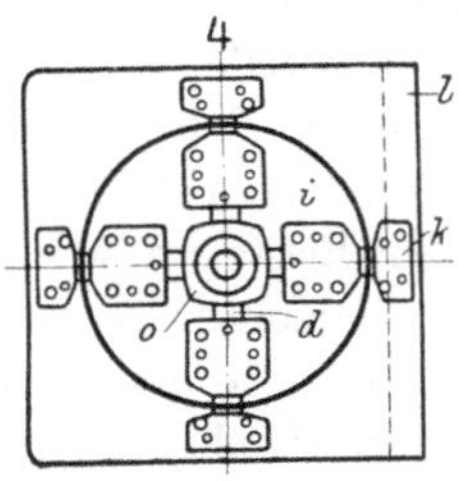

Fig. 79.

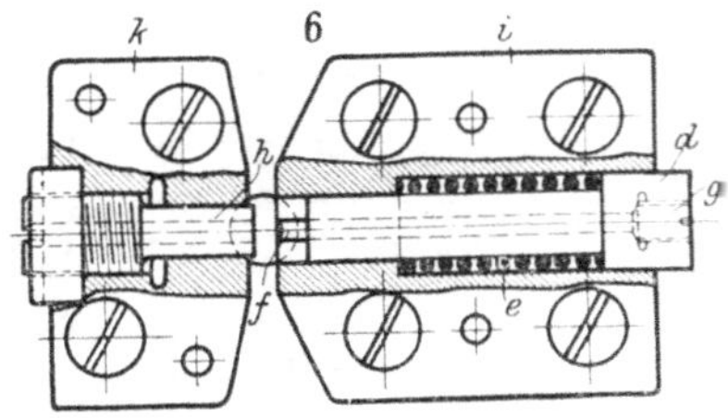

Fig. 80.

Die eigentlichen Lochstempel und Platten sind in gußeiserne Führungsblöcke k und i, die mit Schrauben und Prisonstiften auf dem Winkel l befestigt werden, eingesetzt. Die Lochstempel aus Stubsstahl sind in zylindrischen Führungsbolzen d, die der Länge nach durchbohrt sind, gut eingepaßt und werden darin durch einen kleinen angenieteten Kopf einerseits und die dagegen drückende Schlitzschraube g andrerseits gehalten. Die Führungen sind außen abgesetzt und dementsprechend auch die Bohrungen in den Führungsblöcken i, um Raum für die Rückzugfeder e zu geben. Die Lochplatte h ist ebenfalls ein zylindrischer Bolzen aus Werkzeugstahl, welcher der Länge nach durchbohrt und am äußeren Ende mit aufgeschnittenem Gewinde versehen ist; an dem äußeren Ende sitzt eine Gegenmutter, die beim Nachschleifen der Lochplatte eine Nachstellung gestattet.

Es ist selbstverständlich, daß man bei dieser Anordnung nicht an vier Löcher gebunden ist, auch diese nicht unter rechtem Winkel stehen müssen. Eine obere Grenze wird durch die Größe des Umfanges der Nockenscheibe gegeben, da immer noch genügend Winkelweg vorhanden sein muß, um die Stempel durch das Material zu treiben, andrerseits die Schräge der Erhöhungen ein gewisses Maß nicht überschreiten darf und der Durchmesser der Nockenscheibe von dem Durchmesser der zu lochenden Kappen abhängt.

Soll das Werkzeug für eine andere Kappengröße Verwendung finden, so werden die Stempelteile i ausgewechselt.

Von besonderem Interesse ihres konstruktiven Aufbaues wegen sind die in diese Gruppe gehörigen Werkzeuge mit mehrteiliger Schnittplatte, die im folgenden besprochen werden.

Mehrfachschnitt mit zusammengesetzter Schnittplatte für genaue Arbeiten. Um ausnehmend genaue, austauschbare Teile aus Blech herzustellen, bei denen z. B. die zugelassene Abweichung nicht mehr als 0,012 mm betragen darf, muß man Werkzeuge mit zusammengesetzten Schnittplatten anwenden.

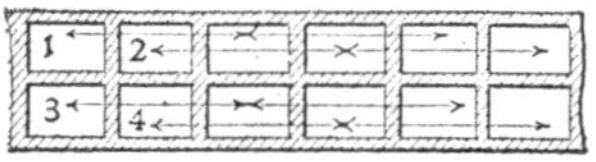

Fig. 81.

Fig. 81 zeigt den Abfallstreifen. Das Blech kommt in langen Streifen von genügender Breite an die Presse, so daß man die Blanketts in zwei Reihen stanzen kann. Der Streifen geht jedoch viermal durch das Werkzeug. Bei dem ersten Durchgang werden die auf der Linie 1 liegenden Blanketts gestanzt, beim zweiten Weg die auf der Linie 2 usw.

Die Werkzeuge, welche in Fig. 82/83 abgebildet sind, dienen zum Stanzen eines kleinen Blanketts aus Werkzeugstahl von 1,6 mm Dicke, das nachher Teil eines ganz besonders genauen Präzisionsartikels ist. Diese Werkzeuge sind sorgfältig durchgebildet und stanzen gleichzeitig drei Blanketts.

Fig. 83 gibt Grundriß, Seitenansicht und einen Schnitt des gesamten Unterteiles. A ist die gußeiserne Grundplatte, in welcher die Teile der Schnittplatte B mittels Schwalbenschwanzes und Zylinderkopfschrauben D und der Prisonstifte L befestigt sind. Der gestrichelte Linienzug E—E zeigt die Stoßlinie der Schnittplattenhälften. Die Art und Weise, wie die gegenseitige Verriegelung der Teile längs E—E durchgebildet ist, sichert eine steife und starke Schnittplatte, die mit den Stempeln dauernd vollständig ausgerichtet bleibt und während ihrer Lebensdauer jeder Möglichkeit eines Durchscherens vorbeugt. Die beiden Teile B sind genaue Wiederholungen, wodurch es leicht und bequem wird, die Teile herzustellen, zu schleifen und, wenn nötig, auszubessern. In dem Grundriß und der Seitenansicht ist auch der Anschlag zur Regelung des Materialvorschubes nach jedem Hub zu sehen. H ist ein Block,

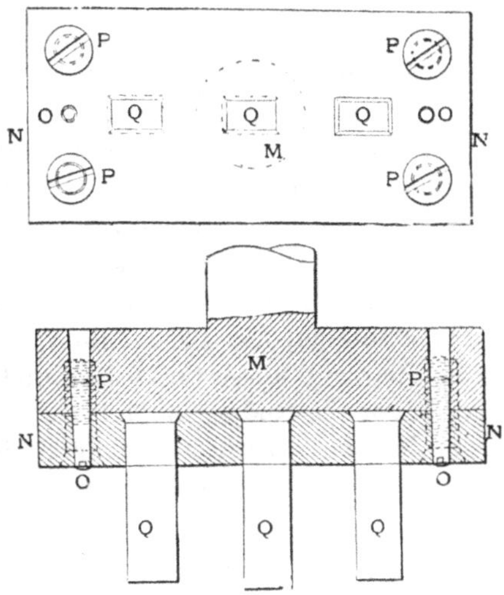

Fig. 82.

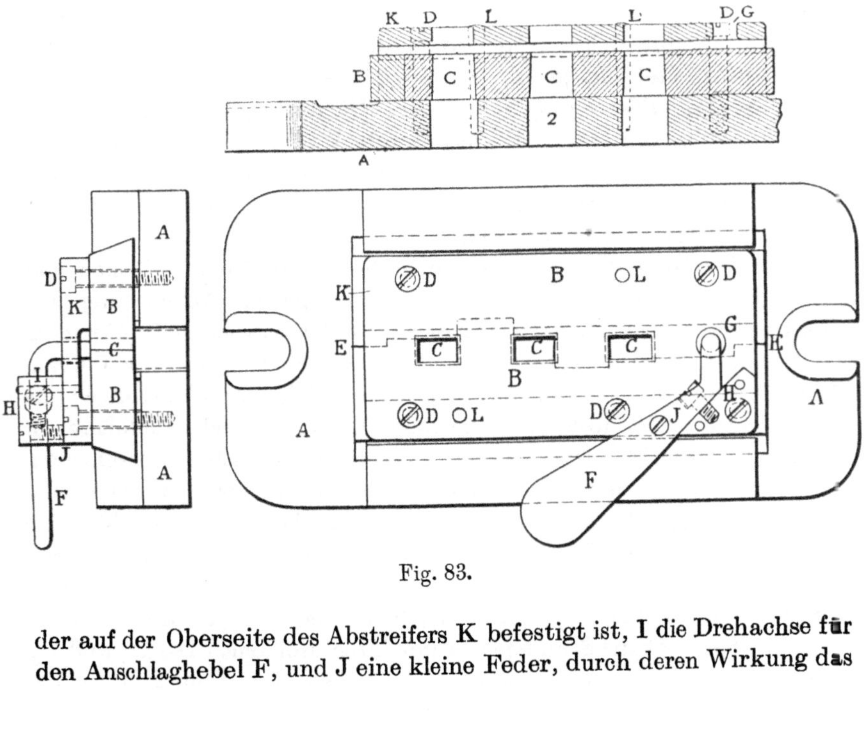

Fig. 83.

der auf der Oberseite des Abstreifers K befestigt ist, I die Drehachse für den Anschlaghebel F, und J eine kleine Feder, durch deren Wirkung das

Materialende bei G auf die Oberfläche der Schnittplatte niedergedrückt wird. Bei Verwendung dieses Anschlages ist es nicht nötig, beim Vorschub das Material nach jedem Hub anzuheben, der Arbeiter hat nur das verlängerte Ende von F niederzudrücken und, nachdem er das Material vorgeschoben hat, loszulassen, wobei der Anschlag das Material in der neuen Arbeitsstellung sichert.

Fig. 82 ist ein Längsschnitt und ein Grundriß des Stempels. Dieser besteht aus dem gewöhnlichen, gußeisernen Stempelkopf M, einer stählernen Kopfplatte N, die durch zwei Prisonstifte O gesichert und durch vier versenkte Kopfschrauben P befestigt ist, und schließlich aus den Stempeln Q aus Werkzeugstahl. Der mittlere Stempel ist um etwas mehr als eine Metalldicke länger als die beiden anderen, so daß die Wirkung auf das Blech allmählich eintritt, da der mittlere Stempel sein Blankett ausgestanzt hat und in die Schnittplatte eingetreten ist, bevor die beiden anderen mit dem Material in Berührung kommen, so daß das Blech in seiner Stellung tatsächlich gesichert und ein Verziehen oder Verschieben ausgeschlossen ist.

Werkzeug mit zweifach zusammengesetzter Schnittplatte. Man hört häufig selbst von erstklassigen Werkzeugmachern, daß zusammengesetzte Schnitte sich nicht bezahlt machen. Nichtsdestoweniger ist es oft von größerem Vorteil, die Schnitte in Teilen herzustellen, als auf andere Weise, wenn die Umrißlinien oder ihre Winkel schwer bearbeitet und gefeilt werden können. Weiter ist der Ersatz eines bei der Bearbeitung oder beim Härten zugrunde gegangenen Werkzeugteils leichter. Zusammengesetzte Werkzeuge werden in vielen Werkstätten an Werkzeugen für doppeltwirkende Ziehpressen verwendet, auch in Verbundwerkzeugen in Verbindung mit der Subpresse, da beide sich leicht dazu einrichten lassen.

Bei der Herstellung von geometrischen Lehrmitteln, Flächen und Körpern, kommt ein Satz solcher Werkzeuge zur Anwendung, die Interesse erregen können. Die Körper werden aus gut getrocknetem, hartem Holz und die Flächen aus Lederpappe von ungefähr 0,6 mm Dicke gemacht. Die im folgenden beschriebenen Werkzeuge (die Zeichnungen sind nicht im gleichen Maßstab) sind vielleicht im einzelnen ziemlich von anderen derartigen Schnitten verschieden.

Fig. 84 zeigt den Schnittstempel, der von der Schnittfläche nach rückwärts ungefähr 1^0 Abschrägung hat, so daß ein kleiner Teil des Materials in der Spindelpresse abgeschert werden kann, wenn man den Stempel an die gehärtete Schnittplatte genau anpaßt. Wenn es sich um das Stanzen von Pappe handelt, bleibt der Stempel immer weich.

Fig. 85 zeigt die Schnittplatte mit ihren drei Teilen D_1, D_2 und D_3. Die vier unabhängigen Positionskeile haben ihre Keilnuten in der

Mitte zwischen Ober- und Unterfläche der Teile D_1 und D_2 eingefräst, ebenso in dem mittleren Teil D_3, wie durch die gestrichelten Linien ersichtlich gemacht ist.

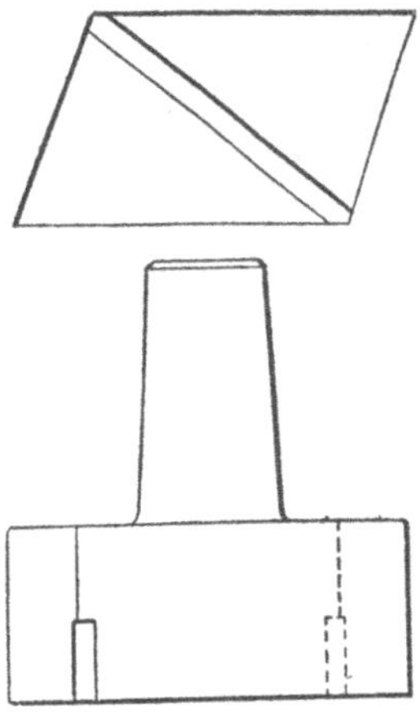

Fig. 84.

Über der Schnittplatte sind die Führungsleisten mit dem darüber befindlichen Abstreifer B angebracht. Sie sind aus Material von 4,7 mm Dicke, der Abstreifer aus kalt gewalztem Material von 2,3 mm Dicke gemacht. Beide werden durch ¼″ Sechskantkopfschrauben gehalten. Die Schnittplatte hat von der Schnittkante nach abwärts eine Schräge von ½°.

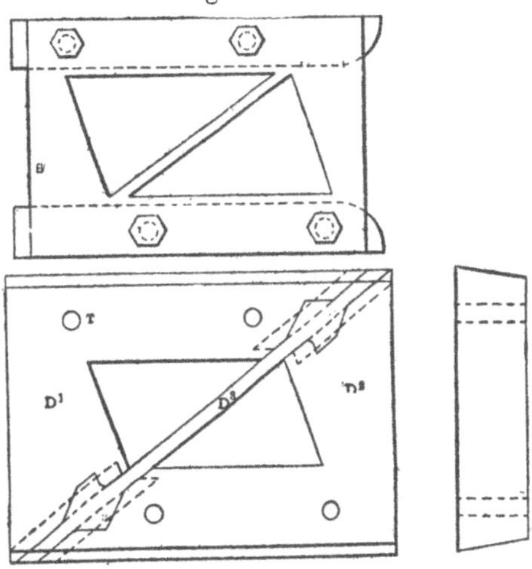

Fig. 85.

Bei der ersten Ausführung war vielleicht der merkwürdigste Teil der ganzen Arbeit, daß die drei Teile der Schnittplatte aus drei verschiedenen Arten Werkzeugstahl gemacht wurden. Dies geschah, um die wirkliche Brauchbarkeit der in Frage stehenden Materialien zu prüfen. Der Stahl für D_3 war ein erprobtes Material, während für die beiden anderen Teile Versuchsmaterial zur Verwendung kam. Die rückwärts offne, einfach wirkende Presse, in der diese Werkzeuge verwendet wurden, hatte eine Sperradschaltung mit einer Reihe von Sperrklinken, so daß man den Vorschub in engen Grenzen ändern konnte.

Werkzeug mit sechsteiliger Schnittplatte für Blanketts aus Werkzeugstahl. Das im folgenden abgebildete Werkzeug ist zur Herstellung des Schlittens für eine genaue Paßarbeit entworfen worden. Da das Material der Schlitten Werkzeugstahl ist und diese Teile später genau bearbeitet werden sollten, mußte man Werkzeuge von genauer und dauerhafter Bauart verwenden.

Aus der Zeichnung, Fig. 86, die einen Teil des Abfallstreifens und die Durchgänge des Streifens selbst angibt, kann die Stellung der gleichzeitig gestanzten Blanketts im Streifen und zueinander leicht eingesehen werden. Es sind nämlich AA in einem Preßhub zu stanzen, BB in dem nächsten, dann CC usw., bis der ganze Streifen aufgearbeitet worden ist. Da alle diese Blanketts in

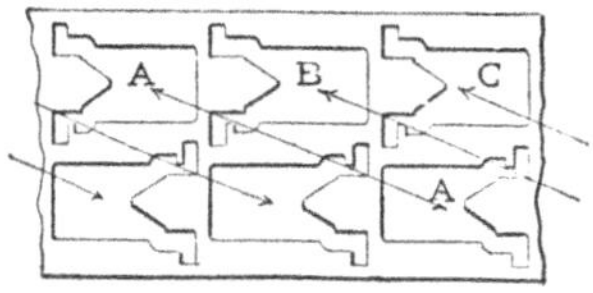

Fig. 86.

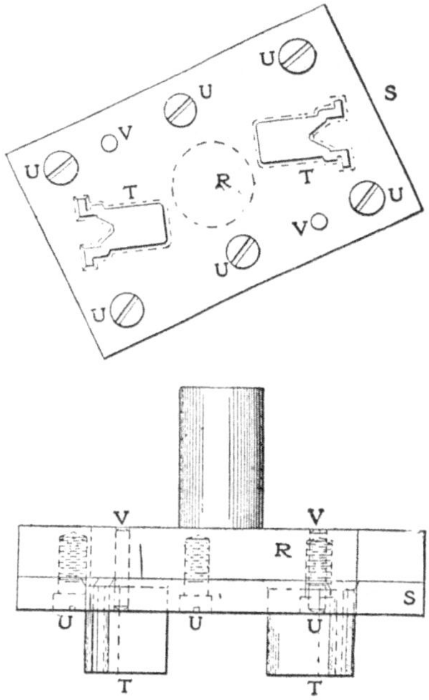

Fig. 87.

ihren Abmessungen ganz ungewöhnlich genau werden müssen und das Material Werkzeugstahl ist, kommt ein Werkzeug mit geteilter Schnittplatte zur Anwendung. Die Art und Weise, wie diese Teile eingestellt und verriegelt sind, die Leichtigkeit, mit der sie ausgearbeitet werden können, und die verhältnismäßig billige Erneuerung derselben wird den Praktiker bestechen und zweifellos die Anwendung ähnlicher Arbeitsweisen für solche genauen Arbeiten anregen.

In dem Schnittkasten, Fig. 88, ist G die Grundplatte aus Gußeisen, mit Schlitzen für die Befestigungsschrauben, in der bei

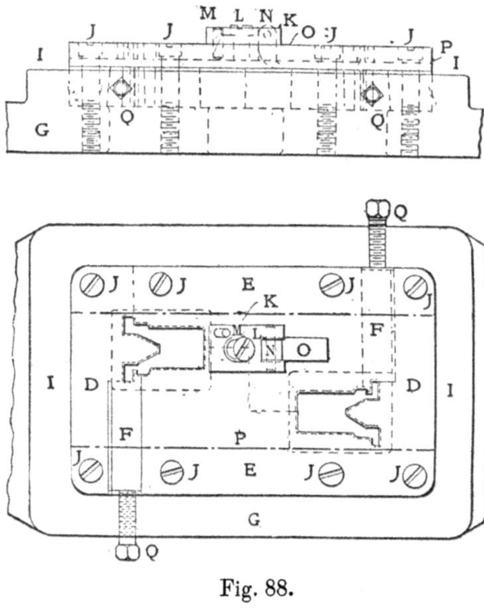

Fig. 88.

I eine Aussparung zur Aufnahme der Teile der Schnittplatte sehr genau ausgearbeitet ist. Die eigentliche Schnittplatte besteht aus sechs Teilen, D, E, F, die in der Aussparung I der Grundplatte durch sechs Zylinderkopfschrauben J und zwei Stellschrauben Q befestigt werden. P ist eine Stahlplatte, in der Abstreif- und Führungsplatte vereint sind, wie durch die strichpunktierten Linien ersichtlich wird. Ein längliches Loch O für den Anschlagfinger N ist in dem Abstreifer vorgesehen. K, L und M sind der Tragarm, Prisonstift und die Befestigungsschraube für den Anschlag. Bei Verwendung eines Anschlages dieser Art wird der Streifen nach jedem Hub vorwärts gestoßen und dann scharf zurückgezogen, so daß das Vorderende des letzten ausgestanzten Loches sich an die gerade Fläche des Anschlagfingers N anlegt. Die Teile F der Schnittplatte werden mit einer schrägen Fläche versehen, so daß sie nach dem Zusammensetzen und Befestigen der übrigen Teile innerhalb des Sitzes I in der Grundplatte mittels der Schrauben J sich unter dem Arbeitsdruck nicht herausheben können.

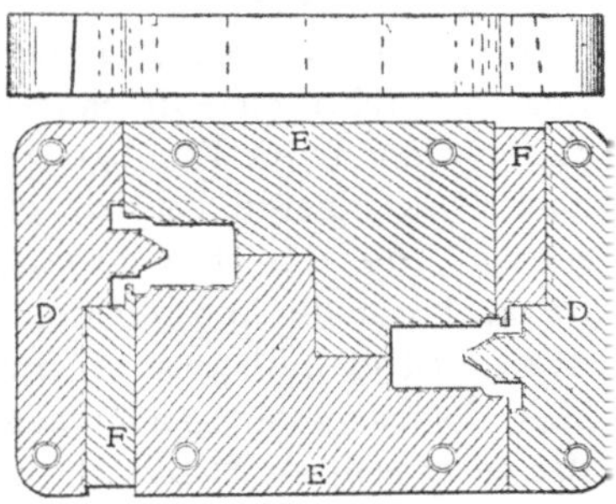

Fig. 89.

Der Stempel, Fig. 87, besteht aus einem gußeisernen Stempelkopf R, einer Kopfplatte S aus Maschinenstahl, sechs Zylinderkopfschrauben U zwei Prisonstiften V und den zwei Schnittstempeln T aus Werkzeugstahl, die geschliffen, gehärtet und auf Strohgelb angelassen werden

Bei der Herstellung der Schnittplattenteile D, E und F, Fig. 89, ist genaue Arbeit und sorgfältiges Einhalten der Maße nötig, z. B. es muß überall an der Schnittkante entlang ein Hinterschliff von 1° gegeben werden; dazu gehört das Schleifen und Einpassen der Verriegelungen und Paßflächen nach dem Härten usw. Schließlich wird man bemerken, daß die Teile D, ebenso wie E und F, genaue Wiederholungen sind. Dadurch, daß das Zusammensetzen der Schnittplatte in dieser Weise erfolgt, wird das Bearbeiten, Schleifen, Polieren und Einpassen der Teile bedeutend vereinfacht und beschleunigt; auch kann die Erneuerung von durchgescherten, gebrochenen oder abgesplitterten Teilen ohne besondere Schwierigkeit ausgeführt werden.

Ein sehr interessanter Mehrfachschnitt, der in die Gruppe der Folgewerkzeuge fällt, ist auf S. 96 beschrieben und dient zur Herstellung von Ketteltaschen; ein zweiter Mehrfachschnitt zur Herstellung endloser Bänder für Kartonnagenklammern mit einem Mehrfachbiegewerkzeug ist auf S. 452 zu finden. Es gehört zu diesem Werkzeug eine eigene Zuführungsvorrichtung, Fig. 643.

Der Folgeschnitt.

Mit den bisher beschriebenen Schnittwerkzeugen wurden solche Stücke hergestellt, bei denen zur Herstellung des für diesen Arbeitsgang gewünschten Ausschnittes eine Umrißlinie in einem Stempelhub genügte. Wenn aber ein Blankett in einem Arbeitsgang gleichzeitig mehrere getrennte bzw. einander einschließende Umrißlinien aufweisen soll, so müßten die betreffenden Stempel ineinander liegen, bzw. die Schnittplatten einander einschließen. Als einfachster Fall sei die Herstellung einer Unterlagscheibe angeführt. Da diese Arbeit verwickelte und teure Werkzeuge ergeben würde, so greift man, wenn es nicht auf eine größtmögliche Genauigkeit ankommt, zu dem Ausweg, daß die beiden oder mehr Umrisse wohl in einem Arbeitsgang, aber in mehreren aufeinanderfolgenden Preßhüben gestanzt werden. Es werden beim Durchgang durch das Werkzeug in den Streifen zuerst die inneren Umrisse gestanzt, dann der Streifen weitertransportiert, bis die eben gestanzten Umrisse genau unter dem zweiten Stempel, der die äußeren Umrisse stanzt, sich befinden, so daß bei dem zweiten Stößelhub unter dem zweiten Stempel das fertige Stück herausfällt. Es ist dabei angenommen worden, daß die gesamte zur Herstellung des gewünschten Blanketts erforderliche Arbeit sich auf zwei Stempelgruppen verteilen läßt.

In dem obenerwähnten Beispiel, der Herstellung einer Unterlagscheibe, mußte man bei Verwendung des Schnittes mit Einlage, Fig. 18, den Blechstreifen zuerst durch eine Presse gehen lassen, um die Scheiben zu stanzen, dann die Scheiben von Hand in eine zweite Presse zum Ausschneiden einlegen. Bei Verwendung des Folgeschnittes werden fast in derselben Zeit, die zum Ausstanzen der Scheiben erforderlich war, die fertigen Unterlagscheiben ausgeworfen. Es wird nur eine Presse und ein Arbeiter benötigt, es gelangt nur ein einziges Werkzeug zur Verwendung, dessen Wirtschaftlichkeit durch Verwendung eines selbsttätigen Vorschubes gegenüber dem Einlegen von Hand noch bedeutend gesteigert werden kann.

Folgeschnitt für Unterlagscheiben. Ein derartiges Werkzeug für Handvorschub zum Stanzen von Unterlagscheiben ist in Fig. 90[1]) als Schulbeispiel wiedergegeben. Der Schnittstempel a und der Lochstempel b sind in der Kopfplatte c vernietet, die wiederum mit vier Zylinderkopfschrauben l an dem Stempelkopf d befestigt ist. Die Stellung der beiden Stempel in bezug auf den Stempelzapfen wird so gewählt, daß das Werkzeug nicht einseitig beansprucht wird. Der Schnittkasten besteht aus der Schnittplatte f, welche deutlich

[1]) W. T. 1909, S. 579.

die beiden um eine Schaltung voneinander entfernten Schnitte für das Loch und den Außenumriß zeigt, und auf welcher die Führungs- bzw. Abstreiferplatte g mit Schrauben k und Prisonstiften i befestigt ist. Die Führungsplatte ist mit einer gehobelten Nut für den Blechstreifen versehen und erhält über dem Aufhängestift eine Aussparung, sowie eine Ausdrehung über den eigentlichen Schnitten. Um nämlich auf der Schnittplatte eine möglichst kleine Schleiffläche

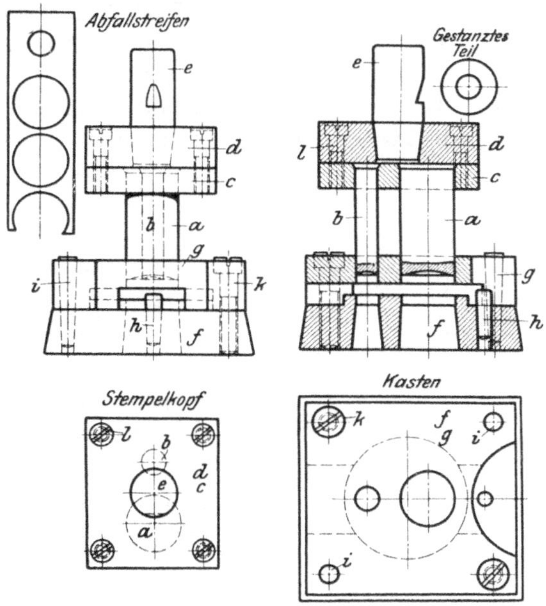

Fig. 90.

zu erhalten, gibt man ihr einen zylindrischen Ansatz von 3—5 mm Höhe, der die Schnitte aufnimmt und dem eine etwas tiefere Ausdrehung der Führungsplatte entspricht. Da die Führungsplatte aus einem Stück gemacht ist, fällt das Einstellen der Führungsschienen für den Blechstreifen vor und nach dem Schleifen weg, und man braucht die konischen Löcher für die Prisonstifte nicht aufzureiben.

Diese Art Werkzeuge haben infolge ihrer Billigkeit und leichten Herstellung eine sehr große Verbreitung gefunden, doch genügen sie nicht, wenn bei der Arbeit größere Genauigkeit oder eine Steigerung der Herstellungsmenge verlangt wird. Man hat infolgedessen verschiedene Wege, diesem Übelstand abzuhelfen, eingeschlagen, von denen einer, der den Charakter des Folgewerkzeuges beibehält, hier angeführt werden soll[1]).

---

[1]) W. T. 1913, S. 39.

In Fig. 91 ist eine einfache Rosette für Baubeschläge nebst dem dazugehörigen Abfallstreifen wiedergegeben, deren Verarbeitung auf den Folgewerkzeugen, Fig. 92/94, mit verschiedener Genauigkeit und in steigender Menge erfolgen kann. In dem Blechstreifen, der nur 2 mm breiter als die Rosette ist, werden zuerst die fünf Löcher gestanzt und der betreffende Teil des Streifens bis unter den Ausschnittstempel transportiert, wo beim Niedergang des Stempels das fertige Stück ausfällt, während gleichzeitig im Streifen daneben die Löcher für das folgende Stück gestanzt werden.

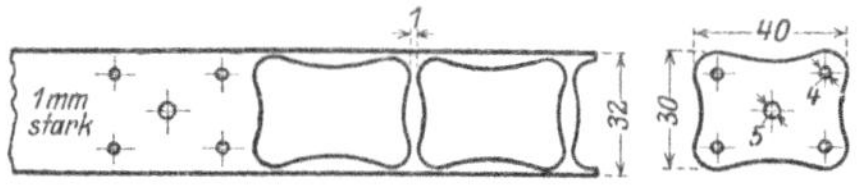

Fig. 91.

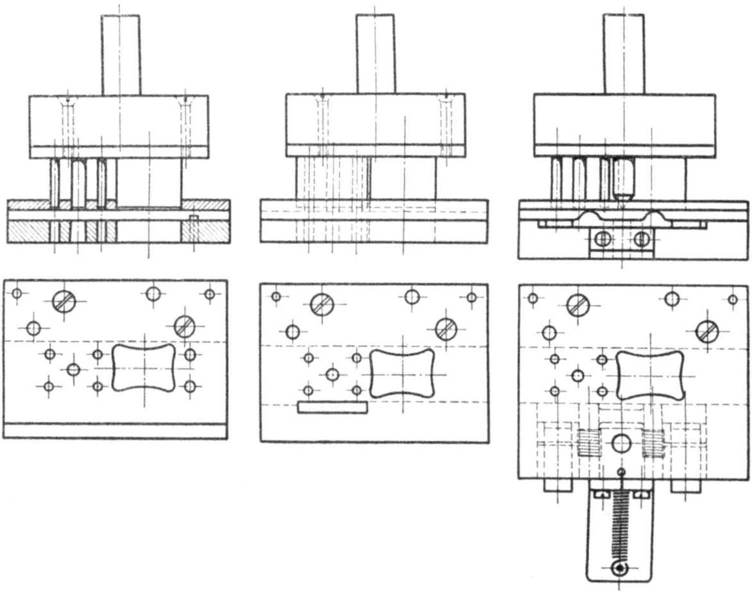

Fig. 92—94.

Bei Verarbeitung[1]) auf dem Werkzeug, Fig. 92, mit Aufhängestift und Handzuführung unter der gewöhnlichen Handspindelpresse kann ein geübter Arbeiter in 10 Stunden aus Bandmaterial 6000 Stück herstellen. Wird dasselbe Werkzeug in gleicher Weise in der Exzenterpresse mit selbsttätiger Zuführung verwendet, so steigert sich die Herstellungsmenge in der gleichen Zeit auf 12 000 Stück. Die Genauigkeit

[1]) Die folgenden Angaben stammen von der Firma Rob. Tümmler in Döbeln.

der Arbeit wird dadurch aber nicht erhöht. Wenn man dagegen neben den Lochstempeln einen besonderen Schnittstempel, den Seitenschneider, Fig. 93, anbringt, der gleichzeitig die eine Kante des Blechstreifens auf genaue Breite zuschneidet, so bildet der im Blech entstehende Ansatz einen genauen Anschlag beim Weiterschieben des Streifens; jedenfalls fällt das lästige Aufhängen des Abfallstreifens auf den Aufhängestift weg. Mit diesem Werkzeug lassen sich in der zuletzt geschilderten Weise 20 000 Stück Rosetten in 10 Stunden herstellen. Auch bei dieser Arbeitsweise entstehen durch das Weiterschieben des Streifens und die Aufeinanderfolge der zu einem Stück gehörigen Arbeiten Ungenauigkeiten, die aber geringer als bei dem erstbeschriebenen Werkzeug sind. Dagegen bedingt die Verwendung eines Seitenschneiders einen größeren Materialverbrauch infolge der größeren Breite des Blechstreifens, sowie einen größeren Kraftbedarf durch die Mehrarbeit eines Stempels.

In Fig. 94 ist schließlich dasselbe Werkzeug dem zwangläufigen Vorschub[1]) in dem Stanzautomaten der Firma Rob. Tümmler in Döbeln angepaßt.

Aufhängestift und Seitenschneider fallen weg, so daß der Streifen frei zwischen den Führungen durchgehen kann, wobei er durch die seitlich ersichtliche Federvorrichtung dauernd gegen die feste Führungsleiste gedrückt wird. Die Genauigkeit wird hier dadurch erreicht, daß der Streifen durch die zwangläufige Zuführungsvorrichtung nach jedem Hub um das gleiche Stück weitertransportiert wird. Es ist also für die Genauigkeit der Arbeit, d. h. das Übereinstimmen der Löcher mit dem äußeren Umriß der Rosette, nicht mehr das Werkzeug, sondern die Genauigkeit des Vorschubapparates maßgebend. Bei 90 Preßhüben in der Minute lassen sich in 10 Stunden ungefähr 48 000 Rosetten herstellen.

Folgewerkzeuge für einen Hauswecker. (C. Lorenz A. G., Berlin.) Zur Darstellung der konstruktiven Einzelheiten, welche bei dem Bau dieser einfachen Werkzeuge auftreten können, sollen eine Anzahl Werkstattzeichnungen solcher Folgeschnitte, die bei dem Bau eines elektrischen Hausweckers Verwendung finden, gebracht werden. Die Verwendung der einzelnen Teile, sowie der gesamte Herstellungsgang derselben, soweit sie nicht auf diesen Werkzeugen fertiggestellt werden, ist aus der Tabelle Nr. 1, I—IV zu entnehmen. Hier soll nur auf die einzelnen Abweichungen der Bauart Bezug genommen werden, da die Werkzeuge sonst mit dem in Fig. 90 angeführten Schulbeispiel übereinstimmen.

[1]) Siehe Fig. 641/2.

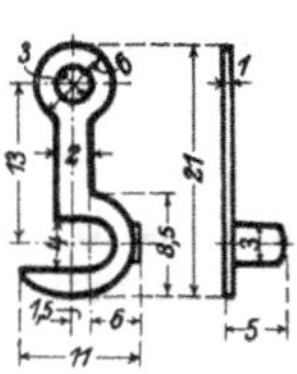

Fig. 95.

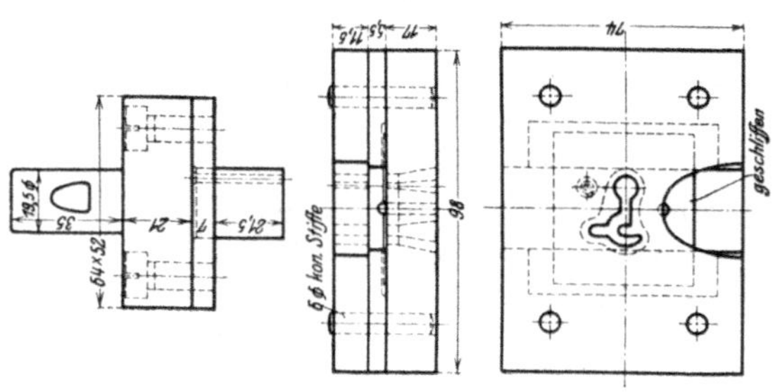

Fig. 96.

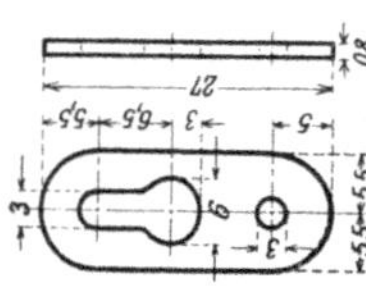

Fig. 97.

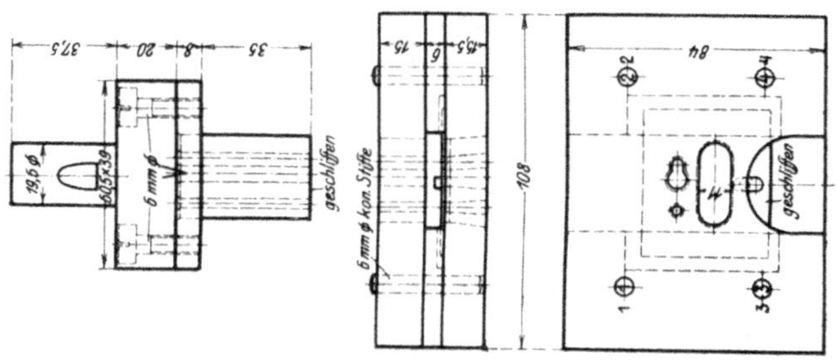

Fig. 98.

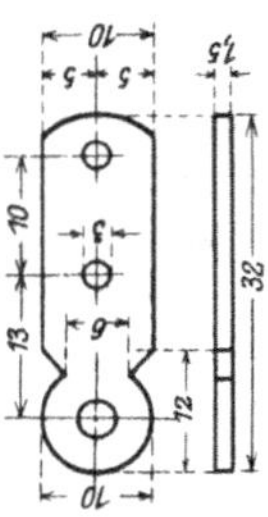

Fig. 99.

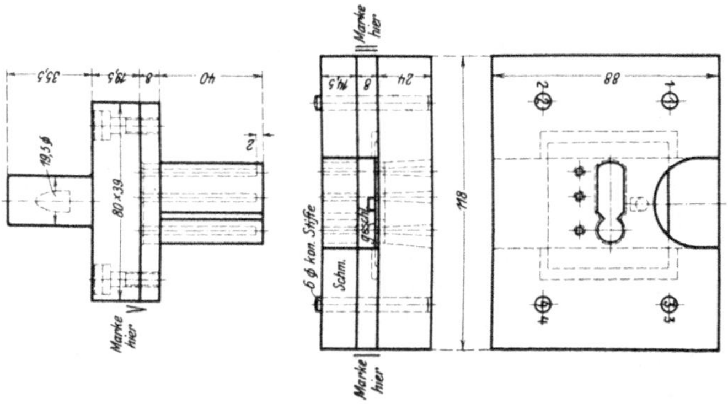

Fig. 100.

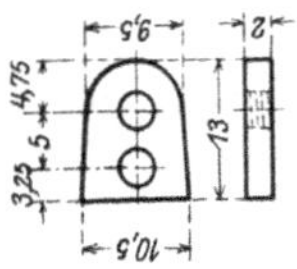

Fig. 101.

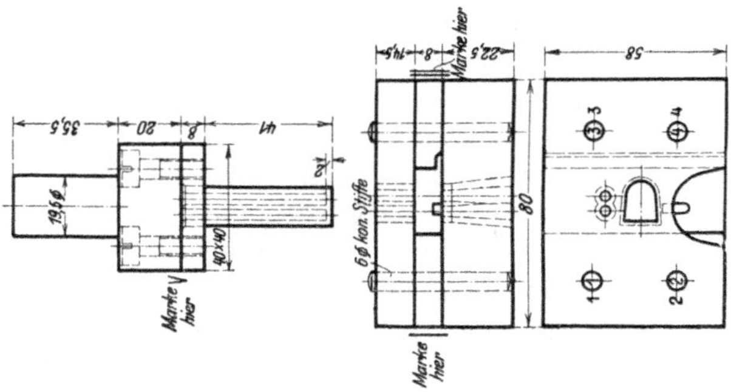

Fig. 102.

Es handelt sich dabei um kleine Teile, die teils aus Federstahl, teils aus starkem Eisenblech von 1,5 und 2 mm Dicke hergestellt werden, deren Genauigkeit jedoch nicht übermäßig groß zu sein braucht.

Vor allem ist ersichtlich, daß alle Stempelzapfen mit gleichem Durchmesser ausgeführt sind, damit das Neueinrichten eines Werkzeuges möglichst schnell vor sich geht. Außerdem sind alle Teile der Stempel und der Schnittkasten nach jeder Richtung gezeichnet, so daß beim Auseinandernehmen der Werkzeuge immer wieder die richtige Stellung der einzelnen Teile erhalten werden kann. Sämtliche Stempel sind vernietet, die Lochstempel fast durchweg kürzer gehalten als die Ausschnittstempel, so daß der Blechstreifen sicher gehalten wird, wenn die dünnen Lochstempel zum Arbeiten kommen, die so gegen Bruch geschützt werden.

Fig. 95 und Werkzeug, Fig. 96, gehören zum Haken, mittels dessen der Holzkasten des Hausweckers an der hölzernen Grundplatte gehalten wird. Der seitliche Fortsatz wird in einem einfachen Biegewerkzeug, S. 352, umgebogen. Bei diesem, wie den zwei folgenden Werkzeugen ist die erhöhte Schliffläche um die Schnitte herumgelegt und in der Führungsplatte die entsprechende Aussparung gelassen worden. Fig. 97 ist die Aufhängeöse für den Wecker aus Schwarzblech von 1,5 mm Dicke, die in dem Werkzeug nach Fig. 98 hergestellt wird. Die eine Klemme für den Zuleitungsdraht ist in Fig. 99 dargestellt. Das dazugehörige Werkzeug ist Fig. 100. Das Material ist nickelplattiertes Eisenblech. In Fig. 101 ist die Mutter für die Befestigung der Schelle gezeichnet, welche in dem entsprechend geformten Schwanzende der Grundplatte[1]) eingepaßt ist. Das Material ist Schwarzblech von 2 mm Dicke, die im Verhältnis zu den Abmessungen des ganzen Stückes bereits sehr groß ist, während der Durchmesser der zu stanzenden Löcher die Blechdicke nur wenig übertrifft.

Die beiden folgenden Werkstücke Fig. 103 und 105, welche ebenfalls zu dem Hauswecker gehören, zeigen Formen, die an die Herstellung des Stempels, Fig. 104 und 106, bedeutende Anforderungen stellen. Die schmale lange Form der zweiten Klemme aus nickelplattiertem Schwarzblech mit der Lagerung für die Kontaktschraube verlangt besondere Sorgfalt beim Härten des im Vergleich zu den beiden dicken Enden dünnen Steges. Außerdem ist die Stellung des Stempelzapfens zwischen Loch- und Ausschnittstempel wegen der großen Entfernung der Teile voneinander sorgfältig mit Rücksicht auf möglichst gleichmäßige Druckverteilung zu wählen. Bei dem Hammer verlangt das schmale Ende, links in Fig. 105, eine besondere Unterstützung

[1]) Vgl. Fig. 587.

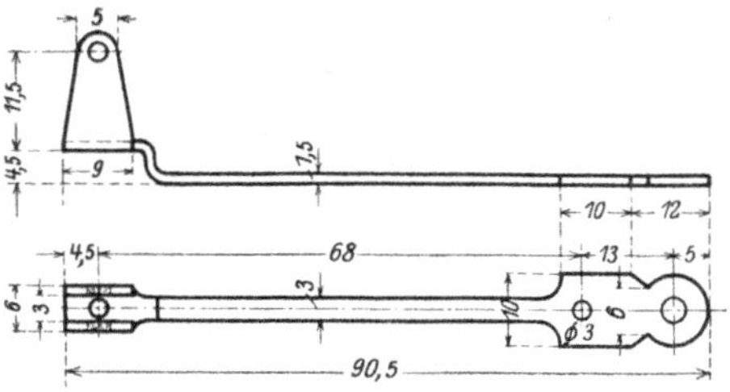

Fig. 103.

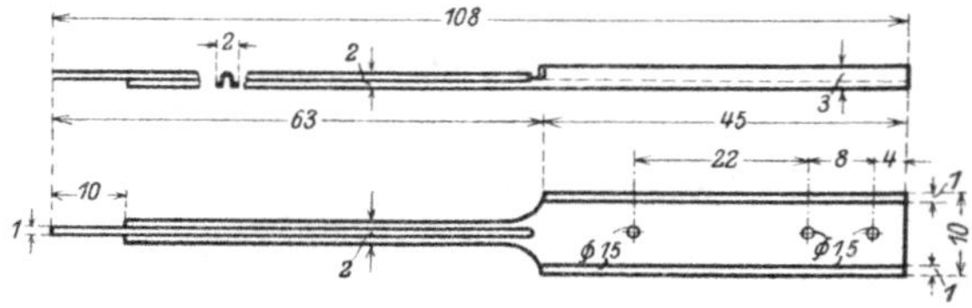

Fig. 105.

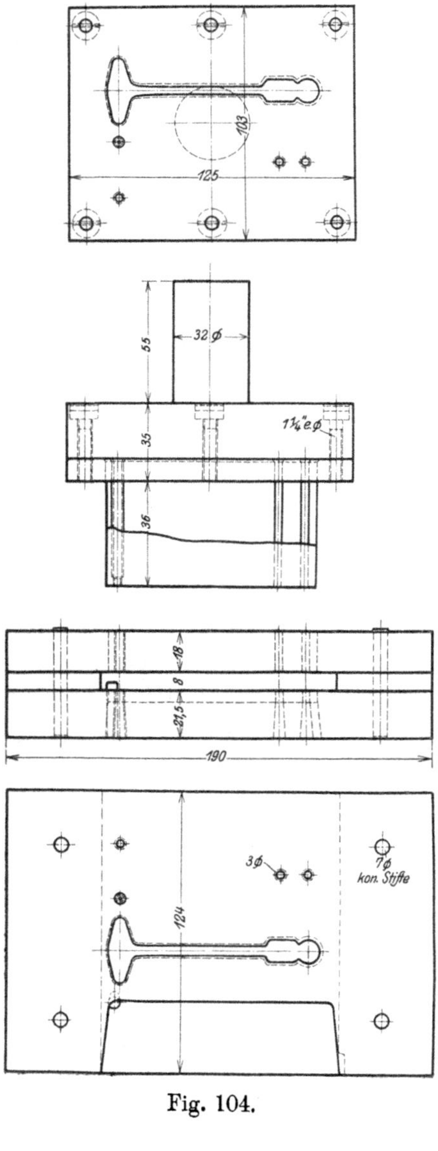

Fig. 104.

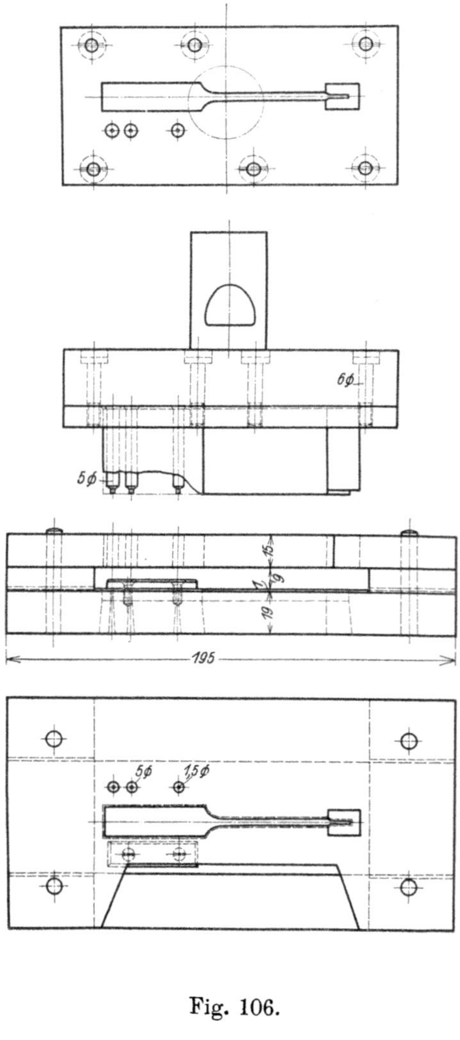

Fig. 106.

im Stempel, damit es gegen Ausbrechen geschützt ist und eine sichere Führung in der Führungsplatte erhält. Dasselbe gilt für die Loch-

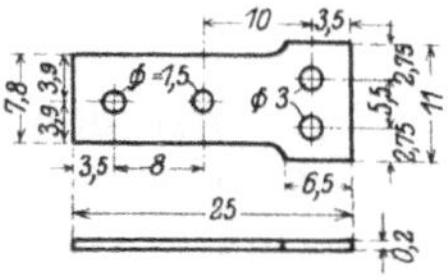

Fig. 107.

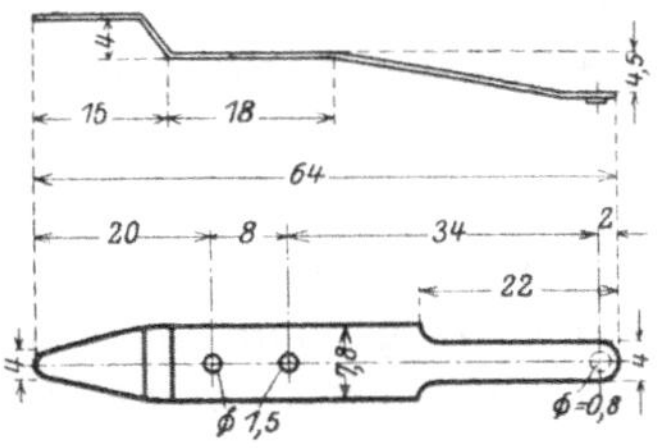

Fig. 109.

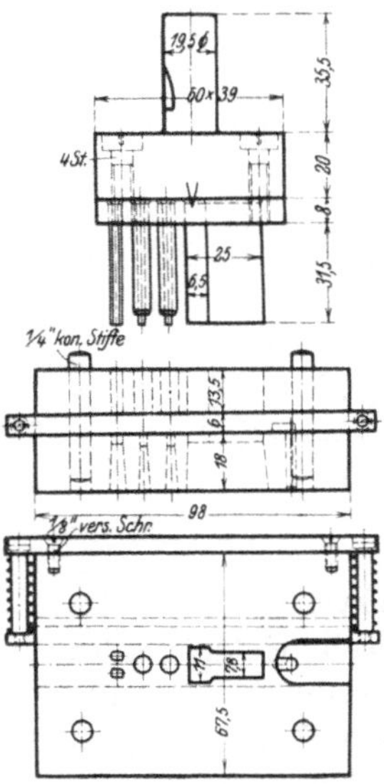

Fig. 108.

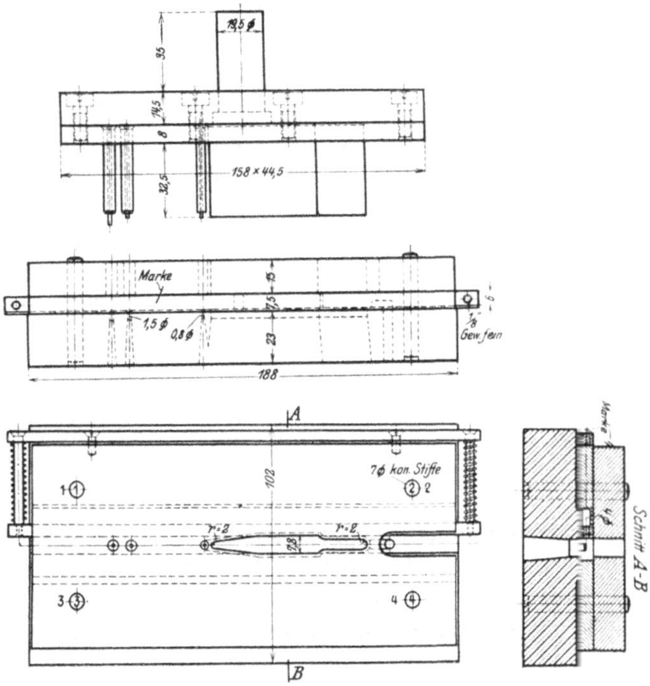

Fig. 110.

stempel, die Löcher von nur 1,5 mm Durchmesser in 1 mm starkes Blech zu stanzen haben. Da das Stück verhältnismäßig schmal ist, wird der Aufhängestift als viereckiges Paßstück, welches dem Ausschnitt im Abfallstreifen entspricht, gebildet, damit eine möglichst gute Mittelstellung der Löcher erhalten wird.

Fig. 107 und 109 sind Teile der Kontaktfeder und werden aus Federstahl hergestellt. Da es sich in beiden Fällen um sehr dünnes Material, 0,2—0,3 mm, handelt, sind die Werkzeuge,

Fig. 108 und 110, mit seitlicher federnder Streifenausrichtung versehen, deren Form von dem schematischen Bild in Fig. 94 abweicht und jedenfalls leichter einzubauen ist. Das letzte Werkzeug schneidet nur das Blankett für die Hammerfeder aus, deren endgültige Form in Fig. 109 erscheint.

Folgeschnitt zur Herstellung Galle'scher Kettenglieder. Bei den bisher besprochenen Folgeschnitten wurde nach dem Lochen der äußere Umriß als ein Ganzes im zweiten Schnitt hergestellt. Daß man aber unter gewissen Umständen durch abweichende Verteilung der Schnitte Vorteile erzielt, soll Fig. 111/12 zeigen. Köhler und Bovenkamp in Barmen legen in diesem Folgewerkzeug den Aus-

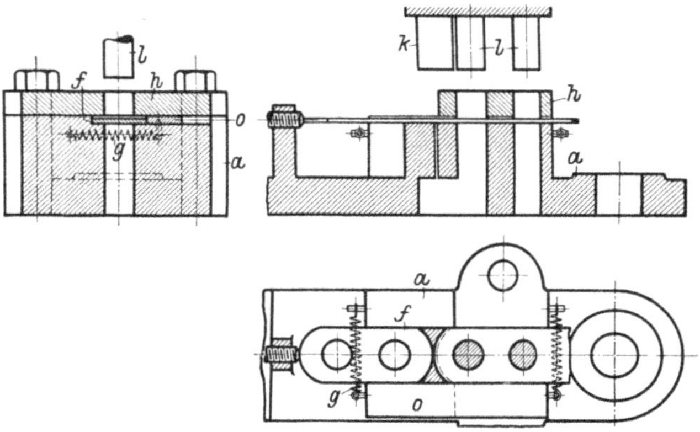

Fig. 111 und 112.

schnittstempel k, Fig. 111, so, daß gleichzeitig das rückwärtige Ende des fertigen und das vordere Ende des vorgeschnittenen Kettengliedes erzeugt wird. Die schematische Zeichnung des Werkzeuges zeigt alle Teile, die Lochstempel l, den Ausschnittstempel k, die Führungsplatte h und den Werkzeugunterteil a deutlich. Der Blechstreifen, welcher die genaue Breite des zu stanzenden Kettengliedes hat, wird von rechts vorgeschoben, bis das ausgeschnittene Ende an die links gezeichnete Anschlagschraube anstößt, wobei eine Federausrichtung g den Streifen gegen den festen Anschlag schiebt. Man kann infolgedessen dauernd die Gelenkkettenglieder stanzen, indem man den Streifen einfach vor und gegen den Anschlag schiebt. Das Aufhängen des vorgelochten Streifens ist dadurch vermieden. Außerdem wird, da die Blechstreifen bereits die Breite der Kettenglieder haben und die beiden Abrundungen sehr nahe stehen können, ein geringerer Materialabfall erzielt. Nach Angabe der Patentschrift wird die Arbeit um das Vierfache verkürzt.

Werkzeug zur Herstellung einer Federschließe. In Tabelle Nr. 31 ist der Herstellungsgang einer Schließe, wie sie für kleine Kästchen oder Schatullen verwendet wird, angegeben. Das fertige Blankett mit den aufgebogenen Enden zeigt Fig. 113. Das Ausschneiden des Blanketts erfolgt auf einem einfachen Folgeschnitt, Fig 114, wie er für die eben besprochenen Gelenkkettenglieder durch Fig. 111 verbessert worden ist. Die Besonderheit dieses Werkzeuges liegt in der Abrundung der Ecken des Ausschnittstempels bei r. Man erhält damit die beiden Enden des Blanketts leicht aufgebogen, so daß das Einrollen der beiden Scharnierteile erleichtert wird. B und D sind Grund- und Seitenansicht der Schnittplatte.

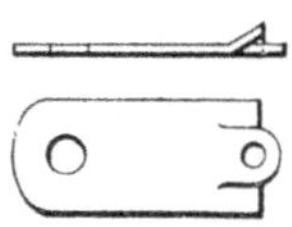

Fig. 113.

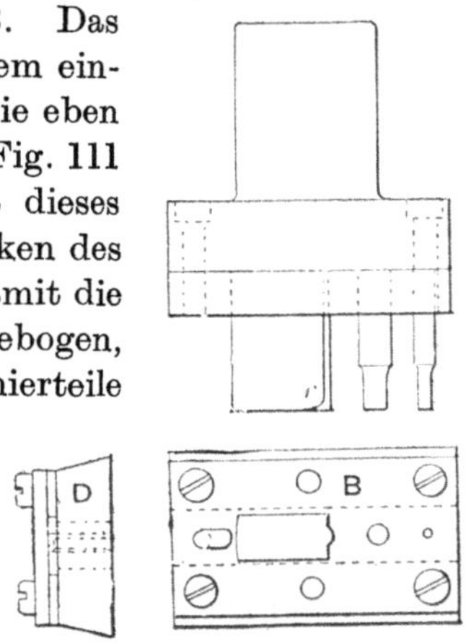

Fig. 114.

Gelochte Kappe mit zwei aufgebogenen Spitzen. Die Kappe in Fig. 115 wird in zwei Arbeitsgängen fertiggestellt, von denen der erste in einem Folgeschnitt geschieht, dessen Schnittplatte der Fig. 116 entspricht. Da die Löcher genau in der Mitte der Kappe stehen müssen, ist im Ausschnittstempel ein Lochsucher vorgesehen, der in das mittlere Loch des Blanketts eintritt und so dieses ausmittelt.

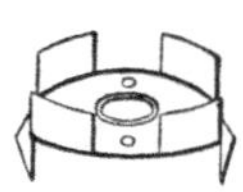

Fig. 115.

Die weitere Arbeit geschieht in einem Verbundwerkzeug nach Fig. 592 S. 481.

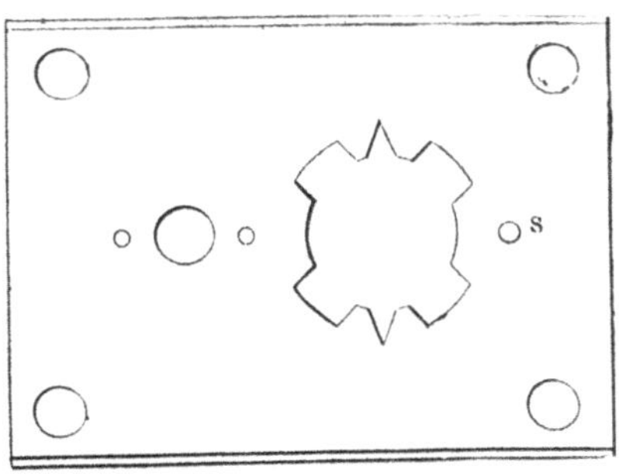

Fig. 116.
S = Anschlagstift.

Folgeschnitt mit Lochsuchern. Das Blankett, Fig. 117, für das auf S. 364 beschriebene Stück wird auf dem in Fig. 119 dargestellten Werkzeug hergestellt. Bemerkenswert dabei sind die Lochsucher bei m und der bewegliche Anschlag bei f, der durch die am Stempelkopf befindliche Anschlagschraube in dem Moment betätigt wird, wenn die Stifte m den Streifen zu fassen beginnen.

Das Rohmaterial ist halbhartes Messingblech, ungefähr 1,5 mm dick. Die Schnittplatte c, Fig. 119, ist aus Werkzeugstahl, wird aus dem Vollen gearbeitet, dann gehärtet und geschliffen. Der Abstreifer d ist aus Eisen, wird mit Schrauben an der Schnittplatte befestigt und durch Prisonstifte gesichert. Der Anschlag f ist von bekannter Ausführung, ähnlich wie in Fig. 83 bei F—G.

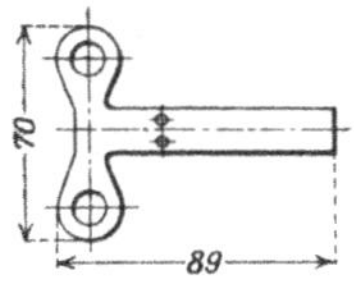

Fig. 117.

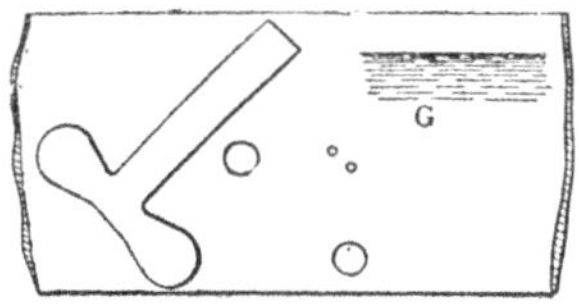

Fig. 118.
G = Walzrichtung.

Der Stempelkopf g ist ein Schmiedestück aus weichem Stahl und bearbeitet; Loch- und Schnittstempel k, j sind darin in der gewöhnlichen Weise befestigt, und die Kopfplatte h in derselben Weise wie der Abstreifer angeschraubt. Besondere Aufmerksamkeit verdient die Austeilung der Stempel bzw. der Nut für den Blechstreifen in dem Schnittkasten.

Der Schnitt ist unter 45° gesetzt, und zwar aus folgendem Grunde: Man wird aus der in den ersten zwei Biegegängen erhaltenen Form, Fig. 471, S. 363, sehen, daß bei xx eine doppelte Biegung

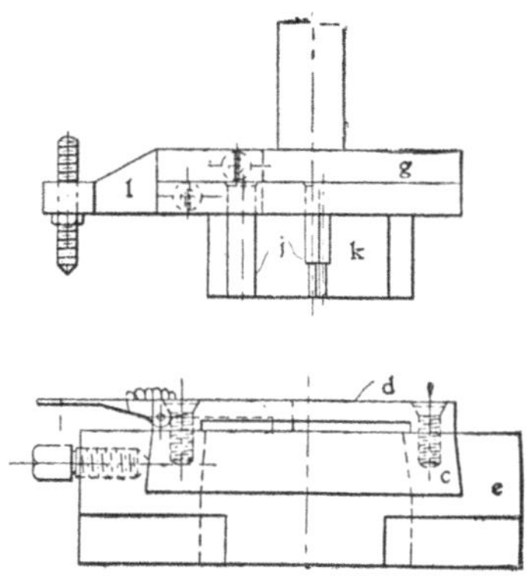

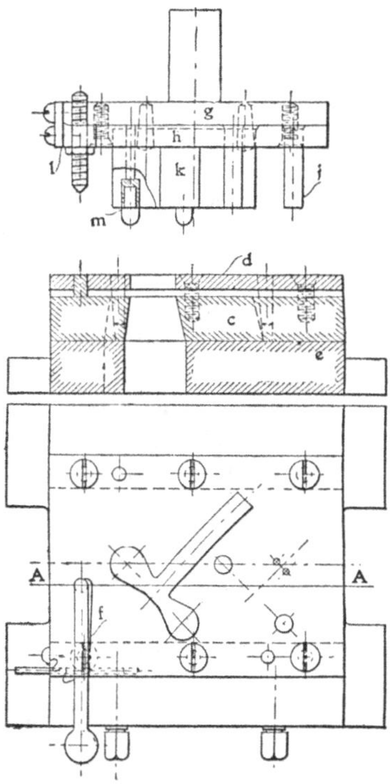

Fig. 119.

vorhanden ist. Wenn nun die Blanketts in oder unter 90° zu der Walzrichtung geschnitten würden, so würden bei der einen oder anderen dieser Biegungen Schwierigkeiten auftreten. Daher wird das Material, dessen Walzrichtung bei G, Fig. 118, angedeutet ist, nach der daselbst angegebenen Richtung gestanzt. Bei m ist ersichtlich, wie die Lochsucher in dem Schnittstempel gehalten sind. Die gewöhnliche Art, diese Stifte unmittelbar in den Stempel einzupassen, ist nicht zufriedenstellend. In diesem Falle ist der Stempel gebohrt, ausgebohrt und dann gehärtet worden, wonach ein Stöpsel aus weichem Stahl in das ausgebohrte Loch eingetrieben wird. Dieser Stöpsel wird dann sorgfältig angerissen und gebohrt, so daß der Lochsucher genau eingepaßt werden kann.

Folgewerkzeug mit gleichzeitigem Massenschnitt[1]). Dieses von der Firma Rob. Tümmler in Döbeln gebaute Werkzeug zur abfallosen Herstellung von Sechskantmuttern wird in Pressen mit dem Vorschubapparat dieser Firma verwendet und ist in Fig. 120 in Ansicht wiedergegeben. Auf dem Werkzeug werden aus Eisenblech von 3 mm Dicke bei jedem Preßhub mittels der Stempel a acht Muttern mit

Fig. 120.

Loch fertiggestellt, während gleichzeitig aus dem übrigbleibenden Streifen durch den durchgehenden Schnittstempel b sieben Muttern abgeschnitten werden, deren übrige Seiten aus dem Abfall der zuerst gestanzten Muttern gebildet werden. Dementsprechend sind im ganzen 15 besondere Lochstempel c vorgesehen, die gleichzeitig die Bohrungen an den nachher in zwei aufeinanderfolgenden Schaltungen hergestellten Muttern lochen.

An dem Werkzeug erkennt man die gußeiserne Gesenkplatte d, die der Länge nach eine rechtwinklige Nut eingehobelt erhält, in welche die Schnittplatte eingepaßt wird. Die Führung für den Streifen ist mit einem seitlichen Federanschlag e nach Fig. 94 zur Streifenausrichtung durchgebildet, über welcher die kräftige Führungsplatte f für sämtliche Loch- und Schnittstempel mittels vier Sechskantkopfschrauben befestigt ist. Das Bemerkenswerte an diesem Werkzeug ist die Teilung der Schnittstempel zum Ausschneiden der Sechskante in zwei einander ergänzende Gruppen und die Vereinfachung der Schnittstempel für die zweite Mutternreihe zu einem geradlinigen Schnittstempel, so daß gleichzeitig die Herstellungskosten des Werkzeuges verringert werden.

[1]) W. T. 1913, S. 105; Kurrein, Folgewerkzeuge und Verbundwerkzeuge.

Folgeschnitt mit Seitenschneider[1]). Für das in Fig. 93 besprochene Prinzip, die Genauigkeit des Folgeschnittes durch Anwendung des Seitenschneiders zu erhöhen, ist in Fig. 121 wiederum ein Schulbeispiel gegeben und daneben das fertige Stück mit dem Abfallstreifen (Abfallriemen), der die Wirkung des Seitenschneiders am deutlichsten zeigt, abgebildet. Lochstempel b, Schnittstempel a und Seitenschneider c sind in der gewöhnlichen Weise durch Vernieten in der Kopfplatte d befestigt,

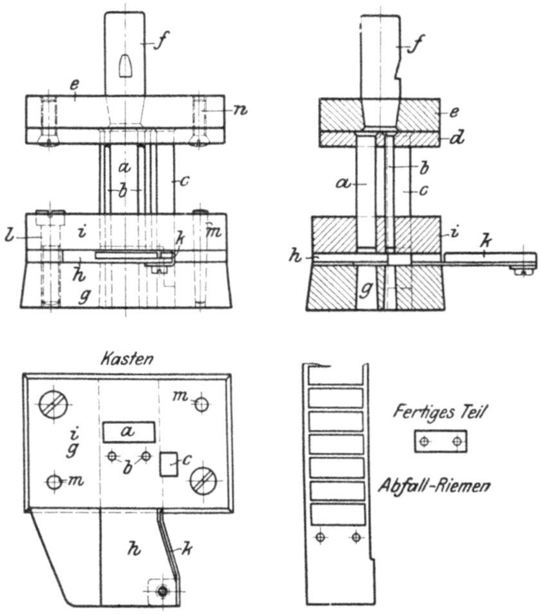

Fig. 121.

die wiederum durch Schrauben n an dem Stempelkopf e angeschraubt ist. Der Stempelzapfen f ist mit Konus in den Stempelkopf eingepaßt und vernietet. Der Seitenschneider c ist außerhalb des Schnittes nach unten verlängert und führt sich mit dieser Verlängerung dauernd in der Schittplatte, so daß einem seitlichen Abrutschen vorgebeugt wird. Der Schnittkasten ist in der gewöhnlichen Weise aufgebaut, nur die Führungsplatte ist mit einem seitlichen Fortsatz h versehen, der eine Feder k trägt, welche den Blechstreifen dauernd gegen den festen Anschlag drückt. Die Entfernung von Mitte Lochstempel bis Mitte Schnittstempel ist gleich der Länge des Seitenschneiders.

Folgeschnittwerkzeug mit doppeltem Seitenschneider. In Fig. 122 wird eine selbsttätige Zuführung für feine Locharbeit angegeben,

[1]) W. T. 1909, S. 580, Fig. 6.

die sich gut verwenden läßt, wenn das Blockwerkzeug zu teuer wird. Fig. 122 zeigt das Stück, den Abstreifer bzw. die Führungsplatte B, die Seitenschneider C, die Lochmatrize D, die Schnittplatte E und die Schrauben F, die den Abstreifer auf A niederhalten. Das Stück in Fig. 122 ist wie das Werkzeug (Unterteil) in $\frac{2}{3}$ natürlicher Größe gezeichnet. Die Löcher in dem fertigen Stück haben 1,36 mm Durchmesser und die Radien der umschriebenen Kreise sind 2,36 mm. Der Arbeitsgang besteht darin, daß der Materialstreifen bei H eintritt, wo er zwischen den Seitenschneidern C bis zu deren innerem Ende an beiden Rändern zugeschnitten wird, so daß er durch den verengten Teil der Führungsplatte B hindurchtreten kann. Es ist vollständig klar, daß das Material von rückwärts, wenn so weit nachgeschoben, daß das unbeschnittene Material sich an dem Absatz in der Führung anlegt, nicht weiter kann und dort gehalten wird, so daß eine genau begrenzte Zuführung erreicht wird. Wenn nun die Rollenzuführung so eingestellt ist, daß sie das Material noch z. B. 1,6 mm weiter, als der zugeschnittene Teil erlaubt, transportieren würde, so werden die Löcher, da das Material an beiden Seiten durch die zugeschnittenen Kanten und ebenfalls am Ende gehalten ist, genau an die richtige Stellung unter den Schnittstempel gebracht.

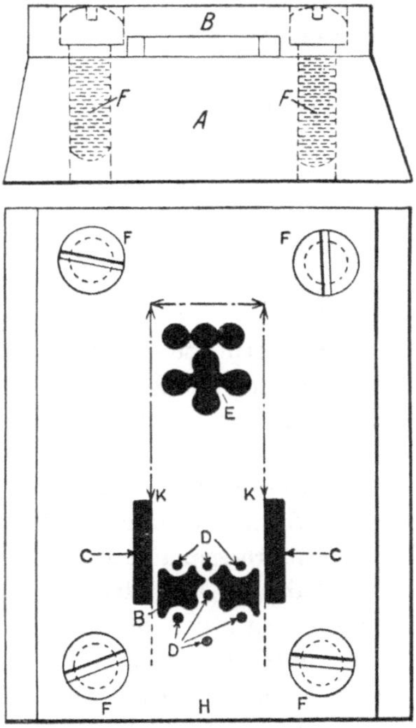

Blankett.

Fig. 122.

Das Werkzeug ist sehr einfach herzustellen und in ähnlicher Form viel bei Prägearbeiten verwendet worden, wo das Stück sehr nahe an der Prägung ausgeschnitten werden und doch kein Grat bleiben sollte; dabei müssen natürlich die Lochstempel in das Material eingetreten sein, bevor der Prägestempel das Material aus der Form gezogen hat.

Schnittwerkzeug zur Herstellung von Schuppentaschen. Der Körper der Handtaschen besteht aus eng nebeneinander liegenden Ringen, welche durch vierzackige Krallen, Fig. 123, zusammengehalten werden (vgl. deren Herstellung S. 392). Die Ringe werden nicht einzeln, sondern als zusammenhängender Streifen in einem Mehrfachfolgeschnitt mit Seitenschneider gestanzt.

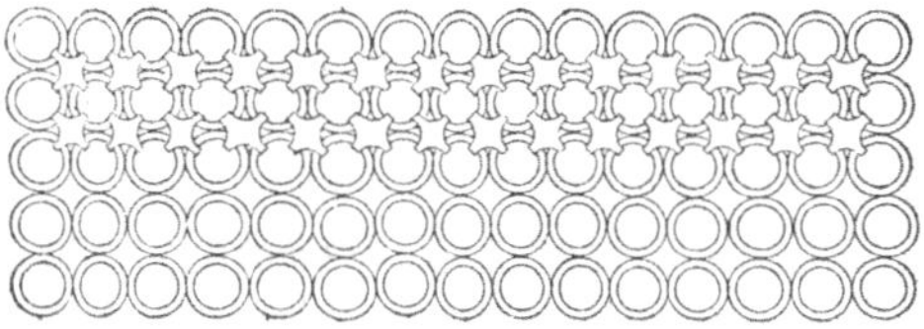

Fig. 123.

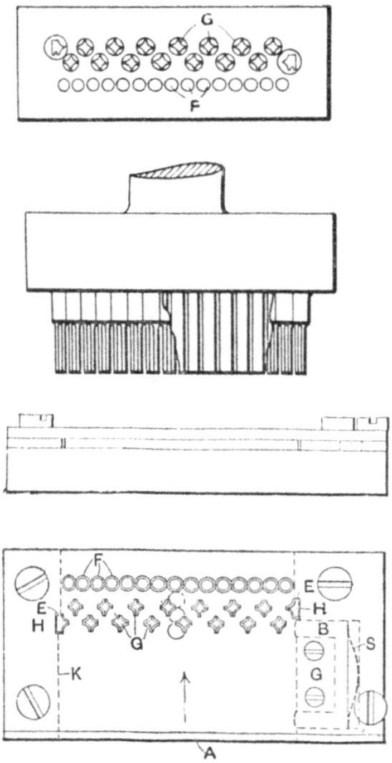

Fig. 124.

Stempel und Schnittplatte zur Herstellung der Ringe sind in Fig. 124 dargestellt und bilden ein verhältnismäßig einfaches Werkzeug. Das Material wird von Hand in Streifen in der Pfeilrichtung von A aus in das Werkzeug eingeführt und durch die Federführungsplatte B—G sicher gegen die Führung K gedrückt. Nach jedem Schnitt wird das Material vorgeschoben, bis seine Kante an die Seitenschneider E kommt, welche sehr genau gefeilt sind. Das Material kommt zuerst unter die Stempel G, die den Außenumfang der Ringe ausschneiden. Jede Lochseite, welche von diesen Stempeln hergestellt wird, beträgt ungefähr 0,07 mm weniger als ein Viertelkreis, so daß zwischen den Ringen ungefähr 0,15 mm Material übrig bleibt, das die Ringe zusammenhält. Hierauf kommt das Material unter die runden Stempel F, die das im Inneren der Ringe befindliche Material entfernen und eine aus Ringen bestehende Platte übrig lassen, wie in Fig. 123 ersichtlich ist. Durch Mädchen werden dann die Krallen zwischen die Ringe eingesetzt und in einer Fußpresse geschlossen, worauf das Ganze mehrere Male in der Hand gerollt wird, so daß das Material zwischen den Ringen gebrochen und ein biegsamer Körper erhalten wird.

Wenn bei Werkzeugen sehr feine Schnitt- oder Prägestempel vorhanden sind, die bei ganz geringer außerachsiger Stempelführung leicht abbrechen können, wird der Stempelkopf mit dem Schnittkasten durch besondere Führungen verbunden, wie es bei den Verbundwerkzeugen, vgl. S. 116, die dann Säulen- oder Blockwerkzeuge genannt werden, allgemein gebräuchlich ist.

Folgewerkzeug mit besonderen Führungssäulen. Gewisse Schnitte haben die Neigung, an den Kanten Grat zu zeigen, was in folgendem Werkzeug vermieden wird:

Das Blankett, Fig. 125, aus kaltgewalztem Material, rd. 2,8 mm dick mit 5 Löchern von 1,8 mm Durchmesser, welche flach und sehr glatt sein müssen, soll hauptsächlich auf die billigste Weise hergestellt werden. Mehrere Versuche in zur Verfügung stehenden Pressen zeigten nicht das verlangte Ergebnis, so daß das in Fig. 126 abgebildete Werkzeug gemacht wurde, das eigentlich stark an ein Säulenwerkzeug erinnert, jedoch für diese Gattung Arbeit besser geeignet ist. Die Grundplatte und der Stempelkopf sind aus Gußeisen, die vier Führungssäulen von 22 mm Durchmesser aus Werk-

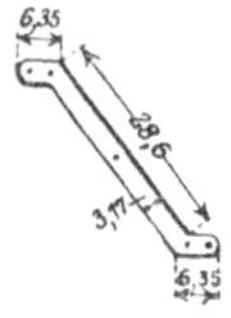

Fig. 125.

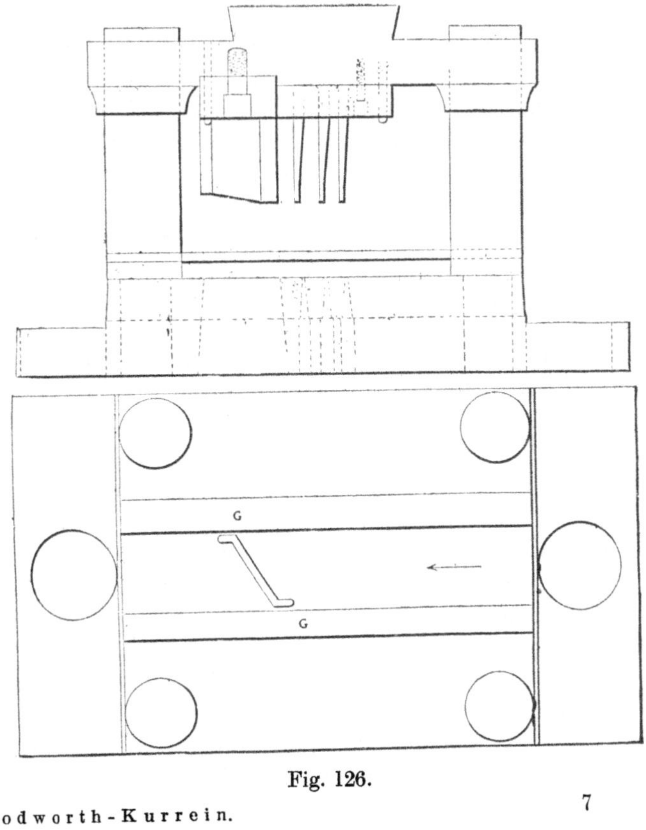

Fig. 126.

zeugstahl, gehärtet, geschliffen und gut in die Kopfplatte eingepaßt, deren Löcher gebohrt werden, nachdem alle Teile zusammengespannt worden sind.

Der Schnittkasten wird in der Presse befestigt, der Stempelkopf nach den Führungen ausgerichtet, und die Kopfplatte angeschraubt und mit Prisonstiften gesichert. Der Schnittstempel wird eingepaßt, während der Stempelkopf an den Führungen sitzt, dann gehärtet; die kleinen Stempel werden eingeschliffen. Der Stempel wird mit Keil in einer Presse befestigt, die in der tiefsten Stellung die Führungssäulen noch frei läßt. Mit einer solchen Presse, die mit 250 Umdr./min. lief und Greiferzuführung hatte, schnitt ein Junge 450 000 Stück, ohne dem Werkzeug Schaden zuzufügen und hätte ohne Nachschärfen der Werkzeuge noch weit mehr herstellen können. Im Vergleich mit den erzielten besseren Ergebnissen fallen die Mehrauslagen nicht ins Gewicht; der Werkzeugmacher kann außerdem die Stempel bei der Montage keinesfalls durchscheren.

Der Verbundschnitt.

Das sicherste Mittel zur Erzielung einer hohen Genauigkeit der gestanzten Stücke ist die Vereinigung sämtlicher Schnitte, welche das fertige Stück bilden, in einem einzigen Werkzeug, so daß bei einem Preßhub ein fertiges Stück ausfällt. Ein solches Werkzeug setzt sich naturgemäß aus mehreren Einzelteilen zusammen, welche beim Arbeiten Relativbewegungen gegeneinander ausführen, die teils durch Federkraft, teils zwangläufig durch Hebel bewirkt werden. Die Bauart mit Federn ist jedoch heute die gebräuchlichere. Die einzelnen Teile müssen auf das sorgfältigste ineinander eingepaßt werden und ohne Spiel in einander gehen. Da bei diesen Werkzeugen der Stempel nur sehr wenig, besonders bei den feinsten Arbeiten nur ungefähr 0,05 mm, in die Schnittplatte eintritt, kann damit eine Massenherstellung der feinsten Teile ohne Bruch- und Biegungsgefahr erfolgen, und da diese Werkzeuge als Säulen- und Blockwerkzeuge meistens eine besondere Führung für den Stempel erhalten, auch die Austauschbarkeit der Stücke gewährleistet werden.

Diese Werkzeuge, auch Subpressen genannt, werden entweder mit Säulenführung, welche die Schweizer zuerst angewendet haben, oder mit Zylinderführung, die mehr von den Amerikanern gebraucht wird, gebaut. Die erste Konstruktion, das Säulenwerkzeug, eignet sich mehr für den Großwerkzeugbau, während die Zylinderführung, das Blockwerkzeug, mehr für die allerfeinsten Arbeiten verwendet wird, obwohl nach Stock die Amerikaner solche Führungen bis 500 mm Durchmesser ausgeführt haben. Bei Verwendung dieser Werkzeuge braucht der

Arbeiter nur den Hub der Presse einzustellen, so daß viel Zeit für das Ausrichten der Werkzeuge gespart wird und die Bruchgefahr bedeutend verringert wird. Dagegen sind die Herstellungs- und Reparaturkosten dieser Werkzeuge sehr hoch, so daß in vielen Fällen ein billigeres Folgewerkzeug mit einem zwangläufigen Vorschub, vgl. Fig. 94, mit der Subpresse hat ernstlich in Wettbewerb treten können. Mit Rücksicht aber auf die Güte und Menge der Arbeit — man kann bis zu 50 000 Stück kleiner Gegenstände ohne Nachschleifen des Werkzeuges herstellen — ist heute das Säulen- oder Blockwerkzeug für die feinsten Arbeiten immer noch an erste Stelle zu setzen, wobei gleichzeitig die hohen Herstellungskosten aufgewogen werden.

Da aber bei den gewöhnlichen Werkzeugen dieser Art ausnahmslos Bandmaterial verarbeitet wird, tritt dabei die Frage nach einer zuverlässigen und billigen Materialzuführung[1]) in den Vordergrund. Da diese im allgemeinen teuer sind und besondere Auf- und Abwickelvorrichtungen benötigen, stehen kleinere Fabriken der Verwendung der Säulen- und Blockwerkzeuge nicht sehr wohlwollend gegenüber. Es wird deshalb auch ein solches Werkzeug angeführt werden, welches die Verwendung von Abfall- und Plattenmaterial zuläßt.

Ich möchte aber auch hier[2]) den Gedanken vertreten, daß es vom wirtschaftlichen Standpunkt vorteilhafter sein muß, die Mehrkosten, welche die erreichbare Genauigkeit eines Stückes verlangt, auf die Maschine und nicht auf das Werkzeug zu verteilen. Wenn man imstande sein wird, mittels eines der billigen Werkzeuge und eines genau und sicher arbeitenden Vorschubes dieselbe Arbeit zu leisten wie mit einem teuren Blockwerkzeug, so ist dies die bessere Lösung. Es kommen dann die Mehrkosten des Vorschubes allen auf der Presse verwendeten Werkzeugen zu gute, die darauf entfallenden Unkosten verteilen sich auf alle in der Presse hergestellten Stücke, während bei den Blockwerkzeugen die Gestehungskosten, sowie die Verzinsung während der Zeit, wo das Werkzeug unbenutzt am Lager liegt, ein einziges Stück belasten.

Es sollen zuerst Verbundschnitte besprochen werden, welche nach Art der bisher besprochenen Werkzeuge in die Pressen eingesetzt werden, also keine besonderen Geradführungen haben, welchen sich die Werkzeuge mit Zwei- und Mehrsäulenführungen anschließen werden, während zum Schluß die Zylinderführungen der amerikanischen Bauart aufgeführt werden sollen.

---

[1]) Gugel, Materialzuführungsvorrichtungen an Exzenter- und Ziehpressen.

[2]) Vgl. Kurrein, Folgewerkzeuge und Verbundwerkzeuge, W. T. 1913 S. 39ff.

α) Verbundschnittwerkzeuge ohne Führungen.

Verbundlochwerkzeug. Fig. 128 zeigt ein Verbundlochwerkzeug, mit dem mehrere Löcher am Umfang und eines in der Mitte des Blanketts, Fig. 127, hergestellt werden. Das Zieh- und Prägewerkzeug zur Herstellung des Blanketts entspricht einigen, die später beschrieben werden. Das Blankett ist ein Aluminiumdeckel von 47,7 mm Durchmesser und 6,35 mm Tiefe, der einen sechsstrahligen Stern erhaben in den Boden eingeprägt hat. Beim Entwurf dieses Werkzeuges mußte man auf Genauigkeit und lange Lebensdauer Rücksicht nehmen, da die Ausbesserung teuer und umständlich ist. Man wird deshalb in der Beschreibung finden, daß alle Schnittkanten bei Abnutzung nachgestellt werden können. Auf dem Umfang des Blanketts sind fünf Löcher herzustellen, was durch Seitenschieber bewirkt wird. D ist ein abgeschrägter Ring, Fig. 128, der an der Kopfplatte befestigt ist und die Seitenschieber betätigt. Die Ringform ist gewählt worden, da sie beim Ausrichten des Werkzeuges eine größere Genauigkeit gewährleistet und zugleich als Führung für den mittleren Stempel K dient. Dieser Ring paßt gleitend über die Führungen B und ist so eingestellt, daß der Stempel K gerade in das Metall eintritt, wenn die Seitenschieber C voll eingetrieben sind.

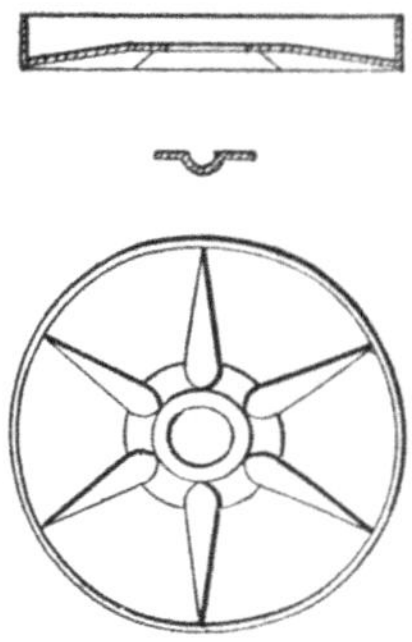

Fig. 127.

E ist einer der Stempel, die leicht herausgenommen, nachgeschliffen und wieder eingesetzt werden können, da sie in der Längsrichtung mittels eingeschnittenen Gewindes nachgestellt und durch eine Druckschraube gehalten werden. G ist eine der schwalbenschwanzförmigen Lochplatten, die in die Gesenkplatte eingelassen sind. Diese ist größer als die übrigen, da sie zwei Löcher enthält. Die anderen drei sind von gleicher Größe. Die Löcher werden von rückwärts ausgerieben, damit die Abfälle leicht durch die Löcher J fallen können. Obzwar sich bei der Ausführung keine Schwierigkeiten durch Steckenbleiben der Abfälle ergaben, wurden doch gerade über diesen Löchern Bolzen mit einer abgeschrägten Fläche eingetrieben, so daß jeder derartigen Möglichkeit vorgebeugt wurde. H ist eine zentrale Büchse bzw. die Schnittplatte für den Stempel K und wird von einer Schraube mit innerem Loch I getragen.

Wenn sich die Büchse H an der Schnittkante abnützt, kann sie herausgenommen, nachgeschliffen und mittels der Schraube I nachgestellt werden. Hierbei schiebt H mit der Abschrägung am oberen Ende die schwalbenschwanzförmigen Lochplatten G heraus. Nun können diese

Teile auch leicht nachgeschliffen werden, so daß das ganze Werkzeug wieder im Arbeitszustand ist.

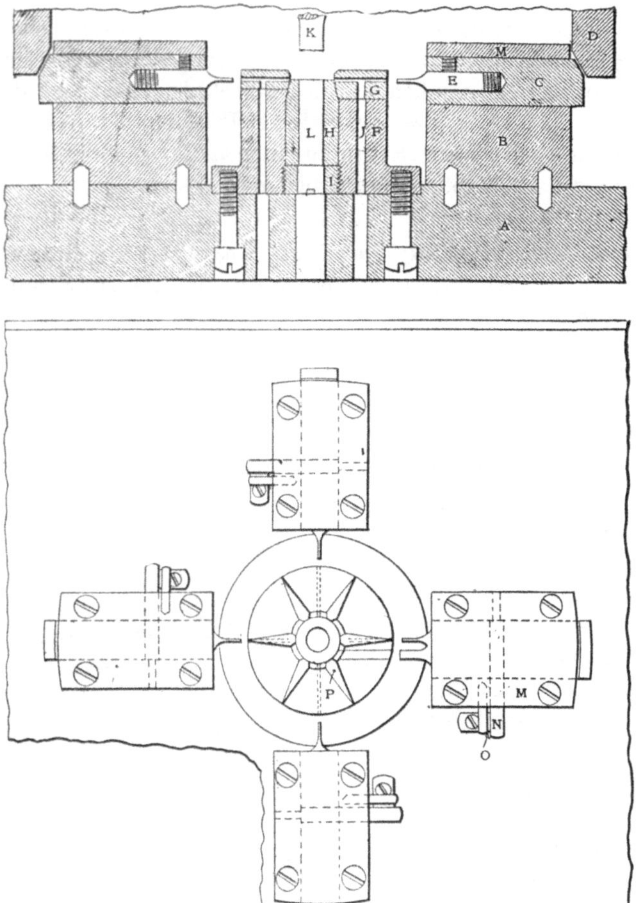

Fig. 128.

Im Grundriß ist ersichtlich, daß in die Führungen M ein kleiner Stahlbolzen eingetrieben ist, und ebenfalls ein Bolzen N in die Seitenschieber C. Wenn der Stempel E in die Lochplatte G eintritt, geht der Bolzen N mit dem Schlitten vor, wofür in der Führung ein Langloch ausgeschnitten ist. Dadurch wird die Feder O gegen den Bolzen in der Führung gebogen, so daß sie genügend gespannt wird, um den Schlitten beim Rückgang des Stempels in seine Anfangslage zurückzuziehen. Diese

Feder ist ein einfaches Stück Stahlblech, das während des Biegens angelassen und an die gußeiserne Grundplatte in der richtigen Stellung angeschraubt worden ist.

Das eben besprochene Werkzeug ist nur insoweit als Verbundwerkzeug zu bezeichnen, als es bei einem Stempelhub zwei verschiedenartige Locharbeiten, nämlich die der Mittelstempel K und der Seitenstempel E, ausführt. Das eigentliche Bild der Verbundwerkzeuge, daß nämlich verschiedene Schnitte ineinander gleichzeitig arbeiten, zeigen erst die im folgenden besprochenen Werkzeuge.

Verbund-Schnittwerkzeug für Lederringe. Fig. 130 gibt einen Verbundschnitt für Dichtungsringe aus Leder, Fig. 129, im Schnitt wieder, die in einem besonderen Fall für das Füllventil eines Benzinbehälters verwendet wurden.

Stempel und Schnitt sind in halber Größe dargestellt. A, B, C, F, G, I und K sind aus Maschinenstahl, D, H und J aus Werkzeugstahl und L aus Gußeisen hergestellt. Der Stempelzapfen A ist zur Aufnahme der Feder R ausgebohrt, die durch den eingeschraubten Stöpsel I in Stellung erhalten wird. Dieser letztere ist durchbohrt und führt den

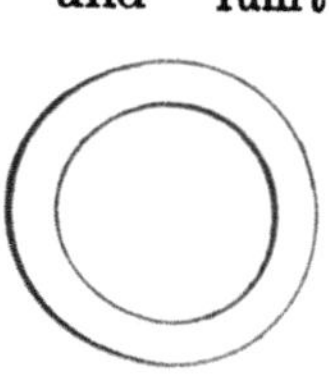

Fig. 129.

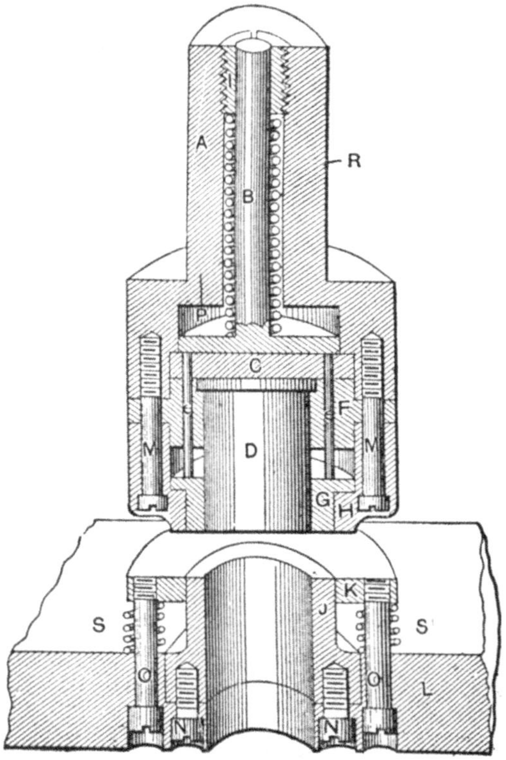

Fig. 130.

Zapfen des Auswerferstempels B, parallel mit dem Teil, der in P niedergeht. Der untere Teil von A wird, wo C eingepaßt ist, ein wenig größer ausgebohrt als P. C dient als Anlage für den Stempel D und ist gut in A eingepaßt, während F, das als Kopfplatte für den

Stempel D dient, streng in A und H paßt. Der Stempel D paßt gut in F. Der Abstreifer G wird durch vier Stifte e betätigt, die durch in F und C gebohrte Löcher gehen und gegen B drücken. Beim Aufwärtsgang des Stempels schiebt die Feder R den Lederring vom Stempel ab und läßt ihn auf der Oberfläche der Schnittplatte liegen, so daß er, sobald Stempel und Schnittplatte sich voneinander trennen, abgleiten kann, wenn dieses Werkzeug in einer schrägen Presse arbeitet; sonst muß er von Hand weggenommen werden. Der Schnittring H für den Außenumfang des Lederringes wird von vier Zylinderkopfschrauben M gehalten, ohne daß, da H, F und A gut eingepaßt sind, eine Gefahr besteht, daß er seine Lage ändert.

Die Schnittplatte wird in folgender Weise hergestellt:

L, aus Gußeisen, paßt in eine 100 mm Froschplatte und ist zur Aufnahme der Schnittplatte J, die durch vier Zylinderkopfschrauben N gehalten wird, ausgebohrt. J wird gut in L eingepaßt. Der Auswerfer K wird ebenfalls durch vier Zylinderkopfschrauben O gehalten und durch vier Federn S betätigt.

Der mit diesem Werkzeug hergestellte Lederring ist in Fig. 129 ersichtlich. Bei der ganzen Arbeit war kein Spiel in den Abmessungen vorgesehen worden, da der Ring ein ganz bestimmtes Maß haben sollte. In einem solchen Fall ist nicht zu befürchten, daß sich die Abmessungen der Werkzeugteile nach dem Nachschleifen ändern; da der Stempel in die Schnittplatte nicht mehr als 1,6 mm eintrat, arbeitete das Werkzeug zur vollsten Zufriedenheit.

Bei der Herstellung von Verbundschnitten dieser Art wird man finden, daß man durch Weglassen eines jeden Spieles in den Abmessungen der Teile gute Ergebnisse erhält, solange man dünnes Material verarbeitet, da auf diese Weise Grat vermieden wird, und eine Veränderung der Abmessungen des Blanketts nach dem Schleifen nicht zu befürchten ist.

Verbund-Schnittwerkzeug für Anhänger. Das Verbundwerkzeug in Fig. 134 dient zur Herstellung von Anhängern nach Fig. 131/32 und wird in einer einfach wirkenden Presse verwendet.

Wie aus Fig. 131 ersichtlich ist, ist es notwendig, das Blankett auszuschneiden, dreimal zu lochen, und das Material zwischen den zwei kleinen Löchern mit einem Radius von 4 mm aufzuschneiden, damit das zwischenliegende Material zur Aufnahme eines Drahtringes umgeschlagen werden kann. Diese Klappe wird dann in ihre frühere Stellung zurückgebogen und mit einer kleinen gummierten Etikette, die über die Verbindungsstelle geklebt wird, gesichert.

In dem Unterteil, Fig. 134, wird die Schnittplatte A zuerst gemacht Die gewellte Schnittkante wird in der Weise hergestellt, daß der Umfang in eine gerade Anzahl gleicher Teile geteilt wird und jeder zweite Teil gebohrt, während das dazwischen liegende Material fast ohne Spielraum kreisförmig ausgearbeitet wird, so daß die konvexen Kreise dem Durchmesser des gebohrten Loches entsprechen, wodurch sich der Umfang

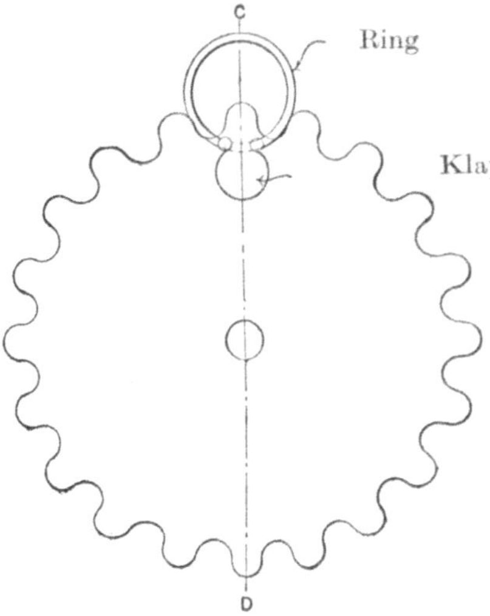

Fig. 131.

Fig. 132. Schnitt C—D.

nach Fig. 133 ergibt. Die Schnittplatte wird dann gehärtet und an der Grundfläche, die ein wenig ausgedreht ist, sowie am äußeren Umfang, soweit er in die Grundplatte eintritt, geschliffen. Ebenso wird die obere Fläche auf ungefähr 15⁰ angeschliffen.

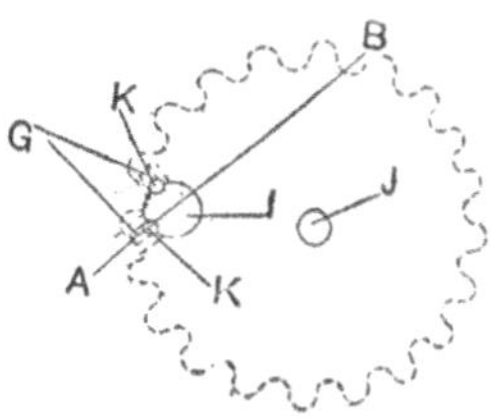

Fig. 133.

Der Hauptstempel E, Fig. 134/35, wird nun vorgearbeitet und schließlich durch die Schnittplatte durchgeschert.

Der Auswerfer B, Fig. 134, wird mit einer 15,8 mm breiten Ringnut, die 3,2 mm von der Unterkante beginnt und so tief ist, daß die innern Spitzen der Schnittplatte A vorbeigehen, gedreht. Hierauf wird er von unten in A eingesetzt, die Schnittkante darauf angerissen und das Material oberhalb der Nut entsprechend ausgearbeitet.

Die Kopfplatte C, eine Stahlscheibe von 3,2 mm Dicke, wird zunächst an der Platte B befestigt. Nachdem die zwei größten Löcher durch beide durchgebohrt und ausgerieben worden sind, wird die Platte als Lehre zum Bohren des Stempels E verwendet, wonach das größte Loch bei I mit der gleichen Stahlsorte gestöpselt wird, so daß sich der Bohrer nicht verlaufen kann. Dann werden die Löcher für die beiden kleinen

Stempel K gebohrt. Der Auswerfer B wird nun wieder benutzt, um die kleinen Löcher im Stempel E zu bohren, wozu der letztere einen Stöpsel wie B erhalten hatte. Der kurze Stöpsel, der im Auswerfer B verwendet worden ist, wird als Stempel F, Fig. 134/35, verwendet, und der Stöpsel aus dem Stempel E wird nun der lange Stempel I in Fig. 134.

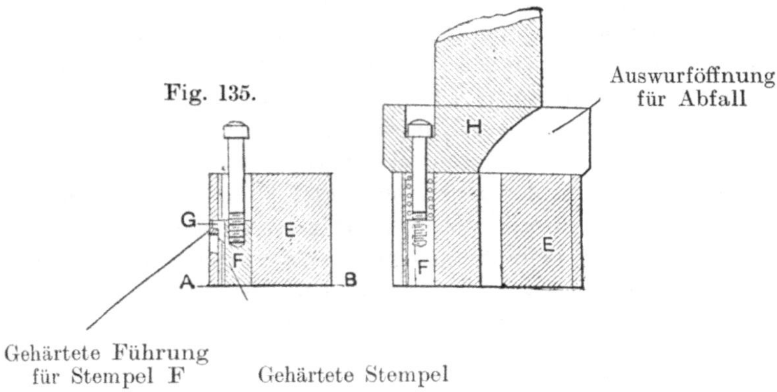

Fig. 135.

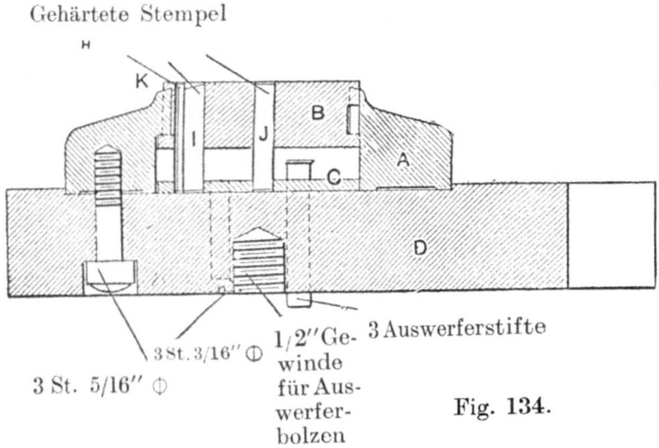

Fig. 134.

Seine Stirnfläche wird abgeschrägt, so daß er in den Oberstempel E, der jetzt als Schnittplatte arbeitet, eintreten, und das Material bis zu den vorher von den Stempeln K hergestellten, kleinen Löchern aufschneiden kann. Der Stempel F ist unter dem gleichen Winkel wie der Stempel I abgeschrägt, jedoch nicht ganz bis an die Kante der kleinen Löcher, welche die Lochplatte bilden. Die Abfälle werden aus den Oberstempeln E durch die abgeschrägten Führungen G, Fig. 135, seitwärts entfernt.

Der Stempel F, Fig. 135, wird durch zwei rechtwinklige Führungen bei G, die in kleine Löcher über den Auswurföffnungen im Körper des Oberstempels E von innen eingesetzt werden, an der Drehung verhindert, wobei der Stempel durch seine Stellung selbst ein Sichlosarbeiten der Führungen verhindert.

Die Stempel K, K, I und J werden in der Kopfplatte C, Fig. 134, vernietet, und die in letzterer befindlichen Gewinde mit Benützung der Gesenkplatte D als Lehre hergestellt.

Da sich die Abfälle von den Stempeln I und J durch den Oberstempel E heraufarbeiten müssen, ist man gegen die gewöhnliche Abstreifplatte im Vorteil, da man den Abfall nicht aus dem fertigen Stück entfernen muß, der sonst durch den Auswerfer in dasselbe zurückgedrückt wird.

Es wird vielleicht neu sein, daß in diesem Falle Druckluft zur Entfernung der fertigen Stücke aus dem Werkzeug verwendet und so die Arbeit der Schwerkraft in der geneigten Presse ersetzt wurde. Diese Arbeitsweise wurde mit diesem Werkzeug in einer senkrechten Presse mit Erfolg durchgeführt. Beim Abwärtsgang wurde die Luft zusammengepreßt und auf die Oberfläche der Schnittplatte beim Aufwärtsgang der Presse ausgeblasen[1]), so daß der in diesem Falle natürlich leichte Gegenstand entfernt wurde.

Verbundschnitt zum Stanzen und Lochen von Ankerscheiben. Die Ankerscheibe, Fig. 136, ist aus Stahlblech von 0,8 mm Dicke hergestellt mit einem Durchmesser von 197 mm und einer Bohrung

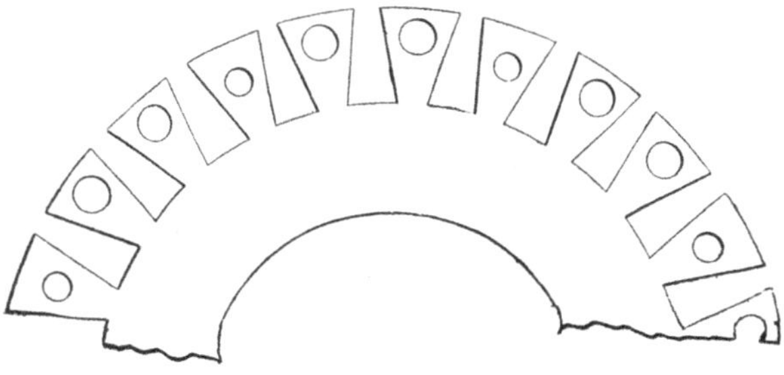

Fig. 136.

von 92 mm. Die Schnittplatte, Fig. 137, wird an dem Pressenstößel befestigt und der Stempel, Fig. 138, auf der Aufspannplatte. Die Schnittplatte besteht aus einem Stahlring a, der auf seiner Innenfläche dem Umriß der Ankerscheibe entspricht, während in der Mitte des Ringes ein runder Stahlstempel b und nahe dem Rande eine Anzahl kleiner, runder Stempel c angeordnet sind. Alle diese Teile sind an die Gußeisenplatte d angepaßt und, wie ersichtlich, mit Schrauben und Muttern befestigt. Ein Abstreifer aus Stahl e paßt lose in die Schnittplatte und wird dort von den Schrauben f und den Federn g gehalten.

[1]) Vergl. Fig. 290, S. 229.

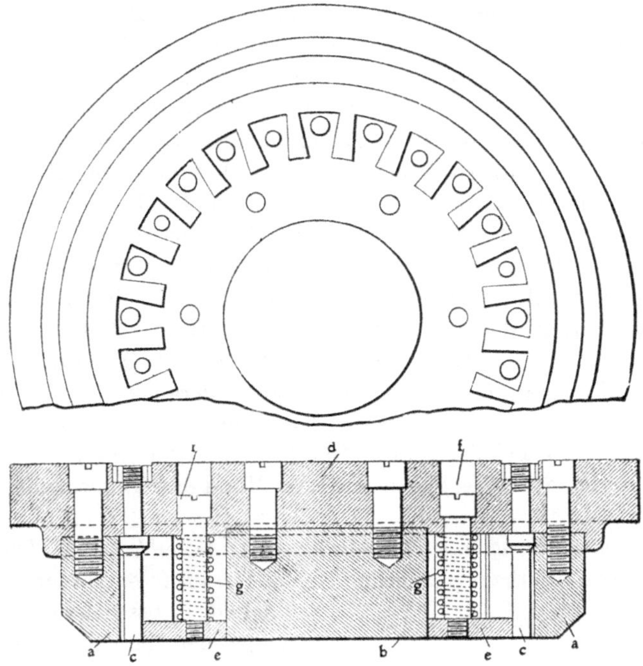

Fig. 137.

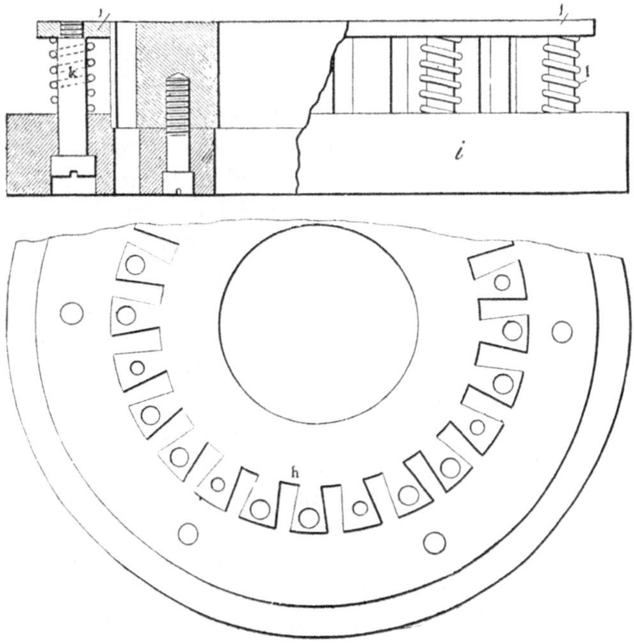

Fig. 138.

Der Stempel, Fig. 138, besteht aus einem Stahlring, der an der äußeren Umfläche, wie bei h ersichtlich, entsprechend der Schnittplatte genutet, in der Mitte ausgebohrt und nahe dem Umfange mit einer Reihe kleiner Löcher für die kleinen Stempel c des Ringes a versehen ist. Der eigentliche Stempel ist mit Schrauben, wie gezeichnet, auf der gußeisernen Grundplatte i befestigt und hat einen Stahlabstreifer j, der durch die Schrauben k und die Federn l gehalten wird. Beim Abwärtsgang des Stößels wird die Schnittplatte über die äußere Schnittkante des Stempels, der auf der Aufspannplatte der Presse befestigt ist, geführt und dabei der Außenumfang des Blanketts gestanzt; gleichzeitig stellt der runde Stempel b innen die Bohrung und die kleinen Stempel c am Rande des Blanketts die betreffenden Löcher her. Beim Aufwärtsgang des Stößels wirft der Federauswerfer das Blankett aus der Schnittplatte und streift den Abfall an der Außenseite des Stempels ab, so daß das fertige Blankett und der Abfall lose auf der Oberfläche des Stempels liegen und von Hand entfernt werden können.

Verbundwerkzeug zur Herstellung von Ankerblechen. Dem eben beschriebenen Werkzeug amerikanischer Herkunft entsprechen die im folgenden beschriebenen Werkzeuge der Firma L. Schuler in Göppingen.

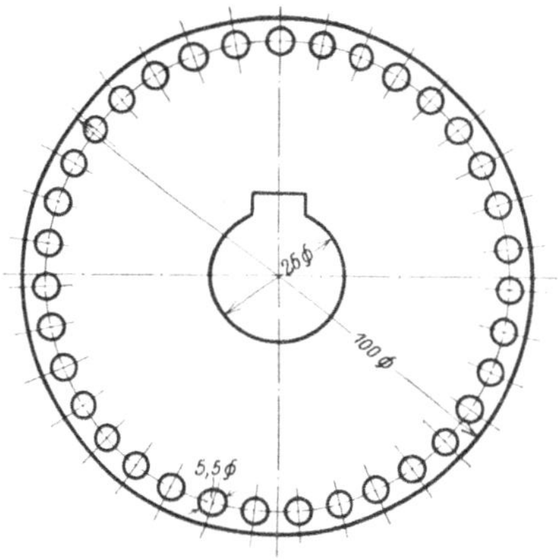

Fig. 139.

Fig. 140 stellt ein Werkzeug dar, das zur Massenherstellung von Scheiben kleinen Durchmessers in senkrechten Exzenterpressen mit durchbrochenem Ständer vorteilhaft zur Verwendung kommt. Die Maschinen sind im allgemeinen mit selbsttätigen Auswerfern in Tisch und Stößel ausgestattet. Das Ankerblech, Fig. 139, wird gleichzeitig ausgeschnitten, genutet und mit dem Mittelloch und der Keilnut versehen. Der Durchmesser der Scheiben beträgt 100 mm, der Bohrung 26 mm und der Nuten 5,5 mm. In diesem Fall wird nur der untere Auswerfer von der Presse betätigt, während der obere mittels dreier Wurstfedern von 19 mm Außendurchmesser, 5 mm Drahtstärke und einer ungespannten Länge von 68 mm bei 9 vollen Windungen abgefedert ist. Der Stempel hat einen zylindrischen Zapfen aus Schmiedeeisen, der mit zölligem, fest eingetriebenem Gewinde im Stempelkopf a befestigt ist. Dieser wird aus Grauguß hergestellt und mit einer Aus-

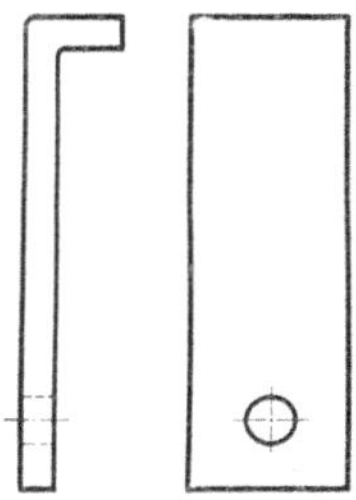

Fig. 140 a.

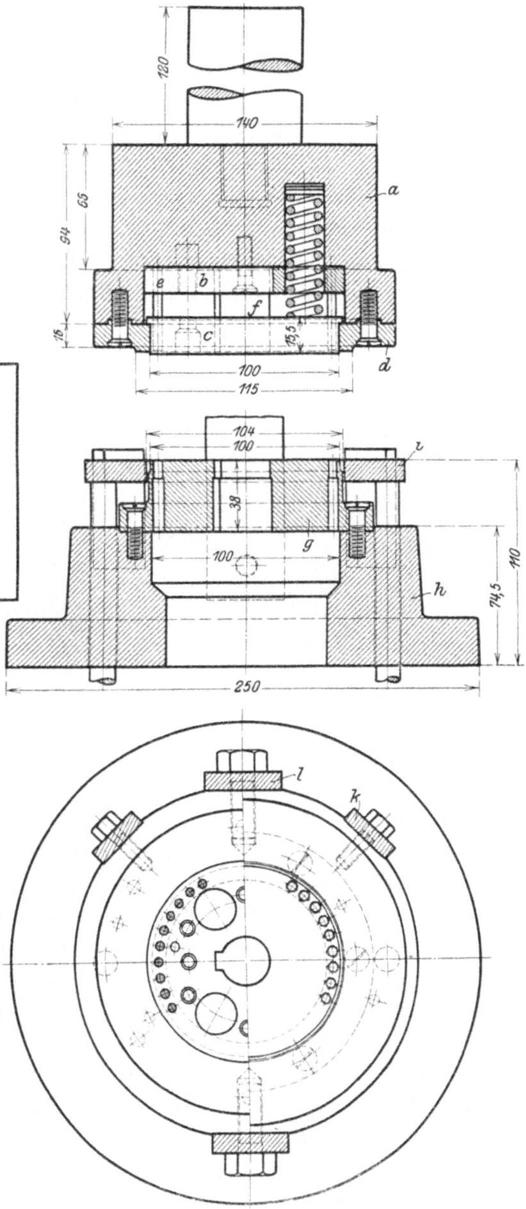

Fig. 140.

drehung für die Kopfplatte b und den Auswerfer c versehen. Außerdem sind drei Löcher von 20 mm Durchmesser für die Auswerferfedern gebohrt, sowie je sechs $^5/_{16}$'' Gewinde für die Kopfplatte und den Schnittring d. Die Unterfläche des Stempelkopfes a erhält eine Ringnut von 12 mm Breite und 2,5 mm Tiefe eingedreht, in welcher sich der Schnittring d ausmittelt. Der Stempelkopf ist außen von 160 auf 140 mm Durchmesser abgesetzt und bildet so einen Anschlag für die beiden Winkel l, Fig. 140a.

An der Unterfläche des Stempelkopfes ist der Schnittring d aus Werkzeugstahl mit seiner Andrehung zentrisch eingepaßt und mit sechs versenkten Kopfschrauben angeschraubt. In ihm führt sich der Auswerfer c, der sich mit einem angedrehten Bund rückwärts an die Schnitt-

Fig. 141. Ansicht des Werkzeuges in Fig. 140.

platte anlegt. Andererseits wird er durch drei Entlastungsschrauben von $^5/_{16}$'' Durchmesser, die in ihm tief versenkt sind, von der Kopfplatte aus getragen. Diese ist aus Schmiedeeisen hergestellt. Die Nut- und Lochstempel e und f sind aus Werkzeugstahl in der üblichen Weise gemacht und in die schmiedeeiserne Kopfplatte eingenietet; ihre Enden stehen 0,5 mm oberhalb des Schnittringes, so daß zuerst der Außenumfang und dann die Nuten und das mittlere Loch gestanzt werden. Die Kopfplatte von 107 mm Durchmesser ist in die Ausdrehung des Stempelkopfes genau eingepaßt und mit sechs versenkten $^5/_{16}$'' Kopfschrauben befestigt.

Der Werkzeugunterteil g bildet den Schnittstempel für den Außenumfang und die Schnittplatte für die Nut- und Lochstempel wird aus Werkzeugstahl hergestellt und mittels eines angedrehten Flansches und sechs Stück versenkter $^5/_{16}$'' Kopfschrauben auf der gußeisernen Gesenkplatte h befestigt. Diese erhält eine Ausdrehung von 2,5 mm Tiefe und 135 mm Durchmesser, in der sich die Schnittplatte g ausmittelt. Außerdem ist die Gesenkplatte mit einem großen Mittelloch von 100 mm Durchmesser versehen, durch welches die Abfälle, die von den Loch- und Nutstempeln herrühren, ausfallen können.

An der Matrize führt sich mit 1 mm Spiel im Durchmesser der Auswerfer i, der aus Schmiedeeisen von 12 mm Dicke hergestellt ist und von zwei Stangen von 12 mm Durchmesser, welche durch die Gesenkplatte durchgehen, durch den Pressenantrieb betätigt wird. Zum Einlegen der ausgeschnittenen Scheiben dienen zwei Anschlagleisten k, die ebenfalls an der Gesenkplatte angeschraubt sind. Ein runder breiter Flansch dient zur Befestigung der Gesenkplatte am Preßtisch.

Die Schnittplatte g ist mittels je zweier Prisonstifte von 5 mm Durchmesser in ihrer Stellung gesichert.

Verbundwerkzeug für Ringe mit Innennuten (Statorbleche). Fig. 143 stellt einen nach den gleichen Grundsätzen von L. Schuler-Göppingen gebauten Verbundschnitt zur Herstellung von Ringen mit Innennuten, Fig. 142, dar. Dieses Werkzeug kommt, wie das eben beschriebene, für die Massenherstellung von Ringen kleinen Durchmessers zur Verwendung. Das innere Loch in den Scheiben wird auf geeigneten Exzenterpressen gestanzt, worauf im zweiten Gang die Nuten und der äußere Durchmesser auf dem im folgenden beschriebenen Werkzeug fertiggestellt werden. Auch hier ist die Presse mit selbsttätigem Auswerferantrieb für den Unterteil zu versehen, während der obere Abstreifer durch fünf Federn von 19 mm Durchmesser, 5 mm Drahtstärke und 68 mm ungespannter Länge bei 9 Windungen betätigt wird.

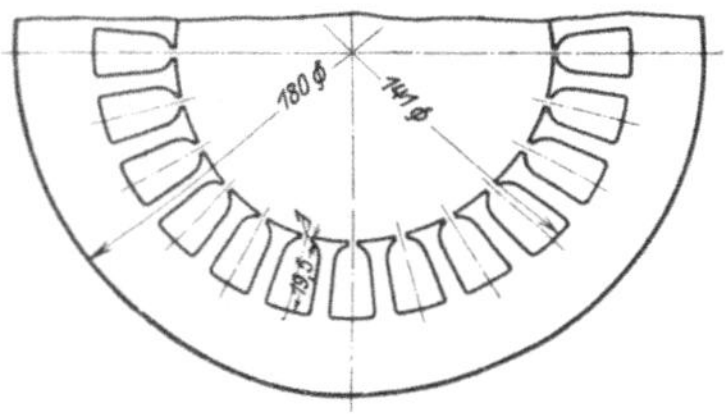

Fig. 142.

Der Stempelkopf a aus Gußeisen trägt, mittels Eindrehung ausgemittelt und mit 10 versenkten Kopfschrauben von $^5/_{16}$″ Durchmesser befestigt, den Schnittring b aus Werkzeugstahl, der außerdem durch zwei Prisonstifte von 5 mm Durchmesser gesichert ist. In eine Ausdrehung des Stempelkopfes von 186 mm Durchmesser ist die schmiedeeiserne Kopfplatte c eingepaßt, mit 10 Schrauben, wie die Schnittplatte, befestigt und durch Prisonstifte gesichert. In ihr sind die 24 Nutenstempel d eingepaßt und vernietet, welche hier in gleicher Höhe wie der Schnittring abschneiden. Sie führen sich in der Abstreiferplatte e, die um 0,5 mm in der tiefsten Stellung über die Schnittplatte vortritt. In dieser Stellung liegt ihr angedrehter Bund auf der rückwärtigen Oberfläche der Schnittplatte auf, gegen die sie durch die Federn angedrückt wird. Der Stempelzapfen f ist in derselben Weise wie im vorangehenden Werkzeug befestigt.

Der Unterteil setzt sich aus der gußeisernen Gesenkplatte g, die mit einem runden Aufspannflansch versehen ist, dem Schnittstempel h

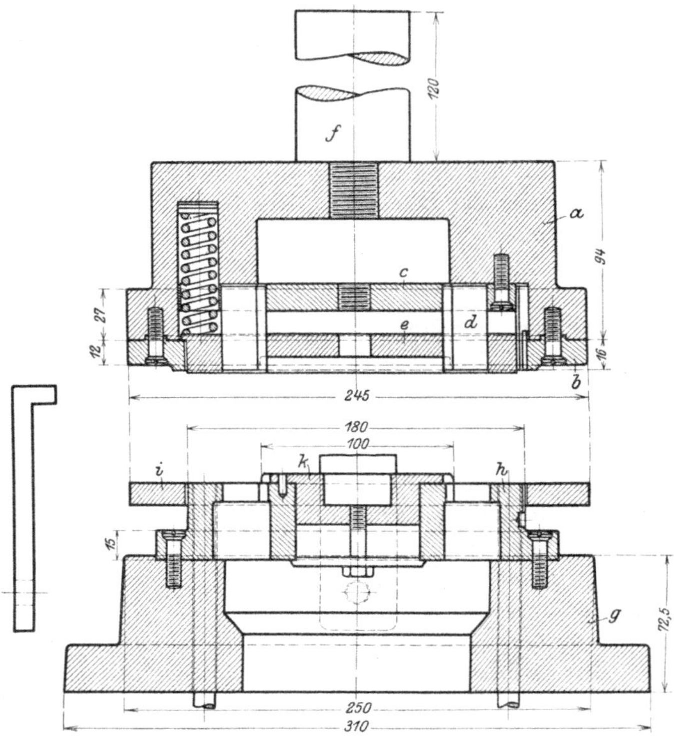

Fig. 144.

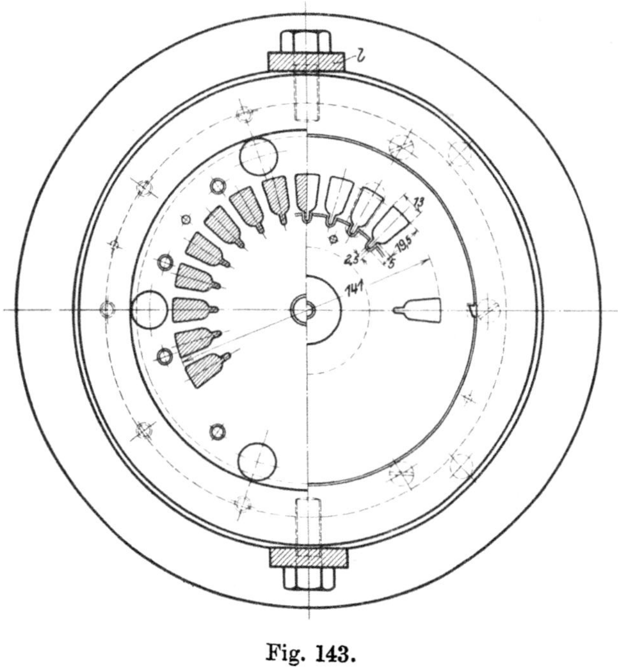

Fig. 143.

aus Werkzeugstahl, der auch die Lochschnitte enthält, dem Auswerfer i und dem Einlegerring k zusammen. Die Gesenkplatte ist wiederum genügend ausgebohrt, um den Abfall nach unten durchfallen zu lassen und oben mit einer 1 mm tiefen Ausdrehung zur Ausmittlung des Schnittstempels versehen. Dieser wird mittels sechs versenkter Kopfschrauben von $^5/_{16}''$ Durchmesser gehalten und durch zwei Stahlstifte gesichert. Der Schnittstempel ist unter den Lochschnitten verhältnismäßig weit ausgespart, um ein leichtes und sicheres Durchfallen des Abfalles zu gewährleisten. In der Mitte erhält der Stempel eine Bohrung von 65 mm

Fig. 145. Ansicht des Werkzeuges in Fig. 143.

Durchmesser, die zur Aufnahme und zentrischen Führung des Einlegeringes k dient, der das vorgelochte Blankett beim Einlegen ausmittelt. Der Einlegering wird mittels Unterlagscheibe und Kopfschraube in dem Schnittstempel befestigt und durch einen Stift am Drehen verhindert. Er ist am Umfang geschlitzt, um die Enden der Nutenstempel durchtreten zu lassen, da die auszustanzenden Nuten nach dem inneren Umfang des Ringes offen sind. Mit 1 mm Spiel im Durchmesser führt sich auf dem Stempel der Auswerfer i, der hier von vier Auswerferstangen durch den Pressenantrieb betätigt wird. Außerdem sind noch zwei Winkel l, Fig. 144, wie im ersten Werkzeug vorgesehen.

Herstellung eines Verbundwerkzeuges mit zusammengesetzter Schnittplatte. Der folgende Weg, einen Verbundschnitt für sehr kleine Ankerscheiben herzustellen, bietet Interesse.

Das in der Fig. 146 dargestellte Werkzeug findet Verwendung, um Ankerscheiben von 38 mm Durchmesser zu stanzen. Von den Scheiben wird verlangt, daß sie genau ausgeteilt und eben seien, daß die Nuten genaue Abmessungen haben und frei von Grat seien. Man erreicht dies leichter, wenn man die Schnittplatte aus einzelnen Teilen aufbaut, wie Fig. 146/47 zeigt. Dieser Teil ist aus den Zeichnungen genügend deutlich zu entnehmen.

Bei Herstellung dieser Werkzeuge wird zuerst der Stempel und der Auswerfer auf Maß gefräst. Ist man jedoch seiner Fräsmaschine

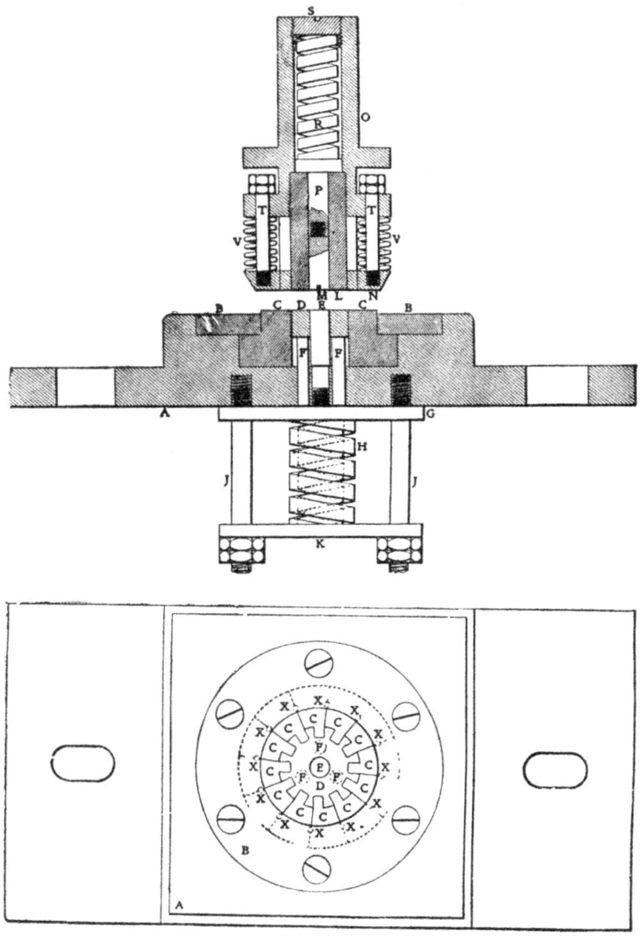

Fig. 146.

und seines Teilkopfes bezüglich Genauigkeit nicht sicher, so ist es am besten, gegenüberliegende Nuten in einem Ausrichten zu fräsen, d. h. man schneidet Ober- und Unterseite ohne Losspannen des Stückes aus dem Futter.

Zur Herstellung der Segmente macht man eine Lehre und Profilfräser A, nach Fig. 148 und 149, hobelt den Stahl für die Segmente auf die Form C und befestigt das Gußeisenstück B am

Tisch der Fräsmaschine und bearbeitet es genau mit einem scharfen Seitenfräser. Dann wird C in der in Fig. 148 gezeichneten Stellung aufgespannt und mit Klammern, die in die Löcher D passen, befestigt. Nachdem die erste Form gefräst worden ist, geht man nach Fig. 149 vor. Es wird ein Gußeisenring D, Fig. 147, von 13 mm Dicke gemacht und mit Gewinde für die dort sichtbaren Schrauben E versehen. Dann wird das Stück für die Segmente auf die Form in Fig. 147 gehobelt und in die richtigen Längen für die einzelnen Segmente geschnitten. Nun wird der Stempel L auf einen Dorn gesteckt und mit den Segmenten in ihrer Stellung mittels des Ringes D und der Schrauben E zusammengehalten und das eine Ende auf Maß abgedreht. Dann befestigt man die Grundplatte A auf der Planscheibe und dreht sie passend zu dem abgedrehten Segmentumfang und zum Ring B aus. Jetzt wird der Dorn aus dem Stempel L ausgetrieben und ein abgesetzter Kupfer- oder Rotgußstöpsel dafür in das Ende von L eingetrieben. Segmente und Stempel werden nun zusammen in die Grundplatte A eingetrieben und fertig abgedreht, worauf die Segmente in ihre endgültige Stellung eingetrieben werden. Schließlich bohrt man die Löcher für die 12 Stifte x, Fig. 146, und setzt diese ein. Dann wird der Ring B angepaßt und mit Schrauben befestigt.

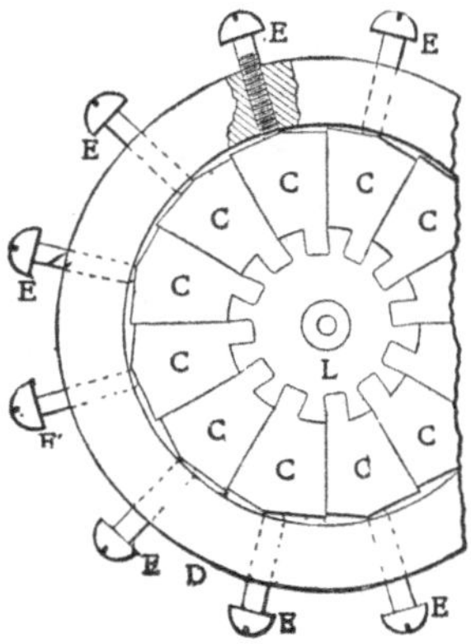

Fig. 147.

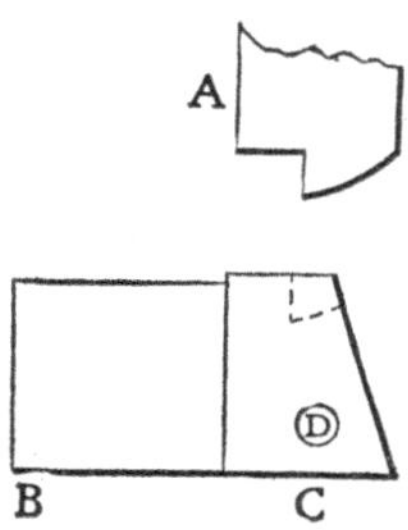

Fig. 148.

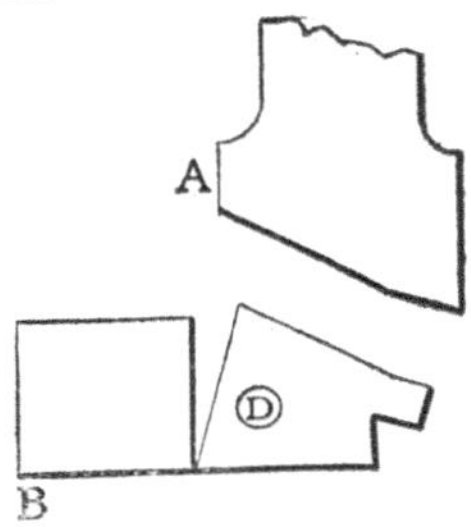

Fig. 149.

Jetzt kann man den Stempel wieder heraustreiben, vorausgesetzt, daß man beim Eintreiben nicht den Rotgußstöpsel vergessen hat. Der Auswerfer M, Fig. 146, ist in den durch eine Feder R betätigten Plunger P eingeschraubt, damit man ihn des leichteren Ausrichtens halber herausnehmen kann. Der Auswerfer D ist noch nicht eingesetzt und wird nach dem Ausrichten einfach auf die Tragstifte F niedergelassen. Durch Anziehen der Bolzen T, Fig. 146, mittels der

Muttern, wird der Abstreiferring gehoben, so daß der Stempel beim Ausrichten in den Unterteil eintreten kann.

H und R sind Federn aus angelassenem Vierkantstahl. Von den Federn V sind acht vorhanden. S ist eine Stellschraube, die die Spannung der Feder R nachzustellen gestattet. Die Schnittplatte muß gehärtet und angelassen werden, letzteres gerade, bis die Farben zu erscheinen beginnen. Der Stempel kann auf ein mittleres Strohgelb angelassen werden. Wenn der Auswerfer D aus weichem Stahl gemacht wird, muß er im Einsatz gehärtet werden.

Für die Abstreifer, besonders wenn sie nicht zuviel Arbeit zu leisten haben, kann mit Vorteil Bandagenstahl verwendet werden, da dieser in vieler Beziehung sehr brauchbar ist, ohne Anlassen genügend hart und dabei zäh und widerstandsfähig genug ist. Er hat jedoch auch seine Fehler, von denen der wichtigste die schwierige Bearbeitung ist.

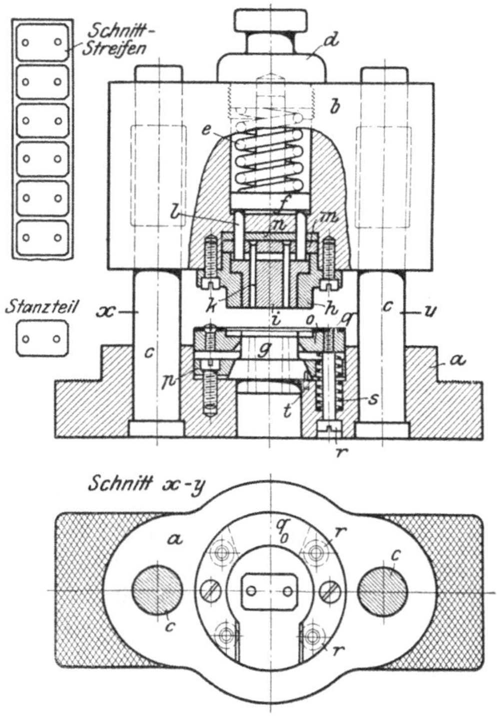

Fig. 150.

β) Verbundwerkzeuge mit Säulenführung (Säulenwerkzeuge).

Die Werkzeuge mit Säulenführung bauen sich im allgemeinen etwas niedriger als die im folgenden besprochenen Werkzeuge mit Zylinderführung und lassen sich, was für kleinere Werkstätten von Vorteil ist, so einrichten, daß innerhalb gewisser Grenzen in dieselbe Führung verschiedene Werkzeuge eingesetzt werden können. Sie werden entweder mit zwei oder mit vier Säulen gebaut.

Fig. 150 gibt wieder ein Schulbeispiel[1]) für ein derartiges Werkzeug, welches die in Fig. 91/94, 121 gegebene Reihe in der Entwicklung der Werkzeuge für denselben Schnitt weiterführt.

[1]) W. T. 1909, S. 581; Stock, Schnittwerkzeuge.

Jedes derartige Werkzeug besteht aus einem Unterteil a und einem Oberteil b, deren gegenseitige Lage durch die Führungssäulen c gesichert ist. Während der Unterteil in gewöhnlicher Weise auf dem Pressentisch festgeschraubt wird, ist die Verbindung des Oberteils mit dem Pressenstößel verschiedenartig. Eine der gebräuchlichen Befestigungen ist bei d gegeben. Der Zapfen ist ausgedreht, so daß er in eine seitlich offene T-Nut des Stößels eingeschoben werden kann. Führen sich die Säulen unmittelbar im Oberteil, so wird man vorteilhaft Ober- und Unterteil zusammengießen, bohren und dann erst teilen. Werden die Führungen dagegen durch besondere Stahlbüchsen, welche auf die Säulen aufgepaßt werden, gebildet, so kann man diese Büchsen, wie Fig. 151 zeigt[1]), mit Weißmetall außen eingießen. Diese Büchsen brauchen nicht auf der ganzen Länge zu führen, sondern können in der Mitte frei gedreht werden, so daß sich im Betriebe dort eine Ölkammer bildet, in welche auch das Schmierloch mündet. Die Muttern aus Maschinenstahl werden im Durchmesser um 5 mm größer als die entsprechende Weißmetallfläche gehalten. Die Löcher für die Weißmetallbüchsen werden konisch gebohrt und mit zwei Längsnuten versehen, die einerseits das Verdrehen der Weißmetallbüchsen zu verhindern haben, andererseits beim Herausschlagen derselben deren richtiges Wiedereinsetzen gewährleisten müssen. Beim Eingießen der Büchsen prüfe man erst, ob der Oberteil genau senkrecht in den Schnittstempel eingedrückt und ob der Abstreifer gerade zurückgeschraubt worden ist. Die Stahlbüchsen werden durch untergelegte Eisenstücke gegen den Oberteil angedrückt und der Zwischenraum mit Weißmetall ausgegossen. Dann zieht man die Muttern fest. Diese Passung hat den Vorteil, daß sie von dem fertigen Schnitt ausgeht also für eine Werkstätte, die nicht auf die Massenherstellung ähnlicher Werkzeuge eingerichtet ist, leichter herzustellen ist als die genau passende gebohrte Führung des ersten Werkzeuges. Natürlich ist die erstere aber haltbarer. Man gibt vorteilhaft beiden Säulen verschiedene Durchmesser, damit der Oberteil bei symmetrischen Schnitten nicht einmal verkehrt aufgesetzt wird. Die Führungen im Oberteil werden auch manchmal geschlitzt und durch Klemmschrauben zusammengezogen, damit man der Abnutzung derselben begegnen kann. Das eigentliche Werkzeug deckt sich in seiner Bauart mit den bereits besprochenen Verbundwerkzeugen. Im Unterteil ist, wie bei allen größeren Schnitten, der Stempel g in einer Ausdrehung zentriert und wird mittels des Über-

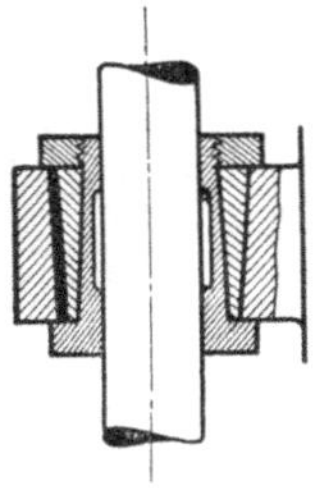

Fig. 151.

[1]) Illustrierte Zeitung für Blechindustrie 1909, S. 2398.

lagringes p und eines angedrehten Konus durch vier versenkte Kopfschrauben niedergehalten und durch den Stift t gegen Drehung gesichert. Auf dem Stempel führt sich der Abstreifer o, der gleichzeitig als Führung q für den Blechstreifen ausgebildet ist; er wird durch vier Federn s und Schrauben r gehalten. Im Oberteil ist der Schnittring h zentriert und durch versenkte Zylinderkopfschrauben mittels eines Flansches gehalten, während der innere Zentrierrand desselben die Kopf- und Gegenplatte m, n für die Lochstempel k festhält. In der inneren Ausdrehung des Schnittringes führt sich der obere Auswerfer i, der durch die Druckstangen l, Druckplatte f und Feder e betätigt wird und gleichzeitig als Führung für die Lochstempel k dient. Diese werden bis knapp über das Blech geführt, so daß Abbrechen und Verlaufen möglichst vermieden ist. Der obere Auswerfer drückt das Stück in den Streifen und prägt diesen gleichzeitig glatt.

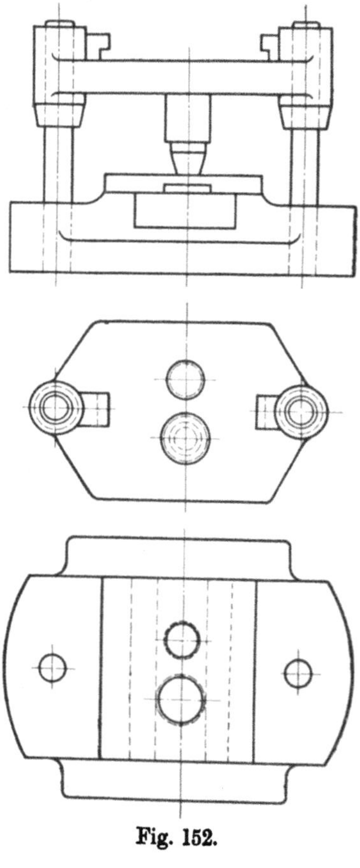

Fig. 152.

Subpresse mit Führungssäulen für ein Folgewerkzeug. (Vgl. Fig. 126.) Die Authomometer Company in St. Louis, Mo. stellt eine Maschine unter dem Namen „Burroughs Addiermaschine" her, deren Konstruktion eine große Menge genauer Stanzarbeiten enthält. Für alle diese Arbeiten werden Subpressen verwendet. Diese Maschinen enthalten eine ganze Reihe Räder, Hebel, Knaggen und Zahnräder, welche alle durch Stanzarbeit aus kaltgewalztem, blankem Stahl hergestellt werden. Die Säulenwerkzeuge, ungefähr 300 Stück, sind alle von gleicher Bauart.

Die Zeichnungen, Fig. 152, zeigen die daselbst regelmäßig verwendete Bauart, in die ein Folgewerkzeug zur Herstellung einer Unterlagscheibe eingesetzt ist. Die Konstruktion ist vollständig verschieden von der als Normalform bekannten Bauart der Firmen Pratt & Whitney und E. W. Bliss; es werden Werkzeuge dieser Bauart für fast alle Stücke verwendet.

Der große Vorteil dieser bei der Authomometer Co. in steter Verwendung befindlichen Werkzeuge ist die Gewißheit, die Stempel und

Schnitte unbedingt genau einstellen zu können. Die Abnutzung der Führungen ist über alle in Gebrauch befindlichen Subpressen verteilt und ist sehr gering, da die Führungen nur den Seitendruck der Verbindungsstange aufzunehmen haben. Weiter ist es unmöglich, die Werkzeuge falsch in die Maschine einzusetzen oder während des Einbaues zu beschädigen. Die wohl höheren Anschaffungskosten werden bei weitem durch die Tatsache aufgewogen, daß die Werkzeuge selbst viel schneller ausgewechselt werden können, so daß man nicht allein unmittelbar Zeit erspart, sondern auch die Pressen eine viel kürzere Zeit leerstehen hat. Bei Anwendung der Subpresse wird die gesamte Einstellung der Werkzeuge ein für allemal auf der Werkbank vorgenommen, wo alle Teile leicht zugänglich und sichtbar sind, während richtig eingesetzte Prisonstifte und Schrauben jederzeit gestatten, Teile, die zum Schleifen auseinandergenommen werden müssen, wieder fehlerlos zusammenzubauen.

Diese Subpressen werden sehr sorgfältig gemacht, besonders was die Parallelführungen und deren Passungen in den Büchsen betrifft. Die hier abgebildeten Stempel sind so geformt, daß sie unmittelbar an dem beweglichen Teil der Subpresse angebracht werden können, doch kommt für die meisten Stempel anderer Form eine Kopfplatte zur Verwendung. In dem obigen Werk wird der Stempel meistens ungehärtet gelassen; falls der Ausschnitt verwendet wird, bleibt die Stirnfläche des Stempels vollständig eben. Das Spiel in der Schnittplatte beträgt auf allen Seiten 5 % der Dicke des verarbeiteten Materials.

Zahlen über die relative Verwendung der Blockwerkzeuge gegenüber unmittelbar in die Presse eingebauten Werkzeugen zu geben ist wohl nicht nötig; es mag genügen, daß — wenigstens in Amerika — sie die Verwendung der letzten um ein Vielfaches übersteigt. Sie sind übrigens beinahe unverwüstlich.

Die in Werkzeugen nach Fig. 152 hergestellten Stücke stellen erstklassige Werkstättenarbeit dar.

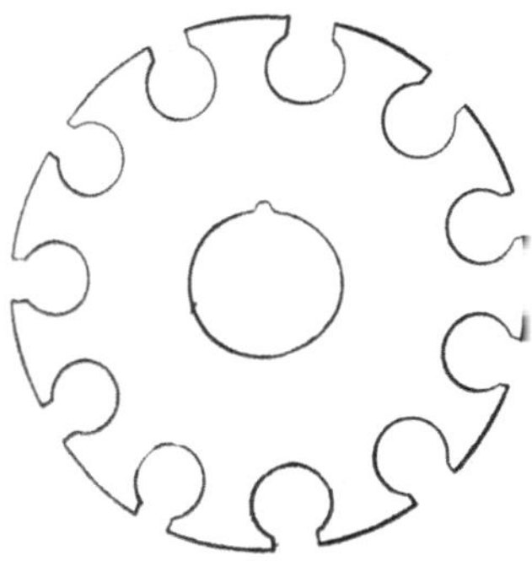

Fig. 153.

Verbundschnitt mit vier Führungssäulen für kleine Ankerscheiben. Im folgenden ist eine Herstellungsweise von Verbundschnitten für kleine Ankerscheiben beschrieben, die von jenen in Fig. 146/49 ziemlich verschieden ist.

Fig. 153 ist das Blankett aus Blech, das mit diesem Werkzeug hergestellt worden ist, in natürlicher Größe, doch hat man viel größere Scheiben auf Werkzeugen dieser Art hergestellt.

Fig. 155 gibt einen Schnitt und einen Grundriß des Unterteiles. A ist der Schnittstempel, der in der Drehbank ausgebohrt und auf dem Dorn abgedreht werden muß, wobei man genug Material für das Fertig-

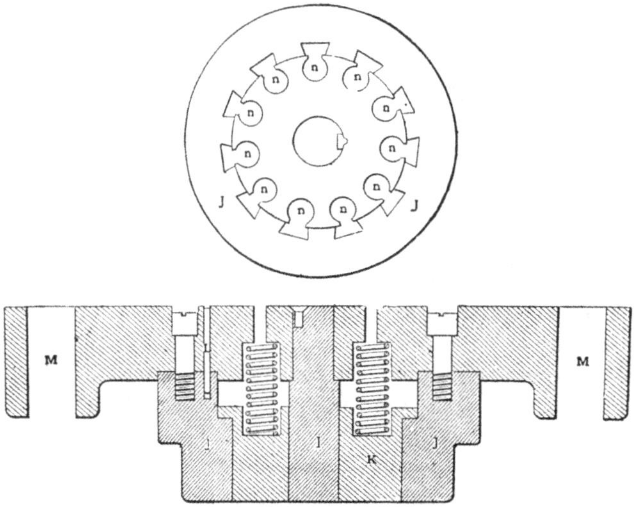

Fig. 154.

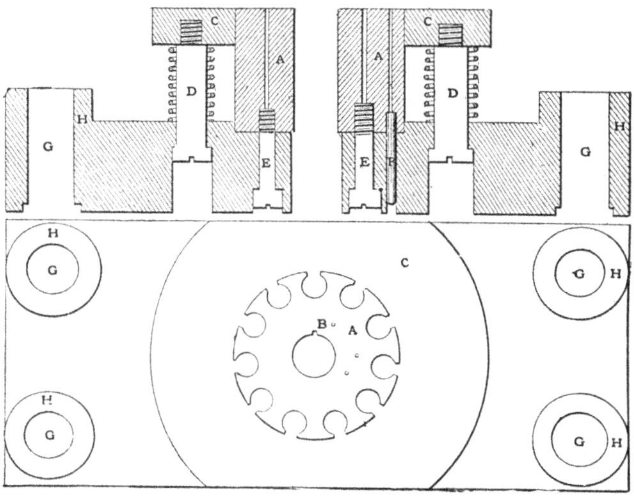

Fig. 155

schleifen stehen läßt. Ein Lehre mit elf gleichmäßig ausgeteilten Löchern und einem Putzen, der in das Loch des Schnittstempels paßt, kann für das Bohren und Ausreiben der Löcher für die Umfangsnuten in Verwendung treten. Dann wird das Stück zwischen Spitzen auf der

Fräsmaschine aufgespannt und die Wand bis zu den Löchern in der verlangten Breite durchgeschnitten. Hierauf wird bei B der Sitz für den Rundkeil eingefeilt und der Schnittstempel gehärtet, auf lichtes Strohgelb angelassen und auf Maß geschliffen.

C ist der Abstreifer, der durch sechs Schrauben D in seiner Stellung gehalten und durch Federn nach aufwärts gedrückt wird. E sind vier Schrauben, welche den Schnittstempel in der gußeisernen Gesenkplatte, hier zugleich Grundplatte, halten. F ist ein Prisonstift, der den Stempel gegen Drehung sichert. G sind vier Löcher von 26 mm Durchmesser, welche die Führungssäulen der Subpresse aufnehmen und mit Ausdrehungen für deren Köpfe versehen sind.

Fig. 154 ist ein Schnitt durch den Oberteil. Bei Herstellung dieses Werkzeuges wird der äußere Schnittring J auf der Drehbank aufgespannt und fast auf Maß abgedreht und ausgebohrt. Die Nuten werden, unter einem Winkel von 10^0 auf einer Seite schwalbenschwanzförmig ausgeschnitten, auf 2^0 abgeschrägt, wobei das größere Maß an der Schnittkante ist. Hierauf wird der Schnittring gehärtet und geschliffen, bis der Stempel eintreten kann, ebenso am Außenumfang, bis er in die Ausdrehung der Kopfplatte paßt. Der innere Stempel I wird gedreht, gehärtet und geschliffen, nachdem die schwalbenschwanzförmige Nut, welche den Keilnutenstempel aufnehmen soll, eingearbeitet worden ist. K ist der Abstreifer, der von dem Schnittring getragen und durch Federn betätigt wird. L ist ein Prisonstift, der den Schnittring bei seiner Arbeit an der Drehung in der Kopfplatte verhindert. M sind wieder die Löcher für die Führungssäulen der Subpresse.

In Fig. 154 sind n Teile des Schnittringes, die nach dem Schleifen des Ringes eingepaßt werden müssen, indem sie in die Schwalbenschwanznuten eingetrieben werden, bis sie 0,4 mm unterhalb der Stirnfläche stehen. Nun läßt man den Schnittring in dem Unterteil andrücken und reißt den Umriß an, bezeichnet alle Teile und entfernt alles Material bis zur angerissenen Linie. Dann setzt man den Stempel wieder ein und kontrolliert die Arbeit. Nach Fertigstellung härtet man und treibt die Stücke in ihre richtige Stellung ein. Da die Nuten abgeschrägt sind, wird natürlich jedesmal ein Preßsitz erreicht werden.

Die nächste Arbeit ist die Befestigung von Schnittring und Stempel ohne Abstreifer in den betreffenden Grundplatten. Zu dem Zweck setzt man sie zusammen, richtet aus, unterbaut mit Prismen und klemmt sie zusammen. Dann bohrt man die vier 26 mm Löcher für die Führungssäulen der Subpresse und reibt sie aus. Diese Säulen müssen aus Werkzeugstahl gemacht, gehärtet und geschliffen werden, bis sie genau in die Kopfplatte passen, und auf ungefähr 50 mm vom Kopf ein wenig größer gelassen werden. Sie werden von unten in die Grundplatte des

Unterteiles eingetrieben, worauf alle Teile des ganzen Werkzeuges zusammengebaut werden, die, falls sorgfältig gearbeitet worden ist, ein für allemal passen müssen.

Ist es nötig, die Schnittplatte wieder anzuschleifen, so treibt man die Führungssäulen sorgfältig heraus und bringt alles wieder in Ordnung. Der Schnittring J und die Außenkante des Schnittstempels A müssen gerade geschliffen werden, die Innenseite soll eine Schräge von $^1/_{200}$ haben.

Säulenwerkzeug für Ankerscheiben mit besonderem Abstreifer. Ankerscheiben nach Fig. 156 werden im ganzen in großen Mengen in einem Verbundschnitt mit besonderem Abstreifer hergestellt; dieses Werkzeug ist in Fig. 157/58 abgebildet. Arbeit dieser Art wird gewöhnlich auf Pressen, die am Ober- und Unterteil besondere Auswerfer besitzen, hergestellt, so daß die Abstreifer im Schnitt entbehrlich werden. Das im folgenden beschriebene Werkzeug sollte jedoch auf einer Presse ohne diese Einrichtung verwendet werden. Fig. 157 ist ein Grundriß und Schnitt des Werkzeugoberteiles, Fig. 158 enthält Grundriß und Schnitt des Unterteiles, sowie eine Seitenansicht des Werkzeuges mit dem Abstreifer. Dieser war besonders mit Rücksicht auf die Schwierigkeiten, die in derartigen Werkzeugen damit entstehen, entworfen worden, d. h. das Blankett soll sicher und richtig abgestreift werden. Dies ist ziemlich schwierig, wenn die Presse mit keinem besonderen Auswerfer versehen ist, da die lange Schnittlinie eine beträchtliche Kraft hinter dem Abstreifer erfordert und nicht genug Platz vorhanden ist, um genügend starke Feder- oder Gummipuffer anzubringen, ohne das Werkzeug unnötig verwickelt zu bauen.

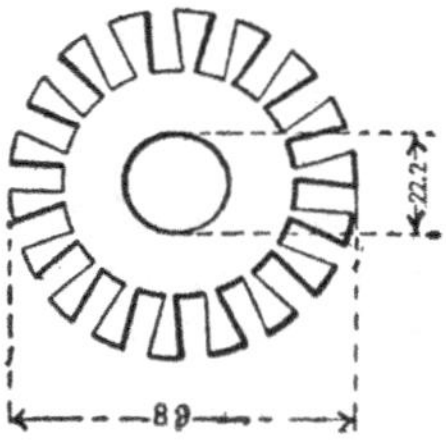

Fig. 156.

Über die beste Art und Weise, den Abstreifer einzurichten, sind die Meinungen geteilt. Manche halten einen feststehenden Abstreifer für eine besondere Arbeit für das beste, andere ziehen die beweglichen vor, während manchmal beide gleich gut arbeiten. Feststehende Abstreifer verbiegen oft das Blankett, während bewegliche es ebnen und niederhalten, bevor die Werkzeuge in Tätigkeit kommen, was oft von Wichtigkeit ist, wenn eine besonders genaue Arbeit erzielt werden soll. Werden Federn oder Gummi verwendet, so wird oft ein großer Teil der Kraft der Presse zum Zusammendrücken dieser verbraucht, bevor die eigentliche Preßarbeit beginnt. Dieser Punkt muß berücksichtigt werden, besonders wenn die Presse bereits bis an die Grenze ihrer Leistungsfähigkeit belastet ist. Man bemerkt, daß der Schnitt für das Mittelloch sich in dem Unterteil befindet, so daß der Abfall durchfallen kann. Das Werkzeug ist als Säulenwerkzeug mit vier Führungssäulen H

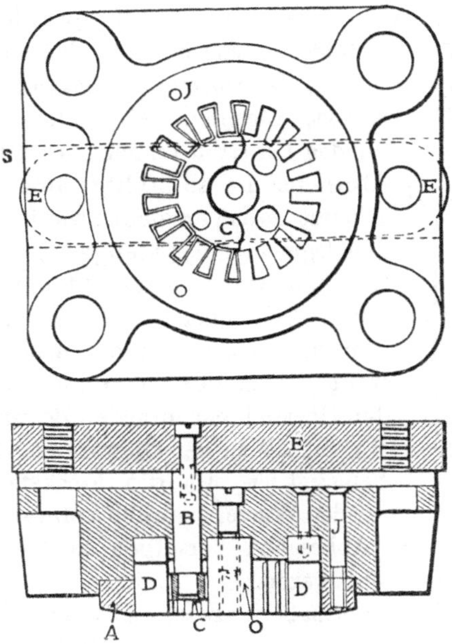

Fig. 157.

gebaut, die im Unterteil befestigt sind und gut gleitend im Oberteil gehen. Der Schnitt A für den Außendurchmesser der Scheibe ist in einer Ausdrehung des Oberteils genau gesichert und durch drei Schrauben J gehalten. Die Schnittplatte für die Nuten ist aus Segmenten, D, aufgebaut, so daß beim Bruch eines Teiles schnell Ersatz geschaffen werden kann, ohne daß der ganze Ring verloren ist. Jedes Segment wird von einer Kopfschraube mit versenktem Kopf gehalten. Der Stempel O stellt das mittlere

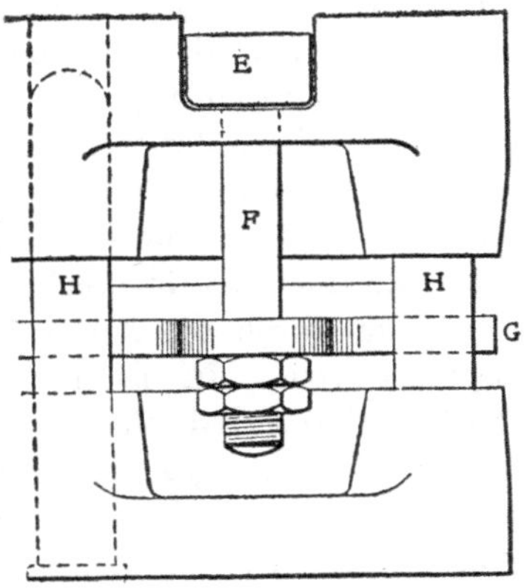

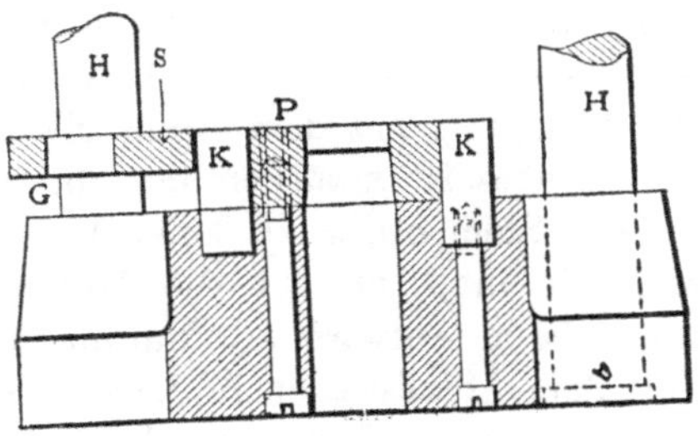

Loch her und arbeitet mit der Schnittplatte P zusammen, Fig. 158. Das Loch ist auf ungefähr 6,5 mm zylindrisch, so daß man, ohne den Durchmesser zu ändern, nachschleifen kann Unterhalb ist es konisch, damit der Abfall frei durchfallen kann. Der Stempel für die Nuten ist aus Segmenten K aufgebaut, die in derselben Weise wie die in der Schnittplatte gehalten und gesichert werden.

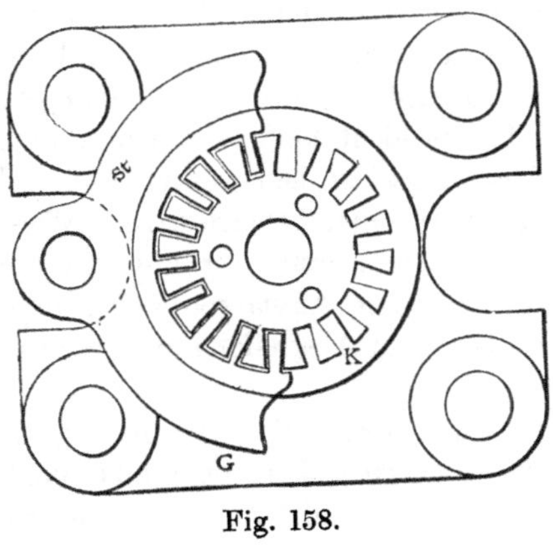

Fig. 158.

Es folgt nun die Einrichtung des Abstreifers, der hier Hauptsache ist. Er ist vollständig in Fig. 159 abgebildet. Eine Stange E ist in eine durchgehende Nut auf der Rückenfläche des Oberteiles eingelegt. An dieser sind zwei Stangen F befestigt, die den Abstreifer G für den Unterteil tragen. Dieser Abstreifer ist sorgfältig ausgearbeitet, so daß er leicht zwischen den Segmenten des Stempels gleitet, und wird in seiner Stellung durch Schrauben und Gegenmuttern gehalten und nachgestellt. C ist der Abstreifer für den Oberteil, der mit vier Bolzen B ebenfalls an der Querstange E befestigt ist. Da die Fortsätze an der Wurzel ziemlich schwach sind, muß der Abstreifer, um ein Abbrechen derselben zu verhindern, dicker als sonst nötig gemacht werden.

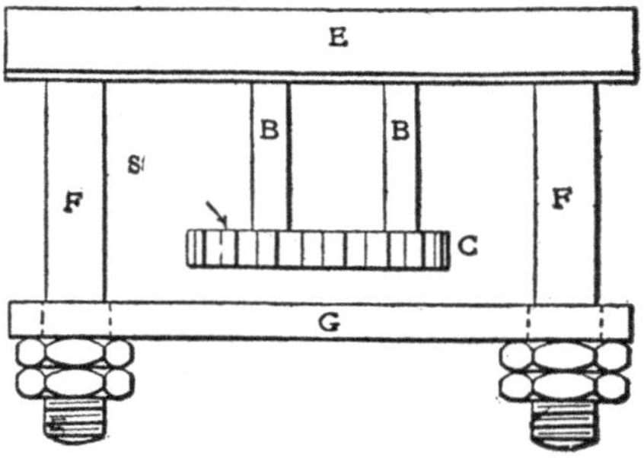

Fig. 159.

Diese Einrichtung arbeitet folgendermaßen: Beim Abwärtshub gehen die Abstreifer mit dem Schnitt herunter, bis der Abstreifer C auf dem Material aufruht. Er steigt dann in dem Oberteil um einen Betrag, der dem Unterschied zwischen der Dicke der Querstange und der Tiefe der Nut, in welcher er ruht, entspricht, bis seine Aufwärtsbewegung durch die Berührung der Stange E mit dem Stempelkopf begrenzt wird. In dieser Stelle des Hubes ist natürlich der untere Abstreifer unterhalb der Stirnfläche des Stempels, so daß der Unterteil des Werkzeuges ausgeschnitten sein muß, um für die Muttern in dieser Stellung Platz zu machen. Beim Aufwärtshub nimmt der Abstreifer G des Unterteiles, der mit dem Abfallstreifen in Berührung kommt, diesen mit und bringt gleichzeitig beim Weiterhub die Querstange E in der Nut zum Aufliegen, so daß durch diese Abwärtsbewegung der obere Abstreifer C die Scheibe auswirft. Der Abfall und die Scheibe liegen nun lose auf der Schnittplatte, so daß sie leicht vom Arbeiter entfernt werden können. Es ist nicht nötig, die Herstellung der verschiedenen Teile dieses Werkzeuges zu beschreiben, da sich diese mit den Teilen der im Vorangegangenen beschriebenen Werkzeuge decken. Die größte Sorgfalt ist der Herstellung von Stempel und Segmenten zuzuwenden. Alle Teile mit Ausnahme der Körper für Ober- und Unterteil, der Schrauben und Abstreifer sind gehärtet.

Säulenwerkzeug für genaue Loch- und Stanzarbeit. Das in Fig. 161 abgebildete Werkzeug dient zum Stanzen eines Blanketts aus blankgewalztem Bessemerstahl von 6,4 mm Breite und 3,2 mm Dicke, von der in Fig. 160 gezeichneten Länge[1]). Aus dem Schnitt und Grundriß des Werkzeuges läßt sich alles leicht entnehmen. Der Oberteil A ist zur Aufnahme der gehärteten und geschliffenen Schnittplatte B, Fig. 164, ausgearbeitet. Die Schnittplatte ist mit Preßsitz eingepaßt und wird durch die Schrauben, die in der Abbildung ersichtlich sind, gehalten; sie ist aus Teilen aufgebaut, deren größter Teil gefräst ist. Sie hat einen Auswerfer C, der durch die Stifte D in Verbindung mit den Federn E niedergedrückt wird und dessen freie Bewegung genügend groß für das Nachschleifen und eine Blankettdicke ist. F sind die Einstell-

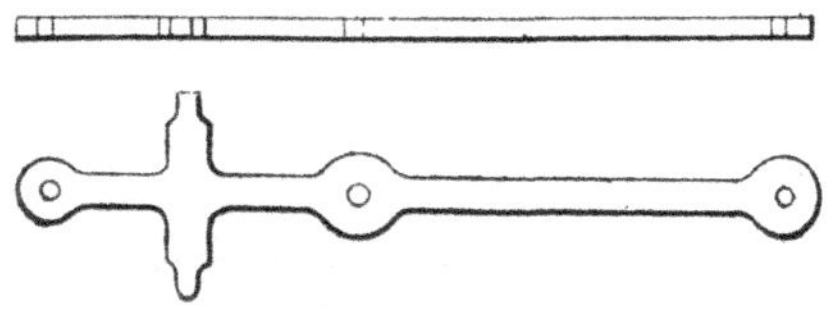

Fig. 160.

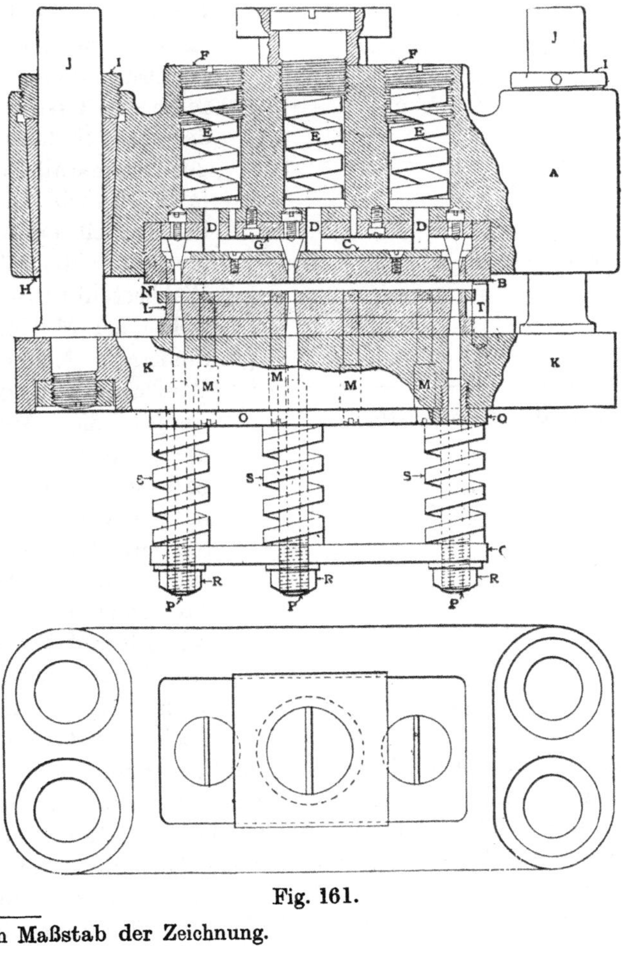

Fig. 161.

[1]) Im Maßstab der Zeichnung.

schrauben, mittels derer die Spannung in den Federn geregelt werden kann. G, Fig. 162, ist die Kopfplatte für die Lochstempel; sie paßt genau in die Ausfräsung der Schnittplatte und ist am Oberteil A durch Schrauben und Prisonstifte gesichert. Die Führungssäulen J werden, nachdem sie geschliffen und gut in den Unterteil K eingepaßt worden sind, durch die Muttern scharf angezogen. Die Einstellmuttern I sind auf die Führungssäulen J gut passend aufgerieben und mitteln gleichzeitig den Oberteil A aus, der jetzt für das Eingießen der Weißmetall-Büchsen H fertig ist. Die Ausdrehung im Oberteil erhält Nuten, die ein Mitdrehen der Weißmetallbüchse beim Anziehen der Muttern I verhindern. Nachdem das Weißmetall eingegossen ist, werden die Muttern I etwas nachgerieben, so daß sie ein wenig Spielraum haben. Der Unterteil, die Grundplatte K, wird dann zur Aufnahme des Stempels, Fig. 165, gehobelt und dieser mittels vier Schrauben daran befestigt. Die anderen acht Löcher von 6,5 mm Durchmesser dienen für die Stifte M, die darin mit Spiel gehen und an den Auswerfer N angeschraubt sind. Sie sind mit einem Absatz versehen, der in die Ausdrehung der Grundplatte paßt, welche gerade tief genug ist, um den Auswerfer so weit steigen zu lassen, bis er bündig mit dem Stempel steht; die anderen Enden der Stifte sind auch auf 6,5 mm Durchmesser ab edreht und mit einem Schlitz für den Schraubenzieher

Fig. 162. Kopfplatte.

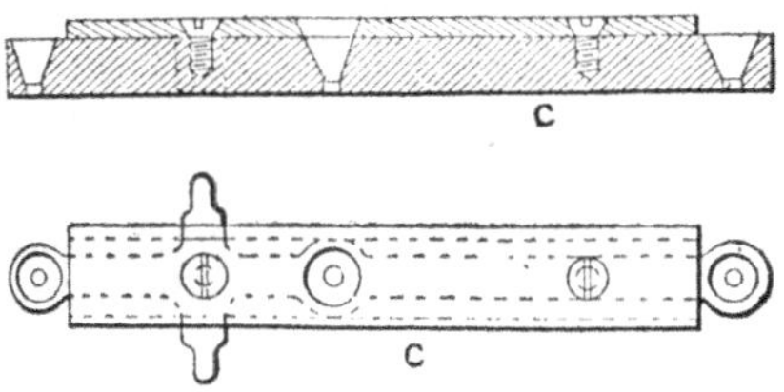

Fig. 163. Auswerfer.

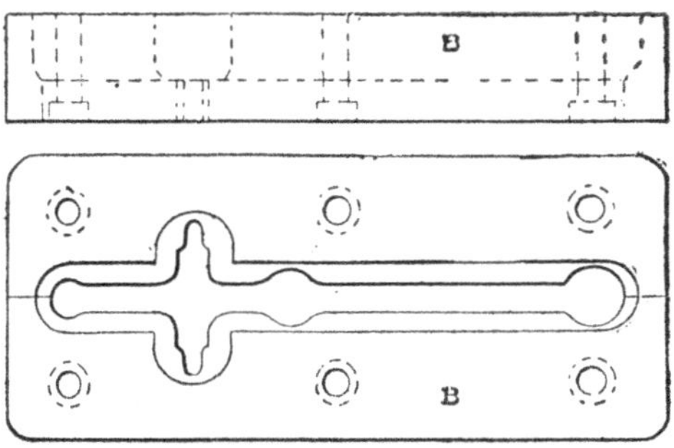

Fig. 164. Schnittplatte.

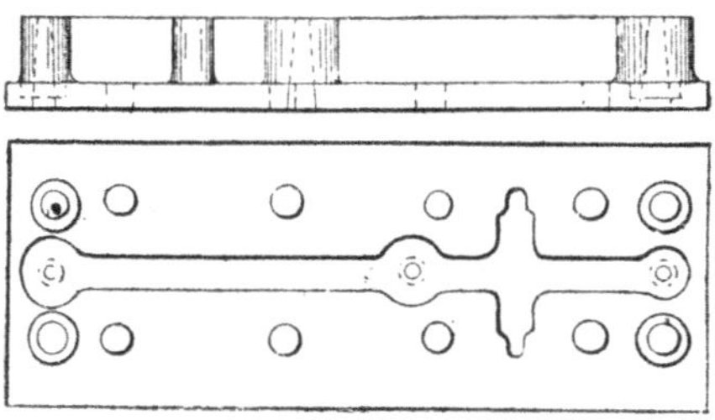

Fig. 165. Stempel.

versehen, so daß die untere Fläche des Absatzes als Anschlag für die Platte O dient. Sie hat drei Löcher von 13 mm Durchmesser, welche den Bolzen P gerade so viel Spiel geben, daß sie leicht an ihnen gleiten kann. Diese Bolzen sind vollständig der Länge nach durchbohrt, so daß die Abfälle der Lochstempel durchfallen können, und sind in die Unterfläche der Grundplatte eingeschraubt. Die Platte Q ist in der gleichen Weise eingepaßt und wird durch die Muttern R niedergepreßt, bis die Federn genügend zusammengedrückt sind.

Die Arbeit des Werkzeuges und die Herstellung der Stücke braucht keine weitere Beschreibung, da aus den vorangegangenen Beispielen und den Zeichnungen alles erklärlich ist.

γ) Verbundschnittwerkzeuge mit Zylinderführung (Blockwerkzeuge).

Die Herstellung der Werkzeuge für Blockwerkzeuge. In dem Buche Woodworths „American Tool-Making and interchangeable Manufacturing", Kap. 28, wird bloß von der Konstruktion und Arbeit der Subpresse und den dazugehörigen Werkzeugen gesprochen, und da die dort erörterten Herstellungsweisen für die Taschenuhren bzw. die Groß-Uhrenfabrikation nicht genügen, bringt Woodworth an dieser Stelle auch etwas über die Herstellung diesbezüglicher Werkzeuge.

Die in obiger Quelle angegebenen Arbeitsweisen beschreiben nicht die Herstellung der Stempel und Schnitte, nehmen aber an, daß der Stempel durchgeschert werden muß. Dies ist nun bei der Herstellung der Uhrräder und anderer genau gelochter, unregelmäßiger Stücke unzulässig, da alle Stempel auf Maß gefräst werden und die Stempel für die Speichenräder innerhalb des Kranzes geschliffen und poliert sind, so daß ein Scheren durch die Schnittplatte zum Einpassen ausgeschlossen ist. Die oben erwähnten Arbeitsweisen sind für unregelmäßige Umrisse dort anwendbar, wo Stempel und Schnittplatte auf Passen gefeilt werden, wenn aber beide Teile vollständig mit der Maschine bearbeitet werden, muß man auf die im folgenden beschriebenen, bewährten Arbeitsweisen zurückgreifen.

Fig. 166 ist ein Längsschnitt durch ein vollständiges Blockwerkzeug mit Stempel und Schnittplatte. Der Ständer A wird auf der Unterseite eben abgedreht und ausgebohrt. In diese Ausbohrung paßt eine besondere Planscheibe, an die ein passender Putzen angedreht ist. An sie wird der Ständer angeschraubt, auf 1 : 24 konisch ausgebohrt und in einem Aufspannen bei festgezogenem Vorgelege vier Längsnuten eingezogen, welche die Weißmetallbüchse gegen Drehung sichern.

Der Ständer wird auf einen Dorn, der in den eben ausgedrehten Konus paßt, gesteckt, der obere, zylindrische Teil angedreht und bei R

Gewinde angeschnitten, an welches später der Deckel C angepaßt wird. Nun wird die Unterfläche des Ständers noch einmal abgedreht und ausgebohrt, solange der Ständer noch am Dorn ist, um so sämtliche Flächen genau parallel, bzw. rechtwinklig zueinander zu erhalten.

Dann wird der Ring C um ungefähr 0,13 mm kleiner als der fertige Durchmesser des Plungers gebohrt, ausgedreht und schließlich ein Gewinde eingeschnitten, das auf den Ständer paßt. Nun wird — nur zu Herstellungszwecken — ein Ring von ungefähr 3,2 mm Dicke gemacht, der in den Ring C paßt und dessen Innendurchmesser um 3,2 mm kleiner als die Ausbohrung des Ständers bei D (rechts) ist. Der Ring C wird nun an den Ständer gegen den Absatz festgeschraubt und auf den richtigen Plungerdurchmesser ausgebohrt.

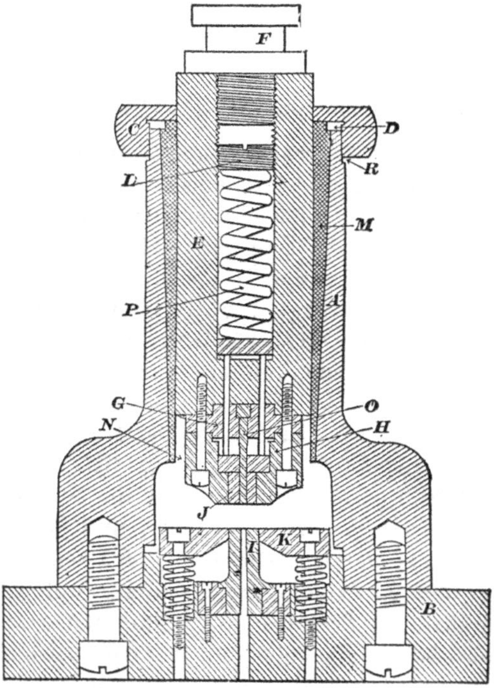

Fig. 166.

Hierauf wird die Grundplatte B gehobelt und an die obenerwähnte Planscheibe angeschraubt, eben und winklig abgedreht und der Putzen passend für die Ausdrehung des Ständers A gedreht; außerdem wird die Mitte für den Fuß des Stempels I ausgedreht. Der Plunger E wird dann zentriert und in der Lünette zur Aufnahme der Abstreiferfeder P ausgebohrt, wobei gleichzeitig das Gewinde für die Einstellschraube L und den Mitnehmer F geschnitten wird. Der letztere — aus Maschinenstahl — wird in den Stempel bis an den Absatz eingeschraubt, der Plunger in der Bank vorgeschruppt und dann zwischen Spitzen auf Maß geschliffen, so daß eine hochpolierte Oberfläche verbleibt, die nicht mit Schmirgel berührt werden darf.

Nun wird die Kopfplatte G für den Schnittring H hergestellt, indem man ein Stück Material in einem Futter einspannt und einen Putzen an einem Ende auf richtiges Maß abdreht und vor dem Abstechen einen zweiten Putzen auf der gegenüberliegenden Seite vor-schruppt. Dann wird der Plunger wieder auf der Drehbank in der Lünette eingespannt und für den eben angedrehten Putzen der Kopfplatte G ausgedreht, der in die Ausdrehung bis an den Absatz eingetrieben wird. So wird der bereits vorgeschruppte Putzen, vom Plunger mitgenommen, abgedreht, bis er in die rückwärtige Ausdrehung der Schnittplatte H paßt, so daß diese, nachher eingesetzt, unbedingt richtig laufen muß. Hierauf werden noch in den Plunger der Länge nach Nuten eingefräst, so daß er sich nicht in dem Weißmetall drehen kann.

Als nächstes wird die Grundplatte B gebohrt, mit Gewinde versehen und zur Aufnahme der Auswerferfedern und Schrauben ausgebohrt, worauf die Löcher für die Prisonstifte und die Schrauben, welche Ständer und Grundplatte verbinden, gebohrt und ausgebohrt werden. Damit sind Grundplatte, Ständer und Plunger fertig.

Die Schnittplatte H wird im Futter eingespannt, in der Lünette gedreht und zur Aufnahme des Abstreifers J und des Putzens der Kopfplatte G ausgedreht; dann wird sie gebohrt und entsprechend der Zahnwurzel im Musterrad ausgedreht. Zum Schluß wird die Schnittplatte abgestochen und im Futter die Vorderfläche parallel mit der Hinterfläche gedreht.

Das Blankett der Schnittplatte wird ausgeräumt, wobei man mit großer Sorgfalt vorgehen muß. Die Räumnadeln[1]) werden entweder mit drei oder vier Zahnstufen, je nach der Größe der Schnittplatte gemacht, wobei die Stufen von 0,038—0,064 mm ansteigen, und die Zähne gefräst werden. Dabei muß aber der Fräser genau zentral und gerade arbeiten. Die Räumnadeln werden gehärtet und angelassen. Nachdem in der Schnittplatte noch die Löcher für die Schrauben und Prisonstifte gebohrt worden sind, wird sie in Knochenmehl gepackt, sorgfältig erwärmt, dann in Öl abgelöscht und auf leichtes Strohgelb angelassen. Dann wird die Unterfläche der Schnittplatte geschliffen und poliert.

Nun wird der Stempel I und der Abstreifer J gefräst, wobei die Fräsmaschine in ihrer Einstellung belassen wird, bis sie in die Schnittplatte passen, dann in der Spitze und Lünette auf den Durchmesser des Mittelloches im Musterrad gebohrt. Das Loch im Stempel wird nach rückwärts konisch ausgerieben, während das Loch im Abstreifer gerade bleibt.

Für den Stempel I wird eine Grundplatte gemacht, die genau in die eigentliche Grundplatte paßt, wo sie mit Schrauben und Prisonstiften gehalten wird, und eine andere für den Abstreifer J, die leicht in der Ausdrehung auf der Rückseite der Schnittplatte sitzt. Der Grund, daß man den Abstreifer mit einer Grundplatte versieht, liegt darin, daß man dann mit dem Fräser gerade durch arbeiten kann.

[1]) Siehe Fig. 679/80.

und doch einen rechtwinkligen Absatz erhält, gegen den sich der Abstreifer bei seiner Arbeit anlegen kann. Als nächstes wird der Stempel in die Grundplatte eingesetzt und fest niedergeschraubt, worauf er leicht an der Stirnfläche abgeschliffen wird, um so eine vollständig rechtwinklige Schnittfläche zu erhalten.

Schnittplatte, Abstreifer und Kopfplatte werden nun zusammengebaut, mit Schrauben und Prisonstiften am Plunger befestigt und die Einstellschraube L für die Spannfeder P so eingestellt, daß der Abstreifer ungefähr 1,6 mm über der Stirnfläche der Schnittplatte zu stehen kommt. Hierauf wird der Ständer auf der Grundplatte aufgebaut, die Schrauben angezogen und der Plunger, nachdem er ein wenig mit einem öligen Tuch abgewischt und mit ein wenig Flockengraphit bestreut worden ist, sorgfältig in den Ständer niedergelassen, wobei man den Stempel I in die Schnittplatte eintreten läßt, bis er durch den Abstreifer aufgehalten wird. Dann wird der Ring C niedergeschraubt, während sich noch der kleinere Ring innerhalb desselben befindet, so daß Stempel, Schnittplatte und Plunger genau in Linie kommen und gleichachsig im Ständer stehen.

Der Ständer A wird nun auf einer kräftigen Richtplatte umgekehrt aufgestellt, leicht mit einem Bunsenbrenner erwärmt und das Weißmetall M von zwei Seiten gleichzeitig eingegossen. Nach dem Abkühlen des Weißmetalls wird der Plunger zurückgezogen, der Ständer von der Grundplatte entfernt und das Weißmetall bei N, wo es beim Eingießen überlief, abgedreht, so daß der Plunger weiter nach abwärts gehen kann. Nach dieser Arbeit werden Ständer und Grundplatte wieder zusammengebaut — der kleine Ring ist aus der Überwurfmutter C herausgenommen worden —, dann wird der Plunger durchgetrieben und ein Blankett aus Messingblech von ungefähr 0,25 mm Dicke ausgeschnitten. Man zieht jetzt den Plunger wieder zurück und untersucht Stempel und Schnittplatte sorgfältig, ob irgendwo ein Scheren stattgefunden hat, da in diesem Fall der Ständer noch einmal mit Weißmetall ausgegossen werden muß. Wenn jedoch Grundplatte und Ständer genau und sorgfältig hergestellt worden sind, und das Weißmetall mit der richtigen Temperatur eingegossen worden ist, wird man ein vollständig ausgerichtetes Werkzeug erhalten haben. Nun wird die Kopfplatte G zur Aufnahme des Lochstempels O gebohrt und ausgebohrt, der andrerseits wieder in den Abstreifer J und den Stempel I passend gedreht sein muß. Dieser hat auch noch den Zweck, den Lochstempel bei der Arbeit zu stützen.

Der untere Abstreifer K wird hergestellt und ähnlich wie die Schnittplatte mit der Räumnadel ausgearbeitet, für die Schrauben und Führungsstifte ausgebohrt, gehärtet und geschliffen. Für Uhrenarbeit werden im allgemeinen unter dem Abstreifer acht Federn nötig sein. Der Abstreifer wird in der Grundplatte so eingestellt, daß er ungefähr 0,125 mm über der Stirnfläche des Stempels steht, der Ständer dann auf die Grundplatte niedergeschraubt und die Abstreiferfeder P auf die richtige Spannung eingestellt, womit das Werkzeug arbeitsfertig ist.

Blockwerkzeug deutscher Konstruktion[1]). In der angegebenen Quelle wird ein ganz ähnliches Blockwerkzeug, welches gleichzeitig als Schulbeispiel die Reihe der Fig. 91—94, 121 und 150 abschließt, mit allen Teilen in Einzeldarstellung beschrieben. Es ist in Fig. 168/9 und das Blankett mit seinem Abfallstreifen in Fig. 167 wiedergegeben. Da die einzelnen Teile mit Rücksicht auf das eben beschriebene amerikanische Werkzeug selbstverständlich sind, sollen nur die Abweichungen näher besprochen werden. Bei A ist der Plunger zur gleichzeitigen Aufnahme von Kopf- und Gegenplatte für die Lochstempel und die Schnittplatte ausgedreht, wodurch eine sicherere Ausmittlung als bei der doppelten Einpassung des amerikanischen Werkzeuges erreicht wird. Die Einpassung des Schnittstempels in die feste Grundplatte und die davon in gewissem Grade unabhängige Befestigung durch den konisch ausgedrehten Überlagring sichert besser die Zentrierung des Stempels, siehe B. Eine besondere Führungsplatte

Fig. 167.

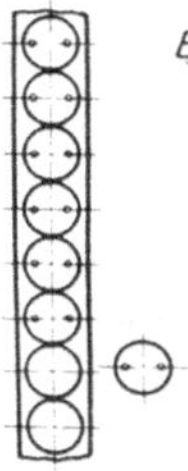

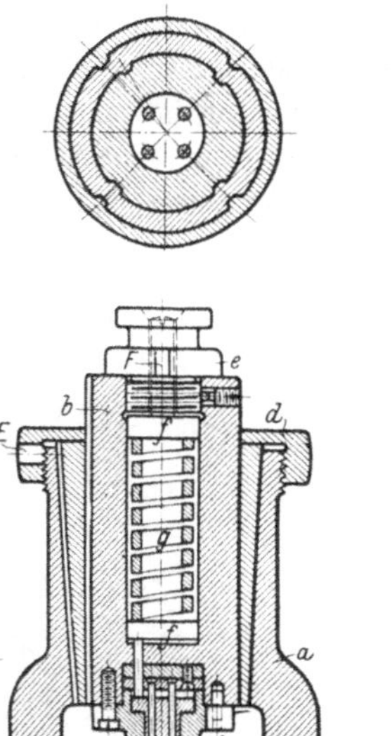
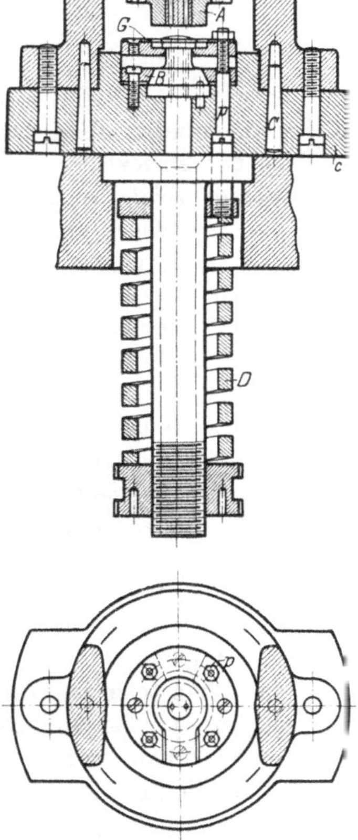

Fig. 168.

[1]) Illustr. Zeitschrift f. Blechindustrie 1909, S. 4; Knopf, Das Blockwerkzeug.

G für das Blech ist mit dem Abstreifer fest verbunden. Dadurch, daß der Federauswerfer für sich allein unter den Werkzeugunterteil verlegt worden ist, baut sich dieser gedrungener und kann die Pressung, ohne an dem Werkzeug etwas zu lösen, geändert werden. An weiteren Verbesserungen sind die Prisonstifte C der Grundplatte, die Anzug-

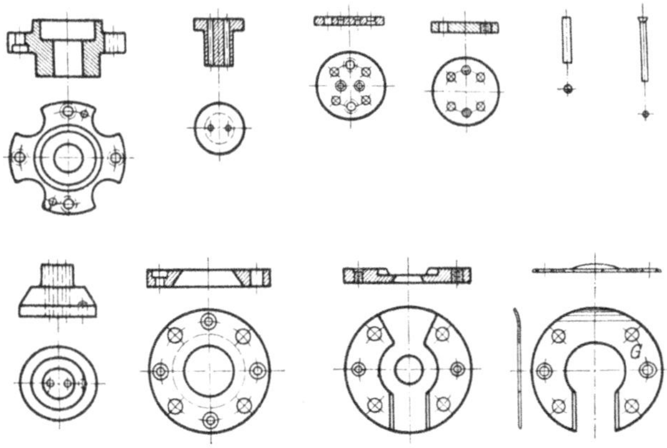

Fig. 169.

löcher im oberen Ring bei E und die von außen mögliche Nachstellung der Druckfeder im Plunger zu erwähnen. Der Schnitt durch den Ständer zeigt die Lage der Nuten für das Weißmetallfutter, von denen eine versetzt ist, damit der Plunger nur in einer Stellung eingeführt werden kann. Fig. 170 zeigt den Notstempel, mittels dessen der Mitnehmer des Blockwerkzeuges in einen gewöhnlichen Stößel eingehängt wird.

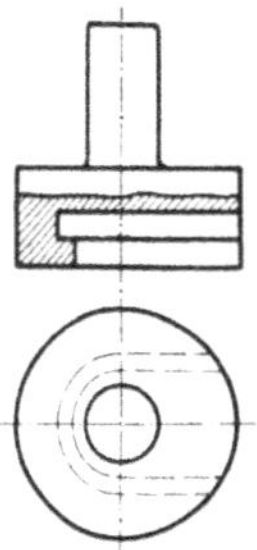

Fig. 170.

Abmessungen der Blockwerkzeuge. Es überrascht einen wirklich, bei einem Gang durch die Werkstätten, die sich mit der Herstellung kleiner Massenartikel beschäftigen, zu sehen, wie gering die Verwendung der Blockwerkzeuge zur Herstellung vieler kleiner Teile aus dünnem Blech ist. Sie sind stabil, genau, verläßlich und handlich und von der Abnutzung der Schlitten und Schienen vollständig unabhängig, wobei sie die Genauigkeit der Teile großer Pressen erreichen.

Die nebenstehenden Zeichnungen, Fig. 171 und 172, sowie die zugehörigen Tabellen sind aus dem Entwurf einer Reihe Blockwerkzeuge entstanden, welche den im folgenden, nach ihrer Wichtigkeit geordneten Bedingungen entsprechen:

1. Genauigkeit.
2. Stärke und Steifigkeit.

3. Geringes Gewicht, soweit dieses mit den beiden ersten Punkten zu vereinen ist.
4. Handlichkeit.
5. Billigkeit und Haltbarkeit.

Der Ständer wird mit einem Kern, dessen Anzug 2° beträgt, gegossen. Der Kern hat vier halbrunde Nuten durch die ganze Länge des Loches, welche das Weißmetall gegen Verdrehen sichern. Das Loch wird konisch gemacht, damit man das Weißmetall, wenn notwendig, leichter entfernen kann. Die Unterfläche wird eben abgedreht und das Loch G nach innen 1° konisch gebohrt. In dieses Loch wird der Zapfen der Grundplatte, der eine Höhe von 19 mm hat, sorgfältig eingepaßt, und diese mittels der beiden Schrauben bb an dem Ständer befestigt. Der Plunger wird gedreht, erhält bei O Gewinde für den Mitnehmer,

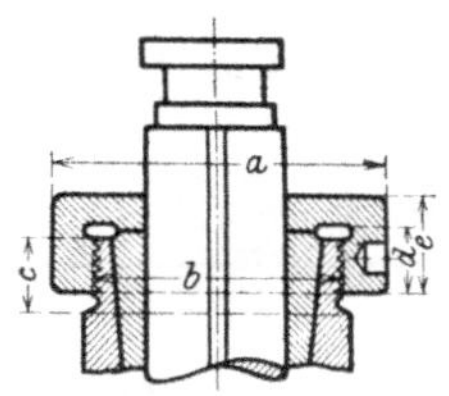

Fig. 172.

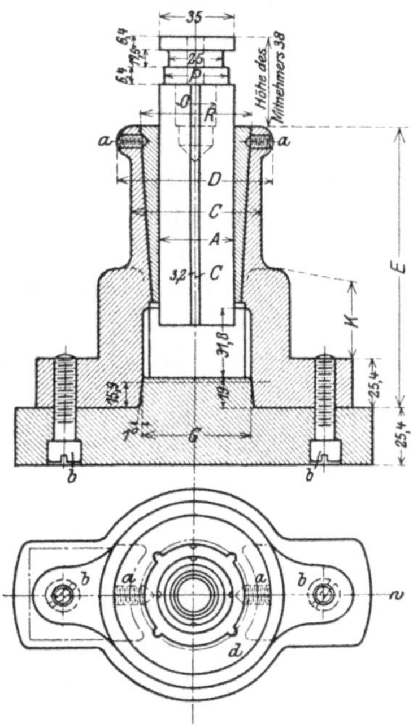

Fig. 171.

20 Gänge/Zoll engl., und wird schließlich auf Maß geschliffen. In den Plunger werden ebenfalls vier halbzylindrische, jedoch ungleich verteilte, Nuten eingefräst, so daß er nur in der richtigen Stellung in die Weißmetallbüchse eingesetzt werden kann. Der Plunger wird dann in geeigneter Weise zentrisch im Ständer gehalten und das Weißmetall eingegossen. Zwei bis vier geschlitzte Spitzschrauben ohne Köpfe werden in den Ständer eingeschraubt, so daß jede zufällige Verschiebung der Weißmetallbüchse verhindert wird. Die Unterseite des Stößels einer größeren Presse ist zur Befestigung des Mitnehmers, der in den Plunger eingeschraubt ist, eingerichtet; dann befestigt man den Ständer am Unterteil der Presse und läßt das Werkzeug einlaufen. Auf der Oberfläche der Weißmetallbüchse ist eine Ölnut, welche den Plunger umgibt, vorgesehen. Die Mitnehmer für die verschiedenen

Blockwerkzeuge werden in ihren Abmessungen einheitlich gehalten, wenigstens was ihre Länge und Einpassung im Pressenstößel der größeren Presse betrifft, so daß nur der Durchmesser des Ansatzes P und das Gewinde O sich ändert. Doch ist die Gangzahl des Gewindes stets die gleiche, nämlich 20 Gänge auf 1″ engl.

Im Gebrauch läßt man die Werkzeugteile am besten dauernd im Ständer und Plunger befestigt. Doch kann man sie natürlich auch leicht entfernen und schnell mit der Gewißheit auswechseln, daß sämtliche Teile unabhängig von allen Veränderungen, die inzwischen in der Presse vorgenommen worden sind, ausgerichtet bleiben.

Eine Konstruktionseinzelheit betreffend, sei erwähnt — wie Woodworth sagt —, daß mehrere Firmen in ihren Blockwerkzeugen statt der runden Kopfform des Ständers die Seiten gerade halten. In das Gußeisen wird dann ein Gewinde geschnitten, auf welches eine im Einsatz gehärtete Überwurfmutter paßt, die gegen die Oberfläche der Weißmetallbüchse angezogen wird, so daß diese mittels des Konus in die Ständerhöhlung gepreßt wird. Woodworth sagt, daß der einzige Punkt, in dem sich diese Konstruktion vor der oben beschriebenen unterscheidet, die höheren Kosten sind. Doch gibt er auch in Fig. 172 diese Einzelheit wieder und eine dazugehörige Tabelle.

Abmessungen verschiedener Größen des Blockwerkzeuges Fig. 171.

<table>
<tr><th>Pressen-Nr.</th><th>A</th><th>B</th><th>C</th><th>D</th><th>E</th><th>G</th><th>K</th><th>O</th><th>P</th><th>R</th></tr>
<tr><td>1</td><td>32</td><td rowspan="6">127</td><td>60</td><td>73</td><td rowspan="6">178</td><td>50</td><td rowspan="3">45</td><td>19</td><td>28</td><td>44</td></tr>
<tr><td>2</td><td>46</td><td>73</td><td>86</td><td>64</td><td>25</td><td>38</td><td>57</td></tr>
<tr><td>3</td><td>57</td><td>86</td><td>98</td><td>76</td><td>32</td><td>44</td><td>70</td></tr>
<tr><td>4</td><td>70</td><td>98</td><td>111</td><td>89</td><td rowspan="3">48</td><td>38</td><td>50</td><td>82</td></tr>
<tr><td>5</td><td>82</td><td>111</td><td>124</td><td>102</td><td>44</td><td>57</td><td>95</td></tr>
<tr><td>6</td><td>95</td><td>124</td><td>136</td><td>115</td><td>51</td><td>64</td><td>108</td></tr>
</table>

Abmessungen verschiedener Größen des Kopfes nach Fig. 172.

<table>
<tr><th>Pressen-Nr.</th><th>a</th><th>b</th><th>c</th><th>d</th><th>e</th></tr>
<tr><td>1</td><td>70</td><td>54</td><td rowspan="2">16</td><td rowspan="2">14</td><td rowspan="2">19</td></tr>
<tr><td>2</td><td>82</td><td>67</td></tr>
<tr><td>3</td><td>95</td><td>80</td><td rowspan="2">17,5</td><td rowspan="2">16</td><td rowspan="2">22</td></tr>
<tr><td>4</td><td>108</td><td>92</td></tr>
<tr><td>5</td><td>120</td><td>105</td><td rowspan="2">19</td><td rowspan="2">17,5</td><td rowspan="2">30</td></tr>
<tr><td>6</td><td>133</td><td>118</td></tr>
</table>

Bemerkung: Alle Maße sind auf ganze Millimeter abgerundet.

Stahl für die Plunger der Blockwerkzeuge. Es bestehen verschiedene Ansichten über das geeignetste Material für die Plunger der Blockwerkzeuge; die einen nehmen Gußeisen, die anderen ziehen Maschinenstahl vor.

Der amerikanische Verfasser gibt aus langjähriger Erfahrung an, daß in jedem Fall, ob es sich nun um einen Plunger von 32 mm oder 127 mm Durchmesser handelt, Stahl das geeignetste Material ist.

Der Stahl kostet weniger als die Hälfte des Gußeisens, so daß man bereits daran ein wenig gewinnt; doch sollte dieser Punkt überhaupt nicht in Frage kommen, wenn er auf Kosten der Dauerhaftigkeit oder Güte des Werkzeuges erreicht würde. In diesem Falle gewinnt man aber gleich auf zwei Seiten. Es ist dabei nicht einmal notwendig, den gelegentlichen Verlust eines schlechten Gusses, der erst nach fast vollständiger Bearbeitung sichtbar wird, einzurechnen. Nach dem Fertigschleifen wird das feine Korn des Stahles sehr bald eine glatte, abgearbeitete Oberfläche zeigen, die natürlich auch die Weißmetallfütterung weniger angreift als das poröse Gußeisen.

Es ist der Einwand gemacht worden, daß der Plunger aus demselben Material wie der gußeiserne Ständer sein soll, damit sich beide infolge der durch die Arbeit eintretenden Erwärmung der Presse gleichmäßig ausdehnen. Der Einwand hat seine Richtigkeit, doch wird dieser Punkt wohl nicht von großem Einfluß sein. Es handelt sich bei dieser Frage um viel wichtigere Punkte. Mehr als einmal enthält der Plunger ein Werkzeug, daß aus einer ganzen Anzahl Stücke aufgebaut ist, welche mit der größtmöglichen Genauigkeit eingepaßt werden müssen. Vergleicht man nun ein Stück Stahl mit seinem durchaus gleichmäßigen Gefüge mit einem Gußstück, dessen Mitte wahrscheinlich von Gußlöchern und Gasblasen durchsetzt ist, so kann man leicht einsehen, daß das Einpassen kleiner Schrauben und Prisonstifte in Gußeisen keineswegs eine einwandfreie Arbeit liefert, so daß dem Stahl der Vorzug zu geben ist.

Registrieren der Subpressen. Die nebenstehenden Tabellen sind Wiedergaben der Indexkarten, die in der amerikanischen Authomometer Company in St. Louis[1]) verwendet werden, um jederzeit Stand und Verwendung eines Werkzeuges aufzufinden. Die Herstellung der Addiermaschine und die dazu gehörigen Reparaturen umfassen im ganzen ungefähr 700 Arbeitsgänge. Alle Stücke erhalten eine Kennummer und alle Blockwerkzeuge eine Standnummer. Mit Bezug auf die Arbeits- und Werkzeugkarte Nr. 212 kann man ersehen, daß es drei Stücke dieser Nummer gibt, 212, 212 N und 212 Fr. Die Arbeitsgänge an dem Stück sind in ihrer Reihenfolge angegeben; gegenüber jedem Arbeitsgang steht die Standnummer des Blockwerkzeuges, auf dem diese

[1]) Vergl. S. 118.

Arbeit gemacht wird; diese Nummer ist auch die Kennummer für die Werkzeugkarte, die keiner weiteren Erklärung bedarf.

Das Werkzeug 3—4 z. B. ist in der Abteilung „3", im Fache „4" des Werkzeugständers zu finden. Sowohl das Werkzeug wie der Ständer tragen Messingschilder mit der gleichen Standnummer, so daß die Nummern übereinstimmen werden, wenn das Werkzeug am richtigen Platze steht. In der oben genannten Werkstätte werden alle Werkzeuge in dem Säulenwerkzeug, Fig. 152, verwendet.

Arbeits- und Werkzeugkarte für Stück Nr. 212.

Kennummer: Nr. 212, 212 N, 212 Fr.

Material: 220 · 222 · 1,6 mm Messing. Arbeitsgänge: 4.

Anzahl Stücke für 1 Maschine .	1	Ersatzwerkzeuge	Säulenwerkzeug
Anzahl Stücke auf 1 m	3,28		
Anzahl Stücke auf 1 kg	3,23		
Arbeitsgänge:	Blockwerkzeug		
212—1 Runde Löcher.	4—12		
212—2 Nuten	3—4	Alte oder Reparaturwerkzeuge	
212—3 Lochen für Nr. 227 . .	7—5		
212—4 Lochen für Nr. 215 . .	7—5		
212—N—1	6—11	212—2 Alte Nuten	V—C—I
212—N—2	8—4	212—N—2	V—C—I
212—N—3	7—5		
212—N—4	7—5		
usf.			

Werkzeugkarte.

Säulenwerkzeug 3—4. 212—2 Nuten.

Besondere Bauart			Bemerkungen
Presse	7—4		Anschläge auswechseln für 212 bis 213.
Froschplatte	1		Stifte auswechseln für 212 N bis 213 N.
Exzenter.	2		Stempel wenig einölen für 2—4.
Bolzen für Plunger . . .	am Werkzeug		Papier auf dem Stück halten.
Zeichnungen	10—2	1160	Säulenwerkzeug V—C für Fr. verwenden.
Modelle	1—97	A—98	
	1—99	A—100	
Sonstige Stücke zum Säulenwerkzeug.			
212—2.	—	—	
212—N—2	—	—	
212—Fr—2.	—	—	

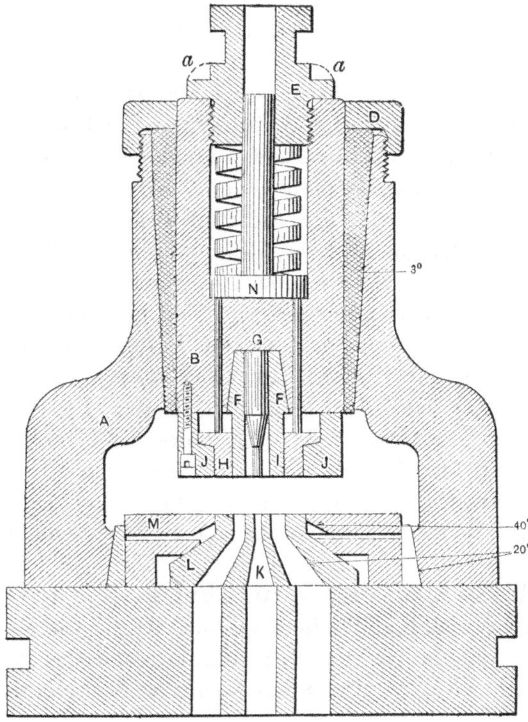

Fig. 173.

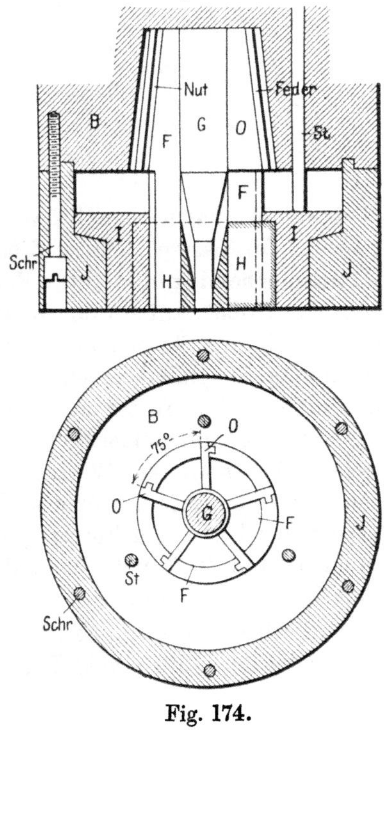

Fig. 174.

Blockwerkzeug zur Herstellung von Uhrrädern. Das Blockwerkzeug zum Stanzen eines Rades, Fig. 173, besteht aus 17 Teilen, einschließlich des oberen und unteren Abstreifers. Dazu gehören, Fig. 174, fünf Teilstempel F für die Segmente, ein Mittellochstempel G, fünf Füllstücke O, ein Stern H, der die fünf Teilstempel F auseinanderhält, ein Ring I, der den äußeren Ringumfang ausfüllt, als Auswerfer und drei runde Stangen St, die eine Verbindung zwischen dem oberen Auswerfer I und der Unterseite des abgefederten Kolbens N herstellen, vorhanden. Alle diese Teile sind an den Plunger angepaßt. Die fünf Teilstempel F, Fig. 176, werden in einer Länge auf den nötigen Winkel, in diesem Fall 75° gehobelt, so daß die Speichen schräg gegen den Rand zulaufen. In eine Seite wird eine Nut für die Keilfüllstücke O, Fig. 177, gehobelt. Diese werden so auf Länge abgeschnitten, daß man sie an den Enden abdrehen kann, worauf sie in einem besonderen Futter auf der Fußdrehbank fertiggemacht werden.

Herstellung des Futters für fünf Teilstempel. Der beste Weg, das Futter nach Fig. 175 herzustellen, ist, für dieses ein Blankett, das ein wenig größer als der Innendurchmesser des zu stanzenden Rades ist, zu machen, in dieses ein Loch von demselben Durchmesser

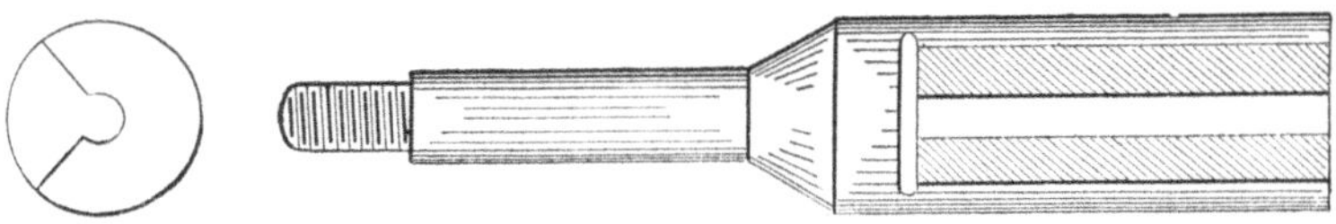

Fig. 175.

wie die Radnabe zu bohren und darin einen Stöpsel einzupassen, der ungefähr 6,4 mm vorsteht. An diesem Ende wird er im Durchmesser der Dicke der Arme entsprechend abgedreht. Man muß vor dem Hobeln den Auslauf für den Hobelstahl auf der Innenseite im Futter schneiden. Außerdem muß ein zur Shapingmaschine passender Halter gemacht werden, an dem man den Spindelstock einer Mechanikerdrehbank befestigen kann, so daß das Futter auf den richtigen Winkel gehobelt werden kann, während es im Spindelstock eingespannt ist. Man kann hierbei so vorgehen, daß man den Werkzeughalter nach jeder Seite unter 37° 30′ einstellt, oder ihn in seiner senkrechten Stellung beläßt, erst die eine Seite der V-Nut im Futter hobelt und dann das Futter mit dem Spindelkopf um 105° dreht, worauf die andere Seite in gleicher Weise gehobelt werden kann.

Herstellung der Teilstempel F. Nach Fertigstellung des Futters wird die Außenfläche der Teilstempel F, Fig. 176, abgedreht, indem man jeden in dem Futter mit einem Ring mit Druckschraube befestigt. Sie werden mit einer Zugabe von 0,25 mm fürs Schleifen zur Hälfte gedreht und dann die andere Hälfte fertig gemacht. Die Innenfläche der Kreisausschnitte wird mit einer Zugabe fürs Schleifen von 0,075 mm ausgebohrt und alle Enden auf eine Länge abgedreht. Man härtet sie nun bei der geringsten Temperatur, bei der der Stahl noch Härte annimmt, und läßt soviel an, als zur Beseitigung des Schwindens notwendig ist. Dann werden die Seiten sorgfältig an einem Gußeisenblock vorpoliert, wobei man genau auf die Einhaltung der richtigen Schräge sehen muß. Die Ausschnitte werden in den Plunger mit 1° Anzug eingesetzt. Nach dem obenerwähnten Polieren werden die Teilstempel einzeln in das Futter eingespannt und jeder am Ende rechtwinklig und in der Länge mittels einer Rundschleifvorrichtung auf der Drehbank mit 1° Anzug geschliffen, wobei man alle genau auf denselben Durchmesser bringt.

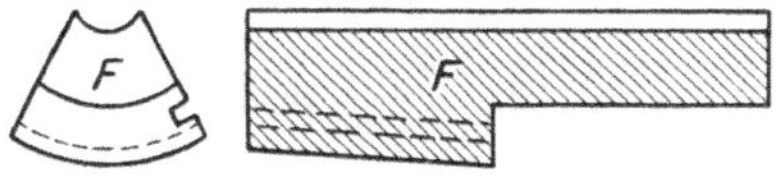

Fig. 176.

Herstellung der Füllstücke O und des Sternes H. Die fünf Füllstücke O nach Fig. 177 werden in einer Länge mit der richtigen, den Radspeichen entsprechenden Schräge im Querschnitt und einer Feder auf der einen Seite, die in die Nut der Teilstempel, Fig. 176, paßt, gehobelt und auf solche Länge abgeschnitten, daß sie bündig mit dem Absatz in F werden. Für den Stern H, Fig. 178, wird ein Blankett hergestellt, das um 25,4 mm im Durchmesser größer als der Innendurchmesser des Radkranzes ist, und so lang wie der Abstreifer I ist. Der Drehbankspindelstock wird wieder an dem auf der Shapingmaschine verwendeten Halter befestigt und unter Verwendung dieses Aufbaues die Speichen mit einem Vorfräser ausgeschruppt und schließlich mit einem besonderen Fräser, der für jede Radgröße besonders gemacht werden muß, fertiggestellt. Nach dem Fräsen müssen die fünf Teilstempel gut zwischen den Stern eingepaßt werden. Dann kommt der Stern H wieder auf die Drehbank, wo das große Loch in der Mitte für den Lochstempel G ausgebohrt und das Stück auf Länge abgestochen wird. Man gibt dabei eine Zugabe für das Einspannen in einem Zangenfutter zum Abdrehen der Stirnfläche und bohrt dann das kleine Loch. Dieses dient als Führung für das dünne Ende des Lochstempels G und muß nach dem Härten ausgeschliffen werden. Der Stern wird dann gehärtet und auf Purpur angelassen; sollte sich nachher noch eine Nacharbeit als notwendig erweisen, so geschieht es mittels eines

Fig. 177.

Kupferreibers unter Verwendung von feinstem Schmirgel. Der Auswerfer I, Fig. 178 und 174, in den der Stern paßt, wird innen ausgedreht und in einem Zangenfutter, das auf den Spindelstock paßt, befestigt, worauf auf der Innenfläche die fünf Nuten so tief eingehobelt werden, daß die Speichen des Sternes gut hineinpassen. Dann wird er gehärtet, angelassen und innen wie außen geschliffen. Nun können die Stoßstellen des Sternes im Auswerfer verlötet werden. Der äußere Schnittring J wird gehärtet, angelassen, innen und außen geschliffen und mit vier bis sechs Stahlschrauben Schr, die Federhärte haben sollen, befestigt. Durch die Schraubenlöcher des Schnittringes wird dann von außen ein feiner Sägeschnitt geführt, der Härterisse verhindern soll.

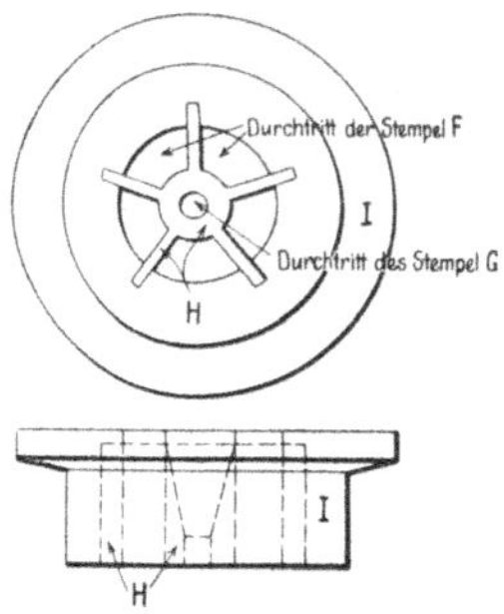

Fig. 178.

Nun wird der Plunger B auf der Drehbank in der Lünette eingespannt, die Ausdrehung für den Schnittring J hergestellt und letzterer darin mit Preßsitz eingepaßt. Jetzt wird das konische Loch für die Teilstempel F mit 1^0 Anzug ausgebohrt und diese darin ebenso eingepaßt.

Während der Plunger noch so in der Bank eingespannt ist, können die Teilstempel F in ihrer Stellung auf ihrer Außenseite, entsprechend dem Innendurchmesser des Radkranzes, und ihr inneres Loch, entsprechend dem Nabendurchmesser des Rades, eingeschliffen werden, so daß auch die Nabe des Sternes hineinpaßt. Selbstverständlich muß die Nabe des Sternes und des zu stanzenden Rades den gleichen Durchmesser haben. Gleichzeitig kann in einem besonderen Futter das kleine Loch im Stern geschliffen werden. Es werden noch die drei Löcher für die Auswerferbolzen St in den Plunger gebohrt, die Werkzeugteile zusammengebaut und die Stirnfläche eben abgeschliffen.

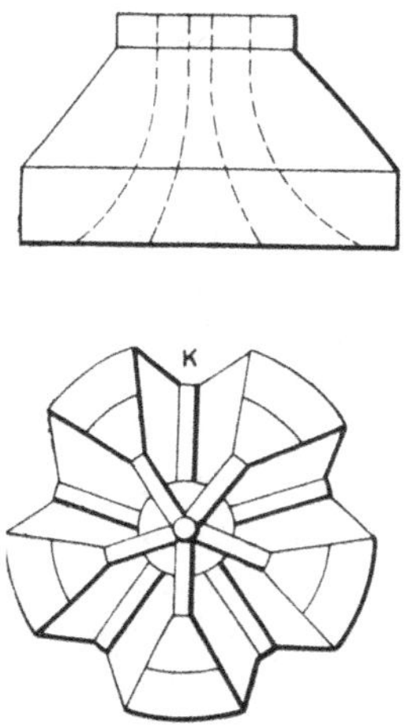

Fig. 179.

Herstellung des unteren Stempels K. Nachdem die Grundplatte gedreht und gebohrt worden ist, wird zunächst der untere Stempel K, Fig. 179, hergestellt. Das betreffende Blankett wird gut in einem Futter eingespannt, vorgefräst und mit demselben Fräser, mit dem der Stern H gemacht wurde, fertiggestellt. Nach dem Fräsen muß es an die fünf Teilstempel F angepaßt werden und, mit diesen als Führung, in seine richtige Stellung in der Grundplatte — diese und der Ständer sind hierbei zusammengespannt — eingepreßt werden, nachdem man ein Zentrierloch durch den Stempel K in die Grundplatte gebohrt hat.

Die obere Stirnfläche des Stempels wird gehärtet, angelassen und die Speichen auf Maß geschliffen, während das kleine mittlere Loch nach rückwärts $\frac{1}{4}^0$ Anzug hat. Der runde Schnittstempel L ist einfache Arbeit, ebenso der untere Auswerfer M. Die Löcher in der Grundplatte sollen mit Spiel gemacht sein, damit die Abfälle durchfallen können. Die ausgeschnittenen Räder werden wieder in den Messingstreifen zurückgedrückt.

Unter den unteren Auswerfer, der durch vier Schrauben gehalten wird, werden acht Federn eingelegt. Vor dem Härten sind noch Löcher für die Führungsstifte in den Abstreifer zu bohren, welcher dann blau angelassen wird.

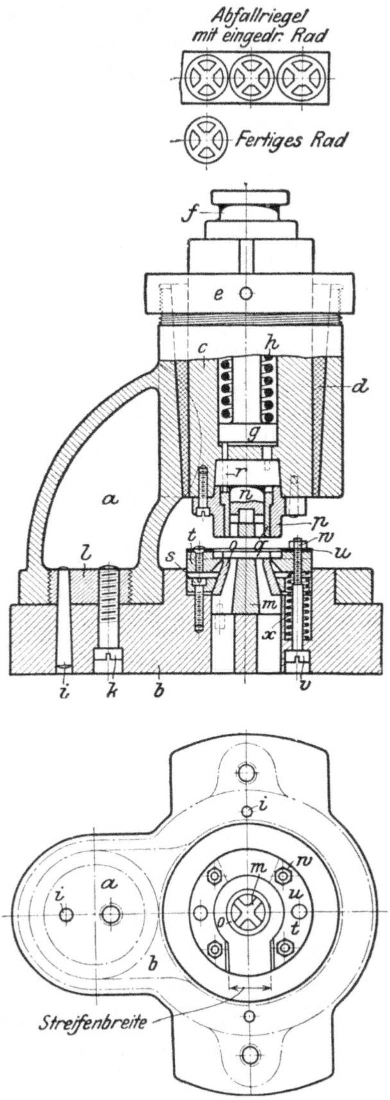

Fig. 180.

Offenes Blockwerkzeug[1]). Während bei dem gewöhnlichen Blockwerkzeug stets Bandmaterial verarbeitet werden muß und infolge dessen selbsttätige Zuführungsvorrichtungen zur vollen Ausnutzung des Werkzeugs gebraucht werden, kann auf dem in Fig. 180 abgebildeten offenen Blockwerkzeug jeder Abfall verarbeitet werden. Diese Form kann deshalb für kleinere Betriebe, wo von einem einzigen Teil dauernd geringe Mengen hergestellt werden, oder von mehreren, wenn das Werkzeug selbst auswechselbar eingerichtet worden ist, mit Vorteil gegenüber dem

[1]) W. T. 1909, S. 582.

geschlossenen Block ausgeführt werden. Auf diesem Werkzeug wird ein Rad mit vier Speichen, Fig. 180, ohne die Nabenbohrung hergestellt. Das Werkzeug selbst ist, wenigstens was den Schnittstempel anbelangt, ganz gleich dem eben besprochenen amerikanischen, dagegen sind abweichende Einzelheiten, wie die Befestigung der Stempel und die Anordnung der Auswerferfedern v, w, x, deutlich zu erkennen. Auch ist eine Führungsplatte u vorgesehen. Der Schnittstempel für Speichen und Nabe n und der Auswerfer q, Fig. 181, sind hier aus dem Vollen herausgearbeitet, so daß der Zusammenbau und etwaige Ausbesserungen leichter zu machen sind. Die Herstellung des Schnittstempels selbst wird wohl etwas mehr Werkstättenarbeit verlangen. Von besonderem Interesse ist die Anordnung des Ständers. Ein kräftiger Hohlgußkörper a wird, wie vorher bei den andern Werkzeugen beschrieben worden ist, zur Führung des Plungers mit Weißmetall ausgegossen. Der Block ist mit einer angegossenen Fußplatte versehen, welche gleichzeitig die Ausmittlung von Ober- und Unterteil gewährleistet. Die Befestigung erfolgt durch drei von unten versenkte Zylinderkopfschrauben k, während die richtige Stellung der beiden Teile gegeneinander durch drei Prisonstifte i, wie im Grundriß, Fig. 180, ersichtlich ist, gewährleistet wird. Die Kernverschraubung l ist aus gußtechnischen Gründen einzusetzen.

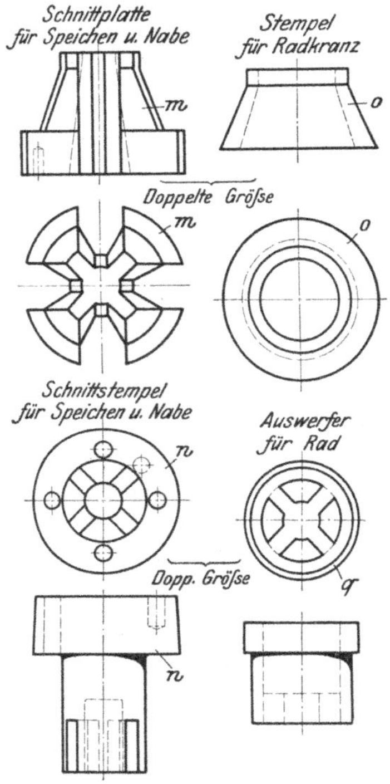

Fig. 181.

Schnittring u für den Radkranz und Schnittplatte m für Speichen und Nabe sind mit besonderen Prisonstiften gesichert.

Herstellung eines Blockwerkzeuges für ein unregelmäßiges Stück. Der beste Weg, die Teile eines Blockwerkzeuges zur Herstellung eines unregelmäßigen Stückes, ähnlich dem, welches im Grundriß des Schnittstempels, Fig. 186, und des Auswerfers, Fig 183, dargestellt ist, anzufertigen, ist der folgende:

Die wichtigste Vorbedingung für eine genaue Arbeit ist die Herstellung einer Bohrlehre, Bo, Fig. 187, nach der die Löcher gebohrt werden müssen. Ein rundes, flaches Stück Werkzeugstahl von genügenden Abmessungen — in diesem Fall 38 mm Durchmesser und 5 mm Dicke — wird abgedreht und vollständig eben und parallel geschliffen, da von der Genauigkeit der Bohrlehre viel abhängt. Eine Seite desselben wird mit „Oben“, die andere mit „Unten“ bezeichnet, um bei den folgenden Arbeiten Verwechslungen zu vermeiden; auch das Muster soll in gleicher Weise bezeichnet werden. Arbeitet man ohne ein solches, so muß die Bohrlehre nach einer Zeichnung oder Maßskizze angerissen werden, worauf der Mittelpunkt angekörnt, die Lehre auf der Planscheibe der Drehbank mit einem Fühlhebel ausgerichtet und befestigt wird; dann werden alle Löcher gebohrt und auf denselben Durchmesser ausgerieben. Arbeitet man hingegen nach einem Muster, so muß dessen Oberseite an die Unterseite der Lehrenplatte angelötet werden, wobei man ein Paar parallele Streifen einlegt, damit man durch die Bohrlehre hindurchbohren kann. Eine Mechanikerdrehbank mit Zangeneinrichtung ist für diese Art Arbeit am besten geeignet. Man verwendet ein Futter mit einem schwachen Konus an einem Ende und dreht einen Führungsstift dafür, der gehärtet und geschliffen wird, bis er in das größte Loch des Musters paßt. Dann wird die Bohrlehre mit dem Muster zusammen auf der Planscheibe der Bank mit dem größten Loche auf den Führungsstift aufgehängt

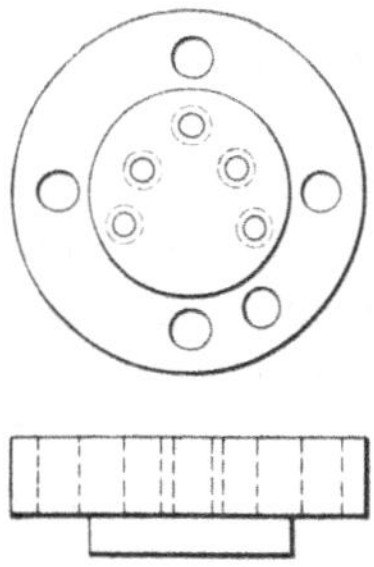

Fig. 182. Kopfplatte.

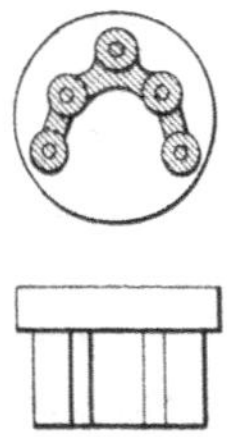

Fig. 183. Oberer Auswerfer.

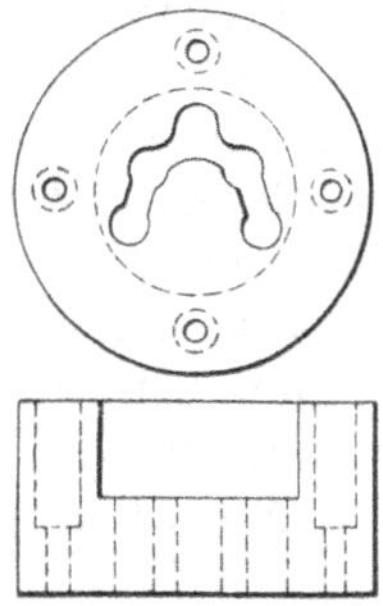

Fig. 184. Schnittplatte.

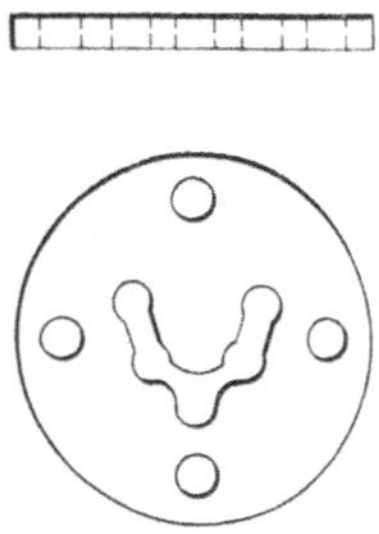

Fig. 185. Unterer Abstreifer.

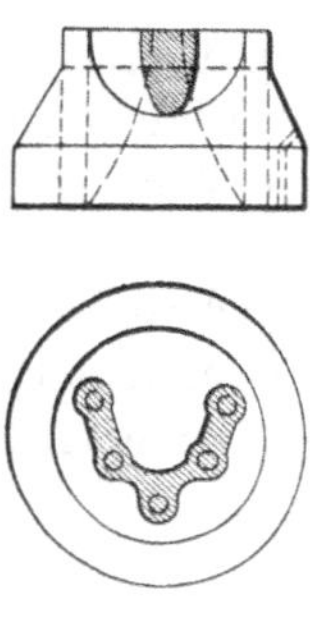

Fig. 186. Schnittstempel.

und festgezogen, wobei man aufpassen muß, daß die Platte nicht verzogen wird. Nun wird das erste Loch gebohrt, ausgearbeitet und auf ungefähr den Durchmesser ausgerieben. Nach Fertigstellung des ersten Loches wird der Führungsstift auf den Durchmesser des nächst kleineren Loches abgeschliffen, und in gleicher Weise vorgegangen, bis alle Löcher fertig sind. In diesem Fall müssen alle Löcher in der Bohrlehre die gleiche Größe haben. Wenn die Bohrlehre fertig ist, wird ein gleichachsiger Einlegering, Dr, Fig. 188, aus Maschinenstahl hergestellt und auf der einen Seite auf den Durchmesser und die Dicke der Bohrlehre ausgedreht; hernach wird der Ring umgedreht und erhält gleichachsige Eindrehungen für jedes der folgenden Stücke: den unteren Stempel, Fig. 186, den unteren Auswerfer, Fig. 185, die obere Schnittplatte, Fig. 184, die Kopfplatte, Fig. 182, und schließlich den oberen Auswerfer, Fig. 183. Diese fünf Stücke, die das Werkzeug bilden, werden vorgeschruppt, auf Maß gedreht und geschliffen, so daß sie genau in die Ausdrehungen des Einlegeringes, Fig. 188, passen.

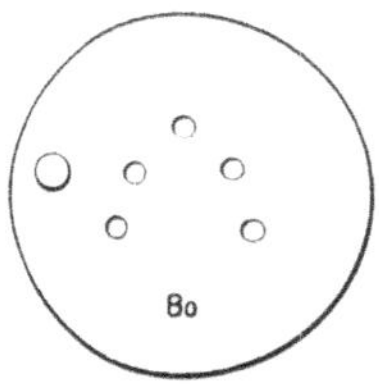

Fig. 187.

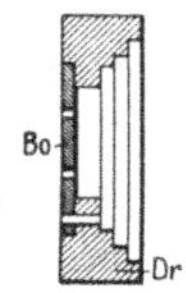

Fig. 188.

Herstellung der oberen Schnittplatte. Das Stahlstück für die obere Schnittplatte, Fig. 184, wird auf die richtige Tiefe mit einer Zugabe von ungefähr 0,15 mm für das Schleifen ausgedreht. Dann bohrt man ein Loch für einen Prisonstift durch das Blankett für die obere Schnittplatte und die Kopfplatte, Fig. 182, welche beide mit demselben Prisonstift gehalten werden sollen, und den Einlegering. Ebenso verfährt man mit dem unteren Stempel, Fig. 186, sowie der Bohrlehre, Fig. 187, wobei jedes Stück in seiner zugehörigen Eindrehung liegt. Die beiden übrigen Stücke brauchen nicht für die Prisonstifte gebohrt zu werden, da sie ungehärtet bleiben. Nun wird die Bohrlehre mit der „Oben“ bezeichneten Seite in den Ring eingelegt, so daß diese Seite nächst der Planscheibe liegt; dann legt man die obere Schnittplatte, Fig. 184, mit ihrem Prisonstift in seiner Stellung in die Ausdrehung ein, hängt das Ganze auf den Führungsstift und befestigt es sicher auf der Planscheibe, wie Fig. 189 erkennen läßt. In dieser Weise bohrt man die fünf Löcher und reibt sie mit einer Zugabe von 0,05 mm für Schleifen nach dem Härten aus, nachdem die vier Schraubenlöcher gebohrt und versenkt worden sind. Der Rest der Schnittplatte kann mit einem kleinen Fingerfräser im Schlitten einer Universal-Werkzeugschleifmaschine bearbeitet werden,

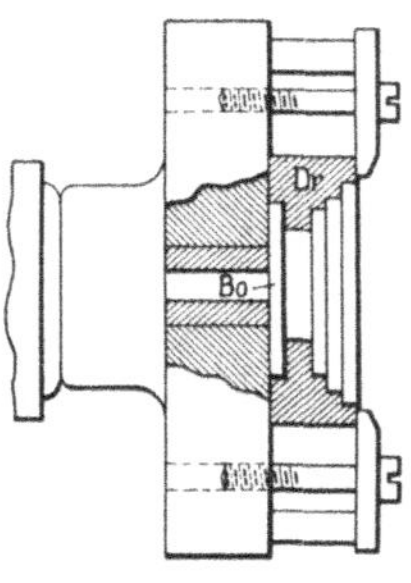

Fig. 189.

deren Umdrehungszahl man verringert. Schließlich wird noch mit einer feinen Feile nachgearbeitet und gehärtet. Der obere Abstreifer soll in derselben Weise wie die obere Schnittplatte gebohrt werden, d. h. die Seite „Oben" der Bohrlehre liegt nächst der Planscheibe, nur erhalten die Löcher 1^0 Anzug; die Führungsbüchsen werden gesondert hergestellt und werden, nachdem sie gehärtet und innen nach dem Durchmesser der Lochstempel geschliffen, sowie außen gut in die obere Schnittplatte eingepaßt worden sind, bis an den Bund in die Abstreiferplatte eingepreßt. Man kann noch Füllstücke in die Platte einsetzen, obwohl dies nicht unbedingt notwendig ist, da die Führungsbüchsen allein bereits den Abfall wieder in den Streifen zurückdrücken. Die Kopfplatte, Fig. 182, soll mit dem Putzen an der Planscheibe liegen, da die Löcher Anzug nach rückwärts haben, und die Bohrlehre mit der mit „Unten" bezeichneten Seite an der Planscheibe anliegen. Diese Platte soll denselben Prisonstift haben wie die obere Schnittplatte, damit die Löcher sich in Linie befinden. Sodann befestigt man sie auf der Planscheibe auf dem Führungsstift, bohrt die Löcher und reibt sie mit einem Konus von 2^0 Anzug aus. Konische Reibahlen darf man aber nur mit der größten Vorsicht verwenden.

Herstellung der kleinen Lochstempel. Die kleinen Stempel werden auf der Stange vorgeschruppt und mit dem Abstechstahl teilweise getrennt, dann gehärtet und passend angelassen. Danach müssen die Stempel gerade und auf Maß geschliffen und ihr Konus der Kopfplatte angepaßt werden, doch müssen alle Arbeiten in einem Schleifen ausgeführt werden. An den Stempeln sind zwei Konen vorgesehen, um sie so stark wie möglich zu halten. Die Führungsbüchsen im Abstreifer müssen an die kurzen Konen der Stempel angepaßt werden, damit der Abstreifer so nahe wie möglich herankommen kann. Nachdem alle Stempel in die Kopfplatte eingepaßt und fest eingetrieben worden sind, wird die Platte in einem Futter, welches so ausgedreht worden ist, daß die Stempel freiliegen, befestigt und die Bruchstellen an den Rückseiten der Stempel glatt und bündig mit der Platte geschliffen. Die untern Enden der kleinen Stempel und die Führungsbüchsen werden nach dem Zusammensetzen auch zusammen geschliffen. Die Schraubenlöcher in der Kopfplatte werden gebohrt, nachdem die obere Schnittplatte geschliffen und an die Kopfplatte angepaßt worden ist, wobei sie mit Hilfe des Prisonstiftes ausgerichtet werden.

Herstellung des unteren Stempels. Der untere Stempel, Fig. 186, und der Einlegering, Dr. werden durch den Prisonstift miteinander verbunden — die Seite „Unten" der Bohrlehre liegt an der Planscheibe — und an dem Führungsstift aufgehängt, worauf die Löcher wie vorher fertiggestellt werden. Sie werden ungefähr 6,5 mm tief und ungefähr 0,075 mm kleiner als die Stempel ausgerieben — als Zugabe für das Schleifen —, wonach man das Stück von der Bank entfernen kann. Ein kleiner Führungsstift, der in die eben gebohrten Löcher paßt, muß hergestellt werden, auf welchem dann der untere Stempel auf der Planscheibe befestigt wird, um die erweiterten Löcher für die Abfälle zu bohren. Da der Stempel mit Konus eingesetzt ist, bedarf er keiner Schraubenlöcher zur Befestigung.

Herstellung des unteren Abstreifers. Der untere Abstreifer, Fig. 183, wird in die Ausdrehung des Einlegeringes eingelegt, wobei die Bohrlehre mit der Seite „Unten" an der Planscheibe liegt, dann auf dem Führungsstift auf der Planscheibe befestigt und gebohrt, ausgearbeitet und nachgerieben, bis die Löcher im Durchmesser denen in dem Muster entsprechen, und dann aus dem Ring entfernt, damit man die Schraubenlöcher bohren und die Gewinde einschneiden kann. Die Löcher, die das Loch für den Stempel bilden, müssen gut erweitert werden, und der Rest gebohrt und leicht über den unteren Stempel eingepaßt werden. Zwei Löcher müssen auch in die Platte für die Führungsstifte des Bandmateriales gebohrt werden, welche in ihrer Stellung den oberen Stempel nicht behindern dürfen. Dann wird gehärtet und angelassen.

Schleifen der oberen Schnittplatte. Nachdem die obere Schnittplatte gehärtet und angelassen worden ist, wird sie in einem Futter von außen eingespannt, die Stirnfläche rechtwinklig abgeschliffen und die Ausdrehung für den oberen Abstreifer passend geschliffen. Nun wird sie umgekehrt und mit der Ausdrehung in dem Futter gehalten, worauf der äußere Durchmesser auf richtiges Maß geschliffen wird. Das Loch für den Prisonstift wird wohl ein wenig Nachschleifen verlangen, und, falls der äußere Durchmesser der Schnittplatte kleiner als die Ausdrehung des Einlegeringes geworden sein sollte, wird die entsprechende Ausdrehung ein wenig zugestaucht, so daß man sie aufs neue ausrichten kann. Nun wird die Schnittplatte mit dem Prisonstift in seiner Stellung in die Ausdrehung eingelegt und die Bohrlehre mit der entsprechenden Seite an die Planscheibe angelegt. Das Ganze wird auf dem Führungsstift an der Planscheibe befestigt und die Löcher mit feinstem Schmirgel auf Maß geschliffen.

Einstellen und Anpassen der Teile. Die Schnittplatte kann nun an den Plunger der Subpresse, deren Konstruktion bereits beschrieben worden ist, angepaßt werden, wobei es nur notwendig sein könnte, die geraden Seiten, die sich vielleicht durch das Härten verzogen haben, mit feinem Schmirgel nachzuschleifen. Hierauf müssen drei kleine Führungsstifte gedreht werden, deren kleinere

Enden in den unteren Stempel und deren größere Enden in die obere Schnittplatte eingepaßt werden müssen, damit der untere Stempel in der Grundplatte der Subpresse seine richtige Stellung erhält. Eine dünne Unterlagscheibe, ungefähr 2,5 mm dick, ist unterhalb des unteren Stempels einzupassen und dieser selbst größer zu lassen, so daß er durch die drei kleinen Führungsstifte und die obere Schnittplatte ungefähr 1 mm in die Grundplatte der Subpresse eingedrückt wird; dann wird der Prisonstift in die Grundplatte eingebohrt, der untere Stempel entfernt und auf seiner Stirnfläche mit einer dünnen Schicht Lot überzogen, um einen scharfen Abdruck der oberen Schnittplatte zu erhalten, wobei man von dem überflüssigen Lot soviel wie möglich abschabt. Dann wird das über den Eindruck hinausstehende Material so nahe wie möglich weggefräst und das Stück wieder in die Subpresse eingesetzt, um nun einen neuen Eindruck auf dem Stahl selbst zu nehmen. Hierauf wird der Stempel gefräst und gefeilt, wobei man ungefähr 0,025 mm — je weniger desto besser — zum Nacharbeiten mit Feile und Schaber übrig läßt. Nun härtet man und läßt das Stück an, entfernt das obere Werkzeug aus dem Plunger und sieht nach, wie der Stempel in die Schnittplatte paßt. Ist dann noch etwas Nacharbeiten erforderlich, so wird mit einem flachen Schmirgelreiber sorgfältig nachpoliert, um die Ecken nicht zu runden.

Ausrichten der Stempel und Schnitte. Man stellt nun fest, daß die Grundflächen der oberen Schnittplatte sicher rechtwinklig zur Achse sind und setzt den unteren Stempel in die Schnittplatte ein, wobei man den Zwischenraum zwischen beiden mit einem parallelen Ring ausfüllt. Dann schleift man die Grundfläche des Stempels ab, stellt die Schräge richtig und paßt ihn in die Grundplatte ein, worauf der Stempel und die Schnittplatte ausgerichtet sein werden. Es bleibt nur noch übrig, den Stempel wieder in die Ausdrehung des Einlegeringes einzulegen, die Bohrlehre in ihrer richtigen Stellung einzusetzen und die Lochschnitte für die kleinen Stempel auf Maß zu schleifen, wobei man ihnen nach rückwärts $\frac{1}{4}^{0}$ Anzug gibt. Dann wird die dünne Unterlagscheibe entfernt, Schnittplatte und Stempel wieder in die Subpresse eingesetzt und geprüft, ob sich beide richtig in Linie befinden. Etwaige Fehler müssen jetzt gefunden werden.

Fig. 189 zeigt die Planscheibe auf der Spindel der Drehbank und die Weise, wie die Bohrlehre auf dem Führungsstift aufzuhängen ist. Es mag noch hinzugefügt werden, daß bei dieser Art Arbeit stete Aufmerksamkeit unbedingt erforderlich ist.

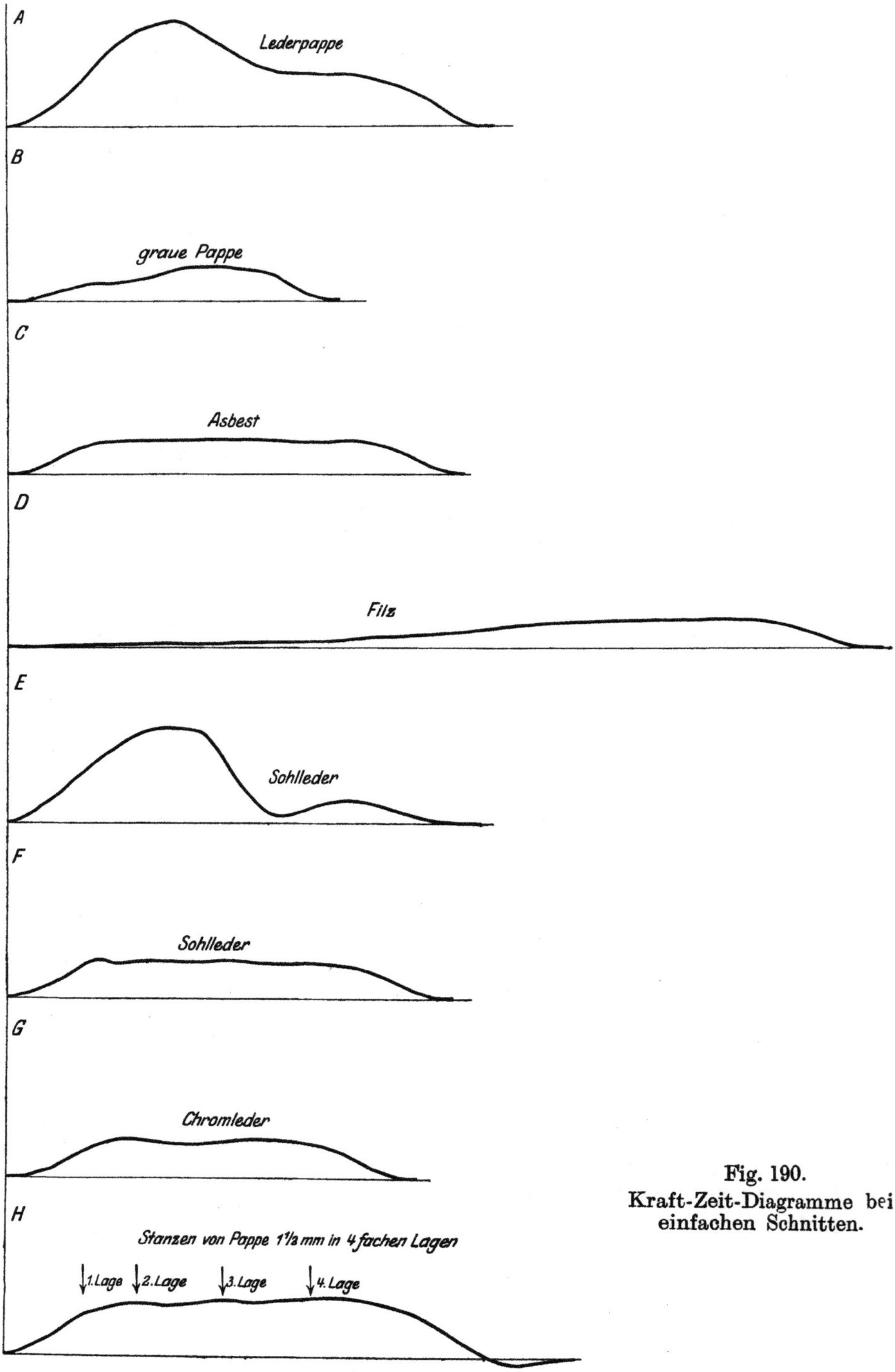

Fig. 190.
Kraft-Zeit-Diagramme bei einfachen Schnitten.

2. Einfacher Schnitt ohne Gegenschnitt.

Einfache Schnitte sind jene, die auf der Keilwirkung einer zugeschärften Kante beliebiger Form beruhen. Sie arbeiten ohne Gegenschnitt, so daß man nicht mehr von einer reinen Scherwirkung im Sinne des Begriffes Scherfestigkeit reden kann. Es ist infolgedessen bei demselben Material die zur Abtrennung notwendige Kraft von der Schärfe des Schnittes oder in anderen Worten von der Breite der zuerst zur Auflage kommenden Keilfläche und weiter von dem Keilwinkel der beiden Seitenflächen des Messers abhängig. Deshalb kommen derartige Werkzeuge nur für weiches Material, Papier, Pappe, Leder usw. zur Anwendung, vielleicht auch noch manchmal für ganz dünnes, weiches Messingblech.

Versuche in dieser Richtung sind noch keine gemacht worden, bis auf die Doktor-Dissertation von Wormser, der die Vorgänge beim Stanzen von Lederpappe, Filz usw. untersucht hat[1]). Es zeigen sich dabei ganz bestimmte Formen der Schaulinien für jedes Material, von denen einige in Fig. 190 als Kraft-Zeitdiagramme und in Fig. 191 als Kraft-Wegdiagramme wiedergegeben sind. Infolge der geringen Dicken der gestanzten Materialien zeigen die letzten den Verlauf der Kräfte nicht so deutlich wie die mit gleichbleibender Trommelgeschwindigkeit geschriebenen, auseinandergezogenen Kraft-Zeitdiagramme. Der allgemeine Verlauf der Schaulinien ist ungefähr derselbe wie beim gleichen Stanzen auf einer langsam arbeitenden Materialprüfmaschine, doch steigen die Ordinaten (Kräfte) mit wachsender Stanzgeschwindigkeit, bis sie bei ungefähr 300 minutlichen Umdrehungen der Versuchsstanze je nach dem Material von ungefähr 70 % bis ungefähr 130 % und in einem Fall 286 % der Werte bei ruhigem Druck erreichen. Die Kräfte steigen zunächst immer rascher, dann langsamer an, bis sie nach Erreichen eines Höchstpunktes

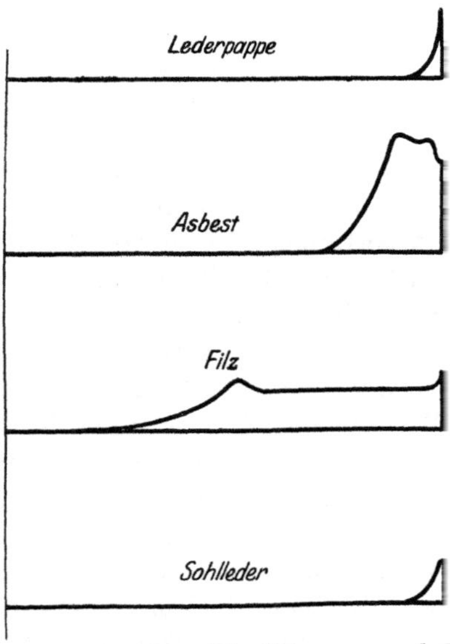

Fig. 191. Kraft-Weg-Diagramme bei einfachen Schnitten.

[1]) A. Wormser, Versuche über den Kräfteverlauf beim Stanzen. Dr.-Ing.-Dissert. Berlin 1913.

mit einem merklichen Knick abfallen. Aus dem Wegdiagramm ist zu ersehen, daß der absteigende Ast der Kurve stark auseinandergezerrt ist, wie auch tatsächlich der Abfall der Kräfte in der letzten, dünnen Schicht des Materiales erfolgt.

Die zweite Erhebung im Diagramm ist auf das Eindringen des Messers in die Unterlage zurückzuführen.

Mit dem Kräfteverlauf ist auch gleichzeitig der Schnitt selbst zur Beurteilung heranzuziehen, wobei man bei der Arbeit zwischen Loch- und Umrißmessern, Fig. 192, unterscheiden muß. Je nachdem, ob das Loch oder der ausgeschnittene Teil zur Verwendung kommt, ist der äußere oder innere Durchmesser des Stanzmessers senkrecht gelassen. Demgemäß ergibt sich das Bild der Materialverschiebung am Putzen oder am Loch. Fig. 193 zeigt die Materiallagerung für den

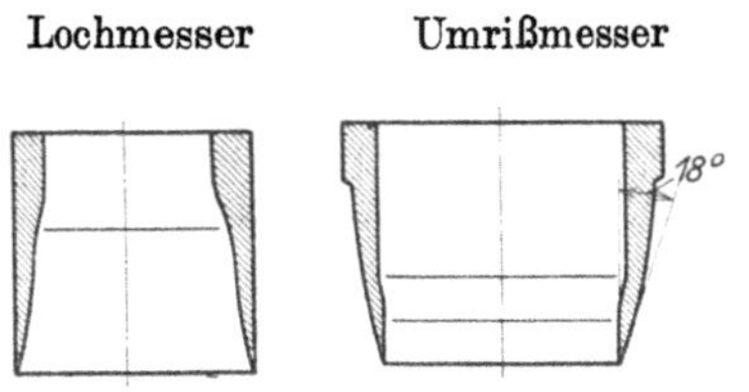

Fig. 192.

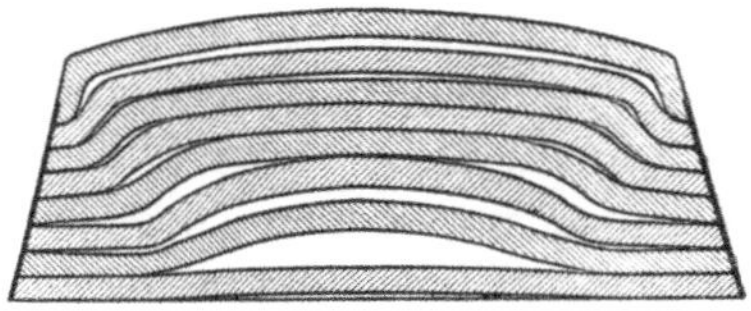

Fig. 193. Schnitt durch Pappe.

ersten Fall, also am Putzen. Das Material erscheint schon bei 0,5 mm Schnittiefe geschnitten, bis bei höheren Drücken sich leichte Aufwerfungen des inneren Schnittrandes bemerkbar machen, welche bei tieferen Schnitten immer stärker werden. Dabei schließt sich die Schnittfuge nicht mehr, sondern bleibt in der vom Messer erzeugten Breite stehen. Es tritt eine Materialverschiebung der Schichten auf, die mit dem Knick im Schaubild zusammenfällt. Dabei kann das Material von einer gewissen Stärke an der Zerreißwirkung des Messers auf die tiefer liegenden Schichten nicht widerstehen, so daß die Verschiebung und der damit verbundene Druckabfall vor dem Durchlaufen der ganzen Materialdicke erfolgen muß. Die übrigen Schichten werden dann aber zerrissen, wie man aus der Betrachtung des Schnittes, der aus einem dunkeln, glänzenden, geschnittenen und einem matten, hellen, gerissenen Teil besteht, erkennt. Aus diesen Materialverschiebungen ist zu erklären, daß der Schneidwiderstand mit dem Ein dringen des Messers in das Material wächst.

Einfache Schnitte für Papier und Pappe. (C. Krause in Leipzig.) Der in Fig. 194 abgebildete Einschieber für Zigarettenschachteln wird mittels des in Fig. 195 in zwei Ansichten dargestellten einfachen Stanzmessers hergestellt. Der äußere Rahmen ist aus einem Stahl von entsprechendem Dreieckquerschnitt geschlossen geschweißt und die einzelnen Eckmesser bei a und b durch Vernieten in ihm befestigt. Die auf der Oberseite sichtbaren Versteifungen werden in eingearbeitete Nuten eingepaßt und durch versenkte Zylinderkopfschrauben dort befestigt, so daß an der Schnittkante das Messer vollständig freitragend ist. Im Gegensatz zu diesem Messer, welches ein geschlossenes Ganzes bildet, gestattet das in Fig. 196 mit seinem zugehörigen Schnitt dargestellte verstellbare Stanzeisen für Kuvertfaltschachteln, eine Einstellung auf verschiedene Kuvertgrößen innerhalb der durch die Gleitschienen gegebenen Abmessungen. Der Rahmen wird durch vier prismatische Schienen gebildet, welche sich in Führungssteinen kreuzen und dort festgestellt werden können. Diese Steine dienen auch zur Befestigung der einzelnen Stanzmesser an den Enden, während sonstige Schnitte in der Mitte auf besonderen Führungssteinen angeschraubt werden. Die Gleitschienen werden mit eingravierten Maßstäben geliefert, so daß eine schnelle Einstellung der einzelnen Messer ermöglicht wird. Um ein Bild der Abmessungen der Apparate zu geben, sei bemerkt, daß sich mit den normalen Apparaten Faltschachteln von 60 · 80 mm bei 50 mm Tiefe bis zu 200 · 220 mm bei 70 mm Tiefe herstellen lassen.

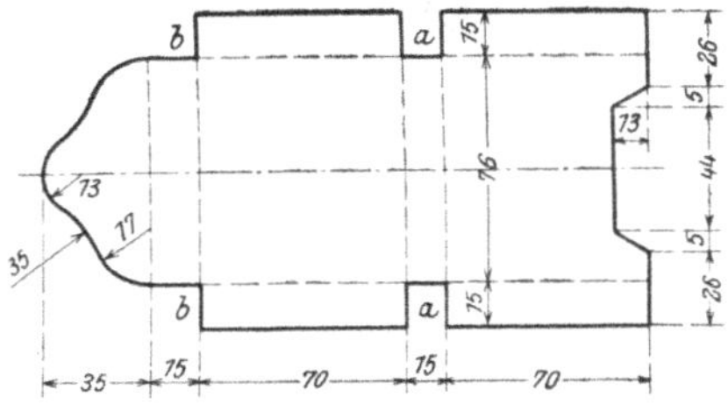

Fig. 194.

Fig. 195.

Der gesamte Herstellungsgang setzt sich aus vier Arbeitsgängen zusammen. Im ersten Gang wird das Material auf der Schneidmaschine auf genaue Größe zugeschnitten, im zweiten werden die Umbiegestellen gerillt, worauf auf dem angeführten Stanzeisen die eigentliche Form bis auf den Zungenschnitt ausgestanzt wird, dessen Herstellung der letzte Arbeitsgang ist.

Ein in gewissem Sinne ähnlich zusammengesetzter Schnitt für Unterlagscheiben, Ringe usw. ist in Fig. 197 mit einem Stanzmesser daneben wiedergegeben. Mit Hilfe eines Messerkopfes a, welcher drei oder vier Einstellschienen trägt, werden die beiden Stanzmesser sicher gleichachsig eingesetzt. Das innere Messer hat Innenfase und einen federnden Innenauswerfer und verbleibt fest mit dem Stempelkopf a

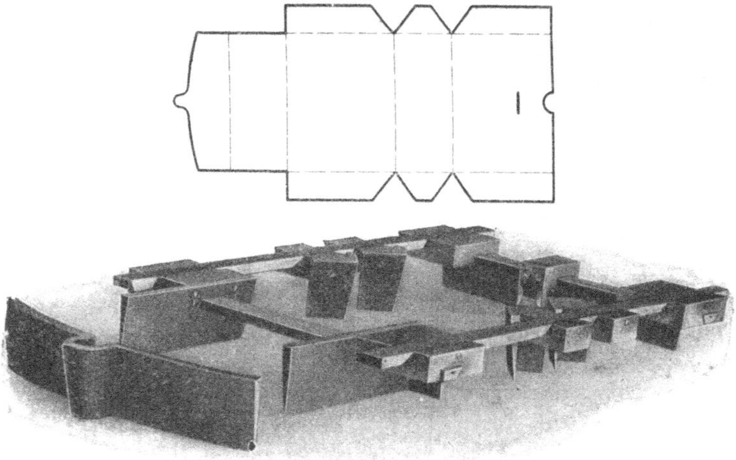

Fig. 196.

Fig. 197.

verschraubt. Das äußere Stanzmesser c führt sich mit seinen Nuten d auf den Leisten des Stempelkopfes a und ist abnehmbar, damit man nach jeder Stanzung das äußere Messer entfernen kann, um den schmalen ausgestanzten Ring nicht zu beschädigen. Ist dagegen der auszustanzende Ring so groß, daß man zwischen beiden Messern noch einen besonderen federnden Auswerfer anbringen kann, so ist das äußere Messer auch festzuschrauben. Am Rande des Messerkopfes sind federnde Niederhalter e angebracht, welche eine gleichmäßige

Auflage des Papieres bewirken und ein Kippen des Apparates bei kleinen Schnitten verhindern sollen. Die Vorteile dieser Einrichtung liegen also in der unbedingt sicheren Ausmittlung der beiden Schnitte, in der großen Verstellbarkeit des Messerkopfes und der dadurch hervorgerufenen Verbilligung des Werkzeuges, da derselbe Messerkopf für eine ganze Reihe Einzelmesser genügt.

Es können aber auch Werkzeuge nach Art der bisher für die Metallverarbeitung angeführten für weichere Stoffe gebaut werden, doch bleibt jedenfalls die Innen- und Außenfase an den Messern anzubringen. Zwei derartige reine Verbundwerkzeuge sind im folgenden beschrieben.

Verbund-Schnittwerkzeug für Unterlagscheiben. Werkzeuge zur Herstellung von Unterlagscheiben nach Fig. 198 können in allen Größen von 3,2 mm Lochdurchmesser und 8 mm Außendurchmesser bis zu 50 mm Loch- und 100 mm-Außendurchmesser hergestellt werden, so daß man damit alle Sorten Material, Gummi, Filz, Papier und Leder ausstanzen kann. Unter Verwendung eines Hammers und Handstempels dieser Art kann man bis zu 800 Unterlagscheiben in einer Stunde ausschlagen.

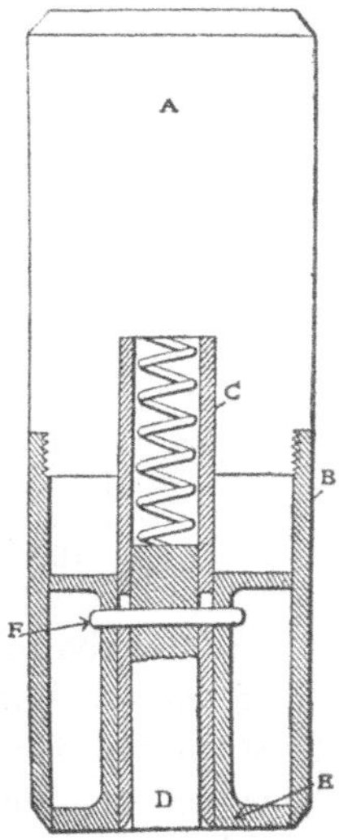

Fig. 198.

Alle Werkzeuge dieser Form sollen aber auch zum Gebrauch in der Kraftpresse, falls dies verlangt wird, eingerichtet sein; doch wird man im allgemeinen finden, daß selbst die großen 100 mm-Unterlagscheiben sauber und glatt mit einem Hammerschlag in 5 mm-Geschirrleder ausgeschlagen werden können. In dem Werkzeug in Fig. 198 ist der Zapfen A abgesetzt und bis an den Absatz mit Gewinde versehen, sowie zur Aufnahme des inneren Lochstempels C ausgebohrt, welcher denselben Außendurchmesser wie das Loch in der Unterlagscheibe hat. Das Stück ist in den Zapfen eingetrieben und hat ein Langloch für den Stift F, der durch die Auswerferstempel D und E geht, so daß diese sich gleichzeitig bewegen. Um den Stempel zusammenzubauen, wird C in A eingesetzt, die Feder in die Ausdrehung von C eingelegt und der Stempel D nachgeschoben. Dann wird der Stempel in seine Stellung gebracht, der Stift F eingeschlagen und zuletzt der äußere Schnittstempel B angeschraubt.

Ausschneiden und Lochen einer Unterlagscheibe aus Filz. Das Werkzeug, welches in Fig. 199 abgebildet ist, ist zum Gebrauch in einer Fußpresse entworfen und mit Ausnahme des aus Maschinenstahl gemachten Stempelkopfes C durchaus aus Stahl hergestellt. Das Werkzeug wird zum Ausstanzen von Filzunterlagscheiben von 20 mm Durchmesser mit einem 8 mm Loch in einem Arbeitsgang verwendet.

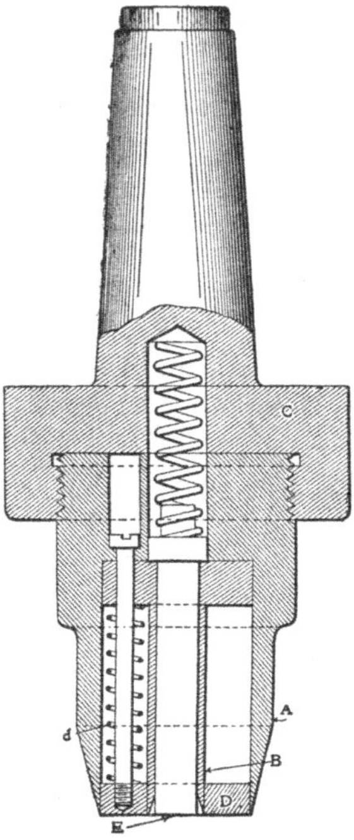

Fig. 199.

A ist der äußere Schnittstempel, dessen Schnittkante messerartig zugeschärft ist, also Außenfase hat, und den äußeren Durchmesser der Unterlagscheibe herstellt. Der Kopf von B, dessen Schnittkante nur 1,2 mm dick ist, ist in A eingetrieben. Schnittstempel A wird in Stempelkopf C eingeschraubt, bis er am Grunde der Ausdrehung aufsitzt. D ist die ringförmige Abstreifplatte, die sich sowohl in A als auch an B passend führt und durch 3 Schlitzkopfschrauben, deren eine gezeichnet ist, von 7/32″ Durchmesser gehalten wird. Der Abstreifer wird durch drei Wurstfedern d niedergedrückt und entfernt so die fertige Unterlagscheibe. E ist ein Auswerfer aus Werkzeugstahl, der in die Bohrung von B passend gedreht und durch einen Bund am oberen Ende am Herausfallen gehindert wird. Darüber liegt in der Ausdrehung von A und C eine Wurstfeder, die durch den auf E ausgeübten Druck den Abfall, der durch den inneren Lochstempel B erzeugt wird, ausstößt.

Zum Arbeiten wird der Stempel in die Fußpresse eingesetzt und ein Stück weiches Ahornholz mit der Stirnseite nach oben auf der Froschplatte der Presse befestigt. Ein Stück Filz wird nun auf diesen Block gelegt und der Stempel durch einen schnellen Fußtritt durchgetrieben, bis er gerade das Holz berührt, was durch den einstellbaren Anschlag auf der Rückseite der Presse erreicht wird, so daß er nicht in das Holz selbst eindringt. Beim Aufwärtsgang der Presse werfen die Abstreifer die Unterlagscheibe und den Abfall aus.

Werkzeuge dieser Art können fast auf jedem weichen Material verwendet werden, wie Leder, dünnes Gummi, Papier, Karton, ja sie sind auch auf Messing von 0,5 mm Dicke gebraucht worden.

Beim Anlassen der Stempel muß man Obacht geben, den Stahl nicht zu überhitzen; Fischöl gibt die besten Erfolge und gibt Anlaßfarben von Dunkelbraun bis Blau. Nach dem Härten und Anlassen müssen die Kanten mit einem feinen Schleifstein so lange abgezogen werden, bis sie scharf sind.

B. Arbeiten mit Formänderung.

Die zweite große Gruppe der Preßwerkzeuge bewirkt die Entstehung des fertigen Stückes durch eine Formänderung einer gewissen Materialmenge, die in der für die betreffende Verarbeitung bestgeeigneten Form unter das Werkzeug gebracht wird. Es sind also beim Arbeiten mit Formänderung vor allem zwei große Gruppen zu unterscheiden; die eine Gruppe arbeitet mit einer Materialverschiebung aller kleinsten Teilchen des zu verarbeitenden Stückes, die andere arbeitet ohne Materialverschiebung, so daß nur die Entfernung der an der Arbeitsstelle nebeneinander liegenden Teilchen eine Änderung erfährt. Während bei den Schnittwerkzeugen eine Materialtrennung zur Erzeugung der gewünschten Form nötig war, findet hier eine Verschiebung der kleinsten Materialteile gegeneinander statt, ohne daß deren Zusammenhang gelöst werden darf. Es muß also in jedem Falle die Fließgrenze des Materiales überwunden werden. Trotzdem die Scherfestigkeit, dasselbe Material vorausgesetzt, höher als diese liegt, wirken doch in diesem Falle größere Gesamtkräfte, da die Schnittwerkzeuge diese Kraft nur auf die eigentliche Schnittfläche auszuüben brauchen, während bei den Werkzeugen, die mit Formänderungen arbeiten, das ganze unter dem Werkzeug liegende Material, bzw. die ganze Oberfläche desselben der Beanspruchung bis über die Fließgrenze ausgesetzt werden muß. Dies würde aber bei verhältnismäßig noch ganz kleinen Arbeitsstücken zu derartig großen Kräften und Maschinen führen, daß eine wirtschaftliche Arbeit ausgeschlossen wäre. Deshalb kommen hier noch zwei Hilfsmittel in Betracht, die Erwärmung und die lebendige Kraft einer bewegten kleinen Masse. Die Erwärmung erhöht die Bildsamkeit der in Betracht kommenden Materialien derartig, daß die Kosten für die Erwärmung eines Stückes durch die dadurch erzielte leichtere und schnellere Bearbeitung nicht nur aufgewogen werden, sondern man noch mit einer großen Arbeitsersparnis rechnen kann. Ebenso ist die Mehrleistung einer bewegten kleinen Masse gegenüber der eines großen, ruhigen Druckes eine so bekannte Sache, daß die einfache Anführung hier genügt.

Innerhalb dieser beiden Gruppen wird von den erwähnten Hilfsmitteln zur Erleichterung der Bearbeitung Gebrauch gemacht, so daß sich die in der Praxis gebräuchlichen Bearbeitungsweisen als Unterabteilungen ergeben.

1. Formänderung mit Materialverschiebung.

Eine Klarlegung der inneren Vorgänge im Material ist bis jetzt noch nicht einwandfrei erfolgt, sowohl was die Behandlung im kalten wie im warmen Zustand betrifft. Bei Betrachtung eines normalen Zug- und Druckdiagramms, Fig. 200, für Formänderung im kalten Zustand fallen alle diese Arbeiten in den Bereich der bleibenden

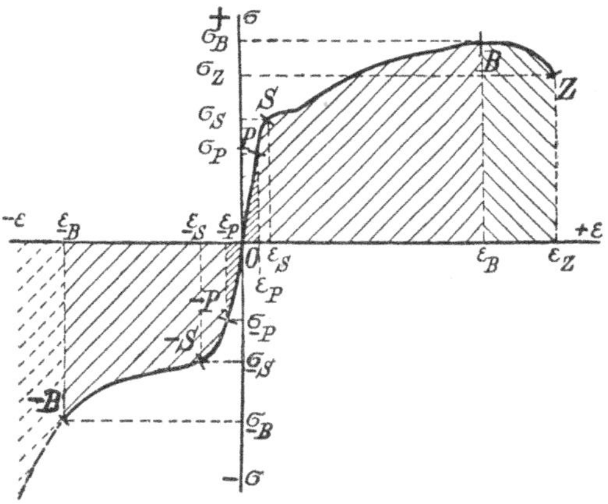

Fig. 200.

Formänderungen nach Überschreiten der Fließgrenze des Materials in das Gebiet -S -B, bzw. S B. Die inneren Vorgänge, nach welchen die bei der Kaltbearbeitung unter Druck[1]) entstehenden Materialverschiebungen vor sich gehen, sind um so schwieriger zu verfolgen, als der Einfluß der an den Druckflächen auftretenden Reibung zwischen Stempel und Material einen ganz bedeutenden Einfluß auf die Materialverschiebung hat. Heyn[2]) hat diese Vorgänge sehr anschaulich an geätzten Schnitten verfolgt, wobei sich das in Fig. 201/2 dargestellte Bild der Materialbewegung ergab. Die früher gerade verlaufenden Fasern krümmen sich nach außen, und der Krümmungsradius wird um so kleiner, je weiter die Faser von der Körperachse entfernt liegt. Steigt die Zusammendrückung über ein von der Höhe des Körpers abhängiges Maß, so tritt die in Fig. 202 wiedergegebene Erscheinung auf. Von den Kanten des Körpers bzw. den Ecken des Schnittes gehen dunkle Diagonalen nach innen, in welchen die Fasern eine doppelte Krümmung, also eine erhöhte

---

[1]) Beim Ziehen kommen ähnliche Einflüsse in Betracht, S. 250.

[2]) Heyn, Z. Ver. deutsch. Ing. 1900, S. 435.

Beanspruchung des Materials zeigen. Es tritt an diesen Stellen bereits eine Materialzerstörung auf. Anders verhalten sich dagegen dieselben Materialien, wenn sie unter allseitigem Druck belastet werden[1]). Unter

Fig. 201.

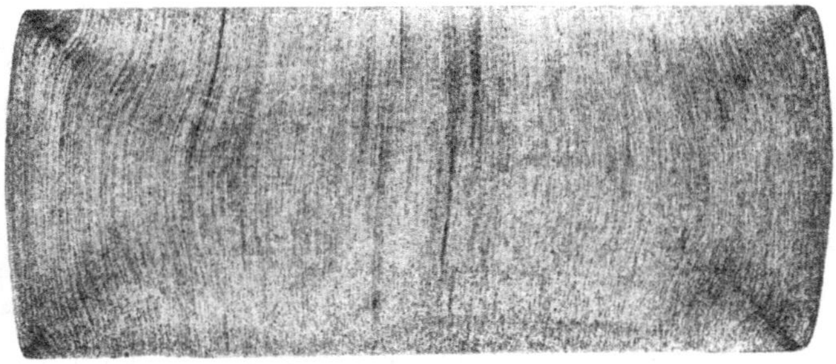

Fig. 202.

diesen Verhältnissen kann man selbst bei spröden Körpern ein Fließen und eine plastische Deformation erzeugen. Beim Gesenkschmieden, Gesenkpressen und Prägen wird das Material anfangs in der ersten Art gedrückt, bis sich beim Schließen der Gesenkhälften nahezu der zweite Zustand einstellt. Eine Anwendung allseitigen Druckes findet beim Huberschen Preßverfahren statt.

[1]) Karman, Festigkeitsversuche unter allseitigem Druck. Z. Ver. deutsch. Ing. 1911, S. 1749.

a) Gesenkschmieden.

α) Warmbearbeitung.

Ebenso wie beim Handschmieden muß man beim Gesenkschmieden zwischen einer bloßen Oberflächenwirkung und einem das ganze Material durchdringenden Druck unterscheiden. Das in Fig. 202 gezeigte Bild ergibt sich im letzteren Falle, während leichte Schläge, die nur auf die Oberfläche des zu schmiedenden Stückes wirken, eine Formänderung nach Fig. 203 hervorrufen. Vergleicht man damit die Materiallagerung[1]) an einem kalt zu einem Vierkant ausgeschmiedeten Rundstab, Fig. 204/5, aus der sich ergibt, daß die größte Korndeformation, also auch die größte Materialdurcharbeitung, an den Stellen der größten Materialverschiebung und in der Richtung dieser, also in den Punkten E und F, erfolgte, während die in der Mitte bei M liegenden Körner ihre Würfelform und nahezu quadratischen Querschnitt beibehalten haben, so erkennt man, daß bei der äußeren Form in

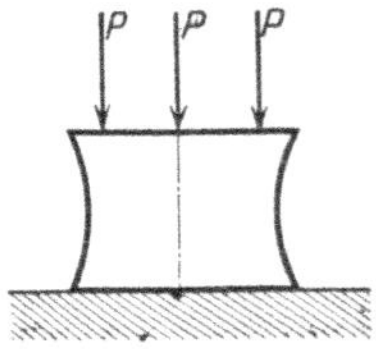

Fig. 203.

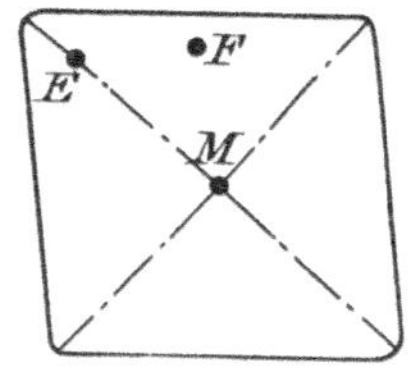

Fig. 204.

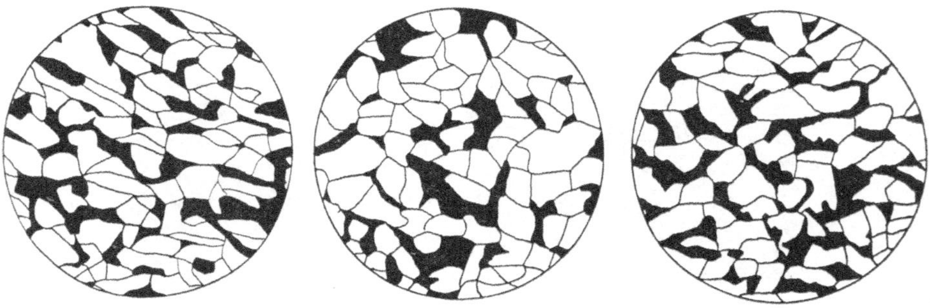

Fig. 205. Kleingefüge des kalt in einem Vierkant ausgeschmiedeten Rundstab, Fig. 204.

Stelle E Stelle M Stelle F

Fig. 204 nur die an den Berührungsflächen von Bahn und Schmiedestück liegenden Teile eine Deformation und Durcharbeitung erfahren haben werden. Wenn dagegen durch möglichst allseitigen Druck und einen Überschuß an die ganze Masse durchdringender Kraft eine durchgängige Materialverschiebung Platz greift, so ist das Bild ein ganz anderes. Fig. 206 a—c[2]) zeigen anschaulich an makroskopischen Ätzungen

[1]) Heyn, Z. d. V. d. I., a. a. O.

[2]) Vgl. Katalog der Firma de Fries, Düsseldorf.

den Unterschied der Materiallagerung bei einem aus dem vollen gedrehten, einem auf die gewöhnliche Weise durch Absetzen und Strecken und einem auf der Schmiedemaschine Ajax-de Fries, also unter möglichst allseitigem Druck geschmiedeten Bolzen. Die Materialbewegung im Kopf, welche der unter freiem Druck erhaltenen sehr ähnlich ist, zeigt — bei c — keine Unterbrechungen der Fasern an dem Übergang vom Kopf in den Schaft, im Gegensatz zu der vollständig scharf abgeschnittenen

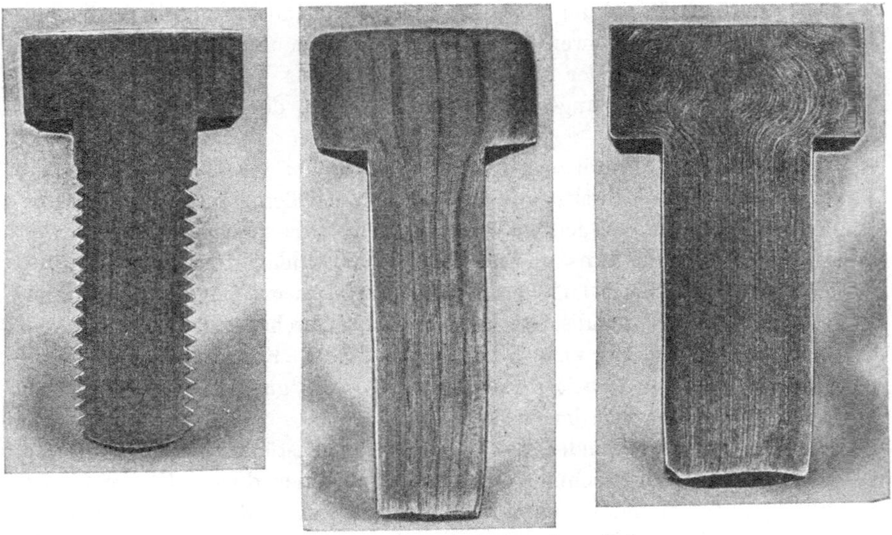

Fig. 206. Geätzter Querschnitt eines Bolzens.

a) Aus dem Vollen gedreht. b) Abgesetzt und gestreckt. c) Auf der Schmiedemaschine Ajax-de Fries gestaucht.

Faser — bei a —, beim gedrehten und der nur unvollkommen der Form des Körpers folgenden Faser bei dem nach dem landläufigen Verfahren b, geschmiedeten Bolzen.

Um diese Wirkung zu erzielen, ist es notwendig, das Material während des Schmiedens in genügend fließbarem Zustand zu erhalten, also bei richtiger Hitze zu arbeiten. Nach Hoyer, Mech. Technologie, S. 144, sind die Druckkräfte zur Formänderung auf 1 qmm Fläche bei den verschiedenen Schmiedetemperaturen:

gelbwarm	5,5 kg,
dunkelgelb . . .	6,4 kg,
rot	6,7 kg,
dunkelrot	16,0 kg,
braun	17,6 kg,
beinahe schwarz	22,3 kg.

Daraus ergibt sich, daß bei Unterschreitung des Rot die Drücke auf den Quadratmillimeter plötzlich sehr stark zunehmen, so daß ein Schlag, der bei einem genügend warmen Werkstück eine volle Durcharbeitung entsprechend Fig. 206 c noch geben würde, bei einer Abkühlung unter Rot nur noch eine Oberflächenwirkung nach Fig. 203 erzielen würde.

Drittens kommt, wie Sobbe[1]) sagt, der notwendige Deformationsdruck auch in Beziehung zum Grad der Formänderung selbst, die bei einem Schlag erzielt werden soll, so daß man auf Drücke kommen kann, welche trotz der starken Erhitzung des Werkstückes die im Diagramm Fig. 200 angegebenen Werte über der Fließgrenze noch übersteigen.

Zu diesem Kapitel sagt nun Woodworth:

Um Schmiedestücke mit Erfolg herzustellen, sind eine ganze Reihe schwieriger Fragen zu beantworten. Die wichtigsten sind: die richtige Hitze, die Anzahl und Stärke der Schläge und der Einfluß der schnellen Schläge auf das Material im Gegensatz zu der langsamen Preßarbeit der hydraulischen Presse. Zur Entscheidung dieser Fragen gehört nicht allein eine gute mechanische Kenntnis sowie Vertrautheit mit den Arbeitsweisen, sondern auch gute Erfahrung in der Gesenkschmiedearbeit im besonderen.

Man hat eingewendet, daß Gesenkschmiedestücke manchmal nicht so stark wie handgeschmiedete sind, doch kann dieser Einwurf nur gelten, wenn er durch eine Prüfung des gleichen Materiales,das auch unter den gleichen Bedingungen verarbeitet worden ist, bewiesen wird. Natürlich ist mehr Arbeit auf dem Material verrichtet worden, wenn ein Stück von Hand geschmiedet worden ist, als wenn dies im Gesenk geschah; das Stück wird weniger gebogen, es werden weniger Schläge gegeben, weniger Fertigarbeit wird aufgewendet, alles Dinge, die eine Qualitätsverbesserung des Stückes herbeiführen, vorausgesetzt, daß die Arbeiten bei der richtigen Hitze ausgeführt worden sind.

Es ist ziemlich allgemein bekannt, daß die Zusammendrückung, der die Stücke beim Gesenkschmieden unterworfen werden, wenn die Arbeit bei einer hohen Temperatur, die zu Anfang oft eine Schweißhitze ist, gemacht wird, das Material verdichtet und sein Gefüge verbessert, gerade wie die vielen leichten Schläge beim Handschmieden. Daraus ergibt sich, daß der wichtigste Punkt die Einhaltung der richtigen Hitze während der ganzen Gesenkschmiedearbeit ist, damit das Material in die verlangte Form fließen kann. Wenn das Material genügend warm und der Hammer genügend kräftig ist, so kommt es auf die zu schmiedende Form wenig an, da das Material im heißen Zustande wie eine

---

[1]) W.T. 1908, S. 436.

bildsame Masse fließt, indem es den Linien des geringsten Widerstandes folgt.

Mit Berücksichtigung dieser Punkte ergeben sich drei Wege, das angestrebte Ziel zu erreichen. Man kann die Temperatur des Materials erhöhen, um die Bildsamkeit des Materials zu erhöhen und so die Bedingungen mehr den Verhältnissen beim Gießen zu nähern — ein Weg, der sehr verführerisch erscheint, immer aber die Gefahr des Überhitzens und Verbrennens mit sich bringt.

Andererseits kann der Druck erhöht werden.

Bei Verwendung des Fallhammers kann man im Gegensatz zur langsamen Preßarbeit eine schnelle Folge leichter Schläge anwenden. Diese eignen sich besser für dünne Arbeitsstücke, während dicke, in die Tiefe zu bearbeitende Gegenstände eher unter der Presse auszuarbeiten sind. Gibt man auf derartige Arbeitsstücke leichte Schläge, so wird das Innere des Materials, vgl. Fig. 204/5, nicht wie bei der Preßarbeit verdichtet. Wenn leichte Schläge auf kleine Schmiedestücke im Gesenk gegeben werden, so kommt die Arbeit der Handschmiederei ziemlich nahe.

Die richtige Schlagzahl und Stärke für ein gegebenes Schmiedestück zu finden, ist reine Erfahrungssache. Die Temperatur wechselt von Schweißhitze am Anfang bis zu einem Dunkelrot, manchmal ist das Stück am Ende der Arbeit bereits dunkel. Im Anfang der Bearbeitung fließt das Material wie ein weicher Körper, später bewirken die Schläge mehr eine Verdichtung und Bearbeitung der Oberfläche. Ein Schmiedestück, das beträchtliche Verschiedenheiten in seinen Abmessungen zeigt, wird nicht so gleichmäßig durchgearbeitet werden können als eines, dessen Abmessungen alle ziemlich gleich sind.

Selbst wenn der Einwand, daß Gesenkschmiedestücke nicht so stark wie handgeschmiedete sind, berechtigt wäre, muß er gegen die Anforderungen moderner Herstellungsmethoden zurücktreten. Es ist heute notwendig, Hunderte gleicher Teile zu einem fast verschwindenden Stücklohn herzustellen. Wenn der Beweis erbracht werden sollte, daß tatsächlich das Gesenkschmieden schwächere Stücke herstellt als das Handschmieden, so müßte man mit Rücksicht auf die Erfordernisse des Verkaufes das gewünschte Endziel auf einem anderen Weg, z. B. durch Verwendung eines besseren Materials, bessere Einstellung und Einhaltung der richtigen Hitze, erreichen.

Eigenheiten der Gesenkschmiedestücke. Eine Eigenheit aller im Gesenk geschmiedeten Teile ist ein kleiner unregelmäßiger Grat, der sich längs des ganzen Umfanges des Stückes hinzieht und dessen Ursache ist, daß das Material bzw. der Materialüberschuß zwischen die beiden Gesenkhälften als den einzigen Ausweg, wohin es unter dem Druck entweichen kann, fließt. Es ist ja richtig, daß dieser Grat gelegentlich nicht erscheint, aber in den meisten Fällen tritt er auf. Vermeiden kann man ihn nur dadurch, daß das Blankett genau in seine richtige Stellung gelegt wird, dort während des Schlages verbleibt, und daß das Blankett auch nur die notwendige Materialmenge enthält. Mehr oder weniger sind diese Grate immer vorhanden, doch werden sie nachher in einer Kraftpresse unter Anwendung eines Abgratwerkzeuges mit ausgehöhltem Stempel, der sich soweit als möglich der oberen Fläche des Stückes anpaßt, abgeschnitten. Man kann natürlich auf diesem Wege nur Stücke herstellen, die leicht aus den Gesenken entfernt werden können, weshalb diese eine entsprechende Schräge und keine hohen, senkrechten Flächen haben.

Anwendung der Gesenkschmiederei. Die allgemeine Anwendung des Schmiedens im Gesenk ist so ausgebreitet, daß es unmöglich ist, alle Gebiete hier aufzuzählen. Auf diesem Wege werden viele Tausende kleiner Teile von Werkzeugen, Maschinen, Eßbestecken und Gebrauchsgegenständen sehr schnell hergestellt, wobei eine Gleichmäßigkeit wie bei der Stanzarbeit, jedoch eine viel glattere Oberfläche und ein dichteres Materialgefüge erzielt wird. Die meisten dieser Gegenstände haben eine erhöht-abgerundete Form, die sich mittels Stanzens oder Prägens aus Blechmaterial überhaupt nicht herstellen läßt. Solche Schmiedestücke sind gewöhnlich besser als die besten handgeschmiedeten Sachen, auch billiger und gleichmäßiger. Sie sind oft sogar billiger als gleiche Gußstücke, da sie weniger Nacharbeit erfordern, leichter austauschbar sind und deshalb für die meisten Zwecke geeigneter als diese sind.

Schmiedegesenke. Die Gesenke für Schmiedearbeit werden in zwei Teilen gemacht. Der eine, der obere, ist an dem Bären des Fallhammers befestigt und bewegt sich mit diesem zwischen den beiden senkrechten Führungen, wobei der Hammerbär mittels Reibrollen, die vom Arbeiter ein- und ausgeschaltet werden, angehoben wird. Der untere Gesenkteil ist an dem Amboß befestigt. Der Hammer wird angehoben und fällt sofort nach der Auslösung, wobei er mit dem Obergesenk auf die glühende Metallstange, die auf dem unteren Gesenk aufliegt, aufschlägt und so die Form in beiden Gesenken mit dem weichen Material ausfüllt. Nach einer Anzahl solcher Schläge ist der Gegenstand fertiggestellt.

Die zur Herstellung von Gesenkschmiedestücken besonderer Form und Größe notwendigen Gesenke werden entweder nach einer Zeichnung oder einem Modell gemacht. Der letztere Weg erleichtert jedoch den Entwurf der Gesenke und die Veranschlagung ihrer Herstellungskosten.

Bei Herstellung dieser Gesenke muß der Gesenkschmied vor allem wissen, ob die Zeichnung oder das Modell die fertigen oder die Schmiedeabmessungen aufweist. Ebenso muß er die notwendige Zugabe für die Bearbeitung kennen. Meistens wird auf jeder zu bearbeitenden Fläche 0,8 mm zugegeben, außer wenn das Stück nur durch Schleifen oder Polieren fertiggestellt wird; dann beträgt die Zugabe nur 0,25 mm. Flächen, die unbearbeitet bleiben, werden so nahe wie möglich auf Maß geschmiedet. Die Schmiedestücke sind in der Dicke ein wenig verschieden — z. B. von 0,25 bis zu 0,8 mm —, was von ihrer Form und dem verwendeten Material abhängt. Man kann sie jedoch auf Maß herstellen, indem man sie in besonderen Gesenken, womit aber auch zusätzliche Kosten verbunden sind, nachschmiedet.

Außer den Schmiedegesenken, deren Kosten und Haltbarkeit von den an sie gestellten Anforderungen abhängen, benötigt man noch Abgratwerkzeuge, mit denen der Grat, der sich durch das Herausquellen des überschüssigen Materiales zwischen den Gesenken bildet, abgeschnitten wird.

Vor der Indienststellung eines neuen Gesenkes wird damit in Blei eine Probe geschlagen, die mit dem Modell verglichen wird. Dieses Probestück ist von dem Modell oder der Zeichnung meistens um einen gewissen Betrag, den Anzug, verschieden. Es ist dies die Schräge, die den Stücken gegeben werden muß, damit man sie nach der Arbeit aus den Gesenken entfernen kann; sie beträgt gewöhnlich 7°. Man kann diese Schräge durch Zugabe an dem Stück oder durch Abarbeiten gegenüber dem Modell erhalten. Gewöhnlich wird Material zugegeben.

Arbeitsweisen beim Gesenkschmieden. Das Gesenkschmieden von Stahl, Eisen oder Kupfer ist eine bedeutend schnellere Arbeitsweise als das Handschmieden und gestattet, nicht allein große Mengen, sondern auch Formen herzustellen, die beim Handschmieden fast unmöglich zu erreichen sind. Es gibt nun zwei Arbeitsweisen: Blankett vorbereiten, Schmieden und Zuschneiden, entweder mittels kombinierten Schmiedegesenkes mit „Überlaufkanal" oder mit besonderen Abgratwerkzeugen, wie es eben das Stück erfordert. Beim Ausschneiden der Blanketts, die nachher im Gesenk verschmiedet werden sollen, ist das Stück sorgfältig zu studieren und, wenn man genug Zeit hat, das Schmiedegesenk zuerst auszuarbeiten, zu härten und fertigzustellen. Dann macht man ein Holzmodell, entsprechend dem gedachten Blankett, gießt dieses im Sand in Blei ab und versucht es im Fallhammer. Nach dem Ergebnis kann man dann die Fehler verbessern und versucht wieder in gleicher Weise. Fast alle Schmiedegesenke müssen wohl mit einem Überlaufkanal[1]) versehen

[1]) Vgl. Gesenkpressen S. 194.

werden, d. h. man muß einen schmalen Kanal rund um die Form im Gesenk anbringen. Dieser Kanal soll das überschüssige, herausquellende Metall an den Kanten abschneiden und andererseits auch den Druck verringern, der entstehen würde, wenn sich dieser Metallüberschuß in Form einer dünnen Fläche oder eines Kranzes zwischen die beiden Gesenkflächen eindrängen würde. Die Form dieser Kanalkante hängt in großem Maße von der Gestalt des Stückes selbst ab.

Um bei der Arbeit von den Gesenken übermäßige Hitze wie Zunder abzuhalten, ist es gut, die Gesenke mittels Preßluft abzublasen, die man entweder von dem Schmiedegebläse oder einer besonderen Leitung abzweigt. Manchmal wird ein sehr kräftiger Strahl notwendig sein, um das Gesenk frei von kleinen Eisenteilchen zu halten.

Bei der Herstellung der Gesenke kann viel Arbeit erspart werden, indem man Stahlmodelle herstellt und nach Einpassen der Schwalbenschwänze die Gesenke erwärmt, worauf das kalte Stahlmodell zwischen ihnen geschlagen wird, so daß man eine vollständige Form des Stückes erhält. Man muß jedoch dabei scharf aufpassen, daß das Modell richtig zwischen den Gesenkteilen liegt, um das Stück nicht zu biegen. Im allgemeinen ist es notwendig, die Gesenke roh vorzuarbeiten und ihnen ihre endgültige Gestalt auf diesem warmen Wege zu geben.

In Gesenken für lange oder sehr genaue Stücke muß man auch eine Zugabe für Schwinden geben — die beiden Schwindmaße für das Gesenk und das zu schmiedende Stück betragen zusammen $^1/_{192}$.

Schwierige Stücke mit mehreren hervorstehenden Teilen werden manchmal wohl unter dem Hammer roh vorgearbeitet und nachher im Gesenk fertiggeschmiedet, doch können sie auch von dem Walzmaterial zugeschnitten werden, wobei dann die vorstehenden Teile des Schmiedestückes durch die ununterbrochenen Flanschen der Stange dargestellt werden. Indem man nun Teile des Flansches wegschneidet, erscheint allmählich der Umriß des Schmiedestückes, das dann in einem oder mehreren Gesenken auf die richtige Form gebracht werden kann. Das Kurbellager eines Fahrrades ist ein gutes Beispiel für solche Arbeit.

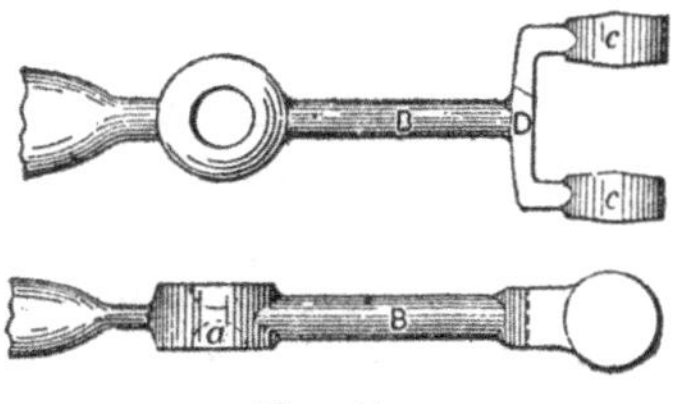

Fig. 207.

Die Herstellung von Schmiedegesenken. Es sei für die in Fig. 207 dargestellte Gabel ein Schmiedegesenk nach Fig. 208 zu bauen.

Die Gesenke werden aus Kohlenstoffstahl (0,45 bis 0,6 % C) hergestellt und sind gewöhnlich 125 bis 200 mm dick. Die Fig. 208 gibt eine Ansicht der fertigen Gesenke. Ober- und Unterseite werden durch eingeschlagene Buchstaben, z. B. T — B, gezeichnet, so daß man sie beim Anreißen nicht verwechseln kann. Die Vorder- und linke Seitenfläche werden unter rechtem Winkel bearbeitet und von diesen Flächen aus die Mittellinien der Eindrücke angerissen und die fertigen Gesenke eingebaut. Das Vorschmiedegesenk befindet sich an der rechten Seite des Gesenkes. Damit wird das warme Rohmaterial, bevor es in das Fertiggesenk kommt, in Formen gebracht, die eine entfernte Ähnlichkeit mit der fertigen Gestalt aufweisen. Manchmal verwendet man ein besonderes Vorschmiedegesenk. Der schwerste Teil des Schmiedestückes liegt immer vorne. In Gesenken für tiefe Formen, mit parallelen Seiten, gibt man an jeder Seite 5° bis 7° Schräge, damit das Schmiedestück nicht im Gesenk stecken bleibt. Als Zugabe für Bearbeitung nimmt man gewöhnlich 0,8 mm und für das Schwinden 0,3—0,38 mm. Wenn die Gesenke fertig sind, wird ein Muster in Blei abgegossen, um festzustellen, ob das Gesenk die richtige Form herstellt.

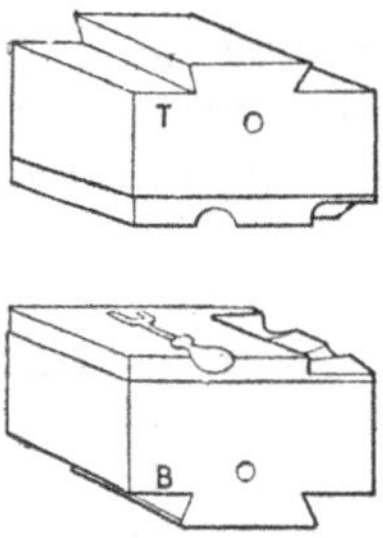

Fig. 208.

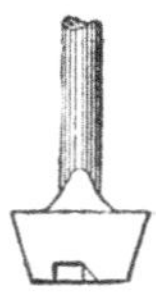

Fig. 209.

Der runde Teil des Eindruckes wird zuerst hergestellt. Hat man viel Material zu entfernen, so arbeitet man das Gesenkblankett auf der Drehbank aus. In der Schmiedewerkstätte der H. Williams Company, Brooklyn, N. Y., wird eine gußeiserne Grundplatte auf der Planscheibe der Drehbank verwendet. Diese Grundplatte hat auf der Rückseite eine Rippe, die in eine Nut der Planscheibe paßt. In die Stirnfläche der Grundplatte ist eine Schwalbenschwanznut eingearbeitet, die mit der im Hammer übereinstimmt und unter rechtem Winkel zu der Rippe auf der Rückseite verläuft. Mittels dieser kann man nun einen Kreis sehr bald ausrichten. Ist der runde Teil des Gesenkes unter 38 mm im Durchmesser, so wird die Arbeit besser auf der Profilfräsmaschine hergestellt. Man schruppt das Stück mit dem zweischneidigen Fräser, Fig. 209, vor und verwendet dann einen halbrunden Fräser zur Fertigarbeit. Der halbrunde Fräser ist sehr nützlich, da er stärker ist, leicht hergestellt und von Hand nachgeschliffen werden kann. Der Stahl in Fig. 209 läßt in der Mitte der Ausdrehung eine Erhöhung stehen, so daß man die Mitte des Putzens an dem Schmiedestück anreißen kann. Ein Gesenk zum Schmieden einer Kugel wird mit einem zweischneidigen Kugelfräser ausgearbeitet. Wenn in das Schmiedestück nachher ein großes Loch zu bohren ist, so läßt man bei der Herstellung im Gesenk einen Putzen stehen oder setzt ihn später ein, so daß das Schmiedestück an dieser Stelle eine Höhlung erhält, wie bei a, Fig. 207, ersichtlich ist. Dieser Putzen muß eine Schräge von 15° an den Seiten und eine gut abgerundete Spitze haben. Bei der Herstellung der Gesenke für das in diesen Figuren abgebildete Stück kann der runde Teil bei B im Gesenk nach der Herstellung des Teiles bei D ausgearbeitet werden.

Die Teile c, Fig. 207, müssen von Hand ausgearbeitet werden. Manchmal werden die kreisförmigen Vertiefungen in der Fräsmaschine bearbeitet, wobei eine lange, halbe Körnerspitze und ein Profilfräser mit dünnem Schaft verwendet werden, indem man zuerst eine Nut ausschneidet, um dem Schaft Platz zu machen. Doch werden die oben angeführten Teile c in den meisten Werkstätten von Hand ausgearbeitet werden müssen. Der Stöckel ist aus gehärtetem Stahl, in Größe und Form der herzustellenden Vertiefung entsprechend, mit ungehärtetem Kopf, so daß der Stahl unter den Hammerschlägen nicht in Stücke fliegt. Teile eines Gesenkes, die nicht mit der Maschine bearbeitet werden können, müssen von Hand ausgearbeitet werden, sei es, daß mangelnde Werkstatteinrichtung oder unregelmäßige Form des Eindruckes der Grund ist. Doch verlangt die Handarbeit ganz besondere Vertrautheit mit Hammer und Meißel, Schaber und Graviernadel, wie auch ein gutes Auge und Formensinn.

Fig. 210.

Fig. 211.

Der Meißel, der zum Ausarbeiten harter Gesenke verwendet wird, muß unter dem richtigen Schnittwinkel angeschliffen werden, um die gekrümmten Oberflächen und schwierigen Ecken abzunehmen. Die Schaber müssen verschiedene Längen und Formen besitzen, so daß sie sich den verschiedensten Anforderungen der Arbeit anpassen. Am nützlichsten unter allen ist der Dreikantschaber, dessen zwei Kanten abgerundet sind; die dritte, die Schnittkante, wird, wie Fig. 210 zeigt, scharf gelassen und rundet sich gegen die Spitze. Ein anderer brauchbarer Schaber ist in Fig. 211 dargestellt. Er ist herz- oder blattförmig und sehr geeignet, kleine Kurven und Ecken auszuarbeiten, besonders wenn sie am Grunde der Vertiefung liegen. Der Stöckel wird mit einer dünnen Schicht Preußisch Blau oder Mennige überzogen und von Zeit zu Zeit in das Gesenk eingetrieben; die so angezeigten Stellen werden nach und nach abgearbeitet, bis man die richtige Form erhält. Für die Abrundungen und kleinen Ecken wird aus Bohrerstahl vom verlangten Durchmesser eine Graviernadel oder ein Schaber gemacht. Um Platz im Werkzeugkasten zu ersparen, kann man einen kleinen Handgriff — 5 bis 6,5 mm — an einem Ende mit Gewinde versehen und alle kleinen Schaber daran anpassen. Die Schaber lassen in ihrer Arbeit einen schmalen Grat, der dann mit Lochfeilen entfernt werden muß. Manchmal werden die Vertiefungen auch mit einem Block aus weichem Tannenholz und Schmirgelpulver auspoliert, doch ist dies nicht der gewöhnliche Gebrauch. Wenn die Eindrücke bis an die Umrißlinien ausgearbeitet worden sind, macht man von dem Gesenk einen Abguß in Blei, um die Stellen, die noch der Verbesserung bedürfen, herauszufinden. Der Bleiabguß wird dann auf seine Abmessungen kontrolliert und, wenn er sich als richtig herausstellt, von dem Obergesenk ein Halbabguß gemacht, der als Lehre beim Anreißen der Abgratwerkzeuge dient, d. h. in jenen Werkstätten, wo nicht nach Blechlehren gearbeitet wird. Ist das Blei überhitzt oder zu oft umgeschmolzen worden, so fließt es nicht leicht und wird starr, bevor der Eindruck ganz gefüllt ist. Wenn man aber die Form mit Kreide bestreut, fließt das Blei leichter.

Das Vorschmiedegesenk auf der rechten Seite des Gesenkes wird ungefähr 1,6 bis 2,4 mm kleiner als der wagrechte Querschnitt des Schmiedestückes, ohne jähe Übergänge oder Krümmungen, gemacht. Man will damit das erwärmte Material schmäler als die fertige Form erhalten, so daß das Material zuerst auf den Grund der Form trifft und so gezwungen wird, in die Breite zu gehen und das Gesenk auszufüllen. Zum Vorschmieden schwerer Stücke werden auch Gußeisengesenke verwendet.

Der Überlaufkanal, der ungefähr 4 bis 6,5 mm Tiefe und 20 mm Breite hat, wird längs des Umrisses der Gesenkform gefräst, so daß der Materialüberschuß aus den Gesenken dorthin abfließen kann. Dieser Grat wird dann in den Abgratwerkzeugen entfernt. Auch das Obergesenk erhält eine Nut von ungefähr 1,5 mm Tiefe, die längs des Umrisses der Form, 6,5 mm von deren Kante entfernt, eingefräst wird. Die Einführungsöffnung für das Material ist von der Form gegen die Vorderseite zu leicht konisch, damit das Gesenk nicht an dieser Stelle zu sehr geschwächt wird.

Bei der Herstellung der Gesenke für die Massenfabrikation kleiner Gesenkschmiedestücke werden mehrere Formen ausgearbeitet, von denen eine als Vorschmiedegesenk Verwendung findet und ungefähr 0,8 mm schmäler und tiefer als das Fertiggesenk ist. Manche Gesenke, besonders wenn es sich um schwierige Formen handelt, müssen verriegelt werden, d. h. die Stirnflächen müssen so geformt sein, daß sie nur in der richtigen Lage aufeinanderpassen. Man muß dabei Sorge tragen, daß die Teile, welche die Verriegelung besorgen, so hoch sind, daß sie nicht klaffen, wenn sie auf das Material auftreffen, denn sonst würde das Gesenk einen unvollständigen Abdruck und ein unbrauchbares Schmiedestück geben. Ist die Stirnfläche der Gesenke gekrümmt, so werden besondere Fräser, ähnlich denen in Fig. 212, für die Bearbeitung der Oberfläche und der Überlaufkanäle gemacht. Um

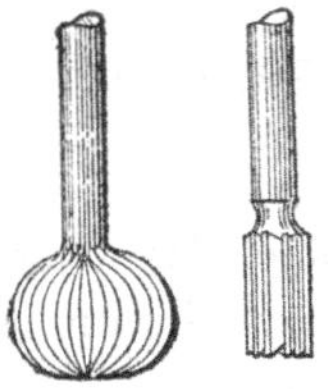

Fig. 212.

eine Führung zur Bearbeitung der gekrümmten Flächen zu haben, werden manchmal die Linien auf die Seitenfläche des Gesenkblanketts übertragen und dort die Kurve angerissen. Dann wird eine Anreißnadel an der Profilfräsmaschine so festgemacht, daß ihre Spitze in Linie mit dem Fräser ist, und nun das Material weggenommen, wobei man mit der Anreißnadel längs der vorgezeichneten Linie fährt. Gesenke zum Schmieden von Zahnrädern oder ähnlichen Stücken werden mit einem Stöckel hergestellt, in dem die Zähne eingefräst worden sind und der dann in die Matrize eingetrieben wird.

Das Härten der Schmiedegesenke. Beim Härten aller Arten Schmiedegesenke, besonders der großen, darf man unter keiner Bedingung das ganze Stück ins Wasser werfen, sondern muß, sobald das Stück die richtige Temperatur erreicht hat, einen Strahl kalten Wassers auf die Stirnfläche des Gesenkes in einer passenden Aufstellung spritzen. Der Strahl soll so groß sein, daß er fast die ganze Oberfläche des Gesenkes bedeckt, wobei man das Wasser so lange fließend erhalten muß, bis die Siedetemperatur erreicht worden ist; dann kann man das Gesenk allmählich abkühlen lassen. Dabei soll man das Gesenk auf einer kleinen Erhöhung so aufstellen, daß das Wasser, welches die Oberfläche abkühlt, nicht gleichzeitig die unteren Teile des Gesenkes abkühlen kann. Man erreicht dadurch, daß nur die Oberfläche des Gesenkes gehärtet wird und Härterisse oder Spannungen vermieden werden. Das Stück kann, wenn verlangt, angelassen werden, indem man von rückwärts anwärmt und, nachdem man die richtige Anlauffarbe erreicht hat, durch einen Strahl kalten Wassers das Gesenk abschreckt.

Vervielfältigung von Schmiedegesenken. Um Schmiedegesenke, die andauernd für dasselbe Stück Verwendung finden sollen und deshalb fortwährend erneuert werden müssen, zu vervielfältigen, macht man Modelle, mittels derer die neuen Gesenke jederzeit geschlagen werden können. Handelt es sich um ein sehr tiefes Gesenk, so schlägt man folgenden Weg ein: Zuerst wird der Block für das Gesenk erhitzt und das Modell eingeschlagen, wobei man durch eine genügende Anzahl von Schlägen eine vollkommene Form herstellt. Dann wird das Gesenk sorgfältig ausgeglüht und der Zunder durch Eintauchen in Salzsäure entfernt.

Nachdem das Gesenk in warmem Wasser abgewaschen, getrocknet und überall geölt worden ist, wird das Modell in die Form eingelegt und untersucht, ob durch das Schwinden nicht passende Stellen entstanden sind. Zeigen sich keine Fehler, so arbeitet man alle geraden Flächen und Kanten ab, kurz stellt das Gesenk nach jeder

Richtung fertig. Dann wird das Stück mit dem Modell im Hammer eingelegt, die Flächen leicht geölt, und das Gesenk durch eine Reihe leichter Schläge, die einigen wenigen, schw ren vorzuziehen sind, fertiggestellt. Nachdem das Gesenk aus dem Hammer herausgenommen worden ist, erhält es die letzten Nacharbeiten und wird, wie oben beschrieben, gehärtet. Auf diese Weise kann man ein Gesenk, das, wenn von Hand ausgearbeitet, nach amerikanischer Angabe 1000—1200 M. kosten würde, leicht in einem Tage mit einem Kostenaufwand von wenigen Mark für das Stück herstellen. Man hat dabei noch den Vorteil, daß jedes Stück eine genaue Wiederholung des Modells ist, was besonders, wenn es sich um verwickelte Einzelheiten im Gesenk handelt, bei Handarbeit äußerst schwierig ist.

Schmiedegesenk für Luft- oder Dampfhammerarbeit. Das erste der beiden im folgenden beschriebenen Gesenke ist für den Gebrauch in einem Lufthammer bestimmt, während das zweite für die angeführte Arbeit unter dem Dampfhammer zu arbeiten hat. Beide stellen ein gutes Beispiel moderner Gesenkschmiederei dar. Das Schmiedestück, dessen aufeinanderfolgende Arbeitsstufen, erste Hitze, zweite Hitze und fertiges Stück, in Fig. 213 dargestellt sind, ist ein Bremsteil, der für die Drehgestell- und vierräderigen Wagen der Neuseeland-Eisenbahnen geliefert wurde.

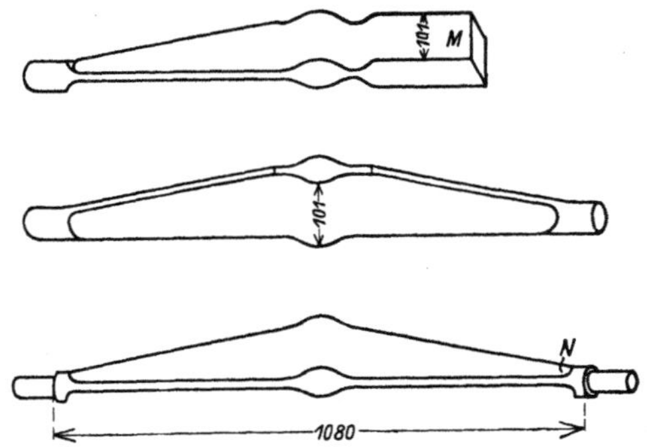

Fig. 213.

Fig. 214 ist das Gesenk für die erste und zweite Hitze und wird in folgender Weise verwendet: Ein Stück weicher Stahl, 57×101 mm im Querschnitt und 438 mm lang, wird fast auf Weißglühhitze gebracht und bei A in die Gesenke eingelegt, wo das Auge in der Mitte vorgeschmiedet wird, und mit Hilfe der parallelen Gesenkfläche bei B und der geneigten Flächen bei C das Material gestreckt wird, bis es die Form des ersten Arbeitsganges, Fig. 213, angenommen hat. Selbstverständlich müssen alle drei Gesenkteile mehrere Male verwendet werden, bis die richtige Form erreicht worden ist. Während der letzten Schläge werden die Gesenkflächen mit Wasser bespritzt, wodurch das Schmiedestück eine sehr glatte Oberfläche erhält.

Der Teil A an dem Gesenk, der zur Herstellung des Auges dient, und die geneigten Flächen bei C sind so gestellt, daß die Flächen B gerade zusammenkommen, wenn die richtigen Abmessungen für das Auge oder die Schräge erreicht worden sind. In demselben Gesenk wird das Stück im zweiten Arbeitsgang nach Fig. 213 umgeformt, enthält jedoch bereits weniger Arbeit, da das Auge bereits hergestellt ist, und nur der in der ersten Figur mit M bezeichnete Teil gestreckt und geformt werden muß.

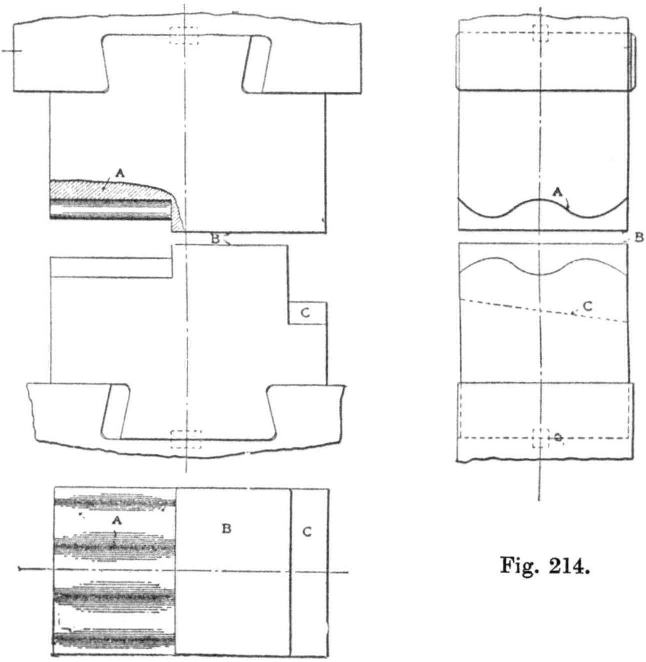

Fig. 214.

Die beiden Balkenenden, die in der Abbildung des fertigen Stückes zu erkennen sind, werden in den in Fig. 215 dargestellten Gesenken mit einem Dampfhammer von 135 kg Bärgewicht, und zwar in einer Hitze, für jedes Ende hergestellt. Die Abrundung N an dem fertigen Schmiedestück, Fig. 213, wird in dem mit D bezeichneten Teil des Gesenkes hergestellt, wobei die richtige Länge des Balkens durch die Lage dieser Abrundung mit Hilfe einer Lehre bestimmt wird. Die Stauchkanten bei E bringen die Enden auf Maß, da Bund und Hals in einem Arbeitsgang gebildet werden. Die geneigte Fläche bei F, die nur in der unteren Gesenkhälfte vorhanden ist, dient dazu, irgendwelche Unregelmäßigkeiten der Kanten, die beim Schmieden der Enden eingetreten sind, auszugleichen.

Das fertige Stück wiegt 18 kg und kann, nach Woodworth, leicht zu ungefähr 8 Pf. für das Kilo Schmiedegewicht hergestellt werden.

Die Gesenke bestehen aus ungehärtetem Werkzeugstahl und gestatten über 4000 Schmiedestücke herzustellen, bevor die Gesenkflächen erneuert werden müssen. Ein Satz der in Fig. 214/15 dargestellten Gesenke wurde aus Stahlguß hergestellt und gab ebenfalls sehr befriedigende Resultate.

Die Bremsbalken wurden früher ohne diese Gesenke hergestellt, nur unter Verwendung einfacher Döper. Dabei hing sehr viel von der Geschicklichkeit des Schmiedes ab; auch die aufgewendete Zeit war aus demselben Grunde sehr verschieden. Während manche Schmiede in der ersten Hitze bloß das Auge in der Mitte herstellen konnten, war ein anderer mit Benützung eines Dampfhammers von 360 kg Bärgewicht mit zwei Hilfsarbeitern zum Erhitzen des Blanketts und Halten der Döper imstande, in einem Arbeitsgang die erste Form in Fig. 213 herzustellen. Mit den beschriebenen Gesenken hingegen konnte ein Schmied von ganz mittelmäßiger Geschicklichkeit nach ein wenig Übung die Schmiedestücke in zufriedenstellender Qualität und Zeit herstellen.

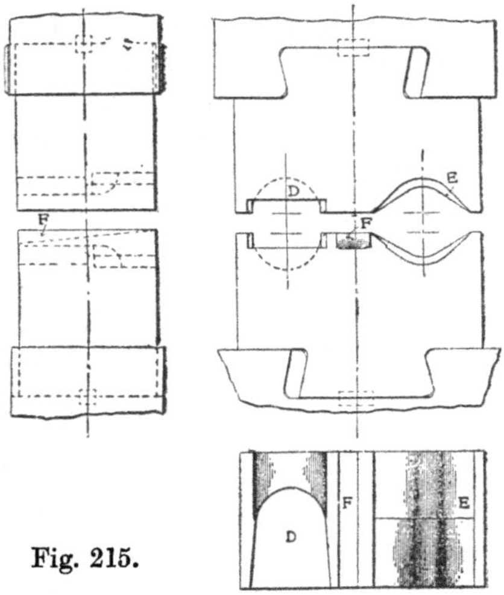

Fig. 215.

Der dabei verwendete Lufthammer, Fabrikat Massey, hatte 360 kg Bärgewicht und wurde von einem 20 PS-Motor angetrieben. An ihm arbeiteten zwei Schmiede abwechselnd in der Weise, daß der eine die Arbeit begann, wenn der zweite mit seinem Stück fertig war, so daß der Hammer fast dauernd lief. Das Schmieden dauerte $4\frac{1}{2}$ Minuten. Ein Wattmesser zeigte die größte Belastung des Hammers an, wenn das Auge ausgeschmiedet wurde, was jedesmal die erste Arbeit war. Zum Schmieden dieses Stückes hatte die Maschine unter Vollast zu arbeiten, während der ersten Schläge sogar mit Überlast, so daß diese Schmiedearbeit für diese Größe und Bauart des Hammers einen Grenzfall darstellt.

Werkzeuge für Schmiedemaschinen. Wenn schon das Gesenkschmieden unter dem Hammer einen großen Fortschritt in bezug auf Herstellungssteigerung, Verbilligung und Gleichheit der Stücke gegenüber dem Handschmieden bedeutet, so tritt die volle Ausnutzung der Gesenkschmiedearbeit erst bei Verwendung der Schmiedemaschine ein. Während bei der Gesenkschmiedearbeit unter dem Hammer in vielen Fällen immer noch die Geschicklichkeit des Arbeiters die Herstellung des Stückes stark beeinflußt und immer beeinflussen wird, solange nicht das Stück mit einem Einlegen ins Gesenk fertig geschmiedet werden kann, geschehen bei der Schmiedemaschine alle Arbeiten selbsttätig. Es müssen wohl die Werkzeuge der Arbeit der Schmiedemaschine angepaßt werden, die einzelnen Arbeitsgänge sind ebenfalls anders als beim Schmieden unter dem Hammer, doch wird mittels der Schmiedemaschinen eine derartige Steigerung der Herstellung und gleichmäßige Form der Stücke, verbunden mit einer so vollständigen Durcharbeitung des Materials wie kaum bei einer anderen Arbeitsweise, erreicht. Selbstverständlich lassen sich nicht alle und besonders verwickelte Formen auf der Schmiedemaschine herstellen, doch findet man beim Durchblättern der Kataloge der verschiedenen Firmen, welche Schmiedemaschinen bauen, so viele und oft auch so verwickelte Schmiedestücke abgebildet, daß man sich eine Massenfabrikation moderner Art ohne die Schmiedemaschine kaum denken kann.

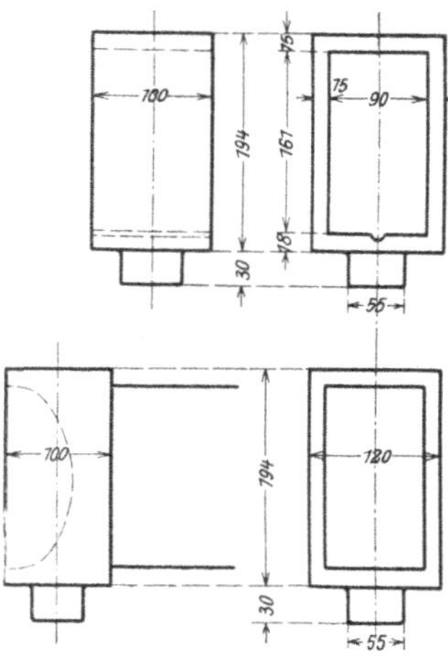

Fig. 216 und 217.

Man darf also die Verbilligung der Kosten nicht bei der Herstellung von einigen wenigen Stück suchen, wo die Auswechslung der Werkzeuge, die Kosten der Sonderwerkzeuge für die Maschine und die Amortisation der Maschine ungleich höher ins Gewicht fallen. Die Firma Hasenclever-Düsseldorf gibt an, daß die Verwendung der Schmiedemaschine schon bei einer Herstellung von 20 Schmiedestücken von Vorteil ist.

Um die Arbeitsweise der Schmiedemaschinen zu erklären, sollen nach Angaben derselben Firma einige Arbeitsgänge beschrieben werden.

Die Herstellung eines Federbundes, Fig 216, erfolgt aus einem Flacheisen von der ungefähren Größe des Loches. Die Stange wird am vorderen Ende angewärmt und von der Maschine in einem Druck so gestaucht, daß der fertige Bund mit dem Zapfen vorn an der Stange sitzt, Fig. 217. In derselben Hitze wird das Stück in dem unter dem Stauchgesenk befindlichen Lochgesenk zurückgelocht, so daß der Lochputzen an der Stange sitzen bleibt und kein Materialverlust eintritt.

Ein anderes Beispiel für eine Schmiedearbeit in drei Gängen ist die in Tabelle Nr. 8 wiedergegebene Ventilbrücke. Hier wird die Stange etwas stärker als der Lochdurchmesser gewählt. Das Material wird in den beiden ersten Gängen vor- und fertiggestaucht, dabei das Material hinter dem Flansch auf den Lochdurchmesser eingekniffen, so daß im dritten Gang zurückgelocht werden kann.

Die Werkzeuge sind natürlich der Schmiedemaschine angepaßt und werden in die Klemmbacken der Maschine[1]) eingesetzt. Sie sind infolgedessen für irgendeine Schmiedepresse oder einen beliebigen

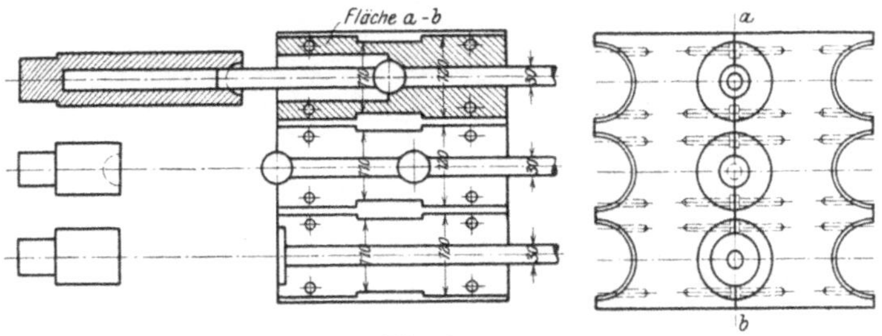

Fig. 218.

Hammer nicht zu verwenden, so daß man sie möglichst wirtschaftlich auszubilden versucht, um diesen Mangel auszugleichen. Eine derartige Form baut die Firma Hasenclever nach Fig. 218. Man kann erstens die Klemmbacken auf zwei, manchmal auch auf vier Seiten mit eingearbeiteten Kalibern versehen oder noch weiter gehen, indem man die Klemmbacken mit — der Herstellung auf der Drehbank wegen — runden Einsatzstücken versieht, in welche die einzelnen Kaliber erst eingearbeitet werden. Dadurch ergeben sich für die einzelnen Gesenkteile verhältnismäßig kleine Stahlstücke. Die Einsätze werden mit angedrehten Bunden versehen, so daß die weitere Befestigung durch kleine, wenig beanspruchte Schrauben geschehen kann. In der Fig. 218 ist ein derartiger Satz Werkzeuge zur Herstellung von Geländersäulen in drei Arbeitsgängen gegeben. Die mittlere Kugel wird zuerst angestaucht, wobei der hohle Stempel eine solche Länge des Stangenmateriales aufnimmt, als zur Herstellung beider Kugeln und der dazwischen liegenden Stangenlänge nötig ist. Hierauf kommt die Stange mit dem angestauchten, mittleren Kugelbund in das mittlere Gesenk, wo die obere Kugel angestaucht wird. Dann wird die Stange umgedreht und im untersten Gesenk der Fuß angestaucht.

[1]) Vgl. Abschnitt IV, Maschinen, Fig. 670.

Im folgenden werden eine Reihe Schmiedegesenke für verschiedene Stücke, welche in einem, zwei oder mehr Arbeitsgängen fertiggestellt werden, beschrieben.

Der Hackerstiel in Fig. 219[1]) wird aus dem Vierkanteisen in einem Arbeitsgang fertiggestellt. Der Kopf wird auf drei Seiten gestaucht, während auf der vierten Flachseite eine schmale Nut ein-

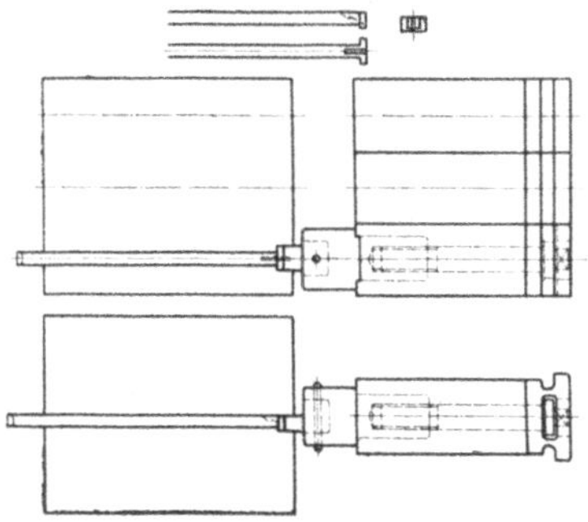

Fig. 219.

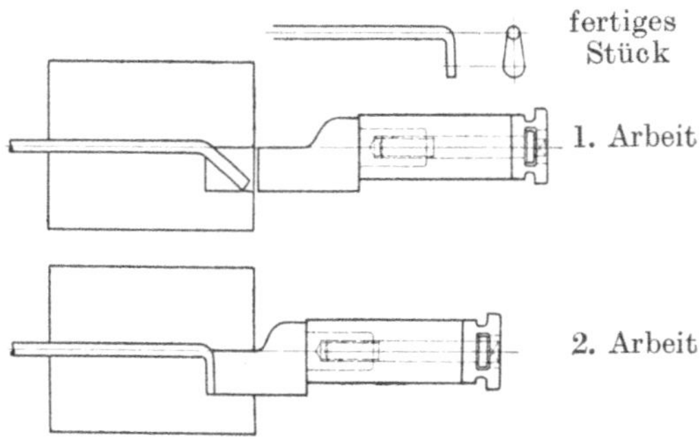

Fig. 220.

gepreßt wird. Der Stempel staucht den Kopf an, während die seitliche Nut durch den Klemmbacken eingepreßt wird. Die Befestigung des Stempels im Stempelkopf und dessen Befestigung im Stößel ist ebenfalls aus Fig. 219 ersichtlich. Es sei dies ein Beispiel, wo die gesamte Schmiedearbeit auf Stauchen beschränkt ist und ein Arbeitsgang das Stück fertig stellt. Die Herstellung eines Handgrifflappens, Fig. 220, geschieht am Ende wohl auch in einem Arbeitsgang, doch wird eine ausgeprägte Biegearbeit an dem ungestauchten Rundmaterial durch die eigentümliche Form der Klemmbacken vorgenommen, bevor der Stempel seine eigentliche Arbeit, das Breitstauchen und Fertigbiegen des Lappens, getan hat. Es bildet also dieses Beispiel den Übergang zu den Schmiedearbeiten in zwei Arbeitsgängen und zeigt den Arbeitsgewinn, der durch richtigen Entwurf der Werkzeuge in der Schmiedemaschine erhalten werden kann, sehr deutlich.

Ein Beispiel einer Verbindung von Staucharbeit und Abschneiden in demselben Werkzeug ist die Herstellung des Sechskantbolzens, Fig. 221 (Gesenk I), in einem Arbeitsgang, während die Herstellung des Bolzens mit Vierkantkopf und Konus in zwei Arbeitsgängen in den Gesenken II, III mit gleichzeitigem Abschneiden dargestellt ist. Die dazugehörigen Werkzeuge sind der Einfachheit halber in einem Schieber zusammen gezeichnet, wobei der Grundriß des

---

[1]) Die Werkzeuge, Fig. 219—221, werden von der Firma Hasenclever-Düsseldorf gebaut.

untersten Gesenkes die Anordnung der Schneidbacken bei a zeigt. Im ersten Gang, Gesenk II, wird der Konus angestaucht und das Material für den Vierkant in einem steilen Konus vorgestaucht, so daß das Material bereits gründlich zum Fließen gebracht worden ist, bevor die weitgehende Formänderung beim Stauchen des Kopfes vorgenommen wird, und gleichzeitig der Kraftbedarf und die Beanspruchung der Werkzeuge und Maschine dafür verringert wird.

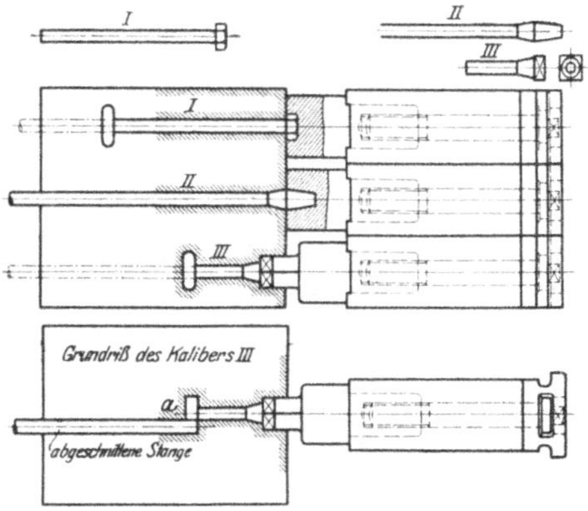

Fig. 221.

Es müssen aber natürlich nicht alle Arbeiten auf der Schmiedemaschine vorgenommen werden, man erspart bereits genug an Kosten, wenn die teuren und verwickelten Arbeiten der Maschine überwiesen und billige Vor- und Endarbeiten auf andere Weise ausgeführt werden. In Tabelle Nr. 6 wird die Herstellung einer gegabelten Zugstange[1]) auf der Schmiedemaschine, Defries-Ajax, in vier Gängen beschrieben, welche deutlich erweist, daß die Hauptarbeit, die Herstellung der Gabel, der Schmiedemaschine zugewiesen ist, so daß die Gestehungskosten des Stückes bedeutend herabgedrückt werden. Das Vierkantmaterial wird zuerst mit der Warmsäge auf 175 mm Tiefe geschlitzt, worauf im Vorschmiedegesenk eine angenäherte Form der Gabel hergestellt wird. Die Gabel ist nunmehr 145 mm lang, dafür aber nur 57 mm breit, und die Maulöffnung auf 25 mm ausgeweitet, also bereits sehr nahe dem Endmaß gebracht. Im zweiten Gang auf der Schmiedemaschine erhält die Gabel ihre endgültige Form mit 117,5 mm Länge und 55/70 mm Breite bei einer Maulöffnung von 27 mm. Es zeigt sich also, daß die Fertigarbeit hauptsächlich in einer Durcharbeitung der beiden Lappen zu finden ist. Hernach wird im vierten Gang das bis jetzt vierkantige Ende unter dem Hammer gereckt und gleichzeitig rund geschmiedet.

Die dazu erforderlichen Werkzeuge der Schmiedemaschine sind in Fig. 222 abgebildet, und zwar der Vorstauchstempel links oben, der Fertigstauchstempel links unten, während die beiden Matrizen in einem zweiteiligen Block auf der rechten Seite ausgearbeitet sind. Es ist dabei zu ersehen, daß 0,4 mm der Stangenbreite zum Festklemmen ver-

---

[1]) Nach Angaben der Firma De Fries & Co., A.-G. in Heerdt-Düsseldorf.

wendet werden, da die Matrize nur 44 mm breit ist, während die Stange 44,4 mm breit ist. Die Stempel stellen ein ziemlich schwieriges Stück Werkzeugarbeit dar, damit durch geeignet gewählte Übergänge von dem schmalen, langen, eigentlichen Formstempel auf dessen Kopf und

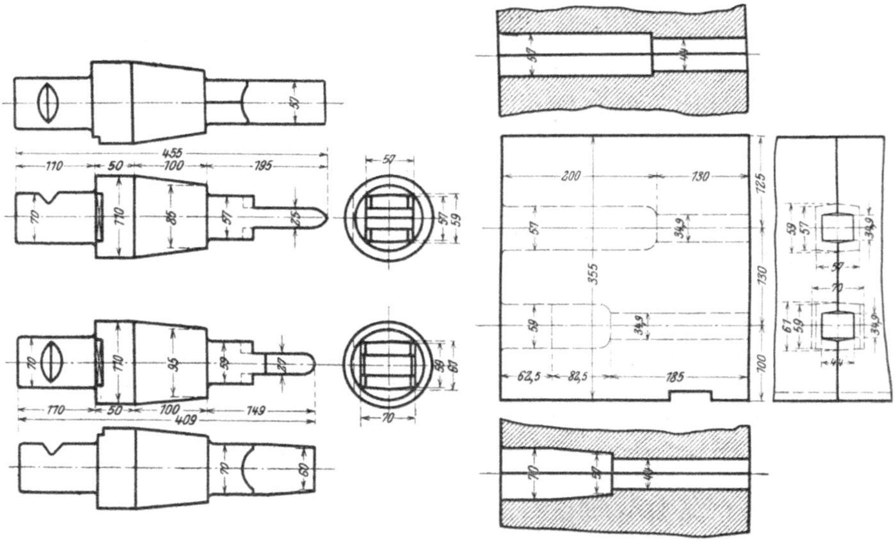

Fig. 222.

Zapfen der Druckausgleich derart erfolgt, daß die Stempel den starken Beanspruchungen gewachsen bleiben. Besonders die scharfen Übergänge zur Herstellung der Gabelenden sind gefährliche Stellen. Die Matrizen sind verhältnismäßig einfach und genügend deutlich aus der Zeichnung zu entnehmen.

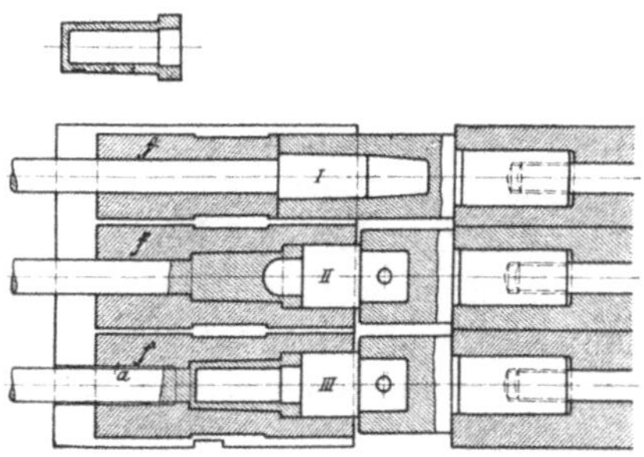

Fig. 223.

Die Herstellung eines Stückes in **drei** Gängen auf der Schmiedemaschine[1]) mit nachfolgendem Abschneiden ist in Tabelle Nr. 7 gegeben. Es ist dies eine **Büchse mit Boden**, Fig. 223, bei welcher die Materialverschiebungen in der Längsrichtung des Stückes verhältnismäßig groß gegenüber dessen Querabmessungen sind. Die drei Satz Werkzeuge können wie Fig. 218 gleichzeitig in den Stößel bzw. die Klemmbacken der Maschine ein-

[1]) Nach Angaben der Firma Hasenclever, A.-G., Düsseldorf.

gesetzt werden, wie die Zeichnung zeigt, und stellen, bis auf das Abschneiden, die fertige Büchse aus dem Rundmaterial her.

Im ersten Arbeitsgang wird das Rundmaterial auf Durchmesser und Länge entsprechend dem Körper der Büchse gestaucht und wiederum ein steiler Konus angestaucht, aus dem im zweiten Gang der Flansch gepreßt wird. Gleichzeitig wird das weitere für den Flansch notwendige Material aus einer Verlochung der Büchse geholt. Im dritten Gang tritt der eigentliche Lochstempel in die auf ihrem Außenumfang fertige Büchse und drängt das ganze Material durch den Boden der Büchse nach außen, so daß die Stange, aus welcher die Büchse geformt wird, dabei nach außen zurückgeschoben wird. Um hier die Reibung zwischen dem zurückfließenden Material und der Matrize möglichst zu vermindern, ist die Bohrung der Matrize bei a nur auf ein kurzes Stück passend und auf den größten Teil der Länge frei gehalten. Die so fertige Büchse hängt an der Stange und wird nachher auf der Warmsäge abgeschnitten.

Wenn man Fig. 217 mit dieser Herstellung vergleicht, so erkennt man, daß die Arbeit des Zurücklochens in beiden Fällen dieselbe ist nur daß man bei der Büchse, um den Boden zu behalten, nicht mit dem Lochstempel durchgehen kann. Auch die Arbeitsweise in Fig. 221 läßt sich nicht anwenden, da das Material nach dem Klemmen erst durch die Klemmstelle unter dem Druck des Lochstempels zurückfließen muß.

Bolzenschmieden und Mutternherstellung. Sobald es sich um Massenfabrikation der Bolzen, also eine reine Sonderherstellung ganz bestimmter Stücke in großen Massen, handelt, können ganz andere Arbeits- und Werkzeugformen zur Verwendung kommen.

Gebr. Hübner in Chemnitz bauen eine besondere Bolzenschmiedemaschine, bei der die Köpfe durch vier Stempel am Umfang und einen Stauchstempel von vorn hergestellt werden. Die Gesamtanordnung der Werkzeuge in der Maschine ist aus Fig. 224 oben ersichtlich, während die einzelnen Werkzeuge selbst mit den Klemmbacken für den Bolzenschaft in Fig. 224a dargestellt sind. Der angewärmte Bolzen wird vom Arbeiter zwischen die Klemmbacken A eingespannt, welche durch Kniehebel bedient werden. Diese Klemmbacken sind vierkantige Stahlstücke, welche an den vier Seitenflächen je eine halbkreisförmige Nut besitzen und somit in vier Stellungen[1]) benutzt werden können. Eine Nacharbeit ist nur an den Quadratseiten erforderlich. Die Hammer B, C und D, welche die Sechskantform des Kopfes herstellen, also geschlossen die Matrize bilden, werden in vier zueinander senkrechten Schlitten gehalten, welche von dem Hauptschlitten aus durch Hebel betätigt werden. Diese Führungen liegen alle

[1]) Vgl. S. 245, Fig. 309a.

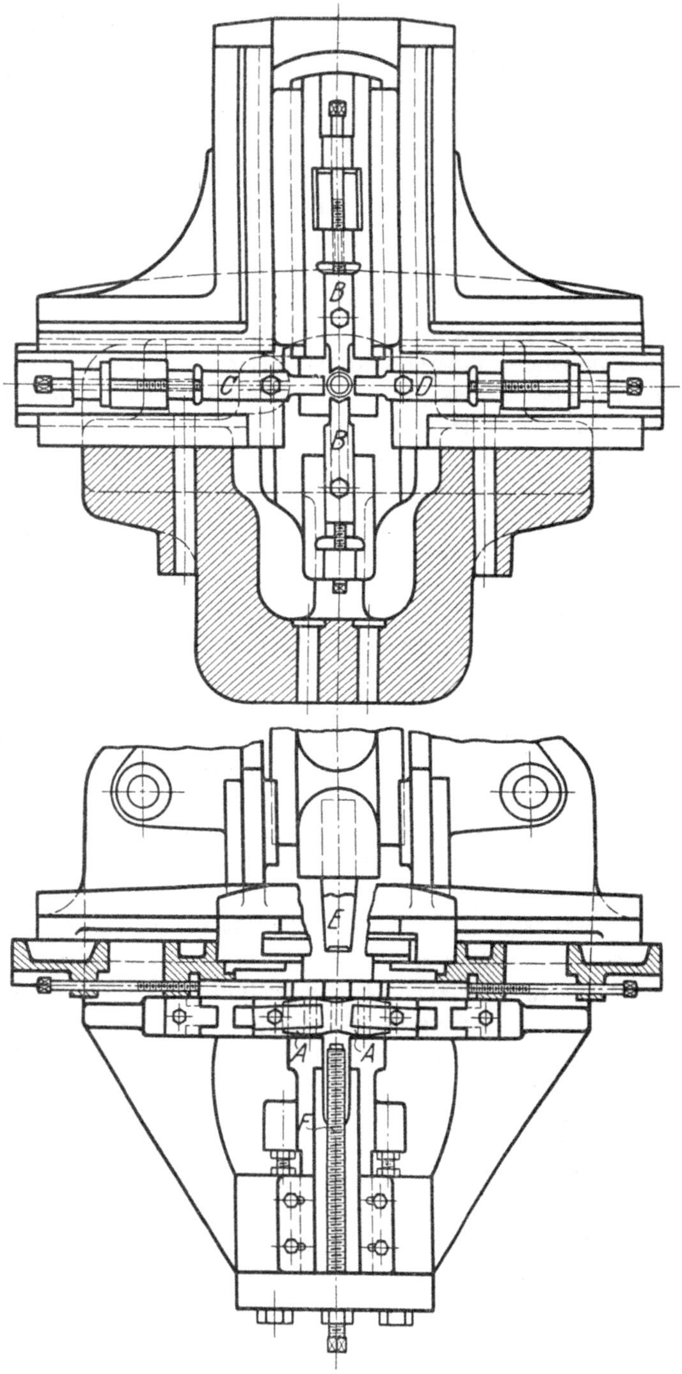

Fig. 224.

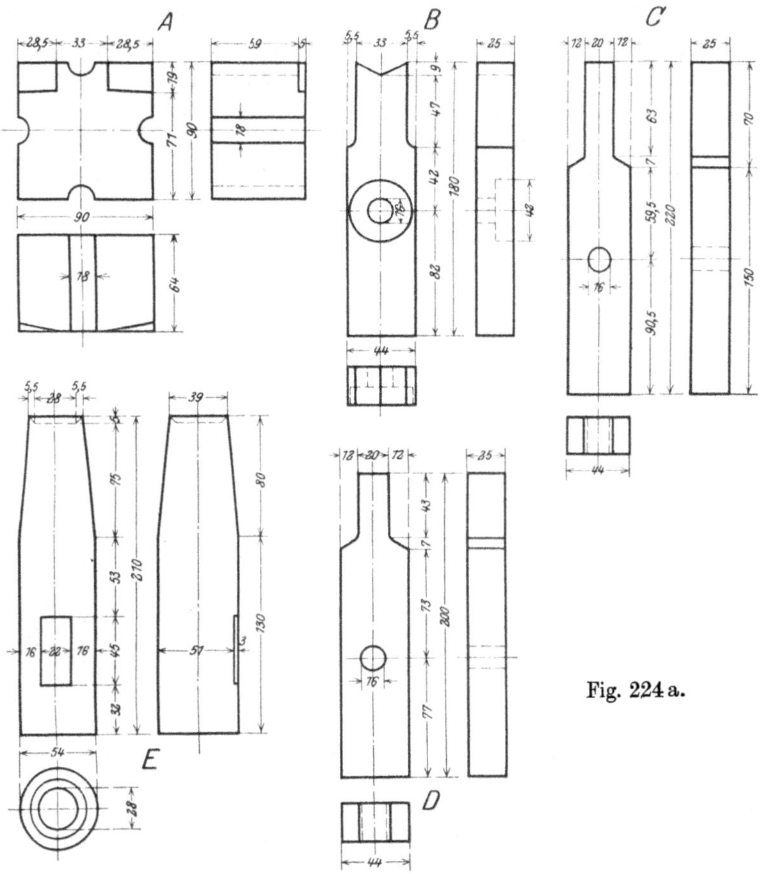

Fig. 224 a.

in einer senkrechten Ebene, während der Stempel E, der die obere Fläche des Bolzenkopfes fertig stellt, unmittelbar vom Hauptschlitten in einer wagerechten Ebene bewegt wird. Der Arbeiter rückt nach dem Einspannen des Bolzens, der durch einen verstellbaren Endanschlag auf genaue Schaftlänge eingestellt wird, mittels eines Handhebels die Maschine ein, die nach Fertigstellung des Bolzenkopfes selbsttätig wieder ausgerückt wird, während der Antrieb der Maschine dauernd läuft. Es werden dadurch die eigentlichen, beim Schmieden benötigten Maschinenteile sehr geschont.

Über amerikanische diesbezügliche Verhältnisse sagt Woodworth:

Besondere Gesenke zum Bolzenschmieden. Die alte Herstellungsweise der Gesenke für die Bolzenschmiedemaschine ist ein ganz eigenes Ding, nicht allein bezüglich der Kosten, sondern auch bezüglich der Ausbesserung und des Ersatzes der alten Gesenke; dies tritt besonders hervor, wenn es sich um die Größen von $\frac{1}{2}$—3 Zoll handelt. Jedes Paar Gesenke, die hier gemeint sind, Fig. 225, bestand aus Blöcken — entsprechend den Schnitten A-B und C-D — von 400×266×225 mm und konnte für zwei Bolzengrößen verwendet werden. Infolgedessen mußte

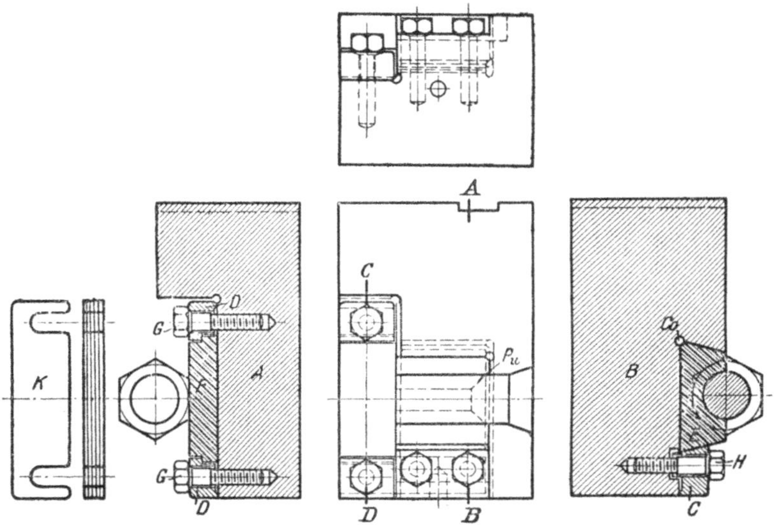

Fig. 226. Schnitt C-D. Fig. 225. Schnitt A-B.

man jedesmal eine große Zahl schwerer Stücke in Bewegung setzen, wenn eine andere Bolzengröße geschmiedet werden sollte. Außerdem wurde zur Herstellung eines Satzes dieser Gesenke nicht allein eine große Menge Stahlmaterial verbraucht, sondern auch unnötige Arbeitskosten in der Werkstätte und Werkzeugmacherei aufgewendet. Da diese Blöcke noch dazu aus weichem Stahl waren, nützten sie sich bald ab, die Ecken rundeten sich, so daß sie zur Herstellung von Bolzen, besonders für Staatslieferungen, untauglich wurden. Das im folgenden beschriebene Gesenk ist das Ergebnis von Versuchen, diese Übelstände zu beseitigen und ein Stück herzustellen, dessen Erneuerung mit geringen Kosten verbunden ist und auch an den Stirnflächen der Abnützung wegen gehärtet werden kann.

Der Halter besteht aus zwei gleichen Stahlgußstücken, A und B, die zur Aufnahme der Gesenke aus Werkzeugstahl, F und E, bearbeitet werden. Nachdem sie einmal in der Maschine befestigt worden sind, werden sie beim Auswechseln der Gesenke weiter nicht mehr berührt.

Die Gesenke bestehen aus zwei abgeschrägten Stücken E aus gehärtetem Werkzeugstahl, die das Stangenmaterial zu fassen haben, während der Pressenstößel zwischen den Flächen der Stücke F den Kopf bildet. Ein Paar Gesenkplatten E kann zur Herstellung zweier Bolzengrößen verwendet werden; es kann z. B. das Gesenk zur Herstellung roher $1^1/_8''$ Bolzen zur Herstellung von einzölligen bearbeiteten Bolzen dienen, da für beide das gleiche Rohmaterial verwendet wird. Die Verschiedenheit der Sechskantköpfe wird ausgeglichen, indem man Beilagstücke K, Fig. 226, unter das Werkzeugstahlstück F legt und es durch die Bolzen G festzieht. Auf diese Weise kann man durch verschiedene Dicke dieser Beilagen jede Schlüsselweite herstellen.

Um ein Gesenk auszuwechseln, braucht man nur die Schrauben G und H zu lösen, worauf die Gesenke am Ende herausgezogen werden können. Ebenso kann man die Beilagen, da deren Löcher nach der einen Seite zu durchgeschnitten sind, von dem einen Ende des Blockes herausnehmen.

Die Herstellung der Bolzen in einer Maschine mit diesen Gesenken ist in keiner Weise von der Arbeitsweise der alten Gesenke verschieden.

Bolzengesenke aus Gußeisen. Beim Arbeiten in der Bolzenstauchmaschine geben die Gesenke oft der Werkzeugmacherei andauernd zu tun, so daß die Unterhaltungskosten dafür ziemlich hoch werden. In einer Schmiedewerkstätte versuchte man die verschiedensten Materialien und Härteverfahren und fand, daß die höchste Verarbeitungsziffer eines Paares Gesenke 30 000 halbzöllige Bolzen (7500 im Tag) war, worauf die Gesenke ausgeglüht und nachgearbeitet werden mußten. Zur Verringerung der Kosten usw. versuchte man es mit der Herstellung eines Modells nach Fig. 227, das selbstverständlich in einem Stück gemacht wird, während die Stahlgesenke aus zwei Stücken gemacht und mit einer Schraube, Fig. 228, zusammengeschraubt werden. Nach diesem Modell wurden Abgüsse aus Hartguß hergestellt. Wenn die Gesenke aus der Gießerei kamen, mußten sie nur ein wenig nachgeschliffen werden und konnten in die Maschine eingesetzt werden. Zu allgemeinem Erstaunen konnte man mit diesen Gesenken 75 000 Stück halbzöllige Bolzen, also zwei und einhalb Mal soviel wie mit den teuren Stahlgesenken, herstellen, wobei hier zu den Modellkosten nur die Gießerei- und Schleifereikosten kamen.

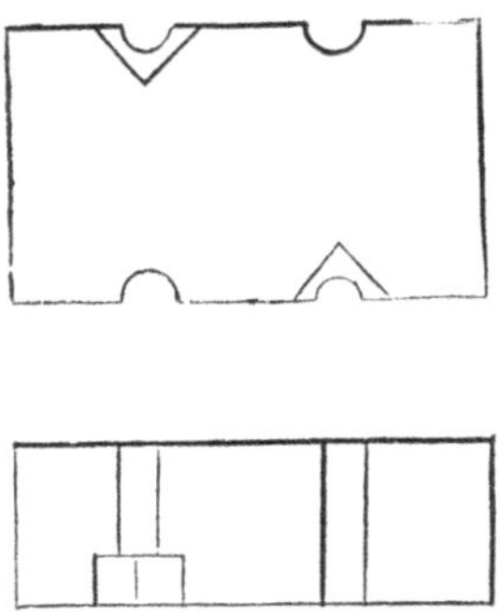

Fig. 227.

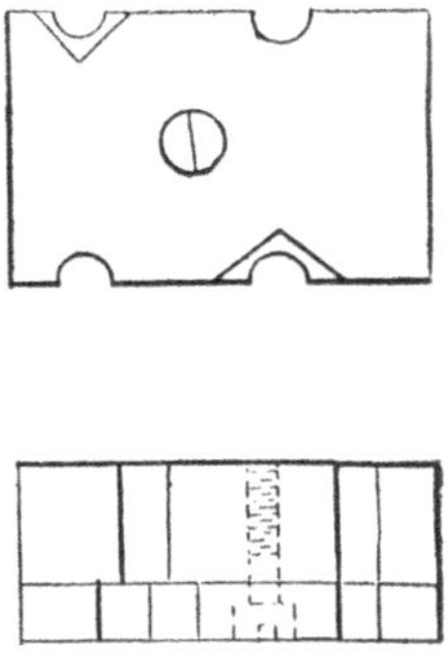

Fig. 228.

β) Kaltbearbeitung.

Wenn auch in den meisten Fällen Schmiedearbeiten mit erwärmtem Material ausgeführt werden, so kann man für Vorbereitungsarbeiten oder bei der Verarbeitung kleiner Stücke oder weicherer Mate-

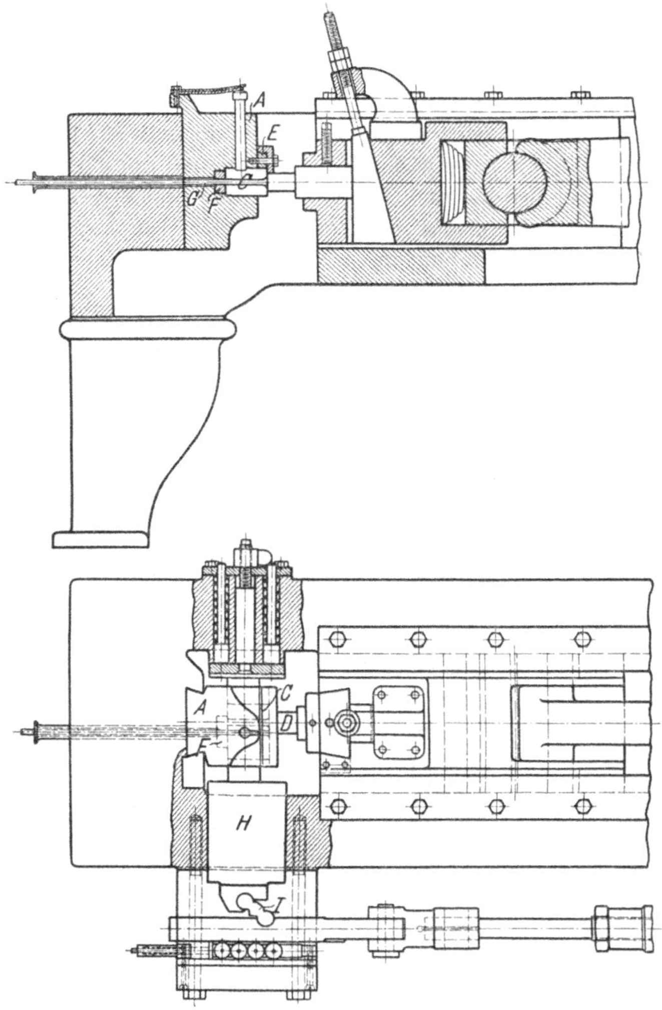

Fig. 229.

rialien auch ohne Erwärmung mit der gleichen Arbeitsweise, also mit Hammer und Gesenk, genügend Formveränderung erreichen, um einen brauchbaren Erfolg zu erzielen.

Eine der einfachsten und dabei doch weitgehendsten Anwendungen des Kaltstauchens geschieht bei dem Herrichten der Bolzen für die

Schmiedemaschine. Da durch das Wegfallen der Vorwärmung unmittelbar vom Stangenmaterial, selbst von der Rolle gearbeitet werden kann, und die einzelnen Bolzen bei jedem Pressenhub vorgestaucht und gleichzeitig abgeschnitten werden, wird die Herstellung der Bolzen derartig verbilligt, daß sich die Kaltvorstauchmaschine bald bezahlt macht. Durch das Kaltvorstauchen wird auch einer Rißbildung zwischen Schaft und Kopf in weitem Maße vorgebeugt und außerdem das Schmieden hoher Köpfe ermöglicht.

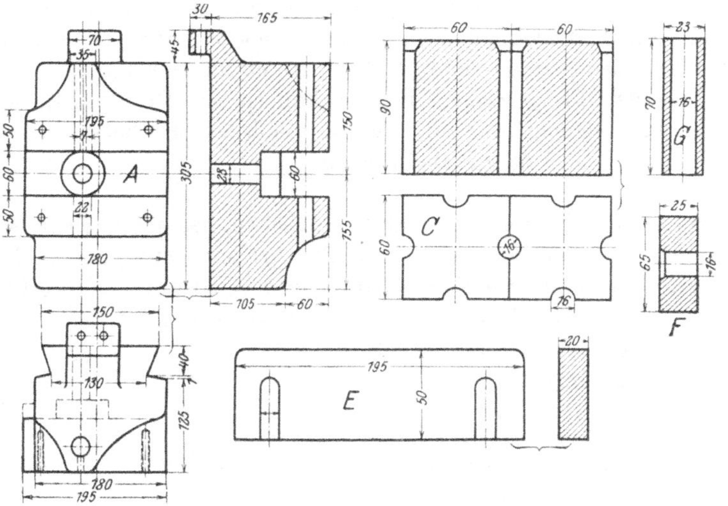

Fig. 229 a.

Die Anordnung der Werkzeuge in der Maschine zeigt Fig. 229[1]), während die Werkzeuge selbst in Fig. 229a in vergrößertem Maßstab dargestellt sind. In dem Körper der Bolzenstauchmaschine ist ein Werkzeughalter A mit Schwalbenschwanz und Führungsleiste eingesetzt, der rückwärts eine Bohrung hat, durch welche das Stangenmaterial zugeführt wird. Diese Bohrung ist für den größten zu verarbeitenden Materialdurchmesser gemacht und wird durch Einsatzbüchsen G verkleinert, wenn kleinere Durchmesser zur Verarbeitung kommen. In den Werkzeughalter ist rückwärts der feststehende Scherring F eingesetzt, während an der Vorderseite eine durchgehobelte Nut die beiden Gesenkhälften C aufnimmt. Diese werden durch Überlagplatten und Zwischenbeilagen E gehalten. Die Stauchgesenke C werden aus Werkzeugstahl hergestellt und erhalten an den vier Seitenflächen die Gesenke eingearbeitet. Die rückwärtige Fläche dient als zweiter Scherbacken und arbeitet mit dem Ring F zusammen.

[1]) Gebr. Hübner in Chemnitz.

Sobald die richtige Materiallänge durch den Vorschub zwischen die Gesenkhälften vorgezogen worden ist, werden diese — gleichzeitig Klemmbacken — durch die Wirkung des Kniehebels I geschlossen, beim Weitergehen des Kniehebels I unter Überwindung der Federkraft des festen Widerlagers auf der anderen Maschinenseite gegen dieses vorgeschoben und dabei zwischen dem Ring F und den Backen C das Rundmaterial abgeschert. Da dabei eine reine Scherwirkung auftritt, werden die Bolzenenden gerade abgeschnitten, was für die weitere Verarbeitung von Wichtigkeit ist. In der Endstellung dieser Bewegung beginnt der Stauchstempel D zu arbeiten und stellt den vorgestauchten Kopf in einem Schlage her. Nach Angaben der Firma Gebr. Hübner können in 10 Stunden 40 000 Bolzen vorgestaucht werden und gewöhnliches Flußeisen von $^5/_8{}''$ um 30 mm leicht kalt gestaucht werden.

Oft genügt aber für diese Gattung Arbeiten, besonders wenn es sich um weniger genaue oder kleine Stücke handelt, ein ganz einfaches Hammergesenk und ein einfacher Hammerschlag, während bei größeren Stücken die gebräuchliche Hammerarbeit notwendig ist.

Im folgenden sind Belege dafür nach Woodworth gegeben:

Gesenk zum Konischformen hohler Schraubenköpfe. Bei Wagen, die an Stelle der Gewichte durch ungleicharmige Hebel von einer Anzahl gleichartiger Stücke ein Dutzend, Gros, usw. auswiegen, sind

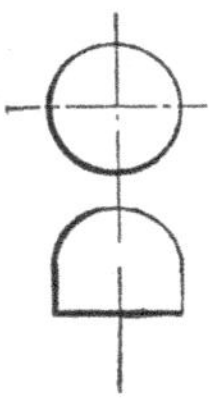
Fig. 230.

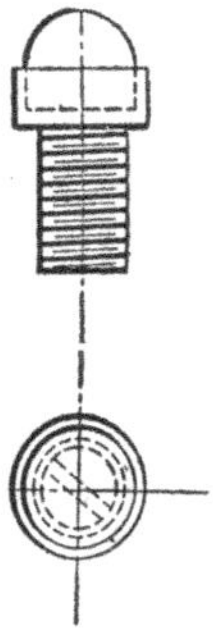
Fig. 231.

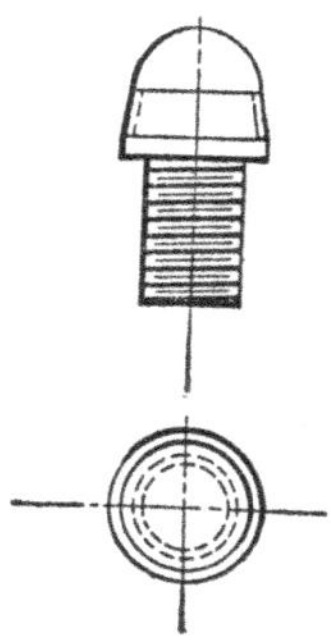
Fig. 232.

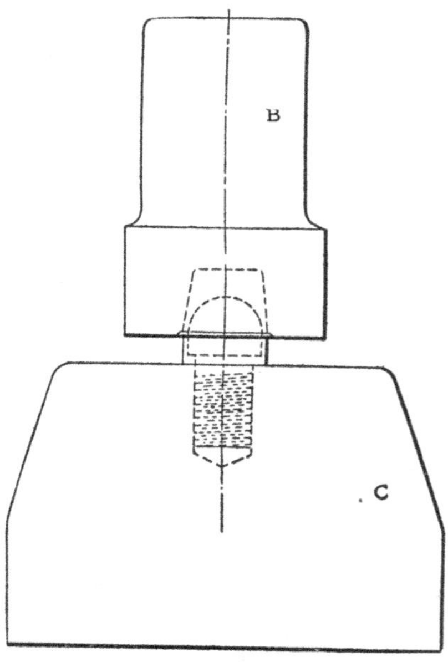

Fig. 233.

häufig Drehscheiben vorgesehen, so daß die Wage von jeder Seite des Ladentisches verwendet werden kann. Um die Tischplatte zu schonen, werden dabei oft mit Gummispitzen versehene Fußschrauben, die zugleich als Einstellschrauben dienen, verwendet. In Fig. 230 ist Grund und Seitenansicht der Gummispitze gegeben, die genau so geformt ist, daß sie in die Ausdrehung des Schraubenkopfes paßt. Die Spitzen wurden früher nach Fig. 231 einzementiert; da sie dann jedoch häufig lose wurden, so entschied man sich, sie durch Stauchen zu befestigen. Die in Fig. 233 abgebildeten Werkzeuge dazu wurden auf der Werkbank verwendet, da man keine Presse zur Verfügung hatte. In Fig. 233 ist B der Stauchstempel aus Werkzeugstahl, der ein konisches Loch hat, das gerade groß genug für den Schraubenkopf ist, wenn die Spitze eingesetzt ist, und C ein gußeisernes Auflager mit einem Loch, in das die Schraube lose paßt. Ein leichter Hammerschlag auf das obere Ende von B formt den Kopf nach Fig. 232.

Schlagen der Prägewerkzeuggesenke für Druckknöpfe. Bei dem allgemeinen Gebrauch der Druckknöpfe denkt man wenig daran, wieviel Mühe die Herstellung der wirklich gebrauchsfertigen Stücke

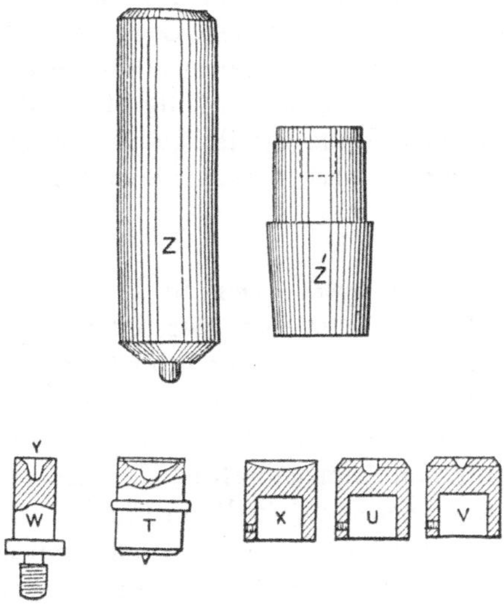

Fig. 234.

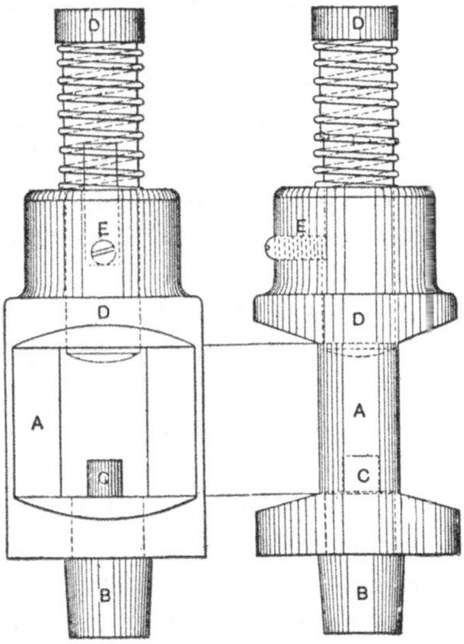

Fig. 235.

kostet. Auf den ersten Blick erscheint es, als ob man sie nur in einer der bekannten Weisen zu verbinden hat, wenn sie schließlich nur halten. Wenn man jedoch solche Druckknöpfe, die zu fest einschnappen, an dünnem Material befestigt, so zerreißt man beim Öffnen derselben sicherlich den betreffenden Gegenstand.

Man gab nun einem Werkzeugschlosser den Auftrag, die Prägewerkzeuge, Fig. 234, für die Druckknöpfe herzustellen. Sie wurden wohl alle mit Fassonwerkzeugen nach Lehren hergestellt, mußten jedoch nachher von Lehrlingen abgeschmirgelt und poliert werden. Wenn nun einer derselben die Schmirgelleinwand etwas zu lange an einen Teil anhielt, so arbeitete das Werkzeug bei der Inbetriebsetzung nicht richtig.

Man schlug dann einen anderen Weg ein, der obigem Übelstand abhalf. Fig. 235 stellt ein kleines Blockwerkzeug bzw. eine Subpresse dar, die in einem Fuß-Fallhammer verwendet wird. A ist der Pressenkörper, B ein konischer Zapfen, der einerseits streng in A, andererseits mit dem konischen Ende in den Amboß des Fallhammers paßt. C nimmt die Matrize auf, während sie durch den Schlagstempel D geschlagen wird. Ein Gesenk nach Fig. 234 wird auf den Zapfen C aufgesetzt, der Hammerbär auf die verlangte Höhe angehoben und auf D fallen gelassen, so daß der Schlagstempel D auf der Matrize die genaue Gesenkform herstellt. Die Feder hebt den Schlagstempel D an, so daß man den fertigen Werkzeugteil entfernen und einen neuen einlegen kann. Mittels der Schraube E wird D in der richtigen Höhe gehalten. Die zur Herstellung des Teiles W, Fig. 234, notwendige Schlagstärke ist so gering, daß man hierfür keinen Fallhammer zu verwenden brauchte. Es wurde einfach bei Y ein Loch gebohrt, worauf ein leichter Hammerschlag auf den Preßstempel zur Fertigstellung genügte. Mit ein wenig Erfahrung kann man die Stärke des Hammerschlages so einrichten, daß alle Werkzeugteile gleich werden. Für die verschiedenen Matrizen wurden auch verschiedene Schlagstempel nach Fig. 234 gemacht.

Gesenk zum Schlagen von Nummerplatten. Der in Fig. 237/38 dargestellte Stempel mit seiner Matrize dient zur Herstellung von Nummerplatten, die von einer Firma zur Bezeichnung ihrer Öfen und verschiedenen Herde verwendet werden. Diese Werkzeuge können wohl als neu für die Herstellung dieser Nummerplatten bezeichnet werden.

Fig. 238 zeigt die Matrize und den Nummerhalter, der aus gehärtetem Stahl hergestellt wird. Das Loch in J hält die in Fig. 236 abgebildeten, auswechselbaren Nummernblöcke und kann gleichzeitig drei Stück derselben aufnehmen. Sie sind aus Messing und messen 17,5 × 17,5 mm; unter Verwendung zweier Zwischenstücke von 8,75 mm Dicke kann man zweistellige Zahlen auf der Nummerplatte erhalten. Der Nummerhalter J wird auf der gußeisernen Grundplatte I, die in eine gewöhnliche Froschplatte paßt, durch $^3/_8''$ Kopfschrauben befestigt. Das Metall darf während des Schlagens keine Falten machen. Der Niederhalter L ist entsprechend der Matrize J gelocht und liegt auf den Druckstäben A so, daß er ungefähr 1,6 mm über das Untergesenk herausragt. Sobald der Stößel herabkommt, drückt er Platte L und dadurch die untere Druckplatte H nieder, so daß die Spannung der Gummifedern die Faltenbildung das Metalles verhindert.

Fig. 237 ist ein Schnitt durch die obere Gesenkhälfte, deren Kopf und Zapfen E aus Gußeisen hergestellt und durch dessen Zapfen zur Aufnahme der Feder N ein Loch von 19 mm gebohrt ist. Die Gesenkplatte G entspricht in ihrer Form der Matrizenplatte J, nur ist sie um die doppelte Metalldicke größer. In Fig. 237 sind auch die Negative D der Nummernblöcke in Stellung gezeichnet. Sie werden durch die Stange C, die durch ein Loch von 8 mm Durchmesser P, Fig. 236, in den Blöcken geschoben wird, in ihrer Stellung gehalten. Die beiden Schrauben K sichern die Stange C. Da die Blöcke sich gegen den Stempelkopf E bewegen müssen, sind die Löcher für die Stange C in der Gesenkplatte G größer gebohrt. Die Platte T liegt lose auf den Nummernblöcken D, damit sich die Pressung gleichmäßig auf diese verteilt. Auf der Platte T ruht wieder der Druckstempel B. Da die Breite der Blöcke nur 17,5 mm beträgt, mußte man zu diesem Zwischenglied im Auswerfer greifen, da sonst der Auswerferkolben von 19 mm Durchmesser sich auf die vorstehenden Kanten der Gesenkplatte G auflegen würde, so daß die Feder nicht auf das Stück T drücken könnte. Die Gesenkplatte G ist an dem Stempelkopf mit vier Kopfschrauben F befestigt. O ist Einstellschraube und Gegenhalt für die Feder.

Fig. 236.

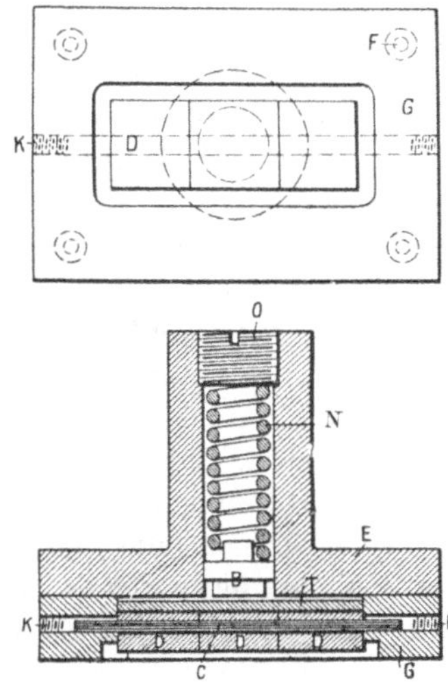

Fig. 237.

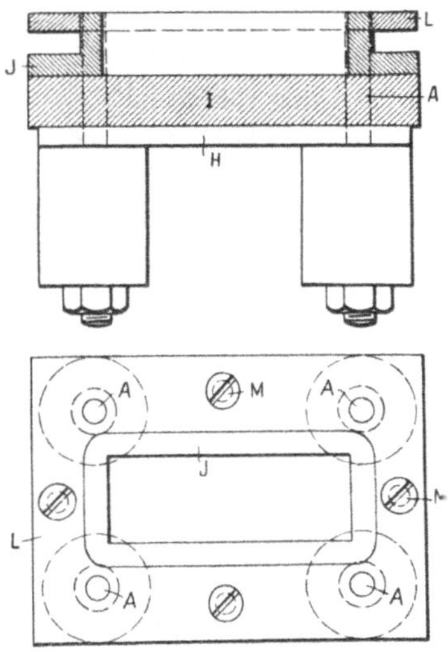

Fig. 238.

Der Vorteil dieser Gesenke ist, daß man die Nummern auswechseln kann. Man braucht nur einige wenige Nummernblöcke, während sonst jede Nummer ihren besonderen Block haben müßte.

In dieser Zusammenstellung gibt das Gesenk ein gutes brauchbares Werkzeug ab.

Arbeitsweisen bei der Herstellung von Juweliergesenken. Bei der Herstellung von Gesenken für Juwelierarbeiten gibt es verschiedene Arbeitsweisen, die alle einander ziemlich ähnlich sind, deren jede aber irgendeinen kleinen Vorteil besitzt, so daß einmal diese, das andere Mal jene zur Herstellung kleiner Gesenke vorgezogen wird. —

Im allgemeinen ist es viel leichter, auf der Außenseite irgend eines Gegenstandes, als auf seiner Innenseite zu arbeiten, da im ersten Fall jeder Punkt leicht zugänglich ist, besonders wenn es sich um kleine Werkzeuge handelt. Wie beim Schneiden eines Gewindes der Gewindebohrer als Werkzeug hergestellt und damit in ein Loch ein Gewinde geschnitten wird, so ist es auch bequemer, den Stempel zu machen, und damit das Gesenk herzustellen, als das Gesenk mit kleinen Fräsern, Sticheln usw. auszuarbeiten. Man greift auch sehr selten auf diese Hilfsmittel zurück, um irgendeinen Fehler des erhaltenen Gesenkes zu verbessern, außer man kann den Fehler im Gesenk mit weniger Arbeit verbessern als im Stempel.

Ein Gesenk kann auf diese oder eine andere Weise bis auf das Härten fertiggestellt werden; vor dem Härten ist es jedoch gut, einige Versuchsblanketts in weichem Blei zu prägen, um Maße abzunehmen, Versuche zu machen usw.; dann werden die gewöhnlichen Abänderungen manchmal im Gesenk, meistens aber im Stempel gemacht, weil es so leichter ist, und das Gesenk ein zweites Mal oder öfter, bis der beabsichtigte Erfolg erzielt wird, geschlagen werden kann. Es ist auch manchmal bequemer, eine kleine Änderung im Gesenk vorzunehmen als im Stempel, weil man die Änderung nur an einer Stelle des Gesenkes zu machen hat, während der ganze Stempel bis auf diese Stelle würde nachgearbeitet werden müssen.

Bei dem Nacharbeiten der Gesenke und manchmal auch der Stempel erweisen sich runde Stangen von Arkansasstein als sehr brauchbar. Um den Stein für diese Arbeit geeignet zu formen, läßt man ihn langsam in der Bank oder Bohrmaschine laufen und feilt auf Form; zum Schleifen gibt man höhere Umdrehungszahlen und benützt raffiniertes Petroleum, um ein Verschmieren zu verhindern.

Es gibt viele Teile an Zwicker- und Brillenfassungen, die sehr genau auf Maß und Form ausgeführt werden müssen, doch verlangt man im allgemeinen von geprägten Teilen mehr Symmetrie und schönes Aussehen als Genauigkeit der Abmessungen. Bei der Verarbeitung von vollem Gold ist es jedoch der Materialkosten wegen nicht wünschenswert, wenn die Teile ein bestimmtes Gewicht überschreiten, doch kann man dies leicht verhindern, indem man die Dicke der Teile oder die Tiefe der Prägung regelt.

Gesenke für diese Arbeit werden oft durch den Zunder der ausgeglühten Stücke abgenutzt. Es sind Sätze von 7000 bis 12000 Prägestücken gezählt worden, was jedoch von dem verarbeiteten Material, der Sauberkeit usw. der Stücke abhängt; goldplattierte Gegenstände sind am besten zu bearbeiten. Die arbeitenden Oberflächen dieser Gesenke werden bald stumpf, so daß die Prägungen nicht besser werden als die Gesenkfläche, und das Gesenk, besonders für goldplattiertes Material, wieder neu poliert werden muß. Hat ein Gesenk eine gewisse Vergrößerung durch das Nacharbeiten, welches jedoch nur wenig die ursprüngliche Größe überschreiten darf, erfahren, so wird es nicht weiter benützt, da die Prägungen aus einem derartigen Gesenk in der nachfolgenden Verarbeitung, in den Einspannfuttern usw. Schwierigkeiten machen würden.

Es hilft auch nicht viel, die Fläche eines solchen abgenützten Gesenkes abzuschleifen, da die Prägung nur weiter und dünner wird.

Um zu beweisen, daß nur eine begrenzte Anzahl Stücke kalt geschlagen werden kann, sei eine Klammer mit V-Nuten, die über die ganze Länge der Klammer laufen, angeführt. Diese erfordert Gesenke, welche auf andere Weise hergestellt werden müssen. Bedingung ist, daß die Nuten einen scharfen Winkel bilden sollen. Von mancher Seite wird behauptet, daß dies nicht auf kaltem Wege erzielt werden kann, gleichzeitig aber zugegeben, daß es auch nicht mit einem warmen Gesenkblankett gemacht werden kann. Es ist nun interessant, eine erstklassige Arbeit dieser Art durch kalte Bearbeitung vorzuführen, wobei alle feinen Einzelheiten vorhanden sind, und viel weniger Arbeit aufgewendet wird als bei Herstellung eines großen Teils der Prägung von Hand. Man kann sagen, daß fast alle Prägegesenke kalt geschlagen und alle feinen Einzelheiten nur in dieser Weise hergestellt werden sollen[1]).

Es wäre dagegen falsch, zu behaupten, daß alle Gesenke, große und kleine, sowohl kalt wie warm geschlagen werden können, obwohl die Unmöglichkeit hier nicht zugegeben werden soll. Wenn jemand einen offenbaren Beweis der hier vorhandenen Möglichkeiten finden wollte, so möge er die Aufschlagstelle eines Geschosses in einer Panzerplatte ansehen und beobachten, wie genau die Form des Eindruckes mit der Form des Geschosses übereinstimmt. Die Herstellung großer Gesenke auf diese Art und Weise wird wohl aus wirtschaftlichen Gründen nie zur Anwendung kommen, doch handelt es sich in unserem Falle nur um kleine Gesenke.

Es soll für einen Augenblick die Arbeitsweise eines Stempels bei der Herstellung einer Arbeit nach Fig. 239/40 betrachtet werden. Angenommen, P in Fig. 239 stellte einen Schnitt eines Stempels, der ein genutetes Stück mit sehr scharfen Kanten herstellen soll, dar. D ist die Matrize. Wenn nun bei der Herstellung D mittels P geschlagen wird,

[1]) Vgl. S. 224.

so schneiden die Spitzen p in die Oberfläche des Stahlblanketts, indem sie nicht unähnlich einem Keil, der in das Material eindringt, wirken. Bei der weiteren Abwärtsbewegung von P muß das bereits zwischen den Kanten p gespaltene Material innerhalb der V-Nuten in diesen nach aufwärts steigen. Da sich aber der Raum dort immer mehr verkleinert, wird es entweder bis in den Scheitel der Nuten steigen oder zusammengedrückt werden, vielleicht auch beides — und dieses tritt tatsächlich ein —, und schließlich den scharfen Winkel in diesen Nuten bilden, was aber selten der Fall ist. In Wirklichkeit beobachtet man, daß dieser Idealzustand nur teilweise erreicht wird, und aus einem ziemlich einfachen Grunde. Das Metall muß in die V-Nuten fließen und wird, da der zu füllende Raum sich verengt, zusammengedrückt, bis der spezifische Druck, der notwendig ist, um diese V-Nuten scharf auszuprägen, größer als die Druckfestigkeit des Materials wird. Infolgedessen wird der Stempel als Ganzes bei jedem Schlage tiefer eingedrückt, so daß durch wiederholte Schläge wenig an der Form der Nuten verbessert wird. Die gekrümmten Scheitel p′ in Fig. 239 sollen diese teilweise gebildeten Nuten darstellen.

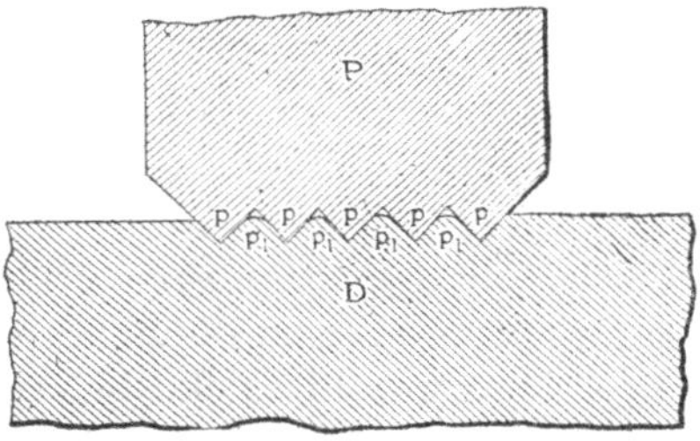

Fig. 239.

Dies sind die Verhältnisse, die man bei der kalten Herstellung der Gesenke antrifft; man umgeht sie auch nicht, wenn man das Blankett warm verarbeitet, weil dann auch die Druckfestigkeit des Materials sinkt, so daß die Verhältnisse dieselben bleiben, wie oben geschildert worden ist. Der einzige Vorteil der warmen Verarbeitung mag sein, daß das Gefüge des Materials nicht sehr verändert wird; da jedoch die Lebensdauer kalt geschlagener Gesenke keinen großen Einfluß in dieser Richtung zeigt, fällt dieser Vorteil weg. Bei der Verwendung eines Fallhammers zum Schlagen der Gesenke erzielt man gewöhnlich bessere Resultate, denn es hat den Anschein, daß mehr Details ausgearbeitet werden, wenn der Schlag schneller erfolgt. Dies ist der Grund, die Verwendung des Fallhammers zu empfehlen.

Woodworth hatte oft Gelegenheit, Greifflächen, wie oben beschrieben, herzustellen, wobei — trotzdem mit Verwendung des Fallhammers allein gewöhnlich keine scharfen Winkel erzielt wurden — diese doch die Hauptbedingung waren. Wenn das Gesenk, wie zuerst erwähnt, bloß geschlagen wurde, waren diese Kanten p′ ganz leicht abgerundet, und wenn ein Stück in diesem Gesenk geprägt wurde, hatte es, selbstverständlich als Spiegelbild, scharfe V-Nuten mit gerundetem Grund, was allen Ansprüchen zu entsprechen schien. Wenn jedoch diese Einzelheiten tatsächlich scharf ausgeprägt werden sollen, so darf man dem natürlichen Widerstand des Materials, diese Form anzunehmen, nicht entgegenarbeiten, sondern muß einen Weg einschlagen, der diese Form in folgender Weise leicht ergibt: Angenommen, wir hätten einen Modellstempel in der verlangten Form und Detailausführung und das Gesenkblankett, beide mit hochpolierten Arbeitsflächen. Wir befestigen nun beide Teile im Fallhammer und geben einen Schlag, der einen genügend tiefen Eindruck und ein ziemlich gutes Resultat gibt. Nun kann man das Gesenk herausnehmen und die Nuten ausschaben, bis man die Kanten p′ scharf hat und der Winkel ein wenig spitzer, als im fertigen Stück erforderlich, ist, so daß in verzerrt-vergrößertem Maßstab das in Fig. 240 gezeichnete Bild entsteht. Dann poliert man das Gesenk fertig, wobei man Kratzer und Abrunden der Kanten zu vermeiden hat; hierauf wird das Blankett wieder in die Presse eingesetzt und ein neuer Schlag gegeben, vielleicht auch mehrere, wenn nötig, bis man die verlangte Tiefe erreicht hat. So wird der gesamte Eindruck tiefer geschlagen, wobei sich die V-Nuten zu demselben Winkel wie die im Stempel umformen und genau so scharf werden.

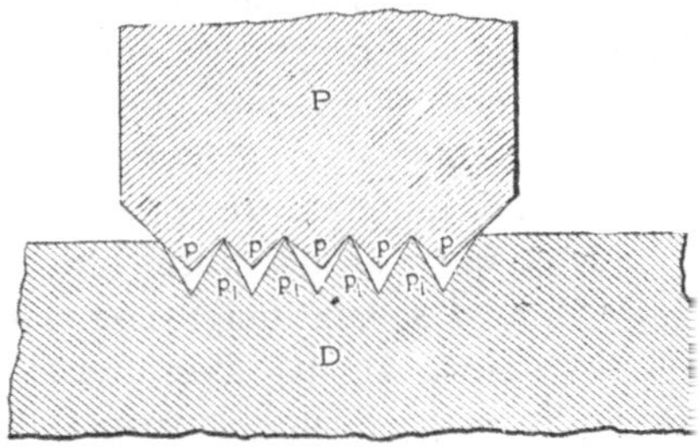

Fig. 240.

Der Vorteil dieser Arbeitsweise mittels kalten Schlagens ist, daß bedeutend weniger genaue Handarbeit erforderlich wird, weil es nur nötig ist, die Kanten der V-Nuten zu schärfen, ohne den anderen Teilen des Eindruckes besondere Aufmerksamkeit zuzuwenden, der doch nach der Fertigstellung des Gesenkes die genaue Wiederholung des Stempels wird. Das ganze Geheimnis liegt darin, daß man den Winkel der Nuten scharf und ein wenig spitzer, als im fertigen Stück erforderlich ist, macht.

b) Gesenkpressen.

Beim Gesenkschmieden erreichen die bewegten Massen und damit die bei der Schlagarbeit auftretenden Verluste bei größeren Arbeitsstücken solche Werte, daß die Güteziffer der ganzen Anlage auf unzulässige Werte herabgedrückt wird. Außerdem ist der Einfluß der Hammerarbeit auf das Fundament, benachbarte Maschinen und nicht zum wenigsten auf die Teile des Hammers selbst ein derartig ungünstiger, daß man von der Verwendung der im Gesenk geschmiedeten Stücke zu den Stahlgußstücken überging und ihre Nachteile gegenüber den ins Unmögliche wachsenden Kosten der Herstellung gleichgroßer Gesenkschmiedestücke unter dem Hammer in den Kauf nahm. Diese Übelstände hat erst die Entwicklung des Schmiedepressens behoben, als man erkannte, wie große Kräfte angewendet werden müssen, um die gleichmäßige Durcharbeitung der großen Schmiedestücke zu erreichen. Das Ergebnis war aber auch derart, daß es in bezug auf Durcharbeitung des Materiales, Vollkommenheit der Formgebung und Schnelligkeit der Herstellung an erste Stelle trat.

Selbstverständlich wurde dies aber erst nach langwierigen Versuchen erreicht: die Form der Stücke mußte dem neuen Arbeitsverfahren angepaßt, die Werkzeuge entsprechend umgebaut und schließlich die richtige Hubzahl für die Presse gefunden werden. Diese mußte bei Einhaltung der der Presse eigentümlichen, langsamen stetigen Druckwirkung doch immerhin hoch genug sein, um die Ausnutzung der Schmiedetemperatur möglichst hoch zu treiben und so eine wirtschaftliche Herstellung zu ermöglichen.

Das Material, das durch die hohe Erwärmung schon sehr bildsam geworden ist, kommt unter dem ruhigen Druck in seiner ganzen Masse — im Gegensatz zu der Hammerwirkung — zum Fließen, so daß die Gesenkformen leichter ausgefüllt werden als bei der Hammerarbeit. Dieser Fließvorgang darf deshalb auch nicht aufgehalten oder unterbrochen werden, solange die Form nicht ausgefüllt ist; es dürfen also dem fließenden Material keine verengten Querschnitte entgegenstehen, hinter welchen es sich wieder auszudehnen hätte. Ebenso sollen starke, einseitige Materialbewegungen in der Form nach Möglichkeit vermieden werden. Rippen z. B. sollen erst nach bereits erfolgter Materialbewegung ausgefüllt werden.

Die Gesenke werden aus Gußeisen, Stahlguß, Flußstahl oder armiertem Stahl hergestellt, sind meistens zweiteilig, manchmal aber auch dreiteilig, das letzte aber nur bei Pressen mit mehreren Preßkolben.

Bei der Bemessung der Gesenke ist auf die Erwärmung der Gesenke und das Schwindmaß, bei gegossenen Gesenken auf das doppelte Schwindmaß Rücksicht zu nehmen.

Bei gegossenen Vorgesenken entfällt die Bearbeitung oder wird auf ein ganz geringes Maß beschränkt, während Stahlgesenke aus dem Vollen gearbeitet oder nach einem Stahlmodell geschlagen werden, dem nur geringes Nacharbeiten folgt.

Bei Werkstücken, welche sich im Gesenk festsetzen können, ist eine Bohrung anzubringen, durch welche das fertige Schmiedestück leicht aus dem Gesenk herausgeschlagen werden kann.

Im übrigen gilt das bereits über die Hammergesenke Gesagte bezüglich der Herstellung der Gesenke auch hier.

Die Gesenke lassen sich im allgemeinen in drei Gruppen einteilen[1]), ähnlich den Formen der Gießerei.

Man unterscheidet das offene Gesenk, dessen obere Fläche durch die Pressenbahn geschlossen wird, ungefähr der Herdformerei entsprechend, das geteilte Gesenk, welches mit der gewöhnlichen zweiteiligen Kastenform Ähnlichkeit hat, und schließlich das geschlossene Gesenk, bei dem die ganze Form des Gegenstandes in einen Stahlblock eingearbeitet ist, in welchen durch einen Stempel das Material eingepreßt wird. Dazu müssen auch die Strangpressen gerechnet werden.

Das offene Gesenk kommt nur für die allereinfachsten Formen, die auch in den Dickenabmessungen keine großen Unterschiede aufweisen, zur Verwendung, da das überschüssige Material oberhalb des Gesenkes abfließen muß. Das geteilte Gesenk, Fig. 241, nimmt seiner leichten Herstellung und verhältnismäßig geringen Beanspruchung bei der Preßarbeit wegen wohl den breitesten Raum ein und findet die größte Verwendung, wenn auch in den meisten Fällen ein Abgraten der Preßteile nicht zu umgehen ist. Durch die Teilfuge, welche die beiden Gesenkhälften trennt, wird dem bildsamen Material ein Ausweg unter dem Pressendruck gegeben, der sich beim erkalteten Preßteil als Grat, Finne usw. unangenehm bemerkbar macht. Dieser Rand muß dann in besonderen Abgratgesenken, die meistens unmittelbar neben der Presse stehen, entfernt werden. Es ist daher die Lage der Teilfuge und die Ausbildung des Abflusses für das überschüssige Material von großer Wichtigkeit. Das Gesenk wird vorteilhaft in seiner senkrechten Mittellinie geteilt, so daß Putzen u. dergl. in der Druckrichtung liegen. Eine Teilung nach der wagerechten Mittellinie, bei welcher diese durch seitliches Verdrängen des Materials ausgebildet werden müßte, ist weniger gut, da dann auch der Grat längs der gekrümmten Umfläche entfernt werden müßte. In Fig. 241 ist an einer Stelle ein Kanal zur Ableitung des überschüssigen Materials vorgesehen. Da das Material hier durch einen im Verhältnis zum Stück engen Kanal austreten muß, wird man diese Form vorteilhaft dann wählen, wenn es sich um zylindrische Stücke handelt, deren Durchmesser nicht allzu groß ist, damit die Materialbewegung nicht zuviel Kraft erfordert. In Fig. 241a ist eine längs der ganzen Trennungsfuge umlaufende Rinne r vorgesehen, in welche sich das Material leicht einlegt, dann aber infolge seiner geringen Dicke rasch erkaltet und beim Fertigpressen einen unverhältnismäßig großen Widerstand erzeugt. Dieser Rand muß nach dem Pressen abgegratet werden. Die beste Wirkung hat die Ausführung nach Fig. 241b, bei welcher ein umlaufender Kanal o in 3 bis 12 mm Entfernung von

[1]) Vgl. Uhland, Techn. Rundschau 1910, S. 21, 29, woher auch Fig. 241/2 entnommen sind.

der inneren Gesenkkante angebracht ist, der den Materialüberschuß aufzunehmen hat. Die Kante k erfährt aber naturgemäß eine rasche Abnutzung, wodurch diese Gesenkform verteuert wird.

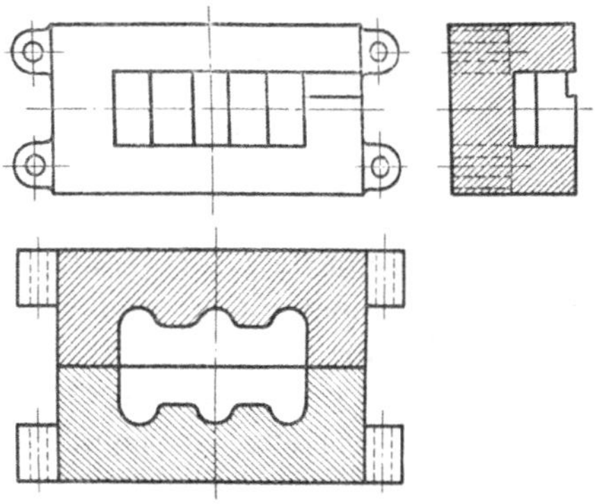

Fig. 241.

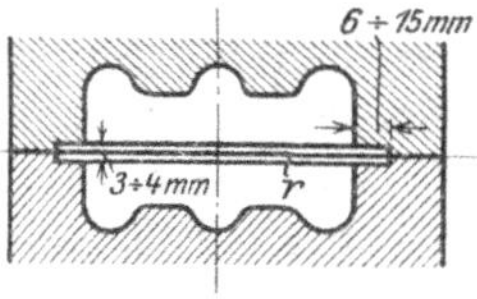

Fig. 241a.

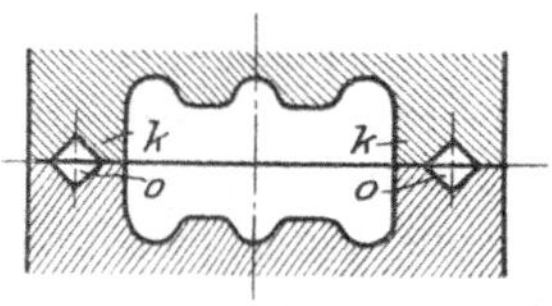

Fig. 241b.

Die vollkommenste Ausführung, sowohl was Gesenk, wie Preßgut betrifft, weist ohne Zweifel das geschlossene Gesenk auf, Fig. 242.

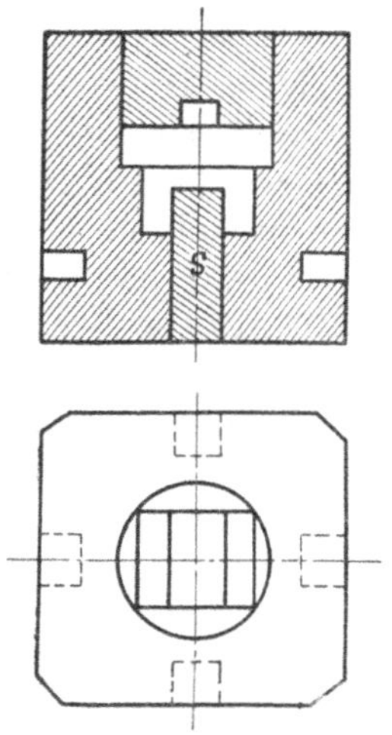

Fig. 242.

Bei diesen Gesenken muß die Masse des Blanketts (Rohlings usw.) genau der fertigen Form entsprechen. Die größte Öffnung des Gesenkes wird durch den Preßstempel, der mit an der Formgebung beteiligt ist, ausgefüllt und dient zum Einlegen des Blanketts. Man kann infolge des allseitig geschlossenen Arbeitsraumes mit den Preßdrücken viel höher als bei den erstbesprochenen Werkzeugen gehen und erhält infolgedessen viel schärfer ausgearbeitete Werkstücke. Folglich wird diese Bauart nur für Fertiggesenke verwendet, so daß sich die Mehrkosten der teureren Herstellung und kräftigeren Ausbildung wieder durch geringere Nacharbeit bezahlt machen.

Bei allen diesen Gesenkformen hat sich im modernen Schnellbetrieb die Bauart in armiertem Stahl oder Stahl als die beste herausgestellt, da die Werkzeuge viel kleiner und trotzdem haltbarer werden, so daß sich die Mehrkosten der Herstellung durch die Herstellungssteigerung bezahlt machen.

Die größten Gesenke, z. B. für Kesselböden, Geschützlafetten u. ä, werden auf hydraulischen oder dampfhydraulischen Pressen verwendet. Für mittlere Größen kommen die Reibradpressen und für kleine Arbeiten, z. B. bei der Messingwarmpresserei, auch Exzenterpressen zur Verwendung.

Gesenk für Sechskantbolzen. (Gebr. Hübner in Chemnitz.) Zur Herstellung von Bolzenköpfen auf der Reibradpresse[1]) derselben Firma dient das in Fig. 243 abgebildete Gesenk. Die eigentlichen formgebenden Teile sind die Matrize E und der Stempel D aus Werkzeugstahl, welche mittels eines konischen Ansatzes ineinander eingeführt werden. Der Stempel wird im Pressenstößel mittels zylindrischen Zapfens und Druckschraube befestigt. Die Matrize wird mit einer konisch ausgedrehten Zentrier- und Niederhaltbüchse B aus Stahlguß durch zwei $^7/_8$" Kopfschrauben auf die Froschplatte A niedergezogen. Die Froschplatte ist rechteckig 380 × 225 aus Schmiedeeisen und wird mit vier Kopfschrauben am Pressentisch, durch eine Andrehung ausgemittelt, befestigt. Sie ist mit einer durchgehenden Bohrung versehen, durch welche die Auswerferstange H nach unten geführt wird. Diese wird unterhalb durch eine Führungsbüchse G gehalten. So sind alle Teile, ineinander ausgemittelt, leicht auszuwechseln, wobei die Zugänglichkeit des Werkzeuges in der Presse immer gewahrt bleibt.

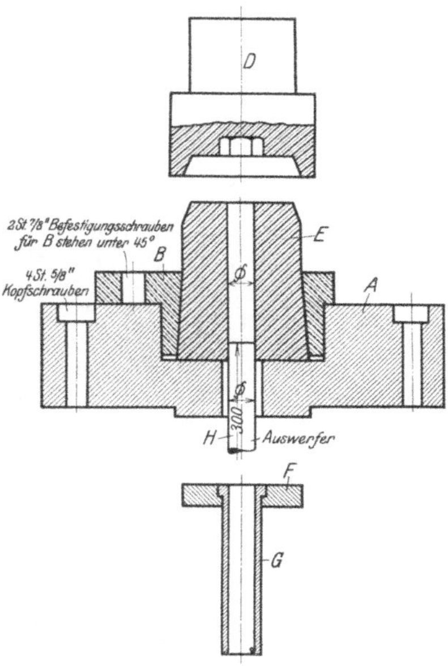

Fig. 243.

Gesenk für einen schmiedeeisernen Deckel. (Maschinenfabrik Hasenclever in Düsseldorf.) Eine etwas verwickeltere Preßarbeit, die gleichzeitig mit einem Lochen des Stückes in einem Arbeitsgang verbunden wird, erfordert der in Fig. 244 dargestellte Deckel. Das Werkzeug

[1]) Siehe Fig. 661.

war nach einem vorliegenden Muster herzustellen. Der Deckel erhält eine beiderseits gekrümmte Begrenzungsfläche mit vorstehenden Nasen und Einkerbungen im unteren Ansatz, sowie ein konisches Loch von

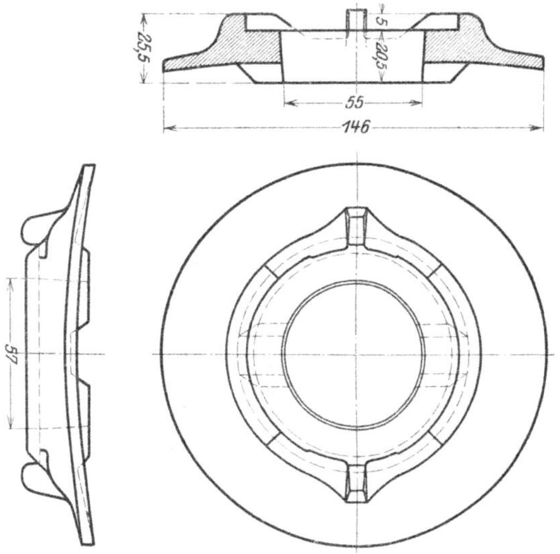

Fig. 244.

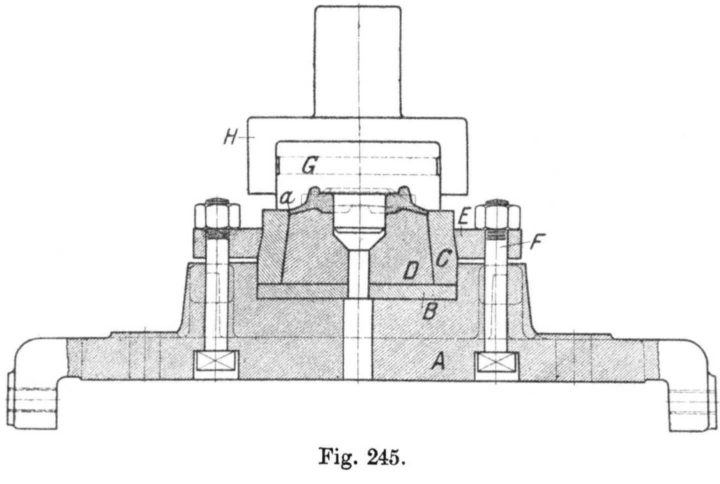

Fig. 245.

55 mm Durchmesser. Aus diesem Grunde kommt ein nach dem Umfang geteiltes Gesenk zur Anwendung, in dem der Lochstempel vor dem Schließen der beiden Gesenkhälften das Loch im Blankett herstellt, so daß erst nach seinem Eintritt in die Matrize das Gesenk nach innen vollständig geschlossen erscheint.

Die Matrize, Fig. 245, baut sich auf einer gußeisernen Froschplatte auf, die mit Löchern für die Befestigungsschrauben und mit Putzen für seitliche Stellschrauben ausgestattet ist, so daß die Ausmittlung beider Gesenkteile untereinander erfolgen kann. Die gußeiserne Froschplatte A erhält eine zylindrische Ausdrehung von 210 mm Durchmesser und 37 mm Tiefe, in welche die eigentlichen Matrizenteile eingepaßt werden. In dieser Ausdrehung liegt die Unterplatte B aus Stahl, auf welche die beiden Matrizenteile C und D niedergezogen werden. Sie sind mit Konus aufeinandergepaßt und teilen sich in die Formgebung so, daß der äußere Ring C die senkrechte Umfläche des Deckels bei a formt, während der innere Teil D die Unterfläche herstellt. Im Teil D ist ebenfalls ein mittleres Loch für den Ausstoßbolzen vorgesehen. Der äußere Matrizenteil, der durch seinen Konus den inneren Matrizenteil D niederhält, wird durch den Überlagring E, ebenfalls mit Konus, und zöllige Kopfschrauben niedergepreßt. Die Formgebung des oberen Teiles von der Teilfuge bei a angefangen, einschließlich des inneren Loches, übernimmt der Stempel G, der in einen Stempelkopf H eingesetzt ist.

Gesenk zum Kümpeln von Kesselböden. (Deutsche Niles-Werkzeugmaschinenfabrik.) Die moderne Herstellung der Kesselböden und ähnlicher Teile, die sämtlich aus dickem Blech bis zu den gößten Abmessungen im Durchmesser bei verhältnismäßig niedrigem Mantel in einem zwischen Preß- und Zieharbeit liegenden Arbeitsverfahren hergestellt werden, erfolgt heute ausnahmslos in großen hydraulischen Pressen. Die Grundlagen dieser Werkzeuge sind wohl die gleichen wie für die gewöhnlichen Werkzeuge der Blechverarbeitung, doch erfordern die großen Kräfte und die Warmverarbeitung der Bleche, verbunden mit den großen zu bewegenden Massen der Arbeitsstücke, eine konstruktiv vollständig verschiedene Lösung der Werkzeuge. Sie müssen aus den einfachsten Formen aufgebaut sein, ihre Verbindung mit der Presse möglichst leicht herzustellen und dabei doch sicher

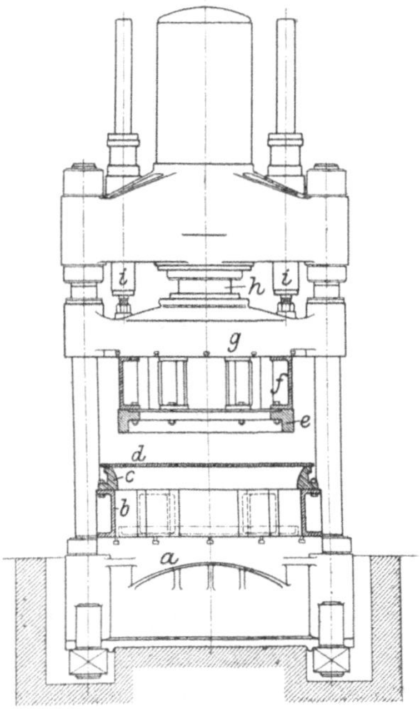

Fig. 246.

genug für die großen Stücke und Drücke ausgeführt sein. Ein derartiges Werkzeug zum Kümpeln eines Kesselbodens von 1870 mm Durchmesser, eingebaut in eine hydraulische Presse derselben Firma[1]), zeigt Fig. 246. Auf dem unteren feststehenden Tisch a, der Amboßbahn, ist auf sechs Zwischenstücken b aus Gußeisen das Gesenk c verschraubt. Auf diesem liegt das zu kümpelnde Blech d. Der Preß- und Ziehstempel e ist mittels sechs ähnlicher Zwischenstücke f mit dem Kolbenquerhaupt, der Pressenbahn g, durch Stahlbolzen von 36 mm Durchmesser verschraubt. Der Druckkolben h bewirkt den Arbeits-(Abwärts-) Gang des Stempels, während zwei Rückzugzylinder i, die ebenfalls hydraulisch betätigt werden, das Querhaupt und den Stempel e wieder nach aufwärts bringen.

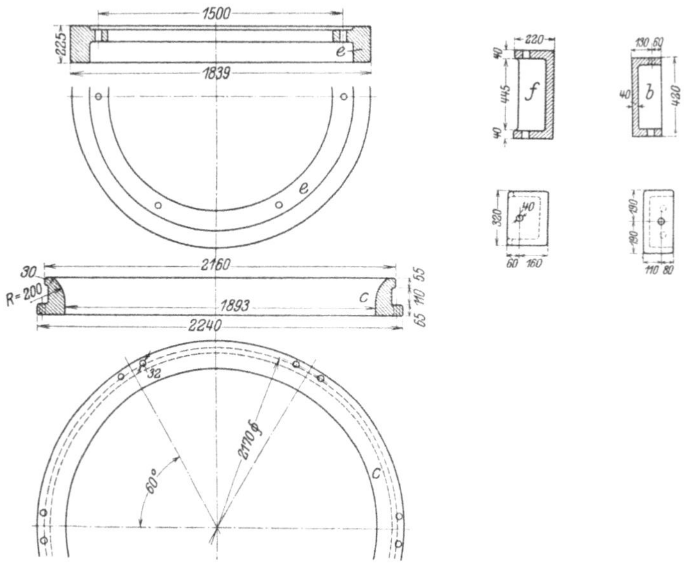

Fig. 247.

Die Einzelheiten des Werkzeuges sind maßstäblich in Fig. 247 wiedergegeben. Das Gesenk c wird aus Stahlguß hergestellt, ist im ganzen 230 mm hoch und wird auf der Unter- und Innenfläche bearbeitet. Während die Unterfläche aber nur als Auflagerfläche eben gedreht sein muß, ist die Innenfläche als Ziehfläche entsprechend sorgfältig zu glätten. Die Innenfläche ist mit Krümmungsradien von 30 mm an der Einlaufstelle und einem anschließenden Radius von 200 mm, der dann in den geraden Ziehring von 1893 mm Durchmesser übergeht, ausgedreht. Die gußeisernen Zwischenstücke sitzen natur-

[1]) Siehe Fig. 668/9.

gemäß auf der Außenseite und werden alle gleichzeitig gehobelt, damit die Auflage des Ziehrings auch wirklich allseitig erfolgt. Die Verbindung der Zwischenstücke mit dem Gesenk erfolgt durch 12 geschmiedete Stahlkopfschrauben, von denen je zwei an einem Fußzwischenstück sitzen, während die Befestigung in den Schlitzen der Amboßbahn durch je eine gleiche Schraube erfolgt. Der Preß- und Ziehstempel e wird aus Gußeisen mittels Schablonenformerei hergestellt und wiegt rd. 1350 kg. Er ist zylindrisch auf 1839 mm Durchmesser und 225 mm Höhe gedreht, auf der Oberfläche als Auflagerfläche eben abgedreht, während die Außen- und Unterfläche sorgfältig geglättet sind. Die Befestigung an der Pressenbahn erfolgt wieder, unter Zwischenstellung der Gußstücke f, die aber jetzt auf der Innenseite stehen, mittels je sechs geschmiedeter Kopfschrauben von 30 mm Durchmesser aus Stahl. Diese sitzen am Stempel auf einem innen umlaufenden Flansch.

Der Einbau der Werkzeuge muß sehr sorgfältig erfolgen, damit sie tatsächlich ausgerichtet sind, wenn auch die Pressen im allgemeinen so gebaut sind, daß sie geringen außerachsigen Beanspruchungen, die gelegentlich nicht zu vermeiden sind, Widerstand leisten können.

Werkzeuge für Metallpresserei[1]). (Kabelwerk Oberspree der A. E. G.) Das Material für die Metallpresserei ist meistens Messing, in einzelnen

Fig. 248.

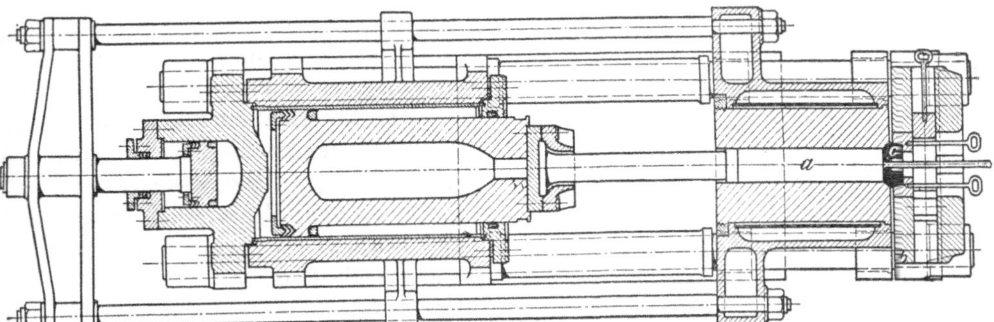

Fig. 249.

Fällen auch Kupfer, wo es sich um elektrische Leitfähigkeit handelt. Das gegossene Material wird auf eigenen Strangpressen in verschieden profilierte Stäbe, die das Ausgangsmaterial für die Warmpresserei dar-

[1]) Z. V. d. I. 1913, S. 1377ff., Adler, Metallpreßteile.

stellen, gepreßt. Die Querschnitte sind sehr mannigfach, von dem einfachen runden Stab bis zu dem unsymmetrischen dreistrahligen Stern u. dgl. in allen Formen vorhanden, vgl. Fig. 248. Die Barren von ungefähr 150 kg Gewicht werden hocherhitzt in den Aufnehmer a der Strangpresse, Fig. 249, eingelegt, worauf mittels Schiebers b, der die Matrize c trägt, eine Abdichtung des Preßraumes unter Druck erfolgt. Dann wird der Stempel d mit einem Druck von 5—8000 kg/qcm vorgeschoben und treibt das Material durch die Matrize. Die so erhaltenen Stangen werden in Stücke zersägt, deren Länge dem zu verarbeitenden Blankett entspricht. Die Blanketts werden in besonderen Glühöfen, die neben den Pressen stehen, auf etwa 750° erwärmt und von einem hinter der Presse stehenden Arbeiter mit einer Zange in die Matrize eingelegt, während ein zweiter vor der Presse befindlicher Arbeiter die Form schmiert und die Presse einschaltet. Das fertige Stück wird selbsttätig ausgeworfen. Das Gesenk wird durch Druckluft gekühlt.

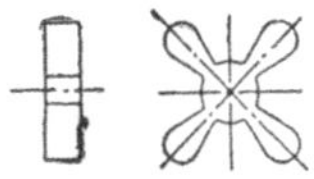

Fig. 250.

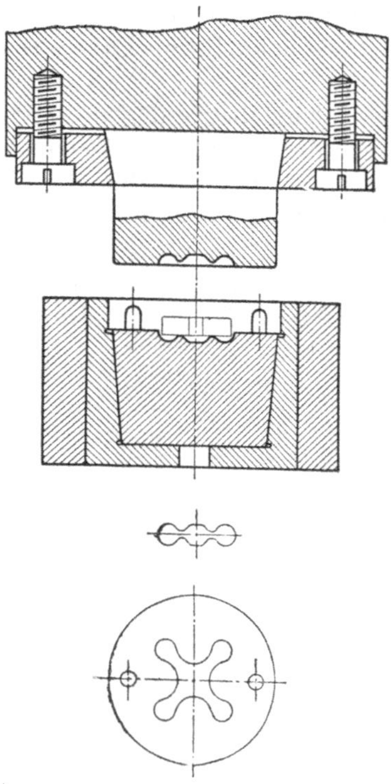

Fig. 251.

Adler teilt die Arbeitsverfahren der Warmpresserei in drei Grundverfahren, das Quetsch verfahren, das Stauch verfahren und das Spritz verfahren, nach der Hauptbewegung, welche das unter Druck fließende Material ausführt. Die Verfahren sind in Fig. 251/5 mit den Werkzeugen dargestellt. Bei dem Quetschverfahren beschränkt sich die Formgebung auf eine Materialverschiebung an Ort und Stelle, so daß das Blankett so eng wie möglich dem fertigen Gegenstande angepaßt wird. Es erfolgt z. B. die Herstellung des Kreuzknebels, Fig. 251, aus einem Abschnitt des vierstrahligen Stangenmaterials, Fig. 250. Das Material für jeden Teil, die Arme oder den Körper, liegt größtenteils schon an der Stelle, wo es beim Arbeitsgang umgeformt wird. Infolgedessen werden meistens im Verhältnis zur Breite dünne Formen beim Quetschverfahren verwendet, jedenfalls ist die innerhalb des Materiales vor sich gehende Materialverschiebung die geringste von den drei angeführten Arbeitsverfahren.

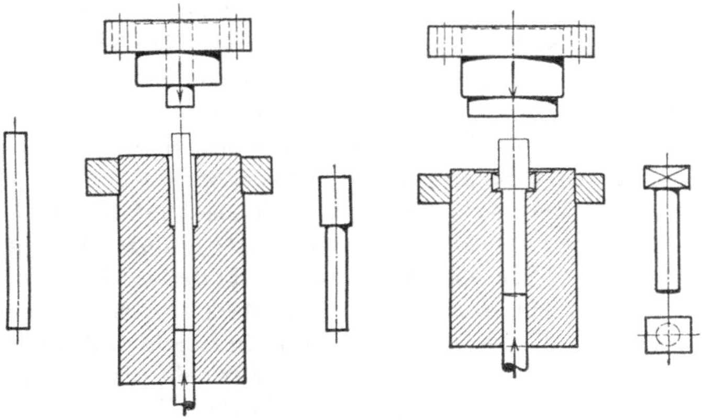

Fig. 252.

Bei dem zweiten, dem Stauchverfahren, kommt hauptsächlich eine Materialbewegung in der Längsrichtung der gepreßten Stange zur Anwendung, welche das Material in einen Körper von geringerer Höhe und größerem Querschnitt umformt. Man kann dabei die volle Materialbewegung nicht immer in einem Arbeitsgang erreichen, wie die Herstellung des Bolzens mit Vierkantkopf, Fig. 252, zeigt. Die Rundstange von der Dicke des Bolzenschaftes wird zuerst auf eine gewisse Höhe rund vorgestaucht und erst im zweiten Gang der Zylinderkopf in einen niedrigeren, aber breiteren Vierkantkopf fertig gestaucht.

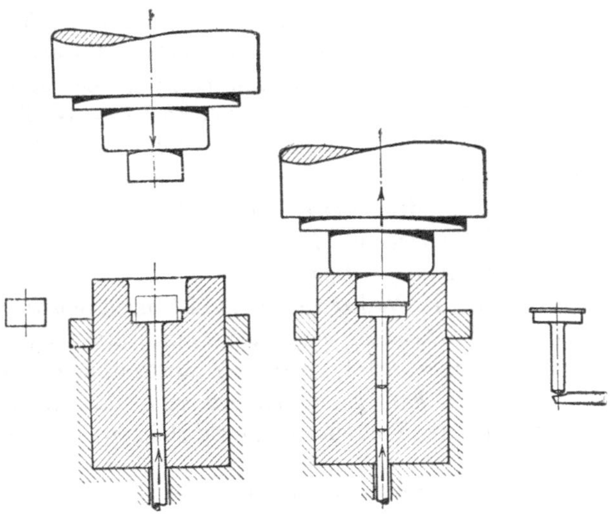

Fig. 253.

Bei dem Spritzverfahren I, Fig. 253, wird der umgekehrte Weg eingeschlagen. Aus einem kurzen, dicken Metallstück wird beim Niedergang des Stempels der breite Teller gestaucht und das übrige Material in die Bohrung des Untergesenkes herausgedrückt, so daß der Zapfen des Ventiltellers erzeugt wird. Grundsätzlich ließe sich ein derartiges Stück auch nach dem Stauchverfahren herstellen, doch wäre dies unwirtschaftlich, da mehrere Stauchungen nötig wären, um den im Durchmesser so stark vom Zapfen abweichenden Teller aus einem dünnen, der Zapfendicke entsprechenden Stangenmaterial zu stauchen. Dagegen läßt sich beim Spritzverfahren die Endform in einem einzigen Arbeitsgange herstellen.

Das Material muß aber nicht notwendigerweise in der Mitte der Form vereinigt sein, sondern kann auch am Umfang der Form erscheinen, wie die nach Spritzverfahren II hergestellten Teile zeigen, Fig. 254. In dieser Abbildung ist in allgemeinen Zügen das Werkzeug zum Pressen einer Kapsel skizziert. Es ist in seinem vollständigen Aufbau ungleich verwickelter[1]) als die bisher besprochenen Gesenke, obgleich die einzelnen Teile im Vergleich zu einem Blockwerkzeug noch aus verhältnismäßig einfachen Einzelgliedern bestehen. Der eigentliche Preßstempel a wird aus einem glatten zylindrischen Stück gebildet, damit er bei dem großen Verschleiß, dem diese Werkzeuge unterworfen sind, leicht und billig ausgewechselt werden kann. Der Stempel bildet nur die oberste Fläche der Kapsel. Der äußere Umfang der Kapsel wird durch den Mittelteil b des Gesenkes gebildet, welcher am Pressenstößel angehängt ist, so daß er von diesem beim Aufwärtsgang nach oben mitgenommen werden kann. Auch dieser, der eigentliche formgebende Teil, ist in der geometrisch einfachsten Form als Zylinder ausgeführt, der an seinem oberen Ende mit einem kegelförmigen Ansatz versehen ist, mittels dessen das Gesenk in dem Befestigungsring getragen wird. Die ganze Innen- und Unterfläche der Kapsel wird endlich durch den im Unterteil sitzenden Stempel c fertiggestellt.

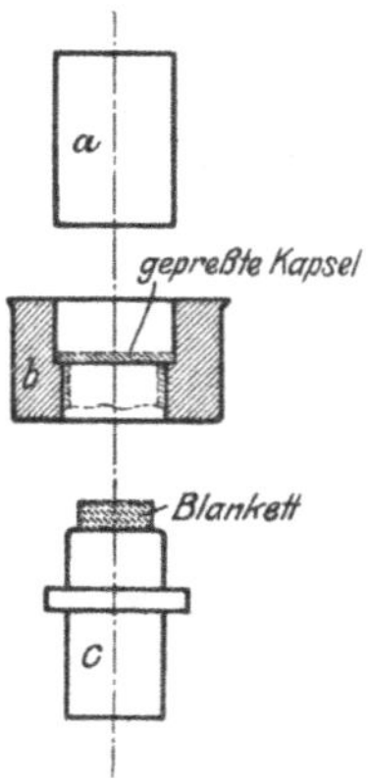

Fig. 254.

Für die Werkzeuge eignet sich nach obiger Quelle schwach legierter Chrom-Nickelstahl, der nach dem Härten innen fest und zäh, außen aber glashart wird, so daß er unter dem Preßdruck nicht zerspringt, aber auch hart genug ist, um eine saubere Oberfläche und richtiges Maß für die Preßteile zu gewährleisten.

Herstellung einer Flügelmutter (Tabelle Nr. 9). Von der warmgepreßten Rundstange von 21,2 mm Durchmesser wird ein Blankett von 27,4 mm Länge abgeschnitten. Aus diesem werden in einem Arbeitsgang zwei an ihren Grundflächen zusammenhängende Muttern in dem

[1]) Siehe Z. d. V. d. I. 1913, S. 1385, Fig. 40/1.

Gesenk, Fig. 255, warm gepreßt. Die Gesenke sind nach der Mittellinie der Lappen geteilt und stellen in der dazu senkrechten Trennungsebene der beiden Muttern eine kräftige Nut her, um das spätere Entzweischneiden zu erleichtern. Das Untergesenk ist mit zwei kräftigen Führungsstiften versehen, damit unabhängig von den Pressenführungen ein genaues Aufeinanderpassen der beiden Gesenkhälften erreicht wird. Die eigentlichen formgebenden Gesenkteile sind in ihren äußeren Abmessungen normalisiert, damit sie bei Verbrauch leichter, schneller und mit geringeren Kosten ausgewechselt werden können. Die Gesenke werden entweder von Hand oder auf der Kopierfräsmaschine ausgearbeitet, gegebenenfalls auf der Maschine nur vor- und von Hand fertiggearbeitet. Zu beachten ist die Ausführung der äußeren Form als Kegel, von denen sich der eine unter dem Pressendruck fester in seinen Sitz zieht, während der andere das Bestreben loszukommen hat.

Nach dem Pressen werden die Teile geglüht, so daß die entstehende Oxydschicht durch Beizen mit Salpetersäure entfernt werden muß.

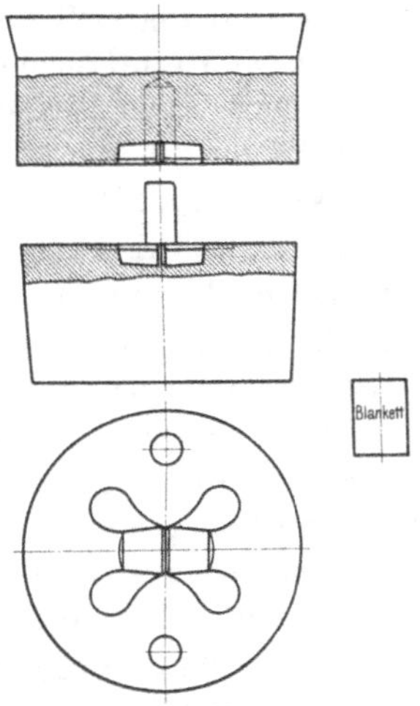

Fig. 255.

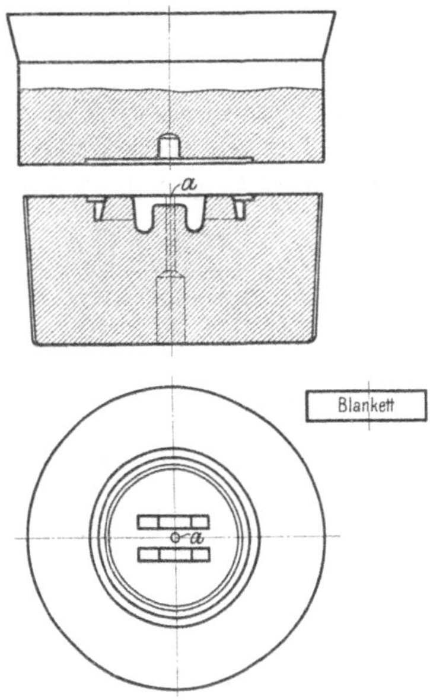

Fig. 256.

Dann kommen die Teile in die Abgratpresse, wo mit einem gewöhnlichen Abgratwerkzeug der Grat entfernt wird. In dem beschriebenen Fall einer Flügelmutter kommt noch der Kaltschnitt zur Trennung der beiden Muttern dazu. Stücke dieser Art verlangen keinen besonderen Auswerfer im Gesenk.

Deckel für ein Schwimmergefäß (Fig. 256). Dieser Deckel von 58 mm Durchmesser und ungefähr 2,5 mm Wandstärke stellt bereits eine viel größere Materialbewegung dar. Das Ausgangsstück ist eine Scheibe von 40 mm Durchmesser und 10 mm Dicke. Der Deckel erhält einen umlaufenden Rand von 8 mm Höhe und 3 mm Dicke im Mittel, auf der Unterseite vier durch zwei Rippen verbundene Lappen von ungefähr 12 mm Höhe und ungefähr 3 mm Dicke, während auf der Oberseite ein mittlerer Zapfen von 10 mm Höhe und 8 mm Durchmesser ausgeprägt werden muß. Das Verfahren nähert sich bereits dem Spritzverfahren. Das Werkzeug ist zweiteilig, die Trennungsfuge in die Mitte des oberen Flansches verlegt, so daß nach dem Abgraten die Fertigarbeit sehr leicht geschehen kann. Da das Stück infolge seiner dünnen Wandstärken und verhältnismäßig hohen Wände, wie der Lappen, nicht leicht von selbst aus dem Untergesenk herausgeht, ist bei a ein Auswerferstift vorgesehen, dessen Wirkung natürlich auf der Innenseite des Deckels zu sehen ist.

Das Stück wird in einem Arbeitsgang fertiggestellt, wobei der Verlust durch Grat usw. ca. 17% beträgt.

Herstellung eines Heizungsgriffes in zwei Preßgängen (Tabelle Nr. 10). Der im fertigen Zustand gebogene und gelochte Heizungsgriff wird aus einer Rundstange von 14 mm Durchmesser und 95,5 mm Länge in zwei Pressungen mit einer zwischengeschalteten Biegung hergestellt. Das Vorpreßgesenk, Fig. 257 in fünf Ansichten, ist verhältnismäßig einfach und arbeitet nach dem Quetschverfahren. Der größte Teil der Form ist in dem Untergesenk enthalten, so daß nur die im Gesenk oben liegende Fußfläche des ungebogenen Griffes durch den Oberstempel hergestellt wird. Die Form im Untergesenk ist am Grunde eines tiefen, durch die ganze Länge des Untergesenkes gehenden Schlitzes eingearbeitet, der dem schmalen, langen Oberstempel eine gute seitliche Führung gibt, so daß er nicht weggedrückt werden kann. Außerdem sind noch zwei kräftige Sucherstifte am Untergesenk vorgesehen, die in der Kopfplatte des Stempels in entsprechende Löcher eingreifen. Da auch hier die ganze Tiefe des Stückes im Untergesenk liegt, ist ein Auswerferstift in der Mitte unter dem tiefsten Teile vorgesehen.

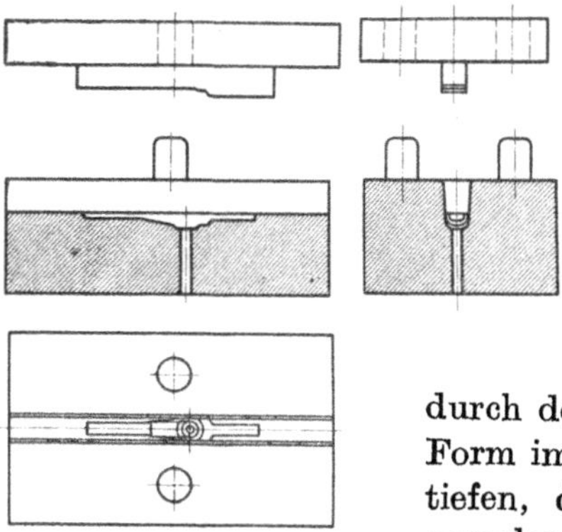

Fig. 257.

Nachdem der auf dem Gesenk, Fig. 257, vorgepreßte Griff abgegratet worden ist, wird das eine Ende desselben unter rechtem Winkel abgebogen und in einem geteilten Gesenk, Fig. 258, nach dem Quetschverfahren fertiggestellt. Das Gesenk ist nach der Mittellinie des Stückes geteilt, wobei die beiden formgebenden Gesenkteile wieder den Normalien angepaßt sind. Zu bemerken ist die Lage der unsymmetrischen Gesenkform in dem kreisförmigen Gesenk, die so gewählt ist, daß die Druckverteilung möglichst gleichmäßig erfolgt. Zwei Sucherstifte sind ebenfalls vorgesehen, damit beide Gesenkteile in der richtigen Stellung aufeinandertreffen. Nach dem nun folgenden Abgraten zeigt das Stück einen Gewichtsverlust von ungefähr 63 % bezogen auf das Fertiggewicht und 39 % bezogen auf das Blankettgewicht.

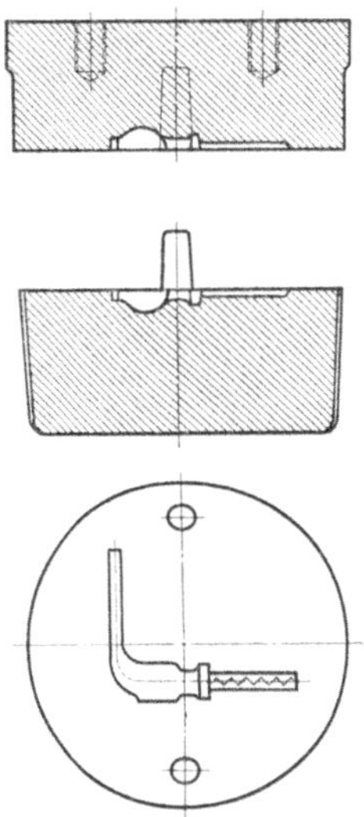

Fig. 258.

Die Gewichtsabnahmen werden von Interesse sein

Gewicht des Blanketts . . . . 124 g
Gewicht des gebeizten Griffes, nicht abgegratet . . . . 123 g
Gewicht des gepreßten Griffes, abgegratet . . . . .110—100 g
Gewicht des gebogenen, fertiggepreßten Teiles, nicht abgegratet. . . . . . . 108 g
Gewicht des abgegrateten, gelochten, gebogenen Griffes . . . . 76 g

Alle Teile sind blank gewogen, und sind Abweichungen bis ungefähr 10 % möglich.

Warmpressung mit Hilfe eines vierteiligen Gesenkes (Teile a, b, c, d nach Zeichnung des Kabelwerkes der A. E. G). In Tabelle Nr. 11 ist der Herstellungsgang eines Stückes angegeben, zu dessen letzter Arbeit ein vierfach geteiltes Gesenk verwendet wird. Das Blankett ist ein Abschnitt einer Rundstange von 45 mm Durchmesser und ist 32,5 mm hoch. Dieses Stück wird auf einem zweiteiligen Gesenk, Fig. 259, dessen Teilfuge in der Mittelebene des Stückes liegt, vorgepreßt. Der entstehende Grat ist rund 1 mm dick. Da das Stück mit seinen stark wechselnden Abmessungen auch ziemlich tief im Gesenk liegt, ist ein Auswerferstift vorgesehen. Auch zwei Sucherstifte mit abgerundeten Köpfen

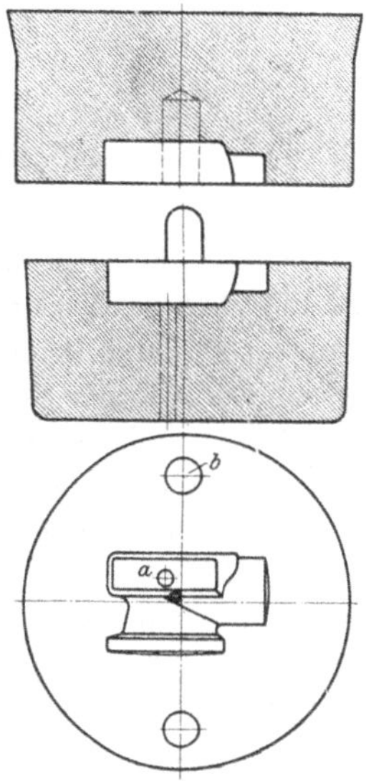

Fig. 259.

sind vorhanden und den ganzen Abmessungen des Werkzeuges entsprechend kräftig bemessen. Nach dem Abgraten, welches in diesem Fall keine Schwierigkeiten aufweist, kommt das Stück in das Fertiggesenk, Fig. 260, dessen Aufbau ziemlich verwickelt ist. Durch die nach der jetzt notwendigen Druckrichtung unsymmetrische Form des Stückes ist die außerachsige Lage des Oberstempels bedingt, während die Teilfuge des Untergesenkes, die durch die Mittellinie des Fußes zu legen ist, bedingt, daß die Form im Untergesenk noch weiter seitlich verschoben werden muß. Dazu kommt, daß die Unterschneidungen im Stücke in zwei zueinander senkrechten Ebenen liegen, so daß man mit dem einfacheren, dreiteiligen Gesenk nach Fig. 254 nicht auskommt und das Untergesenk nach der Linie xx wieder teilen muß. Daraus ergibt sich die Arbeit der einzelnen Stempelteile. Der Oberstempel a bildet die innere Aussparung bei c, Tabelle Nr. 11 G. 4, und die dazu gehörige obere Fläche des Preßstückes. Er ist mit seinem vernieteten Kopf in eine normalisierte Kopfplatte eingesetzt, die am Pressenstößel in der gezeichneten Weise gehalten werden kann. In dem Mittelteil ist b der eigentliche Gesenkteil, der an seiner Mantelfläche eine Führung für den Oberstempel abgibt, die in dem gegebenen Falle wegen der außerachsigen Lage des Stempels notwendig ist. An seiner Unterfläche hilft Teil b an der Formgebung des zylindrischen Zapfens des Stückes mit und trägt gleichzeitig die beiden Sucherstifte c, welche in ihm, weil er ungeteilt ist, angebracht sind. Von besonderem Interesse ist der Unterteil d des Gesenkes. Er besorgt die Formgebung der Unterfläche des vierkantigen Oberteiles des Preßstückes, so daß der beim Pressen herausgequetschte Grat gerade noch in die ebenen Seitenflächen fällt, außerdem der unteren Hälfte des zylindrischen Zapfens und des unterschnittenen Fußes. Die beiden Gesenkhälften, die nach der

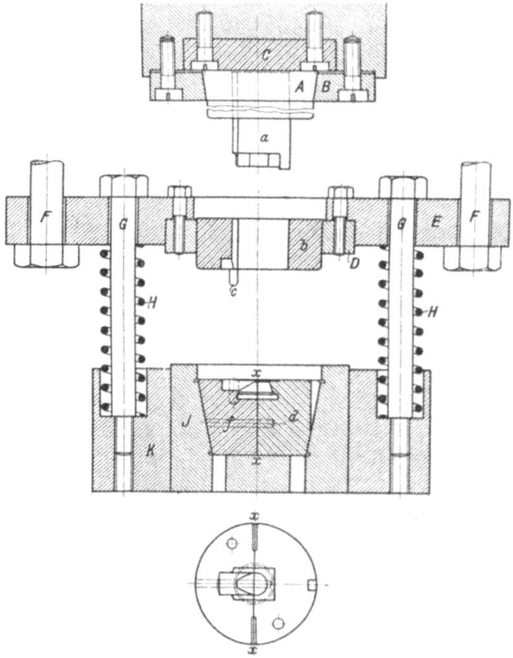

Fig. 260.

Linie xx in einer senkrechten Ebene zusammenstoßen, nehmen je eine Hälfte des Fußes auf. Wegen dieser Stoßfläche sind die Löcher für die Sucherstifte unter einem Winkel von 30° dazu gelegt worden. Um das fertige Stück aus dem Untergesenk zu entfernen, sind bei f senkrecht zur Trennungsfläche der Gesenkhälften verlaufende Bohrungen vorgesehen, in welche man Stifte hineinstößt, um die beiden Teile auseinanderzutreiben. Andrerseits erleichtern diese Stifte auch das richtige Einlegen der beiden Gesenkhälften. Die äußere Form des zusammengesetzten Unterteils entspricht wieder der Normalform. An dem fertigen Stück ist natürlich auch der feine Grat, den diese Trennungslinie hervorruft, zu sehen.

Die Gewichte der einzelnen Stücke ergeben sich wie folgt:

Gewicht des Blanketts	435 g
Gewicht des gebeizten, nicht abgegrateten Teils	430 g
Gewicht des vorgepreßten, abgegrateten Teils	400 g
Gewicht des fertiggepreßten, nicht abgegrateten Teils	384 g
Gewicht des fertiggepreßten, abgegrateten Teils	375 g

Daraus ergibt sich ein totaler Abfall, bezogen auf das fertige Stück, von 16%. In dem übrigen Teil der Fig. 260 ist eine Art des Einbaues derartiger Werkzeuge gegeben, die einerseits erkennen läßt, wie die nicht unmittelbar formgebenden Teile eines Werkzeuges normalisiert werden können, andrerseits die verhältnismäßig großen Abmessungen dieser Teile gegenüber den eigentlichen Gesenkteilen zeigt, damit sie der schweren gleichzeitigen Beanspruchung durch Wärme und Druck gewachsen sind. Der Oberstempel a ist in seiner Kopfplatte A vernietet, welche mit einem Kegel in dem Tragring B ruht und durch die Zylinderkopfschrauben, die gleichzeitig den Ring tragen, fest gegen die Druckplatte C im Stößel gepreßt wird. Da der Tragring in einer Ausdrehung des Stößels ausgemittelt wird, wird gleichzeitig der Stempel in dem Stößel gleichachsig eingespannt. Der Mittelteil b wird ähnlich wie der Stempel durch einen schmalen Bund in seinem Tragring D gehalten und mit diesem durch von oben eingeschraubte Sechskantkopfschrauben in einer Ausdrehung des Mitnehmers E befestigt. Während beim Preßgang der Mitnehmer durch den auf ihn drückenden Stempel unter Ausspannung der Tragfedern H, die auf den Führungsbolzen G gleiten, gleichachsig nach unten bewegt wird, nehmen die Köpfe der Schrauben F denselben schließlich beim Aufwärtsgang mit. Beim Aufwärtsgang des Stößels trennen sich Ober- und Mittelteil voneinander. Die zweiteilige Matrize sitzt in einem zylindrischen Unterkasten, der auf der Unterseite mit Löchern zum Herausstoßen der Gesenkhälften versehen ist. Dieser Teil hat beim Pressen die ganze seitliche Komponente des Preßdruckes aufzunehmen und soll dementsprechend bemessen sein. Da er in dem Grundplattenteil des Werkzeuges außerdem gut eingepaßt ist, übernimmt dieser, sobald die elastische Formänderung des Teiles J ein bestimmtes Maß überschreitet, einen Teil des Druckes, so daß jedenfalls genügend Sicherheit gegen Aufreißen vorhanden ist.

Die Bearbeitung von goldplattiertem Material. Das Geheimnis der Herstellung guter Fassungen aus goldplattiertem Material liegt darin, daß man das Material so verarbeitet, daß das Gold, welches als eine verhältnismäßig dünne Schicht das Messing oder Grundmaterial bedeckt, soweit wie möglich als vollständige Decke ohne Risse oder Falten auf dem Grundmaterial erhalten wird, so daß dieses in keiner Weise zutage tritt. Denn selbst die kleinsten Löcher usw. dieser Art werden sehr bald durch den Schweiß, der bei manchen stark ätzend wirkt, schwarz oder grün und beweisen so, daß ein fehlerhaftes Stück

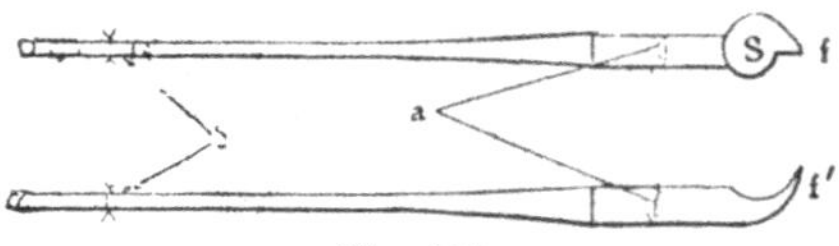

Fig. 261.

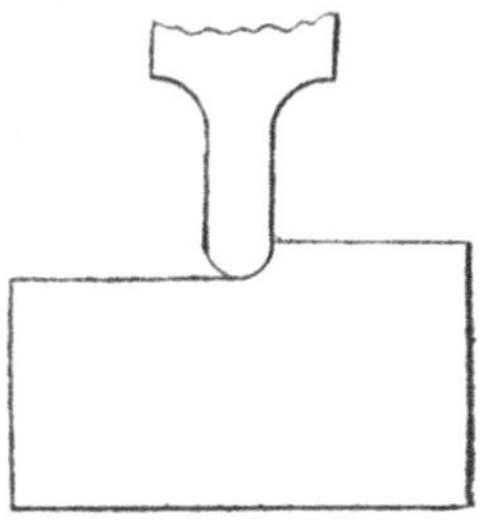
Fig. 262.

aus der Fabrik herausgegangen ist. Bei Herstellung der Ohrenbügel für Brillen — ein Teil derselben ist in Fig. 261 abgebildet — gehen folgende Arbeiten vor sich. Der goldplattierte Draht, der gewöhnlich von Spezialfabriken bezogen wird, hat den Durchmesser a, Fig 261, während das Ende, das den Bogen hinter dem Ohre des Trägers bildet, auf den Durchmesser b gestreckt wird. Diese Arbeit verringert selbstverständlich die Dicke des Goldüberzuges, doch wird derselbe dabei in keiner Weise gebrochen oder verletzt. Volles Gold, Neusilber, goldähnliche Legierungen, Stahl usw. werden sämtlich auf Streckmaschinen verarbeitet, wobei die einen mit Handzuführung, die anderen mit selbsttätiger Zuführung versehen sind. Wenn es sich um volles Metall handelt, werden die Enden der Bügel, die mit dem an die Linsenfassung angelöteten Scharnierstück ein Gelenk bilden, auf die richtige Dicke ausgebreitet und nachher in einem Schnittwerkzeug nach den in Fig. 261 gezeichneten Umrissen ausgeschnitten. Falls man es hingegen mit goldplattiertem Material zu tun hat, würde das Ausschneiden das Grundmaterial an der Kante des ausgeflachten Stückes freilegen. Das im folgenden beschriebene Gesenk dient zum Ausflachen dieser Enden auf die richtige Form, ohne die Goldplattierung zu verletzen, und gibt zugleich einen kleinen Kunstgriff, der es ermöglicht, auch die Schneide f auszubilden.

Nach dem Strecken wird das Arbeitsstück in einem Gesenk entsprechend Fig. 262 ungefähr auf die Form f′ gebracht. Da das Gold dehnbarer als das Grundmetall ist, bleibt es flach und dünn und behält eine saubere Oberfläche, — es scheinen sich die beiden Goldstärken an dieser Stelle zusammenzuschweißen. Würde man diese Enden abschneiden, so würde wieder das Grundmaterial zutage treten. Die Teile werden nun auf die richtige Form in einem Gesenk abgeflacht, Fig. 263.

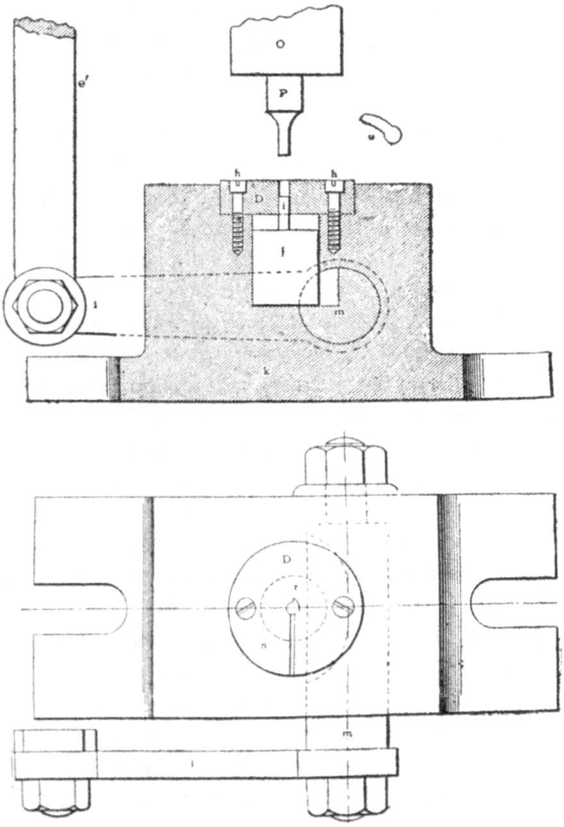

Fig. 263.

Das Werkzeug besteht aus einer Froschplatte k, deren Unterseite gehobelt ist und auf den Tisch der Presse zu liegen kommt. Von oben wird in diese Platte ein Loch gebohrt, dessen Grund eben ist so daß der Kolben j dort aufruht, und welches gleichzeitig für die Matrize D ausgebohrt ist, die mit zwei Schrauben h befestigt ist. Zur Aufnahme von m wird durch die Froschplatte ein Loch, rechtwinklig zur Vorderfläche und parallel zur Grundfläche gebohrt, dessen Achse in der Ebene der Bodenfläche des Loches für j liegt. Ein Stück Werkzeugstahl m wird dann, wie im Schnitt ersichtlich, an seiner oberen Seite ausgefräst, so daß eine Fläche desselben mit der Auflagerfläche für j abschneidet, und leicht drehbar in das Loch eingepaßt. Das vordere, vor der Zeichenebene heraustretende Ende von m trägt einen Hebel l, wie punktiert gezeichnet, und einen nicht gezeichneten Stift, der l an der Drehung auf seiner Achse verhindert. Am linken Ende des Hebels l ist ein Auge vorgesehen, an dem eine Zugstange e′ angelenkt ist, die mit dem oberen Ende des Pressenstößels verbunden ist. Die Matrize D wird nach der bei u und r sichtbaren Linie ausgearbeitet, ebenso der Stempel P und Auswerfer i. Dieser ist so lang, daß er, wenn der Stempel am oberen Hubende angelangt ist, fast bis an die Oberfläche von D kommt. Die Matrize D ist nahe der Oberfläche ein wenig erweitert, so daß das fertige Arbeitsstück, sobald es durch den Auswerfer bis nach oben gedrückt worden ist, lose liegt und leicht entfernt werden kann. Der Auswerfer paßt wohl gut in das glatte Loch der Matrize, muß sich jedoch darin leicht bewegen lassen. Um die bestmöglichen Ergebnisse an dem fertigen Werkstück zu erhalten, muß das Loch in D auf der Innenseite besonders glatt gehalten werden.

Bei der Herstellung dieses Gesenkes, das doppelt gemacht werden muß, geht man folgendermaßen vor: Die Platte D, die von einer bereits auf den richtigen Durchmesser abgedrehten Stange abgesägt wird, wird an den beiden Seiten eben gehobelt und die Gesenkform n angenähert mittels Bohrer und Feile ausgearbeitet, worauf ein Schnittstempel, der etwas kleiner als das Fertigmaß ist, in einer Schraubenpresse durch D durchgedrückt wird; dann werden noch zwei oder drei andere Stempel durch das Loch getrieben. Diese Stempel sind in Form genau gleich dem Schnittstempel, erhalten jedoch am unteren Ende an Stelle der scharfen Kanten des ersteren abgerundete und geglättete Kanten, da sie zu glätten und nicht zu schneiden haben. Sie werden in zwei oder drei Größen mit vielleicht 0,05 mm Unterschied gemacht und sauber geglättet, wobei die Arbeitsrichtung dafür längs des Umfanges, also rechtwinklig zur schließlichen Arbeitsbewegung liegt. Sie werden selbstverständlich gehärtet und auf dunkles Strohgelb angelassen.

Zuerst wird nun Stempel Nr. 1, der am kleinsten und gerade ein wenig größer als der oben erwähnte Schnittstempel ist, durchgestoßen. Dem folgt der etwas größere Stempel Nr. 2, dann Nr. 3 und schließlich derjenige mit dem Fertigmaaß. Man benutzt dazu am besten eine Schraubenpresse, wobei Stempel und Matrize so befestigt sind, daß keines von beiden ausweichen kann. Beide werden eingefettet und dann der Stempel Nr. 1 auf- und abwärts bewegt, wobei man ihn beim ersten Hub nur wenig eintreten läßt; dann zieht man zurück, vergrößert den Hub ein wenig usw., bis man ihn ganz durchgetrieben hat. Wenn man alle drei Stempel in dieser Weise durcharbeiten läßt, erhält man ein sehr gutes, glattes Loch. Das Metall wird wohl an den Umrißlinien ein wenig aufgeworfen, doch läßt sich dies nachträglich leicht entfernen. Nun wird noch, wie oben erwähnt, an der Oberfläche das Loch zur leichteren Entfernung des Arbeitsstückes ein wenig vergrößert, das Gesenk ausgeglüht und der Stempel Nr. 3 noch einmal durchgetrieben.

Das Pressen von Hartgummi und Isoliermaterialien (Siemens-Schuckertwerke, Kabelwerk Gartenfeld). Die ungeheure, weite Verbreitung, die die Verwendung der Isoliermaterialien mit dem Anwachsen der elektrischen Anlagen gefunden hat, läßt es berechtigt erscheinen, die Arbeitsweisen auf diesem Gebiete, wenigstens soweit Preßarbeit in Frage kommt, auch zu besprechen. Die Frage gewinnt noch dadurch an Interesse, weil mit den steigenden Gummipreisen die elektrische Industrie eifrigst Ersatzstoffe gesucht und, wie die weiter unten angeführten Beispiele der Werkzeuge aus dem Betrieb des Kabelwerkes der Siemens-Schuckertwerke beweisen, auch gefunden hat, welche dann in ähnlicher Weise wie das Grundmaterial verarbeitet werden. Wenn es sich auch in allen Fällen um Gesenkarbeit handelt und die Herstellung der Gesenke sich im allgemeinen an die bereits besprochenen Werkzeuge anlehnt, so ist die Anpassung an die besonderen Erfordernisse der drei in Betracht kommenden Materialien immerhin von so einschneidendem Einfluß auf die endgültige Ausführung der Werkzeuge, daß unmöglich eines derselben für Arbeit auf dem andern Material verwendet werden kann.

Es handelt sich um drei Stoffe: Hartgummi, warm gepreßte Isoliermaterialien, in welchen Schellack, Kopale, Asphalte usw. das Bindemittel abgeben, und kalt gepreßte Isoliermaterialien mit oxydierenden Ölen oder Teeren als Bindemittel. Der Wert dieser drei Materialien in bezug auf Verarbeitungseigenschaften entspricht ungefähr der oberen Reihenfolge; im Preis steht Hartgummi am höchsten.

Hartgummi-Umpressung. Bei vielen Teilen, die neben ihrer elektrischen Isolierfähigkeit auch einen gewissen Widerstand gegen mechanische Beanspruchung besitzen sollen, genügt die Festigkeit des Hartgummis allein nicht, andrerseits erspart man durch das in Fig. 264 und 265 angegebene Verfahren an diesem Material selbst. Da das Hartgummi nach seiner Verarbeitung vom Naturgummi zur Platte immer noch im weichen Zustand vorliegt, kann man durch gegossene Bleigesenke genügend genaue und scharfe Pressungen erhalten, die dann durch Vulkanisieren in den bekannten, festen Zustand übergeführt werden. Man erreicht dadurch den Vorteil, daß die Originalform, nach welcher das Gesenk gegossen wird, fast gar nicht abgenutzt wird und überhaupt nicht in die Werkstätte kommt. Da auch das Blei der Form selbst unter gewissen Vorsichtsmaßregeln und mit einem durch das Umschmelzen bedingten Abbrand wieder verwendet werden kann, ist die Materialersparnis augenfällig. Dazu kommt die leichte Bearbeitung der Bleiform, die sich

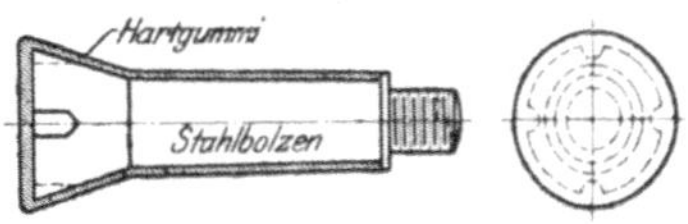

Fig. 264.

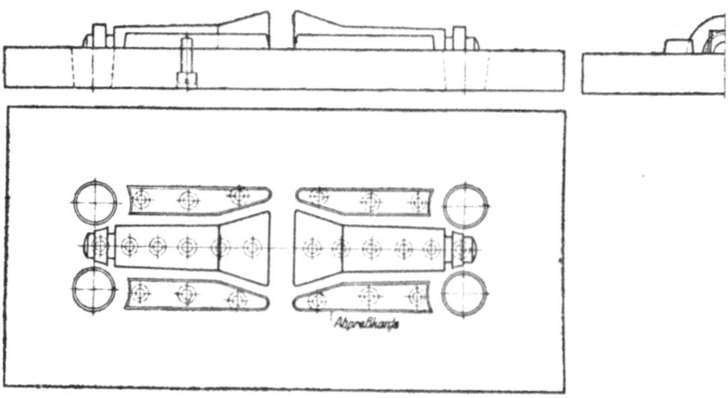

Fig. 265.

auf ein Hobeln der Auflagerflächen beschränken kann. Eine derartige Gießform für ein Gesenk, welches zum Umpressen von Stahlbolzen, Fig. 264, Verwendung findet, zeigt Fig. 265. Das Modell ist auf einer Grundplatte aus Schmiedeeisen aufgebaut und ist zum gleichzeitigen Pressen von zwei Bolzen eingerichtet. Die eigentlichen Modelle, Kernhälften, die in der Gesenkform den Mantel aus Hartgummi bilden, werden aus Bessemerstahl hergestellt und sehr sauber poliert, so daß an dem abgegossenen Teil, soweit er für die Formgebung des Gesenkes in Frage kommt, keine Nacharbeit mehr nötig ist. Zeigen sich am Abguß irgendwelche Gußfehler, so wird man bei den geringen Herstellungskosten des eigentlichen Gesenkes zu keiner Nacharbeit greifen, sondern einen neuen Abguß machen. Die „Kernhälften" werden mit je sechs versenkten Kopfschrauben von $\frac{1}{4}$ Zoll Durchmesser auf der Grundplatte befestigt. Zu beiden Seiten der Kernhälfte sind Auflagen vorgesehen, die als „Abpreßkanten" in dem Gesenk vertiefte Kanäle vorstellen, in welchen sich das überschüssige Material sammeln kann, so daß bei dem verarbeiteten weichen Material nur ein verschwindend dünner Grat übrig bleiben wird. Außerdem sind in der Gußform vier konische Löcher nach Kaliber eingearbeitet, die dazu dienen, mit geeigneten Bolzen in den Gesenken nach dem Guß die beiden Gesenkhälften genau aufeinanderzupassen.

Im Gegensatz zu diesen Gesenken sind die Werkzeuge für Pressungen in dem zweiten Material entsprechend den dabei auftretenden Drücken sehr kräftig gehalten. Ein derartiges Preßstück, einen Fuß für einen Stechkontakt, zeigt Fig. 266. Das Stück ist auf der Unterseite mit drei ziemlich hohen Rippen versehen, die stark abgeschrägt worden sind, damit das Stück leichter aus der Form entfernt werden kann. Außerdem sind zwei große, zylindrische und drei kleinere, konisch abgesetzte Löcher in dem Stück herzustellen.

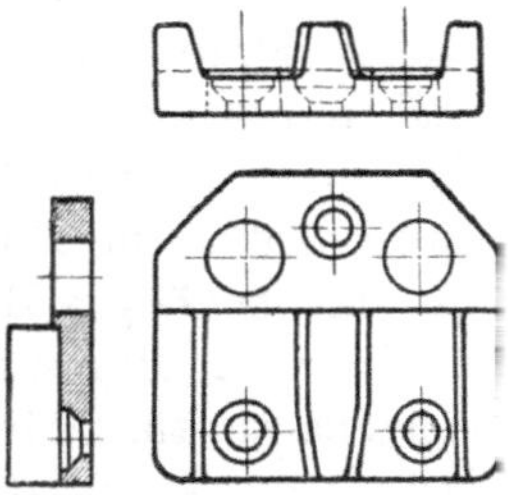

Fig. 266.

Infolge der eigenartigen Herstellung des Preßgutes aus einem pulverförmigen Rohmaterial und der notwendigen Erwärmung des

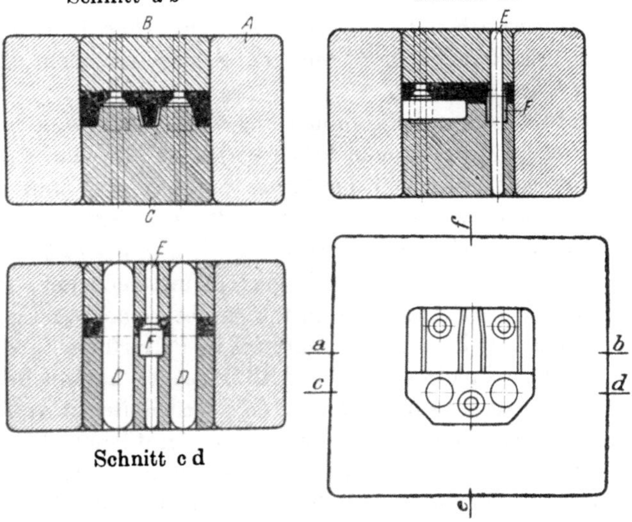

Fig. 267.

Gesenkes. teils allein, teils mit der aufgefüllten Preßmasse, ist das Werkzeug aus losen Einzelstücken hergestellt, die vom Arbeiter vor dem Auffüllen der Form mit Preßmasse erst zusammengesetzt werden. Das zu obigem Stück gehörige Werkzeug ist in vier Schnitten in Fig. 267 dargestellt. Ähnlich wie bei dem Werkzeug für Warmpressung von

Messing, Fig. 254 und 260, wird die vollständige Form durch Zusammenarbeit dreier gesonderter Hauptteile erzeugt, eines Oberstempels, eines Unterstempels und des Mantels. Dieser hat hier, wo das Material in Pulverform aufgefüllt wird, zuerst als Füllgefäß zu dienen, worauf er bei Beginn der Pressung als seitliches Widerlager für die gepreßte Masse arbeitet, gleichzeitig die Führung für Ober- und Unterstempel abgibt und schließlich am Ende der Pressung die Hubbegrenzung für den Preßweg und mit seiner inneren Mantelfläche einen Teil der Gesenkform selbst darstellt. Weitere Teile des Werkzeuges sind zwei zylindrische Kerne D aus extrahartem zähen Werkzeugstahl und zwei kleinere Kerne E mit den dazugehörigen Büchsen F aus demselben Material. Der Mantel A wird seiner Beanspruchung entsprechend sehr kräftig gehalten, aus Siemens-Martinstahl hergestellt und allseitig bearbeitet. Die innere Mantelfläche ist poliert. Der Oberstempel B paßt gut gleitend in den Mantel und bildet bei der Pressung eine glatte Unterfläche des Preßstückes, in ihm führen sich sämtliche fünf Kerne D, E in sauber ausgeriebenen Bohrungen. Der Oberstempel wird wie der Unterstempel aus mittelhartem Werkzeugstahl gefertigt und bleibt ungehärtet. Der Unterstempel übernimmt den größten Teil der Formgebung, die auf der Unterfläche des Preßstückes liegt, und nimmt alle fünf Kerne sowie die Büchsen F auf, die hier zwar gut, aber doch so eingepaßt sind, daß sie durch einen leichten Schlag aus dem Untergesenk entfernt werden können.

Die gefüllten Gesenke kommen angewärmt unter eine hydraulische Presse, welche die Preßstücke fertig verlassen. Da das ganze Gesenk mit dem fertigen Stück von der Presse entfernt wird und das Gesenk zur Herausnahme des Stückes in seine Teile zerlegt werden muß, so wird wenigstens mit zwei gleichen Gesenken gearbeitet.

Das dritte Material wird kalt gepreßt und nach dem Pressen und Herausheben aus der Form gebrannt, damit es die notwendige Festigkeit erhält. Auch hier kommt das Material in Pulverform in das Gesenk, so daß der Mantel gleichzeitig das Maß bildet, bis zu welcher Höhe der Arbeiter das Gesenk zu füllen hat. Die Arbeitsweise ist von den bisher besprochenen verschieden und mehr der eigentlichen Preßarbeit auf Metall ähnlich. Die Werkzeugteile bleiben in der Presse befestigt, der Oberstempel am Pressenstößel, Unterstempel und Mantel am Pressentisch. Die Arbeit kann auf irgendeiner Pressenart — Hand-, Reibungs-, Kniehebel- oder hydraulische Presse — geschehen. Zum Entfernen der gepreßten Stücke aus dem Gesenk ist ein Auswerfer vorgesehen, der nach Beendigung der Pressung von Hand oder durch einen Fußtritt betätigt wird. Dagegen ist die Zusammensetzung des Gesenkes unter Zuhilfenahme eines Mantels, der den vollen seitlichen Druck aufzunehmen hat, beibehalten worden.

Fig. 268 stellt ein derartiges Werkzeug zum Pressen einer Kappe, die in der Zeichnung im Schnitt bei A eng schraffiert gezeichnet ist, dar. Der Oberstempel wird mit seinem zylindrischen Zapfen von 20 mm Durchmesser in der gewöhnlichen Weise im Pressenstößel befestigt, ist aus

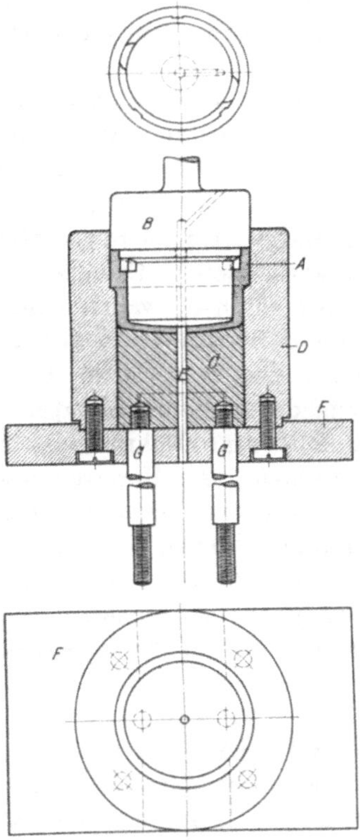

Fig. 268.

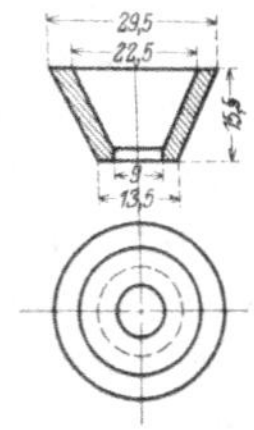

Fig. 269.

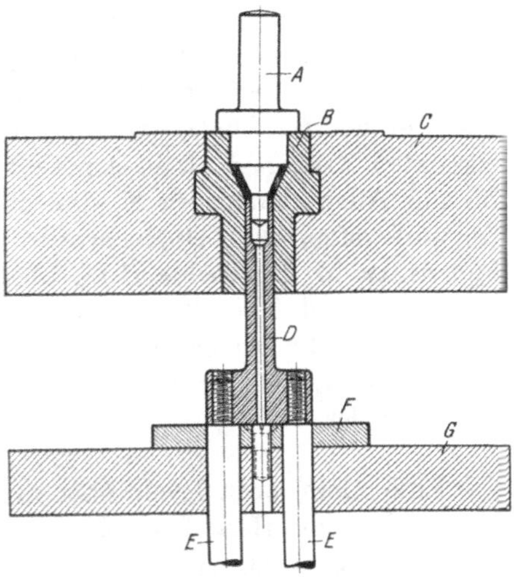

Fig. 270.

Gußstahl hergestellt und gehärtet. Der Zapfen selbst bleibt weich. Der Oberstempel erhält innen eine Bohrung zur Aufnahme des Kernes von 5,5 mm Durchmesser, in welcher der Kern mit 0,5 mm Spiel geht. Der Kern ist aus Gußstahl, gehärtet und in der Grundplatte selbst verschraubt. Der Oberstempel bildet die gesamte Innenfläche und die Grundfläche der Kappe aus. Die Umfläche der Kappe wird in der Endstellung der Preßwerkzeuge durch den Mantel D gebildet, dessen Innenfläche daher wie bei den früher besprochenen Werkzeugen poliert ist. Der Mantel wird aus Siemens-Martinstahl hergestellt und mit einer Andrehung versehen, mittels der er in einer Ausdrehung der Grundplatte F ausgemittelt wird. Er wird durch vier halbzöllige versenkte Zylinderkopfschrauben von der Unterseite der Grundplatte aus in ihr gehalten. Die Oberseite des

Preßstückes wird von dem Unterstempel C hergestellt, der gleichzeitig als Auswerfer dient und durch die Stifte G für diese Arbeit betätigt wird. Der Unterstempel ist aus Gußstahl und gehärtet. Er hat außerdem eine durchgehende Bohrung, in welcher sich der Kern passend führt. Die Auswerferstangen G aus Siemens-Martinstahl sind in den Unterstempel bis an den Bund niedergeschraubt und führen sich mit ihrem zylindrischen Teil in der Grundplatte.

Ein anderes Werkzeug zur Herstellung einer kleinen konischen Büchse, Fig. 269, ist in Fig. 270 abgebildet. Da es im ganzen und großen nach denselben Grundsätzen wie das vorangehende Werkzeug gebaut ist, wird eine kurze Besprechung neben der Zeichnung genügen. Der Stempel ist aus mittelhartem Werkzeugstahl gefertigt, wird gehärtet und blau angelassen, während der Zapfen A weich bleibt. Er bildet die Innenfläche und die breitere Grundfläche der Büchse, während ihre äußere Umfläche von der zweiteiligen Büchse B und die kleine Grundfläche vom Auswerferstempel D gebildet wird. Das Material für B ist dasselbe wie für den Oberstempel A, während der Auswerferstempel D aus extra zähhartem Stahl gemacht wird. F ist eine Führungsplatte für die Auswerferstangen E, die in dem verbreiterten Unterteil des Auswerferstempels D eingeschraubt sind. Zu bemerken wäre, daß der Füllansatz in der Büchse B im obersten Teil mit 0,5 mm konisch gemacht ist, so daß der Stempel A seinen Weg finden kann. Um über die auftretenden Drücke einigen Anhaltspunkt zu geben, sei bemerkt, daß der Mantel C, der für dieses Gesenk verwendet, für 25000 kg Druck berechnet ist.

Eine dem eben beschriebenen Arbeitsverfahren ähnliche Arbeit kommt bei der Herstellung der Pastillen und chemischen Tabletten wie bei der Farbenfabrikation vor. In beiden Fällen ist das Rohmaterial entweder pulverförmig oder ein aus Pulver angemachter Teig, der ebenfalls große Pressendrücke zur Fertigstellung der Form verlangt.

Stempel und Gesenk für Pastillen. Die im folgenden beschriebenen Werkzeuge sind in einer Vielfachpresse, Patent Richards, zur gleichzeitigen Herstellung von vier Pastillen, nach Fig. 271, benutzt worden.

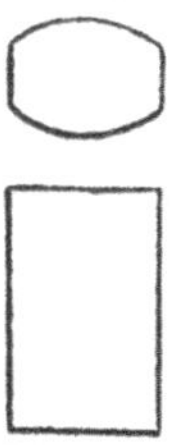

Fig. 271.

Diese Pastille enthält 2,6 g des Pulvers und hat stumpf-sechseckigen Querschnitt mit nur zwei flachen Seitenflächen. Die Pastille mußte nämlich durch den Flaschenhals hindurch, der aber nicht groß genug war, um einen vollständigen Sechskantquerschnitt durchzulassen; außerdem geht diese Form leichter vom Stempel ab, und schließlich kann der Stempel, Fig. 272, bei dieser Formgebung verstärkt werden. Die Arbeit verlangt acht Stempel nach Fig. 272, von denen vier länger als die anderen gemacht wurden, ein Gesenk nach Fig. 272, zwei Kopfplatten nach Fig. 273 und einen Ring dazu nach Fig. 274. Das Gesenk muß so hart wie nur möglich sein, ohne Risse zu zeigen, und die Stempel gehärtet und auf helles Strohgelb angelassen werden, da die gepulverten Chemikalien sehr stark auf Stempel und Gesenk wirken; die Stempel müssen gut eingepaßt und hoch poliert werden.

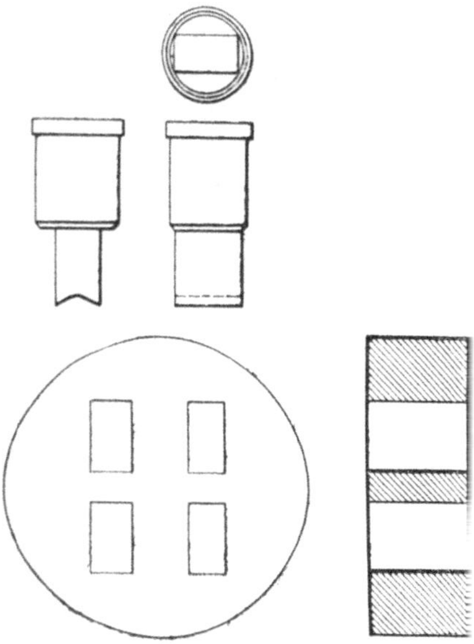

Fig. 272.

Als vorläufiges Futter dient ein Stück nach Fig. 275. Es wird auf beiden Seiten eben bearbeitet, worauf die drei Löcher a, b und c gebohrt werden. Unter Benutzung der Löcher a und c wird das Gußstück auf der Planscheibe befestigt, der Kreis für den Durchmesser der Matrize angerissen und passend auf 25 mm Tiefe ausgebohrt; der Rest des Loches wird dann auf den kleinen Durchmesser der Kopfplatte, Fig. 273, fertiggedreht. Hieraufwird auf der Planscheibe ein möglichst großer Kreis angerissen und vier Punkte auf derselben unter gleichem Abstand angezeichnet, die Blanketts für Fig. 272, 273 und 274 in das Futter eingespannt und auf beiden Stirnflächen abgedreht, wobei sie mit der Klemmschraube d festgehalten werden.

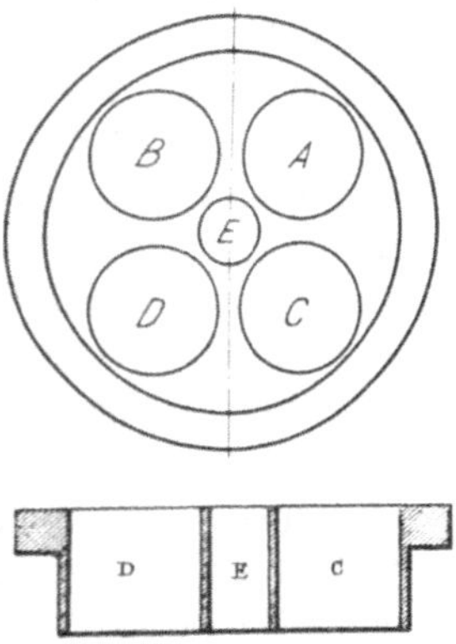

Fig. 273.

Nun wird in die Mitte der beiden Stücke, Fig. 273 und 274, ein Loch E von 9,5 mm Durchmesser gebohrt, das letzte Stück in das Futter, Fig. 275, eingespannt und ein Stangenzirkel ge-

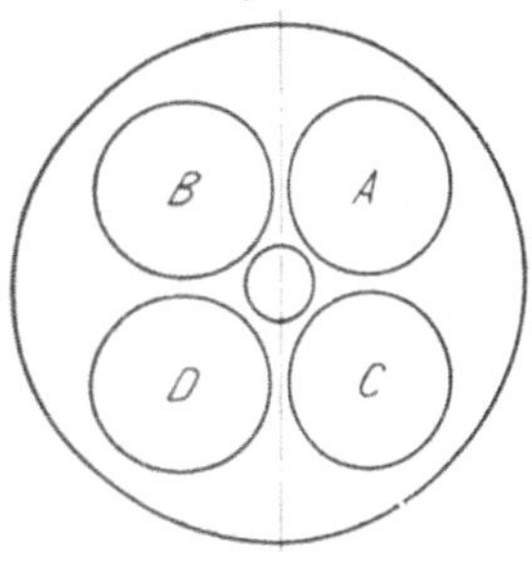

Fig. 274.

macht. Die eine Spitze desselben setzt man in einen leichten Körner, der vorher ins Drehbankbett eingeschlagen worden ist, ein und die andere Spitze in eine der vier Teilungen der Planscheibe und zieht eine Linie durch den Mittelpunkt des Stückes in Fig. 274 mit einer Anreißnadel, die am Support befestigt wird. Hierauf wird die Planscheibe bis zur nächsten Teilung gedreht und eine zweite Linie angerissen, bis auf dem Ring, Fig. 274, alle vier Linien angerissen sind. Die letzte Linie wird auch noch am Futter bei X angerissen und dann das Futter um 119 mm verschoben. Dieses Maß ist nämlich die Mittelentfernung der Löcher A, B, C, D von der Mitte des Loches E in der Kopfplatte, Fig. 273, sowie im Ring, Fig. 274. Zu dieser Arbeit benutzt man die gebohrten Löcher a und b

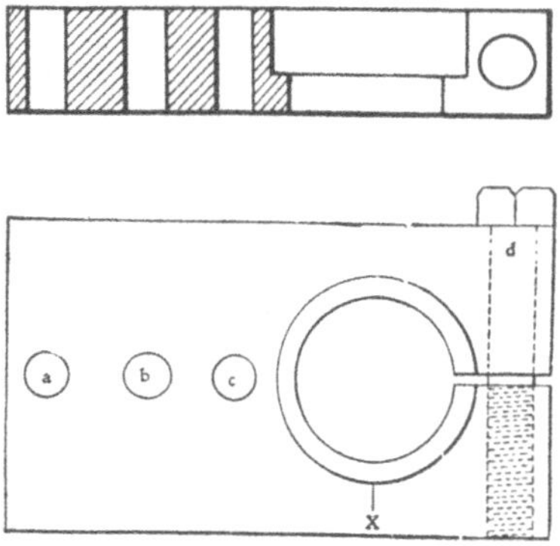

Fig. 275

und bohrt und reibt das erste Loch für den Stempel aus. Nun wird die Klemmschraube d gelüftet und der Ring gedreht, bis die nächste Mittellinie auf die Linie X zu stehen kommt, das nächste Loch fertiggestellt usw., bis alle vier Löcher für die Stempel fertig sind. Hierauf werden die Kopfplatten mit je einem Loch versehen und ein Dorn, der in die beiden Löcher der Kopfplatten und des Ringes paßt, und ebenso ein Dorn für das 9,5 mm Loch in deren Mitten gemacht. Dann wird eine Kopfplatte und der Ring zusammen auf den Dorn gesteckt und in das Futter, Fig. 275, eingespannt, so daß eine der Linien am Ring mit der Linie X zusammenfällt. Nun wird das nächste Loch in die Kopfplatte gebohrt, die Klemmschraube gelöst, und die Kopfplatte ebenso wie vorher der Ring fertiggemacht. Als nächstes werden beide Stücke auf einem 9,5 mm Dorn aufgebracht und am Umfange fertiggedreht. Das Futter wird von der Drehbank entfernt und die Matrize, Fig. 272, in die eine Seite und eine Kopfplatte in die andere Seite eingesetzt und alles zusammengeklemmt. Nun bohrt man ein Stück auf 9,3 mm, dreht es so ab, daß es in die Löcher der Kopfplatte paßt, härtet es ein wenig und bohrt die Matrize mit einem entsprechenden Bohrer von 9,3 mm Durchmesser. Dem folgt die Herstellung einer Räumnadel[1]) nach Fig. 276, die um 0,15 mm kleiner als die Stempel ist, und an die ein Ansatz angedreht wird, der in das Loch der Matrize paßt. Dieser wird in das Loch der Matrize eingesteckt und nach dem Ende der Räumnadel die Löcher für die Stempel angerissen,

[1]) Vgl. Fig. 679/80.

wonach die Matrize gebohrt, das übrige Material ausgestoßen und weggefeilt wird, so daß die Räumnadel eintreten kann. Dann nimmt man ein kleines Stück Stahl, paßt es genau an die Räumnadel und dreht es außen ab, bis es in die Kopfplatte paßt. Nun wird die Matrize und Kopfplatte in dem Futter fest zusammengeklemmt, die Führung für die Räumnadel in die Kopfplatte eingesetzt, worauf die Schnittplatte mit der Räumnadel ausgearbeitet und durch weniges Nachfeilen fertiggestellt wird. Nach dem Härten zeigten sich wohl noch einige Schwierigkeiten, doch hatten diese wenig Einfluß.

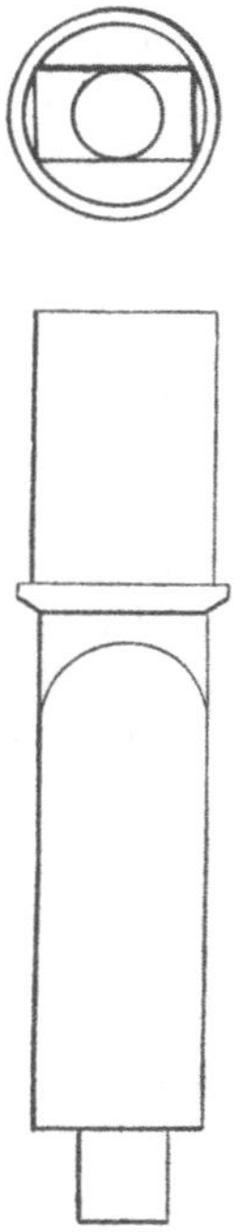

Fig. 276.

Die Stempel werden so abgeschnitten, daß zwei aus einem Stück gemacht werden können, und zwar vier Gruppen, je ein langer und ein kurzer, nach Fig. 272. Die Stempel werden an den Stirnflächen abgedreht, ungefähr 0,4 mm gerade gedreht und hierauf nach dem Mikrometer auf Maß fertig gemacht. Sie werden auf der Fräsmaschine entzweigeschnitten. Die oberen Stempel werden in die Matrize eingepaßt und in die Kopfplatte eingesetzt, die in eine Mutter paßt und an dem Stempelkopf festgeklemmt ist; dieser Arbeit folgt das Abdrehen auf gleiche Länge. Hierauf werden die unteren Stempel in den Ring und in die Kopfplatte gesteckt. Nachdem das Ganze in den Stempelkopf eingebracht ist, wird die Matrize über die Stempel gesteckt, so daß sie sich nicht verdrehen können. Dann werden sie abgedreht. Man kann die oberen Stempel nicht in derselben Weise abdrehen, da sie nicht durch die Matrize treten. Die oberen Stempel haben nun alle die gleiche Länge, ebenso die unteren Stempel. Schließlich wird ein Kopffräser entsprechend der Ober- und Unterfläche der Pastillen gemacht, mit dem die Enden der Stempel ausgefräst werden, wonach sie poliert, gehärtet, angelassen und wieder poliert werden.

Die Herstellung eines solchen Gesenkes dauerte 19 Tage.

Pressen der Farben. Dauernd findet man neue Gebiete für die Anwendung der Kraftpressen, die ursprünglich nur für die Bearbeitung von Blech entworfen worden sind und jetzt in einer großen Reihe anderer Fabrikationen mit Vorteil Verwendung finden. Eine der letzten ist die Herstellung der Wasserfarben in den bekannten Formen. Diese Pressen arbeiten vollständig selbsttätig. Vor der Beschreibung der Pressen soll die Vorbereitung des dabei verarbeiteten Materiales besprochen werden.

Nachdem die Farbe gründlich gemischt worden ist und die Steifheit von Glaserkitt angenommen hat, wird sie aus dem Trog genommen und wie Teig bis auf ungefähr 7 mm Dicke ausgewalzt. Dann werden daraus von Hand lange Streifen von ungefähr 38 mm Breite mittels eines Rollenschneiders geschnitten, dessen dünne, runde Stahlscheiben in der angegebenen Entfernung stehen. Die folgende Arbeit ist gleichartig, nur haben die Schneidrollen jetzt eine Entfernung von 20 mm. Dieser Schnitt wird quer über die Streifen geführt und zerschneidet diese in kleine Fleckchen, 38 × 20 mm, die man dann eine Zeitlang trocknen läßt.

Nun tritt die Kraftpresse in Tätigkeit. Die halbharten Stückchen werden auf die Teilscheibe über die Taschen gelegt, die regelmäßig am Umfange der Scheibe verteilt sind. Da diese Stückchen meistens eine etwas unregelmäßige Form haben, muß man sie in die Taschen der Scheibe eindrücken; dies geschieht vermittels eines der drei Stempel, welcher zur äußersten Linken des Hauptstempels steht. Die Matrize ist unterhalb der Teilscheibe, genau unter dem Stempel in der Mitte des Stößels, angebracht, so daß er die Stückchen beim Durchgang durch die Taschen in die Formplatte stößt und genügend Druck auf sie ausübt, um die Feuchtigkeit zu entfernen, das Material zu verdichten und jedem Stückchen eine gleiche Form mit passender Prägung zu geben.

Die soweit fertigen Farben neigen zum Haften an der Matrize, die deshalb mit einem abnehmbaren Boden versehen ist. Mit Hilfe eines durch eine Nocke betätigten Aufwerfers, der unterhalb der Teilplatte und ihres Antriebes angebracht ist, wird der Boden der Formplatte beim Aufgang des Stößels gehoben und die gebildete Farbe gezwungen, wieder in die Taschen einzutreten. Der dritte Stempel, der zur Rechten des Stößels steht, befindet sich an einem Arm beträchtlich vor der Mitte des Schlittens und drückt die fertige Farbe aus den Taschen in einen unterhalb befindlichen Behälter aus. Die Teilvorrichtung arbeitet in bekannter Weise. Derartige Pressen machen bei 10stündiger Arbeitszeit in einem Tag 35000 Farben und wiegen gebrauchsfertig ungefähr 360 kg.

Das Ehrhardtsche Pressverfahren[1]). Das **Ehrhardt**sche Verfahren zur Herstellung nahtloser Hohlkörper aus Metall wird auf zwei verschiedene Arten ausgeübt. Die eine umfaßt die Herstellung von im Verhältnis zu ihrem Durchmesser dickwandigen Hohlkörpern, die andere, die nicht durch das deutsche Patent geschützt ist, stellt die dünnwandigen Hohlkörper her.

Die dem Hauptverfahren zugrunde liegende Idee ist, wie **Sobbe** an der angegebenen Stelle sagt, das Bestreben, eines in der Periode des „intensiven Fließens“ befindlichen Metallblockes beim Eintreiben eines

[1]) W. T. 1908, Heft 8, 9, 10, **Sobbe**, Technologie des Schmiedepressens.

Dornes in das bildsame Material eine Form anzunehmen, deren Umfläche bei möglichst geringer Fläche den größten Inhalt einschließt. Wenn dergestalt in ein Prisma P mit scharfen Kanten ein Dorn D, Fig. 277, eingetrieben wird, immer vorausgesetzt, daß sich das Material in dem bildsamen Zustand und unter den Bedingungen befindet, die den Zustand des intensiven Fließens hervorrufen, so wird die Umfläche sich einem Kreiszylinder nähern, an dem die scharfen Kanten noch erkennbar sind. Werden dagegen die Kanten abgerundet, so wird das Prisma vollständig in die Kreiszylinderform übergehen, wenn man noch Vorsorge trifft, daß eine Matrize vorhanden ist, welche als Hohlzylinder gerade die Kanten des ursprünglichen Prismas berührt. Dabei wird vorausgesetzt, daß die Fläche des Dornes D, gleich der Summe der vier Kreisabschnitte a, Fig. 277, ist. In diesem Fall wird an der Höhe des Prismas nach der Pressung nichts geändert, es fließt das Material, den Linien des geringsten Widerstandes folgend, in die freien Kreisabschnitte. Daraus ergibt sich, daß dieses Verfahren ohne Abänderungen auf bestimmte Dickenverhältnisse der Wandungen zum Durchmesser beschränkt ist.

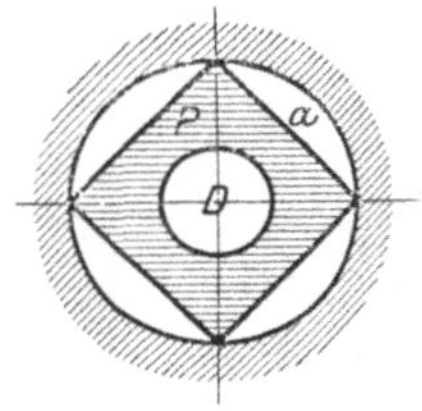

Fig. 277.

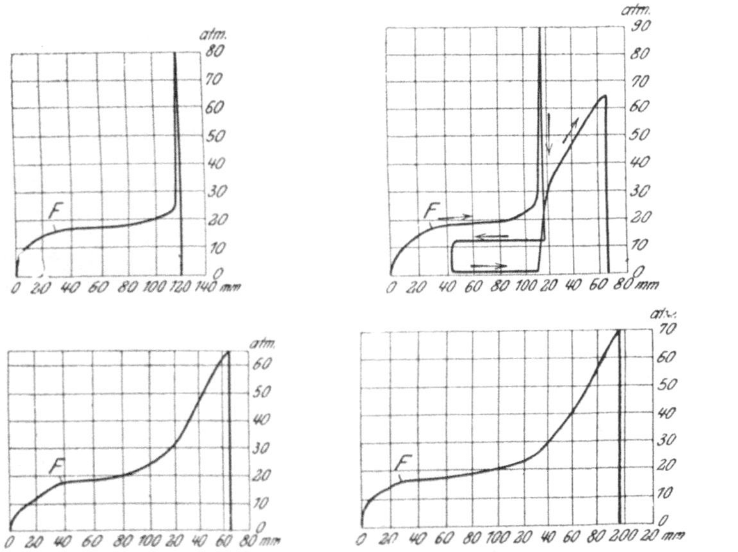

Fig. 278.

Sobbe hat an Versuchen — vgl. die angegebene Quelle — gezeigt, daß beim Zusammendrücken eines Eisenblockes von genügender Hitze der Druck bei verhältnismäßig geringer Zusammendrückung des Körpers schnell auf eine Höhe ansteigt, bei welcher ein starkes Fließen des Metalles bei nahezu gleichbleibendem Druck einsetzt — Punkt F im

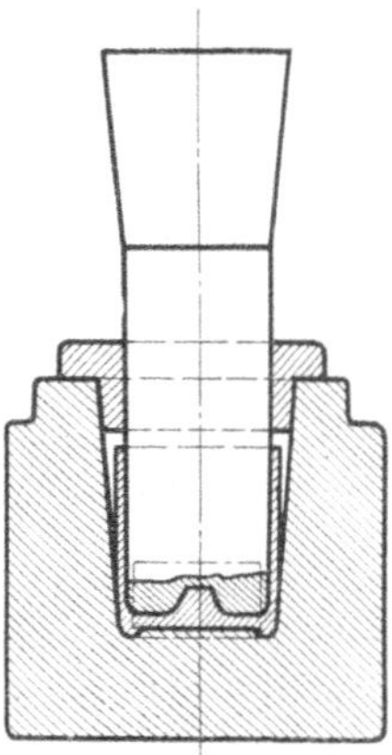

Fig. 279.

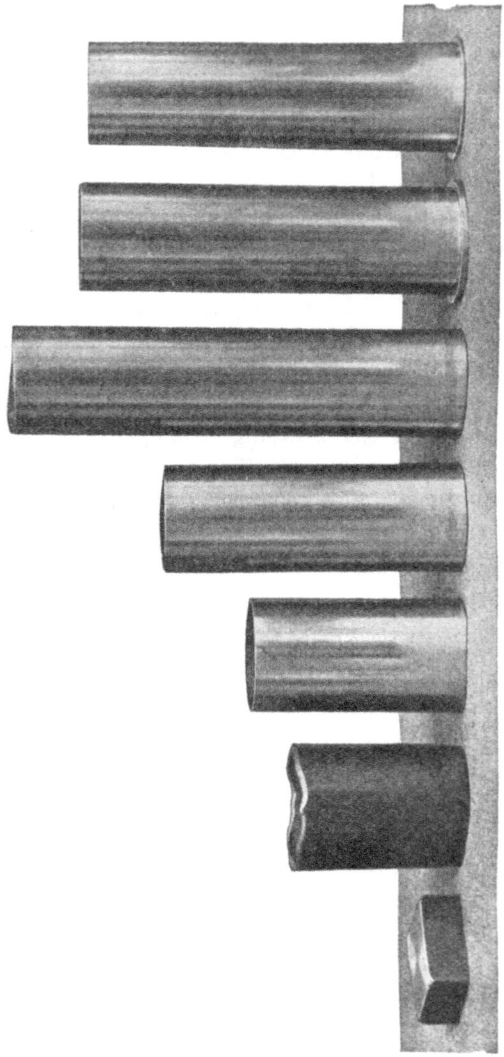

Fig. 280.

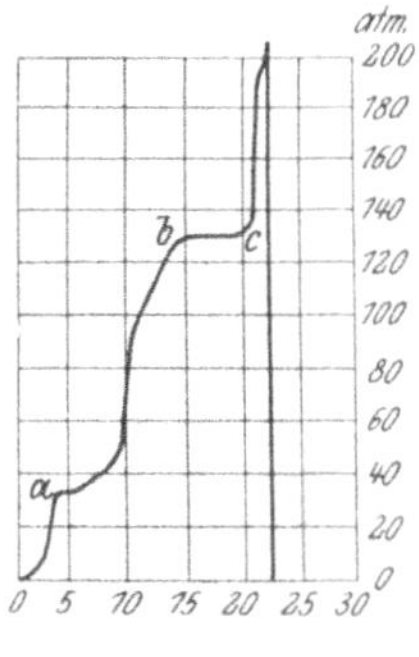

Fig. 281.

Schaubild, Fig. 278, der erst gegen das Ende der Zusammendrückung wieder schneller ansteigt. Diese Periode nannte Sobbe das „intensive Fließen". Weiter zeigte er, daß ein gedrückter Körper unter diesen Bedingungen, sei es nun ein Zylinder oder ein Prisma, im Querschnitt senkrecht zur Druckrichtung kreisförmige Begrenzungslinien annimmt, sich also das Material nach den Linien des größten Inhaltes, bei kleinster Umfläche einstellt.

Wenn dagegen Hohlkörper mit dünnen Wandungen, wie Patronenhülsen usw., hergestellt werden sollen, so tritt die Wirkung der Matrize in vollem Umfang ein. Das verwendete Gesenk ist in Fig. 279 und die erhaltenen Arbeitsgänge in Fig. 280 dargestellt. Der Ausgangspunkt, das Blankett, wird von einem Metallprisma, Fig. 280, gebildet, das beim Aufsetzen des Stempels zuerst die Vertiefungen des Stempels und der Matrize ausfüllt und beim Weitergang des Stempels gezwungen wird, zuerst die runde Form der Matrize auszufüllen, während der folgende Druck das Material zwischen Matrize und Stempel in die Höhe schiebt, so daß das Material eine der Stempelbewegung gerade entgegengesetzte Bewegung machen muß. Bei dieser letzten Arbeit tritt eine bedeutende Drucksteigerung ein, wie das Diagramm, Fig. 281, der betreffenden hydraulischen Presse zeigt. In Fig. 281 ist bei a die Stelle, wo das Material in die Vertiefungen von Stempel und Matrize gepreßt wird. Dieser Druck bleibt während der nachfolgenden Periode des intensiven Fließens, bis das Blankett die runde Matrize ausgefüllt hat, gleich, um erst wieder, und zwar sehr beträchtlich anzusteigen, wenn die Aufwärtsbewegung des Materials zwischen Stempel und Matrize begonnen hat. Dieses tritt bei dem Punkte b ein und dauert so lange, bis das Material zu der verlangten Höhe angestiegen ist, wobei der Druck, da das Material jetzt gleichmäßig fließt; ungeändert bleibt, bis zu dem Punkte c, bei welchem der Stempel auf die Aufsatzstücke aufsetzt, so daß der Druck seinen Größtwert erreicht.

Das in diesem Fall verwendete Material ist eine Legierung von 72% Cu und 28% Zn, die als in der Hitze zwar nicht schmiedbar angesehen wird. Dadurch aber, daß das Material bis in die Periode des intensiven Fließens gebracht wird, ist eine fehlerfreie Umformung in der Hitze möglich.

c) Das Prägen.

Das Prägen ist eine dem Gesenkpressen ganz ähnliche Arbeit, unterscheidet sich nur durch die geringere Tiefe des Eindruckes, der erhaben oder vertieft sein kann. In den meisten Fällen handelt es sich dabei um dünnes Material, bei welchem durch die Prägung die Form aus der ebenen, bzw. glatten Oberfläche herausgedrückt wird. In dem Fall zeigt die Unterseite des Bleches das Negativ der geprägten Form. Das Charakteristische ist jedenfalls, daß eine Materialverschiebung bei dem Prägen stattfindet, nur nicht in so weitgehendem Maße als beim Schmieden oder Pressen. Auch wird die Arbeit gewöhnlich im kalten Zustand des Materiales gemacht.

Andrerseits wird oft die Grenze zwischen diesen Arbeitsvorgängen schwer zu ziehen sein, ebenso wie sich manche Stücke leicht als gebogene Arbeiten werden bezeichnen lassen. Während aber bei einer Biegung die Materialbewegung keine Verschiebung der kleinsten Teilchen gegeneinander bedingt und sich auf eine gesetzmäßige Entfernung der Teilchen an der gezogenen Faser und Näherung derselben an der gedrückten Faser, beide Male längs der Biegekante, beschränkt, tritt beim Prägen eine Materialverschiebung nach der Form der Matrize auf, die außerdem die Teilchen auch aus der ursprünglichen Materialebene entfernt.

Eine Prägearbeit, die eben noch einen Übergang vom Pressen darstellt, sei den anderen Prägewerkzeugen vorangestellt.

Prägewerkzeug für Glasklammern an Augengläserfassungen aus Gold. Auf S. 47 und in Fig. 282 ist der Herstellungsgang für diesen Teil gegeben. Hierbei wird das Prägewerkzeug zuerst fertiggestellt.

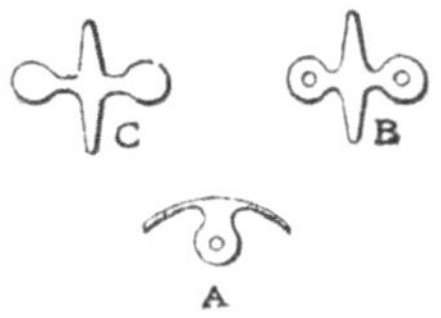

Fig. 282.

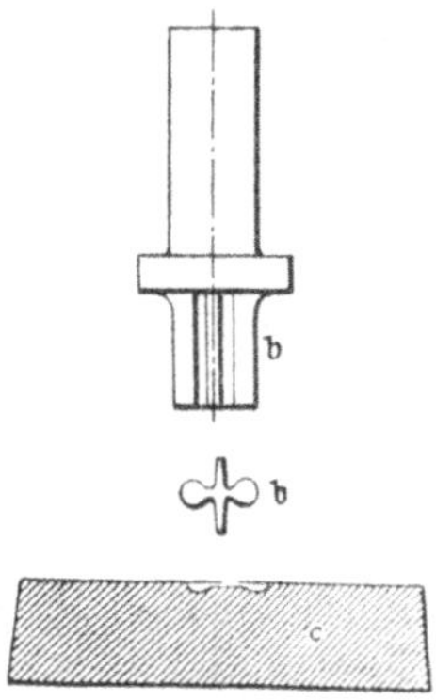

Fig. 283.

Von einem Streifen weichen Neusilbers werden zwei Musterstücke gefeilt und eines derselben vollständig fertiggestellt, wie A, Fig. 282, zeigt, und an eine Linse angeschraubt, während das andere, B, an die Stirnfläche eines Stempels angelötet wird. Dieser Stempel b, Fig. 283, wird als Schlagstempel für die Prägematrize verwendet. Er wird auf der Fräsmaschine vorgearbeitet und nachher auf die genauen Abmessungen des Musterstückes, das an seiner Stirnfläche angelötet ist, gefeilt. Bei der Herstellung eines Schlagstempels dieser Art muß man ihn so kurz wie möglich machen, damit er sich nicht biegt, wenn man die Matrize schlägt. Alle Stempel, die hier zur Verwendung kommen, werden aus Achtkantstahl von 25,4 mm Durchmesser gedreht, und zwar immer zwei Stempel zusammen. Dann wird die Matrize ganz lose in die Froschplatte passend gehobelt. Beim Schlagen der Matrize geht man folgendermaßen vor: Die Matrize wird kalt in die Froschplatte eingesetzt und der Stempel im Stößel befestigt. Nun befestigt man die Froschplatte ebenfalls in der Presse und läßt den Stößel von Hand so weit herabgehen, bis der Schlagstempel ein wenig über der Stelle in der Matrize steht, wo die Form geschlagen werden soll. Nun wird die Matrize herausgenommen und der Stößel so weit heruntergeführt, daß er genügend tief in die Matrize eindringen kann, was durch Seitwärts-Heranschieben dieser bestimmt wird. Dann wird der Stößel um seinen vollen Hub angehoben, die Matrize herausgenommen und erwärmt. Es ist von Vorteil, eine zweite Matrize neben der zu schlagenden zu befestigen, um einen Anschlag für das erwärmte Stück zu haben. Selbstverständlich kann man keine Zeit verlieren, um das erwärmte Stück erst auszurichten. Ein Kreidestrich oder ähnliche Marken genügen kaum, da es manchmal notwendig ist, den Schlag zu wiederholen, wobei sich die Matrize, wenn nicht ordentlich eingespannt, sehr leicht verschiebt. Es ist auch nicht anzuraten, das warme Stück mittels Preßschrauben in der Froschplatte zu befestigen, da man dann immer unschöne Marken an der Matrize erhält, und die Schrauben auch meistens nicht passen. Ist das Stück nun warm genug, so wird es mit Zangen aufgenommen, schnell zur Presse gebracht und in die Froschplatte eingelegt. Das Schwungrad der Presse wird von Hand gedreht und der Stößel nach abwärts bewegt. Sollte der Eindruck nicht tief genug sein, so kann die Einstellmutter schnell nachgestellt werden, so daß man beim zweiten Schlag das gewünschte Ergebnis erhält.

Sobald die Matrize genügend abgekühlt ist, wird sie in die Shapingmaschine eingespannt und einige leichte Späne genommen. Dies ist nötig, um die Kanten der Vertiefung, die durch das Schlagen eingezogen worden sind, in eine Flucht mit der übrigen Matrizenfläche zu bringen. Um dann in den runden Teilen größere Übereinstimmung zu erhalten, macht man aus Bohrerstahl zwei Endfräser, feilt die Enden auf die verlangte Form und mit einer Dreikantfeile die Zähne an. Man erhält so auch den genauen Durchmesser der runden Teile, worauf die Vertiefung mit dem Schaber fertiggestellt und hochglanzpoliert wird Dann wird der Prägestempel gemacht und glashart gehärtet. Meistens würde man nun auch die Matrize härten, doch wird hier ein anderer Weg eingeschlagen.

Nun soll das mit Gewinde versehene runde Ende des Stückes schwerer sein als das andere, obwohl beide den gleichen Durchmesser haben. Dies bedingt, daß eine der runden Seiten im Blankets tiefer sein muß als die andere, so daß man, um die verschiedenen Abmessungen zu erhalten, ein wenig probieren muß. Um die Prägematrize zu prüfen, benutzt man einen dünnen Streifen Blei, bis man ein tadelloses Stück erhält. Dann wird die Matrize so hoch wie nur möglich gehärtet. Wenn man ein wenig Schmierseife vor dem Anwärmen in die Vertiefung einreibt, so kommt das Stück aus dem Härtebad sehr hart und blank an dieser Stelle.

Das Prägen von kleinen Medaillons und das Schlagen von Medaillen. Die kleinen Medaillen, die als Heiligenbilder Verwendung finden, werden aus Aluminium, Messing, Kupfer, Gold oder Silber, wie eben verlangt, gemacht und fallen, soweit ihre Herstellung in Frage kommt, in die Gruppe „Münzen", da gleiche Maschinen und Werkzeuge wie für diese verwendet werden.

Die gewöhnliche Form, wie sie im Abfallstreifen, Fig. 284, ersichtlich ist, wird meistens aus Aluminium, manchmal auch aus Silber und Gold gemacht. Das Aluminium kommt in Bändern von 61 m Länge herein, die gerade breit genug sind, um zwei Reihen Blanketts zu stanzen und gleichzeitig zu lochen.

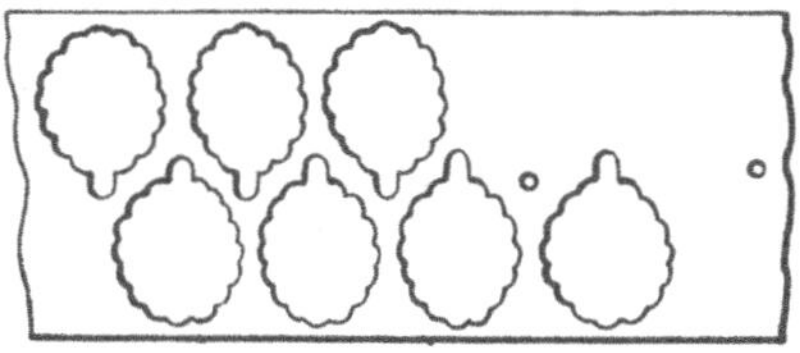

Fig. 284.

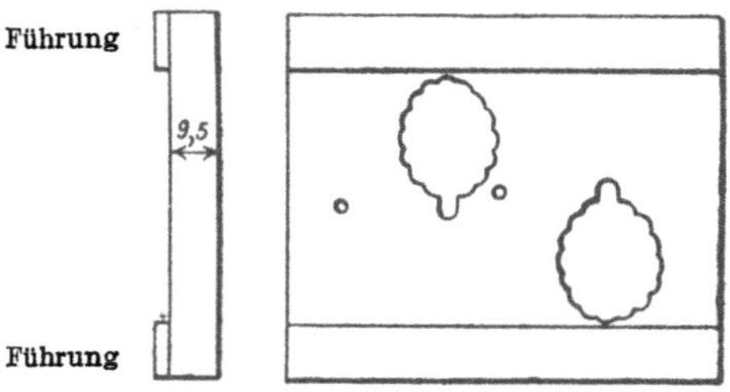

Fig. 285.

Die Schnittplatte wird nach Fig. 285 hergestellt und meistens ungefähr 10 mm dick gehalten, um eine gleichmäßige Härtung zu erzielen. Sie erhält kaum irgendwelche Abschrägung und wird an der Gesenkplatte mit Schrauben und Prisonstiften befestigt. Die Stempel werden, mit Ausnahme der kleineren Lochstempel, weich gelassen und in der gewöhnlichen Weise in der Kopfplatte befestigt, die nachher mittels Schrauben und Prisonstiften an dem Stempelkopf befestigt wird. Der gewöhnliche Federabstreifer, der am Stempel befestigt ist und die Schnittplatte vollständig frei läßt, wird hier verwendet, so daß am Unterteil nur Führungen für die verarbeiteten Bänder nötig sind.

Für diese Arbeit wurde eine Presse Nr. 19 von Bliß verwendet mit einem Sperrad-Walzenapparat, der von der Kurbelwelle betätigt wird und leicht für die verschiedenen Größen einzustellen ist. Auf dieser Presse wurden ungefähr 175 000 Stücke im Tag ausgeschnitten. Die Blanketts werden dann, um sie von Öl und Schmutz zu reinigen, gewaschen, worauf sie in die Prägeabteilung, die mehrere Pressen enthält, kommen.

Anfangs zeigten sich ziemlich viel Schwierigkeiten durch die Abnutzung und ganz besonders durch das Nachsinken der Gesenkplatte infolge der fortwährenden Zusammendrückung unterhalb der Prägung. Nach vielen Versuchen wurde die Froschplatte aus Stahlguß gemacht und zur Aufnahme eines gehärteten und geschliffenen Druckstückes aus Stahl mit fünf Absätzen, wie Fig. 291 zeigt, ausgebohrt, was sich schließlich als zufriedenstellend erwies. Die Gesenke, und zwar zwei für jedes Medaillon, werden aus dem besten, gehämmerten Werkzeugstahl gemacht und auf 41,3 mm Durchmesser bei 44,5 mm ganzer Länge in folgender Weise hergestellt: Die Stange wird in Stücke von 47,6 mm Länge geschnitten und an der Grundfläche vollständig eben und rechtwinklig abgedreht, während die andere Endfläche in der Mitte leicht erhaben gehalten und dann hochpoliert wird, worauf die Prägung in das Gesenkblankett geschlagen werden kann.

Der Schlagstempel wird sehr flach von Hand graviert, sorgfältig gehärtet und danach auf volle Tiefe in die Matrize geschlagen. Dafür wurde anfangs ein Fallhammer verwendet, doch ohne ausreichenden Erfolg, da es unmöglich ist, den Grund des Eindruckes bei mehreren Schlägen vollständig flach zu erhalten. Schließlich wurde diese Arbeit in der Prägepresse auf kaltem Wege und immer mit vollem Erfolg ausgeführt.

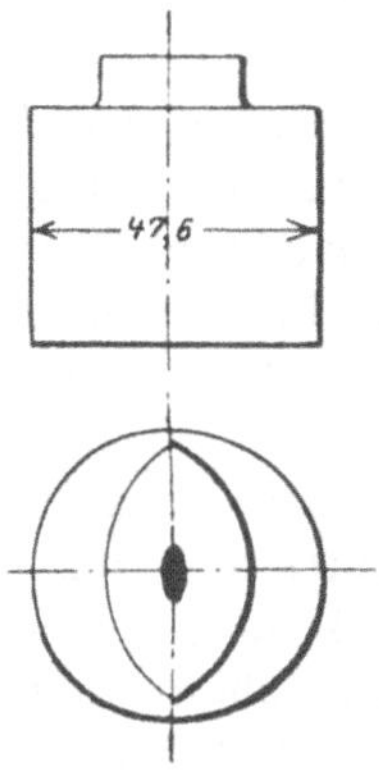

Fig. 286.

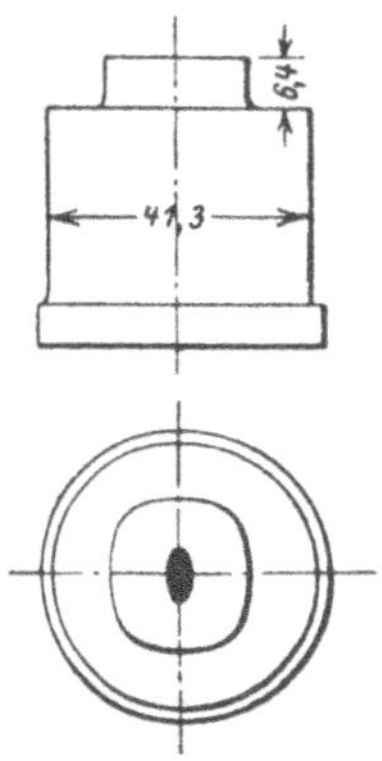

Fig. 287.

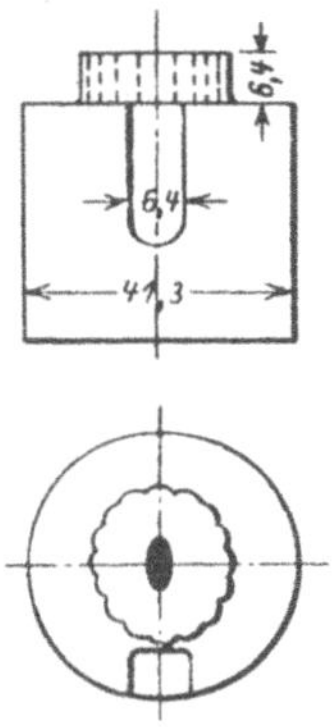

Fig. 288.

Das in dieser Weise mit der Prägung versehene Gesenk wird sehr sorgfältig in einer besonderen Spannvorrichtung für sämtliche ovalen Formen eingespannt und erhält zu beiden Seiten der Stirnfläche nach Fig. 286 einen Absatz von 6,4 mm Höhe angedreht, so daß die Prägung, angedeutet durch die kleine schwarze Ellipse, genau in der Mitte steht. Hiernach wird die Matrize wieder in einem gewöhnlichen Backenfutter eingespannt und die Stirnfläche oder der Grund der Prägung genau in Mitte und winkelrecht ausgerichtet, ohne Rücksicht auf die Umfläche der Matrize. Dann wird die Stirnfläche abgedreht, bis die Prägung die richtige Tiefe zeigt, die Umfläche sehr sorgfältig auf Maß gedreht und die beiden Scheitel des Ovals an der Stirnfläche bis an den Absatz herunter abgedreht, so daß eine Art Zapfen, ähnlich wie Fig. 287, übrig bleibt. Nun ist das Gesenk im allgemeinen für das Einschneiden des Wellenumrisses fertig. Diese Wellen werden vollständig gerade bis an den Absatz ausgearbeitet, so daß am Absatz selbst eine kleine Abrundung, die nach einer Lehre gemacht wird, Fig. 288, verbleibt, und die Kanten der Stirnfläche leicht abgerundet, so daß das fertige Medaillon eine erhöhte Kante erhält. Das Ende der Matrize, welches früher im Futter eingespannt war, wird sodann in einem besonderen Futter genau auf Durchmesser und Länge abgedreht.

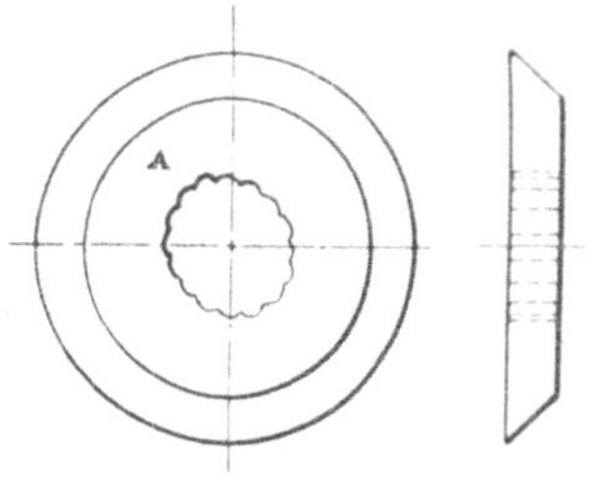

Fig. 289.

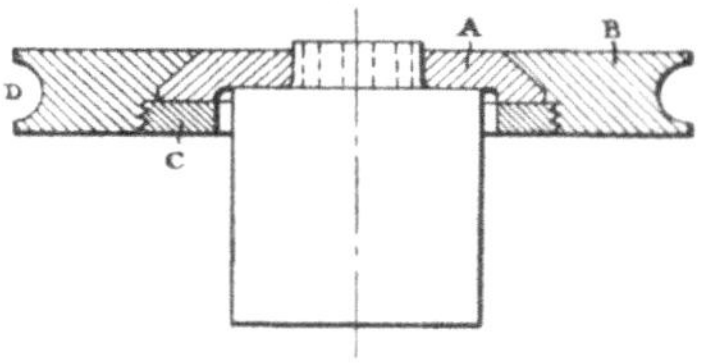

Fig. 290.

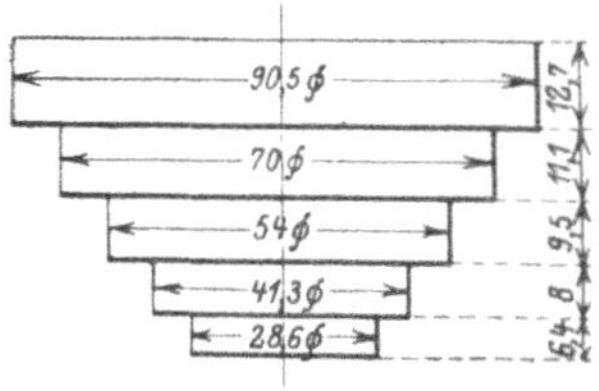

Fig. 291.

Dieser Arbeit folgt das Einfräsen einer Nut in die Umfläche der Matrize, die in der Mittellinie der Matrize und der eingeprägten Figur liegt, und in welche ein Stück zum Prägen des Auges eingepaßt, siehe Fig. 288, und mittels Schraube und Prisonstift befestigt wird. Dies ist keine besondere Arbeit.

Es mag auf den ersten Blick eigentümlich erscheinen, daß das Auge nicht in einem Stück mit der übrigen Matrize hergestellt wird. Tatsächlich wurde es beim ersten Versuch so gemacht, aber ohne Erfolg, da sich dieses Stück beim Härten vom übrigen Matrizenkörper absplitterte, so daß man diesen zur Ausbesserung ausglühen mußte.

Die Löcher für die Schraube und für die Prisonstifte des Auges werden sorgfältig mit Lehm gefüllt, dann härtet man die Matrize und läßt sie langsam abkühlen. Sie wird dann mit feinstem Schmirgel und einem Kork an allen erhöhten Teilen hochpoliert, ihre Grundfläche eben und parallel zur Stirnfläche geschliffen, wonach sie in die Presse eingesetzt werden kann.

Die Matrizen werden mittels besonderer Klammern in den Gesenkplatten gehalten, wodurch die Grundfläche einer jeden Matrize vollständig flach gegen das gehärtete Stahldruckstück niedergepreßt wird, vgl. Fig. 634.

Eine gehärtete Ringlehre A, Fig. 289, die verhindert, daß das Material über einen bestimmten Betrag hinaus ausgedehnt wird, wird in einem zweiten Stahlring B befestigt. In diesem, der ebenfalls gehärtet und geschliffen ist, wird Ring A mittels einer Ringmutter C, die ungehärtet bleibt und zwei Löcher zum Anziehen hat, befestigt. Die Platte B erhält an gegenüberliegenden Seiten zwei zylindrische Nuten D, in welche die Mitnehmer der Klammern — nicht gezeichnet[1]) — eingesetzt werden, die, nachdem das Blankett geprägt worden ist und die Presse sich wieder auf dem Rückwärtsgang befindet, die Ringlehre bis auf den Absatz und ein wenig unter die Oberfläche der Matrize niederdrücken, wie Fig. 290 zeigt, so daß das geprägte Blankett ausgeworfen werden kann. In diesem Falle wurde mit einer Abblasvorrichtung gearbeitet, die zur geeigneten Zeit ein Ventil öffnete und mittels eines scharfen Luftstromes das Blankett in die Abfallrinne und in den unterhalb befindlichen Kasten blies.

Die Zuführung der Blanketts zum Gesenk wird auf Seite 522 beschrieben und erfolgt mittels eines durch den Pressenhub betätigten Schlittens, so daß der Arbeiter nur den Zuführtrichter mit Blanketts gefüllt zu halten und aufzupassen hat, daß das Auge sich in der richtigen Lage befindet, da der Schlitten am Ende des Rückweges jedesmal nur ein Blankett faßt und vorwärts in die Ringlehre transportiert, die sich ausgenommen beim Abstreifen immer in einer Ebene mit der Schlittenführung befindet. Dabei ist das Loch im Schlitten, wenn die Presse steht, genau über der Ringlehre oder der Matrize, geht beim Beginn der Bewegung der Presse nach rückwärts aus dem Wege und erfaßt ein Blankett, wie oben näher beschrieben worden ist.

Die Herstellung von Löffeln, Gabeln und Zierlöffeln. Die im folgenden beschriebenen Arbeitsweisen und Zeichnungen beziehen sich auf die Herstellung (vgl. Tabelle Nr. 12) der gewöhnlichen Löffel. Die meisten gestreckten und gepreßten Löffel werden aus Neusilber gemacht, da dieses ein für diese Zwecke genügend kräftiges, steifes Material ist.

Das Material wird in Rollen von bestimmter Breite und Dicke angeliefert — die Abmessungen werden erfahrungsgemäß festgelegt —, aus denen (S. 45) die Blanketts ohne Abfall ausgeschnitten werden.

Nach dem Ausschneiden geht das Stück zu den Querstreckwalzen, die das Material für die eigentlichen Streckwalzen vorbereiten. Bei der letzteren Arbeit werden die Blanketts, die erst ungefähr die halbe Länge des fertigen Löffels und eine gleichmäßige Dicke haben, auf die richtige Länge gestreckt. Es wird sowohl das Ende für den Handgriff wie das für die Pfanne breitgewalzt, wobei gewöhnlich jedes Ende für sich bearbeitet wird; manche Löffel sind jedoch so geformt, daß beide Enden in einem Durchgang gewalzt werden können. Hier sei der erste Fall betrachtet. Zuerst wird der Pfannenteil quergewalzt, wobei das Blankett die Form II annimmt, dann im dritten Gang der Handgriff, Form III. Nun sind beide Enden fast auf die richtige Dicke breit-

[1]) Vgl. dazu Fig. 634.

gewalzt, so daß das Stück jetzt in den eigentlichen Streckwalzen bearbeitet werden kann.

Diese Arbeit ist der erste Gang durch die Längsstreckwalzen, nach dem das Material die Form IV angenommen hat. Hierauf wird es zum erstenmal ausgeglüht, dann gebeizt, gereinigt und für die letzten Walzen vorbereitet, die das Blankett auf die notwendige Länge strecken und dem Stück überall die richtige Dicke geben. Man erhält Form V. Nachdem die Blanketts noch einmal ausgeglüht worden sind, sind sie für das Zuschneiden des äußeren Umrisses fertig und kommen wieder zur Presse zurück. Diese Arbeit wird wieder mit Schnittwerkzeugen ausgeführt, vgl. Form VI, die das zugeschnittene Blankett, umgeben von dem Abfall, zeigt.

Das soweit fertige Blankett kommt nun auf die Tuchpolierscheibe, wo es geglättet, von Grat und sonstigen Unebenheiten befreit wird, da der folgende Arbeitsgang, der im Prägen des Handgriffes allein besteht, eine sehr glatte Oberfläche verlangt. Es werden nämlich aus den gestreckten Blanketts, Form V, viele verschiedene Formen und Arten zugeschnitten und auf dem Handgriff die verschiedensten Muster geprägt, so daß man deshalb schon vorzieht, nicht den ganzen Löffel, Griff und Pfanne, in einem Werkzeug zu prägen. Verschiedene Muster, um den Löffel in einem Schlag zu prägen, sind entworfen worden, doch nur bei sehr einfachen Formen. Wenn dagegen der Griff in einem besonderen Werkzeug geprägt wird, so kann man ein einziges Gesenk für die Pfanne mit den verschiedensten Handgriffen verwenden und spart auf diese Weise eine Menge teurer Werkzeuge, ohne die Zahl seiner Muster zu verringern. Der Löffel, der jetzt der Form VII entspricht, wird jetzt an eine ziemlich feste Polierscheibe gebracht und mit ihr die scharfen, vom Prägen herrührenden Grate entfernt. Diese Grate sind bei Neusilber sehr gering, da die Blanketts so genau ausgeschnitten werden, daß sie gerade in die Matrizen passen und die Form ausfüllen, so daß kaum Grat bleibt, jedenfalls nicht genug, um ein zweites Mal in einem Schnittwerkzeug abgraten zu müssen. Ein in dieser Arbeit erfahrener Arbeiter kann das Stück sehr fein zuschneiden und polieren, so daß der Löffel am Ende dieses Arbeitsganges sehr glatt bleibt.

Nach dem Polieren kommt der Löffel an eine kleine Fußpresse, die ein geeignetes Werkzeug zum Biegen der Handgriffenden hat. Damit erhält dieses Ende die mehr für die weitere Prägearbeit an der Pfanne als aus einem anderen Grunde notwendige Form. Dadurch kann nämlich das geprägte Muster am Handgriff immer richtig in das Gesenk zur Herstellung der Pfanne eingelegt werden; ohne diese Hilfe, vgl. Form VIII, würden oft eine ganze Anzahl Stücke auf der falschen Seite ausgeprägt und verdorben werden.

Nun kann zur Herstellung der Pfanne geschritten werden; dies geschieht unter dem Fallhammer, wobei auch ein Teil des Stieles nach Form IX geformt wird. Damit sind die Löffel zum letzten Polieren und Glätten fertig, welches eine Industrie für sich ist und große Erfahrung verlangt, da gewöhnlich an Neusilberlöffeln kein Nachfeilen oder Nacharbeiten zugelassen wird. Nach diesem letzten Polieren kommen die Löffel in ein Prägegesenk, das oft aus hartem Holz ist, manchmal auch mit Kupfer überzogen wird und die genaue Gestalt des fertigen Löffels besitzt. Man verwendet hartes Holz, um zu verhindern, daß die Musterung durch diese Arbeit beschädigt oder vernichtet wird, was bei Stahl- oder Eisengesenken unausbleiblich wäre. Diese Gesenke kommen meistens in einer Fußpresse oder einem leichten Tischfallhammer zur Verwendung. Manchmal werden auch besondere Bankhämmer — für Handbetrieb — dazu gemacht, wobei jeder Hammer seine besondere Form hat. So wird das Werkzeug für den Gebrauch sehr bequem, da man keinerlei Gesenke auszuwechseln hat. Form X zeigt den fertigen Teelöffel. Es bleiben nur noch die verschiedenen Marken, Zeichen usw. einzuprägen, wonach die Löffel versilbert, poliert, kontrolliert und gepackt werden können.

Löffel aus gestrecktem Stahl werden ungefähr in gleicher Weise wie die Neusilberlöffel hergestellt. Es gibt natürlich auch eine ganze Anzahl, die nicht aus gestrecktem Material hergestellt, sondern gleichmäßig dick sind, in welche zur Verstärkung starke Rippen eingepreßt werden. Exzenterpresse und Fallhammer sowie die Streckwalzen sind die wichtigen Maschinen der Löffelherstellung.

Die Herstellung von Neusilbergabeln. Neusilbergabeln werden auf fast die gleiche Weise hergestellt, nur tritt an Stelle des Prägens der Pfanne das Schneiden der Zinken, wie in Fig. 22, S. 46 ersichtlich ist.

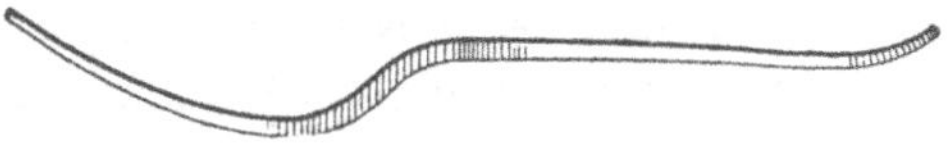

Fig. 292.

Manchmal, wenn es sich um ganz glatte, einfache Muster handelt, werden die Gabeln mittels eines schweren Fallhammers, Handgriff und Zinken in einem Gesenk, mit einem einzigen Schlage fertiggestellt und sehen dann wie Fig. 292 aus.

Die abgebildete Form gibt die Stücke so wieder, wie sie aus den Gesenken, in denen sie gestanzt oder geprägt werden, genommen werden. Die Herstellung von Gabeln und Löffeln aus Silber erfolgt in derselben Weise, nur wird mehr Handarbeit darauf aufgewendet; es werden noch mehrere Präge- und Zuschneidearbeiten eingeschoben und viel von Hand gefeilt.

Die Herstellung von Zierlöffeln. Die größte Zahl der Zierlöffel wird aus Silber gemacht, das in Ingots nach Gewicht gekauft wird. Das Silber wird, da eine große Hitze notwendig ist, gewöhnlich in Gasöfen im Tiegel geschmolzen. Wenn das Silber in den Tiegeln schmilzt, wird eine kleine Handvoll Holzkohle darüber gestreut; dadurch werden alle Verunreinigungen ausgeschieden und an die Oberfläche gebracht. Das geschmolzene Silber wird in eine Form gegossen, die in ihrer Größe verschieden ist, meistens aber ungefähr 76 mm breit, 19 mm dick und 305 mm lang ist. Die Gußform wird poliert und muß frei von allen Luftblasen und kleinen Löchern sein, die gewöhnlich im Gußeisen vorkommen. Die Form erhält immer eine leichte Abschrägung und einen keilförmigen Kern aus Stahl, der im Boden eingelegt wird. Wenn man den Ingot aus der Form entfernen will, treibt man diesen Keil einfach hinein. Gewöhnlich wird die Form erwärmt und mit etwas öliger Putzwolle ausgewischt, um ein Ankleben des Silbers zu verhindern. Nachdem sich das Silber in der Form bis zu einem bestimmten Grad abgekühlt hat, wird der Ingot entfernt und, falls er frei von Luftblasen und Löchern ist, mit heißem Wasser, Seife und einer harten Bürste abgerieben, um so alle anhaftenden Fremdkörper zu entfernen.

Der gewaschene Ingot kommt in ein Walzwerk mit Walzen aus Werkzeugstahl, die geschliffen, gehärtet und poliert sind. Die Walzen haben rd. 150 mm Durchmesser und sind rd. 250 mm lang, doch bestimmt sich ihre Größe meistens nach dem zu walzenden Ingot. Das Silber geht mehrere Male durch diese Walzen, wird dabei immer dünner, breiter und länger, bis schließlich die Kanten einzureißen beginnen. Dies ist das Zeichen für den Arbeiter, daß das Material vor weiterem Walzen erst ausgeglüht werden muß. Der Ingot ist jetzt ungefähr 9,5 mm dick, 127 mm breit und ungefähr 508 mm lang. Zum Ausglühen faßt der Arbeiter den Ingot an einem Ende mit der Zange und zieht ihn durch den Gasofen hin und her, bis er rotglühend geworden ist; dann läßt er ihn abkühlen. Nachher wird der Ingot in ein Säurebad getaucht, wodurch alles Oxyd und aller Zunder entfernt wird. Aus dem Säurebad kommt das Silber in kochendes Wasser und wird hier abgebürstet, so daß alle Spuren der Säure entfernt werden, damit die Walzen nicht durch die Säure angegriffen werden. Nun trocknet man das Silber in heißen Sägespänen, wonach es weiter ausgewalzt und zwischendurch, sooft wie nötig, ausgeglüht wird, was man jedesmal am Einreißen der Kanten erkennt.

Beim Auswalzen muß man die beiden Walzen äußerst sorgfältig parallel stellen; sonst beginnt das Material zu rollen und walzt sich nicht gerade aus, wobei es sich auf der einen Seite mehr streckt als auf der anderen und bricht. Fig. 293 zeigt die verwendeten Walzen, die von Zeit zu Zeit geschliffen und poliert werden müssen, da sie sich ungleichmäßig abnutzen und dann das Silber ungleichmäßig walzen. Auch zeigen sich manchmal an der Oberfläche der Walzen Unregelmäßigkeiten, die man entfernen muß, da sie sonst immer am Silber sichtbar werden.

Nachdem das Silber auf die verlangte Dicke ausgewalzt worden ist, wird es zum letzten Male geglüht, gescheuert und getrocknet, worauf es auf die Fertigwalzen, die grundsätzlich denen in Fig. 293 ähnlich, aber kleiner im Durchmesser, genauer geschliffen und besser poliert sind,

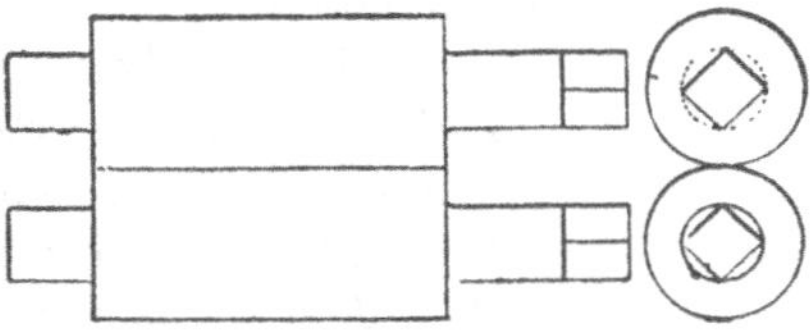

Fig. 293.

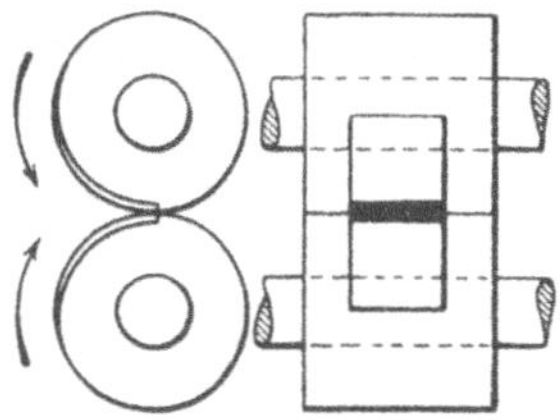

Fig. 294.

kommt. Auf diesem Walzwerk wird keine starke Streckung vorgenommen, sondern bloß eine gleichmäßige Materialdicke angestrebt. Nach dem Fertigwalzen werden die gewalzten Silberbleche, deren Dicke ungefähr von 0,64 mm bis 0,9 mm wechselt, einem Paar Kreisscheren zugeführt, auf denen die unregelmäßigen Kanten abgeschnitten und das Silber auf die richtige Breite zugeschnitten wird. Die Rollen Silberblech sowie die Abschnitte von den Scheren und der Schmelzrest aus dem Tiegel werden dann im Büro nachgewogen und so die Verluste bestimmt. Dann ist das Silber für die weitere Verarbeitung zu Löffeln bereit.

Die Rollen werden nun zu den Scheren gebracht und in die der Löffelgröße entsprechenden Längen zerschnitten. Hierauf werden die Enden für den Griff und die Pfanne keilförmig gestreckt. Die Silberbleche werden immer kürzer als der fertige Löffel abgeschnitten, da sie sich beim Strecken längen. Dieses Strecken geschieht in dem in Fig. 294 abgebildeten Streckwalzwerk. Die Walzen sind in der Mitte auf der einen Seite mit einer ungefähr 3,2 mm tiefen Nut versehen, deren Tiefe sich allmählich vermindert, bis sie in die Oberfläche der Walze übergeht. Dieser Teil der Walzen ist meistens 50 oder 75 mm breit, und seine Länge beträgt ungefähr ein Viertel des Umfanges. Die Walzen drehen sich immer in der Richtung der Keilnut, so daß an einem bestimmten Punkte die beiden tiefsten Stellen der Nut zusammentreffen und so eine gemeinsame Öffnung von 6,4 mm geben. An dieser Stelle wird das Stück zum Verjüngen und Strecken eingelegt. Da der Griff und die

Pfanne verschiedene Abschrägungen erhalten, befinden sich in jeder Walze zwei Nuten. Man faßt die Silberstücke in einer Zange, deren Maulbreite angibt, wie weit das Silberblech zwischen die Walzen gesteckt werden soll. Wenn die Walzen so stehen, daß die Tiefe der Nut am größten ist, stößt der Arbeiter das in der Zange richtig gefaßte Blankett so weit in die Öffnung, bis die Zange mit den Walzen in Berührung kommt, und hält es hier so lange, bis die Walzen das Stück fassen und den Keil ausstrecken. Die Walzen sind in Ständern, die auf jede Blechdicke einstellbar sind, gelagert.

Nach dem Ausschneiden des Griffes und der Pfanne (S. 45) wird zum Einprägen des Musters auf den Handgriffen geschritten.

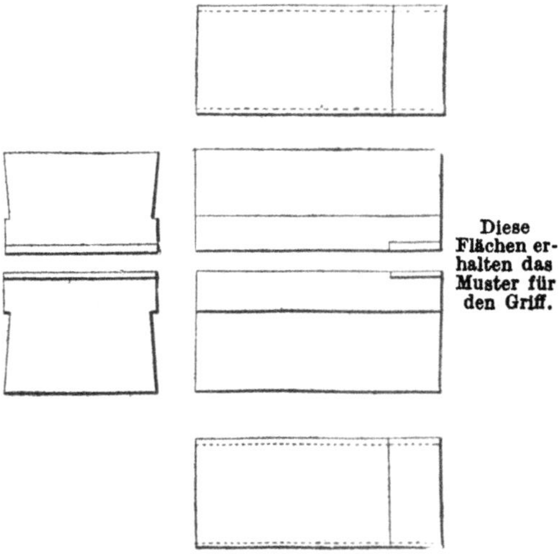

Fig. 295.

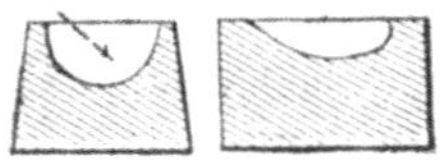

Fig. 296.

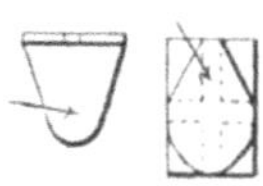

Fig. 296 a.

Diese Arbeit geschieht in einem Fallhammer zwischen den in Fig. 295 abgebildeten Gesenken, in welche das Muster, das sich auf der Ober- und Unterseite des Handgriffes befinden soll, eingearbeitet ist. Es sind dies zusammengepaßte Gesenke. Die Tiefe der Prägung bestimmt immer die Dicke des Silbers, je tiefer das Muster, desto dicker das Material. Dann kommen die Blanketts in ein Abgratwerkzeug, in dem die Umrisse des Löffels sauber zugeschnitten werden. Das Werkzeug ist dem in Fig. 23/4 ähnlich, nur ist der Stempel nicht flach, sondern in der Mitte ausgeschnitten, so daß am Rande eine Schnittkante bleibt. Infolgedessen läßt der Stempel die Prägung unversehrt und schneidet gleichzeitig eine saubere Kante ab. Nach dem Zuschneiden des Umrisses wird die Pfanne hohl geprägt.

Dies geschieht im Fallhammer zwischen dem Gesenk Fig. 296 und dem Prägestempel Fig. 296 a. Das Gesenk wird zwischen den Führungen am Fallhammerbett befestigt und der Prägestempel am Hammer angebracht. Der Arbeiter legt nun ein Blankett auf die Oberfläche des Gesenkes, läßt den Hammer fallen, der mit dem Formstempel die Pfanne ausprägt. Ein Fallhammer mit selbsttätiger Auslösung ist das Beste für diese Arbeit, da dann jeder Schlag mit gleicher Stärke geführt wird. Soll nun noch der Name einer Stadt oder sonst ein Muster in die Pfanne eingeprägt werden, so wird dieses, wie Fig. 296 andeutet, in den Formstempel eingraviert. Dadurch werden zwei Arbeitsgänge vereinigt, da das Muster zugleich mit der Pfanne geprägt wird.

Die Löffel kommen nun an die Werkbank, wo sie sauber gefeilt werden. Dabei werden alle Kratzer und Grate von der Pfanne entfernt, worauf die Löffel wieder zur Presse zurückkommen. Dort werden die Handgriffe zwischen zwei Blöcken aus hartem Holz auf die verlangte Form gebogen. Von da wandern die Löffel in den Polierraum; hier wird auf einer gewöhnlichen Tuchpolierscheibe mit Polierkalk, der von Zeit zu Zeit erneuert werden muß, gearbeitet. Dadurch werden wieder alle scharfen Kratzer und Kanten vom Silber entfernt, worauf noch ein zweites Polieren auf einer Tuchscheibe mit Safran folgt. Dann kommen die Löffel in den „Färberaum", wo sie galvanisch vergoldet, andere oxydiert werden; andere wieder erhalten eine Mattierung durch ein Sandstrahlgebläse. Die vergoldeten Löffel gehen in den Polierraum zurück, werden dort noch einmal poliert und dann in heißem Salmiakwasser gewaschen, wodurch alle Flecken entfernt werden, und schließlich in heißen Sägespänen getrocknet, wonach sie zum Versand fertig sind.

Die Fabrikation dieser Zierlöffel ist sehr groß, es hat z. B. eine amerikanische Firma über 1000 Muster für die Pfannen, unter denen fast jede größere Stadt der Vereinigten Staaten und Kanadas sowie vieler anderer Länder vertreten ist.

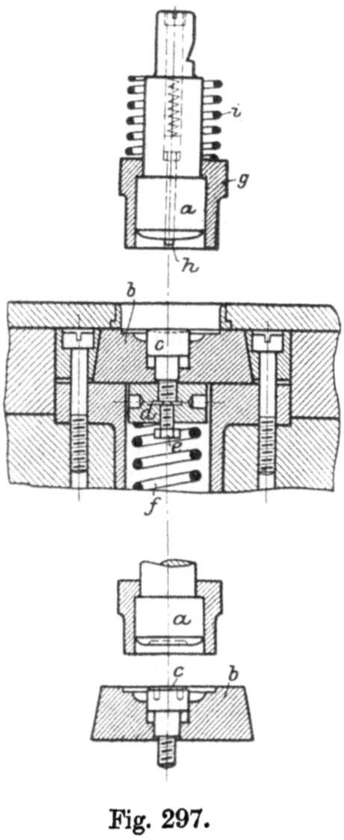

Fig. 297.

Einfaches Prägewerkzeug deutscher Bauart. Bei der Herstellung der Gehäuse für Taschenuhren[1]) werden die einzelnen Teile für Glasring, Deckel, Zwischenring u. dgl. in einem einfachen Werkzeug hohl geprägt, Fig. 297 oben, während in dem unterhalb gezeichneten, abge-

[1]) WT. 1910, S. 406.

änderten Werkzeug der Boden hohl geprägt und gleichzeitig eine Erhöhung eingeprägt wird.

Die Blanketts werden auf irgend einem bekannten Wege, z. B. Mehrfachschnitt mit zwangläufigem Bandvorschub, ausgeschnitten und kommen dann in eine einfache Revolverpresse, auf welcher die Werkzeuge nach Fig. 297 das Blankett hohl prägen. In Fig. 297 ist a der Prägestempel, der mit der Matrize b und Auswerfer c zusammen den Boden des Blanketts hohl prägt und den Rand hochzieht. An dem Stempel führt sich der Blechhalter g, der durch eine Feder auf das Blankett niedergepreßt wird; in seiner tiefsten Stellung legt er sich mit einer Ausdrehung an einen Absatz des Stempels an. Der Stempel ist der Länge nach durchbohrt und nimmt einen Abstreiferstift h, der durch eine in der Bohrung gelagerte und mit einer Schlitzschraube gehaltene Feder betätigt wird, auf. Im Unterteil erscheint oberhalb des eigentlichen Werkzeuges der Revolverteller mit der dem Stück entsprechenden eingesetzten Büchse.. In dem im Schnitt angedeuteten Pressentisch ist die Prägematrize b gelagert und wird mittels des konisch ausgedrehten Überlagringes und versenkter Zylinderkopfschrauben, welche durch die Führungs- und Zentrierbüchse durchgehen, gehalten. Die Prägematrize ist zur Aufnahme des Auswerfers c ausgebohrt und mit einer mittleren Bohrung für dessen Zapfen versehen, der in einer von der Feder f getragenen Mutter d verschraubt ist. Damit wird der Auswerfer so eingestellt, daß das geprägte Stück wieder richtig in den Revolverteller kommt, um von diesem weiter transportiert zu werden.

Gesenk zur Fertigstellung von Knöpfen mit Zelluloidkappen. Fig. 298 stellt einen Knopf dar, dessen Oberseite aus einer gemusterten Metallkappe mit einem durchsichtigen Zelluloidüberzug besteht. Die Rückseite wird von einem gewöhnlichen Knopfunterteil gebildet, der von der Kappe, die mit ihrem Rand den Zelluloidüberzug festhält, umschlossen ist.

Fig. 298.

Hier soll der letzte Arbeitsgang, das Schließen, beschrieben werden.

Das Untergesenk Fig. 299 besteht aus einer Grundplatte, die eine Säule trägt, auf der eine Büchse durch eine starke Feder in ihrer Stellung erhalten wird. Die Kappe mit einer flachen Zelluloidplatte darüber wird nun, wie aus der Abbildung ersichtlich ist, in diese Büchse eingelegt. Der Stempel trägt ebenfalls eine Büchse und führt bei seinem Abwärtsgange diese weit genug herunter, so daß die Kappe mit der Zelluloidplatte in diese eintreten kann, wodurch beide beim Aufgang des Stempels mit nach oben genommen werden. Stempel und Hülse werden durch einen Gasbrenner warm genug erhalten, damit das Zelluloid, ohne zu brechen oder Falten zu erhalten, die neue Form annehmen kann. Der innere Durchmesser der Hülse am Stempel muß selbstverständlich um die doppelte Dicke der Zelluloidplatte größer sein als der der Kappe.

Jetzt kommt das in Fig. 300 dargestellte Schließgesenk zur Anwendung. Der Arbeiter hält es in der linken Hand und legt den Knopfunterteil hinein, wie gezeichnet. Das Schließgesenk wird dann auf die Hülse des Unterteils in Fig. 299 gesetzt und dort bis zum Niedergang des Stempels gehalten. Dann wird der Haken, der die obere Hülse in Stellung erhält, beiseite gedrückt, so daß die Hülse am Bund des Stempels zum Anliegen kommt. Der Rand der Kappe kommt beim Schließen des Gesenkes zwischen die Innenfläche der Büchse und der Hülse im Schließgesenk,

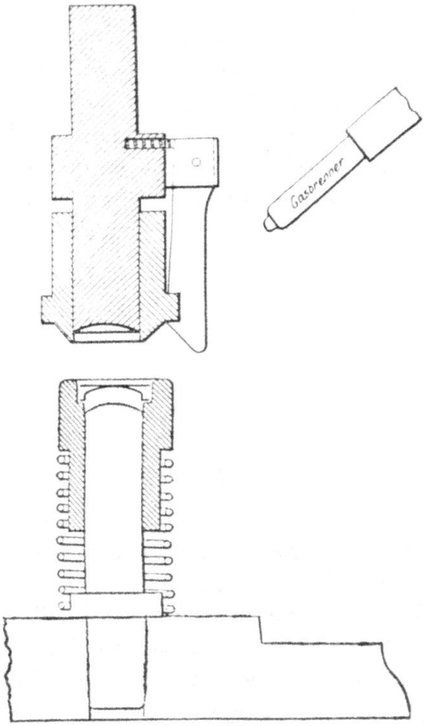

Fig. 299.

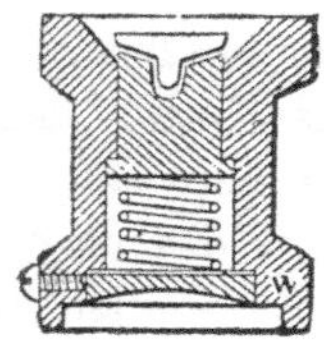

Fig. 300.

wodurch er umgeschlagen und gegen das Zelluloid gedrückt wird, worauf beide zusammen gegen den Knopfunterteil gepreßt werden. Bei diesem letzten Hube wird das von Hand gehaltene Schließgesenk niedergedrückt, bis die Unterlagscheibe W fest auf der inneren Säule des Untergesenkes sitzt, so daß für das Schließen ein harter Druck erhalten wird. Da der Knopfunterteil lose in die Kappe paßt, kann er in sie eintreten, bevor das Schließen vor sich geht.

Das Knopfunterteil ruht auf einem leicht abgefederten Zylinder innerhalb des Schließgesenkes, dessen Unterlagscheibe W eine Druckfeder trägt, damit der Zylinder während der Schließarbeit nach unten ausweichen kann.

Beim folgenden Aufwärtsgange des Stempels wird der fertige Knopf entfernt, während sich die Hülse am Stempel selbsttätig einhakt, so daß der nächste Knopf eingelegt werden kann.

Das Ausflachen von ausgeschnittenen Scheiben. Gelegentlich wird die Lieferung einer größeren Zahl Scheiben, die vollständig eben und frei von Buckeln sein sollen, verlangt. Um diese Arbeit mit Erfolg durchzuführen, benutzt man am besten eine der folgenden Arbeitsweisen.

In einem Falle hatte man einen Satz Scheiben von 200 mm Durchmesser und 8 mm Dicke, die eben sein und gleichmäßige Dicke haben sollten. Das Material für die Scheiben, die mit einem Loch von 50 mm Durchmesser zu versehen waren, war Stahl von 4 % C.

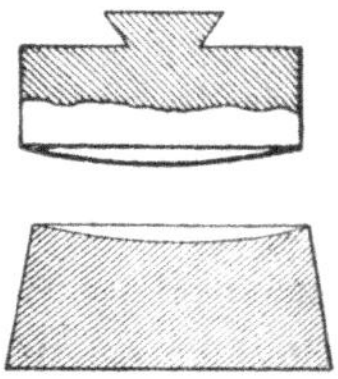
Fig. 301.

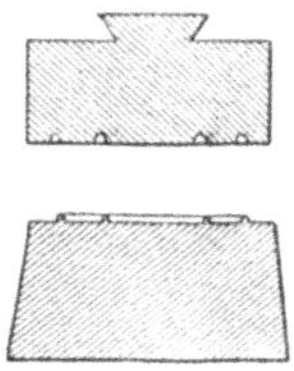
Fig. 302.

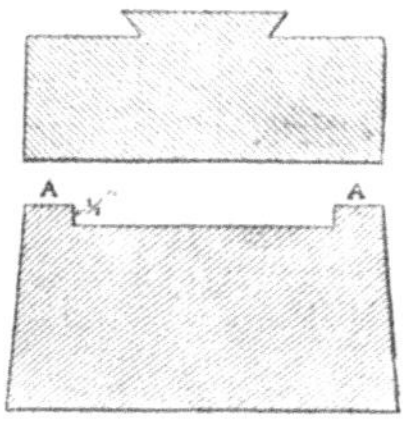

Fig. 303.

Zwei bis dreimaliges Schlagen derselben zwischen einem Paar ebener Gesenke hatte keinen Einfluß, so daß ein Paar Gesenke nach Fig. 301 hergestellt wurde, von denen das untere mit 1,6 mm Pfeilhöhe hohl und das obere in gleichem Maße erhaben gemacht wurde. Die Blanketts wurden durch einen kräftigen Schlag gut gekrümmt, dann umgedreht und noch einmal gerade genug geschlagen, um sie eben zu machen. Nachdem man mit dieser Arbeitsweise einige Übung erlangt hatte, stellten sich keine weiteren Schwierigkeiten ein.

Die nächste Aufgabe war die Herstellung von Scheiben aus Federkupfer von 178 mm Durchmesser und 0,9 mm Dicke. Obwohl vorher bekannt war, daß die Scheiben flach sein mußten, erwartete man in dieser Richtung keine Schwierigkeiten. Sowie man jedoch an das Ausflachen der Scheiben kam, häuften sich die Schwierigkeiten. Die erst beschriebenen Gesenke ergaben wohl eine richtige Höhlung der Kupferscheiben, doch konnte keine Art Schlag sie nachher wieder eben machen. Der Grund schien in der Federung des Materials zu liegen, da die Scheiben stets einen Buckel aufwiesen. Schließlich wurde ein Paar Gesenke nach Fig. 302 mit zwei Ringnuten hergestellt, so daß die Erhöhungen des einen Gesenkes mit den Vertiefungen des andern übereinstimmten, wozu man mit vieler Mühe wirklich erstklassige Werkzeuge herstellte. Damit wurden die Scheiben ziemlich gut, so daß man schließlich durch Verändern der Ringnuten — bis sie groß genug waren, um die Federung in den Scheiben aufzunehmen — die Scheiben wirklich eben erhielt. Die Ringe versteiften die Blanketts außerdem und hielten sie in der Form, in der sie die Ausflachgesenke verließen.

Die dritte Aufgabe bestand im Ausflachen einer Reihe viereckig ausgeschnittener Platten, die auf eine verlangte Dicke gebracht werden sollten und, falls dies in der Hitze besser erreicht werden konnte, auch so verarbeitet werden durften. Ein altes Paar Ausflachgesenke wurde ausgeglüht, eines derselben nach Fig. 303 ausgehobelt, so daß die Seiten A 6,5 mm hoch blieben, und dann nach dem Härten auf den Seiten und der Stirnfläche geschliffen. Das Ausflachen geschah mit einem leichten Schmiede-Fallhammer; man erwärmte das Material bis auf Dunkelrot oder heiß genug, daß sich kein Zunder bildete, und gab einen Schlag. Die Stücke kamen aus den Gesenken eben und mit der verlangten Dicke, doch überstiegen die Kosten die vorkalkulierten Preise.

Wenn sehr dünnes Material ausgeflacht werden muß, so gibt es keinen besseren Weg, als die Gesenke eben und parallel zu schleifen, sie mit ein wenig Schmirgel und Öl aufeinanderzupassen, so daß ein Stück Seidenpapier, wenn die Gesenke in die Presse eingebaut worden sind, gleichzeitig an allen vier Ecken gefaßt wird. Unter Benutzung eines leichten, durch Maschinenkraft angetriebenen Fallhammers dürfen die Kosten des Ausflachens von dünnem Material nicht 25 bis 34 Pfennig für das Tausend — amerikanische Angabe — überschreiten. Woodworth gibt einen Fall an, wo bei Verwendung eines Hand-Fallhammers in 10stündiger Arbeitszeit 44 000 Stück hergestellt wurden.

Huberpreßverfahren. Das eine der bekannten Hilfsmittel beim Ausbauchen vorgezogener Hohlkörper, Sand und Wasser, die zur Druckübertragung von einem nicht der Matrize entsprechenden Stempel auf das Arbeitsstück dienen, das Wasser, wird in dem Huberverfahren als grundlegender Teil der Preßarbeit angewendet. Bei diesem Arbeitsverfahren wird der vorgezogene und gegebenenfalls auch gebauchte Hohlkörper in ein mehrteiliges Gesenk, das die Entfernung des fertig gepreßten Körpers durch Zerlegen des Gesenkes gestattet, eingelegt und in den Innenraum desselben hoher Wasserdruck von bis zu 5000—6000 Atm. eingeleitet. Unter diesem allseitigen Druck auf die Innenfläche fließt das Metall, ohne seinen Zusammenhang aufzugeben, nach außen und in die Vertiefungen des Gesenkes hinein.

In der Praxis zeigte sich aber, daß die Ausprägung der feinsten Formen trotz des hohen Druckes immer noch zu wünschen übrig ließ, so daß die zwischen Hohlkörper und Gesenk verbleibende, eingeschlossene Luft entfernt werden mußte. In dem D. R. P. Nr. 201 136 sind einige grundsätzliche Anwendungsmöglichkeiten gezeigt, von denen Fig. 304 die Anwendung auf ein Werkzeug für Huberpressung an einem ausgebauchten Hohlkörper zeigt. Der zu prägende Hohlkörper o ist in die mehrteilige Matrize m_1 eingesetzt. Die Matrize wird von einem Mantel g, der die seitlichen Drücke beim Prägen aufzunehmen hat, umschlossen, während dieser mit einem in das Innere des zu prägenden Hohlkörpers hineinragenden Gummibeutel g_1 gegen die Außenluft abgeschlossen wird. Durch das seitlich angebrachte Rohr wird der Raum zwischen Gummibeutel und Mantel durch Öffnen des Hahnes mit einer Luftpumpe in Verbindung gebracht, so daß die zwischen Hohlkörper und Matrize befindliche Luft abgesaugt und der Gummibeutel fest an die Innenwand des Hohlkörpers angepreßt wird. Die weitere Prägung erfolgt in derselben Weise wie beim gewöhnlichen Huberpreßverfahren.

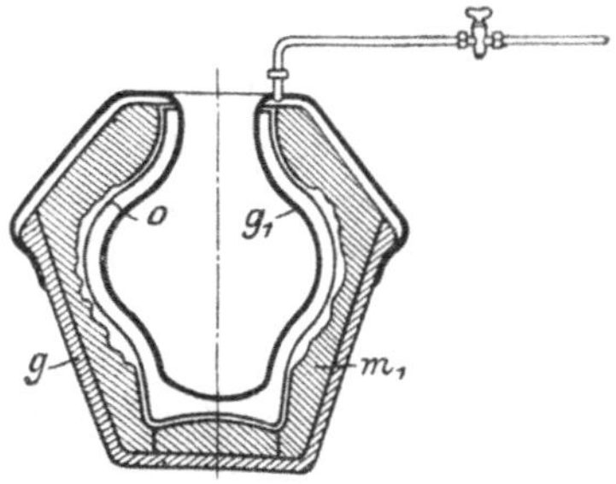

Fig. 304.

Prägen von Blechplakaten[1]). Für das Prägen der großen Blechplakate, die bis 2 m × 1 m messen, würden die Herstellungskosten der Matrizen aus Stahl derartig hoch werden, daß die Möglichkeit eines Wettbewerbes vollständig in Frage gestellt würde. Es sind nämlich die Materialkosten dieser Werkzeuge fast, und die Arbeitslöhne ganz verloren, d. h. sie belasten den einen Artikel derartig, daß die Preise für derartige Plakate viel zu hoch werden. In der angegebenen Quelle ist ein Weg zur Herstellung solcher Matrizen angegeben, der den Vorzug der Billigkeit hat und die teure Handgravüre umgeht. Die Prägung wird wohl etwas weniger scharf als bei einer Stahlmatrize, doch spielt dies bei den großen Abmessungen und den auf Fernwirkung berechneten Plakaten erheblich weniger Rolle.

Die Herstellung dieser Matrizen ist in großen Zügen — in der angegebenen Quelle ist eine genaue Einzeldarstellung gegeben — folgende: Man hobelt zwei Stahlgußplatten von 4—7 cm Stärke auf beiden Seiten eben und klebt das zu prägende Plakat mit Fischleim glatt auf eine der Platten. Dann schneidet man mit der Laubsäge die Buchstaben und Verzierungen des zu prägenden Plakates aus Messingblech aus, dessen Dicke mit der Höhe der Buchstaben in der Prägung

[1]) Illustr. Zeit. f. Blechind. 1908, S. 1247 ff.

übereinstimmt, bricht die Kanten ein wenig und klebt die Buchstaben und Verzierungen mit dick angerührtem Fischleim auf die Stahlgußplatte dort auf, wo sie im aufgezogenen Plakat erscheinen sollen. Inzwischen ist eine zweite der unteren gleiche Stahlplatte am Pressenstößel befestigt und mit dieser genau und parallel ausgerichtet worden. Dann legt man ein Stück graue Pappe, die etwas dünner als die Prägung hoch ist, auf die untere Stahlplatte, wobei man des besseren Erkennens halber die Oberfläche der Prägung mit Tusche bestreichen kann, und läßt die Spindel niedergehen, bis sich die Prägung auf der Pappe deutlich markiert; dann läßt man die Presse 3—4 Stunden unter Druck stehen, damit der Fischleim hart wird und die aufgeklebten Buchstaben festsitzen. Der ausgequollene Leim muß dann sorgfältig neben den Buchstaben weggekratzt werden, da er sich sonst beim Prägen markiert. Die angeprägte Pappe wird nun sorgfältig mit der Laubsäge ausgeschnitten, so daß ein Negativ des zu prägenden Plakates erscheint. Legt man dieses dann auf die untere Stahlplatte, so müssen die Messingbuchstaben ungefähr ein Drittel der Höhe der Prägung über die Pappe hervorragen. Diese Negativplatte wird nun mit andern Pappen derselben Qualität mittels Tischlerleims zusammengeleimt, bis die ganze Dicke, je nach der Tiefe der Prägung, 2 bis 5 cm beträgt. Das Ganze läßt man zwischen zwei Eisenplatten am besten über Nacht unter Druck trocknen. Man muß auch hier darauf achten, daß kein Leim in den ausgesägten Raum kommt. Die so erhaltene Pappmatrize wird unter der Presse fertig ausgeprägt und gibt ein genaues Negativ der unteren Preßplatte, da, wie oben gesagt, die ausgesägte Tiefe niedriger als die Messingbuchstaben ist und der Rest durch die Prägung selbst erhalten wird, wobei gleich die genaue Form geprägt wird. Dann wird die Pappmatrize auf der Rückseite mit „Gelbzug" bestrichen, auf die untere Stahlplatte aufgelegt und durch Niedergehen der Spindel an der oberen Stahlplatte festgeklebt. Die bedruckten, lackierten und getrockneten Plakate werden auf der bedruckten Seite mit Speckstein eingerieben und in der gewöhnlichen Weise geprägt. Handelt es sich um sehr große Lieferungen, Zehntausende und mehr, so empfiehlt es sich, einen Überzug aus schwachem Schwarzblech auf der unteren Tafel im Druck festzuprägen. Diese Platte bleibt auf der Stahlplatte liegen, so daß über ihr alle Bleche geprägt werden. Dadurch kommen die Übergänge noch besser heraus. Selbstverständlich muß die Pappmatrize dementsprechend erweitert werden.

Für dauernde Lieferungen kann man die Messingbuchstaben auf der unteren Stahlplatte auflöten, statt sie anzukleben.

Gesenk zum Rundprägen von Ofenrahmen. In Fig. 305 ist ein Gesenk zum Rundprägen von Ofenrahmen dargestellt. Der Blechstreifen wird zuerst um die Ecken des Untergesenkes gebogen und dann in den Schlitz gesteckt, so daß seine untere Kante auf der Stahlleiste ruht, während die obere Kante ein wenig höher als der Beginn der Abrundung im Untergesenk zu stehen kommt. Das Obergesenk geht nun nieder und bleibt eine Metalldicke vom Untergesenk entfernt stehen, worauf beide zusammen weiter nach abwärts gehen, so daß der Blechstreifen gezwungen wird, in den gekrümmten Raum zwischen Ober- und Untergesenk einzutreten, ohne daß er Falten bilden kann.

Prägegesenke für Ofenecken. Das im folgenden beschriebene Gesenk dient zur Herstellung von Ecken für Herde nach Fig. 306, ohne das Rohmaterial zu

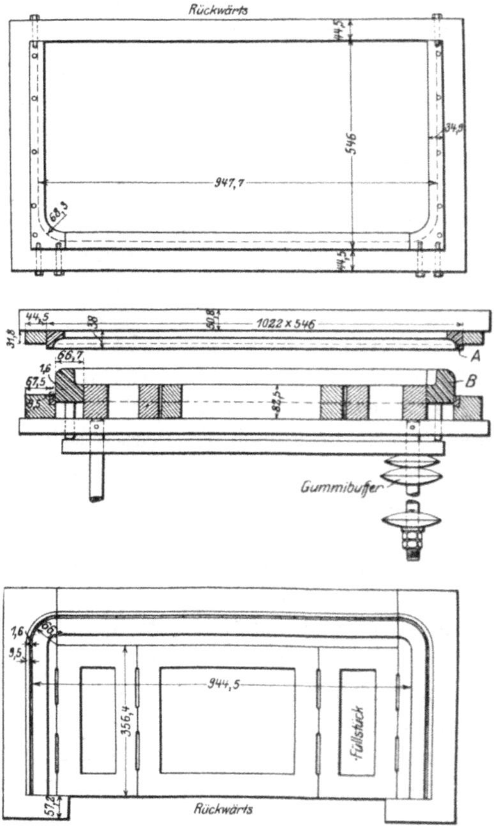

Fig. 305.

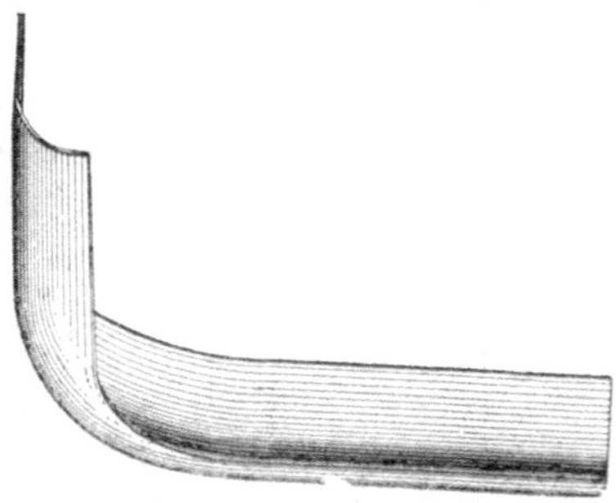

Fig. 306.

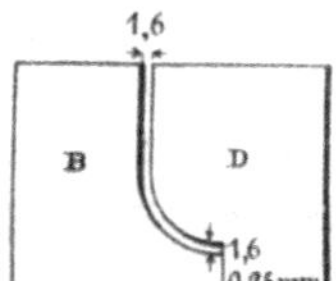

Fig. 307.

fälteln oder zu beschädigen. Nach dem Prägen werden die Ecken vernickelt und so verwendet.

In Fig. 308 ist A eine Grundplatte aus Gußeisen, die stark genug gemacht wird, um die beträchtliche Beanspruchung während des Arbeitsvorganges auszuhalten. B ist ein Winkelstück aus Stahl, das von unten und seitwärts mit der Grundplatte verschraubt ist. Dieser Winkel wird aus Werkzeugstahl hergestellt, gehärtet und geschliffen.

C ist der gußeiserne Stößel, an welchem der Prägestempel D, der auch aus Werkzeugstahl hergestellt, gehärtet und geschliffen ist, mit Zylinderkopfschrauben befestigt ist. Der Stößel wird von Hand mittels Kniehebel bewegt.

Die Grundplatte wird zuerst auf der Unterseite gehobelt und dann die Führung für den Stößel ausgefräst. Der aufrechtstehende Teil und die Auflagerfläche für B werden ebenfalls ausgefräst und daran der Winkel B befestigt, so daß die Winkelhalbierende mit der Mittellinie des Stößels

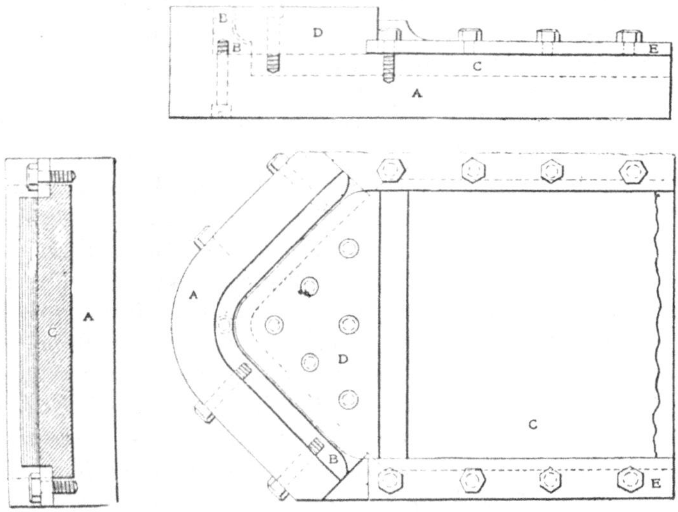

Fig. 308.

zusammenfällt. Hierauf wird der Stößel gehobelt und mit Gleitsitz in die Grundplatte eingepaßt. Er wird durch aufgeschraubte Führungen E festgehalten. Das Schmiedestück für den Winkel B aus Werzkeugstahl wird, nachdem es roh vorgeschruppt worden ist, mit einem besonderen Fräser, der genau dem herzustellenden Profil entspricht, bearbeitet. Der Winkel wird jedoch nicht auf 90⁰, sondern nur auf 87⁰ ausgefräst, so daß er sich nach dem Härten auf 90⁰ öffnet. Zum Fräsen wird der Winkel in einem Schraubstock festgespannt. Nach dem Härten wird der Winkel allseitig geschliffen und, wie oben erwähnt, in der Grundplatte befestigt. Der Prägestempel D wird aus einem einzigen Stück Werkzeugstahl herausgearbeitet. Nach dem Vorschruppen wird eine besondere Aufspannplatte gemacht, die auf den Teilkopf der Fräsmaschine paßt, und ein besonderer Fräser hergestellt, der genau ein Negativ des für den Winkel verwendeten ist und nur um die Blechdicke, 1,6 mm, kleiner ist. Der Stempel wird dann auf 90⁰ gefräst, wobei der Scheitel gleich die richtige Krümmung erhält, indem man den Fräser genau über den Mittelpunkt des Teilkopfes stellt, dann denselben um 90° schwenkt, worauf der Fräser wieder in der Geraden vorrücken kann. Dann wird der Stempel gehärtet und geschliffen, wobei an der Unterseite so viel weggenommen wird, daß der Zwischenraum am Grunde 0,25 mm kleiner wird, als die Dicke des Rohmaterials beträgt (siehe Fig. 307), da das Blech beim Verarbeiten die Neigung hat, den Biegestempel und Stößel abzuheben. An den Enden werden Stempel und Gesenk abgerundet, so daß sich das Profil nach und nach verliert, damit sich, nachdem die Ecken an beiden Enden des Blechstreifens geformt sind, beim Biegen und Formen des dazwischenliegenden Teiles keine Stoßstelle ergibt.

Als Rohmaterial werden Blechstreifen von 40 mm Breite und 1,6 mm Dicke verwendet. Diese werden auf die richtige Länge zugeschnitten, so daß man die beiden Enden biegen kann und ein langes gerades Stück zwischen ihnen behält. Diese Stücke werden zuerst an jedem Ende in der richtigen Entfernung voneinander flach gebogen, bevor sie in dieses Gesenk kommen. Ein Stück wird dann mittels des Prägestempels an dem Gesenk in der richtigen Stellung festgehalten, worauf der Stößel durch den Kniehebel nach vorne durchgedrückt und durch Einschalten der Presse das Blech in die Form hineingedrückt wird. Nachdem durch Anziehen des Kniehebels das Blech freigeworden ist, wird mit dem anderen Ende des Streifens in gleicher Weise verfahren. Der dazwischenliegende Teil des Bleches ist immer noch flach. Er wird in einem zweiten Gesenk mit demselben Profil geformt. Der Biegestempel zum Anbiegen der Ecken ist ganz flach, da er nur den Zweck hat, beim Abwärtsgang der Presse gegen das Blech zu drücken und dieses in die Form hineinzuschieben.

Gewindedrücken. Beim Gewindedrücken wird der mit Gewinde zu versehende Bolzen zwischen zwei Backen A, B, Fig. 309, die mit den dem Gewinde entsprechenden Riefen versehen sind, unter kräftigem Druck gerollt. Diese Riefen liegen schräg, entsprechend dem Steigungswinkel des Gewindes. Die Backe B ist fest, während die Backe A eine hin- und hergehende Bewegung macht, aber nicht verstellbar ist, während die Backe B sowohl in der Höhe wie in der

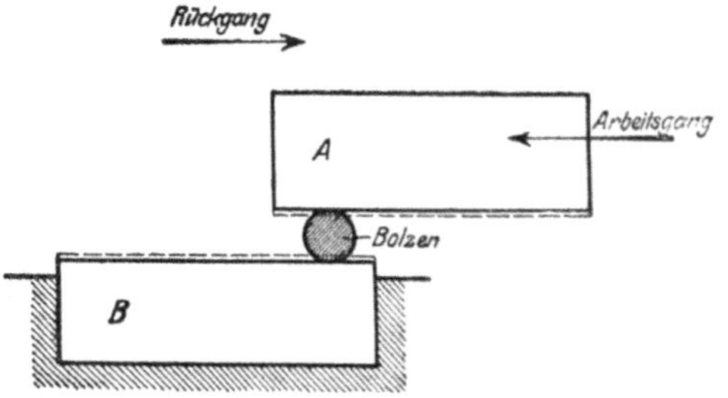

Fig. 309.

Schräge verstellbar ist. Zum Drücken des Gewindes wird der Bolzen in eine Zuführung vor der Backe B gesteckt, worauf die ankommende bewegliche Backe A den Bolzen mitnimmt und zwischen den beiden Werkzeugteilen durchrollt. Am anderen Ende der Backe fällt der fertige Bolzen aus. Die Höhe der Backe ist gleich der Gewindelänge + 10 mm, wobei die hin- und hergehende Backe manchmal aus Zweckmäßigkeitsgründen 15 mm länger gemacht wird.

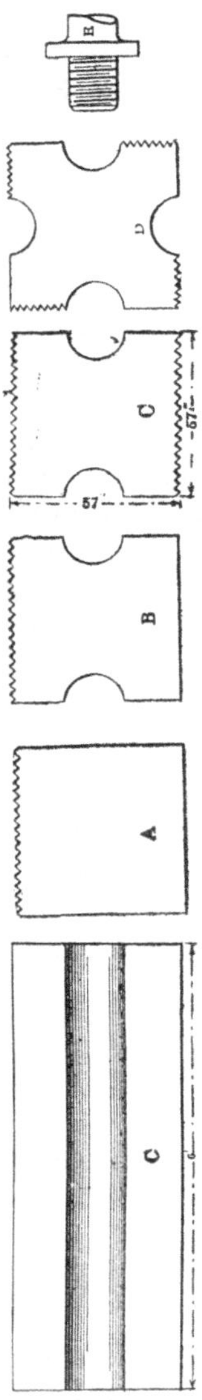

Fig. 309 a.

Selbstverständlich entspricht die Schrägstellung der Riefen dem Steigungswinkel der Schraube.

Man walzt Eisengewinde bis zu 32 mm Durchmesser und Holzgewinde bis zu 16 mm 15 bis 40 Stück minutlich.

In den Figuren 309 a ist ein Werkzeug zum Gewindedrücken oder Walzen dargestellt. Die ersten Werkzeuge wurden mit einer Schnittkante wie A oder B gemacht, doch fand man nach dem Härten, daß die Schnittfläche konvex geworden war, und zwar wenigstens um 0,25 mm. Es erscheint dies wohl auf den ersten Blick wenig, doch genügt es, um das Gewinde unbrauchbar zu machen. Die Gewinde werden nämlich mit 24 Gängen auf den Zoll hergestellt, und 0,25 mm Pfeil im Werkzeug gibt einen in der Mitte um 0,5 mm kleineren Gewindedurchmesser als an den Enden, so daß eine Mutter, die an den Enden stramm geht, in der Mitte lose sein wird. Um über diese Schwierigkeit hinwegzukommen, wurden auf dem Werkzeug an zwei gegenüberliegenden Seiten Gewinde angeschnitten und Nuten angehobelt, wie bei C, Fig. 309 a, ersichtlich ist. Auf diese Weise erhielt man ein tadelloses Werkzeug und gleichzeitig zwei Werkzeuge in einem Stahlblock.

Von mancher Seite wird der Einwand erhoben werden, daß die geschnittenen Flächen nicht festgeklemmt werden sollen, da die Werkzeuge sich verschieben und die geschnittenen Zähne Nuten in die Klemmbacken der Maschine einarbeiten werden. Dies ist jedoch nicht der Fall, da die klemmenden Oberflächen viel größer als die Berührungsflächen des zu drückenden Materials sind, so daß keine Gefahr eines Verschiebens der Gesenke während der Drückarbeit besteht. D in Fig. 309 a zeigt eine andere Werkzeugform für vier Schnittflächem auf einen Stahlblock und dient zum Schneiden **kurzer** Gewinde auf **kurzem** Material. Man kann nun nicht alle acht Flächen mit Gewinde versehen, wie man vielleicht anzunehmen geneigt wäre, da das Werkzeug D an mit Bund versehenen Stücken (vgl. E Fig. 309a) verwendet werden soll und beim Einschneiden von Gewinde an den acht freien Flächen die Gewinde an den gemeinsamen Kanten nicht vollkommen ausfallen und beim Härten abbrechen würden. Wenn jedoch die Gewinde nur bis auf ungefähr 3 mm von dem Kopf geschnitten werden sollen, kann man alle acht Flächen mit Gewinde versehen und die Kanten abrunden. Man beachte, daß im Werkzeug D das Material gleichmäßig verteilt ist und so beim Härten gleichmäßiges Schwinden erzielt wird.

Herstellung der Werkzeuge zum Gewindedrücken. Die wichtigste Bedingung bei der Herstellung der Werkzeuge zum Gewindedrücken ist, daß die Gewinde oder die Schneidzähne vollständig richtig auf dem Blankett geschnitten sind, so daß sie auf dem Arbeitsstück ein richtiges Gewinde herstellen. Manchmal verwendet man ein paar Werkzeuge, die für ein 5/16″ Gewinde mit 26 Gängen auf den Zoll geschnitten sind,

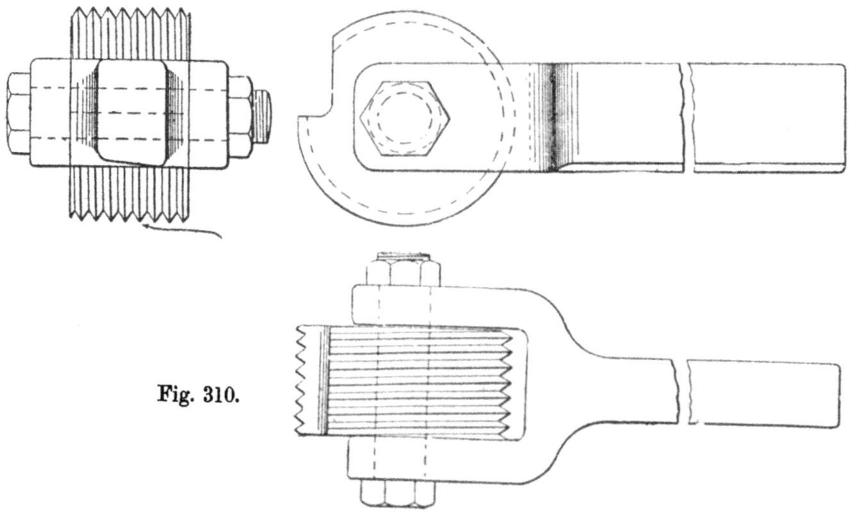

Fig. 310.

auch für ¾″ mit 20 Gängen, was für rohe Arbeit wohl ausreicht, am Ende sich jedoch nicht bezahlt macht. Soll ein Werkzeug für verschiedene Materialstärken Verwendung finden, so muß es einstellbar sein; ein solches Werkzeug ist jedoch nicht leicht herzustellen und läßt sich schwer in die Maschine einbauen. Erfahrung und viele Versuche haben gezeigt, daß es am billigsten ist, für jede Bolzenstärke ein eignes Werkzeug zu machen. Diese sind dann Normalien, einfach in ihrer Herstellung, und gestatten jedem Arbeiter die Auswechselung und Einstellung, ohne daß der Werkzeugmacher dabei notwendig ist. Diese Werkzeuge werden auf der Wagerecht-Stoßmaschine unter Verwendung eines runden Messers hergestellt. Das Messer wird auf der Drehbank wie ein gewöhnliches Gewinde geschnitten und muß ebenso breit oder ein wenig breiter als das Blankett sein.

Die Abbildungen Fig. 310 zeigen das Messer und seinen Halter. Es ist sehr wichtig, daß die Unterseite des Halters rechtwinklig zur Steigung des abgebildeten Gewindes gehobelt wird, sonst ist das Messer auf der einen Seite nicht frei.

Fig. 311 stellt ein besonderes Spannfutter für die Blanketts zum Schneiden der Gewinde dar. Das Futter ist auf einer gußeisernen Grundplatte drehbar, die am Tisch der Wagerecht-Stoßmaschine festgeklemmt wird. Nach Einstellung des Winkels für das zu schneidende Werkzeug wird das Futter festgeschraubt, an seine Außenfläche eine Marke angerissen und dabei Größe und Zahl der Gänge angeschrieben, so daß man im Wiederholungsfall das Futter in kurzer Zeit wieder einstellen kann. Eine Seite wird „rechts“, die andere „links“ gezeichnet, so daß man auch beide Steigungen gleich einstellen kann.

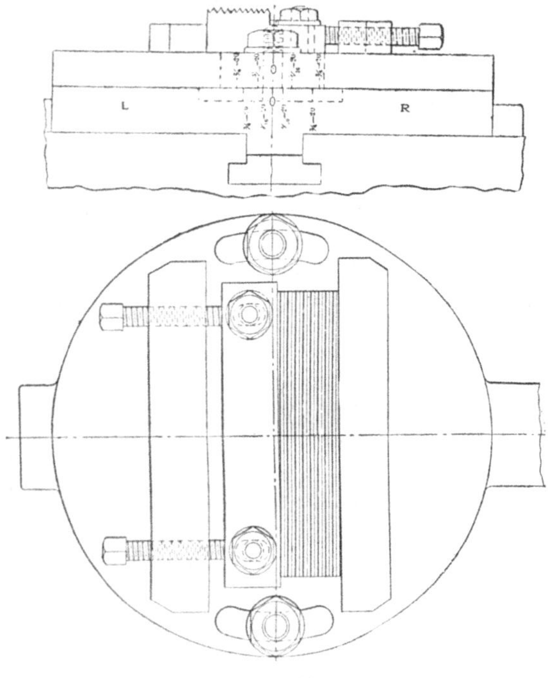

Fig. 311.

Gewindedrückwerkzeuge (Gebr. Hübner, Chemnitz). Für Schrauben bis zu 25 mm Durchmesser ist das Kaltdrücken (Kaltwalzen) der Gewinde manchmal billiger als das Schneiden, um so mehr, als die Kaltwalzwerkzeuge durch ihre große Leistung den Gewindeschneidemaschinen überlegen werden. Obwohl diese Maschinen in ihrem mechanischen Aufbau verhältnismäßig einfach sind, soll doch eine Abbildung einer der-

Fig. 312.

artigen deutschen Maschine, Fig. 312, gegeben werden, um einen Überblick über die Abmessungen zu geben und einen Rückschluß auf die bei dieser Arbeit auftretenden Kräfte zu erlauben. Von einer rückwärts gelagerten Riemenscheibe wird ein Triebling aus Stahl unmittelbar angetrieben, der in das eigentliche Antriebsrad, das den Kurbelzapfen trägt, eingreift. Von diesem Kurbelzapfen wird die nachstellbare Pleuelstange, die mit Bronzelagern versehen ist, angetrieben und versetzt den Schlitten, der den beweglichen Gewindewalzbacken trägt, in hin- und hergehende Bewegung. Während das größere Modell der Maschine für Bolzen von 25 mm Durchmesser lang-

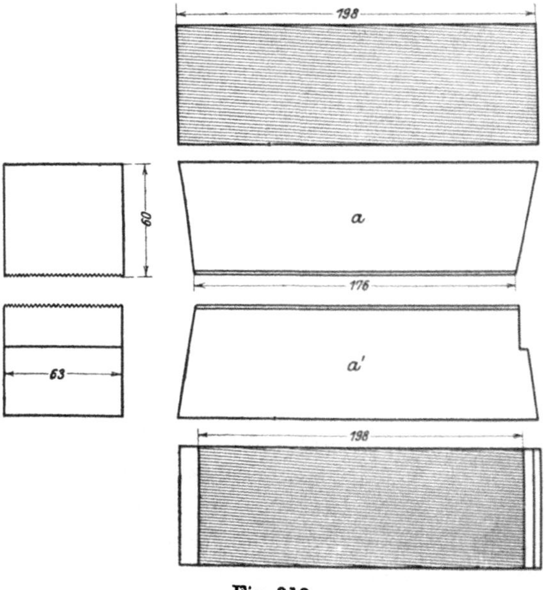

Fig. 313.

samen Arbeitsgang und beschleunigten Rücklauf besitzt, ist bei dem kleinen Modell für bis zu 16 mm Bolzendurchmesser mit Rücksicht auf die höheren Umdrehungszahlen gleichmäßiger Hin- und Rücklauf vorgesehen, wodurch die Schrauben genügend Zeit haben, herauszufallen.

Die Gleitflächen des Schlittens erhalten aufgeschraubte Platten aus Phosphorbronze. Gegen Beschädigung durch Unachtsamkeit ist eine Brechbolzenkupplung vorgesehen. Die Werkzeuge sind nach Fig. 313 ausgebildet und nur an einer Fläche mit Gewinde versehen, im Gegensatz zu dem vorher besprochenen amerikanischen Vorschlag, wofür einerseits die Formgebung für die Einspannung, andererseits aber die Einstellvorrichtung für die Backen der Grund ist. Die Querschnittsabmessungen decken sich mit denen der amerikanischen Teile.

Selbsttätige Graviermaschinen. Da es beim Prägen meistens auf künstlerische Durcharbeitung des Musters und folglich auch des Gesenkes sehr ankommt, so sind die Kosten der Prägewerkzeuge sehr hoch, was um so mehr ins Gewicht fällt, als die Muster in den betreffenden Industriezweigen der Mode unterworfen sind und sehr stark wechseln. Infolgedessen sind die Bestrebungen, die teure Handarbeit durch Maschinenarbeit zu ersetzen, hier nicht zur Ruhe gekommen. Bereits die Herstellung der Gesenke für das Warmpressen von Messing erfolgt teilweise[1]) auf der Maschine. Die Formen sind aber dort verhältnismäßig einfach, während die verwickelten künstlerischen Werkzeuge, wie sie bei der Herstellung der Bijouteriewaren, Beschläge für Möbel usw. vorkommen, auf anderen Maschinen ausgeführt werden, die unter anderen von Hälbig & Sohn in Berlin[2]) gebaut werden. Diese Firma hat die Graviermaschinen als rundarbeitende Kopiermaschinen oder als selbsttätige Reduziermaschinen gebaut. Die ersteren arbeiten nur in Originalgröße von einem runden Modell, während die zweite Gruppe die Gravierung in Naturgröße, vergrößert oder verkleinert wiedergibt.

Die selbsttätige Reduziermaschine besteht aus einem Bett, welches zwei Planscheiben trägt, die auf parallelen, wagrechten Wellen senkrecht zum Bett laufen. Eine derselben trägt das Modell, die andere die zu gravierende Stahlplatte. Gegenüber diesen Scheiben sitzt auf einer Längsführung je ein Support, der gegenüber der Modellscheibe einen Tasterstift und gegenüber der zu gravierenden Platte einen Fräser trägt, der mit 2000 Uml. in der Minute von einem besonderen Motor angetrieben wird. Der Taster und der Fräser gehen während der Arbeit gleichmäßig langsam nach unten, so daß nach und nach das ganze Modell abgetastet wird.

[1]) Z. d. V. d. I. 1913 S. 1385/6.
[2]) Vgl. auch Ill. Zeit. f. Blechind. 1908 Nr. 12, S. 612/3.

Die Maschine wird von einem zweiten Motor angetrieben, der den Planscheiben für gewöhnlich eine gleichsinnige Bewegung erteilt. Wenn beide Planscheiben im entgegengesetzten Sinn umlaufen, so wird die gravierte Stahlplatte ein Spiegelbild des Modelles darstellen, was in manchen Fällen als angenehmer Vorteil empfunden werden wird.

Die Außenbewegung des Tasterstiftes erfolgt zwangläufig von dem Modell aus, während seine Innenbewegung durch eine Gewichtsbelastung bewirkt wird. Diese Bewegungen werden durch die Führungswelle, an welcher, wie oben gesagt, beide Supporte sitzen, zwangläufig auf den Fräsersupport übertragen. Das Verhältnis der Übertragung wird durch die Schräglage dieser Welle bestimmt, welche durch eine an der Seite der Maschine angeordnete Hebelvorrichtung eingestellt wird. Die Firma gibt an, daß das Halbmodell einer Kanne von ungefähr 190 × 130 × 50 mm Abmessungen in vier Tagen ununterbrochener Arbeit hergestellt wurde.

d) Das Ziehen.

Ziehen im Sinne der Preßtechnik ist eine Formänderungsarbeit, bei der ein Materialkörper gezwungen wird, unter der Einwirkung einer an diesem Körper selbst angreifenden und durch seinen Querschnitt als kraftübertragenden Teil gehenden, äußeren Kraft durch einen festen Ring zu gehen, der einen kleineren Querschnitt als der Körper selbst hat. Die kleinsten Teilchen des gezogenen Körpers müssen dabei eine Verschiebung zulassen, ohne ihren Zusammenhang aufzugeben. Handelt es sich um das Ziehen eines vollen Körpers, der auch nach dem Ziehen voll bleibt, wie Draht und Stangenmaterial, oder Rohre von einer solchen Wandstärke, daß sie ohne Anwendung eines die Höhlung ausfüllenden Dornes gezogen werden können, so erscheint es selbstverständlich, daß die Änderung der Querschnittsfläche eine Vergrößerung der Länge bedingt, so daß der Körperinhalt vor und nach dem Ziehen der gleiche ist. Es muß an der Stelle, an der das Material den Ziehring berührt, ein Fließen des Materiales im kalten Zustand eintreten. Für die Materialbewegung als solche ist es gleichgültig, ob der in Fig. 314 beim Drahtziehen oder der in Fig. 315 beim Weiterziehen eines Hohlkörpers abgebildete Vorgang stattfindet, da der Stempel den Druck des im Innern fehlenden Materiales ersetzt. In beiden Fällen aber ist das Fließen ein zwangläufiges, da die Querschnittsverminderung und die Ziehgeschwindigkeit

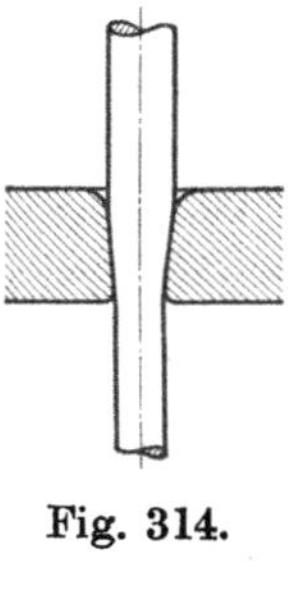

Fig. 314.

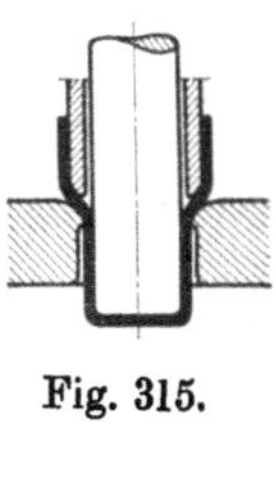

Fig. 315.

durch die betreffenden Werkzeuge bestimmt sind. Wenn man dem Material jedoch gestattet, unter der Einwirkung einer äußeren Kraft im kalten Zustand ins Fließen zu geraten, so tritt dabei eine ganz bestimmte Längenänderung und Querschnittsverminderung ein, die sich nur wenig ändert, wenn die Zeit bei einem gewöhnlichen Zugversuch an einem Normalstab von $1/2$ Minute auf vielleicht 20 Minuten ausgedehnt wird.

Es ist nun festgestellt worden von Kirsch[1]) und auf mikrographischem Wege vom Verfasser[2]), daß das Fließen von einem Ende des Stabes seinen Anfang nimmt und nicht, wie man bei oberflächlicher Überlegung glauben sollte, gleichzeitig über den ganzen Körper sich erstreckt. Verfasser hat an Stäben, die gerade bis in die Fließgrenze belastet wurden, mit großer Wahrscheinlichkeit festgestellt, daß diese Erscheinung der nur teilweise beginnenden Formänderung auch an den Ferritkörnern des Probestabes zu verfolgen ist. Bei weiterer Belastung geht dann die Deformation des ganzen Stabes und die Korndeformation gleichmäßig vor sich, bis erst wieder beim Beginn der Zerstörung, verkennbar durch den Beginn der Kontraktion, örtliche besonders starke Stab- und Korndeformationen auftreten.

Vergleicht man damit die Tatsache, daß bei sehr schneller Beanspruchung, wie sie durch Schlagzerreißversuche bewirkt wird, die Dehnung und die Kontraktion nur auf eine kurze Stelle des Stabes beschränkt bleiben und der übrige Stab im Gegensatz zum langsamen Zugversuch ungeändert bleibt, so ist damit das Einreißen eines Stückes bei zu großer Ziehgeechwindigkeit srklärt, während bei einer anderen, der notwendigen Materialdeformation entsprechenden Ziehgeschwindigkeit das Stück anstandslos durch das Werkzeug geht.

Nun folgen aber beim Fließen die kleinsten Materialteilchen genau bestimmten Gesetzen, die wohl zum Teil erkannt und teils auch theoretisch behandelt worden[3]), aber noch lange nicht vollkommen so weit geklärt sind, daß auf Grund dieser Erkenntnis eine rein theoretische Klarlegung der Ziehvorgänge erfolgen könnte.

Bei gleichmäßiger Beanspruchung eines Materiales entstehen auf der Oberfläche des Körpers regelmäßige Liniensysteme, die, wie Ewing und Rosenhain durch unmittelbare Verfolgung des Vorganges unter dem Mikroskop gefunden haben, durch ein Heraustreten einzelner Materialschichten aus der ebenen Oberfläche des Stückes infolge eines

[1]) Kirsch, Beiträge z. Studium des Fließens; Mitt. Berlin 1887, 1888, 1889.

[2]) Kurrein, Die Gefügeänderung in Flußeisen etc. Baumaterialienkunde 1904, 13/6.

[3]) Tresca, Sur l'écoulement des corps solides. — Hartmann, Distributions des deformations dans les métaux etc. 1896. — Ewing und Rosenhain, Baumaterialienkunde, 1899, S. 365. — Rejtö, Über die Entstehung der Kraftlinien usw. Baumaterialienkunde. — Ludwik, Elemente der technologischen Mechanik. Springer, Berlin 1909.

Gleitens der kleinsten Teile aneinander erzeugt werden. Bei flachen Blechen sind dies zwei geradlinige Liniensysteme, die unter einem Winkel, der größer als 45° ist, gegeneinander geneigt sind, wie Fig. 316[1])

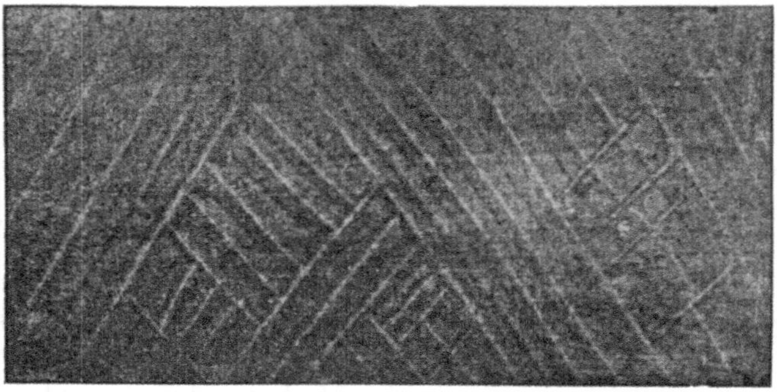

Fig. 316.

Fig. 317.

zeigt. Diese Linien gehen, Fig. 317[2]), auf der Mantelfläche eines zylindrischen Körpers in zwei Schraubensysteme über. Dasselbe Ergebnis erhält man bei Druckbeanspruchung eines Vollkörpers oder eines Rohres auf inneren oder äußeren Überdruck.

Beim Ziehen eines Hohlkörpers treten alle diese Fälle gleichzeitig auf, da die Grundfläche des Blanketts unter dem Stempel einer Druckbeanspruchung, die Mantelfläche einer Zugbeanspruchung und die unter dem Niederhalter liegende Fläche einer aus beiden zusammengesetzten Beanspruchung unterworfen ist, wozu noch im letzten Fall die äußere Reibung zwischen Blech und Niederhalter tritt. Außerdem ist, wie Hartmann gezeigt hat, die Beanspruchung der Bleche auf der Ober- und Unterseite, Fig. 318/19, eine verschiedene — wulstartige Linien oben bis auf $1/3$ des Halbmessers und Furchen unten. Sobald der runde Stempel, dessen Kraftlinienbild radial erscheint, durch einen viereckigen Stempel ausgewechselt wird, ändert sich die Kraftverteilung so, Fig. 320,

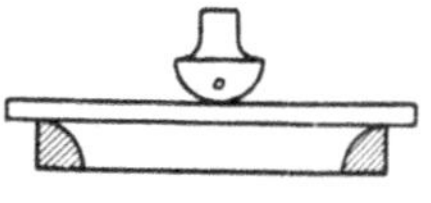

Fig. 318.

[1]) Martens, Materialienkunde, Tafel I, Fig. 16.

[2]) Rejtö, Über Entstehung der Kraftlinien. Baumaterialienkunde, Sonderdruck, S. 2.

daß jetzt die Kraftlinien von den Ecken des vierkantigen Stempels ausgehen, also die Stellen der größten Beanspruchung anzeigen. Aus der Form der Kraftlinien, je nachdem sie wulstig oder furchig erscheinen, läßt sich schließen, ob die Beanspruchung des Materiales Druck oder Zug ist. Bei der Beanspruchung durch den runden Stempel, Fig. 318, ergaben sich drei Stufen, innerhalb deren das Material sich verschieden

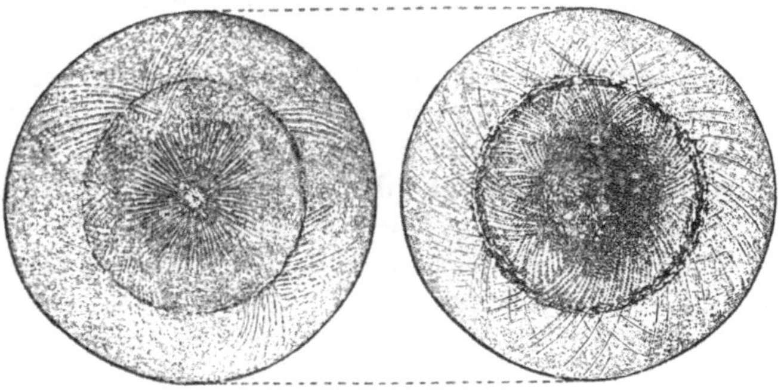

Fig. 319.

verhielt. In der ersten Stufe entstehen auf der oberen Fläche radialewulstige Kraftlinien über eine Länge gleich einem Drittel des Halbmessers, während anf der Unterseite furchenartige Linien bis zu der Hälfte des Halbmessers gehen. Daraus ist ersichtlich, daß die untere Fläche auf Zug, die obere auf Druck beansprucht wird. Dieser Stufe folgt eine andere, bei der von dem Grenzkreis der radialen Kraftlinien logarithmische Linien ausgehen, die wieder auf der Oberfläche gedrücktes, auf der Unterfläche gezogenes Material andeuten. In der dritten Stufe bildet sich auf der oberen Fläche um das Werkzeug ein Wulst, doch erscheinen jetzt auf der Oberfläche des Bleches auch Anzeichen, daß die Beanspruchung auf Zug erfolgt, während die untere Fläche ihren Charakter beibehält. Dieser Zustand ist wohl derjenige, in welchem die in der Praxis geübten Ziehvorgänge stattfinden, wenn also ein gleichmäßiges Fließen über den ganzen Körper begonnen hat. Ist die Ziehgeschwindigkeit gering genug, daß dieser Zustand über den ganzen Körper eintreten kann, bevor eine örtliche Kontraktion

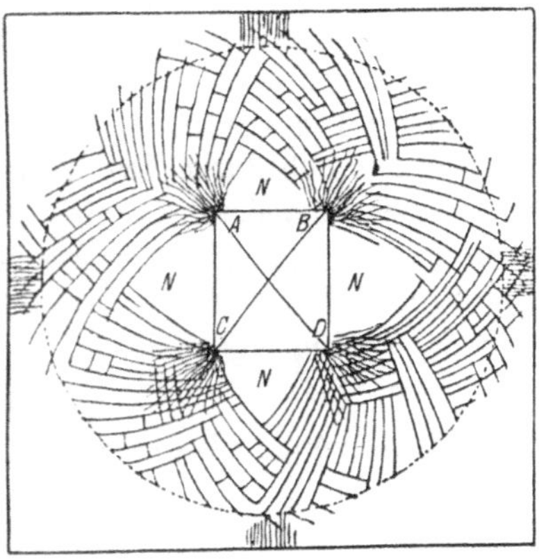

Fig. 320.

einsetzt, wie es beim gewöhnlichen Zugversuch im Gegensatz zum Schlagzerreißversuch der Fall ist, so wird der Ziehvorgang anstandslos durchgeführt werden können, bis die dem Material eigene Formänderung erreicht worden ist. Ist aber der Weiterschlag mit einer so großen Querschnittsverminderung verbunden, daß diese bei der zwangläufigen Bewegung zwischen Ziehwerkzeug und Ziehgut die der freien Querschnittsveränderung entsprechende Größe überschreitet, so muß auch bei dem langsamen Ziehvorgang ein Reißen eintreten, wie beim freien Zerreißversuch nach Überschreitung der dem Material eigenen Dehnung und Querschnittsverminderung die örtliche Kontraktion und damit die Zerstörung des Materiales eintritt.

Nun hat Ludwik[1]) gezeigt, daß der Einfluß der Dehnung oder „spezifischen Schiebung" — welchen Ausdruck er einführt — auf die innere Reibung — jene Schubspannung, die eine bleibende Verschiebung einleitet, — die Formänderungsvorgänge beherrscht und man aus der gewöhnlichen Schaulinie eines Zerreißversuches eine Beziehung zwischen den beiden Größen, die „Fließkurve", ableiten kann, welche für das Verhalten des Materiales im Gebiet der bleibenden Formveränderungen erst die charakteristischen Merkmale angibt.

Es lassen sich demnach derartige Vorgänge, also auch Zieharbeiten, nicht mittels der allgemein bisher für Berechnungen verwendeten Zahlen der zulässigen Beanspruchung oder Bruchdehnung berechnen, ebenso wie alle Formeln, die auf reiner Erfahrung beruhen, kaum etwas mehr als für den einzelnen Fall gültige Werte darstellen. Aus dem Grunde scheint mir der von Musiol eingeschlagene Weg, die Kalibrierung der Ziehwerkzeuge nur auf die Dehnung des zu verarbeitenden Materiales zurückzuführen, wenigstens für praktische Zwecke am aussichtsreichsten.

Es sei[2]) in Fig. 321 A B C D ein Stück der Mantelfläche vor dem Ziehen, das sich durch das Ziehen in A B′ C′ D′ nach den drei Koordinatenachsen ändert. Die drei entsprechenden spezifischen Dehnungen seien φ_x, φ_y, φ_z, und die spezifische Dehnung der Diagonale sei φ_d. Die Durchmesser vor und nach dem Ziehen seien d und d_1, die Blechstärken δ und δ_1. Dann ergibt sich

$$E B_1 = E B (1 - \varphi_d).$$

Nach Einsetzung der betreffenden Bezeichnungen erhält man, da außerdem

$$\delta_1 = \delta (1 + \varphi_z) \quad \text{und} \quad \frac{1}{\pi^2} \sim 0{,}1 \text{ ist,}$$

$$\frac{d_1}{d} = \sqrt{(1 - \varphi_d)^2 - \frac{\delta^2}{10\, d^2} [(1 + \varphi_z)^2 - (1 - \varphi_d)^2]}\,.$$

[1]) Ludwik, Elemente der technologischen Mechanik. Berlin 1909. Verlag J. Springer.

[2]) Musiol, Die Kalibririerung der Ziehwerkzeuge. Stahl und Eisen 1907, S. 477 ff.

Durch Einführung der Blechdicke vor dem Ziehen und Berücksichtigung, daß es sich um Blechdicken von ungefähr 1 mm handelt, kann man schließlich die Glieder 0,1 δ^2 unter dem Wurzelzeichen, die sich bei dem Umwerten ergeben, vernachlässigen und erhält

$$\varphi_d = 1 - \frac{d_1}{d} \,.$$

Aus denselben Gründen kann man die spezifische Dehnung $\varphi_d = \varphi_x$ setzen und kommt schließlich auf die Formel

$$\frac{d_1}{d} = \sqrt{m^2 - \frac{\delta^2 \cdot n^2}{d^2 \cdot 10}}, \quad \text{worin } m^2 = (1 - \varphi_x)^2$$

$$n^2 = [(1 + \varphi_z)^2 - (1 - \varphi_x)^2]$$

Infolge dieser Annäherungen sind zwar die an einem ähnlichen Ziehkörper gefundenen Werte der spezifischen Dehnungen notwendige Vorbestimmungen für die Gültigkeit der Formel, berücksichtigen aber gerade dadurch die unfreie Formänderung des Materiales, ohne sich um deren wirkliche, vorderhand noch unbekannte, allgemeine Gesetze zu kümmern.

Es handelt sich hier wohl auch nur um eine Annäherungsmethode, die aber in ihren Formeln — das ist der große Fortschritt gegenüber den von allen Praktikern angegebenen Tabellen — Konstanten verwendet, die aus gleichartigen Versuchen gewonnen werden müssen und die Längs- und Querdehnung des Materiales, d. h. in der Richtung des Umfanges und des Halbmessers, enthalten.

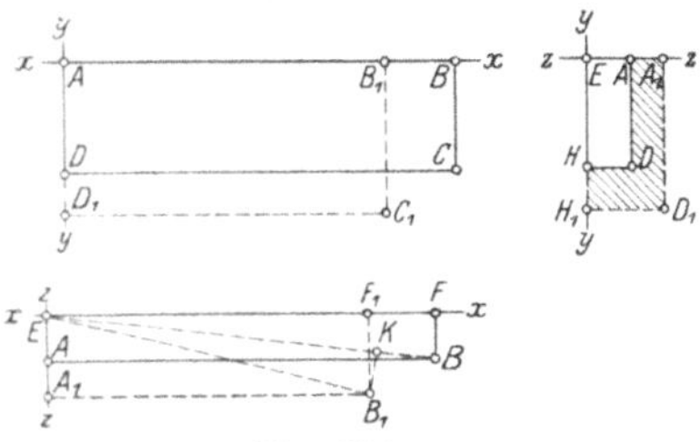

Fig. 321.

Ermittlung der Scheibendurchmesser für Hülsen. Im Gegensatz zu diesen Erwägungen gibt Woodworth ein Schaubild, welches auf der Inhaltsgleichheit der Scheibe und der daraus gezogenen Hülse beruht. Man findet in der Literatur verschiedene Formeln veröffentlicht, die den Durchmesser der Scheibe für eine gegebene Hülsengröße zu finden versuchen. Die meisten Formeln beruhen auf der Annahme, daß die Metalldicke an den Seiten der gezogenen Hülse die gleiche wie im ursprünglichen Material ist, d. h. daß der Boden und die Seiten der Hülse gleiche Materialstärke aufweisen.

Außerdem gehen die meisten Formeln von der Oberfläche der Hülse an Stelle des Inhaltes aus, so daß sie eine Fehlerquelle enthalten, die bei dünnem Material wohl vernachlässigt werden kann, bei dickem jedoch beträchtlich ins Gewicht fällt. Nun neigt man gerade jetzt mehr zu derartigen Arbeiten aus dickem Material.

Die Formel 1 ist eine rein theoretische Ableitung aus der Inhaltsgleichheit vor und nach dem Ziehen, setzt also runde Hülsen mit scharfen, nicht gerundeten Kanten voraus, enthält ebenso keine Zugabe für das Zuschneiden der oberen Kante. Ist

D der Durchmesser der Scheibe
d „ äußere Durchmesser der Hülse
d_1 „ innere „ „ „
h die „ Tiefe „ „
t „ Dicke des Bodens,

so ergibt sich aus der Inhaltsgleichung

1) $$D = \sqrt{d^2 + (d^2 - d_1^2)\frac{h}{t}}.$$

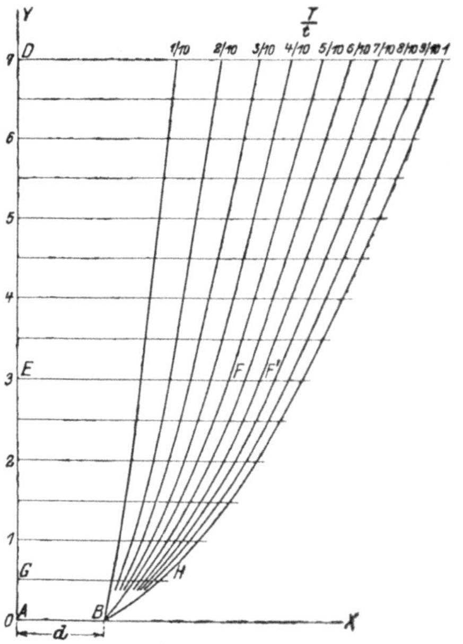

Fig. 322.

Führt man $d - d_1 = 2\,T$ als Materialdicke der Seitenfläche ein und setzt dies in die obige Gleichung ein, so erhält man unter Vernachlässigung des Gliedes mit T in der zweiten Potenz

2) $$D = \sqrt{d^2 + 4\,dh\frac{T}{t}}.$$

Diese Formel ist hyperbolisch und gibt für dünne Materialstärken genügend genaue Werte. Das Schaubild in Fig. 322 ist danach entworfen, so daß jede Linie einem bestimmten Verhältnis $\frac{T}{t}$ entspricht; die dazwischenliegenden Werte können leicht interpoliert werden. Ist A B (im Schaubild = 1 cm) das Maß des Durchmessers d der zu ziehenden Hülse, so ist die Abszisse von einem Punkt der Linie A D, der der gewünschten Hülsentiefe d, 2 d, 3 d usw. — natürlich im gleichen Maßstab — entspricht, bis zu der dem Dicken-Verhältnis entsprechenden Schaubildlinie F bzw. F′ usw. der Durchmesser der Scheibe für diese

Hülsentiefe. Setzt man in 1) für $d - T = d_a$, so erhält man eine Formel, die für dickes Material den in Formel 2) gemachten Fehler aufhebt.

$$3) \qquad D = \sqrt{d^2 + 4\, d_a h \frac{T}{t}}.$$

Diese letzte Formel wird auch mit Vorteil angewendet, wenn die Dicke der Seitenfläche der Hülse nicht gleichmäßig ist; man setzt dann für T die mittlere Dicke ein.

Bestimmung des Scheibendurchmessers nach einer Musterhülse. Man hatte eine doppelt wirkende hydraulische Presse zur Herstellung der Hülsen nach Fig. 323 zu bauen, welche den ersten Arbeitsgang für die Stahl-Nickel-Mantelgeschosse (vgl. Tabelle Nr. 16) darstellt. Nach dem Muster und dem Rohmaterial von der Dicke d in cm hatte man den Scheibendurchmesser D in cm zu bestimmen. Man bestimmte dazu zuerst das Gewicht der Hülse G, dann durch Wägen in Wasser den Gewichtsverlust G_1 in Gramm und erhält für den Scheibendurchmesser D

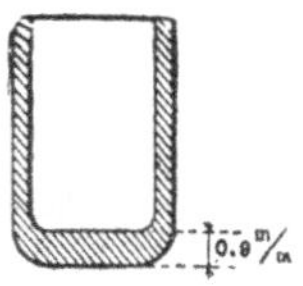

Fig. 323.

$$4) \qquad D = \sqrt{\frac{4\, G_1}{3{,}14\, d}}.$$

Abmessungen der Scheiben für Ziehen in Hülsen. Beim Ziehen des Mantels einer Hülse aus einer Scheibe wird das Material gezwungen, sich von dem größeren auf einen kleineren Durchmesser zu verschieben, so daß der daraus folgende Materialüberschuß sich entweder in einer Verdickung des Ringes unter dem Niederhalter oder Verlängerung des Mantels wiederfinden muß. Im allgemeinen hält man jedoch bei dem Ziehen die Dicke gleich oder nahezu gleich der ursprünglichen, so daß die Mantelhöhe größer als die ursprüngliche Ringbreite wird.

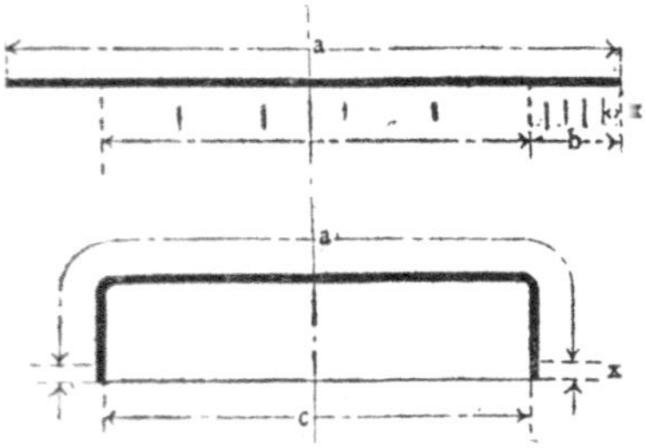

Fig. 324.

Um nun bei der Herstellung einer Hülse von einer gegebenen Scheibe die zu erreichende Mantelhöhe zu bestimmen, wendet man unter Annahme gleichbleibender Materialdicke in Ring und Mantel wieder das Gesetz der Inhaltsgleichheit an. Es muß in Fig. 324 das im Ring von der Breite b enthaltene Material zur Mantelbildung verwendet werden, so daß sich ergibt, wenn

a der Durchmesser des Blanketts,
c = a — 2 b der innere Durchmesser der gebildeten Hülse,
x der Höhenzuwachs des Flansches über b

ist:

$$\frac{\pi}{4}(a^2 - c^2) = \pi c (b + x);$$

durch Elimination von a und Reduzieren der Gleichung

5) $$c : b = b : x.$$

Man kann dabei dann auch graphisch vorgehen, indem man nach Fig. 325 über c einen Kreis schlägt, von einem Ende mit b eine Sehne abträgt, worauf die Projektion dieser Sehne auf c das gesuchte Stück x ist.

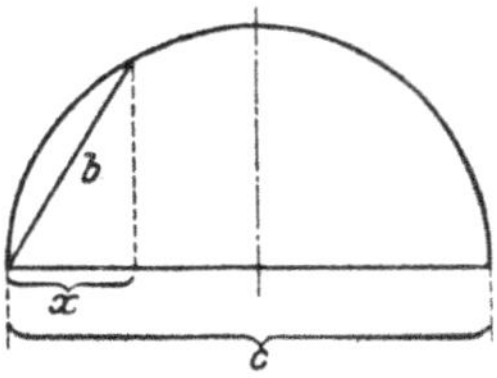

Fig. 325.

Wie gesagt, bei Entwicklung dieser Regel ist angenommen worden, daß die Dicke des Mantels dieselbe wie die des Blanketts ist. Wird jedoch der Flansch um 2 bis 5 % dünner, so wird er auch im Verhältnis tiefer.

1. Ziehwerkzeuge für Metalle.

Bevor auf die einzelnen besonderen Fälle der Ziehtechnik eingegangen wird, sollen die beim Ziehen verwendeten Werkzeuge in schematischen Bildern angeführt und die tatsächlichen Abstufungs-

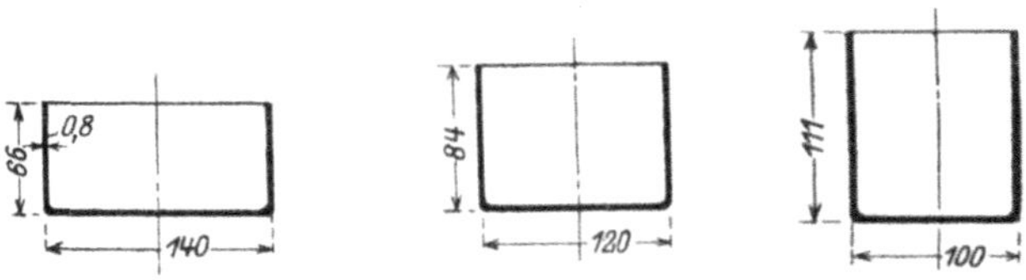

Fig. 326.

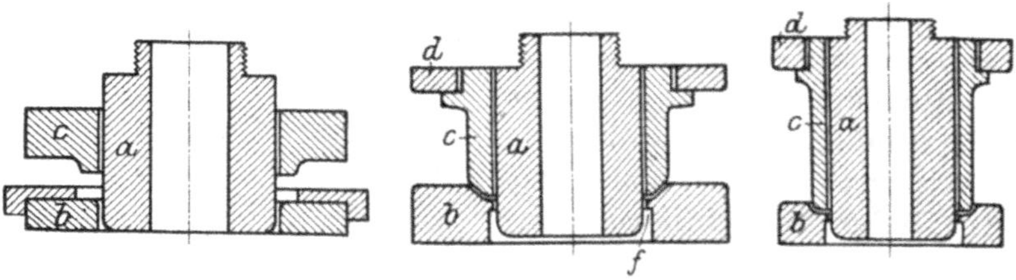

a b c
Fig. 327.

verhältnisse, Fig. 326, an einem auf ähnlichen Werkzeugen in drei Gängen erhaltener Becher (Aluminium) gezeigt werden. Die Werkzeuge, Fig. 327 a, b, c, bestehen aus drei Teilen, die unbedingt erforderlich sind, und zwar dem Stempel a, der die innere Form des gezogenen Gegenstandes bestimmt, dem Ziehring (Matrize) b, der die äußere Form des Gegenstandes festlegt, und dem Niederhalter c, der das ausgeschnittene Blech (Blankett, Platine, Scheibe) während des Ziehens auf die Oberfläche des Ziehringes b niederdrückt, um Faltenbildung zu verhindern. Es ist leicht einzusehen, daß der Kreisringausschnitt a b c d, Fig. 328, frei keine derartige Deformation bzw. Verschiebung seiner kleinsten Teilchen unter der von dem Zylindermantel auf ihn wirkenden Zugkraft machen würde, daß er in einen immer kleiner werdenden Kreisring übergehen und schließlich in einen Teil des Zylindermantels a′ b′ c′ d′ überginge. Wenn der Teil des Blanketts, der jetzt unter dem Niederhalter liegt, sich bei gleicher Zieharbeit frei deformieren könnte, so würden sich die einzelnen Materialschichten mit dem Fortschreiten des Ziehens allmählich übereinanderschieben und Falten bilden. Um diese Faltenbildung zu vermeiden, wird der Niederhalter mit solcher Kraft niedergepreßt, daß das Material unter dem Drucke noch fließen, ohne zu reißen, aber nicht Falten bilden kann. Die richtige Einstellung dieses Druckes gehört mit der Abstufungsbestimmung zwischen den einzelnen Zügen heute, wo richtige, allgemein anwendbare Versuchsergebnisse noch nicht vorliegen, zu den schwierigsten Erfahrungsarbeiten dieser Fabrikation. Außerdem ist an dem ersten Werkzeug noch ein Einlegering für das Blankett vorgesehen, der die Scheibe beim Einlegen gegen Stempel und Ziehring ausmittelt. Die Spielräume zwischen Stempel, Niederhalter und Ziehring sind übermäßig breit gezeichnet. Der Stempel wird je nach der Art der Arbeit, ob auf hartem oder weichem Material gearbeitet wird, der Materialdicke entsprechend oder ein wenig kleiner gemacht. Das Material für die eigentlich arbeitenden Teile ist Stahl, der wieder je nach dem bearbeiteten Material entweder gehärteter Werkzeugstahl oder im Einsatz gehärteter Maschinenstahl ist. Im Stempel ist ein richtig bemessenes Luftloch, das der Luft unter allen Umständen den Zutritt gestattet, anzubringen, damit das Abstreifen des gezogenen Teiles anstandslos vor sich gehen kann. Selbstverständlich müssen alle Teile des Werkzeuges in der Presse ganz genau ausgerichtet und so befestigt werden, daß ein Lockerwerden der einzelnen Teile beim Arbeiten ausgeschlossen ist, da sonst bei den großen Kräften unweigerlich das Werkzeug oder ein Teil der Presse zu Bruche geht, da die bei den Führungsschnitten vorhandene feste Führung des Stempels in dem Schnittkasten fehlt

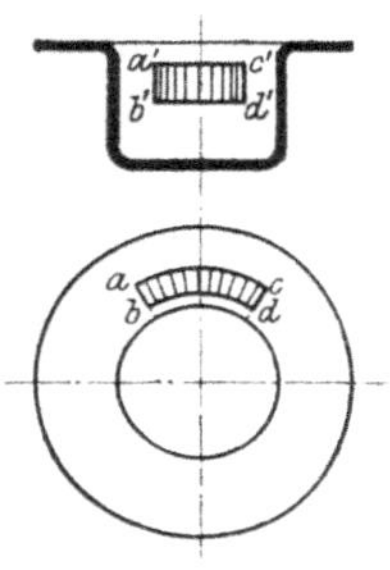

Fig. 328.

Das Werkzeug a, Fig. 327, wird in der Praxis das Anschlagwerkzeug genannt und zieht aus dem Blankett das Näpfchen.

Das vorgezogene Näpfchen kommt in das nächste Ziehwerkzeug b, den Weiterschlag, auch Tieferzieh- oder Nachziehwerkzeug genannt. Hier kann das Material bereits stärker beansprucht werden als im ersten Gang, da die Materialverschiebung eine mehr der früheren ähnliche Formänderung bewirkt. Außerdem wird, um den Übergang zu erleichtern, die Niederhalterfläche konisch mit abgerundeten Übergängen ausgeführt. Unterhalb des Ziehringes ist bei f eine scharfe Kante vorgesehen, die die gezogene Hülse abfängt, sobald sie den Ziehring verlassen hat und sich dabei ein wenig öffnet, so daß sie hier beim Aufgang des Stempels abgestreift wird.

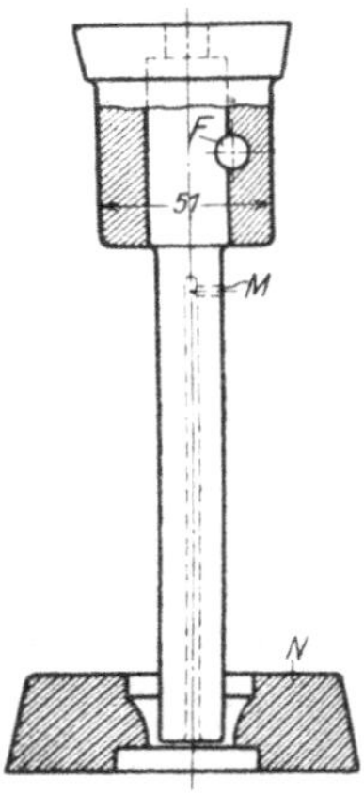

Wenn nun, was bei reinen Zieharbeiten meistens der Fall ist, die gezogenen Stücke einer weiteren Formveränderung bedürfen, so erfolgt das Weiterziehen in ähnlich geformten Werkzeugen, vgl. c, Fig. 327, die sich weder im Material noch im Aufbau besonders unterscheiden, sondern eben nur in ihren Abmessungen der gewünschten Formänderung des verarbeiteten Gegenstandes angepaßt sind. Fig. 326 zeigt einen derartigen Herstellungsgang aus drei

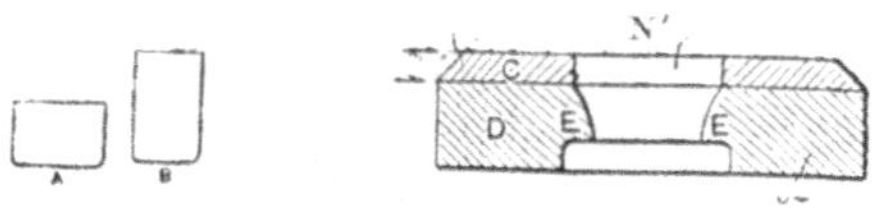

Fig. 329.

Ziehstufen in Aluminium, deren Abstufungsverhältnisse $d_1 : d_2$ der Reihe nach betragen: 0,86; 0,835.

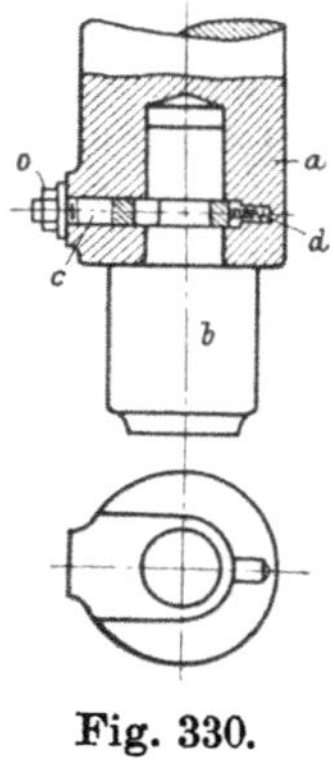

Fig. 330.

Vorteilhafte Konstruktion von Werkzeugen zum Nachziehen. Eine große Anzahl Ziehplatten wird wie N, Fig. 329, aus dem Vollen hergestellt und in der Froschplatte mittels vier Druckschrauben befestigt, die an den abgeschrägten Seitenflächen der Ziehplatte angreifen. Nach einer anderen Arbeitsweise hingegen wird die Ziehplatte aus zwei Teilen wie bei N' gemacht, deren Einlegering C für das vorgezogene Näpfchen A paßt. Die Einlegeringe werden in verschiedenen Größen gemacht und mit verschiedenen Ziehplatten D verwendet. Es kommt manchmal vor, daß ein Einlegering mit nicht weniger als 10 verschiedenen Ziehplatten verwendet wird. Es ist nun bekannt, daß eine abgenützte Ziehplatte gehämmert, wieder gehärtet, geschliffen und auf

Maß poliert werden kann. Wenn man nun die Ziehplatte aus dem Vollen herstellt, wie N zeigt, so muß der Einlegering auch geschliffen werden, nachdem die Ziehplatte gestaucht worden ist, so daß man offenbar Material und Zeit spart, wenn man diese Werkzeuge aus zwei Teilen macht. Eine Froschplatte und Einspannung für derartige Werkzeuge ist in Fig. 520 gezeichnet.

Es ist vielleicht hier am Platze, etwas über die Art und Weise, den Stempel in dem Stößel zu halten, zu sagen. In mehreren Fällen hat man die gewöhnliche Befestigung mittels Druckschraube und Gegenmutter aufgegeben und sie durch andere mit besserem Erfolg ersetzt, vgl. Fig. 329/30.

Der gußeiserne Stempelkopf paßt in eine Schwalbenschwanznut im Stößel der Presse und wird durch einen Keil, der nicht abgebildet ist, gehalten. Der Stift bei F paßt in die halbrunde Nut des Ziehstempels G, wenn der Stempel in den Kopf eingesetzt ist, so daß der Stempel beim Abstreifen der Hülse an der scharfen Kante E nicht herausgezogen werden kann. In obiger Figur erscheint der Stempel länger, als zum Ziehen der Hülse B, Fig. 329, nötig ist; er wird aber so gemacht, um Hülsen desselben Innendurchmessers, aber verschiedener Länge zu ziehen. Eine Befestigung mit eingedrehtem Hals zeigt Fig. 330.

Um nun eine wirtschaftlichere Ausnutzung der Zieharbeit zu ermöglichen, erlaubt die Doppelziehpresse der Bonner Maschinenfabrik Mönkemöller in Bonn eine unmittelbare Aufeinanderfolge zweier Ziehgänge in einem Kurbelhube der Presse. Die Arbeitsgänge und die Ausbildung dieser Werkzeuge sind schematisch in Fig. 331 dargestellt. In der Stellung I arbeitet das Werkzeug als Anschlagwerkzeug, indem der Stempel a und der innere Niederhalter b gleichzeitig als ein einziger Stempel nach abwärts gehen und aus dem Blech, welches unter dem

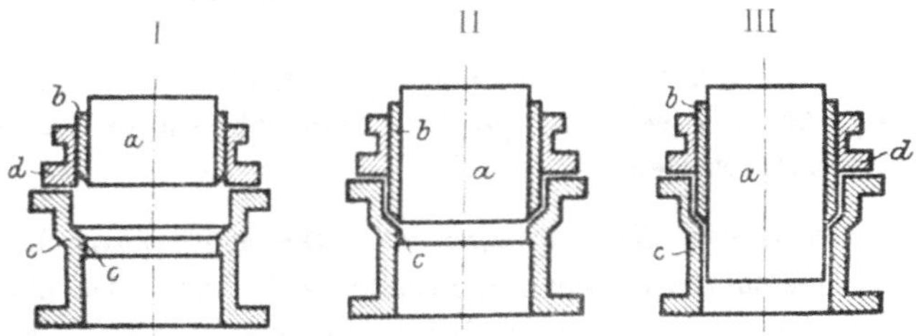

Fig. 331.

äußeren Niederhalter d und dem Ziehring c gehalten ist, das Näpfchen ziehen. Sobald die erste Ziehtiefe erreicht ist, Fig. 331—II, arbeitet der Niederhalter b nicht mehr als Ziehstempel, sondern als Niederhalter des Weiterschlagwerkzeuges, dessen Stempel jetzt nur mehr a allein ist, und dessen Ziehring der innere zylindrische Teil bei c ist. Der Stempel a geht nun allein nach abwärts, III, und zieht die vorgezogene Hülse weiter.

Die Vorteile dieser Werkzeuge, die in besonderen Pressen[1]) verwendet werden, bestehen in der Zeitersparnis und den verringerten Bedienungskosten, so daß man auf ungefähr die doppelte Leistung der entsprechenden einfachen Presse kommt, oder bei gleicher Leistung die halben Kosten an Zeit, Lohn und Platzaufwand hat.

Ziehen eines Messingbechers mit Rand. In Tabelle Nr. 13 ist ein Herstellungsgang, bestehend aus drei Zügen und einem Umlegen des Flansches, Fig. 425, gegeben. Es sind dies die Arbeiten, die nötig sind, um einen Becher von 20,6 mm Durchmesser und 23,8 mm Höhe mit einem Randflansch von 27 mm Durchmesser aus einem Blankett von 47,6 mm Durchmesser und 0,8 mm Dicke aus weichem Messing herzustellen. Tabelle Nr. 13, G. 5 gibt einen Schnitt durch den fertigen Becher. Bei dem Entwurf von Werkzeugen zum Ziehen gerader Becher lassen sich meistens nur angenäherte Regeln geben, um die Anzahl Gänge und Durchmesser der für jeden Gang zu verwendenden Werkzeuge zu geben. Praxis und lange Erfahrung sind bei der Herstellung der Werkzeuge erforderlich, um unnötige Arbeitsgänge zu vermeiden. Glatte Politur, Schmierung und keine zu große Tiefe der Ziehringe sind Punkte. die unbedingt beachtet werden müssen.

Die Ziehringe sind einfach rund, mit rechteckigen Kanten auf der Unterseite, die beim Durchzug als Abstreifer wirken. Die beiden Stempel für die ersten zwei Züge haben einen Anzug von $^1/_{500}$; der dritte ist vollständig zylindrisch. Alle haben Luftlöcher, um das Abstreifen der Becher zu erleichtern. Nach jedem Zug muß ausgeglüht werden. Nach dem Ziehen werden die Enden auf der Drehbank auf die zur Bildung der Flanschen nötige Länge abgestochen, worauf die Herstellung des Flansches folgt. Das dazu gehörige Werkzeug wird zu Fig. 425/6 beschrieben werden.

Eine der weitgehendsten Anwendungen der Zieherei ist die Patronenherstellung, bei der selbsttätige Zuführvorrichtungen in weitestem Maße verwendet werden.

Das Ziehen von Messingpatronen. Die amerikanischen 30—30 und 0,303″ Geschosse verwenden rauchloses Pulver. Da sich bei der Verwendung des rauchlosen Pulvers eine unerklärliche Brüchigkeit der Hülse nach der ersten Entladung einstellt, werden nur sehr wenige derselben wieder geladen. Infolgedessen erreicht der jährliche Bedarf nur dieser beiden Größen eine beinahe unglaubliche Ziffer. Diese beiden Größen sind sowohl in bezug auf äußeren Durchmesser wie auf ihr allgemeines Aussehen wenig voneinander verschieden, so daß,

[1]) Vgl. Fig. 650/1 Beschreibung dieser Doppelziehpressen.

da auch die Werkzeuge fast gleich sind, in der folgenden Beschreibung nur die Werkzeuge für das 0,303″ Geschoß beschrieben werden sollen.

Obwohl die Genauigkeit des Durchmessers und der Form aller Gewehrpatronen bis ins einzelne vorgeschrieben ist, damit man beliebige Fabrikate verwenden kann, und die zur Herstellung angewendeten Arbeitsgänge sehr zahlreich sind, so betragen doch die Kosten einer Hülse obigen Kalibers nur ungefähr 3 Pf.

Fig. 332a—m gibt die Arbeitsgänge bei der Herstellung der Patrone wieder, die sich mit den in Tabelle Nr. 14 gegebenen Schnitten größtenteils decken, so daß diese zum Verständnis mit benutzt werden kann.

Bei der Geschoßfabrikation wird, wo immer nur möglich, selbsttätige Zu- und Abführung der Stücke eingerichtet, wozu entweder Reibungs- oder Sperradvorrichtungen verwendet werden. Nach jedem Ziehen werden die Stücke ausgeglüht, wobei die Hülsen in eiserne Trommeln mit beinahe luftdicht schließenden Deckeln verpackt werden. Nachher läßt man sie ohne Verwendung von Wasser abkühlen.

Dann werden die Stücke in offene Bottiche mit schräger Achse gelegt, die langsam gedreht werden, während verdünnte Schwefelsäure über die Hülsen gegossen wird. Nach Entfernung des Glühspanes wird die Säure von den Stücken durch Übergießen mit Wasser abgewaschen.

Der Hals der gezogenen Hülse wird vor dem Einziehen in einer Maschine ausgeglüht, die mittels eines Reibungsvorschubes eines der Stücke nach dem andern an einer Gasflamme vorbeiführt und in einem Behälter absetzt.

Ziehen und Nachziehen der Hülse. Im ersten Arbeitsgange wird aus Messingblech von 0,23 mm Dicke in einer gewöhnlichen doppeltwirkenden Ziehpresse die Scheibe ausgeschnitten und das Näpfchen gezogen, Fig. 332a. Die Figuren 332a — m stellen den vollständigen Arbeitsgang vom flachen Blankett bis zum fertigen Geschoß dar.

Alle Ziehringe sind Gesenkschmiedestücke, die auf der Außenseite unbearbeitet bleiben. Sie werden in einem Futter gehalten, gebohrt, vor- und nachgerieben bis auf ungefähr 0,8 mm. Hierauf wird der Ziehring auf der Unterseite unter Fertigmaß genügend ausgebohrt, so daß eine bearbeitete Oberfläche von 1,6 mm Tiefe auf der Unterseite vorhanden ist, und die gleiche Höhe in dem Loch für die Ziehfläche bleibt.

Nach dem Bohren werden die Ziehringe gehärtet und auf Maß poliert. Wenn die kleinen Durchmesser abgenutzt sind, werden sie ausgeglüht und auf das nächstgrößere Maß nachgerieben, bis sie schließlich im ersten Zug vollständig abgenutzt sind.

Für das erste und zweite Nachziehen, Fig. 332b, verwendet man zwei Ziehringe übereinander, so daß im ganzen sechs besondere Zieh-

Arbeitsgang	1	2	3	4	5	6	7	8	9	10	11
Gegenstand	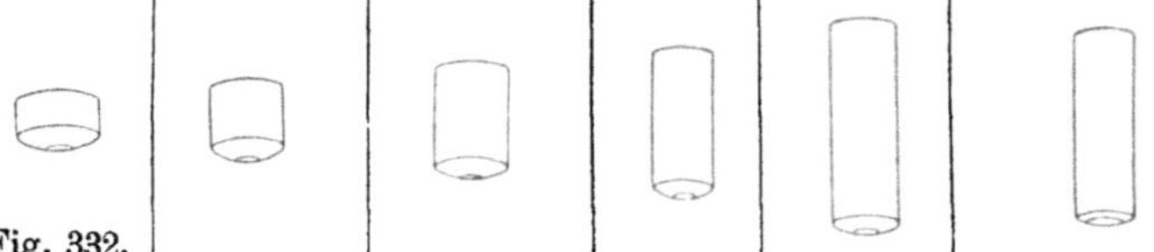 Fig. 332. a			b	c		d		e		f
Arbeitsverfahren	Scheibe ausschneiden Näpfchen ziehen		Ausglühen, Beizen, Waschen	1. Ziehen und 2. Ziehen der Hülse in einem Gang		Ausglühen, B. W.	3. Ziehen	Ausglühen, B. W.	4. Ziehen	Ausglühen, B. W. Abschneiden	Vorpressen des Bodens
Werkzeug	Verbund-Schnitt- und Ziehwerkzeug			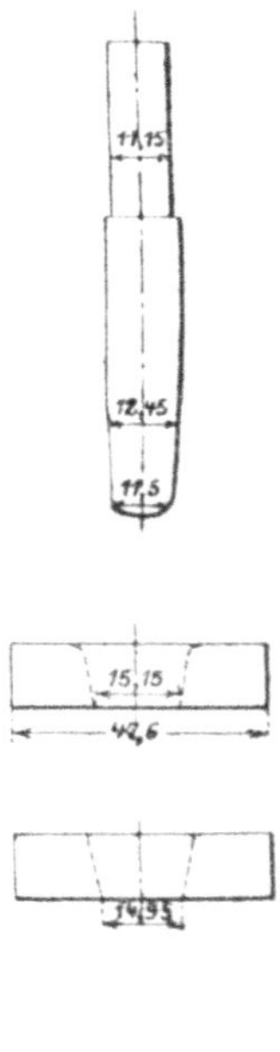 Fig. 333.			Einfache Zieh-Werkzeuge wie Fig. 335, bzw. 333				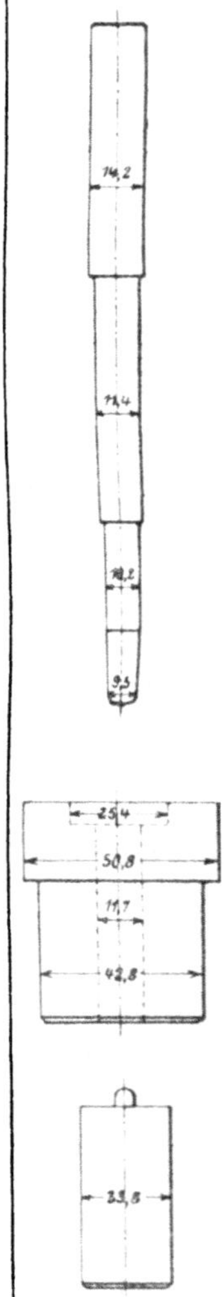 Fig. 334.

12	13	14	15	16	17	18	19	20	21	22	23
g			h	i		k		l			m
5. Ziehen	Ausglühen, B. W.	Abschneiden	Fertigpressen des Bodens	Lochen	Ausglühen	1. Einziehen	2. Einziehen	Zuschneiden auf Länge	Öffnung fertig bearbeiten	Kalibrieren	Fertigarbeiten und zusammensetzen

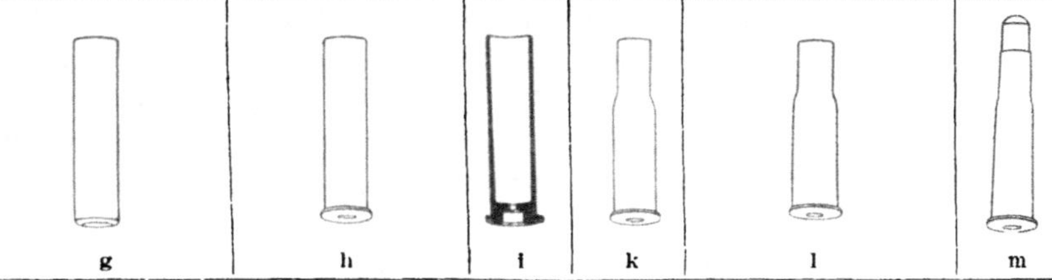

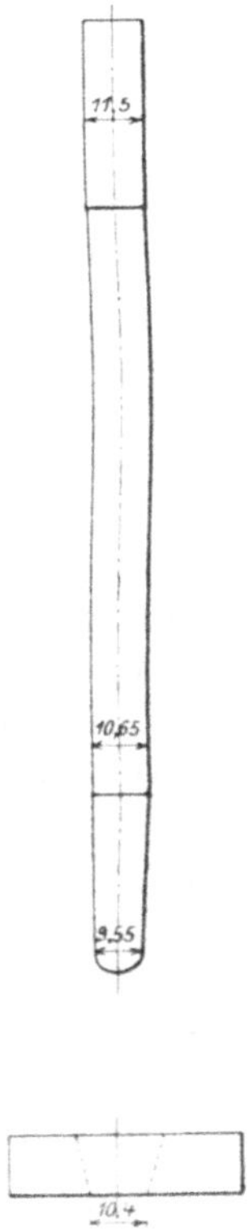

Fig. 335.

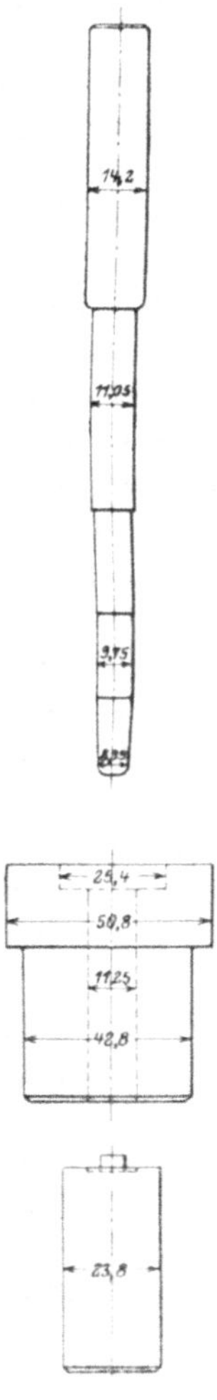

Fig. 336.

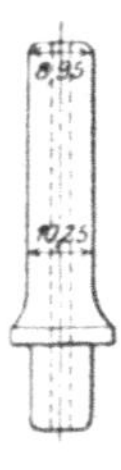

Fig. 337.

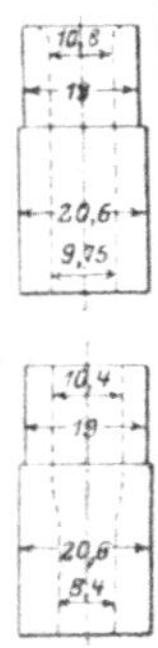

Fig. 338.

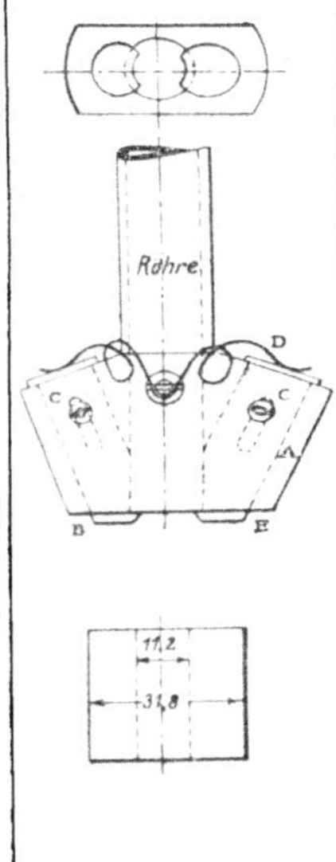

Fig. 339.

ringe und fünf Stempel beim Nachziehen verwendet werden. Die Ziehringe werden auf der Grundplatte in einer Froschplatte gehalten, die sie sicher niederhält und eine kleine seitliche Verschiebung zur Selbsteinstellung gestattet. Die Stempel werden gedreht, ihre Enden mit einem Kastenwerkzeug mit besonderem Formstahl für jede Größe bearbeitet, dann gehärtet und auf den verlangten Durchmesser geschliffen.

Der erste und fünfte Stempel für das Nachziehen, b und g, sowie der obere und untere Ziehring zu b und der Ring zu g sind in Fig. 333/335 abgebildet. Das zweite, dritte und vierte Ziehwerkzeug c, d, e sind ausgelassen, da sie in ihren allgemeinen Formen und Abmessungen mit den abgebildeten übereinstimmen und nur die Loch- und Stempeldurchmesser sich ändern.

Nach dem fünften Arbeitsgang werden die Hülsen auf Länge in einer selbsttätigen Maschine, „Abringelmaschine", zugeschnitten, in deren Zuführungsrinne sie alle im gleichen Sinne eingelegt werden. Dadurch werden alle unregelmäßigen Kanten entfernt und alle Stücke gleich lang, so daß beim fünften Nachziehen g der Boden weniger leicht ausgestoßen wird. Von der Abringelmaschine kommen die Stücke in eine andere selbsttätige Maschine, wo der Sitz für den Zünder roh vorgepreßt wird. Die betreffenden Werkzeuge sind Gesenk, Stempel und Prägestempel, Fig. 334. Die Hülsen gehen wie im vorangehenden Falle durch eine Zuführungsrinne, worauf der Stempel eine derselben erfaßt, in den Ring schiebt und dort hält, bis der Prägestempel herankommt und die Vertiefung einprägt. Infolge der selbsttätigen Zuführung wird hier sehr schnell gearbeitet.

Nun werden die Hülsen zum fünften Male nachgezogen, g, wieder auf Länge abgeschnitten und der Boden fertiggepreßt. Die Werkzeuge für diese Arbeit sind in Fig. 336 abgebildet. Die Arbeitsweise ist ähnlich dem Vorpressen der Vertiefung für den Zünder, weshalb sie auf der gleichen Maschine ausgeführt wird. Bei dieser Arbeit wird der Rand gepreßt, die Schrift eingeprägt und die Vertiefung für den Zünder angebracht.

Zur Herstellung des Loches für den Zünder wird am Gestell der Presse ein genuteter Revolverteller eingebaut, der eine zwangläufige, genaue Schaltung durch Schalt- und Sperrklinken erhält. Fest auf dem Revolverteller werden 16 säulenförmige Lochschnitte, deren Außenform der Innenform der Hülse entspricht, mit einem entsprechenden Loch, in das der eigentliche Lochstempel eintritt, befestigt. Von unten ist, wie aus Fig. 337 ersichtlich ist, das große Loch erweitert, damit die Abfälle herausfallen können. Der Stempel besteht aus einem Stück Bohrerstahl, welches passend angelassen wird. Die Einrichtung zur Entfernung der Hülsen, Fig. 339, von dem Lochwerkzeug ist am Ende der Beschreibung angegeben.

Nach dem Ausglühen des Halses wird die Hülse eingezogen. Dies geschieht in einem Arbeitsgang in zwei Werkzeugen, die nebeneinander an dem Stößel der Presse befestigt sind, und deren Form und Abmessungen aus Fig. 338 zu ersehen sind. Auch hier kann der Revolverteller mit Vorteil verwendet werden. Die Hülsen werden mit dem Rande in den Teller eingehängt und so an den ersten Ziehring, der sie konisch vorzieht, gebracht und dann im zweiten Fertig-Werkzeug auf die richtige, eingezogene Form gebracht. Bei jedem Preßhub wird eine Hülse fertig und fällt beim weiteren Drehen des Revolvertellers nach unten aus. Von da gehen die Stücke zu einer anderen selbsttätigen Maschine, ähnlich jener, welche die Hülsen auf Länge zuschneidet, wo dann der Rand mit einem runden Formfräser auf den verlangten Durchmesser und die nötige Dicke gebracht wird. Die Hülsen gehen nun durch einen sechsten Automaten, wieder ähnlich der ersten Abringelmaschine, in welcher sie auf genaue Länge abgeschnitten werden.

Es mag hier erwähnt werden, daß die Vorrichtungen zum Festhalten der Hülsen Einzug- oder Ausstoßwerkzeuge irgendeiner bekannten Bauart sind. Meistenteils treten die Hülsen, bevor die Köpfe gepreßt werden, eine nach der andern selbsttätig in die Zuführungshülsen ein, gehen durch die Spindel und kommen am rückwärtigen Ende heraus. Alle diese Maschinen sind mit selbsttätigen Zuführungen für das Werkstück ausgestattet.

Nachdem die Stücke durch alle diese Arbeitsstufen gegangen sind, sind sie oft leicht verbogen oder ein wenig aufgeweitet, so daß besondere Kalibrierwerkzeuge angewendet werden, um etwaige Abweichungen richtig zu stellen und alle Hülsen auf gleiches Maß zu bringen. Ein derartiges Werkzeug ist in Fig. 339 abgebildet.

Zwölf Werkzeuge dieser Art sind auf einem Revolverteller von 25 oder mehr mm Dicke aufgebaut. Während dieser Teller sich dreht, werden die Hülsen leicht und schnell von dem Arbeiter mit dem eingezogenen Ende in die Gesenke gesteckt. Ein Stempel mit flachem Ende stößt dann die Hülse bis an den Rand in das Gesenk, worauf beim Weiterdrehen des Tellers das Arbeitsstück an dem Stempel vorbeigeführt und beim nächsten Preßhub um ungefähr 6 mm aus dem Gesenk gehoben wird; in diesem Augenblick wird die Hülse von einem Greifer nach Fig. 339 gefaßt und entfernt.

Der Greifer ist an dem Stößel der Presse sicher befestigt, so daß er genau über ein Gesenk zu stehen kommt, wenn der Revolverteller unter den Stempel geschaltet wird. Die Röhre am Kopfe des Greifers geht ein Stück gerade nach aufwärts, biegt dann U-förmig nach außen um und wirft die Stücke in einen Kasten neben der Presse aus.

Der Körper A des Greifers ist ein Schmiedestück aus Stahl, auf der Außenseite bearbeitet, und erhält ein mittleres Loch, welches ein wenig größer als der größte Durchmesser der Hülse ist. B sind zwei Stücke aus Werkzeugstahl, die mit einem Schlitz für die Schrauben C versehen sind, so daß eine Drehbewegung verhindert ist, dagegen eine Auf- und Abwärtsbewegung derselben leicht erfolgen kann. Die Drahtfeder D hat die Stücke B nach unten zu drücken und so die Hülsen festzuhalten. Beim Abwärtsgang des Stößels kommen die runden Ausdrehungen bei E mit dem Rand der Hülse in Berührung, werden seitwärts nach oben gedrückt, so daß sie sich öffnen und den Rand eintreten lassen; da nun die Federn andauernd nach abwärts drücken, wird die Hülse unterhalb des Kopfes gepackt und in senkrechter Stellung gehalten. Beim folgenden Niedergang des Stempels kommt das vorstehende untere Ende der Hülse mit dem Oberteil eines der Werkzeuge am Revolverteller in Berührung und wird dadurch nach oben in die Ablaufröhre geschoben, so daß die Hülsen, da diese Arbeit ununterbrochen vor sich geht, allmählich durch die ganze Röhre nach außen befördert werden. Die letzte Arbeit, das Abschrägen der Innenkante der Öffnung, wird von Hand besorgt, indem man die Hülse gegen einen schnellaufenden Versenker andrückt.

Deutsche Werkzeuge zur Herstellung der Patronenhülsen (Fritz Werner, Berlin W, Lützowstr.). Wie aus der Tabelle Nr. 14 und der vorangehenden Beschreibung der amerikanischen 0,303″ Patronenhülsen zu ersehen ist, handelt es sich bei allen drei Hülsen um ein ziemlich gleiches Stück, bei welchem nur die Anzahl der Ziehgänge und die Herstellung des Bodens mit dem Zünderloch und des Randes eine Verschiedenheit in den Arbeitsgängen und Maschinen verursacht.

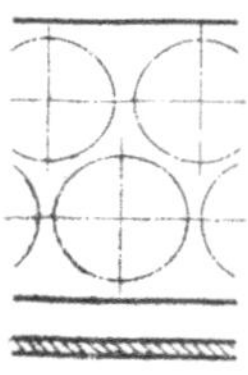

Fig. 340.

Während nach der deutschen Angabe die einzelnen Ziehgänge und das Pressen des Bodens mit seiner Vertiefung für das Loch des Zünders mit den amerikanischen Arbeitsgängen übereinstimmen, wird bei dem deutschen Verfahren das Einziehen und Kalibrieren in einen Arbeitsgang verlegt, die Zündlöcher nicht gelocht, sondern gebohrt und die Ausziehernut, die bei der amerikanischen Patrone angepreßt worden war, hier eingedreht. Das ist auch der augenfälligste Unterschied beider Hülsen, abgesehen von dem nachherigen Verwendungszweck. Aus Fig. 340 ist ersichtlich, daß das Rohmaterial in Streifen, die das Ausschneiden zweier Blanketts gestatten, verwendet wird. Daß hier Rollenmaterial mit selbsttätiger Auf- und Abwickelvorrichtung, verbunden mit einem selbsttätigen Vorschub, verwendet wird, ist wohl selbstverständlich. Aus dem Streifen wird im ersten Arbeitsgang wie bei der amerikanischen Hülse in einem Verbund-Schnitt- und Ziehwerkzeug das Näpfchen, Fig. 341 a, hergestellt, welches in den nächsten drei Arbeitsgängen weitergezogen wird. Zwischen jedem Ziehen wird die Hülse wieder ausgeglüht, gebeizt und gewaschen. Dem dritten Ziehen folgt das Abringeln, Fig. 342, welches auf selbsttätigen Maschinen vorgenommen wird. Die zugeführten Hülsen werden von dem umlaufenden Dorn gefaßt, von dem herankommenden Gegenhalter in ihrer Stellung gesichert, worauf der Dorn in die Hülse eintritt. Sobald er die richtige Tiefe erreicht hat, bewegt sich der Arm, der das Kreismesser trägt, heran

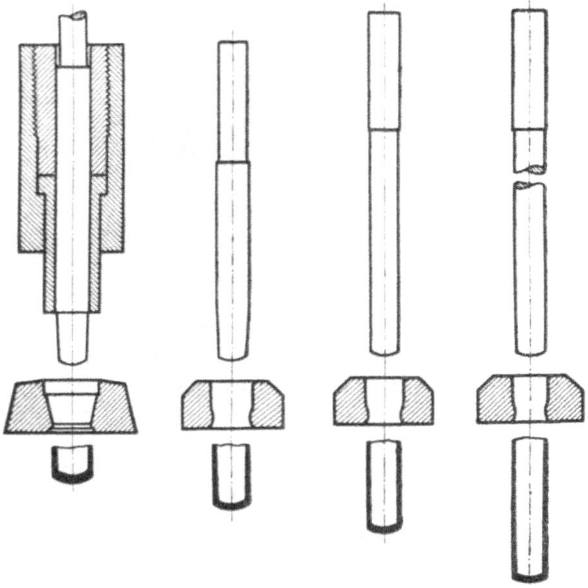

Fig. 341.

und schneidet an der Hülse die übrige Länge weg. Dann geht das Kreismesser zurück, ebenso der Gegenhalter, worauf der Abstreifer Abfall und Hülse von dem Dorn abschiebt. Dieser Arbeit folgt das Vorpressen des Bodens, Fig. 343, dessen Werkzeuge denen in Fig. 334 entsprechen. Bei diesem Werkzeug erfolgt die Pressung in einem allseits geschlossenen Preßgesenk, so daß eine schärfere Arbeit erzielt wird. Aus diesem Grund ist die Führungsbüchse für die Hülse von dem eigentlichen Gesenk getrennt hergestellt, in das der Stempel, der die Vertiefung für die Zündlöcher preßt, geführt eintritt. Deutlich ersichtlich ist die Befestigung der Gesenkteile so durch Verschraubung, daß man jeden derselben leicht und schnell auswechseln kann, ohne die richtige Ausmittlung der Werkzeuge zu stören. Nach dem Vorpressen werden die Hülsen zum vierten Male gezogen, Fig. 344, wieder abgeschnitten, Fig. 345, worauf der Boden in einem Werkzeug nach Fig. 346, das

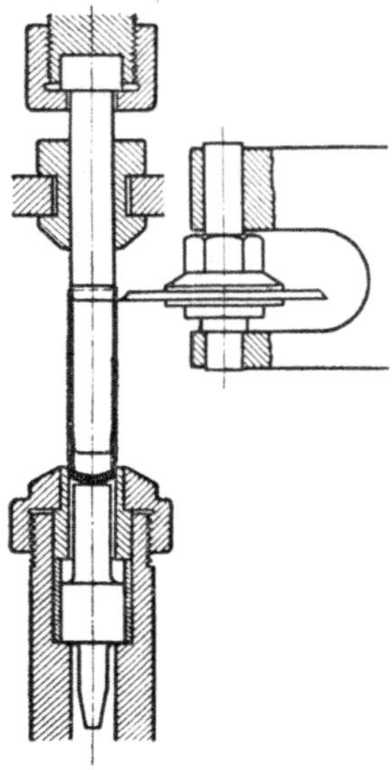

Fig. 342.

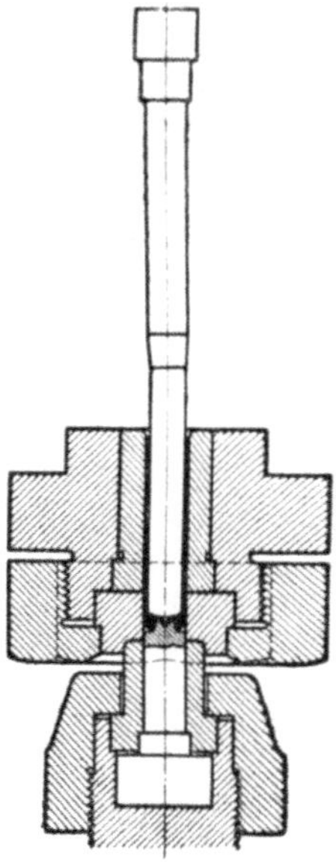

Fig. 343.

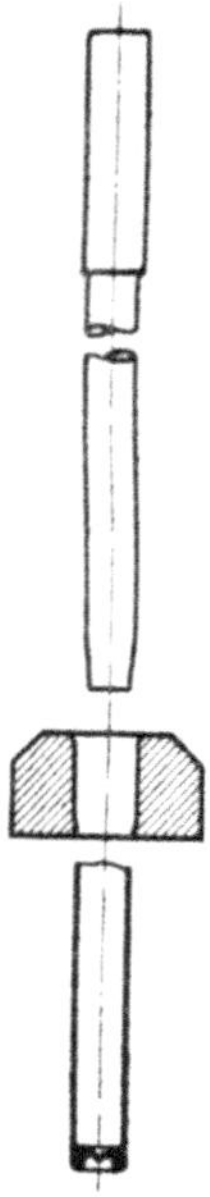

Fig. 344.

Fig. 345.

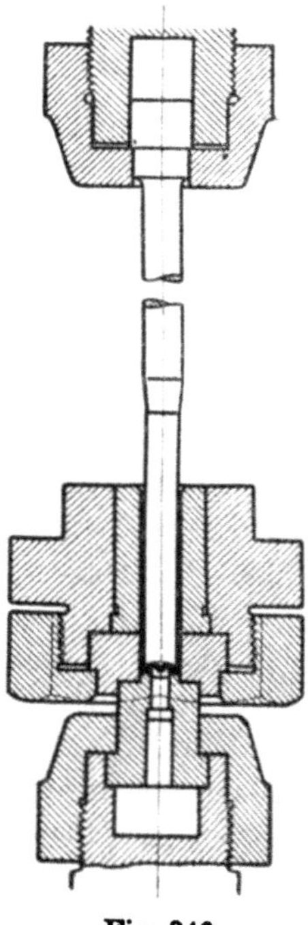

Fig. 346.

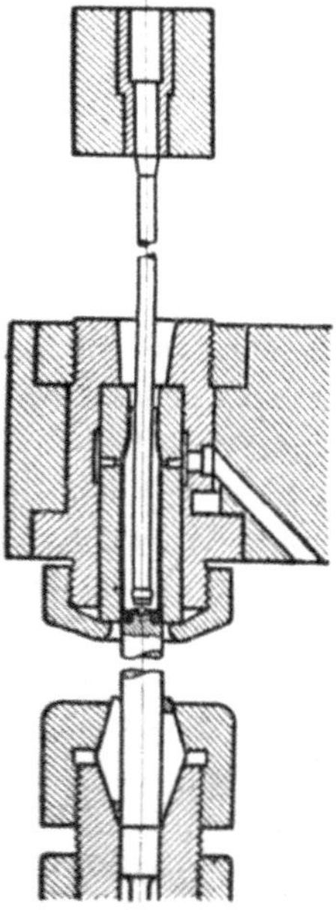

Fig. 347.

nach denselben Grundsätzen wie das nach Fig. 343 gebaut ist, fertiggepreßt wird. Die folgenden Arbeitsgänge sind aber von dem amerikanischen Verfahren verschieden. Es erfolgt das Einziehen und Kalibrieren gleichzeitig in einem Arbeitsgang auf einem Werkzeug nach Fig. 347. Das eigentliche Einziehgesenk ist aus einem Stück hergestellt und wird mit einer Überwurfmutter in einer Büchse gehalten, durch die eine

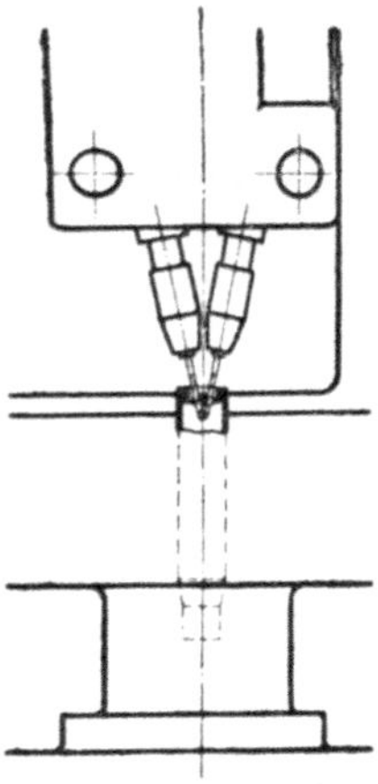

Fig. 348.

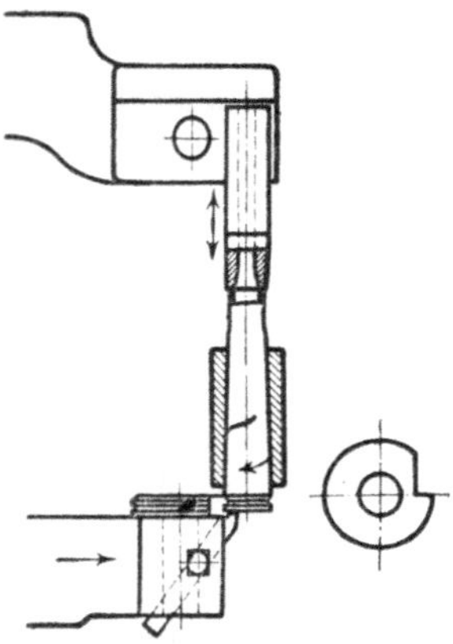

Fig. 349.

Druckschmierung für die Einziehstelle vorgesehen ist. Die Hülse wird zwischen den unteren Stempel, der den Boden kalibriert, und das Einziehgesenk eingeschoben, wobei der obere Gegenstempel, der gleichzeitig die Stelle eines Auswerfers vertritt, die Hülse festhält. Hierauf werden, Fig. 348, die Zündlöcher in einer eignen zweispindligen Bohrvorrichtung gebohrt, bei der die schnellaufenden Bohrer in ihren Führungshülsen gegen die feststehende, im Futter gehaltene Hülse vorgeschoben werden. Die gebohrten Hülsen kommen dann in einen Automaten, in welchem mit einem kreisförmigen Formstahl, Fig. 349, die Ausziehernut eingedreht wird, während ein im gleichen Halter befestigter Drehstahl den Boden gerade dreht und ein geradlinig an die Hülse herangeführter Fräser das Stück auf genaue Länge abfräst. Sämtliche Werkzeuge erhalten die Vorschub-, die Hülse die Drehbewegung.

Im übrigen sei auf Tabelle Nr. 15 verwiesen.

Die Herstellung der Nickelkupfermantel-Geschosse. Der Geschoßmantel der 30-Kaliber-Patrone, die in den Militärwerkstätten der Vereinigten Staaten hergestellt wird, ist eine Legierung von Kupfer und Nickel. Das Mantelblankett wird ausgeschnitten und in einer Reihe Arbeitsgänge fertig gezogen und gepreßt, die in Tabelle Nr. 16 wiedergegeben sind. Das Rohmaterial, aus welchem auf einer doppelt wirkenden Presse die Scheibe von ungefähr 25 mm Durchmesser ausgeschnitten wird, ist 0,9 mm dick. Nach drei aufeinanderfolgenden Ziehgängen wird das übrige Material an der Öffnung abgestochen und die Hülse noch einmal gezogen, die man dann immer noch ein wenig zu lang läßt, so daß man sie schließlich auf die genaue Länge abstechen kann. Sämtliche hier verwendeten Ziehpressen sind mit selbsttätigen Zuführungen versehen. Die Geschosse selbst, zur Füllung der Mäntel, werden aus einer Legierung von Blei und Zinn in der Bleigießerei zu Frankford, Ill. gegossen. Diese Gießerei enthält sieben

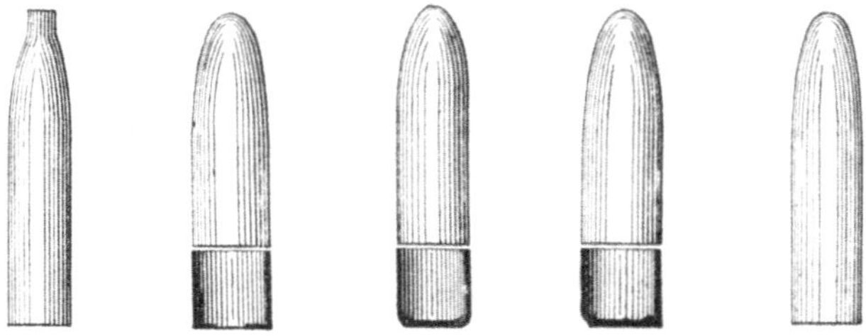

Fig. 350.

mit Öl geheizte Schmelzöfen mit je vier Tiegeln. Die Formen fassen je 16 Geschosse, von denen ein Mann täglich 225 kg gießt. Von der Bleigießerei kommen die Bleikerne zu den Zusammensetzmaschinen, wo die Mäntel, nachdem sie in der Scheuertrommel gereinigt worden sind, geformt und mit Blei gefülllt werden. In Fig. 350 sind die letzten Herstellungsstufen gegeben, wie die Geschosse die Zusammensetzmaschinen verlassen.

Diese Pressen werden von zwei Arbeitern bedient und vollführen gleichzeitig sechs Arbeitsgänge an ebensoviel Mänteln. Zwei Revolverteller, die sich in einer wagerechten Ebene drehen, bringen die Arbeitsstücke unter die Stempel. Die Revolverteller haben am Umfang eine Reihe von Kammern und stehen so, daß der Teller, der die Bleigeschosse trägt, den Teller mit den Mänteln an einer Stelle übergreift, so daß dort die Bleifüllung aus ihrer Kammer durch den oberen Revolverteller und in den Mantel im unteren Revolverteller gedrückt werden kann. Diese Revolverteller erhalten einen Sperrad-Vorschub, bei dem durch jede Schaltung alle Stempel einmal zum Arbeiten kommen. Sobald der Mantelteller den ersten Mantel unter den ersten Stempel bringt, wird die Spitze des Mantels vorgepreßt und beim folgenden Hub unter dem nächsten Stempel fertiggeformt. Bei der dritten Schaltung wird der Bleikern in den Mantel gepreßt und gleichzeitig das offene Ende desselben leicht umgebörtelt, während unter dem nächsten Stempel diese Arbeit zu Ende geführt wird. Der letzte Stempel drückt nun das ummantelte Geschoß durch eine Kalibrieröffnung und aus der Maschine heraus. In einer zweiten Kalibrierung wird das rückwärtige Ende des Geschosses leicht konisch geformt und danach am flachen Ende ein O eingeprägt.

Die Einrichtung der Maschinen zum Laden der Geschosse besteht hauptsächlich aus zwei wagerechten Revolvertellern, einer für die Patrone und einer für das Geschoß, einem Pulvermagazin mit Ladevorrichtung und einer Reihe Stempel, die von einem senkrechten, hin- und hergehenden Stößel betätigt werden. Der Teller zur Rechten ist mit Nuten zur Aufnahme der Patronen versehen, die diesen Nuten durch eine dauernd sich drehende Scheibe zugeführt werden, auf deren offenes Ende sie vom Arbeiter mit der rechten Hand niedergelegt werden. Die Scheibe zur Linken ist durch Räderübersetzung mit dem ersten Revolverteller verbunden und enthält Löcher, in welche der Arbeiter mit der linken Hand die Geschosse einzeln fallen läßt. Die Schaltung der Teller erfolgt wieder durch einen Sperradantrieb, dessen Hubpause so lang ist, daß die Pulverladevorrichtung und die Stempel ihre Arbeiten ausführen können.

Die Ladevorrichtung besteht aus einem hin- und hergehenden Schlitten, der eine Kammer für die genaue Pulvermenge enthält. Dieser Schlitten wird durch eine Führung unter dem Magazin hin- und herbewegt und geht bei jeder Schaltungspause so weit vor, bis die Kammer gerade über der leeren Patrone steht, wo sie ihre Füllung abgibt. Hierauf kehrt der Schlitten wieder zum frischen Füllen unter das Magazin zurück. Die Ladevorrichtung erhält eine Schüttelbewegung, um ein Festsetzen des Pulvers beim Ablauf zu verhindern und das Ablegen der vollen Ladung zu gewährleisten. Sobald die gefüllte Patrone in die nächste Stellung weitergeschaltet worden ist, drückt ein Stempel das Pulver darin fest, während ein zweiter, mit einer Glocke in Verbindung stehender Stempel ein Warnungszeichen gibt, wenn aus irgendeinem Grunde eine Patrone ohne richtige Füllung an dem Magazin vorbeigegangen sein sollte. Sobald nun die Patrone unter den übergreifenden Teil des Revolvertellers kommt, drückt ein Stempel das Geschoß hinein, worauf das fertiggeladene Geschoß aus der Maschine entfernt wird. Vor dem Laden werden die Geschosse in eine Lösung von japanischem Wachs und Graphit getaucht und die Patronen sorgfältig gereinigt. Auf dieser Maschine können sowohl Platz- wie scharfe Patronen geladen werden, und zwar ungefähr 16 500 Stück in einem achtstündigen Arbeitstag.

Das schlotförmige Magazin für das Pulver aus Stahlblech sieht wohl äußerlich ziemlich unförmig aus, enthält jedoch kaum mehr als ungefähr 1 bis 1,5 kg Pulver, und ist auch genügend hoch, um den Arbeiter im Falle einer Explosion zu schützen. Das Pulver wird tatsächlich in einer geraden Röhre aus Pappe von zirka 50 mm im Durchmesser gehalten, die mit einem sich erweiternden Ende bis an die Metallumhüllung reicht. Zur Anzeige des Pulverstandes in der Röhre dient ein Kugelgewicht, das auf dem Pulver liegt und durch eine Schnur über eine feste Scheibe ein leichteres Gegengewicht trägt, das außen beim Stande des Arbeiters sichtbar ist und durch sein Sinken den Stand des Pulvers in der Röhre anzeigt.

Einsetzen der Patronen in die Magazine. Nach der Kontrolle werden scharfe wie Platzpatronen in Magazine eingesetzt, Fig. 351, was ebenfalls auf Maschinen geschieht. Die Magazine und die Federn werden zuerst auf besonderen Maschinen zusammengesetzt und dann in das Zuführungsmagazin der ersten oben erwähnten Maschine, System

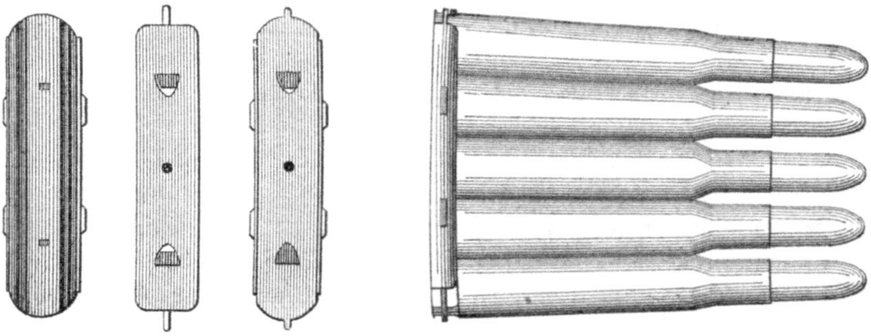

Fig. 351. Patronenmagazin.

Lissak, eingelegt. In jeden Schlitz des Revolverkopfes dieser Maschine werden beim Schalten fünf Patronen eingelegt, über welche, sobald der betreffende Schlitz bis an das Magazin geschaltet worden ist, das Magazin gezogen wird. Wenn das gefüllte Magazin der Fallrinne gegenübersteht, wird es aus dem Revolverkopf gehoben und durch einen in senkrechter Richtung hin- und hergehenden Arm mit federnden Greifern aus der Maschine entfernt.

Scharfe Patronen sind 0,25 mm länger als Platzpatronen, damit nicht zufällig eine scharfe Patrone in ein Magazin mit Platzpatronen eingesetzt wird. Es würde dann die scharfe Patrone über die andern und die Magazinführung so weit herausstehen, daß ein Vorwärtsziehen des Magazins verhindert wäre und die Maschine stehenbleiben würde. Von diesen Zusammensetzmaschinen sind eine Anzahl im Gebrauch, deren jede ungefähr 100 Patronen in der Minute einsetzt.

Das Ziehen der Patronenhülsen für Schnellfeuergeschütze. Die Entwicklung der Schnellfeuergeschütze hat es sofort mit sich gebracht und ist eigentlich dadurch erst richtig möglich geworden, daß bedeutende Verbesserungen in bezug auf schnelle Ladefähigkeit der Munition durchgeführt worden sind. Schnellfeuergeschütze unterscheiden sich von den gewöhnlichen dadurch, daß die Ladung und der Zünder in einer Metallhülse untergebracht sind. Das Geschoß ist entweder mit dieser Hülse in Verbindung oder nicht, was von der Größe des Geschützes abhängt. Bei den gewöhnlichen Geschützen sind Geschoß, Ladung und Zünder getrennt, wobei die Ladung meistens in einem verbrennbaren Seiden- oder Sergesäckchen untergebracht ist. Die Vorteile der Metallhülsen gegenüber den verbrennbaren Säckchen sind folgende: Sie lassen ein schnelleres Laden zu, da nur zwei Teile einzusetzen sind. Außerdem werden Fehlzündungen eher vermieden. Das Reinigen des Geschützrohres, das nötig ist, um eine Vorzündung der nächsten Ladung durch glimmende Überreste der Säckchen zu verhindern, fällt weg. Die Ausdehnung der Hülse beim Feuern ersetzt die Liderungsringe, weil die Gase von selbst abgedichtet werden.

Gleichzeitiges Laden mit fester Munition, wobei das Geschoß mit der Ladungshülse fest verbunden ist, wird bei Schnellfeuergeschützen bis ungefähr 75 mm Durchmesser angewandt, darüber hinaus würden die Geschosse dieser Art zu unhandlich werden. Deshalb werden Geschütze von 75 bis 150 mm Durchmesser gesondert geladen, das Geschoß für sich und die in einer Metallhülse gelagerte Pulverladung auch für sich. Über 150 mm Durchmesser spricht man nicht mehr von Schnellfeuergeschützen und verwendet bei ihnen gesonderte Ladung in verbrennbaren Säckchen. Für die größeren Schnellfeuergeschütze ist die gesonderte Ladung wünschenswert, nicht allein des bedeutenden Gewichtes des ganzen Geschosses wegen, sondern auch wegen der Gefahr, geladene und mit Zünder versehene Geschosse zusammen mit geladenen Hülsen aufzubewahren. Bei getrennter Ladung hingegen können die Geschosse nach Belieben nahe dem Geschütz aufgestapelt werden, während die Ladung erst im letzten Augenblick aus dem Magazin geholt wird. Die Bedingungen für Aufbewahrung, Beförderung und Gebrauch der Geschosse und der Ladungen sind so verschieden, daß die beiden Teile besser getrennt bleiben.

Die Hülsen für die Ladung solcher Schnellfeuergeschütze werden im allgemeinen aus Messing gemacht, da man fand, daß dieses Material den auftretenden Arbeitsbedingungen am besten entspricht. Von all den verschiedenen Legierungen nimmt diese Kupfer-Zinklegierung, das Messing, den ersten Rang ein. Das käufliche Messing besteht aus zwei Teilen Kupfer und einem Teil Zink und wird in fast allen Ländern für Patronenhülsen für Gewehre und Schnellfeuergeschütze benutzt. Die genaue Zusammensetzung ist 67 % Kupfer, 33 % Zink mit einer zulässigen Abweichung von ± 5 % für jedes Metall.

Die gesamte Herstellung dieser Metallhülsen besteht mit Ausnahme von zwei oder drei Arbeitsgängen aus Zieharbeiten im kalten Zustand. Das Messing kann kalt in hohem Grade gestreckt werden, heiß jedoch nicht. Nach dem Ziehen des Näpfchens folgen verschiedene weitere Ziehgänge, um Durchmesser und Dicke der Hülse zu verringern und die Länge zu vergrößern, wobei sich das Volumen kaum merklich ändert. Bei jedesmaligem Ziehen wird das Material bis weit über die Streckgrenze beansprucht, so daß es nachher ausgeglüht werden muß. Die ersten Ziehgänge, solange die Hülse noch kurz ist, werden auf einer senkrechten Ziehpresse ausgeführt; sobald jedoch das Stück so lang wird, daß das Zurückziehen des Stempels Schwierigkeiten zu machen beginnt, geht man zur liegenden Presse über. Der Ziehring ist ein gehärteter, angelassener Stahlring, dessen Innenfläche die Form eines Kegelstumpfes hat und dessen Achse in einer Linie mit dem Stempel liegt. Der Arbeitsgang geht so vor sich, daß man den vorgezogenen Gegenstand richtig ausgemittelt in die weite Öffnung des Ziehringes einlegt und den Stempel vorgehen läßt, bis er den Boden des Näpfchens berührt. Der hydraulische Druck preßt nun die Hülse durch den engeren Teil des Ziehringes, so daß hierbei der Durchmesser und die Wandstärke verringert, die Länge der Hülse hingegen vergrößert wird, was mit einem beträchtlichen Fließen des Materials verbunden sein muß.

Herstellung der Hülsen. (Tabelle Nr. 17.) Der erste Gang ist das Ausschneiden der Blanketts und die Herstellung der Näpfchen aus denselben; dies geschieht in zwei Stufen, Fig. 352a und b. Vor Beginn ist es notwendig, den Ziehring gegen den Stempel auszumitteln, indem man die Entfernung zwischen Ring und Stempel an drei Stellen mißt. Dann wird der Hub des Ziehstempels eingestellt, so daß er am Ende des Hubes gerade nur so tief geht, daß das gebildete Näpfchen aus dem Ziehring ausgestoßen wird, um Zeit und Kraft nicht zu verschwenden. Bei Beginn des Hubes wird 100 bis 200 mm mehr als der notwendige Hub nach oben eingestellt, so daß der Arbeiter Zeit hat, das Blankett einzulegen, solange die Presse frei ist. Dann werden Werkzeug und Blankett gut geschmiert und diese in die Presse eingelegt. Die größte Kraft im hydraulischen Zylinder, die durch ein an den Zylinder angeschlossenes Manometer angezeigt wurde, überstieg in diesem Falle

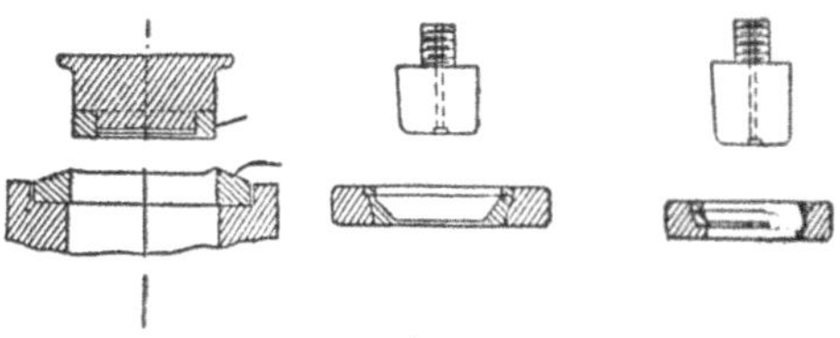

a b c

Fig. 352.

nicht 70 at. Dann wird das Näpfchen, das mit einer Stahlklammer umfaßt wird, 28 Minuten lang bei 740° C ausgeglüht. Der Glühspan an der Oberfläche wird durch Beizen in einem bleigefütterten Trog, der im Verhältnis 1:4 mit Wasser verdünnte Schwefelsäure enthält, während 8 bis 15 Minuten je nach der Stärke des Bades entfernt. Dann werden die Näpfchen in gleichen Trögen in fließendem Wasser gewaschen, wodurch bald jede Spur der Säure entfernt wird. Das Nachziehen des Näpfchens geschieht ganz in der gleichen Weise, nur tritt das Werkzeug nach Fig. 352c an Stelle des vorigen. Auch werden dieselben Vorsichtsmaßregeln betreffs Ausmittlung und Schmierung der Werkzeuge befolgt.

In diesem Falle betrug die größte Wasserpressung ungefähr 80 at, während das folgende Ausglühen bei 650° C 20 Minuten dauerte. Das Scheuern und Waschen nach dem Nachziehen ist das gleiche wie im ersten Gang. Das Verhalten des Materials während des Ziehens ist die beste Qualitätsprüfung, da sich jede Unreinigkeit oder unvollständiges Ausglühen sofort durch Einreißen oder Rauhwerden der Oberfläche zeigt.

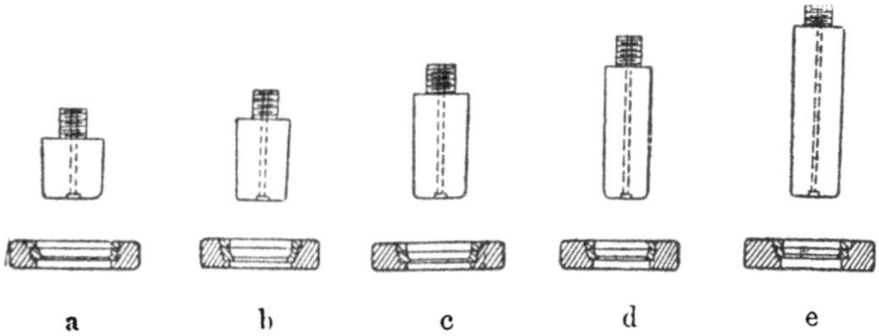

Fig. 353.

1. Ziehen. Fig. 353a zeigt das Werkzeug, Tabelle Nr. 17 IV die gezogene Hülse. Größter Wasserdruck ungefähr 91,4 at. Bei 650° C 28 Minuten ausglühen.

2. Ziehen. Fig. 353b, Tabelle Nr. 17 V, die gezogene Hülse. Wasserpressung ungefähr 95 at. Bei 650° C 26 Minuten ausglühen.

3. Ziehen. Fig. 353c das Werkzeug, Tabelle Nr. 17 VI die gezogene Hülse. Wasserdruck ungefähr 93 at. Bei 640° C 25 Minuten ausglühen.

4. Ziehen. Vor dem Ziehen wird der Boden der Hülse abgeflacht, bevor man die Vertiefung für den Zünder einprägt. Das Abflachen geschieht durch Pressen der Hülse zwischen dem Stempel und einer flachen Stahlscheibe, die auf dem Ziehring aufliegt. Die Scheibe wird dann entfernt und das Ziehen geht in der gewöhnlichen Weise vor sich. Fig. 353d das Werkzeug, Tabelle Nr. 17 VIII die gezogene Hülse. Wasserpressung 70 at. Bei 630° C 22 Minuten ausglühen.

5. Ziehen. Dies ist der letzte Arbeitsgang auf der senkrechten Presse. Fig. 353e das Werkzeug, Tabelle Nr. 17 IX die gezogene Hülse. Wasserpressung 49 at. Bei 630° C 20 Minuten ausglühen.

Einprägen der Vertiefung für den Zünder.

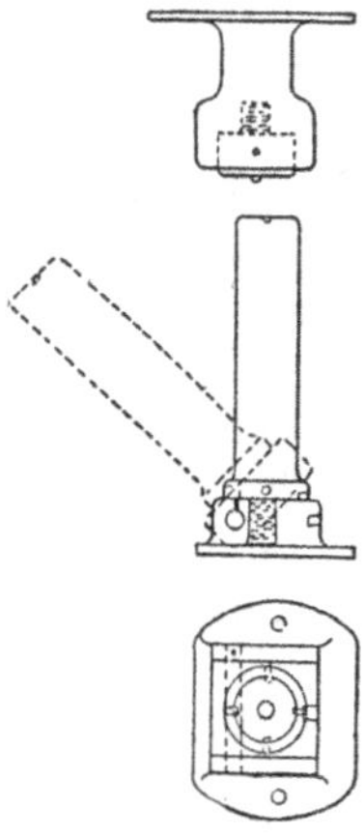
Fig. 354.

Diese Arbeit wird auf einer senkrechten 1000-t-Presse mit den in Fig. 354 gezeichneten Werkzeugen ausgeführt. An dem Plunger der Presse, der sich nach oben bewegt, ist eine Druckplatte befestigt, an die ein Stahlstempel angelenkt ist, dessen Außenform der Innenform der Hülse nach dem 5. Ziehen entspricht. Dieser Stempel kann einseitig herausgeschwungen werden, so daß man die Hülse leicht einlegen und entfernen kann. An der Unterseite des oberen Preßkopfes ist ein Halter befestigt, an den ein flaches, angelassenes Stahlstück angeschraubt ist, das in der Mitte eine Erhöhung trägt, die der Form der einzuprägenden Vertiefung für den Zünder entspricht. Am Stempel befindet sich die entsprechende Vertiefung. Der Druck zwischen den Arbeitsflächen beträgt 314 t und die Wasserpressung 175 at. Ein folgendes Ausglühen findet nicht statt.

6. Ziehen. Für diese und die folgenden Arbeiten werden größere, liegende Pressen verwendet, weil die nunmehrige Länge der Hülsen

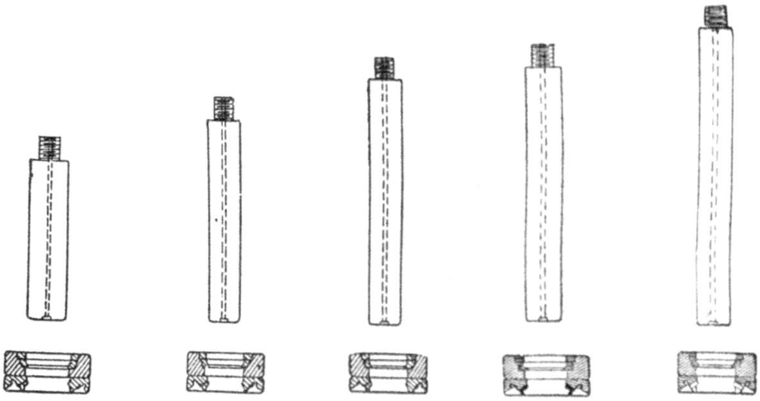
a b c d e

Fig. 355.

die Verwendung der stehenden Presse mit ihrem kurzen Hub nicht gestattet. Die Werkzeuge für diese Arbeit erscheinen in Fig. 355a—e. Bis jetzt konnte das Abstreifen der Hülsen durch die Abstreifkante im Ziehring erfolgen. Wenn die Hülsen von dem beträchtlichen Seitendruck befreit werden, verhindert ihr Ausfedern nach dem Ziehen, daß sie von dem rückkehrenden Stempel noch einmal durch den Ziehring zurückgezogen werden. Nach dem 6. Zug hingegen ist die seitliche Pressung geringer, so daß man andere Abstreifer anwenden muß. Unter jedem Ziehring sind acht Greiferfinger [1]) eingebaut, die durch Federn nach innen gedrückt werden. Während des Ziehens weichen sie vor der vorgehenden Hülse in Aussparungen zurück, treten jedoch sofort wieder heraus, wenn der Zug vollendet ist, und halten die Hülse während des Stempelrückganges zurück, wobei ihre Arbeit durch die Keilwirkung der Auflagerflächen in den Ausschnitten unterstützt wird. Der sechste Zug geschieht entweder auf 18- oder 16zölligen, liegenden Pressen, die beide genügend stark sind. Tabelle Nr. 17 XI die gezogene Hülse. Bei 630° C 18 Minuten ausglühen.

7. Ziehen. Fig. 355b das Werkzeug, Tabelle Nr. 17 XII die gezogene Hülse. Bei 630° C ausglühen.

8. Ziehen. Fig. 355c das Werkzeug, Tabelle Nr. 17 XIII die gezogene Hülse. Bei 600° C 14 Minuten ausglühen.

9. Ziehen. Fig. 355d das Werkzeug, Tabelle Nr. 17 XIV die gezogene Hülse. Bei 570° C ausglühen.

10. Ziehen. Fig. 355e das Werkzeug. Tabelle Nr. 17 XV die gezogene Hülse. Dies ist der letzte Zug, nach dem das Stück nicht mehr ausgeglüht wird.

Vorpressen des Bodens. Die Herstellung des Patronenbodens und Randes ist vielleicht die interessanteste Arbeit in der ganzen Fabrikation. Der Gesamtdruck, welchen diese Fläche beim Feuern auszuhalten hat, ist sehr hoch; bei dem 150-mm-Schnellfeuergeschütz der spanischen Armee, für die diese Hülse bestimmt ist, beträgt der durch die Explosion hervorgerufene Gasdruck ungefähr 2680 kg/cm². Aber selbst dieser Druck wird bei der Erprobung der Geschütze überschritten, da hierbei drei Entladungen mit 3150 kg/cm² Gasdruck vorgenommen werden. Daraus kann leicht die große Beanspruchung des Patronenbodens ersehen werden. Da für eine richtige Arbeit der Geschütze beim Feuern größere Formänderungen der Patronen unbedingt unzulässig sind, muß während der Herstellung der Patronen der Boden einem Druck, der zwei bis dreimal den wahrscheinlichen Arbeitsdruck übersteigt, ausgesetzt werden. Das Vorpressen des Bodens geschieht auf einer stehenden 2500-t-Presse in drei Stufen. Für den ersten Gang kommen die Werkzeuge in Fig. 356a zur Verwendung. Ein gußeiserner Ständer A auf dem Stößel der Presse trägt die Grundplatte und die Matrize B, die den Boden außen formt. Innerhalb des Ständers A ist ein Preßstempel C eingesetzt, über den die Hülse nach dem 10. Zug ge-

[1]) Siehe Fig. 358.

stülpt wird. Dieser Stempel, der einem Gesamtdruck von 1650 t Widerstand leisten muß, ist aus bestem gehärteten Stahl und zweifelsohne das empfindlichste Werkzeug in der ganzen Fabrikation. Zuerst wird die Hülse zwischen Ständer A und Stempel C eingelegt. Darüber wird der Gegenstempel D aus hartem Stahl gelegt, der in der Mitte eine Vertiefung hat, so daß nicht die ganze Bodenfläche den Druck überträgt. Unter dem Drucke von 1600 t wird an dem Boden eine innere und äußere Erhöhung angepreßt und gleichzeitig das Material nach auswärts zur Bildung des Flansches gezwängt.

In Fig. 356b ist das Werkzeug für den zweiten Arbeitsgang gegeben. Es ist dem im ersten Gange gleich, nur wird ein kleinerer Gegenstempel von 75 mm Durchmesser über die Hülse an Stelle des Stempels D gestellt. Unter einem Gesamtdruck von 600 t wird nun die äußere Er-

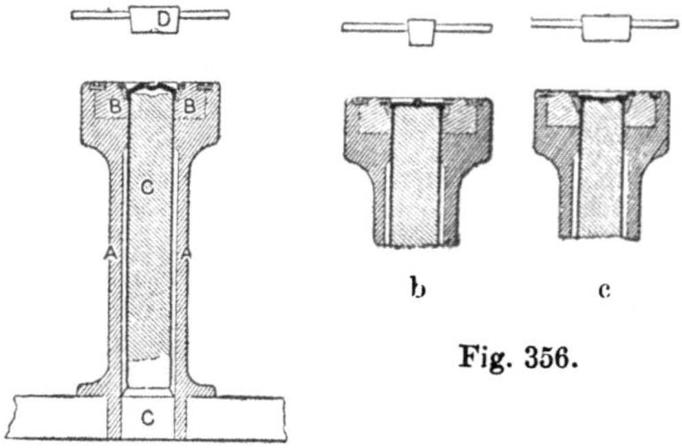

a b c

Fig. 356.

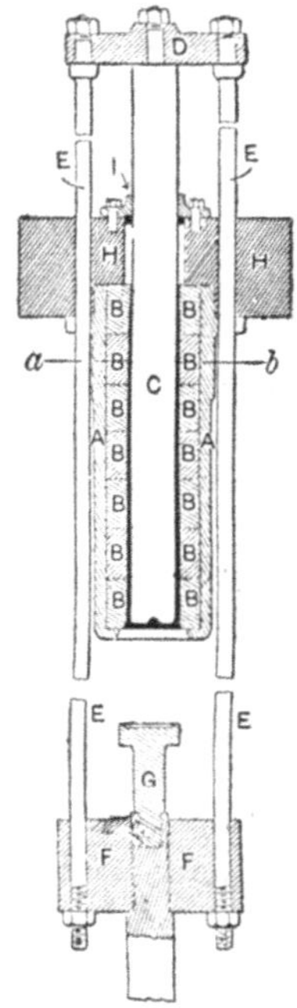

Fig. 357.

höhung abgeflacht und alles Material aus ihr nach innen getrieben, so daß rings um das Zündloch genügend Material übrigbleibt. Schließlich wird im dritten Gange mit dem Werkzeug in Fig. 356c unter einem Gesamtdruck von 1650 t der Boden flachgepreßt und endgültig geformt.

Einziehen. In diesem Arbeitsgang erhält die Hülse ihre endgültige äußere Form, so daß sie leicht in die Geschoßkammer gehen kann. Die Arbeit geht auf einer liegenden Presse vor sich, damit man den langen Hub derselben ausnutzen kann. In das feste Querhaupt H der Presse, Fig. 357, ist ein gußeiserner, säulenförmiger Ständer A eingeschraubt, innerhalb dessen sieben Ringe aus angelassenem Stahl B eingesetz, sind, deren innere Länge zusammen genau der Länge der Geschoßkammet entspricht. Die Hülse wird durch die Presse in diesen Raum einger drückt, muß aber durch einen besonderen Auswerfer zwangläufig entfernt werden. Der zylindrische Auswerferstempel C, dessen Kopf der Innenseite des Patronenbodens angepaßt ist, wird durch die Querhäupter D und F, sowie die Zugstangen E mit dem Plunger der Presse verbunden, so daß er der Bewegung folgt, wobei er durch die Brille I geführt wird. Der Stempel G, der in den Plunger der Presse eingeschraubt ist, drückt die Hülse bei seinem Vorwärtsgang in die Ringe, während der Auswerferstempel C sie beim Rückgang herausdrückt. In manchen Fällen wird das Einziehen in zwei Gängen durchgeführt, zwischen denen ein Ausglühen stattfindet. Vor dem ersten Einziehen wird die Hülse bei 560° C in einem kleinen senkrechten Ofen ausgeglüht, wobei man den Boden der Hülse außerhalb in der Luft stehen läßt. Die Hülse kommt dann wieder in die Presse und wird ungefähr bis zur Hälfte in die Ringkammer eingedrückt, wobei man als Vorsichtsmaßregel den Pressenhub auf die Hälfte einstellt. Beim Rückzug wird ein hölzernes Zwischenstück zwischen Ausstoßstempel und Patronenboden eingelegt und so die Patrone herausgedrückt. Die Hülse kommt dann wieder in den Glühofen, wobei man, wie vorher, den Boden außerhalb des Ofens beläßt, und glüht bei 500° C aus. Dann wird die Hülse fertig eingezogen, indem man sie vollständig in die Ringkammer eindrückt.

Die übrigen Arbeiten sind rein mechanischer Natur, und zwar Abdrehen der Enden, des Bodens, der Nut, des Zünderloches, Abstechen auf genaue Länge usw., die ähnlich wie in Tabelle Nr. 14 ausgeführt werden. Es ist jedoch notwendig, zu erwähnen, daß während der ganzen Herstellung Dicke und Durchmesser der Hülsen auf das sorgfältigste mittels Kaliber und Lehren kontrolliert werden, und zwar besonders die ersten zwei oder drei Stücke in jedem Satz um die Genauigkeit der Werkzeuge und ihre Einstellung zu prüfen Die Enden der Hülsen werden des öftern zwischen den einzelnen Ziehstufen auf Länge abgestochen, da die Stücke sich leicht durch unregelmäßiges Ausglühen oder Ungleichmäßigkeiten des Metalles ungleichmäßig ziehen oder an den Kanten einreißen. Ebenso wird die Dicke des Hülsenendes sehr genau mittels Grenzlehren geprüft. Die Schmierung der Werkzeuge geschieht entweder mit Olivenöl oder Seifenwasser.

Werkzeug zum Nachziehen großer Hülsen. Fig. 358 zeigt ein Werkzeug für den dritten und vierten Zug einer großen Hülse. Es ist nur der Ziehring für den dritten Gang abgebildet, während zum vierten Zuge Ziehring und Abstreifer ausgewechselt werden. Alle Teile des gezeichneten Werkzeuges sind rund, so daß Ziehringe, Anschläge, Abstreifer für andere Formen leicht angebracht werden können.

A ist die Hülse nach dem zweiten Zug in Stellung für den dritten Zug. B ist ein Führungsring aus Gußeisen oder Stahl; das erste Material wurde benutzt, da es sich bewährte und leicht zu bearbeiten war.

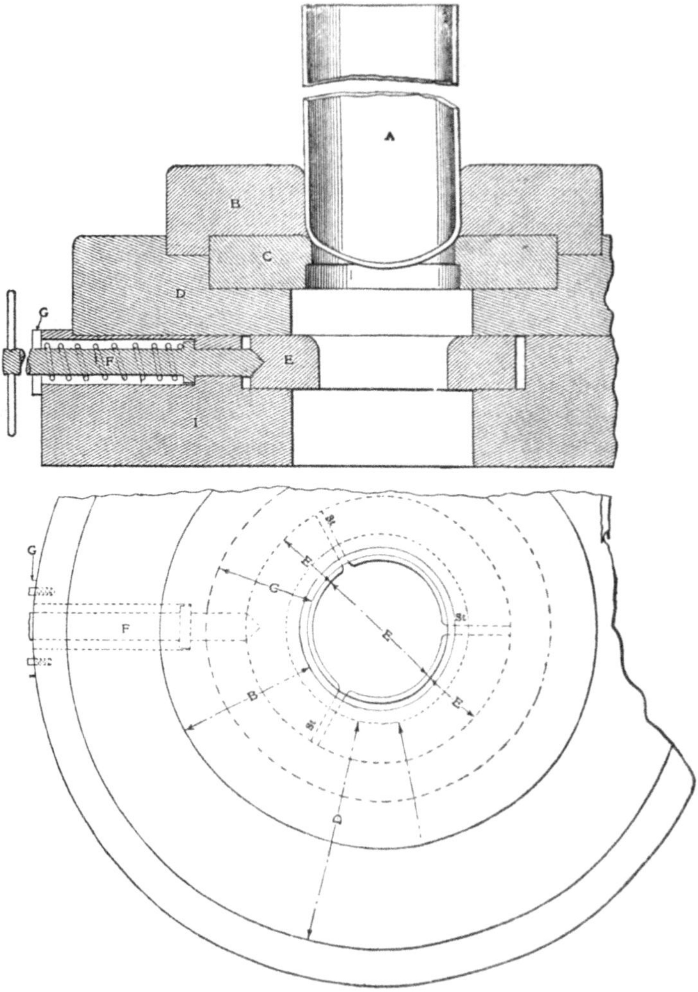

Fig. 358.

C ist der Ziehring aus Stahl; seine oberen oder Ziehkanten sind abgerundet und nach dem Härten so glatt wie möglich poliert. Man kann die Teile, die beim Ziehen mit der Hülse in Berührung kommen, mit nicht genug Sorgfalt fertigstellen, da sich die kleinste Rauheit an der fertigen Hülse wieder zeigt. Das Ziehloch wird hinterdreht, um die Reibung soviel wie möglich zu vermindern. D ist eine gußeiserne Gesenkplatte, in die der Ziehring C und die Führung B eingepaßt werden, und die selbst wiederum in die Froschplatte I ungefähr 3,2 mm eingelassen ist. E ist der Abstreifer, bei dem die oberen Kanten abgerundet und die unteren scharf gelassen sind; er wird aus Werkzeugstahl gemacht und so hart wie möglich angelassen. Das Loch wird auf denselben Durchmesser wie das im Zieheisen gebohrt und das Stück nachher in drei Teile zerschnitten. Man darf von den Teilen nicht mehr wegschneiden, als daß sie sich gerade genug für die Abstreifarbeit schließen oder, in anderen Worten, den Stempel einschließen, nachdem die Hülse durchgegangen ist.

Beim Abwärtsgang des Stempels wird der Abstreifer zwangläufig geöffnet, so daß die Hülse durchgehen kann, und dann der Abstreifer vermittels der Federn, die auf die drei Stifte F drücken, rings um den Stempel geschlossen; nun ist der Abstreifer in der richtigen Stellung, um beim Rückgang des Stempels die Hülse abzustreifen. Es wird wohl von Vorteil sein, auf einen Punkt in der Konstruktion hinzuweisen, der bei der Herstellung eines solchen Werkzeuges besondere Schwierigkeiten ergab. Der äußere Durchmesser des Abstreifers war nämlich fast gleich dem Durchmesser des Ziehringes gemacht worden, so daß in der Froschplatte I eine sehr große Ausdrehung nötig wurde. Infolgedessen hatten C und die Gesenkplatte D nicht genügend Auflage, so daß sich letztere durchbog und den Abstreifer E an seiner Bewegung verhinderte. Ein weiterer wichtiger Punkt ist das Abrunden und Polieren aller Innenkanten am Abstreifer nach dem Zerschneiden in drei Teile; sonst werden an dem gezogenen Stück Marken sichtbar. Die Grundplatte wird an dem Preßtisch durch Bolzen, die durch gebohrte Löcher in D und I gehen, befestigt. Die Stempel für den dritten und vierten Zug werden infolge ihres kleinen Durchmessers in einem Stück hergestellt. Bei dem Stempel für den ersten und zweiten Zug ist der Durchmesser viel größer, so daß die Herstellung aus einem Stück ein unverhältnismäßig größeres Stück Stahl verlangen würde und die Kosten der Erneuerung bedeutend größer würden.

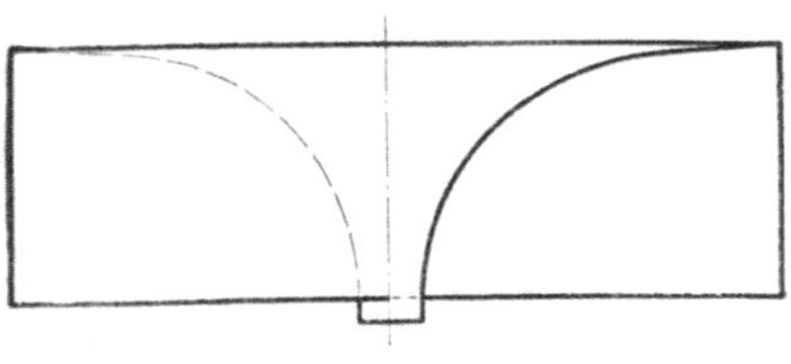
Fig. 359

Ziehen eines Zylinders mit aus dem Boden sich erhebenden Konus[1]). Die Herstellung eines Stückes nach Fig. 359 ist nur für geringe Höhen auf Ziehwerkzeugen, für größere auf der Drückbank vorgenommen worden. Da diese Arbeit durch Um-

[1]) W.T. 1910, S. 499. Pickersgill, Fachausstellung Stuttgart.

spannen und Nachdrücken ziemlich viel Zeit in Anspruch nimmt, hat die **Erzgebirgische Schnittwerkzeug- und Maschinenfabrik in Schwarzenberg** einen Satz Werkzeuge hergestellt, die gestatten, dieses Stück auf Kurbelpressen fertigzustellen. Zuerst wird das Blankett ausgeschnitten und dann auf dem in Fig. 360 abgebildeten Ziehwerkzeug hochgezogen. Nach Fertigstellung des äußeren Umfanges wird der Boden in mehreren Arbeitsgängen konisch herausgezogen. Sobald die gewünschte Höhe erreicht worden ist, wird der hochgezogene Teil in einem nach Fig. 361 ausgebildeten Kalibrierwerkzeug zur Erzielung eines glatten Aussehens fertiggedrückt.

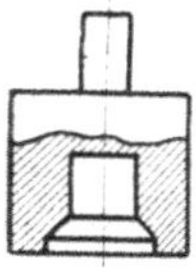

Fig. 360.

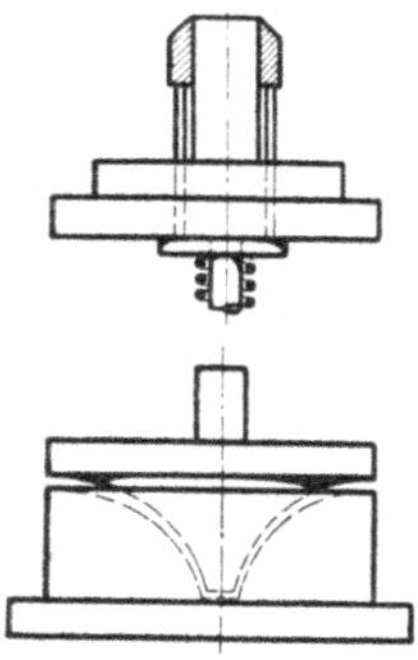

Fig. 361.

Ziehen eines konischen Lampenschirmes. (Bonner Maschinenfabrik Mönkemöller, G. m. b. H.) Zur Herstellung eines konischen Lampenschirmes in einem Zuge verwendet die Firma ihre Doppelziehpressen[1]), deren Werkzeuge in grundsätzlicher Bauart bereits in Fig. 331 abgebildet worden sind.

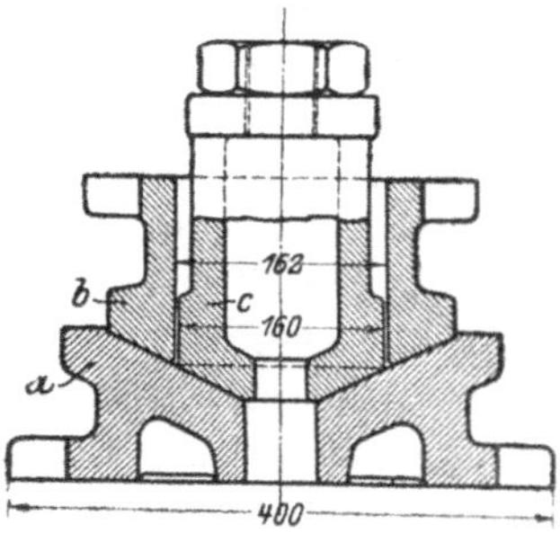

Fig. 362.

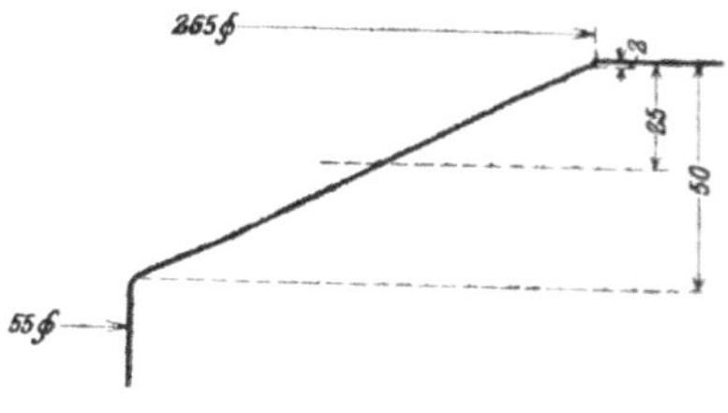

Fig. 363.

Zur Herstellung des Lampenschirmes wird zuerst eine Scheibe von 265 mm Durchmesser ausgeschnitten und in die Ausdrehung der Matrize a, Fig. 362, vom gleichen Durchmesser eingelegt. Die Linie der Fig. 363 stellt die Erzeugende der Matrize dar und zeigt diese Ausdrehung von 2 mm Tiefe deutlicher. Hierauf gehen Blechhalter b und Ziehstempel c nach abwärts und ziehen den ersten Kegelstumpf auf 25 mm Tiefe, Fig. 363, bis zu der gestrichelten Linie in Fig. 362 vor. Von hier arbeitet der Teil b als Blechhalter, während der innere Ziehstempel c den Kegel auf die volle Tiefe auszieht. Gewöhnlich folgt ein Einrollen des äußeren

[1]) Vgl. Beschreibung der Presse Fig. 650/1.

Randes auf Werkzeugen gewöhnlicher Bauart, vgl. Fig. 432. Die eingerollten und gezogenen Teile werden dann auf dem Lochwerkzeug in Fig. 18 mit dem mittleren Loch versehen und gleichzeitig dessen innerer Rand hochgezogen.

Die Ziehwerkzeuge sind sämtlich aus Gußeisen hergestellt, so daß ihre Bearbeitung leicht ist.

Ziehwerkzeug für einen Becher mit Rippen. Der im folgenden beschriebene und abgebildete Satz Werkzeuge besitzt manche wichtigen Punkte für alle derartige Arbeiten; ebenso wird die Herstellung des Stempels für den letzten Gang jedem von Hilfe sein, der ähnliche Stempel zu machen oder einen ähnlichen Becher zu ziehen hat. Die Tabelle Nr. 19 gibt die aufeinanderfolgenden Arbeitsgänge der Herstellung des Bechers ohne Boden nach Fig. 364. Es ist ein konischer Becher mit eingeprägten Rippen, der aus einem runden Blankett von 58,7 mm Durchmesser und 2,4 mm Dicke aus weichem Messing gezogen wird. Die Rippen gehen nach innen und nach außen von sechs gleichmäßig verteilten, viereckigen Nuten aus, die nach innen zusammenlaufen und ungefähr 5 mm vom Grunde verschwinden. Die Lieferbedingungen verlangten in diesem Fall, daß am Grunde des konischen Teiles ein scharf ausgeprägter Absatz vorhanden sein sollte, der aber nur außen an den Rippen selbst ausgebildet sein sollte. Auf der Innenseite sollte kein Absatz sichtbar sein, sondern es sollten die Nuten ohne Unterbrechung nahe dem Grunde verschwinden. Man sieht, daß der Druck beim letzten Zug viel größer als sonst sein muß, um diese Forderungen zu erfüllen, denn das Material war an der Stelle, wo der gerade und der konische Teil zusammenstoßen, ganz dünn gepreßt, um den Absatz auszubilden.

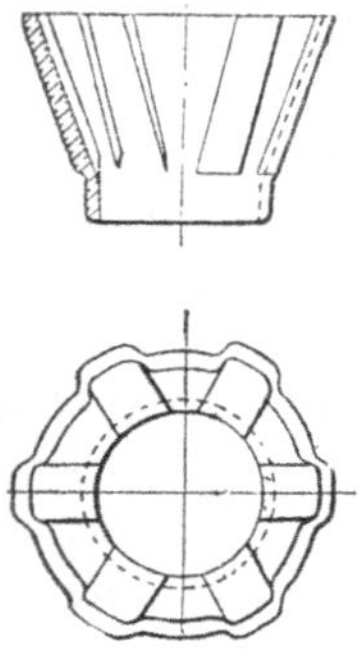

Fig. 364.

Die Ziehringe und die Zapfen aller Stempel werden gleichmäßig gemacht, so daß nur zwei Kopfplatten, eine für die Stempel und eine für die Ringe, benötigt werden. Die Stempel werden durch eine Kopfschraube gesichert, die Ziehringe in die Grundplatte eingelassen und durch vier Kopfschrauben gehalten, die gleichmäßig auf der Umfläche verteilt sind und mit ihren Spitzen in die Umfläche des Ringes eintreten. Um von einem Gang zum anderen überzugehen, ist es nur nötig, die Schrauben zu lösen, das gerade im Gebrauch befindliche Werkzeug zu entfernen und das nächste einzusetzen. Jedem Zug folgt ein sorgfältiges Ausglühen, um ein gleichmäßiges Materialgefüge zu sichern und die Möglichkeit eines Einreißens auf das geringste Maß zu beschränken.

Die Hauptarbeit ist, daß der Stempel für den letzten Zug so billig wie möglich hergestellt wird und trotzdem der starken Beanspruchung, der er unterworfen ist, gewachsen ist. Er wurde zuerst in Teilen hergestellt, auf Form abgedreht und dann mit den Schwalbenschwänzen versehen, die zur Aufnahme der Nutenstücke dienen. Diese wurden gesondert hergestellt und in die Schwalbenschwänze bis an den Absatz eingepreßt. Dann wurde das Ende gebohrt, mit Gewinde versehen und der gerade Zapfen angeschraubt. Nachdem die Teile sorgfältig zusammengepaßt worden waren, wurden sie gezeichnet, auseinandergenommen, gehärtet und dann wieder wie zuvor zusammengebaut. Der Stempel, der in dieser Weise gemacht worden war, hielt jedoch im Gebrauch nur kurze Zeit, da die Nuten an den Kanten, wo die Beanspruchung am größten war, abzusplittern und Risse zu zeigen anfingen, wobei oft 100 und mehr Becher durch die Presse gingen, bevor der Fehler entdeckt wurde. Deshalb wurden die gehärteten Teile durch weiche ersetzt, die aber nach kurzer Zeit abflachten und dauernd Ausbesserungen erforderten. Um über diese Schwierigkeiten hinwegzukommen, wurde der Stempel schließlich, wie in Fig. 365 dargestellt ist, aus einem einzigen Stück Werkzeugstahl gemacht. Dieses wurde gefräst, gemeißelt und fertig gefeilt, sodann gehärtet und auf lichtes Strohgelb angelassen, wonach man keine Anstände mehr hatte. Die Ziehringe für den ersten

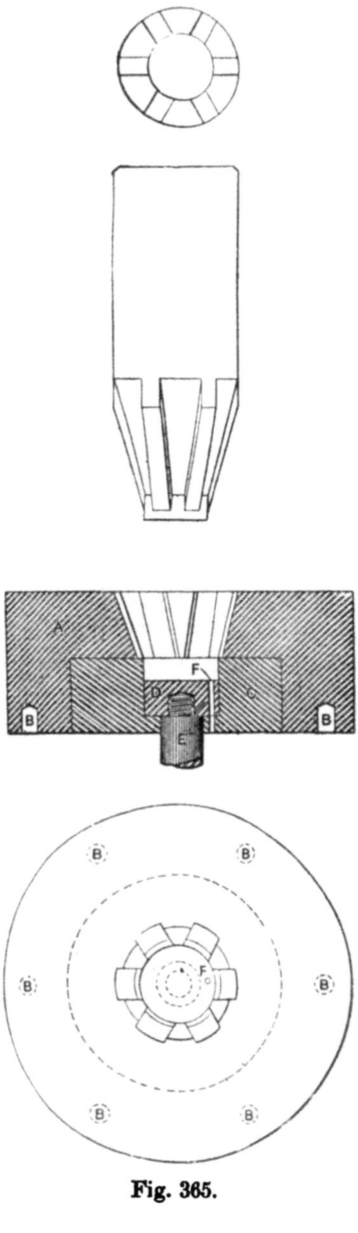

Fig. 365.

und zweiten Zug werden aus Scheiben mit ungefähr 125 mm Durchmesser und 51 mm Höhe aus Werkzeugstahl gemacht, und mit Löchern versehen, um die Näpfchen nach Tabelle Nr. 19 zu ziehen. Es sind einfache Ziehringe mit einer scharfen Kante auf der Unterseite, um das Näpfchen vom Stempel abzustreifen, der leicht konisch ist und einen dem Innendurchmesser des Näpfchens entsprechenden Durchmesser hat. In diesem Fall wird der Stempel um die doppelte Dicke des Materials 0,2 mm kleiner im Durchmesser als das Loch im Ziehring gemacht. Vor dem letzten Gang ist es gebräuchlich, die Stempel mehr als die doppelte Materialdicke kleiner als den Ziehring zu machen und ihnen einen Anzug von ungefähr $^1/_{100}$ zu geben, um das Stück leichter abstreifen zu können; meistens sind jedoch die genauen Maße gezogener Arbeit von geringerer Wichtigkeit bis zum letzten Gang. Infolgedessen wird ein konischer und unter Maß gehaltener Stempel die Fertigstellung des Werkstückes in keiner Weise behindern, da diese im letzten Gang durch ein Werkzeug geschieht, das genau den verlangten Abmessungen des Stückes entspricht. Eine allgemeine Regel ist, den Stempel für kurze Näpfchen aus dickem Material kleiner und konisch und bei dünnem Material nahezu auf Maß zu machen, da das letztere mehr dazu neigt, Falten zu bilden, wenn es nicht ordentlich zwischen den Werkzeugteilen ausgepreßt wird. Das Werkzeug für den dritten Gang wird nach dem Fertigmaß, aber mit glatten Wandungen gemacht. Das Werkzeug für den letzten Zug ist in Fig. 365 dargestellt. Der Teil A wird zuerst auf der Drehbank gedreht und ausgebohrt und dann auf einem besonderen Dorn an dem Teilkopf einer Fräsmaschine befestigt. Sechs Löcher von 6,5 mm Durchmesser werden bei B gebohrt und ausgerieben; dabei wird das Stück so geschaltet, daß diese gleichmäßig verteilt werden. Dann wird ein besonderer Eisenwinkel gemacht, um das Stück unter dem der Schräge des Bechers entsprechenden Winkel zu halten, und zwei Löcher sorgfältig so in seine Stirnfläche gebohrt, daß sie mit zwei einander gegenüberstehenden Löchern auf der Rückseite des Ringes übereinstimmen. In diese Löcher werden Führungsstifte von 6,5 mm ⌀ eingetrieben, die ungefähr 5 mm über die Oberfläche des Winkels hervorragen. Der Winkel wird dann an den Tisch einer Shapingmaschine angeschraubt, der Ring auf der Stirnfläche des Winkels mittels der Führungsstifte eingestellt und in richtiger Stellung festgezogen. Mittels der sechs gebohrten Löcher kann man den Ring zum Hobeln der Nuten genau ausrichten, das mit einem Profilstahl, der an das Ende eines Werkzeughalters für Innenarbeit angeschraubt ist, geschieht. Nach dem Hobeln werden die Nuten gefeilt und vollständig glatt poliert und der Ring gehärtet. Das Stück C wird gedreht und nahezu auf Preßsitz in der Unterfläche des Ringes geschliffen und beide dann auf der

Unterseite überschliffen. D ist der Gegenstempel und Auswerfer aus Stahl und E der Auswerferbolzen, die vom Pressenstößel beim Aufwärtsgang betätigt werden. F ist ein Luftloch. Nach dem Ziehen wird der Boden des Bechers in einem einfachen Schnittwerkzeug ausgeschnitten.

Ziehwerkzeuge zur Herstellung von Brennerteilen für Petroleumsturmlaternen. In Tabelle Nr. 20 sind die Herstellungsstufen eines Ober- und eines Unterteiles eines gewöhnlichen Petroleum-Flachbrenners abgebildet[1]). Der Arbeitsgang ist ganz ähnlich wie bei dem vorher besprochenen Stück, doch verlangte die bedeutend weitergehende Umformung des Blanketts sowie das in diesem Fall verarbeitete Material — Weißblech —, welches kein Ausglühen zwischen den einzelnen Zügen gestattet, eine besonders sorgfältige Formverteilung auf die einzelnen Werkzeuge.

Die Herstellung erfolgte in diesem Falle[2]) auf einer liegenden Ziehpresse, einem sogenannten Stoßwerk, so daß die Anordnung des Matrizenteiles im Pressenstößel und des Ziehstempels am Pressentisch erfolgt, wo auch der Stempel auf Federn gelagert ist. Der zwangläufige Auswerfer ist im Matrizenteil vorgesehen und wird durch einen durchgehenden Keil, der beim Rückgang der Presse gegen feststehende, einstellbare Anschläge stößt, betätigt. Der Niederhalter ist auf Federn gelagert. Alle Matrizen und Stempel sind gleicher Bauart, so daß eine allgemeine Besprechung genügen dürfte und nur die Abweichungen der einzelnen Werkzeugteile, Fig. 366/73, bei den einzelnen Zügen erwähnt werden sollen.

Die Matrizenteile A werden aus einem zylindrischen Stück Stahl, das oben mit einem Zapfen und $1\frac{3}{4}$ zölligem Gewinde zum Einschrauben in den Pressenstößel versehen ist, gedreht und gehärtet. Sie sind in der Mitte durchbohrt, damit der Zapfen des Auswerfers B, der durch einen Stift am vollständigen Durchfallen gehindert wird, durchtreten und von dem im Stößel geführten, durchgehenden Keil betätigt werden kann. Auf der Unterseite ist der Körper entsprechend der Form des Auswerfers oder des Brennerteiles ausgedreht und auf dem arbeitenden Umfang poliert. An der Formgebung nehmen auch die Auswerfer B teil, deren Innenseiten als Matrizen ausgebildet sind, und die infolgedessen aus Stahl gehärtet und poliert sind. Die Unterflächen der Matrizenteile A sind Niederhalter und werden deshalb ebenfalls geschliffen und poliert. An der Umfläche sind noch zwei Löcher von ungefähr 17 mm

[1]) Angaben und Werkzeuge stammen von der Firma Ed. Sommerfeld, Metallwarenfabrik, Berlin SO 26.

[2]) Bei der jetzigen Herstellung wird auf der Revolverpresse, Fig. 646 gearbeitet.

Durchmesser vorgesehen, mittels derer die Matrizen in den Pressenstößel eingeschraubt werden. Die eigentlichen Ziehstempel C sind aus

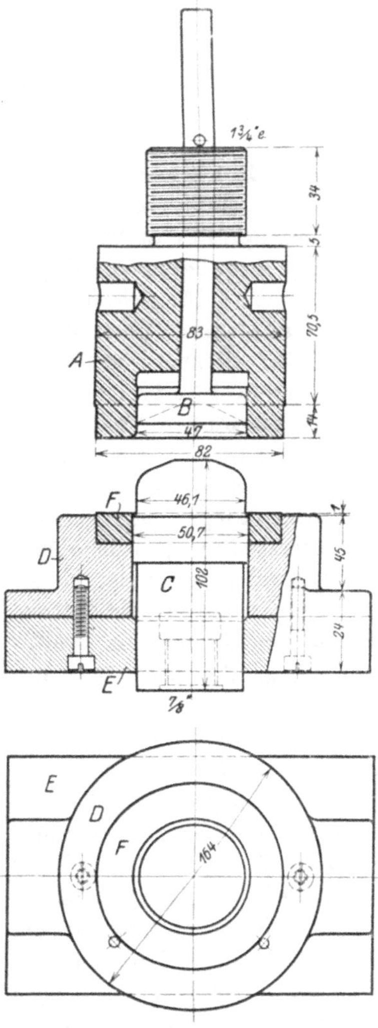

Fig. 366.

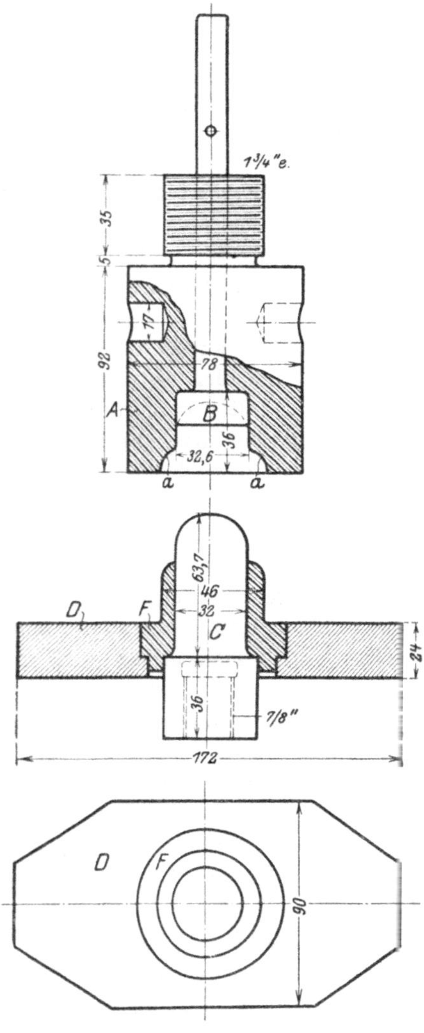

Fig. 367.

Stahl, gehärtet, geschliffen und an den formgebenden Stellen poliert. Auf der Unterseite erhalten sie $^7/_8$'' Gewinde, mittels deren sie mit dem Federhebel der Presse verbunden sind. Die Führungsringe D für den Stempel sind meistens aus Gußeisen und mittels versenkter Zylinderkopfschrauben auf schmiedeeisernen Froschplatten E gehalten

Die Auflage- und Stoßflächen beider Teile sind gedreht, so daß sie genau senkrecht zu der Stempelführung sind. Blechhalterringe F aus gehärtetem Stahl, die je nach der Arbeit in der Form verschieden sind, werden in eine Ausdrehung der Führungsplatte eingepaßt und auf der Oberseite geschliffen. Die Innenausdrehung derselben bildet die eine Führung für den Stempel, während in der Grundplatte tiefer unten eine zweite vorgesehen ist.

Fig. 366 zeigt das Werkzeug, auf dem aus dem flachen Blankett das Näpfchen vorgezogen wird. Der Auswerfer B ist hier, wie oben beschrieben, ein einfacher, dem Näpfchen angepaßter Stahlstempel. Im folgenden Gang, Fig. 367, wird aus dem Näpfchen bereits die Kappe in ihrer Grundform gezogen. Während der untere Flansch und ein Stück des Mantels unberührt bleibt, wird aus dem Oberteil die Kappe und der erste Absatz vorgezogen. Zu diesem Zwecke ist der Blechhalter entsprechend geformt und arbeitet mit einem entsprechend hohl ausgedrehten Teil der Matrize A zusammen. Aus diesem Grunde ist er auch nicht mehr als einfacher Ring ausgebildet, sondern trägt einen Zylinder, der als Führung für das im ersten Gang gezogene Näpfchen dient und oben, dem Teil a im Matrizenkörper entsprechend, ballig gedreht ist. Im vierten Gang beginnt bereits die eigentliche Ausbildung der scharfen Kanten des oberen Teiles und zugleich das Tieferziehen der Kappe, so daß der selbsttätige Auswerfer als Teil der Matrize eine solche Tiefe erhält, daß zur Entfernung des Stückes ein zweiter Federauswerfer G eingebaut werden muß. Fig. 368 zeigt, wie der Teil des selbsttätigen Auswerfers, der als Matrize dient, mittels einer Eindrehung im Kopf derselben ausgemittelt und durch zwei versenkte Kopfschrauben gehalten wird. Er ist durchbohrt und dient dem dritten Teil der Matrize, dem Federauswerfer G, zur Führung und zur Auflage. Dieser wird wiederum durch eine in einer Ausbohrung des Kopfteiles getragene Feder gegen einen Absatz in B gepreßt, so daß auf diese Weise die gebildete Kappe aus dem selbsttätigen Auswerfer entfernt wird. Selbstverständlich sind die beiden Matrizenteile im selbsttätigen Auswerfer aus Stahl, gehärtet und geschliffen, sowie an den arbeitenden Flächen poliert. Im Federauswerfer und im Stempel ist ein Luftloch H vorgesehen, das bei allen folgenden Werkzeugen wiederkehrt, da die Tiefe des gezogenen Stückes es verlangt. Fig. 369 stellt das Werkzeug für den 5. Gang dar, das im wesentlichen dem eben besprochenen entspricht und an dem Stück den unteren Absatz scharf herstellt.

Diesen Arbeiten, bei welchen der Brenner-Oberteil im ganzen bereits seine Form erhalten hat, folgt das Ausschneiden des Schlitzes in der Kappe für den Docht und eines kleinen seitlichen Ausschnittes am unteren Umfang. Zugleich wird der Rand abgeschnitten. Danach kommt das in Fig. 370 abgebildete Werkzeug zur Anwendung, das die

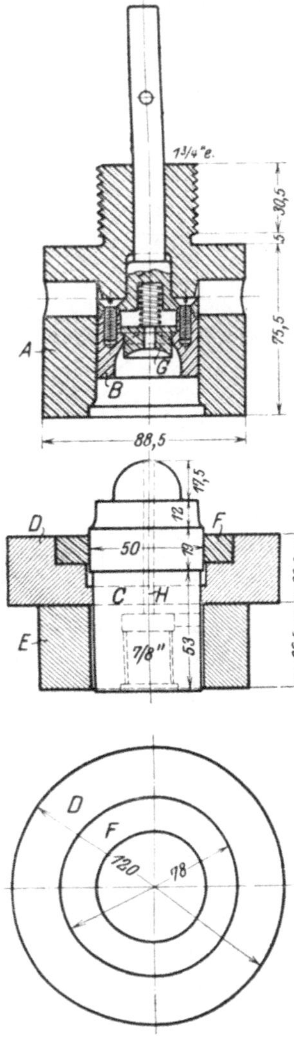

Fig. 368.

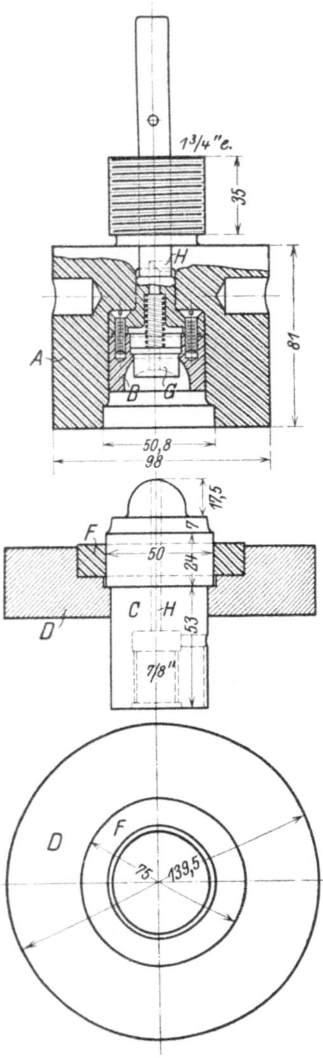

Fig. 369.

Teile auf genaues Endmaß bringt. Hier ist nur der Lochsucher G zu bemerken.

Die drei in Tabelle Nr. 20 b angeführten Züge des Unterteils werden auf Werkzeugen nach Fig. 371—373 hergestellt, die im wesentlichen mit den eben beschriebenen Werkzeugen für die Brenner-Oberteile übereinstimmen, so daß eine besondere Beschreibung der Zeichnungen unterbleiben kann. Da die Tiefe des Stückes geringer ist und

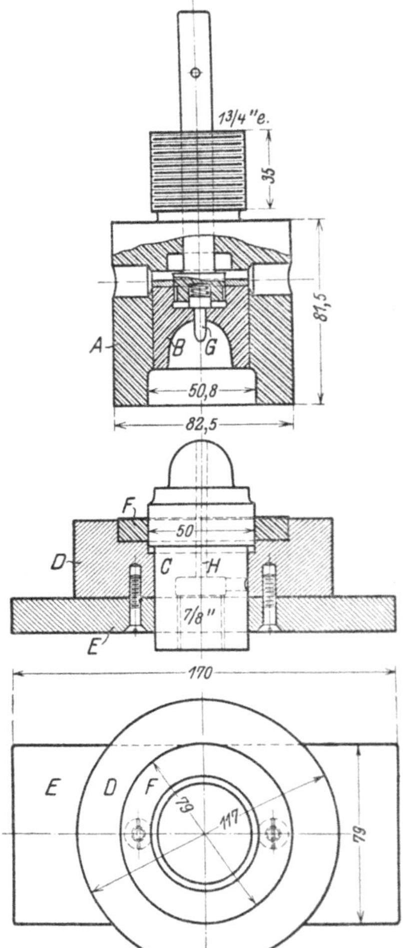

Fig. 370.

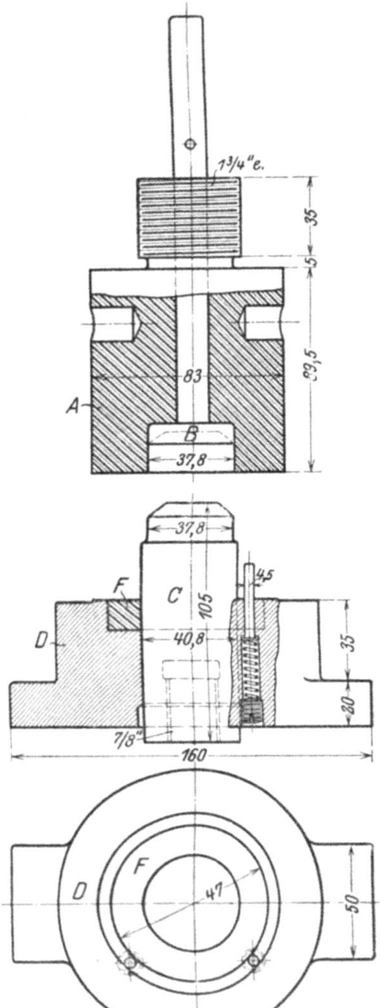

Fig. 371.

die Absätze mehr ineinander übergehen, so kommt man mit drei Zügen aus. Die Werkzeuge, die in einer gleichen Presse wie oben verwendet werden, sind auch, was Material und Ausführung betrifft, ganz gleich den früheren; dagegen sollen die beiden Reihen zeigen, wie weit die gesamten Ziehwerkzeuge für ein Stück in den Einzelheiten abweichen oder übereinstimmen können. Es soll nur auf den Einbau

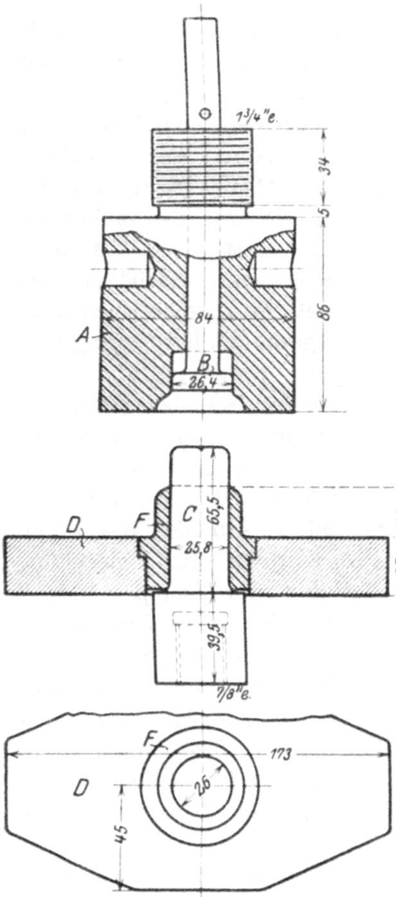

Fig. 372.

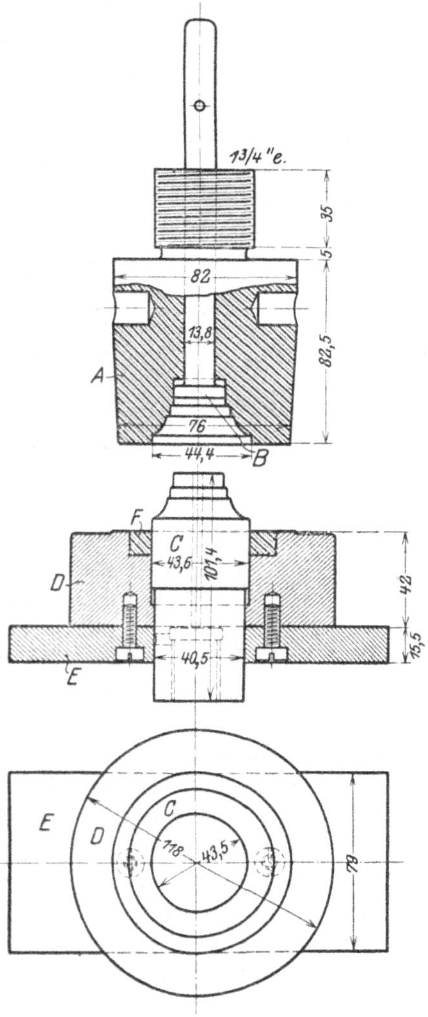

Fig. 373.

der federnden Anschlagstifte in Fig. 371 aufmerksam gemacht werden, deren zwei vorhanden sind, die beim Niedergang des Matrizenteiles nach innen ausweichen.

Das Ziehen einer Nabe aus einer starken Blechscheibe. Die Beschreibung einer Arbeitsweise und die Abbildung der dazugehörigen Werkzeuge zur Herstellung des in Fig. 374 abgebildeten Gegenstandes wird in jenen Fällen Interesse bieten, wo man bei der Herstellung ähnlicher Stücke auf Schwierigkeiten gestoßen ist. Nach der Herstellung der Nabe wird der Rand in derselben Richtung wie die Nabe in einem gewöhnlichen Ziehwerkzeug hochgezogen; hier sei jedoch nur das Werkzeug zur Fertigstellung der Nabe beschrieben. Der Außendurchmesser des fertigen Gegenstandes beträgt ungefähr 67 mm. Nach dem Auftrag mußte bei dem Stück die Metalldicke durchweg 2,8 mm betragen. Schließlich wurde auf der Rückseite der Nabe eine Abweichung von 0,4 mm gestattet, da keine Presse stark genug war, das Stück unter Einhaltung des Maßes fertigzustellen. Die größte Schwierigkeit war das Ziehen der kleinen Nabe b in der Mitte, Fig. 374, ohne das Material zu strecken. Nachdem man einige Zeit erfolglos mit den gewöhnlichen Ziehwerkzeugen Versuche gemacht hatte, ging man den folgenden Weg:

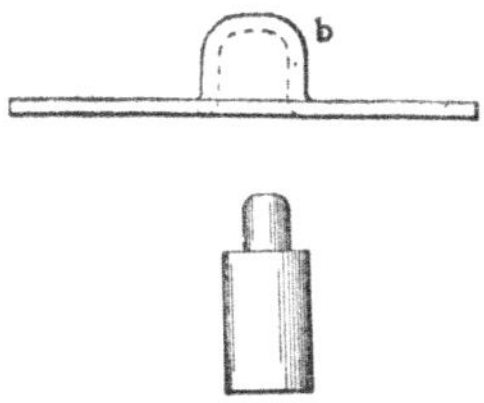

Fig. 374.

Das Ausschneiden der Scheiben aus Stahlblech von 2,8 mm Dicke geschah in der bekannten Weise und kann hier übergangen werden. In Fig. 375 ist B die Matrize, C das Arbeitsstück und A ein Teil des Stempels. Die Scheibe wird in eine Eindrehung der Matrize gelegt und der Stempel A nach abwärts gebracht, bis der Außendurchmesser von C um ungefähr 0,05 bis 0,07 mm kleiner als der Auswerfer J in Fig. 376 ist. Die Tiefe von C muß groß genug sein, um noch das Material für das Durchdrücken der Nabe herzugeben, deren erste Herstellungsstufe bei C in Fig. 376 und deren letzte Form in Fig. 374 erscheint. Die Teile A und B sind aus weichem Stahl und in Knochenmehl gehärtet, so daß sie leicht zufriedenstellend bearbeitet werden können.

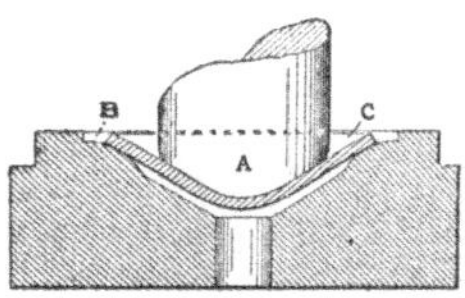

Fig. 375.

Fig. 376 stellt Stempel und Matrize für das Ziehen der Nabe dar. D ist die Kopfplatte, die in den Stößel der Presse eingepaßt ist — man zieht hier die umgekehrte Anordnung, der Matrize im Ober- und des Stempels im Unterteil, vor — E ist eine Überwurfmutter, welche die Matrize F gleichachsig mit der Stirnfläche von D hält. L ist die Gesenkplatte für den Stempel, die an der Oberfläche für den Auswerfer J ausgedreht ist. Der Stempel K wird in einem Loch am Boden dieser Ausdrehung gehalten und durch die Druckschraube N gesichert. Der Stempel geht durch J und arbeitet mit F zusammen. O ist der Zapfen irgendeines zwangläufigen Auswerferantriebes und betätigt durch die Stifte M den Auswerfer J.

Dies ist eine Umkehr der gewöhnlichen Anordnung und wird durch die Tatsache notwendig, daß das Material aus der Nabe in den Flansch ausfließt, wenn man C in Fig. 376 in einem zweiten gleichartigen Arbeitsgang in Werkzeugen der gewöhnlichen Bauart weiter bearbeitet, so daß das Material bei Erreichen des letzten Ganges gedehnt und unbrauchbar geworden ist. Bei dem beschriebenen Wege hingegen wird C, Fig. 375, auf den Auswerfer J mit der erhabenen Seite nach oben gelegt; beim Abwärtsgang der Matrize F fließt, da C rundherum festgehalten ist, nur wenig Material in den Flansch und wird mit einer geeigneten Presse in die Ecke bei a, Fig. 376, gedrückt. Im ganzen

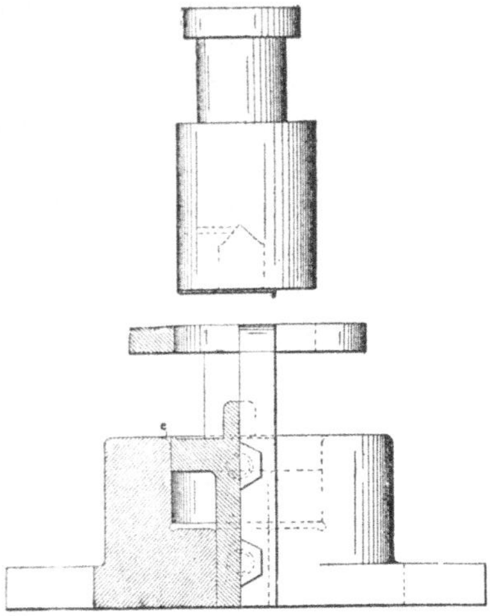

Fig. 377.

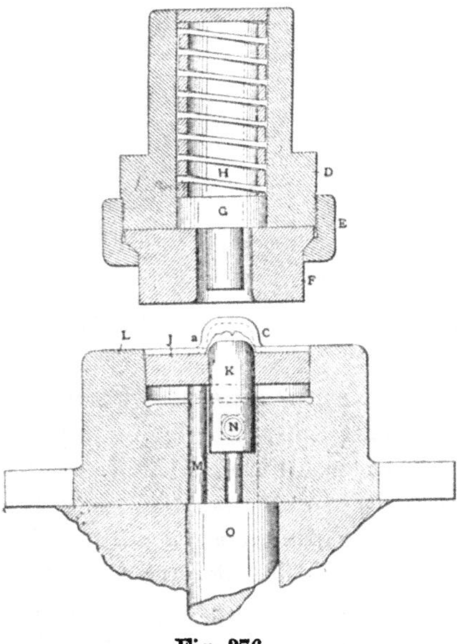

Fig. 376.

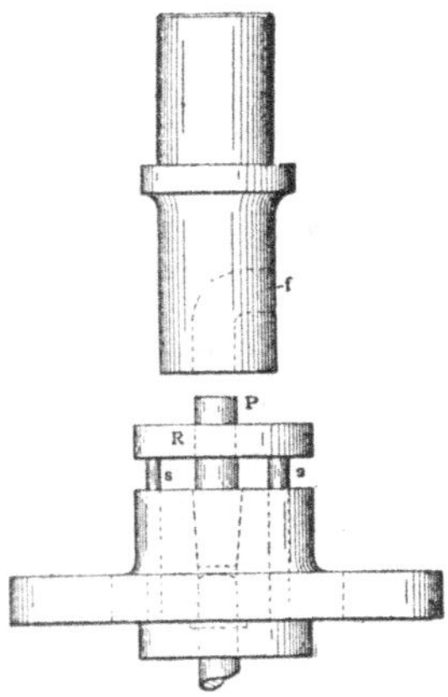

Fig. 378.

sind fünf Ziehstufen nötig, deren Stempel und Matrizen in ihren Haltern auswechselbar eingepaßt werden. Bei diesem Satz Werkzeuge soll man nur gutes Material verwenden und eine Presse von mindestens 125 t Druckkraft. D muß an der auf F liegenden Fläche gehärtet und geschliffen werden und in dem Stößel am Ende und Absatz gut aufliegen. Auch die Ausdrehung in L soll gehärtet und geschliffen sein.

Fig. 374 (unten) zeigt den Stempel zu dem fertigen Stück, Fig. 377 das Kalibrierwerkzeug. Des Zusammenhanges wegen ist hier auch das Lochwerkzeug in Fig. 378 abgebildet, das den Boden der Nabe ausschneidet. Es kann als ein weiteres, einfaches Beispiel für einen offenen Rundschnitt, neben Fig. 18, betrachtet werden.

Ziehen von Feldkesseln aus Aluminium. (L. Schuler, Göppingen.) In den Figuren 379—384 ist ein mustergültiger Satz Ziehwerkzeuge nach den Angaben obiger Firma wiedergegeben, der zur Herstellung des deutschen Armee-Feldkessels Verwendung findet. Die bisher be-

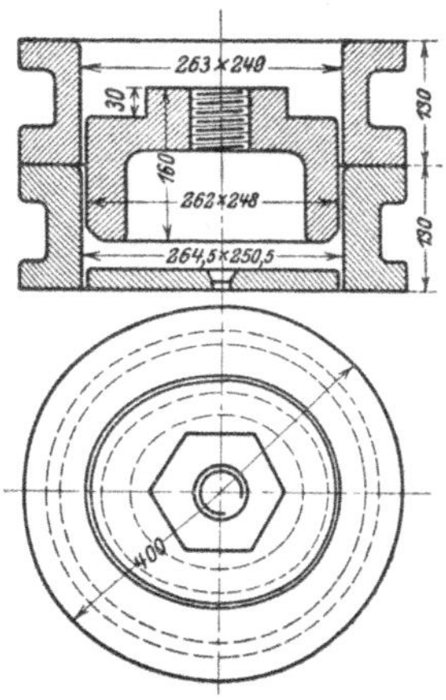

Fig. 379.

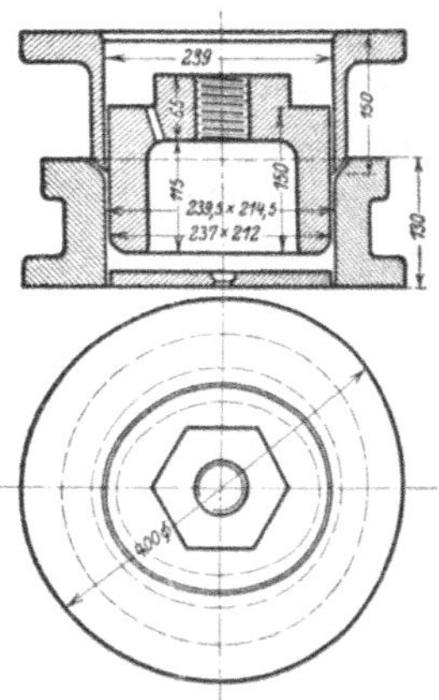

Fig. 380.

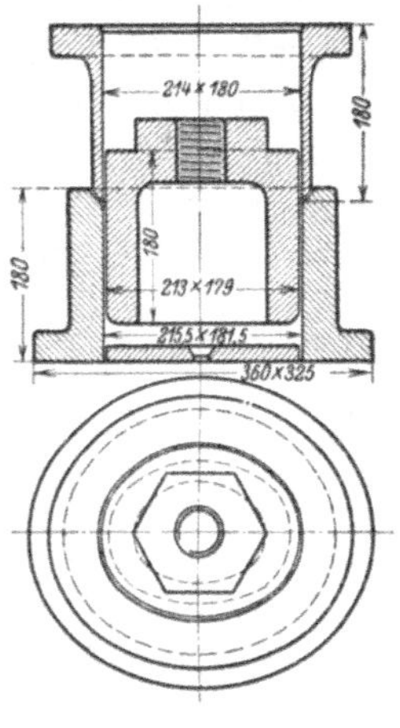

Fig. 381.

sprochenen Werkzeuge hatten ausnahmslos symmetrische Querschnitte, deren Spannungs- und Dehnungsverteilung beim Ziehen naturgemäß günstiger als jene der unsymmetrischen Querschnitte ist. Außerdem hat das Aluminium eine zu geringe Festigkeit, so daß man für eine gerade ausreichende Spannung unter dem Niederhalter sorgen muß, indem man ihn auf der Innenseite etwas mehr tragen läßt. Bei den Feldkesseln handelt es sich fast ausschließlich um Militärlieferungen, die also meistens große Mengen umfassen und nur nach sorgfältiger Kontrolle abgenommen zu werden pflegen, so daß Fehler an Werkzeugen für derartige Stücke, seien es solche Fehler, die von vornherein vorhanden sind, seien es solche, die im Verlaufe der Arbeit erst auftreten, großen Schaden mit sich bringen.

Die abgebildeten Werkzeuge sind sämtlich, vom Anschlag- bis zum Kalibrierwerkzeug, in allen Teilen aus Gußeisen hergestellt und konstruktiv vollständig gleichartig durchgebildet, was Befestigung der

Stempel, Matrizen, Niederhalter usw. betrifft. Besondere Sorgfalt ist auf die Abrundungen an den Stempel- und Matrizenkanten, die Übergangsstellen an den Niederhaltern verwendet worden, da gerade bei derartiger Arbeit sich oft die kleinste, scharfe Kante durch Ausschußstücke später unliebsam bemerkbar macht.

An einer fast gleichen Abstufung[1]) wird an der in der Fußnote angegebenen Stelle die Wichtigkeit des „Schwärzens" bei der Herstellung dieser Werkzeuge, deren Teile vollständig genau zusammenarbeiten müssen, ausgeführt. Um nämlich bei dem Entwurf der Abstufungswerkzeuge eine richtige Prüfung des mit ihnen erhaltenen Zuges zu erhalten, ist es notwendig, sich zuerst von der einwandfreien Arbeit derselben

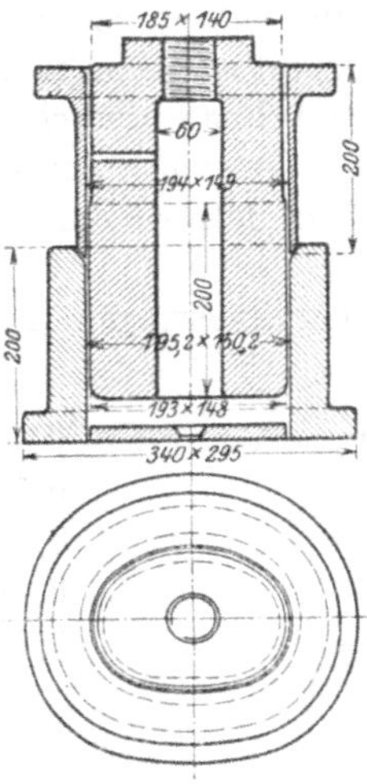

Fig. 382.

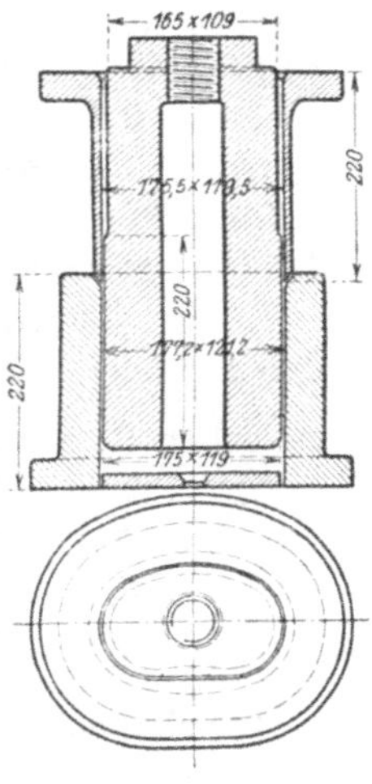

Fig. 383.

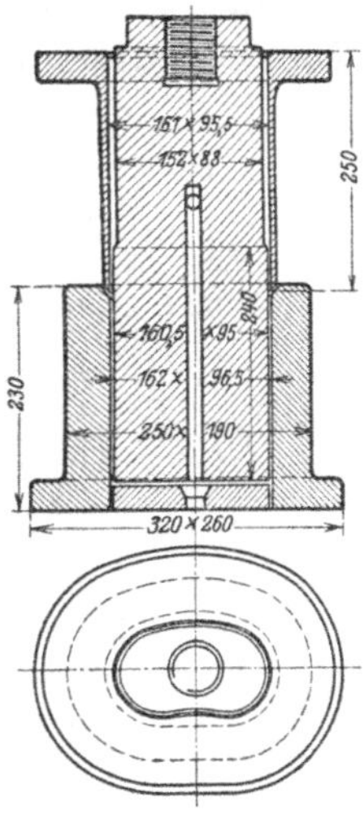

Fig. 384.

zu überzeugen und dann erst die Schuld an Ausschußstellen im verarbeiteten Material zu suchen. Zu diesem Zweck werden die Werkzeuge mit geschwärzten Blanketten geprüft. Die Schwärze, die aus Kienruß und Öl in Form eines dicken Breies hergestellt wird, wird dann ganz dünn auf das Blech aufgetragen. Wenn das Blech in die Werkzeuge eingelegt und diese ein wenig angezogen werden, so werden sich auf dem Blech hellere und dunklere Stellen markieren. Die helleren Stellen auf dem Blech entsprechen jenen, die am Werkzeug geschwärzt erscheinen, dort also zu stark angreifen, so daß das Werkzeug an diesen Stellen nachgearbeitet werden muß. Dies ist nun mit sorgfältig auf gleichmäßige Dicke hin nachgemessenen Blechen an dem Werkzeug

[1]) Z. f. Werkzeugm. 1913, S. 534.

solange fortzusetzen, bis nicht die geringste Ungleichmäßigkeit zu finden ist. Dann kann man das Werkzeug der eigentlichen Arbeit zuführen. Entstehen dann noch Ausschußstücke, so werden diese auf Rechnung eines ungenügend gleichmäßigen Blechmateriales zu setzen sein.

An den abgebildeten Werkzeugen ist zu erkennen, daß im ersten Zug ein nach beiden Achsen symmetrischer Querschnitt gezogen wird, der im zweiten und dritten Zug weiter verkleinert wird, wobei bereits nur mehr eine Symmetrieebene vorhanden ist, immer aber noch beide Hälften Ellipsenform haben. Im vierten Zug beginnt das Abflachen der einen Querschnitthälfte, dem im fünften und sechsten Zug ein deutliches Einziehen dieser Wandung folgt, das aber immer von einer Querschnittsverringerung und entsprechenden Verlängerung der Hülse begleitet ist. Das verarbeitete Aluminiumblech ist 0,7 mm stark und dementsprechend zwischen den Werkzeugteilen ein Spiel von 2,5 mm bis herunter beim letzten Zug auf 1,5 mm gelassen worden.

Die Querschnittsverminderungen f_2/f_1 der Stempel betragen

zwischen	2.	und	1.	Zug	. . . . . .	0,835
„	3.	„	2.	„	. . . . . .	0,745
„	4.	„	3.	„	. . . . . .	0,77
„	5.	„	4.	„	. . . . . .	0,737
„	6.	„	5.	„	. . . . . .	0,775

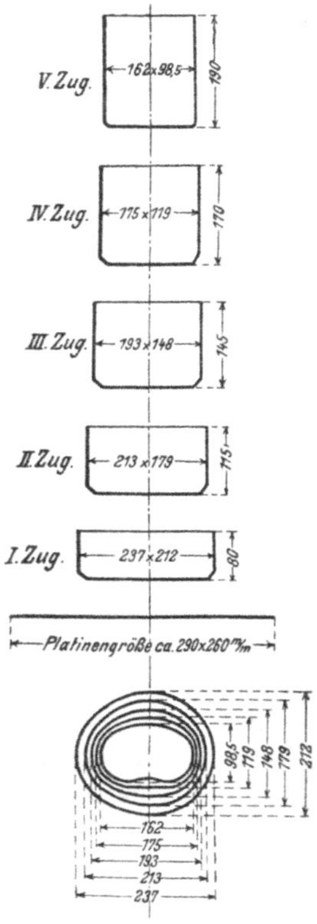

Fig. 385.

Sie bleiben also mit Ausnahme des ersten Zuges fast gleich.

In der oben angegebenen Quelle ist ein ähnlicher Satz Züge für einen Feldkessel, Fig. 385, gegeben, der in fünf Zügen vollendet ist und mit den übrigen Maßen die Verlängerung der einzelnen Züge enthält, deren Verhältnis von rd. 0,7 auf 0,9 im letzten Zug allmählich fällt.

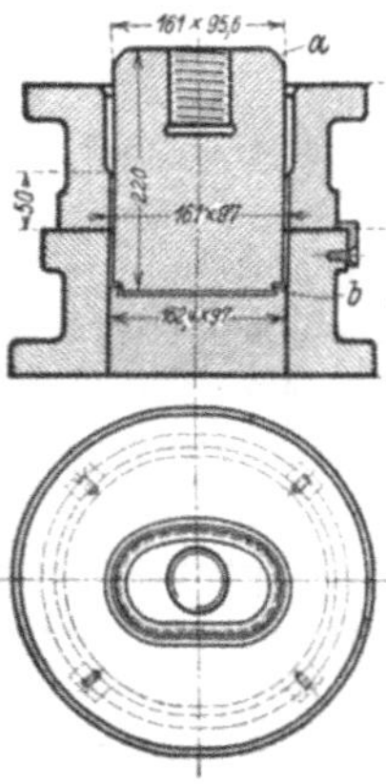

Fig. 386.

In Fig. 386 ist das Werkzeug zur Herstellung des Deckels wiedergegeben, das insoweit bemerkenswert ist, als dazu zwei Stempel gehören, einer wie a als Anschlagwerkzeug und einer wie b als Kalibrierwerkzeug. Außerdem sind vier Anschläge zum Einlegen des Blanketts vorgesehen, das ein Oval von 260 × 230 mm ist. Der gezogene Deckel ist 59 mm hoch, erhält einen 10 mm breiten Rand und wird wie der Kessel aus 0,7 mm starkem Aluminiumblech gemacht.

Mit Ausnahme der Anschläge sind auch hier alle Werkzeugteile aus Gußeisen gemacht und werden in gleicher Weise wie die Kesselwerkzeuge eingepaßt. Die Werkzeuge sind für eine Ziehpresse mit beweglichem Tisch und beweglicher Stempelführung, Größe TE, der Firma Schuler gebaut. Die Maschine zieht bis zu einem Durchmesser von 400 mm bei einer Ziehtiefe von 220 mm und 12 Niedergängen in der Minute. Der Kraftverbrauch stellt sich auf ungefähr 4 PS.

Wie bekannt, nutzt das gleichzeitige Niedergehen des Tisches die Leerzeit der Presse günstiger aus, da bei gleicher Hubzahl der Presse Zeit zum Einlegen der Blanketts gewonnen wird. Die Einstellbarkeit des Niederhalters unter Anwendung des mechanischen Antriebes verringert die toten Zeiten beträchtlich; seine Selbsteinstellung durch Lagerung desselben auf einer Kugelfläche bewirkt, daß kleine Unterschiede in der Blechdicke ausgeglichen werden.

Ziehen viereckiger Hülsen. Die Werkzeuge zum Ziehen viereckiger Hülsen sind, soweit die konstruktive Durchbildung in Betracht kommt, denen für runde Hülsen fast gleich. Dagegen ist die Abstufung und die Form der einzelnen Stufen mit Rücksicht auf die auszubildenden Kanten besonders zu entwerfen.

Wenn nämlich aus einem rechteckigen Blankett, Fig. 387, eine rechteckige Hülse gezogen werden sollte, so kann man sich vorstellen, daß das Material der Seitenflächen a b c d einfach umgeklappt wird, während die Zwickel bei x x als Falten herausgedrückt oder, wenn dies nicht geht, als verlängerte Hülsenecken nach Fig. 388 erscheinen würden. Dabei ist klar, daß die Materialbeanspruchung an den Seitenmitten sehr gering ist, während sie in den Ecken und den anliegenden, durch die Ziehwirkung mitgenommenen Teilen der Seitenflächen unverhältnismäßig groß wird, wodurch dort ein Einreißen und Aufspringen bis auf den Boden sehr häufig ist. Es sei noch bemerkt, daß bei einer zu starken Eckenbildung nach Fig. 388 die Hülse auch beim Abstreifen aufreißen kann, nachdem die eigentliche Zieharbeit anstandslos durchgeführt worden ist. Aus diesen Gründen wird man beim Anschlagwerkzeug die Abrundungen der Ecken so groß, wie nur angängig, halten, ebenso diejenigen an

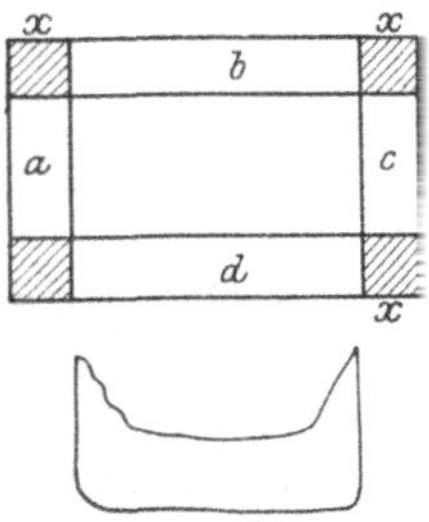

Fig. 387 und 388.

Boden, und die oberen Abrundungen der Ecken am Ziehring größer als bei einem derartigen Werkzeug für runde Hülsen lassen. In den weiteren Zügen werden diese Abrundungen allmählich kleiner, bis sie beim letzten Zug, wenn nötig, in die scharfen Kanten übergehen können. Ein vollständiger Satz Ziehstufen, deren Endergebnis eine scharfkantige, rechteckige Büchse ist, ist in Tabelle Nr. 21 dargestellt. Das betreffende Anschlagwerkzeug, das gleichzeitig den Ausschnitt aus dem Blech herstellt, ist auf S. 413 beschrieben.

Man muß also hier dem Zuschnitt der Ecken große Sorgfalt zuwenden. Das einfache Probieren ist, wie ein Praktiker in der Ill. Zeitschrift für Blechindustrie zugesteht, eine äußerst mühselige Sache, während von einer anderen Seite[1]) aus der Praxis eine Näherungskonstruktion gegeben wird, die infolge ihrer Einfachheit tatsächlich einen Teil dieser Arbeit erspart. In das volle Rechteck des Blanketts, Fig. 389, ABC wird der Boden des ersten Zuges A′ B′ C′ eingezeichnet, so daß eine Randhöhe a ringsherum verbleibt. Dann schlägt man aus der Ecke B mit a als Radius einen Kreis, an welchen man aus den Punkten 1 und 2, die auf einem Kreis aus B mit dem Radius 2 a liegen, die beiden Berührungskreise mit dem gleichen Radius a zieht. Der schraffierte Ausschnitt gibt dann die Form des Blanketts. Ein bei weitem besserer, wenn auch durch die Vorversuche etwas längerer Weg wird nach ähnlichen Grundsätzen, wie auf S. 254 für die Ermittlung der einzelnen Züge angegeben, von Musiol[2]) an einem Beispiel für eine Weißblechdose, die in fünf Zügen nach Fig. 390[3]) hergestellt wird, eingeschlagen. Es handelt sich um eine ziemlich tiefe Dose mit einem kleinen Rand und mehreren Ausprägungen im Boden.

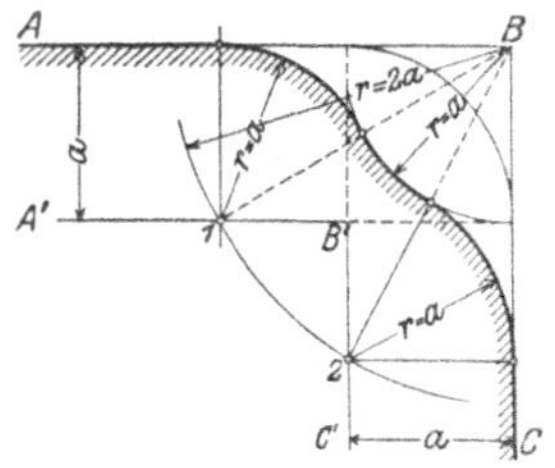

Fig. 389.

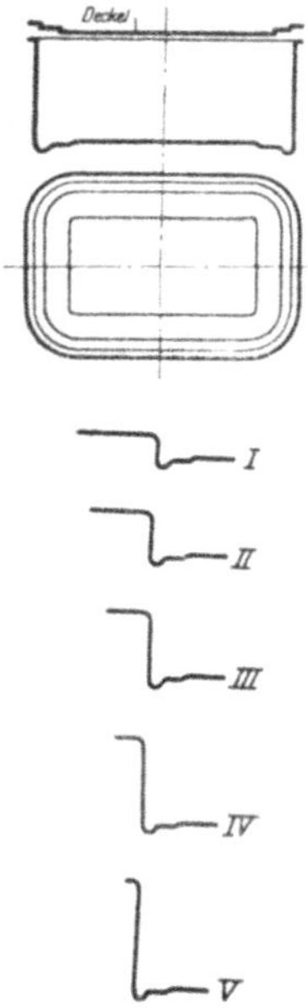

Fig. 390.

Zur Ermittlung des zugehörigen Blanketts geht Musiol[4]) wieder von der Dehnungsziffer des betreffenden Rohmateriales aus, die er auf

[1]) Zeitschr. f. Werkzeugm. 1913, S. 374.

[2]) Musiol, Rechnerische und zeichnerische Methode der Zuschnittsermittlung in der Ziehpressentechnik.

[3]) Zeitschr. f. Werkzeugm. 1911, S. 199.

[4]) Illustr. Zeitschr. f. Blechindustrie 1911, S. 1136.

einem gewöhnlichen Ziehwerkzeug für eine runde Hülse vom Durchmesser der einen Rechteckseite des Bodens bestimmt. Dann teilt er den Hohlkörper, Fig. 391, in vier Zylinderteile, die den Ecken der rechteckigen Hülse entsprechen, und die übrigbleibenden prismatischen Teile. Die Zylinderteile entsprechen einem Zylinderdurchmesser von 44 mm, zu dem er die entsprechende Scheibe nach seinen Formeln aus den nach obigen Versuchen erhaltenen Dehnungsziffern zu 113 mm Durchmesser bestimmt. Zur Bestimmung der beiden Hauptachsen des Blanketts wickelt er die beiden Schnitte des Hohlkörpers nach den Hauptachsen I und II ab, multipliziert diese Länge von 270 bzw. 219 mm mit den betreffenden linearen Dehnungsziffern, so daß sich für die beiden Achsen Längen von 251 bzw. 210 mm ergeben. Das Blankett wird mit den gefundenen Werten in folgender Weise, Fig. 392, angerissen: Man zeichnet sich den Umriß des Bodens an mit den beiden Hauptachsen I, II und zieht aus den Krümmungsmittelpunkten der Ecken A B usw. Kreise mit den oben gefundenen Radien von 56,5 mm. Dann trägt man die Längen 125,5 und 105 mm auf den bezüglichen Hauptachsen nach beiden Richtungen auf, zieht die Parallelen in den Endpunkten zu den Hauptachsen und Tangenten an die Kreise aus den Krümmungsmittelpunkten. Wenn man nun noch in dieses Sechseck an den Scheiteln eingeschriebene Kreise in der gezeichneten Weise beschreibt, bildet diese Linie den Blankettumriß.

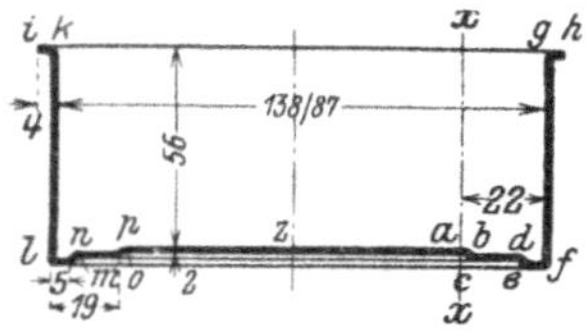

Fig. 391.

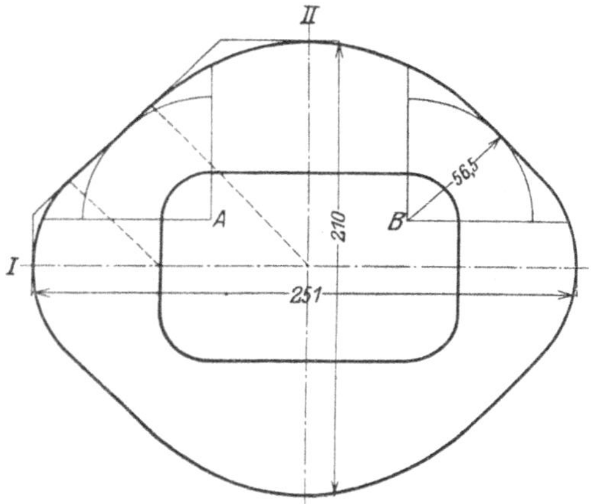

Fig. 392.

Werkzeug zum Einziehen von Messingrohren. Im Gegensatz zu den verwickelten Vorrichtungen zum Einziehen der an einem Ende geschlossenen Patronenhülsen ist in Fig. 393 ein Werkzeug zum Einziehen der Enden von Messingrohren wiedergegeben, das aus einem

Bedürfnis der Praxis entstanden ist. Dieses Werkzeug besteht aus Stempel und Matrize mit zwangläufigen Schaltungen, kann jedoch nur in einer Presse mit großer Bauhöhe unter dem Stößel

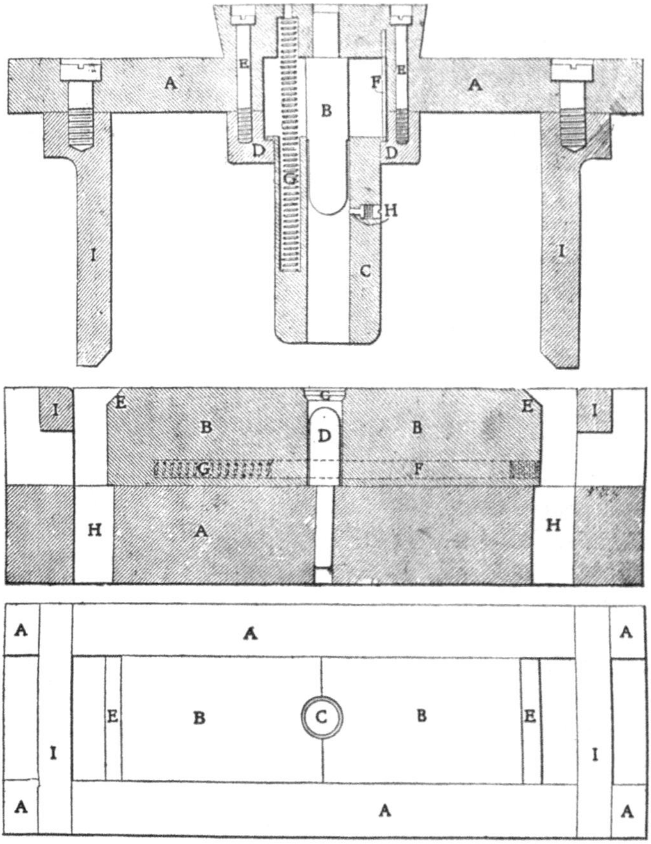

Fig. 393.

und mit ungewöhnlich langem Hub verwendet werden. In dem besprochenen Falle muß ein Arbeitsstück von 82,5 mm Länge leicht entfernt werden können. Das verwendete Rohrmaterial hat einen Außendurchmesser von ungefähr 14,6 mm und wird an einem Ende auf einer Länge von 25 mm auf 12 mm Durchmesser eingezogen. Die Wandstärke wird dabei im kleineren Durchmesser ein wenig verstärkt, auch die Dichte des Materials ein wenig durch die Bearbeitung vergrößert, während die Länge ungeändert bleibt. Die Arbeitsstücke aus einem derartigen Werkzeug sind, was Außendurchmesser und Länge betrifft, gleich, während sich die Unterschiede nur auf der Innenseite des eingezogenen Teiles zeigen.

Fig. 393 stellt einen Grundriß und Schnitt durch den Unterteil und einen Schnitt durch den Stempel dar. A ist die gußeiserne Grundplatte, die allseitig bearbeitet und mit einer Längsnut versehen ist, in der die gehärteten und geschliffenen Schlitten B gleiten. Vor dem Härten werden die Stücke vorläufig zusammengespannt und auf der Planscheibe einer Drehbank angezeichnet, so daß die Mitte des Loches C in die Stoßfläche beider Schlitten fällt, und dann dieses bearbeitet und auf Maß poliert. Nachdem die Dreharbeit fertig ist, werden die Schlitten so hart wie möglich im Salzbad gehärtet und überhaupt nicht angelassen. Die inneren Stoßflächen der Schlitten werden leicht auf einer Flächenschleifmaschine überschliffen, um jede vom Härten herrührende Unebenheit zu beseitigen, so daß sie, und zwar besonders in der Nähe des Loches C, statt aneinander anliegen, da sonst jeder Fehler am fertigen Stück sichtbar würde. Die Stücke werden nun wieder zusammengeklemmt und der zurückgesetzte Durchmesser des Loches C auf Maß poliert. Unten wird er bis auf 3,2 mm von der Oberkante um 0,05 mm größer gehalten, um die Reibung zu vermindern, wenn die Röhre durchgedrückt wird, da sich die Arbeit unmöglich zufriedenstellend mit einem langen, geraden Loche ausführen ließ. D ist ein gehärteter Bolzen, der in die Grundplatte A eingetrieben ist. Der Raum zwischen der Lochwand und dem Bolzen beträgt 0,1 mm mehr als die Dicke der Röhre.

Die Außenkanten der Schlitten B werden bei E unter 45° abgeschrägt

Die Konstruktion der übrigen Teile des Werkzeuges sowie ihre Arbeit und die Arbeitsweise des ganzen Werkzeuges beim Einziehen der Rohre bedürfen keiner weiteren Erklärung.

Ziehwerkzeug mit Blanketthalter. Fig. 394 zeigt den Werkzeugunterteil für den ersten Zug einer Hülse aus Messingblech von 1 mm Dicke. Der Durchmesser B (Grundriß) des Blanketts beträgt 133 mm.

Der Ziehring A ist aus Werkzeugstahl gemacht und hat bei K und H zwei besondere Merkmale. Die Abrundung bei K hat einen Radius von 12,7 mm; ist sie kleiner, so reißt der Boden aus, ist sie größer so bilden sich Falten, und das Material reißt oben ein. Für diese Blankettgröße ist ein Stempel von 76 mm Durchmesser die kleinste Größe die mit Erfolg für den ersten Zug verwendet werden kann; kleinere Stempel reißen den Boden des Metalls aus. Die Ausdrehung H vermindert die Reibung und dient gleichzeitig als Abstreifkante, D ist ein gußeiserner Führungsring und gleichzeitig Befestigungsplatte, B ein gehärteter Stahlring, der in D eingesetzt ist und dem Durchmesser des Blanketts entspricht. E sind Befestigungsbolzen, die durch die Hebel F betätigt werden. G ist eine der Federn, die die Platte D hochhalten, während das Material zugeführt wird, da D während des Ziebens als

Blechhalter dient, um **Faltenbildung** zu verhindern. Der Ring B wird wegen der starken Abnutzung beim Ziehen eingesetzt. Der Hebel F wird an der Schraube E mit einem Klemmschlitz befestigt, damit

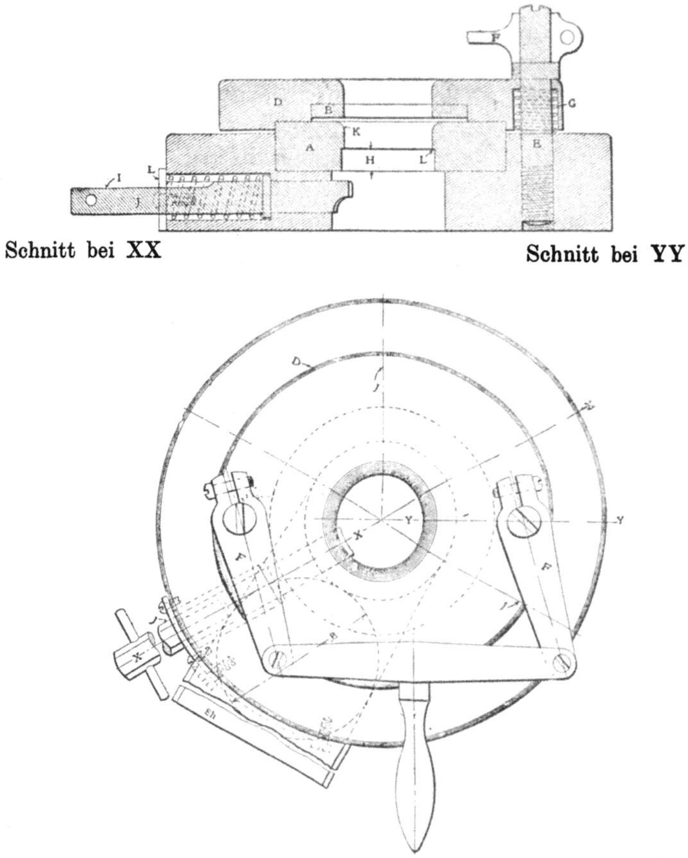

Fig. 394.

man, nachdem D auf das Blankett niedergepreßt worden ist, F einstellen und in einer für die rechte Hand des Arbeiters bequemen Stellung festziehen kann, so daß die linke Hand frei bleibt, um das Material auf dem Zuführungstisch Sh vorzuschieben.

Dieses Werkzeug war in eine große Presse mit Zahnräderübersetzung, die ununterbrochen läuft, eingebaut worden.

Hilfsmittel beim Abstreifen im Ziehgesenk. Fig. 395 zeigt ein Ziehwerkzeug zur Herstellung eines Näpfchens aus Stahlblech von ungefähr 2 bis 0,7 mm Dicke. Beim Abstreifen vom Stempel oder Entfernen desselben aus der Matrize hatte man dabei oft Schwierigkeiten, kurz, es war ein Stück, das schwer zu verarbeiten war. Während man aber das Abstreifen vom Stempel verhältnismäßig leicht erreichte, verursachte das Entfernen des Stückes aus der Matrize größere Schwierigkeiten. Manche dieser Näpfchen hatten ein kleines Loch in der Mitte, andere einen vollen Boden; doch mußte der Boden in beiden Fällen so flach wie irgend möglich sein. Die Bodenfläche des Ziehringes hatte ein kleines Loch für den Auswerferstift. Die Hülse mit dem Loch im Boden konnte jedoch damit nicht ausgeworfen werden, während die Hülse mit vollem Boden manchmal ausgeworfen und ein anderes Mal wieder der Boden durch den Auswerferstift aufgebogen wurde, wie links unten angedeutet ist.

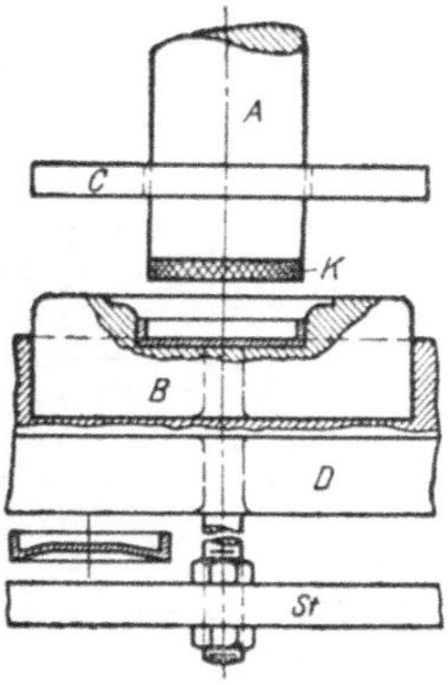

A = Stempel
B = Ziehring
C = Abstreifer
D = Froschplatte
St = Auswerfer
K = Kordierung am unteren Ende von A

Fig. 395.

Alle diese Schwierigkeiten wurden durch Kordieren des Stempelendes beseitigt, wonach man den unteren Auswerfer nicht mehr benötigte. Der Stempel wurde jetzt ungefähr 0,075 mm kleiner im Durchmesser als das verlangte Maß gemacht, dann sein Ende auf ungefähr 5 mm Länge kordiert und die scharfe Kante mit Schmirgelleinwand gerundet. Durch das Kordieren erhielt der Stempel dann die richtige Abmessung,

Die Reibung war so groß, daß man vorher sehr oft Hülsen mit dem Meißel aus dem Ziehring entfernen mußte; durch das Kordieren hingegen konnte der Stempel diese Reibung überwinden und die Hülse regelmäßig mit nach oben nehmen, wo sie vom Abstreifer C richtig abgestreift wurde.

2. Das Ziehen von Pappschachteln.

(Sächsische Kartonnagen-A.-G., Dresden-A.) Die allgemein gebräuchlichen runden Pappschachteln bestehen entweder aus zwei Teilen, dem Deckel und dem mit Boden versehenen Körper, oder aus drei Stücken, nämlich zwei Deckeln und einem zylindrischen Mantel ohne Boden.

Die Herstellung der Deckel und Körper mit Boden erfolgt in vielen Fällen durch Ziehen, das sich jedoch, sowohl was die Art der Arbeit und Vorbereitung als auch was das Werkzeug betrifft, von den in der Metallzieherei verwendeten Verfahren unterscheidet.

In Fig. 396 ist das Werkzeug und drei verschieden hohe Deckel abgebildet, die sämtlich mit demselben hergestellt werden. Für Rund- oder Fassonschachteln kommen der Hauptsache nach gleichgebaute Werkzeuge zur Anwendung, die in die Gruppe der Verbundwerkzeuge fallen, da sie gleichzeitig ziehen und abschneiden. Die Ausbildung und Vereinigung der bezüglichen arbeitenden Teile rechtfertigt jedoch ihre Anführung hier. Der Stempel besteht entweder aus einem einzigen Stück Werkzeugstahl, links in der Figur, oder er ist aus mehreren Stücken zusammengesetzt, rechts in der Figur. In diesem Fall besteht er aus dem eigentlichen Ziehstempel a, der auf der Unterfläche die Prägung usw. angearbeitet und in der Mitte ein Luftloch erhält, das durch einen Ringkanal und eine Bohrung im Schnittring c mit der Außenluft in Verbindung steht. Auf den Stempelzapfen werden Schnittring c und Beilagringe d nach Bedarf aufgeschoben und von oben durch ein am Stempelzapfen aufgeschnittenes Gewinde mittels einer kräftigen Mutter b gehalten. Der Stempel a und der Schnittring c sind aus bestem Werkzeugstahl hergestellt. Die Beilagringe d können entweder, wie in der Zeichnung ersichtlich ist, ober- und unterhalb des Schnittringes eingelegt werden, bzw. beide oberhalb oder unterhalb desselben. Je nach der Stellung derselben ergibt sich eine verschiedene Deckelhöhe, wie aus der Figur ersichtlich ist. Der stählerne Ziehring e, hier „Zugbüchse" genannt, ist in eine gußeiserne Grundplatte f, „Zugmantel", eingepreßt und legt sich mit einem eingedrehten Absatz gegen eine entsprechende Kante der Grundplatte f. Der Auswerfer g arbeitet mit der Unterseite des Stempels zusammen, um die Prägung des Deckels herzustellen, und wird von unten in der üblichen Weise durch Federbuffer betätigt; er ist ebenfalls aus Stahl. Zur Verhinderung der Faltenbildung ist die Faltenplatte h vorgesehen und mit Zylinderkopfschrauben befestigt, unterhalb welcher das Blankett von vorn eingeschoben wird.

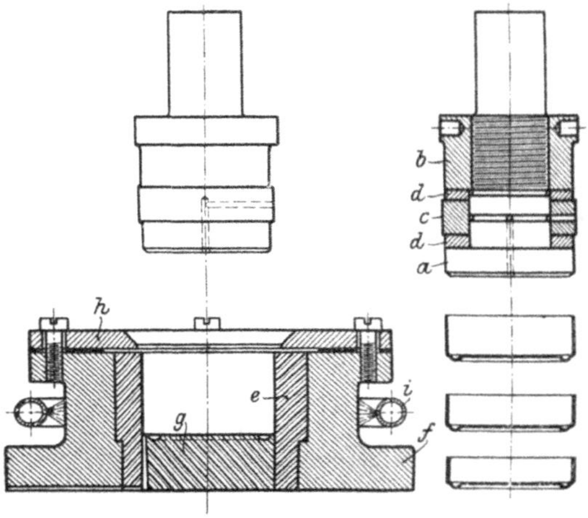

Fig. 396.

Sollen bloß geringe Mengen eines Stückes hergestellt werden, so ist ein Stempel aus einem Stück, links in der Figur, anzuwenden, im anderen Falle der zusammengesetzte Stempel, da man leicht, wenn die eine Schnittkante stumpf geworden ist, den Schnittring umkehren und gegebenenfalls ganz auswechseln kann.

Bei der Einspannung des fertigen Werkzeuges in der Presse ist große Sorgfalt anzuwenden, da damit die Haltbarkeit desselben Hand in Hand geht. Man setzt zuerst den Stempel in den Unterteil ein und befestigt ihn mit der angefeilten Fläche mittels der Druckschraube im Stößel, nachdem man diesen so tief wie möglich über den Zapfen heruntergeschoben hat. Dann wärmt man den Unterteil an, bis auch der Stempel warm geworden ist, läßt den Stößel wieder hochgehen und schiebt, ohne den Unterteil zu verschieben, ein Blankett vorsichtig ein. Hierauf läßt man den Stempel nach abwärts gehen, bis er aufsitzt, und zieht die Spannschrauben an. Zwischen den Unterteil oder bei mittelbarer Heizung zwischen die Wärmeplatte und den Pressenkörper ist stets eine Isolierschicht einzulegen, damit die Wärme nicht auf den Pressentisch übergeht.

Das Anwärmen ist genau der zu verwendenden Pappe anzupassen; zeigen sich nach dem Ziehen Blasen auf der Pappe, so ist dies ein Zeichen, daß die Form zu heiß ist. Die Heizung geschieht mit Leuchtgas, Spiritus oder Petroleum. Um gute Erfolge zu erzielen, ist eine richtige Pappendicke von größter Bedeutung. Bei Verwendung von zu schwacher Pappe werden die Falten nicht genügend ausgepreßt, bei zu starker Pappe reißen die Ränder ein. Stark satinierte Pappe ist schwächer zu wählen als weiche und schwach satinierte.

Um einen schönen Glanz des gezogenen Randes zu erhalten und zu vermeiden, daß die Schachteln beim Ziehen reißen, ist die Pappe vorteilhaft vorher mit warmem Seifenschaum zu behandeln, der mit einem Schwamme leicht auf die später äußere Fläche der Pappe aufzutragen ist. Am besten bestreicht man die Pappstreifen vor dem Ausschneiden. Den Seifenschaum setzt man warm mit guter weißer Kernseife, etwa 30 g auf 1 Liter Wasser, an. Statt mit Seifenschaum kann man die Pappe auch mit einem trockenen Stück Kernseife leicht bestreichen.

Die Herstellung der zylindrischen Schachtelteile ohne Boden geschieht auf einer Ziehbank, die im allgemeinen ganz gleich der bei dem Drahtziehen verwendeten ist. Das Ziehloch dagegen ist ebenso wie im eben besprochenen Werkzeug angeheizt. Der ganze Bogen Papier, der ein Vielfaches der Schachtellänge beträgt, wird nun um einen hochglanzpolierten Stahldorn gewickelt und gleichzeitig geleimt, worauf der Stahldorn an die Ziehkette angehängt und das Ganze durch das Ziehloch gezogen wird. Nach dem Abstreifen wird das fertige lange Rohr auf die benötigten Längen zerschnitten.

3. Werkzeuge zur Herstellung von Lederpackungen[1]).

Nutzeffekt der Lederpackung. Wenn man Druckwasser, besonders für hohen Druck, zur Kraftübertragung verwendet, so ist eine dichte Kolbenpackung das wichtigste und teuerste Glied der Anlage. Je höher der Druck, um so wichtiger wird diese Frage, in manchen Fällen hängt geradezu der größtmögliche Druck von der Güte der Lederpackung ab. Es gibt nun eine ganze Reihe anderer Packungen, die alle ihre Vorteile haben, doch wird man viele Fälle finden, in denen die gewöhnliche, selbstdichtende Lederpackung weitaus am besten, vielleicht sogar unersetzlich ist. Ihr größter Nachteil sind jedoch die im Verhältnis zu anderen Packungen hohen Anschaffungskosten. Wenn auch heutzutage verschiedene Firmen diese Leder in hervorragender Qualität fabriksmäßig herstellen, wird man doch bei einzelnen Unternehmungen, die hydraulische Maschinen haben, von dem Maschinisten erwarten, daß er sich seine Leder selbst macht.

Formen der Ledermanschetten. Im folgenden seien die notwendigen Werkzeuge, die Herstellungsweise beschrieben und eine Reihe Winke zur erfolgreichen Herstellung der verschiedenen Manschetten gegeben.

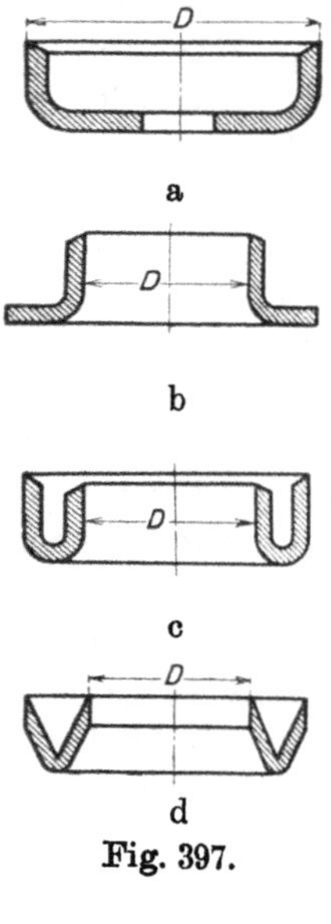

Fig. 397.

Obwohl die Herstellung einer guten brauchbaren Manschette eine ziemlich einfache Arbeit ist, so verlangt sie doch viel Erfahrung und Sorgfalt. Die verschiedenen Formen sind in Fig. 397 abgebildet. Die ersten drei Formen, die Napfform a, die Hutform b und der Nutring c, sind allgemein bekannt, die in d abgebildete Dachform baut die Firma Gehrckens mit Erfolg mehrfach übereinander ein. Dabei kommt in die Manschettendachform erst ein Lederring und darauf die zweite Dachform.

Im allgemeinen ist bekannt, daß die Höhe des Leders praktisch fast keinen Einfluß auf seine

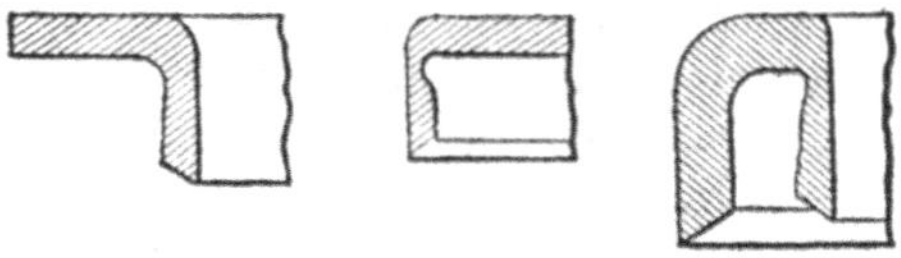
Fig. 398.

Reibung und Dichtungsfähigkeit hat, da die Dichtung nur an einer einzigen Stelle der Anlagefläche erfolgt. Dies ist am besten aus Fig. 398 er-

[1]) Die Anmerkungen in diesem Absatz entstammen Angaben der Firma C. Otto Gehrckens, Hamburg.

sichtlich, die Schnitte aller drei Lederformen in abgenutztem Zustand zeigt. Wie man auch später erkennen wird, erfährt auch die Anlagestelle die größte Abnutzung und ist gleichzeitig auch die Stelle der größten Beanspruchung beim Ziehen. Weiter wächst die Beanspruchung an diesem Punkt mit der Höhe des Leders. Man muß also um so mehr Sorgfalt bei der Herstellung aufwenden, um am fertigen Leder überall gleiche Dicke zu erhalten, so daß es offenbar anzuraten und sicher billiger ist, die Höhe des Leders so gering wie möglich zu halten.

Ein Packungsleder muß genau in die dazu bestimmte Ausdrehung passen, ausgenommen in der Höhe, und soll, wenn irgend möglich, mit einem Tragring versehen werden, gegen den es sich anlegt und der an der Auflagerfläche des Leders genau die gleiche Krümmung wie das Leder hat.

Im Falle des U-Leders ist dieser Ring sehr notwendig und muß Löcher erhalten, durch die das Druckwasser unter das Leder eintreten kann. Die verdrückte Form des U-Leders in Fig. 398 ist auf das Fehlen eines solchen Ringes zurückzuführen. Wenn man auf diese Einzelheiten achtet, so wird das Leder seine Gestalt viel länger behalten und deshalb auch länger halten; auch öffnet sich das Leder viel schneller, wenn man Druckwasser einläßt.

Auswahl und Vorbereitung des Leders. Für diesen Zweck wird von den Händlern besonders vorbereitetes Leder verkauft. Es ist weicher als das gewöhnliche Leder und muß von bester Qualität sein; Rückenstücke sind am besten. Dann geht man folgendermaßen vor[1]): Man wählt ein Stück gesundes, fehlerloses Leder von dem verlangten Durchmesser; besonders gefährlich sind die „Engerlinge“ (engl. Nadellöcher), die sich erst öffnen und besser zu sehen sind, wenn das Leder naß gestreckt wird. Findet man eines dieser feinen Löcher, so verwendet man das Stück für eine kleinere Größe, da man sicher sein kann, daß sich das Loch während der Verarbeitung öffnet und das fertige Leder undicht macht. Hat man ein gesundes Stück ausgewählt, so schabt man mit einem Messer alles weiche Material auf der Fleischseite weg und macht das Leder blank und überall gleichdick[2]), je nach der Größe

[1]) Außerordentlich schwierig ist die Auswahl des Leders, namentlich um die sogen. U-Leder herzustellen; es muß sehr geschmeidig und ein Ia eichenlohgruben, nach altem Verfahren gegerbtes Leder oder Chromleder sein. Das Leder sollte vorher nur ganz leicht gefettet und frei von überflüssigen Gerbstoffen sein, es darf in trocknem Zustande nicht brechen. Auch muß es frei von Engerlingen und von groben Schnitten, die die Schlachter verursachen, sein.

[2]) Über das Wachstum der Haut muß man genau orientiert sein, um an Punkten, wo sich das Leder mehr streckt als an anderen, nicht zu viel von der Dicke desselben fortzunehmen, denn der Wert der Manschetten liegt darin, daß das Leder nach der Pressung eine gleichmäßige Dicke aufweist; wenn man vor

Woodworths Tabellen der Blankett-

Form des Leders	Dichtungsdurchmesser*) D in Zoll	$^1/_2$"	$^3/_4$"	1"	1"	$1^1/_4$"
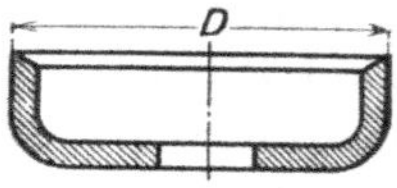	Außendurchmesser D in mm		19	25,5		32
	Blankett . . . . . . in mm		35	47,5		54
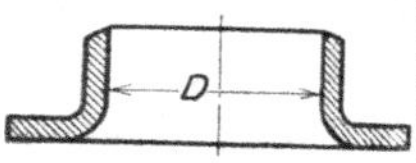	Innendurchmesser D in mm	12,5	19	25,5	25,5	38
	Außendurchmesser . in mm	39,5	39,5	41	54	66,5
	Blankett . . . . . in mm	51	54	60	76	92
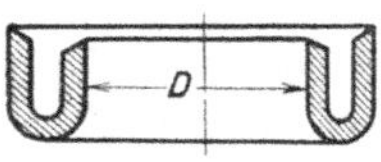	Innendurchmesser D in mm			25,5		
	Außendurchmesser . in mm			51		
	Blankett . . . . . . in mm			89		

*) Die Millimetermaße sind auf $^1/_2$ mm abgerundet.

des Leders 5—8 mm. Die Dicke muß sorgfältig und genau von den Formen abgemessen werden, da sich bei zu dickem Leder große Schwierigkeiten ergeben. Nun weicht man das Lederstück in Wasser ein, wobei man es zwischen Daumen und Finger gut verarbeitet, bis es überall weich und biegsam geworden ist; dann wird es mit Talg eingerieben, worauf es zum Ziehen fertig ist. Welche Seite der Haut, die glatte Haar- oder die rauhe Fleischseite, sich weniger abnutzt, darüber gehen die Meinungen sehr auseinander. Die Haarseite hat ein feineres Korn und ist deshalb eher wasserdicht als die Fleischseite. Wenn man daher die Haarseite innen und die Fleischseite an der Reibungsfläche verwendet, so wird das Leder fast so lange wasserdicht bleiben, als es überhaupt hält; wenn man andererseits die Haarseite als Reibungsfläche verwendet, so nutzt sich das Leder weniger ab und sieht besser aus. Es ist jedoch immer das beste, die glatte Seite nach innen zu legen und die rauhe Seite die Abnutzung tragen zu lassen.

Herstellung der Napfform. Diese werden selten bei Kolben über 76 mm Durchmesser verwendet[1]) und sollen für 25 mm und weniger Durchmesser ungefähr 10 mm Höhe und bei 76 mm Durchmesser bis zu 16 mm Höhe haben. Das vorbereitete Blankett wird mit der glatten

dem Pressen das Leder ganz gleichmäßig hat, so kann man niemals eine ganz gleichmäßige Manschette erhalten.

[1]) Die Preisliste der Firma Gehrckens enthält für alle Formen Durchmesser bis zu 1000 mm mit Höhen bis zu 45 mm.

Durchmesser für Lederpackungen.

1½"	1¾"	2"	2¼"	2¾"	3"	3¾"	4½"	6"	8"	8½"	11"	14½"	18"
38		51		70	76								
63,5		79,5		98,5	108								
38,5	44,5		57	70									
66,5	73		92	95									
92	92		117,5	127									
38		51		70		95	114,5	152,5	203	216	279,5	368	457
63,5		79,5		95		120,5	152,5	190,5	241,5	254	317,5	466,5	501,5
105		133,5		165		190,5	235	279,5	343	355,5	419	508	654

Seite nach oben ausgemittelt auf den Ziehring gelegt und der Stempel gleichachsig aufgesetzt, wie Fig. 399a zeigt. Dann wird der Stempel allmählich und in Absätzen hineingedrückt, bis das Leder mit dem Grunde des Ziehringes gleichsteht, Fig. 399b; man schneidet dann die Oberkante des Leders bündig mit der Oberkante des Ziehringes ab. Nun wird das Leder noch ein wenig weiter eingedrückt und der Ring nach Fig. 399c eingesetzt. Dadurch verhindert man, daß der Ring das Leder an der Abrundung bei A verdrückt und verbreitert. Dann wird das Ganze auf die Unterplatte gesetzt und der Ring noch ein wenig weiter hineingedrückt, der Ziehring bis auf die Grundplatte heruntergezogen oder heruntergeschlagen und in der Stellung, wie Fig. 399c zeigt, festgeklemmt[1]). Nach einiger Zeit entfernt man Form und Leder zusammen und stellt das Ganze an einen warmen Ort, z. B. einen Kesselraum oder bei kleinen Größen auf einen Dampfzylinder und läßt trocknen. Das dauert je nach der Größe des Leders gewöhnlich mehrere Stunden bis zu einem Tage. Man muß aufpassen, daß der Platz nicht zu heiß ist, da sonst das Leder ernstlich beschädigt werden kann. Wenn man das Leder aus der Form nimmt, soll es vollständig hart geworden

[1]) Nach der Pressung sollten die Manschetten besonders mit animalischem Fett behandelt werden, um dem Leder eine längere Lebensdauer zu geben. Zu erwähnen ist hierbei die patentierte „Cog"-Ledermanschette der Firma Gehrckens. Während der Pressung dieser Manschette wird Graphit in das Leder hineingepreßt, der bewirkt, daß die Manschetten sich durch leichteres Arbeiten und geringeren Verschleiß auszeichnen.

sein; Tatsache ist, sagt **Woodworth**, daß das Leder, je besser es ist[1]), um so härter wird. Nun kann man die abgeschrägte Kante mit einem gewöhnlichen Messer anarbeiten und mit feinem Glaspapier glätten. Dann ist das Leder fertig. Die Abschrägung soll ungefähr 30° betragen und — dies ist sehr wichtig — bis hinauf an die Kante des Leders gehen, so daß eine scharfe Kante ringsherum bleibt.

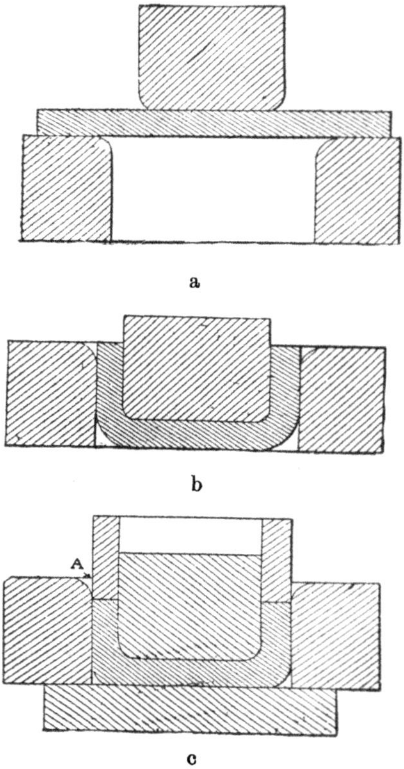

Fig. 399.

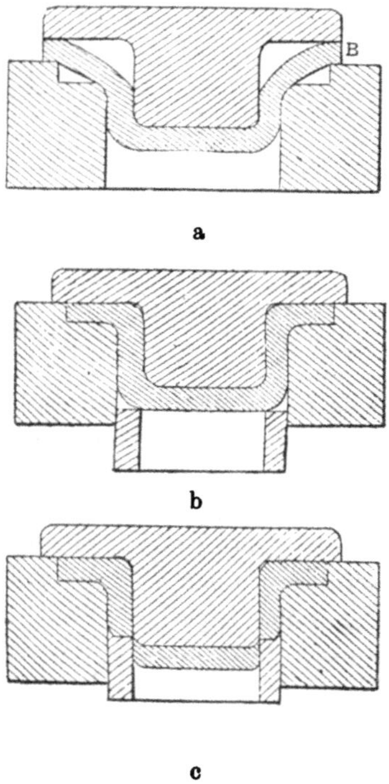

Fig. 400.

Herstellung der Hutform. Was für die Größe der Napfform gesagt worden ist, gilt auch hier. Die ganze Höhe wechselt zwischen 12 und 20 mm. Die Herstellungsstufen sind folgende: Zuerst wird, mit der rauhen Seite nach oben, Fig. 400a, soviel als möglich des Teiles B in

[1]) Auf das Trocknen der Ledermanschetten ist große Sorgfalt zu verwenden. Dieser Behauptung: „Je besser das Leder ist, um so härter wird es", widerspricht die Firma Gehrckens; denn eine ganz feine Eichenlohgrubengerbung, die sich speziell für die Manschetten eignet, wird nicht steinhart, während eine ganz billige, norddeutsche Gerbung, die vielleicht 30% niedriger im Preise steht, sicherlich steinhart wird. Das spätere Imprägnieren mit warmem, animalischem Fett darf nur geschehen, wenn die Ledermanschetten vollständig trocken sind, da sonst das nasse Fett die Fasern verbrennt.

den Ziehring hineingedrückt; das übrige kann von der scharfen Kante abgeschert werden. Dann wird nach Fig. 400b der Ring eingesetzt und der Boden ausgeschnitten. Der Ring wird ein wenig nachgedrückt und in dieser Stellung, Fig. 400c, festgeklemmt. Dann kann das Leder getrocknet und fertiggemacht werden wie das vorangehende; man kann dafür auch denselben Ring verwenden.

Herstellung des Nutringes (*U-Leder*). Die U-Leder sind die verbreitetste Form der Lederdichtung. Sie sind als Dichtung den beiden vorgenannten überlegen und haben auch weniger Reibung, da sie nachgiebiger sind. Sie sind jedoch teurer und verlangen größere Sorgfalt bei ihrer Herstellung, so daß immerhin möglichst die beiden ersten verwendet werden. Das U-Leder kann für jeden Kolben von 25 mm aufwärts verwendet werden, mit folgenden Höhen:

			Kolben				Lederhöhe	
			bis	26	mm	Durchm.	12	mm
über	26	mm	,,	76	,,	,, ...	12—16	,,
,,	75	,,	,,	500	,,	,, ...	20	,,
,,	500	,,				,, [1])	25—38	,,

Die Herstellungsweise ergibt sich aus Fig. 401; a zeigt die erste Stellung, mit der glatten Lederseite nach innen. In dieser Stellung bleibt das Leder von ½ bis zu mehreren Stunden, je nach der Größe. Dann dreht man das Ganze um und setzt den mittleren Stempel ein, mit dem man die Mitte herausdrückt, wie in Fig. 401b ersichtlich ist. In dieser Stellung läßt man das Leder trocknen, worauf man es aus der Form nimmt und eine Linie am Umfang anreißt, die die Höhe des Leders angibt, Fig. 401c. Nun befestigt man das Ganze in einem Schraubstock und sägt den Überschuß mit einer gewöhnlichen Holzsäge längs dieser Linie ab.

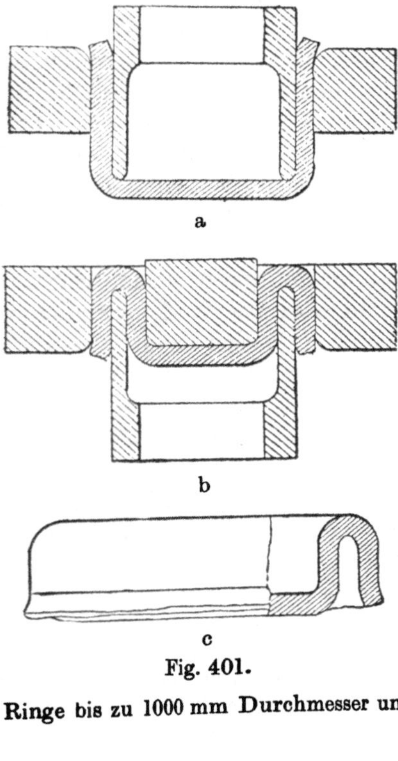

c
Fig. 401.

[1]) Die Firma Gehrckens stellt die Ringe bis zu 1000 mm Durchmesser und 45 mm Höhe her.

Dann wird das Leder eben abgeschliffen, indem man ein Stück Glaspapier auf eine glatte Unterlage legt und den Ring darauf abreibt. Hierauf wird die Schräge angearbeitet und wieder mit Glaspapier geglättet. Sollen sehr viele Ringe gemacht werden, so wird sich folgender Vorgang empfehlen. Wenn das Leder trocken ist, nimmt man es aus dem Ziehring heraus und befestigt es auf der hölzernen Planscheibe einer kleinen Drehbank. Mit einem Holz-Drehmeißel sticht man dann die verlangte Höhe ab, arbeitet die Schräge an und poliert. Die Politur, die manche Leder aufweisen, erhält man in der Drehbank mittels eines Stockes aus Hartholz und Schmierseife; dadurch wird aber nur das Aussehen, keinesfalls die Qualität des Leders als Packungsmaterial gehoben.

Praktische Winke für die Herstellung. Bei allen Arbeiten muß man alle Stempel und Ringe, sowie die Blanketts genau ausmitteln, so daß kein Teil übermäßig gestreckt wird. Man muß jedoch immer eine volle Lederscheibe verwenden und sollte gar nicht erst versuchen, aus einem Ring eine Manschette formen zu wollen. Da man den aus der Mitte herausgeschnittenen Teil leicht für ein kleineres Leder verwenden kann, so braucht man nichts wegwerfen. Bezüglich des auszuübenden Druckes sei bemerkt, daß der oft geübte Gebrauch, ein Loch in die Mitte des Blanketts zu schneiden und den Stempel mit Bolzen und Mutter niederzuziehen, nicht zu empfehlen ist, da sich das Leder unvermeidlich immer von dem Loch aus wegzieht und manchmal auch einreißt; außerdem wird das Loch oft größer als notwendig geschnitten. Hat man eine große Zahl Leder zu machen, so erweist sich eine kleine hydraulische Presse[1]) als sehr nützlich, meistens aber lohnt es sich nicht. Für die kleineren Leder, bis zu ungefähr 50 mm Durchmesser, genügt ein gewöhnlicher Schraubstock. Für größere Abmessungen kann man eine Werkbank mit zwei T-Nuten versehen, die sich unter rechtem Winkel kreuzen und zur Aufnahme von zölligen Bolzen dienen; die Klammern werden selbstverständlich in gewohnter Weise verwendet. Man hat diese einfache Vorrichtung sogar für Leder bis zu 660 mm Durchmesser und 35 mm Höhe verwendet.

Einsetzen der Leder in die Maschine. Man soll immer einen vollen Satz Leder auf Lager haben, um ein abgenutztes Leder sofort erneuern zu können. Es ist dies praktischer als die Lagerung des rohen Leder-

---

[1]) Von der Verwendung einer hydraulischen Presse rät die Firma Gehrckens dringend ab; die Firma hält eine gewöhnliche Handpresse für vorteilhafter, da man mit derselben fühlen kann, welchen Druck das Leder erleidet, während dies bei der hydraulischen Presse nicht möglich ist, und dadurch viel mehr Leder beim Pressen verdorben wird als sonst.

stückes, da die fertigen Leder weniger dem Verderben ausgesetzt sind. Man erspart sich auch unangenehme Überraschungen, wenn eine Dichtung plötzlich nachgibt. Bevor man ein Leder in die Maschine einsetzt, muß man es gut mit bestem Talg einfetten, bis es weich ist; auch Klauenöl[1]) ist sehr gut, doch teuer. Wenn trotzdem das Leder nicht weich werden will, so gibt man es auf einige Minuten in warmes Wasser, doch muß man es nachher sofort in die Maschine einsetzen, sonst verliert es seine Form, geht auf und muß noch einmal in das Ziehwerkzeug zurück. Ebenso muß man beim Reinigen der Maschine das Leder sofort, nachdem es herausgenommen worden ist, in die Form legen, sonst läßt es sich, nachdem es trocken geworden ist, nicht wieder einsetzen. Man muß immer die Höhe der Ausdrehung für das Leder und die Höhe des Leders nachmessen, um sicher zu sein, daß man in dieser Richtung genügend Spiel hat; denn, wenn das Leder in dieser Richtung geklemmt wird, kann es sich unter dem Wasserdruck nicht öffnen. Ebenso ist auf die Abschrägung zu achten; es ist dies eine Kleinigkeit, aber doch sehr wichtig und wird oft übersehen. Das Leder dichtet sonst nicht gut; in Maschinen, wo Leder ohne Abschrägung eingesetzt wurden, hat man damit viel Unannehmlichkeiten gehabt; nach dem Ausbessern der Leder arbeiteten diese Maschinen richtig.

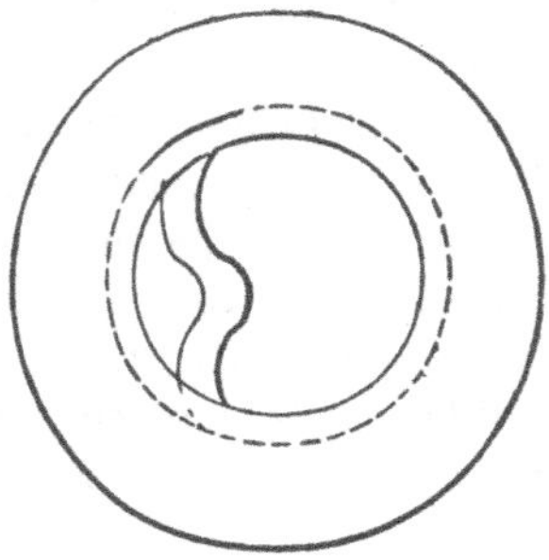

Fig. 402.

Nur beim Einsetzen eines U-Leders in einen Zylinder ohne Stopfbüchse[2]), in dem bloß die Ausdrehung für das Leder vorgesehen ist,

[1]) Mineralöl darf auf keinen Fall verwendet werden, da es das Leder angreift.

[2]) Vorschrift der Firma C. Otto Gehrckens: Um ein Brechen des Nutringes (U-Leder) beim Einsetzen in die Zylindernute zu vermeiden, falls das Leder durch Lagern etwas hart geworden, empfiehlt es sich, den Nutring durch mäßiges Anweichen in lauwarmem Wasser auf der zuletzt hineinzudrückenden Stelle etwas geschmeidig zu machen, nur darf man nicht übertreiben, damit der Nutring nicht die Form verliert. Unmittelbar vor dem Einsetzen, nach dem eben beschriebenen mäßigen Anweichen, fette man den Nutring mit erwärmtem reinen Rindstalg ein.

Der Nutring wird mit beiden Händen langsam und vorsichtig in den Zylinder eingeführt, dann etwas eingebogen und in die Zylindernute hineingedrückt. Hierzu gehört ein Griff, welcher sich mit der Zeit erlernt. (Einige lernen es nie!)

Die Knickstelle ist darauf sorgfältig wieder herauszuarbeiten und zu glätten, damit auch nicht der kleinste Teil in den Zylinderraum hineinragt; andernfalls wird der Preßkolben den vorstehenden Teil der Manschette mitreißen, sie dadurch beschädigen und schnell unbrauchbar machen.

findet man Schwierigkeiten. Wenn das Leder einen großen Durchmesser hat, so genügt dessen eigene Elastizität beim Einsetzen, wenn man allmählich und mit einiger Sorgfalt vorgeht. Hat das Leder jedoch nur einen kleinen Durchmesser, so muß man soviel wie möglich davon in die Ausdrehung hineindrücken und dann mit einem Hammerstiel die übrige Länge gerade hinunterstoßen. Das Leder erhält dann eine Form nach Fig. 402 und kann leicht vollends in die Ausdrehung hineingedrückt werden; es darf dabei aber nicht zu weich sein, sonst verliert es bei dieser Behandlung seine Form.

Im Gebrauch soll man den Plunger blank und gut geschmiert erhalten. Wenn möglich, soll der Teil des Zylinders oder der Teil des Plungers, auf dem das Leder arbeitet, mit Bronze oder, in mehr moderner Ausführung, mit Kupfer überzogen sein; das Leder hält dann viel länger und braucht weniger Schmierung. Wenn das Leder auf Gußeisen mit geringer Schmierung arbeitet, so scheint es tatsächlich kleine Eisenteile mit fortzureißen. Dies ist selbst bei langsamer Bewegung unter hohem Druck, z. B. in einer Schmiedepresse, ganz deutlich zu sehen. Wenn man nach längerem Gebrauch ein solches Dichtungsleder herausnimmt und trocknen läßt, so sind selbst mit bloßem Auge kleine Körnchen Eisen, die in der arbeitenden Lederfläche eingebettet sind, sichtbar. Schließlich soll man möglichst reines Wasser, das frei von Sand usw. ist, benutzen.

4. Drahtziehen.

Unter Drahtziehen versteht man eine Durchmesserverkleinerung eines vollen, stangenartigen Metallkörpers — Draht —, der durch ein Ziehloch, das einen kleineren Durchmesser als der zu ziehende Körper hat, mittels einer an dem Draht selbst angreifenden Kraft im kalten Zustand gezogen wird. Der an den Flächen des Ziehloches auftretende Widerstand bringt den dort befindlichen Teil des zu ziehenden Drahtes zum kalten Fließen, so daß das Material, ohne seinen Zusammenhang aufzugeben, auf den kleineren Querschnitt verringert

---

Stulpen (Manschetten) von kleinem Durchmesser, sehr dickwandige oder solche, die schon längere Zeit als Vorrat gelagert haben und dadurch etwas hart geworden oder eingekrochen sind, behandle man, wie oben angegeben; auch empfiehlt es sich, derartige, namentlich Chromleder-Manschetten, vor dem Einsetzen einige Minuten über heißem Wasserdampf zu halten, wodurch dieselben geschmeidig werden und etwas aufquellen. Genaue Regeln lassen sich hierüber aber nicht feststellen, weil hierbei die jeweilig vorliegenden Umstände, wie Größe und Wandstärke der Manschetten, Alter, Ledersorte usw., mitsprechen. Es muß der Erfahrung des Maschinisten überlassen bleiben, den richtigen Zeitpunkt zu erkennen, in welchem die Manschette die erforderliche Geschmeidigkeit besitzt, auch kommt es sehr auf den geschickten Handgriff beim Einsetzen an.

wird, während seine Länge dementsprechend zunimmt. Die beim Ziehen zu überwindende Kraft muß kleiner als die Zugfestigkeit des Drahtquerschnitts sein, und wird von der Festigkeit und folglich auch der Härte des gezogenen Materiales, der Durchmesserverkleinerung beim Ziehen, dem Winkel des Ziehloches und der Oberflächenbeschaffenheit, sowohl des Ziehloches wie des Drahtes, der Ziehgeschwindigkeit und der Schmierung abhängen.

Über die Größe der Abstufung sind Versuche mit Kupferdrähten gemacht worden, weil dieses Material am gleichmäßigsten herzustellen und am dehnbarsten ist, von den Edelmetallen, die hier weniger in Frage kommen, abgesehen.

Die ziffermäßigen Werte für die praktisch brauchbaren Querschnittsabnahmen lassen sich nach den vorliegenden Versuchen nicht berechnen, so daß man auf die Angaben der Praxis angewiesen ist.

Für Stahldraht finden sich[1]) neben anderen folgende Werte für die Abstufungen:

	2,5 mm Durchm.	2 mm Durchm.	1,5 mm Durchm.	1 mm Durchm.
Walzdraht:	6	5,5	5,5	4,8
	5 bei 475° Ausglühen	4,9 bei 465°	4,8 bei 480°	4,2
		4,4 Ausglühen	4,2 Ausglühen	3,6
			3,7	2,7
Nach dem Glühen:	4,4	4,4	3,7	2,7
	3,9	3,9	3,2	2,3
	3,4	3,4	2,7	1,9
	2,9	2,9	2,25	1,55
	2,5	2,4	1,85	1,25
		2	1,5	1

Für Kupfer wird angegeben, daß die Durchmesser von 7 bis 5 mm um einen Millimeter, von 5 bis 2 mm um 0,5 mm und von da auf 1,5 mm um 0,2 mm fallen sollen. Für feinere Drähte von 1 mm bis 0,5 mm abwärts, die durch Diamanten gezogen werden, gibt eine englische Firma Abstufungen von ungefähr 0,1 mm an.

Woodworth gibt allgemein, ohne das Material anzugeben, an, daß jeder Drahtzieher seine eigenen Gedanken über die Querschnittsabnahme hat, und sagt:

Bei großen Durchmessern, z. B. 5 mm Durchmesser, können mehrere Nummern in einem Durchzug gemacht werden. Von da bis ungefähr 0,08 mm Durchmesser wird gewöhnlich für jede Nummer ein Durchzug gemacht. Von hier ab wird um $^1/_{1000}$ Zoll engl. jedesmal abgestuft, und um $^1/_{10\,000}$ Zoll engl., wenn ein Durchmesser von $^1/_{1000}$ Zoll

[1]) „Aus der Praxis für die Praxis". Claryscher Verlag.

engl. erreicht worden ist. Soweit die Ziffern des amerikanischen Autors. Diese angegebenen Ziffern hängen natürlich vom Material ab, und sind manchmal viel kleinere Unterschiede notwendig.

Mit allen derartigen Erfahrungsangaben wird man aber höchstens ein Erfahrungsgesetz für unregelmäßige Querschnitte der gezogenen Drähte ableiten können, während die meisten gezogenen Drähte Kreisquerschnitt haben, für welchen die Aufstellung eines Gesetzes von bedeutend weniger Größen abhängig ist. Versuche, die an Kupferdraht vorgenommen worden sind[1]), haben gezeigt, daß die einfache Annahme einer Formel für die Zugkraft K

$$K = c\,(\operatorname{tg}\alpha + \mu)\,(Q - q),$$

worin

c eine Konstante, die von der Ziehgeschwindigkeit, dem Material und der Beschaffenheit des Ziehloches abhängig ist, bedeutet,

α der Winkel des Ziehloches,

μ der Reibungskoeffizient zwischen Wandung und Material,

Q, q die Querschnitte des Drahtes vor und nach dem Ziehen

sind, keineswegs ein Gleichbleiben des Reibungskoeffizienten ergibt, wenn der Winkel α des Ziehloches abgeändert wird.

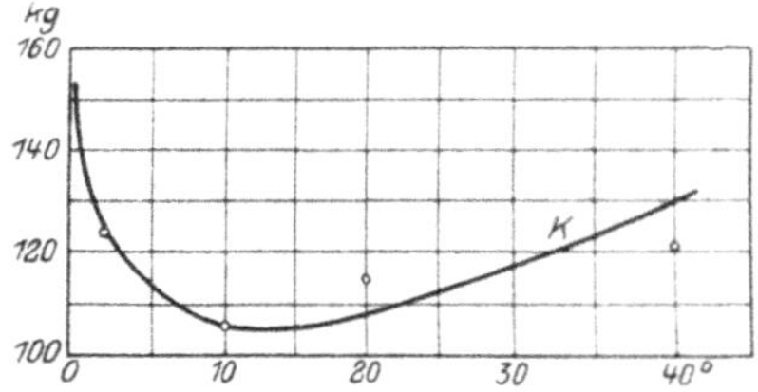

Fig. 403.

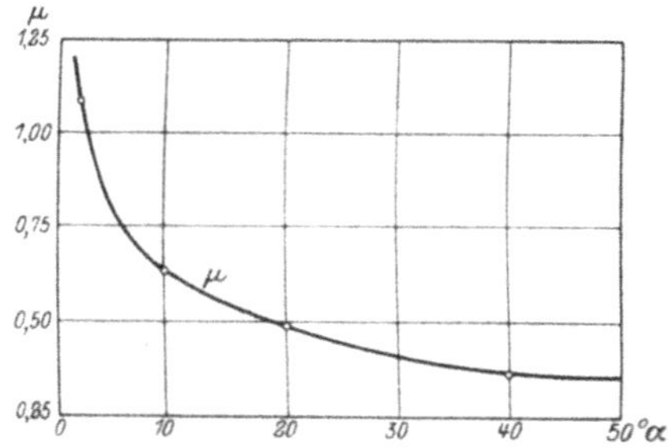

Fig. 404.

Bei einer Querschnittsabnahme von 32,3 % und Winkel im Ziehloch von 2—40° ergaben sich Zugkräfte nach Fig. 403, aus denen durch eine etwas verwickelte Rechnung eine Beziehung zwischen μ und α, Fig. 404, abgeleitet wurde, die ein deutliches Abfallen des Reibungskoeffizienten mit steigendem Winkel α zeigt. Wenn auch diese Versuche nicht umfassend genug waren, um ein endgültiges Gesetz auf ihnen zu gründen, so ist doch die Tatsache an und für sich wichtig genug, um angeführt zu werden.

Daß aber nicht allein die geometrische Form des Ziehloches für dieses Verhalten maßgebend ist, zeigen die Mikrophotographien zweier Drähte, die nicht vollständig durch das Ziehloch gegangen sind, Fig. 405.

---

[1]) Gewecke, Versuche über Drahtziehen. Dr.-Diss. Darmstadt.

An der Stelle, wo der Draht im Ziehloch steht, ist eine ganz scharf begrenzte Zone zu sehen, die von dem Gefüge des Materials vor und hinter dem Ziehloch vollständig abweicht. Während vor und hinter dem Ziehloch die Längsstreifen, welche von den in die Länge gezogenen Kupferkristallen herrühren, deutlich erkennbar sind, enthält diese Stelle ein feinkörniges, gleichmäßiges Gefüge ohne ausgesprochene Richtung der Materiallagerung. Wenn man damit das

Fig. 405.

Bild, Fig. 406, an der Kontraktionsstelle eines nichtgerissenen Stabes[1]) vergleicht, so erscheint die Längung der einzelnen Ferritkörner ganz deutlich und gibt dasselbe Bild wie die dem Ziehloch benachbarten Teile des Drahtes in bezug auf den Materialaufbau, abgesehen von den durch die Kontraktion hervorgerufenen größeren Querschnittsveränderungen. Beim Ziehen haben wir es aber offenbar mit einer vollständig durchgehenden, vorübergehenden Gefügeänderung des Materiales zu tun.

Fig. 406.

[1]) Kurrein, Gefügeänderungen in Flußeisen. Baumaterialienkunde 1904. Tafel I, Bild 12, 5.

Die Zieheisen sind natürlich die wichtigste Bedingung für eine erfolgreiche Arbeit. Sie müssen aus einem Material hergestellt werden, das hart genug ist, um sich nicht zu schnell oder ungleichmäßig abzunutzen; ihre Herstellung muß, mit Rücksicht auf eine möglichst lange Verwendungsdauer, so billig wie möglich sein.

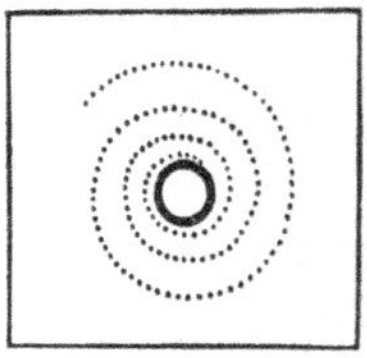
Fig. 407. Zuhämmern eines Zieheisens.

Für Eisen, Stahl, Bronze und dergl. härtere Materialien verwendet man am besten Zieheisen aus besonderem Gußstahl, d. i. naturhartem Chromstahl. Sie sind entweder rechteckig mit mehreren Ziehlöchern in einer oder mehreren Reihen, oder rund mit einem Loche in der Mitte. Die Platten werden, wenn die Löcher sich durch den Gebrauch ausgeweitet haben, auf Kirschrot erwärmt, und die Vorderfläche, an der der Draht das Zieheisen verläßt, mit Hammer und Punze gestaucht, Fig. 407, worauf das Loch auf Maß nachgerieben wird.

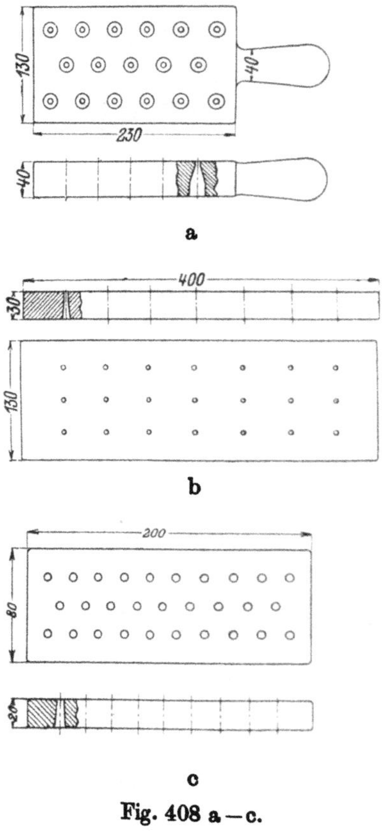

Fig. 408 a—c.

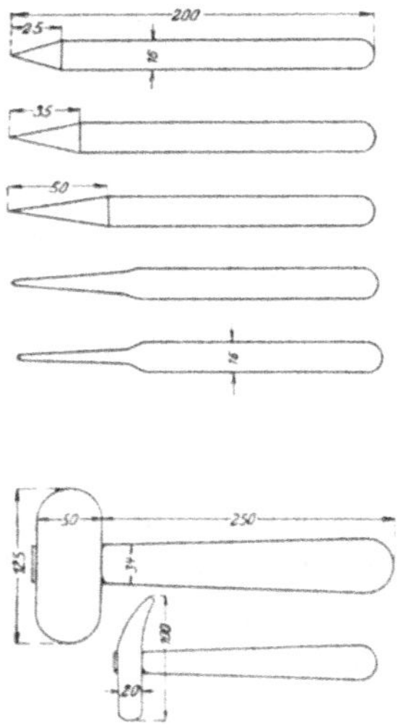

Fig. 409.

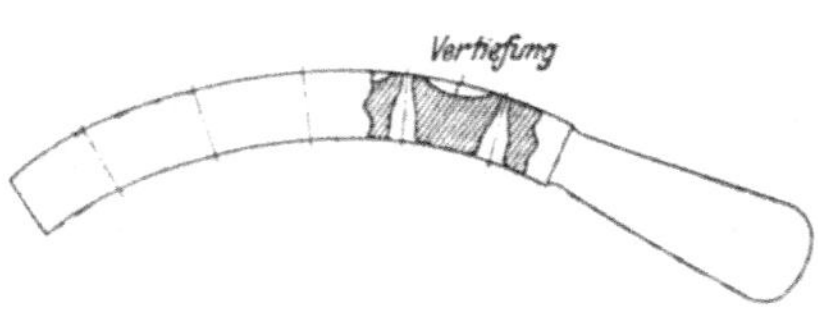

Fig. 410.

Kohlenstoffstahl wird zum Ziehen von Draht unter 14 mm Durchmesser überhaupt nicht verwendet, da er sich beim Härten und Abschrecken wirft und die Ziehlöcher unrund werden und nicht gerade bleiben, was ein Nachschleifen bedingen würde, das aber viel zu teuer wird. Handelt es sich bloß um eine schnelle Verkleinerung des Durchmessers, so braucht das Loch nicht genau zu sein; soll der fertige Draht hingegen genaues Maß haben, so muß das Loch sorgfältig poliert sein, da jede Einkerbung eine Kratzlinie am Draht gibt.

Nach der Form sind drei Arten Zieheisen im Gebrauch, das englische, deutsche und Wiener Zieheisen, Fig. 408 a—c [1]). Am besten haben sich die Zieheisen bewährt, die aus selbsthärtendem Stahl bestehen und nach Abnutzung durch einfaches Aufdornen mit Treibdornen nach Fig. 409 auf die nächste Nummer aufgeweitet werden können. Die Eisen aus Kohlenstoffstahl müssen ausgeglüht, gebohrt, auf die nächste Nummer geschliffen und wieder gehärtet werden. Das Wiederaufstellen des Loches durch kaltes oder warmes Zuhämmern wobei das Material in naturhartem Eisen durch Hämmern nach Fig. 407 von außen beginnend nach dem Loch getrieben wird, bewirkt, daß die englischen Eisen mit der Zeit eine gekrümmte Form, Fig. 410, annehmen. Die dabei verwendeten Hämmer, oben für englische, unten für deutsche Eisen, sind im unteren Teil der Figur 409 dargestellt. Die Hämmer für die Wiener Zieheisen haben dieselbe Form wie die deutschen, sie sind nur kleiner.

Die verschiedenen Schwierigkeiten, die das Nacharbeiten der ausgeweiteten Zieheisen mit sich bringt, haben mehrmals zu Versuchen, nachstellbare Zieheisen zu bauen, geführt. Unterstützt wurden diese

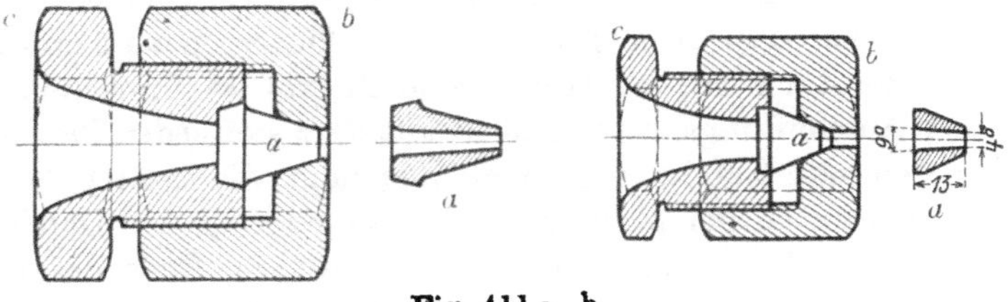

Fig. 411 a, b.

Versuche durch die Kosten des beinahe täglich notwendigen Neueinrichtens der Zieheisen und durch den Umstand, daß man dabei gerade die durch das Ziehen vergütete Metalloberfläche im Ziehloch beim Wiederherrichten entfernen muß.

Ein solches nachstellbares Zieheisen, Patent Glover, das nach vorliegenden Versuchen offenbar gute Erfolge gibt, zeigt Fig. 411 a b.

[1]) Z. Ver. deutsch. Ing. 1906, S. 1895.

Dieses Zieheisen wird für Stahl (a) und in etwas veränderter Ausführung für Kupfer (b) verwendet. Das Zieheisen besteht aus drei Teilen, dem eigentlichen Zieheisen a aus einem naturharten Spezialstahl, einem Halter b dafür aus Stahlguß, der mit Innengewinde versehen ist, in welchen der dritte Teil, die Befestigungs- und Nachstellschraube c, paßt. Wenn durch den Gebrauch die Mündung des eigentlichen Zieheisens ausgeweitet ist, wird die Nachstellschraube angezogen und dabei das in dem kegeligen Loche des Halters c aufliegende Zieheisen a nach vorn gedrückt, wobei der dünne Teil des Kegels zusammengedrückt wird. Da hierbei ein genau axialer Druck durch den kegeligen Innenumfang der Nachstellschraube und des eigentlichen Zieheisens ganz gleichmäßig radial übertragen wird, wird die innere Öffnung, das Ziehloch, sich vollständig gleichmäßig rund schließen und so das Ziehloch in der einfachsten und schnellsten Weise wieder eingerichtet, „wiederaufgestellt", werden. Der Winkel im Ziehloch beträgt 9°. Infolge der eigentümlichen Materialbewegung im Zieheisen, durch die das Material desselben durch das Loch im Halter, wie in einer Strangpresse, bei gleichzeitigem radialen Druck herausgepreßt wird, ist ersichtlich, daß man gehärteten Stahl für diese Zieheisen nicht verwenden kann. Außerdem wird aber gerade die vergütete Oberfläche des Ziehloches durch das Nachstellen nicht zerstört, sondern bleibt in ihrem Zustand im ganzen und großen erhalten. Ein weiterer wichtiger Vorteil ist die Möglichkeit, die Nachstellung durch ungelernte Arbeiter und nicht durch besondere Werkzeugschlosser machen zu lassen.

Ziehwerkzeuge aus Diamanten, Saphiren und Achat. Für Durchmesser unter 1,5 mm und genaue Durchmesser des gezogenen Drahtes ist zweifellos der Diamant das beste Material für Ziehlöcher. Woodworth gibt an, daß er manchmal für Ziehlöcher bis zu 2 mm Durchmesser verwendet wird, was bereits ziemlich groß ist. Immerhin werden die hohen Anschaffungskosten und die mit dem steigenden Durchmesser rascher steigende Gefahr, daß der Stein springt, die meisten davon abhalten, Diamanten bei Ziehlöchern, die mehr als 1,5 mm Durchmesser haben, anzuwenden. Sicher ist der Diamant am Ende das billigste Material für Ziehwerkzeuge, besonders wenn es sich um das Ziehen großer Mengen handelt, da man den Stein mit verhältnismäßig geringen Kosten für die nächste Größe und noch weiter ausreiben kann.

Die Herstellung der Diamantwerkzeuge erfolgt entweder durch Umgießen des Steines mit Bronze oder zuerst mit Stahl und dann durch nochmaliges Umgießen des in Stahl eingehüllten Steines mit Bronze. Man hat auch versucht, den Stein auf der Drückbank unmittelbar in Metall zu fassen, doch ist die Gefahr eines Sprengens des Steines dabei sehr groß, da das Gefühl für den richtigen Druck beim Einpressen leicht täuscht. Eine gebräuchliche Befestigung zeigt Fig. 412[1]). Der Stein wird in die Messingscheiben t eingegossen, die wiederum mit den Stahlbolzen s vergossen werden. Es ist für die Lebensdauer des Steines von großer Wichtigkeit, daß die Messingscheibe fest im Stahlbolzen sitzt, da sonst der Draht beim Ziehen den Stein leicht zersprengt. Um eine gleichmäßige Abnutzung des Steines herbeizuführen, wird der Stein während des Betriebes zwangläufig gedreht, indem die Bolzen s durch die Zahnräder Z von der Antriebswelle der Maschine aus dauernd gedreht werden.

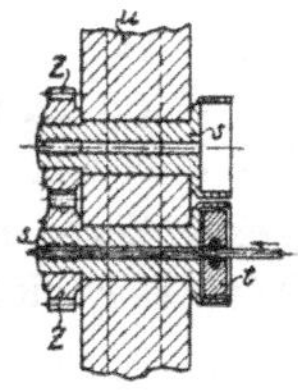
Fig. 412.

Herstellung von Ziehplatten durch Zusammengießen von Diamanten mit Gußeisen und Stahl. Woodworth gibt ein Patent von Fr. Krause in Jersey City N. J. über das Eingießen des Diamanten in Stahl an, das sich offenbar mit dem D. R. G. M. 182 015 deckt. Fig. 413 zeigt die Gußform mit dem Diamanten darin; Fig. 414 zeigt den in Stahl oder Eisen eingegossenen Diamanten. Wie nun weiter Material um die soweit fertiggestellte Ziehplatte gegossen wird, zeigt Fig. 415, und den daraus erhaltenen Guß Fig. 416. Eine Abänderung dieses Vorganges ist in Fig. 417 angegeben. In den Zeichnungen hat die Gußform ein Loch im Boden. Der bereits gebohrte Diamant wird in die Gußform gesetzt und vorläufig durch einen Draht D, der durch das Loch in der Form und dem Diamanten geht, in seiner Stellung gehalten. Das geschmolzene Metall, Stahl oder Gußeisen, wird dann in die Form gegossen und umgibt den Diamanten, der auf einer kleinen Erhöhung über dem Boden der Form sitzt. Nach dem Erkalten wird das Metall aus der Form entfernt, Fig. 414. Auf der Unterseite des Steines, der auf der Auflage in der Gußform ruht, läßt der Metallmantel G das Loch im Diamanten B frei.

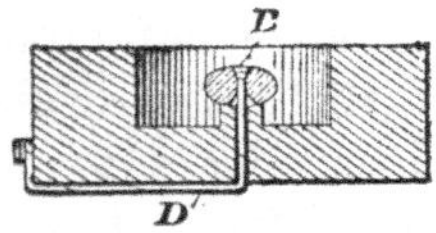

Fig. 413.

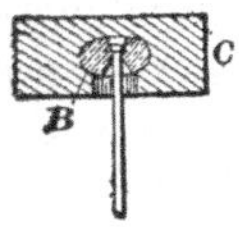

Fig. 414.

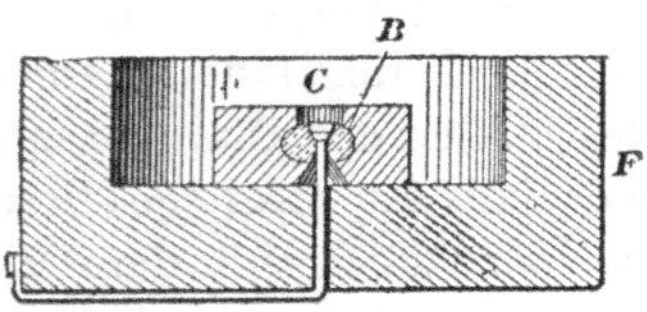

Fig. 415.

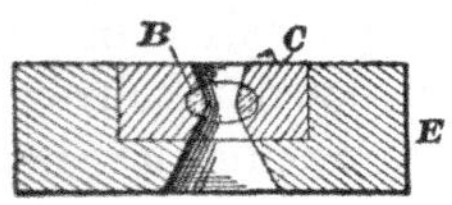

Fig. 416.

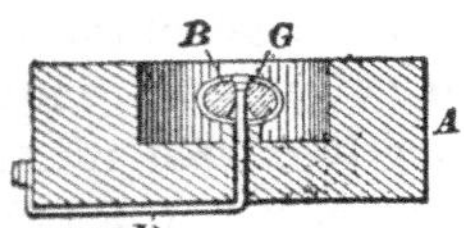

Fig. 417.

[1]) Z. Ver. deutsch. Ing. 1906, S. 2024.

Auf der gegenüberliegenden Seite wird der Mantel G ausgebohrt, um die Bohrung des Diamanten freizulegen. Der Draht D, der während des Gießens den Diamanten in seiner Stellung hält, wird beim Bohren entfernt.

Nun kann die Platte zum Drahtziehen verwendet werden; man kann jedoch das Werkzeug der leichteren Handhabung wegen dadurch vergrößern, daß man in gleicher Weise um den Mantel C einen zweiten Mantel E gießt. Dazu bringt man das Gußstück C, wie in Fig. 415 zu ersehen ist, in eine Form F und hält es dort wieder durch einen Draht fest. Die Bohrung im Diamanten muß, nachdem der Draht eingesetzt ist, mit Graphit, Lehm oder ähnlichem Material verstopft werden, um zu verhindern, daß beim Guß das Material in die Bohrung fließt. Diese zweite Hülle E wird ebenfalls ausgebohrt, um die Öffnung freizulegen wie zuvor, Fig. 416.

Die Gefahr, den Diamanten während dieser Arbeiten zu beschädigen, ist jedem klar, der schon mit diesen so leicht splitternden Teilchen Kohlenstoff für industrielle Zwecke zu tun hatte. Auch der Erfinder übersieht diese Gefahr nicht, ja, sagt sogar, daß die Hitze während des Gießens den Diamanten verletzen könnte. Er schlägt deshalb vor, den Diamanten in Blech einzupacken, wie in Fig. 417 dargestellt ist. Hier hat der Haltedraht einen Kopf, so daß er nicht durch die Öffnung des Steines schlüpfen kann, wenn Stein und Blech aufgefädelt werden. Während des Gusses soll die Umhüllung des Bleches den Diamanten vor dem geschmolzenen Material schützen; jedenfalls ist sofortige Berührung verhindert, und das Blech legiert sich allmählich mit dem Mantel, so daß nach Vollendung der Ziehplatte der Diamant wieder wie in Fig. 414 erscheint.

Der zu den Diamantziehplatten verwendete Stein ist der Bort, der hauptsächlich in Brasilien und Südamerika gefunden wird. Er ist so hart, daß das Bohren eines Loches eine der langwierigsten und schwierigsten Arbeiten ist, besonders, weil es sich fast immer um kleine oder sehr feine Löcher handelt, und man wieder nur mit Diamanten oder Diamantpulver bohren kann. Die Löcher werden mit fein zugespitzten Stahlnadeln gebohrt, auf welche feiner in Öl angeriebener Diamantstaub aufgebracht wird. Die Bohrspindel läuft, nach Woodworth, mit ungefähr 5000 minutl. Umdrehungen und braucht zum Durchbohren eines Stückes von 3,175 mm Dicke bei täglich 10 stündiger Arbeitszeit zwei bis drei Wochen.

Eine etwas vollkommenere Maschine zeigt Fig. 418[1]) in Zeichnung. Der Spindelstock a trägt eine kleine Planscheibe mit dem Diamanten. Der Support c für die Bohrspindel d ist auf einem Halter verschiebbar gelagert, der außerdem noch eine Schlagvorrichtung, be-

[1]) Z. d. Ver. deutsch. Ing. 1906, S. 2027/8.

stehend aus Rolle g und Nockenscheibe h, trägt. Durch die Feder f wird die Bohrspindel stets gegen den Diamanten angedrückt, wobei die Spindel der leichteren Beweglichkeit halber auf Rollen e_1, e_2 gelagert ist. Je nach der Stellung des Hebels i, der um seinen Drehpunkt l schwingt, wird die Bohrspindel mehr oder weniger von dem Stein zurückgezogen, gegen den sie dann durch die Federkraft beim Weiterdrehen der Nockenscheibe h wieder angeschlagen wird. Die Spindel

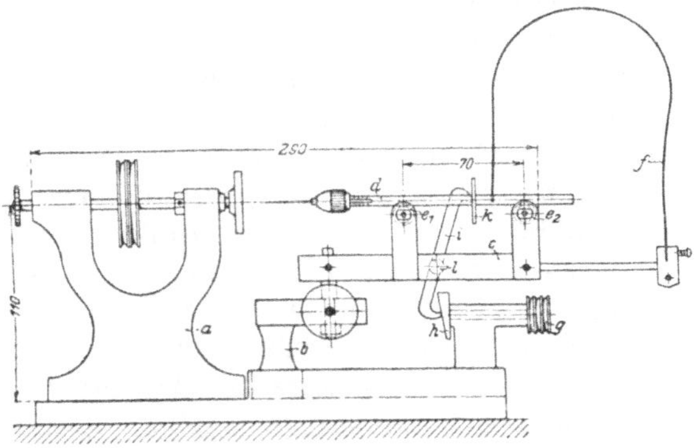

Fig. 418.

macht 1500—2500 minutl. Umdrehungen, während die Spindel 200 bis 300 Schläge ausführt. Die Nadel ist aus Stubsdraht hergestellt und wird in den Stahlhalter mit Schlacke eingekittet oder in einem Futter festgehalten. Der Schlaghub ist so einzustellen, daß die Nadel bei jedem Hub einen Augenblick im Bohrloch stillsteht, so daß eine genau runde Bohrung gewährleistet wird. Die gebohrten Diamanten müssen justiert und das Kaliber eingeschliffen werden.

Der fertige Stein wird in einer Messingplatte von ungefähr 32 mm Durchmesser und 6,5 mm Dicke gefaßt und die Größe und das Gewicht des Steines darauf eingeschlagen.

Bei einer vergleichenden Prüfung bezüglich der Härte des Diamants und des Achates wurde ein ungefähr 12,7 mm dickes Stück des letzten in der von Woodworth angegebenen Weise angebohrt. Der Bohrer hatte in 17 Minuten durchgearbeitet.

Auch Saphire werden zu Gesenken verwendet und sind für manche Zwecke hart genug, während ihr verhältnismäßig niedriger Preis sie in manchen Fällen in den Vordergrund stellt; wo es jedoch auf Dauerhaftigkeit und Genauigkeit als erste Bedingung ankommt, bleibt der Diamant das beste bis jetzt bekannte Material.

Zieheisen aus Hartguß. Gesenke aus Hartguß werden gewöhnlich für Größen unter 20 mm bis 2,5 mm Durchmesser verwendet; sie werden aus feinkörnigem Eisen in einer gußeisernen Form gegossen, wodurch das Material abgeschreckt wird. Die Löcher werden konisch gegossen, wobei die große Öffnung an der Eintrittsstelle des Drahtes ist. Diese Löcher werden jedoch nicht durchgegossen, sondern man läßt an dem kleinen Ende eine dünne Schicht Metall stehen, die nachher ausgestoßen und auf Maß ausgerieben wird. Das Ausreiben geschieht mittels einer Vierkant-Maschinenreibahle, welche sehr langsam umläuft, so daß ein Loch von durchweg gleichem Durchmesser erhalten wird, das je nach der Größe von 1,6 bis 4,8 mm lang ist. Diese Hartguß-Zieheisen erhalten gewöhnlich vier bis acht Löcher, die ungefähr 32 mm Abstand haben; die Dicke der Platte wechselt von 25 bis zu 38 mm, ihre Länge hängt von der Zahl der eingegossenen Löcher ab. Diese Ziehplatten sind in ihren Anschaffungskosten sehr billig und können, ohne an ihrer Nützlichkeit einzubüßen, auf die nächste Größe ausgerieben werden. Hunderte von ihnen werden in jeder Drahtzieherei verwendet, doch ist ihre Herstellung eine eigene Kunst, die mit viel Geduld und Geschicklichkeit gelernt werden muß. Hartgußzieheisen halten ungefähr zehnmal solange als solche aus Werkzeugstahl; der Grund dafür ist, daß ungleiche Materialien sich aufeinander weniger abnutzen als gleichartige. Die Anzahl Meter, die man durch ein Zieheisen von 32 mm Durchmesser ziehen kann, ist sehr verschieden, man kann jedoch 1200 m ohne Neueinrichten des Loches als guten Mittelwert annehmen.

Ziehen von runden und rechteckigen Stahlstangen. Rundes und anderes Stangenmaterial wird in folgender Weise hergestellt: Das Material ist Stahl mit 0,1—0,25 $^0/_0$ C, der gebeizt und gewaschen wird, wobei aller Zunder von den Stangen entfernt werden muß. Die runden Stangen werden durch Gesenke gezogen, die 1,6 mm kleiner im Durchmesser sind als das gewalzte Material, andere Querschnitte im gleichen Verhältnis. Sie gehen dann durch eine Richtmaschine, wo sie geradegerichtet werden; dann werden sie von Hand nachgerichtet, auf Länge abgeschnitten und eingefettet für Lagerung oder Verschiffung.

Das Beizen. Das Beizen besteht darin, daß 10 bis 15 t Rohmaterial in verdünnter Schwefelsäure vier bis fünf Stunden gelassen, dann in Wasser, hierauf in Kalkwasser gewaschen werden, wobei das Waschen ohne Unterbrechung geschieht, indem eine Stange nach der anderen durch die Behälter geht. Es ist Sorge zu tragen, daß nichts an der Stange haften bleibt, was beim Durchgang durch das Zieheisen Marken verursachen kann. Um die Notwendigkeit einer gründlichen Reinlichkeit in dieser Hinsicht zu beweisen, sei gesagt, daß eine Kreidemarke, ein Haar oder ein Faden, an der Stange angebracht, sichtbare Spuren zurücklassen, nachdem die Stange durch das Zieheisen gegangen ist, deren Tiefe wohl schwer zu schätzen sein würde, die aber sichtbar genug sind, um als Schönheitsfehler in Betracht zu kommen.

Ziehen von Wellen. Gezogene Wellen, die im Handel als 3zöllig verkauft werden, haben, so genau wie man sie machen kann, 2,995 Zoll Durchmesser. Der zylindrische Teil des Loches in der Ziehplatte ist 2,994 Zoll, also $\frac{1}{1000}$" (0,0254 mm) kleiner im Durchmesser. Eine meßbare Vergrößerung der äußeren Abmessungen der Ziehplatte ist nicht nachweisbar, und an die Innenseite kann man nicht heran. Die vergrößerte Dichte der Stange, die auf das Ziehen zurückzuführen ist, kann nur durch sehr genaue Messungen, keinesfalls aber im Werkstättenbetrieb bestimmt werden, da die Änderungen, die z. B. bei Kupfer[1]) 0,45—1,44 ‰ betragen, infolge der anfänglichen Rauhheit und sonstiger Unregelmäßigkeiten vollständig verschleiert werden. Außerdem strecken sich nicht alle Stangen gleich. Die deutlich ausgeprägten Kennzeichen des Ziehens sind erhöhte Zugfestigkeit und Härte neben einer großen Verminderung der Dehnbarkeit. Gelegentlich bricht eine Stange, während sie durchs Ziehloch geht, wobei die Bruchstelle mit Ausnahme von Farbe und Textur an hartes Gußeisen erinnert, vgl. Fig. 405, obwohl das Material vor dem Ziehen eine Dehnung von 25 %, ohne zu reißen, ertragen hätte. Je größer die Stange ist, desto mehr herrscht unter sonst gleichen Umständen die Neigung vor beim Ziehen, wenn die Durchmesserverkleinerung die gleiche ist, nämlich 1,58 mm für alle Größen von 20 mm bis 125 mm, zu brechen, da bei der größten Stange auch die größte prozentuale Querschnittsverringerung vorhanden ist. Der zum Ziehen einer Stange von 27 mm Durchmesser durch ein Loch von 25,4 mm nötige Zug beträgt ungefähr 6800 kg, während ungefähr 13 600 kg nötig sind, um eine Stange von 52,4 mm Durchmesser durch ein Loch von 50,8 mm zu ziehen, woraus ersichtlich ist, daß der Zug mit der Querschnittsabnahme wächst. Stangen werden öfter mit einer viel größeren Durchmesserverkleinerung als 1,6 mm im Durchmesser gezogen, und die Veränderungen infolge des Ziehens wechseln in gewissem Maße mit der Größe der Verkleinerung. Stangen aus hochgekohltem Stahl gehen plötzlich durch das Zieheisen und lassen wohl ausgeprägte Marken dort, wo sie im Gesenk für einen Moment angehalten haben. Stangen von 20 mm Durchmesser und weniger werden vier zugleich durch das Eisen gezogen, mit einer Verkleinerung von 0,8 mm. Über 20 mm und unter 48 mm werden zwei Stangen gleichzeitig gezogen. Beim Ziehen werden alle Stangen mehr oder weniger gebogen und nehmen, trotzdem der Zug sie geradezu

[1]) Gewecke, Strukturänderungen. Dr.-Diss. Darmstadt 1909, S. 88.

halten versucht, wenn sie aus dem Zieheisen fallen, alle möglichen Krümmungen an. Wenn man mehrere Stangen gleichzeitig zieht, kann man fühlen, wie einige voreilen, andere zurückbleiben, obwohl die Halter nicht viel Abweichung in dieser Richtung gestatten. Eine Besonderheit kann man beim Ziehen von Stangen bemerken, nämlich die Leichtigkeit, mit der sie gedreht werden können. Man kann eine Stange von 25 mm Durchmesser mit der Hand merklich, wenigstens soweit es die Halter zulassen, im Zieheisen verdrehen. Andererseits kann auch beim Drehen einer Stange im Zieheisen ein Zug von ½ kg mit der Zeit genügen, um die Stange zu reißen oder durchzuscheuern. Stangen werden durch das Zieheisen mit ungefähr 5 m Ziehgeschwindigkeit in der Minute gezogen; man braucht bei dieser Geschwindigkeit ungefähr eine Pferdekraft, um eine Stange von 32 mm Durchmesser zu ziehen. Während des Ziehens werden Stangen und Zieheisen mit dickem Öl geschmiert.

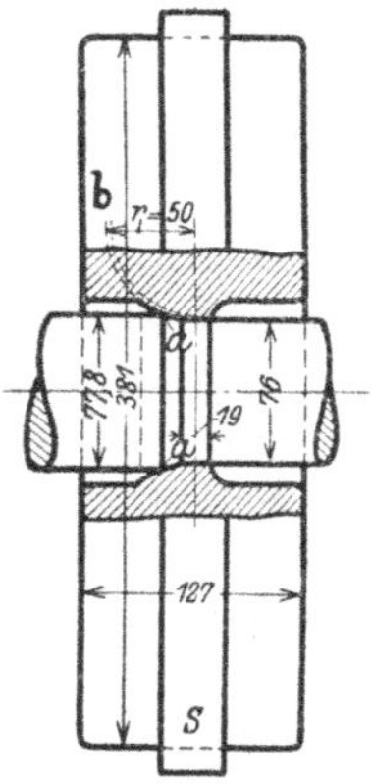

Fig. 419.

Zieheisen aus Eisenbahnrädern. Für die beschriebenen Verhältnisse beim Ziehen von Wellenmaterial verwenden manche Firmen die folgenden Ziehwerkzeuge aus Waggonrädern und schrecken das Material rings um das Ziehloch ab. Fig. 419 zeigt ein derartiges Zieheisen von 76 mm Durchmesser. Auf 19 mm ist das Loch parallel, dann öffnet es sich mit einem Radius von rund 50 mm glockenförmig. Wenn der Absatz bei a stumpf wird, wird er nach der gestrichelten Linie bei b scharf geschliffen oder ausgerieben. Wenn nunmehr ungefähr 3,2 mm im zylindrischen Teile des Loches übrig sind, wird das Zieheisen auf die nächste Größe ausgearbeitet und so weiter, bis der ganze abgeschreckte Teil verarbeitet ist. Ein Schrumpfring S aus Eisen oder Stahl wird auf das Rad aufgezogen, um ihm eine größere Festigkeit zu geben.

Einstellbares Ziehwerkzeug für alle Querschnitte. Fig. 420 zeigt ein Universalwerkzeug zum Ziehen von flachen oder rechteckigen Stäben. Die Rahmenstücke A, B und C werden in der angedeuteten Weise miteinander verschraubt, wobei die Außenteile A und C in der Mitte Löcher zum Durchgang des Stabes haben. Das eigentliche Ziehloch wird aus den Vollzylindern a, b, c, d gebildet, die alle den gleichen Durchmesser, nämlich 97 mm, haben. Die Grundflächen von b, c, d sind rechtwinklig zu den Zylinderachsen mit einem Halbmesser von 48,5 mm ausgebohrt, so daß die Grundfläche des einen Zylinders auf den Umfang des anderen paßt, und sie in der Zusammenstellung nach Fig. 420 ein Ziehloch von 50,8 × 12,7 mm zwischen sich freilassen. Die Länge des Zylinders c bestimmt die Dicke des gezogenen Materials; die Länge der anderen kommt nicht in Betracht. Man sieht daraus, daß der Zylinder a leicht zwanzig oder mehr Arbeitsstellen hat, während Zylinder b und d durch Umdrehen und Wenden jeder sechzehn und c acht Arbeitsstellen, sämtlich ohne Nachschleifen, aufweisen. Wenn das Ziehloch auf einer Seite stumpf wird, werden die Zylinder herumgedreht und die Stange von der anderen Seite durchgezogen. Das Ziehloch wird mittels Stellschrauben und Doppelstellkeilen f und e, die sich hinter jedem Zylinder befinden,

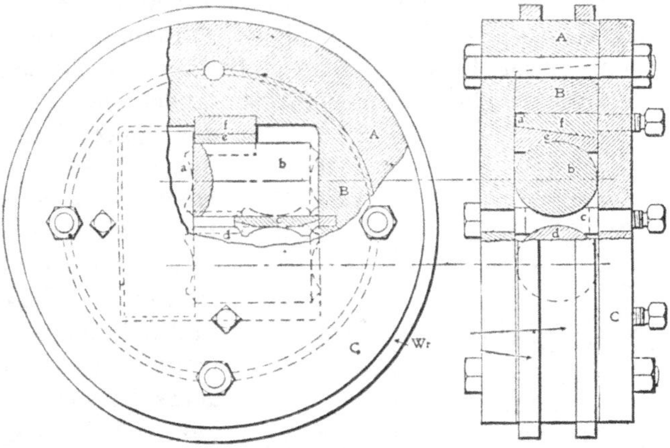

Fig. 420.

nachgestellt. In der Zeichnung sind sie nur für den Zylinder b angegeben. Dieses Ziehwerkzeug liefert vorzüglich gezogenes Material, dessen Oberflächen glatt und richtig und dessen Ecken sauber und scharf sind. Man kann erkennen, daß man durch geeignete Eindrehungen der Zylinder a und c sechs-, achteckige oder elliptische Stangen in diesem Werkzeug ziehen kann, jedoch nicht so vorteilhaft wie die rechteckigen.

Die Ziehbank dafür besteht aus zwei wagerechten Säulen, die aus fünf T-Eisen in einer gewissen Entfernung voneinander aufgebaut sind, um den Wagen zwischen sich aufzunehmen. Ein Ende dieser Doppelsäule nimmt das Zieheisen auf und trägt ein loses Kettenrad mit Welle, während das treibende Kettenrad mit seiner Welle am anderen Ende der Säule befestigt ist. Diese Kettenräder tragen eine Gliederkette, die aus Gliedern von 152 × 20 × 254 mm Mittelentfernung mit 52 mm Nieten aufgebaut ist. Der Wagen hat einen Haken, an dem die Kette angreift, mit einer Verbindung, die sich beim Aufhören des Zuges auch löst. Außerdem hat der Wagen Beißkeile mit einem Anzug von 1 : 6, die das angespitzte Ende der Stange fassen und durch das Zieheisen durchziehen und, wenn die

Stange das Ziehloch verläßt, infolge der Erschütterung und des Rückstoßes diese loslassen.

Ziehwerkzeug aus Umdrehungskörpern. Auf dem gleichen Grundsatz wie das eben beschriebene Werkzeug beruht auch ein Patent der Firma Brown, Boveri & Co., Nr. 230 234 und 230 530, das in Fig. 421 wiedergegeben ist. Hier soll das Werkzeug hauptsächlich zur Herstellung von profilierten Stäben verwendet werden.

Die Konstruktion weicht von dem oben erwähnten Werkzeug insoweit ab, als sie hauptsächlich auf weiteste Ein- und Nachstellung der Werkzeugteile eingerichtet ist. Die eigentlichen zwei Ziehwerkzeuge C und D sind wieder als Umdrehungskörper ausgeführt, die, nachdem der ganze Umfang abgenutzt worden ist, durch Nachdrehen neu verwendbar gemacht werden. Diese Teile werden lose von Achsen — Vierkant J K und Schrauben A B — getragen, die aus einem Stück gemacht sind. Die Vierkante J K tragen feste Nasen O P, in welche die entsprechenden Nuten E F der Werkzeuge durch die im Schlitten verschraubten Gewindebüchsen G H und Muttern L O_1 eingedrückt werden. Entsprechend der Nutenzahl lassen sich verschiedene Teile des Werkzeugumfanges als Arbeitsstellen verwenden. Sollen diese Stellen noch vermehrt werden, so wird nach dem Zusatzpatent der Vierkant J bzw. K aus zwei Stücken gemacht, die auf gleiche Weise gegeneinander versetzt werden können. Außerdem können die Vierkante J K selbst noch in ihrer Lagerung im Ständer verdreht werden, wodurch die Anzahl Arbeitsstellen noch einmal verdoppelt wird. Die ganze Anordnung wird in je einem Schlitten getragen, der durch die vier Stellschrauben U V in Verbindung mit der Gegenschraube und der Führungsstange Z auf irgendeine gewünschte Profilhöhe eingestellt werden kann.

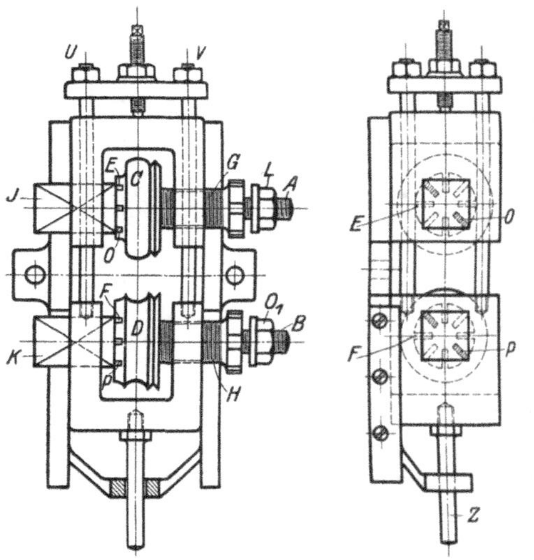

Fig. 421.

Selbstverständlich lassen sich auch mehr als zwei Umdrehungskörper zur Herstellung des Profils verwenden, so daß wie bei dem Werkzeug in Fig. 420 ein geschlossenes Profil erreicht werden kann.

e) Das Einrollen, Einziehen und Ausbauchen, Falzen.

Die Arbeiten des Einrollens, des Einziehens und Ausbauchens und des Falzens haben mit den vorher besprochenen Arbeiten gemein, daß die gewünschte Formänderung des Körpers durch eine Verschiebung der kleinsten Teilchen erfolgt. Während aber beim Ziehen die Formveränderung und Bewegung der Teilchen in dem ganzen Material vor sich geht, ist bei den oben angeführten Arbeiten nur ein mehr oder weniger großer Teil des Materials an der augenblicklichen Materialverschiebung beteiligt, z. B. bei dem Einrollen des äußeren Randes des nach Fig. 362 und 18 hergestellten Lampenschirmes nur der äußere Rand, beim Ausbauchen immer nur der unter der Rolle liegende Teil oder bei einer mit Falz zu versehenden Zarge eben nur jene Stelle des Materiales, die für die Bildung des Falzes gebraucht wird.

1. Das Einrollen.

Beim Einrollen wird unter Mithilfe einer Matrize und der dem Bleche innewohnenden Steifigkeit das Blech durch Druck gezwungen, sich nach der Form der Matrize allmählich zu biegen. Diese Arbeitsweise wird am besten durch die Herstellung des Auges für das in Fig. 12/4 abgebildete Scharnier klar, wofür das in Fig. 422 abgebildete Werkzeug Verwendung findet.

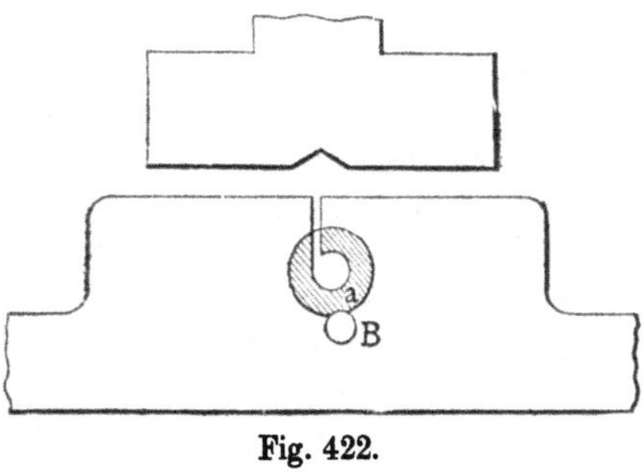

Fig. 422.

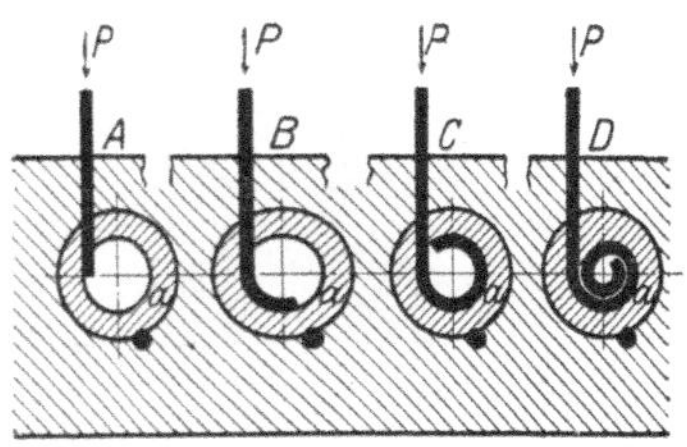

Fig. 423.

Die ausgeschnittene und abgetrennte Scharnierhälfte wird mit dem Ende, das die Öse für den Scharnierbolzen bilden soll, in den Schlitz des Einrollwerkzeuges a eingeschoben. Fig. 423 A. Unter dem durch den Stempel in Fig. 422 ausgeübten Druck auf die senkrecht stehende Kante wird das Blech gezwungen, sich längs des Kreisumfanges, Fig. 423 B, vorzuschieben, bis es sich allmählich ganz herumlegt, C, und sich, wenn der Druck noch weiter ausgeübt wird, schließlich ganz aufrollt, wie D zeigt.

Die Gesenkplatte, Fig. 422, ist aus Gußeisen; der Zylinder a aus Werkzeugstahl, gehärtet und poliert, um das Loch zu glätten. Stift B sichert dauernd die ausgerichtete Stellung von a gegen Verdrehen. Der Stempel ist in Arbeitsstellung darüber dargestellt. In einem derartigen Werkzeug werden die Arbeitsstücke durch einen Stift, der mittels Handhebels betätigt wird, herausgeschoben. Es kann jedoch dazu auch der Aufwärtshub der Presse benutzt werden, wenn es sich um große Herstellungsmengen handelt.

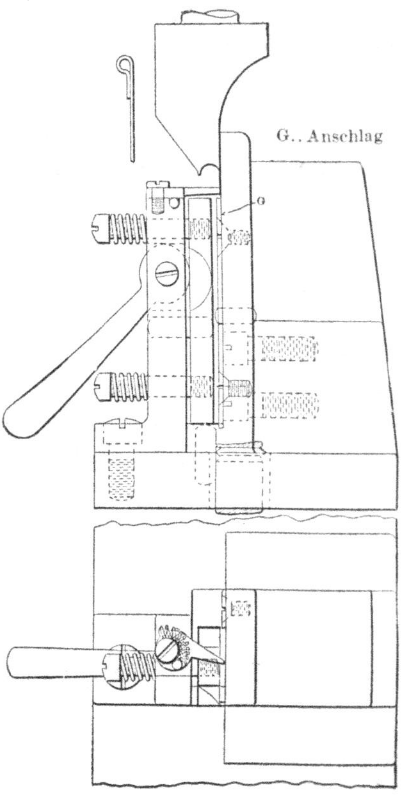

Fig. 424.

Die umgekehrte Arbeitsweise, bei der das Werkstück festgehalten wird und der Einrollbacken nach abwärts geht, wird bei dem dritten Arbeitsgang in Tabelle Nr. 32 verwendet.

Es ist nämlich im zweiten Arbeitsgang das Auge bereits vorgebogen worden, da das gerade Ende desselben nach dem Einrollen erhalten bleiben muß.

Hierbei wird das Blankett im dritten Gang gegen die Anschlagplatte G gedrückt, Fig. 424, wobei der kleine federnde Finger zurückgedrückt wird. Dann wird der Handhebel angehoben und das Blankett dadurch an dem Ständer festgehalten, worauf der Stempel auf seinem Abwärtswege das Auge schließt. Hierbei wird die Befestigungsplatte geöffnet, so daß sich das Ende vollständig umlegen und andererseits der Stempel das Blankett nicht nach vorwärts drücken kann. Dann wird der Handhebel niedergedrückt, wodurch der Federfinger das Blankett auswerfen kann. Das Ausfallen der Blankette wird durch die abgeschrägte Unterfläche des Gesenkes ermöglicht.

Der letzte Arbeitsgang in Tabelle Nr. 13, in dem der Flansch des gezogenen Bechers zu einem breiten flachen Rand umgelegt wird, ist die beste Einführung in die beim Einrollen eines geschlossenen Ringes vor sich gehende Arbeit. Während bei dem bisher besprochenen Einrollen eines Streifens die Arbeit aus einzelnen, unmittelbar einander folgenden Biegungen zusammengesetzt war, erfolgt bei dem Einrollen eines geschlossenen Randes die umgekehrte Arbeit, die beim Ziehen des Näpfchens aus der Platine vor sich geht. Sobald der Rand der Hülse in die Abrundung bei E, Fig. 425, eintritt, wird er nach außen gedrängt, so daß das Material in der Richtung des Umfanges gestreckt wird. Diese Bewegung setzt sich auf die in der Hülse weiter unten liegenden Streifen fort, solange der Stempel nach abwärts geht, während das bereits umgebörtelte Material durch das nachfließende Material gezwungen wird, unter der Stempelfläche F längs dieser nach außen zu fließen. Diese Materialbewegung ist die erste Hälfte der vollständigen, beim Einrollen vor sich gehenden Materialverschiebung.

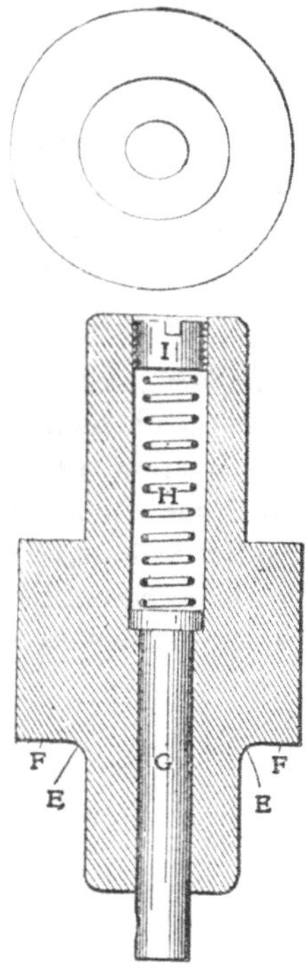

Fig. 425.

Der Becher wird in die Matrize, Fig. 426, die 0,05 mm größer im Durchmesser als der letzte Ziehring gehalten ist, eingesetzt und durch den Stempel eingedrückt. Hierbei geht der Auswerfer a in der Matrize hinab und legt sich auf den Boden auf; inzwischen wird der Flansch durch die Abrundung im Stempel bei E nach außen umgelegt und durch den Ansatz F, sobald er in Berührung mit der Stirnfläche der Matrize kommt, ausgeflacht. Sobald der Stempel wieder so weit emporgegangen ist, daß die Matrize frei ist, wird der Becher durch den Auswerfer a genügend angehoben, daß der Arbeiter ihn entfernen kann. Der Auswerfer wird durch eine Wurstfeder aus 3 mm Draht, die durch Stellmuttern B eingestellt wird, betätigt. D sind Luftlöcher. Der Federbolzen G kommt weit genug unter das Ende des Stempels, um

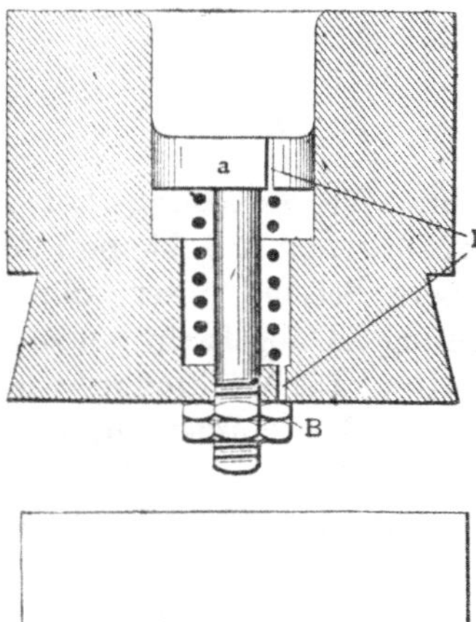

Fig. 426.

das Stück abzustreifen und während der Flanschbildung den Boden des Bechers niederzupressen. Die Bewegung, welche der Rand der Hülse in dem eben besprochenen Werkzeug macht, kann entweder nach außen oder nach innen erfolgen, wobei das Material im zweiten Falle zusammengestaucht wird. Grundsätzlich haben die Werkzeuge dieselbe Form und das Material erhält dieselbe Verschiebung.

Schließwerkzeug. Das in Fig. 428 dargestellte Werkzeug zwingt den aufrechtstehenden Rand des flachen Deckels, unter dem Schließdruck nach einwärts zu gehen und sich mit dem eingerollten Rand auf dem Bodenblech festzulegen.

Die Herstellung der Knallkapseln, Tab. Nr. 33, die bei Nebel als Gefahrsignale im Eisenbahnverkehr dienen, erfolgte früher mittels anderer Werkzeuge als des in Fig. 428 abgebildeten, und zwar nach Fig. 427

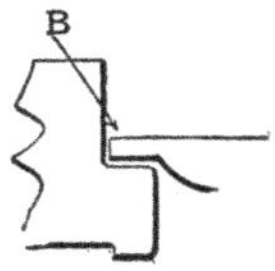

Fig. 427.

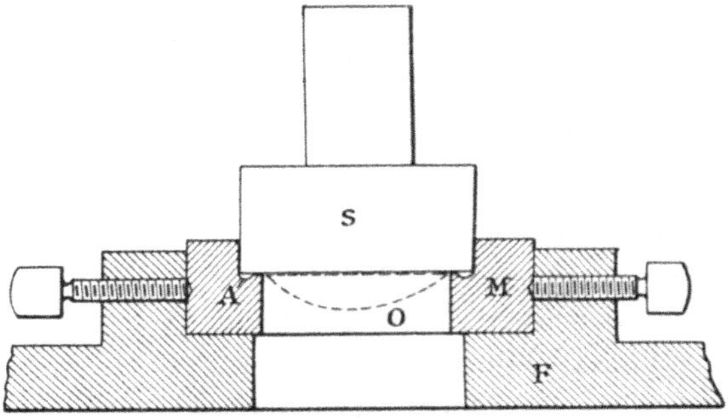

Fig. 428.

ohne die Vertiefung bei A, sondern bloß mit einer rechteckigen Ausdrehung und kleinen Abrundung bei B. Man erwartete, daß die obere Scheibe unter dem Druck unter die untere kriechen und genügend abgeflacht würde, um die Unterplatte und das Pulver zu halten. Die Werkzeuge wurden aber nachher nach Fig. 428 abgeändert und gaben ausgezeichnete Erfolge. Das Material zur Herstellung dieser Knallkapseln besteht aus dünnem Weißblech. Die Vertiefung bei A entspricht einer Blechdicke. Die arbeitenden Teile, Stempel S und Matrize M, sind gehärtet und angelassen. Die Tabelle zeigt das Blankett für den Oberteil vor dem Schließen, für den Unterteil und endlich beide Teile, wie sie in einem Werkzeug nach Fig. 428 geschlossen werden. Unterhalb des mit Pulver gefüllten Bodens ist in der Matrize M und der Froschplatte F eine genügend weite Öffnung O vorgesehen, damit im Falle einer Explosion der Pulverladung beim Einrollen der Arbeiter keinen Schaden leiden kann.

Einziehwerkzeug für eine Linsenfassung. Das flache Näpfchen, das später die Linsenfassung bildet, wird auf dem Verbundwerkzeug nach Fig. 538 ausgeschnitten, geprägt und gezogen. Im zweiten Gang wird der senkrecht gezogene Rand in dem einfachen Einrollwerkzeug, Fig. 429, nach innen eingerollt. Die beiden Herstellungsstufen der Fassung sind in der Matrize und darüber gezeichnet. Die Matrize ist auf der Stirnfläche zur Aufnahme des Blanketts, von denen eines in Arbeitsstellung gezeichnet ist, ausgedreht. In dem Stempel wird die Ausdrehung Z auf den geforderten Durchmesser und im gleichen Aufspannen das Loch für den Auswerfer S und seinen Zapfen R ausgebohrt. Die Ausdrehung Z ist ein wenig abgeschrägt, so daß das Näpfchen leicht seine richtige Stellung findet. Die Arbeit, die durch diesen Stempel verrichtet wird, ist ähnlich der im ersten Arbeitsgang, Fig. 538, S. 419. Das Stück wird vom Stempel so weit in die Höhe mitgenommen, bis der Auswerfer es abstreift, worauf es aus der geneigten Presse in den unterhalb befindlichen Kasten fällt.

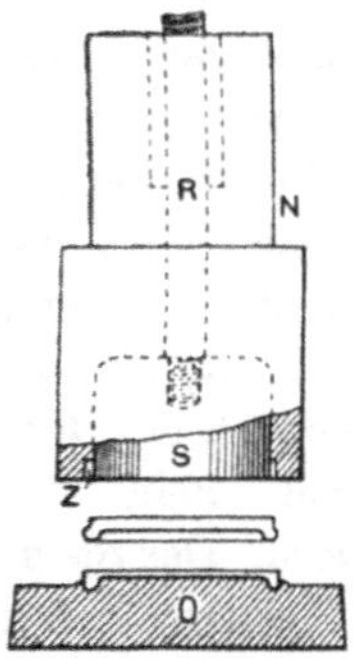

Fig. 429.

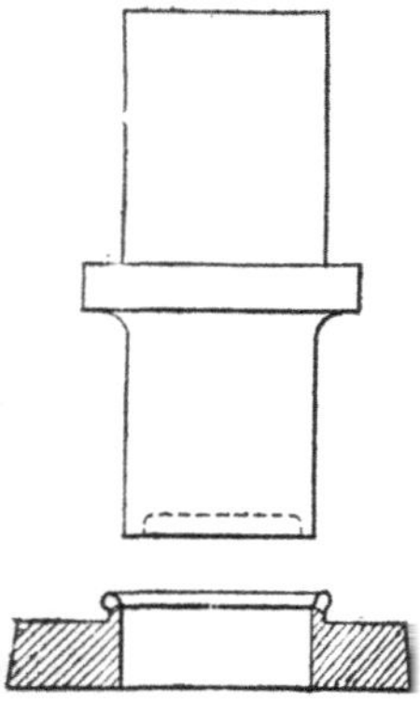
Fig. 429 a.

Fig. 429a stellt das Werkzeug zum Ausschneiden des Bodens dar, so daß das Stück in Gestalt des Ringes, Fig. 537c, übrig bleibt. Vorher jedoch wird der Ring an seinem Umfang noch beträchtlich abgedreht, wie L[1]) zeigt; dies geschieht mittels eines gewöhnlichen Abstechstahles auf der Drehbank, wobei die Ringe in einem Federfutter eingespannt werden. Der abgedrehte Teil wird dann über der Linse abgeschrägt.

Einrollen einer Linsenfassung. Die hier besprochenen Fassungen dienen beim Auswählen und Anpassen der Gläser an das Auge des Kunden zum Halten der zu prüfenden Linse und bestehen aus einem kleinen Ring mit Handgriff. Von ihnen ist eine ganze Anzahl für Linsen verschiedener Brennweite in ständigem Gebrauch, so daß man sie bei der Prüfung der Brille leicht in besondere Fassungen einsetzen kann. Im ersten Gang wird aus dem Blechstreifen auf einem Verbundschnitt- und Ziehwerkzeug, Fig. 523, ein flacher Deckel gezogen, dessen aufrechtstehender Rand auf dem Werkzeug, Fig. 430, eingerollt wird Hier wird bereits das Material auf ¾ eines vollen Kreises eingerollt. Das innerhalb des Ringes liegende Material wird dann auf einem Rundschnitt mit besonderem Niederhalter, Fig. 430a, ausgeschnitten. Das Werkzeug ist, wie Fig. 429a, an dieser Stelle mitangeführt, damit

[1]) S. 419.

der Vergleich mit den später besprochenen Verbundwerkzeugen, die dieselbe Arbeit liefern, erleichtert wird. Während die in der Presse hergestellten Ringe schneller gemacht werden und billiger sind, haben die in den Automaten hergestellten ein besseres Aussehen und stellen eine bessere Qualität dar, so daß man sich leicht bewegen läßt, diese selbst zu einem höheren Preis zu kaufen.

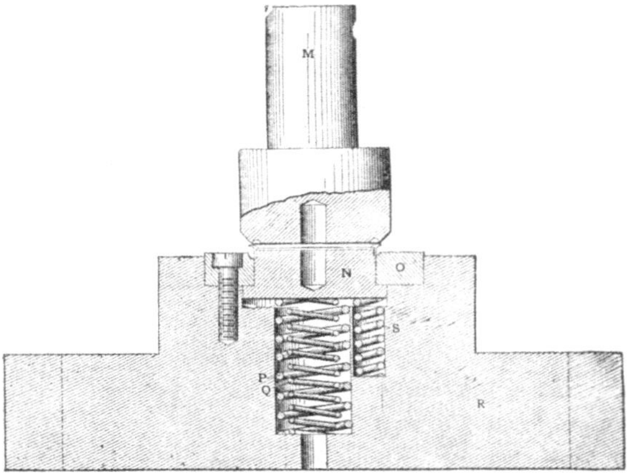

Fig. 430.

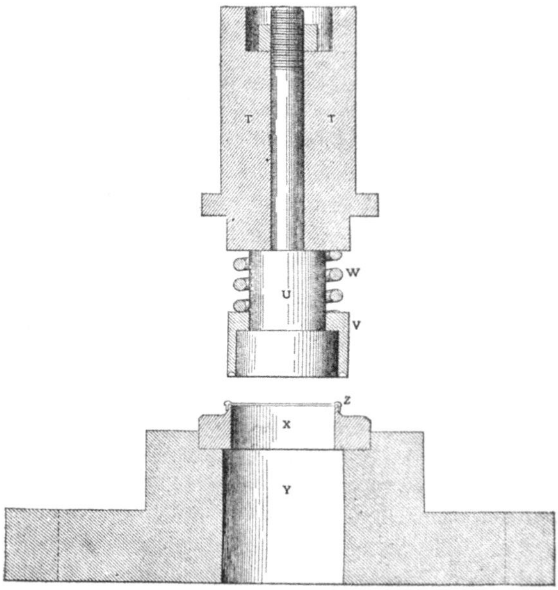

Fig. 430a.

Fig. 430 zeigt das Einrollwerkzeug, in welchem der Rand der Fassung durch den Einrollstempel hohl gedrückt wird. Der Stempel M hat einen Zapfen, der in einer größeren Büchse, die für ähnliche kleine Stempel verwendet wird, gehalten ist, damit man deren Zapfen nicht in den Abmessungen der großen Stempel herstellen muß.

Der im ersten Gang hergestellte Deckel wird zuerst auf die Platte N gelegt, die mit dem Arbeitsstück niedergeht und auf dem Grunde bei S zum Aufsitzen kommt. Das noch übrige im Näpfchen vorhandene Material fließt nun in die Aussparungen von M und O. Durch die Federn P, Q und S wird N zurückgeführt, sobald sich die Werkzeugteile voneinander trennen. Die Feder Q innerhalb P ist eine oft geübte Verstärkung der Kraft des Abstreifers, dessen Wirkung noch durch die drei Federn S weiter unterstützt wird.

Das letzte Werkzeug, Fig. 430 a, hat in T einen gußeisernen Stempelkopf mit dem Stahlstempel U zum Ausschneiden der Mitte des Blanketts. Die Schnittplatte X hat eine sehr dünne Kante, die in Verbindung mit dem oberen Abstreifer V beim Niedergang des Stempels die Form der Fassung bei Z sichert, während U bei seinem weiteren Abwärtsgang das Loch schneidet. Die Feder W ist aus Draht von 3,2 mm Durchmesser.

Einroll- und Ausschnittwerkzeug deutscher Bauart. Bei der Herstellung der verschiedenen Teile eines Taschenuhrgehäuses[1]) kommen ähnliche Arbeiten, wie eben für Linsenfassungen beschrieben worden sind, vor. Der vollständige Herstellungsgang solcher Teile ist in Tabelle Nr. 25/6 gegeben, aus dem hier die Werkzeuge für das Einrollen eines Mittelringes, Fig. 431, und das Aushauen eines Glasringes besprochen werden. Es sei vorausgeschickt, daß diese Werkzeuge für eine Materialzuführung mittels Revolvertellers entworfen sind. Die Anordnung des Einrollwerkzeuges (Einziehwerkzeuges), Fig. 432, ist der in Fig. 429 entgegengesetzt, soweit die Stellung des federnden Niederhalter- und Auswerferstempels c in Frage kommt. Der Revolverteller bringt den gezogenen Mittelring bis unter das Werkzeug, wo er sich selbsttätig in die Matrize a einlegt. Dann geht der Stempel nach abwärts, wobei der federnde Niederhalter c den Mittelring in der Matrize festhält, gegebenenfalls erst richtig legt. Beim Weitergang des Stößels faßt der Einrollstempel b mit seiner konischen Ausdrehung bei d den aufrechtstehenden Rand des Mittelringes und zieht ihn nach innen ein, so daß er die Form Fig. 431 annimmt. Bei starken Blechen und kleinen Rollradien muß man zweimal rollen.

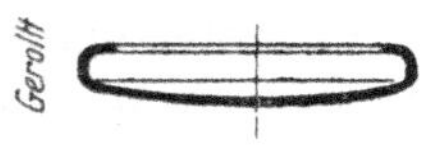

Fig. 431.

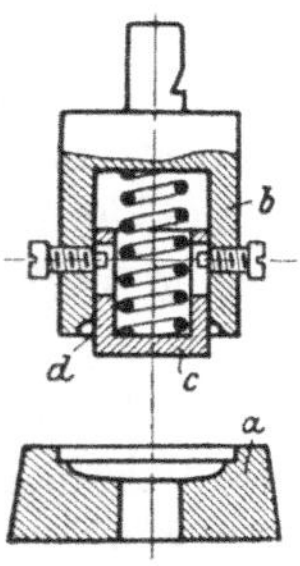

Fig. 432.

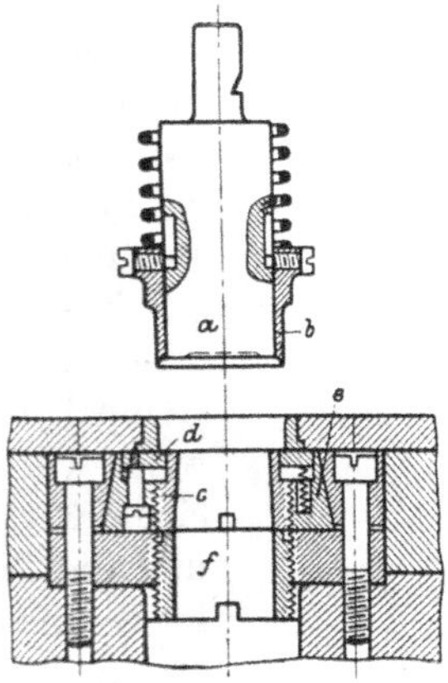

Fig. 433.

Wenn ein derartiger Ring ausgeschnitten werden soll, so muß der Niederhalter, wie in Fig. 429 a u. 430 a gezeigt und in Fig. 433 bei b ersichtlich ist, eine Aussparung für den eingerollten Teil haben. Die Arbeitsweise ist dieselbe wie in Fig. 430, doch ist die Ausbildung der

[1]) W. T. 1910, S. 406/7.

Schnittplatte mit dem besonderen Auswerfer sorgfältiger durchgebildet. Die Zuführung erfolgt wieder mittels des Revolvertellers, unter dem erst die Schnittplatte beginnt. Der zu lochende Ring liegt auf der Schnittplatte c, die hier mit Gewinde in die Gesenkplatte e eingeschraubt ist, so daß sie nach dem Nachschleifen wieder auf richtige Höhe eingestellt werden kann. In der Grundplatte ist eine Gegenschraube f eingesetzt, die das unbeabsichtigte Herausdrehen der Schnittplatte verhindert. Der Abfall fällt durch das große mittlere Loch aus, während der ausgeschnittene Ring durch den federnden Auswerfer d in den Revolverteller zurückgedrückt wird. Die Gesenkplatte mit dem Schnittring wird mittels eines konischen Überlagringes und versenkter Zylinderkopfschrauben in dem Unterteil ausgemittelt und festgehalten. Der Stempel a bildet die Führung für den Niederhalter b, der durch Federkraft nach abwärts gedrückt und durch zwei Schrauben, die in seitlichen Schlitzen des Stempels gleiten, geführt ist. Das Einrollen des ausgeschnittenen Deckels erfolgt auf einem ganz ähnlichen Werkzeug wie Fig. 432, nur liegt der Deckel umgekehrt auf der Matrize.

Es ist aber nicht immer notwendig, nach diesen allgemeinen Formen zu arbeiten, man kann, besonders wenn es sich um größere Mengen besonderer Teile handelt, Sonderbauarten von Werkzeugen mit Erfolg verwenden, wie die folgenden Beschreibungen angeben.

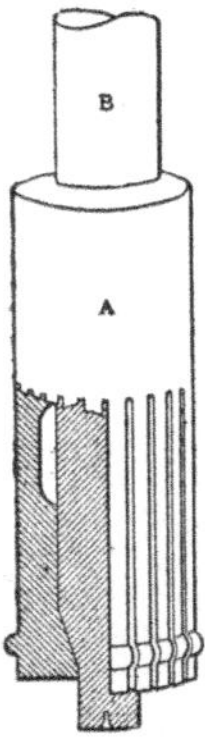

Fig. 434.

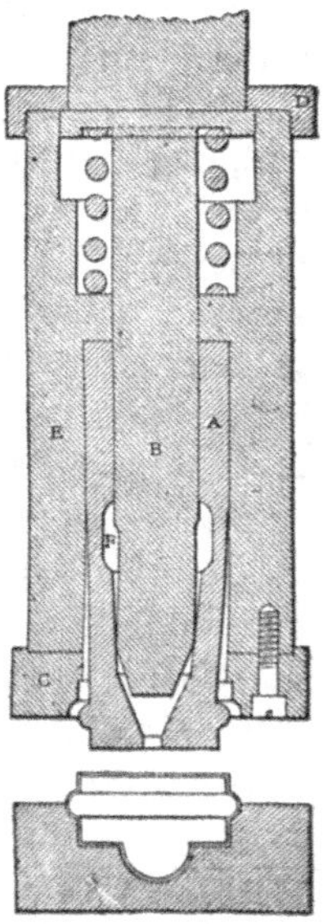

Fig. 435.

Spreizdorn zum Bördeln. Stempel und Matrize, wie in Fig. 434/5 abgebildet, dienen zur Herstellung eines halbrunden Wulstes an einer Hülse, die in der Matrize, Fig. 435, in Arbeitsstellung abgebildet ist.

Zuerst wird die Stempelhülse A hergestellt, dann der innere Spreizstempel B. Dann werden A und B weich verlötet und die Schlitze in A eingefräst, wobei B als Dorn dient und die Schnitte ein wenig durch A in B eintreten, so daß sich an der Innenseite von A kein Grat bilden kann. Die Teile werden dann durch Ausschmelzen des Lots getrennt und B so weit abgedreht, daß sich A im Gebrauch zusammenziehen kann, wobei gleichzeitig die Frässchnitte in B entfernt werden.

Wenn jedoch für diesen Stempel keine entsprechende Matrize vorgesehen wird, kann man scharf ausgeprägte oder sehr regelmäßige

Wulste nicht erwarten; außerdem wird durch diesen Vorgang der Durchmesser des Arbeitsstückes vergrößert. Um wirklich gute Arbeit zu liefern, muß man eine Matrize nach Fig. 435 anwenden, in der das vollständige Werkzeug mit dem zusammengezogenen Stempel abgebildet ist. Bei dieser Bauart des ganzen Werkzeuges ist keine Möglichkeit vorhanden, daß die einzelnen Teile in Unordnung geraten. A und B entsprechen den Teilen in Fig. 434, während E eine Hülse aus weichem Material ist, in die A eingetrieben wird und die zugleich die obere Hälfte C der Matrize trägt. Der Stempel A und die Matrize C gehen zusammen nach abwärts, wobei sie durch eine im oberen Teil befindliche Spiralfeder vorwärts gedrückt werden, die stark genug sein muß, um C niederhalten zu können, während der Wulst gebildet wird. A und C kommen gleichzeitig zur Auflage, während B weiter nach abwärts geht, bis sich A genügend geöffnet hat, um einen guten Wulst herzustellen. Die Ausdrehung F ist vorgesehen, damit A, welches an dieser Stelle Federhärte haben soll, genügend federt. Alle anderen gehärteten Teile werden nur auf mittleres Strohgelb angelassen. Nur die eigentlichen arbeitenden Teile und das Ende des Stempels B müssen durchgehärtet werden. D ist ein Ring aus weichem Material, der den Druck der Feder auf die Hülse E aufnimmt.

Herstellung eines doppelten Wulstes in einer Messinghülse. Bei der Herstellung der Messinghülse nach Fig. 436 b kommen ziemlich ungewöhnliche Werkzeuge zur Anwendung, die in Fig. 437 dargestellt sind. Die Herstellung geschieht in zwei Gängen: zuerst wird die Hülse nach Fig. 436a in gewöhnlicher Weise gezogen und, nachdem sie an der Kante a auf Länge abgestochen worden ist, in dem in Fig. 437 abgebildeten Werkzeug fertiggestellt. Um das fertige Stück aus dem Werkzeug zu entfernen, muß die Matrize zweiteilig sein. Dazu werden zwei Stücke Stahl gehobelt, mit Prisonstiften versehen und miteinander verspannt fertiggestellt, so daß das Loch mit den Nuten durch die Schnittlinie genau halbiert wird. Der eine Teil, Block B, wird auf der Gußeisenplatte D festgeschraubt, während der andere C auf ihr gleitet und in dem Block E aus Maschinenstahl geführt wird. Zwischen den Blöcken B und C werden die vier starken Wurstfedern F in Bohrungen eingelegt. An einer in E eingeschraubten Drehachse G ist der Nockenhebel H aus Maschinenstahl drehbar gelagert, mit dem man unter dem Druck der Federn die Matrize bequem öffnen und schließen kann.

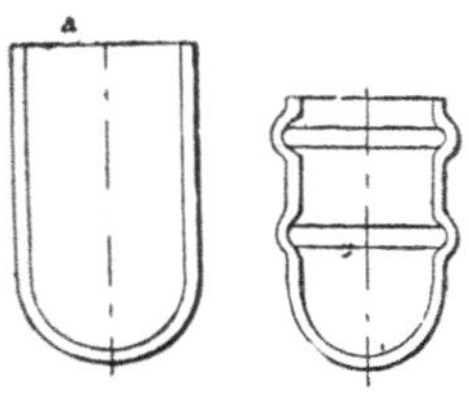

Fig. 436 a, b.

Die Arbeitsweise ist in kurzem wie folgt: Nachdem die Hülse in die Matrize, die selbstverständlich geschlossen ist, eingelegt worden ist, geht der Stempel I, dessen Durchmesser genau mit dem inneren und äußeren Durchmesser des Näpfchens übereinstimmt, herab, bis er, ohne bis dahin irgendeine eigentliche Arbeit geleistet zu haben, in die in Fig. 438 gezeichnete Stellung kommt. Da nun der Stempel genau in die Hülse paßt und diese am Grunde aufruht, so muß das Material beim weiteren Abwärtsgang des Stempels in die Nuten fließen. Dann wird

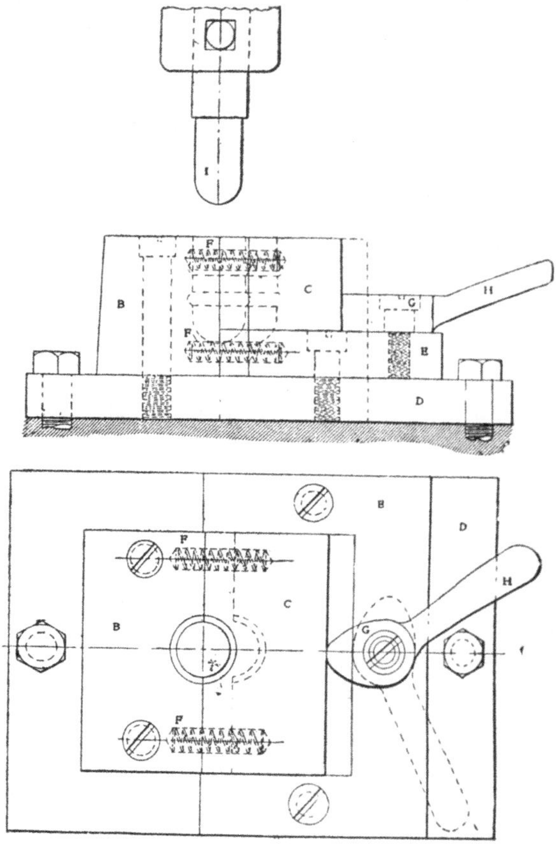

Fig. 437.

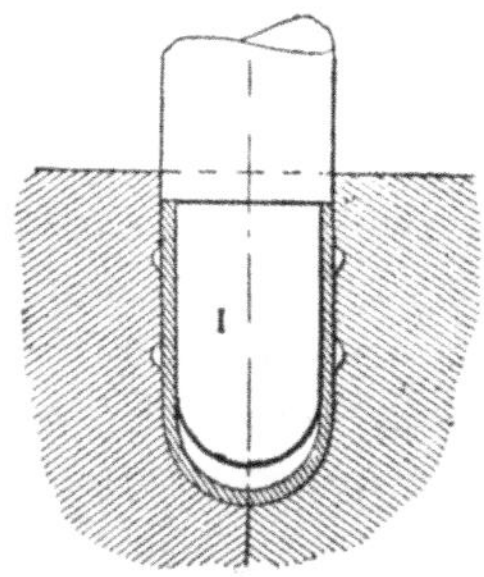

Fig. 438.

der Stempel zurückgezogen, der Nockenhebel geöffnet und die Hülse herausgenommen.

2. Das Falzen.

Das Falzen besteht in dem Umbiegen der Kanten bei geraden Blechen und dem Einrollen der Kanten bei runden Blechteilen, derart, daß zwei Teile auf diese Weise miteinander verbunden werden und durch den auf die zusammengefügten Bleche ausgeübten Schließdruck auch luft- und wasserdicht vereinigt werden können. Diese Arbeitsweise, die zuerst maschinell von Kircheis in Aue, D. R. P. 66860 (bereits abgelaufen), ausgeführt worden ist, hat eine durchgreifende Umwälzung in der Herstellung von Blechemballagen, von Blechgegenständen für den Hausgebrauch usw. hervorgerufen und das früher übliche Löten oder Nieten fast vollständig verdrängt. Wenn das Ergebnis auch das gleiche ist, so sind doch die Werkzeuge und die Reihenfolge der Arbeiten verschieden, je nachdem es sich um das Falzen einer geraden Zarge oder das Schließen einer runden oder rundeckigen Dose handelt. Auch die Form der Falze ist verschieden, Fig. 439, je nachdem es sich um einen innen oder außen liegenden Falz handelt. Es ist dies der sogenannte Doppelfalz.

Eine Maschine zur Herstellung derartiger Falze an geraden Zargen zeigt Fig. 440[1]). Die vorgebogene Blechzarge I, Fig. 439, wird mit den für den Falz nötigen Anbiegungen in einem Stößelhub versehen und der Falz nach II oder III geschlossen. An der schrägen Seitenfläche des Stößels a arbeitet der Schieber mit der Falzschiene b, die von der Exzenterwelle aus durch die Zugstange d angetrieben werden. An der Falzschiene ist ein federnder Anschlag c angebracht, der als Anschlag für die Zarge dient. Der unterhalb befindliche Hohldorn e trägt die zweite Falzschiene f, die sich gegen den inneren, drehbaren Dorn g stützt. Man legt die Zarge wie gezeichnet ein, so daß sich das eine Ende an den federnden Anschlag c anlegt und das andere Ende in die Hohlkehle der unteren Falzschiene f einspringt. Beim Einschalten der Maschine wird zuerst Schiene b nach abwärts bewegt, wobei sie die Blechkanten um die Kanten der Falzschiene f einwärts und auswärts biegt und gleichzeitig ineinander fügt.

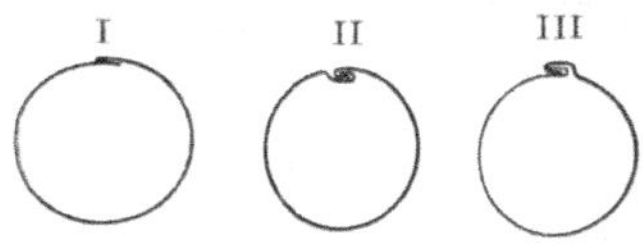

Fig. 439.

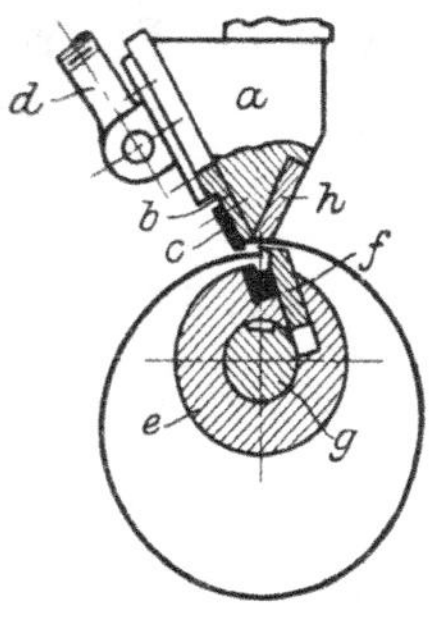

Fig. 440.

Gleichzeitig dreht sich der innere Kern so weit, daß die Aussparung unter die untere Falzschiene kommt, so daß sie zurückgehen und dem Druck der Stößelschiene h ausweichen kann. Je nach der Form des Dornes wird ein Falz nach II oder III, Fig. 439, gebildet.

Um den bei langen Zargen stark überhängenden Dorn gegen Durchfedern abzustützen, hat Kircheis eine von der Exzenterwelle betätigte Unterstützung vorgesehen, die das Zapfenende umschließt und nach vollbrachter Arbeit wieder freigibt.

Im Anschluß daran sei in Fig. 441/2 die von Woodworth beschriebene Einrichtung wiedergegeben, die zum ununterbrochenen Falzen von runden Zargen für Blecheimer, Kannen u. dgl. dient.

---

[1]) Adler, Die Werkzeugmaschinen auf der Brüsseler Ausstellung 1910. Z. Ver. deutsch. Ing. 1910, S. 1854, Fig. 47.

In einem Zubringen des Werkstückes werden in zwei Preßhüben zuerst die Kanten umgebrochen, ineinandergeschoben und schließlich geschlossen. Fig. 441 gibt eine Vorder- und Seitenansicht des Stempels und des Dornes in ihrer Stellung in der Presse wieder. Der Stempel A ist übrigens ein gutes Beispiel für schwierige Hobelarbeit. Der gußeiserne Körper trägt drei Stücke aus Werkzeugstahl, B, D und E. Die zugeschärften Kanten von D und E sind um eine Metalldicke dünner gehobelt als die entsprechenden Nuten im Stück B. Dieses Stück bewegt sich bei seiner Arbeit in der Richtung des Pfeiles bis in die gestrichelte Stellung F, in der es in der Vorderansicht gezeichnet ist. Wenn dessen Unterfläche an die Oberfläche des Dornes stößt, wird B gegen die Federn bei C gedrückt, wobei die Kanten der Zarge nach

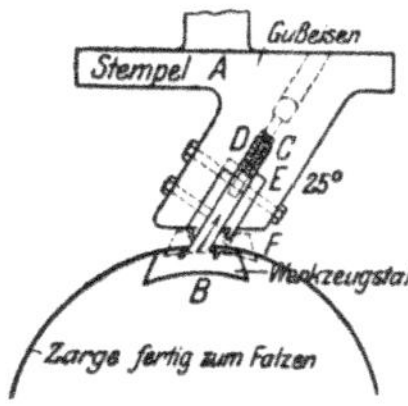

Fig. 441.

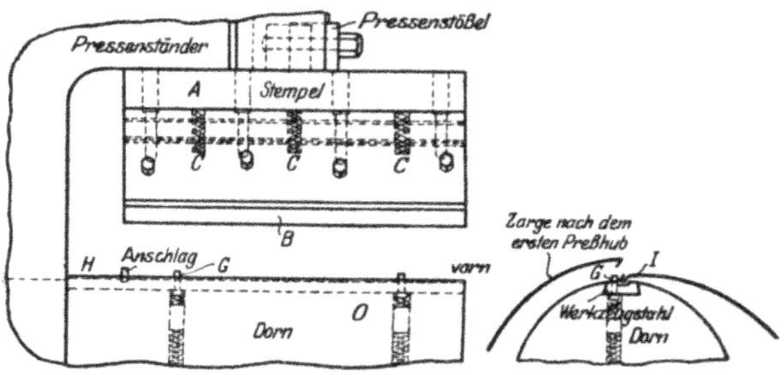

Fig. 442.

Fig. 442 umgebogen werden. Wenn der Teil B beim Aufwärtsgang der Presse den Dorn verläßt, so drücken ihn die Federn C wieder in seine Anfangsstellung, in der er durch Zylinderkopfschrauben, die in der Seitenansicht deutlicher zu sehen sind, gehalten wird. Zur Fertigstellung des Falzes werden nun die Kanten der Zarge ineinandergesteckt und direkt auf den Dorn und gegen die Federanschläge G und den festen Endanschlag H gehalten. Die Presse wird dann eingerückt, wobei die Unterfläche von B die Kanten des Falzes flachdrückt; vgl. den fertigen Falz in Fig. 439. Dieses Werkzeug zum Falzen, das in Verbindung mit dem Dorn gebraucht wird, kann für Stücke bis ungefähr 300 mm Länge verwendet werden. Für größere Längen wird der Dorn durch einen einsetzbaren Bock von unten abgestützt, welcher augenblicklich und leicht zur Seite ausgeschwungen werden kann, um eine neue Zarge auf den Dorn aufzuschieben. Auf diese Art können innere oder äußere Falze hergestellt werden. Die hier beschriebene Arbeitsweise ist für den ersten Falz; beim letzten wird die Nut I in B gehobelt, und der Stahleinsatz im Dorn bleibt glatt.

Das Falzen der Böden und Schließen der Deckel runder, ovaler oder eckiger Blechkörper mit runden Ecken geschieht in einer von Fig. 440 insofern abweichenden Weise, als die ineinandergelegten Flansche von

Zarge und Deckel bzw. Boden nach Fig. 443[1]) eingerollt werden, bis ein geschlossener Rundfalz erreicht worden ist. Dieser würde für die Verbindung genügen, doch drückt man den Rundfalz meistens, teils aus Schönheitsrücksichten, teils um ein im Falz eingeschlossenes, organisches Dichtungsmaterial gleichmäßig nach allen Stellen zu pressen, flach. In Fig. 443 ist der durch eine innere Rille gegen Verziehen versteifte Deckel in die aufrechtstehende Zarge, die in der in Fig. 439 gezeichneten Weise geschlossen worden ist, eingelegt und wird dort durch einen Stempel gehalten. Wie ersichtlich, ist der Deckelflansch etwas länger als der Zargenflansch, so daß er sich zuerst einrollt und beim Näherrücken der Rolle allmählich den Zargenflansch mitnimmt. Für diese Arbeiten sind zwei Arten Maschinen im Gebrauch, solche, in denen die Dose umläuft, die ältere Bauart, und solche, die mit feststehender Dose, die neuere Bauart, arbeiten. Im letzten Teil der Figur ist das Andrücken des Rundfalzes mit Hilfe einer glatten Druckrolle gezeigt. Es sei noch erwähnt, daß an den Stellen, wo der Falz in der Zarge mit dem Bodenfalz zusammentreffen würde, der Bodenflansch weggeschnitten sein muß, damit die Druckrollen an der Stelle nicht eine vierfache Materialdicke einzurollen haben.

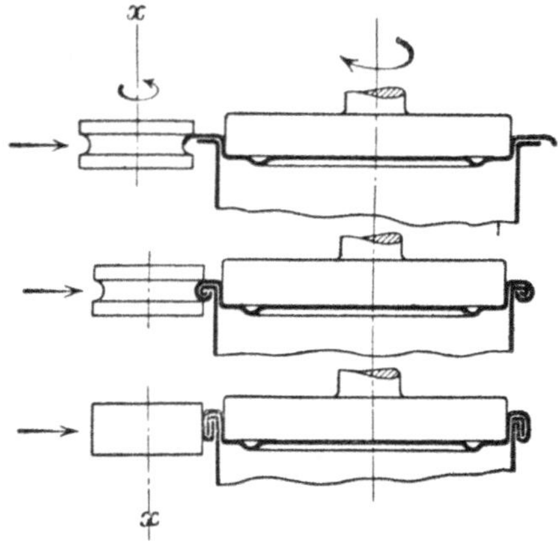

Fig. 443.

3. Das Ausbauchen und Einziehen.

Unter Ausbauchen oder Einziehen versteht man eine derartige Materialverschiebung an gezogenen Umdrehungskörpern, welche durch eine Reihe unmittelbar aufeinanderfolgender, örtlicher Einzeldeformationen das Material der gezogenen Hülse derartig streckt oder staucht, daß aus Hohlkörpern, nach Fig. 444a, gebauchte oder eingezogene Hohlkörper nach Fig. 444b entstehen. Es ist dies dieselbe Arbeit wie das Treiben, bei dem früher mit Treibstock und Hammer aus dem ebenen Blech durch einzelne örtliche Streckungen schließlich der Hohlkörper erhalten wurde. Jetzt tritt dafür der ruhige Druck

[1]) Zeitschr. f. Werkzeugm. 1911, S. 334.

der Druckrolle A, Fig. 445, gegen die Formwalze B an Stelle des Schlages des Hammers auf den Treibstock. Die einzelnen Schläge des Treibhammers streckten die getroffene Stelle des Bleches, so daß sie sich ein wenig über das unbearbeitete Blech erhob. Dann wurde sorgfältig Schlag neben Schlag gesetzt, bis die ganze Oberfläche des Bleches allmählich die gewünschte Hohlform annahm. Die Wirkung eines ruhigen Druckes der Druckrolle A würde bei den zur Verfügung stehenden Kräften allein nicht ausreichen, um die gleiche Wirkung zu erzielen, weshalb man den Druck mit einem Walzvorgang vereinigt. Es erfolgt also bei umlaufenden Werkzeugen, Fig. 445, ein Auswalzen, d. h. ein Strekken oder Stauchen des Materials zwischen Druck- und Formwalze, auf einem Ringe des Hohlkörpers, der in seiner Breite der Breite der Berührungsstelle zwischen Druck- und Formwalze entspricht. Wenn nun gleichzeitig mit der Drehbewegung der drei Stücke, Hülse, Form- und Druckwalze, eine Verschiebung der Druckwalze in der Längsachse der Hülse vor sich geht, so werden nach und nach alle Teile der Hülse in eng aneinanderliegenden Schraubenlinien gestreckt, und so wie beim Treiben nach und nach der ganze Körper ausgebaucht.

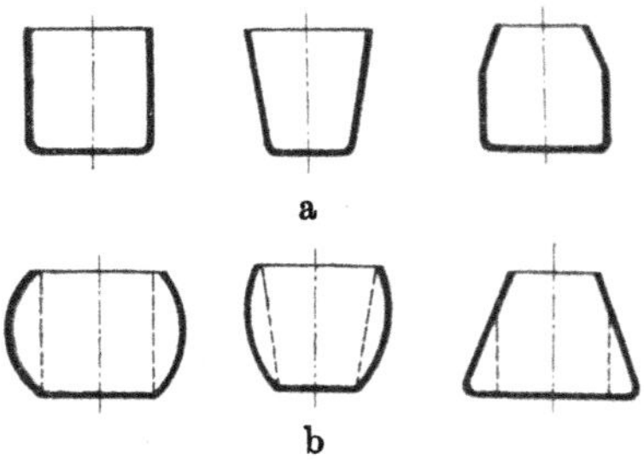

Fig. 444.

Für diese Arbeiten sind zwei grundsätzlich verschiedene Verfahren üblich, die schematisch in Fig. 445 a und b dargestellt sind. Im ersten Fall wird eine geschlossene Form aus Gußeisen, Stahlguß oder Stahl, die zweiteilig, wenn nötig auch mehrteilig ist, an der Spindel der Ausbauchbank befestigt und läuft mit ihr um. In diese Form, deren Innenfläche der Form des auszubauchenden Gegenstandes entspricht, wird die Hülse eingelegt und die Druckrolle A eingeführt, die am Support befestigt ist. Indem man die Druckrolle gegen das Werkstück und die Formrolle anstellt, nimmt diese die Hülse und die Druckrolle mit und streckt das Material an der betreffenden Stelle. Allmählich wird dann die Druckrolle vorgeschoben, so daß schließlich die ganze Hülse ausgebaucht wird. Die Formrolle B, Teilkluppe genannt, hat den Vorteil, daß das Arbeitsstück gezwungen wird, gleichachsig zu laufen, und gleichzeitig sehr genau in den Außenabmessungen erhalten wird. Die Nachteile bestehen in der teuren Herstellung und dem großen Gewicht der Teilkluppe, das einseitig an der Bankspindel weit ausladet und infolgedessen schwere Spindelstöcke bedingt. Das Ein- und Ausbringen des Werkstückes dauert verhältnismäßig lange Zeit, so daß die Arbeit teuer wird. Da sich außerdem bei den billigeren Teilformen aus Gußeisen leicht Poren und unganze Stellen zeigen, die die Form zum Gebrauch ungeeignet machen, ist man auf die bedeutend teureren Formen aus besonderem Stahlguß oder Werkzeugstahl angewiesen. Infolgedessen ging man, wo immer nur möglich, auf den in Fig. 445 b gezeichneten Apparat über. Dieser besteht aus einer außenliegenden Formrolle B, die durch Reibung von dem Arbeitsstück mitgenommen wird. Sie ist in einem besonderen Halter am Support befestigt und wird je nach der Form des zu drückenden Gegenstandes ausgewechselt. Die Druckrolle A bleibt wie früher. Das Arbeitsstück wird von Hand in ein umlaufendes Spannfutter eingelegt und mit einer gleichachsigen Einspannvorrichtung von der Pinole des Reitstockes aus festgeklemmt. In beiden Fällen drückt der Arbeiter die Druckrolle so lange an, bis er die Wandungen des gedrückten Gefäßes allseitig an der Formrolle anliegen hat. Dann stellt er im ersten Fall die Bank ab, öffnet das Teilfutter und nimmt das Stück heraus. Im zweiten Fall ist es nicht notwendig, die Bank stillzusetzen, sondern der Arbeiter zieht einfach die Pinole mit der Haltevorrichtung zurück und nimmt bei umlaufender Bank das Stück von Hand aus ab.

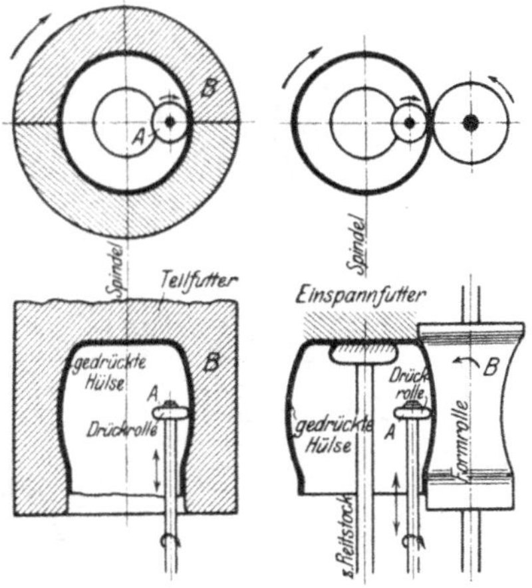

Fig. 445.

Für die Arbeit ist selbstverständlich die erstere Arbeitsweise, bei der die Druckrolle den Formrollenkreis von innen berührt, vorteilhafter als die Anordnung in Fig. 445 b, wo die beiden Kreise Außenberührung haben.

Eine Sonderbauart[1]) gibt Musiol, D. R. P. 159 576, nach Fig. 446 an. Diese Bauart vereinigt einen großen Teil der Vorzüge beider oben besprochenen Arbeitsweisen, doch wird sie wohl teurer als jede der beiden werden. Sobald es sich aber um eine große Zahl gleichartiger Gegenstände handelt, wird die Schnelligkeit der Herstellung in Verbindung mit der größeren Genauigkeit der gedrückten Stücke sicher die einmaligen Anschaffungskosten ausgleichen.

Hier ist das Futter selbst als Spindel ausgebildet, indem die Außenseite des zweiteiligen Futters durchweg als Umdrehungskörper ausge-

[1]) Illustrierte Zeitschr. f. Blechindustrie 1906, S. 504.

bildet ist, der in der Mitte die Riemenscheibe und an beiden Enden die Zapfen der Spindel hergibt. Wie ersichtlich, wird der Durchmesser der Lagerstellen bedeutend größer, da innerhalb derselben der Durchmesser des geformten Gegenstandes aufgenommen werden muß, doch werden dabei die Lager günstiger beansprucht, da die Druckwirkung zwischen beide Lager fällt. Das Futter ist geschlossen ausgeführt und die Teilfuge nach dem größten Umfang gelegt, so daß der Teil J immer in den Lagern verbleibt, solange dasselbe Stück gedrückt wird. Der Vorsatz N, der wieder von der Pinole des Reitstockes angedrückt wird, ist im Innern nach dem zu bauchenden Gegenstande ausgearbeitet und bildet mit dem Teil J, wenn geschlossen, den vollständigen äußeren Umriß des Arbeitsstückes. Es können alle Arbeiten bei umlaufender Spindel, Futterteil J, ausgeführt werden, da der Teil N stehen bleibt, sobald der vom Reitstock ausgeübte Druck entfernt und N zurückgezogen wird.

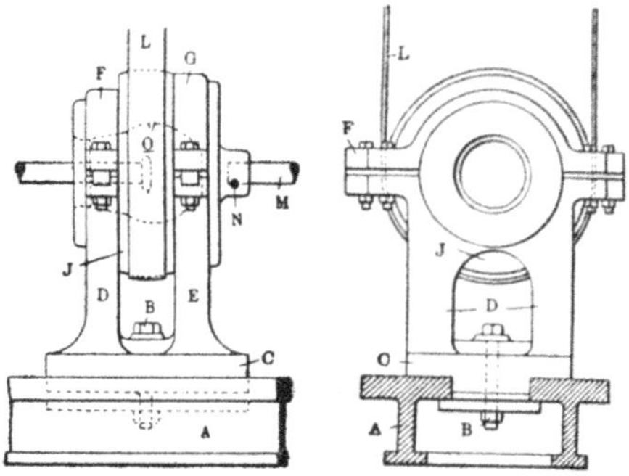

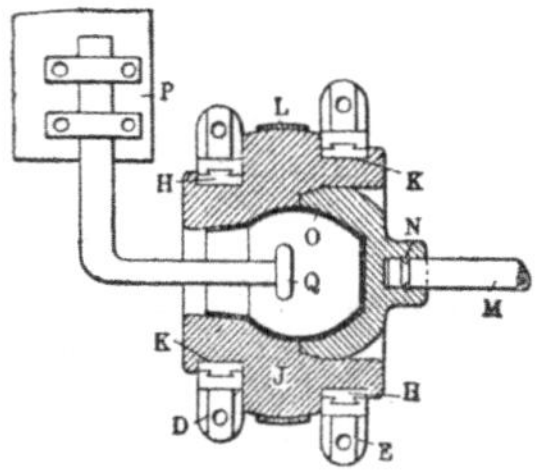

Fig. 446.

Verkehrt-Drücken. Das Umkehren gezogener Blechformen durch Drücken erlaubt Formen herzustellen, die eine der gewöhnlichen Arbeitsweisen nicht fertig bringt. Als Beispiel sei der Winkel einer Automobilhupe gewählt, bei dem die Tiefe 1,6 mm größer als der Halbmesser sein soll. Es ist unmöglich, eine Hülse nach Fig. 447 zu ziehen, so daß man ohne Kenntnis des Verkehrtdrückens (auch Verkehrtstanzen genannt) eine Hülse nach Fig. 448 durch Ziehen (vgl. Fig. 359/61) und Drücken herstellen und dann den äußeren Rand auf einem nach dem fertigen Stück geformten Futter drücken würde. Während dieser Weg bei Messing oder anderem Metall, das ausgeglüht werden kann, gangbar ist, versagt er bei Weißblech und kann auf folgende einfache Weise ersetzt werden.

Fig. 447.

Fig. 448.

Auf einem Futter nach Fig. 449 wird eine gezogene Hülse so gedrückt, daß, wie ersichtlich, das ganze Material in der dem fertigen

Stück entgegengesetzten Richtung liegt. Unter Verwendung des Drückfutters nach Fig. 451 bewegt sich das Material beim Verkehrtdrücken in der in Fig. 450 angedeuteten Weise nach innen. Das Futter muß man so bemessen, daß die Hülse gerade etwas weniger Material enthält, als notwendig ist, um beim Umkehren bis auf den Grund des Futters zu kommen, da sich das Material beim Umkehren dehnt, so daß das Arbeitsstück glatt, ohne Falten zu machen, herunterkommt, bevor es den Boden des Futters ereicht.

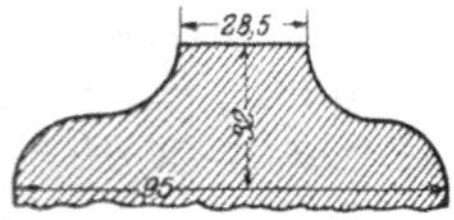

Fig. 449.

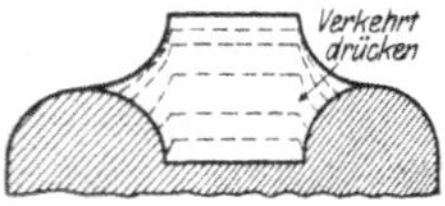

Fig. 450.

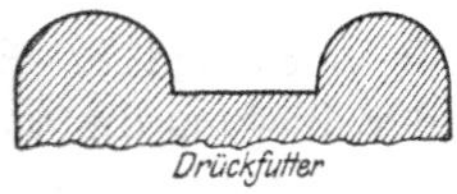

Fig. 451.

Bis vor kurzem wurde ein Einfülltrichter aus Weißblech gezogen, gedrückt und schließlich innen mit Rollen glatt gedrückt, um die Falten zu glätten. Innendrücken ist jedoch eine unangenehme Arbeit und soll möglichst vermieden werden; denn wenn es sich darum handelt, so starke Falten, wie eben erwähnt, herauszudrücken, so erhält man nachher beim Vernickeln eine minderwertige Arbeit. Deshalb entschloß man sich, die oben beschriebene Arbeitsweise zu versuchen. Die Hülse wurde zuerst nach Fig. 452 gedrückt, worauf man mit dem alten Innendrückfutter als Umkehrfutter arbeitete, Fig. 453. Die Einfachheit und Sauberkeit der Arbeit übertraf alles bisher Gemachte. Das hölzerne Drückfutter wurde nun in Stahl wiederholt, daran ein Flansch angesetzt, so daß der eingerollte Rand daran anliegen konnte, und dann zum Umkehren verwendet. Das Umkehren in dieser Art verbilligte nicht allein die Herstellungskosten, indem vier Arbeitsgänge überflüssig wurden, sondern beseitigte auch die Schwierigkeit beim Ausdrücken der Falten und das Innendrücken, das immer teuer ist.

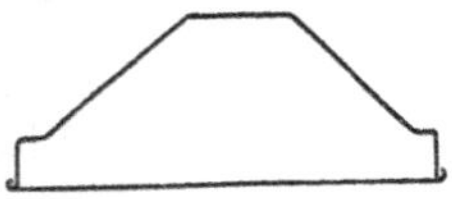
Fig. 452.

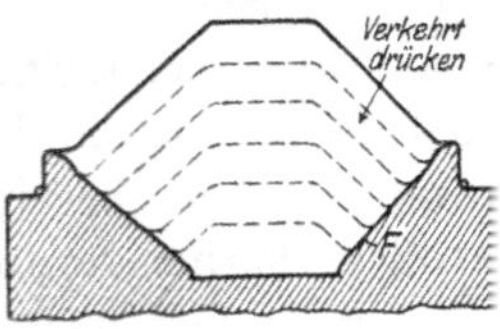

Fig. 453.

Einer der Vorteile des Umkehrens, besonders bei Weißblech, ist, daß die obere oder gedrückte Oberfläche meistens vernickelt oder poliert werden muß; wenn nun das Drückfutter richtig gemacht ist, so ist in der Hülse immer etwas weniger Material, als unbedingt nötig, um das Futter auszufüllen, vorhanden, da sich das Material beim Umkehren, anstatt zu fälteln, dehnt und die obere Fläche glatt und sauber bleibt.

2. Formänderung ohne Materialverschiebung.

Das Biegen.

Beim Biegen wirken alle äußeren Kräfte, also auch die Auflagerdrücke, winkelrecht zur Schwerachse des Balkens, es kommen also äußere Kräfte in der Balkenachse nicht vor. Es liegen auch alle Kräfte und Auflagerdrücke in einer Ebene durch die Balkenachse. Wenn nun der Balken senkrecht zu dieser Ebene eine beliebige Abmessung hat, so ändert diese nichts an der Kraft- und Spannungsverteilung.

Diese Erklärung der Mechanik für das Biegen gilt auch in der Preßtechnik. Indessen gelten die Formeln der Mechanik, die für innerhalb der Elastizitätsgrenze liegende Belastungen aufgestellt sind, hier nur im Beginn der Bearbeitungsvorgänge. Sie werden im weiteren Verlauf durch bleibende Formänderungen innerhalb des Gebietes $S\,B\,\varepsilon_S\,\varepsilon_B$ und $-S\,-B\,-\varepsilon_S\,-\varepsilon_B$ im Schaubild, Fig. 200, ersetzt, da bei jeder Biegung die auf der einen Seite der neutralen Linie liegenden Fasern gezogen und die auf der andern gedrückt werden. Beim Biegen im Sinne der Preßtechnik wird ein an einem oder an mehreren Punkten (auf der Matrize) gelagerter „Stab" (das Blech) durch eine äußere Kraft (den Biegestempel) an einem oder mehreren Punkten, bzw. auf seiner ganzen Länge belastet, bis er bleibende Durchbiegungen aufweist. Ihre endgültige Form ist aber nicht wie in der Mechanik die elastische Linie oder die frei unter den Kräften sich einstellende bleibende Durchbiegungslinie, sondern wird durch die Form des Biegestempels und der Matrize erhalten, die das zu biegende Blech zwischen sich fassen und ihm beim Aufsitzen ihre Form geben. Dabei kann die Biegung örtlich, also nur an einer Kante, oder über die ganze Breite des Bleches, also mit einer allmählichen Abrundung, vor sich gehen. In den meisten Fällen, besonders aber, wenn es sich um härteres Material handelt, wird nach dem Aufhören des Pressendruckes ein Zurückfedern des gebogenen Bleches stattfinden, so daß man, um eine bestimmte, bleibende Form zu erhalten, eine etwas größere Formänderung als die verlangte durchzuführen hat, damit sich das Stück nach dem Zurückfedern auf die richtige Weite öffnet. Wenn z. B. aus einem harten Draht ein U gebogen werden soll, so wird man die beiden Schenkel unter Winkeln, die kleiner als 90° sind, zusammenbiegen.

Bei dieser Art der Verarbeitung werden die kleinsten Teilchen des Materiales insoweit nicht gegeneinander verschoben, als die Länge der neutralen Faser gleichbleibt, während bei den im vorhergehenden Absatz besprochenen Arbeiten, dem Ziehen, Prägen und Börteln, diese Länge durch die Verarbeitung geändert wird.

Diese ist auch der Unterschied zwischen dem Pressen im Biegegesenk nach vollendeter Biegung, wenn der Stempel in der Matrize zum Aufsitzen kommt, und dem Pressen, wie es vorher besprochen wurde, wo unter dem Pressendruck eine Umformung des Materiales vor sich geht.

Die Gleiterscheinungen, die als Grundlage jeder bleibenden Formänderung angenommen worden sind, werden dadurch aber nicht berührt.

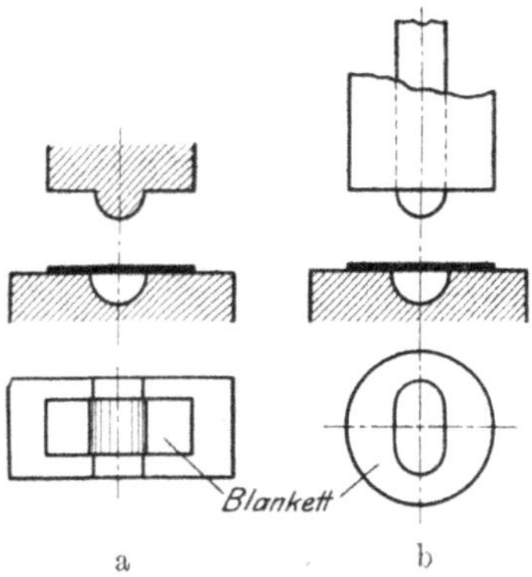

Fig. 454.

Der entwickelte Unterschied wird am besten verständlich, wenn man die Biegung eines U-förmigen Bleches, Fig. 454 a, mit dem Ausprägen einer Rippe aus einer Platte, Fig. 454 b, vergleicht. Im Schnitt sehen beide Werkzeuge fast gleich aus, auch ergibt ein Schnitt durch die beiden fertigen Körper dasselbe Bild. Während jedoch in Fig. a beim Biegen das Material, von der Verlängerung der Außen- und Verkürzung der Innenfaser an der gebogenen Stelle abgesehen, ohne Änderung seiner ganzen Länge in einem einfachen Biegungsvorgang dem Stempel folgt, wird in Fig. b die Platte festgehalten und unter der Einwirkung des nachfolgenden Prägestempels das darunter befindliche Material gezwungen, sich in diesem Teil zu strecken, so daß nachher die Länge eines Mittelschnittes um $(d + \delta)\left(\frac{\pi}{2} - 1\right)$ größer sein wird, wenn d der Durchmesser der Abrundung des Prägestempels und δ die Blechstärke ist.

a) Das Einfachbiegen.

Die einfachste Arbeit dieser Art ist das Biegen eines gleichseitigen Winkels unter 90°, wie in Fig. 455, 1—3, dargestellt ist.

Das Biegen des Stückes an und für sich ist einfach genug, es ist jedoch notwendig, daß der Scheitel des Winkels genau in die Mitte des Bleches fällt, und daß die Stücke austauschbar werden. Das in 2 dargestellte Werkzeug wird so lange gebraucht werden können, als die Biegekante des Stempels scharf und die arbeitenden Kanten der Matrize glattbleiben werden. Sobald jedoch der Stempel stumpf wird und die Biegekanten sich abgenutzt haben, besteht keine Sicherheit, daß ein Werkzeug dieser Form dauernd zufriedenstellend arbeitet

In Fig. 455-3 ist eine Matrize dargestellt, die tatsächlich obige Schwierigkeiten zum großen Teil überwindet. Die größeren Anschaffungskosten werden durch die Austauschbarkeit der Arbeitsstücke und die Lebensdauer des Werkzeuges wettgemacht. Der Stempel A, Fig. 455-2, ist in beiden Werkzeugen gleich, C sind die Anschläge, D das zu biegende Blech und Fig. 455-1 der fertige Winkel. Die Biegematrize ist, wie Fig. 455-3 zeigt, mit einer durchgehenden Nut versehen, in der der Auswerfer E leicht gleitet. Dieser Auswerfer hat einen zylindrischen Zapfen L, der durch die Matrize geht. Am unteren Ende dieses Zapfens sind zwei Anschlagmuttern K aufgeschraubt, um die Hubhöhe des Auswerfers zu begrenzen. Die Grundplatte J ist unterhalb des Auswerfers durchbohrt und dort mit einem eingeschraubten Messing- oder Eisenrohr versehen. Der Auswerfer wird von einer Feder G nach oben gedrückt, bis die Mutter K die Unterfläche der Biegematrize berührt. Die Oberkanten des in den Auswerfer eingeschnittenen Winkels müssen dann mit der Oberfläche der Matrize bündig stehen, wie aus der Zeichnung ersichtlich ist.

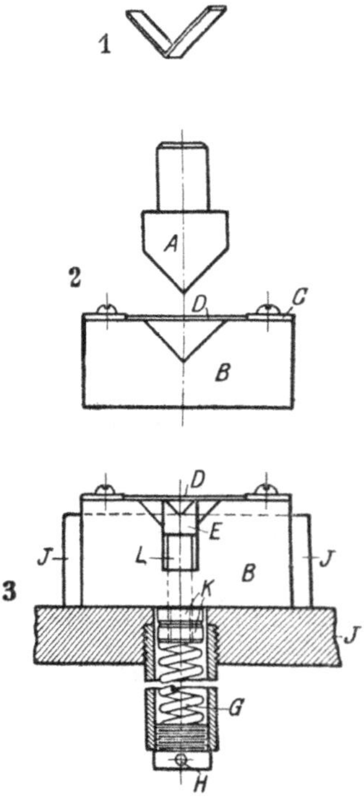

Fig. 455.

Die Wirkungsweise ist selbstverständlich. Betreffs der Herstellung der Matrize ist zu beachten, daß die Nut gestoßen und das Loch für den Auswerfer gebohrt werden soll; auch soll der Auswerfer mäßig hart eingetrieben werden, bevor der Winkel gehobelt wird.

Dieses Hilfsmittel, einen Auswerfer in einem Biegegesenk zu verwenden, ist sehr vorteilhaft und sollte überall dort angewendet werden, wo es das herzustellende Stück erlaubt. In dem eben beschriebenen Stück befindet sich der Winkelscheitel in der Mitte des Stückes, so daß die beiden Winkelschenkel unter 45° zur Oberfläche der Matrize liegen können.

Fig. 456 stellt einen Winkel dar, dessen Schenkel im Verhältnis 1 : 4 ungleich lang sind. Dies bedingt, daß die Winkel in der Matrize von 45° verschieden sein müssen. Es ist klar, daß ein Werkzeug nach Fig. 455-2 in diesem Falle nicht verwendet werden kann. Obwohl man ein Werkzeug nach Fig. 455-3 benutzen könnte, würden sich dabei doch Schwierigkeiten bei der Einstellung der Anschläge ergeben, so daß man besser den Winkel von 45° aufgibt.

Ein der ersten Anordnung 2 entsprechendes, aber den geänderten Verhältnissen angepaßtes Werkzeug zeigt Fig. 456-5, doch erlaubt die Verwendung des Auswerfers eine bessere Lösung. Der Stempel ist wieder in beiden Fällen derselbe. Die Winkel zwischen den Schenkeln des Ausschnittes und der Matrizenoberfläche hängen von dem Verhältnis der Schenkellängen ab und sind in dem vorliegenden Fall 80° und 10°. Die beiden Löcher, die in dem längeren Schenkel ersichtlich sind, werden vorteilhaft als Anschläge ausgenutzt, wie Fig. 456-6 zeigt, wo das Stück vor dem Biegen auf dem Auswerfer liegt. Entsprechend den zwei Löchern in dem zu biegenden Stück werden zwei Prisonstifte in dem Auswerfer befestigt, deren Stellung selbstverständlich genau ermittelt werden muß. Gleichzeitig müssen in dem Stempel entsprechende Aussparungen vorgesehen werden, da die Stifte durch das Blech hindurchtreten müssen. Die Stifte werden oberhalb der Blechdicke zugespitzt, um das Einlegen des Bleches zu erleichtern.

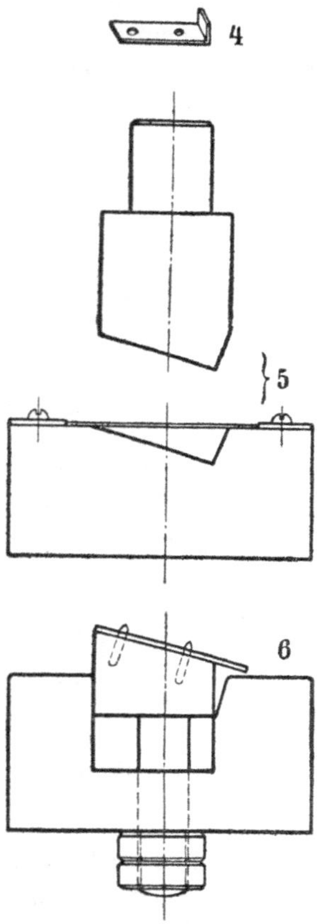

Fig. 456.

Bei Verwendung eines Werkzeuges dieser Form wird vorausgesetzt daß das Blech bereits auf Maß geschnitten und gelocht ist. In Fig. 568 ist jedoch ein einziges Werkzeug für alle drei Arbeiten, Schneiden, Lochen und Biegen, wiedergegeben.

Biegewerkzeuge für die Grundplatte eines Hausweckers. (C. Lorenz, Berlin SO.) Die in Tabelle Nr. 1—I angegebene Grundplatte wird im 3. und 4. Arbeitsgang in den beiden einfachen Biegewerkzeugen nach Fig. 457/8 bearbeitet. Der lange rechteckige Streifen links im Blankett (1. Gang) muß zusammengebogen und dadurch verstärkt werden, da er den Hammer zu tragen hat. Da die volle Biegung der Hälfte dieses

Endes um 180° nicht leicht in einem Werkzeug an dem für diese Arbeit etwas unhandlichen Stück ausführbar ist, wird das Ende im Werkzeug, Fig 457, auf einen spitzen Winkel vorgebogen und dann mit einem flachen Stempel im zweiten Werkzeug flach umgelegt, so daß das linke Ende jetzt zwei Materialdicken enthält, in die man das Gewinde für die

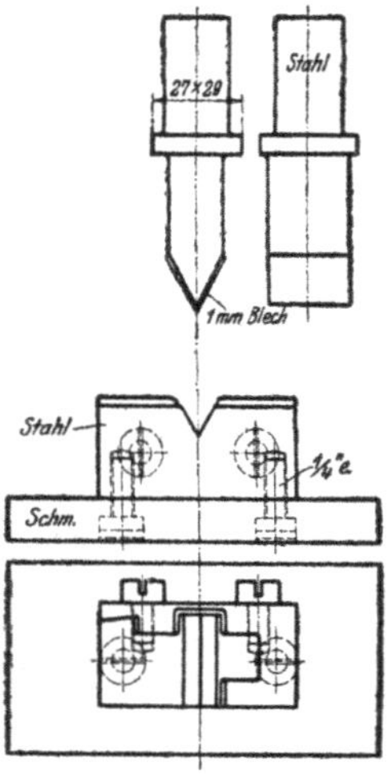

Fig. 457.

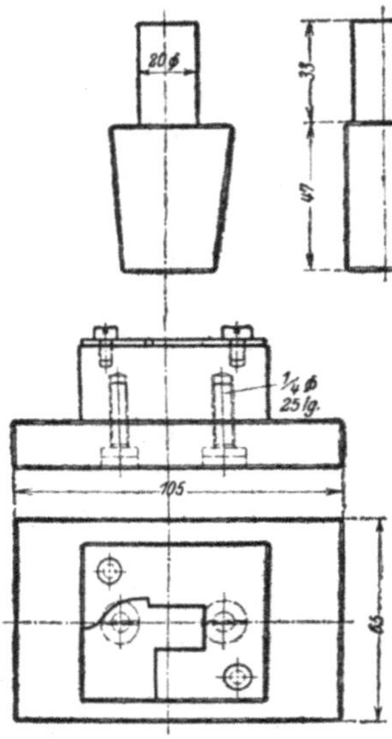

Fig. 458.

Befestigungs- und Kontaktschrauben gut einschneiden kann. Die Ausführung der Werkzeuge entspricht der einfachen Bauart in Fig. 455-2, doch ist die eigene Form der Anschlagplatte, die durch die Form des Blanketts bestimmt ist, bemerkenswert. Da die Biegung in der Hälfte des Blechendes und rechtwinklig zu dessen Länge sein muß, damit später beim Aufstellen das umgebogene Ende nicht mit einer Ecke an die Grundplatte stößt, sind die benachbarten Teile des Blanketts in beiden Werkzeugen mit als Anschlag verwendet worden.

Ebenfalls zum Hauswecker gehört ein einfaches Biegewerkzeug, Fig. 459, das den kleinen linken Fortsatz des Hakens nach dem Ausschneiden umbiegt. Das Werkzeug besteht aus einer Grundplatte, in welche ein schmaler Schlitz, der in seinen Abmessungen dem Querschnitt dieses Fortsatzes entspricht, eingearbeitet ist. Im oberen Teile ist der Ausschnitt rechteckig, so daß der Haken sicher gehalten wird, im unteren Teil jedoch wird das Loch auf einen Durchmesser, der der größten Breite des Fortsatzes entspricht, ausgebohrt, so daß man den umgebogenen Haken, wenn nötig, leicht von unten herausstoßen kann. Unmittelbar neben diesem Schlitz ist ein Hebel angelenkt, der beim Herunterschlagen den mit seinem Fortsatz in dem kleinen Schlitz steckenden, aufrechtstehenden Haken umbiegt, so daß aus dem neben der Zeichnung abgebildeten flachen Blankett der fertige Haken nach Fig. 95 wird.

Ein selbsttätiges Biegewerkzeug. An dem Blankett A, Fig. 460, sollen in einem Arbeitsgang die beiden Fortsätze nach rechts und links unter rechtem Winkel umgebogen werden, Fig. B. Obwohl es zwei ganz einfache Biegearbeiten sind, bedingt das gleichzeitige Biegen beider Enden ein verwickelteres Werkzeug, Fig. 461, als die bisher besprochenen einfachen Anordnungen. Es handelt sich um ein Zwischenstück, das bei der Herstellung von Armaturen verwendet wird, um die Ventilationsöffnungen festzulegen. In Fig. 461 ist eine Seitenansicht dieses Werkzeuges zum Teil im Schnitt in dem Punkte des Abwärtsganges gezeichnet, wo die Greifer das Blech zu fassen beginnen. Die Hauptteile des Werkzeuges sind die Matrize E aus Werkzeugstahl und der Stempel D, der in einer Nut des gußeisernen Stempelkopfes C mit Schrauben gehalten wird. In die Matrize wird der Länge nach eine Nut mit einer rechtwinkligen und einer schrägen Seitenfläche gefräst, wie aus der Figur ersichtlich ist. In diese Nut wird ein Keilstück f eingepaßt, das an der senkrechten Seitenfläche Platz läßt, um das Blech stramm zu halten. In dem Keil sind kleine Stifte befestigt, an die die Enden der Federn g′ eingehängt werden. Diese Federn werden um die Schrauben j gewickelt und mit ihren äußeren Enden unter kleinen Kopfschrauben befestigt. Diese Federn heben den Keil auf und entfernen ihn von dem Blech, wodurch es nach dem Biegen frei wird und ein neues eingelegt werden kann. Die Bewegung des Keiles wird durch die Schrauben h h′ begrenzt, die lose in der Matrize E gehen und im Keil f eingeschraubt sind.

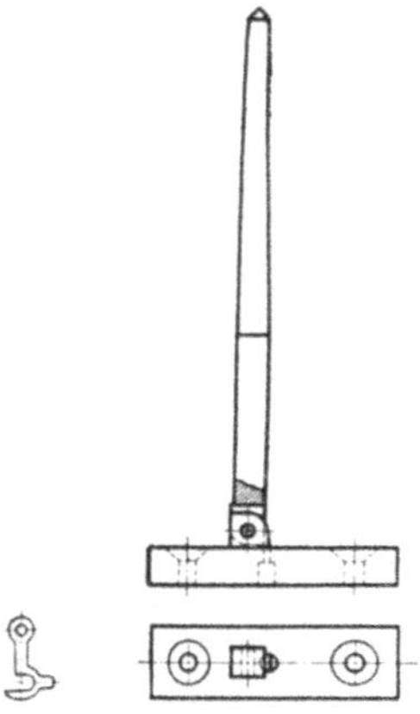

Fig. 459.

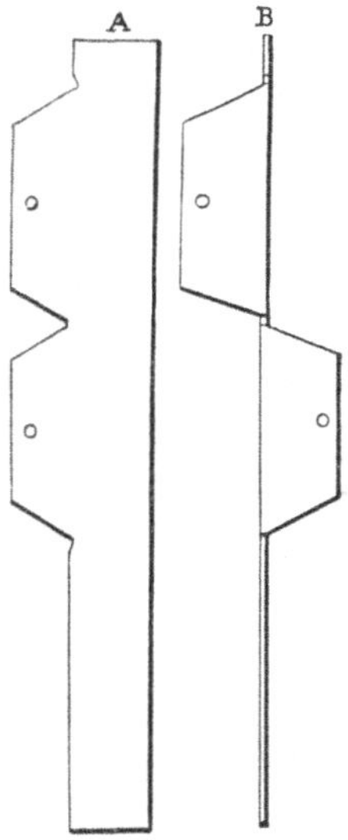

Fig. 460.

Das flache Blankett wird nun zwischen Keil und Matrize eingeschoben, bis es an einen Anschlag (der nicht gezeichnet ist) stößt, durch den seine richtige Stellung gesichert wird. Beim Niedergange des Stößels drücken die Stempel e e′ den Keil in die Matrize und spannen so das Blech fest. Die Stempel müssen lang genug sein, um dies zu tun, bevor die Greifer a a′ das Blech fassen, ebenso müssen die Federn i i′ dieser Stempel stark genug sein. Der Greifer a dreht sich auf dem Bolzen b, der auf der der Biegung entgegengesetzten Seite des Bleches

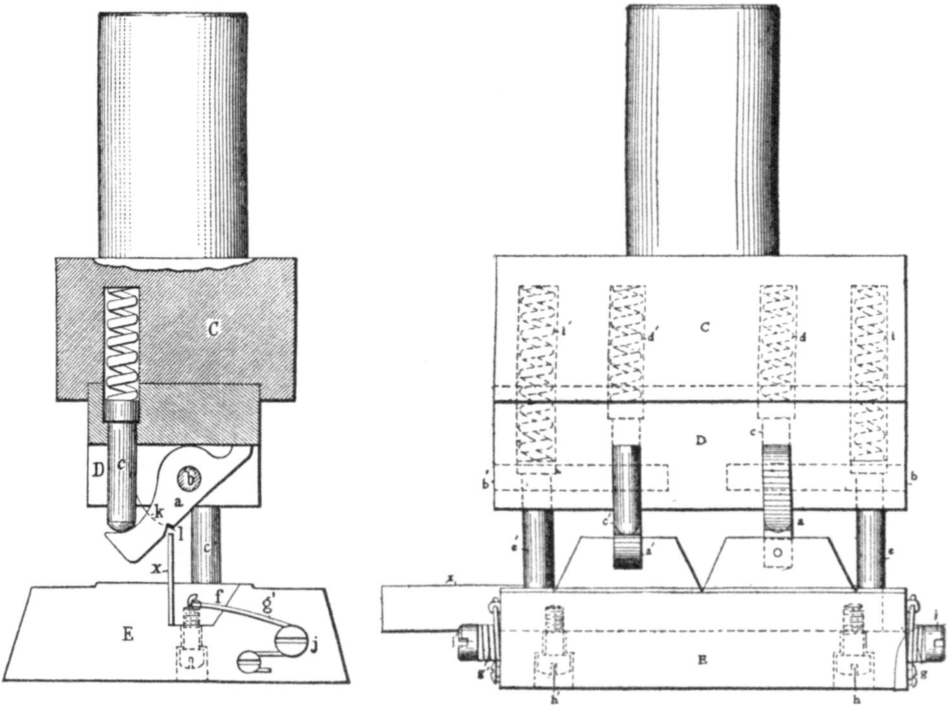

Fig. 461.

liegt. Der kleine Arm am entgegengesetzten Ende dient als Anschlag, um die Nut bei l genau über die Blechkante zu stellen, während die Stempel c c′, die durch schwache Federn d d′ betätigt werden, auf die Oberseite des unteren Endes drücken, so daß die Greifer a a′ nach jedem Pressenhub in ihrer Anfangsstellung gehalten werden. Die obere Blechkante wird von der Nut im Greifer beim Abwärtshub erfaßt und in der Richtung der gestrichelten Linie bei k seitlich abgebogen, bis sie die Stirnfläche des Stempels D erreicht, der den oberen

Lappen weiter niederbiegt, bis er durch den Schließdruck rechtwinklig zu dem in der Matrize festgehaltenen Teil fertiggebogen wird. Da die beiden oberen Blechlappen in entgegengesetzten Richtungen abgebogen werden müssen, so haben die beiden Greifer a a′ jeder in anderer Richtung zu arbeiten, so daß die Stifte b b′ sowie die Stempel c c′ auf entgegengesetzten Seiten der Mittellinie stehen. Alle Stempelteile sind gehärtet, die Greifer müssen leicht auf den Zapfen b b′ und in den Nuten des Stempels gehen, damit sie unter Druck nicht klemmen. Früher wurde das Blech in zwei Arbeitsgängen gebogen.

Biegewerkzeug für die Grundplatte eines Hausweckers. (C. Lorenz, Berlin SO.) Die Grundplatte des Hausweckers (Tabelle Nr. 1-I, G. 6) wurde im zweiten Arbeitsgang in der Mitte rechteckig ausgeschnitten, doch blieb der Ausschnitt mit einer seiner Längsseiten in Verbindung mit dem Stück. Dieser Ausschnitt wird nach dem Ziehen der Grundplatte im 6. Gang aufgebogen, so daß er senkrecht zur Grundplatte steht, und dient dann als Joch und Träger für die beiden Magnetkerne, die darin vernietet werden. Wegen des eigentümlich geformten Umrisses der Grundplatte und der längs dieses Umrisses hochgezogenen Kanten kann ein einfaches Biegewerkzeug nicht verwendet werden, obwohl die Biegung als solche einfach genug ist. Tatsächlich ist der Arbeitsvorgang mehr ein Ziehen einer geraden Fläche, ohne daß eine Materialverschiebung dabei auftritt. Es ist eine Umkehr der gebräuchlichen Anordnung, da der Stempel in den Unterteil und die Matrize in den Oberteil verlegt worden sind. Die gezogene Grundplatte wird auf den im Schnitt gezeichneten Federauswerfer aufgelegt, dessen Umfang dem inneren Umfang der Grundplatte entspricht, worauf der Werkzeug-

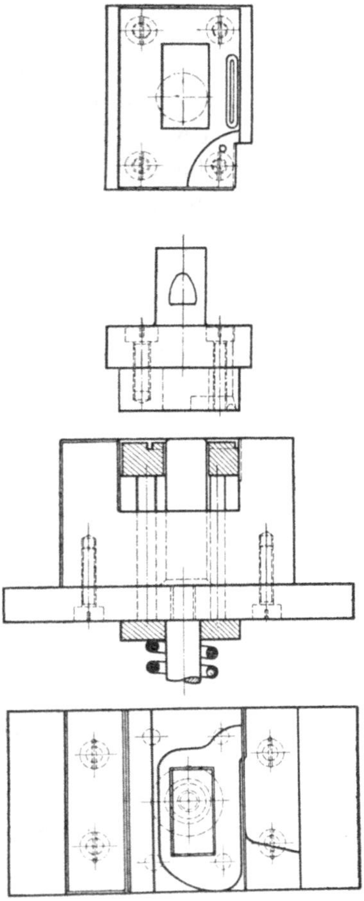

Fig. 462.

oberteil niedergeht, das Arbeitsstück auf dem Auswerfer festhält und mit diesem gleichzeitig weiter nach abwärts geht und in die Grundplatte eintritt. Dabei tritt der feststehende Stempel unter den auf drei Seiten ausgeschnittenen Teil der Grundplatte, biegt diesen hoch, wobei beide zusammen in einen inneren entsprechenden Ausschnitt des Oberteiles eintreten. Gleichzeitig wird der im Werkzeug nach Fig. 458 flach umgebogene Fortsatz ebenfalls rechtwinklig zur Grundplatte aufgestellt, indem er zwischen die äußere Seitenwand des Unterteils und eine rechts oben im Grundriß ersichtliche Aussparung im Oberteil hineingebogen wird. Stempelkopf und Grundplatte sind aus Maschinenstahl, der „Stempel", also hier die Matrize, der Auswerfer und der Stempel im Unterteil sind gehärtet. Es scheint, daß die Stempelkante, die den seitlichen Fortsatz bei e, Tabelle Nr. 1-I, G. 6, aufbiegt, der gefährlichste Teil ist, wie das eingesetzte Stück an dieser Stelle beweist.

Rundbiegen. Die bisher behandelten Werkzeuge haben sämtlich durch Biegen eine mehr oder weniger scharfe Kante in dem Blech erzeugt. Wenn aber statt des scharfkantigen ein runder Stempel mit einer gleichen Matrize zusammenarbeitet, so werden durch Biegen auch Zylinderflächen und weiterhin zylindrische Flächen mit einer beliebigen krummen Linie als Leitlinie erzeugt, deren Achse senkrecht zur Bewegungsrichtung des Biegestempels liegt.

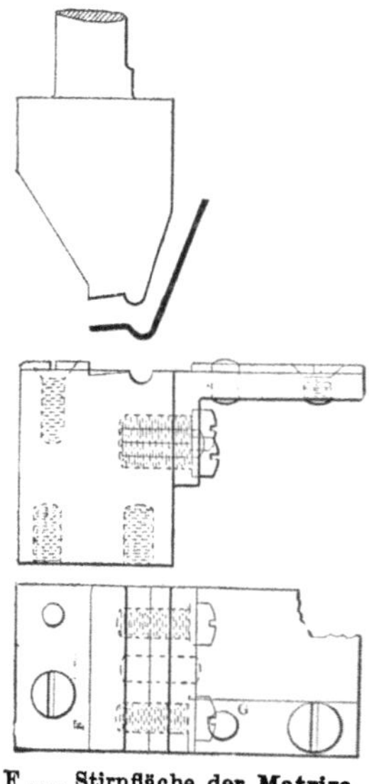

F ... Stirnfläche der Matrize.
G ... Anschlag.

Fig. 463.

Ein einfaches derartiges Werkzeug ist in Fig. 463 abgebildet, welches den zweiten Arbeitsgang des in Tabelle Nr. 32 dargestellten Stückes herstellt. Das Material für dieses Stück ist blank gewalzt und 1 mm dick. Bei sämtlichen Arbeiten an diesem Stück mußte man sehr sorgfältig darauf achten, die vorgeschriebenen Maße und Krümmungen einzuhalten.

Nach dem Ausschneiden und Lochen, das in einem normalen Folge- oder Verbundwerkzeug vor sich gehen kann, wird das rechte Ende des Blanketts in dem Biegewerkzeug, Fig. 463, zur Hälfte vorgebogen, damit man auf dem Einrollwerkzeug nach Fig. 424 in der einfachsten Weise arbeiten kann. Auch das Biegewerkzeug wird in der einfachsten Form gebaut; das Blech wird gegen F und den Anschlag G gelegt, worauf der halbrunde Stempelfortsatz das Blech in die entsprechende Vertiefung der Matrize eindrückt. Das

Ende des Bleches bleibt gerade und wird beim Biegen von der entsprechenden Stempelfläche gerade gehalten. Dieses gerade Ende ist die Ursache, daß das Einrollen in zwei Stufen, einem Vorbiegen und einem Einrollen, gemacht werden muß.

Rundbiegewerkzeug für kupferne Schellen. In Fig. 464/5 sind die Werkzeuge zur Herstellung von Schellen aus Kupferblech, wie bei A gezeichnet, dargestellt. Diese Schellen müssen genau kreisförmig sein und die Abmessungen A und D genau eingehalten werden, so daß sie genau auf die verschiedenen Pfosten passen. Das in Streifen von der richtigen Breite vorhandene Material wird auf Länge abgeschnitten und dann in die Nute der Matrize E, Fig. 464, eingelegt, worauf der heruntergehende Stempel die erste Form herstellt. Im zweiten Arbeitsgang wird das vorgebogene wellenförmige Blech mit den Enden unter die Stifte I, Fig. 465, und an die Anschläge K gelegt, so daß die Mitte des Stückes auf dem Bolzen H aufruht. Beim Heruntergehen des Stempels J wird das Stück um den Bolzen H gebogen und so in seine endgültige Form A, Fig. 464, gebracht. Alle Teile der Werkzeuge sind aus Flußeisen zu machen; die Stempel C, J, die Matrize E und der Bolzen H sind im Einsatz zu härten.

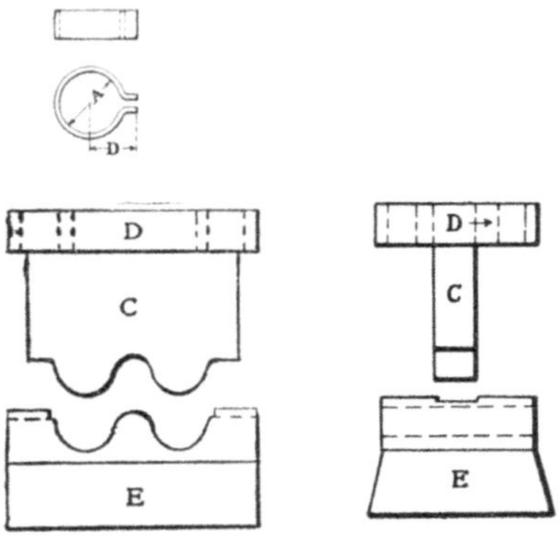

Fig. 464.

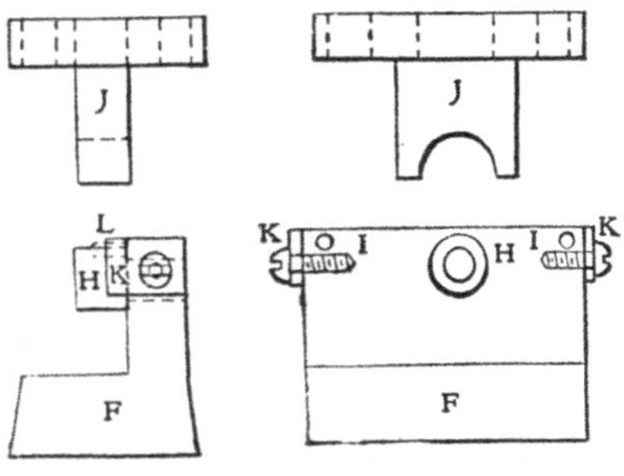

Fig. 465.

Warmbiegen einer Stahlfeder. Die beiden Biegewerkzeuge in Fig. 466/7 dienen, vgl. Tabelle Nr. 27, zur Herstellung einer Feder aus Stahlblech von 11 mm $\times$ 2,4 mm Abmessung in zwei Gängen, in einer Hitze für jeden Gang. Die erste Biegung geschieht im Werkzeug nach Fig. 466 auf die dort durch die Linie x angedeutete Form, die infolge der starken Formveränderung offenbar eine große Hebelübersetzung erfordern muß, falls sie von Hand werden soll. Die erhaltenen Formen müssen überdies nach Lehre stimmen und austauschbar sein. Das Werkzeug soll auch gestatten, durch Auswechslung der Biegebacken verschiedene Krümmungen herzustellen.

Der Körper a besteht aus Grauguß, hat an jedem Ende ein Loch für die Befestigungsschrauben und wird so aufgestellt, daß der Handhebel rechts vom Arbeiter zu liegen kommt. In die Grundplatte ist an der Seite des Gesenkes ein Trog für das Fett bzw. Öl eingegossen, das beim Arbeiten auf die Biegebacken aufgetragen wird; der Arbeiter

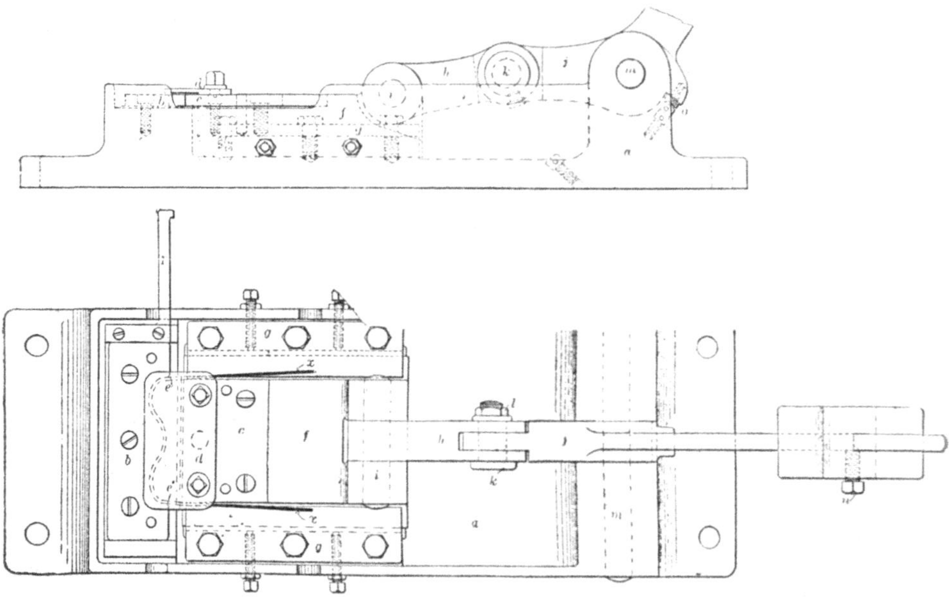

Fig. 466.

hat in seiner linken Hand eine Bürste, mit der er etwas Öl über die Arbeitsfläche des äußeren Backens vor dem jedesmaligen Biegen streicht. Der Stahl darf nur bis gerade auf dunkle Rotglut erwärmt werden.

Die Biegebacken b und c werden aus Werkzeugstahl hergestellt und durch drei Schrauben und zwei Prisonstifte in ihrer Stellung gehalten. Die inneren Ecken von b sind bei e abgerundet, damit das Material nicht eingeschnitten wird, wenn es in den Biegebacken gepreßt wird. Vor der Herstellung der Biegebacken muß ihre genaue Form ermittelt werden, da die Feder nach dem Biegen gehärtet wird. Die Federnlänge wurde in einem Falle errechnet, jedoch kann ihre Form auch durch den Versuch gefunden werden, indem man so lange Stücke biegt und härtet, bis eines die richtige Form erhält. Nach dem einen Biegebacken wird dann der zweite hergestellt. Die Platte d oberhalb der Biegebacken, die im Stempel c durch zwei Schrauben gehalten wird, hat die beiden Enden des Stahles während des Biegens auszurichten; gleichzeitig verhindert sie das Aufwärtsspritzen des Öles, das den Arbeiter verbrennen kann. Der T-förmige Stößel f, der den Biegebacken c trägt, ist aus Gußeisen und allseitig bearbeitet. Er wird durch zwei Stahlschienen g niedergehalten, deren jede mit drei Schrauben in a befestigt ist. Die beiden Stelleisten aus Stahl erhalten Stellschrauben mit Gegenmuttern, um den Stößel zentrisch zu führen und bei Abnützung Nachstellung zu ermöglichen. Am hinteren Ende des Stößels ist ein Kniegelenk vorgesehen, das aus den Hebeln h und j und den drei Bolzen i, k und m besteht. Der Bolzen i ist aus Werkzeugstahl und sitzt im Stößel f mit Preßsitz und im Hebel h mit Gleitsitz, der Zylinderkopfbolzen k dreht sich in h und j und hat am einen Ende Gewinde für die Mutter l. Der Handhebel j schwingt um die Achse m, die mit Preßsitz in j und mit Gleitsitz in a sitzt. Am äußersten Ende des Handhebels j ist ein Belastungsgewicht aus Gußeisen angebracht, welches mittels der Schraube n einstellbar befestigt ist. Dieses Gewicht wiegt ungefähr 20 kg und gestattet in Verbindung mit dem Kniehebel eine leichte Bedienung der Maschine, obgleich die Biegearbeit einen bedeutenden Druck verlangt. Zur Begrenzung der Druckwirkung auf das Metall ist je eine Anschlagschraube o vorgesehen, die die Bewegung des Handhebels j begrenzt. Diese Schrauben aus Werkzeugstahl sind auf Federhärte angelassen. Auch für die Rückzugbewegung des Biegestempels sind Anschläge vorgesehen, wie aus der Zeichnung ersichtlich ist. Der Anschlag p ist mit zwei Schrauben an dem Gesenk festgeklemmt und dient zur genauen Einstellung des Arbeitsstückes vor dem Biegen, so daß beide Enden der gebogenen Feder genau gleichlang werden. Die Wände der Grundplatte sind genügend ausgeschnitten, um die Enden des Stahles durchtreten können.

Der Arbeiter nimmt mit einer Zange, die er in der linken Hand hält, den Stahlstreifen aus dem Feuer und legt ihn gegen den Anschlag p und den Biegebacken. Mit der rechten Hand bewegt er den Handhebel j, wobei durch die erste Bewegung von d das Stück sogleich in einer wagerechten Ebene festgelegt wird, da die Entfernung zwischen Platte d und der Grundplatte dem Arbeitsstück gerade den Durchgang gestattet. Das eigentliche Biegen geschieht sehr schnell.

Das Fertigbiegen der Federn geschieht auf einem Werkzeug nach Fig. 467, das auf einer gußeisernen Grundplatte a aufgebaut ist, die an jeder Ecke ein Loch für die Befestigungsschrauben und an der dem Arbeiter abgewendeten Seite die Handkurbel für das Festklemmen der Feder trägt. Durch eine Umdrehung dieser Kurbel klemmt der

Arbeiter das Arbeitsstück x zwischen e und b und biegt dann durch eine gleichzeitige Einwärtsbewegung der Gleithebel l mit der rechten und linken Hand die Enden von x nach der Form, die durch die dicke Linie angedeutet ist. Der entsprechende Biegebacken b wird in der Mitte der Grundplatte mittels vier Schrauben und zwei Prisonstiften befestigt. Die Herstellung dieses Biegebackens geschieht unter Be-

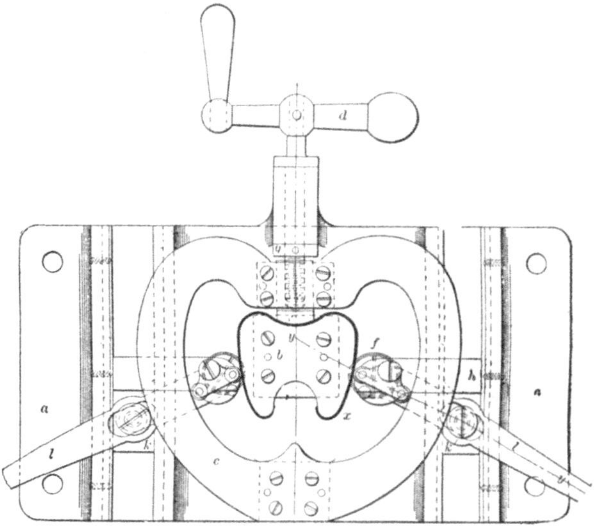

Fig. 467.

obachtung derselben Vorsichtsmaßregeln wie bei dem vorangehenden Werkzeug und in derselben Weise aus Werkzeugstahl, worauf das Stück auf Strohgelb angelassen wird. Der Gleitstein e aus Werkzeugstahl wird in eine in die Grundplatte eingehobelte Nut eingepaßt und mit einer passenden Deckplatte versehen. Er erhält in der Mitte das Gewinde für die Schraube der Handkurbel d. Der feste Biegebacken b ist an die Stirnfläche des Gleitsteins gepaßt, so daß ihr Profil mit der im ersten Arbeitsgang erhaltenen Feder übereinstimmt. An beiden Seiten der Grundplatte a sind Schwalbenschwanznuten unter 45^0 eingehobelt, die Stelleisten und Nachstellschrauben haben. In diesen Nuten bewegen sich die Gleitsteine h, die die Rollenhalter tragen. Sie sind auf ihren eingeschraubten Achsen drehbar und tragen jeder zwei kleine Rollen, deren eine sich auf der Innenseite der Führung c abrollt, während die andere gegen das zu biegende Stück x drückt. Eine Blattfeder f an der einen Seite drückt die Rollenhalter dauernd gegen die Führung c. Rollen, Rollenhalter und Führung sind aus Stahl und im Einsatz gehärtet. Die Handgriffe l sind an den Gleitsteinen h so gelagert, daß sie der Längsbewegung folgen zu können, und zum gleichen Zwecke mit Langlöchern für die Drehachse k versehen.

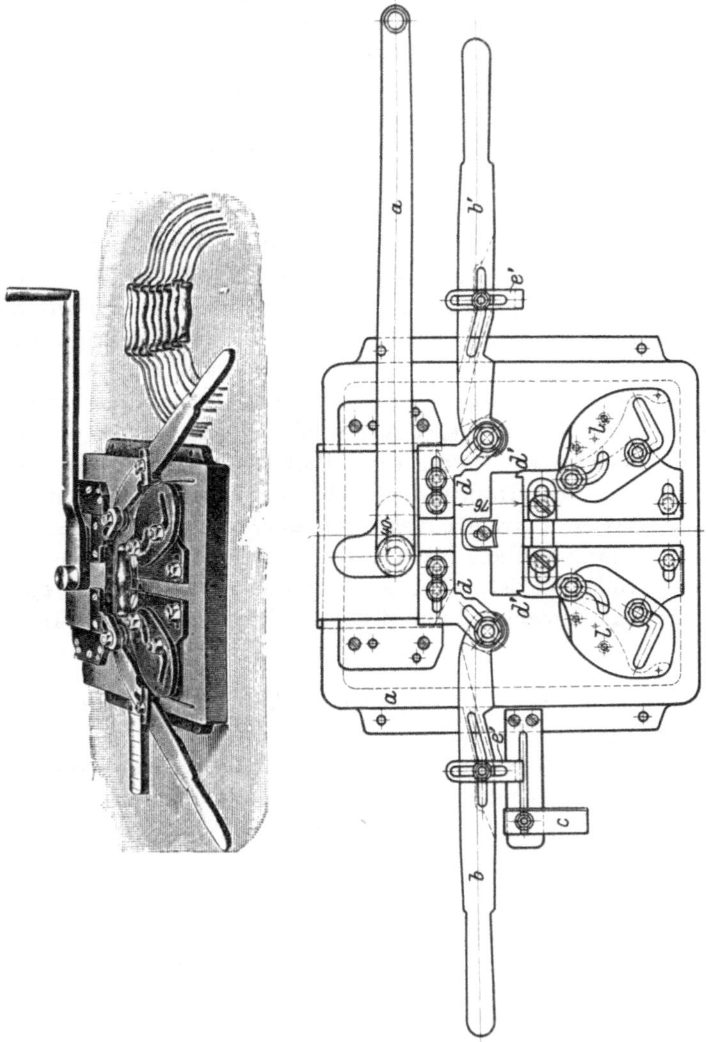

Fig. 468. Eimerbügel-Biegeapparat.

Eimerbügel-Biegeapparat (Erzgebirgische Schnittwerkzeug- und Maschinenfabrik). In Zweck und Ausführung dem eben besprochenen Werkzeug sehr ähnlich, gestattet jedoch dieser Apparat eine weitergehende Einstellung, um verschiedene Bügelbreiten herzustellen, während der mittlere Teil, der den Handgriff aufnimmt, ungeändert bleibt. Das Biegen geschieht hier kalt. Während in der einfachsten, früher gebräuchlichen Arbeitsweise zur Herstellung des Bügels mit eingeklemmtem Handgriff mittels Schablone 5 Arbeitsgänge notwendig waren, genügt mit der in Fig 468 angegebenen Einrichtung ein einziger Arbeitsgang. Mit der nötigen Geschicklichkeit kann ein Arbeiter nach Angaben des Fabrikanten bis zu 600 Bügel in der Stunde biegen. Der ganze Apparat[1]) ist auf einer gußeisernen Grundplatte a, die mit Löchern für die Befestigungsschrauben versehen ist, aufgebaut. Es sind zwei Sätze fester Biegebacken vorhanden, von denen der eine, rückwärtige, d′, das Einbiegen des Drahtes zu beiden Seiten des Holzgriffes besorgt, während die zwei symmetrischen vorne liegenden Teile l den Seiten des Bügels die Form geben. Alle Teile sind entsprechend der zu verarbeitenden Drahtstärke und Griffgröße einstellbar, wofür am Apparat Marken vorgesehen sind. Der bewegliche Backen d des rückwärtigen Werkzeuges ist in einer auf der Grundplatte aufgeschraubten Schlittenführung gehalten und wird darin mittels des Handhebels g nach vorne gegen den festen Backen bewegt, wodurch der Draht zu beiden Seiten des Griffes eingeknickt wird. Hierauf werden die beiden Hebel b b′, welche die beweglichen Biegebacken e′ zur Herstellung der Krümmung der Drahtenden nach der Form der beiden festen vorderen Backen l tragen, von ihrer der vorderen Werkzeugkante parallelen Anfangsstellung nach vorne gezogen, womit das Biegen des Bügels vollendet ist. Ein der Bügellänge entsprechend einstellbarer Anschlag c sichert beim Einlegen die richtige Drahtlänge.

Biegewerkzeug mit Anschlagstiften. An dem Schließteil, dessen vollständiger Herstellungsgang in Tabelle Nr. 31 gegeben ist, sollen im zweiten Arbeitsgang die bereits beim Ausschneiden angebogenen Enden, welche nach dem Einrollen das Scharnier bilden, unter rechtem Winkel abgebogen werden. Da das Stück bereits vorgelocht worden ist und die Scharnierhälften genau rechtwinklig zur Längsachse des Teiles stehen müssen, so werden hier in der Matrize Stifte angebracht, die das Blankett vor dem Biegen ausmitteln. Bei dem Abwärtsgang des Biegestempels L treten die Stifte allmählich in ihn ein, so daß das schmale Blankett nicht herausspringen kann. Während der Stempel sich auf die Matrize D aufsetzt, biegt er

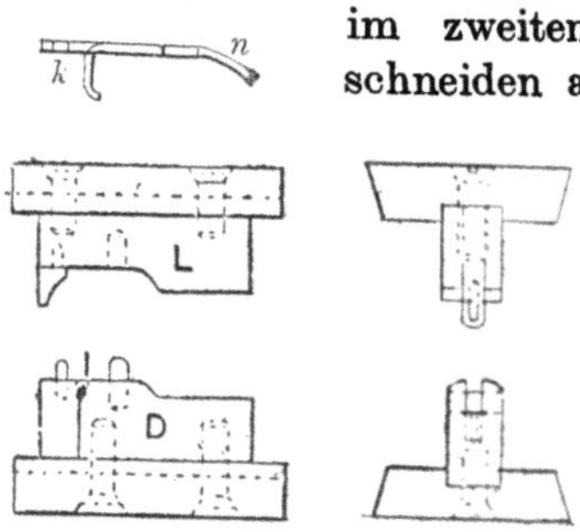

Fig. 469.

[1]) Vgl. W. T. 1910, S. 500.

das eine Ende bei n rund ab und die beiden seitlichen Fortsätze, die Scharnierenden k, nach abwärts unter rechtem Winkel.

Beiderseitiges Aufbiegen eines Bleches. Das Blankett für den eigentümlich geformten Teil, der nach Tabelle Nr. 32 hergestellt wird, wird nach dem Einrollen des Auges, Fig. 424, im vierten Gang auf beiden Seiten hochgebogen, so daß ein U-förmiger Querschnitt entsteht. Dazu dient ein Werkzeug nach Fig. 470. In diesem Arbeitsgang müssen die Blechkanten, die im letzten Arbeitsgang scharf umgelegt werden sollen, bereits gebrochen werden. Das Blankett wird gegen die Anschlagplatte eingelegt und zwischen den Federstiften im Stempel und dem Auswerfer, der hier auf einer Gummifeder gelagert ist, gehalten. Beim Niedergang drückt

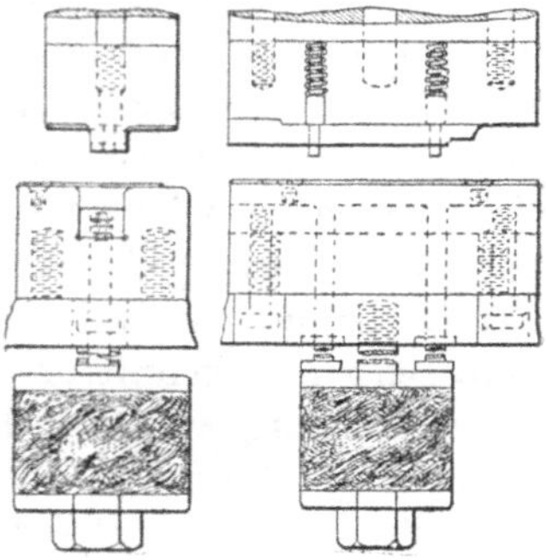

Fig. 470.

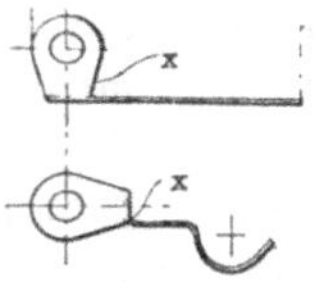

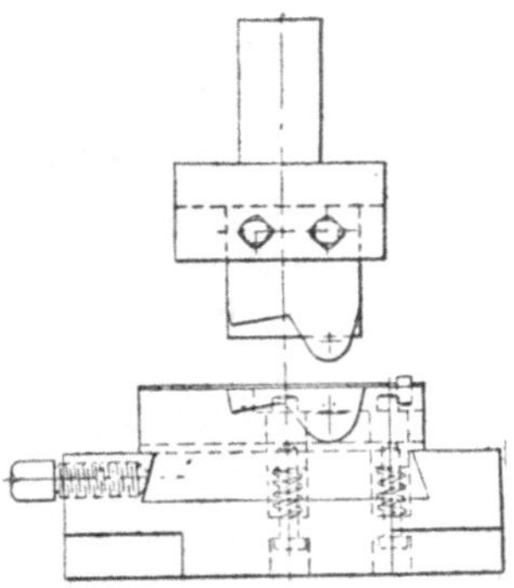

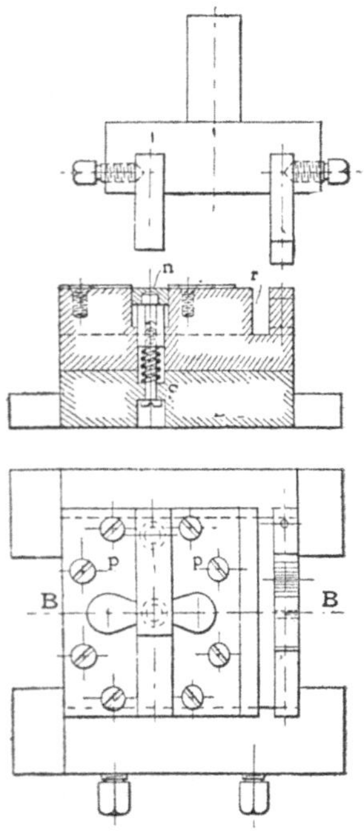

Fig. 471.

der Stempel das Stück in die Vertiefung der Matrize, wobei gleichzeitig die Gummifeder zusammengepreßt wird, während das Blech zwischen dem mittleren Stempelfortsatz und dem Ausschnitt in der Matrize scharf rechtwinklig umgelegt wird. Nach Rückgang des Stempels drückt die Gummifeder unter dem Auswerfer das Stück aus dem Werkzeug heraus.

Doppeltes Biegewerkzeug. Das Werkzeug in Fig. 471 vereinigt eigentlich zwei vollständig getrennte Einzelwerkzeuge in einem Stempel und einer Matrize. Es dient zur Herstellung der beiden Biegungen an dem in Fig. 117/8 mit seinem Abfallstreifen dargestellten Stück, das auf dem in Fig. 119 abgebildeten Folgewerkzeug ausgeschnitten worden ist. Hier sollen erstens die beiden seitlichen Augen unter einem rechten Winkel aufgestellt werden und dann die Grundfläche des Stückes bei x rechtwinklig abgeknickt und an der mit dem Kreuz bezeichneten Stelle rund gebogen werden. Das Werkzeug ist an und für sich einfach. Zum Biegen der beiden seitlichen Lappen dient der linke Teil des Werkzeuges mit dem Federauswerfer n und den Anschlägen p, die auf die Stirnfläche der Matrize aufgeschraubt sind und mit den Lappen entsprechenden Ausschnitten versehen sind. Die Matrize für die zweite Biegearbeit ist glatt nach der entsprechenden Linie ausgeschnitten und durch einen Schlitz r, der das Einlegen und Entfernen des Stückes erleichtert, von dem ersten Teil getrennt. Zu bemerken wäre, daß die beiden Biegestempel in den Stempelkopf eingepaßt und durch Klemmschrauben darin gehalten werden.

Stanzen, Biegen und Lochen von Goldmaterial. In Fig. 29 ist der Herstellungsgang eines kleinen U-förmigen Stückes aus Gold dargestellt, das sowohl mit großer Genauigkeit geformt als auch, wie bei B sichtbar, mit scharfen Ecken versehen werden sollte. An dem Stück sind drei Arbeitsgänge auszuführen, und zwar: zuerst das Ausschneiden des Blanketts A, zweitens das Biegen nach B und schließlich die Herstellung des kleinen Loches im Grund bei C. Das Material ist 0,5 mm dick und wird nach dem Anlöten an ein Stück Schmuck poliert. Man versuchte mehrere Wege beim Biegen, die alle nicht den gewünschten Erfolg gaben, da man zuviel nacharbeiten mußte, um das erforderliche Fertigaussehen zu erhalten Schließlich gelangte man auf die im folgenden beschriebene Weise zum Ziele.

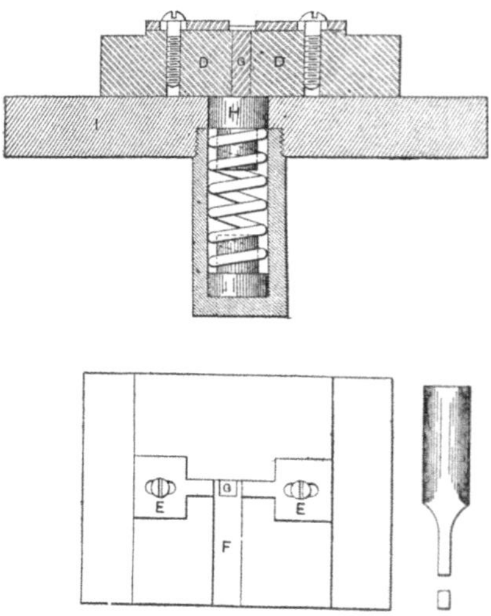

Fig. 472.

Das Stück wurde durch die oben angeführten Arbeiten fertiggestellt; um jedoch die genauen Abmessungen des Blanketts zu erhalten, wurde das Stück zuerst gebogen. Das Werkzeug dafür ist in Fig. 472 abgebildet und offenbar sehr einfacher Bauart. Die Matrize ist zur Befestigung der Anschlagplatten auf der Oberfläche gehobelt und mit einer Nut F versehen, um das Einschieben der Blanketts zu erleichtern. Eine zweite Nut, unter rechtem Winkel zu F, dient in Verbindung mit den Enden von E zur richtigen Einstellung des Blanketts für das Biegen.

Um nun ein viereckiges Loch von den genauen Außenabmessungen des gebogenen Stückes zu erhalten, wird auf der Fräsmaschine ein Räumstempel gemacht und die Öffnung, von der Unterseite beginnend, durch die Matrize durchgearbeitet. Beim Räumen eines Loches dieser Art darf man keinen zu großen Span nehmen, sonst erhält man eine rauhe Oberfläche ohne scharf ausgeprägte Kanten. In dem Maße, wie der Räumstempel der Oberfläche der Matrize näherkommt, muß die zu entfernende Materialmenge abnehmen, bis endlich in den Ecken gerade nur noch berührt wird. Nach dem Räumen wird die Matrize mit einem Polierstock poliert und gehärtet, darf jedoch nur wenig angelassen werden. Hierauf wird sie mit einem Messingpolierstock und feinstem Schmirgelpulver hochglanzpoliert. In das viereckige Loch der Matrize wird der Stempel G eingepaßt, der während des Biegens das Auflager für das Arbeitsstück abgibt und nach der Fertigstellung des Stückes als Auswerfer arbeitet. Er ruht auf dem eigentlichen Auswerfer, der aus dem Stempel H und der darunter befindlichen Wurstfeder besteht. I ist die gußeiserne Gesenkplatte, in welche die Messinghülse für den Auswerfer eingeschraubt ist. Der Biegestempel ist in der einen Richtung um die doppelte Metalldicke schmäler als die Öffnung des Loches und in der dazu senkrechten Richtung von gleicher Breite. Mit einem derartigen Werkzeug kann man das Metall mit scharfen Kanten biegen.

Biegewerkzeug für die Hammerfeder eines Hausweckers (C. Lorenz, Berlin SO). In Tabelle Nr. 1-II ist der Herstellungsgang der Hammerfeder (Fig. 109) gegeben, die aus 0,3 mm starkem Federstahlblech gebogen wird. Die Maßzeichnung des dazugehörigen Biegewerkzeuges ist in Fig. 473 gegeben. Stempel und

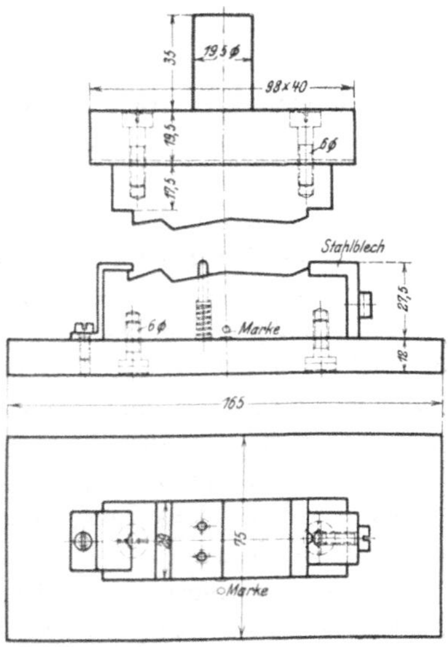

Fig. 473.

Matrize sind aus Stahl, gehärtet und an den arbeitenden Flächen poliert. Sie werden mit versenkten Kopfschrauben im Stempelkopf bzw. der Gesenkplatte befestigt. Da das Stück durch die vielfache Biegung vor dem Biegen nur auf den einzelnen Kanten der Matrize aufliegt, ist eine gute Ausrichtung vorgesehen, die das Stück so lange hält, bis der Stempel an einer Stelle gefaßt hat, so daß eine Verschiebung des Bleches nicht mehr eintreten kann. Das Blankett wird mit den beiden Abrundungen in den aufgeschraubten Anschlagblechen der Länge nach gehalten, während die beiden auf Federn gelagerten Anschlagstifte in der Mitte der Matrize durch den Stempel bei seinem Abwärtsgange niedergedrückt werden, so daß sie bis zum letzten Augenblick das Blankett zwischen sich fassen. Wegen des Zurückfederns des Stahlbleches nach dem Biegen werden in Stempel und Matrize die einzelnen Winkel steiler ausgeführt, als sie im fertigen Stück sein sollen.

Herstellung eines Schiffchenträgers einer Nähmaschine. Der Herstellungsgang dieses Trägers ist in Fig. 474 dargestellt und besteht aus sechs Einzelarbeiten, nämlich Ausschneiden, Prägen, Bohren, Gewindeschneiden und zwei Biegearbeiten. Das Material für das Stück ist kaltgewalztes Stahlblech von 1,6 mm Dicke. In das Loch bei D wird das Gewinde geschnitten, mit dem der Träger am Hebel befestigt wird.

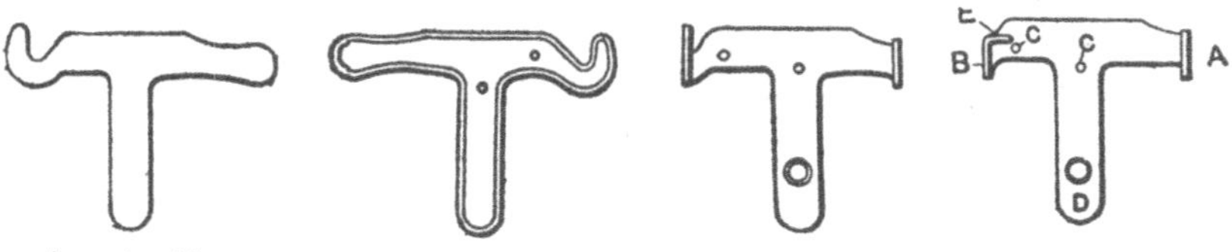

Fig. 474.

Das Ausschneiden und Prägen geschieht mittels der bekannten Arbeitsweisen, es wäre nur zu bemerken, daß Ausschnitt- und Prägestempel hier vorteilhaft aus dem vollen Block geschnitten werden, da das harte, verhältnismäßig dicke Material große Anforderungen an die schmalen Arbeitsflächen der Stempel stellt. Beim Prägen werden die scharfen Schnittkanten nach außen gerundet. Hierauf werden die Träger gebeizt und gescheuert, um allen Grat zu entfernen und die Stücke blank zu machen, worauf die beiden kleinen Löcher in einer eigenen Bohrlehre gebohrt werden. Das Loch D wird dann gesondert gelocht und mit Gewinde versehen.

Das erste Biegen, durch das die Enden A und B rechtwinklig abgebogen werden, erfolgt in dem Werkzeug nach Fig. 475. Man sieht aus der Zeichnung, daß die Kante bei X steiler als unter einem rechten

Winkel abgebogen ist, damit das Blankett nach dem Biegen durch seine Elastizität auf den richtigen, rechten Winkel zurückfedert. Der **Matrizenkörper W ist aus gehärtetem Werkzeugstahl hergestellt und wird in der Nut der Grundplatte Y durch eine versenkte Kopfschraube Z befestigt. Die Stirnflächen des Biegestempels werden bei a gehärtet und geschliffen, wobei die Entfernung der senkrechten Flächen um die doppelte Blechdicke größer als die Matrize ist. Die Anschlagplatte wird aus dem vollen Block herausgefräst. Die zweite Biegung besteht darin, daß an der Ecke B, Fig. 474, das Ende des bereits gebogenen Teiles noch einmal rechtwinklig, bei E, umgebogen**

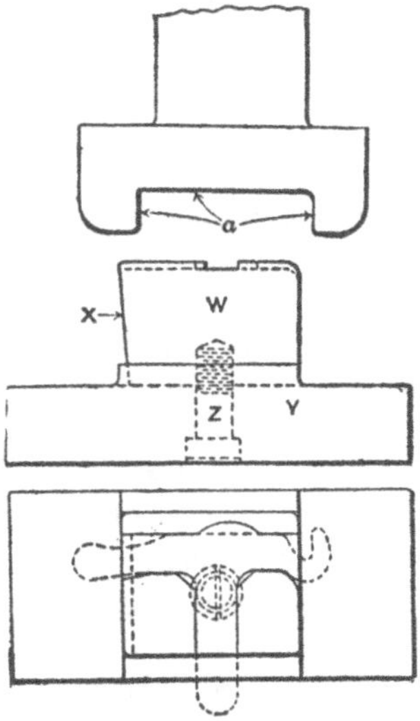

Fig. 475.

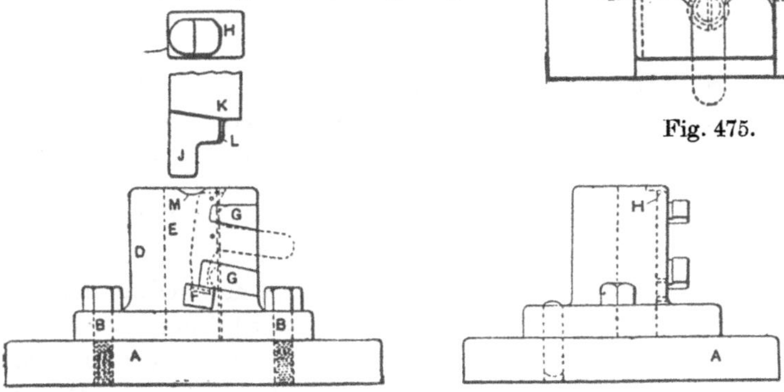

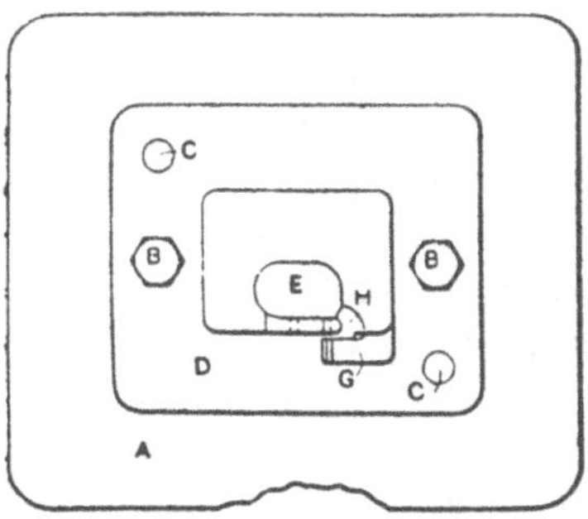

Fig. 476.

wird. **Dabei darf aber die Biegung des übrigen Teiles in keiner Weise verändert werden.** Das **Werkzeug dafür ist in Fig. 476 abgebildet und** macht im **fertigen Entwurf einen** sehr **einfachen Eindruck.** Wenn **es sich aber darum handelt, irgendeine** derartige **Arbeit auf dem einfachsten,** billigsten und **dabei doch sichersten** Wege **auszuführen,** so **liegt fast im**mer **die Schwierigkeit in dem Entwurf eines derartigen Werkzeuges und seltener in der Ausführung.**

Die Grundplatte A aus Gußeisen trägt, mit Schrauben B und Prisonstiften C befestigt, den Matrizenblock D, in den ein senkrechtes

Loch E eingearbeitet ist. An der einen Seite, fast am Grunde des Loches E, ist ein viereckiges Loch F bis in das Loch E durchgearbeitet. Bei G sind zwei Anschläge mit Nasen vorgesehen und eine Ausfräsung bei H und M an der Oberfläche der Matrize. Der Schiffchenträger wird, wie die gestrichelten Linien zeigen, gegen die senkrechte Seite des Matrizenblockes angelegt, wobei sich die bereits umgebogenen Kanten in die Vertiefungen H und F einlegen. Die beiden Anschläge halten mit ihren Nasen den Schiffchenträger fest, wenn der Stempel beim Abwärtsgang mit der Kante L die Ecke E umbiegt. Der Stempel hat eine verlängerte Führung J, die in die eine Seite der Öffnung E eintritt, bevor die Kante L das Blech berührt, so daß der Stempel nicht ausweichen kann. Der Teil K des Stempels glättet die abgebogene Seite und preßt sie in Stellung, worauf der Stempel wieder nach aufwärts geht.

Man muß das fertig gebogene Stück beim Herausnehmen ein wenig anheben und gleichzeitig nach vorwärts schieben, damit das Stück aus den Anschlägen kommt; wenn die Spitze des Stückes über der Öffnung M steht, kann es entfernt werden.

Biegen von Rohren auf verbessertem Wege. Die Hauptschwierigkeit beim Biegen von Rohren ist deren Bestreben, einzuknicken und Falten zu bilden, was bei der im folgenden beschriebenen Arbeitsweise vermieden wird.

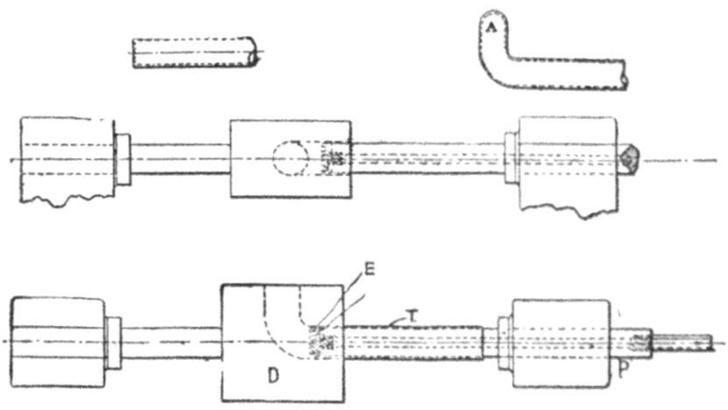

Fig. 477.

Das Rohr in Fig. 477 z. B. sei aus Messing von 45 mm Durchmesser und 1,6 mm Wandstärke und soll unter rechtem Winkel ohne Falten oder andere Beschädigungen gebogen werden; ebenso soll der Krümmungsradius genau eingehalten werden. Zu diesem Zweck wird die Biegematrize aus zwei Hälften hergestellt, nach der verlangten Krümmung und dem Rohrdurchmesser ausgefräst, gehärtet und die Stoßfläche beider Hälften sehr glatt poliert.

Das Rohr A, das an einem Ende geschlossen wird, wird, wie Fig. 477 zeigt, in die untere Matrizenhälfte so eingelegt, daß das geschlossene Ende E am Anfang der Krümmung liegt; dann wird die obere Matrizenhälfte D in ihrer Stellung festgeklemmt und ein Plunger in das offene Ende der Röhre eingeführt. Der Plunger und die Matrize liegen gegen feste Widerlager an. Man läßt nun Druckwasser von ungefähr 400 at ein, so daß das Rohr, welches am vorderen Ende geschlossen ist, durch den Wasserdruck vorwärts und gleichzeitig durch die Krümmung der Matrize geschoben wird. Dabei bleiben die Rohre rund, und der Wasserdruck von innen verhindert die Faltenbildung. Das Rohr wird dann herausgenommen, und der durch die Pressung gerundete Boden bei A abgeschnitten, so daß man einen tadellosen Krümmer erhält. Die Matrize muß aber sehr gut geschmiert werden, da die Reibung an den Seiten sehr groß ist.

b) Das Mehrfachbiegen.

Wie bei den Schnittwerkzeugen gleichzeitig mehrere voneinander vollständig getrennte Schneidvorgänge mit gesonderten Stempeln und Schnittplatten, die aber alle in einem einzigen Werkzeug vereinigt sind, stattfinden können, so ist die gleiche Arbeitsweise auch beim Biegen möglich. Es wird also das in Fig. 473 abgebildete Werkzeug nicht als ein Mehrfachbiegewerkzeug zu bezeichnen sein, da die verschiedenen Biegungen der Feder gleichsinnig in einem einzigen Stempel liegen, dessen Unterfläche dem fortlaufenden Linienzug aller Biegungen entsprechend geformt ist. Sonst müßte man ein Werkzeug nach Fig. 464 auch als ein Mehrfachbiegewerkzeug bezeichnen, da man sich jede runde Biegung aus einer unendlich großen Anzahl Einzelbiegungen entstanden denken könnte.

Ein Werkzeug aber, das eine halbrunde Öse nach Fig. 493 mit geschlossenen Enden biegt, bedarf zu dieser Arbeit eines halbrunden und zweier gerader Seitenstempel, so daß drei vollständig getrennte Biegungen auszuführen sind, deren sämtliche Einzelwerkzeuge in einem Werkzeug mit einem Antrieb nach Fig. 494 vereinigt sind.

Mehrfachbiegewerkzeug zum Hauswecker (C. Lorenz, Berlin SO). Der Ständer für die Unterbrecherschraube des Hausweckers wird im zweiten Arbeitsgang (Tabelle Nr. 1—IV) aus dem ausgeschnittenen und gelochten Blankett auf die in Fig. 103 gezeichnete Form gebogen. Die beiden seitlichen Lappen, welche das Gewinde für die Unterbrecherschraube aufnehmen, sind rechtwinklig gegen die Grundfläche abzubiegen und diese selbst knapp vor den Lappen stufenförmig einzuknicken. Diese drei Biegungen werden in einem Werkzeug nach Fig. 478 in einem Pressenhub fertiggestellt. Die Arbeiten sind so verteilt, daß der Stempel die beiden Lappen gefaßt und zwischen dem Auswerfer und der Matrize hochgebogen hat, bevor das Abbiegen der Stufe in der Grundplatte beginnt. Die betreffenden Flächen sind schräg angearbeitet, so daß die Federung die Winkel wieder zu rechten macht. Der Stempel besteht aus einem Stempelkopf von Maschinenstahl, in den der eigentliche Biegestempel mit zwei konischen Stiften in einer

Nut eingepaßt und festgezogen ist. Die Stifte sind, einer mit Anzug von links und einer von rechts, eingepaßt, so daß der Anzug des Stempels gleichmäßig erfolgt. Der Stempel für beide Biegungen ist aus dem Vollen gearbeitet, gehärtet und angelassen. Die Matrize ist etwas verwickelterer Bauart und auf einer Grundplatte aus Maschinenstahl mit zwei versenkten, versetzten Kopfschrauben von unten befestigt. Die linke Hälfte ist mit einem Schlitz zum Aufbiegen der Lappen versehen, in dem der schmale Teil des Stempels arbeitet, und in den von unten der abgesetzte Teil des Auswerfers eintritt. Auf der Unterseite geht dieser Schlitz auf die volle Breite durch, da der Auswerfer dort mit seiner vollen Breite von den Auswerferstiften, die in bekannter Weise von der Federplatte getragen werden, abgestützt wird. Auf der Oberfläche links ist auch die Anschlagplatte aufgelegt, in die die Lappen des Blanketts eingelegt werden. Die Anschlagplatte ist aus vollem Blech herausgearbeitet und mit zwei seitlichen Kopfschrauben an den senkrechten Seitenwänden der Matrize festgeschraubt. Dadurch richtet man sie gut aus und spart gleichzeitig auf der Oberfläche der Matrize an Platz. Der Schlitz für das Umbiegen der Lappen geht so tief in die Matrize hinein, daß die Lappen vollständig aufgebogen und zwischen Stempel Matrize und Auswerfer vollständig gefaßt werden, bevor das Abbiegen der Stufe im rechten Teil der Matrize stattfindet. Dadurch wird ein sicheres Festhalten des kürzeren Blechteiles im Werkzeug erreicht und ein Ausweichen bei der zweiten Biegung, die das Stück zu Ausschuß machen würde, verhindert. Die Matrize ist selbstverständlich aus Werkzeugstahl, gehärtet und angelassen.

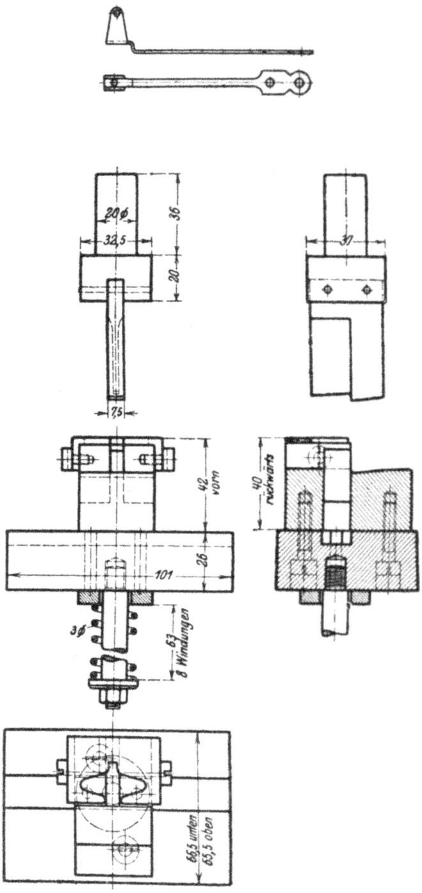

Fig. 478.

Biegewerkzeug für Augenglasklammern aus Gold. Das in Fig. 479 dargestellte Biegewerkzeug vollführt den letzten Arbeitsgang bei der Herstellung einer Augenglasklammer aus Gold. Die Glasklammer ist auf Werkzeugen nach Fig. 26 ausgeschnitten worden; dann wurden auf dem Prägewerkzeug, Fig. 283, die beiden runden Lappen ausgeprägt und auf dem Lochwerkzeug, Fig. 27, gelocht. Als letzte Arbeit folgt nun das Abbiegen dieser

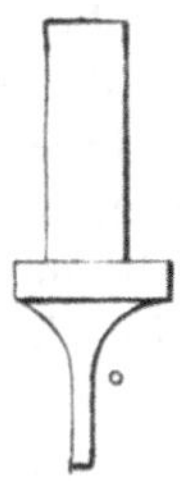

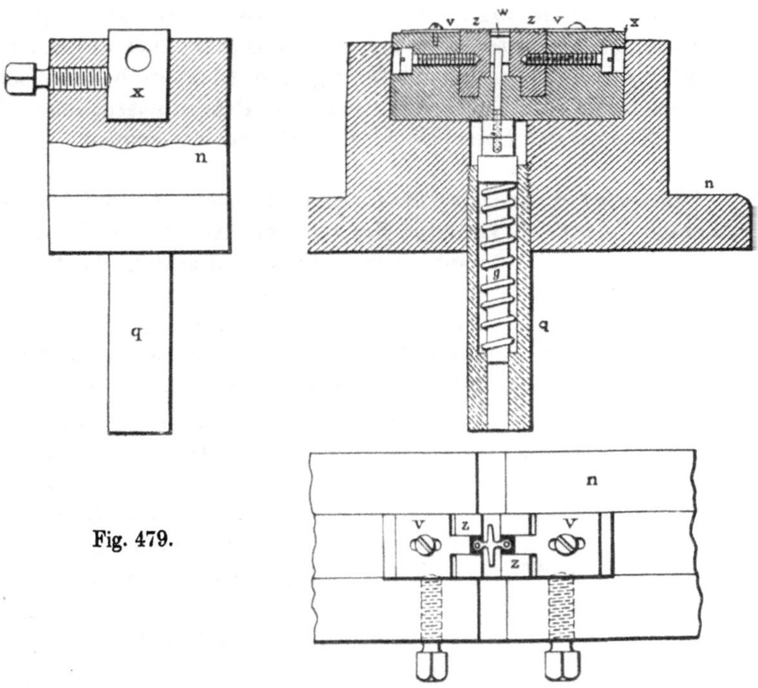

Fig. 479.

Lappen unter rechtem Winkel mit gleichzeitigem Rundbiegen der Mittelrippe.

Die gußeiserne Froschplatte n ist der Länge nach so genutet, daß die Matrize x aus weichem Stahl gut paßt. In diese sind die gehärteten Backen z und der gehärtete Auswerferstempel w, der mittels Stift und Gegenmutter auf dem Plunger g aufruht, eingepaßt. Die Backen z erhalten Schlitze entsprechend der Breite der runden Lappen der Glasklammer, und richten so das Blankett nach der Seite aus, während die mit Zungen versehenen, einstellbaren Anschlagplatten v es auf die richtige Mittellinie einstellen. Die weitere Konstruktion ergibt sich aus der Zeichnung. Um das Blankett zu biegen, wird es in die Nut der Backen z eingelegt, wobei nur die runden Teile gehalten werden. Beim Niedergang der Presse biegt der Stempel o, Fig. 479, die Klammerlappen einwärts, bis die Mittelrippe auf dem Auswerferstempel w aufruht. Dann geht dieser mit nach abwärts, bis er aufsitzt, wodurch die Mittelrippe der Klammer nach A, Fig. 25, gebogen wird. Gleichzeitig drückt der Stift in w oder vielmehr die Gegenmuttern an seinem Ende die Feder in der Büchse q zusammen. Beim Aufwärtsgang des Stempels o wird der Auswerferstempel durch diese Feder nach oben geschoben und hält das fertige Stück am Stempel, von dem es mit den Fingern abgenommen werden kann. Die Messingbüchse q ist in die Froschplatte n so tief eingeschraubt, daß der Auswerferstempel das in der Anschlagplatte liegende Blankett beinahe berührt. Es ist anzuraten, die Klammern nach dem Prägen auszuglühen, da die scharfen Kanten des Stempels o sie leicht brechen können.

Biegewerkzeug für Ofengrundplatten aus Stahl. Das im folgenden beschriebene Werkzeug vollführt den zweiten Arbeitsgang bei der Herstellung von Ofengrundplatten, die bereits auf die Form J, Fig. 480, gebracht worden sind. Die Größe dieser Grundplatten ist — entsprechend den drei üblichen Ofengrößen — 736 × 508, 787 × 508 und 838 × 508 mm. Fig. 481 zeigt das erste Biegen und die fertige Grundplatte.

Fig. 480.

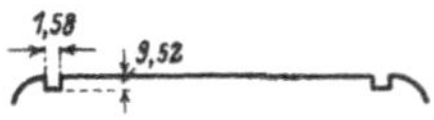

Fig. 481.

Fig. 482 ist ein Schnitt durch das geschlossene Werkzeug, dessen einzelne Teile in Fig. 483/4 abgebildet sind. A, Fig. 483, ist die Grundplatte der Matrize, auf welcher das Stahlblech während der Arbeit ruht. Die Matrize ist aus mehreren Teilen aufgebaut, indem die vier Leisten G an den Lang- und Querseiten angebracht werden, wie der Schnitt F in Fig. 483 durch die Matrize mit den Teilen G in Stellung zeigt. In die Grundplatte der Matrize A sind zwei Gruppen Schwalbenschwanznuten rechtwinklig zueinander eingehobelt, in die die entsprechenden Schwalbenschwänze auf der Unterseite der Leisten G passen. Die Löcher an den vier senkrechten Wänden in A dienen zur Aufnahme von Federn, die die Leisten G nach beendigtem Biegen beim Aufwärtsgang des Stempels zurückdrücken. Der Körper des Biegestempels ist in Fig. 484, B dargestellt. An

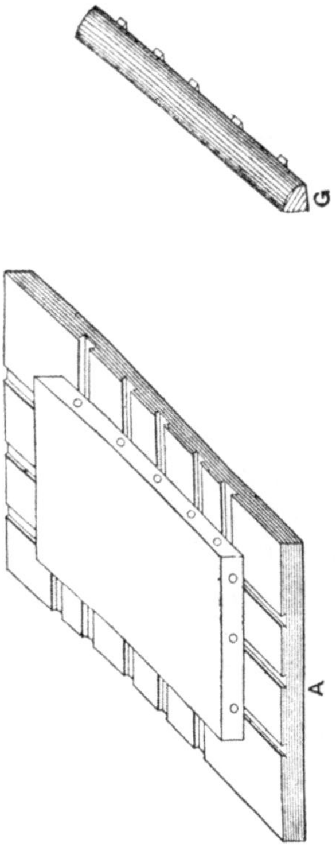

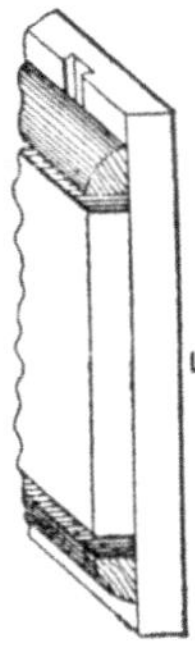

Fig. 483.

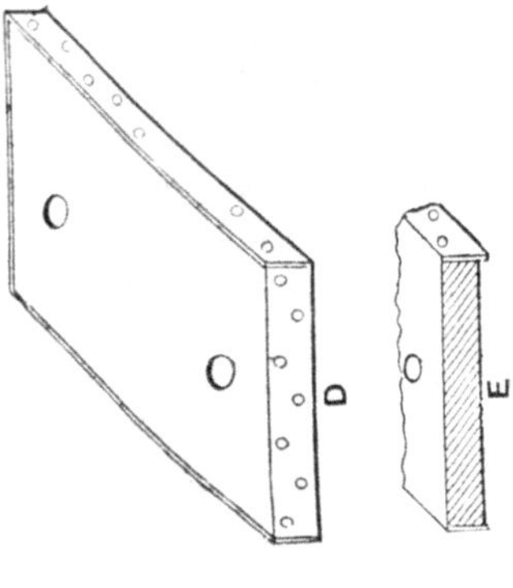

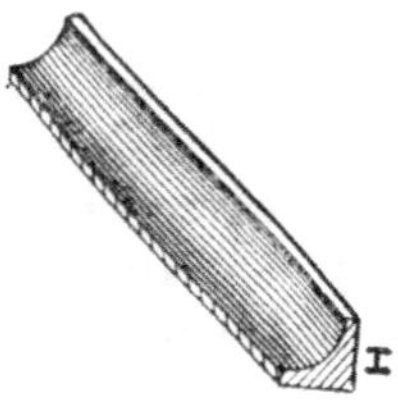

Fig. 484.

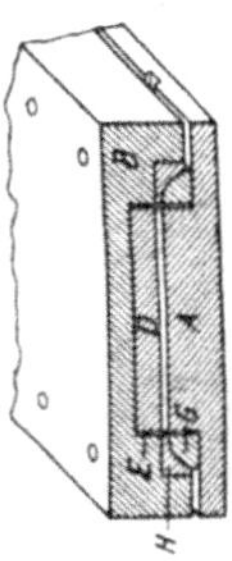

Fig. 482.

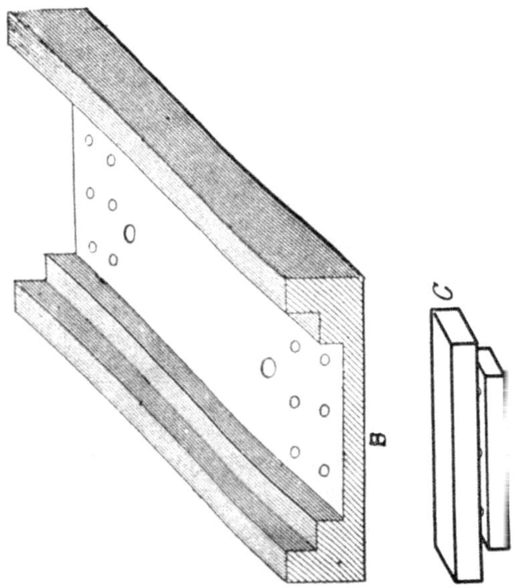

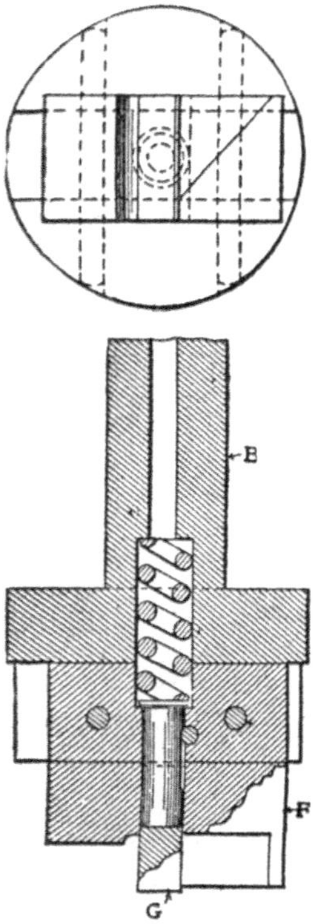

Fig. 486.

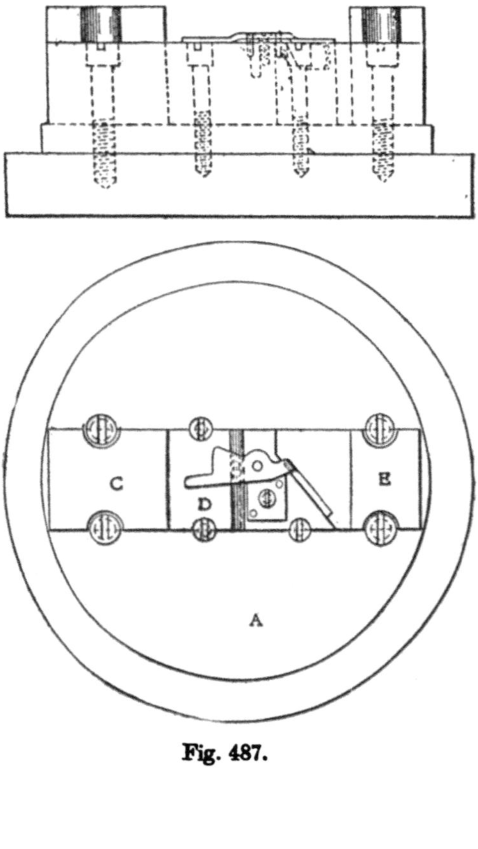

Fig. 487.

ihm sind Stahlleisten H befestigt, die in Übereinstimmung mit den Stücken G der Matrize die Lang- und Querseiten des Stempels bilden. Die Stücke C werden in den Stempelkörper an den beiden Querseiten eingepaßt, um die Endleisten H zu halten, so daß der Biegestempel geschlossen ist. Die Druckplatte D/E erhält angepaßte Seitenstreifen, die ungefähr 6,5 mm unter der Unterfläche vorstehen, wie im Schnitt bei E ersichtlich ist. Diese Streifen an der Druckplatte treten bei dem zweiten Biegen in diesem Werkzeug in die beim ersten Biegen erzeugten Nuten in J und über die Innenkanten der Matrize, so daß sich das Blech nicht aufbiegen kann, wenn die Seitenteile beim Biegen nach innen vorrücken.

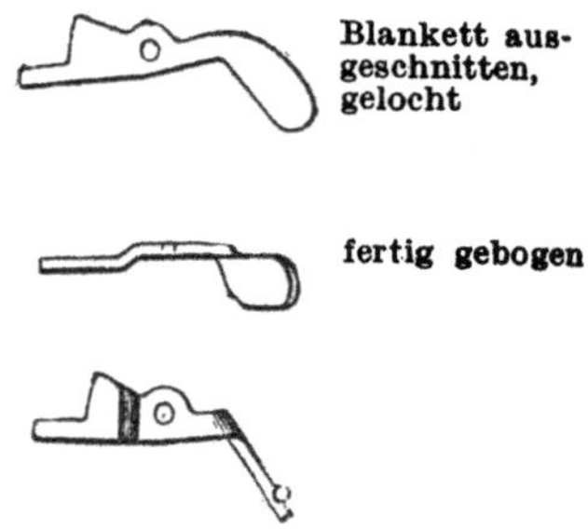

Fig. 485.

Biegen eines eigenartig geformten Stückes. Fig. 485 zeigt den Herstellungsgang eines eigenartig geformten Stückes und Fig. 486/7 das Biegewerkzeug dafür.

In Fig. 487 ist A eine gewöhnliche, runde Grundplatte aus Gußeisen. Sie wird zur Aufnahme der Teile C, D und E ausgefräst. Diese Stücke werden allseitig bearbeitet, gut eingepaßt und befestigt.

Das Arbeitsstück ist in dem Werkzeug in Fig. 487 in der fertig gebogenen Stellung gezeichnet. Das Loch im Blankett und das Anschlagblech auf der Vorderseite sichern die richtige Stellung des Arbeitsstückes beim Biegen.

Der Stempelkopf, Fig. 486, aus Gußeisen ist ebenfalls rund und wird zur Aufnahme des Stempels F ausgefräst, der gut in jenen eingepaßt wird. Hierauf wird dieser für den Niederhaltstempel G gebohrt und gestoßen. Der letzte kommt zuerst mit dem Blech in Berührung und hält es nieder, während der Winkel angebogen wird.

Die Stücke C und E wurden höher als D gemacht, um den Stempel zu führen und den Seitendruck aufzunehmen.

Biegewerkzeuge für zwei aufeinanderfolgende Arbeiten. In Fig. 488 ist der Herstellungsgang eines Blechgegenstandes abgebildet, der in einem Ausschneiden und zwei Biegearbeiten hergestellt wird.

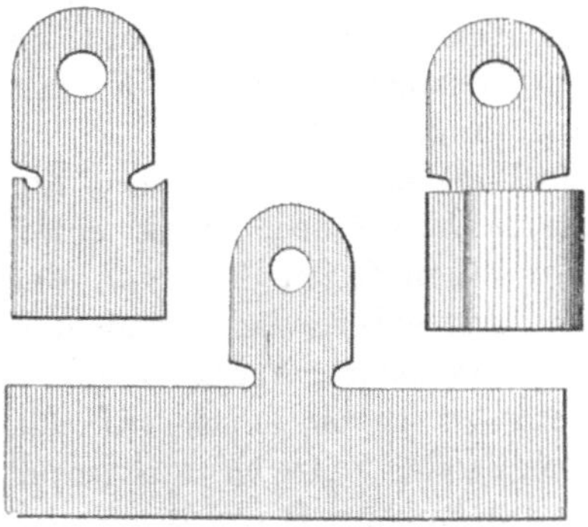

Fig. 488.

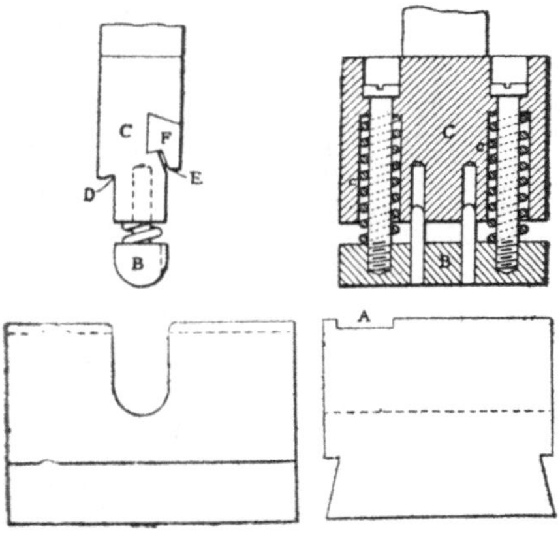

Fig. 489.

Fig. 489 und 490 stellen zwei Werkzeuge dar, um ein flaches Stück Kupfer, ungefähr 54 mm lang, 12 mm breit und 0,64 mm dick, in Kreisform mit unregelmäßigem Umriß und übergreifenden Enden zu biegen, von denen das eine zuletzt nach außen abgebogen wird. Auf der einen Seite, nahezu in der Mitte des Bleches, befindet sich ein vorstehendes Auge, das dem Auftrage entsprechend eben bleiben muß und beim Biegen nicht beeinflußt werden darf, und dessen Mittellinie mit der Mittellinie des gebildeten Ringes zusammenfallen soll.

Nachdem in gewöhnlicher Weise Maße und Umriß des Blanketts erhalten worden sind, wird das Schnittwerkzeug hergestellt, das weiter keiner Erklärung bedarf.

Stempel und Matrize für die erste Biegung sind in Fig. 489 gezeichnet; in diesem Werkzeug wird das Stück U-förmig gebogen und das höhere Ende, ungefähr 1,6 mm unterhalb des oberen Endes, nach innen gebogen. Das Blech wird in den Ausschnitt A eingelegt, der als Seitenanschlag dient, während Stifte von ungefähr 3 mm Durchmesser, die nicht gezeichnet sind, die Enden festlegen. Das Blech wird dann so weit vorgeschoben, daß das Auge über die Vorderseite der Matrize überhängt, so daß es während der Biegung des Ringes eben bleibt. Der Unterteil B des Stempels wird durch die Schraubenfedern c in seiner Stellung gehalten, bis er mit dem Grunde der Matrize in Berührung kommt. Beim weiteren Abwärtsgang des Stößels stößt die Nut im Stempel bei D an dieser Seite auf das Blech und drückt es herunter, so daß beim Hubende dieses Blechende bei allen Stücken die richtige Länge erhält. Gleichzeitig tritt das andere Ende des Bleches in den Schlitz E und wird dadurch einwärts gebogen. Um die Herstellung dieses Schlitzes zu erleichtern, wird im Stempel C an dieser Stelle ein besonderes Stück F mit Schwalbenschwanz eingesetzt.

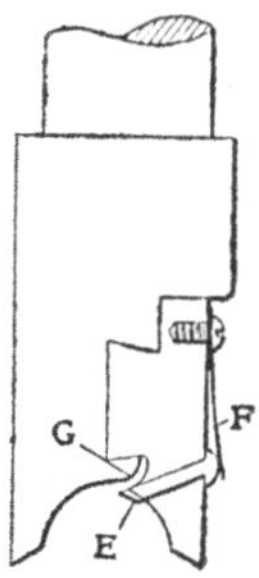

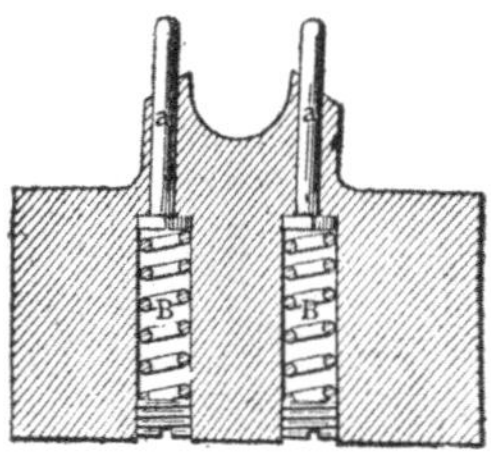

Fig. 490.

In diesem Fall ist es empfehlenswert, am Grunde der Matrize hinter dem Stück einen Blechstreifen oder einen festen Stift von der gleichen Dicke wie das Arbeitsstück einzulegen, damit der Stempel darauf zur Auflage kommt. Dadurch wird ein zu harter Druck auf das Arbeitsstück vermieden, so daß es gleichzeitig unter dem Stempel gleiten kann, ohne einzuknicken, wenn die Nut bei D das Blech vor sich her schiebt.

Fig. 490 zeigt das Schließwerkzeug. Nach dem ersten Biegen wird das Arbeitsstück auf die Matrize zwischen die beiden Stifte a gelegt, die es in der richtigen Stellung halten. Diese Stifte werden durch zwei leichte Schraubenfedern B gehalten, deren Druck durch zwei Stellschrauben von unten geregelt werden kann. In der Mitte der Vorderseite der Matrize ist mittels Schrauben und Schwalbenschwanz ein Gußeisenblock befestigt — in der Zeichnung nicht sichtbar — mit einem Stift an jeder Seite, der während der Schließarbeit das Auge des Arbeitsstückes zu führen hat. Beim Niedergang des Stempels übt der

Stift E einen leichten Druck auf das Arbeitsstück aus und bewirkt, daß das an diesem Ende vorher deshalb abgebogene Stück in den Schlitz tritt. Der Stift wird durch eine flache Feder F in seiner Stellung gehalten. Während das eine Ende des Arbeitsstückes auf der rechten Seite nach abwärts geht, folgt das linke dem Stempel, bis es in Berührung mit der Oberseite des Stiftes E kommt, wodurch es in die Nut G geleitet und so auswärts gebogen wird. Ein zylindrischer Bolzen ist notwendig, um bei dieser Arbeit das richtige Innenmaß des Auges zu erhalten.

Mehrfachbiegewerkzeug. Dieses Werkzeug stellt den Übergang zu jenen Biegewerkzeugen dar, die mit Seitenstempeln arbeiten.

Die gestellte Aufgabe verlangte, daß das Blankett nach Fig. 491 aus Messingblech von ungefähr 1,6 mm Dicke doppelt U-förmig und selbstverständlich möglichst in einem Gang gebogen werden sollte.

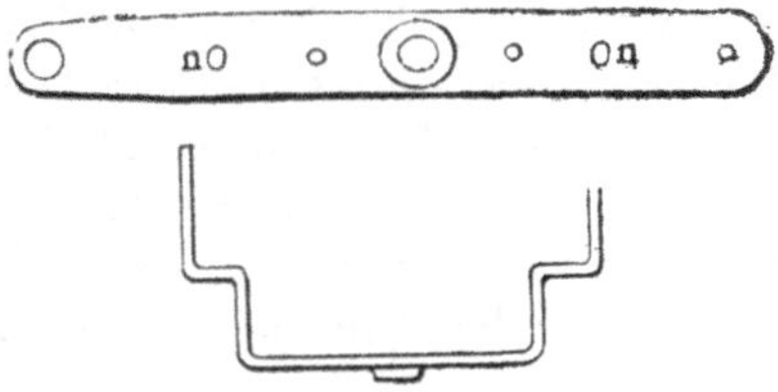

Fig. 491.

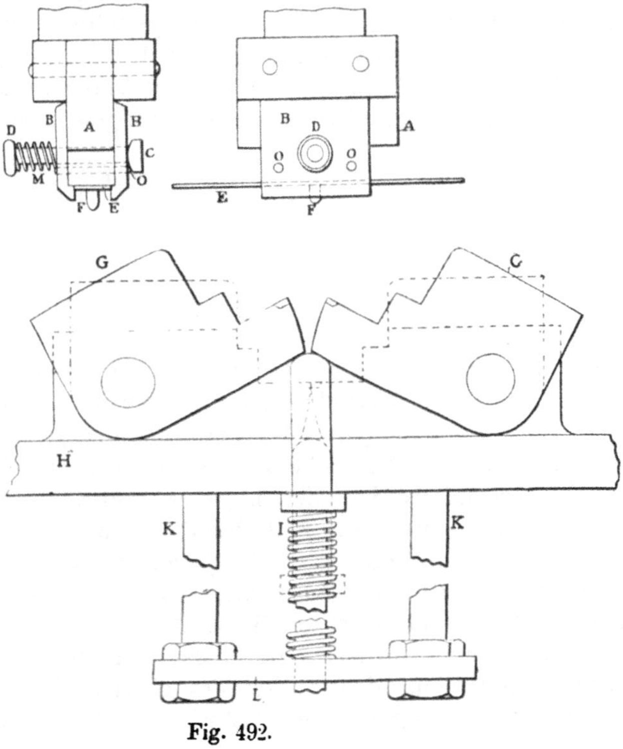

Fig. 492.

Das Blankett wurde in gewohnter Weise gelocht und ausgeschnitten. Das Biegewerkzeug ist in Fig. 492 abgebildet. Das Blankett wird auf dem Stempel mittels der Klammern B festgehalten. Die beiden Teile G der Matrize sind um ihre Bolzen drehbar und werden durch den Federbolzen I in der gehobenen Stellung gehalten. Wenn nun das Blech durch den Stempel nach abwärts gebracht wird, stellen sich diese Teile G in die durch die gestrichelten Linien angedeutete Stellung, so daß das Biegen des Blanketts auf die verlangte Form mit möglichst geringer Streckung des Bleches vor sich geht. Die Breitenabmessung des Bleches wurde nicht bemerkenswert geändert, wenn es sich auch als notwendig herausstellte, die Löcher n in dem Blech länglich zu lochen, damit sie am fertigen Stück rund werden. F ist ein Sucherstift, der in das mittlere Loch des Bleches paßt.

Der Bolzen C, der eine mit Fischhaut versehene Mutter D und eine Feder M trägt, erlaubt die Klammern B so weit zu öffnen, daß das Blankett sicher von dem Stempel gehalten wird. O sind Führungsstifte, die in dem Stempel fest und in B lose sitzen.

Werkzeug mit Seitenstempeln zum Biegen verschiedener Draht augen. Fig 494 zeigt eine Werkzeuganordnung für die Kraftpresse, um dünne Drähte in die in Fig. 493 dargestellten Formen zu biegen. Mittels des abgebildeten Werkzeuges wird die in Fig. 493 an erster Stelle abgebildete D-Form hergestellt. Obwohl die Bewegung der Presse wie gewöhnlich senkrecht ist, geht das Biegen in einer wagerechten

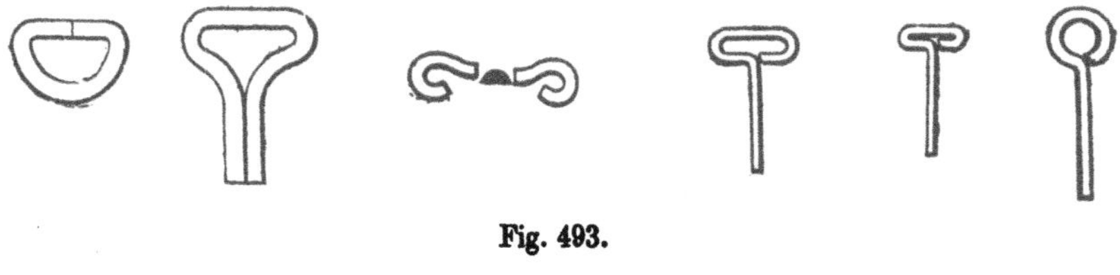

Fig. 493.

Ebene vor sich. In Fig. 494 ist eine Vorderansicht, eine Seitenansicht der ganzen Vorrichtung und ein Grundriß des unteren Teiles gegeben. In die Grundplatte A ist eine kreuzförmige Schwalbenschwanznut gehobelt, in der das Gleitstück B arbeitet, dessen Oberfläche bündig mit der Grundplatte ist oder ein geringes darüber steht. Das Gleitstück B wird durch ein Keilstück C in Bewegung gesetzt, dessen unteres, schmales und paralleles Ende so lang ist, daß es sich immer in einer Nut der Grundplatte A passend führt. Das Keilstück geht durch den Schlitz E im Gleitstück B und hat an der Keilfläche eine vorstehende Leiste a, die in der Nut bei E arbeitet, so daß beim Aufwärtsgang des Stößels das Gleitstück B in seine Ruhestellung zurückgezogen wird. Außerdem sind noch zwei andere Gleitstücke G und H vorhanden, die unter rechten Winkeln zu B arbeiten und durch die Keilstücke I und J in Tätigkeit versetzt werden. Die geneigten Kanten K und L dieser Mitnehmer liegen höher als die von C, so daß die Gleitstücke G und H später bewegt werden als Gleitstück B. An der Stirnseite von B ist der Biegestempel M und an G und H die Seitenstempel N und O befestigt.

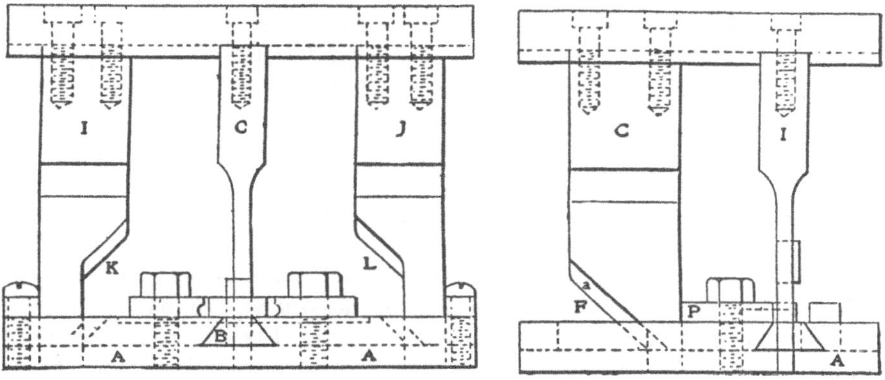

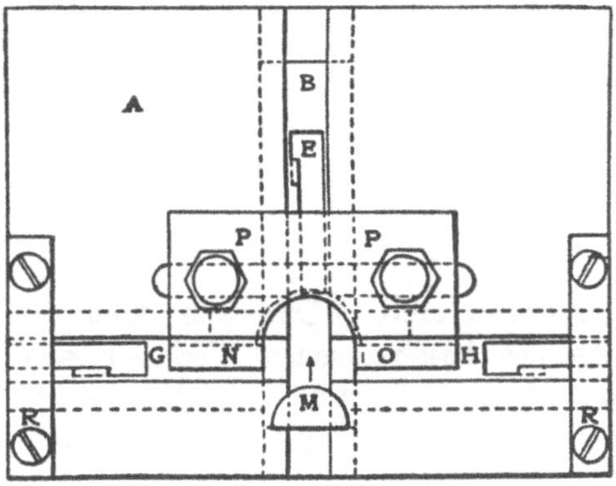

Fig. 494.

Der zu biegende Draht von 6,3 mm Durchmesser wird auf Länge geschnitten, geglüht und ein Stück in die Mitte vor den Biegestempel M gelegt. Wenn nun der Stößel der Presse niedergeht, zieht die schräge Kante des Keilstückes C das Gleitstück B in der Pfeilrichtung vor, wobei der Stempel M den Draht in den halbkreisförmigen Ausschnitt der Platte P eindrückt und so U-förmig biegt, worauf die Seitenstempel N und O vorwärts gehen, die Enden umbiegen und derart das „D“ schließen.

R sind Stahlstreifen, die an der Grundplatte A mit Schrauben befestigt sind und als Führungen für die Rückenfläche der Keilstücke I und J, ebenso wie P für C, dienen. Die Überlagplatte P hat überdies an der Unterseite einen Keil, der in einen entsprechenden Schlitz der Grundplatte paßt, in dem P mit zwei Kopfschrauben befestigt ist. Die

Zeichnung zeigt das Werkzeug in seinem ursprünglichen Entwurf, so daß die Befestigungsteile für die betreffende Presse, in der es zur Verwendung kommen sollte, nicht sichtbar sind.

Die letzten drei Augenformen in Fig. 493 waren aus weichem, geradegerichtetem Messingdraht gemacht worden, der von rechts dem Werkzeug zugeführt wurde, das in diesem Fall mit einer Abschneidevorrichtung ausgestattet war. Der innere Stempel des Auges wurde vor Schluß des Pressenhubes selbsttätig herausgezogen, so daß der Ring flach zusammengedrückt werden kann. Bei der Herstellung des doppelten Auges, Form 3, war der innere Formstempel fest, während der äußere bewegt wurde, da in diesem Falle der erste zu schwach war.

Biegewerkzeug mit Seitenstempeln. Eine besonders nützliche Form eines Biegewerkzeuges, wenn es sich um schnelle Arbeit handelt, stellt Fig. 496 dar. Das Material wird bis auf den Grund der Matrize niedergedrückt, worauf es durch die Seitenstempel zusammengebogen wird. Der Stempel entspricht in seinen Abmessungen gewöhnlich den Innenabmessungen des fertigen Gegenstandes. Fig. 495 zeigt Stücke, die nach diesem Verfahren hergestellt worden und in Form und Ab-

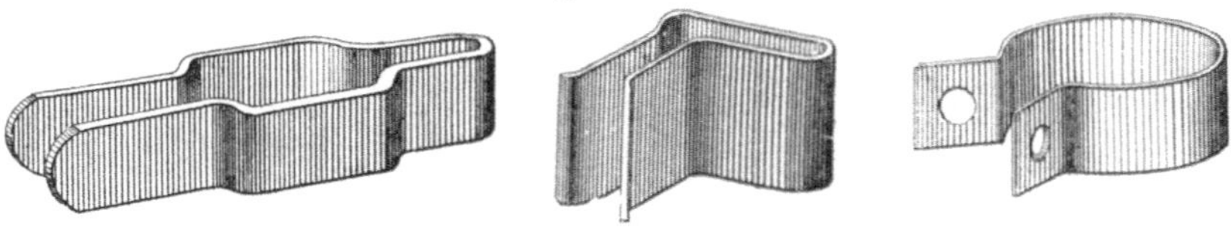

a b c

Fig. 495.

messungen ganz gleichmäßig ausgefallen sind. Das erste Stück in Fig. 495 wird auf dem im folgenden beschriebenen Werkzeug aus hartem Messingblech, 1,6 mm dick und 8 mm breit, hergestellt. Nach mehreren Versuchen wurde der folgende Weg als der beste beibehalten:

In Fig. 496 ist ein Längsschnitt durch das ganze Werkzeug sowie Seitenansicht und Grundriß der Matrize wiedergegeben. A ist eine gußeiserne Grundplatte, die die Seitenstempel B trägt, welche durch Führungen aus 6,5 mm dicken, auf die Oberfläche der Grundplatte festgeschraubten Stahlstreifen festgehalten werden. C sind Nuten, die in die Oberflächen der Seitenstempel eingehobelt werden, um das Metall seitlich zu führen. D sind Endanschläge und E Stifte, die in die Grundplatte der Matrize eingepaßt sind und die Bewegung der Seitenstempel nach außen begrenzen. Diese erhalten die dem notwendigen Gleitweg entsprechenden Aussparungen. Wenn der Stempel nach aufwärts geht, drücken die beiden Wurstspiralfedern F die beiden Seitenstempel nach auswärts. Die Stifte G in ihnen übertragen den Federdruck.

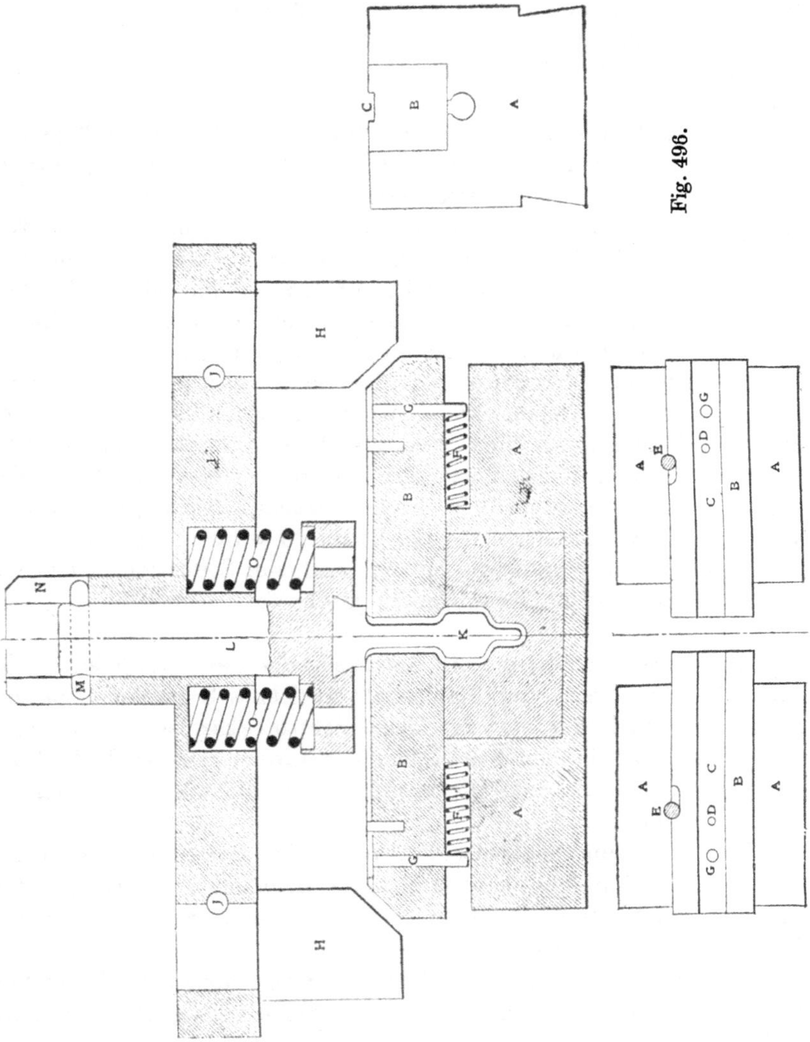

Fig. 496.

Die Schnittzeichnung läßt den Arbeitsvorgang deutlich erkennen. H sind Keilstücke, deren Enden gedreht und in den Stempelkopf I eingepaßt sind, wo sie durch konische Stahlstifte J gesichert werden. Das untere Ende K ist der Biegestempel, der mittels Schwalbenschwanz in der Kopfplatte befestigt wird, die ihrerseits in einem ausgeriebenen Loche des Stempelkopfes gleitet und durch einen Stift M in ihrer Stellung gehalten wird. Der äußere Zapfen erhält bei N einen Schlitz, in dem der Stift M während der Arbeit gleiten kann. Die Federn O halten den Stempel K nieder, bis die erste Biegung vollendet ist. Ihre Zahl und Größe hängt von der zur ersten Biegung nötigen Kraft ab. Wenn nötig, können sie in größerer Anzahl um den Stempel herum angeordnet werden.

In allen Werkzeugen dieser Art müssen die Federn stark genug sein, um durch Niederdrücken des Blanketts die erste Biegung auszuführen, bevor irgendeine bemerkenswerte Verziehung eintritt. Der Stempel und die Matrize sollten so entworfen sein, daß die abgeschrägten Flächen der Keilstücke H die Seitenstempel B berühren, wenn der Biegestempel im Grunde der Matrize aufruht. Beim weiteren Niedergange des Stößels werden die Seitenstempel in die Matrize hineingedrückt, wodurch die seitlichen Biegungen vor sich gehen.

Die Teile von Stempel und Matrize ändern sich je nach der Form des herzustellenden Gegenstandes. Wenn an beiden Seiten des Stückes dieselbe Biegung ausgeführt werden soll, stehen die abgeschrägten Flächen an H und B unter 45°. Bei unsymmetrischer Arbeit sind die Winkel verschieden, so daß der eine Seitenstempel einen größeren Weg als der andere zu machen hat. Ein Beispiel hierfür ist Fig. 495 b.

Es sind sogar Werkzeuge mit drei und vier Seitenstempeln schon gebaut worden und haben erfolgreich gearbeitet.

Mehrfachbiegewerkzeug mit Seitenstempeln für Stahlfedern. In Fig. 497 ist ein mit einer Stahlfeder versehenes Scharnier offen und geschlossen abgebildet, das für kleine Schmuckkästchen Verwendung findet, damit der Deckel beim Öffnen mit einem gewissen Federdruck aufspringt und nach dem Öffnen nicht über die wagerechte Stellung hinaus aufklappt.

Die bei diesem Scharnier verwendete Feder wird in dem Biegewerkzeug nach Fig. 498 aus dem Blankett in einem Arbeitsgang fertiggestellt. Das Material ist Federstahl von 0,3 mm Dicke, der nachher federhart gehärtet wird.

Die Schwierigkeiten solcher Biegewerkzeuge werden durch das hier verarbeitete Material noch erhöht, das verhältnismäßig schwer im ganz gleichen Zustand zu bekommen ist, wenn es sich um Massenlieferungen handelt, bei denen immer genau dasselbe Stück in gleicher Form und Qualität geliefert werden soll. Kurzum, Woodworth sagt, daß es in solchen Fällen beim Probieren eines solchen Werkzeuges billiger ist, ein neues Stück zu machen, wenn aus irgendeinem Grund das Versuchswerkzeug gebrochen ist. Auch sind die Keilstücke zur Bewegung der Seitenstempel den Federn vorzuziehen, da jene zwangläufig arbeiten und dadurch die Gefahr eines Stecken-

bleibens und Bruches vermeiden. Der Winkel der Keilstücke kann bis zu 45° betragen.

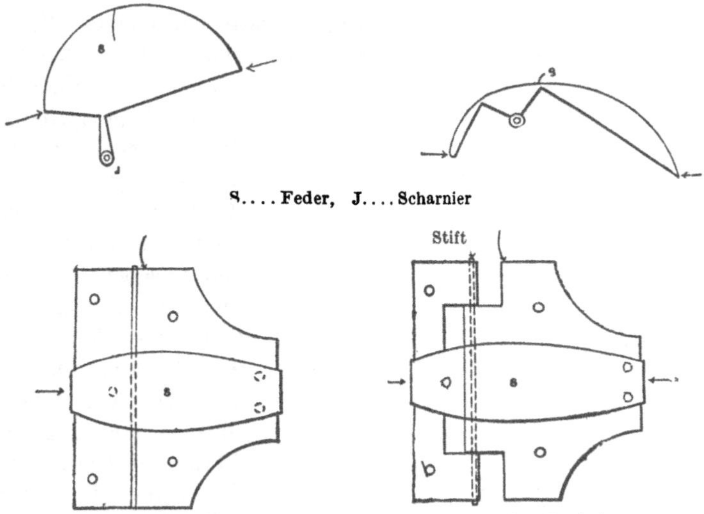

Stellung der Feder und des Scharniers, wenn der Deckel geschlossen / offen

Fig. 497.

Das Werkzeug, Fig. 498, besteht aus einem Stempelteil, der neben dem eigentlichen Biegestempel b noch vier Keilstücke zur Bewegung der Seitenstempel trägt, und einem Unterteil. Die Grundplatte A ist aus Gußeisen und trägt die Seitenstempel B von rechteckigem Querschnitt aus gehärtetem Stahl, die in Führungen C gleiten, die durch vier $^5/_{16}''$ Zylinderkopfschrauben D befestigt werden. Links sind die Schrauben der Deutlichkeit wegen ausgelassen. Die Vorderseiten der Führungen C erhalten Langschlitze, durch welche gehärtete Stahlstifte E von 6,5 mm Durchmesser durchtreten, die unter dem Druck der Keilstücke die Seitenstempel nach außen schieben. Diese Stifte werden in den Seitenstempeln wieder durch fest eingetriebene Prisonstifte gehalten. Außerdem sichert der Stift in dem linken Seitenstempel die Feder F und den Gleitstift G in ihren richtigen Stellungen. Die Grundplatte ist bei H freigeschnitten, um die Anschläge für das Blankett herzugeben. Beim Abwärtsgang des Stößels wird das Blankett zwischen den Biegestempel b und den rechteckigen Auswerfstempel I, der durch die Feder J und die Stellschraube K gehalten wird, geklemmt. Um einen toten Gang, der leicht durch die geneigte Arbeitslage dieser Stempel herbeigeführt wird, in den Seitenstempeln B aufzuheben, sind, wie links gezeichnet, unter diesen Federn L eingebaut.

Der eigentliche Biegestempel b wird in dem Stempelkopf a gehalten, in dem er sich gleitend führt. Eine Schraube c und eine Feder d halten ihn in der tiefsten Stellung, bis der Druck beim Aufsitzen ihn nach oben zurückschiebt. Die Feder c muß selbstverständlich stark genug sein, um die erste Biegung zu bewirken. Durch den Stift e, dessen oberes Ende frei im Stempelkopf gleitet, wird der Stempel gegen Verdrehung gesichert. Die Keilstücke P, die am Stempelkopf mit Schrauben und Prisonstiften befestigt sind, pressen im letzten Teil des Abwärtsganges des Stößels die Seitenstempel nach innen und biegen die Enden des Blanketts scharf um. An der Vorderseite des Stempelkopfes, genau parallel und unter dem richtigen Winkel ausgerichtet, sind zwei weitere Keilstücke aus Stahlblech (6,5 mm dick) angeschraubt, deren arbeitende Kanten gehärtet sind. Diese stehen gerade so weit über die Vorderseite der Matrize heraus, daß sie an dieser vorbeigehen und mit den Stiften E zusammenarbeiten können. In der höchsten Stellung des Stempels stehen die unteren geraden Enden der Keilstücke in Verbindung mit den Stiften, so daß sie nie aus dem Eingriff mit der Matrize kommen. Hat die Presse einen längeren Hub als notwendig, so werden die unteren Enden der Keilstücke meistens etwas abgeschrägt, wie die gestrichelte Linie links zeigt, damit die Stifte nicht bei einem zufälligen Einschalten der Presse beim Auftreffen der Keilstücke abgebrochen werden.

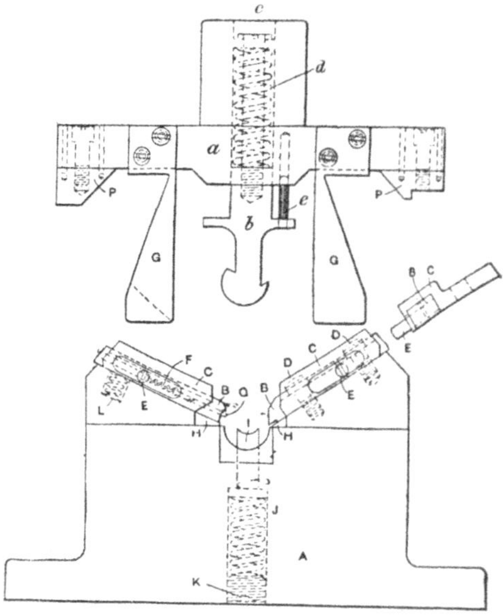

Fig. 498.

Alle arbeitenden Teile sind gehärtet und angelassen; das Werkzeug wird in eine schnellaufende, schräge Presse eingebaut, aus der die Stücke von selbst ausfallen können.

Werkzeuge für den Herstellungsgang in Tabelle Nr. 30. Die Herstellung des Stückes in Tabelle Nr. 30, G. 4 erfolgt fortlaufend aus einem Streifen weichen Kupferbleches, der 0,8 mm dick und 15,8 mm breit ist.

Das Schnittwerkzeug, das als einfacher Führungsschnitt ausgebildet ist, schneidet die Streifenenden G. 1 in ununterbrochener Folge ohne Materialverlust richtig aus. Das Kupferblech liegt in einer Nut und wird bei jedem Aufwärtsgang des Stempels bis an den Anschlag vorgeschoben. Als Führungsplatte dienen zwei aufgeschraubte Leisten, die entsprechend ausgeschnitten sind und gegen die sich der Stempel beim Schneiden anlegt, so daß ein Ausfedern desselben verhindert wird.

1. Biegen. Fig. 499 stellt das Werkzeug für das erste Biegen dar, das die in der Tabelle G. 2 abgebildete Form liefert. In diesem Werkzeug sind die Seitenstempel in Form von Rollen c ausgeführt. Sie sind ebenso breit wie die Matrize und leicht drehbar in sie eingepaßt. Kopfschrauben sichern sie in ihrer bündigen Stellung. An jeder Seitenfläche befindet sich eine Stellschraube mit Gegenmutter d. Die Spitzen dieser Schrauben treten in Nuten der Rollen ein und begrenzen deren Drehbewegung. Wenn der Stempel außerhalb der Matrize steht, so ziehen die Federn die Rollen in die Ruhestellung zurück, so daß die oberen Kanten des Ausschnittes fast mit den senkrechten Flächen der Matrize zusammenfallen. Die Stellung der Rollen in der Figur entspricht der tiefsten Stellung des Stempels, nachdem das Arbeitsstück über ihm zusammengebogen worden ist. Beim Aufwärtsgang des Stempels nimmt dieser das Arbeitsstück mit, wobei sich die Rollen zurückdrehen und das Arbeitsstück freigeben. Das Stück wird dann mit der Hand vom Stempel abgezogen. In der Figur ist übrigens das Arbeitsstück an die vordere Kante der Matrize vorgeschoben gezeichnet, um seine Lage mehr zu verdeutlichen. Das Biegen vollzieht sich dagegen in der Mitte der Matrize, wobei das Stück auf die Flächen b zwischen die einstellbaren Anschläge a gelegt wird.

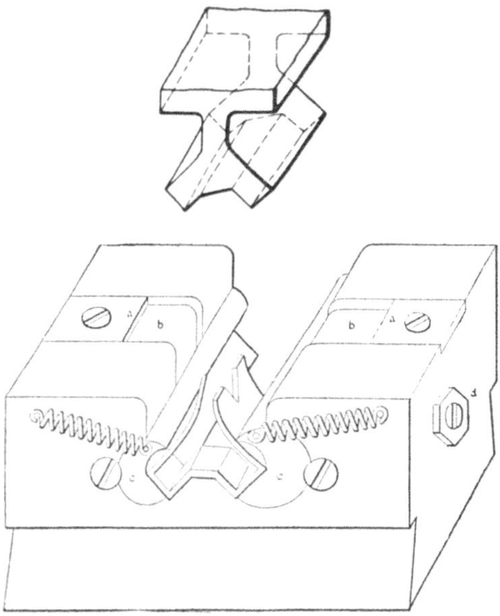

Fig. 499.

Durch die Anwendung der Rollen wird zum mindesten ein Arbeitsgang gespart, denn so kann das Biegen viel weiter getrieben werden, als mit andern Einrichtungen möglich ist.

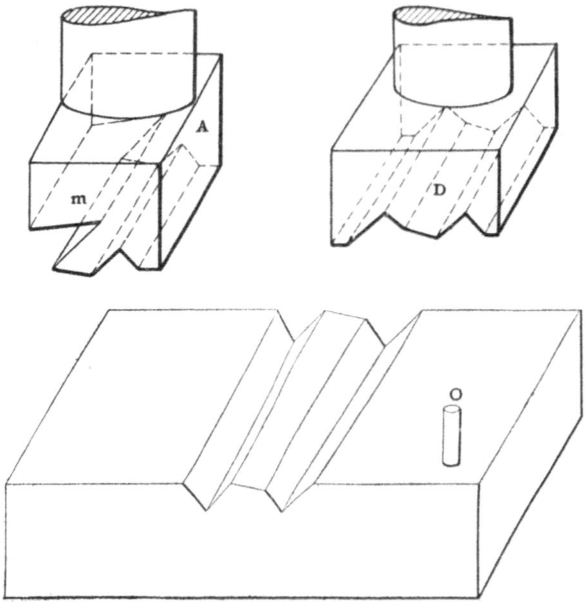

Fig. 500.

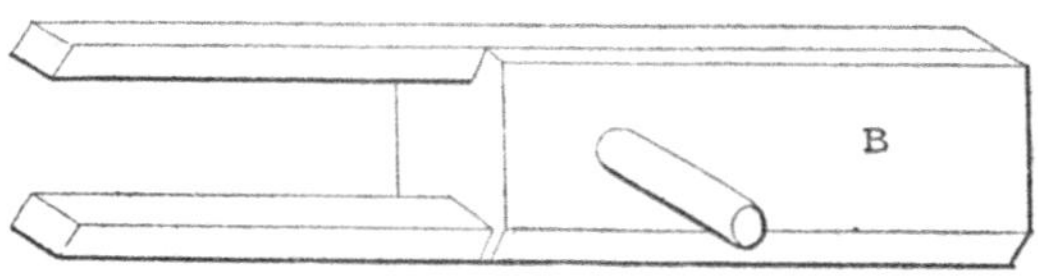

Fig. 501.

2. Biegen. Die Form nach G. 3, Tabelle Nr. 30, wird zwischen der Matrize, Fig. 500, und dem Stempel A erhalten, wobei der Dorn B, Fig. 501, zur Verwendung kommt. Die mit m bezeichnete Fläche des Stempels dient dazu, das Ende n so weit als möglich zu biegen.

Das 3. Biegen erfolgt auf derselben Matrize, Fig. 500, mit dem Stempel D und dem oben erwähnten Dorn. Der kleine Stift o in der Ecke der Matrize dient zum Abziehen des fertigen Stückes von dem Dorn.

Dreifaches Biegewerkzeug (Gebrüder Bing A.-G., Nürnberg). Der Blechwinkel in Fig. 502 enthält im ganzen drei rechtwinklige Biegungen, A, B, C, deren eine, nämlich die runde Fläche B, bereits einem Ziehen sehr ähnlich ist, und eine U-förmig aufgebogene Rinne D an der Längsseite. Das entsprechend ausgeschnittene Blankett wird in einem Arbeitsgang in dem in Fig. 503/4 abgebildeten Werkzeug hergestellt.

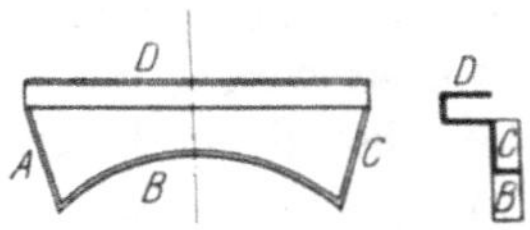

Fig. 502.

Der Stempel Fig. 503 besteht aus einem Stempelkopf a mit

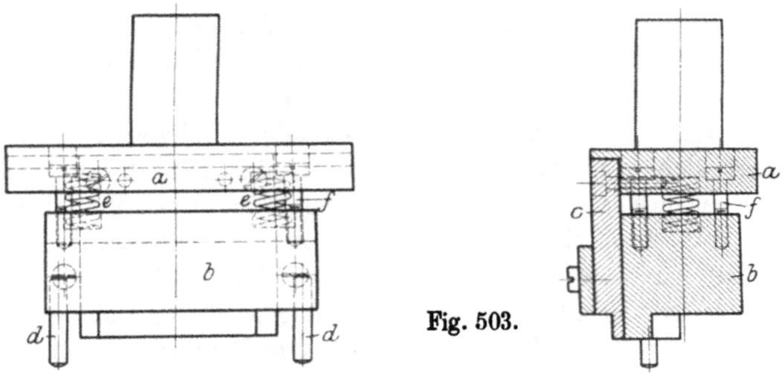

Fig. 503.

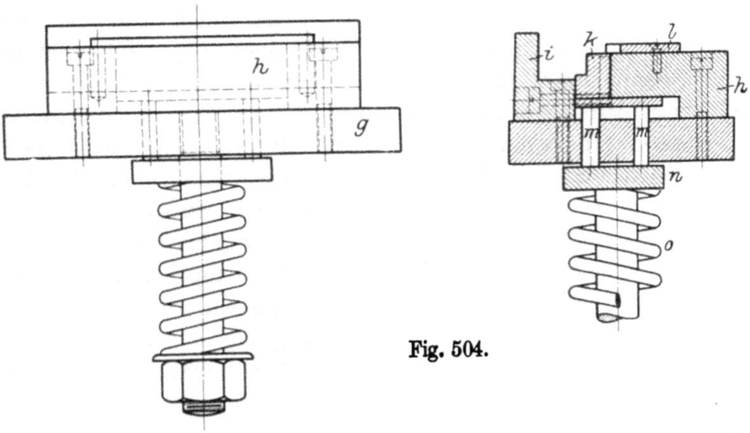

Fig. 504.

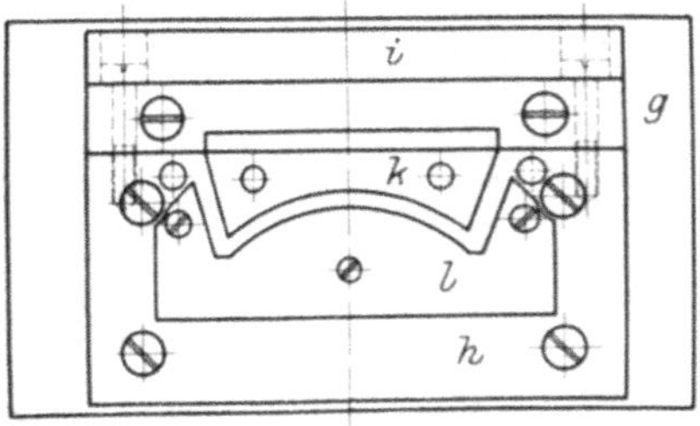

einem zylindrischen Zapfen aus Maschinenstahl, an dem der Biegestempel c für D fest mit zwei Zylinderkopfschrauben und Prisonstiften befestigt ist. Dieser Stempel dient gleichzeitig dem Biegestempel b, der die drei Biegungen A, B, C herstellt, als Führung, wozu eine Überlagplatte an den beiden überstehenden Seiten des Stempels b angeschraubt ist, die den Stempel c umgreift. Der Stempel b ist an den vier Zylinderkopfschrauben f, die im Stempelkopf lose gehen, angehängt und wird durch die zwei Wurstfedern e, die stark genug sein müssen, um die Biegung des Bleches bei D an der ersten Kante auszuführen, stets in seine tiefste Stellung gedrückt. Um ein richtiges Aufsetzen des Oberteiles zu sichern, sind zwei Führungsstifte d in dem beweglichen Biegestempel b eingepaßt, die in entsprechende Löcher im festen Matrizenteil eintreten.

Die Matrize ist auf einer Grundplatte g, Fig. 504, aus zwei Hauptteilen h und i, die mit versenkten Zylinderkopfschrauben von der Oberseite aus befestigt sind, aufgebaut. Diese beiden Teile sind außerdem noch durch Längsschrauben miteinander verbunden. Zwischen beiden führt sich der Auswerferstempel k, der mittels der Stifte m und der Druckplatte n auf dem Federauswerfer o gewöhnlicher Bauart ruht. Auf dem Matrizenteil h ist eine Anschlagplatte l aufgeschraubt, die im Verein mit der erhöhten Kante der anderen Matrizenhälfte i das Blankett in seiner richtigen Stellung sichert.

Beim Abwärtsgang des Stößels wird das Blankett zwischen dem Stempel b und dem unteren Auswerfer k gefaßt und, da infolge des geringeren Druckes der Federn im Stempel der Teil b zuerst zurückgedrückt wird, der erste Winkel der U-förmigen Biegung bei D angebogen. Sobald dann beim weiteren Abwärtsgang des Stößels der Stempel b zum Aufsitzen am Stempelkopf kommt, wird der Auswerferstempel k unter Überwindung der Federkraft des Auswerfers niedergedrückt, dessen Feder wiederum stark genug sein muß, um die dreifache Biegung A, B und C auszuführen. Vor dem Hubende tritt aber, nachdem bereits diese drei Kanten angebogen worden sind, der Fortsatz des festen Stempels c zwischen Auswerferstempel und Matrize i, wodurch die U-förmige Biegung bei D fertiggestellt wird.

Die Materialien der einzelnen Werkzeugteile, deren Bearbeitung usw. decken sich mit den in den früheren Beschreibungen gemachten Angaben.

C. Vereinigung mehrerer verschiedener Arbeitsvorgänge in einem Werkzeug.

1. Abtrennung und Formänderung mit Materialverschiebung.

Die bisher beschriebenen Werkzeuge verrichteten entweder nur einen einzigen Arbeitsvorgang oder mehrere, die stets derselben Gruppe angehörten.

Wenn nun, wie im folgenden besprochen werden wird, Arbeitsvorgänge, die verschiedenen Gruppen angehören, in einem Werkzeug vereinigt werden, so ist bei der Ausbildung dieser Werkzeuge auf die jeder einzelnen Gruppe zugehörigen Merkmale des einfachen Werkzeuges besondere Rücksicht zu nehmen, wenn es auch manchmal vorkommen wird, daß ein Teil des neuen Werkzeuges die Arbeiten zweier Teile der in diesem Werkzeug vereinigten Einzelarbeiten ausführt.

Diese Werkzeuge können entweder nach den Grundsätzen der Folgewerkzeuge, wo eine Einzelarbeit erst nach der anderen zur Ausführung kommt, oder der Verbundwerkzeuge ausgestaltet sein, bei denen in einem Pressenhub alle Einzelwerkzeuge an demselben Stück gleichzeitig zur Arbeit gelangen.

In beiden Fällen wird aber bei jedem Pressenhub ein fertiges Arbeitsstück ausgeworfen.

a) Folgewerkzeuge.

Werkzeuge, die mehrere verschiedenartige Arbeiten ausführen, sind, als Folgewerkzeuge ausgebildet, einfacher in der Durchbildung der betreffenden Einzelwerkzeuge, da genügend Platz vorhanden ist, um die zu jedem Werkzeug gehörigen Einzelteile unterzubringen. Dagegen ist die Zuführung des Materials zu den Einzelwerkzeugen verwickelter und verlangt manchmal ganz besondere Einrichtungen, die nur an dem einen Werkzeug gebraucht werden können. In dieser Beziehung wird diese Werkzeugart teurer.

Schnitt- und Ziehwerkzeug mit selbsttätigem Vorschub. In Fig. 506 ist ein Werkzeug dargestellt, mit dem aus Bandmaterial durch Lochen, Ausschneiden und Ziehen ein Näpfchen, Fig. 505, mit unregelmäßiger Oberkante hergestellt wird.

In Fig. 505 ist der Streifen — 0,28 mm dick und $^5/_8$" (15,875 mm) breit — mit den zwei Ausschnitten, die durch die Stempel I hergestellt werden, und das fertige Stück in drei Ansichten abgebildet. Das Material in dem Blechstreifen zur Linken der gestrichelten Linie ist das Blankett für ein Näpfchen.

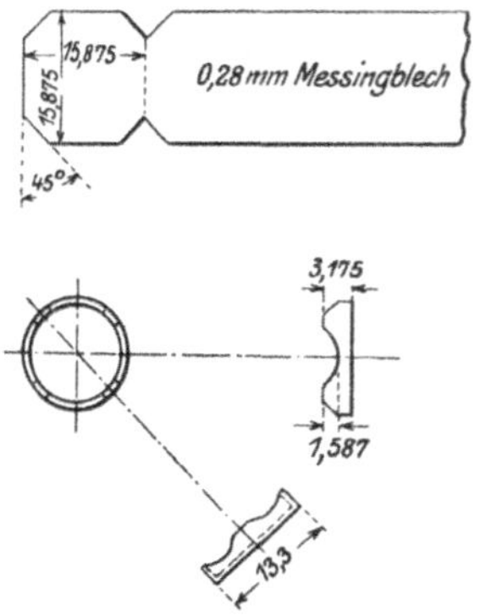

Fig. 505.

In Fig. 506 ist eine Vorderansicht des Werkzeuges, eine Seitenansicht und ein Grundriß der Grundplatte und des Schlittens gegeben, in denen der Stempel der Deutlichkeit wegen ausgelassen worden ist. A, ist das gußeiserne Bett, welches den Schlitten B trägt. Der Arm C für den Schlittenantrieb, seine Verbindung mit der Presse und seine Arbeitsweise stimmen mit Fig. 572 überein.

Hier muß der Teil des Streifens, der das Blankett bildet, abgetrennt werden, bevor eine der formgebenden Arbeiten stattfindet. Die Schnittplatte D liegt infolgedessen über der Oberfläche des Ziehringes E und ist auf der Unterseite mit einer Nut für den Zubringerschlitten F versehen, der mittels der Stellschraube G einstellbar ist.

Während der Blechstreifen zwischen Schnittplatte D und Abstreiferplatte H durchgeht, werden durch die Schnittstempel I die Winkel ausgeschnitten. Hierauf wird eine Blankettlänge vorgeschoben und durch den Schnittstempel J das Blankett abgeschnitten. Der Stempel ist so ausgebildet, daß er das Blankett nach dem Abtrennen von dem Streifen sicher vor den Zubringerschlitten F legt. Die folgende Aufwärtsbewegung des Stößels führt der Presse eine neue Blankettlänge des Bandmaterials zu, während das abgeschnittene Blankett über den Ziehring E befördert wird.

Die überhängenden Teile M_2 der Anschlagplatte M sind vor der Schnittplatte D nach oben abgeschrägt, so daß das Blankett seine richtige Stellung einnehmen muß. Die Größe des Vorschubes wird durch die Schraube G in Verbindung mit dem Schlitten F und dem Anschlag L geregelt, der in einer Nut der Anschlagplatte M liegt und durch die Schraube M' einstellbar ist. Die seitliche Einstellung des Streifens liegt vollständig fest, da sie von der Breite des verarbeiteten Materials abhängig ist, die gleich bleiben muß, um gute

Ergebnisse zu erzielen. Die Knagge X soll verhindern, daß der Streifen bei der Auswärtsbewegung des Schlittens zurückgezogen wird. Der Einbau des Auswerferstempels P, der durch die Gummifeder Q

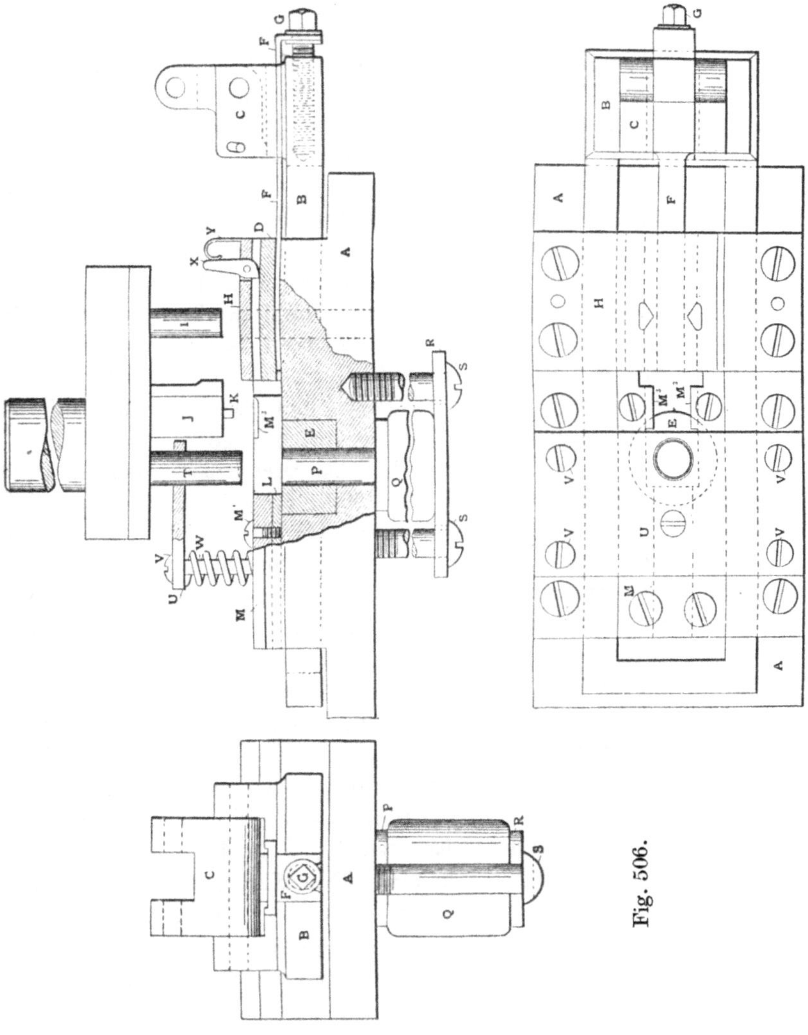

Fig. 506.

betätigt wird, erweist sich als nötig, da der Grund des Näpfchens leicht hohl sein soll; hätte das Näpfchen einen flachen Boden, so genügte ein gewöhnlicher Durchzug. Da durch die Wirkung des Auswerfers die Näpfchen an dem Ziehstempel hängen bleiben, muß ein Abstreifer U eingefügt werden, der sich an den Stiften V entlang bewegt und durch die Spiralfedern W stets nach oben gedrückt wird. Da der Weg des

Stempels größer ist als der Weg des Abstreifers, wird das Näpfchen beim weiteren Aufwärtsgehen des Stempels sicher von ihm abgezogen.

Werkzeug zur Herstellung von Schuppentaschen. Die Werkzeuge zur Herstellung der Schuppentaschen sind besonders interessant. Die weitere Beschreibung der dazugehörigen Werkzeuge (vgl. Fig. 124) wird zeigen, wieviel Genauigkeit und Erfindungsgabe beim Entwurf und der Ausführung derartiger Werkzeuge aufgewendet werden muß. Der Körper dieser Taschen besteht aus Reihen von Ringen, die durch vierzackige Krallen zusammengehalten werden. Ein Teil eines solchen Schuppenkörpers ist in Fig. 123 und die Kralle in Fig. 507 dargestellt. Die zwei unteren Reihen zeigen die Ringe, wie sie aus dem Schnittwerkzeug in Streifen kommen, während die oberen Reihen bereits durch die Krallen zusammengehalten werden. Sowohl die Ringe, wie die Krallen werden auf der Kraftpresse hergestellt. Hier sei das Werkzeug zur Herstellung der Krallen beschrieben.

Fig. 507.

Der Stempel, Fig. 508a, besteht aus drei Teilen, dem Schnittstempel D, der im Schnitt in Fig. 508c ersichtlich ist, dem Ziehstempel F und dem Keilstück G. Durch die Mitte des Schnittstempels geht ein kleines Stück Draht mit einem Ansatz am oberen Ende, gegen das eine Spiralfeder drückt, die in dem Körper des Stempels durch eine Stellschraube ohne Kopf gehalten wird. Dieser Federbolzen hat den Zweck, das Blankett sicher in die Vertiefung Y der Scheibe Z zu drücken und zu verhindern, daß es am Stempel hängen bleibt, wozu ohne diese Vorkehrung Neigung vorhanden wäre. Der Ziehstempel F wird durch eine Kopfschraube K im Stempelkopf gehalten, doch ist das Loch im Kopf soviel tiefer als das Ende des Stempels, daß, falls die Schaltung nicht richtig arbeiten und der Stempel gegen die Scheibe stoßen würde, F in dem Loche aufwärts gleitet und nicht zu Bruche kommt. Das Keilstück G wird so eingestellt, daß es mit der Rolle U arbeitet und beim Abwärtsgang die Scheibe Z schaltet. Bei T ist ein Loch vorgesehen, das zur Entfernung der Abfälle dient, da sich sonst beim Arbeiten des Werkzeuges Schwierigkeiten ergeben würden.

In der unteren Figur ist D die eigentliche Schnittplatte aus einem Stück Werkzeugstahl, die in die Gesenkplatte C eingelassen ist und mit der Oberfläche derselben abschneidet. Darüber ist die Abstreifplatte B befestigt, die eine Nut A zur Einführung des Rohmaterials enthält. Die kreisrunde Platte Z ist sehr sorgfältig in die Grundplatte A eingelassen und dreht sich um einen Bolzen in ihrer Mitte. In die Umfläche dieser Scheibe sind die Nuten für die Sperr- und Schaltklinke geschnitten, so daß die Scheibe bei einer Umdrehung achtmal festgestellt wird. Zur Linken dieser Scheibe befindet sich ein Schlitten P, der die Schaltklinke M trägt und mit einer Rückzugfeder S und einem einstellbaren Anschlag W versehen ist. Die Sperrklinke Q, die in einer Aussparung der Grundplatte gelagert ist, wird durch eine Feder R betätigt, die stark genug ist, um die Scheibe in ihre

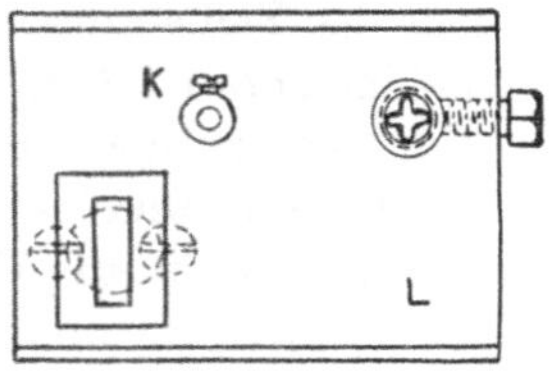

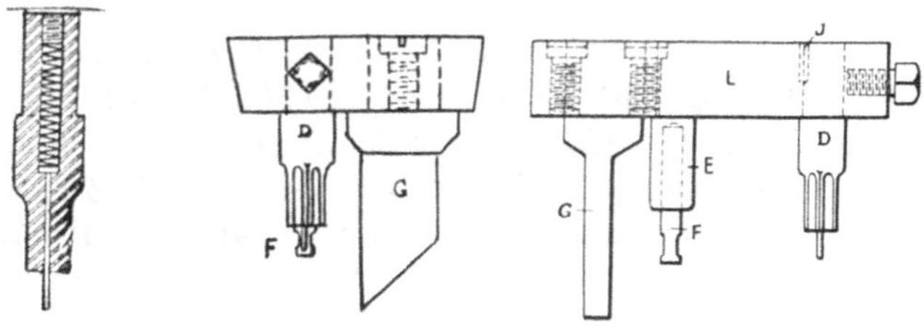

c a

richtige Stellung zu bringen, falls die Schaltung unregelmäßig arbeiten sollte. Die acht Löcher X sind in der Scheibe Z so gebohrt, daß sie durch die Schaltvorrichtung genau zentral unter die Schnittplatte gebracht werden. Diese Löcher werden bei Y ein wenig größer als das Blankett ausgebohrt, damit diese leicht auf den Grund der Ausbohrung niedergedrückt werden kann, wo es liegen bleibt, bis die Schaltvorrichtung es unter den Ziehstempel bei H bringt. Der Stempel F zieht das Blankett in die Form nach Fig. 507 und drückt es durch das Loch H aus der Maschine.

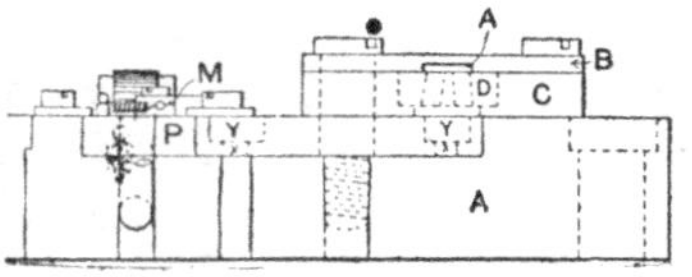

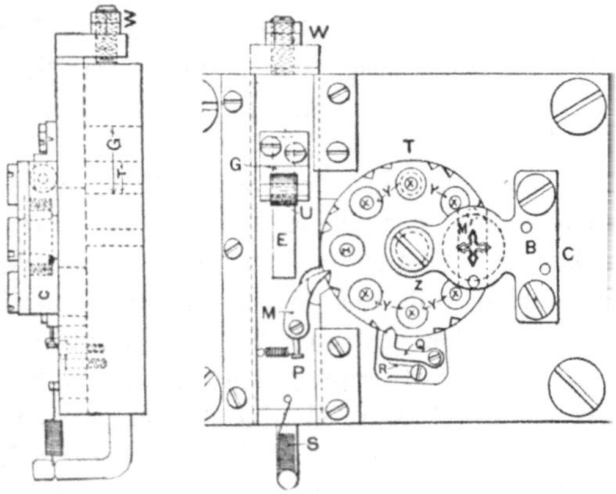

b

Fig. 508.

Die Arbeitsweise ist also folgende: Das Bandmaterial wird mittels eines gewöhnlichen Walzenapparates der Nute A zugeführt.

Nachdem die Blanketts ausgeschnitten worden sind, werden sie durch den Federbolzen im Stempel D in die Vertiefung Y gedrückt, in der sie bleiben, bis der Vorschub sie unter den Ziehstempel bei H bringt, wo sie durch den Ziehstempel F, Fig. 508, gezogen werden, um dann aus der Maschine herausfallen. Bei jedem Abwärtsgange der Presse bewegt sich der Schlitten P durch das Keilstück G und Rolle U so weit, daß die Scheibe Z um einen Zahn weiter geschaltet wird. Nachdem das Keilstück wieder in die Höhe gegangen ist, wird der Schlitten durch die Feder S in seine ursprüngliche Stellung zurückgeführt.

Folgewerkzeug zum Lochen, Ausschneiden und Ziehen mit Zubringerschlitten. In Fig. 510 ist eine verbesserte Bauart eines Werkzeugs zur Herstellung der kleinen Schloßgehäuse an den Rahmen billiger Ledertäschchen durch Lochen, Ausschneiden und Ziehen dargestellt. Vor Verwendung solcher Folgewerkzeuge waren zur Herstellung dieser Gehäuse zwei Arbeitsgänge notwendig: der erste bestand im Lochen und Ausschneiden, der zweite im Ziehen. Bei diesen Werkzeugen wird das Material hingegen ununterbrochen zugeführt und bei jedem Hub ein fertiger Gegenstand hergestellt. Fig. 509 zeigt das flache Blankett und das fertige Schloßgehäuse. Die Enden des Blanketts sind halbrund, mit Ausnahme der vorstehenden Nasen S, mit denen das Gehäuse an dem Rahmen des Täschchens befestigt wird. In die zwei Nuten bei T treten die Schloßhaken ein, U ist das Loch für den Fingerdruckknopf und V das Schlüsselloch.

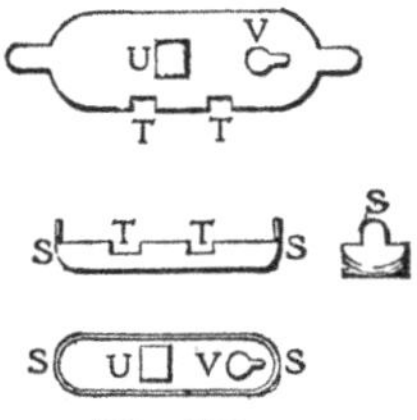

Fig. 509.

Fig. 510 zeigt das ganze Werkzeug in drei Ansichten, mit Ausnahme des Exzenterantriebes, der den Zubringerschlitten für das Blankett betätigt. A ist die Gesenkplatte, auf der sämtliche arbeitenden Teile des Werkzeuges angebracht sind. C ist ein erhöhter Teil derselben, der zur Aufnahme der Schnittplatte bearbeitet und mit einer Längsnut zur Aufnahme des Zubringerschlittens M versehen wird. Bei F und G sind die Ausschnitte für die Lochstempel, bei H der Umrißschnitt und bei I ein Loch für den selbsttätigen Anschlagstift gezeichnet. J sind zwei Führungsplatten für das Bandmaterial, E die Abstreifplatte und K vier Kopfschrauben, die die Schnittplatte auf der Gesenkplatte befestigen und in ihrer Stellung sichern. L ist ein Fortsatz des Abstreifers mit einem Loch für den Ziehstempel. Aus der Figur ist ersichtlich, daß die Unterseite der Schnittplatte mit einer Nut versehen ist, welche mit der in der Gesenkplatte übereinstimmt. Dadurch wird der Schlitten M benützt, um die Teile des Werkzeuges vollkommen gegeneinander auszurichten.

Der Zubringerschlitten M ist aus Werkzeugstahl, gehärtet und geschliffen, mit einer Ausfräsung für das Blankett und einem Loch bei U, welches den Ziehring bildet. Der Schlitten erhält eine hin- und hergehende Längsbewegung durch die Mitnehmerstangen N an jedem Ende und die runde Stange Q, die sich in den Lagern R durch einen Exzenterantrieb von der Hauptwelle der Presse aus verschiebt. Durch die Stellschrauben PP ist der Schlitten auf der runden Stange Q in der Längsrichtung einstellbar. Der Stempelkopf mit den einzelnen Stempeln ist ohne Beschreibung verständlich.

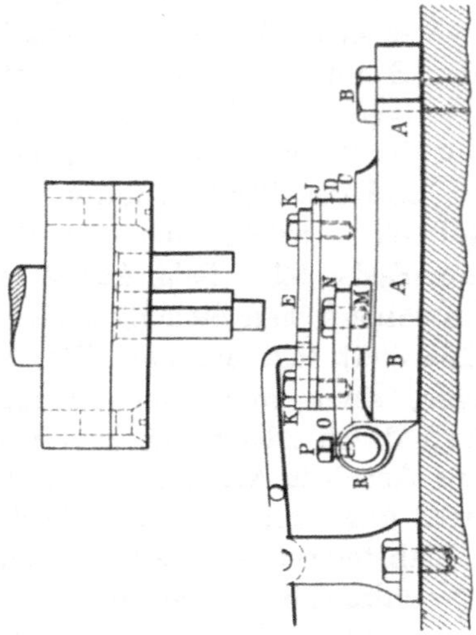

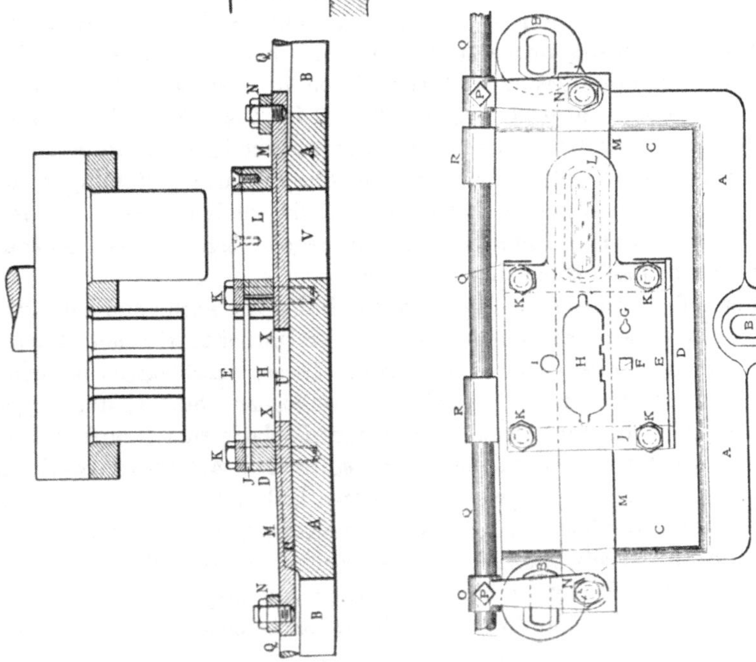

Fig. 510.

Bei der Arbeit mit diesem Werkzeug hat der Arbeiter das Blech oder den Streifen zwischen die Führungsplatten J bis gegen den selbsttätigen Anschlag einzuführen, wobei die Presse ununterbrochen läuft. Zuerst werden die Löcher U und V und beim nächsten Hub das Blankett ausgeschnitten und durch die Schnittplatte H gedrückt — der Stempel geht weit genug nach abwärts —, bis das Blankett auf die Oberfläche des Schlittens fällt, wo es sich selbst auf dem Sitz X einstellt. Beim nächsten Hub wird der Schlitten mit dem Blankett durch den Exzenterantrieb selbsttätig vorgeschoben, bis dieses genau unter dem Ziehstempel liegt. Dann bleibt der Schlitten stehen, während der Stempel weiter nach abwärts geht und das Blankett durch das Ziehloch U zieht, von wo es durch das Loch V, welches weiter als das gezogene Gehäuse ist, in einen Kasten am Boden fällt. Beim Aufwärtsgang des Stempels kehrt der Schlitten in seine frühere Stellung zurück, und das inzwischen ausgeschnittene Blankett fällt auf den Sitz X.

Die Einfachheit dieses Werkzeuges, seine große Herstellungsmenge, wie die geringe Menge Ausschuß durch Abnützung desselben sollten diese Werkzeuge auch für andere Arbeit in Aufnahme bringen. Da auch diese Konstruktion verhindert, daß zwei Blanketts gleichzeitig durch den Schlitten unter den Ziehstempel befördert werden, können Unglücksfälle nur durch grobe Nachlässigkeit entstehen.

Fast jede Art von Kraftpressen kann mit der Exzenterbewegung für dieses Werkzeug ausgestattet werden. Bei richtiger Arbeit der Presse können mit diesem Werkzeug 60 bis 75 Gehäuse in der Minute hergestellt werden, also bis ungefähr 40 000 im Tag. Mit selbsttätiger Zuführung von der Rolle, die gleich in der notwendigen Breite geliefert wird, können eine ganze Reihe kleiner Gegenstände viel billiger als durch irgendwelche andere Mittel hergestellt werden. Dazu würden Stifte, Krampen, Riemenhaken, kleine unregelmäßige Gehäuse, Metallknöpfe, Ösen, Schloßteile usw. gehören.

Der selbsttätige Anschlagstift kann bei allen Loch- und Schnittwerkzeugen verwendet werden, wo die Fertigarbeiten für den Gegenstand auf einer Seite aufeinander folgen. Dieser Anschlagstift besteht, wie ersichtlich, aus einer einstellbaren, gebogenen Stange, deren gebogenes Ende auf der Oberfläche der Schnittplatte ruht, während der Stempel arbeitet. Der Anschlag wird selbsttätig aus dieser Stellung beim Aufgang des Stempels nach jedem Hub gehoben, so daß das Material für den nächsten Arbeitsgang weiter zugeführt werden kann. Da er dann durch das Loch im Abstreifer in das eben gelochte Blech einfällt, arbeitet er als ein genauer Anschlag- bzw. Aufhängestift, der weder die Zuführung des Materials behindert, noch ein Aufheben desselben nach jedem Hub verlangt, wie es sonst nötig wäre.

Folgewerkzeug mit Zubringereinrichtung zur Herstellung von Unterlagscheiben für Fahrradfelgen. Das Werkzeug in Fig. 512 ist als Folgewerkzeug für Ausschneide-, Präge- und Locharbeit gebaut. Auf diesem

Werkzeug werden die gewöhnlichen käuflichen Unterlagscheiben für Fahrradfelgen hergestellt, die in zwei verschiedenen Herstellungsstufen vergrößert in Fig. 511 dargestellt sind. Diese Unterlagscheiben müssen alle den gleichen Außendurchmesser haben, während für das Loch eine Abweichung von ungefähr 0,07 bis 0,1 mm zugelassen wird. Die Zähne müssen gleiche Länge haben, die Vertiefung genau in der Mitte sein und die richtige Tiefe für den Kopf des Nippels haben. Da nun zwei Sätze Prägewerkzeuge vorhanden sind, kann ein Fehler an wenigstens acht Stellen auftreten. Das verarbeitete Rohmaterial sind Metallbänder von ungefähr 30 m Länge, 16 mm Breite und 0,8 mm Dicke. Abfallmaterial in diesem Werkzeug zu verarbeiten, wurde gar nicht erst versucht.

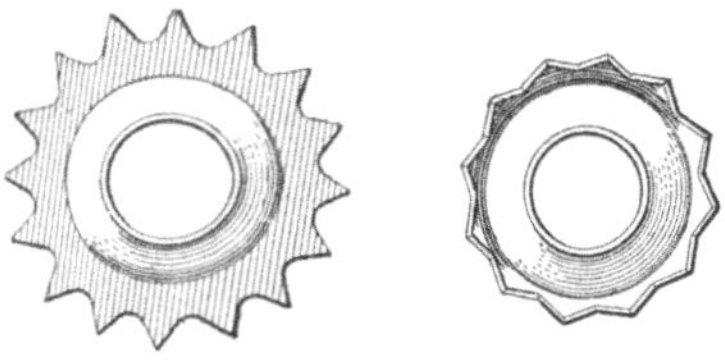

Fig. 511.

Das Werkzeug arbeitet in folgender Weise: Das Stahlband wird mittels einer doppelten Walzenzuführung in das Werkzeug gebracht und durch den kleinen Stempel H, Fig. 512, gelocht. Der Schlitten geht vor und führt dabei eine Stücklänge zu. Da aber der kleine und große Stempel um zwei Vorschube auseinander stehen, geschieht bei diesem Hube nichts weiter, als daß der kleine Stempel H den Streifen noch einmal locht. Beim nächsten Hub wird das Stahlband unter den großen Stempel G gebracht, wo die Scheibe ausgeschnitten und in der Ausdrehung des Schlittens bei S abgelegt wird, und zwar nicht durch ihr eigenes Gewicht, sondern zwangläufig durch den Stempel. Der Gleitstein in der Pleuelstange geht jedoch eine halbe Weglänge weiter, bevor die Bewegung des Schlittens D beginnt, so daß während der Bewegung von D aus einer Stellung in die andere der Gleitstein die obere Hälfte seines Hubes beendigt und ungefähr die Hälfte des Rückweges macht, bevor der Schlitten zur Ruhe kommt. Aus dem Grundriß ist der Winkel der Nockenscheibe von ungefähr 105° zu ersehen. Bei dem vierten Niedergange des Stößels wird — angenommen, der Schlitten D ist in der gezeichneten Stellung — der große Stempel G ein Blankett ausschneiden und in dem linken der beiden großen Löcher in D ablegen. Der Schlitten D bringt dabei das erste Blankett unter den Prägestempel FF, der es aus D in die Matrize niederdrückt, wobei die Zähne des Blanketts aufgebogen werden. Am Ende des Hubes kommt der Stempel FF mit dem Blankett gegen den inneren Ring [1]) und den Gegenstempel K, der die innere Prägung herstellt — in der Zeichnung ist der Stempelkopf BB im Schnitt in einer tieferen Stellung

[1]) In der Zeichnung nicht schraffiert, oberhalb K.

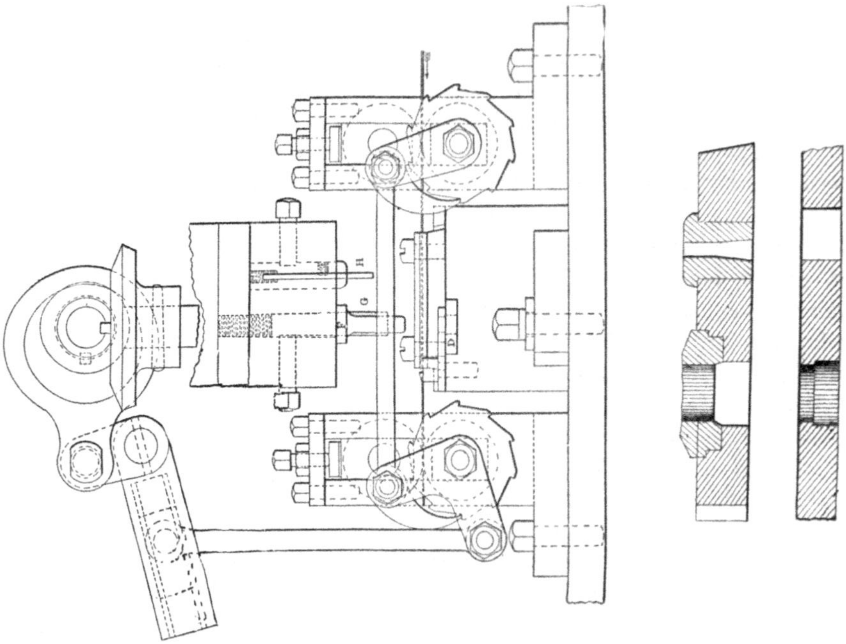

Fig. 513.

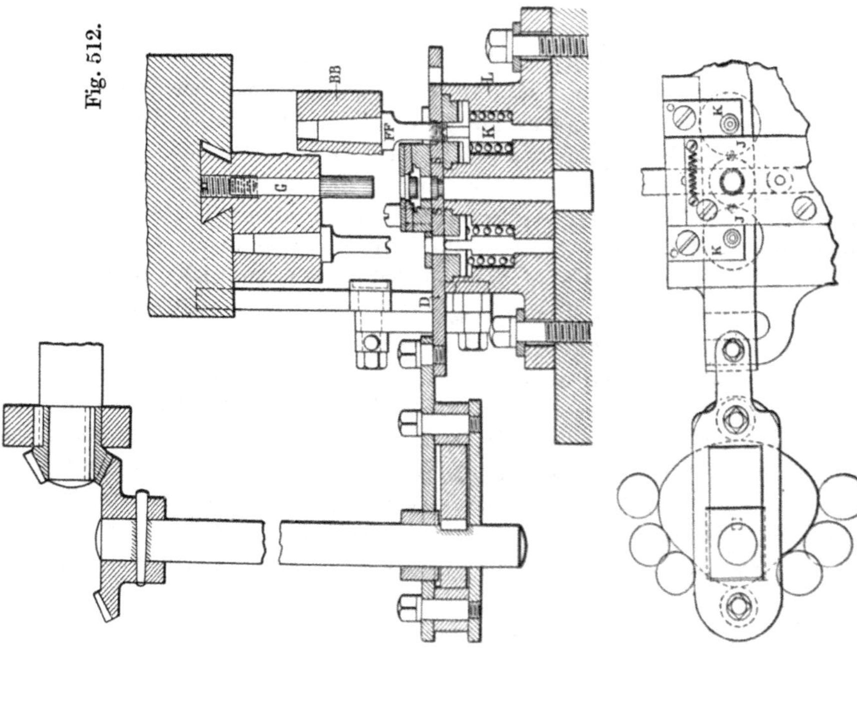

Fig. 512.

gezeichnet, um diese Arbeit deutlich zu zeigen. In diesem Hub geht der linke Prägestempel leer und kommt erst beim nächsten Abwärtshub zur Wirkung.

Wenn der Schlitten D nun zurückgeht, drückt die Feder L den Ring oberhalb K hinauf und damit die eben geformte Unterlagscheibe in den Schlitten zurück, so daß sie bei dem folgenden Linksgang des Schlittens durch das Loch unterhalb G in einen Kasten unter der Presse fällt. Von nun an wird, da jetzt erst das Werkzeug voll ist, bei jedem Hub eine fertige Unterlagscheibe ausgeworfen, so daß bei 10stündiger Arbeit, da die Presse leicht mit hundert minutlichen Hüben laufen kann, 60000 Stück im Tag hergestellt werden. Ein solches Werkzeug lief, um einen Rekord aufzustellen, $6\frac{1}{2}$ Tage und stellte 390000 Unterlagscheiben her, ohne nachgeschliffen zu werden. Fig. 513 zeigt Einzelheiten der Werkzeuge.

Ein fünffaches Folgewerkzeug. In dem Werkzeug in Fig. 515 werden fünf getrennte Arbeiten nach Tabelle Nr. 4 ausgeführt, bevor der fertige Gegenstand, Fig. 514, die Presse verläßt. In vielen Werkstätten würden drei Pressen und drei Sätze Werkzeuge verwendet werden, um dieselbe Arbeit zu verrichten, die hier ein einziges Werkzeug ausführt.

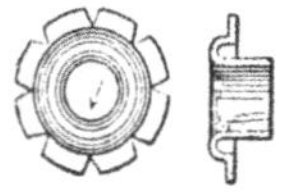
Fig. 514.

Es erscheint nicht angezeigt, die Schnittplatte ganz aus einem Stück zu machen, da dann der kleinste Härteriß das ganze Werkzeug unbrauchbar machen würde. Deshalb wurde eine Gesenkplatte aus Stahl verwendet, in die Vertiefungen für die einzelnen Matrizen gebohrt wurden. Diese Löcher können in folgender Weise ausgeteilt werden, um eine größtmögliche Genauigkeit zu erhalten: Die eine Seite und das eine Ende der Gesenkplatte werden genau unter rechtem Winkel bearbeitet und auf der Oberfläche der Platte eine Mittellinie angerissen. Die Stellung der Löcher wird dann ungefähr mit einem Bleistift markiert und das erste Loch in der Drehbank gebohrt, wozu der Block auf der Planscheibe befestigt wird. Nach Fertigstellung des ersten Loches werden, bevor man die Aufspannung des Blockes löst, auf der Planscheibe Winkel festgespannt, die an den bearbeiteten Flächen des Blockes anliegen. Dann werden die Befestigungsklammern des Blockes gelöst und derselbe parallel zu sich selbst so weit vorgeschoben, daß man ein Zwischenstück, dessen Dicke der verlangten Entfernung der beiden Löcher entspricht, zwischen Winkel und Block einschieben kann. Dann wird der Block für die Gesenkplatte wieder festgespannt und das zweite Loch gebohrt. Indem man diese Arbeitsweise wiederholt und ein Zwischenstück nach dem anderen zufügt, bis alle Löcher gebohrt sind, kann man ihre Austeilung auf der Gesenkplatte in der genauesten Weise vornehmen. Der Stempelkopf und der Abstreifer werden mit Benutzung derselben Zwischenstücke gebohrt.

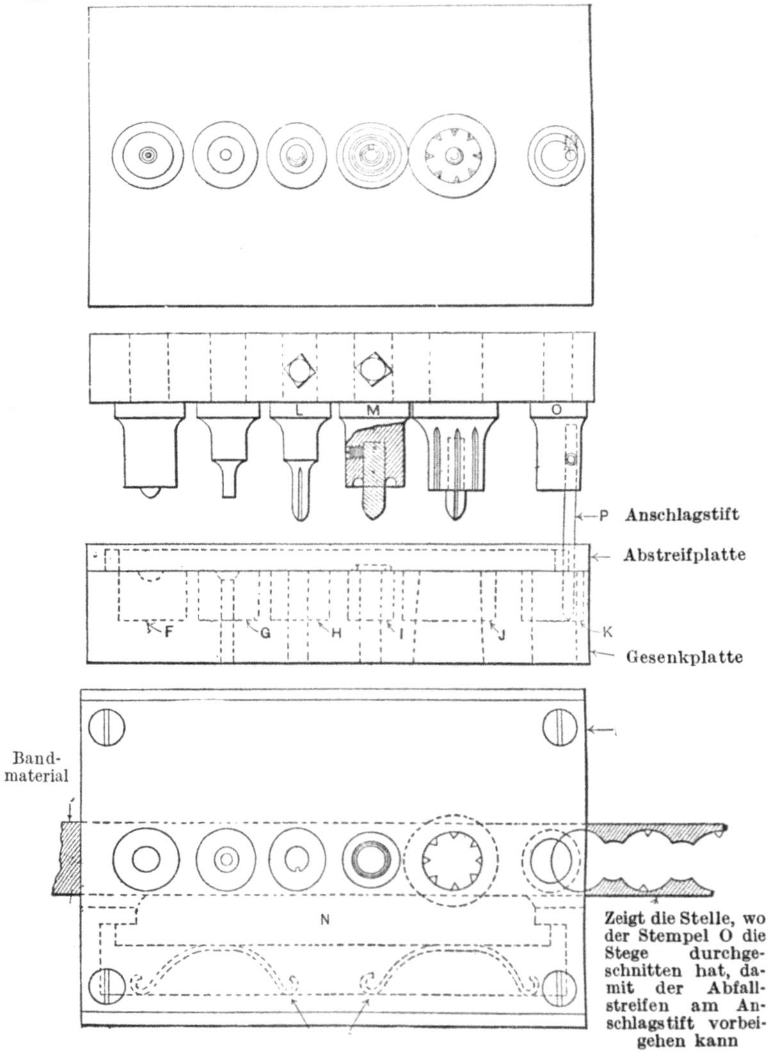

Fig. 515.

Hierauf werden die Büchsen F, G, H, I, J und K gemacht, gehärtet und auf Maß poliert. Um den Außenumfang dieser Teile gleichachsig mit dem Loch zu erhalten, wird die Büchse auf ein Stück weichen Stahles, das in einem Futter gehalten wird, aufgezogen und abgedreht. Nun werden diese Büchsen in ihre Sitze in der Gesenkplatte eingepreßt, womit der Werkzeugunterteil fertig ist. Dann werden die Stempel allseitig geschliffen, um sie sicher gerade zu erhalten, und in den Stempelkopf eingepreßt. Die Zieh- und Prägestempel L und M werden mit Druckschrauben gehalten, um ein Herausziehen derselben aus dem Stempelkopf zu verhindern. Wenn man ein Werkzeug mit zwei oder mehr Stempeln verwendet, so ergeben sich häufig Schwierigkeiten, wenn die Breite des verarbeiteten Materials nicht gleichmäßig ist. Paßt ein Streifen aus der Lieferung in die Nut des Abstreifers, so ist zu erwarten, daß der nächste überhaupt nicht hineingeht. Man wendet deshalb einen Federanschlag N an, wie bereits in Fig. 108 und 110 angegeben wurde. Die Nut im Abstreifer wird 1,5 mm breiter als die Sollbreite des Bandmaterials gehobelt und der Abstreifer rückwärts ausgeschnitten, damit der Federanschlag N leicht darin gleiten kann, wenn der Abstreifer in seiner Stellung im Werkzeug befestigt ist. Aus der Figur ist leicht zu ersehen, wie die Federn den Anschlag und den Streifen gegen die feste Führung im Abstreifer pressen. Der Stempel O hat keine Arbeit auszuführen, bevor die Unterlagscheibe nicht vollständig fertig ist. Er dient zum Durchschneiden des übrigbleibenden Steges in dem Streifen, so daß dieser weiter vorgeschoben werden kann, bis der Anschlagstift an den nächsten Steg kommt. Da der Anschlagstift nicht aus dem Streifen heraustritt, kann bei der Zuführung auch kein Hängenbleiben desselben usw. eintreten.

Nachdem das Werkzeug vollständig in die Presse eingebaut worden ist, muß natürlich der Stößel fünf Hübe machen, bevor das erste fertige Stück ausgeworfen wird, doch wird nachher bei jedem Preßhub ein fertiges Stück hergestellt.

Der erste Stempel, zur Linken beginnend, prägt die Vertiefung in den Streifen, wobei die Stirnfläche des Stempels den Streifen beim Auftreffen gerade richtet. Der zweite Stempel stellt in dieser Vertiefung das Loch her, der nächste zieht die Hülse und bildet gleichzeitig die Keilnut, die am fertigen Stück zu sehen ist, der vierte Stempel ist der Prägestempel und der vorletzte schneidet das Stück aus.

b) Verbundwerkzeuge.

Werkzeuge, die mehrere verschiedene Einzelarbeiten ausführen, können auch so gebaut werden, daß sämtliche Arbeiten gleichzeitig an demselben Stück in einem Pressenhub fertiggestellt werden. Diese Werkzeuge seien entsprechend den Verbundschnittwerkzeugen allgemein als Verbundwerkzeuge bezeichnet.

Sie werden für einfach- und doppeltwirkende Pressen gebaut und können auch mit Vorteil in Revolverpressen verwendet werden. Naturgemäß wird dasselbe Werkzeug einfacher, wenn es als Verbundwerkzeug für eine doppeltwirkende Presse gebaut ist als zur Verwendung in einer einfachwirkenden, da dann alle eingebauten Federn usw. wegfallen.

Die einfachste Form, die auch die weiteste Verbreitung gefunden hat, ist das Verbundwerkzeug zum gleichzeitigen Ausschneiden und Ziehen, da es die Verwendung von Bandmaterial für Zieharbeiten gestattet. Während aber bei einem Folgewerkzeug für ähnliche Arbeiten das Ausschneiden die letzte Arbeit bildet, die erst den Gegenstand von dem Streifen trennt, sobald alle an diesem Gegenstand auszuführenden Arbeiten beendet sind, wird in dem Verbundwerkzeug zuerst die Scheibe ausgeschnitten, worauf das Ziehen stattfindet. Aus diesem Grunde findet man auch häufiger als sonst die umgekehrte Anordnung, daß der Ziehstempel in dem Unterteil, der Schnittplatte, sitzt und der Ziehring im Oberteil, am Pressenstößel, befestigt ist. Dann ist auch ein einfacher Durchzug nicht möglich, und das Stück muß nach oben von dem Stempel durch einen besonderen Abstreifer entfernt werden, während im ersten Fall die einfache Abstreifkante im Ziehring wie bei einem einfachen Ziehwerkzeug genügt.

Ziehwerkzeug für eine doppeltwirkende Exzenterpresse. Das im folgenden beschriebene Ziehwerkzeug wurde auf einer doppeltwirkenden Exzenterpresse mit selbsttätigem Walzenvorschub verwendet. In Fig. 516 stellt A den Schnittstempel und B den Ziehstempel vor. Der erste hat einen geraden Zapfen S', der am Ende mit Gewinde versehen ist und in dem Stempelkopf durch eine Gegenmutter gesichert gehalten wird. Der Ziehstempel B wird an dem inneren Stößel der Presse mittels der Ausdrehung S befestigt. Bei der Herstellung werden Schnittplatte und Ziehring gesondert gehalten, da die Ziehplatte D viel öfter als die Schnittplatte C erneuert werden muß.

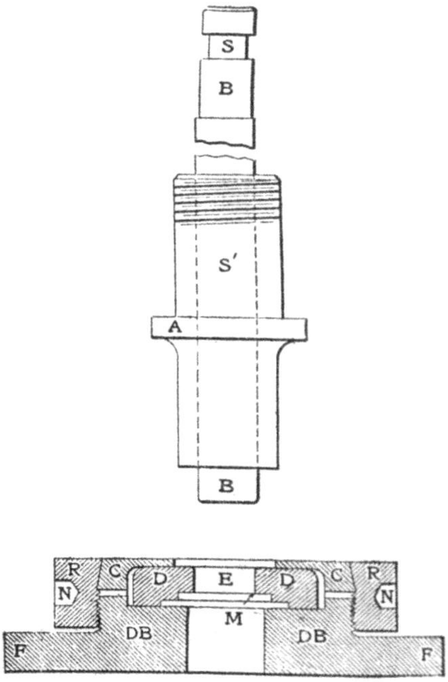

Fig. 516.

Bei derartigen Verbundwerkzeugen schneidet der Stempel A die Scheibe aus und hält sie auf der Oberfläche des Ziehringes D sicher fest, während der Stempel B niedergeht und das

Blankett durch den Ziehring D durchzieht. Wenn nun der Stößel wieder in die Höhe geht, wird die Hülse an der Kante M vom Stempel B abgestreift.

Die untere oder Ziehplatte D ist seitlich frei beweglich und mittelt sich deshalb selbst aus, wird jedoch in der gußeisernen Gesenkplatte DB durch die Schnittplatte C niedergehalten, welche mit der Überwurfmutter R darauf niedergezogen wird. Zum Anziehen der Mutter dienen die Löcher N. Der Schnittstempel A bleibt ungehärtet, damit seine Schnittkante leichter an die Schnittplatte C angepaßt werden kann. Der Ziehring D wird gehärtet, angelassen und geschliffen, während das Ziehloch E sorgfältig auf Maß poliert wird. Der Ziehstempel B ist gehärtet und geschliffen, aber nicht angelassen. Die Schnittplatte C ist gehärtet und angelassen und wird an der Schnittkante innen unter einem Winkel von $1^1/_2{}^0$ angeschliffen.

Es ist sehr wichtig, daß sowohl die Schnittplatte wie der Ziehring gehärtet und mit einer richtigen und ebenen Auflage abgeschliffen werden, da sonst die Hülsen beim Ziehen Falten machen. Mit einem derartigen Satz Werkzeuge hat man ungefähr 3000 kg Messinghülsen aus Material von 0,38 mm Dicke in drei Tagen mit einmaliger Einstellung des Werkzeuges ausgeschnitten und gezogen. Es ist natürlich nicht zu leugnen, daß die Anschaffungskosten dieses Werkzeuges etwas höher als die einer gangbaren Form sind, doch wurde der Unterschied durch das Ergebnis mehr als aufgewogen.

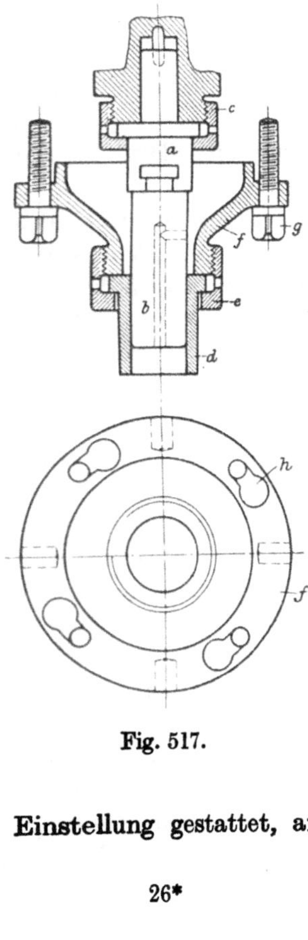

Fig. 517.

Stempelbefestigung für doppeltwirkende Pressen (deutsche Bauart). In Fig. 517[1]) ist eine Stempelbefestigung angegeben, die ein schnelles Einspannen und Lösen des Ziehstempels und des Niederhalters gestattet. Unleugbar wird sie teurer sein als die vorher und im folgenden angegebene Befestigung, doch vermeidet sie jedes Zwängen der beiden aufeinandergleitenden Teile, da sich der Niederhalter nur nach dem Stempel ausmittelt. Der Niederhalter d ist nämlich mit der Überwurfmutter e, die ihm eine seitliche Einstellung gestattet, an

[1]) W. T. 1910, S. 409.

dem Deckel f des Stößels befestigt; sie wird erst angezogen, wenn der Niederhalter sich im ausgerichteten Werkzeug am Stempel ausgemittelt hat. Die Verbindung des Stempels mit dem Stößel erfolgt durch einen Notstempel a mit T-Nut, während dieser mit Zylinderzapfen und Überwurfmutter fest an dem Stößel befestigt wird. Zum Auswechseln des Stempels braucht man nur die vier Schrauben g zu lösen und kann Deckel und Niederhalter nach unten abziehen.

Verbundwerkzeug zum Ausschneiden, Ziehen und Rändeln. Die Fig. 519/20 zeigen ein Verbundwerkzeug, das zur Herstellung eines gerändelten Deckels, Fig. 518, in einer doppeltwirkenden Kraftpresse in einem Gang dient. Auch die Froschplatte, die in Verbindung mit diesen Werkzeugen verwendet wird, spart viel Zeit. Diese Werkzeuge schneiden und ziehen einen Messingdeckel von 76 mm Durchmesser, 13 mm Tiefe und 1,3 mm Blechdicke. Der Deckel, Fig. 518, gehört zu einer zylindrischen Schachtel und ist an der Außenseite gerändelt, um das Öffnen und Schließen zu erleichtern.

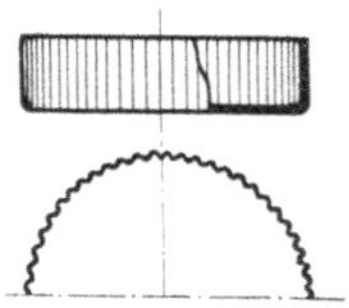

Fig. 518.

Fig. 519 zeigt den Schnittstempel A, der an der Schnittkante 102 mm Durchmesser hat. Er hat einen konischen Zapfen und paßt hart in den Stempelkopf D, der an der Presse mit zwei $^5/_8{}''$ Schrauben befestigt ist.

Der Ziehstempel B wird aus zwei Teilen hergestellt. Der eigentliche Ziehstempel ist aus Werkzeugstahl gemacht, 82,5 mm lang und mit einem 1″ Gewinde versehen, mit dem er in seinem Stempelkopf B′ verschraubt ist. Der Stempel ist gehärtet, geschliffen und auf Maß poliert. Der angedeutete Luftkanal macht nicht allein das Ziehen leichter, sondern erleichtert auch das Abziehen des Stückes vom Stempel. Der Stempelkopf B′ wird für verschiedene Größen des Ziehstempels verwendet und wird deshalb in der Presse durch einen eingedrehten Zapfen gehalten, der in die Nut des Stößels paßt, wie in Fig. 517 gezeichnet ist.

Der Werkzeugunterteil besteht aus zwei Teilen, der Schnittplatte und dem Ziehring. Die erste, F, ist aus Werkzeugstahl, gehärtet, oben und unten geschliffen und an der Oberkante abgeschrägt, so daß sie mit der entsprechenden Abschrägung der Überwurfmutter K in Fig. 520 zentrisch angezogen werden kann. Das Loch von 102 mm Durchmesser ist mit sehr wenig Spiel ausgeschliffen, da sich sonst das Blankett nach dem Ausschneiden leicht von einer Seite zur anderen schiebt, wodurch der Deckel ungleich hoch ausfallen würde.

L ist der Ziehring. Seine Ausdrehungen bei E und G dienen dazu, aus dem Blankett den gerändelten Deckel zu ziehen. Die

Einziehöffnung des Ziehringes ist bei E abgerundet, so daß das Blankett leichter durch den Ring gedrückt werden kann. Der Durchmesser

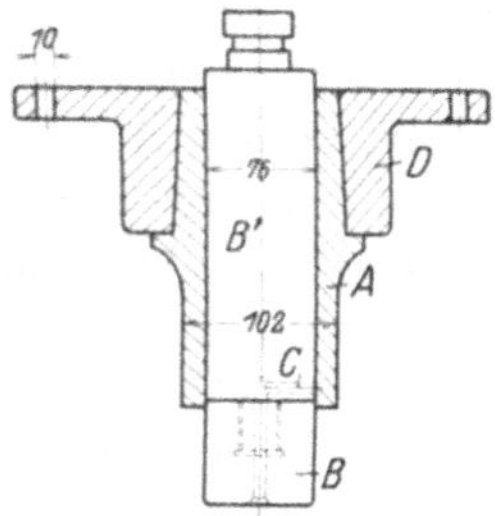

Fig. 519.

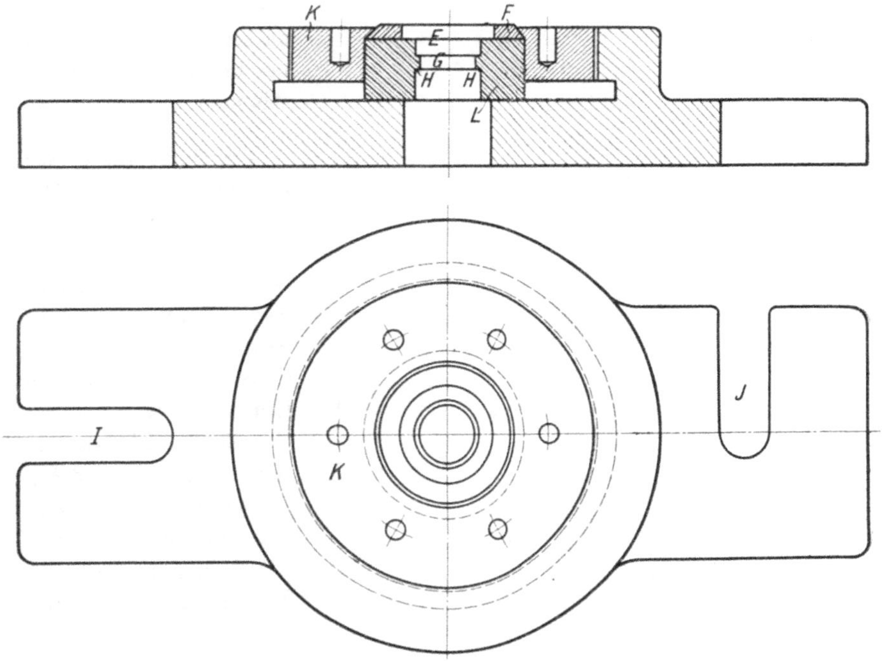

Fig. 520.

bei E ist um 0,5 mm größer als bei G, so daß der Deckel gezogen wird, bevor die Rändelung durch G beginnt. Der Durchmesser von G ist wiederum etwas kleiner als das Endmaß, da sich der Deckel nach dem Durchgang durch den Ring ein wenig ausdehnt. Die Ecken bei H sind scharf und streifen beim Aufwärtsgang der Presse den Deckel vom Stempel ab. G wird mit dem Rändelwerkzeug, Fig. 521, gerändelt.

Kurz besprochen sei noch die Froschplatte. Der Schnitt- und Ziehring werden durch die Überwurfmutter K in ihrer Stellung gehalten, welche mit einem Aufsteckschlüssel, der mit Stiften in die gezeichneten Löcher greift, angezogen wird. Die Froschplatte ist von den sonst gebräuchlichen verschieden, da sie die Schraubenschlitze I und J unter rechtem Winkel zueinander trägt. Wenn man die Froschplatte aus der Presse nimmt oder einsetzt, so braucht man nur die Schrauben zu lösen, die man sonst gänzlich entfernen muß, bevor die Froschplatte entfernt werden kann. Beim Einsetzen dieser Froschplatte wird der Schlitz I an die linke Schraube gesetzt und die Platte nach rechts herumgedreht, bis die andere Schraube in den Schlitz J eintritt. Die Zuführung sowie der Abstreifer sind nicht abgebildet, da sie mit andern hier beschriebenen übereinstimmen.

Fig. 521.

Verbundwerkzeug mit Schutzvorrichtung. Die Konstruktion eines solchen Verbundwerkzeuges zum Schneiden und Ziehen der Näpfchen für Edisongewinde, deren Herstellungsgang in Tabelle Nr. 22 angegeben ist, zeigt Fig. 522[1]).

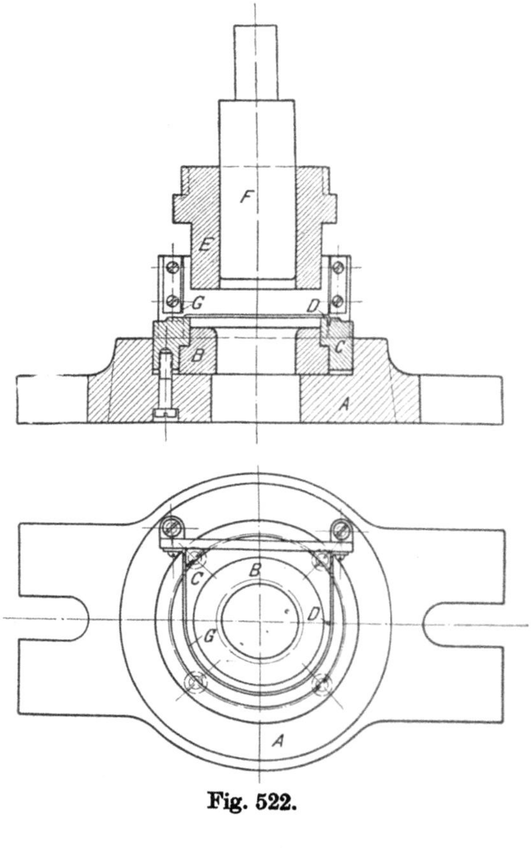

Fig. 522.

Der Oberteil besteht aus dem Schnittstempel E, in dem sich der Ziehstempel F führt, die beide in den Stößeln einer doppeltwirkenden Ziehpresse befestigt werden. Der Unterteil wird von der gußeisernen Froschplatte A gebildet, die in einer Ausdrehung den Schnittring C trägt, der von unten durch vier versenkte Zylinderkopfschrauben auf den Ziehring B niedergezogen wird. Er ist abgesetzt und dient

[1]) W T. 1909, S. 16. Perls, Massenfabrikation der Edisongewinde.

so gleichzeitig als Zentrier- und Haltevorrichtung für den Ziehring B; beide werden aus Werkzeugstahl in der üblichen Weise hergestellt. Die untere scharfe Kante des Ziehringes B bildet den Abstreifer für das gezogene Näpfchen, während ein aus einem gelochten Blech gebildeter Schutzring G gleichzeitig als Abstreifer für das Blech dient.

Beim Niedergang des Preßenstößels wird das Blankett vom Schnittstempel E ausgeschnitten, der sich jetzt aufsetzt und als Niederhalter für den folgenden Ziehvorgang dient, bei dem der Ziehstempel F das Näpfchen durch den Ziehring nach unten durchzieht.

In dem Schnittring C ist ein Anschlagstift D vorgesehen.

In der angezogenen Quelle wird auch über den oben von Woodworth angeführten Wert der Vereinigung mehrerer solcher Werkzeuge in einem Mehrfachwerkzeug geurteilt. Doch wird die beinahe entgegengesetzte Schlußfolgerung gezogen, daß es sich wohl noch lohnen kann, mehrere einfache Schnittwerkzeuge zu einem Mehrfachschnitt zu verbinden, keineswegs gälte dies aber von Verbundwerkzeugen für Zieharbeiten obiger Art. Es wüchsen nämlich die Herstellungs- und Reparaturkosten sowie die Kosten für das Einspannen unverhältnismäßig an, außer wenn es sich um kleine und niedrige Stücke handelt.

Verbundwerkzeug für die einfachwirkende Presse. Bei der Herstellung von Linsenfassungen, S. 335, wird im ersten Gang das Blankett durch ein Verbundschnitt- und Ziehwerkzeug nach Fig. 523 ausgeschnitten und gleichzeitig hochgezogen. Dieses Werkzeug kann als Beispiel für ein Verbundwerkzeug, das in einer einfachwirkenden Presse arbeitet, angesehen werden. Auf den ersten Blick ist die ungleich verwickeltere Bauart gegenüber den bis jetzt besprochenen Werkzeugen erkennbar, da durch den notwendigen Einbau der Federn auch die umgekehrte Werkzeuganordnung — Stempel im Unterteil und Ziehring im Oberteil — angewendet werden muß. Das verarbeitete Material ist Messingblech von 0,38 mm Dicke.

A in Fig. 523 ist der Stahlstempel mit dem Befestigungszapfen am oberen Ende, während der untere Teil von 36,8 mm Durchmesser mit dem Schnitt I die Scheibe ausschneidet. Hierauf zieht der innere Umfang von A mit 32 mm Durchmesser den Umfang der Scheibe über den Ziehstempel F, der 31,2 mm Durchmesser hat. Wenn der Stempel A über F heruntergeht, geht auch der Abstreifer H, der auf drei Federn J ruht, nieder, um dann bei seinem Aufgang das Näpfchen von F abzuheben. Diese Federn sind aus 1,6 mm Stahldraht gemacht. Der Stempel B mit seiner Feder E wirft das Stück aus A aus, während der äußere Abstreifer C das Bandmaterial von der Außenseite von A abstreift. Dieser äußere Abstreifer hat drei Schlitze mit

Anschlagschrauben b und zwei Nuten bei a, um den Anschlagstiften G auszuweichen. Die Feder D ist aus 3,2 mm Draht gewickelt. Die Schnittplatte I ist selbstverständlich fest in der Grundplatte L gelagert. Das Mittelloch in B dient zum Ausrichten der Werkzeuge mittels des Stiftes G.

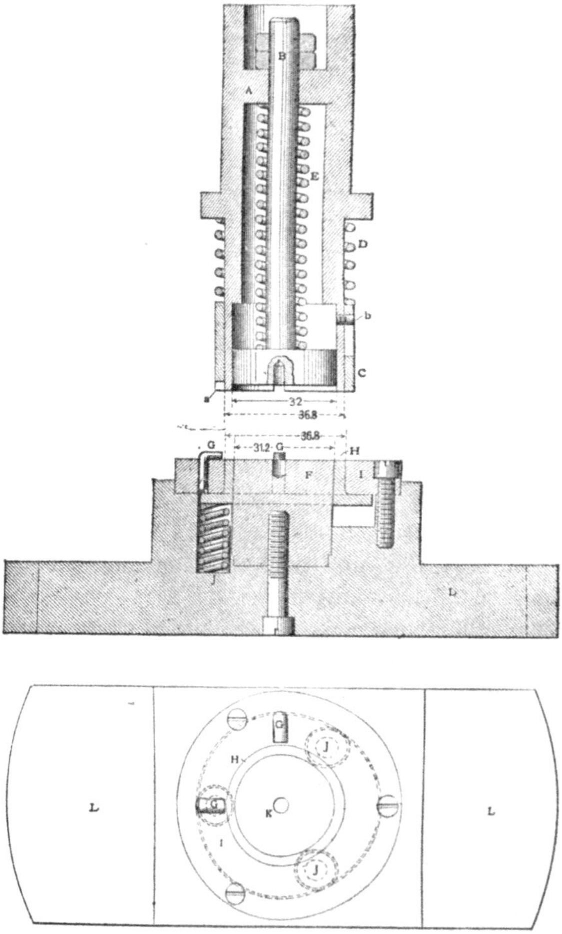

Fig. 523.

Verbundwerkzeuge für einfachwirkende Pressen zur Herstellung von Kannenschnauzen. In Tabelle Nr. 24 ist der Herstellungsgang der Schnauzen für große Blechkannen zum Schiffstransport von Flüssigkeiten angegeben. In den Fig. 524/5 sind die bezüglichen Werkzeuge dargestellt. Es sind Verbundwerkzeuge, die in der einfachwirkenden Presse gebraucht werden und bei jedem Preßhub eine bis drei Arbeiten vollführen. Mit diesen Werkzeugen kann man im Tag je nach der Gewandheit des Arbeiters 12000—15000 fertige Schnauzen herstellen.

Das erste Werkzeug, Fig. 524, besteht aus fünf Hauptteilen: A ist die gußeiserne Froschplatte, die nach der Linie a b geteilt ist, damit man das Werkzeug bei Ausbesserungen leicht auseinandernehmen kann. B ist der Schnittring, der in die obere Hälfte der Froschplatte eingelassen und durch vier — nicht gezeichnete — Zylinderkopfschrauben gehalten wird. C ist der Ziehstempel, der in die untere Hälfte der Gesenkplatte eingepaßt ist und ebenfalls durch Zylinderkopfschrauben, die nicht gezeichnet sind, gehalten wird. D ist der Blechhalter bzw. Auswerfer, der auf drei Stiften liegt, von denen einer gezeichnet ist, die von der Federdruckplatte E getragen werden. Diese ruht auf einer Gummifeder, die sich auf dem Bolzen F führt und durch eine Unterlagscheibe mit Mutter, die in der Zeichnung nicht sichtbar ist, gespannt werden kann, um den Druck des Niederhalters während des Ziehens zu regeln. G ist der Schnittstempel und zugleich Ziehring, dessen Außendurchmesser in den Schnittring B paßt.

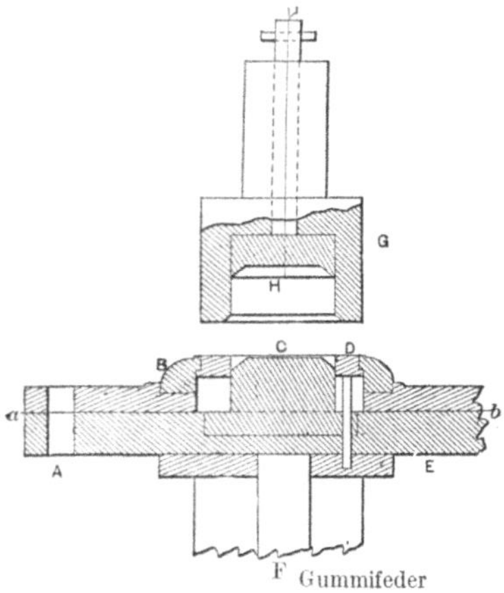

Fig. 524.

Der innere Durchmesser entspricht dem Durchmesser des Ziehstempels C, vermehrt um die doppelte Metalldicke. H ist der obere Auswerfer, der gleichzeitig mit dem Ziehstempel C zusammen arbeitet. Er bildet die Bodenfläche des Näpfchens beim Hubende.

Diese Werkzeuge werden in eine schräge Presse eingebaut. Sobald der Stempel mit der Schnittplatte B in Berührung tritt, schneidet er die Scheibe aus, die durch den Blechhalter D gegen die Stirnfläche des Stempels G gedrückt wird. Beim weiteren Abwärtsgang des Stempels G wird das Blech über den Ziehstempel C gezogen, worauf beim folgenden Aufwärtsgang des Stempels der Bolzen I, der auf der Oberseite des Stempelzapfens austritt, an eine Stange in der Presse anstößt[1]), so daß der Auswerfer H niedergedrückt und das in Tabelle Nr. 24 G. 2 abgebildete Näpfchen ausgestoßen wird und nach rückwärts abgleitet.

Die nun folgende Umformung im zweiten Ziehgang, Tabelle Nr. 24 G. 3 geschieht in einem Verbundwerkzeug nach Fig. 601, S. 488.

---

[1]) Vgl. Fig. 366/73.

Im dritten Gang G. 4 folgt das Ausschneiden des Bodens und das Umlegen des unteren Flansches, als Vorbereitung für das Einrollen im letzten Gang G. 5.

Fig. 525 stellt das Werkzeug für den dritten Gang dar, der aus drei Einzelarbeiten besteht, A ist die Froschplatte, B die Schnittplatte, C die Matrize, D der Niederhalter und E der untere Schnittring. Die Druckplatte F nimmt die Tragstifte für den Niederhalter auf, während G eine Abfallröhre für die ausgeschnittenen Böden ist. Diese finden beim Legen von Dachpappe Verwendung, so daß man tatsächlich in einem Arbeitsgang zwei Gegenstände[1]) herstellt. Die Werkzeuge werden in einer schrägen Presse verwendet. Beim Abwärtsgang des Stempels durchschneidet der Stempel I den Boden, während gleichzeitig der Stempel H die umgebördelte Kante zuschneidet und beim weiteren Abwärtsgang die Hülse über die Kante der Matrize C zieht. Beim Aufwärtsgang des Stempels kommt die Stange mit dem Bolzen M in Berührung, so daß der Abstreifer J durch den Bolzen K nach abwärts gestoßen wird und die in Tabelle Nr. 24 G. 4 abgebildete Schnauze auswirft.

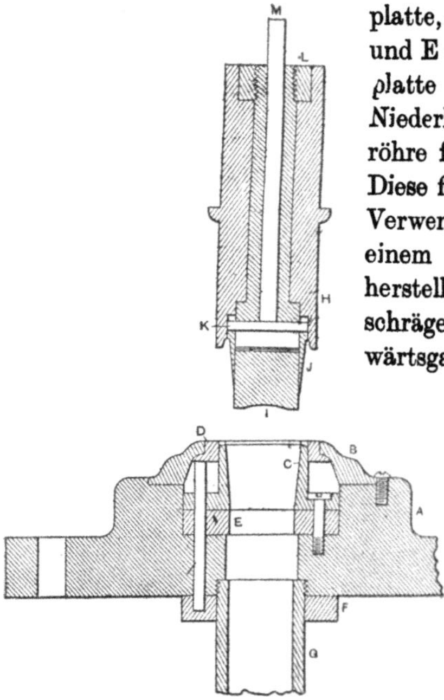

Fig. 525.

Fig. 602, S. 489, gibt schließlich das Werkzeug für den vierten und letzten Gang an.

Verbundwerkzeug für die einfachwirkende Presse zur Herstellung einer Blechzwinge. In Fig. 527 ist eine Schnittzeichnung eines Verbundwerkzeuges zur Herstellung einer Blechzwinge in einem Hub abgebildet. Darüber ist eine Ansicht der Stirnfläche des Stempels und in Fig. 526 Schnitt und Grundriß der herzustellenden Zwinge abgebildet. Der Schnittstempel C ist aus ungehärtetem Werkzeugstahl und erhält den Befestigungszapfen c angedreht. E ist der innere Schnittstempel mit seinem Zapfen e, der in den Schnittstempel C eingetrieben und leicht vernietet ist. Die gußeiserne Gesenkplatte B ist mit Putzen zur Befestigung an der Presse versehen. K ist die Schnittplatte, die, aus gehärtetem Werkzeugstahl gemacht, innen, außen und am Grunde

[1]) Vgl. Werkzeug f. Anhänger. WT. 1913, S. 106.

geschliffen ist. J ist der Ziehstempel, der auch den kleinen, inneren Schnitt bildet; er ist aus Werkzeugstahl gemacht, gehärtet und innen poliert. H ist der Niederhalter, der durch eine große Feder, die in der Zeichnung nicht sichtbar ist, mittels dreier Stifte I fest gegen die Stirnfläche des Stempels C gepreßt wird. Dieser Ring H dient auch als Abstreifer und schiebt die fertige Zwinge vom Stempel J ab. Um sie dann aus dem Schnittstempel C zu entfernen, tritt ein Auswerfer in Tätigkeit, der aus drei Stiften a besteht, die an ihren unteren Enden Segmentköpfe D tragen. Die Köpfe passen in den Ringraum zwischen den Stempeln C und E, während die Stifte a selbst fest in die Scheibe M eingetrieben sind, die beim Rückgang des Stößels an den fest-

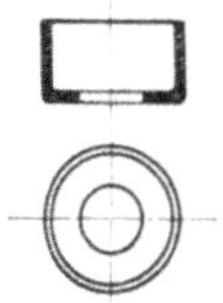

Fig. 526.

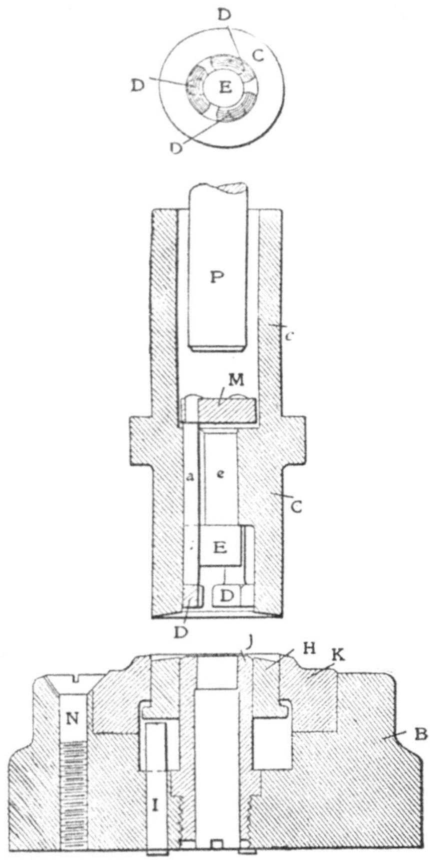

Fig. 527.

stehenden Auswerferstift P anstößt. Die Löcher für die Stifte a sind etwas größer gebohrt als der Ringraum zwischen den Stempeln, wie in Fig. 527 ersichtlich ist, und die Stifte an der Innenseite abgeflacht.

Verbundwerkzeug mit zwangläufigem Auswerfer. Die Zeichnung Fig. 528 zeigt ein Verbundschnitt- und Ziehwerkzeug, das zur Herstellung des in das Werkzeug eingezeichneten Näpfchens dient. Das Näpfchen wird in einer doppeltwirkenden Presse in einem Gang ausgeschnitten und gezogen, kann aber nicht nach unten abgestreift werden, so daß ein besonderer Auswerfer nötig wird.

A ist der Ziehstempel, der an seinem oberen Ende mit Gewinde versehen ist, mit dem er in den inneren Stößel der Presse eingeschraubt wird. Das untere Ende des Stempels ist nach dem Näpfchen geformt. Der Schnittstempel B ist ebenfalls mit Gewinde versehen und wird

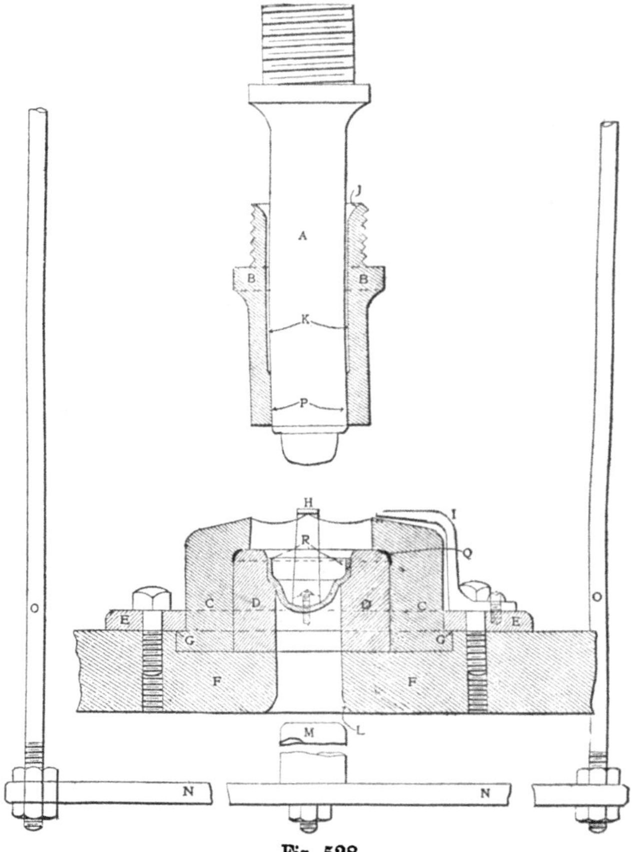

Fig. 528.

in den äußeren Stößel der Presse eingeschraubt. Er wird bei J abgerundet, um den Stempel A frei zu lassen, und bei K ausgedreht, damit nicht seine ganze innere Länge geschliffen werden muß. Außen wird er auf den verlangten Durchmesser geschliffen und hat bei P einen guten Gleitsitz auf A.

Die Schnittplatte C, die dem Durchmesser des Blanketts entspricht, ist innen auf 2^0 konisch hinterdreht und hat ungefähr 1,6 mm Scherfläche an der Oberkante. Bei Q sind die Ecken weggeschnitten, um ein Einschleifen zu vermeiden, und bei G ein Absatz zur Befestigung vorgesehen. Dieser Teil ist nur auf der Innenfläche geschliffen. Der

Ziehring D paßt in C und wird von C niedergehalten. E ist ein Überlagring aus Flußeisen, der mit der Gesenkplatte F verschraubt ist und die beiden Teile C und D niederhält. Die gußeiserne Gesenkplatte, F, wird auf dem Pressentisch befestigt. H ist ein Anschlag und I eine seitliche Führung für das Material; jener erstreckt sich längs der Rückseite, dieser entlang der Seite der Schnittplatte.

M ist ein Auswerfer, der an der Querstange N befestigt wird, die durch die Zugstangen O mit dem Pressenstößel verbunden ist. Die Gewinde werden lang genug zur Nachstellung von M geschnitten. Aus dem Bandmaterial wird das Blankett von B ausgeschnitten, worauf es beim Weitergang des Stempels B nach abwärts auf den Ziehring D niedergedrückt und dort gehalten wird, während der Stempel A es fertig zieht.

Ausschneiden und Ziehen einer rechteckigen Dose (vgl. Fig. 390, S. 300, Ziehen von Weißblechdosen). Die im folgenden beschriebene Herstellung eines Verbundschnitt- und Ziehwerkzeuges für ovale und viereckige Dosen hat verschiedene Vorteile.

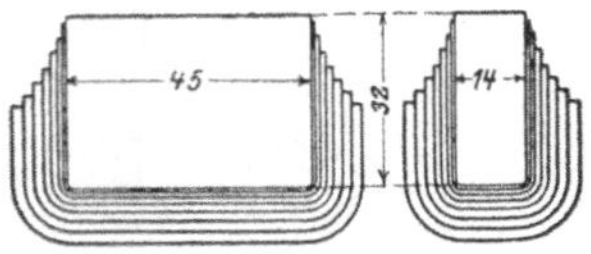

Fig. 529.

In Tabelle Nr. 21 ist der Herstellungsgang mit den Abmessungen der einzelnen Züge und in Fig. 529 sind die übereinander gezeichneten Züge gegeben.

Fig. 530 zeigt die Gesenkplatte E, die zur Aufnahme der unteren oder Ziehplatte D von vorn nach rückwärts und von rechts nach links

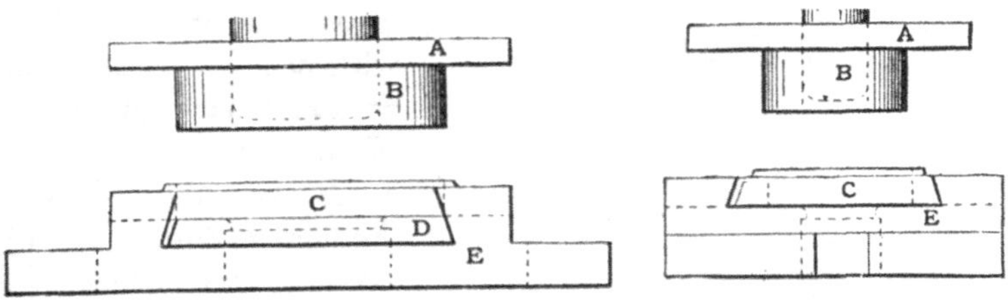

Fig. 530.

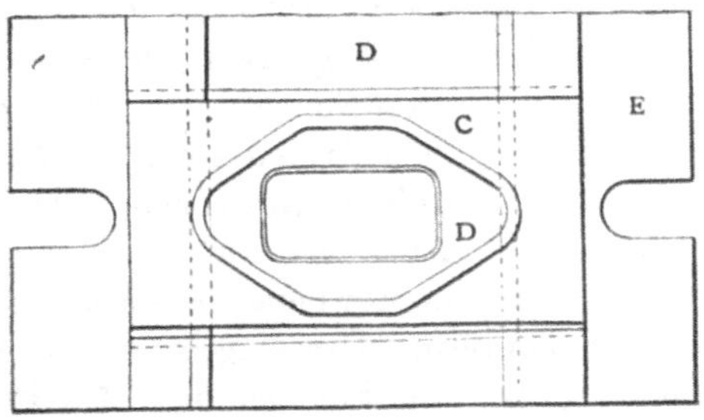

für die obere, die Schnittplatte C, ausgehobelt ist. Beide Teile sind mit 10^0 seitlicher Schräge gehobelt und mit Keilen in der Gesenkplatte befestigt. Damit ist erreicht, daß man das Gesenk aus einzelnen Teilen, die leicht zu bearbeiten sind, herstellen, die Schnittflächen beider Matrizen nach dem Härten schleifen und im Falle eines Bruches jede der beiden, ohne die andere zu verstellen, entfernen kann. Man kann auch die beiden Teile gegeneinander, Schnittkante gegen Ziehkante, einstellen, was bei andern Verbundwerkzeugen nicht so leicht ist. Der Niederhalter und Schnittstempel B werden an der Kopfplatte A mit Schrauben und Prisonstiften in gewohnter Weise befestigt. Fig. 529 und Tabelle Nr. 21 geben die aufeinanderfolgenden Gänge, acht an der Zahl, die nötig sind, um eine rechteckige Dose, 45 × 14 × 32 mm tief, herzustellen. Die Vorschrift verlangte eine Dose aus Kupferblech, von obigen Abmessungen, deren Kanten voll und scharf ausgebildet sein mußten, während am Boden eine Krümmung derselben mit 0,8 mm Halbmesser erlaubt war. Das Material war in der fertigen Dose 2,2 mm dick zu halten.

Fig. 530 gibt drei Ansichten des Werkzeuges für den ersten Gang, während Tabelle Nr. 21 G. 1 Form und Abmessung des Blanketts zeigt. Die folgenden Züge werden mit nach jeder Richtung hin, mit Ausnahme der Form, gleichen Werkzeugen ausgeführt wie runde Stücke.

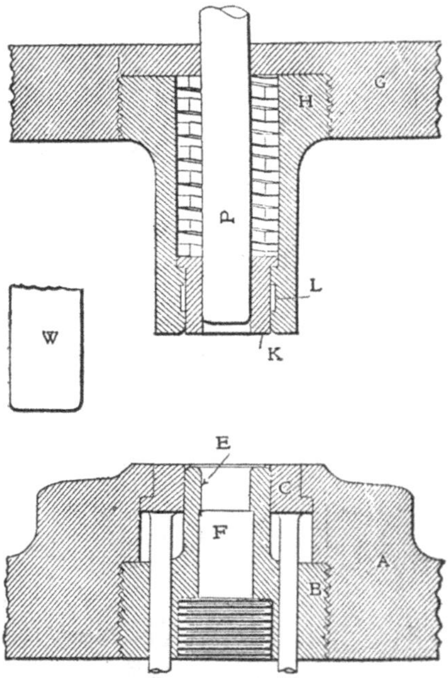

Fig. 531.

Kombiniertziehen von Hülsen. Fig. 531 gibt eine Arbeitsweise, die in der doppeltwirkenden Exzenterpresse zum Ziehen und gleichzeitigen Nachziehen einer Hülse, wie z. B. zur Herstellung von Aluminiumhülsen in einem Gang, Verwendung finden kann. Man kann dieselbe Einrichtung auch zur schnellen Herstellung von Flaschendeckeln und ähnlichen Blechhülsen verwenden, die mit Rücksicht auf ihren Durchmesser ohne Nachziehen nicht auf die volle Tiefe gezogen werden können. Der eingeschlagene Weg besteht in der bekannten Vereinigung von Stempel und Matrize in doppelter Ausführung ineinander. Die Aluminiumhülsen W werden aus halbhartem Blech von ungefähr 0,5 mm Dicke auf 14,3 mm Länge und 9,5 mm Durchmesser in einem Gang gezogen. Die

Scheiben dazu haben 27 mm Durchmesser und werden aus der Tafel geschnitten.

Die Gesenkplatte A ist gleichzeitig Schnittring und besteht aus einem Schmiedestück, das aus Stahl und Eisen zusammengeschweißt wird, dessen Schnittkante gehärtet, angelassen und geschliffen wird, während der Grundkörper weich bleibt, damit man den äußeren Ziehstempel B, der aus Werkzeugstahl hergestellt und gehärtet ist, einschrauben und den Niederhalter C einsetzen kann. Dieser wird durch die Stifte D und eine Gummifeder, die mit Unterlagscheiben an einer in B eingeschraubten Führungsröhre gehalten wird, getragen. Die Hülsen fallen durch die Führungsröhre nach ihrer Fertigstellung heraus. Der Ziehstempel B arbeitet aber auch als Ziehring für den inneren Stempel P und wird, nachdem er zum Ziehen der ersten Hülse passend abgedreht worden ist, auf die genaue Größe der Fertighülse ausgebohrt; die Ziehkante ist bei E abgerundet und eine scharfe Abstreifkante bei F angedreht.

Der Stempelkopf ist aus Gußeisen; in ihm wird der Stahlstempel H bei J fest niedergeschraubt; gleichzeitig dient diese Fläche als Widerlager für die Feder, die den oberen Niederhalter K betätigt. Der Schnittstempel H ist aus Werkzeugstahl gemacht und als Ziehring für den ersten Zug ausgebohrt. Die Bohrung ist jedoch reichlich groß, d. h. sie ist um mehr als die Metalldicke größer.

Das Zusammenbauen, die Arbeitsweise des Werkzeuges usw. kann leicht aus den Figuren ersehen werden.

Ausschneiden und Halbrundprägen. Im folgenden wird das Rundprägen einer zylindrischen Hülse beschrieben. Die Hülse wird nicht vollständig eingerollt, sondern nur mit einer halbkreisförmigen Wulst am Boden versehen, um in diesem Teil, der sich am Grunde der Hülse befindet, eine Isolation aufzunehmen. Hier werden zwei Arbeiten gleichzeitig ausgeführt, wodurch man viel Zeit erspart, nämlich das Ausschneiden („Aushauen") des Bodens der Hülse und, sobald der Stößel an das Hubende kommt, das Prägen.

Die Konstruktion des Stempels und der Matrize ist folgende: Fig. 532 stellt die fertige Hülse dar, Fig. 533 eine Ansicht des vollständigen Werkzeuges und einen Schnitt durch dasselbe. A ist der Schnittstempel zum Ausschneiden des Bodens der Hülse; er ist in der Grundplatte B vernietet, die auf den Pressentisch paßt. C ist der Prägestempel, in den oben die verlangte Abrundung eingearbeitet wird. Dieser Teil wird von der Grundplatte aus durch Federn sicher oben gehalten, um die Hülse, während A seine Arbeit vollführt, festzuhalten. D ist die Schnittplatte und zugleich die Matrize, die beim

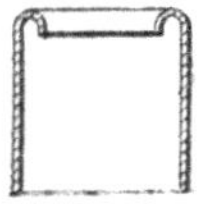
Fig. 532.

Niedergang die Hülse über C ausprägt. Wenn der Oberteil seinen Hub beendet hat, hat der Stempel A das aus dem Boden der Hülse ausgeschnittene Blechstück durch einen Schlitz in der oberen Kopfplatte ausgestoßen. Die Matrize D hat nach Fertigstellung der Hülse das Metall durch die Federwirkung genügend weit vom Stempel abgezogen, so daß der Arbeiter die fertige Hülse leicht entfernen kann.

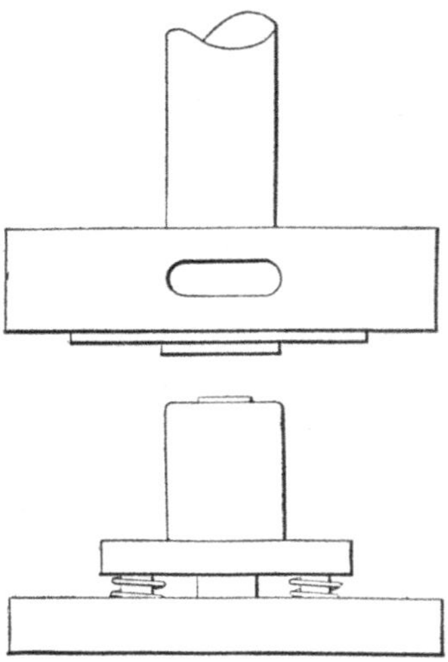

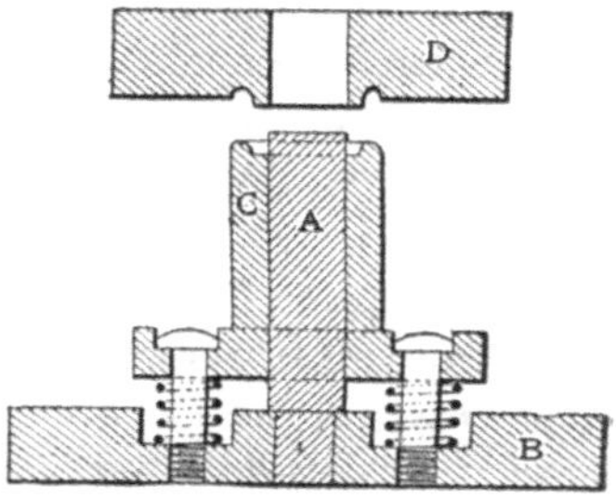

Fig. 533.

Verbundschnitt- und Prägewerkzeug. Das in Fig. 534 abgebildete Stück wird auf einem Verbundwerkzeug zum Ausschneiden, Prägen und Lochen in einem Gang aus galvanisch verzinktem Eisen hergestellt. Es hat 120 mm Durchmesser und ein Mittelloch von 54 mm Durchmesser, sowie drei gleichmäßig verteilte Lappen und eine halbkreisförmige Rippe auf einer Seite.

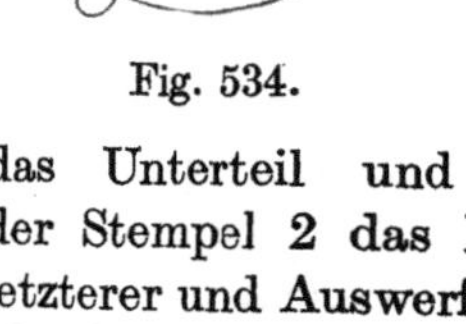

Fig. 534.

Der Unterteil, Fig. 536, enthält den äußeren Schnittring 1, den inneren Prägestempel 2, den inneren Schnittring 3 und den Auswerfer, der aus den Teilen 4, 5, 6 und 7 besteht. Die Stirnfläche des Stempels 2, Fig. 535, berührt zuerst mit der Schnittkante das Unterteil und schneidet das Blankett aus. Dann drückt der Stempel 2 das Blankett auf den Prägestempel 2 nieder, wobei letzterer und Auswerfer 4 niedergedrückt werden, bis sie auf der Bodenplatte 12 ruhen. Die Pressung zwischen den Stirnflächen von Prägestempel und Matrize prägt aus dem Blankett die halbkreisförmige Rippe heraus, während der innere Schnittstempel 3 das Mittelloch ausschneidet. Hierauf geht der Stempel in die Höhe, läßt das Blankett los, das nun durch den Federauswerfer 4 nach oben ausgeworfen wird. Der Abstreifer 5 im Oberteil entfernt den Abfall von dem Stempel.

Bei der Herstellung der verschiedenen Teile des Unterteiles wird der äußere Schnittring 1 aus einem vollen Block Werkzeugstahl von 178 mm Durchmesser und 51 mm Dicke gemacht. Er kann jedoch auch aus einem geschmiedeten Ring aus Werkzeug- oder Maschinenstahl[1]) gemacht werden. Der Schnittring wird allseitig bearbeitet, wobei die Innenseiten beim Feilen ungefähr 3 mm im Durchmesser größer als das fertige Blankett gelassen werden, um für das Ausprägen der Rippe Material vorzusehen. Die Schnittkante wird ohne jeden Hinterschliff gerade nach abwärts geführt und die drei Aussparungen für die Lappen des Blanketts ausgearbeitet. Die Schnittplatte wird ungefähr 20 mm unterhalb der Stirnfläche auf der Innenseite 12,7 mm im Durchmesser größer ausgedreht, so daß ein rechtwinkeliger Absatz gebildet wird. Am Grunde ist die Schnittplatte noch einmal abgesetzt und auf 8 mm um 9,5 mm größer im Durchmesser ausgedreht. Hier wird Gewinde eingeschnitten und die runde Bodenplatte 12, die in der Ausdrehung und bündig mit der Unterfläche steht, eingepaßt. In Platte 12 werden vier Löcher von 6,5 mm Durchmesser für die Stahlstifte 6 gebohrt.

Fig. 535.

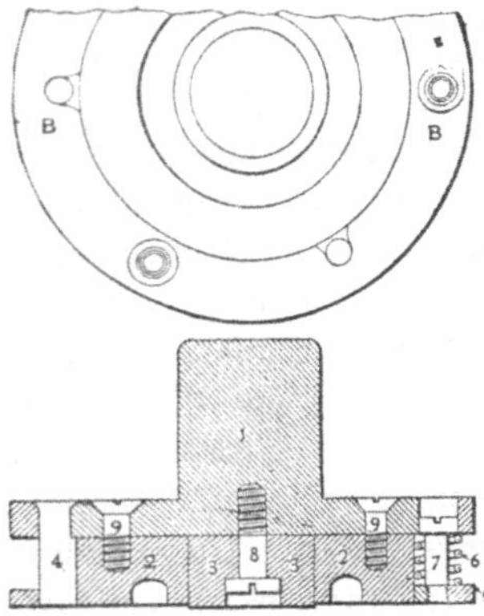

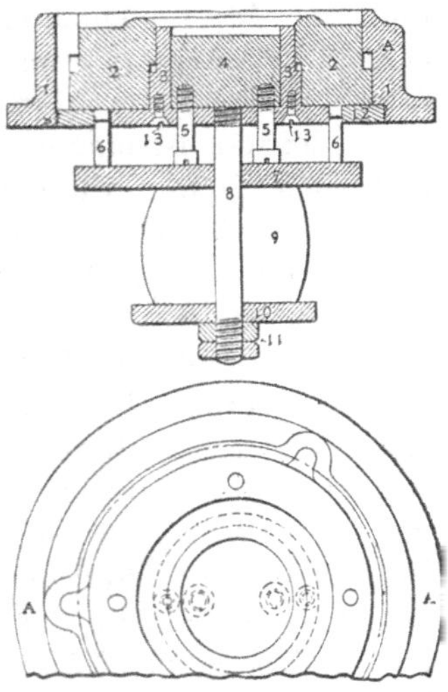

Fig. 536.

Der innere Prägestempel 2 wird am besten aus Werkzeugstahl gemacht und bleibt ungehärtet; seine Stirnfläche wird nach der Innenfläche des Blanketts gedreht, und sein Außenumfang in den Schnittring so eingepaßt, daß er sich leicht und doch ohne Spiel in demselben nach oben und unten bewegen kann. Der ebene Teil der Stirnfläche bleibt, wenn Auswerfer 4 am Grunde aufruht, ungefähr 6,5 mm unter der Schnittkante oder Stirnfläche des Schnittringes, und steht bündig mit derselben, wenn der Block gegen den inneren Absatz der Schnittplatte emporgeschoben wird. Das Loch in der Mitte des Prägestempels

[1]) Vgl. S. 58.

ist auf den inneren Durchmesser der Rippe ausgebohrt. Ungefähr 20 mm von der Oberkante ist es 3,2 mm zurückgesetzt, so daß eine rechtwinklige Ausdrehung gebildet wird, in die der Absatz des inneren Schnittringes 3 genau paßt, so daß der Prägestempel 2 sich gut nach beiden Richtungen bewegen kann. Der innere Schnittring 3 ist aus Werkzeugstahl gemacht, gehärtet und allseitig geschliffen, und erhält an der Unterfläche vier $1/4''$ Gewinde, mit denen er mit versenkten Schrauben 13 an der Bodenplatte 12 befestigt wird. Die Oberkante des inneren Schnittringes steht in gleicher Höhe wie der ebene Teil des inneren Prägestempels 2, wenn derselbe auf der Bodenplatte 12 aufruht. Der Auswerfer 4 wird aus Maschinenstahl gemacht und paßt so in den inneren Schnittring, daß er leicht seine Bewegung nach beiden Richtungen machen kann. Wenn er auf der Grundfläche aufruht, steht seine Oberfläche 6,5 mm unter der Oberfläche des inneren Schnittringes. Im Auswerfer sind auf der Unterseite zwei Stahlschrauben 5 von $3/8''$ Durchmesser gut eingepaßt, die aber in der Bodenplatte mit Spiel gehen. Diese Schrauben werden so lang gemacht, daß der Abstand zwischen ihren Köpfen, wenn der Auswerfer am Grunde aufruht, und der Bodenplatte gleich dem Abstand zwischen der Oberfläche des Auswerfers 4 und der des Schnittringes 1 ist. Die vier Stahlstifte 6 sind 6,5 mm kürzer als der Abstand zwischen den Platten 12 und 7; letztere ist eine runde Scheibe aus Maschinenstahl von 9,5 mm Dicke und 220 mm Durchmesser mit einem Loch von 16 mm Durchmesser in der Mitte, in das der Federbolzen 8, der in die Bodenplatte 12 eingeschraubt ist, paßt. Eine Gummifeder 9, ungefähr 75 × 75 mm, wird durch eine Unterlagscheibe 10 und zwei $5/8''$ Muttern 11 gehalten. Damit sind sämtliche Teile des Unterteiles angeführt worden.

Im Oberteil, Fig. 535, wird der Stempelkopf 1 am besten aus einem Schmiedestück aus Stahl gemacht, doch kann auch Gußeisen Verwendung finden, wobei aber seine Abmessungen größer gewählt werden müssen. Daran befestigt ist der runde Schnittstempel 2, zugleich Prägematrize, aus Werkzeugstahl, der, gehärtet und allseitig geschliffen, mit seiner Außenseite gut in den Schnittring paßt. Die Ausdrehung für die Rippe ist um 6,5 mm tiefer ausgedreht, als die Rippe im Blankett hoch ist, damit man den Stempel nachschleifen kann; die Breite der Ausdrehung stimmt mit der Breite der Rippe am Blankett überein. Das Loch in der Mitte des Stempels ist groß genug, um den inneren Schnittstempel 3, der mit einer Schraube 8 in der Mitte an dem Stempelkopf befestigt ist, aufzunehmen. Der innere Schnittstempel wird ungefähr 1,6 mm länger als die Prägematrize bzw. Schnittstempel 2 gemacht. Er paßt gut in den inneren Schnittring, der ungefähr 1,6 mm im Durchmesser kleiner als der Lochdurchmesser im fertigen Blankett zu halten ist, da man für das Ausprägen der Rippe Material zugeben muß.

Die drei kleineren Stempel 4, die die Lappen am Blankett ausschneiden, sind aus Werkzeugstahl und werden durch Drehen oder Schleifen an den äußeren Kreisumfang des Stempels angepaßt. Die zylindrischen Enden derselben passen in entsprechende Löcher des Stempelkopfes und sind rückwärts leicht vernietet, damit sie sich nicht herausziehen. Die äußeren Umflächen passen in die Ausschnitte des Schnittringes. Der Abstreifer 5, an der Außenseite des Oberteils, ist aus flachem, 6,5 mm dickem Maschinenstahl gemacht und wird durch die drei Schrauben 7 so in seiner Stellung gehalten, daß er ein wenig über der Stirnfläche des Stempels steht. Er wird durch drei Stahlfedern angespannt. Beim Nachschleifen von Stempel und Schnitt infolge der Abnützung werden der innere Schnittring und Stempel unterlegt, um ihre richtige Stellung zu behalten.

Verbundwerkzeug zur Herstellung von Linsenfassungen. Die in Fig. 537 dargestellte Herstellung eines Ringes in drei aufeinanderfolgenden Gängen bedient sich der Werkzeuge in Fig. 538 und 429. Dieser Ring findet als Linsenfassung Verwendung, so daß die Konstruktion der Werkzeuge für die Herstellung ähnlicher Stücke (Tabelle Nr. 26) [1]) von Interesse sein dürfte. Die Ringe werden im ersten Arbeitsgang auf die Form A, Fig. 537, aus halbhartem Messingblech geprägt. Der Prägestempel C, Fig. 538, wird nach dem Außendurchmesser der Linse abgedreht, wobei man mit Hilfe einer Lehre aus weichem Draht, die dem Querschnitt der Fassung entspricht, die richtige Gestalt der Stirnfläche erhält. Mit dem Härten des Prägestempels C wartet man, bis alle Teile zusammengebaut worden sind. Dann wird der Stempel M gedreht, der ein wenig größer als das Fertigmaß gelassen wird, und inzwischen auf der Bank die Ausdrehung hergestellt und das Loch für den Auswerfer gebohrt. Nun wird die Lehre auf der Stirnfläche von C angelötet und mittels ein wenig Mennige die hervortretenden Stellen ermittelt und dann die Ausdrehung auf richtiges Maß fertiggemacht. Um den Durchmesser des Schnittstempels und Schnittringes zu finden, entfernt man durch Abschmelzen des Lotes die Drahtlehre und streckt sie gerade. Nachdem die Außenseite von M auf diesen Durchmesser gedreht worden ist, wird das Stück im Gasofen erwärmt und gehärtet. In diesem Falle hatte man keine

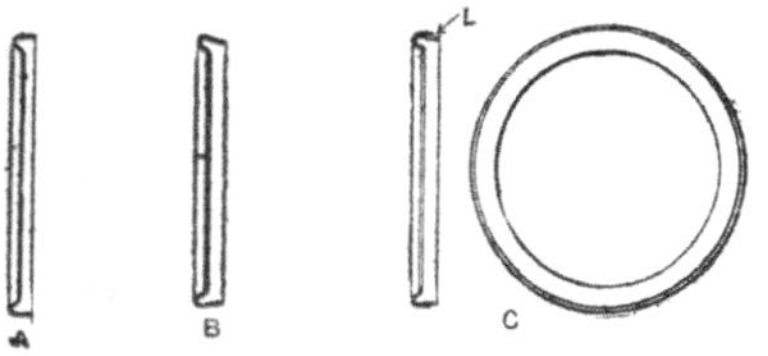

Fig. 537.

[1]) Vgl. W. T. 1910, S. 405.

Schleifmaschine zur Verfügung, so daß man den Stempel auf das sorgfältigste langsam und gleichmäßig erhitzte und ihn tatsächlich aus dem Härtebad ohne Veränderung, als wenn er von der Bank gekommen wäre, erhielt. Den Schnittring B bohrt man von rückwärts so, daß das Loch an der Schnittkante ungefähr 0,1 mm kleiner als auf der Rückseite wird. Die Gesenkplatte A aus Gußeisen wird zu dem Schnittring passend ausgebohrt, und nach dem Einsetzen desselben drei Löcher, halb in den Schnittring, halb in die Grundplatte gebohrt und mit Gewinde versehen, wie bei E ersichtlich ist. In diese Löcher werden dann drei Befestigungsschrauben eingepaßt. Beim Härten des Schnittringes wurde ebenso vorsichtig wie beim Härten des Stempels vorgegangen. Als nächstes wird die Büchse D aus Werkzeugstahl, die zwischen dem Prägestempel und dem Schnittring gleitet, fertiggestellt. Nachdem die Teile des Werkzeuges zusammengebaut worden sind, werden die zwei Löcher für die Schrauben G von der Rückseite der Gesenkplatte in den Prägestempel gebohrt und mit Gewinde versehen, so daß der Prägestempel C jetzt zum Härten fertig ist. F sind zwei Stahlstifte, die durch die Wirkung der weichen Gummifeder I gegen die Büchse D drücken.

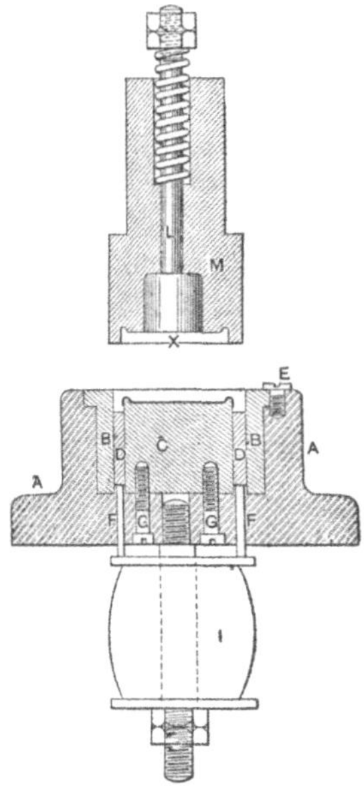

Fig. 538.

Das Werkzeug arbeitet in folgender Weise: Der Schnittstempel M tritt in den Schnittring B ein und schneidet damit die Scheibe auf den richtigen Durchmesser aus. Während der Stempel weiter nach abwärts geht, biegt er die Scheibe über den Prägestempel C und bildet, sobald er aufsitzt, den bei A, Fig. 537, ersichtlichen Ring. Beim Aufgang des Stempels streift die Büchse D, die vorher die Gummifeder zusammengedrückt hat, den Ring von dem Prägestempel C ab, worauf der Stempel M ihn nach oben mitnimmt, bis der Auswerfer beim Hubende an den Anschlagarm in der Presse kommt. Da die Presse geneigt ist, so fällt das Blankett an dem Werkzeug vorbei in einen Kasten.

Besonderes Verbundschnitt- und Prägewerkzeug für Schnurscheibenhälften aus Stahlblech. In Fig. 539a—d ist eine besondere Pressenausführung nebst Werkzeugbefestigung gegeben. Da die Presse weder geneigt gebaut, noch der Ständer nach rückwärts offen ist, man jedoch die fertigen Stücke durch die Wirkung der Schwerkraft entfernen wollte, griff man zu dem Ausweg, die ganze Presse nach der Seite zu kippen was sich in jeder Beziehung bewährte. Natürlich war das Aussehen ohne Zweifel sehr ungewöhnlich. Dazu wurden an Stelle der gewöhnlichen Füße neue eingebaut, die der Presse eine seitliche Neigung von 30^0 gaben. Trotzdem stand die Presse, da der Schwerpunkt ihres oberen Teiles ziemlich tief lag, sicher auf diesen Füßen, und man konnte, obwohl der kürzere Fuß den weitaus größeren Teil des Gewichtes zu tragen hatte, beim Arbeiten keine Erschütterungen bemerken.

Die besondere Arbeit für diese Presse bestand in der Herstellung der Seitenhälften einer kleinen Schnurscheibe aus weichem Stahlblech von ungefähr 50 mm Durchmesser und 0,1 mm Dicke. Es waren davon ungefähr 40000 Stück im Tag zu liefern. Ursprünglich wurden sie in drei Arbeitsgängen hergestellt, und zwar wurde die Scheibe zuerst ausgeschnitten, dann gelocht und schließlich in Prägewerkzeugen auf die verlangte Form geprägt; mit diesen Arbeiten waren drei Pressen und, da ohne selbsttätigen Vorschub gearbeitet wurde, auch drei Arbeiter dauernd beschäftigt. Nachdem die Blanketts die Werkzeuge verlassen haben, werden sie, wie gewöhnlich, in Trommeln mit Sägespänen gescheuert, um alle Grate und scharfen Kanten zu entfernen, ihnen eine Art Politur zu geben oder sie wenigstens blank zu machen und von allem Öl zu befreien, das bei der Arbeit im Verbundwerkzeug reichlich verwendet wird.

Obwohl die Konstruktion auf den ersten Blick ziemlich kostspielig aussieht, wird bald klar, daß sie sich schnell bezahlt macht. Es wurden von allen Teilen, die brechen könnten oder sich leicht abnutzen würden, Ersatzteile angefertigt, damit die Presse so ununterbrochen wie nur möglich laufen konnte und man nur zum Auswechseln, aber nicht zum Ausbessern anhalten mußte. Die Änderungen an der Presse bestanden, außer in den geneigten Füßen, in einem besonderen Stößel, einem neuen Exzenter für dessen Antrieb, einem besonderen zweiten Exzenter mit Bügel, das innerhalb des Schlittens arbeitete, mit einer Führung am Kopfe der Presse und einer passenden schweren Platte, die in sich Froschplatte, Gesenkplatte und Grundplatte für die selbsttätige Rollenzuführung, die ebenfalls zugefügt wurde, vereinigte.

In der folgenden Beschreibung der Einzelheiten der Zeichnung sind alle Teile mit Zahlen bezeichnet. 12 ist der Stößel, der senkrecht im Preßständer geführt wird und seinerseits eine Führung (äußerer Schlitten) 13 trägt, die der Zugänglichkeit halber aus zwei Teilen hergestellt wird, die zusammen eine zylindrische Führung für den Präge-

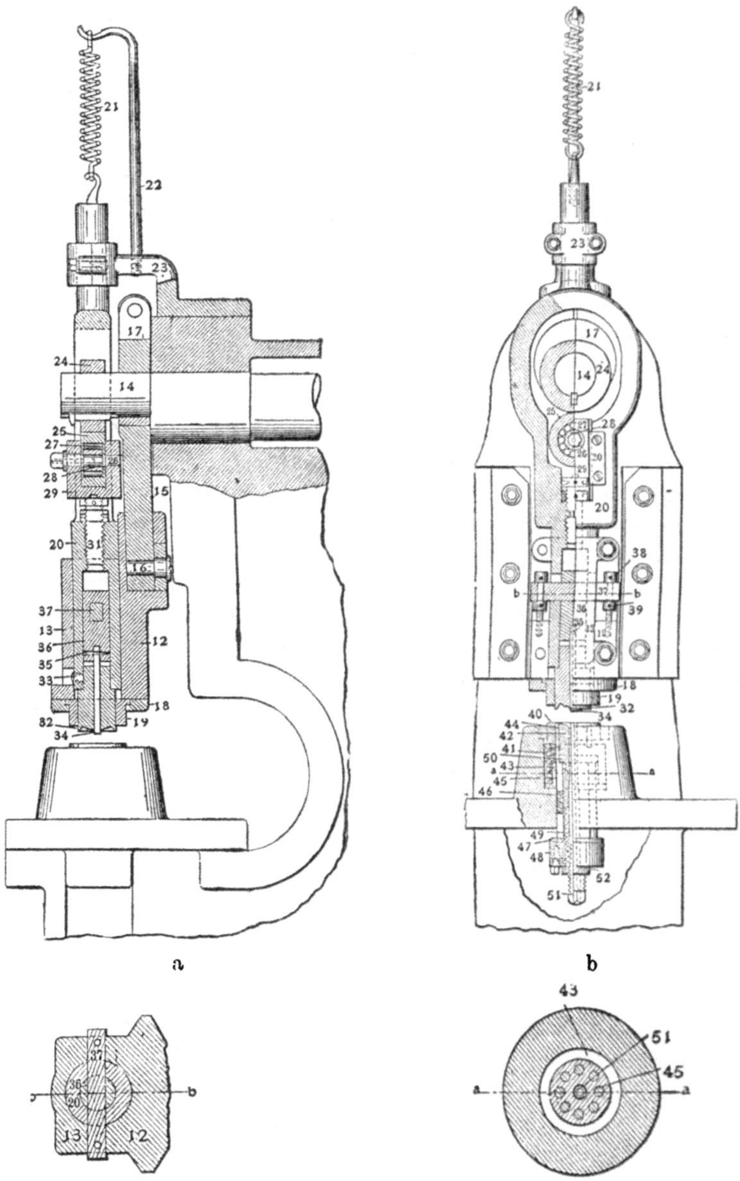

a b

c
Schnitt bei b b.

d
Schnitt bei a a.

Fig. 539.

stempel bilden. Der Schlitten 12 ist mit der Kurbelwelle durch eine Pleuelstange 15 verbunden, deren eines Ende im Schlitten an den

Bolzen 16 angehängt ist, und deren anderes Ende den Exzenterring zu dem Exzenter 17 bildet, das einstellbar aufgeklemmt ist, so daß man jede gewünschte Voreilung zur Kurbelwelle erhalten kann. Am unteren Ende des Schlittens ist mittels des Ringes 18 der Blechhalter 19 befestigt, der zugleich Schnittplatte ist. Innerhalb der Führung 13 gleitet der Schlitten 20, der für gewöhnlich durch die Feder 21 in seiner höchsten Stellung gehalten wird, die den oberen Teil dieses Schlittens mit einem Ständer 22 auf dem Pressenkopf verbindet. Ein Teil dieses Ständers 23 bildet zugleich die Führung für das obere Ende des Schlittens 20. Auf der Kurbelwelle ist ein Exzenter 24 aufgekeilt, das den Schlitten 20 bewegt, während die Feder 21 ein auf Rollen im Schlitten gelagertes, gehärtetes Stahlrad 25 ständig dagegen drückt. Die Drehachse 26 für das Rad 25 ist mit einer eingepaßten, gehärteten Stahlbüchse versehen, auf der die Rollen erst laufen. Um auch hier eine Einstellung zu ermöglichen, ist diese Drehachse in einer Gabel 29 gelagert, die in einer Führung 30 senkrecht verschiebbar ist und durch eine Einstellschraube 31 bewegt wird.

Das obere Prägewerkzeug 32 ist an dem unteren Ende des inneren Schlittens 20 durch eine Druckschraube 33 befestigt, die durch eine Aussparung an der Vorderseite des äußeren Schlittens 13 hindurchgeht. Ein Lochstempel zum Lochen der Mitte des Blanketts geht durch ein mittleres Loch im Prägestempel 32 und wird an dem oberen Ende durch einen Stift 35 in dem einstellbaren Plunger 36 gehalten, der von einer Aussparung des Schlittens 12 aus getragen wird. Durch den Plunger 36 geht nämlich eine Querstange 37, deren Enden einstellbar in der Führung 13 durch Schrauben 38 und Gegenmuttern 39 so gehalten werden, daß der Stempel 34 senkrecht zum Schlitten 12 eingestellt werden kann.

Um den Materialstreifen während des Prägens zu halten, ist bei 40 ein unterer Niederhalter vorgesehen, der in Linie mit dem oberen Ring 19 steht und durch eine starke Stahlfeder 41 fest gegen den Ring angedrückt wird, sobald dieser niederzugehen beginnt, so daß das Material unter starker Spannung festgehalten wird. Die untere Prägematrize 42 ruht fest auf dem Ring 43, der wieder durch die gußeiserne Gesenkplatte getragen wird, welche auf dem Pressenständer sitzt.

Um den Teil, der die Scheibe bildet, auszuschneiden oder nach dem Prägen von dem Abfallstreifen zu trennen, ist ein unterer Schnittring 44 vorgesehen, der die Prägematrize einschließt und von einer Reihe einstellbarer Stahlstifte getragen wird. Zwei von diesen, einer bei 45 in der Zeichnung, Fig. 539d und 540a, der andere gerade gegenüber, werden mit dem Stempel durch eine Wurstfeder 46 in Berührung gehalten, so daß der Ring gleichzeitig die geprägte Scheibe aus der

Prägematrize auswirft. Die anderen Stifte ruhen auf einer Stahlunterlagscheibe 47, die durch die geteilte Mutter 48 getragen wird, so daß 47 sicher in jeder Stellung festgezogen werden kann.

Die Stifte sind so lang, daß zwischen der Unterseite des Stempels und der Stirnfläche der Stifte so viel Platz bleibt, wie die Zusammendrückung der Federn 46 beträgt. In dem Maße, wie sich der Schnittring abnutzt oder nachgeschliffen wird, kann er durch die Mutter 48 und Schraube 49 nachgestellt werden. Der Niederhalter 19 arbeitet auch als Schnittstempel, da das Material zwischen den Kanten der Ringe 19 und 44 abgeschert wird, wie aus der Schnittzeichnung, Fig. 540a, zu ersehen ist. Es sei bemerkt, daß der Streifen erst ausgeschnitten wird, wenn das Blankett vollständig geprägt worden ist.

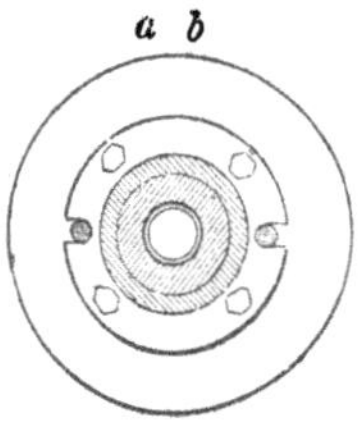

Fig. 540 a.

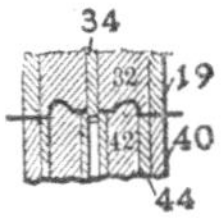

Fig. 540 b.

Infolge der hin- und hergehenden Bewegung des Stempels 44 hat die Prägematrize 42 das Bestreben, aus dem Halter nach oben zu kriechen; diesem Bestreben wird dadurch entgegengearbeitet, daß eine Büchse 50 an der Unterfläche der Prägematrize und mit Gewinde an der Büchse 43 befestigt ist. Damit das mittlere Loch, Lochstempel 34, gleichzeitig mit dem Ausschneiden des Blanketts hergestellt werden kann, ist eine runde Schnittplatte, Fig. 540b, vorgesehen, die von der hohlen Schraube 51 getragen und nachgestellt wird. Diese Schraube sitzt mit Gewinde in der Büchse 43, Fig. 539d, und wird in ihrer Stellung durch die Gegenmutter 52 gehalten.

Die Arbeitsweise ist kurz folgende: Der Metallstreifen wird zwischen die Prägestempel eingelegt oder selbsttätig zugeführt, worauf die Umdrehung der Kurbelwelle die Schlitten 12 und 20 zusammen nach abwärts bewegt. Sobald die Blechhalter 19 und 40 mit dem Arbeitsstück in Berührung kommen, wird der untere Teil niedergedrückt, und die Prägestempel 32 und 42 beginnen das Material auszuprägen, während der Blechstreifen durch die Niederhalter 19 und 40 festgehalten wird. Wenn der Prägestempel 32 am Ende seines Abwärtshubes angelangt ist und das Blankett fertig geprägt ist, bewirkt die gegenseitige Einstellung der Exzenter, die die Schlitten antreiben, daß der Schlitten 12 und mit ihm der Niederhalter 40 und Stempel 34 ihre Abwärtsbewegung fortsetzen. In diesem Augenblick sind aber die Federn 46 bereits so zusammengedrückt worden, daß der Ring 44 fest auf den Stiften aufruht und bei der weiteren Abwärtsbewegung des Stempels 19, der nun gegen die feste Kante des Schnittringes 44 zu arbeiten kommt, das Material abgeschert und die Mitte durch den Stempel 34 gelocht wird. Sobald bei der weiteren Bewegung der Kurbelwelle der Schlitten und

Stempel (bzw. Niederhalter) 19 zurückgehen, wird der untere Niederhalter 40 durch die Federn 46 und 41 aus der Matrize 42 nach oben gedrückt, worauf, da sich inzwischen das Exzenter 24 weitergedreht hat, der Prägestempel 32 nach abwärts zu gehen beginnt und das gebildete Blankett wie den Streifen mit sich nimmt. Infolgedessen wird, da der Streifen gleichzeitig vorwärts transportiert wird, das gebildete Blankett aus dem Bereich der Werkzeuge entfernt und abgeliefert, womit der Streifen für den nächsten Arbeitsgang bereit ist.

c) Folgewerkzeuge gegenüber Verbundwerkzeugen[1]).

In dem Bestreben um eine wirtschaftlich günstigere und qualitativ bessere Herstellung der in der Blechbearbeitung erzeugten Stücke haben die Amerikaner die Ausbildung des Verbundwerkzeuges vorgezogen. Man verlangt heute nicht mehr ähnliche, sondern austauschbare Stücke, die auch in hohem Maße genau sein sollen. Wenn auch nicht geleugnet werden soll, daß zur Herstellung der genauesten und verwickeltsten Arbeiten die Subpresse oder das Verbundwerkzeug mit besonderen Führungen immer noch an erster Stelle steht, so sind eine Reihe Bestrebungen der Fabrikanten sorgfältig zu verfolgen, die versuchen, die Genauigkeit und Herstellungssteigerung der obigen Werkzeuge mit einfacheren zu erreichen. Dafür aber teilen die zusätzlichen Kosten für die bei diesen Werkzeugen notwendigen Mehreinrichtungen der Presse selbst zu, so daß die Vorteile und die dafür aufgewendeten Kosten sich nicht mehr auf ein einzelnes Werkzeug, sondern auf alle in dieser Presse verwendeten Werkzeuge verteilen. Wenn sich nun ein großer Teil der bis jetzt auf Block- oder Verbundwerkzeugen hergestellten Gegenstände durch Wahl geeigneter Hilfsmittel an der Maschine selbst mittels der einfacheren Werkzeuge herstellen läßt, ohne daß die Wirtschaftlichkeit und die Genauigkeit der Herstellung Schaden leiden, so muß man dies wohl als die empfehlenswertere Lösung ansehen. In dem einen Fall kommen nämlich die Vorteile des Blockwerkzeuges nur dem einen damit hergestellten Gegenstande zugute, und die ganzen dafür aufgewendeten hohen Anschaffungs- und Unterhaltungsmehrkosten gegenüber den einfachen Werkzeugen belasten die Gestehungskosten des einen Stückes, wozu noch die Verzinsung derselben in der ganzen Zeit, während der das Werkzeug unbenutzt bleibt, kommt. Im andern Fall verteilen sich die Mehrkosten der besonderen Pressenkonstruktion auf die verschiedenen in ihr verwendeten Werkzeuge, die Presse wird ausreichender beschäftigt sein — kurz, man verlegt die Vorteile, die Genauigkeit

[1]) W. T. 1914, S. 39, 82, 105.

und Herstellungssteigerung außerhalb des Werkzeuges selbst, das immer nur für den einzelnen Fall gilt, in die Pressenantriebe, deren Vorteile jedem in die Presse eingebauten Werkzeuge zugute kommen. Wenn dieser Gedanke systematisch durchgeführt würde, so würden langsam die für jedes Stück von der betreffenden Werkstätte „neu" zu machenden Erfindungen einer neuen Verbesserung eines Werkzeuges, alle die kleinen und großen, streng gehüteten Geheimnisse einer nach geregelten Grundsätzen arbeitenden und nach anerkannten Gesichtspunkten entwickelten Werkzeugherstellung weichen. Die im folgenden beschriebenen Werkzeuge deutscher Bauart sollen diesen Gedanken ausführen und die Wege zeigen, die bisher zur Lösung dieser Frage versucht worden sind.

Blockwerkzeug zum Ausschneiden, Prägen und Lochen mit Schutzgitter. Im folgenden sollen zwei Werkzeuge deutscher Herkunft besprochen werden, die beide ganz ähnliche Stücke herstellen wie die in Fig. 535/536 und 539a, b abgebildeten Verbund- und Einzelwerkzeuge.

Zuerst, Fig. 541[1]), wird ein Blockwerkzeug unmittelbar den amerikanischen Beispielen zum Vergleich gegenübergestellt werden, dann ein Satz Einzelwerkzeuge zur Herstellung eines ähnlichen Stückes in einer besonderen Revolverpresse[2]) angeführt werden, deren Ausführung die Frage der Vor- und Nachteile, die beiden Arbeitsweisen anhaften, am besten beleuchten wird.

Das Verbundwerkzeug, Fig. 541, ist für die einfachwirkende Presse mit Federauswerfer gebaut und besteht aus dem Stempel, der mittels Druckschraube in dem Pressenstößel befestigt wird, dem Unterteil und einem daran mittels Kopfschrauben befestigten Schutzgitter aus gelochtem Blech, aus dem der Stempel in seiner höchsten Stellung nicht heraustritt. Das Werkzeug ist ein sehr interessantes Beispiel für die konstruktive Durchbildung und Vereinigung einer besonders großen Anzahl Einzelarbeiten in einem einzigen Werkzeug, um so mehr, als das hergestellte Stück verhältnismäßig kleine Abmessungen aufweist.

Der Stempelkopf a aus Gußeisen trägt fest verschraubt den Schnittstempel b. Der Stempelkopf erhält eine mittlere Ausdrehung, in welche die Kopfplatte c für die Loch- und Ausschnittstempel eingesetzt ist. Unterhalb dieser ist die Ausdrehung erweitert und gibt dem Prägestempel d Raum, der durchbohrt ist und so die Führung für die Loch- und Ausschnittstempel abgibt. Er ist, damit die Verbindung der ein-

---

[1]) Dieses Werkzeug stammt von der Firma Gebr. Bing, Nürnberg.
[2]) Angegeben von der Firma Reiß & Martin, Berlin.

zelnen Teile deutlicher wird, in angehobener Stellung gezeichnet, in der die Federn g, die auf ihn wirken, zusammengedrückt sind. Auf seiner Unterseite ist der Schnittstempel ebenfalls ausgedreht und nimmt an dieser Stelle den Abstreifer e, der zugleich als Niederhalter beim Ausschneiden des Blanketts dient und mit Stiften im Schnittstempel aufgehängt ist, auf. Im Schnittstempel b führt sich wiederum der

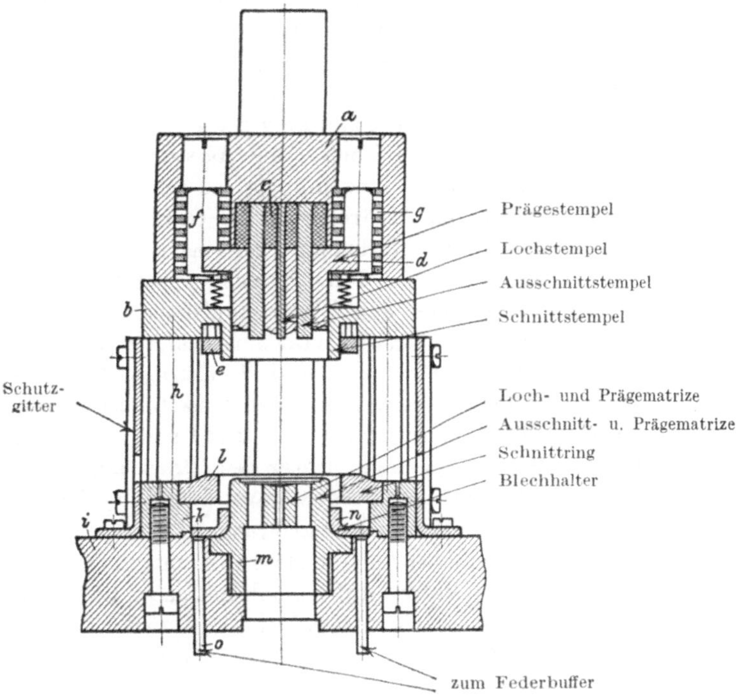

Fig. 541.

Prägestempel d, der durch zwei kräftige Spiralfedern g aus Vierkantstahl nach abwärts gedrückt wird. Er ist in den Schnittstempel gleitend eingepaßt. Die Spannung der Federn kann mit Hilfe zweier Schlitzkopfschrauben f, die als Führungen für die Federn dienen, geregelt werden.

So sind alle Einzelwerkzeuge des Oberteils aneinander genau zentrisch geführt, so daß die einzelnen Teile bei ihrer Herstellung einander als Lehren dienen können. Der Oberteil führt sich an vier runden Säulen h, die im Unterteil befestigt sind und innerhalb des Schutzgitters stehen. Der Unterteil ist auf der gußeisernen Froschplatte i aufgebaut, auf welcher, mit einem angedrehten Rand ausgemittelt, die Gesenkplatte k mittels versenkter Kopfschrauben befestigt ist.

In die Gesenkplatte k ist ein Sitz für den Schnittring l eingedreht; da dieser aus konstruktiven Gründen auf dem größten Teil seiner Unterfläche nicht aufliegt, ist er in der Mitte bedeutend verstärkt worden. Der durch diese Konstruktion unterhalb der Schnittplatte freiwerdende Raum dient zur Aufnahme des unteren Auswerfers und Niederhalterringes n, der durch durchgehende Stifte o von einer Feder nach oben gepreßt wird und so im Verein mit dem Schnittstempel das Blankett festhält, während in der Matrize m vom Prägestempel d die Rippen und die Mitte ausgeprägt werden. In der Matrize sind auch die Schnitte für Ausschnitt- und Lochstempel untergebracht. Auf der Unterseite ist sie auf einen ziemlich großen Durchmesser ausgebohrt, so daß alle Abfälle der Loch- und Ausschnittstempel nach unten durchfallen können. Sie ist unabhängig von der Schnittplatte in einer verhältnismäßig breiten Ausdrehung der Gesenkplatte ausgemittelt. Die Arbeitsweise des Werkzeuges ist folgende:

Der Blechstreifen wird von dem Vorschubapparat durch den Schlitz in dem Schutzgitter über den Unterteil gebracht, worauf beim Niedergang des Stempels der Streifen von dem Schnittstempel b und Blechhalter n sowie dem Abstreifer e und Schnittring l festgehalten wird. Beim Abwärtsgang des Stempels wird das Blankett ausgeschnitten. Dabei bleiben der Blechstreifen zwischen e und l und das Blankett zwischen b und n gehalten, während unter dem Druck der Federn g die Rippe und die Mitte von dem Prägestempel d und der Matrize m ausgeprägt werden. Beim weiteren Abwärtsgang des Stempels weichen die Abstreifer e und Niederhalter n nach oben bzw. unten aus, während der Prägestempel auf der Matrize aufliegend und unter Überwindung des Federdruckes in den Schnittstempel weiter zurücktritt, so daß die inneren Schnitt- und Lochstempel zur Wirkung kommen können. Beim Rückgang des Stempels streift der Blechhalter n das Stück von der Matrize ab, während der Abstreifer e den Blechstreifen von dem Schnittstempel entfernt.

Vereinigung von einfachen Verbund- und Einzelwerkzeugen in der Revolverpresse (Reiß & Martin, Berlin). Eine Reihe ganz ähnlicher Arbeiten wird nun teils hintereinander, teils gleichzeitig an dem in Fig. 542 abgebildeten Unterkasten eines Kosmos-Petroleumbrenners ausgeführt. Der Herstellungsgang ist in Tabelle Nr. 23 gegeben.

Um in bezug auf Schnelligkeit der Herstellung mit dem Verbundwerkzeug gleichen Schritt halten zu können, sind diese Werkzeuge in eine mehrstempelige Revolverpresse dieser Firma eingebaut und, um die notwendige Genauigkeit zu erhalten, mit Sucherstiften und selbsttätigen Ausmittelvorrichtungen für die von dem Revolverteller unter das Werkzeug gebrachten Stücke versehen worden.

Die Herstellung des Unterkastens für den Petroleumbrenner (Fig. 542) aus 0,5 mm starkem Messingblech erfordert im ganzen sechs Arbeitsgänge.

Für den späteren Zusammenbau in den Brennern ist es notwendig, daß die Ausschnitte in den Stücken genau sitzen und keinerlei Grat aufweisen, so daß die in den drei letzten Gängen erzielte Genauigkeit für die Brauchbarkeit des Stückes ausschlaggebend ist.

Das Rohmaterial wird in Rollen von solcher Breite angeliefert, daß je nach dem verwendeten Werkzeug mittels Massenschnittes 5 oder 7 Scheiben (Blanketts) auf einmal gestanzt werden können. Da hierbei auch mit selbsttätiger Zuführung gearbeitet wird, stellt sich das Ausschneiden der Platten so billig, daß es mit den Verbundwerkzeugen zum Stanzen und Ziehen in Wettbewerb treten kann. Bei 80 Preßhüben in der Minute werden 400 bzw. 560 Stück minutlich fertiggestellt. Als Arbeitsverlust kommt nur die Zeit zum Einlegen eines neuen Blechstreifens in die Presse in Betracht.

Fig. 542.

Im zweiten Gang werden die Scheiben auf einer schnellaufenden, schräg stellbaren Kurbelpresse gezogen, wobei eine Gummifeder zur Bedienung des Blechniederhalters Verwendung findet. Die Maschine wird schräg gestellt, um das Einfallen der Scheiben in das Werkzeug zu erleichtern, während sie durch die Wirkung der Schwerkraft nach dem Ziehen aus der Presse entfernt werden. Die Presse macht ungefähr 100 Hübe minutlich, womit man mit einem Arbeiter bei neunstündiger Arbeitszeit 22000—25000 Näpfchen täglich erhält.

Das Gewindedrücken erfolgt im dritten Gang auf einer mit selbsttätiger Zuführung ausgestatteten Maschine[1]), so daß der Arbeiter bloß die Näpfchen in den Zuführungskanal einzulegen hat, da die mit Gewinde versehenen Näpfchen auch selbsttätig ausgeworfen werden. Die selbsttätige Zuführung wird auf S. 526 beschrieben.

Für die drei letzten Arbeitsgänge werden Werkzeuge nach Fig. 543 verwendet. Das fertig gezogene, mit Gewinde versehene Stück wird auf dem Revolverteller von dem Arbeiter vorn eingelegt und kommt von links an die Werkzeuge heran, wo zuerst die beiden Ringnuten eingeprägt werden. Beim Aufgang des Stempels wird es unter der vereinigten Wirkung des Federauswerfers und des Abstreifers lose im Loch des Revolvertellers abgelegt, der es nun unter

[1]) W. T. 1909, S. 16.

den mittleren Stempel, den Innenschneider, befördert. Hier wird es mittels des federnden Niederhalters c beim Niedergang des Stempels zuerst ausgemittelt, da dieser der Innenform des Stückes entsprechend gebildet ist, und zugleich an die Schnittplatte angedrückt. Beim weiteren Abwärtsgang des Stempels treten der Lochstempel a und der Schnittstempel b gleichzeitig in Tätigkeit. Beim Rückgang des Oberstempels streift der Niederhalter das Stück von dem Stempel ab und hält es so im Revolverteller zurück, während der Abfall durch die Schnittplatte nach unten ausfällt.

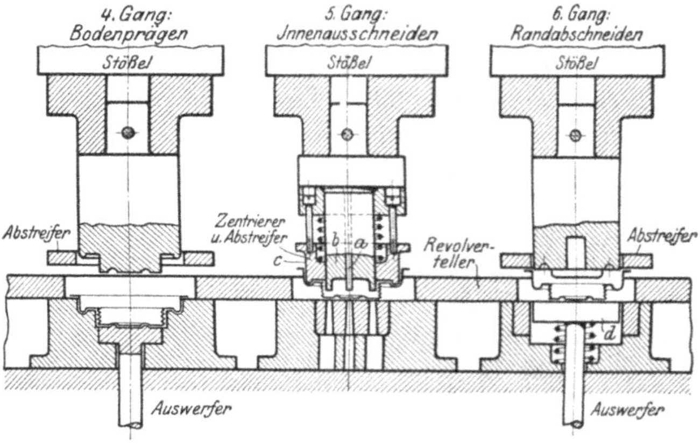

Fig. 543.

Der letzte, sechste Gang ist das Randabschneiden. Damit dieser Schnitt wieder gleichachsig mit dem vorhergehenden wird, findet das Arbeitsstück, das jetzt von dem Revolverteller auf dem im Unterteil befindlichen Federauswerfer d abgesetzt worden ist, an dem Schnittwerkzeug eine doppelte, innere und äußere Führung und Ausmittlung. Es zeigt sich dabei, daß die nötigen Hublängen der einzelnen Stempel ziemlich verschieden sind, und auch die Wege, während derer die größte Kraftäußerung von jedem Stempel verlangt wird, wohl gleichzeitig enden, nicht aber gleichzeitig beginnen. Während bei den Schnittstempeln die größte Kraftäußerung ungefähr nach Durchdringung der Mitte der Blechdicke überschritten ist, dauert sie beim Prägen durch die ganze Tiefe der zu prägenden Nut oder, genauer gesagt, bleibt nach Erreichung des Höchstdruckes stehen, bis die volle Nutentiefe erreicht ist, also bis zum Aufsetzen.

Bei Verwendung von Einzelwerkzeugen in der Revolverpresse ist es sehr schwer, alle in dem Stößel befestigten Werkzeuge so genau einzustellen, daß sie bei dem Niedergang des Schiebers gleichzeitig die ihnen zugewiesene Arbeit mit der verlangten Genauigkeit ohne kleine Abweichungen ausführen. Bei nicht sehr gut geführten, kräftigen Stößeln werden sich die Einwirkungen des einzelnen Stempels infolge kleiner elastischer Formänderungen oder Spieles in der Führung auf die Arbeit des nebenan liegenden geltend machen. Andrerseits bedingen die notwendigen gleichen Hübe aller Stempel öfter eine längere Führung, als dem notwendigen Hub für diesen Arbeitsgang entsprechen würde, d. h. man muß die Werkzeuge besonders bauen. Solange die verschiedenen Stempel gleiche Werkzeuge darstellen, wird sich der Fehler infolge Beeinflussung der einzelnen Stempel aufeinander in bezug auf das Arbeitsstück verringern. Wenn es sich jedoch darum handelt, mit diesen Pressen die Arbeit der Verbundwerkzeuge zu ersetzen oder die Genauigkeit der Arbeit eines Folgewerkzeuges zu erhöhen, so treten alle oben angeführten Punkte in vollem Maße auf.

Hier setzt nun die der Firma Reiß & Martin patentierte Pressenkonstruktion[1]) ein, die zur Herstellung der drei letzten Gänge verwendet wird.

d) Mehrfachwerkzeuge.

Verbindung von Einzelwerkzeugen mit einem genauen Vorschub. (Rob. Tümmler, Döbeln.) Ein andrer Weg, die Genauigkeit und Herstellungsmenge der Verbundwerkzeuge zu erreichen, wird von Tümmler-Döbeln eingeschlagen, der durch eine größtmögliche Genauigkeit des Vorschubes die zur Herstellung notwendigen Einzelarbeiten ganz einfachen Einzelwerkzeugen zuweist. Die Herstellungssteigerung erreicht er durch eine Mehrfachanordnung der Werkzeuge, die gerade bei den kleinen Teilen, für die sich dieser Vorschub besonders eignet, mit den Verbundwerkzeugen ernstlich in Wettbewerb tritt, da diese um so schwieriger herzustellen sind und um so teurer werden, je weiter sich der herzustellende Gegenstand nach oben und nach unten von einer gewissen Mittelgröße abweicht.

Charakteristisch für diese Arbeit ist das Verbleiben des Blanketts in dem Streifen bis zu dem letzten Gang, wozu oft besondere Stellen zwischen Streifen und Blankett stehen bleiben, die erst im letzten Gang entfernt werden und nur zu dem Zweck vorhanden sind, um eine unbedingt sichere Mitnahme des Stückes durch den Vorschub zu sichern.

Ein Beispiel für eine derartige Arbeit ist bereits in der abfallfreien Herstellung von Sechskantmuttern, Fig. 120, gegeben worden Hier soll an einem Beispiel, das wie Fig. 541 und 542 Ausschneiden, Prägen und Ziehen enthält und gleichzeitig als Mehrfachschnitt aus-

---

[1]) Siehe Fig. 647/8.

geführt ist, eine andere Arbeitsteilung für denselben Endzweck, den Ersatz des teuren Verbundwerkzeuges, gezeigt werden.

In Fig. 544/5 ist eine Blechplombe und das Werkzeug zu ihrer Herstellung gegeben, woraus die Lösung eines Mehrfachwerkzeuges für jene Fälle ersichtlich ist, in denen nicht bloß Abtrennung des Materials oder einfache Formänderung desselben bewirkt wird wie in den eben angeführten, sondern auch eine Formänderung mit Materialverschiebung notwendig ist. Mehr noch als bei den Werkzeugen nach Fig. 120 und 543 tritt hier die Bedingung hervor, das Stück bis zur Fertigstellung in dem Streifen zu halten, was in diesem Falle, wo der letzte Gang ein Ziehen ist, bemerkenswert gelöst ist.

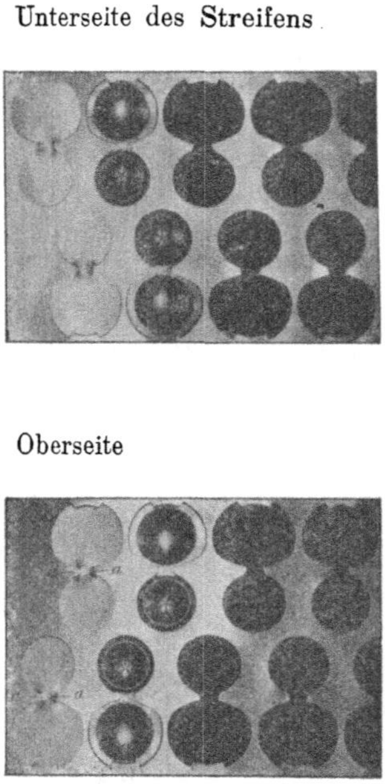

Fig. 544.

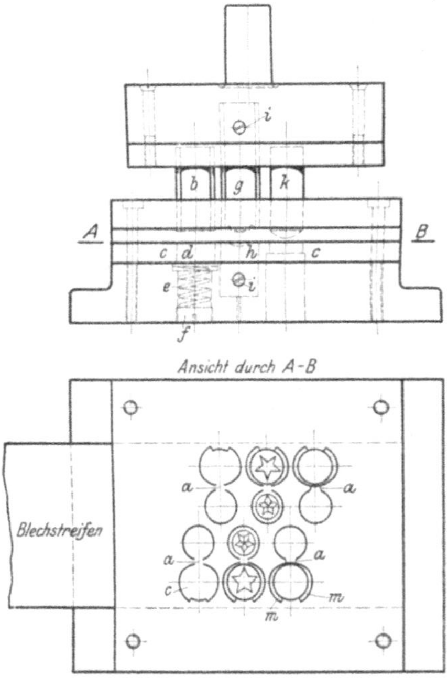

Fig. 545.

Die Blechplomben werden aus endlosem, verzinntem Stahlband oder zugeschnittenen Weißblechstreifen hergestellt, indem von links nach rechts (Fig. 544/5) zuerst die Blanketts bis auf die Stellen a ausgeschnitten und dann durch den Federauswerfer d wieder zum Weitertransport in den Streifen zurückgedrückt werden. Beim nächsten Hub wird irgendein Zeichen, Firma usw. und gleichzeitig die Form der Plombe geprägt. Diesem folgt im letzten Gang das Abschneiden der Stellen a, mit denen das Stück im Streifen hängt, und gleichzeitig das Hochziehen des Randes, so daß hier die fertige Plombe nach unten ausfällt.

Die Konstruktion des Werkzeuges sei nur insofern besprochen, als Abweichungen von der gewöhnlichen Bauart auftreten. Der Prägestempel g im Stempelkopf, wie die Prägematrize h sind nicht wie sonst vernietet, sondern mittels je einer Preßschraube i befestigt, damit die Prägewerkzeuge nach Bedarf ausgewechselt werden können. Der Schnittstempel b und die Schnittplatte c sind an den Stellen a, in Fig. 545 links, abgerundet, damit der Streifen an diesen Stellen nicht durchschnitten wird. Der Auswerfer d wird durch die mittels Schraube f einstellbare Feder e betätigt und ist durch einen angedrehten Rand gezwungen, in der obersten Stellung mit der Oberfläche der Schnittplatte bündig abzuschneiden. Der Ziehstempel k ist gleichzeitig Schnittstempel für die Stellen a und zieht, nachdem diese Stellen durchschnitten sind, den Rand hoch, während die Unterkante des Ziehringes als Abstreifer scharf bleibt.

Ein anderes Werkzeug derselben Firma, bei dem diese Punkte noch deutlicher hervortreten, wird bei der Herstellung von Maßstabkappen, Fig. 560, S. 450, beschrieben werden.

Werkzeug zum schnellen Ausschneiden und Ziehen kleiner Näpfchen. Fig. 546 stellt ein mehrfaches Verbundwerkzeug zum gleichzeitigen Ausschneiden und Ziehen von acht Näpfchen dar.

Die Schnittplatte wird aus einer Platte aus Tiegelgußstahl hergestellt. Nachdem diese auf 152 mm Durchmesser und 25,4 mm Dicke abgedreht worden ist, werden die acht Schnitte C angerissen, jedes Loch sorgfältig auf der Planscheibe der Drehbank ausgemittelt und auf ungefähr 0,125 mm weniger als der Durchmesser des Blanketts ausgebohrt, damit man nach dem Härten noch Material zum Schleifen hat. Nachdem die acht Löcher der Reihe nach ausgebohrt worden sind, werden vier Löcher von 6,35 mm Durchmesser gebohrt und für die entsprechenden Zylinderkopfschrauben Fl versenkt. Zwei weitere Löcher für Prisonstifte Do desselben Durchmessers und ein Loch für einen Anschlagstift Ga folgen, wonach die Schnittplatte gehärtet wird. Alle Löcher, mit Ausnahme der acht Schnitte, werden vor dem Härten mit Lehm gefüllt. Nach dem Härten und Anlassen wird die Schnittplatte auf der Ober- und Unterseite eben geschliffen und die Seitenfläche gerundet. Bevor man den Unterteil weiter bearbeitet, wird die Kopfplatte aus Maschinenstahl als eine genaue Wiederholung der Schnittplatte bezüglich Durchmesser und Dicke, wie bezüglich der Löcher für Schrauben und Prisonstifte gemacht. Hierauf wird eine gußeiserne Gesenkplatte B genau auf Durchmesser und Höhe der Schnittplatte ausgedreht und mit Löchern für die Prisonstifte Do und Gewinde für die versenkten Kopfschrauben Fl versehen. Nachdem nun die Schnittplatte mittels Schrauben und Prisonstifte in der Gesenkplatte befestigt worden ist, wird das Ganze auf der Planscheibe der Drehbank aufgespannt und so ausgerichtet, daß einer der acht Schnitte genau rund läuft, worauf er auf den genauen Durchmesser der Scheibe ausgeschliffen wird. Dann wird die Schnittplatte aus der Gesenkplatte entfernt, ohne daß man diese verschiebt oder ihre Stellung auf der Planscheibe ändert, und die Gesenkplatte zur Aufnahme des Ziehstempels Po ausgedreht. Danach wird die Kopfplatte mit der Stirnfläche nach unten in derselben Weise und mit den gleichen Mitteln wie die Schnittplatte befestigt und zur Aufnahme der Schnittstempel ausgebohrt. Die Gesenkplatte wird dann von der Planscheibe entfernt und die Kopfplatte herausgenommen. Nun wird die Schnittplatte wieder eingesetzt und die übrigen sieben Schnitte geschliffen, die Gesenkplatte für die Ziehstempel ausgedreht und die Kopfplatte der Reihe nach, wie bereits beschrieben, ausgebohrt. Dadurch, daß man in dieser Weise alle Teile herstellt, erhält man ein vollständig ausgerichtetes Werkzeug mit genau passenden Teilen.

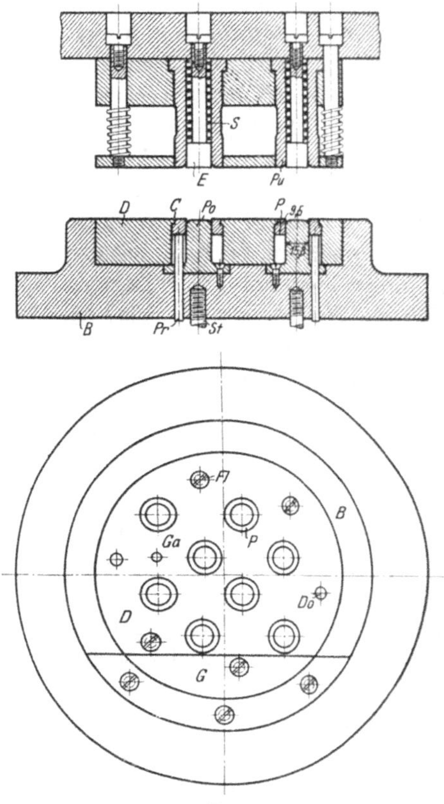

Fig. 546.

Dann wird ein Ziehstempel Po aus Werkzeugstahl für jede Ausdrehung der Gesenkplatte B gemacht und auf den genauen Durchmesser des verlangten Näpfchens abgedreht. Sein unteres Ende paßt in die Ausdrehung der Gesenkplatte, in der der Ziehstempel mit drei kleinen Schrauben befestigt wird. Drei Tragstifte Pr, auch aus Werkzeugstahl, gehen durch den Fuß eines jeden Ziehstempels und durch die Gesenkplatte und tragen die Niederhalter P, wie aus der Zeichnung zu ersehen ist. In der Mitte unter jeder Ausdrehung ist ein $^3/_8$″ Gewinde zur Aufnahme des Bolzens St für Gummifeder und Unterlagscheibe geschnitten. Die Stempel aus Werkzeugstahl werden nach ihrer Fertigstellung in Öl gehärtet und dann auf 0,025 mm im Durchmesser kleiner als die Schnittkante C geschliffen.

Jeder Stempel erhält einen durch eine Feder S betätigten Auswerfer E, der auf der Rückseite des Kopfes mit einer Zylinderkopfschraube gehalten wird. Die Kopfplatte ist aus Maschinenstahl und, statt mit dem gewöhnlichen 38-mm-Zapfen befestigt, an die Putzen des Pressenstößels angeschraubt. Ein ebenfalls durch Federn betätigter Abstreifer, der den Abfall nach dem Ausschneiden von den Stempeln abstreift, ist ebenso wie die Auswerfer E in den Stempeln, durch lange Zylinderkopfschrauben, die durch die Kopfplatte gehen, befestigt und in seiner Stellung gehalten. Eine Anschlagplatte G für das Blech, bestehend aus einem Stahlstreifen von ungefähr 3,2 mm Dicke, ist an der Gesenkplatte auf der dem Arbeiter zugekehrten Seite befestigt.

Dieses Mehrfachverbundwerkzeug fand auf einer schrägen Presse Verwendung, so daß die Näpfchen, nachdem sie gezogen und abgestreift worden sind, frei herausfallen konnten. Es wurden davon 120000 im Tage hergestellt. Man sieht aus der Stellung des Aufhängestiftes Ga, der, nebenbei bemerkt, so abgebogen ist, daß das Anschlagende des Streifens beinahe an die nächste Schnittkante rechts von Ga, Fig. 546, kommt, daß bei dem ersten Preßhub bloß sechs Näpfchen fertiggestellt werden, bei jedem folgenden jedoch acht; der Vorschub beträgt abwechselnd immer eine Teilung und drei Teilungen.

Ein fünffaches Verbundwerkzeug zum Ausschneiden, Ziehen und Prägen von fünf Blechschalen in einem Gang. Fig. 547/48 gibt Grundriß und Schnitt des oberen und unteren Teiles eines fünffachen Verbundwerkzeuges wieder. Damit werden in einem Gang fünf Schalen von 32 mm Durchmesser und 11 mm Höhe aus dekoriertem Blech durch Ausschneiden, Ziehen und Prägen hergestellt. Die Grundsätze des Aufbaues sind im wesentlichen dieselben wie bei jedem einfachen Verbundwerkzeug, weshalb auch keine ins einzelne gehende Beschreibung der Herstellung der verschiedenen Teile mehr gegeben sei, da sich der hierbei zu befolgende Weg aus den bis jetzt beschriebenen Werkzeugen entnehmen läßt. Eine allgemeine Beschreibung des Zusammenbaues der Teile und der Arbeitsweise wird für alle Zwecke genügen und im Verein mit den Zeichnungen zeigen, wie ähnliche Werkzeuge für derartige Arbeiten, gegebenenfalls noch in größerer Zahl, einzurichten sind. Es liegt nach Woodworth kein Grund vor, diesen Grundsatz auch auf Werkzeuge auszudehnen, mit denen eine

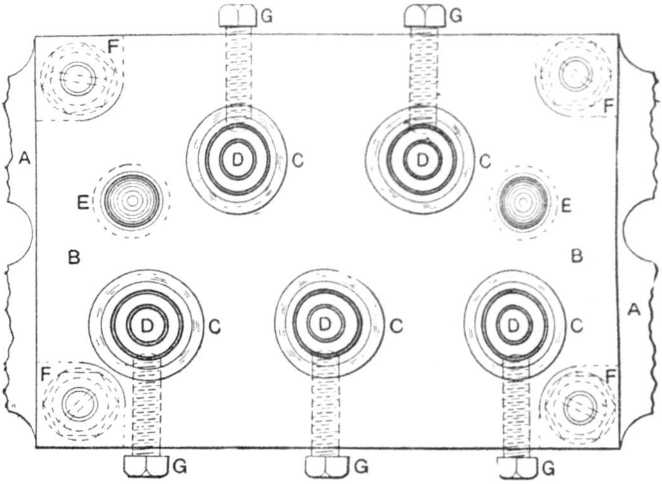

Fig. 547.

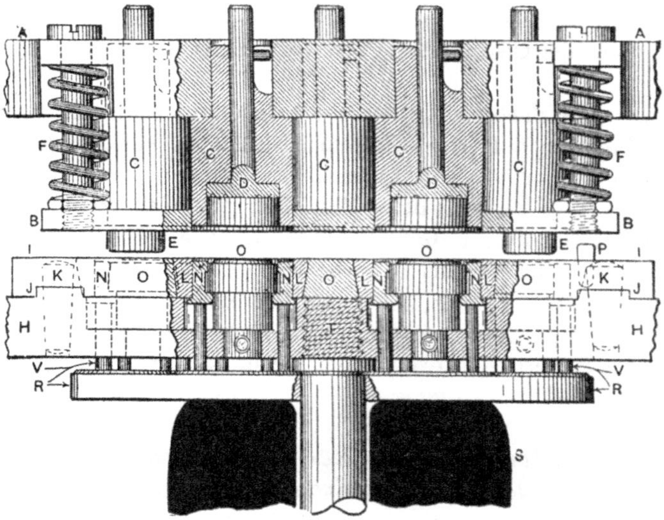

Fig. 548.

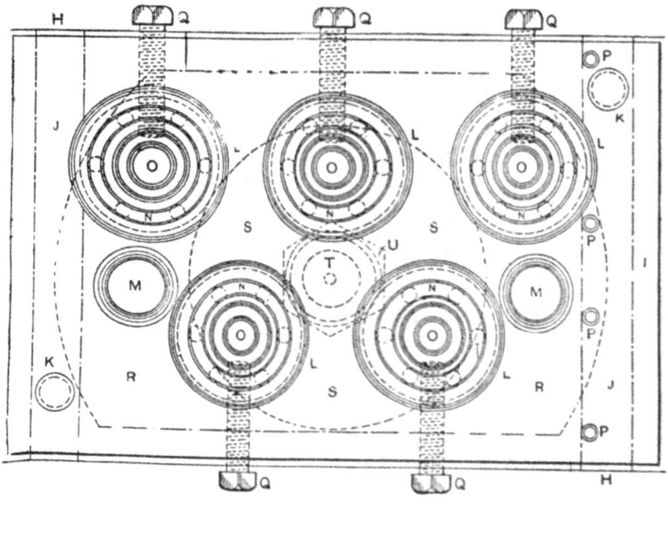

noch größere Anzahl Stücke hergestellt werden kann, wenn nur eine Presse von genügender Größe und Stärke zur Verfügung steht[1]).

Aus Fig. 547, der Ansicht und dem Schnitt durch die erste Reihe Werkzeuge, ist deren Konstruktion deutlich zu ersehen. Das gleiche gilt für Fig. 548, den Grundriß und Schnitt des Unterteils, während in der Stückliste, S. 438/9, die Einzelheiten gesondert aufgeführt werden.

Die Stellung der in der Stückliste abgebildeten Teile in dem fertigen Werkzeug kann bei Verfolgung der gleichen Buchstaben leicht gefunden werden.

Im Oberteile, Fig. 547, haben wir zuerst die Kopfplatte A, die die fünf Verbundschnittstempel und Ziehringe C mit den Auswerfern D, die gleichzeitig Prägematrizen sind, den Federabstreifer B für das Blech und die zwei Führungsstifte E für den Oberteil enthält. In der Kopfplatte werden die Stempel durch die Schrauben G gehalten. Die Kopfplatte ist aus Gußeisen, während der Abstreifer B aus Stahl hergestellt ist. Die vereinigten Schnittstempel und Ziehringe C sind aus Werkzeugstahl, gehärtet und angelassen; sie werden sorgfältig poliert und geschliffen, ebenso die fünf vereinigten Auswerfer und Prägematrizen C. Die Führungsstifte E werden möglichst genau hergestellt.

Der Unterteil besteht aus der Grund- oder Froschplatte H, in der die fünf Zieh- und Prägestempel O, die fünf Niederhalter N mit ihren 30 Tragstiften — je sechs für jeden Schnitt — der Federbolzen T und fünf Preßschrauben Q zur Befestigung der Zieh- und Prägestempel angebracht sind. Dazu kommt die obere Gesenkplatte I, in der die fünf Schnitte L und die zwei Führungsbüchsen M für die Stifte E sowie die vier Anschlagstifte P befestigt sind. Diese beiden Platten sind aus Maschinenstahl und werden durch die beiden gehärteten, konischen Prisonstifte K gesichert und an ihren Enden durch Schrauben, die nicht gezeichnet sind, gehalten. R ist die obere Federplatte aus Gußeisen und S die Gummifeder. Die Bearbeitung und die Fertigstellung der Sitze für die verschiedenen Schnitte und Gesenke in den Gesenkplatten bedingte manche sehr genaue und sorgfältige Arbeit, um alle arbeitenden Teile vollständig auszurichten.

Man wird bemerken, daß die beiden Platten H und I gegeneinander vollständig genau mittels Feder und Nut J verriegelt sind. Die fünf Schnitte L werden in die konischen Löcher der Platte I eingepreßt, während die fünf Niederhalter N gegen die Absätze der Ziehstempel O zum Anliegen kommen, so daß deren Bestreben, die Schnitte lose zu arbeiten, beseitigt wird. Die 30 Druckstifte werden aus Bohrerstahl,

[1]) Vgl. dazu S. 407, die gegenteilige Ansicht in W.T. 1909, S. 16: Perls, Die Massenfabrikation der Edisongewinde.

Stückliste zum Werkzeug Fig. 547–548.

Nr.	Gegenstand	Bezeichnung	Stückzahl	Bezeichn. in Fig.	Material
1	**A. Oberteil.** 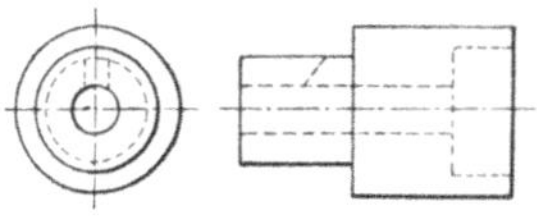	Schnittstempel und Ziehring	5	C	W. Z. St. gehärtet, angelassen, poliert, geschliffen
2	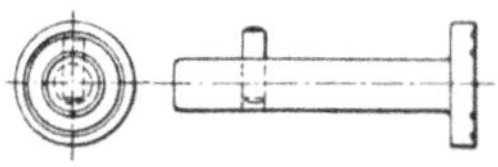	Prägematrize und Auswerfer	5	D	W. Z. St. wie Nr. 1
3		Federn z. Abstreifer	4	F	Masch.-Stahl.
4	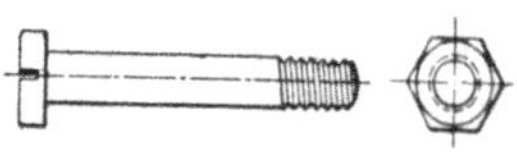	Bolzen und Muttern z. Abstreifer	4 4	F	W. Z. St.
5		Führungsstifte	2	E	W. Z. St.
6	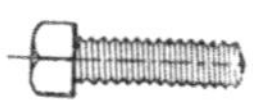	Druckschrauben z. Schnittstempel	5	G	Masch.-Stahl
7	siehe Fig. 547	Kopfplatte	1	A	Gußeisen
8	siehe Fig. 547	Abstreifer	1	B	Masch.-Stahl
9	**B. Unterteil.** 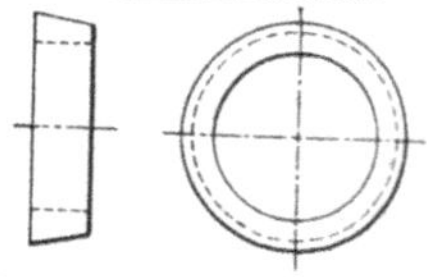	Schnittring	5	L	W. Z. St. wie Nr. 1
10	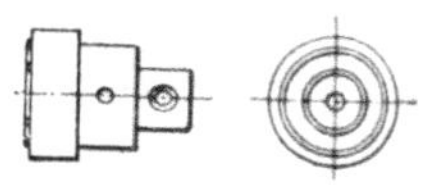	Zieh- und Prägestempel	5	O	W. Z. St. wie Nr. 1

Stückliste der Teile zum Werkzeug Fig. 547—548.

Nr.	Gegenstand	Bezeichnung	Stückzahl	Bezeichn. in Fig.	Material
11	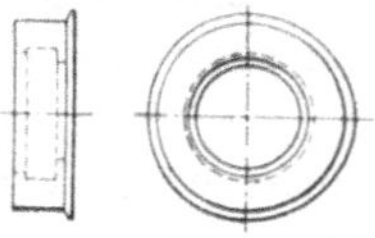	Blechhalter	5	N	W. Z. St.
12	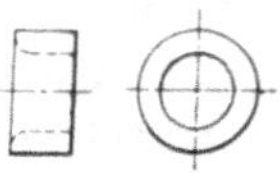	Führungsbüchsen zu Nr. 5	2	M	W. Z. St.
13		Druckschraube für Nr. 10	5	Q	Masch.-Stahl
14		Anschlagstifte für Blech	4	P	Bohrer-Stahl
15		Tragstifte für Blechhalter N	30	V	Bohrer-Stahl
16		Prisonstifte für Gesenkplatte	2	K	Stahl gehärtet
17	siehe Fig. 548	Grund- oder Froschplatte	1	H	Masch.-Stahl
18	siehe Fig. 548	Gesenkplatte	1	I	Masch.-Stahl
19	siehe Fig. 548	Federplatte	1	R	Gußeisen
20	siehe Fig. 548	Feder	1	S	Gummi
21	nicht gezeichnet	Befestigungsschrauben	6	—	Masch.-Stahl
22	nicht gezeichnet	Untere Federplatte	1	—	Schm.
23	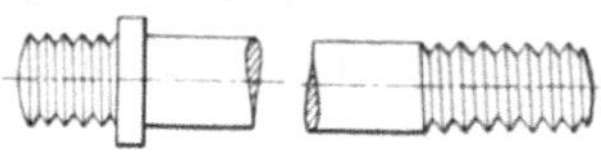	Federschraube	1	T	Masch.-Stahl
24	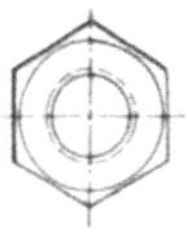	Mutter zu Nr. 23	2		Masch.-Stahl

in ihrer Länge genau auf Maß mit einer Toleranz von nur 0,025 mm gearbeitet. So unscheinbar diese Stifte erscheinen, so wichtig sind sie für die richtige Arbeit des Werkzeuges. Wenn sie nicht genau gleiche Länge haben, so kann man nie das Material gleichmäßig, frei von Falten und ohne Beschädigung des Dekors ziehen, da die Niederhalter durch die ungleiche Länge der Stifte ungleichmäßig angedrückt werden. Die Arbeitsweise des Werkzeuges ist aus folgender Beschreibung zu ersehen:

Das Werkzeug wird in eine schräge Presse eingebaut und diese mit einem von zwei Stangen betätigten Auswerfer am Stößel versehen, damit von diesem aus die Auswerfer und Prägematrizen betätigt werden können. Der Oberteil wird an den Stößel angeschraubt und der Unterteil auf der Tischplatte der Presse mittels großer Kopfschrauben befestigt. Die Werkzeuge sind in einer Presse von ungefähr 1600 kg Gewicht mit sehr kräftigem Antrieb verwendet worden, die einen normalen Hub von 51 mm, einen größten Hub von 89 mm und eine kreisrunde Öffnung in der Tischplatte von 330 mm hatte. Beide Teile wurden so ausgerichtet, daß die Presse beim Abwärtsgang die Stifte E in die Büchsen M eintreten ließ. Hierauf wurde der Hub eingestellt und die Stärke der Federpressung geregelt.

Da die Presse geneigt ist, wird das Material von links nach rechts zugeführt, wobei dessen eine Kante an die Anschlagstifte P kommt und die andere bis zur Kante der hinteren Schnittreihe geht. Beim Einschalten der Presse geht der Oberteil herab, wobei der Abstreifer B das Material flach legt und sicher auf der Oberfläche von I hält, während die Blanketts in den Schnitten N ausgeschnitten, weiter gezogen und geprägt werden. Beim Aufgang des Schlittens schiebt der Abstreifer B den Abfall von den Stempeln, während der Stößelauswerfer die fünf Abstreif- und Prägestempel D betätigt, die beim Abwärtshub die fertigen Näpfchen von den Stempeln abschieben, so daß sie auf die Oberfläche der Gesenkplatte und von da durch die Schwerkraftwirkung nach rückwärts fallen.

Wenn man überlegt, daß die Arbeit dieses Werkzeuges die fünffache eines einzelnen ist, so kann man dessen Leistung bei 10stündiger Arbeit ermessen und die Ersparnis an Arbeit und Kosten schätzen. Man kann, die Materialkosten nicht gerechnet, derartige runde, geprägte Schalen, dekoriert, mit ungefähr 8 Pf. für 100 Stück — nach Angabe Woodworths — und, wenn die bestellte Stückzahl in die Millionen geht, noch billiger verkaufen.

2. Abtrennung und Formänderung ohne Materialverschiebung.

a) Folgewerkzeuge.

Auch für die Vereinigung von Schnittarbeiten mit Biegearbeiten, d. h. solchen, bei denen eine Formänderung ohne eine Verschiebung der kleinsten Teile, wie auf S. 348 erörtert wurde, stattfindet, gelten dieselben Grundsätze, wie auf S. 389 für die Vereinigung von Schnitt- und Zieh- oder Prägearbeiten beschrieben worden sind.

Werkzeug zur Herstellung eines Drahtbügels. Vermutlich eine der einfachsten Arbeiten ist die Herstellung des Drahtbügels in Tabelle Nr. 31 G. 3 auf dem in Fig. 549 abgebildeten Folgewerkzeug.

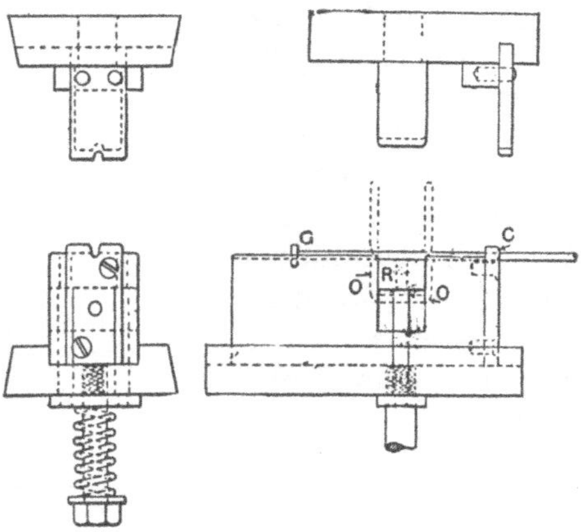

Fig. 549.

Der Draht wird, wie in der Figur ersichtlich, in das Untergesenk eingelegt, und beim Abwärtsgang des Stößels bei C von einem besonderen Schnittstempel abgeschnitten. Der Biegestempel, der ungefähr 3,5 mm kürzer als der Schnittstempel ist, drückt den halb in der Nut des Auswerfers gegen den Anschlag G liegenden Draht in die Nut O. Beim Aufgang des Biegestempels folgt der Auswerfer R durch Federkraft und entfernt den U-förmig gebogenen Drahthaken, worauf mit dem Vorschieben des Drahtes der Kreislauf von neuem beginnt. Dieses Werkzeug wird in einer geneigten Presse verwendet, um die Entfernung der Haken aus dem Werkzeug mittels Schwerkraft allein zu ermöglichen. Fig. 549 zeigt auch die Seitenansicht des Werkzeuges.

Klammern für Kistenecken. Zum Verstärken der Ecken von Holzkisten werden häufig gewellte Blechklammern verwendet, die nach dem Zusammensetzen der Kiste in die Stirnfläche der Ecke eingeschlagen werden, wie Fig. 550 zeigt. Da diese Nieten in großen Mengen gebraucht werden und demgemäß sehr billig hergestellt werden müssen, werden dazu Pressen mit selbsttätiger Streifenzuführung verwendet.

Fig. 551 zeigt eine Vorderansicht, zum Teil im Schnitt, des Werkzeuges und eine Ansicht des Stempels von unten. Es ist ein Folgewerkzeug für Streifenzuführung und stellt bei jedem Preßhub ein fertiges Stück her. Auf der Grundplatte A sind der Zuführungstisch E und die Biegematrize B befestigt. Jener ruht auf Federbolzen, dieser ist mit Schwalbenschwanz in der Grundplatte befestigt und enthält gleichzeitig die untere Schnittkante. Der Stempel besteht aus dem gewöhnlichen, gußeisernen Stempelkopf I, dem Schnittstempel G und dem Biegestempel J. Der Schnittstempel ist mittels Schwalbenschwanzes im Stempelkopf gehalten und durch die große Kopf-

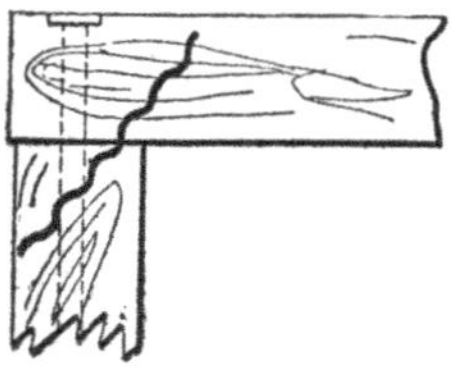
Fig. 550.

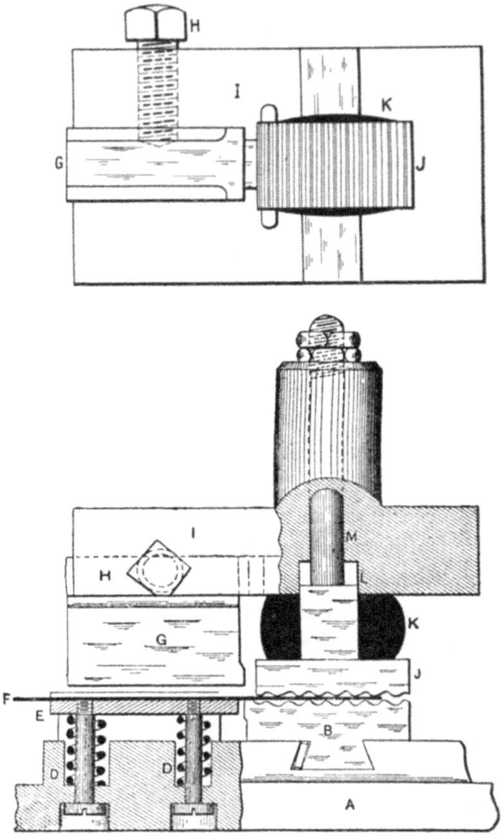

Fig. 551.

schraube H gesichert. So können beide Teile, Schnittstempel und Platte, nach dem Nachschleifen stets wieder eingestellt werden. Der Biegestempel J besitzt einen zylindrischen Zapfen M, der in ein ausgeriebenes Loch des Stempelkopfes paßt, während der Stempel sich in der durchgehenden Nut L auf der Unterseite des Stempelkopfes führt. Zwischen Biegestempel und Kopf ist eine harte Gummifeder K eingelegt. Schnitt- und Biegestempel, G und J, sind gehärtet und auf dunkles Strohgelb, die Matrize B auf lichtes Strohgelb angelassen.

In der Figur steht das Werkzeug in Arbeitsstellung. Der Blechstreifen wird über den Führungstisch zugeführt, wobei selbsttätig bei jedem Hub eine bestimmte Streifenlänge über die Oberfläche der Matrize vorgeschoben wird. Beim Niedergange des Stempels schneidet der Schnittstempel G jenen Teil des Streifens, der auf der Biegematrize B liegt, ab. Dieser wird nun beim weiteren Niedergang des Stempels dort vom Biegestempel J festgehalten, der unter Zusammendrückung der Gummifeder stehen bleibt, während der Schnittstempel mit dem Führungstisch weiter nach abwärts geht, bis der Stempel J im Stempelkopf zur Auflage kommt, wenn das Biegen der Matrize beginnt. Beim Wiederaufgang des Stempels fällt das fertige Stück von der Matrize seitwärts ab, da die Presse geneigt ist.

Ausschneiden und Biegen einer Metallrolle. Das im folgenden beschriebene Folgewerkzeug dient zur Herstellung, vgl. Tabelle Nr. 28,

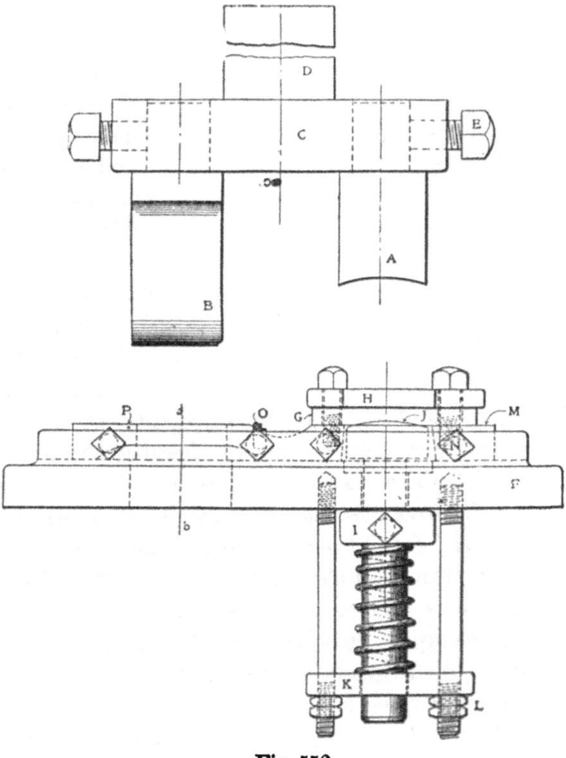

Fig. 552.

einer Blechrolle aus Material von ca. 1 mm Dicke. Im allgemeinen wird das Ausschneiden in einem Gang und das Biegen in einem zweiten vorgenommen, wobei das Blankett von Hand oder mittels kleiner Pinzetten in das Biegegesenk eingelegt wird. Beide Wege sind langsam und gefährlich, da ja bekanntlich mehr als ein Arbeiter bei dieser Art Arbeit seine Finger verloren hat. Mit einem Werkzeug wie das im folgenden beschriebene hat ein Junge über 250 000 Stück in kürzerer Zeit hergestellt als früher drei mit den alten Werkzeugen. Der Schnittstempel ist bei A in Fig. 552 und der Biegestempel bei B in Fig. 553 ersichtlich. C ist der gußeiserne Stempelkopf mit seinem Zapfen D, der in den Pressenstößel paßt. Die Stempel werden mittels der Kopfschrauben E und die Matrizen in der Gesenkplatte F durch Schwalbenschwanz und Schrauben gehalten. G ist eine Unterlage für den Abstreifer selbst. I ist ein einstellbarer Anschlagring, der auf dem Auswerfer J befestigt ist, welcher sich in der Platte K führt, die auf den Einstellmuttern L ruht. Die Schnittplatte wird bei M eingesetzt und durch die Schrauben N gehalten; der Anschlagstift ist O. Das Biegewerkzeug, das in Fig. 553 besonders herausgezeichnet ist, liegt bei P. Die Schnittplatte besteht aus einem Stück Werkzeugstahl und ist eben; der Stempel aber ist hohl, damit ein allmählicher Scherschnitt erhalten wird. Das Blankett, wie in Tabelle Nr. 28, G. 1 angedeutet, wird wohl durchgestoßen, jedoch nicht von dem Materialstreifen getrennt, da es an den Seiten gehalten und auf diese Weise[1]) unter den Biegestempel befördert wird, so daß gleichzeitig an der einen Stelle ein Blankett ausgeschnitten und an der anderen Stelle eines gebogen werden kann.

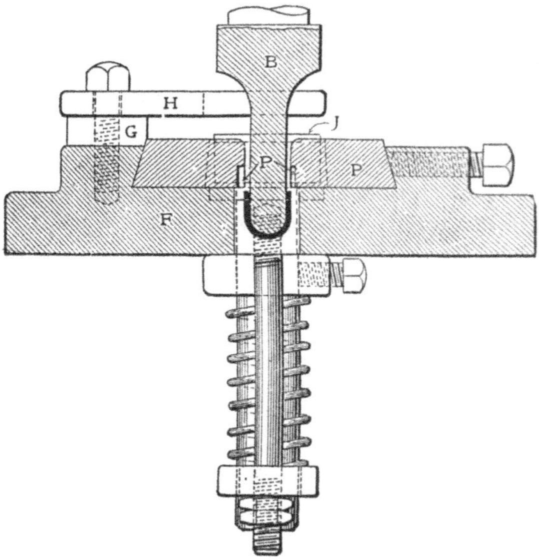

Fig. 553.

Infolge der Form des Stempels wird das Blankett gewölbt und seine Kante Q ungefähr 1,5 mm niedriger als die Kante R, so daß sie an den Anschlagstift O stoßen muß, da die Schnittplatte an dieser Stelle zu diesem Zwecke hohl gemacht wird.

[1]) Vgl. Fig. 544/545 und 559/560.

Zuweilen könnte jedoch das durchgeschnittene Blankett an den Seiten nicht haften und würde durchfallen, wenn nicht der Auswerfer J vorgesehen wäre, der bei seinem Aufwärtsgang das Blankett durch die Schnittplatte wieder in den Abfall hineindrückt. Der Kopf des Auswerfers J wird durch eine harte Feder, die auf der Platte K ruht und sich gegen den Stellring I preßt, der am Auswerfer mittels der Kopfschraube befestigt ist, gegen das Blankett gedrückt.

Die Biegematrize in Fig. 554 wird aus vier Teilen hergestellt, und zwar aus 2 Stücken P aus Werkzeugstahl mit den Zwischenstücken S aus Maschinenstahl. Bei T ist eine Aussparung angedeutet, deren Kante gleichzeitig als Abstreifer wirkt. Die Matrizenteile P werden so hart wie möglich gemacht und nachher über einem offenen Feuer ein wenig anlassen. Wenn die Matrize abgenutzt ist, kann sie in der Mitte nachgeschliffen, an den Kanten gerichtet und sehr schnell und billig wieder auf Maß gebracht werden.

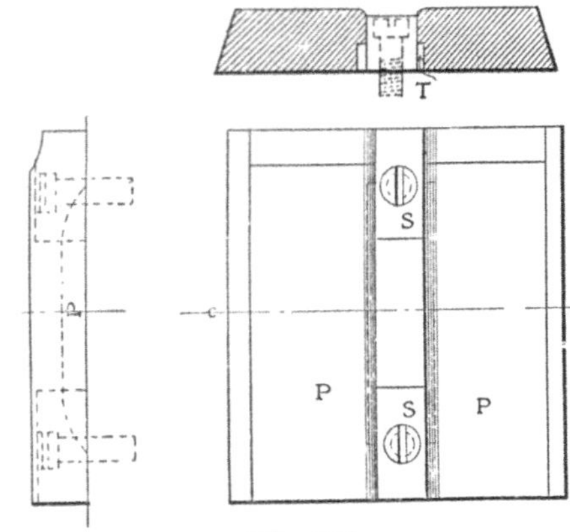

Fig. 554.

Ein dreifaches Folgewerkzeug für dünnes Material. Das in Fig. 555 in zwei Bearbeitungsstufen abgebildete Stück wird auf einem Folgewerkzeug, das in vier Ansichten in Fig. 556 abgebildet ist, hergestellt. Das Stück wird aus halbhartem Kupfer, 0,18 mm dick und 6,5 mm breit, mit einem Innendurchmesser von 5,5 mm im zylindrischen Teile hergestellt. Das hierzu verwendete Bandmaterial wird auf eine Achse aufgerollt angeliefert, von der das Band leicht abrollen kann, wenn es von Hand zwischen den Führungsstiften A A zugeführt wird; es wird so weit vorgeschoben, daß der Schnittstempel B das Ende abschneidet. Der abgefederte Niederhalter C fängt das Band und hält es fest, während das Ende abgeschnitten wird und der bei E sichtbare Ausschnitt im Stempel das abgeschnittene Ende auf seinem weiteren Abwärtsgang freigibt. Nun übernimmt der Biegestempel D das Band und bildet zusammen mit der Matrize F die untere Hälfte der Krümmung, wobei genug Material für die obere Hälfte übrig bleibt. Das Band wird dann vorgeschoben und der bereits gebogene Halbkreis, Fig. 555, in die zweite Matrize G eingelegt. Beim nächsten Abwärtsgange der Presse stößt das Keilstück H den Stift I ein und

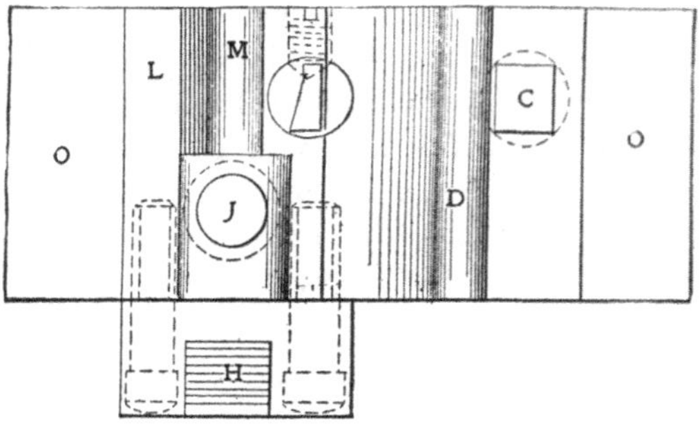

Fig. 555.

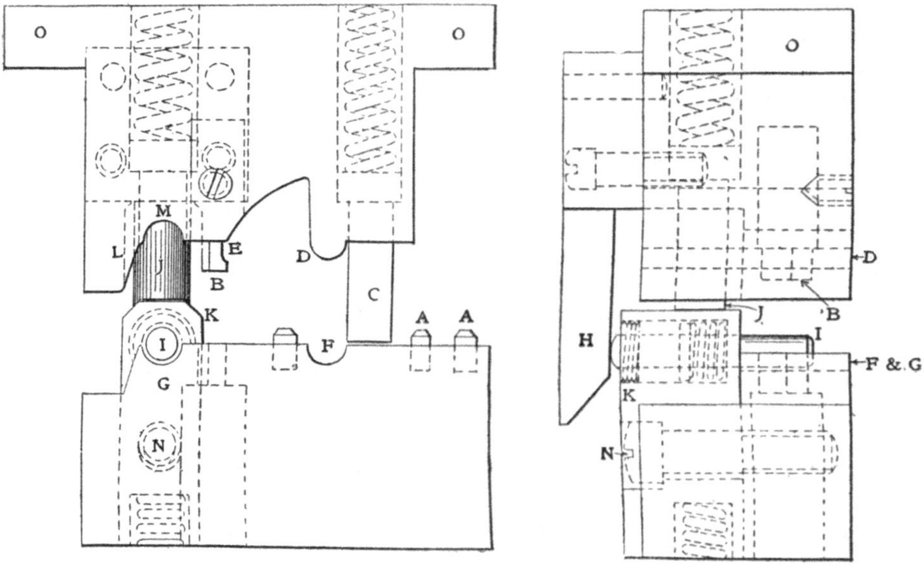

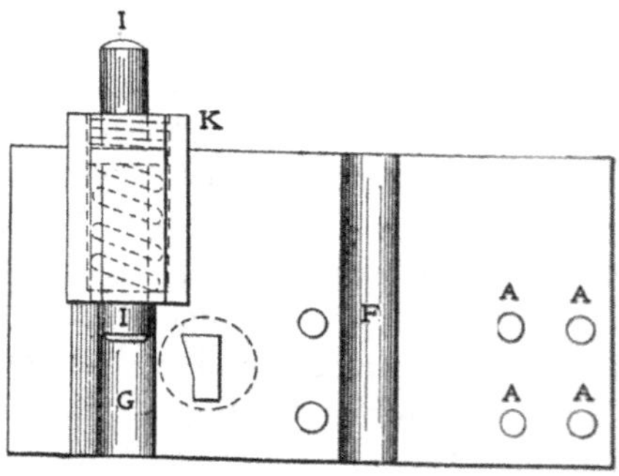

Fig. 556.

über das Band weg, worauf der Stempel J, der eine stärkere Feder besitzt, den Block K, den Stift I und alle mit ihm in Verbindung stehenden Teile herunter stößt und so das Band festhält, während die obere Hälfte des Bleches gebogen und die umgelegten Enden gleichzeitig auf die richtige Form zugeschnitten werden, womit das Stück fertig ist. Gleichzeitig wird jedoch auch ein neuer Halbkreis an dem unter D befindlichen

Teil des Bandes, das noch mit der Rolle zusammenhängt, gebogen. Der überragende Teil L hat das aufrechtstehende Ende des Bandes in Fig. 555 zu erfassen und die Biegung nach M in dem Augenblick einzuleiten, wo die Presse nach abwärts geht und das Band durch I, J und K niedergehalten wird.

Zu der bisherigen Beschreibung dieses Werkzeuges ist zu erwähnen, daß die Feder über C nicht so stark sein darf, daß das Bandmaterial verletzt wird. Keilstück H soll tief genug heruntergehen, um den Stift I einzuschieben, bevor der Stempel J faßt, und dieser wiederum lang genug sein, um das Band festzuhalten, bevor das freie Ende desselben von L gefaßt wird. Der Schnittstempel B wird aus einem zylindrischen Stück herausgearbeitet, durch eine Kopfschraube in seiner Stellung gehalten und gegen Drehung gesichert, wie die Figur zeigt. Auch muß der Ausschnitt bei E so weit heruntergehen, daß das Ende des Bandes freigegeben wird, bevor der Biegestempel D das Band berührt. Die zu B gehörige Schnittplatte wird von oben durchgearbeitet, indem von unten ein Loch in die Grundplatte bis auf 6,5 mm von der Oberfläche gebohrt wird. Die Schraube N hält nicht allein den Gegenstempel K, sondern begrenzt auch seine Aufwärtsbewegung. Die Lappen O an der gemeinsamen Kopfplatte dienen zur Befestigung mit Schrauben und Prisonstiften am Stempelkopf, während der Unterteil in der verschiedenartigsten Weise gehalten werden kann. In der Seitenansicht ist der Niederhalter C und seine Feder ausgelassen, um die Deutlichkeit zu erhöhen; er würde jedoch, wenn eingezeichnet, hinter B erscheinen.

Folgewerkzeug zum Abschneiden und Biegen. Fig. 557 zeigt ein Folgewerkzeug zur Herstellung eines Stückes nach Tabelle Nr. 29. Es ist wohl etwas verwickelter Bauart, weil es mit Rücksicht auf möglichst billige Herstellung des Gegenstandes entworfen worden ist. Der Unterteil enthält die Stücke I, J, K und G aus Werkzeugstahl, während S aus Maschinenstahl und X aus Gußeisen hergestellt ist. Das Winkeleisen O ist an dem Stück K mit zwei Schrauben befestigt. Der Oberteil wird später beschrieben werden.

Das Material wird in der Schere auf die gewünschte Breite zugeschnitten und so in die Zuführung O eingelegt, die das Stück während des ununterbrochenen Biegens in der richtigen Stellung erhält, während es dem ersten Biegewerkzeug K zugeführt wird. Dieses ist auf die verlangte Form jedoch ein wenig tiefer als J ausgearbeitet, so daß der Auswerfer N darin seinen Platz finden kann, ohne die für das Blankett notwendige Höhe zu verkleinern. Das in der Zuführungsleiste O liegende Material wird bis an die Kante des Biegewerkzeuges K vorgeschoben, bis ungefähr 6 mm — um soviel zieht sich das Material ein — überstehen. Nach Einschalten der Presse erhält man die erste Form, Tabelle Nr. 29 G. 1. Das Stück wird dann weiter geschoben, bis es an die mit L bezeichneten Seite des Blocks J anstößt, wodurch die genaue Länge — 38 mm – für das nächste Blankett bestimmt wird. Wenn die Presse wieder eingeschaltet wird, so erhält man ein zweites Blankett, also eine ununterbrochene Formgebung, wie Tabelle Nr. 29 G. 1 zeigt. Hierauf wird das Material

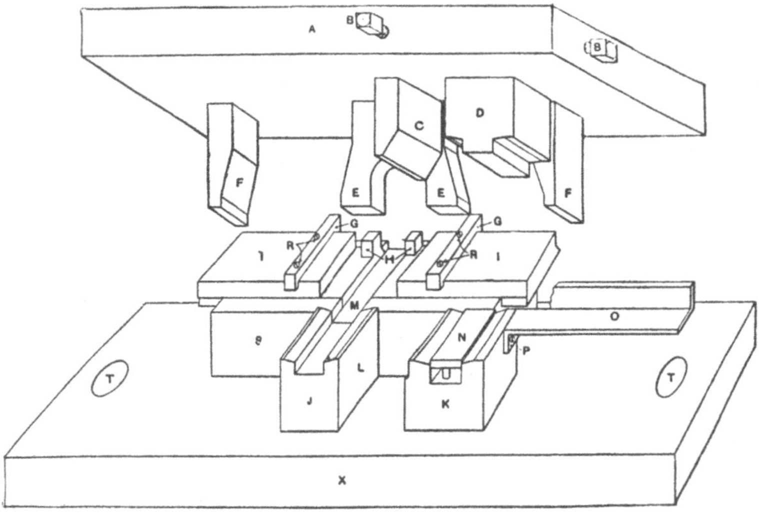

Fig. 557.

dem Block J zugeführt, dessen Form mit den bereits hergestellten Stücken übereinstimmt, so daß man beim dritten Preßhub wieder ein neues Blankett in K erhält und gleichzeitig das erstgeformte Stück in J abschneidet, vgl. Tabelle Nr. 29 G. 2. Da dieses Werkzeug in eine schräge Presse eingebaut ist, so fällt das abgeschnittene Stück durch sein Gewicht in das Schließgesenk M. Die Kante L ist die Schnittkante der Matrize und C der zugehörige Schnittstempel. Der Block J ist ein wenig abgeschrägt, damit das Metall möglichst gleichmäßig von beiden Seiten eingezogen werden kann. Der Schnittstempel C ist ungefähr 1,6 mm länger als der Biegestempel D, so daß er das Material abschneidet, bevor das Biegen beginnt. Er ist ebenfalls abgeschrägt, damit er nicht den Teil des Materials, der gebogen wird, behindert.

Das abgeschnittene Stück gleitet beim Aufwärtsgang des Stößels in den Teil des Werkzeuges, der mit M bezeichnet ist, und stößt dort gegen die Anschläge H, die verhindern, daß das Stück durchfällt, bevor es fertiggeformt ist. Der Block S enthält eine Schwalbenschwanznut, in der die Schlitten I gut eingepaßt gleiten;

die Keilstücke F, deren Abschrägung so gewählt ist, daß sie dem Schlitten den erforderlichen Weg erteilen, sind aus Werkzeugstahl, gehärtet und haben eine Abschrägung von 6,5 mm an den Keilflächen. Die Leisten G sind ebenfalls aus Werkzeugstahl, gehärtet und werden durch die Schrauben R gehalten. Auch die Keilstücke E sind aus Werkzeugstahl hergestellt und haben dieselbe Abschrägung wie die Keilstücke F. Das Stück wird nun in der oben beschriebenen Weise zugeführt, worauf beim Abwärtsgang des Stößels die Keilstücke F die Schlitten I nach innen drücken und das Stück auf die verlangte Form zusammenbiegen. Die Keilstücke E arbeiten mit den Leisten G zusammen und bringen beim Aufwärtsgang der Presse die Schlitten in ihre Anfangsstellung zurück, so daß das Stück, da die Presse geneigt ist, nach rückwärts ausfallen und ein neues eintreten kann. In der Weise erhält man einen ununterbrochenen Arbeitsgang, bis der Streifen zu Ende ist.

Es sei bemerkt, daß die Keilstücke E und F unten ungefähr auf 6,5 mm gerade gelassen sind, so daß sie mit ihren zusammenarbeitenden Teilen stets in Berührung bleiben. Ein Teil der Keilstücke E ist weggeschnitten, damit das fertige Stück ausfallen kann.

Das Werkzeug wird mittels der Löcher T in einer Presse mit 45 mm Hub befestigt. Die Teile F, E und C werden durch Druckschrauben B gehalten, während die anderen Teile durch Zylinderkopfschrauben und Prisonstifte gesichert sind, da Schrauben allein nicht genügen, um die Teile unbedingt genau in ihrer Lage zu erhalten. Die Keilstücke F und E müssen zusammenarbeiten und deshalb gleiche Abschrägung besitzen. A ist aus Gußeisen, während I, J, K und G aus gehärtetem Werkzeugstahl auf ein dunkles Strohgelb angelassen werden. Der Oberteil ist mit einem Befestigungszapfen versehen, der nicht gezeichnet ist.

Aus der Tabelle ist ersichtlich, daß das Stück auf etwas über 90° abgebogen wird, damit die Enden beim Schließen nicht auseinandergehen und das Stück oben offen lassen. Auch muß man die zweifache Metalldicke entweder unter den Schlitten I oder im Grunde des Werkzeuges M frei lassen, um das Stück während des Schließens nicht zu klemmen. Nach dem dritten Hub der Presse wird bei jedem Hub ein fertiger Gegenstand ausgeworfen. Die Kosten für 100 Stück gibt Woodworth mit 12,6 Pf. an.

Mehrfachfolgewerkzeug für genauen Vorschub[1]). (Rob. Tümmler Döbeln.) Dieselben Verhältnisse, die bei Gegenüberstellung der Folge- und Verbundwerkzeuge und bei Beschreibung des Werkzeuges für Blechplomben, Fig. 544/5, zur Sprache kamen, treten bei dem Werk-

---

[1]) Bezüglich des Vorschubes siehe S. 532/3.

zeug zur Herstellung von **Maßstabkappen**, Fig. 558, in erhöhtem Maße auf.

Wegen der Kleinheit dieses Stückes tritt hier die Frage der Genauigkeit bei der Aufeinanderfolge der einzelnen Arbeitsgänge, die jedesmal genau an die richtige Stelle kommen müssen, in viel höherem Maße in den Vordergrund. Die Zahl der Arbeitsgänge ist erhöht, und

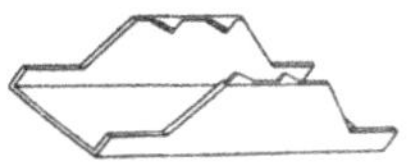

Fig. 558.

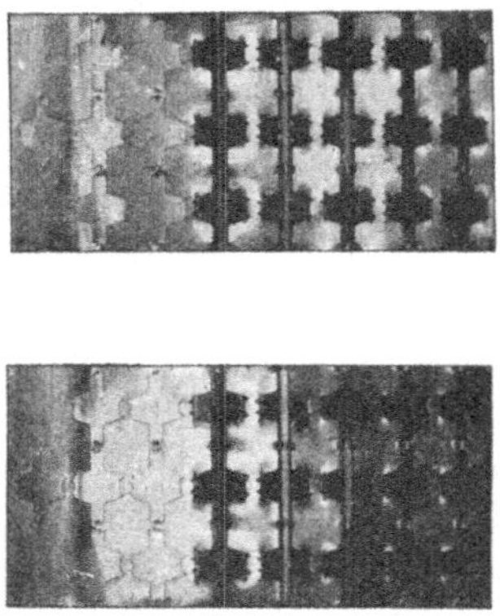

Fig. 559.

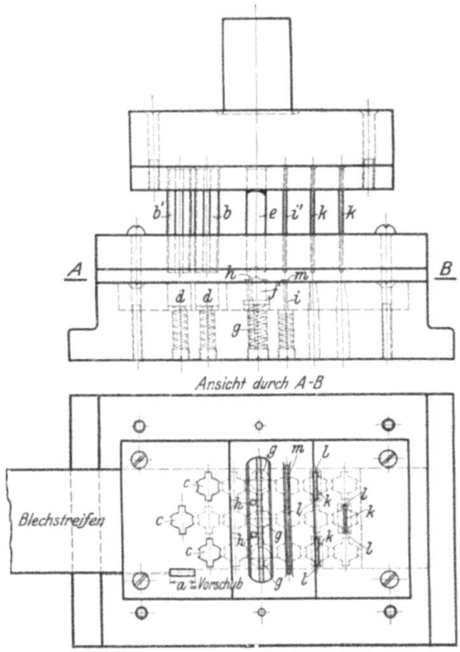

Fig. 560.

auch die Zahl der in einem Hub hergestellten Stücke. An dem Werkzeug ist ein Seitenschneider vorgesehen, damit es auch in der gewöhnlichen Presse benutzt werden kann. Das Material der Maßstabkappen ist Messing- oder Stahlblech.

Die einzelnen Arbeiten folgen in nachstehender Weise aufeinander:

Zuerst wird das Blankett ausgeschnitten, so daß immer noch zwei Haltepunkte, wie bei dem Werkzeug für Plomben, im Streifen verbleiben. Im zweiten Gang werden die Befestigungsspitzen der Kappen im spitzen Winkel abgebogen, zugleich wird eine Biegung durch die Mitte der Kappe vorgenommen, die der endgültigen Biegung entgegengesetzt ist. Im dritten Gang folgt das Fertigbiegen. Damit jetzt die hochzustellenden Seiten der Kappen nicht im Abfallstreifen, Fig. 559, hängen bleiben, wurde im zweiten Gang die entgegengesetzte Biegung durchgeführt. Im vierten Gang endlich wird die fertiggebogene Kappe an den Haltepunkten aus dem Abfallstreifen ausgeschnitten und daraus entfernt.

Das Werkzeug, Fig. 560, unterscheidet sich in dem allgemeinen Aufbau scheinbar in nichts von einem gewöhnlichen Folgewerkzeug. Im Stempelteil sind in der mittels versenkter Kopfschrauben an dem Stempelkopf befestigten Kopfplatte die Stempel b, b′, e, i′ und k in gewohnter Weise eingesetzt und vernietet. Dagegen ist deren Stellung und Arbeit auf den betreffenden Blanketts sehr verschieden. Die drei Schnittstempel b, b′ liegen nicht in einer Linie, sondern sind auf zwei Vorschübe — vgl. die Länge des Seitenschneiders a = dem Vorschub — verteilt, und zwar sowohl aus Rücksicht auf die Befestigung der Schnittstempel als auch auf die gleichzeitige Beanspruchung des Bleches. In der Gesenkplatte ist die aus drei Teilen bestehende Schnittplatte gut eingepaßt und mittels versenkter Kopfschrauben befestigt. In ihrem ersten Teil sind die drei Schnitte c so mit dem Seitenschneider a angeordnet, daß beim ersten Hub der mittlere Schnittstempel b′ und der Seitenschneider und beim zweiten Hub — nach einer Schaltung — die beiden seitlichen Schnittstempel b arbeiten. Man wird bemerken, daß die Führungsplatte verhältnismäßig sehr dick ist, damit die schmalen, feinen Stempel eine gute Führung erhalten. Nachdem die Blanketts ausgeschnitten worden sind, werden sie durch die Federauswerfer d wieder in den Streifen zurückgedrückt, wo sie noch durch zwei Haltepunkte, wie die Plomben in Fig. 544, gehalten werden. Nun wird der Streifen einen Vorschub leer geschaltet, damit das nächste Werkzeug über die Teilfuge der Schnittplatte hinüberkommt. Nach dem dritten Vorschub kommt der Streifen unter den Biegestempel e, der die Mittelrippe durchdrückt und die Befestigungsspitzen unter spitzem Winkel aufbiegt. Dieses geschieht zwischen dem Federauswerfer f, der an seiner Stirnfläche als Matrize ausgebildet ist, und den beiden seitlichen aus der Schnittplatte ausgearbeiteten Rippen. Stempel und Auswerfer sind über die ganze Breite des Bleches in einem Stück geführt, so daß der Verbindungssteg und die Haltepunkte mit durchgebogen werden. Der Auswerfer, der wiederum durch einen unterhalb befindlichen Anschlag in seiner Höchststellung gesichert ist, wird von drei Federn g getragen, die so stark sein müssen, daß sie das Biegen der Rippen aushalten. Dann werden beim weiteren Niedergang des Stempels die Befestigungsspitzen abgebogen. Zwei kleine Auswerferstifte h, die ebenfalls auf Federn gelagert sind, heben den Streifen nach dem Biegen so hoch über die Rippen weg, daß er ungehindert weiterbefördert werden kann. In der nächsten Schaltung erfolgt das Fertigbiegen der Kappen, und zwar wiederum dreier in einer Linie, weshalb der bezügliche Werkzeugteil aus einem Stempel i' und einer Matrize m, i besteht, die sich wieder über die ganze Blechbreite erstrecken. Es sind wieder zwei seitliche Rippen m in der Matrize vorhanden, zwischen welchen und dem Stempel die Seitenteile der Maßstabkappen hochgestellt werden, während unter der Einwirkung des federnden Auswerfers i die im Gang vorher durchgedrückte Rippe wieder ausgeflacht wird. Dieser Auswerfer drückt dann den Streifen aus der Matrize und hebt ihn über die Rippen, so daß er weiterbefördert werden kann. Die Kappen sind nun bis auf das Abschneiden der Haltepunkte l fertig; dies geht während der beiden nächsten Schaltungen vor sich, und zwar wiederum durch drei einzelne Schnittstempel k, von denen der mittlere um eine Schaltung versetzt ist. Es wird also das beim ersten Hube gestanzte Blankett nach der sechsten Schaltung aus der Maschine herausfallen. Danach werden jedoch bei jedem Pressenhube drei fertige Kappen ausgeworfen. Die Schnitte für die Stempel k sind im dritten Teil der Schnittplatte angebracht, so daß die Endkante des Mittelteiles noch als eine Schnittkante für die beiden äußeren Schnitte dient. Es ist dies eine bedeutende Erleichterung bei der Herstellung dieses Teiles. Die Führungsplatte ist mit den Führungsleisten und der Gesenkplatte, in welche die dreiteilige Schnittplatte versenkt ist, durch Schrauben und Prisonstifte zu einem Ganzen verbunden.

Mehrfach-Schnitt- und Biegewerkzeug zur gleichzeitigen Herstellung von 11 Stücken. (Rob. Tümmler, Döbeln.) Zur Herstellung von Kartonnagenklammern in „endlosen" Bändern, d. h. bis zu der Länge des angelieferten Bandmateriales, dient das im folgenden beschriebene Werkzeug, das die Firma seit 15 Jahren baut. Das Bandmaterial wird ohne Abfall verarbeitet und dabei mittels selbsttätigen Vorschubes von der Rolle ab- und nach Fertigstellung wieder auf die Rolle aufgewickelt.

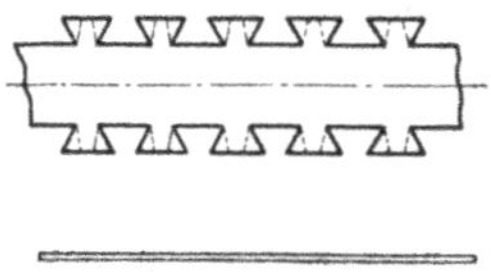

Fig. 561.

Bei den vielen, gleichzeitig arbeitenden Teilen im Werkzeug und den kleinen Abmessungen der ausgeschnittenen Klammerzacken, Fig. 561, ist eine große Genauigkeit des Vorschubes notwendig, da sonst beim Eintritt in das Biegewerkzeug, wo die Zacken, die ungefähr $1{,}5 \times 2{,}5$ mm messen, seitlich abgebogen werden, leicht Fehlstellen im Bande entstehen können. Aus diesem Grund ist eine besondere Transporteinrichtung, Fig. 643, vorgesehen, durch die nach dem Ausschneiden jede Klammer für sich gezogen wird. Der einzige Abfall sind die beiden äußeren Streifen des Bandes, die während des Schneidens an den Aussparungen im Oberteil unter g herausfallen, so daß sie nicht an das Biegewerkzeug gelangen können.

Mit dem vorliegenden Werkzeuge werden 11 Streifen geschnitten und gebogen und zu Kartonnagenklammern fertiggestellt. Fig. 562 und Fig. 563 I ist das Schnittwerkzeug, Fig. 563 II das Biegewerkzeug.

Schnittwerkzeug. Dieses ist für Ober- und Unterteil gleich aus lauter zusammengesetzten kleinen Schnittstempeln e gefertigt, die in einem Stempelkopf c zusammengespannt werden. Diese einzelnen Stempel e sind durch genaue Durchzugsschnitte einheitlich geschaffen, wodurch eine ganz genaue Zusammensetzung möglich ist und auch keinerlei Zwischenfugen sichtbar werden. Durch diese kleinen Schnitt stempel ist einmal eine derartig verwickelte Schnitteinrichtung zu schaffen möglich, außerdem kann jeder Schaden sofort durch Auswechseln der einzelnen Stempel auf billigste Art beseitigt werden.

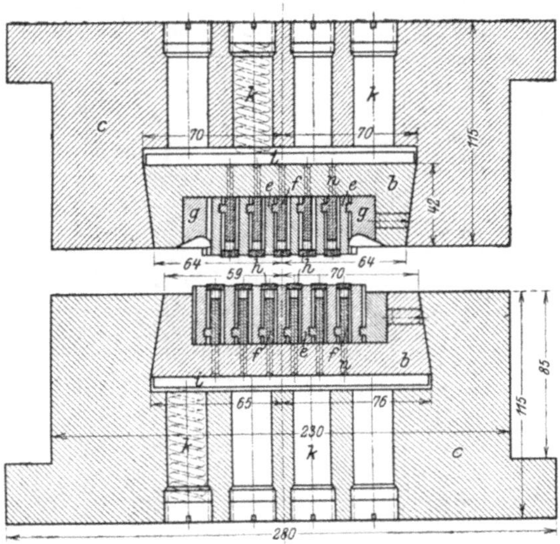

Fig. 562. Schnitt A—B.

Um ein ganz genaues Ineinandergreifen des Ober- und Unterteiles zu ermöglichen, sind noch Führungsstifte (im Grundriß ersichtlich) eingesetzt. Die beiden Teile des Werkzeuges, Oberteil und Unterteil, haben ganz gleiches Aussehen. Der Oberteil enthält 6 Reihen Einzel-Stempel e. Ferner enthält der Oberteil zwischen den Schnittstempeln 5 Zwischenschienen f und 2 Seitenschienen g. Die Zwischen- und Seitenschienen haben einen Ansatz, der zur Befestigung der einzelnen Stempel dient. Im übrigen sind die Stempel noch durch in der Kopfplatte b versenkte Schrauben vorn und seitlich fest zusammengespannt. Über den Zwischenschienen f sind Auswerfer h angebracht und durch 4 Schrauben, welche durch die Zwischenschiene mit der Kopfplatte b verbunden sind, gehalten. Unter der Kopfplatte b ist der Stempelkopf c ausgefräst. In dieser Ausfräsung liegt die Druckplatte i, welche den Federdruck durch die kleinen Stifte auf die Auswerfer h überträgt. Durch die Federn k werden die Auswerfer h in der

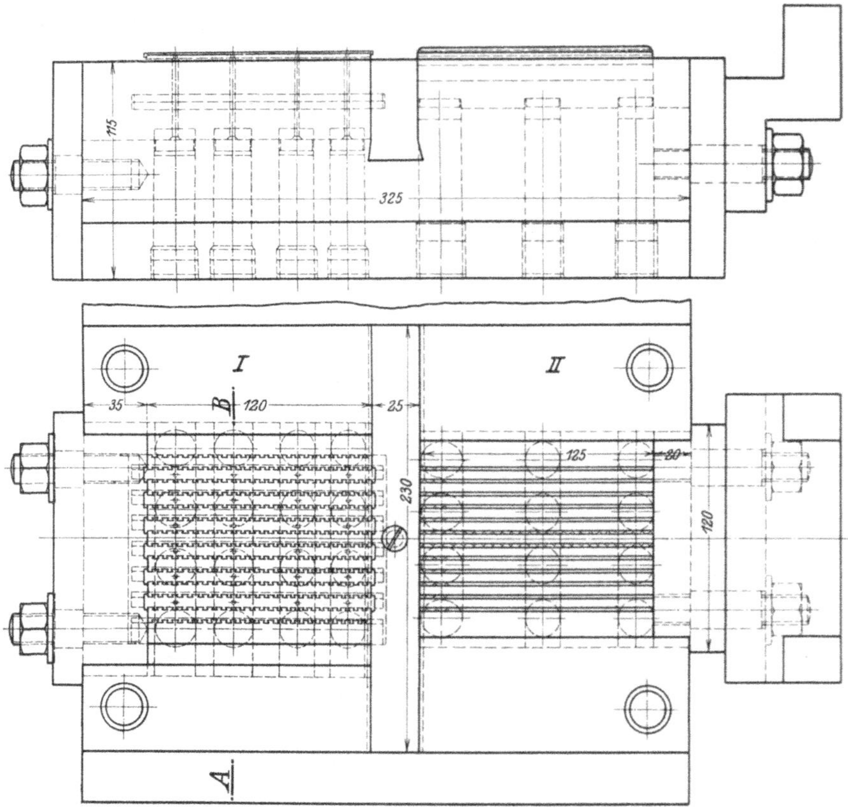

Fig. 563.

richtigen Lage gehalten, so daß die geschnittenen Streifen emporgehoben und zum Biegestempel weiterbefördert werden können. Die beiden Reihen der Seitenstempel g sind an der Unterseite ausgefräst, damit der Abfall herausfallen kann. Der Werkzeugunterteil ist genau in gleicher Weise gefertigt wie der Oberteil, nur daß er 7 Reihen Stempel und 6 Reihen Auswerfer enthält.

Biegewerkzeug. Dieses besteht aus einzelnen Biegestempeln, die in einem oberen und unteren Halter befestigt sind, die abwechselnd im Stempelkopf und in der Gesenkplatte eingespannt werden. Wie beim Schnittwerkzeug sind auch hier Auswerferschienen von gleicher Bauart vorhanden, um die fertiggeprägten Klammern aus den Tiefen der Biegewerkzeuge herauszuheben.

Zur Herstellung dieser Klammern aus Rollenmaterial dient eine Kurbelpresse. Das Bandmaterial wird meist in Rollen von rd. 300 Meter angeliefert und auf einen Aufwickelapparat aufgelegt. Die fertig geschnittenen und gebogenen Klammern werden ebenfalls durch einen Aufwickelapparat aufgerollt, und zwar dergestalt, daß der Aufwickelapparat diese 11 Klammern jede für sich auf eine Vorrichtung aufrollt, so daß die Klammern gleich versandbereit von der Presse kommen. Der Aufwickelapparat ist derartig gebaut, daß 100 Meter Klammern abgenommen werden können, worauf neue Holzachsen zum Aufwickeln einzulegen sind.

Die Presse arbeitet selbsttätig, ebenso Zuführung und Aufwickelapparat, vgl. Fig. 645.

b) Verbundwerkzeuge.

Werkzeug für Rohrschellen. Rohrschellen nach Fig. 564 werden in einem Verbund-, Schnitt-, Loch- und Biegewerkzeug nach Fig. 565 hergestellt, und zwar ein Stück bei jedem Pressenhub. Das Blech wird gebogen, abgeschnitten und die Schraubenlöcher gelocht. Das Werkzeug ist einstellbar gebaut, damit man Stempel für 4 verschiedene Rohrdurchmesser einbringen kann, wobei in allen Fällen die Füße und Schraubenlöcher der Schellen gleichbleiben. Der Stempel in Fig. 565 entspricht einem Rohr von 38 mm (1½″ engl.) Durchmesser; außerdem wurden Stempel für 32, 26, 13 mm (1¼, 1 und ½″ engl.) Rohrdurchmesser vorgesehen. Zur Einstellung werden die Klötze F, gegeneinander verschoben und Zwischenstücke eingelegt, die dann durch die Schrauben M entsprechend den verschiedenen Stempeln festgestellt werden. Hierauf werden die Blöcke F mit den im Grundrisse sichtbaren Schrauben auf dem Bett befestigt. Keile K, die mit Schwalbenschwänzen in den Klötzen befestigt sind und in einer gefrästen Nut des Bettes gleiten, sichern die winkelrechte Stellung der Werkzeugteile. Die Grundplatte ist aus Gußeisen und hat auf der Unterseite gefräste Nuten zum Versenken der Schraubenköpfe. Zwei Ständer, die seitlich an die Grundplatte angeschraubt und durch Keile gesichert sind, tragen die Schrauben M, die sich gegen die Gesenkplatte oder im anderen Fall gegen die Beilagplatte stemmen und so den Gegendruck während der Arbeit aufnehmen. Der dargestellte Anschlag I dient nur für drei Größen Schellen, während für die größten, für die der Stempel gezeichnet ist, ein anderer eingesetzt werden muß. Die Biegebacken D sind aus Werkzeugstahl, gehärtet und werden durch je zwölf Federn gegen das Blech gedrückt, sobald der Stempel dasselbe in die Matrize einzupressen beginnt. Diese Federn sind in

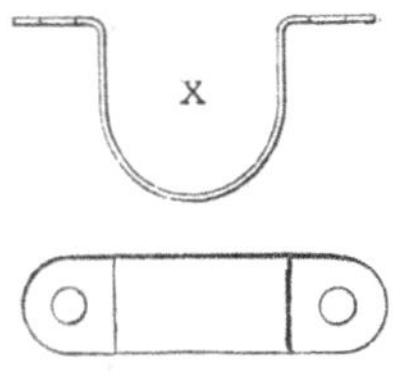

Fig. 564.

entsprechende Löcher auf der Unterseite der Matrizenteile D eingelegt. Diese Spiralfedern haben einen lichten Durchmesser gleich dem Durchmesser des Federdrahtes, sind aus gezogenem Stahldraht hergestellt und werden nach dem Wickeln angelassen. Sie arbeiten zufriedenstellend, da die ganze notwendige Zusammendrückung nur rd. 1,5 mm ist.

Die Matrizenteile D werden durch die Stifte E geführt, die gleichzeitig als Lochstempel dienen, um die Schraubenlöcher in den Schellen herzustellen. Der Biegestempel A, an dem die Schnittstempel

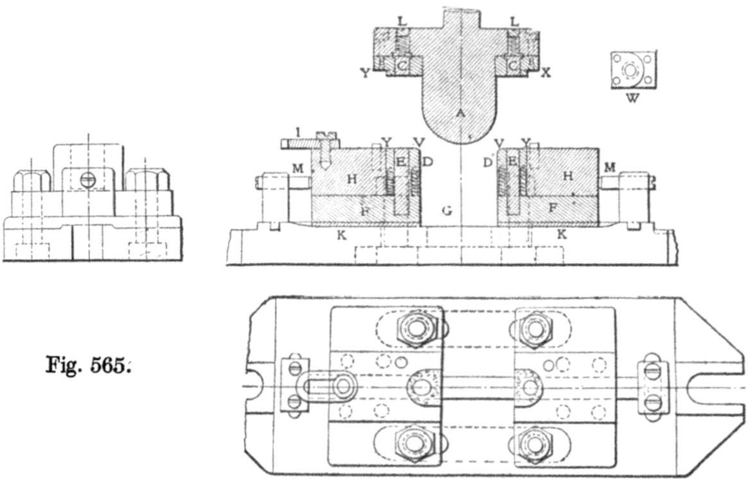

Fig. 565.

B angeschraubt und durch Prisonstifte gesichert sind, ist aus Stahl. Andrerseits sind diese Schnittstempel zugleich die Schnittplatten für die bereits erwähnten Lochstempel E. Die von ihnen ausgeschnittenen Scheiben werden sofort beim Aufgang des Stempels durch die federnden Auswerfer C, die durch die Stellschrauben L eingestellt werden können, ausgeworfen. Die seitlichen Lappen X am Stempel sind eingefügt, um zu verhindern, daß das Blech sich an den Schnittkanten reibt, während es in die Matrize gezogen wird. Tatsächlich kommt dieses Reiben vor, jedoch nur in so geringem Maße, daß weder das Arbeitsstück noch das Werkzeug Schaden leiden. Diese Fortsätze bieten auch Platz für die Befestigungsschrauben der Schnittstempel. Ein Grundriß derselben ist bei W ersichtlich. Das Rohmaterial für die Schellen ist weiches Messingblech in Streifen von der richtigen Breite. Die Dicke beträgt 1,6 mm.

Verbund-Schnitt- und Biegewerkzeug zur Herstellung der Grundplatte eines Hausweckers. (C. Lorenz, Berlin, SO.) Die Grundplatte des Hausweckers, vgl. Tabelle Nr. 1, ist im ersten Gang auf dem in Fig. 20

abgebildeten Werkzeug ausgeschnitten worden. Das ausgeschnittene Blankett wird ausgeglüht und kommt dann in das in Fig. 566 abgebildete Verbundwerkzeug, mittels dessen die in Tabelle Nr. 1 - I G. 2 ersichtlichen Löcher gelocht und der dort im Schnitt A—B erscheinende aufgebogene Rechtecklappen ausgeschnitten, an einer Seite aufgebogen und an der einen Kante scharf umgebogen wird.

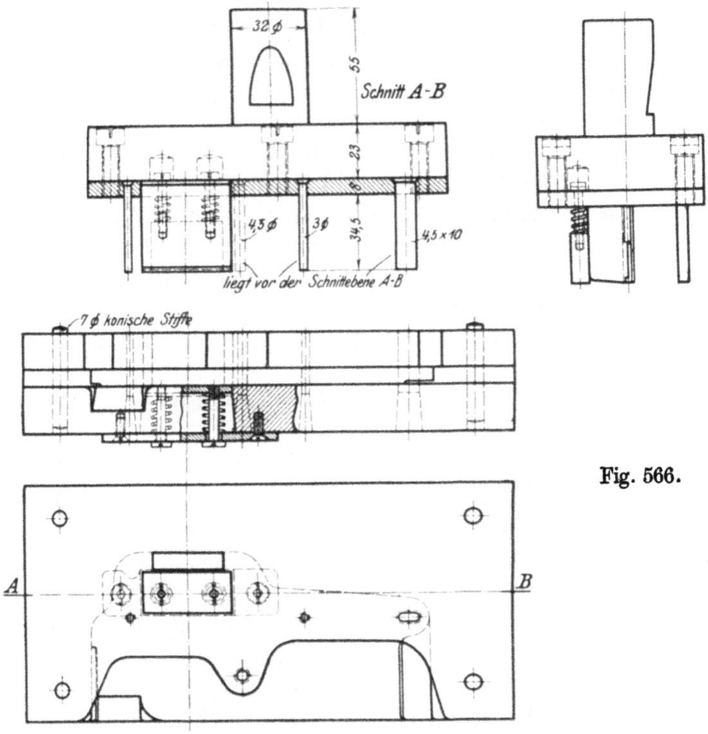

Fig. 566.

Der Aufbau des Werkzeuges deckt sich mit den verschiedenen andern Werkzeugen dieser Firma, die im Anschluß an die in Tabelle Nr. 1, I—IV angegebenen Arbeitsgänge besprochen werden. In der Kopfplatte werden die vier Lochstempel in der gebräuchlichen Weise vernietet, ebenso der rechteckige Ausschnittstempel. Die Kopfplatte wird dann mit vier versenkten Zylinderkopfschrauben von oben im Stempelkopf befestigt. Ebenso sind die Schnitte in der Schnittplatte in gewöhnlicher Weise ausgearbeitet, die dann mit einer verhältnismäßig dicken Abstreiferplatte durch vier konische Stifte zu dem Schnittkasten vereinigt ist. Von Interesse ist, wie der rechteckige Ausschnittstempel gleichzeitig als Biegestempel verwendet wird. Der Stempel ist an drei Kanten, an denen er auch als Schnittstempel arbeiten soll, scharfkantig, während die eine Langseite, an der der Lappen mit dem Blankett zusammenhängen soll, abgerundet ist. An der anderen Langseite ist der Stempel zurückgesetzt, so daß das Material, das bereits gelocht und an den Schmalseiten des Rechteckes durchgeschnitten ist, sich um die untere Kante und um die abgerundete des Stempels abbiegen muß, bevor die obere Kante es abschneidet. Man müßte eigentlich diese Arbeit als Prägen ansprechen, da der Lappen noch an zwei Seiten mit dem übrigen Blankett zusammenhängt, wenn die doppelte Biegung ausgeführt wird, die also nicht ohne eine Streckung des Materiales vor sich gehen kann. Aus diesem Grunde ist auch der in der Seitenansicht des Stempels sichtbare federnde Niederhalter vorgesehen, der verhindert, daß die schmale Blankettfläche hinter dieser Kante beim Abbiegen Falten wirft. Er wird von zwei versenkten Kopfschrauben getragen, die in Ausbohrungen des Stempelkopfes sitzen und gleichzeitig als Führungen für die Druckfedern dienen. In dem entsprechenden Teil der Schnittplatte ist ein einfacher Federauswerfer vorgesehen, dessen Schrauben sich gegen eine auf der Unterseite der Schnittplatte angeschraubte schmale Blechplatte anlegen.

Die Arbeitsweise ist also folgende: Das Blankett wird von Hand in die auf der Oberseite der Schnittplatte eingefräste Aussparung eingelegt, wo es gleichzeitig mit dem Auftreffen der Lochstempel von dem Niederhalter festgehalten wird. In diesem Augenblick beginnen auch die tiefsten Ecken der schmalen Schnittkanten des Rechteckausschnittstempels die Kanten im Blankett einzuschneiden und die stumpfe Längskante des Stempels das Blech an dieser Stelle hochzubiegen, bzw. zu prägen. Diese Arbeit ist eben ein Mittelding zwischen den beiden genannten Grundarbeiten. Mit dem weiteren Fortschreiten des Stößels nach abwärts wird allmählich die ganze Länge der beiden Schmalseiten durchgeschnitten und im gleichen Maße die Doppelbiegung durchgeführt, bis in der tiefsten Stellung des Stößels die obere scharfe Schnittkante des Rechteckstempels, rechts in der Seitenansicht, die eine Langseite durchschneidet. Infolge dieser eigentümlichen Arbeitsweise ist auch die Führungsplatte so stark gemacht, damit die kleinen Lochstempel von 3 mm Durchmesser nicht durch ein Nachgeben oder Verziehen des Bleches brechen.

c) Folge- und Verbundwerkzeuge.

Einfachste Form einer Vereinigung von Folge- und Verbundwerkzeugen. Der in Fig. 456 und 567 dargestellte ungleichseitige Winkel, der auf dem Werkzeug Fig. 456 in zwei Gängen erhalten wird, kann auch vorteilhaft in einem Verbundwerkzeug mit Vorlocher hergestellt werden. In Fig. 568 ist ein solches Werkzeug im Schnitt dargestellt, bei dem das Schneiden, Lochen und Biegen in einem Gang vor sich geht. Hier ist A die Schnittplatte, die eine Führungs- und Abstreiferplatte B für die Lochstempel H trägt und weit genug geschlitzt sein muß, um den Blechstreifen leicht und doch ohne seitliches Spiel einführen zu können. Der Blechstreifen wird nun durch den Schlitz zuerst bis an die Schnittkante a vorgeschoben, worauf der Stempelkopf F, der den Schnitt- und Biegestempel D und die mittels Kopfplatte E befestigten Lochstempel H trägt, niedergeht und das Blech locht. Während der Stempel in die Höhe geht, wird der Blechstreifen bis an den Anschlag G vorgeschoben, so daß von jetzt ab die Arbeitsgänge ununterbrochen aufeinanderfolgen und bei jedem Preßhub ein fertiges Stück hergestellt wird, wie es unterhalb des Stempels in punktierten Linien angedeutet ist.

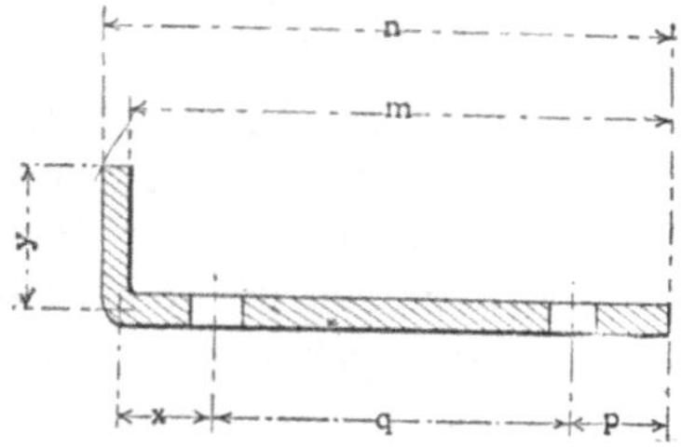

Fig. 567.

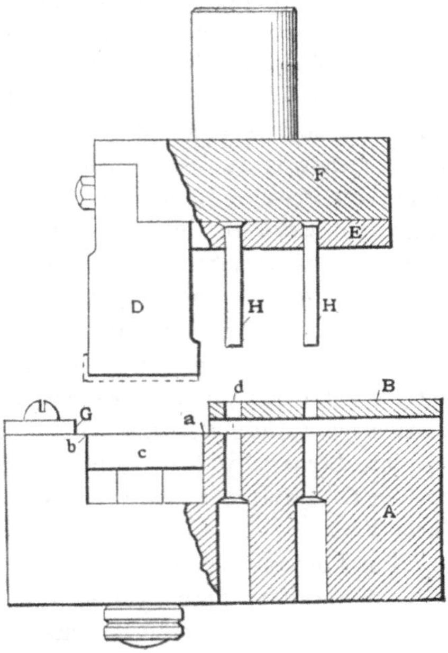

Fig. 568.

Die Feder unter dem Auswerfer c kann so lange verwendet werden, als das zu verarbeitende Material nicht zu stark ist. Ist das Material dicker als 1,25 mm, so wird an Stelle der Feder ein Puffer aus weichem Gummi verwendet. Es hängt natürlich von dem Grade des Ausglühens ab, wieviel Druck zwischen Auswerfer und Stempel nötig ist. Der Druck muß selbstverständlich groß genug sein, um ein Gleiten des Materials zu verhindern, während der Stempel an der Kante b vorbeigeht. Das Bandmaterial für ein Werkzeug dieser Art muß gerade gerichtet sein und die genaue Breite haben. Löcher, die in die Stücke gestanzt werden, sind, soweit als möglich, als Anschläge für die folgenden Arbeitsgänge zu benutzen. Dies erzeugt gleichmäßige Arbeit.

In diesen Verbundwerkzeugen muß die ganze Länge des Bleches, sowie die Entfernung zwischen der Schnittkante der Matrize a und dem kleinen Loch d genau bekannt sein, bevor man die kleinen Löcher oder den Anschlag G festlegt. Wenn man nach dem Lochen der beiden kleinen Löcher d die Länge des Blechstückes ändert, indem man den Anschlag G verschiebt, so bringt man die Löcher in dem fertigen Stück an eine falsche Stelle. Bei der Herstellung ist es daher vorteilhafter, Matrize und Stempel ohne die kleinen Löcher in der Führungsplatte und ohne die dazugehörigen Stempel H fertigzumachen. Alle übrigen Teile können bis auf das Härten und Anlassen fertiggestellt werden.

Der auf dem besprochenen Werkzeug hergestellte Winkel ist in Fig. 567 in vergrößertem Maßstabe im Schnitt dargestellt. Von dessen Abmessungen sind bekannt: m, das Maß für den Stempel, n dasselbe + der Blechdicke und gibt das Maß für die Länge a—b, Fig. 568, ab. Die Strecken x und y sind dagegen unbekannt und notwendig, um die kleinen Löcher d gegen den Anschlag G festzulegen. Hier führt nun der Versuchsweg am besten zum Ziel. Das soweit halbfertige Werkzeug wird in eine Presse eingebracht, ein passendes Stück Blech von bekannter Länge, mit zwei in der richtigen Entfernung voneinander gebohrten Löchern, zwischen Stempel und Auswerfer gelegt und gebogen. Die Länge p + q ist bekannt, entweder aus der Zeichnung oder dem Musterstück, ebenso die Länge des Bleches vor dem Biegen. Zieht man nun von der Länge des Versuchsbleches den Unterschied zwischen y und der entsprechenden Länge des Versuchswinkels ab, so erhält man die richtige Länge des Blanketts. Diese Länge vermindert um die bekannte Strecke p + q ergibt die Entfernung des kleinen Loches d von der Schnittkante bei a. Diese Arbeitsweise ergibt die Lösung in einem Versuch.

Nun können die kleinen Löcher d in der Schnittplatte A fertiggemacht werden; dann werden die Gewinde für die Schrauben der Führungsplatte B geschnitten, die Löcher für die Prisonstifte gebohrt und die Schnittplatte gehärtet, bevor die Führungsplatte fertiggemacht wird, bzw. bevor die Löcher für die kleinen Stempel übertragen werden.

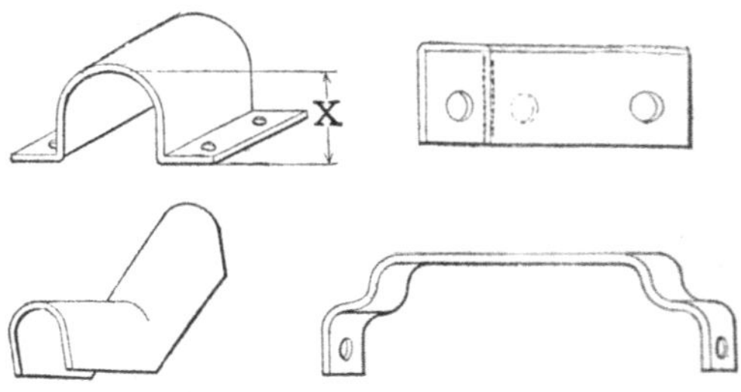

Fig. 569 a—d.

Folge- und Verbundwerkzeug zum Schneiden, Lochen und Biegen. Die in Fig. 569a—d abgebildeten Stücke werden auf einem Verbundwerkzeug vorteilhaft hergestellt. Fig. 570 stellt ein solches Schnitt-

und Biegewerkzeug mit Vorlocher dar, mittels dessen in einem Gang das in Fig. 569a abgebildete Stück hergestellt werden kann.

Die Froschplatte A ist eine flache Platte, am besten aus Stahlguß, mit einer eingehobelten Nut zur Aufnahme des Biegestempels D und der Schnitt- und Lochplatte B, die mittels Schrauben daran befestigt werden. Unterhalb des Biegestempels D wird Gewinde in die Aufspannplatte A zur Aufnahme des Auswerferbolzens L geschnitten.

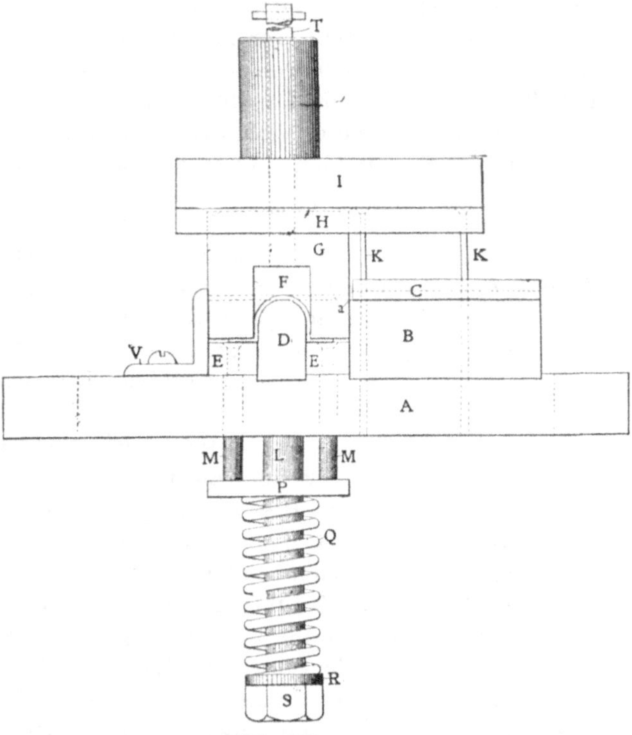

Fig. 570.

Die Auswerfer E tragen vier eingenietete Stifte M, die durch die Froschplatte A hindurchgehen und auf die Platte P drücken, so daß die Schraubenfeder Q zusammengepreßt wird. Zur Herstellung einer Form nach Fig. 569a müssen die Auswerfer E so hoch gemacht werden, daß sie gerade auf der Froschplatte aufliegen, wenn die Entfernung ihrer oberen Fläche vom Scheitel des Biegestempels D gleich der Strecke X in Fig. 569a ist. Andrerseits müssen die Stifte M so lang sein, daß die oberen Flächen der Auswerfer E sich genau in einer Ebene mit dem Scheitel des Biegestempels D und der Oberfläche der Schnittplatte befinden, wenn die Platte P gegen die Unterfläche der Froschplatte gepreßt wird.

Der Schnittstempel und Biegematrize G muß unmittelbar unter dem Stempelzapfen angebracht werden, damit der Auswerfer F beim Aufwärtsgang der Presse, der mittels der Stange T[1]) von dem Hebel der Presse bedient wird, den fertiggestellten Gegenstand aus dem Oberstempel entfernen kann. Die Presse steht geneigt, so daß die fertigen Stücke von selbst herausfallen können und ein ununterbrochenes Arbeiten ermöglichen.

Der Scheitel des Biegestempels D muß sich in gleicher Höhe wie die Schnittplatte B befinden, damit der Schnittstempel G das Blech niederhält, bevor das eigentliche Scheren stattfindet. Die gestrichelten Linien oberhalb D stellen das Blech dar, die Schnittkante von B ist a.

Die Herstellungsziffer eines gut ausgeführten derartigen Werkzeuges ist sehr hoch, die Arbeit zufriedenstellend und gleichmäßig, ohne Materialverlust.

Verbund- und Folgewerkzeug zum Schneiden, Lochen und Biegen mit selbsttätiger Zuführung. Die Grundzüge des im folgenden beschriebenen Folgewerkzeuges lassen sich auf eine ganze Reihe Preßarbeiten übertragen. Das Werkzeug in Fig. 572 ist mit einer selbsttätigen Zuführung[2]) versehen, die einfach und vollständig in sich geschlossen ist und zwangläufig arbeitet, auch keine besonderen verwickelten Verbindungen mit der Presse verlangt. In Fällen, wo die Herstellungsmenge nicht groß genug ist, um die Auslagen dafür hereinzubringen, kann von ihr abgesehen werden und die Zuführung von Hand erfolgen, ohne dadurch die Vorteile des Werkzeuges zu vermindern.

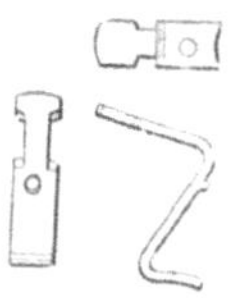

Fig. 571 a, b.

Aus der Zeichnung Fig. 571 a b, die den Streifen und das herzustellende Stück darstellt, läßt sich leicht erkennen, daß man es hier mit einer Biegearbeit zu tun hat, deren Schwierigkeiten in der genauen Einstellung für die einzelnen Herstellungsstufen des Stückes liegen. In dem abgebildeten Werkzeug wird bei jedem Preßhub ein Stück fertiggestellt, wobei die Stücke austauschbar sind.

Fig. 572 gibt Vorderansicht, Seitenansicht des Werkzeuges und einen Grundriß der Gesenkplatte mit entfernt gedachtem Abstreifer wieder. M ist die Gesenkplatte, die eine Nut zur Aufnahme des Schlittens P hat, K die Loch- und Schnittplatte und J die Biegematrize. Die Breite derselben muß dem zu biegenden Stück entsprechen, K dagegen mindestens 6 mm breiter als der Streifen sein, aus dem das

[1]) Vergl. Fig. 366/73, 524/5.

[2]) Dieser Antrieb gilt auch für Fig. 506.

Blankett ausgeschnitten wird. Das Schneiden, Lochen und Biegen der kleinen Fortsätze, die in Fig. 571 b ersichtlich sind, geschieht durch die Schnitt- bzw. Biegestempel C, D und E. Die Schnittplatte zur Linken des Stempels E ist mit einem abgeschrägten Kanal versehen, so daß der Vorwärtsbewegung des Streifens, nachdem die seitlichen Fortsätze gebogen worden sind, kein Widerstand entgegengesetzt wird. Der Kanal darf sich jedoch nicht bis an die Schnittkante der Schnittplatte G′ unterhalb G, erstrecken, da diese die volle Breite des Streifens haben muß.

Die gestrichelten Linien an den Grundflächen der Stempel C und E, Fig. 572, stellen die Schnitt- und Biegeflächen dar, da die Stempel,

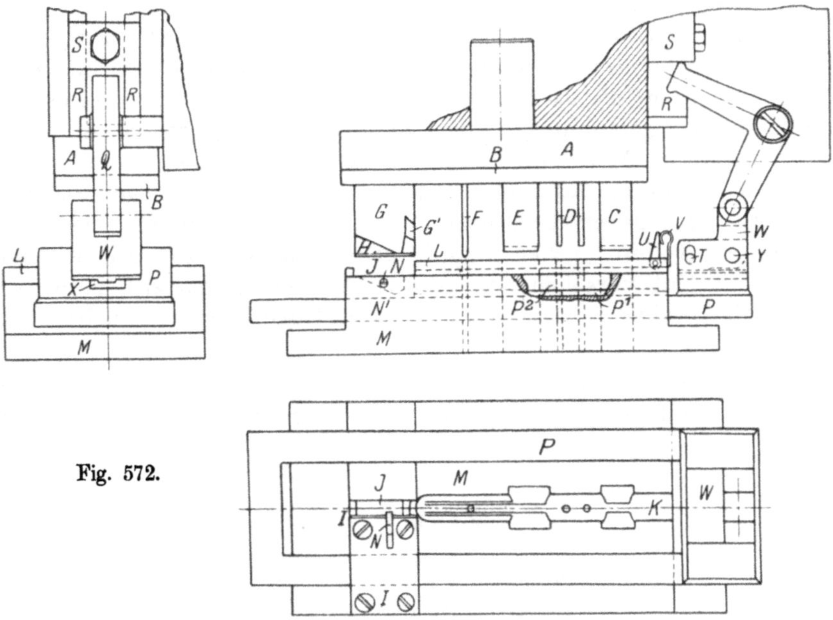

Fig. 572.

um zufriedenstellend zu arbeiten, in den Schnittkasten eingetreten und in ihm geführt sein sollen, bevor das Schneiden und Biegen stattfindet.

Zwischen dem Biegestempel E und der Schnittkante G′ muß eine Blankettlänge freibleiben, um den unbedingt notwendigen Anschlagstift F auszunutzen. Der Biegestempel G hat eine Führungsplatte H, die am Grunde abgeschrägt ist, und die Biegematrize J eine entsprechende Platte I an der entgegengesetzten Seite. Da der Biegestempel und die Matrize dieselbe Breite wie der fertige Gegenstand besitzen, wird dieser an seinen Platz gebracht und dort in seiner Stellung gesichert, so daß nach dieser Richtung hin alle Stücke gleich werden. Das Stück G_1, das die Schnittkante des Stempels G bildet, ist besonders eingesetzt, um es schnell ausbessern zu können.

Der Auswerferstift N ist von der Oberfläche der Matrize J nach aufwärts abgeschrägt und wird durch eine Spiralfeder, welche durch die Schraube N' in ihrer Stellung gehalten wird, herausgedrückt. Beim Eindrücken des Streifens in die Matrize wird der Auswerfer in die Nut zurückgedrängt. Nach Aufhören des Druckes wird das fertige Stück durch den Auswerfer entfernt und fällt, da die Presse geneigt steht, durch sein eigenes Gewicht heraus.

In einem derartigen Werkzeug wird vollständige Austauschbarkeit der Stücke dadurch erzielt, daß der Streifen durch den Anschlagstift so lange gehalten wird, bis die Biegearbeit fast vollendet ist, da das Abschneiden des fertigen Gegenstandes beinahe der letzte sich abspielende Vorgang ist.

Die selbsttätige Zuführung ist, wie Woodworth sagt, wohl neuartig. Es sind verschiedene Werkzeuge mit derartigen Zuführungen gebaut worden, die sehr gut gearbeitet haben, wobei der Streifen von der Rolle abgenommen wurde. Diese Werkzeuge stellen in einem zehnstündigen Arbeitstag 18 000 bis 60 000 Stück her.

An der Seite des Stößels der Presse ist ein U-förmiges Stück R befestigt, das durch die Klemme S, nach oben und unten einstellbar, in seiner Stellung gehalten wird. Diese ist auch zur Nachstellung mit einer Nut versehen. Mit dieser Einrichtung läßt sich eine ganz feine Nachstellung, besser als mit Sperrad und Klinke, erreichen.

Folge- und Verbundwerkzeug mit Säulenführungen. An den Momenthebelausschaltern, Fig. 573, befindet sich innerhalb des Hebels eine

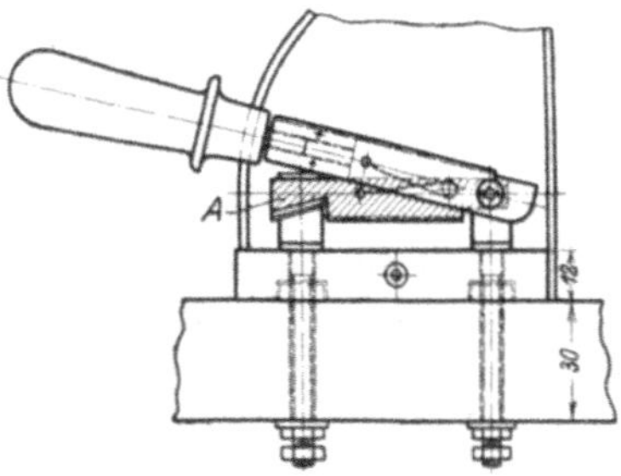

Fig. 573.

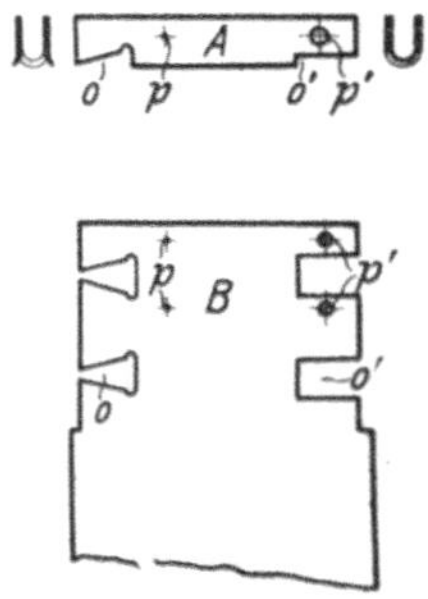

Fig. 574.

durch eine Feder mit dem Hebel verbundene U-förmige Schiene, die entweder aus Kupfer oder Messing hergestellt wird. Die Herstellung dieses Teiles, der in Fig. 574 mit dem Blechstreifen wiedergegeben ist, aus dem der Teil ohne Abfall ausgeschnitten wird, erfolgt in einem Säulenwerkzeug, das als Folgewerkzeug arbeitet, jedoch durch die Verbindung und den Antrieb der arbeitenden Teile mit dem Aufbau eines Verbundwerkzeuges viel gemein hat.

Die Arbeitsweise des Werkzeuges, Fig. 575/6, ist folgende: Der Streifen wird von der Vorderseite bei o o′ ein- und bis gegen die Anschläge e geschoben, worauf beim folgenden Niedergang des Stempels die mit o und o′ bezeichneten Stempel, die gleichzeitig als Seitenschneider arbeiten, die seitlichen Schlitze ausschneiden und das Blech

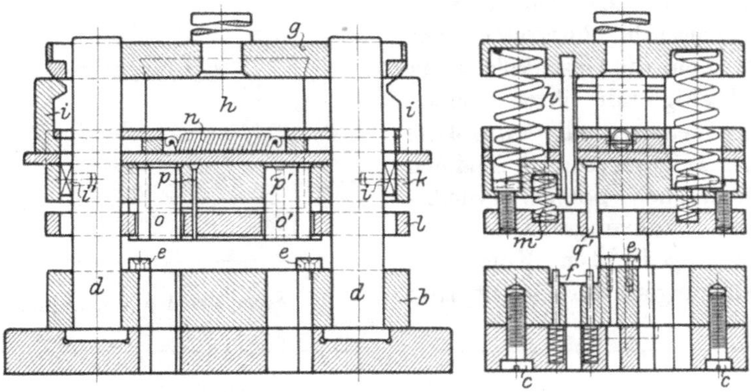

Fig. 575.

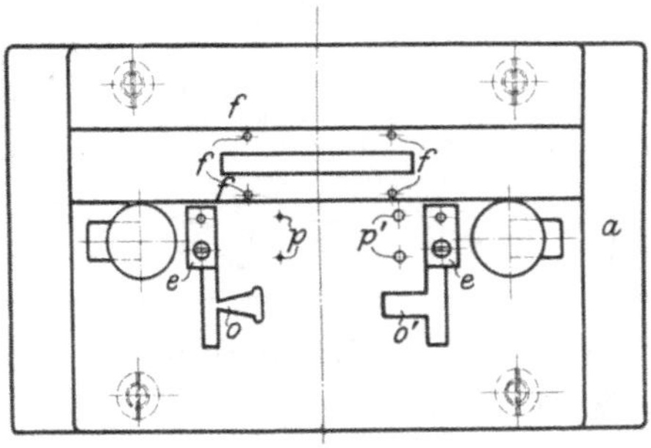

auf die richtige Breite zuschneiden. Dann kann das Blech beim folgenden Stempelaufgang so weit zwischen die Anschläge e vorgeschoben werden, bis die von den Seitenschneidern ausgeschnittenen Ecken, Fig. 574, an die Anschläge stoßen. Damit ist jetzt die einem Blankett entsprechende Blechbreite unter die Lochstempel p p′ gebracht worden, die bei dem zweiten Stößelniedergang diese Löcher herstellen; gleichzeitig ist natürlich eine Blankettlänge zurück auf jeder Seite wieder ein Schlitz ausgeschnitten worden. Beim dritten Stößelaufgang wird das Blech wieder um eine Blankettlänge vorgeschoben, so daß das zuerst ausgeschnittene Blankett jetzt unter dem Schnittstempel q′ und unter dem Biegestempel h liegt. Beim Abwärtsgang schneidet der Stempel q die volle Blechbreite ab, so daß ein Blankett jetzt auf den Federstiften f liegt, wo es von dem federnden Niederhalter gehalten und beim weiteren Abwärtsgehen des Stößels von dem Biegestempel h U-förmig durchgebogen wird.

Das Werkzeug hat einen einfachen Unterteil, an dem die beiden Führungssäulen d befestigt sind, an denen sich der verwickelter aufgebaute Oberteil führt. Es wird in eine gewöhnliche einfachwirkende Presse eingebaut, deren Hub nur so groß wie der Weg des Schnittstempels + dem des Biegestempels sein muß. In diesem Fall ist dies ungefähr 35—40 mm.

Schnittplatte und Biegematrize b aus einem Stück Werkzeugstahl hergestellt und werden auf die Gesenkplatte, die gleichzeitig mit ihren verlängerten Seiten a als Grundplatte dient, aufgeschraubt. Dadurch werden die beiden zylindrischen Führungssäulen mit ihren angedrehten Köpfen in Ausdrehungen der Gesenkplatte

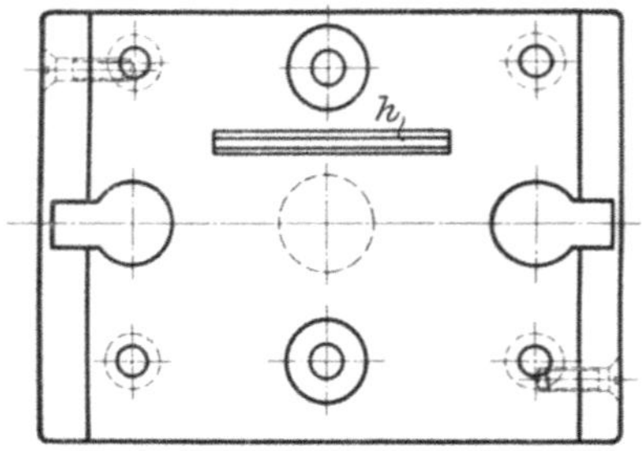

Stempelkopf g von unten.

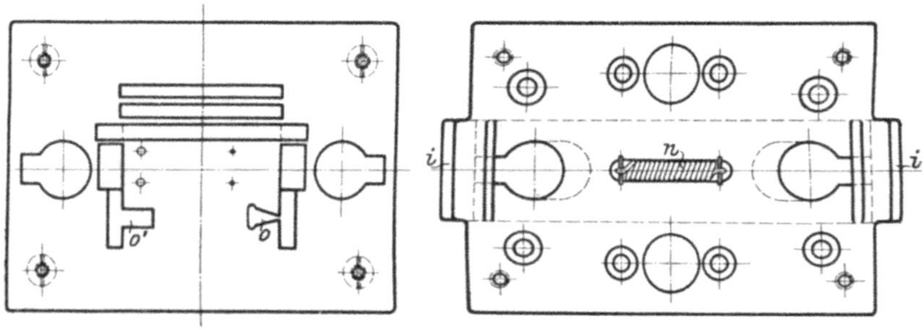

Führungs- und Abstreifplatte l

Zwischenplatte k

Fig. 576 a—c.

nieder- und festgezogen. Auf der Schnittplatte sind zwei Streifen e mittels Schrauben und Prisonstiften befestigt, die dem Streifen als Führung dienen. Gegen diese Anschläge legt sich der durch die Seitenschneider, wie bei o o′ in Fig. 574 ersichtlich ist, auf die richtige Breite zugeschnittene Blechstreifen.

Der Oberteil des Werkzeuges ist ungleich verwickelter zusammengesetzt. Fest mit dem Stempelzapfen ist nur der Stempelkopf g mit dem Biegestempel h verbunden, während die Zwischenplatte k an den nur im Grundriß, Fig. 576 c, des Stempelkopfes g angedeuteten Schrauben aufgehängt und durch zwei Federn abgestützt wird. An diese Zwischenplatte k ist die Kopfplatte für die Schnitt- und Lochstempel o o′ und p p′ angeschraubt, während der Biegestempel h sich in ihr führt. In der Zwischenplatte sind zwei Schlitten mit Nocken i gleitend eingepaßt, die mit entsprechenden Aussparungen in dem Stempelkopf g zusammenarbeiten. Die Führungs- und Abstreiferplatte l für die Loch- und Schnittstempel ist mit vier Kopfschrauben an k angehängt und durch Federn m dagegen abgestützt.

Beim Aufwärtsgang des Stempels geht der ganze Oberteil als ein einheitliches Stück nach abwärts, so daß die Loch- und Schnittstempel ihre Arbeit ausführen können, da die Nocken i sich gegen den Stempelkopf g anlegen. Ist der Stößel ein Stück seines Weges nach abwärts gegangen und hat der Schnittstempel q den Streifen und das Blankett getrennt, so drücken die an den Führungssäulen befestigten Anschläge i' die Schlitten mit den Nocken seitwärts, so daß der Stempelkopf mit dem Biegestempel unter Zusammendrückung der beiden großen Spiralfedern an den Anschlägen vorbeigeht und das Blankett U-förmig durchbiegen kann. Geht nun der Stößel wieder nach aufwärts, so halten die zwischen Stempelkopf g und Kopfplatte eingelegten Druckfedern die Teile so lange auseinander, bis die Ausschnitte im Stempelkopf an den Anschlägen der Führungssäulen vorbeigegangen sind, worauf die Feder n die beiden Schlitten wieder zusammenzieht, worauf der ganze Stempel wieder als ein Stück in die Höhe geht.

Die Loch-, Schnittstempel und der Biegestempel sind aus Werkzeugstahl hergestellt, gehärtet und geschliffen, desgleichen die Nocken i und die Führungsplatten. Die anderen Teile werden aus Siemens-Martinstahl gemacht.

Folge- und Verbundwerkzeug zum Abschneiden, Biegen und Einschlagen von Krampen. Werkzeuge nach Fig. 577/8 wurden in einer Maschine zur Herstellung und zum Einschlagen von siebenundzwanzig Krampen in einer Reihe bei jedem Hub verwendet. Diese werden aus einer Rolle Flachdraht gemacht und durch einen Stoffriemen in ein schmales Eichenbrettchen eingetrieben, wobei die Spitzen durch das Brettchen dringen und glatt aufhören. Diese so mit Eichenbrettchen armierten Riemen werden an landwirtschaftlichen Maschinen verwendet. Der Hauptteil der Maschine ist das Werkzeug zur Herstellung und zum Eintreiben der Krampen. Zwei verschiedene Riemenbreiten und dem entsprechend zwei Leistenlängen werden verwendet. Deshalb stellen die Werkzeuge am Ende jeder Reihe zwei Krampen her und treiben sie in einer Entfernung von ungefähr 14,3 mm ein, die übrigen dreiundzwanzig Werkzeuge dagegen machen nur je eine Krampe.

Vor Herstellung der vollen Ausrüstung von fünfundzwanzig Werkzeugen wurde ein Versuchswerkzeug gemacht, mit einem billigen Antrieb versehen und durch Herstellung von drei Millionen Krampen auf Dauerhaftigkeit und zufriedenstellendes Arbeiten untersucht.

wobei verschiedene Male mehrere tausend Krampen in die Stoffriemen und Eichenleisten eingetrieben wurden. Am Schlusse dieser Versuche schien das Werkzeug noch zur Herstellung von mindestens weiteren drei Millionen Krampen gut genug, ohne daß ein Nachschärfen notwendig gewesen wäre. Mit Ausnahme einiger weniger Fehlstücke, die durch kurze Schleifen im Draht verursacht wurden, wurden keine fehlerhaften Stücke geliefert.

Der vollständige Apparat besteht der Hauptsache nach aus der vorderen und hinteren Schnittplatte, den gleitenden Schnitt- und Biegestempeln b, dem gleitenden Biegestempel c, den Schwingbiegestempeln d und

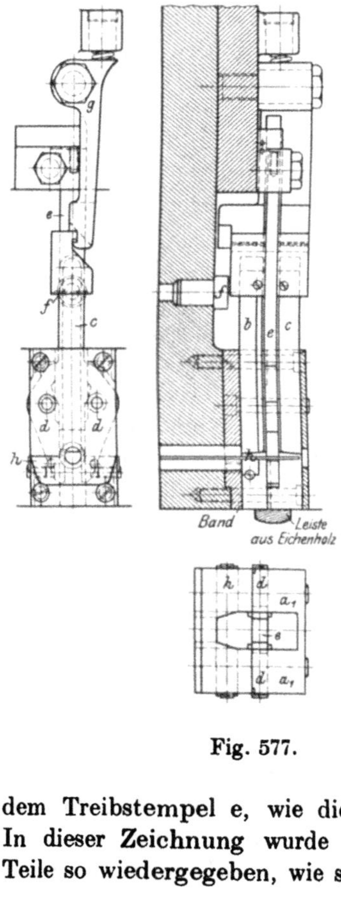

Fig. 577.

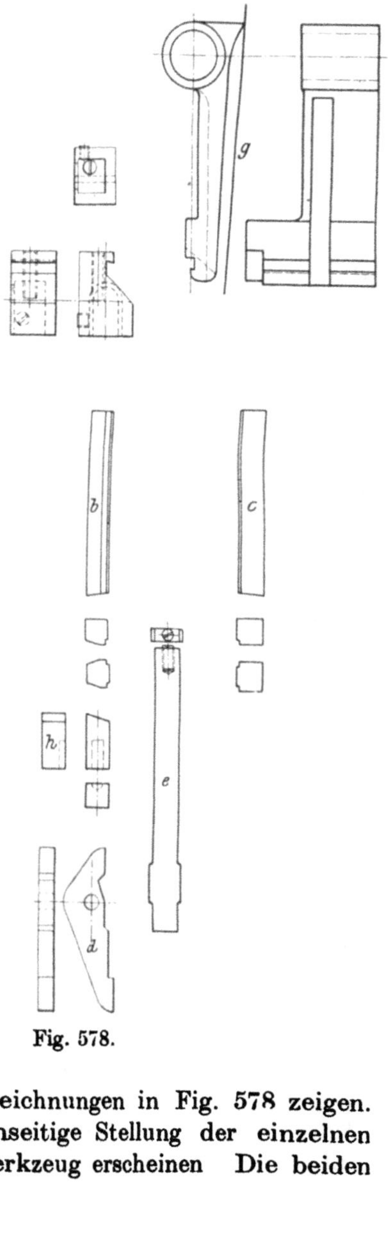

Fig. 578.

dem Treibstempel e, wie die Einzelzeichnungen in Fig. 578 zeigen. In dieser Zeichnung wurde die gegenseitige Stellung der einzelnen Teile so wiedergegeben, wie sie im Werkzeug erscheinen Die beiden

Flachdrähte, die von links zugeführt werden, liegen über den Schwingbiegestempeln d, so daß die Enden, die der Schnittstempel bei seinem Abwärtsgange geschnitten hat, durch den Biegestempel c und den Schnitt- und Biegestempel b über den Schwingbiegestempel d gebogen werden, da die beiden ersteren durch den Schalthaken g am Querhaupt der Maschine zusammen nach abwärts gedrückt werden. Wenn diese beiden Biegestempel am Ende ihres Weges angelangt sind und ihre unteren Endflächen bündig mit der Gesenkunterfläche stehen, stößt ein Anschlagbolzen f an die schräge Fläche des Schalthakens, der dadurch von seiner Verbindung mit den Stempelköpfen gelöst wird, so daß beim weiteren Abwärtsgange des Stößels der Presse die beiden Stempel b und e stehen bleiben.

Um diese Zeit stößt auch das untere Ende des Treibstempels e, der fest mit dem Querhaupt der Maschine verbunden ist, die fertigen Krampen nach abwärts. Der breitere Teil des Treibstempels hält in seiner oberen Stellung, wie aus Fig. 578 ersichtlich ist, die oberen Enden der Schwingbiegestempel d auswärts und in der zum Biegen der Krampen nötigen Stellung. Nachdem nun Schnitt- und Biegestempel b und c die Stifte fertiggestellt haben, drückt dieser breitere Teil des Treibstempels die unteren Enden der Schwingbiegestempel d aus dem Bereich der Krampen und des abwärtsgehenden Treibstempels. Die Nuten an den Kanten der Schnitt- und Biegestempel b und c, Fig. 578, geben eine Zwangführung für die Krampen während des Biegens und Eintreibens, bis sie im Holz gegriffen haben. Die Führung ist so sicher und gut, daß die Stifte selbst durch Knoten in den Eichenleisten durchgetrieben werden, ohne daß sie brechen oder sich verbiegen. Beim Rückgang des Querhauptes faßt der Schalthaken die Stempelköpfe und zieht beide Biegestempel zurück, so daß der Draht zugeführt werden kann, während der Treibstempel e die oberen Enden der Schwing-Biegestempel d nach außen drückt, so daß die unteren Enden derselben sich in der zum Biegen der Stifte nötigen Stellung befinden. Die festen Schnittplatten h werden im rückwärtigen Teil der Gesenkplatte am Ende durch eine Schraube in einer halbrunden Nut gehalten. Die rückwärtige Gesenkplatte trägt dazu passende Gewinde, während der Schnittbacken ohne Gewinde über die Schraube paßt. Seitwärts wird der Schnittbacken durch eine Schraube im rückwärtigen Teil der Gesenkplatte gehalten.

Dies sind die Doppelwerkzeuge, die an den Enden des Bandes je zwei Krampen herstellen. Die einfachen Werkzeuge in der Mitte zur Herstellung einer Krampe erhalten nur einen Schnittstempel und eine Schnittplatte, einen Schwingbiegestempel; außerdem hatte der Treibstempel nur eine einseitige Verbreiterung zum Antrieb des Schwingstempels. Die rückwärtige Gesenkplatte ist dementsprechend ausgebildet. Die Maschine macht jedesmal nur eine Umdrehung, worauf das Werkzeug am oberen Hubende mit allen Teilen in einer solchen Stellung stehen bleibt, daß eine neue Drahtlänge zugeführt werden kann. Stoffriemen und Eichenleiste werden mittels einer Stange aus angelassenem Stahl, die nach der Leiste gekrümmt ist, fest gegen die Unterfläche der Gesenkplatte gepreßt. Alle wichtigen Teile, die einer Abnutzung unterworfen sind, werden aus steirischem Werkzeugstahl gefertigt, sorgfältig gehärtet und angelassen.

3. Abtrennung und Formänderung mit und ohne Materialverschiebung.

Verbund-Schnitt-, Präge- und Biegewerkzeug. Das im folgenden beschriebene Werkzeug dient zum Ausschneiden des Bleches aus dem Streifen, zum Biegen der beiden Seiten und zum Hohlprägen des so vorbereiteten Stückes. Der Unterteil des Werkzeuges enthält ohne die Schrauben neun Teile, und selbst der Stempel besteht aus vier Teilen. Das Werkzeug dient zur Herstellung eines besonderen Teiles nach Fig. 579 aus Messingblech von 2,5 mm Dicke.

Fig. 579.

Die erste Arbeit bei der Herstellung des Werkzeuges ist die Fertigstellung der Grundplatte A, Fig. 581, die aus Werkzeugstahl, 64 · 51 · 16 mm, gemacht und an den beiden kürzeren Seiten unter 5⁰ gehobelt wird, damit sie in die Aufspannplatte paßt. Hierauf wird das Loch B in der Platte A hergestellt, das in seiner Breite dem Stück, in seiner Länge dem Schnittstempel entspricht, so daß noch zu beiden Seiten des Prägestempels Material übrigbleibt. Hierauf wird das Loch bis zu den Linien C ausgeschnitten, so daß die Tiefe des Ausschnittes der Dicke der Leisten an den Biegeplatten entspricht.

Die Grundplatte kommt dann wieder auf die Shapingmaschine, auf der die drei Nuten, D für den Anschlag, E für die Schnittplatte und F für die Biegeplatten, eingearbeitet werden. Die beiden ersten Nuten sind 4,7 mm, die letztere 7,9 mm tief. Die Biegeplatten werden in die Nut F passend gehobelt und erhalten Leisten von 6,3 mm Tiefe und 3,95 mm Dicke, die durch die Gesenkplatte bis zur punktierten Linie G reichen. Die obere Kante dieser Leiste wird dann nach der punktierten Linie H, entsprechend der Krümmung des Stückes, bis zur Linie C nachgearbeitet. Die Stücke werden dann gehärtet und angelassen.

Die Schnittplatte von 2,35 mm Dicke wird in die Nut E so eingepaßt, daß der mittlere Teil bis zur Kante des Loches für den Schnittstempel in der Gesenkplatte reicht, während das entgegengesetzte Ende mit der Kante der Grundplatte abschließt. Der Anschlag J wird in die Nut D, genau wie auf der entgegengesetzten Seite die Schnittplatte eingepaßt, nur beträgt seine Dicke entsprechend der vollen Nutentiefe 4,7 mm. Dieser Anschlag begrenzt die Zuführung des Blechstreifens.

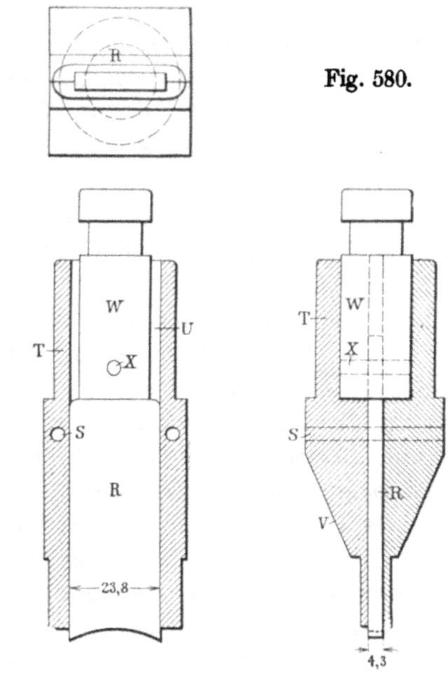

Fig. 580.

Die beiden Führungsschienen K und L werden auch in die Nut E eingepaßt und über der Schnittplatte befestigt. Die Schiene K besteht aus einem Stück mit zwei Ansätzen, von denen der eine fast bis an das Ende der Schnittplatte, der andere bis an die Kante der Grundplatte reicht. Die gegenüberliegende Schiene L wird anders gemacht. Ein Stück wird gehobelt, in die Nut E (rechts) eingepaßt und zwei Löcher für die Stifte M gebohrt. Hierauf wird ein schmaler Streifen entsprechend der Länge über den beiden Ansätzen der Schiene K geschnitten und in ihn zwei Löcher gebohrt, in welche die Stifte M eingetrieben werden. Zwei kleine zylindrische Schraubenfedern werden in die Löcher des feststehenden Teiles eingelegt. Dann werden die Innenflächen dieser Führungsschienen auf 0,8 mm unterschnitten, damit der Blechstreifen auf der Schnittplatte niedergehalten wird. Schließlich erhalten beide Teile Langlöcher für die Befestigungsschrauben. Die beiden Widerlager oder Einstellblöcke N werden auf einen

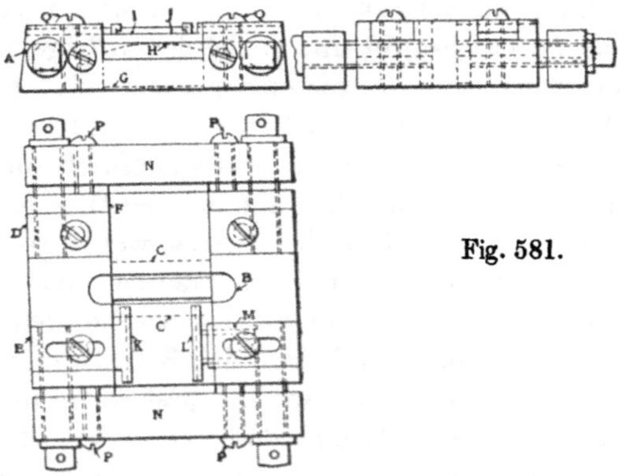

Fig. 581.

Querschnitt von 12,7 · 9,5 mm gehobelt, in der Länge gleich der Breite der Grundplatte und auch an den Enden abgeschrägt. Hierauf werden sie mittels der Schrauben O an der Grundplatte befestigt, während die Schrauben P als Druckschrauben dienen und ihr Muttergewinde in den Teilen N haben. Mit Hilfe dieser Widerlager werden die Biegeplatten eingestellt, da deren Enden über die Kanten der Grundplatte bis an jene reichen müssen. Die Schrauben Q dienen zur Befestigung der Führungen K und L, sowie der Schnittplatte in der Grundplatte. Nach Anlassen der Schnittplatte wären die Teile zum Aufbau des Werkzeuges fertig, jedoch muß zuerst der Stempel fertiggestellt werden.

Der Stempel, Fig. 580, aus Schnitt- und Biegestempel bestehend, wird in seinen beiden Hauptteilen gleichzeitig hergestellt. Der Schnittstempel ist aus Werkzeugstahl (19 · 38 · 203 mm) und erhält in der Mitte eine Nut entsprechend der halben Dicke des Prägestempels, von 23,8 mm Breite und 2,15 mm Tiefe. Dieses Stück wird dann in zwei gleich lange Stücke zerschnitten. Nun wird zur Herstellung des Prägestempels R geschritten. Das Stempelblatt R wird 4,3 mm dick und 23,8 mm breit gehobelt, fertigbearbeitet und in die oben erwähnten Hälften des Schnittstempels eingepaßt. Hierauf werden die beiden Hälften mit Klammern zusammengezogen und die Bolzenlöcher S gebohrt. In diese werden dann Bolzen streng eingepaßt, da dies die einzige Möglichkeit, die beiden Hälften zusammenzuhalten, ist, und die Druckschraube im Stößel das übrige besorgen soll. Der Zapfen des Stempels wird passend in das Loch im Stößel der Presse gedreht, das Loch U etwas größer als der Zapfen des Biegestempels gehalten, die Seiten V abgeschrägt und schließlich die Schnittkante des Stempels so genau wie möglich auf die richtige Größe gebracht. Nun muß der Biegestempel noch vollendet werden. Das untere Ende desselben wird nach dem Krümmungsradius des Stückes ausgearbeitet, der Zapfen W passend in den entsprechenden Pressenstößel gedreht und schließlich das Ende für die Befestigungsvorrichtung genutet. Am anderen Ende wird eine Nut gestoßen, in die der eigentliche Biegestempel eingepaßt und dort durch den Bolzen X gesichert wird.

Nachdem die Biegeplatten in die Nut F eingelegt, der Anschlag J in Nut D mittels der Schrauben Q befestigt worden ist, wird auch die Schnittplatte I in die Nut E gelegt und über ihr die beiden Führungsschienen K und L und die Widerlager N befestigt. Nun kommt der Biegestempel in Verwendung. Zwei Stücke des zu verwendenden Materials werden seitwärts an den Biegestempel angelegt, das Ganze zwischen die Biegeplatten gestellt und diese durch Lösen und Anziehen der Schrauben P und Q eingestellt. Zum Schluß werden noch die Führungsschienen K und L entsprechend der Breite der Streifen eingestellt.

Verbund-Schnitt-, Biege- und Prägewerkzeug. Das Werkzeug in Fig. 583 dient zur Herstellung des in Fig. 582 in Vorder- und Seitenansicht dargestellten Gegenstandes in einem Arbeitsgang. Der Gegenstand wird aus dünnem Eisenblech oder Zinkblech hergestellt und mißt, innen gemessen, 63,5 mm in der Länge und 35 mm in der Breite. Die Kanten werden an den Enden auf 3,1 mm Länge umgebogen und eine halbrunde Nut in der Mitte nach oben und seitlich nach unten ausge-

Fig. 582.

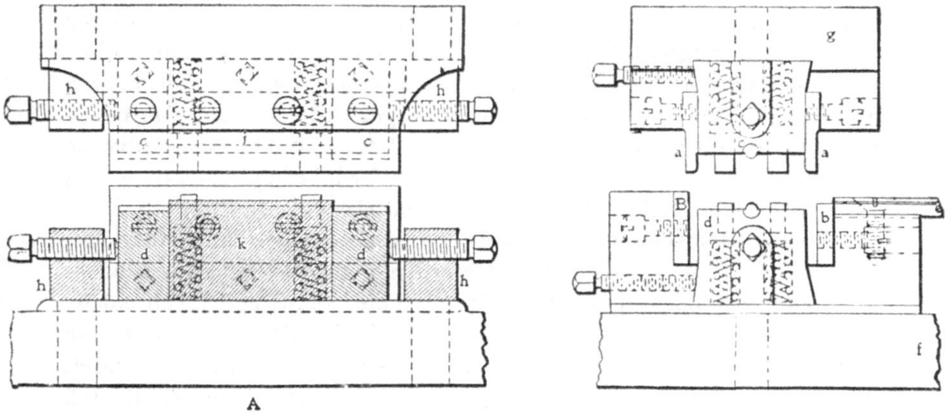

Fig. 583.

prägt, damit der fertige Gegenstand sich auf einem durch diese Nut gezogenen Draht von ungefähr 2 mm Durchmesser drehen kann.

Bei der Herstellung des Werkzeuges wird Stempelkopf g und Grundplatte f aus Gußeisen gemacht. Der Teil g wird zuerst auf der Ober- und Unterseite fertigbearbeitet und zur Aufnahme der Teile a ausgehobelt; außerdem wird er zur Aufnahme von c und j unter 15⁰ ausgehobelt. Dann werden zwei Löcher von 11 mm Durchmesser zur Aufnahme der Widerlager h gebohrt, die mit Preßsitz eingepaßt werden. Hierauf werden die Teile c und j ebenfalls mit Preßsitz in g eingepaßt und in ihrer Stellung durch Druckschrauben in h und g gehalten. Die Rippe auf der Oberfläche von c wird in folgender Weise hergestellt: Die Oberflächen zu beiden Seiten der Mitte werden bis auf die nötige Dicke des Stückes abgehobelt, so daß in der Mitte genügend Material für die Rippe stehen bleibt. Nun wird ein Stück Werkzeugstahl, 25 · 12 mm, passend in den Werkzeughalter der Shapingmaschine, auf der 12 mm Oberfläche fertigbearbeitet, durch diese Fläche ein Loch von 1,2 mm Durchmesser gebohrt und von rückwärts mit einer konischen Reibahle ausgerieben, bis es ungefähr 0,07 mm mehr als das verlangte Maß hat. Dieses Werkzeug wird dann abgeschnitten und auf ungefähr 0,5 mm, die Blechdicke, unterhalb der Lochmitte fertigbearbeitet, während die Seiten des Loches am weitesten Ende parallel gefeilt werden. Das Werkzeug erhält noch einen geringen Hinterschliff und wird nach dem Härten zum Hobeln der Rippen verwendet. Der Teil j wird auf dieselbe Dicke wie c gehobelt, worauf in der Mitte eine halbrunde Nut, 2 mm mehr 0,13 mm im Durchmesser, eingehobelt wird.

Dieser Teil wird dann zur Aufnahme der vier Federbolzen gebohrt und versenkt. Hierauf werden die Teile c und j gehärtet, auf lichtes Strohgelb angelassen, an den Endflächen rechtwinklig geschliffen und in die richtige Stellung in g eingetrieben, worauf nach Eintreiben der Widerlager h die Kopfschrauben angezogen werden. Die Teile a werden 4,75 mm dick gemacht und ragen 6,35 mm unterhalb c j hervor; ihre inneren Kanten sind abgerundet, damit eine Beschädigung des Arbeitsstückes vermieden wird. Vier Zylinderkopfschrauben und zwei Prisonstifte halten die Teile fest und sichern ihre Stellung.

Bei der Herstellung des Unterteiles wird derselbe Plan befolgt, bis auf die Teile B b, die erst gemacht werden, wenn die Breite des Bleches ermittelt worden ist. In diesem Falle sind die halbrunden Nuten natürlich in d und die Rippe auf k. Die Führung für das Blech wird aus einem Stück Flachstahl e, 203 mm lang, 6 mm dick und 25 mm breit, gebildet, das auf einer Seite eine 6 mm tiefe Nut von solcher Breite erhält, daß man das Blech leicht durchschieben kann. Dieses Stück wird auf der einen Seite der Gesenkplatte rechtwinklig zur Mittellinie befestigt. Der Arbeiter drückt beim Arbeiten das Blech gegen den Grund der Nut.

Dann wird ein Muster in das Werkzeug eingelegt und der Stempel niedergedrückt. Die vier Federbolzen, von denen die oberen 6,5 mm, die unteren 4,7 mm im Durchmesser haben, halten das Blech und beugen einem Verziehen desselben vor. Bei der Abwärtsbewegung des Stempels biegen die Stempel a die Blechseiten über d k herunter, während die Rippen und Nuten in c j und d k die Mittelpartie prägen und schneiden.

Nachdem auf diese Weise die notwendige Blechbreite bestimmt worden ist, werden die Teile a außen auf Maß geschliffen und sorgfältig auf gleiche Dicke gebracht. Dann erst werden die Teile B b gemacht, wobei B um 3,2 mm höher gehalten wird als b. Die Teile werden dann so geschliffen, daß die arbeitenden Kanten gleichweit von der Mittelnut entfernt sind und sauber über die Teile a passen. Sie werden in gleicher Weise wie die Teile a gehalten, ebenfalls gehärtet und auf lichtes Strohgelb angelassen. Die Führung wird noch an g befestigt, womit das Werkzeug gebrauchsfertig ist.

Beim Arbeiten mit diesem Werkzeug wird der Blechstreifen, der 63,5 mm breit abgeschnitten worden ist, vom Arbeiter in die Nut bei e eingeschoben, bis er an B anstößt, worauf die Presse eingeschaltet wird. Das Material wird dann bei a b abgeschnitten, gebogen und mit der Nut versehen, wie oben beschrieben worden ist. Beim Aufwärtsgang der Presse entfernen die oberen Federbolzen das Material von dem Stempel, während die unteren es hochheben, wodurch es frei liegt und vom Arbeiter weggestoßen werden kann, so daß die Presse für einen neuen Arbeitsgang frei ist.

Wenn die Schnittkanten c und k stumpf geworden sind, werden die Stücke an den Stirnflächen nachgeschliffen, und zwar k an beiden Flächen, worauf j auf die gleiche Dicke geschliffen wird. Die Teile c werden nur an den Schnittflächen geschliffen und dann die Teile wieder zusammengebaut. Die Teile a und b werden auf der Schleifmaschine, ohne sie aus dem Werkzeug zu entfernen, geschliffen, wobei die inneren Kanten von a mit einer kleinen Schleifscheibe gerundet und mit Schmirgelleinwand geglättet werden. Das Werkzeug arbeitete sehr gut und konnte an den Schnittkanten dauernd nachgeschliffen werden, da die Entfernung zwischen den Teilen c nicht von besonderer Wichtigkeit war.

Verbundwerkzeug zum Ausschneiden, Lochen, Ziehen und Biegen mit zwangläufigem Abstreifer. Die Figuren 585/6 stellen ein eigenartiges Verbundwerkzeug dar, das in einem Gang einen Untersatz aus Weißblech herstellt. Fig. 584 gibt die Form und die Abmessungen des Stückes.

Fig. 585 ist ein Schnitt durch den Stempel in der Richtung HO in Fig. 586 und zeigt den zwangläufigen Abstreifer. Der innere Lochstempel B aus Werkzeugstahl wird durch den gehärteten Stahlkeil C in seiner Stellung gehalten, während der Abstreifer D für den inneren Lochstempel B auf Schiebesitz ausgebohrt ist und bei E als Führung für den Stempel gut, bei F dagegen lose paßt, damit der Abstreifer frei arbeiten kann. Der Keil C wird in den Stempelzapfen eingetrieben und geht lose im Schlitz des Lochstempels B und Abstreifers D. Der Schlitz in D ist lang genug, um dem Abstreifer ein Hinabgehen bis an die Stirnfläche des Stempels zu ermöglichen; beim Aufwärtsgang des Stempels geht D solange mit, bis es an den Ausrückhebel anstößt.

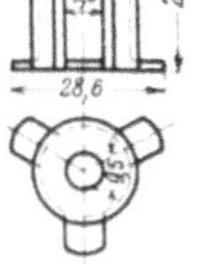

Fig. 584.

Da diese Untersätze verschieden hoch — 6,35 bis 28,6 mm — sein sollen, muß man den inneren Lochstempel B, wie in Fig. 585 ersichtlich ist, für den niedrigsten Untersatz lang genug machen.

Es ist allgemein bekannt, daß eine Feder in einem Stanzwerkzeug eine Reihe Unzukömmlichkeiten mit sich bringt, weshalb

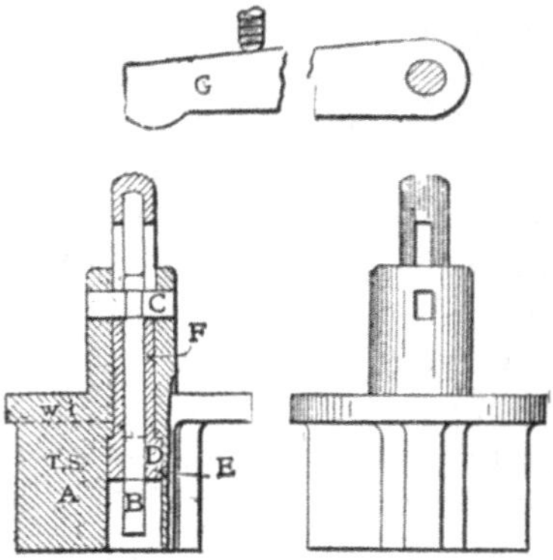

Fig. 585.

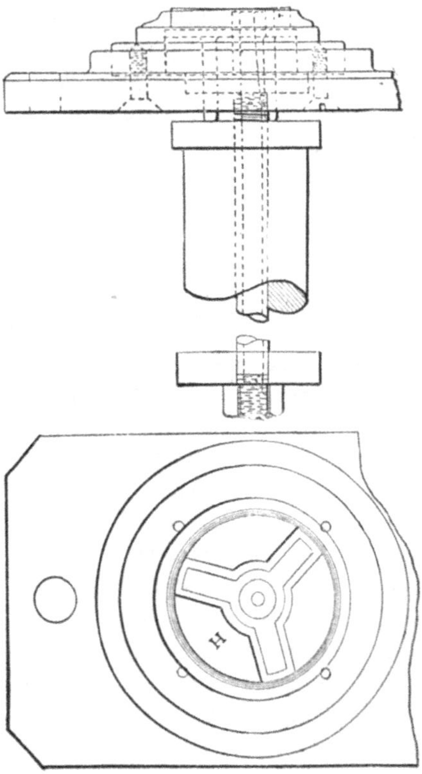

Fig. 586.

man Federn, wo irgend möglich, durch eine zwangläufig wirkende Vorrichtung zu ersetzen sucht, besonders in Verbundwerkzeugen, die bei großer Ziehlänge zur Unterbringung der Feder wenig Raum bieten. Ein andrer gegen die Anwendung der Feder sprechender Punkt ist in diesem Fall, daß eine Feder, die stark genug ist, das Blankett auszuwerfen, das Blankett oder die drei Fortsätze, Fig. 584, verbiegen würde, eine zu schwache hingegen ab und zu versagten könnte, ohne daß der Arbeiter es sieht, so daß plötzlich zwei Blanketts im Stempel hätte.

Fig. 586 ist ein Grundriß des Unterteiles, das mit Rücksicht auf möglichst wenig Abfall entworfen wurde. Darüber ist eine Vorderansicht des Unterteiles.

Der Gummipuffer ist an einem Rohr statt einer Stange befestigt, damit der Abfall durch das Loch in der Mitte nach außen fallen kann. In diesem Falle mußte man eine lange Gummifeder verwenden; sonst ist es im allgemeinen besser die Pufferfeder aus einzelnen kurzen Stücken zusammenzusetzen, da jedes Stück sich dann unabhängig von den andern strecken kann.

4. Formänderung mit und ohne Materialverschiebung.

Diese Gruppe umfaßt Werkzeuge, bei denen Zieh- oder Prägearbeiten und Biegearbeiten gleichzeitig in einem Werkzeug ausgeführt werden. Da hier die einzelnen Arbeiten bereits bedingen, daß der Gegenstand in einer oder der anderen Weise aus der Ebene des Bleches heraustritt, so wird die Anwendung von Folgewerkzeugen bei diesen Vereinigungen mehrerer Arbeiten wenig oder gar nicht zur Ausführung kommen, so daß meistens Verbundwerkzeuge angeführt sind.

Zieh- und Prägewerkzeug für die Grundplatte des Hausweckers. (C. Lorenz, A.-G. Berlin.) Im fünften Arbeitsgang der Tabelle 1—I wird das geschnittene und gelochte Blankett, dessen Fortsatz bereits flach umgebogen worden ist, gezogen, wodurch ein ungefähr 4 mm hoher, umlaufender Rand entsteht, der der Grundplatte die nötige Steifigkeit

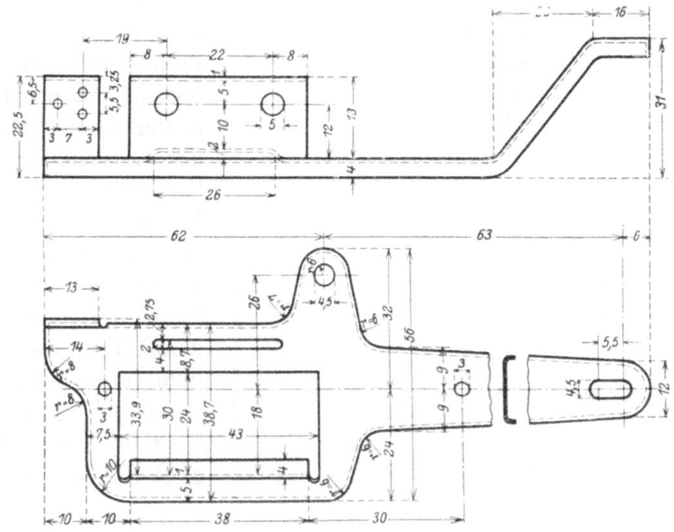

Fig. 587.

gibt. Gleichzeitig wird in der Mitte der Grundplatte neben dem rechteckigen Ausschnitt eine kurze, gerade Rippe eingeprägt. Die vollständige fertige Grundplatte ist in Fig. 587 mit allen Maßen abgebildet.

Das Verbund-Zieh- und Prägewerkzeug ist in Fig. 588 wiedergegeben. Der Stempel ist, wie bei allen Ziehwerkzeugen, sehr einfach, da die Prägung keine besonderen Vorkehrungen erfordert. Er besteht aus einem Stempelkopf aus Schmiedeeisen, der mit dem Stempelzapfen aus einem Stück hergestellt ist. Der eigentliche Ziehstempel ist mittels versenkter Zylinderkopfschrauben an der ebenen Unterfläche des Stempelkopfes befestigt und entspricht in seinem Grundriß dem im Unterteil gezeichneten Auswerfer, in den Abmessungen ist er um die Materialdicke 1,1 mm + dem notwendigen Spiel kleiner als der Ausschnitt in der Ziehplatte. Die Rippe ist durch Herunterarbeiten des Stempels um deren Höhe erhalten worden. An der Stelle, wo der Fortsatz, der nachher in dem in Fig. 462 beschriebenen Werkzeug hochgestellt wird, flach in der Blechebene über den zu ziehenden Rand hinüberragt und in der Matrize rechts ein entsprechender Ausschnitt erkennbar ist, ist der Stempel um rd. 3 mm über die Umrißlinie verlängert, so daß dieses Ende gegen einen Einfluß beim Ziehen geschützt ist und flach bleibt.

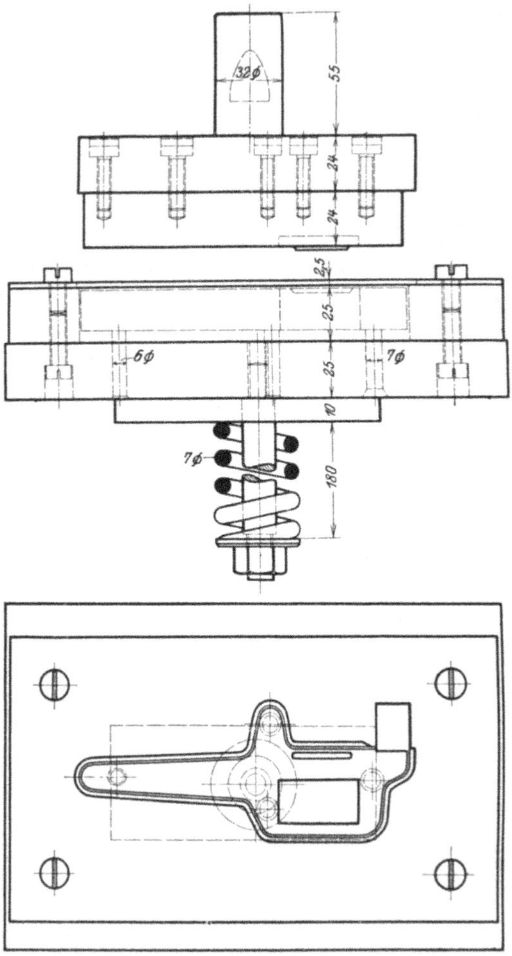

Fig. 588.

Der Werkzeugunterteil besteht aus vier Teilen. Die Gesenkplatte ist aus Schmiedeeisen und die Matrize aus gehärtetem Werkzeugstahl auf ihr mit von unten versenkten Zylinderkopfschrauben befestigt. Auf ihr ist weiter in denselben Gewinden die Einlage für das Blankett aufgeschraubt, die aus Blech von 2,5 mm Dicke gearbeitet ist. Außerdem ist ein Auswerfer vorgesehen, der auf der Oberfläche das Prägegesenk für die Rippe bildet. Er ruht auf vier Auswerferstiften, die auf der unterhalb der Gesenkplatte befindlichen Druckplatte des Federauswerfers aufstehen. Die Feder ist so stark, daß das Ausprägen der Rippe stattfinden kann, bevor das Ziehen beginnt, wodurch gleichzeitig das Blech während des Ziehens gehalten wird. Auch für den inneren Ausschnitt ist sowohl im Stempel wie im Auswerfer eine Aussparung vorgesehen, damit die im zweiten Gang erfolgte Biegung nicht beschädigt wird.

Vereinigtes Biege- und Prägewerkzeug für den Hauswecker. Nachdem nach Tabelle No. 1—I G. 6 die beiden Lappen der Grundplatte rechtwinklig aufgebogen worden sind, wird der Arm für die Glocke mit einer

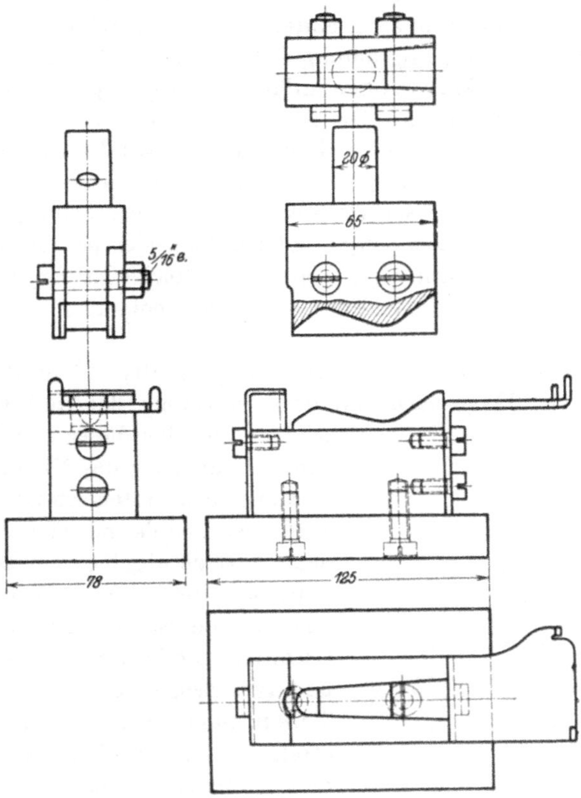

Fig. 589.

doppelten Biegung versehen, so daß sie die richtige Höhe über der Grundplatte bekommt. Da diese doppelte Biegung auch an den gezogenen Seitenflächen ausgeführt werden muß, ist diese Arbeit als ein Prägen zu bezeichnen. Die Ausführung erfolgt mit einem

Werkzeug sehr einfacher Form, Fig. 589, das gerade dadurch an Interesse gewinnt. Infolge der eigenartigen Arbeit wird der Hauptstempel und der Stempelzapfen aus gehärtetem Werkzeugstahl aus einem Stück gemacht. Der Stempel besteht aus drei Teilen, dem mittleren Hauptteil, der dieBiegung ausführt, und den beiden Seitenteilen die in Aussparungen des Mittelteiles eingepaßt sind und durch eingepaßte, zylindrische Kopfschrauben gehalten werden. Diese Teile werden blau angelassen, während der mittlere Teil, der eigentliche Biegestempel, auf Strohgelb angelassen wird.

Den Unterteil des Werkzeuges bildet eine Gesenkplatte aus Maschinenstahl, auf der die Biegematrize mit zwei versenkten Zylinderkopfschrauben befestigt ist. Sie ist aus dem vollen Block herausgearbeitet und an den arbeitenden Flächen geschliffen. Die Anschläge sind sehr sorgfältig durchgebildet, so daß das Stück zwischen ihnen und den senkrechten Seitenflächen des Stempels gehalten wird, bevor das Biegen beginnt. Die Anschläge selbst sind aus Stahlblech hergestellt und mit Kopfschrauben an den Stirnflächen der Matrize angeschraubt. Als Anschlag ist an der linken Seite ein gehärteter Stahlblock mit dem Blech verlötet. Die beiden Schrägen auf der Biegematrize, die an dem fertigen Stück die beiden parallelen Flächen, Fig. 587, ergeben sollen, sind wegen der Rückfederung des Stückes wieder unter spitzeren Winkeln als 180^0 angeordnet.

Biegen und Prägen. Der letzte Gang der Herstellung des Stückes in Tabelle Nr. 32 geschieht auf dem Werkzeug, das in Fig. 590 dargestellt ist. Bei dieser Arbeit wird die lange Krümmung des Mittelteiles hergestellt, das kastenförmige Ende geschlossen und das mit einem Auge versehene Ende abgebogen. Das Werkzeug arbeitet in der einfachwirkenden Presse und ist mit einem kräftigen Gummiauswerfer versehen. Der Aufbau des Werkzeuges erfolgt aus einzelnen Teilen, damit man diese leicht nachstellen und schleifen kann, wenn eine Abnutzung stattgefunden hat. In seiner Herstellung und seinem Aufbau ähnelt das Werkzeug sehr dem Biegewerkzeug für den 4. Gang in Fig. 470.

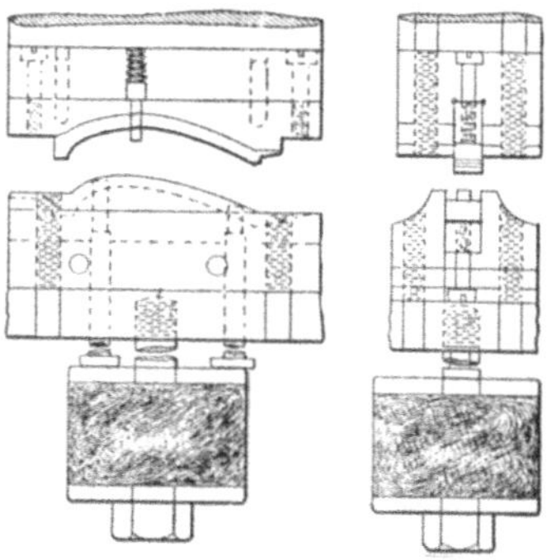

Fig. 590.

Werkzeug zur Herstellung eines Messingknopfes. Das ausgeschnittene Blech ist in Fig. 591 bei A dargestellt und muß nach B in derselben

Figur umgeformt werden. Das Blech wird auf die Gesenkplatte, Fig. 592, gelegt und seine Stellung durch die versenkbaren Anschlagstifte 1, 2, 3 sowie die Anschlagblöcke 4 und 5 gesichert. Der Stempel A geht dem Keilstück B ein wenig voraus, so daß das Ziehen des Knopfes durch den Stempel A beginnen kann, bevor B überhaupt zur Wirkung kommt. Beim Niedergange stößt B an den Seitenstempel C und schiebt ihn vorwärts gegen den Stempelkopf von A, der entsprechend der Form von B ausgeschnitten ist. Beim Rückgange des Stößels ziehen die Federn D den Seitenstempel C zurück. Das Werkzeug ist bei E mit einem Auswerfer, der den fertigen Knopf ausstößt, und mit einem Abstreifer bei F versehen, der durch die oben im Schnitt gezeichnete Feder betätigt wird. In der gezeichneten Stellung ist gerade ein Stück ausgeworfen worden.

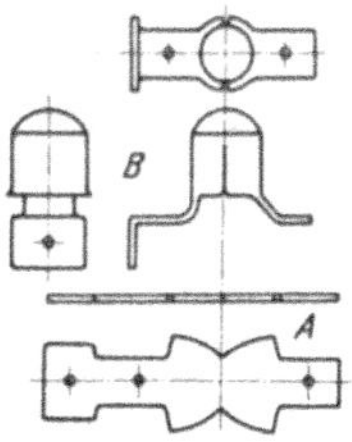

Fig. 591.

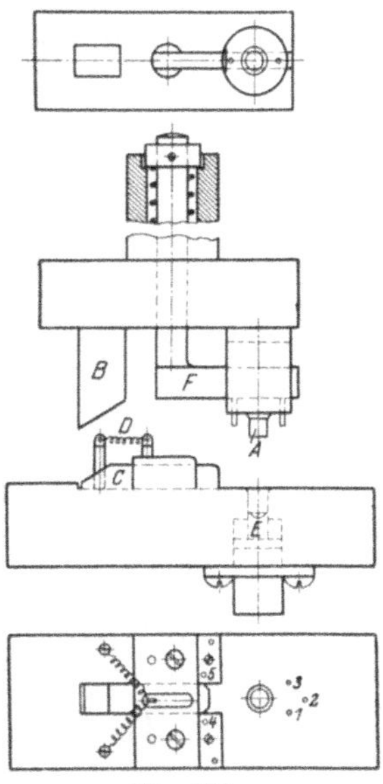

Fig. 592.

Verbundwerkzeug zum Biegen, Ziehen und Prägen. Das in Fig. 115 dargestellte Stück, das aus dem im Folgeschnitt Fig. 116 ausgeschnittenen Blankett hergestellt ist, wird in einem weiteren Gang fertiggestellt. Der Stempel zum Biegen, Ziehen und Prägen mit dem dazugehörigen Untergesenk, die in einem Gang das Stück fertigstellen, werden nach Fig. 593 hergestellt. Das Blankett wird in eine eingearbeitete Einlage A, die zwei Anschlagplatten B enthält, eingelegt, wodurch eine genaue Ausmittlung des Stückes erzielt wird.

Die überragenden Fortsätze C des Stempels biegen zuerst die scharfen Spitzen des Blanketts über einen Auswerfer, der sich in der Mitte der Gesenkplatte befindet und von einer kräftigen Feder E getragen wird. Bei dem weiteren Abwärtsgang zieht der runde Teil des Stempels die vier Lappen des Blanketts in den runden Teil D des Untergesenkes, dessen Durchmesser um die doppelte Metalldicke größer als der Durchmesser des Stempels ist. Dieser Teil hat leicht abgerundete Kanten, so daß die vier Fortsätze des Blanketts in entgegengesetzter Richtung wie die Spitzen gebogen und gleichzeitig in eine zylindrische Form gezogen werden, bis schließlich der Stempel, wenn der Auswerfer am Grunde aufsitzt, zusammen mit der Ausdrehung auf dessen Oberfläche eine Vertiefung in den Boden des Blanketts einprägt. Beim Rückgang des Stempels wird das Blankett durch die Feder E, die wegen ihrer bedeutend größeren Lebensdauer einem Gummipuffer vorzuziehen ist, ausgeworfen.

Da der für die Feder zur Verfügung stehende Raum begrenzt ist, wird ein kleines Stück Bohrerstahl, das gerade in den Auswerfer paßt, in die Drehbank eingespannt. Hierauf wird in dieses ein Flachgewinde mit der richtigen Steigung geschnitten, wobei man ca. 3 mm Material für die Feder übrig läßt und das Material in der Mitte der Feder mit einem kleinen Bohrer entfernt. Dann wird das Stück ausgebohrt, auf richtige Länge abgeschnitten und in Öl gehärtet.

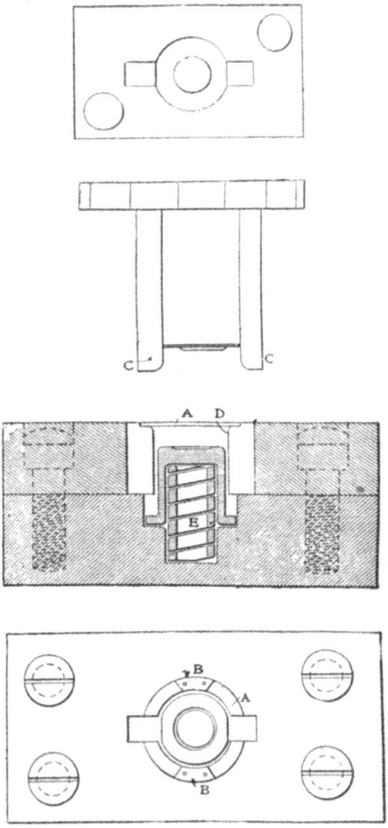

Fig. 593.

Biege- und Prägewerkzeug für den Hammer eines Hausweckers. (C. Lorenz, Berlin.) Der Hammer für den Hauswecker, Tabelle Nr. 1—III, wird im zweiten Arbeitsgang in einem Verbund-Biege- und Prägewerkzeug aus dem Blankett, vgl. S. 88, Fig. 105, fertiggestellt. Das Stück ist wegen der großen Verschiedenheit seiner Abmessungen für die Verarbeitung bemerkenswert, da es auch aus verhältnismäßig starkem — 1 mm — blankgewalzten Eisenblech hergestellt ist. Das dünne Ende von 1 mm Breite, an dem die Kugel des Hammers vernietet wird, darf durch die Prägung des Stieles nicht geschwächt werden. Dieser soll zur Verstärkung auf seiner ganzen Länge rd. 2 mm hoch rundgeprägt werden, wenn auch die Breite des ungeprägten Stieles nur rd. 3,3 mm beträgt. Das Aufbiegen der Seiten des breiten Teiles bereitet keine Schwierigkiten. Das Werkzeug, Fig. 594, ist als Verbundwerkzeug ohne eigene Führungen ausgebildet, so daß der Stempel erst im letzten Augenblick unmittelbar vor seiner Arbeit an beiden Enden eine kurze Führung in den Anschlagplatten des Unterteiles erhält. Der Oberteil ist einfach, er besteht aus einem schmiedeeisernen Stempelkopf, der in einem Stück mit dem Zapfen hergestellt ist, und einem angeschraubten Stempel. Dieser ist aus Werkzeugstahl, gehärtet und an den arbeitenden Flächen dunkel strohgelb angelassen. Auf der rechten Seite, wo das dünne Ende des Hammers niedergehalten und gleichzeitig eine Führung erhalten werden soll, ist ein durchgehärtetes Stahlstück eingesetzt, damit der Ersatz dieses dünnen, leicht dem Bruch ausgesetzten Stückes ohne allzu große Kosten geschehen kann.

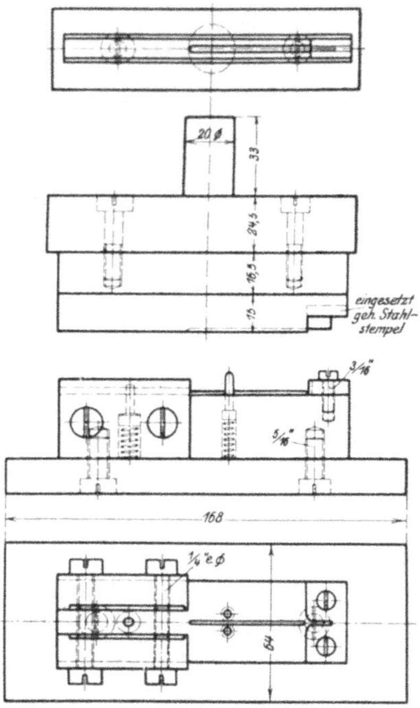

Fig. 594.

Der Unterteil besteht aus einer flachen Grundplatte und der Matrize, die mit zwei versenkten Zylinderkopfschrauben auf ihr befestigt ist. Die Matrize selbst ist aus vier Stücken zusammengeschraubt, von denen der Hauptkörper das Prägen der Rippe besorgt und gleichzeitig der Träger der übrigen Teile ist, da nur er mit der Grundplatte verschraubt ist. An ihm sind die Anschläge und Führungen mit Zylinderkopfschrauben befestigt.

Die Führungen für den linken Teil des Stempels sind gleichzeitig als Anschläge für den breiten Teil des Blanketts ausgebildet und sind die Seitenteile der Biegematrize. Der schmale Teil des Hammers wird durch zwei federnde Anschlagstifte, die unter dem Stempel nach unten ausweichen, in seiner Arbeitsstellung gehalten, während unter dem breiten, linken Teil des Hammers ein federnder Auswerferstift vorgesehen ist.

Verbund-Biege- und Prägewerkzeug für eine Federklammer. Die Federklammer nach Fig. 595 wird in dem in Fig. 596/7 abgebildeten Werkzeug hergestellt.

Fig. 596 ist die Vorder- und Seitenansicht des Ober- und Unterteiles, während Fig. 597 die Matrize allein darstellt. Die Grundplatte

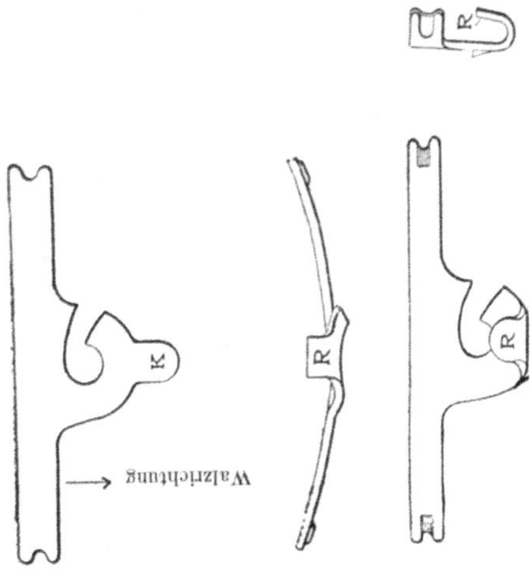

Fig. 595.

Fig. 596.

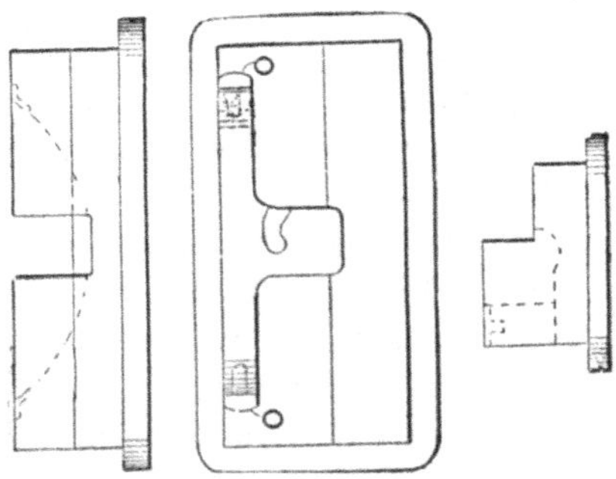

Fig. 597.

ist aus Gußeisen, A ist die Matrize, in der das Blankett geprägt, auf die richtige Krümmung gebogen und sein mit K bezeichnetes Ende unter rechtem Winkel abgebogen wird. Dieser Teil besteht aus gehärtetem Werkzeugstahl. Das Stück B, ebenfalls aus Werkzeugstahl, aber nicht gehärtet, bildet einen Rahmen für die Matrize A und wird durch einen Stift an jedem Ende in der Mitte der Grundplatte in seiner Lage gesichert.

Die Löcher D sind die Führungen für die Druckstifte zwischen der Matrize und der Auswerferfeder; das Gewinde in der Mitte der Gesenkplatte dient zur Aufnahme des Federbolzens.

C ist ein festes Keilstück aus gehärtetem Werkzeugstahl, das dem Seitenstempel E seine Bewegung erteilt. F sind Führungen für diesen. Der Seitenstempel E stößt beim Abwärtsgang des Stempels an die Schräge L des Keilstückes C und beginnt so langsam gegen den Stempel G vorzurücken. Während der Seitenstempel längs der Schräge L nach abwärts geht, formt der Stempel das Blankett in dem Gesenk und biegt den Haken R rechtwinklig um. Wenn E bei M ankommt, hat der Stempel das Blankett bis auf das Ausprägen der Enden und der Mitte und das Umbiegen des Hakens fertiggestellt. Die Biegegesenk A ruht nun auf der Grundplatte. Das Umbiegen des Hakens erfolgt an der Stelle des Stempels G, wo die punktierte Linie P angedeutet ist. Der Ausschnitt ist gerade weit genug, um Raum für die Rippe Q und eine Metallstärke, hier hartes Messing von ungefähr 1 mm Dicke, zu geben. An der kurzen Schräge bei M tritt dazu der Seitenstempel E tatsächlich in Tätigkeit, während er auf seinem Abwärtsgang längs L bloß leer mitgeht, so daß er das Blankett erst im geeigneten Augenblick berührt. Nachdem E die 45^0 Schräge bei M passiert hat, geht der Seitenstempel E mit dem ganzen Stempel längs N herab. Während dieser Bewegung führt der Stempel G das Ausprägen der Enden und der Mitte aus. Beim folgenden Aufgang des Stößels hängt der Haken R in der Öffnung bei P, so daß der Arbeiter leicht das fertige Stück abnehmen kann. Der Krümmungshalbmesser des großen Bogens am Oberstempel ist 44,5 mm, am fertigen Stück 81 mm. Bei der ersten Ausführung dieses Werkzeuges zeigte es sich, daß man für das Federn des Blanketts durch das Prägen keine besondere Zugabe geben mußte.

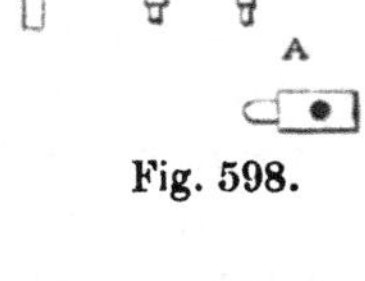

Fig. 598.

Biege- und Prägewerkzeug für Juwelierarbeit. Die Zeichnungen in Fig. 599 stellen ein Werkzeug dar, in dem ein kurzes Stück Neusilberdraht auf die Form A, Fig. 598, gebracht wird. Links in der Figur ist der Stift gezeichnet, wie er vom Automaten mit einem kleinen angedrehten Zapfen am einen Ende kommt. Der Stift wird in dieser Form zwischen die Klemmbacken a in das Werkzeug eingespannt und darin auf die in den drei andern Ansichten abgebildete Form gebracht. In dem Gesenk wird der Stift gebogen, ein Teil desselben ausgeflacht, und darauf eine Erhöhung ausgeprägt. Die Arbeit geschah früher in zwei Gängen, Biegen und Prägen, doch war dieses Vorgehen zu langsam und zeitigte keine zufriedenstellenden Ergebnisse.

In dem Gesenk, Fig. 599, wird nun folgendermaßen gearbeitet: Der Stift wird zwischen die gehärteten Stahlbacken a eingelegt, die mit Schwalbenschwanz in die Schlitten b aus ungehärtetem Werkzeugstahl eingesetzt sind. Diese wiederum sind in gleicher Weise in der

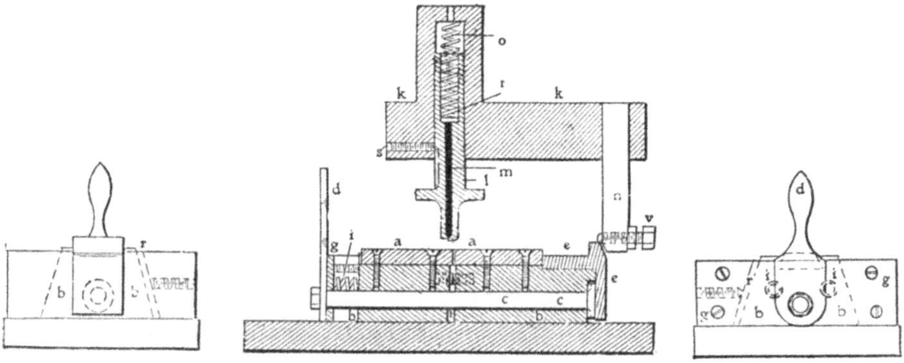

Fig. 599.

gußeisernen Froschplatte geführt. Dann wird der Handgriff d nach dem Arbeiter zu umgeschlagen und schließt durch das Exzenter c, welches die Backen a anhebt, diese leicht. Wenn nun der Stößel hinabgeht, stößt der Prägestempel l mit seiner abgeschrägten Vorderseite an den Stift. Gleichzeitig stößt der Anschlag n an das Keilstück e und schiebt die Schlitten b und mit ihnen die Backen a unter dem Stempel l vorwärts, wodurch der Stift abgebogen wird. Dabei werden durch die Arbeit von n zwei starke Federn i zusammengepreßt und erhöhen den Druck zwischen den Backen a, der so lange anwächst, bis der Anschlag n den höchsten Punkt des Keilstückes e hinter sich hat. Ein wenig später während des Abwärtsganges legt sich der Stempel l mit seinem Absatz an den gußeisernen Stempelkopf k, so daß jetzt zwangläufig ein harter Druck erzielt wird, durch den der Stift ausgeprägt, d. h. ausgeflacht, und mit der Erhöhung versehen wird.

Beim Abwärtsgang des Stempels übt die Feder o zwischen Kopf und Stempel einen harten Druck auf den langen Stahlstift m aus, der an seinem unteren Ende auf die Lochabmessung im Stempel abgefeilt ist, so daß das Material für die Erhöhung in den Stempel eintreten kann. Zwischen der Feder o und Stahlstift m ist eine Unterlagscheibe r eingelegt. Beim Aufwärtsgang arbeitet dieser Stift auch als Auswerfer auf die ausgeprägte Erhöhung und stößt diese aus dem Stempel, bevor der Anschlag n den höchsten Punkt des Keilstückes e verläßt. In diesem Augenblick gehen die Schlitten b wieder in ihre Anfangsstellung zurück, und der Handgriff d liegt an der Platte g an. Hierauf wird das Exzenter durch Aufwärtsdrehen des Handgriffes freigelegt und der fertige Stift aus den Backen herausgenommen.

Man kann sehen, daß die beiden Schlitten b durch zwei Federn nur so weit von einander gehalten werden, um das Einlegen und Entfernen des Stiftes zu ermöglichen. Die Stifte könnten leicht in die Öffnung zwischen die beiden Schlitten hineinfallen, wenn die Entfernung derselben größer als die Dicke des Stiftes wäre. Die Schraube s im gußeisernen Stempelkopf paßt in eine in den Stempel l gefräste Nut, so daß derselbe gegen Drehung gesichert und gleichzeitig der Abwärtsweg der Feder o begrenzt wird.

Der Handhebel d ist aus weichem Stahl gemacht und im Einsatz gehärtet. Die Platte g aus weichem Stahl wird mit vier Schrauben an der Froschplatte befestigt. Das Keilstück e ist aus Werkzeugstahl hergestellt, gehärtet und nicht angelassen; es wird mit zwei Schrauben, die nicht gezeichnet sind, am Schlitten b befestigt. Auch der Anschlag n ist aus Werkzeugstahl gemacht und gehärtet.

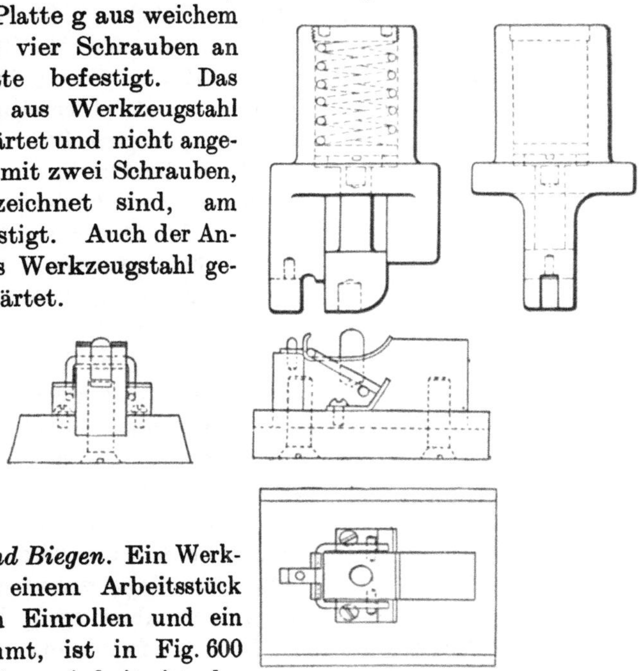

Fig. 600.

Einrollen und Biegen. Ein Werkzeug, das an einem Arbeitsstück gleichzeitig ein Einrollen und ein Biegen vornimmt, ist in Fig. 600 abgebildet. Diese Arbeit ist der letzte Gang bei der Herstellung des Schließteiles in Tabelle Nr. 31. Der im dritten Gang fertiggestellte U-förmige Drahthaken muß in die beiden vorgebogenen Enden eingelegt und durch Einrollen derselben gehalten werden. Gleichzeitig wird dem Hauptkörper die endgültige Krümmung erteilt.

Das im zweiten Gang vorgebogene Stück wird so auf den Unterteil gelegt, daß die beiden vorgebogenen Scharnierenden beiderseits auf den Stützflächen aufliegen. Dann wird der Haken eingelegt, der sich, wie die Zeichnung erkennen läßt, auf Stifte auflegt und gegen die Anschläge abstützt. Der Aufhängestift tritt in das größere Loch im Schließteil ein und sichert so dessen Stellung, worauf beim Abwärtsgang des Stempels der Biegestempel, gleichzeitig als Niederhalter arbeitend, das Stück formt. Beim weiteren Abwärtsgang des Stößels tritt der Biegestempel unter Überwindung des Federdruckes in den Stempelkopf zurück, hält aber das Stück immer noch fest, so daß der Einrollstempel die aufrechtstehenden Scharnierenden fassen und über dem Drahthaken umbiegen kann. Im letzten Teil des Hubes wird die Zunge zwischen dem erhöhten Matrizenteil rechts und dem entsprechenden Teil des Stempelkopfes scharf abgebogen, womit die in G. 5 abgebildete endgültige Form des Stückes erreicht ist.

Nachziehen und Prägen. Die in Tabelle Nr. 24 in ihrem ganzen Herstellungsgang wiedergegebene Schnauze für eine Blechkanne wird im dritten Gang auf dem in Fig. 601 abgebildeten Werkzeug nachgezogen und gleichzeitig mit einem vertieft geprägten Boden versehen.

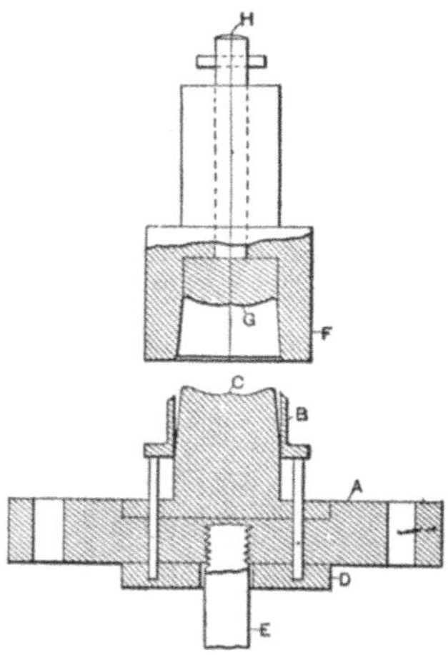

Fig. 601.

A ist die Froschplatte, B ein Blechhaltering, der durch Stifte und eine Gummifeder, ähnlich wie beim Werkzeug in Fig. 524, getragen wird und in seiner tiefsten Stellung im Verein mit dem Ziehstempel C die Hohlkehle formt. Der Ziehstempel C in diesem Werkzeug ist konisch und dementsprechend der Stempel F konisch ausgedreht. Der Auswerfer im Stempel F hat eine besondere Form, da der Boden später ausgeschnitten und für andere Zwecke verwendet werden soll. Die Schnauze wird auf den Blechhaltering gelegt, worauf der Stempel sie bei seinem Abwärtsgang herabzieht und nach dem Stempel C formt. Beim Aufwärtsgang wird die Hülse durch den Zapfen H, wie im ersten Gang, ausgeworfen. Das gezogene Stück sieht jetzt wie G. 3 in Tabelle Nr. 24 aus.

Kalibrieren und Einrollen. Nachdem im vierten Gang das eben beschriebene Stück nachgezogen und ausgehauen worden ist, kommt es im letzten Gang in das Werkzeug in Fig. 602. Obwohl dieses sehr einfach gebaut ist und nur aus vier Teilen besteht, hängt doch sehr viel von ihm ab, da alle diese Schnauzen an der fertigen Kanne ein genaues Maß haben müssen, um die Verschlußkappe aufzunehmen, die übrigens, wenn an der Kanne angebracht, wasserdicht sein muß. Das Werkzeug besteht aus der Froschplatte A und einem Formstempel B aus Werkzeugstahl, der gehärtet, angelassen und auf Maß geschliffen ist. Auch der Oberteil, die dazugehörige Matrize, ist gehärtet, angelassen und auf Maß geschliffen. Nachdem das Werkzeug in die Presse eingebaut worden ist, wird die Schnauze auf den Formstempel B aufgesetzt, worauf der Oberstempel D beim Abwärtsgang mit der inneren Ausdrehung F zuerst das Stück berührt. Beim weiteren Abwärtsgang wird diese Kante umgerollt und auf die Form in G. 5 der Tabelle zusammengepreßt. Beim Aufgang des Stempels wird die Schnauze wie in den vorangehenden Arbeitsgängen ausgeworfen.

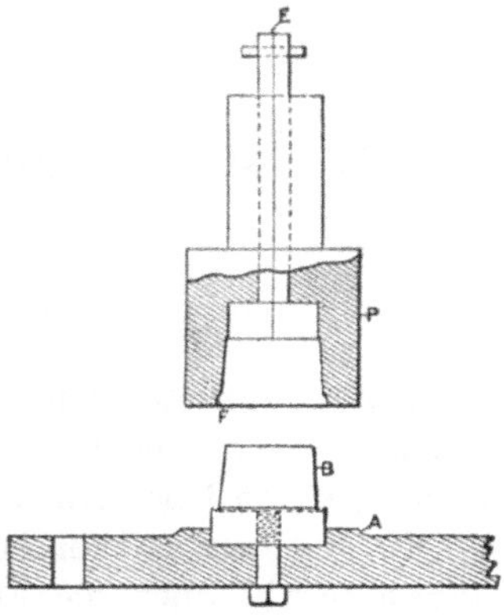

Fig. 602.

Manchmal besteht die Ansicht, man brauche keine wirklich guten Gesenke, um Blechgefäße herzustellen, während gerade diese Werkzeuge die äußerste Genauigkeit bei dem Entwurf erfordern und außerdem Weißblech in seinem Verhalten beim Ziehen ein ganz eigentümliches Material ist.

III. Besondere Herstellungsverfahren.

Herstellung der Stahlfedern.

Der eigentliche Erfinder der Stahlfeder ist unbekannt geblieben. Frankreich, England und Amerika haben Namen, die darauf Anspruch machen, so daß es schwer ist, einem die Berechtigung zuzusprechen.

Nach einer ausführlichen, geschichtlichen Ausführung spricht Woodworth über die Einführung und Bedeutung der Feder mit stumpfer Spitze als Geschäftsfeder gegenüber der spitzen Schönschreibfeder von früher. Als Stahl für die Federnerzeugung gibt er den besten englischen und schwedischen mit seiner feinkörnigen Struktur und besonderen Dichte an und sagt, daß amerikanische Erze derartiges Material nicht herzustellen gestatten.

Vorarbeiten der Federnfabrikation. Der Stahl wird in Blechen von verschiedener Länge, Breite und Dicke geliefert, diese von den Fabrikanten in Streifen passender Breite geschnitten und dann in längiiche Eisenkästen gepackt. Diese werden mit der Öffnung nach unten ln einen zweiten gleichartigen Kasten gepackt und der Zwischenraum luftdicht ausgefüllt. Nun erwärmt man diese Kästen im Glühofen allmählich, bis sie außen dunkelrot erscheinen, und läßt sie dann langsam abkühlen. Um den dabei entstehenden schwachen Glühspan von den Streifen zu entfernen, werden sie in ein Bad von verdünnter Schwefelsäure getaucht, das den Glühspan lose macht, der dann in Scheuertrommeln durch Wasser vollständig entfernt wird. Aus den Scheuertrommeln kommen die Streifen mit silbergrauer Oberfläche und gehen dann durch mehrere Walzwerke, bis sie die richtige Dicke, im allgemeinen ungefähr 0,15 mm, erhalten haben. Diese Arbeit verlangt große Sorgfalt und Übung, da eine Abweichung von nur 0,025 mm in der Dicke des Streifens genügen würde, um die Biegsamkeit der Feder ernstlich zu beeinträchtigen. Die Streifen sind nun dreimal so lang wie anfangs und haben eine glänzende Oberfläche.

Formgebung der Feder. Diese Vorarbeiten werden von Männern und Knaben ausgeführt, die weiteren Arbeiten jedoch durch weibliche

Kräfte, die für diese Arbeit geeigneter sind. Das Ausschneiden der Blanketts erfolgt durch ein gewöhnliches Schnittwerkzeug. Woodworth beschreibt nun diese Arbeit mittels einfacher Werkzeuge und Handvorschub des Streifens mit Benutzung des Aufhängestiftes. Fig. 603 zeigt dagegen eine mehr moderne Arbeit, wie sie auf selbsttätigen Exzenterpressen (vgl. S. 540, Stanzautomat mit selbsttätigem Bandvorschub von Rob. Tümmler in Döbeln) bedeutend schneller und genauer erhalten wird. Beim Ausschneiden der Blanketts wird ein kleines V an der oberen Kante, die in den Federhalter gesteckt wird, eingeprägt, das man bei sorgfältiger Untersuchung finden kann und welches für die nachfolgenden Arbeiten von Wichtigkeit ist, da die Arbeiterin danach die rauhe und glatte Seite des Blanketts unterscheiden kann.

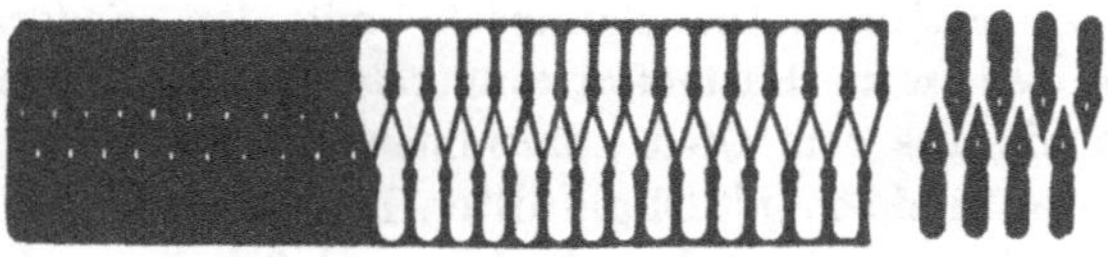

Fig. 603.

Im nächsten Arbeitsgang wird die Beschriftung eingeprägt. Dies geschieht in einem kleinen Fallwerk, das vom Fuß aus betätigt wird. Die Arbeiterin nimmt eine Handvoll Blanketts in ihre linke Hand und ordnet sie durch eine Handbewegung in eine kleine Reihe zwischen Daumen und Zeigefinger, so daß das erste Blankett in passender Stellung von der rechten Hand übernommen werden kann. Mit der rechten Hand wird das Blankett mit seiner Spitze gegen die Arbeiterin in die Führung des Fallwerkes gelegt, worauf die Arbeiterin den Stempel fallen läßt. Manche Arbeiterinnen sind so geschickt, daß sie zwischen 30000 und 35000 Federn im Tag prägen können. Ist jedoch die Prägung ungewöhnlich groß, so wird sie auf einen späteren Arbeitsgang verschoben, vgl. Tabelle Nr. 5 G. 5.

Nun wird die Feder gelocht, wodurch sie die gewünschte Elastizität bekommt und später die Tinte hält. Das Lochen geschieht auf einer Spindelpresse, die mit geeigneten Zuführungsvorrichtungen versehen ist. In dem oben angegebenen Werkzeug, Fig. 603, geschieht das Lochen („Ausstechen“) gleich beim Ausschneiden des Blanketts mittels eines Folgeschnittes. Um die Blanketts für das Hohlprägen („Stampfen“) weicher zu machen, werden sie ausgeglüht. Dazu werden sie sorgfältig von Staub und anhaftendem Fett befreit, mit Holzkohlepulver in runde, eiserne Töpfe gepackt, die wieder von größeren Kästen umschlossen werden, um den Eintritt der Verbrennungsgase zu verhindern. Dann werden die Kästen in den

Glühofen eingesetzt und bis auf ein dunkles Rot erhitzt, wonach man sie allmählich abkühlen läßt.

Nach dem Ausglühen sind die Blanketts so weich und biegsam, daß sie leicht die verschiedenen Formen des folgenden Arbeitsganges („Stampfen") annehmen. Dies geschieht wieder in der Presse, in deren Stößel der Prägestempel befestigt ist. Auf dem Pressentisch ist in einer stählernen Froschplatte das Prägegesenk befestigt, das mit vier Führungsleisten aus Stahl versehen ist. Die Arbeiterin läßt das Blankett zwischen die Führungen gleiten, zwischen denen es bis zum Ausprägen gehalten wird. Hierauf werden die geprägten Blanketts in dünnen Schichten in runde Schüsseln mit Deckeln gelegt und gehärtet.

Härten der Federn. Die Schüsseln mit den Federn werden 20 bis 30 Minuten in die Härteöfen eingesetzt, bis sie hellrot werden. Dann werden sie aus dem Ofen genommen und ihr Inhalt in einen großen Korb, der in einem mit Öl gefüllten Trog versenkt ist, geworfen. Der Korb ist selbstverständlich gelocht, so daß beim Ausheben das Öl in den Trog zurück abfließt und die Federn im Korb zurückgehalten werden. Das anhaftende Öl wird dann in sich drehenden Sieben entfernt. Nun sind die Federn sehr fettig und glashart. Zur Entfernung der letzten Fettspuren kommen sie wieder in Siebe, mit denen sie in einen Trog mit kochendem Sodawasser eingetaucht werden.

Anlassen der Federn. Nachdem die Federn gründlich gereinigt worden sind, werden sie in einen eisernen Zylinder gebracht, der über einem Holzkohlenfeuer in Drehung erhalten wird, bis die Federn die verlangte Anlaßfarbe erreicht haben. Danach sind die Federn dunkel und rauh an der Spitze. Nun werden sie blank gescheuert, indem man sie in ein Bad von verdünnter Schwefelsäure taucht, das alle anhaftenden Spuren des Härtens und Anlassens entfernt. Doch ist dabei große Sorgfalt anzuwenden, da der Stahl leicht selbst angegriffen werden kann. Dann kommen die Federn in eiserne Trommeln, die mit einer Mischung von Wasser und feingemahlener, gesiebter Tiegelmasse gefüllt sind. Diese Trommeln läßt man fünf bis acht Stunden laufen, wonach die Federn für dieselbe Zeit in Trommeln mit Löschkalk und dann in anderen mit trockenen Sägespänen gescheuert werden. Hierauf werden sie zwischen der Mitte und Spitze geschliffen. Dies geschieht auf Holzscheiben aus Erlenholz von 267 mm Durchmesser und 13 mm Breite, die mit Leder überzogen sind, auf dem Schmirgel aufgetragen wird. Die Schleifräder werden durch Riemen angetrieben. Die Arbeiterin hält die Feder fest und schleift mit einer leichten Berührung die Oberfläche ab.

Spalten der Federn. Die letzte und weitaus wichtigste Arbeit, die auf Maschinen ausgeführt wird, ist das Spalten der Federn. Das dazu benützte Werkzeug besteht aus zwei länglichen Stücken Stahl von ungefähr 38 × 9,5 × 32 mm Abmessungen. Die Schneidkanten dieser Werkzeuge sind ebenso fein und empfindlich wie die Schneide eines Rasiermessers. Die Arbeiterin schiebt die Feder in das Werkzeug mit der Spitze gegen die Führungsanschläge, worauf beim Einschalten der Maschine der Schnittstempel die Feder spaltet.

Die Firma Soennecken verwendet zum Spalten der Federn das Werkzeug in Fig. 604, bei dem der Schnittstempel a abgeschrägt ist, so daß er an der Spitze der Feder zu schneiden beginnt und allmählich bis zum Ende seines Weges die ganze Länge durchschneidet. Die Schnittplatte ist nur zur Hälfte, bis zur Mittellinie der Feder, vorhanden, so daß beim Auftreffen des Schnittstempels die nicht aufliegende Federhälfte, von der Spitze beginnend, heruntergedrückt und gespalten wird. Nach dem Rückgang des Schnittstempels federt diese Hälfte in ihre Anfangsstellung zurück. Die Arbeit wird auf einer kleinen Spindelpresse gemacht. Auf dieser Herstellungsstufe sind die äußeren Kanten jeder Spitze glatt, die inneren jedoch scharf und rauh. Dieser Mangel wird wieder in Scheuertrommeln beseitigt, in denen die Federn fünf bis sechs Stunden verbleiben, wonach sie herausgenommen und mit Sägespänen poliert werden. Hierauf werden die Federn gefärbt, indem man sie in über einem Koksfeuer sich drehende Trommeln bringt, worin sie so lange bleiben, bis sie die verlangte Färbung angenommen haben. Sollen sie noch lackiert werden, so legt man sie in eine Lösung von Schellack in Alkohol. Man läßt diese Lösung nachher abtropfen und bringt die Federn in sich drehende Trockenzylinder. Dann werden die Federn auf eiserne Bleche aufgestreut und in einem Ofen erwärmt, bis sich der Lack gleichmäßig über die ganze Oberfläche der Federn ausgebreitet hat. Der Lack gibt den Federn den Glanz und verhindert das Rosten. Schließlich werden die Federn noch sorgfältig kontrolliert, bevor sie in den Handel gebracht werden.

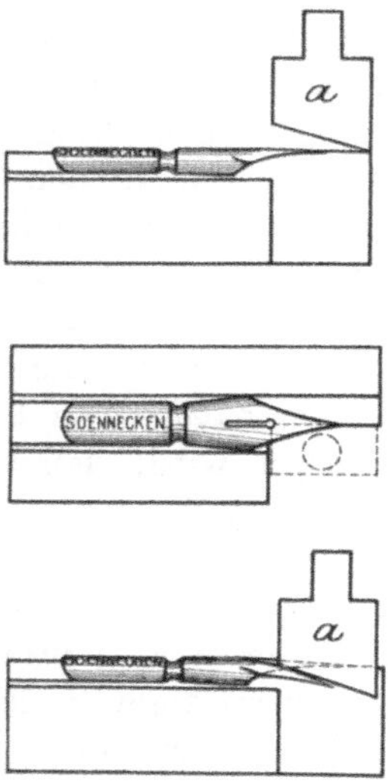

Fig. 604.

Herstellung von amerikanischen Goldfedern.

Die moderne Goldfeder ist durch den amerikanischen Fabrikanten zu ihrer heutigen Vollkommenheit gebracht worden, wie sich auch diese ganze Industrie von Beginn an durch Anwendung amerikanischer Arbeitsweisen kennzeichnet. Da zu der Herstellung der Goldfeder eine große Geschicklichkeit notwendig ist, werden in diesen Werkstätten nur die erfahrensten Leute verwendet.

Das verwendete Gold kommt aus der amerikanischen Münze. Dann wird es geschmolzen und auf ungefähr 16 Karat Feingehalt legiert und in lange, schmale Bänder ausgewalzt, aus denen die Blanketts, die beträchtlich dicker als die fertige Feder sind, auf der Exzenterpresse ausgeschnitten werden. Die stumpfe Spitze des Blanketts wird gespalten oder mit einer Vertiefung zur Aufnahme des Iridiums versehen, das die harte Spitze aller guten Goldfedern bildet. Das Iridium wird mit einem Teige aus Borax, der in Wasser gemahlen worden ist, bedeckt und in den Schlitz am Ende des Blanketts eingelegt. Es wird dann dort durch eine Art Einschmelzen des umgebenden Goldes gehalten, das sich dadurch völlig mit dem Iridium verbindet. Hierauf geht das Blankett durch ein Streckwalzwerk, das die Dicke der Feder von der Spitze nach rückwärts allmählich vermindert. Diese Walzen haben eine kleine Aushöhlung, in welche die Iridiumspitze eingelegt wird, damit sie durch das Walzen nicht beschädigt wird. Nach dem Walzen wird die Spitze jeder Feder gehämmert und steif gemacht. Es ist dies die wichtigste Arbeit in der Herstellung, da die Elastizität der Spitze nur von dieser Arbeit abhängt. Nun wird die Feder zugeschnitten, was wohl meistens auf einer Exzenterpresse mit selbsttätiger Zuführung geschieht. Nach dem Zuschneiden wird die Feder gestempelt, wobei Name des Fabrikanten und die Nummer der Feder eingeprägt wird. Dieser Arbeit folgt das Stampfen, durch das in einem geeigneten Werkzeug das Blankett hohlgeprägt wird. Dazu ist sehr wenig Kraft notwendig. Die endgültige Krümmung erhält die Feder in einem Werkzeug, das mit zwei Backen die gegenüberliegenden Kanten der Feder erfaßt und zusammendrückt. Im nächsten Arbeitsgang wird die Iridiumspitze gespalten, indem sie gegen eine schnell umlaufende, sehr dünne Kupferscheibe, auf die Schmirgel und Öl aufgetragen wird, gehalten wird. Dann wird die Feder in der Presse gespalten und der Schlitz mit einer feinen Kreissäge gesäubert. Nun werden die Spitzen durch Hämmern zusammengebracht und die Federinnenseite in einer hohlen und die Außenseite in einer erhabenen Form poliert. Auf diese Weise erhält die Feder eine gleichmäßige Oberfläche und größere Elastizität. Hierauf werden die Federspitzen von Hand gerichtet und dann die Feder auf einer Bank geschliffen, die mit einer dünnen Stahlscheibe und einem Kupferzylinder, die mit feinem Schmirgel und Öl bestrichen sind, ausgestattet ist. Mit dem Polieren der Feder auf der Tuchscheibe ist die Herstellung der Feder beendet. Bevor die Feder in den Handel kommt, wird sie gründlich auf ihre Elastizität, Schliff und Gewicht untersucht und einer Kontrollperson übergeben, die die Feder prüft und nachwiegt.

Herstellung der Stecknadeln.

Die frühere Herstellung der Stecknadel durch Handarbeit war sehr langwierig und mühevoll, da jede Nadel durch die Hände von vierzehn bis achtzehn Personen gehen mußte. Jetzt wird die Stecknadel in den Vereinigten Staaten mit den verbesserten selbsttätigen Maschinen der Firma Atwood & Fowler hergestellt.

Der Gang der Herstellung ist kurz folgender: Die Drahtbunde werden auf Rollen aufgelegt, von denen der Draht selbsttätig mittels Zangen zwischen festen Bolzen durchgezogen wird, wodurch er gerade gerichtet wird. Ein Stück Draht, ungefähr der Länge einer Nadel entsprechend, wird von zwei seitlichen Backen so erfaßt, daß ein Stück desselben vorsteht und durch einen dagegen vorrückenden Stauchstempel zu einem Kopf vorgestaucht wird. Dann wird der Draht freigegeben und ungefähr 1 mm vorgeschoben, worauf der Kopf noch einmal von demselben Werkzeug gestaucht wird. Dadurch ist der Kopf fertiggestellt und der Draht in der gewünschten Länge abgeschnitten. Zur Bildung des Kopfes ist eine Drahtlänge von ungefähr 3,2 mm nötig. Wollte man versuchen, den Kopf in einem Gang zu stauchen, so würde wahrscheinlich der Draht umknicken, statt sich zu stauchen.

Die mit Kopf versehenen Drähte fallen in einen Aufnehmer und ordnen sich dort in einem schmalen Schlitz an, der von zwei zugeschärften, geneigten Schienen gebildet wird. Die Entfernung beider Schienen ist gerade groß genug, daß der Nadelschaft durchfallen kann, während der Kopf auf den Schienen liegen bleibt, so daß die Nadeln zwischen denselben in einer geordneten Reihe hängen. Wenn sie nun in dieser Stellung das untere Ende der geneigten Führung erreicht haben, so werden sie zwischen zwei Teilen der Maschine gefaßt und weiter befördert, wobei sie sich drehen, bis sie vor einen Zylinderfräser kommen, der die Spitzen anschärft. Die fertigen Nadeln werden von der Maschine ausgeworfen und, wenn es Messingnadeln sind, durch Kochen in dünnem, saurem Bier gereinigt. Hiernach werden sie verzinnt. Dies geschieht so, daß man in einer Kupferschale abwechselnd Lagen von Nadeln und körnigem Zinn aufeinanderschichtet, worauf man eine wässerige Lösung von doppeltweinsaurem

Kali darübergießt. Durch Erwärmung wird das Zinn aufgelöst und setzt sich auf den Nadeln ab. Nach dieser Arbeit werden die Nadeln in sich drehenden Scheuertrommeln mit Kleie oder Sägespänen gereinigt und blank gemacht. Zum Schluß müssen sie noch in die Papiere eingesteckt werden, wie man sie dann im Handel erhält. Dies geschieht auf einer selbsttätigen Maschine folgendermaßen: Die Nadeln kommen in einen Zuführtrichter, mit dem eine Stahlplatte verbunden ist, welche lange schmale Schlitze, entsprechend der Anzahl Nadeln in einer Reihe des Papieres, trägt. Die Nadeln werden durch ein kammartiges Werkzeug so aufgerührt, daß deren Schäfte durch die Schlitze in der Stahlplatte fallen und die Nadeln an ihren Köpfen aufgehängt erscheinen. Der Arbeiter schiebt nun lange schmale Papierstreifen in die Maschine, die zwei erhöhte Falten anfalzt, in die nun von der Maschine die Nadeln in Reihen eingestochen werden. Diese Arbeit wiederholt sich, bis die richtige Anzahl Nadelreihen in jedem Papierstreifen steckt.

Herstellung von Nadeln für Wirkereimaschinen.

Die Nadeln für die Wirkerei sind mit einer Stahlfeder oder einem Schloß versehen. Das arbeitende Ende wird nahezu auf eine Spitze ausgezogen, dieser Teil dann zu einem offenen Haken so umgebogen, daß eine Nut, die vorher in die Nadel eingearbeitet worden ist, dieser Spitze gegenüber kommt. Wenn die Nadel arbeitet, so tritt diese Spitze zur rechten Zeit vor und erfaßt das Garn, das durchgezogen wird, um den Stich zu bilden. Bei der weiteren Vorwärtsbewegung wird die Spitze in die Nut dadurch eingedrückt, daß sie zu diesem Zwecke mit einer besonderen Vorrichtung in Berührung kommt, so daß die Nadel durch die Fadenschlinge ohne sich zu fangen treten kann.

Die Nadel mit Schloß hat an Stelle des Federhakens einen kurzen, steifen Haken, der durch Ausziehen des arbeitenden Endes der Nadel fast zu einer Spitze und Biegen desselben, damit er in das Schloß paßt, hergestellt wird. Das Schloß ist in einer Nut, die in den Nadelkörper eingefräst wird, verlagert und dreht sich auf einer feinen Niete, die durch die Wandungen der Nut hindurchgeht. Da sowohl das Schloß wie die Wände der Nut, durch die die Achse vernietet ist, sehr fein sind (jeder Teil ist nur 0,25 mm dick), muß selbstverständlich äußerste Sorgfalt und Übung bei der Herstellung der Nadel obwalten. Das Schloß hat den Zweck, bei der Herstellung und dem Abstreifen der Fadenschlinge, des Stiches, zu verhindern, daß das Garn anders als zur rechten Zeit unter dem Haken gefangen wird.

Herstellung der Nähmaschinen-Nadel.

Anfänglich war die Herstellungsweise der Nähmaschinennadel in Amerika dieselbe, die in England zur Herstellung der Handnähnadel in Gebrauch war und sehr viel Handarbeit erforderte. Das Ausziehen des Schaftes auf die richtigen Abmessungen, das Nuten der Seiten wurde durch Prägen im Gesenk erreicht. Dabei wurde das überschüssige Material als Grat nach beiden Seiten ausgepreßt und mit Handscheren entfernt, worauf noch ein Abgraten im Gesenk folgte. Hierauf wurden die Nadeln gerundet und durch Feilen angespitzt. Allmählich wurden die obigen Arbeiten durch Walzen, Schleifen, Drehen und Fräsen ersetzt, bis schließlich die ganze Arbeit auf Automaten erfolgte.

Bei der Herstellung der Nähnadeln geht das Stück durch die folgenden Arbeitsgänge: Blankett, 1. Ausziehen, 2. Ausziehen und Anspitzen, Nuten, Öhr stanzen, Härten und Anlassen, Grat entfernen, Schleifen, Scheuern, Öhr polieren, 1. Inspektion, Geraderichten von Hand, Spitze fertig machen, Fertigmachen.

Bei der Fabrikation der Nadeln werden zwei Wege befolgt, deren Arbeiten in den meisten Punkten die gleichen sind, bis auf die Herstellung des breiten Teiles, bei der sich eine Verschiedenheit ergibt. Bei der einen Arbeitsweise wird der breite Teil bereits beim Abschneiden des Drahtes auf Länge gebildet, während bei der anderen Herstellungsweise der Draht in Stücke von ungefähr $1/3$ der fertigen Nadellänge geschnitten und dann durch Kaltstrecken auf seine richtige Länge gebracht wird.

Da die dabei verwendeten Maschinen größtenteils Geheimnis des betreffenden Fabrikanten sind, kann hier nur eine allgemeine Beschreibung erfolgen, doch kann man aus der folgenden einige Jahre zurückliegenden amerikanischen Arbeitsweise immerhin genug ersehen.

Früher wurden die Nadeln aus dem besten Tiegelgußstahl gemacht, der in Bunden angeliefert und auf einer selbsttätig arbeitenden Maschine geradegerichtet wurde. Dann kam er in eine Maschine, die das Kopfende der Nadel anstauchte und den Draht in Stücke geeigneter Länge abschnitt. Die Blanketts kamen nun an drei Schleifmaschinen, von denen die beiden ersten mit grober und feiner Schmirgelscheibe arbeiteten, während die dritte einen Schleifriemen mit Schmirgel hatte. Von den Zuführtrichtern dieser Maschinen kamen die Nadeln zu einem genuteten, endlosen Transportschlitten, der den Teil der Nadelschäfte, der im Durchmesser verjüngt werden sollte, an die Schleifscheibe brachte; der Teil des Schaftes, der seinen ursprünglichen Durchmesser behält, wird nachher in der Nähmaschine befestigt. Während die Nadeln an der Schleifscheibe vorbeigehen, werden sie durch zwei hin- und hergehende Platten gedreht, so daß sie auf allen Seiten geschliffen werden. Nach dem Feinschleifen auf dem Schleifriemen der dritten Maschine werden die Nadeln durch selbsttätigen Vorschub zwei Kreissägen zugeführt, die sie mit Nuten versehen. Die Sägen werden solange an die Nadeln angedrückt, bis die Nuten die verlangte Tiefe und Form zeigen. In einer Presse mit Riemenantrieb wird dann das Öhr gestanzt, worauf die Nadeln in einem Flammofen mit Holzkohlenfeuerung auf Kirschrot erwärmt und nachher in Walfischtran abgeschreckt werden. Dann werden die Nadeln in Blechschalen, die an den Armen einer sich drehenden Welle hängen, gelegt und in einem Ofen, der durch die Abgase der Härteöfen betrieben wird, angelassen. Dieser Arbeit folgt das Reinigen auf Schmirgelleinen, wobei die Nadeln in Bündeln zu zwanzig Stück zwischen Finger und Daumen gehalten und unter Druck auf dem Leinen gerollt werden. Die gereinigten Nadeln werden in Flachzangen, ungefähr 70 auf einmal, mit den Nuten nach oben gepackt und gegen eine Messingbürste gehalten, die mit 8000 Umdrehungen pro Minute umläuft, wodurch die Nuten poliert werden. Dieselbe Arbeit wird mit einer Borstenbürste wiederholt, womit das Polieren der Nuten beendet erscheint. Während die Nadeln noch in diesen Zangen gehalten werden, wird ein mit Schmirgel und Öl getränkter Baumwollfaden in verschiedenen, schrägen Richtungen durch die Öhre hin- und hergezogen, damit das Poliermaterial an allen Stellen des Öhres zur Wirkung kommt. Nachdem die Nadeln von diesem Polierfaden entfernt worden sind, werden sie durch eine rotierende Haarbürste gereinigt und Öhr, Spitze und Schaft genau kontrolliert. Die guten Nadeln werden von Hand gerade gerichtet, wobei man sie auf einem Richtamboß in Augenhöhe hin- und herrollt und dabei jede kleine Krümmung mit einem Schlage eines kleinen Hammers ausgleichen kann. Die Endarbeiten sind das Fertigspitzen auf einer feinen Schmirgelscheibe und Fertigpolieren mit einer rotierenden Haarbürste unter Anwendung von Safran und Alkohol.

Kaltstrecken der Nadelspitzen. Bei der zweiten Herstellungsweise wird der Draht in eine Richt- und Abschneidemaschine geführt, in der er geradegerichtet und in Stücke von ungefähr $1/3$ der fertigen Nadellänge abgeschnitten wird. Dann kommen die Blanketts in kleine umlaufende eiserne Scheuertrommeln, wo Grat, Zunder und Schmutz entfernt werden. Nach einer weiteren Reinigung sind die Nadeln zum Kaltstrecken vorbereitet. Sie werden nun in den Zuführtrichter eingelegt, von dem immer ein Blankett auf einmal abgenommen wird, dessen Ende dann an einen Satz sich drehender Stahlgesenke gebracht wird. Durch das andauernde Öffnen und Schließen der sich drehenden

Werkzeuge werden die Drahtenden zusammengepreßt und auf die Schaftform und Länge ausgestreckt. Dieser Arbeit folgt das Stempeln der Blanketts. Da infolge des Kaltstreckens eine geringe Ungleichheit der Nadellängen unvermeidlich ist, werden sie auf Scheren beschnitten und wieder geradegerichtet. Eine besondere Einrichtung dieser Maschinen ist eine Schraubenzuführung, durch welche die Nadeln einerseits quer transportiert werden, wobei die Enden der Schäfte gegen eine Anschlagfläche ausgerichtet werden, und andrerseits gleichzeitig vorwärts, so daß die Spitzen an ein Messer gebracht werden, das alle Nadeln genau auf gleiche Länge abschneidet. Hierauf gehen die Nadeln durch ein Gesenk, in dem die Nadelnummer auf den Schaft eingeschlagen wird. Die folgenden Arbeiten sind die gleichen, wie oben bereits beschrieben worden ist.

Seitdem für alle diese Arbeiten selbsttätige Maschinen eingeführt worden sind, wird der ganze Transport von einer Arbeit zur nächsten vollständig ohne Handarbeit durchgeführt. Der Werdegang dieser Einrichtungen ist ein typisches Beispiel für die Entwicklung dieser Industrie in den letzten zwanzig Jahren.

IV. Zuführungen und Maschinen.

Der Bau einzelner Werkzeugformen zeigte deutlich den Einfluß einer besonderen Zuführung und einer besonderen Presse. Es ist deshalb notwendig, bei der Konstruktion der Werkzeuge gegebenenfalls die vorhandenen Zuführungen oder Pressen zu berücksichtigen, so daß einige Vertreter für die einzelnen Bauarten in diesem Abschnitt angeführt werden sollen.

A. Zuführungen[1].

Die Zuführungen der Pressen lassen sich nach ihrer Arbeit in drei Gruppen einteilen: die eine Gruppe bringt das Blech in Platten oder in Streifen zur Presse, die zweite Gruppe bringt die bereits vorgearbeiteten Stücke einzeln an das Werkzeug, und in einer dritten Gruppe schließlich werden gleichzeitig der Streifen und die vorgearbeiteten Stücke in die Maschinen, durch die Werkzeuge und aus der Presse befördert.

1. Zuführungen für Blech.

Die Zuführungsvorrichtungen der ersten Gruppe umfassen die Plattenzuführungen von Schuler[2]) und Bliss, die aber infolge des verwickelten Aufbaues nur zum Massenausschneiden von Scheiben, gegebenenfalls auch zum Ziehen der Näpfchen Verwendung finden. Sie arbeiten so, daß der Tafel durch den Träger eine Verschiebung in der Längsrichtung gegeben wird, bis eine volle Länge ausgeschnitten worden ist, worauf der Träger durch eine Querschaltung um eine Schaltbreite senkrecht zur ersten Richtung verschoben wird, worauf mittels der Längsschaltung wieder eine volle Länge ausgeschnitten

---

[1]) Eine ausführliche Behandlung dieses Gebietes ist in Gugel, Materialzuführungsvorrichtungen an Exzenter- und Ziehpressen, Springer, Berlin 1912, zu finden.

[2]) Z. d. V. d. I. 1910, S. 1851.

werden kann. Es kann aber auch die Schaltung zickzackförmig vor sich gehen, so daß gleichzeitig zwei Lochreihen fertiggestellt werden.

Die zweite Form, die Zuführung von Blechstreifen durch den Walzenapparat, wird ungleich häufiger angewendet; sie stellt die in verschiedenen Abänderungen meist gebaute Form der Zuführung für Blech dar.

Die übrigen Formen können hier übergangen werden.

Der Walzenapparat, dessen gewöhnlich Form Fig. 605 darstellt, besteht im allgemeinen aus zwei Walzenpaaren, rechts und links von dem Werkzeug, die das zu verarbeitende Band zwischen sich fassen und während des Aufganges der Presse um einen gewissen Betrag vorwärtsschieben, zum Stillstand kommen, bevor der Stempel in das Blech eintritt, und, nachdem der Stempel gearbeitet hat, das Blech wieder weiter befördern.

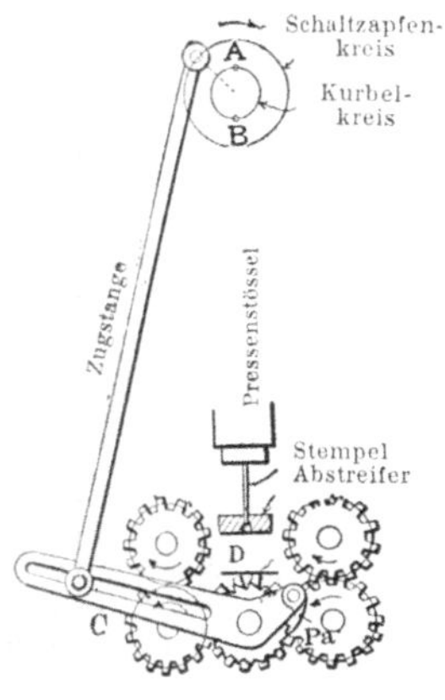

Fig. 605.

Diese ruckweise Bewegung der Walzen erfolgt meistens von der Kurbelwelle A der Presse durch eine auf dieser befestigte Kurbelscheibe B, deren Zapfen mittels einer Zugstange den Schalthebel C und die Schaltklinke Pa bewegt. Diese Schaltklinke greift in das Schaltrad des Walzenapparates ein, mit dem die Antriebsräder des Walzenapparates dauernd in Verbindung stehen. Um eine Überschaltung zu vermeiden, sind Sperrklinken vorgesehen, die zwangläufig oder durch Federkraft nach Beendigung des Schaltweges in die Zähne des Schaltrades einfallen und dessen Bewegung begrenzen.

Die Zuführung muß beginnen, nachdem das Material von dem Stempel abgestreift worden ist, und beendet sein, bevor der Stempel das Material wieder berührt. Diese Stellungen des Stößels sind auf die bezüglichen Kurbelwinkel zu übertragen, so daß der Winkel zwischen a und b, Fig 606, den Teil der Drehung der Kurbelwelle darstellt, der für die Zuführung in Betracht kommt. Der Schaltzapfen muß auf seiner Kurbelscheibe dann einen derartigen Voreilwinkel erhalten, daß mit Berücksichtigung der endlichen Stangenlängen sein Schaltweg in diesen Winkel des Kurbelzapfens fällt. Die weitere Einstellung auf die richtige Zähnezahl am Schaltrad erfolgt durch Verschiebung der Zugstange in dem Schlitz C, Fig. 605. Betreffs einer genauen

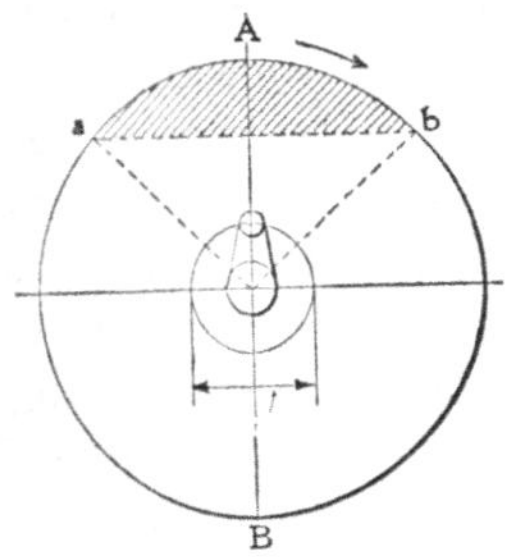

Fig. 606.

kinematischen Behandlung sei auf die angegebene Sonderliteratur verwiesen.

In Fig. 607 ist eine Zuführung ersichtlich, die eine abweichende Konstruktion aufweist und für feine Arbeit geeigneter ist. In diesem Fall erfolgt die Einstellung der Zähnezahl durch Nachstellen der Schraube F unter dem Daumenende und Verschieben des Bolzens in dem Schlitz E. Während das Zuführungsexzenter von A nach B weitergeht, stößt die Gabel den Bolzen nieder und treibt so das Schaltrad in der Pfeilrichtung weiter. Von B nach D hebt sich die Gabel wieder, während der Hebel durch die Feder zurückgezogen wird. Das Zuführungsexzenter eilt der Kurbelwelle um 90° voraus.

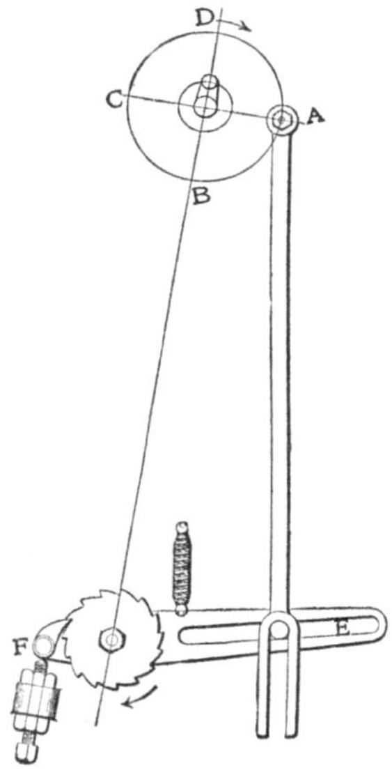

Fig. 607.

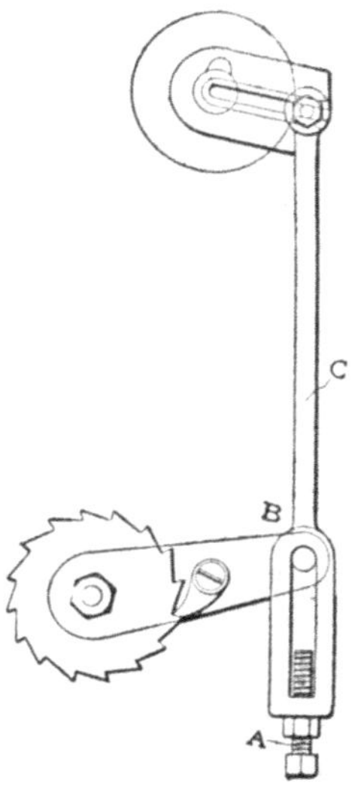

Fig. 608.

Fig. 608 zeigt eine kräftige Bauart für schwere Arbeit. Die Neigung der Schaltradvorrichtungen zu einer Überschaltung wird durch die Anwendung einer Bremse, die an einer passenden Stelle angebracht wird, aufgehoben. Die Bremse wird gewöhnlich auf den Walzen oder an der Stirnfläche eines der Zahnräder der Zuführung angebracht. Diese Einrichtung ist sehr einfach und kann mit Vorteil beim Lochen schwerer Bleche verwendet werden. Die Einstellung der Zähnezahl des Schaltrades hängt von der Einstellung der Schraube A und der Stellung des Bolzens in dem T-Schlitz der Schaltscheibe ab. Die geschlitzte Stange stößt den Bolzen B nieder, während beim Aufgang der Stange die Spitze der Schraube an den Bolzen anstößt und ihn in seine Anfangstellung zurückbringt.

Bei den obigen Einrichtungen wird die Bewegung des Schaltrades durch einen Satz Stirnräder auf die Walzen übertragen, je nach der Bewegungsrichtung der Zuführungswalzen und anderen Umständen.

Die Abbildungen sind natürlich nur schematisch, während man bei der tatsächlichen Ausführung die konstruktive Ausbildung der Einzelheiten sorgfältig überlegen muß. Im allgemeinen ist es vorteilhaft, große getriebene Zahnräder an der Schaltradwelle zu verwenden, da diese die Belastung beim Anlaufen beträchtlich verringern.

Die Sperrklinke der Walzenapparate. Die Sperrklinke darf wegen ihrer Kleinheit nicht übersehen werden. Die im allgemeinen einfache Form derselben gestattet mehr Aufmerksamkeit ihrer Anwendung als ihrer Konstruktion zu schenken. Fig. 609 zeigt ein Schaltrad, das sich um eine Mittelachse dreht. C ist der hin- und hergehende Klinkenhebel, der die Schaltklinke A trägt. Wenn der Klinkenhebel nach rechts geht, hebt sich die Schaltklinke und gleitet über die Spitzen der Zähne; beim Rückgang fällt sie gegen eine Zahnflanke ein und nimmt so das Rad mit. Die leergehende Sperrklinke B ist so gestellt, daß sie hinter jedem Zahn einfällt und dadurch eine Rückdrehung des Schaltrades verhindert.

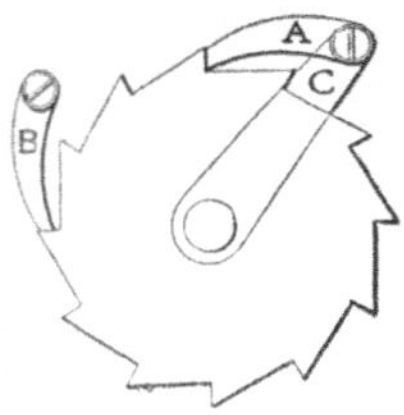

Fig. 609.

Wenn am Rade eine große Anzahl feiner Zähne vorhanden ist, wird durch den dauernden Gebrauch die Zahnfläche abgenützt und der Vorschub ungenau. Da es wenig wahrscheinlich ist, daß gerade gegenüberstehende Zähne beschädigt werden, kann man die in Fig. 610a gezeichnete Anordnung verwenden. Die Sperrzähne sind gleichlang und arbeiten so zusammen, daß von jedem Zahn und jeder Sperrklinke, solange das Schaltrad in guter Verfassung ist, die halbe Belastung aufgenommen wird und, falls ein Zahn fehlt, der andere ziemlich sicher den Druck überträgt; man erhält also durch die zweite Sperrklinke eine Sicherheitsvorrichtung.

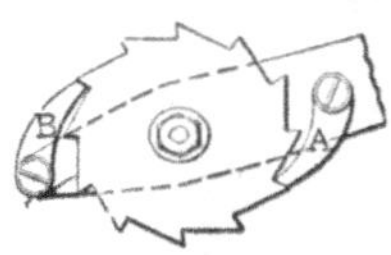

Fig. 610a.

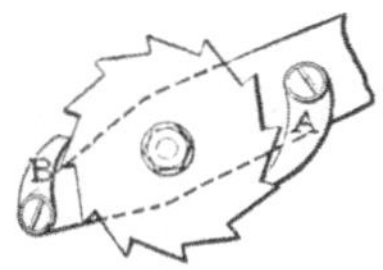

Fig. 610b.

Angenommen nun, man hätte mit einem Schaltrad von bestimmter Zähnezahl Schaltungen von der halben Zahnteilung, also die eines Rades mit doppelt soviel Zähnen zu erreichen, so kann man, statt ein neues Rad einzubauen, eine Schaltklinke gegen eine kürzere auswechseln, die in der halben Teilung einklinkt, wie Fig. 610b zeigt. Die beiden Klinken arbeiten abwechselnd, indem jede das Rad jedesmal um einen halben Zahn vorschiebt. Ebenso kann man die Schaltung durch Verwendung dreier Klinken auf ein Drittel verkleinern.

Fig. 611 zeigt eine andere sehr bequeme Schaltvorrichtung, bei der die Einstellung der für den Hub notwendigen Zähnezahl leicht erreicht wird. Die

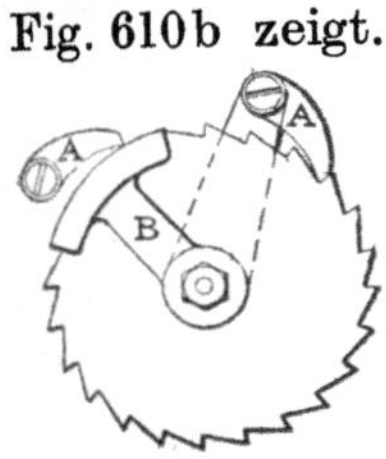

Fig. 611.

Sperrklinke A ist breit genug, um über die Kappe B, die um die Mittelachse drehbar ist, zu gleiten. Wenn der Vorschub im Sinne des Uhrzeigers erfolgt, so sind, wie ersichtlich, nur vier Zähne frei, die arbeiten werden, und die Klinke kommt beim Rückweg wieder auf der Kappe zur Ruhe. Durch Verlängern der Kappe kann man die Schaltung ganz aufheben.

Ausführungsformen. Um die Verschiedenheiten der Ausführung dieser Konstruktionsgrundlagen zu zeigen, sind im folgenden drei verschiedene Bauarten bekannter Firmen angeführt worden.

Fig. 612. L. Schuler, Göppingen.

Fig. 612 ist die von Schuler-Göppingen gebaute Form des Walzenapparates, der mit selbsttätiger Walzenabhebung bei jedem Stößelniedergang versehen ist. Der Apparat kann als Ganzes von dem Pressentisch entfernt werden. Trotzdem ist zum Einspannen und Ausrichten eines Werkzeuges nur nötig, die Walzenpaare abzuklappen. Die Schaltung erfolgt, wie in früheren schematischen Zeichnungen angegeben worden ist, mittels Zugstange und Schaltklinke von der Hauptwelle der Maschine aus auf das an der Vorderseite sichtbare Schaltrad, von dem es durch gefräste Räder auf die Walzen übertragen wird. Beim Rückgang des Schalthebels hebt sich die Klinke von den Zähnen ab, wodurch die Abnützung beider Teile verringert und die Bewegung geräuschlos wird. An dem Apparat ist auf der Einlaufseite, rechts, ein Auflagertisch mit Führungsleisten vorgesehen und außerdem ist die Oberwalze auf dieser Seite zum Abheben eingerichtet, damit die

Streifen schnell eingelegt werden können. Der zur Weiterbeförderung des Blechstreifens notwendige Druck wird durch verstellbare Federhebel, die auf die oberen Walzen wirken, erzeugt.

Eine konstruktiv davon grundsätzlich verschiedene Ausführung wird von der Bonner Maschinenfabrik Mönkemöller & Co. nach Fig. 613 gebaut. Der Antrieb erfolgt hier von der Exzenterwelle durch eine seitlich sitzende Zugstange auf die Schalträder des rechten Apparates, während die Übertragung auf den linken Teil durch Kettentrieb erfolgt.

Fig. 613. Bonner Maschinenfabrik Mönkemöller & Co.

Die Belastung der Walzen erzeugen einarmige Gewichtshebel, die auf die Lagerbüchsen der Oberwalzen drücken und gemeinsam beim Niedergang der Presse durch Kettenzug vom Stößel aus abgehoben werden. Die Verstellung der Gewichte auf den Belastungshebeln gestattet eine Regelung des Druckes auf die Walzen. Die beiden Teile des Apparates können rechts und links nach Lösen der Verbindungen heruntergeklappt werden, so daß der Pressentisch zum Aufspannen des Werkzeugunterteiles frei wird.

Walzenapparat mit Kugelschaltung (Reiß & Martin, Berlin SO). In Fig. 614 ist eine Exzenterpresse der obigen Firma wiedergegeben,

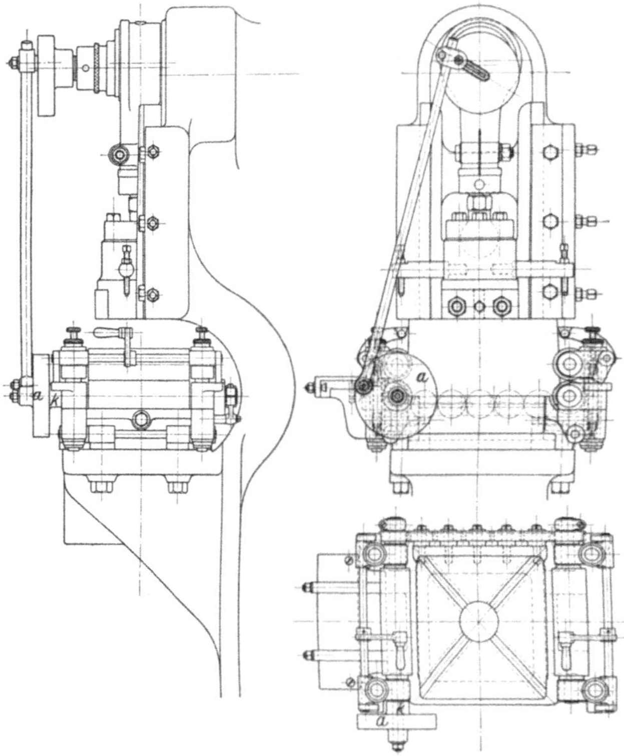

Fig. 614. Reiß & Martin, Berlin SO.

die mit einem eigenartigen Schaltapparat für die Walzen versehen ist, dessen Einzelheiten in Fig. 614a gezeichnet sind. Der Walzenapparat ist in bekannter Ausführung mit Räderübertragung von den linken zu den rechten Walzen versehen und ist um seinen Zapfen ausschwenkbar. Die Belastung der Walzen erfolgt hier durch Federn. Der Antrieb des Apparates erfolgt durch eine Schlitzkurbel auf der Kurbelwelle, in der ein Gleitstein entsprechend der gewünschten Vorschublänge eingestellt werden kann. Während nun die Übertragung von hier aus in den meisten Fällen durch Schaltklinke und Schaltrad erfolgt, ist hier ein Kugelschaltgetriebe mit der Achse der Antriebswalze verbunden, das nach Mitteilungen des Fabrikanten eine sehr genaue Schaltung ergibt. In Fig. 614a ist diese Schaltscheibe gezeichnet. Sie besteht aus einer lose auf der Achse der Walze laufenden Mitnehmerscheibe a, in der die Mitnehmerzapfen b eingepaßt und durch Spitzschräubchen c gesichert sind. An diese Zapfen b ist an der Innenseite eine schräge Fläche d angearbeitet, gegen die die Kugel e durch die Wurstfeder f dauernd gepreßt wird. Diese

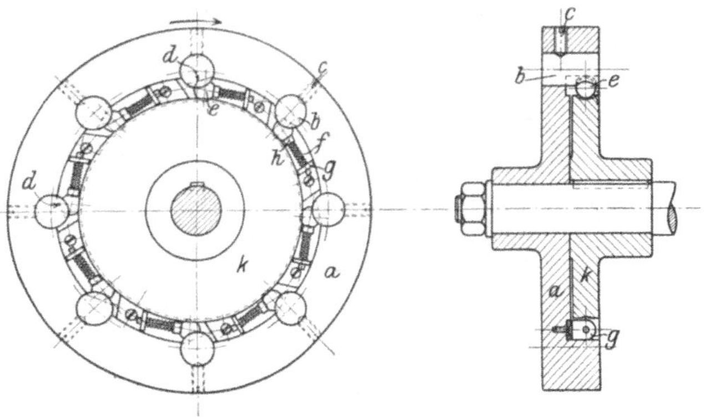

Fig. 614 a.

Federn f sind an kleinen Böckchen g, die an der Mitnehmerscheibe a verschraubt sind, befestigt und drücken mit einem kugelig ausgearbeiteten Druckstück h gegen die Kugel. Wenn nun durch die Bewegung der Schubstange die Mitnehmerscheibe a eine Schaltung in der Pfeilrichtung erhält, so werden durch den Eigenwiderstand des Walzenapparates die Kugeln, die sich in den zwischen b und der Schaltscheibe k entstehenden Keilraum einlegen, relativ gegen die Scheibe in der Pfeilrichtung unter Zusammendrückung der Federn bewegt, bis diese Kraft groß genug ist bzw. die Federn vollständig zusammengedrückt sind, so daß die Schaltscheibe k mitgenommen wird. Diese Scheibe ist auf der Achse der Walze aufgekeilt. Beim Rückgang der Mitnehmerscheibe a treten die Kugeln nach Entlastung der Federn sofort aus den Keilräumen heraus, da durch geeignete Bemessung der Winkel eine Klemmung nicht stattfinden kann.

2. Zuführungen für vorgearbeitetes Material.

Vorgearbeitetes Material kann entweder in der Form der nächsten Presse zugebracht werden, daß ein ganzer Teil, z. B. ein Ankerblech, ausgeschnitten und gelocht, zum Nuten weiterverarbeitet werden soll, oder es können die Blanketts, deren jedes einen für sich weiter zu verarbeitenden Teil vorstellt, an die nächste Presse gebracht werden.

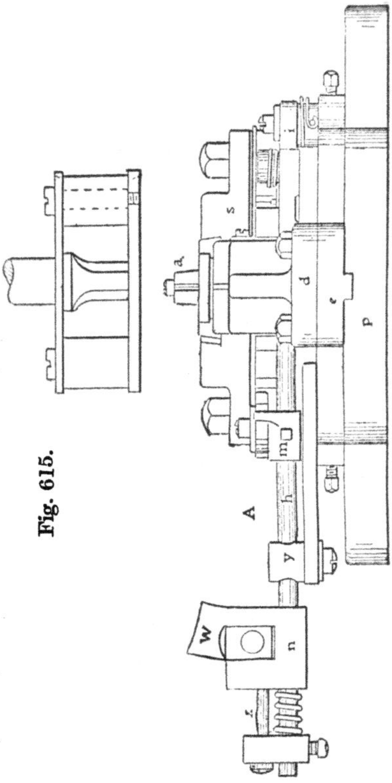

Fig. 615.

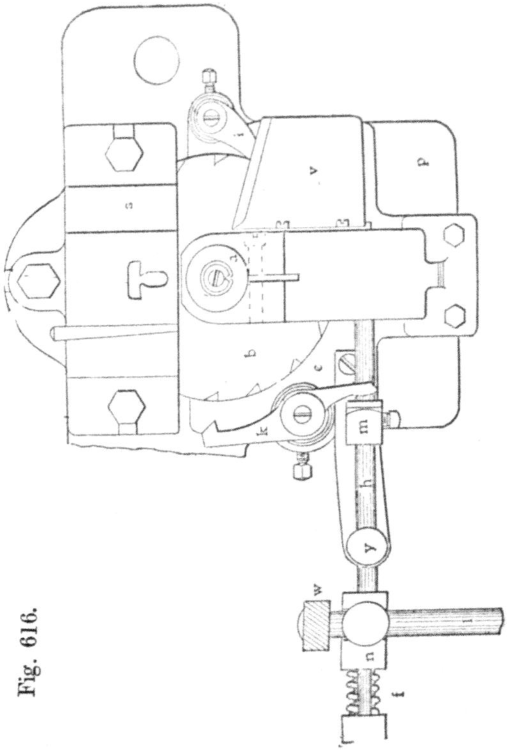

Fig. 616.

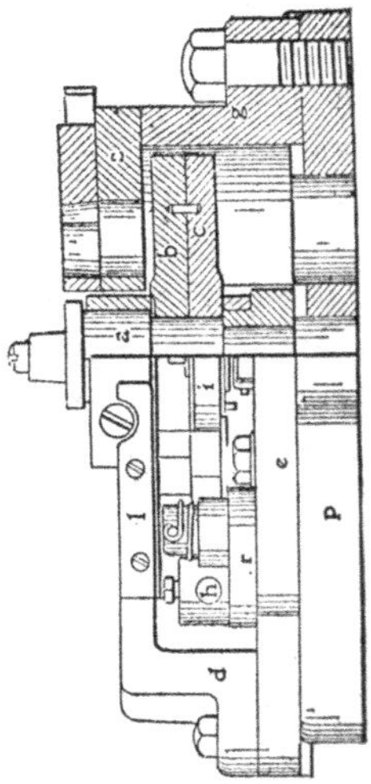

Die für beide Zuführungen verwendeten kinematischen Glieder haben viel Ähnlichkeit, wenn auch die Zuführung der Teile an die Apparate verschieden ist.

a) Zuführung von Teilen für Mehrfacharbeit.

Besondere Einrichtung zum Nuten von kleinen Ankerscheiben. Die Zeichnungen Fig. 615—618 geben ein in tatsächlichem Gebrauche stehendes Werkzeug wieder, das in jeder Presse mit genügend Zwischenraum zwischen Stößel und Aufspannplatte zu verwenden ist. Es kann leicht und schnell auf jeden Scheibendurchmesser zwischen 76 und 356 mm eingestellt werden und kostet nach Schätzung der amerikanischen Quelle rund 325 M. Die Einrichtung stammt aus einer Werkstätte, die verschiedene elektrische Apparate, darunter zwölf verschiedene Anker, baut, von denen keine Größe in Gruppen von über 50 Stück auf einmal hergestellt wird. Die Schnittplatten sind aus Schnellstahl hergestellt, 51 mm breit und 13 mm dick. Man kann sich leicht ohne weitere Erklärung irgendeine Übertragung der Bewegung des Stößels bzw. Schlittens auf die Schaltwelle vorstellen. Auf der Rückseite des oberen Endes des im Schnitt gezeichneten langen Hebels W, Fig. 616, befindet sich ein Zapfen mit einer Rolle, die durch eine Wurstfeder gegen die schräge Kante einer Führungsschiene gedrückt wird. Diese Führungsschiene ist ungefähr 10 mm dick und wird unter den Klemmuttern des Schlittens gehalten; sie ist mit Langlöchern zur Nachstellung versehen. Mit dem Auf- und Abwärtsgang des Schlittens wird so dem Hebel W eine schwingende Bewegung gegeben.

In Zeichnung Fig. 616 ist ein Grundriß, eine Vorderansicht und eine Seitenansicht und in Fig. 617 ein Schnitt unterhalb A mit dem Teil hinter der Achse a im Schnitt gegeben. Diese ist am unteren Ende in der Platte e und mit ihrem oberen Ende im Arm d gelagert. Sie trägt zwei Radscheiben, b und c, die durch Prisonstifte verbunden sich gleichzeitig drehen, Fig. 618. Das obere Rad ist auf der Achse aufgekeilt; die beiden Räder tragen entgegengesetzt gerichtete Sperrzähne, wie in Fig. 617 ersichtlich ist. Die obere Scheibe b dient

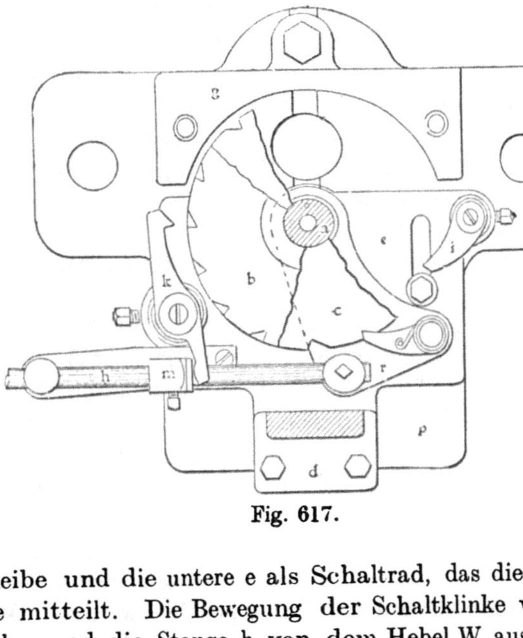

Fig. 617.

als Teilscheibe und die untere e als Schaltrad, das die Bewegung der Teilscheibe mitteilt. Die Bewegung der Schaltklinke wird durch den Schalthebel r und die Stange h von dem Hebel W aus, der teilweise im Schnitt in Fig. 616 angedeutet ist, besorgt. Der Schalthebel r dreht sich nicht unmittelbar auf der Achse a, sondern auf einem konzentrischen Putzen der Platte e, die das Halslager umgibt. Diese Platte trägt auch die Sperrklinke i, die eine Rückbewegung der Schaltscheibe verhindert, und den Sperrhaken k, der mit Nuten der Teilscheibe im Eingriff steht. In diese wird er gewöhnlich durch eine Feder eingedrückt und aus ihnen durch einen einstellbaren Stein m herausgezogen, wenn dieser beim Rückwärtsgang der Stange h mit der rückwärtigen Verlängerung des Sperrhakens k in Berührung kommt und ihn soweit zurückzieht, daß er die Zähne freigibt. Diese Bewegung dauert jedoch an, so daß sich beim nächsten Vorwärtsgang der Stange h das Rad ein wenig weiter nach rechts dreht, bevor der Sperrhaken wieder einklinkt und die Feder ihn in die nächste ankommende Nut eindrücken kann. Die radiale Fläche der Nut am Rad wird gegen den Sperrhaken mit Druck vorgeschoben; um dies sicher zu erreichen und immer gleichen Druck zu haben, gibt man der Teilscheibe dadurch eine geringe Überschaltung, daß man die Führungsplatte über dem Schlitten so einstellt, daß die Stange h eine etwas größere Bewegung erhält als notwendig, um die Scheibe eine Teilung weiter zu schalten, und dieses Mehr durch die Wurstfeder f aufnehmen läßt. Zu dem Zwecke kann der Block n auf der Stange gleiten und wird in seiner senkrechten Stellung durch den Stift x erhalten, der fest in n ist und durch den stehenden Block links leicht gleiten kann. Der Stein n ist durch ein Universalgelenk mittels des langen Bolzens l mit dem Hebel W verbunden.

Die Schnittplatte ist mit einem Keil in der stählernen Gesenkplatte s einstellbar auf der Platte g, Fig. 617, befestigt, die wiederum auf der Grundplatte p unverrückbar befestigt ist. Es ist zu bemerken, daß die ganze Schaltung auf der Platte e aufgebaut ist, die mit einer Feder in eine Nut der Grundplatte p paßt und mit einem Langloch für die Befestigungsschraube versehen ist, so daß man die ganze Vorrichtung nach dem Durchmesser der zu nutenden Ankerscheibe einstellen kann.

Fig. 618 ist ein Schnitt durch die Achse a. Der Stift o paßt in eine radiale Keilnut in der oberen Scheibe. Der Stift t paßt leicht in ein radiales Loch und kann durch Drehen der senkrechten Stellschraube nach außen gedrückt werden, wodurch das Rad mit der Achse fest verbunden wird. Das obere Ende der Achse ist ausgebohrt für die Zapfen der gehärteten Stahlfutter, die für die verschiedenen Bohrungen der zu nutenden Ankerscheiben auf Lager gehalten werden. Diese werden durch eine axiale Schraube und Prisonstift in ihrer Stellung gehalten.

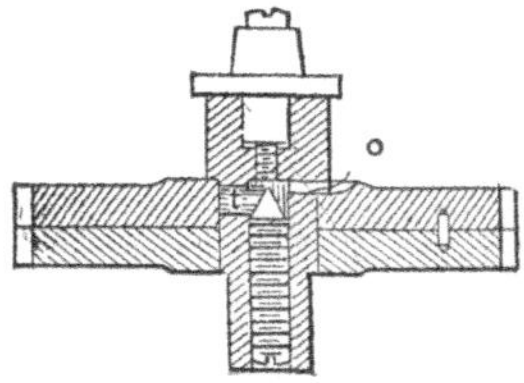

Fig. 618.

Um Öl in das untere Halslager zu bringen, muß von oben ein Loch gebohrt werden, das an dem unteren Absatz an einer Stelle heraustritt, wo es nicht mit dem Keil und Befestigungsstift zusammenkommt.

Der Stempel, Fig. 615, ist mit einem Abstreifer versehen und hat zwei Gummipuffer als Federn.

Ist der Ring für Größen über 150 mm im Durchmesser eingestellt, so fallen die Abfälle unmittelbar durch, bei den kleineren Durchmessern hingegen fallen sie auf die Teilscheibe und werden mitgenommen, bis sie von der Messerplatte v, Fig. 616, abgestreift werden, von wo man sie abbürsten kann.

Man wird bemerken, daß das Loch in der Schnittplatte T-Form besitzt, damit man viereckige oder unregelmäßig geformte Bleche, die bereits mit der Bohrung versehen sind, gleichzeitig nuten und abschneiden kann. Die Abschnitte können mit einem Holzstäbchen abgestrichen werden, ohne daß man die Presse abstellen muß.

Eine ganz ähnliche Konstruktion findet sich beim Revolverteller der Presse von Hiltmann & Lorenz [1]). Die Stelle der mit dem

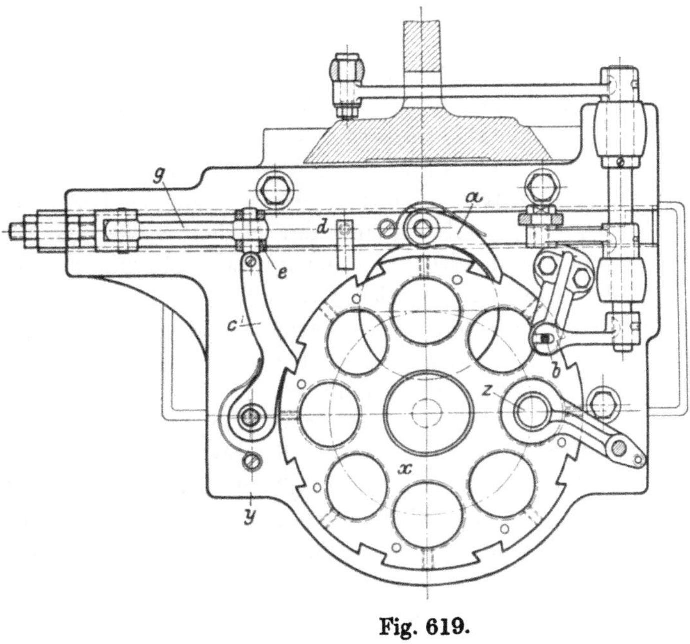

Fig. 619.

Sperrad gekuppelten Teilscheibe der eben besprochenen Konstruktion vertritt hier der Revolverteller x, Fig. 619, der durch die Schalt- und

[1]) W.T. 1910 S. 407.

Sperrhaken a und c weitergeschaltet bzw. abgefangen wird. Die genaue Einstellung erfolgt hier jedoch durch einen Stift b an Stelle der dritten Sperrklinke der vorher besprochenen Konstruktion. Die Bewegung der Klinke a erfolgt gleich jener von r in Fig. 616 durch ein Gestänge, das jedoch in diesem Fall mit einer Sicherung, Fig. 620, gegen Bruch bei falscher Schaltung ausgerüstet ist. Der Hebel e wird nämlich von einer federnd nachgiebigen Stange a b betätigt. Da der Drehpunkt der beiden Teile a und b um die Strecke x (2 bis 5 mm) aus der Stoßlinie liegt, wird bei Überschreitung einer gewissen Druckkraft die Feder c in der Pfeilrichtung so weit durchgebogen, daß beim Weitergang der Presse die Schubstange einfach zusammenknickt, wenn der Revolverteller an der Weiterdrehung verhindert ist.

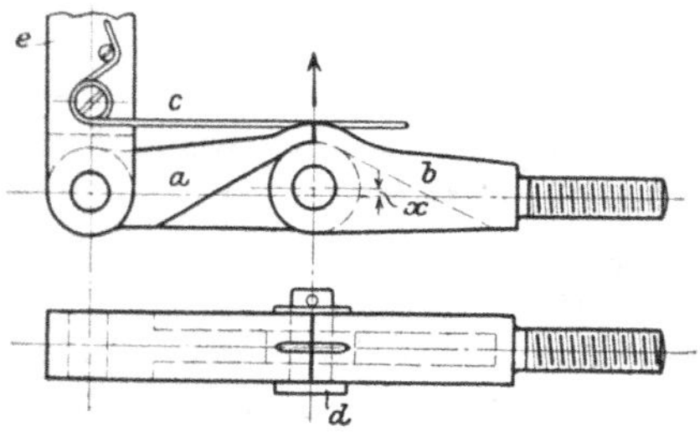

Fig. 620.

Teilapparat zum Nuten von Ankerscheiben. Fig. 622 stellt einen Teilapparat dar, der beim Nuten von Ankerscheiben Verwendung findet. In diesem Werkzeug werden bloß die Nuten ausgestanzt, da die vorhergehenden Arbeiten, nämlich die Herstellung des Loches für die Bohrung und der Keilnut sowie das Ausschneiden des äußeren Durchmessers in einem großen Rundschnitt erledigt ist, bevor die Scheibe in dieses Werkzeug kommt, vgl. Fig. 621. Eine nähere Beschreibung dieser Arbeiten kann hier fortfallen.

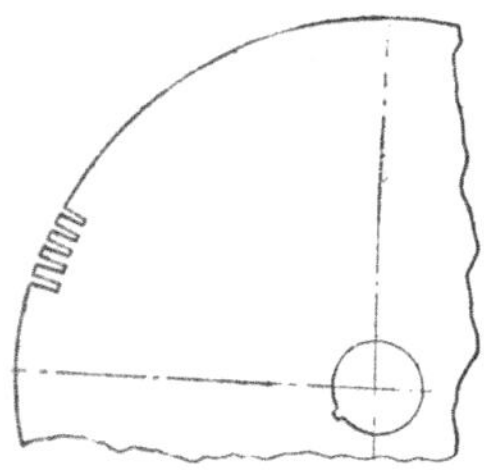
Fig. 621.

In Fig. 622 ist A die Schnittplatte aus Werkzeugstahl mit fünf Nuten, die gebohrt und auf Maß gefeilt werden und deren Seiten einen Hinterschliff von $\frac{1}{2}^0$ erhalten, damit die Abfälle leicht durchfallen können. Diese Platte wird gehärtet und oben und unten genau geschliffen. B ist die gußeiserne Grundplatte, die gehobelt und der Zeichnung entsprechend angepaßt wird. C, die Teilscheibe, ist aus weichem Stahl, passend gedreht und mit einem Loch mit Gewinde zur Aufnahme der in Fig. 623 ersichtlichen Bronzebüchse H versehen. In diesem Fall hat die Ankerscheibe fünfundsiebzig Nuten; da die Schnittplatte fünf Löcher enthält, muß die Teilscheibe 15 Nuten erhalten. Die Mitte der Platte ist entsprechend dem Loch in der Ankerscheibe gedreht und erhält einen Keil, der in die Keilnut der

Ankerscheibe paßt. Der Arm D aus Flußeisen, der die Teilscheibe trägt, wird mittels Schrauben und Prisonstiften an der gußeisernen Grundplatte B befestigt. Er ist ebenfalls ausgebohrt und mit Gewinde versehen, um den in Fig. 623 dargestellten Drehzapfen aufzunehmen. Außerdem ist er auf ungefähr ein Drittel seiner Länge durch

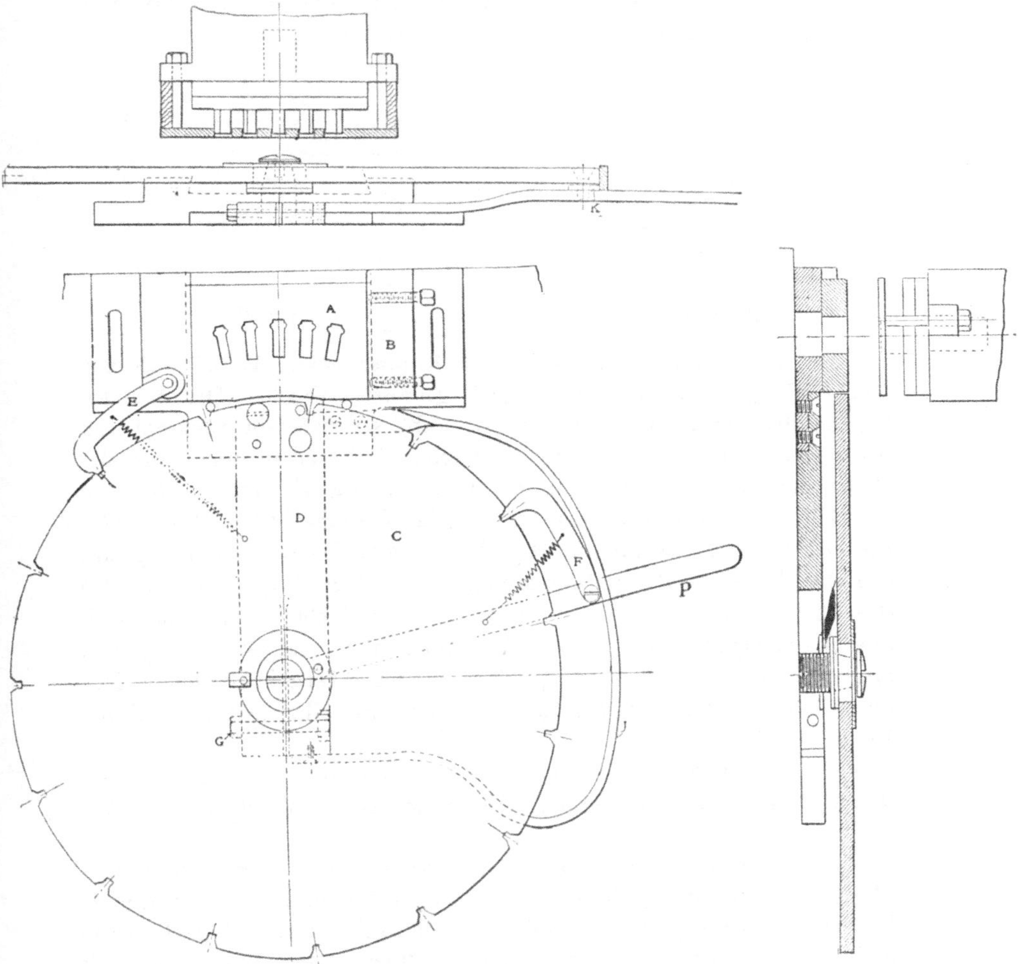

Fig. 622.

einen Sägeschnitt geschlitzt und mit einer Klemmschraube G für den Drehzapfen I versehen. Der Hebel E arbeitet als Sperrklinke; seine Zugfeder ist stark genug, um ihn in der Nut zu halten und so die Teilscheibe in ihrer Stellung zu sichern. P ist der Handhebel für die Schaltung und J seine Tragschiene.

Die Arbeitsweise ist folgende: Die Ankerscheibe, Fig. 621, wird in richtiger Stellung auf der Teilscheibe C, Fig. 622, befestigt und die fünf ersten Nuten gestanzt. Ein langsamer Zug am Hebel P gibt dann die Sperrklinke E frei und dreht durch die Schaltklinke F die Platte, bis die Sperrklinke E in die nächste Nut einfällt, worauf die nächsten fünf Löcher gestanzt werden können. Der Vorgang wiederholt sich, bis der ganze Kreis vollendet ist.

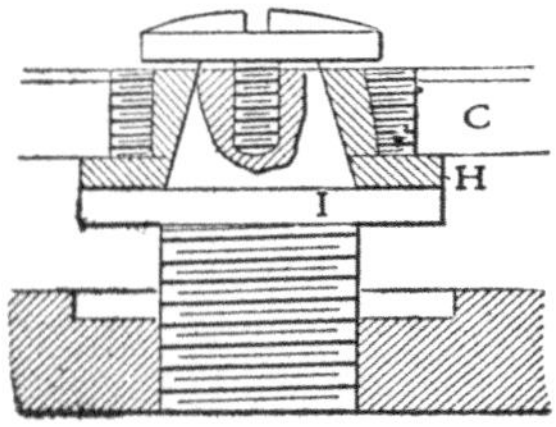

Fig. 623.

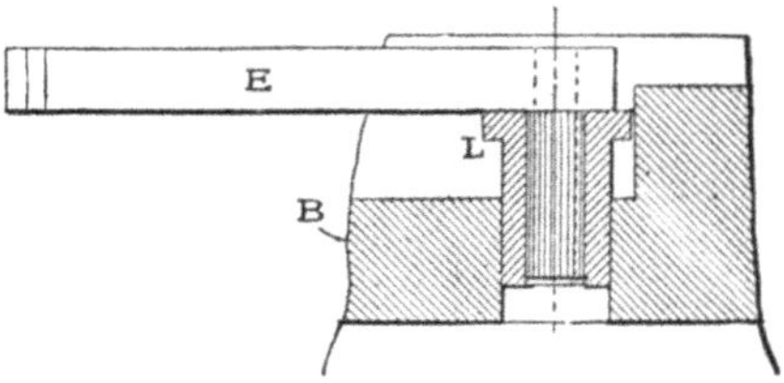

Fig. 624.

Es sei bemerkt, daß in dieser Konstruktion Vorkehrung zur Nachstellung der verschiedenen Teile getroffen wurde. Die Arbeitskanten der Schnittplatte nützen sich ab, so daß ein Nachschleifen ihrer Oberfläche nötig wird. Dadurch wird sie niedriger als die Teilscheibe C und auch niedriger als die Hebel E und F. Um nun die Teilscheibe C nachzustellen, löst man die Klemmschraube G, Fig. 622, und dreht den Drehzapfen I, Fig. 623, so lange, bis die Teilscheibe und die Oberfläche der Schnittplatte in gleicher Höhe liegen, worauf man die Klemmschraube G wieder anzieht. Der Hebel E wird dadurch gesenkt, daß man die Büchse L in die gußeiserne Grundplatte B, Fig. 624, eintreibt, bis sie die richtige Stellung zur Teilscheibe hat. Der Schalthaken F wird von P abgenommen, um ihn in seine richtige Lage zu bringen.

Diese Werkzeuge, die aus England stammen, werden in verschiedenen Größen bis zu einem Scheibenhalbmesser von 508 mm gemacht.

Schnellaufende selbsttätige Nutenmaschine (L. Schuler, Göppingen). D. R. P. angemeldet. Bei diesen mit bis zu 400 minutl. Umdrehungen laufenden Exzenterpressen, Fig. 625—628, mit selbsttätigem Vorschub und selbsttätiger Ausrückung nach Fertigstellung eines Bleches handelt es sich darum, die Geschwindigkeiten den bewegten Massen derart anzupassen, daß die Beschleunigungs- und Verzögerungskräfte nicht allzugroß werden. Vor allem wurde für den Vorschub an Stelle des Kurbelantriebes eine offene Hubkurve[1]) verwendet. Der aufsteigende Teil der Kurve wird zum Vorschalten ausgenützt und beträgt nicht, wie bei

[1]) Vgl. dazu Fig. 641.

dem gewöhnlichen Kurbelantrieb, 180° sondern 210°, wodurch es möglich ist, größere Zeiteinheiten für die zu beschleunigenden und zu verzögernden Hubgestänge des Teilrades, sowie für die zu stanzenden Bleche zu gewinnen. Auf dem absteigenden Teil der Kurve mit 140° ist nur der Leerweg (Rückweg) des leichten Hubgestänges zu vollziehen. Die Hubrolle des Hubhebels wird mittels einer Feder ständig an die Hubkurve angepreßt, deren Form so gewählt ist, daß die Geschwindigkeiten für den Vorschub in einer viel günstigeren Kurve verlaufen als bei dem Kurbelvorschub. Durch diesen Vorschub wird nicht nur ein unbedingt ruhiger Gang, sondern auch eine genaue Schaltung erreicht. Zwischen dem auf- und absteigenden Teil der Kurve ist noch eine Strecke von 10° Länge übrig. Dieser Teil verläuft konzentrisch, damit alle mit dem Vorschub bewegten Teile während dieser Periode den Beharrungszustand erreichen; dieser Zustand wird noch durch Abbremsen mittels zweiteiliger Ringbremsen sowohl an dem Teilrad als an dem Hubhebel kräftig unterstützt. Wie bekannt, verlangen die Ankerbleche in der Teilung peinlichste Genauigkeit. Um daher während der Stanzarbeit den Vorschub genau zu halten, wird eine Sperrklinke durch eine weitere Kurve gesteuert. Die beiden Kurven sind nun so zueinander gestellt, daß bei Beginn des Rückwärtsganges der Schaltklinke die Sperrklinke in eine Zahnlücke der Teilscheibe eintritt und so die ganze Schalt- und Teilvorrichtung auf das genaueste regelt und vor Beginn der Stanzarbeit sperrt.

Die Hubregelung erfolgt durch Verkürzung oder Verlängerung von zwei Hubhebeln, welche mittels einer Mutter und Klemmringen auf einem gemeinschaftlichen Bolzen eingestellt werden können. Die Ausrückung nach dem Fertigschalten und Stanzen eines Bleches erstreckt sich auf den Stößel mit Werkzeug und auf das Hubgestänge mit Schalt- bzw. Sperrgestänge, während Kurbel, Pleuelkopf und Schwungrad sich ständig bewegen. Der in langen Prismen geführte Stößel ist in der Höhe verstellbar, während sich der Umsteuerhebel für den Verkupplungsdrehschieber zwischen Pleuelstange und Stößel in einer Kulisse bewegt. Diese in wagrechter Richtung geführte Kulisse ist auf einer Seite mit dem Ein- und Ausrückgestänge verbunden, während die entgegengesetzte Seite das Gestänge für die Ausrückung des Hubgestänges trägt. Der Sattelstock, der durch eine Spindel verstellbar ist, ist auf Prismenführungen gelagert und trägt die in langem Halslager geführte Sattelstockachse, auf der zur Befestigung der zu stanzenden Bleche das Klemmplattenpaar (D.R.P.) mit abnehmbarer Oberplatte und Exzenterzugspannung Verwendung findet. Die Unterplatte sitzt mit Doppelkonus und mit Gegenmutter in der Höhe einstellbar auf dem zylindrischen Ansatz der Sattelstockachse.

Das Teilrad ist auf dem konischen Endzapfen der Sattelstockachse aufgepreßt und verschraubt. Der Ausrückhebel, von einem auf dem Teilrad sitzenden Anschlag betätigt, ist auf der vierkantigen Ausrückwelle aufgeklemmt. Derselbe läßt sich daher einem größeren oder

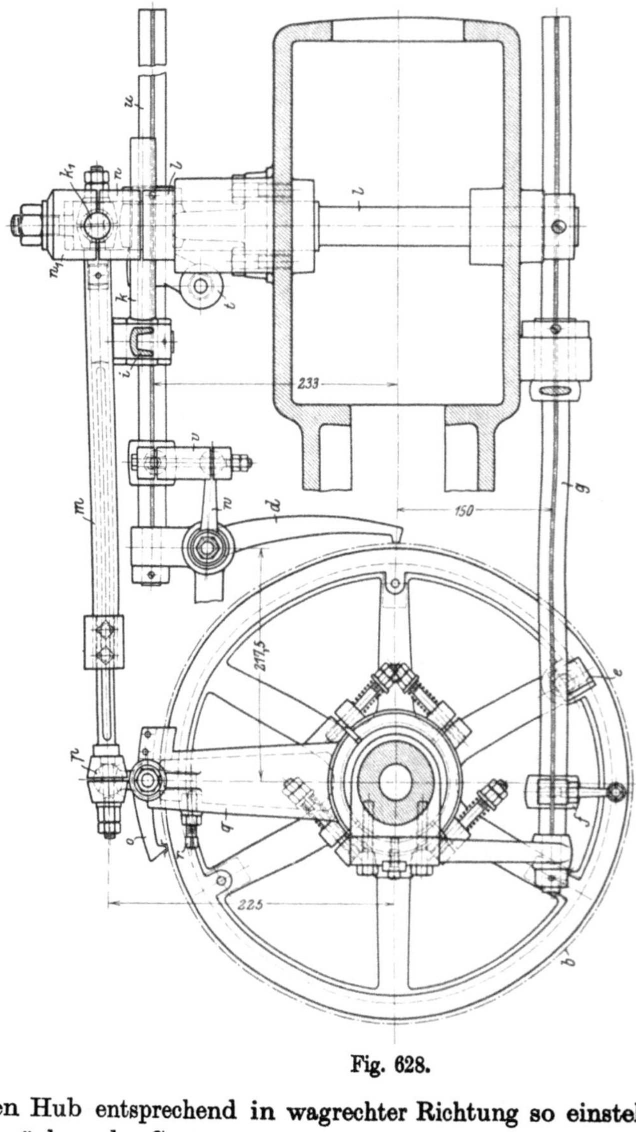

Fig. 628.

kleineren Hub entsprechend in wagrechter Richtung so einstellen, daß das Ausrücken des Gestänges erfolgt, bevor der Stößel seine Höchstlage erreicht hat. Die Teilringe sind in naturhartem Stahl ausgeführt und leicht auswechselbar. Ein gehärteter Einlegekeil bürgt für die

Additional material from *Die Werkzeuge und Arbeitsverfahren der Pressen,*
ISBN 978-3-662-23911-7, is available at http://extras.springer.com

EXTRA
MATERIALS
extras.springer.com

richtige Lage, während das Festspannen des Ringes entsprechend der Maschinengröße mittels 3 oder 4 Spannpratzen erfolgt. Die Maschinen werden am besten unmittelbar mit regelbarem Motor angetrieben, zu welchem Zwecke an der Rückseite des sehr kräftig ausgebildeten Maschinengestelles eine Konsole angebracht werden kann.

Beschreibung der einzelnen Bewegungsvorgänge. In Bewegung befinden sich nur das Schwungrad mit Kurbelwelle und der Pleuelkopf E mit Pleuelkopfführung F. Das Einrücken erfolgt durch Aufwärtsziehen des Handhebels A, der durch die Welle B mit dem Federhebel C in Verbindung steht. Dadurch wird die Feder D in der Zeit, in der der Pleuelkopf E mit Führung F bereits über den oberen Totpunkt hinaus und über den unteren Totpunkt zurück bis zum oberen Totpunkt gelangt, und infolgedessen auch das senkrechte Gestänge G und durch Übertragung der Hebel H, H_1 auch das wagerechte Gestänge J mit dem Führungsstück K gespannt. Ist nun der obere Totpunkt erreicht, so wird die Spannung der Feder von dem Stück K mit der Kulisse auf die Führungsrolle L des Hebels M, welche mit dem Verkupplungsdrehschieber N verbunden ist, übertragen, und der Drehschieber N findet freien Raum zur Verdrehung, um den Stößel O mit der in Bewegung befindlichen Pleuelkopfführung F zu kuppeln. Nach der Verkupplung wird der Handhebel A freigegeben und die Feder D entlastet. Die Momentausrückstange P wird durch die gespannte Feder Q nach vorn gedrückt und dabei das Gestänge G gegen eine Rückwärtsbewegung durch die Stellschraube R abgestützt. Die Rückwärtsbewegung wird durch die Feder S hervorgerufen, deren Spannung ungeführ halb so stark wie die der Feder D ist.

Die Hubrolle des Doppelwinkelhebels U, die den Vorschub betätigt, wird durch die Regulierschraube V so eingestellt, daß sie von dem dauernd umlaufenden Hubexzenter W stets etwas angehoben wird.

Während dieses Anhebens wird der Hebel X etwas frei und durch die Feder Y beiseite gedrückt. Infolgedessen ist nun auch der ausgerückte Zustand des Vorschubgestänges behoben, indem jetzt die Hubrolle des Hebels U sich an der ganzen Kurve abwälzt. Der Hubhebel Z, welcher durch die Kurve a betätigt wird und für die Sperrung bei einer Überschaltung der Teilscheibe b dient, sitzt in ausgerücktem Zustande gestützt durch die Einstellschrauben C auf dem Hubhebel U auf und bleibt somit auch ausgerückt. Da nun die Hubexzenter derart gegeneinandergestellt sind, daß die absteigende Kurve von W gegen die absteigende Kurve von a etwas voreilt, entsteht zwischen der Schraube C und ihrem Auflagepunkt ein solcher Spielraum, daß der Sperrklinke d genügend Zeit zur Verfügung steht, in eine Zahnlücke des Teilrades einzufallen, bevor die Stanzarbeit vor sich geht. Ist eine ganze Runde Stanzarbeit beendet, so drückt der auf dem Teilrad befestigte Anschlag e mit seiner Keilfläche unter den Hebel f.

Die dadurch entstehende Drehbewegung der Vierkantwelle g wird auf den Hebel h übertragen, dieser und mit ihm die Momentausrückungsstange P so weit nach links geschoben, daß das nach unten unter Federdruck stehende Gestänge G freigegeben wird.

Diese Kraftäußerung, hervorgerufen durch die Feder D, überträgt sich weiter durch die Hebel H, H_1 auf das wagrechte Gestänge J, Führungsrolle L, Hebel M auf den Verkupplungsdrehschieber N. Dieser stellt sich somit in der oberen Totlage in die ursprüngliche Vertikallage wieder ein, und der Stößel bleibt stehen, da der Pleuelkopfführung F der freigewordene Leerweg zur Verfügung steht. Inzwischen hat das Hubexzenter W den höchsten Hub erreicht; zwischen der Stellschraube V des Doppelwinkelhebels U und dem mit dem Horizontalgestänge J verbundenen Hebel X entsteht der früher erwähnte kleine Spielraum, und der Hebel tritt vor die Stellschraube, wodurch die Bewegung des Hubgestänges aufgehoben wird. Ebenso kommt das Gestänge für die Sperrung zur Ruhe, da die Schraube C des Hebels Z wieder auf dem Hebel U aufruht.

Der Doppelwinkelhebel U überträgt die Hubbewegung durch das Gestänge i, Fig. 628, und Hebel k auf die drehbar gelagerte Welle l, sowie Hebel k_1 auf das Schubgestänge m. Die Hebel k, k_1 sind für kleineren und größeren Hubeinstellungen verschiebbar angeordnet und werden durch Klemmbüchsen n, n_1 auf der Welle l gehalten. Die Schaltklinke o ist mittels Kugelgelenk p mit dem Schubgestänge m verbunden und ist in dem auf der Nabe des Sattelstockes mit Federdruck abgebremsten Hebel q drehbar angeordnet. Der Schaltklinkenzahn paßt genau in die Zahnlücken der Teilscheibe. Bei der Hubrückwärtsbewegung hebt sich zuerst der Klinkenkopf aus der Zahnlücke, und erst beim Anschlag des nach innen verlängerten Teiles an der Stellschraube r wird der schwingende Hebel q mitgenommen. Ebenso verhält es sich bei der Hubvorwärtsbewegung. Zuerst legt sich der Klinkenkopf zwischen die Zahnflanken, und dann erst erfolgt der eigentliche Vorschub. Das Teilrad wird gleichzeitig durch eine Ringbremse mit regelbarem Federdruck abgebremst. In dem Augenblick, in dem der Schaltkegelkopf bei der Hubrückwärtsbewegung die Zahnlücke verläßt, erfolgt vor Beginn der Stanzarbeit ein genauestes Verriegeln des jeweiligen Vorschubes durch die mit dem Gestänge s t n v w besonders gesteuerte Sperrklinke d.

Ein Teilkopf für eine Lochstanze. In den Zeichnungen, Fig. 629 bis 630, sind verschiedene Ansichten einer Einrichtung für eine wagrechte Lochstanze gegeben, mit der Kesselböden und ähnliche mit Flanschen versehene Teile genau geteilt und gelocht werden können, ohne daß man die Stücke vorher von Hand anreißen muß. Es sei angenommen, daß das zu lochende Stück eine Mittelöffnung habe, durch die ein Bolzen hindurchgesteckt werden kann, der mittels einer darübergesteckten Klammer das Werkstück fest auf der senkrechten Säule h hält. Der Schlitten, der diese Säule trägt, kann mit Hilfe eines vom Handrad auf der Vorderseite der Maschine aus betätigten Zahnstangentriebes in wagrechter Richtung verschoben werden. Die Achse der senkrechten Säule ist so immer

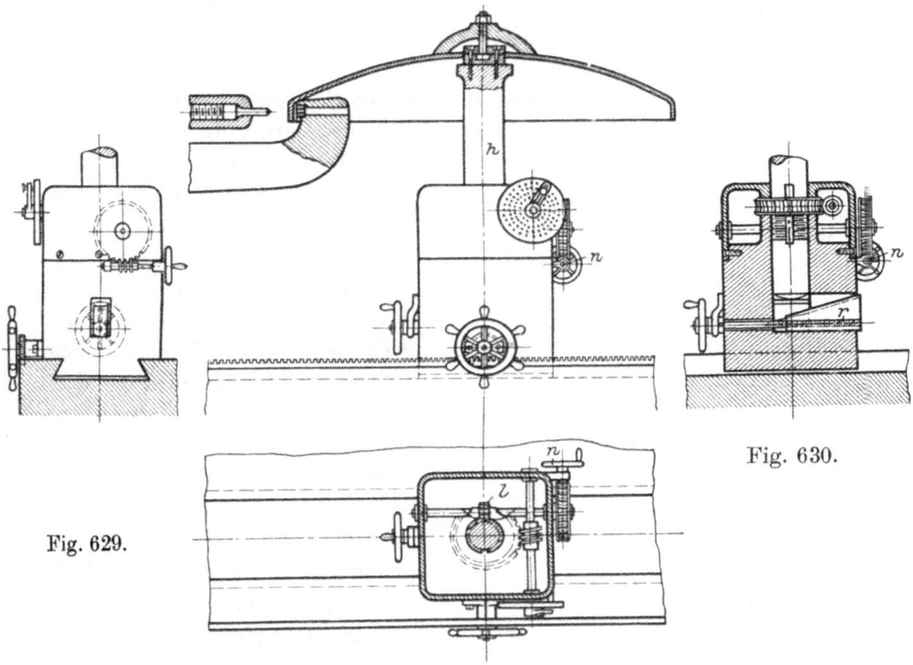

Fig. 630.

Fig. 629.

in einer Linie mit dem Lochstempel. Außerdem sind in die Säule Zähne eingeschnitten, die in Verbindung mit dem Zahnrad l, welches durch eine Schneckenradübersetzung von der Handkurbel n aus betätigt wird, das Heben und Senken der Säule bewirken. Wenn die Säule und das Arbeitsstück in der Höhe richtig eingestellt sind, wird das Gewicht der Säule von dem Keile r, der durch Schraube und Handkurbel links, Fig. 630, vorgeschoben wird, aufgenommen. Die Teilscheibe in Verbindung mit dem Schneckenradantrieb bewirken die richtige Rundschaltung des Stückes, entsprechend der Zahl der zu stanzenden Löcher. Das Schneckenrad ist mit Keil und Langnut befestigt, damit die Auf- und Abwärtsbewegung der Säule nicht gehindert ist.

Diese Einrichtung läßt sich natürlich in gleicher Weise an senkrechten Stanzen zum Teilen und Stanzen flacher kreisrunder Scheiben usw., wie für Bohrmaschinen verwenden.

b) Zuführung von Teilen für Einzelarbeit.

Von den Zuführungen für einzelne Blanketts kommen entweder Schieber in Schlittenform oder Revolverteller in Betracht. Von den ersten sind bei Besprechung der Folgewerkzeuge bereits verschiedene Bauarten, die unmittelbar mit dem Werkzeug zusammenhängen, angeführt worden.

Vor der Beschreibung einiger Sonderbauarten wird es vielleicht von Interesse sein, eine Einführungsvorrichtung, die also mehr Führung als Zuführung ist, zu besprechen.

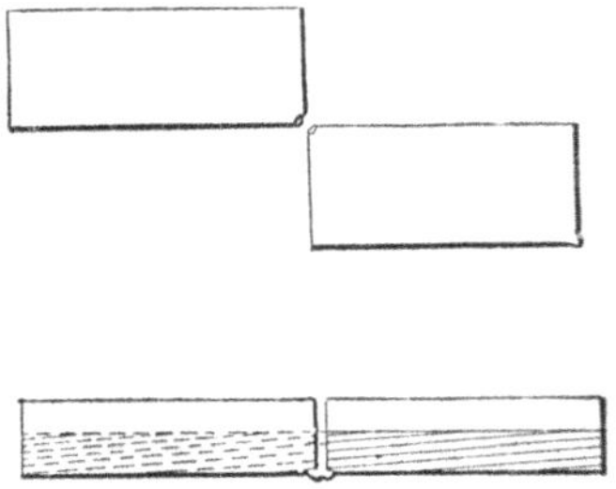

Fig. 631.

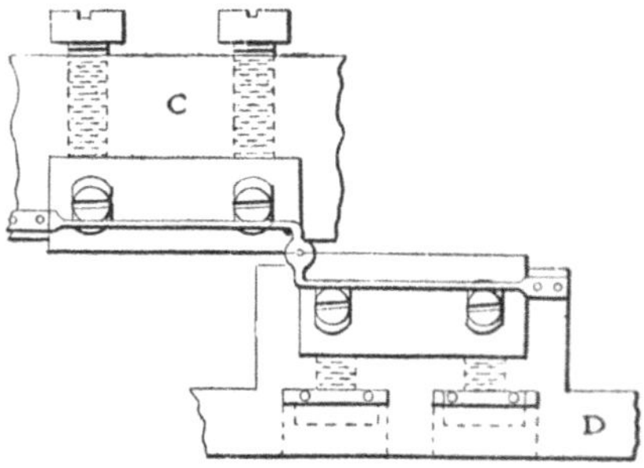

Fig. 632.

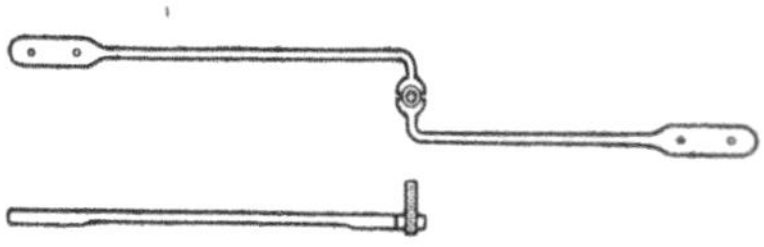

Fig. 633.

Einführungsvorrichtung für Schraubenbolzen in eine Gewindedrückmaschine. Die Fig. 631—633 stellen eine Vorrichtung dar, mittels der kleine Schraubenblanketts beim Eintritt in die Gewindedrückwerkzeuge kontrolliert werden können. Diese Schrauben sind im ganzen ungefähr 4 mm lang und aus Golddraht von 1 mm Durchmesser mit 90 Gängen auf den Zoll hergestellt. Bei der Arbeit ergab sich der Übelstand, daß die Schrauben sich infolge ihrer geringen Länge beim Eintritt in die Werkzeuge verdrehten und so die Maschine zum Stillstand brachten.

Fig. 631 gibt eine Ansicht und einen Grundriß der Werkzeuge ohne die Vorrichtung, während Fig. 632 eine Ansicht des Werkzeuges mit der Einführungsvorrichtung in ihrer richtigen Stellung zeigt. Die Federn sind nach Fig. 633 hergestellt und die eine an dem Schlitten C, die andere an dem Arm D, wo die Werkzeuge befestigt sind, angebracht. Das Schraubenblankett wird durch den Plunger zwischen die beiden Federn eingeschoben, die dann das äußere Ende des Bolzens umschließen, wie im Grundriß sichtbar ist.

Die Vorrichtung ist gut verwendbar und kann für ähnliche Arbeiten benutzt werden.

Selbsttätige Schlittenzuführung für ein Werkzeug zum Prägen von Medaillen. Auf S. 225—29 ist ein Prägewerkzeug für Medaillen beschrieben worden, dem hier die Beschreibung der selbsttätigen Schlittenzuführung folgt.

Fig. 634 gibt eine Vorderansicht der oberen und unteren Gesenkplatte und einen Grundriß der letzten mit den Werkzeugen in Arbeitsstellung beim Prägen eines Blanketts; gleichzeitig sind die Anschlag- und Abstreifplatten, ein Abstreifanschlag mit seinem Antrieb (nur auf einer Seite) und die Kugellagereinrichtung zur Betätigung des Zahnstangenvorschubes dargestellt.

Wie bekannt, ist die Arbeit der Prägepresse verschieden von der einer gewöhnlichen Exzenterpresse, in welcher der untere Werkzeugteil fest steht und der Stempel oder die Matrize, welche am Stößel befestigt sind, nach abwärts gehen, während in der Prägepresse das obere Werkzeug festbleibt und das untere nach oben geht.

Die Gesenkplatten sind, wie bereits erklärt, aus Stahlguß, nach der Zeichnung bearbeitet und mit gehärteten und geschliffenen Stahldruckstücken A, vgl. Fig. 291, für die Werkzeuge, Fig. 635, versehen, die fest von der Oberseite aus eingetrieben werden. Die Werkzeuge, Abstreifer und Anschlagplatten sind bereits an obiger Stelle beschrieben worden.

Die zwei Abstreifanschläge, Fig. 634 oben, — einer derselben ist bei B gezeichnet — sind aus zwei Stücken hergestellt; es sind Schmiedestücke aus Werkzeugstahl, die in einem Augenlager C — auch aus Werkzeugstahl — gehalten werden, das in die Gesenkplatte eingelassen und mittels zweier versenkter Schrauben gehalten wird. Die beiden Teile der Anschläge, die gut aneinander gepaßt sind, werden außen durch eine Feder verbunden, so daß der Anschlag B, wenn er beim Aufwärtsgang der Presse mit der Nase D in Berührung kommt, gerade weit genug öffnet, um an der Nase vorbei zu gleiten, worauf er, wenn die Presse am Ende des Aufwärtsganges angelangt ist, wieder in die gezeichnete Stellung zurückschnellt; beim Abwärtsgang faßt der Anschlag die innere Ecke der Nase und drückt so den Abstreifer und die Anschlagplatte auf den Absatz der Matrize nieder, damit das geprägte Medaillon, wie auch bereits erwähnt, durch einen starken Luftstrom weggeblasen werden kann. Das innere Ende des Anschlages paßt in eine Nut des Abstreifers und ist gehärtet. Sein äußeres Ende

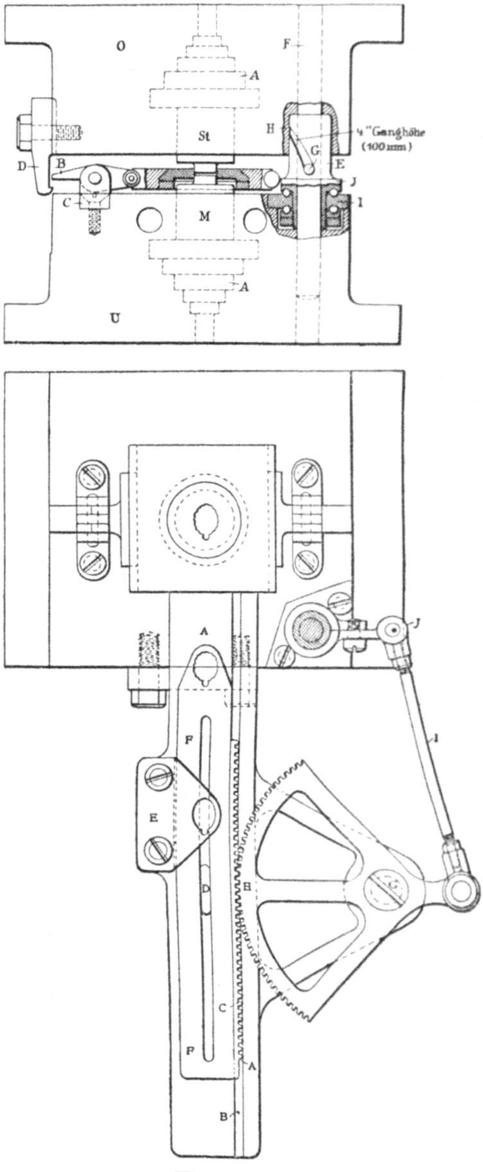

Fig. 634.

ist an der Stelle, wo er mit der Nase in Berührung kommt, ebenfalls gehärtet.

Die Büchse E, welche die Zuführungseinrichtung betätigt, ist aus Werkzeugstahl hergestellt, innen in der Mitte etwas größer ausgebohrt, so daß sie sich nur ungefähr 20 mm an beiden Enden führt, und außen wie gezeichnet abgedreht. Der untere Absatz ist mit einer Ringnut für das Kugellager versehen; der anschließende Zapfen erhält Gewinde — 24 Gänge auf den Zoll —, auf das sich die Einstellmuttern, unterhalb I, schrauben, von denen die obere mit einer Laufrinne für das Kugellager versehen und gehärtet ist. In diese Büchse wird dann eine Spiralnut H eingefräst, die breit genug ist, um die Rolle G, die auf einem Stift sitzt, aufzunehmen, wobei die Steigung der Nut gleich einem Viertel Umdrehung auf 1 Zoll oder entsprechend dem vollen Preßhub ist. Nachdem die Büchse in Walratöl gehärtet worden ist, wird sie poliert, damit sie leicht auf dem ebenfalls gehärteten Stift F gleitet, der fast ganz durch beide Gesenkplatten hindurchgeht und in der oberen durch eine Preßschraube, die in der Zeichnung nicht ersichtlich ist, gehalten wird.

Die Platte I ist auch aus Werkzeugstahl gemacht und gehärtet;

sie wird mittels zweier versenkter Schrauben an der Gesenkplatte befestigt. Beide Seiten dieser Platte erhalten eine Ringnut für das Kugellager J.

Bei der erstmaligen Herstellung der Büchse E konnte man kein befriedigendes Arbeiten erzielen, man versuchte alles mögliche, Vulkanfiber usw., um die Reibung auf der Platte I während der notwendigen Viertelumdrehung zu überwinden. Nachdem mehrere Platten gebrochen waren, setzte man versuchsweise ein Kugellager ein, das zur vollsten Zufriedenheit arbeitete.

Der Grundriß der unteren Gesenkplatte, Fig. 634 unten, mit allen ihren Teilen zeigt, wie die Schlitteneinrichtung sich darstellen würde, wenn die Presse den halben Hub nach oben zurückgelegt hätte.

Der Arm A ist aus Gußeisen und an seiner Oberseite wie an der Fläche, mit der er an der Gesenkplatte befestigt ist, gehobelt. Die Nut B bildet eine Führung für die Zahnstange C und verhindert eine Klemmung und Reibung und infolgedessen übermäßige Abnützung des Anschlages D, der gerade hinter dem Einführtrichter E angebracht ist; der Anschlag kann im Falle einer Störung unter die Oberfläche des Schlittens niedergelassen werden und der Schlitten F leicht entfernt werden, indem man auf einen Federhebel unterhalb des Armes drückt, ohne daß man dabei den Einführtrichter entfernen muß. Die Auflagerfläche des Putzens für das Segment H ist gefräst, gebohrt und mit Gewinde für die große Kopfschraube G, die Drehachse für H, versehen. Der ganze Arm ist mit zwei Kopfschrauben an der Gesenkplatte befestigt und hat außerdem zwei mit Gewinde versehene Löcher für die Befestigungsschrauben des Einführtrichters.

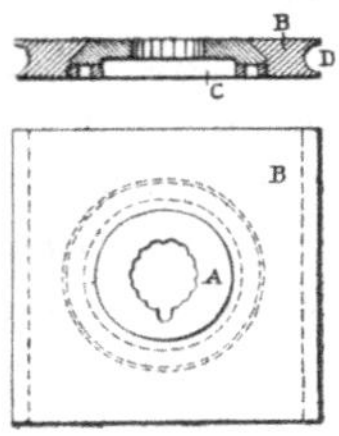

Fig. 635.

Das Segment ist aus Messing hergestellt, auf den richtigen Radius gedreht und mit einer Zahnteilung, Mod. 2, versehen. In das Ende, das mit der Zugstange I verbunden ist, ist von unten ein Bolzen eingeschraubt, der ungefähr 12 mm über die obere Stirnfläche heraussteht und über den das Auge der Zugstange geschoben wird, wonach man das Segment leicht loskuppeln und einstellen kann, ohne die ganze Vorrichtung in Stücke zu zerlegen.

Der geteilte Ring bei J, der an der Büchse befestigt ist und sie mit der Zuführung verbindet, kann auf der Büchse eingestellt und in jeder Stellung festgeklemmt werden.

Zuführungsplatte und Zahnstange werden aus Messing hergestellt, die erste, ein wenig dünner als das Blankett, ist an die Zahnstange angenietet, die auf der ganzen Länge einen eingefrästen Schlitz besitzt, der die Platte aufnimmt und gerade in gleiche Höhe mit der Oberfläche des Armes bringt. Der Schlitz, der durch die Mitte und fast über die ganze Länge der Platte geschnitten ist, begrenzt die Bewegung der Zahnstange an beiden Hubenden und ist von großem Vorteil, wenn der Einführtrichter in seine richtige Stellung eingestellt werden soll.

Der Einführtrichter ist ein Messingrohr von dem gleichen Querschnitt wie das Medaillon und im ganzen ungefähr 125 mm hoch. Ein Schlitz, der an der Spitze ungefähr 12 mm breit ist, erleichtert das Füllen der Röhre mit den zu prägenden Blanketts und verengt sich allmählich auf dieselbe Größe wie das Auge des Medaillons, damit das Blankett auch sicher in das Loch der Zuführungsplatte bei deren Rückbewegung fällt. Diese Röhre ist an eine Messingplatte von ungefähr 3,5 mm Dicke angelötet, die wiederum an den sogenannten Fuß, auch aus Messing und ungefähr 10 mm dick, angelötet und genietet ist, der dann mit zwei Schrauben an dem Arm befestigt wird.

Die ganze Tätigkeit des Arbeiters beschränkt sich auf das Bedienen des Fußtrittes der Presse und Füllen des Trichters mit Blanketts, da diese ununterbrochen von dem Zuführschlitten entfernt werden.

Die Zuführung wird durch die Spiralnut in der Büchse E, Fig. 634 oben, folgendermaßen geregelt: Bei der ersten Bewegung der Presse, die, wie gesagt, nach oben vor sich geht, veranlaßt die Rolle in der Büchse diese und den geteilten Ring, sich nach links zu drehen, so daß die Verbindungsstange I, Fig. 634 unten, das Segment dreht und dadurch den Zuführschlitten gegen den Einführtrichter vorschiebt; am Ende des Aufwärtshubes hat die Rolle G den Grund der Spiralnut in der Büchse E erreicht, und das Loch in dem Zuführschlitten ist unmittelbar unter dem Einführtrichter E, Fig. 634 unten, angekommen, so daß ein Blankett hineinfallen kann. Beim Niedergang der Presse arbeiten alle Teile in umgekehrter Richtung, der Zuführschlitten und das Blankett werden unter das Gesenk befördert, so daß das Blankett beim Hubende unmittelbar über der Anschlagplatte zu stehen kommt und durch sie auf die Oberfläche des Gesenkes fällt. Das Loch in der Zuführungsplatte befindet sich, wenn die Presse in Ruhestellung ist, immer in dieser Stellung. Während des Aufwärtshubes wird der Zuführschlitten gegen den Trichter zu bewegt, und die Abstreifanschläge B, wie bereits beschrieben, gleiten federnd an den Nasen D vorbei und schnellen in ihre ursprüngliche Stellung zurück; beim Abwärtsgang kommen die Anschläge mit den Nasen in Eingriff, wodurch der Abstreifer und die Anschlagplatte unter die Stirnfläche des Gesenkes niedergedrückt werden, das Blankett ausgeworfen und die Abstreifanschläge und -platte wieder in ihre ursprüngliche Stellung, Fig. 634 oben, gebracht werden. Alle Teile sind so genau gegeneinander ausgerichtet und eingestellt, daß sie allerdings erst nach langer Mühe und vielen Versuchen gut miteinander arbeiteten.

Revolverapparate. Die Revolverapparate werden in zwei Formen verwendet, je nachdem sie fest mit der für verschiedene Arbeiten verwendeten Presse verbunden sind, oder ob sie für Sonderarbeiten an Sondermaschinen angebracht sind.

Im ersten Falle handelt es sich um den gewöhnlichen Revolverteller, der mit verschieden großen Einlagen versehen wird, entsprechend der Größe der eben zu verarbeitenden Teile. Die Grundlagen aller von den verschiedenen Fabrikanten gebauten Formen sind dieselben; sie unterscheiden sich nur in konstruktiven Einzelheiten, die eine mehr oder weniger genaue Schaltung, schnellere Hubzahl usw. zulassen. Eine Ausführungsform ist in Fig. 619 gegeben, eine andere in Verbindung mit einer besonderen Presse in Fig. 646. Betreffs dieser Formen und der kinematischen Behandlung sei wieder auf die jüngste Literatur verwiesen, wo gerade dieses Gebiet sehr eingehend behandelt worden ist.

Eine von dem gebräuchlichen Bilde vollständig abweichende Ausführung soll zeigen, wie weit sich Bestrebungen der Praxis von der Einfachheit entfernen können, um einen zweifelhaften Vorteil zu erzielen. In Fig. 636 ist ein einstellbarer Revolverteller der Firma Wilh. Sondermann in Niederseßmar, Rheinland, abgebildet, der das Auswechseln der Einlagen für die verschiedenen Arbeiten unnötig macht. An Stelle der gewöhnlichen Einlegebüchsen sind in jeder Einlage 6 Schieber h vorgesehen, die am äußeren Ende die eingesetzten Greifer i tragen. Diese Schieber h gleiten in Nuten g, die in der eigentlichen Tischplatte des Revolvertisches so ausgespart sind, daß die Schieber und mit ihnen die Greifer nur eine radiale Bewegung machen können. Nun tragen die Schieber kleine, senkrechte Stifte k, die in Schlitzen m einer kreisförmigen Deckplatte n gleiten. Da diese Schlitze schräg zu den Nuten g

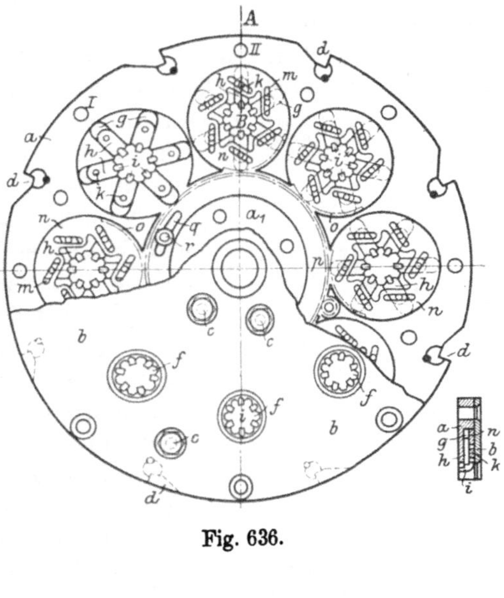

Fig. 636.

stehen, so sind die Schieber bei einer Drehbewegung der Deckplatten n gezwungen, sich radial nach innen oder außen zu bewegen. Zu diesem Zweck sind die Deckplatten n auf einem Teile ihres Umfanges mit einer Verzahnung versehen, die in ein auf der Achse des Revolvertisches verlagertes Stirnrad p eingreifen, das durch Vermittlung des rechts ersichtlichen kleinen Trieblings gedreht werden kann. Links ist bei q ein kreisförmiger Schlitz vorgesehen, durch den die Klemmschraube bis an die Oberseite der großen Deckplatte des Revolvertellers geht. Durch Anziehen dieser Schraube wird der ganze Antrieb der Verstellung der in den Öffnungen des Tellers b sichtbaren Greifer i gesperrt, worauf der eigentliche Tisch a mit der aufgeschraubten oberen Deckplatte und den dazwischen eingebauten Schaltteilen als ein Ganzes arbeiten kann. Bei d sind die üblichen Schaltnuten des Revolvertellers ersichtlich.

Revolverapparate für Sonderzwecke. In Fig. 637 ist ein Revolverapparat für eine Gewindedrückmaschine[1]) (Reiß & Martin, Berlin) abgebildet, der in einer senkrechten Ebene arbeitet und die gezogenen Näpfchen, die hintereinander in den Einwurf eingelegt werden, einzeln an die Drückrollen bringt. Die Einrichtung ist für eine vollständig selbsttätig arbeitende Gewindedrückmaschine gedacht, bei der der Arbeiter die Näpfchen nur in den Einwurf einzulegen hat. Bei der Schaltung des Revolvertellers nimmt eine der sechs Nuten ein Näpfchen aus der Zuführungsröhre mit und bringt es bei der nächsten Schaltung an die feststehende Drückrolle, worauf die bewegliche Rolle herunter gegen das Arbeitsstück geschwungen wird und nach Fertigstellung des Gewindes aus dem Bereich des Revolvertellers zurückgedreht wird. Dann wird durch eine Kurvenscheibe die feststehende Drückrolle zurückgezogen und der Revolverteller weitergeschaltet, wodurch das mit Gewinde versehene Näpfchen vor den Auswerfer kommt, der es aus der Nut im Revolverteller entfernt und in den Abfallkanal ausstößt.

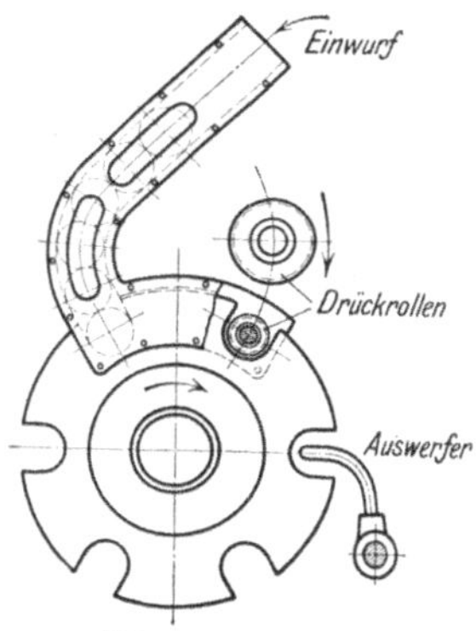

Fig. 637.

Statt des umlaufenden Revolvertellers baut die Firma auch einen schwingenden Zuführapparat. Die Gewindefutter werden für die verschiedenen Arbeitsstücke ausgewechselt, ebenso Revolverteller und Zuführung. Die Maschine ist für Gewindedurchmesser von 25 bis

[1]) Vgl. W.T. 1909, S. 19.

55 mm eingerichtet. Ihr Kraftbedarf beträgt ungefähr 0,5 PS. Sie arbeitet mit 60 Umdrehungen in der Minute. Das ergibt bei neunstündiger Arbeitszeit eine tägliche Leistung von ungefähr 20—22000 Stück.

Während bei den größeren Gegenständen, die in der Presse verarbeitet werden, Zuführungsvorrichtungen zur Beschleunigung der Herstellung oder aus Sicherheitsgründen angewendet werden, sind sie bei den kleinen nach Millionen herzustellenden Stücken, wie Patronen, Nadeln, Federn, Druckknöpfen und dgl. unbedingte Notwendigkeit, weshalb deren Ausbildung von seiten der Fabrikanten die größte Aufmerksamkeit zugewendet worden ist. Die größte Zahl derselben ist wohl nur für den eigenen Betrieb des betreffenden Werkes gebaut worden. Bei diesen Vorrichtungen kommen deshalb auch die verschiedensten Hilfsmittel zur Anwendung, so daß die drei nächsten Beschreibungen nur mögliche Anwendungen und Ausführungsformen vorstellen sollen.

Nadelspitzensortiervorrichtung (Jos. Drummen, Aachen-Laurensberg). Eine Vorrichtung, die das Sortieren der Nadeln nach Spitze und Kopf selbsttätig vornimmt, ist nach der Patentzeichnung in Fig. 638

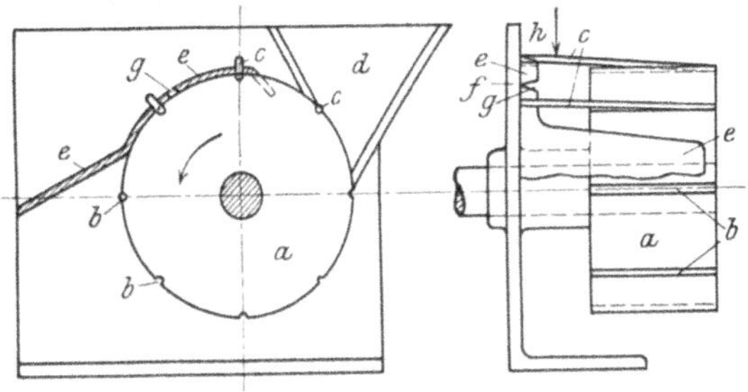

Fig. 638.

gegeben. Das Sortieren erfolgt nach dem Dickenunterschied zwischen Spitze und Kopf. Es wird zu diesem Zweck eine Walze, Trommel, Scheibe oder dgl. a mit Nuten b zur Aufnahme der zu sortierenden Körper c versehen. Die Walze a wird unter einem trichterförmigen Behälter d in der Pfeilrichtung gedreht, wobei jede Nut nur eine Nadel mitnimmt. An der Seitenwand f ist eine kurze und der Spitze der Nadeln entsprechende schmale Leiste e angebracht. Über diese werden sämtliche Nadeln mit dem an Seitenwand f anliegenden Ende hinweggeführt. Die Leiste e hat an einer Stelle einen Spalt oder Ausschnitt g, durch welchen die Spitzen der Nadeln c hindurchgehen können, dagegen die Kopfenden nicht. Durch das Auflaufen der Nadeln c auf die Leiste e werden sie an dieser Seite etwas gehoben, jedoch durch ein über der Walze a liegendes Band fest in ihren Nuten gehalten. Hierdurch drücken sie auf die Leiste e in der Pfeilrichtung h, also zur Walzenachse hin. Kommen nun Nadeln mit spitzem Ende am Ausschnitt g an, so schlüpfen diese durch und werden von der Walze a unter der Leiste e weitergeführt.

Die Nadeln jedoch, welche mit ihrem dicken Ende über der Leiste e ankommen, können nicht durch den Spalt g und werden sodann von der nach unten zu breiter werdenden Leiste e ganz aus den Nuten herausgehoben und rollen ab.

Es laufen also alle von der Walze a geförderten Nadeln über die Leiste des Abnehmers e bis zu der Öffnung g, wo die Nadeln mit den Spitzen wieder in die Nuten der Walze zurückfallen, die anderen entgegengesetzt liegenden Nadeln dagegen gehen über den Schlitz g hinweg zum Abnehmer.

Durch die regelbare Zunge ist es möglich, auf ein und derselben Vorrichtung Körper verschiedener Dicke zu sortieren, da der Schlitz g größer oder kleiner gemacht werden kann.

Mit dieser Vorrichtung können auch durchgebogene Nadeln heraussortiert werden, wobei es gleichgültig ist, ob die Nadeln schon spitzenrecht liegen oder nicht. Die Öffnung g muß dann so groß gestellt werden, daß auch die dicken Enden durchgehen können.

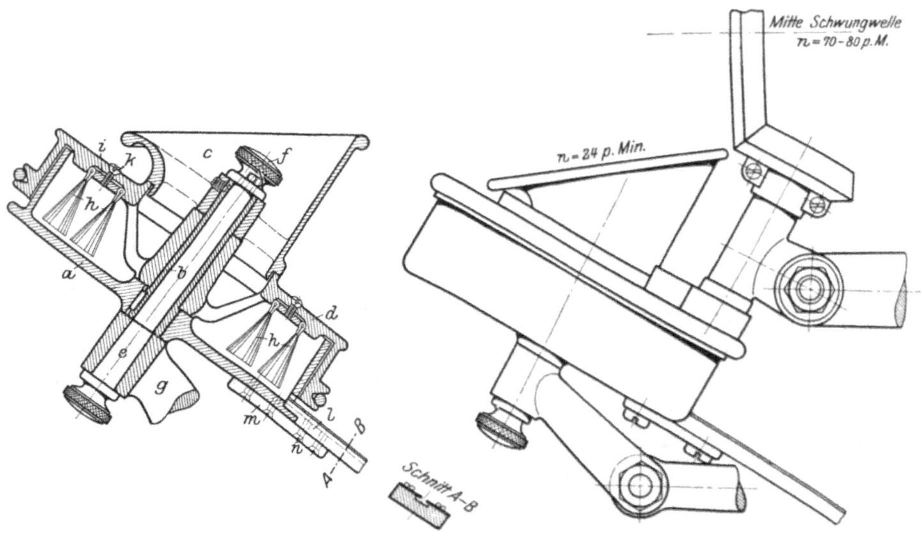

Fig. 639.

Zuführungsapparat für Druckknöpfe[1]). In der angegebenen Quelle veröffentlicht Stock einen Zuführungsapparat für Druckknöpfe, der

[1]) W.T. 1910 S. 660. Fig. 7/8.

in Fig. 639 im Schnitt und in einer etwas veränderten Ausführung in Ansicht abgebildet ist. Der Apparat beruht auf der Wirkung der Fliehkraft und gleichzeitig der Schwerkraft. Er besteht aus einer feststehenden Trommel a, die unter 30° auf einer Büchse b aufgekeilt ist und mit dieser durch eine Klemmschraube f auf der Achse e festgestellt werden kann. Auf derselben Büchse ist der Zuführtrichter c befestigt. Zwischen beiden läuft, die Trommel a am Umfang überdeckend, der Bürstenhalter d, der in dem einen Fall durch einen Schnurtrieb, im anderen Fall durch einen Kegelradantrieb mit ungefähr 24 Umdrehungen pro Minute angetrieben wird. Diese Drahtbürsten von denen vier Stück am Umfang sitzen, sind ungefähr 20 mm breit.

Am tiefsten Punkt der Trommel ist eine Ausfallöffnung, an die sich die Zuführrinne, die bis an die betreffende Arbeitsmaschine geht, anschließt. Diese Rinne bildet eine T-Nut, durch die die Druckknöpfe hindurchgleiten.

Die Druckknöpfe werden in den Zuführungstrichter eingeschüttet, von den umlaufenden Bürsten h erfaßt, durch die Fliehkraft nach außen geschleudert und gelangen durch allmähliches Hinabgleiten an die Rinne, von wo sie nach außen befördert werden.

Um ein Bild über die Abmessungen des Apparates zu geben, sei bemerkt, daß die Trommel 180 mm inneren Durchmesser und rund 40 mm Tiefe hat.

Selbsttätiger Zuführungsapparat für Patronenhülsen (Fritz Werner, Berlin W 35). Eine in der äußeren Ausbildung an die eben besprochene Konstruktion erinnernde Zuführungsvorrichtung ist in Fig. 640 in Schnitt und Ansicht dargestellt. Dieser Apparat hat vor dem eben besprochenen die zwangläufige Bewegung und die selbsttätige Abstellvorrichtung voraus.

Der Zuführungsapparat ruht auf einem Ständer 1, an dem der Boden 2 einer schräg gestellten, oben offenen Trommel 3 befestigt ist. Über dem Trommelboden dreht sich ein Teller 4, der seine Drehbewegung durch eine von der betreffenden Arbeitsmaschine angetriebene Schnurscheibe 5 unter Vermittelung des Schneckenradgetriebes 6, 7 erhält. In der Nähe seines Umfanges besitzt der Teller eine größere Zahl Durchbrechungen 8, die sämtlich auf einem zur Drehachse konzentrischen Kreis liegen und so bemessen sind, daß eine Hülse bequem hindurchfallen kann. Am Scheitelpunkt der Trommel ist unterhalb der jeweils obersten Durchbrechung 8 ein Greifer 9 schwingend angeordnet, der durch die Hebel 10, 11, 12 vermittels einer auf der Antriebswelle sitzenden Kurvenscheibe 13 entgegen der Wirkung einer Zugfeder 14 in die Öffnungs- oder Schließlage bewegt wird (die Zeichnung zeigt die Schließlage). Der Hebel 11,

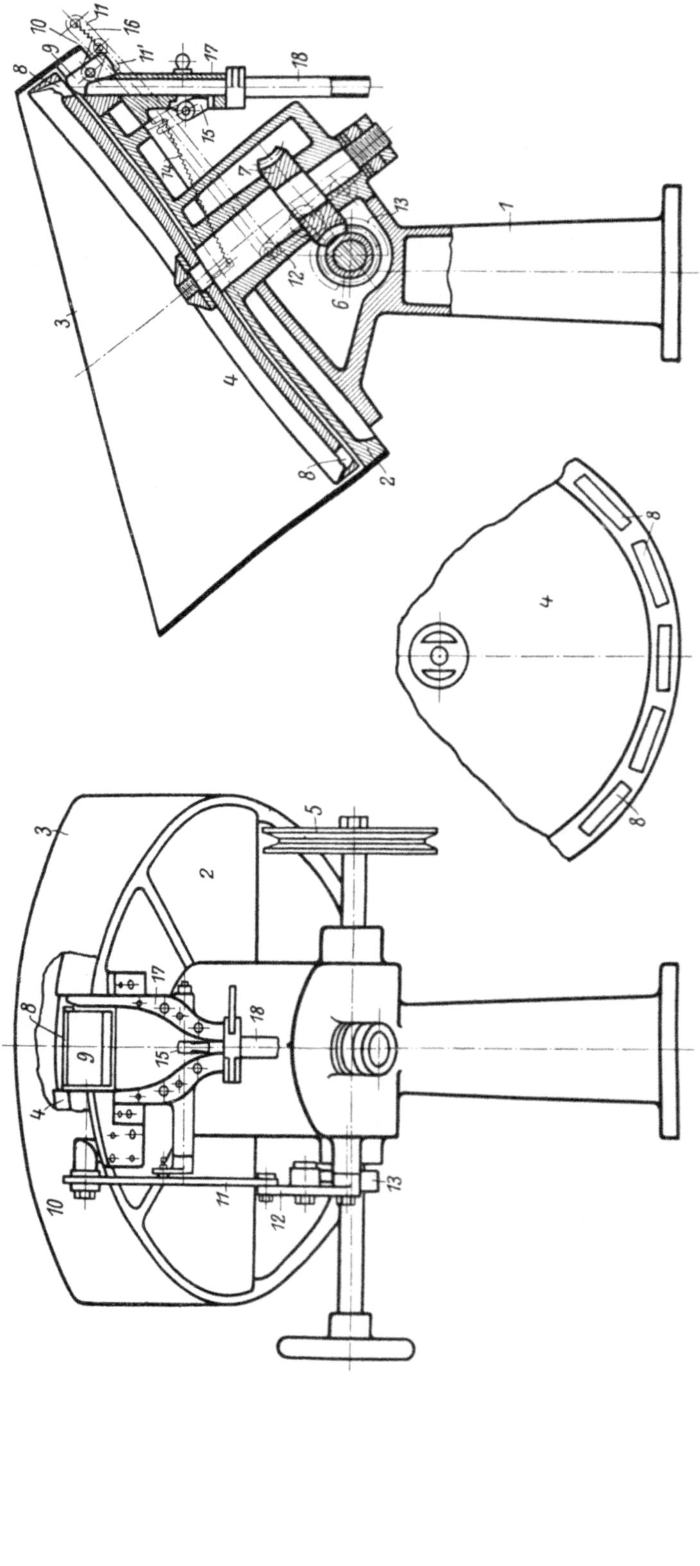

Fig. 640.

welcher bei 11′ geschlitzt und durch eine Feder 16 noch weiterhin mit dem Verbindungsbolzen des Hebels 10 gekuppelt ist, steuert gleichzeitig einen Fühlhebel 15, der in den Ausfallkasten 17 eintreten kann, an welchen sich das nach der Bearbeitungsstelle der Werkstücke führende Ausfallrohr 18 anschließt.

Die Arbeitsweise des Apparates ist folgende: Die Hülsen werden in die schräge Trommel einfach eingeschüttet, bis diese ungefähr bis zur Mitte des Tellers 4 gefüllt ist. Die tieferliegenden Durchbrechungen 8 des Tellers nehmen hierbei je eine Hülse auf und führen diese nacheinander bis zum Scheitelpunkt der Trommel, wo der Greifer 9 jedesmal öffnet, sobald das Werkstück in der Ausfallstellung angelangt ist. Eine im Ausfallkasten 17 angebrachte Gleichrichtevorrichtung bewirkt, daß sämtliche Werkstücke mit dem gleichen Ende nach unten abfallen, gleichgültig, ob sie von den Durchbrechungen 8 in der richtigen oder verkehrten Lage aufgenommen wurden.

Da der Teller des Apparates mehr Werkstücke zuführen soll, als an der Arbeitsstelle verbraucht werden, so würde das Ausfallrohr bald gänzlich mit Werkstücken angefüllt sein und es könnten hierdurch Störungen eintreten. Dies wird durch den Fühlhebel 15 verhütet, welcher bei jeder Bewegung des Greifers in den Ausfallkasten eintritt und der, falls seine Bewegungsbahn durch die bis an diese Stelle reichende Werkstücksreihe im Ausfallrohr gesperrt ist, die Zurückbewegung des Hebels 11 und somit auch die Öffnung des Greifers 9 verhindert. Dies geschieht so lange, bis die nachsinkenden Werkstücke im Ausfallrohr genügend Platz für die Bewegung des Hebels 15 und damit für den Eintritt weiterer Werkstücke schaffen.

Infolge der schrägen Lagerung der Trommel bleibt in der Nähe des Ausfalles in jedem Durchbruch des rotierenden Tellers immer nur ein Werkstück haften, während etwa mit hochgenommene weitere Werkstücke stets wieder nach dem Werkstückvorrat an der tiefsten Stelle der Trommel zurückrollen. Diese Anordnung bürgt somit bei größtmöglicher Leistungsfähigkeit für ein unbedingt einwandfreies Arbeiten.

3. Zuführungen für die Streifen und die vorgearbeiteten Stücke.

Zu dieser Gruppe gehören die zwangläufig bewegten Vorschubvorrichtungen, da diese allein imstande sind, die Streifen mit den darin noch gehaltenen, vorgearbeiteten Formen so genau weiterzubefördern, daß in einzelnen Fällen mit diesem Vorschub ausgestattete Folgewerkzeuge mit dem Verbundwerkzeug in Wettbewerb treten können.

Hierzu gehört der Vorschubapparat von Tümmler-Döbeln, dessen Antrieb durch das „ruhende“ Gesperre von Hundhausen erfolgt. Werkzeuge, die mit diesen Vorschüben arbeiten, sind in Fig. 120, 545, 560 wiedergegeben worden. Ihre Besonderheit liegt darin, daß sämtliche Stücke bis zum letzten Ausschnitt im Streifen verbleiben und mit demselben weiterbefördert werden, was auch für die Sicherheit des Arbeiters einen nicht zu unterschätzenden Vorteil bedeutet.

Der Unterschied gegenüber den bekannten Schaltwerken besteht darin, daß die Bewegung der Sperrklinken zwangläufig von unrunden Scheiben aus nach ganz bestimmten Gesetzen erfolgt. Der Grundgedanke der Wirkungsweise ergibt sich am besten aus der Originalzeichnung[1]), Fig. 641. Das zu schaltende Rad S wird von dem Schieber a in der Pfeilrichtung geschaltet, wenn die Kurvenscheibe I, die am Ende des Hebels A sitzende Rolle A_1 nach auswärts drückt und gleichzeitig die kleinere innenliegende Kurvenscheibe II, die am Ende des Hebels C sitzende Rolle C_1 im gleichen Sinn so bewegt, daß der Schieber a, der sich in seiner am Hebel A festen Führung A_2 verschieben kann, in die Zähne des Rades S eingedrückt wird. Während dieser Umdrehung der Kurvenscheiben steht der Sperrschieber, der in der am Ständer G festen Führung G_2 gleitet, außer Eingriff mit den Zähnen des Rades S. Die Kurvenscheiben sind ungleich geteilt[2]), um für die eigentliche Schaltung I, während der außer dem Antriebshebel A und dem Schaltschieber a das große Rad S vom Stillstande aus allmählich beschleunigt und von seiner höchsten Geschwindigkeit wieder bis auf 0 verzögert werden muß, eine halbe Umdrehung der Kurvenscheibe auszunutzen, während von der anderen Hälfte je ein Drittel für den leeren Rückgang des Hebels A sowie die Umsteuerungen der Schieber ausgenutzt werden.

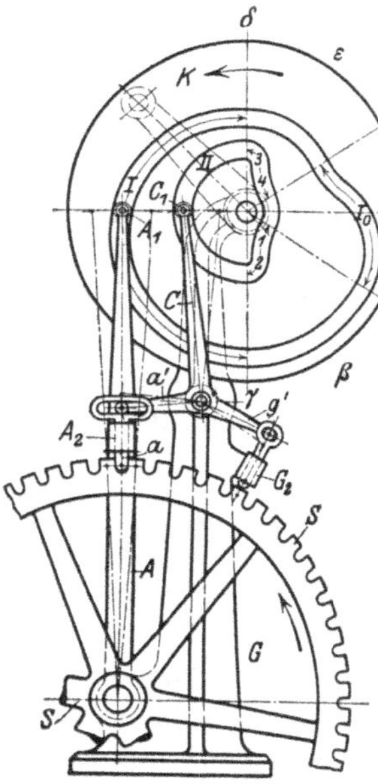

Fig. 641.

In dem Stanzautomaten von Tümmler ist dieses Prinzip für den Bandvorschub verwendet worden; an Stelle des Schaltschiebers treten

[1]) Hundhausen, E.T.Z. 1902, Heft 51/52.
[2]) Vgl. Gugel, Materialzuführungsvorrichtungen, S. 58.

die Vorschubbacken und an Stelle des Sperrschiebers die am Pressentisch festen Klemmbacken f, Fig. 642. Das Ein- und Ausheben der Schieber in Fig. 641 wird bei dem Apparat von Tümmler durch das Öffnen und Schließen der Klemm- und Vorschubbacken ersetzt. Die eigentliche Arbeit des Vorschubapparates geht bei dem Stanzautomaten in folgender Weise vor sich: Die auf der Exzenterwelle sitzende Hauptkurvenscheibe a bewirkt durch die Vermittlung zweier Schieber b und n die Verschiebung und die Klemmbewegung der Backen e und f. Es schließt sich Backe e links und gleichzeitig rechts, wodurch der Blechstreifen gehalten und durch den jetzt einsetzenden Vorschub um eine Schaltlänge vorgeschoben wird. Die inneren Backen f, die während dieser Bewegung offen waren, schließen sich, nachdem der Streifen zur Ruhe gekommen ist, und halten das Blech während der Preßarbeit fest. In dieser Zeit öffnen sich die Vorschubbacken e und gehen in ihre Anfangsstellung zurück.

Die Schaltbewegung der Backen e erfolgt, wie bereits gesagt, von der Hauptkurvenscheibe a, die einen Schieber b in wagrechter Richtung hin- und herbewegt. Diese Bewegung wird durch den Steuerhebel c, der mittels Kugelgelenkes mit dem Hebel d verbunden ist, und eine kurze Zugstange auf einen durch die beiden Vorschubbacken e und die Stangen e' gebildeten Rahmen übertragen, der durch diese Stangen am Pressentisch gut geführt ist. Die genaue Einstellung des Hubes bzw. des Blechvorschubes erfolgt durch die Mikrometerverstellung g am Steuerhebel c. Dazu löst man die beiden Schrauben i, die zur Befestigung des am Pressenkörper verschiebbaren Böckchens dienen, sowie die Schrauben h, wonach man ohne weiteres den ganzen Apparat am Steuerhebel c herauf- und herunterschieben kann. Hierauf zieht man die Schrauben i des verstellbaren Böckchens am Pressenständer und die Schraube des unteren Böckchens am Steuerhebel fest. Nun verschiebt man mit Hilfe der kleinen Kurbel k den Mikrometeroberteil soweit, bis der Stift l ein Loch findet, löst die untere Schraube h des Mikrometeraufsatzes und zieht die obere fest, worauf man den Mikrometeroberteil wieder mit Hilfe der Kurbel k auf das gewünschte Maß einstellt. Jetzt müssen alle Schrauben fest angezogen werden. Die an den beweglichen Vorschubbacken e angebrachten Anschlagschrauben m müssen so eingestellt werden, daß sie beim Gang der Presse leicht rechts und links anschlagen.

Das Öffnen und Schließen der Backen wird von derselben Kurvenscheibe a durch den Schieber n bewirkt, der sich in senkrechter Richtung auf- und niederbewegt. Von ihm aus werden durch einen Schubstangenmechanismus 1—4 die unrunden Scheiben o in Pendelbewegung versetzt. Die Köpfe der Zugstangen, welche diese Kurvenscheiben umgreifen, sind an die Schieber p angelenkt, die wiederum im Pressenständer geführt sind und mit Hilfe der Kniegelenke q die Backen öffnen nud schließen. Die Enden der Schieber p sind soweit gegabelt, daß die Vorschubbewegung der Backen e unbeschadet der Kniehebelsteuerung erfolgen kann. Die unrunden Scheiben o sind so auf ihren Achsen

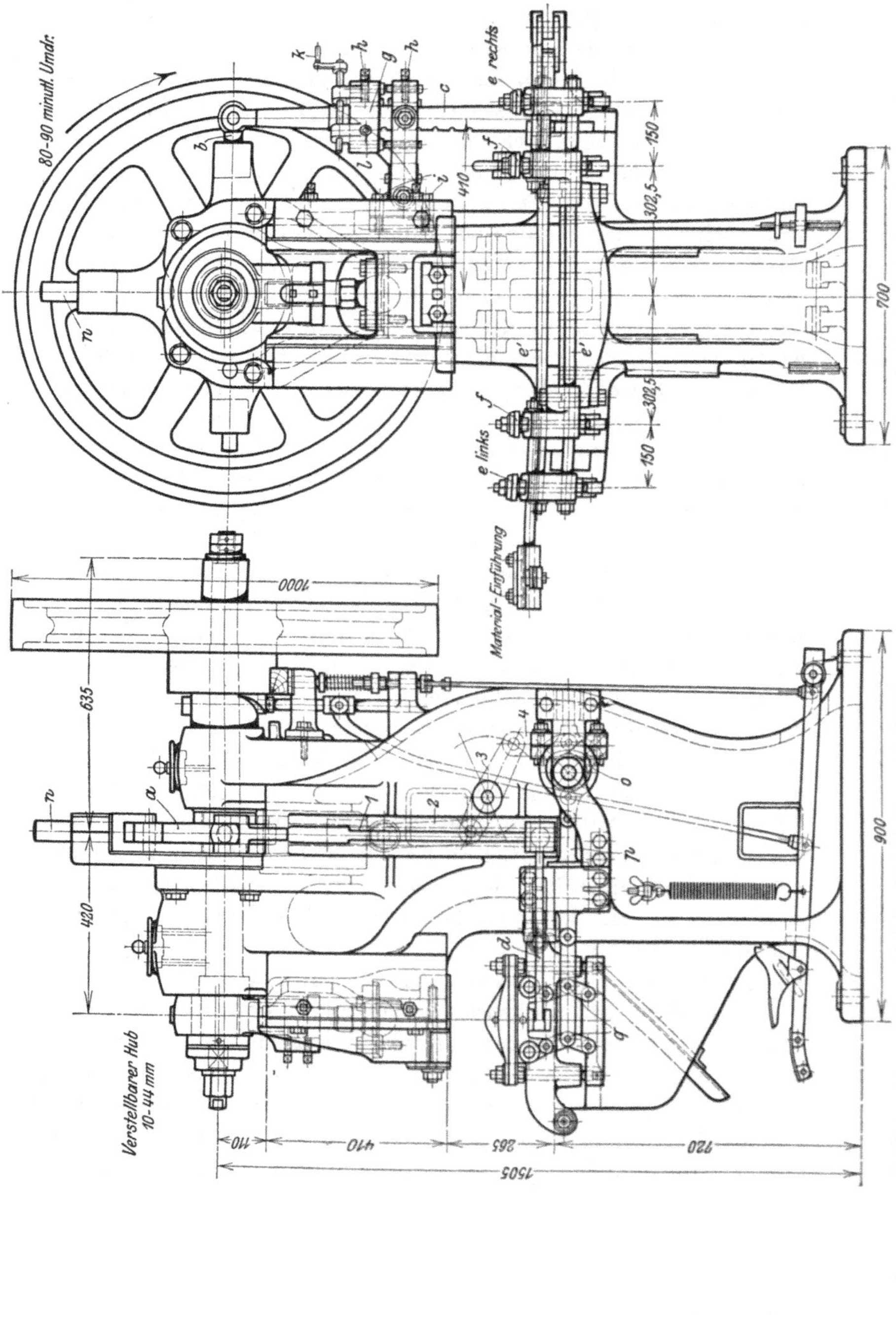

versetzt, daß ein Paar Backen sich stets schließt, bevor das andere sich öffnet.

Die in Fig. 642 abgebildete Pressenart ist hauptsächlich zur Herstellung flacher Stücke durch Ausschneiden, Prägen, Biegen und Durchziehen bestimmt, wobei diese Arbeitsgänge außerdem in einem Werkzeug mehrfach wiederholt werden können. Diese „Stanzautomaten" sind feststehend in senkrechter Bauart konstruiert, damit der Abfall und das fertige Stück nach unten entfernt werden kann, und sind für Folgewerkzeuge oder für Verbundwerkzeuge verwendbar. Der Vollständigkeit halber möge erwähnt werden, daß bei diesen Maschinen eine Einrichtung zur Ausschaltung der Vorschubbacken e in Zusammenhang mit einem selbsttätig bewegten Streifenanschlag vorhanden ist, durch die ein gleichmäßiger, abfalloser Anschnitt erreicht wird, ferner eine Hubverstellung mittels exzentrischer Büchse und Zahnkranz, die innerhalb der normalen Hubgrenzen den Hub von 2 zu 2 mm zu verstellen gestattet.

Besondere Vorschubeinrichtung für Kartonnageklammern (Rob. Tümmler, Döbeln). In Fig. 643 ist ein besonderer ebenfalls zwangläufig arbeitender Vorschub für das in Fig. 562/3 abgebildete Werkzeug zur abfallfreien Herstellung von Kartonnageklammern abgebildet. Da die Klammern mit den aufgebogenen Zacken nicht unter den Greifern des eben beschriebenen Apparates durchgehen könnten, so wird auf der Austrittsseite des Werkzeuges ein besonderer Zuführungsapparat eingebaut, der insoweit auch zwangläufig ist, als die Streifen beim Vorwärtsgang von eigenen Greifern vorgeschoben werden, während bei der Ausschaltung dieser jeder Streifen durch eine eigene Bremsvorrichtung festgehalten wird.

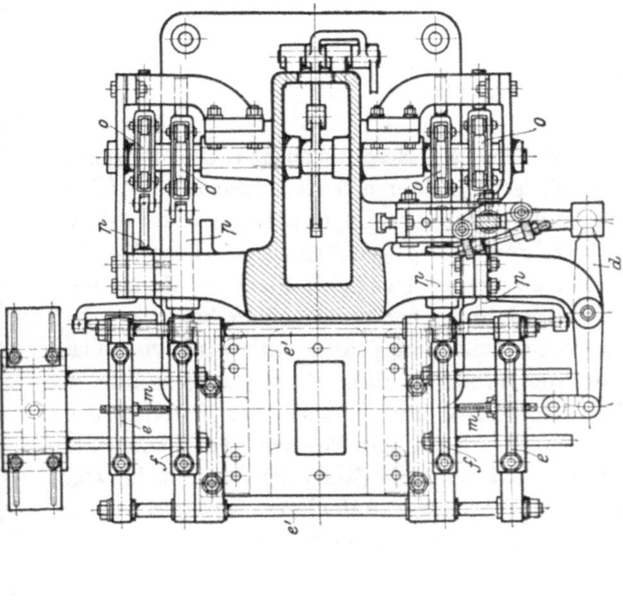

Fig. 642. Rob. Tümmler-Döbeln.

Außerdem muß der Vorschub wegen der Kleinheit der einzelnen Schnitte und Biegungen, sowie des vielfachen Mehrfachschnittes in diesem Werkzeug so genau wie nur möglich sein. Diesen Bedingungen genügt der abgebildete Vorschubapparat.

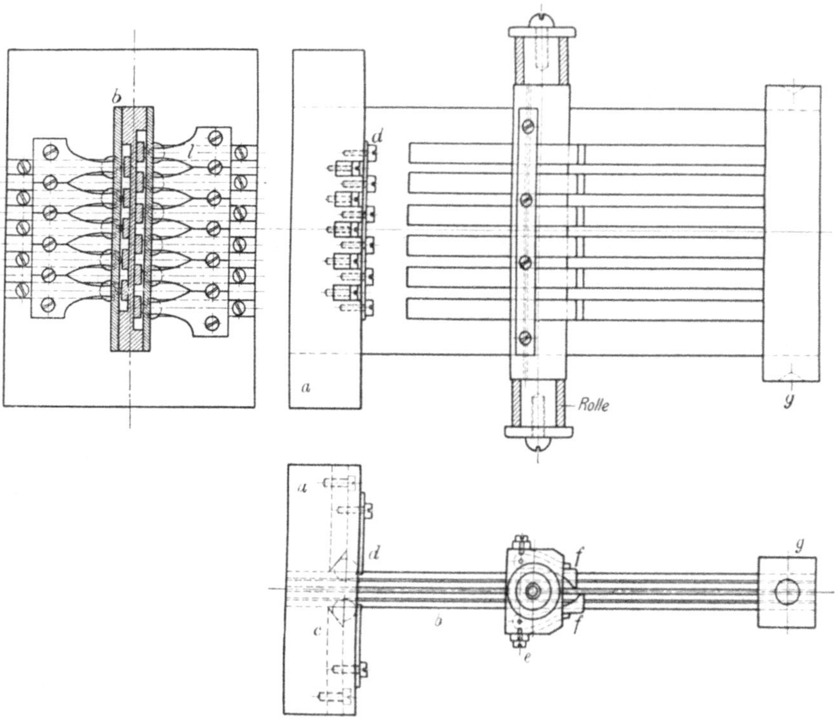

Fig. 643.

Teil a wird an dem Werkzeug befestigt, ebenso die einzelnen Transportschienen b, welche aus mehreren Teilen bestehen. Über jeder einzelnen Transportschiene ist eine Kugel c eingebaut, welche durch je eine Feder d in der richtigen Lage gehalten wird. Diese Kugeln dienen als Klemme für den zu transportierenden Streifen, damit der Streifen sich beim Transport nicht wieder zurückziehen kann.

Teil e ist der eigentliche Transportapparat und wird vor- und rückwärts von der Presse aus betätigt. Beim Vorwärtsgehen faßt er mit seinen Greifern f in die Zähne bzw. Spitzen der Streifen ein, wodurch der Transport erfolgt. Beim Zurückgehen des Transports heben sich die Greifer von selbst über die Zähne der Streifen hinweg.

Teil g ist genau wie Teil a zur Aufnahme der Transportschienen sowie zur Befestigung an dem Werkzeug bestimmt.

B. Maschinen.

Die Beschreibung einzelner Maschinen als Vertreter der verschiedenen Bauarten soll nach dem Antrieb und der Verwendung geordnet erfolgen, und zwar

1. Exzenterpressen.
2. Spindelpressen.
3. Hydraulische Pressen.
4. Schmiedemaschinen.
5. Hilfsmaschinen.

1. Exzenterpressen.

Bei den gewöhnlichen Exzenterpressen ohne Druckregler oder Sicherheitsglied, die immer noch die Mehrzahl der im Gebrauch befindlichen Maschinen bilden, ist der Stößelhub nach oben und unten hart begrenzt, so daß die Presse mit ihren Führungen und Werkzeugen auf das sorgfältigste ausgerichtet werden muß. Ebenso ist ein gelegentliches Nachsehen der Führungen notwendig, was dann meistens ein Wiederausrichten nach sich zieht. Dieses kann nach denselben Grundsätzen wie im allgemeinen Werkzeugmaschinenbau erfolgen. Derartige Angaben finden sich unter anderen:

Berichte des Versuchsfeldes für Werkzeugmaschinen an der Techn. Hochschule Berlin, Heft I.

Werkstattstechnik 1909: Schlesinger, Prüfung einer Universalfräsmaschine. S. 459. — W. Pockrandt, Beispiele für das Ausrichten der Werkzeugmaschinen. S. 557.

Werkstattstechnik 1913: Ausrichten von Wellen. S. 317.

Finkelstein, Prüfung der Arbeitsgenauigkeit von Werkzeugmaschinen, Berlin, Springer.

Eine Sonderarbeitsweise, die in manchen Fällen mit Erfolg bzw. in ähnlicher Weise wird angewendet werden können, gibt Woodworth an:

Ausrichten einer Presse. Wenn man findet, daß die Laufflächen des Stößels sich kaum noch in brauchbarer Verfassung befinden, und man keine Maschine besitzt, um sie neu zu hobeln, muß man sie gut passend schaben und dann die Sitze der Stempelzapfen parallel mit den Laufflächen hobeln, worauf die Unterfläche des Stößels genau rechtwinklig hergerichtet wird. Sollen die Pressen Stempel mit vierkantigen Zapfen aufnehmen, so können sie ohne Schwierigkeit dafür hergerichtet werden.

Ein Hebel A, Fig. 644, und ein runder Zapfen B werden hergestellt und so zusammengepaßt, daß kein bemerkbarer toter Gang darin ist. Der Zapfen wird dann sicher an dem neueingepaßten Pressenstößel befestigt und eine Mikrometerschraube am Ende des Hebels A festgemacht. Die Froschplatte C muß überall gleich dick gehobelt werden und eine ihrer Seiten als Richtplatte verwendet werden, nach der die Oberfläche des Pressentisches gefeilt und geschabt wird.

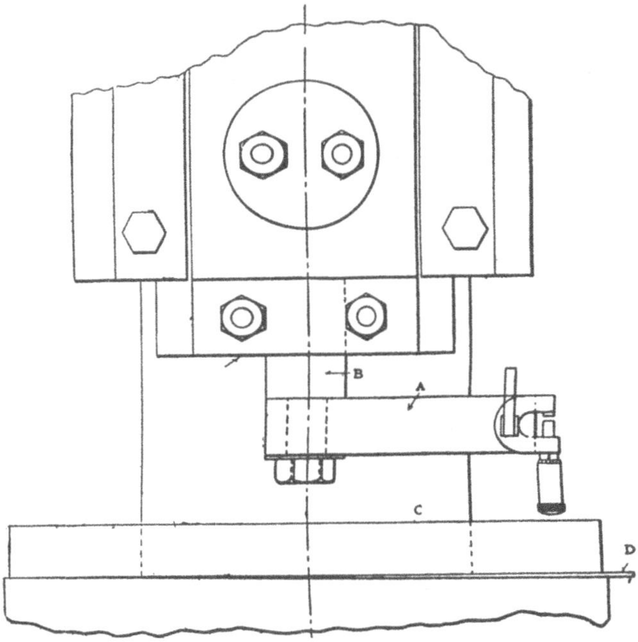

Fig. 644.

Hierauf wird die Mikrometerschraube niedergeschraubt und an jeder Ecke der Froschplatte eine Ablesung gemacht, wobei man nur Sorge tragen muß, daß die Mikrometerschraube auf die Froschplatte ohne Druck niedergeschraubt wird. Ein Vergleich der Ablesungen wird zeigen, daß das eine Ende höher als das andere ist, worauf man durch Unterlegen von Streifen Seidenpapieres, wie vergrößert in Fig. 644 bei D ersichtlich ist, das niedrigere Ende hebt. Hat man schließlich in den vier Ecken gleiche Ablesungen erhalten, so kann man die Froschplatte entfernen und in der Hobelmaschine einspannen. Die Unterlagen müssen mit den entsprechenden Kanten der Froschplatte gleich gezeichnet werden und beim Einspannen in der Hobelmaschine vertauscht werden, worauf ein Span über die ganze Breite der Froschplatte genommen wird. Man erhält dann beim Wiedereinsetzen ihre Oberfläche rechtwinklig zur Stößelführung.

Man darf aber diesen Weg nicht einschlagen, wenn die Froschplatten auf verschiedenen Pressen verwendet werden sollen. Die Platten müssen vielmehr deutlich mit der Nummer der dazu gehörigen Presse und an ihrer dem Arbeiter zugekehrten Seite mit „Vorne" bezeichnet werden.

Da verschiedene einfache Formen der Exzenterpressen in Fig. 612 bis 614 abgebildet worden sind, soll hier eine Exzenterpresse, die mit allen möglichen Bewegungen und Hilfsmitteln ausgestattet ist, besprochen werden.

Einarmiger doppeltgelagerter Stanzautomat (Rob. Tümmler, Döbeln). In Fig. 645 ist eine Exzenterpresse mit dem in Fig. 642 beschriebenen Greiferspeiseapparat und Auf- und Abwickelvorrichtung für das Band abgebildet. Die Bezeichnungen für den Schaltantrieb decken sich mit denen in Fig. 642. Bei A ist die Maschine mit einem selbsttätigen Blechrichtapparat versehen. Das Bemerkenswerte dieser Ausführung liegt in der Ausbildung des Ständers, der auf seinem feststehenden Untersatz um 30° nach rückwärts schräg gestellt werden kann, wodurch Teile, die von den selbsttätigen Auswerfervorrichtungen wieder auf der Werkzeugoberfläche abgelegt werden, durch ihr Gewicht abgleiten und so aus der Presse entfernt werden. Durch eine besondere Stellvorrichtung können diese Pressen aber auch auf ihrem Untersatz ganz herumgelegt werden, bis sie wagrecht liegen. Das bedingt, daß alle Antriebsteile für den Vorschub zwangläufig arbeiten. Infolgedessen lassen sich diese Pressen für Arbeiten, die mit einem Ziehen oder Prägen verbunden sind, ebenfalls verwenden, wobei man entweder nur die Schwerkraft oder zwangläufige Auswerfer verwendet.

Die allgemeinen Hauptteile der Presse decken sich mit den an anderen Stellen besprochenen; zu erwähnen ist der in Rippenguß aus besonderem Eisen hergestellte Ständer, der fest angeordnete Tisch, die beide mit Rücksicht auf möglichst geringe Federung bemessen sind. Der Stößel ist durch eine Kugelschraube in weiten Grenzen verstellbar, so daß man verschieden hohe Werkzeuge einspannen kann. Er ist, um möglichst viele Werkzeuge einspannen zu können, mit einer breiten Auflagerfläche, Klemmdeckel und Zapfenloch versehen und erhält außerdem zwei äußere Augen mit durchgehenden Bolzenlöchern, damit auch ganz große Werkzeuge eingespannt werden können. Die Feststellung des Kugelzapfens erfolgt mittels eines in den Pleuelkopf eingeschnittenen Gewindekeiles und zweier Druckschrauben. Die Maschinen sind mit Momenteinrückung und selbsttätiger Auslösung im höchsten Stößelstand, sowie mit einer Bremseinrichtung in dieser Lage versehen, die durch Fußtritte ein- bzw. ausgerückt werden.

Fig. 645. Rob. Tümmler-Döbeln.

Mehrstempelige Revolverpresse (Reiß & Martin, Berlin SO). Bei der Besprechung der Herstellungsgänge in Tabelle Nr. 23, der Herstellung des Unterkastens für Kosmos - Petroleumbrenner, Fig. 542, wurden verschiedene Schwierigkeiten bei der Stempelanordnung der Revolverpressen angeführt, deren Beseitigung die Firma Reiß & Martin zur Konstruktion ihrer mehrstempeligen Revolverpresse, Fig. 646—648, geführt hat. Die Presse, die den auf S. 425 besprochenen Übelständen in interessanter und einfacher Weise begegnet, ist in Fig. 646 abgebildet. Es werden gewissermaßen drei Einzelpressen nebeneinander gestellt und von einem Antrieb aus betätigt. Zu jedem Werkzeug gehört ein besonderer Stößel, der auch in eigenen, nachstellbaren Schwalbenschwanzführungen gleitet. Diese Stößel sind, wie aus der Figur ersichtlich ist, mittels Gewindemuttern an den Pleuelstangen einstellbar und werden jeder auch von einem besonderen Exzenter angetrieben. Der Hub eines jeden der Stempel kann leicht unabhängig von dem andern eingestellt werden. Die Achsen sind gut in dem kräftigen, gemeinsamen Ständer gelagert und werden rückwärts durch Stirnräder von einer Stelle aus, die zu dem Hauptantrieb hinüberleitet, angetrieben. Es kann infolgedessen jedes für sich allein niederbewegte Werkzeug selbst das Arbeitsstück genau in die erforderliche Lage bringen. Dies kann z. B. eintreten, wenn der Ziehstempel des einen Werkzeuges den äußeren Umfang des Arbeitsstückes so weit verringert hat, daß es im Revolverteller nicht mehr zentrisch geführt ist. Es tritt dann eine im Oberstempel oder am Stößel angebrachte besondere Zentriervorrichtung in Tätigkeit, wie in Fig. 543 zu erkennen ist.

Fig. 646. Reiss & Martin-Berlin.

Ein weiterer Vorteil der getrennten Anordnung liegt darin, daß man irgendeinen Stempel ausschalten kann, ohne ihn aus seiner Befestigung herauszunehmen, da man seine Antriebswelle ausschalten kann.

Die Einrichtung des Antriebes für einen einfach- und einen doppeltwirkenden Stößel für eine derartige Presse zeigen Fig. 647—648. Ob nun der Antrieb mittels Stirnräder, wie in der Zeichnung dargestellt, oder mittels Kegelräder gelöst ist, ist grundsätzlich für die Ausführung wohl gleichgültig. Die Antriebsräder sind mit der Exzenterwelle nicht fest, sondern durch eine lösbare Kupplung verbunden.

Die Presse läuft mit 55 Hüben in der Minute und stellt bei neunstündiger Arbeitszeit nach Angaben des Fabrikanten 25000 Teile her, mit drei Arbeitsgängen an jedem Stück. Werden die Arbeiten auf Hand- oder Exzenterpressen ausgeführt, so kann man bei einem guten Arbeiter im Mittel auf 7—8000 Teile im Tag rechnen.

Die bisher besprochenen Exzenterpressen arbeiteten mit in der Höhe nicht verstellbarem Tisch, so daß die für das Einlegen und Entfernen der Arbeitsstücke verfügbare Zeit nur durch die Stößelbewegung bestimmt wurde. Da für diese Arbeiten immer ein gewisses geringstes Maß an Zeit verfügbar sein muß, und außerdem erst bei einer gewissen Entfernung beider Werkzeugteile voneinander diese Arbeiten sicher ausgeführt werden können, so ist die Hubzahl der Presse und damit die stündliche Erzeugungsmenge durch diese Zeiten nach oben begrenzt. Wenn man nun gleichzeitig während dieses verfügbaren Stößelweges auch den Pressentisch nach abwärts gehen läßt, so vergrößert sich die Entfernung beider Werkzeugteile doppelt so schnell, so daß die Erzeugungsmenge der Presse bereits allein durch eine beschleunigte Hubzahl vergrößert werden kann. Eine derartige Presse der Firma L. Schuler in Göppingen wurde bereits mit Rücksicht auf die dieser Bauart innewohnenden Vorteile beim Ziehen von Feldkesseln aus Aluminium auf S. 299 besprochen.

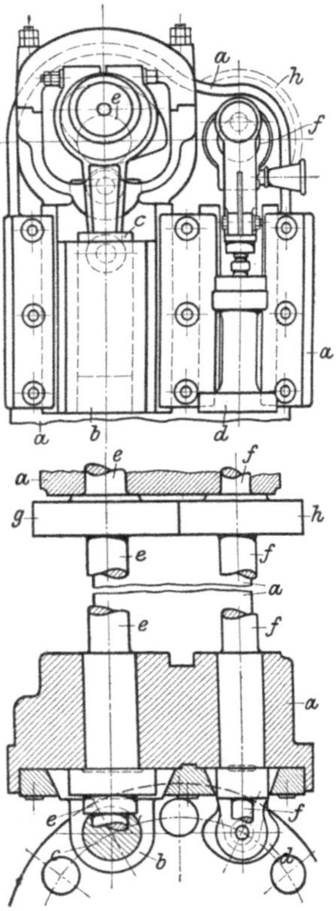

Fig. 647.

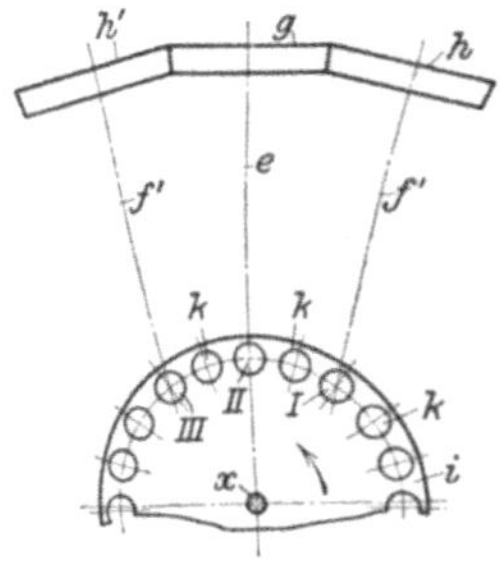

Fig. 648.

In Fig. 649 ist die Zeichnung einer Räder- oder Kurbelziehpresse mit beweglichem Tisch der Firma Bonner Maschinenfabrik Mönkemöller wiedergegeben. Die Maschine besteht aus drei Teilen, dem Bett A, das den Antrieb enthält, den beiden seitlichen Ständern B,

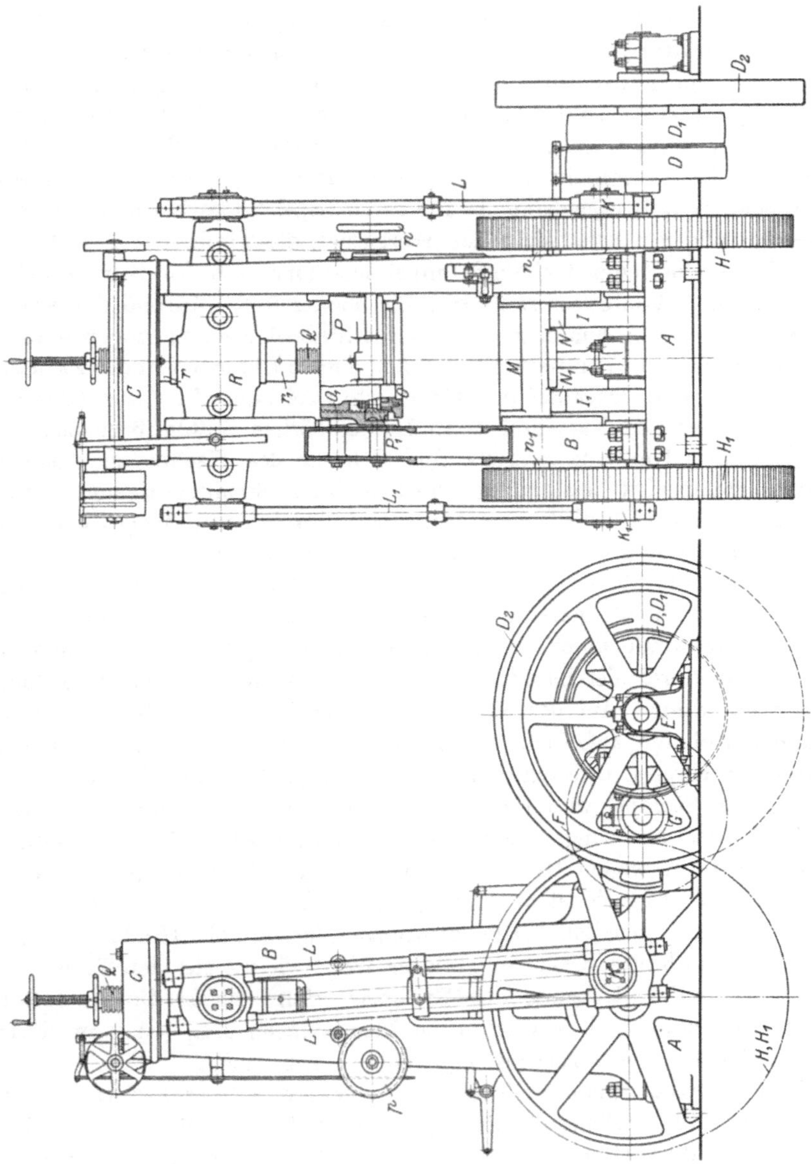

Fig. 649. Bonner Maschinenfabrik Mönkemöller.

die die Führung und Befestigung für die Halter der einzelnen Werkzeugteile tragen, und einem Kopfstück C, an dem der Riemenantrieb für die Blechhalterverstellung angebracht ist und das gleichzeitig die beiden Seitenständer gegeneinander absteift. Der Antrieb erfolgt von den beiden Riemenscheiben, Fest- und Losscheibe $D D_1$, deren Durchzugkraft durch ein schweres Schwungrad D_2 verstärkt wird, mittels der Stirnräder E, F und G auf die eigentlichen Antriebsräder $H H_1$ der Presse. Diese sitzen auf einer durchgehenden, in drei Lagern gelagerten Welle, die zwischen den Lagern die beiden Kurvenscheiben $I I_1$ für die Tischbewegung trägt. Die beiden Räder $H H_1$ tragen außen zwei Kurbelzapfen $K K_1$, von denen durch lange, zweiteilig ausgeführte Pleuelstangen $L L_1$ das Querhaupt R für den Stempel betätigt wird. Die drei Hauptteile zur Befestigung der Werkzeuge, der Tisch, der Blechhalter und das Querhaupt für den Stempel, sind innerhalb der Seitenständer eingebaut. Von diesen steht der Blechhalter fest, während die beiden anderen Teile in Führungen, die durch Stelleisten nachgezogen werden können, laufen. Der Tisch M, auf dem mittels Untersätzen der Ziehring befestigt wird, trägt auf der Unterseite zwei Laufrollen aus Hartguß N_1, die mit den Kurvenscheiben II_1 aus hartem Stahlguß zusammenarbeiten. Um eine zwangläufige Führung des Tisches zu erhalten und gleichzeitig ein Hängenbleiben beim Stempelaufgang auszuschließen, sind seitlich zwei Führungsrollen bei $n n_1$ vorgesehen, die innerhalb der großen Zahnräder in Leitkurven laufen.

Der Blechhalter O ist mit einer Kalotte einstellbar und wird, durch Federbolzen getragen, an dem Gewindekörper O_1 befestigt, dessen Mutterngewinde durch einen Schneckenantrieb $p P_1$ verdreht werden kann, wodurch der Blechhalter von außen leicht und schnell verstellt werden kann. Auf der rechten Seite der Schneckenwelle p ist eine Riemenscheibe aufgekeilt, damit man die Verstellung auch mittels des am Kopf C befestigten Riemenvorgeleges bewirken kann.

Der Stempel wird mittels des am Kopf der Spindel sichtbaren Handrades und langer Schraube in der Stempelspindel Q befestigt, die dann in weiten Grenzen mittels des Handkreuzes in der Höhe verstellbar ist. Sie ist in dem Querhaupt R gelagert und durch zwei Muttern $r r_1$ gehalten. Die Führungen für das als Hohlguß hergestellte Querhaupt R sind in gleicher Weise wie die für den Tisch ausgeführt. Die Bewegung des Stempels erfolgt durch die an Zapfen des Querhauptes R angreifenden Pleuelstangen $L L_1$, also in einem festgelegten Voreilungswinkel gegenüber dem Tisch.

Die Arbeitsweise der Presse ist demnach in kurzem folgende: In der höchsten Stellung des Stempels und tiefsten Stellung des Tisches wird das Blankett eingelegt, worauf der Tisch hochgeht, bis er gegen den feststehenden Blechhalter kommt, der das Blech faßt. Dann geht der Stempel nach abwärts und zieht den Hohlkörper. Beim Hochgehen des Stempels wird der Hohlkörper von dem Stempel abgestreift, worauf der Ziehring mit dem Tisch niedergeht die Hülse zwangläufig ausgeworfen und vom Arbeiter vom Tisch entfernt wird.

Doppelziehpresse (Bonner Maschinenfabrik Mönkemöller G. m. b. H.). Eine noch weitergehende Ausnützung der toten Zeiten erreicht die

Fig. 650.

Bonner Maschinenfabrik in ihren Doppelziehpressen, Fig. 650—651. Die Richtlinien dieser Arbeit und die erzielten Vorteile sind bereits auf S. 261 nebst einer grundsätzlichen Darstellung des Arbeitsvorganges gegeben worden. Es werden bei dieser Pressenkonstruktion zwei Züge hintereinander ausgeführt, indem der erste Ziehstempel beim zweiten Zug als Blechhalter arbeitet. Infolgedessen ist dieser Blechhalter also der Ziehstempel des ersten Zuges in einer gesonderten Führung an das Querhaupt des Ziehstempels angehängt.

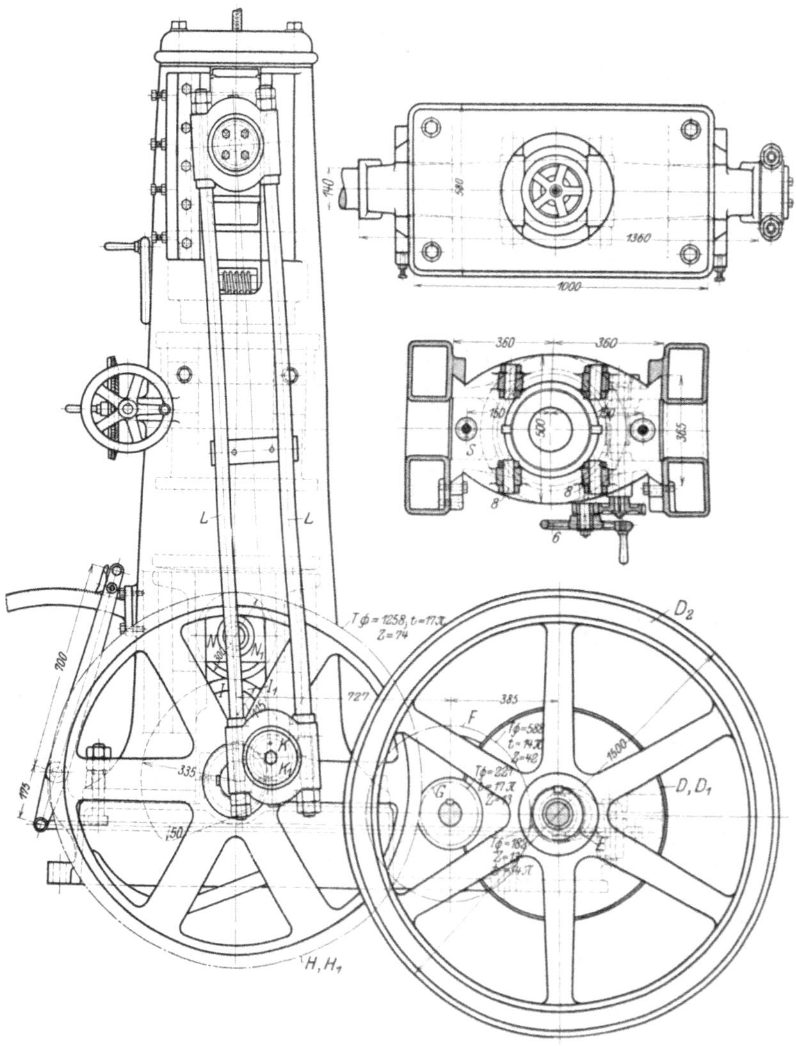

Fig. 651.

Da die Einzelheiten des Gesamtaufbaues sich mit denen der vorherbeschriebenen Presse decken, sollen hier nur die durch die neue Arbeit bedingten Abweichungen besprochen werden. Die Teile, die sich mit der einfachen Ziehpresse derselben Firma decken,

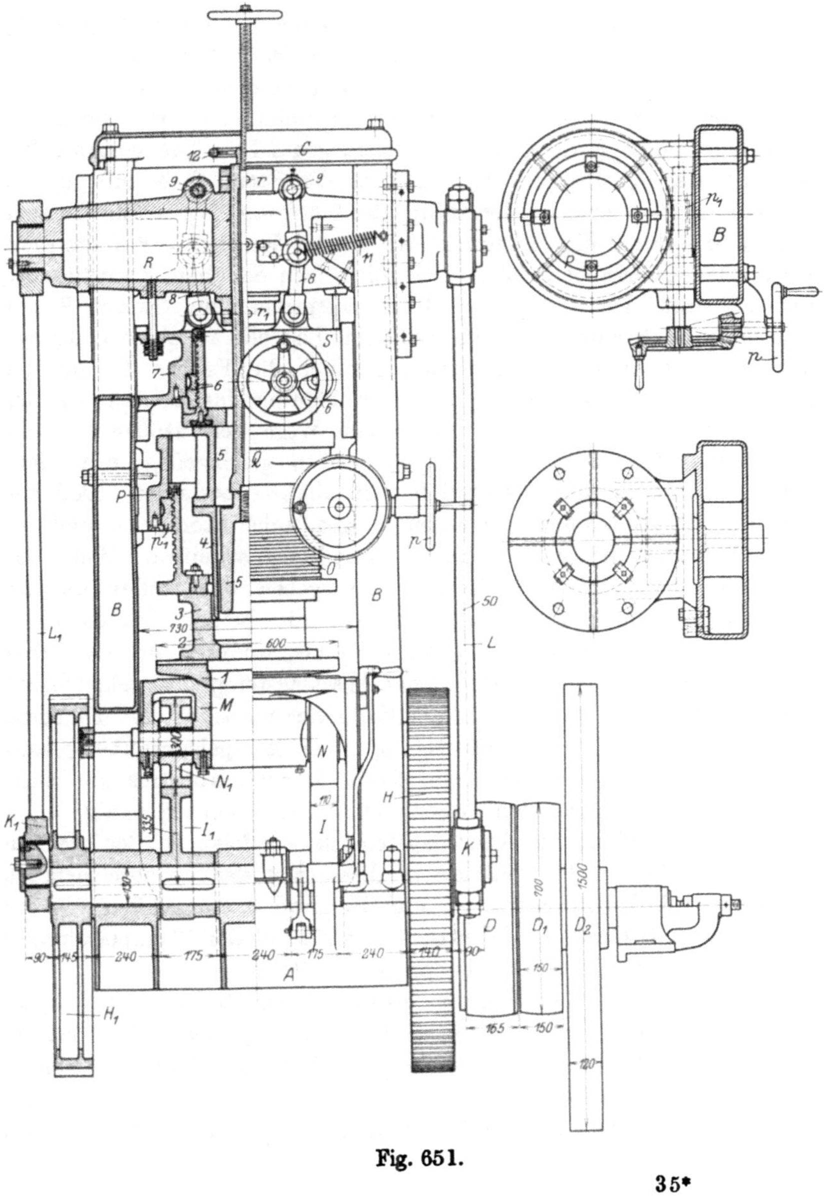

Fig. 651.

sind auch mit den gleichen Buchstaben bezeichnet wie in Fig. 649 ohne jedoch eine gesonderte Besprechung zu erfahren.

Auf dem Pressentisch M ist auf einer Kalottenplatte 1 der Ziehring 2 befestigt, gegen den der Blechhalter 3 sich legt und das Blech festhält. Dieser liegt wie in der vorhergehenden Maschine fest und wird in derselben Weise mittels Handrades p und Schneckengetriebes auf der Einstellmutter p_1 nachgestellt, bis die genügende Anpressung auf dem Blech vorhanden ist. Damit ist das Blech für den ersten Zug fertig. Nun gehen Ziehstempel 4 und 5 gemeinsam herunter, wobei aber nur 4 der ziehende Stempel ist. Er erhält seine Bewegung von dem Querhaupt R durch die Pleuelstangen LL_1. Die gesamte Bewegung des Querhauptes S, an dem der Stempel 4 mittels des Zwischenstückes und der Verstellmutter 6 befestigt ist, erfolgt in zwei Stufen, die durch die Bewegung der beiden Kniehebel 8 9 geregelt wird. Das Querhaupt S führt sich mit seinen Schuhen 7 am Ständer und ist mit den Kniehebeln 8 an entsprechenden Bolzen 9 des Querhauptes R aufgehängt, während eine Leitrolle an einer Leitkurve des Ständers abrollt und gleichzeitig durch die Federn 11 dauernd gegen diese gepreßt wird. Da diese Führung für den ersten Teil des Weges senkrecht also parallel zum Wege des Stößels ist, so wird sich Querhaupt S mit Stempel 4 gleichzeitig mit dem Querhaupt R und dem Stempel 5 nach abwärts bewegen. Dadurch wird aus dem Blech die erste Form gezogen, wobei Stempel 4 und der obere Teil der Matrize 2 als Anschlagwerkzeug arbeiten. Wenn der Stempel 4 seinen Weg vollendet hat, gehen die Rollen an der Ecke der Führung vorbei und knicken die beiden Kniehebel 8 unter dem Zug der Federn 11 seitwärts aus, wodurch beim weiteren Abwärtsgang des Querhauptes R der Anschlagstempel 4 stehen bleibt und für den jetzt folgenden Zug als Blechhalter arbeitet. Nun geht der Ziehstempel 5, dessen Befestigung in der Stempelschraube Q deutlich sichtbar ist, mit dem Querhaupt R nach abwärts und zieht das Arbeitsstück durch den engen Teil des Ziehringes 2 fertig.

Die Einstellung des Anschlagstempels 4 erfolgt in der gleichen Weise wie die des Blechhalters in dieser und der vorbeschriebenen Presse mittels Schneckenantrieb und Handrad 6 von außen, die des Ziehstempels 5 durch das Handrad 12 am Kopf der Stempelschraube Q, wiederum wie in der erstbesprochenen Presse.

Der Antrieb der Presse erfolgt durch Riemen auf Fest- und Losscheibe, wozu noch eine Reibungskupplung mit Scheibenbremse tritt, damit die Presse in jeder Stellung leicht und stoßfrei eingerückt werden kann. Selbstverständlich kann der Antrieb auch mittels Elektromotor auf das Schwungrad oder unmittelbar gekuppelt erfolgen.

Die Maschinen werden von 220—680 mm Stempeldurchmesser bei 325—900 mm Blankettdurchmesser für eine Ziehtiefe von 140—360 mm gebaut. Dementsprechend macht die Presse 12 bis 5 Stößelhübe minutlich. Der Kraftbedarf steigt von 3 bis 8 PS und das Gewicht der Presse von 6500 bis 32000 kg.

Die abgebildete Presse nimmt eine größte Blechscheibe von 460 mm Durchmesser auf für einen größten Ziehstempel von 320 mm Durchmesser bei einer Gesamtziehtiefe von 190 mm. Die Ziehtiefe des äußeren Stempels beträgt 90 mm.

Liegende Ziehpressen (L. Schuler, Göppingen). Beim Ziehen von Brennerteilen, S. 288, wurde bereits die Verwendung einer liegenden Kurbelpresse angeführt. Sie werden schon lange verwendet, da sie ein bequemes Einlegen der Scheiben und gleichfalls eine leichte Entfernung der fertigen Stücke gestatten. Da sich auch die Werkzeuge verhältnismäßig leicht und schnell umspannen lassen, andrerseits auch nicht teurer, ja eher billiger als die für senkrechte Pressen sind, dürfte sich diese Form Pressen für kleinere Betriebe mit wechselnden Aufträgen empfehlen.

Sie werden hauptsächlich zum Einziehen, Nachziehen und Konischziehen vorgezogener Näpfchen verwendet und finden deshalb in der Lampenfabrikation, zur Herstellung von Schalen, Sieben, Bleistift- und Federhalterhülsen, Zwingen usw. Verwendung. Eine wagrechte Kurbelpresse mit selbsttätiger Zuführung mittels Füllkanals und beweglichem Schieber, selbsttätigem Auswerfer und der verbesserten Drehkeilkupplung[1]) nach Schuler ist in Fig. 652—654 wiedergegeben. Die Maschine besteht aus einem kräftigen, verrippten Bett A, das auf zwei Füßen BB_1 aus Rippenguß ruht. Auf der linken Seite, unmittelbar über dem einen Fuß, ist der Antrieb für den Stößel untergebracht. Er erfolgt von einer Fest- und Losscheibe mit Schwungrad CC_1C_2 über eine in zwei mittleren Lagern gelagerte Welle auf ein Ritzel D, das mit dem eigentlichen Antriebsstirnrad E auf der Kurbelwelle kämmt. An dieser Stelle ist die Sicherheitskupplung, die Schuler als Drehkeilkupplung F ausführt, eingebaut. Sie verhindert, daß der Stößel ein zweites Mal niedergeht, wenn der Arbeiter vergißt, nach beendigtem Arbeitshub den Fußtritt G freizugeben. Durch diese Kupplung wird der Stößel am linken Hubende nach jeder Umdrehung stillgesetzt, gleichgültig, in welcher Stellung sich der Fußtritt G befindet. Wird ein zweiter Stößelniedergang verlangt, so muß der Fußtritt losgelassen und ein zweites Mal nach abwärts gedrückt werden. Arbeitet

---

[1]) Z. d. V. d. I. 1910, S. 1849, Fig. 24/25 mit genauer Beschreibung der Kupplung.

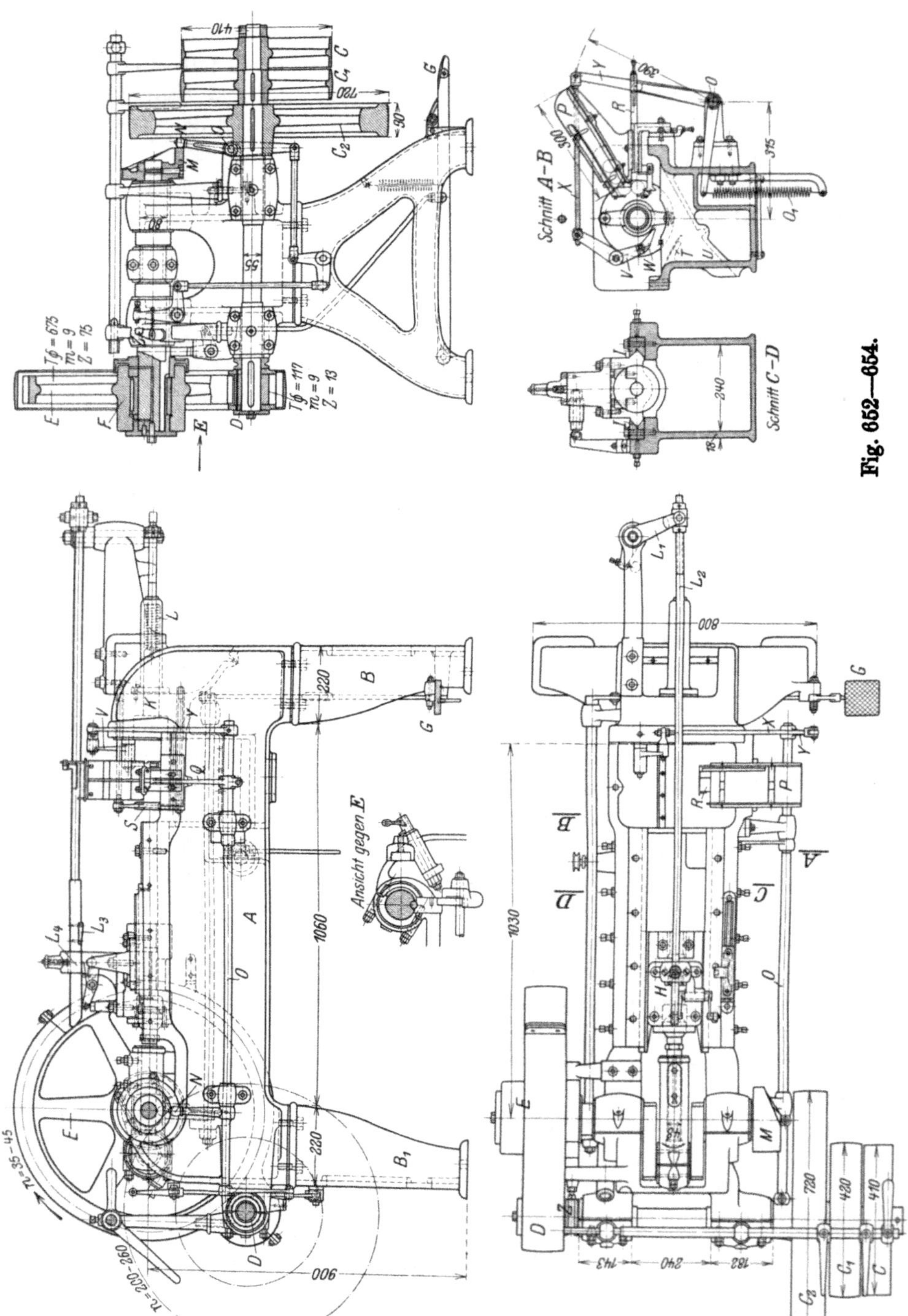

Fig. 652—654.

dagegen eine mit dieser Sicherheitsvorkehrung ausgestattete Presse mit selbsttätigem Materialvorschub, so läßt sich die Kupplung feststellen und arbeitet dann als gewöhnliche feste Kupplung.

Der Stößel H läuft in langen nachstellbaren Prismen I und ist einstellbar. Am Pressentisch K ist der zwangläufige Auswerfer L zu erwähnen, der durch das Gestänge $L_1 L_2$ gesteuert wird, indem beim Stößelrückgang der Keil L_3 von dem Mitnehmer L_4 und dadurch die Auswerferstange mitgenommen wird. Eine im Auswerfer L eingebaute Feder zieht das Auswerfergestänge nach Freigabe des Keiles L_3 wieder in die Anfangslage zurück.

Besondere Aufmerksamkeit beansprucht die selbsttätige Zuführung mit Füllkanal und beweglichem Schieber, Fig. 654, die mit einer besonderen Sicherheitsvorrichtung ausgestattet ist, falls ein Röhrchen während des Ziehens abreißt oder im Ziehring stecken bleibt. Die selbsttätige Zuführung wird von einer Leitkurve M auf der Kurbelwelle der Presse betätigt. Auf dieser Kurve schleift eine Rolle N auf einem einarmigen Hebel, der durch seine Schwingbewegung die Welle O verdreht und durch die Feder O_1 immer in Berührung mit der Leitkurve gehalten wird. Von der Welle O wird dann die gesamte Bewegung der Zuführungsvorrichtung abgeleitet, wie Fig. 654 zeigt. Die Näpfchen werden in die Zuführungsrinne P, die in gewissen Grenzen entsprechend dem Durchmesser und der Länge der zu ziehenden Näpfchen verstellbar ist, eingelegt. Sie rollen, wie gezeichnet, die Zuführungsrinne herab und werden von dem Schieber R, der durch den Hebel Q von der Welle O bewegt wird, nach der Achse der Maschine vor den Ziehring befördert. Die Öffnung in dem Schieber ist so groß, daß immer nur ein Näpfchen vor das Werkzeug kommen kann. Die gezogene Hülse, die mit dem Stempel nach links genommen wird, wird dann von einem Federabstreifer S vom Stempel entfernt und fällt über das Blech T auf eine schräge Fläche U, von der sie aus der Maschine entfernt wird. Die Sicherheitsabstellvorrichtung besteht aus einem schwingenden Hebel V mit angelenkter Greifernase W, die durch die Zugstange X von dem Hebel Y, der auf der Welle O verkeilt ist, aus in schwingende Bewegung versetzt wird. Bleibt nun eine Hülse im Werkzeug stecken, so geht die Greifernase W, indem sie sich um ihren Bolzen aufbiegt, an der Hülse vorbei, schnappt unter ihr aus und verhindert so den Hebel V und damit die Welle O an der Rückdrehung, wodurch der Anschlag bei Z an der Drehkeilkupplung nicht zur Berührung kommt und die Presse stillgesetzt wird.

Pressen der Kartonnagenindustrie (Karl Krause, Leipzig). Ebenso wie die Werkzeuge der Kartonnagenindustrie, die auf S. 59, 150 besprochen worden sind, in ihrem Aufbau von den Werkzeugen der

Metallverarbeitung sehr verschieden sind, so erfolgt auch in den meisten Fällen die konstruktive Durchbildung der bezüglichen Pressen nach ganz anderen Gesichtspunkten. Infolge des verhältnismäßig geringen Kraftbedarfes können die Pressen leichter, der Antrieb einfacher werden, wohingegen die Form des verarbeiteten Materiales, das meistens in Bogen angeliefert wird, eine breite Ausladung und einen breiten Durchgang an den Werkzeugen verlangt. Deshalb werden auch so oft als angängig mehrere Schnitte gleichzeitig geführt, so daß sich der Einfluß der Durchbildung der Werkzeuge wieder in der besonderen

Fig. 655.

Ausführung des Stößels und des Tisches zeigt. Da ausserdem die geringe Steifigkeit der Bogen das freihändige Einlegen unmöglich macht, sind besondere Anlegetische zu beiden Seiten der Werkzeuge vorgesehen. Eine derartige Presse für den Einbau der Werkzeuge nach Fig. 44—53 zur Herstellung von Faltschachteln, Einschiebern für Zigarettenschachteln und dgl. zeigt Fig. 655. Die Maschine baut sich in einem aus zwei seitlichen Ständern aus Rippenguß gebildeten Gestell auf, das im Unterteil die durchgehende Antriebswelle und im oberen Teil die beiden nachstellbaren Führungen für den Stößel enthält. Um die volle Breite des Stößels möglichst ausnutzen zu können, liegt der Antrieb des Stößels durch Exzenter, Pleuelstangen und Kreuz-kopf außen an den beiden Seiten des Ständers, so daß die volle Breite zwischen den Seitenständern für den Papierdurchgang freibleibt. Der Antrieb erfolgt von einer Riemenscheibe mit eingebauter Reibungskupplung, von der durch ein Paar eingebauter Stirnräder die Exzenterwelle angetrieben wird. Der Stößel ist hier mit einem durchgehenden Schlitz zur Befestigung der Stempel versehen, doch wird an Stelle dessen auch eine an der Vorderseite des Stößels angebrachte Nut zur Befestigung der Werkzeuge nach Fig. 44—53 vorgesehen. Auf

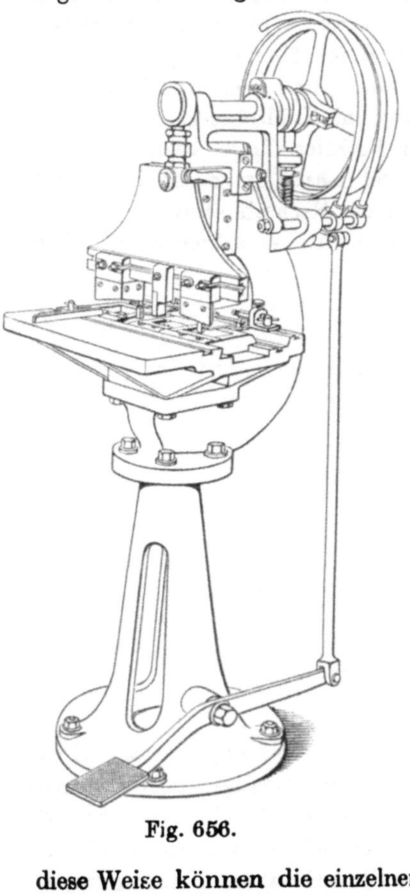

Fig. 656.

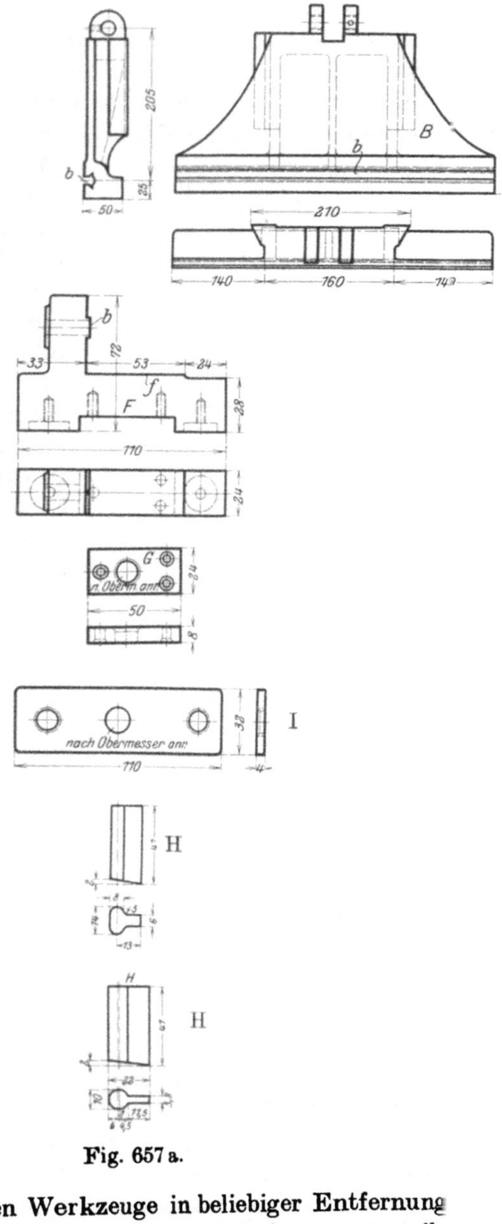

Fig. 657 a.

diese Weise können die einzelnen Werkzeuge in beliebiger Entfernung voneinander eingesetzt werden. Die Befestigung der Werkzeugunterteile erfolgt auf zwei durchgehenden Schienen am Tisch in einer Schwalbenschwanznut, wodurch die gleiche Einstellungsmöglichkeit wie beim Oberteil erhalten wird.

Kleine Universalstanzmaschine (Karl Krause, Leipzig). Eine kleine, durch Riemen angetriebene, einständrige Schnittpresse zeigt Fig. 656, deren Einzelteile, Tisch, Stößel und Werkzeuge einzeln in maßstäblichen Zeichnungen in Fig. 657 wiedergegeben sind. Die Maschine dient zur

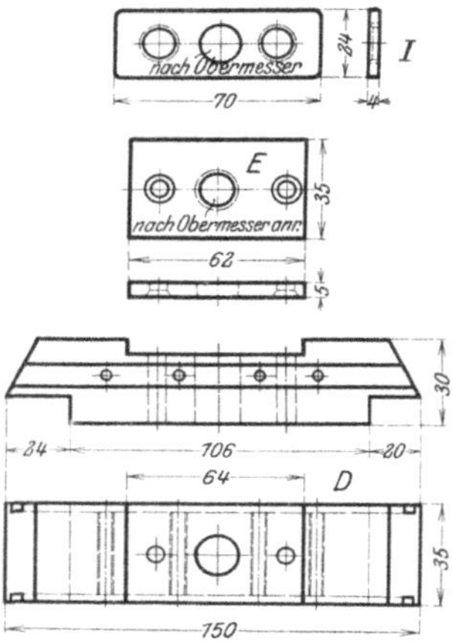

Massenherstellung kleiner Schnitte, wie Automaten- und Zigarettenschachteln, Kartothekschnitten usw., die 7 cm Tiefe nicht überschreiten und bis zu 40 cm Breite haben können. Die Maschine ist für Riemen- oder Elektromotorantrieb eingerichtet und wird mit Momentkupplung, Fußeinrückung und selbsttätiger Ausrückung mit Stillsetzung des Stößels in der höchsten Lage versehen. Der Stößel ist einstellbar gemacht, damit man beim Nachschleifen der Werkzeuge den Materialverlust ausgleichen

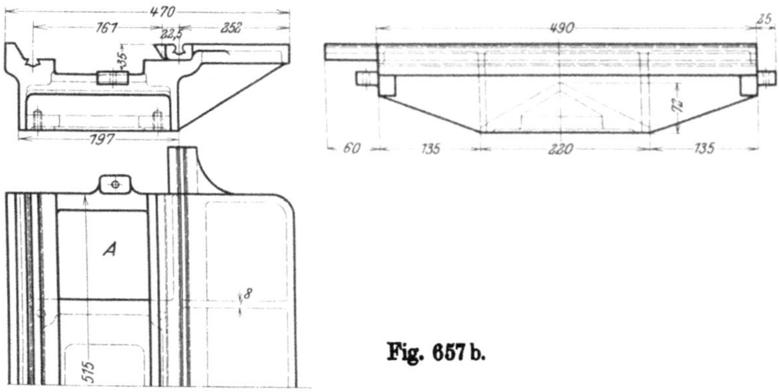

Fig. 657 b.

kann. Eine Anlegeschiene für die Kartonbogen, deren Größe übrigens nicht begrenzt ist, wird am Tisch befestigt. Die wichtigsten arbeitenden Teile der Maschine sind einzeln in Fig. 657 herausgezeichnet. Der eigentliche Pressentisch A, der die Werkzeugunterteile in der nachstellbaren Schwalbenschwanznut aufnimmt, ist ein verripptes Gußstück mit einem breiten Anlegetisch, der auf dem

Zwischenstück des Ständers verschraubt wird. Der Stößel B wird in Schwalbenschwanzführungen mit aufgeschraubten Leisten gehalten und verbreitert sich nach unten auf 440 mm, also auf das Doppelte der Führungsbreite. Er ist mit der obenerwähnten, vorderen Schwalbenschwanznut bei b zur Befestigung der Stempelteile versehen. Außerdem ist noch ein Winkel vorgesehen. Auf der Tischplatte A wird in den gezeichneten Schwalbenschwanznuten die Gesenkplatte (Untereinsatz) D befestigt, in der die Schnittplatte E verschraubt wird. Als Schnitt ist der Einfachheit halber ein kreisförmiges Loch angegeben, während andere Formen, z. B. für Registerkartenschnitte, an den Stempeln H zu sehen sind. An dem Stößel B wird der Stempelkopf F mit einer Kopfschraube befestigt und durch die durchgehende Feder und Nut b in seiner Stellung gesichert; gleichzeitig nimmt eine bearbeitete Leiste auf der Oberseite bei f den Arbeitsdruck des Schnittes auf. In dem Stempelkopf F wird die Kopfplatte G, in der der Stempel H in gewohnter Weise vernietet ist, mittels dreier versenkter Flachkopfschrauben befestigt. Bei I sind schließlich die beiden Abstreifer noch abgebildet. Die übrigen Teile sind aus der Zusammenstellungszeichnung Fig. 656 deutlich zu ersehen.

Prägepresse für Karton (Karl Krause, Leipzig). Als Beispiel für eine Prägepresse der Kartonnagenindustrie sei eine Kniehebelpresse mit Riemenantrieb, Fig. 658, angeführt. Dadurch, daß auch hier wie beim Ziehen von Schachteln, S. 305—307, die Werkzeuge angeheizt werden müssen, erhält die Presse ihre besondere Form. Auf einem Gußbett A sind durch zwei durchgehende Keile die beiden Hauptsäulen der Presse aus Schmiedeeisen $B_1 B_2$ befestigt, die das schwere Querhaupt C tragen. An dem Querhaupt ist oberhalb der Matrize der Heizkörper Y befestigt. Dieser Teil der Maschine steht fest und trägt auch den gesamten Antrieb. Außerdem sind an den Säulen der Maschine die Führungen $D_1 D_2$ für den beweglichen Tisch befestigt, die gleichzeitig die Lagerung für die erste Welle R der Maschine enthalten. Auf dieser Welle sitzen links die beiden Antriebsriemenscheiben T, Fest- und Losscheibe mit dem Schwungrad, hinter denen außerdem noch eine Reibungskupplung U zur augenblicklichen Stillsetzung der Presse in jeder Stellung eingebaut ist. Von dieser Welle wird die Bewegung durch ein Paar Stirnräder auf die Zwischenwelle V_1 und von dieser durch ein weiteres Stirnradvorgelege auf die eigentliche Antriebswelle N übertragen. Von dieser Welle wird mittels unrunder Scheiben die Kniehebelstelze JK bewegt, die einerseits in der festen Pfanne v aufruht, andrerseits mit ihrer oberen Pfanne den Tisch F auf- und abwärts bewegt. Zwangläufig mit dieser Bewegung wird der Werkzeugunterteil h von den Exzentern H, Zugstangen H_1 und Schwingen $i_1 i_2$ ein-

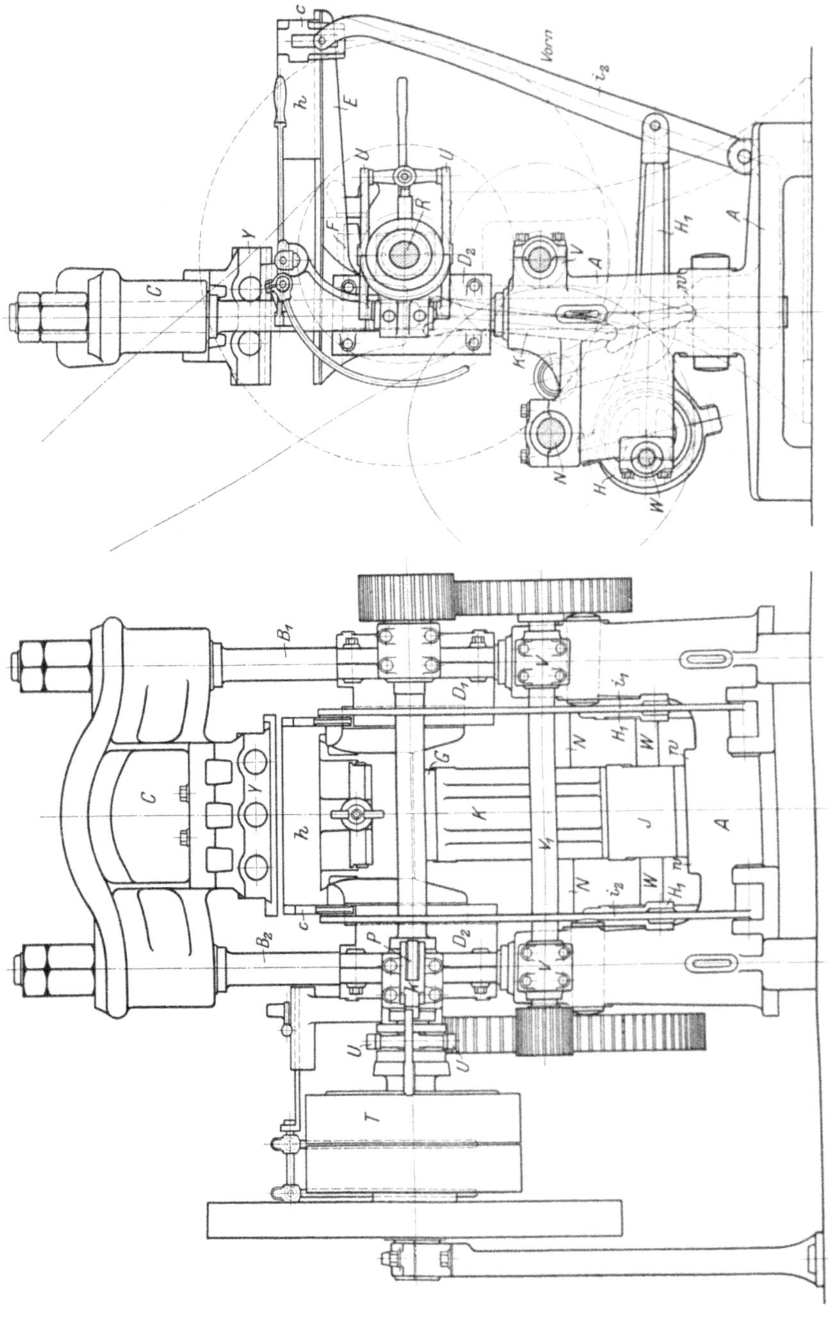

Fig. 658.

und ausgeschoben, wozu sich die Führungszapfen bei c in zwei senkrechten Kulissen führen, wodurch die Schwingbewegung ohne Behinderung des Auf- und Abwärtshubes ausgeführt werden kann. Die abgebildete Presse hat eine Druckfläche von 33 × 44 cm bei einem senkrechten Hub von 16 mm. Es ist dies die kleinste Größe dieser Bauart. Die Maschine ist für einen Arbeitsdruck von rund 95000 kg gebaut und wird mit einem Druck von 110000 kg geprüft. Die Presse macht 8—11 Hübe in der Minute und wiegt rund 2500 kg.

Die Presse dient hauptsächlich für Buchbinderarbeiten zum Vergolden und Prägen und kann auch mit besonderen Farbwerken verbunden werden. Ein Druckanzeiger mit Kontrollvorrichtung kann geliefert werden.

Indiziervorrichtung für Pressen[1]). Peiseler beschreibt an der angegebenen Stelle einen Versuchsapparat, der bei der Firma Krause in Leipzig in Gebrauch ist und dort für das Indizieren von Schnitt-

Fig. 659.

pressen verwendet wird. Der Apparat ist in Fig. 659 abgebildet, eingebaut in eine Schnellschneidemaschine. Der Apparat besteht aus einem gemeinsamen Tragkörper, der eine Anzahl nebeneinanderliegender Bohrungen, in die öldicht geschliffene Kolben eingepaßt sind, enthält.

[1]) W.T. 1913, S. 710, Fig 10—16.

Diese Zylinder stehen durch eine gemeinsame Bohrung in dem Tragkörper miteinander und gleichzeitig mit einem Indikator, in Verbindung der seine Bewegung mittels eines Hubreduktors von dem Pressenstößel erhält. Außerdem ist ein Anzeigemanometer gewöhnlicher Bauart angeschlossen. Es erscheinen also in den Schaubildern, Fig. 660, die Arbeitsdrücke in Funktion der Eindringtiefen, der Messerwege, aufgetragen. Mit dem gewöhnlichen Manometer allein sind keine richtigen Zahlenwerte zu erhalten, da bei der plötzlich einsetzenden Drucksteigerung der Zeiger des Manometers die richtige Angabe weit überfliegt, was man auch bei anderen derartigen Belastungen, z. B. bei Bremsungen feststellen kann. Die drei Schaubilder in Fig. 660, die den Einfluß verschiedener Arbeitsgeschwindigkeit zeigen, ergeben mit abnehmender Geschwindigkeit auch einen abnehmenden Arbeitsdruck. Wie ersichtlich, ist der Abfall von 10100 kg bei 14,5 Schnitten minutlich auf 9500 kg bei 10,5 Schnitten minutlich ziemlich beträchtlich und steigert sich bis zu 4570 kg beim Durchdrehen von Hand. Dagegen bleibt der Abfall des Druckes vom Beginn bis zum Ende des Schnittes fast gleich mit rd. 1500 kg und fällt erst im dritten Fall auf rd. 1000 kg.

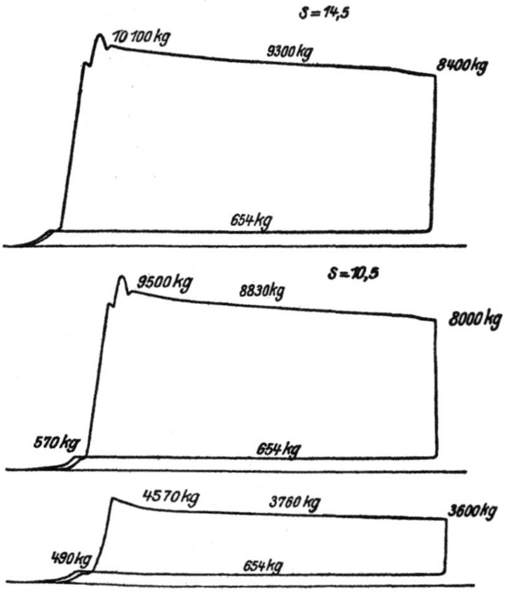

Fig. 660.

2. Spindelpressen.

Die Spindelpresse als solche ist wohl die älteste Form der Presse und war ursprünglich nur für den Handantrieb eingerichtet. Sie hat infolgedessen die größte Wandlung bei der Einrichtung für Kraftantrieb erfahren. Sie ist hauptsächlich da am Platze, wo neben der Druckwirkung eine Schlagwirkung unerläßlich ist, wie bei schweren Prägearbeiten für Münzen, Medaillen, Besteckgriffe und dgl. Aber gerade dieser Vorzug birgt auch den größten Nachteil in sich. Da nämlich beim Kraftantrieb die Verbindung des Antriebes mit der Spindel keine zwangläufige ist und infolgedessen die Spindel mit der größten während des Abwärtshubes durch die Schwungradwirkung erreichten lebendigen Kraft auf das Arbeitsstück auftrifft, so ist die Gefahr eines Ständer- oder Spindelbruches im Falle, daß kein Arbeitsstück im Werkzeug liegt, sehr groß. Es werden infolgedessen in letzter Zeit die größten Vorsichtsmaßregeln bei dem Entwurf des Ständers angewendet, um so mehr, als die genaue, zahlenmäßige Berechnung der der Schlagwirkung[1]) entsprechenden Beanspruchung des Ständers auf große, rechnerische Schwierigkeiten stößt und man die Abhängigkeit dieser Drucksteigerung von der Hubzahl bzw. Geschwindigkeit des niedergehenden Oberstempels noch nicht gesetzmäßig festgelegt hat.

Um diesen Punkt zu umgehen, werden jetzt die Reibungsspindelpressen mit Sicherheitsvorrichtungen im Schwungrad ausgestattet, oder der Ständer durch schmiedeeiserne Ankerbolzen zusammengehalten, die leichter innerhalb bestimmter Sicherheitsgrenzen zu bemessen sind und außerdem nicht der Gefahr eines plötzlichen Bruches unterliegen. Auch wird der Ständer aus Stahlguß hergestellt.

Die Spindelpressen werden entweder — und zwar in der größten Zahl — zweiständrig oder einständrig gebaut. Für die größten Drücke wird aus den obenerwähnten Gründen meistens eine verhältnismäßig kleine Ausladung verwendet, wie es auch meistens für Prägen genügt. Wenn die Pressen dagegen zum Ausschneiden oder Ziehen flacher Formen verwendet werden, ist die Beanspruchung geringer und die Ausladung kann größer werden.

Reibungsspindelpresse für Bolzenstauchen (Gebr. Hübner, Chemnitz). Die Reibungspresse, Fig. 661, ist im eigenen Betrieb der genannten Firma entstanden und deshalb hauptsächlich den Arbeiten der Bolzenherstellung angepaßt, weshalb auch die Ausladung des Ständers klein ist. Trotzdem ist die Maschine mit Rücksicht auf die bei Bolzenarbeiten manchmal vorkommenden sehr schweren Arbeiten in allen Teilen sehr kräftig bemessen. Der Pressenkörper besteht aus zwei getrennten Teilen, dem Ständer A und dem Querhaupt B, die durch schmiedeeiserne Anker mit Gegenmuttern zusammengezogen werden, die die Zugbeanspruchungen aufnehmen. Das Querhaupt B der Presse bildet ein Stück für sich und trägt die lange Bronzemutter D für die Spindel und zu beiden Seiten die Lager CC_1 für den Reibungsantrieb. Die Spindel E ist aus Tiegelgußstahl von 80 kg/mm² Festigkeit hergestellt und erhält durch die Reibscheiben F_1F_2 beim Abwärtsgang eine beschleunigte Bewegung, da das Spindelschwungrad F mit der

[1]) Seehase, Die experimentelle Ermittlung des Verlaufes der Stoßkraft und die Bestimmung der Deformationsarbeit beim Stauchversuch. Dr.-Diss. Berlin 1914.

Spindel nach abwärts geht. Die Belederung erfolgt durch geschichtete Lederringe, die mit den Stirnflächen auf den Reibscheiben laufen. Man erreicht damit eine längere Dauer des Materiales, obwohl diese aus einem Stück geschnittenen Ringe bedeutend teurer als die umgelegten flachen Lederriemen sind. Die Spindelmutter D aus Phosphorbronze erhält ungefähr die vierfache Länge des Spindeldurchmessers, wodurch eine gute Führung und Lagerung gegen den seitlichen Druck der Reibscheiben erhalten wird. Der Stößel G ist aus geschmiedetem Stahl und gleitet auf Verschleißleisten aus Phosphorbronze, so daß durch diese Führungen die Spindel außerdem noch gegen Verbiegen geschützt wird.

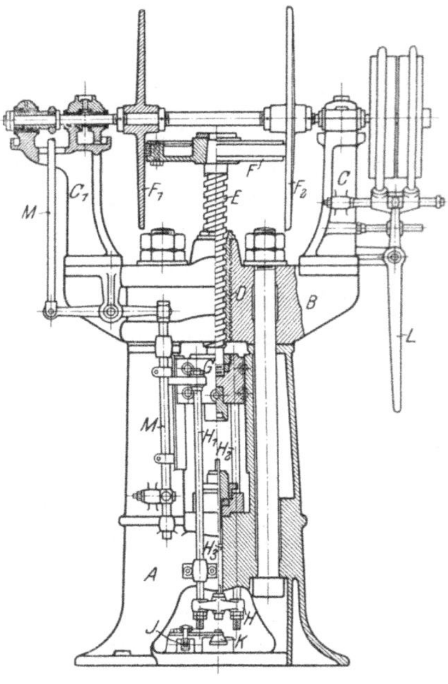

Fig. 661.

Die Firma gibt an, daß ein Lederbezug bei täglichem Betriebe und täglichem Einfetten ungefähr 5 Jahre hält.

Das Auswerfergestänge H ist außerhalb an der Pressebefestigt und an dem Stößel angehängt. Es besteht aus einer Brücke H und zwei Gehängestangen $H_1 H_2$, die den Gegenstempel bzw. Auswerferstift H_3 bewegen. Diese Einrichtung wird gleichzeitig ausgenutzt, um große Bolzen, die nicht in einer Stauchung fertiggestellt werden können, in zwei aufeinander folgenden Schlägen in einer Hitze und einem Einlegen zu verarbeiten. Zu diesem Zweck ist bei gleichem Werkzeug der Gegenstempel in seiner Höhe verstellbar gemacht, indem der Arbeiter nach dem ersten Schlag mittels des Hebels J den Schlitten K unter die Brücke H schiebt, so daß sich der Gegenstempel H_3 auf den Untersatz auflegt, weiter in das Werkzeug hineingeschoben und dadurch der vorgestauchte Bolzen ein größeres Stück aus dem Werkzeug herausgeschoben wird, so daß beim nächsten Schlag die volle Materialmenge für das Fertigstauchen des Kopfes aus dem Werkzeugunterteil herausragt.

Die Einrückung des Antriebes erfolgt mittels der Riemengabel L, während von dem Gestänge M aus das Spindelschwungrad an die Reibscheiben angelegt wird. Mittels einer eigenen Einrichtung lassen sich die Werkzeuge, Fig. 243, leicht und schnell zentrisch einspannen.

Spindelpresse mit Druckreglerschwungrad (Paten Schull). Die Überbeanspruchung der Spindeln und des Ständers bei einem Hartaufschlagen des Stempels auf dünne Bleche oder beim Fehlen eines Werkstückes im Werkzeug bedingt, daß die gesamte im Schwungrad aufgespeicherte Energie in einem ganz kurzen Wege zu vernichten ist und infolgedessen sehr hohe Drücke und große Beanspruchungen dieser Teile die Folge sind. Das Patent Schull (Bauart Mönkemöller & Co.) verbindet nun das Antriebsrad A der Spindel mit dem Schwungring D

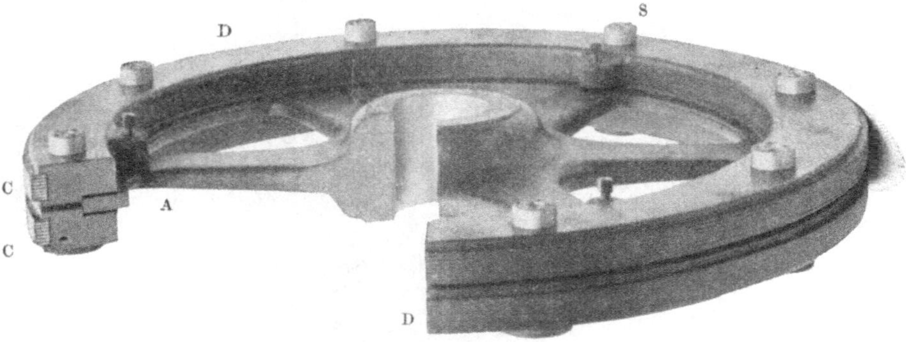

Fig. 662.

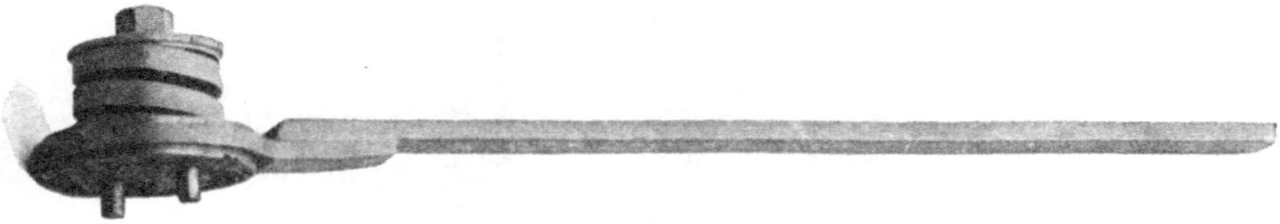

Fig. 663.

nicht fest, sondern durch Reibung. Der Schwungring D, der auch die Belederung C für den Antrieb durch die Reibscheiben trägt, ist aus zwei gleichen Teilen hergestellt, die mittels einer Anzahl am Umfang verteilter Schrauben S gegen den Kranz des Rades A gepreßt werden. Um die genügende Reibung zu erzeugen, werden an den Berührungsflächen Ringe aus Vulkanfiber eingelegt. Es ist klar, daß man durch ein übermäßiges Anziehen der Schrauben S die Wirkung dieser Reibungskupplung vollständig aufheben kann. Es soll ja doch beim Auftreffen des Stößels auf die Unterlage die dem Schwungring innewohnende zu große lebendige Kraft ein Gleiten des Schwungringes D auf den Fiberringen des Rades A unter Überwindung der Reibung die durch das Anziehen der Schrauben erzeugt wird, hervorbringen.

Infolgedessen wird durch diese Reibungsarbeit ein regelbarer Teil der lebendigen Kraft aufgezehrt, so daß der Ständer oder die Spindel nur eine bestimmte Beanspruchung erfahren kann. Fig. 662 zeigt ein derartiges Spindelschwungrad, das die eigenartige Ausbildung der Schraubenköpfe erkennen läßt. Damit man nämlich beim Anziehen der Schrauben nicht

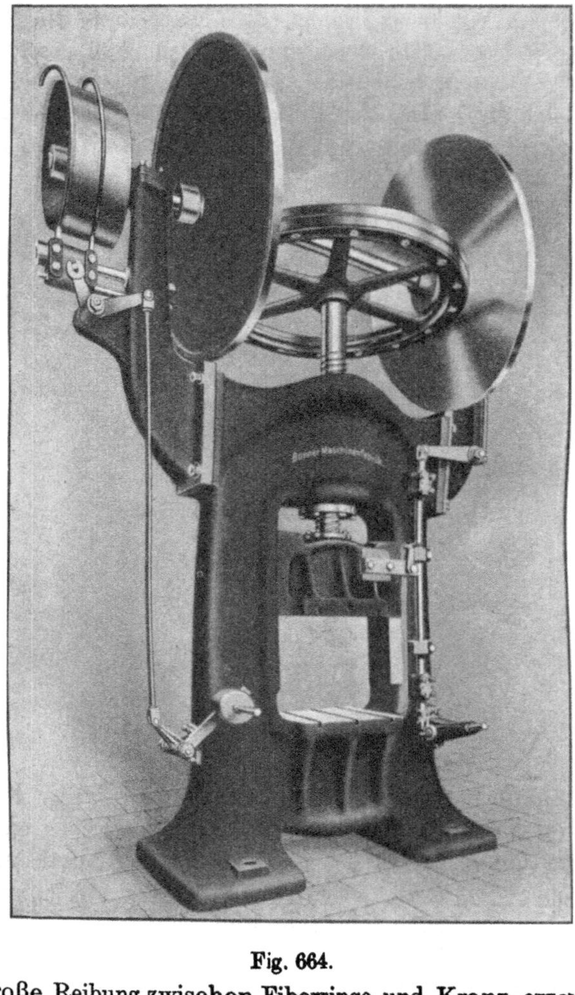

Fig. 664.

eine zu große Reibung zwischen Fiberringe und Kranz erzeugt, erfolgt das Anziehen der Muttern mittels eines Federschlüssels, dessen Feder sich beim Überschreiten eines gewissen Drehmomentes beim Anziehen einfach verdreht, ohne die Schrauben des Ringes weiter anzuziehen.

Fig. 663 zeigt den Schlüssel und Fig. 664 eine mit einem derartigen Schwungrad ausgerüsteten Spindelpresse der Firma Bonner

Maschinenfabrik Mönkemöller & Co. Das Bemerkenswerte und der große Vorteil dieser Einrichtung ist der leichte Einbau, der die Presse, wie die Abbildung zeigt, so wenig verändert, daß äußerlich kaum ein Unterschied gegen die gewöhnliche Bauart wahrzunehmen ist. Durch diese Sicherheitsvorrichtung wird es möglich, die Presse ohne Gefahr mit einer höheren Umdrehungszahl laufen zu lassen bzw. die volle Leistung des Schwungrades auszunutzen. Andrerseits kann man auch bei Zieharbeiten, die eine verhältnismäßig kleine Kraft, aber auf einem langen Weg erfordern, die dem Rade innewohnende lebendige Kraft vollständig ausnutzen und erzielt laut Angabe Leistungssteigerungen bis über 100 %.

3. Hydraulische Pressen.

Die bisher angeführten, mechanischen Pressen sind teils durch ihre arbeitenden Teile, wie die Exzenterpressen, teils durch den innewohnenden Bewegungsvorgang in ihrer Anwendung begrenzt, was sich besonders bei großen Zieh- und Preßarbeiten bemerkbar macht. Nach Überschreitung bestimmter Ziehtiefen oder bei stärkeren Blechen tritt bald derselbe Fall ein, der bereits bei der Entwicklung des Gesenkpressens S. 191 angeführt wurde: die Pressen arbeiten nicht mehr wirtschaftlich.

Ohne näher auf die Vor- und Nachteile der hydraulischen Pressen hier einzugehen, sollen an drei Formen solcher Pressen Anwendungsgebiete derselben gegeben werden, die eine Ergänzung zu dem bereits früher über die diesbezüglichen Werkzeuge Gesagten bilden.

Einfache hydraulische Presse für die Herstellung von Juweliergesenken. Wie bereits häufiger bei der Besprechung der Prägewerkzeuge angeführt worden ist, werden die bezüglichen Matrizen nach einem nach jeder Richtung hin fertiggestellten Musterstempel geschlagen, d. h. unter starkem Druck oder unter dem Fallwerk nach und nach in eine Stahlplatte eingetrieben, bis der Abdruck in der Platte die richtige Tiefe für das Arbeitsstück hat. Woodworth gibt für solche Arbeiten, hauptsächlich Gesenke für Gold- und Silberwaren, eine sehr einfache hydraulische Presse, Fig. 665, an.

Die eigentliche Presse besteht aus nur drei Teilen, dem Ständer A, dem Kolben B und der Gegenplatte C; die gewöhnlichen Verbindungsstangen, Muttern usw. sind gänzlich weggelassen worden und alle diese Teile in der Konstruktion zu einem Ganzen vereinigt worden. An der Oberseite erhält der Ständer eine Ausdrehung zur Aufnahme der Gegenplatte C und an der Unterseite die Bohrung für den Kolben.

Außerdem ist e an der Oberseite eben gedreht, so daß sich dort eine Auflagerfläche ergibt, wie der Schnitt zeigt. Die Platte auf der Oberseite des Ständers dient nur zum Halten der Gegenplatte, wenn der Druck abgelassen wird. Der Zapfen am unteren Teil des Kolbens ist in die Ausbohrung im Ständer passend gedreht, wodurch der Kolben eine genaue Führung erhält. Demnach zeigt die Zeichnung alle Teile der Presse, die zum Gebrauch nur an eine passende Druckpumpe angeschlossen werden muß. Es ist einleuchtend, daß diese Konstruktion eine große Steifigkeit besitzt und besonders für die Herstellung der Gesenke in der Juwelierwerkstätte und ähnliche Zwecke, wo ein gehärteter Musterstempel kalt unter großem Druck langsam in die Matrize eingedrückt wird, verwendet werden kann.

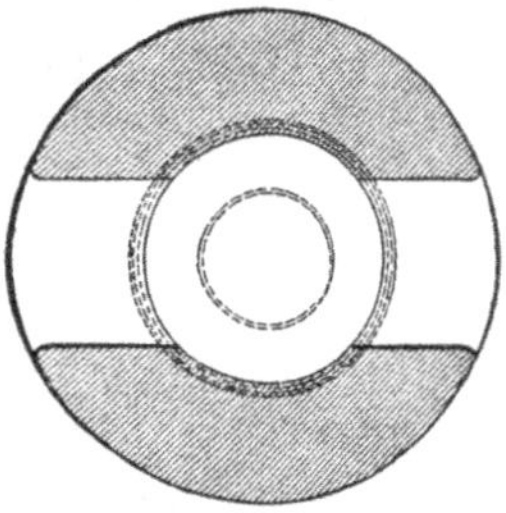

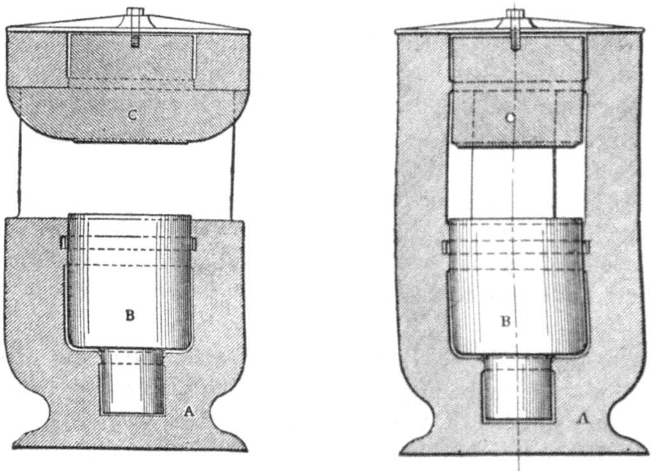

Fig. 665.

Doppeltwirkende hydraulische Ziehpresse[1]). Die Abbildung und Zeichnungen, Fig. 666—667, stellen eine kräftige Presse dar, die für schwere Zieharbeiten entworfen und mit einer eigenartigen Kolben- und Zylinderanordnung zur Betätigung des Ziehstempels und der Niederhalter versehen ist.

---

[1]) Die Maschine ist von L. Schuler, Göppingen (im amerik. Original nicht angegeben), vgl. W. T. 1912, S. 458.

In der Grundplatte der Presse bei A, die auf einem geeigneten Fundament verankert ist, sind vier kräftige Schraubenspindeln B verlagert, die das obere feste Querhaupt C mit der Matrize D tragen.

Fig. 666.

C kann gehoben und gesenkt werden, wozu die vier Muttern an den Hauptschrauben B durch das Zahnrädergetriebe mittels der Riemenscheiben am Kopf der Presse nach vor- und rückwärts gedreht werden können. Das bewegliche Querhaupt, der Tisch E, trägt den Blechhalter und kann durch den hohlen Plunger G im Zylinder H, geführt an den vier Schrauben B, gehoben und gesenkt werden. Dieser Plunger ist selbst wiederum ein Zylinder, in dem ein Scheibenkolben I, der in der gezeichneten Weise mit dem Ziehstempel J verbunden ist,

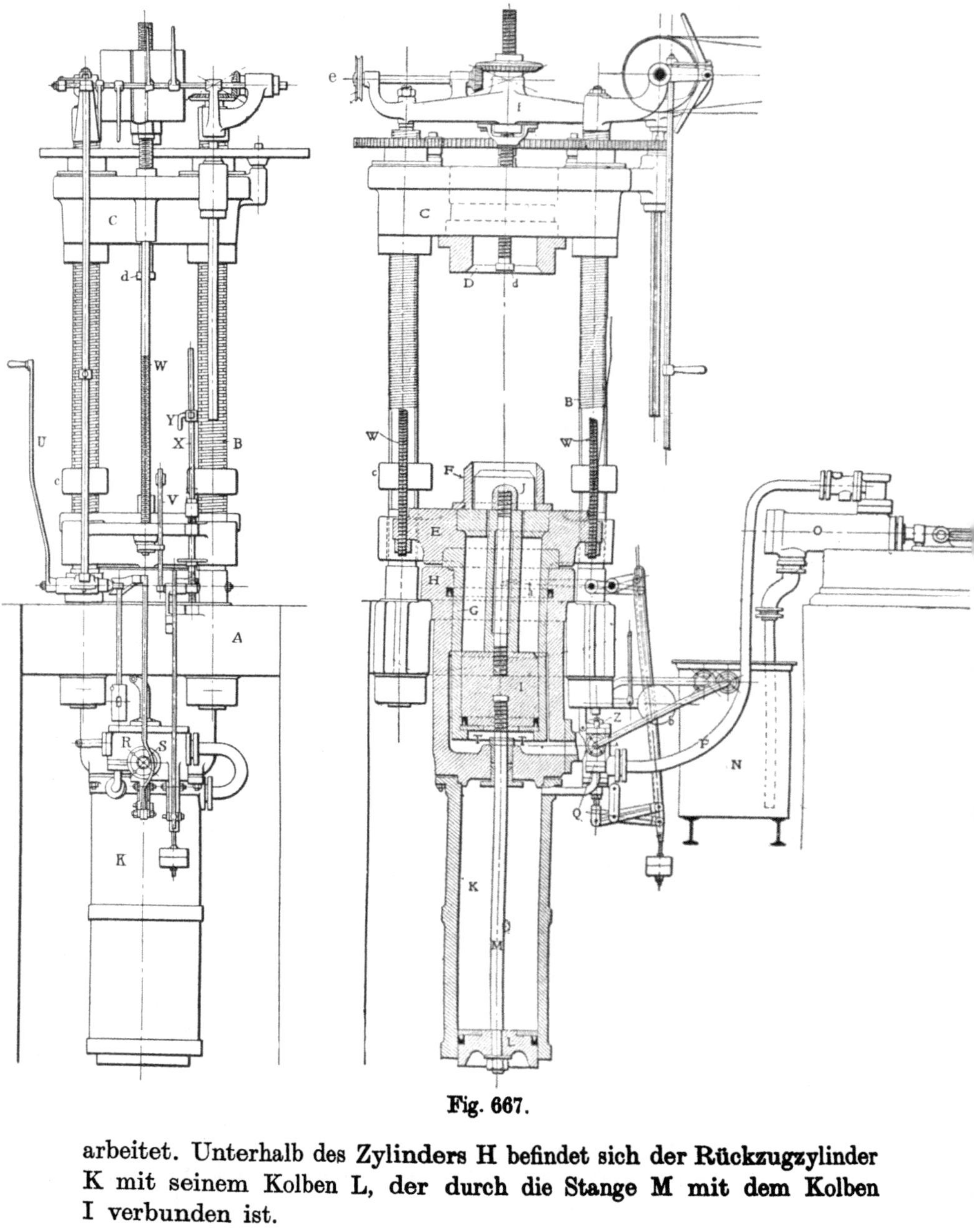

Fig. 667.

arbeitet. Unterhalb des Zylinders H befindet sich der Rückzugzylinder K mit seinem Kolben L, der durch die Stange M mit dem Kolben I verbunden ist.

Bei N ist der Flüssigkeitsbehälter und bei O eine zweizylindrige[1]) Riemenpumpe, die die Flüssigkeit durch das Rohr P zum Ventilgehäuse Q und von da in die Zylinder drückt. In dem Ventilgehäuse Q befinden sich die Ventile R und S, von denen das erste die Zu- und Ableitung der Flüssigkeit zu und von dem Raum T unterhalb der Kolben G und I besorgt und durch den Hebel U gesteuert wird, während das Ventil S die Flüssigkeitsbewegung zu und von dem Zylinder K regelt und durch ein Handrad eingestellt wird. Dieses Ventil wird nach der Einstellung durch ein Hebelsystem selbsttätig gesteuert, das durch eine Stellstange V und die einstellbaren Anschläge an den Stangen W sowie eine Stange X, die einen Anschlagdaumen Y trägt, bewegt wird. Bei Z befindet sich ein Ventil, das den Druck auf den Rückzugkolben L regelt und zu diesem Zweck durch den Fußtritt a bedient wird. Da es gleichzeitig als Sicherheitsventil für den Rückzugzylinder dient, ist es auch mit einem Gewichtshebel b versehen. Durch dieses Ventil Z wird der Druck auf den Blechhalter nach Wunsch vergrößert, ohne daß man die Arbeit der Maschine unterbrechen muß.

Nachdem die Presse für die vorliegende Arbeit eingestellt worden ist, und alle drei Stempel sich in ihrer tiefsten Stellung befinden, wird der Handhebel U in Mittelstellung gestellt und die Pumpe angelassen, wodurch die Flüssigkeit durch das Ventilgehäuse Q und zurück in den Behälter N gedrückt wird. Der Steuerhebel wird dann nach rechts herübergeschlagen, worauf der Zylinder K — aber nur bei Inbetriebsetzung der Maschine — sich zuerst füllt und alle Kolben in die Höhe gehen. Der Kolben L hebt die Flüssigkeit im Zylinder K und treibt sie unter die anderen Kolben, so daß deren Bewegung beschleunigt wird. Die Kolben gehen in die Höhe, bis das Blankett hart zwischen dem Halter F und der Matrize D gefaßt wird, wobei der Tisch E an die Anschlagmuttern c stößt, wodurch das Blankett vor einem zu großen Druck bewahrt wird. Gleichzeitig hebt der Anschlagdaumen Y die Stange X und schließt das Ventil S, wodurch die Verbindung zwischen dem Zylinder K und dem Raum T geschlossen wird, während der Kolben I unter einem durch die selbsttätige Regelung der Pumpe erhöhten Druck weiter steigt.

Während der Stempel das Arbeitsstück auf die verlangte Tiefe in der Matrize zieht, stößt der Boden des Stückes an eine Abstellstange d, die durch ein geeignetes Getriebe und eine Kette über das Rad e so eingestellt werden kann, daß der Rahmen f und die Stangen W angehoben und dadurch das Ventil S geöffnet wird. Dadurch wird eine Verbindung zwischen dem Zylinder K und dem Raum T hergestellt. Gleichzeitig wird der Hebel U wieder auf Mittelstellung gestellt, wodurch die Kolben

[1]) Vierfach wirkende Differentialpumpe.

schnell abwärts gehen, die Flüssigkeit K füllt und der Rest derselben in den Behälter zurückfließt.

Wenn aus irgend einem Grund der Stempel an dem Arbeitsstück hängen bleibt und so die Kolben am Zurückgehen verhindert, kann der Kontrollhebel nach links herüber geschlagen werden. Dann lassen die Ventile Druck nach K eintreten und die Kolben werden in ihre ursprüngliche Stellung zurückgezogen.

Fig. 668.

Schwere Kümpelpresse (Deutsche Niles-Werkzeugmaschinenfabrik, Oberschöneweide). Die Photographie, Fig. 668, und die Zeichnungen, Fig. 669, stellen eine hydraulische Presse für Arbeiten auf starkem

Kesselblech und ähnlichen Materialien dar. Sie ist mit einem Arbeits- und zwei Rückzugzylindern versehen und wird durch ein Steuerventil mit Verteiler, die auf der Photographie seitwärts neben der Presse aufgestellt sind, betätigt. Werkzeuge und Arbeiten für diese Presse sind auf S. 197, Fig. 246/7, zu finden.

Die Presse wird außer zur Herstellung von Kesselböden auch zu allen sonstigen Kümpel-, Bördel-, Stanz- und Gesenkpreßarbeiten verwendet.

Die Presse ist als 4-Säulen-Presse gebaut. Der Unterteil A der Presse besteht entweder aus ausgeglühtem Stahlformguß oder besonderem Grauguß. Der obere Querholm B, in welchen der Preßzylinder C und die beiden Rückzugzylinder $D_1 D_2$ besonders eingesetzt sind, und das bewegliche Querhaupt E sind aus ausgeglühtem Stahlformguß.

Das bewegliche Querhaupt E kann entweder als ein Stück oder dreiteilig ausgeführt werden. Diese Anordnung hat den Vorteil, daß die Seitenteile abgenommen werden können, Fig. 669, um bei Gesenkpreßarbeiten besser an die Presse zu gelangen. Die Befestigung der Seitenteile erfolgt dann in besonders sorgfältiger Weise mittels Feder und Nut und Verschraubung, so daß dieselbe Festigkeit erzielt wird, wie wenn ein geschlossenes Stück vorhanden wäre. Die Säulenführungen des beweglichen Querhauptes sind lang gehalten, um eine gute, kräftige Führung zu erzielen, damit dieselben auch bei geringen außerachsigen Arbeiten den erhöhten Spannungen genügend Widerstand zu leisten vermögen.

Säulen F, Plunger G, Preß- und Rückzugzylinder sind aus S.M.-Stahl gefertigt und die Zylinder aus dem Vollen gebohrt. Die Abdichtung sämtlicher Kolben erfolgt durch eine besondere Lederpackung (D.R.G.M.). Diese hat eine sehr lange Lebensdauer und hat sich im Betriebe sehr gut bewährt. Auf schnelle und bequeme Auswechselbarkeit der von außen zugänglichen Manschetten und Dichtungen ist bei der Konstruktion besonders Rücksicht genommen worden.

Diese Auswechselung der Manschetten ist ohne weitere Störung der übrigen Teile der Presse dadurch ermöglicht, daß der Kolben G mit dem Querhaupt durch einen zweiteiligen Ring E_1 verbunden ist und bis an den Ring zylindrisch mit seinem Arbeitsdurchmesser verläuft. Unterhalb dieses Ringes ist der Plunger kugelförmig abgedreht und arbeitet mit einer in das bewegliche Querhaupt eingelegten Kalotte E_2 aus Stahl, so daß der Plunger und das Querhaupt ohne Zwang ihren Führungen folgen können. Auch die Plunger für die Rückzugzylinder gehen nach beiden Seiten glatt durch, so daß die beiderseitigen Packungen anstandslos eingesetzt und entfernt werden können.

Unterhalb der Grundplatte A ist eine hydraulisch angetriebene

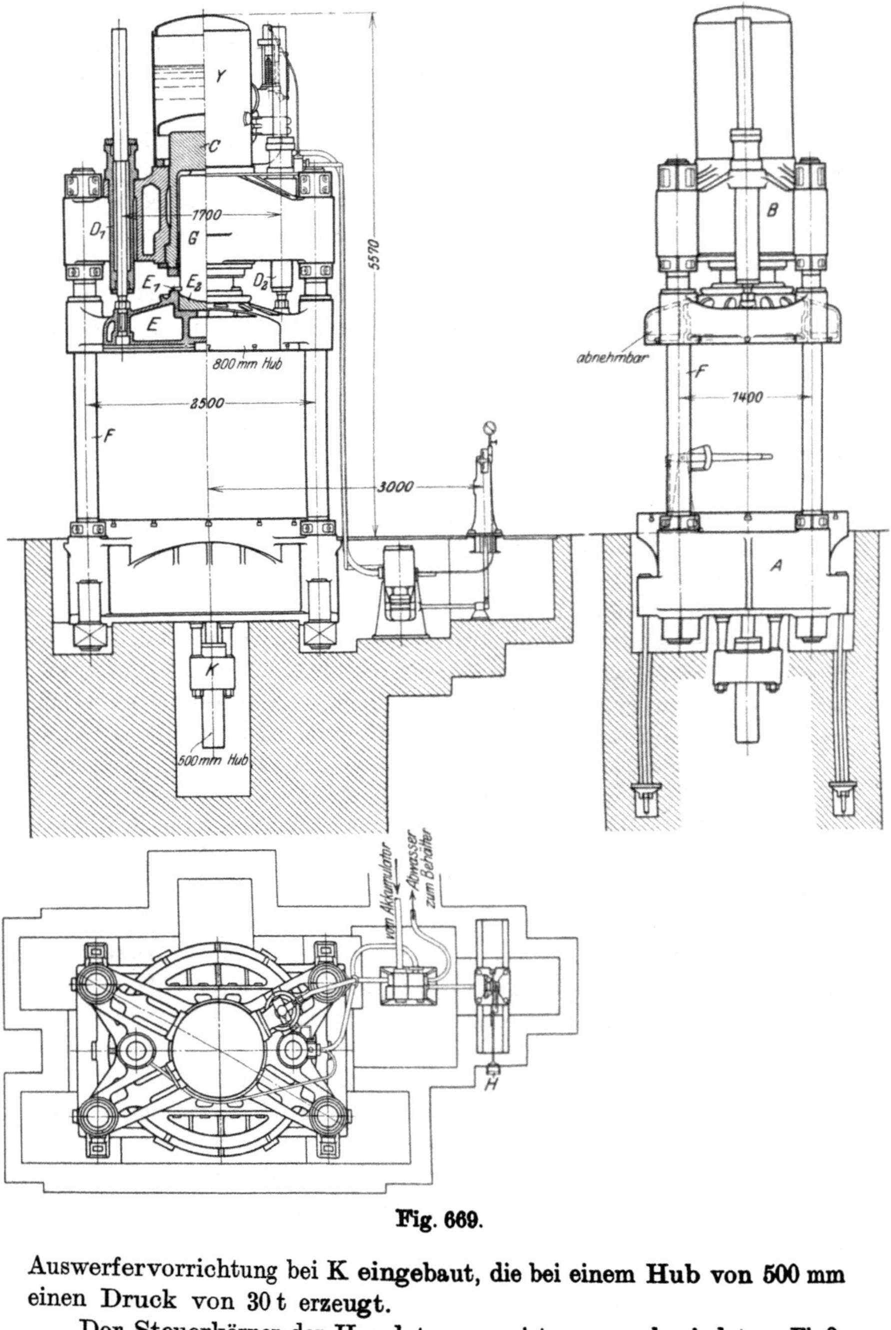

Fig. 669.

Auswerfervorrichtung bei K eingebaut, die bei einem Hub von 500 mm einen Druck von 30 t erzeugt.

Der Steuerkörper der Handsteuerung ist aus geschmiedetem Flußstahl hergestellt und aus dem Vollen gebohrt. Die Innenteile sind aus hochprozentigem Nickelstahl gefertigt. Für die Hochdruckleitung wird nahtloses Stahlrohr, für die Niederdruckleitung schmiedeeisernes, patentgeschweißtes Rohr verwandt.

Die Arbeitsweise der Presse ist folgende:

Zuerst sinken, nachdem der Pressenführer durch eine geringe Abwärtsbewegung des Steuerhebels H den Anstoss hierzu gegeben, die beweglichen Pressenteile G und E durch ihr Gewicht schnell herab, bis das obere Gesenk auf dem Arbeitsstück aufruht. Während dieser Bewegung ist durch eine Füllsteuerung Wasser aus dem über der Presse angeordneten Füllbehälter Y in den Preßzylinder übergetreten und hat diesen vollständig angefüllt. Nach einem weiteren Impuls des Pressenführers schließt sich die Füllsteuerung wieder, und es wird im Preßzylinder C Preßdruck erzeugt. Ist auf die vorgeschriebene Weise die eigentliche Preßarbeit vollführt worden, so wird der Steuerhebel H in seine oberste Stellung gebracht und dadurch den Rückzugzylindern D_1D_2 Druckwasser zugeführt, gleichzeitig die Füllsteuerung aber wieder derart betätigt, daß der Übertritt des im Preßzylinder C befindlichen Wassers in den Füllbehälter Y möglich ist. Hierdurch werden die beweglichen Pressenteile so lange wieder aufwärts bewegt, bis der Pressenführer den Steuerhebel in die Stillstand- bzw. zu einem weiteren Arbeitshub in die Niedergangstellung bringt.

Die Abmessungen der Füllsteuerung wie auch der von Hand zu betätigenden hydraulischen Steuerung sind derart gehalten, daß sowohl die Leerwege als auch der Preß- und Rückzug der Presse äußerst schnell von statten gehen. Die zur Verwendung kommenden, hydraulischen, Steuerteile sind derart ausgebildet, daß eine sehr leichte Beweglichkeit, wie auch leichte Regelung der Preßgeschwindigkeit seitens des Pressenführers ermöglicht wird.

Die Presse hat einen Preßdruck von 500 t; der Betriebswasserdruck beträgt 200 Atm., der Hub 800 mm, Säulenstellung von Mitte zu Mitte 2500 × 1400 mm, freie Höhe zwischen den Aufspannflächen 1800 mm.

4. Schmiedemaschinen.

Die selbsttätigen Schmiedemaschinen stellen liegende Kurbelpressen dar, bei denen durch große Massenanhäufung und gleichzeitig im Verhältnis dazu große Hubzahl (je nach der Größe von 80—30 minutlich), auf das zwischen den Klemmbacken gefaßte Material neben der Druckwirkung gleichzeitig eine Schlagwirkung ausgeübt wird, so daß eine dem Schmieden ähnliche Arbeit geleistet wird. Die Vorteile der Schmiedemaschine gegenüber dem Handschmieden liegen in der beschleunigten Herstellung, der infolge des größeren Arbeitsdruckes gleichmäßigeren, durch die ganze Materialtiefe gehenden Stauchung und der Verarbeitung bei höherer Temperatur infolge der schnelleren Arbeitsweise. Da außerdem die Schmiedemaschinen bis zu drei Satz Gesenke aufnehmen können, so können auch Stücke, zu deren Fertigstellung sonst mehrere Gänge notwendig sind, mit denselben Vorteilen bearbeitet werden, man kann also an Hitzen ersparen. Durch die selbsttätige Arbeitsweise dieser Maschinen kann man an Arbeitskräften nicht allein dem Handschmieden, sondern auch der Dampfhammerarbeit gegenüber sparen. Bei den zugehörigen Werkzeugen, S. 171, ist dieses näher ausgeführt worden.

Die Maschinen werden entweder mit feststehender oder mit beweglicher, rechter Klemmbacke gebaut. Für die gewöhnlichen Arbeiten reicht die erste Form aus; wenn aber die zu stauchenden Bunde zu groß werden, so lassen sich die Teile aus einer der Schmiedearbeit entsprechenden Maschinengröße nicht entfernen, so daß man auf die unwirtschaftliche Verwendung einer größeren Maschine angewiesen wäre. Ebenso ist es bei Schmiedeteilen, bei denen die überstehenden Enden länger als der Stößelhub sind. Im allgemeinen werden von den in Betracht kommenden Firmen beide Formen gebaut, weshalb im Anschluß an das über die dazugehörigen Werkzeuge, S. 172—177, Gesagte die Beschreibung je einer Bauart genügen dürfte.

Schmiedemaschine mit fester rechter Klemmbacke (Hasenclever A.-G., Düsseldorf). Die Maschine besteht aus einem, wie bei allen derartigen Maschinen, sehr schwer und kräftig ausgebildeten Bett, in dem der Stößel A aus Siemens-Martin-Stahl in Führungen mit auswechselbaren Verschleißplatten gleitet. In der Verlängerung der Stößelachse sind die Gesenke B in den Backen befestigt, von denen die rechte Backe feststeht, während die linke Backe durch eine mehrfache Hebelübersetzung C in einem bestimmten Verhältnis zum Stößelhub nach innen, bzw. nach außen bewegt wird. Da die Klemmbacken das Stück fassen müssen, bevor der Stauchstempel im Stößel seine Arbeit beginnt, muß notgedrungen ein Teil des Stößelweges für die nutzbare Stauchlänge bei derartigen Maschinen verloren gehen. Je sorgfältiger nun bei einer derartigen Maschine die konstruktive Durchbildung in dieser Richtung erfolgt, desto geringer ist dieser Verlust. Besondere Sorgfalt wird bei den Schmiedemaschinen dem Antrieb und den Sicherheitsvorrichtungen zugewendet, damit nicht eine Überlastung der Maschine einen Bruch des Bettes mit sich bringt. Der Antrieb erfolgt durch Riemen über eine als Schwungrad ausgebildete Riemenscheibe D auf der Kurbelwelle E. Bei den größeren Maschinen ist ein Zahnradvorgelege und hinter diesem die Abscherbolzen eingeschaltet, die im Fall einer Überbeanspruchung der Maschine abgeschert werden und so den Stößel stillsetzen. Die Übertragung des Arbeitsdruckes und Weges auf den Stößel erfolgt durch eine eigenartig ausgebildete Pleuelstange F, G, deren Bolzen beim Arbeiten nur auf Druck und nicht auf Biegung

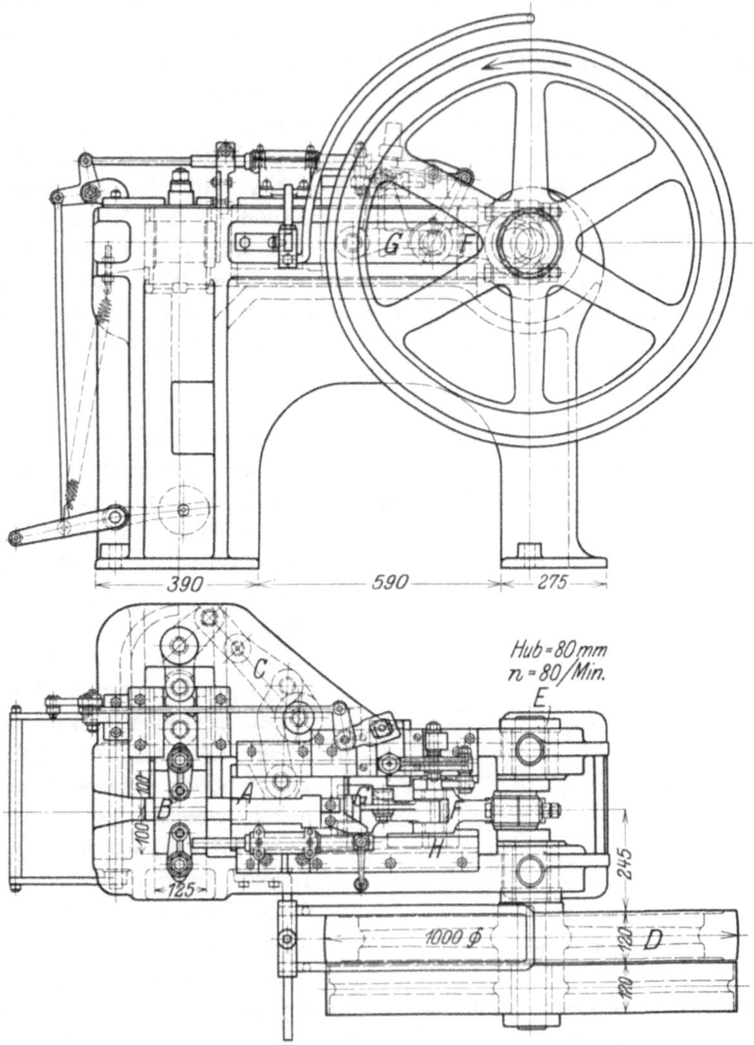

Fig. 670.

beansprucht werden. An dieser Stelle ist auch Vorkehrung getroffen, daß die Maschine entweder ununterbrochen oder nur mit einem Schlag arbeiten kann. Der Pleuelstangenteil G kann nämlich herausgeklappt werden, so daß der mittlere Bolzen H der Pleuelstange, der sich in einem beiderseitigen Langschlitz des Stößels führt, den Stößel unter allen Umständen nach rechts in die Totlage mitnimmt. Wird dagegen durch eine mit dem Teil G verbundene Führungsrolle, die auf einer Leitkurve abläuft, dieser aus seiner Verbindung mit dem Stößel gehoben, so geht der Bolzen H mit dem Teil der Pleuelstange leer hin und her, der Stößel bleibt also nach einem Hub in der rechten Totlage bei geöffneten Backen stehen, worauf ein frisches Arbeitsstück eingelegt werden kann.

Die Maschinen werden für einen größten Arbeitsdruck von 16000 bis 750000 kg gebaut und nehmen dementsprechend Bolzen von 16 bis 125 mm Durchmesser auf. Bei der kleinsten Größe ist der gesamte Stößelhub 80 mm, von denen 42 mm nach Schluß der Klemmbacken noch zur Verfügung stehen; dabei ist der Weg der Klemmbacken 30 mm. Für die größte Maschine sind diese Ziffern 300 bzw. 180 bzw. 120 mm, die Hubzahlen 85 minutlich bei der kleinsten und 40 minutlich bei der größten Maschine. Der Kraftbedarf beträgt bei Einzelantrieb rd. 6—40 PS.

Schmiedemaschine mit beweglicher rechter Klemmbacke (de Fries & Co. A.-G., Düsseldorf). Diese Firma baut ihre Schmiedemaschinen nach dem System Ajax-de Fries und legt ihre Sicherheitsvorrichtung[1]) in die Pleuelstange. Diese wird zweiteilig gemacht und der Drehbolzen beider Hälften ein wenig mit seinem Mittelpunkt über die Verbindungslinie der beiden Hauptbolzen gestellt[2]). Da diese beiden Hälften nur durch einen Zerreißbolzen zusammengehalten werden, so wird bei Überschreitung eines gewissen Druckes in der Pleuelstange dieser Bolzen abreißen und die beiden Hälften der Pleuelstange bei fortschreitender Kurbelwellenbewegung sich aus ihrer Lage nach oben herausdrehen; die Pleuelstange knickt nach dem Zerreißen des Bolzens zusammen und der Stößel bleibt stehen.

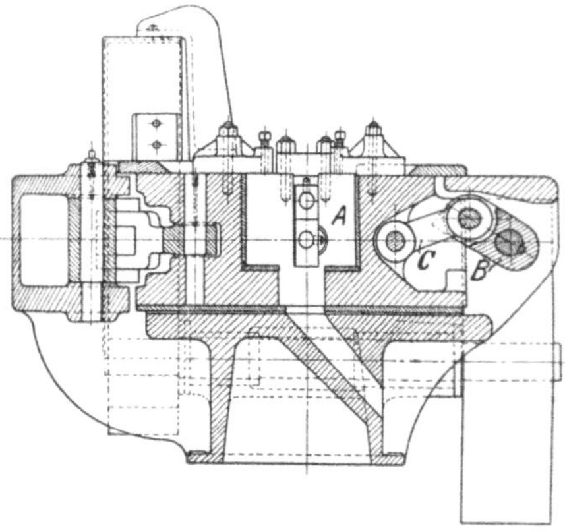

Fig. 671.

In Fig. 671—672 ist die Einrichtung der rechten, beweglichen Klemmbacke A im Schnitt und in Ansicht wiedergegeben. Die Bewegung erfolgt hier von Hand, kann aber auch von der Kurbelwelle

[1]) W.T. 1912, S. 111.
[2]) Vgl. auch S. 512, Fig. 620.

aus betätigt werden. Zur Bewegung dient ein Kniehebelsystem B, C, das durch den Schließdruck im Gesenk in der Arbeitsstellung gehalten wird, da während des Arbeitshubes des Stößels die Strecklage des Kniehebels nach unten durchgedrückt wird, so daß der Kniehebel gegen einen festen Anschlag im Ständer stößt, der dann den Gegendruck sicher aufnimmt. Durch Drehen der in Fig. 672 sichtbaren Hand-

Fig. 672.

kurbel wird das Zahnradsegment verdreht und dadurch das auf dem gleichen Bolzen sitzende Kniehebelsystem zurückgeführt und die rechte Klemmbacke A wieder geöffnet. Durch eine derartige Einrichtung erreicht man eine bedeutend größere Klemmbackenöffnung, wodurch man entsprechend größere Stücke auf den kleineren Maschinen verarbeiten kann. So kann man bei geschlossener, rechter Klemmbacke von 45—175 mm Raum verwenden, während bei geöffneter, größter Weite derselben von 70—315 mm zur Verfügung stehen. Natürlich ist nicht diese ganze Weite für das Schmieden nutzbar zu machen, sondern nur für das Entfernen der Stücke. Das zu verarbeitende Material kann von 25 bis zu 150 mm Durchmesser oder Vierkant haben. Der Kraftbedarf der Maschinen beträgt 8 — 50 PS, die Stößelbewegung nach Schluß der Klemmbacken 60 — 240 mm.

5. Hilfsmaschinen.

Gesenknutmaschine mit hin- und hergehender Bewegung. Das Ausschneiden von Vertiefungen in Stahl oder anderen Metallen bei der Herstellung von Gesenken geschieht auf verschiedene, wohlbekannte Weisen, mit dem Meißel, Bohrer, auf der Fräs- oder Stoßmaschine oder mit der Feile, wie es sich eben ergibt. Jede Abweichung von diesen Arbeitsweisen ist etwas ausgesprochen Neues. Es gibt viele Möglichkeiten, das Metall nach den verwickelten Formen der Fig. 673 auszuarbeiten. Die Maschine nach Fig. 674 ist patentiert und läßt sich in keine der obenerwähnten Gruppen einreihen, sondern nimmt eine Sonderstellung ein. Die Arbeitsweise der Maschine beruht darauf, daß ein schwingendes Werkzeug in das Material vorgeschoben wird und die abgetrennten Splitter mit dem gezahnten Ende des Formstahles „auslöffelt". Ein hin- und hergehender Messerkopf ist zur Auf-

Fig. 673.

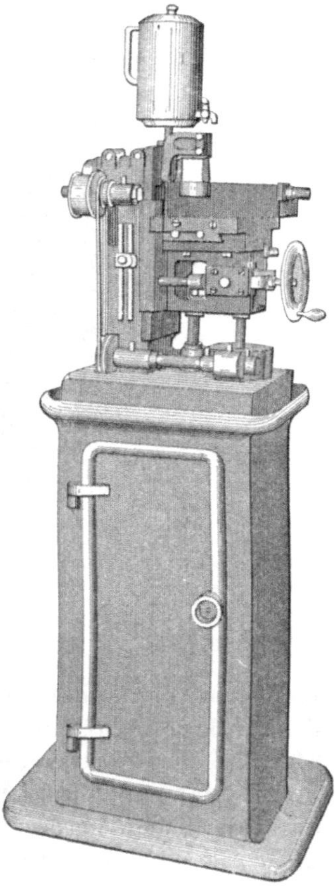

Fig. 674.

nahme des Messerschaftes geteilt und wird durch eine wagrechte Welle am Kopfe der Maschine angetrieben. Der Messerkopf, der nach jeder Richtung einstellbar ist, sowie der Vorschub können, wie aus der Figur ersichtlich, von Hand oder mechanisch betrieben werden.

Das Arbeitsmuster in Fig. 673 dürfte besonderes Interesse erwecken. Es ist ein viereckiger Stahlblock 51 × 30 × 30 mm, in den beliebige, sauber geschnittene Löcher eingearbeitet sind. Drei dieser Löcher gehen vollständig durch den Block und stehen, was Sauberkeit und Schärfe des Schnittes betrifft, den mit Räumnadeln gearbeiteten Löchern gleich. Das viereckige Loch am Ende hingegen verändert seine Form in der halben Blocktiefe zu einem sechseckigen Loch, während das mittlere der drei Löcher am anderen Ende nur bis zur halben Tiefe durchgeht und einen halbkreisförmigen Grund hat. Das viereckige Loch ist 16 × 16 mm; der enge Schlitz 3,2 × 22 mm; der Schnitt in der Mitte 8 × 25 mm und der mit den halbrunden Enden 6,4 × 22 mm.

Das viereckige Loch und der 3,2 × 22 mm-Schlitz zeigen die Verwendungsmöglichkeit dieses Werkzeuges für unregelmäßige Querschnitte, die für andere Werkzeuge sehr schwierig sind. Ein Formmesser kann für jeden Umriß gemacht werden und damit das Material durchgearbeitet werden. Hier kann ein Einwand einsetzen: Bei einer oberflächlichen Betrachtung könnte es scheinen, als ob die Arbeitsweise mehr einem Auskrümeln des Materials als einem richtigen Schneiden entspricht. Dagegen spricht, daß die Zeit für jeden Schnitt an dem in Fig. 673 dargestellten Musterstück nur ungefähr drei Minuten beträgt. Bei Berücksichtigung dieses Umstandes ergibt sich eine derartige Arbeitsgeschwindigkeit für das Ausarbeiten der Gesenke und ähnlicher Stücke, daß Meißel und Feile weit zurücktreten müssen.

Feilmaschinen (Gebrüder Thiel, Ruhla i. Thür.). Bereits bei Besprechung der Herstellung der Werkzeuge für Fig. 28, S. 50, wurde von Seite Woodworths die senkrechte, von der üblichen wagrechten abweichende Feilarbeit als besonders empfehlenswert gelobt. Derselbe Grundsatz, verbunden mit einer Steigerung der Erzeugung durch Anwendung der Maschinenarbeit statt der Handarbeit, liegt den Feilmaschinen zugrunde. Bei der Herstellung von Schnittplatten und Gesenken, wo die genauen Umrißlinien dauernd eingehalten werden müssen, treten die Übelstände des Handfeilens, die die Firma Gebrüder Thiel sehr anschaulich in der Fig. 675 darstellt, auf. Das Arbeitsstück ist in den meisten Fällen sehr ungünstig für das Auge des Arbeiters im Schraubstock eingespannt, so daß man es zur genauen Kontrolle öfters ausspannen muß, wodurch die Genauigkeit der Arbeit wahrscheinlich, sicher aber die Arbeitszeit leidet. Das einfallende Licht wird in den seltensten Fällen die Umrißlinie so beleuchten, daß der Arbeiter ohne Hilfsmittel, wie Spiegel oder Drehen des Schraubstockes, was auch nicht immer möglich ist, seine Arbeit genau kontrollieren kann. Bei der Feilmaschine liegt das Arbeitsstück dagegen wagrecht so auf dem Tisch, daß man nach Wegblasen der Feilspäne, ohne das

Stück aus der Einspannung zu nehmen, im auffallenden Licht die Arbeit genau kontrollieren kann. Außerdem ist die Feile durch den Gegenhalter stets gerade geführt, wodurch die bearbeiteten Kanten auch wirklich winkelrecht werden. Da die Feilmaschine auch leicht in eine Sägemaschine umgeändert werden kann, so wird sich für Vorarbeiten diese

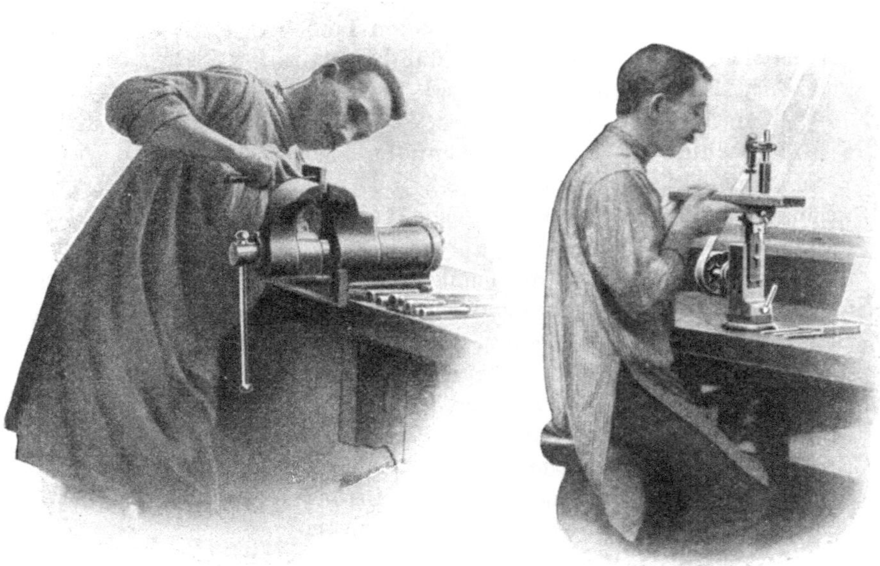

Fig. 675.

Arbeitsweise an Stelle des Ausbohrens durch Setzen eines Loches an das andere längs der auszuschneidenden Umrisse sehr empfehlen. Man kann auch, um nach dem Härten verzogene Gegenstände herzurichten, Schmirgel- und Diamantfeilen in die Maschine einspannen, wobei man im ersten Fall Kupfer, im zweiten Fall weichen Maschinenstahl als Träger des Schleifmittels verwendet.

Die Feilmaschine wird in zwei Größen gebaut, Fig. 676 zeigt die größere Maschine.

Die Maschine wird mit dem Fuß a des gußeisernen Ständers b auf der Feilbank festgeschraubt und erhält ihren Antrieb durch die Riemenscheibe I über die Stirnräder II und III auf die Welle der Kulissenscheibe c, deren umlaufende Bewegung durch die Schwingstange r in eine hin- und hergehende verwandelt und auf den Feilenhalter e übertragen wird. Die Hubhöhe ist verstellbar und ebenso in doppelter Weise der Feilenhalter, so daß man Feilen verschiedener Länge verwenden kann. Die Feilen können schnell und leicht durch die seitlichen Klemmschrauben i ausgewechselt werden. Das Arbeitsstück ruht auf dem drehbaren Tisch d, der auch in gewissen Grenzen schräg zur Feilrichtung einstellbar und dazu mit zwei entsprechenden Skalen versehen ist. Das Arbeitsstück wird auf dem Tisch von Hand bewegt und dadurch allmählich alle Umrißlinien an die Feile gebracht.

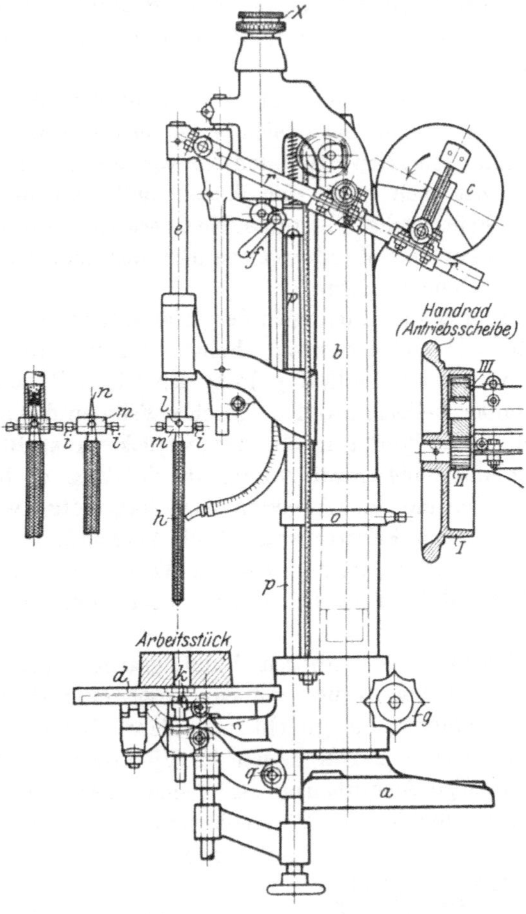

Fig. 676.

Der Feilmaschine sind 3 Spannringe beigegeben, die einen Wechsel der verschiedenen Feilenprofile augenblicklich gestatten, wie es bei der Bearbeitung von Schnitten und derartigen Werkzeugen oft erforderlich ist.

Das Einspannen der einzelnen Feilen erfolgt auf folgende Weise: Durch Drehen des Handrades (Antriebscheibe) wird die Feilenspindel e möglichst weit hochgezogen, dann löst man die Klemmschraube f und das Klemmrädchen g und senkt den Tisch d der Feilmaschine soweit nach unten, bis man die Feile h in die Feilenspindel e einführen kann. Dann klemmt man mittels der Schrauben i die Feile h etwas fest und bewegt den Tisch d wieder soweit nach oben, bis die Körnerspitze der Feile h in die Mitte des Hohlkörners k auftrifft. Durch entsprechendes Anziehen der beiden Schrauben i wird die Feile so ausgerichtet, daß sie genau läuft. Bild n zeigt eine Feile h mit aufgespanntem Ring m.

So liegt jede Art Feilen in dem Ring, wie beim Handfeilen im Heft.

Auf diese Art kann man sich eine Anzahl der verschiedenen Feilenprofile zum schnellen Auswechseln vor Beginn der Arbeit vorrichten. Dies hat auch noch den Vorteil, daß man nach Abnutzung der vorderen Feilenfläche — ohne Neuausrichten — durch bloßes Umspannen (Lösen und Wiederfestspannen der Schraube l) auch die hintere Fläche benutzen kann.

Beim Einführen und Entfernen der Arbeitsstücke (bei Bearbeitung von Innenflächen) verfährt man ähnlich, indem man die Klemmschraube f und Klemmrädchen g löst, den Tisch d senkt, wodurch bei hochstehender Feilenspindel e zwischen Feile und Tisch ein freier Raum entsteht. Man führt nun das Arbeitsstück ein, stellt den Tisch d, dessen Höheneinstellung man vorher durch Ring o begrenzt hat, wieder richtig ein, und das Arbeiten kann fortgesetzt werden.

Der vorstehenden Erläuterung ist die Verwendung von 250 mm langen Feilen zugrunde gelegt. Um die Verwendung auch kürzerer Feilen von 200 und 150 mm Länge zu ermöglichen, ist wie folgt zu verfahren:

Klemmschraube f des oberen Führungsarmes an der hinteren Führungsstange p bleibt immer auf den Strich eingestellt.

Den Längenunterschied gleicht man dann durch Verstellen des unteren Führungsarmes durch Lösen der Vierkantschraube q aus.

Die vollständige Ausnutzung der Feilen jeder Länge und jeden Querschnittes ist durch die Verstellbarkeit des Hohlkörners k gewährleistet.

Fig. 677.

Das Räumen. Verschiedentlich wurden bei der Herstellung der einzelnen Werkzeuge, die bis jetzt besprochen worden sind, zur Fertigstellung durchgehender Löcher, die ein genaues Maß haben sollten und gleichzeitig so sauber wie möglich ausgeführt sein sollten, Räumnadeln verwendet. Besonders[1]) wenn es sich um unregelmäßige

[1]) Vergl. S. 129.

Löcher oder sehr verwickelte Querschnitte handelt, wie sie als Arbeitsbeispiele in Fig. 677 angegeben sind, ist in der Massenfabrikation die Räumarbeit das weitaus billigste und schnellste Arbeitsverfahren. Die Arbeit, Fig. 678, besteht darin, daß eine Räumnadel genau gerad-

Fig. 678.

linig geführt, langsam durch das fast auf Maß vorgebohrte Loch durchzogen wird und mit äußerst feinen Spänen, aber dafür am ganzen Umfang gleichzeitig schneidend, den verlangten Querschnitt in mehreren aufeinanderfolgenden Zügen ausarbeitet. Dadurch, daß die gleichzeitig arbeitenden Zähne der Räumnadel einander genau gegenüberliegen, kann ein Verdrücken oder Ausweichen der Nadel nicht stattfinden, da die folgenden etwas größeren Zähne auch immer genau in der Mittellinie weiterarbeiten müssen. Wenn ein Loch mit einer gewöhnlichen Reibahle ausgerieben wird, so macht jede Stelle der Schnittkante einen Weg, der dem Durchmesser des Loches × der Anzahl Umdrehungen der Reibahle bei ihrem Wege durch das Arbeitsstück entspricht. Wird dagegen dasselbe Loch ausgeräumt, so ist der Weg eines jeden Punktes der Schnittkante gleich der Länge des Loches. Man arbeitet also bedeutend schneller, besonders wenn man der Räumnadel eine Schnittgeschwindigkeit von mindestens 1,25 m minutlich gibt. Ferner behält eine Räumnadel ihre Abmessungen bedeutend länger als eine Reibahle, weil das Fertigende der Räumnadel eine Anzahl Zähne gleichen Durchmessers hat, die, da die Räumarbeit nur sehr kleine Späne nimmt, auch nur einer sehr geringen Abnutzung unterworfen sind. Fig. 679/80 zeigt Formen und Abmessungen solcher Räumnadeln.

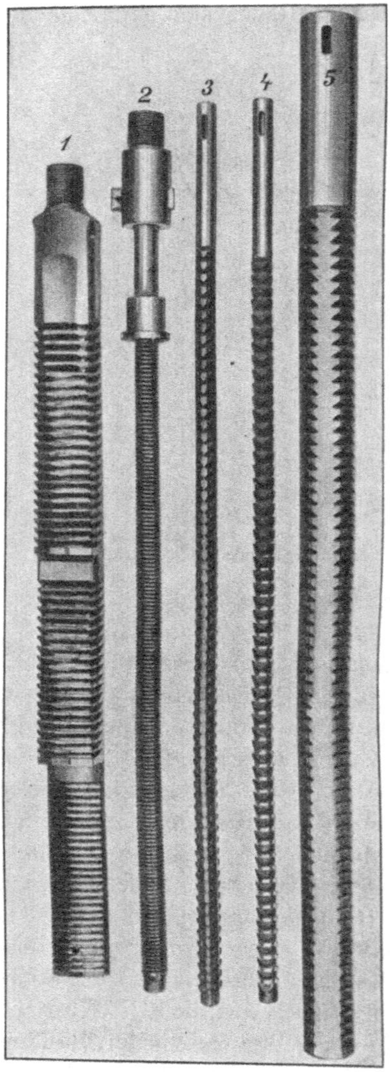

Fig. 679.

Die betreffenden Löcher, Fig. 681, werden z. B. bis auf 0,4 bis 0,45 mm unter das Fertigmaß ausgebohrt, worauf die Räumarbeit vor sich gehen kann.

Die glatten, runden Teile an diesen Räumnadeln sollen das Verlaufen der Räumnadel verhindern. Indem man zwischen die Zähne der Räumnadel gerade, zylindrische Teile einschaltet, wird die Räumnadel ausgerichtet im Loch gehalten, da immer ein Teil der geraden Zylinderfläche mit dem Arbeitsstück in Berührung ist, während mehrere Zähne an einer anderen Stelle schneiden.

Die Räumnadel wird durch das Arbeitsstück gesteckt und mit Hilfe eines durchgehenden Keiles an dem Schlitten der Maschine befestigt, der für diesen Zweck in geeigneter Weise geschlitzt ist.

Die Vorarbeit bei nicht gebohrten Löchern kann in verschiedener Weise, Fig. 681, erfolgen. Das Loch in dem Pleuelstangenkopf, der in Fig. 678 bei der Bearbeitung in der Räumnadelmaschine gezeigt ist, kann entweder durch Ausbohren mittels nebeneinander gesetzter Löcher und nachheriges Ausmeißeln vorgearbeitet werden oder auf der Stoßmaschine so nahe wie möglich an das Fertigmaß bearbeitet werden. Die rechte Hälfte der Fig. 681 zeigt die Form eines rohen Loches, entweder in einem Gesenkschmiedestück oder in einem Gußstück.

Die Räumarbeit selbst erfolgt auf besonderen Maschinen[1]) nach

[1]) Schuchhardt & Schütte, Berlin.

Fig. 682, bei denen, wie Fig. 678 zeigt, das zu räumende Arbeitsstück auf der linken Maschinenseite gegen den Arbeitstisch angelegt wird, worauf die Räumnadel durch die Schraubenspindel der Maschine langsam und stetig durch das vorgeschruppte Loch durchgezogen wird. In Fig. 682 ist eine neuere Ausführung der Maschine abgebildet,

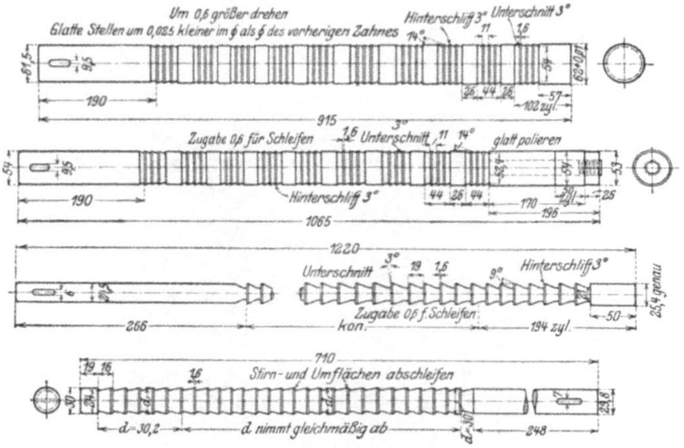

Fig. 680.

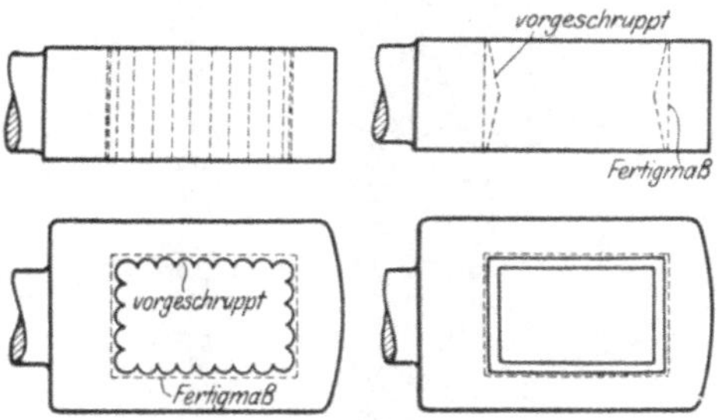

Fig. 681.

bei der zwei nebeneinander liegende Spindeln gleichzeitig zwei Löcher ausräumen können, wodurch die Herstellungsmenge verdoppelt wird. Die Antriebsspindeln werden in dem Trog, der gleichzeitig den Arbeitsdruck auf den Hauptständer der Maschine überträgt, geführt und durch den in Fig. 683 abgebildeten Räderantrieb bewegt. Der Antrieb erfolgt durch Riemen auf drei Scheiben, für Vor- und Rückgang der Maschine, sowie für Leerlauf. Die Rücklaufscheibe ist kleiner im Durchmesser, so daß die Maschine beim Rücklauf mit erhöhter Geschwindigkeit arbeitet. Bei der Maschine mit zwei Räumnadeln, Fig. 682 ist der Antrieb insofern anders, als die Kupplung die eine Spindel vorwärts treibt, während die andere rückwärts geht, so daß die Leerzeiten fortfallen. Es kann auch die eine Spindel ganz ausgeschaltet werden. Die Spindeln A, Fig. 683, erhalten Flachgewinde und werden durch eine Bronzemutter B vorgetrieben. Diese Mutter

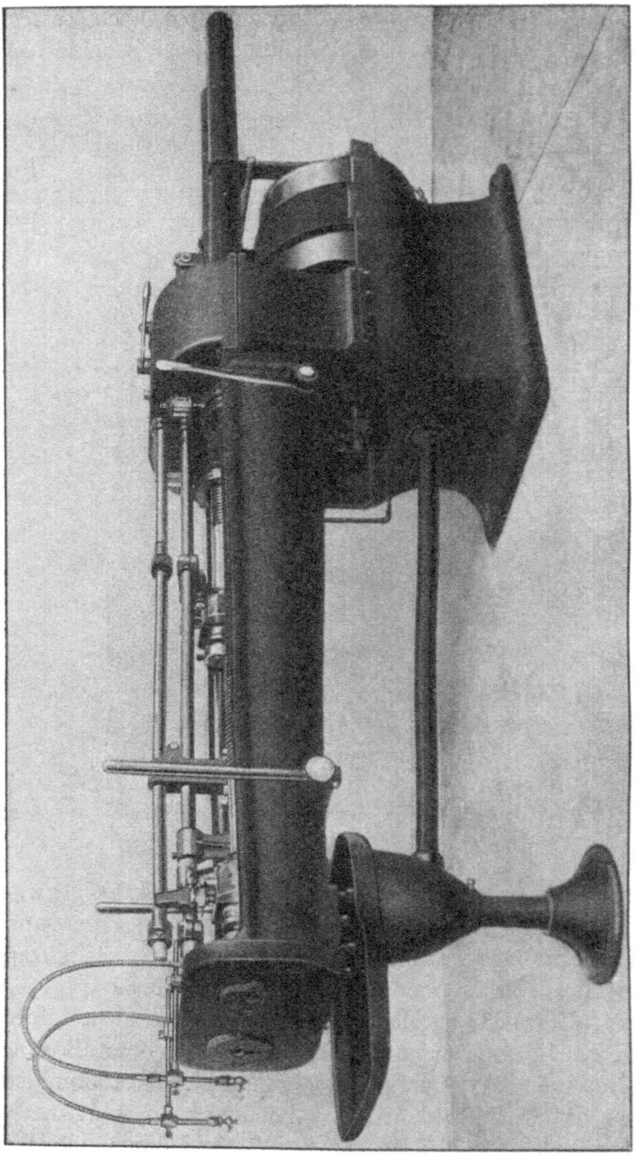

Fig. 682.

ist in der langen, als Hauptlager ausgebildeten Nabe des Rades C mit Schrauben und Keil so befestigt, daß bei Kupplung dieses Zahnrades mit dem Antrieb die Spindel den langsamen Vorschub erhält. Auf der Nabe des Rades C sitzt außen das kleinere Stirnrad D, das den schnellen Vorschub vermittelt. Die angeführten vier Teile werden in Laufbüchsen in dem gußeisernen, alsgeschlossener Kasten ausgebildeten Ständer der Maschine gehalten und durch ein Druckkugellager gegen den Ständer in achsialer Richtung abgestützt, wodurch die bei dem hohen Arbeitsdruck auftretenden Reibungsverluste beträchtlich, nach einer Angabe des Fabrikanten bis zu 30% vermindert werden.

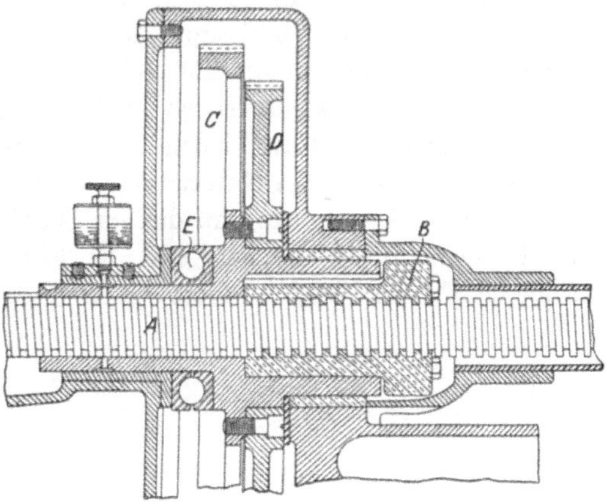

Fig. 683.

Die Verbindung der Räumnadeln mit der Spindel erfolgt entweder nach Fig. 679/80 durch Keil oder mittels Gewinde. Die Zugspindel erhält im letzten Fall vorn ein Innengewinde, in das sich der Zapfen des Kopfes einschraubt. Der Kopf ist vorn geschlitzt und mit Innengewinde versehen, so daß der Gewindezapfen der Räumnadel dort eingeschraubt und gleichzeitig durch die Klemmutter von außen festgezogen werden kann. Außerdem kann man an dem Spindelkopf einen senkrechten Schlitten verstellen, wodurch man eine genaue Einstellung der Nadel für das Arbeitsstück in der Höhe vornehmen kann.

Das Ausschalten der Räumnadeln geschieht selbsttätig durch einstellbare Anschläge, während ein Handhebel unmittelbar neben dem Arbeitsstück das Ein- und Ausrücken der Maschine vom Stande des Arbeiters ermöglicht.

Bezüglich der Abmessungen sei bemerkt, daß man mit der größten Maschine Löcher bis zu 75 × 75 mm und Keilnuten bis zu 250 × 38 mm ausräumen kann. Die Spindel hat rund 70 mm Durchmesser bei 1300 mm Länge mit zwei Gängen auf 1 Zoll engl. und läuft in einer Mutter von 225 mm Länge. Die beiden Schnittgeschwindigkeiten betragen 915 und 1830 mm in der Minute. Die Maschine wiegt rund 1600 kg.

Das Arbeitsstück braucht nur insofern festgespannt zu werden, als zum Halten notwendig ist, da die zylindrischen Teile der Räumnadel das Arbeitsstück ausgerichtet halten. Wenn man diese Räumnadeln in Gußeisen gebraucht, so wird Seifenlösung mit Vorteil als Schmiermittel verwendet, da dann die geräumte Fläche ein hochpoliertes Aussehen erhält. Für Chrom-Nickelstahl gibt irgendein gutes Bohröl zufriedenstellende Ergebnisse. An manchen Arbeitsstücken wird vor dem Räumen überhaupt nicht gebohrt und sehr oft nur eine einzige Räumnadel verwendet. Wenn aber das Arbeitsstück länger als beispielsweise 50 mm ist, so wird gewöhnlich in zwei Stufen mit zwei verschiedenen Räumnadeln gearbeitet. Selbstverständlich kann das Räumen im rohen Material nur dann erfolgen, wenn die Räumarbeit die erste Bearbeitung an diesem Stück ist, da sonst die Räumnadel, die dem rohen Loche folgt, die ganze übrige Bearbeitung der anderen Flächen unbrauchbar machen würde.

Sachregister.

P

R

S

T

U

V

W

Z

GPSR Compliance

The European Union's (EU) General Product Safety Regulation (GPSR) is a set of rules that requires consumer products to be safe and our obligations to ensure this.

If you have any concerns about our products, you can contact us on ProductSafety@springernature.com

In case Publisher is established outside the EU, the EU authorized representative is:

Springer Nature Customer Service Center GmbH
Europaplatz 3
69115 Heidelberg, Germany

Batch number: 10682325

Printed by Printforce, the Netherlands

AF844549

Graduate Texts in Mathematics 40

John G. Kemeny
J. Laurie Snell *Anthony W. Knapp*

Denumerable Markov Chains

with a chapter of Markov Random Fields by David Griffeath

Springer-Verlag
New York Berlin Heidelberg London Paris
Tokyo Hong Kong Barcelona Budapest

John G. Kemeny

President Emeritus
Dartmouth College
Hanover, NH 03755

J. Laurie Snell

Department of Mathematics
Dartmouth College
Hanover, NH 03755

Anthony W. Knapp

Department of Mathematics
State University of New York
Stony Brook, New York 11794

David S. Griffeath

Department of Mathematics
University of Wisconsin
Madison, Wisconsin 53706

Mathematics Subject Classifications
60J05, 60J10

Library of Congress Cataloging in Publishing Data

Main entry under title:
Denumerable Markov Chains.
(Graduate texts in mathematics; 40)
First ed. by J. G. Kemeny, J. L. Snell, and
A. W. Knapp is entered under Kemeny, John G.
Bibliography: p. 471
Includes indexes.
1. Markov processes. I. Kemeny, John G.
II. Series.
QA274.7.K45 1976 519.2'33 76-3535
Second Edition

Softcover reprint of the hardcover 2nd edition 1976
Originally published in the University Series in Higher Mathematics
(D. Van Nostrand Company); edited by M. H. Stone, L. Nirenberg,
and S. S. Chern.

R.R. Donnelley & Sons Company.
Manufactured in Mexico.

ISBN 978-1-4684-9457-0 ISBN 978-1-4684-9455-6 (eBook)
DOI 10.1007/978-1-4684-9455-6

PREFACE TO THE SECOND EDITION

With the first edition out of print, we decided to arrange for republication of *Denumerable Markov Chains* with additional bibliographic material. The new edition contains a section Additional Notes that indicates some of the developments in Markov chain theory over the last ten years. As in the first edition and for the same reasons, we have resisted the temptation to follow the theory in directions that deal with uncountable state spaces or continuous time. A section entitled Additional References complements the Additional Notes.

J. W. Pitman pointed out an error in Theorem 9-53 of the first edition, which we have corrected. More detail about the correction appears in the Additional Notes. Aside from this change, we have left intact the text of the first eleven chapters.

The second edition contains a twelfth chapter, written by David Griffeath, on Markov random fields. We are grateful to Ted Cox for his help in preparing this material. Notes for the chapter appear in the section Additional Notes.

J.G.K., J.L.S., A.W.K.
March, 1976

PREFACE TO THE FIRST EDITION

Our purpose in writing this monograph has been to provide a systematic treatment of denumerable Markov chains, covering both the foundations of the subject and some topics in potential theory and boundary theory. Much of the material included is now available only in recent research papers. The book's theme is a discussion of relations among what might be called the descriptive quantities associated with Markov chains—probabilities of events and means of random variables that give insight into the behavior of the chains.

We make no pretense of being complete. Indeed, we have omitted many results which we feel are not directly related to the main theme, especially when they are available in easily accessible sources. Thus, for example, we have only touched on independent trials processes, sums of independent random variables, and limit theorems. On the other hand, we have made an attempt to see that the book is self-contained, in order that a mathematician can read it without continually referring to outside sources. It may therefore prove useful in graduate seminars.

Denumerable Markov chains are in a peculiar position in that the methods of functional analysis which are used in handling more general chains apply only to a relatively small class of denumerable chains. Instead, another approach has been necessary, and we have chosen to use infinite matrices. They simplify the notation, shorten statements and proofs of theorems, and often suggest new results. They also enable one to exploit the duality between measures and functions to the fullest.

The monograph divides naturally into four parts, the first three consisting of three chapters each and the fourth containing the last two chapters.

Part I provides background material for the theory of Markov chains. It is included to help make the book self-contained and should facilitate the use of the book in advanced seminars. Part II contains basic results on denumerable Markov chains, and Part III deals with discrete potential theory. Part IV treats boundary theory for both transient and recurrent chains. The analytical prerequisites for the two chapters in this last part exceed those for the earlier parts of the book and are not all included in Part I. Primarily, Part IV presumes that the reader is familiar with the topology and measure theory of compact metric spaces, in addition to the contents of Part I.

Two chapters—Chapters 1 and 7—require special comments. Chapter 1 contains prerequisites from the theory of infinite matrices and some other topics in analysis. In it Sections 1 and 5 are the most important for an understanding of the later chapters. Chapter 7, entitled "Introduction to Potential Theory," is a chapter of motivation and should be read as such. Its intent is to point out why classical potential theory and Markov chains should be at all related.

The book contains 239 problems, some at the end of each chapter except Chapters 1 and 7.

For the most part, historical references do not appear in the text but are collected in one segment at the end of the book.

Some remarks about notation may be helpful. We use sparingly the word "Theorem" to indicate the most significant results of the monograph; other results are labeled "Lemma," "Proposition," and "Corollary" in accordance with common usage. The end of each proof is indicated by a blank line. Several examples of Markov chains are worked out in detail and recur at intervals; although there is normally little interdependence between distinct examples, different instances of the same example may be expected to build on one another.

A complete list of symbols used in the book appears in a list separate from the index.

We wish to thank Susan Knapp for typing and proof-reading the manuscript.

We are doubly indebted to the National Science Foundation: First, a number of original results and simplified proofs of known results were developed as part of a research project supported by the Foundation. And second, we are grateful for the support provided toward the preparation of this manuscript.

J. G. K.
J. L. S.
A. W. K.

Dartmouth College
Massachusetts Institute of Technology

RELATIONSHIPS AMONG MARKOV CHAINS

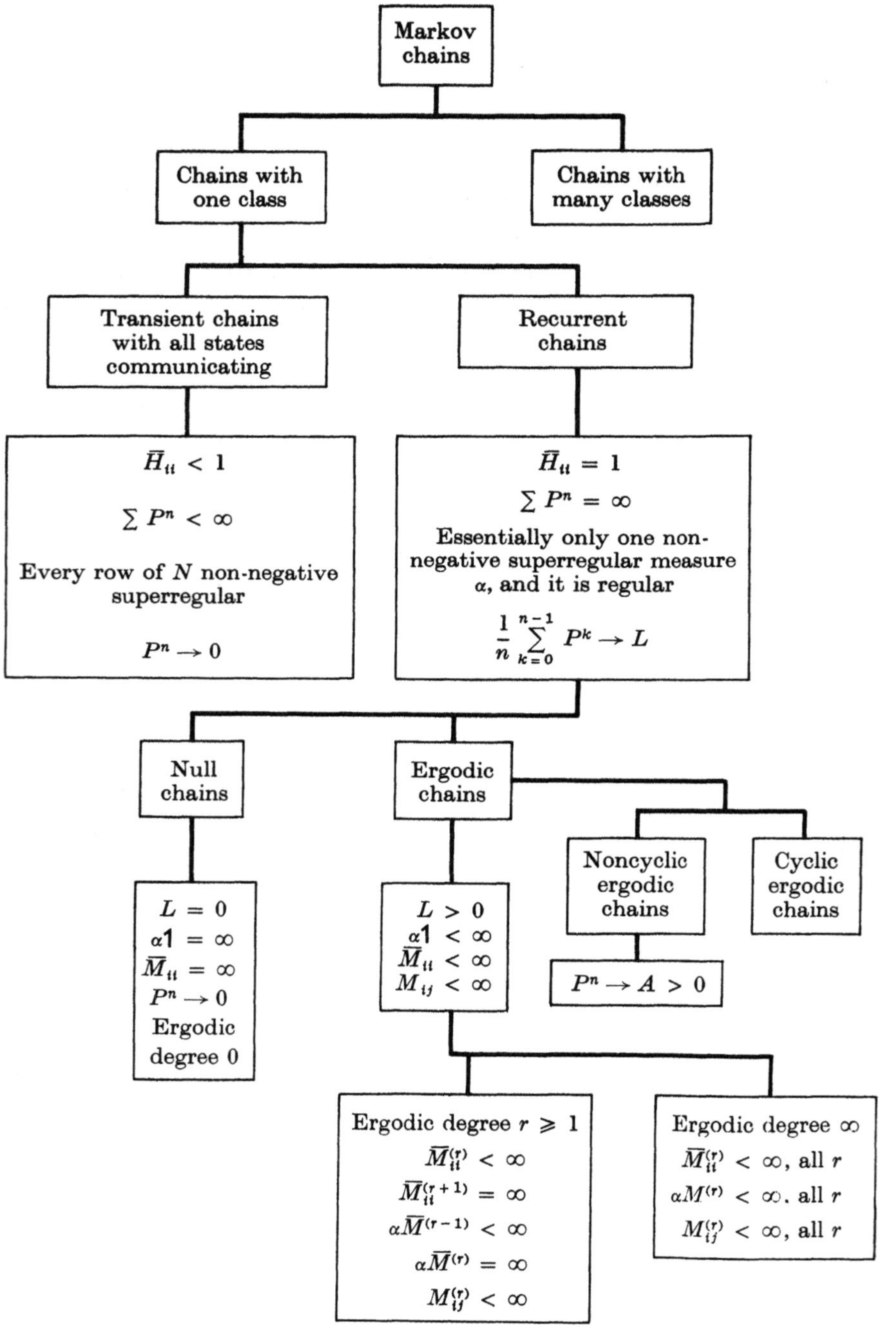

TABLE OF CONTENTS

Chapter 5: TRANSIENT CHAINS

Chapter 6: RECURRENT CHAINS

Chapter 7: INTRODUCTION TO POTENTIAL THEORY

Chapter 8: TRANSIENT POTENTIAL THEORY

CHAPTER 9: RECURRENT POTENTIAL THEORY

CHAPTER 10: TRANSIENT BOUNDARY THEORY

CHAPTER 11: RECURRENT BOUNDARY THEORY

CHAPTER 12: INTRODUCTION TO RANDOM FIELDS

CHAPTER 1

PREREQUISITES FROM ANALYSIS

1. Denumerable matrices

The word **denumerable** in the sequel means finite or countably infinite. Let M and N be two non-empty denumerable sets. A **matrix** is a function with domain the set of ordered pairs (m, n), where $m \in M$ and $n \in N$, and with range a subset of the extended real number system—the reals with $+\infty$ and $-\infty$ adjoined. We call the sets M and N **index sets**. The matrix is called a **finite matrix** if both M and N are finite sets.

To say that the m-nth **entry** of the matrix is x or is equal to x, we mean that the value of the function on the pair (m, n) is x. A matrix is said to be **non-negative** if all of its entries are non-negative, and it is said to be **positive** if all of its entries are positive. We agree to use upper-case italic letters to stand for matrices. If A is a matrix, we denote the m-nth entry of A by A_{mn}. Some examples of matrices are as follows:

(1) If all entries of a matrix are equal to zero, we say that the matrix is the **zero matrix**, denoted by 0.

(2) A matrix for which M and N are the same set is called a **square matrix**. The entries corresponding to $m = n$ are **diagonal entries**; other entries are **off-diagonal entries**.

(3) A square matrix whose off-diagonal entries all equal zero is a **diagonal matrix**. The diagonal matrix obtained from a square matrix A by setting all of its off-diagonal entries equal to zero is denoted A_{dg}.

(4) A diagonal matrix whose diagonal entries are all equal to one is called the **identity matrix**, denoted by I.

(5) A matrix whose second index set contains only one element is called a **column vector**. If we wish to distinguish a column vector from an arbitrary matrix, we shall denote the former by a lower-case italic letter.

(6) A matrix whose first index set contains only one element is called a **row vector.** If we wish to distinguish a row vector from an arbitrary matrix, we shall denote the former by a lower-case Greek letter.

(7) If A is a matrix defined on index sets M and N, define a matrix A^T, called the **transpose** of A, to have index sets N and M and to have entries given by $(A^T)_{nm} = A_{mn}$. The transpose of the transpose of A is simply A.

(8) The column vector all of whose entries are equal to one is denoted $\mathbf{1}$; the row vector with all entries one is $\mathbf{1}^T$. A matrix other than a row or column vector which has all entries equal to one is denoted by E.

(9) If A is an arbitrary matrix and c is a real number, cA is the matrix whose entries are given by $(cA)_{mn} = cA_{mn}$.

(10) The matrix $-A$ is defined to be the matrix $(-1)A$.

(11) A **constant** (column) **vectór** is a vector of the form $c\mathbf{1}$ for some extended real number c.

(12) A **bounded vector** is a vector all of whose entries are less than or equal in absolute value to some finite real number c.

Two matrices A and B are equal, written $A = B$, if they have the same index sets and if $A_{mn} = B_{mn}$ for every m and n. Inequalities are defined similarly. For example, $A > B$ if A and B have the same index sets and if $A_{mn} > B_{mn}$ for every m and n. In particular, non-negative matrices are those for which $A \geq 0$, and positive matrices are those for which $A > 0$.

Addition of matrices is defined for matrices A and B having the same index sets M and N. Their **sum** $C = A + B$ has the same index sets, and addition is defined entry-by-entry:

$$C_{mn} = A_{mn} + B_{mn}.$$

The sum $C = A + B$ is **well defined** if no entry of C is given by $\infty - \infty$ or by $-\infty + \infty$. We leave the verification of the following properties of matrices with index sets M and N to the reader:

(1) $A + 0 = A$ for every A.

(2) For every A having all entries finite, $A + (-A) = 0$.

(3) For any matrices A, B, and C,

$$A + (B + C) = (A + B) + C$$

if the indicated sums on at least one side of the equality are well defined.

Up to now, we have imposed no orderings on our index sets, and in fact nothing we have done so far necessitates doing so. We shall define even matrix multiplication shortly in a way that requires no ordering.

There is, however, a standard way of representing matrices as rectangular arrays, and for this purpose one normally orders the index sets with the usual ordering on the non-negative integers. The elements of the index sets are thus numbered 0, 1, 2, . . . either up to some integer r if the index set is finite or indefinitely if the index set is infinite. Under such orderings of its index sets, a matrix A is represented as

$$A = \begin{pmatrix} A_{00} & A_{01} & A_{02} & \cdots \\ A_{10} & A_{11} & A_{12} & \cdots \\ A_{20} & A_{21} & A_{22} & \cdots \\ \vdots & & & \ddots \end{pmatrix}.$$

We note that other representations are possible if at least one of the index sets is infinite; such representations come from ordering the index sets with an order type other than that of the non-negative integers. We shall meet another order type with its corresponding representation at the end of this section. We point out, however, that orderings are completely irrelevant as far as the fundamental properties of matrices are concerned, and we shall have little occasion to refer to them again.

For any real number a_m, define $a_m{}^+$ and $a_m{}^-$ by

$$\begin{aligned} a_m{}^+ &= \max(a_m, 0) \\ a_m{}^- &= -\min(a_m, 0). \end{aligned}$$

The sum of denumerably many non-negative terms $\sum_{m\in M} a_m{}^+$ or $\sum_{m\in M} a_m{}^-$ always exists independently of any ordering on M. Therefore, we say that $\sum_{m\in M} a_m = \sum_{m\in M} a_m{}^+ - \sum_{m\in M} a_m{}^-$ is **well defined** if not both of $\sum_{m\in M} a_m{}^+$ and $\sum_{m\in M} a_m{}^-$ are infinite.

Definition 1-1: Let A be a matrix with index sets K and M, and let B be a matrix with index sets M and N. Suppose the sums

$$\sum_{m\in M} A_{km}B_{mn}$$

are well defined for every k and every n. Then the **matrix product** $C = AB$ is said to be well defined; its index sets are K and N, and its entries are given by $C_{kn} = \sum_{m\in M} A_{km}B_{mn}$. Matrix multiplication is not defined unless all of these properties hold.

Most of the propositions and theorems about matrices that we shall deal with are statements of equality of matrices $A = B$. Such statements are really just assertions about the equality of corresponding entries of A and B, and a proof that A equals B need only contain an argument that an arbitrary entry of A equals the corresponding entry of B. With this understanding, we see that the proof of the additive properties of matrices is reduced to a trivial repetition of the properties of real numbers. Propositions about multiplication, however, when looked at entry-by-entry involve a new idea.

Let A be a matrix with index sets M and N and let m and n be fixed elements of M and N, respectively. The mth **row** of A is defined to be the restriction of the function A to the domain of pairs (m, s), where s runs through the set N. Similarly the nth **column** of A is defined to be the restriction of the function A to the domain of pairs (t, n), where t runs through the elements of the set M. We note that the mth row of a matrix is a row vector and that the nth column is a column vector. With these conventions matrices can be thought of as sets of rows or as sets of columns, and addition of matrices is simply addition of corresponding rows or columns of the matrices involved. Furthermore, the k-nth entry in the matrix product of A and B is the product of the kth row of A by the nth column of B and is of the form $\sum_{m\in M} \pi_m f_m$, where π is a row vector and f is a column vector. That is, propositions about matrix multiplication, when proved entry-by-entry, may sometimes be proved by considering only the product of a row vector and a column vector.

Because of the correspondence of row vectors to rows and column vectors to columns, we shall agree to call the domain of a row vector or a column vector the elements of a single index set.

Connected with any definition of multiplication are five properties which may or may not be valid for the structure being considered. All five of the properties do hold for the real numbers, and we state them in this context:

(1) Existence and uniqueness of a multiplicative identity. The real number 1 satisfies $c1 = 1c = c$ for every c.

(2) Commutativity: $ab = ba$

(3) Distributivity: $a(b + c) = ab + ac$
$(a + b)c = ac + bc$

(4) Associativity: $a(bc) = (ab)c$

(5) Existence and uniqueness of multiplicative inverses of all non-zero elements.

We can easily settle whether the first two properties hold for matrix multiplication. First, the identity matrix I plays the role of the multiplicative identity, and the identity is clearly unique. Second, commutativity can be expected to fail except in special cases because it is not even necessary for the index sets of two matrices to agree properly after the order of multiplication has been reversed.

The validity of the third property, that of distributivity, is the content of the next proposition.

Proposition 1-2: If A, B, and C are matrices and if AB, AC, and $AB + AC$ are well defined, then $A(B + C) = AB + AC$. Similarly $(D + E)F = DF + EF$ if DF, EF, and $DF + EF$ are all well defined.

PROOF: We prove only the first assertion. We may assume that A is a row vector π and that B and C are column vectors f and g. Then

$$\begin{aligned}\pi f + \pi g &= \sum_{m\in M} \pi_m f_m + \sum_{m\in M} \pi_m g_m \\ &= \sum_{m\in M} (\pi_m f_m + \pi_m g_m) \\ &= \sum_{m\in M} \pi_m (f_m + g_m) \\ &= \pi(f + g).\end{aligned}$$

The fourth and fifth properties are related and nontrivial. Associativity does not always hold, but useful sufficient criteria for its validity are known. For an example of how associativity may fail, let A be a matrix whose index sets are the non-negative integers and whose entries are given by

$$A = \begin{pmatrix} 1 & -1 & 0 & 0 & 0 & \cdots \\ 0 & 1 & -1 & 0 & 0 & \cdots \\ 0 & 0 & 1 & -1 & 0 & \cdots \\ 0 & 0 & 0 & 1 & -1 & \cdots \\ 0 & 0 & 0 & 0 & 1 & \cdots \\ \vdots & & & & & \ddots \end{pmatrix}$$

Then

$$\mathbf{1}^T(A\mathbf{1}) = 0,$$

whereas

$$(\mathbf{1}^T A)\mathbf{1} = 1.$$

All the products involved are well defined, but the multiplications do not associate.

We shall not consider the problem of existence of inverses, but uniqueness rests upon associativity. For suppose $AB = BA = AC = CA = I$. Since $AC = I$, we have $B(AC) = B$, and since $BA = I$, we have $(BA)C = C$. Therefore, $B = C$ if and only if $B(AC) = (BA)C$. With this note we proceed with some sufficient conditions for associativity.

Lemma 1-3: Let b_{ij} be a sequence of real numbers nondecreasing with i and with j. Then $\lim_i \lim_j b_{ij} = \lim_j \lim_i b_{ij}$, both possibly infinite.

PROOF: In the extended sense $\lim_i b_{ij} = L_j$ exists and so does $\lim_j b_{ij} = L_i^*$. Now $\{L_j\}$ is nondecreasing, for if $L_j > L_{j+k}$, then for i sufficiently large $b_{ij} > L_{j+k} \geq b_{i,j+k}$, which is impossible. Similarly $\{L_i^*\}$ is nondecreasing, so that $\lim_j L_j = L$ and $\lim_i L_i^* = L^*$ exist in the extended sense. If $L \neq L^*$, we may assume $L^* > L$ and hence L is finite. Then there exists an i such that $L_i^* > L$. Hence

$$\begin{aligned} L_i^* &> L \\ &\geq L_j \quad \text{for all } j \\ &\geq b_{ij} \quad \text{for all } j. \end{aligned}$$

Thus b_{ij} is bounded away from its limit on j, a contradiction.

Following the example of Lemma 1-3, we agree that all limits referred to in the future are on the **extended real line**.

Proposition 1-4: Non-negative matrices associate under multiplication.

PROOF: Since we are interested in each entry separately of a triple product, we may assume that we are to show that $\pi(Af) = (\pi A)f$, where $\pi \geq 0$, $A \geq 0$, $f \geq 0$, π is a row vector, f is a column vector, and the index sets are subsets of the non-negative integers. Then

$$\pi(Af) = \sum_m \sum_n \pi_m A_{mn} f_n$$

and

$$(\pi A)f = \sum_n \sum_m \pi_m A_{mn} f_n.$$

Set $b_{ij} = \sum_{m=0}^{i} \sum_{n=0}^{j} \pi_m A_{mn} f_n = \sum_{n=0}^{j} \sum_{m=0}^{i} \pi_m A_{mn} f_n$ and apply Lemma 1-3 to complete the proof.

If A is an arbitrary matrix we define A^+ and A^- by the equations

$$\begin{aligned} (A^+)_{mn} &= \max \{A_{mn}, 0\} \\ (A^-)_{mn} &= -\min \{A_{mn}, 0\}. \end{aligned}$$

Then $A = A^+ - A^-$, $A^+ \geq 0$, and $A^- \geq 0$. For row and column vectors, the matrices π^+, π^-, f^+, and f^- are defined analogously. We note that if Af is well defined, then so are Af^+ and Af^-. Powers of matrices are defined inductively by $A^0 = I$, $A^n = A(A^{n-1})$. The absolute value of a matrix A is $|A| = A^+ + A^-$. Proposition 1-4 now gives us five corollaries.

Corollary 1-5: Matrices associate if the product of their absolute values has all finite entries.

PROOF: We are again to prove that $\pi(Af) = (\pi A)f$, and we do so by setting $\pi = \pi^+ - \pi^-$, $A = A^+ - A^-$, and $f = f^+ - f^-$, applying distributivity, and using Proposition 1-4 on the resulting non-negative matrices.

Corollary 1-6: Finite matrices with finite entries associate.

PROOF: The result follows from Corollary 1-5.

Corollary 1-7: If A and B are non-negative matrices and f is a column vector such that $A(Bf)$ and $(AB)f$ are both well defined, then $A(Bf) = (AB)f$. In particular, if C is a non-negative matrix, if $n > 0$, and if C^nf and $C(C^{n-1}f)$ are well defined, then $C^nf = C(C^{n-1}f)$.

PROOF: Consider f^+ and f^- separately and apply Proposition 1-4. For the second assertion, set $A = C$ and $B = C^{n-1}$.

Similarly one proves two final corollaries.

Corollary 1-8: If A, B, C, and D are non-negative matrices such that either

(1) ABD, AB, and BD, or
(2) ACD, AC, and CD

are finite-valued, then $(A(B - C))D = A((B - C)D)$.

Corollary 1-9: If A, B, and C are matrices such that either

(1) A has only finitely many non-zero entries in each row,
(2) C has only finitely many non-zero entries in each column, or
(3) B has only finitely many non-zero entries,

and if $(AB)C$ and $A(BC)$ are well defined, then

$$(AB)C = A(BC).$$

Some of these conditions are cumbersome to check, but there is a simple sufficient condition. Suppose that we write a general product as $\prod_{i=1}^{n} (A_i - B_i)$, with $A_i \geq 0$ and $B_i \geq 0$. If all the 2^n products

$$A_1A_2 \ldots A_n,\ B_1A_2 \ldots A_n, \ldots, B_1B_2 \ldots B_n$$

are finite, then we see from Proposition 1-2 and Corollary 1-5 that we may freely use distributivity and associativity.

The effect of matrix multiplication on matrix inequalities is summarized by the next proposition, whose proof is left to the reader.

Proposition 1-10: Matrix inequalities of the form $A \geq B$ or $B \leq A$ are preserved when both sides of the inequality are multiplied by a non-negative matrix. Inequalities of the form $A > B$ or $B < A$ are preserved when both sides are multiplied by a positive matrix, provided the products have all entries finite.

Next we consider the problem of "block multiplication" of matrices. The picture we have in mind is the following decomposition of the matrices involved in a product:

$$\begin{pmatrix} A_1 & A_2 \\ A_3 & A_4 \end{pmatrix}\begin{pmatrix} B_1 & B_2 \\ B_3 & B_4 \end{pmatrix} = \begin{pmatrix} C_1 & C_2 \\ C_3 & C_4 \end{pmatrix}.$$

More specifically, let K, M, and N be index sets and let K', M', and N', respectively, be non-empty subsets of the index sets. Impose orderings on K, M, and N so that the elements of K', M', and N' precede the other elements, which comprise the complementary sets $\tilde{K}'$, $\tilde{M}'$, and $\tilde{N}'$. Let A, B, and C be matrices such that

(1) A is defined on K and M,
(2) B is defined on M and N, and
(3) $AB = C$ is well defined.

Let matrices A_1, A_2, A_3, and A_4 be defined as the restriction of the function A to the sets

(1) K' and M' for A_1
(2) K' and $\tilde{M}'$ for A_2
(3) $\tilde{K}'$ and M' for A_3
(4) $\tilde{K}'$ and $\tilde{M}'$ for A_4.

Pictorially what we are doing is writing A as four submatrices with

$$A = \begin{pmatrix} A_1 & A_2 \\ A_3 & A_4 \end{pmatrix}.$$

We perform the same kind of decomposition for B and C and obtain

$$\begin{pmatrix} A_1 & A_2 \\ A_3 & A_4 \end{pmatrix}\begin{pmatrix} B_1 & B_2 \\ B_3 & B_4 \end{pmatrix} = \begin{pmatrix} C_1 & C_2 \\ C_3 & C_4 \end{pmatrix}.$$

The proposition to follow asserts that the submatrices of A, B, and C

multiply as if they were entries themselves. Its proof depends on the fact that matrix multiplication is defined independently of any ordering on the index sets.

Proposition 1-11:

$$A_1B_1 + A_2B_3 = C_1$$
$$A_1B_2 + A_2B_4 = C_2$$
$$A_3B_1 + A_4B_3 = C_3$$
$$A_3B_2 + A_4B_4 = C_4.$$

PROOF: We prove only the first identity since the others are similar.

$$(C_1)_{ij} = C_{ij} = \sum_{m \in M} A_{im}B_{mj} = \sum_{m \in M'} A_{im}B_{mj} + \sum_{m \in \bar{M}'} A_{im}B_{mj}$$
$$= (A_1B_1)_{ij} + (A_2B_3)_{ij}.$$

Notice that if the submatrix A_1 has at least one infinite index set, then the representation of A by

$$A = \begin{pmatrix} A_1 & A_2 \\ A_3 & A_4 \end{pmatrix}$$

is not the standard one

$$A = \begin{pmatrix} A_{00} & A_{01} & \cdots \\ A_{10} & A_{11} & \cdots \\ \vdots & & \ddots \end{pmatrix}.$$

The ordering on the index sets of A is not of the same type as that of the non-negative integers. We recall once more, however, that the fundamental properties of matrices are independent of any orderings on the index sets. It is only the representation of a matrix as an array which requires these orderings.

Limits of matrices play an important part in the study of denumerable Markov chains. We shall touch only briefly at this time on the problems involved.

Definition 1-12: Let $\{A^{(k)}\}$ be a sequence of matrices. We say that $A = \lim_{k\to\infty} A^{(k)}$ exists if $A_{mn} = \lim_{k\to\infty} (A^{(k)})_{mn}$ exists for every m and n.

Notice that limits of matrices are defined entry-by-entry. No uniformity of convergence to the limiting matrix is assumed.

The type of problem that arises is as follows. Let π be a row vector and let $\{f^{(k)}\}$ be a sequence of column vectors converging to a column vector f. Is it true that $\{\pi f^{(k)}\}$ necessarily converges to πf? The answer to this question is in the negative unless some additional hypothesis is added. What is being attempted is an interchange of the order of two limit operations—one from the series which defines $\{\pi f^{(k)}\}$ and the other from the limit as k tends to ∞. Such an interchange can be justified only under special circumstances, and we shall obtain later in this chapter some sufficient conditions as special cases of theorems of measure theory.

2. Measure theory

Let X be an arbitrary non-empty set of points and let $\mathscr{F}$ be a family of subsets of X. We say that $\mathscr{F}$ is a **field of sets** if

(1) the empty set $\varnothing$ is in $\mathscr{F}$,
(2) whenever A is a set of $\mathscr{F}$, the complement of A, denoted $\tilde{A}$, is in $\mathscr{F}$, and
(3) whenever A and B are sets of $\mathscr{F}$, so is their union, denoted $A \cup B$.

A field of sets $\mathscr{F}$ is called a **Borel field** if it has the additional property that whenever $A_n \in \mathscr{F}$ for $n = 1, 2, 3, \ldots$, so is $\bigcup_{n=1}^{\infty} A_n$.

The intersection of sets A and B is indicated by $A \cap B$, and the difference $A \cap \tilde{B}$ is denoted $A - B$. From the above definitions the reader can easily establish the following result.

Proposition 1-13: If $\mathscr{F}$ is a field of sets, then $\mathscr{F}$ contains $\varnothing$ and X and is closed under complementation, finite unions, finite intersections, and differences. If $\mathscr{F}$ is a Borel field, then $\mathscr{F}$ is closed under denumerable intersections.

Proposition 1-14: For any class of sets $\mathscr{C}$ of the points of a set X, there exists a unique smallest Borel field containing $\mathscr{C}$.

PROOF: The family of all subsets of X forms a Borel field containing $\mathscr{C}$. Form the intersection of all Borel fields which contain $\mathscr{C}$ and call the resulting family of sets $\mathscr{F}$. Let A be in $\mathscr{F}$; then A is in all Borel fields containing $\mathscr{C}$ and so is $\tilde{A}$. Hence $\tilde{A}$ is in $\mathscr{F}$. A similar argument applies to intersections and denumerable unions. Thus $\mathscr{F}$ is the smallest Borel field containing $\mathscr{C}$.

Definition 1-15: A function ρ from a field of sets $\mathscr{F}$ to the extended real number system is called a **set function**. If $\rho(A) \geq 0$ for every A in $\mathscr{F}$, ρ is said to be **non-negative**. If $\rho(A \cup B) = \rho(A) + \rho(B)$ whenever A and B are in $\mathscr{F}$ and $A \cap B = \varnothing$, ρ is said to be **additive**. Supposc A_n is in $\mathscr{F}$ for $n = 1, 2, 3, \ldots$, and suppose $A_i \cap A_j = \varnothing$

whenever $i \neq j$. If $\rho(\bigcup_{n=1}^{\infty} A_n) = \sum_{n=1}^{\infty} \rho(A_n)$ holds whenever $\bigcup_{n=1}^{\infty} A_n$ is a set of $\mathscr{F}$, then ρ is said to be **completely additive**. In discussing set functions, we shall assume that there are no two sets A and B in $\mathscr{F}$ such that $\rho(A) = +\infty$ and $\rho(B) = -\infty$, and we shall assume that ρ is not identically infinite.

An additive set function ρ has the properties that

(1) $\rho(\varnothing) = 0$,
(2) $\rho(\bigcup_{n=1}^{N} A_n) = \sum_{n=1}^{N} \rho(A_n)$ for disjoint sets $\{A_n\}$, and
(3) $\rho(A \cup B) + \rho(A \cap B) = \rho(A) + \rho(B)$.

If ρ is non-negative and additive and if A is contained in B, then $\rho(A) \leq \rho(B)$. To see this, set $C = B \cap \tilde{A}$ so that A and C are disjoint and $A \cup C = B$. Then $\rho(A) + \rho(C) = \rho(B)$ by additivity, and the result follows at once. We shall now establish two facts about completely additive set functions.

Proposition 1-16: Let ρ be an additive set function defined on a field of sets $\mathscr{F}$. Let $\{A_n\}$ be a sequence of sets in $\mathscr{F}$ such that $A_1 \subset A_2 \subset \cdots$, and suppose $A = \bigcup_{n=1}^{\infty} A_n$ is in $\mathscr{F}$. If ρ is completely additive, then $\lim_{n\to\infty} \rho(A_n) = \rho(A)$. Conversely, if $\lim_{n\to\infty} \rho(A_n) = \rho(A)$ for all such sequences, ρ is completely additive.

PROOF: Set $B_1 = A_1$ and $B_n = A_n \cap \tilde{A}_{n-1}$. Then $A_n = \bigcup_{k=1}^{n} B_k$ disjointly, and by additivity $\rho(A_n) = \sum_{k=1}^{n} \rho(B_k)$. But $A = \bigcup_{k=1}^{\infty} B_k$ and by complete additivity $\rho(A) = \sum_{k=1}^{\infty} \rho(B_k)$. The proof of the converse is left to the reader.

A consequence of this proposition is the following:

Corollary 1-17: Let ρ be an additive set function defined on a field of sets $\mathscr{F}$ in such a way that $\rho(A) < \infty$ for every A. Let $\{A_n\}$ be a sequence of sets in $\mathscr{F}$ such that $A_1 \supset A_2 \supset A_3 \supset \cdots$ and $\bigcap_{n=1}^{\infty} A_n = \varnothing$. If ρ is completely additive, then $\lim_{n\to\infty} \rho(A_n) = 0$. Conversely, if $\lim_{n\to\infty} \rho(A_n) = 0$ for all such sequences, then ρ is completely additive.

A non-negative completely additive set function on a field of sets $\mathscr{F}$ is called a **measure**. The set of points X with a measure defined on its field $\mathscr{F}$ is called a **measure space**. We shall usually denote measures by μ or ν. If there is no ambiguity about what measure is involved, we shall frequently refer to X by itself as the measure space.

If X is a measure space with field of sets $\mathscr{F}$ and measure μ, then X is a set in $\mathscr{F}$, and we define $\mu(X)$ to be the **total measure** of the space. A **probability space** is a measure space of total measure one.

We give four examples of measure spaces.

(1) Let X be any set, let $\mathscr{F} = \{\varnothing, X\}$, and define $\mu(\varnothing) = 0$ and $\mu(X) = a \geq 0$. Then X is the trivial measure space.

(2) Let X be Euclidean n-space, let $\mathscr{F}$ be the Lebesgue measurable sets, and let μ be Lebesgue measure (the natural generalization of length, area, or volume).

(3) Let X be the set of six possible outcomes for tossing a die. Assign weight $\frac{1}{6}$ to each of the six points in the space, and for any subset of X assign as a measure the sum of the weights of the points in the set. Then $\mathscr{F}$ is the family of all subsets of X, and X is a probability space.

(4) Let X be a denumerable index set, and let π be a non-negative row vector with X as its index set. Assign as a weight to each point of X the value of the corresponding entry of π. For any subset of X assign as a measure the sum of the weights of the points in the set. Then $\mathscr{F}$ is the family of all subsets of X, and X is a measure space with total measure $\pi\mathbf{1}$.

The sets of a field on which a measure μ is defined are called the **μ-measurable** or simply the **measurable** subsets of X. In the construction of a measure on a field, it is possible for a non-empty set A to be assigned measure zero. In example (2) above, for instance, every denumerable set and even certain uncountable sets are sets of measure zero. Suppose B is a subset of such a set A. If B is measurable, then $\mu(B) \geq 0$ since μ is a measure. But

$$\mu(B) \leq \mu(A) = 0$$

since $B \subset A$ and A is of measure zero. Thus, a measurable subset of a set of measure zero is of measure zero. But there is no reason why such a set B has to be measurable. However, one can agree to add all subsets of sets of measure zero to a field and extend the resulting family of sets to the smallest field containing the family. Such an extended field is called an **augmented field**. It consists precisely of all sets of the form $(C - D) \cup E$, where C is a set in the original field and D and E are subsets of a set of measure zero. Therefore the augmented field of a Borel field is again a Borel field. Note that in any augmented field every subset of a set of measure zero is measurable and has measure zero. In later chapters of this book all fields will be augmented.

If a statement about the points of a measure space X fails to be true only for a set of points which is a subset of a set of measure zero, we say that the statement holds for **almost all** points of X or that it is true **almost everywhere** (abbreviated **a.e.**).

Proposition 1-18: Let μ be a measure defined on a field of sets $\mathscr{F}$. If $\{A_n\}$ is a sequence of sets in $\mathscr{F}$, if A is in $\mathscr{F}$, and if $A \subset \bigcup_n A_n$, then

$$\mu(A) \leq \sum_n \mu(A_n).$$

PROOF: Write $B_n = A_n - (\bigcup_{k=1}^{n-1} A_k)$. The sets B_n are disjoint in pairs, and consequently the sets $A \cap B_n$ are also disjoint. Furthermore, $\bigcup_n B_n = \bigcup_n A_n$ so that

$$\begin{aligned} A &= A \cap \Big(\bigcup_n A_n\Big) \\ &= A \cap \Big(\bigcup_n B_n\Big) \\ &= \bigcup_n (A \cap B_n). \end{aligned}$$

By hypothesis μ is a measure. It is therefore completely additive and

$$\begin{aligned} \mu(A) &= \sum_n \mu(A \cap B_n) \\ &\leq \sum_n \mu(B_n) \quad \text{since } A \cap B_n \subset B_n \\ &\leq \sum_n \mu(A_n) \quad \text{since } B_n \subset A_n. \end{aligned}$$

To conclude this section we shall establish a result known as the Extension Theorem. The proof follows the proof of Rudin [1953].

Theorem 1-19: Let $\mathscr{F}$ be a field of sets in a space X and let ν be a measure defined on $\mathscr{F}$. Suppose X can be written as the denumerable union of sets in $\mathscr{F}$ of finite measure. If $\mathscr{G}$ is the smallest Borel field containing $\mathscr{F}$, then ν can be extended in one and only one way to a measure defined on all of $\mathscr{G}$ which agrees with ν on sets of $\mathscr{F}$.

Before proving the theorem, we need some preliminary lemmas and definitions. The property in the statement of the theorem that X is the denumerable union of sets of finite measure is summarized by saying that ν is **sigma-finite**.

Let ν be a measure defined on a field of sets $\mathscr{F}$ in a space X, and suppose $X = \bigcup_{n=1}^{\infty} A_n$ with $A_n \in \mathscr{F}$ and $\nu(A_n) < \infty$. For each subset B of X, define $\mu(B) = \inf \{\sum \nu(B_n)\}$, where the infimum is taken over all denumerable coverings of B by sets $\{B_n\}$ of $\mathscr{F}$.

Lemma 1-20: The set function μ is non-negative. If A and B are subsets of X such that $A \subset B$, then $\mu(A) \leq \mu(B)$. If C is a set in $\mathscr{F}$, then $\mu(C) = \nu(C)$.

Proof: We see that μ is non-negative because μ is the limit of non-negative quantities. If $A \subset B$, then $\mu(A) \leq \mu(B)$ because every covering of B is a covering of A. Let C be in $\mathscr{F}$. Then $\{C\}$ is a covering of C and $\mu(C) \leq \nu(C)$. And for any covering $\{C_n\}$,

$$\nu(C) \leq \sum_n \nu(C_n)$$

by Proposition 1-18. Therefore,

$$\nu(C) \leq \inf \sum_n \nu(C_n) = \mu(C).$$

Lemma 1-21: If $\{A_n\}$ is an arbitrary sequence of subsets of X and if $A = \bigcup_n A_n$, then $\mu(A) \leq \sum_n \mu(A_n)$.

Proof: Let $\epsilon > 0$ be given. Let $\{B_k^{(n)}\}$ with $k = 1, 2, 3, \ldots$ be a denumerable covering of A_n such that $B_k^{(n)}$ is in $\mathscr{F}$ and $\sum_k \nu(B_k^{(n)}) \leq \mu(A_n) + \epsilon/2^n$. This choice is possible by the definition of μ. Then since all the B's form a covering of A, we have

$$\begin{aligned}\mu(A) &\leq \sum_n \sum_k \nu(B_k^{(n)}) \\ &\leq \sum_n \mu(A_n) + \epsilon\end{aligned}$$

and the assertion follows.

We define a set theoretic operation $\oplus$ for subsets of X by

$$A \oplus B = (A \cap \tilde{B}) \cup (B \cap \tilde{A}).$$

The set $A \oplus B$ is called the **symmetric difference** of A and B. A point is in $A \oplus B$ if it is in A or B but not both. We leave the details of the proof of the next lemma to the reader.

Lemma 1-22: The subsets of a space X form a ring under the operations $\oplus$ and $\cap$ with additive identity $\varnothing$ and multiplicative identity X. Every set is its own additive inverse.

Define a distance d between subsets of X by $d(A, B) = \mu(A \oplus B)$. We note that d has the properties

$$d(A, A) = \mu(\varnothing) = 0$$

and

$$d(A, B) = \mu(A \oplus B) = d(B, A).$$

Since

$$A \cup B = (A \oplus B) \cup B,$$

we have

$$A \subset (A \oplus B) \cup B$$

and by Lemmas 1-20 and 1-21

$$\mu(A) \leq d(A, B) + \mu(B).$$

Replacing A by $A \oplus B$ and B by $C \oplus B$, we obtain the triangle inequality

$$d(A, B) \leq d(A, C) + d(C, B).$$

Lemma 1-23: For any subsets A_1, A_2, B_1, B_2, A, and B, of X,

$$d((A_1 \cup A_2), (B_1 \cup B_2)) \leq d(A_1, B_1) + d(A_2, B_2)$$

$$d((A_1 \cap A_2), (B_1 \cap B_2)) \leq d(A_1, B_1) + d(A_2, B_2)$$

$$d(B, A) = d(\tilde{B}, \tilde{A}).$$

PROOF: We prove only the first and third assertions. First we observe that $(A_1 \cup A_2) \oplus (B_1 \cup B_2) \subset (A_1 \oplus B_1) \cup (A_2 \oplus B_2)$. For suppose $x \in (A_1 \cup A_2) \oplus (B_1 \cup B_2)$. We may assume without loss of generality that $x \in A_1 \cup A_2$ but $x \notin B_1 \cup B_2$. If $x \in A_1$, then $x \notin B_1$ so that $x \in A_1 \oplus B_1$. Similarly if $x \in A_2$, then $x \in A_2 \oplus B_2$ and the containment is established. The first assertion of the lemma now follows by applying Lemmas 1-20 and 1-21. For the third part, we have

$$B \oplus A = (A \cap \tilde{B}) \cup (\tilde{A} \cap B)$$

and

$$\tilde{B} \oplus \tilde{A} = (\tilde{A} \cap B) \cup (A \cap \tilde{B})$$

so that

$$B \oplus A = \tilde{B} \oplus \tilde{A}.$$

Definition 1-24: Convergence of sets in measure is defined by saying that $A_n \to A$ if $\lim_{n \to \infty} d(A_n, A) = 0$. Let $\mathscr{F}^*$ be the collection of all subsets A of X for which there exists a sequence $\{A_n\}$ of sets in $\mathscr{F}$ having the property that $A_n \to A$. Let $\mathscr{G}^*$ be the family of denumerable unions of sets in $\mathscr{F}^*$.

Lemma 1-25: If $\{A_n\}$ and $\{B_n\}$ are sequences of sets in $\mathscr{F}$ such that $A_n \to A$ and $B_n \to B$, then $A_n \cup B_n \to A \cup B$, $A_n \cap B_n \to A \cap B$, and $\tilde{A}_n \to \tilde{A}$. Therefore, $\mathscr{F}^*$ is a field of sets. For any $C_n \to C$, $\lim_n \mu(C_n) = \mu(C)$.

PROOF: Since by Lemma 1-23

$$d((A_n \cup B_n), (A \cup B)) \leq d(A_n, A) + d(B_n, B)$$

$$d((A_n \cap B_n), (A \cap B)) \leq d(A_n, A) + d(B_n, B)$$

$$d(A_n, A) = d(\tilde{A}_n, \tilde{A}),$$

we have $A_n \cup B_n \to A \cup B$, $A_n \cap B_n \to A \cap B$, and $\tilde{A}_n \to \tilde{A}$. The limit of $\mu(C_n)$ is established by the inequalities

$$\mu(C_n) \leq d(C_n, C) + \mu(C)$$

and

$$\mu(C) \leq d(C_n, C) + \mu(C_n).$$

Lemma 1-26: μ is additive on $\mathscr{F}^*$.

PROOF: Let A and B be disjoint sets in $\mathscr{F}^*$ and pick $\{A_n\}$ and $\{B_n\}$ in $\mathscr{F}$ such that $A_n \to A$ and $B_n \to B$. Then since ν is additive on $\mathscr{F}$ and since μ agrees with ν on sets of $\mathscr{F}$, we have

$$\mu(A_n \cup B_n) + \mu(A_n \cap B_n) = \mu(A_n) + \mu(B_n).$$

By Lemma 1-25,

$$\mu(A \cup B) + \mu(A \cap B) = \mu(A) + \mu(B)$$

or

$$\mu(A \cup B) = \mu(A) + \mu(B).$$

Lemma 1-27: If $A = \bigcup_n A_n$ with A in $\mathscr{G}^*$ and $\{A_n\}$ a sequence of disjoint sets in $\mathscr{F}^*$, then $\mu(A) = \sum_n \mu(A_n)$.

PROOF: Since $A \supset (A_1 \cup A_2 \cup A_3 \cup \cdots \cup A_k)$, we have, by Lemma 1-20, $\mu(A) \geq \mu(A_1 \cup A_2 \cup \cdots \cup A_k)$, and by Lemma 1-26 the right side equals $\sum_{n=1}^{k} \mu(A_n)$ for each k. Hence

$$\mu(A) \geq \sum_n \mu(A_n),$$

and equality holds by Lemma 1-21.

Lemma 1-28: If A is in $\mathscr{G}^*$ and if $\mu(A)$ is finite, then A is in $\mathscr{F}^*$.

REMARK: If A is in $\mathscr{F}^*$, then $\mu(A)$ is not necessarily finite.

PROOF OF LEMMA: Write $A = \bigcup_n A_n$ with $\{A_n\}$ in $\mathscr{F}^*$, A_n disjoint sets, and set $B_n = \bigcup_{k=1}^{n} A_k$. Then

$$\begin{aligned} d(A, B_n) &= \mu([A \cap \tilde{B}_n] \cup [\tilde{A} \cap B_n]) \\ &= \mu\left(\bigcup_{k=n+1}^{\infty} A_k\right), \end{aligned}$$

which by Lemma 1-27

$$= \sum_{k=n+1}^{\infty} \mu(A_k).$$

Since the last expression on the right is the tail of a convergent series, we have $B_n \to A$. Since $B_n \in \mathscr{F}^*$, we can find L_n in $\mathscr{F}$ such that $d(L_n, B_n) \leq 1/n$. Then $d(L_n, A) \leq 1/n + d(B_n, A)$, and hence $L_n \to A$. Thus A is in $\mathscr{F}^*$.

REMARK: If $A \in \mathscr{G}^*$ with $\mu(A)$ finite, then A is in $\mathscr{F}^*$; hence for every $\epsilon > 0$ there is a B in $\mathscr{F}$ such that $\mu(A \oplus B) \leq \epsilon$. Conversely, if there exists such a B_ϵ for any ϵ, then $B_{1/n} \to A$ so that $A \in \mathscr{F}^*$ and, *a fortiori*, $A \in \mathscr{G}^*$. These observations give a characterization of the sets A in $\mathscr{G}^*$ for which $\mu(A)$ is finite.

Lemma 1-29: μ is completely additive on $\mathscr{G}^*$, and $\mathscr{G}^*$ is a Borel field.

PROOF: Suppose

$$A = \bigcup_n A_n$$

is the union of disjoint sets $\{A_n\}$ in $\mathscr{G}^*$. Then $\mu(A) \geq \mu(A_n)$ for every n by Lemma 1-20, so that we may assume $\mu(A_n) < \infty$ for every n. The complete additivity of μ now follows from Lemmas 1-28 and 1-27. For the proof that $\mathscr{G}^*$ is a Borel field, we see clearly that $\mathscr{G}^*$ is closed under denumerable unions. It remains to be proved that $\mathscr{G}^*$ is closed under complementation. Since ν is sigma-finite, let

$$X = \bigcup_n A_n$$

with A_n in $\mathscr{F}$ and $\mu(A_n) = \nu(A_n) < \infty$. Let B in $\mathscr{G}^*$ be given and suppose $B = \bigcup_k B_k$ with B_k in $\mathscr{F}^*$. Since

$$A_n \cap B = \bigcup_k (A_n \cap B_k)$$

and since $A_n \cap B_k$ is in $\mathscr{F}^*$, $A_n \cap B$ is in $\mathscr{G}^*$. But by Lemma 1-20,

$$\mu(A_n \cap B) \leq \mu(A_n)$$

and we have assumed that $\mu(A_n) < \infty$. Thus by Lemma 1-28, $A_n \cap B \in \mathscr{F}^*$, and since $\mathscr{F}^*$ is a field,

$$A_n \cap (X - (A_n \cap B)) = A_n \cap \tilde{B}$$

is in $\mathscr{F}^*$. Therefore

$$\tilde{B} = X \cap \tilde{B} = \bigcup_n (A_n \cap \tilde{B})$$

is in $\mathscr{G}^*$, and the proof is complete.

We are now in a position to prove the Extension Theorem.

PROOF OF THEOREM 1-19: Existence of the extension of ν to a measure μ defined on $\mathscr{G}^*$ is proved by Lemmas 1-20 and 1-29. Since, by

Lemma 1-29, $\mathscr{G}^*$ is a Borel field containing $\mathscr{F}$, $\mathscr{G}^*$ contains $\mathscr{G}$. The extended measure restricted to sets of $\mathscr{G}$ has the desired properties. For uniqueness, suppose μ' is another measure on $\mathscr{G}$ that agrees with ν on $\mathscr{F}$. Since, by sigma-finiteness, X is the union of sets A_n in $\mathscr{F}$ of finite ν-measure, we may assume that X is a disjoint union of sets of finite measure by letting $B_n = A_n - \bigcup_{k<n} A_k$. Let C be any set in $\mathscr{G}$; we want to show $\mu'(C) = \mu(C)$. By definition,

$$\mu(C) = \inf\left\{\sum_n \nu(C_n)\right\}$$

with the infimum taken over covers $\{C_n\}$, where C_n is in $\mathscr{F}$. For any fixed cover $\{C_n\}$, we have

$$\mu'(C) \leq \mu'\left(\bigcup_n C_n\right) \leq \sum_n \mu'(C_n) = \sum_n \nu(C_n).$$

Therefore

$$\mu'(C) \leq \inf\left\{\sum_n \nu(C_n)\right\} = \mu(C).$$

Writing

$$\mu'(C) = \sum_n \mu'(C \cap B_n),$$

we see that it is sufficient to show that

$$\mu'(C \cap B_n) \geq \mu(C \cap B_n).$$

But

$$\mu'(C \cap B_n) + \mu'(\tilde{C} \cap B_n) = \nu(B_n) = \mu(C \cap B_n) + \mu(\tilde{C} \cap B_n).$$

Now we know that μ' is dominated by μ:

$$\mu'(\tilde{C} \cap B_n) \leq \mu(\tilde{C} \cap B_n) \leq \mu(B_n) < \infty.$$

If

$$\mu'(C \cap B_n) < \mu(C \cap B_n),$$

we obtain the contradiction

$$\mu'(C \cap B_n) + \mu'(\tilde{C} \cap B_n) < \mu(C \cap B_n) + \mu(\tilde{C} \cap B_n).$$

3. Measurable functions and Lebesgue integration

Let $\mathscr{F}$ be a Borel field of sets in a set X. The **measurable sets** of X are the sets of $\mathscr{F}$.

Definition 1-30: Let f be a function with domain X and with range the extended real number system. The function f is said to be a **measurable function** if for each real number c the set $\{x \mid f(x) < c\}$ is measurable.

The content of the next proposition is that the property $f(x) < c$ may be replaced by any of the conditions $f(x) \leq c$, $f(x) > c$, or $f(x) \geq c$. Therefore, if f is a measurable function, then the set

$$\{x \mid c \leq f(x) \leq d\}$$

is measurable; either or both of the signs $\leq$ may be replaced by $<$, and the set is still measurable.

Proposition 1-31: The following four conditions are equivalent:

(1) $\{x \mid f(x) < c\}$ is measurable for every c.
(2) $\{x \mid f(x) \leq c\}$ is measurable for every c.
(3) $\{x \mid f(x) > c\}$ is measurable for every c.
(4) $\{x \mid f(x) \geq c\}$ is measurable for every c.

PROOF: From

$$\{x \mid f(x) \leq c\} = \bigcap_{n=1}^{\infty} \left\{x \mid f(x) < c + \frac{1}{n}\right\},$$

$$\{x \mid f(x) > c\} = X - \{x \mid f(x) \leq c\},$$

$$\{x \mid f(x) \geq c\} = \bigcap_{n=1}^{\infty} \left\{x \mid f(x) > c - \frac{1}{n}\right\},$$

and

$$\{x \mid f(x) < c\} = X - \{x \mid f(x) \geq c\},$$

we see that (1) implies (2), that (2) implies (3), that (3) implies (4), and that (4) implies (1).

Proposition 1-32: Every constant function is measurable.

PROOF: If $f(x) = a$ identically, then $\{x \in X \mid f(x) < c\}$ is either $\varnothing$ or X.

In analogy with our procedure for matrices in Section 1, we define f^+ and f^- by

$$f^+(x) = \max\{f(x), 0\}$$

$$f^-(x) = -\min\{f(x), 0\}.$$

Proposition 1-33: If f is measurable, then so are f^+, f^-, and $|f|$.

PROOF:
$$\{x \mid f^+ \geq c\} = \{x \mid f \geq c\} \cup \{x \mid 0 \geq c\}$$
$$\{x \mid f^- \geq c\} = \{x \mid f \leq -c\} \cup \{x \mid 0 \geq c\}.$$

The set $\{x \mid 0 \geq c\}$ is either $\varnothing$ or X and is therefore measurable. For $|f|$ we have

$$\{x \mid |f| \leq c\} = \{x \mid -c \leq f(x) \leq c\}.$$

Proposition 1-34: Let f and g be measurable functions whose values are finite at all points. Then $f + g$ and $f \cdot g$ are measurable.

PROOF: We prove only the assertion about $f + g$. Order the rational numbers and call the nth one r_n. Then

$$\{x \mid (f + g)(x) > c\} = \bigcup_{n=1}^{\infty} (\{x \mid f(x) > c + r_n\} \cap \{x \mid g(x) > -r_n\}),$$

so that $f + g$ is measurable.

Corollary 1-35: If f is measurable, then so is cf for every constant c.

Proposition 1-36: Let $\{f_n\}$ be a sequence of measurable functions. Then the functions

(1) $\sup_n f_n(x)$
(2) $\inf_n f_n(x)$
(3) $\limsup_n f_n(x)$
(4) $\liminf_n f_n(x)$

are all measurable.

PROOF: The assertions follow from the observations that

$$\{x \mid \sup f_n(x) > c\} = \bigcup_{n=1}^{\infty} \{x \mid f_n(x) > c\},$$

$$\{x \mid \inf f_n(x) < c\} = \bigcup_{n=1}^{\infty} \{x \mid f_n(x) < c\},$$

$$\limsup_n f_n(x) = \inf_n \sup_{m>n} f_m(x)$$

and

$$\liminf_n f_n(x) = \sup_n \inf_{m>n} f_m(x).$$

The supremum of finitely many functions is their pointwise maximum. Therefore the maximum and minimum of finitely many measurable functions are both measurable.

Corollary 1-37: If $\{f_n\}$ is a sequence of measurable functions and if $f = \lim_n f_n$ exists at all points, then f is measurable.

We shall give three examples of measurable functions.

(1) Let $\mathscr{F}$ be the family of sets on the real line which are either finite unions of open and closed intervals or complements of such sets. Then $\mathscr{F}$ is a field of sets. Let $\mathscr{G}$ be the smallest Borel field containing $\mathscr{F}$. All continuous real-valued functions are measurable with respect to the Borel field $\mathscr{G}$.

(2) Let X be a space for which $\mathscr{B}$ is the family of all subsets of X. Then every function f defined on X is measurable.

(3) Let X be the union of a sequence of disjoint sets $\{A_n\}$, and let $\mathscr{B}$ be the family of all sets which are unions of sets in the sequence. Then a function f is measurable if and only if its restriction to the domain A_n is a constant function for each n. In particular, if $\mathscr{B} = \{X, \varnothing\}$, then the measurable functions are the constant functions.

Let A be any subset of X. The **characteristic function** of A, denoted $\chi_A(x)$, is defined by

$$\chi_A(x) = \begin{cases} 1 & \text{if } x \in A \\ 0 & \text{otherwise.} \end{cases}$$

A function that takes on only a finite number of values is called a **simple function**. It may be represented, uniquely, in the form

$$(*) \qquad s = \sum_{n=1}^{N} c_n \chi_{A_n}$$

where the c_n are the distinct values the function takes on and the sets A_n are disjoint. The simple function is measurable if and only if all of the sets $A_1, A_2, \ldots, A_N$ are measurable.

Proposition 1-38: For any non-negative function f defined on X, there exists a sequence of non-negative simple functions $\{s_n\}$ with the property that for each $x \in X$, $\{s_n(x)\}$ is a monotonically increasing sequence converging to $f(x)$. If f is measurable, the $\{s_n\}$ may be taken to be measurable.

Proof: For every n and for $1 \le j \le n2^n$ set

$$A_{nj} = \left\{x \,\middle|\, \frac{j-1}{2^n} \le f(x) < \frac{j}{2^n}\right\}$$

and

$$B_n = \{x \mid f(x) \ge n\}.$$

Then

$$s_n = \sum_{j=1}^{n2^n} \frac{j-1}{2^n} \chi_{A_{nj}} + n\chi_{B_n}$$

increases monotonically with n to f. If f is measurable, then so are $\{A_{nj}\}$ and B_n; thus s_n is measurable.

If μ is a measure defined on a Borel field $\mathscr{B}$ of subsets of X, we denote the measure space by the ordered triple $(X, \mathscr{B}, \mu)$. In $(X, \mathscr{B}, \mu)$ let E be a set of the family $\mathscr{B}$, and suppose s is a non-negative, measurable simple function, represented as in (*) above. Since s is measurable, A_n is measurable and $\mu(A_n \cap E)$ is defined for every n. Set

$$I_E(s) = \sum_{n=1}^{N} c_n\mu(A_n \cap E).$$

For any non-negative measurable function f, define the **Lebesgue integral** of f on the set E with respect to the measure μ by

$$\int_E f d\mu = \sup I_E(s),$$

where the supremum is taken over all simple functions s satisfying $0 \leq s \leq f$. We note that the value of the integral is non-negative and possibly infinite. It can be verified that if s is a non-negative measurable simple function, then

$$\int_E s d\mu = I_E(s).$$

If f is an arbitrary measurable function, then by Proposition 1-33, $\int_E f^+ d\mu$ and $\int_E f^- d\mu$ are both defined. If the integrals of f^+ and f^- are not both infinite, we define the **integral** of f by $\int_E f d\mu = \int_E f^+ d\mu - \int_E f^- d\mu$. The function f is said to be **integrable** on the set E if $\int_E f^+ d\mu$ and $\int_E f^- d\mu$ are both finite.

Following our examples of measure spaces and measurable functions, we give three examples of integration. A fourth example will arise in Chapter 2.

(1) Let $\mathscr{B} = \{\varnothing, X\}$ and suppose $\mu(X) = 1$. Only the constant functions are measurable and

$$\int_\varnothing c d\mu = 0; \qquad \int_X c d\mu = c.$$

(2) Let X be the real line, let $\mathscr{G}$ be the Borel field of sets constructed in the first example of measurable functions, and let μ be Lebesgue measure. Continuous functions are measurable, and it can be shown that the value of the Lebesgue integral of a continuous function on a closed interval agrees with the value of the Riemann integral. More generally one finds that every Riemann integrable function is Lebesgue integrable, but not conversely.

(3) Let X be the denumerable set of points described in Example 4 of measure spaces, and let π be a non-negative row vector defined on X. Then π defines a measure on X. If f is an arbitrary column vector defined on X, then f is a function on the points of X. Furthermore, f is measurable since all subsets of X are measurable sets. The reader should verify that the integral of f over the whole space X with respect to the measure π is the matrix product πf and that the condition for the integral to be defined is precisely the condition for the matrix product to be well defined. Because of this application of Lebesgue integration, we often speak of column vectors as **functions** and non-negative row vectors as **measures**. We shall return to this example in Section 5 of this chapter. The proof of the next proposition is left to the reader.

Proposition 1-39: The Lebesgue integral satisfies these seven properties:

(1) If c is a constant function,

$$\int_E c\,d\mu = c\mu(E).$$

(2) If f and g are measurable functions whose integrals are defined on E and if $f(x) \leq g(x)$ for all $x \in E$, then

$$\int_E f\,d\mu \leq \int_E g\,d\mu.$$

(3) If f is integrable on E and if c is a real number, then cf is integrable and $\int_E cf\,d\mu = c\int_E f\,d\mu$.

(4) If f is measurable and $\mu(E) = 0$, then $\int_E f\,d\mu = 0$.

(5) If E' and E are measurable sets with $E' \subset E$ and if f is a function for which $\int_E f\,d\mu$ is defined, then $\int_{E'} f\,d\mu$ is defined. In particular,

$$\int_{E'} f^+\,d\mu \leq \int_E f^+\,d\mu$$

and

$$\int_{E'} f^-\,d\mu \leq \int_E f^-\,d\mu.$$

(6) If $|f(x)| \leq c$ for all $x \in E$, if $\mu(E) < \infty$, and if f is measurable, then f is integrable on E.

(7) If f is measurable on E and if $|f| \leq g$ for a function g integrable on E, then f is integrable on E.

Corollary 1-40: If f is a non-negative measurable function with $\int_E f\,d\mu = 0$, then $f = 0$ a.e. on E.

PROOF: The subset of E where $f(x) \geq 1/n$ must have measure zero since otherwise f would have positive integral by (1) and (2) of Proposition 1-39. The set where $f \neq 0$ on E is the countable union on n of these sets.

4. Integration theorems

We shall make frequent use of four important facts about the Lebesgue integral. We develop these results as the four theorems of this section.

Theorem 1-41: Let f be a fixed measurable function and suppose that $\int_X f d\mu$ is defined. Then the set function $\rho(E) = \int_E f d\mu$ is completely additive.

PROOF: If we can prove the theorem for non-negative functions, we can write $f = f^+ - f^-$ and apply our result separately to f^+ and f^-. We therefore assume that f is non-negative. We must show that if $E = \bigcup_{n=1}^{\infty} E_n$ disjointly, then $\rho(E) = \sum_{n=1}^{\infty} \rho(E_n)$. If f is a characteristic function χ_A, then $\rho(E) = \int_E \chi_A d\mu = \mu(A \cap E)$ and the complete additivity of ρ is a consequence of the complete additivity of μ. If f is a simple function, the complete additivity of ρ is a consequence of the result for characteristic functions and of the fact that the limit of a sum is the sum of the limits. Thus, for general f we have for every simple function s satisfying $0 \leq s \leq f$,

$$\int_E s d\mu = \sum_{n=1}^{\infty} \int_{E_n} s d\mu \leq \sum_{n=1}^{\infty} \rho(E_n)$$

by property (2) of Proposition 1-39. Hence

$$\rho(E) = \sup_s \int_E s d\mu \leq \sum_{n=1}^{\infty} \rho(E_n).$$

We now prove the inequality in the other direction. By property (5) of Proposition 1-39, $\rho(E) \geq \rho(E_n)$ for every n since $f = f^+$. Thus if $\rho(E_n) = +\infty$ for any n, the desired result is proved. We therefore assume $\rho(E_n) < \infty$ for every n. Let $\epsilon > 0$ be given and choose a measurable simple function s satisfying $0 \leq s \leq f$,

$$\int_{E_1} s d\mu \geq \int_{E_1} f d\mu - \frac{\epsilon}{2},$$

and

$$\int_{E_2} s d\mu \geq \int_{E_2} f d\mu - \frac{\epsilon}{2}.$$

This choice is possible by the definition of the integral as a limit. Then

$$\rho(E_1 \cup E_2) = \int_{E_1 \cup E_2} f d\mu \geq \int_{E_1 \cup E_2} s d\mu = \int_{E_1} s d\mu + \int_{E_2} s d\mu$$

$$\geq \int_{E_1} f d\mu + \int_{E_2} f d\mu - \epsilon$$

$$= \rho(E_1) + \rho(E_2) - \epsilon.$$

Hence

$$\rho(E_1 \cup E_2) \geq \rho(E_1) + \rho(E_2).$$

By induction, we obtain

$$\rho(E_1 \cup E_2 \cup \cdots \cup E_n) \geq \rho(E_1) + \cdots + \rho(E_n)$$

and

$$\rho(E) \geq \rho(E_1) + \cdots + \rho(E_n) \quad \text{for every } n.$$

Hence

$$\rho(E) \geq \sum_{n=1}^{\infty} \rho(E_n).$$

The proofs of two corollaries of Theorem 1-41 are left as exercises. These results use only the additivity of the integral and not the complete additivity.

Corollary 1-42: If f is measurable, if $\int_E f d\mu$ is defined, and if

$$\mu(E \oplus F) = 0,$$

then $\int_F f d\mu$ is defined and $\int_E f d\mu = \int_F f d\mu$.

Corollary 1-43: If $\int_E f d\mu$ is defined, then

$$\left| \int_E f d\mu \right| \leq \int_E |f| d\mu.$$

If f is integrable on E, then so is $|f|$.

Let f and g be two functions whose integrals on E are defined. Suppose the set $A = \{x \in E \mid f(x) \neq g(x)\}$ is of measure zero; that is, suppose f and g are equal almost everywhere on E. Writing $E = A \cup (E \cap \tilde{A})$, we find by applying Corollary 1-42 twice,

$$\int_E f d\mu = \int_{E \cap \tilde{A}} f d\mu = \int_{E \cap \tilde{A}} g d\mu = \int_E g d\mu.$$

Functions which differ on a set of measure zero thus have equal integrals. Therefore, when we are thinking of a function f defined on X in terms of integration, it is sufficient that f be defined at almost all points of X. And, if we agree to augment Borel fields of sets by adjoining subsets of sets of measure zero, we see that if f and g differ on a set of measure zero and if f is measurable, then g is measurable. With the convention of augmenting the field, we obtain from Corollary 1-37, for example, the result that if $\{f_n\}$ is a sequence of measurable functions such that

$$f(x) = \lim_{n\to\infty} f_n(x)$$

almost everywhere, then f is measurable. Similarly the theorems to follow would be valid with convergence almost everywhere if the underlying Borel field were augmented; the necessary modifications in the proofs are easy.

We now state and prove the Monotone Convergence Theorem, which is due to B. Levi.

Theorem 1-44: Let E be a measurable set, and suppose $\{f_n\}$ is a sequence of measurable functions such that

$$0 \leq f_1 \leq f_2 \leq \cdots$$

and

$$f(x) = \lim_{n\to\infty} f_n(x).$$

Then

$$\int_E f d\mu = \lim_{n\to\infty} \int_E f_n d\mu.$$

PROOF: Since the $\{f_n\}$ increase with n, so do the $\left\{\int_E f_n d\mu\right\}$. Therefore

$$k = \lim_{n\to\infty} \int_E f_n d\mu$$

exists. Since f is non-negative and is the limit of measurable functions, we know that $\int_E f d\mu$ exists, and since $f_n \leq f$, we have

$$\int_E f_n d\mu \leq \int_E f d\mu$$

for every n. Therefore, $k \leq \int_E f d\mu$. Let c be a real number satisfying $0 < c < 1$, and let s be a measurable simple function such that $0 \leq s \leq f$. Set

$$E_n = \{x \in E \mid f_n(x) \geq cs(x)\}$$

so that $E_1 \subset E_2 \subset E_3 \subset \cdots$. Then $E = \bigcup_{n=1}^{\infty} E_n$. For any n we have

$$k \geq \int_E f_n d\mu \geq \int_{E_n} f_n d\mu \geq c \int_{E_n} s d\mu.$$

Since the integral is a completely additive set function (Theorem 1-41), we have by Proposition 1-16

$$\lim_{n \to \infty} c \int_{E_n} s d\mu = c \int_E s d\mu.$$

Thus, as $n \to \infty$,

$$k \geq c \int_E s d\mu.$$

Letting $c \to 1$, we have $k \geq \int_E s d\mu$, and taking the supremum over all s, we find $k \geq \int_E f d\mu$.

Proposition 1-45: Suppose $h = f + g$ with f and g integrable on E. Then h is integrable on E and

$$\int_E h d\mu = \int_E f d\mu + \int_E g d\mu.$$

Proof: We first prove the assertion for f and g non-negative. For simple functions the assertion is trivial. If f and g are not both simple, apply Proposition 1-38 to find monotone sequences of non-negative measurable simple functions $\{t_n\}$ and $\{u_n\}$ converging to f and g. Then $\{s_n = t_n + u_n\}$ converges to h, and since

$$\int_E s_n d\mu = \int_E t_n d\mu + \int_E u_n d\mu,$$

the result follows from Theorem 1-44. Next, if $f \geq 0$, $g \leq 0$, and $h = f + g \geq 0$, we have $f = h + (-g)$ with $h \geq 0$ and $(-g) \geq 0$, so that

$$\int_E f d\mu = \int_E h d\mu + \int_E (-g) d\mu$$

or

$$\int_E h d\mu = \int_E f d\mu + \int_E g d\mu.$$

Since the right side is finite and since $h \geq 0$, h is integrable. For an arbitrary $h \geq 0$, decompose E into the disjoint union of three sets, one where $f \geq 0$ and $g \geq 0$, one where $f \geq 0$ and $g < 0$, and one where

$f < 0$ and $g \geq 0$. Theorem 1-41 then gives the desired result. Finally, for general h, write $h = h^+ - h^-$ and consider h^+ and h^- separately.

Corollary 1-46: Let E be a measurable set, and suppose $\{f_n\}$ is a sequence of non-negative measurable functions with

$$f(x) = \sum_{n=1}^{\infty} f_n(x).$$

Then

$$\int_E f d\mu = \sum_{n=1}^{\infty} \int_E f_n d\mu.$$

Proof: The result follows from Proposition 1-45 and Theorem 1-44.

Proposition 1-47: If f is a non-negative integrable function, then for every $\epsilon > 0$ there is a $\delta > 0$ such that, if $\mu(E) < \delta$, then

$$\int_E f d\mu < \epsilon.$$

Proof: Set $f_n = \min(f, n)$. By Theorem 1-44,

$$\lim_n \int_X f_n d\mu = \int_X f d\mu.$$

Since f is integrable, we may find an N such that

$$\int_X (f - f_N) d\mu < \frac{\epsilon}{2}$$

by Proposition 1-45. Let $\delta = \epsilon/(2N)$. If $\mu(E) < \delta$, then

$$\begin{aligned}
\int_E f d\mu &= \int_E (f - f_N) d\mu + \int_E f_N d\mu \quad \text{by Proposition 1-45} \\
&\leq \int_X (f - f_N) d\mu + N\mu(E) \quad \text{by Proposition 1-39} \\
&< \epsilon.
\end{aligned}$$

Our third theorem for this section is known as Fatou's Theorem.

Theorem 1-48: Let E be a measurable set, and let $\{f_n\}$ be a sequence of non-negative measurable functions. Then

$$\int_E \liminf_n f_n d\mu \leq \liminf_n \int_E f_n d\mu.$$

In particular, if $f(x) = \lim_n f_n(x)$,

$$\int_E f d\mu \le \liminf_n \int_E f_n d\mu.$$

PROOF: Set $g_n(x) = \inf_{k \ge n} f_k(x)$. Then $0 \le g_1(x) \le g_2(x) \le \cdots$, and g_n is measurable on E. We have $g_n(x) \le f_n(x)$ and

$$\lim_n g_n(x) = \liminf_n f_n(x).$$

By Theorem 1-44,

$$\int_E \liminf_n f_n d\mu = \int_E \lim_n g_n d\mu = \lim_n \int_E g_n d\mu.$$

The result now follows from the inequality $\int_E g_n d\mu \le \int_E f_n d\mu$.

The fourth basic integration theorem is the Lebesgue Dominated Convergence Theorem.

Theorem 1-49: Let E be a measurable set, and suppose $\{f_n\}$ is a sequence of measurable functions such that for some *integrable* g, $|f_n| \le g$ for all n. If $f(x) = \lim_n f_n(x)$, then $\lim_n \int_E f_n d\mu$ exists and

$$\int_E f d\mu = \lim_n \int_E f_n d\mu.$$

PROOF: By property (7) of Proposition 1-39, f is integrable and so is f_n for every n. Apply Fatou's Theorem first to $f_n + g \ge 0$ to obtain $\int_E (f + g) d\mu \le \liminf \int_E (f_n + g) d\mu$ or

$$\int_E f d\mu \le \liminf_n \int_E f_n d\mu.$$

Apply the theorem once more to $g - f_n \ge 0$ to obtain

$$\int_E (g - f) d\mu \le \liminf_n \int_E (g - f_n) d\mu$$

or

$$-\int_E f d\mu \le \liminf_n \int_E (-f_n) d\mu$$

or

$$\int_E f d\mu \ge \limsup_n \int_E f_n d\mu.$$

Therefore $\lim_n \int_E f_n d\mu$ exists and has the value asserted.

Corollary 1-50: Let E be a set of finite measure and suppose $\{f_n\}$ is a sequence of measurable functions such that $|f_n| \leq c$ for all n. If $f(x) = \lim_n f_n(x)$, then $\int_E f d\mu = \lim_n \int_E f_n d\mu$.

Much of the discussion of this section has dealt with the following problem: A sequence of integrable functions f_n converges a.e. to a function f; we want to be able to conclude that $\int f_n d\mu$ tends to $\int f d\mu$. First we should note that at almost all points $f_n^+ \to f^+$ and $f_n^- \to f^-$, and hence it is sufficient to check the convergence of the integral separately for these two sequences. Thus we may consider the case $f_n \geq 0$ alone. For non-negative functions Fatou's inequality is the only general result; one cannot conclude equality without a further hypothesis. Two sufficient conditions are given by monotone and dominated convergence.

But if we restrict our attention to a space of finite total measure, then we can give a necessary and sufficient condition.

Definition 1-51: A sequence $\{f_n\}$ of non-negative integrable functions is said to be **uniformly integrable** if for each $\epsilon > 0$ there is a number k such that

$$\int_{\{x|f_n(x)>k\}} f_n d\mu < \epsilon$$

holds for every n.

Equivalently we may require that the inequality holds for all sufficiently large n. For suppose it holds for $n > N$ and for the number c. Then since each f_n is integrable, there is a k_n depending on n (and, of course, ϵ) such that

$$\int_{\{x|f_n(x)>k_n\}} f_n d\mu < \epsilon$$

and we may choose $k = \sup\{k_1, k_2, \ldots, k_N, c\}$.

Proposition 1-52: If $\{f_n\}$ is a sequence of non-negative integrable functions tending to f and if $\mu(E) < \infty$, then $\int_E f_n d\mu \to \int_E f d\mu$ if and only if the $\{f_n\}$ are uniformly integrable.

REMARK: The sequence $\{f_n\}$ need converge to f only almost everywhere for the proposition to be valid, provided f is assumed measurable. This measurability condition is always satisfied if the underlying Borel field is augmented.

Proof: We shall use the notation $f^{(k)}$ for the function truncated at k; that is, $f^{(k)}(x) = \inf(f(x), k)$. Let

$$A_n = \int_E f d\mu - \int_E f_n d\mu$$

$$B^k = \int_E f d\mu - \int_E f^{(k)} d\mu$$

$$C_n^k = \int_E f^{(k)} d\mu - \int_E f_n^{(k)} d\mu$$

$$D_n^k = \int_E f_n d\mu - \int_E f_n^{(k)} d\mu = \int_{\{x | f_n > k\}} (f_n - k) d\mu.$$

We have

$$A_n = B^k + C_n^k - D_n^k.$$

Clearly $f^{(k)}$ increases monotonically to f, so that B^k tends to 0 by monotone convergence. Since $f_n \to f$, $f_n^{(k)} \to f^{(k)}$. But $f_n^{(k)}$ is bounded by k. Thus, on the totally finite measure space E, $\lim_n C_n^k = 0$ by Corollary 1-50. Hence by choosing a large k and then a sufficiently large n (depending on k), we can make the first two terms on the right side as small as desired. If the functions are uniformly integrable, then we can find a large k (perhaps larger than the one already chosen) for which D_n^k will be small for all n. Hence, for all sufficiently large n, the left side is small. Thus the integrals converge.

Conversely, suppose that $A_n \to 0$. Then there is a k such that for all sufficiently large n, $D_n^k < \epsilon/2$. Thus for all n sufficiently large, we have

$$\begin{aligned} \epsilon/2 &> \int_{\{x|f_n > k\}} (f_n - k) d\mu \\ &\geq \int_{\{x|f_n > 2k\}} (f_n - k) d\mu \\ &\geq \frac{1}{2} \int_{\{x|f_n > 2k\}} f_n d\mu, \end{aligned}$$

since $f_n - k > \frac{1}{2} f_n$ on the set in the last two integrals. Taking $2k$ as the number in the equivalent definition, we see that we have uniform integrability.

5. Limit theorems for matrices

We have already said that if π is a row vector and if $\{f^{(k)}\}$ is a sequence of column vectors converging to f, then it is not necessarily true that

$\pi f^{(k)}$ converges to πf. Our object in this section is to obtain sufficient criteria to justify saying that $\pi f = \lim_{k\to\infty} \pi f^{(k)}$.

In the examples of Lebesgue integration, we noted that non-negative row vectors are measures and column vectors are functions. Functions are integrated by forming the matrix product of the measure and the function. Thus, the theorems of Section 4 immediately give us the following four results. In each of them it should be borne in mind that:

(1) There is a corresponding result if row vectors are thought of as functions and column vectors are thought of as measures.
(2) These results imply corresponding results about matrices which are not just row or column vectors. (Recall the discussion in Section 1 about proving matrix equalities entry-by-entry.)

Proposition 1-53: Let $\pi \geq 0$ be a row vector and suppose $\{f^{(k)}\}$ is a sequence of column vectors converging entry-by-entry to f and satisfying

$$0 \leq f^{(1)} \leq f^{(2)} \leq \dots .$$

Then $\pi f = \lim_k \pi f^{(k)}$.

PROOF: This result is a restatement of the Monotone Convergence Theorem as it applies to matrices.

Corollary 1-54: Let $\pi \geq 0$ be a row vector and suppose $\{f^{(k)}\}$ is a sequence of non-negative column vectors such that

$$f = \sum_{k=1}^{\infty} f^{(k)}.$$

Then

$$\pi f = \sum_{k=1}^{\infty} \pi f^{(k)}.$$

PROOF: This corollary is immediate from Corollary 1-46.

Proposition 1-55: Let $\pi \geq 0$ be a row vector and suppose $\{f^{(k)}\}$ is a sequence of non-negative column vectors. Then

$$\pi(\liminf_k f^{(k)}) \leq \liminf_k (\pi f^{(k)}).$$

If $f = \lim_k f^{(k)}$ exists, then $\pi f \leq \liminf_k (\pi f^{(k)})$.

PROOF: This is Fatou's Theorem.

Proposition 1-56: Let $\pi \geq 0$ be a row vector such that $\pi 1$ is finite. If $\{f^{(k)}\}$ is a sequence of column vectors such that $|f^{(k)}| \leq c1$ and $f = \lim_k f^{(k)}$ exists, then

$$\pi f = \lim_k \pi f^{(k)}.$$

PROOF: The result follows from Corollary 1-50.

A harder problem arises with a sequence of non-negative row vectors $\pi^{(k)}$ converging to a row vector π. It is not sufficient for $\pi^{(k)}1 \leq M$ and $|f| \leq c1$ in order for $\pi f = \lim_k \pi^{(k)}f$. For set

$$\begin{aligned} \pi^{(1)} &= (1 \quad 0 \quad 0 \quad 0 \quad \ldots) \\ \pi^{(2)} &= (0 \quad 1 \quad 0 \quad 0 \quad \ldots) \\ \pi^{(3)} &= (0 \quad 0 \quad 1 \quad 0 \quad \ldots) \\ &\vdots \end{aligned}$$

and take $f = 1$. Then $\pi = 0$ so that $\pi f = 0$, while $\lim_k \pi^{(k)}f = 1$. We give two sufficient conditions for

$$\pi f = \lim_k \pi^{(k)}f;$$

our integration theorems do not provide us with quick proofs, however. The second proposition is closely related to the Silverman–Toeplitz Theorem on summability methods.

Proposition 1-57: If $\{\pi^{(k)}\}$ is a sequence of non-negative row vectors converging to π, if f is a column vector such that $0 \leq f \leq c1$ for some c, and if $\pi 1 = \lim_k \pi^{(k)}1$ with $\pi 1$ finite, then

$$\pi f = \lim_k \pi^{(k)}f.$$

PROOF: Take f as a measure and $\{\pi^{(k)}\}$ as a sequence of non-negative functions and apply Fatou's Theorem. We have

$$\pi f \leq \liminf \pi^{(k)}f.$$

With $c1 - f$ as a measure and $\{\pi^{(k)}\}$ as a sequence of functions, Fatou's Theorem gives

$$\pi(c1 - f) \leq \liminf \pi^{(k)}(c1 - f).$$

Since $\pi 1$ is finite and $\lim \pi^{(k)}1 = \pi 1$, we find

$$-\pi f \leq \liminf (-\pi^{(k)}f)$$

or

$$\pi f \geq \limsup \pi^{(k)}f.$$

Proposition 1-58: Let $\{\pi^{(k)}\}$ be a sequence of row vectors converging to π and satisfying $|\pi^{(k)}|1 \leq M$. Suppose f is a column vector with the property that for any $\delta > 0$ only finitely many entries of f have absolute value greater than δ. Then

$$\pi f = \lim_k \pi^{(k)} f.$$

PROOF: The entries of f are clearly bounded, say by c. Numbering the entries, we have for every N

$$|\pi f - \pi^{(k)} f| \leq \sum_{j=1}^{N} |\pi_j - \pi_j^{(k)}|\,|f_j| + \sum_{j>N} (|\pi_j| + |\pi_j^{(k)}|)|f_j|.$$

Let $\epsilon > 0$ be given. Choose N sufficiently large that $|f_j| < \epsilon/4M$ for $j > N$. Choose k sufficiently large that $|\pi_j - \pi_j^{(k)}| < \epsilon/2cN$ for $j \leq N$. Then $|\pi f - \pi^{(k)} f| < \epsilon$, and the result is established.

As we noted in Section 1, results about general denumerable matrices can be reduced to results about row and column vectors. In particular, the propositions of the present section apply equally well to sequences of the forms $\{A^{(k)} f\}$ and $\{\pi A^{(k)}\}$.

6. Some general theorems from analysis

In this section we collect a variety of results from analysis which we shall need in later chapters. We prove only some of them. At first reading the reader may find it wise to skip this section, returning to it later as the material is required.

a. Stirling's formula. Stirling's formula gives an asymptotic expression for $m!$ The approximation is

$$m! \sim \frac{m^m}{e^m} \sqrt{2\pi m},$$

where the symbol $\sim$ indicates that the ratio of the two quantities tends to one as m increases. For a proof, see Courant and Hilbert [1953], pp. 522–524. Stirling's formula immediately gives an approximation for the **binomial coefficient** $\binom{rn}{n}$ for large n. The coefficient $\binom{a}{b}$ is defined as $\dfrac{a!}{b!(a-b)!}$.

Lemma 1-59: For $r > 1$,

$$\binom{rn}{n} \sim c \frac{1}{\sqrt{n}} \left(\frac{r^r}{(r-1)^{r-1}}\right)^n,$$

where c is a constant depending on r but not on n.

The multinomial coefficient $\binom{a}{b, c, \ldots, d}$ is defined to be $\frac{a!}{b!c! \cdots d!}$.

b. Difference equations. An nth order linear difference equation with constant coefficients is an expression of the form

$$y_{k+n} + c_{n-1}y_{k+n-1} + \cdots + c_1 y_{k+1} + c_0 y_k = r_k,$$

where y_k and r_k are functions defined on the integers and where the $c_{n-1}, \ldots, c_0$ are complex numbers. The equation is **homogeneous** if $r_k = 0$ and **nonhomogeneous** otherwise. For a nonhomogeneous solution, we refer to any single function y_k satisfying the equation as a **particular solution**, and we call the set of functions satisfying the same equation with $r_k = 0$ the **homogeneous solution**. The totality of solutions to any difference equation is known as the **general solution**.

Proposition 1-60: Every solution of the difference equation

$$y_{k+n} + c_{n-1}y_{k+n-1} + \cdots + c_0 y_k = 0$$

is a linear combination of n fixed functions, obtained as follows: If a is a root of multiplicity q of the **characteristic equation**

$$x^n + c_{n-1}x^{n-1} + \cdots + c_1 x + c_0 = 0,$$

then q of the functions are

$$a^k, ka^k, k^2a^k, \ldots, k^{q-1}a^k.$$

Conversely, each of these functions is a solution of the difference equation. Furthermore, each solution of the equation

$$y_{k+n} + c_{n-1}y_{k+n-1} + \cdots + c_0 y_k = r_k$$

is the sum of a fixed particular solution and some solution of

$$y_{k+n} + c_{n-1}y_{k+n-1} + \cdots + c_0 y_k = 0.$$

Conversely, every such sum is a solution of the nonhomogeneous equation.

For a proof of this proposition, see Hildebrand [1956], pp. 202–203.

c. Cesaro summability and Abel summability. Let $\{a_n\}$ be a sequence of real numbers. Define b_n to be the arithmetic mean of the first n terms of the sequence $\{a_n\}$. The sequence $\{b_n\}$ is called the sequence of **Cesaro averages** of the sequence $\{a_n\}$. The sequence $\{a_n\}$ is said to be **Cesaro summable** if its sequence, $\{b_n\}$, of Cesaro averages has a limit. If $A^{(n)}$ is a sequence of matrices, the sequence of Cesaro averages $B^{(n)}$

is defined entry-by-entry: $B_{ij}^{(n)}$ is the Cesaro average of $A_{ij}^{(1)}, A_{ij}^{(2)}, \ldots, A_{ij}^{(n)}$. The basic fact about Cesaro summability is the following proposition.

Proposition 1-61: If a sequence $\{a_n\}$ converges to a limit L, then the sequence of Cesaro averages $\{b_n\}$ converges and its limit is L.

PROOF: Let f be the column vector whose nth entry is $a_n - L$, and let $\pi^{(n)}$ be the row vector defined by

$$\pi_j^{(n)} = \begin{cases} 1/n & \text{for } 1 \leq j \leq n \\ 0 & \text{for } n > j. \end{cases}$$

Then

$$b_n = \pi^{(n)}f + L.$$

Now $|\pi^{(n)}|1 = 1$, $\lim_n \pi^{(n)} = 0$, and f is a column vector with entries tending to 0. Hence by Proposition 1-58, $\lim_n \pi^{(n)}f = 0$. Therefore $\lim_n b_n = L$.

The converse of Proposition 1-61 is false. The sequence $\{a_n\}$ defined by $a_{2n} = 0$, $a_{2n+1} = 1$ does not converge, but it is Cesaro summable.

Let $\{c_n\}$ be a sequence of real numbers. (In most applications the partial sums $c_0 + \cdots + c_n$ are assumed bounded.) If the limit

$$\lim_{t \uparrow 1} \sum_{n=0}^{\infty} c_n t^n$$

exists, the limit is called the **Abel sum** of the series $\sum c_n$ and the series is said to be **Abel summable**. Abel's Theorem is the following result.

Proposition 1-62: If the series $\sum c_n$ converges to a finite limit L, then it is Abel summable and its Abel sum is L.

PROOF: Since the partial sums converge to a finite limit, the c_n are bounded and $\sum c_n t^n$ converges absolutely for $t < 1$. Let $\{t_k\}$ be any sequence of positive reals less than one and increasing to one. Let f be the column vector whose nth entry is $(c_0 + \cdots + c_n) - L$, and let $\pi^{(k)}$ be the row vector defined by

$$\pi_j^{(k)} = (1 - t_k)t_k^{\,j}.$$

Then

$$\pi^{(k)}f = \sum c_n t_k^{\,n} - L.$$

Now $|\pi^{(k)}|1 = 1$, $\lim_k \pi^{(k)} = 0$, and f is a column vector with entries tending to 0. Hence by Proposition 1-58, $\lim \pi^{(k)}f = 0$. That is,

$$\lim_{k \to \infty} \sum c_n t_k^{\,n} = L.$$

This equality for every such sequence $\{t_k\}$ implies that

$$\lim_{t \uparrow 1} \sum c_n t^n = L.$$

d. Convergent subsequences of matrices. A bounded sequence of real numbers has a subsequence which converges to a finite limit. We shall obtain a generalization of this result to matrices.

Proposition 1-63: Let $\{A^{(n)}\}$ be a sequence of matrices with the property that for some pair of real numbers c and d, $cE \le A^{(n)} \le dE$ for all n. Then there exists a subsequence of matrices $\{A^{(n_\nu)}\}$ which converges in every entry.

PROOF: Since there are only denumerably many entries in each matrix, they can be numbered by a subset of the positive integers. Select a subsequence $\{A_1^{(n)}\}$ which converges in the first entry. Let $A_2^{(2)}, A_2^{(3)}, A_2^{(4)}, \ldots$ be a subsequence of $A_1^{(2)}, A_1^{(3)}, \ldots$ which converges in the second entry. In general, let $A_m^{(m)}, A_m^{(m+1)}, \ldots$ be a subsequence of $A_{m-1}^{(m)}, A_{m-1}^{(m+1)}, \ldots$ which converges in the mth entry. Finally set

$$A^{(n_\nu)} = A_\nu^{(\nu)}.$$

Then $\{A^{(n_\nu)}\}$ converges in every entry.

Corollary 1-64: Let $\{A^{(n)}\}$ be a sequence of matrices with the property that $cE \le A^{(n)} \le dE$ for all n. Then $\lim_n A^{(n)} = A$ exists if and only if $\lim_\nu A^{(n_\nu)} = A$ for every convergent subsequence $\{A^{(n_\nu)}\}$.

PROOF: The necessity of the condition is trivial. For the sufficiency suppose $\lim A_{ij}^{(n)}$ does not exist. Then $\liminf A_{ij}^{(n)} < \limsup A_{ij}^{(n)}$. Pick a subsequence of $\{L^{(n)}\}$ that converges in the i-jth entry to $\limsup A_{ij}^{(n)}$, and do the same for $\liminf A_{ij}^{(n)}$. Apply Proposition 1-63 to extract subsequences convergent in all entries from these sequences, and the result follows at once.

e. Sets of positive integers closed under addition. The greatest common divisor of a non-empty set of positive integers is the largest integer which divides all of them. If the set consists of $\{n_1, n_2, \ldots\}$, its greatest common divisor is denoted $(n_1, n_2, \ldots)$.

Lemma 1-65: If T is a set of positive integers with greatest common divisor d, then there exists a finite subset of T for which d is the greatest common divisor.

PROOF: Let $n_1 = k_1 d$ be an element of T. If $k_1 = 1$, $\{n_1\}$ is the required set. If not, choose n_2 such that $n_1 \nmid n_2$. Then $(n_1, n_2) = k_2 d$ with $k_2 < k_1$. If $k_2 = 1$, $\{n_1, n_2\}$ is the required set. Otherwise, find n_3 such that $k_2 d \nmid n_3$, and set $(n_1, n_2, n_3) = k_3 d$ with $k_3 < k_2$. Continuing in this way, we obtain a decreasing sequence of integers $k_1, k_2, \ldots$ bounded below by 1. It must terminate, and then we have constructed the finite set.

Lemma 1-66: Let T be a non-empty set of positive integers which is closed under addition and which has the greatest common divisor d. Then all sufficiently large multiples of d are in the set T.

PROOF: If $d \neq 1$, divide all the elements in T by d and reduce the problem to the case $d = 1$. By Lemma 1-65 there is a finite subset $\{n_1, \ldots, n_s\}$ of T with greatest common divisor 1. Then there exist integers $c_1, \ldots, c_s$ with the property

$$c_1 n_1 + \cdots + c_s n_s = 1.$$

If we collect the positive terms and the negative terms and note that T is closed under addition, we find that T contains non-negative integers m and n with $m - n = 1$. Suppose $q \geq n(n-1)$. Then $q = an + b$ with $a \geq n - 1$ and $0 \leq b \leq n - 1$. Thus

$$q = (a - b)n + bm$$

and q is in T.

f. Renewal theorem. The Renewal Theorem, one of the important results in the elementary theory of probability, can be stated purely in terms of analysis.

Theorem 1-67: Let $\{f_n\}$ be a sequence of non-negative real numbers such that $\sum f_n = 1$ and $f_0 = 0$, and suppose the greatest common divisor of those indices k for which $f_k > 0$ is 1. Set $\mu = \sum n f_n$, $u_0 = 1$, and $u_n = \sum_{k=0}^{n-1} u_k f_{n-k}$. If μ is infinite, then $\lim_n u_n = 0$, and if μ is finite, then $\lim_n u_n = 1/\mu$.

For a proof, see Feller [1957], pp. 306–307.

g. Central Limit Theorem. Identically distributed independent random variables are defined in Sections 3-2 and 6-4. The mean of a random variable is its integral over its domain, and its variance is the integral of its square minus the square of its integral, a quantity which is always non-negative.

Theorem 1-68: Let $\{y_n\}$ be a sequence of identically distributed independent random variables with common mean μ and with common finite variance $\sigma^2 > 0$. Set $s_m = y_1 + \cdots + y_m$. Then for all real α and β with $\alpha < \beta$,

$$\lim_{n \to \infty} \Pr\left[\alpha < \frac{s_m - m\mu}{\sigma\sqrt{m}} < \beta\right] = \Phi(\beta) - \Phi(\alpha),$$

where

$$\Phi(x) = \frac{1}{\sqrt{2\pi}} \int_{-\infty}^{x} e^{-u^2/2} du.$$

For a proof, see Doob [1953], p. 140.

CHAPTER 2

STOCHASTIC PROCESSES

1. Sequence spaces

We shall introduce the concept of a stochastic process in this chapter and develop the basic tools needed to treat the processes. Before the formal development, we shall indicate the intuitive ideas underlying the formal definitions.

We imagine that a sequence of experiments is performed. The outcomes may be arbitrary elements of a specified set, such as the set {"yes," "no," "no opinion"}, the set {heads, tails}, the set {fair, cloudy, rainy, snowy}, or a set of numbers. The experiments may be quite general in nature, but we impose some natural restrictions:

(1) The set of possible outcomes is denumerable. (This restriction is natural for the present book. It would be removed in a more general treatment of stochastic processes.)

(2) The probability of an outcome for the nth experiment is completely determined by a knowledge of the outcomes of earlier experiments. Here "probability" is used heuristically, to motivate the later precise definition.

(3) The experimenter is, at each stage, aware of the outcomes of earlier experiments.

We shall first consider a sequence of n experiments, where n is specified at the beginning. Later we shall consider a denumerably infinite sequence of experiments. In each case we assume that the experiments do not stop earlier. However, this is an unimportant restriction; a process that terminates may be represented in our framework by allowing the outcome "stopped." The following are examples of such sequences of experiments:

(1) A sociologist wishes to find out whether people feel that television is turning us into a nation of illiterates. He asks a carefully selected sample of subjects and receives the answer "yes," "no," or "no opinion" in each case.

(2) A gambler flips a coin repeatedly, recording "heads" or "tails."

(3) A meteorologist records the daily weather for 23 years, classifying each day as "fair," "cloudy," "rainy," or "snowy."

(4) A physicist tries to determine the speed of light by making a series of measurements. (Since each measurement is recorded only to a certain number of decimal places, the possible outcomes are rational numbers, and hence the set is denumerable.)

(5) A physicist makes a count of the number of radioactive particles given off by an ounce of uranium. A measurement is made every second, and the outcome of the nth measurement is the total number of particles given off until then.

The exact way in which probabilities are determined from an experiment is a deep problem in the philosophy of science, and it will not concern us here. We will assume that the nature of the experiment yields us certain probabilities, namely the probability that the nth experiment results in an outcome c, given that the previous experiments resulted in outcomes $c_0, c_1, \ldots, c_{n-1}$. We then design a probability space in which one can compute the probability of a wide variety of statements concerning the experiments and in which the specified (conditional) probabilities turn out as given.

The elements of our probability space Ω are sequences of possible outcomes for the experiments (either sequences of length n, for a finite series of experiments, or infinite sequences). The elements of the Borel field $\mathscr{B}$ of measurable sets will be the truth sets of statements to which probabilities are to be assigned. (The **truth set** of a statement about the experiments is the set of all those sequences in Ω for which the statement is true.) A measure μ is constructed, and the probability of a statement is the measure of its truth set. In particular, $\mu(\Omega) = 1$ in a probability space, and hence the probability of a logically true statement is 1.

Let us first consider the case where n experiments are performed. The possible outcomes are conveniently represented by a tree, with each path through the tree representing a sequence of possible outcomes. In the diagram n equals 3, Ω has 8 elements, and $\mathscr{B}$ consists of all subsets of Ω.

The numbers on the branches, known as **branch weights**, represent the conditional probabilities mentioned above. For example, p_2 is the probability of heads, given that the first toss came up heads, while $1 - p_3$ is the probability that tails is the outcome that follows two heads. The weight assigned to the path HHT is taken to be the product of the branch weights, $p_1p_2(1 - p_3)$. The measure $\mu(A)$ of a set of paths A is the sum of the weights of the paths in A. In the usual

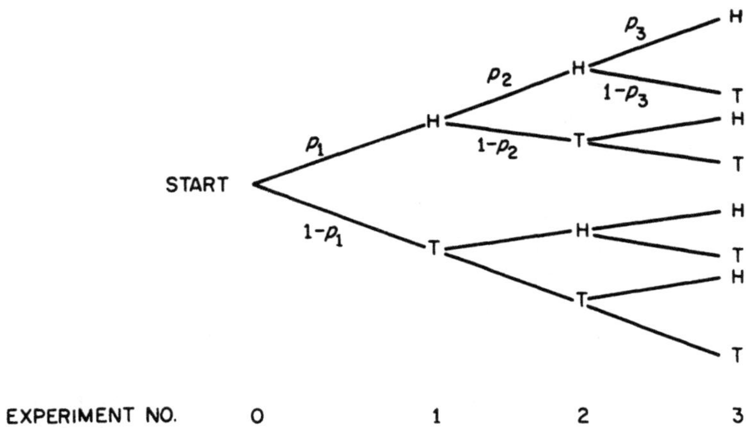

setup for coin tossing, each p is $\frac{1}{2}$, and the weights of the paths are each $\frac{1}{8}$.

The branch weights may be arbitrary non-negative numbers, but the sum of the weights of branches starting from a given branch-point must be one.

After we define conditional probabilities, it will be easy to verify that the numbers written on the branches do indeed turn out to be the desired conditional probabilities (see Kemeny, Mirkil, *et al.* [1959], Chapter 3).

Let us suppose that we have constructed a tree Ω_n for a series of n experiments. We consider k additional experiments, obtaining a tree Ω_{n+k}. We wish our probabilities to be consistent in the sense that a statement p about Ω_n has the same probability when computed in either measure space. Our method of computing measures has this consistency property. This assertion is easily verified for $k = 1$, and the result follows by induction.

In constructing an infinite tree Ω for a sequence of experiments, our measure is required to have the property that a statement about the first n outcomes has the same probability as if computed with respect to the finite tree Ω_n. (Of course, the same probability may be computed with respect to Ω_{n+k}, but the result is the same by consistency.) This convention assigns probabilities to many simple statements. We can then show that the probability of a much larger class of statements is uniquely determined. We will now consider the problem abstractly.

Let S be a denumerable set; S is called the **state space**. Let Ω be the set of all infinite sequences of elements of S. A typical element ω of Ω is represented as

$$\omega = (c_0, c_1, c_2, \ldots),$$

where $c_0, c_1, c_2, \ldots$ are elements of S. The points ω of Ω are called

paths, the whole space Ω is called a **sequence space**, and the value c_n on a path ω is called the nth **outcome** on ω. The function $x_n(\omega)$, defined from Ω to S by

$$x_n(c_0, c_1, c_2, \ldots, c_n, \ldots) = c_n$$

is called the nth **outcome function**, and the nth outcome is said to occur **at time** n.

Let $\mathscr{F}_n$ be the family of all unions of sets in Ω of the form

$$\{\omega \mid x_0(\omega) \in S_0 \wedge x_1(\omega) \in S_1 \wedge \cdots \wedge x_n(\omega) \in S_n\},$$

where $S_0, S_1, \ldots, S_n$ are subsets of the state space S. (Notice that the sets of $\mathscr{F}_n$ arise from the class of all subsets of the tree Ω_n described above.) It is clear that for each n, $\mathscr{F}_n$ is a Borel field. Let $\mathscr{F}$ be the family of sets defined by

$$\mathscr{F} = \bigcup_{n=0}^{\infty} \mathscr{F}_n.$$

Each set in $\mathscr{F}$ is a set of paths for which a finite number of outcomes are restricted to lie in certain subsets of S. All other outcomes are unrestricted. The reader should verify that $\mathscr{F}$ is a field of sets. In Section 3 we shall see that $\mathscr{F}$ is not a Borel field; in the meantime, we let $\mathscr{G}$ be the smallest Borel field containing $\mathscr{F}$ (Proposition 1-14). After we have defined a measure on $\mathscr{G}$, the Borel field $\mathscr{B}$ which we are seeking will be the augmented field obtained from $\mathscr{G}$ by adding subsets of sets of measure zero.

The sets of $\mathscr{F}$ are known as **cylinder sets**. If a cylinder set C is a set in $\mathscr{F}_n$, we note that C may be written as the denumerable union

$$C = \bigcup_i B_i^{(n)}$$

of (disjoint) **basic cylinder sets**

$$B_i^{(n)} = \{\omega \mid x_0(\omega) = c_0 \wedge x_1(\omega) = c_1 \wedge \cdots \wedge x_n(\omega) = c_n\}.$$

A basic cylinder set in $\mathscr{F}_n$ is the set of all possible continuations of a single path in Ω_n. We let $\nu(B_i^{(n)})$ equal the product of the branch weights on this path in Ω_n.

Recalling that the probability measures we assigned to Ω_n were defined consistently, we can show that ν is uniquely defined on the sets of $\mathscr{F}$. It has the properties that $\nu(\Omega) = 1$ and that the restriction of ν to $\mathscr{F}_n$ is a measure for every n.

We will next show that ν can be extended to a measure μ on the smallest Borel field containing $\mathscr{F}$. This result will be a consequence of Theorem 2-4. First we prove a series of lemmas. In each case $\mathscr{F}_n$, $\mathscr{F}$, and ν are as defined above.

Lemma 2-1: Let ν be a set function defined on $\mathscr{F} = \bigcup \mathscr{F}_n$ in such a way that the restriction of ν to $\mathscr{F}_n$ is a measure for every n. Then ν is non-negative and additive.

PROOF: Non-negativity is trivial. For additivity, let A and B be disjoint sets in $\mathscr{F}$. Then $A \in \mathscr{F}_m$ and $B \in \mathscr{F}_n$, say. Since the $\mathscr{F}_j$ are nested, A and B are both in $\mathscr{F}_r$, where $r = \sup(m, n)$. Since ν is a measure when restricted to $\mathscr{F}_r$,

$$\nu(A \cup B) = \nu(A) + \nu(B).$$

We shall in fact establish that ν is completely additive, a result due to Kolmogorov.

Lemma 2-2: Suppose $C_0 \supset C_1 \supset C_2 \supset \cdots$ is a sequence of sets in $\mathscr{F}$ such that $C_n \in \mathscr{F}_n$ and $\lim_n \nu(C_n) > 0$. Then for every m there is a basic cylinder set $B_i^{(m)}$ in $\mathscr{F}_m$ such that

(1) $\lim\limits_{n} \nu(C_n \cap B_i^{(m)}) > 0$
(2) $B_i^{(m)} \subset C_m$.

PROOF: By complete additivity of ν on $\mathscr{F}_r$, where $r = \sup(m, n)$, we have $\nu(C_n) = \sum_j \nu(C_n \cap B_j^{(m)})$ with $\nu(C_n \cap B_j^{(m)})$ monotonically decreasing in n. Then

$$0 < \lim_n \nu(C_n) = \lim_n \sum_j \nu(C_n \cap B_j^{(m)}) = \sum_j \lim_n \nu(C_n \cap B_j^{(m)}).$$

The interchange of limit and sum is justified by dominated convergence as follows: The functions of j, namely $\nu(C_n \cap B_j^{(m)})$, satisfy

$$\nu(C_n \cap B_j^{(m)}) \le \nu(C_0 \cap B_j^{(m)}),$$

and we know that $\sum_j \nu(C_0 \cap B_j^{(m)}) = \nu(C_0)$ is finite since $\nu(\Omega) = 1$.

Thus, since a denumerable sum cannot be positive unless one of the terms is positive, we have for some $j = i$

$$\lim_n \nu(C_n \cap B_i^{(m)}) > 0.$$

Hence (1) is satisfied. But the terms in the sequence $\nu(C_n \cap B_i^{(m)})$ monotonically decrease to a positive limit, so that

$$\nu(C_m \cap B_i^{(m)}) > 0.$$

Now $C_m \in \mathscr{F}_m$ and is thus the union of basic cylinder sets. Since $C_m \cap B_i^{(m)}$ cannot be empty, we must have $B_i^{(m)} \subset C_m$.

Lemma 2-3: Suppose $A_0 \supset A_1 \ldots$ is a sequence of sets in $\mathscr{F}$ such that $A_n \in \mathscr{F}_n$, and $\lim_n \nu(A_n) = L > 0$. Then there exists a sequence $\{B^{(n)}\}$ of basic cylinder sets such that

(1′) $B^{(n)}$ is a basic cylinder set of $\mathscr{F}_n$.
(2′) $\lim_m \nu(A_m \cap B^{(n)}) > 0$.
(3′) For $n > 0$, $B^{(n)} \subset B^{(n-1)}$.
(4′) $B^{(n)} \subset A_n$.

Remark: Property (3′) indicates that we are actually constructing a single path by adjoining branches one at a time.

Proof of Lemma: The construction is by induction. For $n = 0$ apply Lemma 2-2 to the sequence $\{A_n\}$ and the integer $m = 0$. The $B_i^{(0)}$ that the lemma gives is $B^{(0)}$. Property (2′) follows from (1), (3′) is vacuous, and (4′) follows from (2). Suppose that we have constructed $B^{(0)}, \ldots, B^{(n)}$; we want $B^{(n+1)}$. Let

$$C_k = \begin{cases} A_k & \text{for } k < n \\ A_k \cap B^{(n)} & \text{for } k \geq n. \end{cases}$$

The sequence of sets $\{C_k\}$ is decreasing and $C_k \in \mathscr{F}_k$; we have

$$\lim_k \nu(C_k) > 0$$

by (2′) of the inductive hypothesis. Applying Lemma 2-2 to $\{C_k\}$ and the number $m = n + 1$, we obtain a basic cylinder set $B_i^{(n+1)}$, which we take as $B^{(n+1)}$. By (2),

$$B^{(n+1)} \subset C_{n+1} = A_{n+1} \cap B^{(n)}.$$

Hence (1′), (3′), and (4′) hold. By (1),

$$\begin{aligned} 0 < \lim_k \nu(C_k \cap B^{(n+1)}) &= \lim_k \nu(A_k \cap B^{(n)} \cap B^{(n+1)}) \\ &= \lim_k \nu(A_k \cap B^{(n+1)}) \qquad \text{by (3′)}. \end{aligned}$$

Hence (2′) holds.

Theorem 2-4: Let Ω be a sequence space, let $\mathscr{F}_n$ be the Borel field of sets generated by all statements

$$x_0 = c_0 \wedge \cdots \wedge x_n = c_n,$$

and let $\mathscr{F} = \bigcup_n \mathscr{F}_n$. Suppose for every n there is a probability measure ν_n defined on $\mathscr{F}_n$ with the property that the restriction of ν_{n+1} to $\mathscr{F}_n$ is ν_n. Let ν be the set function on $\mathscr{F}$ whose restriction to $\mathscr{F}_n$ is ν_n for all n. Then ν is a measure.

PROOF: By Lemma 2-1 and the contrapositive of the second half of Corollary 1-17, it is sufficient to show that if $A_0 \supset A_1 \supset A_2 \supset \cdots$ is a decreasing sequence of sets in $\mathscr{F}$ with $\lim_n \nu(A_n) = L > 0$, then $\bigcap_n A_n \neq \varnothing$. Since every set in $\mathscr{F}_n$ is a set in $\mathscr{F}_{n+1}$, we may assume that $A_n \in \mathscr{F}_n$ by repeating the same set several times in the sequence and by adding, if necessary, the set Ω a finite number of times at the beginning of the sequence. The hypotheses of Lemma 2-3 are satisfied. We then have $\bigcap_n B^{(n)} = \{\omega\}$, where ω is a single path of Ω. For every n, $\omega \in B^{(n)} \subset A_n$; hence $\omega \in \bigcap_n A_n$ and $\bigcap_n A_n \neq \varnothing$.

Applying Theorem 1-19 we extend the measure ν defined on $\mathscr{F}$ uniquely to a measure μ defined on the Borel field $\mathscr{G}$. Augmenting the field $\mathscr{G}$, we obtain the Borel field $\mathscr{B}$ with which we shall work. The extended measure μ is called **tree measure** and satisfies $\mu(\Omega) = 1$. This completes our construction of the sequence space $(\Omega, \mathscr{B}, \mu)$. From now on when we refer to a sequence space we shall mean $(\Omega, \mathscr{B}, \mu)$ constructed in the indicated manner.

2. Denumerable stochastic processes

We turn now to the definition of a denumerable stochastic process. After the definition we shall show that every sequence space defines in a natural way a stochastic process and that every denumerable stochastic process can, in a way to be described shortly, be considered as a sequence space.

Let S be a denumerable set, which will be called the **state space**, and let $(\Omega, \mathscr{B}, \mu)$ be a probability space. Points of Ω will be denoted ω.

Definition 2-5: Let $\{f_n\}$ be a sequence of functions with domain Ω and range in S, and let $\{\mathscr{B}_n\}$ be a sequence of Borel fields. The pair $(f_n, \mathscr{B}_n)$ is called a **denumerable stochastic process** on Ω if these two conditions are satisfied:

(1) $\mathscr{B}_0 \subset \mathscr{B}_1 \subset \mathscr{B}_2 \subset \cdots$; and for every n, $\mathscr{B}_n \subset \mathscr{B}$.
(2) For every fixed n and for each $s \in S$, the set $\{\omega \mid f_n(\omega) = s\}$ is a set in $\mathscr{B}_n$.

The second condition in the definition is a measurability requirement on f_n. If we were to think of the family $\mathscr{S}$ of all subsets of S as a Borel field, our condition would be equivalent to the demand that the inverse image under f_n of any set in $\mathscr{S}$ be a set in $\mathscr{B}_n$.

First we shall show that every sequence space defines a stochastic process in a natural way. Let $(\Omega, \mathscr{B}, \mu)$ be a sequence space. We take the outcome functions x_n as the sequence of functions and $\{\mathscr{F}_n\}$ as the sequence of Borel fields. It is clear that

$$\{\omega \mid x_n(\omega) = c\}$$

is a set of $\mathscr{F}_n$ and that $\mathscr{F}_0 \subset \mathscr{F}_1 \subset \cdots \subset \mathscr{F} \subset \mathscr{G} \subset \mathscr{B}$. The pair $(x_n, \mathscr{F}_n)$ therefore is a stochastic process and is referred to simply as $\{x_n\}$. We have thus shown that the outcomes of a sequence of experiments with a denumerable range form a denumerable stochastic process.

When we begin to discuss Markov chains, we shall wish to confine ourselves to stochastic processes defined on a sequence space. Therefore, our second task is to show that the restriction to sequence space is actually no restriction at all; this will indicate that our treatment of denumerable Markov chains is completely general.

Let $(f_n, \mathscr{B}_n)$ be a denumerable stochastic process on Ω' with state space S. Let μ' be a measure on Ω'. We shall construct a sequence space $(\Omega, \mathscr{B}, \mu)$ in such a way that the behavior of the process $(f_n, \mathscr{B}_n)$ on Ω' may be studied completely by studying the stochastic process $\{x_n\}$ on Ω.

Define Ω to be the space of all sequences of elements of S, and let $\mathscr{G}$ be the Borel field obtained by the construction in Section 2-1. We shall establish a correspondence between paths of Ω and subsets of Ω', and we define a measure μ on $\mathscr{G}$.

The correspondence we choose is

$$\omega \leftrightarrow \{\omega' \mid f_0(\omega') = x_0(\omega) \wedge \cdots \wedge f_n(\omega') = x_n(\omega) \wedge \cdots\}.$$

To assign a measure to Ω, we must assign a measure to cylinder sets, such as

$$A = \{\omega \mid x_0(\omega) \in S_0 \wedge \cdots \wedge x_n(\omega) \in S_n\}.$$

Noticing that the set A', defined by

$$A' = \{\omega' \mid f_0(\omega') \in S_0 \wedge \cdots \wedge f_n(\omega') \in S_n\},$$

is a set in $\mathscr{B}_n$ and is therefore μ'-measurable, we define

$$\mu(A) = \mu'(A').$$

The measure μ can then be extended to a measure defined on all of $\mathscr{B}$, and the construction of the space $(\Omega, \mathscr{B}, \mu)$ is complete. We thus see that an arbitrary stochastic process defined on Ω' may be considered as a process on a suitable sequence space Ω in which the f_n are outcome functions. The probability of any statement concerning the f_n can be computed in the sequence space.

3. Borel fields in stochastic processes

Probabilities are numbers assigned to statements about stochastic processes. We may now formally define the **probability** of a statement to be the measure of the statement's truth set. In symbols $\Pr[p] = \mu(P)$, where $P = \{\omega \mid p\}$. If the set P is not a set in the Borel field on which μ is defined, then $\Pr[p]$ is undefined. Statements for which $\Pr[p]$ is defined are called **measurable statements.**

In a stochastic process we see that a Borel field of sets represents a state of knowledge. The more sets there are in the Borel field, the more statements there are that we know how to assign probabilities to. Let us analyze briefly what this fact implies about Definition 2-5.

In a denumerable stochastic process we are given an increasing sequence of Borel fields $\mathscr{B}_n$ such that $\mathscr{B}_n \subset \mathscr{B}$ for every n. The field $\mathscr{B}_n$ represents the state of knowledge of the process up to time n. The fact that the Borel fields are increasing means that our knowledge of the process never decreases as time goes on, and the fact that all $\mathscr{B}_n$ are contained in $\mathscr{B}$ means that our total knowledge of the process necessarily includes knowledge of what happens in a finite number of steps.

Similarly, condition (2) in the definition is an abstract formulation of the requirement that in a stochastic process the present does not depend upon the future. We conclude, therefore, that our definition satisfies the conditions imposed at the beginning of Section 1.

We shall apply this insight about the role of Borel fields to a specific example to show that the field $\mathscr{F}$ in Section 1 is not the same as the Borel field $\mathscr{G}$. Let Ω be the sequence space constructed when S is taken as a two-element set. Measures $2^{-(n+1)}$ are assigned to each set B of paths of the form

$$B = \{\omega \mid x_0 = c_0 \wedge \cdots \wedge x_n = c_n\}.$$

This measure is eventually extended to a measure μ on the Borel field $\mathscr{B}$, and we obtain $\mu(\Omega) = 1$. The state space for this example is often taken as $S = \{\text{heads, tails}\}$, and the model for the stochastic process is the tossing of a balanced coin.

For the coin-tossing process a well-known example of a statement whose truth set is not in the field $\mathscr{F}$ but is in the Borel field $\mathscr{G}$ is involved in the Strong Law of Large Numbers (which we shall consider in more detail in Chapter 3). Let k and n be integers and let r_n be the fraction of the first n outcomes which are heads. Let p be the statement about r_n that

$$\frac{1}{2} - \frac{1}{k} < r_n < \frac{1}{2} + \frac{1}{k}.$$

Consider the statement q about $k > 0$ that

$$(\forall k)(\exists N)(\forall n)(n > N \rightarrow p).$$

We write the statement in this form to demonstrate that its truth set is in $\mathscr{G}$. In words, the statement q asserts that for any $k > 0$ there exists an N such that if $n > N$, then $|r_n - \frac{1}{2}| < 1/k$. That is, q says that $\lim_{n\to\infty} r_n = \frac{1}{2}$. The truth set of the statement q is in $\mathscr{G}$ but not $\mathscr{F}$, because the notion of limit cannot be expressed in terms of finitely many of the r_n. The Strong Law of Large Numbers asserts that

$$\Pr[q] = 1.$$

4. Statements of probability zero or one

In a probability space $(\Omega, \mathscr{B}, \mu)$ a statement with truth set Ω is logically true, whereas a statement with truth set $\varnothing$ is logically false. However, the Strong Law of Large Numbers asserts not that a certain set is Ω but that it has measure one. A statement whose truth set has measure one is said to be **almost always, almost everywhere, almost surely** true, or **true a.e.** Correspondingly, a statement with truth set of measure zero is almost surely false, and the negation $\sim p$ of a statement p which is almost surely false is almost surely true.

Two useful propositions are related to the subject of almost sure statements.

Proposition 2-6: Let $\{p_n\}$ be a denumerable set of statements and let q be the statement that p_n holds for all n. If $\Pr[p_n] = 1$ for all n, then $\Pr[q] = 1$.

PROOF: Let $\{P_n\}$ be the truth sets of the statements $\{p_n\}$. Applying Proposition 1-18, we have

$$1 - \Pr[q] = \mu\Big(X - \bigcap_n P_n\Big) = \mu\Big(\bigcup_n \tilde{P}_n\Big) \le \sum_n \mu(\tilde{P}_n) = 0$$

so that

$$\Pr[q] = 1.$$

The second result is one of the Borel–Cantelli Lemmas.

Proposition 2-7: If $\{p_n\}$ is a sequence of statements for which $\sum \Pr[p_n]$ is finite, then with probability one only finitely many of the p_n are true.

Proof: Let q_n be the statement that all of the statements $p_n, p_{n+1}, \ldots$ are false. Let $\epsilon > 0$ be given. Choose N large enough so that $\sum_{n=N}^{\infty} \Pr[p_n] < \epsilon$. Then

$$\begin{aligned} 1 - \Pr[q_N] &= \Pr[\text{at least one of } p_N, p_{N+1}, \ldots \text{ occurs}] \\ &= \Pr[p_N \vee p_{N+1} \vee \cdots]. \end{aligned}$$

By Proposition 1-18 the right-side is

$$\leq \sum_{n=N}^{\infty} \Pr[p_n] < \epsilon.$$

Hence,

$$\begin{aligned} \Pr[\text{finitely many } p_n \text{ are true}] &= \Pr\left[\bigvee_{n=1}^{\infty} q_n\right] \\ &\geq \Pr[q_N] \\ &> 1 - \epsilon. \end{aligned}$$

Since this inequality holds for every $\epsilon > 0$, the probability must be 1.

5. Conditional probabilities

Let $(\Omega, \mathscr{B}, \mu)$ be a probability space. If p and q are statements such that $\Pr[q] \neq 0$, the **conditional probability of** p **given** q, written $\Pr[p \mid q]$, is defined by

$$\Pr[p \mid q] = \Pr[p \wedge q]/\Pr[q].$$

If $\Pr[q] = 0$, we shall normally agree that $\Pr[p \mid q] = 0$. (Alternatively, we might leave $\Pr[p \mid q]$ undefined if $\Pr[q] = 0$. Such a convention would be adopted in a more general context.)

The case $\Pr[q] = 0$ is not very interesting. Suppose $\Pr[q] \neq 0$, and let $Q \subset \Omega$ be the truth set of q. If Q is taken as a space of points, then $\mathscr{B}' = \{B' \mid B' = Q \cap B, B \in \mathscr{B}\}$ is a Borel field of sets in Q. For any set B' in $\mathscr{B}'$ we define a set function ν by

$$\nu(B') = \frac{\mu(B')}{\mu(Q)}.$$

The reader should verify that ν is a measure on $\mathscr{B}'$. Furthermore, $\nu(Q) = \mu(Q)/\mu(Q) = 1$. Therefore, $(Q, \mathscr{B}', \nu)$ is a probability space. We may thus speak of the probabilities of statements relative to Q, and we see that their values coincide with conditional probabilities given q and relative to Ω. That is, conditional probabilities possess the same properties as ordinary probabilities.

We shall apply this notion of conditional probability to the sequence space considered in the preceding sections. In this space the sets of $\mathscr{B}_n$ are denumerable disjoint unions of sets of the form

$$B_n = \{\omega \mid x_0 \in S_0 \wedge \cdots \wedge x_n \in S_n\}.$$

Since the state space S is denumerable, each of the sets $S_0, S_1, \ldots, S_n$ is denumerable and B_n is the denumerable union of disjoint sets of the form $\{\omega \mid x_0 = c_0 \wedge \cdots \wedge x_n = c_n\}$. By definition of conditional probability,

$$\Pr[x_0 = c_0 \wedge \cdots \wedge x_{n-1} = c_{n-1} \wedge x_n = c_n] = \Pr[x_n = c_n \mid x_0 = c_0 \wedge \cdots \wedge x_{n-1} = c_{n-1}] \times \Pr[x_0 = c_0 \wedge \cdots \wedge x_{n-1} = c_{n-1}].$$

By induction, we find

$$\begin{aligned}\Pr[x_0 = c_0 \wedge \cdots \wedge x_n = c_n] = \Pr[x_0 = c_0]\cdot\Pr[x_1 = c_1 \mid x_0 = c_0] \\ \times \Pr[x_2 = c_2 \mid x_0 = c_0 \wedge x_1 = c_1] \\ \times \Pr[x_3 = c_3 \mid x_0 = c_0 \wedge x_1 = c_1 \wedge x_2 = c_2] \\ \times \cdots \times \Pr[x_n = c_n \mid x_0 = c_0 \wedge \cdots \wedge x_{n-1} = c_{n-1}].\end{aligned}$$

We have established the following result.

Proposition 2-8: The measure on a sequence space is completely determined by

(1) the starting probabilities, $\Pr[x_0 = c_0]$, and
(2) the transition probabilities,

$$\Pr[x_n = c_n \mid x_0 = c_0 \wedge \cdots \wedge x_{n-1} = c_{n-1}].$$

Conditional probabilities, as we anticipated, have their place in tree diagrams. To each branch we may assign a conditional probability. We see now the abstract formulation of the fact that the probabilities of statements like $x_0 = c_0 \wedge x_1 = c_1 \wedge \cdots \wedge x_n = c_n$ are computed simply by multiplying together the appropriate branch weights.

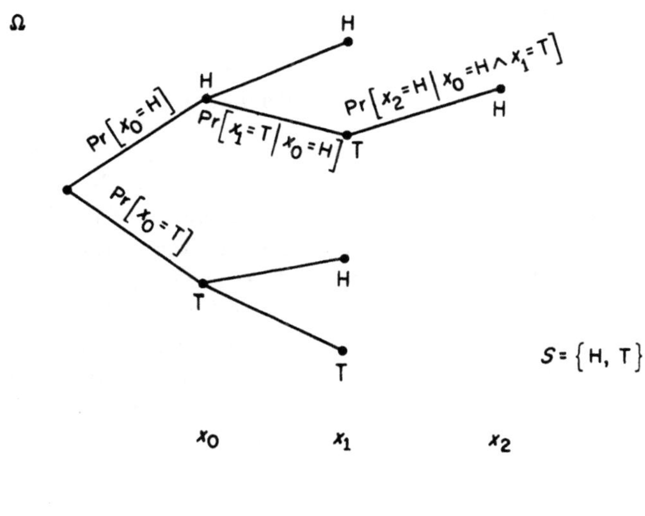

Two statements p and q are said to be **probabilistically independent** if $\Pr[p \wedge q] = \Pr[p]\Pr[q]$. A stochastic process defined on sequence space is called an **independent process** if the statements

$$x_n = c_n$$

and

$$x_0 = c_0 \wedge \cdots \wedge x_{n-1} = c_{n-1}$$

are independent for all $n \geq 1$ and for all $c_0, c_1, \ldots, c_n$. Coin tossing is an example. We shall see that an independent process is a special kind of Markov process. If, in addition, for each c the probability that $x_n = c$ does not depend on n, then the process is called an **independent trials process.**

6. Random variables and means

Let $(\Omega, \mathscr{B}, \mu)$ be a probability space. A measurable function $\mathbf{f}$ with domain Ω and range in the extended real-valued number system is called a **random variable.** We may apply all the properties of the measurable functions in Section 3 of Chapter 1.

If $\mathbf{f}(\omega)$ is an extended-real-valued function defined on the space Ω of a stochastic process and if Ω has the property that for some n, $\mathbf{f}(\omega)$ is measurable with respect to the Borel field $\mathscr{B}_n$, then $\mathbf{f}$ is a random variable because $\mathscr{B}_n \subset \mathscr{B}$. Such a function is said to be $\mathscr{B}_n$-measurable. For the special case in which Ω is sequence space, a function $\mathbf{f}$ is $\mathscr{B}_n$-measurable for some n if its values depend only on a bounded number of outcomes.

In terms of sequence space, we give two examples of random variables.

(1) Define a function $\mathbf{u}_j^{(n)}$ by

$$\mathbf{u}_j^{(n)} = \begin{cases} 1 & \text{if } x_n = j \\ 0 & \text{otherwise.} \end{cases}$$

Since the value of $\mathbf{u}_j^{(n)}$ depends only on outcome n, the function $\mathbf{u}_j^{(n)}$ is $\mathscr{B}_n$-measurable and is therefore a random variable. Letting $\mathbf{n}_j = \sum_{n=0}^{\infty} \mathbf{u}_j^{(n)}$ and noting that the limit of a sequence of random variables is a random variable, we see that $\mathbf{n}_j$ is a random variable. The function $\mathbf{n}_j(\omega)$ counts the number of times that the outcome j occurs on the path ω; it will appear again after we introduce Markov chains.

(2) Let $\mathbf{t}_j$ be the time to reach j. That is, define $\mathbf{t}_j(\omega)$ to be the first time n along path ω such that $x_n(\omega) = j$. If j is never reached, set $\mathbf{t}_j = +\infty$. Then $\mathbf{t}_j$ is a random variable because it is the limit on n of the function $\mathbf{t}_j^{(n)} = \inf(\mathbf{t}_j, n)$, which is $\mathscr{B}_n$-measurable.

Definition 2-9: A **random time** $\mathbf{t}$ is a random variable satisfying these two properties:

(1) Its range is in the non-negative integers with $\{+\infty\}$ adjoined.
(2) For each integer n the set $\{\omega \mid \mathbf{t}(\omega) = n\}$ is a set of $\mathscr{B}_n$.

The random variable $\mathbf{t}_j$ defined in the second example above is an example of a random time.

The **mean** of a random variable $\mathbf{f}$, denoted $\mathrm{M}[\mathbf{f}]$, is defined by $\mathrm{M}[\mathbf{f}] = \int_\Omega \mathbf{f}d\mu$, where $\mathrm{M}[\mathbf{f}]$ exists if and only if $\int_\Omega \mathbf{f}d\mu$ exists and where μ is the measure associated with the probability space. Since means are Lebesgue integrals, they satisfy all the properties of Lebesgue integrals. In particular, if $\{\mathbf{f}_n\}$ is a sequence of non-negative random variables, then by Corollary 1-46,

$$\mathrm{M}\left[\sum_{n=0}^{\infty} \mathbf{f}_n\right] = \sum_{n=0}^{\infty} \mathrm{M}[\mathbf{f}_n].$$

In addition, if $\mathbf{g}_n$ is a sequence of random variables with the properties that $|\mathbf{g}_n| \leq c$ for every n and that $\mathbf{g}_n$ converges to $\mathbf{g}$, then

$$\mathrm{M}[\mathbf{g}] = \lim_{n\to\infty} \mathrm{M}[\mathbf{g}_n]$$

by Corollary 1-50. An important application of these facts is the following result.

Proposition 2-10: In a sequence space,

$$\mathrm{M}[\mathbf{n}_j] = \sum_{n=0}^{\infty} \Pr[x_n = j].$$

PROOF:

$$\begin{aligned}\mathrm{M}[\mathbf{n}_j] &= \mathrm{M}\left[\sum_n \mathbf{u}_j^{(n)}\right] \\ &= \sum_n \mathrm{M}[\mathbf{u}_j^{(n)}] \\ &= \sum_n \int_\Omega \mathbf{u}_j^{(n)} d\mu \\ &= \sum_n \int_{\{\omega \mid x_n(\omega) = j\}} 1 d\mu \\ &= \sum_n \Pr[x_n = j].\end{aligned}$$

7. Means conditional on statements

Suppose $(\Omega, \mathscr{B}, \mu)$ is a probability space on which a denumerable stochastic process is defined. We say that a denumerable family of subsets $\mathscr{R} = \{R_1, R_2, \ldots\}$ is a **partition** of Ω if the sets R_j are disjoint, exhaustive, and measurable. Each such subset R_j is called a **cell** of the partition; we allow the possibility that a cell may have measure zero. The reader should notice that the sets $\{\varnothing, R_1, R_2, \ldots\}$ together with all possible denumerable unions form a Borel field which we shall call $\mathscr{R}^*$. Since the sets in $\mathscr{R}$ are measurable, we see that $\mathscr{R}^* \subset \mathscr{B}$.

Two examples of partitions are typical.

(1) Let the process be coin tossing, and define a partition by

$$R_1 = \{\omega \mid x_0(\omega) \text{ is a head}\}, \qquad R_2 = \{\omega \mid x_0(\omega) \text{ is a tail}\}.$$

More generally, let $\mathscr{R}$ consist of disjoint exhaustive measurable sets which are in $\mathscr{B}_n$ for some fixed n.

(2) Suppose $\mathbf{f}$ is a random variable whose range is denumerable. If the range is $\{a_j\}$, define $R_j = \{\omega \mid \mathbf{f}(\omega) = a_j\}$. Then $\{R_j\}$ is a partition.

A denumerable set of statements $\{q_i\}$ about a stochastic process is said to be a **set of alternatives** if the truth sets Q_i of the statements form a partition. Since the integral is a completely additive set function, we obtain, for every random variable $\mathbf{f}$ whose mean exists, the relation

$$\mathrm{M}[\mathbf{f}] = \sum_i \int_{Q_i} \mathbf{f}\, d\mu.$$

Let p be a statement with measurable truth set P. If $\mathbf{f}$ is a random variable and if $\Pr[p] \neq 0$, then the **conditional mean** of $\mathbf{f}$ given p, written $\mathrm{M}[\mathbf{f} \mid p]$, is defined by

$$\mathrm{M}[\mathbf{f} \mid p] = \frac{1}{\Pr[p]} \int_P \mathbf{f}\, d\mu.$$

If $\Pr[p] = 0$, then $\mathrm{M}[\mathbf{f} \mid p]$ is defined to be zero. (In a more general setting, $\mathrm{M}[\mathbf{f} \mid p]$ is not defined when $\Pr[p] = 0$.)

Proposition 2-11: If $\mathrm{M}[\mathbf{f}]$ exists and $\{q_i\}$ is a set of alternatives with truth sets $\{Q_i\}$, then

$$\mathrm{M}[\mathbf{f}] = \sum_i \Pr[q_i] \cdot \mathrm{M}[\mathbf{f} \mid q_i].$$

PROOF: By definition of conditional mean, we have

$$\int_{Q_i} \mathbf{f}\, d\mu = \Pr[q_i] \cdot \mathrm{M}[\mathbf{f} \mid q_i].$$

Summing both sides of the equation on i gives the result immediately.

Corollary 2-12: Let p be a statement with measurable truth set P, and let $\{q_i\}$ be a set of alternatives with truth sets Q_i. Then

$$\Pr[p] = \sum_i \Pr[q_i] \cdot \Pr[p \mid q_i] = \sum_i \Pr[p \wedge q_i].$$

PROOF: Let $\mathbf{f}$ be the characteristic function of the set P and apply Proposition 2-11.

8. Problems

1. For coin tossing, show that
   (a) the probability of getting only finitely many "heads" is 0;
   (b) there will be infinitely many "heads" a.e.

2. Consider the experiment of selecting "1" with probability $\frac{1}{2}$, "2" with probability $\frac{1}{3}$, or "3" with probability $\frac{1}{6}$. If this experiment is repeated infinitely often, show that the probability of selecting "1" only finitely often is 0.

3. In Problem 2, let $\mathscr{B}_n = \mathscr{F}_n$. Which of the following f_n form a denumerable stochastic process $(f_n, \mathscr{B}_n)$?
   (a) f_n = the nth outcome.
   (b) f_n = time at which nth "3" is selected.
   (c) $f_n = s_n$ = sum of the first n numbers selected.
   (d) $f_n = \dfrac{s_n - a}{\sqrt{cn}}$, where a and c are constants.

4. Show that the following converse of the Borel–Cantelli Lemma is false: If $\sum \Pr[p_n] = +\infty$, then the probability that only finitely many p_n are true is less than 1.

5. Start with 4 jacks and 4 queens from a bridge deck. Find the probability of drawing two cards, both of which are jacks. Then compute the conditional probability of the same on each of the following conditions:
   (a) One card drawn is a jack.
   (b) One card drawn is a red jack.
   (c) One card drawn is the jack of hearts.
   (d) The first card drawn is the jack of hearts.

6. For coin tossing, define three random variables which are not measurable with respect to any finite tree Ω_n.

7. Let ${}^i\mathbf{n}_j(\omega)$ be the number of times on path ω that a j occurs before the first i occurs (if $i = j$, take ${}^i\mathbf{n}_j(\omega) = 0$). Prove that ${}^i\mathbf{n}_j$ is a random variable, and develop an infinite series representation for $\mathrm{M}[\mathbf{t}_i]$ in terms of it.

8. In coin tossing, let $\mathbf{t}$ be the first time that "heads" comes up.
   (a) Is $\mathbf{t}$ a random time?
   (b) Find $\mathrm{M}[\mathbf{t}]$.
   (c) Find $\mathrm{M}[\mathbf{t} \mid \text{first outcome is "tails"}]$.

9. In a randomly selected two-child family, let **f** = the number of boys. Find
   (a) M[**f**].
   (b) M[**f** | first child is a boy].
   (c) M[**f** | there is at least one boy].

10. For coin tossing, let **f** = the number of "heads" in the first three tosses. Let q_i be the statement that there were exactly i "heads" in the first two tosses ($i = 0, 1, 2$). Find M[**f**], M[**f** | q_0], M[**f** | q_1], and M[**f** | q_2]. Verify that the first of these is a linear combination of the last three with appropriate probabilities as weights.

11. Let $\{q_i\}$ be a set of alternatives and let **f** be a non-negative random variable. Prove that if M[**f** | q_1] > M[**f**], then there is an alternative q_j such that M[**f** | q_j] < M[**f**].

12. Let X be the unit interval. Let $\mathscr{F}$ consist of all finite unions of finite intervals (with or without endpoints). We classify a point as an interval of length 0. The measure to be constructed in this problem is called **Lebesgue measure**, and the Borel field is the class of all **Borel sets**.
    (a) Show that $\mathscr{F}$ is a field.
    (b) Show that every set A in $\mathscr{F}$ can be written as a finite *disjoint* union of intervals $A_1, A_2, \ldots, A_n$.
    (c) If A is decomposed as in part (b), we define

$$\nu(A) = \sum_{k=1}^{n} l(A_k),$$

    where l denotes length. Show that ν is consistently defined and that ν is a non-negative additive set function.
    (d) Show that if A is a finite union of intervals and if $\epsilon > 0$ is given, then there is a finite union K of closed intervals and a finite union G of open intervals such that $K \subset A$, $A \subset G$, and

$$\nu(K) + \epsilon \geq \nu(A) \geq \nu(G) - \epsilon.$$

    [*Note:* An open interval here means a set which is the intersection of X with an open interval of the real line.]
    (e) Let A and A_n, $n = 1, 2, \ldots$, be finite unions of intervals with the A_n disjoint and with $\cup A_n = A$. Use parts (c) and (d) and the Heine–Borel Theorem to prove that, for any $\epsilon > 0$,

$$\nu(A) \leq \sum \nu(A_n) + \epsilon.$$

    (f) Deduce from part (e) that ν is completely additive on $\mathscr{F}$.
    (g) Apply Theorem 1-19 and describe the resulting measure space.
    (h) Using complete additivity, prove that a denumerable set of points has measure 0.
    (i) Why does the proof of (h) not show that every set has measure 0?
    (j) Show directly from the definition

$$\mu(B) = \inf \left[\sum \nu(A_n)\right]$$

    that every denumerable set of points has measure 0.

13. Let $(\Omega, \mathscr{B}, \mu)$ be the unit interval with Lebesgue measure. If $x \in \Omega$, let $f(x) = x^2$. Let p be the statement "$x \leq \frac{1}{2}$."
(a) Show that f is a random variable.
(b) Find $\mathrm{M}[f]$.
(c) Find $\mathrm{M}[f \mid p]$ and $\mathrm{M}[f \mid {\sim}p]$.
(d) Relate your answer in (b) to the solution of (c).

CHAPTER 3

MARTINGALES

1. Means conditional on partitions and functions

In this chapter we consider a natural abstraction of the idea of a fair game in gambling. We shall give several applications of the basic result, the Martingale Convergence Theorem.

We begin by defining what we mean by the conditional mean of $\mathbf{f}$ given a partition $\mathscr{R}$ of the domain. Let $(\Omega, \mathscr{B}, \mu)$ be a measure space. We shall normally assume $\mu(\Omega) \leq 1$, but such an assumption is not necessary as long as μ is finite.

Definition 3-1: Let $\omega \in \Omega$ and let r_j be the statement that ω is in a cell R_j of $\mathscr{R}$. If $\mathbf{f}$ is a random variable, then the **conditional mean of f given** $\mathscr{R}$, written $M[\mathbf{f} \mid \mathscr{R}]$, is defined to be a function of ω whose value at every point in the cell R_j is the constant $M[\mathbf{f} \mid r_j]$.

Next, we observe that if $M[\mathbf{f}]$ exists and is finite, then $M[\mathbf{f} \mid \mathscr{R}]$ exists and is finite. For, on cells of measure zero, the conditional mean is defined to be zero. If $\mu(R_j) > 0$, then

$$M[|\mathbf{f}| \mid \mathscr{R}] = M[|\mathbf{f}| \mid R_j] = \frac{1}{\mu(R_j)} \int_{R_j} |\mathbf{f}| d\mu \leq \frac{1}{\mu(R_j)} M[|\mathbf{f}|] < \infty.$$

The next proposition provides an equivalent definition of the mean of $\mathbf{f}$ given $\mathscr{R}$.

Proposition 3-2: $M[\mathbf{f} \mid \mathscr{R}]$ is characterized almost everywhere by these two properties:

(1) $M[\mathbf{f} \mid \mathscr{R}]$ is constant on each cell of $\mathscr{R}$.
(2) $\int_{R_j} \mathbf{f} d\mu = \int_{R_j} M[\mathbf{f} \mid \mathscr{R}] d\mu$.

Proof: We must show that $M[\mathbf{f} \mid \mathscr{R}]$ has these two properties and that any random variable satisfying these properties is equal to the conditional mean of **f** given $\mathscr{R}$ a.e. Set $\mathbf{g} = \mathrm{M}[\mathbf{f} \mid \mathscr{R}]$, and suppose first that **g** is known to be a conditional mean. Then **g** satisfies (1) by definition. If $\mu(R_j) = 0$, both sides of (2), are 0. If not, equality again follows from the definition of $\mathrm{M}[\mathbf{f} \mid \mathscr{R}]$.

Conversely, if a function **g** satisfies (1) and (2), then it agrees with $\mathrm{M}[\mathbf{f} \mid \mathscr{R}]$ on all paths in cells of positive measure; that is, it agrees with $\mathrm{M}[\mathbf{f} \mid \mathscr{R}]$ a.e.

We give two examples of conditional means.

(1) Let the stochastic process be coin tossing and let

$$R_1 = \{\omega \mid x_0(\omega) \text{ is a head}\}$$
$$R_2 = \{\omega \mid x_0(\omega) \text{ is a tail}\}.$$

Let **f** be the total number of heads on the zeroth and first tosses. Then

$$\mathrm{M}[\mathbf{f} \mid \mathscr{R}] = \begin{cases} \frac{3}{2} & \text{on } R_1 \\ \frac{1}{2} & \text{on } R_2 \end{cases}$$

and

$$\mathrm{M}[\mathbf{f}] = (\tfrac{1}{2})(\tfrac{3}{2}) + (\tfrac{1}{2})(\tfrac{1}{2}) = 1.$$

(2) For any stochastic process and for any denumerable-valued random variable **f**, let $\mathscr{R}$ be the trivial partition $\{\Omega\}$. Then

$$\mathrm{M}[\mathbf{f} \mid \mathscr{R}] = \mathrm{M}[\mathbf{f}]$$

on every path ω.

A partition $\mathscr{R}$ is said to be **contained** in $\mathscr{S}$, written $\mathscr{R} \subset \mathscr{S}$, if every cell of $\mathscr{R}$ is a union of cells of $\mathscr{S}$. If $\mathscr{R} \subset \mathscr{S}$, then $\mathscr{R}$ is the "coarser" subdivision, and $\mathscr{S}$ is called a **refinement** of $\mathscr{R}$.

Proposition 3-3: Conditional means satisfy these properties:

(1) $\mathrm{M}[\mathrm{M}[\mathbf{f} \mid \mathscr{S}] \mid \mathscr{R}] = \mathrm{M}[\mathbf{f} \mid \mathscr{R}]$ if $\mathscr{R} \subset \mathscr{S}$.
(2) $\mathrm{M}[\mathrm{M}[\mathbf{f} \mid \mathscr{S}]] = \mathrm{M}[\mathbf{f}]$ for any $\mathscr{S}$.
(3) If **g** is constant on each cell of $\mathscr{R}$, then $\mathrm{M}[\mathbf{g} \mid \mathscr{R}] = \mathbf{g}$ a.e., and if $\mathrm{M}[\mathbf{fg}]$ exists and is finite, then $\mathrm{M}[\mathbf{fg} \mid \mathscr{R}] = \mathbf{g}\mathrm{M}[\mathbf{f} \mid \mathscr{R}]$.
(4) If **f** and **g** assume only finite range values and if $\mathrm{M}[\mathbf{f} \mid \mathscr{R}]$ and $\mathrm{M}[\mathbf{g} \mid \mathscr{R}]$ both exist and are finite, then $\mathrm{M}[\mathbf{f} + \mathbf{g} \mid \mathscr{R}]$ exists, is finite, and is given by

$$\mathrm{M}[\mathbf{f} + \mathbf{g} \mid \mathscr{R}] = \mathrm{M}[\mathbf{f} \mid \mathscr{R}] + \mathrm{M}[\mathbf{g} \mid \mathscr{R}].$$

PROOF: For (1) it is sufficient to show that

$$\int_{R_j} \mathrm{M}[\mathrm{M}[\mathbf{f} \mid \mathscr{S}] \mid \mathscr{R}]d\mu = \int_{R_j} \mathrm{M}[\mathbf{f} \mid \mathscr{R}]d\mu$$

since both functions are constant on cells of $\mathscr{R}$. Applying property (2) of Proposition 3-2 three times, we have

$$\begin{aligned}\int_{R_j} \mathrm{M}[\mathrm{M}[\mathbf{f} \mid \mathscr{S}] \mid \mathscr{R}]d\mu &= \int_{R_j} \mathrm{M}[\mathbf{f} \mid \mathscr{S}]d\mu \\ &= \sum_i \int_{S_i \subset R_j} \mathrm{M}[\mathbf{f} \mid \mathscr{S}]d\mu \\ &= \sum_i \int_{S_i \subset R_j} \mathbf{f}d\mu \\ &= \int_{R_j} \mathbf{f}d\mu \\ &= \int_{R_j} \mathrm{M}[\mathbf{f} \mid \mathscr{R}]d\mu.\end{aligned}$$

The proofs of (3) and (4) use the same technique and are left to the reader. Property (2) follows from (1) with $\mathscr{R}$ taken as the trivial partition $\{\Omega\}$.

Definition 3-4: The **cross partition** $\mathscr{R} \otimes \mathscr{S}$ of two partitions $\mathscr{R}$ and $\mathscr{S}$ is the family of sets defined by

$$\mathscr{R} \otimes \mathscr{S} = \{R_i \cap S_j \mid R_i \in \mathscr{R}, S_j \in \mathscr{S}, R_i \cap S_j \neq \varnothing\}.$$

For example,

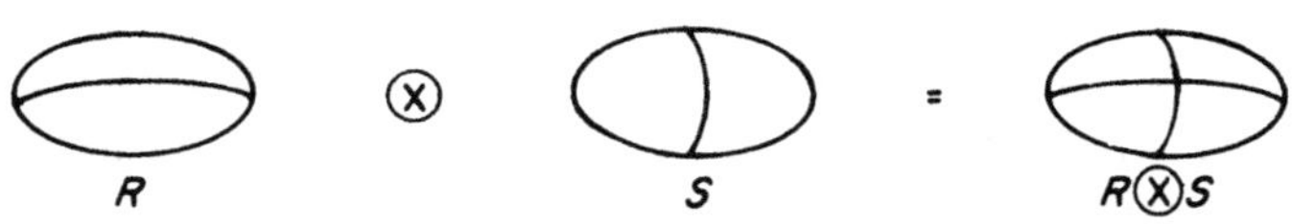

Since the intersection of measurable sets is measurable and since the sets $\{R_i \cap S_j\}$ are disjoint and exhaustive, a cross partition is a partition.

In Example 2 of Section 2-7 we noted that every denumerable-valued random variable determines a natural partition of Ω. We call the partition associated with the denumerable-valued random variable $\mathbf{g}$, $\mathscr{R}^g$. In terms of the natural partition induced by $\mathbf{g}$, we define the **conditional mean of f given g**, by

$$\mathrm{M}[\mathbf{f} \mid \mathbf{g}] = \mathrm{M}[\mathbf{f} \mid \mathscr{R}^g].$$

Then $\mathrm{M}[\mathbf{f} \mid \mathbf{g}]$ is constant on sets where $\mathbf{g}$ is constant, and on the set where $\mathbf{g} = c$ for some constant c, $\mathrm{M}[\mathbf{f} \mid \mathbf{g}]$ has the value $\mathrm{M}[\mathbf{f} \mid \mathbf{g} = c]$.

Observing that the operation of forming the cross partition of two partitions is both associative and commutative, we define more generally

$$\mathrm{M}[\mathbf{f}_{n+1} \mid \mathbf{f}_0 \wedge \cdots \wedge \mathbf{f}_n] = \mathrm{M}[\mathbf{f}_{n+1} \mid \mathscr{R}^{f_0} \otimes \cdots \otimes \mathscr{R}^{f_n}].$$

If p is a statement with truth set P, we define

$$\Pr[p \mid \mathbf{f}_0 \wedge \cdots \wedge \mathbf{f}_n] = \mathrm{M}[\chi_P \mid \mathbf{f}_0 \wedge \cdots \wedge \mathbf{f}_n],$$

where χ_P is the characteristic function of the set P.

A sequence of random variables $\{\mathbf{f}_n\}$ is said to be **independent** if, for every n and for every A,

$$\Pr[\mathbf{f}_{n+1} \in A \mid \mathbf{f}_0 \wedge \cdots \wedge \mathbf{f}_n] = \Pr[\mathbf{f}_{n+1} \in A]$$

almost everywhere. The reader should show that if $\{\mathbf{f}_n\}$ is a sequence of independent random variables, then

$$\mathrm{M}[\mathbf{f}_{n+1} \mid \mathbf{f}_0 \wedge \cdots \wedge \mathbf{f}_n] = \mathrm{M}[\mathbf{f}_{n+1}].$$

In the special case in which, for each A, $\Pr[\mathbf{f}_n \in A]$ does not depend on n, the random variables are said to be **identically distributed**.

2. Properties of martingales

With the background of Section 3-1, we proceed to define martingales and to give several examples of them. We still work with the probability space $(\Omega, \mathscr{B}, \mu)$ and a denumerable set of states S. We assume, however, that the set S is a subset of the extended real number system.

Definition 3-5: Let $\{f_n\}$ be a sequence of denumerable-valued random variables, and suppose $\mathscr{R}_0 \subset \mathscr{R}_1 \subset \cdots$ is an increasing sequence of partitions of Ω. The pair $(\mathbf{f}_n, \mathscr{R}_n)$ is called a **martingale** if three conditions are satisfied:

(1) $\mathrm{M}[|\mathbf{f}_n|]$ is finite for each n.
(2) $\mathbf{f}_n$ is constant on cells of $\mathscr{R}_n$.
(3) $\mathbf{f}_n = \mathrm{M}[\mathbf{f}_{n+1} \mid \mathscr{R}_n]$.

If (1) and (2) are satisfied and (3) is replaced by $\mathbf{f}_n \geq \mathrm{M}[\mathbf{f}_{n+1} \mid \mathscr{R}_n]$, then $(\mathbf{f}_n, \mathscr{R}_n)$ is a **supermartingale**. If (1) and (2) are satisfied and (3) is replaced by $\mathbf{f}_n \leq \mathrm{M}[\mathbf{f}_{n+1} \mid \mathscr{R}_n]$, then $(\mathbf{f}_n, \mathscr{R}_n)$ is a **submartingale**.

For a martingale, condition (3) in Definition 3-5 implies condition (2), but for a supermartingale or a submartingale it does not.

When we defined partitions in Section 2-7, we noted that every partition $\mathscr{R}_n$ determines a Borel field $\mathscr{R}_n^*$ and that the Borel field satisfied $\mathscr{R}_n^* \subset \mathscr{B}$. Since $\mathscr{R}_n \subset \mathscr{R}_{n+1}$ implies $\mathscr{R}_n^* \subset \mathscr{R}_{n+1}^*$, we see that a martingale is a stochastic process. The reader should notice that the condition that $\mathbf{f}_n$ is constant on cells of $\mathscr{R}_n$ is equivalent to the condition that $\mathbf{f}_n$ is measurable with respect to $\mathscr{R}_n^*$.

Throughout the discussion of martingales, it is convenient to keep in mind the idea of a fair game, which we shall introduce as our first example below. We shall see that the fair game is a special case of the following situation. Let $\mathbf{f}_0, \mathbf{f}_1, \ldots$ be a sequence of denumerable-valued random variables defined on a probability space such that $\mathrm{M}[|\mathbf{f}_j|]$ is finite. Setting $\mathscr{R}_n = \mathscr{R}^{f_0} \otimes \cdots \otimes \mathscr{R}^{f_n}$, we see that the sequence $\{\mathscr{R}_n\}$ is clearly increasing. Therefore, only condition (3) of Definition 3-5 need be satisfied for $(\mathbf{f}_n, \mathscr{R}_n)$ to be a martingale. In particular, we see that such a sequence of random variables forms a martingale if and only if

$$\mathrm{M}[\mathbf{f}_{n+1} \mid \mathbf{f}_0 \wedge \mathbf{f}_1 \wedge \cdots \wedge \mathbf{f}_n] = \mathbf{f}_n.$$

When the partitions $\mathscr{R}_n$ are obtained from the $\mathscr{R}^{f_n}$ in the way we have just described, we agree to refer to the pair $(\mathbf{f}_n, \mathscr{R}_n)$ simply as $\{\mathbf{f}_n\}$.

We shall give three examples of martingales at this time. More examples will appear after we introduce Markov chains.

(1) Let $\{\mathbf{y}_n\}$ be a sequence of independent random variables with denumerable range and let $\mathbf{s}_n = \mathbf{y}_0 + \cdots + \mathbf{y}_n$ represent the nth partial sum. Then $\{\mathbf{s}_n\}$ with its partition obtained from the $\mathscr{R}^{s_n}$ in the natural way is a martingale if and only if $\mathrm{M}[\mathbf{y}_k] = 0$ for every k. We have

$$\begin{aligned}
\mathrm{M}[\mathbf{s}_{n+1} \mid \mathbf{s}_0 \wedge \cdots \wedge \mathbf{s}_n] &= \mathrm{M}[\mathbf{y}_{n+1} + \mathbf{s}_n \mid \mathbf{s}_0 \wedge \cdots \wedge \mathbf{s}_n] \\
&= \mathrm{M}[\mathbf{y}_{n+1} \mid \mathbf{s}_0 \wedge \cdots \wedge \mathbf{s}_n] \\
&\quad + \mathrm{M}[\mathbf{s}_n \mid \mathbf{s}_0 \wedge \cdots \wedge \mathbf{s}_n] \\
&= \mathrm{M}[\mathbf{y}_{n+1} \mid \mathbf{s}_0 \wedge \cdots \wedge \mathbf{s}_n] + \mathbf{s}_n \\
&= \mathrm{M}[\mathbf{y}_{n+1} \mid \mathbf{y}_0 \wedge \cdots \wedge \mathbf{y}_n] + \mathbf{s}_n \\
&= \mathrm{M}[\mathbf{y}_{n+1}] + \mathbf{s}_n \quad \text{by independence} \\
&= \mathbf{s}_n \ \text{ if and only if } \mathrm{M}[\mathbf{y}_{n+1}] = 0.
\end{aligned}$$

A special case in which the $\mathbf{y}_k$ have identical distributions is the sequence of plays of a game of chance. A fair game is one in which the expected fortune $\mathrm{M}[\mathbf{s}_{n+1}]$ at any time $n + 1$ is equal to the actual fortune $\mathbf{s}_n$ at time n. Matching pennies is a fair game, whereas roulette is unfair. A game like roulette that is favorable to the house is a supermartingale. From the calculation above, we see that a game is fair if and only if the mean amount won in each round is zero.

(2) A particle moves on a line stopping at points whose coordinates are integers. At each step it moves n units to the right with probability $\{p_n\}$, where $n = \ldots, -2, -1, 0, 1, 2, \ldots$, only finitely many of the p_n are different from zero, $p_n \neq 0$ for some negative value and some positive value of n, and $\sum p_n = 1$. The particle's position after j steps is x_j. Set $f(s) = \sum_k p_k s^k$, and consider the positive roots of the equation $f(s) = 1$. Since $f''(s) > 0$ for $s > 0$, there are at most two roots of the equation, and since $f(1) = 1$, either one or two roots exist.

As $s \to 0$ or $\infty, f(s) \to \infty$. Hence either 1 is a minimum point or there are two positive roots. If 1 is a minimum point, then $f'(1) = 0$, and hence $\{x_j\}$ is a martingale, and if r is a root other than 1, then $\{r^{x_j}\}$ is a martingale. The details of verifying these assertions are left to the reader. We shall need to use these results later.

(3) Let Ω be the closed unit interval [0, 1] on the real line, let the measure of an interval be its length, and let $\mathscr{R}_n$ be the partition $\{[0, 2^{-n}], (2^{-n}, 2 \cdot 2^{-n}], \ldots, ((2^n - 1)2^{-n}, 1]\}$. Let f be a monotone increasing function on [0, 1] and let $\mathbf{f}_n$ be a function which is constant on the interval $(c2^{-n}, (c + 1)2^{-n}]$ and whose value at any point in the interval is

$$2^n(f((c + 1)2^{-n}) - f(c2^{-n})).$$

Thus $\mathbf{f}_n$ is an approximation to the derivative f', if it exists. The reader should verify that $(\mathbf{f}_n, \mathscr{R}_n)$ is a martingale.

3. A first martingale systems theorem

In the first example in the preceding section, we saw that martingales bear some relation to gambling. A fair game is a martingale, a game favorable to the house is a supermartingale, and a game favorable to the player is a submartingale. A gambling system is a device to take advantage of the nature of a game of chance in order to increase the player's expected fortune. Systems theorems are theorems which prove that certain classes of gambling systems do not work. For our first systems theorem, which we shall need in the proof in the next section, we require a lemma.

Lemma 3-6: Let $\{\mathscr{R}_n\}$ be an increasing sequence of partitions and let $\{\mathbf{f}_n\}$ be a sequence of random variables. Suppose R_k is a set in the Borel field $\mathscr{R}_k^*$.

If $(\mathbf{f}_n, \mathscr{R}_n)$ is a submartingale, then $\int_{R_k} \mathbf{f}_{k+1} d\mu \geq \int_{R_k} \mathbf{f}_k d\mu$.
If $(\mathbf{f}_n, \mathscr{R}_n)$ is a martingale, then $\int_{R_k} \mathbf{f}_{k+1} d\mu = \int_{R_k} \mathbf{f}_k d\mu$.
If $(\mathbf{f}_n, \mathscr{R}_n)$ is a supermartingale, then $\int_{R_k} \mathbf{f}_{k+1} d\mu \leq \int_{R_k} \mathbf{f}_k d\mu$.

Proof: We shall prove only the first assertion. By Proposition 3-2, $\int_{R_k} \mathbf{f}_{k+1} d\mu = \int_{R_k} \mathrm{M}[\mathbf{f}_{k+1} \mid \mathscr{R}_k] d\mu$, and since $\{\mathbf{f}_n, \mathscr{R}_n\}$ is a submartingale, we know that

$$\mathrm{M}[\mathbf{f}_{k+1} \mid \mathscr{R}_k] \geq \mathbf{f}_k.$$

The result follows immediately from integrating this inequality.

Proposition 3-7: Let $(\mathbf{f}_n, \mathscr{R}_n)$ be a submartingale, and suppose that $\{\boldsymbol{\epsilon}_n\}$ is a sequence of random variables such that $\boldsymbol{\epsilon}_n(\omega) = 1$ or 0 for every

ω and such that the set $\{\omega \mid \boldsymbol{\epsilon}_n(\omega) = 1\}$ is a set in $\mathscr{R}_n^*$, the Borel field generated by $\mathscr{R}_n$. Define

$$\hat{\mathbf{f}}_n = \mathbf{f}_0 + \boldsymbol{\epsilon}_0(\mathbf{f}_1 - \mathbf{f}_0) + \boldsymbol{\epsilon}_1(\mathbf{f}_2 - \mathbf{f}_1) + \cdots + \boldsymbol{\epsilon}_{n-1}(\mathbf{f}_n - \mathbf{f}_{n-1}).$$

Then $(\hat{\mathbf{f}}_n, \mathscr{R}_n)$ is a submartingale and $\mathrm{M}[\hat{\mathbf{f}}_n] \le \mathrm{M}[\mathbf{f}_n]$.

REMARK: Analogous results hold for martingales and for supermartingales, but we need only what is proved here.

PROOF OF PROPOSITION: We first show that $\mathrm{M}[\hat{\mathbf{f}}_{n+1} \mid \mathscr{R}_n] \ge \hat{\mathbf{f}}_n$. We have

$$\begin{aligned}\mathrm{M}[\hat{\mathbf{f}}_{n+1} \mid \mathscr{R}_n] &= \mathrm{M}[\hat{\mathbf{f}}_n + \boldsymbol{\epsilon}_n(\mathbf{f}_{n+1} - \mathbf{f}_n) \mid \mathscr{R}_n] \\ &= \mathrm{M}[\hat{\mathbf{f}}_n \mid \mathscr{R}_n] + \mathrm{M}[\boldsymbol{\epsilon}_n(\mathbf{f}_{n+1} - \mathbf{f}_n) \mid \mathscr{R}_n] \\ &= \hat{\mathbf{f}}_n + \mathrm{M}[\boldsymbol{\epsilon}_n(\mathbf{f}_{n+1} - \mathbf{f}_n) \mid \mathscr{R}_n].\end{aligned}$$

Since $\{\omega \mid \boldsymbol{\epsilon}_n(\omega) = 1\}$ is the union of cells of $\mathscr{R}_n$, $\boldsymbol{\epsilon}_n$ is constant on cells of $\mathscr{R}_n$. Thus by Proposition 3-3, the above expression

$$\begin{aligned}&= \hat{\mathbf{f}}_n + \boldsymbol{\epsilon}_n \mathrm{M}[(\mathbf{f}_{n+1} - \mathbf{f}_n) \mid \mathscr{R}_n] \\ &= \hat{\mathbf{f}}_n + \boldsymbol{\epsilon}_n(\mathrm{M}[\mathbf{f}_{n+1} \mid \mathscr{R}_n] - \mathbf{f}_n),\end{aligned}$$

which is

$$\ge \hat{\mathbf{f}}_n$$

because $(\mathbf{f}_n, \mathscr{R}_n)$ is a submartingale and $\boldsymbol{\epsilon}_n$ is non-negative. It remains to be shown that $\mathrm{M}[\mathbf{f}_n - \hat{\mathbf{f}}_n] \ge 0$; we prove the result by induction on n. For $n = 0$, $\mathbf{f}_0 = \hat{\mathbf{f}}_0$ and $\mathrm{M}[\mathbf{f}_0 - \hat{\mathbf{f}}_0] = 0$. Suppose we have proved that $\int_\Omega (\mathbf{f}_k - \hat{\mathbf{f}}_k)d\mu \ge 0$. Then we have $\hat{\mathbf{f}}_{k+1} = \hat{\mathbf{f}}_k + \boldsymbol{\epsilon}_k(\mathbf{f}_{k+1} - \mathbf{f}_k)$, and when we subtract both sides from $\mathbf{f}_{k+1}$, we get

$$\begin{aligned}\mathbf{f}_{k+1} - \hat{\mathbf{f}}_{k+1} &= \mathbf{f}_{k+1} - \hat{\mathbf{f}}_k - \boldsymbol{\epsilon}_k(\mathbf{f}_{k+1} - \mathbf{f}_k) \\ &= (1 - \boldsymbol{\epsilon}_k)(\mathbf{f}_{k+1} - \mathbf{f}_k) + (\mathbf{f}_k - \hat{\mathbf{f}}_k).\end{aligned}$$

Thus

$$\begin{aligned}\int_\Omega (\mathbf{f}_{k+1} - \hat{\mathbf{f}}_{k+1})d\mu &\ge \int_\Omega (1 - \boldsymbol{\epsilon}_k)(\mathbf{f}_{k+1} - \mathbf{f}_k)d\mu \quad \text{by hypothesis} \\ &= \int_{\{\boldsymbol{\epsilon}_k(\omega) = 0\}} (\mathbf{f}_{k+1} - \mathbf{f}_k)d\mu \\ &\ge 0 \quad \text{by Lemma 3-6.}\end{aligned}$$

To see the connection of Proposition 3-7 to gambling systems, we again consider Example 1 of Section 3-2. We take the partial sums $\mathbf{s}_n$ as the $\mathbf{f}_n$, and we observe that the differences $\mathbf{s}_k - \mathbf{s}_{k-1}$ in the definition of $\hat{\mathbf{s}}_n$ are simply the random variables $\mathbf{y}_k$. The $\hat{\mathbf{s}}_n$ become modified fortunes, fortunes changed by not playing in every round of the game. When $\epsilon_k = 1$, the player participates in the $k + 1$st game; and when $\epsilon_k = 0$, he does not. The whole set of ϵ's, therefore, represents a gambling system; the condition that the set of paths for which $\epsilon_n(\omega) = 1$ be the union of cells of $\mathscr{R}_n$ is the condition that the system not depend on any knowledge of the future. For the special case in which the process is a submartingale, the expected fortune after time $n + 1$ is greater than or equal to the fortune at time n and the game is favorable for the player. The content of Proposition 3-7 is that the player's expected fortune after time $n + 1$ would not have been increased by a system which caused him not to play in certain rounds.

4. Martingale convergence and a second systems theorem

In this section we shall prove two theorems which will be of great use in our treatment of Markov chains. The two results will indicate the value of recognizing martingales when they appear in our later work.

Definition 3-8: Let $\{\mathbf{f}_n\}$ be a sequence of random variables defined on points ω, and let r and s be numbers with $r < s$. An **upcrossing** of $[r, s]$ is said to occur between $n - k$ and n for the point ω if these conditions are satisfied:

(1) $\mathbf{f}_{n-k}(\omega) \leq r$
(2) $r < \mathbf{f}_{n-k+m}(\omega) < s \quad$ for $0 < m < k$
(3) $\mathbf{f}_n(\omega) \geq s$.

The reader should notice that no two upcrossings overlap.

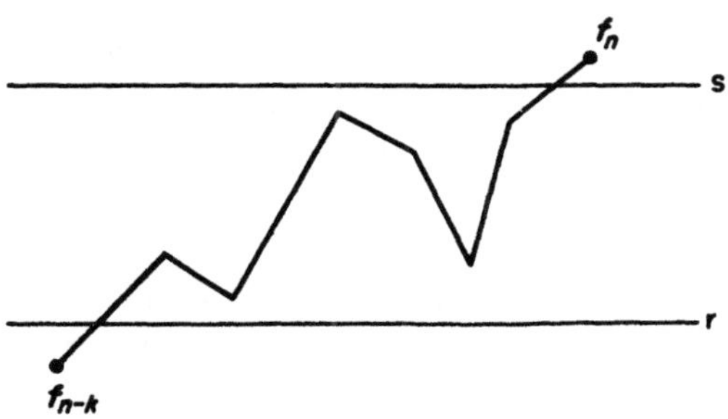

After three preliminary results, we proceed with a proof of the Martingale Convergence Theorem. Proposition 3-11 is known as the Upcrossing Lemma.

Lemma 3-9: If $(\mathbf{f}_n, \mathscr{R}_n)$ and $(\mathbf{g}_n, \mathscr{R}_n)$ are submartingales, then so is $(\sup(\mathbf{f}_n, \mathbf{g}_n), \mathscr{R}_n)$.

PROOF:

$$M[|\sup(\mathbf{f}_n, \mathbf{g}_n)|] \le M[\sup(|\mathbf{f}_n|, |\mathbf{g}_n|)]$$

$$= \int_{|\mathbf{f}_n| \ge |\mathbf{g}_n|} |\mathbf{f}_n| d\mu + \int_{|\mathbf{f}_n| < |\mathbf{g}_n|} |\mathbf{g}_n| d\mu < \infty.$$

The function $\sup(\mathbf{f}_n, \mathbf{g}_n)$ is clearly constant on cells of $\mathscr{R}_n$ if $\mathbf{f}_n$ and $\mathbf{g}_n$ each are. Furthermore,

$$M[\sup(\mathbf{f}_{n+1}, \mathbf{g}_{n+1}) \mid \mathscr{R}_n] \ge M[\mathbf{f}_{n+1} \mid \mathscr{R}_n] \ge \mathbf{f}_n$$

and

$$M[\sup(\mathbf{f}_{n+1}, \mathbf{g}_{n+1}) \mid \mathscr{R}_n] \ge M[\mathbf{g}_{n+1} \mid \mathscr{R}_n] \ge \mathbf{g}_n$$

so that

$$M[\sup(\mathbf{f}_{n+1}, \mathbf{g}_{n+1}) \mid \mathscr{R}_n] \ge \sup(\mathbf{f}_n, \mathbf{g}_n).$$

Lemma 3-10: If $(\mathbf{f}_n, \mathscr{R}_n)$ is a martingale, $M[\mathbf{f}_n] = M[\mathbf{f}_{n-1}]$. If $(\mathbf{f}_n, \mathscr{R}_n)$ is a supermartingale, $M[\mathbf{f}_n] \le M[\mathbf{f}_{n-1}]$. If $(\mathbf{f}_n, \mathscr{R}_n)$ is a submartingale, $M[\mathbf{f}_n] \ge M[\mathbf{f}_{n-1}]$.

PROOF: The result is immediate from Lemma 3-6 when R_k is taken as Ω.

Proposition 3-11: Let $(\mathbf{f}_j, \mathscr{R}_j)$ be a submartingale, and let $\boldsymbol{\beta}(\omega)$ be the number of upcrossings of $[r, s]$ between times 0 and n. Then

$$M[\boldsymbol{\beta}] \le \frac{M[(\mathbf{f}_n - r)^+]}{s - r} \le \frac{M[|\mathbf{f}_n|] + |r|}{s - r}.$$

PROOF: Consider first the special case $\mathbf{f}_k \ge 0$ for $0 \le k \le n$ and $r = 0$. Let $\hat{\mathbf{f}}_k$ be defined as in Proposition 3-7 with the $\boldsymbol{\epsilon}$'s to be specified. For a given path ω, define, by induction on m, $\boldsymbol{\epsilon}_m(\omega) = 1$ whenever $\mathbf{f}_m(\omega) = 0$, and let $\boldsymbol{\epsilon}_{m+k}(\omega) = 1$ as long as $\mathbf{f}_{m+k}(\omega) < s$. As soon as $\mathbf{f}_m(\omega) \ge s$, require that $\boldsymbol{\epsilon}_m(\omega) = 0$ until m is large enough so that $\mathbf{f}_m(\omega) = 0$ again. Then $\hat{\mathbf{f}}_m$ measures the increase in the sequence $\mathbf{f}_0(\omega), \ldots, \mathbf{f}_n(\omega)$ during upcrossings (plus a possible "partial upcrossing" at the end) and is greater than or equal to the minimum increase in each upcrossing multiplied by the number of upcrossings. That is,

$$s \cdot \boldsymbol{\beta}(\omega) \le \hat{\mathbf{f}}_n(\omega).$$

Hence

$$s \cdot M[\boldsymbol{\beta}] \le M[\hat{\mathbf{f}}_n],$$

which by Proposition 3-7 is

$$\le M[\mathbf{f}_n].$$

Therefore,

$$M[\boldsymbol{\beta}] \leq (1/s)M[\mathbf{f}_n]$$

and the special case is proved. For a general sequence $\{\mathbf{f}_n\}$ and general r, consider the function $(\mathbf{f}_k - r)^+$, which is the supremum of the zero function and the function $\mathbf{f}_k - r$. It is readily verified that constant functions are martingales and that the difference of a submartingale and a martingale is a submartingale. Thus $(\mathbf{f}_k - r, \mathscr{R}_k)$ is a submartingale and by Lemma 3-9 $((\mathbf{f}_k - r)^+, \mathscr{R}_k)$ is a submartingale. Applying the special case proved above to $(\mathbf{f}_k - r)^+$ and upcrossings of $[0, s - r]$, we find

$$\begin{aligned} (s - r)M[\boldsymbol{\beta}] &\leq M[(\mathbf{f}_n - r)^+] \\ &\leq M[|\mathbf{f}_n - r|] \\ &\leq M[|\mathbf{f}_n| + |r|] \\ &= M[|\mathbf{f}_n|] + |r| \end{aligned}$$

and the proof of the Upcrossing Lemma is complete.

Theorem 3-12: If $(\mathbf{f}_n, \mathscr{R}_n)$ is a submartingale with the property that $M[|\mathbf{f}_n|] \leq K < \infty$ for all n, then

$$\lim_{n \to \infty} \mathbf{f}_n(\omega)$$

exists *and is finite* for almost all points ω.

PROOF: Failure of almost-everywhere convergence means that there exists a set of points ω of positive measure for which the sequence diverges. At least one of two things must happen. Either $\mathbf{f}_n(\omega)$ for each fixed ω in a set of positive measure oscillates infinitely often above and below rationals $r(\omega)$ and $s(\omega)$ with $r(\omega) < s(\omega)$, or else $|\mathbf{f}_n(\omega)|$ diverges to $+\infty$ on a set of positive measure. We consider the cases separately.

(1) Suppose $|\mathbf{f}_n(\omega)|$ diverges to $+\infty$ on a set E of positive measure m. Then by Fatou's Theorem

$$\begin{aligned} \int_E \liminf_n |\mathbf{f}_n(\omega)| d\mu &\leq \liminf_n \int_E |\mathbf{f}_n(\omega)| d\mu \\ &\leq \liminf_n \int_\Omega |\mathbf{f}_n(\omega)| d\mu \\ &= \liminf_n M[|\mathbf{f}_n|] \\ &\leq K. \end{aligned}$$

But lim inf $|\mathbf{f}_n(\omega)| = +\infty$ on E, and E has positive measure m. Thus the left side of the inequality is infinite, and we have arrived at a contradiction.

(2) Suppose $\mathbf{f}_n(\omega)$ oscillates infinitely often above and below rationals $r(\omega)$ and $s(\omega)$ on a set of positive measure m. Order the set of all pairs of rationals (which is a denumerable set) and call the kth pair q_k. Consider the denumerable family of sets A_k defined by $A_k = \{\omega \mid \mathbf{f}_n(\omega)$ oscillates infinitely often above and below the rationals of the pair $q_k\}$. It is possible for more than one set to have the same point in it, but, on the other hand, every point ω for which $\mathbf{f}_n(\omega)$ oscillates infinitely often is in some A_k. Therefore,

$$\sum \mu(A_k) \geq \mu(\bigcup A_k) = m > 0$$

and there must exist a t for which $\mu(A_t) > 0$. That is, for every ω in a set A_t of positive measure, $\mathbf{f}_n(\omega)$ oscillates infinitely often above and below fixed rationals r and s with $r < s$. Let $\boldsymbol{\beta}_n(\omega)$ be the number of upcrossings of $[r, s]$ by $\mathbf{f}_0(\omega), \ldots, \mathbf{f}_n(\omega)$. By Proposition 3-11,

$$\begin{aligned} \mathrm{M}[\boldsymbol{\beta}_n] &\leq \frac{\mathrm{M}[|\mathbf{f}_n|] + |r|}{s - r} \\ &\leq \frac{K + |r|}{s - r} \\ &= c \quad \text{for every } n. \end{aligned}$$

Furthermore, the $\boldsymbol{\beta}_n$ are non-negative and increasing with n to a function $\boldsymbol{\beta}$, so that $\mathrm{M}[\boldsymbol{\beta}] = \lim \mathrm{M}[\boldsymbol{\beta}_n] \leq c$ by the Monotone Convergence Theorem. But $\mathrm{M}[\boldsymbol{\beta}] = +\infty$ since there are infinitely many upcrossings on a set of positive measure. This contradiction establishes the Martingale Convergence Theorem.

Corollary 3-13: Every non-negative supermartingale converges to a finite-valued function almost everywhere. In particular, every non-negative martingale converges almost everywhere.

PROOF: If $(\mathbf{f}_n, \mathscr{R}_n)$ is a non-negative supermartingale, then $(-\mathbf{f}_n, \mathscr{R}_n)$ is a submartingale. Since $\mathbf{f}_n \geq 0$, $|-\mathbf{f}_n| = \mathbf{f}_n$ and hence $\mathrm{M}[|-\mathbf{f}_n|] \leq \mathrm{M}[\mathbf{f}_0]$ by Lemma 3-10. Therefore, $\{-\mathbf{f}_n\}$ converges almost everywhere by Theorem 3-12, and so does $\{\mathbf{f}_n\}$.

We postpone a discussion of applications of Theorem 3-12 and Corollary 3-13 to the next section. We shall find that the corollary is used more frequently than the theorem itself.

A random time which is finite almost everywhere is called a **random stopping time** or simply a **stopping time**. If $\mathbf{t}$ is a stopping time, then the set $\bigcap_{n=1}^{\infty} \{\omega \mid \mathbf{t}(\omega) \geq n\}$ has measure zero. If $\{\mathbf{f}_n\}$ is a sequence of random variables, we define a function $\mathbf{f}_t$ almost everywhere by

$$\mathbf{f}_t(\omega) = \mathbf{f}_n(\omega) \quad \text{if} \quad \mathbf{t}(\omega) = n.$$

Since

$$\{\omega | \mathbf{f}_t(\omega) < c\} = \bigcup_{n=1}^{\infty} (\{\omega \mid \mathbf{t}(\omega) = n\} \cap \{\omega \mid \mathbf{f}_n(\omega) < c\}),$$

$\mathbf{f}_t$ is a random variable.

Lemma 3-14: If $(\mathbf{f}_n, \mathscr{R}_n)$ is a martingale and if $\mathbf{t}$ is a stopping time for which $\int_\Omega \mathbf{f}_t d\mu$ exists, then for any n

$$\int_\Omega \mathbf{f}_t d\mu = \int_{\{\mathbf{t} \leq n\}} \mathbf{f}_n d\mu + \int_{\{\mathbf{t} > n\}} \mathbf{f}_t d\mu.$$

PROOF: We have

$$\int_\Omega \mathbf{f}_t d\mu = \sum_{k=0}^{n} \int_{\{\mathbf{t} = k\}} \mathbf{f}_t d\mu + \int_{\{\mathbf{t} > n\}} \mathbf{f}_t d\mu$$

$$= \sum_{k=0}^{n} \int_{\{\mathbf{t} = k\}} \mathbf{f}_k d\mu + \int_{\{\mathbf{t} > n\}} \mathbf{f}_t d\mu,$$

which by Lemma 3-6

$$= \sum_{k=0}^{n} \int_{\{\mathbf{t} = k\}} \mathbf{f}_n d\mu + \int_{\{\mathbf{t} > n\}} \mathbf{f}_t d\mu$$

$$= \int_{\{\mathbf{t} \leq n\}} \mathbf{f}_n d\mu + \int_{\{\mathbf{t} > n\}} \mathbf{f}_t d\mu.$$

Theorem 3-15: If $(\mathbf{f}_n, \mathscr{R}_n)$ is a martingale and if $\mathbf{t}$ is a stopping time, then $\mathrm{M}[\mathbf{f}_t] = \mathrm{M}[\mathbf{f}_0]$ if

(1) $\mathrm{M}[|\mathbf{f}_t|] < \infty$, and

(2) $\lim_n \int_{\{\mathbf{t} > n\}} \mathbf{f}_n d\mu = 0.$

REMARK: Analogous results hold for submartingales and for supermartingales. Inequalities replace the equality in the conclusion.

Proof of theorem: By (1), $\int_\Omega \mathbf{f}_\mathbf{t} d\mu$ exists, so that Lemma 3-14 applies. Thus, for any n,

$$\int_\Omega \mathbf{f}_\mathbf{t} d\mu = \int_{\{\mathbf{t} \le n\}} \mathbf{f}_n d\mu + \int_{\{\mathbf{t} > n\}} \mathbf{f}_\mathbf{t} d\mu$$

$$= \int_\Omega \mathbf{f}_n d\mu - \int_{\{\mathbf{t} > n\}} \mathbf{f}_n d\mu + \int_{\{\mathbf{t} > n\}} \mathbf{f}_\mathbf{t} d\mu,$$

which by Lemma 3-10

$$= \int_\Omega \mathbf{f}_0 d\mu - \int_{\{\mathbf{t} > n\}} \mathbf{f}_n d\mu + \int_{\{\mathbf{t} > n\}} \mathbf{f}_\mathbf{t} d\mu.$$

Using condition (1) together with Corollary 1-17 and the complete additivity of the integral as a set function, we see that

$$\lim_n \int_{\{\mathbf{t} > n\}} \mathbf{f}_\mathbf{t} d\mu = 0.$$

Since $\int_{\{\mathbf{t} > n\}} \mathbf{f}_n d\mu \to 0$ by hypothesis, we have $\int_\Omega \mathbf{f}_\mathbf{t} d\mu = \int_\Omega \mathbf{f}_0 d\mu$.

Corollary 3-16: Suppose $(\mathbf{f}_n, \mathscr{R}_n)$ is a martingale defined on a space Ω of finite total measure and $\mathbf{t}$ is a stopping time. If $|\mathbf{f}_n| \le K$ for all n, then $M[\mathbf{f}_\mathbf{t}] = M[\mathbf{f}_0]$.

Proof: We must show the two conditions of Theorem 3-15 are satisfied. For (1) we have $|\mathbf{f}_\mathbf{t}| \le K$ by definition, and hence $\mathbf{f}_\mathbf{t}$ is integrable. For (2) we have

$$\left| \int_{\{\mathbf{t} > n\}} \mathbf{f}_n d\mu \right| \le \int_{\{\mathbf{t} > n\}} |\mathbf{f}_n| d\mu$$

$$\le \int_{\{\mathbf{t} > n\}} K d\mu$$

$$= K\mu(\{\omega \mid \mathbf{t}(\omega) > n\})$$

$$\to 0.$$

In terms of gambling systems, the result of Lemma 3-10, namely that for martingales $M[\mathbf{f}_n] = M[\mathbf{f}_0]$, states that the expected fortune at any fixed stopping time is equal to the initial fortune if the game is fair. That is, the fairness of a game is not altered by deciding in advance to stop playing at some fixed time. But what about a system where the player stops according to how the game is going? The system he adopts is represented by the random time $\mathbf{t}$, and Theorem 3-15 and

Corollary 3-16 give sufficient criteria for the game still to be fair. Corollary 3-16 by itself is a general result; it covers the situation, for example, where the game stops if either the player or the house goes bankrupt. If the game does stop under such circumstances, the corollary states that the fairness of the game is not altered by any gambling system whatsoever. Similar remarks apply to supermartingales. If the amount of money that a player has is limited, no system that he adopts for stopping according to how the game is going will make an unfair game favorable.

The following proposition is useful in proving that certain random times are stopping times.

Proposition 3-17: Let $(\mathbf{f}_n, \mathscr{R}_n)$ be a martingale, let $\mathbf{t}$ be a random time, and let $\hat{\mathbf{f}}_n$ be the stopped process with

$$\hat{\mathbf{f}}_n(\omega) = \mathbf{f}_{\min(n,\mathbf{t}(\omega))}(\omega).$$

Then $(\hat{\mathbf{f}}_n, \mathscr{R}_n)$ is a martingale.

PROOF: We first note that $M[|\hat{\mathbf{f}}_n|] < \infty$ since $|\hat{\mathbf{f}}_n| \leq \sum_{j=0}^{n} |\mathbf{f}_j|$ and each $\mathbf{f}_j$ is integrable. Next, let R be a cell in $\mathscr{R}_n$ with $\mu(R) \neq 0$. In R we have

$$\begin{aligned}
M[\hat{\mathbf{f}}_{n+1} \mid \mathscr{R}_n] &= \frac{1}{\mu(R)} \int_R \hat{\mathbf{f}}_{n+1} d\mu \\
&= \frac{1}{\mu(R)} \left[\int_{R \cap \{\mathbf{t} \leq n\}} \hat{\mathbf{f}}_{n+1} d\mu + \int_{R \cap \{\mathbf{t} > n\}} \hat{\mathbf{f}}_{n+1} d\mu \right] \\
&= \frac{1}{\mu(R)} \left[\int_{R \cap \{\mathbf{t} \leq n\}} \hat{\mathbf{f}}_n d\mu + \int_{R \cap \{\mathbf{t} > n\}} \mathbf{f}_{n+1} d\mu \right]
\end{aligned}$$

by definition of $\hat{\mathbf{f}}_n$. Since $(\mathbf{f}_n, \mathscr{R}_n)$ is a martingale and $\{\mathbf{t} > n\}$ is in $\mathscr{R}_n$, the above expression by Lemma 3-6 is

$$\begin{aligned}
&= \frac{1}{\mu(R)} \left[\int_{R \cap \{\mathbf{t} \leq n\}} \hat{\mathbf{f}}_n d\mu + \int_{R \cap \{\mathbf{t} > n\}} \mathbf{f}_n d\mu \right] \\
&= \frac{1}{\mu(R)} \left[\int_{R \cap \{\mathbf{t} \leq n\}} \hat{\mathbf{f}}_n d\mu + \int_{R \cap \{\mathbf{t} > n\}} \hat{\mathbf{f}}_n d\mu \right] \\
&= \frac{1}{\mu(R)} \int_R \hat{\mathbf{f}}_n d\mu.
\end{aligned}$$

The last expression equals $\hat{\mathbf{f}}_n$ since $\hat{\mathbf{f}}_n$ is constant on R.

The application of Proposition 3-17 is this: Let $(\mathbf{f}_n, \mathscr{R}_n)$ be an integer-valued martingale, and let S be a set of integers. Let the martingale almost surely have the property that it can be constant from some time on only for values in S (and possibly for no values). We stop $\mathbf{f}_n$ the first time it takes on a value in S. That is, we let $\mathbf{t}$ be the first time that $\mathbf{f}_n$ is in S, and we introduce the stopped process $\hat{\mathbf{f}}_n$. If the values of $\hat{\mathbf{f}}_n$ are bounded from below or from above, then the "stopped process," is almost sure to stop. The proof proceeds as follows.

First, assume that $\hat{\mathbf{f}}_n \geq 0$. Then $(\hat{\mathbf{f}}_n, \mathscr{R}_n)$ is a non-negative martingale, by Proposition 3-17, which must converge a.e. to a finite value depending on ω. Since by hypothesis these values must be in S for a.e. ω, the process $\{\hat{\mathbf{f}}_n\}$ almost surely stops. Next, if $\{\hat{\mathbf{f}}_n\}$ is bounded below, apply the result for non-negative martingales to $\hat{\mathbf{f}}_n$ plus a suitable constant. Finally, if $\hat{\mathbf{f}}_n \leq c$, apply the result for $\{\hat{\mathbf{f}}_n\}$ bounded below to $\{-\hat{\mathbf{f}}_n\}$.

These results are used in the next section in Examples 1 and 4. In Example 1 a fair game is stopped when it leaves a certain finite set, whereas in Example 4 it is stopped when a positive value is reached for the first time. By the above argument these random times are stopping times.

Proposition 3-18: Let $\mathscr{R}_0 \subset \mathscr{R}_1 \subset \cdots$ be an increasing sequence of partitions and let $\mathscr{R}^*$ be the smallest augmented Borel field containing the field $\bigcup \mathscr{R}_n{}^*$. Let $\mathbf{f}$ be a random variable measurable with respect to $\mathscr{R}^*$ and having finite mean, and set $\mathbf{g}_n = \mathrm{M}[\mathbf{f} \mid \mathscr{R}_n]$. Then $(\mathbf{g}_n, \mathscr{R}_n)$ is a martingale, and

$$\lim_{n\to\infty} \mathbf{g}_n = \mathbf{f}$$

almost everywhere.

PROOF: We may assume that $\mathbf{f} \geq 0$ since the general case follows by considering $\mathbf{f}^+$ and $\mathbf{f}^-$ separately. Then $\mathbf{g}_n \geq 0$ and $\mathrm{M}[|\mathbf{g}_n|] = \mathrm{M}[\mathbf{f}] < \infty$ by conclusion (2) of Proposition 3-3. Since, in addition,

$$\begin{aligned} \mathrm{M}[\mathbf{g}_{n+1} \mid \mathscr{R}_n] &= \mathrm{M}[\mathrm{M}[\mathbf{f} \mid \mathscr{R}_{n+1}] \mid \mathscr{R}_n] \\ &= \mathrm{M}[\mathbf{f} \mid \mathscr{R}_n] \quad \text{by (1) of Proposition 3-3} \\ &= \mathbf{g}_n, \end{aligned}$$

we see that $(\mathbf{g}_n, \mathscr{R}_n)$ is a non-negative martingale. By Corollary 3-13, $\mathbf{g} = \lim \mathbf{g}_n$ exists a.e. We shall prove that $\mathbf{g} = \mathbf{f}$ a.e.

First we prove that the $\mathbf{g}_n$ are uniformly integrable. Let $\epsilon > 0$ be given. Choose, by Proposition 1-47, $\delta > 0$ small enough so that $\mu(E) < \delta$ implies $\int_E \mathbf{f}d\mu < \epsilon$. Now

$$N\mu(\{\mathbf{g}_n \geq N\}) \leq \mathrm{M}[\mathbf{g}_n] = \mathrm{M}[\mathbf{f}]$$

or

$$\mu(\{\mathbf{g}_n \geq N\}) \leq \frac{\mathrm{M}[\mathbf{f}]}{N}.$$

Choose N large enough so that the right side is less than δ. Then for all n we have

$$\int_{\{\mathbf{g}_n \geq N\}} \mathbf{f} d\mu < \epsilon.$$

Since

$$\int_{\{\mathbf{g}_n \geq N\}} \mathbf{g}_n d\mu = \int_{\{\mathbf{g}_n \geq N\}} \mathbf{f} d\mu$$

by Lemma 3-6, we conclude the $\mathbf{g}_n$ are uniformly integrable.

Let E be any subset of $\mathscr{R}_n{}^*$. For $m \geq n$,

$$\int_E \mathbf{g}_m d\mu = \int_E \mathbf{f} d\mu.$$

By uniform integrability and Proposition 1-52,

$$\lim_m \int_E \mathbf{g}_m d\mu = \int_E \mathbf{g} d\mu.$$

Therefore

$$\int_E \mathbf{f} d\mu = \int_E \mathbf{g} d\mu$$

for all E in $\bigcup \mathscr{R}_n{}^*$. The two sides of this last equation, considered as set functions, are equal on $\bigcup \mathscr{R}_n{}^*$. By the uniqueness half of Theorem 1-19 they must be equal on $\mathscr{R}^*$. That is, $\mathbf{f}$ and $\mathbf{g}$ are measurable with respect to $\mathscr{R}^*$ and satisfy

$$\int_E \mathbf{f} d\mu = \int_E \mathbf{g} d\mu.$$

Taking E successively to be the set where $\mathbf{f} \geq \mathbf{g}$ and the set where $\mathbf{f} \leq \mathbf{g}$ and applying Corollary 1-40, we find that $\mathbf{f} = \mathbf{g}$ a.e.

5. Examples of convergent martingales

Four examples will serve at present to illustrate Theorems 3-12 and 3-15. Each of the first three refers to the correspondingly numbered example in Section 2.

Example 1: The sequence $\{\mathbf{s}_n\}$ of nth partial sums of the independent random variables $\mathbf{y}_n$ forms a martingale if $\mathrm{M}[\mathbf{y}_n] = 0$. Suppose the $\mathbf{y}_n$ have identical distributions and mean zero, suppose they assume only the values 0, 1, and -1, and suppose that the process is stopped whenever $\mathbf{s}_n(\omega) = M$ or $\mathbf{s}_n(\omega) = -N$. (Mean zero implies that the outcomes $+1$ and -1 are equally likely.) The player of the fair game is ruined if $s_n(\omega)$ ever equals $-N$ and he breaks the bank if $s_n(\omega) = M$. Set

$$p = \text{Pr[player breaks the bank]}$$

and

$$q = \text{Pr[player is eventually ruined]}.$$

By the remarks following Proposition 3-17, $p + q = 1$. In this situation, Corollary 3-16 applies and

$$\begin{aligned} 0 = M[s_0] = M[s_t] &= p \cdot M + q(-N) \\ &= pM + (1 - p)(-N) \end{aligned}$$

or

$$p = \frac{N}{M + N}$$

and

$$q = \frac{M}{M + N}.$$

EXAMPLE 2: With the particle moving on the line, suppose there is an $r > 0$ which is a root of the equation $f(s) = 1$. We shall assume that $r < 1$. Then $\{r^{x_n}\}$ is a non-negative martingale, and by Corollary 3-13, $\{r^{x_n}\}$ converges to a limiting function a.e. Since x_n is integer-valued, this convergence means either

(1) for almost all ω, there is an N such that if $n \geq N$, then $x_n = x_N$, or
(2) $\lim x_n = +\infty$.

Now $x_N = x_{N+1} = x_{N+2} = \cdots = x_{N+k}$ means that the particle fails to move for k consecutive steps. Since such a thing happens with probability $p_0{}^k < 1$, the probability that $x_n = x_N$ for all $n \geq N$ is zero. Thus case 1 is eliminated, and we have established that $\lim x_n = +\infty$ a.e. That is, for any k and for almost all ω, $x_n(\omega) = k$ for only finitely many n.

EXAMPLE 3: When the f_n's are the difference quotients of a monotone function f, the pair $(f_n, \mathscr{R}_n)$ forms a non-negative martingale. By Corollary 3-13, f_n converges to a limiting function at almost all points. The limiting function will turn out to be the derivative of f. However, our argument considered only nested partitions, and hence it provides only part of the proof of the existence of f' a.e.

Next we consider an example where Theorem 3-15 is *not* applicable.

EXAMPLE 4: Suppose that a player plays the fair game of Example 1 and that he stops the first time that he is ahead. The process is stopped when $s_n(\omega) = 1$. We have already seen that this is a stopping time. Then $s_0 = 0$, and $s_t = 1$ a.e. Hence $M[s_0] = 0 \neq 1 = M[s_t]$. Why is the theorem not applicable? Condition (1) is clearly satisfied. However,

$$0 = \int s_n d\mu = \int_{\{t \le n\}} s_n d\mu + \int_{\{t > n\}} s_n d\mu.$$

The first term equals the probability that 1 has been reached and tends to 1. Hence the second term tends to -1, not to 0. Thus condition (2) is violated.

In practice this gambling strategy cannot be implemented, since the gambler would need infinite capital to be able to absorb arbitrarily large losses.

6. Law of Large Numbers

The Strong Law of Large Numbers, which may be derived from the Martingale Convergence Theorem, is formulated as follows.

Theorem 3-19: Let $\{y_n\}$ be a sequence of independent identically distributed random variables with finite mean $a = M[y_k]$. If $s_n = y_1 + \cdots + y_n$ and $s^* = s_n/n$, then

$$\Pr[\lim_n s_n^* = a] = 1.$$

We shall prove the theorem for the special case where the random variables have finite range; say, $\Pr[y_k = j] = p_j$ for a finite number of j's. For more generally applicable proofs the reader is referred to the bibliography. (See Feller [1957], pp. 244–245, for an elementary proof in the case y_n is denumerable-valued; and see Doob [1953], pp. 334–342, for a general proof using the Martingale Convergence Theorem.)

We introduce a useful tool, the generating function

$$\varphi(t) = \sum_j p_j \cdot t^j.$$

It is a well-behaved function satisfying $\varphi(1) = 1$ and $\varphi'(1) = a$. Let

$$f(k, n) = \frac{t^k}{\varphi(t)^n}$$

for some $t > 0$ to be specified. We shall show that $\{f(s_n, n)\}$ is a martingale. The conditional mean of $f(s_{n+1}, n + 1)$ given $s_n = k$ is

$$\frac{\sum_j p_j \cdot t^{k+j}}{\varphi(t)^{n+1}} = f(k, n) \cdot \frac{\sum_j p_j \cdot t^j}{\varphi(t)} = f(k, n) = f(s_n, n).$$

Since $M[f(s_0, 0)] = 1 < \infty$, $\{f(s_n, n)\}$ is a martingale, and it is clearly non-negative. Thus, by the Martingale Convergence Theorem, $f(s_n, n)$ converges to a finite limit a.e., where

$$f(s_n, n) = t^{s_n}/\varphi(t)^n = [t^{s_n^*}/\varphi(t)]^n.$$

Fix $\epsilon > 0$, let $b = a + \epsilon$, and form the function $g(t) = t^b/\varphi(t)$. Since $g(1) = 1$ and $g'(1) = b - a > 0$, we have $g(t_0) > 1$ for some sufficiently small t_0 greater than 1. If $s_n^*(\omega) \geq b$, we have

$$[t_0^{s_n^*(\omega)}/\varphi(t_0)] \geq g(t_0) > 1,$$

and hence if $s_n^*(\omega) \geq b$ for infinitely many n, then $f(s_n(\omega), n)$ has a subsequence tending to $+\infty$. By the convergence of f, we conclude that

$$\limsup s_n^*(\omega) \leq b = a + \epsilon$$

a.e. Similarly, by choosing a suitable $t < 1$ we would find that

$$\liminf s_n^*(\omega) \geq a - \epsilon$$

a.e. Since ϵ is arbitrary, s_n^* converges to a with probability one.

7. Problems

1. Show that if $\{f_n\}$ are denumerable-valued independent random variables, then
$$M[f_{n+1} \mid f_0 \wedge \cdots \wedge f_n] = M[f_{n+1}].$$
Show also that
$$\Pr[f_0 \in A_0 \wedge f_1 \in A_1 \wedge \cdots \wedge f_n \in A_n] = \Pr[f_0 \in A_0] \cdot \cdots \cdot \Pr[f_n \in A_n].$$

2. Let $\{f_n\}$ be a sequence of denumerable-valued independent random variables and let $\{g_n\}$ be a sequence of Borel-measurable functions defined on the real line. Show that $\{g_n(f_n)\}$ is a sequence of independent random variables. If the f_n are identically distributed and all the g_n's are equal to g, show that the $g(f_n)$'s are identically distributed.

3. Verify that Examples 2 and 3 of Section 2 are martingales.

4. Prove that if $(f_n, \mathcal{R}_n)$ and $(g_n, \mathcal{R}_n)$ are martingales, then so is $(f_n + g_n, \mathcal{R}_n)$. Does the same hold for $(f_n g_n, \mathcal{R}_n)$?

5. Prove that if $(f_n, \mathcal{R}_n)$ and $(g_n, \mathcal{R}_n)$ are non-negative supermartingales, then so is $(\min(f_n, g_n), \mathcal{R}_n)$.

6. Prove that if $(f_n, \mathcal{R}_n)$ is a martingale, then $(|f_n|, \mathcal{R}_n)$ is a submartingale. If the f_n form a martingale on their cross partitions, do the $|f_n|$ form a submartingale on their own cross partitions?

7. Let $(f_n, \mathcal{R}_n)$ be a submartingale. Prove that, for any $c > 0$,
$$c \Pr\left[\max_{i \leq n} f_i \geq c\right] \leq M[|f_n|].$$
[*Hint:* Take as stopping time the first time c is surpassed, or n, whichever comes first.]

8. Prove that every submartingale can be written as the sum of a martingale and an increasing submartingale. [*Hint:* If $(\mathbf{x}_n, \mathscr{R}_n)$ is given, put $\mathbf{f}_0 = 0$, $\mathbf{f}_n = \mathrm{M}[\mathbf{x}_n \mid \mathscr{R}_{n-1}] - \mathbf{x}_{n-1}$, $\mathbf{z}_n = \mathbf{f}_0 + \cdots + \mathbf{f}_n$, and $\mathbf{y}_n = \mathbf{x}_n - \mathbf{z}_n$.]
9. Consider the following stochastic process: A white and a black ball are placed in an urn. One ball is drawn, and this ball is replaced by two of the same color. Let $\mathbf{f}_n$ be the fraction of white balls after n experiments.
   (a) Prove that $\{\mathbf{f}_n\}$ is a martingale.
   (b) Prove that it converges.
   (c) Prove that the limiting distribution has mean $\frac{1}{2}$.
   (d) Prove that the probability of ever reaching a fraction $\frac{3}{4}$ of white balls is at most $\frac{2}{3}$. [*Hint:* Use Problem 7.]
10. We consider an experiment with each outcome one of two possible outcomes H or T. We have two different hypotheses A and B as to how the underlying measure for a stochastic process should be assigned. For a given sequence HHT...H we denote by $p_n(\text{HHT}\ldots\text{H})$ the assignment under hypothesis A and by $r_n(\text{HHT}\ldots\text{H})$ the assignment under hypothesis B. Let f_n be defined by

$$f_n(\text{HHT}\ldots\text{H}) = \frac{p_n(\text{HHT}\ldots\text{H})}{r_n(\text{HHT}\ldots\text{H})}.$$

   (a) Show that if the measure is defined by hypothesis B, then $\{f_n\}$ is a martingale and hence converges a.e.
   (b) Specialize to the case of tossing a biased coin. Let hypotheses A and B be that the probability of heads is, respectively, p and r. Show that if the measure is defined by hypothesis B, then $\{f_r\}$ converges to 0 a.e. if $p \neq r$.

Problems 11 to 13 concern a type of stochastic process employed by psychologists in learning theory. The state space consists of the rational points on the unit interval, and we are given two rational parameters, $0 < b \leq a < 1$. From a point x the process moves to $bx + (1 - a)$ with probability x, or to bx with probability $1 - x$. It is started at an interior point x_0.

11. Show that if $b = a$, then $\{x_n\}$ is a martingale.
12. Prove that the process converges either to 0 or to 1, and compute the probability of going to 1 as a function of the starting position x_0.
13. Show that if $b < a$, then $\{x_n\}$ is a supermartingale and the process converges to 0 a.e.

Problems 14 to 18 concern the notion of conditional mean given a Borel field and show how it generalizes the notion of conditional mean given a partition. It will be necessary to know the Radon–Nikodym Theorem to solve Problem 14. Let $(\Omega, \mathscr{B}, \mu)$ be a probability space in which every subset of a set of measure zero is measurable.

14. If $\mathbf{f}$ is a random variable such that $\mathrm{M}[|\mathbf{f}|] < \infty$ and if $\mathscr{G}$ is a Borel field contained in $\mathscr{B}$, show that there exists a function $\mathrm{M}[\mathbf{f} \mid \mathscr{G}]$ defined on Ω satisfying
   (i) $\mathrm{M}[\mathbf{f} \mid \mathscr{G}]$ is measurable with respect to $\mathscr{G}'$, the augmented field obtained from $\mathscr{G}$.
   (ii) If G is a set in $\mathscr{G}$, then $\int_G \mathbf{f}\,d\mu = \int_G \mathrm{M}[\mathbf{f} \mid \mathscr{G}]\,d\mu$.

Show that $M[f \mid \mathscr{G}]$ is unique in this sense: Any two functions satisfying (i) and (ii) differ only on a set of μ-measure zero. We can therefore define any determination of $M[f \mid \mathscr{G}]$ to be the conditional mean of f given $\mathscr{G}$.

15. Show that if $\mathscr{G}$ is the Borel field generated by a partition $\mathscr{R}$, then $M[f \mid \mathscr{G}] = M[f \mid \mathscr{R}]$ a.e. Show that if $\mathscr{G} = \mathscr{B}$, then $M[f \mid \mathscr{G}] = f$ a.e.

16. State and prove a result for these conditional means in analogy with Proposition 3-3. [*Hint:* In (3), the condition that g be constant on cells of $\mathscr{R}$ should be replaced with the condition that g be measurable with respect to $\mathscr{G}$.]

17. Generalize the definition of martingale in Definition 3-5, using these conditional means. Verify that the statements and proofs of Lemma 3-6, Proposition 3-7, Lemmas 3-9 and 3-10, Proposition 3-11, Theorem 3-12, and Corollary 3-13 apply with only minor changes to this generalized notion of martingale. Perform the same verification for Lemma 3-14, Theorem 3-15, Corollary 3-16, and Proposition 3-17.

18. Prove the following generalization of Proposition 3-18: Let $\mathscr{G}_0 \subset \mathscr{G}_1 \subset \cdots$ be an increasing sequence of Borel fields in $\mathscr{B}$, let $\mathscr{G}$ be the least Borel field containing $\bigcup \mathscr{G}_n$, and let f be a random variable with $M[|f|] < \infty$. Then $(M[f \mid \mathscr{G}_n], \mathscr{G}_n)$ is a martingale, and

$$\lim_{n \to \infty} M[f \mid \mathscr{G}_n] = M[f \mid \mathscr{G}]$$

a.e.

CHAPTER 4

PROPERTIES OF MARKOV CHAINS

1. Markov chains

During all of our discussion of Markov chains, we shall wish to confine ourselves to stochastic processes defined on a sequence space. We have shown that an arbitrary stochastic process may be considered as a process on a suitable Ω in which the outcome functions f_n are coordinate functions. We see, therefore, that in a sense no generality is lost by discussing Markov chains in terms of sequence space.

Definition 4-1: Let $(\Omega, \mathscr{B}, \mu)$ be a sequence space with a denumerable stochastic process $\{x_n\}$ defined from Ω to a denumerable state space $\mathscr{S}$ of more than one element. The process is called a **denumerable Markov process** if

$$\Pr[x_{n+1} = c_{n+1} \mid x_0 = c_0 \wedge \cdots \wedge x_{n-1} = c_{n-1} \wedge x_n = c_n] = \Pr[x_{n+1} = c_{n+1} \mid x_n = c_n]$$

for any n and for any $c_0, \ldots, c_{n+1}$ such that

$$\Pr[x_0 = c_0 \wedge \cdots \wedge x_n = c_n] > 0.$$

The condition that defines a Markov process is known as the **Markov property**. If a denumerable Markov process has the property that for any m and n and for any c_n, c_{n+1} such that $\Pr[x_m = c_n] > 0$ and $\Pr[x_n = c_n] > 0$,

$$\Pr[x_{n+1} = c_{n+1} \mid x_n = c_n] = \Pr[x_{m+1} = c_{n+1} \mid x_m = c_n]$$

holds, then the process is called a **denumerable Markov chain**. The condition that defines a Markov chain is called the **Markov chain property**.

All Markov chains that we shall discuss will be denumerable. From Proposition 2-8 we immediately have the following result.

Proposition 4-2: The measure on the space for a Markov chain is completely determined by

(1) the starting probabilities, $\Pr[x_0 = i]$, and
(2) the one-step transition probabilities, the common value of $\Pr[x_{n+1} = j \mid x_n = i]$ for all n such that $\Pr[x_n = i] > 0$.

If S is the set of states for a Markov chain, we customarily denote representative elements of S by $i, j, k, \ldots$, and 0. For any Markov chain we define on the set S a row vector π and a square matrix P by

$$\pi_i = \Pr[x_0 = i],$$
$$P_{ij} = \Pr[x_{n+1} = j \mid x_n = i], \quad \text{where} \quad \Pr[x_n = i] > 0.$$

The vector π is the **starting distribution**, and the matrix P is the **transition matrix** for the chain. They satisfy the properties $\pi \geq 0$, $\pi\mathbf{1} = 1$, and $P \geq 0$. If $\Pr[x_n = i] = 0$ for all n, then the ith row of P is not covered by the above definition, and we shall agree to take $P_{ij} = 0$ for all j in this case.

The definition of P implies that, for each i, $(P\mathbf{1})_i = 1$ or 0. It will be convenient, however, to think of Markov chains from a point of view which allows P to be any matrix with $P \geq 0$ and $P\mathbf{1} \leq \mathbf{1}$. To do so, we shall admit the possibility that some of the paths in the sequence space are of finite length. Intuitively a path of finite length is one along which the Markov chain can "disappear"; the process disappears from state i with probability $1 - (P\mathbf{1})_i$. Mathematically paths of finite length can be introduced as follows: Suppose a Markov chain with state space S has a distinguished state 0 for which $\pi_0 = 0$ and $P_{00} = 1$. We shall sometimes identify entry to state 0 with the act of disappearing in a process with state space $S - \{0\}$ which also will be called a Markov chain. The transition matrix for the new Markov chain is the same as the original one except that the 0th row and column are omitted. Any path in the original process which has 0 as an outcome is now thought of as a path of finite length which terminates before the first occurrence of 0. The original process can be recovered from the new process by re-introducing state 0 to the state space and by requiring that the transition probabilities to state 0 in the original process be the same as the probabilities of disappearing in the new process. With these conventions a Markov chain determines a vector π and a matrix P with $\pi \geq 0$, $\pi\mathbf{1} = 1$, $P \geq 0$, and $P\mathbf{1} \leq \mathbf{1}$.

Conversely, if π is a row vector defined on S for which $\pi \geq 0$ and $\pi\mathbf{1} = 1$ and if P is a square matrix defined on S for which $P \geq 0$ and $P\mathbf{1} \leq \mathbf{1}$, then π and P define a unique Markov chain with state space S by Theorem 2-4. If $(P\mathbf{1})_j < 1$, then the process has positive probability equal to $1 - (P\mathbf{1})_j$ of disappearing each time it is in state j. Whenever convenient, the act of disappearing may be thought of as entry to an ideal state adjoined to S.

Any state i of a Markov chain P for which $P_{ii} = 1$ is said to be an **absorbing state**. If outcome i occurs at some time, the process is said to enter the absorbing state and to become **absorbed**. It is easily seen that once the process has been absorbed, it is impossible for it to leave the absorbing state.

If P is a Markov chain with starting distribution π and if q is a statement about the process, we denote the probability of q by $\Pr_\pi[q]$. If

$$\pi_k = \begin{cases} 1 & \text{when } k = i \\ 0 & \text{otherwise,} \end{cases}$$

we may alternatively write $\Pr_i[q]$. Similarly if $\mathbf{f}$ is a random variable, we write $M_\pi[\mathbf{f}]$ or $M_i[\mathbf{f}]$, depending on the starting distribution. With this notational convention, we are free to discuss a whole class of Markov chains at once. The class contains all chains whose transition matrices are some fixed matrix P, and two chains of the class differ only in their starting distributions. Most of our treatment of Markov chains will be on this more natural level, where a matrix P, but no distribution π, is specified.

We conclude this section with a simple but useful proposition. Its proof is left to the reader.

Proposition 4-3: If P is a Markov chain, then for $n \geq 0$,

$$\Pr_i[x_n = j] = (P^n)_{ij}$$

and

$$\Pr_\pi[x_n = j] = (\pi P^n)_j.$$

We shall use the notation $P_{ij}^{(n)}$ for $(P^n)_{ij}$, the n-step probability from i to j.

2. Examples of Markov chains

We give ten examples of Markov chains; we shall refer to all of them from time to time.

EXAMPLE 1: Weather in the Land of Oz.

The Land of Oz is blessed by many things, but good weather is not one of them. They never have two nice days in a row. If they have a nice day, they are just as likely to have snow as rain the next day. If they have snow (or rain), they have an even chance of having the same the next day. If there is a change from snow or rain, only half of the time is this a change to a nice day.

The weather is conveniently represented as a Markov chain with the three states S = {Rain, Nice, Snow}. The transition matrix becomes

$$P = \begin{array}{c} \\ \text{R} \\ \text{N} \\ \text{S} \end{array} \begin{array}{c} \begin{array}{ccc} \text{R} & \text{N} & \text{S} \end{array} \\ \begin{pmatrix} \frac{1}{2} & \frac{1}{4} & \frac{1}{4} \\ \frac{1}{2} & 0 & \frac{1}{2} \\ \frac{1}{4} & \frac{1}{4} & \frac{1}{2} \end{pmatrix} \end{array}.$$

Example 2: Chain with a set of states E made absorbing.

Let P be an arbitrary Markov chain with S the set of states. Let a subset E of S be specified. We modify the original process by requiring that if the process is ever in a state j of E, it does not leave that state. The new process is also a Markov chain; its transition matrix P' differs from the P-matrix in that $P'_{jj} = 1$ and $P'_{ji} = 0$ for every $j \in E$ and for every $i \neq j$. The new process is called the chain with E made absorbing.

Example 3: Finite drunkard's walk.

A drunkard walks randomly on a street between his house and a lake, starting at a bar in the middle. He has some idea of which way is home. The steps along the way are labeled by the integers from 0 to n; the bar, some integer i between 0 and n, is the starting state, and the drunkard moves one step toward home (state n) with probability p and one step toward the lake (state 0) with probability $q = 1 - p$. States 0 and n are absorbing. We assume that $p \neq 0$ and $p \neq 1$. The transition matrix is

$$P = \begin{array}{c} \\ 0 \\ 1 \\ 2 \\ \\ n-1 \\ n \end{array} \begin{array}{c} \begin{array}{cccccccc} 0 & 1 & 2 & 3 & & n-1 & n \end{array} \\ \begin{pmatrix} 1 & 0 & 0 & 0 & \cdots & 0 & 0 \\ q & 0 & p & 0 & \cdots & 0 & 0 \\ 0 & q & 0 & p & \cdots & 0 & 0 \\ & \cdots & & & & & \cdots \\ 0 & 0 & 0 & 0 & \cdots & 0 & p \\ 0 & 0 & 0 & 0 & \cdots & 0 & 1 \end{pmatrix} \end{array}.$$

The reader should verify that if $p = \frac{1}{2}$, then $\{x_n\}$ is a martingale and that if $p \neq \frac{1}{2}$, then $\{(q/p)^{x_n}\}$ is a martingale.

Example 4: Infinite drunkard's walk.

For this process, which is an extension of the one in Example 3, the states are the non-negative integers, and state 0 is absorbing. For each $i > 0$, we have

$$P_{i,i+1} = p, \quad P_{i,i-1} = q; \quad \text{and} \quad p + q = 1.$$

We assume $p \neq 0$ and $q \neq 0$. The transition matrix is

$$P = \begin{matrix} & \begin{matrix} 0 & 1 & 2 & 3 & \end{matrix} \\ \begin{matrix} 0 \\ 1 \\ 2 \\ \end{matrix} & \begin{pmatrix} 1 & 0 & 0 & 0 & \cdots \\ q & 0 & p & 0 & \cdots \\ 0 & q & 0 & p & \cdots \\ \vdots & & & & \ddots \end{pmatrix} \end{matrix}.$$

If $p = \frac{1}{2}$, then $\{x_n\}$ is a martingale, and if $p \neq \frac{1}{2}$, then $\{(q/p)^{x_n}\}$ is a martingale.

Example 5: Basic example.

A sequence of tasks is to be performed in a certain order, each with its own probability of success. Success means that the process goes to the next state; failure means that the process must start over at state 0. Thus the states are the non-negative integers, and with each positive integer i we associate two probabilities p_i and q_i such that $p_i + q_i = 1$. The value p_i is the transition probability from state $i - 1$ to state i, and q_i is the transition probability from state $i - 1$ to 0. Thus p_i is the probability of succeeding in the ith task. We assume that $p_i < 1$ for infinitely many i, and we normally assume that $p_i > 0$ for every i. The transition matrix is

$$P = \begin{matrix} & \begin{matrix} 0 & 1 & 2 & 3 & \end{matrix} \\ \begin{matrix} 0 \\ 1 \\ 2 \\ \end{matrix} & \begin{pmatrix} q_1 & p_1 & 0 & 0 & \cdots \\ q_2 & 0 & p_2 & 0 & \cdots \\ q_3 & 0 & 0 & p_3 & \cdots \\ \vdots & & & & \ddots \end{pmatrix} \end{matrix}.$$

In connection with this example, we define a row vector β by

$$\beta_0 = 1$$

$$\beta_i = \prod_{k=1}^{i} p_k.$$

Then β_i is the probability of i successes in a row after the process starts at 0. The reader should verify that a necessary and sufficient condition for $\beta = \beta P$ is that $\lim_{i \to \infty} \beta_i = 0$. This Markov chain will be referred to hereafter as "the basic example."

Example 6: Sums of independent random variables.

The states of a Markov chain P are the elements of an index set I on which an operation of addition is defined in such a way that I becomes an abelian group. A probability distribution $\{p_i\}$ defined on I satisfies $p_i \geq 0$ and $\sum p_i = 1$. The Markov chain P is defined by $p_{ij} = p_{j-i}$.

The name of this Markov chain is derived from thinking of performing independent experiments which have probability p_i of outcome i. The states of the chain are the partial sums of these results, and the sum changes from i to j with probability $P_{i,i+k} = p_k$ if $j = i + k$.

For the case in which the index set I is the set of integers with the usual concept of addition defined on them, martingales arise as in Example 1 of Section 3-2. We shall apply these ideas in Chapter 5.

EXAMPLE 7: Two classes of random walks.

We shall be concerned especially with two kinds of random walks. The symmetric random walk in n-dimensions is defined to be a sums of independent random variables process on the lattice of integer points in n-dimensional Euclidean space. The transition probability from one lattice point to another is $(2n)^{-1}$ if the two points are a Euclidean distance of one unit apart; the transition probability is zero otherwise. Thus, from each point the process moves to one of $2n$ neighboring points with probability $(2n)^{-1}$.

A second kind of random walk with which we shall be concerned is a sums of independent random variables process on the integers with $P_{i,i+1} = p$ and $P_{i,i-1} = q$ for every i. We shall call this process the p-q random walk. If $p = q = \frac{1}{2}$, then $\{x_n\}$ is a martingale, and if $p \neq \frac{1}{2}$, then $\{(q/p)^{x_n}\}$ is a martingale.

EXAMPLE 8: General random walks on the line.

The state space for a random walk on the line is the set of integers, and for each integer i, three probabilities p_i, q_i, and r_i with $p_i + q_i + r_i = 1$ are specified. A Markov chain is defined by

$$\begin{aligned} P_{i,i+1} &= p_i \\ P_{i,i-1} &= q_i \\ P_{i,i} &= r_i. \end{aligned}$$

The drunkard's walk and the p-q random walk are both special cases.

An important case of random walks on the line which we have not discussed yet is the reflecting random walk. For this chain the process is started at a state which is a non-negative integer, and the assumption is made that $q_0 = 0$. The process never reaches the negative integers, and the state space may just as well be taken as $\{0, 1, 2, \ldots\}$.

EXAMPLE 9: Branching process.

The state space is the set of non-negative integers, and a fixed probability distribution $p = \{p_0, p_1, p_2, \ldots\}$ is specified. Suppose the mean $\sum kp_k$ of p is m. Let $\{\mathbf{y}_n\}$ be a sequence of non-negative integer-valued independent random variables with common distribution p, and set $\mathbf{s}_n = \mathbf{y}_1 + \cdots + \mathbf{y}_n$. Let $p_j^{(n)} = \Pr[\mathbf{s}_n = j]$. Then the branching process is defined to be a Markov chain with transition probabilities $P_{ij} = p_j^{(i)}$.

The usual model is the following. A species of bacteria has the distribution p representative of the number of offspring one such bacterium has before it dies. The value of $\mathbf{y}_k$ represents the number of offspring the kth bacterium has while it is alive, and $\mathbf{s}_n$ represents the total number of bacteria produced by n bacteria in one generation of the colony. The rth position, x_r, of the stochastic process is the number of bacteria in the rth offspring generation.

As we have noted, the branching process is a Markov chain. Let $\{x_n\}$ be the outcome functions for the chain started in state 1 (that is, with one bacterium in the colony initially), and suppose the mean m is finite. Then $\{x_n/m^n\}$ with its natural partition forms a martingale. The reader should verify that $\mathrm{M}[|x_n/m^n|]$ is finite; we shall show that $\mathrm{M}[x_{n+1}/m^{n+1} \mid x_1/m \wedge \cdots \wedge x_n/m^n] = x_n/m^n$. First we note that

$$\mathrm{M}[\mathbf{s}_n] = \sum \mathrm{M}[\mathbf{y}_i] = nm,$$

so that if we know that the process is in state r, then the mean state that it is in after the next step is rm. Then

$$\begin{aligned}
\mathrm{M}[x_{n+1}/m^{n+1} \mid x_1/m \wedge \cdots \wedge x_n/m^n]& \\
&= \mathrm{M}[x_{n+1}/m^{n+1} \mid x_n/m^n] \quad \text{by the Markov property} \\
&= \mathrm{M}[x_{n+1}/m^{n+1} \mid x_n] \\
&= (1/m^{n+1})\mathrm{M}[x_{n+1} \mid x_n] \\
&= (1/m^{n+1})x_n m \quad \text{by the remarks above} \\
&= x_n/m^n.
\end{aligned}$$

EXAMPLE 10: Tree process.

Let $\{x_n\}$ be a denumerable stochastic process defined on sequence space, and let S be the set of states. Define a set T to be the set of all finite sequences of elements in S. Define a new stochastic process as follows: If t and u are elements of T for which

$$t = (c_0, c_1, c_2, \ldots, c_n)$$

and

$$u = (c_0, c_1, c_2, \ldots, c_n, c_{n+1}),$$

define

$$\Pr[y_{n+1} = u \mid y_n = t] = \Pr[x_{n+1} = c_{n+1} \mid x_0 = c_0 \wedge \cdots \wedge x_n = c_n].$$

The process $\{y_n\}$ defined from the same space to the set T is a Markov chain; the entire history of the original process up to time n is contained in the knowledge of the value of the nth outcome function y_n for the new process.

An example of a tree process is obtained by considering an individual's voting history in successive years. Letting D and R represent the political parties, we see that his possible histories can be conveniently represented as a tree:

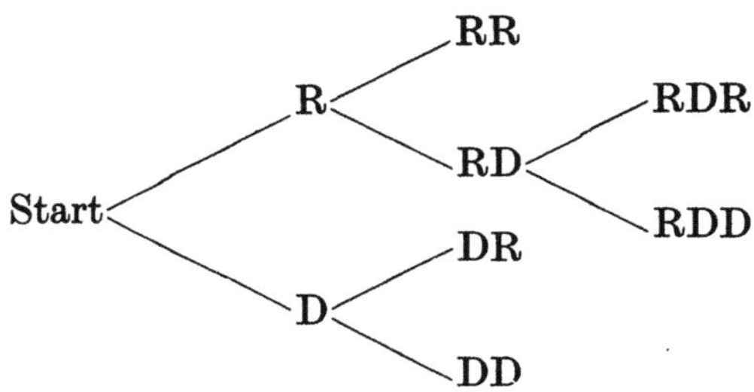

The chain is in each state—D, R, DD, DRR, etc.—at most once.

3. Applications of martingale ideas

Let P be the transition matrix for a Markov chain. A column vector f is said to be a P-**regular function**, or simply a **regular function**, if $f = Pf$. The function is **superregular** if $f \geq Pf$; it is **subregular** if $f \leq Pf$.

The reader should convince himself that the regularity of a function is a condition of the following form: At each point of the domain, the value of the function is equal to the average value of the function at neighboring points. By neighboring points we mean those states that it is possible for the process to reach in one step, and by average value we mean the average obtained by weighting the function values at neighboring points by the transition probabilities to those states. A function f is said to be **regular at a point** j if $f_j = (Pf)_j$.

Regular measures may be defined analogously with regular functions. A non-negative row vector π is a **regular measure** if $\pi = \pi P$; it is **superregular** if $\pi \geq \pi P$ and **subregular** if $\pi \leq \pi P$.

Let h be a P-regular function and let $h(x_n)$ denote h_j if $x_n = j$. Suppose $\mathrm{M}[|h(x_n)|]$ is finite. We shall show that $(h(x_n), \mathscr{R}_n)$ is a martingale, where $\mathscr{R}_n$ is the cross partition $\mathscr{R}^{x_0} \otimes \mathscr{R}^{x_1} \otimes \cdots \otimes \mathscr{R}^{x_n}$ determined by the outcome functions $x_0, x_1, \ldots, x_n$ for the Markov chain. It is sufficient to show that

$$\mathrm{M}[h(x_{n+1}) \mid x_0 \wedge \cdots \wedge x_n] = h(x_n).$$

On the cell of the cross-partition where $x_0 = i, \ldots, x_n = j$ and where $\Pr[x_0 = i \wedge \cdots \wedge x_n = j] > 0$,

$$\begin{aligned} M[h(x_{n+1}) \mid x_0 \wedge \cdots \wedge x_n] &= M[h(x_{n+1}) \mid x_0 = i \wedge \cdots \wedge x_n = j] \\ &= \sum_k \Pr[x_{n+1} = k \mid x_0 = i \wedge \cdots \wedge x_n = j] \cdot h_k \\ &= \sum_k P_{jk} h_k \quad \text{by the Markov chain property} \\ &= h_j \quad \text{since } h \text{ is regular} \\ &= h(x_n). \end{aligned}$$

Thus $(h(x_n), \mathscr{R}_n)$ is a martingale. Similarly, superregular functions are associated naturally with supermartingales and subregular functions correspond to submartingales. The proofs differ from the above proof only by insertion of the appropriate inequality sign in the next to the last step.

Most of our applications of martingale ideas we shall leave to the next few chapters. We shall, however, settle some things about branching processes at this time. Let $\{x_n\}$ be the outcome functions for a branching process started in state 1, and suppose the mean $m = \sum kp_k$ is finite. As we noted in Section 2, $\{x_n/m^n\}$ forms a non-negative martingale, which by Corollary 3-13 converges almost everywhere to a finite limiting function g. One can show that g is not a constant function; that is, the value of the limit of $\{x_n(\omega)/m^n\}$ very much depends upon the early history of the path. The exact distribution of g, however, is an unsolved problem.

On the other hand, information about whether the process dies out (by being absorbed at state 0) is not hard to obtain. Let $\varphi(s) = \sum_j p_j s^j$ and suppose $r = \varphi(r)$, $r \geq 0$, and $r \neq 1$. Then $\{r^{x_n}\}$ is a non-negative martingale. First suppose $r > 1$. Since $\{r^{x_n}\}$ is a non-negative martingale, it converges to a finite limit almost everywhere, and since $r > 1$, x_n itself converges with probability one. Since x_n is integer-valued, x_n is constant on almost all paths from some point on. It is left as an exercise to show that the constant must be zero and that the process therefore dies out with probability one. Next suppose $r < 1$. Then $\{r^{x_n}\}$ is bounded and converges almost everywhere to a limiting function r^{x_∞}, which must be 0 or 1 (that is, $x_\infty = \infty$ or 0) almost everywhere. By dominated convergence we have

$$\begin{aligned} r = M[r^{x_0}] = M[r^{x_\infty}] = {} & 1 \cdot \Pr[\text{process dies out}] \\ & + 0 \cdot \Pr[\text{process does not die out}], \end{aligned}$$

so that r is the probability that the process dies out in the long run.

Finally, suppose $r = 1$ is the only non-negative root of $s = \varphi(s)$. Then $m = 1$ and $\{x_n\}$ is a non-negative martingale. Once again we must have $x_n \to 0$ with probability one, and the process is almost certain to die out.

The reader should notice that the case $r = 1$ has the property that $\mathrm{M}[x_n] = 1$ for all n, whereas $\mathrm{M}[\lim x_n] = 0$. The process in this case is an example of a fair game whose final expected fortune is strictly less than the starting fortune.

4. Strong Markov property

The **strong Markov property** is a rigorous formulation of the following assertion about a Markov chain: If the present is known, then the probability of any statement depending on the future is independent of what additional information about the past is known. In this section we shall state and prove this result; our procedure will be first to prove a conceptually simpler special case and then to obtain the general theorem as an easy consequence. In the special case the time of the present will be a fixed time n, whereas in the general case the time of the present will be allowed to depend on the past history of the process. That is, the time of the present will be a random time. Knowledge of the present, then, means knowledge of the outcome at the time of the present.

If $\omega = (c_0, c_1, c_2, \ldots, c_{n-1}, c_n, c_{n+1}, \ldots)$ is a point in a sequence space, we agree to call the path

$$(c_n, c_{n+1}, \ldots)$$

by the name ω_n.

Lemma 4-4: Let $\{p_k\}$ be a sequence of statements whose truth sets are disjoint in pairs, let $\bigvee p_k$ be their disjunction, and suppose $\Pr[\bigvee p_k] > 0$. If p is a statement for which $\Pr[p \mid p_k] = c$ whenever $\Pr[p_k] > 0$, then $\Pr[p \mid \bigvee p_k] = c$.

PROOF: For each k,

$$c \cdot \Pr[p_k] = \Pr[p \wedge p_k].$$

Thus

$$c \sum_k \Pr[p_k] = \sum_k \Pr[p \wedge p_k].$$

Since the p_k are disjoint statements, it follows from complete additivity that $\sum \Pr[p_k] = \Pr[\bigvee p_k]$ and that $\sum \Pr[p \wedge p_k] = \Pr[p \wedge (\bigvee p_k)]$. Thus $c = \Pr[p \wedge (\bigvee p_k)]/\Pr[\bigvee p_k]$ and the lemma follows.

Throughout the remainder of this section let $(\Omega, \mathscr{B}, \mu)$ be a fixed sequence space and $\{x_n\}$ a fixed denumerable Markov chain defined on Ω. The field of cylinder sets will be denoted by $\mathscr{F}$, as in Section 2-1, and the smallest Borel field containing $\mathscr{F}$ will be called $\mathscr{G}$.

Definition 4-5: The **tail-field** $\mathscr{T}_n$ is the smallest augmented Borel field containing all truth sets of statements $x_n = c_n \wedge \cdots \wedge x_m = c_m$, $m \geq n$. [Thus $\mathscr{T}_0 = \mathscr{B}$ and $\mathscr{T}_n \subset \mathscr{T}_{n-1}$.]

A statement relative to the field $\mathscr{F}_n$ defined in Section 2-1 is one whose truth set depends only on outcomes $x_0, \ldots, x_n$, whereas a statement relative to $\mathscr{T}_n$ is one whose truth set does not depend on outcomes $x_0, \ldots, x_{n-1}$. Specifically, a set R in $\mathscr{B}$ is in $\mathscr{T}_n$ if and only if, whenever $\omega \in R$ and ω' is such that $\omega'_n = \omega_n$, then $\omega' \in R$.

We note that the class of sets $\mathscr{T}_n \cap \mathscr{F}$, being the intersection of fields, is again a field. Moreover, $\mathscr{T}_n$ is the smallest augmented Borel field containing $\mathscr{T}_n \cap \mathscr{F}$, so that the uniqueness statement of Theorem 1-19 applies: A probability measure on $\mathscr{T}_n$ is completely determined by its values on $\mathscr{T}_n \cap \mathscr{F}$.

Lemma 4-6: Let $\{x_n\}$ be a Markov chain with starting distribution π, let q be a statement relative to $\mathscr{F}_{n-1}$, let r be a statement relative to $\mathscr{T}_{n+1} \cap \mathscr{F}$, and suppose $\Pr_\pi[q \wedge x_n = i] > 0$. Then

$$\Pr_\pi[r \mid q \wedge x_n = i] = \Pr_\pi[r \mid x_n = i] = \Pr_i[r'],$$

where r' is so chosen that $\omega \in R$ if and only if $\omega_n \in R'$.

Remark: Such an r' exists (and is unique), since r is a statement relative to $\mathscr{T}_{n+1} \cap \mathscr{F}$.

Proof:

Case 1: r is of the form $x_{n+1} = j$. Write q as a disjunction $q = \bigvee q_m$, where

$$q_m\colon \quad x_0 = c_0^{(m)} \wedge \cdots \wedge x_{n-1} = c_{n-1}^{(m)}.$$

For each m such that $\Pr_\pi[q_m \wedge x_n = i] > 0$,

$$\Pr_\pi[r \mid q_m \wedge x_n = i] = \Pr_\pi[x_{n+1} = j \mid q_m \wedge x_n = i] = P_{ij}$$

by Definition 4-1. Hence, since $\Pr_\pi[q \wedge x_n = i] > 0$,

$$\Pr_\pi[r \mid q \wedge x_n = i] = P_{ij} = \Pr_\pi[r \mid x_n = i]$$

by Lemma 4-4. Taking r' as $x_1 = j$, we have

$$P_{ij} = \Pr_i[x_1 = j] = \Pr_i[r'].$$

Case 2: r is of the form $x_{n+1} = c_{n+1} \wedge \cdots \wedge x_m = c_m$, $m > n$. We have

$$\begin{aligned}\Pr_\pi[r \mid q \wedge x_n = i] \\ &= \Pr_\pi[x_{n+1} = c_{n+1} \mid q \wedge x_n = i] \\ &\quad \times \Pr_\pi[x_{n+2} = c_{n+2} \mid q \wedge x_n = i \wedge x_{n+1} = c_{n+1}] \\ &\quad \times \cdots \times \Pr_\pi[x_m = c_m \mid q \wedge x_n = i \wedge \cdots \wedge x_{m-1} = c_{m-1}].\end{aligned}$$

The general factor on the right is

$$\Pr_\pi[x_{n+k+1} = c_{n+k+1} \mid q \wedge x_n = i \wedge \cdots \wedge x_{n+k} = c_{n+k}].$$

First, suppose that none of these factors is zero. Then we may apply Case 1 with $n + k$ in place of n and $q \wedge x_n = i \wedge \cdots \wedge x_{n+k-1} = c_{n+k-1}$ in place of q. The q's drop out of the conditions, and the product of the new conditional probabilities is $\Pr_\pi[r \mid x_n = i]$.

Next, suppose that at least one of the factors is zero; let the first such factor from the left be

$$\Pr_\pi[x_{n+k+1} = c_{n+k+1} \mid q \wedge x_n = i \wedge \cdots \wedge x_{n+k} = c_{n+k}].$$

We must show that $\Pr_\pi[r \mid x_n = i] = 0$. If $k = 0$, then by Case 1

$$0 = \Pr_\pi[x_{n+1} = c_{n+1} \mid q \wedge x_n = i] = \Pr_\pi[x_{n+1} = c_{n+1} \mid x_n = i],$$

and hence $\Pr_\pi[r \mid x_n = i] = 0$. If $k > 0$, then

$$\Pr_\pi[q \wedge x_n = i \wedge \cdots \wedge x_{n+k-1} = c_{n+k-1}] > 0,$$

and Case 1 gives

$$\begin{aligned}0 &= \Pr_\pi[x_{n+k+1} = c_{n+k+1} \mid q \wedge x_n = i \wedge \cdots \wedge x_{n+k} = c_{n+k}] \\ &= \Pr_\pi[x_{n+k+1} = c_{n+k+1} \mid x_n = i \wedge \cdots \wedge x_{n+k} = c_{n+k}].\end{aligned}$$

Hence $\Pr_\pi[r \mid x_n = i] = 0$.

Finally r' is the statement $x_1 = c_{n+1} \wedge \cdots \wedge x_{m-n} = c_m$, and, since $\Pr_\pi[x_n = i] \geq \Pr_\pi[q \wedge x_n = i] > 0$, we have

$$\Pr_\pi[r \mid x_n = i] = P_{i,c_{n+1}} \cdot P_{c_{n+1},c_{n+2}} \cdot \cdots \cdot P_{c_{m-1},c_m} = \Pr_i[r'].$$

Case 3: r is arbitrary in $\mathscr{T}_{n+1} \cap \mathscr{F}$. A general statement r reduces to the denumerable union of the type statements in Case 2, and the result follows from the complete additivity of the probability measure.

The lemma to follow is the strong Markov property for the case in which the time of the present is a fixed time n.

Lemma 4-7: Let $\{x_n\}$ be a Markov chain with starting distribution π, let q be a statement relative to $\mathscr{F}_n$, let r be a statement relative to $\mathscr{T}_n$, and suppose $\Pr_\pi[q \wedge x_n = i] > 0$. Then

$$\Pr_\pi[r \mid q \wedge x_n = i] = \Pr_\pi[r \mid x_n = i] = \Pr_i[r'],$$

where $\omega \in R$ if and only if $\omega_n \in R'$.

Proof: Write

$$q = \bigvee_m (x_0 = c_0^{(m)} \wedge \cdots \wedge x_{n-1} = c_{n-1}^{(m)} \wedge x_n = c_n^{(m)}).$$

If we set $q^* = \bigvee_m (x_0 = c_0^{(m)} \wedge \cdots \wedge x_{n-1} = c_{n-1}^{(m)})$, where the disjunction is taken over just those m such that $c_n^{(m)} = i$, then

$$(q^* \wedge x_n = i) \equiv (q \wedge x_n = i)$$

and q^* is a statement relative to $\mathscr{F}_{n-1}$. In the special case where r is relative to $\mathscr{T}_n \cap \mathscr{F}$, we may write

$$r = \bigvee_m (x_n = c_n^{(m)} \wedge \cdots \wedge x_N = c_N^{(m)}) \qquad (N \text{ fixed})$$

and

$$r^* = \bigvee_m (x_{n+1} = c_{n-1} \wedge \cdots \wedge x_N = c_N^{(m)})$$

with the second disjunction taken over only those m such that $c_n^{(m)} = i$. Then

$$\Pr_\pi[r \mid q \wedge x_n = i] = \Pr_\pi[r^* \mid q^* \wedge x_n = i],$$

$$\Pr_\pi[r \mid x_n = i] = \Pr_\pi[r^* \mid x_n = i],$$

and

$$\Pr_i[r'] = \Pr_i[r^{*\prime}].$$

By Lemma 4-6,

$$\Pr_\pi[r^* \mid q^* \wedge x_n = i] = \Pr_\pi[r^* \mid x_n = i] = \Pr_i[r^{*\prime}].$$

Hence

$$\Pr_\pi[r \mid q \wedge x_n = i] = \Pr_\pi[r \mid x_n = i] = \Pr_i[r'].$$

We have thus established the lemma for every r measurable with respect to $\mathscr{T}_n \cap \mathscr{F}$. But $\mathscr{T}_n$ is the smallest augmented Borel field containing $\mathscr{T}_n \cap \mathscr{F}$, and by Theorem 1-19 any two measures on $\mathscr{T}_n$ which agree on $\mathscr{T}_n \cap \mathscr{F}$ must agree on all of $\mathscr{T}_n$. Thus

$$\Pr_\pi[r \mid q \wedge x_n = i], \quad \Pr_\pi[r \mid x_n = i], \quad \text{and} \quad \Pr_i[r'],$$

which define such measures as r varies, are equal for every r measurable with respect to $\mathscr{T}_n$.

Turning to the general case of the strong Markov property, let $\mathbf{t}$ be a random time. We define ω_t pointwise to be ω_n at all points where

$\mathbf{t}(\omega) = n$. We do not define ω_t if $\mathbf{t}(\omega) = \infty$. Similarly the outcome function x_t is defined to be $x_n(\omega)$ if $\mathbf{t}(\omega) = n$, and it is not defined for $\mathbf{t}(\omega) = \infty$.

Definition 4-8: The field $\mathscr{F}_t$ is the Borel field of all sets A such that, for each n, $A \cap \{\omega \mid \mathbf{t}(\omega) = n\}$ is in $\mathscr{F}_n$. The tail-field $\mathscr{T}_t$ is the smallest augmented Borel field containing all truth sets of statements

$$x_t = c_t \wedge \cdots \wedge x_{t+k} = c_{t+k}, \qquad k \geq 0.$$

A statement q relative to $\mathscr{F}_t$ is one such that, for each n, the statement $q \wedge \mathbf{t} = n$ depends only on outcomes $x_0, \ldots, x_n$. A statement r relative to $\mathscr{T}_t$ is one whose truth set does not depend on outcomes before time $\mathbf{t}$. Specifically, a set R in $\mathscr{B}$ is in $\mathscr{T}_t$ if and only if whenever $\omega \in R$ and ω' is such that $\omega'_t = \omega_t$, then $\omega' \in R$.

We state the strong Markov property as the next theorem.

Theorem 4-9: Let $\{x_n\}$ be a Markov chain with starting distribution π, let $\mathbf{t}$ be a random time, let q be a statement relative to $\mathscr{F}_t$, let r be a statement relative to $\mathscr{T}_t$, and suppose $\Pr_\pi[q \wedge x_t = i] > 0$. Then

$$\Pr_\pi[r \mid q \wedge (x_t = i)] = \Pr_\pi[r \mid x_t = i] = \Pr_i[r'],$$

where $\omega \in R$ if and only if $\omega_t \in R'$.

PROOF: We shall prove the theorem for any statement r measurable with respect to $\mathscr{T}_t \cap \mathscr{F}$. The theorem for general r will then follow, as in the proof of Lemma 4-7, from the uniqueness half of Theorem 1-19. Since $x_t \neq i$ when $\mathbf{t} = \infty$, we have

$$(q \wedge x_t = i) \equiv \bigvee_{n=0}^{\infty} (q \wedge x_n = i \wedge \mathbf{t} = n).$$

We are going to apply Lemma 4-4 with p the statement r, with p_n the statement $q \wedge x_n = i \wedge \mathbf{t} = n$, and with c the constant $\Pr_i[r']$. To do so, we must show that

$$\Pr_\pi[r \mid q \wedge x_n = i \wedge \mathbf{t} = n] = \Pr_i[r']$$

whenever $\Pr_\pi[q \wedge x_n = i \wedge \mathbf{t} = n] > 0$, and we will have proved that

$$\Pr_\pi[r \mid q \wedge x_t = i] = \Pr_i[r'].$$

The fact that $\Pr_\pi[r \mid x_t = i]$ equals both of these quantities will follow by taking q to be a tautology.

Thus we first note that $q \wedge \mathbf{t} = n$ is measurable with respect to $\mathscr{F}_n$. In addition there exists a statement $\hat{r}$ measurable with respect to $\mathscr{T}_n$ such that

$$(r \wedge \mathbf{t} = n) \equiv (\hat{r} \wedge \mathbf{t} = n);$$

this is so because r is the denumerable union of statements

$$x_t = c_t \wedge \cdots \wedge x_{t+N} = c_{t+N},$$

and we may take $\hat{r}$ to be the same union over the statements

$$x_n = c_t \wedge \cdots \wedge x_{n+N} = c_{t+N}.$$

In this notation the statement r' is the union of the statements

$$x_0 = c_t \wedge \cdots \wedge x_N = c_{t+N},$$

and we have that ω is in the truth set of $\hat{r}$ if and only if ω_n is in the truth set of r'. Hence

$$\begin{aligned}\Pr_\pi[r \mid q \wedge x_n = i \wedge \mathbf{t} = n] &= \Pr_\pi[r \wedge \mathbf{t} = n \mid q \wedge x_n = i \wedge \mathbf{t} = n] \\ &= \Pr_\pi[\hat{r} \wedge \mathbf{t} = n \mid q \wedge x_n = i \wedge \mathbf{t} = n] \\ &= \Pr_\pi[\hat{r} \mid (q \wedge \mathbf{t} = n) \wedge x_n = i] \\ &= \Pr_\pi[\hat{r} \mid x_n = i] \\ &= \Pr_i[r'],\end{aligned}$$

the last two equalities following from Lemma 4-7.

An equivalent way of stating the first equality of the conclusion of the preceding theorem is

$$\Pr_\pi[q \wedge r \mid x_\mathbf{t} = i] = \Pr_\pi[q \mid x_\mathbf{t} = i] \Pr_\pi[r \mid x_\mathbf{t} = i].$$

This is the form in which the theorem asserts that if the present is known, then the past and future are independent.

5. Systems theorems for Markov chains

As immediate consequences of the strong Markov property, we can prove two systems theorems for Markov chains. The first states that if p is a statement depending on outcomes only beyond some random time $\mathbf{t}$, then one may compute $\Pr_\pi[p]$ as if the chain were started with the initial distribution $\Pr_\pi[x_\mathbf{t} = j]$.

Theorem 4-10: Let $\{x_n\}$ be a Markov chain, and let p be a measurable statement with truth set P satisfying

(1) $\Pr_\pi[p \wedge (\mathbf{t} = \infty)] = 0$, and
(2) there exists a statement p' with truth set P' such that if $\mathbf{t}(\omega) < +\infty$, then $\omega \in P$ if and only if $\omega_\mathbf{t} \in P'$.

Then

$$\Pr_\pi[p] = \sum_{k \in S} \Pr_\pi[x_\mathbf{t} = k] \Pr_k[p'].$$

PROOF: By (1) we have

$$\begin{aligned}\Pr_\pi[p] &= \sum_k \Pr_\pi[\omega \in P \wedge x_t = k] \\ &= \sum_k \Pr_\pi[\omega_t \in P' \wedge x_t = k] \\ &= \sum_k \Pr_\pi[x_t = k] \Pr_\pi[\omega_t \in P' \mid x_t = k] \\ &= \sum_k \Pr_\pi[x_t = k] \Pr_k[p'] \quad \text{by Theorem 4-9.}\end{aligned}$$

Theorem 4-10, which is a result about probabilities of statements, can also be thought of as a result about means of characteristic functions. Then Theorem 4-11 to follow becomes a straightforward generalization to arbitrary functions.

Theorem 4-11: Let $\{x_n\}$ be a Markov chain, and let $\mathbf{f}$ be a random variable satisfying

(1) $\Pr_\pi[\mathbf{f} \neq 0 \wedge \mathbf{t} = \infty] = 0$, and
(2) there exists a random variable $\mathbf{f}'$ such that if $\mathbf{t}(\omega) < \infty$, then $\mathbf{f}(\omega) = \mathbf{f}'(\omega_t)$.

Then

$$\mathrm{M}_\pi[\mathbf{f}] = \sum_k \Pr_\pi[x_t = k]\, \mathrm{M}_k[\mathbf{f}'].$$

PROOF: If $\mathbf{f}$ assumes negative values, we may prove the result for $\mathbf{f}^+$ and $\mathbf{f}^-$ separately. We therefore assume $\mathbf{f} \geq 0$. Let $p_j^{(m)}$ be the statement $j/2^m \leq \mathbf{f} < (j+1)/2^m$, for $1 \leq j < m\cdot 2^m$, let $p_0^{(m)}$ be the statement $0 < \mathbf{f} < 1/2^m$, and let $p_{m2^m}^{(m)}$ be the statement $m \leq \mathbf{f}$. Define statements $p_j^{(m)'}$ similarly for $\mathbf{f}'$. Then $p_j^{(m)}$ and $p_j^{(m)'}$ satisfy the hypotheses of Theorem 4-10, so that

$$\Pr_\pi[p_j^{(m)}] = \sum_k \Pr_\pi[x_t = k] \Pr_k[p_j^{(m)'}].$$

Hence

$$\begin{aligned}\mathrm{M}_\pi[\mathbf{f}] &= \lim_m \sum_{j=0}^{m2^m} \frac{j}{2^m} \Pr_\pi[p_j^{(m)}] \\ &= \lim_m \sum_k \Pr_\pi[x_t = k] \sum_{j=0}^{m2^m} \frac{j}{2^m} \Pr_k[p_j^{(m)'}] \\ &= \sum_k \Pr_\pi[x_t = k] \lim_m \sum_{j=0}^{m2^m} \frac{j}{2^m} \Pr_k[p_j^{(m)'}] \quad \text{by monotone convergence} \\ &= \sum_k \Pr_\pi[x_t = k]\, \mathrm{M}_k[\mathbf{f}'].\end{aligned}$$

6. Applications of systems theorems

The theorems of Section 5 will play an important role in our study of denumerable Markov chains. At this time we shall not illustrate the full power of the theorems but shall be content instead to use them in developing some of the machinery needed for the classification of states in Section 7.

We begin by introducing some notation. Define

$$\delta_{ij} = \begin{cases} 1 & \text{if } i = j \\ 0 & \text{otherwise.} \end{cases}$$

Let h_j be the statement about a Markov chain that state j is eventually reached. We have already defined the random variables $\mathbf{n}_j$ and $\mathbf{t}_j$ for general stochastic processes (see Section 2-6); $\mathbf{n}_j$ is the number of times in state j, and $\mathbf{t}_j$ is the time to reach state j. Let $f_j^{(k)}$ be the statement that $\mathbf{t}_j(\omega) = k$.

Confining ourselves to Markov chains, we associate the quantities $\bar{h}_j$, $\bar{\mathbf{n}}_j$, $\bar{\mathbf{t}}_j$, and $\bar{f}_j^{(k)}$ with h_j, $\mathbf{n}_j$, $\mathbf{t}_j$, and $f_j^{(k)}$. They are defined as follows:

$$\begin{aligned} \bar{h}_j&: \quad h_j \text{ is true for } \omega_1 \\ \bar{\mathbf{n}}_j(\omega) &= \mathbf{n}_j(\omega_1) \\ \bar{\mathbf{t}}_j(\omega) &= \mathbf{t}_j(\omega_1) + 1 \\ \bar{f}_j^{(k)}&: \quad \bar{\mathbf{t}}_j(\omega) = k. \end{aligned}$$

In terms of these quantities, we define a collection of matrices. We note that, in general, an expression of the form $\{\mathrm{M}_i[\mathbf{g}_j]\}$ stands for a matrix.

$$\begin{aligned} H_{ij} &= \mathrm{Pr}_i[h_j] \\ N_{ij} &= \mathrm{M}_i[\mathbf{n}_j] \\ F_{ij}^{(k)} &= \mathrm{Pr}_i[f_j^{(k)}] \\ \bar{H}_{ij} &= \mathrm{Pr}_i[\bar{h}_j] \\ \bar{N}_{ij} &= \mathrm{M}_i[\bar{\mathbf{n}}_j] \\ \bar{F}_{ij}^{(k)} &= \mathrm{Pr}_i[\bar{f}_j^{(k)}]. \end{aligned}$$

It is trivial to verify that $H_{ii} = 1$, that $F^{(0)} = I$, that $F^{(1)} = P - P_{dg}$, and that $N = I + \bar{N}$.

Proposition 4-12: If P is a Markov chain, then

$$N = \sum_{k=0}^{\infty} P^k.$$

PROOF: The result follows immediately from Propositions 2-10 and 4-3.

Proposition 4-13: If P is a Markov chain, then $\bar{N} = PN$ and $N = I + PN$.

Proof: The second identity follows from the first by adding I to both sides. To obtain the first one, we apply Theorem 4-11 with $\mathbf{f} = \bar{\mathbf{n}}_j$ and the random time identically one. Then $\mathbf{f}' = \mathbf{n}_j$ since $\bar{\mathbf{n}}_j(\omega) = \mathbf{n}_j(\omega_1)$ by definition, and thus

$$\begin{aligned} \mathrm{M}_i[\bar{\mathbf{n}}_j] &= \sum_k \mathrm{Pr}_i[x_1 = k]\, \mathrm{M}_k[\mathbf{n}_j] \\ &= \sum_k P_{ik}\, \mathrm{M}_k[\mathbf{n}_j] \end{aligned}$$

or

$$\bar{N} = PN.$$

Proposition 4-14: If P is a Markov chain, then

$$H = \sum_{k=0}^{\infty} F^{(k)}, \quad \bar{H} = \sum_{k=1}^{\infty} \bar{F}^{(k)}, \quad \text{and} \quad \bar{H} = PH.$$

Proof: The first two assertions follow from the complete additivity of μ; we have $h_j \equiv \bigvee f_j^{(k)}$ and $\bar{h}_j \equiv \bigvee \bar{f}_j^{(k)}$ disjointly. For the third assertion we apply Theorem 4-10 with $p = \bar{h}_j$ and the random time identically one. Then p' is the statement h_j and

$$\begin{aligned} \bar{H}_{ij} = \mathrm{Pr}_i[p] &= \sum_k \mathrm{Pr}_i[x_1 = k]\, \mathrm{Pr}_k[p'] \\ &= \sum_k P_{ik} H_{kj}. \end{aligned}$$

Proposition 4-15: If P is a Markov chain, then

$$\begin{aligned} N_{ij} &= H_{ij} N_{jj} \\ \bar{N}_{ij} &= \bar{H}_{ij} N_{jj} \\ N_{ii} &= 1 + \bar{H}_{ii} N_{ii}. \end{aligned}$$

Proof: The third assertion follows from the second and the identity $N = I + \bar{N}$ with $j = i$. For the first assertion we apply Theorem 4-11 with $\mathbf{f} = \mathbf{n}_j$ and the random time equal to $\mathbf{t}_j$. It is clear that $\mathbf{n}_j(\omega) = \mathbf{n}_j(\omega_{t_j})$ if $\mathbf{t}_j(\omega) < \infty$. Therefore $\mathbf{f}' = \mathbf{n}_j$, and

$$\begin{aligned} N_{ij} = \mathrm{M}_i[\mathbf{n}_j] &= \sum_k \mathrm{Pr}_i[x_{t_j} = k]\, \mathrm{M}_k[\mathbf{n}_j] \\ &= \mathrm{Pr}_i[x_{t_j} = j]\, \mathrm{M}_j[\mathbf{n}_j] \\ &= \mathrm{Pr}_i[\mathbf{t}_j < \infty]\, \mathrm{M}_j[\mathbf{n}_j] \\ &= H_{ij} N_{jj}. \end{aligned}$$

Similarly, for the second assertion we apply Theorem 4-11 with $\mathbf{f} = \bar{\mathbf{n}}_j$, $\mathbf{f}' = \mathbf{n}_j$, and the random time equal to $\bar{\mathbf{t}}_j$. By the same kind of argument, we find

$$\mathrm{M}_i[\bar{\mathbf{n}}_j] = \bar{H}_{ij}\,\mathrm{M}_j[\mathbf{n}_j].$$

Proposition 4-16: Let p be the statement that a Markov chain reaches state j and then state k with $j \neq k$. Then $\Pr_i[p] = H_{ij}H_{jk}$.

PROOF: In the notation of Theorem 4-10, if $\mathbf{t}_j$ is taken as the random time, then p' is the statement h_k. The theorem applies and

$$\begin{aligned}\Pr_i[p] &= \sum_m \Pr_i[x_{\mathbf{t}_j} = m]\Pr_m[p'] \\ &= \Pr_i[x_{\mathbf{t}_j} = j]\Pr_j[h_k] \\ &= H_{ij}H_{jk}.\end{aligned}$$

In our discussion of Markov chains, we shall make frequent use of the following notational devices. Let k and j be states of a Markov chain. By ${}^k\mathbf{n}_j(\omega)$ is meant the number of times on the path ω that the process is in state j before (and not including) the first time that the process is in k. We define ${}^k\bar{\mathbf{n}}_j$ as the number of visits to j before the process reaches k *after* time 0. Notice that ${}^j\mathbf{n}_j(\omega) = 0$, but ${}^j\bar{\mathbf{n}}_j(\omega)$ is 1 if ω starts with j. For fixed k we introduce the corresponding matrices ${}^kN$ and ${}^k\bar{N}$ by

$$\begin{aligned}{}^kN_{ij} &= \mathrm{M}_i[{}^k\mathbf{n}_j(\omega)], \\ {}^k\bar{N}_{ij} &= \mathrm{M}_i[{}^k\bar{\mathbf{n}}_j(\omega)].\end{aligned}$$

They are related as follows:

$${}^k\bar{N}_{ij} = \delta_{ij} + \mathrm{M}_i[{}^k\mathbf{n}_j(\omega_1)].$$

We further define ${}^kH_{ij}$ to be the probability of hitting j before k having started in i; ${}^k\bar{H}_{ij}$ is the probability of hitting j before hitting k after time 0, having started in i.

We will later want a more general notation than ${}^k\mathbf{n}_j(\omega)$. By ${}^E\mathbf{n}_j(\omega)$ we shall mean the number of times on the path ω that the process is in j before it is in any state of the set E. It is sometimes convenient, in this connection, to think in terms of the modified chain in which the states of E have been made absorbing. Again we have matrices ${}^EN$, and we also introduce the matrices ${}^EH$ and ${}^E\bar{H}$ analogously.

If E is a subset of the set of states S for which neither E nor $\tilde{E}$ is empty, we shall decompose the P matrix into

$$P = \begin{array}{c} \\ E \\ \tilde{E} \end{array}\!\!\begin{array}{c} \begin{array}{cc} E & \tilde{E} \end{array} \\ \begin{pmatrix} T & U \\ R & Q \end{pmatrix} \end{array}$$

according to the method discussed at the end of Section 1-1. If A is an arbitrary matrix indexed by the set S, we write A_E for the restriction of A to a matrix indexed only by E. As an example, we note that $P_E = T$.

7. Classification of states

We introduce a partial ordering on the states S of a Markov chain P. Two states i and j are said to be R-related, written $R(i, j)$ if $H_{ij} > 0$, that is, if it is possible to reach j from i. If $R(i, j)$ and $R(j, i)$, we say that i and j **communicate** and write $i \sim j$. To see that R is a partial ordering, we note that

(1) $H_{ii} = 1 > 0$ so that $R(i, i)$.
(2) If $R(i, j)$ and $R(j, k)$, then $R(i, k)$ because $H_{ik} \geq H_{ij}H_{jk} > 0$ by Proposition 4-16.

The reader should verify that $\sim$ is an equivalence relation.

The relation $\sim$ therefore partitions the states of S into equivalence classes within the ordering, and movement from state to state is within a class or upward through the ordering. We do not assert the existence of maximal classes; we shall see an example later where no maximal classes are present. (The reader should then be able to exhibit an example of a chain having no minimal classes.)

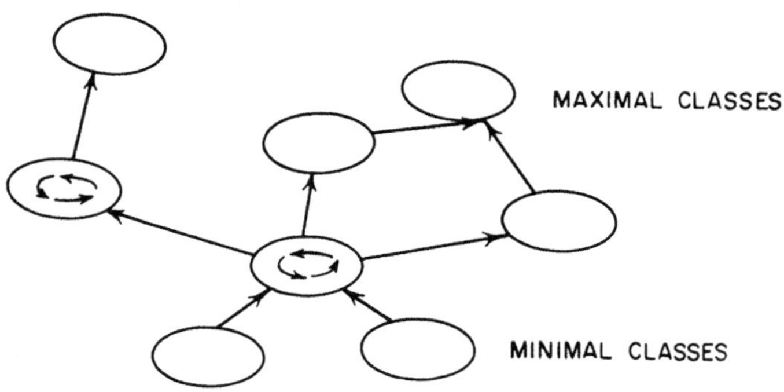

FLOW IN A MARKOV CHAIN

Proposition 4-17: States i and j are R-related if and only if there exists an $n \geq 0$ for which $(P^n)_{ij} > 0$.

PROOF: Suppose $n \geq 0$ is the smallest exponent for which $(P^n)_{ij} > 0$. Then $F_{ij}^{(n)} = (P^n)_{ij} > 0$, and since $H_{ij} = \sum_n F_{ij}^{(n)}$, we have $R(i, j)$. Conversely, if $R(i, j)$, then $H_{ij} > 0$ and it must be true that $F_{ij}^{(n)} > 0$ for some n. Thus $(P^n)_{ij} > 0$.

Definition 4-18: A state i is said to be **recurrent** if $\bar{H}_{ii} = 1$; it is said to be **transient** if $\bar{H}_{ii} < 1$.

The lemma to follow contains some identities connecting H and $\bar{H}$ which will be used in the next few propositions. The reader should study these examples of the use of the strong Markov property in order to develop his intuition.

Lemma 4-19: The following statements hold:

(1) The probability starting in i of returning to state i at least k times is $(\bar{H}_{ii})^k$. (Use the convention $0^0 = 1$.)

(2) The probability starting in i of returning at least k times to i before hitting j is $({}^j\bar{H}_{ii})^k$, provided $i \neq j$.

(3) The probability starting in i of returning to i *via* j is ${}^i\bar{H}_{ij}H_{ji}$, provided $i \neq j$.

(4) The probability starting in i of reaching j for the first time after n returns to i is $({}^j\bar{H}_{ii})^n \, {}^i\bar{H}_{ij}$, provided $i \neq j$.

(5) The probability starting in i of being in state j at least n times is $H_{ij}(\bar{H}_{jj})^{n-1}$.

PROOF: The proofs are all by Theorem 4-10.

(1) Use induction on k. For $k = 0$ the result is trivial; assume that it holds for $k - 1$. Let p be the statement that the process returns to i at least k times, and let $\mathbf{t} = \bar{\mathbf{t}}_i$. Then p' is the statement that the process returns to i at least $k - 1$ times.

$$\begin{aligned}\Pr_i[p] &= \sum_j \Pr_i[x_{\bar{t}_i} = j] \Pr_j[p'] \\ &= \Pr_i[x_{\bar{t}_i} = i] \Pr_i[p'] \\ &= \Pr_i[\bar{\mathbf{t}}_i < \infty] \Pr_i[p'] \\ &= \bar{H}_{ii}(\bar{H}_{ii})^{k-1} \quad \text{by inductive hypothesis.}\end{aligned}$$

(2) If $i \neq j$, the result is the same as (1) for the chain in which the single state j has been made absorbing (see Example 2, Section 4-2).

(3) Let p be the statement that the process returns to i *via* j, and let $\mathbf{t}$ be the time that j is reached if j is reached before a return to i, or $+\infty$ if j is not reached before i. Then p' is the statement that i is reached, and

$$\begin{aligned}\Pr_i[p] &= \sum_k \Pr_i[x_{\mathbf{t}} = k] \Pr_k[p'] \\ &= \Pr_i[x_{\mathbf{t}} = j] H_{ji} \\ &= \Pr_i[\mathbf{t} < \infty] H_{ji} \\ &= {}^i\bar{H}_{ij}H_{ji}.\end{aligned}$$

(4) The argument is the same as in (3). Use the systems theorem with $\mathbf{t}$ equal to the time of the nth return to i if the return occurs before j is reached, or $+\infty$ otherwise.

(5) The proof is by induction on n and is the same as in (1) except that the random time becomes $\mathbf{t} = \bar{\mathbf{t}}_j$.

Proposition 4-20: State i is transient if and only if $N_{ii} < +\infty$. Then $N_{ii} = 1/(1 - \bar{H}_{ii})$.

PROOF:

$$N_{ii} = \sum_{k=1}^{\infty} k \Pr_i[\mathbf{n}_i = k],$$

which upon rearrangement of terms becomes

$$= \sum_{k=1}^{\infty} \sum_{m=k}^{\infty} \Pr_i[\mathbf{n}_i = m],$$

which by complete additivity is

$$= \sum_{k=1}^{\infty} \Pr_i[\mathbf{n}_i \geq k]$$

$$= \sum_{k=1}^{\infty} (\bar{H}_{ii})^{k-1} \quad \text{by conclusion (1) of Lemma 4-19.}$$

The right side is finite if and only if $\bar{H}_{ii} < 1$.

Corollary 4-21: If j is a transient state, then $N_{ij} < \infty$ for all states i in the chain, and $N_{ij} = H_{ij}N_{jj}$.

PROOF: From Proposition 4-15 we have

$$N_{ij} = H_{ij}N_{jj} \leq N_{jj}.$$

The result now follows from Proposition 4-20.

We are now in a position to put together the ideas of recurrence and transience with the partial ordering R and the equivalence relation $\sim$. We need two lemmas before we can prove our fundamental result—that all states in an equivalence class are of the same type, recurrent or transient.

Lemma 4-22: If $i \neq j$ and $R(i, j)$, then ${}^j\bar{H}_{ii} < 1$ and ${}^jN_{ii} < +\infty$.

PROOF: Suppose ${}^j\bar{H}_{ii} = 1$. By conclusion (2) of Lemma 4-19 the probability of returning n times before hitting j is $({}^j\bar{H}_{ii})^n = 1$. Hence, by Proposition 2-6, there is probability one for returning infinitely often before hitting j, in contradiction to the relation $R(i, j)$. For the second half of the lemma, we have, as in Proposition 4-20,

$$^{j}N_{ii} = \sum_{n=0}^{\infty} (^{j}\bar{H}_{ii})^{n} < +\infty.$$

Lemma 4-23: If i is recurrent and $R(i, j)$, then $H_{ji} = 1$ and $H_{ij} = 1$.

Proof: The result is obvious if $i = j$. If $i \neq j$, consider returning to i with and without first reaching j. By conclusion (3) of Lemma 4-19,

$$1 = \bar{H}_{ii} = {}^{i}\bar{H}_{ij}H_{ji} + {}^{j}\bar{H}_{ii}$$

Since ${}^{i}\bar{H}_{ij} + {}^{j}\bar{H}_{ii} \leq 1$, this equation is a contradiction unless ${}^{j}\bar{H}_{ii} = 1$ or $H_{ji} = 1$. The first alternative is ruled out by Lemma 4-22. Thus $H_{ji} = 1$ and ${}^{i}\bar{H}_{ij} = 1 - {}^{j}\bar{H}_{ii}$. Next, since i is recurrent, one may compute H_{ij} by summing the probabilities of reaching j for the first time after n returns to i, where $n = 0, 1, 2, \ldots$. By conclusion (4) of Lemma 4-19,

$$H_{ij} = \sum_{n=0}^{\infty} ({}^{j}\bar{H}_{ii})^{n}\, {}^{i}\bar{H}_{ij} = (1 - {}^{j}\bar{H}_{ii}) \sum_{n=0}^{\infty} ({}^{j}\bar{H}_{ii})^{n} = 1.$$

The last equality holds, since ${}^{j}\bar{H}_{ii} < 1$ by Lemma 4-22.

Proposition 4-24: All states in an equivalence class are of the same type, recurrent or transient.

Proof: It is sufficient to show that if one state in an equivalence class is recurrent, so are all others. Let i be a recurrent state, and suppose $j \sim i$, $j \neq i$. Then $\bar{H}_{jj} \geq H_{ji}H_{ij}$, since the probability of returning is at least as great as the probability of returning via i. (We have used an argument familiar from Proposition 4-16 to compute the latter.) Hence $\bar{H}_{jj} = 1$ by Lemma 4-23.

Corollary 4-25: If i is recurrent and $i \sim j$, then $H_{ij} = H_{ji} = 1$.

Proof: The corollary follows from Lemma 4-23.

Because of Proposition 4-24 we are free to speak of transient and recurrent classes of states. We shall mention a few simple results about classes of states. By a **closed class** we mean one that it is impossible to leave. A process cannot disappear when it is in a closed class.

Proposition 4-26: Recurrent classes are closed and maximal with respect to the partial ordering R.

PROOF: It is sufficient to prove that a recurrent class S' is closed, since closed classes are clearly maximal. Suppose the class can be left, say from a state $j \in S'$. If k is a state outside S' for which $P_{jk} > 0$, then it is not true that $R(k, j)$ because j and k do not communicate. Thus $\bar{H}_{jj} \leq 1 - P_{jk} < 1$, and j is not recurrent.

Proposition 4-27: If a Markov chain is started in a recurrent class S', then the chain is in every state of S' infinitely often with probability one. In particular, if i and j are in S', then $N_{ij} = +\infty$.

PROOF: Suppose the chain is started in state i. Then, by conclusion (5) of Lemma 4-19, the probability of being in state j at least n times is $H_{ij}(\bar{H}_{jj})^{n-1} = 1$. By Proposition 2-6 the chain is in state j infinitely often with probability one. Again by Proposition 2-6 it is in *every* state infinitely often with probability one.

Proposition 4-28: A Markov chain is in a finite subset of transient states only finitely often, with probability one.

PROOF: If the chain were in a finite set S' infinitely often with positive probability, it would be in one state j of S' infinitely often with positive probability. Such an occurrence would imply that N_{ij} is infinite for some i, in contradiction to Corollary 4-21 if j is transient.

We single out two kinds of Markov chains for special attention. We note that every absorbing state forms a one-element recurrent class, and conversely.

Definition 4-29: A Markov chain is said to be a **recurrent chain** if its states comprise a single equivalence class and if that class is recurrent. A chain is called a **transient chain** if all of its recurrent states are absorbing.

If P is an arbitrary Markov chain with r recurrent classes, then all properties of P can be deduced from the properties of one transient and r recurrent chains. This assertion follows from the observations:

(1) If the process P starts in a recurrent state j, movement from state to state is confined to the single equivalence class to which j belongs. The properties of the chain started in j are the properties of a chain while it is in one recurrent class; they are thus the properties of a recurrent chain.

(2) If the process P starts in a transient state, its behavior while in transient states is the same as the behavior of the transient chain P' obtained from P by making all recurrent states absorbing. If P enters a recurrent state, then P' becomes absorbed. And after P has entered a recurrent state, its properties are those of a recurrent chain. Thus the properties of P may be studied by considering the one transient chain P' and the r separate recurrent chains.

Because of these observations, we shall restrict our discussion in subsequent chapters to Markov chains which are either transient or recurrent.

The reader should notice that every chain whose states form only one equivalence class is either a transient chain or a recurrent chain. Shortly we shall examine the basic example, in which all pairs of states communicate, to determine when it is transient and when it is recurrent.

First we discuss some properties of maximal classes for a moment. Not every chain has maximal classes; a tree process, for example, consists of infinitely many transient classes of one state each. None of the classes is maximal. Even if a chain does have a maximal class, that class does not have to be closed. The process may have a positive probability of disappearing from some state in the maximal class.

Nor is it true that all closed classes are recurrent. An additional condition is needed.

Proposition 4-30: All closed equivalence classes consisting of finitely many states are recurrent.

PROOF: Let the states be the first n positive integers, and suppose the class is transient. Then N_{ij} is finite for every i and j in the class. Therefore

$$c = M_i\left[\sum_{j=1}^{n} \mathbf{n}_j\right] = \sum_{j=1}^{n} N_{ij}$$

is finite. But c is the mean total number of steps taken in the class, and c is infinite because the class is closed. This contradiction establishes the proposition.

To see that infinite closed equivalence classes need not be recurrent, we consider the basic example, whose states form a single equivalence class. Let $\bar{H}_{00}^{(n)}$ be the probability that the chain, started in state 0, returns to 0 at some time up to and including time n. Then

$$\bar{H}_{00} = \lim_n \bar{H}_{00}^{(n)}.$$

But

$$1 - \bar{H}_{00}^{(n)} = p_1 p_2 \cdots p_n = \beta_n,$$

since a single step other than away from zero returns the process to zero at once. Now in order for the process to be recurrent it is necessary and sufficient that $\bar{H}_{00} = 1$ or that

$$\lim_n \beta_n = \lim_n (1 - \bar{H}_{00}^{(n)}) = 0.$$

The reader should be able to construct examples where $\lim_n \beta_n = 0$ and where $\lim_n \beta_n \neq 0$. Thus the basic example may be either transient or recurrent.

8. Problems

1. Find an expression analogous to that in Proposition 4-3 for $\Pr_i[x_n = j]$ in a Markov process.
2. Let p_m be the Poisson distribution with mean m on the non-negative integers. A game is played as follows: A random integer n_1 is selected with probabilities determined by p_1. A second random integer n_2 is selected with probabilities determined by p_{n_1}. The ith random integer is selected with probabilities determined by $p_{n_{i-1}}$. Prove that with probability one the integer 0 is eventually selected.
3. Show that if $h \geq 0$ is a column vector for which $P^n h$ converges, then the limit function is non-negative superregular.
4. Let j be an absorbing state. Prove that the probability starting at i of ever reaching j is a regular function.
5. Show that an independent trials process is a Markov chain in which P_{ij} is independent of i. Let 0 be any fixed state and let $\mathbf{t}$ be any stopping time. Show that $\Pr_\pi[x_{\mathbf{t}+1} = j] = P_{0j}$, and give an example to show that $\Pr_\pi[x_{\mathbf{t}} = j]$ does not have to equal P_{0j}.
6. If the symmetric random walk in 3 dimensions is started at the origin, the probability of being at the origin after n steps is 0 if n is odd and is of the order of magnitude of $n^{-3/2}$ for n even. Prove that the probability of returning to the origin is less than 1.
7. Consider the following random walk in the plane. If the process is not on an axis, it is equally likely to move to any of the four neighboring states. If it is at the origin, it stays at the origin. Otherwise, on the x-axis it takes a step away from the origin, whereas on the y-axis it takes a step toward the origin. Give a complete classification of the states.
8. Let j be a transient state in a closed class. Prove that there must be a state i in the class such that $H_{ij} < 1$.
9. Prove that every tree-process is a transient chain and that each equivalence class of states is a unit-set.
10. Prove or disprove: In a chain with a minimal class and with no closed class, there is no non-zero non-negative regular measure.

Problems 11 to 14 refer to a reflecting random walk, that is, a random walk on the non-negative integers with $q_0 = 0$.

11. Prove that the only regular functions are constants.

12. Let

$$\beta_i = p_0 p_1 \cdots p_{i-1}, \qquad \gamma_i = q_1 q_2 \cdots q_i, \qquad \alpha_i = c\beta_i/\gamma_i.$$

Show that, for any choice of the constant c, α is a regular measure.

13. Show that all regular measures are of the form given in the previous problem.

14. Show that $\alpha_j P_{ji}/\alpha_i = P_{ij}$ for all i and j.

Problems 15 to 18 refer to a branching process and use the notation of Sections 2 and 3 of the text.

15. Show that the roots of the equation $\varphi(r) = r$ satisfy the following conditions:
(a) There is an $r < 1$ if $m > 1$.
(b) There is an $r > 1$ if $m < 1$.
(c) $r = 1$ is the only root if $m = 1$.

16. Show that $\{r^{x_n}\}$ is a martingale if and only if $\varphi(r) = r$.

17. Show that $\{x_n\}$ is a martingale if and only if $m = 1$.

18. What condition on m will assure that the branching process has positive probability of survival (of not dying out)?

Problems 19 to 24 concern space-time processes and martingales. If P is a Markov chain with state space S, we define the **space-time process** to be a Markov chain whose states are pairs (i, n), where i is in S and n is a non-negative integer, and which moves from (i, n) to $(j, n + 1)$ with probability P_{ij}.

19. Prove that any space-time process is transient. What can be said about classification of states?

20. Prove that if $f(i, n)$ is a finite-valued non-negative regular function for the space-time process, then $f(x_n, n)$ is a martingale for the process P started at a given state 0.

21. Specialize to the case of sums of independent random variables on the integers with $p_k = 0$ for $k < 0$. Define $\varphi(t) = \sum_k p_k t^k$ for all $t \geq 0$ for which the right side is finite. Show that $\varphi(t)$ is defined at least for $0 \leq t \leq 1$. Fix a t for which $\varphi(t)$ is defined and put

$$f(i, n) = \frac{t^i}{[\varphi(t)]^n}.$$

Prove that $f(i, n)$ is regular for the space-time process.

22. In Problem 21 show that $f(x_n, n)$ converges a.e. if the process is started at 0.

23. Specialize further to the case where $p_0 = p_1 = \frac{1}{2}$, and define, for $0 \leq t \leq 1$,

$$g(i, n) = 2^n t^i (1 - t)^{n-i}.$$

Show by change of variable in Problem 22 that $g(x_n, n)$ converges almost everywhere in the process started at 0.

24. Using only the result of Problem 23, prove that if p is any number between 0 and 1, not equal to $\frac{1}{2}$, then the probability that $x_n = [np]$ for infinitely many n is 0. Here $[np]$ is the nearest integer to np.

CHAPTER 5

TRANSIENT CHAINS

1. Properties of transient chains

Recall that a transient Markov chain is a Markov chain all of whose recurrent states are absorbing. Its transition matrix satisfies $P\mathbf{1} \leq \mathbf{1}$. For any transient state j in the chain, we have seen that $\bar{H}_{jj} < 1$ and $N_{ij} < +\infty$ for every i. If E is any set of states, we can put the transition matrix in the canonical form

$$P = \begin{array}{c} \\ E \\ \tilde{E} \end{array} \begin{array}{c} \begin{array}{cc} E & \tilde{E} \end{array} \\ \begin{pmatrix} T & U \\ R & Q \end{pmatrix} \end{array}.$$

In the special case in which P is a transient chain and E is the set of absorbing states, we find that $T = I$ and $U = 0$. (If there are no absorbing states, we agree to write $P = Q$. We shall assume that not all states are absorbing, however.) Thus, for a transient chain,

$$P = \begin{pmatrix} I & 0 \\ R & Q \end{pmatrix}.$$

The matrices R and Q for a transient chain will always be associated with this standard decomposition. We observe that Q itself is the transition matrix for a transient chain and that this chain has only transient states. Some authors actually define a transient chain to be one with all states transient. However, in the study of these chains, it is often convenient to add absorbing states to ensure $P\mathbf{1} = \mathbf{1}$. And as we saw in Chapter 4, the decomposition of general Markov chains into transient and recurrent chains depends on allowing absorbing states in transient chains. For these reasons we have adopted the slightly more general definition of transient chain which permits absorbing states.

Let P be the transition matrix of a transient chain, and consider the quantity N_{ij}, the mean number of times in state j when the process is started in state i. If j is an absorbing state, then this quantity is infinite if j can be reached from i with positive probability and 0 otherwise. If i is absorbing, it is 0 unless $i = j$, and then it is infinite. Hence N_{ij} is of interest only when i and j are transient. Thus we shall agree to restrict N to these entries: The matrix so restricted is called the **fundamental matrix** for the chain. We shall show that the restricted matrix is the matrix $\{N_{ij}\}$ for the chain determined by Q. In what follows, N always denotes the restricted matrix associated with P.

Lemma 5-1: If P is a transient chain and if $\tilde{E}$ is the set of transient states, then $(P^k)_{\tilde{E}} = Q^k$.

PROOF: We readily verify by induction that

$$P^k = \begin{array}{c} \\ E \\ \tilde{E} \end{array} \begin{array}{c} \begin{array}{cc} E & \tilde{E} \end{array} \\ \begin{pmatrix} I & 0 \\ (I + Q + \cdots + Q^{k-1})R & Q^k \end{pmatrix} \end{array},$$

and the result follows at once.

Proposition 5-2:

$$N = \sum_{k=0}^{\infty} Q^k.$$

PROOF: For transient states i and j, we have in the P-process

$$N_{ij} = \sum_k (P^k)_{ij} = \sum_k (Q^k)_{ij},$$

by Proposition 4-12 and Lemma 5-1.

Proposition 5-3: N is finite-valued, and $\lim_k Q^k = 0$.

PROOF: N_{ij} in the P-process is finite when j is transient; hence N is finite-valued. Therefore $\lim_k (Q^k) = 0$ by Proposition 5-2.

We recall that $\bar{N}_{ij} = \mathrm{M}_i[\bar{\mathbf{n}}_j]$, and $N = I + \bar{N}$. Hence $\bar{N} = \sum_{k=1}^{\infty} Q^k$.

Proposition 5-4: If P is a transient chain, then

$$\begin{aligned} \bar{N} &= QN \\ N &= I + QN \\ N_{ij} &= H_{ij}N_{jj} \\ \bar{N}_{ij} &= \bar{H}_{ij}N_{jj} \\ N_{ii} &= 1 + \bar{H}_{ii}N_{ii} \\ N_{ii} &= 1/(1 - \bar{H}_{ii}). \end{aligned}$$

PROOF: The first two assertions are restatements of Proposition 4-13 for the case where Q is our transition matrix. The last four results are a restatement of Proposition 4-15.

Note that the conclusions of Proposition 5-4 show how to compute N from H and $\bar{H}$. Our next result establishes a method of finding N without using the H-matrix: For finite matrices the knowledge that N is a (two-sided) inverse of $I - Q$ is sufficient to determine N uniquely, but for infinite matrices it is not. For if r is a Q-regular column vector and β is a Q-regular row vector, then $N + r\beta$ is a second two-sided inverse of $I - Q$. We shall see that such regular vectors r and β often exist.

In Section 2 we shall obtain a refinement of Proposition 5-5 by proving that N is the unique *minimum* non-negative inverse of $I - Q$ on each side.

Proposition 5-5: $N(I - Q) = (I - Q)N = I$ and $QN = NQ \leq N$. In particular, every row of N is a Q-superregular measure, and every column of N is a Q-superregular function.

PROOF: The second and third assertions follow from the first, and $QN = N - I$ by Proposition 5-4. Also $NQ = N - I$ by Proposition 5-2 and monotone convergence. Since N has finite entries, the first assertion follows.

If P is a transient chain with a non-empty set E of absorbing states, we define the **absorption matrix** B to have index sets $\tilde{E}$ and E and to have entries

$$B_{ij} = \Pr_i[\text{process is absorbed at } j].$$

The B-matrix is not square; it has the same index sets as the R-matrix.

Proposition 5-6: If P is a transient chain with a non-empty set of absorbing states, then $B = NR$.

PROOF: Let i be transient and let j be absorbing. By Theorem 4-10 with the random time equal to the constant n and with the statement p taken as the assertion that the process is absorbed at j on the $n + 1$st step, we have

$$\begin{aligned}\Pr_i[p] &= \sum_k (P^n)_{ik}R_{kj} \\ &= \sum_k (Q^n)_{ik}R_{kj}.\end{aligned}$$

Summing on n, we find

$$B = \sum_{n=0}^{\infty} (Q^n R),$$

which by monotone convergence

$$= \left(\sum_{n=0}^{\infty} Q^n\right) R$$
$$= NR.$$

As a result, we see from the proof of Lemma 5-1 that if P is a transient chain, then

$$\lim_k P^k = \begin{pmatrix} I & 0 \\ B & 0 \end{pmatrix}.$$

Let P be an arbitrary Markov chain, let E be a subset of the set of states, and let s_E be the statement that the process is in states of E infinitely often. Define s^E by $s_i^E = \Pr_i[s_E]$.

Proposition 5-7: For any subset E of states in a Markov chain P s^E is a P-regular function.

PROOF: Letting p be the statement s_E and taking the random time to be identically one, we see that p' in Theorem 4-10 is also s_E and that

$$\Pr_i[s_E] = \sum_k P_{ik} \Pr_k[s_E]$$

or

$$s^E = Ps^E.$$

For any Markov chain P we define a **hitting vector** h^E and an **escape vector** e^E by

$h_i^E = \Pr_i[\text{process eventually reaches } E]$

and

$e_i^E = \Pr_i[\text{process goes on first step from } E \text{ to } \tilde{E} \text{ and then never returns to } E]$.

We notice that if $i \in E$, then $h_i^E = 1$, and that if $j \in \tilde{E}$ then $e_j^E = 0$.

The **absorption matrix** B^E for the set E is defined to be a square matrix with index set the set of all states and with entries defined by

$B_{ij}^E = \Pr_i[\text{process at some time enters } E \text{ and first entry is at state } j]$.

We see that the B^E-matrix is computed by finding the entries of the B-matrix for the process with states of E made absorbing. Specifically, if E is the set of all absorbing states, then

$$B^E = \begin{matrix} & E & \tilde{E} \\ E & I & 0 \\ \tilde{E} & B & 0 \end{matrix}.$$

The matrices s^E, h^E, e^E, and B^E are interrelated as in the following proposition, whose proof is left to the reader.

Proposition 5-8: Let P be an arbitrary Markov chain. Then

(1) $h^E = B^E\mathbf{1}$.
(2) $h^E = e^E + Ph^E$ and hence $e^E = (I - P)h^E$.
(3) $s^E = \mathbf{1}$ if and only if $h^E = \mathbf{1}$ and $P\mathbf{1} = \mathbf{1}$.
(4) If $E \subset F$, then $h^E \le h^F$ and $s^E \le s^F$.
(5) $s^E = B^E s^E$.
(6) If $E \subset F$, then $B^F B^E = B^E$.
(7) $s^E = \lim P^n h^E$.

2. Superregular functions

Superregular measures and functions were defined in Section 4-3; a vector is P-superregular if $h \ge Ph$. Let P be a transient chain, and let Q be the restriction of P to transient states. As we have seen before, Q is a transition matrix. Our object in this section is to obtain a standard decomposition of non-negative Q-superregular functions and to use it in a consideration of the solutions to the equation $(I - Q)x = f$. Our results will hold equally well for Q-superregular measures, but we shall not supply the proofs. A way of transforming rigorously theorems and proofs about functions into theorems and proofs about measures will emerge later when we discuss duality. Generalizations of the present results will arise in the study of potential theory.

The transformation later of theorems about functions into theorems about measures by duality will require the existence of a positive finite-valued Q-superregular measure. Any row of N will suffice if all pairs of transient states communicate, but if not, we proceed as follows: Number the states, beginning with 1, and take

$$\beta = \sum_i 2^{-i} N^{(i)},$$

where $N^{(i)}$ is the ith row of N. It is clear that β is superregular because it is the sum of non-negative superregular measures; β is positive because $\beta_j = \sum_i 2^{-i}N_{ij} \geq 2^{-j}N_{jj} \geq 2^{-j} > 0$. Finally β is finite-valued because

$$\beta_j = \sum_i 2^{-i}N_{ij} = \sum_i 2^{-i}H_{ij}N_{jj} \leq \sum_i 2^{-i}N_{jj} \leq N_{jj} < \infty.$$

The lemma and theorem to follow hold for arbitrary Markov chains. In the transient case they will most often be applied to the chain Q. The theorem has an analog in classical potential theory, but we postpone a discussion of this point until the end of Section 8-1 after we have introduced Markov chain potentials.

Lemma 5-9: Let P be any Markov chain and let $N = \sum P^n$. If Nf is well defined and finite-valued, then $(I - P)(Nf) = f$.

PROOF: Write $f = f^+ - f^-$. Then Nf^+ and Nf^- are both finite-valued by hypothesis. Since $PN + I = N$, we have $PN \leq N$ and hence $PNf^+ \leq Nf^+$ and $PNf^- \leq Nf^-$. Therefore, by Corollary 1-5,

$$\begin{aligned}(I - P)(Nf) = Nf - P(Nf) &= Nf - (PN)f \\ &= Nf - (N - I)f \\ &= f.\end{aligned}$$

Theorem 5-10: Let P be any Markov chain and let $N = \sum P^n$. Any non-negative P-superregular finite-valued function h has a unique representation $h = Nf + r$, where r is regular. In the representation f and r are both non-negative, and $f = (I - P)h$.

PROOF: Since h is P-superregular,

$$h \geq Ph \geq P^2h \geq \cdots \geq 0.$$

Thus P^nh converges to a non-negative function r. By the Dominated Convergence Theorem,

$$Pr = P(\lim P^nh) = \lim P^{n+1}h = r.$$

Hence r is regular. Also

$$h = P^{n+1}h + (I + P + \cdots + P^n)(h - Ph).$$

Since $h - Ph \geq 0$, we may apply monotone convergence in passing to the limit on n; we obtain

$$h = r + N(h - Ph).$$

Set $f = h - Ph$, and existence follows.

For uniqueness, suppose that $h = r' + Nf'$ with r' regular. Then Nf and Nf' are finite-valued since h is. Multiplying the equation

$$r + Nf = r' + Nf'$$

through by $I - P$ and applying Lemma 5-9, we obtain

$$f = f'.$$

Hence also $r = r'$.

We return now to the special case of transient chains, where $N = \sum Q^n$. A solution g to an equation is the **minimum non-negative solution** if whenever h is a non-negative solution, we have $h \geq g \geq 0$.

Proposition 5-11: If $f \geq 0$ and if Nf is finite, then Nf is the minimum non-negative solution of $(I - Q)x = f$.

PROOF: By Lemma 5-9, Nf is a solution. Let x be any non-negative solution. Then x is finite-valued and superregular. By Theorem 5-10, $x = Nf + r$ where $r \geq 0$. Hence $x \geq Nf$.

It follows that N is the minimum non-negative right inverse of $(I - Q)$. To prove that the jth column of N is minimum, define f by $f_i = \delta_{ij}$ and then apply Proposition 5-11. After the analog of Proposition 5-11 for measures has been established, we find similarly that N is the minimum non-negative left inverse of $(I - Q)$.

3. Absorbing chains

A class of Markov chains of special interest is the class of absorbing chains. We shall use the material developed in the two preceding sections to establish the basic facts about absorbing chains.

Definition 5-12: A Markov chain P is said to be **absorbing** if, for every starting state, the probability of ending in an absorbing state is one.

If P is a Markov chain containing a recurrent nonabsorbing state i, then the process cannot be absorbed if it is started in state i. That is, all absorbing chains are transient chains. It is not true, however, that all transient chains are absorbing. The property $P1 = 1$ is a necessary condition. But even it is not sufficient, since the basic example can be transient but is never absorbing.

The proposition to follow is the special case of the identity $B^E 1 = h^E$ in which E is the set of all absorbing states.

Proposition 5-13: If P is a transient chain, then P is absorbing if and only if $B\mathbf{1} = \mathbf{1}$.

The next two propositions give two ways in which absorbing chains arise.

Proposition 5-14: If P is a finite transient chain such that $P\mathbf{1} = \mathbf{1}$, then P is absorbing.

PROOF:

$$\begin{aligned} B\mathbf{1} &= (NR)\mathbf{1} = N(R\mathbf{1}) \\ &= N[(I - Q)\mathbf{1}] \quad \text{since } (R\mathbf{1} + Q\mathbf{1})_i = (P\mathbf{1})_i = 1 \\ &= [N(I - Q)]\mathbf{1} \quad \text{by Corollary 1-6} \\ &= \mathbf{1}. \end{aligned}$$

Let $\mathbf{a}(\omega)$ be the time on the path ω of a chain P that absorption takes place. If the process is not absorbed along ω, define $\mathbf{a}(\omega) = +\infty$. Since $\mathbf{a}(\omega) = \sum_j \mathbf{n}_j(\omega)$, where the sum is taken over the transient states, we see that $\mathbf{a}$ is measurable and we conclude that $\mathbf{a}$ is a random time. Define the column vector a by $a_i = M_i[\mathbf{a}]$. The vector a is indexed by the transient states. It is clear that the chain P is absorbing if and only if $\mathbf{a}$ is finite a.e.

Proposition 5-15: If P is a recurrent chain and if ${}^EP$ is the Markov chain obtained by making a non-empty set E of states absorbing, then ${}^EP$ is absorbing.

PROOF: Let $j \in E$. Since $H_{ij} = 1$ for every i, $\mathbf{t}_j(\omega)$ is finite almost everywhere. But $\mathbf{a}(\omega) \leq \mathbf{t}_j(\omega)$, and ${}^EP$ is thus absorbing.

The notation ${}^EP$ will be used in later sections to refer either to the chain P with the states made absorbing or to the chain P made so that it disappears instead of entering E. If

$$P = \begin{pmatrix} T & U \\ R & Q \end{pmatrix},$$

then these two chains are, respectively,

$$\begin{pmatrix} I & 0 \\ R & Q \end{pmatrix} \quad \text{and} \quad \begin{pmatrix} 0 & 0 \\ 0 & Q \end{pmatrix}.$$

It will be clear from the context which one is meant.

There is one more important way in which absorbing chains arise. Suppose that $P\mathbf{1} \neq \mathbf{1}$. If we add an absorbing state 0, the state "stopped," to the state space S and define $\bar{P}$ by

$$\begin{aligned} \bar{P}_{ij} &= P_{ij} \quad \text{if } i \neq 0 \text{ and } j \neq 0 \\ P_{0j} &= \delta_{0j} \\ \bar{P}_{i0} &= 1 - \sum_{k \in S} P_{ik} \quad \text{if } i \neq 0, \end{aligned}$$

then $\bar{P}$, called the **enlarged chain**, may be absorbing. If P is a finite transient chain, then $\bar{P}$ necessarily will be absorbing by Proposition 5-14.

With this set of propositions to indicate how absorbing chains arise, we conclude with an investigation of the properties of the vector a.

Proposition 5-16: If P is a transient chain, then $a = N\mathbf{1}$.

PROOF:

$$\begin{aligned} (N\mathbf{1})_i &= \sum_j N_{ij} \quad \text{summed over the transient states} \\ &= \sum_j M_i[\mathbf{n}_j] \\ &= \mathrm{M}_i\Big[\sum_j \mathbf{n}_j\Big] \quad \text{by monotone convergence.} \end{aligned}$$

But $\mathbf{a} = \sum_j \mathbf{n}_j$, where the sum is taken over transient states j. Thus $(N\mathbf{1})_i = \mathrm{M}_i[\mathbf{a}] = a_i$.

Corollary 5-17: If P is a transient chain for which $P\mathbf{1} = \mathbf{1}$ and if a has only finite entries, then $x = a$ is the unique minimum non-negative solution of the equation $(I - Q)x = \mathbf{1}$.

PROOF: It is the unique minimum non-negative solution by Proposition 5-16 and Proposition 5-11.

4. Finite drunkard's walk

The finite drunkard's walk is a Markov chain defined on the integers $\{0, 1, \ldots, n\}$ with states 0 and n absorbing and with transition probabilities

$$P_{i,i+1} = p$$

and

$$P_{i,i-1} = q = 1 - p \quad \text{for } 0 < i < n.$$

If we set $r = q/p$, two cases arise. Either $r = 1$ and $\{x_n\}$ is a martingale or $r \neq 1$ and $\{r^{x_n}\}$ is a martingale.

We shall use the second martingale systems theorem (Theorem 3-15) to compute the entries of B, H, N, and a for the case $r = 1$.

To compute the entries of the B matrix when $r = 1$, we note that $B\mathbf{1} = \mathbf{1}$ by Proposition 5-14; therefore $B_{i0} = 1 - B_{in}$ for each transient state i. Since $\{x_n\}$ is a bounded martingale, Corollary 3-16 applies with the time taken as the time of absorption (which is a stopping time because P is absorbing if and only if $\mathbf{a}$ is finite a.e.). Then

$$i = 0B_{i0} + nB_{in}$$

so that

$$B_{in} = i/n$$

and

$$B_{i0} = 1 - i/n.$$

To find the entry H_{ij} of the H-matrix, we make state j absorbing and consider the resulting process. If $i \le j$, the modified process is the drunkard's walk on the integers $\{0, \ldots, j\}$ with j absorbing. Hence $H_{ij} = i/j$. If $i \ge j$, the modified process is the drunkard's walk on $\{j, \ldots, n\}$. Renumbering the states, we can consider the process as starting at $i - j$ and taking place on the states $\{0, \ldots, n - j\}$. Thus,

$$H_{ij} = 1 - \frac{i - j}{n - j} = \frac{n - i}{n - j}.$$

To get $\bar{H}_{ii}$, where i is transient, we use the fact that $\bar{H} = PH$, so that

$$\begin{aligned}\bar{H}_{ii} &= pH_{i+1,i} + qH_{i-1,i}\\ &= 1 - \frac{n}{2i(n - i)} \quad \text{since } p = q = \tfrac{1}{2}.\end{aligned}$$

The N-matrix is determined as a function of $\bar{H}$ and H by

$$N_{jj} = \frac{1}{1 - \bar{H}_{jj}}$$

and

$$N_{ij} = H_{ij}N_{jj}.$$

We find

$$N_{ij} = \begin{cases} \dfrac{2}{n}\,(i)(n - j) & \text{for } i \le j \\[2ex] \dfrac{2}{n}\,(j)(n - i) & \text{for } i \ge j. \end{cases}$$

Finally, we have

$$a_i = \mathrm{M}_i[\mathbf{a}] = (N\mathbf{1})_i = \sum_{j=1}^{n-1} N_{ij} = i(n - i).$$

The case $r \neq 1$ proceeds in the same way. By the systems theorem

$$r^i = r^0 B_{i0} + r^n B_{in}.$$

From this equation one easily deduces that

$$B_{i0} = \frac{r^i - r^n}{1 - r^n}.$$

After first computing H and N, we find

$$a_i = \mathrm{M}_i[\mathbf{a}] = \frac{1}{q - p}\left[i - n\frac{1 - r^i}{1 - r^n}\right].$$

When $r > 1$, this process is sometimes known as "gambler's ruin" because of the following interpretation. A gambler walks into a gambling house with i dollars in his pocket, and the house has $n - i$ dollars to bet against him. In a given game the gambler has probability p of winning. Since the house fixes the odds, we have $p < \frac{1}{2}$ and therefore $r > 1$. If the game is played repeatedly, x_k in the above Markov chain represents the gambler's cash after k games, and B_{i0} is the probability of his eventual ruin. Since $r > 1$,

$$B_{i0} = \frac{1 - r^{-(n-i)}}{1 - r^{-n}} > 1 - r^{-(n-i)}$$

is nearly 1 when $n - i$ (the house's capital) is large. Thus the gambler is nearly sure to be ruined, no matter how rich he is. However, a_i is approximately $i/(q - p)$, which is very large if i is substantial and p is near to $\frac{1}{2}$. Thus the gambler is likely to have a long run for his money.

5. Infinite drunkard's walk

Extending the finite drunkard's walk to a process P defined on all of the non-negative integers, we set

$$\begin{aligned} P_{0i} &= \delta_{0i} & \\ P_{i,i+1} &= p & \text{for } 0 < i < \infty \\ P_{i,i-1} &= q = 1 - p & \text{for } 0 < i < \infty. \end{aligned}$$

Again we take $r = q/p$.

Our first problem is to establish the sense in which the infinite drunkard's walk P is the limiting case of the finite drunkard's walk.

Let ${}^nB$, ${}^nN$, and ${}^na$ denote, respectively, the absorption matrix, the fundamental matrix, and the mean time to absorption vector for the finite drunkard's walk on the integers $\{0, \ldots, n\}$. Define in connection with the infinite drunkard's walk the random variable ${}^k\mathbf{n}_j$ to be the number of times the process is in state j up to the first time it is in state k. Let p be the statement that the process is not absorbed at state 0, and let p_k be the statement that the process is not absorbed at state 0 at any time up to the time it reaches state k.

Proposition 5-18: In the infinite drunkard's walk

$$B_{i0} = \lim_n {}^nB_{i0},$$

$$N_{ij} = \lim_n {}^nN_{ij},$$

and

$$a_i = \lim_n {}^na_i.$$

PROOF: We have $\Pr[p] = 1 - B_{i0}$ and $\Pr[p_n] = 1 - {}^nB_{i0}$. Since the union of the truth sets of the p_n is the truth set of p, we have, by Proposition 1-16,

$$1 - B_{i0} = \lim_n (1 - {}^nB_{i0}).$$

For the N-matrix we note that

$${}^nN_{ij} = \mathrm{M}_i[{}^n\mathbf{n}_j]$$

and

$$\mathbf{n}_j = \lim_n {}^n\mathbf{n}_j \quad \text{monotonically.}$$

The result for N therefore follows from the Monotone Convergence Theorem. Since $a_i = \mathrm{M}_i[\mathbf{a}] = \sum_j N_{ij}$, the assertion about a_i is also a consequence of monotone convergence.

Taking the limits of some of the quantities computed for the finite drunkard's walk we find that

$$B_{i0} = \begin{cases} 1 & \text{if } r \geq 1 \\ r^i & \text{if } r \leq 1 \end{cases}$$

and

$$a_i = \begin{cases} \dfrac{i}{q - p} & \text{if } r > 1 \\ +\infty & \text{if } r \leq 1. \end{cases}$$

The value of B_{i0} shows that the chain is absorbing when $p \leq q$; that is, $\mathbf{a}$ is finite almost everywhere when $p \leq q$. However, $\mathrm{M}_i[\mathbf{a}]$ is finite only when $p < q$.

If we have calculated B_{i0} and have seen that **a** is a stopping time when $p \leq q$, we may compute $\mathrm{M}_i[\mathbf{a}]$ directly from martingales without any knowledge of the N-matrix. Let $p \leq q$, and for $0 < s \leq 1$ define

$$f(k, n) = \frac{s^k}{(ps + qs^{-1})^n},$$

where n represents time and k represents position. Now

$$f(k, n) = s^k(ps + qs^{-1})^{-n} \leq (ps + qs^{-1})^{-n},$$

which is maximized when $s = \sqrt{q/p}$. Thus

$$\begin{aligned} f(k, n) &\leq (p/q)^{n/2} \\ &\leq 1 \quad \text{since } p \leq q. \end{aligned}$$

Hence f is bounded. It is easily seen that $\{f(x_n, n)\}$ is a martingale: Since f is bounded, $\mathrm{M}[f]$ is finite; the reader may verify the regularity property by showing that

$$p \cdot f(x_n + 1, n + 1) + q \cdot f(x_n - 1, n + 1) = f(x_n, n).$$

Let **a** be the stopping time of Corollary 3-16. Taking i as the starting state, we have

$$\frac{s^i}{(ps + qs^{-1})^0} = \sum \Pr[\mathbf{a} = n] \frac{s^0}{(ps + qs^{-1})^n}.$$

Set $1/u = ps + qs^{-1}$. Then

$$s = \frac{1 - \sqrt{1 - 4pqu^2}}{2pu}$$

and

$$\sum \Pr[\mathbf{a} = n]u^n = \left(\frac{1 - \sqrt{1 - 4pqu^2}}{2pu}\right)^i.$$

Defining

$$\varphi(u) = \sum \Pr[\mathbf{a} = n]u^n,$$

we note that

$$\varphi'(u) = \sum n \Pr[\mathbf{a} = n]u^{n-1}$$

and that

$$\varphi'(1) = \sum n \Pr[\mathbf{a} = n] = \mathrm{M}_i[\mathbf{a}] = a_i.$$

Using the fact that $\sqrt{1 - 4pq} = q - p$ to calculate $\varphi'(1)$, we find that $a_i = i/(q - p)$, in agreement with the result obtained by the longer method.

The present method further allows us to find the probability distribution of Pr[**a** $= n$] by expanding $[(1 - \sqrt{1 - 4pqu^2})/(2pu)]^i$ as a power series in u. We thus find that

$$\Pr[\mathbf{a} = n] = 0 \quad \text{for } n < i$$
$$\Pr[\mathbf{a} = i] = q^i$$
$$\Pr[\mathbf{a} = i + 1] = 0$$
$$\Pr[\mathbf{a} = i + 2] = ipq^{i+1}$$

and so on.

6. A zero-one law for sums of independent random variables

Historically, the first infinite Markov chain that was studied was the sums of independent random variables process. We gather some of the results in the next few sections, beginning with two propositions and a corollary of rather general applicability.

Proposition 5-19: If P is a Markov chain for which the only bounded P-regular functions are constant vectors, then, for each subset of states E, $\Pr_i[s_E] = 0$ or 1, independently of the starting state i.

PROOF: By Proposition 5-7, s^E is regular and it is clearly bounded; therefore $s^E = c\mathbf{1}$. On the other hand, by Proposition 5-8,

$$s^E = B^E s^E$$

so that

$$c\mathbf{1} = cB^E\mathbf{1} = ch^E.$$

Therefore, either $c = 0$ or $h^E = \mathbf{1}$. In the latter case, $s^E = \mathbf{1}$ by Proposition 5-8.

As in Example 6 of Section 4-6, we let $p_k = P_{i,i+k}$, which, by assumption, is independent of i.

Proposition 5-20: Let P be the transition matrix of a Markov chain obtained from sums of independent random variables. If, for each pair of states q and r, there is a state s such that q can be reached from s or s can be reached from q *and* such that r can be reached from s or s can be reached from r, then the only bounded regular functions are constant functions. In particular, the hypothesis is satisfied if all pairs of states communicate.

PROOF: Let f be non-constant regular and suppose $f_q \neq f_r$. We shall assume that s can be reached from both q and r; the proof in the other cases is completely analogous. Let $q, q + a_1, q + a_1 + a_2, \ldots,$

$q + a_1 + \cdots + a_m$ and $r, r + b_1, \ldots, r + b_1 + \cdots + b_n$ be respective paths of positive probability leading from q to s and from r to s. Then

$$f_{q+a_1+\cdots+a_m} = f_s = f_{r+b_1+\cdots+b_n},$$

so that at least one of the equalities in the two chains

$$f_q = f_{q+a_1} = f_{q+a_1+a_2} = \cdots = f_{q+a_1+\cdots+a_m}$$

and

$$f_r = f_{r+b_1} = f_{r+b_1+b_2} = \cdots = f_{r+b_1+\cdots+b_n}$$

must be false, since otherwise $f_q = f_r$. Without loss of generality, let $f_{q+\cdots+a_{k-1}} \neq f_{q+\cdots+a_k}$ and let $a = a_k$. Then $p_a > 0$. Let $g_i = f_{i+a} - f_i$. Then g is not identically 0. Further, g is regular because

$$\begin{aligned}\sum_j P_{ij} g_j &= \sum_j P_{ij}(f_{j+a} - f_j)\\ &= \sum_j P_{ij} f_{j+a} - \sum_j P_{ij} f_j\\ &= \sum_j P_{i+a,j+a} f_{j+a} - \sum_j P_{ij} f_j\\ &= f_{i+a} - f_i\\ &= g_i.\end{aligned}$$

Suppose that for all i, $|f_i| \leq c$. Then $|g_i|$ is bounded by $2c$. Since multiplying g by -1 affects neither its regularity nor its boundedness, we may assume $b = \sup_i g_i$ is positive and finite. For any i and any $m > 0$,

$$\left|\sum_{k=0}^{m-1} g_{i+ka}\right| = |f_{i+ma} - f_i| \leq 2c.$$

Choose N so that $N \cdot b/2 > 2c$. Let $p^{(n)} = P^{(n)}_{i,i+na} \geq (p_a)^n > 0$, let $p = \min_{n<N} p^{(n)}$, and let t be a state such that $g_t > b(1 - p/2)$. A choice for t exists since b is finite and since $p > 0$. Then for $n < N$,

$$\begin{aligned}b\left(1 - \frac{p^{(n)}}{2}\right) \leq b\left(1 - \frac{p}{2}\right) &< g_t\\ &= \sum_k P^{(n)}_{tk} g_k\\ &= P^{(n)}_{t,t+na} g_{t+na} + \sum_{k \neq t+na} P^{(n)}_{tk} g_k\\ &\leq p^{(n)} g_{t+na} + (1 - p^{(n)})b.\end{aligned}$$

Thus

$$g_{t+na} > b/2 \quad \text{for } n < N.$$

Hence

$$\sum_{k=0}^{N-1} g_{t+ka} > 2c,$$

a contradiction.

Corollary 5-21: If P is a sums of independent random variables Markov chain in which all pairs of states communicate, then for each subset of states E, $\Pr_i[s_E] = 0$ or 1, independently of the starting state i.

7. Sums of independent random variables on the line

Let P be a sums of independent random variables Markov chain indexed by the integers and defined by the probability distribution $\{p_k\}$. The set of integers k for which $p_k > 0$ we shall call the set of k-values associated with the chain. We shall assume that the greatest common divisor of the k-values is one. Thus, if both positive and negative k-values exist, we see (from Lemma 1-66, for example) that all pairs of states communicate.

The mean m for the process is defined by $m = \sum_k kp_k$ and is said to exist if and only if the positive and negative parts of the sum are not both infinite. In this section we shall establish the following result.

Proposition 5-22: If P is a Markov chain representing sums of independent random variables on the line, if there are finitely many k-values and if they have greatest common divisor one, if $\sum p_k = 1$, and if $m = \sum kp_k$, then in order for the chain to be recurrent it is necessary and sufficient that $m = 0$.

Before we come to the proof, two comments are in order. The first is that the proposition can be generalized to the case where there are infinitely many k-values as long as the mean m still exists and the k-values still have greatest common divisor one. The same condition $m = 0$ is necessary and sufficient for P to be recurrent. The second comment is that the necessity of the condition $m = 0$ is an immediate consequence of the Strong Law of Large Numbers (Theorem 3-19) and that the special added assumption we used in the proof of that theorem translates exactly into the condition that there are only finitely many k-values. Nevertheless, we give a different proof.

For the proof we may assume that both positive and negative k-values exist. Otherwise, the chain is obviously transient. Recalling our discussion in Example 2 of Section 3-2, we observe that if both positive and negative k-values exist, then there are either two distinct real roots or one double real root of the equation

$$f(s) = \sum_k p_k s^k = 1.$$

If there is a root r other than $s = 1$, then $\{r^{x_n}\}$ is a non-negative martingale. And if $s = 1$ is a double root, then $\{x_n\}$ is a martingale.

But $f'(1) = (\sum kp_k s^k)_{s=1} = \sum kp_k = m$, and $f(s) = 1$ has a double root at $s = 1$ if and only if $m = 0$. In the case $m \neq 0$, $\{r^{x_n}\}$ converges a.e. and must converge to the zero function. Thus, according as $r < 1$ or $r > 1$, we have $\lim x_n = +\infty$ or $\lim x_n = -\infty$. Hence, the process returns to each state only finitely often with probability one. But if the chain were recurrent, each state would be reached infinitely often with probability one. Therefore, if $m \neq 0$, the chain is transient.

In the case in which $m = 0$, let $-u$ be the smallest k-value and let v be the largest k-value. Let E be the set of states $\{-u, \ldots, -2, -1\}$ and let E' be the set of states $\{j, j + 1, \ldots, j + v - 1\}$ for some fixed j. Start the process in state i with $0 \leq i < j$, and let $\mathbf{t}$ be the time to reach the set $E \cup E'$. The chain stopped at time $\mathbf{t}$ is absorbing by Proposition 5-14, and $\mathbf{t}$ is therefore a stopping time. Since $\{x_n\}$ is a bounded martingale before time $\mathbf{t}$, Corollary 3-16 applies. Therefore, $\mathrm{M}[x_0] = \mathrm{M}[x_t]$, and for $0 \leq i < j$, we have

$$\begin{aligned} i = \mathrm{M}[x_0] = \mathrm{M}[x_t] &= \sum_{k \in E} B^E_{ik} k + \sum_{k \in E'} B^{E'}_{ik} k \\ &\geq -u \sum_{k \in E} B^E_{ik} + j \sum_{k \in E'} B^{E'}_{ik} \\ &= -uh^E_i + jh^{E'}_i, \end{aligned}$$

which, by Proposition 5-13,

$$= -uh^E_i + j(1 - h^E_i).$$

Then

$$(j + u)h^E_i \geq j - i$$

and

$$h^E_i \geq \frac{j - i}{j + u}.$$

Letting $j \to \infty$, we find that $h^E_i = 1$ for all $i \geq 0$.

Reversing the argument for $i \leq 0$ and $F = \{1, \ldots, v\}$, we find similarly that $h^F_i = 1$ for all $i \leq 0$. Thus, for any state i, $h^{E \cup F}_i = 1$. By Proposition 5-8, $s^{E \cup F}_i = 1$. Since $E \cup F$ is a finite set, Proposition 4-28 applies, and the chain is recurrent.

8. Examples of sums of independent random variables

Calculations with sums of independent random variables on the line normally involve either martingales or difference equations. We shall illustrate in this section each of these methods with an example.

EXAMPLE 1: Let the defining distribution for a Markov chain P representing sums of independent random variables on the integers be $\{p_1 = q, p_2 = p\}$. The process is obviously transient since $\bar{H}_{jj} = 0$

and $N_{jj} = 1$ for all j. Since $N_{ij} = H_{ij}N_{jj} = H_{ij} = H_{0,j-i}$, we will have determined the H-matrix and the N-matrix completely by finding the value of H_{0k} for all k. We note first that $H_{0k} = 0$ if k is negative. For the case $k \geq 0$, $\{r^{x_n}\}$ is a nonconstant martingale if r is a nonzero root other than one of the equation $\sum p_k s^k = qs + ps^2 = 1$. Thus $\{(-1/p)^{x_n}\}$ is a martingale. Taking the stopping time $\mathbf{t}$ as the time when the process reaches or passes state k, we find from Corollary 3-16 that for the process started at state 0

$$(-1/p)^0 = H_{0k}(-1/p)^k + (1 - H_{0k})(-1/p)^{k+1}.$$

Therefore,

$$H_{0k} = \frac{1}{1+p}(1 - (-p)^{k+1}).$$

It is interesting to note that

$$\lim_{k\to\infty} H_{0k} = \frac{1}{1+p}.$$

This result can also be obtained from the Renewal Theorem of Section 1-6 if we observe that

$$m = \sum kp_k = q + 2p = 1 + p$$

so that

$$\frac{1}{1+p} = \frac{1}{m}.$$

EXAMPLE 2: Let $p_{-1} = \frac{2}{9}$, $p_1 = \frac{3}{9}$, and $p_2 = \frac{4}{9}$. Then

$$m = \sum kp_k = 1,$$

and the process is transient by Proposition 5-22. Let the transition matrix be called P.

If g is a P-regular vector at state i, then $g_i = (Pg)_i$ and

$$g_i = \tfrac{2}{9}g_{i-1} + \tfrac{3}{9}g_{i+1} + \tfrac{4}{9}g_{i+2}.$$

We shall need a characterization of such vectors in the calculation of the H-matrix. The difference equation we have just formed may be rewritten as

$$4g_{i+2} + 3g_{i+1} - 9g_i + 2g_{i-1} = 0.$$

Its characteristic equation (see Section 1-6b) is

$$4k^3 + 3k^2 - 9k + 2 = 0,$$

whose solutions are $k = 1, \frac{1}{4}$, and -2. Thus,

$$g_i = A + B(\tfrac{1}{4})^i + C(-2)^i.$$

In finding the entries of the H-matrix, we know that $H_{ij} = H_{i-j,0}$; therefore, it is sufficient to consider only the entries H_{i0}. We shall look at the cases $i \geq 0$ and $i \leq 0$ separately.

Suppose $i \geq 0$. Since the only negative k-value is -1, it is impossible for the process to start in state $i + 1$ and reach state 0 without first passing through state i. Thus, for $i \geq 0$,

$$\begin{aligned} H_{i+1,0} &= H_{i+1,i}H_{i,0} \\ &= H_{1,0}H_{i,0} \end{aligned}$$

with $H_{00} = 1$. The result is a first-order difference equation whose solution is $H_{i,0} = c(H_{1,0})^i$. Setting $i = 0$ shows that $c = 1$. Thus $H_{i,0}$ is exponential.

But $\bar{H} = PH$, so that $H_{i,0} = (PH)_{i,0}$ for all $i > 0$. Thus $H_{i,0}$ satisfies the difference equation

$$g_i = \tfrac{2}{9}g_{i-1} + \tfrac{3}{9}g_{i+1} + \tfrac{4}{9}g_{i+2}$$

for $i > 0$. It therefore satisfies

$$g_{i+1} = \tfrac{2}{9}g_i + \tfrac{3}{9}g_{i+2} + \tfrac{4}{9}g_{i+3}$$

for $i \geq 0$. Hence $H_{i,0} = A + B(\tfrac{1}{4})^i + C(-2)^i$ for all $i \geq 0$. Since $H_{i,0}$ is known to be exponential, two of the coefficients A, B, and C are zero and the other is one. The alternatives $C = 1$ and $A = 1$ are eliminated, respectively, by the facts that -2 is not a probability and that P drifts to the right a.e. Thus,

$$H_{i,0} = (\tfrac{1}{4})^i \quad \text{for } i \geq 0.$$

For $i \leq 0$ we again use the fact that $\bar{H} = PH$, and we find that $H_{i,0}$ is a solution of the equation

$$g_{i-2} = \tfrac{2}{9}g_{i-3} + \tfrac{3}{9}g_{i-1} + \tfrac{4}{9}g_i$$

for all $i \leq 1$. Therefore,

$$H_{i,0} = A + B(\tfrac{1}{4})^i + C(-2)^i$$

for all $i \leq 1$. Known values for $H_{i,0}$ when $i = 0$ and $i = 1$ give us two conditions on the three unknowns A, B, and C. The fact that $H_{i,0} \leq 1$ as $i \to -\infty$ tells us that $B = 0$. We have as a result

$$H_{i,0} = \begin{cases} (\tfrac{1}{4})^i & \text{for } i \geq 0 \\ \tfrac{3}{4} + \tfrac{1}{4}(-2)^i & \text{for } i \leq 1. \end{cases}$$

From a knowledge of the H-matrix, we can compute $\bar{H}$ by $\bar{H} = P\bar{H}$. The entries of the N-matrix follow from

$$N_{jj} = 1/(1 - \bar{H}_{jj})$$

and

$$N_{ij} = H_{ij}N_{jj}.$$

9. Ladder process for sums of independent random variables

For a sums of independent random variables Markov chain defined on the integers, we define a sequence $\mathbf{s}_i(\omega)$ of **positive step times** inductively as follows: $\mathbf{s}_0(\omega)$ is the least n such that $x_0(\omega) > 0$, and $\mathbf{s}_i(\omega)$ is the least n such that $x_n(\omega) > x_{\mathbf{s}_{i-1}(\omega)}(\omega)$. If we construct a stochastic process by watching the old Markov chain only at the positive step times—that is, by calling the nth outcome in the new process the $\mathbf{s}_n$th outcome in the old process—then the strong Markov property as formulated in Theorem 4-9 implies that the new process is a Markov chain. We shall go through this implication in detail.

Proposition 5-23: If P is a sums of independent random variables Markov chain defined on the integers, then the stochastic process whose nth outcome is the $\mathbf{s}_n$th outcome in P is a Markov chain P^+. Moreover, $P^+_{ij} = P^+_{0,j-i}$.

PROOF: The times $\mathbf{s}_n$ are random times. Applying Theorem 4-9 to the time $\mathbf{s}_n$ and the statement $r \equiv (x_{\mathbf{s}_{n+1}} = c_{n+1})$, we find that, if $\Pr_\pi[x_{\mathbf{s}_0} = c_0 \wedge \cdots \wedge x_{\mathbf{s}_n} = c_n] > 0$, then

$$\Pr_\pi[x_{\mathbf{s}_{n+1}} = c_{n+1} \mid x_{\mathbf{s}_0} = c_0 \wedge \cdots \wedge x_{\mathbf{s}_n} = c_n] = \Pr_\pi[x_{\mathbf{s}_{n+1}} = c_{n+1} \mid x_{\mathbf{s}_n} = c_n].$$

And if $\Pr_\pi[x_{\mathbf{s}_n} = c_n] > 0$, then

$$\Pr_\pi[x_{\mathbf{s}_{n+1}} = c_{n+1} \mid x_{\mathbf{s}_n} = c_n] = \Pr_{c_n}[x_{\mathbf{s}_1} = c_{n+1}].$$

Thus the process is a Markov chain P^+. The fact that $P^+_{ij} = P^+_{0,j-i}$ follows from the fact that P represents sums of independent random variables.

The chain P^+ is called the **ladder process** for P. The ladder process moves from i to j if j is the first state greater than i that is reached in the original process. If the mean step m in P is positive, then the process reaches or passes any given positive state with probability one, so that the $\mathbf{s}_i$ are finite a.e. Hence $P^+\mathbf{1} = \mathbf{1}$, and the ladder process represents sums of independent random variables.

As an example, we shall compute explicitly the ladder process associated with Example 2 of the preceding section. For the given chain we have $p_{-1} = \frac{2}{9}$, $p_1 = \frac{3}{9}$, and $p_2 = \frac{4}{9}$. The ladder process has two k-values, namely 1 and 2, and thus has a distribution $\{p_1^+, p_2^+\}$. Since the positive step times are finite a.e., we have $\sum p_k^+ = 1$ or

$$p_1^+ + p_2^+ = 1.$$

To find the values of p_1^+ and p_2^+, we note that

$$\begin{aligned} p_2^+ &= {}^1H_{02} \\ &= P_{02} + P_{0,-1}p_1^+p_2^+ \\ &= p_2 + p_{-1}p_1^+p_2^+. \end{aligned}$$

Putting in the known values for p_{-1} and p_2, we find

$$p_1^+ = p_2^+ = \tfrac{1}{2}.$$

The ladder process for our Example 2 is therefore an instance of Example 1 in the same section.

10. The basic example

The basic example is a Markov chain with state space the non-negative integers and with transition probabilities

$$\begin{aligned} P_{i-1,i} &= p_i, \quad i > 0 \\ P_{i-1,0} &= q_i = 1 - p_i. \end{aligned}$$

We normally assume that none of the p_i's is 0. A row vector β is defined by

$$\begin{aligned} \beta_0 &= 1 \\ \beta_i &= p_i\beta_{i-1} \quad \text{for } i > 0; \end{aligned}$$

it is regular if and only if $\lim_{i\to\infty} \beta_i = 0$, and the process is recurrent or transient according as the limit is or is not 0.

In this section we shall compute the H and N matrices for the basic example when it is transient, and we shall show that a transient basic example has no non-zero regular row vector.

The process cannot leave the set $\{0, 1, \ldots, j\}$ without hitting j. Hence $H_{ij} = 1$ if $i \leq j$. If $i > j$, then j can be reached only via 0, so that

$$H_{ij} = H_{i0}H_{0j} = H_{i0}$$

by Proposition 4-16. Thus we need only find H_{i0}. The only way the process can fail to reach 0 is to continue moving to the right from i.

Let $\beta_\infty = \lim_{i\to\infty} \beta_i = \prod_{i=1}^{\infty} p_i$. Then

$$1 - H_{i0} = \prod_{j>i} p_i = \frac{\beta_\infty}{\beta_i},$$

and we find

$$H_{ij} = \begin{cases} 1 & \text{if } i \leq j \\ 1 - \dfrac{\beta_\infty}{\beta_i} & \text{if } i > j. \end{cases}$$

Then

$$\begin{aligned} \bar{H}_{jj} &= p_{j+1}H_{j+1,j} + q_{j+1}H_{0j} \\ &= p_{j+1}\left(1 - \frac{\beta_\infty}{\beta_{j+1}}\right) + q_{j+1} \\ &= 1 - \frac{\beta_\infty}{\beta_j}. \end{aligned}$$

Suppose now that the process is transient—that is, that $\beta_\infty > 0$. Then

$$N_{jj} = \frac{1}{1 - \bar{H}_{jj}} = \frac{\beta_j}{\beta_\infty},$$

so that

$$N_{ij} = H_{ij}N_{jj} = \begin{cases} \dfrac{\beta_j}{\beta_\infty} & \text{if } i \leq j \\ \dfrac{\beta_j}{\beta_\infty} - \dfrac{\beta_j}{\beta_i} & \text{if } i > j. \end{cases}$$

If $\beta_\infty > 0$, we know that β is not regular. Indeed, a transient basic example has no non-zero regular row vector. For if α is regular, then

$$\sum_i \alpha_i q_{i+1} = \alpha_0$$

and

$$\alpha_{j-1}p_j = \alpha_j \quad \text{for } j > 0.$$

From the second condition we find by induction that $\alpha_j = \alpha_0\beta_j$. Then the first condition yields

$$\alpha_0 = \alpha_0 \sum_i \beta_i q_{i+1} = \alpha_0 \sum_i (\beta_i - \beta_{i+1}) = \alpha_0(1 - \beta_\infty).$$

Thus $\alpha_0 = 0$ and $\alpha = 0$.

11. Problems

1. Consider the finite Markov chain with states

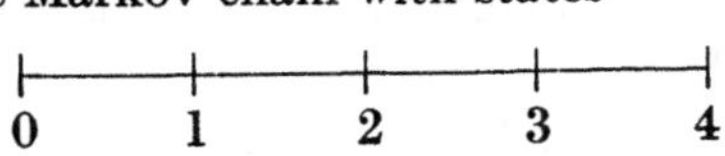

States 0 and 4 are absorbing. At each of the other states the process takes a step to the right with probability $\frac{2}{3}$, or a step to the left with probability $\frac{1}{3}$. Compute P, N, and B by means of Propositions 5-5 and 5-6.

2. If the states of a transient chain form a single closed set, show that each column of N is a non-constant positive superregular function. [*Note:* We shall see later that there are no such functions for recurrent chains; hence their existence is a necessary and sufficient condition for a closed set to be transient.]

3. Prove Proposition 5-8. Prove also that $h^E = Ne^E + s^E$. Interpret each result.

4. In the basic example, let $E = \{0, 1, 2\}$. Compute B^E, h^E, e^E, and s^E. Check formulas (1), (2), and (3) in Proposition 5-8.

5. Prove an analog of Theorem 5-10 for row vectors. Use it to show that if $\pi \geq \pi P \geq 0$ in a transient basic example, then there is a measure μ such that $\pi = \mu N$.

6. For a transient chain let

$$x_i = \mathrm{M}_i[\mathbf{n}_E],$$

where $\mathbf{n}_E$ is the number of times the chain is in the finite set of states E. Use a systems theorem to find an equation of the form

$$(I - P)x = y,$$

and prove that x is the minimum non-negative solution.

7. Find the probability in the p-q random walk started at 0 of reaching $+n$ before $-n$. [*Hint:* Use the results obtained for the finite drunkard's walk.] If $p > q$, what happens to this probability as n increases?

8. The one-dimensional symmetric random walk is a process to which Corollary 5-21 applies. If E is the set of primes, is s^E equal to 1 or is it equal to 0?

9. Let $x_0, x_1, x_2, \ldots$ be the outcome functions for the symmetric random walk on the integers started at 0. Show that there is no non-constant non-negative function $f(n)$ defined on the integers such that $f(x_0)$, $f(x_1), \ldots$ is a martingale.

10. Show by direct computation that the sums of independent random variables process on the integers with $p_2 = \frac{1}{3}$ and $p_{-1} = \frac{2}{3}$ is recurrent.

11. Find H and N for sums of independent random variables on the integers with $p_{-1} = p_2 = \frac{1}{2}$.

Problems 12 to 19 refer to sums of independent random variables on the integers with $p_{-1} = \frac{1}{3}$ and $p_1 = \frac{2}{3}$.

12. Find H and N.

13. Describe the long-range behavior of the chain.

14. Give two examples of infinite sets E to illustrate the two possibilities $s^E = 1$ and $s^E = 0$.

15. Find all non-negative regular functions.

16. Give a necessary and sufficient condition on a non-negative function f that Nf be finite-valued.
17. In the previous problem, let $g = Nf$. Choose an f satisfying your condition, and show that $(I - P)g = f$.
18. Let

$$h_i = \begin{cases} 2 + (\frac{1}{2})^i & \text{if } i < 0 \\ (3 + i)(\frac{1}{2})^i & \text{if } i \geq 0. \end{cases}$$

Show that h is superregular, and decompose h as in Theorem 5-10.
19. Use the function of Problem 18 together with the Martingale Convergence Theorem to prove that the process is to the left of 0 only finitely often a.e.

Problems 20 to 22 refer to the game of tennis. It will be necessary to know how one keeps score in tennis. A match is being played between A and B, and A has probability p of winning any one point.

20. Set up a single game as a transient chain with the two absorbing states "A wins" and "B wins." [Minimize the number of states, e.g., identify "30–30" with "deuce."] Compute the probability that A wins the game as a function of p.
21. Suppose that A has probability p' of winning a game. What is the probability that he wins a set? What of winning the match (if he is required to win three sets)?
22. What is the probability that A wins the match if $p = 0.6$? What if $p = 0.51$?

CHAPTER 6

RECURRENT CHAINS

1. Mean ergodic theorem for Markov chains

Recurrent chains are Markov chains such that the set of states is a single recurrent class. They have the properties that $P\mathbf{1} = \mathbf{1}$, $\bar{H} = E$, and $\mathrm{M}_i[\mathbf{n}_j] = \infty$. The study of recurrent chains begins with a characterization of finite-valued non-negative superregular measures and functions; the reader should turn back to Sections 1-6c and 1-6d for the terms referred to in what follows.

We shall apply Proposition 1-63 and Corollary 1-64 to the sequence of matrices obtained as the Cesaro sums of the powers of a recurrent chain P. Define

$$L^{(n)} = \frac{1}{n}\sum_{k=0}^{n-1} P^k.$$

Then

$$0 \le L^{(n)} \le E \quad \text{for all } n.$$

Theorem 6-1: If P is the matrix of a recurrent chain, then the sequence of powers of P is Cesaro summable to a limiting matrix L with the properties $L \ge 0$ and $LP = L = PL = L^2$.

Proof: We shall show that every convergent subsequence converges to the same limit L. The proof proceeds in four steps.

(1) Since

$$L^{(n)} = \frac{1}{n}(I + P + \cdots + P^{n-1}),$$

we have

$$PL^{(n)} = \frac{1}{n}(P + P^2 + \cdots + P^n) = L^{(n)}P$$

$$= L^{(n)} + \frac{1}{n}(P^n - I).$$

Let $\{L^{(n_\nu)}\}$ be a convergent subsequence; such a sequence exists by Proposition 1-63. Set $L = \lim_\nu L^{(n_\nu)}$. Then $L \geq 0$. Since

$$\lim_n \frac{1}{n}(P^n - I) = 0,$$

we have $\lim_\nu PL^{(n_\nu)} = L = \lim_\nu L^{(n_\nu)}P$.

(2) By Proposition 1-56 (dominated convergence), we have

$$\lim_\nu PL^{(n_\nu)} = P \lim_\nu L^{(n_\nu)} = PL$$

and thus

$$PL = L.$$

By Proposition 1-55 (Fatou's Theorem), we may further conclude

$$(\lim_\nu L^{(n_\nu)})P \leq \lim_\nu (L^{(n_\nu)}P) = L$$

and

$$LP \leq L.$$

(3) Suppose LP is not equal to L. Then for some i and j, $(LP)_{ij} < L_{ij}$. Summing the inequalities $(LP)_{ik} \leq L_{ik}$ on k, we obtain

$$\sum_k (LP)_{ik} < \sum_k L_{ik}$$

since strict inequality holds in the jth entry. Thus $[(LP)\mathbf{1}]_i < (L\mathbf{1})_i$. Since L, P, and $\mathbf{1}$ are non-negative, associativity holds and $(LP)\mathbf{1} = L(P\mathbf{1}) = L\mathbf{1}$. Therefore, $[(LP)\mathbf{1}]_i = (L\mathbf{1})_i$, and we have a contradiction. Hence $LP = L = PL$. By induction, we readily see that $LP^n = L = P^nL$ for every n. Adding these results, we obtain finally

$$LL^{(n)} = L = L^{(n)}L.$$

(4) Let $\{L^{(n_\mu)}\}$ be a convergent subsequence with limit $\bar{L}$. It is sufficient to show that $L = \bar{L}$. From step (3) we have $L = LL^{(n_\mu)}$ for any μ, and by Fatou's Theorem $L\mathbf{1} \leq \mathbf{1}$ and $\bar{L}\mathbf{1} \leq \mathbf{1}$. Thus, by dominated convergence,

$$L = \lim_\mu (LL^{(n_\mu)}) = L\bar{L}.$$

Interchanging the roles of L and $\bar{L}$, we find $\bar{L} = \bar{L}L$. But by Fatou's Theorem $\bar{L}L \leq L$ and $L\bar{L} \leq \bar{L}$. Therefore,

$$\bar{L} = \bar{L}L \leq L \quad \text{and} \quad L = L\bar{L} \leq \bar{L}.$$

Hence $L = \bar{L}$ and $L = L^2$.

Definition 6-2: If P is a recurrent chain, P is said to be a **null chain** if $L = 0$. If $L \neq 0$, P is said to be an **ergodic chain** and the limit matrix L is called A.

Proposition 6-3: If P is a recurrent chain, every constant function is regular and the only non-negative (finite-valued) superregular functions are constants.

PROOF: Constant functions are trivially regular since $P1 = 1$. Let h be a finite-valued non-negative superregular function, and let the chain be started in any fixed state i. Since $M[|h(x_0)|] = h_i < \infty$, $(h(x_n), \mathscr{R}_n)$ is a non-negative supermartingale (see Section 4-3). Thus $\lim_n h(x_n)$ exists and is finite with probability one by Corollary 3-13. If h is not a constant function, then $h_j \neq h_k$ for some j and k. Since the chain is in states j and k infinitely often a.e., $h(x_n) = h_j$ and $h(x_n) = h_k$ for infinitely many n with probability one. Thus $h(x_n)$ diverges a.e., a contradiction.

To prove the corresponding result for measures, we introduce the dual matrix $\hat{P}$, defined whenever a positive finite-valued P-superregular measure α exists. The entries of $\hat{P}$ are $\hat{P}_{ij} = \alpha_j P_{ji}/\alpha_i$. Although we shall investigate $\hat{P}$ more fully in the next section, we mention some of its properties here. Suppose P is recurrent. Since $\hat{P}_{ij} \geq 0$ and since

$$\sum_j \hat{P}_{ij} = \frac{1}{\alpha_i} \sum_j \alpha_j P_{ji} \leq \frac{1}{\alpha_i} \alpha_i = 1,$$

$\hat{P}$ is a transition matrix. Since all pairs of states communicate in P, they do in $\hat{P}$. Now, using induction on n, we note that if

$$(\hat{P}^{n-1})_{ij} = \frac{\alpha_j (P^{n-1})_{ji}}{\alpha_i} \quad \text{for all } i \text{ and } j,$$

then

$$(\hat{P}^n)_{ij} = \sum_k \hat{P}_{ik} (\hat{P}^{n-1})_{kj} = \frac{\alpha_j}{\alpha_i} \sum_k (P^{n-1})_{jk} P_{ki} = \frac{\alpha_j (P^n)_{ji}}{\alpha_i}.$$

Summing on n, we see that $M_i[\hat{\mathbf{n}}_i] = +\infty$ because $M_i[\mathbf{n}_i] = \infty$. Hence $\hat{P}$ is recurrent.

Proposition 6-4: If P is a recurrent chain, all (finite-valued) non-negative superregular measures are regular and are uniquely determined up to multiplication by a constant. A non-zero non-negative superregular measure is positive.

PROOF: We prove the second assertion first. Suppose $\alpha \geq \alpha P$. Then $\alpha \geq \alpha P^n$ for every n. If $\alpha_i = 0$ and $\alpha_j > 0$, find n such that $(P^n)_{ji} > 0$. We have

$$\alpha_i \geq \sum_m \alpha_m (P^n)_{mi} \geq \alpha_j (P^n)_{ji} > 0,$$

a contradiction. Hence $\alpha > 0$. For the first assertion of the proposition, let α and β be non-zero non-negative finite-valued superregular measures. Then α and β are positive. Use α to form the recurrent chain $\hat{P}$. Then $\hat{P}1 = 1$ since $\hat{P}$ is recurrent. Therefore,

$$1 = \sum_j \hat{P}_{ij} = \sum_j \frac{\alpha_j P_{ji}}{\alpha_i} = \frac{1}{\alpha_i} \sum_j \alpha_j P_{ji}.$$

Thus $\alpha_i = \sum_j \alpha_j P_{ji}$ and α is regular. If we can show that $\{\beta_j/\alpha_j\}$ is a superregular function for $\hat{P}$, we will have shown that $\beta = c\alpha$, and the proof will be complete. We have

$$\sum_j \hat{P}_{ij} \frac{\beta_j}{\alpha_j} = \sum_j \frac{\alpha_j}{\alpha_i} P_{ji} \frac{\beta_j}{\alpha_j} = \frac{1}{\alpha_i} \sum_j \beta_j P_{ji} \leq \frac{\beta_i}{\alpha_i}.$$

Proposition 6-5: If P is ergodic, then $A = 1\alpha$, $\alpha 1 = 1$, and α is regular.

PROOF: We have $PA = A$. Thus every column of A is regular and must be constant by Proposition 6-3. Hence $A = 1\alpha$. Since $AP = A$, every row of A is regular and α must be regular. It therefore remains to be shown that $\alpha 1 = 1$. Now $A^2 = A$ so that $(1\alpha)(1\alpha) = (1\alpha)$. By associativity $1(\alpha(1\alpha)) = 1\alpha$ so that $\alpha(1\alpha) = \alpha$. But $\alpha(1\alpha) = (\alpha 1)\alpha$ so that $(\alpha 1)\alpha = \alpha$. If $\alpha_j \neq 0$, then from $(\alpha 1)\alpha_j = \alpha_j$ we may conclude $\alpha 1 = 1$.

The existence of a positive regular measure for ergodic chains is thus an easy matter to prove. For null chains, however, the proof is harder since the limiting matrix $L = 0$ is no help. The technique we shall use is to watch the recurrent chain P only while it is in a subset E of the set of states.

Let E be a subset of states and let P^E be the stochastic process whose nth outcome is the outcome of P the nth time the process P is in the set E. We shall see in Lemma 6-6 that P^E is a Markov chain. From its interpretation it is clear that P^E is recurrent if P is recurrent. Moreover, if $E \subset F$, then $(P^F)^E = P^E$.

The index set for the matrix P^E is taken to be E. Writing P as

$$P = \begin{array}{c} \\ E \\ \tilde{E} \end{array} \begin{array}{c} \begin{array}{cc} E & \tilde{E} \end{array} \\ \begin{pmatrix} T & U \\ R & Q \end{pmatrix} \end{array},$$

we have the following relationship between P^E and P.

Lemma 6-6: For an arbitrary Markov chain P, P^E is a Markov chain and

$$P^E = T + UNR,$$

where $N = \sum_{k=0}^{\infty} Q^k$.

REMARK: The lemma holds even if N has infinite entries, provided we agree as usual that $0 \cdot \infty = 0$.

PROOF: Let y_n be the nth outcome in the P^E-process. If

$$\Pr_\pi[y_0 = c_0 \wedge \cdots \wedge y_{n-1} = c_{n-1}] > 0,$$

let $\mathbf{t}$ be the random time of outcome $n - 1$ and apply Theorem 4-9. Then

$$\begin{aligned}\Pr_\pi[y_n &= c_n \mid y_0 = c_0 \wedge \cdots \wedge y_{n-1} = c_{n-1}] \\ &= \Pr_\pi[y_n = c_n \mid y_0 = c_0 \wedge \cdots \wedge y_{n-1} = c_{n-1} \wedge x_{\mathbf{t}} = c_{n-1}] \\ &= \Pr_{c_{n-1}}[y_1 = c_n],\end{aligned}$$

and it follows that P^E is a Markov chain. Now let i and j be in E. Applying Theorem 4-10 with the random time identically one and with the statement that E is hit after time 0 first at state j, we have

$$\begin{aligned}P_{ij}^E &= \sum_k P_{ik} B_{kj}^E \\ &= \sum_{k \in E} P_{ik} B_{kj}^E + \sum_{k \notin E} P_{ik} B_{kj}^E \\ &= P_{ij} + \sum_{k \notin E} P_{ik} B_{kj}^E.\end{aligned}$$

The result then follows from Proposition 5-6.

Lemma 6-7: For an arbitrary Markov chain P, if E is a subset of states and β is a finite-valued non-negative P-superregular measure, then β_E is P^E-superregular.

PROOF: Since $\beta \geq \beta P$, multiplication of the submatrices of β by the submatrices of P gives the two relations

$$\beta_E \geq \beta_E T + \beta_{\tilde{E}} R$$

and

$$\beta_{\tilde{E}} \geq \beta_E U + \beta_{\tilde{E}} Q.$$

We may rewrite the second relation as

$$\beta_{\tilde{E}}(I - Q) \geq \beta_E U \geq 0.$$

The proof of Theorem 5-10 translates directly into a proof for row vectors. From it we find

$$\beta_{\tilde{E}} = \gamma N + \rho,$$

where $\gamma = \beta_{\bar{E}}(I - Q)$, $N = \sum Q^n$, and ρ is non-negative and Q-regular. Hence

$$\beta_{\bar{E}} \geq \gamma N \geq (\beta_E U)N.$$

Thus $\beta_E P^E = \beta_E T + \beta_E UNR \leq \beta_E T + \beta_{\bar{E}} R \leq \beta_E$.

Lemma 6-8: No finite null chains exist.

PROOF: We have $L^{(n)}1 = 1$ or $\sum_j L_{ij}^{(n)} = 1$ for every n. Since the limit of a finite sum is the sum of the limits,

$$(L1)_i = \sum_j L_{ij} = \sum_j \lim_n L_{ij}^{(n)} = \lim_n \sum_j L_{ij}^{(n)} = 1.$$

Theorem 6-9: Every recurrent chain P has a positive finite-valued regular measure α which is unique up to multiplication by a scalar. Furthermore, $\alpha 1 < \infty$ if and only if P is ergodic.

PROOF: Order the states by the positive integers, let E be the first n of the states, and let F be the first $n + 1$. Then P^E and P^F are ergodic chains and have regular measures α^E and α^F. Also $(P^F)^E = P^E$. Thus α_E^F is P^E-regular by Lemma 6-7, and we may choose α^F such that $\alpha_E^F = \alpha^E$ by the uniqueness part of Proposition 6-4. The procedure of adding a single state to F may be continued by induction, and we set $\alpha = \lim_{E \to S} (\alpha^E \quad 0)$. Now for any of these sets E we have

$$\alpha_E T \leq \alpha_E T + \alpha_E UNR = \alpha_E P^E = \alpha_E$$

or

$$\alpha_E T \leq \alpha_E.$$

Thus,

$$\alpha \begin{pmatrix} T & 0 \\ 0 & 0 \end{pmatrix} = (\alpha_E T \quad 0) \leq (\alpha_E \quad 0) \leq \alpha.$$

As $E \to S$, the entries of $\begin{pmatrix} T & 0 \\ 0 & 0 \end{pmatrix}$ increase monotonically from zero to the entries of P. Hence, by monotone convergence, $\alpha P \leq \alpha$ and, by Proposition 6-4, α is regular. Clearly $\alpha > 0$, and we know that if P is ergodic then $\alpha 1 < \infty$. Conversely, suppose $\alpha 1 < \infty$. Then, by dominated convergence,

$$\begin{aligned} \alpha L &= \lim_n \alpha L^{(n)} \\ &= \lim_n \frac{1}{n} \alpha(I + \cdots + P^{n-1}) \\ &= \alpha \\ &\neq 0 \end{aligned}$$

and $L \neq 0$.

2. Duality

The proof of Proposition 6-4 is somewhat artificial without further explanation. What was used was a standard method of converting proofs about functions into proofs about measures. We had proved uniqueness for non-negative finite-valued superregular functions, and the idea was to take advantage of this fact in the result for measures.

The isomorphism that exists between row vectors and column vectors is known as **duality**. Not only does duality make rigorous the correspondence between row and column vectors, but also it provides easy proofs of some new results.

Definition 6-10: Let P be an arbitrary Markov chain transition matrix and suppose there exists a positive finite-valued P-superregular measure α. The α-**dual matrix** of P is a matrix $\hat{P}$ defined by

$$\hat{P}_{ij} = \frac{\alpha_j P_{ji}}{\alpha_i}.$$

Let D be a diagonal matrix with diagonal entries $1/\alpha_i$.

We note that $\hat{P} = DP^T D^{-1}$.

We cannot define duality in general, because we are not always assured of the existence of a positive superregular measure. However, there are only two important special cases, and we know that a superregular measure α exists for each of them:

(1) P is recurrent. Then there exists a unique α-dual of P. We call $\hat{P}$ the **dual** of P or the **reverse chain.** We shall investigate the properties of the reverse chain in some detail in Section 8.

(2) P has only transient states. Then, as we saw in Section 5-2, a positive superregular α exists. All duality statements are relative to such a vector α, but there is no assurance that α is unique.

Proposition 6-11: If P is a transition matrix, then so is $\hat{P}$. If all pairs of states in P communicate, then all pairs of states in $\hat{P}$ communicate.

PROOF: It is clear that $\hat{P} \geq 0$. For $\hat{P}\mathbf{1} \leq \mathbf{1}$ we have

$$\sum_j \hat{P}_{ij} = \sum_j \frac{\alpha_j P_{ji}}{\alpha_i} = \frac{1}{\alpha_i} \sum_j \alpha_j P_{ji} \leq \frac{1}{\alpha_i} \alpha_i = 1.$$

If i and j communicate in P by the routes

$$i, m_1, m_2, \ldots, m_r, j$$

and

$$j, n_1, n_2, \ldots, n_s, i,$$

then they communicate in $\hat{P}$ by

$$i, n_s, \ldots, n_1, j,$$
$$j, m_r, \ldots, m_1, i.$$

Proposition 6-12: If P is a transition matrix, then

$$\hat{P}^n = D(P^n)^T D^{-1} \quad \text{and} \quad \{\mathrm{M}_i[\hat{\mathbf{n}}_j]\} = D\{\mathrm{M}_i[\mathbf{n}_j]\}^T D^{-1}.$$

If P is either (1) recurrent or (2) transient with only transient states then $\hat{P}$ is of the same type. If P is of the second type, then

$$\hat{N} = DN^T D^{-1}.$$

Proof: The proof of the first assertion is by induction on n. The case $n = 1$ is Definition 6-10. Suppose that

$$\hat{P}^{k-1} = D(P^{k-1})^T D^{-1}.$$

Then

$$\hat{P}^k = \hat{P}\hat{P}^{k-1} = (DP^T D^{-1})(D(P^{k-1})^T D^{-1}) = D(P^k)^T D^{-1}.$$

Associativity holds because all the matrices are non-negative. Now

$$\{\mathrm{M}_i[\hat{\mathbf{n}}_j]\} = \sum_k \hat{P}^k = D \sum_k ((P^k)^T) D^{-1} = D\{\mathrm{M}_i[\mathbf{n}_j]\}^T D^{-1}.$$

In particular, if $\mathrm{M}_j[\mathbf{n}_j]$ is infinite, then so is $\mathrm{M}_j[\hat{\mathbf{n}}_j]$. Hence, by Proposition 6-11, if P is recurrent, so is $\hat{P}$.

Definition 6-13: Let P be a transition matrix, and let α be a positive finite-valued superregular measure. Let Y be any square matrix, let β be any row vector, and let f be any column vector all indexed by the set of states. Define D to be a diagonal matrix whose diagonal entries are $1/\alpha_i$. The **duals** of Y, β, and f are defined by

$$\begin{aligned} \text{dual } Y &= DY^T D^{-1} \\ \text{dual } \beta &= D\beta^T \\ \text{dual } f &= f^T D^{-1}. \end{aligned}$$

The **dual** of a number is that number.

We see that the dual of a row vector is a column vector and that the dual of a column vector is a row vector. The reader should note that $\hat{P}$ is identical with dual P and that part of the content of Proposition 6-12 is that $\mathrm{M}_i[\mathbf{n}_j]$ transforms to the $\hat{P}$ chain in the same way that P_{ij} does:

$$\{\mathrm{M}_i[\hat{\mathbf{n}}_j]\} = \text{dual } \{\mathrm{M}_i[\mathbf{n}_j]\}.$$

The fundamental properties of duals are listed in the next proposition.

Proposition 6-14: Let X and Y be square matrices, row vectors, or column vectors indexed by the set of states for a Markov chain P. Suppose $\hat{P}$ is the α-dual of P. Then

(1) dual dual $X = X$.
(2) dual $(X + Y) =$ dual $X +$ dual Y.
(3) dual $(cX) = c$ dual X.
(4) dual $(XY) =$ dual Y dual X.
(5) dual $I = I$ and dual $0 = 0$.
(6) If $X \geq 0$, then dual $X \geq 0$; and if $X > 0$, then dual $X > 0$.
(7) If $X \geq Y$, then dual $X \geq$ dual Y; and if $X > Y$, then dual $X >$ dual Y.
(8) If f is a P-superregular (or subregular) column vector, then dual f is a $\hat{P}$-superregular (or subregular) row vector; and if β is a P-superregular (or subregular) row vector, then dual β is a $\hat{P}$-superregular (or subregular) column vector.
(9) dual $\mathbf{1} = \alpha$ and dual $\alpha = \mathbf{1}$. The measure α is $\hat{P}$-superregular.
(10) If $\lim_n X^{(n)} = X$, then $\lim_n$ dual $X^{(n)} =$ dual X.

PROOF: We shall prove only (1) and (4); the rest of the proof is left to the reader. For (1) we have

$$\begin{aligned}\text{dual dual } X &= \text{dual } (DX^TD^{-1}) \\ &= D(DX^TD^{-1})^TD^{-1} \\ &= DD^{-1}X^{TT}DD^{-1} \\ &= X.\end{aligned}$$

Associativity holds because D and D^{-1} are diagonal matrices.

For (4) we have

$$\begin{aligned}\text{dual } XY &= D(XY)^TD^{-1} \\ &= DY^TX^TD^{-1} \\ &= (DY^TD^{-1})(DX^TD^{-1}) \\ &= \text{dual } Y \text{ dual } X.\end{aligned}$$

We may summarize Proposition 6-14 by saying that the operation dual is its own inverse, it reverses products, and it preserves sums, equalities, inequalities, regularity, and limits. We know, for example, that the dual of a recurrent chain is recurrent, and since dual is one-one, a dual recurrent chain is the most general recurrent chain. Hence a proof "for all recurrent chains $\hat{P}$" is a proof for all recurrent chains.

The essential feature of duality lies in this last statement; we shall apply it to the proof of Proposition 6-4. We start with a recurrent chain P and two positive superregular measures α and β. Forming $\hat{P}$, the

α-dual of P, we observe that since $\hat{P}$ is the most general recurrent chain and since **1** is $\hat{P}$-regular, the dual of **1** must be P-regular. Thus α is regular. Now since β is P-superregular and non-negative, dual β is $\hat{P}$-superregular and non-negative. Hence dual β is a constant vector, and the proof that β is a constant multiple of α is complete.

To form the α-dual of the restriction of a matrix, we use the appropriate restrictions of D and D^{-1} which make the matrix products defined. For example, write

$$P = \begin{array}{c} \\ E \\ \tilde{E} \end{array}\begin{array}{c} \begin{array}{cc} E & \tilde{E} \end{array} \\ \begin{pmatrix} T & U \\ R & Q \end{pmatrix} \end{array}.$$

By definition,

$$\text{dual } T = D_E T^T D_E^{-1},$$

$$\text{dual } U = D_{\tilde{E}} U^T D_E^{-1},$$

$$\text{dual } R = D_E R^T D_{\tilde{E}}^{-1},$$

and

$$\text{dual } Q = D_{\tilde{E}} Q^T D_{\tilde{E}}^{-1}.$$

Note that

$$\begin{pmatrix} \hat{T} & \hat{U} \\ \hat{R} & \hat{Q} \end{pmatrix} = \hat{P} = DP^T D^{-1}$$

$$= \begin{pmatrix} D_E & 0 \\ 0 & D_{\tilde{E}} \end{pmatrix}\begin{pmatrix} T^T & R^T \\ U^T & Q^T \end{pmatrix}\begin{pmatrix} D_E^{-1} & 0 \\ 0 & D_{\tilde{E}}^{-1} \end{pmatrix}$$

$$= \begin{pmatrix} D_E T^T D_E^{-1} & D_E R^T D_{\tilde{E}}^{-1} \\ D_{\tilde{E}} U^T D_E^{-1} & D_{\tilde{E}} Q^T D_{\tilde{E}}^{-1} \end{pmatrix},$$

so that

$$\text{dual } T = \hat{T},$$

$$\text{dual } U = \hat{R},$$

$$\text{dual } R = \hat{U},$$

and

$$\text{dual } Q = \hat{Q}.$$

To make effective use of duality, it is convenient to know what interpretation, if any, the duals of the matrices associated with P have in terms of the $\hat{P}$-process. At this time we shall calculate the duals of ${}^EP$, P^E, and B^E.

If we let ${}^{E}P$ be the process P watched until it enters E, then ${}^{E}P$ has transition matrix

$$ {}^{E}P = \begin{pmatrix} 0 & 0 \\ 0 & Q \end{pmatrix} $$

and fundamental matrix

$$ {}^{E}N = \begin{pmatrix} 0 & 0 \\ 0 & \sum Q^n \end{pmatrix}. $$

Hence dual ${}^{E}P = {}^{E}\hat{P}$ and dual ${}^{E}N = {}^{E}\hat{N}$.

The duals of P^E and B^E are not so trivial to settle, and we shall state what they are as the next two propositions.

Proposition 6-15: The dual of P^E is $\hat{P}^E$.

PROOF: By Lemma 6-6,

$$ P^E = T + U\left(\sum Q^n\right)R. $$

Hence

$$ \begin{aligned} \text{dual } P^E &= \text{dual } T + (\text{dual } R)\left(\sum (\text{dual } Q)^n\right)(\text{dual } U) \\ &= \hat{T} + \hat{U}\left(\sum \hat{Q}^n\right)\hat{R} \\ &= \hat{P}^E. \end{aligned} $$

Proposition 6-16:

$$ (\text{dual } B^E)_{ij} = \begin{cases} {}^{E}\hat{\hat{N}}_{ij} & \text{if } i \in E \\ 0 & \text{if } i \notin E, \end{cases} $$

where ${}^{E}\hat{\hat{N}}_{ij}$ is the number of times that the α-dual process started at i is in j before returning to E.

PROOF: Let $N = \sum Q^n$. Then

$$ B^E = \begin{pmatrix} I & 0 \\ NR & 0 \end{pmatrix}, $$

so that

$$ \text{dual } B^E = \begin{pmatrix} I & \hat{U}\hat{N} \\ 0 & 0 \end{pmatrix}. $$

Thus if $i \notin E$, then $(\text{dual } B^E)_{ij} = 0$, and if i and j are in E, then

$$ (\text{dual } B^E)_{ij} = \delta_{ij} = {}^{E}\hat{\hat{N}}_{ij}. $$

If $i \in E$ and $j \notin E$, then the result that

$$ (\hat{U}\hat{N})_{ij} = {}^{E}\hat{\hat{N}}_{ij} $$

follows from Theorem 4-11 with the random time identically one.

If we define ${}^E\bar{N}$ to be a matrix indexed by E and S whose i-jth entry is the mean number of times starting in i that the process is in j before returning to E, then we may rewrite the result of Proposition 6-16 as

$$\text{dual } B^E = \begin{pmatrix} {}^E\hat{N} \\ 0 \end{pmatrix}$$

The rest of this section contains applications of Propositions 6-15 and 6-16. We begin by deriving two identities relating B^E to other matrices, and we shall then dualize the first identity to obtain a result which will be used in Chapters 8 and 9. Finally we shall apply Proposition 6-16 in a different way to get a probabilistic interpretation for α_j/α_i.

Proposition 6-17: For any set E,

$$(I - P)B^E = \begin{pmatrix} I - P^E & 0 \\ 0 & 0 \end{pmatrix}.$$

If $N^{(n)} = I + P + \cdots + P^n$, then

$$B^E N^{(n)} = N^{(n)} + {}^E N(P^{n+1} - I).$$

Proof: Set $N = \sum Q^n$. For the first identity we have

$$(I - P)B^E = \begin{pmatrix} I - T & -U \\ -R & I - Q \end{pmatrix}\begin{pmatrix} I & 0 \\ NR & 0 \end{pmatrix} = \begin{pmatrix} I - (T + UNR) & 0 \\ -R + (I - Q)NR & 0 \end{pmatrix}$$

$$= \begin{pmatrix} I - P^E & 0 \\ 0 & 0 \end{pmatrix} \quad \text{since } (I - Q)N = I$$

For the second identity, we have

$${}^E N = \begin{pmatrix} 0 & 0 \\ 0 & N \end{pmatrix}$$

and hence

$${}^E NP = \begin{pmatrix} 0 & 0 \\ NR & N - I \end{pmatrix}$$

and

$${}^E N(P - I) = \begin{pmatrix} 0 & 0 \\ NR & -I \end{pmatrix} = B^E - I.$$

Therefore ${}^E N(P^{n+1} - I) = [{}^E N(P - I)]N^{(n)} = (B^E - I)N^{(n)}$.

Proposition 6-18: For any set E

$$\begin{pmatrix} {}^E\bar{N} \\ 0 \end{pmatrix}(I - P) = \begin{pmatrix} I - P^E & 0 \\ 0 & 0 \end{pmatrix}.$$

PROOF: Apply duality to the first identity of Proposition 6-17, using Propositions 6-15 and 6-16. Then

$$\begin{pmatrix} {}^E\hat{\bar{N}} \\ 0 \end{pmatrix}(I - \hat{P}) = \begin{pmatrix} I - \hat{P}^E & 0 \\ 0 & 0 \end{pmatrix}.$$

Since this identity holds for all reverse processes $\hat{P}$, it holds for all processes.

Using Proposition 6-16, we can obtain a simple interpretation for the ratio α_j/α_i. The case in which P is recurrent is of special importance because α is unique up to multiplication by a constant. But first we prove a more general result.

Proposition 6-19: Let α be a positive finite-valued superregular measure for P, and let $\hat{P}$ be the α-dual for P. Then for any set E,

$$\sum_{i \in E} \alpha_i \, {}^E\bar{N}_{ij} = \alpha_j \hat{h}_j^E.$$

PROOF: By Proposition 6-16,

$$(\text{dual } \hat{B}^E)_{ij} = \begin{cases} {}^E\bar{N}_{ij} & \text{for } i \in E \\ 0 & \text{for } i \notin E. \end{cases}$$

Therefore

$$\begin{aligned} \text{dual } \hat{h}^E = \text{dual } (\hat{B}^E 1) &= \alpha(\text{dual } \hat{B}^E) \\ &= \sum_{i \in E} \alpha_i \, {}^E\bar{N}_{ij}. \end{aligned}$$

Corollary 6-20: Let α be a positive finite-valued superregular measure for P, and let $\hat{P}$ be the α-dual of P. Then

$${}^i\bar{N}_{ij} = \frac{\alpha_j}{\alpha_i} \hat{H}_{ji}.$$

PROOF: Set $E = \{i\}$ in Proposition 6-19.

In particular, ${}^i\bar{N}_{ij} \leq \alpha_j/\alpha_i$ for any such α.

Corollary 6-21: Let α be a positive finite-valued regular measure for a recurrent chain P. Then

$$ {}^{i}\bar{N}_{ij} = \alpha_j/\alpha_i. $$

PROOF: Since P is recurrent, so is $\hat{P}$. Thus $\hat{H}_{ji} = 1$, and we may apply Corollary 6-20.

Corollary 6-22: Let α be a positive finite-valued regular measure for a recurrent chain P. Then

$$ \sum_{i \in E} \alpha_i \, {}^{E}\bar{N}_{ij} = \alpha_j \qquad \text{for all } j. $$

PROOF: Apply Proposition 6-19. Then $\hat{h}_j^E = 1$.

Definition 6-23: Let P be a recurrent chain. Set

$$ M_{ij} = \mathrm{M}_i[\mathbf{t}_j] $$

and

$$ \bar{M}_{ij} = \mathrm{M}_i[\bar{\mathbf{t}}_j]. $$

The matrix M is called the **mean first passage time matrix**. Similarly M_{iE} is the mean time from i until E is reached, and $\bar{M}_{iE}$ is the mean time from i to return to E.

Proposition 6-24: If P is a recurrent chain with positive regular measure α, then

$$ \sum_{i \in E} \alpha_i \bar{M}_{iE} = \alpha \mathbf{1}. $$

PROOF: We have

$$ \sum_{i \in E} \alpha_i \bar{M}_{iE} = \sum_{i \in E} \sum_j \alpha_i \, {}^{E}\bar{N}_{ij} = \sum_j \sum_{i \in E} \alpha_i \, {}^{E}\bar{N}_{ij} $$
$$ = \sum_j \alpha_j = \alpha \mathbf{1}, $$

the next to last equality following from Corollary 6-22.

Proposition 6-25: If P is a recurrent chain with positive regular measure α, then

$$ \bar{M}_{ii} = \begin{cases} \dfrac{1}{\alpha_i} & \text{if } P \text{ is ergodic and } \alpha\mathbf{1} = 1 \\ +\infty & \text{if } P \text{ is null.} \end{cases} $$

PROOF: Set $E = \{i\}$ in Proposition 6-24.

3. Cyclicity

Let P be a recurrent Markov chain and let i be a fixed state of the chain. Define a set of positive integers T by

$$T = \{k \mid (P^k)_{ii} > 0,\ k > 0\}.$$

Let d be the greatest common divisor of the integers in T.

Lemma 6-26: T is non-empty and is closed under addition.

PROOF: T is clearly non-empty since P is recurrent. Suppose m and n are integers in T. Then $(P^m)_{ii} > 0$ and $(P^n)_{ii} > 0$, so that

$$\begin{aligned}(P^{m+n})_{ii} &= \sum_k (P^m)_{ik}(P^n)_{ki} \\ &\geq (P^m)_{ii}(P^n)_{ii} \\ &> 0.\end{aligned}$$

Hence $m + n$ is in T.

Noting the discussion in Section 1-6e, we arrive at the following result, using Lemma 1-66.

Lemma 6-27: T contains all sufficiently large multiples of its greatest common divisor d.

The integer d we shall call the **period of the chain for the state** i.

Proposition 6-28: The period of a recurrent chain for the state i is a constant independent of the state i.

PROOF: Let i and j be any two states in the chain. Since the chain is recurrent, i and j communicate. Let d be the period associated with state i and let $\bar{d}$ be the period associated with state j. Suppose the minimum possible time for the process to go from state i to state j is s, and suppose the minimum time for the process to go from j to i is t. By Lemma 6-27 let N be large enough so that the process can return to j in $n\bar{d}$ steps for all $n \geq N$. Then the process can go from i to j in s steps, return to j in $N\bar{d}$ steps, and go back to i in t steps. Hence $d \mid (s + N\bar{d} + t)$. Similarly, $d \mid (s + (N + 1)\bar{d} + t)$. Thus d divides the difference, or $d \mid \bar{d}$. Reversing the roles of i and j, we find that $\bar{d} \mid d$. Therefore, $d = \bar{d}$.

We may thus speak of the **period of a recurrent chain** without ambiguity. Every recurrent chain has a period, and that period is finite. If, for example, P is a recurrent chain in which $P_{ii} > 0$ for some i, then P has period one.

Definition 6-29: A recurrent chain is said to be **non-cyclic** if its period is one and **cyclic** if its period is greater than one.

Let P be a recurrent chain of period d. Define a relation R on the states of P by the following: We say that $i\ R\ j$ if and only if, starting at i, the process can reach j in md steps for some m. From the definition of the period d, it follows that $i\ R\ i$. The symmetry of R follows from the fact md plus the time to return from j to i must be a multiple of d. To see that R is transitive, we note that if j can be reached from i in md steps and if k can be reached from j in nd steps, then k can be reached from i in $(m + n)d$ steps.

Thus R partitions the states into cyclic subclasses. The reader may verify that there are d distinct subclasses and that the nth class contains all those states which it is possible to reach from the starting state only at times which are congruent to n modulo d. The process moves cyclically through the classes in the specified order. Furthermore, if the chain is watched after every dth step, the resulting process is again a Markov chain (by the strong Markov property), and its behavior will be noncyclic. The transition matrix for the new process is P^d, and its form is that of d separate recurrent chains:

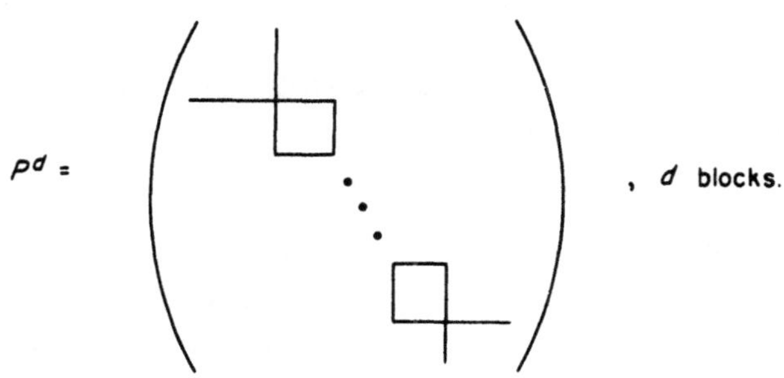

The entries in each block are the entries of a recurrent noncyclic chain, and the entries which are not in any block are all zeros.

The observation that P^d is really d separate recurrent noncyclic chains enables us to study representatively the properties of all recurrent chains by considering only noncyclic chains. Thus, it is to noncyclic chains that we now turn our attention. The main tool in their study will be chains representing sums of independent random variables.

4. Sums of independent random variables

We have already investigated some of the properties of sums of independent random variables Markov chains. Such processes are especially important because of how they arise from general recurrent chains (see Proposition 6-32), and it is for this reason that we now discuss their origin.

For concreteness we shall confine ourselves to sums of independent random variables chains defined on the integers. Before recalling the definition of independent random variables, we remark that if $\mathbf{y}$ is a real-valued function defined on a probability space and having a denumerable range, then a necessary and sufficient condition for $\mathbf{y}$ to be measurable (and hence to be a random variable) is that the inverse image under $\mathbf{y}$ of every one-point set be measurable. The condition is necessary because $\{\omega \mid \mathbf{y}(\omega) = c\} = \{\omega \mid c \leq \mathbf{y}(\omega) \leq c\}$ must be measurable, and it is sufficient because $\{\omega \mid \mathbf{y}(\omega) < c\}$ is a countable union of such sets. Therefore, if $\mathbf{y}$ is a denumerable-valued random variable and if E is an arbitrary set of real numbers, the set $\{\omega \mid \mathbf{y}(\omega) \in E\}$ is measurable.

Definition 6-30: The denumerable-valued random variables $\mathbf{y}_1, \mathbf{y}_2, \mathbf{y}_3, \ldots$ defined on Ω are **independent** if, for every finite collection of sets $E_1, E_2, \ldots, E_m$ of reals, it is true that

$$\Pr[\mathbf{y}_{n_k}(\omega) \in E_k \text{ for } k = 1, \ldots, m] = \prod_{k=1}^{m} \Pr[\mathbf{y}_{n_k}(\omega) \in E_k].$$

The random variables are **identically distributed** if, for any m and n and for any set E of reals, it is true that

$$\Pr[\mathbf{y}_m(\omega) \in E] = \Pr[\mathbf{y}_n(\omega) \in E].$$

An independent process $\{y_n\}$ was defined in Section 2-5 as one in which the statements $y_0 = c_0 \wedge \cdots \wedge y_{n-1} = c_{n-1}$ and $y_n = c_n$ are probabilistically independent for every $n > 0$ and for every choice of the c's. We see that an independent process is that special case of a Markov process in which $\Pr_\pi[y_{n+1} = j \mid y_n = i]$ is independent of i. Moreover, an independent process is a Markov chain if and only if it is an independent trials process.

Proposition 6-31: Let $\{\mathbf{y}_n\}$ be a stochastic process defined from a sequence space Ω to a denumerable set of real numbers S. The stochastic process is an independent process if and only if the $\{\mathbf{y}_n\}$ are independent random variables. It is an independent trials process if and only if the $\{\mathbf{y}_n\}$ are independent and identically distributed.

PROOF: We are to prove first that the $\{y_n\}$ are independent if and only if the statements $y_0 = c_0 \wedge \cdots \wedge y_{n-1} = c_{n-1}$ and $y_n = c_n$ are probabilistically independent for any $n \geq 1$ and for any choice of the c's. Independence of the y's means

$$\Pr[y_0 = c_0 \wedge \cdots \wedge y_n = c_n] = \prod_{k=1}^{n} \Pr[y_k = c_k]$$

and

$$\Pr[y_0 = c_0 \wedge \cdots \wedge y_{n-1} = c_{n-1}] = \prod_{k=1}^{n-1} \Pr[y_k = c_k]$$

for all n. This statement holds if and only if

$$\Pr[y_0 = c_0 \wedge \cdots \wedge y_n = c_n] = \Pr[y_0 = c_0 \wedge \cdots \wedge y_{n-1} = c_{n-1}]\Pr[y_n = c_n]$$

for all n. Second, we are to prove that the $\{y_n\}$ are also identically distributed if and only if

$$\Pr[y_n = c_n] = \Pr[y_m = c_m]$$

for any n and m. But this assertion is clear from Definition 6-30.

Let $\{y_n\}$ for $n \geq 0$ be a sequence of independent random variables which are identically distributed for $n \geq 1$ and which have range in the union of the integers and $\{-\infty, +\infty\}$, and define inductively

$$x_0 = y_0$$

and

$$x_{n+1} = y_{n+1} + x_n \quad \text{for } n \geq 0.$$

If the y_n are finite-valued a.e., we claim that the random variables x_n are the outcome functions for a sums of independent random variables process on the integers with starting distribution $\pi_i = \Pr[y_0 = i]$. Setting

$$p_k = \Pr[y_n = k], \quad n > 0,$$

which is a constant not depending on any other function in the sequence (by independence) and not depending on n (by identical distributions) we see that $\sum p_k = 1$ since y_n is finite-valued a.e. Moreover, if $\Pr[x_0 = a \wedge \cdots \wedge x_{n-1} = i] > 0$ with $n > 0$, then

$$\begin{aligned}\Pr[x_n = j \mid x_0 = a \wedge x_1 = b \wedge \cdots \wedge x_{n-1} = i] &\\ = \Pr[y_n = j - i \mid y_0 = a \wedge y_1 = b - a \wedge \cdots \wedge y_{n-1} = i - h] &\\ = \Pr[y_n = j - i] &\\ = p_{j-i}. &\end{aligned}$$

Hence the process is a Markov chain representing sums of independent random variables.

Conversely, let P be the transition matrix for a sums of independent random variables Markov chain with state space the integers and with outcome functions x_n. Let $\mathbf{y}_n = x_n - x_{n-1}$ for $n > 0$. We shall show that x_0 and the $\mathbf{y}_n$ are independent and that the $\mathbf{y}_n$ are identically distributed; it is clear that x_0 and the $\mathbf{y}_n$ are finite-valued a.e. In fact, we have

$$\begin{aligned}\Pr_\pi[x_0 = c_0 \wedge (\mathbf{y}_k = c_k \text{ for } 1 \leq k \leq n)]&\\ &= \pi_{c_0} P_{c_0, c_0 + c_1} \cdot \cdots \cdot P_{c_0 + \cdots + c_{n-1}, c_0 + \cdots + c_n}\\ &= \pi_{c_0} p_{c_1} \cdots p_{c_n}\\ &= \Pr_\pi[x_0 = c_0] \cdot \prod_{k=1}^{n} \Pr_\pi[\mathbf{y}_k = c_k],\end{aligned}$$

and independence follows by taking countable disjoint unions of such statements; since

$$\Pr_\pi[\mathbf{y}_n = j] = p_j,$$

the $\mathbf{y}_n$ are identically distributed.

Sums of independent random variables appear in a natural way in the study of recurrent chains. The result to follow associates to every recurrent chain P a sums of independent random variables chain P^* with state space the integers.

Proposition 6-32: Let P be a recurrent chain with outcome functions x_n. For a fixed state s let $\mathbf{t}_n(\omega)$ be the $(n + 1)$st time on the path ω that state s is reached. Then the random times $\mathbf{t}_n$ for $n \geq 0$ are the outcome functions for a sums of independent random variables ladder process P^* with state space the integers.

PROOF: If $\Pr_\pi[\mathbf{t}_0 = c_0 \wedge \cdots \wedge \mathbf{t}_{n-1} = c_{n-1}] > 0$, then

$$\begin{aligned}\Pr_\pi[\mathbf{t}_n = c_n \mid \mathbf{t}_0 = c_0 \wedge \cdots \wedge \mathbf{t}_{n-1} = c_{n-1}]&\\ &= \Pr_\pi[x_{c_{n-1}+1} \neq s \wedge \cdots \wedge x_{c_n - 1} \neq s \wedge x_{c_n} = s \mid x_0 \neq s\\ &\qquad \wedge \cdots \wedge x_{c_0 - 1} \neq s \wedge x_{c_0} = s \wedge x_{c_0+1} \neq s\\ &\qquad \wedge \cdots \wedge x_{c_{n-1}} = s]\\ &= \Pr_s[x_1 \neq s \wedge \cdots \wedge x_{c_n - c_{n-1}} = s] \quad \text{by Theorem 4-9}\\ &= \Pr_s[\bar{\mathbf{t}}_s = c_n - c_{n-1}],\end{aligned}$$

where $\bar{\mathbf{t}}_s$ is the time to return to state s. Hence

$$\Pr_\pi[\mathbf{t}_n = j \mid \mathbf{t}_0 = a \wedge \cdots \wedge \mathbf{t}_{n-1} = i] = \bar{F}_{ss}^{(j-i)}.$$

Set $P^*_{ij} = \bar{F}^{(j-i)}_{ss}$. Since by recurrence of P

$$(P^*1)_k = \sum_j P^*_{kj} = \sum_j \bar{F}^{(j-k)}_{ss} = \bar{H}_{ss} = 1,$$

P^* is a sums of independent random variables Markov chain.

The following is a converse to the preceding result.

Proposition 6-33: Let P' be a sums of independent random variables ladder process on the integers. There exists a recurrent chain P and a state s such that the times to return to s are the outcome functions for P'.

PROOF: Let $p'_k = P'_{0k}$ for $k \geq 1$; then $\sum_{k=1}^{\infty} p'_k = 1$. We take P to be a basic example and s to be state 0; the values of p_i and q_i in the basic example are yet to be specified. Define recursively the q_i's by the relations

$$p'_1 = q_1 = \beta_0 - \beta_1$$

and

$$p'_n = p_1 \cdots p_{n-1} q_n = \beta_{n-1} q_n = \beta_{n-1} - \beta_n.$$

In P we have $\Pr[\mathbf{t}_1 - \mathbf{t}_0 = k] = p'_k$ as required; it remains to be proved that P is recurrent. We have

$$\sum_{k=1}^{n} p'_k = \sum_{k=1}^{n} (\beta_{k-1} - \beta_k) = \beta_0 - \beta_n = 1 - \beta_n,$$

and since $\sum_{k=1}^{\infty} p'_k = 1$, we must have $\lim_n \beta_n = 0$. Hence P is recurrent.

We close this section with two remarks about sums of independent random variables and their relation to recurrence. First, we have seen in Proposition 5-22 that a sums of independent random variables process on the integers with finitely many k-values is a recurrent chain if and only if the k-values have mean zero and their greatest common divisor is one. Second, we note that an infinite recurrent chain representing sums of independent random variables must be null, since $\alpha = 1^T$ is regular and $1^T 1 = \infty$.

5. Convergence theorem for noncyclic chains

By restricting our attention to noncyclic recurrent chains, we can prove a stronger result than the Mean Ergodic Theorem, namely that P^n itself converges with n to a limiting matrix. We shall give two proofs of this convergence theorem—the first a matrix proof using sums of independent random variables and the second the classical proof using the Renewal Theorem of Section 1-6f. We shall further show that, conversely, the truth of the convergence theorem just when P is the basic example implies the full validity of the Renewal Theorem.

We begin by proving two lemmas needed in both proofs of the convergence theorem; their effect is to formulate noncyclicity in a number-theoretic way.

Lemma 6-34: For any Markov chain and any states i and j,

$$P_{ij}^{(0)} = \delta_{ij}$$

$$\bar{F}_{ij}^{(0)} = 0,$$

and

$$P_{ij}^{(n)} = \sum_{k=1}^{n} \bar{F}_{ij}^{(k)} P_{jj}^{(n-k)} = \sum_{k=0}^{n-1} P_{jj}^{(k)} \bar{F}_{ij}^{(n-k)} \quad \text{for } n > 0.$$

Proof: The first two statements are obvious; for the third we first note that if $\Pr_i[\mathfrak{t}_j = k] > 0$, then for $n \geq k$,

$$\begin{aligned}
\Pr_i[x_n = j \mid \mathfrak{t}_j = k] &= \Pr_i[x_n = j \mid x_k = j \wedge x_{k-1} \neq j \wedge \cdots \wedge x_1 \neq j] \\
&= \Pr_i[x_n = j \mid x_k = j] \quad \text{by Lemma 4-6} \\
&= \Pr_j[x_{n-k} = j] \quad \text{by Lemma 4-6} \\
&= P_{jj}^{(n-k)}.
\end{aligned}$$

Hence no matter what the value of $\Pr_i[\mathfrak{t}_j = k]$, it is true that, for $n \geq k$,

$$\Pr_i[\mathfrak{t}_j = k] \Pr_i[x_n = k \mid \mathfrak{t}_j = k] = \bar{F}_{ij}^{(k)} \cdot P_{jj}^{(n-k)}.$$

Using $x_n = j \wedge \mathfrak{t}_j = k$, $1 \leq k \leq n$, as a set of alternatives for $x_n = j$, we have

$$\begin{aligned}
P_{ij}^{(n)} &= \sum_{k=1}^{n} \Pr_i[\mathfrak{t}_j = k] \cdot \Pr_i[x_n = j \mid \mathfrak{t}_j = k] \\
&= \sum_{k=1}^{n} \bar{F}_{ij}^{(k)} P_{jj}^{(n-k)} \\
&= \sum_{k=0}^{n-1} P_{jj}^{(k)} \bar{F}_{ij}^{(n-k)} \quad \text{by a change of variable.}
\end{aligned}$$

Lemma 6-35: A recurrent chain is noncyclic if and only if the set $Z = \{k \mid \bar{F}_{ii}^{(k)} > 0\}$ has greatest common divisor one.

Proof: If Z has greatest common divisor one, then the period for the state i is one. Conversely, suppose that the greatest common divisor is c. We shall show that c divides d, the period of the chain. Hence, if $c > 1$, the chain is cyclic. Let n be the smallest integer for which $P_{ii}^{(n)} > 0$ and $c \nmid n$, and write $n = qc + r$ with $0 < r < c$. By Lemma 6-34,

$$P_{ii}^{(n)} = \sum_{k=1}^{n} \bar{F}_{ii}^{(k)} P_{ii}^{(n-k)}$$

$$= \sum_{j=1}^{q} \bar{F}_{ii}^{(jc)} P_{ii}^{((q-j)c+r)}.$$

Then the right side is zero since every term $P_{ii}^{((q-j)c+r)}$ is zero, a contradiction. Thus $P_{ii}^{(n)} > 0$ only if $c \mid n$.

The next two lemmas lead to the convergence theorem; the first one is a consequence of Proposition 6-32 and the zero-one law for sums of independent random variables.

Lemma 6-36: Let P be a noncyclic recurrent chain, and for a fixed state s let E and F be any two sets of integers whose union is the set of all non-negative integers. Then either

$$\Pr_\pi[x_n = s \quad \text{for infinitely many } n \in E] = 1$$

or

$$\Pr_\pi[x_n = s \quad \text{for infinitely many } n \in F] = 1 \quad \text{(or both)},$$

and whichever alternative holds is independent of the starting distribution π.

PROOF: Form the process P^* of Proposition 6-32. We shall first show that for any two states i and j there is a state k which it is possible to reach from both i and j; for this purpose it is sufficient to show that from state 0 it is possible to reach all sufficiently large states, since P^* represents sums of independent random variables. Now the set of states which can be reached from 0 is non-empty and is closed under addition (since P^* represents sums of independent random variables); its greatest common divisor is one by Lemma 6-35. Hence by Lemma 1-66 all sufficiently large states can be reached.

By the zero-one law, which is Propositions 5-19 and 5-20,

$$\Pr_i[\mathbf{t}_n \in E \text{ infinitely often}] = \Pr_i[x_n = s \text{ for infinitely many } n \in E]$$

is zero or one and is independent of i. Thus

$$\Pr_\pi[x_n = s \text{ for infinitely many } n \in E]$$

is zero or one and is independent of π. If

$$\Pr_\pi[x_n = s \text{ for infinitely many } n \in E] = 0$$

and

$$\Pr_\pi[x_n = s \text{ for infinitely many } n \in F] = 0,$$

then

$$\Pr_\pi[x_n = s \text{ infinitely often}] = 0,$$

in contradiction to the recurrence of P.

The following lemma uses the notation $\|\beta\| = \sum_i |\beta_i|$.

Lemma 6-37: Let P be a noncyclic recurrent chain, and let β and γ be probability vectors. Then $\lim_{n\to\infty} \|(\beta - \gamma)P^n\| = 0$.

PROOF: Let $E = \{n \mid (\beta P^n)_s \leq (\gamma P^n)_s\}$ and $F = \{n \mid (\beta P^n)_s \geq (\gamma P^n)_s\}$. By Lemma 6-36, either

$$\Pr_\beta[x_n = s \text{ for infinitely many } n \in E] = 1$$

or

$$\Pr_\gamma[x_n = s \text{ for infinitely many } n \in F] = 1,$$

and by symmetry we may assume that the former alternative holds.

Let $h^{(n)}$ be the statement that $x_m = s$ for some $m \in E$ with $m < n$, and let

$$\beta_k^{(n)} = \Pr_\beta[x_n = k \wedge \sim h^{(n)}].$$

Then

$$\|\beta^{(n)}\| = \beta^{(n)}\mathbf{1} = \Pr_\beta[\sim h^{(n)}] \to 0$$

by the above assumption. Also

$$\beta^{(0)} = \beta$$

and

$$\beta_k^{(n+1)} = \begin{cases} \sum_j \beta_j^{(n)} P_{jk} & \text{if } n \notin E \\ \sum_{j \neq s} \beta_j^{(n)} P_{jk} & \text{if } n \in E. \end{cases}$$

We may represent this last identity conveniently by using ϵ, a row vector such that $\epsilon_i = \delta_{is}$, and by defining

$$\delta^{(n)} = \begin{cases} \beta_s^{(n)}\epsilon & \text{if } n \in E \\ 0 & \text{otherwise.} \end{cases}$$

Then

$$\beta^{(n+1)} = (\beta^{(n)} - \delta^{(n)})P.$$

Next, we define

$$\gamma^{(n)} = \beta^{(n)} + (\gamma - \beta)P^n.$$

From this relation,

$$\gamma^{(n-1)}P = \beta^{(n-1)}P + (\gamma - \beta)P^n$$

and

$$\gamma^{(0)} = \gamma$$

and hence

$$\gamma^{(n+1)} = \gamma^{(n)}P + \beta^{(n+1)} - \beta^{(n)}P = (\gamma^{(n)} - \delta^{(n)})P.$$

We shall show by induction that $\gamma^{(n)} \geq \delta^{(n)}$. First, $\gamma^{(0)} = \gamma \geq 0$. If $0 \notin E$, then $\delta^{(0)} = 0 \leq \gamma^{(0)}$. If $0 \in E$, then $\gamma_s^{(0)} = \gamma_s \geq \beta_s = \beta_s^{(0)}$ by the definition of E, and $\beta_s^{(0)} = \delta_s^{(0)}$. Thus $\gamma^{(0)} \geq \delta^{(0)}$ in either case. Suppose $\gamma^{(n-1)} \geq \delta^{(n-1)}$. Then

$$\gamma^{(n)} = (\gamma^{(n-1)} - \delta^{(n-1)})P \geq 0.$$

If $n \notin E$, then $\delta^{(n)} = 0$ and $\gamma^{(n)} \geq \delta^{(n)}$. If $n \in E$, then $[(\gamma - \beta)P^n]_s \geq 0$ by the definition of E, and hence

$$\gamma_s^{(n)} \geq \beta_s^{(n)} = \delta_s^{(n)}$$

by the definition of $\gamma^{(n)}$. Thus $\gamma^{(n)} \geq \delta^{(n)}$ for every n.

In particular, we have $\gamma^{(n)} \geq 0$. Thus

$$\begin{aligned}
\|\gamma^{(n)}\| &= \gamma^{(n)}\mathbf{1} \\
&= \beta^{(n)}\mathbf{1} + [(\gamma - \beta)P^n]\mathbf{1} \\
&= \beta^{(n)}\mathbf{1} + (\gamma - \beta)(P^n\mathbf{1}) \\
&= \beta^{(n)}\mathbf{1} + (\gamma - \beta)\mathbf{1} \\
&= \beta^{(n)}\mathbf{1} \\
&= \|\beta^{(n)}\| \to 0.
\end{aligned}$$

Finally

$$\|(\beta - \gamma)P^n\| = \|\beta^{(n)} - \gamma^{(n)}\|$$

by the definition of $\gamma^{(n)}$, and the right side is

$$\leq \|\beta^{(n)}\| + \|\gamma^{(n)}\| \to 0.$$

Theorem 6-38: If P is a noncyclic recurrent chain, then $\lim_{n\to\infty} P^n$ exists. If P is ergodic, then $\lim P^n = A = \mathbf{1}\alpha$ and $\lim_n \|\pi P^n - \alpha\| = 0$ for every probability vector π. If P is null, then $\lim_n (\pi P^n) = 0$ for every probability vector π.

PROOF: Every recurrent chain is either ergodic or null; taking π to be a vector with 1 in the ith entry and zeros elsewhere, we see that the existence of lim P^n follows from the other assertions of the theorem.

Let P be ergodic and let $A = 1\alpha$ be the Cesaro limit of the powers of P. We have $\alpha P^n = \alpha$ for every n. Letting $\beta = \pi$ and $\gamma = \alpha$ in Lemma 6-37, we obtain the desired result.

Let P be null and suppose the assertion of the theorem is false. Then by Corollary 1-64, for some probability vector π, there is an increasing sequence $\{n_k\}$ of positive integers and there is a row vector $\rho \neq 0$ such that

$$\lim_k (\pi P^{n_k})_i = \rho_i \quad \text{for every state } i.$$

Certainly $\rho_i \geq 0$. Summing on i, we obtain

$$\rho 1 = \sum_i \rho_i = \sum_i \lim_k (\pi P^{n_k})_i \leq \lim_k \sum_i (\pi P^{n_k})_i = 1,$$

the inequality following from Fatou's Theorem. Applying Lemma 6-37 with $\beta = \pi$ and $\gamma = \pi P$, we see that

$$\lim_k (\pi P^{n_k + 1})_i = \rho_i \quad \text{for each } i.$$

By Fatou's Theorem,

$$\rho P = (\lim \pi P^{n_k})P \leq \lim \pi P^{n_k + 1} = \rho.$$

Hence ρ is non-negative superregular and satisfies $\rho 1 < \infty$; ρ must be regular by Proposition 6-4, and the fact that P is null then contradicts Theorem 6-9.

Corollary 6-39: If P is a null chain (not necessarily noncyclic), then lim $P^n = 0$.

PROOF: Let P have period d. By Theorem 6-38

$$\lim_n P^{nd} = 0.$$

By dominated convergence, $\lim_n P^{nd+r} = 0$ for each $r = 0, 1, 2, \ldots, d - 1$. Hence,

$$\lim_n P^n = 0.$$

The classical proof of Theorem 6-38 that follows proves only that lim P^n exists.

SECOND PROOF OF THEOREM 6-38: We first prove the theorem for the ith diagonal entry. Set $f_n = \bar{F}_{ii}^{(n)}$, $u_n = (P^n)_{ii}$, and

$$\mu = \sum_n nf_n = \sum_n n\bar{F}_{ii}^{(n)} = \sum_n n \Pr_i[\mathfrak{t}_i = n] = \mathrm{M}_i[\mathfrak{t}_i] = \bar{M}_{ii}.$$

Lemmas 6-34 and 6-35 establish all the hypotheses of Theorem 1-67 except for the fact that $\sum_n f_n = 1$:

$$\sum_n f_n = \sum_n \bar{F}_{ii}^{(n)} = \bar{H}_{ii} = 1.$$

Therefore $u_n \to 1/\mu$ or 0 according as μ is finite or infinite, and the value of the limit for the diagonal entries follows from Proposition 6-25. For the off-diagonal entries, let i and j be any two distinct states. Define a row vector β and a sequence of column vectors $\{g^{(n)}\}$ by

$$\beta_m = \bar{F}_{ij}^{(m)}$$

$$g_m^{(n)} = \begin{cases} (P^{n-m})_{jj} & \text{if } n \geq m \\ 0 & \text{if } n < m. \end{cases}$$

Then $\lim_n g_m^{(n)} = L_{jj}$ exists since we have proved the theorem for diagonal entries. Furthermore, by Lemma 6-34,

$$\begin{aligned}(P^n)_{ij} &= \sum_{m=1}^{n} \beta_m (P^{n-m})_{jj} \\ &= \sum_{m=1}^{\infty} \beta_m g_m^{(n)} \\ &= \beta g^{(n)}.\end{aligned}$$

Since $\beta 1 = 1$ and $g^{(n)} \leq 1$, the Dominated Convergence Theorem applies and

$$\begin{aligned}\lim_n (P^n)_{ij} &= \lim_n \beta g^{(n)} \\ &= \beta \lim_n g^{(n)} \\ &= \beta(L_{jj} 1) \\ &= L_{jj}.\end{aligned}$$

As a converse to the second proof of Theorem 6-38, we shall show that the convergence of P^n for every noncyclic recurrent chain implies the truth of the Renewal Theorem. This result is of particular interest because all that is needed is convergence of the diagonal entries of P^n, when P is a noncyclic recurrent case of the basic example.

Proposition 6-40: If every noncyclic recurrent chain converges to a limiting matrix, then the Renewal Theorem holds.

PROOF: Let the sequence $\{f_n\}$ be given. Define $r_i = \sum_{k>i} f_k$; the r_i tend to 0 because $\sum_k f_k = 1$. As long as $r_i > 0$, define

$$p_{i+1} = \frac{r_{i+1}}{r_i} \quad \text{for } i = 0, 1, 2, \ldots.$$

If $r_i = 0$ for some i, then $p_i = 0$ and the p_k for $k > i$ are irrelevant. Set $q_i = 1 - p_i$ and let the p_i and the q_i represent the transition probabilities associated with the basic example. We have

$$\beta_i = p_1 p_2 \ldots p_i = \frac{r_1}{r_0}\frac{r_2}{r_1} \cdots \frac{r_i}{r_{i-1}} = r_i.$$

That is, $\beta_i = r_i$. Since $r_i \to 0$, $\beta_i \to 0$ and the chain is recurrent. Now

$$\begin{aligned} \bar{F}_{00}^{(n)} &= \beta_{n-1}(1 - p_n) \\ &= \beta_{n-1} - \beta_n \\ &= \sum_{k>n-1} f_k - \sum_{k>n} f_k \\ &= f_n. \end{aligned}$$

Thus $\mu = \sum_n n f_n = \bar{M}_{00}$. By Lemma 6-34 we see that $u_n = P_{00}^{(n)}$. The Markov chain is noncyclic by Lemma 6-35 because the greatest common divisor of the k's for which $f_k > 0$ is 1. Therefore $\lim u_n = \lim P_{00}^{(n)}$ exists. On the other hand, by Proposition 6-25 the Cesaro limit of $P_{00}^{(n)}$ is $1/\bar{M}_{00} = 1/\mu$ if $\bar{M}_{00} < \infty$ or 0 if $\bar{M}_{00} = +\infty$. Hence by Proposition 1-61

$$\lim u_n = \begin{cases} 1/\mu & \text{if } \mu < \infty \\ 0 & \text{if } \mu = +\infty. \end{cases}$$

6. Mean first passage time matrix

The matrices M and $\bar{M}$ have already been defined by

$$\begin{aligned} M_{ij} &= \mathrm{M}_i[\mathbf{t}_j] \\ \bar{M}_{ij} &= \mathrm{M}_i[\bar{\mathbf{t}}_j]. \end{aligned}$$

In Proposition 6-25 we saw that

$$\bar{M}_{ii} = \begin{cases} 1/\alpha_i & \text{if } P \text{ is ergodic and } \alpha 1 = 1. \\ \infty & \text{if } P \text{ is null.} \end{cases}$$

Proposition 6-41: In any recurrent chain, $\bar{M} = E + PM$.

PROOF: Apply Theorem 4-11 with the random time equal to one Then

$$\begin{aligned} \mathrm{M}_i[\bar{\mathfrak{t}}_j] &= \sum_k P_{ik}\mathrm{M}_k[\mathfrak{t}_j + 1] \\ &= \sum_k P_{ik}(M_{kj} + 1) \\ &= \sum_k P_{ik}M_{kj} + \sum_k P_{ik} \\ &= (PM)_{ij} + 1. \end{aligned}$$

For an ergodic chain P we define D to be the diagonal matrix whose diagonal entries are $1/\alpha_i$, where $\alpha 1 = 1$. From Proposition 6-25 we see that

$$\bar{M} = D + M.$$

Proposition 6-42: If P is an ergodic chain, the mean first passage time matrix M satisfies these properties:

(1) $M_{dg} = 0$ and $M \geq 0$.
(2) M is finite-valued.
(3) $(I - P)M = E - D$.

PROOF: The first statement is obvious, and the third follows immediately from Proposition 6-41 and the identity $\bar{M} = D + M$ if we can show that M is finite-valued. The problem thus reduces to proving (2). We know that $M_{ii} = 0$; therefore let i and j be distinct states. We shall show that M_{ij} is finite. Let $\mathfrak{t} = \min(\bar{\mathfrak{t}}_i, \bar{\mathfrak{t}}_j)$. Then

$$\begin{aligned} \frac{1}{\alpha_j} = \bar{M}_{jj} &= \mathrm{M}_j[\bar{\mathfrak{t}}_j] \\ &\geq \Pr_j[x_t = i]\mathrm{M}_j[\bar{\mathfrak{t}}_j \mid x_t = i] \\ &\geq \Pr_j[x_t = i]M_{ij} \quad \text{by Theorem 4-11} \\ &= {}^j\bar{H}_{ji}M_{ij}. \end{aligned}$$

If we can show that ${}^j\bar{H}_{ji} > 0$, we will have

$$M_{ij} \leq \frac{\bar{M}_{jj}}{{}^j\bar{H}_{ji}} < \infty.$$

But $0 < \alpha_i/\alpha_j = {}^j\bar{N}_{ji} = {}^j\bar{H}_{ji}\,{}^jN_{ii}$ by Corollary 6-21 and Proposition 4-15, so that ${}^j\bar{H}_{ji} > 0$.

The remarkable fact about the mean first passage time matrix M for ergodic chains is that the converse of Proposition 6-42—namely

Theorem 6-43—is also true. Thus once a candidate for M has been found, even by guessing, we need check only that it satisfies (1), (2), and (3).

Theorem 6-43: If P is an ergodic chain, the mean first passage time matrix M is characterized by these properties:

(1) $M_{dg} = 0$ and $M \geq 0$.
(2) M is finite valued.
(3) $(I - P)M = E - D$.

PROOF: Proposition 6-42 shows that M satisfies these properties. Let Y be any matrix for which (1), (2), and (3) hold. Let 0 be an arbitrary fixed state of the chain. It is sufficient to show that y, the zeroth column of Y, is equal to m, the zeroth column of M. Forming the chain ${}^0P$, in which state 0 has been made absorbing, and writing

$$^0P = \begin{pmatrix} 1 & 0 & 0 & \cdots \\ \vdots & & Q & \end{pmatrix} \quad \text{and} \quad y = \begin{pmatrix} y_0 \\ \bar{y} \end{pmatrix} \quad \text{and} \quad m = \begin{pmatrix} m_0 \\ \bar{m} \end{pmatrix},$$

we see that $m = \{M_i[a]\}$ and by Corollary 5-17 that $\bar{m}$ is the minimum finite-valued non-negative solution of the equation $(I - Q)\bar{x} = 1$. We first show that $\bar{y}$ is another finite-valued non-negative solution. We know that $\bar{y}$ is finite-valued and non-negative by hypothesis. The identity $(I - P)Y = E - D$ yields in the zeroth column

$$\begin{pmatrix} 1 - P_{00} & \cdots \\ \vdots & I - Q \end{pmatrix} \begin{pmatrix} y_0 \\ \bar{y} \end{pmatrix} = \begin{pmatrix} 1 - 1/\alpha_0 \\ 1 \end{pmatrix}.$$

But $y_0 = 0$ so that $(I - Q)\bar{y} = 1$. We conclude that $\bar{y} \geq \bar{m}$. Since $y_0 = m_0 = 0$, we have $y \geq m$. Hence

$$(I - P)(y - m) = (I - P)y - (I - P)m = 0$$

and $y - m$ is a finite-valued non-negative P-regular function. Thus $y - m = c1$ by Proposition 6-3. Looking at the zeroth entries, we see that $0 = y_0 - m_0 = c$. Therefore, $y = m$.

7. Examples of the mean first passage time matrix

In this section we shall compute the mean first passage time matrix associated with two infinite recurrent chains. The first example is a reflecting random walk, and the second is the basic example.

EXAMPLE 1: Reflecting random walk.

A random walk on the non-negative integers is defined by the transition probabilities

$$P_{00} = q, \quad \text{where } q \neq 0 \text{ and } q \neq 1,$$
$$P_{i,i+1} = p = 1 - q \quad \text{for } i \geq 0$$
$$P_{i,i-1} = q \quad \text{for } i > 0.$$

We note that the process P with 0 made absorbing is the infinite drunkard's walk P'. For the present chain we have

$$\bar{H}_{00} = pH_{10} + qH_{00} = pH_{10} + q.$$

But H_{10} is the absorption probability B_{10} for the infinite drunkard's walk. And $B_{10} = 1$ if $p \leq q$, and $B_{10} < 1$ if $p > q$. Therefore

$$\bar{H}_{00}\begin{cases} = 1 & \text{if } p \leq q \\ < 1 & \text{if } p > q, \end{cases}$$

and P is recurrent if and only if $p \leq q$.

A similar relation holds for $\bar{M}_{00}$; we have

$$\bar{M}_{00} = 1 + pM_{10} + qM_{00} = 1 + pM_{10}.$$

Since M_{10} is the mean time to absorption $M_1[\mathbf{a}]$ in the P' chain, we see that $\bar{M}_{00}$ is finite if and only if $p < q$. That is,

$$P \text{ is} \begin{cases} \text{transient} & \text{if } p > q \\ \text{null} & \text{if } p = q \\ \text{ergodic} & \text{if } p < q. \end{cases}$$

The chain is never cyclic, since $P_{00} > 0$.

We shall compute M for the ergodic case. Let $r = q/p > 1$. A P-regular measure α must satisfy $\alpha = \alpha P$, or

$$\alpha_0 = \alpha_0 q + \alpha_1 q$$
$$\alpha_i = \alpha_{i-1}p + \alpha_{i+1}q \quad \text{for } i > 0.$$

From the first equation we obtain $\alpha_1 = \alpha_0/r$, and from the second, which is a second-order difference equation, we obtain

$$\alpha_i = A + Br^{-i} \quad \text{for } i \geq 0.$$

From the two equations we find $\alpha_0 = A + B$ and $\alpha_0/r = \alpha_1 = A + B/r$. Therefore, $A/r = A$, and since $r > 1$, we must have $A = 0$. Choosing B so that $\alpha 1 = 1$, we arrive at the result

$$\alpha_i = (1 - 1/r)r^{-i}.$$

The entries M_{i0} of the mean first passage time matrix are easily found once the value of α_0 is known. Letting m be the zeroth column of the M matrix, we see from Proposition 6-42 that

$$(I - P)m = \mathbf{1} - \{\delta_{i0}/\alpha_0\}$$

or that

$$m_0 - qm_0 - pm_1 = 1 - 1/\alpha_0$$

and

$$m_i - qm_{i-1} - pm_{i+1} = 1 \quad \text{for } i > 0.$$

Since $\alpha_0 = 1 - p/q$, we have $1 - 1/\alpha_0 = -p/(q - p)$. The fact that $m_0 = 0$ then reduces the first relation to

$$-pm_1 = -p/(q - p),$$

so that

$$m_1 = 1/(q - p).$$

The difference equation for m_i has as a general solution

$$m_i = A + B(q/p)^i + i/(q - p) \quad \text{for } i \geq 0.$$

The boundary conditions on m_0 and m_1 imply that $A = B = 0$ and that

$$M_{i0} = i/(q - p).$$

The computation of M_{0i} uses the same general methods. First, we note that if $i < j$, then the process must pass through i from 0 to get to j. Hence

$$M_{0i} + M_{ij} = M_{0j}$$

or

$$M_{ij} = M_{0j} - M_{0i}.$$

Now

$$M_{01} = p + q(1 + M_{01})$$

so that

$$M_{01} = 1/p.$$

For $0 < i < j$,

$$M_{ij} = p(1 + M_{i+1,j}) + q(1 + M_{i-1,j})$$

or

$$M_{0j} - M_{0i} = 1 + p(M_{0j} - M_{0,i+1}) + q(M_{0j} - M_{0,i-1}).$$

Thus for $i > 0$,

$$pM_{0,i+1} - M_{0i} + qM_{0,i-1} = 1,$$

and for $i \geq 0$,

$$pM_{0,i+2} - M_{0,i+1} + qM_{0i} = 1.$$

Solving this equation and using the relations $M_{00} = 0$ and $M_{01} = 1/p$, we find

$$M_{0i} = \frac{q}{(q-p)^2}(r^i - 1) - \frac{i}{q-p}.$$

Algebraic manipulation yields the alternate formula

$$M_{0i} = \frac{1}{p}\sum_{k=0}^{i-1} r^k(i-k).$$

Since $M_{i0} = M_{ij} + M_{j0}$ when $i > j$ and since $M_{0i} + M_{ij} = M_{0j}$ when $i < j$, we may summarize our results as follows:

$$M_{ij} = \begin{cases} \dfrac{i-j}{q-p} & \text{if } i \geq j \\ \dfrac{q}{(q-p)^2}(r^j - r^i) - \dfrac{j-i}{q-p} & \text{if } i \leq j. \end{cases}$$

Example 2: Basic Example.

The vector β defined for the basic example has the property that $\beta P = \beta$ if and only if $\lim_{i\to\infty} \beta_i = 0$. Furthermore, P is recurrent if and only if $\lim_{i\to\infty} \beta_i = 0$. When P is recurrent, it is null if $\sum_i \beta_i$ is infinite and ergodic if $\sum_i \beta_i$ is finite. In the ergodic case the regular measure α for which $\alpha 1 = 1$ has entries

$$\alpha_i = \frac{\beta_i}{\sum_i \beta_i}.$$

Entries M_{ij} of the mean first passage time matrix for the basic example satisfy the equations

$$M_{0i} + M_{ij} = M_{0j} \qquad \text{for } i < j$$

and

$$M_{ij} = M_{i0} + M_{0j} \quad \text{for } i > j.$$

Since

$$M_{i0} + M_{0i} = \bar{M}_{ii} = \frac{\sum \beta_i}{\beta_i} \quad \text{for } i > 0,$$

it is sufficient to compute M_{0i} for the chain. Taking the statements {the process moves from 0 to $k \leq i - 1$ and then to zero, the process moves directly to i} as a set of alternatives, we find that

$$M_{0i} = \beta_i i + \sum_{k=0}^{i-1} \beta_k q_{k+1}(k + 1 + M_{0i}).$$

Solving this equation with the aid of the relation $\beta_k q_{k+1} = \beta_k - \beta_{k+1}$, we obtain

$$M_{0i} = \frac{1}{\beta_i} \sum_{k<i} \beta_k.$$

Therefore, for $i > 0$,

$$M_{i0} = \frac{1}{\beta_i} \sum_{k \geq i} \beta_k.$$

The general entries may be computed from

$$\begin{aligned} M_{ij} &= M_{0j} - M_{0i} \quad \text{if } i < j \\ M_{ij} &= M_{i0} + M_{0j} \quad \text{if } i > j. \end{aligned}$$

8. Reverse Markov chains

Let $\{x_n\}$ be the outcome functions for a denumerable Markov process defined on a space Ω and with range in S. The outcome functions appear in a certain order and represent the forward passage of time. One may well wonder, however, if, when the functions are looked at in reverse order, the process is in any sense still Markovian. It is the purpose of this section to discuss this question; as a by-product of the discussion, we shall gain an interpretation for the dual of an ergodic Markov chain.

The sense in which a Markov process reversed in time is still a Markov process is the following.

Proposition 6-44: Let $\{x_n\}$ be a denumerable Markov process and let N be a fixed positive integer. Define $y_n = x_{N-n}$ for $0 \leq n \leq N$ and $y_n =$ "stop" for $n > N$. Then the functions y_n are the outcome functions for a denumerable Markov process with the same state space with "stop" adjoined.

PROOF: We shall show that the functions y_n satisfy the Markov property. Clearly, this needs to be checked only for $n \leq N$. If $\Pr[y_0 = c_0 \wedge \cdots \wedge y_{n-1} = c_{n-1}] > 0$, then

$$\begin{aligned} \Pr[y_n = c_n \mid y_0 = c_0 \wedge \cdots \wedge y_{n-1} = c_{n-1}] \\ = \Pr[x_{N-n} = c_n \mid x_N = c_0 \wedge \cdots \wedge x_{N-n+1} = c_{n-1}] \\ = \frac{\Pr[x_{N-n} = c_n \wedge x_{N-n+1} = c_{n-1} \wedge \cdots \wedge x_N = c_0]}{\Pr[x_{N-n+1} = c_{n-1} \wedge \cdots \wedge x_N = c_0]}. \end{aligned}$$

The numerator is

$$\begin{aligned} \Pr[x_{N-n} = c_n \wedge x_{N-n+1} = c_{n-1}] \\ \times \Pr[x_{N-n+2} = c_{n-2} \mid x_{N-n} = c_n \wedge x_{N-n+1} = c_{n-1}] \\ \times \cdots \times \Pr[x_N = c_0 \mid x_{N-n} = c_n \wedge \cdots \wedge x_{N-1} = c_1], \end{aligned}$$

which by the Markov property is

$$\Pr[x_{N-n} = c_n \wedge x_{N-n+1} = c_{n-1}] \cdot P_{c_{n-1}c_{n-2}} \cdot \cdots \cdot P_{c_1 c_0}.$$

The denominator similarly reduces to

$$\Pr[x_{N-n+1} = c_{n-1}] \cdot P_{c_{n-1}c_{n-2}} \cdot \cdots \cdot P_{c_1 c_0}.$$

Dividing, we obtain

$$\begin{aligned}\Pr[y_n = c_n \mid y_0 = c_0 \wedge \cdots \wedge y_{n-1} = c_{n-1}] \\ &= \frac{\Pr[x_{N-n} = c_n \wedge x_{N-n+1} = c_{n-1}]}{\Pr[x_{N-n+1} = c_{n-1}]} \\ &= \Pr[x_{N-n} = c_n \mid x_{N-n+1} = c_{n-1}] \\ &= \Pr[y_n = c_n \mid y_{n-1} = c_{n-1}].\end{aligned}$$

It is not true in general that, if the original process is a Markov chain P, then the new process is also a Markov chain. Let P be started with distribution π. We then have, if $\Pr[y_{n-1} = i] > 0$,

$$\begin{aligned}\Pr[y_n = j \mid y_{n-1} = i] &= \Pr_\pi[x_{N-n} = j \mid x_{N-n+1} = i] \\ &= \frac{\Pr_\pi[x_{N-n} = j \wedge x_{N-n+1} = i]}{\Pr_\pi[x_{N-n+1} = i]} \\ &= \frac{(\pi P^{N-n})_j \cdot P_{ji}}{(\pi P^{N-n+1})_i}.\end{aligned}$$

The last quantity on the right need not be independent of n. Nevertheless, if P is ergodic, there is a case where we *can* state a positive result—a result which gives us an interpretation for the dual of P.

Proposition 6-45: Let $\{x_n\}$ be the outcome functions for an ergodic chain P, let N be a fixed positive integer, and let α be the unique P-regular probability measure. If P is started with distribution α, then the process $\{y_n = x_{N-n}, 0 \le n \le N\}$ is an initial segment of the Markov chain with transition matrix $\hat{P}$ and with starting distribution α.

PROOF: If $\Pr[y_{n-1} = i] > 0$, then $\sigma = \alpha P^N = \alpha$ and

$$\begin{aligned}\Pr_\alpha[y_n = j \mid y_{n-1} = i] &= \frac{(\alpha P^{N-n})_j \cdot P_{ji}}{(\alpha P^{N-n+1})_i} \\ &= \frac{\alpha_j P_{ji}}{\alpha_i} \\ &= \hat{P}_{ij}\end{aligned}$$

independently of n for $n \le N$.

The motivation for calling $\hat{P}$ the reverse chain when P is recurrent now becomes clear.

9. Problems

1. Compute for the Land of Oz example P^2, P^4, and P^8. What is $A = \lim P^n$? Show that each row of A is a regular measure and that each column is a regular function.
2. Let

$$P = \begin{matrix} & 1 & 2 & 3 & 4 & 5 & 6 \\ 1 & \frac{1}{2} & 0 & 0 & \frac{1}{4} & \frac{1}{4} & 0 \\ 2 & 0 & 1 & 0 & 0 & 0 & 0 \\ 3 & 0 & 0 & \frac{3}{5} & 0 & \frac{2}{5} & 0 \\ 4 & \frac{1}{2} & 0 & 0 & 0 & 0 & \frac{1}{2} \\ 5 & 0 & 0 & \frac{3}{5} & 0 & \frac{2}{5} & 0 \\ 6 & 0 & \frac{1}{4} & \frac{1}{4} & \frac{1}{4} & \frac{1}{4} & 0 \end{matrix}.$$

The process is started in state 1. Find the probability of being in the various states in the long run.

3. In the basic example, let

$$p_i = \left(\frac{i}{i+1}\right)^2.$$

Is the chain transient, null, or ergodic?

4. Prove that $\alpha = 1^T$ is regular for any sums of independent random variables process. Give a careful statement as to the existence of transient, null, and ergodic examples.
5. Establish the following relationships between a chain with transition matrix P and one with matrix P^E:
   (a) If P is transient, then P^E is transient.
   (b) If P is recurrent, then P^E is recurrent and $\alpha^E = c\alpha_E$.
   (c) If P is ergodic, then P^E is ergodic, but the converse is not always true.
6. Prove that if a recurrent P has column-sums equal to 1, then $\hat{P} = P^T$.
7. Consider sums of independent random variables on the integers with $p_{-1} = \frac{1}{3}$ and $p_1 = \frac{2}{3}$. Choose two essentially different positive regular measures α, and show that each gives a correct expression for ${}^i\bar{N}_{ij}$ in Corollary 6-20.
8. Show that if $P_{ii} > 0$ for a single state in a recurrent chain, then the chain is noncyclic.
9. Show by an example that $\hat{M}_{ij} = \alpha_j M_{ji}/\alpha_i$ need not be true.
10. Show that in an ergodic chain αM may be either finite-valued or infinite-valued.
11. Determine whether the following chain is transient or recurrent:

$$P = \begin{pmatrix} \frac{1}{2} & 0 & \frac{1}{4} & \frac{1}{4} \\ 0 & 1 & 0 & 0 \\ \frac{1}{4} & \frac{1}{4} & \frac{1}{4} & \frac{1}{4} \\ 0 & 0 & 0 & 1 \end{pmatrix}.$$

If transient: Put into standard form, and find N, B, and a.
If recurrent: Is it cyclic? Find α, M, $\hat{P}$, $\hat{M}$.

12. Do the same for

$$P = \begin{pmatrix} 1 - c & c \\ 1 & 0 \end{pmatrix}, \qquad 0 < c \leq 1.$$

13. Find α and M for an independent trials process by the methods of this chapter, and check your answers by a direct computation. [See Problem 5 in Chapter 4.]

14. (a) Complete the work of finding M for the basic example.
(b) Find the reverse of the basic example (when recurrent), and compute M for this chain.

15. A light bulb in a fixture lasts j time units with probability f_j. It is replaced with a similar bulb when it burns out. Assume that $\sum f_j = 1$, $f_0 = 0$, and $f_1 > 0$. Let x_n be the length of time that the bulb in use at time n has lasted (taken to be 0 if there is a replacement at time n). Show that $\{x_n\}$ is the set of outcome functions for a Markov chain and discuss the connection with the basic example. Show that the probability that a bulb is replaced at time n tends to a limit as $n \to \infty$.

Problems 16 to 26 refer to sums of independent random variables on the circle. Mark n ($n \geq 3$) points on a circle, labeled $i = 0, 1, \ldots, n - 1$ clockwise. The process either moves one step clockwise with probability $\frac{2}{3}$, or it moves one step counterclockwise with probability $\frac{1}{3}$.

16. Prove that the chain is ergodic. Is it cyclic?

17. Find a positive regular measure α with $\alpha\mathbf{1} = 1$. Interpret it.

18. Construct the reverse chain.

19. Compute M by means of Theorem 6-43. [It is sufficient to find M_{i0}.]

20. Show that for large n,

$$M_{i0} \sim 3(n - i - n(\tfrac{1}{2})^i).$$

Compare this result with the absorption times of a drunkard's walk on $\{0, 1, \ldots, n\}$ with $p = \frac{2}{3}$.

21. Show that the approximation in the previous problem is excellent for $n = 50$.

22. Use the approximation of Problem 20 to show that the maximum value of M_{i0} occurs approximately at

$$i \doteq \frac{\log n}{\log 2}.$$

Check this conclusion for $n = 50$.

23. Let n be even, and let E be the set of even-numbered states. Compute P^E.

24. For $n = 3$, compute P^2, P^4, P^8, and P^9. What is the limit of P^n?

25. Repeat Problem 24 for $n = 4$.

26. Show that $\alpha M = c_n \mathbf{1}^T$, and find an asymptotic expression for c_n.

CHAPTER 7

INTRODUCTION TO POTENTIAL THEORY

1. Brownian motion

One of the fruitful achievements of probability theory in recent years has been the recognition that two seemingly unrelated theories in physics—one for Brownian motion and one for potentials—are mathematically equivalent. That is, the results of the two theories are in one-to-one correspondence and any proof of a result in one theory can be translated directly into a proof of the corresponding result in the other theory. In this chapter we shall sketch how this equivalence comes about, and we shall see that Brownian motion is a process which is like a Markov chain except that it does not have a denumerable state space and time does not proceed in discrete steps. The details of this equivalence can be found in Knapp [1965]. The important thing to notice will be that the definitions of potential-theoretic concepts in terms of Brownian motion do not depend on isolated specific properties of the process but depend only on the Markovian character of Brownian motion. In other words, there is reasonable hope of defining for an arbitrary Markov chain a potential theory in which analogs of the classical theorems hold.

We begin by describing Brownian motion. In 1826 the botanist Robert Brown observed that microscopic particles, when left alone in a liquid, are seen to move constantly in the fluid along erratic paths. Much later Albert Einstein investigated this movement of particles from a theoretical point of view. Einstein was able to derive statistical laws which estimate how a large number of particles spread over a period of time, and his predictions were verified.

In setting up a probabilistic model for this so-called Brownian motion, we simply replace Einstein's estimate of what happens to a large number of particles by a probability for what happens to one particle. We are then to require that

$$\Pr[\text{particle started at } u \text{ is in } E \text{ at time } t] = \int_E \frac{1}{(2\pi t)^{3/2}} e^{-|u-y|^2/2t} dy,$$

where E is a Borel set in three-dimensional (Euclidean) space R^3 and $|u - y|$ denotes the Euclidean distance from u to y. If we use the notation $\Pr_u[x_t \in E]$ for the left side and the notation $p^t(u, E)$ for the right side, we have

$$\Pr_u[x_t \in E] = p^t(u, E).$$

By Theorem 1-41, $p^t(u, F)$ is a measure depending on t and u and defined on the smallest Borel field containing the open sets of R^3. Therefore, we may write

$$\Pr_u[x_t \in E] = \int_E p^t(u, dy).$$

The physical theory also makes us require that if $t_1 < t_2 < \cdots < t_n$, then $\{x_{t_1}, x_{t_2} - x_{t_1}, \ldots, x_{t_n} - x_{t_{n-1}}\}$ should behave like a set of independent random variables with $x_{t+s} - x_t$ having the same distribution as x_s. That is, we require that

$$\begin{aligned}\Pr_u[x_{t_1} \in E_1 \wedge \cdots \wedge (x_{t_n} - x_{t_{n-1}}) \in E_n] \\ &= \Pr_u[x_{t_1} \in E_1] \cdot \cdots \cdot \Pr_u[x_{t_n} - x_{t_{n-1}} \in E_n] \\ &= \Pr_u[x_{t_1} \in E_1] \cdot \cdots \cdot \Pr_u[x_{(t_n - t_{n-1})} \in E_n].\end{aligned}$$

This identity implies that we must have

$$\begin{aligned}\Pr_u[x_q \in E \wedge x_r \in F \wedge \cdots \wedge x_s \in G \wedge x_t \in H] \\ &= \int_E p^q(u, dw) \int_F p^{r-q}(w, dx) \int \cdots \int_H p^{t-s}(y, dz).\end{aligned}$$

We note that with these various requirements we have given more than one definition for $\Pr_u[x_t \in E]$ and that we must check, for example, that

$$\Pr_u[x_s \in R^3 \wedge x_t \in E] = \Pr_u[x_t \in E]$$

and

$$\Pr_u[x_s \in F \wedge x_t \in R^3] = \Pr_u[x_s \in F].$$

Such identities can be verified by direct calculation. It should not be too surprising that such consistency conditions arise since they arose with denumerable stochastic processes earlier: In the proof of the Kolmogorov Extension Theorem in Chapter 2 we required that the measures on cylinder sets all be consistent.

Now for any denumerable Markov chain P we have

(1) $$0 \leq P_{ij} = \Pr_i[x_1 = j],$$

(2) $$\Pr_i[x_1 \in S] = \sum_j P_{ij} \leq 1,$$

(3a) $$\Pr_i[x_1 = j \wedge x_2 = k \wedge \cdots \wedge x_{n-1} = r \wedge x_n = s] = P_{ij}P_{jk} \cdots P_{rs}.$$

The last equality (3a) implies and is implied by

(3b) $\Pr_i[x_1 \in E \wedge x_2 \in F \wedge \cdots \wedge x_{n-1} \in G \wedge x_n \in H]$

$$= \sum_{j \in E} P_{ij} \sum_{k \in F} P_{jk} \cdots \sum_{s \in H} P_{rs}.$$

It is easy to prove both the Markov property and the Markov chain property from (3a), and hence (1), (2), and (3b) give an equivalent definition of Markov chain. The analogous statements for Brownian motion are

(1') $$0 \le \int_E p^t(u, dy) = \Pr_u[x_t \in E],$$

(2') $$\Pr_u[x_t \in R^3] = \int_{R^3} p^t(u, dy) = 1,$$

and

(3') $\Pr_u[x_q \in E \wedge x_r \in F \wedge \cdots \wedge x_t \in H]$

$$= \int_E p^q(u, dw) \int_F p^{r-q}(w, dx) \int \cdots \int_H p^{t-s}(y, dz).$$

As expected, these statements imply that the position of a particle at time $t + s$ depends only on s and the particle's position at time t, and not on the value of t or what happened to the particle before time t. (This assertion can be formulated precisely in terms of means of functions given a Borel field, which are a technical generalization of conditional means of functions given a partition.)

As with denumerable Markov chains we need not require that a Brownian motion particle be started deterministically at a state u. If we start the particle according to probabilities assigned by a measure μ on R^3, then we have

$$\Pr_\mu[x_t \in E] = \int_{R^3} \int_E \frac{1}{(2\pi t)^{3/2}} e^{-|u-y|^2/2t} dy d\mu(u)$$
$$= \int_{R^3} \int_E p^t(u, dy) d\mu(u).$$

A similar expression holds for the probability of being in a finite sequence of sets at specified times.

In Section 3 we shall need a formal definition of Brownian motion, and we use the formula for $\Pr_\mu[x_t \in E]$ to motivate it. We define a transformation P^t of the measures μ on R^3 with $\mu(R^3) = 1$ into themselves by

$$(\mu P^t)(E) = \Pr_\mu[x_t \in E] = \int_E \left[\int_{R^3} \frac{1}{(2\pi t)^{3/2}} e^{-|x-y|^2/2t} d\mu(x) \right] dy.$$

Later we shall replace the expression in brackets by $f(y, t)$ for simplicity of notation. That μP^t is a measure follows from Theorem 1-41, and that $(\mu P^t)(R^3) = 1$ follows from the identity

$$(\mu P^t)(R^3) = \int_{R^3} \left[\int_{R^3} \frac{1}{(2\pi t)^{3/2}} e^{-|x-y|^2/2t} dy\right] d\mu(x)$$
$$= \int_{R^3} \left[\frac{1}{(2\pi)^{3/2}} \int_{R^3} e^{-|u|^2/2} du\right] d\mu(x) = 1$$

after a change of variable.

Definition 7-1: The totality of theorems about the operators P^t, the measures μ on R^3 with $\mu(R^3) = 1$, and quantities definable in terms of them and properties of R^3 is called **Brownian motion theory**.

We immediately extend P^t by linearity to be defined on all finite measures on R^3 and all differences of two finite measures.

2. Potential theory

Classical potential theory begins as a study of Coulomb's law of attraction of electrical charges in physics. This law states that every two charges in the universe attract (or repel) each other with a force whose direction is the line connecting them and whose magnitude is proportional to the magnitude of each of them and inversely proportional to the square of the distance between them. That is,

$$F = \epsilon_0 \frac{Qq}{r^2},$$

where ϵ_0 is a constant depending on the units. As an aid in the study, one introduces the notion of potential: The potential at a point x due to a charge q is the work (or energy) required to bring a unit charge from infinity to the point x. It can be shown that this potential is independent of the path along which the charge is brought to the point x and that its value is

$$\frac{1}{2\pi} \frac{q}{|x - x_0|},$$

where x_0 is the position of the charge and where the constant $1/2\pi$ has been fixed after a certain choice of units.

More generally one defines a **charge distribution** to be the difference of any two finite measures defined on the Borel sets of R^3, that is, the smallest Borel field in R^3 containing all open sets. The potential at x due to the charge distribution is again the work required to bring a unit charge from infinity to the point x. Since force (and hence work) are additive, the potential due to a charge distribution consisting of charges $q_1, \ldots, q_n$ at points $x_1, \ldots, x_n$ is

$$\frac{1}{2\pi} \sum_{i=1}^{n} \frac{q_i}{|x - x_i|}.$$

Passing to the limit in an appropriate sense, we would expect the potential due to an arbitrary charge distribution μ to be

$$\frac{1}{2\pi} \int_{R^3} \frac{d\mu(y)}{|x - y|}.$$

After checking that such an expression is always well defined, we shall define a potential to be any function of this form.

Lemma 7-2: If μ is a charge distribution, then

$$g(x) = \frac{1}{2\pi} \int_{R^3} \frac{1}{|x - y|} \, d\mu(y)$$

is finite a.e. with respect to Lebesgue measure.

PROOF: It suffices to prove the lemma for the case where μ is a measure, since the general case follows by taking differences. Let K_n denote the closed ball about the origin of radius n, and form

$$\int_{K_n} g(x)dx = \frac{1}{2\pi} \int_{K_n} \int_{R^3} \frac{1}{|x - y|} \, d\mu(y)dx.$$

By Fubini's Theorem we may interchange the order of integration to get

$$\int_{K_n} g(x)dx = \frac{1}{2\pi} \int_{R^3} \left[\int_{K_n} \frac{1}{|x - y|} \, dx \right] d\mu(y).$$

The inside integral on the right is bounded by its value at the origin, which is some finite number c. Thus the right side does not exceed

$$\frac{1}{2\pi} c\mu(R^3) < \infty,$$

and g must be finite a.e. on K_n. Since the countable union of the sets K_n is R^3, we conclude that g is finite a.e.

Definition 7-3: The function

$$\frac{1}{2\pi} \int_{R^3} \frac{1}{|x - y|} \, d\mu(y)$$

for μ a charge distribution will be called the **potential** of μ.

The class of theorems relating charges and potentials and quantities definable in terms of them and properties of R^3 is called **classical potential theory**. The operator transforming a charge into its potential is called the **potential operator.** The kernel of the potential operator is called the **Green's function**.

As we have defined it, potential theory contains the subject known in physics as electrostatics. Our definition includes the notions of distance, charge, and potential, and all the quantities commonly arising in electrostatics are definable in terms of these three notions. As an illustration, Table 7-1 shows how some of the quantities arising in electrostatics are related dimensionally to distance, charge, and potential. The table uses the notation

distance	$= x$		distance	$= x$
time	$= t$	and	charge	$= q$
mass	$= m$		potential	$= V$
charge	$= q$			

TABLE 7-1. DIMENSIONS OF ELECTROSTATIC CONCEPTS

Concept	Dimensions	Potential-Theoretic Dimensions
Capacity	q^2t^2/mx^2	q/V
Charge	q	q
Energy	mx^2/t^2	Vq
Field	mx/t^2q	V/x
Force	mx/t^2	Vq/x
Potential	mx^2/t^2q	V

We give four examples to illustrate how concepts may be defined explicitly in terms of distance, charge, and potential.

(1) We can reasonably ask what the total amount of work required to assemble a charge distribution is if only an "infinitesimal" amount of charge is brought into position at one time. The way to compute this amount of work is to integrate the potential function against the charge distribution, provided the integral exists. Thus we define the **energy of a charge distribution** to be the integral of its potential with respect to the charge, provided the integral exists.

(2) The **total charge** of a charge distribution μ is $\mu(R^3)$.

(3) If a total amount of charge q is put on a piece of conducting metal in R^3, the charge will redistribute itself in such a way that the potential is a constant on the set where the metal is. The situation where the potential is constant on the metal is the one which minimizes energy

among all charges μ with total charge q and with μ vanishing away from the set where the metal is, and this situation is referred to as equilibrium. We define an **equilibrium potential** for a closed set E to be a potential which is 1 on E and which comes from a charge distribution which has all its charge on E. An **equilibrium set** is a set which has an equilibrium potential.

(4) The capacity of a conductor in R^3 is defined as the total amount of charge needed to produce a unit potential on the set where the conductor is. We thus define the **capacity** of any equilibrium set to be the total charge of a charge distribution producing an equilibrium potential.

To indicate the directions in which classical potential theory leads, we shall state without proof some of the theorems in the subject. The **support** of a charge is defined to be the complement of the union of all open sets U with the property that the charge vanishes on U and every measurable subset of U.

(1) *Uniqueness of charge:* A potential uniquely determines its charge.

(2) *Determination of potential:* A potential is completely determined by its values on the support of its charge.

(3) *Uniqueness of equilibrium potential:* A set E has at most one equilibrium potential. (This assertion is a corollary of (2).)

(4) *Characterization of equilibrium potential:* The equilibrium potential for an equilibrium set E is equivalent to the pointwise infimum of all potentials which have non-negative charges and which dominate the constant function 1 on E.

(5) *Principle of domination:* Let h and g be potentials arising from non-negative charges $\bar{\mu}$ and μ, respectively. If h dominates g on the support of μ, then h dominates g everywhere and $\bar{\mu}(R^3) \geq \mu(R^3)$.

(6) *Principle of balayage:* If g is a potential with a non-negative charge and if E is a closed set in R^3, then there is a unique potential $\bar{g}$ with a non-negative charge with support in E such that $\bar{g} = g$ on E. The potential $\bar{g}$ (called the balayage potential) satisfies $\bar{g} \leq g$ everywhere, and its total charge does not exceed the total charge of g. The balayage potential is equivalent to the pointwise infimum of all potentials which have non-negative charges and which dominate g on E. It is equivalent to the supremum of all potentials which are dominated by g on E and whose charges have support in E.

(7) *Principle of lower envelope:* The pointwise infimum of potentials with non-negative charges is equivalent to a potential with a non-negative charge.

(8) *Non-negative potentials:* The charge distribution of a non-negative potential has non-negative total charge.

(9) *Energy of equilibrium potential:* If E is an equilibrium set of finite energy, then the equilibrium potential minimizes energy among all potentials whose charges have support in E and whose total charge is equal to the capacity of E.

3. Equivalence of Brownian motion and potential theory

Kakutani [1944] observed that several of the basic quantities of potential theory, like equilibrium potential, had simple probabilistic interpretations in terms of Brownian motion. If E is an equilibrium set, the value of the equilibrium potential at x is the probability that a Brownian motion process started at x ever hits the set E. Doob and Hunt extended Kakutani's work, and it gradually became clear that in a certain sense Brownian motion and potential theory were really the same subject.

To say that they are exactly the same would be to say that every theorem about Brownian motion should interest a person studying potential theory, and conversely. Although it is doubtful that this situation is the case at present, it is certainly true that modern developments in the two subjects are moving more and more in this direction.

We shall now show that there is a natural way in terms of Brownian motion of obtaining the operator mapping charges into potentials, and that, conversely, from the potential operator it is possible to recover the family $\{P^t\}$ of transition operators for Brownian motion. These facts make it clear that in a technical sense the two theories are identical.

The proof in the first direction is easy and is completed by Proposition 7-4. We recall from the definition of μP^t that

$$(\mu P^t)(E) = \int_E f(x, t)dx$$

for a certain function $f(x, t)$.

Proposition 7-4: Every theorem about potentials can be formulated as a theorem about Brownian motion. Specifically, if μ is a charge, then the potential g of μ satisfies

$$g(x) = \lim_{T \to \infty} \int_0^T f(x, t)dt,$$

where

$$(\mu P^t)(E) = \int_E f(x, t)dx.$$

PROOF: We may assume that $\mu \geq 0$ without loss of generality. Then

$$\int_0^T f(x, t)dt = \int_0^T \frac{1}{(2\pi t)^{3/2}} \left[\int_{R^3} e^{-|x-y|^2/2t} d\mu(y)\right] dt.$$

By Fubini's Theorem we may interchange the order of integration. The above expression is

$$= \int_{R^3} \left[\int_0^T \frac{1}{(2\pi t)^{3/2}} e^{-|x-y|^2/2t} dt \right] d\mu(y).$$

We make the change of variable on t which sends $|x - y|^2/t$ into u^2. The expression becomes

$$= \int_{R^3} \left[\int_{|x-y|/\sqrt{T}}^{\infty} \frac{2}{(2\pi)^{3/2}} |x - y|^{-1} e^{-u^2/2} du \right] d\mu(y).$$

By the Monotone Convergence Theorem,

$$\begin{aligned} \lim_{T\to\infty} \int_0^T f(x, t)dt &= \int_{R^3} \left[\int_0^{\infty} \frac{2}{(2\pi)^{3/2}} |x - y|^{-1} e^{-u^2/2} du \right] d\mu(y) \\ &= \int_{R^3} \frac{1}{|x - y|} d\mu(y) \left[\frac{1}{2\pi} \int_{-\infty}^{\infty} \frac{1}{\sqrt{2\pi}} e^{-u^2/2} du \right] \\ &= \frac{1}{2\pi} \int_{R^3} \frac{1}{|x - y|} d\mu(y) \\ &= g(x). \end{aligned}$$

Proposition 7-4 is a precise statement of the connection between Brownian motion and potential theory in one direction. We see that formally the potential operator is

$$\lim_{T\to\infty} \int_0^T P^t dt.$$

Thus to complete the proof of the equivalence of the two theories, what we need to do essentially is recover a sequence from its limit. Of course, we cannot do so unless we know some other properties of the sequence, and it is the isolation of these properties that makes this half of the equivalence difficult. We shall not go into the details here, but we can indicate the general approach to the problem.

Let C_0 denote the set of continuous real-valued functions f on R^3 which vanish at infinity; that is, which are such that for any $\epsilon > 0$ there is a ball of finite radius in R^3 outside of which f is everywhere less than ϵ in absolute value. We define Q^t on C_0 by

$$(Q^t f)(y) = \int_{R^3} \frac{1}{(2\pi t)^{3/2}} e^{-|x-y|^2/2t} f(x)dx.$$

The following facts can be checked:

(1) If f is in C_0, then so is $Q^t f$.
(2) $\sup_y |(Q^t f)(y)| \le \sup_y |f(y)|$.

(3) $Q^{t+s} = Q^tQ^s$.
(4) $(Q^tf)(y)$ converges uniformly to $f(y)$ as t decreases to 0.
(5) $(Q^tf)(y)$ converges uniformly to 0 as t increases to ∞.

Now any set function in the class M of differences of finite measures is completely determined by its effect on all the functions in C_0, and a direct calculation shows that

$$\int_{R^3} fd(\mu P^t) = \int_{R^3} (Q^tf)d\mu$$

for every $\mu \in M$ and $f \in C_0$. It follows that Q^t and this equation completely determine P^t. Therefore, it is enough to recover Q^t from the potential operator in order to prove our result.

For every $f \in C_0$ such that $(1/t)[(Q^tf)(y) - f(y)]$ converges uniformly as t decreases to 0, we define

$$Af = \lim_{t \downarrow 0} \frac{1}{t}[Q^tf - f].$$

It turns out that if A is known on its entire domain of definition, then Q^t is completely determined by the definition of A and by the first four properties of Q^t listed above. Thus if A could be defined within potential theory, then so could each of the operators Q^t: They are the unique family of operators such that the definition of A and properties (1), (2), (3), and (4) hold. The actual proof of this existence and uniqueness consists in writing down a concrete formula for Q^t in terms of A and t; we reproduce it in order to show that nothing appears in the formula except A, t, and the identity operator I:

$$Q^tf = \lim_{\lambda \to \infty} \sum_{k=0}^{\infty} \frac{1}{k!} t^k[\lambda^2(\lambda I - A)^{-1} - \lambda I]^k f.$$

For every f in C_0 such that $\int_0^T (Q^tf)(y)dt$ converges uniformly as $T \to \infty$, we define

$$Gf = \lim_{T \to \infty} \int_0^T (Q^tf)dt.$$

It can be shown readily from the five properties of Q^t that G and $-A$ are inverse operators on their respective domains. Thus each uniquely determines the other. Finally (and here is where some work is required) G looks sufficiently like the potential operator when its definition is compared with the formulas of Proposition 7-4 that the potential operator determines G. Thus the potential operator determines G, G determines A, A determines Q^t, and Q^t determines P^t. Hence every theorem of Brownian motion theory is a theorem of potential theory.

4. Brownian motion and potential theory in *n* dimensions

In the mathematical formulation of Brownian motion and potential theory there is no need to restrict the underlying space to be three-dimensional. We can define an n-dimensional Brownian motion operator P^t by

$$(\mu P^t)(E) = \int_E \left[\frac{1}{(2\pi t)^{n/2}} \int_{R^n} e^{-|x-y|^2/2t} d\mu(y)\right] dx.$$

The potential operator differs in appearance from dimension to dimension more than the Brownian motion operator does, but its kernel is still a constant multiple of the integral of $|x|^{-(n-1)}$. The potential $g(x)$ of μ, the difference of two finite measures, is defined by

$$g(x) = \begin{cases} -\displaystyle\int_{R^1} |x - y| d\mu(y) & \text{in dimension 1} \\ 2\displaystyle\int_{R^2} \log |x - y| d\mu(y) & \text{in dimension 2} \\ c_n \displaystyle\int_{R^n} \frac{1}{|x - y|^{n-2}} d\mu(y) & \text{in dimension } n \geq 3, \end{cases}$$

where

$$c_n = \tfrac{1}{2}\pi^{-n/2}\Gamma(\tfrac{1}{2}(n - 2)).$$

In dimension $n \geq 3$, $g(x)$ is necessarily finite a.e., but in dimensions 1 and 2 we shall need to assume that g is finite a.e.

The fact that the kernel $1/|x|^{n-2}$ tends to zero at infinity in dimension $n \geq 3$ but the kernels $|x - y|$ and $\log |x - y|$ do not tend to zero in dimensions 1 and 2 gives us a clue that the potential theory or dimensions 1 and 2 will differ sharply from that in higher dimensions. We shall discuss the reason for this difference shortly.

In dimension $n \geq 3$, Brownian motion theory and potential theory are again equivalent. The formula

$$g(x) = \lim_{T\to\infty} \int_0^T \frac{1}{(2\pi t)^{n/2}} \int_{R^n} e^{-|x-y|^2/2t} d\mu(y) dt$$

generalized from Proposition 7-4 is still valid, and the discussion of Section 3 goes over with little change to establish the equivalence.

But in dimensions 1 and 2, it does not. The above formula is not true for these dimensions, and the argument in Section 3 fails after the operator G is introduced. The reason for this failure is the following. We recall that in dimension $n \geq 3$ the potential operator is formally $\lim_{T\to\infty} \int_0^T P^t dt$. In dimensions greater than or equal to 3, this quantity is finite, whereas in dimensions 1 and 2 it is infinite. Now

$\lim_{T\to\infty}\int_0^T P^t dt$ plays much the same role for Brownian motion that $\sum_{n=0}^{\infty} P^n$ plays for denumerable Markov chains. It is finite if the process is transient, and infinite if the process is recurrent. In fact, the distinction between transience and recurrence is what is relevant for Brownian motion here: In dimensions 3 and greater a Brownian motion particle after leaving the unit ball of R^n returns to it with probability less than one, whereas in dimensions 1 and 2 it returns with probability equal to one.

The potential operator in dimensions 1 and 2 arises in a different way. Specifically, the formula generalized from Proposition 7-4 is not valid in general, but it is valid if μ has total charge zero and if a mild additional condition is satisfied. The exact formulation of this result will interest us later, and we give it as the next proposition.

Proposition 7-5: In R^n for $n = 1$ or 2, let $\mu = \mu^+ - \mu^-$ be the difference of two finite measures, and suppose that

$$\int_{R^1} |x - y| d\mu^+(y) < \infty \quad \text{and} \quad \int_{R^1} |x - y| d\mu^-(y) < \infty \quad \text{a.e. if } n = 1$$

or

$$\int_{R^2} |\log |x - y|| d\mu^+(y) < \infty \quad \text{and} \quad \int_{R^2} |\log |x - y|| d\mu^-(y) < \infty \quad \text{a.e. if } n = 2.$$

If $\mu(R^n) \neq 0$, then

$$\lim_{T\to\infty} \int_0^T \frac{1}{(2\pi t)^{n/2}} \int_{R^n} e^{-|x-y|^2/2t} d\mu(y)dt = +\infty \quad \text{or} \quad -\infty \quad \text{a.e.}$$

If $\mu(R^n) = 0$, then

$$g(x) = \lim_{T\to\infty} \int_0^T \frac{1}{(2\pi t)^{n/2}} \int_{R^n} e^{-|x-y|^2/2t} d\mu(y)dt$$

exists, is finite a.e., and satisfies

$$g(x) = \begin{cases} -\int_{R^1} |x - y| d\mu(y) & \text{if } n = 1 \\ 2\int_{R^2} \log |x - y| d\mu(y) & \text{if } n = 2. \end{cases}$$

Proof: We prove the result for $n = 1$; the ideas in the proof for $n = 2$ are similar. The same calculation as at the beginning of the proof of Proposition 7-4 shows that for $n = 1$

$$\int_0^T \frac{1}{\sqrt{2\pi t}} \int_{-\infty}^{\infty} e^{-|x-y|^2/2t} d\mu(y)dt = \int_{-\infty}^{\infty} |x - y| \left[\int_{|x-y|/\sqrt{T}}^{\infty} \frac{2}{\sqrt{2\pi}} u^{-2} e^{-u^2/2} du \right] d\mu(y).$$

If we use integration by parts on the terms in the brackets, differentiating the exponential and integrating u^{-2}, we find that the right side is

$$= \int_{-\infty}^{\infty} |x - y| \frac{2}{\sqrt{2\pi}} \left[-u^{-1} e^{-u^2/2} \Big|_{\frac{|x-y|}{\sqrt{T}}}^{\infty} - \int_{|x-y|/\sqrt{T}}^{\infty} e^{-u^2/2} du \right] d\mu(y)$$

$$= -\int_{-\infty}^{\infty} |x - y| \left[\int_{|x-y|/\sqrt{T}}^{\infty} \frac{2}{\sqrt{2\pi}} e^{-u^2/2} du \right] d\mu(y) + \frac{2}{\sqrt{2\pi}} \int_{-\infty}^{\infty} \sqrt{T} e^{-|x-y|^2/2T} d\mu(y).$$

We let T tend to ∞ and consider each term separately. In the first term the expression in brackets increases to 1. If we write the integral as the difference of one with respect to μ^+ and one with respect to μ^- and use the fact that

$$\int_{-\infty}^{\infty} |x - y| d\mu^+(y) < \infty \quad \text{and} \quad \int_{-\infty}^{\infty} |x - y| d\mu^-(y) < \infty \qquad \text{a.e.,}$$

we see by the Dominated Convergence Theorem that the first term tends a.e. to

$$-\int_{-\infty}^{\infty} |x - y| d\mu(y).$$

Next we consider the second term. Suppose first that $\mu(R^1) \neq 0$. The second term tends, as $T \to \infty$, to

$$\lim_{T\to\infty} \sqrt{T} \lim_{T\to\infty} \frac{2}{\sqrt{2\pi}} \int_{-\infty}^{\infty} e^{-|x-y|^2/2T} d\mu(y),$$

and the integral and the second limit may be interchanged by dominated convergence to become

$$= \left(\lim_{T\to\infty} \sqrt{T} \right) \left(\frac{2}{\sqrt{2\pi}} \int_{-\infty}^{\infty} d\mu(y) \right)$$
$$= \pm\infty.$$

To complete the proof we shall show that if $\mu(R^1) = 0$, then

$$\int_{-\infty}^{\infty} \sqrt{T} e^{-|x-y|^2/2T} d\mu(y)$$

tends to zero a.e. as $T \to \infty$. Since $\mu(R^1) = 0$, we have

$$\int_{-\infty}^{\infty} \sqrt{T} e^{-|x-y|^2/2T} d\mu(y) = \int_{-\infty}^{\infty} \sqrt{T}(e^{-|x-y|^2/2T} - 1) d\mu(y).$$

We shall prove that the right side even tends to zero when μ is replaced by μ^+ or μ^- Let us do so for μ^+. First we show that

$$|\sqrt{T}(e^{-|z|^2/2T} - 1)| \leq k|z|$$

for a fixed constant k and for all T. By differential calculus methods we find that

$$\frac{e^{-|z|^2/2T} - 1}{|z|}$$

assumes its maximum value as a function of $|z|$ when $|z|$ satisfies

$$\frac{|z|^2}{T} = e^{|z|^2/2T} - 1.$$

The unique positive solution occurs for $2 \leq |z|^2/T \leq 3$. Let $b = |z|^2/T$ with $2 \leq b \leq 3$ be the point at which the solution occurs. Then

$$\left|\sqrt{T}\,\frac{e^{-|z|^2/2T} - 1}{|z|}\right| \leq \left|\sqrt{T}\,\frac{e^{-b/2} - 1}{\sqrt{bT}}\right| = \frac{1 - e^{-b/2}}{\sqrt{b}} = k$$

and

$$|\sqrt{T}(e^{-|z|^2/2T} - 1)| \leq k|z|.$$

Put $|z| = |x - y|$. Since

$$\int_{-\infty}^{\infty} k|x - y| d\mu^+(y) < \infty \qquad \text{a.e.,}$$

we have by dominated convergence

$$\lim_{T\to\infty} \int_{-\infty}^{\infty} \sqrt{T}(e^{-|x-y|^2/2T} - 1) d\mu^+(y)$$
$$= \int_{-\infty}^{\infty} \lim_{T\to\infty} [\sqrt{T}(e^{-|x-y|^2/2T} - 1)] d\mu^+(y)$$

for almost every x. The integrand on the right side is identically zero.

The hypotheses of Propositions 7-4 and 7-5 are worth reviewing and comparing. In the transient case, Proposition 7-4, we started with any element μ of M and we were able to conclude both that the potential operator was defined on μ and that its value was the Brownian motion limit. In the recurrent case, Proposition 7-5, we started with any element μ of M and we had to assume that the potential operator was defined on μ; we then concluded that the potential of μ was equal to the Brownian motion limit if and only if μ had total charge zero. We shall see that the same thing happens in potential theory defined for denumerable Markov chains.

5. Potential theory for denumerable Markov chains

We turn our attention now to those properties of Brownian motion which relate it to potential theory. In this section we shall answer the following questions:

(1) How can the connection between Brownian motion and classical potential theory be used to define a potential theory for denumerable Markov chains?

(2) How does potential theory differ in the transient and recurrent cases, and what form does the potential operator take?

(3) What is the nature of the inverse operator that transforms potentials into charges?

(4) What other Markov chain concepts play a role in potential theory?

For definiteness, let P be a denumerable Markov chain which either is recurrent or is transient with no absorbing states, and let α be a positive finite-valued P-superregular measure.

Before defining a potential theory for denumerable Markov chains, we should discuss some properties of the operators P^t and Q^t. The operators P^t and Q^t act respectively on differences of finite measures and on functions in C_0 according to the equations

$$(\mu P^t)(E) = \int_{R^n} \left[\int_E \frac{1}{(2\pi t)^{n/2}} e^{-|x-y|^2/2t} dy \right] d\mu(x)$$

and

$$(Q^t f)(y) = \int_{R^n} \frac{1}{(2\pi t)^{n/2}} e^{-|x-y|^2/2t} f(x)dx,$$

and they are related by the identity

$$\int_{R^n} (Q^t f) d\mu = \int_{R^n} f d(\mu P^t).$$

The linearity properties

$$(\mu + \nu)P^t = \mu P^t + \nu P^t$$

and

$$(c\mu)P^t = c(\mu P^t),$$

together with a certain property of continuity, imply that the action of P^t on differences of finite measures is analogous to the action of a matrix on row vectors. Similarly, the corresponding properties for Q^t imply that the action of Q^t on functions in C_0 is like that of a matrix on column vectors. But the real insight into P^t and Q^t comes in realizing that these matrices are identical. To be more specific, we must reformulate the assertion for a countable space.

Let P^* be a continuous linear operator on row vectors μ which have a finite sum, and let Q^* be a continuous linear operator on column vectors f whose components tend to 0. Suppose that P^* and Q^* are related by the identity

$$\mu \cdot (Q^* f) = (\mu P^*) \cdot f$$

for all μ and f of the type specified, where $\cdot$ stands for vector multiplication. Let $\delta^{(i)}$ and $d^{(i)}$ be the row and column vectors having ith component equal to 1 and all other components equal to 0. The vector $\delta^{(i)}$ is in the domain of P^*, and $d^{(i)}$ is in the domain of Q^*. If we define a square matrix $\{P^*_{ij}\}$ by

$$P^*_{ij} = (\delta^{(i)} P^*) \cdot d^{(j)} = (\delta^{(i)} P^*)_j,$$

then

$$(\mu P^*)_j = \sum_k \mu_k (\delta^{(k)} P^*)_j = \sum_k \mu_k P^*_{kj}.$$

Hence the operator P^* may be represented by the matrix $\{P^*_{ij}\}$. But by the identity relating P^* and Q^*,

$$P^*_{ij} = \delta^{(i)} \cdot (Q^* d^{(j)}).$$

Hence by a similar argument, the operator Q^* is also representable by the same matrix $\{P^*_{ij}\}$.

Thus the denumerable Markov chain analog of the pair of operators P^t and Q^t can be expected to be a single matrix depending on t. For $t = 1$, this matrix can be taken to be the transition matrix P of the Markov chain. Then the relations $P^{t+s} = P^t P^s$ and $Q^{t+s} = Q^t Q^s$ for integers t and s imply that the analog of P^t and Q^t for any other integer value of t is a power of the matrix P.

Lebesgue measure has a special property with respect to Brownian motion which is summed up in the equation

$$\int_E dx = \int_{R^n} \left[\int_E \frac{1}{(2\pi t)^{n/2}} e^{-|x-y|^2/2t} dy \right] dx.$$

If we call Lebesgue measure σ and use notation that earlier was reserved for finite measures, this equation becomes

$$\sigma = \sigma P^t.$$

That is, σ is regular for P^t. Thus the analog of σ for the Markov chain P should be a P-regular measure. But if P is transient, then P need not have a regular measure. We therefore relax our requirement and ask only that the analog of σ be a P-superregular measure. We can then decree that the specified measure α is to be the analog of σ.

The problem of defining potentials for Markov chains becomes a problem of translating notions about Brownian motion into notions about Markov chains. Following Propositions 7-4 and 7-5, we recall that a potential $g(x)$ is obtained from a charge μ in this way: If we abbreviate the equation

$$(\mu P^t)(E) = \int_{R^n} \left[\int_E \frac{1}{(2\pi t)^{n/2}} e^{-|x-y|^2/2t} dy\right] d\mu(x)$$
$$= \int_E \left[\int_{R^n} \frac{1}{(2\pi t)^{n/2}} e^{-|x-y|^2/2t} d\mu(x)\right] dy$$

as

$$(\mu P^t)(E) = \int_E f(x, t) dx,$$

then

$$g(x) = \lim_{T\to\infty} \int_0^T f(x, t) dt.$$

Translating the relation for $(\mu P^t)(E)$ into notions about Markov chains, we write

$$\sum_{i\in E} (\mu P^n)_i = \sum_{i\in E} f_i^{(n)} \alpha_i.$$

If E is the one-point set $\{i\}$, we find that

$$f_i^{(n)} = \frac{1}{\alpha_i} (\mu P^n)_i$$

or

$$f^{(n)} = \text{dual } (\mu P^n).$$

The equation defining $g(x)$ translates into

$$g = \lim_{n\to\infty} [f^{(0)} + f^{(1)} + \cdots + f^{(n)}]$$

or

$$\boxed{g = \text{dual} \lim_{n\to\infty} [\mu(I + P + \cdots + P^n)]}$$

Classically, potentials are left as point functions and are never transformed into set functions because such a transformation is frequently impossible. In Markov chain potential theory, however, every column vector can be transformed into a row vector by the duality mapping.

If we take the dual of both sides of the boxed equation, we get

$$\text{dual } g = \lim_{n\to\infty} [\mu(I + P + \cdots + P^n)].$$

For simplicity in notation we shall adopt the convention that dual g, and not g, is the potential of μ. We can at last formulate a definition.

Definition 7-6: Any row vector μ with $\mu 1$ well defined and finite for which the limit

$$\nu = \lim_{n\to\infty} [\mu(I + P + \cdots + P^n)]$$

exists and is finite-valued is called a **left charge** with **potential measure** ν.

The condition that $\mu 1$ be well defined and finite is the analog of the condition that a charge in R^n be the difference of two finite measures.

The boxed equation for g yields an alternate possibility, namely

$$g = \lim_{n\to\infty} [(I + \hat{P} + \cdots + \hat{P}^n)(\text{dual } \mu)]$$

with $\alpha(\text{dual } \mu) = \mu 1$ finite. If we had gone through the same argument for the process $\hat{P}$, we would have obtained the same equation with the carets removed. We therefore complement Definition 7-6 as follows.

Definition 7-7: Any column vector f with αf well defined and finite for which the limit

$$g = \lim_{n\to\infty} [(I + P + \cdots + P^n)f]$$

exists and is finite-valued is called a **right charge** with **potential function** g.

From our knowledge of what happens with Brownian motion, we should expect that the Markov chain potential operators will arise in different ways in the transient and recurrent cases. Consequently, we shall treat the different kinds of processes separately, handling the transient case in Chapter 8 and the recurrent case in Chapter 9.

In the transient case of Brownian motion the operator was formally

$$\lim \int_0^T P^t dt = \int_0^\infty P^t dt,$$

and it is no surprise that for Markov chains it is the matrix $N = \sum_{n=0}^{\infty} P^n$ which turns out to be the potential operator both for left charges and right charges (see Theorem 8-3). Once we have the potential operator, it will not be difficult to develop a full theory in analogy with classical potential theory.

In the recurrent case, however, the problem of finding the potential operator is not so easy. The information that we will find, just as in

Proposition 7-5, is that for any charge of total charge zero on which the potential operator is defined the potential operator should agree with the operator which is formally

$$\lim \int_0^T P^t dt.$$

It will turn out that there are many possible potential operators for left charges and many others for right charges. Of these the matrices $-C$ and $-G$, respectively, will be representative (see Definition 9-24). But if we ask that the same matrix work for both left and right charges, then we shall see that there is a matrix K such that all such potential operators are of the form $-K + c\mathbf{1}\alpha$, where c is a constant (see Theorem 9-84). With $-K$ as our operator, we have some hope of imitating classical potential theory if we redefine charge and potential in terms of K: The column vector f is a charge, for instance, with potential g if

$$g = -Kf.$$

From this new definition of charges and potentials, we shall be able, just as in the transient case, to prove theorems which are analogs of some of the main results of classical potential theory.

In discussing the relation of Brownian motion to potential theory in Section 3, we mentioned that the operator inverse to $-A$, where

$$Af = \lim_{t \downarrow 0} \frac{1}{t} (Q^t f - f),$$

was of the same form as the potential operator. It is thus quite believable that $-A$ should be essentially the inverse operator that transforms potentials into charges. Now the definition of A involves a derivative, and when concepts in Brownian motion are translated into concepts in Markov chains, derivatives transform into differences. Therefore, the proper analog of Af for Markov chains is $Pf - f = (P - I)f$. That is, $I - P$ plays the role of $-A$. In Theorems 8-4 and 9-15 we shall see that $I - P$ is indeed the operator that transforms potentials in the sense of Definitions 7-6 and 7-7 into charges.

With Brownian motion the operator A is a constant multiple of the Laplacian operator Δ for smooth enough functions, where

$$\Delta f = \left(\frac{\partial^2}{\partial x_1^2} + \cdots + \frac{\partial^2}{\partial x_n^2}\right) f.$$

If a function f satisfies the equation $\Delta f = 0$ in a neighborhood of a point x, then f is said to be **harmonic** in a neighborhood of x. The analog in the case of denumerable Markov chains is that if a column vector f satisfies $(P - I)f = 0$ at the point i, then f is **regular** at i. Thus we can expect that regular functions will have some of the same behavior for Markov chains that harmonic functions have classically. As an example, a function harmonic on a connected open set in R^n cannot assume its maximum value inside that open set unless the function is constant. We shall see in Corollary 8-44 that an analogous result holds for Markov chains.

A twice continuously differentiable function f is said to be **superharmonic** if $\Delta f \leq 0$. The analog of this property is the condition that $(P - I)f \leq 0$ or $f \geq Pf$. Thus the analog of a superharmonic function is a superregular function.

TABLE 7-2. MARKOV CHAIN ANALOGS OF POTENTIAL THEORY CONCEPTS

Classical Notion	Markov Chain Analog
R^n	State space S
P^t and Q^t	P^n
Lebesgue measure	α
Potential	$\lim \mu(I + P + \cdots + P^n)$ or $\lim (I + P + \cdots + P^n)f$
Total charge	$\mu 1$ or αf
Potential operator	Transient: N Recurrent: $-K$
Inverse operator	$I - P$
Harmonic function	Regular function
Superharmonic function	Superregular function
Connected set	Communicating set

6. Brownian motion as a limit of the symmetric random walk

The symmetric random walk in n dimensions was defined in Chapter 4 as a Markov chain obtained from sums of independent random variables on the integer lattice in R^n with the probability of going from any state to any of the $2n$ neighboring states equal to $1/(2n)$. In potential theory for Markov chains this process assumes the role of the "classical case," exhibiting in its potential theory much of the special behavior of the theory in Section 2. For instance, the matrix of the potential operator for this process has the same asymptotic behavior at infinity as the potential kernel has classically: $\log |x|$ in two dimensions, $1/|x|$ in three dimensions, and so on.

The reason for this coincidence is that Brownian motion is in a precise sense the limit of the symmetric random walk. Specifically if the random walk is considered first on the integer lattice, then on the half-integer lattice, then on the quarter-integer lattice, and so on, then the probabilities in the kth process of being in a fixed ball in R^n after time $4^k t$ converge to the probability in Brownian motion of being in that ball after time t. We shall prove this result only in the one-dimensional case, and we make use of the Central Limit Theorem (Theorem 1-68).

Consider for fixed k the random walk on the line having as states the points of the form $j2^{-k}$, for j any integer, and having transition probabilities $\frac{1}{2}$ from any state to each of its two neighbors. This process is the symmetric random walk with a change in scale. Let $x_n^{(k)}$ be the nth outcome function, and let $x_0^{(k)} = 0$. As in Section 1, we let x_t denote the position in Brownian motion.

Proposition 7-8: Brownian motion in one dimension is a limit of the symmetric random walk in this sense: If t is a diadic rational and if α and β are real numbers, then

$$\lim_{k\to\infty} \Pr_0[x_{4^k t}^{(k)} \in (\alpha, \beta)] = \Pr_0[x_t \in (\alpha, \beta)].$$

PROOF: The random variables $x_{n+1}^{(k)} - x_n^{(k)}$ are independent and identically distributed and have mean 0 and variance

$$\sigma^2 = \frac{1}{2}\left(\frac{1}{2^k}\right)^2 + \frac{1}{2}\left(-\frac{1}{2^k}\right)^2 = \frac{1}{4^k}.$$

Let $m = 4^k t$ be an integer. Since

$$x_m^{(k)} = \sum_{n=1}^{m} [x_n^{(k)} - x_{n-1}^{(k)}],$$

$x_m^{(k)}$ has mean 0 and variance $m/4^k = t$. Hence, by the Central Limit Theorem,

$$\lim_{k\to\infty} \Pr_0\left[\frac{\alpha}{\sqrt{t}} < \frac{x_m^{(k)}}{\sqrt{t}} < \frac{\beta}{\sqrt{t}}\right] = \Phi\left(\frac{\beta}{\sqrt{t}}\right) - \Phi\left(\frac{\alpha}{\sqrt{t}}\right)$$

or

$$\lim_{k\to\infty} \Pr_0[\alpha < x_m^{(k)} < \beta] = \Phi\left(\frac{\beta}{\sqrt{t}}\right) - \Phi\left(\frac{\alpha}{\sqrt{t}}\right).$$

On the other hand, by definition,

$$\begin{aligned}\Pr_0[x_t \in (\alpha, \beta)] &= \int_\alpha^\beta \frac{1}{\sqrt{2\pi t}} e^{-y^2/2t} dy \\ &= \int_{\alpha/\sqrt{t}}^{\beta/\sqrt{t}} \frac{1}{\sqrt{2\pi}} e^{-u^2/2} du \\ &= \Phi\left(\frac{\beta}{\sqrt{t}}\right) - \Phi\left(\frac{\alpha}{\sqrt{t}}\right).\end{aligned}$$

Therefore,

$$\lim_{k\to\infty} \Pr_0[x_{4^k t}^{(k)} \in (\alpha, \beta)] = \Pr_0[x_t \in (\alpha, \beta)].$$

7. Symmetric random walk in *n* dimensions

As was mentioned in Section 6, the n-dimensional symmetric random walk and n-dimensional Brownian motion share a number of properties because the second process is the limit of the first. Of these we shall prove for the random walk just two—that the process is recurrent in dimensions 1 and 2 and transient in dimension $n \geq 3$ and that in the transient case the columns of the N-matrix tend to zero. The second result is in analogy with the behavior of the potential kernel $1/|x|^{n-2}$ in Brownian motion.

For the first problem we note immediately that all states communicate in the random walks of all dimensions; hence each of them is either transient or recurrent. In one dimension the state space is the integers, and

$$P_{i,i+1} = P_{i,i-1} = \tfrac{1}{2}.$$

Since the mean step in zero, the process is recurrent by Proposition 5-22. A more direct proof of the recurrence proceeds as follows. It is impossible to get from state 2 to state 0 without going through 1 This fact, together with the translation invariance of the hitting probabilities, implies that

$$H_{20} = H_{21}H_{10} = (H_{10})^2.$$

But

$$\bar{H}_{00} = \tfrac{1}{2}H_{10} + \tfrac{1}{2}H_{-1,0} = \tfrac{1}{2}H_{10} + \tfrac{1}{2}H_{01} = H_{10}$$

since $H_{01} = H_{10}$. Therefore, the identity

$$H_{10} = \tfrac{1}{2}H_{00} + \tfrac{1}{2}H_{20} = \tfrac{1}{2} + \tfrac{1}{2}(H_{10})^2$$

implies that $H_{10} = 1$ and hence $\bar{H}_{00} = 1$. Consequently, the process is recurrent. Still a third proof can be based on a calculation of N_{00}. In fact, we have

$$P_{00}^{(2n)} = 2^{-2n}\binom{2n}{n}$$

because in order for the process to return in $2n$ steps to 0, it must make n steps to the right and n to the left; each such possibility has probability 2^{-2n}. By Lemma 1-59,

$$\binom{2n}{n} \sim c2^{2n}\frac{1}{\sqrt{n}}.$$

Hence the tail of the series $N_{00} = \sum P_{00}^{(2n)}$ dominates a constant multiple of the tail of the series $\sum 1/\sqrt{n}$. Therefore, N_{00} is infinite and the process is recurrent.

In the two-dimensional case there are two simultaneous processes going on in perpendicular directions, and for the process to return to the origin, both of them must return to their zero positions at the same time. Letting k be the number of steps to the right and $n - k$ the number of steps up in $2n$ steps, we have

$$P_{00}^{(2n)} = 4^{-2n} \sum_{k=0}^{n} \binom{2n}{k, k, n-k, n-k}.$$

If we multiply numerator and denominator of the multinomial coefficient by $(n!)^2$, we see that it equals $\binom{2n}{n}\binom{n}{k}^2$. Thus

$$P_{00}^{(2n)} = 4^{-2n}\binom{2n}{n} \sum_{k=0}^{n} \binom{n}{k}^2.$$

The identity

$$\sum_{k=0}^{n} \binom{n}{k}^2 = \binom{2n}{n}$$

shows that

$$P_{00}^{(2n)} = 4^{-2n}\binom{2n}{n}^2.$$

Since

$$\binom{2n}{n} \sim c2^{2n}\frac{1}{\sqrt{n}},$$

we have

$$P_{00}^{(2n)} \sim c^2\frac{1}{n}.$$

Thus the series $N_{00} = \sum P_{00}^{(2n)}$ dominates a multiple of $\sum 1/n$, N_{00} must be infinite, and the process is recurrent.

An alternate proof that the process is recurrent in two dimensions proceeds as follows. If we introduce the new coordinates

$$u = x + y$$
$$v = x - y,$$

then the two-dimensional symmetric random walk described in the coordinates (u, v) executes two one-dimensional symmetric random walks independently of each other. Hence $P_{(0,0)(0,0)}^{(2n)}$ is the probability that $u = 0$ and $v = 0$ after $2n$ steps, which is $(P_{00}^{(2n)})^2 \sim c^2/n$.

In three dimensions we calculate N_{00}. We have

$$P_{00}^{(2n)} = 6^{-2n} \sum_{\substack{j,k \\ j+k\leq n}} \binom{2n}{j, j, k, k, n-j-k, n-j-k}$$
$$= 2^{-2n}\binom{2n}{n} \sum_{j,k} 3^{-2n}\binom{n}{j, k, n-j-k}^2.$$

The coefficients $\binom{n}{j, k, n-j-k}$ are dominated by the central term

$$\binom{n}{n/3,\ n/3,\ n/3},$$

where the gamma function may be used for $(n/3)!$ if 3 does not divide n. This fact and the observation that the coefficients $\binom{n}{j, k, n-j-k}$ sum to 3^n implies that

$$P_{00}^{(2n)} \leq 2^{-2n}\binom{2n}{n}3^{-n}\binom{n}{n/3,\ n/3,\ n/3}.$$

Summing on n and using the approximations of Section 1-6a, we see that the series $N_{00} = \sum P_{00}^{(2n)}$ is dominated by a multiple of $\sum 1/n^{3/2}$. Therefore N_{00} is finite, and the process is transient.

If any higher-dimensional random walk were recurrent, then the process projected to a three-dimensional set would be recurrent, and the latter process watched only when it changes state would also be recurrent. But this last process is exactly the three-dimensional symmetric walk. We conclude therefore that the random walk in all dimensions greater than three is transient, and we have completed the proof of the following proposition.

Proposition 7-9: The symmetric random walk is recurrent in dimensions one and two and is transient in all dimensions greater than or equal to three.

In the transient case of dimension $n \geq 3$, the jth entry in the 0th column of the N-matrix is of the order of a constant times $|j|^{-(n-2)}$. We conclude this chapter by proving the weaker result that the entries of that column tend to zero, but our proof will be for a more general situation.

Proposition 7-10: Let P be a Markov chain with an infinite state space such that

(1) P is transient.
(2) $P = P^T$.
(3) P has only finitely many non-zero entries in each row.

Then

$$\lim_{j \to \infty} H_{0j} = 0.$$

If, in addition, P represents sums of independent random variables, then

$$\lim_{j\to\infty} N_{0j} = 0.$$

In particular, the conclusions apply to the symmetric random walk in any dimension $n \geq 3$.

PROOF: We note that hypothesis (3) is equivalent to

(3′) For any state 0 and for any given m, there exist only finitely many states j for which $F_{0j}^{(k)} \neq 0$ for some $k \leq m$.

If the conclusion about H_{0j} is false, then for some $\epsilon > 0$ and for infinitely many j we have $H_{0j} > \epsilon$. By (1), N is finite-valued. Therefore,

$$\lim_{k\to\infty} P^k N = \lim_{k\to\infty} [N - (I + \cdots + P^{k-1})] = 0$$

and

$$\lim_{k\to\infty} P^k \bar{H} = 0$$

since $\bar{H} \leq H \leq N$. Choose m large enough so that $(P^m\bar{H})_{00} < \epsilon^2$. Since there are infinitely many j such that $H_{0j} > \epsilon$, we can find by (3′) such a j with $F_{0j}^{(k)} = 0$ for all $k \leq m$. Then

$$\begin{aligned}\Pr_0[\text{hit } 0 \text{ after time } m] &\geq \Pr_0[\text{ever hit } j \text{ and return to } 0]\\ &= H_{0j}H_{j0}\\ &= H_{0j}H_{0j} \quad \text{by (2)}\\ &> \epsilon^2.\end{aligned}$$

But

$$\Pr_0[\text{hit } 0 \text{ after time } m] = (P^m\bar{H})_{00} < \epsilon^2,$$

a contradiction. Therefore $H_{0j} \to 0$.

Finally if P represents sums of independent random variables,

$$N_{0j} = H_{0j}N_{jj} = H_{0j}N_{00}.$$

Hence $N_{0j} \to 0$. (Note that we really need only that N_{jj} is bounded.)

As Markov chains the symmetric random walks have some special properties, reflecting corresponding special properties of Brownian motion. For instance, α is a constant for the random walk, and $P = P^T$. Consequently $P = \hat{P}$. We shall see that although many results of classical potential theory generalize to all transient and most recurrent chains, some will require further assumptions which happen to be true for symmetric random walks.

CHAPTER 8

TRANSIENT POTENTIAL THEORY

1. Potentials

In this chapter P denotes a Markov chain all of whose states are transient—that is, a transient chain with no absorbing states. Every such chain has at least one (strictly) positive superregular measure, as we saw in Chapter 5; for example, the sum of 2^{-i} times the ith row of N is such a measure.

We select one such positive superregular measure, to be fixed throughout the chapter, and call it α. All of transient potential theory will be relative to the distinguished vector α.

Let $\hat{P}$ be the α-dual of P. Since all states are transient in $\hat{P}$ and since $\hat{\hat{P}} = P$, we see that $\hat{P}$ is the most general chain of the type we consider. The distinguished measure for $\hat{P}$ is taken to be *the same* α.

As an example, let P be a transient Markov chain whose states communicate. Then $P\mathbf{1} \leq \mathbf{1}$ and $0 < N = \sum_{k=0}^{\infty} P^k < \infty$. Every non-negative non-zero superregular row vector β is positive, for if $\beta_j \neq 0$, then for every state i and integer k

$$\beta_i \geq (\beta P^k)_i = \sum_m \beta_m P_{mi}^{(k)} \geq \beta_j P_{ji}^{(k)}.$$

The right side must be positive for some k, since j communicates with i. Thus in this special case any non-negative superregular row vector may be taken as α; in particular, α may be taken as a row of N.

In the general case, if $P\mathbf{1} \neq \mathbf{1}$, we have defined the enlarged chain $\bar{P}$ by adding an absorbing state a to P and by setting

$$\begin{aligned} \bar{P}_{ij} &= P_{ij} \qquad \text{if } i \neq a \text{ and } j \neq a \\ \bar{P}_{ia} &= 1 - \sum_k P_{ik} \\ P_{aj} &= \delta_{aj}. \end{aligned}$$

If $P\mathbf{1} = \mathbf{1}$, we shall agree that P is its own enlarged chain.

It will be convenient to say that the product of a row vector and a column vector is **finite** when we mean that it is well defined and finite. We recall the terminology **function** for column vector and **signed measure** for row vector.

Definition 8-1: If μ is a signed measure with $\mu 1$ finite and if

$$\nu = \lim_n [\mu(I + P + \cdots + P^{n-1})]$$

exists and is finite-valued, then μ is called a **left charge** with **potential measure** ν. If f is a function with αf finite and if

$$g = \lim_n [(I + P + \cdots + P^{n-1})f]$$

exists and is finite-valued, then f is called a **right charge** with **potential function** g. In either case a **pure potential** is a potential of a non-negative charge.

The condition that αf be finite is the natural analog of the classical theory as described in Chapter 7. It states that f is integrable with respect to the distinguished superregular measure α. Similarly, the condition on μ is that the distinguished superregular function 1 be integrable with respect to μ.

Potential functions have a simple probabilistic interpretation in terms of games. If f denotes a payment function in which f_j is the payment a player receives each time he is in state j, then $P^n f$ denotes the expected payment on the nth step. Thus $(I + P + \cdots + P^{n-1})f$ is the total payment before time n, and the potential g is the expected total payment in the long run. It is clear intuitively that g_i should equal $\sum_j N_{ij} f_j$, and we now prove this result.

Lemma 8-2: If $\mu 1$ is finite, then μN is finite-valued. If αf is finite, then Nf is finite-valued.

PROOF: We have

$$N_{ij} = H_{ij} N_{jj} \leq N_{jj}.$$

Thus

$$|(\mu N)_j| \leq \sum_i |\mu_i| N_{ij} \leq \sum_i |\mu_i| N_{jj} = (|\mu| 1) N_{jj} < \infty.$$

For the second half let $\mu = \text{dual}\, f$ and apply the first result to $\hat{P}$, noting that $\mu 1 = \alpha f$. Then

$$\infty > |\mu \hat{N}|_j = \left| \sum_i f_i \alpha_i (\alpha_j N_{ji} / \alpha_i) \right| = \left| \alpha_j \sum_i N_{ji} f_i \right| = \alpha_j |Nf|_j.$$

Since $\alpha_j \neq 0$, Nf is finite-valued.

Theorem 8-3: If αf is finite, then f is a charge and its potential is $g = Nf$.

Proof: By Lemma 8-2, Nf is well defined and finite-valued. Hence so are both Nf^+ and Nf^-. By monotone convergence,

$$\lim_n [(I + P + \cdots + P^{n-1})f^+] = Nf^+$$

and

$$\lim_n [(I + P + \cdots + P^{n-1})f^-] = Nf^-.$$

Thus

$$\lim_n [(I + P + \cdots + P^{n-1})f] = Nf^+ - Nf^- = Nf.$$

Thus f is a charge if and only if it is integrable with respect to α, and N is the potential operator that transforms a charge into its potential. In particular, f is a charge for P if and only if it is a charge for $\hat{P}$. We shall now show that $I - P$ is the inverse operator.

Theorem 8-4: If g is a potential, then $(I - P)g$ is its charge.

Proof: Let f be a charge with potential g. By Theorem 8-3, $g = Nf$. Hence, by Lemma 5-9, $(I - P)g = f$.

Therefore, there is a one-to-one correspondence between charges and potentials. Note that Theorem 8-4 implies that a potential is regular at all states where the charge is zero.

The method used to derive the second half of Lemma 8-2 is of general importance. We prove a result for all our P's for signed measures (or functions). We apply the result to $\hat{P}$ and obtain a corresponding result for functions (or signed measures). Then since $\hat{P}$ is the form of the most general transient chain being considered, the new result holds for all P's. Such results will loosely be described as **duals**.

The duals of Theorems 8-3 and 8-4 state that a signed measure μ is a charge if and only if $\mu 1$ is finite. Its potential is $\nu = \mu N$, and $\mu = \nu(I - P)$.

From now on we shall prove theorems only for functions; the dual results for signed measures can always be proved by the indicated method. The key to the success of the method is that the dual of a right charge for P is a left charge for $\hat{P}$, and the dual of a potential function for P is a potential measure for $\hat{P}$.

From Theorem 8-3 we see immediately that the class of potentials is quite extensive. We can even prove that there exists a strictly positive pure potential, a result we shall need later on.

Proposition 8-5: There exists a strictly positive pure potential.

PROOF: Number the states and let $f_j = 2^{-j}/\alpha_j$. Then

$$\alpha f = \sum_j \alpha_j(2^{-j}/\alpha_j) = 1,$$

so that f is the charge of a pure potential g, by Theorem 8-3. Furthermore,

$$g_i = \sum_j N_{ij}f_j \geq N_{ii}f_i \geq f_i > 0,$$

so that g is strictly positive.

For many purposes it is sufficient in studying potentials to consider only pure potentials. The reason for this simplification is the following.

Proposition 8-6: Any potential may be represented as the difference of two pure potentials.

PROOF: Write $g = Nf = Nf^+ - Nf^-$.

Note by Theorem 8-4 that a potential is superregular if and only if it is a pure potential.

We recall from Theorem 5-10 that a non-negative superregular function h is uniquely representable as $h = Nf + r$ with r regular. In the representation, $r = \lim_n P^n h \geq 0$ and $f = (I - P)h \geq 0$. (The dual of this result allows the unique representation of a non-negative superregular measure π as $\pi = \mu N + \rho$ with ρ regular. In this representation, $\rho = \lim_n \pi P^n \geq 0$ and $\mu = \pi(I - P)$.) This result is the analog of a classical theorem due to F. Riesz: In any open set of Euclidean space which corresponds to a transient version of Brownian motion, any non-negative superharmonic function is uniquely the sum of a pure potential for the region and a non-negative harmonic function. The pure potential may have infinite total charge. We now generalize the Markov chain result, and in so doing we obtain a useful necessary condition that potentials must satisfy.

Proposition 8-7: If $(I - P)h = f$, if h is finite-valued, and if Nf is finite-valued, then h has a representation in the form

$$h = Nf + r$$

with r regular. The vector r satisfies $r = \lim_n P^n h$.

PROOF: Set $r = h - Nf$. Then

$$\begin{aligned}(I - P)(h - Nf) &= (I - P)h - (I - P)(Nf) \\ &= (I - P)h - f \quad \text{by Lemma 5-9} \\ &= f - f = 0.\end{aligned}$$

Hence r is regular. Now

$$f = (I - P)h = h - Ph$$

or

$$h = Ph + f.$$

Since Nf is finite-valued and since $P^n f^+ \leq Nf^+$ and $P^n f^- \leq Nf^-$, $P^n f$ is finite-valued for every n. By induction we see that

$$P^{k-1}h = P^k h + P^{k-1}f$$

and that $P^k h$ is finite-valued. Summing for $k = 1, \ldots, n$, we obtain

$$h = P^n h + (I + P + \cdots + P^{n-1})f.$$

By dominated convergence the second term tends to Nf. Hence

$$h = \lim_n P^n h + Nf.$$

Corollary 8-8: If $(I - P)h = f$, if h is finite-valued, and if αf is finite then h is a potential if and only if $\lim_n P^n h = 0$.

PROOF: By Theorem 8-3, f is a charge and Nf is its potential. Apply Proposition 8-7 and write

$$h = Nf + \lim P^n h.$$

If $\lim P^n h = 0$, then h is the potential of f. Conversely, if $\lim P^n h \neq 0$, then h cannot be a potential because, by Theorem 8-4, it would have to have f as its charge.

Corollary 8-9: If g is a potential, then $\lim_n P^n g = 0$.

PROOF: Take $h = g$ in Corollary 8-8.

In the discrete analog of the classical case—three-dimensional symmetric random walk—every potential g is bounded and satisfies $\lim_i g_i = 0$. In our theory we obtain only the weaker result, Corollary 8-9; that g may be unbounded will be shown in Section 7.

The stronger results of the classical theory are due to special features, as the next proposition shows. In the classical case α is chosen as 1^T

and N_{ii} is independent of i. Hence, $\alpha_i \geq kN_{ii}$ for some positive constant k. Furthermore, $\lim_i N_{ij} = 0$ by Proposition 7-10.

Proposition 8-10: If α is chosen so that $\alpha_i \geq kN_{ii}$ for all i with k a positive constant, then all potentials are bounded. If, in addition, $\lim_i N_{ij} = 0$, then $\lim_i g_i = 0$.

PROOF: The dual of $N_{ij} \leq N_{jj}$ is $N_{ij} \leq (N_{ii}/\alpha_i)\alpha_j$. If $\alpha_i \geq kN_{ii}$, then

$$|g_i| \leq \sum_j N_{ij}|f_j| \leq \frac{N_{ii}}{\alpha_i} \sum_j \alpha_j|f_j| \leq \frac{1}{k} \sum_j \alpha_j|f_j|.$$

Thus g is bounded. Now suppose $\lim_i N_{ij} = 0$. By Proposition 8-6 we may assume that g is a pure potential. Define a sequence of functions $h_j^{(i)} = N_{ij}/\alpha_j$ and a measure $\mu_j = \alpha_j f_j$. Then μ is a finite measure, and

$$h_j^{(i)} \leq \frac{N_{jj}}{\alpha_j} \leq \frac{1}{k}$$

is bounded independently of i and j. Hence, by dominated convergence,

$$\lim_i g_i = \lim_i \sum_j h_j^{(i)}\mu_j = \sum_j (\lim_i N_{ij})f_j = 0.$$

Both conditions of Proposition 8-10 hold for the basic example with α chosen as β. However, only the first condition holds for the reverse of the basic example (see Section 6). In Section 7 we shall see an example where both conditions fail.

2. The *h*-process and some applications

Duality is a transformation which interchanges the roles of row and column vectors. Our purpose now is to describe a useful transformation of transient chains into new transient chains in which row and column vectors are transformed into vectors of the same type.

Definition 8-11: Let h be a positive finite-valued superregular function for a transient chain P. The ***h*-process** is a Markov chain P^* with transition probabilities

$$P^*_{ij} = \frac{P_{ij}h_j}{h_i}.$$

It is left to the reader to verify that P^* is a transition matrix, that all states are transient, and that if the states of P communicate, the same is true for P^*. Let U be a diagonal matrix with diagonal entries $1/h_i$. Then $P^* = UPU^{-1}$.

Definition 8-12: The **h-process transformation** is a transformation defined on square matrices Y, row vectors π, and column vectors f by

$$Y^* = UYU^{-1}$$

$$\pi^* = \pi U^{-1}$$

$$f^* = Uf.$$

The h-process transformation yields results similar to those with duality. If P is a transient chain, then P^* also is transient. Moreover, powers of P transform to the P^* process the same way that P does, and the fundamental matrix for P^* is N^*. Sums and products are preserved in their given order; and equalities, inequalities, and limits are preserved entry-by-entry. Any superregular function (or signed measure) for P transforms into a superregular function (or signed measure) for P^*.

If α is the distinguished superregular measure for P, we select α^* as the distinguished superregular measure for P^*. Then $\alpha^* f^* = \alpha f$ and $N^* f^* = (Nf)^*$. Hence if f is a right charge with potential g in P, then f^* is a right charge for P^* with potential g^*.

If we decompose P as

$$P = \begin{array}{c} \\ E \\ \tilde{E} \end{array} \begin{array}{c} \begin{array}{cc} E & \tilde{E} \end{array} \\ \begin{pmatrix} T & U \\ R & Q \end{pmatrix} \end{array},$$

then

$$B^E = \begin{pmatrix} I & 0 \\ (\sum Q^n)R & 0 \end{pmatrix}.$$

From this decomposition we see that $(B^E)^*$ is the B^E-matrix of the P^*-chain because Q and R transform into Q^* and R^*, because products are preserved, and because $I^* = I$.

We shall now give some applications of the h-process.

Definition 8-13: The **support of a charge** is the set on which the charge is not 0; the **support of a potential** is the support of its charge. A charge or potential is said to have **support in E** if its support is a subset of E.

The function $\mathbf{1}$ is always superregular, and hence by the representation theorem $\mathbf{1} = Nf + r$, where $f = (I - P)\mathbf{1}$ and r is regular. That is,

$$f_i = 1 - (P\mathbf{1})_i = \bar{P}_{ia}$$

in the enlarged chain. Moreover,

$$(Nf)_i = \sum_j N_{ij}\bar{P}_{ja} = B_{ia}.$$

Thus $r_i = 1 - B_{ia}$ is the probability that the P-process, started at i, continues indefinitely. The enlarged chain is absorbing if and only if $\mathbf{1} = Nf$.

Proposition 8-14: Let $g > 0$ be a pure potential with support in E, and let P^* be the h-process for $h = g$. Then $g^* = \mathbf{1}$ is a potential, $\overline{P^*}$ is absorbing, and $(B^E)^*\mathbf{1} = \mathbf{1}$.

PROOF: Potentials transform into potentials; hence $g^* = \mathbf{1}$ is a potential. Since $\mathbf{1}$ is then of the form N^*f^*, $\overline{P^*}$ is absorbing. The absorbing state a can be reached only from a state i such that $(P^*\mathbf{1})_i < 1$; for such a state i,

$$0 < [(I - P^*)\mathbf{1}]_i = [(I - P^*)g^*]_i = f_i^*.$$

Hence $f_i > 0$ and i must be in E. Thus the P^*-process with probability one reaches E from all states, and $(B^E)^*\mathbf{1} = \mathbf{1}$.

What underlies Proposition 8-14 is this: The h-process tends to follow paths along which h is large. But since potentials tend to zero on the average ($P^n h \to 0$ for a potential), if h is a potential, then the paths in the h-process disappear. See Chapter 10 for details.

Proposition 8-15: If h is a non-negative finite-valued superregular function, then $B^E h \le h$ for any set E.

PROOF: First suppose that $h > 0$. Form the h-process; then $h^* = \mathbf{1}$. Since $(B^E)^*\mathbf{1} \le \mathbf{1}$, we have $B^E h \le h$. (The conclusion that an inequality for the h-process implies an inequality for the original process is one we shall draw frequently. If it were false, then the inequality $B^E h \le h$ would fail in some entry. But the h-process transformation preserves inequalities entry-by-entry.)

Now suppose that h has some zero entries. Apply the special case above to the function $h + \epsilon\mathbf{1}$. Then

$$B^E(h + \epsilon\mathbf{1}) \le h + \epsilon\mathbf{1}.$$

Letting ϵ tend to zero, we obtain $B^E h \le h$.

Proposition 8-16: If h is a non-negative superregular function and if E is any set of states, then $\bar{h} = B^E h$ satisfies the following:

(1) $\bar{h} \le h$ and $\bar{h}_E = h_E$.

(2) $\bar{h}$ is the pointwise infimum of all non-negative superregular functions which dominate h on E.

(3) If $\bar{f} = (I - P)\bar{h}$, then
$$\bar{f}_E = (I - P^E)h_E \geq 0$$
and
$$\bar{f}_{\tilde{E}} = 0.$$
Therefore, $\bar{h}$ is regular on $\tilde{E}$ and is superregular everywhere.

(4) If $E \subset F$, then $B^E h \leq B^F h$.

Proof: Statement (1) follows from Proposition 8-15 and the fact that $B_{ij}^E = \delta_{ij}$ for i in E. For (2) let x be a non-negative superregular function such that $x_E \geq h_E$. Since the $\tilde{E}$ columns of B^E are zero,
$$x \geq B^E x \geq B^E h = \bar{h},$$
and (2) holds provided we show in (3) that $\bar{h}$ is superregular. For (3), since $P\bar{h} \leq Ph \leq h$ is finite-valued,
$$\bar{f} = (I - P)(B^E h) = [(I - P)B^E]h = \begin{pmatrix} I - P^E & 0 \\ 0 & 0 \end{pmatrix}\begin{pmatrix} h_E \\ h_{\tilde{E}} \end{pmatrix}.$$
But h_E is P^E-superregular by the dual of Lemma 6-7, and (3) follows. Finally for (4), if $E \subset F$, then by conclusion (6) of Proposition 5-8,
$$B^E h = B^F(B^E h) \leq B^F h.$$

We now prove two lemmas and a proposition which conclude that a charge and its potential may both be computed from a knowledge of the values of the potential on the support. The first lemma is interesting in itself because of its game interpretation, which we shall discuss after proving the result.

Lemma 8-17: For any set of states E,
$$N = B^E N + {}^E N.$$
If g is a potential with charge f, then
$$g = B^E g + {}^E N f.$$

Proof: In Theorem 4-11, let $\mathbf{f}_j$ be the number of times in j when and after E is reached (or 0 if E is not reached), and let $\mathbf{t}$ be the time when E is reached (or $+\infty$ if E is not reached). Then Theorem 4-11 yields
$$\begin{aligned} \mathrm{M}_i[\mathbf{f}_j] &= \sum_k \Pr_i[x_{\mathbf{t}} = k]\, \mathrm{M}_k[\mathbf{n}_j] \\ &= \sum_k B_{ik}^E N_{kj}. \end{aligned}$$

But $\mathbf{f}_j$ is the difference of the total number of times the process is in j and the number of times it is in j before reaching E. Hence

$$\mathrm{M}_i[\mathbf{f}_j] = N_{ij} - {}^E N_{ij},$$

and the first equation follows. To get the second equation we multiply through by f; associativity in $B^E N f$ holds because $B^E N|f|$ is finite-valued.

In the game interpretation of potentials, f_j is a payment received each time the process is in state j, and g is the expected total gain if the process is started in i. The second equation of Lemma 8-17 states that g is the expected gain when and after E is entered *plus* the expected gain before E is entered. If all the payments are non-negative, then it is obvious from this interpretation that $g \geq B^E g$. If the support of f is in E, then all non-zero payments occur in E, and the expected gain before reaching E is zero. Hence, as we shall see formally in Proposition 8-19, $g = B^E g$.

Lemma 8-18: The fundamental matrix for P^E is N_E.

PROOF: The assertion is probabilistically clear because the number of times the process P is in a state of E when watched only in states of E is the same as the number of times the process P is in a state of E.

Proposition 8-19: If g is a potential with support in E, then g_E determines g, $g = B^E g$, $f_E = (I - P^E)g_E$, and $g_E = N_E f_E$.

PROOF: The fact that $g = B^E g$ is immediate from Lemma 8-17. Hence g_E determines g. Since $g = B^E g$, we have $f_E = (I - P^E)g_E$ by conclusion (3) of Proposition 8-16. Finally $g_E = N_E f_E$ either by Lemma 8-18 or by direct calculation:

$$\begin{pmatrix} g_E \\ g_{\tilde{E}} \end{pmatrix} = \begin{pmatrix} N_E & N_2 \\ N_3 & N_4 \end{pmatrix} \begin{pmatrix} f_E \\ 0 \end{pmatrix} = \begin{pmatrix} N_E f_E \\ N_3 f_E \end{pmatrix}.$$

Next we shall prove that the columns of B^E are always potentials.

Proposition 8-20: For any set of states E the columns of B^E are potentials with support in E, and

$$B^E = N\begin{pmatrix} I - P^E & 0 \\ 0 & 0 \end{pmatrix}.$$

Proof: Since

$$\alpha\begin{pmatrix} I - P^E & 0 \\ 0 & 0 \end{pmatrix} = (\alpha_E(I - P^E) \quad 0)$$

is finite-valued, each column of $N\begin{pmatrix} I - P^E & 0 \\ 0 & 0 \end{pmatrix}$ is a potential with support in E. Thus by Proposition 8-19,

$$\begin{aligned} N\begin{pmatrix} I - P^E & 0 \\ 0 & 0 \end{pmatrix} &= B^E N\begin{pmatrix} I - P^E & 0 \\ 0 & 0 \end{pmatrix} \\ &= B^E\begin{pmatrix} N_E(I - P^E) & 0 \\ N_3(I - P^E) & 0 \end{pmatrix} \quad \text{with } N = \begin{pmatrix} N_E & N_2 \\ N_3 & N_4 \end{pmatrix} \\ &= B^E\begin{pmatrix} I & 0 \\ N_3(I - P^E) & 0 \end{pmatrix} \quad \text{by Lemma 8-18} \\ &= B^E. \end{aligned}$$

Corollary 8-21: For any set E, $\lim_n P^n B^E = 0$.

Proof: Apply Proposition 8-20 and Corollary 8-9.

Finally we work toward a proof that a non-negative superregular function dominated by a potential is a potential, a result we state as Proposition 8-25.

Lemma 8-22: If E is a finite set and if h is non-negative superregular, then $B^E h$ is a pure potential of finite support.

Proof: $B^E h$ is a finite linear combination of columns of B^E and is therefore by Proposition 8-20 a potential with support in the finite set E. Since h is non-negative superregular, $B^E h$ is non-negative superregular by conclusion (3) of Proposition 8-16. Hence, $B^E h$ is a pure potential.

Proposition 8-23: Every non-negative superregular function is the limit of an increasing sequence of pure potentials of finite support.

Proof: Let $E_1 \subset E_2 \subset E_3 \subset \cdots$ be an increasing sequence of finite sets with union the set of all states S, and let $h^{(n)} = B^{E_n} h$. Then $h^{(n)}$ is an increasing sequence of pure potentials of finite support by

Lemma 8-22 and conclusion (4) of Proposition 8-16. If i is in E_n, then $h_i^{(n)} = h_i$, so that $\lim h^{(n)} = h$.

If g is a potential with charge f, then the **total charge** of g (or f) is defined to be αf (see Section 7-5).

Lemma 8-24: If $N\bar{f} \leq Nf$ with $\bar{f} \geq 0$ and αf finite, then

$$0 \leq \alpha\bar{f} \leq \alpha f < \infty.$$

PROOF: By the dual of Proposition 8-23, we may find a sequence of finite measures $\pi^{(n)}$ such that α is the monotone limit of $\pi^{(n)}N$. Since $N\bar{f} \leq Nf$, we have $\pi^{(n)}(N\bar{f}) \leq \pi^{(n)}(Nf)$ and

$$\lim_n \pi^{(n)}(N\bar{f}) \leq \lim_n \pi^{(n)}(Nf).$$

Since $\bar{f} \geq 0$,

$$\pi^{(n)}(N\bar{f}) = (\pi^{(n)}N)\bar{f},$$

and

$$\lim_n \pi^{(n)}(N\bar{f}) = (\lim_n \pi^{(n)}N)\bar{f} = \alpha\bar{f}$$

by monotone convergence with $\bar{f}$ as the measure. And since $\pi^{(n)}Nf^+ \leq \alpha f^+ < \infty$ and $\pi^{(n)}Nf^- \leq \alpha f^- < \infty$,

$$\pi^{(n)}(Nf) = \pi^{(n)}Nf^+ - \pi^{(n)}Nf^-.$$

Hence

$$\lim_n \pi^{(n)}(Nf) = \alpha f^+ - \alpha f^- = \alpha f$$

by monotone convergence for each term. Thus $\alpha\bar{f} \leq \alpha f$; $\alpha\bar{f} \geq 0$ since $\bar{f} \geq 0$, and $\alpha f < \infty$ by hypothesis.

Proposition 8-25: If h is a non-negative superregular function dominated by a potential g, then h is a potential and its total charge is no greater than the total charge of g.

PROOF: Let $g = Nf$. Write $h = N\bar{f} + \lim P^n h$ with $\bar{f} \geq 0$. Since $0 \leq h \leq g$, we have $0 \leq P^n h \leq P^n g$. But $P^n g \to 0$ by Corollary 8-9, so that $P^n h \to 0$ and $h = N\bar{f}$. Since $\alpha|f| < \infty$, we have, by Lemma 8-24; $0 \leq \alpha\bar{f} \leq \alpha f < \infty$. Hence h is a potential and $\alpha\bar{f} \leq \alpha f$.

Corollary 8-26: A non-negative potential $g = Nf$ has non-negative total charge.

PROOF: Let $g = Nf \geq 0$, and set $\bar{f} = 0$. Since αf is finite, $\alpha f \geq \alpha\bar{f} = 0$ by Lemma 8-24.

Proposition 8-25 has an interesting interpretation in terms of the enlarged chain. The jth column of N is a potential with charge $\{\delta_{ij}\}$. If the infimum of the column is positive, then the column dominates a constant function $k\mathbf{1}$, so that $\mathbf{1}$ is a potential by Proposition 8-25. Hence, the infimum of every column of N is zero unless the extended chain is absorbing. In particular, if $P\mathbf{1} = \mathbf{1}$, then the infimum of every column of N is 0.

In the case of the symmetric random walk in three dimensions, $P\mathbf{1} = \mathbf{1}$. Thus the infimum of every column of N is zero, and since P is symmetric, the infimum of every row is zero. This fact, although not providing a proof of Proposition 7-10, does give us more insight into that result.

3. Equilibrium sets and capacities

In proving analogs in the next section to the classical potential principles, we shall need to restrict the supports of the charges involved. The notion we shall need is that of an equilibrium set.

Definition 8-27: A set E is an **equilibrium set** for P if there is a pure potential which assumes the value 1 at every point of E and which has support in E. Such a potential is called an **equilibrium potential** for E A set E is a **dual equilibrium set** for P if there is a pure potential measure with support in E which equals α on E.

We proceed to give two characterizations of equilibrium potentials.

Proposition 8-28: A set E is an equilibrium set if and only if both

(1) $\alpha e^E < \infty$ and
(2) for any starting distribution the set E is entered only finitely often a.e.

When E is an equilibrium set, the hitting vector h^E is the *unique* equilibrium potential and its charge is the escape vector e^E.

PROOF: Suppose E is an equilibrium set. If x is an equilibrium potential for E, then $B^E x = x$ by Proposition 8-19 and $x_E = \mathbf{1}$ by definition. Since $x_{\tilde{E}}$ does not affect the value of $B^E x$, we have

$$x = B^E x = B^E \mathbf{1} = h^E,$$

and h^E must be the equilibrium potential. Its charge is

$$(I - P)h^E = e^E$$

by Theorem 8-4 and Proposition 5-8. Thus E is an equilibrium set if and only if h^E is a potential or if and only if $\alpha e^E < \infty$ and $\lim P^n h^E = 0$. But $s^E = \lim P^n h^E$ by conclusion (7) of Proposition 5-8.

Corollary 8-29: All finite sets are equilibrium sets.

PROOF: Apply Propositions 4-28 and 8-28.

Proposition 8-30: If E is an equilibrium set, then the equilibrium potential is the pointwise infimum of all pure potentials which dominate $\mathbf{1}$ on E.

PROOF: Since $h^E = B^E\mathbf{1}$, the result follows from conclusion (2) of Proposition 8-16.

We shall use the notation η^E for dual e^E.

Definition 8-31: If E is an equilibrium set, the **capacity** of E is defined by $C(E) = \alpha e^E = \eta^E\mathbf{1}$.

In terms of total charge, Definition 8-31 states that the capacity of an equilibrium set is to be the total charge of the equilibrium potential.

Lemma 8-32: A set E is an equilibrium set if and only if both $(P^E)^n\mathbf{1} \to 0$ and $\alpha_E[(I - P^E)\mathbf{1}] < \infty$. If E is an equilibrium set, then $C(E) = \alpha_E[(I - P^E)\mathbf{1}] = [\alpha_E(I - \hat{P}^E)]\mathbf{1}$.

PROOF: We shall apply Proposition 8-28. $[(P^E)^n\mathbf{1}]_i$ is the probability starting in $i \in E$ of returning to E at least n times. Thus $(P^E)^n\mathbf{1} \to 0$ is a necessary and sufficient condition for being in E only finitely often a.e. for any starting distribution. Secondly $(I - P^E)\mathbf{1} = e_E^E$ and $\alpha_E e_E^E = \alpha e^E$. Hence αe^E is finite if and only if $\alpha_E[(I - P^E)\mathbf{1}] < \infty$. And if E is an equilibrium set, then

$$C(E) = \alpha_E[(I - P^E)\mathbf{1}].$$

Under duality a number is transformed into itself. Hence

$$C(E) = [(\text{dual } \mathbf{1})(\text{dual } (I - P^E))](\text{dual } \alpha_E)$$
$$= [\alpha_E(I - \hat{P}^E)]\mathbf{1}.$$

Proposition 8-33: E is an equilibrium set if and only if $\mathbf{1}$ is a potential for P^E with α_E as the distinguished measure. Also $C(E)$ is the same computed for P as for P^E.

PROOF: 1 is always superregular for P^E. The two conditions given by Lemma 8-32 are precisely the conditions that 1 be a potential. Also $\alpha_E[(I - P^E)1]$ is the capacity of E in P^E.

Proposition 8-34: E is an equilibrium set for $\hat{P}$ if and only if it is a dual equilibrium set for P. When E is such a set, $\alpha_E = \hat{\eta}_E^E N_E$.

PROOF: E is an equilibrium set for $\hat{P}$ if and only if 1 is a potential for $\hat{P}^E$ with $1 = \hat{N}_E \hat{e}_E^E$. By duality, this condition is equivalent to the assertion that α_E is a potential measure for P^E with $\alpha_E = \hat{\eta}_E^E N_E$. The result then follows from the dual of Proposition 8-33.

We would like the result that $C(E) = \hat{C}(E)$. However, an equilibrium set for P need not be an equilibrium set for $\hat{P}$ (see Section 6). Therefore, the following is the best possible result:

Proposition 8-35: If E is an equilibrium set for both P and $\hat{P}$, then $C(E) = \hat{C}(E)$.

PROOF: By Proposition 8-34, we have $\alpha_E = \hat{\eta}_E^E N_E$, so that $\hat{\eta}_E^E = \alpha_E(I - P^E)$ by Lemma 8-18. By Lemma 8-32 applied to $\hat{P}$,

$$C(E) = \alpha_E e_E^E = (\hat{\eta}_E^E N_E) e_E^E = \hat{\eta}_E^E (N_E e_E^E) = \hat{\eta}_E^E h_E^E$$
$$= \hat{\eta}_E^E 1 = [\alpha_E(I - P^E)]1 = \hat{C}(E).$$

Proposition 8-36: If F is a dual equilibrium set and $E \subset F$, then E is a dual equilibrium set and $\hat{C}(E) = \hat{\eta}^F h^E$.

PROOF: We shall use Proposition 8-28 to prove that E is a dual equilibrium set. By Proposition 8-34, $\alpha_F = (\hat{\eta}^F N)_F$ and $\hat{C}(F) = \hat{\eta}^F 1 < \infty$. Hence $\alpha_E = (\hat{\eta}^F N)_E$ and

$$\alpha_E(I - P^E) = \left[\hat{\eta}^F N \begin{pmatrix} I - P^E & 0 \\ 0 & 0 \end{pmatrix}\right]_E$$
$$= (\hat{\eta}^F B^E)_E$$

by Proposition 8-20. Then

$$\alpha \hat{e}^E = \alpha_E[(I - \hat{P}^E)1] = [\alpha_E(I - P^E)]1 = (\hat{\eta}^F B^E)_E 1 = \hat{\eta}^F B^E 1 = \hat{\eta}^F h^E.$$

Since $\hat{\eta}^F h^E \leq \hat{\eta}^F 1 < \infty$, we have just verified the first condition of Proposition 8-28—that $\alpha \hat{e}^E < \infty$. The second condition is trivial for a subset of an equilibrium set. Hence E is a dual equilibrium set and $\hat{C}(E) = \alpha \hat{e}^E = \hat{\eta}^F h^E$.

The dual of this result states that any subset E of an equilibrium set F is an equilibrium set, and $C(E) = \eta^F \hat{h}^E$.

Proposition 8-37: The union of a finite number of equilibrium sets is an equilibrium set.

PROOF: Let $E_1, \ldots, E_n$ be equilibrium sets and let $E = \bigcup_{k=1}^n E_k$. Then

$$\begin{aligned}\sum_{i \in E} \alpha_i e_i^E &\le \sum_{k=1}^n \sum_{i \in E_k} \alpha_i e_i^E \\ &\le \sum_{k=1}^n \sum_{i \in E_k} \alpha_i e_i^{E_k} \\ &= \sum_{k=1}^n C(E_k) < \infty,\end{aligned}$$

and if the process is in each E_k only finitely often a.e., then it is in E only finitely often a.e. Hence, by Proposition 8-28, E is an equilibrium set.

Some of the classical results hold only if the support of a potential is a reasonably small set. It will always be satisfactory to have a finite set as support. A more general assumption is that the support is an equilibrium set. Since equilibrium sets include all finite sets, since a subset of an equilibrium set is an equilibrium set, and since finite unions of equilibrium sets are equilibrium sets, we may think of equilibrium sets as a class of "reasonably small" sets.

Choquet has introduced a generalized notion of capacity. In our case his definition takes the following form.

Definition 8-38: A **Choquet capacity** is a non-negative monotone increasing set function such that, for any sets $A_1, A_2, \ldots, A_n$,

$$\begin{aligned}C(A_1 \cap A_2 \cap \cdots \cap A_n) \le \sum_i C(A_i) &- \sum_{i \ne j} C(A_i \cup A_j) \\ &+ \sum_{i \ne j \ne k} C(A_i \cup A_j \cup A_k) \\ &- \cdots - (-1)^n C(A_1 \cup \cdots \cup A_n).\end{aligned}$$

A simple way of constructing one of these capacities is to let π be a fixed starting distribution and to take $C(E)$ to be the probability of ever entering E. That is, $C(E) = \pi h^E$. This set function is monotone because h^E is. The right side of the inequality in the definition of capacity is the probability that all sets are entered. The left side is the probability that the intersection of the sets is entered, which is one way of entering all sets, though in general not the only way. Hence Choquet's definition is satisfied. Since we may clearly also replace $C(E)$ by $kC(E)$ with $k > 0$, π may be any non-negative finite measure.

We shall show from this construction that our Definition 8-31 yields a Choquet capacity on equilibrium sets. For convenience, we will give a proof for $\hat{P}$. For any fixed equilibrium set F of $\hat{P}$, Proposition 8-36 tells us that the capacity of any subset E is $\hat{\eta}^F h^E$. Hence the above argument applies with $\pi = \hat{\eta}^F$. Thus the Choquet conditions hold for all subsets of F and, since F is any equilibrium set, they hold for all equilibrium sets.

A more general method of obtaining a Choquet capacity within our framework is as follows. Let π be a measure, and let h be a strictly positive superregular function such that $\pi h < \infty$. Define $C(E) = \pi B^E h$. Forming the h-process, we see that $C(E) = \pi^*(B^E)^*1 = \pi^*(h^E)^*$ with $\pi^*1 = \pi h < \infty$. Thus by the special case $C(E)$ is a Choquet capacity for the h-process and hence satisfies the same axioms in the original chain. Moreover, we see that the situation in the earlier case is just the present case with $h = 1$. On the other hand, this more general method includes a second interesting case: If g is a pure potential for which πg is finite, then by Lemma 8-17,

$$\pi B^E g = \pi g - \pi\, {}^E N f;$$

$\pi B^E g$ is a Choquet capacity which assumes its maximum value πg on all sets E containing the support of f. If $\pi 1 = 1$, then $\pi B^E g$ in the game interpretation is the expected gain when and after E is reached.

Definition 8-31 is reasonable only for equilibrium sets, since otherwise it is possible to have $e^E = 0$. We could instead restrict the definition to finite sets and define the capacity of an infinite set as the supremum of the capacities of its finite subsets. We will show, under an additional assumption, that this new approach agrees with Definition 8-31 on equilibrium sets and assigns infinite capacity to all other sets.

Proposition 8-39: If $\hat{N}$ has columns which tend to zero, then a set E for which the supremum of the capacities of its finite subsets is finite is an equilibrium set, and the supremum is the capacity of the set.

PROOF: Let $E_1 \subset E_2 \subset \ldots$ be an increasing sequence of finite sets whose union is E. We must prove that if $\sup C(E_n)$ is finite, then h^E is a potential and $C(E) = \sup C(E_n)$. First we note that h^E is the monotone limit of h^{E_n}. If $i \in E_m$, then the ith component of e^{E_n} decreases for $n \geq m$. Thus $\lim e^{E_n} = \bar{e}$ exists. Since $\hat{N}$ has columns that tend to zero and since $\eta^{E_n}1 = C(E_n) \leq \sup C(E_n) < \infty$,

$$\eta^{E_n}\hat{N} \to (\text{dual } \bar{e})\hat{N}$$

by Proposition 1-58. By duality,

$$h^{E_n} = Ne^{E_n} \to N\bar{e},$$

and

$$\alpha\bar{e} = (\text{dual } \bar{e})1 = \sum_i \lim_n \eta_i^{E_n},$$

which by Fatou's Theorem is

$$\leq \lim_n (\eta^{E_n}1) = \lim C(E_n) = \sup C(E_n).$$

Hence $h^E = N\bar{e}$ and $\alpha\bar{e} < \infty$. Thus E is an equilibrium set. Also

$$C(E) = \alpha\bar{e} \leq \sup C(E_n).$$

But $C(E_n) \leq C(E)$ for every n, so that $\sup C(E_n) \leq C(E)$. Thus

$$C(E) = \sup C(E_n).$$

The converse, that an equilibrium set has the property that the supremum of the capacities of its finite subsets is finite, follows trivially from the monotonicity and the finiteness of capacity on equilibrium sets.

4. Potential principles

We shall now derive analogs of several of the fundamental theorems of classical potential theory. The first is the solution to the Dirichlet problem; in the uniqueness statement we shall need a lemma, for which we shall give two proofs.

Lemma 8-40: If $\bar{P}$ is an absorbing chain, then P has no bounded non-zero regular function.

PROOF 1: Suppose $Ph = h$ with $|h| \leq c1$. Since $h_i = \sum_i P_{ij}h_j$, we have $|h_i| \leq \sum_i P_{ij}|h_j|$ or $|h| \leq P|h|$. Therefore, $|h| \leq P^n|h| \leq P^n(c1)$. But P^n1 is the probability that the process continues at least until time n, which tends to zero as n tends to infinity because $\bar{P}$ is absorbing. Hence $h = 0$.

PROOF 2: Let a be the absorbing state of $\bar{P}$ and let $\mathbf{t} = \mathbf{a}$ be the time to absorption; $\mathbf{t}$ is a stopping time since $\bar{P}$ is absorbing. If $h = Ph$, set

$$h' = \begin{matrix} a \\ S \end{matrix}\begin{pmatrix} 0 \\ h \end{pmatrix}.$$

Then $h'(x_n(\omega))$ in the $\bar{P}$-process forms a bounded martingale. By Corollary 3-16 to the second martingale systems theorem,

$$0 = h'_a = \mathrm{M}[h'(x_t(\omega))] = \mathrm{M}[h'(x_0(\omega))].$$

But $x_0(\omega)$ is arbitrary, so that $h' = 0$ and $h = 0$.

The method in the second proof is of some importance, and we shall meet it again later.

Theorem 8-41: Let E be an arbitrary set of states, and suppose that ${}^EP$, the chain P with E made absorbing, is an absorbing chain. If h_E is any bounded function defined on E, then there exists a unique bounded function $\bar{h}$ whose restriction to E is h_E and which is regular on $\tilde{E}$. The function is

$$\bar{h} = B^E\begin{pmatrix} h_E \\ 0 \end{pmatrix}.$$

PROOF: For existence, set

$$\bar{h} = B^E\begin{pmatrix} h_E \\ 0 \end{pmatrix}.$$

The product is defined, since h_E is bounded and B^E has row sums one Then the restriction of $\bar{h}$ to E is h_E because $(B^E)_{ij} = \delta_{ij}$ for i and j in E. Moreover,

$$(I - P)\bar{h} = (I - P)B^E\begin{pmatrix} h_E \\ 0 \end{pmatrix} = \begin{pmatrix} I - P^E & 0 \\ 0 & 0 \end{pmatrix}\begin{pmatrix} h_E \\ 0 \end{pmatrix} = \begin{pmatrix} (I - P^E)h_E \\ 0 \end{pmatrix},$$

so that $\bar{h}$ is regular outside of E; associativity is justified in the triple product because $(I + P)B^E\begin{pmatrix} |h_E| \\ 0 \end{pmatrix}$ is finite-valued.

For uniqueness, let

$$k = \begin{pmatrix} h_E \\ \bar{k} \end{pmatrix}$$

be another such bounded function. Then $\bar{h} - k$ is a bounded function which is zero on E and regular outside E. If Q is the transition matrix for the transient states of ${}^EP$, then $(\bar{h} - k)_{\tilde{E}}$ is a non-zero bounded Q-regular function, in contradiction to Lemma 8-40, since $\bar{Q}$ is absorbing.

Next we prove the Maximum Principle.

Theorem 8-42: Let E be an arbitrary set of states and suppose that h is a finite-valued function such that $h = B^E h$. Then the supremum of the values of h is equal to the supremum of the values of h on E. If the states of $\tilde{E}$ communicate in ${}^E P$, if ${}^E P$ is absorbing, and if h assumes its maximum on $\tilde{E}$, then h is constant on $\tilde{E}$.

REMARK: Corresponding results hold for infima by replacing h by $-h$.

PROOF:

$$h_i = \sum_{j \in E} B^E_{ij} h_j$$
$$\leq \sum_{j \in E} B^E_{ij} \left(\sup_{j \in E} h_j\right) \leq \sup_{j \in E} h_j.$$

Suppose that h assumes its maximum on $\tilde{E}$, that ${}^E P$ is absorbing, and that the states of $\tilde{E}$ communicate in ${}^E P$. Let i be a state where the maximum is assumed, and let k be any state of E that can be reached in ${}^E P$ from $\tilde{E}$. Since the transient states of ${}^E P$ communicate, we have $B^E_{ik} > 0$. Moreover,

$$\begin{aligned} h_i &= B^E_{ik} h_k + \sum_{j \neq k} B^E_{ij} h_j \\ &\leq B^E_{ik} h_k + h_i \sum_{j \neq k} B^E_{ij} \qquad \text{since } h_j \leq h_i \\ &= B^E_{ik} h_k + h_i(1 - B^E_{ik}) \quad \text{since } B^E 1 = 1 \\ &= h_i - B^E_{ik}(h_i - h_k). \end{aligned}$$

Therefore, $h_i = h_k$ for all such k. Then for any $m \in \tilde{E}$, $B^E_{mj} > 0$ precisely for those j for which $B^E_{ij} > 0$, and $h_j = h_i$ for those j. Thus

$$h_m = \sum_{j \in E} B^E_{mj} h_j = \sum_{j \in E} B^E_{mj} h_i = h_i.$$

Corollary 8-43: If g is a potential with support in a finite set, then g is bounded.

PROOF: Since $g = B^E g$ for any potential, we may apply Theorem 8-42. The supremum in E is over a finite number of values.

Corollary 8-44: Let E be an arbitrary set of states, and suppose that

(1) ${}^E P$ is an absorbing chain.
(2) the states of $\tilde{E}$ communicate in ${}^E P$.
(3) every state of E can be reached in ${}^E P$ from $\tilde{E}$.

If h is a bounded function regular outside of E, then h cannot assume its maximum on $\tilde{E}$ unless h is constant everywhere.

PROOF: By Theorem 8-41, h is the unique solution to the Dirichlet problem for the function h_E. Hence

$$h = B^E\begin{pmatrix} h_E \\ 0 \end{pmatrix}.$$

Multiplying through by B^E and applying Proposition 5-8, we have

$$B^E h = B^E B^E\begin{pmatrix} h_E \\ 0 \end{pmatrix} = B^E\begin{pmatrix} h_E \\ 0 \end{pmatrix} = h.$$

By Theorem 8-42, h is constant on $\tilde{E}$. As shown in the proof of that theorem, h assumes the same constant value at every state of E which can be reached in ${}^E P$ from $\tilde{E}$.

The result that follows is the Principle of Domination.

Theorem 8-45: Let h be a finite-valued non-negative superregular function, and let $g = Nf$ be a potential. If h dominates g on the support of f^+, then h dominates g everywhere. If, in addition, h is a potential $N\bar{f}$, then $\alpha f \leq \alpha\bar{f}$.

PROOF: If g is a pure potential supported in E, then $g = B^E g$ by Proposition 8-19. But by Proposition 8-16, $B^E g$ is the pointwise infimum of all non-negative superregular functions which dominate g on E. Thus the first half is proved if g is a pure potential. For arbitrary g, write $g = Nf^+ - Nf^-$. We have $Nf^+ - Nf^- \leq h$ on the support of f^+, so that

$$Nf^+ \leq h + Nf^-$$

on the support of f^+. Applying the special case to the superregular function $h + Nf^-$ and the potential Nf^+, we have

$$Nf^+ \leq h + Nf^- \quad \text{or} \quad g \leq h$$

everywhere. Finally, if $h = N\bar{f}$, then

$$Nf^+ \leq N(\bar{f} + f^-)$$

implies

$$\alpha f^+ \leq \alpha(\bar{f} + f^-)$$

by Lemma 8-24. Hence $\alpha f \leq \alpha\bar{f}$.

Next we prove the Principle of Balayage.

Theorem 8-46: If g is a pure potential and if E is any set of states, then there is a unique pure potential $\bar{g}$ with support in E such that $\bar{g} = g$ on E. The potential $\bar{g}$ satisfies $\bar{g} \leq g$ everywhere, and its total charge does not exceed the total charge of g.

PROOF: For existence, let $\bar{g} = B^E g$. Then $\bar{g} \leq g$, $\bar{g}_E = g_E$, and $\bar{g}$ is superregular by conclusions (1) and (3) of Proposition 8-16. By Proposition 8-25, $\bar{g}$ is a potential, and the total charge of $\bar{g}$ is less than or equal to that of g.

For uniqueness, if h were another such potential, we would have

$$\bar{g} = B^E \bar{g} = B^E h = h$$

by Proposition 8-15 and the fact that $\bar{g}_E = g_E = h_E$.

If E is any set and if g is a pure potential, we refer to the potential $\bar{g}$ of Theorem 8-46 as the **balayage potential** of g on E.

Corollary 8-47: The balayage potential $\bar{g} = B^E g$ of g on E is the pointwise infimum of all pure potentials which dominate g on E.

PROOF: Apply conclusion (2) of Proposition 8-16.

Corollary 8-48: The balayage potential of g on E is the supremum of all pure potentials with support in E which are dominated by g on E.

PROOF: Certainly the balayage potential does have the stated property. Thus let $\bar{g} = B^E g$ and let h be a potential with support in E and with $h_E \leq g_E$. Then by Proposition 8-19, $h = B^E h$ and

$$\bar{g} = B^E g \geq B^E h = h.$$

If g has support in E, then g itself is the balayage potential of g on E. In particular, h^E is the balayage potential of h^E on E for E an equilibrium set.

Next we prove the Principle of Lower Envelope.

Lemma 8-49: The pointwise infimum of non-negative superregular functions is non-negative superregular.

PROOF: It is clearly non-negative. If

$$h_\beta \geq P h_\beta$$

for all β, then

$$P(\inf h_\beta) \le Ph_\beta \le h_\beta$$

for all β, so that

$$P(\inf h_\beta) \le \inf h_\beta.$$

Theorem 8-50: The pointwise infimum of pure potentials is a pure potential.

PROOF: Apply Lemma 8-49 and Proposition 8-25.

Finally we prove the Principle of Condensers. We are to think of two sets E and F as the two plates of a condenser with a positive charge placed on E and a negative charge placed on F in such a way as to produce a unit voltage drop. Since in equilibrium there should be a uniform voltage on each plate and since the 0-value of voltage is arbitrary, we will require that the potential be 1 on E and 0 on F. The theorem is proved for an equilibrium set E with **finite boundary** $\bar{\bar{E}}$, that is, a set E that can be entered or left only through the finite set $\bar{\bar{E}}$.

Theorem 8-51: Let E be an equilibrium set with finite boundary, and let F be any disjoint set of states. Then there is a potential $g = N_f^r$ which is 1 on E and 0 on F and which is such that f^+ has support in E f^- has support in F, and $\alpha f \ge 0$.

PROOF: Let $g_i = {}^F H_{iE}$, the probability starting at i that E is reached before F. Clearly, g is 1 on E and 0 on F. Furthermore, $0 \le g \le h^E$. Since $P^n h^E \to 0$ for the equilibrium set E, we have $P^n g \to 0$.

Let $f = (I - P)g$. We are going to apply Corollary 8-8 to conclude that g is a potential with charge f, but to do so we must show that α_f^r is finite. If i is in the complement of $E \cup F$, then $g_i = (Pg)_i$; hence f has support in $E \cup F$. Write

$$P^{E \cup F} = \begin{matrix} & \begin{matrix} E & F \end{matrix} \\ \begin{matrix} E \\ F \end{matrix} & \begin{pmatrix} X & Y \\ Z & W \end{pmatrix} \end{matrix}.$$

Noting that if $i \in E \cup F$, then $(Pg)_i$ is the probability that the next entry to $E \cup F$ is in E, we have

$$f_i = \begin{cases} g_i - (Pg)_i = 1 - (Pg)_i = 1 - (X\mathbf{1})_i & \text{if } i \in E \\ 0 - (Pg)_i = -(Z\mathbf{1})_i & \text{if } i \in F. \end{cases}$$

Hence

$$f_{E\cup F} = \begin{cases} 1 - X1 & \text{on } E. \\ -Z1 & \text{on } F. \end{cases}$$

Thus f^+ has support in E and f^- has support in F. Furthermore $X1 = 1$ except for the boundary of E; hence f^+ has finite support and $\alpha f^+ < \infty$. Moreover

$$\alpha f^- = \alpha_F Z1 = \sum_{j\in \tilde{E}} (\alpha_F Z)_j < \infty$$

since $Z_{ij} = 0$ except when j is on the boundary of E. Thus $\alpha|f| < \infty$. Hence g is a potential. Finally $\alpha f \geq 0$ by Corollary 8-26.

5. Energy

Classically, energy is the integral of the potential with respect to the charge, and we shall adopt the obvious analog of this definition. Throughout this section we shall write

$$\mu = \text{dual } f$$
$$\nu = \text{dual } g.$$

Definition 8-52: If $g = Nf$ is a potential and $|\mu|\ N\ |f| < \infty$, then its energy is defined to be $\mathbf{I}(g) = \mu g = \nu f$, and g is said to have **finite energy**.

If all potentials are bounded, then all of them have finite energy. For if $\bar{g} = N|f|$, then

$$|\mu|\ N\ |f| = \sum_i \alpha_i |f_i| \bar{g}_i$$
$$\leq \sup \bar{g}_i(\alpha|f|)$$
$$< \infty.$$

In any case a potential of finite support has finite energy.

We can write energy either purely in terms of the charge or purely in terms of the potential:

$$\mathbf{I}(g) = \mu(Nf) = \nu[(I - P)g].$$

Since the dual of a number is the same number, we also have

$$\mathbf{I}(g) = (\mu\hat{N})f = [\nu(I - \hat{P})]g.$$

If f is a charge for P, then, as noted after Theorem 8-3, f is also a charge for $\hat{P}$. In the two processes we have

$$\mathbf{I}(Nf) = \mu(Nf) = (\mu\hat{N})f$$

and

$$\mathbf{I}(\hat{N}f) = (\mu N)f = \mu(\hat{N}f),$$

and the energies are equal since the matrices associate by Corollary 1-5.

The expression for $\mathbf{I}(g)$ in Definition 8-52 disguises the fact that energy depends only on the values of the potential on the support E. We shall derive a simple dependence of $\mathbf{I}(g)$ on P^E.

Proposition 8-53: If g is a potential of finite energy with support in E, then

$$\mathbf{I}(g) = \nu_E[(I - P^E)g_E] = [\nu_E(I - \hat{P}^E)]g_E.$$

PROOF: Since f has support in E,

$$(I - P^E)g_E = f_E$$

by Proposition 8-19, and

$$\nu_E[(I - P^E)g_E] = \nu_E f_E = \nu f = \mathbf{I}(g).$$

The other half of the proposition is the dual of the first half.

Classically energy is non-negative. We shall prove shortly that the energy of a potential is non-negative provided it is finite. To do so, we first introduce a definition. If $g = Nf$ and $\bar{g} = N\bar{f}$ are potentials of finite energy, we define

$$(g, \bar{g}) = \tfrac{1}{2}(\mu\bar{g} + \bar{\mu}g),$$

provided the matrix products are well defined. (We shall show soon that this condition is always satisfied.)

Note that $(g, g) = \mathbf{I}(g)$. We wish to show that $(g, \bar{g})$ is an inner product. The reader should verify that $(g, \bar{g})$ satisfies (1), (2), and (4) in general and (3) when all the potentials have finite support. We shall prove (5) and the general case of (3) below.

(1) $(g, \bar{g}) = (\bar{g}, g)$.
(2) For every real number c, $(cg, \bar{g}) = c(g, \bar{g})$.
(3) $(g + g', \bar{g}) = (g, \bar{g}) + (g', \bar{g})$.
(4) If g is a pure potential for which $(g, g) = 0$, then $g = 0$.
(5) $(g, g) \geq 0$ for all g.

Lemma 8-54: If g has support in a finite set E, then

$$\mathbf{I}(g) = \tfrac{1}{2} \sum_{i \in E} \left[(\alpha_i m_i + \pi_i)g_i^2 + \sum_{j \in E} \alpha_i P_{ij}^E (g_i - g_j)^2\right] \geq 0,$$

where

$$m_i = 1 - \sum_{j \in E} P_{ij}^E \geq 0 \quad \text{and} \quad \pi_i = \alpha_i - \sum_{k \in E} \alpha_k P_{ki}^E \geq 0.$$

PROOF: We shall apply Proposition 8-53. The matrices involved are finite matrices so that distributivity and associativity hold. Hence

$$\begin{aligned}\mathbf{I}(g) &= \nu_E g_E - \nu_E P^E g_E \\ &= \tfrac{1}{2}\sum_{i\in E}\Big[\alpha_i g_i^2 + \alpha_i g_i^2 + \sum_{j\in E}(-2\alpha_i P_{ij}^E g_i g_j)\Big] \\ &= \tfrac{1}{2}\sum_{i\in E}\Big[\alpha_i\Big(1 - \sum_{j\in E} P_{ij}^E\Big)g_i^2 + \Big(\alpha_i - \sum_{k\in E}\alpha_k P_{ki}^E\Big)g_i^2 \\ &\qquad + \sum_{j\in E}(\alpha_i P_{ij}^E g_i^2 - 2\alpha_i P_{ij}^E g_i g_j + \alpha_i P_{ij}^E g_j^2)\Big] \\ &= \tfrac{1}{2}\sum_{i\in E}\Big[(\alpha_i m_i + \pi_i)g_i^2 + \sum_{j\in E}\alpha_i P_{ij}^E (g_i - g_j)^2\Big].\end{aligned}$$

Since P^E is a transition matrix and α_E is P^E-superregular, m and π are non-negative. Hence $\mathbf{I}(g) \geq 0$.

From properties (1), (2), and (3), we can prove that Schwarz's inequality holds for g and $\bar{g}$ whenever they have finite support.

Lemma 8-55: If $g = Nf$ and $\bar{g} = N\bar{f}$ are two potentials of finite support, then

$$(g, \bar{g})^2 \leq \mathbf{I}(g)\mathbf{I}(\bar{g}).$$

PROOF: By Lemma 8-54 we have

$$\mathbf{I}(xg - \bar{g}) = (xg - \bar{g}, xg - \bar{g}) \geq 0$$

for all real x. Hence by properties (1), (2), and (3), we find that

$$x^2(g, g) - 2x(g, \bar{g}) + (\bar{g}, \bar{g}) \geq 0$$

for all real x. If $(g, g) = 0$, then, for $-2x(g, \bar{g}) + (\bar{g}, \bar{g})$ to be non-negative for all x, it must be true that $(g, \bar{g}) = 0$, and the lemma is trivial. Otherwise, the discriminant of the quadratic equation in x must be non-positive, so that

$$4(g, \bar{g})^2 - 4(g, g)(\bar{g}, \bar{g}) \leq 0$$

or

$$\begin{aligned}(g, \bar{g})^2 &\leq (g, g)(\bar{g}, \bar{g}) \\ &= \mathbf{I}(g)\mathbf{I}(\bar{g}).\end{aligned}$$

Lemma 8-56: Let $g = Nf$ and $\bar{g} = N\bar{f}$ be pure potentials of finite energy, let

$$E_1 \subset E_2 \subset E_3 \cdots$$

be an increasing sequence of finite sets with union the set of all states S, and let

$$f^{(n)} = \begin{pmatrix} f_{E_n} \\ 0 \end{pmatrix}, \qquad \bar{f}^{(n)} = \begin{pmatrix} \bar{f}_{E_n} \\ 0 \end{pmatrix}.$$

$$g^{(n)} = Nf^{(n)}, \quad \text{and} \quad \bar{g}^{(n)} = N\bar{f}^{(n)}.$$

Then $(g, \bar{g}) = \lim_{n\to\infty} (g^{(n)}, \bar{g}^{(n)})$.

PROOF: We have $(g, \bar{g}) = \frac{1}{2}(\mu\bar{g} + \bar{\mu}g)$, and by symmetry it is enough to show that $\bar{\mu}^{(n)}g^{(n)}$ converges to $\bar{\mu}g$. By monotone convergence, we have $\lim g^{(n)} = g$, since $0 \le f^{(1)} \le f^{(2)} \le \cdots$. Let

$$h_i^{(n)} = \begin{cases} g_i^{(n)} & \text{if } i \in E_n \\ 0 & \text{otherwise.} \end{cases}$$

Then

$$\bar{\mu}^{(n)}g^{(n)} = \bar{\mu}h^{(n)}$$

and

$$\lim h^{(n)} = \lim g^{(n)} = g.$$

The functions $h^{(n)}$ are non-negative and increasing; also $\bar{\mu}$ is non-negative. Thus by monotone convergence,

$$\lim \bar{\mu}^{(n)}g^{(n)} = \lim \bar{\mu}h^{(n)} = \bar{\mu} \lim h^{(n)} = \bar{\mu}g.$$

Lemma 8-57: If g and $\bar{g}$ are pure potentials of finite energy, then

$$(g, \bar{g})^2 \le (g, g)(\bar{g}, \bar{g}).$$

Consequently $(g, \bar{g}) < \infty$.

PROOF: Form the approximations to g and $\bar{g}$ as in the statement of Lemma 8-56. By Lemmas 8-54 and 8-55,

$$(g^{(n)}, \bar{g}^{(n)})^2 \le (g^{(n)}, g^{(n)})(\bar{g}^{(n)}, \bar{g}^{(n)}).$$

Applying Lemma 8-56 to each factor, we obtain

$$(g, \bar{g})^2 \le (g, g)(\bar{g}, \bar{g}).$$

If g and $\bar{g}$ are any potentials of finite energy, then

$$|(g, \bar{g})| \le (N|f|, N|\bar{f}|) \le \sqrt{\mathbf{I}(N|f|)\mathbf{I}(N|\bar{f}|)} < \infty.$$

Therefore $(g, \bar{g})$ is always well defined. We can now prove (3) in general by breaking charges into positive and negative parts.

Proposition 8-58: If g has finite energy, then $\mathbf{I}(g) \ge 0$.

PROOF: Write $g = Nf = Nf^+ - Nf^-$, where Nf^+ and Nf^- are pure potentials. Then

$$\begin{aligned} \mathbf{I}(g) &= (g, g) = (Nf^+ - Nf^-, Nf^+ - Nf^-) \\ &= \mathbf{I}(Nf^+) - 2(Nf^+, Nf^-) + \mathbf{I}(Nf^-) \\ &\geq \mathbf{I}(Nf^+) - 2\sqrt{\mathbf{I}(Nf^+)\mathbf{I}(Nf^-)} + \mathbf{I}(Nf^-) \quad \text{by Lemma 8-57} \\ &= (\sqrt{I(Nf^+)} - \sqrt{I(Nf^-)})^2 \\ &\geq 0. \end{aligned}$$

We can finally prove Schwarz's inequality for all potentials of finite energy by proceeding just as in the proof of Lemma 8-55.

We now begin the proof of the fundamental result about energy, the theorem that justifies the name "equilibrium potential." Unfortunately the result fails for the most general transient chain, so that some extra hypothesis is needed. We shall prove the theorem—Theorem 8-61—under the hypothesis $P = \hat{P}$, a condition that *is* satisfied in the classical case of the three-dimensional symmetric random walk with $\alpha = 1^T$.

Lemma 8-59: The energy of the equilibrium potential on E is the capacity $C(E)$.

PROOF: $\mathbf{I}(h^E) = \eta^E h^E = \eta^E_E h^E_E = \eta^E_E 1 = \eta^E 1 = C(E)$.

Lemma 8-60: If E is an equilibrium set, if $g = Nf$ is a potential with support in E, and if $P = \hat{P}$, then $(g, h^E) = \alpha f$.

PROOF:

$$\begin{aligned} 2(g, h^E) &= \mu h^E + \eta^E g \\ &= \mu h^E + \eta^E Nf \\ &= \mu h^E + \mu \hat{N} e^E && \text{by duality} \\ &= \mu h^E + \mu N e^E && \text{since } N = \hat{N} \\ &= 2\mu h^E && \text{since } h^E = Ne^E \\ &= 2\mu 1 && \text{since } f \text{ has support in } E \\ &= 2\alpha f. \end{aligned}$$

Theorem 8-61: Suppose that P is a chain in which $P = \hat{P}$. If E is an equilibrium set, then the equilibrium potential for E minimizes energy among all potentials of finite energy whose support is in E and whose total charge is $C(E)$.

PROOF: Let $g = Nf$ be a potential with $\alpha f = C(E)$ and with support in E. By Lemma 8-60, $(g, h^E) = C(E)$, and by Lemma 8-59, $\mathbf{I}(h^E) = C(E)$. Furthermore, $\mathbf{I}(h^E) \neq 0$ by property (4). Therefore, by Schwarz's inequality, we have

$$\mathbf{I}(g) \geq \frac{(g, h^E)^2}{\mathbf{I}(h^E)} = \frac{C(E)^2}{C(E)} = C(E) = \mathbf{I}(h^E).$$

We shall see in the next section that the theorem need not be true if $P \neq \hat{P}$.

6. The basic example

In this section we shall work out what the results of the preceding five sections mean in terms of the basic example.

First we compute $\hat{P}$, the β-dual of P.

$$\hat{P}_{ij} = \frac{\beta_j P_{ji}}{\beta_i} = \begin{cases} \beta_{i-1} p_i / \beta_i = 1 & \text{if } j = i - 1 \\ \dfrac{\beta_j q_{j+1}}{\beta_0} = \beta_j - \beta_{j+1} & \text{if } i = 0, \end{cases}$$

Thus the reverse process proceeds deterministically a step at a time to the left until it reaches 0. From 0 it may step into any state and does so with probability

$$\hat{P}_{0j} = \beta_j - \beta_{j+1}.$$

Since $\sum_j \hat{P}_{0j} = 1 - \beta_\infty < 1$ and since 0 is reached from all states with probability one, the extended chain for $\hat{P}$ is absorbing. We saw in Section 5-10 that P has no non-zero regular measure; on the other hand β is regular for $\hat{P}$ since

$$\sum_i \beta_i \hat{P}_{ij} = \beta_0(\beta_j - \beta_{j+1}) + \beta_{j+1} 1 = \beta_j.$$

From Section 5-10 we know that

$$N_{ij} = \begin{cases} \dfrac{\beta_j}{\beta_\infty} & \text{if } i \leq j \\ \dfrac{\beta_j}{\beta_\infty} - \dfrac{\beta_j}{\beta_i} & \text{if } i > j. \end{cases}$$

Hence

$$\hat{N}_{ij} = \frac{\beta_j N_{ji}}{\beta_i} = \begin{cases} \dfrac{\beta_j}{\beta_\infty} & \text{if } i \geq j \\ \dfrac{\beta_j}{\beta_\infty} - 1 & \text{if } i < j. \end{cases}$$

We note that N has columns tending to 0, but $\hat{N}$ has columns bounded away from 0. The latter fact by itself implies that the extended chain for $\hat{P}$ is absorbing.

Next we find the general form of potentials. In P if $g = Nf$, then

$$g_i = (Nf)_i = \frac{1}{\beta_\infty} \sum_{j=0}^{\infty} \beta_j f_j - \frac{1}{\beta_i} \sum_{j=0}^{i-1} \beta_j f_j. \tag{1}$$

In $\hat{P}$ if $g = \hat{N}f$, then

$$g_i = \frac{1}{\beta_\infty} \sum_{j=0}^{\infty} \beta_j f_j - \sum_{j=i+1}^{\infty} f_j. \tag{2}$$

In either case, g is finite-valued if $\beta|f| < \infty$, in agreement with Lemma 8-2. (For the reverse chain we have

$$\infty > \sum_j \beta_j |f_j| \geq \beta_\infty \sum_j |f_j|,$$

and hence $\sum_j |f_j| < \infty$.)

Let $\mu = \text{dual} f$ and $\nu = \text{dual} g$. Then as required by duality

$$\begin{aligned} \nu_i &= \frac{\beta_i}{\beta_\infty} \sum_{j=0}^{\infty} \beta_j f_j - \sum_{j=0}^{i-1} \beta_j f_j \\ &= \frac{\beta_i}{\beta_\infty} \sum_{j=0}^{\infty} \mu_j - \sum_{j=0}^{i-1} \mu_j = (\mu \hat{N})_i. \end{aligned}$$

Thus μ is a left charge with potential measure ν for $\hat{P}$.

Theorem 8-4 demands that $f = (I - P)g$ when g is a potential with charge f. We have from (1)

$$\begin{aligned} (Pg)_i &= p_{i+1} g_{i+1} + q_{i+1} g_0 \\ &= \frac{p_{i+1}}{\beta_\infty} (\beta f) - \frac{p_{i+1}}{\beta_{i+1}} \sum_{j=0}^{i} \beta_j f_j + \frac{q_{i+1}}{\beta_\infty} (\beta f) \\ &= \frac{1}{\beta_\infty} (\beta f) - \frac{1}{\beta_i} \sum_{j=0}^{i} \beta_j f_j. \end{aligned}$$

Hence $g_i - (Pg)_i = f_i$ and $(I - P)g = f$.

For both P and $\hat{P}$, we have

$$N_{ii} = \hat{N}_{ii} = \frac{\beta_i}{\beta_\infty},$$

and thus the condition $\beta_i \geq kN_{ii}$ of Proposition 8-10 is satisfied with $k = \beta_\infty$. Hence all potentials in both P and $\hat{P}$ are bounded. We can see directly the boundedness from (1) and (2). In (1) and (2) we have

$$|g_i| \leq \frac{2\beta|f|}{\beta_\infty}.$$

The estimate

$$|g_i| \leq \frac{\beta|f|}{\beta_\infty}$$

of Proposition 8-10 is better; it takes into account the cancellation of the first and second terms in each expression. If $f \geq 0$ in (2), then the g_i increase monotonically to $\beta|f|/\beta_\infty$, so that the proposition gives the best possible bound. In (1) we have $\lim_i g_i = 0$, in agreement with the second half of Proposition 8-10 since the columns of N tend to zero. However, $\hat{N}$ does not have columns tending to zero, and $\lim_i g_i$ in (2) is not necessarily 0.

We determine the regular functions and signed measures for P and $\hat{P}$ as follows. If r is a P-regular function, then

$$r_i = (Pr)_i = p_{i+1}r_{i+1} + q_{i+1}r_0.$$

Thus

$$r_0 = p_1r_1 + q_1r_0, \quad p_1r_0 = p_1r_1,$$

and

$$r_0 = r_1.$$

Hence only the constant functions are P-regular. Dually only multiples of β are regular signed measures for $\hat{P}$. Since P has no regular signed measures, $\hat{P}$ has no regular functions. (Recall that $\hat{P}1 \neq 1$.) Therefore, the non-negative superregular functions of P are pure potentials plus non-negative constants, whereas only pure potentials are non-negative superregular functions for $\hat{P}$.

Next, we determine the equilibrium sets for P and $\hat{P}$. It is clear that P will be in any infinite set infinitely often a.e. Hence only finite sets are equilibrium sets. Let us verify this fact in terms of equilibrium potentials.

Let E be a finite set with m as last element. Then

$$e_i^E = \begin{cases} \dfrac{\beta_\infty}{\beta_m} & \text{if } i = m \\ 0 & \text{otherwise} \end{cases}$$

and

$$\beta e^E = \beta_m e_m^E = \beta_\infty.$$

By (1),

$$h_i^E = \frac{1}{\beta_\infty}(\beta e^E) - \frac{1}{\beta_i}\sum_{j=0}^{i-1} \beta_j e_j^E$$

or

$$h_i^E = \begin{cases} 1 & \text{if } i \leq m \\ 1 - \dfrac{\beta_\infty}{\beta_i} & \text{if } i > m. \end{cases}$$

It is clear probabilistically that these are the correct values for h^E; when $i > m$ the only way to avoid hitting E from i is to march straight out to the right. The probability of doing so is β_∞/β_i. Thus we see that h^E satisfies the conditions of an equilibrium potential: All non-empty finite sets have capacity β_∞.

If E is an infinite set, then $h^E = 1$ and $e^E = (I - P)h^E = 0$. Hence h^E is not a potential, and E is not an equilibrium set. The fact that the supremum of the capacities of finite subsets of an infinite set E is β_∞ does not contradict Proposition 8-39, since $\hat{N}$ does not have columns tending to zero.

For the reverse chain $\hat{P}$ let E be any set (finite or infinite) with least element m. Then $\hat{h}_i^E = 1$ for $i \geq m$. For $i < m$,

$$\hat{h}_i^E = \hat{h}_0^E = \hat{H}_{0m} = \frac{\hat{N}_{0m}}{\hat{N}_{mm}} = \frac{(\beta_m/\beta_\infty) - 1}{\beta_m/\beta_\infty} = 1 - \frac{\beta_\infty}{\beta_m}.$$

The next to last equality follows from the fact that $m = 0$ is incompatible with $i < m$. The process can escape from E only via m, and for $m > 0$

$$\hat{e}_m^E = [(I - \hat{P})\hat{h}^E]_m = \hat{h}_m^E - \hat{h}_{m-1}^E = \frac{\beta_\infty}{\beta_m}.$$

If $m = 0$,

$$\hat{e}_0^E = 1 - \sum_j \hat{P}_{0j} = \beta_\infty = \frac{\beta_\infty}{\beta_0}.$$

Hence, in either case,

$$\beta e^E = \beta_m e_m^E = \beta_\infty.$$

Thus *all* sets are equilibrium sets, and every set has capacity β_∞. We note that only the finite sets are equilibrium sets for both P and $\hat{P}$, and their capacity β_∞ is the same in both, as predicted by Proposition 8-35.

The dual of Proposition 8-23 is that every non-negative superregular measure is the increasing limit of pure potential measures of finite support. We shall produce the charges for $\hat{P}$ which give rise to the potential measures which increase to β. Let E_m be the set $\{0, \ldots, m\}$. The functions $h^{E_m} = Ne^{E_m}$ are the functions which Proposition 8-23 gives as increasing to 1. Therefore, by duality the measures $\eta^{E_m}\hat{N}$ should increase to β, and the charges we seek in $\hat{P}$ are the η^{E_m}. From our above calculations we have

$$\eta_i^{E_m} = \begin{cases} \beta_\infty & \text{if } i = m \\ 0 & \text{otherwise,} \end{cases}$$

and

$$(\eta^{E_m}\hat{N})_i = \begin{cases} \beta_i & \text{if } i \leq m \\ \beta_i - \beta_\infty & \text{if } i > m. \end{cases}$$

Next suppose as in Lemma 8-24 that $N\bar{f} \leq Nf$ with $\bar{f} \geq 0$ and βf finite. Since by (1)

$$(N\bar{f})_0 = \frac{1}{\beta_\infty}(\beta\bar{f})$$

and

$$(Nf)_0 = \frac{1}{\beta_\infty}(\beta f),$$

we must have $\beta\bar{f} \leq \beta f$. The proof for the reverse chain is not so simple, however. If $\hat{N}\bar{f} \leq \hat{N}f$ with $\bar{f} \geq 0$ and βf finite, then

$$\frac{1}{\beta_\infty}(\beta\bar{f}) - \sum_{j=i+1}^{\infty} \bar{f}_j \leq \frac{1}{\beta_\infty}(\beta f) - \sum_{j=i+1}^{\infty} f_j$$

for all i. According to the proof of Lemma 8-24, we multiply through by η^{E_m} and take the limit on m. We have

$$\beta\bar{f} - \beta_\infty \sum_{j=m+1}^{\infty} \bar{f}_j \leq \beta f - \beta_\infty \sum_{j=m+1}^{\infty} f_j.$$

As $m \to \infty$, we obtain $\beta\bar{f} \leq \beta f$.

Turning to the potential principles, we shall first illustrate the Principle of Domination (Theorem 8-45). We do so in $\hat{P}$. Let $f \geq 0$ and $\bar{f} \geq 0$ be given, let E be the support of $\bar{f}$, and suppose that $g_i \geq \bar{g}_i$ for $i \in E$. For convenience, suppose $0 \in E$. From (2) we have

$$\frac{\beta f}{\beta_\infty} - \sum_{j>i} f_j \geq \frac{\beta\bar{f}}{\beta_\infty} - \sum_{j>i} \bar{f}_j \quad \text{for } i \in E.$$

Let k be in $\tilde{E}$, and let i be the largest state of E for which $i < k$ (i exists since $0 \in E$). Then

$$\begin{aligned} g_k = \frac{\beta f}{\beta_\infty} - \sum_{j>k} f_j \geq \frac{\beta f}{\beta_\infty} - \sum_{j>i} f_j \geq \frac{\beta\bar{f}}{\beta_\infty} - \sum_{j>i} \bar{f}_j \\ = \frac{\beta\bar{f}}{\beta_\infty} - \sum_{j>k} \bar{f}_j = \bar{g}_k, \end{aligned}$$

the next to last equality holding since $\bar{f}_j = 0$ for $i < j \leq k$. Hence $g \geq \bar{g}$.

Next we examine the Principle of Balayage (Theorem 8-46) for $\hat{P}$. Let $f \geq 0$ be given, and for convenience let E be an infinite set containing 0. We wish to choose $\bar{f}$ with support in E so that $\bar{g}_E = g_E$. For $i \in E$, we must have

$$\frac{\beta\bar{f}}{\beta_\infty} - \sum_{j>i} \bar{f}_j = \frac{\beta f}{\beta_\infty} - \sum_{j>i} f_j.$$

Let k be the next element of E greater than i. We must also have

$$\frac{\beta\bar{f}}{\beta_\infty} - \sum_{j>k} \bar{f}_j = \frac{\beta f}{\beta_\infty} - \sum_{j>k} f_j.$$

Subtracting, we find that $\bar{f}_k = \sum_{j=i+1}^{k} f_j$. This equation determines all of $\bar{f}$ except for $\bar{f}_0$. Adding the relations for $\bar{f}_k$ with $k > 0$, we obtain $\sum_{j>0} \bar{f}_j = \sum_{j>0} f_j$.

Since for $g_0 = \bar{g}_0$ we need

$$\frac{\beta\bar{f}}{\beta_\infty} - \sum_{j>0} \bar{f}_j = \frac{\beta f}{\beta_\infty} - \sum_{j>0} f_j,$$

we must choose $\beta\bar{f} = \beta f$. Thus set

$$\bar{f}_0 = f_0 + \sum_{j>0} \beta_j(f_j - \bar{f}_j)$$

$$\bar{f}_k = \sum_{j=i+1}^{k} f_j \quad \text{for } i, k \in E \text{ and } j \notin E \text{ when } i < j < k.$$

To see that we have actually chosen $\bar{f}_0 \geq 0$, we note that β_j decreases with j, so that if $i, k \in E$ with no $j \in E$ for $i < j < k$, then

$$\begin{aligned} \beta_{i+1}(f_{i+1} - \bar{f}_{i+1}) + \beta_{i+2}(f_{i+2} - \bar{f}_{i+2}) + \cdots + \beta_k(f_k - \bar{f}_k) \\ &= \beta_{i+1}f_{i+1} + \beta_{i+2}f_{i+2} + \cdots + \beta_k f_k - \beta_k\bar{f}_k \\ &= \beta_{i+1}f_{i+1} + \cdots + \beta_k f_k - \beta_k(f_{i+1} + \cdots + f_k) \\ &= (\beta_{i+1} - \beta_k)f_{i+1} + \cdots + (\beta_{k-1} - \beta_k)f_{k-1} \\ &\geq 0. \end{aligned}$$

We shall illustrate the Principle of Condensers (Theorem 8-51) for P in the case $E = \{0, 1, \ldots, a\}$ and $F = \{b, b+1, \ldots\}$ with $0 < a < b$. We have

$$g_i = {}^F H_{iE} = \begin{cases} 1 & \text{for } i \leq a \\ 1 - \dfrac{\beta_b}{\beta_i} & \text{for } a < i < b \\ 0 & \text{for } i \geq b. \end{cases}$$

Then

$$f_i = g_i - p_{i+1}g_{i+1} - q_{i+1}g_0 = \begin{cases} \dfrac{\beta_b}{\beta_a} & \text{if } i = a \\ -q_{i+1} & \text{if } i \geq b \\ 0 & \text{otherwise.} \end{cases}$$

Hence

$$\beta f = \beta_b - \sum_{i \ge b} \beta_i q_{i+1} = \beta_\infty.$$

We can verify from (1) that g is the potential of f, and we see that f^+ has support in E, f^- has support in F, and that $\beta f \ge 0$.

We can show by example that the Principle of Condensers does not hold for all equilibrium sets E. In $\hat{P}$ let E be the set of even states and let F be the set of odd states. All sets are equilibrium sets for $\hat{P}$, but E does not have a finite boundary. If

$$g_i = \begin{cases} 1 & \text{for } i \in E \\ 0 & \text{for } i \in F, \end{cases}$$

the theorem requires that g be a potential. But if $i \in F$,

$$f_i = 0 - (\hat{P}g)_i = -1.$$

Hence

$$(\hat{N}f^-)_0 = -\sum_{j \text{ odd}} \hat{N}_{0j} = -\sum_{j \text{ odd}} \left(\frac{\beta_j}{\beta_\infty} - 1\right).$$

The expression on the right side may be infinite if the β's are chosen properly. Let

$$\beta_i = \frac{1}{i+2} + \frac{1}{2}$$

$$\beta_\infty = \tfrac{1}{2}.$$

The β's determine the transition probabilities uniquely, and for this choice $(\hat{N}f^-)_0$ is infinite, a contradiction.

Equation (1) gives us the following relation for energy in the P-process.

$$\mathbf{I}(g) = \frac{1}{\beta_\infty}(\beta f)^2 - \sum_{i=0}^{\infty} f_i \sum_{j=0}^{i-1} \beta_j f_j.$$

It is not difficult to see that (2) yields the same value for the same f. But it is not easy to see that $\mathbf{I}(g) \ge 0$ if g is not a pure potential. We shall now show that Theorem 8-61 fails if the assumption $P = \hat{P}$ is dropped. In P let $E = \{0, 1, \ldots, m\}$. We have seen that $C(E) = \beta_\infty$. Thus any potential with total charge $\beta f = \beta_\infty$ equal to that of the equilibrium potential has energy

$$\beta_\infty - \sum_{i=0}^{\infty} f_i \sum_{j=0}^{i-1} \beta_j f_j.$$

The equilibrium potential has energy β_∞, which is a *maximum* (not a minimum) among pure potentials. For example, let

$$f_i = \begin{cases} \frac{1}{2}\beta_\infty & \text{if } i = 0 \\ \dfrac{1}{2\beta_1}\beta_\infty & \text{if } i = 1 \\ 0 & \text{otherwise.} \end{cases}$$

Then $\beta f = \beta_\infty$ and

$$\mathbf{I}(g) = \beta_\infty - f_0(\beta_0 f_1) = \beta_\infty - \frac{1}{4\beta_1}\beta_\infty^2 < \beta_\infty.$$

7. An unbounded potential

In Proposition 8-10 we saw that a sufficient condition for all potentials to be bounded is that α be chosen so that $\alpha_i \geq kN_{ii}$ for all i. The purpose of this section is to show that unbounded potentials may exist when this hypothesis is not satisfied; the potential we exhibit will have a bounded charge whose support is at the same time an equilibrium set and a dual equilibrium set.

The chain P will be a modification of sums of independent random variables on the line with $p_1 = \frac{1}{3}$ and $p_{-1} = \frac{2}{3}$. Let

$$\left.\begin{aligned} P_{i,i-1} &= \tfrac{2}{3} \\ P_{i,i+1} &= \tfrac{1}{3} \end{aligned}\right\} \quad \text{for } i \leq 0$$

and

$$\left.\begin{aligned} P_{ii} &= 1 - \frac{1}{i} \\ P_{i,i-1} &= \frac{2}{3i} \\ P_{i,i+1} &= \frac{1}{3i} \end{aligned}\right\} \quad \text{for } i > 0.$$

If the process is watched only when it changes states, it becomes the $p_1 = \frac{1}{3}$, $p_{-1} = \frac{2}{3}$ process, so that we may compute H from the latter chain.

$$H_{ij} = \begin{cases} 1 & \text{if } i \geq j \\ (\frac{1}{2})^{j-i} & \text{if } i < j. \end{cases}$$

Therefore,

$$\bar{H}_{ii} = P_{i,i+1}\cdot 1 + P_{i,i-1}\cdot\tfrac{1}{2} + P_{ii}\cdot 1 = 1 - \tfrac{1}{2}P_{i,i-1},$$

and

$$N_{jj} = \frac{1}{1 - \bar{H}_{jj}} = \frac{2}{P_{j,j-1}}.$$

Hence

$$N_{ij} = \begin{cases} 3 & \text{if } j \le 0, \quad i \ge j \\ 3(\frac{1}{2})^{j-i} & \text{if } j \le 0, \quad i < j \\ 3j & \text{if } j > 0, \quad i \ge j \\ 3j(\frac{1}{2})^{j-i} & \text{if } j > 0, \quad i < j. \end{cases}$$

Let $E = \{1, 2, 3, \ldots\}$. Since the process goes toward $-\infty$ with probability one, it can be in E only finitely often a.e. Moreover, $e_i^E = 0$ unless $i = 1$, so that $\alpha e^E < \infty$ for any choice of α. Thus E is an equilibrium set.

We shall take α to be the zeroth row of N. Then

$$\alpha_j = N_{0j} = \begin{cases} 3 & \text{if } j \le 0 \\ 3j(\frac{1}{2})^j & \text{if } j > 0. \end{cases}$$

If we calculate $\hat{P}$, we find that

$$\hat{P}_{ij} = P_{ij} \quad \text{for } i > 0,$$

$$\left.\begin{aligned} \hat{P}_{i,i+1} &= \tfrac{2}{3} \\ \hat{P}_{i,i-1} &= \tfrac{1}{3} \end{aligned}\right\} \quad \text{for } i < 0,$$

and

$$\hat{P}_{01} = \hat{P}_{0,-1} = \tfrac{1}{3}.$$

With probability one the $\hat{P}$ process reaches 0 from all states, and from there it can disappear. Hence the extended chain for $\hat{P}$ is absorbing, and $\hat{P}$ is in any set only finitely often a.e. As before, $\alpha\hat{e}^E < \infty$, and E is therefore an equilibrium set for $\hat{P}$.

Thus E is both an equilibrium set and a dual equilibrium set. We shall choose a bounded charge with support in E. Let

$$f_i = \begin{cases} 1 & \text{if } i \in E \\ 0 & \text{otherwise.} \end{cases}$$

Then

$$\alpha f = \sum_{j>0} 3j(\tfrac{1}{2})^j = \frac{3(\frac{1}{2})}{(1 - \frac{1}{2})^2} = 6.$$

Thus f is a charge. Its potential is

$$g_i = \sum_{j>0} N_{ij} = \begin{cases} \displaystyle\sum_{j>0} 3j(\tfrac{1}{2})^{j-i} & \text{if } i \le 0 \\ \displaystyle\sum_{j=1}^{i} 3j + \sum_{j>i} 3j(\tfrac{1}{2})^{j-i} & \text{if } i > 0. \end{cases}$$

Summing these expressions, we find that

$$g_i = \begin{cases} 6(\frac{1}{2})^{|i|} & \text{for } i \le 0 \\ \frac{3}{2}(i^2 + 3i + 4) & \text{for } i > 0. \end{cases}$$

Thus $\lim_{i \to +\infty} g_i = +\infty$, and g is unbounded.

We note that $P^n g$ for large n is a weighted sum of g values along paths that the process is likely to take. Thus the fact that $\lim_{i \to +\infty} g_i = +\infty$ does not contradict $P^n g \to 0$ since the process moves toward $+\infty$ with probability zero. On the other hand, in the direction that the process does go, namely $-\infty$, we do have $\lim_{i \to -\infty} g_i = 0$.

8. Applications of potential-theoretic methods

Many useful quantities for transient chains arise as means of non-negative random variables: $h_i = \mathrm{M}_i[\mathbf{z}]$. To compute h we can often use a systems theorem argument (Theorem 4-11 with the random time identically one) to obtain an expression of the form $h = Ph + f$. Under appropriate circumstances we may write $(I - P)h = f$, and if h can be shown to be a potential, we conclude $h = Nf$. The purpose of this section is to give some sufficient conditions under which all these steps are valid and to apply the results.

We first restrict our attention to the case of a bounded non-negative random variable $\mathbf{z}$. Later in this section we extend our results to obtain Theorem 8-67, which is a powerful tool applicable even if the vector h is not necessarily finite-valued.

To maximize the number of potentials, we choose a row of N as α. Then all finite-valued functions of the form Nf are potentials, and a set E is an equilibrium set if and only if the process is in E only finitely often a.e. Let $\mathbf{z}$ be a bounded non-negative random variable, and let $\mathbf{z}^{(n)}(\omega) = \mathbf{z}(\omega_n)$. We shall assume that $\mathbf{z}^{(1)} \le \mathbf{z}$. Then $\mathrm{M}_i[\mathbf{z}]$ is finite for all i, and $\mathrm{M}_i[\mathbf{z}^{(1)}] \le \mathrm{M}_i[\mathbf{z}]$. Define

$$h_i = \mathrm{M}_i[\mathbf{z}]$$

and

$$f_i = \mathrm{M}_i[\mathbf{z} - \mathbf{z}^{(1)}] \ge 0.$$

Lemma 8-62: The column vector h is superregular and satisfies $(I - P)h = f$. Furthermore, $\mathbf{z}^{(\infty)} = \lim_{n \to \infty} \mathbf{z}^{(n)}$ exists and is finite.

PROOF: By Theorem 4-11 with the random time taken to be identically one, we have

$$\mathrm{M}_i[\mathbf{z}^{(1)}] = \sum_k P_{ik} \mathrm{M}_k[\mathbf{z}] = (Ph)_i.$$

Therefore,

$$[(I - P)h]_i = \mathrm{M}_i[\mathbf{z}] - \mathrm{M}_i[\mathbf{z}^{(1)}] = \mathrm{M}_i[\mathbf{z} - \mathbf{z}^{(1)}]$$
$$= f_i.$$

Since $f \geq 0$, h is superregular; and since $\mathbf{z}^{(1)} \leq \mathbf{z}$, we have

$$\mathbf{z} \geq \mathbf{z}^{(1)} \geq \mathbf{z}^{(2)} \geq \cdots \geq 0.$$

Therefore $\mathbf{z}^{(\infty)}$ exists and is finite.

Lemma 8-63: The function h satisfies $h = Nf$ if and only if $\mathbf{z}^{(\infty)} = 0$ almost everywhere.

PROOF: By dominated convergence,

$$\mathrm{M}_i[\mathbf{z}^{(\infty)}] = \lim \mathrm{M}_i[\mathbf{z}^{(n)}],$$

and by Theorem 4-11 with the random time n,

$$\{\mathrm{M}_i[\mathbf{z}^{(n)}]\} = P^n\{\mathrm{M}_i[\mathbf{z}]\} = P^n h.$$

Hence

$$\mathrm{M}_i[\mathbf{z}^{(\infty)}] = \lim P^n h.$$

By Theorem 5-10,

$$h_i = (Nf)_i + \mathrm{M}_i[\mathbf{z}^{(\infty)}].$$

Thus $h = Nf$ if and only if $\mathrm{M}_i[\mathbf{z}^{(\infty)}] = 0$ for every i. Since $\mathbf{z}^{(\infty)} \geq 0$, $\mathrm{M}_i[\mathbf{z}^{(\infty)}] = 0$ for all i if and only if $\mathbf{z}^{(\infty)} = 0$ a.e.

Lemma 8-64: If either of these conditions is satisfied, then $h = Nf$:

(1) There exists an equilibrium set E such that $\mathbf{z}(\omega) = 0$ for every path ω which does not go through E.
(2) The enlarged chain is absorbing and $\mathbf{z}(\omega) = 0$ for every path ω which begins in the absorbing state.

PROOF: The second condition is just the first for E, the set of all transient states. For the first condition, on every path which does not pass through E, $\mathbf{z}(\omega) = 0$, so that $\mathbf{z}^{(n)}(\omega) = 0$ for every $n \geq 1$ on such paths. On almost all paths which do pass through E, there is an n which is a function of the path and which denotes the last time the process passes through E on that path. Therefore, on almost all paths there is an n depending on the path such that $\mathbf{z}^{(n)}(\omega) = 0$. Hence $\mathbf{z}^{(\infty)} = 0$ a.e., and the result follows by Lemma 8-63.

We shall now generalize our considerations. Let $\mathbf{z}(\omega)$ be a non-negative random variable with $h_i = \mathrm{M}_i[\mathbf{z}]$ not necessarily finite. Suppose $\mathbf{z} \geq \mathbf{z}^{(1)}$, and define ${}^m\mathbf{z}(\omega) = \min(m, \mathbf{z}(\omega))$. We agree to set $\mathbf{z} - \mathbf{z}^{(1)} = 0$ at all points where $\mathbf{z}^{(1)}$ is infinite. Then

$$\mathbf{z}(\omega) = \lim_{m \to \infty} {}^m\mathbf{z}(\omega).$$

The crucial property of the functions ${}^m\mathbf{z}$ is that

$$({}^m\mathbf{z})^{(1)} = {}^m(\mathbf{z}^{(1)}).$$

We denote the common value of $({}^m\mathbf{z})^{(1)}$ and ${}^m(\mathbf{z}^{(1)})$ by ${}^m\mathbf{z}^{(1)}$, and we define ${}^m\mathbf{z}^{(n)}$ analogously.

Lemma 8-65: If $\mathbf{z} \geq \mathbf{z}^{(1)} \geq 0$, then $\mathbf{z} - \mathbf{z}^{(1)} \geq {}^m\mathbf{z} - {}^m\mathbf{z}^{(1)} \geq 0$.

PROOF:

$${}^m\mathbf{z} - {}^m\mathbf{z}^{(1)} = \begin{cases} \mathbf{z} - \mathbf{z}^{(1)} & \text{if } \mathbf{z} \leq m \\ 0 & \text{if } \mathbf{z}^{(1)} > m \\ m - \mathbf{z}^{(1)} & \text{if } \mathbf{z} > m \geq \mathbf{z}^{(1)}. \end{cases}$$

Define vectors ${}^mh$ and ${}^mf$ by

$${}^mh_i = \mathrm{M}_i[{}^m\mathbf{z}]$$

and

$${}^mf_i = \mathrm{M}_i[{}^m\mathbf{z} - {}^m\mathbf{z}^{(1)}].$$

Lemma 8-66: ${}^{m+1}f \geq {}^mf$ and $\lim_m {}^mf = f$.

PROOF: We note that ${}^m({}^{m+1}\mathbf{z}) = {}^m\mathbf{z}$. Applying Lemma 8-65 to ${}^{m+1}\mathbf{z}$, we find that ${}^{m+1}\mathbf{z} - {}^{m+1}\mathbf{z}^{(1)} \geq {}^m\mathbf{z} - {}^m\mathbf{z}^{(1)}$. Then ${}^{m+1}f_i \geq {}^mf_i$. Since $\lim ({}^m\mathbf{z} - {}^m\mathbf{z}^{(1)}) = \mathbf{z} - \mathbf{z}^{(1)}$ a.e., $\lim {}^mf = f$ by the Monotone Convergence Theorem.

Theorem 8-67: Let $\mathbf{z}$ be a non-negative random variable. Define

$$\mathbf{z}^{(1)}(\omega) = \mathbf{z}(\omega_1)$$
$$\mathbf{z}(\omega) - \mathbf{z}^{(1)}(\omega) = 0 \quad \text{when } \mathbf{z}^{(1)}(\omega) = +\infty$$
$$h_i = \mathrm{M}_i[\mathbf{z}]$$
$$f_i = \mathrm{M}_i[\mathbf{z} - \mathbf{z}^{(1)}],$$

and suppose that $\mathbf{z} \geq \mathbf{z}^{(1)}$. If $\mathbf{z}$ satisfies either one of the following conditions, then $h = Nf$:

(1) There exists an equilibrium set E such that $\mathbf{z}(\omega) = 0$ for every path ω which does not go through E.

(2) The extended chain is absorbing and $\mathbf{z}(\omega) = 0$ for every path ω which begins in the absorbing state.

Proof: For each m, ${}^m\mathbf{z}$ is bounded. If one of the conditions applies to $\mathbf{z}$, then it applies to ${}^m\mathbf{z}$ because $0 \le {}^m\mathbf{z} \le \mathbf{z}$. Hence by Lemma 8-64,

$$ {}^m h = N\,{}^m f. $$

By Lemma 8-66, ${}^m f$ increases to f, and by monotone convergence ${}^m h$ increases to h. Therefore $h = Nf$ by the Monotone Convergence Theorem.

We now apply Theorem 8-67 in four special cases.

First let P be an absorbing chain with fundamental matrix N. We define $a_i^{(r)} = \mathrm{M}_i[\mathbf{a}(\omega)^r]$, where $\mathbf{a}(\omega)$ is the absorption time defined in Chapter 5. The column vector $a^{(r)}$ is indexed by the transient states of P.

Proposition 8-68: If P is an absorbing chain, then

$$ a^{(r)} = \sum_{m=1}^{r-1} \binom{r}{m} NQa^{(m)} + N1. $$

Proof: Start the process in a transient state i, and let $\mathbf{a}$ be the absorption time. Since i is transient,

$$ \mathbf{a}^r(\omega) = (\mathbf{a} + 1)^r(\omega_1) $$

so that

$$ \mathbf{a}^r(\omega) - \mathbf{a}^r(\omega_1) = (\mathbf{a} + 1)^r(\omega_1) - \mathbf{a}^r(\omega_1) $$

or

$$ (\mathbf{a}^r - (\mathbf{a}^r)^{(1)})(\omega) = ((\mathbf{a} + 1)^r - \mathbf{a}^r)(\omega_1). $$

By Theorem 4-11 with the random time identically one,

$$ \begin{aligned} \mathrm{M}_i[\mathbf{a}^r - (\mathbf{a}^r)^{(1)}] &= \sum_k P_{ik}\mathrm{M}_k[(\mathbf{a} + 1)^r - \mathbf{a}^r] \\ &= \sum_{k\text{ abs.}} R_{ik}\mathrm{M}_k[1] + \sum_{k\text{ trans.}} Q_{ik}\mathrm{M}_k\left[\sum_{m=0}^{r-1} \binom{r}{m} \mathbf{a}^m\right] \\ &= \sum_{m=1}^{r-1} \binom{r}{m} \sum_k Q_{ik}a_k^{(m)} + 1 \end{aligned} $$

since $\sum R_{ik} + \sum Q_{ik} = 1$. In Theorem 8-67 let $\mathbf{z} = \mathbf{a}^r$. Then $\mathbf{z}$

satisfies $\mathbf{z} \geq \mathbf{z}^{(1)}$, and condition (2) of the theorem holds for $\mathbf{z}$. As we have just shown,

$$f = \sum_{m=1}^{r-1} \binom{r}{m} Qa^{(m)} + 1,$$

and we also have $h = a^{(r)}$. Hence $h = Nf$ by the theorem.

Corollary 8-69: Let P be an absorbing chain. Then there exist real numbers c and d such that

$$a^{(r)} \leq cNa^{(r-1)} \leq da^{(r)}.$$

In particular, $a^{(r)}$ is finite-valued if and only if $Na^{(r-1)}$ is finite-valued.

PROOF: By Proposition 8-68,

$$a^{(r)} = N \sum_{m=1}^{r-1} \binom{r}{m} Qa^{(m)} + N\mathbf{1},$$

even if both sides are infinite. Hence

$$\begin{aligned} a^{(r)} &\leq (2^r - 2)NQa^{(r-1)} + N\mathbf{1} && \text{since } a^{(m)} \leq a^{(r-1)} \\ &= (2^r - 2)(N - I)a^{(r-1)} + N\mathbf{1} && \text{since } N - I = NQ \\ &\leq (2^r - 2)Na^{(r-1)} + N\mathbf{1} \\ &\leq (2^r - 1)Na^{(r-1)} && \text{since } \mathbf{1} \leq a^{(r-1)}. \end{aligned}$$

For the other inequality,

$$\begin{aligned} a^{(r)} &\geq NQa^{(r-1)} \\ &= Na^{(r-1)} - a^{(r-1)} \\ &\geq Na^{(r-1)} - a^{(r)} \quad \text{since } a^{(r-1)} \leq a^{(r)}. \end{aligned}$$

Hence $Na^{(r-1)} \leq 2a^{(r)}$.

As a second example, let P be a transient chain, and for any two transient states i and j define

$$W_{ij} = \mathrm{M}_i[\mathbf{n}_j^2].$$

The reader may verify with the aid of Theorem 4-11 that if $\mathbf{z} = \mathbf{n}_j^2$, then

$$\mathrm{M}_i[\mathbf{z} - \mathbf{z}^{(1)}] = (2N_{dg} - I)_{ij}.$$

Now $\{j\}$ is an equilibrium set, and $\mathbf{n}_j^2 = 0$ for all paths not going through this set. Hence, for fixed j, the column vector h with $h_i = W_{ij}$ satisfies condition (1) of Theorem 8-67. Therefore,

$$W = N(2N_{dg} - I).$$

We note that W is finite-valued; similarly one shows that $\mathrm{M}_i[\mathbf{n}_j{}^r] < \infty$ for $r > 2$.

Next let E be an equilibrium set and let $\mathbf{z}$ be the time at which the process is in E for the last time (or 0 if E is never reached). Set $v^E = \mathrm{M}_i[\mathbf{z}]$. Then

$$\mathbf{z} - \mathbf{z}^{(1)} = \begin{cases} 1 & \text{if } E \text{ is ever reached after time } 0 \\ 0 & \text{otherwise.} \end{cases}$$

Hence $\mathrm{M}_i[\mathbf{z} - \mathbf{z}^{(1)}] = \bar{h}_i^E$, the probability that, from i, E is reached after time 0. The random variable $\mathbf{z}$ satisfies condition (1) of the theorem. Hence

$$v^E = N\bar{h}^E.$$

Finally, let E be an equilibrium set and let j be any state. Let $\mathbf{z}_j$ be the number of times in j before E is left for the last time (or 0 if E is never reached). Define

$$N_{ij}^E = \mathrm{M}_i[\mathbf{z}_j].$$

Then $\mathbf{z}_j$ satisfies condition (1) of the theorem, and we have

$$\mathbf{z} - \mathbf{z}^{(1)} = \begin{cases} 1 & \text{if } x_0(\omega) = j \text{ and } E \text{ is ever reached} \\ 0 & \text{otherwise.} \end{cases}$$

Hence

$$\mathrm{M}_i[\mathbf{z} - \mathbf{z}^{(1)}] = \delta_{ij}h_j^E,$$

and

$$N_{ij}^E = N\{\delta_{ij}h_j^E\} = N_{ij}h_j^E.$$

9. General denumerable stochastic processes

We shall show that any denumerable stochastic process can be represented within a transient Markov chain in such a manner that potential theory applied to the chain yields corresponding results for the stochastic process.

Throughout this section we shall deal with a probability space Ω with measure μ, and a fixed sequence $\{\mathscr{R}_i\}$ of partitions of Ω. Each $\mathscr{R}_i$ has a denumerable number of cells, and $\mathscr{R}_i \subset \mathscr{R}_{i+1}$. For convenience, we assume that $\mathscr{R}_0 = \{\Omega\}$. We recall that $(f_n, \mathscr{R}_n)$ is a stochastic process if f_n is constant on each cell of $\mathscr{R}_n$. (This condition is Definition 2-5 expressed in terms of partitions.) If $U \in \mathscr{R}_n$, define $f_n(U)$ to be this constant value.

Definition 8-70: If $\{\mathscr{R}_i\}$ is a sequence of partitions, the **space-time Markov chain** for $\{\mathscr{R}_i\}$ is defined to be a Markov chain whose states are all ordered pairs $\langle U, n\rangle$, where $U \in \mathscr{R}_n$ and $\mu(U) > 0$, and whose transition probabilities are

$$P_{\langle U,n\rangle,\langle V,m\rangle} = \begin{cases} \dfrac{\mu(V)}{\mu(U)} & \text{if } m = n + 1 \text{ and } V \subset U \\ 0 & \text{otherwise.} \end{cases}$$

The chain is started in the state $\langle \Omega, 0\rangle$, which will be called state 0.

Proposition 8-71: The space-time chain is transient, and if $i = \langle U, n\rangle$, then

$$N_{0i} = H_{0i} = P_{0i}^{(n)} = \mu(U).$$

PROOF: State i can be entered only on the nth step, as is clear from Definition 8-70. Hence the chain is in i at most once, and $N_{0i} = H_{0i} = P_{0i}^{(n)}$. Along the path from 0 to i there is a unique sequence of cells

$$U \subset U_{n-1} \subset U_{n-2} \subset \cdots \subset U_1 \subset \Omega \quad \text{with } U_k \in \mathscr{R}_k.$$

Then

$$\begin{aligned} P_{0i}^{(n)} &= P_{0\langle U_1,1\rangle} P_{\langle U_1,1\rangle,\langle U_2,2\rangle} \cdots P_{\langle U_{n-1},n-1\rangle,\langle U,n\rangle} \\ &= \frac{\mu(U_1)}{\mu(\Omega)} \frac{\mu(U_2)}{\mu(U_1)} \cdots \frac{\mu(U)}{\mu(U_{n-1})} = \mu(U). \end{aligned}$$

For example, let Ω be a sequence space with some probability measure μ, and let $\mathscr{R}_n$ be the partition such that $\mathscr{R}_n^* = \mathscr{F}_n$. As usual, we may think of a cell of $\mathscr{F}_n$ as a path $\langle i_1, i_2, \ldots, i_n\rangle$ of length n in Ω. A state of the space-time chain may also be thought of as such a path, and the chain moves from $\langle i_1, i_2, \ldots, i_n\rangle$ to $\langle j_1, j_2, \ldots, j_{n+1}\rangle$ only if $j_k = i_k$ for $1 \le k \le n$. The probability of such a transition is

$$\Pr[x_{n+1} = j_{n+1} \mid x_1 = i_1 \wedge \cdots \wedge x_n = i_n].$$

The starting state 0 may be thought of as the empty sequence.

Definition 8-72: If $(f_n, \mathscr{R}_n)$ is a stochastic process, then the function f defined on the states of a space-time chain by

$$f(\langle U, n\rangle) = f_n(U)$$

is said to **correspond to** the process $(f_n, \mathscr{R}_n)$.

We write $f \sim (f_n, \mathscr{R}_n)$ when f corresponds to $(f_n, \mathscr{R}_n)$. If we identify two stochastic processes $(f_n, \mathscr{R}_n)$ and $(g_n, \mathscr{R}_n)$ when $f_n = g_n$ a.e. for every n, then the correspondence between all stochastic processes on $\{\mathscr{R}_n\}$ and all functions on the states of the space-time chain is one-one. We now restrict ourselves to the case where the f_n are real-valued functions. Under the following definition the correspondence preserves inequalities, linear combinations, and limits.

Definition 8-73: Operations on stochastic processes are defined by

(1) $(f_n, \mathscr{R}_n) \leq (g_n, \mathscr{R}_n)$ if $f_n \leq g_n$ a.e. for all n.
(2) $a(f_n, \mathscr{R}_n) + b(g_n, \mathscr{R}_n) = (af_n + bg_n, \mathscr{R}_n)$.
(3) $\lim_k (f_n{}^{(k)}, \mathscr{R}_n) = (f_n, \mathscr{R}_n)$ if $\lim f_n{}^{(k)} = f_n$ a.e. for all n.

Lemma 8-74: If $(f_n, \mathscr{R}_n) \sim f$, then

$$(\mathrm{M}[f_{n+k} \mid \mathscr{R}_n], \mathscr{R}_n) \sim P^k f$$

in the sense that if either quantity is well defined, then so is the other, and if they are both well-defined, then they correspond.

Proof: We shall proceed by induction on k. If $k = 0$, the result is trivial. Suppose that both quantities exist for some k and that they correspond. Then

$$(P^{k+1}f)_i = \sum_j P_{ij}(P^k f)_j$$

or

$$(P^{k+1}f)_{\langle U,n\rangle} = \sum_{\substack{V \subset U \\ V \in \mathscr{R}_{n+1}}} \frac{\mu(V)}{\mu(U)} (P^k f)_{\langle V,n+1\rangle}.$$

By inductive hypothesis, $(\mathrm{M}[f_{n+k} \mid \mathscr{R}_n], \mathscr{R}_n) \sim P^k f$. Hence, by definition of the correspondence,

$$\begin{aligned}(P^k f)_{\langle V,n+1\rangle} &= \mathrm{M}[f_{n+1+k} \mid \mathscr{R}_{n+1}](V) \\ &= \frac{1}{\mu(V)} \sum_{\substack{W \subset V \\ W \in \mathscr{R}_{n+1+k}}} \mu(W) f_{n+1+k}(W),\end{aligned}$$

and

$$\begin{aligned}(P^{k+1}f)_{\langle U,n\rangle} &= \sum_{\substack{V \subset U \\ V \in \mathscr{R}_{n+1}}} \frac{\mu(V)}{\mu(U)} \sum_{\substack{W \subset V \\ W \in \mathscr{R}_{n+1+k}}} \frac{\mu(W)}{\mu(V)} f_{n+1+k}(W) \\ &= \frac{1}{\mu(U)} \sum_{\substack{W \subset U \\ W \in \mathscr{R}_{n+1+k}}} \mu(W) f_{n+1+k}(W) \\ &= \mathrm{M}[f_{n+1+k} \mid \mathscr{R}_n](U).\end{aligned}$$

That is, $P^{k+1}f$ exists if and only if $\mathrm{M}[f_{n+(k+1)} \mid \mathscr{R}_n]$ does, and if they exist, then they correspond.

Proposition 8-75: If $(h_n, \mathscr{R}_n) \sim h$, then $(h_n, \mathscr{R}_n)$ is a supermartingale (martingale) if and only if h is superregular (regular) and $P^k h$ is finite-valued for all k.

PROOF: If h is superregular, then $Ph \leq h$. Hence, by Lemma 8-74, $\mathrm{M}[h_{n+1} \mid \mathscr{R}_n] \leq h_n$ a.e. for all n. If $(P^k h)$ is finite-valued, then

$$(P^k h)_0 = \sum_{U \in \mathscr{R}_k} \mu(U) h_k(U) = \mathrm{M}[h_k]$$

is finite. Since h_n is constant on cells of $\mathscr{R}_n$, Definition 3-5 is satisfied.

Conversely, if $(h_n, \mathscr{R}_n)$ is a supermartingale, then $\mathrm{M}[h_{n+1} \mid \mathscr{R}_n] \leq h_n$, and hence $Ph \leq h$ by Lemma 8-74. Moreover, $\mathrm{M}[h_{k+n}] = (P^{k+n}h)_0$ is finite. If $i = \langle U, n \rangle$, then $P^{(n)}_{0i} = \mu(U) > 0$. From

$$(P^{k+n}h)_0 = \sum_j P^{(n)}_{0j} (P^k h)_j,$$

we see that $(P^k h)_i$ must be finite. The proof for martingales simply replaces $Ph \leq h$ by $Ph = h$.

Definition 8-76: $(f_n, \mathscr{R}_n)$ is a **stochastic process charge** with **potential** $(g_n, \mathscr{R}_n)$ if $\sum_n \mathrm{M}[|f_n|] < \infty$ and if $(g_n, \mathscr{R}_n) = \sum_k (\mathrm{M}[f_{n+k} \mid \mathscr{R}_n], \mathscr{R}_n)$. If, in addition, $f_n \geq 0$ then the potential is called a **pure potential.**

We shall make use of potential theory results for the space-time chain P. As the distinguished measure α, we select $\alpha_j = N_{0j} > 0$. As usual, if αf is finite, then f is a charge.

Proposition 8-77: Charge functions correspond to charge stochastic processes, and their potentials also correspond.

PROOF: If $(f_n, \mathscr{R}_n) \sim f$, then by Proposition 8-71

$$\begin{aligned} \alpha|f| &= \sum_{\langle U, n \rangle} \mu(U) \cdot \mathrm{M}[|f_n| \mid \mathscr{R}_n](U) \\ &= \sum_n \sum_{U \in \mathscr{R}_n} \mu(U) \cdot |f_n(U)| \\ &= \sum_n \mathrm{M}[|f_n|]. \end{aligned}$$

Since sums and limits are preserved, we have by Proposition 8-74

$$g = Nf = \sum_k P^k f \sim \sum_k (\mathrm{M}[f_{n+k} \mid \mathscr{R}_n], \mathscr{R}_n).$$

Thus g is the potential of f if and only if the two conditions of Definition 8-76 are fulfilled by $(f_n, \mathscr{R}_n)$ and $(g_n, \mathscr{R}_n)$.

We give two applications of the correspondence. The first is a decomposition of non-negative supermartingales.

Proposition 8-78: If $(h_n, \mathscr{R}_n)$ is a non-negative supermartingale, then there is a unique representation

$$(h_n, \mathscr{R}_n) = (r_n, \mathscr{R}_n) + (g_n, \mathscr{R}_n)$$

of $(h_n, \mathscr{R}_n)$ as the sum of a martingale and a potential. In the representation the martingale is a non-negative martingale, the potential is a pure potential, and r_n satisfies $r_n = \lim_k \mathrm{M}[h_{n+k} \mid \mathscr{R}_n]$. Moreover, $(g_n, \mathscr{R}_n)$ is the difference between a martingale and a process consisting of an increasing sequence of random variables.

PROOF: Existence and uniqueness of the representation follows immediately from Theorem 5-10 and Propositions 8-75 and 8-77. Then $r_n = \lim_k \mathrm{M}[h_{n+k} \mid \mathscr{R}_n]$ by Lemma 8-74. For the last part, let $(f_n, \mathscr{R}_n)$ be the charge of $(g_n, \mathscr{R}_n)$. Set

$$s_n = f_0 + \cdots + f_{n-1}$$

and

$$s = \lim s_n.$$

Then s_n increases monotonically to s, and

$$\mathrm{M}[s] \leq \sum_n \mathrm{M}[f_n \mid \mathscr{R}_n] < \infty$$

by monotone convergence. Since $f_n \geq 0$,

$$\begin{aligned} g_n &= \sum_k \mathrm{M}[f_{n+k} \mid \mathscr{R}_n] = \mathrm{M}\Big[\sum_k f_{n+k} \mid \mathscr{R}_n\Big] \\ &= \mathrm{M}[s - s_n \mid \mathscr{R}_n] = \mathrm{M}[s \mid \mathscr{R}_n] - s_n. \end{aligned}$$

Hence $\{g_n\}$ is the difference between the martingale $\{\mathrm{M}[s \mid \mathscr{R}_n]\}$ and the increasing sequence s_n.

As the second application, we give a proof of the Upcrossing Lemma, Proposition 3-11, as it applies to non-negative supermartingales. The present estimate is better than the one in Chapter 3.

Proposition 8-79: Let r and s be real numbers with $0 \leq r < s$. Let $\boldsymbol{\beta}(\omega)$ be the number of upcrossings on ω of $[r, s]$ by the non-negative supermartingale $(\mathbf{f}_k, \mathscr{R}_k)$ up to time n. Then

$$\mathrm{M}[\boldsymbol{\beta}] \leq \frac{r}{s - r}.$$

Proof: Let $f \sim (\mathbf{f}_k, \mathcal{R}_n)$; f is non-negative superregular. Let E and F be the sets of states in the space-time chain defined by

$$E = \{\langle U, m\rangle \mid m \le n \quad \text{and} \quad \mathbf{f}_m(U) \le r\}$$

$$F = \{\langle U, m\rangle \mid m \le n \quad \text{and} \quad \mathbf{f}_m(U) \ge s\}.$$

Hence $f_i \le r$ for $i \in E$ and $f_j \ge s$ for $j \in F$. For any other state, $i = \langle U, m\rangle$ with $m \le n$, $r < f_i < s$. Now $h_0^F = (B^F 1)_0$ is the probability that the random variables $\mathbf{f}_0, \ldots, \mathbf{f}_n$ ever take on a value greater than or equal to s. Similarly, $(B^E B^F 1)_0$ is the probability of at least one upcrossing, and in general $[(B^E B^F)^k 1]_0$ is the probability of at least k upcrossings by $\mathbf{f}_0, \mathbf{f}_1, \ldots, \mathbf{f}_n$. Hence

$$\mathrm{M}[\boldsymbol{\beta}] = \sum_{k=1}^{\infty} \Pr[\boldsymbol{\beta} \ge k] = \sum_{k=1}^{\infty} [(B^E B^F)^k 1]_0.$$

Since the chain cannot be in F after time n, F is an equilibrium set, and $h^F = B^F 1$ is a potential. For $i \in F$

$$(B^F 1)_i = 1 \le \frac{1}{s} f_i.$$

By the Principle of Domination (Theorem 8-45),

$$(B^F 1) \le \frac{1}{s} f$$

everywhere. Hence $B^E B^F 1 \le (1/s) B^E f$. For $i \in E$,

$$(B^E f)_i \le r.$$

Since f is superregular and since $r1$ is superregular,

$$B^E f \le r1$$

everywhere by conclusion (2) of Proposition 8-16. Thus

$$B^E B^F 1 \le \frac{r}{s} 1.$$

By induction,

$$(B^E B^F)^k 1 \le \left(\frac{r}{s}\right)^k 1,$$

and hence

$$\mathrm{M}[\boldsymbol{\beta}] \le \sum_{k=1}^{\infty} \left(\frac{r}{s}\right)^k = \frac{r}{s - r}.$$

10. Problems

1. If P is a transient chain whose states communicate and if $\alpha > 0$ is superregular, show that
   (a) $\hat{P}$ stops (disappears) a.e. if and only if α is a potential measure.
   (b) The mean stopping time is finite if and only if α is a charge. [*Hint:* Adjoin an absorbing state to $\hat{P}$.]

2. Let P be a recurrent chain, and let E be a finite set. Show that for any specified values of h_E there is a unique bounded function h with the specified values which is regular on $\tilde{E}$. Show that h takes on its maximum on E.

3. Illustrate the result of the previous problem for the symmetric random walk on the integers with the specified values $h_0 = 0$ and $h_1 = 1$.

4. Let P have only transient states, let π be a specified probability vector and let $\alpha = \pi N$. Let $f \geq 0$ be a charge, and define the random variable

$$\mathbf{s} = \sum_{n=0}^{\infty} f(x_n).$$

   Show that $\mathbf{s}$ is finite a.e. and that $\mathrm{M}_\pi[\mathbf{s}] = \alpha f$. What is the value of $\mathrm{M}_i[\mathbf{s}]$? Give a game-interpretation.

5. In the framework of Problem 4, introduce a second charge $\bar{f}$, its potential $\bar{g}$, and $\bar{\mathbf{s}}$. Prove that if $\mu = \text{dual } f$, then

$$(g, \bar{g}) = \tfrac{1}{2}\mathrm{M}_\pi[\mathbf{s}\bar{\mathbf{s}}] + \tfrac{1}{2}\mu\bar{f}.$$

   Find the corresponding expression for $\mathbf{I}(g)$.

6. If P is a chain with only transient states, let Var_{ij} be the variance of $\mathbf{n}_j$ for the process started at i. Show that

$$\mathrm{Var}_{ij} = N_{ij}(2N_{jj} - N_{ij} - 1).$$

7. In the framework of Problem 6, if $P\mathbf{1} = \mathbf{1}$, prove that for each j there is an i such that $\mathrm{Var}_{ij} \geq N_{ij}$.

Problems 8 to 19 refer to the following Markov chain: The states are the non-negative integers. From state i either the process moves one step to the right with probability $p_i > 0$, or it remains at i.

8. Find H and N.

9. Give a *simple* characterization of
   (a) the regular functions,
   (b) the non-negative superregular functions,
   (c) the pure potentials, where $\alpha_j = N_{0j}$,
   (d) the potentials, where $\alpha_j = N_{0j}$.

10. What does Theorem 5-10 say about this chain?

11. If g is a potential with charge f, give a simple characterization in terms of g of the support of f. Of the support of f^+.

12. Use Problems 9 and 11 to verify that Theorem 8-45 holds for this chain [*Hint:* Distinguish the cases where the support of f^+ is finite and where it is infinite.]

13. Use Problems 9 and 11 to construct a counterexample to Theorem 8-45 if the assumption $h \geq 0$ is omitted.
14. If $\alpha_j = N_{0j}$, form $\hat{P}$ and compute $\hat{N}$.
15. For $E = \{0, 1, 2\}$, find $\bar{h}^E$. [See the end of Section 8.] Show that $v^E = N\bar{h}^E$ has the desired interpretation. [Remember that the chain starts at time 0, not at time 1.]
16. Show that $C(E) = 1$ for all equilibrium sets. Show that both the hypothesis and the conclusion of Proposition 8-39 are false for P.
17. Show that $\hat{C}(E) = 1$ for all sets E. Show that both the hypothesis and the conclusion of Proposition 8-39 hold for $\hat{P}$.
18. Let $E = \{0, 1, \ldots, n\}$. Find B^E. If g is a potential, what is the form of its balayage potential on E? Show that Theorem 8-46 is satisfied.
19. Show that if $\mu = \text{dual } f$, then

$$\mathbf{I}(g) = \frac{1}{2}\left(\sum_i \mu_i\right)^2 + \frac{1}{2}\sum_i \mu_i^2.$$

Problems 20 to 30 refer to sums of independent random variables on the integers with $p_{-1} = \frac{1}{3}$ and $p_1 = \frac{2}{3}$. Use the results of Problems 12 to 19 in Chapter 5.

20. Show that there are two essentially different positive regular measures and that all regular measures are linear combinations of the two basic measures.
21. Show that if $f \geq 0$ and $\alpha f < +\infty$ for either α, then $g = Nf$ is finite-valued. Show also that $\lim P^n g = 0$.
22. Let $E = \{0, 1, \ldots, n\}$. Compute B^E. Choose a non-negative super-regular function h (not a constant), and verify the various parts of Proposition 8-16.
23. For E as above, compute e^E, and verify that $Ne^E = h^E$.
24. For E as above, compute $C(E) = \alpha e^E$ for each of the two basic measures. What happens as n increases?
25. Form $\hat{P}$ and compute $\hat{N}$ for each of the two basic measures. In each case, does $\hat{N}$ have columns tending to 0?
26. Use the results of the last two problems to show that each assignment of capacities is consistent with Proposition 8-39, even though $\lim_n C(E)$ is finite in one case and not in the other.
27. Show that there are infinite equilibrium sets for this chain.
28. Choose $\alpha = \mathbf{1}^T$. Prove that if $\lim_{n\to+\infty} h_n = 0$ and $\sum |h_i - h_{i-1}| < \infty$, then h is a potential.
29. Choose a function h satisfying the conditions of Problem 28, compute its charge f, and check that $h = Nf$.
30. Let $E = \{0, 1, 2\}$. Compute P^E and the fundamental matrix of this finite chain. Verify that the latter is N_E.

CHAPTER 9

RECURRENT POTENTIAL THEORY

1. Potentials

Throughout this chapter P is a recurrent chain which is either null or noncyclic ergodic. For such a chain, $\lim_n P^n$ always exists; we let $L = \lim P^n$.

In a recurrent chain the non-negative finite-valued superregular measures are uniquely determined up to multiplication by a constant, and the non-zero ones are positive and regular. We choose one such non-zero regular measure and call it α.

If P is noncyclic ergodic, then $L_{ij} = \alpha_j / \sum_k \alpha_k$; whereas if P is null, then $L_{ij} = 0$. In either case, $L_{ik} - L_{ij}\alpha_k/\alpha_j = 0$.

Duality for P is defined with respect to the regular measure α. The dual $\hat{P}$ of a null chain is null, and the dual of a noncyclic ergodic chain is noncyclic ergodic. In general, if two results are duals, we shall prove only one of the pair. As usual, the key to the proof by duality of the second result is that $\hat{P}$ is the most general chain of the type we consider in this chapter.

As Definition 9-1 suggests, we define charges and potentials in the same way as in transient potential theory.

Definition 9-1: If μ is a signed measure with $\mu 1$ finite and if

$$\nu = \lim_n [\mu(I + P + \cdots + P^{n-1})]$$

exists and is finite-valued, then μ is called a **left charge** with **potential measure** ν and **total charge** $\mu 1$. If f is a function with αf finite and if $g = \lim_n [(I + P + \cdots + P^{n-1})f]$ exists and is finite-valued, then f is called a **right charge** with **potential function** g and **total charge** αf. The **support** of a charge is the set on which the charge is not zero; the support of a potential is the support of its charge.

The definition of the support of a potential is not justified until we prove a uniqueness theorem for the charge of a potential, but such a result will follow directly from conclusion (2) of Theorem 9-15.

We note that the dual of a left (right) charge for P is a right (left) charge for $\hat{P}$ and that their total charges are the same (since the dual of a number is the same number).

Although we adopt the same definitions as with transient chains, the results are sometimes significantly different. For example, the only pure potential is the zero potential: If $f \geq 0$, then by monotone convergence $\lim_n [(I + P + \cdots + P^{n-1})f]_i = \sum_i N_{ij}f_j$, where $N_{ij} = +\infty$ for every i and j. Thus the limit is finite-valued only if $f = 0$.

On the other hand, every row (or column) of $I - P$ is a charge, and the potential of the ith row (column) of $I - P$ is the ith row (column) of $I - L$. For if μ is the ith row of $I - P$, then $|\mu|1 \leq 2 < \infty$ and

$$\begin{aligned} \nu_j &= \lim_n [\mu(I + P + \cdots + P^{n-1})]_j = \lim_n (I - P^n)_{ij} \\ &= (I - L)_{ij}; \end{aligned}$$

the assertion for columns is dual.

Our first potential theory result will be that every charge has total charge zero. To prove this fact, we require the Doeblin Ratio Limit Theorem, Theorem 9-4. We recall that $H_{ij}^{(n)}$ is the probability starting in i of reaching j before or at time n and that $N_{ij}^{(n)}$ is the mean number of times the process started at i is in j up to and including time n. Hence $H_{ij}^{(n)} = \sum_{k=0}^{n} F_{ij}^{(k)}$ and $N_{ij}^{(n)} = \sum_{k=0}^{n} (P^k)_{ij}$; from the latter relation we see that $\hat{N}_{ij}^{(n)} = (\alpha_j/\alpha_i)N_{ji}^{(n)}$. In terms of this notation the Doeblin Ratio Limit Theorem states that in any Markov chain with a positive superregular measure

$$\lim_n \frac{N_{ij}^{(n)}}{N_{i'j'}^{(n)}}$$

exists and is finite for any states i, j, i', and j' which communicate. We shall give a simple proof of this important result.

Lemma 9-2: Let P be any Markov chain with a positive superregular measure α. If i and j communicate, then the quantities

$$N_{jj}^{(n)} - N_{ij}^{(n)}$$

and

$$N_{ii}^{(n)} \frac{\alpha_j}{\alpha_i} - N_{ij}^{(n)}$$

are non-negative and bounded. In particular, $|N_{jj}^{(n)} - N_{ij}^{(n)}| \leq {}^iN_{jj}$.

PROOF: If i and j communicate, then ${}^i\bar{H}_{jj} < 1$, so that ${}^iN_{jj} = 1/(1 - {}^i\bar{H}_{jj}) < \infty$. It is clear that

$$N_{jj}^{(n)} - N_{ij}^{(n)} \geq 0$$

and that

$$N_{jj}^{(n)} \leq {}^iN_{jj} + N_{ij}^{(n)},$$

and hence the first expression is non-negative and bounded by ${}^iN_{jj} < \infty$. By duality, $N_{jj}^{(n)} - N_{ji}^{(n)}(\alpha_j/\alpha_i)$ is non-negative and bounded; if we multiply by α_i/α_j and interchange j and i, we obtain the second result.

Lemma 9-3: Let P be any Markov chain with a positive superregular measure α. If i and j communicate, then for all n

$$\frac{N_{ij}^{(n)}}{N_{jj}^{(n)}} \leq 1 \quad \text{and} \quad \frac{N_{ij}^{(n)}}{N_{ii}^{(n)}} \leq \frac{\alpha_j}{\alpha_i},$$

and

$$\lim_n \frac{N_{ij}^{(n)}}{N_{jj}^{(n)}} = H_{ij} \quad \text{and} \quad \lim_n \frac{N_{ij}^{(n)}}{N_{ii}^{(n)}} = \frac{\alpha_j}{\alpha_i}\hat{H}_{ji} = {}^i\bar{N}_{ij}.$$

PROOF: By Lemma 9-2,

$$0 \leq N_{jj}^{(n)} - N_{ij}^{(n)} \leq c \quad \text{for all } n.$$

Hence

$$0 \leq 1 - \frac{N_{ij}^{(n)}}{N_{jj}^{(n)}} \leq \frac{c}{N_{jj}^{(n)}}.$$

Therefore $N_{ij}^{(n)}/N_{jj}^{(n)} \leq 1$ for all n. If j is recurrent, then $N_{jj}^{(n)} \to +\infty$ and the ratio $N_{ij}^{(n)}/N_{jj}^{(n)}$ must tend to 1; we have $H_{ij} = 1$ since i must be recurrent if i and j communicate. Hence

$$\frac{N_{ij}^{(n)}}{N_{jj}^{(n)}} \to H_{ij}.$$

If j is transient, then

$$\frac{N_{ij}^{(n)}}{N_{jj}^{(n)}} \to \frac{N_{ij}}{N_{jj}} = \frac{H_{ij}N_{jj}}{N_{jj}} = H_{ij}.$$

The other results are duals, and the assertion about ${}^i\bar{N}_{ij}$ follows from Corollary 6-20.

The following is the Ratio Limit Theorem.

Theorem 9-4: Let P be any Markov chain with a positive superregular measure α, and let i, j, i', and j' be any states which communicate. Then

$$\lim_n \frac{N_{ij}^{(n)}}{N_{i'j'}^{(n)}}$$

exists. If all four states are recurrent, the limit is $\alpha_j/\alpha_{j'}$. If the states are transient, the limit is

$$\frac{H_{ij}\hat{H}_{jj'}\alpha_j}{H_{i'j'}H_{j'j}\alpha_{j'}}.$$

REMARK: Since the states in question communicate, they must be either all recurrent or all transient.

PROOF: Write

$$\frac{N_{ij}^{(n)}}{N_{i'j'}^{(n)}} = \left[\frac{N_{i'j'}^{(n)}}{N_{j'j'}^{(n)}}\right]^{-1}\left[\frac{N_{j'j}^{(n)}}{N_{j'j'}^{(n)}}\right]\left[\frac{N_{j'j}^{(n)}}{N_{jj}^{(n)}}\right]^{-1}\left[\frac{N_{ij}^{(n)}}{N_{jj}^{(n)}}\right],$$

and apply Lemma 9-3 to each factor.

Proposition 9-5: Every charge has total charge zero.

PROOF: If μ is a left charge, then $\sum |\mu_k| < \infty$ and

$$\nu_j = \lim_n \sum_k \mu_k N_{kj}^{(n)}$$

is finite. Therefore

$$\sum_k \mu_k \left(\frac{N_{kj}^{(n)}}{N_{jj}^{(n)}}\right) \to 0.$$

Since by Lemma 9-3, $N_{kj}^{(n)}/N_{jj}^{(n)} \leq 1$, dominated convergence gives

$$0 = \lim \sum_k \mu_k \left(\frac{N_{kj}^{(n)}}{N_{jj}^{(n)}}\right) = \sum_k \mu_k \lim \left(\frac{N_{kj}^{(n)}}{N_{jj}^{(n)}}\right) = \sum \mu_k = \mu 1.$$

The result for functions is dual.

The condition that a function f satisfy $\alpha f = 0$ is a strong necessary condition for it to be a charge, but it is by no means sufficient. In fact, it is not even sufficient in general if f also has finite support. We shall return to discuss this point at length in Section 2.

We now establish as Theorem 9-7 an identity which will play a fundamental role when we develop an operator which transforms charges into potentials.

Lemma 9-6: Let $\{a_n\}$ and $\{b_n\}$ be two sequences of real numbers such that $a_n \geq 0$, $\sum a_n = a < \infty$, $|b_n| < B$, and $|b_n - b_{n-1}| \to 0$. Then

$$\lim_{n\to\infty} \sum_{k=0}^{n} a_k(b_n - b_{n-k}) = 0.$$

PROOF: Let $\epsilon > 0$ be given. Choose N sufficiently large that $\sum_{k>N} a_k < \epsilon/(4B)$ and pick N' large enough so that for all $n \geq N'$

$$|b_n - b_{n-1}| \leq \frac{\epsilon}{2aN}.$$

Then for $n \geq N + N'$, we have

$$\begin{aligned}\left|\sum_{k=0}^{n} a_k(b_n - b_{n-k})\right| &\leq \sum_{k=0}^{N} a_k|b_n - b_{n-k}| + \sum_{k=N+1}^{n} a_k(|b_n| + |b_{n-k}|)\\ &\leq \sum_{k=0}^{N} a_k \sum_{j=0}^{k-1} |b_{n-j} - b_{n-j-1}| + 2B \sum_{k=N+1}^{n} a_k\\ &\leq \sum_{k=0}^{N} a_k \sum_{j=0}^{k-1} \frac{\epsilon}{2aN} + \frac{\epsilon}{4B}\cdot 2B, \quad \text{since } n - j \geq N'\\ &\leq \frac{\epsilon}{2a}\sum_{k=0}^{N} a_k + \frac{\epsilon}{2}\\ &\leq \epsilon.\end{aligned}$$

Theorem 9-7: Let i, j, and k be arbitrary states in a recurrent Markov chain which is either null or noncyclic ergodic. Then

$$\lim_{n\to\infty} [(N_{kk}^{(n)} - N_{ik}^{(n)})\alpha_j/\alpha_k + N_{ij}^{(n)} - N_{kj}^{(n)}] = {}^kN_{ij}.$$

PROOF: We may assume that neither i nor j equals k, since otherwise both sides are clearly zero. We begin by establishing four equations:

$$(1) \qquad N_{kk}^{(n)} = \sum_{\nu=0}^{\infty} F_{ik}^{(\nu)} N_{kk}^{(n)}$$

$$(2) \qquad N_{ik}^{(n)} = \sum_{\nu=0}^{n} F_{ik}^{(\nu)} N_{kk}^{(n-\nu)}$$

$$(3) \qquad N_{kj}^{(n)} = \sum_{\nu=0}^{\infty} F_{ik}^{(\nu)} N_{kj}^{(n)}$$

$$(4) \qquad N_{ij}^{(n)} = \sum_{\nu=0}^{n} F_{ik}^{(\nu)} N_{kj}^{(n-\nu)} + {}^kN_{ij}^{(n)}.$$

Equations (1) and (3) follow from the fact that $\sum F_{ik}^{(\nu)} = H_{ik} = 1$. Equation (2) comes from Theorem 4-11 with the random time $\mathbf{t} = \min(\mathbf{t}_k, n)$, and equation (4) is a similar result, except that the sum has been broken into two parts representing what happens after and before state k is reached for the first time.

Multiply (1) by α_j/α_k, (2) by $-\alpha_j/\alpha_k$, (3) by -1, and (4) by 1, and add. We obtain

$$(N_{kk}^{(n)} - N_{ik}^{(n)})\alpha_j/\alpha_k + (N_{ij}^{(n)} - N_{kj}^{(n)})$$
$$= {}^kN_{ij}^{(n)} + \sum_{\nu=0}^{n} F_{ik}^{(\nu)}(b_n - b_{n-\nu}) + \sum_{\nu=n+1}^{\infty} F_{ik}^{(\nu)}b_n,$$

where $b_n = N_{kk}^{(n)}\alpha_j/\alpha_k - N_{kj}^{(n)}$, and $\{b_n\}$ is a bounded sequence by Lemma 9-2. The first term ${}^kN_{ij}^{(n)}$ on the right side tends to ${}^kN_{ij}$, and the third term tends to zero since $\{b_n\}$ is bounded and $\sum F_{ik}^{(\nu)}$ is finite. It is thus sufficient to show that

$$\lim_n \sum_{\nu=0}^{n} a_\nu(b_n - b_{n-\nu}) = 0,$$

where $a_\nu = F_{ik}^{(\nu)}$. Since $a_n \geq 0$, $\sum a_n = 1$, and $\{b_n\}$ is bounded, we need show only that $(b_n - b_{n-1}) \to 0$ to apply Lemma 9-6. But

$$\begin{aligned} b_n - b_{n-1} &= (N_{kk}^{(n)} - N_{kk}^{(n-1)})\frac{\alpha_j}{\alpha_k} - (N_{kj}^{(n)} - N_{kj}^{(n-1)}) \\ &= P_{kk}^{(n)}\frac{\alpha_j}{\alpha_k} - P_{kj}^{(n)} \\ &\to L_{kk}\frac{\alpha_j}{\alpha_k} - L_{kj} \\ &= 0. \end{aligned}$$

If in the recurrent chain P we make a set of states E absorbing, then ${}^EP$ is an absorbing chain and results about transient chains may be applied to it. For example, the result $N_{ij} = H_{ij}N_{jj}$ yields ${}^EN_{ij} = {}^EH_{ij}\,{}^EN_{jj}$. We shall also make frequent use of the fact noted in Section 6-2 that ${}^E\hat{N}_{ij} = (\alpha_j/\alpha_i)\,{}^EN_{ji}$.

At this point we begin developing the machinery needed for the main result of this section, Theorem 9-15. We first need two preliminary identities, which we establish as Propositions 9-12 and 9-13.

Lemma 9-8: For any pair of states i and j,

$$\frac{{}^jN_{ii}}{\alpha_i} = \frac{{}^iN_{jj}}{\alpha_j}.$$

PROOF 1: If $i = j$, both sides are zero. If $i \neq j$, then from $\bar{H}_{jj} = 1$ and (3) of Lemma 4-19, we find

$${}^i\bar{H}_{jj} + {}^j\bar{H}_{ji}H_{ij} = \bar{H}_{jj} = 1,$$

so that

$$1 - {}^i\bar{H}_{jj} = {}^j\bar{H}_{ji}.$$

Hence

$${}^{i}N_{jj} = 1/(1 - {}^{i}\bar{H}_{jj}) = 1/{}^{j}\bar{H}_{ji}.$$

Therefore, by Proposition 5-4 and Corollary 6-21,

$$\frac{{}^{j}N_{ii}}{{}^{i}N_{jj}} = {}^{j}\bar{H}_{ji}\,{}^{j}N_{ii} = {}^{j}\bar{N}_{ji} = \frac{\alpha_i}{\alpha_j}.$$

PROOF 2: Set $i = j$ in Theorem 9-7. Then

$$\lim\,[(N_{kk}^{(n)} - N_{ik}^{(n)})\alpha_i + (N_{ii}^{(n)} - N_{ki}^{(n)})\alpha_k] = \alpha_k\,{}^{k}N_{ii}.$$

Interchange i and k, and the left side stays the same. The right side becomes $\alpha_i\,{}^{i}N_{kk}$.

Lemma 9-9: For any states i, j, k with $i \neq j$,

$${}^{j}N_{ki} + {}^{i}N_{kj}\,\frac{\alpha_i}{\alpha_j} = {}^{j}N_{ii}.$$

PROOF:

$$\begin{aligned} {}^{j}N_{ki} + {}^{i}N_{kj}\,\frac{\alpha_i}{\alpha_j} &= {}^{j}H_{ki}\,{}^{j}N_{ii} + {}^{i}H_{kj}\,{}^{i}N_{jj}\,\frac{\alpha_i}{\alpha_j} \\ &= ({}^{j}H_{ki} + {}^{i}H_{kj})\,{}^{j}N_{ii} \quad \text{by Lemma 9-8} \\ &= {}^{j}N_{ii}. \end{aligned}$$

Lemma 9-10:

$$\left| N_{ki}^{(n)} - N_{k0}^{(n)}\,\frac{\alpha_i}{\alpha_0} \right| \leq {}^{0}N_{ii}.$$

PROOF: If $i = 0$, both sides are zero. Otherwise we have

$$\hat{N}_{ik}^{(n)} \leq {}^{0}\hat{N}_{ik} + \hat{N}_{0k}^{(n)}.$$

If we multiply through by α_i/α_k, we obtain

$$N_{ki}^{(n)} \leq {}^{0}N_{ki} + \frac{\alpha_i}{\alpha_0}\left(\frac{\alpha_0}{\alpha_k}\,\hat{N}_{0k}^{(n)}\right) = {}^{0}N_{ki} + \frac{\alpha_i}{\alpha_0}\,N_{k0}^{(n)}.$$

Hence

$$N_{ki}^{(n)} - \frac{\alpha_i}{\alpha_0}\,N_{k0}^{(n)} \leq {}^{0}N_{ki}.$$

Interchanging i and 0 and multiplying by α_i/α_0 gives

$$\frac{\alpha_i}{\alpha_0}\,N_{k0}^{(n)} - N_{ki}^{(n)} \leq {}^{i}N_{k0}\,\frac{\alpha_i}{\alpha_0}.$$

Therefore,

$$\left| N_{ki}^{(n)} - \frac{\alpha_i}{\alpha_0} N_{k0}^{(n)} \right| \le {}^0N_{ki} + {}^iN_{k0} \frac{\alpha_i}{\alpha_0} = {}^0N_{ii} \quad \text{by Lemma 9-9.}$$

As usual, we agree that ${}^EN$, and in particular ${}^0N$, is a square matrix indexed by the set of all states; the entries of ${}^EN$ on the rows or columns indexed by E are all zeros.

Lemma 9-11: If $\mu 1$ is finite, then $\mu\, {}^0N$ is finite-valued. Dually, if αf is finite, then ${}^0Nf$ is finite-valued.

PROOF: If $\mu 1$ is finite, then

$$\sum_i |\mu_i|\, {}^0N_{ij} = \sum_i |\mu_i|\, {}^0H_{ij}\, {}^0N_{jj} \le {}^0N_{jj} \sum_i |\mu_i| < \infty.$$

Proposition 9-12: If $\mu 1 = 0$, then

$$\lim_{n\to\infty} \sum_k \mu_k \left[N_{ki}^{(n)} - \frac{\alpha_i}{\alpha_0} N_{k0}^{(n)} \right] = \sum_k \mu_k\, {}^0N_{ki}.$$

Dually, if $\alpha f = 0$, then

$$\lim_{n\to\infty} \sum_k [N_{ik}^{(n)} - N_{0k}^{(n)}] f_k = \sum_k {}^0N_{ik} f_k.$$

PROOF: Let

$$S_k^{(n)} = \left[(N_{00}^{(n)} - N_{k0}^{(n)}) \frac{\alpha_i}{\alpha_0} + (N_{ki}^{(n)} - N_{0i}^{(n)}) \right].$$

Since $\mu 1 = 0$, we have

$$\sum_k \mu_k \left[N_{ki}^{(n)} - \frac{\alpha_i}{\alpha_0} N_{k0}^{(n)} \right] = \sum_k \mu_k S_k^{(n)}.$$

By Theorem 9-7, $\lim_n S_k^{(n)} = {}^0N_{ki}$. Hence if we can prove that $S_k^{(n)}$ is bounded independently of n and k, then, since $\mu 1$ is finite, the result of the proposition follows by dominated convergence. (Note that $\sum_k \mu_k\, {}^0N_{ki}$ is absolutely convergent by Lemma 9-11.) We have

$$|S_k^{(n)}| \le \left| N_{00}^{(n)} \frac{\alpha_i}{\alpha_0} - N_{0i}^{(n)} \right| + \left| N_{ki}^{(n)} - N_{k0}^{(n)} \frac{\alpha_i}{\alpha_0} \right|.$$

The first term on the right is bounded according to Lemma 9-2, and the second term is bounded according to Lemma 9-10.

Proposition 9-13: If $\alpha f = 0$, then f is a charge if and only if

$$\lim_n [P^n({}^0Nf)]$$

exists and is finite-valued. If f is a charge, then its potential g satisfies

$$g = {}^0Nf - \lim_n [P^n({}^0Nf)].$$

PROOF: We have

$$(P\,{}^0N)_{ij} = \sum_k P_{ik}\,{}^0N_{kj} = \begin{cases} 0 & \text{if } j = 0 \text{ since } {}^0N_{k0} = 0 \\ {}^0\bar{N}_{0j} = \dfrac{\alpha_j}{\alpha_0} & \text{if } i = 0 \text{ and } j \neq 0 \\ {}^0N_{ij} - \delta_{ij} & \text{if } i \neq 0 \text{ and } j \neq 0 \end{cases}$$

$$((I - P)\,{}^0N)_{ij} = \begin{cases} 0 & \text{if } j = 0 \\ -\alpha_j/\alpha_0 & \text{if } i = 0 \text{ and } j \neq 0 \\ \delta_{ij} & \text{if } i \neq 0 \text{ and } j \neq 0. \end{cases}$$

By Lemma 9-11, ${}^0Nf$ is finite-valued, and hence associativity holds in the triple product $(I - P)\,{}^0Nf$. We find

$$((I - P)\,{}^0Nf)_i = \begin{cases} \sum_{j \neq 0} \delta_{ij} f_j = f_i & \text{for } i \neq 0 \\ \sum_{j \neq 0} \left(-\dfrac{\alpha_j}{\alpha_0}\right) f_j = \dfrac{1}{\alpha_0}(\alpha_0 f_0) = f_0 & \text{for } i = 0. \end{cases}$$

The next to last equality uses the fact that $\alpha f = 0$. We see therefore that $(I - P)\,{}^0Nf = f$, so that

$$\begin{aligned} &\lim_{n\to\infty} (I + P + \cdots + P^{n-1})f \\ &\qquad = \lim_{n\to\infty} (I + P + \cdots + P^{n-1})[(I - P)\,{}^0Nf] \\ &\qquad = \lim_{n\to\infty} (I - P^n)\,{}^0Nf. \end{aligned}$$

Associativity where required in the last equation follows from the distributive property, which holds because ${}^0Nf$ is finite-valued.

In the discussion that follows, E and F denote non-empty subsets of states.

Lemma 9-14: $B^E\,{}^FN = {}^FN - {}^EN$, provided $F \subset E$.

PROOF: By Theorem 4-11 with the time to reach E as the random time, we find that $(B^E\,{}^FN)_{ij} = \sum_k B^E_{ik}\,{}^FN_{kj}$ is the mean of the total number of times that the process is in state j before F, counting from the time when E is first entered. This mean is the difference of the number of times in j before F and the number of times in j before E, since F cannot be entered unless E is entered.

Theorem 9-15: If f is a function with $\alpha f = 0$ and if

$$\lim\ [(I + P + \cdots + P^{n-1})f]_0$$

exists and is finite for some state 0, then f is a charge. If g is its potential, then

(1) $g = {}^0Nf + g_0 1$.
(2) $f = (I - P)g$.
(3) $P^n g \to 0$.
(4) $B^E g = g$ if the support is contained in E.
(5) $f_E = (I - P^E)g_E$ if the support is contained in E.

Proof: By Proposition 9-12, if $\lim_n \sum_k N_{0k}^{(n)} f_k$ exists, then

$$\lim_n \sum_k [N_{ik}^{(n)} - N_{0k}^{(n)}]f_k = \sum_n {}^0N_{ik} f_k$$

or

$$\left(\lim_n \sum_k N_{ik}^{(n)} f_k\right) - \left(\lim_n \sum_k N_{0k}^{(n)} f_k\right) = ({}^0Nf)_i.$$

Hence $\lim_n \sum_k N_{ik}^{(n)} f_k$ exists, and f is a charge.

(1) Therefore $g_i - g_0 = ({}^0Nf)_i$ and (1) follows.

(2) From the proof of Proposition 9-13 we have $(I - P)\,{}^0Nf = f$, since $\alpha f = 0$. Thus if we multiply (1) by $I - P$, we get (2).

(3) By (2) we have $g = Pg + f$, and, since $P^k f$ is finite-valued, we see by induction that $P^k g$ is finite-valued and that

$$P^{k-1}g = P^k g + P^{k-1}f.$$

Adding these relations for $k = 1, \ldots, n$, we obtain

$$g = P^n g + (I + P + \cdots + P^{n-1})f.$$

Hence $P^n g \to 0$.

(4) Let $0 \in E$. By Lemma 9-14 with $F = \{0\}$,

$$B^E\,{}^0N = {}^0N - {}^EN,$$

and by (1)

$$\begin{aligned} B^E g &= B^E\,{}^0Nf + B^E g_0 1 \\ &= {}^0Nf - {}^ENf + g_0 1 \quad \text{since } B^E 1 = 1 \\ &= g - {}^ENf \quad \text{by (1).} \end{aligned}$$

Since f has support in E, ${}^{E}Nf = 0$ because ${}^{E}N_{ij} = 0$ when $j \in E$.

(5) $f = (I - P)g = (I - P)(B^E g) = [(I - P)B^E]g$

$$= \begin{pmatrix} I - P^E & 0 \\ 0 & 0 \end{pmatrix}\begin{pmatrix} g_E \\ g_{\tilde{E}} \end{pmatrix} \quad \text{by Proposition 6-17}$$

$$= \begin{pmatrix} (I - P^E)g_E \\ 0 \end{pmatrix}.$$

We note from conclusion (2) that a potential uniquely determines its charge.

Corollary 9-16: If $(I - P)g = f$, then g is a potential if and only if αf is finite and $P^n g \to 0$.

PROOF: If g is a potential, then $P^n g \to 0$ by conclusion (3) of Theorem 9-15 and f is the charge by conclusion (2). Hence αf is finite by definition. Conversely, if $(I - P)g = f$, we have by induction

$$g = P^n g + (I + \cdots + P^{n-1})f.$$

If $P^n g \to 0$, then $g = \lim [(I + P + \cdots + P^{n-1})f]$; if αf is finite, then g is a potential by definition.

We already know that the columns of $I - L$ are potentials whose charges are the corresponding columns of $I - P$. Corollary 9-16 allows us to enlarge this result as follows.

Corollary 9-17: If P is a null chain and if g is a function for which αg is finite, then g is a potential.

PROOF: We shall apply Corollary 9-16. By writing $g = g^+ - g^-$, we may assume that $g \geq 0$. Then

$$(P^n g)_i = \sum_j P_{ij}^{(n)} g_j = \frac{1}{\alpha_i} \sum_j (\alpha_j g_j)\hat{P}_{ji}^{(n)}.$$

Since αg is finite and since $\hat{P}_{ji}^{(n)}$ is bounded by one, we have by dominated convergence

$$\lim_n (P^n g)_i = \frac{1}{\alpha_i} \sum_j (\alpha_j g_j) \lim_n \hat{P}_{ji}^{(n)} = 0.$$

Therefore $P^n g \to 0$. Set $f = (I - P)g$. Then

$$\alpha|f| \leq \alpha(g + Pg) = \alpha g + \alpha(Pg) = \alpha g + (\alpha P)g = \alpha g + \alpha g < \infty.$$

A corresponding result for noncyclic ergodic chains will be proved in Section 3.

2. Normal chains

The condition at the end of Section 1 that g is a potential in a null chain if αg is finite does not provide a sufficiently large class of potentials for a satisfactory potential theory. In this section we shall impose the condition on P that any function f of finite support with $\alpha f = 0$ be a charge and that the corresponding result for signed measures hold; such a chain will be called **normal**. The justification for considering normal chains will consist in our showing that the class of normal chains is quite extensive; we shall see in Section 3, for example, that all noncyclic ergodic chains are normal.

Our procedure in introducing normal chains will be not to define them as above but to give a definition which is computationally simpler to check. The point of departure is the identity

$$g = {}^0Nf - \lim_n [P^n({}^0Nf)]$$

of Proposition 9-13.

Proposition 9-18: If for each j there is some i such that $\lim (P^n\, {}^0N)_{ij}$ exists, then $\lim (P^n\, {}^0N)$ exists and has constant columns which are finite-valued.

PROOF: Let j be a fixed state and let

$$f_k = \begin{cases} \alpha_j/\alpha_0 & \text{for } k = 0 \\ -1 & \text{for } k = j \\ 0 & \text{otherwise.} \end{cases}$$

Since $\alpha f = 0$, we have $(I - P)\, {}^0Nf = f$. Hence

$$\begin{aligned} \lim\, [(I + P + \cdots + P^{n-1})f]_i & \\ &= \lim\, [(I + P + \cdots + P^{n-1})(I - P)\, {}^0Nf]_i \\ &= \lim\, [(I - P^n)\, {}^0Nf]_i \\ &= ({}^0Nf)_i + \lim\, (P^n\, {}^0N)_{ij}. \end{aligned}$$

This limit exists for some i by hypothesis, and hence, by Theorem 9-15, f is a charge and $\lim (P^n\, {}^0N)_{kj}$ exists for all k. Hence $\lim (P^n\, {}^0N)$ exists. By Fatou's Theorem,

$$P \lim\, (P^n\, {}^0N) \leq \lim P\, P^n\, {}^0N = \lim\, (P^n\, {}^0N),$$

so that each column of the limit is non-negative superregular. Hence the columns of the limit are constants by Proposition 6-3.

Definition 9-19: If the indicated limits exist, then

$$^i\nu_j = \lim_n \sum_k (P^n)_{mk}\, {}^iN_{kj}$$

and

$$^i\lambda_j = {}^i\nu_j/{}^iN_{jj} = \lim_n \sum_k (P^n)_{mk}\, {}^iH_{kj}.$$

Notation independent of m in Definition 9-19 is justified by Proposition 9-18. We note that $^i\nu_j$ exists if and only if $^i\lambda_j$ exists, that $^i\nu_j$ is finite, and that $0 \le {}^i\lambda_j \le 1$. Furthermore, $^i\lambda_j$ is the probability that j is entered in the long run before i, and $^i\nu_j$ is the mean number of times in j, in the long run, before reaching i.

Definition 9-20: A chain is **normal** if for some fixed state 0 and for all j, $^0\lambda_j$ and $^0\hat{\lambda}_j$ both exist.

We note the important fact that the dual of a normal chain is normal.

Proposition 9-21: If P is a normal chain and if, for a given function f with $\alpha f = 0$, $\sum_k {}^0N_{kk}f_k$ is finite, then its potential g exists, is bounded, and satisfies $g = [{}^0N - \mathbf{1}\,{}^0\nu]f$.

PROOF: Since P is normal, $P^n\,{}^0N \to \mathbf{1}\,{}^0\nu$. Furthermore,

$$\sum_k (P^n)_{ik}\,{}^0N_{kj} \le \sum_k (P^n)_{ik}\,{}^0N_{jj} = {}^0N_{jj}$$

and ${}^0Nf^+$ and ${}^0Nf^-$ are both finite-valued. Thus we have dominated convergence in Proposition 9-13, f is a charge, and $g = [{}^0N - \mathbf{1}\,{}^0\nu]f$. Since by Theorem 9-15 $|g_i - g_0| = |\sum_k {}^0N_{ik}f_k| \le \sum_k {}^0N_{kk}|f_k|$, g is bounded.

Corollary 9-22: If P is normal and if f is a function of finite support with $\alpha f = 0$, then f is a charge.

PROOF: We have $\sum_k {}^0N_{kk}|f_k| < \infty$. Apply Proposition 9-21.

The converse to Corollary 9-22 is the following.

Lemma 9-23: If the function f defined by

$$f_i = \begin{cases} \alpha_j/\alpha_0 & \text{if } i = 0 \\ -1 & \text{if } i = j \\ 0 & \text{otherwise} \end{cases}$$

is a charge with potential g, then ${}^0\lambda_j$ exists and $g_0 = {}^0\nu_j$. If all functions f of support $\{0, j\}$ with $\alpha f = 0$ are charges and if all signed measures μ of support $\{0, j\}$ with $\mu 1 = 0$ are charges, then P is normal.

PROOF: For the first assertion we have, by Proposition 9-13,

$$\begin{aligned} g_0 &= ({}^0Nf)_0 - \lim (P^n\, {}^0Nf)_0 \\ &= -\lim (P^n\, {}^0Nf)_0 \quad \text{since } {}^0N_{0k} = 0 \text{ for all } k \\ &= \lim (P^n\, {}^0N)_{0j} \qquad \text{since } {}^0N_{k0} = 0 \text{ for all } k \\ &= {}^0\nu_j. \end{aligned}$$

Hence ${}^0\lambda_j$ exists. For the second assertion we see from the hypothesis about functions that ${}^0\lambda_j$ exists for all j. Dually we obtain from the hypothesis about signed measures that ${}^0\hat{\lambda}_j$ exists for all j. Hence P is normal.

Definition 9-24: Matrices C and G are defined by

$$C_{ij} = \lim_n (N_{jj}^{(n)} - N_{ij}^{(n)})$$

$$G_{ij} = \lim_n \left(N_{ii}^{(n)} \frac{\alpha_j}{\alpha_i} - N_{ij}^{(n)}\right)$$

whenever the indicated limits exist.

According to Lemma 9-2, the quantities defining C_{ij} and G_{ij} are bounded and non-negative. Thus all entries of C and G which exist are finite and non-negative. We have further that $C_{ii} = G_{ii} = 0$ for every i and that

$$\hat{G}_{ij} = \frac{\alpha_j C_{ji}}{\alpha_i}.$$

Hence $\hat{G} = \text{dual } C$ and $\hat{C} = \text{dual } G$.

Lemma 9-25: G_{0j} exists if and only if ${}^0\nu_j$ exists. If they both exist, then $G_{0j} = {}^0\nu_j$. Dually $C_{j0} = (\alpha_0/\alpha_j)\, {}^0\hat{\nu}_j$.

PROOF: We need only note that for the potential defined in Lemma 9-23

$$g_0 = \lim_n \left[N_{00}^{(n)} \frac{\alpha_j}{\alpha_0} - N_{0j}^{(n)}\right] = G_{0j}.$$

Hence G_{0j} exists if and only if $g_0 = {}^0\nu_j$ exists, and then they are equal. The other result follows by duality.

Theorem 9-26: If for some fixed state 0 and all other states j,

(1) ${}^0\lambda_j$ and ${}^0\hat{\lambda}_j$ both exist,
or (2) G_{0j} and C_{j0} both exist,
or (3) all functions and signed measures with support $\{0, j\}$ and total charge 0 are potential charges,

then P is normal. Conversely, if P is normal, then C, G, ${}^i\lambda_j$, ${}^i\nu_j$ all exist (for both P and $\hat{P}$), $G_{ij} = {}^i\nu_j = {}^i\lambda_j\, {}^iN_{jj}$ and

$$C_{ij} = \frac{\alpha_j}{\alpha_i}\, {}^j\hat{\nu}_i = {}^j\hat{\lambda}_i\, {}^iN_{jj},$$

and all functions and signed measures of finite support and total charge 0 are potential charges.

Proof: (1) is the definition of normality. If (2) holds, then ${}^0\nu_j$ and ${}^0\hat{\nu}_j$ exist by Lemma 9-25, and (1) holds. The sufficiency of (3) was shown in Lemma 9-23.

Conversely, suppose that P is normal. Then Corollary 9-22 assures that all f with finite support and $\alpha f = 0$ are potential charges. Consider the charge

$$f_k = \begin{cases} \alpha_j/\alpha_i & \text{if } k = i \\ -1 & \text{if } k = j \\ 0 & \text{otherwise.} \end{cases}$$

By Lemmas 9-23 and 9-25, its potential has as ith component $G_{ij} = {}^i\nu_j$, and ${}^i\nu_j = {}^i\lambda_j\, {}^iN_{jj}$ by definition. The remaining assertions follow by duality.

Corollary 9-27: If P is normal, then

$${}^0N_{ij} - {}^0\nu_j = -[G_{ij} - G_{i0}(\alpha_j/\alpha_0)].$$

Proof: The potential of Lemma 9-23 has 0th component $g_0 = {}^0\nu_j$. By Theorem 9-15,

$$g_i = g_0 + ({}^0Nf)_i = {}^0\nu_j - {}^0N_{ij}.$$

By definition,

$$g_i = \lim \left[N_{i0}^{(n)} \frac{\alpha_j}{\alpha_0} - N_{ij}^{(n)}\right] = G_{ij} - G_{i0}\frac{\alpha_j}{\alpha_0}.$$

Corollary 9-28: If P is a normal chain and if f is a function with $\alpha f = 0$ and $\sum_k {}^0N_{kk} f_k$ finite, then f is a charge and its potential g satisfies $g = -Gf$.

PROOF: By Proposition 9-21,

$$\begin{aligned} g &= [{}^0N - 1\,{}^0\nu]f \\ &= \left[-G + \frac{1}{\alpha_0}\{G_{i0}\alpha_j\}\right]f \quad \text{by Corollary 9-27} \\ &= -Gf \quad \text{since } \alpha f = 0. \end{aligned}$$

The dual of Corollary 9-28 is that in a normal chain if μ is a signed measure with $\mu 1 = 0$ and $\sum_k \mu_k\,{}^0N_{kk}$ finite, then μ is a charge and its potential ν satisfies $\nu = -\mu C$.

We can use Theorem 9-26 to conclude that all symmetric sums of independent random variables processes are normal; in particular, the one-dimensional and two-dimensional symmetric random walks are normal.

Corollary 9-29: If P is a null or noncyclic ergodic sums of independent random variables process with $P = P^T$, then P is normal,

$$C_{ij} = G_{ij} = \tfrac{1}{2}\,{}^jN_{ii} \quad \text{and} \quad {}^i\lambda_j = \tfrac{1}{2}.$$

PROOF: If we put j in for k and i in for i and j in Theorem 9-7, we obtain

$$\lim_{n\to\infty}\left[(N_{jj}^{(n)} - N_{ij}^{(n)})\frac{\alpha_i}{\alpha_j} + (N_{ii}^{(n)} - N_{ji}^{(n)})\right] = {}^jN_{ii}.$$

Now $\alpha_i = \alpha_j$ and $N_{ii}^{(n)} = N_{jj}^{(n)}$ in sums of independent random variables, and $N_{ij}^{(n)} = N_{ji}^{(n)}$ in a symmetric process. Hence

$$\lim\,[2(N_{jj}^{(n)} - N_{ij}^{(n)})] = {}^jN_{ii}$$

or $C_{ij} = \frac{1}{2}\,{}^jN_{ii}$. Alternatively

$$\lim\left[2(N_{ii}^{(n)}\frac{\alpha_j}{\alpha_i} - N_{ij}^{(n)})\right] = {}^jN_{ii}$$

or $G_{ij} = \frac{1}{2}\,{}^jN_{ii} = \frac{1}{2}\,{}^iN_{jj}$. Hence ${}^i\lambda_j = \frac{1}{2}$.

The strongest known result for concluding that a function f is in the domain of the potential operator G is Corollary 9-28, but that result involves a condition that is hard to check. The definition and theorem to follow give a more useful condition.

Definition 9-30: A function f in a normal chain is said to be a **weak charge** if $\alpha f = 0$ and if Gf and $\hat{C}f$ are both finite-valued. A signed measure μ is called a weak charge if $\mu 1 = 0$ and if μC and $\mu\hat{G}$ are both finite-valued.

The dual of a weak charge for P is a weak charge for $\hat{P}$.

Theorem 9-31: If P is a normal chain and if f is a weak charge, then f is a charge and its potential g is bounded and satisfies $g = -Gf$.

PROOF: Set i and j both equal to k in Corollary 9-27. Then

$$^0N_{kk} - {}^0\nu_k = -[G_{kk} - G_{k0}(\alpha_k/\alpha_0)] = G_{k0}(\alpha_k/\alpha_0).$$

Since $^0\nu_k = G_{0k}$, $^0N_{kk} = G_{k0}(\alpha_k/\alpha_0) + G_{0k}$. Thus

$$\sum_k {}^0N_{kk}f_k = \sum_k \left(\frac{\alpha_k G_{k0}}{\alpha_0} + G_{0k}\right)f_k = (\hat{C}f)_0 + (Gf)_0,$$

and $\sum_k {}^0N_{kk}f_k$ must be finite. The result follows by Proposition 9-21 and Corollary 9-28.

Dually if P is normal and μ is a weak charge, then ν exists, is bounded by a multiple of α, and satisfies $\nu = -\mu C$.

For the symmetric random walks in one and two dimensions, we have $C = G = \hat{C} = \hat{G}$ by Corollary 9-29. Therefore, Theorem 9-31 states for these cases that if $\alpha f = 0$ and if Gf is finite-valued, then f is a charge and its potential is bounded and satisfies $g = -Gf$; this is the analog of the Brownian motion result.

We shall now introduce the recurrent analog of the equilibrium set for transient chains—namely, the small ergodic set. Potentials with supports in small ergodic sets will be found to have special properties.

Lemma 9-32: If h is a non-negative bounded column vector such that $\lim_n P^n h$ exists, then the limit is a constant vector.

PROOF: By Fatou's Theorem

$$P \lim (P^n h) \leq \lim P^{n+1} h.$$

Thus the limit is finite-valued, non-negative, and superregular. Hence it is constant.

In particular, if $\lim P^n B^E$ exists, then it has constant columns.

Definition 9-33: If the indicated limit exists, then

$$\lim_{n \to \infty} P^n B^E = 1\lambda^E.$$

The entry λ_i^E is the probability in the long run of entering E at i. In the special case where E is a two-point set, we have $^0\lambda_i = \lambda_i^{\{i,0\}}$.

Proposition 9-34: If λ^E exists, then $\lambda^E \geq 0$, $\lambda^E_{\tilde{E}} = 0$, and $\lambda^E 1 \leq 1$.

Proof: The first two assertions follow from the facts that $P^n B^E \geq 0$ for every n and that $(P^n B^E)_{ij} = 0$ if j is not in E. Also,

$$(\lambda^E 1)1 = (1\lambda^E)1 = (\lim P^n B^E)1 \leq \lim (P^n B^E 1) = 1,$$

so that $\lambda^E 1 \leq 1$.

Definition 9-35: If λ^E exists and is such that $\lambda^E 1 = 1$, then E is said to be a **small set**.

To justify the name small set and to prove existence of small sets, we shall show that subsets of small sets are small and that in a normal chain all finite sets are small.

Proposition 9-36: If E is a subset of a small set F, then E is small and $\lambda^E = \lambda^F B^E$.

Proof: By Proposition 5-8, we have $B^E = B^F B^E$. Since each row of $P^n B^F$ tends to the probability vector λ^F, it follows from Proposition 1-57 that

$$P^n B^E = P^n B^F B^E \to 1\lambda^F B^E.$$

Thus λ^E exists and $\lambda^E = \lambda^F B^E$. In addition, $\lambda^E 1 = \lambda^F B^E 1 = \lambda^F 1 = 1$.

Lemma 9-37: If P is a normal chain and if $\sum_{k \in E} {}^0N_{kk} P^E_{kj}$ is finite for every $j \in E$, then λ^E exists and the columns of $B^E - 1\lambda^E$ are potentials with support in E.

Proof:

$$(I - P)B^E = \begin{pmatrix} I - P^E & 0 \\ 0 & 0 \end{pmatrix},$$

and the columns on the right are charges for a bounded potential by Proposition 9-21. Thus the limits

$$\lim_n (I + P + \cdots + P^{n-1}) \begin{pmatrix} I - P^E & 0 \\ 0 & 0 \end{pmatrix} = \lim_n (I - P^n)B^E$$

exist, and the resulting potentials are the columns of $B^E - 1\lambda^E$.

Proposition 9-38: In a normal chain all finite sets are small.

Proof: Let E be a finite set. By Lemma 9-37, λ^E exists. Moreover

$$\begin{aligned} \lambda^E 1 &= \sum_{j \in E} \lambda_j^E = \sum_{j \in E} \lim\, (P^n B^E)_{ij} \\ &= \lim \sum_{j \in E} (P^n B^E)_{ij} \\ &= \lim\, (P^n B^E 1)_i \\ &= 1. \end{aligned}$$

Corollary 9-39: In a normal chain ${}^i\lambda_j + {}^j\lambda_i = 1$.

Proof: Since ${}^i\lambda_j = \lambda_j^{\{i,j\}}$ and ${}^j\lambda_i = \lambda_i^{\{i,j\}}$, we may apply Proposition 9-38.

Definition 9-40: A set E of states is an **ergodic set** if P^E is an ergodic chain.

Proposition 9-41: E is ergodic if and only if $\alpha_E 1 < \infty$. All finite sets are ergodic, a subset of an ergodic set is ergodic, and the union of two ergodic sets is ergodic.

Proof: $\alpha_E P^E = \alpha_E$. Hence P^E is ergodic if and only if $\alpha_E 1 < \infty$. Thus the ergodic sets are the sets of finite α-measure; the remaining statements follow from this observation.

Small sets and ergodic sets might seem to be related notions, but we shall see later that they are actually independent.

Proposition 9-42: If E is a small set and if g is a bounded potential with support in E, then g is regular at points of $\tilde{E}$ and $\lambda^E g = 0$. Conversely, if E is small and ergodic and if g is a bounded function which is regular at points of $\tilde{E}$, and which satisfies $\lambda^E g = 0$, then g is a bounded potential with support in E.

Proof: Let g be a bounded potential with support in E. Since $(I - P)g$ is the charge, g is regular in $\tilde{E}$. By Theorem 9-15, $B^E g = g$ and hence $P^n B^E g = P^n g$. Since $P^n g \to 0$, $P^n B^E g \to 0$. But by Proposition 1-57, since E is small, $P^n B^E g \to 1(\lambda^E g)$. Hence $\lambda^E g = 0$.

For the converse $g(x_n)$ is a bounded martingale at points of $\tilde{E}$, since g is bounded and regular in $\tilde{E}$. By Corollary 3-16 with the stopping time the time to reach E, we have $g = B^E g$. Now, since E is small,

$$P^n g = P^n B^E g \to 1(\lambda^E g) = 0 \quad \text{by Proposition 1-57.}$$

In addition,

$$\alpha[(I - P)g] = \alpha[(I - P)B^E g] = \alpha\begin{pmatrix}(I - P^E)g_E \\ 0\end{pmatrix} = \alpha_E(g_E - P^E g_E).$$

Since E is ergodic and g is bounded, $\alpha_E g_E$ is finite and hence

$$\alpha_E(g_E - P^E g_E) = \alpha_E g_E - (\alpha_E P^E)g_E = 0.$$

Therefore g is a potential by Corollary 9-16, and it is bounded by hypothesis. Its charge $\begin{pmatrix}(I - P^E)g_E \\ 0\end{pmatrix}$ has support in E.

For potentials of total charge zero in recurrent chains, the Principle of Balayage takes the following form.

Proposition 9-43: Let E be a small ergodic set. If x is a function bounded on E, then there is a unique potential g with support in E which differs from x on E by a constant function. The potential is $g = B^E x - (\lambda^E x)\mathbf{1}$.

PROOF: Let $g = B^E x - (\lambda^E x)\mathbf{1}$. Then g is bounded, and

$$(I - P)g = \begin{pmatrix}(I - P^E)x_E \\ 0\end{pmatrix};$$

therefore g is regular on $\tilde{E}$. Since $\lambda^E g = 0$, g is a potential with support in E, by Proposition 9-42; and g differs from x by the constant $\lambda^E x$ on E.

For uniqueness, let g' be another such potential. Then $g_E - g'_E = k\mathbf{1}$ and $g - g' = B^E g - B^E g'$. Then $g - g' = k\mathbf{1}$. Since $P^n(g - g') \to 0$, we must have $k = 0$. Hence $g = g'$.

The result to follow is the total-charge-zero version of the Principle of Condensers.

Proposition 9-44: Suppose that E and F are disjoint sets such that $E \cup F$ is small and ergodic. Then there exists a function h such that:

(1) $h_i = 1$ if $i \in E$ and $h_j = 0$ if $j \in F$.
(2) $f = (I - P)h$ has its positive values in E and its negative values in F.
(3) h is the sum of a potential and a constant function.

PROOF: For existence, let x be a function which is 1 on E and 0 on F, and set $h = B^{E \cup F}x$. By Proposition 9-43, (1) and (3) are satisfied and f has support in $E \cup F$. For (2) we note that, if $i \in E$, then f_i is the probability of return to F before E and, if $i \in F$, then $-f_i$ is the probability of return to E before F.

For uniqueness, let x be a function satisfying (1), (2), and (3). Then $x = g + c\mathbf{1}$, where g is a potential, by (3). By (2),

$$(I - P)g = (I - P)x$$

vanishes outside of $E \cup F$ so that g has support in $E \cup F$. Hence $g = B^{E \cup F}g$. Therefore,

$$B^{E \cup F}x = B^{E \cup F}g + B^{E \cup F}(c\mathbf{1}) = g + c\mathbf{1} = x,$$

and x is uniquely determined by its values on $E \cup F$, which are fixed by (1).

We conclude this section with some results about normal chains which will be needed later.

Proposition 9-45: In a normal chain

$$G_{ik}\frac{\alpha_j}{\alpha_k} + G_{kj} - G_{ij} = {}^kN_{ij}$$

$$C_{ik}\frac{\alpha_j}{\alpha_k} + C_{kj} - C_{ij} = {}^kN_{ij}$$

$$G_{ik}\frac{\alpha_i}{\alpha_k} + G_{ki} = {}^kN_{ii}$$

$$C_{ik}\frac{\alpha_i}{\alpha_k} + C_{ki} = {}^kN_{ii}.$$

PROOF: The first two expressions follow from Theorem 9-7 and Definition 9-24. For the other two, set $i = j$.

By Lemma 9-32, if $\lim (P^n\,{}^EN)$ exists, then it is finite-valued and has constant columns.

Definition 9-46: ${}^E\nu_j = \lim_n \sum_k (P^n)_{ik}\,{}^EN_{kj}$, provided the limit exists.

Proposition 9-47: ${}^E\nu_j$ exists if and only if $\lambda_j^{E \cup \{j\}}$ exists. If they both exist, then ${}^E\nu_j = \lambda_j^{E \cup \{j\}}\,{}^EN_{jj}$. Hence in a normal chain ${}^E\nu$ exists for all finite sets E.

Proof: ${}^E\nu_j = \lim \sum_k (P^n)_{ik} \, {}^EN_{kj}$ and

$$\begin{aligned}\lambda_j^{E\cup\{j\}} \, {}^EN_{jj} &= \lim \sum_k (P^n)_{ik} B_{kj}^{E\cup\{j\}} \, {}^EN_{jj} \\ &= \lim \sum_k (P^n)_{ik} \, {}^EH_{kj} \, {}^EN_{jj} \\ &= \lim \sum_k (P^n)_{ik} \, {}^EN_{kj}.\end{aligned}$$

Remark: Actually, it can be shown that if E is small, then $E \cup \{j\}$ is small. Hence ${}^E\nu$ exists for all small sets. (See the problems.)

3. Ergodic chains

For this section let P be a noncyclic ergodic chain, and choose α so that $\alpha\mathbf{1} = 1$. Then $L = \lim P^n = A = \mathbf{1}\alpha$. The dual of P, namely $\hat{P}$, is also noncyclic ergodic, and the mean first passage time matrix for P has all finite entries (see Proposition 6-42).

We begin by proving that all noncyclic ergodic chains are normal and by giving an existence theorem for potentials.

Lemma 9-48: For every i and j,

$$0 \leq [N_{jj}^{(n)} - N_{ij}^{(n)}] \leq M_{ij} < \infty \quad \text{and} \quad \lim\,[N_{jj}^{(n)} - N_{ij}^{(n)}] = M_{ij}\alpha_j.$$

Proof: Summing over the powers of P in Lemma 6-34, we obtain

$$N_{ij}^{(n)} = \sum_{k=0}^{\infty} F_{ij}^{(k)} N_{jj}^{(n-k)},$$

where we use the convention $N_{jj}^{(m)} = 0$ if $m < 0$. Since $\sum_{k=0}^{\infty} F_{ij}^{(k)} = H_{ij} = 1$, we have

$$N_{jj}^{(n)} - N_{ij}^{(n)} = \sum_{k=0}^{\infty} F_{ij}^{(k)}[N_{jj}^{(n)} - N_{jj}^{(n-k)}].$$

As $n \to \infty$, we obtain

$$\begin{aligned}\lim_n\,[N_{jj}^{(n)} - N_{ij}^{(n)}] &= \lim_n \sum_{k=0}^{\infty} F_{ij}^{(k)}[N_{jj}^{(n)} - N_{jj}^{(n-k)}] \\ &= \sum_{k=0}^{\infty} F_{ij}^{(k)} \lim_n\,[N_{jj}^{(n)} - N_{jj}^{(n-k)}]\end{aligned}$$

by dominated convergence, since $N_{jj}^{(n)} - N_{jj}^{(n-k)} \leq k$ and $\sum F_{ij}^{(k)}k = M_{ij} < \infty$. Thus

$$\begin{aligned}\lim\,[N_{jj}^{(n)} - N_{ij}^{(n)}] &= \sum_{k=0}^{\infty} F_{ij}^{(k)} \lim_n \sum_{m=n-k+1}^{n} P_{jj}^{(m)} \\ &= \alpha_j \sum_{k=0}^{\infty} kF_{ij}^{(k)} \\ &= M_{ij}\alpha_j.\end{aligned}$$

Theorem 9-49: Every noncyclic ergodic chain is normal, and $C_{ij} = M_{ij}\alpha_j$. In matrix form $C = MD^{-1}$.

PROOF: By Lemma 9-48, C exists and satisfies $C_{ij} = M_{ij}\alpha_j$. Since $\hat{P}$ is noncyclic ergodic, $\hat{C}$ exists and hence G exists. Thus P is normal by Theorem 9-26.

Dually, $G_{ij} = \alpha_j \hat{M}_{ji}$ or $G = \hat{M}^T D^{-1}$.

For noncyclic ergodic chains we can prove a stronger result than Theorem 9-31 about the existence of potentials.

Theorem 9-50: If f is a function with $\alpha f = 0$ and Gf finite-valued, then f is a charge and its potential g is such that $g = -Gf$ and αg is finite. If, in addition, $\hat{C}f$ is finite-valued, then g is bounded. Dually, if μ is a signed measure with $\mu 1 = 0$ and μC finite-valued, then μ is a charge and its potential ν is such that $\nu = -\mu C$ and $\nu 1$ is finite; if, in addition, $\mu\hat{G}$ is finite-valued, then ν is bounded by a multiple of α.

PROOF: We shall prove the dual statements.

$$[\mu N^{(n)}]_i = \sum_k \mu_k N_{ki}^{(n)} = -\sum_k \mu_k [N_{ii}^{(n)} - N_{ki}^{(n)}]$$

since $\sum_k \mu_k N_{ii}^{(n)} = N_{ii}^{(n)}(\mu 1) = 0$. Now $N_{ii}^{(n)} - N_{ki}^{(n)} \le M_{ki}$ by Lemma 9-48, and $\alpha_i \sum_k \mu_k M_{ki} = (\mu C)_i$ is finite by hypothesis. Hence, by dominated convergence, μ is a charge and

$$\nu_i = -\sum_k \mu_k C_{ki}.$$

By the dual of Theorem 9-15,

$$\nu = \mu\,{}^0N + \frac{\nu_0}{\alpha_0}\alpha.$$

To show that $\nu 1$ is finite, it suffices to show that $|\mu|\,{}^0N1 < \infty$. But

$$\begin{aligned} |\mu|\,{}^0N1 &= \sum_{i,j} |\mu_i|\,{}^0N_{ij} \\ &= \sum_i |\mu_i| M_{i0} \\ &= \frac{1}{\alpha_0}\sum |\mu_i| C_{i0} < \infty \quad \text{by hypothesis.} \end{aligned}$$

If $\mu\hat{G}$ is finite-valued, then ν is bounded by a multiple of α, according to Theorem 9-31.

Corollary 9-51:

$$(I - P)(-C) = I - A$$

and

$$(-G)(I - P) = I - A.$$

PROOF: A row of $I - P$ is a charge whose potential is the corresponding row of $I - A$. Since

$$\begin{aligned} 0 \le (PC)_{ij} &= \sum_k P_{ik}C_{kj} \\ &= \alpha_j \sum_k P_{ik}M_{kj} \\ &= \alpha_j(\bar{M}_{ij} - 1) \quad \text{by Proposition 6-41} \\ &< \infty, \end{aligned}$$

$(I - P)(-C) = PC - C$ is finite-valued. Hence $(I - P)(-C) = I - A$ by Theorem 9-50. The second result is dual.

From Theorem 9-50 we have a sufficient condition on charges for their potentials to exist. We turn now to conditions on functions to ensure that they are potentials, the same problem that we touched on for null chains at the end of Section 1. We shall prove as Theorem 9-53 the result for noncyclic ergodic chains that corresponds to Corollary 9-17 for null chains.

Lemma 9-52: If h is a function for which αh is finite, then $P^n h \to Ah$.

PROOF:

$$\begin{aligned} \sum_j P_{ij}^{(n)} h_j &= \frac{1}{\alpha_i} \sum_j (\alpha_j h_j) \hat{P}_{ji}^{(n)} \\ &\to \frac{1}{\alpha_i} \sum_j (\alpha_j h_j) \alpha_i \quad \text{by dominated convergence} \\ &= \alpha h. \end{aligned}$$

Theorem 9-53: If g is a function for which αg is finite, then g is a potential if and only if $\alpha g = 0$. If g is such a potential and if f is its charge, then $g = (I - A)\, {}^0Nf$.

PROOF: Set $f = (I - P)g$. Then $\alpha f = \alpha[(I - P)g] = \alpha g - \alpha P g = 0$. By Lemma 9-52, $P^n g \to Ag = \mathbf{1}(\alpha g)$. Therefore, by Corollary 9-16, g is a potential if and only if $\alpha g = 0$. If g is such a potential, then by Theorem 9-15

$$g = {}^0Nf + g_0\mathbf{1}.$$

If we multiply through by α, we obtain

$$0 = \alpha\, {}^0Nf + g_0.$$

Hence

$$g = {}^0Nf - A\, {}^0Nf = (I - A)\, {}^0Nf.$$

Thus integrable potentials, those for which αg is finite, are precisely those integrable functions whose integral is zero. In particular, every bounded potential has integral zero.

Examples show that associativity of $A({}^0Nf)$ in Theorem 9-53 may fail. However, if it does hold, then Gf is finite-valued and $g = -Gf$. In fact, since the columns of ${}^0N$ are bounded, $\alpha\, {}^0N$ is finite-valued. Hence, by Lemma 9-52,

$$\lim P^n\, {}^0N = A\, {}^0N.$$

On the other hand, by definition

$$\lim P^n\, {}^0N = \mathbf{1}\, {}^0\nu.$$

Hence by associativity

$$g = {}^0Nf - \mathbf{1}\, {}^0\nu f = [{}^0N - \mathbf{1}\, {}^0\nu]f.$$

Therefore, by Corollary 9-27,

$$\begin{aligned} g_i &= -(Gf)_i + G_{i0} \frac{\alpha f}{\alpha_0} \\ &= -(Gf)_i \quad \text{since } \alpha f = 0. \end{aligned}$$

For a further discussion of this point, see the Additional Notes.

The existence of small sets for noncyclic ergodic chains is settled by the following proposition. Obviously all sets in such chains are ergodic sets.

Proposition 9-55: In a noncyclic ergodic chain all sets are small, and $\lambda^E = \alpha B^E$ and ${}^E\nu = \alpha\, {}^EN$. In particular, for the set of all states, $\lambda^S = \alpha$.

PROOF: By Proposition 1-57, $P^nB^E \to AB^E$. Hence $\lambda^E = \alpha B^E$ and $\lambda^E\mathbf{1} = \alpha B^E \mathbf{1} = \alpha \mathbf{1} = 1$, so that every set E is small. For $E = S$, $B^S = I$ and thus $\lambda^S = \alpha$. The assertion about ${}^E\nu$ is proved similarly.

Corollary 9-56: If αh is finite and if $(I - P)h = f$, then $B^E h$ is finite-valued and $B^E h = h - {}^E N f$. If f has support in E, then $B^E h = h$. If in addition h is bounded, then $\lambda^E h = \alpha h$.

PROOF: Let $g = h - (\alpha h)\mathbf{1}$. Then g is a potential by Theorem 9-53, and $(I - P)g = (I - P)h = f$. The proof of conclusion (4) of Theorem 9-15 shows that $B^E g$ is finite-valued and $B^E g = g - {}^E N f$. Therefore $B^E h$ is finite-valued and $B^E h = h - {}^E N f$. If f has support in E, then ${}^E N f = 0$ since ${}^E N_{ij} = 0$ for j in E. Hence $B^E h = h$ and $\alpha(B^E h) = \alpha h$. If h is bounded, associativity holds, and we conclude from Proposition 9-55 that $\lambda^E h = \alpha h$.

For total-charge-zero potentials in noncyclic ergodic chains, the following is the form that the Principle of Balayage takes.

Proposition 9-57: If αh is finite and $\lambda^E h$ is finite, then there is a unique potential g with support in E which differs from h by a constant on E.

PROOF: For existence $B^E h$ is finite-valued by Corollary 9-56, and we let

$$g = B^E h - (\lambda^E h)\mathbf{1}.$$

Then g differs from h by the constant $\lambda^E h$ on E. Since $\alpha g = 0$, g is a potential by Theorem 9-53. Moreover,

$$f = (I - P)g = (I - P)B^E h = \begin{pmatrix} (I - P^E)h_E \\ 0 \end{pmatrix},$$

so that the support of f is in E. For uniqueness, let g' be another such potential. Then $g - g'$ is constant on E. Since $B^E(g - g') = g - g'$, $g - g'$ is constant everywhere. But if $P^n(g - g') \to 0$, then $g - g'$ must be 0. Hence g is unique.

The following summary of the results for ergodic chains may be helpful. We consider the set of all states as a denumerable measure space of finite total measure $\alpha\mathbf{1}$; the measure assigned to state i is α_i. A function h is integrable if αh is finite; we restrict our attention to integrable functions.

We know that an integrable function is uniquely representable as the sum of a constant and a potential. The constant is the integral of the function. Hence an integrable function is a potential if and only if its integral is zero.

We conclude this section with some results about M and λ^E which hold for ergodic chains.

Proposition 9-58:

$$M_{ik} + M_{kj} - M_{ij} = \frac{{}^kN_{ij}}{\alpha_j},$$

$$M_{ik} + M_{ki} = \frac{{}^kN_{ii}}{\alpha_i}.$$

PROOF: In Proposition 9-45, substitute $M_{ij}\alpha_j$ for C_{ij} in the second and fourth equations.

Lemma 9-59: In any infinite ergodic chain, for fixed i and k

$$\lim_j {}^iN_{kj} = \lim_j {}^iH_{kj} = 0.$$

PROOF: Since ${}^iH_{kj} \leq {}^iN_{kj}$, it suffices to prove the result for ${}^iN_{kj}$. But $\sum_j {}^iN_{kj} = M_{ki} < \infty$, so that $\lim_j {}^iN_{kj} = 0$.

Proposition 9-60: In any infinite noncyclic ergodic chain, for fixed i

$$\lim_j {}^i\lambda_j = 0.$$

PROOF:

$$\begin{aligned} \lim_j {}^i\lambda_j &= \lim_j \sum_k \alpha_k\, {}^iH_{kj} && \text{by Proposition 9-55} \\ &= \sum_k \alpha_k \lim_j {}^iH_{kj} && \text{by dominated convergence} \\ &= 0 && \text{by Lemma 9-59.} \end{aligned}$$

Proposition 9-61: $\hat{\lambda}_i^E = \alpha_i \bar{M}_{iE}$ for i in E.

PROOF: By Proposition 6-16,

$$\alpha_j \hat{B}_{ji}^E = \alpha_i\, {}^E\bar{N}_{ij} \quad \text{for } i \text{ in } E.$$

Summing on j gives

$$\hat{\lambda}_i^E = \sum_j \alpha_j \hat{B}_{ji}^E = \alpha_i \sum_j {}^E\bar{N}_{ij} = \alpha_i \bar{M}_{iE}.$$

Thus Proposition 6-24 is the assertion that all sets are small in an ergodic chain.

Lemma 9-62: If $S_{ij} = M_{ij} + M_{ji}$, then

$$S_{ij} = \frac{{}^{i}N_{jj}}{\alpha_j},$$

$$S_{ij} = S_{ji} = \hat{S}_{ij} = \hat{S}_{ji},$$

$$M_{ij} = {}^{j}\hat{\lambda}_i S_{ji},$$

and

$$\frac{M_{ji}}{M_{ij}} = \frac{{}^{i}\hat{\lambda}_j}{{}^{j}\hat{\lambda}_i}.$$

PROOF: The first result is a restatement of Proposition 9-58. Since ${}^{i}\hat{N}_{jj} = {}^{i}N_{jj}$, we have $S_{ij} = \hat{S}_{ij}$; $S_{ij} = S_{ji}$ by definition. Finally, the last two results follow from the identities

$$M_{ij} = \frac{C_{ij}}{\alpha_j} = \frac{1}{\alpha_i}\,{}^{j}\hat{\nu}_i = {}^{j}\hat{\lambda}_i S_{ji},$$

which are consequences of Theorem 9-49 and Lemma 9-25.

From this lemma we see that

$$\hat{M}_{ij} = {}^{j}\lambda_i S_{ji} = {}^{j}\lambda_i \hat{S}_{ij} = {}^{j}\lambda_i (M_{ij} + M_{ji}),$$

a formula which gives a means of computing $\hat{M}$ from quantities in P. Moreover, from Proposition 9-61 we have

$${}^{j}\hat{\lambda}_i = \alpha_i \bar{M}_{i,\{i,j\}},$$

so that the lemma gives, on multiplication by S_{ji},

$$M_{ij} = {}^{j}\hat{\lambda}_i S_{ji} = \alpha_i \frac{{}^{j}N_{ii}}{\alpha_i} \bar{M}_{i,\{i,j\}}$$

or

$$M_{ij} = {}^{j}N_{ii} \bar{M}_{i,\{i,j\}}.$$

Proposition 9-63: In any infinite noncyclic ergodic chain

$$\lim_j \frac{M_{ji}}{M_{ij}} = 0,$$

$$\lim_j M_{ij} = +\infty$$

and

$$\lim_j \frac{C_{ij} - G_{ij}}{\alpha_j} = +\infty.$$

PROOF: By Lemma 9-62,

$$\frac{M_{ji}}{M_{ij}} = \frac{{}^i\hat{\lambda}_j}{{}^j\hat{\lambda}_i}.$$

By Proposition 9-60 the numerator tends to 0 and the denominator tends to 1. Hence the first assertion follows. Therefore, for all but finite many states j,

$$2M_{ij} \geq M_{ji} + M_{ij} \geq \bar{M}_{jj} = \frac{1}{\alpha_j}.$$

Since $\sum \alpha_j < \infty$, $\alpha_j \to 0$ and $M_{ij} \to \infty$. Finally

$$\begin{aligned}(C_{ij} - G_{ij})/\alpha_j &= M_{ij} - \frac{{}^i\nu_j}{\alpha_j} = M_{ij} - \frac{1}{\alpha_j}\,{}^i\lambda_j\,{}^iN_{jj} \\ &= M_{ij} - {}^i\lambda_j S_{ij} = M_{ij} - {}^i\lambda_j(M_{ij} + M_{ji}) \\ &= M_{ij}\left[1 - {}^i\lambda_j\left(1 + \frac{M_{ji}}{M_{ij}}\right)\right].\end{aligned}$$

The factor in brackets tends to 1 since ${}^i\lambda_j \to 0$ and $M_{ji}/M_{ij} \to 0$. Thus the third assertion of the proposition follows from the fact that $M_{ij} \to \infty$.

Corollary 9-64: In any infinite noncyclic ergodic chain, $C \neq G$.

PROOF: If $C = G$, then $(C_{ij} - G_{ij})/\alpha_j = 0$ for every j.

On the other hand, there are many finite ergodic chains with $C = G$ An example is

$$P = \begin{pmatrix} 1 - a & a \\ a & 1 - a \end{pmatrix}$$

for $0 < a < 1$.

4. Classes of ergodic chains

Let P be a recurrent chain. In this section we shall investigate the finiteness of the rth moments of certain random times and obtain formulas for these moments. Let

$$\begin{aligned} M_{ij}^{(r)} &= \mathrm{M}_i[\mathbf{t}_j{}^r], \\ b_j^{(r)} &= \mathrm{M}_j[\bar{\mathbf{t}}_j{}^r], \\ c_j^{(r)} = \sum_k \alpha_k\, \mathrm{M}_k[\mathbf{t}_j{}^r] &\quad \text{or} \quad c^{(r)} = \alpha M^{(r)}. \end{aligned}$$

These quantities are rth moments of the first passage times, the return times, and the equilibrium first passage times, respectively.

Let 0 be a distinguished state and write

$$P = \begin{matrix} & \{0\} & \widetilde{\{0\}} \\ \{0\} & P_{00} & U \\ \widetilde{\{0\}} & R & Q \end{matrix} \quad \text{and} \quad \alpha = \begin{matrix} \{0\} & \widetilde{\{0\}} \\ (\alpha_0 & \bar{\alpha}). \end{matrix}$$

The chain ${}^0P$ obtained from P by making 0 absorbing is an absorbing chain and has $N = \sum Q^k$ as its fundamental matrix. The time to absorption $\mathbf{a}$ in the chain ${}^0P$ is the same as the time $\mathbf{t}_0$ to reach 0 in the chain P. As in Section 8-8 we let $a^{(r)}$ be a column vector indexed by the transient states of ${}^0P$ and satisfying

$$a_i^{(r)} = \mathrm{M}_i[\mathbf{t}_0{}^r] = M_{i0}^{(r)}.$$

Since $M_{00}^{(r)} = 0$, Proposition 8-68 enables us to compute any column of $M^{(r)}$. From the relation $c_0^{(r)} = \bar{\alpha}a^{(r)}$ we also have a formula for the computation of $c^{(r)}$. An easy calculation gives $b_0^{(r)}$ in terms of $a^{(m)}$ with $m \le r$:

$$\begin{aligned} b_0^{(r)} = \mathrm{M}_0[\bar{\mathbf{t}}_0{}^r] &= \sum_k P_{0k}\, \mathrm{M}_k[(\mathbf{t}_0 + 1)^r] \\ &= P_{00} + \sum_{k \ne 0} P_{0k} \sum_{m=0}^{r} \binom{r}{m} \mathrm{M}_k[\mathbf{t}_0{}^m] \\ &= P_{00} + \sum_{m=0}^{r} \binom{r}{m} Ua^{(m)}. \end{aligned}$$

The first three propositions to follow give conditions for the finiteness of $M^{(r)}$, $b^{(r)}$, and $c^{(r)}$.

Proposition 9-65: If $r > 0$, then $b_0^{(r)} < \infty$ if and only if $c_0^{(r-1)} < \infty$.

PROOF: Since $a^{(m)} \le a^{(r)}$ for $m \le r$, $b_0^{(r)} < \infty$ if and only if $Ua^{(r)} < \infty$. Multiplying the inequalities of Corollary 8-69 through by U gives

$$Ua^{(r)} \le cUNa^{(r-1)} \le dUa^{(r)}.$$

For $j \ne 0$,

$$(UN)_j = \sum_{k \ne 0} P_{0k}\, {}^0N_{kj} = {}^0\bar{N}_{0j} = \frac{\alpha_j}{\alpha_0}.$$

Thus

$$UNa^{(r-1)} = \frac{1}{\alpha_0} \sum_{j \ne 0} \alpha_j\, \mathrm{M}_j[\mathbf{t}_0^{r-1}] = \frac{1}{\alpha_0} c_0^{(r-1)},$$

and $c_0^{(r-1)} < \infty$ if and only if $b_0^{(r)} < \infty$.

Lemma 9-66: For any two distinct states i and j, let $\mathbf{t}_{i,j}(\omega) = \mathbf{t}_j(\omega_{t_i})$ be the additional time needed to reach j after i. Then for any powers r and s

$$M_\pi[\mathbf{t}_i{}^r\, \mathbf{t}_{i,j}^s] = M_\pi[\mathbf{t}_i{}^r]\, M_i[\mathbf{t}_j{}^s].$$

PROOF:

$$\begin{aligned} M_\pi[\mathbf{t}_i{}^r\, \mathbf{t}_{i,j}^s] &= \sum_{m,n} \Pr_\pi[\mathbf{t}_i = m \wedge \mathbf{t}_{i,j} = n]\, m^r n^s \\ &= \sum_{m,n} \Pr_\pi[\mathbf{t}_i = m]\, \Pr_\pi[\mathbf{t}_{i,j} = n \mid \mathbf{t}_i = m]\, m^r n^s \\ &= \sum_{m,n} \Pr_\pi[\mathbf{t}_i = m]\, \Pr_i[\mathbf{t}_j = n]\, m^r n^s \\ &\qquad \text{by the strong Markov property} \\ &= \sum_m \Pr_\pi[\mathbf{t}_i = m]\, m^r \Big(\sum_n \Pr_i[\mathbf{t}_j = n] n^s\Big) \\ &= M_\pi[\mathbf{t}_i{}^r]\, M_i[\mathbf{t}_j{}^s]. \end{aligned}$$

Proposition 9-67: The vector $b^{(r)}$ is finite-valued if and only if $M^{(r)}$ is finite-valued.

PROOF: By definition $b_j^{(r)} = M_j[\bar{\mathbf{t}}_j{}^r]$. Let $\mathbf{t} = \min(\mathbf{t}_i, \bar{\mathbf{t}}_j)$ for $i \neq j$, and let $\mathbf{u} = \bar{\mathbf{t}}_j - \mathbf{t}$ (or 0 if $\mathbf{t} = \infty$). Then $\bar{\mathbf{t}}_j = \mathbf{t} + \mathbf{u} \geq \mathbf{u}$, so that

$$\begin{aligned} b_j^{(r)} &\geq M_j[\mathbf{u}^r] \\ &= \sum_k \Pr_j[x_t = k]\, M_k[\bar{\mathbf{t}}_j{}^r] \quad \text{by Theorem 4-11} \\ &\geq \Pr_j[x_t = i]\, M_i[\bar{\mathbf{t}}_j{}^r] \\ &= {}^j\bar{H}_{ji} M_{ij}^{(r)}. \end{aligned}$$

Since ${}^j\bar{H}_{ji} > 0$, if $b_j^{(r)} < \infty$, then $M_{ij}^{(r)} < \infty$ for all i.

Conversely, suppose that $M^{(r)}$ is finite-valued. Since $\bar{\mathbf{t}}_j \leq \mathbf{t}_i + \mathbf{t}_{i,j}$ for any state $i \neq j$,

$$\begin{aligned} b_j^{(r)} &\leq M_j[(\mathbf{t}_i + \mathbf{t}_{i,j})^r] \\ &= \sum_{m=0}^{r} \binom{r}{m} M_j[\mathbf{t}_i{}^m\, \mathbf{t}_{i,j}^{r-m}] \\ &= \sum_{m=0}^{r} \binom{r}{m} M_{ji}^{(m)} M_{ij}^{(r-m)} \quad \text{by Lemma 9-66.} \end{aligned}$$

Hence $b^{(r)}$ is finite-valued.

Proposition 9-68: If $b_0^{(r)} < \infty$ for some state 0, then $b_j^{(r)} < \infty$ for all j.

Proof: The proof is by induction on r. For $r = 0$, $b_j^{(0)} = 1$, and the result is trivial. Suppose that the result holds for $r \le n$ and that $b_0^{(n+1)} < \infty$. Then $b_0^{(r)} < \infty$ for $r \le n$, so that $b_j^{(r)} < \infty$ for $r \le n$ by inductive assumption. By Proposition 9-67, $M^{(r)}$ is finite-valued for $r \le n$, and by Proposition 9-65, $c_0^{(r)} < \infty$ for $r \le n$. Thus for $j \ne 0$

$$\begin{aligned} c_j^{(n)} &= \sum_k \alpha_k \, \mathrm{M}_k[\mathbf{t}_j{}^n] \\ &\le \sum_k \alpha_k \, \mathrm{M}_k[(\mathbf{t}_0 + \mathbf{t}_{0,j})^n] \quad \text{since } \mathbf{t}_j \le \mathbf{t}_0 + \mathbf{t}_{0,j} \\ &= \sum_k \alpha_k \sum_{r=0}^{n} \binom{n}{r} \mathrm{M}_k[\mathbf{t}_0{}^r \, \mathbf{t}_{0,j}^{n-r}] \\ &= \sum_k \alpha_k \sum_{r=0}^{n} \binom{n}{r} M_{k0}^{(r)} M_{0j}^{(n-r)} \\ &= \sum_{r=0}^{n} \binom{n}{r} c_0^{(r)} M_{0j}^{(n-r)} < \infty. \end{aligned}$$

Therefore $b^{(n+1)} < \infty$ by Proposition 9-65.

Finally we show the connection between the moments in P and those in $\hat{P}$.

Lemma 9-69: $\bar{F}_{00}^{(n)} = \hat{\bar{F}}_{00}^{(n)}$ in a recurrent chain.

Proof: By Lemma 6-34, for $n \ge 1$,

$$P_{00}^{(n)} = \sum_{k=1}^{n} \bar{F}_{00}^{(k)} P_{00}^{(n-k)}.$$

By induction we see that $\bar{F}_{00}^{(n)}$ is a function of $P_{00}^{(k)}$ for $k \le n$; similarly $\hat{\bar{F}}_{00}^{(n)}$ is the same function of $\hat{P}_{00}^{(k)}$. Since $P_{00}^{(k)} = \hat{P}_{00}^{(k)}$, $\bar{F}_{00}^{(n)} = \hat{\bar{F}}_{00}^{(n)}$.

Proposition 9-70: $b^{(r)} = \hat{b}^{(r)}$ and $c^{(r)} = \hat{c}^{(r)}$.

Proof: By Lemma 9-69,

$$b_0^{(r)} = \mathrm{M}_0[\bar{\mathbf{t}}_0{}^r] = \sum_n n^r \bar{F}_{00}^{(n)} = \sum_n n^r \hat{\bar{F}}_{00}^{(n)} = \mathrm{M}_0[\hat{\bar{\mathbf{t}}}_0{}^r] = \hat{b}_0^{(r)}.$$

For the second assertion,

$$\alpha_i \Pr_i[\mathbf{t}_0 = m] = \sum \alpha_i P_{ik_1} P_{k_1 k_2} \dots P_{k_{m-1}0}$$
$$= \sum \alpha_0 \hat{P}_{0k_{m-1}} \dots \hat{P}_{k_2 k_1} \hat{P}_{k_1 i},$$

where each of the sums is taken over the denumerable number of sequences $k_1, \dots, k_{m-1}$ with each k_j different from 0. Hence for $m > 0$,

$$\sum_i \alpha_i \Pr_i[\mathbf{t}_0 = m] = \alpha_0 \sum \hat{P}_{0k_{n-1}} \dots \hat{P}_{k_2 k_1}$$
$$= \alpha_0(1 - \hat{\bar{H}}_{00}^{(m-1)})$$
$$= \alpha_0(1 - \bar{H}_{00}^{(m-1)}) \quad \text{by Lemma 9-69}$$
$$= \sum_i \alpha_i \Pr_i[\hat{\mathbf{t}}_0 = m] \quad \text{by symmetry.}$$

If we multiply through by m^r and sum on m, we obtain $c_0^{(r)} = \hat{c}_0^{(r)}$.

We thus arrive at a hierarchy of recurrent chains:

Definition 9-71: A recurrent chain P has **ergodic degree** r if $b_0^{(r)} < \infty$ but $b_0^{(r+1)} = \infty$. If $b_0^{(r)} < \infty$ for every r, then P is said to have **infinite ergodic degree**. [It should be noted that $b_0^{(0)} = 1$; hence the degree is always defined.]

We may summarize our previous results as follows:

(1) Ergodic degree does not depend on the choice of the state 0.
(2) P is of ergodic degree $r > 0$ if and only if $c_0^{(r-1)} < \infty$ but $c_0^{(r)} = \infty$. (The choice of 0 is immaterial.)
(3) P is of ergodic degree r if and only if $M^{(r)}$ is finite-valued but $\alpha M^{(r)}$ is infinite-valued.
(4) $\hat{P}$ has the same ergodic degree as P.
(5) If P is of infinite ergodic degree, then $M^{(r)}$, $b^{(r)}$, and $c^{(r)}$ are finite-valued for all r.

For example, null chains have ergodic degree 0 and ergodic chains have ergodic degree at least 1. We shall see in Section 6 that the basic example may be of any degree $r = 0, 1, 2, \dots, \infty$.

Proposition 9-72: Every finite recurrent chain has infinite ergodic degree.

PROOF: For a fixed state 0 the result follows by induction on r from Propositions 9-68, 9-67, and 9-65.

5. Strong ergodic chains

A **strong ergodic chain** is a recurrent Markov chain of ergodic degree 2 or greater. Every strong ergodic chain is ergodic, and by Proposition 9-72 all finite recurrent chains are strong ergodic. By Proposition 9-65, P is strong ergodic if and only if αM is finite-valued. If P is strong ergodic, then so is $\hat{P}$. By Proposition 9-70, $\alpha M = \alpha \hat{M}$.

In this section P is a noncyclic strong ergodic chain and α is chosen so that $\alpha 1 = 1$.

Proposition 9-73: If P is strong ergodic and if f is a bounded function, then Gf is finite-valued. If, in addition, $\alpha f = 0$, then f is a charge and its potential g satisfies $g = -Gf$.

PROOF: If $|f| \leq k1$, then

$$G|f| \leq kG1 = k(\hat{M}^T D^{-1})1 = k\hat{M}^T \alpha^T = k(\alpha \hat{M})^T < \infty.$$

Hence $G|f|$ is finite-valued. If $\alpha f = 0$, then f is a charge with potential $-Gf$ by Theorem 9-50.

Definition 9-74: For any noncyclic ergodic chain P, a matrix Z is defined by $Z = \sum_{n=0}^{\infty} (P - A)^n$ whenever that sum exists *and is finite-valued.*

Proposition 9-75: Z exists if and only if P is strong ergodic. If P is strong ergodic, then $Z = A - G(I - A)$.

PROOF: Suppose P is strong ergodic. Since $\alpha(I - A) = 0$, each column of $I - A$ is a charge with potential

$$-G(I - A) = \sum_{n=0}^{\infty} P^n(I - A)$$

by Proposition 9-73. By induction we verify that for $n > 0$

$$P^n - A = (P - A)^n$$

and hence $P^n(I - A) = P^n - A = (P - A)^n$. Therefore

$$-G(I - A) = I - A + \sum_{n=1}^{\infty} P^n(I - A) = I - A + \sum_{n=1}^{\infty} (P - A)^n$$
$$= -A + \sum_{n=0}^{\infty} (P - A)^n.$$

Hence Z exists and equals $A - G(I - A)$.

Conversely, suppose Z exists. Then

$$\begin{aligned}(\alpha M)_j\alpha_j &= (\alpha M D^{-1})_j = (\alpha C)_j \\ &\leq \liminf_n \sum_k [\alpha_k(N_{jj}^{(n-1)} - N_{kj}^{(n-1)})] \quad \text{by Fatou's Theorem} \\ &= \liminf_n (N_{jj}^{(n-1)} - n\alpha_j) \\ &= \liminf_n \sum_{m=0}^{n-1} (P^m - A)_{jj} \\ &= \liminf_n \sum_{m=0}^{n-1} [(P - A)^m]_{jj} - A_{jj} \\ &= Z_{jj} - \alpha_j.\end{aligned}$$

Hence αM is finite-valued, and P is strong ergodic.

Proposition 9-76: If P is strong ergodic, then

(1) dual $Z = \hat{Z}$.
(2) $Z\mathbf{1} = \mathbf{1}$ and $\alpha Z = \alpha$.
(3) $Z(I - P) = I - A = (I - P)Z$.
(4) $Z(I - P + A) = I = (I - P + A)Z$.
(5) $Z = A - G(I - A) = A - (I - A)C$.

Proof: For (1) we have

$$\text{dual } Z = \text{dual}\left(\sum_{n=0}^{\infty} (P - A)^n\right) = \sum_{n=0}^{\infty} [\text{dual } (P - A)]^n$$
$$= \sum_{n=0}^{\infty} (\hat{P} - A)^n = \hat{Z}.$$

Hence in conclusions (2) through (5) the second result is always the dual of the first, and we need verify only the first. Conclusion (5) comes from Proposition 9-75. For (2), we have

$$\begin{aligned}Z\mathbf{1} &= [A - G(I - A)]\mathbf{1} = A\mathbf{1} - (G - GA)\mathbf{1} \\ &= \mathbf{1} - G\mathbf{1} + GA\mathbf{1} \quad \text{since } G\mathbf{1} < \infty \text{ by Proposition 9-73} \\ &= \mathbf{1} - G\mathbf{1} + G\mathbf{1} \\ &= \mathbf{1}.\end{aligned}$$

For (3),

$$\begin{aligned}Z(I - P) &= [A - G(I - A)](I - P) \\ &= (-G + GA)(I - P) \\ &= -G + GA + GP - GAP,\end{aligned}$$

since by Proposition 9-73 all terms are finite-valued,

$$\begin{aligned} &= -G + GA + GP - GA \\ &= -G(I - P) \\ &= I - A \quad \text{by Corollary 9-51.} \end{aligned}$$

Conclusion (4) follows directly from (2) and (3).

Proposition 9-77: In a strong ergodic chain

$$C = EZ_{dg} - Z$$

and

$$M = (EZ_{dg} - Z)D.$$

PROOF: The second assertion follows from the first, since $C = MD^{-1}$. For the first we have

$$\begin{aligned} Z_{ij} - \alpha_j &= \lim_n \sum_{m=0}^{n} (P^m - A)_{ij} \\ &= \lim [N_{ij}^{(n)} - (n + 1)\alpha_j] \end{aligned}$$

and similarly

$$Z_{jj} - \alpha_j = \lim [N_{jj}^{(n)} - (n + 1)\alpha_j]$$

Hence

$$Z_{jj} - Z_{ij} = \lim [N_{jj}^{(n)} - N_{ij}^{(n)}] = C_{ij}.$$

The next proposition shows that Z may be used as a single potential operator in place of both $-C$ and $-G$. Since (dual Z) $= \hat{Z}$, duality takes as simple a form as for transient potentials. Beginning in Section 8 we shall develop an operator $-K$ which exists for all normal chains and which has properties similar to those of Z.

Proposition 9-78: In a strong ergodic chain, Z may be used as either a right or a left potential operator.

PROOF: If $g = -Gf$ and $\alpha f = 0$, then

$$g = [A - G(I - A)]f = Zf$$

by Proposition 9-75. The result for signed measures is dual.

The operator Z is used in Kemeny and Snell [1960] in the analysis of finite recurrent chains, which are all strong ergodic. A number of quantities associated with recurrent chains are computed in that book in terms of Z. The proposition to follow is a sample.

Proposition 9-79: If P is strong ergodic, then

(1) $M_{\alpha j} = \dfrac{Z_{jj}}{\alpha_j} - 1,$
(2) $Z_{jj} = \bar{M}_{\alpha j}\alpha_j,$
(3) $M_{\alpha j} - M_{ij} = \hat{M}_{\alpha i} - \hat{M}_{ji}.$

PROOF: From Proposition 9-77 and 9-76,

$$\alpha M = (\alpha E Z_{dg} - \alpha Z)D = (EZ_{dg} - \alpha)D = EZ_{dg}D - 1^T,$$

and (1) follows. For (2),

$$\bar{M}_{\alpha j} = M_{\alpha j} + \alpha_j \bar{M}_{jj} = M_{\alpha j} + 1 = \frac{Z_{jj}}{\alpha_j}.$$

Finally by (1) and Proposition 9-77,

$$M_{\alpha j} - M_{ij} = \frac{Z_{ij}}{\alpha_j} - 1 = \frac{\hat{Z}_{ji}}{\alpha_i} - 1 = \hat{M}_{\alpha i} - \hat{M}_{ji}.$$

6. The basic example

The basic example P is recurrent if and only if $\beta_n \to 0$. Then β is regular, and we choose $\alpha = \beta$. First we consider the case where P is null, that is, where $\sum_i \beta_i = +\infty$.

Since a null chain with high probability will be outside a given finite set after a long time, the finite set must be re-entered from the left. Hence if E is finite,

$$\lambda_i^E = \begin{cases} 1 & \text{if } i \text{ is the first state of } E \\ 0 & \text{otherwise.} \end{cases}$$

In particular,

$$^i\lambda_j = \begin{cases} 1 & \text{if } j < i \\ 0 & \text{if } j \geq i. \end{cases}$$

Since

$$^iN_{jj} = \frac{1}{1 - {}^i\bar{H}_{jj}} = \frac{1}{{}^j\bar{H}_{ji}} = \begin{cases} \dfrac{\beta_j}{\beta_i} & \text{if } j < i \\ 1 & \text{if } j \geq i, \end{cases}$$

we have

$$G_{ij} = {}^i\nu_j = {}^i\lambda_j\, {}^iN_{jj} = \begin{cases} \dfrac{\beta_j}{\beta_i} & \text{if } j < i \\ 0 & \text{if } j \geq i. \end{cases}$$

To compute C, we note that the reverse chain $\hat{P}$ enters finite sets from the right. If E is finite,

$$\hat{\lambda}_i^E = \begin{cases} 1 & \text{if } i \text{ is the last state of } E \\ 0 & \text{otherwise.} \end{cases}$$

Hence

$${}^i\hat{\lambda}_j = \begin{cases} 1 & \text{if } j > i \\ 0 & \text{if } j \le i. \end{cases}$$

Since ${}^i\hat{N}_{jj} = {}^iN_{jj}$,

$$\hat{G}_{ij} = {}^i\hat{\lambda}_j\,{}^iN_{jj} = \begin{cases} 1 & \text{if } j > i \\ 0 & \text{if } j \le i \end{cases}$$

and

$$C_{ij} = \frac{\beta_j}{\beta_i}\hat{G}_{ji} = \begin{cases} \dfrac{\beta_j}{\beta_i} & \text{if } j < i \\ 0 & \text{if } j \ge i. \end{cases}$$

Note that $C = G$ and that $-C$ or $-G$ is obtained if, in the transient case, we let $\beta_\infty = +\infty$ in the formula for N.

Let us consider λ^E for infinite sets. For convenience we assume that 0 is in E. The probability of entering E in the long run at a state $j > 0$ is no greater than the probability of being between 0 and j in the long run, and hence $\lambda_j^E = 0$ for $j > 0$. For any state k, let k' be the next state in E (with the convention $k = k'$ if $k \in E$). Then

$$\lambda_0^E = \lim_n \sum_k P_{0k}^{(n)} B_{k0}^E = \lim_n \left[\sum_{k>0} P_{0k}^{(n)}\left(1 - \frac{\beta_{k'}}{\beta_k}\right) + P_{00}^{(n)}\right]$$

$$= 1 - \lim_n \sum_k P_{0k}^{(n)} \frac{\beta_{k'}}{\beta_k}.$$

Therefore the last term must be 0 for a small set.

We shall use this criterion to give an example of an ergodic set which is not a small set. Let $\frac{1}{2} \le p < 1$ and

$$P_{i,i+1} = \begin{cases} p & \text{if } i \text{ is a power of 2} \\ 1 & \text{otherwise,} \end{cases}$$

and let E be the set of all powers of 2 together with 0. Then

$$\beta_i = \begin{cases} 1 & \text{if } i = 0 \text{ or } 1 \\ p^n & \text{if } i > 1 \text{ and } 2^{n-1} < i \le 2^n. \end{cases}$$

Since $\sum \beta_i = 2 + \sum_{n=1}^{\infty} 2^{n-1}p^n = \infty$ if $p \ge \frac{1}{2}$, the process is null.

On the other hand, $\sum_{i\in E}\beta_i = 1 + \sum_{k=0}^{\infty} p^k < \infty$, since $p < 1$; hence E is an ergodic set. Now

$$\sum_k P_{0k}^{(n)}\frac{\beta_{k'}}{\beta_k} = \sum_k P_{0k}^{(n)}\cdot 1 = 1,$$

so that $\lambda^E = 0$ and E is not small.

For this example the bounded function $g = 1$ is regular outside of E and has $\lambda^E g = 0$, but $P^n g$ does not tend to 0 and hence g is not a potential. Thus Proposition 9-42 fails if we require that E be ergodic but not necessarily small.

Moreover, the same process and the same set give an example of a function f with support in E and having $\alpha f = 0$ and Gf finite-valued which is such that $-Gf$ is not a potential. For any function f,

$$(-Gf)_i = -\frac{1}{\beta_i}\sum_{j=0}^{i-1}\beta_j f_j.$$

If we let

$$f_i = \begin{cases} 0 & \text{for } i = 0 \text{ and for } i \notin E. \\ -1 & \text{for } i = 1 \\ 1 & \text{for other } i \in E, \end{cases}$$

and specialize to the case $p = \frac{1}{2}$, then

$$\alpha f = -1 + \sum_{n>0} 2^{-n} = 0$$

and

$$(-Gf)_i = \begin{cases} 0 & \text{for } i = 0 \text{ or } 1 \\ 2 & \text{otherwise.} \end{cases}$$

By Corollary 9-17 the function

$$g_i = \begin{cases} 2 & \text{for } i = 0 \text{ or } 1 \\ 0 & \text{otherwise} \end{cases}$$

is a potential. Since the sum of $-Gf$ and the potential g is a constant vector and not a potential, $-Gf$ cannot be a potential. We should note that f is not a weak charge since

$$(\hat{C}f)_i = (\hat{G}f)_i = \sum_{j>i} f_j = +\infty.$$

With minor modifications of the above example, we can make $\lambda^E 1 = \lambda_0^E$ assume any value between 0 and 1. For example, to get $\lambda_0^E = p$ with $\frac{1}{2} \le p < 1$, redefine E to consist of 0 and all states of the form $2^k + 1$. To get values of $\lambda_0^E < \frac{1}{2}$, redefine the process (and the set) so that the process can move from i to 0 only when i is a power of n.

Now we consider the case where P is ergodic. Let

$$\alpha_j = \beta_j \Big/ \sum_k \beta_k$$

and define

$$\sigma_i = \sum_{k=0}^{i-1} \beta_k.$$

Then $\sigma_0 = 0$ and $\sigma_\infty = \sum_k \beta_k$; σ_∞ is finite for all ergodic basic examples, and $\alpha_j = \beta_j/\sigma_\infty$.

The formula for ${}^iN_{jj}$ still applies, but $\lambda^E = \alpha B^E$ and λ^E is positive on all of E. In particular, if $i < j$,

$${}^i\lambda_j = \sum_k \alpha_k\, {}^iH_{kj} = \sum_{k=i+1}^{j} \alpha_k \frac{\beta_j}{\beta_k} = (j - i)\alpha_j.$$

If $i > j$, then

$${}^i\lambda_j = 1 - {}^j\lambda_i = 1 - (i - j)\alpha_i.$$

Therefore

$$G_{ij} = {}^i\lambda_j\, {}^iN_{jj} = \begin{cases} (j - i)\alpha_j & \text{if } j \geq i \\[2ex] \dfrac{\alpha_j}{\alpha_i} + (j - i)\alpha_j & \text{if } j < i. \end{cases}$$

To get C, we first compute $\hat{G}$. If $i > j$,

$$\begin{aligned} {}^i\hat{\lambda}_j &= \sum_k \alpha_k\, {}^i\hat{H}_{kj} = \sum_{k=j}^{i-1} \alpha_k + \sum_{k<j} \alpha_k\, {}^i\hat{H}_{0j} \\ &= \frac{1}{\sigma_\infty}\left[\sum_{k=j}^{i-1} \beta_k + \sum_{k<j} \beta_k\left(1 - \frac{\alpha_i}{\alpha_j}\right)\right] \\ &= \frac{1}{\sigma_\infty}\left[(\sigma_i - \sigma_j) + \sigma_j\left(1 - \frac{\alpha_i}{\alpha_j}\right)\right] \\ &= \frac{\sigma_i}{\sigma_\infty} - \frac{\alpha_i}{\alpha_j}\frac{\sigma_j}{\sigma_\infty}. \end{aligned}$$

If $i < j$,

$${}^i\hat{\lambda}_j = 1 - {}^j\hat{\lambda}_i = 1 - \frac{\sigma_j}{\sigma_\infty} + \frac{\alpha_j}{\alpha_i}\frac{\sigma_i}{\sigma_\infty}.$$

Thus

$$\hat{G}_{ij} = \begin{cases} \dfrac{\alpha_j}{\alpha_i}\dfrac{\sigma_i}{\sigma_\infty} - \dfrac{\sigma_j}{\sigma_\infty} & \text{if } j \leq i \\[3ex] 1 + \dfrac{\alpha_j}{\alpha_i}\dfrac{\sigma_i}{\sigma_\infty} - \dfrac{\sigma_j}{\sigma_\infty} & \text{if } j > i, \end{cases}$$

and

$$C_{ij} = \frac{\alpha_j}{\alpha_i}\hat{G}_{ji} = \begin{cases} \dfrac{\sigma_j}{\sigma_\infty} - \dfrac{\alpha_j}{\alpha_i}\dfrac{\sigma_i}{\sigma_\infty} & \text{if } j \geq i \\[2ex] \dfrac{\sigma_j}{\sigma_\infty} - \dfrac{\alpha_j}{\alpha_i}\dfrac{\sigma_i}{\sigma_\infty} + \dfrac{\alpha_j}{\alpha_i} & \text{if } j < i. \end{cases}$$

Also

$$\hat{C}_{ij} = \frac{\alpha_j}{\alpha_i}G_{ji} = \begin{cases} (i-j)\alpha_j & \text{if } j \leq i \\ (i-j)\alpha_j + 1 & \text{if } j > i. \end{cases}$$

In an ergodic basic example, C is never equal to G. For suppose $C_{0j} = G_{0j}$ for all j. Then

$$\frac{\sigma_j}{\sigma_\infty} = j\alpha_j$$

or

$$\sigma_j = j\beta_j$$

for all j. By induction, $\beta_j = \beta_{j+1}$ for every j, in contradiction to the fact that β_j must tend to 0 in a recurrent basic example.

In an ergodic basic example, we have

$$M_{ij} = \begin{cases} \dfrac{1}{\alpha_j}\dfrac{\sigma_j}{\sigma_\infty} - \dfrac{1}{\alpha_i}\dfrac{\sigma_i}{\sigma_\infty} & \text{if } j \geq i \\[2ex] \dfrac{1}{\alpha_j}\dfrac{\sigma_j}{\sigma_\infty} - \dfrac{1}{\alpha_i}\dfrac{\sigma_i}{\sigma_\infty} + \dfrac{1}{\alpha_i} & \text{if } j < i, \end{cases}$$

since $M_{ij} = C_{ij}/\alpha_j$.

It is clear that $\mathrm{M}_i[\mathfrak{t}_i^n]$ is finite if and only if $\mathrm{M}_0[\mathfrak{t}_0^n]$ is finite. In other words, ergodic degree is independent of the state. On the other hand, $\mathrm{M}_0[\mathfrak{t}_0^n]$ is finite if and only if $\mathrm{M}_0[\hat{\mathfrak{t}}_0^n]$ is finite and if and only if $\mathrm{M}_\alpha[\hat{\mathfrak{t}}_0^{n-1}]$ is finite. But trivially, $\mathrm{M}_\alpha[\hat{\mathfrak{t}}_0^{n-1}] = \sum_k \alpha_k k^{n-1} = (1/\sigma_\infty)\sum_k \beta_k k^{n-1}$. Hence P and $\hat{P}$ are of ergodic degree at least n if and only if $\sum_k \beta_k k^{n-1}$ is finite. The chain with $p_i = (i/(i+1))^{n+1}$ has $\beta_k = (k+1)^{-n-1}$; for this chain $\sum_k k^{n-1}\beta_k < \infty$ while $\sum_k k^n\beta_k = \infty$, and the chain is of degree n. To obtain a chain of infinite degree, let $p_i = p$ for every $i > 0$; such a chain represents "repetition of a single task." Then $\beta_k = p^k$ and $\sum_k k^n p^k < \infty$ for all n.

Turning to ergodic potentials, we know that if $\alpha f = 0$ and if Gf is finite-valued, then f is a charge with potential g and

$$g_i = -(Gf)_i = -\sum_j j\alpha_j f_j - \frac{1}{\alpha_i}\sum_{j<i}\alpha_j f_j.$$

Hence if $\alpha f = 0$ and if $\sum_j j\alpha_j f_j$ is finite, then f is a charge and its potential g satisfies $g = -Gf$.

We can show as follows that $\alpha f = 0$ is not a sufficient condition for f to be a charge, even in an ergodic chain. Let h be a function such that $h \geq 0$, $\alpha h = +\infty$, $f = (I - P)h$, and $\alpha f = 0$. For example, choose $p_i = (i/(i + 1))^{3/2}$, so that $\beta_i = (i + 1)^{-3/2}$ and the chain is ergodic; then $h_i = \sqrt{i + 1}$ has all the required properties. For any such function h

$$(I + P + \cdots + P^{n-1})f = (I - P^n)h$$

and, by Fatou's Theorem,

$$\liminf_n (P^n h) \geq Ah = +\infty.$$

Hence f is not a charge.

For ergodic chains, potentials among all integrable functions are characterized by the fact that they have integral zero—that is, $\alpha g = 0$ or $\nu 1 = 0$. We shall conclude this section by constructing a non-integrable potential. The chain we use will be the reverse of the fixed-task example with $p = \frac{1}{2}$. Then $\alpha_k = (\frac{1}{2})^{k+1}$ and

$$P_{i+1,i} = 1$$
$$P_{0i} = (\tfrac{1}{2})^{i+1} = \alpha_i.$$

Let ν be a row vector; the necessary and sufficient condition on ν that ν be a potential is that $\nu P^n \to 0$ and that $[\nu(I - P)]1 = 0$. Now

$$P_{0j}^{(0)} = \delta_{0j} \quad \text{and} \quad P_{0j}^{(n)} = \sum_k P_{0k} P_{kj}^{(n-1)} = \sum_k \alpha_k P_{kj}^{(n-1)}$$
$$= \alpha_j.$$

If $i \geq n$, then $P_{ij}^{(n)} = \delta_{i,j+n}$, and if $i < n$, then

$$P_{ij}^{(n)} = \sum_k P_{ik}^{(i)} P_{kj}^{(n-i)} = P_{0j}^{(n-i)} = \alpha_j.$$

Therefore

$$P_{ij}^{(n)} = \begin{cases} \delta_{i,j+n} & \text{if } i \geq n \\ \alpha_j & \text{if } i < n. \end{cases}$$

Now $(\nu P^n)_j = \sum_i \nu_i P_{ij}^{(n)} = \alpha_j \sum_{i<n} \nu_i + \nu_{j+n}$. Thus $\nu P^n \to 0$ if and only if

$$\lim_{n\to\infty} \sum_{i=0}^{n} \nu_i = 0.$$

On the other hand, we are trying to construct ν so that $|\nu|1 = +\infty$, and hence the series $\sum_{i=0}^{\infty} \nu_i$ must converge conditionally to 0.

The condition $[\nu(I - P)]1 = 0$ imposes more restrictions on ν. We have

$$[\nu(I - P)]_j = (\nu_j - \nu_{j+1}) - \alpha_j \nu_0,$$

so that $[\nu(I - P)]1$ is well defined if $\sum |\nu_j - \nu_{j+1}| < +\infty$. And if $\sum |\nu_j - \nu_{j+1}| < +\infty$, then

$$[\nu(I - P)]1 = \nu_0 - \lim_j \nu_j - \sum_j \alpha_j \nu_0 = -\lim_j \nu_j.$$

We conclude that ν is a nonintegrable potential if:

(1) $\sum_{i=0}^{\infty} \nu_i$ converges conditionally to 0.
(2) $\sum |\nu_j - \nu_{j+1}| < +\infty$.

If $\nu_0 = 0$, these conditions are necessary and sufficient. It is easily verified that the sequence

$$0, 1, -1, \tfrac{1}{4}, \tfrac{1}{4}, -\tfrac{1}{4}, -\tfrac{1}{4}, \tfrac{1}{9}, \tfrac{1}{9}, \tfrac{1}{9}, -\tfrac{1}{9}, -\tfrac{1}{9}, -\tfrac{1}{9}, \tfrac{1}{16}, \tfrac{1}{16}, \tfrac{1}{16}, \tfrac{1}{16}, -\tfrac{1}{16}, \ldots$$

satisfies conditions (1) and (2).

7. Further examples

EXAMPLE 1: Independent trials process.

In an independent trials process $P_{ij} = p_j$ independently of i, where $p_j > 0$ and $\sum p_j = 1$. In such a chain $P^n = P$ since

$$P_{ij}^{(n)} = \sum_k P_{ik}^{(n-1)} P_{kj} = \sum_k P_{ik}^{(n-1)} p_j = p_j = P_{ij}.$$

It follows that P is recurrent and that $A = \lim P^n = P$; the chain is noncyclic ergodic and $\alpha_j = p_j$. In addition,

$$Z = I + \sum_{n=1}^{\infty} (P^n - A) = I,$$

so that P is strong ergodic.

If $\alpha f = 0$, then f is a charge with potential $g = Zf = f$. If $\mu 1 = 0$ then μ is a charge with potential $\nu = \mu Z = \mu$.

We have

$$\Pr_0[\bar{\mathfrak{t}}_0 = k] = (1 - p_0)^{k-1} p_0.$$

Hence P is of infinite ergodic degree. To compute the M-matrix, we note that

$$\Pr_i[\bar{\mathfrak{t}}_j = k] = (1 - p_j)^{k-1} p_j$$

and therefore

$$\bar{M}_{ij} = p_j \sum_k k(1 - p_j)^{k-1} = p_j\left(\frac{1}{p_j^2}\right) = \frac{1}{p_j}.$$

Example 2: Reflecting random walk.

We return to the reflecting p-q random walk of Example 1 of Section 6-7. We first determine its ergodic degree. From 0 the chain returns to 0 either in one step or in an even number of steps, say $2n$, and

$$\Pr_0[\bar{\mathbf{t}}_0 = 1] = q.$$

Each path returning to 0 in $2n$ steps has n steps to the right and n to the left. Thus the probability of such a path is $(pq)^n$. From Feller [1957], p. 71, we see that the number of such paths is $\binom{2n-2}{n-1}\frac{1}{n}$. Therefore

$$\Pr_0[\bar{\mathbf{t}}_0 = 2n] = \binom{2n-2}{n-1}\frac{1}{n}(pq)^n.$$

We know from earlier results that the chain is recurrent if and only if $p \leq \frac{1}{2}$. The moments of the return time to 0 in this case are

$$\mathrm{M}_0[\bar{\mathbf{t}}_0^r] = q + \sum_{n=1}^{\infty}(2n)^r\binom{2n-2}{n-1}\frac{1}{n}(pq)^n.$$

By Stirling's formula, for large n

$$(2n)^r\binom{2n-2}{n-1}\frac{1}{n}(pq)^n \sim c(4pq)^n \cdot n^{r-3/2}$$

where c is a constant. If $p = \frac{1}{2}$, then $4pq = 1$, the series converges only for $r = 0$, and the chain is null. But if $p < \frac{1}{2}$, then $4pq < 1$ and all moments are finite. We may summarize as follows:

$$\text{If}\begin{Bmatrix} p > \frac{1}{2} \\ p = \frac{1}{2} \\ p < \frac{1}{2}\end{Bmatrix},\quad \text{then } P \text{ is}\begin{cases}\text{transient}\\ \text{null}\\ \text{ergodic of infinite degree.}\end{cases}$$

If $p < \frac{1}{2}$, we can find $C_{ij} = M_{ij}\alpha_j$ from the calculation of M_{ij} in Section 6-7. The reflecting random walk satisfies $P = \hat{P}$, and hence

$$G_{ij} = \hat{G}_{ij} = \frac{\alpha_j C_{ji}}{\alpha_i} = \alpha_j M_{ji}$$

If $p = \frac{1}{2}$, the process is with high probability far from state 0 after a long time. Hence ${}^i\lambda_j = 1$ if $j > i$. Thus

$$G_{ij} = \begin{cases} {}^iN_{jj} & \text{if } j > i \\ 0 & \text{if } j \leq i.\end{cases}$$

We may compute ${}^iN_{jj}$ for $j > i$ as follows. For $j > 0$,

$$\begin{aligned} {}^0\bar{H}_{jj} &= \tfrac{1}{2}\,{}^0H_{j+1,j} + \tfrac{1}{2}\,{}^0H_{j-1,j} \\ &= \tfrac{1}{2} + \tfrac{1}{2}\,{}^0H_{j-1,j} \\ &= \frac{1}{2} + \frac{1}{2}\frac{j-1}{j} \quad \text{from Section 5-4} \\ &= 1 - \frac{1}{2j}. \end{aligned}$$

Therefore, ${}^0N_{jj} = 2j$ and ${}^iN_{jj} = {}^0N_{j-i,j-i} = 2(j - i)$. Thus

$$G_{ij} = \begin{cases} 2(j-i) & \text{if } j > i \\ 0 & \text{if } j \le i. \end{cases}$$

To find the C matrix, we note that

$$\begin{aligned} C_{ij} &= \frac{\alpha_j}{\alpha_i} G_{ji} \quad \text{since } P = \hat{P} \\ &= G_{ji} \quad\quad \text{since } \alpha = 1^T. \end{aligned}$$

In the case $p = \frac{1}{2}$, the condition $\alpha f = 0$ is the condition $\sum_j f_j = 0$. Then

$$\begin{aligned} (Gf)_i &= 2 \sum_{j>i} (j-i)f_j \\ &= 2 \sum_j (j-i)f_j + 2 \sum_{j \le i} (i-j)f_j \\ &= 2 \sum_j jf_j + 2 \sum_{j<i} (i-j)f_j. \end{aligned}$$

Now

$$(\hat{C}f)_j = (G^Tf)_j = (f^TG)_j = 2 \sum_{i<j} f_i(j-i),$$

and the right side is always finite. Thus if Gf is finite-valued and $\alpha f = 0$, f is a weak charge. That is, if $\sum_j f_j = 0$ and if $\sum_j j|f_j| < \infty$, then

$$g_i = -2 \sum_j jf_j - 2 \sum_{j<i} (i-j)f_j$$

is a bounded potential.

Example 3: Sums of independent random variables on the line.

We consider one of the chains covered in Proposition 5-22; they represent sums of independent random variables on the line with finitely many k-values. If the mean of the k-values is zero, the process is recurrent. And since $\alpha = 1^T$, such a chain must be null.

In a recurrent process of this type

$$\left[N_{ii}^{(n)}\frac{\alpha_j}{\alpha_i} - N_{ij}^{(n)}\right] = [N_{ii}^{(n)} - N_{ij}^{(n)}] = [N_{jj}^{(n)} - N_{ij}^{(n)}].$$

It follows that G exists if and only if C exists, and if they both exist, then $C = G$. We are going to show that for this process ${}^i\lambda_j$ does indeed exist for all i and j and hence the process is normal.

Noting that

$${}^i\lambda_j = \lim_n \sum_k P_{0k}^{(n)}\,{}^iH_{kj},$$

we write for fixed N

$$\sum_k P_{0k}^{(n)}\,{}^iH_{kj} = \sum_{k=-N}^{N} P_{0k}^{(n)}\,{}^iH_{kj} + \sum_{k<-N} P_{0k}^{(n)}\,{}^iH_{kj} + \sum_{k>N} P_{0k}^{(n)}\,{}^iH_{kj}.$$

Suppose for the moment that the two limits

$${}^iH_{+\infty,j} = \lim_{k\to+\infty} {}^iH_{kj}$$

and

$${}^iH_{-\infty,j} = \lim_{k\to-\infty} {}^iH_{kj}$$

exist. Choose N large enough so that

$$|{}^iH_{kj} - {}^iH_{+\infty,j}| < \epsilon$$

and

$$|{}^iH_{-k,j} - {}^iH_{-\infty,j}| < \epsilon$$

for $k > N$. Then

$$\begin{aligned}\sum_k P_{0k}^{(n)}\,{}^iH_{kj} &= \sum_{k=-N}^{N} P_{0k}^{(n)}\,{}^iH_{kj} + {}^iH_{-\infty,j}\Big(\sum_{k<-N} P_{0k}^{(n)}\Big)\\ &\quad + {}^iH_{+\infty,j}\Big(\sum_{k>N} P_{0k}^{(n)}\Big) + \sum_k P_{0k}^{(n)}\epsilon_k,\end{aligned}$$

where

$$\epsilon_k = \begin{cases} 0 & \text{for } -N \le k \le N \\ {}^iH_{kj} - {}^iH_{\pm\infty,j} & \text{otherwise.}\end{cases}$$

Since $|\epsilon_k| < \epsilon$, the last term on the right is less than ϵ in absolute value for every n. Moreover, the first term on the right tends to zero with n, since P is null. Finally by the Central Limit Theorem (Theorem 1-68)

$$\lim_n \sum_{k<-N} P_{0k}^{(n)} = \lim_n \sum_{k>N} P_{0k}^{(n)} = \frac{1}{2}.$$

Therefore ${}^i\lambda_j$ exists and satisfies

$${}^i\lambda_j = \tfrac{1}{2}\,{}^iH_{-\infty,j} + \tfrac{1}{2}\,{}^iH_{+\infty,j}.$$

The missing step in the argument is the proof that ${}^iH_{-\infty,j}$ and ${}^iH_{+\infty,j}$ exist. We shall now show that ${}^iH_{-\infty,j}$ exists; the other proof is similar. Since ${}^i\lambda_j = {}^0\lambda_{j-i}$ and ${}^0\lambda_{-j} = {}^j\lambda_0 = 1 - {}^0\lambda_j$, it suffices to prove that ${}^0H_{-\infty,j}$ exists for $j > 0$. If v is the largest k-value and $E = \{0, \ldots, v\}$, then

$${}^0H_{-k,j} = \sum_{s \in E} B^E_{-k,s}\, {}^0H_{sj},$$

since neither 0 nor j can be reached from the negative side except through E. If we can show that $B^E_{-\infty,s} = \lim_{k \to -\infty} B^E_{k,s}$ exists, then since E is finite, we will have

$${}^0H_{-\infty,j} = \sum_{s \in E} B^E_{-\infty,s}\, {}^0H_{sj}.$$

Thus form the ladder process P^+. Then $p_j^+ = 0$ unless $0 < j \leq v$. By the special choice of E we have $B^E_{-k,s} = (B^+)^E_{-k,s}$. Letting $f_j = p_j^+$ and $u_n = H^+_{-n,0}$, we see that $u_0 = 1$ and

$$u_n = \sum_{k=1}^{n} p_k^+ H^+_{-(n-k),0} = \sum_{k=0}^{n-1} p^+_{n-k} H_{-k,0} = \sum_{k=0}^{n-1} f_{n-k} u_k.$$

Hence by the Renewal Theorem (Theorem 1-67), we have

$$\lim_{n \to \infty} H^+_{-n,0} = \frac{1}{\mu^+},$$

where $\mu^+ = \sum j p_j^+$. But $(B^+)^E_{-k,0} = H^+_{-k,0}$, and

$$H^+_{-k,s} = (B^+)^E_{-k,s} + \sum_{j<s} (B^+)^E_{-k,j} H^+_{j,s}.$$

Since $\lim_{k \to \infty} H^+_{-k,s} = 1/\mu^+$, we can prove by induction on s that $(B^+)^E_{-\infty,s}$ exists. Therefore ${}^0H_{-\infty,j}$ exists, and P is normal.

We conclude this section by treating the case of the one-dimensional symmetric random walk, in which $p_{-1} = p_{+1} = \frac{1}{2}$. For $j > 0$, it is clear that ${}^0H_{-\infty,j} = 0$ and ${}^0H_{+\infty,j} = 1$. Hence ${}^i\lambda_j = \frac{1}{2}$ and $G_{ij} = \frac{1}{2}\, {}^iN_{jj}$, in agreement with Corollary 9-29. For $j > 0$, ${}^0N_{jj}$ is the same as for Example 2 with $p = \frac{1}{2}$, since the two processes stopped at 0 are identical. Hence ${}^0N_{jj} = 2j$. If $j > i$, then

$${}^iN_{jj} = {}^0N_{j-i,j-i} = 2(j - i),$$

whereas if $j < i$, then ${}^iN_{jj} = {}^jN_{ii} = 2(i - j)$. Hence

$$G_{ij} = \tfrac{1}{2}\, {}^iN_{jj} = |i - j|.$$

Thus the potential operator is the absolute value of the distance, just as in classical one-dimensional potential theory. The conditions

$\alpha f = 0$ and Gf finite are the conditions $\sum_j f_j = 0$ and $\sum_j j|f_j| < \infty$. Since $\hat{C}f = Gf$, we see that if $\sum f_j = 0$ and $\sum j|f_j| < \infty$, then

$$g_i = -\sum_j |i - j| f_j$$

is a bounded potential.

8. The operator K

We introduce a new matrix to serve as potential operator for recurrent chains. The operator K will combine many of the properties of C and G and will have the single drawback that some of its entries may be negative. Our procedure will be first to tie the K-operator into our present notion of potential, then to define in terms of K the recurrent version of capacity, and finally to introduce so-called generalized potentials, in which we do not require total charge zero. With the generalized potentials we shall be able to prove analogs of the classical potential principles.

Let a distinguished state 0 be specified.

Definition 9-80: The K-matrix is defined by

$$K_{ij} = \lim_{n \to \infty} \left[N_{00}^{(n)} \frac{\alpha_j}{\alpha_0} - N_{ij}^{(n)} \right],$$

whenever the limit exists.

Lemma 9-81: If the indicated entries of C and G exist, then

$$K_{ij} = C_{ij} + (G_{0j} - C_{0j})$$

and

$$K_{ij} = G_{ij} + (C_{i0} - G_{i0}) \frac{\alpha_j}{\alpha_0}.$$

PROOF:

$$\begin{aligned} K_{ij} &= \lim \left[N_{00}^{(n)} \frac{\alpha_j}{\alpha_0} - N_{ij}^{(n)} \right] \\ &= \lim \left[N_{00}^{(n)} \frac{\alpha_j}{\alpha_0} - N_{0j}^{(n)} \right] - \lim \left[N_{jj}^{(n)} - N_{0j}^{(n)} \right] + \lim \left[N_{jj}^{(n)} - N_{ij}^{(n)} \right] \\ &= (G_{0j} - C_{0j}) + C_{ij}. \end{aligned}$$

By Proposition 9-45,

$$G_{i0} \frac{\alpha_j}{\alpha_0} + G_{0j} - G_{ij} = {}^0N_{ij} = C_{i0} \frac{\alpha_j}{\alpha_0} + C_{0j} - C_{ij},$$

and hence

$$C_{ij} + (G_{0j} - C_{0j}) = G_{ij} + (C_{i0} - G_{i0}) \frac{\alpha_j}{\alpha_0}.$$

Lemma 9-82: K exists if and only if P is normal.

PROOF: If P is normal, then C and G exist, and hence K exists by Lemma 9-81. Conversely, suppose K exists. Since $K_{i0} = C_{i0}$ and $K_{0j} = G_{0j}$, C_{i0} and G_{0j} exist for all i and j. Thus P is normal by Theorem 9-26.

Lemma 9-83: $\hat{K} = \text{dual } K$.

PROOF: Since $\hat{G} = \text{dual } C$ and $\hat{C} = \text{dual } G$, we have

$$(\alpha_j/\alpha_i)\hat{K}_{ji} = (\alpha_j/\alpha_i)\left[\hat{G}_{ji} + (\hat{C}_{j0} - \hat{G}_{j0})\frac{\alpha_i}{\alpha_0}\right]$$
$$= C_{ij} + (G_{0j} - C_{0j}) = K_{ij}.$$

The fact that $\hat{K} = \text{dual } K$ is the key property that the K matrix has and the C and G matrices lack. It is what is behind our first important result.

Theorem 9-84: Let

$$\nu = \lim\,[\mu(I + P + \cdots + P^n)]$$

and

$$g = \lim\,[(I + P + \cdots + P^n)f]$$

be potentials of weak charges in a normal chain P. Then $\nu = -\mu K$ and $g = -Kf$. If Y is any matrix such that for all potentials ν and g of finite support $\nu = \mu Y$ and $g = Yf$, then $Y = -(K + k(1\alpha))$, where k is a constant.

PROOF: By Theorem 9-31, $\nu = -\mu C$ and $g = -Gf$. But from Lemma 9-81 and the fact that $\mu 1 = \alpha f = 0$, we see that $-\mu K = -\mu C$ and $-Kf = -Gf$. Hence K has the desired property. If Y is an operator that serves for charges μ, then $\mu Y = -\mu K$, so that $\mu(Y + K) = 0$ Taking μ to be the row vector with 1 in the ith entry and -1 in the jth entry, we see that μ is a charge and that the ith and jth rows of $Y + K$ are equal. Hence $Y + K$ has constant columns. A similar argument with potential functions shows that $Y + K$ has rows proportional to α Therefore $Y + K = -k(1\alpha)$ for some k.

Actually at any time when $\nu = -\mu C$ or $g = -Gf$, we may use K in place of C and G. Thus K serves for both functions and measures. And the theorem shows that the only other two-sided potential operators differ from $-K$ by a multiple of 1α.

We note from the definition of K that if $C = G$, then $K = C = G$. Thus in the classical cases (the symmetric random walks), the three operators coincide.

We introduce the notation

$$^E d_i = {}^E\hat{v}_i/\alpha_i, \quad l_i^E = \hat{\lambda}_i^E/\alpha_i \quad \text{and} \quad W^E = \begin{pmatrix} I - P^E & 0 \\ 0 & 0 \end{pmatrix}.$$

Thus $^E d$ = dual $^E\hat{v}$ and l^E = dual $\hat{\lambda}^E$.

Proposition 9-85: If P is normal and E is a finite set, then

$$(-K)W^E = B^E - 1\lambda^E$$

and

$$W^E(-K) = \begin{pmatrix} ^E\bar{N} \\ 0 \end{pmatrix} - l^E\alpha.$$

PROOF: Lemma 9-37 proves that a column of $B^E - 1\lambda^E$ is a potential with charge the corresponding column of W^E. The first result then follows from Theorem 9-84. The second result follows by duality from Proposition 6-16.

Corollary 9-86: If P is normal and E is a finite set, then

$$-K_E(I - P^E) = I - 1\lambda_E^E$$

and

$$(I - P^E)(-K_E) = I - l_E^E\alpha_E.$$

PROOF: Restrict the equations of Proposition 9-85 to square matrices indexed by the states of E.

We shall see shortly how Corollary 9-86 may be used to compute P^E from K_E. Although we have the formula $P^E = T + UNR$ for P^E, this expression is not of practical value for infinite chains, since N is indexed by the states of $\tilde{E}$. On the other hand, K_E *can* be computed without finding all of K, and hence P^E can be calculated from K_E for finite sets E by using only finite matrices.

From the fact that $-K$ is a two-sided potential operator, we see from the proofs of Proposition 9-85 and Corollary 9-86 that

$$(I - P^E)C_E = -I + l_E^E\alpha_E$$

and

$$G_E(I - P^E) = -I + 1\lambda^E.$$

We turn now to a discussion of capacity. Throughout the remainder of the section we shall assume that P is a normal chain.

Proposition 9-87: For any finite set E there is a constant $k(E)$ such that

$$\lambda_E^E K_E = k(E)\alpha_E \quad\text{and}\quad K_E l_E^E = k(E)\mathbf{1}.$$

Furthermore,

$$k(E) = \lambda_E^E K_E l_E^E \quad\text{and}\quad k(E) = \hat{k}(E).$$

PROOF: In Corollary 9-86, multiply the first equation by K_E on the right and the second by K_E on the left and equate the results. Then

$$\mathbf{1}(\lambda_E^E K_E) = (K_E l_E^E)\alpha_E.$$

Thus for some constant $k(E)$, we have $K_E l_E^E = k(E)\mathbf{1}$ and $\lambda_E^E K_E = k(E)\alpha_E$. Multiplication of $\lambda_E^E K_E = k(E)\alpha_E$ on the right by l_E^E gives $k(E) = \lambda_E^E K_E l_E^E$, since $\alpha_E l_E^E = 1$. The dual of this equation is $k(E) = \hat{\lambda}_E^E \hat{K}_E \hat{l}_E^E$, and thus $k(E) = \hat{k}(E)$.

Definition 9-88: For a finite set E the constant $k(E)$ such that $K_E l_E^E = k(E)\mathbf{1}$ is called the **capacity** of E.

Just as the K-matrix in general depends upon the state 0 selected, so capacity in general is a function of the distinguished state. If $E = \{i\}$, then $\lambda_j^E = \delta_{ij}$, and from Proposition 9-87 we see that

$$K_{ii} = k(\{i\})\alpha_i$$

or

$$k(\{i\}) = (G_{0i} - C_{0i})/\alpha_i.$$

In particular, $k(\{0\}) = 0$.

If we form a K' matrix by using a distinguished state $0'$, then

$$\begin{aligned} K'_{ij} - K_{ij} &= (G_{0'j} - C_{0'j}) - (G_{0j} - C_{0j}) \\ &= (C_{00'} - G_{00'})\alpha_j/\alpha_{0'} \end{aligned}$$

by Proposition 9-45. But $K = K' + k\mathbf{1}\alpha$ by Theorem 9-84. Thus

$$k = \frac{G_{00'} - C_{00'}}{\alpha_{0'}} = k(\{0'\})$$

and

$$K = K' + k(\{0'\})\mathbf{1}\alpha.$$

Therefore, from Proposition 9-87 we see that

$$k(E) = k'(E) + k(\{0'\})$$

for every finite set E. Thus capacity is determined up to an additive constant. If we let $E = \{0\}$, we find that $0 = k'(\{0\}) + k(\{0'\})$ or that $k'(\{0\}) = -k(\{0'\})$. Note that since $k(\{0'\}) = (G_{00'} - C_{00'})/\alpha_{0'}$, when $C = G$ the capacity does not depend upon the choice of 0.

In general, there is no reason why $k(E)$ should be positive. However, if 0 is in E, then

$$k(E) = (Kl^E)_0 = (Gl^E)_0 \geq 0,$$

since Kl^E is constant on E. The dual expression is

$$\alpha_0 k(E) = (\lambda^E C)_0.$$

In the following discussions we shall restrict ourselves to sets which contain the reference point. We accordingly denote by $\mathscr{L}_0$ the collection of all finite sets which contain 0.

Proposition 9-89: If E is in $\mathscr{L}_0$, then $k(E) > 0$ if and only if there is a j in E such that ${}^0\lambda_j > 0$ and $\hat{\lambda}_j^E > 0$ (or, equivalently, there is a j in E such that ${}^0\hat{\lambda}_j > 0$ and $\lambda_j^E > 0$).

PROOF: For $E \in \mathscr{L}_0$,

$$\begin{aligned} k(E) &= \sum_{j \in E} G_{0j} l_j^E \\ &= \sum_{j \in E} \frac{{}^0\nu_j \hat{\lambda}_j^E}{\alpha_j} \\ &= \sum_{j \in E} {}^0\lambda_j \left(\frac{{}^0N_{jj}}{\alpha_j}\right) \hat{\lambda}_j^E. \end{aligned}$$

Thus $k(E) > 0$ if and only if there is a non-zero term. Since ${}^0N_{jj}/\alpha_j > 0$, the first assertion of the proposition follows. The equivalent condition follows by duality.

Proposition 9-90: If P is noncyclic ergodic or if P is a symmetric sums of independent random variables process, then $k(E) > 0$ for all sets E in $\mathscr{L}_0$ with more than one point.

PROOF: If P is ergodic, then $\lambda^E = \alpha B^E$ by Proposition 9-55. Hence $\lambda_j^E \geq \alpha_j > 0$ for $j \in E$. Since $\hat{P}$ is also ergodic, ${}^0\hat{\lambda}_j > 0$ for $j \neq 0$. Therefore, if E contains a state other than 0, then $k(E) > 0$ by Proposition 9-89.

If P is a symmetric sums of independent random variables process, then $K_{ij} = \frac{1}{2}\, {}^jN_{ii} \geq \frac{1}{2}$ for $j \neq i$, by Corollary 9-29. If $l_0^E > 0$, then $k(E) \geq K_{j0} l_0^E > 0$ for any other state $j \in E$. If $l_0^E = 0$, then $l_j^E > 0$ for some $j \in E$, and $k(E) \geq K_{0j} l_j^E > 0$.

Proposition 9-91: For any set E in $\mathscr{L}_0$, $K_E{}^{-1}$ exists if and only if $k(E) > 0$. If $k(E) > 0$, then

$$K_E = \left(P^E + \frac{1}{k(E)}\, l_E^E \lambda_E^E - I\right)^{-1}.$$

Proof: If $k(E) = 0$, then $\lambda_E^E K_E = 0$ and K_E^{-1} does not exist. If $k(E) > 0$, then by Corollary 9-86,

$$\begin{aligned} K_E\left(P^E + \frac{1}{k(E)}\, l_E^E \lambda_E^E - I\right) &= I - 1\lambda_E^E + K_E\left(\frac{1}{k(E)}\, l_E^E \lambda_E^E\right) \\ &= I - 1\lambda_E^E + 1\lambda_E^E \quad \text{by Proposition 9-87} \\ &= I. \end{aligned}$$

From Proposition 9-91 we obtain the following expression for P^E in terms of K_E^{-1} and α_E.

Corollary 9-92: If E is a set in $\mathscr{L}_0$ with $k(E) > 0$, then

$$P^E = I + K_E^{-1} - \frac{(K_E^{-1}1)(\alpha_E K_E^{-1})}{\alpha_E K_E^{-1}1}.$$

Proof: By Proposition 9-91,

$$P^E = I + K_E^{-1} - \frac{1}{k(E)}\, l_E^E \lambda_E^E.$$

Multiplying through on the right by 1 gives

$$K_E^{-1}1 = \frac{1}{k(E)}\, l_E^E.$$

If we multiply both sides by α_E, we get

$$k(E) = \frac{1}{\alpha_E K_E^{-1}1}.$$

Hence

$$l_E^E = \frac{K_E^{-1}1}{\alpha_E K_E^{-1}1}.$$

By duality,

$$\lambda_E^E = \frac{\alpha_E K_E^{-1}}{\alpha_E K_E^{-1}1},$$

and substitution for $k(E)^{-1} l_E^E \lambda_E^E$ gives the desired result.

We now begin a series of lemmas which lead to the fact that $k(E)$ is a Choquet capacity in the sense of Definition 8-38.

Lemma 9-93: If E is a finite set, then

$$\lim_n \,(B^E N^{(n)} - N^{(n)}) = {}^E d\alpha - {}^E N.$$

Proof: By Proposition 6-17,

$$B^E N^{(n)} - N^{(n)} = {}^E N P^{n+1} - {}^E N.$$

Now

$$\text{dual } ({}^E N P^{n+1}) = \hat{P}^{n+1}\, {}^E\hat{N} \to 1\, {}^E\hat{\nu} \quad \text{by Proposition 9-47,}$$

so that

$${}^E N P^{n+1} \to {}^E d\alpha.$$

Lemma 9-94: If E is a finite set, then

$$B^E K = K + {}^E N - {}^E d\alpha$$

and

$$K\begin{pmatrix} {}^E\bar{N} \\ 0 \end{pmatrix} = K + {}^E N - 1\, {}^E\nu.$$

Proof: Each row of $B^E - I$ is a weak charge. By Theorem 9-84 and Lemma 9-93,

$$(B^E - I)K = -\lim_n\, (B^E - I)N^{(n)} = {}^E N - {}^E d\alpha.$$

The second result is dual.

Lemma 9-95: If E is a finite set, then

$$\lambda^E K = k(E)\alpha + {}^E\nu$$

and

$$K l^E = k(E)1 + {}^E d.$$

Proof: Multiply the equation of Lemma 9-94 on the right by l^E to obtain

$$\begin{aligned} B^E(K l^E) &= K l^E + {}^E N l^E - {}^E d(\alpha l^E) \\ &= K l^E - {}^E d. \end{aligned}$$

But $(K l^E)_E = K_E l^E_E$, which by Proposition 9-87 equals $k(E)1$. Hence

$$K l^E = k(E)1 + {}^E d.$$

The other result is dual.

Lemma 9-96: If E, F, and L are finite sets with E and F both contained in L, then

$$[k(E) - k(F)] = \lambda^L({}^F d - {}^E d) = ({}^F\nu - {}^E\nu)l^L.$$

PROOF: Multiplying the equation of Lemma 9-95 through by λ^L, we obtain

$$\lambda^L K l^E = k(E) + \lambda^{L\ E}d$$

and similarly

$$\lambda^L K l^F = k(F) + \lambda^{L\ F}d.$$

Subtracting, we see that it is sufficient to prove that $\lambda^L K l^E = \lambda^L K l^F$ But

$$\lambda^L K l^E = (\lambda^L_L K_L) l^E_L = k(L)\alpha_L l^E_L = k(L)$$

since the support of l^E is in L. Similarly, $\lambda^L K l^F = k(L)$. The other equality is dual.

Lemma 9-97: For any sets $A_1, A_2, \ldots, A_r$ which have at least one point in common,

$$^{A_1 \cap A_2 \cap \cdots \cap A_r}N_{ij} \geq \sum_s {}^{A_s}N_{ij} - \sum_{s \neq t} {}^{A_s \cup A_t}N_{ij} + \sum_{s \neq t \neq u} {}^{A_s \cup A_t \cup A_u}N_{ij} - \cdots + (-1)^{r+1}\, {}^{A_1 \cup A_2 \cup \cdots \cup A_r}N_{ij}.$$

PROOF: The left side is the mean number of times in state j, starting in i, before the intersection of sets is reached. We shall show that the right side is the mean number of times in j before all of the sets have been reached. The former is clearly at least as large as the latter. Let $\mathbf{n}_j(\omega)$ be the number of times on ω that the process is in j before all sets are entered. On the path ω let S_k be the set of times at which the process is in j before A_k is entered. Then $\mathbf{n}_j(\omega)$ is the cardinal number of $S_1 \cup S_2 \cup \cdots \cup S_r$ and equals

$$\sum_s n(S_s) - \sum_{s \neq t} n(S_s \cap S_t) \pm \cdots,$$

where $n(A)$ is the cardinal number of A. Since $n(S_s \cap \cdots \cap S_t)$ is the number of times in j before $A_s \cup \cdots \cup A_t$ is entered, the result follows.

Proposition 9-98: Capacity is a monotone increasing set function, and for any sets $A_1, A_2, \ldots, A_r$ in $\mathscr{L}_0$,

$$k(A_1 \cap \cdots \cap A_r) \leq \sum_i k(A_i) - \sum_{i \neq j} k(A_i \cup A_j) + \sum_{i \neq j \neq k} k(A_i \cup A_j \cup A_k) - \cdots + (-1)^{r+1} k(A_1 \cup \cdots \cup A_r).$$

PROOF: We first prove monotonicity. Let $A \subset B$; then

$$^{A}N \geq {}^{B}N.$$

Thus

$$P^n\, {}^{A}N \geq P^n\, {}^{B}N.$$

As $n \to \infty$,

$$^{A}\nu \geq {}^{B}\nu.$$

Therefore

$$({}^{A}\nu - {}^{B}\nu)l^L \geq 0.$$

If L contains A and B, then by Lemma 9-96, $k(B) - k(A) \geq 0$ or

$$k(B) \geq k(A).$$

Therefore k is monotone. The other inequality of the proposition follows in exactly the same manner starting from the result of Lemma 9-97.

In the next section we shall prove potential principles for sets of positive capacity. It is therefore particularly annoying when all sets in a chain have capacity zero, and we treat this case now.

Definition 9-99: A normal chain is **degenerate** if, for every choice of the reference state 0, $k(E) = 0$ for all finite sets E.

Lemma 9-100: If, for a single reference point 0, $k(E) = 0$ for all $E \in \mathscr{L}_0$ and for all one-point sets, then the chain is degenerate.

PROOF: If $0'$ is any other reference point, then

$$k(E) = k'(E) + k(\{0'\}) = k'(E).$$

Hence it suffices to show that $k(E) = 0$. If $0 \in E$, $k(E) = 0$ by hypothesis. If $0 \notin E$, let $0'$ be any point of E, and $0 = k'(\{0'\}) \leq k'(E) = k(E) \leq k(E \cup \{0\}) = 0$. Both inequalities follow by the monotonicity proved in Proposition 9-98.

Lemma 9-101: A normal chain is degenerate if and only if for every pair of states i and j either ${}^{i}\lambda_j = 0$ and ${}^{i}\hat{\lambda}_j = 1$ *or* ${}^{i}\lambda_j = 1$ and ${}^{i}\hat{\lambda}_j = 0$.

PROOF: If the chain is degenerate, then $k(\{i, j\}) = 0$ if i is chosen as the reference point. Hence ${}^{i}\lambda_j = 0$ or ${}^{i}\hat{\lambda}_j = 0$ by Proposition 9-89. By symmetry ${}^{j}\lambda_i = 0$ or ${}^{j}\hat{\lambda}_i = 0$, and hence ${}^{i}\lambda_j = 1$ or ${}^{i}\hat{\lambda}_j = 1$.

Conversely, if ${}^0\lambda_j = 0$ or ${}^0\hat{\lambda}_j = 0$ for all states j, then $k(E) = 0$ for all $E \in \mathscr{L}_0$ (by Proposition 9-89 since $\lambda_j^E \le {}^0\lambda_j$). And

$$k(\{j\}) = \frac{1}{\alpha_j}(G_{0j} - C_{0j}) = \frac{{}^0N_{jj}}{\alpha_0}({}^0\lambda_j - {}^j\hat{\lambda}_0).$$

But ${}^0\lambda_j$ and ${}^j\hat{\lambda}_0$ are either both 0 or both 1. Hence $k(\{j\}) = 0$. Thus the converse follows from Lemma 9-100.

The basic example and its reverse are degenerate when null. It can be shown that degenerate chains are all of a type similar to the basic example. (See the problems.)

Proposition 9-102: If P is not degenerate, then there are two states 0 and 1 such that, with 0 as reference point, $k(E) > 0$ for all sets containing both states.

Proof: Choose an arbitrary point as the reference state 0. Then, by Lemma 9-100, there is either a one-point set or a set in $\mathscr{L}_0$ which has non-zero capacity.

If $k(\{1\}) > 0$ for some state, then we may choose 0 and 1 as indicated, since $k(E) \ge k(\{1\}) > 0$ if $1 \in E$. If $k(\{1\}) < 0$ for some state, then use 1 as a new reference point, and $k'(\{0\}) = -k(\{1\}) > 0$. Hence we simply interchange the roles of 0 and 1.

Otherwise $k(E) \ne 0$ for some $E \in \mathscr{L}_0$. Then $k(E) > 0$. Let E be a set in $\mathscr{L}_0$ with positive capacity and containing as few states as possible, let 1 be a state of E other than 0, and let $F = E - \{1\}$. By minimality, $k(F) = 0$. Thus, by Proposition 9-98,

$$0 = k(\{0\}) = k(F \cap \{0, 1\}) \le k(F) + k(\{0, 1\}) - k(E),$$

and

$$k(\{0, 1\}) \ge k(E) > 0.$$

Hence any set containing 0 and 1 also has positive capacity.

We conclude this section by applying our results to ergodic chains.

Proposition 9-103: If P is ergodic, $\sum_{k \in E} \lambda_k^E M_{k0} = k(E)$.

Proof: Since $\alpha_0 k(E) = (\lambda^E C)_0$, we have $k(E) = (\lambda^E M)_0$.

This proposition enables us to give an interpretation to $k(E)$. Since $\lambda^E = \alpha B^E$, $k(E) = \alpha(B^E M)_0 = \mathrm{M}_\alpha[\text{time to reach 0 after } E \text{ is entered}] = \mathrm{M}_\alpha[\mathbf{t}_0 - \mathbf{t}_E]$. This function is monotone in E since $\mathbf{t}_0 - \mathbf{t}_E$ is monotone on each path. The capacity inequalities follow from the fact that the time to enter all of n sets is no greater than the time to enter the intersection; hence the time to reach 0 from the intersection is no greater than the time to reach 0 after all n sets are entered. We also see why $k(E) > 0$, if E has more than one state.

Corollary 9-104: $(\lambda^E M)_j = (\hat{\lambda}^E \hat{M})_j$ for all $j \in E$.

PROOF: Choose $j \in E$ as reference point. Then

$$(\lambda^E M)_j = k(E) = \hat{k}(E) = (\hat{\lambda}^E \hat{M})_j.$$

Proposition 9-105: In an infinite ergodic chain $K + k1\alpha$ has negative entries for each k.

PROOF:

$$\begin{aligned}(K + k1\alpha)_{jj} &= G_{0j} - C_{0j} + k\alpha_j \\ &= \left(-\frac{C_{0j} - G_{0j}}{\alpha_j} + k\right)\alpha_j.\end{aligned}$$

Hence

$$\lim_j \frac{(K + k1\alpha)_{jj}}{\alpha_j} = -\infty \quad \text{by Proposition 9-63.}$$

Proposition 9-106: Let $\{E_n\}$ be an increasing sequence of finite sets with union the set of all states S. Then $\lim_{n\to\infty} k(E_n) < +\infty$ if and only if the chain is strong ergodic. In a strong ergodic chain, $k(E) = M_{\alpha 0} - M_{\alpha E}$, and hence $\hat{M}_{\alpha E} = M_{\alpha E}$.

PROOF: $k(E) = \mathrm{M}_\alpha[\mathbf{t}_0 - \mathbf{t}_E]$ and therefore

$$\sum_{i \in E} \alpha_i M_{i0} \le k(E) \le M_{\alpha 0}.$$

Hence as $E_n \to S$, $k(E_n) \to M_{\alpha 0}$, which is finite only in a strong ergodic chain. Then

$$M_{\alpha 0} - M_{\alpha E} = k(E) = \hat{k}(E) = \hat{M}_{\alpha 0} - \hat{M}_{\alpha E}.$$

Hence $\hat{M}_{\alpha E} = M_{\alpha E}$, since $\alpha M = \alpha \hat{M}$.

For strong ergodic chains the concept of capacity can be extended to infinite sets, and Proposition 9-103, Corollary 9-104, and Proposition 9-106 hold for all sets E.

9. Potential principles

In this section we shall assume that P is a nondegenerate normal chain. We shall assume that 0 and 1 are chosen according to Proposition 9-102, and we denote the collection of all finite sets containing both 0 and 1 as $\mathscr{L}$. Thus capacity is uniquely defined, $k(E) > 0$ for $E \in \mathscr{L}$ and K_E^{-1} exists by Proposition 9-91.

Definition 9-107: Let f be any function whose non-zero components are on a finite set E. Then f is the **charge** of the **generalized potential** $g = -Kf$, with **support** E and **total charge** αf. If $f \geq 0$, then g is a **pure potential**. Let μ be any signed measure whose non-zero components are on E. Then μ is the **charge** of the **generalized potential measure** $\nu = -\mu K$, with **support** E and **total charge** $\mu 1$.

In this section g and f will be used only for generalized potentials and their charges.

Proposition 9-108: g determines f, and if the support is in $E \in \mathscr{L}$, then g_E determines g.

PROOF: Since $E \in \mathscr{L}$, $k(E) > 0$, K_E^{-1} exists, and

$$f_E = -K_E^{-1} g_E, \qquad g = -K\begin{pmatrix} f_E \\ 0 \end{pmatrix}.$$

To see that g determines f even if the support F fails to contain 0 or 1, we simply let $E = F \cup \{0, 1\}$.

Proposition 9-109: For any g with support in E, $E \in \mathscr{L}$,

$$\begin{aligned} \lambda^E g &= -(\alpha f)k(E) \\ g &= B^E g - (\alpha f)\,{}^E d \\ (I - P^E)g_E &= f_E - (\alpha f)l_E^E. \end{aligned}$$

PROOF: Since $\lambda_E^E K_E = k(E)\alpha_E$ by Proposition 9-87, $-\lambda^E g = (\alpha f)k(E)$. Since $B^E K = K + {}^E N - {}^E d\alpha$ by Lemma 9-94, $-B^E g = -g - {}^E d(\alpha f)$. Since $(I - P^E)(-K_E) = I - l_E^E \alpha_E$ by Corollary 9-86,

$$(I - P^E)g_E = f_E - l_E^E(\alpha f).$$

Proposition 9-110: If g is a pure potential and E is any set in $\mathscr{L}$, not necessarily the support of g, then

$$\begin{aligned} -\lambda^E g &\geq (\alpha f)k(E) \\ g &\geq B^E g - (\alpha f)\,{}^E d \\ (I - P^E)g_E &\geq f_E - (\alpha f)l_E^E. \end{aligned}$$

PROOF:

$$\begin{aligned} -\lambda^E g &= \lambda^E K f \\ &= (k(E)\alpha + {}^E\nu)f \quad \text{by Lemma 9-95} \\ &\geq k(E)(\alpha f) \qquad \text{since } {}^E\nu \geq 0 \text{ and } f \geq 0. \end{aligned}$$

$$\begin{aligned} -B^E g &= B^E K f \\ &= Kf + {}^E N f - (\alpha f)\, {}^E d \quad \text{by Lemma 9-94} \\ &\geq Kf - (\alpha f)\, {}^E d \end{aligned}$$

or

$$B^E g \leq g + (\alpha f)\, {}^E d.$$

By Proposition 9-85,

$$W^E(-K) = \begin{pmatrix} {}^E\bar{N} \\ 0 \end{pmatrix} - l^E \alpha.$$

Hence

$$\begin{aligned} (I - P^E) g_E &= {}^E\bar{N} f - (\alpha f) l^E_E \\ &\geq f_E - (\alpha f) l^E_E, \end{aligned}$$

since ${}^E\bar{N}_{ii} = 1$ for $i \in E$ and $f \geq 0$.

The recurrent Maximum Principle is the following corollary.

Corollary 9-111: A generalized potential of non-negative total charge assumes its maximum on the support.

PROOF: By Proposition 9-109, $g = B^E g - (\alpha f)\, {}^E d$. But αf and ${}^E d$ are both non-negative. Hence $g \leq B^E g$, and $g_i \leq \max_{k \in E} g_k$.

Lemma 9-112: If g' is a generalized potential with support in $E \in \mathscr{L}$ and if g is any pure potential such that $g_E \leq g'_E$, then $\alpha f' \geq \alpha f$.

PROOF: By hypothesis and by Propositions 9-109 and 9-110,

$$0 \geq \lambda^E(g' - g) \geq (-\alpha f')k(E) + (\alpha f)k(E).$$

From this it follows that $\alpha f' \geq \alpha f$, since $k(E) > 0$.

Definition 9-113: An **equilibrium potential** for E is a potential with support in E which has total charge one and has constant values on E.

This definition of equilibrium potential agrees with the one for transient chains (Definition 8-27) only up to a constant multiple, but it will be a more convenient definition for recurrent chains. We shall discuss the effect of the change after proving that equilibrium potentials exist.

Theorem 9-114: Any set E in $\mathscr{L}$ has a unique equilibrium potential. The equilibrium charge is l^E and the value of the equilibrium potential on the set is $-k(E)$. For j not in E it has the value $-k(E) - {}^E d_j$.

Proof: Lemma 9-95, together with the fact that $\alpha l^E = 1$, shows that $-Kl^E$ is an equilibrium potential and has the values specified in the theorem.

Conversely, if f is an equilibrium charge, then by Proposition 9-109 $f_E = (I - P^E)g_E + (\alpha f)l_E^E = (I - P^E)g_E + l_E^E$. Since g_E is constant and $P^E 1 = 1$, $(I - P^E)g_E = 0$. Thus $f_E = l_E^E$, and the equilibrium potential is unique.

If we renormalize the equilibrium potential for E so that its value is one on E, then its total charge becomes $-1/k(E)$. Thus if we were following the definitions of transient potential theory, we would define $-1/k(E)$ to be the capacity of E. The set function $-1/k(E)$ is a monotone function, as is $k(E)$, but it does not have as nice a probabilistic interpretation as our choice.

We shall now prove the recurrent Principle of Balayage.

Theorem 9-115: If g is a pure potential and if $E \in \mathscr{L}$ then there is a unique pure potential g' with support in E such that

(1) $g'_E = g_E$,
(2) $g' \leq g$,
(3) $\alpha f' \geq \alpha f$,
(4) $\alpha f' = -k(E)^{-1}\lambda^E g$,

and this unique potential is $B^E g + (\lambda^E g/k(E))\, {}^E d$.

Proof: Since g' is determined by its values on E, there is only one potential satisfying (1), and its charge is

$$\begin{aligned} f'_E &= -K_E^{-1} g_E \\ &= (I - P^E)g_E - \frac{\lambda^E g}{k(E)}\, l_E^E \quad \text{by Proposition 9-91} \\ &\geq (I - P^E)g_E + (\alpha f)l_E^E \\ &\geq f_E. \end{aligned}$$

The last two steps are by Proposition 9-110. Thus g' is pure. Hence

$\alpha f' \geq \alpha f$, by Lemma 9-112, and we have (3). Result (4) follows by multiplying the second equation above by α. Furthermore

$$\begin{aligned} g' &= B^E g' - (\alpha f')^E d && \text{by Proposition 9-109} \\ &\leq B^E g - (\alpha f)^E d && \text{by (1) and (3)} \\ &\leq g && \text{by Proposition 9-110.} \end{aligned}$$

Hence we have (2). The formula for g' follows from (4) and the second part of Proposition 9-109.

The potential g' is, as in the transient case, referred to as the **balayage potential** of g on E.

Next, we prove the recurrent Principle of Domination.

Theorem 9-116: If g and g' are pure potentials, if g' has its support in $E \in \mathscr{L}$, and if $g_E \geq g'_E$, then $g \geq g'$.

PROOF:

$$\begin{aligned} g &\geq B^E g - (\alpha f)^E d && \text{by Proposition 9-110} \\ &\geq B^E g' - (\alpha f')^E d && \text{by hypothesis and by Lemma 9-112} \\ &= g' && \text{by Proposition 9-109.} \end{aligned}$$

Corollary 9-117: The balayage potential of g on E is the infimum of all pure potentials that dominate g on E.

PROOF: If $\bar{g}$ is a pure potential and $\bar{g}_E \geq g_E$, then $\bar{g}_E \geq g'_E$; hence $\bar{g} \geq g'$ by domination.

Proposition 9-118: If g is a pure potential with support in $E \in \mathscr{L}$ and total charge 1, then

$$\min_{i \in E} g_i \leq -k(E) \leq \max_{i \in E} g_i.$$

PROOF: Assume that the first inequality is false. Then $-(Kf)_i > -k(E)$ for all $i \in E$. Hence we may choose a $c > 1$ such that

$$-K(cf) \geq -k(E)1 = -Kl^E$$

on E. Thus $c = c(\alpha f) \leq \alpha l^E = 1$, by Lemma 9-112, which is a contradiction. The other inequality is proved similarly.

We retain the same definition of energy as in transient theory.

Definition 9-119: If $g = -Kf$ is a potential, the **energy** of g, denoted $\mathbf{I}(g)$, is

$$\mathbf{I}(g) = \mu g = \nu f.$$

Lemma 9-120: For any potential g with support in a set $E \in \mathscr{L}$, $\mathbf{I}(g) = \nu_E(I - P^E)g_E + (\alpha f)(\hat{\lambda}^E g)$. If $P = \hat{P}$, then

$$\mathbf{I}(g) = \nu_E(I - P^E)g_E - k(E)(\alpha f)^2.$$

PROOF: By Proposition 9-109,

$$(I - P^E)g_E = f_E - (\alpha f)l_E^E.$$

Hence

$$\nu_E(I - P^E)g_E = \mathbf{I}(g) - (\alpha f)(\nu_E l_E^E) = \mathbf{I}(g) - (\alpha f)(\hat{\lambda}^E g).$$

If $P = \hat{P}$, then $\hat{\lambda}^E g = \lambda^E g = -(\alpha f)k(E)$ by Proposition 9-109.

Lemma 9-121: $\nu_E(I - P^E)g_E = \frac{1}{2}\sum_{i,j\in E} \alpha_i P_{ij}^E(g_i - g_j)^2 \geq 0$, and the value is 0 only if g is constant on E.

PROOF: The proof proceeds as in Lemma 8-54, but $P^E 1 = 1$ and $\alpha_E P^E = \alpha_E$; hence $m_i = 0$ and $\pi_i = 0$. Let F be the subset of all states of E on which $g = g_0$. If $F \neq E$, there are states $i \in F$ and $j \in \tilde{F}$ such that $P_{ij}^E > 0$, since the states of E communicate. Then

$$\nu_E(I - P^E)g_E \geq \tfrac{1}{2}\alpha_i P_{ij}^E(g_i - g_j)^2 > 0.$$

Lemma 9-122: If $E \in \mathscr{L}$, the energy of the equilibrium potential of E is $-k(E)$.

PROOF: Since $g = -Kl^E$ is constant on E, $(I - P^E)g_E = 0$. Since $\alpha f = 1$, $\hat{\lambda}^E g = \hat{\lambda}^E(-k(E)1) = -k(E)$. Hence by Lemma 9-120, the result follows.

Theorem 9-123: If $P = \hat{P}$ and $E \in \mathscr{L}$, then among all potentials with support in E and total charge 1 the equilibrium potential alone minimizes energy.

PROOF: If g has support in E and $\alpha f = 1$, $\mathbf{I}(g) \geq -k(E)$, by Lemmas 9-120 and 9-121. Equality holds only if g is constant on E, in which case g is the equilibrium potential.

10. A model for potential theory

By an **electric circuit** we shall mean a denumerable number of terminals, some of which are connected by wires. The wire connecting terminals i and j has resistance r_{ij} and conductance $c_{ij} = 1/r_{ij}$. If there is no connection between i and j, we let $c_{ij} = 0$; also we define $c_{ii} = 0$. We shall assume that:

(1) For each i, $\sum_j c_{ij} < \infty$. This condition is satisfied, for example, if each terminal is connected to only finitely many other terminals.
(2) The circuit is connected.

From physics we have the following two laws. In the present context the first one may be taken as a definition of current in terms of voltage:

(1) (Ohm) If i is at voltage v_i and j is at voltage v_j, the current flowing from j to i is $(v_i - v_j)c_{ij}$.
(2) (Kirchhoff) The sum of all currents flowing into a given terminal is 0.

If an outside source is attached to a certain terminal, then the Kirchhoff Law (2) does not apply unless account is taken of current flowing from the outside source. But for all terminals i which are not attached to any outside sources, Ohm's Law and Kirchhoff's Law imply:

(3) $\sum_k (v_i - v_k)c_{ik} = 0$.

If a finite set E of terminals is kept at prescribed non-zero voltages by an outside source and if there is a finite set F with $E \subset F$ such that all points in the set $\tilde{F}$ are grounded (kept at 0 voltage), then we shall call the problem of finding voltages at the points i of $F - E$ in such a way that (3) holds a **standard voltage problem**. Note that for finite circuits the voltages may be prescribed at an arbitrary subset E of terminals.

Definition 9-124: A Markov chain P with $P1 = 1$ is said to **represent** some given electric circuit if each state corresponds to a terminal and if any standard voltage problem can be solved in such a way that the voltage vector is P-regular at points of $F - E$.

It follows from Theorem 8-41 that if P represents a circuit, then the solution v to a standard voltage problem is unique and satisfies

$$v = B^{E \cup \tilde{F}} v.$$

Thus the voltage at a point i of $F - E$ can be interpreted as the expected final payment if the chain is started at i and stopped at $E \cup \tilde{F}$ and if a payment of v_j is received if the process reaches state j of E.

Proposition 9-125: For any electric circuit there is a unique Markov chain P such that $P_{ii} = 0$ and P represents the circuit.

Proof: We first prove uniqueness; let P represent the circuit. Let i and j be any two distinct states, and let $E = \{j\}$ and $F = \{i, j\}$. Put a unit voltage at j and ground $\tilde{F}$. Then by (3),

$$\sum_k (v_i - v_k)c_{ik} = 0.$$

Since $v_k = 0$ except when k is i or j and since $c_{ii} = 0$, we have

$$v_i \sum_k c_{ik} = v_i c_{ii} + v_j c_{ij} = c_{ij}.$$

Hence

$$v_i = \frac{c_{ij}}{\sum_k c_{ik}}.$$

(The denominator is not zero since the circuit is connected.) Now since P represents the circuit,

$$v_i = (Pv)_i = \sum_k P_{ik}v_k = P_{ii}v_i + P_{ij}v_j.$$

Since $P_{ii} = 0$ and since $v_j = 1$, we have

$$v_i = P_{ij}.$$

Therefore

$$P_{ij} = \frac{c_{ij}}{\sum_k c_{ik}},$$

and P is unique.

Next we prove existence. Let the circuit be given, and define

$$P_{ij} = \frac{c_{ij}}{\sum_k c_{ik}}.$$

Then $P_{ij} \geq 0$, $P_{ii} = 0$, and

$$\sum_j P_{ij} = \frac{\sum_j c_{ij}}{\sum_k c_{ik}} = 1.$$

Hence P is a transition matrix with $P_{ii} = 0$ and $P\mathbf{1} = \mathbf{1}$. Thus let E and F be finite sets with $E \subset F$, let v_E be specified, and let $v_{\tilde{F}} = 0$ We are to show the standard voltage problem has a solution. Define v by

$$v = B^{E\cup\tilde{F}}\begin{pmatrix} v_E \\ v_{\tilde{F}} \\ 0 \end{pmatrix}.$$

Then v is regular on $F - E$, or

$$v_i - \sum_k P_{ik}v_k = [(I - P)v]_i = 0$$

for i in $F - E$. Since $P\mathbf{1} = \mathbf{1}$,

$$\sum_k (v_i - v_k)P_{ik} = 0,$$

or

$$\sum_k (v_i - v_k)\frac{c_{ik}}{\sum_m c_{im}} = 0.$$

Thus

$$\sum_k (v_i - v_k)c_{ik} = 0,$$

and v is a solution to the standard voltage problem.

Corollary 9-126: Every standard voltage problem has a solution, and that solution is unique.

PROOF: Existence follows from Proposition 9-125 and uniqueness follows from Theorem 8-41.

Shortly we shall show exactly how general the class of chains that represent circuits is. But first we shall exhibit the connection between the currents and voltages of this section and the charges and potentials of Markov chain potential theory. In so doing, what we will be showing is that electric circuits provide a model for the discrete potential theory associated with the class of chains that represent circuits.

In physics, current is the time rate at which charge flows past a point—that is, the derivative of charge with respect to time. Markov chains, however, have a time scale that is discrete and not continuous, and the proper analog of the time rate at which charge flows past a point is the amount of charge that moves past the point in unit time. With discrete time the charge moves to some point, stays for unit time, and then moves to another point. Hence the magnitude of the current at a point is equal to the magnitude of the accumulated charge at that point.

Now in a standard voltage problem, current flows in and out of the circuit through the terminals which are attached to the outside source. The above considerations lead us to define the **charge at terminal** i to be the current μ_i which flows into the circuit; μ_i is taken to be negative if the current flows out. By Kirchhoff's Law (2), the charge will be zero on the set $F - E$. For i in $E \cup \tilde{F}$ the charge is given by

$$\mu_i = \sum_k (v_i - v_k)c_{ik}.$$

Before we can connect μ and v in terms of the representing chain, we need one preliminary result.

Proposition 9-127: Let P be a Markov chain which represents an electric circuit with conductances c_{ij}. Then the row vector α defined by

$$\alpha_i = \sum_k c_{ik}$$

is P-regular, and the α-dual of P is P.

PROOF: We have

$$\alpha_i P_{ij} = \alpha_i \frac{c_{ij}}{\sum_k c_{ik}} = \alpha_i \frac{c_{ij}}{\alpha_i} = c_{ij} = c_{ji} = \alpha_j P_{ji},$$

and hence

$$\sum_i \alpha_i P_{ij} = \sum_i \alpha_j P_{ji} = \alpha_j.$$

Therefore α is regular. Since $P_{ij} = (\alpha_j P_{ji})/\alpha_i$, the α-dual of P is P.

In terms of Proposition 9-127 we can transform the equation $\mu_i = \sum_k (v_i - v_k)c_{ik}$ as follows:

$$\begin{aligned}\mu_i &= v_i\alpha_i - \sum_k \alpha_i P_{ik} v_k \\ &= v_i\alpha_i - \sum_k v_k \alpha_k P_{ki} \\ &= [(\text{dual } v)(I - P)]_i.\end{aligned}$$

Hence $\mu = (\text{dual } v)(I - P)$. Since $P = \hat{P}$,

$$\text{dual } \mu = (I - P)v.$$

Let $f_i = \mu_i/\alpha_i$, that is, f is the dual of the vector of charges at the various terminals. Then

$$f = (I - P)v.$$

We note that all pairs of states communicate in P since the circuit is connected; hence P is either transient or recurrent. But v has only finitely many non-zero entries, so that αv is finite. It follows that if P

is transient or null, then v is a potential in the Markov chain sense. And if P is ergodic, then $v - (\alpha f)\mathbf{1}$ is a potential. In any case, f is the charge.

Conversely, if, in a chain which represents a circuit, g is a potential vanishing outside a finite set F with a charge f such that f has total charge 0 and f has support in $E \cup \tilde{F}$, where $E \subset F$, then g solves the standard voltage problem for E and F with the specified values g_E on E. It is in this sense that electric circuits form a model for potential theory.

We turn to the problem of classifying all Markov chains which represent circuits. A chain P is said to be **α-reversible** if $\hat{P}$, the α-dual of P, equals P.

Proposition 9-128: A Markov chain with $P_{ii} = 0$ represents a circuit if and only if its states communicate and it has a positive regular measure α with respect to which it is α-reversible.

PROOF: If P represents a circuit, then it has a regular measure α and is α-reversible by Proposition 9-127. Its states communicate since the circuit is connected.

Conversely, suppose that P is a transition matrix with the stated properties. Introduce the electric circuit with the states of P as terminals and with $c_{ij} = \alpha_i P_{ij}$. The circuit is well defined since

$$c_{ii} = \alpha_i P_{ii} = 0$$

and since

$$c_{ij} = \alpha_i P_{ij} = \alpha_j P_{ji} = c_{ji}.$$

Since the states communicate, the circuit is connected. To see that P represents the circuit, we note that

$$\alpha_i = \sum_k \alpha_i P_{ik} = \sum_k c_{ik}.$$

Thus

$$P_{ij} = \frac{c_{ij}}{\alpha_i} = \frac{c_{ij}}{\sum_k c_{ik}}.$$

Finally we consider the problem of when the chain representing a circuit is recurrent and when it is transient.

Lemma 9-129: Let P be a chain which represents an electric circuit. Let a unit voltage be put at 0, let F be a finite set containing 0, and let $\tilde{F}$ be kept at voltage 0. The charge at the terminal 0 is

$$\mu_0 = \alpha_0 \cdot {}^0\bar{H}_{0\tilde{F}}.$$

Proof:

$$
\begin{aligned}
{}^0\bar{H}_{0\bar{F}} &= \sum_k P_{0k}\, {}^0H_{k\bar{F}} \\
&= \sum_k P_{0k}(1 - {}^{\bar{F}}H_{k0}) \quad \text{since } {}^{\bar{F}}P \text{ is absorbing} \\
&= \sum_k P_{0k}(1 - B_{k0}^{\{0\}\cup\bar{F}}) \\
&= \sum_k P_{0k}(v_0 - v_k) \\
&= \frac{1}{\alpha_0} \sum_k c_{0k}(v_0 - v_k) \\
&= \frac{\mu_0}{\alpha_0}.
\end{aligned}
$$

Lemma 9-130: In any Markov chain a state 0 is recurrent if and only if ${}^0\bar{H}_{0\bar{F}} \to 0$ for some (or every) increasing sequence of finite sets F with union the set of all states.

Proof: If 0 is transient, then there is a positive probability $1 - \bar{H}_{00}$ that the chain never returns to 0. Hence

$$
{}^0\bar{H}_{0\bar{F}} \geq 1 - \bar{H}_{00} > 0
$$

for every finite set F, and ${}^0\bar{H}_{0\bar{F}}$ cannot approach 0.

Conversely, let 0 be recurrent. Choose N sufficiently large that $\bar{H}_{00}^{(N)} > 1 - \epsilon$. Choose δ close enough to 1 so that

$$
1 - \epsilon < \delta^N < 1.
$$

Then construct an increasing sequence of finite sets $A_0, A_1, \ldots, A_N$ such that $A_0 = \{0\}$ and such that the probability of stepping from any state of A_k to a state of A_{k+1} is greater than δ. Let F be any finite set containing A_N. The probability that the process started at 0 is, for every $n \leq N$, in A_n after n steps is greater than δ^N. Hence

$$
{}^0\bar{H}_{0\bar{F}}^{(N)} < 1 - \delta^N < \epsilon.
$$

Since

$$
1 - \epsilon < \bar{H}_{00}^{(N)} \leq {}^{\bar{F}}\bar{H}_{00}^{(N)} + {}^0\bar{H}_{0\bar{F}}^{(N)},
$$

we have

$$
{}^{\bar{F}}\bar{H}_{00} \geq {}^{\bar{F}}\bar{H}_{00}^{(N)} > 1 - 2\epsilon.
$$

But

$$
{}^{\bar{F}}\bar{H}_{00} + {}^0\bar{H}_{0\bar{F}} = 1,
$$

so that

$$ {}^0\bar{H}_{0\tilde{F}} < 2\epsilon $$

for all F containing A_N.

Proposition 9-131: Let P be a chain which represents an electric circuit. Then the following is a necessary and sufficient condition for P to be recurrent: If terminal 0 is kept at a unit voltage, if all terminals outside a finite set F are grounded, and if F is allowed to increase to S, then the current at terminal 0 tends to zero. Furthermore a necessary and sufficient condition for P to be ergodic is that $\sum_{i,j} c_{ij} < \infty$.

PROOF: The first assertion follows directly from Lemmas 9-129 and 9-130. The second assertion follows from the fact that P is ergodic if and only if $\alpha\mathbf{1} < \infty$, where $\alpha_i = \sum_j c_{ij}$.

11. A nonnormal chain and other examples

We shall show by examples in this section that all three of the following conjectures are false:

(1) Small sets and ergodic sets are identical concepts, or at least one of the notions includes the other.
(2) All null chains are normal.
(3) The existence of either C or G implies the existence of both.

First, we settle the independence of the notions of small sets and ergodic sets. We saw in Section 6 an example of an ergodic set which is not small, and we shall now produce a small set which is not ergodic.

Let P be a chain with states the non-negative integers and with transition probabilities

$$ P_{0i} = p_i = P_{i0} \quad \text{for } i > 0 $$

$$ P_{ii} = q_i = 1 - p_i. $$

All other entries of P are 0. We impose no requirements on the p_i yet except that $p_i > 0$ and $\sum p_i = 1$. It is clear that $\bar{H}_{00} = 1$ and hence P is recurrent. Since $P = P^T$, $\alpha = \mathbf{1}^T$ is regular and P is null. Only finite sets are ergodic, and thus it is sufficient to exhibit an infinite small set. For any set E containing 0,

$$ \lambda_j^E = \lim_n \sum_k P_{0k}^{(n)} B_{kj}^E = \lim_n P_{0j}^{(n)} = 0 \quad \text{for } j > 0, $$

whereas

$$\begin{aligned}\lambda_0^E &= \lim_n \sum_k P_{0k}^{(n)} B_{k0}^E \\ &= \lim_n P_{00}^{(n)} + \lim_n \sum_{k \in \tilde{E}} P_{0k}^{(n)} \\ &= 1 - \lim_n P_{0\tilde{E}}^{(n)}.\end{aligned}$$

Thus any set containing 0 such that $P_{0\tilde{E}}^{(n)} \to 0$ satisfies $\lambda_0^E = 1$ and is therefore a small set. Let

$$p_i = 4^{-k} \quad \text{for } 2^k - 1 \le i \le 2^{k+1} - 2.$$

The p_i assume the constant value 4^{-k} on a block of length 2^k. Thus

$$\sum p_i = \sum_{k=1}^{\infty} 2^k \cdot 4^{-k} = 1.$$

Let E consist of 0 and one state, such as 2^k, from each block, and label the representative of the kth block in E as e_k. Then

$$P_{0E}^{(n)} = P_{00}^{(n)} + \sum_{k=1}^{N-1} P_{0e_k}^{(n)} + P_{0T_N}^{(n)},$$

where $T_N = \{e_N, e_{N+1}, e_{N+2}, \ldots\}$. We can form 2^N disjoint sets like T_N each differing from it in every representative selected from the Nth block on. By symmetry $P_{0T'_N}^{(n)} = P_{0T_N}^{(n)}$ for all such sets T'_N. Hence $P_{0T_N}^{(n)} \le 1/2^N$ for all n. Since $P_{0j}^{(n)} \to 0$ in a null chain,

$$\limsup_n P_{0E}^{(n)} \le \frac{1}{2^N},$$

and we must have $P_{0E}^{(n)} \to 0$.

Second, we shall construct a nonnormal null chain. We shall use generating functions and require the following facts:

(1) If $F(t) = \sum_n f_n t^n$ and $G(t) = \sum_n g_n t^n$, then

$$F(t) \cdot G(t) = \sum_n \left(\sum_{k=0}^{n} f_k g_{n-k} \right) t^n.$$

That is, the coefficients of the series for $F(t) \cdot G(t)$ are the convolutions $\sum f_k g_{n-k}$.

(2) The Abel sum of the series $\sum_n f_n$ is $\lim_{t \to 1^-} F(t)$. If the series $\sum f_n$ converges, then its Abel sum exists and has the same value (Proposition 1-62).

Let P be any recurrent chain; we begin by deriving a necessary and sufficient condition for the series $\sum_n (P_{11}^{(n)} - P_{01}^{(n)})$ defining C_{01} to be

Abel summable. Let $E = \{0, 1\}$ and define generating functions as follows:

$$A_{00}(t) = \sum_n {}^1\bar{F}_{00}^{(n)}t^n, \qquad A_{01}(t) = \sum_n {}^0\bar{F}_{01}^{(n)}t^n$$

$$A_{10}(t) = \sum_n {}^1\bar{F}_{10}^{(n)}t^n, \qquad A_{11}(t) = \sum_n {}^0\bar{F}_{11}^{(n)}t^n$$

$$P_{ij}(t) = \sum_n P_{ij}^{(n)}t^n$$

$$H_0(t) = A_{00}(t) + A_{01}(t) = \sum_n \bar{F}_{0E}^{(n)}t^n$$

$$H_1(t) = A_{10}(t) + A_{11}(t) = \sum_n \bar{F}_{1E}^{(n)}t^n$$

$$Q(t) = \sum_n (P_{11}^{(n)} - P_{01}^{(n)})t^n = P_{11}(t) - P_{01}(t)$$

$$R(t) = \frac{1 - H_1(t)}{1 - H_0(t)}.$$

We note that the series defining C_{01} is Abel summable if and only if $\lim_{t\to 1^-} Q(t)$ exists. We shall prove that this limit exists if and only if $\lim_{t\to 1^-} R(t)$ exists. We have

$$P_{01}^{(n)} = \sum_k ({}^1\bar{F}_{00}^{(k)}P_{01}^{(n-k)} + {}^0\bar{F}_{01}^{(k)}P_{11}^{(n-k)})$$

or

$$P_{01}(t) = A_{00}(t)P_{01}(t) + A_{01}(t)P_{11}(t).$$

Also

$$P_{11}^{(n)} = \delta_{n0} + \sum_k ({}^1\bar{F}_{10}^{(k)}P_{01}^{(n-k)} + {}^0\bar{F}_{11}^{(k)}P_{11}^{(n-k)})$$

or

$$P_{11}(t) = 1 + A_{10}(t)P_{01}(t) + A_{11}(t)P_{11}(t).$$

Solving these equations for $P_{01}(t)$ and $P_{11}(t)$, we find

$$\begin{aligned} Q(t) &= P_{11}(t) - P_{01}(t) \\ &= \frac{1 - A_{00}(t) - A_{01}(t)}{(1 - A_{00}(t))(1 - A_{11}(t)) - A_{01}(t)A_{10}(t)} \\ &= \frac{1}{(1 - A_{00}(t))R(t) + A_{10}(t)}. \end{aligned}$$

Since $A_{00}(1) = {}^1\bar{H}_{00} < 1$ and since $A_{10}(1) = {}^1\bar{H}_{10} > 0$, $\lim_{t\to 1^-} Q(t)$ exists if and only if $\lim_{t\to 1^-} R(t)$ exists.

The example of a nonnormal chain will be like the earlier example in this section, only "doubled." The states are

$$0, a_1, a_2, a_3, \ldots$$

$$1, b_1, b_2, b_3, \ldots,$$

and if $p_i > 0$, $p_i' > 0$, $\sum p_i = 1$, and $\sum p_i' = 1$, then the transition probabilities are

$$\begin{aligned} P_{0a_i} &= p_i = P_{a_i 1} \\ P_{a_i a_i} &= q_i = 1 - p_i \\ P_{1b_i} &= p_i' = P_{b_i 0} \\ P_{b_i b_i} &= q_i' = 1 - p_i'. \end{aligned}$$

All other entries are 0. We see easily that $\bar{H}_{00} = 1$ and that $\alpha = 1^T$ is regular, hence the chain is recurrent and null. For $E = \{0, 1\}$,

$$\bar{F}_{0E}^{(n)} = F_{01}^{(n)} = \sum_i p_i q_i^{n-2} p_i \quad \text{for } n \geq 2.$$

Therefore

$$H_0(t) = \sum_i p_i^2 \sum_{n=2}^{\infty} q_i^{n-2} t^n = \sum_i \frac{p_i^2 t^2}{1 - q_i t}.$$

Similarly,

$$H_1(t) = \sum_i \frac{(p_i')^2 t^2}{1 - q_i' t},$$

and we have defined $R(t)$ by

$$R(t) = \frac{1 - H_1(t)}{1 - H_0(t)}.$$

We again choose the p_i's in blocks as follows. Let $\{n_k\}$ and $\{n_k'\}$ be two rapidly increasing sequences (with magnitude specified later) such that $n_k < n_k' < n_{k+1}$. Let there be n_k consecutive p_i's equal to $\epsilon_k = 1/(2^k n_k)$, and let there be n_k' consecutive (p_i')'s equal to $\epsilon_k' = 1/(2^k n_k')$. The remainder of the proof consists in showing that for suitably chosen $\{n_k\}$ and $\{n_k'\}$

$$\lim_{n \to \infty} R(1 - \epsilon_n) \neq \lim_{n \to \infty} R(1 - \epsilon_n').$$

We shall only sketch the argument.

We begin by estimating $R(1 - \epsilon_n)$ for large n. We have

$$\begin{aligned} H_0(1 - \epsilon_n) &= \sum_i \frac{p_i^2 (1 - \epsilon_n)^2}{1 - (1 - p_i)(1 - \epsilon_n)} \sim \sum_i \frac{p_i^2}{p_i + \epsilon_n} \\ &= \sum_{k=1}^{\infty} n_k \frac{\epsilon_k^2}{\epsilon_k + \epsilon_n} = \sum_{k=1}^{\infty} \frac{1}{2^k} \frac{\epsilon_k}{\epsilon_k + \epsilon_n}. \end{aligned}$$

For $k = n$, $\epsilon_k/(\epsilon_k + \epsilon_n) = \frac{1}{2}$. Choose the sequence n_k so that ϵ_n is negligible compared to ϵ_k when $k < n$. Then

$$H_0(1 - \epsilon_n) \sim \sum_{k=1}^{n-1} 2^{-k} \cdot 1 + 2^{-n} \cdot 2^{-1} = 1 - 2^{-n+1} + 2^{-n-1}.$$

Similarly,

$$H_1(1 - \epsilon_n) \sim \sum_{k=1}^{\infty} 2^{-k} \frac{\epsilon_k'}{\epsilon_k' + \epsilon_n} \sim \sum_{k=0}^{n-1} 2^{-k} = 1 - 2^{-n+1},$$

which differs from $H_0(1 - \epsilon_n)$ if ϵ_n is chosen to be negligible compared with ϵ'_n but much bigger than ϵ'_{n-1}.

$$R(1 - \epsilon_n) \sim \frac{2^{-n+1}}{2^{-n+1} - 2^{-n-1}} = \frac{4}{3}.$$

If $t = 1 - \epsilon'_n$ and n is large we obtain, similarly,

$$H_0(1 - \epsilon'_n) \sim \sum_{k=1}^{\infty} 2^{-k} \frac{\epsilon_k}{\epsilon_k + \epsilon'_n} \sim \sum_{k=1}^{n} 2^{-k} = 1 - 2^{-n}$$

and

$$H_1(1 - \epsilon'_n) \sim \sum_{k=1}^{\infty} 2^{-k} \frac{\epsilon'_k}{\epsilon'_k + \epsilon'_n} \sim \sum_{k=1}^{n-1} 2^{-k} + 2^{-n}.2^{-1}$$
$$= 1 - 2^{-n+1} + 2^{-n-1}.$$

The asymmetry in $H_0(1 - \epsilon'_n)$ and $H_1(1 - \epsilon_n)$ arises because the condition $n_k < n'_k < n_{k+1}$ is not symmetric. We thus find

$$R(1 - \epsilon'_n) \sim \frac{2^{-n+1} - 2^{-n-1}}{2^{-n}} = \frac{3}{2}.$$

Therefore, $\lim_{t \to 1^-} R(t)$ does not exist and C_{01} cannot exist. In particular, P is not normal.

This example has the property that neither C nor G exists. The reverse process has transition probabilities $\hat{P}_{0b_i} = P_{b_i0} = p'_i = \hat{P}_{b_i1}$ and $\hat{P}_{1a_i} = p_i = \hat{P}_{a_i0}$. Thus $\hat{P}$ is the same as P except that the roles of 0 and 1 have been interchanged. The above argument therefore shows that $\hat{C}_{10}$ does not exist, and since $\hat{C}_{10} = G_{01}$ if either exists, G cannot exist.

Not even reversibility ($P = \hat{P}$) is a sufficient condition for a null chain to be normal. A slight modification of the above example provides a counterexample. Let the states be as before, and define p_i and p'_i as above. Let

$$P_{0a_i} = P_{a_i0} = \tfrac{1}{2}p_i, \qquad P_{1b_i} = P_{b_i1} = \tfrac{1}{2}p'_i,$$

$$P_{a_ia_i} = 1 - \tfrac{1}{2}p_i, \qquad P_{b_ib_i} = 1 - \tfrac{1}{2}p'_i,$$

and

$$P_{01} = P_{10} = \tfrac{1}{2}.$$

Set all other entries of P equal to 0. Then $P = P^T$, so that $\alpha = 1^T$ and $\hat{P} = IP^TI = P$. But the same kind of computation as for the preceding example shows that C_{01} does not exist. Thus we see that even a symmetric P need not be normal.

Third, we show that the existence of one of C and G does not imply the existence of the other. Again we modify the first example of a nonnormal chain. Let

$$P_{0a_i} = p_i, \quad P_{a_i a_i} = q_i, \quad P_{1b_i} = p_i', \quad \text{and} \quad P_{b_i b_i} = q_i'$$

as before, and use the same p's. But set

$$P_{a_i 0} = P_{a_i 1} = \tfrac{1}{2}p_i \quad \text{and} \quad P_{b_i 0} = P_{b_i 1} = \tfrac{1}{2}p_i'.$$

It is clear that $\bar{F}^{(n)}_{0,\{0,1\}}$ and $\bar{F}^{(n)}_{1,\{0,1\}}$ are the same as before, so that $\lim_{t\to 1^-} R(t)$ does not exist and C_{01} does not exist. On the other hand, the reverse chain is no longer of the same type and the argument for the nonexistence of G fails. In fact,

$$\begin{aligned} {}^0\lambda_1 &= \lim_n \left[P^{(n)}_{01} + \sum_i P^{(n)}_{0a_i}\,{}^0H_{a_i 1} + \sum_i P^{(n)}_{0b_i}\,{}^0H_{b_i 1}\right] \\ &= \lim_n \left[P^{(n)}_{01} + \frac{1}{2}\sum_i (P^{(n)}_{0a_i} + P^{(n)}_{0b_i})\right] \\ &= \frac{1}{2}. \end{aligned}$$

Hence $G_{01} = \frac{1}{2}\,{}^0N_{11}$ exists. Moreover, if x and y are any two states other than 0 and 1,

$${}^x\lambda_y = \tfrac{1}{2}\,{}^xH_{0y} + \tfrac{1}{2}\,{}^xH_{1y}.$$

Therefore all of G exists. The reverse chain is an example in which C exists but G does not.

12. Two-dimensional symmetric random walk

The purpose of this section is to show how the results of Section 8 may be used to work out some of the matrices associated with the two-dimensional symmetric random walk.

First we shall find the operator K. In this example, $K = C = G$ and

$$K_{(x,y),(x',y')} = K_{(x-x',y-y'),(0,0)}.$$

Hence it suffices to compute one column of K. We let

$$k(x, y) = K_{(x,y),(0,0)}.$$

A row of $I - P$ is a charge whose potential is the corresponding row of I. For this process a row of $I - P$ has only finitely many non-zero entries and is therefore a weak charge. By Theorem 9-84, $(I - P)K = -I$. Thus $k(x, y)$ is the average of its values at the four neighboring points, except that at the origin the average is one larger. We know also that $k(0, 0) = 0$. The high degree of symmetry of the chain implies that

$$k(x, y) = k(-x, y) = k(y, x).$$

In particular, the values at the four points neighboring the origin must be the same. Since their average is one, $k(0, 1) = 1$.

We shall need one more result. It can be shown (see Spitzer [1964]) that

$$k(x, x) = \frac{4}{\pi}\left(\frac{1}{1} + \frac{1}{3} + \cdots + \frac{1}{2x - 1}\right) \quad \text{for } x > 0.$$

This identity, together with the properties above, determines the function k.

In fact, it suffices to restrict attention to $0 \le x \le y$. We know that $k(0, 0) = 0$, $k(0, 1) = 1$, and $k(1, 1) = 4/\pi$. If we know the values of $k(x, y)$ up to a given y_0 for all x such that $0 \le x \le y$, then we can find the values of $k(x, y_0 + 1)$ for $0 \le x \le y_0 + 1$. The averaging and symmetry properties give

$$k(0, y_0 + 1) = 4k(0, y_0) - 2k(1, y_0) - k(0, y_0 - 1)$$

$$k(x, y_0 + 1) = 4k(x, y_0) - k(x + 1, y_0) - k(x - 1, y_0) - k(x, y_0 - 1) \quad \text{for } 0 < x < y_0$$

$$k(y_0, y_0 + 1) = 2k(y_0, y_0) - k(y_0 - 1, y_0).$$

And $k(y_0 + 1, y_0 + 1)$ is given by the identity for $k(x, x)$.

The equations above thus are recursion equations for $k(x, y)$. Actually these equations are so simple that we apparently have a very rapid method of computing K_E for large finite sets E. Unfortunately the recursion is highly sensitive to rounding errors.

In Table 9-1 we give values of $k(x, y)$ for a wedge in the plane. The computations were carried out to 9-place accuracy, but by $y = 10$ the effect of rounding errors was noticeable. Any larger table would require much more accuracy of computation.

TABLE 9-1. $k(x, y)$ FOR A WEDGE-SHAPED REGION

$y = 9$	2.429	2.431	2.444	2.461	2.486	2.514	2.546	2.579	2.614	2.649
$y = 8$	2.353	2.357	2.372	2.395	2.424	2.459	2.496	2.535	2.574	
$y = 7$	2.267	2.274	2.293	2.322	2.359	2.400	2.444	2.489		
$y = 6$	2.168	2.178	2.203	2.241	2.288	2.339	2.391			
$y = 5$	2.052	2.065	2.101	2.153	2.213	2.276				
$y = 4$	1.908	1.930	1.984	2.056	2.134					
$y = 3$	1.721	1.762	1.849	1.952						
$y = 2$	1.454	1.546	1.698							
$y = 1$	1.000	1.273								
$y = 0$	0									
	$x = 0$	$x = 1$	$x = 2$	$x = 3$	$x = 4$	$x = 5$	$x = 6$	$x = 7$	$x = 8$	$x = 9$

If we wish to find $k(E)$, λ^E, and P^E for a finite set of points E, we first construct K_E and then compute the inverse K_E^{-1}. Let $z = K_E^{-1}\mathbf{1}$. Since $\alpha = \mathbf{1}^T$ and since K is symmetric, the results in the statement and proof of Corollary 9-92 simplify to

$$k(E) = 1/(\mathbf{1}^T z)$$

$$\lambda_E^E = k(E)z^T$$

$$P^E = I + K_E^{-1} - k(E)zz^T.$$

Calculations for various sets E appear in Tables 9-2, 9-3, and 9-4.

TABLE 9-2. $k(E)$, λ^E, AND P^E WHEN E CONSISTS OF THREE POINTS ON A LINE; $E = \{(0, 0), (a, 0), (2a, 0)\}$

$a = 1$			
$k(E) = 0.785$			
λ^E	0.393	0.215	0.393
P^E	0.460	0.393	0.148
	0.393	0.215	0.393
	0.148	0.393	0.460

$a = 2$			
$k(E) = 1.082$			
λ^E	0.372	0.256	0.372
P^E	0.610	0.256	0.134
	0.256	0.488	0.256
	0.134	0.256	0.610

$a = 3$			
$k(E) = 1.256$			
λ^E	0.365	0.270	0.365
P^E	0.663	0.212	0.125
	0.212	0.576	0.212
	0.125	0.212	0.663

$a = 4$			
$k(E) = 1.379$			
λ^E	0.361	0.277	0.361
P^E	0.693	0.189	0.118
	0.189	0.621	0.189
	0.118	0.189	0.693

TABLE 9-3. $k(E)$, λ^E, AND P^E WHEN E CONSISTS OF THREE POINTS ON A DIAGONAL; $E = \{(0, 0), (a, a), (2a, 2a)\}$

$a = 1$			
$k(E) = 0.955$			
λ^E	0.375	0.250	0.375
P^E	0.558	0.295	0.147
	0.295	0.411	0.295
	0.147	0.295	0.558

$a = 3$			
$k(E) = 1.407$			
λ^E	0.360	0.279	0.360
P^E	0.699	0.185	0.117
	0.185	0.631	0.185
	0.117	0.185	0.699

$a = 5$			
$k(E) = 1.622$			
λ^E	0.356	0.287	0.356
P^E	0.738	0.157	0.106
	0.157	0.687	0.157
	0.106	0.157	0.738

TABLE 9-4. $k(E)$ AND λ^E FOR THREE SETS E

21 points arranged in an isosceles right triangle of base 6

$k(E) = 1.611$

λ^E

0.162					
0.046	0.059				
0.041	0	0.051			
0.041	0	0	0.051		
0.044	0	0	0	0.059	
0.109	0.044	0.041	0.041	0.046	0.162

13 points arranged in a figure ×

$k(E) = 1.778$

λ^E

0.153						0.153
	0.064				0.064	
		0.030		0.030		
			0.015			
		0.030		0.030		
	0.064				0.064	
0.153						0.153

30 points in a 5-by-6 rectangle

$k(E) = 1.670$

λ^E

0.105	0.042	0.038	0.038	0.042	0.105
0.044	0	0	0	0	0.044
0.041	0	0	0	0	0.041
0.044	0	0	0	0	0.044
0.105	0.042	0.038	0.038	0.042	0.105

It is hard to acquire an intuition for the capacities of sets aside from their monotonicity. However, the values of P^E and λ^E are quite intuitive. The latter, in this random walk, may be thought of as the entrance probabilities to E if the chain is started near ∞. For example, in the case of the 5-by-6 rectangle in Table 9-4, it is clear that the corner positions should be considerably more probable than the points on the side. Points on the short sides are more probable than points on the long ones, and the rectangle cannot be entered at an interior point. Equally instructive are the values of λ^E for three-point sets in Tables 9-2 and 9-3. The middle point is least likely to be hit first, but the difference decreases as the points are spread farther apart.

13. Problems

1. Prove that ${}^iN_{jj} = {}^jN_{ii}$ for recurrent sums of independent random variables.
2. Prove that for any null chain and for any states i and j there is a finite set E such that
$$\limsup_n \frac{N_{ij}^{(n)}}{N_{iE}^{(n)}} < \epsilon.$$
3. Let a recurrent chain P be started in state 0. Let α_i^E be the mean number of times in state i in the time required to reach the set E and then return to 0. Thus, for example, if E is the set of all states, then $\alpha^E = (1/\alpha_0)\alpha$
   (a) Prove that for any set E there is a constant k_E such that $\alpha^E = k_E\alpha$
   (b) Let Q be the transition matrix for the transient states when 0 is made absorbing and let $\bar{\alpha}$ be the restriction of α to the transient states. Show that if E is a set which does not contain 0, then
$$1 - {}^E\bar{H}_{00} = \frac{1}{\alpha_0}\bar{\alpha}[(I - Q)B^E\mathbf{1}],$$
   where B^E is restricted to the transient states.
   (c) Conclude that if E is a set which does not contain 0, then $1/k_E$ is the transient capacity of the set E in the chain Q, provided the distinguished superregular measure is taken as $\bar{\alpha}$.
4. Prove that for a recurrent chain
$${}^jN_{ik} + {}^iN_{jk} = {}^jN_{ii}\frac{\alpha_k}{\alpha_i}.$$
5. For the symmetric random walk in two dimensions, verify that the function whose (x, y)th entry is $(|x| + |y| + 1)^{-1}$ is a potential, using only Corollary 9-16. Show that its charge f satisfies $\alpha f = 0$.
6. Let P be the one-dimensional symmetric random walk, and let $E = \{0, 1, 2, 3\}$.
   (a) Find B^E, λ^E, P^E.
   (b) Find all potential functions with support in E, and find their charges.
   (c) Compute αf and αg for each.
7. Let P be the symmetric random walk in two dimensions, and let a, b, and c be three distinguished states. We play a game as follows: The process is started in 0. Each time it is in a or b we win a dollar, and each time it is in c we lose two dollars.
   (a) Let $g_0^{(n)} = \mathrm{M}_0$ [expected gain to time n]. Prove that $\lim g_0^{(n)}$ exists, and find a *computable* expression for it.
   (b) What happens if the game is changed so that we lose only one dollar when the process is in state c?

Problems 8 to 10 lead to the fact that in a normal chain the union of a small set and a finite set is small.

8. Let E be a small set, and let $F = E \cup \{k\}$ have one more point. Show that
$$B_{ij}^E = B_{ij}^F + B_{ik}^F B_{kj}^E \quad \text{for all } j \in E.$$
   From this equation show that if λ_k^F exists, then λ^F exists and F is a small set.

9. In Problem 8 show that

$$^kH_{ij} = \sum_{m \in E} B^F_{im}\, {}^kH_{mj} \quad \text{for all } j \in E.$$

Use the identity of Problem 8 to eliminate the factors B^F_{im}, and solve the resulting identity for B^F_{ik}. Prove from this result that λ^F_k exists, provided P is normal.

10. Use the results of the two previous problems to prove that the union of a small set and a finite set is small in a normal chain.

Problems 11 to 22 develop a new null example and use it to illustrate results in the chapter. The state space consists of all points $z = (x, y)$ in the plane with integer coordinates ≥ 1. It will be convenient to let $n = x + y$. We let $(1, 1)$ be our state 0. Define

$$P_{(x,y),(x+1,y)} = \frac{x}{x + y + 1}, \quad P_{(x,y),(x,y+1)} = \frac{y}{x + y + 1},$$

and

$$P_{(x,y),0} = \frac{1}{x + y + 1}.$$

11. Verify that P is null recurrent and that $\alpha_z = 2/[(n - 1)n]$.
12. Compute $\hat{P}$.
13. Prove that $P^{(m)}_{0z}$ is the same for all z with $n = x + y$ fixed. Do the same for $\hat{P}^{(m)}_{0z}$.
14. Let $E = \{z \mid x + y \leq n_0\}$. Find λ^E and $\hat{\lambda}^E$.
15. Show that

$$^0N_{zz'} = \begin{cases} \binom{x' - 1}{x - 1}\binom{y' - 1}{y - 1}\Big/\binom{n'}{n}, & \text{if } x \leq x',\ y \leq y' \\ 0 & \text{otherwise.} \end{cases}$$

16. Let f be defined by $f_0 = -1$, $f_{(3,2)} = 10$, and $f_z = 0$ otherwise. Show that f is a charge, and use parts (1) and (3) of Theorem 9-15 to find the potential g.
17. Check that $f = (I - P)g$ for the functions of Problem 16. Does $\alpha g = 0$? Verify that $\lambda^E g = 0$ for all finite sets containing the support.
18. Show that $G_{0z} = 0$ and $G_{z0} = (n - 1)n/2$.
19. Use Problem 18 and Proposition 9-45 to show that

$$G_{zz'} = \frac{(n - 1)n}{(n' - 1)n'} - {}^0N_{zz'}.$$

20. Verify that the potential g found in Problem 16 is $-Gf$.
21. Find C, and compute a potential measure of finite support in two different ways (in analogy with Problems 16 and 20).
22. Let E be the triangular set of Problem 14, and let x be its characteristic function. Verify Proposition 9-43.

Problems 23 to 26 develop some theoretical results for an ergodic chain in terms of the operator K.

23. Express K in terms of M and $\hat{M}$.
24. Prove that $M_{ij} = (K_{ij} - K_{jj})/\alpha_j$.
25. Show that $\hat{M}_{ij} - M_{ji} = k(\{i\}) - k(\{j\})$.
26. What happens to the formulas in Problems 24 and 25 if the reference point 0 is changed?

Problems 27 to 32 carry this development further for finite recurrent chains.

27. Show that $k(S) = M_{\alpha 0} = \hat{M}_{\alpha 0}$.
28. Prove that $K1 = k(S)1$ and $\alpha K = k(S)\alpha$.
29. Prove that $M\alpha^T = c1$, where $c = k(S) - \sum_i k(\{i\})\alpha_i$.
30. Prove that
$$K = \left(\frac{1}{k(S)} 1\alpha + P - I\right)^{-1}.$$
31. Prove that the set of charges is the same as the set of potentials.
32. To what extent do the results of Problems 27 to 31 generalize to strong ergodic chains?

Problems 33 to 39 are intended to illustrate Problems 23 to 32 for the Land of Oz example, which was defined in Chapter 4. [See also Chapter 6, Problem 1.] We choose the middle state (nice weather) to be the distinguished state 0.

33. Show that $P = \hat{P}$ (the chain is reversible).
34. Find M.
35. Find K, using the result of Problem 23.
36. Find $k(S)$, using the result of Problem 27.
37. Find K, using the result of Problem 30, and compare with the value of K found in Problem 35.
38. Check the results given in Problems 24, 25, 28, and 29.
39. Find the most general charge and the most general potential function. Verify that the set of charges is the same as the set of potentials.

Problems 40 to 48 work out the probabilistic solution of the so-called Second Boundary Value Problem in the sense that Theorem 8-41 presented the solution of the First Boundary Value Problem. Let P be an absorbing chain whose transient states communicate and whose absorbing states form a finite set B. B is thought of as our "boundary." To each state k in B we associate a "neighboring" transient state k'. For a given function h, we define its normal derivative d_k at k to be $h_k - h_{k'}$. The problem is to find a function h which is regular on the transient states and has specified normal derivatives.

40. Prove that $h_{\bar{B}} = NRh_B$ for any solution.
41. Modify the original chain so that instead of stopping at an absorbing state k, it moves to the neighboring k' with probability 1. Show that this new chain is recurrent.

42. Let P^* be the transition matrix of the modified chain watched in B. Prove that this is an ergodic chain.

43. Show that the requirement that h have the specified normal derivatives can be written as $(I - P^*)h_B = d$, where d has the given values d_k as components.

44. Prove that $\alpha^* d = 0$ is a necessary condition for a solution to exist.

45. Show that $\alpha^* d = 0$ is also sufficient by showing that d is a charge, that its potential will serve as h_B, and that the h supplied by Problem 40 is a solution.

46. Prove that the most general solution differs from the given one only by a constant.

47. Prove that if the modified chain (indexed on all states) is a normal chain, then the most general solution is

$$h = -K\begin{pmatrix} d \\ 0 \end{pmatrix} + c\mathbf{1}.$$

48. Show that if the transient states are the lattice points in a bounded convex set in n-dimensional Euclidean space and if the process moves as a symmetric random walk which is stopped when it moves out of the convex set, then we can apply the previous results.

Problems 49 to 53 give a complete characterization of degenerate chains.

49. Prove that if P is degenerate, so is $\hat{P}$.

50. Show that if P is degenerate and if we let $i < j$ stand for ${}^j\lambda_i = 1$, then $<$ is a simple ordering.

51. Prove that if $k < i < j$, then ${}^jH_{ki} = 1$. Deduce from this fact that, in moving to the right, the process can move at most one step at a time. [*Hint:* Consider λ^E for $E = \{k, i, j\}$.]

52. Prove that the ordering of states must be that of the integers, the positive integers, or the negative integers.

53. Show that the basic example and its reverse illustrate two of the possible orderings, and construct an example of the third.

CHAPTER 10

TRANSIENT BOUNDARY THEORY

1. Motivation for Martin boundary theory

For purposes of motivation it is convenient to think of the state space of a Markov chain P with only transient states as being similar to the open unit disk of two-dimensional Euclidean space. In two-space the boundary of the disk—namely the circle S^1—has the property that there is a one-one correspondence between the non-negative harmonic functions $h(re^{i\theta})$ in the disk and the non-negative Borel measures μ^h on the circle. The correspondence is

$$h(re^{i\theta}) = \int_{S^1} P(re^{i\theta}, t)d\mu^h(t), \qquad (*)$$

where $P(re^{i\theta}, t)$ is the Poisson kernel

$$\frac{1 - r^2}{1 - 2r\cos(\theta - t) + r^2}.$$

Transient boundary theory seeks an analogous representation theorem for all non-negative P-regular functions defined on the state space.

The first problem that arises is to find what the analogs of the circle (the boundary) and the kernel should be. We would like a representation

$$h(i) = \int K(i, x)d\mu^h(x).$$

In the case of the disk, a calculation with Green's identities shows that any kernel $P(re^{i\theta}, t)$ giving rise to the correspondence (*) and satisfying

$$\frac{1}{2\pi}\int_0^{2\pi} P(re^{i\theta}, t)dt = 1$$

must be the normal derivative at t of the Green's function for the disk relative to the point $re^{i\theta}$. That is,

$$P(re^{i\theta}, t) = \left[\frac{\partial}{\partial n} G(\cdot, re^{i\theta})\right]_t.$$

Application of l'Hospital's rule shows that

$$\begin{aligned}\lim_{\substack{z\to t\\ \text{radially}}} \frac{G(z, re^{i\theta})}{G(z, p)} &= \left[\frac{\partial}{\partial n} G(\cdot, re^{i\theta}) \Big/ \frac{\partial}{\partial n} G(\cdot, p)\right]_t \\ &= P(re^{i\theta}, t)/P(p, t),\end{aligned}$$

where p is any fixed reference point in the disk. Hence, except for the positive factor $P(p, t)$ which depends on t but not on $re^{i\theta}$, the Poisson kernel is equal to

$$\lim_{\substack{z\to t\\ \text{radially}}} \frac{G(z, re^{i\theta})}{G(z, p)}. \tag{**}$$

Therefore this last function may be used in the representation (*) in place of the Poisson kernel; the distinction between the kernels is just a normalizing factor (depending on t) which can be absorbed by changing the measures.

Two comments are in order. First, the limit in (**) need not be taken radially. Any method of approach of z to t, as long as z stays in the interior of the disk, will give the same value. Second, the considerations above apply equally well to any domain in n-dimensional space with a sufficiently smooth boundary. Although the explicit form of the kernel will vary from region to region, it will always be connected to the Green's function in the way we have just described.

R. S. Martin [1941] made use of these observations to define an ideal boundary for an *arbitrary* domain in Euclidean space. If the Green's function for the region is denoted $G(z, y)$, he noted that points t on the ordinary topological boundary of the region did not necessarily have the property that

$$\lim_{z\to t} \frac{G(z, y)}{G(z, p)}$$

exists. He suggested that distinct ideal boundary points u should be associated to subsequences $\{z_n\}$ which yield distinct values for the limits

$$\lim_{z_n\to t} \frac{G(z_n, y)}{G(z_n, p)} = K(y, u).$$

He went on to show that the desired representation theorem is indeed obtained in terms of this boundary and the kernel $K(y, u)$.

Doob [1959], taking advantage of the fact that the N-matrix for a transient Markov chain is the analog of the Green's function (see Proposition 7-4), showed that Martin's approach could be used to obtain a boundary for Markov chains. (We remark that N_{ij} corresponds to $G(j, i)$ with the indices in the reversed order.) As the analog of Martin's kernel he used limits on j of expressions of the form N_{ij}/N_{0j}.

There is a minor restriction imposed by the Doob approach, namely that N_{0j} is assumed non-zero for all j. For a more general chain in which it is not possible to get from state 0 to every other state, what Doob did was to consider only those states that could be reached from 0. We shall not follow him in this respect. We simply use limits of $N_{ij}/(\pi N)_j$ instead, where π is a probability vector such that πN is strictly positive. In terms of this kernel there is a natural space to try as the one corresponding to the closed unit disk. The space should consist of one point for each possible limit of $N_{ij}/N_{\pi j}$. Actually we shall find that this space is too large—that the space has to be cut down a bit for the representation to be unique. The price of uniqueness is that the cut-down space is not compact.

The introduction of π in place of 0 itself leads to a problem. The representation will have to be restricted to π-integrable regular functions h, those for which πh is finite. This requirement evaporated in Martin's or Doob's treatment because for them π assigned unit mass to a point 0 and $h(0)$ was automatically finite.

Hunt [1960] gave a new approach to Martin boundary theory for Markov chains which was more probabilistic in nature than Doob's. We follow Hunt's probabilistic approach, except that we use a different metric and get a boundary which is more like Doob's.

2. Extended chains

We begin by introducing the machinery which we shall use in later sections to develop Martin boundary theory for Markov chains. We are going to use a broader notion of Markov chain than we have been considering so far—namely, a process whose time index starts at $-\infty$ and whose behavior is Markovian only after it has entered certain sets.

That is, we extend the concept of Markov chain in two ways. First, we shall allow any finite measure π as a starting distribution. This is only a minor modification in the theory and is a convenience in that it removes the necessity of normalization in certain constructions. Second, we shall extend the Markov chain to the past, that is, to a stochastic process $\{x_n\}$ where n runs through all the integers (including negative integers). This second extension is an essential one and will be the main topic of this section.

First we must extend the concept of a stochastic process. The **state space** will be a denumerable set S (with at least two elements) and two distinguished other states a and b. The underlying set Ω of the measure space will consist of all doubly infinite sequences

$$\omega = (\ldots, c_{-2}, c_{-1}, c_0, c_1, c_2, \ldots)$$

such that

(1) $c_n \in S$ or $c_n = a$ or $c_n = b$.
(2) If $c_n = a$, then $c_m = a$ for all $m < n$.
(3) If $c_n = b$, then $c_m = b$ for all $m > n$.
(4) $c_n \in S$ for at least one n.

The interpretation of (2) and (3) is that state a stands for "not yet started" and state b stands for "stopped." Thus (4) has the meaning that each path in Ω represents some nontrivial possibility for the process. We shall refer to Ω as a **double sequence space**.

We define the outcome functions x_n as usual except that n may be any integer. A **basic cylinder set** is any truth set in Ω of a statement of the form

$$x_m = c_m \wedge x_{m+1} = c_{m+1} \wedge \cdots \wedge x_n = c_n,$$

where at least one c_i is an element of S. The field generated by the basic cylinder sets is denoted $\mathscr{F}$, and the smallest Borel field containing $\mathscr{F}$ is $\mathscr{G}$.

Definition 10-1: An extended stochastic process $\{x_n\}$ is the set of outcome functions for a measure space $(\Omega, \mathscr{G}, \Pr)$ such that

(1) Ω is a double sequence space with state space $S \cup \{a, b\}$.
(2) $\mathscr{G}$ is the smallest Borel field containing the field of cylinder sets of Ω.
(3) $\Pr[\{\omega \mid x_n = i\}] < \infty$ for every integer n and every i in S.

Note that we do not augment the measure space $(\Omega, \mathscr{G}, \Pr)$ by allowing all subsets of sets of measure zero to be measurable.

We shall use interchangeably the notations $\Pr[P]$ and $\Pr[p]$, where P is the truth set of the statement p. Thus the third condition may be replaced by the condition $\Pr[x_n = i] < \infty$ for all n and for all i in S. From it and from condition (4) in the definition of Ω, we find that Pr must be sigma-finite. However, the measure Pr need not be finite.

As promised, Definition 10-1 extends the definition of stochastic process in two ways: The time index n runs through all the integers, and the measure Pr need not be a probability measure or even a finite measure.

In examples it is necessary to have a method of constructing the measures for extended stochastic processes. If the measure Pr has already been defined consistently on basic cylinder sets, we need a version of the Kolmogorov Theorem to prove that Pr is completely additive on $\mathscr{F}$. Theorem 1-19 then will give the extension of Pr to all of $\mathscr{G}$. At this point, therefore, we stop to outline a proof that Pr is completely additive on $\mathscr{F}$.

Now the argument of Lemma 2-1 easily shows that Pr is non-negative and additive. Extend Ω to include the set of *all* doubly infinite sequences of a's, b's, and elements of S, and define Pr to be zero on all cylinder subsets of the set added. Then if Pr were a finite measure, we could temporarily rearrange the time scale and then conclude complete additivity by Theorem 2-4. But, in general, Pr is merely sigma-finite and therefore we shall write it as the countable sum of totally finite non-negative additive set functions, each of which is a measure on cylinder sets depending on a bounded time interval. Each of the summands is completely additive by the above argument, and therefore Pr is completely additive by Lemma 1-3. Thus all we need to do is decompose Pr as such a sum. The countable family of statements indexed by $i \in S$ and by $n \geq 0$, consisting of

$$x_0 = i$$

$$x_{-n-1} = i \wedge x_{-n} = b$$

$$x_n = a \wedge x_{n+1} = i,$$

is a disjoint exhaustive family in the original double sequence space, and each statement is assigned finite Pr-measure. For each of these statements q, define

$$\Pr^{(q)}(E) = \Pr[E \cap \{\omega \mid q\}]$$

for E in the field of cylinder sets. Then the family $\{\Pr^{(q)}\}$ is the required family of set functions.

We now fix our attention on a single extended stochastic process $\{x_n\}$.

Although we may be dealing with an infinite measure space, the conditional probability

$$\Pr[p \mid q] = \Pr[p \wedge q]/\Pr[q]$$

is still well defined as long as $0 < \Pr[q] < \infty$. We define $\Pr[p \mid q]$ to be zero if $\Pr[q] = 0$.

Definition 10-2: For $E \subset S$ and for any $\omega \in \Omega$ such that $x_n(\omega) \in E$ for some n, let $\mathbf{u}_E(\omega)$ be the infimum of all n such that $x_n(\omega) \in E$ and let $\mathbf{v}_E(\omega)$ be the supremum of all such n. Define $\mathbf{u} = \mathbf{u}_S$ and $\mathbf{v} = \mathbf{v}_S$; $\mathbf{u}$ and $\mathbf{v}$ are called the initial time and the final time, respectively.

By condition (4) of the definition of Ω, we see that $\mathbf{u}(\omega)$ and $\mathbf{v}(\omega)$ are defined for all ω. Moreover, $\mathbf{u}_E(\omega) \leq \mathbf{v}_E(\omega)$ whenever $\mathbf{u}_E(\omega)$ and $\mathbf{v}_E(\omega)$ are defined. The values $\mathbf{u}_E(\omega) = -\infty$ and $\mathbf{v}_E(\omega) = +\infty$ are possible. If $x_n(\omega) \in E$ for some n, we have

$$x_n(\omega) = \begin{cases} a & \text{if } n < \mathbf{u}(\omega) \\ \text{element of } S - E & \text{if } \mathbf{u}(\omega) \leq n < \mathbf{u}_E(\omega) \\ \text{element of } S - E & \text{if } \mathbf{v}_E(\omega) < n \leq \mathbf{v}(\omega) \\ b & \text{if } \mathbf{v}(\omega) < n. \end{cases}$$

Proposition 10-3: The functions $\mathbf{u}_E$ and $\mathbf{v}_E$ have a $\mathscr{G}$-measurable subset of Ω as domain and are each $\mathscr{G}$-measurable.

PROOF: We prove the result for $\mathbf{u}_E$. We have

$$\{\omega \mid \mathbf{u}_E(\omega) \leq k\} = \bigcup_{i \in E} \bigcup_{n=-\infty}^{k} \{\omega \mid x_n(\omega) = i\},$$

and the union of these sets on k is the domain of $\mathbf{u}_E$.

Definition 10-4: Let E be a subset of S and define

$$y_n(\omega) = x_{(n+\mathbf{u}_E(\omega))}(\omega)$$

for all $n \geq 0$ and for all ω such that $\mathbf{u}_E(\omega) > -\infty$. Let $\bar{\Omega}$ be the ordinary sequence space with state space $S \cup \{b\}$ with the measure of each measurable set $A \subset \bar{\Omega}$ defined to be

$$\Pr[\{\omega \mid (y_0(\omega), y_1(\omega), \ldots) \in A\}].$$

The measure space $\bar{\Omega}$ and its outcome functions together are called the **process watched starting in E.**

The process watched starting in a set E is an ordinary stochastic process, except that the starting distribution need not be a probability measure and can possibly be infinite.

Let $j \in S$. The mean number ν_j of times that the process $\{x_n\}$ is in j can, as usual, be computed by

$$\nu_j = \sum_n \Pr[x_n = j],$$

except the sum is over all integers n. By Definition 10-1, each summand is finite, but the sum may be infinite.

Definition 10-5: An extended stochastic process $\{x_n\}$ is an **extended chain** with transition matrix P if the following conditions hold for each finite subset E of S:

(1) The domain of $\mathbf{u}_E$ has positive measure.
(2) $\Pr[\mathbf{u}_E = -\infty] = 0$.
(3) The process watched starting in E is a Markov chain with transition matrix P and finite starting measure (but not necessarily a probability measure).
(4) For all $j \in S$, $\nu_j < \infty$.

Note that the transition matrix P of an extended chain necessarily satisfies $P\mathbf{1} = \mathbf{1}$. The state space of P never needs to be any bigger than $S \cup \{b\}$, but as we shall see shortly it must contain S. If b is in the state space, it clearly must be an absorbing state.

From the definition of the process watched starting in E, we see that the total starting measure for the process is equal to the measure of the set of paths on which there is a first entry to E. That is, it is the measure of the set where $\mathbf{u}_E > -\infty$. By conditions (1) and (2), this measure is positive. Hence the process watched starting in E has a starting measure which is not identically zero.

Applying this observation to the one-point set $\{j\}$, we see that j must be included in the state space of P.

Let $\{x_n\}$ be an extended stochastic process satisfying (1), (2), and (3). We shall derive as Proposition 10-6 a necessary and sufficient condition for (4) to hold. Let E be a finite set of S. We introduce the abbreviations

$$\mu_i^{E,m} = \Pr[\mathbf{u}_E = m \wedge x_m = i]$$

and

$$\mu^E = \sum_m \mu^{E,m}.$$

Then μ^E is the starting measure of the process watched starting in E and is a finite measure with support in E by (3). Our remarks above showed that μ^E is not identically zero.

Since the process watched starting in E is a Markov chain and since (2) holds, the following computation is justified: For $j \in E$,

$$\Pr[x_n = j \wedge x_{n+1} = k \wedge x_{n+2} = s] = \sum_{m,i} \mu_i^{E,m} P_{ij}^{(n-m)} P_{jk} P_{ks},$$

where $P^{(n-m)} = 0$ if $n < m$. A similar computation of $\Pr[p]$ is possible for any p such that p is false if E is never entered and p depends only on outcomes after E is entered.

As an application of this calculation, we can relate condition (4) to properties of P.

Proposition 10-6: Let $\{x_n\}$ be an extended stochastic process satisfying conditions (1), (2), and (3) for an extended chain. Then for $j \in E$,

$$\nu_j = (\mu^E N)_j.$$

Moreover, the process is an extended chain if and only if P is the transition matrix of a transient chain whose transient states are S and which may have b as an absorbing state.

PROOF: We have

$$\nu_j = \sum_n \Pr[x_n = j] = \sum_{n,m,i} \mu_i^{E,m} P_{ij}^{(n-m)} = \sum_{m,i} \mu_i^{E,m} N_{ij} = \sum_i \mu_i^E N_{ij}$$

or

$$\nu_j = (\mu^E N)_j.$$

Taking $E = \{j\}$, we find

$$\nu_j = \mu_j^E N_{jj}.$$

Now μ_j^E is assumed finite by (3) and it is strictly positive by (1). Hence ν_j is finite for all $j \in S$ if and only if all elements of S are transient for a Markov chain with transition matrix P. Furthermore, a cannot occur as a state, and if b occurs, it must be an absorbing state.

Corollary 10-7: Let $\{x_n\}$ be an extended stochastic process satisfying conditions (1), (2), and (3) for an extended chain. Then $\nu_j > 0$ for all j.

PROOF: We have $\nu_j = \mu_j^{\{j\}} N_{jj}$, and each factor on the right side is positive.

An important example, but by no means the most general example, of an extended chain is obtained as follows. Let P be a Markov chain with all states transient and let π be a starting distribution such that πN is strictly positive. (For instance, let π assign weight 2^{-n} to the nth state.) Let $\bar{P}$ be the enlarged chain obtained by adding the absorbing state b. Form an extended stochastic process by defining a measure on cylinder sets as follows: Every basic statement containing the assertion $x_n = i$ for $n < 0$ and $i \neq a$ or the assertion $x_m = a$ for $m \geq 0$ gets probability zero. The statement

$$x_{-m} = a \wedge \cdots \wedge x_{-1} = a \wedge x_0 = i \wedge x_1 = j \wedge x_2 = k$$
$$\wedge \cdots \wedge x_{n-1} = r \wedge x_n = s$$

gets probability $\pi_i \bar{P}_{ij} \bar{P}_{jk} \ldots \bar{P}_{rs}$. The probabilities of all other basic cylinder statements can be obtained from these by adding the probabilities of a suitable number of statements of the form just described. The claim is that this extended stochastic process $\{x_n\}$ is an extended chain with transition matrix $\bar{P}$. Property (1) follows from the fact that $\pi N > 0$. Property (2) is an immediate consequence of the definition of the process. Property (3) follows from Theorem 4-9; the starting distribution for the process watched starting in E is πB^E. Finally Property (4) comes from Proposition 10-6; alternatively we could compute directly that the mean number of times in state j is $(\pi N)_j$ We shall call this process the **extended chain associated with** π **and** P.

If $\{x_n\}$ is an extended chain and E is a finite set in S, we define $\nu^E = \mu^E N$. We know that the process watched starting in E is a Markov chain with starting distribution μ^E. Hence ν_j^E is the mean number of times in j in this Markov chain. That is, it is the mean number of times in j *after entering* E in the extended chain. From this interpretation we see that ν^E is monotone increasing in E and that $\nu_j^E = \nu_j$ for $j \in E$. In order to get an interpretation of ν^E and μ^E in terms of ν, we shall generalize the notion of balayage potential as defined in Chapter 8.

If P is a Markov chain with all states transient, if h is a non-negative finite-valued superregular function, and if E is a finite set of states, we define the **balayage potential** of h on E to be the function $B^E h$. By Lemma 8-22, $B^E h$ is a pure potential with support in E. Now $B^E h$ is the unique pure potential with support in E which agrees with h on E, since if $\bar{h}$ is another such potential, $\bar{h}$ must be the unique balayage potential of $B^E h$ on E (Theorem 8-46) and hence must equal $B^E h$. Moreover, if E_n is an increasing sequence of finite sets with union the set of all states, then the charges of $B^{E_n} h$, namely $(I - P)(B^{E_n} h)$, converge to $(I - P)h$, since

$$\lim_n PB^{E_n} h = Ph$$

by part (4) of Proposition 8-16 and by monotone convergence.

Let us dualize these results. Let γ be a non-negative finite-valued superregular measure, and let E be a finite set. Then there is a unique pure potential measure with support in E which agrees with γ on E. This potential is defined to be the balayage potential of γ on E. It has the property that if E_n is an increasing sequence of sets with union the set of all states, then the balayage charges of γ on E_n converge to $\gamma(I - P)$. By Proposition 6-16, the balayage potential of γ on E is

$$(\gamma_E \quad \gamma_E U({}^E N)_{\tilde{E}}),$$

where

$$P = \begin{pmatrix} T & U \\ R & Q \end{pmatrix}$$

and $({}^E N)_{\tilde{E}} = \sum Q^n$.

If we let γ^E be the balayage potential of γ on E and if β^E is the charge of γ^E, then $E \subset F$ implies $\beta^E = \beta^F B^E$. To see this equality, we note that β^E and $\beta^F B^E$ are both pure charges with support in E and that the potential of $\beta^F B^E$, when restricted to E, is

$$(\beta^F B^E N)_E = (\beta^F N - \beta^F {}^E N)_E = (\beta^F N)_E = \gamma_E = (\beta^E N)_E$$

by Lemma 8-17. Hence the potentials of β^E and $\beta^F B^E$ agree on E, and they must therefore agree everywhere. Thus

$$\beta^E = \beta^F B^E$$

by Theorem 8-4.

Our characterization of ν^E and μ^E in terms of balayage potentials is the content of the next proposition.

Proposition 10-8: For every extended chain with transition matrix P,

(1) ν is a superregular measure for P_S.
(2) $\nu(I - P_S) = \mu$, where $\mu = \lim_E \mu^E$ as E increases to S.
(3) ν^E and μ^E are the balayage potential and charge, respectively, for ν on E.

PROOF: We have

$$\nu^E P_S = \mu^E N P_S = \mu^E(N - I) = \nu^E - \mu^E.$$

Along any increasing sequence of sets E_n with union S, ν^E increases to ν. Hence by monotone convergence

$$\nu P_S = \nu - \lim_n \mu^{E_n}.$$

This equality implies that

$$\nu P_S = \nu - \lim_{E \uparrow S} \mu^E$$

and proves (1) and (2). To prove (3) we need only remark that ν^E is a pure potential with support in E which agrees with ν on E. Hence it must be the balayage potential.

Proposition 10-8 has as a converse the following theorem, which asserts roughly that any superregular measure can be represented as the vector of mean times in the states of some extended chain. This result will not be used until Section 11, and its proof, which is quite long, will not be given until after that section.

Theorem 10-9: Let P^* be the transition matrix of a transient chain with $P^*1 = 1$ and with at most one absorbing state b, and let ν be a non-negative finite-valued measure defined on the transient states and superregular for the restriction of P^* to the transient states. Let S be the support of ν, and suppose S has at least two elements. Then there is an extended chain $\{x_n\}$ having $P^*_{S\cup\{b\}}$ as transition matrix and having ν as its vector of mean times in the states of S. Furthermore, if

$$\nu_S = \mu N + \rho$$

is the unique decomposition of ν_S with ρ regular for P^*_S, then μN is contributed by the paths ω with $\mathbf{u}(\omega) > -\infty$ and ρ is contributed by the paths with $\mathbf{u}(\omega) = -\infty$.

To conclude this section we shall define what we mean by the reverse of an extended stochastic process, and we shall prove that the reverse of an extended chain is an extended chain. The transition matrix of the reverse, at least when restricted to S, will turn out to be the ν-dual of the transition matrix, restricted to S, of the original process. We need the following lemma, whose proof uses the calculation preceding Proposition 10-6.

Lemma 10-10: If $\{x_n\}$ is an extended chain with transition matrix P and if k is in a finite set E in S, then

$$\Pr[x_{v_E-2} = i \wedge x_{v_E-1} = j \wedge x_{v_E} = k] = \nu_i P_{ij} P_{jk} e^E_k,$$

where e^E is the escape vector for P.

PROOF:

$$\begin{aligned}
\Pr[x_{v_E-2} &= i \wedge x_{v_E-1} = j \wedge x_{v_E} = k] \\
&= \sum_n \Pr[x_n = i \wedge x_{n+1} = j \wedge x_{n+2} = k \\
&\qquad\qquad \wedge \{x_n\} \text{ not in } E \text{ after time } (n+2)] \\
&= \sum_{m,s,n} \mu_s^{E,m} P_{si}^{(n-m)} P_{ij} P_{jk} e_k^E \quad \text{by the calculation} \\
&= \sum_n \Pr[x_n = i] P_{ij} P_{jk} e_k^E \quad \text{by the calculation again} \\
&= \nu_i P_{ij} P_{jk} e_k^E.
\end{aligned}$$

For any point ω in Ω we define ω' to be the point in Ω with

$$x_n(\omega') = \begin{cases} x_{-n}(\omega) & \text{if } x_{-n}(\omega) \in S \\ a & \text{if } x_{-n}(\omega) = b \\ b & \text{if } x_{-n}(\omega) = a. \end{cases}$$

Definition 10-11: Let $\{x_n\}$ be an extended chain defined on Ω with measure μ. Set

$$\Pr'[\omega \in A] = \Pr[\omega' \in A]$$

for all sets A for which the right side is defined. The extended stochastic process on Ω defined by the measure Pr′ is called the **reverse** of the extended chain.

Proposition 10-12: If $\{x_n\}$ is an extended chain with transition matrix P, then its reverse is also an extended chain and its transition matrix $\hat{P}$ satisfies

$$\hat{P}_{ji} = \nu_i P_{ij}/\nu_j$$

for all states i and j in S.

PROOF: From the definition of ω', we see that

$$\hat{\mathbf{u}}_E(\omega) = -\mathbf{v}_E(\omega').$$

Hence

$$\Pr'[\omega \in \text{domain } \hat{\mathbf{u}}_E] = \Pr[\omega \in \text{domain } \mathbf{v}_E] = \Pr[\omega \in \text{domain } \mathbf{u}_E] > 0$$

and (1) holds. Since the chain watched starting in E is in the finite set E infinitely often with probability 0 (second half of Proposition 10-6),

$$0 = \Pr[\mathbf{v}_E = +\infty] = \Pr'[\hat{\mathbf{u}}_E = -\infty].$$

Thus (2) holds.

Next, we show that the reverse process watched starting in E is a Markov chain with transition matrix $\hat{P}$. We shall compute only a typical conditional probability: Let $k \in E$ and first suppose $i \neq b$. Since

$$\begin{aligned}\Pr'[x_{\hat{u}_E} = k \wedge x_{\hat{u}_E+1} = j \wedge x_{\hat{u}_E+2} = i] \\ &= \Pr[x_{v_E-2} = i \wedge x_{v_E-1} = j \wedge x_{v_E} = k] \\ &= \nu_i P_{ij} P_{jk} e_k^E\end{aligned}$$

by Lemma 10-10, we have, provided the condition has positive Pr′-measure,

$$\begin{aligned}\Pr'[x_{\hat{u}_E+2} = i \mid x_{\hat{u}_E} = k \wedge x_{\hat{u}_E+1} = j] &= (\nu_i P_{ij} P_{jk} e_k^E)/(\nu_j P_{jk} e_k^E) \\ &= \nu_i P_{ij}/\nu_j \\ &= \hat{P}_{ji}.\end{aligned}$$

Next we compute the typical probability

$$\Pr'[x_{\hat{u}_E} = k \wedge x_{\hat{u}_E+1} = j \wedge x_{\hat{u}_E+2} = b].$$

Since we know b is absorbing, we may assume $j \neq b$. Then this probability is

$$\Pr'[x_{\hat{u}_E} = k \wedge x_{\hat{u}_E+1} = j] - \Pr'[x_{\hat{u}_E} = k \wedge x_{\hat{u}_E+1} = j \wedge x_{\hat{u}_E+2} \in S]$$
$$= \nu_j P_{jk} e_k^E - \sum_{i \in S} \nu_i P_{ij} P_{jk} e_k^E.$$

Hence if $\Pr'[x_{\hat{u}_E} = k \wedge x_{\hat{u}_E+1} = j] > 0$, then

$$\Pr'[x_{\hat{u}_E+2} = b \mid x_{\hat{u}_E} = k \wedge x_{\hat{u}_E+1} = j] = (\nu_j - \sum_{i \in S} \nu_i P_{ij})/\nu_j.$$

(Notice this probability is non-negative because ν is P_S-superregular.) Therefore the reverse process watched starting in E is a Markov chain.

The total starting measure is finite for the reverse process watched starting in E because

$$\sum_i \Pr'[x_{\hat{u}_E} = i] = \sum_i \Pr[x_{v_E} = i]$$
$$= \sum_i \nu_i e_i^E \quad \text{by Lemma 10-10}$$
$$< \infty$$

by (4) for the given process and by the finiteness of the set E. Hence (3) holds.

Finally we prove (4) for the reverse process. The same argument as in Proposition 6-12 shows that

$$\hat{N}_{ji} = \nu_i N_{ij}/\nu_j.$$

Hence $\hat{P}$ is transient and (4) holds by Proposition 10-6.

3. Martin exit boundary

We now define the Martin exit boundary of a transient chain with respect to a given starting distribution. With this boundary we shall be able to describe the long-range behavior of the process and we shall obtain a Poisson integral-type representation for all finite-valued non-negative superregular functions which are integrable with respect to the starting distribution.

Let P be a Markov chain with all states transient and let π be a starting vector ($\pi \geq 0$ and $\pi 1 = 1$). Throughout our discussion P and π will be fixed. The vector πN is non-negative, finite-valued, and superregular. Ordinarily we shall assume that π has been chosen so that πN is strictly positive; that is, so that there is positive probability of reaching any state eventually. But for technical reasons which will arise when we consider h-processes, it will be convenient to adopt conventions about what to do when πN has some zero entries. These conventions we shall discuss at the end of this section, and except when stated otherwise we shall assume that πN is everywhere positive. Let S be the state space of P.

Since $N_{\pi j} = (\pi N)_j$ has been assumed positive, we may define

$$K(i, j) = N_{ij}/N_{\pi j}.$$

The notation $K(\cdot, j)$ will mean $K(i, j)$ considered as a function of the first variable with j fixed. Then for each j, $K(\cdot, j)$ is a non-negative finite-valued superregular function with $\pi K(\cdot, j) = 1$. It is regular everywhere except at j, where it is strictly superregular.

For fixed i the function $K(i, \cdot)$ is bounded, since

$$K(i, j) = \frac{N_{ij}}{N_{\pi j}} = \frac{1}{N_{\pi i}}\left(\frac{N_{\pi i}N_{ij}}{N_{\pi j}}\right) = \frac{1}{N_{\pi i}}\hat{N}_{ji} \leq \frac{1}{N_{\pi i}}\hat{N}_{ii} = \frac{N_{ii}}{N_{\pi i}},$$

where carets denote duality with respect to πN.

The real-valued functions $d_i(j, j')$ defined on $S \times S$ by

$$d_i(j, j') = |K(i, j) - K(i, j')|$$

have the property that $\{K(i, j_n)\}$ is a Cauchy sequence for all i in S if and only if $\lim_{m,n\to\infty} d_i(j_m, j_n) = 0$ for all i. According to the bound we just computed for $K(i, j)$, we may lump the functions d_i into the single finite-valued function d defined by

$$d(j, j') = \sum_{i\in S} w_i N_{\pi i}|K(i, j) - K(i, j')|,$$

where the w_i are positive weights such that $\sum w_i N_{ii}$ is finite.

We shall show that d is a metric for S. Clearly d satisfies all the conditions of a metric except possibly that $d(j, j') = 0$ implies $j = j'$. But if $d(j, j') = 0$, then, since $w_i N_{\pi i} > 0$ for all i, we must have

$$K(\cdot, j) = K(\cdot, j').$$

Multiplying through by P and supposing that $j \neq j'$, we obtain

$$K(j', j) = \sum_i P_{j'i}K(i, j) = \sum_i P_{j'i}K(i, j') < K(j', j'),$$

the strict inequality holding, since $K(\cdot, j')$ is not regular at j'. We conclude that $j = j'$ and that d is a metric.

Proposition 10-13: A sequence $\{j_n\}$ in the metric space (S, d) is Cauchy if and only if the sequence of real numbers $\{K(i, j_n)\}$ is Cauchy for every i.

PROOF: If $\{j_n\}$ is Cauchy, then certainly $\{w_i N_{\pi i} K(i, j_n)\}$ is Cauchy. Since $w_i N_{\pi i} > 0$, $\{K(i, j_n)\}$ is Cauchy.

Conversely, let $\{K(i, j_n)\}$ be Cauchy for all i, and let $\epsilon > 0$ be given. Choose a finite set of states E such that

$$\sum_{k \in S - E} w_k N_{kk} < \epsilon/2,$$

and choose M sufficiently large that

$$|K(i, j_m) - K(i, j_n)| < \frac{\epsilon}{2 \sum w_k N_{kk}}$$

for $i \in E$ and for all $n, m \geq M$. Then $d(j_m, j_n) < \epsilon$.

We define S^* to be the Cauchy completion of the metric space (S, d), and we let $B = S^* - S$. The set B is the **Martin exit boundary** for the chain P started with distribution π.

The set B is not necessarily a boundary in the topological sense, since there are examples in which it is not a closed set in S^*, but the abuse of notation will not disturb us.

From Proposition 10-13 we see that $K(i, \cdot)$ is a uniformly continuous function on S. Hence it extends uniquely to a continuous function on S^*. We shall use the same notation $K(i, \cdot)$ for the function on S^*, but will normally denote points of $B = S^* - S$ by x or y.

The characterization of Cauchy sequences given in Proposition 10-13 shows that the nature of the space S^* does not depend upon the choice of the weights w_i. That is, the Cauchy completions of S corresponding to two different choices of weights are homeomorphic.

Since $K(i, \cdot)$ is continuous on S^*, it follows that the extension of d to S^* is simply

$$d(x, y) = \sum_{i \in S} w_i N_{\pi i} |K(i, x) - K(i, y)|.$$

A repetition of the argument in Proposition 10-13 then shows that $\{x_n\}$ is Cauchy in S^* if and only if $\{K(i, x_n)\}$ is Cauchy for each i. Applying this result to the sequence whose terms are alternately x and then y, we find that $x = y$ if and only if $K(i, x) = K(i, y)$ for all i. We state this conclusion as a proposition.

Proposition 10-14: $K(i, x) = K(i, y)$ for all i if and only if $x = y$.

Proposition 10-15: The space S^* is compact.

PROOF: Since S^* is a metric space, it is enough to prove that any sequence $\{x_n\}$ has a convergent subsequence. Now

$$K(i, x_n) \leq \sup_{j \in S} K(i, j) \leq N_{ii}/N_{\pi i} < \infty.$$

Thus by a diagonal process we may choose a subsequence $\{x_{n_k}\}$ such that $\{K(i, x_{n_k})\}$ is Cauchy for all i. Then $\{x_{n_k}\}$ is Cauchy in S^*. Since S^* is complete, $\{x_{n_k}\}$ is convergent.

The sets in the smallest Borel field containing the open sets of S^* are called **Borel sets**. The boundary B is a Borel set, since it is the complement of a countable set. Finite measures defined on the Borel sets are called **Borel measures**.

We conclude this section by agreeing on what conventions we shall adopt in case πN has some zero entries. If S is the set of all states, let

$$W = \{i \in S \mid (\pi N)_i > 0\}.$$

The special nature of W implies that $P_W = P^W$, and by Lemma 8-18 we see that the fundamental matrix for P_W is N_W. Thus if we agree that boundary theory for P and π is to be interpreted as boundary theory for P_W and π_W, we find that for i and j in W

$$K(i, j) = N_{ij}/N_{\pi j},$$

and $K(i, j)$ is not defined otherwise. Hence the metric is

$$d(j, j') = \sum_{i \in W} w_i N_{\pi i} |K(i, j) - K(i, j')|$$

for j and j' in W. Boundary theory is then done relative to the Cauchy completion of (W, d).

4. Convergence to the boundary

We continue to assume that P is a Markov chain with all states transient and that π is a starting distribution with $\pi N > 0$. Let $\bar{P}$ denote the enlarged chain obtained from P by adding the absorbing state b.

The main theorem of this section will be that with probability one every path ω has the property that either $x_n(\omega)$ converges in S^* or the process along ω disappears in finite time. From the results of the next sections we shall be able to sharpen the theorem by concluding that, when convergence takes place, it a.e. is to a nice subset of the boundary.

Lemma 10-16: Let g be a pure potential for P with charge f. If $\{x_n\}$ is an extended chain of total measure one with transition matrix $\bar{P}$ for which νf is finite, then the limit of $g(x_n)$ as n decreases to $\mathbf{u}$ exists a.e.

PROOF: Form the process watched starting in the finite set E. The claim is that for $n \geq 0$, $\{g(x_{u_E+n})\}$ is a non-negative supermartingale. It satisfies the supermartingale inequality because g is non-negative superregular. Thus, to show that the means are finite, it is sufficient to consider $\mathrm{M}[g(x_{u_E})]$. We have

$$\mathrm{M}[g(x_{u_E})] = \mu^E g = \mu^E N f = \nu^E f \leq \nu f < \infty,$$

since f is non-negative and $\nu^E \leq \nu$. Hence $\{g(x_{u_E+n})\}$ is a non-negative supermartingale.

If $0 \leq r < s$, then Proposition 3-11 applied to $-g$ (or Proposition 8-79) shows that the mean number of downcrossings of $[r, s]$ by $\{g(x_{u_E+j})\}$ up to time n is bounded by $s/(s - r)$ independently of n and of E. Let $n \to \infty$ and then let E increase so that $\mathbf{u}_E$ approaches $\mathbf{u}$. The mean number of downcrossings of $[r, s]$ remains bounded, and by monotone convergence the mean number of downcrossings after time $\mathbf{u}$ is finite. By the argument in the proof of Theorem 3-12, $g(x_n)$ converges a.e. as n decreases to $\mathbf{u}$. It can be shown that the limit is finite a.e., but this fact will not be needed.

Lemma 10-17: Let $\mathscr{G}$ be a Borel field of subsets of a set Ω, let S^* be a compact metric space, and for each n let $f_n: \Omega \to S^*$ be a function with the property that $f_n^{-1}(E)$ is in $\mathscr{G}$ for all Borel sets E. If $f_n(\omega) \to f(\omega)$ for all ω, then $f^{-1}(E)$ is in $\mathscr{G}$ for all Borel sets E.

PROOF: First consider the case of a compact set C. Let $N_\epsilon(C)$ be the open set of all points at a distance less than ϵ from C. Then

$$f^{-1}(C) = \bigcap_{k=1}^{\infty} \bigcup_{n=1}^{\infty} \bigcap_{m=n}^{\infty} f_m^{-1}(N_{1/k}(C)).$$

Let $\mathscr{C}$ be the class of all Borel sets E for which $f^{-1}(E) \in \mathscr{G}$. Then $\mathscr{C}$ is clearly closed under countable unions and complements. Since $\mathscr{C}$ contains all compact sets, it contains all Borel sets.

Theorem 10-18: Let the chain P with all states transient be started with a distribution π such that $\pi N > 0$. For each path ω let $\mathbf{v}(\omega)$ be the supremum of the n such that $x_n(\omega)$ is in S. Then a.e. either $\mathbf{v}(\omega) < +\infty$ and $x_{\mathbf{v}(\omega)}(\omega) \in S$ or $\mathbf{v}(\omega) = +\infty$ and $x_n(\omega)$ converges to a point $x_{\mathbf{v}(\omega)}(\omega) \in S^*$ as n tends to infinity. Furthermore, if $\mathscr{G}$ is the least Borel field containing the cylinder sets for P, then the set of ω where $x_{\mathbf{v}}$ is defined is in $\mathscr{G}$, and the inverse image under $x_{\mathbf{v}}$ of any Borel set in S^* is a set of $\mathscr{G}$.

Proof: First we prove the measurability. The set where $x_{\mathbf{v}}$ is defined is the countable union of the sets $\{\mathbf{v} = 1\}, \{\mathbf{v} = 2\}, \ldots$ and the set $\{\mathbf{v} = \infty \wedge x_n(\omega) \text{ converges}\}$. All of these except possibly the last are certainly in $\mathscr{G}$. The last set is

$$A = \{\omega \mid \mathbf{v}(\omega) = \infty \wedge \limsup_n K(i, x_n(\omega)) = \liminf_n K(i, x_n(\omega)) \text{ for all } i\}$$

and is therefore in $\mathscr{G}$. Now the intersection of $\{\mathbf{v} = n\}$ with the inverse image under $x_{\mathbf{v}}$ of a Borel set E is certainly in $\mathscr{G}$. Therefore to complete the proof of the measurability part of the theorem it is sufficient to prove that the intersection of the set A defined above with $x_{\mathbf{v}}^{-1}(E)$ is in $\mathscr{G}$ for every Borel set E. In Lemma 10-17 let Ω be the set A and let the field be the class of sets $A \cap G$, where $G \in \mathscr{G}$. Since $A \cap x_n^{-1}(E)$ is in the field for all E, the lemma applies and gives the result immediately.

Next we are to prove the almost-everywhere statement. Form the extended chain associated with π and P, as described in Section 2 after Corollary 10-7. All statements about this extended chain after time $n \geq 0$ have the same probabilities as the corresponding statements about P, and the vector ν of mean times in the various states is πN. It is therefore sufficient to show in the extended chain the convergence of $K(i, x_n)$ for all i.

Let $\{\hat{x}_n\}$ be the reverse of this extended chain. Since

$$K(i, j) = \hat{N}_{ji}/N_{\pi i},$$

it suffices to show for each i that in the reverse process $\hat{N}_{x_n, i}$ converges a.e. as n decreases to $\hat{\mathbf{u}}$. But $\hat{N}_{\cdot i}$ is the $\hat{P}$-potential of a unit charge at i. Since the charge has finite support, $\hat{\nu}$ times it must be finite. Therefore the theorem follows by applying Lemma 10-16 to the potential $\hat{N}_{\cdot i}$ for the reverse process $\{\hat{x}_n\}$.

By Theorem 10-18 the statement that $x_{\mathbf{v}}$ exists (or equivalently that $x_{\mathbf{v}} \in S^*$) and the statement that $x_{\mathbf{v}} \in E$, where E is a Borel set in S^*, are both measurable with respect to the least Borel field containing all cylinder sets. But $\Pr_i$ is defined on all such statements. Hence

$$\Pr_i[x_{\mathbf{v}} \in E]$$

is defined if E is a Borel set in S^*.

From now on, we use the notation of Chapter 2 that $\mathscr{F}$ is the field of cylinder sets for P and $\mathscr{G}$ is the least Borel field containing $\mathscr{F}$.

Corollary 10-19: $\Pr_i[x_v \in S^*] = 1.$

PROOF: As in the proof of Theorem 10-18, form the extended chain associated with P and π. By Theorem 10-18 almost every path in the process P (started according to π) satisfies $x_v \in S^*$. Hence the same is true of the extended chain, and hence it is true of those paths in the extended chain which pass through i. On any path ω which passes through i, $x_v(\omega) \in S^*$ if and only if $x_v(\omega_{u_{(i)}}) \in S^*$. Therefore by Definition 10-4, the extended chain watched starting in $\{i\}$ satisfies

$$\Pr_{\mu^{(i)}}[x_v \in S^*] = \mu_i^{\{i\}}.$$

On the other hand,

$$\Pr_{\mu^{(i)}}[x_v \in S^*] = \mu_i^{\{i\}} \Pr_i[x_v \in S^*].$$

Since $\mu_i^{\{i\}} \neq 0$, we must have $\Pr_i[x_v \in S^*] = 1$.

It is to be emphasized that S^* has been constructed for the fixed starting distribution π and that Corollary 10-19 is not the same as Theorem 10-18 restated for the case where π assigns measure one to the state i: *The boundaries for different starting distributions may be different.*

5. Poisson–Martin Representation Theorem

The notation P, π, $K(i, x)$, S^*, $\mathscr{F}$, and $\mathscr{G}$ of Sections 3 and 4 is still in force. We shall use Pr to mean $\Pr_\pi$.

We recall that $\pi K(\cdot, j) = 1$ for all j in S. If $j_n \to x$ in S^*, then $K(\cdot, j_n) \to K(\cdot, x)$ and, for all n, $\pi K(\cdot, j_n) = 1$. Hence $\pi K(\cdot, x) \leq 1$ by Fatou's Theorem. Moreover, we know that $K(\cdot, j)$ is P-superregular for all j in S. If $j_n \to x$, then again by Fatou's Theorem,

$$\begin{aligned} PK(\cdot, x) &= P \lim K(\cdot, j_n) \leq \liminf PK(\cdot, j_n) \\ &\leq \liminf K(\cdot, j_n) = K(\cdot, x). \end{aligned}$$

That is, $K(\cdot, x)$ is P-superregular for all x in S^*. These remarks enable us to prove the following proposition.

Proposition 10-20: If ν is any Borel measure on S^* with $\nu(S^*) = 1$, then the function h, defined by

$$h_i = \int_{S^*} K(i, x) d\nu(x),$$

is finite-valued non-negative superregular and satisfies $\pi h \leq 1$.

PROOF: It is clearly non-negative and is finite-valued because $K(i, \cdot)$ is bounded. By Fubini's Theorem,

$$\pi h = \sum_i \int_{S^*} \pi_i K(i, x) d\nu(x) = \int_{S^*} \left[\sum_i \pi_i K(i, x)\right] d\nu(x) \leq \int_{S^*} d\nu(x) = 1$$

and

$$(Ph)_i = \sum_j \int_{S^*} P_{ij} K(j, x) d\nu(x) = \int_{S^*} \left[\sum_j P_{ij} K(j, x)\right] d\nu(x)$$
$$\leq \int_{S^*} K(i, x) d\nu(x) = h_i.$$

Thus Borel measures on S^* give rise to π-integrable non-negative superregular functions h. Our goal in this section will be to prove conversely that every non-negative (finite-valued) superregular function h arises as the integral over S^* of $K(i, x)$ with respect to some measure. We postpone the uniqueness question to Sections 6 and 7.

Throughout the remainder of this chapter we shall use "superregular" to mean "finite-valued superregular."

Harmonic measure μ is defined on the Borel sets E of S^* by

$$\mu(E) = \Pr[x_v \in E].$$

By Theorem 10-18 the definition of μ makes sense and $\mu(S^*) = 1$. The complete additivity of μ is a consequence of the complete additivity of Pr. Thus μ is a Borel measure. The proposition to follow gives a formula for $\Pr_i[x_v \in E]$ in terms of harmonic measure.

Proposition 10-21: For every Borel set E of S^*,

$$\Pr_i[x_v \in E] = \int_E K(i, x) d\mu(x).$$

PROOF: Let E_n be a fixed increasing sequence of finite sets of S with union S. Let $\mathrm{v}_n(\omega)$ be the last time (possible $+\infty$) that an outcome on the path ω is in E_n, and let $\mathrm{v}_n(\omega) = 0$ if no outcome on ω is in E_n. For any starting distribution γ, Proposition 4-28 implies that

$$\Pr[x_{v_n} = j] = \sum_{k=0}^{\infty} \Pr_\gamma[x_k = j \wedge x_m \notin E_n \text{ after time } k]$$
$$= \sum_{k=0}^{\infty} (\gamma P^k)_j e_j^{E_n}$$
$$= (\gamma N)_j e_j^{E_n}.$$

Hence

$$\Pr_i[x_{v_n} = j] = N_{ij} e_j^{E_n} = K(i, j) N_{\pi j} e_j^{E_n} = K(i, j) \Pr_\pi[x_{v_n} = j].$$

For Borel sets E of S^* define measures by

$$\mu_{in}(E) = \Pr_i[x_{v_n} \in E],$$
$$\mu_i(E) = \Pr_i[x_v \in E],$$
$$\mu_{\pi n}(E) = \Pr_\pi[x_{v_n} \in E],$$

and

$$\mu_\pi(E) = \Pr_\pi[x_v \in E] = \mu(E).$$

What we have just shown is that

$$\mu_{in}(E) = \int_E K(i, x) d\mu_{\pi n}(x).$$

Now if $f \geq 0$ is a Borel measurable function on S^*, the claim is that

$$\int_{S^*} f(x) d\mu_{in}(x) = \int_\Omega f(x_{v_n}(\omega))\, d\Pr_i(\omega).$$

The result for characteristic functions is just the definition of μ_{in}, and for general $f \geq 0$ it follows from the result for simple functions by monotone convergence. Then the result holds for continuous $f \geq 0$ and hence for all continuous f. Similarly for continuous f,

$$\int_{S^*} f(x) d\mu_i(x) = \int_\Omega f(x_v(\omega))\, d\Pr_i(\omega),$$

where we set $x_v(\omega) = 0$ when v is not defined.

As $n \to \infty$, $x_{v_n}(\omega) \to x_v(\omega)$ a.e. $[\Pr_i]$ by Corollary 10-19. When f is continuous, $f(x_{v_n}(\omega)) \to f(x_v(\omega))$ a.e. $[\Pr_i]$. Since continuous functions are bounded, we have

$$\lim_n \int_\Omega f(x_{v_n}(\omega))\, d\Pr_i(\omega) = \int_\Omega f(x_v(\omega))\, d\Pr_i(\omega)$$

by dominated convergence. Hence

$$\lim_n \int_{S^*} f(x) d\mu_{in}(x) = \int_{S^*} f(x) d\mu_i(x)$$

for all continuous f. Similarly

$$\lim_n \int_{S^*} f(x) d\mu_{\pi n}(x) = \int_{S^*} f(x) d\mu_\pi(x).$$

Since $K(i, \cdot)$ is continuous, so is $f(\cdot)K(i, \cdot)$. Therefore

$$\lim_n \int_{S^*} f(x) K(i, x) d\mu_{\pi n}(x) = \int_{S^*} f(x) K(i, x) d\mu_\pi(x)$$

for all continuous f. Since $\mu_{in}(E) = \int_E K(i, x) d\mu_{\pi n}(x)$, we obtain

$$\int_{S^*} f(x) d\mu_i(x) = \int_{S^*} f(x) K(i, x) d\mu_\pi(x)$$

for all continuous f. Therefore

$$\mu_i(E) = \int_E K(i, x) d\mu_\pi(x).$$

Corollary 10-22: $\int_{S^*} K(i, x) d\mu(x) = 1$ for all $i \in S$.

PROOF: Since $\Pr_i[x_v \in S^*] = 1$ by Corollary 10-19, the result follows from Proposition 10-21.

The corollary we have just proved is the representation theorem as it applies to the column vector which is identically one. We shall be able to get the general case by applying the corollary to a suitable modification of the h-process introduced in Chapter 8. We now re-define the h-process in such a way that we allow h to have some entries equal to zero.

Definition 10-23: If $h \geq 0$ is a finite-valued P-superregular function such that $\pi h = 1$, then h-process is defined to be the Markov chain with state space S and with the measure $\Pr^h$ defined by

$$\Pr{}^h[x_0 = c \wedge x_1 = d \wedge x_2 = e \wedge \cdots \wedge x_{n-1} = i \wedge x_n = j] = \pi_c P_{cd} P_{de} \ldots P_{ij} h_j.$$

We readily check that the h-process is indeed a Markov chain. If we define S^h by

$$S^h = \{i \in S \mid h_i > 0\},$$

then the transition matrix P^h of the h-process satisfies

$$P_{ij}^h = \begin{cases} \dfrac{P_{ij} h_j}{h_i} & \text{for } i \text{ and } j \text{ in } S^h, \\ 0 & \text{for } i \in S^h \text{ and } j \in S - S^h \end{cases}$$

and the starting vector π^h satisfies

$$\pi_i^h = \pi_i h_i \quad \text{for all } i.$$

If i is in $S - S^h$, then P_{ij}^h is not defined and we shall agree to take it to be zero. With this definition we compute directly that the fundamental matrix N^h satisfies

$$N_{ij}^h = \begin{cases} \dfrac{N_{ij} h_j}{h_i} & \text{if } i \text{ and } j \text{ are in } S^h \\ \delta_{ij} & \text{otherwise.} \end{cases}$$

Hence P^h has only transient states.

Lemma 10-24: If $h \geq 0$ is a P-superregular function and if $h_i = 0$ and $h_j > 0$, then $N_{ij} = 0$. If, in addition, $\pi h = 1$, then $(\pi^h N^h)_j = h_j(\pi N)_j$, and $(\pi^h N^h)_j > 0$ if and only if j is in S^h. For i and j in S^h,

$$K^h(i, j) = K(i, j)/h_i.$$

PROOF: If $h_i = 0$ and $h_j > 0$, then for every n

$$h_i \geq \sum_k P_{ik}^{(n)} h_k \geq P_{ij}^{(n)} h_j$$

and hence $P_{ij}^{(n)} = 0$. Therefore $N_{ij} = \sum_n P_{ij}^{(n)} = 0$. Consequently,

$$(\pi^h N^h)_j = \sum_{i \in S^h} \pi_i h_i \left(\frac{N_{ij} h_j}{h_i} \right) = h_j \sum_{i \in S^h} \pi_i N_{ij} = h_j \sum_{i \in S} \pi_i N_{ij} = h_j(\pi N)_j.$$

By assumption π is a vector such that πN is strictly positive. Therefore $(\pi^h N^h)_j = 0$ if and only if $h_j = 0$.

Finally, according to the convention at the end of Section 3 and the calculation just completed, $K^h(i, j)$ is defined if i and j are in S^h. We have

$$K^h(i, j) = N_{ij}^h/(\pi^h N^h)_j = \frac{N_{ij} h_j / h_i}{h_j(\pi N)_j} = K(i, j)/h_i.$$

Since the h-process has the property that $(\pi^h N^h)_j$ is positive exactly when j is in S^h, we can, as noted at the end of Section 3, define a metric d^h on $S^h \times S^h$ and we can form the Cauchy completion S^{h*} with boundary B^h. We shall agree to use the same weights in defining d^h that were used in defining d.

Lemma 10-25: The identity map from (S^h, d^h) into (S, d) is an isometry.

PROOF: Let j and j' be in S^h. Then

$$\begin{aligned} d^h(j, j') &= \sum_{i \in S^h} w_i (\pi^h N^h)_i |K^h(i, j) - K^h(i, j')| \\ &= \sum_{i \in S^h} w_i N_{\pi i} h_i |K(i, j) - K(i, j')|/h_i \\ &= \sum_{i \in S^h} w_i N_{\pi i} |K(i, j) - K(i, j')| \\ &= \sum_{i \in S} w_i N_{\pi i} |K(i, j) - K(i, j')| \\ &= d(j, j'). \end{aligned}$$

It follows from Lemma 10-25 that S^{h*} can be canonically identified with a compact subset of S^*. Thus by continuity, $K^h(i, x) = K(i, x)/h_i$ for all i in S^h and x in S^{h*}. Harmonic measure for the h-process will be denoted μ^h. We can consider it to be defined on S^* (as well as on S^{h*}) if we set

$$\mu^h(E) = \mu^h(E \cap S^{h*})$$

for Borel sets E in S^*.

We are finally in a position to state and prove the existence half of the Markov chain analog of the Poisson–Martin Representation Theorem.

Theorem 10-26: If $h \geq 0$ is a finite-valued P-superregular function such that $\pi h = 1$, then

$$h_i = \int_{S^*} K(i, x) d\mu^h(x).$$

Proof: Applying Corollary 10-22 to the h-process, we have

$$\int_{S^{h*}} K^h(i, x) d\mu^h(x) = 1$$

for i in S^h. That is, for i in S^h

$$h_i = \int_{S^{h*}} K(i, x) d\mu^h(x) = \int_{S^*} K(i, x) d\mu^h(x).$$

Now if $i \in S - S^h$, $N_{ij} = 0$ for all $j \in S^h$ by Lemma 10-24. Thus $K(i, j) = 0$ for such i and j. Since $K(i, x)$ is continuous on S^{h*}, $K(i, x) = 0$ for $i \in S - S^h$ and $x \in S^{h*}$. Therefore for such i,

$$h_i = 0 = \int_{S^{h*}} K(i, x) d\mu^h(x) = \int_{S^*} K(i, x) d\mu^h(x).$$

Of course, the representation theorem immediately extends to cover all P-superregular functions $h \geq 0$ for which πh is positive and finite. However, the probabilistic interpretation of the measure μ^h is lost. We shall return to this point in Theorem 10-41 of Section 7 after proving the uniqueness theorem.

6. Extreme points of the boundary

The measure μ^h is not necessarily the unique Borel measure which represents h in the sense of Theorem 10-26, and we consequently need another hypothesis to get uniqueness. What we shall do in this section is to define the set B_e of extreme points of the boundary and the subset $\bar{S} = S \cup B_e$ of S^*. In Section 7 we shall see that μ^h has all its mass on $\bar{S}$ and that μ^h is the unique measure with all its mass on $\bar{S}$ for which the representation in Theorem 10-26 holds.

There are three kinds of behavior of points of the boundary that we shall want to exclude:

(1) x has the property that $\pi K(\cdot, x) < 1$.
(2) x has the property that $K(\cdot, x)$ is not regular.
(3) x has the property that $K(\cdot, x)$ is regular but is a nontrivial convex combination of other non-negative regular functions.

The first two of these possibilities are the topic of the two lemmas to follow. The third possibility will require more of our attention, and we discuss it beginning with Definition 10-29 and Lemma 10-30.

We recall that $\pi K(\cdot, x) \leq 1$ for all x in S^*.

Lemma 10-27: For almost every $x\,[\mu]$ in S^*, $\pi K(\cdot, x) = 1$. The set where the equality holds is a Borel set.

PROOF: For each i, the function $K(i, x)$ is continuous. Hence the countable sum $\pi K(\cdot, x)$ is Borel measurable. Therefore the set where it equals one is a Borel set.

By Corollary 10-22,

$$\int_{S^*} K(i, x) d\mu(x) = 1.$$

Thus by Fubini's Theorem,

$$1 = \pi \mathbf{1} = \pi \int_{S^*} K(\cdot, x) d\mu(x) = \int_{S^*} \pi K(\cdot, x) d\mu(x).$$

But $\int_{S^*} 1 d\mu(x) = 1$ also, and since $1 - \pi K(\cdot, x) \geq 0$, we conclude that $\pi K(\cdot, x) = 1$ a.e. by Corollary 1-40.

We say that a function h is **normalized** if $\pi h = 1$. By Lemma 10-27, $K(\cdot, x)$ is normalized for a.e. $x\,[\mu]$.

We recall that $K(\cdot, x)$ is P-superregular for all $x \in S^*$.

Lemma 10-28: For almost every $x\,[\mu]$ in the boundary B of S^*, the function $K(\cdot, x)$ is regular. The set where it is regular is a Borel set.

PROOF: The set where $P_{i\cdot}K(\cdot, x) = K(i, x)$ is a Borel set since it is the set where a Borel measurable function takes on the value $K(i, x)$. The set where $K(\cdot, x)$ is regular is the countable intersection of these sets.

By Theorem 4-10 with the random time identically equal to one, we see that the column vector whose ith component is (see Proposition 10-21)

$$\int_B K(i, x)d\mu(x) = \Pr_i[x_v \in B]$$

is P-regular. By this observation and by Fubini's Theorem, we have

$$\int_B PK(\cdot, x)d\mu(x) = P\int_B K(\cdot, x)d\mu(x) = \int_B K(\cdot, x)d\mu(x).$$

Since $PK(\cdot, x) \leq K(\cdot, x)$, we must have $PK(\cdot, x) = K(\cdot, x)$ a.e. by Corollary 1-40.

Definition 10-29: A finite-valued function $h \geq 0$ is **minimal** if

(1) h is regular, and
(2) whenever $0 \leq h' \leq h$ with h' regular, then $h' = ch$.

Lemma 10-30: A normalized finite-valued regular function $h \geq 0$ is minimal if and only if it cannot be written as a nontrivial convex combination of two distinct normalized non-negative regular functions.

PROOF: If $h = c_1h_1 + c_2h_2$ is such a convex combination, then either h_1 or h_2, say h_1, is not equal to h. Since $h \geq c_1h_1$, we must have $c_1h_1 = ch$ if h is minimal. Multiplying through by π, we obtain $c_1 = c$. Since $c_1 \neq 0$, we conclude $h_1 = h$, contradiction.

Conversely, if $h \geq h' \geq 0$ with h' regular and h' not equal to 0 or h, then

$$h = (\pi h')\frac{h'}{(\pi h')} + [\pi(h - h')]\frac{h - h'}{\pi(h - h')}$$

exhibits h as a nontrivial convex combination of normalized regular functions, provided we can prove $0 < \pi h' < 1$. If so, then by hypothesis the two normalized functions must be equal to each other and hence equal to h. That is, $h' = (\pi h')h$. Thus we are to prove $0 < \pi h' < 1$. Let $h'_j > 0$. Since $(\pi N)_j > 0$, choose n so that $(\pi P^n)_j > 0$. Since h' is superregular, $h' \geq P^nh'$ and hence $\pi h' \geq \pi P^nh' \geq (\pi P^n)_jh_j > 0$. A similar argument applied to $h - h'$ shows that $\pi h' < 1$.

Definition 10-31: A point x in S^* is an **extreme point** of S^* if the function $K(\cdot, x)$ is minimal and normalized. The set of extreme points is denoted B_e. Let $\bar{S} = S \cup B_e$.

Since $K(\cdot, j)$ is not regular when $j \in S$, no point of S can be in B_e, and B_e must be entirely contained in the boundary B. The set $\bar{S}$ is the subset of S^* with respect to which the uniqueness theorem will be stated. We shall see eventually that $\bar{S}$ is a Borel set and that $\mu(\bar{S}) = 1$ (compare with Lemmas 10-27 and 10-28).

If we form an h-process, we know that $\overline{S^h} \subset S^{h*} \subset S^*$. The following lemma strengthens this conclusion and shows that actually $\overline{S^h} \subset \bar{S}$.

Lemma 10-32: Let $h \geq 0$ be a finite-valued normalized P-superregular function. If x is in S^{h*}, then

(1) $K^h(\cdot, x)$ is normalized if and only if $K(\cdot, x)$ is normalized.
(2) $K^h(\cdot, x)$ is regular for P^h restricted to S^h if and only if $K(\cdot, x)$ is P-regular.
(3) $K^h(\cdot, x)$ is minimal for P^h restricted to S^h if and only if $K(\cdot, x)$ is minimal for P.

Hence $B_e^h = B^h \cap B_e$ and $\overline{S^h} \subset \bar{S}$.

PROOF: Conclusions (1) and (2) follow from the identities

$$\sum_{i \in S^h} \pi_i^h K^h(i, x) = \pi K(\cdot, x)$$

$$\sum_{j \in S^h} P_{\cdot j}^h K^h(j, x) = PK(\cdot, x),$$

both of which use the fact that $K(i, x) = 0$ if i is not in S^h (see the proof of Theorem 10-26).

Thus in (3) we may assume that $K^h(\cdot, x)$ and $K(\cdot, x)$ are both regular. Multiplying both by the same constant, if necessary, we may assume for the purposes of this proof that they are normalized. We shall use Lemma 10-30 and show that a nontrivial decomposition exists for $K(\cdot, x)$ if and only if a nontrivial decomposition exists for $K^h(\cdot, x)$. In fact, if for $i \in S$

$$K(i, x) = c_1 h_i^{(1)} + c_2 h_i^{(2)}$$

nontrivially, then for $i \in S^h$,

$$K^h(i, x) = K(i, x)/h_i = c_1 \frac{h_i^{(1)}}{h_i} + c_2 \frac{h_i^{(2)}}{h_i}.$$

We have

$$\sum_{j \in S^h} P_{ij}^h \frac{h_j^{(1)}}{h_j} = \sum_{j \in S^h} \frac{1}{h_i} P_{ij} h_j^{(1)} = \sum_{j \in S} \frac{1}{h_i} P_{ij} h_j^{(1)} = \frac{h_i^{(1)}}{h_i}$$

and

$$\sum_{i \in S^h} \pi_i^h \frac{h_i^{(1)}}{h_i} = \sum_{i \in S^h} \pi_i h_i^{(1)} = \sum_{i \in S} \pi_i h_i^{(1)} = 1,$$

where the sums over S^h can be replaced by sums over S because $i \in S - S^h$ implies $0 \leq h_i^{(1)} \leq K(i, x)/c_1 = 0$ or $h_i^{(1)} = 0$. Consequently we may assume that

$$\frac{h_i^{(1)}}{h_i} = \frac{h_i^{(2)}}{h_i}$$

for $i \in S^h$. That is, $h_i^{(1)} = h_i^{(2)}$ for $i \in S^h$. But $h_i^{(1)} = h_i^{(2)} = 0$ for $i \in S - S^h$. Hence $h^{(1)} = h^{(2)}$.

Conversely, if for $i \in S^h$

$$K^h(i, x) = c_3 h_i^{(3)} + c_4 h_i^{(4)},$$

then

$$K(i, x) = c_3 h_i^{(3)} h_i + c_4 h_i^{(4)} h_i$$

for $i \in S^h$. Extend $\{h_i^{(3)} h_i\}$ and $\{h_i^{(4)} h_i\}$ to be defined for all $i \in S$ by setting them equal to zero for $i \in S - S^h$. The convex sum of them is still $K(i, x)$, and they are both regular normalized functions, since

$$\sum_j P_{ij} h_j^{(3)} h_j = \sum_{j \in S^h} P_{ij} h_j^{(3)} h_j = h_i^{(3)} h_i$$

and

$$\sum_i \pi_i h_i^{(3)} h_i = \sum_{i \in S^h} \pi_i h_i^{(3)} h_i = \pi^h h^{(3)} = 1.$$

Consequently we may assume that $h_i^{(3)} h_i = h_i^{(4)} h_i$ for all $i \in S$. That is, $h_i^{(3)} = h_i^{(4)}$ for all $i \in S^h$.

We now begin to derive properties of the set B_e of extreme points.

Lemma 10-33: If $h \geq 0$ is a normalized minimal function such that

$$h_i = \int_{S^*} K(i, x) d\nu(x)$$

for a Borel measure ν with $\nu(S^*) = 1$, then ν is concentrated at a single point and that point is extreme.

PROOF: Consider the functions

$$h^A = \frac{1}{\nu(A)} \int_A K(\cdot, x) d\nu(x)$$

as A ranges through the Borel sets with $0 < \nu(A) < 1$. For any such A, h^A and $h^{\tilde{A}}$ are superregular and satisfy $\pi h^A \leq 1$ and $\pi h^{\tilde{A}} \leq 1$ by Proposition 10-20. But

$$h = \nu(A) h^A + \nu(\tilde{A}) h^{\tilde{A}}$$

with h regular and normalized and with $\nu(A) + \nu(\tilde{A}) = 1$. Hence h^A

and $h^{\bar{A}}$ must both be regular and normalized. By Lemma 10-30, $h = h^A = h^{\bar{A}}$. Thus, if $0 < \nu(A) < 1$,

$$\nu(A)h_i = \int_A K(i, x)d\nu(x)$$

for each fixed i. The same is trivially true if $\nu(A) = 0$, and it is true by hypothesis if $\nu(A) = 1$. Hence it is true of all Borel sets. Therefore for fixed i,

$$K(i, x) = h_i \quad \text{a.e. } [\nu].$$

Thus $K(i, x) = h_i$ for *all* i almost everywhere $[\nu]$. Since $\nu(S^*) > 0$, there is at least one point x_0 where it is true. We have

$$K(i, x_0) = h_i$$

for all i. If there were another such point x', then we would have $K(\cdot, x_0) = K(\cdot, x')$, and hence $x_0 = x'$ by Proposition 10-14. Therefore the complement of $\{x_0\}$ has measure zero, or ν is concentrated at x_0. Now, we know that $K(\cdot, x_0) = h$ and h is normalized and minimal. Hence x_0 is extreme.

Lemma 10-34: If h, $h^{(1)}$, and $h^{(2)}$ are normalized non-negative superregular functions with

$$h = c_1 h^{(1)} + c_2 h^{(2)},$$

where $c_1 \geq 0$ and $c_2 \geq 0$, then

$$\mu^h = c_1 \mu^{h^{(1)}} + c_2 \mu^{h^{(2)}}.$$

PROOF: For a typical basic statement $x_0 = i \wedge x_1 = j \wedge x_2 = k$, we have, by Definition 10-23,

$$\Pr{}^h[x_0 = i \wedge x_1 = j \wedge x_2 = k] = \pi_i P_{ij} P_{jk} h_k.$$

Using the analogous identities for $h^{(1)}$ and $h^{(2)}$ and breaking up h_k as $c_1 h_k^{(1)} + c_2 h_k^{(2)}$, we obtain

$$\Pr{}^h[x_0 = i \wedge x_1 = j \wedge x_2 = k] = c_1 \Pr{}^{h^{(1)}}[x_0 = i \wedge x_1 = j \wedge x_2 = k] + c_2 \Pr{}^{h^{(2)}}[x_0 = i \wedge x_1 = j \wedge x_2 = k].$$

Hence the same is true of all statements in $\mathscr{F}$, and by the uniqueness half of Theorem 1-19 we find that

$$\Pr{}^h[p] = c_1 \Pr{}^{h^{(1)}}[p] + c_2 \Pr{}^{h^{(2)}}[p]$$

for all statements p measurable relative to $\mathscr{G}$.

Take p to be the statement $x_v \in E$, where E is a Borel set of S^*. The claim is that $x_v \in E$ if and only if $x_v^h \in E$, and similarly for $h^{(1)}$ and $h^{(2)}$. For each n we certainly have $x_n = x_n^h$. Hence $x_v = x_v^h$ when $\mathbf{v}$ is finite. When $\mathbf{v}$ is infinite, we have $x_v = x_v^h$ because $x_n = x_n^h$ for all n and because S^{h*} is isometric with a subset of S^*. Therefore

$$\Pr^h[x_v \in E] = c_1 \Pr^{h^{(1)}}[x_v \in E] + c_2 \Pr^{h^{(2)}}[x_v \in E]$$

or

$$\mu^h = c_1\mu^{h^{(1)}} + c_2\mu^{h^{(2)}}.$$

Proposition 10-35: Let $h = K(\cdot, x)$. Then $x \in \bar{S}$ if and only if h is normalized and $\mu^h(\{x\}) = 1$.

PROOF: First let $x \in S$. Then h is certainly normalized, and hence the h-process is defined. Suppose we can prove that the h-process disappears with probability one. Then by definition of μ^h, $\mu^h(S^h) = 1$. Hence by Theorem 10-26,

$$K(\cdot, x) = \int_{S^*} K(\cdot, y)d\mu^h(y) = \int_S K(\cdot, y)d\mu^h(y) = \sum_j N_{ij}\left(\frac{\mu^h(\{j\})}{N_{\pi j}}\right).$$

But $K(\cdot, x) = N_{ix}/N_{\pi x} = \sum_j N_{ij}(\delta_{jx}/N_{\pi j})$. By Theorem 8-4,

$$\frac{\delta_{jx}}{N_{\pi j}} = \frac{\mu^h(\{j\})}{N_{\pi j}} \quad \text{for all } j.$$

That is, $\mu^h(\{x\}) = 1$. Thus we are to prove that the h-process disappears with probability one. By the remarks following Definition 8-13, it suffices to show that $\sum_{j \in S^h} N_{ij}^h f_j = 1$ for some f. Take f to be a single mass $1/(N_{\pi x}h_x)$ at x, and the equality follows.

Next let $x \in B_e$. Then h is normalized by definition. By Theorem 10-26,

$$K(\cdot, x) = \int_{S^*} K(\cdot, y)d\mu^h(y).$$

In Lemma 10-33 take ν to be μ^h. Then $\mu^h(\{x_0\}) = 1$ for some x_0. But then $K(\cdot, x) = K(\cdot, x_0)$, and hence $x = x_0$ by Proposition 10-14.

Conversely, suppose h is normalized and $\mu^h(\{x\}) = 1$. If $x \notin S$, then $x \in B$, and by Lemmas 10-28 and 10-32 (conclusion 2), h must be regular. It remains to show that h is minimal. If $h = c_1h^{(1)} + c_2h^{(2)}$, then by Lemma 10-34

$$\mu^h = c_1\mu^{h^{(1)}} + c_2\mu^{h^{(2)}}.$$

Hence $\mu^{h^{(1)}}$ must put all its weight on x, and therefore $h = h^{(1)}$. By Lemma 10-30 we conclude that h is minimal.

7. Uniqueness of the representation

In this section we retain all of the notations of Sections 3 through 6. We are going to prove that μ^h is the unique Borel measure concentrated on $\bar{S}$ for which the representation in Theorem 10-26 holds. The first step will be to show that $\bar{S}$ is a Borel set of harmonic measure one.

We denote by S_N the set of points x in S^* for which $K(\cdot, x)$ is normalized. These are exactly the points for which the h-process with $h = K(\cdot, x)$ has been defined. By Lemma 10-27, S_N is a Borel set of harmonic measure one. Since S_N is a Borel subset of S^*, the notion of a Borel measurable function on S_N is well defined.

Lemma 10-36: If A is a fixed $\mathscr{G}$-measurable set in Ω, then $\Pr^{K(\cdot,x)}[A]$ is a Borel measurable function of x in S_N.

PROOF: By Definition 10-23,

$$\Pr^{K(\cdot,x)}[x_0 = i \wedge x_1 = j \wedge x_2 = k] = \pi_i P_{ij} P_{jk} K(k, x),$$

and the right side is continuous even for x in S^*. Since any cylinder set is the countable disjoint union of basic cylinder sets, the function $\Pr^{K(\cdot,x)}[\omega \in A]$ for $A \in \mathscr{F}$ is a denumerable sum of such functions and hence is Borel measurable.

Let $\mathscr{C}$ be the class of all sets in Ω for which $\Pr^{K(\cdot,x)}[\omega \in A]$ is Borel measurable. If $\{A_n\}$ and $\{B_n\}$ are, respectively, increasing and decreasing sequences of such sets, then

$$\Pr^{K(\cdot,x)}[\bigcup A_n] = \lim_n \Pr^{K(\cdot,x)}[A_n]$$

and

$$\Pr^{K(\cdot,x)}[\bigcap B_n] = \lim_n \Pr^{K(\cdot,x)}[B_n],$$

the latter equality holding since $\Pr^{K(\cdot,x)}$ is a finite measure. Hence $\bigcup A_n$ and $\bigcap B_n$ are both in $\mathscr{C}$. By the Monotone Class Lemma (see Halmos [1950], pp. 27–28), $\mathscr{C}$ contains $\mathscr{G}$.

Lemma 10-37: If A is in $\mathscr{G}$ and if C is a Borel set in S_N, then

$$\Pr[\omega \in A \wedge x_v \in C] = \int_C \Pr^{K(\cdot,x)}[A] d\mu(x).$$

PROOF: If p is the typical basic statement $x_0 = i \wedge x_1 = j \wedge x_2 = k$ then

$$\begin{aligned}\Pr[p \wedge x_v \in C] &= \Pr[p] \cdot \Pr[x_v \in C \mid p] \\ &= \pi_i P_{ij} P_{jk} \Pr_k[x_v \in C]\end{aligned}$$

by Theorem 4-9. By Proposition 10-21 and by Definition 10-23, this expression is equal to

$$= \pi_i P_{ij} P_{jk} \int_C K(k, x) d\mu(x)$$

$$= \int_C \pi_i P_{ij} P_{jk} K(k, x) d\mu(x)$$

$$= \int_C \Pr{}^{K(\cdot,x)}[p] d\mu(x).$$

By Lemma 10-36 the function $\Pr^{K(\cdot,x)}[A]$ is a Borel measurable function of x if A is in $\mathscr{G}$. Fixing C, we may therefore define a set function σ by

$$\sigma(A) = \int_C \Pr{}^{K(\cdot,x)}[A] d\mu(x).$$

Then σ is certainly non-negative, and it is completely additive by the Monotone Convergence Theorem. The calculation above shows that the two set functions $\sigma(A)$ and $\Pr[\omega \in A \wedge x_v \in C]$ agree on basic cylinder sets and hence on all of the field $\mathscr{F}$. By Theorem 1-19 they must agree on all of $\mathscr{G}$.

Proposition 10-38: The set $\bar{S}$ is a Borel set with $\mu(\bar{S}) = 1$.

PROOF: Applying Lemma 10-37 to the statement $x_v \in D$, where D is a Borel set of S^*, we have for any Borel subset C of S_N

$$\Pr[x_v \in D \wedge x_v \in C] = \int_C \Pr{}^{K(\cdot,x)}[x_v \in D] d\mu(x) = \int_C \mu^{K(\cdot,x)}(D) d\mu(x).$$

But by definition

$$\Pr[x_v \in D \wedge x_v \in C] = \mu(D \cap C) = \int_C \chi_D(x) d\mu(x).$$

The set on which two Borel measurable functions agree is a Borel set. Hence for fixed D the set of x's on which

$$\mu^{K(\cdot,x)}(D) = \chi_D(x)$$

holds is a Borel set in S_N. Since these two functions have the same μ-integral over all Borel sets C, they must be equal a.e. $[\mu]$.

We shall let D range over the intersection with S_N of all balls with centers in S and with rational radii. Let $\mathscr{T}$ be the collection of such balls and let

$$T = \{x \in S_N \mid \mu^{K(\cdot,x)}(D) = \chi_D(x) \quad \text{for all } D \text{ in } \mathscr{T}\}.$$

Remembering that S_N is a Borel set of harmonic measure one, we see that T is the denumerable intersection of Borel sets of measure one and is therefore a Borel set with $\mu(T) = 1$. We show $T = \bar{S}$.

First let $x \in T$. Choose s_n in S with $d(x, s_n) < 1/n$, and let D_n be the intersection with S_N of the ball with center s_n and radius $1/n$. By assumption

$$\mu^{K(\cdot,x)}(D_n) = \chi_{D_n}(x) = 1$$

for all n. Hence

$$\mu^{K(\cdot,x)}(\{x\}) = \mu^{K(\cdot,x)}(\cap D_n) = 1.$$

Therefore $x \in \bar{S}$ by Proposition 10-35.

Conversely, if $x \in \bar{S}$, then

$$\mu^{K(\cdot,x)}(\{x\}) = 1$$

by Proposition 10-35, and so

$$\mu^{K(\cdot,x)}(D) = \chi_D(x)$$

for all Borel sets D in S_N. Therefore $x \in T$, and $T = \bar{S}$.

Lemma 10-39: Let h be defined by

$$h_i = \int_S K(i, x)d\nu(x),$$

where ν is a measure with $\nu(\bar{S}) = 1$. Then the h-process is well defined and for any A in $\mathscr{G}$

$$\Pr{}^h[A] = \int_S \Pr{}^{K(\cdot,x)}[A]d\nu(x).$$

Proof: By Proposition 10-20, h is non-negative superregular. By Fubini's Theorem $\pi h = 1$, since $\pi K(\cdot, x) = 1$ for all x in $\bar{S}$. Hence the h-process is well defined. If p is a typical basic statement,

$$x_0 = i \wedge x_1 = j \wedge x_2 = k,$$

$$\begin{aligned}\Pr{}^h[p] &= \pi_i P_{ij} P_{jk} h_k \\ &= \int_{\bar{S}} \pi_i P_{ij} P_{jk} K(k, x)d\nu(x) \\ &= \int_{\bar{S}} \Pr{}^{K(\cdot,x)}[p]d\nu(x).\end{aligned}$$

Proceeding as in Lemma 10-37, we define σ by

$$\sigma(A) = \int_S \Pr^{K(\cdot,x)}[A]d\nu(x).$$

Then σ is a finite measure on $\mathscr{G}$ which agrees with the measure $\Pr^h[A]$ on $\mathscr{F}$. By Theorem 1-19,

$$\Pr^h[A] = \sigma(A)$$

for all A in $\mathscr{G}$.

The theorem to follow is the uniqueness theorem mentioned at the beginning of this section.

Theorem 10-40: If $h \geq 0$ is a normalized superregular function such that

$$h_i = \int_S K(i, x)d\nu(x)$$

for some measure ν on S^* with $\nu(\bar{S}) = \nu(S^*) = 1$, then $\nu = \mu^h$.

PROOF: By Lemma 10-39,

$$\mu^h(C \cap \bar{S}) = \Pr^h[x_v \in C \cap \bar{S}] = \int_S \Pr^{K(\cdot,x)}[x_v \in C \cap \bar{S}]d\nu(x).$$

But by Proposition 10-35,

$$\Pr^{K(\cdot,x)}[x_v \in C \cap \bar{S}] = \mu^{K(\cdot,x)}(C \cap \bar{S}) = \chi_{C\cap\bar{S}}(x).$$

Hence for all Borel sets $C \cap \bar{S}$

$$\mu^h(C \cap \bar{S}) = \int_S \chi_{C\cap\bar{S}}(x)d\nu(x) = \nu(C \cap \bar{S}).$$

Since $\nu(\bar{S}) = 1$, we must have $\mu^h(\bar{S}) = 1$ and hence $\mu^h(C) = \nu(C)$ for all Borel sets C in S^*.

We have not yet proved that the representation in Theorem 10-40 does hold for the measure μ^h, but this fact is a consequence of the theorem to follow, which will summarize the results of the past three sections.

Theorem 10-41: The π-integrable non-negative P-superregular functions h stand in one-one correspondence with the non-negative finite measures ν on the Borel sets of $\bar{S}$, the correspondence of h to ν being

$$h_i = \int_{\bar{S}} K(i, x)d\nu(x)$$

and satisfying $\pi h = \nu(\bar{S})$. The unique representation $h = Nf + r$ with

r regular arises by decomposing the integral over $\bar{S}$ into a part over S and a part over B_e. If h is normalized, then the measure that corresponds to h is μ^h.

PROOF: By Fubini's Theorem any function of the form

$$h_i = \int_{\bar{S}} K(i, x) d\nu(x)$$

with ν finite is non-negative superregular and has $\pi h = \nu(\bar{S})$.

Conversely, let h be given. If $\pi h = 0$, then the superregularity of h implies that

$$0 = \pi h \geq \pi P^n h \geq 0$$

for all n and hence $\pi P^n h = 0$ for all n. Thus $(\pi N)h = 0$. Since πN is everywhere positive, $h = 0$. Thus existence and uniqueness of ν follow if $\pi h = 0$. Next, let πh be positive. Since h and ν must be related by $\pi h = \nu(\bar{S})$, we may, for both existence and uniqueness, divide h by an appropriate positive constant to obtain $\pi h = 1$. Uniqueness of ν and the fact that $\nu = \mu^h$ then follow from Theorem 10-40. Existence of ν follows from Theorem 10-26 provided we can show that $\mu^h(S^* - \bar{S}) = 0$. By Lemma 10-32,

$$\overline{S^h} \subset S^{h*} \cap \bar{S}.$$

Hence

$$\mu^h(\bar{S}) = \mu^h(S^{h*} \cap \bar{S}) \geq \mu^h(\overline{S^h}).$$

But the right side equals one by Proposition 10-38. Thus $\mu^h(\bar{S}) = 1$ and $\mu^h(S^* - \bar{S}) = 0$.

Finally we have $\bar{S} = S \cup B_e$ disjointly, and an application of Fubini's Theorem shows that

$$\int_{B_e} K(i, x) d\nu$$

is regular. Since

$$\int_S K(i, x) d\nu = \sum_{j \in S} N_{ij} \left(\frac{\nu(\{j\})}{N_{\pi j}} \right),$$

the representation $h = Nf + r$ is as asserted.

8. Analog of Fatou's Theorem

In the classical case of the disk, which was discussed in Section 1, normalized Lebesgue measure m on the circle has the distinguishing property that it corresponds to the function 1. If h is the non-negative harmonic function corresponding to a measure ν and if $\nu = fm + \mu_s$ is the Lebesgue decomposition of ν with respect to m, then Fatou's Theorem asserts that for almost every x $[m]$ on the circle, $h(re^{i\theta}) \to f(x)$ whenever $re^{i\theta}$ converges to x nontangentially. In this section we shall prove a Markov chain analog of this theorem in terms of the Martin boundary and the measures given by Theorem 10-41. As expected, harmonic measure μ will play the role of Lebesgue measure.

Our procedure will be first to derive an almost everywhere statement in terms of the measure on the probability space and then to translate this statement into a result in terms of harmonic measure. As a preliminary to the first step, we consider a special case in the lemma below.

We shall be dealing with expressions of the form $\lim_{n\to\infty} h(x_n(\omega))$ in this section, where h is non-negative and P-regular, and we shall adopt the convention that $h(x_n(\omega)) = 0$ if $n > \mathbf{v}$. This definition is motivated by the following consideration: If $\bar{P}$ is the enlarged chain for P, then a P-regular function h extends to be regular for $\bar{P}$ if h is defined to be zero at the absorbing state; consequently if πh is finite, $\{h(x_n)\}$ is a martingale with $\mathrm{M}[h(x_0)] = \pi h$.

Lemma 10-42: If $h \geq 0$ is a normalized bounded regular function, then μ^h is absolutely continuous with respect to μ and

$$h_i = \int_S K(i, x)f(x)d\mu(x) = \int_{B_e} K(i, x)f(x)d\mu(x),$$

where f is the Radon–Nikodym derivative of μ^h with respect to μ. The function f may be taken to be zero on S, and if it is, then

$$\Pr[\lim_{n\to\infty} h(x_n) = f(x_v)] = 1.$$

PROOF: Since h is bounded, $h \leq c\mathbf{1}$ for some constant $c > 1$. Noting that

$$\mathbf{1} = \frac{c-1}{c}\left[\frac{1}{c-1}(c\mathbf{1} - h)\right] + \frac{1}{c}h,$$

set $g = (c - 1)^{-1}(c\mathbf{1} - h)$. Then g and h are non-negative normalized superregular functions, and Lemma 10-34 shows that

$$\mu = \frac{c-1}{c}\mu^g + \frac{1}{c}\mu^h.$$

Thus $\mu^h \leq c\mu$ and μ^h is absolutely continuous with respect to μ. By the Radon–Nikodym Theorem there is a Borel function f such that

$$\mu^h(C) = \int_C f(x)d\mu(x)$$

for all Borel sets C. Since h is regular, $\mu^h(S) = 0$ and we may take f to vanish on S.

By Theorem 10-41 we have

$$\begin{aligned} h_i &= \int_S K(i, x) d\mu^h(x) \\ &= \int_S K(i, x) f(x) d\mu(x) \\ &= \int_{S^*} K(i, x) f(x) d\mu(x). \end{aligned}$$

Now the argument in the proof of Proposition 10-21 shows that

$$\int_{S^*} K(i, x) f(x) d\mu(x) = \int_{\Omega} f(x_v(\omega))\, d\mathrm{Pr}_i(\omega).$$

Thus if $\mathrm{Pr}_\pi[x_0 = j \wedge \cdots \wedge x_n = i] > 0$,

$$\mathrm{M}_\pi[f(x_v) \mid x_0 = j \wedge \cdots \wedge x_n = i] = \mathrm{M}_i[f(x_v)] = h_i.$$

Similarly, if b denotes the absorbing state in the enlarged chain, then

$$\mathrm{M}_\pi[f(x_v) \mid x_0 = j \wedge \cdots \wedge x_n = b] = 0 = h_b.$$

That is, if $\mathscr{R}_n$ is the partition generated by $\{x_0, \ldots, x_n\}$, then

$$\mathrm{M}_\pi[f(x_v) \mid \mathscr{R}_n] = h(x_n).$$

On one hand, the Borel field generated by the $\mathscr{R}_n$ is $\mathscr{G}$, and on the other hand $f(x_v)$ is measurable over $\mathscr{G}$, since it is the composition of a Borel function and a function for which the inverse image of every Borel set is in $\mathscr{G}$. By Proposition 3-18,

$$\mathrm{Pr}[\lim_{n \to \infty} h(x_n) = f(x_v)] = 1.$$

The general case of the almost everywhere statement in terms of the measure on the probability space is covered by the following theorem.

Theorem 10-43: Let $h \geq 0$ be a normalized regular function and let $\mu^h = f\mu + \mu_s$ be the Lebesgue decomposition of μ^h with respect to μ (where f is taken to be zero on the state space S); then

$$\mathrm{Pr}[\lim_{n \to \infty} h(x_n) = f(x_v)] = 1.$$

PROOF: If we let $k = \frac{1}{2}1 + \frac{1}{2}h$, then k is a non-negative normalized superregular function. Since k is strictly positive, we have $S^k = S$

and $S^{k*} = S^*$. By Lemma 10-32, $B_e^k = B_e$. The function h/k is a bounded regular function for the k-process with

$$\sum_i \pi_i^k \left(\frac{h_i}{k_i}\right) = \sum_i \pi_i h_i = 1,$$

and hence Lemma 10-42 yields a function g with

$$\left(\frac{h}{k}\right)_i = \int_S K^k(i, x) g(x) d\mu^k(x)$$

and

$$\Pr{}^k\left[\lim \frac{h(x_n)}{k(x_n)} = g(x_v^k)\right] = 1.$$

As was pointed out in the proof of Lemma 10-34, x_v^k is identical with x_v and also

$$\Pr{}^k[p] = \tfrac{1}{2}\Pr{}^1[p] + \tfrac{1}{2}\Pr{}^h[p]$$

for all p. Thus $\Pr^k[p]$ cannot be one unless $\Pr^1(p)$ and $\Pr^h[p]$ both equal one, and we conclude that

$$\Pr\left[\lim \frac{h(x_n)}{k(x_n)} = g(x_v)\right] = 1$$

or

$$\Pr\left[\lim \frac{h(x_n)}{\frac{1}{2} + \frac{1}{2}h(x_n)} = g(x_v)\right] = 1.$$

Since $\{h(x_n)\}$ is a non-negative martingale, $\lim h(x_n)$ exists a.e. [Pr] and is finite. Therefore the above identity implies that

$$\Pr\left[\lim h(x_n) = \frac{g(x_v)}{2 - g(x_v)}\right] = 1.$$

Thus to complete the proof, it suffices to prove that

$$f(x) = \frac{g(x)}{2 - g(x)} \quad \text{a.e. } [\mu].$$

First we identify g as the Radon–Nikodym derivative of μ^h with respect to μ^k. On one hand, we have

$$\begin{aligned} h_i &= k_i \int_S K^k(i, x) g(x) d\mu^k(x) \\ &= \int_S K(i, x) g(x) d\mu^k(x). \end{aligned}$$

On the other hand, Theorem 10-41 gives

$$h_i = \int_S K(i, x) d\mu^h(x),$$

and thus the uniqueness part of Theorem 10-41 gives $\mu^h = g\mu^k$.

Now, by Lemma 10-34, we have $\mu^k = \frac{1}{2}\mu + \frac{1}{2}\mu^h$. Hence

$$f\mu + \mu_s = \mu^h = g\mu^k = \tfrac{1}{2}g\mu + \tfrac{1}{2}g\mu^h = \tfrac{1}{2}g\mu + \tfrac{1}{2}gf\mu + \tfrac{1}{2}g\mu_s$$

or

$$(2f - g - fg)\mu = (g - 2)\mu_s.$$

Since μ and μ_s are singular with respect to each other, each side is the zero measure. For the left side, this statement means that

$$2f - g - fg = 0 \quad \text{a.e. } [\mu]$$

or

$$f = \frac{g}{2 - g} \quad \text{a.e. } [\mu].$$

The corollary to this theorem is the analog of Fatou's Theorem; it is a translation of the theorem into a result in terms of harmonic measure. The statement of the corollary needs a way of singling out for attention a single point x of S^*. One way of proceeding is to use the $K(\cdot, x)$ process, at least if x is in $\bar{S}$; for in that case Proposition 10-35 shows that

$$\Pr{}^{K(\cdot,x)}[x_v(\omega) = x] = 1.$$

Corollary 10-44: Let $h \geq 0$ be a normalized regular function, and let $\mu^h = f\mu + \mu_s$ (with f equal to zero on S) be the Lebesgue decomposition of μ^h with respect to μ. Then for almost every x $[\mu]$ for which $K(\cdot, x)$ is normalized,

$$\Pr{}^{K(\cdot,x)}[\lim_{n\to\infty} h(x_n) = f(x)] = 1.$$

PROOF: By Theorem 10-43,

$$\Pr[\lim h(x_n) = f(x_v)] = 1.$$

Then by Lemma 10-39,

$$1 = \Pr[\lim h(x_n) = f(x_v)] = \int_{\bar{S}} \Pr{}^{K(\cdot,x)}[\lim h(x_n) = f(x_v)]d\mu(x).$$

Since

$$\Pr{}^{K(\cdot,x)}[\lim h(x_n) = f(x_v)] \leq 1$$

for every x, equality must hold for almost every x $[\mu]$. But

$$\Pr{}^{K(\cdot,x)}[x_v = x] = 1$$

or

$$\Pr{}^{K(\cdot,x)}[f(x_v) = f(x)] = 1$$

for almost all x in $\bar{S}$. Since $\mu(\bar{S}) = 1$, the corollary follows.

The results above clearly extend to all π-integrable regular $h \geq 0$ provided we replace μ^h everywhere by the unique measure ν associated to h by Theorem 10-41.

Since the function f in all three of the above results is equal to zero on S, we may think of f as a function on B_e. If f is so restricted, it is called the **fine boundary function** of h.

9. Fine boundary functions

The results of Section 8 may be extended to non-negative π-integrable superregular functions with the help of the proposition below. Our convention that $h(x_n(\omega)) = 0$ if $n > \mathbf{v}(\omega)$ is still in force.

Proposition 10-45: If g is a π-integrable function of the form Nf with $f \geq 0$, then

$$\Pr[\lim_{n \to \infty} g(x_n) = 0] = 1.$$

For almost every x $[\mu]$ for which $K(\cdot, x)$ is normalized,

$$\Pr^{K(\cdot, x)}[\lim_{n \to \infty} g(x_n) = 0] = 1.$$

PROOF: In the first statement, g is non-negative superregular and πg is finite; hence $\{g(x_n)\}$ is a non-negative supermartingale and $g(x_n) \to z \geq 0$ a.e. [Pr]. Now $g \geq P^n g$ for all n, and $P^n g \to 0$. By dominated convergence $\pi P^n g \to 0$. Thus

$$\mathrm{M}_\pi[z] \leq \lim \mathrm{M}_\pi[g(x_n)] = \lim (\pi P^n g) = 0$$

and $z = 0$ a.e. [Pr]. The proof of the second statement in the proposition is the same as the proof of Corollary 10-44.

Thus if $h \geq 0$ is π-integrable and superregular, we may write, according to Theorem 5-10, $h = Nf + r$, with r regular and Nf and r both π-integrable and non-negative. Corollary 10-44 and Proposition 10-45 may therefore be combined into a single result whenever necessary.

The fine boundary function f of a normalized minimal regular function h takes on an especially simple form. By Lemma 10-33, we must have $h = K(\cdot, y)$ for some y in B_e, and, by Proposition 10-35,

$$\mu^h(\{y\}) = 1.$$

There are two cases. First, if $\mu(\{y\}) = 0$, then μ^h is singular with respect to μ and the fine boundary function of h is zero. By Lemma

10-42, h is unbounded. Second, if $\mu(\{y\}) = a > 0$, then the fine boundary function may be taken as $1/a$ at y and 0 elsewhere, since

$$\mu^h(C) = \chi_C(y) = \int_C f(x)d\mu(x)$$

for all Borel sets C. Moreover, h is bounded by $1/a$ since

$$1 \geq \Pr_i[x_v(\omega) = y] = \int_{\{y\}} K(i, x)d\mu(x) = aK(i, y) = ah_i.$$

The class of π-integrable regular functions $h \geq 0$ with a given μ-integrable non-negative function f as fine boundary function is exactly the class of functions h for which

$$\nu = f\mu + \mu_s$$

is the Lebesgue decomposition with respect to μ of the measure ν associated to h. Thus the class of such functions h is exactly the class of functions of the form

$$h_i = \int_{B_e} K(i, x)f(x)d\mu(x) + \int_{B_e} K(i, x)d\mu_s(x),$$

where μ_s is any Borel measure singular with respect to μ. In this class there is a unique smallest such function

$$\bar{h}_i = \int_{B_e} K(i, x)f(x)d\mu(x).$$

On one hand, Theorem 10-43 gives $\lim h(x_n) = f(x_v)$ a.e. [Pr], and hence

$$M_\pi[\lim h(x_n)] = M_\pi[f(x_v)] = \int_{B_e} f(x)d\mu(x) = \pi\bar{h}.$$

On the other hand,

$$\lim M_\pi[h(x_n)] = \lim \pi P^n h = \pi h.$$

If $h_j > \bar{h}_j$ for some j, choose n so that $(\pi P^n)_j > 0$. Then

$$\pi h = \pi P^n h > \pi P^n \bar{h} = \pi\bar{h}.$$

Thus by Proposition 1-52, $\{h(x_n)\}$ is uniformly integrable if and only if $h = \bar{h}$.

There is a different topology for S^* which is occasionally referred to in connection with fine boundary functions. The **fine topology** for S^* is defined in terms of its neighborhood system as follows: For any point x in $S^* - B_e$ every set in S^* containing x is to be a neighborhood of x. For x in B_e, the neighborhoods of x are the sets A in S^* such that x is in A and

$$\Pr^{K(\cdot,x)}[x_n \in A \text{ from some time on}] = 1.$$

Evidently x is in each of its neighborhoods, the intersection of two neighborhoods is a neighborhood, and a superset of a neighborhood is a neighborhood. We must check that S^* is a neighborhood of x; that is, that the $K(\cdot, x)$-process disappears with probability zero. But this fact is a consequence of Proposition 10-35.

If x is in B_e and if A is an open set containing x in the metric topology of S^*, then

$$\begin{aligned} 1 &\geq \Pr^{K(\cdot,x)}[x_n \in A \text{ from some time on}] \\ &\geq \Pr^{K(\cdot,x)}[x_v \in A] = \mu^{K(\cdot,x)}(A) \geq \mu^{K(\cdot,x)}(\{x\}) = 1. \end{aligned}$$

Therefore A is open in the fine topology, and the fine topology is stronger than the metric topology.

The next lemma and proposition show that a zero-one law holds for the probabilities which define the fine topology. The lemma by itself is of value in checking whether a non-negative regular function is minimal.

Lemma 10-46: A non-negative normalized regular function h is minimal for P if and only if the only bounded regular functions for P^h (restricted to S^h) are constants.

PROOF: If h is minimal, then the regularity of h implies that $\mathbf{1}$ is regular for P^h restricted to S^h. By adding a suitable multiple of $\mathbf{1}$ to a given bounded regular function for P^h, we may assume that it is a non-negative bounded regular function for P^h. Thus let $\bar{h}$ with $0 \leq \bar{h} \leq c\mathbf{1}$ be a regular function for P^h restricted to S^h. Set

$$k_i = \begin{cases} \bar{h}_i h_i & \text{if } i \in S^h \\ 0 & \text{otherwise.} \end{cases}$$

Then k is P-regular and satisfies $0 \leq k_i \leq ch_i$. Since h is minimal, $k_i = c'h_i$ for all i and hence $\bar{h}_i = c'$ for i in S^h. That is, $\bar{h} = c'\mathbf{1}$.

Conversely, if h is not minimal, find a regular function k with $0 \leq k \leq h$ and $k \neq ch$. Set

$$\bar{h}_i = k_i/h_i \quad \text{for } i \in S^h.$$

Then $\bar{h}$ is a non-constant regular function for P^h restricted to S^h.

Proposition 10-47: If x is in B_e, then, for any subset A of S,

$$\Pr^{K(\cdot,x)}[x_n \in A \text{ from some time on}] = 0 \text{ or } 1.$$

Proof: The h-process for $h = K(\cdot, x)$ disappears with probability zero, and thus

$$\Pr^{K(\cdot,x)}[x_n \in S] = 1.$$

Therefore

$$\begin{aligned}\Pr^{K(\cdot,x)}[x_n \in A \text{ from some time on}] &= \Pr^{K(\cdot,x)}[x_n \in S - A \text{ only finitely often}] \\ &= 1 - \Pr^{K(\cdot,x)}[x_n \in S - A \text{ infinitely often}].\end{aligned}$$

By Lemma 10-46, the only bounded regular functions for this process are the constants, and thus, by Proposition 5-19, this last expression must equal zero or one.

Proposition 10-47 shows that fine neighborhoods of x in B_e are those sets A in S^* for which x is in A and

$$\Pr^{K(\cdot,x)}[x_n \in A \text{ from some time on}] > 0.$$

The complement of a fine neighborhood of x is called **thin** at x. Such sets A are characterized by the property

$$\Pr^{K(\cdot,x)}[x_n \in A \text{ infinitely often}] = 0.$$

By Proposition 10-47 this probability must again be zero or one.

Let $h \geq 0$ be a π-integrable regular function with fine boundary function f. The fine topology for S^* has the property that the function $h \cup f$ obtained by extending h to S^* by f is continuous at almost every $[\mu]$ point x in S^*. In fact, the statement is trivial for x not in B_e, and it therefore suffices by Corollary 10-44 to prove the result for every x in B_e for which

$$\Pr^{K(\cdot,x)}[\lim h(x_n) = f(x)] = 1.$$

Let $a < f(x) < b$. We are to produce a fine neighborhood of x such that

$$a < (h \cup f)(y) < b$$

for all y in that neighborhood. Let A be the set of points of S for which $a < h < b$ and form the set $A \cup \{x\}$. We shall show this is a neighborhood of x. The convergence of $h(x_n)$ to $f(x)$ implies that

$$\Pr^{K(\cdot,x)}[a < h(x_n) < b \text{ from some time on}] = 1$$

or

$$\Pr^{K(\cdot,x)}[x_n \in A \text{ from some time on}] = 1.$$

Therefore

$$\Pr^{K(\cdot,x)}[x_n \in A \cup \{x\} \text{ from some time on}] = 1$$

and $A \cup \{x\}$ is a fine neighborhood of x.

10. Martin entrance boundary

The Martin exit boundary was defined in terms of P and a vector $\pi \geq 0$ such that πN is strictly positive. The completion S^* of S in the metric d served for a representation of all π-integrable superregular functions $h \geq 0$ in terms of finite measures on this space.

In this section we shall introduce a different completion $^*S$ of S which will serve for the representation of P-superregular measures. As the analog of π, we fix once and for all a function $f \geq 0$ such that Nf is everywhere positive and finite-valued. The representation theorem will be for superregular measures $\sigma \geq 0$ for which σf is finite.

The formalism is as follows: For i and j in S we define

$$J(i, j) = \frac{N_{ij}}{(Nf)_i}.$$

Then the measure $J(i, \cdot)$ is P-regular everywhere except at i, where it is strictly superregular, and it satisfies $J(i, \cdot)f = 1$ for all i. For each j the function is bounded by $N_{jj}/(Nf)_j$ because, if g is defined as Nf, then

$$J(i, j) = \frac{N_{ij}}{g_i} = \frac{N^g_{ij}}{g_j} \leq \frac{N^g_{jj}}{g_j} = \frac{N_{jj}}{g_j}.$$

We define

$$d'(j, j') = \sum_i w_i(Nf)_i |J(j, i) - J(j', i)|,$$

where the w_i are positive weights such that $\sum w_i N_{ii}$ is finite. The bound we have just computed shows that d' is finite-valued, and d' is a metric if we can show that $d'(j, j') = 0$ implies $j = j'$. But if $d'(j, j') = 0$, then

$$J(j, \cdot) = J(j', \cdot).$$

Multiplying through by P and supposing that $j \neq j'$, we obtain

$$J(j, j') = \sum_i J(j, i)P_{ij'} = \sum_i J(j', i)P_{ij'} < J(j', j'),$$

the strict inequality holding since $J(j', \cdot)$ is not regular at j'. Thus $j = j'$ and d' is a metric.

We define $^*S$ to be the Cauchy completion of S in the metric d'; the set $^*S - S$ is the **Martin entrance boundary**. A sequence $\{j_n\}$ is Cauchy in S if and only if the sequence $\{J(j_n, i)\}$ is Cauchy for every i. Consequently $J(\cdot, i)$ extends to a continuous function on $^*S$. Then $\{x_n\}$ is Cauchy in $^*S$ if and only if $\{J(x_n, i)\}$ is Cauchy for every i. Two points x and y are equal if and only if $J(x, \cdot) = J(y, \cdot)$, and the space $^*S$ is compact.

A P-superregular measure σ is **normalized** if $\sigma f = 1$. It is **minimal** if it is regular and if $0 \leq \bar{\sigma} \leq \sigma$ with $\bar{\sigma}$ regular implies $\bar{\sigma} = c\sigma$. Application of Fatou's Theorem shows that $J(x, \cdot)$ is a P-superregular measure for each x and that it satisfies $J(x, \cdot)f \leq 1$. A point x of $*S$ is **extreme** if $J(x, \cdot)$ is minimal and normalized, and the set of extreme points is denoted B^e. Then $B^e \cap S$ is empty.

The next theorem is the Markov chain analog for the Poisson–Martin Representation Theorem for P-superregular measures.

Theorem 10-48: The sets S and B^e are Borel subsets of $*S$. The non-negative P-superregular measures σ with σf finite stand in one-one correspondence with the non-negative finite measures ν on the Borel sets of $S \cup B^e$, the correspondence of σ to ν being

$$\sigma_i = \int_{S \cup B^e} J(x, i) d\nu(x)$$

and satisfying $\sigma f = \nu(S \cup B^e)$. The unique representation $\sigma = \gamma N + \rho$ with ρ regular arises by decomposing the integral over $S \cup B^e$ into a part over S and a part over B^e.

The proof will be accomplished by using duality, but we shall isolate several of the steps beforehand. Let $\alpha > 0$ be a superregular measure such that $\alpha f = 1$. Duality in the remainder of this section will mean α-duality. Let $\hat{P} = \text{dual } P$ and $\hat{\pi} = \text{dual} f$. Since

$$0 < \text{dual } g = \text{dual } (Nf) = \hat{\pi}\hat{N}$$

and

$$1 = \alpha f = (\text{dual} f)(\text{dual } \alpha) = \hat{\pi}\mathbf{1},$$

the exit boundary of $\hat{P}$ relative to $\hat{\pi}$ is defined. Let $\hat{d}$ be the defining metric of the exit boundary, and let $\hat{S}^*$ be the completion of S under $\hat{d}$. We have

$$\text{dual } J(j, \cdot) = \frac{J(j, \cdot)}{\alpha.} = \frac{N_{j.}}{(Nf)_j\alpha.} = \frac{\hat{N}_{.j}}{(\hat{\pi}\hat{N})_j} = \hat{K}(\cdot, j).$$

Lemma 10-49: If the same weights w_i are used in defining $\hat{d}$ and d', then the identity map on S extends to an isometry of $(*S, d')$ onto $(\hat{S}^*, \hat{d})$.

PROOF: We have

$$\begin{aligned} d'(j, j') &= \sum_i w_i (Nf)_i |J(j, i) - J(j', i)| \\ &= \sum_i w_i \alpha_i (Nf)_i \left| \frac{J(j, i)}{\alpha_i} - \frac{J(j', i)}{\alpha_i} \right| \\ &= \sum_i w_i (\hat{\pi}\hat{N})_i |\hat{K}(i, j) - \hat{K}(i, j')| \\ &= \hat{d}(j, j'). \end{aligned}$$

Under the identification of $^*S$ and $\hat{S}^*$,

$$J(x, \cdot) = \text{dual } \hat{K}(\cdot, x)$$

by the continuity of the functions $J(\cdot, i)$ and $\hat{K}(i, \cdot)$. Under duality a $\hat{P}$-superregular function h corresponds to a P-superregular row vector $\sigma = \text{dual } h$. Since $\hat{P} = \text{dual } P$ and $\hat{\pi} = \text{dual } f$, regularity and normalization are both preserved. But minimality is also preserved since duality preserves inequalities. Therefore $B^e = \hat{B}_e$.

PROOF OF THEOREM 10-48: Let $h = \text{dual } \sigma$, and apply Theorem 10-41. Then

$$h_i = \int_{S \cup \hat{B}_e} \hat{K}(i, x) d\nu(x)$$

with $\hat{\pi}h = \nu(S \cup \hat{B}_e)$. Therefore

$$\frac{\sigma_i}{\alpha_i} = \int_{S \cup \hat{B}_e} \frac{1}{\alpha_i} J(x, i) d\nu(x)$$

and $\sigma f = \nu(S \cup \hat{B}_e)$. If we identify B^e and $\hat{B}_e$ and call ν by the same name on B^e as on $\hat{B}_e$, then we have

$$\sigma_i = \int_{S \cup B^e} J(x, i) d\nu(x)$$

as required. Conversely, if σ is given by

$$\sigma_i = \int_{S \cup B^e} J(x, i) d\nu(x),$$

then

$$(\text{dual } \sigma)_i = \int_{S \cup \hat{B}_e} \hat{K}(i, x) d\nu(x).$$

Therefore dual σ is $\hat{P}$-superregular and σ must be superregular. Since there is only one measure which represents dual σ, there can be only one measure which represents σ. The last statement of the theorem follows from the fact that the unique decomposition $\sigma = \gamma N + \rho$ is transformed under duality into the unique decomposition of dual σ in the $\hat{P}$-process.

If $\sigma > 0$, we can take $\alpha = \sigma$ and then the proof of Theorem 10-48 gives us the following interpretation for $\nu : \nu$ is harmonic measure for the chain with transition matrix σ-dual P and starting distribution σ-dual f.

11. Application to extended chains

In order to assign a probabilistic interpretation to the entrance boundary, we must pass to extended chains. Except where noted, we therefore assume throughout this section that $\{z_n\}$ is an extended chain with state space $S \cup \{a\} \cup \{b\}$, with (enlarged) transition matrix $\bar{P}$, and with vector ν of mean times in the various states of S. Let π and f be row and column vectors, respectively, such that $\pi \geq 0$ $f \geq 0$, $\pi 1 = 1$, $\nu f = 1$, $\pi N > 0$, and $Nf > 0$. By Theorem 8-3, Nf is finite-valued. We continue to use the notations of Sections 3 through 10 for boundary theory of P relative to π and f, and we shall use the vectors μ^E and ν^E defined for extended chains in Section 2.

Our first pair of propositions will give the "final" and "starting" distributions for the extended chain. If x is in $*S$ and if E ranges over finite sets in S, we define

$$p(x) = \lim_{E \uparrow S} J(x, \cdot)e^E.$$

Similarly, if y is in S^*, we define

$$q^\nu(y) = \lim_{E \uparrow S} \mu^E K(\cdot, y).$$

We first show that these limits always exist.

Lemma 10-50: For every x in $*S$ and y in S^*, both $p(x)$ and $q^\nu(y)$ exist, possibly being infinite. Their defining limits are increasing limits as $E \uparrow S$. Each is a Borel measurable function on its domain, and the functions satisfy

$$p(x) = \lim_{E \uparrow S} \lim_{j \to x} \frac{h_j^E}{(Nf)_j}$$

and

$$q^\nu(y) = \lim_{E \uparrow S} \lim_{j \to y} \frac{\nu_j^E}{(\pi N)_j}.$$

PROOF: For any finite set E,

$$h_j^E = \sum_{k \in E} N_{jk} e_k^E$$

and

$$\frac{h_j^E}{(Nf)_j} = \sum_{k \in E} J(j, k)e_k^E.$$

Since E is finite, we may pass to the limit as $j \to x$ to get

$$\lim_{j \to x} \frac{h_j^E}{(Nf)_j} = J(x, \cdot)e^E.$$

For each fixed j, the expression $h_j^E/(Nf)_j$ increases with E, and hence so does the limit as $j \to x$. Therefore $p(x)$ exists, is defined by an increasing limit, and has the asserted value. If we choose an increasing sequence of finite sets E_n with union S, then $p(x)$ is exhibited as the limit of a *sequence* of continuous functions and is therefore Borel measurable.

Similarly we have

$$\nu_j^E = \sum_{k \in E} \mu_k^E N_{kj}$$

and hence

$$\lim_{j \to y} \frac{\nu_j^E}{(\pi N)_j} = \mu^E K(\cdot, y).$$

Thus the same argument applies to $q^{\nu}(y)$, since ν^E increases with E.

Proposition 10-51: On almost every path,

$$z_{\mathbf{v}}(\omega) = \lim_{n \uparrow \mathbf{v}} z_n(\omega)$$

exists in S^*. On almost every path for which the limit exists, either $\mathbf{v} < +\infty$ and $z_{\mathbf{v}} \in S$ or else $\mathbf{v} = +\infty$ and $z_{\mathbf{v}} \in B_e$. Moreover, for any Borel set C in S^*,

$$\Pr[z_{\mathbf{v}} \in C] = \int_C q^{\nu}(y) d\mu(y).$$

PROOF: The process watched starting in E is a Markov chain with transition matrix P and starting distribution μ^E if E is a finite set in S. By Proposition 10-21, $z_{\mathbf{v}}$ exists and has the required values for almost every path in this chain, and in this chain

$$\begin{aligned}
\Pr_{\mu^E}[z_{\mathbf{v}} \in C] &= \sum_{i \in E} \mu_i^E \Pr_i[z_{\mathbf{v}} \in C] \\
&= \sum_{i \in E} \mu_i^E \int_C K(i, x) d\mu(x) \\
&= \int_C \mu^E K(\cdot, x) d\mu(x).
\end{aligned}$$

As E increases to S, we obtain almost all paths of $\{z_n\}$ in this way. Hence $z_{\mathbf{v}}$ exists and is in $S \cup B_e$ a.e., and, by Definition 10-4,

$$\Pr[z_{\mathbf{v}} \in C] = \lim_{E \uparrow S} \int_C \mu^E K(\cdot, y) d\mu(y).$$

By Lemma 10-50, the integrand increases with E. Replacing the sets E by an increasing sequence E_n and applying monotone convergence, we obtain

$$\begin{aligned}
\Pr[z_{\mathbf{v}} \in C] &= \int_C \lim_{E \uparrow S} \mu^E K(\cdot, y) d\mu(y) \\
&= \int_C q^{\nu}(y) d\mu(y).
\end{aligned}$$

Proposition 10-52: On almost every path,

$$z_u(\omega) = \lim_{n \downarrow u} z_n(\omega)$$

exists in $^*S$. On almost every path for which the limit exists, either $\mathbf{u} > -\infty$ and $z_n \in S$ or else $\mathbf{u} = -\infty$ and $z_u \in B^e$. Moreover, for any Borel set C in $^*S$,

$$\Pr[z_u \in C] = \int_C p(x) d\mu^\nu(x),$$

where μ^ν is the unique measure on $S \cup B^e$ representing ν.

PROOF: Form the reverse process and call its measure Pr′. Its transition matrix restricted to the states of S is the ν-dual of P, denoted $\hat{P}$. Form the exit and entrance boundaries of $\hat{P}$ relative to $\hat{\pi} = \nu$-dual f and $\hat{f} = \nu$-dual π. By Proposition 10-51, the limit

$$z_v = \lim_{n \uparrow v} z_n$$

exists in $\hat{S}^*$ a.e. [Pr′], and either $\mathbf{v}$ is finite and $z_v \in S$ or else $\mathbf{v}$ is infinite and a.e. $z_v \in \hat{B}_e$. Therefore

$$z_u = \lim_{n \downarrow u} z_n$$

exists in $\hat{S}^*$ a.e. [Pr], and either $\mathbf{u}$ is finite and $z_u \in S$ or else $\mathbf{u}$ is infinite and a.e. $z_u \in \hat{B}_e$. Since $\hat{S}^*$ and $^*S$ are canonically identified according to Lemma 10-49, the first part of the proposition follows. Moreover we have

$$\Pr[z_u \in C] = \Pr'[z_v \in C] = \int_C \lim_{E \uparrow S} [\mu^E \hat{K}(\cdot, x)] d\mu^\nu(x),$$

where μ^ν is the measure on $\hat{S}^*$ which represents $\mathbf{1}$, that is, the measure on $^*S$ which represents ν. The above expression is

$$= \int_C \lim_{E \uparrow S} [J(x, \cdot)(\text{dual } \mu^E)] d\mu^\nu(x).$$

But μ^E, by Proposition 10-8, is the balayage charge of ν on E and its dual is thus the balayage charge of 1 on E; that is, the charge e^E of h^E

We thus obtain $\int p(x) d\mu^\nu(x)$ and $\int q^\nu(y) d\mu(y)$ as starting and final distributions for the extended chain. The fact that ν is normalized ($\nu f = 1$) implies that μ^ν is a probability measure. Note that $p(x)$ depends only on P, π, and f and not on the extended chain or even the vector ν, but that $q^\nu(y)$ depends on P, π, f, and ν. We shall return to this point shortly.

Conversely, if we select any probability measure μ^σ on $S \cup B^e$, then Theorem 10-48 yields a unique normalized P-superregular measure σ represented by μ^σ. If σ is positive, Theorem 10-9 assures the existence of at least one extended chain with σ as its vector of mean times. Any such chain starts in C with probability

$$\int_C p(x)d\mu^\sigma(x).$$

By varying μ^σ we may change the total measure on $\{z_n\}$, even as to whether it is finite or infinite.

We consider two special cases. First, if 0 is in S, let

$$\sigma_j = J(0, j) = N_{0j}/(Nf)_0.$$

By the uniqueness of the representation we must have $\mu^\sigma(\{0\}) = 1$, and hence we may represent σ (provided it is positive) by an extended chain which has a.e. path starting at 0. The total measure of the process must be

$$\int_{*S} p(x)d\mu^\sigma(x) = p(0)\mu^\sigma(\{0\}) = p(0),$$

and so the interpretation of σ_j as the mean number of times in state j shows that

$$\sigma_j = p(0)N_{0j}.$$

As a check, we can compute $p(0)$ directly. We have

$$p(0) = \lim_{E \uparrow S} \frac{h_0^E}{(Nf)_0} = \frac{1}{(Nf)_0},$$

as required.

Second, choose $\sigma = J(x, \cdot)$, where x is in B^e. Then $\mu^\sigma(\{x\}) = 1$, and (if σ is positive) an extended chain representing σ in the sense of Theorem 10-9 must start a.e. at x with total measure $p(x)$. Therefore, $J(x, \cdot)$ may be interpreted as the vector of mean times for any extended chain which is started almost surely at x, and, if $p(x)$ is finite, then

$$\frac{J(x, \cdot)}{p(x)}$$

is the conditional mean of the number of times in the various states given that the process starts at x.

Returning to the case of a general ν and supposing that $p(x)$ is finite a.e. $[\mu^\nu]$, we see that the identity

$$\nu_j = \int_{*S} \frac{J(x, j)}{p(x)} \cdot p(x)d\mu^\nu(x)$$

may be interpreted as follows: ν_j may be computed by choosing a starting state x according to the starting distribution and then weighting it by the conditional mean of the number of times in j given that the process starts at x.

As we noted earlier, there is an asymmetry between $p(x)$ and $q^\nu(y)$ in that one of these quantities depends on ν and the other does not. This asymmetry is cleared up by a discussion of h-processes for P; we shall see that $p(x)$ depends on h and that $q^\nu(y)$ does not. The special case we have been considering so far will be seen to be the case $h = 1$.

In discussing the entrance boundary for P^h, where $h > 0$ is superregular and normalized, we choose

$$f_i^h = f_i/h_i$$

in analogy with the choice of π^h. Define σ by

$$\sigma_i = \nu_i h_i.$$

Then σ is positive, is P^h-superregular, and satisfies $\sigma f^h = \nu f = 1$. Direct calculation shows that

$$\sigma\text{-dual }(P^h) = \nu\text{-dual } P = \hat{P}$$

and that

$$\sigma\text{-dual }(f^h) = \nu\text{-dual} f = \hat{\pi}.$$

Therefore the entrance boundary for P^h and f^h is the same as that for P and f.

Now let $\{z_n'\}$ be an extended chain representing P^h and σ (existence by Theorem 10-9), and take π^h and f^h as the functions relative to which the exit and entrance boundaries are formed. The process starts in $^*S$ and goes to $S^{h*} = S^*$. To obtain the starting and final distributions, we need the functions p and q. For p we have

$$\begin{aligned} p^h(x) &= \lim_{E \uparrow S} \lim_{j \to x} \frac{(h^E)_j^h}{(N^h f^h)_j} \\ &= \lim_{E \uparrow S} \lim_{j \to x} \frac{\sum_{k \in E} (B^E)_{jk}^h}{(Nf)_j/h_j} \\ &= \lim_{E \uparrow S} \lim_{j \to x} \frac{(B^E h)_j}{g_j}. \end{aligned}$$

For q we have

$$\begin{aligned} q^{\sigma,h}(y) &= \lim_{E \uparrow S} (\mu^E)^h K^h(\cdot, y) \\ &= \lim_{E \uparrow S} \sum_{i \in E} \mu_i^E h_i K(i, y)/h_i \\ &= \lim_{E \uparrow S} \mu^E K(\cdot, y) \\ &= q^\nu(y). \end{aligned}$$

Therefore, in the chain $\{z'_n\}$ we have as starting and final distributions

$$\Pr{}^h[z'_u \in C] = \int_C p^h(x)d\mu^\nu(x)$$

and

$$\Pr{}^h[z'_v \in C] = \int_C q^\nu(y)d\mu^h(y).$$

These relations show the symmetric roles of p and q and give us symmetric interpretations for μ^ν and μ^h.

In the special case of positive minimal ν and h, we must have

$$\nu = J(x_0, \cdot)$$

and

$$h = K(\cdot, y_0),$$

where x_0 and y_0 are in B^e and B_e, respectively. In this case

$$\mu^\nu(\{x_0\}) = \mu^h(\{y_0\}) = 1.$$

Thus the process $\{z'_n\}$ has almost every path starting at x_0 and going to y_0. The paths which start at x_0 have measure $p^h(x_0)$, and the paths which go to b have measure $q^\nu(y_0)$. Therefore we must have

$$p^h(x_0) = q^\nu(y_0).$$

12. Proof of Theorem 10-9

(1) *Construction of the extended stochastic process.* Let

$$\bar{P} = P^*_{S\cup\{b\}} \quad \text{and} \quad P = \bar{P}_S.$$

Redefine ν as being restricted to the domain S. Let ν^E be the balayage potential of ν on a finite set E with μ^E the balayage charge. For each E we shall consider a Markov chain with transition matrix P and starting distribution μ^E, and we shall combine these into a single extended stochastic process by choosing a common time scale. For the sake of convenience we assume that $S = \{0, 1, \ldots\}$. Our measure space will, as usual, be the double sequence space obtained from S, a, and b. To define the process $\{x_n\}$ it is sufficient to assign probabilities consistently to basic statements.

We shall use $\{y_n\}$ to denote the outcome functions for the Markov chain P with various μ^E as starting distributions. These vectors are finite measures, but not necessarily probability measures. Since S is the set of non-negative integers, it has a natural ordering on it. For each path ω in the ordinary sequence space of $\bar{P}$, let $\mathbf{s}(\omega)$ be the smallest numbered state on path ω and let $\mathbf{t}(\omega)$ be the time that $\mathbf{s}(\omega)$ is first entered. (Then $\mathbf{s}$ and $\mathbf{t}$ are defined everywhere except on the one path $(b, b, b, b, \ldots)$.)

There are two kinds of probability assignments to be made on the basic cylinder sets. First, if $i \in S$, define

$$\Pr[x_n = i \wedge x_{n+1} = j \wedge \cdots \wedge x_{n+m} = k]$$
$$= \Pr_{\mu^E}[y_{\mathbf{t}+n} = i \wedge y_{\mathbf{t}+n+1} = j \wedge \cdots \wedge y_{\mathbf{t}+n+m} = k],$$

where $E = \{i\}$ and where the right side is taken as 0 for the set of ω on which $\mathbf{t}(\omega) + n < 0$. Second, define

$$\Pr[x_{-n-r} = a \wedge \cdots \wedge x_{-n-2} = a \wedge x_{-n-1} = a$$
$$\wedge\, x_{-n} = i \wedge x_{-n+1} = j \wedge \cdots \wedge x_{-n+m} = k]$$
$$= \mu_i \Pr_i[y_1 = j \wedge \cdots \wedge y_m = k \wedge \mathbf{t} = n].$$

The effect of these definitions is to fix a time scale with this property: For each path there is a designated state such that the first entry into that state occurs at time 0. Then for $i \in S$, $\Pr[x_n = i]$ is finite for all n. To show that $\{x_n\}$ is an extended stochastic process, we must check that the above definitions on cylinder sets are consistent. That is, we must show typically that

$$\sum_k \Pr[x_n = i \wedge x_{n+1} = j \wedge x_{n+2} = k] = \Pr[x_n = i \wedge x_{n+1} = j],$$

when i and j do not both equal a, and that

$$(*) \quad \sum_i \Pr[x_{-n-1} = i \wedge x_{-n} = j \wedge x_{-n+1} = k]$$
$$= \Pr[x_{-n} = j \wedge x_{-n+1} = k],$$

when j and k do not both equal b. The first identity is immediate from the above definitions. But to prove the second identity, we need some alternate expressions for the left side.

Let $i \in S$, let $E = \{i\}$, let F be a finite set with $E \subset F$, and let q be the statement $y_{\mathbf{t}+n} = i \wedge y_{\mathbf{t}+n+1} = j \wedge y_{\mathbf{t}+n+2} = k$. By the first definition above,

$$\Pr[x_n = i \wedge x_{n+1} = j \wedge x_{n+2} = k] = \Pr_{\mu^E}[q].$$

We shall show that

$$\Pr_{\mu^F}[q] = \Pr_{\mu^E}[q].$$

Since the truth of q depends only on events after the time $\mathbf{t}_E$ when E is reached, Theorem 4-10 gives

$$\Pr_{\mu^F}[q] = \sum_{m \in E} \Pr_{\mu^F}[y_{\mathbf{t}_E} = m] \Pr_m[q]$$
$$= \sum_{m \in E} (\mu^F B^E)_m \Pr_m[q].$$

Now $\mu^F B^E = \mu^E$, since μ^F and μ^E are balayage charges (see Section 2). Therefore

$$\Pr_{\mu^F}[q] = \sum_{m \in E} (\mu^E)_m \Pr_m[q] = \Pr_{\mu^E}[q].$$

It follows that

$$\Pr_{\mu^E}[q] = \lim_{F \uparrow S} \Pr_{\mu^F}[q],$$

which is the first of the two alternate identities we need.

Next, we note that $\nu(I - P) = \mu$ by Theorem 5-10. Hence $\mu_i = \lim_{F \uparrow S} \mu_i^F$ by the property of balayage charges that they converge to the product of the given superregular measure and $I - P$. Thus

$$\begin{aligned}\Pr[x_{-n-2} = a \wedge x_{-n-1} = a \wedge x_{-n} = i \wedge x_{-n+1} = j \wedge x_{-n+2} = k]\\ = \mu_i \Pr_i[y_1 = j \wedge y_2 = k \wedge \mathbf{t} = n]\\ = \lim_F \mu_i^F \Pr_i[y_1 = j \wedge y_2 = k \wedge \mathbf{t} = n]\\ = \lim_F \Pr_{\mu^F}[y_0 = i \wedge y_1 = j \wedge y_2 = k \wedge \mathbf{t} = n],\end{aligned}$$

which is our second alternate identity.

Now we are in a position to prove (*). In the proof we shall use the abbreviation

$$Y = \lim_{F \uparrow S} \sum_{m=1}^{\infty} \sum_{i \in S - F} \Pr_{\mu^F}[y_{m-1} = i \wedge y_m = j \wedge y_{m+1} = k \wedge \mathbf{t} = m + n].$$

We begin by using our two alternate identities and by performing a direct calculation. Recall that i cannot equal b, since $j \neq b$.

$$\begin{aligned}&\sum_{\substack{i \in S \\ \text{or } i = a}} \Pr[x_{-n-1} = i \wedge x_{-n} = j \wedge x_{-n+1} = k]\\ &= \lim_{F \uparrow S} \sum_{i \in F} \Pr_{\mu^F}[y_{t-n-1} = i \wedge y_{t-n} = j \wedge y_{t-n+1} = k]\\ &\quad + \lim_{F \uparrow S} \Pr_{\mu^F}[y_0 = j \wedge y_1 = k \wedge \mathbf{t} = n]\\ &= \lim_F \left\{ \sum_{m=1}^{\infty} \sum_{i \in F} \Pr_{\mu^F}[y_{m-1} = i \wedge y_m = j \wedge y_{m+1} = k \wedge \mathbf{t} = m + n] \right.\\ &\qquad \left. + \Pr_{\mu^F}[y_0 = j \wedge y_1 = k \wedge \mathbf{t} = n] \right\}\\ &= \lim_F \sum_{m=0}^{\infty} \Pr_{\mu^F}[y_m = j \wedge y_{m+1} = k \wedge \mathbf{t} = m + n] - Y\\ &= \lim_F \Pr_{\mu^F}[y_{t-n} = j \wedge y_{t-n+1} = k] - Y\\ &= \Pr[x_{-n} = j \wedge x_{-n+1} = k] - Y.\end{aligned}$$

Thus we are to prove $Y = 0$. Now

$$\begin{aligned}
0 \le Y &\le \lim_F \sum_{m=1}^{\infty} \sum_{i \in S-F} \Pr_{\mu^F}[y_{m-1} = i \wedge y_m = j] \\
&= \lim_F \sum_{m=1}^{\infty} \sum_{i \in S-F} (\mu^F P^{m-1})_i P_{ij} \\
&= \lim_F \sum_{i \in \tilde{F}} (\mu^F N)_i P_{ij} \\
&= \lim_F \sum_{i \in \tilde{F}} \nu_i^F P_{ij} \\
&\le \lim_F \sum_{i \in \tilde{F}} \nu_i P_{ij}.
\end{aligned}$$

Since $\sum_i \nu_i P_{ij}$ is convergent, the right side is zero. Hence $Y = 0$. Therefore $\{x_n\}$ is an extended stochastic process.

(2) *Verification of some of the properties of an extended chain.* We must now check that $\{x_n\}$ satisfies the four properties of an extended chain. First we check (2). Consider the special case in which E is the finite set $\{0, 1, 2, \ldots, i\}$. This set has the property that on any path beginning in a state of E the statement $\mathbf{t} \ge n$ implies the statement that there is an $m \ge n$ with $y_m \in E$. Therefore

$$\begin{aligned}
&\lim_n \Pr[x_{-l} \in E \text{ for some } l \ge n] \\
&\quad = \lim_n \sum_{l=n}^{\infty} \Pr[x_{-l} \in E \wedge x_{-l+1} \notin E \wedge \cdots \wedge x_{-n} \notin E] \\
&\quad = \lim_n \sum_{l=n}^{\infty} \Pr_{\mu^E}[y_{t-l} \in E \wedge \cdots \wedge y_{t-n} \notin E] \\
&\quad = \lim_n \Pr_{\mu^E}[y_{t-l} \in E \text{ for some } l \ge n] \\
&\quad \le \lim_n \Pr_{\mu^E}[\mathbf{t} \ge n] \\
&\quad \le \lim_n \Pr_{\mu^E}[(\exists m) \text{ with } m \ge n \text{ and } y_m \in E].
\end{aligned}$$

Since P is transient and μ^E is finite, the right side is zero. Hence so is the left side. By Corollary 1-17 we conclude that

$$\Pr[\mathbf{u}_E = -\infty] = 0.$$

For the general case of a finite set F, choose a finite set E of the above form which contains it. Then

$$\Pr[\mathbf{u}_F = -\infty] \le \Pr[\mathbf{u}_E = -\infty] = 0.$$

Hence (2) holds.

Now suppose for the moment that we have proved (3) in the form that for any finite set E the process watched starting in E is a Markov chain with starting distribution μ^E and transition matrix P, and suppose we have identified the given measure ν as the vector of mean times in the various states. Then (4) follows, since ν was assumed finite-valued. To prove (1) that the domain of $\mathbf{u}_E$ has positive measure, let $j \in E$. Then

$$0 < \nu_j = \nu_j^E = \sum_{i \in E} \mu_i^E N_{ij}.$$

Hence $\mu_i^E > 0$ for some $i \in E$. But $\mu^E \mathbf{1}$ is the measure of the set on which E is ever entered for the first time. That is, it is the measure of the set where $\mathbf{u}_E > -\infty$. Hence (1) holds.

Thus we are to prove that the process watched starting in the finite set E is a Markov chain with starting distribution μ^E and transition matrix $\bar{P}$, and we are to identify ν. Suppose we have proved this assertion for all finite sets of the form $E = \{0, 1, 2, \ldots, e\}$, and suppose F is an arbitrary finite set. We show first how the assertion follows for F. Choose a set E of the special form containing F. Then watching the process beginning in F is the same as watching the process watched starting in E beginning in F. Thus since the E-process is a Markov chain, so is the F-process by Theorem 4-9. The starting distribution for the F-process is

$$\Pr_{\mu^E}[y_{t_F} = j] = \sum_{i \in S} \mu_i^E \Pr_i[y_{t_F} = j] = \sum_{i \in S} \mu_i^E B_{ij}^F = \mu_j^F.$$

Hence we may assume that E is a set of the form $E = \{0, 1, \ldots, e\}$. Let $\{z_n\}$ be the outcome functions for the process watched starting in E. We shall compute the probability of the typical basic statement p defined as $z_0 = i \wedge z_1 = j \wedge z_2 = k$, where $i \in E$, by considering separately the contributions from paths of $\{x_n\}$ with $\mathbf{u} > -\infty$ and paths with $\mathbf{u} = -\infty$. The notation $\tilde{E}$ will mean $S - E$.

(3) *Contribution to* $\Pr[p]$ *from paths with* $\mathbf{u} > -\infty$. The contribution to $\Pr[p]$ of paths with $\mathbf{u} > -\infty$ is

$$\sum_{n=-\infty}^{\infty} \sum_{m=0}^{\infty} \Pr[x_{-m-n-1} = a \wedge x_{-m-n} \in \tilde{E} \wedge \cdots \wedge x_{-n-1} \in \tilde{E} \wedge x_{-n} = i \wedge x_{-n+1} = j \wedge x_{-n+2} = k]$$

$$= \sum_{n=-\infty}^{\infty} \sum_{m=0}^{\infty} \Pr_{\mu}[y_0 \in \tilde{E} \wedge \cdots \wedge y_{m-1} \in \tilde{E} \wedge y_m = i \wedge y_{m+1} = j \wedge y_{m+2} = k \wedge \mathbf{t} = m + n].$$

The fact that E is of the special form $\{0, 1, \ldots, e\}$ means that if

$$y_0(\omega) \in \tilde{E} \wedge \cdots \wedge y_{m-1}(\omega) \in \tilde{E} \wedge y_m(\omega) = i,$$

then since $i \in E$ we must have $\mathbf{t}(\omega) \geq m$. Hence we need sum on n only over $n \geq 0$. If $n \geq 0$, then for any ω for which $y_0, \ldots, y_{m-1}$ are not in E, we have that $\mathbf{t}(\omega) = m + n$ if and only if $\mathbf{t}(\omega_m) = n$. Therefore we may apply Theorem 4-9 to Pr[—]. In symbols the argument is that the contribution is

$$= \sum_{n \geq 0} \sum_{m \geq 0} \Pr_\mu[y_0 \in \tilde{E} \wedge \cdots \wedge \mathbf{t} = m + n]$$

$$= \sum_{n \geq 0} \sum_{m \geq 0} \Pr_\mu[y_0 \in \tilde{E} \wedge \cdots \wedge y_{m-1} \in \tilde{E} \wedge y_m = i] \times \Pr_i[y_1 = j \wedge y_2 = k \wedge \mathbf{t} = n].$$

Let $B^{E,m}$ denote the matrix of probabilities of entry to E at time m. The above expression is

$$= \sum_{n \geq 0} \sum_{m \geq 0} (\mu B^{E,m})_i \Pr_i[y_1 = j \wedge y_2 = k \wedge \mathbf{t} = n]$$

$$= \sum_{n \geq 0} (\mu B^{E})_i \Pr_i[y_1 = j \wedge y_2 = k \wedge \mathbf{t} = n]$$

$$= (\mu B^{E})_i \Pr_i[y_1 = j \wedge y_2 = k]$$

$$= (\mu B^{E})_i \bar{P}_{ij} \bar{P}_{jk}.$$

Thus the contribution is $(\mu B^{E})_i \bar{P}_{ij} \bar{P}_{jk}$.

(4) *Contribution to* Pr[p] *from paths with* $\mathbf{u} = -\infty$. The contribution to Pr[p] of paths with $\mathbf{u} = -\infty$ is

$$\sum_{n=-\infty}^{\infty} \lim_{m \to \infty} \Pr[x_{-n-m} \in \tilde{E} \wedge \cdots \wedge x_{-n-1} \in \tilde{E} \wedge x_{-n} = i \wedge x_{-n+1} = j \wedge x_{-n+2} = k]$$

$$= \sum_{n} \lim_{m} \sum_{s \in \tilde{E}} \Pr[x_{-n-m} = s \wedge x_{-n-m+1} \in \tilde{E} \wedge \cdots \wedge x_{-n+2} = k]$$

$$= \sum_{n} \lim_{m} \sum_{s \in \tilde{E}} \lim_{F \uparrow S} \Pr_{\mu^F}[y_{t-m-n} = s \wedge y_{t-m-n+1} \in \tilde{E} \wedge \cdots \wedge y_{t-n+2} = k]$$

$$= \sum_{n} \lim_{m} \sum_{s \in \tilde{E}} \lim_{F} \sum_{l=0}^{\infty} \Pr_{\mu^F}[y_l = s \wedge y_{l+1} \in \tilde{E} \wedge \cdots \wedge y_{l+m+2} = k \wedge \mathbf{t} = l + m + n].$$

In the expression $\Pr_{\mu^F}$[—], for fixed n we may assume that m is large, say $m \geq |n|$. Then $\mathbf{t} \geq l$. Now since E is of the special form $\{0, 1, \ldots, e\}$ and since $i \in E$, state i is lower in the ordering on S than any of the states of $\tilde{E}$. Hence

$$y_l(\omega) = s \wedge y_{l+1}(\omega) \in \tilde{E} \wedge \cdots \wedge \mathbf{t}(\omega) = l + m + n$$

for $m \geq n$ if and only if

$$y_0(\omega) \in \tilde{G} \wedge \cdots \wedge y_{l-1}(\omega) \in \tilde{G} \wedge y_l(\omega) = s \\ \wedge y_{l+1}(\omega) \in \tilde{E} \wedge \cdots \wedge \mathfrak{t}(\omega) = l + m + n,$$

where $G = \{0, 1, \ldots, i\}$. In the presence of the information

$$y_0(\omega) \in \tilde{G} \wedge \cdots \wedge y_{l-1}(\omega) \in \tilde{G} \wedge y_l(\omega) = s,$$

it is true that $\mathfrak{t}(\omega) = l + m + n$ if and only if $\mathfrak{t}(\omega_l) = m + n$. Theorem 4-9 now applies. The contribution to $\Pr[p]$ therefore is

$$= \sum_n \lim_m \sum_{s \in \tilde{E}} \lim_F \sum_{l \geq 0} \Pr_{\mu^F}[y_0 \in \tilde{G} \wedge \cdots \wedge y_{l-1} \in \tilde{G} \wedge y_l = s] \\ \times \Pr_s[y_1 \in \tilde{E} \wedge \cdots \wedge y_{m+2} = k \wedge \mathfrak{t} = m + n].$$

Next, we operate on the second factor. In the presence of the information $y_0(\omega) = s \wedge y_1(\omega) \in \tilde{E} \wedge \cdots \wedge y_{m-1}(\omega) \in \tilde{E} \wedge y_m(\omega) = i$, we know by the special form of E that $\mathfrak{t}(\omega) = m + n$ if and only if $\mathfrak{t}(\omega_m) = n$. Hence, by Theorem 4-9, we have

$$\Pr_s[y_1 \in \tilde{E} \wedge \cdots \wedge y_{m+2} = k \wedge \mathfrak{t} = m + n] \\ = \Pr_s[y_1 \in \tilde{E} \wedge \cdots \wedge y_m = i] \cdot \Pr_i[y_1 = j \wedge y_2 = k \wedge \mathfrak{t} = n].$$

The second factor on the right side of this last equation does not depend on m and therefore factors out. We may sum it on n, and we get $\bar{P}_{ij}\bar{P}_{jk}$. Hence the contribution is

$$= \lim_m \sum_{s \in \tilde{E}} \lim_F \sum_{l \geq 0} \Pr_{\mu^F}[y_0 \in \tilde{G} \wedge \cdots \wedge y_{l-1} \in \tilde{G} \wedge y_l = s] \\ \times \Pr_s[y_1 \in \tilde{E} \wedge \cdots \wedge y_m = i]\bar{P}_{ij}\bar{P}_{jk}.$$

The first factor here is

$$\begin{aligned} \sum_{l \geq 0} \Pr_{\mu^F}[y_0 \in \tilde{G} \wedge \cdots \wedge y_{l-1} \in \tilde{G} \wedge y_l = s] &= \mu_s^F + \sum_{l > 0} \Pr_{\mu^F}[\text{—}] \\ &= \mu_s^F + \sum_{l > 0} (\mu^F\, {}^G P^{l-1} P)_s \\ &= \mu_s^F + (\mu^F\, {}^G N P)_s. \end{aligned}$$

Applying the identity ${}^G N = N - B^G N$ of Lemma 8-17, we have

$$\begin{aligned} \mu^F\, {}^G N P &= \mu^F (N - B^G N) P \\ &= \mu^F N P - \mu^F B^G N P \\ &= \mu^F N P - \mu^G N P \quad \text{if } F \supset G \\ &= \nu^F P - \nu^G P. \end{aligned}$$

Since $\nu^F \to \nu$ monotonically and since $\mu_s^F \to \mu_s$, we have

$$\lim_F \sum_{l \geq 0} \Pr_{\mu^F}[y_0 \in \tilde{G} \wedge \cdots \wedge y_{l-1} \in \tilde{G} \wedge y_l = s] = \mu_s + (\nu P)_s - (\nu^G P)_s$$
$$= \nu_s - (\nu^G P)_s,$$

since $\nu(I - P) = \mu$. Thus the contribution is

$$= \lim_m \sum_{s \in \tilde{E}} (\nu - \nu^G P)_s \Pr_s[y_1 \in \tilde{E} \wedge \cdots \wedge y_{m-1} \in \tilde{E} \wedge y_m = i]\bar{P}_{ij}\bar{P}_{jk}.$$

But

$$\Pr_s[y_1 \in \tilde{E} \wedge \cdots \wedge y_{m-1} \in \tilde{E} \wedge y_m = i] = ({}^E P^{m-1} P)_{si}.$$

If

$$P = \begin{pmatrix} T & U \\ R & Q \end{pmatrix}, \quad \text{then} \quad {}^E P = \begin{pmatrix} 0 & 0 \\ 0 & Q \end{pmatrix}$$

and

$${}^E P^{m-1} P = \begin{pmatrix} 0 & 0 \\ Q^{m-1} R & Q^m \end{pmatrix}.$$

From this representation of ${}^E P^{m-1} P$, we see that the sum $\sum_{s \in \tilde{E}}$ may now be extended over all of S, and the contribution is

$$= \lim_m [(\nu - \nu^G P)({}^E P^{m-1} P)]_i \bar{P}_{ij}\bar{P}_{jk}.$$

Now

$$\nu^G P({}^E P^{m-1})P \leq \nu^G P^{m+1} \to 0$$

since ν^G is a potential. Therefore the contribution is

$$= \lim_m (\nu\ {}^E P^{m-1} P)_i (\bar{P}_{ij}\bar{P}_{jk}).$$

Since i is a state of E, this expression is

$$= \lim_m (\nu_{\tilde{E}} Q^{m-1} R)_i (\bar{P}_{ij}\bar{P}_{jk}).$$

From the identity $\nu = \nu P + \mu$, we have

$$\nu_{\tilde{E}} = \nu_E U + \nu_{\tilde{E}} Q + \mu_{\tilde{E}}$$

or

$$\nu_{\tilde{E}} Q = \nu_{\tilde{E}} - (\nu_E U + \mu_{\tilde{E}}).$$

Iterating, we find that

$$\nu_{\tilde{E}} Q^{m-1} = \nu_{\tilde{E}} - (\nu_E U + \mu_{\tilde{E}})(I + Q + \cdots + Q^{m-2})$$

and hence

$$\nu_{\tilde{E}} Q^{m-1} R = \nu_{\tilde{E}} R - (\nu_E U + \mu_{\tilde{E}})(I + Q + \cdots + Q^{m-2})R.$$

By monotone convergence,

$$\lim_m \nu_{\tilde{E}} Q^{m-1} R = \nu_{\tilde{E}} R - (\nu_E U + \mu_{\tilde{E}})^E N_{\tilde{E}} R.$$

As we noted in Section 2, the balayage potential ν^E satisfies

$$\nu^E = (\nu_E \qquad \nu_E U\, {}^E N_{\tilde{E}}).$$

Hence

$$\lim_m \nu_{\tilde{E}} Q^{m-1} R = (\nu - \nu^E)_{\tilde{E}} R - \mu_{\tilde{E}}\, {}^E N_{\tilde{E}} R.$$

If we twice use the fact that ν and ν^E agree on E, we obtain

$$(\nu - \nu^E)_{\tilde{E}} R = [(\nu - \nu^E)P]_E = [(\nu - \mu) - (\nu^E - \mu^E)]_E = \mu_E^E - \mu_E.$$

Thus

$$\lim_m \nu_{\tilde{E}} Q^{m-1} R = \mu_E^E - (\mu_E + \mu_{\tilde{E}}\, {}^E N_{\tilde{E}} R) = (\mu^E - \mu B^E)_E.$$

We conclude that the contribution to $\Pr[p]$ is

$$= (\mu^E - \mu B^E)_i \bar{P}_{ij} \bar{P}_{jk}.$$

(5) *Completion of proof.* From steps 3 and 4, we find that

$$\Pr[z_0 = i \wedge z_1 = j \wedge z_2 = k] = \mu_i^E \bar{P}_{ij} \bar{P}_{jk} \quad \text{for } i \in E.$$

Hence

$$\Pr[z_2 = k \mid z_0 = i \wedge z_1 = j] = \bar{P}_{jk}$$

if $\Pr[z_0 = i \wedge z_1 = j] > 0$. That is, $\{z_n\}$ is a Markov chain with matrix $\bar{P}$ and with starting vector μ^E, provided $E = \{0, 1, \ldots, e\}$. We have seen how the proof for a general finite set follows from this special case.

We must compute the mean number of times in state j in $\{x_n\}$. The mean number starting in E is $(\mu^E N)_j = \nu_j^E$. Hence the mean total number is $\lim_E \nu_j^E = \nu_j$ by monotone convergence.

Finally we show that the contribution to ν of the paths with $\mathbf{u} > -\infty$ is μN. On these paths the process behaves as if it were started with distribution μB^E (see step 3 if $E = \{0, 1, \ldots, e\}$ and use the identity $B^E B^F = B^F$, where $E \supset F$, in the general case). Therefore the contribution is

$$\lim_E (\mu B^E) N = \mu N - \lim_E \mu\, {}^E N$$

by Lemma 8-17. Since ${}^E N \le N$ and $\lim {}^E N = 0$, $\mu\, {}^E N \to 0$ by dominated convergence. Thus the contribution is

$$\lim_E (\mu B^E) N = \mu N.$$

13. Examples

EXAMPLE 1: Basic example.

For the basic example, we recall from Chapter 5 that

$$N_{ij} = \begin{cases} \dfrac{\beta_j}{\beta_\infty} & \text{if } i \leq j \\ \dfrac{\beta_j}{\beta_\infty} - \dfrac{\beta_j}{\beta_i} & \text{if } i > j. \end{cases}$$

In a process of this kind in which all pairs of states communicate, there is no gain in choosing a complicated starting vector π. Indeed, if we choose π to be a unit mass at 0, then *all* finite-valued P-regular functions are certainly π-integrable, and therefore no other choice of π will make the representation theorem yield additional regular functions.

With π chosen as a unit mass at 0, $K(i, j)$ becomes

$$K(i, j) = \frac{N_{ij}}{N_{0j}} = \begin{cases} 1 & \text{if } i \leq j \\ \dfrac{\beta_i - \beta_\infty}{\beta_i} & \text{if } i > j. \end{cases}$$

Since K is of a particularly simple form, it is possible to compute the metric for the exit boundary. If $j < j'$, then

$$\begin{aligned} d(j, j') &= \sum_i w_i N_{0i} |K(i, j) - K(i, j')| \\ &= \sum_{j < i \leq j'} w_i \frac{\beta_i}{\beta_\infty} \left| \left(1 - \frac{\beta_\infty}{\beta_i}\right) - 1 \right| \\ &= \sum_{j < i \leq j'} w_i. \end{aligned}$$

The metric space (S, d) thus is isometric to the subset of the real line consisting of the points $0, w_1, w_1 + w_2, w_1 + w_2 + w_3, \ldots$. Its Cauchy completion contains the one extra point corresponding to $\sum_{i \geq 1} w_i$.

Alternatively we can find S^* directly without using the metric. In fact, if $\{j_n\}$ is a sequence of points in S, then $\{K(i, j_n)\}$ is clearly Cauchy if and only if either $\{j_n\}$ is eventually constant or $\{j_n\}$ tends to infinity.

Either way, there is exactly one boundary point, which we may call $+\infty$, and the relative topology for S is discrete. Moreover,

$$K(i, +\infty) = \lim K(i, j_n) = 1.$$

Since **1** is regular and since there is only one boundary point, the boundary point must be extreme. Moreover, every regular function $h \geq 0$ must be constant. Since the process does not disappear, harmonic measure μ must assign unit mass to $+\infty$.

Next, we consider the entrance boundary. If f is a unit mass at 0, then J is given by

$$J(j, i) = \frac{N_{ji}}{N_{j0}} = \begin{cases} \beta_i/\beta_0 & \text{if } j = 0 \\ \beta_i/\beta_0 & \text{if } j > 0,\ j > i \\ \dfrac{\beta_i}{\beta_\infty}\bigg/\left(\dfrac{\beta_0}{\beta_\infty} - \dfrac{\beta_0}{\beta_i}\right) & \text{if } j > 0,\ j \leq i. \end{cases}$$

The sequence 1, 2, 3, ... again is Cauchy, so that there is necessarily exactly one limit point in $^*S$. But

$$\lim_{n \to \infty} J(n, i) = J(0, i),$$

and consequently $^*S = S$ and the entrance boundary is empty. The state 0 is a limit point and thus S does not have the discrete topology. Since the entrance boundary is empty, there are no non-zero regular measures $\sigma \geq 0$. (We had arrived at this conclusion in Chapter 5, too.)

EXAMPLE 2: The p-q random walk.

The process P is the Markov chain with state space the integers which moves one step to the right with probability p or one step to the left with probability $q = 1 - p$. We shall assume $0 < q < p < 1$. A calculation like that in Section 5-8 shows that

$$H_{ij} = \begin{cases} 1 & \text{if } i \leq j \\ (q/p)^{i-j} & \text{if } i \geq j \end{cases}$$

and then

$$N_{ij} = \begin{cases} (p - q)^{-1} & \text{if } i \leq j \\ (q/p)^{i-j}(p - q)^{-1} & \text{if } i \geq j. \end{cases}$$

Take π and f to be unit masses at state 0; as in Example 1, we may make these choices since all pairs of states communicate. We obtain

$$K(i, j) = \frac{N_{ij}}{N_{0j}} = \begin{cases} 1 & \text{for } j \geq i \text{ and } j \geq 0 \\ (q/p)^i & \text{for } j \leq i \text{ and } j \leq 0. \end{cases}$$

There are two distinct infinite Cauchy sequences, one corresponding to $+\infty$ and the other corresponding to $-\infty$. For these points,

$$K(i, +\infty) = 1$$
$$K(i, -\infty) = (q/p)^i.$$

Both functions are regular and minimal, and thus the boundary contains two points, both extreme. No point of S is a limit point Since the process goes to $+\infty$ with probability one, we have $\mu(\{+\infty\}) = 1$. (The concentration of μ can also be deduced analytically from the representation of $\mathbf{1}$.)

The entrance boundary is treated most quickly by reversing the process with respect to the regular measure $\alpha = \mathbf{1}^T$. Then P is a process of the same type but with the roles of p and q interchanged Consequently

$$J(-\infty, j) = 1$$
$$J(+\infty, j) = (p/q)^j,$$

and $\mu^\alpha(\{-\infty\}) = \hat{\mu}(\{-\infty\}) = 1$.

The function $p(x)$, which does not depend upon α, satisfies

$$p(-\infty) = \lim_{E \uparrow S} \lim_{j \to -\infty} \frac{h_j^E}{N_{j0}} = \lim_{E \uparrow S} \frac{1}{N_{00}} = p - q.$$

For $p(+\infty)$, we note that if j is large, then $h_j^E = H_{jm}$, where m is the last state in E. Also

$$\frac{H_{jm}}{N_{j0}} = \frac{(q/p)^{j-m}}{(p-q)^{-1}(q/p)^j} = \frac{(q/p)^{-m}}{(p-q)^{-1}}.$$

As $E \uparrow S$, $m \to +\infty$ and therefore

$$p(+\infty) = 0.$$

Any extended chain representing $\mathbf{1}^T$ starts at $-\infty$ with probability $p - q$ and goes to $+\infty$.

The p-q random walk can be used to show what happens if π is chosen in such a way that there is a P-regular function which is not π-integrable. In fact, set

$$\pi_i = \begin{cases} (p/q)^i\left(\dfrac{p-q}{p}\right) & \text{for } i \le 0 \\ 0 & \text{for } i > 0. \end{cases}$$

Then $\pi\mathbf{1} = 1$ and the regular function $(q/p)^i$ will not be integrable We must recompute $K(i, j)$. First we note that for $j > 0$,

$$N_{\pi j} = \sum_{i \le 0} \pi_i N_{ij} = (p - q)^{-1} \sum \pi_i = (p - q)^{-1},$$

and that for $j \le 0$,

$$\begin{aligned} N_{\pi j} &= (p-q)^{-1} \sum_{i \le 0} \pi_i H_{ij} = p^{-1} \sum_{i \le 0} (p/q)^i H_{ij} \\ &= p^{-1} \sum_{i \le j} (p/q)^i + p^{-1} \sum_{j < i \le 0} (p/q)^i (q/p)^{i-j} \\ &= (p-q)^{-1}(p/q)^j + p^{-1} j (p/q)^j. \end{aligned}$$

Then

$$K(i,j) = \frac{N_{ij}}{N_{\pi j}} = \begin{cases} 1 & \text{for } j \geq i \text{ and } j \geq 0 \\ \dfrac{(p-q)^{-1}(q/p)^i}{(p-q)^{-1} + p^{-1}j} & \text{for } j \leq i \text{ and } j \leq 0. \end{cases}$$

Again there are two boundary points $+\infty$ and $-\infty$, but this time

$$K(i, +\infty) = 1$$

and

$$K(i, -\infty) = 0.$$

Thus $-\infty$ is not an extreme boundary point. The nonintegrability of $(q/p)^i$ introduced a boundary point whose associated function was identically zero. This example is typical of the general situation, but we shall not pursue the details.

There is still a second way a zero boundary point can arise, and the degenerate case where $p = 1$ and $q = 0$ in the example above illustrates the point. If h is regular for this process, then

$$h_i = P_{i,i+1}h_{i+1} = h_{i+1},$$

and hence h must be constant. On the other hand, it is readily seen that, for any choice of π for which $\pi N > 0$, there are two boundary points $+\infty$ and $-\infty$ and that

$$K(i, +\infty) = 1$$

and

$$K(i, -\infty) = 0.$$

Thus again there is a zero boundary point, but this time no function is missing.

Example 3: Symmetric random walk in three dimensions.

The symmetric random walk in three dimensions is the sums of independent random variables process on the integer lattice in three-dimensional space with transition probability $\frac{1}{6}$ from any point to each of its six neighbors. We recall from Proposition 5-20 that the only bounded regular functions for this process are the constants. We now prove, using in part the methods of transient boundary theory, the deeper result that the only non-negative regular functions are the constants.

In fact, if we choose π to be a unit mass at 0, then every regular function is π-integrable, and the representation theorem assures us that it is enough to prove that $K(\cdot, x) = 1$ for every x in the boundary. That is, it suffices to prove that for every i

$$\lim_{j \to \infty} K(i, j) = 1.$$

We shall prove in a moment the following estimate for N_{0j}: If $j \neq 0$, then

$$N_{0j} = \frac{3}{2\pi|j|} + O(|j|^{-2}).$$

Once we have this estimate, we also have, for $i \neq j$,

$$\begin{aligned} K(i, j) = \frac{N_{ij}}{N_{0j}} &= \frac{\dfrac{3}{2\pi|j - i|} + O(|j - i|^{-2})}{\dfrac{3}{2\pi|j|} + O(|j|^{-2})} \\ &= \frac{|j - i|^{-1} + O(|j|^{-2})}{|j|^{-1} + O(|j|^{-2})} \\ &= \frac{\dfrac{|j|}{|j - i|} + O(|j|^{-1})}{1 + O(|j|^{-1})} \\ &= \frac{|j|}{|j - i|} + O(|j|^{-1}). \end{aligned}$$

Hence, $\lim_j K(i, j) = 1$, as asserted.

Thus we are to prove the estimate for N_{0j}. We first show that

$$N_{0j} = 3 \int_{-1/2}^{1/2} \int_{-1/2}^{1/2} \int_{-1/2}^{1/2} \frac{e^{-2\pi i (j \cdot x)} dx_1 dx_2 dx_3}{3 - \cos 2\pi x_1 - \cos 2\pi x_2 - \cos 2\pi x_3},$$

where $x = (x_1, x_2, x_3)$. Call the right side of the equation h_j. The singularity in the denominator of the integrand is of the order of

$$\frac{1}{3 - (1 - 2\pi^2 x_1^2) - (1 - 2\pi^2 x_2^2) - (1 - 2\pi^2 x_3^2)} = \frac{1}{2\pi^2 |x|^2}$$

near $x = 0$, and $|x|^{-2}$ is integrable in a neighborhood of $x = 0$. Thus h_j is certainly well defined, and moreover $h_j \to 0$ as $j \to \infty$ by the Riemann–Lebesgue Lemma. We claim that

$$(I - P)_{m \cdot} h_{\cdot} = \delta_{m0}.$$

Call the six neighbors of 0 by the names a_s, and call the region

$$-\tfrac{1}{2} < x_1, x_2, x_3 \leq \tfrac{1}{2}$$

by the name E. Then

$$
\begin{aligned}
(I - P)_{m\cdot}h_{\cdot} &= h_m - \tfrac{1}{6}\sum_s h_{m+a_s} \\
&= 3\int_E (3 - \cos 2\pi x_1 - \cos 2\pi x_2 - \cos 2\pi x_3)^{-1} \\
&\qquad\qquad \times \left(e^{-2\pi i(m\cdot x)} - \tfrac{1}{6}\sum_s e^{-2\pi i(m+a_s)\cdot x}\right)dx \\
&= 3\int_E (3 - \cos 2\pi x_1 - \cos 2\pi x_2 - \cos 2\pi x_3)^{-1} \\
&\qquad\qquad \times e^{-2\pi i m\cdot x}\left(1 - \tfrac{1}{6}\sum_s e^{-2\pi i a_s\cdot x}\right)dx \\
&= \int_E e^{-2\pi i m\cdot x}dx \\
&= \delta_{m0}.
\end{aligned}
$$

Now $N_{0\cdot} = N_{\cdot 0}$ is also a function tending to zero (see Proposition 7-10) and satisfying

$$(I - P)_{m\cdot}N_{\cdot 0} = \delta_{m0},$$

and thus $N_{\cdot 0} - h_{\cdot}$ is a bounded regular function. By Proposition 5-20 it must be constant, and that constant must be zero, since both functions vanish at infinity. Thus the formula for N_{0j} is established.

We now have N_{0j} exhibited as the jth Fourier coefficient of a certain periodic function of three variables. The remainder of the proof will presuppose some knowledge of Fourier transforms and tempered distributions.

Let $\varphi(x)$ be an infinitely differentiable function defined on R^3 such that

(1) $0 \le \varphi \le 1$.
(2) $\varphi(x) = 1$ for x in a small neighborhood of 0.
(3) $\varphi(x) = 0$ outside of a slightly larger neighborhood of 0.

For simplicity, let $\|x\|^4 = x_1{}^4 + x_2{}^4 + x_3{}^4$. Denote by $\bar{f}(x)$ the periodic continuation of the function

$$f(x) = \varphi(x)\left[\frac{3}{2\pi^2|x|^2} + \frac{\|x\|^4}{2|x|^4}\right].$$

The difference

$$\bar{f}(x) - \frac{3}{3 - \cos 2\pi x_1 - \cos 2\pi x_2 - \cos 2\pi x_3}$$

is a bounded periodic function which is infinitely differentiable away from 0 and is $O(|x|^2)$ near 0. Moreover, its Laplacian is integrable in $E - \{0\}$. These facts imply that the Fourier coefficients of the difference are $O(|j|^{-2})$ as $j \to \infty$. Hence it is enough to show that the Fourier coefficients of $\tilde{f}$ are equal to

$$\frac{3}{2\pi|j|} + O(|j|^{-2}).$$

It is clear from the definitions that the Fourier coefficients of $\tilde{f}$ satisfy

$$\int_E \tilde{f}(x)e^{-2\pi i j\cdot x}dx = \int_{R^3} f(x)e^{-2\pi i j\cdot x}dx.$$

Here the right side is the Fourier transform of f evaluated at j. We are thus to estimate the Fourier transform of f. Write

$$f(x) = [\varphi(x) - 1]\left[\frac{3}{2\pi^2|x|^2} + \frac{\|x\|^4}{2|x|^4}\right] + \frac{3}{2\pi^2|x|^2} + \frac{\|x\|^4 - \frac{3}{5}|x|^4}{2|x|^4} + \frac{3}{10},$$

and take the Fourier transform of both sides in the sense of distributions. We shall consider the four terms on the right separately.

The transform of the constant $\frac{3}{10}$ is the distribution defined by the measure which assigns weight $\frac{3}{10}$ to the origin. It has the property that it is supported, say, entirely in the unit ball.

In the third term, the numerator is a homogeneous harmonic polynomial, and the transform of the term is known from Fourier analysis to be

$$\frac{15}{8\pi}\,\mathrm{PV}\left[\frac{\|y\|^4 - \frac{3}{5}|y|^4}{2|y|^7}\right],$$

where

$$(\mathrm{PV}[g], \psi) = \lim_{\epsilon\to 0}\int_{|x|\geq\epsilon} g(x)\psi(x)dx.$$

This distribution has the property that it is the sum of a function and a distribution which is supported in the unit ball. The function is

$$\frac{15}{8\pi}\,\frac{\|y\|^4 - \frac{3}{5}|y|^4}{2|y|^7} \quad \text{for } |y| \geq 1,$$

and it is $O(|y|^{-3})$ as $y \to \infty$.

The transform of the second term is known from Fourier analysis to be the distribution arising from the function

$$\frac{3}{2\pi|y|}.$$

We come to the first term. Since the transform of the left side is a function, the transform of the first term must be the sum of a function and a distribution supported in the unit ball. On the other hand, the first term is an infinitely differentiable function. If we iterate taking the Laplacian of it enough times, the resulting function is integrable and has a bounded function as Fourier transform. Since the operation of taking the Laplacian goes into multiplication by $-4\pi^2|y|^2$ under the Fourier transform, we obtain the inequality of functions

$$(4\pi^2|y|^2)^k \text{ FT(first term)} \leq A_k, \qquad k \text{ sufficiently large.}$$

This result implies that the function part of the distribution FT(first term) is $O(|y|^{-2k})$ as $y \to \infty$.

Since the sum of the transforms of the four terms is a function, we must have, outside the unit ball,

$$\text{FT}(f)(y) = O(|y|^{-2k}) + \frac{3}{2\pi|y|} + O(|y|^{-3}).$$

We conclude therefore that

$$\text{FT}(f)(j) = \frac{3}{2\pi|j|} + O(|j|^{-3})$$

and hence that

$$N_{0j} = \frac{3}{2\pi|j|} + O(|j|^{-2}),$$

as required.

EXAMPLE 4:

This example will be a process with an exit boundary point x such that $K(\cdot, x)$ is regular and normalized but is not minimal.

The state space will be

$$S = \{a_i, a_i', b_i \mid i = 0, 1, 2, \ldots\}$$

and the non-zero transition probabilities are

$$P_{a_{i-1}a_i} = p_i, \qquad P_{a_{i-1}b_{i-1}} = q_i = 1 - p_i,$$

$$P_{b_{i-1}b_i} = r_i,$$

$$P_{a'_{i-1}a'_i} = p_i, \qquad P_{a'_{i-1}b_{i-1}} = q_i = 1 - p_i.$$

The picture for this process is as follows.

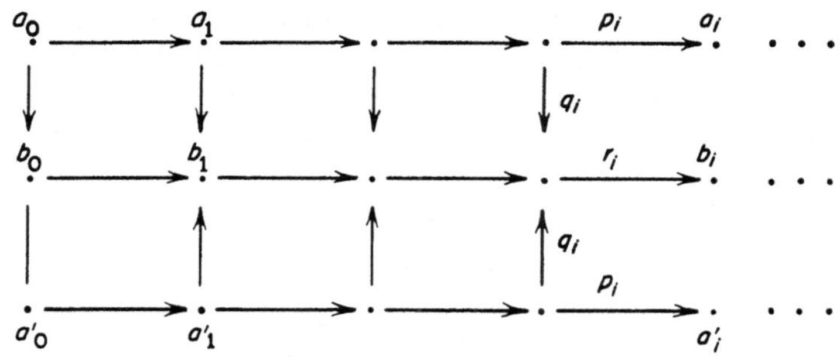

Let

$$\beta_0 = 1, \qquad \beta_i = \prod_{k=1}^{i} p_k, \qquad \beta_\infty = \lim_{i\to\infty} \beta_i,$$

$$\gamma_0 = 1, \qquad \gamma_i = \prod_{k=1}^{i} r_k,$$

$$\sigma_{-1} = 0, \qquad \sigma_i = \sum_{k=0}^{i} \beta_k q_{k+1}/\gamma_k, \qquad \sigma_\infty = \lim_{i\to\infty} \sigma_i.$$

We shall assume for this example that $\sigma_\infty = +\infty$. At the end of the discussion of the example we shall show that this condition is possible.

Any state can be reached either from a_0 or from a'_0. Thus if we set

$$\pi_{a_0} = \pi_{a'_0} = \tfrac{1}{2},$$

then πN is strictly positive. Since the process is in any given state at most once, we must have $H = N$. It is clear that

$$H_{a_i a_j} = H_{a'_i a'_j} = \begin{cases} 0 & \text{if } j < i \\ \beta_j/\beta_i & \text{if } j \geq i, \end{cases}$$

$$H_{b_i b_j} = \begin{cases} 0 & \text{if } j < i \\ \gamma_j/\gamma_i & \text{if } j \geq i, \end{cases}$$

and $H_{a'_i a_j} = H_{a_i a'_j} = 0 = H_{b_i a_j} = H_{b_i a'_j}$. We must compute $H_{a_i b_j}$ (and $H_{a'_i b_j}$). If $j < i$, then $H_{a_i b_j} = 0$. Otherwise we obtain a set of alternatives by considering the first time the process switches from the a-row to the b-row. Then

$$H_{a_i b_j} = \sum_{k=i}^{j} H_{a_i a_k} q_{k+1} H_{b_k b_j}.$$

Substituting, we find

$$H_{a_i b_j} = H_{a'_i b_j} = \begin{cases} 0 & \text{if } j < i \\ \dfrac{\gamma_j}{\beta_i}(\sigma_j - \sigma_{i-1}) & \text{if } j \geq i. \end{cases}$$

In determining the exit boundary we use the observation that whenever finitely many Cauchy sequences exhaust S, then those Cauchy sequences define all of the boundary points. In the present example the claim is that the three sequences $\{a_i\}$, $\{a'_i\}$, and $\{b_i\}$ are each Cauchy sequences.

First consider $K(\cdot, a_j)$ for large j. We have

$$K(a'_i, a_j) = K(b_i, a_j) = 0$$

and, for $i \leq j$,

$$K(a_i, a_j) = \frac{H_{a_i a_j}}{H_{\pi a_j}} = \frac{H_{a_i a_j}}{\frac{1}{2} H_{a_0 a_j}} = \frac{2}{\beta_i}.$$

Hence $a_\infty = \lim a_j$ is a boundary point, and it satisfies

$$K(a_i, a_\infty) = \frac{2}{\beta_i} \quad \text{and} \quad K(a'_i, a_\infty) = K(b_i, a_\infty) = 0.$$

We can check directly that $K(\cdot, a_\infty)$ is a regular function; moreover, it is normalized because

$$\pi K(\cdot, a_\infty) = \tfrac{1}{2} K(a_0, a_\infty) = \tfrac{1}{2} \cdot 2 = 1.$$

Similarly we find that $\{a'_j\}$ is a Cauchy sequence. If we call its limit a'_∞, then

$$K(a'_i, a'_\infty) = \frac{2}{\beta_i} \quad \text{and} \quad K(a_i, a'_\infty) = K(b_i, a'_\infty) = 0.$$

Since $K(\cdot, a_\infty) \neq K(\cdot, a'_\infty)$, a_∞ and a'_∞ are distinct boundary points. We see also that $K(\cdot, a'_\infty)$ is regular and normalized.

Finally we consider the sequence $\{b_j\}$ for large j. For $i \leq j$, we have

$$K(a_i, b_j) = \frac{H_{a_i b_j}}{H_{\pi b_j}} = \frac{H_{a_i b_j}}{H_{a_0 b_j}} = \frac{1}{\beta_i}\left(\frac{\sigma_j - \sigma_{i-1}}{\sigma_j}\right)$$

and

$$K(b_i, b_j) = \frac{H_{b_i b_j}}{H_{\pi b_j}} = \frac{H_{b_i b_j}}{H_{a_0 b_j}} = \frac{1}{\gamma_i \sigma_j}.$$

Since, by hypothesis, $\sigma_\infty = +\infty$, we have

$$\lim_j K(a_i, b_j) = \lim_j K(a'_i, b_j) = \frac{1}{\beta_i}$$

and

$$\lim_j K(b_i, b_j) = 0.$$

Therefore $\{b_j\}$ is Cauchy. Denoting its limit in S^* by b_∞, we conclude

$$K(a_i, b_\infty) = K(a'_i, b_\infty) = 1/\beta_i \quad \text{and} \quad K(b_i, b_\infty) = 0.$$

We can check that $K(\cdot, b_\infty)$ is regular and normalized.

Thus there are exactly three boundary points, and each is associated with a regular normalized function. But

$$K(\cdot, b_\infty) = \tfrac{1}{2}K(\cdot, a_\infty) + \tfrac{1}{2}K(\cdot, a'_\infty)$$

and b_∞ therefore cannot be an extreme point. Then by the representation theorem all non-negative regular functions are generated by $K(\cdot, a_\infty)$ and $K(\cdot, a_{-\infty})$. Hence their linear independence as functions on S implies that they are both minimal. That is, $B_e = \{a_\infty, a'_\infty\}$.

In this example, β_∞ is positive if and only if harmonic measure assigns positive weight to the boundary. In fact,

$$\mu(\{a_\infty\}) = \tfrac{1}{2}\beta_\infty = \mu(\{a'_\infty\}).$$

The remainder of the weight is put on the states from which the process can disappear. We obtain

$$\mu(\{b_j\}) = \Pr_\pi[\text{process disappears from } b_j] = (\gamma_j - \gamma_{j+1})\sigma_j.$$

Note that if $\beta_\infty = 0$, then there exist non-negative regular functions, but none of them is bounded.

We still must show that the condition $\sigma_\infty = +\infty$ puts no restriction on whether β_∞ must be positive. To get $\beta_\infty = 0$, take

$$p_j = q_j = r_j = \tfrac{1}{2}.$$

Then $\beta_j = \gamma_j = 2^{-j}$ and $\sigma_j = \frac{1}{2}(j + 1)$. Hence $\sigma_\infty = +\infty$. To get $\beta_\infty > 0$, take

$$p_j = \frac{j(j+2)}{(j+1)^2}, \qquad q_j = 1 - p_j = \frac{1}{(j+1)^2}, \qquad r_j = \frac{j+1}{j+2}.$$

Then

$$\beta_j = \frac{j+2}{2(j+1)}, \qquad \beta_\infty = \frac{1}{2}, \qquad \gamma_j = \frac{2}{j+2},$$

and

$$\sigma_\infty = \sum_{j=0}^{\infty} \frac{j+2}{2(j+1)} \frac{1}{(j+2)^2} \frac{j+2}{2} = \sum_{j=0}^{\infty} \frac{1}{4(j+1)} = +\infty.$$

EXAMPLE 5:

This example will be a process in which S^* has every point of S as a limit point.

Let S be the set of all finite strings of positive integers of the form

$$(1, k_1, k_2, \ldots, k_n)$$

as n and the k's vary. The transition probability from $(1, k_1, \ldots, k_n)$

to $(1, k_1, \ldots, k_n, m)$ is to be 2^{-m} and all other transition probabilities are zero.

Since each state is entered at most once, we have $N = H$. The hitting probability from

$$I = (1, j_1, \ldots, j_n)$$

to a distinct state

$$J = (1, k_1, \ldots, k_{n'}, m)$$

is

$$H_{IJ} = \begin{cases} 2^{-(k_{n+1} + \cdots + k_{n'} + m)} & \text{if } n \leq n' \text{ and } j_1 = k_1, \ldots, j_n = k_n \\ 0 & \text{otherwise.} \end{cases}$$

Take π to be a unit mass at (1). If

$$0 = (1), \qquad J = (1, k_1, \ldots, k_n), \qquad J_m = (1, k_1, \ldots, k_n, m),$$

the claim is that

$$\lim_{m \to \infty} K(I, J_m) = K(I, J)$$

for all I. It would therefore follow that every point J in S is a limit point of S^*.

For the proof we observe first that $K(I, J) = K(I, J_m) = 0$ unless I is an initial segment of some J_m. Also if $I = J_{m_0}$, then

$$K(I, J) = \lim K(I, J_m) = 0.$$

Thus we may suppose that I is an initial segment of J. In this case, if $I = (1, k_1, \ldots, k_s)$,

$$K(I, J_m) = \frac{H_{IJ_m}}{H_{0J_m}} = 2^{k_1 + \cdots + k_s}$$

and

$$K(I, J) = \frac{H_{IJ}}{H_{0J}} = 2^{k_1 + \cdots + k_s}.$$

Hence $K(I, J) = \lim K(I, J_m)$ always.

EXAMPLE 6: Space-time coin tossing.

Let S be the set of lattice points (n, i) in the plane with $0 \leq i \leq n$. A point of S is to be identified with i heads in n tosses of a fair coin. Thus we take

$$P_{(n,i),(n+1,i)} = P_{(n,i),(n+1,i+1)} = \tfrac{1}{2}$$

with all other transition probabilities equal to zero. In this example, $N = H$ and also

$$H_{(m,i),(n,j)} = \begin{cases} 2^{m-n}\dbinom{n-m}{j-i} & \text{if } n \geq m,\ j \geq i \\ 0 & \text{otherwise.} \end{cases}$$

Let π be a unit mass at (0, 0). Then $\pi N > 0$ since every state can be reached from (0, 0). We obtain

$$K((m, i), (n, j)) = \begin{cases} 2^m \binom{n-m}{j-i} \Big/ \binom{n}{j} & \text{if } n \geq m,\ j \geq i \\ 0 & \text{otherwise.} \end{cases}$$

As a first step in obtaining the exit boundary, we shall prove that if (n_k, j_k) is Cauchy, then $\lim_k (j_k/n_k)$ exists. In fact, set $i = 0$ and $m = 1$ and consider, for $n_k \geq 1$, the expression

$$K((1, 0), (n_k, j_k)) = 2\binom{n_k - 1}{j_k} \Big/ \binom{n_k}{j_k} = \frac{2(n_k - j_k)}{n_k} = 2 - 2\frac{j_k}{n_k}.$$

As $k \to \infty$, the left side converges; hence so does the right side, and $\lim j_k/n_k$ must exist.

Conversely suppose that (n_k, j_k) is an infinite sequence such that $t = \lim j_k/n_k$ exists. We claim that (n_k, j_k) is Cauchy. Fix (m, i) and denote by (n, j) a term of the sequence (n_k, j_k). Then

$$\begin{aligned} K((m, i), (n, j)) &= 2^m \binom{n-m}{j-i} \Big/ \binom{n}{j} \\ &= 2^m \frac{[(j)\ldots(j - i + 1)][(n - j)\ldots(n - j - m + i + 1)]}{(n)\ldots(n - m + 1)}. \end{aligned}$$

Noting that both numerator and denominator have exactly m factors and that m is fixed, divide the numerator and the denominator by n^m and pass to the limit in each factor separately. In each factor we have $j/n \to t$ and, for instance, $(i - 1)/n \to 0$. We get

$$\lim K((m, i), (n, j)) = 2^m t^i (1 - t)^{m-i}.$$

Therefore the classes of infinite Cauchy sequences are in one-to-one correspondence with the rays to infinity, that is, in one-to-one correspondence with points t in [0, 1]. The functions associated to these sequences are

$$K((m, i), t) = 2^m t^i (1 - t)^{m-i}.$$

It is easy to check that all of these functions are regular and normalized, and hence the points in question form the entire boundary B (and nothing more).

Next, we check that the topology of B is the usual topology for the unit interval. The map of the unit interval onto B is continuous since each function $K((m, i), \cdot)$ is continuous in the unit interval topology. Since the map is one-one from a compact space onto a Hausdorff space, it is a homeomorphism.

To see that every point of B is extreme, we assume, on the contrary, that $2^m t_0^i(1 - t_0)^{m-i}$ is not minimal. By Theorem 10-41 there is a measure ν on the Borel sets of $B_e \subset [0, 1]$ with

$$2^m t_0^i(1 - t_0)^{m-i} = \int_{B_e} 2^m t^i(1 - t)^{m-i} d\nu(t).$$

Specializing to the case $m = i$ and extending ν to be defined on $[0, 1]$, we find that

$$t_0^i = \int_0^1 t^i d\nu(t)$$

for all i. But by the Weierstrass Approximation Theorem, ν is completely determined by its integral against all polynomials t^i. Hence ν is a point mass at t_0, and t_0 is extreme.

We can summarize our results so far by saying that no point of S is a limit point, that the boundary B is homeomorphic to the unit interval under the correspondence

$$K((m, i), t) = 2^m t^i(1 - t)^{m-i},$$

and that every point of the boundary is extreme.

The Strong Law of Large Numbers, when applied to coin tossing, is equivalent with the statement that harmonic measure μ concentrates all its mass at $t = \frac{1}{2}$. We can verify this result directly: We do have

$$\int_0^1 2^m(t)^i(1 - t)^{m-i} d\delta_{1/2}(t) = 2^m(\tfrac{1}{2})^i(1 - \tfrac{1}{2})^{m-i} = 1,$$

and, by the uniqueness half of Theorem 10-41, $\delta_{1/2}$ can be the only measure yielding 1. Hence $\delta_{1/2}$ is harmonic measure; that is, harmonic measure concentrates all its mass at $t = \frac{1}{2}$.

By Theorem 10-41 every measure μ^h on the unit interval gives rise to a non-negative regular function h for the process and μ^h is harmonic measure for the h-process. We shall discuss only two special cases.

If μ^h is the unit mass at a point t_0 other than 0 or 1, then

$$h_{(m,i)} = 2^m t_0^i(1 - t_0)^{m-i}$$

and

$$P^h_{(m,i),(m+1,i)} = 1 - t_0,$$

$$P^h_{(m,i),(m+1,i+1)} = t_0.$$

Thus the h-process is space-time coin tossing in which there is probability t_0 of heads and probability $1 - t_0$ of tails. The fact that

$\mu^h(\{t_0\}) = 1$ means that with probability one the ratio of heads to total tosses tends to t_0, again in agreement with the Strong Law of Large Numbers.

If, instead, μ^h is Lebesgue measure, then

$$\begin{aligned} h_{(m,i)} &= 2^m \int_0^1 t^i(1-t)^{m-i}dt \\ &= 2^m \frac{\Gamma(i+1)\Gamma(m-i+1)}{\Gamma((m+1)+1)} \\ &= 2^m \frac{i!(m-i)!}{(m+1)!}. \end{aligned}$$

Consequently, the transition probabilities in the h-process are again computable, and we find

$$P^h_{(m,i),(m+1,i)} = \frac{m-i+1}{m+2}$$

and

$$P^h_{(m,i),(m+1,i+1)} = \frac{i+1}{m+2}.$$

The significance of this process is discussed in Problems 28 and 29.

14. Problems

1. If π assigns positive weight only to finitely many states, show that $\pi K(\cdot, x) = 1$ for every boundary point x. What is the corresponding condition so that $K(\cdot, x)$ is regular for every boundary point?
2. Consider the identity

$$\int_{S^*} PK(\cdot, x)d\mu(x) = P\int_{S^*} K(\cdot, x)d\mu(x) = P\mathbf{1}.$$

What can we conclude about μ if $P\mathbf{1} = \mathbf{1}$? What if $(P\mathbf{1})_i < 1$?

Problems 3 to 12 deal with sums of independent random variables on the integers with $p_{-1} = \frac{2}{9}$, $p_1 = \frac{3}{9}$, $p_2 = \frac{4}{9}$. This example was discussed in Section 5-8, and the results there obtained will be useful.

3. Find N.
4. Let $\pi_0 = 1$, and compute $K(i, j)$.
5. Show that the two boundary points are $\pm\infty$, and find the two minimal functions.
6. From the form of these two functions determine what weight μ must put at each point. Does this agree with your understanding of the long-range behavior of the chain?
7. Let $h_i = (\frac{1}{4})^i$. Find P^h and μ^h. Check the representation of h.

8. Show that the h of Problem 7 is minimal by proving that P^h has only constants for bounded regular functions.
9. Let $f_i = \delta_{i0}$ and $g = Nf$. Find $J(i, j)$.
10. Show that the entrance boundary is "the same" as the exit boundary, and find the two minimal measures.
11. Find $p(-\infty)$ and $p(+\infty)$.
12. Construct extended chains representing the two minimal measures of Problem 10 (in the sense of Theorem 10-9). For each chain, verify that

$$\Pr[z_u \in C] = \int_C p(x)d\mu^v(x).$$

Problems 13 to 20 refer to a "double basic example." Let

$$S = \{0, 1, 2, \ldots, 0', 1', 2', \ldots\}.$$

The non-zero entries of P are

$$P_{i-1,i} = p_i; \quad P_{i-1,0'} = 1 - p_i; \quad P_{(i-1)',i'} = p_{i'}; \quad P_{(i-1)',0} = 1 - p_{i'}.$$

Let

$$\beta_0 = \beta_{0'} = 1, \quad \beta_i = \prod_{k=1}^{i} p_i, \quad \beta_{i'} = \prod_{k=1}^{i} p_{i'},$$

$$\beta_\infty = \lim \beta_i, \qquad \beta_{\infty'} = \lim \beta_{i'}.$$

Assume that $\beta_\infty > 0$ and $\beta_{\infty'} > 0$.

13. Find H. [*Hint:* Use Propositions 4-14 and 4-16.]
14. Let $\pi_0 = 1$, and compute K.
15. Show that there are two boundary points which correspond to ∞ and ∞', and find the minimal functions.
16. From the forms of the two minimal functions, find μ. Give an intuitive interpretation for the two weights.
17. Find the most general non-negative normalized regular function h. Show that h is a convex combination of the two minimal functions.
18. Show that any such h tends to μ^h/μ at each boundary point.
19. Let $h = \frac{1}{5}K(\cdot, \infty) + \frac{4}{5}K(\cdot, -\infty)$. Show that P^h goes to ∞ with probability $\frac{1}{5}$ by computing $\mu^h(\infty)$.
20. For the example of Problem 19, compute P^h explicitly if

$$p_i = \frac{i(i+2)}{(i+1)^2} \quad \text{and} \quad p_{i'} = \frac{(i+1)(i+3)}{(i+2)^2}.$$

Verify the conclusion of Problem 19 by computing h_0^E for

$$E = \{i, i+1, i+2, \ldots\}$$

and by letting i tend to ∞.

Problems 21 to 27 deal with a certain tree-process. The states are all finite sequences of H and T. The empty sequence is the starting state and is denoted state 0. From state $a = (a_1, a_2, \ldots, a_n)$ the process is equally likely to go to $(a_1, a_2, \ldots, a_n, H)$ or to $(a_1, a_2, \ldots, a_n, T)$.

21. Find N.

22. Let $\pi_0 = 1$, and compute K.

23. Show that the boundary is the set of all infinite sequences of H and T.

24. What is the topology of the boundary? [It is *not* that of the unit interval.]

25. Use the identity

$$\Pr_a[x_v \in C] = \int_C K(a, x)d\mu(x)$$

to find the measure μ.

26. Let h be defined by

$$h_a = \begin{cases} 1 & \text{if } a = 0 \\ 2 & \text{if } a = (H) \\ 4 & \text{if } a \text{ begins with } a_1 = a_2 = H \\ 0 & \text{otherwise.} \end{cases}$$

Prove that h is a normalized regular function. Find P^h and μ^h. Show that h is continuous and that $h(x) = d\mu^h/d\mu$ for a.e. x.

27. Let $f_a = 1/m$ if a consists of exactly m H's and m T's, and let f_a equal 0 otherwise. Show that f is a charge, and verify that

$$\Pr[\lim g(x_n) = 0] = 1.$$

Problems 28 and 29 deal with an instance of Polya's urn scheme. An urn contains some white balls and some black balls. A drawing is made with each ball equally likely to be drawn; the ball drawn is then replaced and another of the same color is added to the urn. This scheme is repeated over and over.

28. Let the pair (m, i) stand for $m + 2$ balls in the urn with $i + 1$ of them white. Show that if the outcomes of the Polya urn scheme are taken as such pairs (m, i), then the resulting process is a Markov chain. Note that the transition matrix for this chain is identical with the one for the h-process considered at the very end of Section 13.

29. Let the scheme be started with 1 white ball and 1 black ball, and let r_m be the fraction of white balls at time m. Use the observation in Problem 28 to compute

$$\limsup_m \Pr[|r_m - \tfrac{1}{2}| < \epsilon].$$

Problems 30 to 34 establish the necessity of a necessary and sufficient condition that a transient chain P with all pairs of states communicating have a non-zero non-negative regular measure α. Number the states $0, 1, 2, \ldots$. Let ${}^kL_{ij}$ be the probability that the process started at i reaches j and that the first visit is immediately preceded by a visit to a state $\geq k$. For instance, ${}^0L_{ij} = \bar{H}_{ij}$. The condition on P if α exists is as follows: There must exist an infinite set E of states such that

$$\lim_{k \to \infty} \lim_{\substack{i \to \infty \\ i \in E}} \frac{{}^kL_{ij}}{H_{ij}} = 0 \quad \text{for all } j.$$

30. Prove that

$$\frac{{}^kL_{ij}}{H_{ij}} \le \frac{{}^kL_{ij}}{N_{i0}}\frac{N_{00}}{P_{0j}^{(n)}}$$

for any n with $P_{0j}^{(n)} > 0$.

31. Let

$${}^jP_{ik}^{(n)} = \Pr_i[x_n = k \wedge x_m \ne j \quad \text{if} \quad 0 \le m < n].$$

Show that

$${}^kL_{ij} = \sum_{r=k}^{\infty}\sum_{n=0}^{\infty} {}^jP_{ir}^{(n)}P_{rj},$$

and derive as a consequence that

$${}^kL_{ij} \le N_{ij} - \sum_{r=0}^{k-1} N_{ir}P_{rj}.$$

32. Define f by $f_j = \delta_{j0}$. If an $\alpha > 0$ exists with $\alpha P = \alpha$, show that there is a point x_0 of $^*S$ for which $J(x_0, \cdot)$ is a minimal measure. Choose E to be the set of states in some Cauchy sequence converging to x_0.

33. Prove that

$$\lim_{k\to\infty}\lim_{\substack{i\to\infty\\ i\in E}}\left[\frac{N_{ij}}{N_{i0}} - \sum_{r=0}^{k-1}\frac{N_{ir}}{N_{i0}}P_{rj}\right] = 0.$$

34. Put together the preceding results to obtain a proof of the necessity of the stated condition.

Problems 35 to 38 refer to a transient chain with absorbing states. As usual, we write the transition matrix in the form

$$P = \begin{pmatrix} I & 0 \\ R & Q \end{pmatrix}.$$

35. Put

$$P' = \begin{pmatrix} 0 & 0 \\ R & Q \end{pmatrix}.$$

Show that P' has only transient states. Let π be any starting distribution for P' with $\pi N' > 0$. Prove that if i is an absorbing state of P, then the harmonic measure $\mu(i)$ in P' is equal to the probability in P of absorption in i.

36. If P is the infinite drunkard's walk with $p = \frac{2}{3}$, find harmonic measure for P' when $\pi_1 = 1$.

37. Suppose $P\mathbf{1} = \mathbf{1}$. Find a necessary and sufficient condition on harmonic measure for P' that P should have been an absorbing chain.

38. Let Q be any chain with all states transient and with the enlarged chain $\bar{Q}$ absorbing. Use the condition of Problem 37 to give a new proof that Q has no non-zero bounded regular functions $h \ge 0$.

CHAPTER 11

RECURRENT BOUNDARY THEORY

1. Entrance boundary for recurrent chains

Boundary theory for recurrent chains proceeds along altogether different lines from the approach in Chapter 10. A clue to the difficulty is that every non-negative superregular function is constant, and hence the representation of such functions degenerates. Moreover, since a recurrent chain is in every state infinitely often with probability one, an almost-everywhere convergence theorem is out of the question.

But intuitively, at least for some recurrent chains, there is some limiting behavior going on. For instance, with the one-dimensional symmetric random walk the Central Limit Theorem implies that the probability that the process is on either half-axis after time n tends to $\frac{1}{2}$ as n increases.

If P is a normal chain and E is a finite set, then $(P^nB^E)_{ij}$ is the probability that the process started in i enters E after time n at state j, and its limit λ_j^E is the "long run" probability of entering E at j. As we let E swell to the whole state space S, it is not clear from this interpretation just what happens. Consider therefore the following alternate interpretation: $(P^nB^E)_{ij}$ is the probability that the process started in i at time $-n$ enters E after time 0 at state j, and the limit λ_j^E is the probability that the process started at time $-\infty$ enters E after time 0 at state j. In this interpretation, if we let E swell to S and pass to the limit in the appropriate sense, we can expect λ^E to converge to an entrance distribution for the chain.

Suppose we compute instead the limit of P^nB^E on E followed by the limit on n. By dominated convergence, we have $\lim_{E\to S} P^nB^E = P^n$. Therefore, if we can justify the interchange of limits on E and n, we see that each row of P^n can be expected to converge to an entrance distribution for the chain. This conjecture is in contrast with the situation for transient chains, where P^n converges to an exit distribution.

In other words, the limiting behavior of P^n has something to do with the past history of the process. Our procedure will therefore be as follows: We start with a recurrent chain P, make it disappear after it reaches 0, and apply transient entrance boundary theory to the resulting process. The boundary obtained will be the entrance boundary for P, and it will be suitable in a wide class of chains for describing the limiting behavior of both λ^E and P^n. We shall see that this procedure is canonical in that it does not depend upon the choice of the state 0.

For the remainder of this chapter, let P be a recurrent chain, let 0 be a distinguished state, and let α be a finite-valued positive regular measure; α is unique up to a constant factor.

We define a transient chain Q associated to P and 0 by

$$Q_{ij} = \begin{cases} P_{ij} & \text{if } i \neq 0 \\ 0 & \text{if } i = 0. \end{cases}$$

For this transient chain $N_{ij} = {}^0N_{ij} + \delta_{j0}$. Choose f to be a column vector which places a unit weight at 0. Then $Nf = 1$.

Form the Martin entrance boundary of Q with respect to the reference function f. We have

$$J(i, j) = \frac{N_{ij}}{(Nf)_i} = {}^0N_{ij} + \delta_{j0}.$$

The compact metric space $^*S$ is the completion of S in the metric described in Chapter 10, and we shall see in Proposition 11-1 that $^*S$ does not depend upon the choice of the state 0. The spaces $^*S$ and $^*S - S$ can unambiguously be called the **completed space** and the **recurrent entrance boundary**, respectively, of the chain P. The set B^e of extreme points of $^*S - S$ will also be independent of state 0, and we are therefore free to speak of extreme points of the recurrent boundary.

Since $J(x, j)$ is well defined for x in $^*S$, the above expression for J shows that ${}^0N(x, j)$ is also well defined if we put

$${}^0N(x, j) = \begin{cases} {}^0N_{ij} & \text{if } x = i \in S \\ \lim_n {}^0N_{i_n j} & \text{if } x \in {}^*S - S \text{ and } x = \lim i_n. \end{cases}$$

Proposition 11-1: If $^*S_0$ and $^*S_1$ are the completed spaces of P formed with respect to two distinct states 0 and 1, then the identity map on S extends to a homeomorphism of $^*S_0$ onto $^*S_1$. Under the homeomorphism the extreme parts of the boundary correspond.

PROOF: We first establish the identity

$$ {}^1N_{ij} = {}^0N_{ij} + {}^1N_{0j} - \frac{\alpha_j}{\alpha_1}\, {}^0N_{i1}. \tag{*}$$

In fact, the mean number of times the process started at i visits j from the time of the first visit to 1 until before the first visit to 0 is ${}^0H_{i1}\, {}^0N_{1j}$. Thus if we compute in two ways the mean number of times the process started at i visits j before reaching both 0 and 1, we obtain

$$ {}^1N_{ij} + {}^0H_{i1}\, {}^0N_{1j} = {}^0N_{ij} + {}^1H_{i0}\, {}^1N_{0j}. $$

Substitute ${}^1H_{i0} = 1 - {}^0H_{i1}$ and get

$$ {}^1N_{ij} = {}^0N_{ij} + {}^1N_{0j} - {}^0H_{i1}({}^1N_{0j} + {}^0N_{1j}). $$

Applying Lemma 9-9 to $\hat{P}$ under the identification $i \to 1$, $j \to 0$, and $k \to j$, we find that the term in parentheses equals $(\alpha_j/\alpha_1)\, {}^0N_{11}$. From this relation, (*) follows.

Equation (*) and the expression for the kernel J show immediately that Cauchy sequences of S in the chain relative to state 0 are Cauchy relative to state 1, and by symmetry the converse is also true. Thus the statement about $^*S_0$ and $^*S_1$ follows.

For the assertion about the extreme points, let J_0 and J_1 be the kernels for the two transient chains, and suppose, for instance, that $J_1(y, \cdot)$ is normalized minimal and that $J_0(y, \cdot)$ is not. In any case, $J_0(y, \cdot)$ is normalized since $J_0(y, 0) = 1$. Thus

$$ J_0(y, \cdot) = \int_{^*S} J_0(x, \cdot)\, d\nu(x) $$

by Theorem 10-48, where $\nu(^*S) = 1$ and ν does not concentrate all its mass at one point. Extending equation (*) to $^*S$ and using the connection between the kernels and the N's, we have also

$$ J_1(x, j) = J_0(x, j) + J_1(0, j) - \frac{\alpha_j}{\alpha_1} J_0(x, 1) - \delta_{j0}. $$

Integration of this equation against ν gives

$$ J_1(y, j) = \int_{^*S} J_1(x, j)\, d\nu(x). $$

Since $J_1(y, \cdot)$ is normalized minimal and ν concentrates its weight at more than one point, this last equation contradicts the dual of Lemma 10-33.

2. Measures on the entrance boundary

Let $\delta_{0\cdot}$ and $\delta_{\cdot 0}$ denote the 0th row and column, respectively, of the identity matrix. In this section we consider finite-valued row vectors ν for the Markov chain P with the following three properties:

(1) $\nu \geq 0$.
(2) $\nu_0 = 0$.
(3) $\nu P \leq \nu + \delta_{0\cdot}$.

We shall associate to each such row vector a probability measure β^ν on the entrance boundary and show how β^ν is related to ν probabilistically.

Direct calculation shows that the row vector whose jth entry is $\bar{\nu}_j = \nu_j + \delta_{0j}$ is non-negative and Q-superregular. We introduce a process which, when $\bar{\nu} > 0$, is the $\bar{\nu}$-dual of Q. Let $S^\nu = \{i \mid \nu_i > 0\}$ and define

$$Q^\nu_{ij} = \frac{\nu_j P_{ji}}{\nu_i + \delta_{0i}} \quad \text{for } i, j \in S^\nu \cup \{0\}.$$

All other entries of Q^ν are taken to be 0. Note that

$$N^\nu_{0j} = \frac{\bar{\nu}_j N_{j0}}{\bar{\nu}_0} = \bar{\nu}_j = \nu_j + \delta_{0j}.$$

Proposition 11-2: If ν satisfies (1), (2), and (3), then there is a unique probability measure β^ν on the Borel sets of $S \cup B^e$ such that

$$\nu = \int_{S \cup B^e} {}^0N(x, \cdot) d\beta^\nu(x).$$

PROOF: By Theorem 10-48

$$\bar{\nu} = \int_{S \cup B^e} J(x, \cdot) d\beta^\nu(x)$$

for a unique measure β^ν. Since $\bar{\nu}_0 = 1$, $\bar{\nu}$ is normalized and $\beta^\nu(S \cup B^e) = 1$. Hence

$$\nu = \int_{S \cup B^e} {}^0N(x, \cdot) d\beta^\nu(x).$$

If another such probability measure is given, we can reverse the argument and conclude the measure is β^ν by the uniqueness half of Theorem 10-48.

We define θ^E_j to be the probability in the Q^ν-process started at 0 that there is a last time the process is in E and that this occurrence is at state j. Note that θ^E depends on P, 0, ν, and E and that $\theta^E_j = 0$ for j not in E.

Proposition 11-3: If E is any subset of S containing 0 and if ν satisfies (1), (2), and (3), then

$$\nu_E(I - P^E) = \theta^E_E - \delta_{0\cdot}.$$

PROOF: Using as a set of alternatives the time when the Q^ν-process is last in E, we have for, $j \in E$,

$$\begin{aligned}\theta^E_j &= \sum_{n=0}^{\infty} (Q^\nu)^{(n)}_{0j} \Pr{}^\nu_j[\text{process leaves } E \text{ immediately and never returns}] \\ &= N^\nu_{0j}\left[1 - \sum_{k\in E}\left(Q^\nu_{jk} + \sum_{c,d\in \tilde{E}} Q^\nu_{jc}\, {}^E N^\nu_{cd} Q^\nu_{dk}\right)\right].\end{aligned}$$

Now $N^\nu_{0j} = \nu_j + \delta_{0j}$, and therefore

$$N^\nu_{0j} Q^\nu_{jk} = \nu_k P_{kj}$$

and

$$N^\nu_{0j} Q^\nu_{jc}\, {}^E N^\nu_{cd} Q^\nu_{dk} = \nu_k P_{kd}\, {}^E N_{dc} P_{cj}.$$

for *all* states, not just those in $S^\nu \cup \{0\}$. Substitution and use of Lemma 6-6 give

$$\theta^E_j = \nu_j + \delta_{0j} - \sum_{k\in E} \nu_k P^E_{kj}.$$

In general, θ^E has total measure less than one, since the Q^ν process either may fail to reach E or may return to it infinitely often. If E is finite and $0 \in E$, neither of these alternatives has positive probability and thus $\theta^E 1 = 1$. This conclusion also follows from Proposition 11-3, since finite matrices associate.

The special case $E = S$ yields the following corollary. Let $\bar{\beta}^\nu$ stand for the restriction to S of the measure β^ν defined in Proposition 11-2.

Corollary 11-4: If ν satisfies (1), (2), and (3), then

$$\nu(I - P) = \bar{\beta}^\nu - \delta_{0\cdot}.$$

PROOF: Let $E = S$ in Proposition 11-3. The only way the process can leave S is to disappear, and thus θ^S_j is the probability that the Q^ν-process disappears from state j. But this is just the definition of $\beta^\nu(j)$, since β^ν was defined as harmonic measure for the Q^ν-process started at 0.

We conclude this section by noting the connection between θ^E and β^ν. The proof is contained in the proof of Proposition 10-21.

Proposition 11-5: If ν satisfies (1), (2), and (3) and if $\{E_k\}$ is an increasing sequence of finite sets of states with union S, then the measures θ^{E_k} converge to β^ν weak-star on $^*S$.

3. Harmonic measure for normal chains

We come now to the first convergence theorem. We shall prove that if P is a normal chain, then the measures λ^E converge weak-star to a measure β which will play the role of an entrance distribution for P. This result agrees with the statement in Section 1.

Lemma 11-6: If P is normal, then the row vector $\nu = G_{0\cdot}$ satisfies conditions (1), (2), and (3). Also, for any finite set E containing 0,

$$\nu_E(I - P^E) = \lambda_E^E - \delta_{0\cdot}.$$

PROOF: We know that $G_{0j} \geq 0$ and $G_{00} = 0$. Hence (1) and (2) hold. Condition (3) follows by multiplying the definition of $G_{0\cdot}$ through on the right by P and applying Fatou's Theorem.

Form the K-matrix of Definition 9-80 with respect to the distinguished state 0. By Lemma 9-81, $K_{0j} = G_{0j}$. Hence the formula for $\nu_E(I - P^E)$ is the 0th row of the formula of Corollary 9-86.

Harmonic measure β for a normal chain P is defined to be the measure $\beta = \beta^\nu$ of Proposition 11-2 for $\nu = G_{0\cdot}$. The justification for a name independent of 0 is contained in the following theorem.

Theorem 11-7: If $\{E_k\}$ is an increasing sequence of finite sets with union S, then the measures λ^{E_k} converge weak-star to β on $^*S$. The measure β is independent of the distinguished state 0. Also

$$G_{ij} = \int_{S \cup B^e} {}^iN(x, j)\,d\beta(x)$$

and

$$G(I - P) = 1\beta - I.$$

PROOF: Ultimately the sets E_k contain 0, and Proposition 11-3 and Lemma 11-6 apply. From these results we obtain $\lambda^{E_k} = \theta^{E_k}$, and from Proposition 11-5 we conclude that λ^{E_k} converges to β. Thus β is given a characterization independent of the state 0 (since λ^E does not depend on 0). Since β does not depend on 0, we can use any state i as distinguished state in Proposition 11-2 and Corollary 11-4. The two formulas of the theorem then follow.

4. Continuous and T-continuous functions

The first convergence theorem, which was proved in Section 3, resulted from examining a particular row vector ν satisfying conditions (1), (2), and (3) of Section 2, namely $\nu = G_{0\cdot}$. We now begin to work toward the second convergence theorem, and we do so by examining the vector $\nu = {}^0N(x, \cdot)$. For the present we do not assume that P is normal.

We start by checking that $\nu = {}^0N(x, \cdot)$ does satisfy the three conditions and by identifying θ^E for ν. In fact, $\nu \geq 0$ and $\nu_0 = 0$ because ${}^0N(i, \cdot) \geq 0$ and ${}^0N(i, 0) = 0$. Thus (1) and (2) hold. Moreover we have, for every $i \neq 0$,

$$ {}^0N(i, \cdot)P \leq {}^0N(i, \cdot) + \delta_0. $$

Thus we can let $i \to x$, using Fatou's Theorem, and obtain (3).

The procedure for calculating θ^E is to do so for the approximating vectors ${}^0N(i, \cdot)$ first.

Lemma 11-8: If E is a finite set containing 0 and j, then

$$ \sum_{k \in E} {}^0N_{ik}(I - P^E)_{kj} = B^E_{ij} - \delta_{0j}. $$

PROOF: Apply Lemma 9-37 and conclusion (1) of Theorem 9-15.

REMARK: The lemma remains valid for infinite sets E containing 0 and j, and the proof consists in computing the dual of the left side by means of Propositions 6-16 and 6-17 and Lemma 9-14.

We shall say that a column vector h is **continuous** if it has a (necessarily unique) extension to a continuous function $h(x)$ defined on all of $^*S$.

In this notation the right side of the identity in Lemma 11-8 is continuous for fixed $j \in E$, and hence $B^E_{\cdot j}$ is continuous if $j \in E$. But $B^E_{\cdot j}$ is identically zero for $j \notin E$ and any state can be taken as the reference state 0. Thus we are justified in writing $B^E(x, j)$ for the continuous extension of the jth column of E whenever E is a non-empty finite set. By an **elementary continuous function** is meant a finite linear combination of such functions.

Passing to the limit $i \to x$ in Lemma 11-8, we have

$$ \sum_{k \in E} {}^0N(x, k)(I - P^E)_{kj} = B^E(x, j) - \delta_{0j} $$

whenever $0 \in E$, $j \in E$, and E is finite. Consequently $\theta^E = B^E(x, \cdot)$

by Proposition 11-3, provided E is a finite set containing 0. (Note both sides are zero for states j not in E.)

Let us denote the measure β^{ν} obtained from Proposition 11-2 for $\nu = {}^0N(x, \cdot)$ by β^x. If $\{E_k\}$ is an increasing sequence of finite sets with union S, then Proposition 11-5 asserts that $B^{E_k}(x, \cdot)$ converges weak-star to β^x. We have thus proved the following proposition.

Proposition 11-9: If h is a continuous column vector and if $\{E_k\}$ is an increasing sequence of finite sets with union S, then

$$\lim_k B^{E_k}(x, \cdot)h = \int_{*S} h(y)d\beta^x(y)$$

pointwise for x in $*S$.

If $\nu = {}^0N(x, \cdot)$, then the associated Q-superregular measure is $\bar{\nu} = J(x, \cdot)$. Since β^x is the measure for $J(x, \cdot)$, β^x concentrates its mass at x if x is in S or in B^e. Thus the right side of the identity of Proposition 11-9 equals $h(x)$ for all such x.

Define a linear transformation T of continuous functions to bounded functions on $*S$ by

$$Th(x) = \int_{*S} h(y)d\beta^x(y).$$

A continuous function h such that $Th = h$ is said to be **T-continuous.** (The motivation for this name appears as Problem 2 at the end of the chapter.)

We conclude this section by characterizing the T-continuous functions. Notice that if every boundary point is extreme, then every continuous function is T-continuous.

Lemma 11-10: If h is an elementary continuous function, then $Th = h$.

PROOF: It suffices to consider the function $B^E_{\cdot j}$ where E is a finite set. If $E \subset E_k$, then $B^{E_k}B^E = B^E$ by Proposition 5-8. Passing to the limit as $i \to x$ with k fixed, we have

$$B^{E_k}(x, \cdot)B^E_{\cdot j} = B^E(x, j).$$

Hence the left side of the identity in Proposition 11-9 is $B^E(x, j)$. But the right side is $TB^E(x, j)$.

Lemma 11-11: If h is T-continuous and $\{E_k\}$ is an increasing sequence of finite sets with union S, then for any $\epsilon > 0$ some convex combination of the functions $B^{E_k}(x, \cdot)h$ is within ϵ of h uniformly for x in $*S$.

PROOF: By Proposition 11-9, the functions $B^{E_k}(x, \cdot)h$ converge pointwise to $h(x)$. Thus by dominated convergence their integrals against any Borel measure on $^*S$ converge to the integral of h. That is, the functions converge to h weakly. Thus h is in the weak closure of the set $\{B^{E_k}(x, \cdot)h\}$, is certainly in the weak closure of the convex hull of the set, and must therefore be in the strong closure of the convex hull of the set. (See Dunford and Schwartz [1958], p. 422, Corollary 14.)

Proposition 11-12: The set of T-continuous functions is exactly the uniform closure of the set of elementary continuous functions.

PROOF: Every T-continuous function is contained in the uniform closure according to Lemma 11-11, since $B^{E_k}(x, j)$ vanishes unless j is in E_k. Conversely, every elementary continuous function is T-continuous by Lemma 11-10, and the uniform limit of T-continuous functions is T-continuous, since T has norm no greater than one.

5. Normal chains and convergence to the boundary

As an application of the machinery of Section 4, we can prove the second convergence theorem—that each row of P^n converges weak-star to the harmonic measure β in a suitable class of normal chains. This result was suggested in the discussion in Section 1, and it was pointed out that the key to the proof should be a certain interchange of limits. In fact, this interchange has already taken place and is concealed in the proof of Lemma 11-11.

We begin with a particularly sharp form of the convergence theorem.

Theorem 11-13: If P is normal, if i is any state in S, and if h is T-continuous, then

$$\lim_n (P^n h)_i = \int_{^*S} h(x) d\beta(x).$$

Conversely, if this equation holds for all states i and all T-continuous h, then P is normal.

PROOF: Let $\epsilon > 0$ be given and let $\{E_k\}$ be an increasing sequence of finite sets with union S. Since h is continuous, we can choose k_0 large enough so that

$$\left| \int h(x) d\beta(x) - \lambda^{E_k} h \right| < \epsilon$$

for all $k \geq k_0$ by Theorem 11-7. Truncate the sequence of sets so that it contains only those sets $\{E_k\}$ with $k \geq k_0$. Since h is T-continuous, we can apply Lemma 11-11 to the truncated sequence to obtain a convex combination of the functions $B^{E_k}(x, \cdot)h$ which is uniformly within ϵ of h. If, say,

$$\left\|\sum c_j B^{E_j} h - h\right\| < \epsilon,$$

then also

$$\left|P^n_{i\cdot}\left(\sum c_j B^{E_j} h - h\right)\right| < \epsilon$$

since $P^n 1 = 1$ and $P^n \geq 0$. Consequently,

$$\begin{aligned}\left|\int h(x)d\beta(x) - (P^n h)_i\right| &\leq \left|\int h(x)d\beta(x) - \sum c_j \lambda^{E_j} h\right| \\ &\quad + \left|\sum c_j(\lambda^{E_j} - P^n_{i\cdot} B^{E_j})h\right| \\ &\quad + \left|P^n_{i\cdot}\left(\sum c_j B^{E_j} h - h\right)\right| \\ &< 2\epsilon + \left|\sum c_j(\lambda^{E_j} - P^n_{i\cdot} B^{E_j})h\right|.\end{aligned}$$

The sum on the right is a finite sum and in each summand only finitely many entries of $\lambda^{E_j} - P^n_{i\cdot} B^{E_j}$ can be non-zero. Since $P^n_{i\cdot} B^{E_j} \to \lambda^{E_j}$ pointwise, we conclude that the sum on the right side is less than ϵ for n sufficiently large.

The converse follows by applying the assumption to columns of B^E for two-point sets E.

The convergence theorem is as follows.

Theorem 11-14: If P is normal and if $B = B^e$, then each row of P^n converges weak-star to the harmonic measure β.

PROOF: Apply Theorem 11-13. If $B = B^e$, then every continuous function is T-continuous.

Thus for normal chains with $B = B^e$, the measure β indicates what the chain is "near" in the long run. In the case of null chains with an additional property, we can show that the chain is near the boundary in the long run.

Proposition 11-15: If P is a normal null chain with $B = B^e$ in which every one-point set in S is open, then $\beta(S) = 0$. That is, the measure is entirely on the boundary.

PROOF: If i is given, then the characteristic function of i is continuous. By Theorem 11-14, $\lim P^{(n)}_{0i} = \beta(i)$. But for a null chain, P^n tends to zero.

Under the hypotheses of Theorem 11-15, the number $\beta(X)$ for any Borel subset X of B may be interpreted as the probability that the process is near this part of the boundary in the long run. For example, if x is at a positive distance from other boundary points, then any sufficiently small neighborhood E of x will have a continuous characteristic function. By Theorem 11-14,

$$\lim_n \Pr_i[x_n \in E] = \sum_{j \in E} P_{ij}^{(n)} \to \beta(x).$$

Let us see what our results say for a noncyclic ergodic chain P. Such a chain is necessarily normal, and α may be chosen to have total measure one. If this choice is made, then $G(I - P) = 1\alpha - I$ by Corollary 9-51. Comparison with Theorem 11-7 shows that $\alpha = \bar{\beta}$. Thus the harmonic measure is concentrated entirely on S, in contrast with Proposition 11-15. The measure β is a generalization to all normal chains of the measure α for noncyclic ergodic chains. Thus our results generalize to all normal chains results known for ergodic chains. For example, the representation

$$G_{ij} = \int_{S \cup B^e} {}^iN(x, j)\, d\beta(x)$$

is a generalization of the identity $G_{ij} = \sum_k \alpha_k\, {}^iN_{kj}$, which holds for noncyclic ergodic chains. (Theorem 9-26 gives

$$G_{ij} = {}^i\nu_j = \lim \sum P_{mk}^{(n)}\, {}^iN_{kj},$$

and Proposition 1-57 yields $G_{ij} = \sum \alpha_k\, {}^iN_{kj}$.) As a second example, $(P^n h)_i$ converges to αh for any bounded function h if P is noncyclic ergodic (Lemma 9-52). This result is generalized in Theorem 11-13, but in this theorem we had to make a stronger assumption about h. The difference arises because in a noncyclic ergodic chain each row of P^n actually converges to α in the norm topology of measures, not just in the weak-star topology (see Theorem 6-38).

6. Representation Theorem

Beginning in this section, we connect the results we have obtained so far with the results of Chapter 9 on potential theory. We start by proving a representation theorem. For the moment we do not assume that P is normal.

If μ and ν are row vectors with $\mu = \nu(I - P)$, then μ will be called the **deviation** (from regularity) of ν. If $\mu 1$ is finite, then we say that ν

is of **totally finite deviation.** Dually $f = (I - P)h$ is the deviation of h, and h is of totally finite deviation if αf is finite.

Theorem 11-16: If ν is a non-negative row vector whose deviation μ is totally finite, then $\mu 1 \leq 0$ and there is a probability measure π on B^e such that

$$\nu_j = \nu_0(\alpha_j/\alpha_0) + (\mu\, {}^0N)_j - (\mu 1)\int_{B^e} {}^0N(x, j)d\pi(x).$$

If $\mu 1 \neq 0$, then the probability measure π is uniquely determined.

Proof: We know that

$$N_{ij} = {}^0N_{ij} + \delta_{0j}.$$

If we put $\bar{\mu} = \nu(I - Q)$, we have $\bar{\mu}_i = \mu_i + \nu_0 P_{0i}$ by direct calculation. We therefore get

$$\begin{aligned}(\bar{\mu}N)_j &= (\mu\, {}^0N)_j + (\mu 1)\delta_{0j} + \nu_0\, {}^0\bar{N}_{0j} \\ &= (\mu\, {}^0N)_j + (\mu 1)\delta_{0j} + \nu_0(\alpha_j/\alpha_0).\end{aligned}$$

From Proposition 8-7 applied to the chain Q, we see that $\nu = \bar{\mu}N + \rho$, where ρ is regular and non-negative. The calculation of $\bar{\mu}N$ yields

$$\rho_j = \nu_j - (\bar{\mu}N)_j = \begin{cases} -(\mu 1) & \text{if } j = 0 \\ \nu_j - \nu_0(\alpha_j/\alpha_0) - (\mu\, {}^0N)_j & \text{if } j \neq 0. \end{cases}$$

Since $\rho_0 \geq 0$, $\mu 1 \leq 0$. If $\mu 1 = 0$, then ρ is a Q-regular measure ≥ 0 with $\rho_0 = 0$. Since it is possible to get from any state to 0 eventually in the Q-process, ρ vanishes identically. The representation then follows immediately from the expression for ρ.

Thus suppose $\mu 1 \neq 0$. Define σ by

$$\sigma_j = -(\mu 1)^{-1}[\nu_j - \nu_0(\alpha_j/\alpha_0) - (\mu\, {}^0N)_j].$$

We claim that σ satisfies conditions (1), (2), and (3). It is clear that $\sigma_0 = 0$, and the fact that $\sigma \geq 0$ follows from what we have shown above. For (3), we have

$$(\sigma P)_j = -(\mu 1)^{-1}(\rho Q)_j = -(\mu 1)^{-1}\rho_j = \sigma_j + \delta_{0j},$$

and thus equality holds. Thus, except for the assertion that π is concentrated on B^e, the rest of the theorem is immediate from Proposition 11-2. So simply note from the proof of that proposition that if we have equality in (3), then π is concentrated on B^e.

We define the **exit boundary** of P to be the entrance boundary of $\hat{P}$ That is, $S^* = {}^*\hat{S}$ and $B_e = \hat{B}^e$. We obtain a dual for Theorem 11-16

Theorem 11-17: If h is a non-negative column vector whose deviation f is totally finite, then $\alpha f \leq 0$ and there is a probability measure π on B_e such that

$$h_i = h_0 + ({}^0Nf)_i - \frac{\alpha f}{\alpha_i} \int_{B_e} {}^0\hat{N}(x, i) d\pi(x).$$

If $\alpha f \neq 0$, then the probability measure π is uniquely determined.

We turn to results connected with potential theory. We first apply Theorem 11-17 to give elementary continuous functions a characterization which is valid for all recurrrent chains, normal or not.

Proposition 11-18: If h is an elementary continuous function, then

(1) $(I - P)h = f$ has finite support.
(2) $h = c\mathbf{1} + {}^0Nf$ for some c.
(3) $\alpha f = 0$.
(4) $B^E h = h$ for any set E containing the support of f.

Conversely, if (1) and (2) hold, then h is an elementary continuous function.

PROOF: $(I - P)B^E$ is equal to $I - P^E$ for states in E and is 0 otherwise. Hence $\alpha[(I - P)B^E] = 0$ and (1) and (3) hold for columns of B^E. Since an elementary continuous function is a finite linear combination of such functions for finite sets E, (1) and (3) follow. In particular, any column of B^E is of totally finite deviation and Theorem 11-17 applies. The representation in that theorem establishes (2) for columns of B^E, and the general result follows by linearity.

We shall complete the proof by showing that (2) implies (4) and that (1) and (4) imply that h is an elementary continuous function. Suppose (2) holds. Let $0 \in E$. Then ${}^0N = B^E\, {}^0N + {}^EN$ by Lemma 9-14. If E contains also the support of f, then ${}^ENf = 0$, and we see directly that $B^E h = h$. So (4) holds. If (1) and (4) hold, then (4) holds for some finite set E, and h is exhibited as a finite linear combination of the columns of B^E.

We now assume that P is normal and we shall prove statements for T-continuous functions that look like applications of Proposition 11-18 followed by a passage to the limit.

Proposition 11-19: If P is normal, then a function h of finite deviation is T-continuous if and only if $h = c\mathbf{1} + g$ for a T-continuous potential g. If the representation holds, then

$$c = \int_{*S} h(x)d\beta(x).$$

PROOF: If $(I - P)h = f$ and if h is T-continuous, then

$$(I + P + \cdots + P^n)f = (I - P^{n+1})h \rightarrow h - c\mathbf{1}$$

by Theorem 11-13. If h is of finite deviation, then f is a charge and the limit $h - c\mathbf{1}$ is a potential. This potential is T-continuous, since $\mathbf{1}$ is. (Notice that $\mathbf{1}$ is the 0th column of $B^{(0)}$.) The converse is clear.

Corollary 11-20: If P is normal and if g is a T-continuous potential, then

$$\int_{*S} g(x)d\beta(x) = 0.$$

PROOF: Potentials are of finite deviation by definition. Applying Proposition 11-19 and using the fact that $\mathbf{1}$ is not a potential (since its charge would have to be zero), we see that $c = 0$.

Corollary 11-20 is a recurrent analog of the statement for transient boundary theory that potentials tend to zero along almost all paths of the process.

Corollary 11-21: If P is normal and if h is a T-continuous function whose deviation f is totally finite, then $\alpha f = 0$ and $h = c\mathbf{1} + {}^0Nf$.

PROOF: By Proposition 11-19, h differs from the potential g of f by a constant. But $g = g_0\mathbf{1} + {}^0Nf$ by Theorem 9-15.

Our final proposition enables us to give an interpretation to θ^E for certain infinite sets provided we have an interpretation for finite sets.

Proposition 11-22: If E_k is an increasing sequence of finite sets with union E and if $\theta^E\mathbf{1} = 1$, then $\lim \theta^{E_k} = \theta^E$ pointwise.

PROOF: Since $\theta_j^{E_k}$ is an exit probability, it is decreasing in k from some point on. Hence it tends to a limit, say θ_j. Since $\theta^{E_k}\mathbf{1} = 1$, we have $\theta\mathbf{1} \leq 1$ by Fatou's Theorem. For large k, we have also $\theta_j^E \leq \theta_j^{E_k}$ and hence $\theta_j^E \leq \theta_j$. Thus $1 = \theta^E\mathbf{1} \leq \theta\mathbf{1} \leq 1$. This statement is consistent with the inequality $\theta^E \leq \theta$ only if $\theta^E = \theta$.

As an application of Proposition 11-22, suppose P is normal and we choose $\nu_j = G_{0j}$. Then $\theta^E = \lambda^E$ for finite sets containing 0, and θ^E is thus a limit of λ^E-measures for any set with $\theta^E 1 = 1$, whether or not λ^E exists for the infinite set E. The condition that $\theta^E 1 = 1$ means that the transient chain Q^ν leaves E with probability one; this condition is exactly the statement that E be an equilibrium set for Q^ν. For such a set E and for any reference state $i \in E$, we have

$$[G_E(I - P^E)]_{i\cdot} = \theta^E_E - \delta_{i\cdot}$$

by Proposition 11-3. Since, for two different reference states i in E, we have θ^E as the limit of the same sequence λ^{E_k} (by Proposition 11-22), we may write $G_E(I - P^E) = 1\theta^E_E - I$. Then

$$(\theta^E_E G_E)(I - P^E) = 0$$

and hence $\theta^E_E G_E = k\alpha$ for some constant k. If we form the K-matrix relative to state 0, then Lemma 9-81 gives

$$G_{ij} = K_{ij} + (G_{i0} - C_{i0})(\alpha_j/\alpha_0).$$

Consequently, $\theta^E_E K_E = k'\alpha$, provided

$$\sum_{i \in E} \theta_i G_{i0} < \infty \quad \text{and} \quad \sum_{i \in E} \theta_i C_{i0} < \infty.$$

The conclusion that $\theta^E_E K_E = k'\alpha$ is exactly the statement that k' be the generalized capacity of E and θ^E be the (generalized) recurrent equilibrium charge for E in the sense of Definitions 9-88 and 9-113. However, in the part of Chapter 9 where these points were discussed, we restricted ourselves to finite sets. By means of boundary theory we see we can extend these results to certain infinite sets.

Finally, let us consider the special case where the space $^*S$ for P has only one limit point ∞. Then a neighborhood of ∞ is simply the complement of a finite set not containing ∞. Hence the probability of being within this neighborhood tends to one if P is null. Therefore $P^n h$ tends to $h(\infty)$ for any such null chain and any continuous function h.

If P is null and if P and $\hat{P}$ both have only a single limit point, then P must be normal. For such a chain,

$$G_{ij} = {}^iN(\infty, j) \quad \text{and} \quad C_{ij} = (\alpha_j/\alpha_i)\,{}^j\hat{N}(\hat{\infty}, i),$$

and the representation theorem takes on the simpler form

$$h_i = h_0 + ({}^0Nf)_i - \frac{\alpha f}{\alpha_0} C_{i0}.$$

From Proposition 9-45 we know that

$${}^0N_{ij} - (\alpha_j/\alpha_0)C_{i0} = C_{0j} - C_{ij}.$$

Hence

$$h = h_0 1 + (C_{0\cdot} - C_{i\cdot})f.$$

Thus if Cf is finite-valued, then h differs from Cf by a constant.

7. Sums of independent random variables

The results of this chapter take on an especially neat-looking form where P is a recurrent sums of independent random variables process with state space the lattice in N-dimensional space. Such a chain is always null, and Spitzer [1962] showed it is always normal, has only minimal boundary points, and has no points of S as limit points. In two dimensions there is always a unique boundary point, whereas in one dimension there are one or two, depending on whether the distribution has infinite or finite variance.

In the case of a single boundary point, a continuous function h is one having a limit at infinity, and we have already noted that the convergence of $P^n h$ is trivial. In one dimension with finite variance, the two distinct boundary points correspond to $-\infty$ and $+\infty$, and h, to be continuous, must have limits in both directions. For such a function Theorem 11-14 states that $P^n h$ converges to the average of these two limits. This result also follows from the Central Limit Theorem.

There are such chains in one dimension for which $P^n h$ fails to converge for as nice a function as the characteristic function of the positive integers. However, this behavior can occur only if the variance is infinite; then there is only one boundary point, and h is not continuous.

If $h \geq 0$ is a function whose deviation f is totally finite and if Cf is finite-valued, then the representation theorem takes the form

$$h = -Cf + \text{const.}$$

for the case of one boundary point and

$$h_i = -(Cf)_i + ai + b$$

for the two-boundary-point one-dimensional case. We have already seen that the former identity holds for any chain with a single limit point. The latter follows from the representation theorem together with special knowledge of the nature of C obtained by Spitzer for sums of independent random variables.

8. Examples

EXAMPLE 1: Sums of independent random variables, $p_{-1} = \frac{2}{3}$, $p_2 = \frac{1}{3}$.

This example shows the best possible boundary behavior for a null chain and illustrates the points discussed in Section 7. To determine the recurrent entrance boundary we can work either with the kernel $J(i, j)$ associated to the chain Q or with the matrix ${}^0N_{ij}$ associated to P, since

$$J(i, j) = {}^0N_{ij} + \delta_{j0}.$$

We use the latter.

To compute ${}^0N$, we can proceed in the familiar way—first finding ${}^0H$ and then computing ${}^0N$ from the identities ${}^0\bar{H} = P\,{}^0H$, ${}^0N_{jj} = 1/(1 - {}^0\bar{H}_{jj})$, and ${}^0N_{ij} = {}^0H_{ij}\,{}^0N_{jj}$. As usual, the calculation of ${}^0H$ involves a difference equation valid for certain intervals of i's and j's. In this case, the equation is

$${}^0H_{ij} = \tfrac{2}{3}\,{}^0H_{i-1,j} + \tfrac{1}{3}\,{}^0H_{i+2,j}.$$

The general solution as a function of i is

$${}^0H_{ij} = A + Bi + C(-2)^i$$

and is valid always for a slightly larger interval of i's. For instance, if $0 < j$ and if we are considering i's satisfying $0 \le i \le j$, then the difference equation is valid for $0 < i < j$ and the solution applies for $0 \le i \le j + 1$. The initial conditions in this instance are ${}^0H_{0j} = 0$, ${}^0H_{jj} = 1$ and ${}^0H_{j+1,j} = 1$, and they determine A, B, and C. Similar remarks apply for the other intervals of i's and j's, and the result is:

for $0 < j$

$${}^0N_{ij} = \begin{cases} \tfrac{1}{3}[1 - (-2)^{-j}][1 - (-2)^i] & i \le 0 \\ i - \tfrac{1}{3}(-2)^{-j}[1 - (-2)^i] & 0 \le i \le j \\ j + \tfrac{1}{3}[1 - (-2)^{-j}] & j \le i, \end{cases}$$

for $j < 0$

$${}^0N_{ij} = \begin{cases} \tfrac{1}{3}(-2)^i[(-2)^{-j} - 1] - j & i \le j \\ \tfrac{1}{3}[1 - (-2)^i] - i & j \le i \le 0 \\ 0 & 0 \le i. \end{cases}$$

Therefore there are two boundary points, $+\infty$ and $-\infty$, no point of S is a limit point, and

$${}^0N(+\infty, j) = \begin{cases} j + \tfrac{1}{3}[1 - (-2)^{-j}] & j \ge 0 \\ 0 & j \le 0 \end{cases}$$

$${}^0N(-\infty, j) = \begin{cases} \tfrac{1}{3}[1 - (-2)^{-j}] & j \ge 0 \\ |j| & j \le 0. \end{cases}$$

From the Central Limit Theorem we know that for large n the process is very likely to be far from 0 and equally likely to be on the right or on the left. Hence $\beta(+\infty) = \beta(-\infty) = \frac{1}{2}$. Since β assigns positive mass to each boundary point, both boundary points are extreme and consequently every continuous function is T-continuous. From Theorem 11-7 we have

$$G_{0j} = \tfrac{1}{2}\,{}^0N(+\infty, j) + \tfrac{1}{2}\,{}^0N(-\infty, j),$$

and for sums of independent random variables we have also

$$C_{ij} = G_{ij} = G_{0,j-i}.$$

Therefore

$$C_{ij} = G_{ij} = \begin{cases} \frac{1}{2}|j-i| + \frac{1}{3}[1 - (-2)^{i-j}] & j \geq i \\ \frac{1}{2}|j-i| & j \leq i. \end{cases}$$

Let us see what the representation theorem says for this process. We first need to know ${}^0\hat{N}(x, i)$. We can take $\alpha = \mathbf{1}^T$, and we see that $\hat{P}_{ij} = P_{ji}$ and ${}^0\hat{N}_{ij} = {}^0N_{ji}$. For $\hat{P}$ we again have two boundary points, and

$${}^0\hat{N}(+\infty, i) = \begin{cases} \frac{1}{3}[1 - (-2)^i] & i \leq 0 \\ |i| & 0 \leq i, \end{cases}$$

$${}^0\hat{N}(-\infty, i) = \begin{cases} |i| + \frac{1}{3}[1 - (-2)^i] & i \leq 0 \\ 0 & 0 \leq i. \end{cases}$$

Now let $h \geq 0$ be a function whose deviation f is totally finite, and suppose Cf is finite-valued. The representation theorem gives

$$h_i = h_0 + ({}^0Nf)_i - (\alpha f)[\pi(+\infty)\,{}^0\hat{N}(+\infty, i) + \pi(-\infty)\,{}^0\hat{N}(-\infty, i)].$$

If we use the identity

$${}^0N_{ij} - (\alpha_j/\alpha_0)C_{i0} = C_{0j} - C_{ij},$$

we obtain

$$h_i = -(Cf)_i + (\alpha f)[\pi(-\infty) - \tfrac{1}{2}]i + (h_0 + (Cf)_0).$$

This last equation is an example of the formula

$$h_i = -(Cf)_i + ai + b$$

discussed in Section 7.

EXAMPLE 2: Basic example, null case.
For the basic example we have

$${}^0N_{ij} = \begin{cases} {}^i\bar{N}_{ij} = \alpha_j/\alpha_i & \text{if } j \geq i > 0 \\ 0 & \text{otherwise.} \end{cases}$$

Hence, for fixed j, ${}^0N_{ij} = 0$ if i is sufficiently large. Consequently there is a single limit point p in $*S$, and ${}^0N(p, i) = 0$ for all i. Since ${}^0N(p, i) = {}^0N(0, i)$, we have $p = 0$; the limit point is the state 0. Thus the boundary is empty.

We know from Section 9-6 that in a null basic example $\lambda_0^E = 1$ if E is a finite set containing zero. Therefore the harmonic measure β, which is the weak-star limit of such measures, assigns unit mass to state 0. Thus by Theorem 11-7, $G_{0j} = {}^0N(0, 0) = 0$, in agreement with the result obtained in Chapter 9. This example shows that the condition on one-point sets in Proposition 11-15 cannot be omitted.

We can also check directly, using the results of Section 9-6, that $G_{ij} = {}^iN(0, j)$ and $G(I - P) = 1\delta_{0\cdot} - I$, again in agreement with Theorem 11-7.

The reverse $\hat{P}$ of the basic example has

$$ {}^0\hat{N}_{ij} = \begin{cases} 1 & \text{if } i \geq j > 0 \\ 0 & \text{otherwise.} \end{cases} $$

Again there is one limit point p', but this time we have ${}^0\hat{N}(p', j) = 1 - \delta_{0j}$. The measure $\hat{J}(p', \cdot)$ associated with the transient chain

$$ \hat{Q}_{ij} = \begin{cases} \hat{P}_{ij} & \text{if } i \neq 0 \\ 0 & \text{if } i = 0 \end{cases} $$

satisfies $\hat{J}(p', j) = {}^0\hat{N}(p', j) + \delta_{j0} = 1$ and is easily seen to be $\hat{Q}$-regular, and it follows that p' is a boundary point and is extreme. Put $+\infty = p'$.

It is clear that $\lambda_m^E = 1$ if m is the largest element of the finite set E and if $\hat{P}$ is null. Hence $\beta(+\infty) = 1$, in agreement with Proposition 11-15. Therefore $\hat{G}_{ij} = {}^i\hat{N}(+\infty, j)$, and we find that $\hat{G}(I - \hat{P}) = -I$.

Example 3: Three-line example.

This example is designed to show that P and $\hat{P}$ can have different limiting behavior and to show that in a normal null chain some additional assumption is needed (such as $B = B_e$ in Theorem 11-14) to ensure that each row of P^n converges weak-star to a limiting measure.

Let S consist of three copies of the non-negative integers with typical elements denoted by i, i', and i'', respectively. The process P moves deterministically to the left on the first and third lines and moves from 0 or 0″ to 0′ also with probability one. On the middle line it moves one step to the right or moves to one of the other lines, as shown in the accompanying figure. The quantities on the arrows are transition probabilities, and p and q are positive numbers with sum one. The

p_i and q_i are chosen as in the basic example, except that $q_1 = 0$, β_i is defined as the product of the p_j's up through p_i as in the basic example, and we require that $\beta_i \to 0$ and $\sum \beta_i = +\infty$.

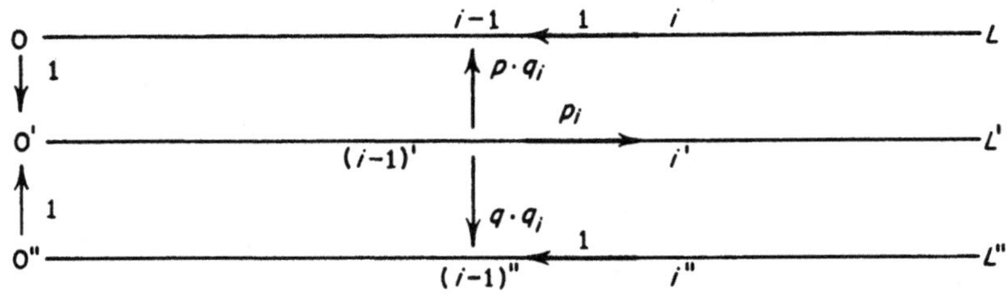

Clearly $\bar{H}_{0'0'} = 1 - \lim \beta_i = 1$, and thus P is recurrent. If α is defined by

$$\alpha_i = p\beta_i, \qquad \alpha_{i'} = \beta_i, \qquad \alpha_{i''} = q\beta_i,$$

then α is a regular measure. Since $\alpha 1 = 2 \sum \beta_i = +\infty$, the chain is null.

To see that P is normal, we compute ${}^{0'}\lambda$ and ${}^{0'}\hat{\lambda}$. Since P is null, the process after a long time is likely to be far to the right of any state we consider. Thus

$${}^{0'}\lambda_a = \begin{cases} p & \text{if } a \in L \\ q & \text{if } a \in L'' \\ 0 & \text{if } a \in L' \text{ and } a \neq 0'. \end{cases}$$

The reverse process $\hat{P}$ moves a step to the right on L (or L'') or switches to L'. On L' it moves deterministically to the left until it reaches $0'$, and then it goes to 0 with probability p and $0''$ with probability q. Thus

$${}^{0'}\hat{\lambda}_a = \begin{cases} 0 & \text{if } a \in L \text{ or } a \in L'' \\ 1 & \text{if } a \in L' \text{ and } a \neq 0'. \end{cases}$$

By Theorem 9-26, P is normal.

We now determine the boundaries of P and $\hat{P}$. If we choose $0'$ as the distinguished state, we see that if a and b are any two states different from $0'$, not necessarily on the same line, then

$${}^{0'}N_{ab} = \begin{cases} 1 & \text{if } a, b \in L \text{ or } a, b \in L'', a \text{ to the right of } b \\ p & \text{if } a \in L', b \in L, a \text{ to the right of } b \\ q & \text{if } a \in L', b \in L'', a \text{ to the right of } b \\ \alpha_b/\alpha_a & \text{if } a \in L', a \text{ to the left of } b \\ 0 & \text{otherwise.} \end{cases}$$

Thus for each b, ${}^{0'}N_{ab}$ tends to a limit along each line, but, as functions of b, the limits are different for each line. The result is three boundary points ∞, ∞', and ∞'', and we have

$$ {}^{0'}N(\infty, b) = \begin{cases} 1 & \text{if } b \in L \\ 0 & \text{otherwise} \end{cases} $$

$$ {}^{0'}N(\infty', b) = \begin{cases} p & \text{if } b \in L \\ q & \text{if } b \in L'' \\ 0 & \text{otherwise} \end{cases} $$

$$ {}^{0'}N(\infty'', b) = \begin{cases} 1 & \text{if } b \in L'' \\ 0 & \text{otherwise.} \end{cases} $$

The measures $J(x, \cdot)$ defined for the associated transient chain satisfy

$$ J(x, b) = {}^{0'}N(x, b) + \delta_{b0'}, $$

and direct calculation shows that

$$ J(\infty', \cdot) = pJ(\infty, \cdot) + qJ(\infty'', \cdot). $$

We conclude that ∞ and ∞'' are extreme boundary points, whereas ∞' is not.

If E is a large finite set of states containing 0 and $0''$, then the chain P is most likely to be to the right of E. Hence $\lambda_k^E = p$ and $\lambda_{m''}^E = q$, where k and m'' are the last elements of E on the first and third lines, respectively. From the form of λ^E, we deduce that

$$ \beta(\infty) = p \quad \text{and} \quad \beta(\infty'') = q. $$

We may interpret this result as follows: L, L', and L'' are neighborhoods of ∞, ∞', and ∞'', respectively. In the long run, the chain is typically far to the right in one of these sets. If it is in L or L'', it must remain there for a long time; but if it is in L', it can leave in one step by switching to L or L''. This behavior is what makes ∞' nonminimal. In other words, far out in L is near ∞, far out in L'' is near ∞'', but far out in L' means near ∞ with probability p and near ∞'' with probability q.

Now let us consider the boundary of $\hat{P}$. We are to look at the limiting behavior of ${}^{0'}\hat{N}_{ab} = \alpha_b\, {}^{0'}N_{ba}/\alpha_a$ along sequences of a's. But for fixed b, this quantity tends to the same limit along all three lines. We thus have just one limit point $\hat{\infty}$, and the corresponding measure is

$$ \hat{J}(\hat{\infty}, b) = {}^{0'}\hat{N}(\hat{\infty}, b) + \delta_{b0'} = \begin{cases} 1 & \text{if } b \in L' \\ 0 & \text{otherwise.} \end{cases} $$

This measure is regular for $\hat{Q}$, and consequently $\hat{\infty}$ is an extreme boundary point. By Proposition 11-15, $\hat{\beta}(\hat{\infty}) = 1$. For $\hat{P}$ the chain either is far out in L' (and hence has to stay there for a long time) or else is in a set from which it can move in one step at any moment to a position far out in L'. The three boundary points for P collapse into one for $\hat{P}$ because a position far out on L or L'' is only one step away from being far out on L'.

We conclude by sketching a proof that the 0'th row of P^n does not converge weak-star as n tends to infinity if the p_i's are chosen appropriately.

Thus some condition such as $B = B_e$ is needed in Theorem 11-14, and some condition such as T-continuity is needed in Theorem 11-13. Let h be the characteristic function of L'. Then h is continuous, since $h(\infty) = h(\infty'') = 0$ and $h(\infty') = 1$; but $(Th)(\infty') = 0$ and thus h is not T-continuous. Let

$$a_n = (P^n h)_{0'} = \Pr_{0'}[x_n \in L'].$$

We shall show that $\{a_n\}$ does not converge if the p_i's are chosen suitably, and hence $P^n_{0'}$ does not converge weak-star. In fact, we shall indicate that $\{a_n\}$ can fail to be even Abel summable. We define

$$A(t) = \sum a_n t^n, \qquad P(t) = \sum P^{(n)}_{0'0'} t^n$$

$$B(t) = \sum \beta_n t^n, \qquad F(t) = \sum \bar{F}^{(n)}_{0'0'} t^n.$$

If we let k be the last time before n that the process is in $0'$, we see that

$$a_n = \sum_{k=0}^{n} P^{(k)}_{0'0'} \beta_{n-k}$$

and hence

$$A(t) = P(t)B(t).$$

For any chain, we have

$$P(t) = \frac{1}{1 - F(t)}.$$

For this chain,

$$\bar{F}^{(2n+2)}_{0'0'} = \beta_n q_{n+1} = \beta_n - \beta_{n+1},$$

whereas $\bar{F}^{(m)}_{0'0'} = 0$ if m is not of the form $2n + 2$. Thus

$$F(t) = 1 - (1 - t^2)B(t^2).$$

Combining our results, we have

$$A(t) = \frac{B(t)}{(1 - t^2)B(t^2)}.$$

The Abel limit of the sequence a_n is

$$\lim_{t\uparrow 1} (1 - t)A(t) = \frac{1}{2} \lim_{t\uparrow 1} \frac{B(t)}{B(t^2)}.$$

Now the sequence $\{\beta_n\}$ which defines $B(t)$ is an arbitrary decreasing sequence of positive numbers subject only to the conditions that $\lim \beta_n = 0$, $\sum \beta_n = +\infty$, and $\beta_0 = \beta_1 = 1$. Such a sequence can be found so that the expression $B(t)/B(t^2)$ oscillates as $t \uparrow 1$.

9. Problems

1. For sums of independent random variables on the integers with $p_{-1} = \frac{2}{3}$ and $p_2 = \frac{1}{3}$, show that if h is continuous and is of finite deviation, then $\alpha f = 0$.
2. Prove that if h and Th are both continuous, then h is T-continuous.
3. Show that for any normal chain $(I - P)C = b\alpha - I$. Identify the vector b.
4. What identities hold for $(I - P)K$ and $K(I - P)$ in a normal chain?
5. Let P be normal and let h be continuous. The balayage potential of h on a small ergodic set E is $B^E h - (\lambda^E h)\mathbf{1}$ (see Proposition 9-43). Let $\{E_k\}$ be an increasing sequence of finite sets with union S. Prove that the balayage potentials of h on E_k converge to $h - (\int h d\beta)\mathbf{1}$ on S.
6. Show that if P is a normal null chain with $B = B_e$ and if x is a point of the boundary with a neighborhood in S^* containing no other limit points, then $\lim_n \Pr[x_n \in E]$ exists for all sufficiently small neighborhoods E and is the same for all such sets E.
7. Prove that in a normal chain the elementary continuous functions are exactly the functions that can be written as the sum of a constant and a potential of finite support. [*Hint:* Use Theorems 9-15 and 11-18.]
8. Prove that if P is a normal null chain and if $P_{\cdot j}$ has only finitely many non-zero entries, then $\bar{\beta}_j = 0$. [*Hint:* Consider columns of $I - P$ as charges.]
9. Show that every T-continuous potential in a normal chain is the uniform limit of potentials of finite support.

Problems 10 to 15 refer to the null chain of Chapter 9, Problems 11 to 22. We shall use the notation and results there developed.

10. Show that the recurrent entrance boundary is empty and that $\beta_0 = 1$.
11. Show that $\hat{B}$ is the unit interval when parametrized by $t = \lim (x/n)$ and that

$$\hat{J}(t, z) = \binom{n-2}{x-1} t^{x-1}(1 - t)^{y-1}.$$

12. What is the form of the most general non-negative function regular for P except at 0?

13. In Problem 12, let $h_0 = 0$ and use Lebesgue measure on $\hat{B}$. Verify that h is regular except at 0.
14. Show that $\hat{\beta}$ is Lebesgue measure on $\hat{B}$.
15. Verify the interpretations of β and $\hat{\beta}$ as limits of rows of P^m and $\hat{P}^m$ by showing this: For either chain n is most probably large in the long run. For $\hat{P}$, the ratio x/n cannot change much in a few steps (if n is large), but for P a transition to 0 is always possible.

CHAPTER 12

INTRODUCTION TO RANDOM FIELDS

David Griffeath

1. Markov fields

One means of generalizing denumerable stochastic processes $\{x_n\}$ with time parameter set $\mathbb{N} = \{0, 1, \ldots\}$ is to consider **random fields** $\{x_t\}$, where t takes on values in an arbitrary countable parameter set T. Roughly, a random field with denumerable state space S is described by a probability measure μ on the space $\Omega = S^T$ of all configurations of values from S on the generalized time set T. In this chapter we discuss certain extensions of Markov chains, called **Markov fields**, which have been important objects of study in the recent development of probability theory. Only some of the highlights of this rich theory will be covered; we concentrate especially on the case $T = \mathbb{Z}$ = the integers, where the connections with classical Markov chain theory are deepest.

Proceeding to the formal definitions, assume as usual that the **state space** S is a countable set of integers including 0, but let the **time parameter set** T be any countable set. The **configuration space** $\Omega = S^T$ is the space of all functions ω from T to S. An element $\omega = \{\omega_t; t \in T\}$ of Ω is called a **configuration**, and is to be thought of as an assignment of values from S to the **sites** t of T. The **outcome function** x_t from Ω to S takes the configuration ω to its value ω_t at site t. Let $\mathscr{B}$ be the minimal complete σ-algebra with respect to which all the outcome functions x_t, $t \in T$, are measurable. In this context, we introduce the following definition.

Definition 12-1: A **random field** is given by $(\Omega, \mathscr{B}, \mu, \{x_t\})$, where μ is a probability measure on $(\Omega, \mathscr{B})$ such that

$$\Pr[x_t = i_t; t \in A] > 0$$

for all finite (non-empty) $A \subset T$ and arbitrary $i_t \in S$. (As always, $\Pr[p] = \mu(\{\omega | p\})$.)

As in the case of stochastic processes, we often identify a random field with its outcome functions $\{x_t\}$ (the remaining structure being understood). At other times it will be more convenient to think of μ as the random field. Our positivity assumption on cylinder probabilities ensures that all conditional probability statements are well-defined. The role of transition probabilities at this new level of generality is played by **characteristics**, which we define next.

Definition 12-2: Given a random field $\{x_t\}$, and finite (non-empty) sets A and Λ such that $A \subset \Lambda$, the (A, Λ)-**characteristic** is the real-valued function on Ω given by

$$\mu_A^\Lambda(\iota) = \Pr[x_t = i_t \text{ for all } t \in A \mid x_t = i_t \text{ for all } t \in \Lambda - A]$$

when evaluated at the configuration $\iota = \{i_t; t \in T\}$. For $a \in T$, we abbreviate $\mu_a^\Lambda = \mu_{\{a\}}^\Lambda$; the collection $\{\mu_a^\Lambda; a \in \Lambda \subset T\}$ is called the **local characteristics** of the random field.

Throughout this chapter A and Λ will *always* be *finite* subsets of T, even when not explicitly identified as such.

Our immediate objective is to formulate the notion of a **Markov field.** As motivation, we return briefly to the setting of Chapter 4.

Definition 12-3: A denumerable stochastic process $\{x_n\}$ satisfies the **two-sided Markov property** if

$$\Pr[x_n = i_n \mid x_k = i_k; k \in \{m, m+1, \ldots, M\} - \{n\}]$$
$$= \begin{cases} \Pr[x_0 = i_0 \mid x_1 = i_1] & \text{if } 0 = m = n < M \\ \Pr[x_n = i_n \mid x_{n-1} = i_{n-1} \wedge x_{n+1} = i_{n+1}] & \\ & \text{if } 0 \le m < n < M \end{cases}$$

whenever $\Pr[x_k = i_k; m \le k \le M] > 0$.

Proposition 12-4: Any Markov chain $\{x_n\}$ satisfies the two-sided Markov property.

PROOF: Let π be the starting distribution, P the transition matrix for $\{x_n\}$. When $\Pr[x_k = i_k; m \le k \le M] > 0$, we consider the quantity $\Pr[x_n = i_n \mid x_k = i_k; k \in \{m, m+1, \ldots, M\} - \{n\}]$. If $0 = m = n < M$ this is simply $\Pr[x_0 = i_0 \mid x_1 = i_1]$ by reversibility of the Markov property (Proposition 6-44). Otherwise the above conditional probability may be evaluated as

$$\frac{(\pi P^m)_{i_m} \prod_{k=m}^{M-1} P_{i_k i_{k+1}}}{(\pi P^m)_{i_m} \left(\prod_{k=m}^{n-2} P_{i_k i_{k+1}}\right) P^{(2)}_{i_{n-1} i_{n+1}} \left(\prod_{k=n+1}^{M-1} P_{i_k i_{k+1}}\right)}$$
$$= P_{i_{n-1} i_n} P_{i_n i_{n+1}} / P^{(2)}_{i_{n-1} i_{n+1}}$$
$$= (\pi P^{n-1})_{i_{n-1}} P_{i_{n-1} i_n} P_{i_n i_{n+1}} / (\pi P^{n-1})_{i_{n-1}} P^{(2)}_{i_{n-1} i_{n+1}}$$
$$= \Pr[x_n = i_n \mid x_{n-1} = i_{n-1} \wedge x_{n+1} = i_{n+1}].$$

The two-sided Markov property, unlike the ordinary Markov property, generalizes to any parameter set T which has a **neighbor system**, i.e., a collection $\partial = \{\partial a; a \in T\}$ of *finite* subsets of T such that (i) $a \notin \partial a$, and (ii) $a \in \partial b$ if and only if $b \in \partial a$, $a, b \in T$. The sites $b \in \partial a$ are called the **neighbors** of a. We write $\bar{a} = \{a\} \cup \partial a$. Also, for $A \subset V$ let

$$\partial A = \{b \in T - A : b \in \partial a \quad \text{for some } a \in A\}; \quad \bar{A} = A \cup \partial A.$$

Definition 12-5: Let T have neighbor system ∂. The random field $\{x_t; t \in T\}$ is a **Markov field** (with respect to ∂) if

$$\mu_a^A = \mu_a^{\bar{a}} \quad \text{whenever } \bar{a} \subset A \subset T,\ A \text{ finite}.$$

We shall usually assume an underlying neighbor system for T, and simply refer to the Markov field $\{x_t\}$. Note that any Markov chain with strictly positive cylinder probabilities is a random field, where $T = \mathbb{N}$. The natural neighbor structure on $\mathbb{N}$ is $\partial 0 = 1$, and for $n \geq 1$, $\partial n = \{n - 1, n + 1\}$. In this case the Markov random field condition is precisely the two-sided Markov property. Proposition 12-4 shows that any Markov chain with positive cylinders may be considered as a Markov field on $\Omega = S^{\mathbb{N}}$. Later we will see that the classes of Markov processes with positive cylinders and Markov fields on $S^{\mathbb{N}}$ actually coincide.

A random field is called **finite** when T is a finite set. Such fields have an elementary theory, which will be developed in the next two sections. First, though, we note that the Markov field property simplifies somewhat when T is finite.

Proposition 12-6: Let $\{x_t\}$ be a finite random field. Then the following three conditions are equivalent:

(1) $\{x_t\}$ is a Markov field.
(2) $\mu_a^T = \mu_a^{\bar{a}}$ for all $a \in T$.
(3) $\mu_a^T(\iota) = \mu_a^T(\iota')$ whenever $a \in T$ and $i_t = i'_t$ for all $t \in \bar{a}$.

Proof: When T is finite, (2) is simply the Markov field property with $\Lambda = T$, while the fact that $\mu_a^{\bar{a}}$ depends only on $t \in \bar{a}$ together with (2) implies $\mu_a^T(\iota) = \mu_a^{\bar{a}}(\iota) = \mu_a^{\bar{a}}(\iota') = \mu_a^T(\iota')$ whenever ι and ι' agree on $\bar{a}$. To see that (3) implies (2), fix $a \in T$ and let $\kappa = \{k_t; t \in T - \bar{a}\}$ be any prescription of values from S on $T - \bar{a}$. Denote by p, q, and r the statements

$$x_a = i_a, \quad x_t = i_t \text{ for all } t \in \partial a, \quad x_t = k_t \text{ for all } t \in T - \bar{a},$$

respectively. Then (3) asserts that

$$\Pr[p \mid q \wedge r_\kappa] = c \quad \text{for all } \kappa,$$

or equivalently,

$$\Pr[p \wedge q \wedge r_\kappa] = c \Pr[q \wedge r_\kappa].$$

Summing over all possible κ, we obtain $\Pr[p|q] = c$. Thus $\Pr[p \mid q \wedge r_\kappa] = \Pr[p|q]$, which is precisely (2) when $k_t = i_t$ for all $t \in T - \bar{a}$. Finally, to show that (2) implies (1) we choose $\bar{a} \subset \Lambda \subset T$ and $\iota = \{i_t\} \in \Omega$. Since $\mu_a^{\bar{a}}$ depends only on sites in $\bar{a}$, (2) yields

$$\begin{aligned}\Pr[x_t = i_t \quad &\text{for all } t \in \Lambda \wedge x_t = k_t \quad \text{for all } t \in T - \Lambda] \\ &= \Pr[x_t = i_t \quad \text{for all } t \in \Lambda - \{a\} \wedge x_t = k_t \quad \text{for all } t \in T - \Lambda]\, \mu_a^{\bar{a}}(\iota),\end{aligned}$$

where $\kappa = \{k_t; t \in T - \Lambda\}$ is any prescription of values from S on $T - \Lambda$. Summing over all possible κ, we conclude that $\mu_a^{\Lambda}(\iota) = \mu_a^{\bar{a}}(\iota)$.

2. Finite Gibbs fields

In this section we introduce an extremely useful representation for the measure μ of an arbitrary *finite* random field. The inspiration behind this approach (and hence most of its terminology) is derived from statistical mechanics, where random fields may be considered as equilibrium distributions for a variety of physical systems.

Definition 12-7: A **potential** U on a finite set T is a family $\{U_A(\iota); A \subset T\}$ of functions from Ω to the real line $\mathbb{R}$ with the property that $U_A(\iota) = U_A(\iota')$ whenever $i_t = i_t'$ for all $t \in A$, and such that $U_\varnothing = 0$. The **energy** H_U of the potential U is given by

$$H_U = \sum_{A \subset T} U_A.$$

U is said to be **normalized** if $U_A(\iota) = 0$ whenever $i_a = 0$ for some $a \in A$. When T has a neighbor system ∂, a set $C \subset T$ is called a **clique** if $b \in \partial a$ whenever $a, b \in C$, $a \neq b$, i.e. if every two distinct sites in C are neighbors. Let $\mathscr{C}$ be the class of all cliques in T. U is called a **neighbor potential** if $U_A = 0$ whenever $A \notin \mathscr{C}$.

Definition 12-8: A finite random field $\{x_t\}$ is a **Gibbs field with potential** U if

$$\mu(\{\iota\}) = z^{-1}e^{H_U(\iota)} \quad \iota \in \Omega,$$

where $z = \sum_{\iota\in\Omega} \exp\{H_U(\iota)\}$ is often called the **partition function.** Implicit is the assumption $z \neq +\infty$, which imposes a condition on U. If U is a neighbor potential, then $\{x_t\}$ is called a **neighbor Gibbs field**, and $H_U = \sum_{C\in\mathscr{C}} U_C$.

We remark that the potential and energy of random field theory should not be confused with those of Markov chain theory presented in earlier chapters. These terms have common origins in classical physics.

Example 12-9: T is sometimes called a **cubic lattice** if no clique in the neighbor system for T has more than two elements. The most important examples are subsets of the d-dimensional integer lattices $\mathbb{Z}^d$, where $\partial a = \{b \in T\colon |a - b| = 1\}$. When T is a finite cubic lattice and U is a neighbor potential, the energy function becomes

$$H_U = \sum_{a\in T} U_{\{a\}} + \sum_{\{a,b\}\in\mathscr{N}} U_{\{a,b\}} \quad \text{where } \mathscr{N} = \{\{a, b\}\colon b \in \partial a\}.$$

In this case U is called a **neighbor pair potential.**

Two lemmas prepare the way for the representation theorem for finite random fields. Given $\iota = \{i_t\}$ and $A \subset T$, the modification $\iota^A = \{i_t^A\}$ of ι has values

$$i_t^A = \begin{cases} i_t & \text{for } t \in A \\ 0 & \text{otherwise.} \end{cases}$$

We abbreviate $\iota^{A+a} = \iota^{A\cup\{a\}}$ when $a \notin A$ and $\iota^{A-a} = \iota^{A-\{a\}}$ when $a \in A$.

Lemma 12-10: If $\{x_t\}$ is a finite random field, $\iota = \{i_t\} \in \Omega$, $A \subset T$ and $a \notin A$, then

$$\frac{\mu_a^T(\iota^A)}{\mu_a^T(\iota^{A+a})} = \frac{\mu(\{\iota^A\})}{\mu(\{\iota^{A+a}\})}.$$

PROOF: $\mu_a^T(\iota^A) = \Pr[x_t = i_t^A \text{ for all } t \in T]/\Pr[x_t = i_t^A \text{ for all } t \in T - \{a\}]$. When we replace ι^A by ι^{A+a} the denominator is unchanged, since $i_t^A = i_t^{A+a}$ for $t \in T - \{a\}$.

Lemma 12-11 (Möbius inversion formula): Let Λ be finite, and let Φ and Ψ be real-valued set functions defined on all subsets of Λ. Then

$$\Phi(A) = \sum_{B\subset A} (-1)^{|A-B|}\Psi(B) \quad \text{for all } A \subset \Lambda$$

if and only if

$$\Psi(A) = \sum_{B \subset A} \Phi(B) \qquad \text{for all } A \subset \Lambda.$$

(Here $|A|$ denotes the cardinality of the set A.)

Proof: Assume that the first condition holds. Then

$$\begin{aligned}\sum_{B\subset A}\Phi(B) &= \sum_{B\subset A}\sum_{D\subset B}(-1)^{|B-D|}\Psi(D)\\ &= \sum_{D\subset A}\Psi(D)\left[\sum_{E\subset A-D}(-1)^{|E|}\right] \qquad (E = B - D)\\ &= \Psi(A),\end{aligned}$$

since the bracketed sum above is 1 if $D = A$ and 0 otherwise. The opposite implication is verified by an analogous computation.

Theorem 12-12: Let $\{x_t\}$ be a finite random field with local characteristics $\{\mu_a^A\}$. Then $\{x_t\}$ is a Gibbs field with **canonical potential** V defined by $V_\varnothing = 0$, and for $A \neq \varnothing$,

$$\begin{aligned}V_A(\iota) &= \sum_{B\subset A}(-1)^{|A-B|}\ln\mu(\{\iota^B\}),\\ &= \sum_{B\subset A}(-1)^{|A-B|}\ln\mu_a^T(\{\iota^B\})\end{aligned}$$

for any fixed $a \in A$. Moreover, V is the unique normalized potential for $\{x_t\}$.

Proof: Let $\mathbf{0}$ denote the configuration with a 0 at every site of T. Fix $\iota \in \Omega$. For $A \subset T$, define $\Psi(A) = \ln[\mu(\{\iota^A\})/\mu(\{\mathbf{0}\})]$ and $\Phi(A) = V_A(\iota)$, where V is the potential given by the first sum in the Theorem. When $A \neq \varnothing$ we have $\sum_{B\subset A}(-1)^{|A-B|} = 0$, and hence

$$\begin{aligned}\Phi(A) = V_A(\iota) &= \left[\sum_{B\subset A}(-1)^{|A-B|}\ln\mu(\{\iota^B\})\right] - \ln\mu(\{\mathbf{0}\})\left[\sum_{B\subset A}(-1)^{|A-B|}\right]\\ &= \sum_{B\subset A}(-1)^{|A-B|}\Psi(B).\end{aligned}$$

When $A = \varnothing$, $\Phi(\varnothing) = V_\varnothing(\iota) = 0 = \ln[\mu(\{\iota^\varnothing\})/\mu(\{\mathbf{0}\})] = \Psi(\varnothing)$, since $\iota^\varnothing = \mathbf{0}$ for any ι. Applying Lemma 12-11 with $\Lambda = T$ we conclude that

$$\ln\frac{\mu(\{\iota\})}{\mu(\{\mathbf{0}\})} = \ln\frac{\mu(\{\iota^T\})}{\mu(\{\mathbf{0}\})} = \sum_{B\subset T} V_B(\iota) = H_V(\iota),$$

and hence

$$\mu(\{\iota\}) = \mu(\{\mathbf{0}\})e^{H_V(\iota)} = z^{-1}e^{H_V(\iota)}.$$

Thus $\{x_t\}$ is a Gibbs field with potential V. For any $a \in A \subset T$ we can write

$$V_A(\iota) = \sum_{B \subset A - a} (-1)^{|A-B|}[\ln \mu(\{\iota^B\}) - \ln \mu(\{\iota^{B+a}\})].$$

This shows that V is normalized, since $\iota^B = \iota^{B+a}$ whenever $i_a = 0$. By Lemma 12-10 the right hand side of the last equation may be rewritten as

$$\sum_{B \subset A - a} (-1)^{|A-B|}[\ln \mu_a^T(\iota^B) - \ln \mu_a^T(\iota^{B+a})]$$
$$= \sum_{B \subset A} (-1)^{|A-B|} \ln \mu_a^T(\iota^B),$$

which establishes the second expression for V in the statement of the Theorem. Finally, suppose that U is any normalized potential for $\{x_t\}$. Then $H_U(\mathbf{0}) = 0$ and $U_D(\iota^B) = 0$ unless $D \subset B$. Therefore

$$\ln \frac{\mu(\{\iota^B\})}{\mu(\{\mathbf{0}\})} = H_U(\iota^B) = \sum_{D \subset B} U_D(\iota^B) = \sum_{D \subset B} U_D(\iota^A)$$

whenever $B \subset A \subset T$. If we apply Lemma 12-11 with $\Lambda = A$, $\Phi(D) = U_D(\iota^A)$ and $\Psi(B) = \ln[\mu(\{\iota^B\})/\mu(\{\mathbf{0}\})]$, the conclusion is

$$U_A(\iota) = \sum_{B \subset A} (-1)^{|A-B|} \ln \frac{\mu(\{\iota^B\})}{\mu(\{\mathbf{0}\})} = V_A(\iota),$$

since the last sum is 0 when $A = \varnothing$, and otherwise

$$\ln \mu(\{\mathbf{0}\}) \textstyle\sum_{B \subset A} (-1)^{|A-B|} = 0.$$

Corollary 12-13: A finite random field is completely determined by its local characteristics $\{\mu_a^A\}$.

PROOF: The second equation in Theorem 12-12 shows that the canonical potential V for μ is determined by the local characteristics, and V determines μ.

Proposition 12-14: Let $\{x_t\}$ be a finite Gibbs field with potential U. Then the canonical potential V for $\{x_t\}$ is related to U by

$$V_A(\iota) = \sum_{B \subset A \subset D \subset T} (-1)^{|A-B|} U_D(\iota^B), \quad A \neq \varnothing.$$

PROOF: Since $\{x_t\}$ is Gibbs,

$$\ln \frac{\mu(\{\iota^B\})}{\mu(\{\iota^{B+a}\})} = \sum_{D \subset T} [U_D(\iota^B) - U_D(\iota^{B+a})]$$

for any $a \in T$ and $B \subset T - \{a\}$. Using the first equation for V in Theorem 12-12, it now follows that whenever $a \in A \subset T$,

$$\begin{aligned} V_A(\iota) &= \sum_{B \subset A - a} (-1)^{|A-B|} \ln \frac{\mu(\{\iota^B\})}{\mu(\{\iota^{B+a}\})} \\ &= \sum_{D \subset T} \sum_{B \subset A} (-1)^{|A-B|} U_D(\iota^{B \cap D}) \\ &= \sum_{D \subset T} \sum_{B_1 \subset D \cap A} \left[(-1)^{|A-B_1|} U_D(\iota^{B_1}) \left\{ \sum_{B_2 \subset \tilde{D} \cap A} (-1)^{|(\tilde{D} \cap A) - B_2|} . \right\} \right] \end{aligned}$$

The inner sum in the last expression is 0 unless $\tilde{D} \cap A = \varnothing$, i.e., unless $A \subset D$.

Corollary 12-15: Given two potentials U' and U'', let $\Delta_A = U'_A - U''_A$. U' and U'' determine the same finite Gibbs field if and only if

$$\sum_{B \subset A \subset D \subset T} (-1)^{|A-B|} \Delta_D(\iota^B) = 0 \quad \text{for every } A \neq \varnothing .$$

PROOF: Letting V' and V'' be the canonical potentials corresponding to U' and U'' respectively, Proposition 12-14 shows that the given equation is equivalent to $V'_A = V''_A$.

3. Equivalence of finite Markov and neighbor Gibbs fields

We now prove an important equivalence theorem which states that the finite Markov fields are precisely those for which the canonical V is a neighbor potential.

Theorem 12-16: Let $\{x_t\}$ be a finite random field with canonical potential V. Then $\{x_t\}$ is a Markov field if and only if V is a neighbor potential.

PROOF: Fix $a \in T$ and $\iota, \iota' \in \Omega$ such that $i_t = i'_t$ whenever $t \in \bar{a}$. Let ${}^s\iota$ and ${}^s\iota'$ be the modifications of ι and ι' respectively obtained by replacing the value at site a with $s \in S$. If V is a neighbor potential, then

$$\begin{aligned} H_V({}^s\iota) &= \sum_{A \subset T - \{a\}} V_A({}^s\iota) + \sum_{a \in A \subset \bar{a}} V_A({}^s\iota) + \sum_{a \in A \not\subset \bar{a}} V_A({}^s\iota) \\ &= \quad \textstyle\sum_1 \quad + \quad \sum_2(s) \quad + \quad 0, \end{aligned}$$

where $\sum_1$ is **independent** of s. Let $\sum'_1$ and $\sum'_2(s)$ be the corresponding sums when ${}^{s}\iota$ is replaced by ${}^{s}\iota'$. Since ${}^{s}\iota$ and ${}^{s}\iota'$ agree on $\bar{a}$, we have $\sum_2(s) = \sum_2'(s)$ for all s. Thus

$$\mu_a^T({}^{s_0}\iota) = \frac{e^{\Sigma_1 + \Sigma_2(s_0)}}{\sum_{s\in S} e^{\Sigma_1 + \Sigma_2(s)}} = \frac{e^{\Sigma_2(s_0)}}{\sum_{s\in S} e^{\Sigma_2(s)}} = \frac{e^{\Sigma_1' + \Sigma_2'(s_0)}}{\sum_{s\in S} e^{\Sigma_1' + \Sigma_2'(s)}} = \mu_a^T({}^{s_0}\iota')$$

for any $s_0 \in S$. Taking $s_0 = i_a = i_a'$ we have verified (3) of Proposition 12-6; this shows that $\{x_t\}$ is Markov. Conversely, if $\{x_t\}$ is Markov, we claim that the canonical potential V is a neighbor potential. To see this, choose $a, b \in A \subset T$ such that $b \notin \bar{a}$. Expand V as

$$V_A(\iota) = \sum_{B \subset A - \{a,b\}} (-1)^{|A-B|} \ln\left[\frac{\mu_a^T(\iota^B)/\mu_a^T(\iota^{B+a})}{\mu_a^T(\iota^{B+b})/\mu_a^T(\iota^{B+a+b})}\right].$$

Since $b \notin \bar{a}$, Proposition 12-6 shows that $\mu_a^T(\iota^{B+b}) = \mu_a^T(\iota^B)$ and $\mu_a^T(\iota^{B+a+b}) = \mu_a^T(\iota^{B+a})$, yielding the desired result.

Corollary 12-17: There is a one-to-one correspondence between the local characteristics $\{\mu_a^T;\, a \in T\}$ for finite Markov fields and normalized (canonical) potentials V for finite neighbor Gibbs fields, given by

$$\begin{aligned} V_A(\iota) &= \sum_{B \subset A} (-1)^{|A-B|} \ln \mu_a^T(\iota^B) \quad && a \in A \in \mathscr{C} \\ &(\equiv 0 && A \notin \mathscr{C}), \end{aligned}$$

and

$$\mu_a^T(\iota) = z^{-1} \exp\left\{\sum_{a\in B \subset \bar{a}} V_B(\iota)\right\},$$

where z is the appropriate normalizing constant.

Proof: If $\{x_t\}$ is Markov, then $V_A(\iota) \equiv 0$ for $A \notin \mathscr{C}$, by the last theorem. The rest of the first equation above was proved in Theorem 12-12, and the second equation was derived in Theorem 12-16.

Another consequence of Theorem 12-16 is an alternative formulation for finite Markov fields.

Corollary 12-18: A finite random field $\{x_t\}$ is Markov if and only if

$$\mu_A^\Lambda(\iota) = \mu_A^{\bar{A}}(\iota) \quad \text{whenever } \bar{A} \subset \Lambda \subset T.$$

Proof: Let $\{x_t\}$ be Markov, with canonical neighbor potential V.

A computation similar to the one in the proof of Theorem 12-16 shows that

$$\mu_A^A(\iota) = z^{-1} \exp\left\{ \sum_{B \subset \bar{A}: B \cap A \neq \varnothing} V_B(\iota) \right\} = \mu_A^A(\iota')$$

whenever $i_t = i_t'$ for all $t \in \bar{A}$. The claim now follows from the straightforward generalization of Proposition 12-6 obtained by replacing $\{a\}$ with A.

Example 12-19: The Markov process case. Let $\{x_n; 0 \leq n \leq N\}$ be a denumerable Markov process viewed from time 0 to time N. Suppose that $\{x_n\}$ has starting distribution π and one-step transition matrix P_n at time n, where

$$P_{n,ij} = \Pr[x_{n+1} = j \mid x_n = i].$$

If π and all the P_n are strictly positive, then $\{x_n\}$ is a Markov field with neighbor system ∂, where $\partial 0 = \{1\}$, $\partial N = \{N - 1\}$, and $\partial n = \{n - 1, n + 1\}$ for $1 \leq n < N$. The local characteristics for $\{x_t\}$ are given by

$$\mu_0^T(\iota) = \pi(i_0)P_{0,i_0i_1} \Big/ \sum_{i \in S} \pi(i)P_{0,ii_1}$$

$$\mu_n^T(\iota) = \frac{P_{n-1,i_{n-1}i_n}P_{n,i_ni_{n+1}}}{\sum_{i \in S} P_{n-1,i_{n-1}i}P_{n,ii_{n+1}}} \qquad 1 \leq n < N,$$

$$\mu_N^T(\iota) = P_{N-1,i_{N-1}i_N}$$

The canonical potential for the process is then given by

$$V_{\{n\}}(\iota) = \ln \mu_n^T(\iota)/\mu_n^T(\mathbf{0}),$$

$$V_A(\iota) = \ln \frac{\mu_n^T(\iota^A)\mu_n^T(\mathbf{0})}{\mu_n^T(\iota^{\{n-1\}})\mu_n^T(\iota^{\{n\}})} \qquad A = \{n - 1, n\},$$

$$V_A(\iota) = 0 \qquad \text{otherwise},$$

$0 \leq n \leq N$, and $\{x_n\}$ is a neighbor Gibbs field with normalized potential V. Conversely, suppose that $\{x_n; 0 \leq n \leq N\}$ is a Markov field with the above neighbor system ∂. A routine calculation using the explicit representation for μ in terms of V shows that

$$\Pr[x_{n+1} = j \mid x_n = i_n, \ldots, x_0 = i_0] = \Pr[x_{n+1} = j \mid x_n = i_n]$$

whenever $0 \leq n < N$, so $\{x_n\}$ is a Markov process. On a finite time parameter set we see that the one and two sided Markov properties are equivalent, so that Markov fields are precisely the Markov processes with positive cylinder probabilities.

4. Markov fields and neighbor Gibbs fields: the infinite case

For the remainder of the chapter, T will be a countably infinite set with some neighbor system ∂. A **neighbor potential** U on T is defined just as in the previous sections. But we can no longer define the probability measure μ for a neighbor Gibbs field with potential U explicitly; instead we must make use of the local characteristics.

Definition 12-20: An infinite random field $\{x_t\}$ is a **neighbor Gibbs field with neighbor potential** U if

$$\mu_a^\Lambda(\iota) = z^{-1}\exp\left\{\sum_{a\in B\subset\bar{a}} U_B(\iota)\right\} \quad \text{for all finite } \Lambda \supset \bar{a},$$

where z is the appropriate normalizing constant. If $s \in S$ and ${}^s\iota$ is the modification obtained by replacing i_a with s, then

$$z = \sum_{s\in S}\exp\left\{\sum_{a\in B\subset\bar{a}} U_B({}^s\iota)\right\}.$$

Theorem 12-21: Let $\{x_t\}$ be an infinite random field. Then $\{x_t\}$ is a neighbor Gibbs field if and only if it is a Markov field.

PROOF: Suppose $\{x_t\}$ is neighbor Gibbs. From the definition we see that $\mu_a^\Lambda(\iota) = \mu_a^\Lambda(\iota')$ whenever $\Lambda \supset \bar{a}$ and ι' agrees with ι on $\bar{a}$. Just as in Proposition 12-6, this implies that $\mu_a^\Lambda(\iota) = \mu_a^{\bar{a}}(\iota)$ for all finite $\Lambda \supset \bar{a}$ so $\{x_t\}$ is Markov. Conversely, if $\{x_t\}$ is Markov we define a potential V by $V_\varnothing = 0$ and

$$V_\Lambda(\iota) = \sum_{B\subset\Lambda}(-1)^{|\Lambda-B|}\ln\mu_a^\Lambda(\iota^B) \quad a \in \Lambda.$$

The argument of Theorem 12-16 shows that V is a neighbor potential and the only normalized potential determined by the $\{\mu_a^\Lambda\}$. An application of the Möbius inversion formula shows that V satisfies the conditions in Definition 12-20, so $\{x_t\}$ is a neighbor Gibbs field. We remark here that Corollary 12–18 also holds for infinite fields; the proof is routine.

Let $\mathscr{G}_V = \mathscr{G}_V(T)$ be the class of neighbor Gibbs fields on the infinite set T with canonical potential V. The bijection of Corollary 12-17, which carries over to the *finite* subsets of T, shows that we may consider equivalently the class of Markov fields with local characteristics $\{\mu_a^\Lambda\}$ corresponding to V. Since we will be considering many fields simultaneously, the elements of $\mathscr{G}_V$ will be thought of as measures μ

governing $\{x_t\}$. In the finite case we have seen that $\mathscr{G}_V$ always contains a single field. When T is infinite this need not be so; indeed, there are exactly three possibilities:

$$\text{(i) } \mathscr{G}_V = \varnothing \qquad \text{(ii) } |\mathscr{G}_V| = 1 \qquad \text{(iii) } |\mathscr{G}_V| = \infty.$$

This follows from the fact that $\mathscr{G}_V$ is **convex**, i.e., if $\mu_1, \mu_2 \in \mathscr{G}_V$ and $0 \le \alpha \le 1$, then $\alpha\mu_1 + (1 - \alpha)\mu_2 \in \mathscr{G}_V$. Examples of (i)–(iii) will be presented in Section 5.

Definition 12-22: When $|\mathscr{G}_V| = \infty$ we say that there is **phase multiplicity** (or **phase transition**) for V. A measure $\mu \in \mathscr{G}_V$ is **extreme** if whenever $\mu = \alpha\mu_1 + (1 - \alpha)\mu_2$, $\mu_1, \mu_2 \in \mathscr{G}_V$, $0 < \alpha < 1$, then $\mu_1 = \mu_2 = \mu$. The class of extreme elements of $\mathscr{G}_V$ is denoted by $\mathscr{E}_V$.

Since $\mathscr{G}_V$ is convex, in the case of phase multiplicity one would hope for an integral representation in terms of $\mathscr{E}_V$. We will obtain such a representation, along with a number of other results on the structure of $\mathscr{G}_V$, by connecting neighbor Gibbs fields with Martin boundaries for certain Markov chains. The remainder of this section is devoted to the study of general structural properties of $\mathscr{G}_V$ with the aid of the boundary theory developed in Chapter 10.

To begin, we fix a neighbor potential V, assume $\mathscr{G}_V \ne \varnothing$, and choose a **reference measure** $\nu \in \mathscr{G}_V$. Also we fix an increasing sequence $\{\Lambda(n), n = 0, 1, \ldots\}$ of finite subsets of T, such that $\overline{\Lambda(n)} \subset \Lambda(n + 1)$ and $\Lambda(n) \uparrow T$ as $n \to \infty$. Write $K(0) = \Lambda(0), K(n) = \Lambda(n) - \Lambda(n - 1)$ for $n \ge 1$. Then any configuration $\iota \in \Omega$ may be thought of as a sequence of subconfigurations $(\kappa^0, \kappa^1, \ldots)$, where $\kappa^n = \{k_t^n; t \in K(n))$ satisfies $k_t^n = i_t$ when $t \in K(n)$. For brevity's sake we denote $\{\omega \mid x_t(\omega) = k_t^n \text{ for all } t \in K(n)\}$ simply as $[\kappa^n]$. Similarly, $[\kappa^0, \kappa^1, \ldots, \kappa^n]$ means $\{\omega \mid x_t(\omega) = k_t^r \text{ for all } t \in K(r), 0 \le r \le n\}$, and so forth. Also, we write $\nu(A|B) = \nu(A \cap B)/\nu(B)$ when convenient (and, of course $\nu(B) > 0$). With these notations in effect, observe the following key property of neighbor Gibbs fields.

Proposition 12-23: If $\iota = (\kappa^0, \kappa^1, \ldots) \in \Omega$, then

$$\nu([\kappa^{m+1}, \ldots, \kappa^n] \mid [\kappa^0, \ldots, \kappa^m]) = \nu([\kappa^{m+1}, \ldots, \kappa^n] \mid [\kappa^m]) \qquad 1 \le m < n < \infty.$$

PROOF: If $\{\nu_A^A\}$ are the characteristics of ν, then the left hand side divided by the right hand side is equal to $\nu_{\Lambda(m-1)}^{\Lambda(n)}(\iota)/\nu_{\Lambda(m-1)}^{\Lambda(m)}(\iota) = 1$, since the numerator and denominator of this last quotient both equal $\nu_{\Lambda(m-1)}^{\overline{\Lambda(m-1)}}(\iota)$ by the Markov field property.

The above result reveals a Markovian structure for ν which can be exhibited explicitly in terms of a Markov chain. The states for the chain are all possible pairs (n, κ^n), $n \geq 0$, $\kappa^n \in S^{K(n)}$. The transition matrix is

$$P_{(n,\kappa^n)(n+1,\kappa^{n+1})} = \nu([\kappa^{n+1}] \mid [\kappa^n]).$$

Proposition 12-23 and a simple induction show that the n-step transition matrix is given by

$$P^{(n)}_{(m,\kappa^m)(m+n,\kappa^{m+n})} = \nu([\kappa^{m+n}] \mid [\kappa^n]).$$

If the initial distribution is $\pi_{(0,\kappa^0)} = \nu([\kappa^0])$, and $\{y_n\}$ denotes the resulting chain, then we obtain the simple relationship

$$\Pr_\pi[y_n = (n, \kappa^n)] = \nu([\kappa^n]).$$

Next, we connect P-regular functions with fields in $\mathscr{G}_V$. A lemma will be useful for this purpose.

Lemma 12-24: If $\mu \in \mathscr{G}_V$, $\iota = (\kappa^0, \kappa^1, \ldots) \in \Omega$, then

$$\frac{\mu([\kappa^0, \kappa^1, \ldots, \kappa^n])}{\nu([\kappa^0, \kappa^1, \ldots, \kappa^n])} = \frac{\mu([\kappa^m, \ldots, \kappa^n])}{\nu([\kappa^m, \ldots, \kappa^n])} \qquad 1 \leq m \leq n < \infty.$$

PROOF: The left hand expression may be rewritten as

$$\frac{\mu([\kappa^m, \ldots, \kappa^n])}{\nu([\kappa^m, \ldots, \kappa^n])} \frac{\mu^{A(n)}_{A(m-1)}(\iota)}{\nu^{A(n)}_{A(m-1)}(\iota)},$$

and the two characteristics agree, being identical functionals of the potential V.

As in Section 10-6, we call a non-negative P-regular function a **normalized** if

$$\pi h = \sum_{\kappa^0 \in S^{K(0)}} \pi_{(0,\kappa^0)} h_{(0,\kappa^0)} = 1,$$

and **minimal normalized** if it cannot be written as a non-trivial convex combination of two distinct non-negative regular functions.

Theorem 12-25: There is a one-to-one correspondence between non-negative normalized P-regular functions h and neighbor Gibbs fields $\mu \in \mathscr{G}_V$ given by

$$h_{(n,\kappa^n)} = \mu([\kappa^n])/\nu([\kappa^n])$$

and

$$\mu([\kappa^0, \ldots, \kappa^n]) = \nu([\kappa^0, \ldots, \kappa^n]) h_{(n,\kappa^n)},$$

$n \geq 0$, $\kappa^r \in S^{K(r)}$. Moreover, h is minimal normalized if and only if μ is in $\mathscr{E}_V$.

Proof: Given $\mu \in \mathscr{G}_V$, define h according to the first equation in the theorem. By Lemma 12-24,

$$\sum_{\kappa^{n+1}} P_{(n,\kappa^n)(n+1,\kappa^{n+1})} h_{(n+1,\kappa^{n+1})} = \sum_{\kappa^{n+1}} \frac{\nu([\kappa^n, \kappa^{n+1}])}{\nu([\kappa^n])} \frac{\mu([\kappa^n, \kappa^{n+1}])}{\nu([\kappa^n, \kappa^{n+1}])}$$
$$= \sum_{\kappa^{n+1}} \frac{\mu([\kappa^n, \kappa^{n+1}])}{\nu([\kappa^n])}$$
$$= \frac{\mu([\kappa^n])}{\nu([\kappa^n])} = h_{(n,\kappa^n)},$$

so h is P-regular. Clearly, h is also non-negative and normalized. Conversely, let h be any non-negative normalized P-regular function. We claim that the second equation of the theorem prescribes cylinder probabilities for a unique measure $\mu \in \mathscr{G}_V$ determined by h. Note first that h must be strictly positive. To see this, write

$$\sum_{\kappa^{n+1}} \nu([\kappa^{n+1}]) h_{(n+1,\kappa^{n+1})} = \pi P^{n+1} h = 1,$$

and

$$h_{(n,\kappa^n)} = (Ph)_{(n,\kappa^n)} = \sum_{\kappa^{n+1}} \frac{\nu([\kappa^n, \kappa^{n+1}])}{\nu([\kappa^n])} h_{(n+1,\kappa^{n+1})}.$$

The first equation above implies that $h_{(n+1,\kappa^{n+1})} > 0$ for some κ^{n+1}, and hence the second shows that $h_{(n,\kappa^n)} > 0$ for all (n, κ^n). Thus all cylinder measures for μ are evidently positive. Next, we use Proposition 12-23 to compute

$$\sum_{\kappa^{n+1}} \mu([\kappa^0, \ldots, \kappa^{n+1}]) = \nu([\kappa^0, \ldots, \kappa^n]) \sum_{\kappa^{n+1}} \frac{\nu([\kappa^0, \ldots, \kappa^{n+1}])}{\nu([\kappa^0, \ldots, \kappa^n])} h_{(n+1,\kappa^{n+1})}$$
$$= \nu([\kappa^0, \ldots, \kappa^n])(Ph)_{(n,\kappa^n)}$$
$$= \nu([\kappa^0, \ldots, \kappa^n]) h_{(n,\kappa^n)} = \mu([\kappa^0, \ldots, \kappa^n]).$$

This shows that the measures $\mu([\kappa^0, \ldots, \kappa^n])$ on $S^{\Lambda(n)}$ are consistent for $n \geq 0$, and also by induction that

$$\sum_{\kappa^0,\ldots,\kappa^n} \sum_{\kappa^{n+1}} \mu([\kappa^0, \ldots, \kappa^{n+1}]) = \sum_{\kappa^0,\ldots,\kappa^n} \mu([\kappa^0, \ldots, \kappa^n]) = 1$$

for all $n \geq 0$, since $\sum_{\kappa^0} \mu([\kappa^0]) = \pi h = 1$. By the usual extension theorems we obtain a unique μ with the desired cylinder probabilities. To see that $\mu \in \mathscr{G}_V$, fix finite A, Λ, with $\bar{A} \subset \Lambda$, and choose n large enough that $\Lambda \subset \Lambda(n)$. Let $\iota = \{i_t; t \in T\} = (\kappa^0, \kappa^1, \ldots) \in \Omega$. Introduce the sets

$$[\iota_A] = \{\omega \mid \omega_t = i_t \text{ for all } t \in A\},$$
$$[\iota_{\Lambda - A}] = \{\omega \mid \omega_t = i_t \text{ for all } t \in \Lambda - A\},$$
$$[\psi] = \{\omega \mid \omega_t = \psi_t \text{ for all } t \in \Lambda(n-1) - \Lambda\} \quad \text{where } \psi \in S^{\Lambda(n-1)-\Lambda}.$$

Then by construction,

$$\mu([\iota_A]) = \sum_{\psi}\sum_{\kappa^n} \nu([\iota_A] \cap [\psi] \cap [\kappa^n])h_{(n,\kappa^n)}$$

and

$$\mu([\iota_{\Lambda - A}]) = \sum_{\psi}\sum_{\kappa^n} \nu([\iota_{\Lambda - A}] \cap [\psi] \cap [\kappa^n])h_{(n,\kappa^n)},$$

where ψ is summed over $S^{\Lambda(n-1)-\Lambda}$, κ^n over $S^{K(n)}$. If ι^{ψ,κ^n} denotes the modification of ι obtained by replacing its values on $\Lambda(n-1) - \Lambda$ with ψ, and its values on $K(n)$ with κ^n, then $\nu_A^{\Lambda(n)}(\iota^{\psi,\kappa^n}) = \nu_A^{\bar{\Lambda}}(\iota)$ for all ψ and κ^n, since $\nu \in \mathscr{G}_V$. Thus

$$\mu([\iota_A]) = \sum_{\psi}\sum_{\kappa^n} \nu_A^{\Lambda(n)}(\iota^{\psi,\kappa^n})\nu([\iota_{\Lambda - A}] \cap [\psi] \cap [\kappa^n])h_{(n,\kappa^n)}$$
$$= \nu_A^{\bar{\Lambda}}(\iota)\mu([\iota_{\Lambda - A}]),$$

and so $\mu_A^{\Lambda}(\iota) = \nu_A^{\bar{\Lambda}}(\iota) = \nu_A^{\Lambda}(\iota)$. Hence $\mu \in \mathscr{G}_V$. To check the one-to-one correspondence, let $\mathscr{H}$ denote the normalized non-negative P-regular functions. We have defined mappings $\rho: \mathscr{G}_V \to \mathscr{H}$ and $\sigma: \mathscr{H} \to \mathscr{G}_V$. One easily verifies that $\sigma(\rho(\mu)) = \mu$ and $\rho(\sigma(h)) = h$, as desired. Finally, ρ is obviously a cone homomorphism in the sense that

$$\rho(\alpha\mu_1 + (1 - \alpha)\mu_2) = \alpha\rho(\mu_1) + (1 - \alpha)\rho(\mu_2),$$

$0 \le \alpha \le 1$, $\mu_1, \mu_2 \in \mathscr{G}_V$. This implies the last assertion in the theorem, and the proof is complete,

Using the Martin boundary theory for the chain P with starting distribution π, as constructed from the reference measure $\nu \in \mathscr{G}_V$, we will now derive several structure properties of $\mathscr{G}_V$.

Definition 12-26: When $\mu \in \mathscr{G}_V$ and $n \ge 1$, set

$$\mu^{\kappa^n}([\kappa^0, \ldots, \kappa^{n-1}]) = \mu_{\Lambda(n-1)}^{\Lambda(n)}(\iota), \quad \iota = (\kappa^0, \kappa^1, \ldots) \in \Omega.$$

For each fixed κ^n on $K(n)$, $\mu^{\kappa^n}(\cdot)$ defines a finite random field on $S^{\Lambda(n-1)}$. The fields μ^{κ^n} are said to have **thermodynamic limit** μ_∞, $t\text{-}\lim_{n\to\infty} \mu^{\kappa^n} = \mu_\infty$ in notation, if there is a measure μ_∞ on Ω such that

$$\lim_{n\to\infty} \mu^{\kappa^n}([\kappa^0, \ldots, \kappa^m]) = \mu_\infty([\kappa^0, \ldots, \kappa^m])$$

for all $m \geq 0$ and all configurations $(\kappa^0, \ldots, \kappa^m)$ on $S^{\Lambda(m)}$.

Theorem 12-27:

$$\nu\left(\left\{\omega \mid t\text{-}\lim_{n\to\infty} \nu^{\kappa^n(\omega)} \in \mathscr{E}_V\right\}\right) = 1,$$

where $\kappa^n(\omega)$ is the configuration of ω restricted to $K(n)$. In particular, $\mathscr{E}_V \neq \varnothing$ whenever $\mathscr{G}_V \neq \varnothing$.

PROOF: Let B_e comprise the extreme points of the Martin boundary for P started with π, and let λ denote harmonic measure. Theorem 10-41 applied to the constant regular function $h = 1$ shows that $\Pr_\pi[y_v \in B_e] = \lambda(B_e) = 1$. Since $\{y_n\}$ visits each state (m, κ^m) at most once, the Martin kernel is given by

$$K((m, \kappa^m), (n, \kappa^n)) = P^{(n-m)}_{(m,\kappa^m)(n,\kappa^n)} \Big/ \sum_{\kappa^0} \pi(\kappa^0) P_{(0,\kappa^0)(n,\kappa^n)}, \quad n \geq m,$$

$(=0$ otherwise$)$. Thus $y_v(\omega) \in B_e$ means that

$$K((m, \kappa^m), x) = \lim_{n\to\infty} \frac{\nu([\kappa^m, \kappa^n(\omega)])}{\nu([\kappa^m])\nu([\kappa^n(\omega)])}$$

exists for every (m, κ^m), and is a minimal regular function of (m, κ^m). By the last theorem, $K(\cdot, x)$ is minimal regular if and only if $\nu([\kappa^0, \ldots, \kappa^m])K((m, \kappa^m), x) = \mu_\infty([\kappa^0, \ldots, \kappa^m])$ for a (unique) $\mu_\infty \in \mathscr{E}_V$. In this case, we deduce from Proposition 12-23 that

$$\frac{\mu_\infty([\kappa^0, \ldots, \kappa^m])}{\nu([\kappa^0, \ldots, \kappa^m])} = \frac{\lim_{n\to\infty} \nu([\kappa^0, \ldots, \kappa^m] \mid [\kappa^n(\omega)])}{\nu([\kappa^0, \ldots, \kappa^m])} = \frac{\lim_{n\to\infty} \nu^{\kappa^n(\omega)}([\kappa^0, \ldots, \kappa^m])}{\nu([\kappa^0, \ldots, \kappa^m])}$$

for all m, κ^r. This shows that $t\text{-}\lim_{n\to\infty} \nu^{\kappa^n(\omega)} \in \mathscr{E}_V$ on $\{\omega: y_v(\omega) \in B_e\}$. Using the reference measure $\nu \in \mathscr{G}_V$, we have produced a set of measures in $\mathscr{E}_V$ which has ν-measure 1.

Theorem 12-28: The elements of $\mathscr{E}_V$ are in one-to-one correspondence with those of B_e. If $\mu^x \in \mathscr{E}_V$ corresponds to $x \in B_e$, then there is a bijection between the probability measures on B_e and the neighbor Gibbs fields in $\mathscr{G}_V$, given by the equations

$$\mu = \int_{x\in B_e} \mu^x d\lambda^\mu(x)$$

and

$$\lambda^{\mu}(E) = \mu\left(\left\{\omega \mid t\text{-}\lim_{n\to\infty} \mu^{\kappa^n(\omega)} = \mu^x \quad \text{for some } x \in E\right\}\right),$$

$$E \quad \text{Borel in } B_e.$$

Proof: We know from Chapter 10 that B_e is in one-to-one correspondence with the class $\mathscr{H}_e$ of minimal normalized regular functions, while Theorem 12-25 gives a bijection between $\mathscr{H}_e$ and $\mathscr{E}_V$. Hence there is a one-to-one correspondence $x \leftrightarrow \mu^x$ between B_e and $\mathscr{E}_V$. For $\mu \in \mathscr{G}_V$, apply Theorem 10-41 to $\rho(\mu) \in \mathscr{H}$, and use Lemma 12-24, to get

$$\frac{\mu([\kappa^0, \ldots, \kappa^m])}{\nu([\kappa^0, \ldots, \kappa^m])} = \int_{x \in B_e} \frac{\mu^x([\kappa^0, \ldots, \kappa^m])}{\nu([\kappa^0, \ldots, \kappa^m])} \, d\lambda^{\rho(\mu)}(x),$$

where λ^h is harmonic measure for the h-process. A routine computation using the explicit form of the Martin kernel, derived in the proof of the previous theorem, shows that

$$\begin{aligned}\lambda^{\rho(\mu)}(E) &= \mu\left(\left\{\omega \mid t\text{-}\lim_{n\to\infty} \nu^{\kappa^n(\omega)} = \mu^x \quad \text{for some } x \in E\right\}\right) \\ &= \mu\left(\left\{\omega \mid t\text{-}\lim_{n\to\infty} \mu^{\kappa^n(\omega)} = \mu^x \quad \text{for some } x \in E\right\}\right),\end{aligned}$$

(The $\rho(\mu)$ process changes the reference measure to μ, and $\mu^{\kappa^n} = \nu^{\kappa^n}$ since both random fields are defined in terms of the same characteristics.) Setting $\lambda^{\mu} = \lambda^{\rho(\mu)}$, the theorem follows.

Corollary 12-29: If $\mu \in \mathscr{E}_V$, then there is an $\iota = (\kappa^0, \kappa^1, \ldots) \in \Omega$ such that

$$\mu = t\text{-}\lim_{n\to\infty} \mu^{\kappa^n}.$$

Proof: The uniqueness of the integral representation implies that if $\mu = \mu^x \in \mathscr{E}_V$, then $\lambda^{\mu}(\{x\}) = \mu(\{\omega: t\text{-}\lim_{n\to\infty} \mu^{\kappa^n(\omega)} = \mu\}) = 1$. The desired ι may therefore be any configuration from a full ω-set with respect to μ.

The entire development of this section has proceeded on the assumption that $\mathscr{G}_V$ is not empty. The problem of determining from the potential V just *when* $\mathscr{G}_V$ is not empty turns out to be a difficult one if the state space S is countably infinite. In the case where T is the integers, we shall have more to say about this later. When S is *finite* on the other hand, it will now be proved that $\mathscr{G}_V$ is *always* non-empty.

Theorem 12-30: If S is finite, then $\mathscr{G}_V \neq \varnothing$. Moreover, the limits

$$\mu^+([\kappa^0, \ldots, \kappa^m]) = \lim_{n \to \infty} \max_{\kappa^n \in S^{K(n)}} \mu^{\kappa^n}([\kappa^0, \ldots, \kappa^m])$$

and

$$\mu^-([\kappa^0, \ldots, \kappa^m]) = \lim_{n \to \infty} \min_{\kappa^n \in S^{K(n)}} \mu^{\kappa^n}([\kappa^0, \ldots, \kappa^m])$$

exist for all $m \geq 0$, $(\kappa^0, \ldots, \kappa^m) \in S^{\Lambda(m)}$, and there is phase multiplicity for V if and only if

$$\mu^+([\kappa^0, \ldots, \kappa^m]) \neq \mu^-([\kappa^0, \ldots, \kappa^m]) \quad \text{for some } m \text{ and } (\kappa^0, \ldots, \kappa^m).$$

(Recall that $\mu^{\kappa^n}([\kappa^0, \ldots, \kappa^m])$ is a certain characteristic which is completely and uniquely determined by V.)

To prove the theorem, we first need the following lemma.

Lemma 12-31: For given $\mu \in \mathscr{G}_V$, $m > 0$ and fixed $(\kappa^0, \ldots, \kappa^m) \in S^{\Lambda(m)}$, abbreviate

$$\mu_n^+ = \max_{\kappa^n \in S^{K(n)}} \mu^{\kappa^n}([\kappa^0, \ldots, \kappa^m]),$$

$$\mu_n^- = \min_{\kappa^n \in S^{K(n)}} \mu^{\kappa^n}([\kappa^0, \ldots, \kappa^m]).$$

Then

(1) $0 < \mu_n^- \leq \mu([\kappa^0, \ldots, \kappa^m]) \leq \mu_n^+$ for each $n > m$, and
(2) μ_n^- is increasing and μ_n^+ is decreasing as $n \to \infty$.

Proof: (1) μ_n^- is a minimum over a finite set of strictly positive probabilities, hence strictly positive. When $n > m$,

$$\mu([\kappa^0, \ldots, \kappa^m]) = \sum_{\kappa^n} \mu^{\kappa^n}([\kappa^0, \ldots, \kappa^m])\mu([\kappa^n])$$

$$\geq \mu_n^- \sum_{\kappa^n} \mu([\kappa^n]) = \mu_n^-,$$

and an analogous estimate establishes the remaining inequality.

(2) For any $n > m$ and $\kappa^{n+1} \in K(n+1)$,

$$\mu^{\kappa^{n+1}}([\kappa^0, \ldots, \kappa^m]) = \sum_{\kappa^n} \mu([\kappa^0, \ldots, \kappa^m] \mid [\kappa^n, \kappa^{n+1}])\mu([\kappa^n] \mid [\kappa^{n+1}])$$

$$= \sum_{\kappa^n} \mu^{\kappa^n}([\kappa^0, \ldots, \kappa^m])\mu([\kappa^n] \mid [\kappa^{n+1}])$$

$$\geq \mu_n^- \sum_{\kappa^n} \mu([\kappa^n] \mid [\kappa^{n+1}]) = \mu_n^-$$

This shows that μ_n^- is increasing with n; a similar estimate proves that μ_n^+ is decreasing.

Proof of Theorem 12-30: The limits μ^+ and μ^- are well-defined by the monotonicity established in the lemma. We now show that for any given configuration $(\kappa^0, \ldots, \kappa^m)$ on $\Lambda(m)$, there is a random field $\underline{\mu} \in \mathscr{G}_V$ such that $\underline{\mu}([\kappa^0, \ldots, \kappa^m]) = \mu^-([\kappa^0, \ldots, \kappa^m])$. Let $\underline{\kappa}^n$ denote the configuration on $K(n)$ for which the minimum value μ_n^- is attained, and define measures $\underline{\mu}_n$, $n \geq 1$, on Ω by

$$\underline{\mu}_n: \begin{cases} \underline{\mu}_n([\iota]) = \mu^{\kappa^n}([\iota_\Lambda]) & \text{whenever } \Lambda \subset \Lambda(n-1) \\ \underline{\mu}_n([\kappa^n]) = 1 \\ \underline{\mu}_n(\{\omega : x_t = 0\}) = 1 & \text{whenever } t \notin \Lambda(n) \end{cases}$$

(The notation $[\iota_\Lambda]$ was introduced in the proof of Theorem 12-25.) These specifications are clearly consistent, so each $\underline{\mu}_n$ is well-defined. Now by the finiteness of S, we can use a diagonal argument (like the one in Proposition 1-63) to choose a subsequence $\{\underline{\mu}_{n'}\}$ such that $\underline{\mu}_{n'}([\iota_\Lambda]) \to \underline{\mu}([\iota_\Lambda])$ for all possible configurations on every finite $\Lambda \subset T$. By the extension theorem, these cylinder limits give rise to a unique $\underline{\mu}$ on Ω. Now observe that

$$\underline{\mu}([\kappa^0, \ldots, \kappa^m]) = \lim_{n' \to \infty} \underline{\mu}_{n'}([\kappa^0, \ldots, \kappa^m]) = \lim_{n' \to \infty} \mu^{\underline{\kappa}^{n'}}([\kappa^0, \ldots, \kappa^m]),$$

the last limit being equal to $\mu^-([\kappa^0, \ldots, \kappa^m])$ by definition. To verify that $\underline{\mu} \in \mathscr{G}_V$, we first note that the $\underline{\mu}_{n'}$ measures of $[\iota_\Lambda]$ are bounded away from 0 for n' sufficiently large, since $\underline{\mu}_{n'}([\iota_\Lambda]) = \mu^{\underline{\kappa}^{n'}}([\iota_\Lambda])$ is strictly positive as soon as $\Lambda \subset \Lambda(n'-1)$ and increases with n' by Lemma 12-31. It follows that $\underline{\mu}$ is strictly positive on finite cylinders, and that the neighbor Gibbs property is inherited from the $\underline{\mu}_{n'}$ (i.e., the limit may be interchanged with the operations defining the characteristics.) This completes the proof that $\mathscr{G}_V$ is always non-empty. By an analogous construction we find a neighbor Gibbs field $\bar{\mu} \in \mathscr{G}_V$ with $\bar{\mu}([\kappa^0, \ldots, \kappa^m]) = \mu^+([\kappa^0, \ldots, \kappa^m])$. If $\mu^+ \neq \mu^-$ (for some $(\kappa^0, \ldots, \kappa^m)$), then evidently $|\mathscr{G}_V| > 1$. If $\mu^+ = \mu^-$, then Lemma 12-31 shows that any $\mu \in \mathscr{G}_V$ is uniquely determined on all events $[\kappa^0, \ldots, \kappa^m]$, $m \geq 0$, hence on all $[\iota_\Lambda]$, finite $\Lambda \subset T$. This shows that $|\mathscr{G}_V| = 1$, completing the proof.

Unfortunately, the general criterion for phase multiplicity just given is often difficult to apply, since μ^+ and μ^- may not be readily computable. A more detailed theory is available for certain "attractive potentials," examples of which will be mentioned in Section 6.

5. Homogeneous Markov fields on the integers

Throughout this section the time parameter set T will be $\mathbb{Z}$ = the integers, $\mathbb{N} = \{0, 1, \ldots\}$ or $-\mathbb{N} = \{\ldots, -1, 0\}$, with the standard neighbor structure ∂ (i.e. $\partial n = \{n' \in T: |n' - n| = 1\}$). Our objective is to treat, in some detail, the classes of Markov fields on these infinite linear lattices.

First let us consider the "one-sided" cases. We show here the previously mentioned fact that the Markov random fields on $\mathbb{N}$ are simply the Markov processes with strictly positive cylinders.

Proposition 12-32: If $T = \mathbb{N}$ and $\mu \in \mathscr{G}_V$, then $\{x_n\}$ is a Markov process. In particular, there are probability measures π_n and transition matrices P_n, $n \geq 0$, such that

$$\pi_{n,i} = \Pr[x_n = i], \quad P_{n,ij} = \Pr[x_{n+1} = j \mid x_n = i],$$

and $\pi_n P_n = \pi_{n+1}$. Similarly, if $T = -\mathbb{N}$ and $\mu \in \mathscr{G}_V$, then $\{x_n\}$ is a Markov process on $-\mathbb{N}$, and there are probability measures π_n and transition matrices P_n, $n < 0$, satisfying the above relations.

Proof: Without loss of generality, assume that V is normalized. It suffices to assume $T = \mathbb{N}$ and check the Markov property (the proof for $T = -\mathbb{N}$ being analogous). Fix $n > 0$, and set $A = \{1, 2, \ldots, n-1\}$. For any $\iota \in \Omega$, define

$$z_n(i_0, i_n) = \sum_{\kappa \in S^A} \exp\left\{\sum_{B \subset \bar{A}: B \cap A \neq \varnothing} V_B(\iota^\kappa)\right\},$$

where ι^κ is the modification obtained by replacing the values of ι on A with those of κ. Now choose a particular $\iota \in \Omega$, and let ι' be the configuration obtained by replacing the value i_0 at site 0 with a 0. Then, using the Markov field property at 0, we have

$$\frac{\Pr[x_0 = i_0 \wedge x_n = i_n]}{\Pr[x_0 = 0 \wedge x_n = i_n]} = \frac{\mu_0^{\bar{A}}(\iota)}{\mu_0^{\bar{A}}(\iota')} \frac{\mu_A^{\bar{A}}(\iota')}{\mu_A^{\bar{A}}(\iota)} = \frac{\mu_0^{\bar{0}}(\iota)}{\mu_0^{\bar{0}}(\iota')} \frac{\mu_A^{\bar{A}}(\iota')}{\mu_A^{\bar{A}}(\iota)}.$$

Writing the characteristics according to Definition 12-20, the last term is

$$\exp\{V_{\{0\}}(\iota) + V_{\{0,1\}}(\iota)\} \frac{z_n(0, i_n)^{-1} \exp\left\{\sum_{B \subset \bar{A}: B \cap A \neq \varnothing} V_B(\iota')\right\}}{z_n(i_0, i_n)^{-1} \exp\left\{\sum_{B \subset \bar{A}: B \cap A \neq \varnothing} V_B(\iota)\right\}}$$

$$= \frac{z_n(i_0, i_n)}{z_n(0, i_n)} e^{V_{\{0\}}(\iota)}$$

after cancellation. Hence

$$\Pr[x_0 = i_0 \wedge \cdots \wedge x_n = i_n]$$
$$= \mu_{A}^{\bar{A}}(\iota) \Pr[x_0 = i_0 \wedge x_n = i_n]$$
$$= z_n(i_0, i_n)^{-1} \exp\left\{ \sum_{B \subset \bar{A}: B \cap A \neq \varnothing} V_B(\iota) \right\} \Pr[x_0 = i_0 \wedge x_n = i_n]$$
$$= z_n(0, i_n)^{-1} \exp\left\{ V_{\{0\}}(\iota) + \sum_{B \subset \bar{A}: B \cap A \neq \varnothing} V_B(\iota) \right\} \Pr[x_0 = 0 \wedge x_n = i_n].$$

Finally, after further cancellation, we obtain

$$\Pr[x_n = i_n \mid x_0 = i_0 \wedge \cdots \wedge x_{n-1} = i_{n-1}]$$
$$= \frac{z_{n-1}(0, i_{n-1})}{z_n(0, i_n)} [e^{V_{\{n-1\}}(\iota) + V_{\{n-1,n\}}(\iota)}] \frac{\Pr[x_0 = 0 \wedge x_n = i_n]}{\Pr[x_0 = 0 \wedge x_{n-1} = i_{n-1}]},$$

a conditional probability depending only on $n - 1$, i_{n-1} and i_n, which we may set equal to $P_{n-1, i_{n-1} i_n}$.

Proposition 12-33: Let $T = \mathbb{N}$, and suppose that $\nu \in \mathscr{G}_V$ has initial measure π_0 and transition matrices P_n. Then any $\mu \in \mathscr{G}_V$ is Markovian with initial measure π'_0 and transition matrices P'_n which satisfy $\pi'_{0,i_0} = \pi_{0,i_0} h_{(0,i_0)}$ and $P'_{n,i_n i_{n+1}} = P_{n,i_n i_{n+1}} h_{(n+1,i_{n+1})}/h_{(n,i_n)}$ for some solution h of the equations

$$h_{(n,i_n)} = \sum_{i_{n+1} \in S} P_{n,i_n i_{n+1}} h_{(n+1,i_{n+1})}.$$

Conversely, any μ arising in this way is in $\mathscr{G}_V$.

PROOF: Apply the construction of the previous section with $T = \mathbb{N}$ and $\Lambda(n) = \{0, 1, \ldots, n\}$. Then the correspondence of Theorem 12-25 is clearly equivalent to the one asserted here, because ν and μ are both Markovian by Proposition 12-32.

Of course, an obvious analogue of the last result holds in case $T = -\mathbb{N}$. A concrete example of phase multiplicity on a half-line will be given later in this section.

We turn our attention now to the "two-sided" case, $T = \mathbb{Z}$. In contrast to the one-sided setting, we shall see shortly that there are Markov fields on $\mathbb{Z}$ which are *not* Markov processes. Since the neighbor structure ∂ on $\mathbb{Z}$ commutes with translation, it is possible to define a class of homogeneous Markov fields.

Definition 12-34: A Markov field $\{x_t\}_{t\in\mathbb{Z}}$ is **homogeneous** if

$$\mu_m^{\overline{m}}(\iota) = \mu_{m+n}^{\overline{m+n}}(\theta^{-n}\iota) \quad \text{for all } m, n \in \mathbb{Z}, \iota \in \Omega,$$

where the value of $\theta^{-n}\iota$ at site t is i_{t-n}. A neighbor potential U is **homogeneous** if

$$U_{\{m\}}(\iota) = u_j \quad \text{and} \quad U_{\{m,m+1\}}(\iota) = u_{jk} \quad \text{whenever } i_m = j, i_{m+1} = k.$$

Two facts follow immediately from the definitions. First, if $\{x_t\}$ is homogeneous Markov, then the canonical V of Theorem 12-21 is a homogeneous neighbor potential. Second, any neighbor Gibbs field with homogeneous neighbor potential U is homogeneous Markov. For such a U, if we set $Q_{jk} = \exp\{\frac{1}{2}u_j + u_{jk} + \frac{1}{2}u_k\}$, then μ is a Gibbs field with potential U if and only if

$$\mu_m^{\overline{m}}(\iota) = \frac{e^{u_j + u_{ij} + u_{jk}}}{\sum_{s\in S} e^{u_s + u_{is} + u_{sk}}} = \frac{Q_{ij}Q_{jk}e^{-(u_i+u_k)/2}}{\sum_{s\in S} Q_{is}Q_{sk}e^{-(u_i+u_k)/2}} = \frac{Q_{ij}Q_{jk}}{(Q^2)_{ik}}$$

whenever $i_{m-1} = i$, $i_m = j$ and $i_{m+1} = k$. On the other hand, any strictly positive Q defines consistent local characteristics by means of the above equation, and these in turn give rise to a homogeneous neighbor Gibbs potential for μ, according to Theorem 12-21. Thus we obtain a multiplicative representation for the characteristics in terms of Q which is more convenient than the one involving the canonical potential V. Let $\mathscr{G}_Q$ denote the class of Markov fields determined by Q in this manner. An immediate requirement for $\mathscr{G}_Q$ to be non empty is $Q^2 < \infty$, and since the local characteristics determine all the characteristics, it follows from this assumption that $Q^n < \infty$ for all n, so that

$$\mu_{[m,n]}^{\Lambda}(\iota) = \mu_{[m,n]}^{[m-1,n+1]}(\iota) = \frac{Q_{i_{m-1}i_m}Q_{i_m i_{m+1}}\cdot \cdots \cdot Q_{i_{n-1}i_n}Q_{i_n i_{n+1}}}{(Q^{n-m+2})_{i_{m-1}i_{n+1}}}$$

whenever $m \le n$ and $[m-1, n+1] \subset \Lambda$. (Here $[m, n]$ denotes $\{m, m+1, \ldots, n\}$.)

We have seen that any homogeneous neighbor Gibbs field is in $\mathscr{G}_Q$ for some Q. But many matrices Q' give rise to the same characteristics, just as in the potential representation. By definition, either $\mathscr{G}_Q = \mathscr{G}_{Q'}$ or $\mathscr{G}_Q \cap \mathscr{G}_{Q'} = \varnothing$; say that Q is **equivalent** to Q' in the former case ($Q \approx Q'$ in notation).

Proposition 12-35: Strictly positive matrices Q' and Q'' are equivalent if and only if

$$\frac{Q''_{ij}}{Q'_{ij}} = c\,\frac{h_j}{h_i} \quad i, j \in S,$$

for some $c > 0$ and strictly positive $h = (h_k)_{k\in S}$.

PROOF: The statements and proofs of Proposition 12-14 and Corollary 12-15 are easily modified to apply to neighbor potentials on an infinite parameter set by replacing T with $\bar{A}$. In our case, if U' and U'' are the homogeneous neighbor potentials such that $u'_k = u''_k = 0$ for all k and $u'_{ij} = \ln Q'_{ij}$, $u''_{ij} = \ln Q''_{ij}$, then U' and U'' determine the same Gibbs fields if and only if

$$\sum_{B \subset A \subset D \subset \bar{A}} (-1)^{|A-B|} \Delta_D(\iota^B) = 0 \quad \text{for every } A = \{m\}, \{m, m+1\},$$

where $\Delta_A = U'_A - U''_A$. Setting $\delta_{ij} = u'_{ij} - u''_{ij}$, these equations become

$$\begin{aligned} &(1) \qquad \delta_{k0} + \delta_{0k} - 2\delta_{00} = 0 \\ &(2) \quad \delta_{ij} - \delta_{i0} - \delta_{0j} + \delta_{00} = 0 \end{aligned} \qquad i, k, j \in S.$$

The combination (2) + $\frac{1}{2}$(1 with $k = i$) + $\frac{1}{2}$(1 with $k = j$) yields

$$(3) \ (u'_{ij} - u''_{ij}) + w_i - w_j - z = 0,$$
$$\text{where } w_k = \tfrac{1}{2}(\delta_{0k} - \delta_{k0}), \quad z = \delta_{00}.$$

Defining $h_k = e^{w_k}$ and $c = e^z$, the desired equation in terms of Q' and Q'' follows. Conversely, if the hypothesis holds then

$$\frac{Q''_{ij}Q''_{jk}}{(Q''^2)_{ik}} = \frac{c^2 Q'_{ij}(h_j/h_i)Q'_{jk}(h_k/h_j)}{c^2 \sum_{s \in S} Q'_{is}(h_s/h_i)Q'_{sk}(h_k/h_s)} = \frac{Q'_{ij}Q'_{jk}}{(Q'^2)_{ik}},$$

so that Q' and Q'' determine the same local characteristics, i.e., $Q' \approx Q''$.

The remainder of this section will be devoted to an analysis of $\mathscr{G}_Q$, the class of homogeneous Markov fields on $\mathbb{Z}$ determined by a strictly positive matrix Q. We let $\mathscr{E}_Q$ comprise the extreme measures in $\mathscr{G}_Q$. It will now be proved, using the Martin boundary arguments of the previous section, that these *extreme* Markov fields on $\mathbb{Z}$ are always Markov processes.

Theorem 12-36: If $\mu \in \mathscr{E}_Q$, then $\{x_n\}_{n \in \mathbb{Z}}$ is a Markov process. Thus μ is determined by measures π_n and transition matrices P_n, $n \in \mathbb{Z}$, where

$$\pi_{n,i} = \Pr[x_n = i], \quad P_{n,ij} = \Pr[x_{n+1} = j \mid x_n = i],$$

and $\pi_n P_n = \pi_{n+1}$.

PROOF: It suffices to check the Markov property. For $n \geq 0$, let $\Lambda(n) = \{-n, -n+1, \ldots, n\}$. By Corollary 12-29, if $\mu \in \mathscr{E}_Q$, then $\mu = t\text{-}\lim_{n \to \infty} \mu^{\kappa^n}$ for some $(\kappa^0, \kappa^1, \ldots) \in \Omega$. This implies that there are states $k_n \in S$ for $n \in \mathbb{Z}$ such that

$$\begin{aligned} &\Pr[x_l = i_l \wedge x_{l+1} = i_{l+1} \wedge \cdots \wedge x_m = i_m] \\ &\quad = \lim_{n \to \infty} \Pr[x_l = i_l \wedge x_{l+1} = i_{l+1} \wedge \cdots \wedge x_m = i_m \mid \\ &\qquad\qquad x_{-n} = k_{-n} \wedge x_n = k_n] \end{aligned}$$

whenever $l \le m$. Therefore

$$\begin{aligned}\Pr[x_m = i_m \mid x_l = i_l \wedge x_{l+1} = i_{l+1} \wedge \cdots \wedge x_{m-1} = i_{m-1}] \\ = \lim_{n\to\infty} \Pr[x_m = i_m \mid x_{-n} = k_{-n} \wedge x_l = i_l \wedge x_{l+1} = i_{l+1} \wedge \\ \cdots \wedge x_{m-1} = i_{m-1} \wedge x_n = k_n].\end{aligned}$$

By the Markov field property, for all n sufficiently large the right hand side becomes

$$\begin{aligned}&\lim_{n\to\infty} \Pr[x_m = i_m \mid x_{-n} = k_{-n} \wedge x_{m-1} = i_{m-1} \wedge x_n = k_n] \\ &= \lim_{n\to\infty} \frac{\Pr[x_{m-1} = i_{m-1} \wedge x_m = i_m \mid x_{-n} = k_{-n} \wedge x_n = k_n]}{\Pr[x_{m-1} = i_{m-1} \mid x_{-n} = k_{-n} \wedge x_n = k_n]} \\ &= \Pr[x_m = i_m \mid x_{m-1} = i_{m-1}] = P_{m-1,i_{m-1}i_m}.\end{aligned}$$

Next we present a useful representation theorem for the extreme homogeneous Gibbs states with matrix Q.

Theorem 12-37: If $\mu \in \mathscr{E}_Q$, then there are strictly positive functions $g_{(n,i)}$ and $h_{(n,i)}$ ($n \in \mathbb{Z}$, $i \in S$) such that

(1) $g_{(n+1,i_{n+1})} = \sum_{i_n\in S} g_{(n,i_n)} Q_{i_n i_{n+1}}$,

(2) $h(n, i_n) = \sum_{i_{n+1}\in S} Q_{i_n i_{n+1}} h_{(n+1,i_{n+1})}$,

and

(3) $\sum_{i_n\in S} g_{(n,i_n)} h_{(n,i_n)} = 1$,

and such that the measures π_n and transition matrices P_n for $\{x_n\}$ are given by

$$\pi_{n,j} = g_{(n,j)} h_{(n,j)} \quad \text{and} \quad P_{n,jk} = Q_{jk} h_{(n+1,k)}/h_{(n,j)}.$$

Moreover, there are constants $c', c'' > 0$ and $k_n \in S$ ($n \in \mathbb{Z}$) such that

$$g_{(m,i_m)} = c' \lim_{n\to-\infty} (Q^{m-n})_{k_n i_m}/(Q^{-n})_{k_n 0} \qquad m < 0,\ i_m \in S$$

$$h_{(m,i_m)} = c'' \lim_{n\to\infty} (Q^{n-m})_{i_m k_n}/(Q^n)_{0k_n} \qquad m > 0,\ i_m \in S$$

Proof: Since $\{x_n\}$ is a Markov process,

$$\mu_{\overline{n+1}}^{\overline{n+1}}(\iota) = \frac{Q_{jk}Q_{k0}}{(Q^2)_{j0}} = \frac{P_{n,jk}P_{n+1,k0}}{(P_nP_{n+1})_{j0}}$$

and

$$\mu_{n+2}^{\overline{n+2}}(\iota) = \frac{Q_{k0}Q_{00}}{(Q^2)_{k0}} = \frac{P_{n+1,k0}P_{n+2,00}}{(P_{n+1}P_{n+2})_{k0}}$$

whenever $i_n = j$, $i_{n+1} = k$, $i_{n+2} = i_{n+3} = 0$. Let $\bar{h}_{(n,j)} = (Q^2)_{j0}/(P_n P_{n+1})_{j0}$, $c_n = Q_{00}/P_{n+1,00}$. Then dividing the first equation by the second and rearranging terms,

$$\frac{P_{n,jk}}{Q_{jk}} = \frac{1}{c_{n+1}} \frac{\bar{h}_{(n+1,k)}}{\bar{h}_{(n,j)}}.$$

Choose $\bar{c}_n$, $n \in \mathbb{Z}$, so that $\bar{c}_0 = 1$ and $\bar{c}_{n-1}/\bar{c}_n = c_n$. Now define $h_{(n,j)} = \bar{c}_n \bar{h}_{(n,j)}$ to get

$$P_{n,jk} = Q_{jk} h_{(n+1,k)}/h_{(n,j)}$$

as desired. Equation (2) follows immediately from the fact that P_n is a transition matrix. If we define $g_{(n,j)} = \pi_{n,j}/h_{(n,j)}$, then (3) holds because π_n is a probability measure, while the equation $\pi_n P_n = \pi_{n+1}$ implies (1). It therefore remains only to derive the representation of g and h as ratio limits of powers of Q. To this end, choose $\Lambda(n)$ and k_n as in the proof of the previous theorem. Then for $m > 0$,

$$\begin{aligned}\Pr[x_m &= i_m \mid x_0 = 0] \\ &= \lim_{n\to\infty} \Pr[x_m = i_m \mid x_{-n} = k_{-n} \wedge x_0 = 0 \wedge x_n = k_n] \\ &= \lim_{n\to\infty} \Pr[x_m = i_m \mid x_0 = 0 \wedge x_n = k_n],\end{aligned}$$

the last since $\{x_n\}$ is a Markov field. In terms of g, h and Q we have

$$\frac{g_{(0,0)}(Q^m)_{0i_m}h_{(m,i_m)}}{g_{(0,0)}h_{(0,0)}} = \lim_{n\to\infty} \frac{g_{(0,0)}(Q^m)_{0i_m}(Q^{n-m})_{i_m k_n}h_{(n,k_n)}}{g_{(0,0)}(Q^n)_{0k_m}h_{(n,k_n)}}$$

so that

$$h_{(m,i_m)} = h_{(0,0)} \lim_{n\to\infty} (Q^{n-m})_{i_m k_n}/(Q^n)_{0k_n}.$$

An analogous computation yields the result for g when $m < 0$.

With the aid of the above representation, cases where $|\mathscr{G}_Q| = 0$, 1 and ∞ will now be discussed.

Example 12-38: Let $S = \mathbb{Z}$, and consider any matrix Q of the form

$$Q_{ij} = q_{j-i} > 0, i, j \in \mathbb{Z}, \quad \text{where} \sum_{i\in\mathbb{Z}} q_i = 1.$$

Suppose that $\mu \in \mathscr{E}_Q$, with g and h the functions of Theorem 12-37. Equations (1) and (2) of that theorem say that $g_{(-m,i)}$ and $h_{(m,i)}$, $m \geq 0$, are space-time harmonic for sums of independent random variables. By analogy to Example 6 of Section 10-13, one can show that the extreme such functions are of the form $c^m t^i$ for some $c > 0$, $t > 0$ ($q_i > 0$ excludes degenerate solutions). Thus,

$$g_{(0,i)}h_{(0,i)} = \int_0^\infty \int_0^\infty [s^i t^i] d\nu(s) d\mu(t)$$

for some measures ν and μ on $(0, \infty)$. It follows that

$$\sum_{i \in \mathbb{Z}} g_{(0,i)}h_{(0,i)} = \int_0^\infty \int_0^\infty \left[1 + \sum_{i=1}^\infty (st)^i + \sum_{i=1}^\infty \frac{1}{(st)^i}\right] d\nu(s) d\mu(t)$$
$$= \infty,$$

since one of the two infinite sums must diverge for any given s and t. This contradicts equation (3) of Theorem 12-37, so $\mathscr{E}_Q$ is empty. $\mathscr{G}_Q$ is therefore empty by Theorem 12-28.

If $Q > 0$ is an ergodic transition matrix, then there is a unique probability measure $\alpha > 0$ such that $\alpha Q = \alpha$. In this case $\mathscr{G}_Q$ clearly contains the stationary process with

$$\pi_{n,i} = \alpha_i \quad \text{and} \quad P_{n,ij} = Q_{ij} \quad \text{for all } n \in \mathbb{Z}.$$

Thus, whenever $Q' \approx Q$ for some ergodic Q, then $|\mathscr{G}_{Q'}| \neq 0$. Our next goal is to show that when S is *finite* $\mathscr{G}_{Q'}$ contains exactly one Markov field for *any* $Q' > 0$, and that this field is a stationary Markov chain on $\mathbb{Z}$. The first step is contained in a lemma.

Lemma 12-39: If S is finite, then any $Q' > 0$ is equivalent to a strictly positive transition matrix Q (with $Q1 = 1$).

PROOF: We show that there is a vector $\bar{h} > 0$ such that $Q'\bar{h} = \bar{c}\bar{h}$ for some constant $\bar{c} > 0$. Then, defining

$$Q_{ij} = \frac{Q'_{ij}\bar{h}_j}{\bar{c}\bar{h}_i},$$

it follows that Q is a transition matrix, while $Q' \approx Q$ by Proposition 12-35. To get $\bar{h}$, let $\mathscr{S} = \{h = (h_i)_{i \in S}: h \geq 0 \text{ and } \sum_{i \in S} h_i = |S|\}$, and define $\bar{c} = \sup\{c: Q'h \geq ch \text{ for some } h \in \mathscr{S}\}$. Easy estimates prove that $0 < \min_{i,j} Q'_{ij} \leq \bar{c} \leq |S| \max_{i,j} Q'_{ij} < \infty$. By definition of $\bar{c}$ and Proposition 1-63, there are constants $c^{(n)}$ and elements $h^{(n)}$ of $\mathscr{S}$, $n = 0, 1, \ldots$, such that $c^{(n)} \to \bar{c}$, $Q'h^{(n)} \geq c^{(n)}h^{(n)}$, and $\lim_{n \to \infty} h^{(n)} = \bar{h}$ for some $\bar{h} \in \mathscr{S}$. This implies that $Q'\bar{h} \geq \bar{c}\bar{h}$. Now if $(Q'\bar{h})_j > \bar{c}\bar{h}_j$ for some j, then $Q'(Q'\bar{h}) > \bar{c}(Q'\bar{h})$ so $Q'(Q'\bar{h}) > (\bar{c} + \epsilon)(Q'\bar{h})$ for small ϵ.

But this contradicts the definition of $\bar{c}$ once we normalize $Q'\bar{h}$. Hence $\bar{c}\bar{h} = Q'\bar{h} > 0$, which shows that $\bar{h} > 0$.

Theorem 12-40: If S is finite, then $|\mathscr{G}_{Q'}| = 1$. $\mathscr{G}_{Q'}$ consists of the Markov chain with transition matrix Q and stationary measure α, where Q is defined as in Lemma 12-39 and α is the regular probability measure for Q.

PROOF: Suppose $\mu \in \mathscr{E}_Q$, and let k_n, $n \in \mathbb{Z}$, be the states in the ratio limit representation of Theorem 12-37 for the functions $g_{(m,i)}$ and $h_{(m,i)}$. Since S is finite, there is some state $j'' \in S$ and an infinite sequence $n'' \to \infty$ such that $k_{n''} \equiv j''$. Hence

$$h_{(m,i_m)} = c'' \frac{\lim_{n''\to\infty} (Q^{n''-m})_{i_m j''}}{\lim_{n''\to\infty} (Q^{n''})_{0j''}} = c'' \frac{\alpha_{j''}}{\alpha_{j''}} = c'',$$

by the convergence theorem for noncyclic ergodic chains. Thus h is constant. Similarly, there is some j' and sequence $n' \to -\infty$ with $k_{n'} \equiv j'$, whereby

$$g_{(m,i_m)} = c' \frac{\lim_{n'\to-\infty} (Q^{m-n'})_{j'i_m}}{\lim_{n'\to-\infty} (Q^{-n'})_{j'0}} = c' \frac{\alpha_{i_m}}{\alpha_0} = c_0\alpha_{i_m}.$$

It follows that $\pi_{n,j} \equiv \alpha_j$ and $P_{n,jk} \equiv Q_{jk}$. In other words, μ is uniquely determined as the process described in the statement of the theorem. For any strictly positive finite matrix Q' with equivalent transition matrix Q, we therefore have $|\mathscr{E}_{Q'}| = |\mathscr{E}_Q| = 1$. Theorem 12-28 now implies $|\mathscr{G}_Q| = 1$.

Next we present a concrete example of a matrix Q with phase multiplicity, for which all the elements of $\mathscr{E}_Q$ can be exhibited explicitly.

Example 12-41: Let $S = \{0, 1, \ldots\}$, and consider the strictly positive matrix Q given by

$$Q_{ij} = \sum_{k=0}^{i\wedge j} \left[\binom{i}{k}(\tfrac{1}{2})^i\right][e^{-1/2}(\tfrac{1}{2})^{j-k}/(j-k)!] \qquad (i \wedge j = \min\{i, j\}).$$

Q may be thought of as describing transition from i to j particles in a population. First particles disappear independently with probability 1/2, and then an independent Poisson distributed population of mean 1/2 is added to those which remain. This interpretation makes it clear that $Q1 = 1$. We remark next that Q is ergodic, with regular measure $\alpha_i = e^{-1}/i!$, $i \in S$. In fact, Q is α-reversible;

$$\alpha_i Q_{ij} = e^{-3/2} \sum_{k=0}^{i \wedge j} \left[\frac{\binom{i}{k}}{i!(j-k)!} \right] (\tfrac{1}{2})^{i+j-k}$$

$$= e^{-3/2} \sum_{k=0}^{i \wedge j} \left[\frac{\binom{j}{k}}{j!(i-k)!} \right] (\tfrac{1}{2})^{i+j-k} = \alpha_j Q_{ji},$$

since the terms in square brackets agree for every i, j and k. Thus $\mathscr{G}_Q$ contains the stationary Markov chain with $\pi_{n,i} = \alpha_i$ and $P_{n,ij} = Q_{ij}$ for every $n \in \mathbb{Z}$. In order to determine the other elements of $\mathscr{G}_Q$, we first compute the powers of Q. For this purpose it is convenient to introduce the generating functions $\gamma_i^n(s) = \sum_{j=0}^{\infty} (Q^n)_{ij} s^j$ $(|s| \leq 1)$, $i \in \mathbb{Z}$. We claim:

$$\gamma_i^n(s) = [1 - (\tfrac{1}{2})^n(1 - s)]^i \exp\{-[1 - (\tfrac{1}{2})^n](1 - s)\}.$$

When $n = 0$ both sides are s^i; the result follows by induction from the following considerations. If we start with i particles at time 0, and S_n denotes the number at time $n + 1$, then $S_{n+1} = X + \sum_{i=1}^{S_n} Y_i$, where X and the Y_i are independent, X Poisson with mean 1/2, the Y_i taking on values 0 and 1 each with probability 1/2. Since X has generating function $e^{-(1-s)/2}$ and each Y_i has generating function $(1 + s)/2$, it follows from the formula for the generating function of a random sum that $\gamma_i^{n+1}(s)$, the generating function of S_{n+1}, equals $e^{-(1-s)/2}\gamma_i^n((1 + s)/2)$ for $n \geq 0$. The desired formula for $\gamma_i^n(s)$ satisfies this recursion relation. Hence $(Q^n)_{ij}$, the coefficient of s^j in the power series expansion of $\gamma_i^n(s)$, is given by

$$(Q^n)_{ij} = \sum_{k=0}^{i \wedge j} \left[\binom{i}{k} (\tfrac{1}{2})^{nk} (1 - (\tfrac{1}{2})^n)^{i-k} \right] [e^{-(1-(1/2)^n)}(1 - (\tfrac{1}{2})^n)^{j-k}/(j - k)!].$$

If $\mu \in \mathscr{E}_Q$, then according to Theorem 12-37, the limits

$$g_{(m,i)} = c' \lim_{n \to \infty} (Q^{n+m})_{k_{-n}i}/(Q^n)_{k_{-n}0}, \; m \in \mathbb{Z}, \; i \in S,$$

exist and are strictly positive, for some $c' > 0$ and fixed sequence k_n, $n \in \mathbb{Z}$. We show next that these limits exist if and only if $\lim_{n \to \infty} k_{-n}/2^n = \theta$ for some θ (≥ 0). Under this assumption

$$(1 - (\tfrac{1}{2})^{n+m})^{k_{-n}} \to e^{-\theta(1/2)^m}$$

and

$$\binom{k_{-n}}{l}((\tfrac{1}{2})^{n+m})^l \to \frac{\theta^l(1/2)^{ml}}{l!},\ l \geq 0.$$

Hence

$$\begin{aligned}\lim_{n\to\infty} (Q^{n+m})_{k_{-n}i} &= \sum_{l=0}^{i}\left[\frac{\theta^l(1/2)^{lm}}{l!} e^{-\theta(1/2)^m}\right][e^{-1}/(i-l)!]\\ &= e^{-(\theta(1/2)^m+1)}\sum_{l=0}^{i}\binom{i}{l}[\theta(1/2)^m]^l/i!\\ &= e^{-(\theta(1/2)^m+1)}(\theta(1/2)^m+1)^i/i!.\end{aligned}$$

We can take this last quantity to be $g_{(m,i)}$ by choosing $c' = e^{-(\theta+1)}$. On the other hand, if $\lim_{n\to\infty} k_{-n}/2^n$ does not exist then we must have $k_{-n}/2^n \to \infty$ as $n \to \infty$, for otherwise the defining ratios would converge to distinct limits along different subsequences. But the ratio $(Q^{n+1})_{k_{-n}i}/(Q^n)_{k_{-n}0}$, for example, is a sum of positive terms including

$$\left[\frac{e^{-(1-(1/2)^{n+1})}}{e^{-(1-(1/2)^n)}}\right]\left[\frac{1-(\frac{1}{2})^{n+1}}{1-(\frac{1}{2})^n}\right]^{k_{-n}}\frac{[1-(\frac{1}{2})^{n+1}]^i}{i!} \geq c_i\left(1+\frac{1}{2^{n+1}}\right)^{k_{-n}},\ c_i > 0,$$

and this last expression tends to ∞ as $n \to \infty$ if $k_{-n}/2^n \to \infty$. We have therefore shown that the limits defining $g_{(m,i)}$ exist if and only if $k_{-n}/2^n \to \theta \in [0, \infty)$ as $n \to \infty$, in which case $g_{(m,i)}$ as a function of i may be taken to be Poisson with mean $\theta(\frac{1}{2})^m + 1$. Since Q is α-reversible, $\alpha_i(Q^n)_{ij} = \alpha_j(Q^n)_{ji}$, and hence the function $h_{(m,i)}$ of Theorem 12-37 can be computed as

$$\begin{aligned}h_{(m,i)} = c''\lim_{n\to\infty}\frac{(Q^{n-m})_{ik_n}}{(Q^n)_{0k_n}} &= c''\lim_{n\to\infty}\frac{\alpha_0(Q^{n-m})_{k_ni}}{\alpha_i(Q^n)_{k_n0}}\\ &= c'''e^{-(\eta 2^m+1)}(\eta 2^m+1)^i\end{aligned}$$

if $k_n(\frac{1}{2})^n \to \eta \in [0, \infty)$; otherwise the limit does not exist. Condition (3) of Theorem 12-37 now dictates $c''' = e^{1-\theta\eta}$. In summary, we have proved that if $\mu \in \mathscr{E}_Q$, then μ is one of the Markov process measures $\mu_{\theta\eta}$, $\theta, \eta \in [0, \infty)$. Here θ determines g, η determines h, and then g and h determine $\mu_{\theta\eta}$ according to Theorem 12-37. Finally, it remains to check that all the $\mu_{\theta\eta}$ are extreme, so that in fact $\mathscr{E}_Q = \{\mu_{\theta\eta};\ \theta, \eta \in [0, \infty)\}$. One first verifies, as in Example 6 of Section 10-13, that the topology of the Martin boundary B associated with $\mathscr{G}_Q$ is the usual topology on $\mathbb{R}^2_+ = \{(\theta, \eta): \theta \geq 0, \eta \geq 0\}$. Then by Theorem 12-28,

$$\mu_{\bar\theta\bar\eta} = \int_0^\infty\int_0^\infty \mu_{\theta\eta}\,d\lambda(\theta, \eta)$$

for a unique probability measure λ on $\mathbb{R}_2^+$ such that $\lambda(\{(\theta, \eta): \mu_{\theta\eta} \in \mathscr{E}_Q\}) = 1$. Evaluating both sides on $\{\omega \mid \omega_n = i, \omega_{n+1} = j\}$, we derive the equation

$$e^{-\bar{\theta}\bar{\eta}}\{e^{-(\bar{\theta}(1/2)^n+1)}(\bar{\theta}(\tfrac{1}{2})^n + 1)^i/i!\}\{e^{-(\bar{\eta}2^{n+1}+1)}(\bar{\eta}2^{n+1} + 1)^j/j!\}$$
$$= \int_0^\infty \int_0^\infty e^{-\theta\eta}\{e^{-\theta(1/2)^n+1}(\theta(\tfrac{1}{2})^n + 1)^i/i!\}$$
$$\times \{e^{-(\eta 2^{n+1}+1)}(\eta 2^{n+1} + 1)^j/j!\}d\lambda(\theta, \eta).$$

Multiply both sides by $u^i v^j$ ($u, v \in \mathbb{R}$), sum over $i, j \in S$, and interchange summation and integration, to obtain

$$e^{-\bar{\theta}\bar{\eta}}\{e^{-\bar{\theta}(1/2)^n(1-u)}\}\{e^{-\bar{\eta}2^{n+1}(1-v)}\}$$
$$= \int_0^\infty \int_0^\infty e^{-\theta\eta}\{e^{-\theta(1/2)^n(1-u)}\}\{e^{-\eta 2^{n+1}(1-v)}\}d\lambda(\theta, \eta).$$

If we make the change of variables: $x = (1 - u)/2^n$, $y = 2^{n+1}(1 - v)$, then for $x \geq 0$ and $y \geq 0$ the right hand side is the double Laplace transform of the measure $e^{-\theta\eta}d\lambda(\theta, \eta)$. Since the equation is satisfied when λ concentrates at $(\bar{\theta}, \bar{\eta})$, the uniqueness theorem for Laplace transforms implies that λ must be this measure. We conclude that $\mu_{\bar{\theta}\bar{\eta}} \in \mathscr{E}_Q$, as desired.

Whenever Q is a transition matrix and $\mu \in \mathscr{E}_Q$ admits a representation with $h_{(n,i)} \equiv 1$, then $\pi_{n,i} = g_{(n,i)}$ and $P_n \equiv Q$. A family of probability distributions π_n, $n \in \mathbb{Z}$, such that $\pi_n Q = \pi_{n+1}$ is called an **entrance law** for Q. For example, the Poisson distributions π_n with mean $\theta(\frac{1}{2})^n + 1$ for fixed $\theta \in [0, \infty)$ constitute an entrance law for the matrix Q of the last example. The case $\theta = 0$ yields the stationary Markov process with regular measure α; when $\theta > 0$ the process $\{x_n\}$ with measure $\mu_{\theta\eta}$ "comes down from infinity" in the sense that $\lim_{n\to-\infty} \Pr[x_n = i] = 0$ for any $i \in S$. A more surprising type of entrance law is described in Problem 8 at the end of the chapter.

The matrix Q of Example 12-41 may also be used to illustrate phase multiplicity on the non-negative half-line. Namely, for $\eta \in [0, \infty)$, let $h^\eta_{(n,i)} = e^{-\eta 2^n}(\eta 2^n + 1)^i$, and define $\pi^\eta_{0,i} = \alpha_i h^\eta_{(0,i)}, P^\eta_{n,jk} = Q_{jk} h^\eta_{(n+1,k)}/h^\eta_{(n,j)}$. If we set $\pi_0 = \alpha$, $P_n \equiv Q$, $\pi'_0 = \pi^\eta_0$ and $P'_n = P^\eta_n$ in Proposition 12-33, then the hypotheses there are clearly satisfied. Thus we have constructed a large family of one-sided Markov fields with the same characteristics as the Markov chain with transition matrix Q and initial distribution α.

As a final application of Example 12-41, we exhibit an element of $\mathscr{G}_Q$ which is not a Markov process. Consider the field $\{x_n\}$ given by the convex combination $\mu = \frac{1}{2}\mu_{00} + \frac{1}{2}\mu_{11}$. Clearly $\mu \in \mathscr{G}_Q$, and to see that μ is not the measure for a Markov process it suffices to check that

$$\Pr[x_2 = 0 \mid x_0 = 0 \wedge x_1 = 0] \neq \Pr[x_2 = 0 \mid x_1 = 0].$$

The π_n and P_n for μ_{11} satisfy $\pi_{0,0} = e^{-4}$, $\pi_{1,0} = e^{-9/2}$, $P_{0,00} = e^{-1}Q_{00}$ and $P_{1,00} = e^{-2}Q_{00}$. Thus

$$\Pr[x_2 = 0 \mid x_0 = 0 \wedge x_1 = 0] = \frac{\frac{1}{2}(e^{-1}Q_{00}Q_{00}) + \frac{1}{2}(e^{-4}e^{-1}Q_{00}e^{-2}Q_{00})}{\frac{1}{2}(e^{-1}Q_{00}) + \frac{1}{2}(e^{-4}e^{-1}Q_{00})}$$

$$= \frac{1 + e^{-6}}{1 + e^{-4}}Q_{00},$$

while

$$\Pr[x_2 = 0 \mid x_1 = 0] = \frac{\frac{1}{2}(e^{-1}Q_{00}) + \frac{1}{2}(e^{-9/2}e^{-2}Q_{00})}{\frac{1}{2}e^{-1} + \frac{1}{2}e^{-9/2}}$$

$$= \frac{1 + e^{-11/2}}{1 + e^{-7/2}}Q_{00}.$$

Hence $\{x_n\}$ has the two-sided Markov property, but *not* the one-sided Markov property.

We conclude this section by mentioning without proof the deepest result to date in the theory of denumerable Markov fields on $\mathbb{Z}$. When T has a group structure it is natural to consider the class $\mathscr{G}_V^0$ consisting of all those $\mu \in \mathscr{G}_V$ which are invariant under translations. The following theorem completely determines the possibilities for $\mathscr{G}_{Q'}^0$, the translation invariant Markov fields on $T = \mathbb{Z}$ with characteristics determined by the strictly positive matrix Q'.

Theorem 12-42: If $\mu \in \mathscr{G}_{Q'}$, then $Q' \approx Q$ for some strictly positive Q such that $Q1 \leq 1$. If $Q1 = 1$ and Q is ergodic, then $|\mathscr{G}_{Q'}^0| = 1$. In this case the unique member of $\mathscr{G}_{Q'}^0$ is the stationary Markov chain on $\mathbb{Z}$ with transition matrix Q and $\pi_n \equiv$ the Q-regular measure α. In all other cases $\mathscr{G}_{Q'}^0 = \varnothing$.

6. Examples of phase multiplicity in higher dimensions

When $T = \mathbb{Z}^d$ for $d \geq 2$, the conclusion of Theorem 12-40 ceases to hold. In other words, there are instances of phase multiplicity for homogeneous potentials U with S finite. This phenomenon is undoubtedly the most important in the theory of random fields, but an adequate treatment is far beyond the scope and purpose of this chapter. Instead, we briefly discuss two examples. Suggestions for further reading are included in the Additional Notes.

Example 12-43: Tree processes. $S = \{0, 1, \ldots, N\}$ and T is a countable **tree** (i.e., T is endowed with a neighbor structure ∂ which defines a connected graph with no loops). Let P be a strictly positive N-state transition matrix with regular measure α, and assume that P is α-reversible. The random field $\{x_t\}$ on S^T is called a **tree process** for (α, P) if μ is defined by the following three properties:

(i) $\Pr[x_t = i] = \alpha_i$ for all $i \in S$, $t \in T$;

(ii) $\Pr[x_{t_0} = i_0 \wedge x_{t_1} = i_1 \wedge \cdots \wedge x_{t_l} = i_l] = \alpha_{i_0} P_{i_0 i_1} \cdot \cdots \cdot P_{i_{l-1} i_l}$ whenever $\{t_0, t_1, \ldots, t_l\}$ is a finite path in the tree T and $i_0, i_1, \ldots, i_l \in S$;

(iii) $\Pr[x_{t_0} = i_0 \wedge \cdots \wedge x_{t_l} = i_l \mid x_{t_r} = i_r \text{ for all } r \in \Lambda]$
$= \Pr[x_{t_0} = i_0 \wedge \cdots \wedge x_{t_l} = i_l \mid x_{t_0} = i_0]$
for any finite $\Lambda \subset T$ and path $\{t_0, \ldots, t_l\} \subset T$ which intersect only at site t_0, and any $i_0, \ldots, i_l \in S$.

For given α and Q, conditions (i)–(iii) determine a well-defined and unique Markov field. According to (i) and (ii) the process behaves like a reversible Markov chain along paths (reversibility ensures that cylinder probabilities are independent of the direction we travel along a path in (ii)), while paths "patch together" because of condition (iii). As a special case, suppose that $S = \{0, 1\}$ and T is the tree with three neighbors for every site (sometimes called the 3-Bethe lattice). Consider the tree processes with measures μ' and μ'' induced by (α', P') and (α'', P'') respectively, where

$$\alpha' = (\tfrac{1}{3}, \tfrac{2}{3}) \qquad P' = \begin{pmatrix} \frac{2}{3} & \frac{1}{3} \\ \frac{1}{6} & \frac{5}{6} \end{pmatrix},$$

$$\alpha'' = (\tfrac{4}{5}, \tfrac{1}{5}) \qquad P'' = \begin{pmatrix} \frac{8}{9} & \frac{1}{9} \\ \frac{4}{9} & \frac{5}{9} \end{pmatrix}.$$

It is not hard to verify that both fields have the same local characteristics, so there is phase transition for the potential V corresponding to these characteristics. For instance,

$$\mu'(\{x_a = 0\} \mid \{x_t = 0 \text{ for all } t \in \partial a\}) = \frac{(\frac{1}{3})(\frac{2}{3})^3}{(\frac{1}{3})(\frac{2}{3})^3 + (\frac{2}{3})(\frac{1}{6})^3} = \tfrac{32}{33},$$

$$\mu''(\{x_a = 0\} \mid \{x_t = 0 \text{ for all } t \in \partial a\}) = \frac{(\frac{4}{5})(\frac{8}{9})^3}{(\frac{4}{5})(\frac{8}{9})^3 + (\frac{1}{5})(\frac{4}{9})^3} = \tfrac{32}{33}.$$

In this setting, if $P = \begin{pmatrix} 1-p & p \\ q & 1-q \end{pmatrix}$ then $\alpha = \left(\frac{q}{p+q}, \frac{p}{p+q}\right)$ and P is always α-reversible. The (α, P)-process is called **attractive** when $p + q \leq 1$. Roughly, attractiveness means that a 1 at site t increases the likelihood of 1's near t (and similarly for 0's). More precisely this implies that in Theorem 12-30 $\mu^{\kappa^n}(\{x_t = 1\})$ is maximized when κ^n consists of "all 1's" on $K(n)$. Using this fact it is possible to show that in the attractive case there is phase multiplicity for the potential corresponding to (α, P) if and only if $(p - q)^2 - 2(p + q) + 1 \geq 0$ and $(p, q) \neq (\frac{1}{4}, \frac{1}{4})$. The (α, P)-process is **repulsive** when $p + q \geq 1$, and in this case one can prove by entirely different methods that phase multiplicity occurs if and only if $p + q > \frac{3}{2}$.

Random fields on countable trees T have the advantage that most physically meaningful quantities can be computed explicitly; the fact that T has no loops enables one to use inductive methods. The natural setting for statistical mechanics is $T = \mathbb{Z}^d$, however, and in this case the theory is immensely more difficult. We summarize some of the leading results for the simplest Markov fields on the two-dimensional integer lattice in our last example.

Example 12-44: Two-dimensional Ising model. $S = \{0, 1\}$, $T = \mathbb{Z}^2$. V is a normalized potential of the form:

$$V_{\{a\}}(\iota) = v_0 \qquad V_{\{a,b\}}(\iota) = v_1$$

whenever $|a - b| = 1$ and $i_a = i_b = 1$, with $V_A(\iota) = 0$ in all other cases. V is **attractive** if $v_1 \geq 0$, **repulsive** otherwise; the intuitive interpretation is the same as in the previous example. When V is attractive there is phase multiplicity if and only if $v_0 + 2v_1 = 0$ and $v_1 > 2\ln(\sqrt{2} + 1)$. For repulsive V there is phase multiplicity in an open neighborhood of the line segment $\{(v_0, v_1): v_0 + 2v_1 = 0$ and $v_1 < K\}$, with K sufficiently negative. Similar results hold in higher dimensions, though less is known.

7. Problems

1. Show that if $S = \{0, 1\}$, T is a finite subset of $\mathbb{Z}^d$, and V is a normalized neighbor potential, then the energy H_V of V may be expressed as

$$H_V(\iota) = \tfrac{1}{2} \sum_{a \in T} \sum_{b \in \bar{a}} v_{ab} i_a i_b \qquad \iota = \{i_l\} \in \Omega,$$

for some $v = \{v_{ab} \in \mathbb{R}, |a - b| \neq 1\}$ satisfying $v_{ab} = v_{ba}$.

2. Give an example of a finite Gibbs field which cannot be represented in terms of a pair potential.

3. Let ∂ be a given neighbor system. The **K-neighbor set** $\partial^K a$ ($K = 1, 2, \ldots$) of $a \in T$ is defined recursively by $\partial^1 a = \partial a$, $\bar{\partial}^1 a = \bar{a}$, and for $K > 1$, $\partial^{K+1} a = \partial(\bar{\partial}^K a)$, $\bar{\partial}^{K+1} a = (\partial^{K+1} a) \cup (\bar{\partial}^K a)$. Thus a **$K$-neighbor** t of a is a site which can be reached from a in K steps to neighboring sites, and no fewer. A random field $\{x_t\}$ is called a **K-Markov field** (with respect to ∂) if $\mu_a^A = \mu_a^{\bar{\partial}^K a}$ whenever $\bar{\partial}^K a \subset A \subset T$, A finite. A potential U is called

a K-**neighbor potential** if $U_A = 0$ whenever A contains two sites which are not K-neighbors. Show that $\{x_t\}$ is K-Markov if and only if the canonical potential V for $\{x_t\}$ is K-neighbor. What does this say when $K = 0$? [*Hint:* Define a new neighbor system.]

4. Suppose that there is a metric d defined on T, and say that $\{x_t\}$ is L-**Markov** ($L \geq 0$) if $\mu_a^A = \mu_a^{B(a,L)}$ whenever $B(a, L) = \{t \in T: d(a, t) \leq L\} \subset A \subset T$. What property for the canonical potential V is equivalent to the L-Markov property for $\{x_t\}$? State and prove a theorem to justify your assertion. Describe the $\sqrt{2}$-Markov fields on $\mathbb{Z}^2$.
5. Show that if $\{x_t\}$ has *any* neighbor potential U, then $\{x_t\}$ is a Markov field. Give an example of a Markov field with potential U, such that U is *not* a neighbor potential. [*Hint:* For the first part, look carefully at the proof of Theorem 12-16.]
6. Let $T = \mathbb{Z}$. Prove that $\mathscr{G}_Q = \varnothing$ if

$$\lim_{n\to\infty} \sup_{i,k\in S} \frac{(Q^n)_{ij}(Q^n)_{jk}}{(Q^{2n})_{ik}} = 0 \quad \text{for some } j \in S.$$

[*Hint:* Show that this condition forces $\Pr[x_0 = j] = 0$.]
7. Give an example of a Markov process $\{x_n\}_{n\in\mathbb{Z}}$ which is *not* extreme in its class of Markov fields. [*Hint:* Use the matrix Q of Example 12-41.]
8. Let $S = \{0, 1, \ldots\}$, $T = \mathbb{Z}$. Define α_i inductively by $\alpha_0 = \frac{1}{2}$, and $\alpha_i = (\alpha_{i-1}//3) + \frac{1}{2}(\frac{1}{4})^i$ for $i \geq 1$. Let

$$Q_{ij} = \begin{cases} \alpha_j & i = 0, j \in S \\ [\frac{1}{2}(\frac{1}{4})^i + \frac{1}{3}\delta_{i-1,j}](\alpha_j/\alpha_i) & i \geq 1, j \in S \end{cases}.$$

Finally, put

$$\pi_{n,i} = \begin{cases} \alpha_i + (\delta_{(-n-1)i} - \alpha_i) \prod_{k=-n-1}^{\infty} (\alpha_k/3\alpha_{k+1}) & n < 0 \\ \alpha_i & n \geq 0 \end{cases}.$$

Show that α and all of the π_n are strictly positive probability vectors on S, and that Q is a strictly positive transition matrix with $Q1 = 1$. Finally, prove that $\alpha Q = \alpha$, $\pi_n Q = \pi_{n+1}$, and $\pi_n \neq \alpha$ for $n < 0$. Thus Q has an entrance law which agrees with the stationary one from time 0 on, but *not* before time 0.
9. Let V be any neighbor potential on $\mathbb{Z}$. Show that if $\mu \in \mathscr{E}_V$, then $\{x_n\}$ is a Markov process.
10. Suppose g and h satisfy (1)–(3) of Theorem 12-37 for some $Q > 0$. Show that $\pi_{n,j}$ and $P_{n,jk}$ as prescribed in that theorem give rise to a well-defined field $\{x_n\}$. Is $\{x_n\}$ in $\mathscr{G}_Q$? Is it in $\mathscr{E}_Q$?

NOTES

Chapter 3:

Stochastic processes with the martingale property were first studied by Lévy [1937]. Lévy considered, in Sections 67 to 70, partial sums of sequences $\{f_n\}$ such that

$$M[f_{n+1} \mid f_0 \wedge \cdots \wedge f_n] = 0.$$

These are a natural generalization of sums of independent random variables with mean 0. He proved theorems such as a central limit theorem suggested by comparison with sums of independent random variables. Ville [1939] recognized the importance of studying processes representing a fair game and for which system theorems should hold. He called these processes martingales. Although he did not prove any convergence theorems, he did prove the inequality given in Problem 7. From this he was able to conclude that non-negative martingales had finite lim sup with probability one. He made application of this to the study of sample paths of coin tossing. In particular, he proved one half of the law of the iterated logarithm. The basic convergence theorem, Theorem 3-12, for martingales was proved by Doob [1940]. In his book on stochastic processes, Doob [1953] introduced submartingales (called semi-martingales in that book) and made a systematic study of the system theorems and convergence theorems for these processes. The proof of Proposition 3-11 for martingales is due to Doob [1940]. The proof given here and the extension to submartingales is due to Snell [1952]. Additional applications of martingale theory to Markov chains may be found in Lamperti [1960a] and [1963a].

Chapter 4:

Markov chains with a finite number of states were introduced by Markov [1907]. Kolmogorov [1936] considered the case of a denumerable number of states. Important contributions in the foundations of Markov chains were made by Doeblin [1938]. There are a number of books devoted to the study of finite Markov chains. Among these are Fréchet [1938], Romanovskiĭ [1949], Kemeny and Snell [1960], Lahres [1964], and Gorden [1965]. The theory of denumerable Markov chains is the subject of a book by Chung [1960].

Finite random walks have been analyzed in some detail in Kemeny and Snell [1960], Chapter 7. See also Kac [1947a]. The books by Spitzer [1964] and Kemperman [1961] give detailed studies of Markov chain problems applied to sums of independent random variables. The class of random walks discussed in Example 8 was introduced by Karlin and McGregor [1959] who made an extensive study of these processes. There is a large literature on branching processes. References to this literature as well as an account of the theory of these processes may be found in a book of Harris [1963]. The process called the basic example in this book is often referred to as a "renewal process."

The recognition of the need for and the importance of system theorems is due to Doob. See Doob [1953], Chapter VII. The strong Markov property which holds for any denumerable Markov chain does not hold for general Markov processes where time is allowed to be continuous and the state space is the real line. For a discussion of this problem in the more general setting, see Blumenthal [1957].

A discussion of system theorems and a rather systematic use of these theorems in Markov chain theory may also be found in Chung [1960].

Chapter 5:

Kemeny and Snell [1960], Chapter III, showed that the fundamental matrix N could be used to obtain moments of many descriptive quantities for finite absorbing chains. The extension of this use of N to denumerable chains was made in Kemeny and Snell [1961b]. Theorem 5-10 is the analog of the Riesz Decomposition Theorem for superregular functions. A systematic discussion of results of this type which exploit the analogy between superregular functions for a Markov chain and classical superharmonic functions may be found in Feller [1956] and Doob [1959]. The proof of Proposition 5-20 is due to Dynkin and Malyutov [1961].

Proposition 5-22 is true even if we drop the hypothesis of finitely many k-values. This theorem is due to Chung and Erdos [1951], and a simplified proof of this result was given by Chung and Ornstein [1962].

Chapter 6:

Theorem 6-9 is due to Derman [1954]. He proved existence by showing that $\alpha_j = {}^i\bar{N}_{ij}$ is a regular measure. His proof of uniqueness is less elementary than ours and uses the Doeblin ratio theorem applied to the chain reversed by α. This ratio theorem proved by Doeblin [1938] states for recurrent chains that $\lim_{n\to\infty} (N_{ij}^{(n)}/N_{kl}^{(n)})$ exists and is independent of i and k. Derman also used the identification of this limit as α_j/α_l.

Proposition 6-24 is due to Kac [1947b].

Doeblin [1938] proved Proposition 6-32 and then applied limit theorems for sums of independent random variables to obtain limit theorems for Markov chains. For details of this technique and resulting limit theorems, see Chung [1960], pp. 75–106. The converse to Proposition 6-32 is due to Yosida and Kakutani [1940].

The fact that $\lim_{n\to\infty} P^n$ exists for a noncyclic recurrent chain is due to Kolmogorov [1936], [1937]. The extension of this result given in Theorem 6-38 is due to Orey [1962], and the first proof (including Lemmas 6-36 and 6-37) is a somewhat simplified version of his proof. The proof using the Renewal Theorem may be found in Feller [1957]. Theorem 6-43 is new.

Chapter 7:

The details of the connection between Brownian motion and classical potential theory are discussed by Knapp [1965]. The recognition of the importance of identifying these two theories started with Kakutuni [1944]. Doob [1954] made significant extensions of the results of Kakutani by further identifying martingale and submartingale theory with the theory of harmonic and superharmonic functions. Important contributions again exploiting connections between Brownian motion and classical potential theory were made by Kac [1951]. The next major contribution was made by Hunt [1957], [1958]. Hunt showed that one could develop a potential theory for essentially the most general Markov process. He considers continuous time and abstract state space. He showed conversely that under rather minimal requirements for a potential theory one can construct a Markov process associated with this theory. His work related to the potential theory that goes with transient processes.

Although many of Hunt's results go over easily to the Markov chain case, even for transient chains new problems arise, and a whole new theory must be developed for the recurrent case. These extensions were made by Kemeny and Snell [1961b] for general Markov chains and by Spitzer [1962] for the important class of Markov chains which arise from sums of lattice-valued independent random variables.

The fact that the symmetric random walk in one and two dimensions is recurrent, whereas in dimension three or greater it is transient was first proved by Polya [1921]. The proof of Proposition 7-10 was supplied by Lamperti.

Chapter 8:

The notion of h-regular function was introduced into the study of potential theory by Brelot [1956] and the corresponding idea of a function regular in the h-process for chains was discussed in Feller [1956] and Doob [1959].

A discussion of equilibrium potential, equilibrium charge, and capacity as they arise in electrostatics and Newtonian potential theory may be found in the book of Kellogg [1929]. A somewhat more modern approach may be found in Brelot [1959]. In the classical theory the Green's function which plays the role of the matrix N is always symmetric. The fact that there is an interesting potential theory even for nonsymmetric operators in probability was first shown by Hunt [1957], [1958].

The results of Sections 1 and 2 of this chapter were for the most part in Doob [1959]. Those of Sections 3 and 4 are specializations to the Markov chain case of results obtained by Hunt [1957], [1958] for more general Markov processes.

Choquet and Deny [1956–57] investigated the problem of the relation between the various potential principles for the case of potentials of the form $g = Gf$, where G is an arbitrary non-negative finite matrix. If G has an inverse they proved that the Principles of Balayage and Domination are equivalent and that each implies the Principle of Lower Envelope. Here every non-negative function f is a charge. They showed further that if G satisfies the Principle of Lower Envelope, then there is a unique permutation of the columns of G such that the resulting matrix satisfies the Principle of Balayage. Also they showed that G satisfies the Principle of Balayage if and only if it is of the form $G = A \sum_{p=0}^{\infty} S^p$, where A is a diagonal matrix with strictly positive diagonal entries and S is a non-negative matrix. Thus the most general operator here is only slightly more general than the class of all matrices of the form $N = (I - Q)^{-1}$, where Q is a finite transient chain. Some investigation of this problem for denumerable matrices was made by Kemeny and Snell [1961b].

The notion of energy seems to be significant only in the case of reversible chains and even here it does not have the nice probabilistic interpretations that the other potential theory concepts have.

The results of Section 8 are taken from Kemeny and Snell [1961b]. The idea of developing a potential theory for supermartingales was suggested by Doob [1961] where he also indicated the proof of Proposition 8-79.

Chapter 9:

The potential theory for recurrent chains discussed in this chapter was introduced by Kemeny and Snell [1961b].

The existence of the limit in Theorem 9-4 was first proved by Doeblin [1938]. The identification of the limit was made by Chung [1950]. The present proof is from Kemeny [1962].

Theorem 9-7 was proved under a mild assumption in Kemeny and Snell [1961b]. The case $i = j$ was proved in general by a method due to Chung in Chung [1961] and Kemeny and Snell [1961c]. The general case was proved by Kemeny [1963].

The fact that all ergodic chains are normal follows from Theorem 4, Chapter 1, Section 11 of Chung [1961]. The remaining results in the first three sections are taken primarily from Kemeny and Snell [1961b].

Proposition 9-65 was first proved by Lamperti [1960b]. Results of the form of Propositions 9-67 and 9-68 may be found in Chung [1960] Chapter 1, Section 11.

The notion of strong ergodic chains introduced here is new. The matrix Z was introduced in Kemeny and Snell [1960]. It was shown in this book that the matrix Z for finite ergodic chains could be used to express the moments of many interesting descriptive quantities and hence played for recurrent chains a role similar to the matrix N for finite absorbing chains.

All sums of independent random variables processes which form aperiodic recurrent Markov chains are normal. This was proved by Kemeny and Snell [1961a] for the case of finite variance and in general by Spitzer [1962].

The operator K was introduced by Kemeny and Snell [1963b]. Most of the results of Sections 8 and 9 are taken from this paper.

The method of associating denumerable chains with electric circuits discussed in Section 10 was carried out by Nash–Williams [1959] under slightly more restrictions on the chain than we impose. He proved also Lemma 9-129.

Chapter 10:

The Martin boundary for Markov chains was introduced independently by Doob [1959] and Watanabe [1960a]. Doob and Watanabe used the methods which were developed in the study of the classical Martin boundary relevant to Newtonian potentials. Details of this approach may be found in Brelot [1956] and Doob [1957] or Watanabe [1960a].

Hunt [1960] gave a new and more probabilistic treatment of Martin boundary theory for Markov chains and completed the work of Doob and Watanabe in several ways. In particular, he introduced a new class of processes called approximate P-chains. These are slightly more general than the processes we have called extended chains. Our treatment of the Martin boundary is for the most part a rewriting of Hunt's paper with more detail supplied, except that we have used a slightly different definition of boundary than that listed by Hunt. The difference is as follows. Doob introduced a metric on the state space, the one we have used if π assigns all its weight to one point, and completed the state space in terms of this metric. A point was called a boundary point by Doob in the completed space if it was a limit point of the original states. It is possible for one of the original states to be such a limit point. To avoid this peculiarity, Hunt modified Doob's metric slightly to make such a point into a new point. Since these new points are always nonminimal points and appear to play no essential role in the theory, we have followed Doob's metric. However, we have chosen to call the boundary simply the new points added by the completion. The observation that $\pi N > 0$ is the only condition on π that is needed appears in Orey [1964].

One can also use G. Choquet's theory of convex cones to develop Martin boundary theory. This approach has been carried out by Neveu [1964]. See also Hennequin and Tortrat [1965].

Brelot [1956] showed that the Martin boundary was ideally suited to the study of the first boundary problem, or the Dirichlet problem. His approach was to generalize the method developed by Perron and Wiener (see Kellogg [1929]) for regions in Euclidean space.

The probabilistic approach to the first boundary problem was first suggested by Kakutani [1945] and done more generally using the Martin boundary by Doob [1958]. The method presented in this book is the probabilistic approach of Kakutani and Doob.

The discussion of fine boundary limits follows that of Doob [1957], who considered these problems for superharmonic functions using Brownian motion theory.

The Martin boundary has now been worked out for several important classes of Markov chains. In particular, Doob, Snell and Williamson [1960] have worked out the boundary for general sums of independent random variables. Related results may be found in Dynkin and Malyutov [1961]. There are close connections between classical moment problems and the Martin boundary for certain of these processes. A discussion of this point may be found in Watanabe [1960b]. Lamperti and Snell [1963] discussed the Martin boundary for the class of random walks introduced by Karlin and McGregor [1959]. This discussion was generalized by Kemeny [forthcoming]. Finally Blackwell and Kendall [1964] have given a discussion of the Martin boundary for the Polya urn scheme.

The result in Example 3 that the only positive regular functions for the symmetric random walk in three dimensions are the constants was proved by Murdoch [1954] by other methods. Murdoch obtained a better estimate of N_{0j} than that given here. The short proof of the estimate for N_{0j} that we give was supplied by E. Stein.

The results in Problems 30 to 34 were discovered by Harris [1957] and Veech [1963].

A point x of S^* is regular for the Dirichlet problem if for each continuous function $f \geq 0$ on S^* the superregular function h with f as boundary values has $\lim_{j \to x} h(j) = f(x)$. An equivalent condition on x is that for each open neighborhood U of x, $\lim_{j \to x} \Pr_j[x_v \in S^* - U] = 0$. Knapp [1966] showed that the set of regular points is a Borel set and gave an example of a chain P with $P1 = 1$ for which the set of regular points was empty.

Chapter 11:

The main results of this chapter were presented in a paper by Kemeny and Snell [1963a]. As noted in this paper, these authors are indebted to W. A. Veech for the important Lemma 11-11. The recurrent boundary was introduced independently by Orey [1964].

ADDITIONAL NOTES

The discussion below deals with some of the developments in the theory since the publication of the first edition. Some papers that predate that publication are mentioned to put matters in context. Citations point to the Additional References except when the bracketed date is followed by "R." The latter citations point to the References section.

Chapter 12:

The foundations for the theory of Markov fields and random fields in general were developed by Dobruschin (e.g., [1968]) in a series of papers. Our treatment of finite random fields and the equivalence theorem for Markov and neighbor Gibbs fields, as presented in Sections 2 and 3, is based on Griffeath [1973]. K. L. Chung and D. Dawson made helpful improvements in the presentation. Theorem 12-16 is due in essence to Averintsev [1970], though he considered only the case $T = \mathbb{Z}^d$. A series of papers, culminating in Grimmett [1973], exploited the Möbius inversion formula to obtain simpler proofs in a more general context. The Martin boundary approach to infinite Gibbs fields is due to Föllmer [1975a], and is based on the work of Dynkin [1971]. Their setting is far more general than the one presented here, so their arguments are not as elementary. The detailed study of countable Markov fields on $\mathbb{Z}$ was initiated by Spitzer [1975a]; much of Section 5 is based on his paper. Föllmer [1975b] has also treated this subject. The proof of Proposition 12-32 was supplied by H. Kesten. Theorems 12-36 and 12-37 were obtained by Spitzer using tail fields rather than the Martin boundary approach. The ratio-limit representation of Theorem 12-37, which does not appear in Spitzer's paper, is cited by Cox [1976]. Example 12-38, due to Spitzer, makes use of Doob, Snell, and Williamson [1960R]. The important Theorem 12-40 was discovered by Dobruschin [1968]. Example 12-41 is a special case of a family of phase transition examples discussed by Cox [1976]. The remarkable Theorem 12-42 is due to Kesten [1976]; the reader is referred to his paper for the proof. Spitzer [1974] gives a very nice exposition of many aspects of random field theory not discussed here. Another useful reference is Dawson [1974]. Tree processes were first studied by Preston [1974], whose book contains a wealth of information on random fields. A more recent reference is Spitzer [1975b]. A lucid exposition of the Ising model may be found in Griffiths [1972]. Problems 7 and 8 are derived from Spitzer [1975a]; Problem 9 is based on a construction of S. Kalikow [1976].

Corrections to the first edition:

Pitman [1974] pointed out that Theorem 9-53 was stated incorrectly in the first edition. It had stated incorrectly that $g = -Gf$, overlooking the possible failure of the relation $A({}^0Nf) = (A\,{}^0N)f$. This theorem and its consequences have been corrected in the second edition.

For general ergodic chains we can give no useful condition under which this associativity holds. However, for a strong ergodic chain associativity holds if f is bounded (and in particular if the potential g is bounded, since $f = (I - P)g$). In fact, it is enough to observe that $\alpha\,{}^0N1 \le \alpha\,{}^0\bar{N}1 = M_{\alpha 0} < \infty$. Thus for a strong ergodic chain, $g = -Gf$ if the charge f is bounded. This conclusion was obtained in a more direct way in Proposition 9-73.

Martingales:

Two books that develop the subject of martingales are those by Meyer [1972] and Neveu [1972a]. Both books begin with the basic material on martingales. Neveu's contains a short chapter on the optimal stopping problem that is discussed later in these notes. The latter part of each book begins to reflect the explosion in the subject of martingales that has centered around integral inequalities.

Early developments were by Burkholder [1966] and Gundy [1967]. Suppose $(f_n, \mathscr{F}_n)$ is a martingale and d_n is the sequence of differences $d_0 = f_0$, $d_n = f_n - f_{n-1}$ for $n \ge 1$, so that $f_n = d_0 + \cdots + d_n$. If v_n is measurable with respect to $\mathscr{F}_{n-1}$, then the sequence g_n with

$$g_n = \sum_{k=0}^{n} v_k d_k$$

is called a *transform* of f_n. It is a martingale if $M[|g_n|] < \infty$ for all n, by imitation of the proof of Proposition 3-7. (The case that the v_n are characteristic functions arises in the proof of the Upcrossing Lemma and is the case of optional sampling.) Burkholder and Gundy deal with questions of convergence and integral boundedness of such transforms. The gambling interpretation is as follows: A gambler playing a sequence of rounds in a fair game can win d_n dollars in round n, and his fortune is then f_n. If $\sup M[|f_n|] < \infty$, his fortune converges to a finite limit f_∞ a.e. and $M[|f_\infty|] \le \sup M[|f_n|]$. In the transformed game he is allowed to vary the stakes according to his past experience; at time n he can win $v_n d_n$ dollars. How can he improve his circumstances by choosing v_n suitably? Burkholder's first theorem is that if $\sup_n |v_n| < \infty$ a.e., then g_n converges a.e. to a finite limit g_∞; however, $M[|g_\infty|]$ may be infinite.

In studying a martingale $\{f_n\}$, Burkholder and Gundy work with the function $(\sum_{n=1}^{\infty} |f_n - f_{n-1}|^2)^{1/2}$ and generalizations. This is called the S-function; some of its properties of convergence and average size are comparable with those of $\{f_n\}$. The prototype for such conclusions is the Khintchine–Kolmogorov Theorem: Let $\{y_n\}$ be independent random variables with $M[y_n] = 0$ and $M[|y_n|^2] = \sigma_n^2$. If $\sum_{n=1}^{\infty} \sigma_n^2 = \sigma^2 < \infty$, then $\sum_{n=1}^{\infty} y_n$ is convergent a.e. and in L^2. Conversely if $\Sigma\, y_n$ is convergent in L^2, then $\Sigma\, \sigma_n^2 = \sigma^2 < \infty$ and $M[|\Sigma\, y_n|^2] = \sigma^2$. There is a corresponding martingale result with $f_n - f_{n-1}$ in place of y_n.

There is a parallel between the theory of martingales and some of the developments in Euclidean Fourier analysis, and some of the theorems in each of the areas motivate theorems in the other. Let Q_0 be the unit cube in n dimensional space $\mathbb{R}^n$, and let $\mathscr{R}_k$ be the partition of Q_0 consisting of cubes of side 2^{-k} with all coordinates of vertices at integral multiples of 2^{-k}. Now let $(f_k, \mathscr{R}_k^*)$ be a martingale; for example, $f_k = M[f \mid \mathscr{R}_k^*]$ is a martingale if f

is integrable on Q_0, and there are other examples. Meanwhile, consider harmonic functions u in $\mathbb{R}_+^{n+1} = \{(x, t) \mid x \in \mathbb{R}^n, t > 0\}$; for example, the Poisson integral of an integrable function on Q_0 (or in all of $\mathbb{R}^n$) is an example, and there are other examples. The parallel is obtained by comparing properties of the function f_k in the special martingale with the function $u(\cdot, 2^{-k})$, where u is harmonic.

This parallel was handled rigorously in one case when the martingale theorem of Burkholder and Gundy [1970] was generalized by Burkholder, Gundy, and Silverstein [1971] to deal with Brownian motion and then to prove a Euclidean theorem.

Except in this one case, however, the idea has been to make the comparison and to proceed by analogy. Under the parallel, the martingale S-function corresponds to the "Lusin area function" of Fourier analysis if the martingale $(f_k, \mathscr{R}_k^*)$ is general, and to the "Littlewood-Paley g-function" of Fourier analysis if the martingale has $f_k = M[f \mid \mathscr{R}_k^*]$. The parallel also gives useful information in dealing with functions of bounded mean oscillation. For an account of the martingale results, see Garsia [1973]. Fefferman [1975] has given a thorough exposition of the Euclidean results and described the parallel in more detail.

Strong ratio limit property:

A noncyclic recurrent chain P has the strong ratio limit property (SRLP) if there are positive numbers $\pi_j > 0$ such that

$$\lim_{n \to \infty} \frac{P_{ij}^{(n+m)}}{P_{kl}^{(n)}} = \frac{\pi_j}{\pi_l}$$

for all i, j, k, l, m. Chung and Erdos [1951 R] showed the SRLP holds for sums of independent random variables on the integers, and an example reproduced in Chung [1960 R] shows the SRLP fails for a certain non-cyclic recurrent P.

Orey [1961] proved the SRLP holds if $P_{00}^{(n+1)}/P_{00}^{(n)}$ tends to 1 and it holds if

$$\limsup_{n \to \infty} \frac{P_{00}^{(m(n+1))}}{P_{00}^{(mn)}} \leq 1.$$

The latter condition holds for a reversible chain since $P_{00}^{(n)}$ is non-increasing in n, and hence the SRLP holds for reversible chains.

The result of Chung and Erdos can be interpreted as showing the SRLP holds if there is spatial homogeneity of the right kind. Kingman and Orey [1964 R] proved the SRLP under a much weaker assumption of spatial homogeneity—that $N_{ii}^{(n)} \geq 1 + \epsilon$ for all i for some $\epsilon > 0$ and some n. Sums of independent random variables clearly have this property.

More recent work has concentrated on transient chains. The SRLP requires reformulation in these cases. Pruitt [1965] gives a definition and proves theorems analogous to those of Orey [1961]. The book by Orey [1971] treats these matters in some detail. See also Freedman [1971].

Applied uses of Markov chains:

Probabilistic functions of finite Markov chains: Let P be a finite Markov chain with state space S and starting vector π. Suppose, for each i in S, that $F_{i\cdot}$ is a probability measure on a finite set Y. We imagine a process in which P takes place unseen in the background and the matrix F is used at each time to produce an outcome in Y. Calling the outcomes y_n, we have

$$(*) \quad \Pr[y_1 = k_1 \wedge \cdots \wedge y_N = k_N] = \sum_{i_0,\ldots,i_N \in S} \pi_{i_0} P_{i_0 i_1} F_{i_1 k_1} P_{i_1 i_2} F_{i_2 k_2} \cdots P_{i_{N-1} i_N} F_{i_N k_N}.$$

For example, a subject in a psychological experiment may have different probabilities for making responses according to his state of mind. If we imagine his frame of mind as the outcome of a Markov chain, then S is the set of frames of mind and Y is the set of responses.

If each row vector $F_{i\cdot}$ has all its mass in one entry, then the situation is that in lumping. The states in S are lumped in some fashion and the lumped states are those in Y. The lumped process need not be a Markov chain. Such processes have been studied extensively for a long time. See Rosenblatt [1971], Chapter III.

The opposite extreme occurs when all F_{ij}, P_{ij}, and π_i are >0. The typical practical problem that arises is to estimate the parameters π, P, and F if a finite sequence of outcomes is all that is known. Specifically one wants values of π, P, and F that make (*) a maximum, given $k_1, \ldots, k_N$. Baum *et al.* [1970] give an iterative procedure in the last paragraph of their paper for passing from one set of values of π, P, and F to another with the property that (*) increases to a critical point. Their theorem that (*) increases to a critical point has been used in modeling letter patterns in English words, in predicting sunspot behavior, and in anticipating the stock market. As indicated above, it also has applications to psychology and sociology.

Optimal stopping problems: Let $(y_n, \mathscr{F}_n)$ be a denumerable stochastic process, and let $x_n = x_n(y_0, y_1, \ldots, y_n)$ be real-valued and measurable. Suppose the x_n are integrable. The problem is to find

$$V = \sup_t M[x_t],$$

where the supremum is taken over all random times t. V is the *value* of the x_n process. If the supremum is attained for some t, t is an optimal strategy, and a further problem is to describe t.

We are to regard the y_n's as some observable outcomes and the x_n's as rewards. We are allowed to choose the time of obtaining our reward, without clairvoyance, and the problem is to maximize the payoff. There is an extensive theory in this generality. See Chow, Robbins, and Siegmund [1971]. Neveu [1972a] treats the martingale case, beginning with "Le problème de Snell," solved in Snell [1952 R].

In many applications the $\{y_n\}$ are the outcome of a Markov chain, and the nth reward function is $x_n(y_0, \ldots, y_n) = f(y_n)$, with f a function on the state space that is independent of n. The value is simply $v_i = \sup_t M_i[f(y_t)]$. If f is bounded, then v is the least nonnegative superregular function $\geq f$. See Dynkin and Yushkevich [1969], Chapter 3, for a treatment of the problem and discussion of strategy. A deeper investigation is the book by Širjaev [1973].

Recurrent potential theory:

Orey [1964 R] developed a recurrent potential theory that avoids the notion of a normal chain and proceeds from axioms for a potential operator.

Neveu [1972b] developed recurrent potential theory from a different perspective, proceeding as follows. For a function h on S with $0 \le h \le 1$, define

$$U_h = \sum_{n=0}^{\infty} (PD_{1-h})^n P = \sum_{n=0}^{\infty} P(D_{1-h}P)^n,$$

where D_h is the diagonal matrix with h_i as ith diagonal entry.

When $h = 1$, $U_h = P$. When $h = 0$, $U_h = P + P^2 + P^3 + \cdots$. In the general case with $f \geq 0$,

$$(U_h f)_i = M_i\left[\sum_{n=0}^{\infty} (1 - h(x_1))(1 - h(x_2)) \cdot \ldots \cdot (1 - h(x_{n-1}))f(x_n)\right].$$

If we write U_E for U_h when h is the characteristic function of E, we have

$$(U_E f)_i = M_i\left[\sum_{1 \le n \le t_E} f(x_n)\right].$$

When f is 1 on E and 0 elsewhere, the right side reduces to the probability that the chain started in state i ever returns to the set E.

Let P be recurrent with positive regular measure α. Neveu proves that there is some h on S with $0 < h \le 1$ such that $U_h h = 1$ and $U_h \ge 1\alpha$. Fix such an h, put $V = U_h - 1\alpha$, and define

$$W = \sum_{n=0}^{\infty} (VD_h)^n V.$$

Clearly W is ≥ 0. The finiteness of W is settled by the facts that $Wh = c1$ and $\alpha D_h W = c\alpha$, where c is the constant $(1 - \alpha h)/(\alpha h)$. The operator $I + W$ is the potential kernel, and Neveu develops an appropriate theory for it.

See also Revuz [1975].

Transient boundary theory:

Transient boundary theory for general denumerable Markov chains stands about where it was in 1966.

Dynkin [1969] gave an account of the theory that does not use extended chains. For the theory of the exit boundary the idea is to use a martingale-upcrossing argument to deal with a superregular measure μ. If $f(i) = \mu_i/(\pi N)_i$, the key result is that $\lim_{n\to\infty} f(x_n(\omega))$ exists a.e. on infinite paths, provided f satisfies a suitable integrability condition. From this result, the result we call Theorem 10-18 follows without reference to any extended chains, and the rest of the theory requires no change.

Athreya and Ney [1972] apply transient boundary theory to branching processes in Chapter II of their book.

Sums of independent random variables:

Sums of independent random variables for a countable group that is not necessarily abelian can be defined just as in the abelian case (see Chapter 4), as long as left and right are distinguished carefully. Let P be the transition matrix for such a chain, and suppose that the states form a single class.

Regarding the question of transience *vs.* recurrence, Kesten [1959] considered a symmetric P as a linear operator on square-summable sequences. He proved that the group admits such a P with spectral radius one if and only if the group is amenable, i.e., there is a non-zero left-invariant positive linear functional on the bounded functions on the group. If the spectral radius is less than one, then not only does N have finite entries but also N is a bounded operator on square-summable sequences; hence recurrence implies spectral radius one. Day [1964] removed the hypothesis of symmetry in Kesten's theorem and found further equivalent conditions. His theorem has been reproved several times by other authors.

For the exact question of transience *vs.* recurrence the results are less decisive. In his book Spitzer [1976] settles the case of processes on the lattice points in Euclidean space. Dudley [1962] proved that a countable *abelian* group has a recurrent P (with one class of states) if and only if the maximum number of linearly independent elements is at most 2. In the non-abelian case Kesten [1967] conjectured that the existence of a recurrent P for a group is related to the growth in n of the number of elements of the group expressible as a product of n generators. Milnor [1968a] showed that this growth function is approximately independent of the set of generators; he pointed out that the existence of a symmetric P with spectral radius less than 1 implies exponential growth, and he gave an example of a solvable group and a symmetric P with spectral radius 1 and with exponential growth. Milnor [1968b] and Wolf [1968] considered classes of countable groups and gave conditions under which the growth function is of polynomial size or of exponential size.

Ney and Spitzer [1966] compute the Martin boundary for transient sums of independent random variables on a lattice with nonzero mean. Kesten and Spitzer [1965] prove the existence of the potential kernel for recurrent sums of independent random variables on countable abelian groups, and they consider the Martin boundary. Kesten [1967] generalized this work to general countable groups, although the extent to which non-trivial non-abelian groups can admit recurrent processes is still not known.

Derriennic [1975] deals with sums of independent random variables on a free group with $n > 1$ generators. It is assumed that the transition matrix has only finitely many nonzero entries in each row and that all states communicate. He shows that such a process is transient and that the boundary consists of all "reduced infinite words." The special case in which the process in one step can move from a word only to the product of that word by a generator or its inverse was considered earlier by other authors. When in the special case all the $2n$ one-step probabilities are equally likely, the resulting example is one that arises in the theory of algebraic groups.

REFERENCES

Blackwell, D., and Kendall, D., "The Martin boundary for Pólya's urn scheme and an application to stochastic population growth," *J. Appl. Probability*, **1**, 284–296 (1964).

Blumenthal, R. M., "An extended Markov property," *Trans. Amer. Math. Soc.*, **85**, 52–72 (1957).

Brelot, M., "Le problème de Dirichlet, Axiomatique et frontière de Martin" *J. Math. Pures Appl.*, **35**, 297–335 (1956).

Brelot, M., *Éléments de la Théorie Classique du Potentiel*, Sorbonne, Paris, 1959.

Choquet, G., "Theory of capacities," *Ann. Inst. Fourier, Grenoble*, **5**, 131–295 (1953–54).

Choquet, G., and Deny, J., "Modèles finis en théorie du potentiel," *J. Anal. Math.*, **5**, 77–135 (1956–57).

Chung, K. L., "An ergodic theorem for stationary Markov chains with a countable number of states," *Proc. Intern. Congs. Math.*, Cambridge, Mass., **1**, 568 (1950).

Chung, K. L., "Contributions to the theory of Markov chains, II," *Trans. Amer. Math. Soc.*, **76**, 397–419 (1954).

Chung, K. L., *Markov Chains with Stationary Transition Probabilities*, Berlin–Göttingen–Heidelberg, Springer-Verlag, 1960.

Chung, K. L., "Some remarks on taboo probabilities," *Illinois J. Math.*, **5**, 431–435 (1961).

Chung, K. L., and Erdös, P., "Probability limit theorems assuming only the first moment I," *Mem. Amer. Math. Soc.*, **6**, 1–19 (1951).

Chung, K. L., and Ornstein, D., "On the recurrence of sums of random variables," *Bull. Amer. Math. Soc.*, **68**, 30–32 (1962).

Courant, R., and Hilbert, D., *Methods of Mathematical Physics*, Vol. I, New York, Interscience Publishers, Inc., 1953.

Derman, C., "A solution to a set of fundamental equations in Markov chains," *Proc. Amer. Math. Soc.*, **5**, 332–334 (1954).

Doeblin, W., "Sur deux problèmes de M. Kolmogoroff concernant les chaines dénombrables," *Bull. Soc. Math. France*, **66**, 210–220 (1938).

Doob, J. L., "Regularity properties of certain families of chance variables," *Trans. Amer. Math. Soc.*, **47**, 455–486 (1940).

Doob, J. L., *Stochastic Processes*, New York, John Wiley & Sons, Inc., 1953.

Doob, J. L., "Semimartingales and subharmonic functions," *Trans. Amer. Math. Soc.*, **77**, 86–121 (1954).

Doob, J. L., "Conditional Brownian motion and the boundary limits of harmonic functions," *Bull. Soc. Math. France*, **85**, 431–458 (1957).

Doob, J. L., "Probability theory and the first boundary value problem," *Illinois J. Math.*, **2**, 19–36 (1958).

Doob, J. L., "Discrete potential theory and boundaries," *J. Math. and Mech.*, **8**, 433–458 (1959).

Doob, J. L., "Notes on martingale theory," *Fourth Berkeley Symposium on Mathematical Statistics and Probability*, Vol. II, Berkeley, Calif., University of California Press, pp. 95–102, 1961.

Doob, J. L., Snell, J. L., and Williamson, R. E., "Application of boundary theory to sums of independent random variables," *Contributions to Probability and Statistics*, Stanford, Calif., Stanford University Press, pp. 182–197, 1960.

Dunford, N., and Schwartz, J. T., *Linear Operators*, Part I, New York, Interscience Publishers, Inc., 1958.

Dynkin, E. B., and Malyutov, M. B., "Random walk on groups with a finite number of generators," *Translation of Dokl. Acad. Sci. of USSR*, **2**, 399–402 (1961) (Original **137**, 1042–1045).

Feller, W., "Boundaries induced by non-negative matrices," *Trans. Amer. Math. Soc.*, **83**, 19–54 (1956).

Feller, W., *An Introduction to Probability Theory and Its Applications*, Vol. I, 2nd ed., New York, John Wiley & Sons, Inc., 1957.

Fréchet, M., "Méthode des fonctions arbitraires;" *Théorie des évenements en chaine dans le cas d'un nombre fini d'états possibles*, 2nd livre, Paris, Gauthiers-Villars, 1938.

Gorden, P., *Théorie des chaînes de Markov finies et ses applications*, Paris, Dunod, 1965.

Halmos, P., *Measure Theory*, Princeton, N.J., D. Van Nostrand Co., Inc., 1950.

Harris, T. E., "Transient Markov chains with stationary measures," *Proc. Amer. Math. Soc.*, **8**, 937–942 (1957).

Harris, T. E., *The Theory of Branching Processes*, Berlin-Göttingen-Heidelberg, Springer-Verlag, 1963.

Hennequin, P.-L., and Tortrat, A., *Théorie des probabilités et quelques applications*, Paris, Masson, 1965.

Hildebrand, F. B., *Introduction to Numerical Analysis*, New York, McGraw-Hill Book Co., 1956.

Hunt, G. A., "Markoff processes and potentials I, II, III," *Illinois J. Math.*, **1**, 44–93, 316–369 (1957); *ibid.*, **2**, 151–213 (1958).

Hunt, G. A., "Markoff chains and Martin boundaries," *Illinois J. Math.*, **4**, 313–340 (1960).

Itô, K., and McKean, H. P., Jr., "Potentials and the random walk," *Illinois J. Math.*, **4**, 119–132 (1960).

Kac, M., "Random walk and the theory of Brownian motion," *Amer. Math. Monthly*, **54**, 369–391 (1947a).

Kac, M., "On the notion of recurrence in discrete stochastic processes," *Bull. Amer. Math. Soc.*, **53**, 1002–1010 (1947b).

Kac, M., "On some connections between probability theory and differential and integral equations," *Proceedings of the Second Berkeley Conference on Mathematical Statistics and Probability*, Berkeley, Calif., University of California Press, pp. 189–215, 1951.

Kakutani, S., "Two-dimensional Brownian motion and harmonic functions," *Proc. Imp. Acad. Tokyo*, **20**, 706–714 (1944).

Kakutani, S., "Markov processes and the Dirichlet problem," *Proc. Jap. Acad.*, **21**, 227–233 (1945).

Karlin, S., and McGregor, J., "Random walks," *Illinois J. Math.*, **3**, 66–81 (1959).

Kellogg, O. D., *Foundations of Potential Theory*, New York, Dover Publications, Inc., 1929.

Kemeny, J. G., "Doeblin's ratio limit theorem," *Notices Amer. Math. Soc.*, **9**, 390–391 (1962).

Kemeny, J. G., "A further note on discrete potential theory," *J. Math. Anal. and Appl.*, **6**, 55–57 (1963).

Kemeny, J. G., "Representation theory for denumerable Markov chains," *Trans. Amer. Math. Soc.*, **125**, 47–62 (1966).

Kemeny, J. G., Mirkil, H., Snell, J. L., and Thompson, G. L., *Finite Mathematical Structures*, Englewood Cliffs, N.J., Prentice-Hall, Inc., 1959.

Kemeny, J. G., and Snell, J. L., *Finite Markov Chains*, Princeton, N.J., D. Van Nostrand Co., Inc., 1960.

Kemeny, J. G., and Snell, J. L., "On Markov chain potentials," *Annals of Math. Stat.*, **32**, 709–715 (1961a).

Kemeny, J. G., and Snell, J. L., "Potentials for denumerable Markov chains," *J. Math. Anal. and Appl.*, **3**, 196–260 (1961b).

Kemeny, J. G., and Snell, J. L., "Notes on discrete potential theory," *J. Math. Anal. and Appl.*, **3**, 117–121 (1961c).

Kemeny, J. G., and Snell, J. L., "Boundary theory for recurrent Markov chains," *Trans. Amer. Math. Soc.*, **106**, 495–520 (1963a).

Kemeny, J. G., and Snell, J. L., "A new potential operator for recurrent Markov chains," *J. London Math. Soc.*, **38**, 359–371 (1963b).

Kemperman, J. H. B., *The Passage Problem for a Stationary Markov Chain*, Chicago, University of Chicago Press, 1961.

Kesten, H., "Ratio theorems for random walks, II," *J. d'Analyse Math.*, **11**, 323–379 (1963).

Kesten, H., and Spitzer, F., "Ratio theorems for random walks, I," *J. d'Analyse Math.*, **11**, 285–322 (1963).

Kingman, J. F. C., and Orey, S., "Ratio limit theorems for Markov chains," *Proc. Amer. Math. Soc.*, **15**, 907–910 (1964).

Knapp, A. W., "Connection between Brownian motion and potential theory," *J. Math. Anal. and Appl.*, **12**, 328-349 (1965).

Knapp, A. W., "Regular boundary points in Markov chains," *Proc. Amer. Math. Soc.*, **17**, 435–440 (1966).

Kolmogorov, A. N., "Anfangsgründe der Theorie der Markoffschen Ketten mit unendlichen vielen möglichen Zuständen," *Mat. Sbornik N.S.*, 607–610 (1936); *Bull. Univ. Moscow*, **1** (1937) (Russian).

Kolmogorov, A. N., *Foundations of the Theory of Probability*, 2nd English ed., New York, Chelsea Publishing Co., 1956.

Lahres, H., *Einführung in die diskreten Markoff-Prozesse und ihre Anwendungen*, Braunschweig, Germany, Friedr. Viewig und Sohn, 1964.

Lamperti, J., "Criteria for the recurrence or transience of stochastic process I," *J. Math. Anal. and Appl.*, **1**, 314–330 (1960a).

Lamperti, J., "The first-passage moments and the invariant measure of a Markov chain," *Annals of Math. Stat.*, **31**, 515–517 (1960b).

Lamperti, J., "Criteria for stochastic processes II: Passage-time moments," *J. Math. Anal. and Appl.*, **7**, 127–145 (1963a).

Lamperti, J., "Wiener's Test and Markov chains," *J. Math. Anal. and Appl.* **6**, 58–66 (1963b).

Lamperti, J., and Snell, J. L., "Martin boundaries for certain Markov chains," *J. Math. Soc. Japan*, **15**, 113–128 (1963).
Lévy, P., *Théorie de l'addition des variables aléatoires*, Paris, Gauthier-Villars, 1937.
Loève, Michel, *Probability Theory*, 3rd ed., Princeton, N.J., D. Van Nostrand, Co., Inc., 1963.
Markov, A. A., "Investigation of an important case of dependent trials," *Izvestia Acad. Nauk SPB*, VI, ser. I (Russian), **61** (1907).
Martin, R. S., "Minimal positive harmonic functions," *Trans. Amer. Math. Soc.*, **49**, 137–172 (1941).
McKean, H. P., Jr., "A problem about prime numbers and the random walk I," *Illinois J. Math.*, **5**, 351 (1961).
Murdoch, B. H., *Preharmonic Functions* (Thesis), Princeton, N.J., Princeton Univ. Library, May 1954.
Naïm, L., "Sur le rôle de la frontière de R. S. Martin dans la théorie du potentiel," *Ann. Inst. Fourier, Grenoble*, **7**, 183–281 (1957).
Nash-Williams, C. St J. A., "Random walk and electric currents in networks," *Proc. Cambridge Phil. Soc.*, **55**, 181–194 (1959).
Neveu, J., "Chaînes de Markov et théorie du potentiel," *Ann. Faculté des Sciences, Clermont*, **3**, 37–65 (1964).
Orey, S., "An ergodic theorem for Markov chains," *Z. Wahrescheinlichkeitstheorie*, **1**, 174–176 (1962).
Orey, S., "Potential kernels for recurrent Markov chains," *J. Math. Anal. and Appl.*, **8**, 104–132 (1964).
Polya, G., "Über eine Aufgabe der Wahrscheinlichkeitsrechnung betriffend die Irrfahrt im Strassennetz," *Math. Ann.*, **84**, 149–160 (1921).
Romanovskii, V. I., *Discrete Markov Chains*, Moscow–Leningrad (Russian), 1949.
Rudin, W., *Principles of Mathematical Analysis*, New York, McGraw-Hill Book Co., 1953.
Snell, J. L., "Applications of martingale system theorems," *Trans. Amer. Math. Soc.*, **73**, 293–312 (1952).
Spitzer, F. L., "Hitting probabilities," *J. Math. and Mech.*, **11**, 593–614 (1962).
Spitzer, F. L., *Principles of Random Walk*, Princeton, N.J., D. Van Nostrand Co., Inc., 1964.
Veech, W., "The necessity of Harris' condition for the existence of a stationary measure," *Proc. Amer. Math. Soc.*, **14**, 856–860 (1963).
Ville, J., *Étude Critique de la notion de collectif*, Paris, Gauthier-Villars, 1939.
Watanabe, T., "On the theory of Martin boundaries induced by countable Markov processes," *Mem. Coll. Science, University Kyoto Series* A, **33**, 39–108 (1960a).
Watanabe, T., "A probabilistic method in Hausdorff moment problem and Laplace–Stieltjes transform," *J. Math. Soc. of Japan*, **12**, 192–206 (1960b).
Yosida, K., and Kakutani, S., "Markov process with an enumerable infinite number of possible states," *Jap. J. Math.*, **16**, 47–55 (1940).

ADDITIONAL REFERENCES

Athreya, K. B., and Ney, P. E., *Branching Processes*, New York, Springer-Verlag, 1972.

Averintsev, M. B., "One method of describing random fields with a discrete argument," *Problems of Information Transmission*, **6**, 169–175 (1970). (Original: *Problemy Peredachi Informatsii*, **6**, 100–108.)

Baum, L. E., Petrie, T., Soules, G., and Weiss, N., "A maximization technique occurring in the statistical analysis of probabilistic functions of Markov chains," *Ann. Math. Stat.*, **41**, 164–171 (1970).

Blackwell, D., and Freedman, D., "The tail σ-field of a Markov chain and a theorem of Orey," *Ann. Math. Stat.*, **35**, 1291–1295 (1964).

Burkholder, D. L., "Martingale transforms," *Ann. Math. Stat.*, **37**, 1494–1504 (1966).

Burkholder, D. L., "Martingale inequalities," *Martingales*, Lecture Notes in Mathematics, **190**, New York, Springer-Verlag, 1–8, 1971.

Burkholder, D. L., and Gundy, R. F., "Extrapolation and interpolation of quasi-linear operators on martingales," *Acta Math.*, **124**, 249–304 (1970).

Burkholder, D. L., Gundy, R. F., and Silverstein, M. L., "A maximal function characterization of H^p," *Trans. Amer. Math. Soc.*, **157**, 137–153 (1971).

Chow, Y.-s., Robbins, H., and Siegmund, D., *Great Expectations: the Theory of Optimal Stopping*, Boston, Houghton Mifflin Company, 1971.

Cox, T., "An example of phase transition in countable one-dimensional Markov random fields," (forthcoming, 1976).

Dawson, D., *Discrete Markov Systems*, Carleton Mathematical Lecture Notes No. 10, Ottawa, Carleton University (1974).

Day, M. M., "Convolutions, means, and spectra," *Illinois J. Math.*, **8**, 100–111 (1964).

Derriennic, Y., "Marche aléatoire sur le groupe libre et frontière de Martin," *Z. Wahrscheinlichkeitstheorie*, **32**, 261–276 (1975).

Dobruschin, P. L., "The description of a random field by means of conditional probabilities and conditions of its regularity," *Th. Prob. and Its Appl.*, **13**, 197–224 (1968). (Original: *Teoriya Veroy. i ee Prim.*, **13**, 201–229.)

Dudley, R. M., "Random walks on abelian groups," *Proc. Amer. Math. Soc.*, **13**, 447–450 (1962).

Dynkin, E. B., *Markov Processes*, 2 vol., Springer-Verlag, New York, 1965.

Dynkin, E. B., "Boundary theory of Markov processes (the discrete case)," *Russian Math. Surveys*, **24**, no. 2, 1–42 (1969). (Original: *Uspekhi Mat. Nauk*, **24**, no. 2, 3–42.)

Dynkin, E. B., "Entrance and exit spaces for a Markov process," *Actes Congrès int. Math., 1970*, **2**, 507–512 (1971).

Dynkin, E. B., and Yushkevich, A. A., *Markov Processes, Theorems and Problems*, New York, Plenum Press, 1969.

Fefferman, C., "Recent progress in classical Fourier analysis," *Proc. International Congress Mathematicians, 1974* vol. 1, Canadian Mathematical Congress, 95–118 (1975).

Flatto, L., and Pitt, J., "Recurrence criteria for random walks on countable abelian groups," *Illinois J. Math.*, **18**, 1–19 (1974).

Föllmer, H., "Phase transition and Martin boundary," *Séminaire de Probabilités IX, Université de Strasbourg*, Lecture Notes in Mathematics, **465**, New York, Springer-Verlag, 305–317 (1975a).

Föllmer, H., "On the potential theory of stochastic fields," *Inter. Stat. Institute*, Warsaw (1975b).

Freedman, D., *Markov Chains*, San Francisco, Holden-Day, 1971.

Garsia, A. M., *Martingale Inequalities*, Reading, Mass., W. A. Benjamin, Inc., 1973.

Griffeath, D., "Markov and Gibbs fields with finite state space on graphs," unpublished manuscript (1973).

Griffeath, D., and Snell, J. L., "Optimal stopping in the stock market," *Ann. of Prob.*, **2**, 1–13 (1974).

Griffiths, R. B., "The Peierls argument for the existence of phase transitions," *Mathematical Aspects of Statistical Mechanics*, J. C. T. Pool (ed.), *SIAM-AMS Proceedings*, Providence, Amer. Math. Soc., **5**, 13–26, 1972.

Grimmett, G. R., "A theorem about random fields," *Bull. London Math. Soc.*, **5**, 81–84 (1973).

Gundy, R. F., "The martingale version of a theorem of Marcinkiewicz and Zygmund," *Ann. Math. Stat.*, **38**, 725–734 (1967).

Kalikow, S., "An entrance law which reaches equilibrium," (forthcoming, 1976).

Kemeny, J. G., "Slowly spreading chains of the first kind," *J. Math. Anal and Appl.*, **15**, 295–310 (1966).

Kemeny, J. G., and Snell, J. L., "Markov chains and summability methods," *Z. Wahrscheinlichkeitstheorie*, **18**, 17–33 (1971).

Kesten, H., "Full Banach mean values on countable groups," *Math. Scand.*, **7**, 146–156 (1959).

Kesten, H., "The Martin boundary of recurrent random walks on countable groups," *Proc. Fifth Berkeley Symposium on Mathematical Statistics and Probability*, Berkeley, University of California Press, vol. 2, part 2, 51–74, 1967.

Kesten, H., "Existence and uniqueness of countable one-dimensional Markov random fields," (forthcoming, 1976).

Kesten, H., and Spitzer, F., "Random walk on countably infinite abelian groups," *Acta Math.*, **114**, 237–265 (1965).

Meyer, P.-A., *Martingales and Stochastic Integrals I*, Lecture Notes in Mathematics, **284**, New York, Springer-Verlag, 1972.

Milnor, J., "A note on curvature and fundamental group," *J. Diff. Geom.*, **2**, 1–7 (1968a).

Milnor, J., "Growth of finitely generated solvable groups," *J. Diff. Geom.*, **2**, 447–449 (1968b).

Neveu, J., *Martingales à temps discret*, Paris, Masson et Cie, 1972a.

Neveu, J., "Potentiel Markovien récurrent des chaines de Harris," *Ann. Inst. Fourier, Grenoble*, **22**, 2, 85–130 (1972b).

Ney, P., and Spitzer, F., "The Martin boundary for random walk," *Trans. Amer. Math. Soc.*, **121**, 116–132 (1966).

Orey, S., "Strong ratio limit property," *Bull. Amer. Math. Soc.*, **67**, 571–574 (1961).
Orey, S., *Limit Theorems for Markov Chain Transition Probabilities*, London, Van Nostrand Reinhold Company, 1971.
Pitman, J. W., "Uniform rates of convergence for Markov chain transition probabilities," *Z. Wahrscheinlichkeitstheorie*, **29**, 193–227 (1974).
Preston, C. J., *Gibbs States on Countable Sets*, Cambridge, Cambridge University Press, 1974.
Pruitt, W. E., "Strong ratio limit property for R-recurrent Markov chains," *Proc. Amer. Math. Soc.*, **16**, 196–200 (1965).
Revuz, D., *Markov Chains*, New York, American Elsevier Publishing Company, Inc., 1975.
Rosenblatt, M., *Markov Processes, Structure and Asymptotic Behavior*, New York, Springer-Verlag, 1971.
Ruelle, D., *Statistical Mechanics, Rigorous Results*, New York, W. A. Benjamin, Inc., 1969.
Şeneta, E., *Non-Negative Matrices*, New York, John Wiley and Sons, 1973.
Širjaev, A. N., *Statistical Sequential Analysis, Optimal Stopping Rules*, Providence, American Mathematical Society, vol. 38, Translations of Mathematical Monographs, 1973.
Spitzer, F., "Introduction aux processus de Markov à paramètre dans Z_ν," *Ecole d'Eté de Probabilités de Saint-Flour III*, Lecture Notes in Math., **390**, New York, Springer-Verlag, 114–189, 1974.
Spitzer, F., "Phase transition in one-dimensional nearest-neighbor systems," *J. Functional Anal.*, **20**, 240–255 (1975a).
Spitzer, F., "Markov random fields on an infinite tree." *Ann. of Prob.*, **3**, 387–398 (1975b).
Spitzer, F., *Principles of Random Walk*, 2nd Edition, New York, Springer-Verlag, 1976.
Wolf, J. A., "Growth of finitely generated solvable groups and curvature of Riemannian manifolds," *J. Diff. Geom.*, **2**, 421–446 (1968).

INDEX OF NOTATION

General Notations

Specific Notations

Specific Notations—(*continued*)

INDEX

GPSR Compliance

The European Union's (EU) General Product Safety Regulation (GPSR) is a set of rules that requires consumer products to be safe and our obligations to ensure this.

If you have any concerns about our products, you can contact us on ProductSafety@springernature.com

In case Publisher is established outside the EU, the EU authorized representative is:

Springer Nature Customer Service Center GmbH
Europaplatz 3
69115 Heidelberg, Germany

Batch number: 10682323

Printed by Printforce, the Netherlands

VOLUME ONE HUNDRED

SEMICONDUCTORS AND SEMIMETALS

Photonic Crystal Metasurface Optoelectronics

SERIES EDITORS

VOLUME ONE HUNDRED

SEMICONDUCTORS AND SEMIMETALS

Photonic Crystal Metasurface Optoelectronics

Edited by

WEIDONG ZHOU

Department of Electrical Engineering, University of Texas at Arlington, Arlington, TX, United States

SHANHUI FAN

Ginzton Laboratory and Department of Electrical Engineering, Stanford University, Stanford, CA, United States

ELSEVIER

ACADEMIC PRESS

An imprint of Elsevier

Academic Press is an imprint of Elsevier
50 Hampshire Street, 5th Floor, Cambridge, MA 02139, United States
525 B Street, Suite 1650, San Diego, CA 92101, United States
The Boulevard, Langford Lane, Kidlington, Oxford OX5 1GB, United Kingdom
125 London Wall, London, EC2Y 5AS, United Kingdom

First edition 2019

ISBN: 978-0-12-817542-2
ISSN: 0080-8784

For information on all Academic Press publications
visit our website at https://www.elsevier.com/books-and-journals

Publisher: Zoe Kruze
Acquisition Editor: Jason Mitchell
Editorial Project Manager: Peter Llewellyn
Production Project Manager: Abdulla Sait
Cover Designer: Greg Harris

Typeset by SPi Global, India

Contents

Contributors

Isabel Barth
The Department of Physics, University of York, York, United Kingdom. (ch5)

Alexander Drayton
The Department of Physics, University of York, York, United Kingdom. (ch5)

Shanhui Fan
Ginzton Laboratory and Department of Electrical Engineering, Stanford University, Stanford, CA, United States. (ch1,4,6)

Cheng Guo
Department of Applied Physics, Stanford University, Stanford, CA, United States. (ch4)

Yu Guo
Electrical Engineering and Ginzton Laboratory, Stanford University, Stanford, CA, United States. (ch3)

Yuri S. Kivshar
Department of Physics and Engineering, ITMO University, St. Petersburg, Russia; Nonlinear Physics Center, Australian National University, Canberra, ACT, Australia. (ch2)

Thomas F. Krauss
The Department of Physics, University of York, York, United Kingdom. (ch5)

Mikhail F. Limonov
Ioffe Institute; Department of Physics and Engineering, ITMO University, St. Petersburg, Russia. (ch2)

Mikhail V. Rybin
Ioffe Institute; Department of Physics and Engineering, ITMO University, St. Petersburg, Russia. (ch2)

Ken Xingze Wang
School of Physics, Huazhong University of Science and Technology, Wuhan, Hubei, China. (ch3)

Zongfu Yu
University of Wisconsin, Department of Electrical and Computer Engineering, Madison, WI, United States. (ch3)

Weidong Zhou
Department of Electrical Engineering, University of Texas at Arlington, Arlington, TX, United States. (ch1,6,7)

Preface

Fano resonance, with its characteristics asymmetric lineshape, differs significantly from conventional resonance with an asymmetric Lorentzian lineshape. In photonics, Fano resonance is first observed from the transmission spectra of gratings and is responsible for one variety of Wood's anomalies. This phenomenon and the underlying mechanism, being common and ubiquitous in many realms of physical sciences, are found in a wide variety of nanophotonic structures and quantum systems, such as quantum dots, photonic crystals (PhCs), plasmonics, and metamaterials. The asymmetric and steep dispersion of the Fano resonance profile promises applicability in a wide range of photonic devices, such as optical filters, switches, sensors, broadband reflectors, lasers, detectors, and slow-light and nonlinear devices. With the advances in nanofabrication processes and the development of integration techniques, Fano resonance-based materials, structures, devices, and systems have been widely investigated and reported over the last decade.

This volume collects the state-of-the-art results on Fano resonance photonic crystal and dielectric metamaterials for photonic and optoelectronic applications. In Chapter 1, the basic principle of Fano resonance in photonic crystal slabs was reviewed, which establishes a basic theoretical framework for the wide varieties of properties observed and to be discussed in this book.

In addition to photonic crystal slabs, dielectric metamaterials also create a low-loss platform for a variety of applications in photonics involving a complex manipulation of the amplitude, polarization, and phase of light. However, for many applications it is important to understand how an artificial periodic system operating as *a photonic crystal* with the properties dominated by the Bragg scattering can transform into *a metamaterial* described by effective parameters. Chapter 2 provides a comprehensive overview of the transitions between photonic crystals and metamaterials. Photonic phase diagram concept was proposed to understand the differences and transition between photonic crystals and dielectric metamaterials.

In Chapter 3, light trapping in photonic crystal structures was introduced for solar cell applications. A statistical temporal coupled mode theory of light trapping was introduced based on a rigorous electromagnetic approach. The upper limit of the angle-integrated light trapping absorption enhancement is proportional to the density of states in any given structure. The theory reveals that the conventional limit can be substantially surpassed when optical modes exhibit deep-subwavelength-scale field confinement.

Chapter 4 discusses the implementation of several isotropic image filters in the wavevector domain using a single photonic crystal slab device. Such a slab is designed so that the guided resonance near the Γ point exhibits an isotropic band structure. Depending on the light frequency and the choice of transmission or reflection mode, the device realizes isotropic high-pass, low-pass, band-reject, and band-pass filtering in wavevector space. These filter functions are important for various image processing tasks, including edge detection, smoothing, white noise suppression, and suppression or extraction of periodic noises.

In Chapter 5, label-free biosensing was discussed the versatility of the photonic crystal guided resonances offers new functionalities, such as multiparameter sensing and imaging, while simultaneously offering competitive sensing performance.

In Chapter 6, Fano resonance filters based on single and couple bi-layer photonic crystal slabs are discussed. These structures possess potentials toward extreme high quality factor filter via the manipulation of dark and bright resonance modes.

In Chapter 7, photonic crystal membrane lasers were discussed for on-chip integration. Based on Fano resonance principles, broadband reflectors were incorporated in a new types of vertical cavity surface-emitting lasers. Based on bandedge mode, photonic crystal surface-emitting lasers were also introduced for on-chip integration based on transfer printing process for hybrid integration.

As seen above, photonic crystals and dielectric metamaterials offer great potential for a wide range of photonic applications. While this book offers a few selected application cases, its potential goes beyond linear photonic applications. Nonlinear and quantum applications are also expected to show much more great progresses. We hope that this volume will be a useful reference book for the scientists and engineers striving for photonic crystal and dielectric metamaterials for new devices and integrated photonic applications.

Weidong Zhou
University of Texas at Arlington
Shanhui Fan
Stanford University

CHAPTER ONE

Fano resonance principles in photonic crystal slabs

Weidong Zhou[a],*, Shanhui Fan[b]
[a]Department of Electrical Engineering, University of Texas at Arlington, Arlington, TX, United States
[b]Department of Electrical Engineering, Stanford University, Stanford, CA, United States
*Corresponding author: e-mail address: wzhou@uta.edu

Contents

1. Introduction

In contrast to a conventional symmetric-shaped Lorentzian resonance, Fano resonance is predominantly used to describe asymmetric resonance (Fano, 1961; Ott et al., 2013), which arises from the constructive and destructive interference of discrete resonance states by broadband continuum states. First observed from the transmission spectra of gratings, Fano resonance is responsible for one variety of Wood's anomalies (Wood, 1902). This phenomenon and the underlying mechanism, being common and ubiquitous in many realms of physical sciences (Fano, 1961; Miroshnichenko et al., 2010), are found in a wide variety of nanophotonic structures and quantum systems, such as quantum dots, photonic crystals (PhCs), plasmonics, and metamaterials. The asymmetric and steep dispersion of the Fano resonance profile promises applicability in a wide range of photonic devices, such as optical filters, switches, sensors, broadband reflectors, lasers, detectors, slow-light, and nonlinear devices. With the advances in nanofabrication processes and the development of integration techniques, Fano resonance-based materials, structures, devices, and systems have been widely investigated and reported over the last few years. There are also excellent reviews written

Semiconductors and Semimetals, Volume 100
ISSN 0080-8784
https://doi.org/10.1016/bs.semsem.2019.04.001

by many experts in the field. Miroshnichenko et al. (2010) presented a detailed review on the principles and origins of Fano resonance in various nanostructures and physical systems. Luk'yanchuk et al. (Luk'yanchuk et al., 2010; Rahmani et al., 2013) and Giannini et al. (Giannini et al., 2011) presented reviews on progress made on Fano resonance in plasmonic nanostructures and metamaterials.

Optical gratings, which can also be viewed as 1D PhCs, are a research topic with a long history, including guided mode resonances (GMRs) in 1D gratings, reported by Magnusson et al. (Kikuta et al., 2003; Magnusson and Shokooh-Saremi, 2008; Magnusson and Wang, 1992). Recently, due to the unique characteristics and potentials for high performance optic and photonic devices, subwavelength high-index-contrast gratings (HCGs) have been proposed and investigated by Chang-Hasnain et al. (Boutami et al., 2007; Carletti et al., 2011; Chang-Hasnain, 2011; Chen et al., 2006) for applications in mirrors, lasers, and focusing elements (Chang-Hasnain, 2011). Here we focus our review on progress made in PhC Fano resonance photonics, with a specific focus on Fano resonances in two-dimensional photonic crystal slabs (2D PCS) (Fan, 2002; Fan and Joannopoulos, 2002; Zhou et al., 2014).

2D PCS structures also support guided resonances (Fan and Joannopoulos, 2002; Tikhodeev et al., 2002). In one of the original papers on this topic, Fan and Joannopoulos (2002) presented a detailed analysis of the guided resonances in 2D PCS structures and investigated the transmission/reflection spectra properties in such systems when these modes are coupled with vertical radiation modes. Subsequently, many theoretical and experimental papers appeared on this topic with investigations on associated physical characteristics of such resonances and related device demonstrations.

2. Principles of Fano resonance in photonic crystal slabs

2.1 2D photonic crystal slabs

2D PCSs are a particularly important class of PhC structures (Joannopoulos et al., 2008; Johnson et al., 1999). A 2D PCS consists of a two-dimensionally periodic index contrast introduced into a high-index guiding layer. As shown in Fig. 1A, PCSs are quasi-3D structures in which the light confinement arises from the in-plane 2D photonic bandgap (PBG) confinement and vertical waveguide index confinement (through the total internal reflection or TIR principle). 2D PCS structures are certainly more amenable to

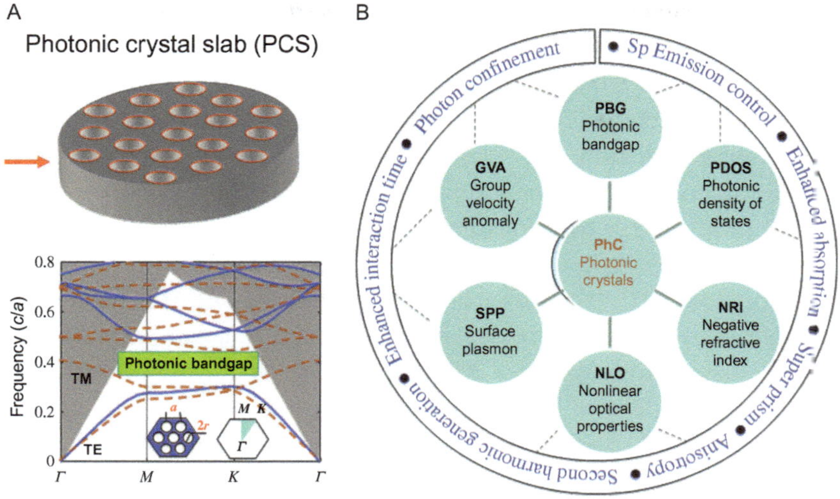

Fig. 1 Two-dimensional photonic crystal slabs (2D PCS): (A) Schematic representation of triangular lattice PCS, with simulated dispersion plot for two (TE/TM) polarizations; and (B) Overview of photonic crystal properties and applications. *Reproduced with permission from Zhou, W., Zhao, D., Shuai, Y.-C., Yang, H., Chuwongin, S., Chadha, A., Seo, J.-H., Wang, K. X., Liu, V., Ma, Z., 2014. Progress in 2D photonic crystal Fano resonance photonics. Prog. Quantum Electron. 38 (1), 1–74.*

fabrication of devices using current semiconductor technology than three-dimensional structures, yet they provide much the same amount of functionality (Fig. 1B) (Zhou, 2011). 2D PCS can be easily integrated with other photonic components for photonic integrated circuits (PICs) using planar micro- and nanofabrication processes (John et al., 2004; Johnson et al., 2001; Mekis et al., 1996; Noda and Baba, 2003; Noda et al., 2002; Notomi, 2011; Seassal et al., 2002). One of the most widely studied structures is the triangular lattice-based air column 2D PCS structure, which possesses a large transverse-electric (TE) and transverse-magnetic (TM) like mode and a large complete TE/TM PBG. A typical dispersion plot is shown in Fig. 1A, where a complete TE/TM bandgap is shown. Note that a light core region is formed due to the vertical confinement. These structures support in-plane-guided modes below the lightlines, which are completely confined by the slab without any coupling to external radiations.

2.2 The presence of Fano resonance in photonic crystal slabs

In addition to in-plane waveguiding, PCS can also interact with external radiation in complex and interesting ways. Of particular importance here is the presence of guided resonance in the structures (Fan and Joannopoulos, 2002; Tikhodeev et al., 2002). Similar to the guided mode, for guided resonance the electro-magnetic power is also strongly confined within the slab. Unlike the guided mode, however, the resonance can be coupled with external radiation (the leaky mode is above the lightline Fig. 1A). Therefore, guided resonance can provide an efficient way to channel light from within the slab to the external environment. A detailed theoretical analysis of guided resonances in 2D PCS's was first presented by Fan and Joannopoulos (2002). Kanskar et al. (1997) presented one of the first experimental reports of guided resonances at optical frequencies in a 2D PCS structure based on a suspended AlGaAs layer.

For light incident on PCS from out of the plane direction (e.g., surface-normal direction), photon transport can take two pathways: a direct or background pathway that does not involve the guided resonance or a resonance-assisted pathway in which photons are coupled into the resonances and subsequently leak out to free space. Assuming that the slab has mirror symmetry in the vertical direction, and taking into consideration the interference between these two pathways, we express the transmission amplitude t as a function of frequency ω

$$t = t_d + f\frac{\gamma}{i(\omega - \omega_0) + \gamma} \quad (1)$$

$$r = r_d \pm f\frac{\gamma}{i(\omega - \omega_0) + \gamma} \quad (2)$$

where t_d and r_d are the direct transmission and reflection coefficients, ω_0 and γ are the center frequency and width of the resonance, and f is the normalized complex amplitude of the resonant mode. The plus and minus sign correspond with even and odd resonant modes for structures with the mirror symmetry. Energy conservation requires

$$f = -(t_d \pm r_d) \quad (3)$$

For $0 < |t_d| < 1$, Eq. (1) gives a Fano resonance lineshape. For example, we consider a system consisting of a square lattice of air holes introduced into a dielectric slab (Fig. 2A). The intensity transmission spectrum is given in Fig. 2B and C, exhibiting Fano resonance lineshapes with strong asymmetric profiles. Using finite-difference time-domain (FDTD) simulations, one can determine the parameters for Eq. (1) and compare the spectrum from this theoretical derivation to that from the FDTD simulation (Fig. 2B and C).

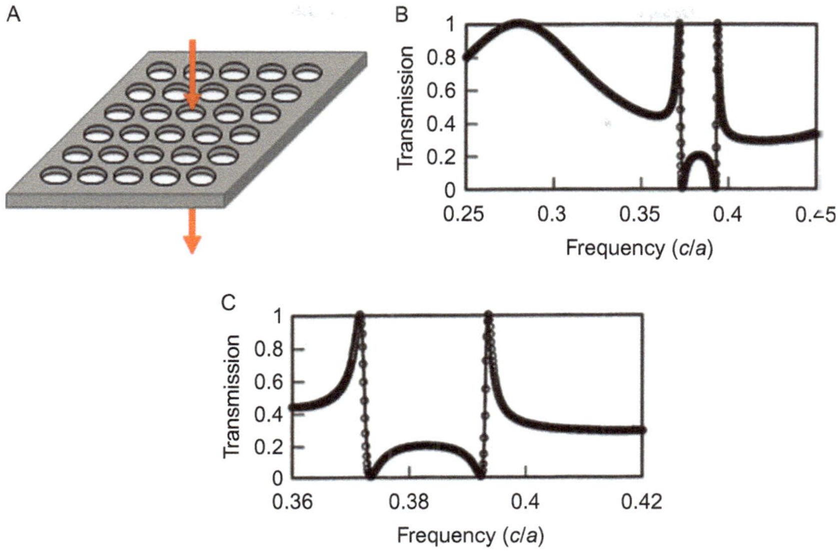

Fig. 2 Fano resonances in 2D photonic crystal slabs: (A) Photonic crystal structure consisting of a square lattice of air holes with a radius of $r=0.2a$ in a dielectric slab with dielectric constant $n_h=12$ and a thickness (t) of $0.5a$, where a is the lattice constant. The *arrow* indicates the direction of the incident light. (B) The intensity transmission spectrum through such a structure. The *circles* are the result of the finite-difference time-domain simulations. The *solid curve* is determined by the analytic theory as represented by Eq. (1). (C) The same plot as (B), except the frequency range is now restricted to [0.36(c/a), 0.42(c/a)] to exhibit further details of the resonance line shape. *Reproduced with permission from Fan, S., Suh, W., Joannopoulos, J., 2003. Temporal coupled-mode theory for the Fano resonance in optical resonators. J. Opt. Soc. Am. 20 (3), 569–572.*

In the vicinity of the Fano resonance, the intensity transmission in Fig. 2B undergoes a transition between 0% and 100%, meaning the constructive and destructive interferences are located close to each other. This phenomenon is due to the rapid variation of the phase along the localized (resonance-assisted) pathway compared to the background pathway.

2.3 Characteristics of Fano resonance in photonic crystal slabs

While Eq. (1) intuitively captures the interference between the background and the resonance-assisted pathways, it can be naturally derived from the more general and phenomenological temporal coupled-mode theory (Joannopoulos et al., 2008), in which a system is modeled as a set of localized and propagating modes. Instead of introducing the general formalism dealing with an arbitrary number of resonances and waveguides (Suh et al., 2004), Fan et al. (2003) focused on the simple case of a single resonance coupled to two ports to better illustrate the essence of the physical process. This two-port configuration could be guaranteed in the structure when the light's wavelength is greater than the lattice constant such that there is no diffraction above or below the slab, as shown in Fig. 2A.

The dynamics of this lossless single-resonance double-port system is described by the coupled-mode equations

$$\frac{\mathrm{d}a}{\mathrm{d}t} = \left(i\omega_0 - \frac{1}{\tau}\right)a + (\kappa_1\, \kappa_2)\begin{pmatrix} s_{1+} \\ s_{2+} \end{pmatrix} \tag{4}$$

$$\begin{pmatrix} s_{1-} \\ s_{2-} \end{pmatrix} = C\begin{pmatrix} s_{1+} \\ s_{2+} \end{pmatrix} + a\begin{pmatrix} d_1 \\ d_2 \end{pmatrix} \tag{5}$$

where a is the amplitude of the resonant mode such that $|a|^2$ represents the electromagnetic energy stored in the resonance. ω_0 and $\gamma = 1/\tau$ are the center frequency and the width of the resonance, respectively. s_{1+} and s_{2+} are the incoming amplitudes of the plane waves on the two ports such that $|s_{1+}|^2$ and $|s_{2+}|^2$ correspond with the intensities. s_{1-} and s_{2-} are the outgoing amplitudes of the plane waves from the two ports such that $|s_{1-}|^2$ and $|s_{2-}|^2$ correspond with the intensities. κ_1 and κ_2 are the coupling constants between the incoming waves and the resonance. d_1 and d_2 are the coupling constants between the outgoing waves and the resonance. C is the background scattering matrix.

The coupling constants are constrained by energy conservation and time-reversal symmetry as follows (Suh et al., 2004):

$$|d_1|^2 + |d_2|^2 = \frac{2}{\tau} \tag{6}$$

$$\kappa_1 = d_1;\ \kappa_2 = d_2 \tag{7}$$

$$wC\begin{pmatrix} d_1^* \\ d_2^* \end{pmatrix} = -\begin{pmatrix} d_1 \\ d_2 \end{pmatrix} \tag{8}$$

By choosing the appropriate positions of the reference planes, without loss of generality, the direct scattering matrix C can be expressed as

$$C = e^{i\phi}\begin{pmatrix} r & it \\ it & r \end{pmatrix} \tag{9}$$

where ϕ is a phase factor, and r and t, both being real, are the reflection and transmission coefficients of the direct process satisfying $r^2 + t^2 = 1$ (Haus, 1984). Combining Eqs. (6–9), one can solve for the coupling constants in terms of the decay rate and the elements in the direct scattering matrix

(Wang et al., 2013), and the theory accounts for asymmetric photonic crystal slabs (Rosenberg et al., 2005). For the structure with mirror symmetry in Fig. 2A, one can further simplify the solution by requiring $d_1 = d_2$, and the intensity reflection coefficient R is therefore

$$R = \frac{r^2(\omega-\omega_0)^2 + t^2(1/\tau)^2 \mp 2rt(\omega-\omega_0)(1/\tau)}{(\omega-\omega_0)^2 + (1/\tau)^2} \tag{10}$$

which agrees with Eq. (2).

Typically there are two techniques for the simulation of the transmission and reflection properties of PCS structures. The first approach is the rigorous coupled wave analysis (RCWA) technique (Moharam and Gaylord, 1981), a relatively straightforward technique for obtaining the exact solution of Maxwell's equations for the electromagnetic diffraction of grating structures. It is a noniterative, deterministic technique utilizing a state-variable method that converges with the proper solution without inherent numerical instabilities. The accuracy of the solution obtained depends solely on the number of terms in the field space-harmonic expansion, ensuring that the energy is always conserved (Moharam et al., 1995). Recently, a Fourier Modal Method-based package was developed by Liu and Fan, and it is now freely available as Stanford Stratified Structure Solver (S^4) software package (Liu and Fan, 2012). Compared to the commercial RCWA package GD-Calc (Johnson, 1981), the S^4 software package offers much higher resolution due to the consideration of higher diffraction orders and also takes much less computation time (Shuai et al., 2013).

The other widely used simulation approach is the three-dimensional finite-difference time domain (3D FDTD) technique (Taflove and Hagness, 2000). In this case, a unit cell is defined by using a periodic boundary condition (PBC) and a perfectly matched layer (PML) in the four lateral and two vertical directions (Qiang et al., 2008), as shown in Fig. 3A. A Gaussian source is launched from the top of the PCS structure, with two power monitors monitoring both the reflected and the transmitted power. For practical structures, where the number of the PhC lattice period is finite, full vectorial 3D FDTD can also offer very precise and accurate simulation results. Depending on the PCS lattice parameters, either Fano resonance filters or broadband reflectors can be realized with different Q values. A set of simulated spectra for transmission and reflection is shown in Fig. 3B and C, respectively. High Q filters with Q greater than 1000 can be easily obtained when a smaller r/a value is used ($r/a = 0.08$, $t/a = 0.34$, $a = 765$ nm for high Q case here).

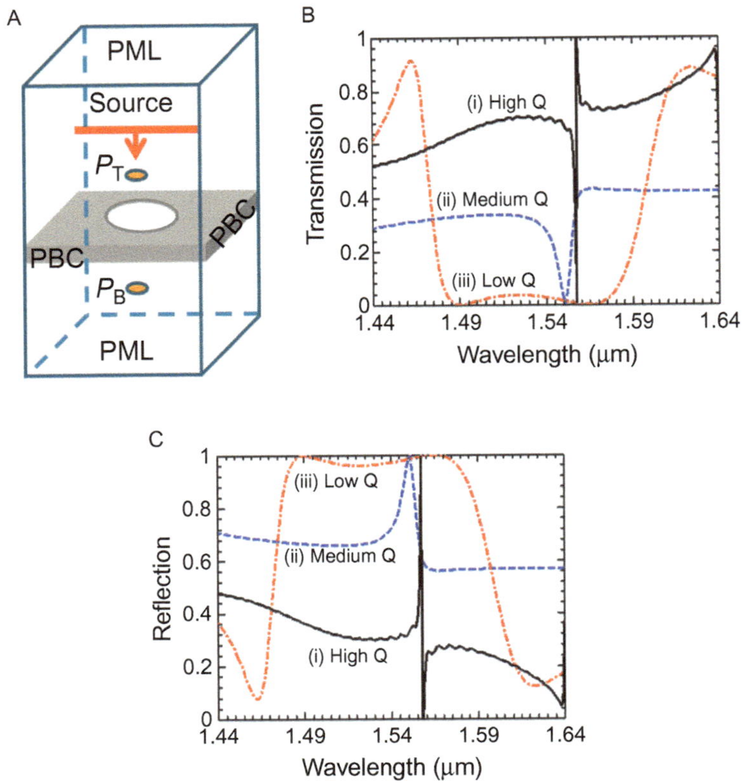

Fig. 3 Simulation and spectral control of Fano resonances in 2D PCS: (A) Schematic of an unit cell used in the simulation based on 3D-FDTD technique; Simulated (B) transmission and (C) reflection spectra for surface-normal Fano filters with different quality factors (Qs) for either highly spectrally selective filters or broadband reflectors. *Reproduced with permission from Zhou, W., Ma, Z., Yang, H., Qiang, Z., Qin, G., Pang, H., Chen, L., Yang, W., Chuwongin, S., Zhao, D., 2009. Flexible photonic-crystal Fano filters based on transferred semiconductor nanomembranes. J. Phys. D Appl. Phys. 42, 234007.*

On the other hand, a larger r/a value can lead to a lower Q filters for broadband reflector design ($r/a = 0.28$, $t/a = 0.347$, $a = 980\,\text{nm}$ for low Q case here). A medium Q design was also presented with $r/a = 0.19$, $t/a = 0.417$, and $a = 600\,\text{nm}$.

The resonance modes were further verified with the field propagation plots based on 3D FDTD simulations. Snapshots of field propagation for the on- and the off-resonance modes are shown in Fig. 4A and B, respectively. Note that, for the on-resonance mode (λ_1), the surface-normal incident light is reflected from the patterned PCS structure due to the coherence (in-phase) reflection, which leads to a dip in the transmission spectra.

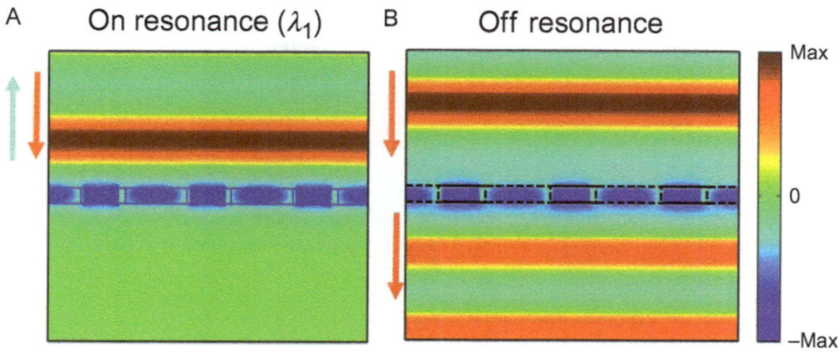

Fig. 4 *Snap shots* of electric field distribution of Fano filters at (A) on-resonance and (B) off-resonance conditions. *Reproduced with permission from Zhou, W., Ma, Z., Yang, H., Qiang, Z., Qin, G., Pang, H., Chen, L., Yang, W., Chuwongin, S., Zhao, D., 2009. Flexible photonic-crystal Fano filters based on transferred semiconductor nanomembranes. J. Phys. D Appl. Phys. 42, 234007.*

On the other hand, light at other spectra locations (off-resonance) can pass through the patterned PCS structure with maximum transmission efficiency.

Shown in Fig. 5 is the simulated dispersion plot for the square lattice PCS structure, based on a three-dimensional (3D) plane wave expansion (PWE) technique (Zhou et al., 2009). The Fano resonance modes operate above the lightline region, denoted as the shaded area. The properties of a few Fano resonance modes (ω_1, ω_2, ω_3) will be discussed in later sections (Qiang et al., 2008). Note also that the existence of PBG is not essential for the design of Fano resonance PCS structures.

In summary, Fano resonances in 2D PCSs arise from the coupling of in-plane discrete guide modes with the vertical continuum free-space radiation mode. Depending on the coupling strength, the output spectral resonance can vary from symmetric to asymmetric, with sharp transition and large phase change between the resonance peak and dip, which can lead to extremely high quality factor (or infinite) optical cavities, or very broadband high reflectivity reflectors. Local field enhancements associated with the Fano resonance cavity can result in strong light-matter interactions for enhanced light emission, detection, sensing, and enhanced optical forces, as well as other nonlinear effects. The existence of the photonic bandgap is not essential in Fano resonance PCS structures. The flexibility in dispersion engineering and the associated slow-light effects can further enhance light-matter interactions (Boutami et al., 2006; Chen et al., 2006; Fan and Joannopoulos, 2002; Fan et al., 2003; Kanamori et al., 2007; Lin et al., 2005; Rosenberg et al., 2005; Suh and Fan, 2004; Taillaert et al., 2002).

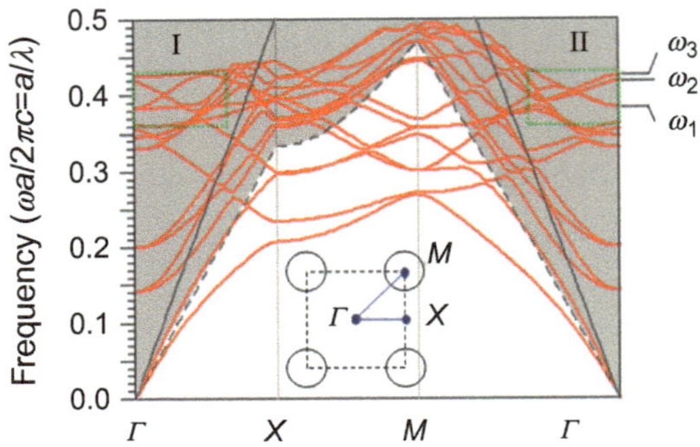

Fig. 5 Simulated dispersion characteristics for the square lattice Si PCS structure on low index glass substrate. The lattice parameters are $r/a=0.19$, $t/a=0.417$, the refractive indices of silicon and glass are 3.48 and 1.5, respectively. *Reproduced with permission from Qiang, Z., Yang, H., Chen, L., Pang, H., Ma, Z., Zhou, W., 2008. Fano filters based on transferred silicon nanomembranes on plastic substrates. Appl. Phys. Lett. 93, 061106.*

References

Boutami, S., Bakir, B.B., Hattori, H., Letartre, X., Leclercq, J.L., Rojo-Romeo, P., Garrigues, M., Seassal, C., Viktorovitch, P., 2006. Broadband and compact 2-D photonic crystal reflectors with controllable polarization dependence. IEEE Photon. Technol. Lett. 18 (7), 835–837.

Boutami, S., Bakir, B., Regreny, P., Leclercq, J., Viktorovitch, P., 2007. Compact 1.55 μm room-temperature optically pumped VCSEL using photonic crystal mirror. Electron. Lett. 43, 282.

Carletti, L., Malureanu, R., Mørk, J., Chung, I.S., 2011. High-index-contrast grating reflector with beam steering ability for the transmitted beam. Opt. Express 19 (23), 23567–23572.

Chang-Hasnain, C.J., 2011. High-contrast gratings as a new platform for integrated optoelectronics. Semicond. Sci. Technol. 26, 014043.

Chen, L., Huang, M.C.Y., Mateus, C.F.R., Chang-Hasnain, C.J., Suzuki, Y., 2006. Fabrication and design of an integrable subwavelength ultrabroadband dielectric mirror. Appl. Phys. Lett. 88, 031102.

Fan, S., 2002. Sharp asymmetric line shapes in side-coupled waveguide-cavity systems. Appl. Phys. Lett. 80 (6), 908.

Fan, S., Joannopoulos, J.D., 2002. Analysis of guided resonances in photonic crystal slabs. Phys. Rev. B 65 (23), 235112.

Fan, S., Suh, W., Joannopoulos, J., 2003. Temporal coupled-mode theory for the Fano resonance in optical resonators. J. Opt. Soc. Am. 20 (3), 569–572.

Fano, U., 1961. Effects of configuration interaction on intensities and phase shifts. Phys. Rev. 124 (6), 1866.

Giannini, V., Francescato, Y., Amrania, H., Phillips, C.C., Maier, S.A., 2011. Fano resonances in nanoscale plasmonic systems: a parameter-free modeling approach. Nano Lett. 11 (7), 2835–2840.

Haus, H.A., 1984. Waves and Fields in Optoelectronics. Prentice-Hall, Englewood Cliffs, NJ.

Joannopoulos, J.D., Johnson, S.G., Winn, J.N., Meade, R.D., 2008. Photonic Crystals: Molding the Flow of Light, second ed. Princeton University Press.

John, S., Toader, O., Chutinan, A., 2004. Photonic band gap architectures for micro-fabrication and diffractionless optical networking. In: IEICE Transactions on Electronics. vol. E87-C(3). The Institute of Electronics, Information and Communication Engineers, pp. 266–273.

Johnson, K., 1981. Coupled scalar wave diffraction theory. Appl. Phys. 24 (3), 249–260.

Johnson, S.G., Shanhui, F., Villeneuve, P.R., Joannopoulos, J.D., Kolodziejski, L.A., 1999. Guided modes in photonic crystal slabs. Phys. Rev. B Condens. Matter 60 (8), 5751–5758.

Johnson, S.G., Povinelli, M.L., Joannopoulos, J.D., 2001. New photonic-crystal system for integrated optics. In: Active and Passive Optical Components for WDM Communication. SPIE-Int. Soc. Opt. Eng, Denver, CO.

Kanamori, Y., Kitani, T., Hane, K., 2007. Control of guided resonance in a photonic crystal slab using microelectromechanical actuators. Appl. Phys. Lett. 90, 031911.

Kanskar, M., Paddon, P., Pacradouni, V., Morin, R., Busch, A., Young, J.F., Johnson, S.R., MacKenzie, J., Tiedje, T., 1997. Observation of leaky slab modes in an air-bridged semiconductor waveguide with a two-dimensional photonic lattice. Appl. Phys. Lett. 70 (11), 1438–1440.

Kikuta, H., Toyota, H., Yu, W., 2003. Optical elements with subwavelength structured surfaces. Opt. Rev. 10 (2), 63–73.

Lin, C., Lu, Z., Shi, S., Jin, G., Prather, D.W., 2005. Experimentally demonstrated filters based on guided resonance of photonic-crystal films. Appl. Phys. Lett. 87, 091102

Liu, V., Fan, S., 2012. S4: a free electromagnetic solver for layered periodic structures. Comput. Phys. Commun. 183 (10), 2233–2244.

Luk'yanchuk, B., Zheludev, N.I., Maier, S.A., Halas, N.J., Nordlander, P., Giessen, H., Chong, C.T., 2010. The Fano resonance in plasmonic nanostructures and metamaterials. Nat. Mater. 9 (9), 707–715.

Magnusson, R., Shokooh-Saremi, M., 2008. Physical basis for wideband resonant reflectors. Opt. Express 16 (5), 3456–3462.

Magnusson, R., Wang, S.S., 1992. New principle for optical filters. Appl. Phys. Lett. 61, 1022.

Mekis, A., Chen, J.C., Kurland, I., Fan, S., Villeneuve, P.R., Joannopoulos, J.D., 1996. High transmission through sharp bends in photonic crystal waveguides. Phys. Rev. Lett. 77 (18), 3787–3790.

Miroshnichenko, A.E., Flach, S., Kivshar, Y.S., 2010. Fano resonances in nanoscale structures. Rev. Mod. Phys. 82 (3), 2257.

Moharam, M., Gaylord, T., 1981. Rigorous coupled-wave analysis of planar-grating diffraction. J. Opt. Soc. Am. 71 (7), 811–818.

Moharam, M., Grann, E.B., Pommet, D.A., Gaylord, T., 1995. Formulation for stable and efficient implementation of the rigorous coupled-wave analysis of binary gratings. J. Opt. Soc. Am. A 12 (5), 1068–1076.

Noda, S., Baba, T., 2003. Roadmap on Photonic Crystals. Springer.

Noda, S., Imada, M., Okano, M., Ogawa, S., Mochizuki, M., Chutinan, A., 2002. Semiconductor three-dimensional and two-dimensional photonic crystals and devices. IEEE J. Quantum Electron. 38 (7), 726–735.

Notomi, M., 2011. Strong light confinement with periodicity. Proc. IEEE 99 (10), 1768–1779.

Ott, C., Kaldun, A., Raith, P., Meyer, K., Laux, M., Evers, J., Keitel, C.H., Greene, C.H., Pfeifer, T., 2013. Lorentz meets Fano in spectral line shapes: a universal phase and its laser control. Science 340 (6133), 716–720.

Qiang, Z., Yang, H., Chen, L., Pang, H., Ma, Z., Zhou, W., 2008. Fano filters based on transferred silicon nanomembranes on plastic substrates. Appl. Phys. Lett. 93, 061106.

Rahmani, M., Luk'yanchuk, B., Hong, M., 2013. Fano resonance in novel plasmonic nanostructures. Laser Photonics Rev. 7 (3), 329–349.
Rosenberg, A., Carter, M., Casey, J., Kim, M., Holm, R., Henry, R., Eddy, C., Shamamian, V., Bussmann, K., Shi, S., Prather, D.W., 2005. Guided resonances in asymmetrical GaN photonic crystal slabs observed in the visible spectrum. Opt. Express 13 (17), 6564–6571.
Seassal, C.D.Y., Letartre, X., Grillet, C., Rojo-Romeo, P., Viktorovitch, P., Benyattou, T., 2002. Optical coupling between a two-dimensional photonic crystal-based microcavity and single-line defect waveguide on InP membranes. IEEE J. Quantum Electron. 38 (7), 811–815.
Shuai, Y., Zhao, D., Tian, Z., Seo, J., Plant, D.V., Ma, Z., Fan, S., Zhou, W., 2013. Double-layer Fano resonance photonic crystal filters. Opt. Express 21 (21), 24582–24589.
Suh, W., Fan, S., 2004. All-pass transmission or flattop reflection filters using a single photonic crystal slab. Appl. Phys. Lett. 84, 4905.
Suh, W., Wang, Z., Fan, S., 2004. Temporal coupled-mode theory and the presence of non-orthogonal modes in lossless multimode cavities. IEEE J. Quantum Electron. 40 (10), 1511–1518.
Taflove, A., Hagness, S.C., 2000. Computational Electrodynamics: The Finite-Difference Time-Domain Method. Artech House.
Taillaert, D., Bogaerts, W., Bienstman, P., Krauss, T., Van Daele, P., Moerman, I., Verstuyft, S., De Mesel, K., Baets, R., 2002. An out-of-plane grating coupler for efficient butt-coupling between compact planar waveguides and single-mode fibers. IEEE J. Quantum Electron. 38 (7), 949–955.
Tikhodeev, S.G., Yablonskii, A.L., Muljarov, E.A., Gippius, N.A., Ishihara, T., 2002. Quasiguided modes and optical properties of photonic crystal slabs. Phys. Rev. B 66 (4), 45102.
Wang, K.X., Yu, Z., Sandhu, S., Fan, S., 2013. Fundamental bounds on decay rates in asymmetric single-mode optical resonators. Opt. Lett. 38 (2), 100–102.
Wood, R., 1902. XLII. On a remarkable case of uneven distribution of light in a diffraction grating spectrum. Lond. Edinb. Phil. Mag. 4 (21), 396–402.
Zhou, W., 2011. Photonic crystal surface-emitting lasers and infrared sensors. In: Nalwa, H.S. (Ed.), Encyclopedia of Nanoscience and Nanotechnology. vol. 20. American Scientific Publishers.
Zhou, W., Ma, Z., Yang, H., Qiang, Z., Qin, G., Pang, H., Chen, L., Yang, W., Chuwongin, S., Zhao, D., 2009. Flexible photonic-crystal Fano filters based on transferred semiconductor nanomembranes. J. Phys. D Appl. Phys. 42, 234007.
Zhou, W., Zhao, D., Shuai, Y.-C., Yang, H., Chuwongin, S., Chadha, A., Seo, J.-H., Wang, K.X., Liu, V., Ma, Z., 2014. Progress in 2D photonic crystal Fano resonance photonics. Prog. Quantum Electron. 38 (1), 1–74.

CHAPTER TWO

Transition from photonic crystals to dielectric metamaterials

Mikhail V. Rybin[a,b,*], Mikhail F. Limonov[a,b], Yuri S. Kivshar[b,c]
[a]Ioffe Institute, St. Petersburg, Russia
[b]Department of Physics and Engineering, ITMO University, St. Petersburg, Russia
[c]Nonlinear Physics Center, Australian National University, Canberra, ACT, Australia
*Corresponding author: e-mail address: m.rybin@mail.ioffe.ru

Contents

1. Introduction

In the late 1980s, Yablonovich (1987) and John (1987) introduced a new class of artificial dielectric media—*photonic crystals*—that are periodic structures having a spatial modulation of the dielectric index with a period of half a wavelength and larger. That time, advanced nanotechnologies became mature enough to support the fabrication of such structures with features of several 100 nm in size. The idea of that concept was to apply the solid-state theory of electrons to the case of electromagnetic waves and to develop a class of "semiconductors for light" materials, which would allow many novel applications. These photonic crystals operate due to spatially extended Bragg resonances that appear as a result of the structure periodicity.

Semiconductors and Semimetals, Volume 100
ISSN 0080-8784
https://doi.org/10.1016/bs.semsem.2019.03.001

At the turn of the century, another class of artificial electromagnetic media—*metamaterials*—was proposed by Pendry et al. (1999) and Smith et al. (2000). Although there are many definitions of metamaterials, here we follow to the seminal papers and define metamaterials as artificial media composed of resonant meta-atoms and described by the effective material parameters (dielectric permittivity ε and magnetic permeability μ) similar to many natural materials. The main feature of metamaterials is a weak spatial dispersion (Simovski, 2018) in contrast to photonic crystals, whose resonant frequencies correspond to the case of strong spatial dispersion because the Bragg resonances are formed by periodically arranged meta-atoms. Metamaterials allow to design media with engineered on demand spatial distribution of ε and μ, which has applications in invisibility (Pendry et al., 2012), hyperlenses (Liu et al., 2007), etc. Unlike dielectric permittivity, natural materials have negligible magnetic response and metamaterials pave the way for achieving negative μ in the optical range. In particular, double negative $\varepsilon(\mathbf{r})$ and $\mu(\mathbf{r})$ metamaterials make it possible to realize Veselago's left-handed media (Shelby et al., 2001; Veselago, 1968).

The first metamaterial was composed of meta-atoms being thin wire RF antennas. Split-ring resonators are widely used to achieve an effective resonant magnetic response of artificial media with some values of negative μ. The split-ring resonators can be much smaller relative to the resonant wavelength that enables an effective media description, however at high frequencies the electric field demonstrates essential Ohmic losses, which is a great disadvantage of metal-based meta-atoms. In 2002, O'Brien and Pendry suggested high-index dielectrics meta-atoms, which support magnetic Mie resonances for the TE polarization (magnetic field oscillates along the rods). They considered the rods with $\varepsilon = 200$ and evaluated an effective magnetic permeability μ. Although the correctness of their effective parameter is a subject of debates (Simovski, 2009), the existence of dielectric metamaterials with negative magnetic response was established.

The main difficulty in the experimental realization of dielectric metamaterials is that the size of meta-atoms (e.g., the diameter of rods) for materials with $\varepsilon = 12$ (the value available for many high-index optical materials) is about 5 times smaller than the resonant wavelength, and this makes the theoretical assumption about the subwavelength operational regime of such structures not applicable. However, numerical simulations suggest that the effective homogenization can work even for relatively large meta-atoms. On the other hand, low-index dielectric periodic structures are presented by the well-studied photonic crystals (see, e.g., Joannopoulos et al., 2008) and thus one may ask a simple question: *When a photonic crystal becomes a metamaterial provided one varies the permittivity from low to high values?* In this chapter, we review the recent studies related to this problem and introduce the concept of *a photonic phase diagram* (see Rybin et al., 2015).

2. Photonic crystals vs metamaterials

Phase transition terminology has been used in a number of articles to discuss effects in artificial photonic structures (Gorlach and Belov, 2014; Mirmoosa et al., 2015; Reyes-Gómez et al., 2014). We briefly remind what the phase transitions in thermodynamics are. First we notice that phase transitions of the first order (i.e., solids become liquids) involving dramatic changes in location of atoms and their interaction has nothing in common with the effects observed in photonics. Other thermodynamic phase transitions are related to second order or continuous phase transitions. In most cases they are associated with changes in electronic properties, which make it possible to observe features, for example, in the temperature dependences of the heat capacity, changes in susceptibility, and so on. The physical origin of the continuous phase transitions is that a change in external parameters (temperature, pressure, etc.) leads to a continuous transformation of the spatial arrangement of atoms that in turn results in a rapid modification of electronic subsystem, in particular, a small displacement of atoms results in a dramatic transformation of band structure.

Unlike conventional materials, photonic structures are fabricated by an arrangement of desired meta-atoms according to a chosen design. In principle, such artificial structures allow one to change the distances between meta-atoms and crystal-lattice symmetry without involving of external influence (e.g., temperature or pressure). However, this modification of system parameters might result in dramatic changes in photonic properties being similar to the continuous phase transitions in the conventional materials. We discuss one important difference in these two cases. The Pauli's exclusive principle prohibits to occupy the same state by different particles does not apply to photons, and hence, Fermi energy is not applicable to photonic structures. Instead, the energy (frequency) of photons is defined by sources of the electromagnetic waves. Thus, frequency is often considered as an additional structure parameter along with lattice spacing and constituent permittivity. For example, a frequency-dependent phase transition between hyperbolic and elliptic regimes of metamaterials was considered (Reyes-Gómez et al., 2014). However, the term “frequency-dependent phase transition” is a confusing point, because a true phase cannot depend on probing parameters, which include the radiation frequency.

Before we give a definition of phase for photonic crystals or metamaterials, let us briefly consider their physical properties and band diagrams assuming the periodic systems. The Bloch theorem makes it possible to classify all solutions by a (quasi) wave vector **k** in the form of exp(i**kr**)·X(**r**), where X(**r**) is a periodic function with the translational symmetry of the crystal lattice. We notice that multiplying the function X(**r**) by the exponent exp(i**gr**) does not change the translational symmetry of the product, given **g** is any vector of the reciprocal lattice. It means that the unique solutions related to different wave vectors are restricted by the Brillouin zone only, which is defined as a region of the reciprocal space containing wave vectors being closer to the origin than to other reciprocal lattice nodes. In the case of a small dielectric contrast, i.e., in a photonic crystal regime, the deviation of the permittivity from its average value can be considered as a small perturbation. The photonic band diagram (a dependence of structure eigenfrequency ω on wave vector **k**) for the unperturbed problem is the light cone defined by the dispersion equation $\omega^2 = c^2|\mathbf{k}|^2\varepsilon^{-1}$ (c is the light velocity constant), but the ambiguity of the choice of the wave vector in the periodic structure results in a set of light cones with apex at each nodes of the reciprocal lattice (Fig. 1A). As long as the dielectric contrast is zero, the modes corresponding to different light

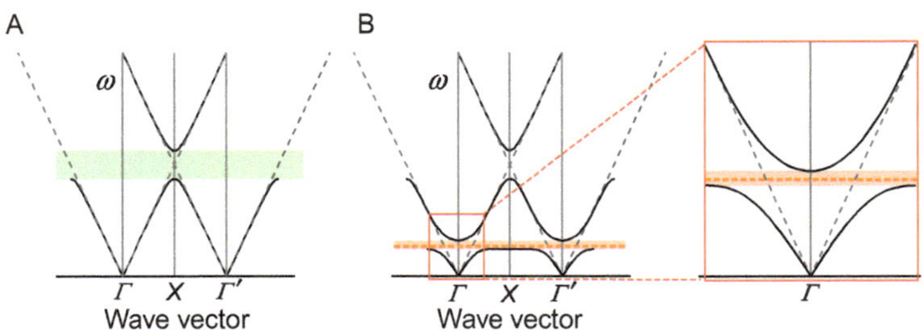

Fig. 1 Difference between the regimes of photonic crystals and metamaterials. (A) Typical bandgap diagram of a photonic crystal: the interaction between two light cones (*gray dashed lines*) results in a frequency gap (*shaded in green*) at the boundary of the Brillouin zone (for example, at the X point). (B) Typical bandgap diagram of a metamaterial: a coupling of the almost flat Mie resonance branch (*orange dashed line*) and the light cone (*gray dashed line*) leads to a polariton-like feature and the Mie gap (*shaded in orange*). *Reproduced with permission from Li, S.V., Kivshar, Y.S., Rybin, M.V., 2018. Toward silicon-based metamaterials. ACS Photonics 5, 4751–4757. Copyright (2018) ACS.*

cones do not mix each other, and only the main light cone at the origin $\mathbf{k}=0$ is enough for a complete description of the electrodynamic problem. However, a nonzero dielectric contrast perturbs the states between different light cones at the boundary of the Brillouin zone that results in opening of bandgaps related to the certain direction of light. These bandgaps arise due to Bragg scattering on the periodic structure. Fig. 1A sketches a pair of unperturbed light cones with vertices at the Γ and Γ' points that cross at the boundary of the Brillouin zone at the X point defining the lowest intersection point for a square lattice. We consider the unique region of wave vectors on the ΓX segment. The branch corresponding to the light cone starts from the Γ point at a certain angle defining the effective refractive index and ends at the X point. Near the X point the dispersion dependence $\omega(\mathbf{k})$ of the lowest branch demonstrates a deviation from a linear function. At high frequencies the second branch appears going from the X point to the Γ point, and the Bragg bandgap is observed between the high- and low-frequency branches.

Now we consider the band diagram of metamaterials, whose meta-atoms support a local resonance at some frequency. Owing to the uncertainty relation, the wave vector of the resonant state is delocalized in the reciprocal space. At the same time, being localized in the real space, the resonance can be considered as a quasi-discrete state with a certain frequency. Fig. 1B exhibits such states by the horizontal orange dotted line, which means that the frequency is well determined and the wave vector is uncertain. Beside this, Fig. 1B demonstrates light cones corresponding to propagating waves (gray dotted lines). A polariton-like feature appears in the band diagram due to a mixing of the propagating waves with a localized state (Kaina et al., 2015; Yves et al., 2017). The branch starting at the Γ point at $\omega=0$ demonstrates a linear dependence at low frequencies and a deviation from the line as it approaches to the eigenfrequency of the localized mode and transforms into a horizontal line merging with an unperturbed local state. With a further increase in frequency, a flat branch emerges from the Γ point, which merges in turn with the continuation of the light cone. As a result a bandgap arises between the high- and low-frequency branches, which is associated with a local resonance supported by the meta-atom. We notice the difference: the spectral position of the Bragg gap is related to the shape of the Brillouin zone and therefore varies depending on the propagation direction in the reciprocal space, whereas the position of the band gap associated with a local resonance does not have dependence on $\mathbf{k}$. It means that in a metamaterial the resonant frequencies correspond to the weak spatial dispersion and the structure can be described by effective material parameters ε and μ, in contrast to the photonic crystals.

Fig. 1B shows the polariton-like feature below the Bragg gap. We remind that the wavelength of Bragg scattering λ_B is governed by the Bragg law $\lambda_B = 2d_{\text{eff}} \cdot \cos(\alpha)$, where d_{eff} is an effective lattice spacing and α is the angle between the wave vector and the normal to the crystal planes. Therefore, there exists the lowest frequency of the Bragg gap, while the high-frequency limit is infinite. It means that in general a range of the weak spatial dispersion is below the lowest Bragg gap. Thus, the criterion for the metamaterial phase is the appearance of the polariton-like feature below the lowest Bragg gap.

We illustrate several problems of dielectric metamaterials by considering one-dimensional structures. Let the first structure consists of a sequence of layers A and B. To achieve resonances on the layers A we set $\varepsilon_A \gg \varepsilon_B$ leading to strong Fabry–Perot resonances at $\omega_{FP} = \pi / \left(w_A \sqrt{\varepsilon_A}\right)$, where w_A is the width of layer A. It is straightforward to demonstrate that the Fabry–Perot frequencies are always higher than the lowest Bragg gap centered at $\omega_B = \pi / \left(w_A \sqrt{\varepsilon_A} + w_B \sqrt{\varepsilon_B}\right)$. Another possibility to obtain resonance response is to use quantum wells as the layer A with the exciton energy (frequency) almost independent on the layer width. An analysis of this system reveals that the Bragg gap width is increasing dramatically when the frequency of exciton resonance is similar to the Bragg frequency (Ivchenko, 2005). Thus, the difference between the Bragg frequency and the frequency of an isolated meta-atom should be large enough to yield a metamaterial with the prototypical polariton-like feature below the lowest Bragg gap.

Let us point out another possibility of achieving the weak spatial dispersion above the lowest Bragg gap. Changes in the structure parameters might lead to an accidental vanishing of spatial dispersion resulting in a photonic crystal with an effective left-handed properties (Decoopman et al., 2006) or a photonic crystal with an effective near-zero refraction index (Huang et al., 2011). Although the accidental vanishing is of some interest, in this chapter we will focus on the regular mechanisms resulting to the weak spatial dispersion around the structure resonances, i.e., dielectric metamaterials.

Here we use the following criterion to distinguish a phase of photonic structure, based on the behavior of the second dispersion branch, which determines the type of the lowest bandgap: (i) in the phase of photonic crystal, the second branch has a minimum frequency at the boundary of the Brillouin zone (see Fig. 1A) and (ii) in the phase of metamaterial, the second branch has a minimum frequency at the center of the Brillouin zone (the Γ point) forming the polariton-like feature (see Fig. 1B).

3. Dielectric rods arranged in a square lattice

In this section, we describe the appearance of metamaterial phase in a square lattice of dielectric rods surrounded by air (Rybin et al., 2015). The rods have a circular profile of radius r and permittivity ε and the lattice constant is a. These three parameters completely determine the photonic structure. However, the scalability of the Maxwell's equations allows us to reduce it to a two parameter set by measuring all distances in the lattice constant a, i.e., we will use a normalized radius r/a and a frequency a/λ, where λ is the vacuum wavelength. We assume that wave vectors are orthogonal to the rod axis, which makes it possible to separate polarization as TE (transverse electric field relative to the rod axis) and TM (transverse magnetic field). The dielectric rods are known to support electric Mie resonances for the TM polarized incident waves and magnetic Mie resonances for the TE polarization (Bohren and Huffman, 1998). First, we consider TE waves because magnetic Mie resonances allow metamaterials with artificial magnetic permeability, which can takes negative values $\mu < 0$.

We have discussed above that the Fabry–Perot resonances in structural elements cannot form the fundamental gap (i.e., the bandgap with the lowest energy in spectrum) in a layered structure. However it is possible for the structures of higher dimensions, for example a periodic system composed of parallel rods. Here we consider two scenarios for the bandgap related to the Mie resonance in rod to become the fundamental gap (Fig. 2). The first one is in reducing the lattice spacing a that increases the Bragg frequency keeping the Mie resonance almost unchanged. At a certain value r/a the order of the Mie and Bragg gaps swaps and the metamaterial phase appears. The second scenario is in increasing the rod permittivity to reduce the Mie resonance frequency, while the lattice geometry (that defines the Bragg bandgaps) are kept. It also might result in the metamaterial phase.

Now, we follow the second scenario to uncover the physics of the metamaterial phase appearance. Three different datasets are computed: (i) Mie scattering spectra on an isolated dielectric rod, (ii) photonic band diagram of a perfect square lattice consisting of homogeneous rods, and (iii) transmission spectra of a structure consisting of 10 layers of rods along the wave incidence direction and being periodic in the orthogonal direction.

First, let us consider the Mie scattering on an isolated rod. Following the Mie theory, the scattering waves are expanded into the multipole cylindrical harmonics $H_n^{(1)}(kr)\exp(in\theta)$ in the far field radiation zone. Here $H_n^{(1)}$ is a Hankel function of order n, and r and θ are the polar coordinates. The harmonic amplitudes are determined by the resonant cylindrical Lorenz–Mie

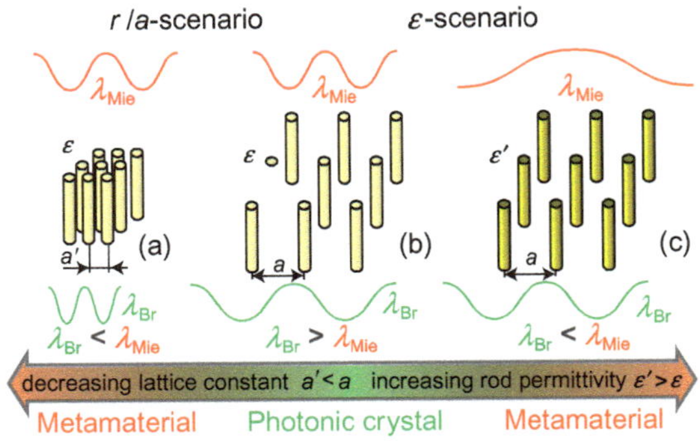

Fig. 2 Transition from photonic crystals to metamaterials. Schematic illustrates two possible transformations of the PhC (B) to MM (A), (C) through the competition between the Bragg and Mie scattering wavelengths. To satisfy the classical homogenization condition in the low-frequency long-wavelength limit ($\lambda_{Bragg} < \lambda_{Mie}$), one may decrease the lattice constant and increase the filling ratio *r*/*a* (*r*/*a*-scenario) or, alternatively, increase the rod permittivity ε the so-called ε-scenario. *Reproduced from the original paper by Rybin, M.V., Filonov, D.S., Samusev, K.B., Belov, P.A., Kivshar, Y.S., Limonov, M.F., 2015. Phase diagram for the transition from photonic crystals to dielectric metamaterials. Nat. Commun. 6, 10102 under a Creative Commons License (http://creativecommons.org/licenses/by/4.0/).*

coefficients a_n and b_n for the TE and the TM polarizations, respectively (see, e.g., Bohren and Huffman, 1998). Fig. 3 shows the calculated Mie scattering efficiency spectra for the dipole cylindrical harmonic with resonances TE_{01}, TE_{02} and higher multipoles with resonances TE_{nm} (n defines the angular behavior and m enumerates resonances with the same n). The spectra reveal a strong decrease of the dimensionless frequencies a/λ and the narrowing of the Mie bands TE_{nm} with increasing ε.

The photonic band diagrams shown in Fig. 3 are calculated numerically by exploiting a plane wave expansion method (Johnson and Joannopoulos, 2001). A large number of plane waves in the expansion (128 × 128) enable a high precision of eigenfrequencies in the photonic diagrams for both photonic crystal and metamaterial phases. We remind that the Brillouin zone of a square lattice has a square shape as well and has three points of high symmetry: Γ at $\mathbf{k}=0$, X at $k=(\pi/a;0)$, M at $k=(\pi/a;\ \pi/a)$, which restrict an irreducible part of the Brillouin zone. Since the lowest frequency of the Bragg resonance in the square lattice corresponds to the ΓX we consider band diagrams for this direction only. For convenience, Fig. 3 shows extended band diagrams for the wave vector in the $\Gamma X \Gamma$ interval with clear extrema of the dispersion branches.

The transmission spectra in Fig. 3 are calculated by using the CST Microwave Studio software for the wave vector of the incident beam parallel to the ΓX direction. We find that transmission spectra of structure with the

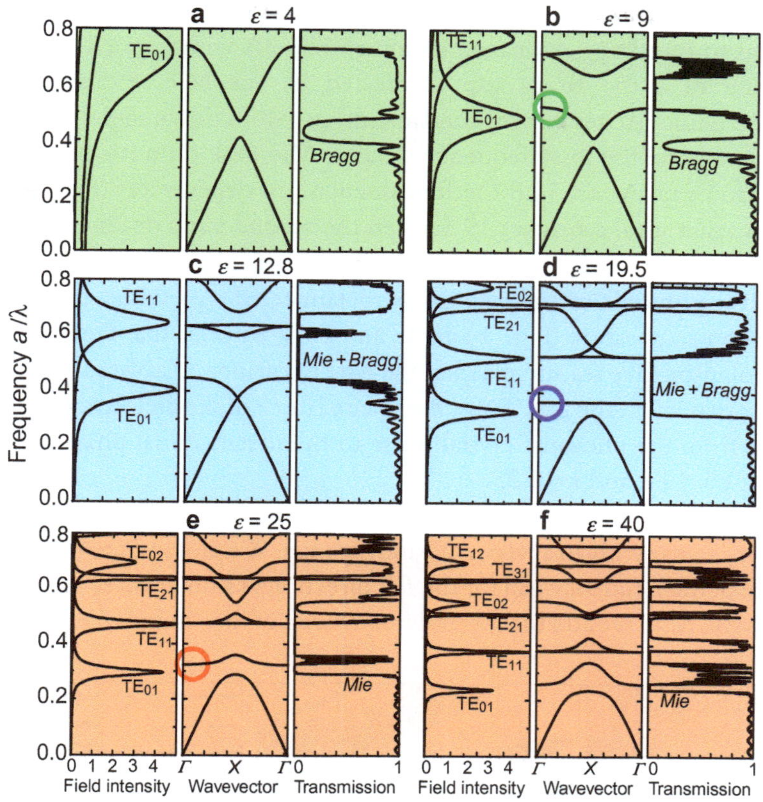

Fig. 3 (Left-hand sections in all panels) Calculated Mie scattering efficiency for an isolated dielectric circular rod for TE_n ($n \geq 0$) modes. (Central sections) The band structure for 2D square lattice of rods with $r = 0.25a$ in air for the TE polarization. The band structure is shown between the Γ point (wave vector $\mathbf{k} = 0$) and the X point ($|\mathbf{k}| = \pi/a$ along the x direction). (Right-hand panels) The transmittance calculated for 10 lattice layers of the 2D square structure of rods in air for the TE polarization. The frequency and wave vector are plotted in dimensionless units a/λ, where λ is vacuum wavelength and a denotes the lattice constant. (A) $\varepsilon = 4$, (B) $\varepsilon = 9$, (C) $\varepsilon = 12.8$, (D) $\varepsilon = 19.5$, (E) $\varepsilon = 25$, and (F) $\varepsilon = 40$. *Reproduced from Rybin, M.V., Filonov, D.S., Samusev, K.B., Belov, P.A., Kivshar, Y.S., Limonov, M.F., 2015. Phase diagram for the transition from photonic crystals to dielectric metamaterials. Nat. Commun. 6, 10102 under a Creative Commons License (http://creativecommons.org/licenses/by/4.0/).*

thickness of 10 rod layers demonstrate an intensive dip for both photonic crystal and metamaterial phases.

The three datasets calculated by independent methods make it possible to analyze an evolution of low-frequency bands of the photonic band diagram being a result of interference between the local Mie resonances and the Bragg resonances extended over the structure. When the permittivity is 4, the Mie resonances are at the higher frequencies and the fundamental gap is related to the Bragg resonance (compare Fig. 3A with Fig. 1A). With the increase of ε the Bragg gap is affected by the Mie resonance TE_{01}. Fig. 3B demonstrates that the fundamental gap shifts down slightly, and both the center and the low-frequency wing of Mie peak do not correspond to any band gap. At $\varepsilon = 12.8$ the fundamental gap degenerates (Fig. 3C) and opens again until ε reaches 19.5 when the second band degenerates into a horizontal line (Fig. 3D). For greater $\varepsilon > 19.5$, the band structure transforms from the photonic crystal type to the metamaterial type with a pronounced polariton-like feature (Figs. 1 and 3E and F). We notice that the bandgap in the polariton-like feature matches the Mie resonance of isolated rod, which means that it is a Mie gap. Thus, the square lattice of the dielectric rods transforms from the photonic crystal phase to the metamaterial phase at critical values $\varepsilon = 19.5$ and $r = 0.25a$.

The picture of metamaterial phase appearance in the photonic structure becomes clear with an analysis of bandgap map. We plot in Fig. 4 the dependence of the frequency gap for the ΓX direction as a function of the rod permittivity. Also, we add the spectral frequencies of low-index Mie resonances

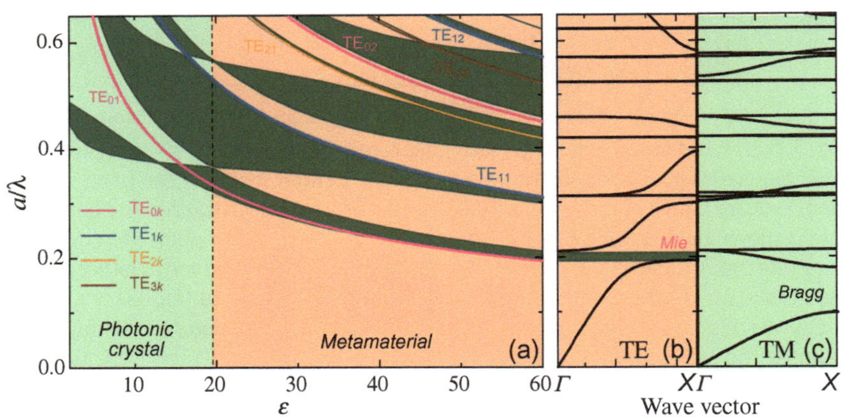

Fig. 4 Bandgap diagram for the 2D square lattice of circular rods. (A) The bandgap diagram obtained from the photonic band structures calculated for $1 \leq \varepsilon \leq 60$ with steps of $\Delta\varepsilon = 1$. The rods are embedded in air, $r/a = 0.25$, and TE polarization is considered. The Bragg and Mie gaps obtained from the band structure calculations are marked by *dark green*. (B and C) The photonic band structure for the 2D square lattice of rods with $\varepsilon = 60$ for TE and TM polarization, respectively, for the ΓX scan of the wave vector **k**. *Reproduced from Rybin, M.V., Filonov, D.S., Samusev, K.B., Belov, P.A., Kivshar, Y.S., Limonov, M.F., 2015. Phase diagram for the transition from photonic crystals to dielectric metamaterials. Nat. Commun. 6, 10102 under a Creative Commons License (http://creativecommons.org/licenses/by/4.0/).*

TE_{nm} to the map. The Bragg gap appears at $a/\lambda=0.5$ when the rod permittivity is low. With the increase of the permittivity the TE_{nm} Mie resonances shift down to $a/\lambda=0.5$, and the Bragg gap becomes broad similar to the case of the resonant photonic crystals (Ivchenko, 2005). The interference between Mie and broaden Bragg resonances results in a nontrivial picture of alternating allowed bands and band gaps. However at $\varepsilon=19.5$ the TE_{01} Mie gap splits off from the broaden Bragg gap and for the higher permittivities the fundamental (lowest) band gap is related to the Mie resonance on the structure elements only.

Recently, Maslova et al. (2018) found conditions for metamaterials with electric response operating for the TM polarization. The TM polarization is known to demonstrate stronger effects with respect to the case of TE polarization. For this reason the effect of the resonant Bragg gap broadening is stronger, which disables metamaterial regime for the structure with and electric metamaterials with $r=0.25a$ (see Fig. 4C). However, sparse structures make is possible to observe metamaterials with electric response (Fig. 5). When the rod permittivity is $\varepsilon=30$ that corresponds to $Ge_2Sb_2Te_5$

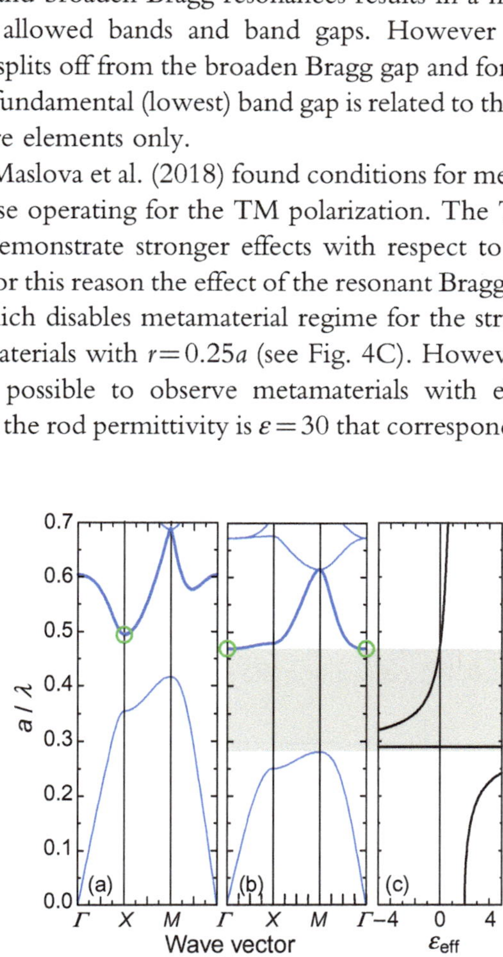

Fig. 5 Band diagrams of dielectric rods arranged in a square lattice for TM polarization. (A) Photonic crystal with $\varepsilon=12$ and $r/a=0.1$. (B) Metamaterial with $\varepsilon=30$ and $r/a=0.1$. The second branches in the diagrams are marked by *thick lines*. *Green circles* correspond to the bottom boundary of the second band, the *red dotted lines* mark frequencies of the maximum field enhancement at $a/\lambda=0.518$ and $a/\lambda=0.472$ in a structure of 5 by 250 rods. (C) Evaluated effective permittivity ε_{eff} of the metamaterial with $\varepsilon=30$ and $r/a=0.1$. *Gray shading* in (B and C) corresponds to the Mie gap. *Reproduced with permission from Maslova, E.E., Limonov, M.F., Rybin, M.V., 2018. Dielectric metamaterials with electric response. Opt. Lett. 43, 5516–5519. Copyright (2018) OSA.*

compound in infrared (Wuttig et al., 2017) the band diagram demonstrates the polariton-like feature.

Metamaterials composed of sparse high-index dielectric rods allow a homogenization procedure by using the approach proposed by O'Brien and Pendry (2002). An effective dielectric index is evaluated as $\varepsilon_{eff} = \langle D \rangle / \langle \varepsilon_0 E \rangle$, where the average of the displacement field D is taken over the 2D unit-cell area and the average of the electric field E is taken along a line at the unit-cell boundary parallel to the rod. The small radius of dielectric rod allows us to perform the averaging over a circle instead of the square unit-cell by setting the circle area to match the area of the unit-cell. This approximation makes it possible to calculate the effective dielectric index analytically by the following formula

$$\varepsilon_{\text{eff}} = \frac{2\pi}{a^2} \frac{\int_0^r \varepsilon_{rod} d_0(k\rho) J_0(\sqrt{\varepsilon_{rod}} k\rho) d\rho + \int_r^R \left(J_0(k\rho) + b_0(k\rho) H_0^{(1)}(k\rho) \right) d\rho}{J_0(kR) + b_0(kR) H_0^{(1)}(kR)},$$

where k is the wave number, b_0 and d_0 are the cylindrical Lorenz–Mie coefficients of the electric field outside and inside the rod, respectively, for the TM polarization (e.g., see Bohren and Huffman, 1998), and $R = a/\sqrt{\pi}$ is the radius of the averaging circle. We plot the effective dielectric index as a function of frequency in Fig. 5C. The interval of the negative ε_{eff} ($0.289 < a/\lambda < 0.471$) is in an excellent agreement with the photonic band gap $0.282 < a/\lambda < 0.471$ confirming the metamaterial nature of this bandgap.

4. Experimental observation of the photonic phase transitions

The aim of the experimental investigations (Rybin et al., 2015) was to confirm the two scenarios of transitions in dielectric structure from photonic crystal to metamaterial shown schematically in Fig. 2. Experiments for the dielectric metamaterials with magnetic response were carried out with the sample designed for the microwave range. This range was chosen because the dielectric index of the rod should be at least 20 for metamaterials, and a sample with a variable lattice constant is easier to create on a centimeter scale. For experiments, it is also useful that distilled water, being an ideal dielectric, has a strong temperature dependence of the permittivity from $\varepsilon = 80$ at 20°C to $\varepsilon = 50$ at 90°C (e.g., see Andryieuski et al., 2015) in the microwave range (1–6 GHz).

The sample called a metacrystal is shown in Fig. 6 composed of 5 by 10 series-connected tubes with outer radius of 20 mm, internal radius of 17 mm, length of 1000 mm and a negligible dielectric index. The tubes were arranged in a square lattice by adjustable holders which make it possible to modify the lattice spacing in the interval $0.65 < r/a < 1.9$. All tubes were connected in a closed series circuit together with a thermostat with heaters and a water pump that support homogeneous temperature of the water inside all tubes of the metacrystal (the difference between the inlet and outlet temperature did not exceed 1°C).

The transmission spectra of the metacrystal in the interval 1–3 GHz were measured in an anechoic chamber. An approximation to the TE-polarized incident plane wave was generated by a rectangular horn antenna (TRIM 0.75–18 GHz; DR) connected to an output port of the vector network analyzer Agilent E8362C. Another similar antenna was connected to an

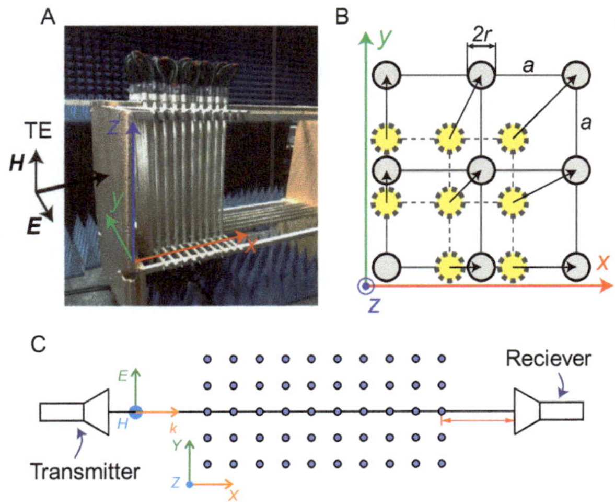

Fig. 6 Experimental setup. (A) An anechoic chamber with a metacrystal composed of 10 tubes in length (*x*-axis) and 5 tubes in width (*y*-axis), 50 tubes in total. Each tube has a length of 1 m (along *z*-axis), an outer radius of 2 cm and an inner radius of 1.7 cm. All tubes are connected in a closed series circuit together with a water heater and a temperature stabilizer. (B) Schematic illustration of the transformation of the 2D square lattice composed of dielectric rods as the lattice constant *a* increase. (C) The scheme shows the square lattice from above. The rods with dielectric permittivity ε and fixed radius *r* are embedded in air. The material is homogeneous along the *z* direction and periodic along *x* and *y*. *Panels (A) and (B) reproduced from Rybin, M.V., Filonov, D.S., Samusev, K.B., Belov, P.A., Kivshar, Y.S., Limonov, M.F., 2015. Phase diagram for the transition from photonic crystals to dielectric metamaterials. Nat. Commun. 6, 10102 under a Creative Commons License (http://creativecommons.org/licenses/by/4.0/).*

input port of the analyzer. The spectra were normalized to the transmission of the free space.

Fig. 7 demonstrates the experimental and theoretical spectra of the metacrystal as a function of the sample volume related to the r/a parameter as well as its temperature, which determines the rod permittivity. During the measurements for the r/a-scenario the temperature of water was stabilized at 90°C while the lattice spacing of the square lattice was reduced from $a = 212$ mm ($r/a = 0.08$) to 89 mm ($r/a = 0.19$). The transmission dip corresponding to the lowest Bragg gap shifts to the higher frequencies and almost disappears at $r/a \approx 0.09$ when it reaches to the frequency of the lowest Mie dipole resonance TE_{01}. For the shorter lattice spacing the Bragg dip appears again at $r/a \approx 0.095$, however at the higher frequency side of the Mie resonance. As a result, the lowest Bragg band and the TE_{01} Mie

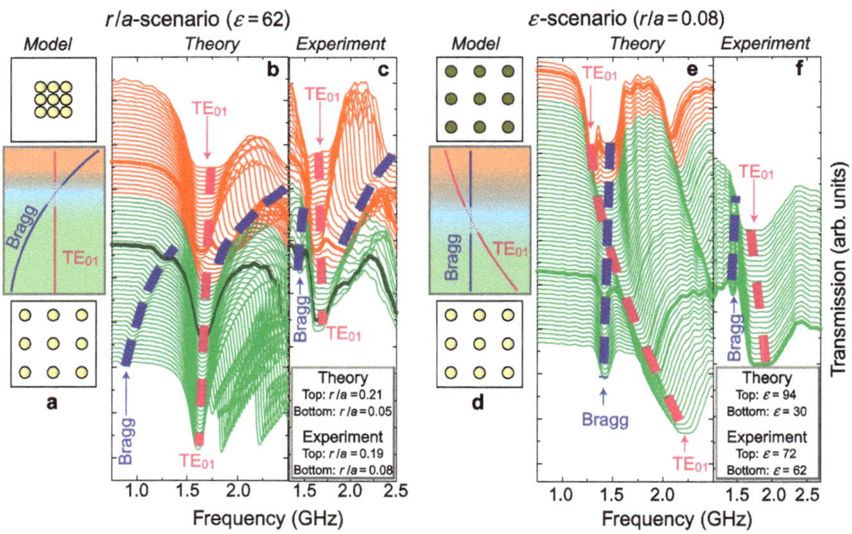

Fig. 7 Transmission spectra for a square lattice of circular rods. TE polarization. (A–C) r/a-scenario, $\varepsilon = 62$. The calculated (B) and experimentally measured (C) transmission spectra of the metacrystal as a function of the filling ratio r/a. The calculated spectra are shown in the range $0.05 \leq r/a \leq 0.21$, the experimental spectra are shown in the range $0.08 \leq r/a \leq 0.19$. (D–F) ε-scenario, $r/a = 0.08$. The calculated (E) and experimentally measured (F) transmission spectra of the metacrystal as a function of the dielectric permittivity. The calculated spectra are shown in the range $30 \leq \varepsilon \leq 94$, the experimental spectra are shown in the range $62 \leq \varepsilon \leq 94$. The spectra are shifted vertically by the constant value. *Reproduced from Rybin, M.V., Filonov, D.S., Samusev, K.B., Belov, P.A., Kivshar, Y.S., Limonov, M.F., 2015. Phase diagram for the transition from photonic crystals to dielectric metamaterials. Nat. Commun. 6, 10102 under a Creative Commons License (http://creativecommons.org/licenses/by/4.0/).*

band change their positions in the frequency scale indicating the transition into the metamaterial phase.

The ε-scenario for the metacrystal with $r/a=0.08$ was realized by varying of the temperature of water leading to modifying the rod permittiv-ity. Increasing of the dielectric permittivity results in the spectral shift of the Mie resonances, and the Bragg dip almost keeps its position. Fig. 7F shows a significant proximity of the Mie and Bragg resonances, although their inter-section was not achieved under the experimental conditions. In theory the Mie resonance becomes the fundamental gap and the metacrystal turns into the metamaterial regime at $r/a=0.08$ and the rod permittivity about 90 (Fig. 7E).

Let us discuss this transition of the metacrystal sample. We remind that the fundamental Mie gap is associated with the negative effective magnetic permeability $\mu_{\text{eff}}<0$ (e.g., see O'Brien and Pendry, 2002) and strong special dispersion above the lowest Bragg gap disables μ_{eff} for the photonic crystal regime. Thus, the continuous change in temperature or sample volume leads to the appearance of the effective magnetic permeability deviating from unity at a certain frequency interval. At the same time thermodynamic phase transitions are known to be associated with dramatic changes in a generalized susceptibility (e.g., Landau et al., 1980). Hence, the observed transition in metacrystal (appearance of the nonzero magnetic susceptibility) owning to the change in temperature or volume can be treated as a phase transition. However, the changes of the thermodynamic quantities (temperature and volume) in such phase transitions manifesting in the photonic response are just means rather than physical causes. The physical reason for the pho-tonic transition is the change of the water permittivity or the lattice spacing. Therefore, the observed transitions are not thermodynamic phase transitions but they are transitions of a new kind that should be referred to as photonic phase transitions.

5. Phase diagram of periodic dielectric photonic structures

Since the transformation between the regimes of photonic crystals and metamaterial is considered as the photonic phase transition, we have to develop corresponding phase diagrams. From Fig. 4 is clearly seen that the TE_{01} Mie gap splits from the Bragg stop-band at $\varepsilon=19.5$ for $r/a=0.25$. For $\varepsilon>19.5$ the band structure has the typical polariton-like dispersion at the low frequencies. Similar maps of the bandgaps vs rod permittivity plotted for a variety of values of r/a reveal the boundary of the metamaterial phase that is shown in Fig. 8 presenting the phase diagram as a function of rod permittivity and the radius normalized by the square lattice constant for the case of TE polarization. The phase diagram unveils the structure parameters required for the metamaterial regime. The left side of the phase boundary corresponds to the r/a- and ε-scenarios described earlier in Fig. 2. However, there exists a competing process resulting in the right side of the boundary. This process is related to the interaction between the neighboring rods and collective modes driving the strong spatial dispersion typical to the photonic crystals.

We also construct the phase diagram for the metamaterials with electric response in TM polarization. The metamaterial phase appears as a narrow area at the left of Fig. 8 corresponding to sparse structures. The reason is in the stronger electric multipole–multipole interaction relative to their magnetic counterparts. Also the minimum rod permittivity for the electric metamaterials with a square lattice is $\varepsilon=27$. We notice that the metamaterials with magnetic response are of interest due to their effective magnetic permeability $\mu_{\text{eff}}\neq 1$, however a variety of applications demand control of the electric response as well as the low-loss dielectric metamaterials with engineered effective permittivity ε_{eff} can significantly improve the light-matter interaction.

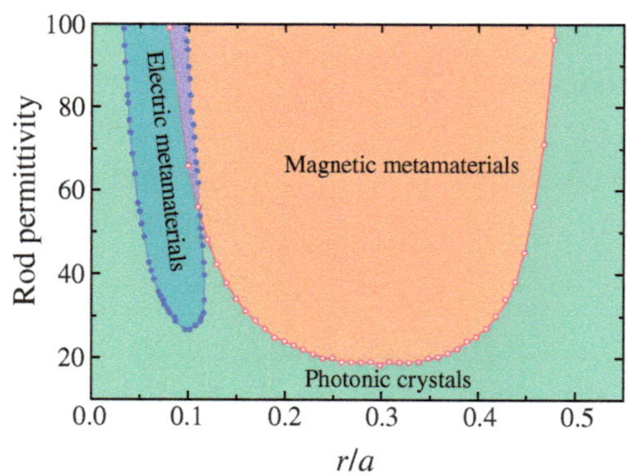

Fig. 8 TM + TE photonic phase diagrams of dielectric rods arranged in a square lattice. The region of the electric metamaterials in the TM polarization is marked by *blue*. The region of the magnetic metamaterials in the TE polarization is marked by *red*. *Circles* are phase boundary obtained from the band diagram analysis. *Solid curves* are guides for eyes only. *Reproduced with permission from Maslova, E.E., Limonov, M.F., Rybin, M.V., 2018. Dielectric metamaterials with electric response. Opt. Lett. 43, 5516–5519. Copyright (2018) OSA.*

Following the intuition one might think that transitions between the photonic crystal and metamaterial regimes are gradual. For example, let us consider the r/a parameter continuously decreasing from a value corresponding to the metamaterial phase (see Fig. 8) through the boundary to a photonic crystal value. The common believe is that the "quantity" of metamaterial is becoming weaker until it disappears completely. However it is not correct description of the transitions. In reality, the field distribution (photonic subsystem) undergoes a dramatic transformation when the structure parameters (ε and r/a) cross the phase boundary. Fig. 9 illustrates the field distribution under the photonic phase transition. We examine the filed profiles in two prism-shaped structures consisting of dielectric rods ($\varepsilon = 25$) arranged in a square lattice. The systems are illuminated by a Gaussian beam directed along the Γ-X (the first prism) and the Γ-M (the second prism) at the lowest frequency of the second dispersion band. We are decreasing the volume of samples by varying the lattice spacing from $r/a = 0.18$ (photonic crystal according to the phase diagram in Fig. 8) to $r/a = 0.20$ (metamaterial).

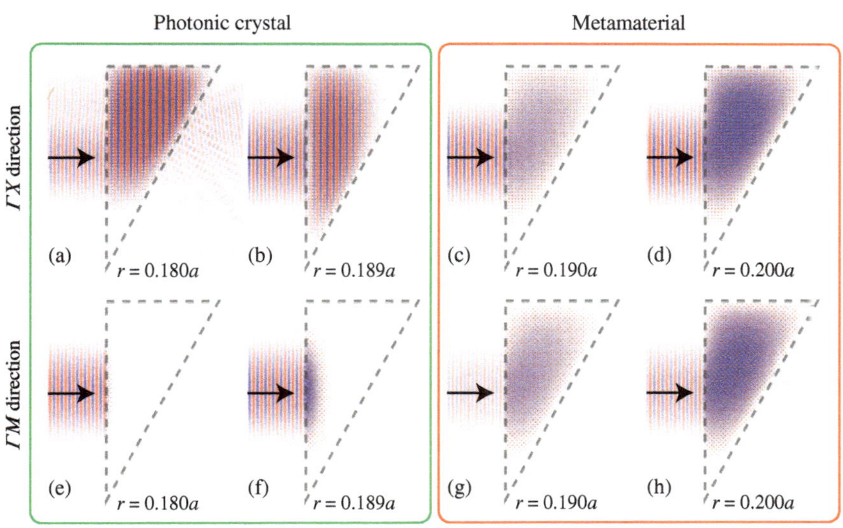

Fig. 9 A dramatic change in the magnetic field pattern depending on the operational photonic crystal or metamaterial regimes (or the corresponding phase). Simulated field patterns for the lowest frequency of the second band vs nanorod radius. An incident Gaussian beam propagates along the ΓX direction (A)–(D) or along the ΓM direction (E)–(H). The structure boundary is marked by a *gray dashed line*; *arrows* show the incident direction, nanorod permittivity $\varepsilon = 25$. Simulations are performed within the multiple scattering theory approach. *Reproduced from Li, S.V., Kivshar, Y.S., Rybin, M.V., 2018. Toward silicon-based metamaterials. ACS Photonics 5, 4751–4757 with permission, ACS.*

The prisms in the photonic crystal regime demonstrate a striped pattern being typical for the Bragg scattering (the neighbor crystalline planes are in anti-phase) for the Γ-X propagation and evanescent wave along the Γ-M (see Fig. 9A and E). In contrast, the metamaterial phase corresponds to almost homogeneous patterns without dependence on the lattice orientation because the structure can be described by a μ-near zero effective parameter (Fig. 9D and H). We notice that the stripped-type pattern is observed for any parameters corresponding to the photonic crystal regime in the phase diagram (Fig. 8), and the homogeneous pattern corresponds to any parameters of metamaterial phase. The pattern is changed right on the phase boundary revealing a sharp transition between the manifestations of these two regimes. The transition takes place at least when the r/a parameter variation is as small as 0.001 (Fig. 9B, C and F, G).

For a deeper insight into the physics of the photonic phase transition we also analyze isofrequency contours for the second dispersion band that plays a crucial role. Fig. 10A demonstrates that in the photonic crystal phase with parameters near the phase boundary ($r/a=0.189$, $\varepsilon=25$) there exist modes with almost a circular dispersion around the Γ point (e.g., at $a/\lambda=0.4242$) due to the Mie resonances of dielectric rods. For higher frequencies the dispersionless contours around the Γ point transforms into a four-pointed star ($a/\lambda=0.4243$) and the effects of spatial dispersion becomes strong.

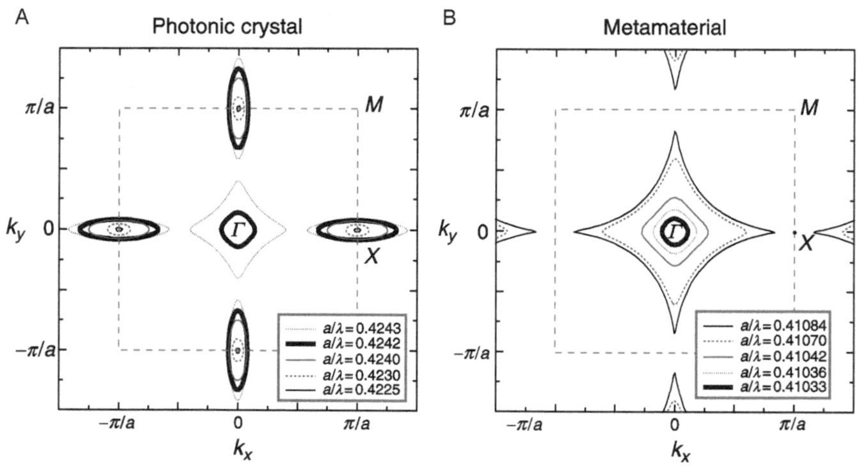

Fig. 10 A dramatic change in isofrequency contours depending on the operational photonic crystal (A) or metamaterial (B) regimes (or the corresponding phase). The isofrequency contours correspond to the second dispersion band. Structure parameters correspond (A) $r/a=0.189$, $\varepsilon=25$ and (B) $r/a=0.190$, $\varepsilon=25$ to structures in Fig. 9B, F, C, and G.

However, event at $a/\lambda=0.4242$ the modes of another type located around the X points coexist to the dimensionless modes around the Γ point that indicates the strong special dispersion regime as well. We notice that the modes around the X points on the Brillouin zone surface appear due to the Bragg scattering on the periodic structure.

Now we consider the metamaterial with ($r/a=0.190$, $\varepsilon=25$) which is described as only one isofrequency contour per frequency. Because the structure parameters are near the phase boundary most frequencies correspond to the four-pointed star revealing the strong spatial dispersion. However for the low-frequency interval below the $a/\lambda=0.41033$ the counters become circles and the structure has the frequency interval with the weak spatial dispersion indicating the metamaterial phase. Now we can describe the transition through the boundary. In the isofrequency representation the Bragg resonances manifest themselves as contours around points on the Brillouin zone surface, whereas the Mie resonances exhibit circles around the Γ point similar to the light cone (dispersion of free photons in homogeneous media). At some frequencies the contours of Bragg modes are superimposed with the dispersionless Mie contours resulting in more the one mode for certain directions of the wave vector. This regime is described by the strong spatial dispersion and we consider it as photonic crystal regime because of the existence of the Bragg scattering. When the frequency of Bragg resonance increases the contours around the Brillouin zone surface disappears and the Mie contours remain the exclusive resonant feature in the structure. Below some frequency the contour becomes circular exhibiting the metamaterial properties of the Mie resonant-driven dielectric structures.

Also we discuss an important consequence from the different nature of the resonances defining the fundamental gap in photonic crystals and metamaterials. In photonic crystals the Bragg scattering requires the periodicity of the structure and the dielectric contrast is not a crucial factor. In contrast, for the metamaterial phase the Mie resonant modes play the decisive role, while the periodicity is important only for the theoretical description of photonic properties by exploiting terms of the band theory. Therefore, a position disorder has to influence on the photonic crystal Bragg bandgaps, unlike the robust Mie bandgaps of the metamaterial regime. We analyze a transmission of the TE-polarized plane wave through a photonic structure in different phases. Fig. 11A shows the light incidence at a central frequency of the Bragg bandgap ($a/\lambda=0.3$, $\varepsilon=4$, $r/a=0.25$) with the striped pattern along the exponential decay that makes the sample opaque. The introduction of the position disorder breaks the periodicity, and the Bragg scattering disappears making the

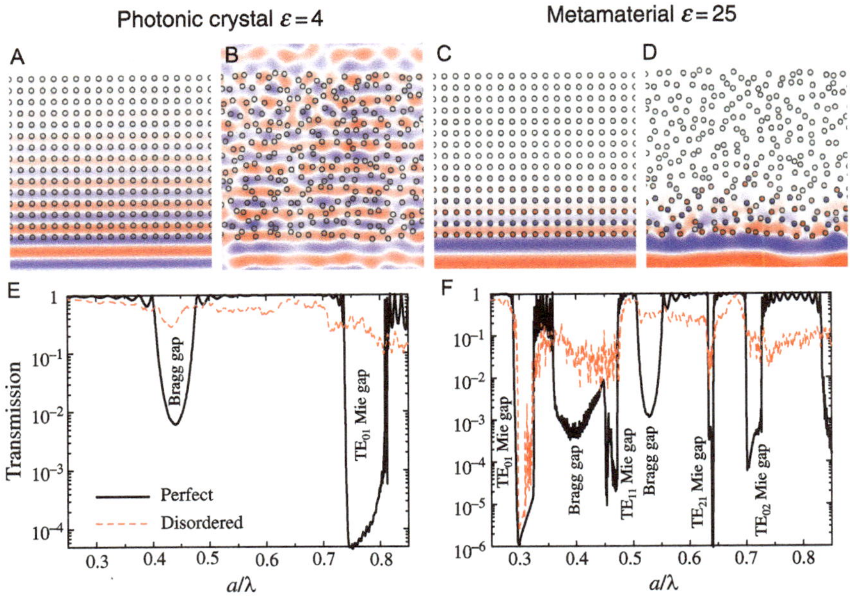

Fig. 11 Transmission of the plane wave through periodic photonic structure with and without position disorder. (A and B) Photonic crystal regime $\varepsilon=4$, $a/\lambda=0.3$. (C and D) Metamaterial regime $\varepsilon=25$, $a/\lambda=0.45$. All structures $r/a=0.25$, TE polarization. The plane wave is incident from the bottom side. Transmission spectra of photonic crystal (E) and metamaterial (F). Spectra of prefect structures are shown by *black solid curve*. Spectra of structures with position disorder are shown by *red dashed curve*.

structure to be transparent (Fig. 11B). The dielectric metamaterial ($a/\lambda=0.45$, $\varepsilon=25$, $r/a=0.25$) operates due to the Mie resonances exhibited as intensive spots inside the rods (Fig. 11C). Similar to the photonic crystal there is an evanescent wave, which dissipates during propagation inside the structure. However, the position disorder of rods has no effect on the propagation length. Fig. 11D demonstrates the robustness of the Mie bandgap with the introduction position disorder that does not make the metamaterial to be transparent, unlike the photonic crystal.

To study this effect in details, we simulated transmission of the structure shown in Fig. 11A–D. The spectra are shown in Fig. 11E and F. The logarithmic scale allows for the curve shapes do not depend on the number of rods along the sample. For the case of photonic crystal without disorder the spectrum exhibit symmetric Bragg gap (at $a/\lambda\approx0.44$) and a knife-blade like dip corresponding to the TE_{01} Mie gap (at $a/\lambda\approx0.47$). The introduction of position disorder leads to the degradation of features corresponding to both gaps by several orders of magnitude. The spectra of metamaterials shown in Fig. 11F reveal a rich structure of dips, which can be identified either as

Bragg gap or as a specific TE_{nk} Mie gaps. Fig. 11F uncovers that after the photonic phase transition to the metamaterial regime the fundamental Mie gap (below all Bragg gaps) is not affected by position disorder in contrast to other gaps Bragg and Mie both. This difference becomes another important feature of photonic structure appearing with the phase transition.

6. Dielectric metamaterials with near-zero effective parameters

First metamaterials were considered as artificial structures with the left-hand behavior owning to the double negative effective ε and μ (Shelby et al., 2001). Another application of metamaterials is in a realization of media with engineered spatial distribution of ε and μ for an invisibility cloaking designed by transformation optics (e.g., see Xu and Chen, 2015). Besides it, metamaterials with near-zero effective parameters ε and/or μ have also attracted a lot of attentions (Liberal and Engheta, 2017). In particular medial with ε-near zero improves the light-matter interaction and enhances nonlinear processes.

Now we discuss the possibility of dielectric ε-near zero metamaterials. The phase diagram for the TM polarization shows that metamaterial regime is allowed for the dielectric rods with $\varepsilon > 27$ (Fig. 8). Materials with such permittivity are not available for the visible range, however at infrared $\varepsilon = 30$ can be achieved by using Ge-Sb-Te (GST) alloy (Wuttig et al., 2017). The most studied compound $Ge_2Sb_2Te_5$ has a crystalline and metastable amorphous phases with $\varepsilon = 30$ and $\varepsilon = 15$, respectively.

Let us demonstrate the metamaterial behavior by existence of the near-zero index regime with a homogeneous field distribution. Fig. 9 reveals that the mode profile does not depend on the orientation of the crystalline lattice. Now we analyze ε-near zero by changing the shape of a metamaterial and by comparing its frequency with the frequency the second band at the Γ point for the infinite structure (Fig. 5B). The structures are composed of about 1050 rods (it depends on specific shape and orientation) with the radius $r/a = 0.1$ having shape of a triangle, a square, and a circle. Fig. 12 demonstrates homogeneous field distributions corresponding to the ε-near zero regime. The distributions do not depend on the lattice orientation. However, the intensities are different is likely because of different boundary conditions. The frequency of the ε-near zero mode $a/\lambda = 0.471$ exactly matches the lowest frequency of the second dispersion band in Fig. 5B.

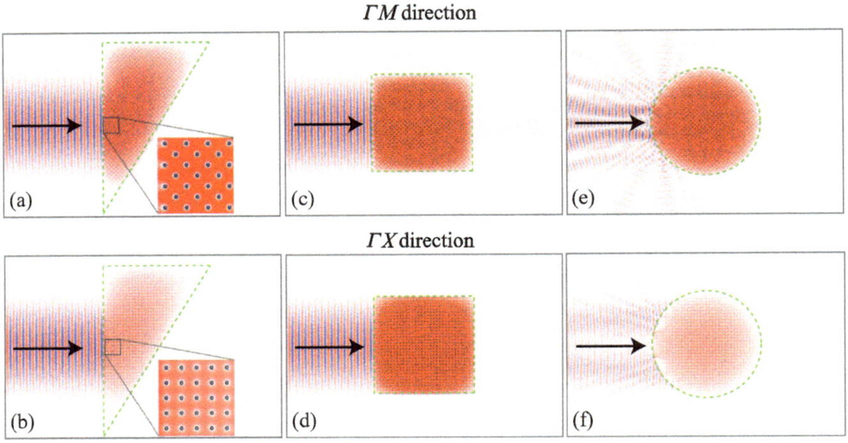

Fig. 12 Distributions of the electric field for the ε-near zero modes in electric metamaterials with different shapes [*triangle* (A and B), *square* (C and D), and *circle* (E and F)] and different lattice orientations [ΓM (A, C, E) and ΓX (B, D, F) directions]. *Green dashed lines* show the metamaterial structure boundaries. *Black arrows* indicate the Gaussian beam incident direction. *Insets* in panels (A) and (B) show the electric field distribution on a larger scale. $a/\lambda = 0.471$, $\varepsilon = 30$, $r/a = 0.1$, TM polarization. *Reproduced with permission from Maslova, E.E., Limonov, M.F., Rybin, M.V., 2018. Dielectric metamaterials with electric response. Opt. Lett. 43, 5516–5519. Copyright (2018) OSA.*

According to the phase diagram in Fig. 8, the dielectric metamaterials with electric response exist for the sparse systems $a > 8.3r$, that is the most of volume is occupied by air. Thus, the ε-near zero can be exploited to enhance light-mater interaction for detecting atoms dissolved in the air by means of an infrared absorbtion line analysis. The sample is illuminated by a Gaussian beam and the modes in the near-zero regime tend to fill all available sample volume because of the infinite wavelength in the metamaterial. Here we use the following figure of merit I/I_0, where I is the electric field intensity integrated over the volume outside the rods, and I_0 is the electric field intensity of the free space integrated over the volume occupied by the sample.

As a reference we consider a photonic crystal having the same geometry as a metamaterial sample (250 by 5 rods, $r/a = 0.1$) but the rod permittivity is 12 (silicon in infrared). It is well-known effect of the electric field localization in the region with lower dielectric index at the low frequencies of the second dispersion band (Joannopoulos et al., 2008). The structure ehibits a well-recognized stripped pattern of the field distribution (Fig. 13A) and the figure of merit demonstrates growth by almost twice (Fig. 13B) with the maximum at $a/\lambda = 0.518$. Now we study the effect of the electric field enhancement by the metamaterial layer composed of the GST rods. In

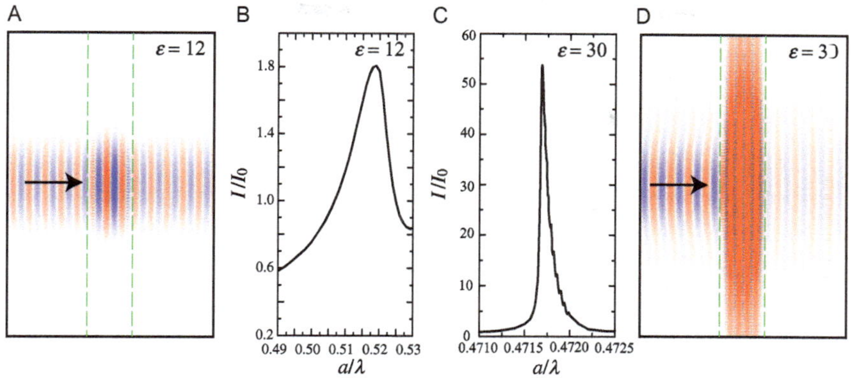

Fig. 13 Electric field localization in the plates being the photonic crystal with $\varepsilon = 12$ and $r/a = 0.10$ or the metamaterial with $\varepsilon = 30$ and $r/a = 0.10$. (A) Distribution of the electric field intensity in the photonic crystal plate. (B) The electric field intensity normalized to the intensity in the free space without account of the field inside the cylinders for the photonic crystal structure. (C) The electric field intensity normalized to the intensity in the free space without account of the field inside the cylinders for the metamaterial regime. (D) Distribution of the electric field intensity in the metamaterial plate. *Green dashed lines* show the plate boundaries. *Black arrows* show the Gaussian beam incident direction. *Reproduced with permission from Maslova, E.E., Limonov, M.F., Rybin, M.V., 2018. Dielectric metamaterials with electric response. Opt. Lett. 43, 5516–5519. Copyright (2018) OSA.*

the near-zero regime, the optical size of the sample is extremely small relative to the wavelength. As a result the Gaussian wave excites a mode extended over the entire structure and the I/I_0 ratio increases over 50 (Fig. 13C and D). The peak frequency $a/\lambda = 0.4717$ differs by a small value from the lowest frequency of the second dispersion band (in Fig. 5) due to the finite size of the sample (it contains only 5 rods along the beam incidence).

7. Toward practical realization of metamaterials in the visible frequency range

In the previous sections, we have demonstrated that the metamaterial phase in a photonic structure with a square lattice demands rod permittivity at least 20 (for magnetic metamaterials in the TE polarization). There are a lot of materials with required permittivity in infrared range, (Te, GeSbTe alloys, etc.) but in the visible range the real part of permittivity growth with its imaginary part. Also we did not optimize the crystalline lattice yet and considered materials with negligible losses only. The aim of this section is to discuss the approaches to yield the metamaterial phase in visible range. We consider here silicon-based structures because of the advanced technologies and the great experience of using silicon for photonics (Staude and Schilling, 2017). In crystalline silicon, electronic transitions between 1.3 and 3.4 eV ($365\,\text{nm} < \lambda < 1130\,\text{nm}$) are indirect, i.e., they take place between states of different wave vectors and only the simultaneous absorption or emission of a photon allows for a weak absorption in that region. For the wavelength shorter than 1000 nm the real part of silicon permittivity grows, while the losses remain relatively weak. In experiments, Mie resonances in silicon nanoparticles are observed almost in entire visible range $\lambda > 465\,\text{nm}$ (Kuznetsov et al., 2016). The complex frequency-dependent dielectric function of silicon does not allow plane wave expansion methods we have used in the previous sections, since the method solves an eigenproblem for square frequency ω^2 and the permittivity-dependent operator (Johnson and Joannopoulos, 2001). Instead, here we exploit the inverse dispersion method, which solves an eigenproblem for the wave numbers k acting as complex eigenvalues (Rybin and Limonov, 2016).

Fig. 14 compares band diagrams of TE modes calculated by the plane wave expansion method $\omega(k)$ and the inverse dispersion methods $k(\omega)$. We consider two structures with the same geometry (square lattice, $r/a = 0.3$) but with different permittivity $\varepsilon = 4$ (photonic crystal according to the phase diagram in Fig. 8) and 30 (metamaterial). Fig. 14A and B exhibits an excellent agreement between the data obtained by the different methods since they solve the same mathematical problem. However the inverse dispersion method finds additional modes with complex wave vectors, including the evanescent waves corresponding to the bandgap frequencies. It appears that the complex band diagrams uncover important information when weak losses are taken into account. Fig. 14C and D shows the diagrams for a dispersion of real part of wave vector as a function of frequency for the structures with permittivity $\varepsilon = 4(1 + i0.01)$ and $30(1 + i0.01)$, respectively. The losses lift the degeneracy of wavevectors (here and below, we imply the real part of wave vectors in the diagram) for frequencies within the bandgap and show that all bands are connected continuously and the second dispersion branch in not well-defined anymore. Overall, the phase determination procedure based on the analysis of the second branch is not applicable to the case of constituents with losses. However, the complex band diagrams unveil another criterion for determination of the metamaterial phase: for the case of Bragg bandgap the modes

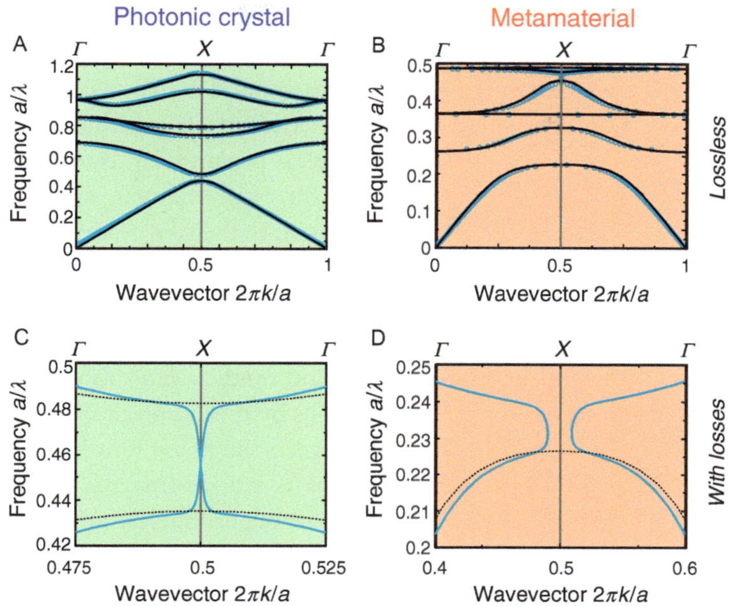

Fig. 14 Phase criteria for the lossless and lossy structures. (A and B) Band diagrams of a square lattice of nanorods with $r/a=0.3$ and lossless permittivity $\varepsilon=4$ (the photonic crystal phase) and 30 (the metamaterial phase). Data obtained by the direct method are shown by *black solid curves*. Results of the inverse dispersion method are shown by *cyan circles*. (C and D) Band diagrams calculated by the inverse dispersion method for the similar structure as in (A and B) but with the nanorod permittivity having an additional imaginary part $\varepsilon''=0.01\varepsilon'$ (*cyan solid curves*). *Black dashed curves* show the results for lossless structures. *Reproduced with permission from Li, S.V., Kivshar, Y.S., Rybin, M.V., 2018. Toward silicon-based metamaterials. ACS Photonics 5, 4751–4757. Copyright (2018) ACS.*

cross each other (Fig. 14C), while for the case of Mie bandgap the modes exhibit avoided crossing (Fig. 14D).

Besides, we optimize the crystalline lattice taking into account two competing conditions preventing the metamaterial phase discussed in Section 5: the lowest Bragg frequency has to be higher than the Mie frequency and the distance between the neighbor rods has to be large enough to preclude from the collective modes formation. Having the densest packing, a hexagonal lattice provides the smallest distance between the crystal planes among other lattices, thus here we consider this type of lattice.

Fig. 15 compares photonic crystal–metamaterial phase diagrams of structures with the hexagonal and square lattices obtained by both considered criteria. First we notice a significant decrease of the minimum permittivity required for the metamaterial phase: the value $\varepsilon = 14$ for the hexagonal lattice is more promising for a practical realization than $\varepsilon = 19.5$ for the square lattice. Besides, we determine another phase boundary (shown by circles in Fig. 15) by the crossing–anticrossing criterion for photonic sutures with losses. For this case we take complex permittivity by the formula $\varepsilon = \varepsilon'(1 + i0.01)$ and the axis of ordinates shows ε' (the real part of ε). The phase diagrams reveal the excellent agreement of the two criteria. The small difference for the denser structures ($r/a > 0.35$) can be explained as different numbers of plane waves involved in calculations. The Hermitian eigenvalue problem solved in the $\omega(k)$ method allows expansion over 128 × 128 plane waves, whereas the inverse dispersion method $k(\omega)$ solves the non-Hermitian eigenvalue problem that limits us by resolution of 25 × 25 plane waves. Another reason for deviation between the phase boundaries evaluated by two approaches is in the permittivity real-to-imaginary parts ratio $\varepsilon''/\varepsilon'$. We find that, when the imaginary part of permittivity grows, an error in the phase boundary is increasing. In particular, for the imaginary part of dielectric permittivity described by $\varepsilon''/\varepsilon' = 0.35$, this deviation is about 5%.

Now we have a complete tool for constructing the phase diagram of periodic photonic structures, which constituents have a complex frequency-dependent permittivity. The phase diagram in Fig. 15 shows that

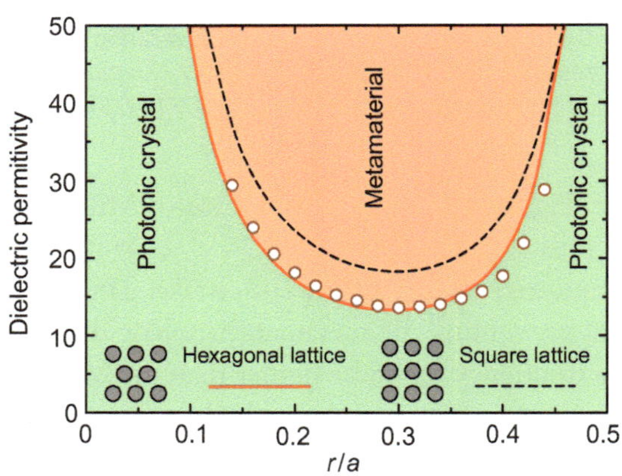

Fig. 15 Phase diagram that allows differentiating metamaterials and photonic crystals. The boundary is calculated for both *hexagonal* (*red line*) and *square* (*black dashed line*) lattices, the *circles* mark the data obtained by the direct and inverse dispersion methods, respectively. *Green and red shadings* correspond to the photonic crystal and metamaterial regimes of a hexagonal lattice, respectively. *Reproduced with permission from Li, S.V., Kivshar, Y.S., Rybin, M.V., 2018. Toward silicon-based metamaterials. ACS Photonics 5, 4751–4757. Copyright (2018) ACS.*

silicon-based metamaterials is possible for wavelength shorter than a certain wavelength that corresponds to the critical value of the frequency-dependent permittivity. By using the crossing/anticrossing criterion and the dispersion of silicon permittivity reported by Green and Keevers (1995), we determine the phase boundary for the hexagonal lattice of silicon nanorods (Fig. 16). We notice that Maxwell's equations are not scalable in this case because of the frequency dependence of permittivity. However it allows an unambiguous mapping of the permittivity to the wavelength, which is used as ordinate axes instead. Since the normalized radius *r*/*a* does not provide exhaustive information about the structure geometry, Fig. 16 shows additional diagrams for radius vs wavelength and lattice constant vs wavelength, as well.

The silicon permittivity grows with shortening of the wavelength that reverses the phase diagram relative to the horizontal (compare the diagrams in Figs. 15 and 16A). Since the real part of the silicon permittivity growths up to 42 the metamaterial phase is allowed even for the square lattice (dashed curves in Fig. 16), however the strong absorption makes such metamaterials impractical. In contrast, the design of silicon-bases structures with the hexagonal lattice allows to achieve the metamaterial phase for the wavelength below 850 nm, where material losses are relatively weak. In practical realizations, silicon metamaterials have to be an array of silicon nanopillars. Advanced nanoimprint lithography, combined with reactive ion etching, has been used to fabricate arrays of silicon nanopillars (see, e.g., a review by Zhang et al., 2017). This method allows the fabrication of well-defined

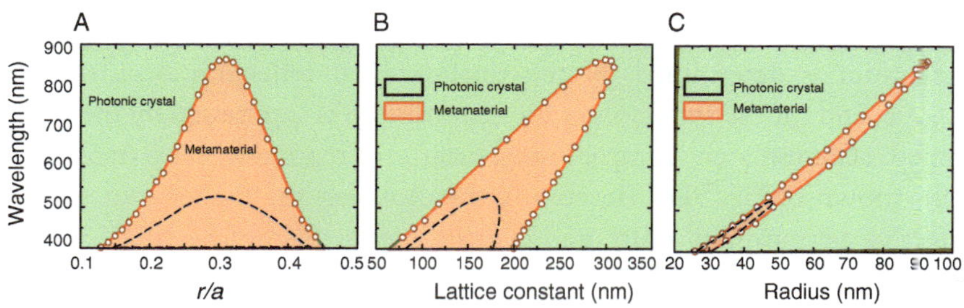

Fig. 16 Phase diagram of a periodic silicon nanorod structure with a hexagonal lattice. *Red circles* show the phase boundary, *solid line* is a guide for eyes only. *Green and red shadings* correspond to the photonic crystal and metamaterial regimes, respectively. The phase boundary for the square lattice is shown by a *black dashed line*. Phase diagram is plotted in the difference axes: (A) ratio *r*/*a* vs wavelength; (B) lattice spacing *a* vs wavelength; and (C) rod radius *r* vs wavelength. *Reproduced with permission from Li, S.V., Kivshar, Y.S., Rybin, M.V., 2018. Toward silicon-based metamaterials. ACS Photonics 5, 4751–4757. Copyright (2018) ACS.*

arrays with precise control in position, density. The reported diameter is as small as 40 nm with the lattice spacing of 200 nm. Fig. 16 shows that those sizes allow silicon-based metamaterials operating almost in entire visible range from 460 to 850 nm. Besides, optimal etching parameters can lead to uniform vertical nanowires with high aspect ratios 60:1, being sufficient for experimental measurements.

Finally, we illustrate metamaterial behavior by considering refraction of a TE-polarized Gaussian beam through a prism made of silicon-based metamaterial. Here we make a number of assumptions: the Snell's law is applicable for description of the beam refraction at the prism surface; the function of effective permeability μ_{eff} is unity at the low frequencies and has a resonant feature around the Mie bandgap TE_{01}, while the effective permittivity ε_{eff} does not depend on frequency; and the effective refractive index reads $n_{\text{eff}} = \sqrt{\varepsilon_{\text{eff}}}\sqrt{\mu_{\text{eff}}}$.

The prism has a shape of right triangle with an acute angle of 30 degree consisting of 1220 silicon nanorods of radius 50 nm arranged in a hexagonal lattice with period $a = 167$ nm. The wide beam (waist is $10a$) is incident on the long side of prism and the transmitted beam is deflected at a certain angle according to the Snell's law. First, we evaluate the effective permittivity of the metamaterial by simulating the beam with the wavelength of $\lambda = 1500$ nm that is far enough from the Mie bandgap TE_{01} (Fig. 17A and B). The transmitted beam propagates at the 44 degree to the normal, which makes it possible to evaluate the effective refractive index $n_{\text{eff}} = 1.39$ and the effective permittivity $\varepsilon_{\text{eff}} = 1.93$ (we assume $\mu_{\text{eff}} = 1$).

The metamaterial prism limits the minimum and maximum values of the permeability can be measured by the simulation of beam refraction. The maximum is determined by the total internal reflection occurring for $n_{\text{eff}} > 2$, and for the small values the transmission is suppressed by the strong impedance mismatch (compare the μ-near zero regime with no transmitted light shown in Fig. 9D). Because of the causality, the higher values of the effective permeability correspond to the modes below the Mie bandgap. In Fig. 17C and D, we show the beam refraction at the wavelength of 700 nm and evaluate the effective refractive index as $n_{\text{eff}} = 1.78$, which yields $\mu_{\text{eff}} = 1.64$ (we assume $\varepsilon_{\text{eff}} = 1.93$). At the wavelength of 495 nm, the prism deviates the Gaussian beam path 3 degree off the incident direction that gives the effective refractive index $n_{\text{eff}} = 0.92$, which yields $\mu_{\text{eff}} = 0.44$ both less than one. Thus, the silicon-based photonic structure exhibits the typical metamaterial behavior.

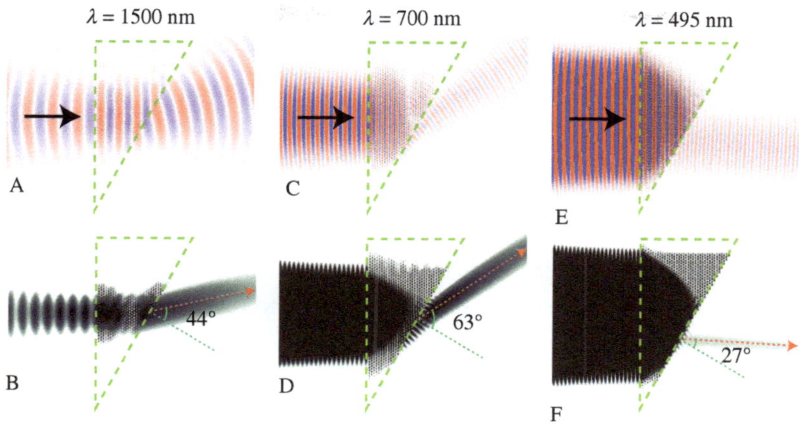

Fig. 17 Refraction of a Gaussian beam on a silicon-based metamaterial prism, for $a=167$ nm and $r=50$ nm. The top panels are the magnetic field profiles (*red* are positive values and *blue* are negative values of H_z); the bottom panels are the electromagnetic field intensities. (A and B) Off-resonance case; $\lambda=1500$ nm. (C and D) the low-frequency edge $\lambda=700$ nm; (E and F) the high-frequency edge $\lambda=495$ nm. *Green dashed triangles* mark the prism. *Black arrows* in the top panels show the propagation direction of the incident beam. The results are for the TE polarization. *Reproduced with permission from Li, S.V., Kivshar, Y.S., Rybin, M.V., 2018. Toward silicon-based metamaterials. ACS Photonics 5, 4751–4757. Copyright (2018) ACS.*

8. Conclusion and outlook

Owing to their unique optical properties, metamaterials have been intensively studied as promising novel optical materials enabling a control over the electromagnetic waves. Although the dielectric metamaterials benefit from their low-losses constituents, the conditions required for the structure to possess the effective parameters characterizing the metamaterial properties have been understood recently. The approach based on the concept of the photonic phase transitions shed light on the underlying physics explaining how the metamaterial properties appear in periodic structures. In this chapter, we have provided a comprehensive overview of the photonic phase transitions in photonic structures between photonic crystal and metamaterial regimes (or "phases"). We have discussed how the corresponding phase diagrams can be constructed. For the lossless case, the appearance of the metamaterial phase is identified by the polariton-like feature formed around the bandgap between the first and second dispersion branches in bandgap diagrams. However, for the realistic case of materials with absorption, the first and second branches are merged into continuous branches and the metamaterial phase can be distinguished by the criterion based on crossing–anticrossing of the branches in complex bandgap diagrams. Although the considered photonic phase transitions are quite dissimilar to conventional thermodynamic phase transitions, the phase transition terminology is feasible because it describes abrupt changes in the photonic patterns observed at the boundary between photonic crystal and metamaterial regimes. Besides, more important is that the developed phase diagrams as a function of structure parameters (permittivity and sizes) become very useful in a design of metamaterials with desired properties. Here we have reviewed several aspects of the metamaterial regimes including near-zero parameters, stability against position disorder, and practical realization of silicon-based metamaterials in the visible frequency range. The study of photonic phase transitions has been started in the recent years, so we anticipate the further development of this concept.

Acknowledgments

This work was supported by the Ministry of Education and Science of the Russian Federation (Grant 3.1500.2017/4.6) and the Strategic Fund of the Australian National University.

References

Andryieuski, A., Kuznetsova, S.M., Zhukovsky, S.V., Kivshar, Y.S., Lavrinenko, A.V., 2015. Water: promising opportunities for tunable all-dielectric electromagnetic metamaterials. Sci. Rep. 5, 13535.

Bohren, C.F., Huffman, D.R., 1998. Absorption and Scattering of Light by Small Particles. Wiley-VCH.

Decoopman, T., Tayeb, G., Enoch, S., Maystre, D., Gralak, B., 2006. Photonic crystal lens: from negative refraction and negative index to negative permittivity and permeability. Phys. Rev. Lett. 97, 073905.

Gorlach, M.A., Belov, P.A., 2014. Effect of spatial dispersion on the topological transition in metamaterials. Phys. Rev. B 90 115136.

Green, M., Keevers, M., 1995. Optical properties of intrinsic silicon at 300 K. Prog. Photovolt. 3, 189–192.

Huang, X., Lai, Y., Hang, Z.H., Zheng, H., Chan, C.T., 2011. Dirac cones induced by accidental degeneracy in photonic crystals and zero-refractive-index materials. Nat. Mater. 10, 582.

Ivchenko, E.L., 2005. Optical Spectroscopy of Semiconductor Nanostructures. Alpha Science, Harrow, UK.

Joannopoulos, J.D., Johnson, S.G., Winn, J.N., Meade, R.D., 2008. Photonic Crystals: Molding the Flow of Light, second ed. Princeton University Press.

John, S., 1987. Strong localization of photons in certain disordered dielectric superlattices. Phys. Rev. Lett. 58, 2486–2489.

Johnson, S.G., Joannopoulos, J.D., 2001. Block-iterative frequency domain methods for Maxwell's equations in a plane-wave basis. Opt. Express 8, 173.

Kaina, N., Lemoult, F., Fink, M., Lerosey, G., 2015. Negative refractive index and acoustic superlens from multiple scattering in single negative metamaterials. Nature 525, 77.
Kuznetsov, A.I., Miroshnichenko, A.E., Brongersma, M.L., Kivshar, Y.S., Luk'yanchuk, B., 2016. Optically resonant dielectric nanostructures. Science 354, aag2472.
Landau, L.D., Lifshitz, E.M., Pitaevskii, L.P., 1980. Statistical Physics, Part I. Pergamon Press, Oxford.
Liberal, I., Engheta, N., 2017. Near-zero refractive index photonics. Nat. Photonics 11, 149.
Liu, Z., Lee, H., Xiong, Y., Sun, C., Zhang, X., 2007. Far-field optical hyperlens magnifying sub-diffraction limited objects. Science 315, 1686.
Maslova, E.E., Limonov, M.F., Rybin, M.V., 2018. Dielectric metamaterials with electric response. Opt. Lett. 43, 5516–5519.
Mirmoosa, M.S., Kosulnikov, S.Y., Simovski, C.R., 2015. Unbounded spatial spectrum of propagating waves in a polaritonic wire medium. Phys. Rev. B 92, 075139.
O'Brien, S., Pendry, J.B., 2002. Photonic bandgap effects and magnetic activity in dielectric composites. J. Phys. Condens. Matter 14, 4035.
Pendry, J.B., Holden, A.J., Robbins, D.J., Stewart, W.J., 1999. Magnetism from conductors and enhanced nonlinear phenomena. IEEE Trans. Microwave Theory Technol. 47, 2075.
Pendry, J.B., Aubry, A., Smith, D.R., Maier, S.A., 2012. Transformation optics and subwavelength control of light. Science 337, 549.
Reyes-Gómez, E., Cavalcanti, S.B., Oliveira, L.E., De Carvalho, C.A.A., 2014. Metric-signature topological transitions in dispersive metamaterials. Phys. Rev. E 89, 033202.
Rybin, M.V., Limonov, M.F., 2016. Inverse dispersion method for calculation of complex photonic band diagram and PT symmetry. Phys. Rev. B 93, 165132.
Rybin, M.V., Filonov, D.S., Samusev, K.B., Belov, P.A., Kivshar, Y.S., Limonov, M.F., 2015. Phase diagram for the transition from photonic crystals to dielectric metamaterials. Nat. Commun. 6, 10102.
Shelby, R.A., Smith, D.R., Schultz, S., 2001. Experimental verification of a negative index of refraction. Science 292, 77–79.
Simovski, C.R., 2009. Material parameters of metamaterials. Opt. Spectrosc. 107, 726.
Simovski, C.R., 2018. Composite Media with Weak Spatial Dispersion. Pan Stanford.
Smith, D.R., Padilla, W.J., Vier, D.C., Nemat-Nasser, S.C., Schultz, S., 2000. Composite medium with simultaneously negative permeability and permittivity. Phys. Rev. Lett. 84, 4184.
Staude, I., Schilling, J., 2017. Metamaterial-inspired silicon nanophotonics. Nat. Photonics 11, 274–284.
Veselago, V.G., 1968. The electrodynamics of substances with simultaneously negative values of ε and μ. Sov. Phys. Usp. 10, 509.
Wuttig, M., Bhaskaran, H., Taubner, T., 2017. Phase-change materials for non-volatile photonic applications. Nat. Photonics 11, 465.
Xu, L., Chen, H., 2015. Conformal transformation optics. Nat. Photonics 9, 15.
Yablonovich, E., 1987. Inhibited spontaneous emission in solid-state physics and electronics. Phys. Rev. Lett. 58, 2059–2062.
Yves, S., Fleury, R., Berthelot, T., Fink, M., Lemoult, F., Lerosey, G., 2017. Crystalline metamaterials for topological properties at subwavelength scales. Nat. Commun. 8, 16023.
Zhang, B., Jie, J., Zhang, X., Ou, X., Zhang, X., 2017. Large-scale fabrication of silicon nanowires for solar energy applications. ACS Appl. Mater. Interfaces 9, 34527–34543.

CHAPTER THREE

Light trapping in photonic structures

Ken Xingze Wang[a,*], Yu Guo[b], Zongfu Yu[c,*]
[a]School of Physics, Huazhong University of Science and Technology, Wuhan, Hubei, China
[b]Electrical Engineering and Ginzton Laboratory, Stanford University, Stanford, CA, United States
[c]University of Wisconsin, Department of Electrical and Computer Engineering, Madison, WI, United States
*Corresponding authors: e-mail address: wxz@hust.edu.cn; zyu54@wisc.edu

Contents

1. Introduction

Photon management strategies, including light trapping, have been tremendously successful in efficiency improvement and cost reduction in practical solar cells (Mokkapati and Catchpole, 2012). Light trapping allows cells to absorb sunlight using an active material layer that is much thinner than the material's intrinsic absorption length. This effect then reduces the amount of materials used in photovoltaic cells, which cuts cell cost in general, and moreover facilitates mass production of photovoltaic cells that are based on less abundant materials. In addition, light trapping can improve

Semiconductors and Semimetals, Volume 100
ISSN 0080-8784
https://doi.org/10.1016/bs.semsem.2019.01.001

cell efficiency, because thinner cells provide better collection of photogenerated charge carriers, and potentially a higher open circuit voltage (Atwater and Polman, 2010; Battaglia et al., 2012a; Bozzola et al., 2012; Callahan et al., 2012; Munday, 2012; Mutitu et al., 2008; Narasimhan and Cui, 2013; Polman and Atwater, 2012; Sandhu et al., 2013; Sheng et al., 2011; Wang et al., 2012, 2013a,b; Wiesendanger et al., 2013; Wong and Morales, 2009; Yu and Fan, 2011; Yu et al., 2010a,b, 2014).

It is known that the light trapping enhancement limit is related to the photonic density of states (DOS) (Sheng et al., 1983). For a bulk silicon cell, this leads to the well known $4n^2$ limit (Deckman et al., 1983; Yablonovitch, 1982; Yablonovitch and Cody, 1982), also referred to as the Yablonovitch limit, the Lambertian limit, the ray-optic limit, or the conventional limit. For thin films, the $4n^2$ limit can be modified depending on the details of guided modes in the film (Agrawal and Peumans, 2008; Haug et al., 2011; Munday et al., 2012; Stuart and Hall, 1997; Yu et al., 2010a). In addition, it has been shown that an elevated local DOS is an important element for improved light trapping (Callahan et al., 2012; Tsakalakos et al., 2007; Wang and Menon, 2014). That description is consistent with our wave optics light trapping theory (Brongersma et al., 2014). The local DOS describes the enhancement of absorption at a given spatial location of the structure. To account for the performance of a solar cell, one needs to integrate over all spatial points in the solar cell where absorption occurs. One thus needs to perform spatial integration of the local DOS, which leads to the DOS picture that we use in this chapter.

Given the importance of DOS in defining the light trapping limit, one naturally should look beyond the relatively simple uniform absorbing layer and consider more complex medium. In particular, the photonic crystal structure is known to be able to drastically influence the DOS (McPhedran et al., 2004), and its ability of dispersion and DOS engineering has been extensively explored and applied to a wide range of applications (Joannopoulos et al., 2008). However, while light trapping in an absorber consisting of photonic crystals has been considered before (Chutinan and John, 2008; Chutinan et al., 2009; Du et al., 2011; Duché et al., 2008, 2011; Eyderman et al., 2013; Gomard et al., 2012; Guldin et al., 2010; John, 2012; Ko et al., 2009; Kuang et al., 2013; Le and John, 2014; Mallick et al., 2010, 2012; Mihi and Míguez, 2005; Oskooi et al., 2014; Park et al., 2009; Tumbleston et al., 2009a,b), the connection between the DOS and the light trapping limit in a photonic crystal has not been explicitly established.

In this chapter, we describe a statistical coupled-mode theory that describes light trapping in general from a rigorous electromagnetic perspective (Yu et al., 2010a). We quantitatively establish the connection between the absorption enhancement and the DOS (Wang et al., 2014a). Applying this relationship, we show that the conventional limit of $4n^2$ is only correct in bulk structures. In the nanophotonic regime, the absorption enhancement factor can go far beyond this limit with proper design.

We further use the connection between the absorption enhancement and the DOS to guide photonic crystal absorber designs. To maximize the absorption by light trapping, one should aim at maximizing the number of accessible modes (used interchangeably with states and resonances) supported by an absorbing structure. Moreover, we show that to take full advantage of the benefit of the photonic DOS, one need to consider additional issues including accessibility to these photonic states, as well as the overlap of the photonic states with the absorber.

As an application of the light trapping theory, we show that the Yablonovitch limit can be reached in ultrathin crystalline silicon (c-Si) solar cells with nanocone gratings. The key to achieve the Yablonovitch limit is to realize that antireflection and light trapping have drastically different structural requirements, and therefore should be optimized differently.

This chapter is organized as follows: In Sections 2 and 3, we briefly review the ray-optic and the wave-optic theory of light trapping, respectively, and derive a direct correspondence between the angle-integrated absorption enhancement and the photonic DOS. In Section 4, we derive the conventional $4n^2$ limit in bulk structures and demonstrate that in the nanophotonic regime the absorption enhancement factor can go far beyond this limit with proper design. In Section 5, we numerically study light trapping in several two-dimensional (2D) photonic crystal absorbers, and account for the numerical results using the theoretical understanding developed in Section 3, highlighting in particular the requirements to observe significant DOS effects for light trapping in photonic crystals. We also briefly discuss three-dimensional (3D) photonic crystals in Section 5. We apply the light trapping theory to enhance absorption in ultrathin crystalline silicon solar cells in Section 6 and conclude in Section 7.

2. Ray optics light trapping theory

The ray optics theory describes conventional light trapping structures in the absence of wave effects. When the front surface is a Lambertian, isotropic surface, the 3D absorption enhancement factor for light incident from any angle is the same and is equal to $4n^2$ (Goetzberger, 1981; Yablonovitch, 1982). This result is commonly referred to as the Yablonovitch limit or the $4n^2$ limit. We refer all such ray optics limits as conventional limits.

The uniform system in Fig. 1 consists of a uniform slab with a refractive index n and an absorption coefficient α. The thickness of the uniform slab, d_{eff}, is chosen such that the amount of dielectric materials per unit area (in 3D) or unit length (in 2D) horizontally is the same as that of the photonic crystal systems considered in Section 5. d_{eff} is therefore typically smaller than the physical thickness, d, of the photonic crystal. They are related by an area (in 2D) or volume (in 3D) fraction η

$$d_{\text{eff}} = \eta d. \tag{1}$$

Except stated explicitly, the thickness of the photonic crystal is chosen or assumed such that d_{eff} is at least equal to a few wavelengths.

We consider two different uniform systems that contain the same amount of absorbing material. Fig. 1 is a slab where both the front and back surfaces are flat, with no mirror at the back surface. We assume perfect antireflection for all frequencies and all angles on both surfaces. The absorption of light incident normally to the top surface of such a slab is defined by the single-pass absorption, αd_{eff}, where $\alpha = 2k_0 \operatorname{Im}\{n\}$ is the absorption coefficient of the material, and k_0 is the wavevector of light in vaccum and $\operatorname{Im}\{n\}$ is the imaginary part of the refractive index.

The goal of any light trapping structure is to achieve absorption greater than the single-pass absorption, while keeping the same amount of the absorbing material. For such light trapping structure, in 3D, we define the absorption enhancement factor as

$$f(\omega, \theta, \varphi) = \frac{A(\omega, \theta, \varphi)}{\alpha d_{\text{eff}}}. \tag{2}$$

Here, ω, θ, and φ are the frequency, the angle of incident light, and the azimuthal angle of light, respectively. A is the absorption coefficient of the

Fig. 1 A dielectric slab with flat surfaces.

Fig. 2 A dielectric slab structure with a randomly textured front surface and a back mirror. The equivalent thickness is the same as in Fig. 1.

structure. There will not be a φ dependency in 2D. The second uniform system represents the conventional light-trapping structure (Fig. 2), where one roughens the front surface of the uniform slab and places a perfect mirror at the back.

Both the isotropic and the anisotropic cases above can be described in a unified fashion by defining an angle-integrated enhancement factor

$$F_{3D}(\omega) = \int_0^{\pi/2} d\theta \int_0^{2\pi} d\varphi f(\omega, \theta, \varphi) \cos\theta \sin\theta \quad (3)$$

where $f(\omega, \theta, \varphi)$ is defined in Eq. (3). For both the isotropic and anisotropic cases, that is, $f(\omega, \theta, \varphi) = 4n^2$ and $f(\omega, \theta, \varphi) = 4n^2 / \sin^2\theta_0 \mathbb{I}_{\theta \le \theta_0}$, respectively, we have $F_{3D} = 4\pi n^2$. Moreover, it has been shown that as an upper limit $F_{3D} = 4\pi n^2$ in fact applies to any light trapping structure of the form of Fig. 2 with an arbitrary angular response of $f(\theta, \varphi)$ (Yu and Fan, 2011).

Since no actual physical light trapping structure has an ideal isotropic angular response, to compare the performance of a physical light trapping structure to the conventional limit, it is imperative that one performs angle integration (Naqavi et al., 2014; Yu and Fan, 2011). Observing an enhancement factor greater than $4n^2$ for a single angle of incidence, for example, should not be taken as the evidence that one has overcome the conventional limit (Ganapati et al., 2014). In this context, the $4\pi n^2$ limit, which is really the conventional limit expressed in an angle-integrated fashion, is very useful when one needs to compare the performance of a physical structure to the ray optics limit.

The ray optics theory also describes the conventional light trapping structure of Fig. 2 in 2D. For the isotropic case the enhancement factor is

$$f(\omega, \theta) = \pi n. \quad (4)$$

The upper limit for the angle-integrated absorption enhancement factor is

$$F_{2D}(\omega) = \int_{-\pi/2}^{\pi/2} d\theta f(\omega, \theta) \cos\theta = 2\pi n \quad (5)$$

which applies to isotropic as well as anisotropic cells in 2D. These 2D results are not nearly as well known as the original $4n^2$ limit in 3D. Moreover, to the best of our knowledge were first derived using the wave optics light trapping theory formalism (Yu et al., 2010b). Therefore below we provide a bit more details on the derivation of the πn limit for the 2D isotropic case using ray optics.

For a uniform 2D structure, it is straightforward to obtain a factor of n from the enhancement of light intensity and another factor of 2 from a back mirror, following the same approach as in the derivation of the $4n^2$ limit (Yablonovitch, 1982). For an oblique ray at an angle θ, the path length is enhanced by $1/\cos\theta$. The angle-integrated enhancement of the path length is obtained by an angular integration over the individual path length enhancement, multiplied by a weighting factor that depends on the surface property and the dimension of the problem (Battaglia et al., 2012b). For a Lambertian surface, the weighting factor is $\cos\theta$ normalized by an angular integration over $\cos\theta$. Therefore, the averaged path length enhancement in 2D is

$$\int_{-\pi/2}^{\pi/2} d\theta \frac{1}{\cos\theta} \frac{\cos\theta}{\int_{-\pi/2}^{\pi/2} d\theta \cos\theta} = \frac{\pi}{2} \tag{6}$$

instead of the 3D factor of 2. Multiplying the three enhancement factors, we obtain the πn enhancement for 2D light trapping.

The different versions of the conventional limits are summarized in Table 1.

3. Wave optics light trapping theory

The understanding of light trapping lies at the heart of optical design for solar cells (Battaglia et al., 2012a; Brongersma et al., 2014; Gomard et al., 2013; Green, 2011; Liang et al., 2012; Mann and Garnett, 2013; Naqavi

Table 1 Conventional light trapping limits in 2D and 3D structures for a single incidence angle and after angular integration. *n* is the refractive index of the weakly absorbing material

Dimension	Isotropic	Angle-integrated
2D	πn	$2\pi n$
3D	$4n^2$	$4\pi n^2$

et al., 2013a; Pala et al., 2013; Peretti et al., 2012; Sheng et al., 2011; Vynck et al., 2012; Wang et al., 2012, 2013a, 2014a,b; Yu et al., 2012; Zeman et al., 2013). Conventional light trapping theory, largely developed in the context of c-Si solar cells in the 1980s, was initially developed using the tools of ray optics. For solar cells with thicknesses or feature sizes that are comparable to the wavelengths of sunlight, ray tracing is no longer applicable. It is therefore essential to develop a light trapping theory entirely from a wave optics perspective.

The development of wave optics light trapping theory has a long history. Yablonovitch, in one of his initial papers that elucidate the $4n^2$ limit (Yablonovitch, 1982; Yablonovitch and Cody, 1982), used a blackbody radiation argument that is directly related to the modal structure of the cell. Sheng, Bloch, and Stepleman argued that the DOS plays an important role and hinted at the possibility of modifying the ray optics limit in thin film structures (Sheng et al., 1983). Stuart and Hall examined light trapping in wavelength-scale thin films and concluded that such a wavelength-scale thin film in fact has an absorption enhancement limit comparable to and somewhat below that of the bulk cell (Stuart and Hall, 1997). All these studies have pointed out the importance of modal structures of the cells in the understanding of the behaviors of light trapping.

As an important step toward a wave optics light trapping theory, Yu, Raman, and Fan recently introduced a statistical temporal coupled mode theory (TCMT) formalism to account for light trapping in the wave optics domain (Yu et al., 2010a,b). This theory has completely reproduced all the key results of conventional ray optics light trapping theory, including the Yablonovitch $4n^2$ limit (Yu et al., 2010a), the correction to the $4n^2$ limit when the single-pass absorption is nonnegligible (Yu et al., 2012), and the correction to the $4n^2$ limit for an absorber with angular selective absorption (Yu and Fan, 2011; Yu et al., 2010a). Moreover, this wave optics theory predicts that one can significantly overcome the conventional light trapping limit in certain nanophotonic structures, a prediction that has been validated with full-field simulations showing that the conventional limit can be overcome for all incident angles and over a wide wavelength ranges (Wang et al., 2014a; Yu et al., 2010a). In this section, we review this wave optics light trapping theory.

A solar cell is modeled as a structure supporting a number of optical modes that may couple to a number of free-space channels and thus become resonances. Light trapping is described by the coupling of incident radiation into the optical modes supported by the solar cell structure. Absorption is enhanced through the aggregate contribution of these resonances. This physics can be captured by a statistical temporal coupled-mode theory (Yu et al., 2010a,b). We assume that free space provides N planewave channels and that the absorber structure supports M resonances. Each planewave channel could guide incoming radiation to each resonance, and each resonance could also leak to each planewave channel. The coupling between the mth resonance to the planewaves with an incident wave from the nth channel is described by the following temporal coupled-mode equation

$$\frac{d}{dt}a_m = \left(j\omega_m - \frac{\sum_{s=1}^{N}\gamma_{m,s} + \gamma_0}{2} \right) a_m + j\sqrt{\gamma_{m,n}}S_n \tag{7}$$

The resonance amplitude a_m is normalized such that $|a_m|^2$ is the electromagnetic energy stored in the resonant mode per unit area, and the incident planewave amplitude S_n is normalized such that $|S_n|^2$ corresponds to its intensity. (n appearing in the subscript denotes the nth channel.) ω_m is the resonant frequency of the mth resonance. $\gamma_{m,\ s}$ is the loss rate from the mth resonance to the sth channel $\forall s \in \{1, 2, \ldots, N\}$. The loss rate is equal to the in-coupling rate, according to energy conservation and time-reversal symmetry. γ_0 is the loss rate of the resonance due to intrinsic material absorption.

In the uniform absorber of Fig. 2, if the thickness is sufficient large that the resonances completely overlap with the absorber in space, $\gamma_0 = \alpha c/n$. In a photonic crystal (for example, in Fig. 6), however, the overlap is generally less than unity (Joannopoulos et al., 2008; Naqavi et al., 2013a). We define Γ to be the overlap, or confinement factor, that characterizes the average overlap of the electric field with the weakly absorbing active material. Hence, the intrinsic loss rate is in general given by

$$\gamma_0 = \alpha \frac{c}{n} \Gamma \tag{8}$$

To rigorously derive Eq. (8), we can treat the absorption as a perturbation to a transparent structure and integrate over the energy loss vs total energy throughout the entire space. This intrinsic loss rate modification generalizes the wave optics light trapping theory over the earlier work (Yu and Fan, 2011).

We calculate the contribution of a single resonance to the absorption spectrum excited by the incident sunlight from a single channel. For the mth resonance and the nth channel, this contribution is

$$A_{m,n}(\omega) = \frac{\tilde{\gamma}_m \gamma_{m,n}}{(\omega - \omega_m)^2 + (\tilde{\gamma}_m + \gamma_m)^2/4}, \quad (9)$$

where ω is the frequency, ω_m is the resonant frequency of the mth mode, $\tilde{\gamma}_m$ is the intrinsic material absorption rate of the mth mode, $\gamma_m = \sum_{n=1}^{N} \gamma_{m,r}$ is the total external coupling rate of the mth mode, $\gamma_{m,n}$ is the decay rate of the mth mode into the nth channel (due to reciprocity, it is also the in-coupling rate of the mth mode from the nth channel), and

$$N = 2\frac{\pi\omega^2}{c^2}\left(\frac{L}{2\pi}\right)^2 \quad (10)$$

is the total number of channels in real space with a period L. For incident sunlight from the nth channel, the overall absorption coefficient, calculated using the concept of spectral absorption cross section (Yu et al., 2010a), is

$$\begin{aligned} A_n(\omega) &= \frac{1}{\Delta\omega}\int_{\omega}^{\omega+\Delta\omega}\sum_{m=1}^{M} A_{m,n}(\omega)\omega \\ &= \frac{1}{\Delta\omega}\sum_{m=1}^{M}\int_{\omega}^{\omega+\Delta\omega} A_{m,n}(\omega)\omega \\ &\approx \frac{1}{\Delta\omega}\sum_{m=1}^{M}\int_{-\infty}^{\infty} A_{m,n}(\omega)\omega, \end{aligned} \quad (11)$$

where

$$M = 2\frac{4\pi n^3 \Delta(\omega^3)}{3c^3}\left(\frac{L}{2\pi}\right)^2\left(\frac{d}{2\pi}\right) \quad (12)$$

is the total number of resonances in the spectral range $(\omega, \omega + \Delta\omega)$ supported by the structure with a thickness d and a refractive index n. The approximation on the limits of integration in the last step of Eq. (11) is justified since $\Delta\omega$ is chosen to be much greater than the linewidths of the resonances, which have high qualify factors, while still being much smaller than ω. It remains to justify the summation in Eq. (11), which holds only if the M modes are orthogonal.

From Eq. (7), we can calculate the broadband absorption enhancement by the mth resonance when light is incident from the nth channel. By summing over the contributions from all M resonances and N channels in the frequency range of $[\omega, \omega + \Delta\omega]$, and by comparing the absorption to the single-pass absorption αd_{eff}, we can calculate the angle-integrated absorption enhancement factor F, which in the k space translates into a summation over all the channels

$$F=\sum_n f_n=\frac{2\pi}{\alpha d_{\text{eff}}\Delta\omega}\sum_m \frac{\sum_n \gamma_{m,n}\gamma_0}{\sum_n \gamma_{m,n}+\gamma_0}\leqslant\frac{2\pi c\Gamma}{n\eta d}\rho(\omega) \tag{13}$$

where $\rho(\omega)=M/\Delta\omega$ is the spectral DOS, and f_m is the contribution to the enhancement factor from the mth resonance. η is the volume or area fraction of the absorbing material. The equality is approached if $\sum_{n=1}^{N}\gamma_{m,n}\gg\gamma_0$ for all m, that is, all the resonances are in the over-coupling regime. Since the material is weakly absorptive, the equality holds as long as all resonances are coupled to external radiation. In other words, all photonic states in the structure contribute to light trapping, and the averaging over the bandwidth results in the appearance of DOS in the enhancement factor F. In next section, we use Eq. (13) to show that the limit of $4n^2$ is only correct in bulk structures and the absorption enhancement factor can go far beyond this limit with proper design in the nanophotonic regime.

4. Fundamental limit of light trapping

Eq. (13) establishes a tight upper limit for absorption enhancement in light-trapping structures. In a bulk structure, $\Gamma=1$, $\eta=1$, the various versions of the conventional limit are readily reproduced by the theory (Table 1) (Yu and Fan, 2011; Yu et al., 2010a,b). In particular, we consider a structure with period L and thickness d that are both much larger than the wavelength. In this case, the resonance can be approximated as propagating plane waves inside the bulk structure. The total number of resonances in the frequency range $(\omega,\omega+\Delta\omega)$ and the number of channels are provided by Eqs. (12) and (10), respectively. With Eq. (13) divided by the number of channels, we find that for the isotropic case

$$F_{\text{iso}}=\frac{2\pi c\dfrac{8\pi n^3\omega^2}{c^3}\left(\dfrac{L}{2\pi}\right)^2\left(\dfrac{d}{2\pi}\right)}{2n\dfrac{\pi\omega^2}{c^2}\left(\dfrac{L}{2\pi}\right)^2 d}=4n^2 \tag{14}$$

which reproduces the $4n^2$ conventional limit, appropriate for the Lambertian emission case with $\sin\theta=1$, where θ is the angle of the emission cone in the medium surrounding the cell. The theory can be generalized to the case of a restricted emission cone and reproduces the standard result of $4n^2/\sin^2\theta$ (Yu et al., 2010a).

The analysis here also points to scenarios where the conventional limit is no longer applicable. Eq. (10) is not applicable when the periodicity is comparable to the wavelength, whereas Eq. (12) is not valid when the film thickness is much smaller than the wavelength. Below, we consider both of these cases.

4.1 Light trapping in structures with wavelength-scale periodicity

We consider a square lattice where the periodicity L is comparable to the wavelength λ. The discrete nature of the channels becomes important (Fig. 3A). To illustrate this effect, we assume that the film has a high refractive index (for example, silicon), such that the wavelength in the material is small compared with the periodicity. We also assume that the film has a thickness of a few wavelengths. In this case, all modes have approximately the same decay rate, and Eq. (12) can still be used to count the number of resonances.

Using Eq. (13), for normally incident light, we calculate the upper limit of the absorption enhancement factor as a function of L/λ (Fig. 3B) when the structure has a square lattice. The discontinuous changes in Fig. 3B correspond to the emergence of new channels. In particular, when $\lambda > L$, there

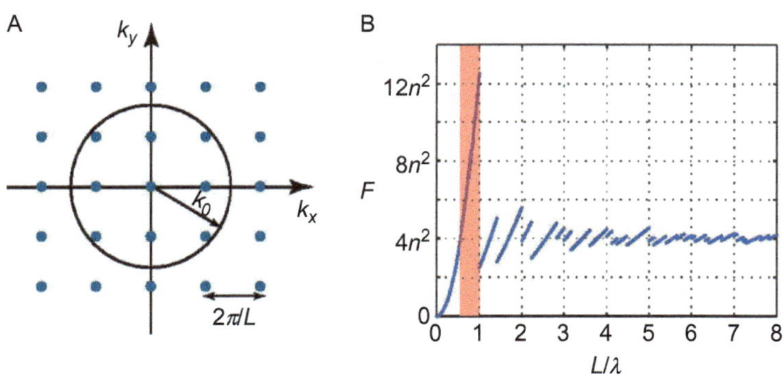

Fig. 3 Light trapping in periodic structures. (A) *Blue dots* represent channels in the k space. Channels in the *circle* correspond to free-space propagating modes. (B) Theoretical upper limit of the absorption enhancement factor using a light-trapping scheme where a square-lattice periodic grating structure is introduced into a thin film. *Red area* represents a spectral range where the upper limit of the absorption enhancement factor F_{iso} is above $4n^2$.

is only a single channel independent of frequency. On the other hand, the number of resonances is frequency dependent. As a result, the maximum enhancement factor increases quadratically as a function of frequency. In order to maximize the absorption, one should choose the periodicity to be slightly smaller than the wavelength range of interest (red region in Fig. 3B). We note that the upper limit for the absorption enhancement factor approaches $4n^2$ for a large period, $L \gg \lambda$.

The above analysis can be used to provide considerable insight into the behavior of grating structures. In particular, one expects that a 2D grating structure is superior to a 1D grating, because a 2D grating can provide access to a significantly larger number of resonances. Also, an asymmetric grating profile should be beneficial, because with a symmetric profile there are resonances that cannot be coupled to incident light due to symmetry constraints. These findings are consistent with existing literature (Heine and Morf, 1995; Tobias et al., 2008).

The use of grating structures on a relatively thick film to enhance optical absorption has been extensively explored (Bermel et al., 2007; Sheng et al., 1983; Tobias et al., 2008). This approach is practically important because it allows one to tailor the device response for specific material parameters and operating conditions such as concentration. From a fundamental perspective, Sheng et al. have argued (Sheng et al., 1983) that the grating may alter the density of state within the structure, leading to enhancement beyond $4n^2$ over particular frequency ranges. However, the cases we consider here involve shallow gratings on the surface of a thick medium. In such a case, the change of density of state in the structure is substantial only in very limited frequency ranges (Fan and Joannopoulos, 2002). Instead, our analysis shows that enhancement beyond $4n^2$ is nevertheless achievable because the grating restricts the number of channels available in free space. Also, in Tobias et al. (2008) and Saeta et al. (2009), enhancement factors above $4n^2$ were predicted using approximate approaches involving a summation of various scattering events in an incoherent fashion. The analysis presented here is more general in the sense that it is based upon electromagnetic analysis. Moreover, our analysis indicates that the potential of significantly exceeding the conventional limit, defined in terms of $4n^2/\sin^2\theta$, is rather limited in these structures; this conclusion arises because, to achieve high-enhancement factors, one needs to use a periodicity comparable to the wavelength of interest, which leads directly to strong angular and spectral dependency, in consistency with previous results (Tobias et al., 2008). Below, we present a strategy that overcomes these issues and exceeds the conventional limit over a large range of angles and frequencies.

4.2 Light trapping in thin films

When the thickness d of the film is comparable to half wavelength in the material, one can reach the single-mode regime where the film supports a single waveguide mode band for each of the two polarizations. In such a case, Eq. (7) is no longer applicable. Instead, the number of resonances in the frequency range of $[\omega, \omega+\delta\omega]$ can be calculated as

$$M = 2 \times \frac{2\pi n_{wg}^2 \omega}{c^2} \left(\frac{L}{2\pi}\right)^2 \delta\omega \tag{15}$$

where the first factor of 2 arises from counting both polarizations. (Here, to facilitate the comparison to the standard conventional limit, for simplicity, we have assumed that the two polarizations have the same group index n_{wg}.) Notice that, in this case, the number of modes no longer explicitly depends upon the thickness d of the film.

In order to highlight the effect of such strong light confinement, we choose the periodicity to be a few wavelengths, in which case the number of channels can still be calculated using Eq. (10). As a result, we obtain the upper limit for the absorption enhancement factor

$$F_{iso} = 2 \times 4n_{wg}^2 \frac{\lambda}{4n_{wg} d} V \tag{16}$$

where the factor $V = \alpha_{wg}/\alpha_0$ characterizes the overlapping between the profile of the guided mode and the absorptive active layer. The absorption coefficient and group index of the waveguide mode are α_{wg} and n_{wg}, respectively.

Eq. (16) in fact becomes $4n^2$ in a dielectric waveguide of $d \approx \lambda/(2n)$. Therefore, reaching the single-mode regime is not sufficient to exceed the conventional limit. Instead, to achieve the full benefit of nanophotonics, one must either ensure that the modes exhibit deep-subwavelength-scale electric-field confinement, or enhance the group index to be substantially larger than the refractive index of the active material, over a substantial wavelength range. Below, using both exact numerical simulations and analytic theory, we will design geometries that simultaneously satisfy both these requirements.

Guided by the theory above, we now numerically demonstrate a nanophotonic scheme with an absorption enhancement factor significantly exceeding the conventional limit. We consider a thin absorbing film with a thickness of 5 nm (Fig. 4A), consisting of a material with a refractive index and a wavelength-independent absorption length of 25 μm. The film is placed on a mirror that is approximated to be a perfect electric conductor (PEC).

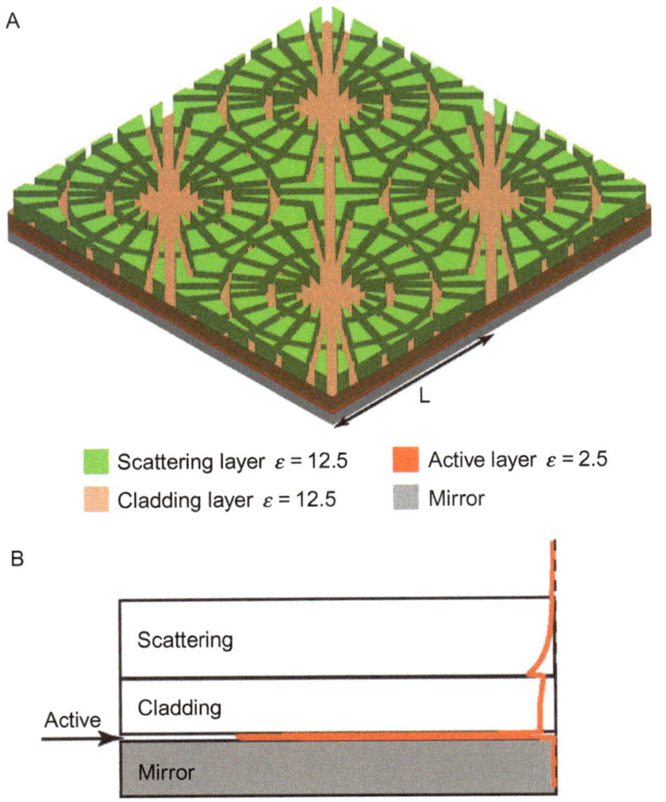

Fig. 4 Structure for overcoming the conventional light-trapping limit. (A) A nanophotonic light-trapping structure. The scattering layer consists of a square lattice of air groove patterns with periodicity $L = 1200$ nm. The thicknesses of the scattering, cladding, and active layers is 80, 60, and 5 nm, respectively. The mirror layer is a perfect electric conductor. (B) The profile of electric-field intensity for the fundamental waveguide mode. Fields are strongly confined in the active layer. To obtain the waveguide mode profile, the scattering layer is modeled by a uniform slab with an averaged dielectric constant.

A PEC mirror is used for simulation convenience. In practice, it can be replaced by a dielectric cladding layer, which produces similar results (Yu et al., 2010a). Our aim here is to highlight the essential physics of nanophotonic absorption enhancement. The choice of material parameters therefore represents a simplification of actual material response. Nevertheless, we note that both the index and the absorption strength here are characteristic of typical organic photovoltaic absorbers in the weakly absorptive regime (Hoppe and Sariciftci, 2004). Furthermore, there is general interest in using thinner absorbers in organic solar cells given their short exciton diffusion lengths of about 3–10 nm (Huynh et al., 2002; Mayer et al., 2007; Yu et al., 1995).

In order to enhance the absorption in the active layer, we place a transparent cladding layer $n_H=\sqrt{12.5}$ on top of the active layer. Such a cladding layer serves two purposes. First, it enhances density of state. The overall structure supports a fundamental mode with group index n_{wg} close to n_H, which is much higher than that of the absorbing material. Second, the index contrast between active and cladding layer provides nanoscale field confinement. Fig. 4B shows the fundamental waveguide mode. The field is highly concentrated in the low-index active layer, due to the well-known slot-waveguide effect (Almeida et al., 2004). Thus, the geometry here allows the creation of a broadband high-index guided mode, with its energy highly concentrated in the active layer, satisfying the requirement in Eq. (16) for high absorption enhancement.

In order to couple incident light into such nanoscale guided modes, we introduce a scattering layer with a periodic pattern on top of the cladding layer, with a periodicity L much larger than our wavelength range of interest. Each unit cell consists of a number of air grooves. These grooves are oriented along different directions to ensure that scattering strength does not strongly depend on the angles and polarizations of the incident light. We emphasize that there is no stringent requirement on these grooves as long as the scattering strength dominates over resonance absorption rates.

We simulate the proposed structure by numerically solving Maxwell's equations (Fig. 5A). The device has a spectrally averaged absorption enhancement factor of $F_{\text{iso}}=119$ (red line) for normally incident light. (All the absorption spectra and enhancement factors are obtained by averaging s and p polarized incident light.) This enhancement factor is well above the conventional limit for both the active material ($4n_L^2=10$) and the cladding material ($4n_H^2=50$). Moreover, the angular response is nearly isotropic (Fig. 5C and D). Thus such enhancement cannot be attributed to the narrowing of angular range in the emission cone, and instead is due entirely to the nanoscale field confinement effect.

Using our theory, we calculate the theoretical upper limit of light-trapping enhancement in this structure. For wavelength $\lambda=500$ nm, we obtain an upper limit of $F_{\text{iso}}=147$. The enhancement factor observed in the simulation is thus consistent with this predicted upper limit. The actual enhancement factor obtained for this structure falls below the calculated theoretical upper limit because some of the resonances are not in the strong overcoupling regime.

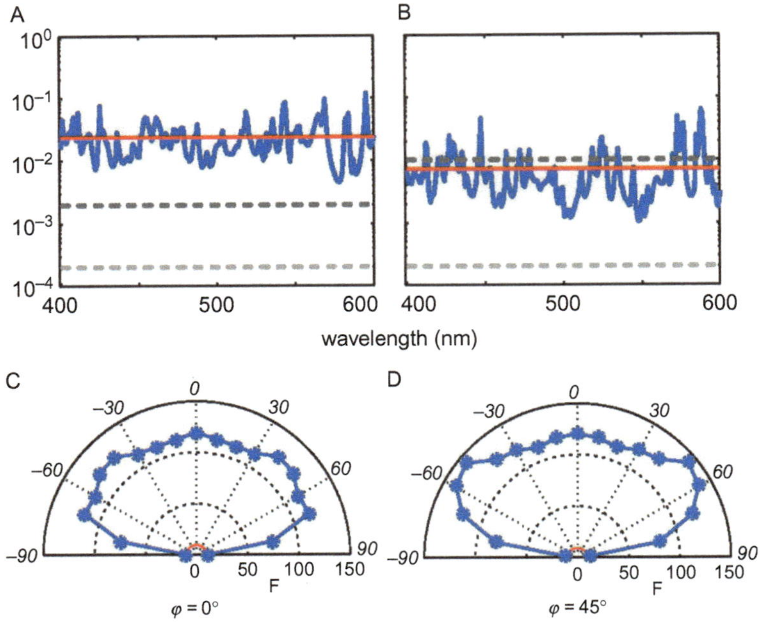

Fig. 5 Absorption with the light-trapping structures. (A) Absorption spectrum for normally incident light for the structure shown in Fig. 4. The spectrally averaged absorption (*red solid line*) is much higher than both the single-pass absorption (*light-gray dashed line*) and the absorption as predicted by the limit of $4n_L^2$ (*dark-gray dashed line*). The vertical axis is the absorption coefficient. (B) Absorption spectrum without nanoscale light confinement. The structure is the same as that of (A) except that the dielectric constant of the active layer is now the same as the cladding layer. The *dark-gray dashed line* represents the absorption as predicted by the limit of $4n_H^2$. (C and D) Angular dependence of the spectrally averaged absorption enhancement factor for the structure in Fig. 4. Incident angles are labeled on top of the semicircles. Incident planes are oriented at 0 (C) and 45 (D) degrees (azimuthal angles) with respect to the [10] direction of the lattice. The *red circles* represent the $4n_L^2$ limit.

To illustrate the importance of nanoscale field confinement enabled by the slot-waveguide effect, we change the index of the material in the absorptive layer to n_H. Such a structure does not exhibit the slot-waveguide effect. The average enhancement in this case is only 37, falling below the conventional limit of 50 (Fig. 5B).

In this section, we have shown that properly designed nanophotonic structures can achieve enhancement factors that far exceed the conventional limit. Our results indicate that substantial opportunities for nanophotonic light trapping exist using only low-loss dielectric components. The basic theory, moreover, is generally applicable to any photonic structure, including nanowire (Kayes et al., 2005; Law et al., 2005) and plasmonic structures (Atwater and Polman, 2010). In plasmonic structures, the presence of nanoscale guided modes may also provide opportunities to overcome the conventional limit.

5. Light trapping in photonic crystals

To highlight the unique physics of light trapping in photonic crystals, and as an illustration of the model system, we consider the structure shown in Fig. 6. It consists of a photonic crystal with a finite number of periods in a vertical direction, with a perfect mirror on its bottom, and with light incident from the top. The crystal is assumed to be of infinite extent in the horizontal directions, and to consist of a periodic array of dielectric elements in air. The dielectric has a real part of refractive index of n, which is assumed to be frequency independent throughout this section, and to be weakly absorbing with a small absorption coefficient $\alpha \ll 1/d$ where d is the thickness of the structure. We assume that the single-pass absorption is negligible. Knowing

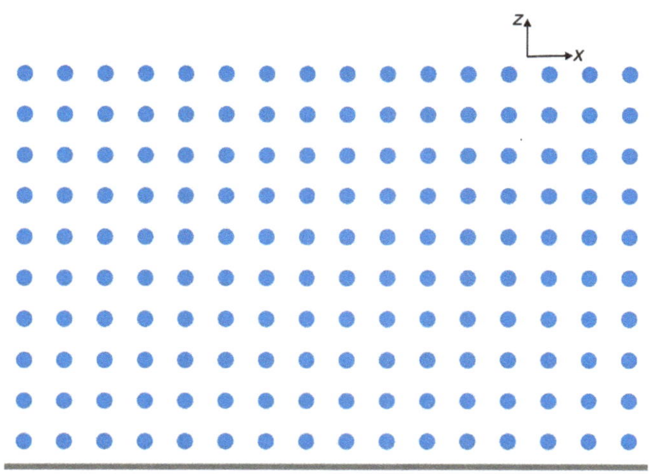

Fig. 6 A 2D photonic crystal structure consisting of a square lattice of dielectric rods in air. The radius of each rod is 0.2a, where a is the lattice constant. The dielectric constant of the rods, colored in *blue*, is 12, with an extremely small imaginary component. In the absorption simulations, the absorption is obtained through 10 layers of such rods with the other dimension being infinite and with a perfect mirror at the *bottom*. Light is incident on *top*. In the band structure and overlap factor calculations, the structure is treated as a lossless 2D photonic crystal infinite in both dimensions. The equivalent thickness of the structure is equal to the thickness of the slab in Fig. 1 so that the two structures contain the same amount of absorbing materials.

the absorption enhancement factor in such a weak absorption limit, one can derive the enhancement factor where the single-pass absorption is no longer negligible (Green, 2002; Yu et al., 2012). Focusing on the weak absorption limit therefore allows us to focus on illustrating the most prominent effects of the DOS engineering in photonic crystals on light trapping absorption enhancement.

Real solar cells are of course 3D structures. To simplify numerical calculations, however, many theoretical literatures on light trapping in solar cells have considered 2D structures (Chutinan and John, 2008; Chutinan et al., 2009; Duché et al., 2011; Gomard et al., 2012; Park et al., 2009; Yu et al., 2010b). By 2D, we mean structures that are uniform in the third dimension. Moreover, we consider only light propagating within a 2D plane perpendicular to the third dimension.

The wave optics theory in Section 3 points to several unique aspects of light trapping in photonic crystals. We examine the most prominent deviations of the optical properties of photonic crystals from those of bulk structures. First, the DOS of a photonic crystal (for example, in Fig. 6) can differ significantly from that of the bulk structure (for example, in Fig. 2). Second, for a bulk structure, if the thickness is greater than a few wavelengths, one can assume complete overlap of the modes with the absorbing material. In contrast, in a photonic crystal, the electromagnetic fields do not completely overlap with the absorbing material, and therefore the modal absorption loss rate is related to the photonic crystal geometries as well as material constants. Third, to access all the modes in a bulk structure, one typically employs surface roughness. In contrast, we will show that a photonic crystal by itself, with appropriate configurations, provides complete access to all its modes without surface disorders.

5.1 Light trapping in 2D photonic crystals

5.1.1 Relevant aspects of the band structure of a 2D photonic crystal

As a model system, we start by considering a lossless and infinite 2D photonic crystal consisting of a square lattice of dielectric rods in air (Fig. 6). The radius of each rod is $r = 0.2a$, where a is the lattice constant. The dielectric material has a nondispersive dielectric constant of 12, which is close to the value of silicon or gallium arsenide at optical frequencies.

We used the MIT Photonic Band (MPB) package to calculate the band structure (Johnson and Joannopoulos, 2001). In Fig. 7, we show the band structure of the photonic crystal along the line segments connecting the high symmetry points. There exist photonic bandgaps for the transverse magnetic

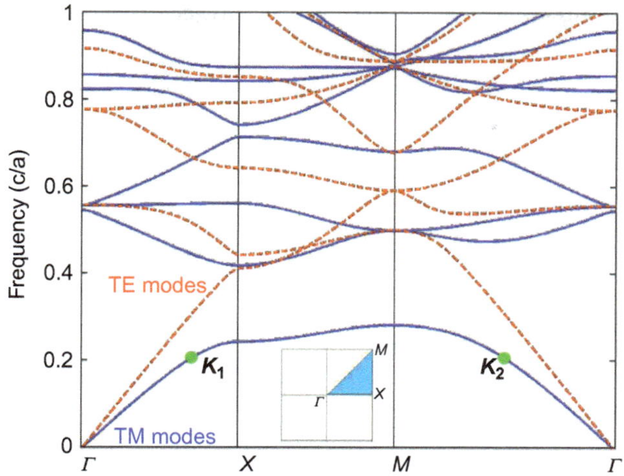

Fig. 7 Band structure of the photonic crystal in Fig. 6. The first Brillouin zone of the square lattice is shown in the *inset*. Γ denotes the *k* point (0,0), *X* denotes (0,π/*a*), and *M* denotes (π/*a*,π/*a*).

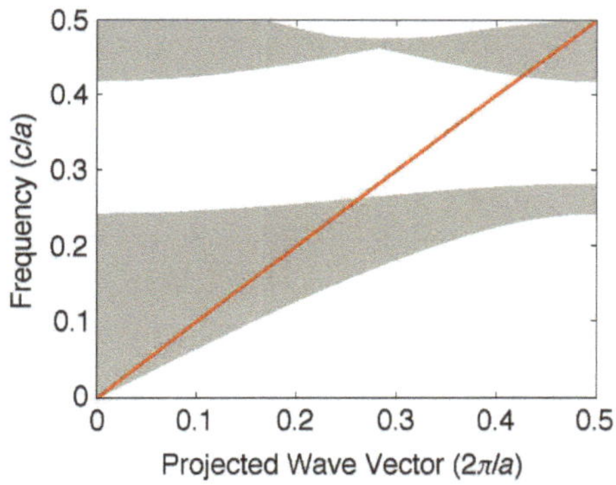

Fig. 8 Projected TM band structure obtained from Fig. 7. The *lightline* is given by $\omega = ck_x$, where k_x is the projected wave vector.

(TM) polarization, but not for the transverse electric (TE) polarization. In order to highlight the features in the DOS of a 2D photonic crystal we choose to work with the TM polarization for all subsequent discussions in 2D structures.

The same photonic band structure in Fig. 7 can alternatively be presented as a projected band diagram (Fig. 8), where we project the 2D band structure $\omega(k_x, k_z)$ onto the $\omega - k_x$ plane. The shaded regions in Fig. 8 corresponds to states in the photonic crystal, the unshaded region is the gap region. We notice that a significant part of the first band and some parts of the second band lie below the light line $\omega = ck_x$. Such a projected band structure will be useful for the discussion of mode coupling issues.

We can calculate the DOS of this photonic crystal by a uniform sampling of all the k-points in the first Brillouin zone. In Fig. 9, we observe two bandgaps at frequencies close to $0.3c/a$ and $0.7c/a$, as well as a number of van Hove singularities where the DOS is divergent (Bassani and Parravicini, 1975; Van Hove, 1953).

For each mode, we can also calculate the overlap factor Γ between the electric field and the dielectric rods using the Dirichlet-to-Neumann (DTN) method (Liu and Fan, 2011). At any given frequency, we select two k points from the segments connecting Γ to X to M and back to Γ. The selected k points correspond to the two leftmost intersections of horizontal cuts with the band structure plot in Fig. 7. Fig. 10 shows the overlap factors for such

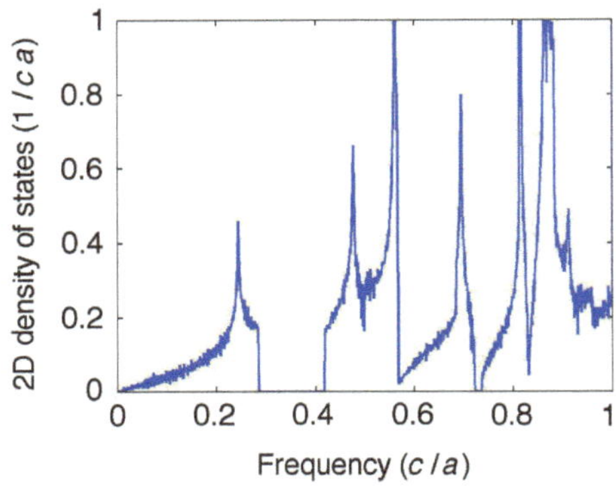

Fig. 9 Density of TM states per area for the 2D photonic crystal of Fig. 6.

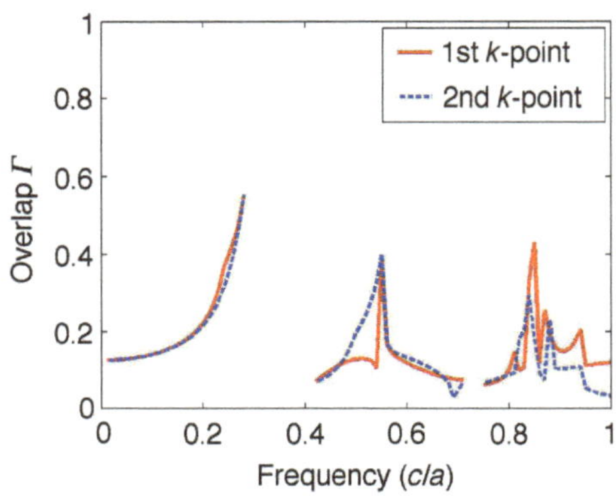

Fig. 10 Overlap factor of the TM mode for two selected k points at each frequency. The k points are selected from *left* to *right*, for example, at a frequency of $0.2c/a$, the first two k points are shown as K_1 and K_2, in Fig. 7. Inside the bandgap, the overlap factor is not defined.

two k points for a range of frequencies. We observe that, at each frequency, the modes at different k points tend to have very similar overlap factors. In the limiting case at zero frequency, the field becomes uniform, and the overlap factors approach the area fraction $\eta = \pi r^2/a^2 \approx 0.126$. The first and second bands have quite different overlap factors, which can be explained by the variational theorem (Joannopoulos et al., 2008). The first band tends to concentrate more of its electromagnetic energy in the dielectric, hence the overlap factors are higher than the area fraction η. Within the first band, there is an increase in overlap factors with respect to frequency, following the trend in the DOS. Γ is not defined in the photonic bandgaps, where light propagation and absorption are forbidden.

5.1.2 Absorption enhancement factor in the 2D photonic crystal and comparison to the density of states

We calculate the absorption in a lossy and finite version of the 2D photonic crystal shown in Fig. 6. The crystal contains a finite number of periods in a vertical direction, with a perfect mirror on its bottom and with light incident from the top. The crystal is assumed to be of infinite extent in the horizontal directions. In the numerical simulations, we chose an extremely weak attenuation coefficient $\alpha = 10^{-8}/d$, and this particular numerical value disappears after taking the ratio between the calculated absorption and the single-pass absorption defined in Eq. (3). The 2D photonic crystal contains 10 layers in the normal direction of incident light. Because the absorption is weak and there are a sufficient number of layers, the DOS and overlap factors can be well approximately by the corresponding lossless and infinite photonic crystal.

We calculate the absorption enhancement factor $f(\theta)$ for the entire range of incident angles θ with the Fourier Modal Method using the Stanford Stratified Structure Solver (S4) package (Liu and Fan, 2012). From $f(\omega, \theta)$, we obtain the angle-integrated absorption enhancement factor F_{2D} using Eq. (5). Numerically, we observe that F_{2D} is not influenced by the choice of number of layers as long as the number is sufficiently large. We plot F_{2D} in Fig. 11 and compare it to both the DOS bound given by Eq. (13) and the conventional limit of $2\pi n$.

To calculate the DOS bound on the right-hand side in Eq. (13), we first convert the summation in the real space as an angular integration in 2D to a summation over channels in the wave vector space. The 2D angle-integrated absorption enhancement factor can be written as

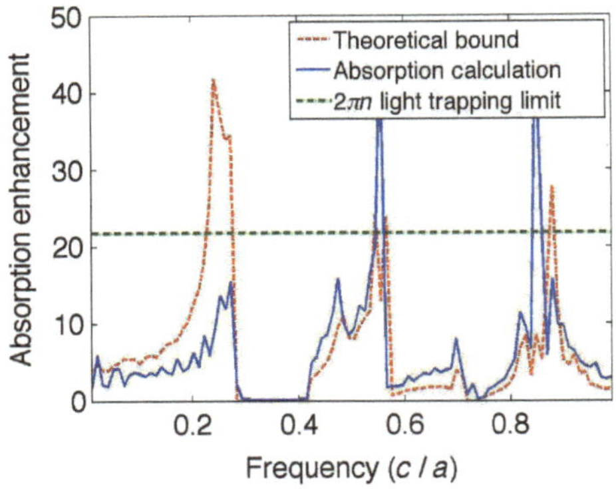

Fig. 11 Comparison of angle-integrated light trapping absorption enhancement factor and its theoretical upper bound for the 2D photonic crystal in Fig. 6.

$$F_{2D} = \int_{-\pi/2}^{\pi/2} d\theta f(\theta)\cos\theta = \frac{1}{k_0}\int_{|k_x|<k_0} dk_x f(k_x) \tag{17}$$

$$= \frac{\Delta k}{k_0}\sum_n f_n \tag{18}$$

where θ is the angle of incidence and $k_x = k_0 \sin\theta$

$$k_0 = \frac{\omega}{c}$$

$$\Delta k = \frac{2\pi}{L}$$

where L is the lattice constant of the photonic crystal or an artificially imposed period of a bulk structure. Using Eqs. (13) and (5), we obtain

$$F_{2d} \leq \frac{(2\pi c)^2 \Gamma}{n\eta\omega}\rho_{2D}(\omega) \tag{19}$$

where $\rho_{2D}(\omega)$ is the DOS per unit area.

We calculated $\rho_{2D}(\omega)$ (Fig. 9), as well as $\Gamma(\omega)$ (Fig. 10) for selected k points. Since the overlap factors Γ are approximately equal for different k points at the same frequency, we simply take an average of the two overlap factors shown in Fig. 10, and use the averaged overlap factor Γ in Eq. (19). With these information and other parameters of the structure, we can obtain the right-hand side expression $(2\pi c)^2 \Gamma \rho_{2D}(\omega)/n\eta\omega$ of Eq. (19), which gives an upper bound for F_{2D}. We plot this bound for a range of frequencies in Fig. 11, referring to it as the "Theoretical Bound" curve.

Fig. 11 verifies our theoretical result on the correspondence between the absorption enhancement and the theoretical bound. We note in particular the following features. First, over a few narrow ranges of frequencies, the Yablonovitch limit is surpassed due to the van Hove singularities in the photonic DOS. Second, overall the absorption enhancement is below the conventional limit. Third, the photonic crystal absorber, without grating or surface roughness, is itself a very efficient scatterer. Over most frequency ranges above $0.3c/a$, there is a close match between the DOS and the absorption curves, implying that most accessible resonances are excited. Unlike the conventional case of uniform slab, there is no need to use a surface grating to achieve light trapping for photonic crystals. However, on the other hand, for the first band from a frequency of 0 to approximately $0.3c/a$, the numerically obtained light trapping absorption enhancement is significantly lower than its upper bound imposed by the DOS. This is because many modes are below the light line and decoupled from external radiation, as we have eluded to in the discussion of Fig. 8.

5.1.3 van Hove singularities in the density of states

For absorption enhancement, we can engineer the photonic crystals and take advantage of the van Hove singularities in the DOS (Bassani and Parravicini, 1975). For practical applications, we might be able to apply the narrowband divergence of DOS to achieve strong absorption improvement, for example, at the bandedge in c-Si where light trapping is crucial (Wang et al., 2012; Yu et al., 2010b).

Van Hove singularities are generally more prominent in lower dimensions (Bassani and Parravicini, 1975). In fact, the DOS can diverge in 2D (Fig. 9, for example) while only its derivative can diverge in 3D (Fig. 19, for example). As a result, in 2D, the angle-integrated absorption enhancement factor indeed surpasses the Yablonovitch limit at van Hove singularities (Fig. 11); however, in 3D, similar absorption enhancement is much more difficult to achieve.

The van Hove singularities in some sense redistribute the optical properties of photonic crystals (Gaponenko, 2010). For real materials, redistribution of the DOS is constrained by the DOS sum rule (Barnett and Loudon, 1996; Scheel, 2008), which implies that an increase in DOS in one frequency region needs to be compensated by a decrease in other frequency regions, limiting the bandwidth and magnitude of the van Hove singularities. In this photonic crystal, the theoretical bound exceeds the conventional limit in the frequency range approximately from $0.22c/a$ to $0.27c/a$, due to the van Hove singularity in DOS. Assuming that the frequency range is centered at a free space wavelength of 1000 nm, this frequency range then corresponds to a wavelength range of 907 to 1114 nm. Therefore, it is possible to use van Hove singularities to achieve light trapping limit above the conventional limit over a bandwidth that is relevant for solar cell light trapping at the crystalline silicon bandedge. We may also stack multiple photonic crystals to use multiple van Hove singularities for an even broader bandwidth of light trapping absorption enhancement.

5.1.4 Long wavelength limit

We observe that away from the van Hove singularities, the absorption enhancement factor in a photonic crystal is in general lower than the $2\pi n$ limit. This is partly due to the fact that the overlap factor Γ is lower than the area or volume fraction, and, more importantly, due to the lower effective DOS of the photonic crystal (Fig. 6) compared to the DOS in the bulk structure (Fig. 2). We can use the effective medium theory (Joannopoulos et al., 2008), which is accurate in the long wavelength limit, to provide a more in depth discussion of the DOS.

In the long wavelength (low frequency) limit, the light does not probe the fine details of the photonic crystal (Lin et al., 2005). At low frequencies in Fig. 9, the DOS of the 2D photonic crystal is linear with respect to frequency, which follows the form of the DOS in a homogeneous dielectric material in 2D. In the long wavelength limit, we can therefore estimate the DOS of the photonic crystal by treating it as a homogeneous effective medium and using the effective dielectric constant of this effective medium. In general, consider a composite material constructed from two constituents with dielectric constants ε_1 and ε_2, and area or volume filling fractions of f_1 and f_2, respectively. The effective dielectric constant ε of the composite is then bounded as follows (Bergman, 1978, 1980; Milton, 1981, 2002):

$$\left(f_1\varepsilon_1^{-1}+f_2\varepsilon_2^{-1}\right)^{-1}\leqslant\varepsilon\leqslant f_1\varepsilon_1+f_2\varepsilon_2 \tag{20}$$

In 2D, imagine that we start with a bulk structure of weakly absorbing material with ε_1 and area A, and compare the structure to a photonic crystal structure for which we use the same amount of absorbing material with ε_1 but dilute it with a transparent material with ε_2 to a total area of A/η, where η is the area fraction of the absorbing material. Then $f_1=\eta$ and $f_2=1-\eta$.

The bound of light trapping enhancement in 2D, as pointed out in Eq. (19), is related to the DOS $\rho(\omega)$, as well as to the area fraction η and the overlap factor Γ. Since the overlap factor $\Gamma \approx \eta$ under the assumption that the electric field is distributed uniformly in space, the area fraction η and the overlap factor Γ cancel out in Eq. (19), and the bound of light trapping enhancement is approximately proportional to $\rho(\omega)$. Furthermore, the 2D DOS is proportional to n^2 or ε:

$$\rho_{2D}(\omega) = \frac{4\varepsilon\omega}{\pi c^2} \tag{21}$$

We further assume that all modes are accessible, which can be made possible by methods discussed in this section. Using Eqs. (20) and (21), the upper bound of the angle-integrated light trapping enhancement factor $F_{2D}/2\pi n$, where $n = \sqrt{\varepsilon_1}$, is then given by

$$\frac{F_{2D}}{2\pi n} \leq \eta + (1-\eta)\frac{\varepsilon_2}{\varepsilon_1} \tag{22}$$

In Fig. 12, we plot this upper bound in Eq. (22) as a function of the area fraction η of the absorbing material. In Fig. 12A we consider the hypothetical case where the absorbing material has a low dielectric constant ε_1 and is embedded in a transparent background of a high dielectric constant with $\varepsilon_2 = 12\varepsilon_1$. In this case, $F_{2D}/2\pi n$ significantly exceeds unity, indicating that such a structure has a light trapping potential that is significantly above conventional light trapping with the same absorbing material in a uniform bulk form. This effect was noted (Yu et al., 2010a) and is due to the enhanced DOS from the high index dielectric material at the background.

In Fig. 12B, we consider the more typical case, where the absorbing material has a high dielectric constant $\varepsilon_1 = 12$ and is embedded in a transparent background with a low dielectric constant $\varepsilon_2 = 1$. In this case, $F_{2D}/2\pi n$ is always less than unity. In other words, in the long wavelength limit, such a structure will always underperform, in terms of its light trapping capability, as compared to a uniform bulk absorbing medium with the same dielectric constant ε_1. The underperformance arises because the presence of the low index material significantly reduces the DOS of the overall structure. The theoretical results illustrated in Fig. 12B are consistent with all numerical results presented in this section. In particular, while these theoretical results were derived using the effective medium theory, which strictly speaking is correct only in the long wavelength limit, all our numerical results indicate that away from the van Hove singularities, the photonic crystal structure that

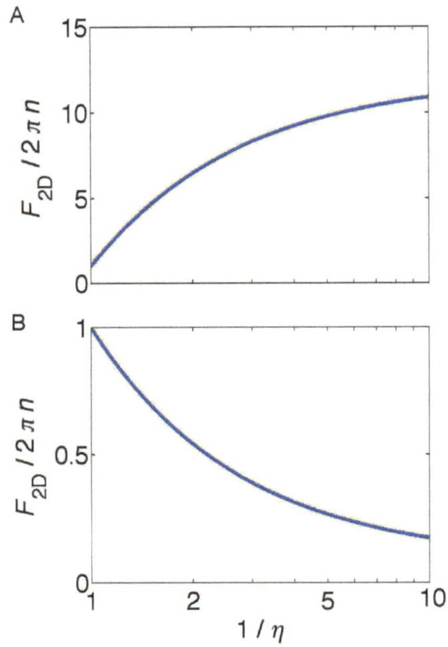

Fig. 12 Upper bounds of 2D angle-integrated absorption enhancement factors by Eq. (22). η is the area fraction of the absorber whose index is n. (A) $\varepsilon_2/\varepsilon_1 = 12$. (B) $\varepsilon_1/\varepsilon_2 = 12$.

we examine has an absorption enhancement that significantly underperforms the conventional limit for the corresponding uniform bulk medium. While the DOS away from the long wavelength limit certainly cannot be estimated from the effective medium theory, the overall trend, which is that the use of a low index material should reduce the average DOS, should in general be valid. This effect also limits the bandwidth over which the use of van Hove singularities can outperform the conventional limit, since it sets a lower floor on the enhancement factor away from van Hove singularities.

Besides the considerations on light trapping, antireflection is also crucial for any practical solar cells (Raut et al., 2011). In particular, optimal antireflection is a prerequisite for optimal light trapping (Yu et al., 2012). Although the DOS of a photonic crystal structure might be lower than that of the bulk structure due to "dilution" of the active material on average, this dilution also lowers the effective refractive index of the photonic crystal structure and provides better optical impedance matching to free space. This mechanism of antireflection adds another dimension to the optimization problem of designing photonic crystal absorbers.

5.1.5 Mode coupling

Eq. (13) provides a tight upper limit for the angle-integrated absorption enhancement factor F. The upper limit can be reached if all the modes are in the over-coupling regime. Since we have assumed that the single-pass absorption is weak, any mode that can couple to the external radiation is over-coupled. Therefore, to reach the upper limit one only needs to ensure that all modes can couple to external radiation. This is evident from Fig. 11, where the absorption enhancement is significantly lower than its limit given by Eq. (13) in the first band, and the limit can be approximately reached only for the second and higher bands. As we noted above, if the modes in the first band were able to couple to external radiation, the $2\pi n$ light trapping limit would also be surpassed over a sizable bandwidth, approximately from $0.22c/a$ to $0.27c/a$ in Fig. 11, at the lowest van Hove singularity. Therefore, it is important to investigate possible methods that would lead to complete accessibility of all the modes.

For the structure shown in Fig. 6, the mode coupling aspect can be understood by considering the projected band diagram shown in Fig 8. Only modes above the light line, that is, modes satisfying $\omega \geq ck_x$, can couple to external radiation. Since $k_x \in [0, \pi/a_x]$, where $a_x = a$ in this case is the periodicity of the truncated structure along the x direction, if follows that significant number of modes below $\omega = 0.5\frac{2\pi c}{a}$, which includes a great portion of the first band and a small portion of the second band, cannot couple to external radiation and thus cannot contribute to light trapping. This effect is seen in Fig. 11, where in the frequency range of the first band, the numerically determined enhancement factor falls far below the theoretical bound of Eq. (5), whereas the numerically determined enhancement factor agrees quite well with the theoretical bound in the upper bands.

For a bulk structure, efficient coupling can be achieved when the light rays are randomized by surface roughness, or when the light is scattered by a grating structure with a period much larger than the wavelength of light (Naqavi et al., 2013b; Yu and Fan, 2011; Yu et al., 2010b). The same underlying physics applies to photonic crystals in the wave optics regime. In the following, we explicitly demonstrate two simple methods to achieve complete coupling. Both methods seek to increase the spatial periodicity a_x of the truncated photonic crystal along the x direction.

In the first method, we simply choose a different crystal truncation. As an example, in Fig. 13 we choose the orientation of the crystal such that the period along the surface $a_x = \sqrt{5}a$. In this case, one can show that all the modes in the crystal are above the light line. Many other different truncations are possible.

In the simulation results shown in Fig. 14, we indeed observe significant improvement of light trapping enhancement in the first band, as compared to the structure in Fig. 6. Also, the light trapping enhancement in the upper bands remain essentially unchanged as we vary the crystal orientation, as expected from the mode coupling argument.

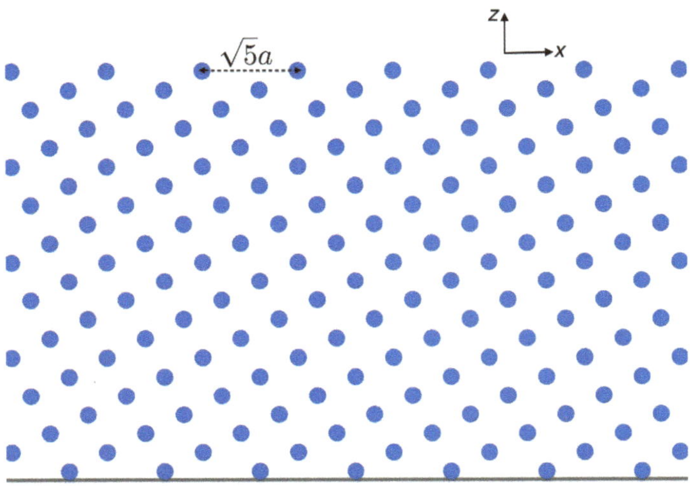

Fig. 13 A different truncation of the 2D photonic crystal in Fig. 6. The structure is rotated by 26.565° from the original structure in Fig. 6. The effectively period is $\sqrt{5}a$, where a is the lattice constant of the square lattice.

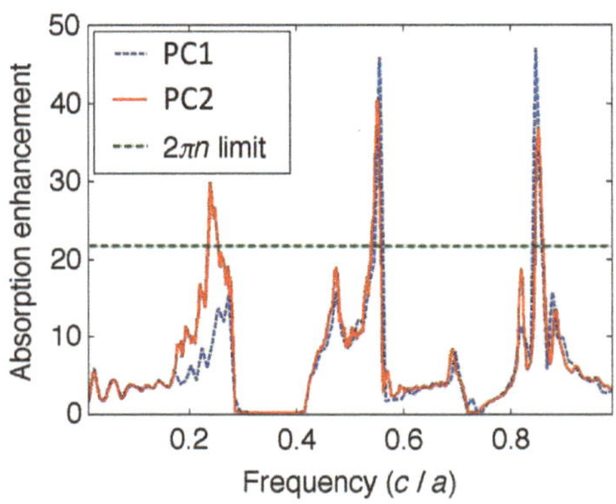

Fig. 14 Comparison of angle-integrated absorption enhancement spectra for the photonic crystals in Fig. 6 (labeled as PC1) and Fig. 13 (labeled as PC2).

In the second method, we add a layer of grating on top of the original structure. In Fig. 15, we reuse the structure from Fig. 6, but add a top layer of transparent rods to keep d_{eff} unchanged. We vary the radii of the five transparent rods in that top layer, therefore creating random scattering of incoming light. Such a layer effectively increases the period to $5a$.

In Fig. 16, we compare the light trapping enhancement without and with the random layer. Again, the second and higher bands are nearly unchanged, but, in the presence of scattering, the absorption in the first band is vastly improved up to the theoretical bound given in Eq. (19).

With either method, we are able to achieve nearly complete coupling of all the modes. Near the frequency range of the first van Hove singularity, roughly between $0.2c/a$ and $0.3c/a$, the calculated light trapping absorption enhancement surpasses the conventional $2\pi n$ limit.

5.2 Light trapping in 3D photonic crystals

The formalism that we have developed for 2D systems can be readily extended to 3D systems. For example, in a 3D system, using Eq. (13), the angle-integrated light trapping enhancement factor can be expressed as

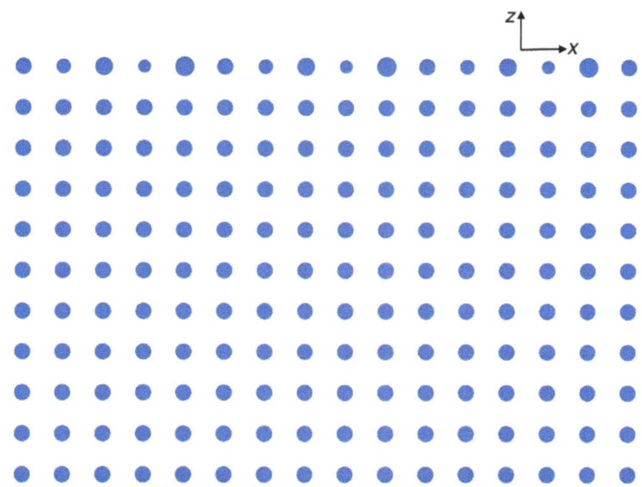

Fig. 15 The photonic crystal in Fig. 6 with one additional scattering layer on *top*. The dielectric rods in the first layer on *top* are transparent while the other rods are weakly absorptive. The real part of the dielectric constant is 12. The radius of each absorptive rod is $0.2a$, where a is the lattice constant. The radii of the transparent rods are alternating in every five rods, being $0.2a$, $0.18a$, $0.22a$, $0.16a$, and $0.24a$, thus forming a scattering layer.

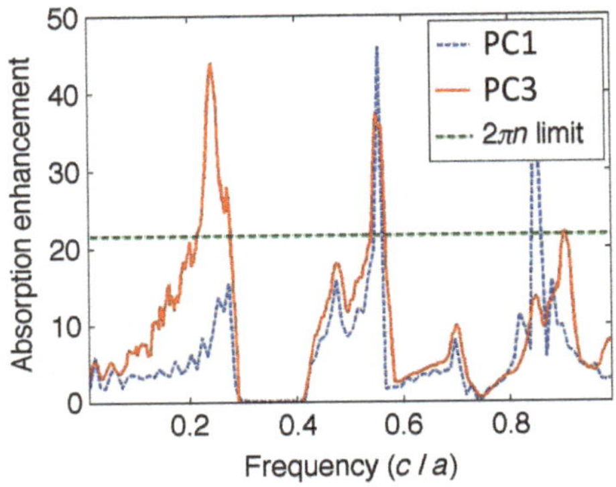

Fig. 16 Comparison of angle-integrated absorption enhancement spectra for the photonic crystals in Fig. 6 (labeled as PC1) and Fig. 15 (labeled as PC3).

$$F_{3D} = \int_0^{\pi/2} d\theta \int_0^{2\pi} d\varphi f(\theta, \varphi) \cos\theta \sin\theta, \\ = \frac{1}{k_0^2} \int \int_{k_x^2 + k_y^2 \leqslant k_0^2} dk_x dk_y f\left(k_x, k_y\right), \tag{23}$$

$$= \frac{\Delta k^2}{k_0^2} \sum_n f_n \tag{24}$$

where θ is the incidence angle, and φ is the azimuthal angle, and

$$k_x = k_0 \sin\theta \cos\varphi, \\ k_y = k_0 \sin\theta \sin\varphi, \\ k_0 = \frac{\omega}{c}, \\ \Delta k = \frac{2\pi}{L}.$$

Similar to Eq. (19), we have

$$F_{3D} \leqslant \frac{(2\pi c)^3 \Gamma}{n\eta\omega^2} \rho_{3D}(\omega) \tag{25}$$

where $\rho_{3D}(\omega)$ is the DOS per unit volume.

Here, we highlight several notable differences between the 3D and 2D photonic crystals. In 3D, one can show that in the long wavelength limit, for a high index absorbing material embedded in a low index transparent background, the same index contrast and volume fraction would cause greater

DOS reduction than in 2D. The result of the effective medium theory for this case gives

$$\frac{F_{3D}}{4\pi n^2} \leqslant \left[\eta + (1-\eta)\frac{\varepsilon_2}{\varepsilon_1}\right]^{\frac{3}{2}} \tag{26}$$

as plotted in Fig. 17 with $\varepsilon_1 = 12$ and $\varepsilon_2 = 1$. As a result, the absorption enhancement is significantly lower than the conventional limit of $4\pi n^2$ (Table 1). In addition, the van Hove singularities are much less prominent since the DOS at van Hove singularities is not divergent (although its derivative is divergent). Therefore, the angle-integrated light trapping enhancement factor does not exceed $4\pi n^2$, although at individual angles and wavelengths the $4n^2$ limit can be surpassed.

We study a 3D woodpile structure (Ho et al., 1994; Sözüer and Dowling, 1994), consisting of the same dielectric with $\varepsilon = 12$ stacked in air as in the 2D simulations, as shown in Fig. 18.

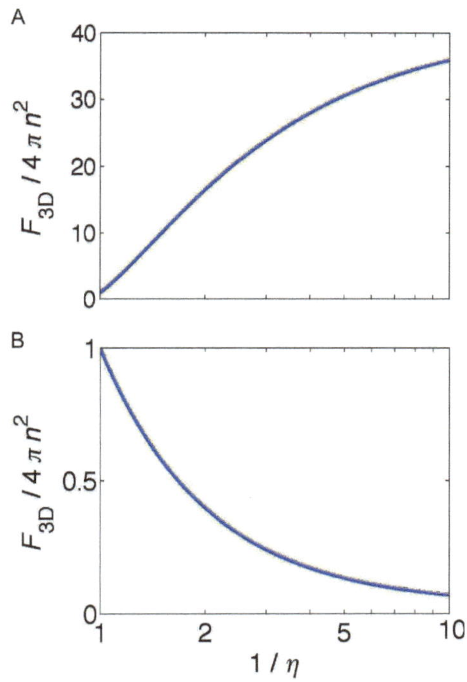

Fig. 17 Upper bounds of 3D angle-integrated absorption enhancement factors by Eq. (26). η is the area fraction of the absorber whose index is n. (A) $\varepsilon_2/\varepsilon_1 = 12$. (B) $\varepsilon_1/\varepsilon_2 = 12$.

Fig. 18 3D photonic crystal. The woodpile structure consists of four alternating layers. Each layer has a thickness of 0.25*a* and consists of a 1D array of infinitely *long square rods* with widths of 0.25*a*, where *a* is the lattice constant. The horizontal locations of the first and the second layers are shifted laterally by 0.5*a* from the horizontal locations of the third and the fourth layers. The lattice orientations of the first and the third layers are orthogonal to the lattice orientations of the second and the fourth layers.

We obtain $\rho_{3D}(\omega)$ by MPB for an infinite and lossless structure, and F_{3D} by S4 for the corresponding finite and lossy structure. We plot the right-hand side expression $(2\pi c)^3\rho_{3D}(\omega)/n\omega^2$ in Eq. (25) as the "Theoretical Bound" curve in Fig. 19, assuming that the overlap factor Γ is equal to the volume fraction η. We plot F_{3D} as the "Absorption Calculation" curve in Fig. 19. The absorption enhancement follows the DOS in the same manner as in the 2D case. We observe that both the theoretical bound and the actual angle-integrated absorption enhancement factor fall significantly below the conventional limit of $4\pi n^2 = 150$ in this case.

To summarize this section, we have considered light trapping in photonic crystals, where the photonic crystals themselves operate as inhomogeneous absorbing media. We show that the DOS of photonic crystals strongly influences the light trapping behaviors. We also note the importance of modal overlap and the need for efficient mode coupling. Our numerical results show that, in a 2D photonic crystal, with the use of van Hove singularities in the DOS, the angle-integrated light trapping absorption enhancement factor can exceed the conventional limit over substantial bandwidths. In a 3D photonic crystal, it is more difficult to use photonic crystals to overcome the conventional limit, due to the weakening of the van Hove singularity and the fact that embedding a high index absorbing material in a low index background reduces the overall DOS away from van Hove singularities.

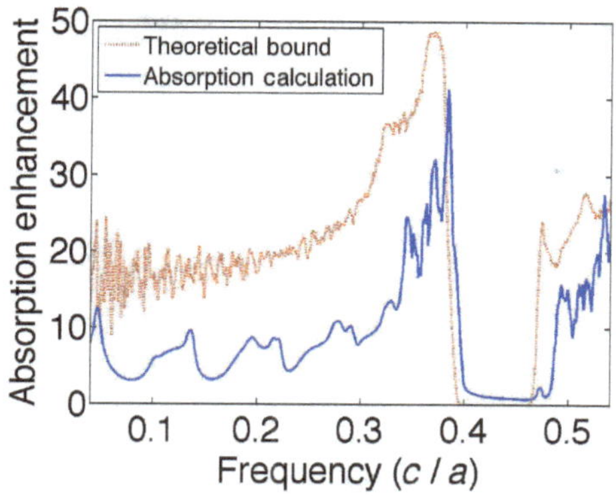

Fig. 19 Comparison of angle-integrated absorption enhancement factor and its theoretical bound calculated by Eq. (25) for the 3D woodpile structure in Fig. 18. The angle-integrated limit $4\pi n^2 = 150$.

6. Absorption enhancement in ultrathin crystalline silicon solar cells with antireflection and light-trapping nanocone gratings

In this section, we apply the light trapping theory to enhance the light absorption in ultrathin-film silicon solar cells for efficiency improvement and cost reduction. We introduce a double-sided grating design, where the front and back surfaces of the cell are separately optimized for antireflection and light trapping, respectively. The optimized structure yields a photocurrent of 34.6 mA/cm^2 at an equivalent thickness of 2 μm, close to the Yablonovitch limit. This approach is applicable to various thicknesses and is robust against metallic loss in the back reflector.

6.1 Motivation

There is significant recent interest in designing ultrathin c-Si solar cells with active layer thickness of a few microns (Biswas and Xu, 2011; Dewan et al., 2011; Fahr et al., 2011; Gjessing et al., 2011; Grandidier et al., 2011; Han and Chen, 2010; Kelzenberg et al., 2010; Li et al., 2010; Mallick et al., 2010; Nagel and Scarpulla, 2010; Paetzold et al., 2011; Park et al., 2009; Sheng et al., 2011; Wang et al., 2010; Zhang et al., 2011). In Fig. 20, we plot the complex dielectric constant of c-Si in comparison with the Air Mass 1.5 (AM 1.5) solar irradiance spectrum. The AM 1.5 spectrum is broadband, hence it is beneficial to increase solar absorption over the entire optical spectrum. The real part of the dielectric constant is much greater than unity, which implies that there is a strong optical impedance mismatch at the air

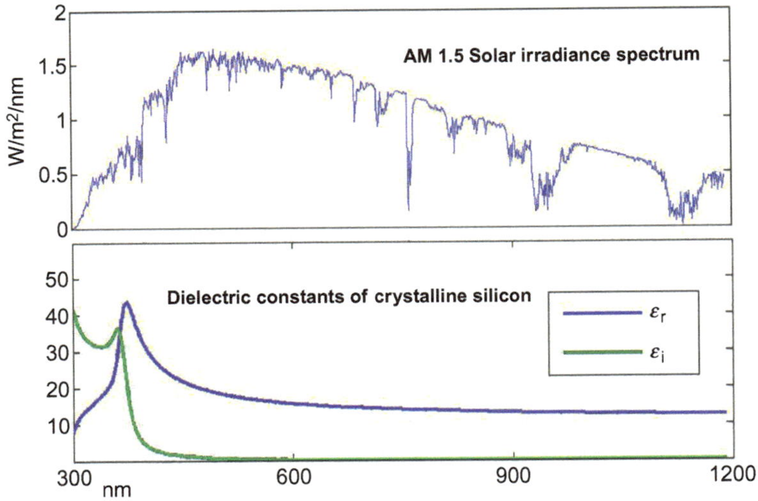

Fig. 20 The AM 1.5 solar irradiance spectrum and the real and imaginary parts of the dielectric constant of c-Si. The *blue curve* is the real part and the *green curve* is the imaginary part.

and c-Si interface, and as a result antireflection is desirable. Close to the c-Si bandedge, the imaginary part of the dielectric constant is negligible, which implies that the absorption depth of infrared photons is extremely large, and as a result light trapping should be performed.

Efficient light absorption in such thin films therefore requires both broadband antireflection coatings and effective light trapping techniques, which often have different design considerations. In this section, we show that by employing a double-sided grating design, we can separately optimize the geometries for antireflection and light trapping purposes to achieve broadband light absorption enhancement. The photocurrent generated by the proposed thin film absorber is able to reach the Yablonovitch limit (Green, 2002; Yablonovitch, 1982; Yablonovitch and Cody, 1982).

We use nanocones as the basic building elements for the grating geometry because of their unique optical properties and compatibility with inexpensive fabrication techniques (Zhu et al., 2009, 2010a,b). The structure we consider, as shown in Fig. 21, contains a c-Si thin film with nanocone gratings also made of c-Si. The circular nanocones form 2D square lattices on both the front and the back surfaces. The film is placed on a mirror. As a starting point we assume the mirror is made of a PEC.

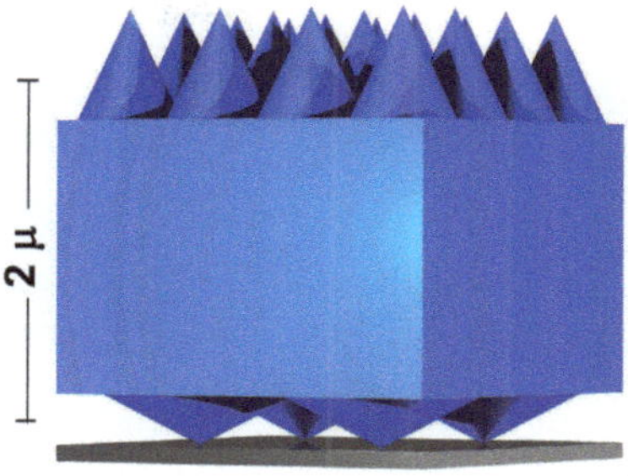

Fig. 21 3D c-Si thin film structures in air. *Blue* represents c-Si, *gray* represents PEC, and *yellow* represents nonabsorbing silicon nitride. The nanocones are made of c-Si, as is the uniform layers, and they are placed in a 2D square lattice on both surfaces of the film.

6.2 Absorption enhancement at Yablonovitch limit

The optimization process is as follows: For a given structure with 2D nanocone gratings, using the rigorous coupled wave analysis (RCWA) (Li, 1997; Liu and Fan, 2012; Tikhodeev et al., 2002), we calculate the absorption spectrum, from which we determine the short circuit current assuming an AM 1.5 incident solar irradiance. In a supercell of period 1000 nm, we optimize the geometry over six parameters—the numbers of primitive cells and the base radii and heights of the nanocones on both sides—for the greatest photocurrent generated from the structure. In the optimization, we adjust those geometrical parameters, as well as the thickness of the uniform layer sandwiched between the top and bottom gratings, while ensuring that the structures always consist of the same amount of silicon as a flat thin film structure with a predetermined thickness. We refer to this thickness as the equivalent thickness of our nanostructured thin film.

Our optimized structure for an equivalent thickness of 2 μm is shown in Fig. 22A. For the top nanocones, the period is 500 nm, the base radius is 250 nm, and the height is 710 nm; for the bottom nanocones, the period is 1000 nm, the base radius is 475 nm, and the height is 330 nm. The thickness of the uniform layer of the thin film is 1736 nm

For this structure, the absorption spectrum is shown in Fig. 23A. This structure generates a short circuit current of 34.6 mA/cm^2 assuming normal incidence from an AM 1.5 solar spectrum.

In Fig. 23, we compare the performance of our structure to both the single-pass absorption and the Yablonovitch limit of 2 μm-thick film. Assuming perfect antireflection but no light trapping, that is, light passing

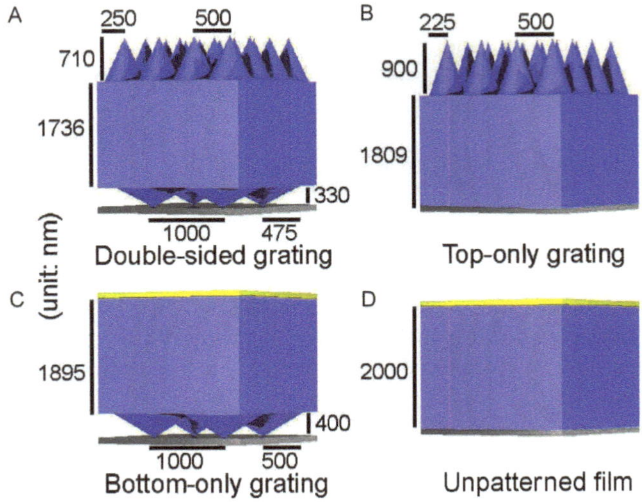

Fig. 22 3D c-Si thin film structures in air. In all panels, *blue* represents silicon, *gray* represents a PEC, and *yellow* represents nonabsorbing silicon nitride. The nanocones are made of silicon, as is the uniform layers, and they are placed in a 2D square lattice either on the front or on the back surface of the film. (A) The optimized double-sided nanostructure, (B) The optimized top-only nanostructure. (C) The optimized bottom-only nanostructure with a thin layer of nonabsorbing silicon nitride on *top*. (D) The flat film with a thin layer of nonabsorbing silicon nitride on *top*.

through the material only once, and assuming normal incidence, the single-pass absorption spectrum (the green curves in Fig. 23) in a thin film with thickness d is given by

$$A_{\text{singlepass}} = 1 - e^{-\alpha d}, \tag{27}$$

where α is the absorption coefficient. Assuming perfect antireflection and perfect light trapping, the absorption spectrum (the red curves in Fig. 23) in a thin film with thickness d is given by the Yablonovitch limit (Green, 2002; Yablonovitch, 1982; Yablonovitch and Cody, 1982)

$$A_{\text{yablonovitch}} = 1 - \frac{1}{1 + 4n^2\alpha d}, \tag{28}$$

where n is the real part of the refractive index. The absorption spectrum of our optimized structure is much higher than the single-pass absorption spectrum given by Eq. (27), and it is very close to the Yablonovitch-limit spectrum given by Eq. (28). The short circuit current of 34.6 mA/cm^2 in our optimized design is very close to 35.5 mA/cm^2, the short-circuit current corresponding to the Yablonovitch limit at the equivalent thickness of 2 μm.

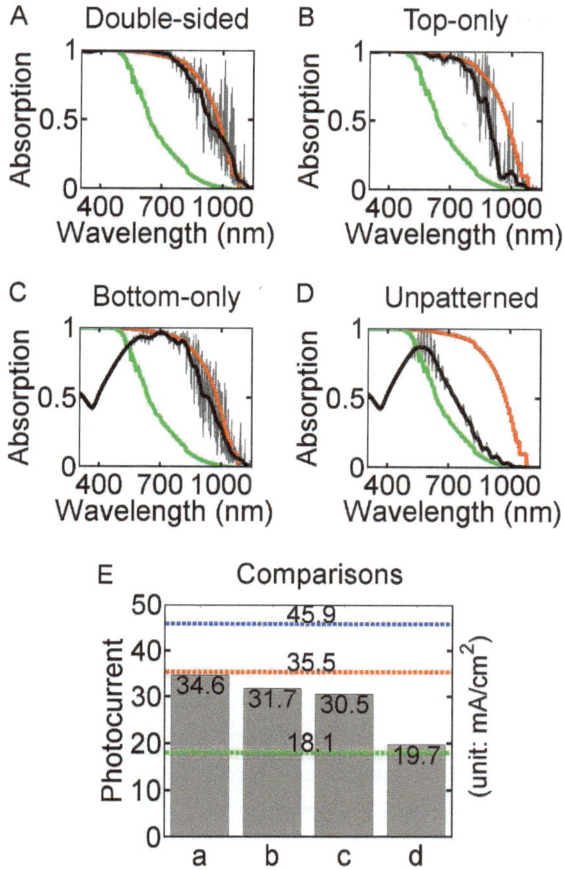

Fig. 23 Absorption spectra under normal incidence from AM 1.5 solar irradiance The *red curves* represent the Yablonovitch limit given by Eq. (28), the *green curves* represent the single-pass absorption spectra given by Eq. (27), and the *thick black curves* are running averages of the absorption spectra for the corresponding structures in Fig. 22. The kinks at the longer wavelength range in the curves are due to the discontinuity of the material constant from the reference book (Palik, 1997). (A) The absorption spectrum of the double-sided structure in Fig. 22A. (B) The absorption spectrum of the "top-only" structure in Fig. 22B. (C) The absorption spectrum of the "bottom-only" structure in Fig. 22C. (D) The absorption spectrum of the flat thin film in Fig. 22D. (E) Comparison of the short-circuit currents generated by the four structures (*gray bars*) the Yablonovitch limit (*red line*), the single-pass absorption (*green line*), and the full absorption (*blue line*).

The device physics for the absorption enhancement strategy in our structure is as follows: The usable solar spectrum for a c-Si cell extends from 300 to 1100 nm. To achieve significant absorption enhancement, one needs broad-band antireflection over the entire usable solar spectrum due to the strong impedance mismatch between silicon and air. One also needs efficient light trapping from roughly 800 to 1100 nm where a silicon cell with an equivalent thickness of a few microns is weakly absorbing. Our strategy is to use the front surface grating for the goal of antireflection, and the back surface grating for the goal of light trapping. Doing so allows us to separately address the different structural requirements in order to achieve these two separate goals. Below, we illustrate this strategy by comparing our optimized structure to optimized "top-only" and "bottom-only" grating structures. All these structures have an equivalent thickness of 2 μm. In the study of these optimized "top-only" or "bottom-only" structures, we will compare their performance to a flat thin film structure with the same equivalent thickness and with a nonabsorbing silicon nitride antireflection coating on top, as shown in Fig. 22D.

The optimized "top-only" structure and its absorption spectrum are shown in Figs. 22B and 23B, respectively. The periodicity of the nanocone grating is 500 nm, and the height is 900 nm, four times the base radius of 225 nm. The planar part of the structure has a thickness of 1809 nm. Comparing the absorption spectrum of such an optimized "top-only" structure (see Fig. 23B) to that of the unpatterned flat thin film in Fig. 23D, we observe substantial absorption enhancement over the entire usable solar spectrum. The short-circuit photocurrent for the "top-only" structure is 31.7 mA/cm^2, compared to 19.7 mA/cm^2 for the flat thin film. From 300 to 700 nm, the absorption curve closely follows the Yablonovitch limit.

The contribution for light absorption enhancement in Fig. 23B, compared to the flat thin film absorption in Fig. 23D, originates mainly from antireflection. Nanocone arrays suppress reflection because the cone geometry provides an averaged, graded index from air to silicon as the radius of its cross section increases from zero to its maximum at the planar film surface. The reflection suppression is broadband since the index-matching is largely independent of wavelength. To achieve effective antireflection, the periodicity of the array has to be in the subwavelength regime for the incoming light to see an effective averaged index. In addition, a high aspect ratio is preferred to provide a smooth index transition from air to silicon. These structural aspects are precisely what we see in the optimized "top-only" structure. However, for longer wavelengths, between 700 to 1100 nm, the absorption of the optimized "top-only" structure falls significantly below the Yablonovitch limit. Therefore, the structural feature of a nanocone that is optimal for antireflection purposes is suboptimal for light trapping. Nevertheless, such a nanocone structure for antireflection has excellent performance in a wide range of light incident angles since the gradual change of refractive index is maintained over this range (Zhu et al., 2009, 2010b).

The optimized "bottom-only" grating structure and its absorption spectrum are shown in Figs. 22C and 23C. The periodicity of the nanocone grating is 1000 nm. For an individual nanocone, the height is 400 nm, and the base radius is 500 nm. The planar part of the structure has a thickness of 1895 nm to ensure the total 2 μm equivalent thickness. We also place a 90 nm thick nonabsorbing silicon nitride antireflection coating on top (Han and Chen, 2010). This structure has a short circuit current of 30.5 mA/cm^2. Comparing the absorption spectrum of the optimized "bottom-only" structure to that of the planar structure in Fig. 22D, we observe significant absorption enhancement beyond 600 nm, where the absorption of the optimized "bottom-only" structure closely follows the Yablonovitch limit. However, in the wavelength range below 600 nm, light absorption is below the Yablonovitch limit and even below the single-pass absorption spectrum due to optical loss from the reflection at the silicon–air interface. The nanocone grating structure on the bottom surface certainly does not contribute to antireflection.

A nanocone structure that is optimal for light trapping, as shown in Fig. 22C, has very different structural characteristics, compared with the optimal nanocone structure for antireflection in Fig. 22B. The spectrum in Fig. 23C shows that the absorption enhancement due to light trapping is primarily in the longer wavelength range where the silicon absorption is weak. Light trapping relies on the excitation of guided resonances (Yu and Fan, 2011; Yu et al., 2010a,b), and therefore a careful choice of periodicity is important. The requirements on periodicity depend on two considerations. On one hand, more guided resonances lead to higher absorption, for which a large period is preferred. On the other hand, each of the guided resonances should not couple and leak to many external channels, that is diffraction directions, for which a small period is preferred. Considering the trade-off between these two requirements, the optimal periodicity for light trapping should be close to the targeted wavelength (Yu et al., 2010a,b). For silicon, light trapping is most important for the 800 to 1100nm wavelength range. We therefore expect that the optimal periodicity is approximately 1000 nm, the wavelength at which the normally incident plane wave is efficiently coupled to guided resonances.

Taking into account the significant difference in the structural requirements for antireflection and light trapping, our optimized design with a double-sided grating structure significantly increases light absorption over the entire solar spectrum and provides a performance that approaches the theoretical limit (see Fig. 23E). Furthermore, the double-sided strategy is a generic approach that can be applied to a range of thicknesses. Fig. 24

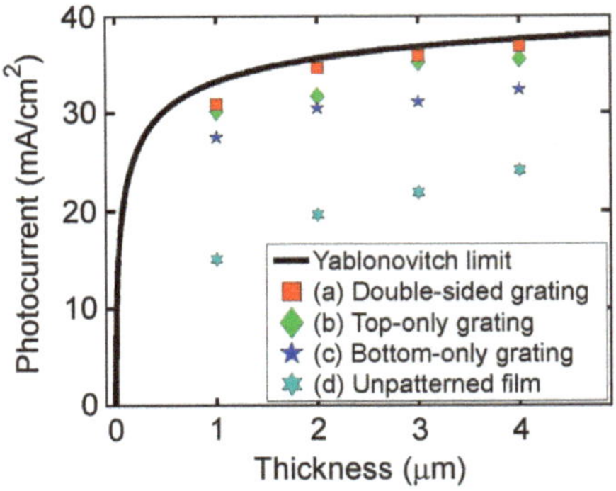

Fig. 24 Photocurrents generated by structures as a function of their equivalent thicknesses. The *black curve* is the Yablonovitch limit calculated by integrating Eq. (28) over the AM 1.5 solar spectrum.

shows that for ultrathin films, our approach can consistently outperform both the "top-only" and "bottom-only" grating designs. The photocurrents from the optimized double-sided grating structures are very close to the theoretical Yablonovitch limit for a range of thicknesses in the few microns range.

To characterize the loss of real mirror, we replace the PEC mirror by a flat silver layer for the optimized structure in Fig. 22A. A thin layer of silicon dioxide is placed between the silicon film and silver as a spacer to reduce metallic loss. We observe only a small reduction in short-circuit photocurrent, from 34.63 to 33.86 mA/cm^2 (see Fig. 25A). Since the grating at the bottom of the silicon layer is sufficiently far from the silver surface, there is no surface plasmon excitation (see Fig. 25B and C). Therefore, our flat silver back-reflector induces no significant local or long-range surface plasmon excitation and its resultant substantial parasitic loss (Biswas and Xu, 2011). When we extend our optimization to include nanocone gratings made of nonabsorbing dielectric materials instead of silicon, we observe similar geometrical configurations and absorption enhancements. Therefore, the strategy has great flexibility in the nanocone grating design regarding material choice for either the grating or the back-reflector.

As a final remark, the doubled-sided structure could be fabricated by applying the Langmuir–Blodgett (LB) assembly method of silica nanospheres and the reactive ion etching (RIE) to each side sequentially (Hsu et al., 2008; Zhu et al., 2009, 2010a). One could use the size of the silica nanoparticles to control the periodicity, and the RIE to control the shapes of the nanocones. This fabrication process could also be applied to

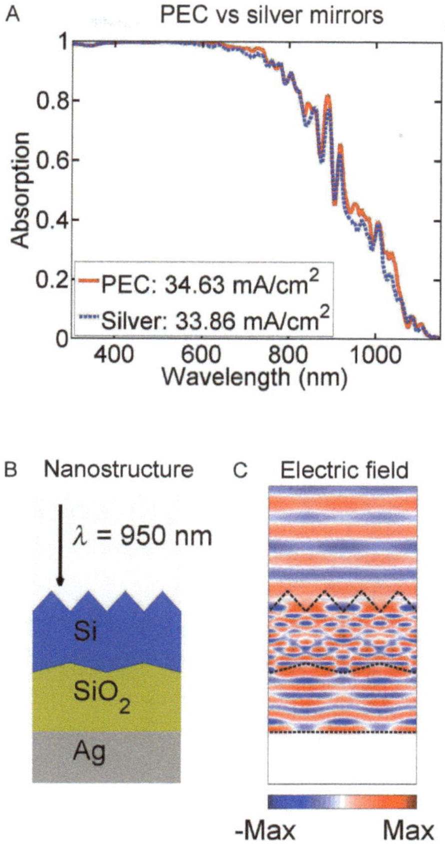

Fig. 25 (A) Comparison of the absorption spectra with PEC and with silver (plus a spacer between the nanostructure and the silver back-reflector). The *red curve* is the absorption spectrum of the structure in Fig. 22A, and the *blue curve* is the absorption spectrum of the same structure, except that the PEC is replaced with real silver (Palik, 1997), and a silicon dioxide layer of thickness 2 μm is placed between silicon and silver as a spacer. (B) and (C) The double-sided grating structure and the electric field profile at the wavelength of 950 nm.

a large scale (Hsu et al., 2008; Zhu et al., 2010a). Since our structure involves only a slight surface modification of a silicon film with nanocones of relatively low aspect ratio, the issues associated with surface recombination and degradation of electronic properties should be less severe than those of other high aspect ratio nanoscale structures. The double-sided nanocone grating design provides an experimentally realistic strategy in efficiency improvement and cost reduction for c-Si solar cells. Although we focus on c-Si, many of the strategies can be applied to other materials (Liang et al., 2012).

7. Conclusion

Light trapping is desirable for weak absorbers because the single-pass absorption is usually too small for a significant percentage of light to be absorbed by the solar absorber. In this chapter, we describe a statistical TCMT formalism to construct a wave optics light trapping theory, which reduces to the conventional light trapping theory in the ray optics regime. We show that the DOS, especially the van Hove singularities, of photonic crystals strongly influences the light trapping behaviors. We validate the wave optics light trapping theory and show that nanophotonic light trapping performance can surpass the conventional benchmarks, such as the Yablonovitch $4n^2$ limit, even after a complete angle integration. We apply light trapping with antireflection to practical photovoltaic cells. Using jointly optimized nanocone gratings on both sides of ultrathin c-Si solar cells, we showed that it is possible to reach the $4n^2$ limit. These results provide a theoretical guidance for the design of optical absorbers that are inhomogeneous in general, including nanostructured photovoltaic and photoelectrochemical cells. We show that to design and experimentally realize high-efficiency photonic crystal solar cells, one should include considerations on the DOS including the effects of the effective medium and van Hove singularities, the overlap factor, the mode coupling, and antireflection designs.

References

Agrawal, M., Peumans, P., 2008. Broadband optical absorption enhancement through coherent light trapping in thin-film photovoltaic cells. Opt. Express 16 (8), 5385–5396.

Almeida, V.R., Xu, Q., Barrios, C.A., Lipson, M., 2004. Guiding and confining light in void nanostructure. Opt. Lett. 29 (11), 1209–1211.

Atwater, H.A., Polman, A., 2010. Plasmonics for improved photovoltaic devices. Nat. Mater. 9, 205–213.

Barnett, S.M., Loudon, R., 1996. Sum rule for modified spontaneous emission rates. Phys. Rev. Lett. 77 (12), 2444–2446.

Bassani, F., Parravicini, G.P., 1975. Electronic States and Optical Transistions in Solids. Pergamon Press.

Battaglia, C., Hsu, C.-M., Söderström, K., Escarré, J., Haug, F.-J., Charrière, M., Boccard, M., Despeisse, M., Alexander, D.T.L., Cantoni, M., Yi, C., Ballif, C., 2012a. Light trapping in solar cells: can periodic beat random? ACS Nano 6 (3), 2790–2797.

Battaglia, C., Boccard, M., Haug, F.-J., Ballif, C., 2012b. Light trapping in solar cells: when does a Lambertian scatterer scatter Lambertianly? J. Appl. Phys. 112, 094504.

Bergman, D.J., 1978. The dielectric constant of a composite material—a problem in classical physics. Phys. Rep. 43 (9), 377–407.

Bergman, D.J., 1980. Exactly solvable microscopic geometries and rigorous bounds for the complex dielectric constant of a two-component composite material. Phys. Rev. Lett. 44, 1285.

Bermel, P., Luo, C., Zeng, L., Kimerling, L.C., Joannopoulos, J.D., 2007. Improving thin-film crystalline silicon solar cell efficiencies with photonic crystals. Opt. Express 15 (25), 16986–17000.

Biswas, R., Xu, C., 2011. Nano-crystalline silicon solar cell architecture with absorption at the classical $4n^2$ limit. Opt. Express 19 (S4), A664–A672.

Bozzola, A., Liscidini, M., Andreani, L.C., 2012. Photonic light-trapping versus Lambertian limits in thin film silicon solar cells with 1D and 2D periodic patterns. Opt. Express 20 (S2), A224–A244.

Brongersma, M.L., Cui, Y., Fan, S., 2014. Light management for photovoltaics using high-index nanostructures. Nat. Mater. 13, 451–460.

Callahan, D.M., Munday, J.N., Atwater, H.A., 2012. Solar cell light trapping beyond the ray optic limit. Nano Lett. 12, 214–218.

Chutinan, A., John, S., 2008. Light trapping and absorption optimization in certain thin-film photonic crystal architectures. Phys. Rev. A 78, 023825.

Chutinan, A., Kherani, N.P., Zukotynski, S., 2009. High-efficiency photonic crystal solar cell architecture. Opt. Express 17 (11), 8871–8878.

Deckman, H.W., Roxlo, C.B., Yablonovitch, E., 1983. Maximum statistical increase of optical absorption in textured semiconductor films. Opt. Lett. 8 (9), 491–493.

Dewan, R., Vasilev, I., Jovanov, V., Knipp, D., 2011. Optical enhancement and losses of pyramid textured thin-film silicon solar cells. J. Appl. Phys. 110, 013101.

Du, Q.G., Kam, C.H., Demir, H.V., Yu, H.Y., Sun, X.W., 2011. Enhanced optical absorption in nanopatterned silicon thin films with a nano-cone-hole structure for photovoltaic applications. Opt. Lett. 36 (9), 1713–1715.

Duché, D., Escoubas, L., Simon, J.-J., Torchio, P., Vervisch, W., Flory, F., 2008. Slow Bloch modes for enhancing the absorption of light in thin films for photovoltaic cells. Appl. Phys. Lett. 92, 193310.

Duché, D., Drouard, E., Simon, J.J., Escoubas, L., Torchio, P., Le Rouzo, J., Vedraine, S., 2011. Light harvesting in organic solar cells. Sol. Energy Mater. Sol. Cells 95, S18–S25.

Eyderman, S., John, S., Deinega, A., 2013. Solar light trapping in slanted conical-pore photonic crystals: beyond statistical ray trapping. J. Appl. Phys. 113, 154315.

Fahr, S., Kirchartz, T., Rockstuhl, C., Lederer, F., 2011. Approaching the Lambertian limit in randomly textured thin-film solar cells. Opt. Express 19 (S4), A865–A874.

Fan, S., Joannopoulos, J.D., 2002. Analysis of guided resonances in photonic crystal slabs. Phys. Rev. B 65, 235112.

Ganapati, V., Miller, O.D., Yablonovitch, E., 2014. Light trapping textures designed by electromagnetic optimization for subwavelength thick solar cells. IEEE J. Photovoltaics 4 (1), 175.

Gaponenko, S.V., 2010. Introduction to Nanophotonics. Cambridge University Press.

Gjessing, J., Sudbø, A.S., Marstein, E.S., 2011. Comparison of periodic light-trapping structures in thin crystalline silicon solar cells. J. Appl. Phys. 110, 033104.

Goetzberger, A., 1981. Optical confinement in thin Si-solar cells by diffuse back reflectors. In: 15th Photovoltaic Specialists Conference. IEEE, pp. 867–870.

Gomard, G., Meng, X., Drouard, E., El Hajjam, K., Gerelli, E., Peretti, R., Fave, A., Orobtchouk, R., Lemiti, M., Seassal, C., 2012. Light harvesting by planar photonic crystals in solar cells: the case of amorphous silicon. J. Opt. 14 (2), 024011.

Gomard, G., Peretti, R., Drouard, E., Meng, X., Seassal, C., 2013. Photonic crystals and optical mode engineering for thin film photovoltaics. Opt. Express 21 (8), A515.

Grandidier, J., Callahan, D.M., Munday, J.N., Atwater, H.A., 2011. Light absorption enhancement in thin-film solar cells using whispering gallery modes in dielectric nanospheres. Adv. Mater. 23, 1272–1276.

Green, M.A., 2002. Lambertian light trapping in textured solar cells and light-emitting diodes: analytical solutions. Prog. Photovolt. Res. Appl. 10 (4), 235–241.
Green, M.A., 2011. Enhanced evanescent mode light trapping in organic solar cells and other low index optoelectronic devices. Prog. Photovolt. Res. Appl. 19 (4), 473–477.
Guldin, S., Hüttner, S., Kolle, M., Welland, M.E., Müller-Buschbaum, P., Friend, R.H., Steiner, U., Tétreault, N., 2010. Dye-sensitized solar cell based on a three-dimensional photonic crystal. Nano Lett. 10 (7), 2303–2309.
Han, S.E., Chen, G., 2010. Toward the Lambertian limit of light trapping in thin nanostructured silicon solar cells. Nano Lett. 10 (11), 4692–4696.
Haug, F.-J., Söderström, K., Naqavi, A., Ballif, C., 2011. Resonances and absorption enhancement in thin film silicon solar cells with periodic interface texture. J. Appl. Phys. 109, 084516.
Heine, C., Morf, R.H., 1995. Submicrometer gratings for solar energy applications. Appl. Opt. 34 (14), 2476–2482.
Ho, K.M., Chan, C.T., Soukoulis, C.M., Biswas, R., Sigalas, M., 1994. Photonic band gaps in three dimensions: new layer-by-layer periodic structures. Solid State Commun. 89 (5), 413–416.
Hoppe, H., Sariciftci, N.S., 2004. Organic solar cells: an overview. J. Mater. Res. 19 (7), 1924–1945.
Hsu, C.-M., Connor, S.T., Tang, M.X., Cui, Y., 2008. Wafer-scale silicon nanopillars and nanocones by Langmuir-Blodgett assembly and etching. Appl. Phys. Lett. 93, 133109.
Huynh, W.U., Dittmer, J.J., Alivisatos, P., 2002. Hybrid nanorod-polymer solar cells. Science 295 (5564), 2425–2427.
Joannopoulos, J.D., Johnson, S.G., Winn, J.N., Meade, R.D., 2008. Photonic Crystals: Molding the Flow of Light, second ed. Princeton University Press.
John, S., 2012. Why trap light? Nat. Mater. 11, 997–999.
Johnson, S.G., Joannopoulos, J.D., 2001. Block-iterative frequency-domain methods for Maxwell's equations in a planewave basis. Opt. Express 8 (3), 173–190.
Kayes, B.M., Atwater, H.A., Lewis, N.S., 2005. Comparison of the device physics principles of planar and radial p-n junction nanorod solar cells. J. Appl. Phys. 97 (11), 114302.
Kelzenberg, M.D., Boettcher, S.W., Petykiewicz, J.A., Turner-Evans, D.B., Putnam, M.C., Warren, E.L., Spurgeon, J.M., Briggs, R.M., Lewis, N.S., Atwater, H.A., 2010. Enhanced absorption and carrier collection in Si wire arrays for photovoltaic applications. Nat. Mater. 9, 239–244.
Ko, D.-H., Tumbleston, J.R., Zhang, L., Williams, S., DeSimone, J.M., Lopez, R., Samulski, E.T., 2009. Photonic crystal geometry for organic solar cells. Nano Lett. 9 (7), 2742–2746.
Kuang, P., Deinega, A., Hsieh, M.-L., John, S., Lin, S.-Y., 2013. Light trapping and near-unity solar absorption in a three-dimensional photonic-crystal. Opt. Lett. 38 (20), 4200–4203.
Law, M., Greene, L.E., Johnson, J.C., Saykally, R., Yang, P., 2005. Nanowire dye-sensitized solar cells. Nat. Mater. 4 (6), 455.
Le, K.Q., John, S., 2014. Synergistic plasmonic and photonic crystal light-trapping: architectures for optical up-conversion in thin-film solar cells. Opt. Express 22 (S1), A1–A12.
Li, L., 1997. New formulation of the Fourier modal method for crossed surface-relief gratings. J. Opt. Soc. Am. A 14 (10), 2758–2767.
Li, J., Yu, H.Y., Wong, S.M., Zhang, G., Lo, G.-Q., Kwong, D.-L., 2010. Si nanocone array optimization on crystalline Si thin films for solar energy harvesting. J. Phys. D. Appl. Phys. 43 (25), 255101.
Liang, D., Huo, Y., Kang, Y., Wang, K.X., Gu, A., Tan, M., Yu, Z., Li, S., Jia, J., Bao, X., Wang, S., Yao, Y., Wong, H.-S.P., Fan, S., Cui, Y., Harris, J.S., 2012. Optical absorption enhancement in freestanding GaAs thin film nanopyramid arrays. Adv. Energy Mater. 2 (10), 1254–1260.

Lin, C., Chen, C., Sharkawy, A., Schneider, G.J., Venkataraman, S., Prather, D.W., 2005. Efficient terahertz coupling lens based on planar photonic crystals on silicon on insulator. Opt. Lett. 30 (11), 1330–1332.

Liu, V., Fan, S., 2011. Efficient computation of equifrequency surfaces and density of states in photonic crystals using Dirichlet-to-Neumann maps. J. Opt. Soc. Am. B 28 (8), 1837.

Liu, V., Fan, S., 2012. S^4: a free electromagnetic solver for layered periodic structures. Comput. Phys. Commun. 183, 2233–2244.

Mallick, S.B., Agrawal, M., Peumans, P., 2010. Optimal light trapping in ultra-thin photonic crystal crystalline silicon solar cells. Opt. Express 18 (6), 5691–5706.

Mallick, S.B., Agrawal, M., Wangperawong, A., Barnard, E.S., Singh, K.K., Visser, R.J., Brongersma, M.L., Peumans, P., 2012. Ultrathin crystalline-silicon solar cells with embedded photonic crystals. Appl. Phys. Lett. 100, 053113.

Mann, S.A., Garnett, E.C., 2013. Extreme light absorption in thin semiconductor films wrapped around metal nanowires. Nano Lett. 13, 3173–3178.

Mayer, A.C., Scully, S.R., Hardin, B.E., Rowell, M.W., McGehee, M.D., 2007. Polymer-based solar cells. Mater. Today 10 (11), 28–33.

McPhedran, R.C., Botten, L.C., McOrist, J., Asatryan, A.A., de Sterke, C.M., Nicorovici, N.A., 2004. Density of states functions for photonic crystals. Phys. Rev. E 69, 016609.

Mihi, A., Míguez, H., 2005. Origin of light-harvesting enhancement in colloidal-photonic-crystal-based dye-sensitized solar cells. J. Phys. Chem. B 109 (33), 15968–15976.

Milton, G.W., 1981. Bounds on the complex permittivity of a two-component composite material. J. Appl. Phys. 52, 5286.

Milton, G.W., 2002. The Theory of Composites. Cambridge University Press.

Mokkapati, S., Catchpole, K.R., 2012. Nanophotonic light trapping in solar cells. J. Appl. Phys. 112, 101101.

Munday, J.N., 2012. The effect of photonic bandgap materials on the shockley-queisser limit. J. Appl. Phys. 112 (6), 064501.

Munday, J.N., Callahan, D.M., Atwater, H.A., 2012. Light trapping beyond the 4n^2 limit in thin waveguides. Appl. Phys. Lett. 100 (121121).

Mutitu, J.G., Shi, S., Chen, C., Creazzo, T., Barnett, A., Honsberg, C., Prather, D.W., 2008. Thin film silicon solar cell design based on photonic crystal and diffractive grating structures. Opt. Express 16 (19), 15238–15248.

Nagel, J.R., Scarpulla, M.A., 2010. Enhanced absorption in optically thin solar cells by scattering from embedded dielectric nanoparticles. Opt. Express 18 (S2), A139–A146

Naqavi, A., Haug, F.-J., Battaglia, C., Herzig, H.P., Ballif, C., 2013a. Light trapping in solar cells at the extreme coupling limit. J. Opt. Soc. Am. B 30 (1), 13.

Naqavi, A., Haug, F.-J., Ballif, C., Scharf, T., Herzig, H.P., 2013b. Limit of light coupling strength in solar cells. Appl. Phys. Lett. 102, 131113.

Naqavi, A., Haug, F.-J., Söderström, K., Battaglia, C., Paeder, V., Scharf, T., Herzig, H.P., Ballif, C., 2014. Angular behavior of the absorption limit in thin film silicon solar cells. Prog. Photovolt. Res. Appl. 22 (11), 1147–1158.

Narasimhan, V.K., Cui, Y., 2013. Nanostructures for photon management in solar cells. Nanophotonics 2 (3), 187–210.

Oskooi, A., Tanaka, Y., Noda, S., 2014. Tandem photonic-crystal thin films surpassing Lambertian light-trapping limit over broad bandwidth and angular range. Appl. Phys. Lett. 104, 091121.

Paetzold, U.W., Moulin, E., Michaelis, D., Böttler, W., Wächter, C., Hagemann, V., Meier, M., Carius, R., Rau, U., 2011. Plasmonic reflection grating back contacts for microcrystalline silicon solar cells. Appl. Phys. Lett. 99, 181105.

Pala, R.A., Liu, J.S.Q., Barnard, E.S., Askarov, D., Garnett, E.C., Fan, S., Brongersma, M.L., 2013. Optimization of non-periodic plasmonic light-trapping layers for thin-film solar cells. Nat. Commun. 4, 2095.

Palik, E.D. (Ed.), 1997. Handbook of Optical Constants of Solids. Elsevier Inc.
Park, Y., Drouard, E., El Daif, O., Letartre, X., Viktorovitch, P., Fave, A., Kaminski, A., Lemiti, M., Seassal, C., 2009. Absorption enhancement using photonic crystals for silicon thin film solar cells. Opt. Express 17 (16), 14312–14321.
Peretti, R., Gomard, G., Seassal, C., Letartre, X., Drouard, E., 2012. Modal approach for tailoring the absorption in a photonic crystal membrane. J. Appl. Phys. 111, 123114.
Polman, A., Atwater, H.A., 2012. Photonic design principles for ultrahigh-efficiency photovoltaics. Nat. Mater. 11, 174–177.
Raut, H.K., Ganesh, V.A., Nair, A.S., Ramakrishna, S., 2011. Anti-reflective coatings: a critical, in-depth review. Energy Environ. Sci. 4, 3779.
Saeta, P.N., Ferry, V.E., Pacifici, D., Munday, J.N., Atwater, H.A., 2009. How much can guided modes enhance absorption in thin solar cells? Opt. Express 17 (23), 20975–20990.
Sandhu, S., Yu, Z., Fan, S., 2013. Detailed balance analysis of nanophotonic solar cells. Opt. Express 21 (1), 1209–1217.
Scheel, S., 2008. Sum rule for local densities of states in absorbing dielectrics. Phys. Rev. A 78 (013841).
Sheng, P., Bloch, A.N., Stepleman, R.S., 1983. Wavelength-selective absorption enhancement in thin-film solar cells. Appl. Phys. Lett. 43 (6), 579.
Sheng, X., Johnson, S.G., Michel, J., Kimerling, L.C., 2011. Optimization-based design of surface textures for thin-film Si solar cells. Opt. Express 19 (S4), A841–A850.
Sözüer, H.S., Dowling, J.P., 1994. Photonic band calculations for woodpile structures. J. Mod. Opt. 41 (2), 231–239.
Stuart, H.R., Hall, D.G., 1997. Thermodynamic limit to light trapping in thin planar structures. J. Opt. Soc. Am. A 14 (11), 3001–3008.
Tikhodeev, S.G., Yablonskii, A.L., Muljarov, E.A., Gippius, N.A., Ishihara, T., 2002. Quasiguided modes and optical properties of photonic crystal slabs. Phys. Rev. B 66, 045102.
Tobias, I., Luque, A., Marti, A., 2008. Light intensity enhancement by diffracting structures in solar cells. J. Appl. Phys. 104, (3), 034502.
Tsakalakos, L., Balch, J., Fronheiser, J., Shih, M.-Y., LeBoeuf, S.F., Pietrzykowski, M., Codella, P.J., Korevaar, B.A., Sulima, O.V., Rand, J., Davuluru, A., Rapol, U., 2007. Strong broadband optical absorption in silicon nanowire films. J. Nanophotonics 1 (1), 013552.
Tumbleston, J.R., Ko, D.-H., Samulski, E.T., Lopez, R., 2009a. Electrophotonic enhancement of bulk heterojunction organic solar cells through photonic crystal photoactive layer. Appl. Phys. Lett. 94, 043305.
Tumbleston, J.R., Ko, D.-H., Samulski, E.T., Lopez, R., 2009b. Absorption and quasiguided mode analysis of organic solar cells with photonic crystal photoactive layers. Opt. Express 17 (9), 7670–7681.
Van Hove, L., 1953. The occurence of singularities in the elastic frequency distribution of a crystal. Phys. Rev. 89, 1189.
Vynck, K., Burresi, M., Riboli, F., Wiersma, D.S., 2012. Photon management in two-dimensional disordered media. Nat. Mater. 11, 1017.
Wang, P., Menon, R., 2014. Optimization of generalized dielectric nanostructures for enhanced light trapping in thin-film photovoltaics via boosting the local density of optical states. Opt. Express 22 (S1), A99–A110.
Wang, W., Wu, S., Reinhardt, K., Lu, Y., Chen, S., 2010. Broadband light absorption enhancement in thin-film silicon solar cells. Nano Lett. 10 (6), 2012–2018.
Wang, K.X., Yu, Z., Liu, V., Cui, Y., Fan, S., 2012. Absorption enhancement in ultrathin crystalline silicon solar cells with antireflection and light-trapping nanocone gratings. Nano Lett. 12 (3), 1616–1619.

Wang, S., Weil, B.D., Li, Y., Wang, K.X., Garnett, E., Fan, S., Yi, C., 2013a. Large-area free-standing ultrathin single-crystal silicon as processable materials. Nano Lett. 13 (9), 4393–4398.

Wang, C., Yu, S., Chen, W., Sun, C., 2013b. Highly efficient light-trapping structure design inspired by natural evolution. Sci. Rep. 3, 1025.

Wang, K.X., Yu, Z., Liu, V., Raman, A., Cui, Y., Fan, S., 2014a. Light trapping in photonic crystals. Energy Environ. Sci. 7, 2725–2738.

Wang, K.X., Zongfu, Y., Liu, V., Brongersma, M.L., Jaramillo, T.F., Fan, S., 2014b. Nearly total solar absorption in ultrathin nanostructured iron oxide for efficient photo-electrochemical water splitting. ACS Photonics 1 (3), 235–240.

Wiesendanger, S., Zilk, M., Pertsch, T., Rockstuhl, C., Lederer, F., 2013. Combining randomly textured surfaces and photonic crystals for the photon management in thin film microcrystalline silicon solar cells. Opt. Express 21 (103), A450–A459.

Wong, B.M., Morales, A.M., 2009. Enhanced photocurrent efficiency of a carbon nanotube p-n junction electromagnetically coupld to a photonic structure. J. Phys. D. Appl. Phys. 42, 055111.

Yablonovitch, E., 1982. Statistical ray optics. J. Opt. Soc. Am. 72 (7), 899–907.

Yablonovitch, E., Cody, G.D., 1982. Intensity enhancement in textured optical sheets for solar cells. IEEE Trans. Electron Devices ED-29 (2), 300.

Yu, Z., Fan, S., 2011. Angular constraint on light-trapping absorption enhancement in solar cells. Appl. Phys. Lett. 98, 011106.

Yu, G., Gao, J., Hummelen, J.C., Wudl, F., JHeeger, A., 1995. Polymer photovoltaic cells: enhanced efficiencies via a network of internal donor-acceptor heterojunctions. Science 270 (5243), 1789–1791.

Yu, Z., Raman, A., Fan, S., 2010a. Fundamental limit of light trapping in solar cells. Proc. Natl. Acad. Sci. U. S. A. 107 (41), 17491–17496.

Yu, Z., Raman, A., Fan, S., 2010b. Fundamental limit of light trapping in grating structures. Opt. Express 18 (S3), A366.

Yu, Z., Raman, A., Fan, S., 2012. Thermodynamic upper bound on broadband light coupling with photonic structures. Phys. Rev. Lett. 109, 173901.

Yu, Z., Sandhu, S., Fan, S., 2014. Efficiency above the Shockley-Queisser limit by using nanophotonic effects to create multiple effective bandgaps with a single semiconductor. Nano Lett. 14 (1), 66–70.

Zeman, M., Isabella, O., Solntsev, S., Jäger, K., 2013. Modeling of thin-film silicon solar cells. Sol. Energy Mater. Sol. Cells 119, 94–111.

Zhang, R.Y., Shao, B., Dong, J.R., Zhang, J.C., Yang, H., 2011. Absorption enhancement analysis of crystallline Si thin film solar cells based on broadband antireflection nanocone grating. J. Appl. Phys. 110, 113105.

Zhu, J., Yu, Z., Burkhard, G.F., Hsu, C.-M., Connor, S.T., Xu, Y., Wang, Q., McGehee, M., Fan, S., Cui, Y., 2009. Optical absorption enhancement in amorphous silicon nanowire and nanocone arrays. Nano Lett. 9 (1), 279–282.

Zhu, J., Yu, Z., Fan, S., Cui, Y., 2010a. Nanostructured photon management for high performance solar cells. Mater. Sci. Eng. R 70, 330–340.

Zhu, J., Hsu, C.-M., Yu, Z., Fan, S., Cui, Y., 2010b. Nanodome solar cells with efficient light management and self-cleaning. Nano Lett. 10 (6), 1979–1984.

CHAPTER FOUR

Optical image processing using photonic crystal slab

Cheng Guo[a], Shanhui Fan[b,*]

[a]Department of Applied Physics, Stanford University, Stanford, CA, United States
[b]Ginzton Laboratory and Department of Electrical Engineering, Stanford University, Stanford, CA, United States
*Corresponding author: e-mail address: shanhui@stanford.edu

Contents

1. Introduction

In this chapter we will discuss the use of photonic crystal slab to accomplish a number of imaging processing tasks, including edge detection, image smoothing, white noise suppression, and suppression or extraction of periodic features. All these tasks involve filtering in the wavevector domain. These tasks are of broad applications ranging from microscopy and medical imaging to industrial inspection and object detection (Abràmoff et al., 2004; Brosnan and Sun, 2004; Dalal and Triggs, 2005; Gonzalez and Woods, 2008; Markham et al., 1963).

All these image processing operations, as mentioned above, can certainly be implemented using conventional electronic computing. However, nowadays there are many big-data applications that require real-time and high-throughput image processing, for which it is of interest to develop

Semiconductors and Semimetals, Volume 100
ISSN 0080-8784
https://doi.org/10.1016/bs.semsem.2019.02.001

techniques to perform these computations with reduced energy consumption and increased speed. Optical image processing is particularly suitable for this purpose, since the images are naturally in the optical domain. Moreover, many of these tasks can be achieved using linear optical devices that have no power consumption except the propagation loss of light through the device, and have very low latency since the latency is determined only by the time for light to propagate through these devices.

Traditionally, optical image processing can be performed using the techniques of Fourier optics. For instance, to achieve filtering in the wavevector domain, one first obtains a Fourier transformation of an image on a Fourier plane by passing the image through a lens. One then performs a spatial filtering on the Fourier plane, followed by a Fourier transformation again through a second lens (Goodman, 2005; Yu et al., 2001). Such technique requires long propagation distance and therefore results in a bulky system (Chang et al., 1990; Eu et al., 1973; Görlitz and Lanzl, 1975; Sirohi and Mohan, 1977).

Thanks to the recent progress in nanophotonics, optical image processing can instead be achieved by compact photonic devices (AbdollahRamezani et al., 2015; Bykov et al., 2014, 2018; Chizari et al., 2016; Dong et al., 2018; Fang and Ruan, 2018; Fang et al., 2017; Golovastikov et al., 2014; Guo et al., 2018b; Hwang et al., 2018; Kwon et al., 2018; Nejad and Fleury, 2018; Pors et al., 2015; Roberts et al., 2018; Silva et al., 2014; Solli and Jalali, 2015; Wu et al., 2017; Youssefi et al., 2016; Zhu et al., 2017). In this chapter, we provide a brief review of optical image processing using the guided resonance in a two-dimensional photonic crystal slab device. In particular, we focus on a particular class of photonic crystal slab structure where the guided resonance near the Γ point exhibits an isotropic band structure. Depending on the light frequency and the choice of transmission or reflection mode, this compact device realizes high-pass, low-pass, band-reject, and band-pass filtering directly in the optical domain. These filtering operations are applied isotropically in the two-dimensional plane, which is particularly attractive for image processing.

The rest of this chapter is organized as follows. In Section 2 we elucidate the underlying physics mechanism of filtering in the wavevector domain using guided resonances. This mechanism underlies the operation of many compact nanophotonic devices for image filtering, including ours. In Section 3 we theoretically analyze how to realize isotropic filters by a photonic crystal slab device using guided resonances with carefully designed isotropic band structures. In Section 4 we numerically demonstrate the isotropic filtering functions of the device. We conclude in Section 5.

2. Filtering in the wavevector domain using guided resonances

We consider a filter in the wavevector domain as characterized by the filter transfer function $H(k_x, k_y)$ (Bracewell, 1986). In general, for a near-normally incident light beam along the z axis with a transverse field profile $S_{in}(x, y)$, upon the operation of the filter, the output beam has a profile $S_{out}(x,y)=\mathcal{F}^{-1}[H(k_x,k_y)\tilde{S}_{in}(k_x,k_y)]$, where $\mathcal{F}^{-1}$ denotes the inverse Fourier transform and $\tilde{S}_{in}(k_x,k_y)$ is the Fourier transform of the input image (Gonzalez and Woods, 2008). If $H(k_x, k_y) = H(|\boldsymbol{k}|)$, i.e., H only depends on the magnitude of the wavevector $|\boldsymbol{k}|$, the filter is isotropic. Here $\boldsymbol{k} = (k_x, k_y)$ refers to the in-plane wavevector.

Filtering in the wavevector domain can be realized using guided resonances hosted by compact photonic device. In general, guided resonances can induce sharp Fano resonance features in the transmission and reflection spectra (Fan and Joannopoulos, 2002; Ochiai and Sakoda, 2001; Tikhodeev et al., 2002; Zhou et al., 2014). Since the band structures of the guided resonances are usually dispersive, i.e., the resonant frequency is wavevector dependent, the transmission and reflection spectra are also wavevector dependent. Filtering in the wavevector domain can thus be realized in both transmission and reflection modes at a specific frequency.

As an illustration, we first show how to use guided resonances to realize second-order differentiation, a specific type of high-pass filter, at transmission along the direction normal to the slab (Guo et al., 2018b). The task is to design an optical system with a transfer function in the wavevector space ($\boldsymbol{k}$-space) of the form $t(\boldsymbol{k}) = \alpha|\boldsymbol{k}|^2$. For one-dimensional differentiation, $\boldsymbol{k} = k_x$; for two-dimensional differentiation, $\boldsymbol{k} = (k_x, k_y)$. We consider the transmission coefficient of a photonic structure which supports a single band of guided resonances as characterized by $\boldsymbol{k}$-dependent resonant frequencies $\omega(\boldsymbol{k})$ and radiative linewidths $\gamma(\boldsymbol{k})$ (Fig. 1A). Near the resonant frequencies, the transmitted amplitude t is expressed as (Fan and Joannopoulos, 2002)

$$t(\omega,\boldsymbol{k}) = t_d + f\frac{\gamma(\boldsymbol{k})}{i[\omega-\omega(\boldsymbol{k})]+\gamma(\boldsymbol{k})}, \tag{1}$$

where ω is the incident light frequency, t_d is the direct transmission coefficient, and f is related to the complex decaying amplitude of the resonance to the transmission side of the slab.

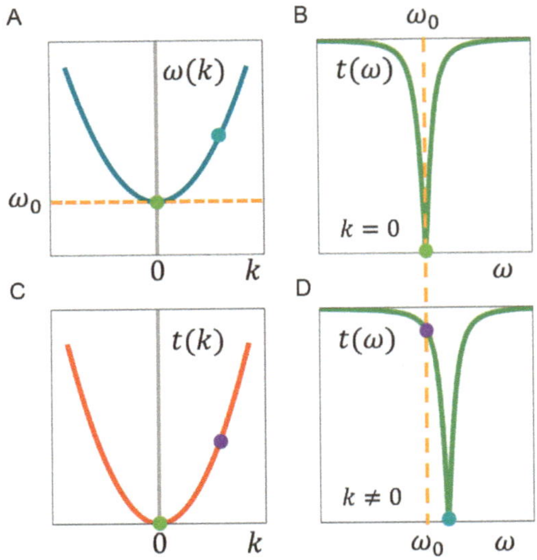

Fig. 1 Operating principle of filtering in the wavevector domain using guided resonances. (A) Dispersion for a single band of guided resonance $\omega(k)$. (B) Transmission spectrum $t(\omega)$ at $k = 0$. (C) k-dependent transmission coefficient $t(k)$ at the resonant frequency at $k = 0$: $\omega = \omega_0$. (D) Transmission spectrum $t(\omega)$ at $k \neq 0$.

In general, f is constrained by the direct process if the system satisfies energy conservation and time-reversal symmetry (Fan and Joannopoulos, 2002; Wang et al., 2013). In particular, if $t_d = 1$, that is, the direct pathway has a 100% transmission coefficient, then

$$f = -t_d = -1 \tag{2}$$

as has been derived in Wang et al. (2013). In this special case,

$$t(\omega, \boldsymbol{k}) = 1 - \frac{\gamma(\boldsymbol{k})}{i[\omega - \omega(\boldsymbol{k})] + \gamma(\boldsymbol{k})}. \tag{3}$$

We plot the spectra of transmission coefficient amplitude $t(\omega)$ at $k = 0$ and $k \neq 0$ in Fig. 1B and D, respectively. Using Eq. (3), zero transmission occurs at the $\boldsymbol{k}$-dependent resonant frequency:

$$\omega = \omega(\boldsymbol{k}). \tag{4}$$

We denote $\omega_0 = \omega(\boldsymbol{k} = \boldsymbol{0})$ and $\gamma_0 = \gamma(\boldsymbol{k} = \boldsymbol{0})$. If we fix the frequency $\omega = \omega_0$, the transmission coefficient $t(\omega_0, \boldsymbol{k})$ is then $\boldsymbol{k}$-dependent. We can perform an expansion of t near $\boldsymbol{k} = \boldsymbol{0}$:

$$t(\omega_0, \boldsymbol{k}) = 0 + \left.\frac{\partial t}{\partial \omega(\boldsymbol{k})}\right|_{\Gamma} \delta\omega(\boldsymbol{k}) + \left.\frac{\partial t}{\partial \gamma(\boldsymbol{k})}\right|_{\Gamma} \delta\gamma(\boldsymbol{k}), \tag{5}$$

where

$$\delta\omega(\boldsymbol{k}) = \omega(\boldsymbol{k}) - \omega_0, \quad \delta\gamma(\boldsymbol{k}) = \gamma(\boldsymbol{k}) - \gamma_0, \tag{6}$$

$$\left.\frac{\partial t}{\partial \omega(\boldsymbol{k})}\right|_{\Gamma} = -\frac{i}{\gamma_0}, \quad \left.\frac{\partial t}{\partial \gamma(\boldsymbol{k})}\right|_{\Gamma} = 0, \tag{7}$$

therefore,

$$t(\omega_0, \boldsymbol{k}) = -\frac{i}{\gamma_0}\delta\omega(\boldsymbol{k}). \tag{8}$$

In this special case, $t(\boldsymbol{k})$ is simply proportional to the band dispersion $\delta\omega(\boldsymbol{k})$ near $\boldsymbol{k} = \boldsymbol{0}$. If $\delta\omega(\boldsymbol{k}) \propto |\boldsymbol{k}|^2$, then $t(\boldsymbol{k}) \propto |\boldsymbol{k}|^2$ as well.

The analysis above indicates that to use guided resonances to perform second-order differentiation, it is sufficient to satisfy the following three conditions:

1. $t_d = 1$.
2. Only one guided resonance band is coupled.
3. The band satisfies the dispersion $(\omega(\boldsymbol{k}) - \omega_0) \propto |\boldsymbol{k}|^2$.

While the discussion above is specific to differentiation, it is relevant for understanding of some of the other filter functionalities as well. Based on the discussion here, in the next section we will illustrate how to use a photonic crystal slab to achieve these filter functionalities.

3. Theoretical analysis of isotropic filters using a photonic crystal slab

In this section, based on the discussion above, we theoretically analyze how to realize isotropic filters using a photonic crystal slab device. Depending on the light frequency and the choice of transmission or reflection mode, this compact device realizes isotropic high-pass, low-pass, band-reject, and band-pass filtering directly in the optical domain (Guo et al., 2018a).

The device is shown in Fig. 2. The dielectric constant of the material for the slabs is $\epsilon = 4.67$, which approximates that of Si_3N_4 in the visible wavelength range. Here we use a dielectric constant that is real, as the extinction coefficient of Si_3N_4 at visible wavelength is negligible. The photonic crystal slab has a thickness of $d = 0.50a$ and a square lattice of air holes with

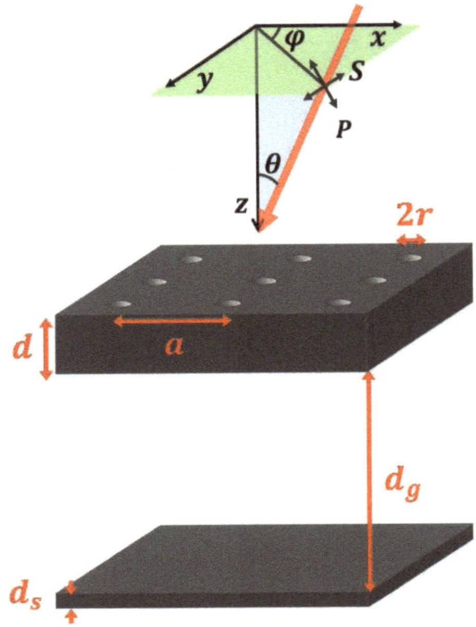

Fig. 2 Geometry of the device, which consists of a photonic crystal slab with a square lattice of air holes, separated from a uniform dielectric slab by an air gap. The slabs have a dielectric constant $\epsilon = 4.67$. The geometry parameters are: $d = 0.50a$, $r = 0.11a$, $d_s = 0.12a$, $d_g = 1.93a$, where a is the lattice constant. The coordinate system is shown above the device. The *red arrow* indicates the direction of the incident light. The electric field directions of the *S* and *P*-polarized light are also indicated. *Reprinted with permission from Guo, C., Xiao, M., Minkov, M., Shi, Y., Fan, S., 2018a. Isotropic wavevector domain image filters by a photonic crystal slab device. J. Opt. Soc. Am. A Opt. Image Sci. Vis. 35 (10), 1685–1691, the Optical Society.*

radii $r = 0.11a$, where a is the lattice constant. The thickness d and radius r are chosen to realize isotropic bands of guided resonances. A uniform dielectric slab with a thickness $d_s = 0.12a$ is placed in the vicinity of the photonic crystal slab. The air gap between the two slabs has a thickness $d_g = 1.93a$. As a side note, the two-layer structure we use in the chapter is quite compact. For a resonant wavelength $\lambda = 500$ nm and $a = 650$ nm the total thickness is $d + d_g + d_s = 1.66$ μm. Such structure can be fabricated using focused ion beam-assisted laser interference lithography (Vogelaar et al., 2001).

To satisfy the first criterion in Section 2, we note that the direct transmission coefficient t_d is related to the nonresonant transmission pathway (Ochiai and Sakoda, 2001). Hence, it is possible to realize $t_d = 1$ by changing the thickness of the slab. In the structure as shown in Fig. 2, we achieve $t_d = 1$ by placing a uniform dielectric slab in the vicinity of the photon crystal slab, and by tuning the distance between the slabs. This has the advantage that we can tune t_d without significantly affecting the band structure of the photonic crystal slab.

To satisfy the second and third criterion in Section 2, one will need to design the band structure of the guided resonances. The design here is in fact quite nontrivial due to the vectorial nature of electromagnetic waves. Since the required filter transfer function $H(\boldsymbol{k})$ is isotropic in $\boldsymbol{k}$-space, it is natural to consider a photonic crystal slab structure that possesses rotational symmetry. As an illustration, here we consider a slab structure with a square lattice of air holes that has C_{4v} symmetry. For concreteness, we consider circular holes; the same analysis will apply to other hole shapes that preserve C_{4v} symmetry. For such a slab, it is known that at the Γ point, which corresponds to $|\boldsymbol{k}| = 0$, the only modes that can couple to external plane waves must be twofold degenerate, belonging to the two-dimensional irreducible representation of the C_{4v} group (Fan and Joannopoulos, 2002; Ochiai and Sakoda, 2001). Near such modes, in the vicinity of the Γ point, from the $k \cdot p$ analysis, the band structure in general can be described by the following 2×2 effective Hamiltonian. (As we consider only small wavevector range near the Γ point, it is sufficient to include only the two modes at the Γ point in our $\boldsymbol{k} \cdot \boldsymbol{p}$ analysis.)

$$\hat{\mathcal{H}}(\boldsymbol{k}) = (\omega_0 - i\gamma_0 + a|\boldsymbol{k}|^2)\hat{\mathbf{I}} + b(k_x^2 - k_y^2)\hat{\boldsymbol{\sigma}}_z + ck_xk_y\hat{\boldsymbol{\sigma}}_x, \tag{9}$$

where a, b, c are three complex coefficients and the $\boldsymbol{\sigma}$'s are the Pauli matrices. This Hamiltonian has two eigenvalues of

$$\omega_\pm(\boldsymbol{k}) - i\gamma_\pm(\boldsymbol{k}) = \omega_0 - i\gamma_0 + a|\boldsymbol{k}|^2 \pm \sqrt{b^2|\boldsymbol{k}|^4 + (c^2 - 4b^2)k_x^2k_y^2}. \tag{10}$$

For a general choice of parameters b and c, the band structure as described by Eq. (10) is not isotropic. On the other hand, Eq. (10) also indicates that when $c = \pm 2b$, both bands will be isotropic:

$$\omega_\pm(\boldsymbol{k}) - i\gamma_\pm(\boldsymbol{k}) = \omega_0 - i\gamma_0 + (a \pm b)|\boldsymbol{k}|^2. \tag{11}$$

To achieve such an isotropic band structure requires detailed tuning of the parameters. Remarkably, with the carefully tuned geometry parameters as above, both bands of the guided resonances that are degenerate at Γ with a frequency $\omega_0 = 0.77098 \times 2\pi c/a$ exhibit almost complete circular symmetry in the real part of the eigenfrequencies:

$$\omega_\pm(\boldsymbol{k}) \approx \omega_0 + A_\pm|\boldsymbol{k}|^2, \tag{12}$$

where $A_+ = 4.35$, $A_- = -1.41$, and the upper (lower) sign corresponds to the upper (lower) band. In Eq. (12), and throughout the rest of the chapter, the wavevector will be in the unit of $2\pi/a$.

The nearly isotropic photonic band structure $(\omega_\pm(\boldsymbol{k}) - \omega_0)$ for the structure shown in Fig. 2 is plotted in Fig. 3A, C, and D. Fig. 3A shows that for both bands, the dispersions along the Γ–M and Γ–X directions have almost identical effective masses. Here the effective mass tensor is defined as $m_{ij}^* = \hbar[\partial^2\omega(\boldsymbol{k})/\partial k_i \partial k_j]^{-1}$. Fig. 3C and D shows that the constant frequency contours for both bands are almost circular. We note that the radiative linewidths $\gamma_\pm(\boldsymbol{k})$, unlike $\omega_\pm(\boldsymbol{k})$, are anisotropic. Nonetheless, $(\gamma_\pm(\boldsymbol{k}) - \gamma_0)$ are much smaller than $(\omega_\pm(\boldsymbol{k}) - \omega_0)$, thus do not affect the circular symmetry of the transfer functions much, as we will show later.

As we mentioned above, for a structure with C_{4v} symmetry, the modes that a normally incident plane wave can couple to at the Γ point are always two-fold degenerate. Consequently, near the Γ point, there are always two bands of guided resonances present. Moreover, off the normal direction, if the direction of incident waves is away from the high symmetry planes, both S and P polarized lights may couple to both bands, leading to complicated polarization conversion effects. Thus, in general, in addition to having an anisotropic band structure near Γ, a photonic crystal slab structure also does not satisfy the criterion 2 above in Section 2 regarding the excitation of a single guided resonance band. Remarkably, however, below we show that once the condition for isotropic bands is satisfied, each of the two bands in fact only couples to one single polarization, *for every direction of incidence.* Below, we refer to this effect, where each polarization excites only a single band away from normal incidence, as the effect of *single-band excitation*. Due to the single-band excitation effect, S and P polarizations are decoupled in this special case.

To illustrate the effect of single-band excitation for every direction when the bands are isotropic, we first show that single-band excitation always occurs when the incident direction is in a mirror symmetry plane of the structure. Using the group theory notation in Sakoda (2005), as our system possesses C_{4v} symmetry, the doubly degenerate states at Γ are E modes. Along the Γ–X direction, this pair of E modes splits into two singly degenerate states of A and B modes which are even and odd, respectively, with respect to the reflection operator associated with the mirror plane $y = 0$. On the other hand, since the S and P polarized lights are odd and even with respect to the same mirror plane, respectively, along the Γ–X direction, the $S(P)$ polarized light can only couple to the $B(A)$ mode. Thus, in general, we

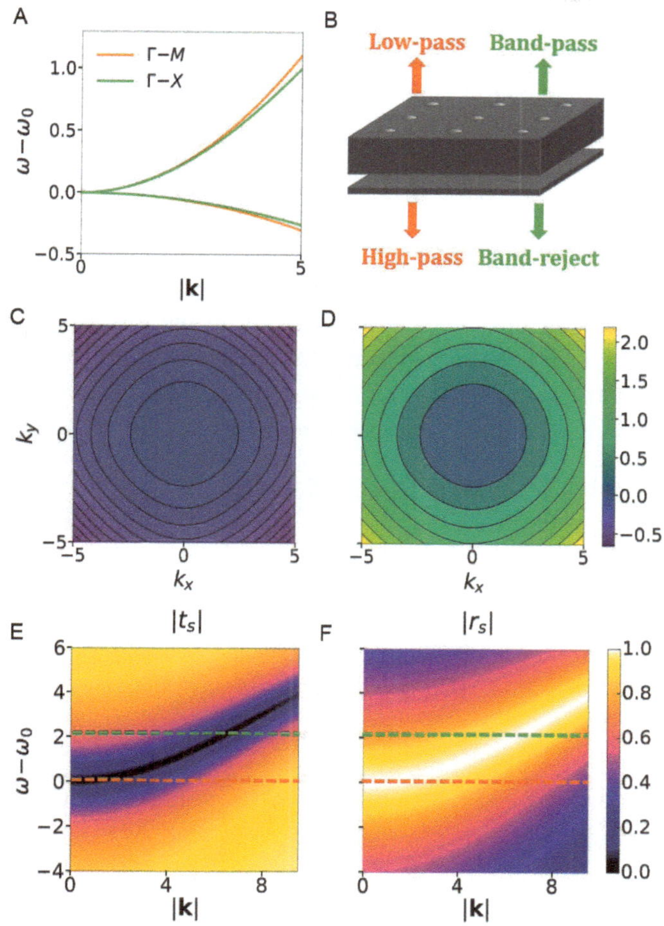

Fig. 3 (A, C, D) Nearly isotropic photonic band structure of the photonic crystal slab shown in Fig. 1 near the frequency $\omega_0 = 0.77098 \times 2\pi c/a$. (A) Band dispersions along Γ–X and Γ–M directions. (C) Constant frequency contours of the lower band. (D) Constant frequency contours of the upper band. (B) Scheme of multiple filtering functions of the device. When the light frequency is on resonance at the normal incidence ($\omega = \omega_0$, labeled by *red arrows* in (B) and *red horizontal lines* in (E,F)), the device realizes isotropic high-pass filter (Laplacian) at transmission and low-pass filter at reflection. When the light frequency is detuned from the resonance at the normal incidence ($\omega = \omega_k$, labeled by *green arrows* in (B) and *green horizontal lines* in (E,F)), the device realizes isotropic band-reject filter at transmission and band-pass filter at reflection. (E) Transmittance $|t|$ for S polarized light as a function of ω and $|\boldsymbol{k}|$ near $\omega_0 = 0.77098 \times 2\pi c/a$ along a general wavevector direction ($\varphi = 14°$). Due to the isotropic band structure, S light only excites the upper band, and the transmission spectra are almost identical along any wavevector direction φ. $|t| = 0$ when $\omega = \omega(\boldsymbol{k})$. (F) Reflectance $|r|$ for S polarized light as a function of ω and $|\boldsymbol{k}|$ near $\omega_0 = 0.77098 \times 2\pi c/a$ along a general wavevector direction ($\varphi = 14°$). The reflection spectra are identical along any wavevector direction. $|r| = 1$ when $\omega = \omega(\boldsymbol{k})$ In all the plots, frequency $(\omega - \omega_0)$ is in the units of $10^{-4} \times 2\pi c/a$, while $|\boldsymbol{k}|$, k_x, and k_y are in the units of $10^{-3} \times 2\pi/a$. *Reprinted with permission from Guo, C., Xiao, M., Minkov, M., Shi, Y., Fan, S., 2018a. Isotropic wavevector domain image filters by a photonic crystal slab device. J. Opt. Soc. Am. A Opt. Image Sci. Vis. 35 (10), 1685–1691, the Optical Society.*

have the effect of single-band excitation when the direction of incidence is in a high symmetry plane such as the $y = 0$ plane (Tikhodeev et al., 2002).

Next, we prove the following statement: if the two-band Hamiltonian is isotropic, that is,

$$\hat{\mathcal{R}}(\varphi)\hat{\mathcal{H}}(\boldsymbol{k})\hat{\mathcal{R}}^{-1}(\varphi) = \hat{\mathcal{H}}(\mathbf{R}(\varphi)\boldsymbol{k}), \tag{13}$$

for every $\varphi \in (0, 2\pi)$, then we have the effect of single-band excitation along all directions. Here, $\mathbf{R}(\varphi)$ and $\hat{\mathcal{R}}(\varphi)$ are the rotation operators that describe the rotation around the z-axis by an angle φ, in the $\boldsymbol{k}$ space and the two-dimensional Hilbert space, respectively. To prove this, we denote the eigenstates of the two bands as $|\boldsymbol{k}, A\rangle$ and $|\boldsymbol{k}, B\rangle$, which connect to the A and B modes along the Γ-X direction, respectively. Eq. (13) implies that:

$$\begin{aligned} |\mathbf{R}(\varphi)\boldsymbol{k}, A\rangle &= \hat{\mathcal{R}}(\varphi)|\boldsymbol{k}, A\rangle, \\ |\mathbf{R}(\varphi)\boldsymbol{k}, B\rangle &= \hat{\mathcal{R}}(\varphi)|\boldsymbol{k}, B\rangle. \end{aligned} \tag{14}$$

We denote the S and P polarized modes as $|\boldsymbol{k}, S\rangle$ and $|\boldsymbol{k}, P\rangle$, respectively. By definition,

$$\begin{aligned} |\mathbf{R}(\varphi)\boldsymbol{k}, S\rangle &= \hat{\mathcal{R}}(\varphi)|\boldsymbol{k}, S\rangle, \\ |\mathbf{R}(\varphi)\boldsymbol{k}, P\rangle &= \hat{\mathcal{R}}(\varphi)|\boldsymbol{k}, P\rangle. \end{aligned} \tag{15}$$

For $\boldsymbol{k} = \boldsymbol{k_x}$ along the Γ–X direction, the $A(B)$ mode does not couple to the $S(P)$ polarization, i.e.,

$$\begin{aligned} \langle \boldsymbol{k_x}, S|\boldsymbol{k_x}, A\rangle &= 0, \\ \langle \boldsymbol{k_x}, P|\boldsymbol{k_x}, B\rangle &= 0. \end{aligned} \tag{16}$$

Then for $\boldsymbol{k} = \mathbf{R}(\varphi)\boldsymbol{k_x}$ along any direction, using Eqs. (14) and (15), we also have

$$\begin{aligned} \langle \boldsymbol{k}, S|\boldsymbol{k}, A\rangle &= \langle \boldsymbol{k_x}, S|\hat{\mathcal{R}}^{-1}(\varphi)\hat{\mathcal{R}}(\varphi)|\boldsymbol{k_x}, A\rangle = 0, \\ \langle \boldsymbol{k}, P|\boldsymbol{k}, B\rangle &= \langle \boldsymbol{k_x}, P|\hat{\mathcal{R}}^{-1}(\varphi)\hat{\mathcal{R}}(\varphi)|\boldsymbol{k_x}, B\rangle = 0. \end{aligned} \tag{17}$$

Due to the single-band excitation effect, as well as the presence of the uniform dielectric slab which sets the background transmission coefficient (Fan and Joannopoulos, 2002) to be unity, for this structure shown in Fig. 2, the transmission and reflection coefficients are:

$$t_{\pm}(\omega, \boldsymbol{k}) = \frac{i[\omega - \omega_{\pm}(\boldsymbol{k})]}{i[\omega - \omega_{\pm}(\boldsymbol{k})] + \gamma_{\pm}(\boldsymbol{k})}, \tag{18}$$

$$r_{\pm}(\omega, \boldsymbol{k}) = -e^{i\phi} \frac{\gamma_{\pm}(\boldsymbol{k})}{i[\omega - \omega_{\pm}(\boldsymbol{k})] + \gamma_{\pm}(\boldsymbol{k})}, \tag{19}$$

where the upper (lower) sign corresponds to $S(P)$ polarized light and upper (lower) band; ω is the incident light frequency. Therefore on resonance,

$$t_{\pm}(\omega_{\pm}(\boldsymbol{k}), \boldsymbol{k}) = 0, \quad r_{\pm}(\omega_{\pm}(\boldsymbol{k}), \boldsymbol{k}) = -e^{i\phi}. \tag{20}$$

The numerically determined transmission and reflection spectra for S polarized light are plotted in Fig. 3E and F. Due to the effect of single-band excitation, S-polarized light only excites the upper band of guided resonances. Moreover, as expected from Eqs. (12), (18), and (19), the resultant transmission and reflection spectra are isotropic, i.e., the spectra are identical along any wavevector direction as defined by the angle φ in Fig. 2. When $\omega = \omega(\boldsymbol{k})$, the transmittance exhibits sharp dips with $|t| = 0$, while the reflectance exhibits peaks with $|r| = 1$, as expected from Eq. (20).

Depending on the operating conditions, the structure as shown in Fig. 2 can be used to perform several very useful image-processing functionalities (Guo et al., 2018a):

1. *Isotropic high-pass filter*

   At the frequency $\omega = \omega_0 \equiv \omega_{\pm}(k=0)$, the $\boldsymbol{k}$-dependent transmittance is:

$$\begin{aligned} |t_{\pm}(\omega_0, \boldsymbol{k})| &= \frac{|\omega_{\pm}(\boldsymbol{k}) - \omega_0|}{\sqrt{[\omega_{\pm}(\boldsymbol{k}) - \omega_0]^2 + \gamma_{\pm}(\boldsymbol{k})^2}} \\ &\approx \frac{|\omega_{\pm}(\boldsymbol{k}) - \omega_0|}{\gamma_0} = \frac{A_{\pm}}{\gamma_0} |\boldsymbol{k}|^2. \end{aligned} \tag{21}$$

   This transmittance realizes the Laplacian, a special isotropic high-pass filter (Guo et al., 2018b).

2. *Isotropic low-pass filter*

   At the frequency $\omega = \omega_0$, if one considers instead the reflected light, the transfer function is:

$$\begin{aligned} |r_{\pm}(\omega_0, \boldsymbol{k})| &= \frac{1}{\sqrt{1 + [\omega_{\pm}(\boldsymbol{k}) - \omega_0]^2 / \gamma_{\pm}(\boldsymbol{k})^2}} \\ &\approx \frac{1}{\sqrt{1 + A_{\pm}^2 |\boldsymbol{k}|^4 / \gamma_0^2}}. \end{aligned} \tag{22}$$

   This transfer function realizes an isotropic low-pass filter with reflection peak $|r_{\pm}| = 1$ at the Γ point.

3. *Isotropic band-reject filter*

Suppose instead we operate away from the frequency ω_0, but at the frequency $\omega = \omega_{\pm,\ q}$, where q is the amplitude of a nonzero in-plane wavevector, the transfer function then becomes

$$|t_\pm(\omega_{\pm,\ q}, \boldsymbol{k})| = \frac{|\omega_\pm(\boldsymbol{k}) - \omega_{\pm,\ q}|}{\sqrt{[\omega_\pm(\boldsymbol{k}) - \omega_{\pm,\ q}]^2 + \gamma_\pm(\boldsymbol{k})^2}} \tag{23}$$

This transfer function realizes an isotropic band-reject filter, with transmission dip $|t_\pm| = 0$ at $|\boldsymbol{k}| = q$.

4. *Isotropic band-pass filter*

At the frequency $\omega = \omega_{\pm,\ q}$, the reflection has a transfer function

$$|r_\pm(\omega_q, \boldsymbol{k})| = \frac{1}{\sqrt{1 + [\omega_\pm(\boldsymbol{k}) - \omega_{\pm,\ q}]^2/\gamma_\pm(\boldsymbol{k})^2}} \tag{24}$$

This transfer function realizes an isotropic band-pass filter, with reflection peak $|r_\pm| = 1$ at $|\boldsymbol{k}| = q$.

Therefore, our device can be used to achieve multiple filtering functions. As schematically shown in Fig. 3B, when the light frequency is on resonance at normal incidence ($\omega = \omega_0$), the device operates as an isotropic high-pass filter (Laplacian) at the transmission mode and an isotropic low-pass filter at the reflection mode. When the light frequency is detuned a bit from the resonance at normal incidence, the device operates as an isotropic band-reject filter at the transmission mode and an isotropic band-pass filter at the reflection mode, where the rejected or passed wavevector components are determined by the light frequency detuning and polarization. For transmission mode, the transmitted image is the required filtered result. For reflection mode, the reflected image is the required filtered result, which can be separated from the incident image by using a beam splitter.

4. Numerical demonstration of the isotropic filters

Now we numerically demonstrate the filtering functions of our device. In all the numerical demonstration below, the incident beam is S polarized. The transmitted image is calculated following the standard way in image processing (Gonzalez and Woods, 2008). (1) Compute the Fourier transform $\tilde{S}_{in}(k_x, k_y)$ of the incident field profile $S_{in}(x, y)$. Note the incident image is $|S_{in}(x, y)|^2$. (2) Compute the Fourier transform of the output field profile, $\tilde{S}_{out}(k_x, k_y) = H(k_x, k_y)\tilde{S}_{in}(k_x, k_y)$, where $H(k_x, k_y)$ is the transfer function. (3) Obtain the output field profile $S_{out}(x, y)$ by inverse Fourier transform. Calculate the output image $|S_{out}(x, y)|^2$.

4.1 Isotropic high-pass filter (the Laplacian)

Fig. 4 illustrates the isotropic high-pass filter (the Laplacian). Fig. 4A plots the filter transfer function: the transmittance for S-polarized light $|t_s|(k_x, k_y)$ at the frequency $\omega_0 = 0.77098 \times 2\pi c/a$. The transfer function is almost isotropic. Fig. 4B plots $|t_s|$ as a function of $|\boldsymbol{k}|$ along a general wavevector direction ($\varphi = 14°$), and the fitting result of Eq. (21). The fitting is very good for $|\boldsymbol{k}|$ up to $0.6 \times 10^{-2} \times 2\pi/a$. These plots confirm that the device indeed operates as an isotropic high-pass filter (the Laplacian) in this case. In Fig. 4A (and also in Fig. 5A as shown later), at larger wavevectors, the transfer function exhibits some anisotropy due to the dependency of the radiative linewidth $\gamma(\boldsymbol{k})$ on the direction of $\boldsymbol{k}$.

The Laplacian enables image sharpening and edge detection (Guo et al., 2018b). Fig. 4E shows an incident image of the Stanford emblem. Fig. 4C plots the Fourier transform of the field profile for this incident image. Fig. 4D shows the calculated Fourier spectrum for the transmitted image, which is obtained by a pointwise product of Fig. 4A and C. The low wavevector components have been filtered out. Fig. 4F is the calculated transmitted image, which contains only the edge of the incident image.

4.2 Isotropic low-pass filter

Fig. 5 illustrates the isotropic low-pass filter. Fig. 5A plots the filter transfer function: the reflectance for S-polarized light $|r_s|(k_x, k_y)$ at the frequency $\omega_0 = 0.77098 \times 2\pi c/a$. The transfer function is almost isotropic. Fig. 5B plots $|r_s|$ as a function of $|\boldsymbol{k}|$ along a general wavevector direction ($\varphi = 14°$), and the fitting result of Eq. (22). The fitting is almost perfect in the wavevector range as shown. These plots confirm that the device indeed operates as an isotropic low-pass filter in this case.

The isotropic low-pass filter accomplishes image smoothing, with applications ranging from character recognition in machine perception, preprocessing functions in printing and publishing industry, to satellite and aerial image processing (Buades et al., 2005; Gonzalez and Woods, 2008). Here we show one specific application of the low-pass filter in white noise reduction. Fig. 5E shows an incident image of the Stanford emblem with substantial white noise components, while Fig. 5C plots the Fourier

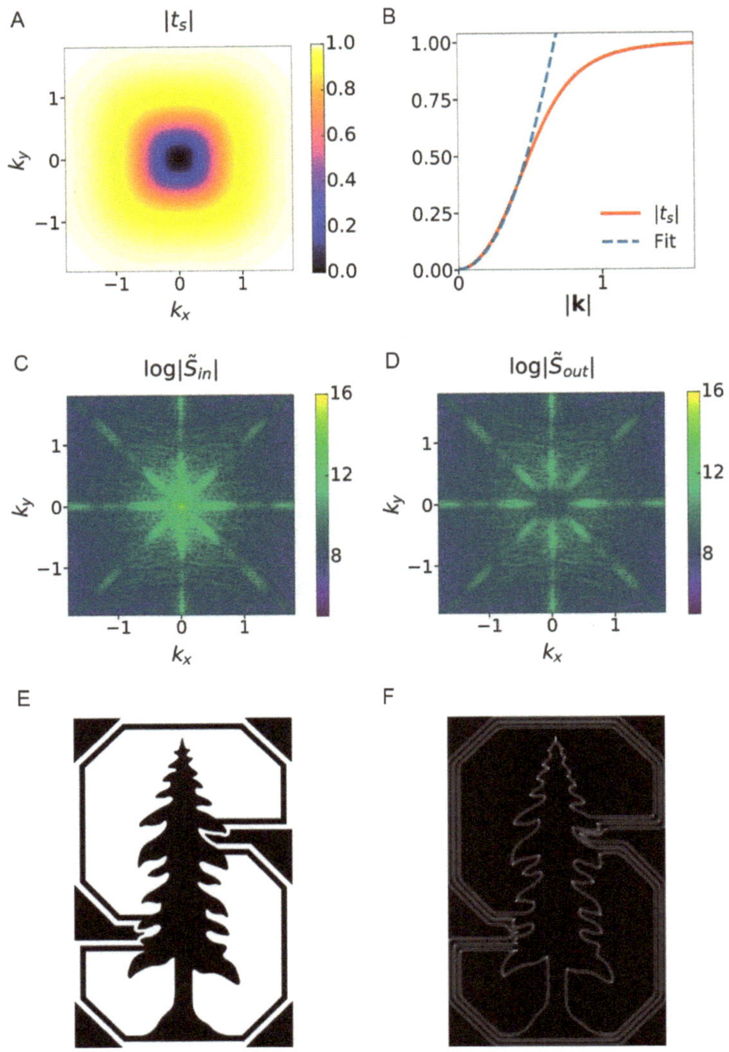

Fig. 4 Isotropic high-pass filter (Laplacian). (A) Transmittance for *S*-polarized light $|t_s|(k_x, k_y)$ at the frequency $\omega_0 = 0.77098 \times 2\pi c/a$. (B) $|t_s|$ as a function of $|\boldsymbol{k}|$ along a general wavevector direction ($\varphi = 14°$), and the fitting result of Eq. (21). The fit is very good for $|\boldsymbol{k}|$ up to $0.6 \times 10^{-2} \times 2\pi/a$. (C) Log plot of the Fourier transform of the field profile for the incident image (E): $\log|\tilde{S}_{in}|(k_x,k_y)$. (D) Log plot of the Fourier transform of the field profile for the reflected image (F): $\log|\tilde{S}_{out}|(k_x,k_y)$. The low wavevector components have been filtered out. (E) Incident image $|S_{in}|^2$ of the Stanford emblem. The image size is $5220a \times 3456a$. (F) Calculated transmitted image $|S_{out}|^2$, which shows the edges with different orientations. $|\boldsymbol{k}|$, k_x, and k_y are in the units of $10^{-2} \times 2\pi/a$. *Reprinted with permission from Guo, C., Xiao, M., Minkov, M., Shi, Y., Fan, S., 2018a. Isotropic wavevector domain image filters by a photonic crystal slab device. J. Opt. Soc. Am. A Opt. Image Sci. Vis. 35 (10), 1685–1691, the Optical Society.*

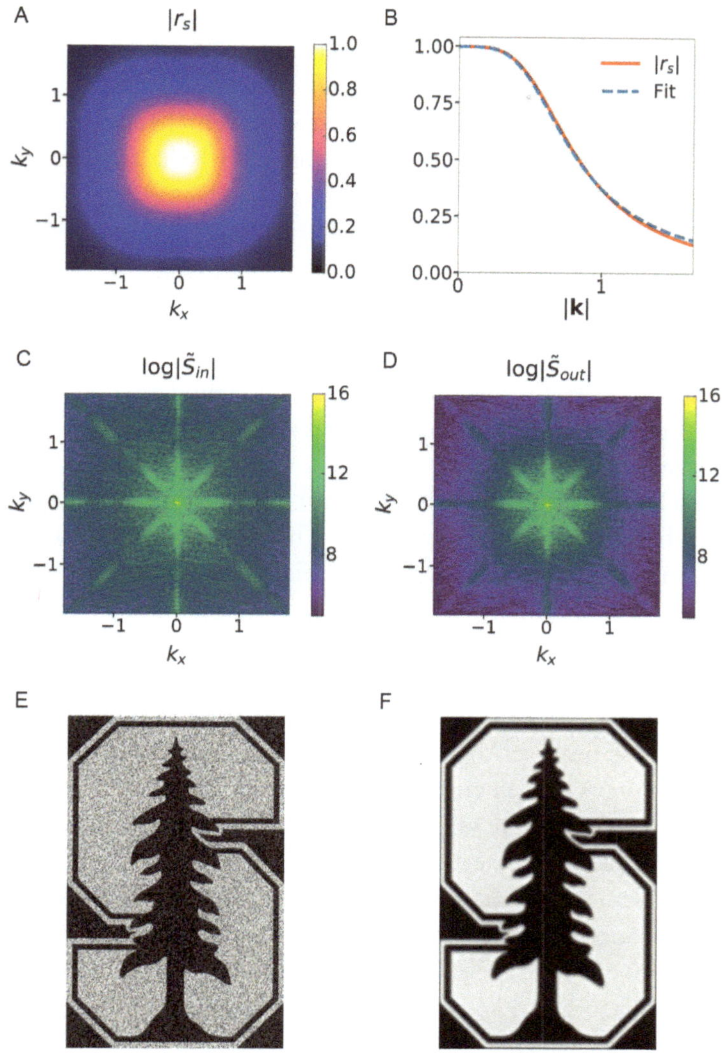

Fig. 5 Isotropic low-pass filter. (A) Reflectance for *S*-polarized light $|r_s|(k_x, k_y)$ at the frequency $\omega_0 = 0.77098 \times 2\pi c/a$. (B) $|r_s|$ as a function of $|\boldsymbol{k}|$ along a general wavevector direction ($\varphi = 14°$), and the fitting result of Eq. (22). (C) Log plot of the Fourier transform of the field profile for the incident image (E): $\log|\widetilde{S}_{in}|(k_x, k_y)$. (D) Log plot of the Fourier transform of the field profile for the reflected image (F): $\log|\widetilde{S}_{out}|(k_x, k_y)$. The high wavevector components have been filtered out. (E) Incident image $|S_{in}|^2$ of the Stanford emblem with substantial white noise component. The image size is $5220a \times 3456a$. (F) Calculated reflected image $|S_{out}|^2$, which reduces the white noise by image smoothing. $|\boldsymbol{k}|$, k_x, and k_y are in the units of $10^{-2} \times 2\pi/a$. *Reprinted with permission from Guo, C., Xiao, M., Minkov, M., Shi, Y., Fan, S., 2018a. Isotropic wavevector domain image filters by a photonic crystal slab device. J. Opt. Soc. Am. A Opt. Image Sci. Vis. 35 (10), 1685–1691, the Optical Society.*

transform of the field profile for this incident image. Fig. 5D shows the calculated Fourier spectrum for the reflected image, which is obtained by a pointwise product of Fig. 5A and C. The high wavevector components have been filtered out. Fig. 5F shows the calculated reflected image, where the noise component has indeed been significantly reduced, demonstrating image smoothing.

4.3 Isotropic band-reject filter

Fig. 6 illustrates the isotropic band-reject filter. Fig. 6A plots the filter transfer function: the transmittance for *S*-polarized light $|t_s|(k_x, k_y)$ at the frequency $\omega_q = 0.77130 \times 2\pi c/a$. The transfer function is almost isotropic. Figure 6B plots $|t_s|$ as a function of $|\boldsymbol{k}|$ along a general wavevector direction ($\varphi = 14°$), which shows $|t_s| = 0$ at $|\boldsymbol{k}| = 0.84 \times 10^{-2} \times 2\pi/a$. We fit the transmittance curve with Eq. (23) together with Eq. (12) and $\gamma(\boldsymbol{k}) \approx \gamma_0$. The fitting is almost perfect in the wavevector range as shown. These plots confirm that the device indeed operates as an isotropic band-reject filter in this case.

The isotropic band-reject filter can effectively eliminate periodic noise, a common type of noise arising typically from electrical or electromechanical interference during image acquisition (Aizenberg and Butakoff, 2002; Banham and Katsaggelos, 1997; Gonzalez and Woods, 2008). As periodic noise appears as peaks in the Fourier transform at locations corresponding to the wavevectors of the periodic interference, it can be isolated and filtered by band-reject filters. Here we show an example of periodic noise reduction with the isotropic band-reject filter. Fig. 6E shows an incident image of the Stanford emblem incorporating a periodic noise component. Fig. 6C plots the Fourier transform of the field profile for this incident image. In the wavevector space, the periodic noise appears as spectral peaks which lie on an approximate circle around the origin. Fig. 6D shows the calculated Fourier spectrum for the transmitted image, which is obtained by a pointwise product of Fig. 6A and C. The spectral peaks corresponding to the periodic noise have been filtered out. Fig. 6F shows the calculated transmitted image, where the periodic noise has indeed been eliminated effectively. Here we note that the rejected wavevector where the maximum rejection occurs in our band-reject filter can be easily tuned by tuning the light frequency.

4.4 Isotropic band-pass filter

Fig. 7 illustrates the isotropic band-pass filter. Fig. 7A plots the filter transfer function: the reflectance for *S*-polarized light $|r_s|(k_x, k_y)$ at the frequency

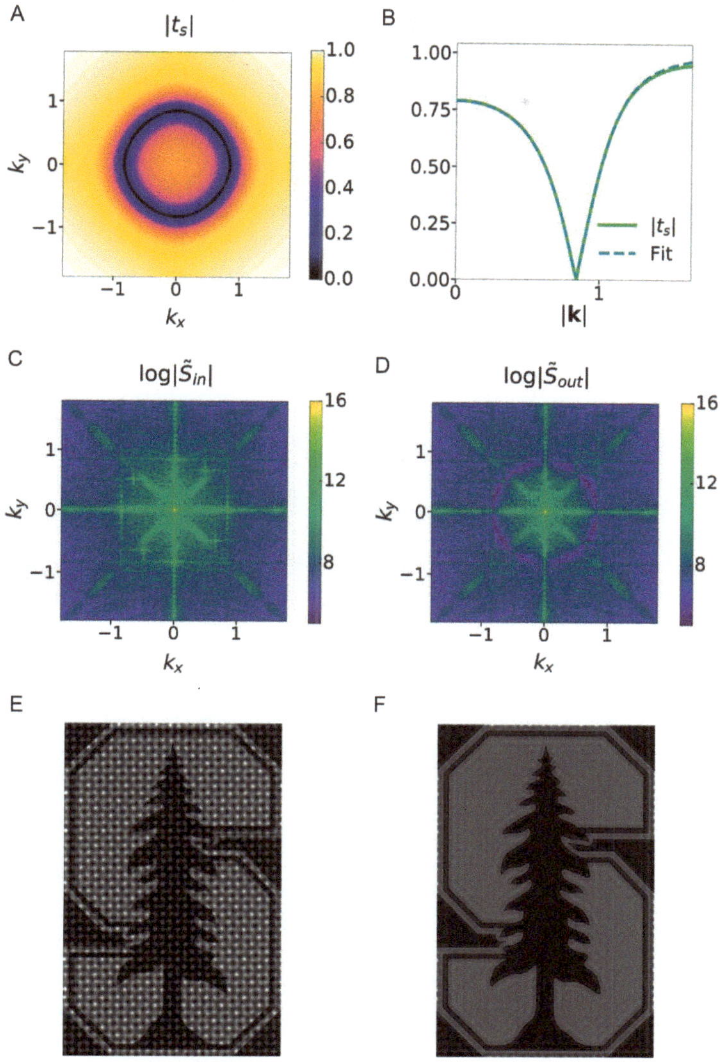

Fig. 6 Isotropic band-reject filter. (A) Transmittance for *S*-polarized light $|t_s|(k_x, k_y)$ at the frequency $\omega_q = 0.77130 \times 2\pi c/a$. (B) $|t_s|$ as a function of $|\boldsymbol{k}|$ along a general wavevector direction ($\varphi = 14°$), and the fitting result of Eq. (23) along with Eq. (12) and $\gamma(\boldsymbol{k}) \approx \gamma_0$. (C) Log plot of the Fourier transform of the field profile for the incident image (E): $\log|\tilde{S}_{in}|(k_x,k_y)$. The sinusoidal noise appears as peaks in the spectrum which lie on an approximate circle around the origin. (D) Log plot of the Fourier transform of the field profile for the transmitted image (F): $\log|\tilde{S}_{out}|(k_x,k_y)$. The wavevector components corresponding to the periodic noise have been filtered out. (E) Incident image $|S_{in}|^2$ of the Stanford emblem with periodic noise. The image size is $5220a \times 3456a$. (F) Calculated transmitted image $|S_{out}|^2$, which eliminates the periodic noise. $|\boldsymbol{k}|$, k_x, and k_y are in the units of $10^{-2} \times 2\pi/a$. *Reprinted with permission from Guo, C., Xiao, M., Minkov, M., Shi, Y., Fan, S., 2018a. Isotropic wavevector domain image filters by a photonic crystal slab device. J. Opt. Soc. Am. A Opt. Image Sci. Vis. 35 (10), 1685–1691, the Optical Society.*

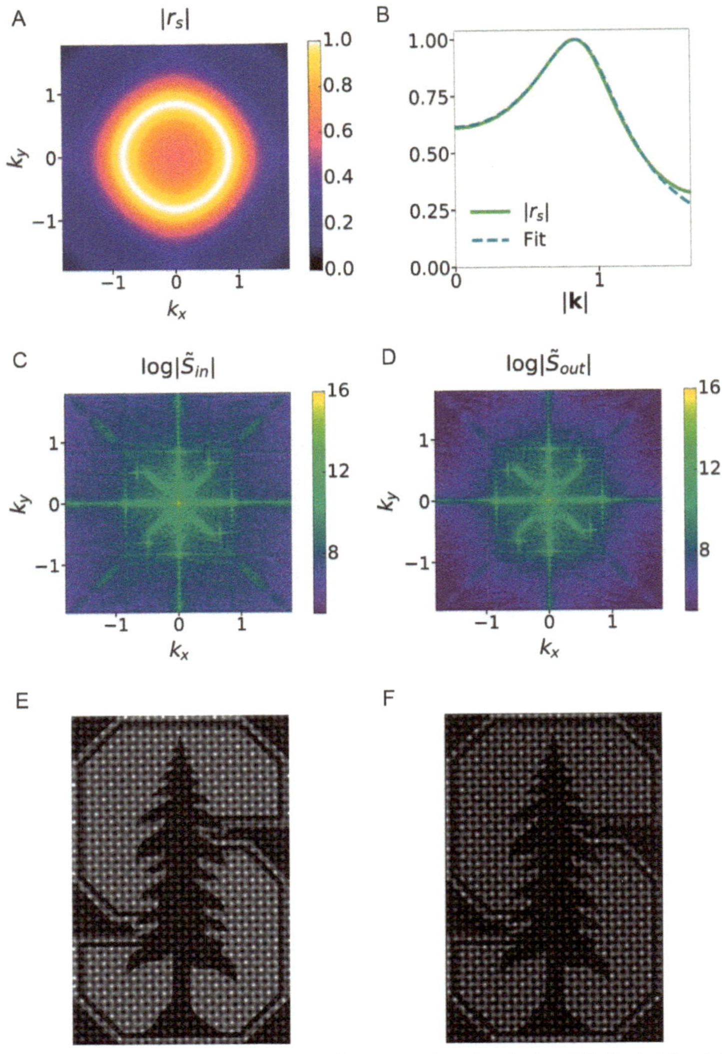

Fig. 7 Isotropic band-pass filter. (A) Reflectance for *S*-polarized light $|r_s|(k_x, k_y)$ at the frequency $\omega_q = 0.77130 \times 2\pi c/a$. (B) $|r_s|$ as a function of $|\boldsymbol{k}|$ along a general wavevector direction ($\varphi = 14°$), and the fitting result of Eq. (24) together with Eq. (12) and $\gamma(\boldsymbol{k}) \approx \gamma_0$. (C) Log plot of the Fourier transform of the field profile for the incident image (E): $\log|\widetilde{S}_{in}|(k_x,k_y)$. The sinusoidal noise appears as impulses in the spectrum which lie on an approximate circle around the origin. (D) Log plot of the Fourier transform of the field profile for the reflected image (F): $\log|\widetilde{S}_{out}|(k_x,k_y)$. (E) Incident image $|S_{in}|^2$ of the Stanford emblem with periodic noise. The image size is 5220*a* × 3456*a*. (F) Calculated reflected image $|S_{out}|^2$, which isolates the periodic noise and simplifies its analysis. $|\boldsymbol{k}|$, k_x, and k_y are in the units of $10^{-2} \times 2\pi/a$. *Reprinted with permission from Guo, C., Xiao, M., Minkov, M., Shi, Y., Fan, S., 2018a. Isotropic wavevector domain image filters by a photonic crystal slab device. J. Opt. Soc. Am. A Opt. Image Sci. Vis. 35 (10), 1685–1691, the Optical Society.*

$\omega_q = 0.77130 \times 2\pi c/a$. The transfer function is almost isotropic. Fig. 7B plots $|r_s|$ as a function of $|\boldsymbol{k}|$ along a general wavevector direction ($\varphi = 14°$), which shows $|r_s| = 1$ at $|\boldsymbol{k}| = 0.84 \times 10^{-2} \times 2\pi/a$. We fit the reflectance curve with Eq. (24) together with Eq. (12) and $\gamma(\boldsymbol{k}) \approx \gamma_0$. The fitting is almost perfect in the wavevector range as shown. These plots confirm that the device indeed operates as an isotropic band-pass filter in this case.

The isotropic band-pass filter performs the opposite operation of the band-reject filter. It is quite useful in isolating the effects on an image caused by selected wavevector bands (Gonzalez and Woods, 2008). Here we show an example of extracting periodic noise patterns with the isotropic band-pass filter. Fig. 7E shows an incident image of the Stanford emblem incorporating periodic noise (same as Fig. 6E), while Fig. 7C plots the Fourier transform of the field profile for this incident image. Fig. 7D shows the calculated Fourier spectrum for the reflected image, which is obtained by a pointwise product of Fig. 7A and C. Fig. 7F shows the calculated reflected image, where the periodic noise pattern is isolated and appears more clearly. This is useful because it simplifies the analysis of the noise, largely independent of the image content (Gonzalez and Woods, 2008).

5. Conclusion

Our design of isotropic wavevector domain image filters using a photonic crystal slab is based on the guided resonances with isotropic band structure. The same idea can extend to other photonic structures that host resonant modes with isotropic band structure. In particular, the phase-shifted Bragg grating in Bykov et al. (2014) can also perform the four filtering functionalities of our device, but with the transmission/reflection mode flipped.

In conclusion, we have shown that isotropic high-pass, low-pass, band-reject, and band-pass filters can be implemented at visible wavelengths using one photonic crystal slab device. Such a simple but multifunctional photonic device may find various applications involving image processing. For instance, it can be used as the first layer of image recognition systems. As such our work indicates the new opportunities for compact, high-throughput, and low-energy-consumption optical analog computing as provided by nanophotonic structures.

Acknowledgments

We acknowledge Dr. Meng Xiao, Dr. Momchil Minkov, and Dr. Yu Shi for collaboration, and Dr. Bo Zhao, Dr. Alexander Cerjan, and Yu Guo for helpful discussion. This work is supported by Samsung Electronics and US Air Force (USAF) (FA9550-17-1-0002).

References

AbdollahRamezani, S., Arik, K., Khavasi, A., Kavehvash, Z., 2015. Analog computing using graphene-based metalines. Opt. Lett. 40 (22), 5239. ISSN 0146-9592. https://doi.org/10.1364/OL.40.005239. https://www.osapublishing.org/abstract.cfm?URI=ol-40-22-5239.

Abràmoff, M.D., Magalhães, P.J., Ram, S.J., 2004. Image processing with ImageJ. Biophotonics Int. 11 (7), 36–42. ISSN 1081-8693.

Aizenberg, I.N., Butakoff, C., 2002. In: Frequency domain medianlike filter for periodic and quasi-periodic noise removal. Proceedings of SPIE, vol. 4667. San Jose, http://proceedings.spiedigitallibrary.org/proceeding.aspx?articleid=876767. http://link.aip.org/link/?PSI/4667/181/1&Agg=doi.

Banham, M.R., Katsaggelos, A.K., 1997. Digital image restoration. IEEE Signal Process. Mag. 14 (2), 24–41. ISSN 10535888. https://doi.org/10.1109/79.581363. http://ieeexplore.ieee.org/document/581363/.

Bracewell, R.N., 1986. The Fourier Transform and Its Applications. McGraw-Hill, New York. ISBN 0073039381.

Brosnan, T., Sun, D.W., 2004. Improving quality inspection of food products by computer vision—a review. J. Food Eng. 61 (1), 3–16. ISSN 02608774. https://doi.org/10.1016/S0260-8774(03)00183-3.

Buades, A., Coll, B., Morel, J.M., 2005. A review of image denoising algorithms, with a new one. Multiscale Model. Simul. 4 (2), 490–530. ISSN 1540-3459. https://doi.org/10.1137/040616024. http://epubs.siam.org/doi/10.1137/040616024.

Bykov, D.A., Doskolovich, L.L., Bezus, E.A., Soifer, V.A., 2014. Optical computation of the Laplace operator using phase-shifted Bragg grating. Opt. Express 22 (21), 25084. ISSN 1094-4087. https://doi.org/10.1364/OE.22.025084. https://www.osapublishing.org/oe/abstract.cfm?uri=oe-22-21-25084.

Bykov, D.A., Doskolovich, L.L., Morozov, A.A., Podlipnov, V.V., Bezus, E.A., Verma, P., Soifer, V.A., 2018. First-order optical spatial differentiator based on a guided-mode resonant grating. Opt. Express 26 (8), 10997–11006. ISSN 1094-4087. https://doi.org/10.1364/OE.26.010997. https://www.osapublishing.org/abstract.cfm?URI=oe-26-8-10997.

Chang, T.Y., Hong, J.H., Yeh, P., 1990. Spatial amplification: an image-processing technique using the selective amplification of spatial frequencies. Opt. Lett. 15 (13), 743. ISSN 0146-9592. https://doi.org/10.1364/OL.15.000743. https://www.osapublishing.org/abstract.cfm?URI=ol-15-13-743.

Chizari, A., Abdollahramezani, S., Jamali, M.V., Salehi, J.A., 2016. Analog optical computing based on a dielectric meta-reflect array. Opt. Lett. 41 (15), 3451. ISSN 0146-9592. https://doi.org/10.1364/OL.41.003451. https://www.osapublishing.org/abstract.cfm?URI=ol-41-15-3451. https://arxiv.org/abs/1605.07150%0Ahttp://doi.org/10.1364/OL.41.003451.

Dalal, N., Triggs, B., 2005. Histograms of oriented gradients for human detection. In: 2005 IEEE Computer Society Conference on Computer Vision and Pattern Recognition (CVPR'05) IEEE, pp. 886–893.

Dong, Z., Si, J., Yu, X., Deng, X., 2018. Optical spatial differentiator based on subwavelength high-contrast gratings. Appl. Phys. Lett. 112 (18), 181102. ISSN 0003–6951. https://doi.org/10.1063/1.5026309. http://aip.scitation.org/toc/apl/112/18. http://aip.scitation.org/doi/10.1063/1.5026309. https://doi.org/10.1063/1.5026309.

Eu, J.K.T., Liu, C.Y.C., Lohmann, A.W., 1973. Spatial filters for differentiation. Opt. Commun. 9 (2), 168–171. ISSN 00304018. https://doi.org/10.1016/0030-4018(73)90251-4. http://linkinghub.elsevier.com/retrieve/pii/0030401873902514.

Fan, S., Joannopoulos, J.D., 2002. Analysis of guided resonances in photonic crystal slabs. Phys. Rev. B 65 (23), 235112. ISSN 0163-1829. https://doi.org/10.1103/PhysRevB.65.235112.

Fang, Y., Ruan, Z., 2018. Optical spatial differentiator for a synthetic three-dimensional optical field. Opt. Lett. 43 (23), 5893–5896.

Fang, Y., Lou, Y., Ruan, Z., 2017. On-grating graphene surface plasmons enabling spatial differentiation in the terahertz region. Opt. Lett. 42 (19), 3840. ISSN 0146-9592. https://doi.org/10.1364/OL.42.003840. https://www.osapublishing.org/abstract.cfm?URI=ol-42-19-3840.

Golovastikov, N.V., Bykov, D.A., Doskolovich, L.L., 2014. Resonant diffraction gratings for spatial differentiation of optical beams. Quantum Electron. 44 (10), 984–988. ISSN 1063-7818. https://doi.org/10.1070/QE2014v044n10ABEH015477. http://stacks.iop.org/1063-7818/44/i=10/a=984?key=crossref.8bee7b769ad2412f3500ff65fffe3a5e%5Cn. http://iopscience.iop.org/1063-7818/44/10/984. http://stacks.iop.org/1063-7818/44/i=10/a=984?key=crossref.8bee7b769ad2412f3500ff65fffe3a5e.

Gonzalez, R., Woods, R., 2008. Digital Image Processing, third ed. Prentice Hall, Upper Saddle River, NJ. ISBN 9780131687288.

Goodman, J.W., 2005. Introduction to Fourier Optics, third ed. Roberts and Company, Englewood. ISBN 0974707724.

Görlitz, D., Lanzl, F., 1975. A holographic spatial filter for direction independent differentiation. Jpn. J. Appl. Phys. 14 (S1), 223. ISSN 0021-4922. https://doi.org/10.7567/JJAPS.14S1.223. http://stacks.iop.org/1347-4065/14/223.

Guo, C., Xiao, M., Minkov, M., Shi, Y., Fan, S., 2018a. Isotropic wavevector domain image filters by a photonic crystal slab device. J. Opt. Soc. Am. A Opt. Image Sci. Vis. 35 (10), 1685–1691.

Guo, C., Xiao, M., Minkov, M., Shi, Y., Fan, S., 2018b. Photonic crystal slab Laplace operator for image differentiation. Optica 5 (3), 251. ISSN 2334-2536. https://doi.org/10.1364/OPTICA.5.000251. https://www.osapublishing.org/abstract.cfm?URI=optica-5-3-251.

Hwang, Y., Davis, T.J., Lin, J., Yuan, X.C., 2018. Plasmonic circuit for second-order spatial differentiation at the subwavelength scale. Opt. Express 26 (6), 7368. ISSN 1094-4087. https://doi.org/10.1364/OE.26.007368. https://www.osapublishing.org/abstract.cfm?URI=oe-26-6-7368.

Kwon, H., Sounas, D., Cordaro, A., Polman, A., Alù, A., 2018. Nonlocal metasurfaces for optical signal processing. Phys. Rev. Lett. 121 (17), 173004.

Markham, R., Frey, S., Hills, G.J., 1963. Methods for the enhancement of image detail and accentuation of structure in electron microscopy. Virology 20 (1), 88–102. ISSN 00426822. https://doi.org/10.1016/0042-6822(63)90143-0. http://linkinghub.elsevier.com/retrieve/pii/0042682263901430.

Nejad, F.Z., Fleury, R., 2018. Performing mathematical operations using high-index acoustic metamaterials. New J. Phys. 20, 073001.

Ochiai, T., Sakoda, K., 2001. Dispersion relation and optical transmittance of a hexagonal photonic crystal slab. Phys. Rev. B 63 (12), 125107. ISSN 0163-1829. https://doi.org/10.1103/PhysRevB.63.125107.

Pors, A., Nielsen, M.G., Bozhevolnyi, S.I., 2015. Analog computing using reflective plasmonic metasurfaces. Nano Lett. 15 (1), 791–797. ISSN 1530-6984. https://doi.org/10.1021/nl5047297.

Roberts, A., Gómez, D.E., Davis, T.J., 2018. Optical image processing with metasurface dark modes. J. Opt. Soc. Am. A Opt. Image Sci. Vis. 35 (9), 1575–1584.

Sakoda, K., 2005. Optical properties of photonic crystals, Springer Series in Optical Sciences, vol. 80. Springer-Verlag, Berlin, Heidelberg. ISBN 3-540-20682-5.

Silva, A., Monticone, F., Castaldi, G., Galdi, V., Alu, A., Engheta, N., 2014. Performing mathematical operations with metamaterials. Science. 343 (6167), 160–163. ISSN 0036-8075. https://doi.org/10.1126/science.1242818.

Sirohi, R.S., Mohan, V.R., 1977. Differentiation by spatial filtering. Opt. Acta 24 (11), 1105–1113. ISSN 0030-3909. https://doi.org/10.1080/713819447. http://www.tandfonline.com/doi/abs/10.1080/713819447.

Solli, D.R., Jalali, B., 2015. Analog optical computing. Nat. Photonics 9 (11), 704–706. ISSN 1749-4885. https://doi.org/10.1038/nphoton.2015.208. http://www.nature.com/doifinder/10.1038/nphoton.2015.208.

Tikhodeev, S.G., Yablonskii, A.L., Muljarov, E.A., Gippius, N.A., Ishihara, T., 2002. Quasiguided modes and optical properties of photonic crystal slabs. Phys. Rev. B 66 (4), 045102. ISSN 0163-1829. https://doi.org/10.1103/PhysRevB.66.045102.

Vogelaar, L., Nijdam, W., van Wolferen, H.A.G.M., de Ridder, R.M., Segerink, F.B., Flück, E., Kuipers, L., van Hulst, N.F., 2001. Large area photonic crystal slabs for visible light with waveguiding defect structures: fabrication with focused ion beam assisted laser interference lithography. Adv. Mater. 13 (20), 1551. ISSN 09359648. https://doi.org/10.1002/1521-4095(200110)13:20¡1551::AID-ADMA1551¿3.0.CO;2-V. http://doi.wiley.com/10.1002/1521-4095%28200110%2913%3A20%3C1551%3A%3AAID-ADMA1551%3E3.0.CO%3B2-V.

Wang, K.X., Yu, Z., Sandhu, S., Fan, S., 2013. Fundamental bounds on decay rates in asymmetric single-mode optical resonators. Opt. Lett. 38 (2), 100. ISSN 0146-9592. https://doi.org/10.1364/OL.38.000100. https://www.osapublishing.org/abstract.cfm?URI=ol-38-2-100.

Wu, W., Jiang, W., Yang, J., Gong, S., Ma, Y., 2017. Multilayered analog optical differentiating device: performance analysis on structural parameters. Opt. Lett. 42 (24), 5270. ISSN 0146-9592. https://doi.org/10.1364/OL.42.005270. https://www.osapublishing.org/abstract.cfm?URI=ol-42-24-5270.

Youssefi, A., Zangeneh-Nejad, F., Abdollahramezani, S., Khavasi, A., 2016. Analog computing by Brewster effect. Opt. Lett. 41 (15), 3467. ISSN 0146-9592. https://doi.org/10.1364/OL.41.003467. https://www.osapublishing.org/abstract.cfm?URI=ol-41-15-3467.

Yu, F.T.S., Jutamulia, S., Yin, S., 2001. Introduction to Information Optics. Academic Press, San Diego. ISBN 0127748113.

Zhou, W., Zhao, D., Shuai, Y.C., Yang, H., Chuwongin, S., Chadha, A., Seo, J.H., Wang, K.X., Liu, V., Ma, Z., Fan, S., 2014. Progress in 2D photonic crystal Fano resonance photonics. Prog. Quantum Electron. 38 (1), 1–74. ISSN 00796727. https://doi.org/10.1016/j.pquantelec.2014.01.001. http://linkinghub.elsevier.com/retrieve/pii/S0079672714000020. https://doi.org/10.1016/j.pquantelec.2014.01.001.

Zhu, T., Zhou, Y., Lou, Y., Ye, H., Qiu, M., Ruan, Z., Fan, S., 2017. Plasmonic computing of spatial differentiation. Nat. Commun. 8, 15391. ISSN 2041-1723. https://doi.org/10.1038/ncomms15391. http://www.nature.com/doifinder/10.1038/ncomms15391. https://doi.org/10.1038/ncomms15391.

CHAPTER FIVE

Guided mode resonances and photonic crystals for biosensing and imaging

Alexander Drayton, Isabel Barth, Thomas F. Krauss*
The Department of Physics, University of York, York, United Kingdom
*Corresponding author: e-mail address: thomas.krauss@york.ac.uk

Contents

Semiconductors and Semimetals, Volume 100
ISSN 0080-8784
https://doi.org/10.1016/bs.semsem.2019.01.002

1. Introduction

This chapter considers applications in the area of sensing. Since the very nature of the resonance is less relevant in this context, we shall refer to it more generally as a guided mode resonance, GMR, which applies whether the Fano nature of the resonance is apparent or not.

The key property of a sensor is its ability to translate a quantity that is difficult to measure into one that can be measured easily, so it acts as a transducer. In the case of GMR sensors, the fact that a GMR is a surface wave that interacts closely with its environment is exploited, so GMR sensors fall into the general class of *surface affinity* sensors. Surface affinity sensors are commonly used in biomedical applications and they detect target molecules by their selective binding to a capture molecule such as an antibody, the antibody being anchored to the surface. The target molecule may be a marker for cancer or some other disease. When the target binds to the antibody, the optical mass on the surface increases and so the refractive index on the surface increases. This local increase in refractive index changes the effective index of the guided mode and thereby its resonance wavelength. By measuring the change in wavelength, one can directly infer back to the change in refractive index and the concentration of target molecules in the biological medium, i.e., the analyte. Ultimately, a GMR sensor is therefore a refractive index sensor and its performance is best described by the smallest refractive index change it can detect; this principle is illustrated in Fig. 1, and the relevant performance metrics will be discussed in Section 3.

A key point to note at the very beginning of this chapter is that GMR sensors are not the most highly performing of all available sensors; other types of sensors offer lower limits of detection. Nevertheless, as the chapter develops, it will become apparent that GMRs offer the most versatile approach toward sensing. They can be readily incorporated into low-cost point-of-care devices because of their ease of coupling, they can perform both the sensing and the imaging function, and they offer a multiparameter approach via the wavelength, the amplitude, the Q-factor, and the phase of the resonance. All of these effects will be discussed in detail in this chapter. The driver behind much of the discussion is therefore not the quest for the highest possible performance, but the desire for versatility, i.e., suitability for a variety of applications, and the highest performance that can be built at the lowest cost, which is much needed by healthcare providers both in the developed and in the developing world.

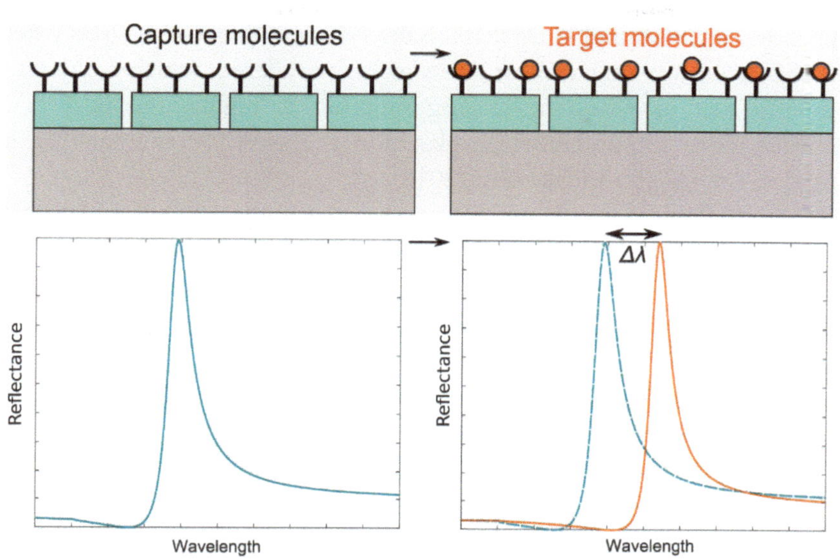

Fig. 1 Sketch of the operation of a GMR surface affinity sensor, which translates the binding of target molecules into a shift of the resonance wavelength.

1.1 A historical note

A guided mode resonance can be understood as the extension of the "Wood anomaly" first described in 1902, when Wood, observing the spectrum of a continuous light source given by a metallic diffraction grating, noticed a surprising phenomenon: "I was astounded to find that under certain conditions, the drop from maximum illumination to minimum, a drop certainly of from 10 to 1, occurred within a range of wavelengths not greater than the distance between the sodium lines" (Wood, 1902). Wood was unable to explain the observation, so he termed it a "singular anomaly." Rayleigh later explained that the anomaly occurs at the wavelength for which a scattered wave emerges tangentially to the grating surface (Rayleigh, 1907), in other words, when the diffracted light lies in the plane of the grating. In the simplest case, i.e., for normal incidence, for the first diffraction order ($m = 1$), the grating equation.

$$\sin(\theta_m) = \frac{m \cdot \lambda}{n \cdot g} \tag{1}$$

Eq. (1) tells us that the diffraction angle $\theta_m = 90$ degrees occurs when the wavelength λ equals the grating period g, which is the point at which the sharp drop in intensity observed by Wood occurs. The refractive index n

in the denominator highlights that it is the wavelength in the material where the diffracted wave travels that needs to be considered, which is the same property that provides the sensing function as highlighted in Fig. 1. A much more comprehensive explanation of the Wood anomaly can be found in the chapter by Maystre (2012).

With the emergence of guided-wave optics in the 1960s, the Anomaly was reconsidered in the context of a guided wave approach. (Hessel and Oliner, 1965). Magnusson, in the 1990s, then recognized that the effect could be extended by creating the grating in a higher refractive index layer that supports guided modes (Wang et al., 1990). The basic physics remained the same, so Eq. (1) still holds, except that one needs to consider the effective index of the guided mode n_{eff}, so the wavelength λ in Eq. (1) becomes λ/n_{eff}.

The idea to track the sharp "drop" observed by Wood as a function of refractive index and to use it for sensing applications was first mentioned in 1992 (Magnusson and Wang, 1992) and the first basic GMR sensor was shown to exhibit sensitive shifts in wavelength to changes in refractive index in the late 1990s. A comprehensive summary of GMR sensor history is available from Magnusson et al. (2011).

2. Figures of merit for sensing

In order to describe the performance of a GMR sensor and to compare it to other sensor modalities, it is essential to define a universal figure of merit (FOM).

Such FOM allows us to benchmark the sensor and to identify routes for optimizing its performance. The FOM then yields the limit of detection (*LOD*), which is the single most important number used to describe a sensor. In terms of the optical performance, the *LOD* is often expressed as the minimum resolvable wavelength shift, e.g., as an *LOD* of $\Delta n = 10^{-6}$. Using the refractive index is convenient in terms of sensor design and assessment and a useful figure to optimize for the photonics engineer. The smallest measurable *LOD* (Armbruster and Pry, 2008) is defined as:

$$\mathrm{LOD} = \mathrm{LOB} + 3\sigma \quad (2)$$

where limit of blank (*LOB*) is the analyte concentration measured when no analyte is present which means *LOB* is merely an offset for the *LOD* hence Eq. (2) can actually be thought of as $LOD \approx 3\sigma$. This consideration specifies the LOD in terms of the refractive index shift that is three times larger than the noise level.

The real purpose of the sensor, however, is to detect the concentration of a biomarker. Therefore, the *LOD* can also be expressed as a biological concentration. For example, the clinically relevant concentration of prostate specific antigen (PSA) in blood is considered to be 4 ng/mL, so a given PSA sensor should have an *LOD* better than this value in order to allow for error-free clinical decisions to be made.

Using the biological concentration as the *LOD* adds a significant level of complication as there are a number of additional factors influencing the operation; devices with very similar optical performance may yield very different biological *LODs*, as the surface chemistry used to functionalize the sample may not always be exactly reproducible when compared to the nanofabrication, which is usually easier to control. Overall, we therefore classify the factors influencing the *LOD* as photonic, electronic and biochemical.

2.1 Photonic factors

Photonic factors include the linewidth or Q-factor of the resonance and its sensitivity to refractive index changes. The Q-factor, typically defined as the ratio of the resonance's central wavelength over its full width half maximum describes the sharpness of the resonance. This is beneficial to sensor operation because a higher Q-factor allows the central wavelength of the resonance to be more accurately defined and can therefore be tracked better. The sensitivity then expresses the shift of the resonance in terms of the wavelength shift $\Delta\lambda$ as a function of external refractive index changes, Δn, thus sensitivity $S = \Delta\lambda/\Delta n$, commonly expressed in nm/RIU, with RIU the dimensionless "refractive index unit." It is clear that the product of Q-factor and sensitivity is a good starting point for the desired FOM. For instance, a GMR sensor exhibits a typical Q-factor of 100–200 and a sensitivity of a similar magnitude, i.e., 100–200 nm/RIU. By comparison, a surface plasmon resonance (SPR) sensor has a sensitivity of 1000 nm/RIU or more, but only a Q of 10 or 20, hence $S \times Q \approx 10^4$–10^5 is similar in both cases (Altintas et al., 2011; Cunningham et al., 2002).

2.2 Electronic factors

Electronic factors mainly describe noise. The light source has instabilities described by its intensity and phase noise, with lasers having additional wavelength noise. The detectors add further thermal noise as well as noise related to the discretization/pixilation in case detector arrays or cameras are being used. Many of these noise sources can be mitigated, e.g., by using advanced algorithms or simply by incrementing the integration time, but it is often difficult to quantify the overall impact of these factors. Several authors have suggested that optical sensing methods are ultimately limited by the photon shot noise, but in practice, all the other sources of noise tend to dominate well before the shot noise limit is reached.

Because noise sources are variable and very much depend on the particular setup being used, including the stability of the laboratory equipment, the FOM is typically *not* expressed as $LOD = k \times Q \times S$, with k some proportionality factor that accounts for the electronic noise factors. In fact, earlier work by White and Fan (2008) attempted to define such a universal FOM, but we note that their figure is not used in the field. Instead, researchers tend to quote Q-factor, sensitivity and *LOD* separately.

As a general rule, high performance sensors achieve *LOD* values of $\Delta n \approx 10^{-5}$–10^{-6}. Putting these values into context with the refractive index variation of water, which is of order $\Delta n = 10^{-4}/°\mathrm{C}$ (Bashkatov and Genina, 2003), we note that achieving an *LOD* of $\Delta n = 10^{-6}$ requires a temperature stability of $\Delta T = 10^{-2}$ degrees. Even though *LODs* of $\Delta n \approx 10^{-7}$–10^{-8} have been reported, it is clear that such values can only be achieved with highly stabilized laboratory equipment and by using liquid analytes that have been preconditioned.

2.3 Biochemical factors

Biochemical factors arguably provide the largest variability. To illustrate this point, we refer to the detection of the carcinoembryonic antigen (CEA) as an example, which is a well-known colorectal cancer marker of mass 180 kDa. Using a nanobeam cavity with a bulk sensitivity of 70 nm/RIU and a Q-factor of 10^4, Liang et al. demonstrated an *LOD* of 0.1 pg/mL (Liang et al., 2013). For the same biomarker, SPR-based devices are only able to reach 3 ng/mL, using the Biacore system (Altintas et al., 2011), which has $Q \approx 10$ and $S \approx 3000$ nm/RIU. The $S \times Q$ product yields an advantage of one order of magnitude in favor of the nanobeam, while the *LOD* in terms of concentration yields an advantage of four orders of magnitude for the same system. Given the limited information available in the respective papers, one can only assume that lower noise and variations in the biochemical functionalization strategies are responsible for these large differences in observed performance.

These biochemical factors include the choice of capture molecule, e.g., as naturally derived antibodies or synthetically manufactured molecules such as affimers or DNA/Peptide aptamers; the choice of surface preparation, e.g., with silane monolayers that act as intermediaries between the sensor surface and the capture molecule; the density of coverage, the pH, the microfluidic environment, but also whether the target molecule is present in a more complex matrix such as blood plasma or urine. The latter point is particularly important as many papers use controlled buffers to determine their limits of detection, which yields somewhat idealized results; if the measurements are carried out in real matrices, nonspecific binding can easily screen the detection of the desired binding events and yield inferior results. It is for these rather complex reasons that sensors with comparable photonic properties may report *LODs* ranging from the ng/mL range into the sub-pg/mL range based on differences in biochemistry alone.

3. Operation of GMR sensors

Having established the most relevant figures of merit, we can consider the operation of GMR sensors in more detail as well as illustrating their operational principle.

First of all, a significant characteristic of a GMR structure is the ease of coupling of light into the grating with out-of-plane illumination. This is in contrast to nonleaky waveguide-based sensors such as microrings (Barrios, 2012) or bimodal waveguides (Herranz et al., 2017). Nonleaky waveguide sensors may achieve higher sensitivity, but they are also more difficult to implement in a practical point-of-care device, because coupling to a waveguide involves careful alignment with tight tolerances, even when a grating coupler is used. A GMR does not require such tolerances and high efficiency coupling can readily be achieved, even with simple collimated LEDs in a rudimentary setup (Block et al., 2008; Triggs et al., 2015).

3.1 Optical signature

As described in Section 1, a characteristic peak in reflection is observed when the diffracted light couples into the waveguide layer. In fact, the thickness of the waveguide layer can be adjusted to yield the inverse, i.e., a peak in transmission, because the leaky mode may interfere constructively or destructively with the directly transmitted light (Rosenblatt et al., 1997). For the sensing application, however, it is more convenient to operate in reflection as it eases the integration of the photonics/electronics, implementing them on one side of the sensor and incorporating the biology/fluidics on the other.

3.2 Polarization dependence

A typical GMR that consists of a 1D grating exhibits birefringence and thus supports two different modes depending on the polarization of the incident light. When the incident light is polarized perpendicular to the grating vector, the resulting mode is called a TE mode, and it is called a TM mode for polarization along the grating vector. TE and TM modes interact differently with the surrounding medium due to the polarization dependent boundary conditions at the interfaces, which yields different distributions of the electric field amplitude, as illustrated by the cross-sectional plots shown in Fig. 2B and C. In particular, the signature of the TM mode exhibits an electric field component in the out-of-plane direction (vertical direction in Fig. 2) because the propagation direction and the electrical field cannot be collinear. Because of these different boundary conditions, the polarization also impacts on the penetration depth of a given mode into the grating, with TM modes propagating for a longer distance before being Bragg-reflected. This longer propagation length means that TM modes experience more grating elements and correspondingly, they exhibit a higher quality factor compared to TE modes.

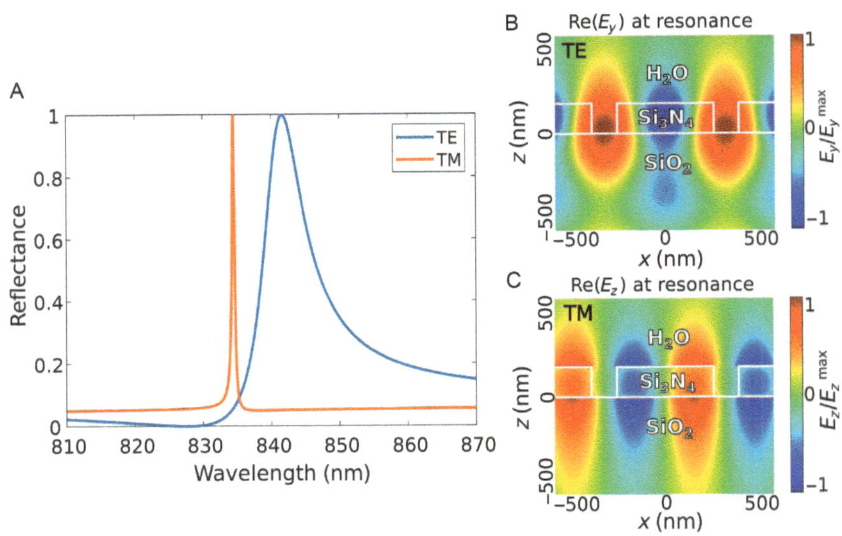

Fig. 2 Examples for the reflectance (A) and spatial distribution of the dominant electric field component at resonance (B, C) for TE and TM modes, respectively. The simulated structure is a Si_3N_4-grating ($n_{Si_3N_4} = 2$) with a thickness of 150 nm, a period of 555 nm and a filling factor of 80% on a SiO_2 substrate ($n_{SiO_2} = 1.45$). Water is assumed to fill the grating grooves and the superstrate ($n_{liquid} = 1.333$) at normal incidence from the substrate layer.

A more detailed explanation for the difference between TE and TM is as follows. The difference lies in their symmetry and the fact that the Q-factor is limited by scattering/radiation loss. As shown in Fig. 2B and C, the field of the TE mode is symmetric with respect to the grating grooves, i.e., the sign of the field is identical on both sides of the groove. This means that light can readily scatter by exciting an out-of-plane radiating wave. The field of the TM mode, however, is asymmetric with respect to the grating grooves, so the sign of the field on one side of the groove is the opposite to the sign on the other. This means that the two scattered components cancel out each other and cannot excite a radiating plane wave (Lalanne et al., 2006). Therefore, the TM mode propagates much further into the grating before eventually scattering out.

3.3 *Q*-factor and sensitivity

The Q-factor difference between the TE and the TM mode is clearly evident in the simulation of Fig. 2, which shows a TM mode with a Q-factor of $Q = 1669$ and a TE mode in the same structure with a much lower Q-factor of $Q = 104$. As the Q-factor is an important element in determining the FOM for sensing, it is clear that the TM mode is advantageous in this respect.

It is useful to understand that GMRs have no intrinsic loss, in contrast to, e.g., the plasmonic structures discussed elsewhere, so the reflectance peaks at 100%. In terms of experimental realization, however, lower reflectance and quality factor are typically being observed, especially for the high-Q TM mode. One obvious limitation is material roughness, which leads to random scattering. The other is the angular dependence of the resonance, i.e., the fact that the resonance sensitively depends on the angle of the incoming beam. While the simulation assumes perfectly collimated light, a real beam tends to cover an angular range, leading to a broadening of the resonance by the nature of the measurement.

The second contributor to the FOM introduced in Section 2 is the sensitivity *S*. The sensitivity describes the change of the resonance wavelength as a function of refractive index, which is directly related to the overlap of the field with the analyte (Block et al., 2008). The ideal situation would be a perfect overlap of the field with the biomolecular binding region in the analyte, but since we are dealing with guided modes that are bound to a high index region, only the tail of the mode overlaps with the analyte, commonly known as evanescence wave sensing, which is discussed in more detail in Section 3.5.

In terms of sensitivity, we need to distinguish between *bulk sensitivity* and *surface sensitivity*. The bulk sensitivity is what is usually quoted, and it measures the overlap of the mode with the analyte. In contrast, the surface sensitivity only measures the overlap between the mode and a thin film on the surface, typically the first 10–20 nm. These numbers are explained by the small size of the relevant molecules; for example, a typical protein (e.g., 40–50 kDa) is of order 5–8 nm in size. Since the purpose of biosensing is usually to measure refractive index changes in a thin film of, e.g., proteins bound to the surface, the surface sensitivity is, among others, the most useful parameter. Unfortunately, surface sensitivity is more difficult to measure and is hence rarely quoted. Instead, most papers quote the bulk sensitivity.

3.4 Bulk sensitivity

The bulk sensitivity measures the response to the refractive index of the entire superstrate (also referred to as the cover). In the case of Fig. 2, the bulk sensitivity of the two modes is similar with 148 nm/RIU (TE) and 154 nm/RIU (TM). The nature of the field confinement is clearly different, however, with 43% of the overall bulk sensitivity of the TE mode relying on the field residing in the grooves, compared to less than 5% for the TM mode.

This means that the analyte has to enter the grooves to be sensed by the TE mode. Given the small size of the grooves (typ. 100 nm) there may be wetting issues in a practical sensing situation, such that the analyte may not be able to diffuse into these grooves rapidly. This consideration therefore suggests that the TE mode is the less favorable choice and that the TM mode is more advantageous, i.e., the same conclusion as in the Q-factor discussion.

Fig. 3 shows the reflectance of the same GMR grating for TM polarization as a function of the superstrate refractive index. It is clear that the dependence is not linear, which is due to variations in mode confinement; if the index of the superstrate layer is low, the mode is strongly confined to the waveguide layer and the overlap between the mode and the analyte is low, resulting in low sensitivity. For higher index, the mode is drawn into the analyte and sensitivity increases. The interesting region for biosensing is a superstrate index around 1.33–1.34, i.e., the index of water or slightly above, since the goal of biosensing is to measure low concentrations of biomarkers in water-based matrices such as blood plasma or urine. In this regime, the resonance shift with respect to the water baseline is sufficiently linear to justify describing it as a linear FOM in terms of nm/RIU.

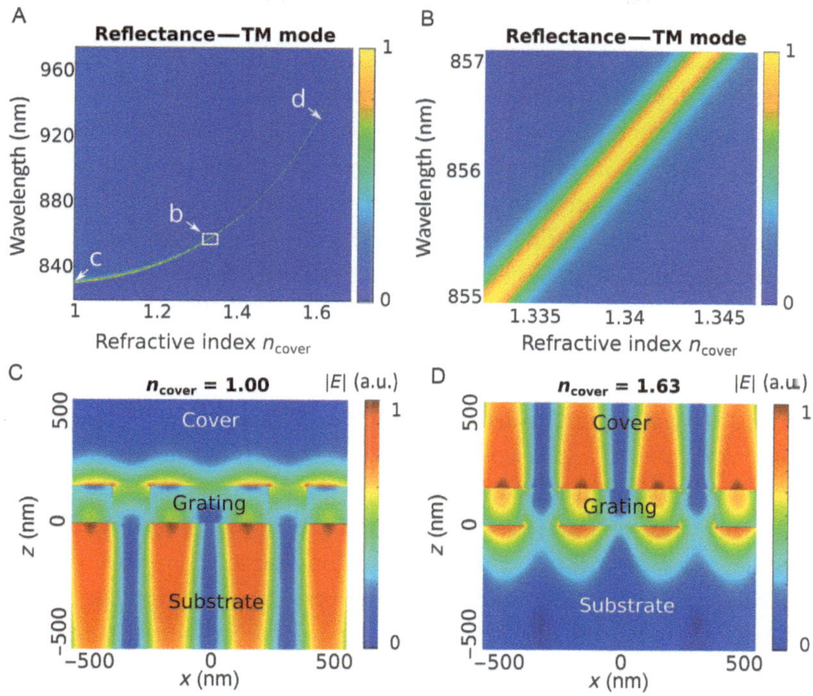

Fig. 3 (A) Resonance map for TM mode showing reflectance as a function of the incidence wavelength and refractive index of the cover layer. Incidence is from the substrate layer. (B) Linear range of sensitivity around the baseline index of water as typically used in biosensing. (C,D) Field plots illustrating the mode confinement for the extremes of lowest and highest refractive index of the cover layer.

3.5 Surface sensitivity

As already stated, the bulk sensitivity is only a proxy for the performance of a sensor and it would be pertinent to quote the surface sensitivity instead. Ultimately, the sensitivity is determined by the overlap of the resonant mode with the analyte. As is well known from guided-mode theory (Yariv, 1991), the evanescent tail of the guided mode can be described in terms of the effective index of the mode n_{eff} and the refractive index of the superstrate n_{sup}.

$$E(z) = E_0 e^{\left(-\frac{z}{p}\right)} \quad p = \frac{2\pi}{\lambda}\sqrt{\left(n_{eff}^2 - n_{sup}^2\right)} \tag{3}$$

with p being the decay-constant in the transverse direction (here: z-direction, using the same convention as in Figs. 2 and 3). Typical values for the decay length $1/p$ in the guided mode systems considered here, i.e., dielectric waveguides, surface plasmons, and GMRs are of order 100–200 nm. This length scale highlights the key issue, namely, that the size of the mode tail tends to be much larger than the biolayer. In order to account for this discrepancy, we need to define another parameter, namely, the surface sensitivity, which describes the overlap integral between the mode and a thin film on the surface of the sensor. Unfortunately, there is no accepted, standardized method for determining the surface sensitivity. Cunningham et al., for example, define a "detection zone" of 25 nm (Block et al., 2008), whereas we believe that it is more realistic to assume a thickness of $t = 10$ nm, which is closer to the size of an antibody-bound protein. To illustrate the impact of the effective index (Eq. 3) we exemplify the mode size and corresponding surface sensitivity on two separate structures, i.e., a GMR realized in silicon nitride ($n_{Si_3N4} = 2$) and one realized in silicon ($n_{Si} = 3.56$), with refractive index being the only variable and the other parameters being kept constant.

The first observation is that the modal field of the Si_3N_4 structure extends much further into the cladding, in direct agreement with Eq. (3), because its effective index is lower. The second observation is that the Q-factor of the silicon grating is much lower; this is because the light propagating into the grating is scattered much more strongly at the silicon interface than at the nitride interface, so the mode in the silicon structure has a much shorter penetration depth L_p, as also discussed in Section 5 (Eq. 4), and a correspondingly lower Q-factor.

Using the definition of a 10 nm thick biolayer, one can then compare the surface sensitivity of the different structures. In terms of refractive index of the biolayer, it is now understood that biomolecules have a refractive index of $n \approx 1.56$ (Young et al., 2018). Using this value would assume that the entire surface layer is densely packed with molecules, however. In reality, packing is not as dense, and one needs to assume some average effective index somewhere between $n = 1.56$ and $n = 1.33$ (water). We chose a value of $n = 1.4$, in part also because it yields a maximum wavelength shift of around $\Delta\lambda \approx 1$ nm (Table 1), which corresponds to experimental observations. In our model, the Si_3N_4 grating shows a surface sensitivity of $\Delta\lambda = 0.9$ nm and the Si grating of $\Delta\lambda = 1.5$ nm. All the relevant parameters are compiled in Table 1 for ease of comparison.

Table 1 clearly highlights the difference between bulk sensitivity and surface sensitivity. While both structures exhibit similar bulk sensitivities of around 180 nm/RIU, the surface sensitivity differs almost by a factor 2.

Table 1 Parameter comparison between Si and Si_3N_4 gratings as displayed in Fig. 4 and their resulting bulk and surface sensitivities

Grating material	Refractive index	Effective index	Bulk sensitivity (nm/RIU)	Surface sensitivity (nm; $n = 1.4$, $t = 10$ nm)
Silicon	3.56	2.63	187	1.50
Silicon nitride	2.00	1.75	176	0.87

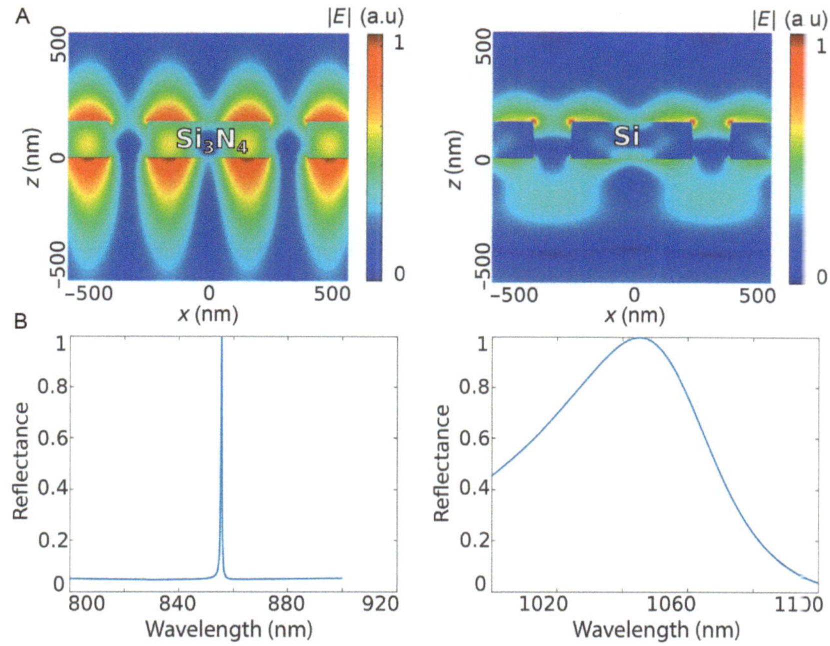

Fig. 4 (A) Electric field profiles of TM modes at resonance for two different dielectric materials and (B) the resulting GMR spectra. All parameters are the same as in Fig. 2 except for the material refractive index.

This difference is a direct consequence of Eq. (3); since the effective index in the silicon structure is higher, the evanescent tail is shorter and the overlap with the biolayer is larger.

An alternative way of highlighting the difference in surface sensitivity is shown in Fig. 5, which is a plot of wavelength shift as a function of biolayer thickness, similar to the study performed in Otte et al. (2009). Using the same structures as in Fig. 4, the graph illustrates the higher surface sensitivity of the higher index silicon grating, especially for a small biolayer thickness. For a large thickness, the two curves converge, as is expected for the similarity in bulk sensitivity.

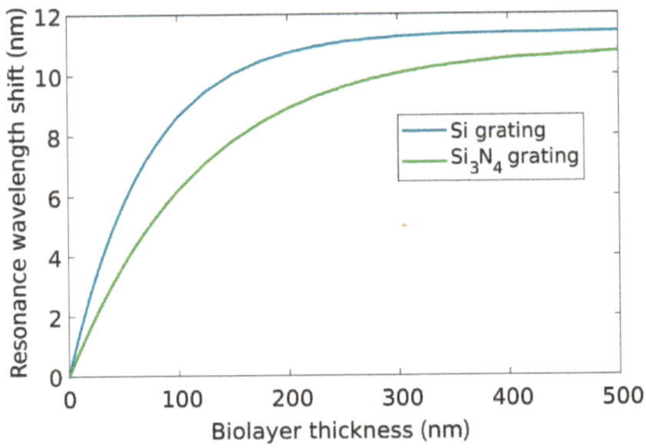

Fig. 5 Simulated resonance wavelength shift for varying biolayer thicknesses assuming a constant refractive index $n_{\text{biolayer}} = 1.4$ on the sensing surface for the two gratings defined in Fig. 4

4. Comparison between GMR sensors and other sensing modalities

Some of the advantages of GMR sensors such as the ease of light in-coupling have already been mentioned. This advantage is traded off against lower Q-factor, so it is very instructive to tension the GMR against competing approaches. The most obvious of the competitors are (a) the widely considered "gold standard" of label-free biosensing, namely, SPR commercialized by Biacore(now GE®), the microring resonator, commercialized by Genalyte®, and the bimodal waveguide interferometric biosensor, pioneered by Lechuga's group at ICN2, Barcelona.

4.1 Surface plasmon resonance

SPR biosensors were developed in the 1980s as one of the first embodiments of optical biosensors (Liedberg et al., 1983). Like GMR, they use out-of-plane excitation, which makes it relatively easy to couple to. The mode that is used for transduction is a surface plasmon mode, however, which is based on a surface charge oscillation at a metal-dielectric interface. The surface plasmon can be understood as a traveling wave and as such, it is excited by *k*-vector matching at the interface; the incident light is directed through a prism or a high-index hemisphere, and by adjusting the angle, its propagation constant can be adjusted to match that of the plasmon mode, illustrated in Fig. 6A. When the *k*-vectors do not match, the light from the prism does not couple into the plasmon and the light is reflected. When the angle varies

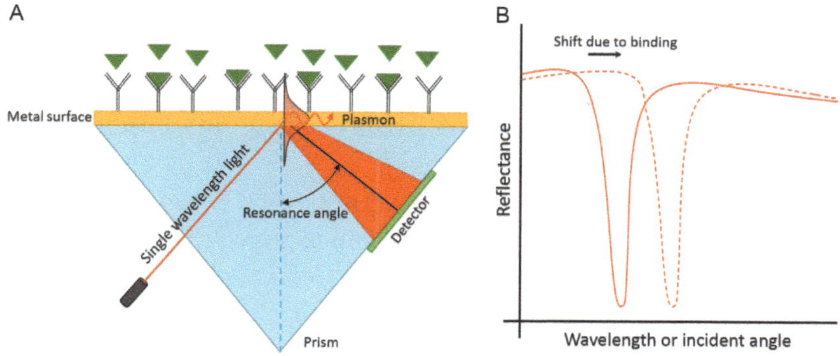

Fig. 6 (A) Operation principle of a surface plasmon resonant sensor and (B) observation of the characteristic dip upon coupling.

and they do match, the light does couple and the plasmon is excited. Since the plasmon is a lossy mode, a lot of the light is dissipated before it couples back, so a characteristic dip is observed in reflection (Fig. 6B). It is the traveling wave nature of the surface plasmon that gives it its exquisite sensitivity, because the smallest changes in effective index of the plasmon mode will lead to a resonance shift (Yu and Fan, 2011). As a result, sensitivities of several 1000 nm/RIU have been reported (Pfeifer et al., 1999).

Because of the lossy nature of the plasmon mode, SPR systems have very low quality factors, however, i.e., of the order of $Q = 10$–20 which is the lowest of the sensing modalities we explore here. The low Q factor is offset by the high sensitivity, however, so the overall FOM $S \times Q \approx 10^4$–10^5 nm/RIU which is comparable to most biosensing platforms including the GMR. As a result, the typical implementation of SPR, such as in Biacor's SPR system, exhibits a LOD in the ng/mL range (Altintas et al., 2011; Homola, 2008), making it useful for immunoassays, while it is not the most sensitive of all available biosensors.

4.2 Nanohole arrays

Nanohole arrays represent a hybrid between SPR and GMR, because they merge plasmon modes with periodic arrays. Nanohole arrays first came to the fore upon the observation of extraordinary optical transmission (EOT). EOT describes the phenomenon whereby light is observed to transmit through an array of subwavelength holes with a transmission that is much higher than expected from geometry alone. EOT is typically explained as a three-step process whereby light first couples to the surface plasmon at the metal interface, the plasmon then propagates through the holes and couples back to a free-space mode on the other side (Genet and Ebbesen, 2007). The conversion to plasmons explains how light can transmit through tiny holes that are well below any classical cutoff (Liu and Lalanne, 2008; Martín-Moreno et al., 2001), which is why the effect is termed "extraordinary" optical transmission.

The recognition that nanohole arrays can be used for sensing came later with key demonstrations by Altug et al. who explored the resonance characteristics and relevant sensing applications (Cetin et al., 2015).

The resonances they studied usually exhibit Q factors of the order of $Q \approx 100$ when working with liquids (Yanik et al., 2011) and sensitivities of up to 650 nm/RIU which yields a FOM of $S \times Q \approx 10^4$–10^5, which is comparable albeit slightly higher than SPR and GMR structures. As a result, nanoholes have exhibited impressive detection limits down to 500 pg/mL for the real-time molecular binding of IgG and 145 pg/mL for vascular endothelial growth factor (VEGF) (Li et al., 2017).

These values demonstrate the strength of the nanohole array method as it combines the benefits of GMR with its ease of coupling with the sensitivity of SPR. The only downside is that the transmission is relatively low ($T \approx 10\%$–20%) which results in weak signals. Since nanoholes only work in transmission, it also means that the collection optics interferes with the fluidic access, i.e., detection electronics and fluidics are on the same side of the device. As stated previously, the fact that GMR and SPR can operate in reflection means that the optics and the electronics can be placed on one side of the device and the fluidics on the other, which is much more elegant for practical implementations.

4.3 Microrings

Microrings are guided-mode devices where light propagates in the form of a whispering gallery mode which is resonant when an integer number of wavelengths fit into the circumference of the ring. The mode is strongly confined to the waveguide (Fig. 7) and is a proper long-lived guided mode (unlike the leaky mode of the GMR or the very lossy mode of the SPR), so very high Q-factors can be achieved. In fact, Q-factors into the $Q \approx 10^8$–10^9 have been achieved with microdisk resonators (Armani et al., 2003) but these are not typically used for sensing as they are impractically high; as explored by Hu et al. (2009), Q-factors above 10^5 are not useful in this context because the wavelength noise of typical lasers then limits the accuracy of determining the resonance. In terms of sensitivity, microrings tend to be

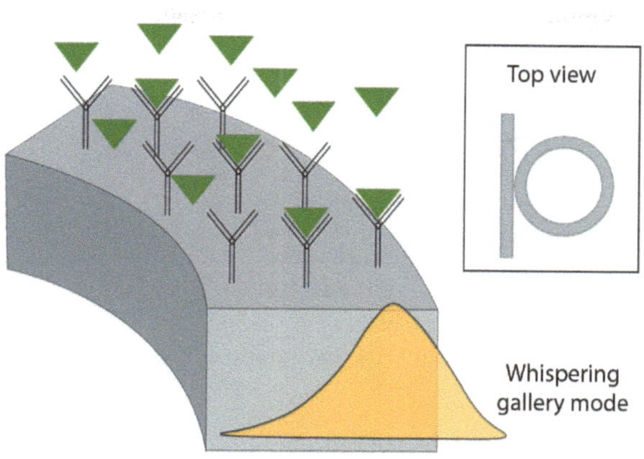

Fig. 7 Operation principle of microring resonator sensor with the *top view* also showing the access waveguide.

worse than the other modalities, because the same strong confinement to the waveguide that is needed for the high Q-factor tends to limit the overlap of the mode with the analyte, such that sensitivities of $S \approx 50$–$80\,\text{nm/RIU}$ are typically being achieved (Bogaerts et al., 2012; Juan-Colás et al., 2017b). Nevertheless, due to the high Q-factor, the overall FOM tends to be quite high, i.e., $S \times Q \approx 10^5$–10^6, which is higher than the other modalities explored so far.

4.4 Bimodal waveguide

Bimodal waveguides represent the class of interferometric waveguide sensors. Bimodal waveguides obtain the sensing information by interfering two modes that coexist in the same waveguide. The advantage of this approach is the "common-path" nature of the interferometer, which ensures high stability and which means that the signal and the reference experience exactly the same external influences. This feature is in contrast to the more commonly employed Mach–Zehnder geometry whereby the signal and the reference experience different paths, so any noise source that is specific to one path does not occur on the other so cannot be normalized out.

The principle of the bimodal interferometer is sketched in Fig. 8. The device starts with a single mode input. The waveguide thickness is then changed at the mode splitter which serves to excite the first order mode. The two modes then propagate with different propagation constants and have a different overlap with the analyte, so they experience a different phase delay on interaction. At the output, the two modes are made to interfere and

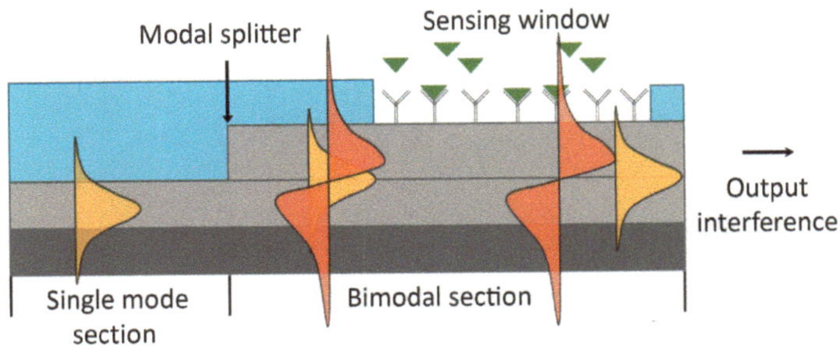

Fig. 8 Operation principle of a bimodal waveguide sensor.

the interference pattern represents a direct measure for the relative phase change between the two modes (Dante et al., 2015). Bimodal waveguides are an elegant solution because they are less complex than conventional interferometers and they have demonstrated detection limits of 2.5×10^{-7} RIU (Herranz et al., 2017).

4.5 Concluding remarks

The various forms of optical biosensor all have benefits and drawbacks, each taking advantage of the modal properties of light. By considering the various modalities, it becomes clear that a higher Q value tends to dominate the $S \times Q$ FOM, however, due to biological considerations, this does not always lead to a low LOD in terms of bioparticle detection. A high Q tends also to reduce the sensitivity, although this reduction is small compared to the gain in Q-factor.

5. Resonant hyperspectral imaging

A key advantage of GMR over most other sensing modalities is its imaging functionality, whereby one can interpret each sensing element as the pixel of an image. By its very nature, the guided mode is spatially confined with a finite lateral extent, so it can be used to detect local changes in refractive index. The lateral extent of the GMR mode can be estimated from Bragg mirror theory (Block et al., 2009) and is expressed as a propagation length L_p, i.e., the distance at which the field has decayed to $1/e$ of its original amplitude,

$$L_p = \frac{\lambda^2}{2\pi\Delta\lambda} \tag{4}$$

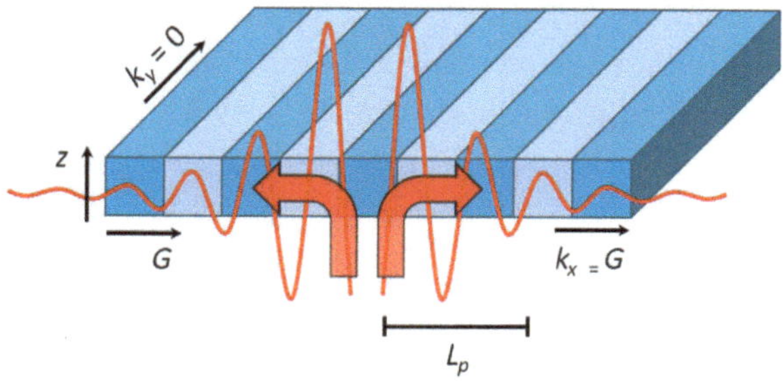

Fig. 9 Illustration of the spatial resolution dependence on orientation.

For the TE mode shown in Fig. 2, i.e., for a center wavelength of $\lambda = 840\,\text{nm}$ and a linewidth of 8 nm, the equation yields a resolution of $L_p = 14\,\mu\text{m}$. In practice, it turns out that the $1/e$ criterion is rather severe and that much better resolution values can be achieved, with 3.5 μm reported in Block et al. (2009) and 6 μm reported in Triggs et al. (2015). A key point to recognize is that the resolution depends on the grating orientation, as illustrated in Fig. 9.

The resolution values typically quoted refer to the resolution along the grating vector, so along the x-direction in Fig. 9, as is also clear from the definition via the propagation length L_p. This configuration assumes the light to be incident in z-direction and to be perfectly collimated, so to have no wavevector components k_x and k_y. The grating then imparts its grating vector G onto the wave, such that $k_x = G$, so the light is deflected into x-direction. Therefore, the spatial resolution is determined by the grating vector and L_p in x-direction. In y-direction, there is no grating vector, so the resolution in y-direction is typically much better and is only limited by the diffraction limit. This observation is somewhat counterintuitive: even though the grating is essential for generating the GMR and for providing the sensing functionality, it actually deteriorates the spatial resolution. Additionally, the resolution depends on the refractive index contrast, with lower contrast yielding poorer resolution (Triggs et al., 2015).

Together, these considerations highlight the trade-off between sensing and imaging. Sensing aims to achieve high Q-factors for a high sensing FOM $S \times Q$, while imaging requires low Q-factors for a high spatial resolution.

Recording an image then requires three-dimensional data, i.e., the x–y coordinates of each pixel together with its wavelength information. Any two of these dimensions can be acquired simultaneously while the third dimension needs to be acquired sequentially for each pixel. The two main recording methods described in the literature are (a) white light illumination in combination with an imaging spectrometer, i.e., a spectrometer that disperses a line of pixels and scans the image line by line (Zhuo et al., 2017) or (b) scanning the illumination wavelength and taking full frame pictures at every wavelength (Triggs et al., 2015). Both methods are in principle equivalent but have different experimental requirements. Arguably, scanning the illumination wavelength (b) is simpler because the imaging optics remains static, while scanning the image (a) either requires scanning the beam or the moving the sample, both of which need to be performed with high accuracy.

This recording method is commonly known as hyperspectral imaging. The resulting dataset, based on its three-dimensional nature, is known as a hyperspectral cube. Hyperspectral imaging was first applied to GMR by Cunningham et al., who termed it photonic crystal enhanced microscopy, PCEM (Chen et al., 2013) although here, we prefer to use the more generic term of resonant hyperspectral imaging.

Valuable examples for the application of resonant hyperspectral imaging can be found in Chen et al. (2013) who demonstrated that it was possible to monitor cell attachment to the surface over extended periods of time (12–24 h) with high stability. This ability to measure over extended periods of time is important when studying stem cell differentiation and related processes. Similarly, drug-induced cell death was monitored over several hours and cell movement, in response to a concentration gradient (chemotaxis), over several days. These examples highlight some of the unique advantages of resonant label-free imaging because alternative methods such as fluorescence have difficulty in maintaining a constant signal over extended periods of time because fluorophores bleach during long exposures.

Similarly, GMR imaging can be used for the early detection of biofilms. The early detection of biofilms is particularly important in an industrial context, where labeled methods such as fluorescence cannot be used. Many commercial biofilm detectors use bacterial autofluorescence, but this requires relatively high surface concentrations to be present; instead, the GMR approach can already see the initial formation of the protein layer which precedes the actual biofilm formation, so it detects biofilms at a much earlier stage (Fischer et al., 2016).

Fig. 10 A representation of how a hyperspectral dataset is used to construct a resonance map from a stack of monochromatic images.

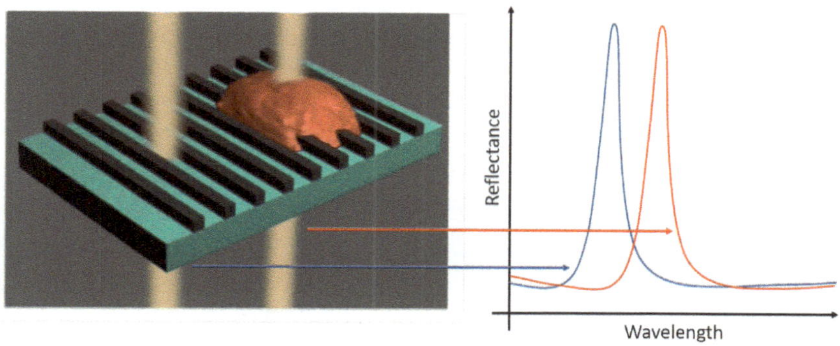

Fig. 11 Illustration of resonant hyperspectral imaging.

5.1 Sensing and imaging

The examples quoted above have illustrated the scope for using guided mode resonances as an alternative imaging modality, mainly for monitoring the behavior of cells. Naturally, this modality can be further enhanced by adding the sensing functionality discussed in Section 3. Combining imaging with sensing therefore offers a unique combination that is difficult to achieve otherwise; other sensing methods cannot provide imaging information, while other imaging methods do not provide sensing information. For example, the well-established imaging method of phase contrast microscopy is sensitive to phase changes, but it is very difficult to detect the presence of a single protein layer. Resonant imaging can combine both functionalities, as illustrated in Figs. 10 and 11.

Fig. 12 provides a schematic and Fig. 13 shows experimental evidence to illustrate the combined imaging and sensing capabilities. The BHK cell in Fig. 13 secrets thrombopoietin, a cytokine essential for maintaining physiological levels of circulating platelets in blood. The purpose of the study was

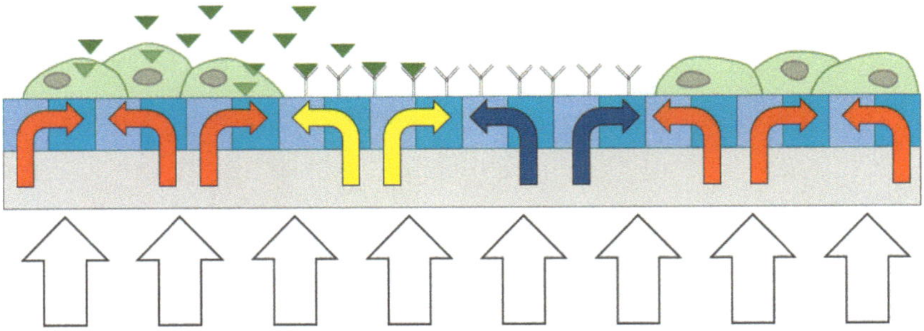

Fig. 12 Schematic of combining both imaging and sensing on the same GMR platform. Cells attached to the surface change the resonance wavelength of the hyperspectral image from *blue* to *red*, as in Fig. 11. When suitably stimulated, they secrete protein biomarkers that are detected via complementary antibodies immobilized on the surface; on binding, the resulting phase change is detected via a change in the hyperspectral image from *blue* to *yellow*.

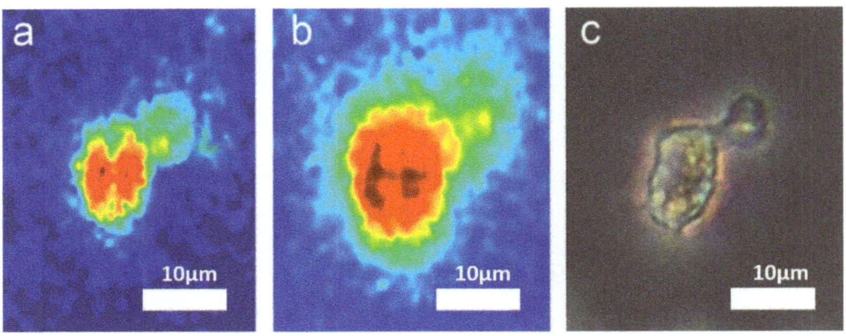

Fig. 13 Comparison of resonant imaging vs phase contrast imaging. All three images show a baby hamster kidney (BHK) cell secreting thrombopoietin (TPO). Panel (A) represents the resonant image at the beginning of the experiment, when only the cell is present ($t = 0$ min). Panel (B) shows a resonant image combining the cell and the secreted proteins at $t = 85$ min. Panel (C) shows a phase contrast micrograph of the cell for reference, which remains unchanged in the presence of the proteins. It is clear that the resonant image can see both the cell and the proteins it secrets while the phase contrast image can only see the cell.

to understand how TPO secretion is regulated in the human body by the concentration of desialylated (aged) platelets in the bloodstream. More detail can be found in Juan-Colas et al. (2018). While the cell can be more clearly distinguished in the phase contrast image of Fig. 13C, TPO secretion cannot be identified. In Fig. 13B, however, the footprint of the cell is significantly larger, and a "halo" of resonance change emerges surrounding the cell due to the antigen-bound TPO, thereby indicating the secretion of proteins.

5.2 Multiparameter imaging

The methods described in the previous sections have enabled the imaging of refractive index changes due to the presence of cells, the binding of proteins, or both. Refractive index changes are detected via the effective index of the guided mode by tracking the central wavelength of the resonance.

The resonance has other properties that can be used for sensing purposes, however. In particular, its height and Q-factor are indicative of optical losses in the system. While the ideal GMR exhibits a reflectance on resonance of 100% (see Figs. 2 and 4), the presence of losses reduces the magnitude of the resonance peak. This reduction allows us to extract more information from the biological system. For example, areas of rapid change of density act as scattering centers that lead to a reduction in the height of the resonance peak. Such rapid changes of density occur around, for instance, cellular "focal adhesions," i.e., the areas at the periphery of a cell associated with adhesion which tend to be higher in molecular density than the cell body itself. The fact that the resonance peak drops around focal adhesions was first recognized by Zhuo et al. (2018) and is illustrated in Fig. 14 taken from Juan-Colás and Krauss (2018).

Fig. 14 makes it clear that the guided mode resonance can detect two separate effects thereby providing more information about the biological system than the wavelength shift alone.

Alternatively, the multiparameter capability of the GMR can provide depth information by a combination of resonance shift and Q-factor. The depth-dependence directly relates to the extent of the evanescent tail of the mode as discussed in Eq. (3), so it applies as long the thickness of the object is within the tail of the resonant mode, while the Q-factor provides information on the refractive index. This capability was first highlighted by Nazirizadeh et al. (2013) where the authors observed that the refractive index of the object impacts on the observed Q-factor of the resonance, similar to the discussion in Figs. 3 and 4; the higher the index of the cover, the lower the index contrast seen by the guided mode, so the further its penetration depth L_p (Eq. 4) and the higher its Q-factor. Using the Q-factor for refractive index information showed an excellent correlation with both phase contrast and atomic force microscopy, thereby providing another impressive example of the multiparameter capability of the GMR approach.

5.3 Plasmonic nanohole arrays

While imaging is an exciting capability of the GMR approach, it is not a unique capability; other platforms offer similar functionality and they are therefore worth mentioning for context. In particular, the plasmonic

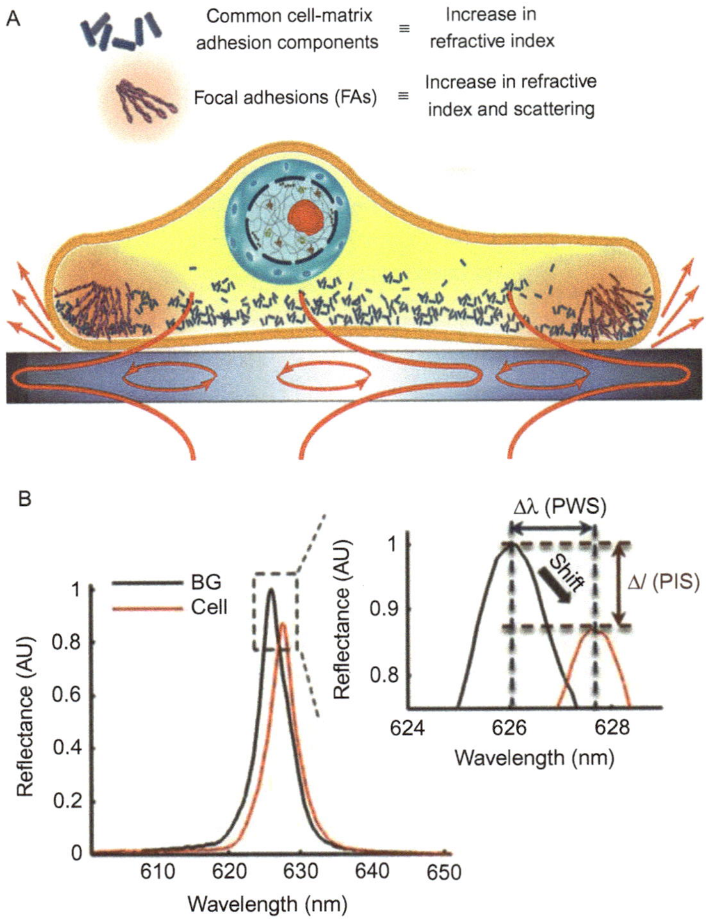

Fig. 14 Illustration of the separate effects of a cell on the optical resonance; while the cell body increases the effective index of the resonant mode and shifts the resonance compared to the background, the additional scattering at the edges of the cell responsible for adhesion (focal adhesions) can be separately identified via a reduction in the amplitude of the resonance peak. (A) Illustration of the scattering at focal adhesions due to their rapid change in density; (B) Corresponding experimental evidence for the refractive index change and additional scattering loss. PWS refers to peak wavelength shift and PIS to peak intensity shift. *Panel (A) taken from Juan-Colás, J. & Krauss, T. F., 2018. Multiparameter resonant imaging for studying cell interactions.* Light Sci. Appl., 7. *Panel (B) taken from fig. 1c of Zhuo, Y. et al., 2018. Quantitative analysis of focal adhesion dynamics using photonic resonator outcoupler microscopy (PROM).* Light Sci. Appl., *7(1), 9.*

nanohole approach has demonstrated imaging with high sensitivity and high spatial resolution. Plasmonic nanohole arrays fabricated in optically thick metal films and they support the EOT phenomenon, which is attributed to the excitation of plasmons by grating coupling of light at normal incidence; these plasmons result in a strong field enhancement in the vicinity of the nanoholes. This strong light confinement enables nanohole arrays to be highly sensitive to minute local refractive index changes, such as those induced by the binding of biomolecules to the sensor surface. In addition, the localization of light around the nanohole provides high spatial resolution. A good example of this capability is provided by Li et al. (2017) where the authors demonstrate high resolution imaging (spatial resolution not quantified) and high sensitivity sensing of VEGF, a signaling protein, directly in complex cell media. The VEGF was secreted from cancer cells attached to the sensor surface, providing a similar functionality as the resonant image shown in Fig. 13.

Nevertheless, the nanohole approach also has its downsides. Intrinsically, even though the transmission is considered "extraordinary," it is relatively low in absolute terms, i.e., around 10%, which limits the signal to noise ratio. In terms of practicality, EOT being observed in transmission also means that the fluidics needs to be integrated into the optical system, i.e., source and detector are on opposite sides of the microfluidic circuitry. In contrast, the GMR, preferentially working in reflection, allows a clean separation between the optical system on one side of the setup and the fluidics on the other; working in reflection therefore mimics the inverted microscope arrangement favored by biologists. In terms of substrate, the nanohole array tends to be fabricated on free-standing silicon nitride membranes (Li et al., 2017), which are fragile and expensive and therefore not suited for low-cost point-of-care applications. Clearly, when identifying the most suitable approach for a given application, it is important to consider all aspects (Juan-Colás et al., 2017a; Pitruzzello and Krauss, 2018) (Table 2).

6. Spatial-spectral equivalence

It is intrinsic to a resonance-based optical sensor that its response is qualified as a function of wavelength; in fact, most of the figures shown in this chapter use wavelength as the parameter to describe a particular aspect of sensor performance. While it is easy to perform spectral characterization in a research laboratory, doing so in the field is more difficult and it means that a spectrometer has to be included into the system in order to provide the readout. Spectrometers are not cheap; despite the efforts of a number of

Table 2 Comparison of different sensing modalities

Structure	Analyte	Q factor	Bulk sensitivity (nm/RIU)	FOM (nm/RIU)	LOD	References
SPR (Biacore)	CEA antigen	~20	$\sim 10^3$	$\sim 2 \times 10^4$	3 ng/mL	Altintas et al. (2011)
GMR	Different Solutions	~85	510	4.3×10^4	—	Huang et al. (2009)
GMR	Streptavi-din	~150	88	1.3×10^4	1 ng/mL	Cunningham et al. (2002)
Microrings	MCP-1	2×10^4	—	—	$0.5\ pg\,mL^{-1}$	Valera et al. (2016) and Washburn et al. (2009)[a]
Bmodal waveguide	HCL Solutions	—	6×10^2 πrad/RIU	—	2.5×10^{-7} RIU	Herranz et al. (2017)
Bimodal Waveguide	hGH	—	6×10^2 πrad/RIU[b]	—	$10\ pg\,mL^{1}$	González-Guerrero et al. (2017)
Nanobeam Cavity	CEA antigen	1×10^4	~70	$\sim 7 \times 10^5$	$0.1\ pg\,mL^{-1}$	Liang et al. (2013)
Nanohole Array	IgG	<100	700	$\sim 7 \times 10^4$	$145\ pg\,mL^{-1}$	Li et al. (2017)

[a]Multiple references used as specific details on Genalyte Maverick™ systems are not openly available.
[b]Assumed to be the same as other work(s) from the same research group due to the same structure being used.

commercial companies who have made major advances in bringing low-cost spectrometers to the market, these devices still represent a significant cost element amounting to US$ 100s or 1000s, depending on the desired spectral resolution. Integrated solutions based on silicon photonics processed in CMOS fabs are emerging that challenge this view (Martens et al., 2018), but nevertheless, adding a spectrometer into the sensor is still an overhead.

The very nature of the GMR approach offers an alternative. As highlighted at the beginning of this chapter, a GMR is based on the grating equation, $\sin\theta = m\lambda/ng$, and the excitation of a resonance requires a particular angle θ, typically close or equal to 90 degrees. In order to obtain the desired value for θ, one can either adjust the wavelength λ or the period g. In combination with the imaging capability, it is then possible to place different gratings in the same field of view, each of which exhibits a different resonance wavelength, so different locations correspond to different wavelengths; this method translates the spectral coordinate into a spatial coordinate. The change in period/resonance wavelength can either be discrete or it can be continuous over the length of the grating, the latter giving rise to the idea of the "chirped GMR" introduced by Triggs et al. (2017) and illustrated in Fig. 15.

When illuminating the chirped grating with a fixed wavelength source such as a laser, the resonance will occur as a horizontal line in the field of view. Upon binding of the target molecules, the refractive index increases as illustrated in Fig. 1, which will lead to the line moving in the field of view; the sensing information is therefore provided by the position of the line.

The same idea was pursued by Tittl et al. (2018), who opted for the discrete translation of the spectral into the spatial information and refer to their approach as "barcoding"; instead of using a 1D GMR grating, they used a

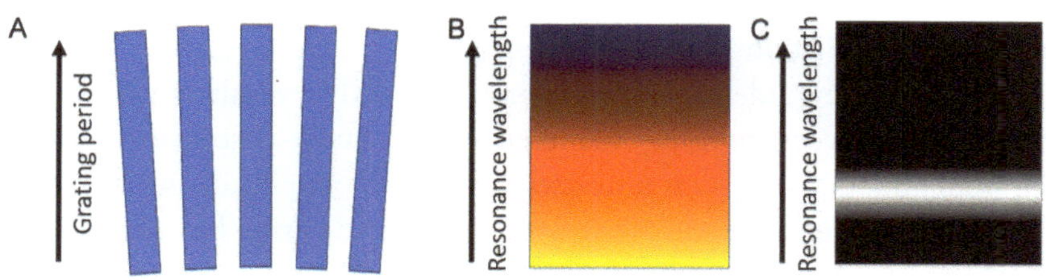

Fig. 15 Illustration of the chirped GMR concept which translates spectral dimension into spatial dimension and combines the sensing and the spectral readout function in a single structure. (A) Design of grating with spatially varying period in one direction. (B) Hyperspectral image illustrating the continuous change of the resonance wavelength as a color code. (C) Illumination with a single wavelength source yields a single line of resonance.

2D pixelated metasurface composed of an array of high-Q resonant metapixels with their resonance frequencies tuned by scaling their size. The idea is the same as in the chirped GMR, namely, that different positions in the field of view map onto different wavelengths. The authors applied this method in the mid-IR, around 6 μm wavelength, to identify various polymers attached to the surface by their vibrational absorption lines. Both the chirped GMR and the pixelated metasurface provide impressive illustrations of the versatility of this approach, which, ultimately only depends on the very simple grating equation everybody learns in High School.

7. Phase-sensitive measurement

All the modalities described so far record the *amplitude* of the light interacting with the biological material, i.e., the cells and proteins. We record the amplitude of the transmission or reflection, the amplitude of the resonance shift or the Q-factor, which is the spectral width of the amplitude response. Nevertheless, amplitude describes only one aspect of the optical interaction and arguably, most advanced photonic devices make use of the phase for advanced functionality.

Recording phase requires an interferometric approach, which can indeed yield excellent results. For example, some of the most sensitive photonic biosensors are based on interferometry, such as the bimodal waveguide (Section 4.3), also (Schmitt et al., 2007) which exploits the interference between two different modes in a waveguide, or the Farfield approach (Cross et al., 2003). These approaches can achieve exquisite *LODs* of $\Delta n < 10^{-8}$ *RIU*. Is it possible to also benefit from the phase approach in the context of GMR?

In a way, the guided mode resonance is already a result of the in-phase interaction between multiply scattered waves, but it is still susceptible to amplitude fluctuations, so is ultimately limited by amplitude noise. Instead, a truly phase-sensitive measurement is limited only by phase noise. Phase noise tends to be significantly lower than amplitude noise for common light sources used in biosensing, including low-cost VCSELs. The phase noise of lasers is inversely proportional to their coherence length l_c:

$$\frac{\Delta\phi}{\phi} = \frac{\lambda}{l_c} \tag{5}$$

Coherence lengths of up to 10s of cm up can be assumed for typical lasers, which results in phase noise of the order of $\Delta\varphi/\varphi = 10^{-5}$–$10^{-6}$ for lasers in the visible or near-infrared range. In contrast, the intensity noise tends to be orders of magnitudes higher with $\Delta I/I \approx 10^{-3}$–$10^{-2}$ typically being observed (Kabashin et al., 2009). For example, a typical HeNe laser exhibits a coherence length of about 30 cm which corresponds to a phase noise of 632.8 nm/30 cm $\approx 2*10^{-6}$ (Thorlabs Inc., 2019a) and an RMS intensity noise of 0.5% $= 5*10^{-3}$ (Thorlabs Inc., 2019b). The performance of a single mode VCSEL is similar with coherence lengths of up to several meters (Dudzik and Rzepka, 2010; Pruijmboom et al., 2008) and RMS noise of 2% typically being reported (Thorlabs Inc., 2019c).

Another argument in favor of the phase approach is illustrated in Fig. 16. The phase changes most rapidly on resonance, where the amplitude curve peaks. At the peak, the amplitude curve is flat, so finding the peak means finding the center of a flat curve, which is difficult. Instead, the slope of the phase curve is largest on resonance and the center is therefore much easier to find.

The third argument in favor of interferometry is slightly more subtle and relates to the ability of minimizing the noise in the measurement. When recording an amplitude spectrum, one can decrease the noise by averaging over multiple spectra, so averaging in time by decreasing the bandwidth of the measurement. When recording phase, one can average over multiple periods of the interferogram without loosing bandwidth, which means that it is possible to reduce noise in postprocessing without loosing time-resolution.

7.1 Realization of phase-sensitive measurements

Phase-sensitive GMR measurements were considered theoretically as early as 2004 (Magnusson et al., 2004). The authors predicted an enhanced detection sensitivity when integrating the GMR into an interferometric experimental setup and discussed the correlation between Q-factor and phase sensitivity. The Q-factor, as discussed in the context of Fig. 4, directly relates to the penetration depth L_p, which is analogous to the propagation length L in an interferometric sensor. Recall that in a traveling wave-type arrangement such as a Mach–Zehnder or a bimodal waveguide, the relation between the active sensing pathlength L and the phase change due to effective index changes in the signal arm Δn_{eff} is given by

$$\Delta\phi = \frac{2\pi L}{\lambda}\Delta n_{\text{eff}} \tag{6}$$

So the phase change is directly proportional to the propagation length L, which is proportional to L_p for a GMR. In the traveling-wave geometry, L can be as long as millimeters or even centimeters, while L_p is restricted to 10s or 100s of micrometers. Even though the GMR is a resonance, which means it is a multipass device whereby the light oscillates back and forth multiple times, the overall interaction length is shorter than in the traveling-wave approach, which explains the higher sensitivity of the interferometric sensor. This higher sensitivity is traded-off against ease of coupling as discussed before. Furthermore, as discussed in Section 2.2, it is questionable whether sensitivities of $\Delta n < 10^{-8}$ RIU are achievable except in the most highly stabilized laboratory environments, given that temperature fluctuations tend to limit the practical sensitivity achievable in the field to $\Delta n \approx 10^{-6}$–10^{-7} RIU. It is therefore not clear whether some of the ultrahigh sensitivities reported with interferometric sensors are practically useful. Nevertheless, it is an interesting question whether interferometry can improve the performance of a GMR sensor.

Realizing a phase-sensitive interferometric measurement requires a reference wave. An obvious way of implementing such a reference wave is to build a classical Mach–Zehnder setup whereby the beam is split into a signal and a reference path with the GMR acting as one of the mirrors in the signal path. This approach was first realized by Sahoo et al. (2017) who highlighted the ability to achieve higher sensitivity by using the phase approach. In Fig. 16, the corresponding reflectance and phase response of a GMR grating are shown, highlighting the strong variation in phase at the point of maximal reflectance corresponding to minimal intensity variation.

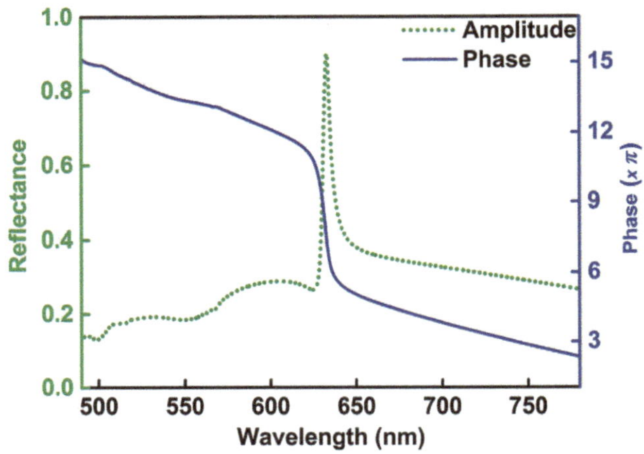

Fig. 16 Reflectance and phase response of GMR grating proposed in Sahoo et al. (2017).

Instead of placing the GMR into an external interferometer, which is very sensitive to alignment tolerances and vibrations, it is preferable to implement a "common-path interferometer" approach, which is much more elegant because the signal and reference beams experience the same geometrical path. Common-path interferometry is similar to the Nomarski differential interference contrast method known in microscopy, whereby the differences in optical pathlength experienced by two orthogonally polarized beams are exploited to enhance imaging contrast interferometrically.

One can learn a lot about common-path interferometry from the related SPR approach. A common-path SPR sensor was first demonstrated in 2005 (Su et al., 2005). This sensor exploits the fact that the SPR is polarization sensitive, i.e., only p-polarized light excites a surface plasmon wave, while s-polarized light is reflected specularly from the gold layer. The s-polarization therefore does not interact with the analyte so can serve as a reference. The idea was developed further by Kabashin et al. (2009) who demonstrated a LOD of $\Delta n \approx 10^{-8}$ RIU, which is comparable to the traveling-wave geometries discussed above, even though the interaction length in an SPR is much shorter.

8. Conclusion

Having now reviewed the various properties of guided mode resonances for sensing applications, one can conclude that GMR arguably provides the most versatile of all optical sensing approaches. It affords high sensitivity, albeit not the ultrahigh sensitivity achieved by other sensor modalities, while also offering a range of other functionalities, such as imaging and multiparameter sensing which are much more difficult or not possible to achieve with other modalities. Similarly, the idea of mapping the spectral response onto the spatial domain as exemplified by the chirped GMR approach is a unique feature that allows combining the sensor element with the wavelength-selective readout element in the same structure; most other optical sensing modalities require an external spectrometer. Together with the ease of coupling to the GMR directly with a collimated beam, the versatility and inherent simplicity make the GMR very suitable for high performance, yet low-cost diagnostic devices that may be deployed in the field and that are much needed by healthcare providers both in the developed and in the developing world.

In terms of scientific developments, the idea of exploiting the phase change on resonance is yet in its infancy and further improvements can be expected, which is further evidence for the versatility of the GMR sensing modality.

References

Altintas, Z., Uludag, Y., Gurbuz, Y., Tothill, I.E., 2011. Surface plasmon resonance based immunosensor for the detection of the cancer biomarker carcinoembryonic antigen. Talanta 86, 377–383. 30 10.

Armani, D.K., Kippenberg, T.J., Spillane, S.M., Vahala, K.J., 2003. Ultra-high-Q toroid microcavity on a chip. Nature 421 (6926), 925.

Armbruster, D.A., Pry, T., 2008. Limit of blank, limit of detection and limit of quantitation. Clin. Biochem. Rev. 29 (Suppl. 1), S49.

Barrios, C.A., 2012. Integrated microring resonator sensor arrays for labs-on-chips. Anal. Bioanal. Chem. 403 (6), 1467–1475.

Bashkatov, A.N., Genina, E.A., 2003. Water refractive index in dependence on temperature and wavelength: a simple approximation. Optical Technologies in Biophysics and Medicine IV. vol. 5068. pp. 393–396.

Block, I.D., Ganesh, N., Lu, M., Cunningham, B.T., 2008. A sensitivity model for predicting photonic crystal biosensor performance. IEEE Sensors J. 8 (3), 274–280.

Block, I.D., et al., 2009. Optimizing the spatial resolution of photonic crystal label-free imaging. Appl. Opt. 48 (34), 6567–6574.

Bogaerts, W., et al., 2012. Silicon microring resonators. Laser Photonics Rev. 6 (1), 47–73.

Cetin, A.E., et al., 2015. Plasmonic nanohole arrays on a robust hybrid substrate for highly sensitive label-free biosensing. ACS Photonics 2 (8), 1167–1174.

Chen, W., et al., 2013. Photonic crystal enhanced microscopy for imaging of live cell adhesion. Analyst 138 (20), 5886–5894.

Cross, G.H., et al., 2003. The metrics of surface adsorbed small molecules on the Young's fringe dual-slab waveguide interferometer. J. Phys. D. Appl. Phys. 37 (1), 74.

Cunningham, B., Li, P., Lin, B., Pepper, J., 2002. Colorimetric resonant reflection as a direct biochemical assay technique. Sensors Actuators B Chem. 81 (2–3), 316–328.

Dante, S., et al., 2015. Linear readout of integrated interferometric biosensors using a periodic wavelength modulation. Laser Photonics Rev. 9 (2), 248–255.

Dudzik, G., Rzepka, J., 2010. Single frequency stable VCSEL as a compact source for interferometry and vibrometry. AIP Conference Proceedings. vol. 1253 (1), pp. 317–320.

Fischer, M., Triggs, G.J., Krauss, T.F., 2016. Optical sensing of microbial life on surfaces. Appl. Environ. Microbiol. 82 (5), 1362–1371.

Genet, C., Ebbesen, T.W., 2007. Light in tiny holes. Nature 7123, 39.

González-Guerrero, A.B., et al., 2017. Direct and label-free detection of the human growth hormone in urine by an ultrasensitive bimodal waveguide biosensor. J. Biophotonics 10 (1), 61–67.

Herranz, S., Gavela, A.F., Lechuga, L.M., 2017. Label-free biosensors based on bimodal waveguide (BiMW) interferometers. In: Methods in Molecular Biology. Springer, pp. 161–185.

Hessel, A., Oliner, A.A., 1965. A new theory of Wood's anomalies on optical gratings. Appl. Opt. 4 (10), 1275–1297.

Homola, J., 2008. Surface plasmon resonance sensors for detection of chemical and biological species. Chem. Rev. 108 (2), 462–493.

Hu, J., Sun, X., Agarwal, A., Kimerling, L.C., 2009. Design guidelines for optical resonator biochemical sensors. J. Opt. Soc. Am. B 25 (5), 1032–1041.

Huang, M., Yanik, A.A., Chang, T.-Y., Altug, H., 2009. Sub-wavelength nanofluidics in photonic crystal sensors. Opt. Express 17 (26), 24224–24233.
Juan-Colás, J., Krauss, T.F., 2018. Multiparameter resonant imaging for studying cell interactions. Light Sci. Appl. 7, 45.
Juan-Colás, J., Johnson, S., Krauss, T.F., 2017a. Dual-mode electro-optical techniques for biosensing applications: a review. Sensors (Switzerland) 17 (9), 1424–8220.
Juan-Colás, J., Krauss, T.F., Johnson, S.D., 2017b. Real-time analysis of molecular conformation using silicon electrophotonic biosensors. ACS Photonics 4 (9), 2320–2325.
Juan-Colas, J., et al., 2018. Quantifying single cell secretion in real time using resonant hyperspectral imaging. Proc. Natl. Acad. Sci. U. S. A. 115 (52), 13204–13209.
Kabashin, A.V., Patskovsky, S., Grigorenko, A.N., 2009. Phase and amplitude sensitivities in surface plasmon resonance bio and chemical sensing. Opt. Express 17 (23), 21191–21204.
Lalanne, P., Hugonin, J.P., Chavel, P., 2006. Optical properties of deep lamellar gratings: a coupled Bloch-mode insight. J. Lightwave Technol. 24 (6), 2442–2449.
Li, X., et al., 2017. Plasmonic nanohole array biosensor for label-free and real-time analysis of live cell secretion. Lab Chip 17 (13), 2208–2217.
Liang, F., et al., 2013. Scalable photonic crystal chips for high sensitivity protein detection. Opt. Express 21 (26), 32306–32312.
Liedberg, B., Nylander, C., Lunström, I., 1983. Surface plasmon resonance for gas detection and biosensing. Sensors Actuators 4, 299–304.
Liu, H., Lalanne, P., 2008. Microscopic theory of the extraordinary optical transmission. Nature 452 (7188), 728.
Magnusson, R., Wang, S.S., 1992. New principle for optical filters. Appl. Phys. Lett. 61 (9), 1022–1024.
Magnusson, R., Lee, K.J., Wawro, D., 2004. Guided- mode resonance biosensors employing phase detection. In: Diffractive Optics and Micro-Optics (p. DTuC2). Optical Society of America.
Magnusson, R., Wawro, D., Zimmerman, S., Ding, Y., 2011. Resonant photonic biosensors with polarization-based multiparametric discrimination in each channel. Sensors 11 (2), 1476–1488.
Martens, D., et al., 2018. Low-cost integrated biosensing platform based on SiN nanophotonics for biomarker detection in urine. Anal. Methods 10, 3066–3073.
Martín-Moreno, L., et al., 2001. Theory of extraordinary optical transmission through subwavelength hole arrays. Phys. Rev. Lett. 86 (6), 1114.
Maystre, D., 2012. Theory of Wood's anomalies. In: Plasmonics. Springer, pp. 39–83.
Nazirizadeh, Y., et al., 2013. Material-based three-dimensional imaging with nanostructured surfaces. Appl. Phys. Lett. 102 (1), 11116.
Otte, M.A., et al., 2009. Identification of the optimal spectral region for plasmonic and nanoplasmonic sensing. ACS Nano 4 (1), 349–357.
Pfeifer, P., et al., 1999. Real time sensing of specific molecular binding using surface plasmon resonance spectroscopy. Sensors Actuators B Chem. 54 (1–2), 166–175.
Pitruzzello, G., Krauss, T.F., 2018. Photonic crystal resonances for sensing and imaging. J. Opt. 20 (7), 73004.
Pruijmboom, A., Schemmann, M., Hellmig, J., Schutte, J., Moench, H., Pankert, J., 2008. VCSEL-based miniature laser-Doppler interferometer. In: Vertical-Cavity Surface-Emitting Lasers XII. vol. 6908. International Society for Optics and Photonics, p. 69080I.
Rayleigh, L., 1907. III. Note on the remarkable case of diffraction spectra described by Prof. Wood. In: The London, Edinburgh, and Dublin Philosophical Magazine and Journal of Science. vol. 14. Taylor & Francis, pp. 60–65. 79.
Rosenblatt, D., Sharon, A., Friesem, A.A., 1997. Resonant grating waveguide structures. IEEE J. Quantum Electron. 33 (11), 2038–2059.

Sahoo, P.K., Sarkar, S., Joseph, J., 2017. High sensitivity guided-mode-resonance optical sensor employing phase detection. Sci. Rep. 7 (1), 7607.

Schmitt, K., et al., 2007. Interferometric biosensor based on planar optical waveguide sensor chips for label-free detection of surface bound bioreactions. Biosens. Bioelectron. 22 (11), 2591–2597.

Su, Y.-D., Chen, S.-J., Yeh, T.-L., 2005. Common-path phase-shift interferometry surface plasmon resonance imaging system. Opt. Lett. 30 (12), 1488–1490.

Thorlabs Inc., 2019a. HeNe Tutorial. [Online] Available at: https://www.thorlabs.com/tutorials.cfm?tabID=9F3258E0-FB6B-46D4-944D-B31A84EF1DAA [Accessed 23 11 2018].

Thorlabs Inc., 2019b. HeNe Lasers: Red. [Online] Available at: https://www.thorlabs.com/newgrouppage9.cfm?objectgroup_id=1516 [Accessed 23 11 2018].

Thorlabs Inc., 2019c. VCSEL Diode Module (CPS850V). [Online] Available at: https://www.thorlabs.com/thorproduct.cfm?partnumber=CPS850V.

Tittl, A., et al., 2018. Imaging-based molecular barcoding with pixelated dielectric metasurfaces. Science 360 (6393), 1105–1109.

Triggs, G.J., et al., 2015. Spatial resolution and refractive index contrast of resonant photonic crystal surfaces for biosensing. IEEE Photonics J. 7 (3), 6801810.

Triggs, G.J., et al., 2017. Chirped guided-mode resonance biosensor. Optica 4 (2), 229.

Valera, E., Shia, W.W., Bailey, R.C., 2016. Development and validation of an immunosensor for monocyte chemotactic protein 1 using a silicon photonic microring resonator biosensing platform. Clin. Biochem. 49 (1), 121–126.

Wang, S.S., Magnusson, R., Bagby, J.S., Moharam, M.G., 1990. Guided-mode resonances in planar dielectric-layer diffraction gratings. J. Opt. Soc. Am. A 7 (8), 1470–1474.

Washburn, A.L., Gunn, L.C., Bailey, R.C., 2009. Label-free quantitation of a cancer biomarker in complex media using silicon photonic microring resonators. Anal. Chem. 81 (22), 9499–9506.

White, I.M., Fan, X., 2008. On the performance quantification of resonant refractive index sensors. Opt. Express 16 (2), 1020.

Wood, R.W., 1902. XLII. On a remarkable case of uneven distribution of light in a diffraction grating spectrum. In: The London, Edinburgh, and Dublin Philosophical Magazine and Journal of Science. vol 4. Taylor & Francis, pp. 396–402. (21).

Yanik, A.A., et al., 2011. Seeing protein monolayers with naked eye through plasmonic Fano resonances. Proc. Natl. Acad. Sci. U. S. A. 108 (29), 11784–11789.

Yariv, A., 1991. Optical Electronics. Saunders College Publishings.

Young, G., et al., 2018. Quantitative mass imaging of single biological macromolecules. Science 360 (6387), 423–427.

Yu, Z., Fan, S., 2011. Extraordinarily high spectral sensitivity in refractive index sensors using multiple optical modes. Opt. Express 19 (11), 10029–10040.

Zhuo, Y., et al., 2017. Label-free imaging of stem cell adhesion and dynamic tracking of boundary evolution using photonic crystal enhanced microscopy (PCEM). Microsc. Microanal. 23 (S1), 1142–1143.

Zhuo, Y., et al., 2018. Quantitative analysis of focal adhesion dynamics using photonic resonator outcoupler microscopy (PROM). Light Sci. Appl. 7 (1), 9.

CHAPTER SIX

Fano resonance photonic crystal filters and modulators

Weidong Zhou[a,*], Shanhui Fan[b]

[a]Department of Electrical Engineering, University of Texas at Arlington, Arlington, TX, United States
[b]Department of Electrical Engineering, Stanford University, Stanford, CA, United States
*Corresponding author: e-mail address: wzhou@uta.edu

Contents

1. Fano resonance photonic crystal filters

1.1 Single-layer filters

As an example of how one might apply the theory in Chapter 1, we can construct single-pole filters using an individual resonance. For $r=1$ and therefore $t=0$, Eq. (1) gives the transmission spectrum

$$T=\frac{(1/\tau)^2}{(\omega-\omega_0)^2+(1/\tau)^2} \tag{1}$$

which exhibits a Lorentzian peak and such a configuration functions as a band-pass filter (Fig. 1A). For $r=0$ and therefore $t=1$, the transmission spectrum

$$T=\frac{(\omega-\omega_0)^2}{(\omega-\omega_0)^2+(1/\tau)^2} \tag{2}$$

Semiconductors and Semimetals, Volume 100
ISSN 0080-8784
https://doi.org/10.1016/bs.semsem.2019.04.003

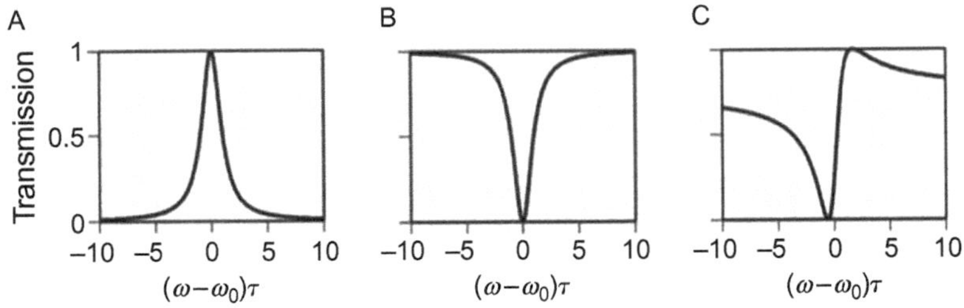

Fig. 1 Single-pole filter lineshapes. ω_0 and $\gamma = 1/\tau$ are the center frequency and width of the resonance, respectively. (A) A bandpass filter constructed by a resonance imposed on a total reflection background. (B) A notch filter constructed by a resonance imposed on a zero reflection background. (C) A Fano filter constructed by a resonance imposed on a partial reflection background. *Reproduced with permission from Zhou, W., Zhao, D., Shuai, Y.-C., Yang, H., Chuwongin, S., Chadha, A., Seo, J.-H., Wang, K. X., Liu, V., Ma, Z., 2014. Progress in 2D photonic crystal Fano resonance photonics. Prog. Quantum Electron. 38 (1), 1–74.*

features a band-rejection filter (Fig. 1B). With all intermediate background parameters being $0 < r < 1$ and $0 < t < 1$, the transmission spectrum given by Eq. (1) exhibits sharp asymmetric Fano lines (Fig. 1C). Many aspects in the spectral response can be controlled if appropriately designed (Thurman and Morris, 2003). Although we have been dealing with planewaves in the temporal coupled-mode theory discussions, similar phenomena can be seen with focused illumination as well (Crozier et al., 2006).

To construct double-pole filters, one should introduce an additional resonance in the photon transport through PCSs. This can be achieved by coupling two PCSs, which we will discuss later, or by using two resonances in a single PCS. A variety of double-pole filters can be constructed depending on, for example, the symmetry of the two resonances and the difference in frequency between the two resonances compared to the width of each individual resonance. In Fig. 2, we show graphs corresponding with different configurations. Fig. 2A is a collection of graphs displaying photon transport through two orthogonal resonances, in addition to the direct pathway. Fig. 2B displays what occurs when the two resonances are nonorthogonal. Because it is relatively difficult to construct two nonorthogonal resonances in a single-layer PCS, we will focus on the orthogonal case in this section and continue the discussion of the nonorthogonal case in Section 1.2, which is on coupled double-layer filters.

Relying on two resonances with opposite mirror symmetries in a single PCS, the structure can serve as an all-pass transmission or a flattop reflection filter (Suh and Fan, 2004). To have two resonances in the vicinity of a signal frequency, one needs to choose appropriate geometrical parameters for the photonic crystal slab. Fig. 3 shows such a configuration.

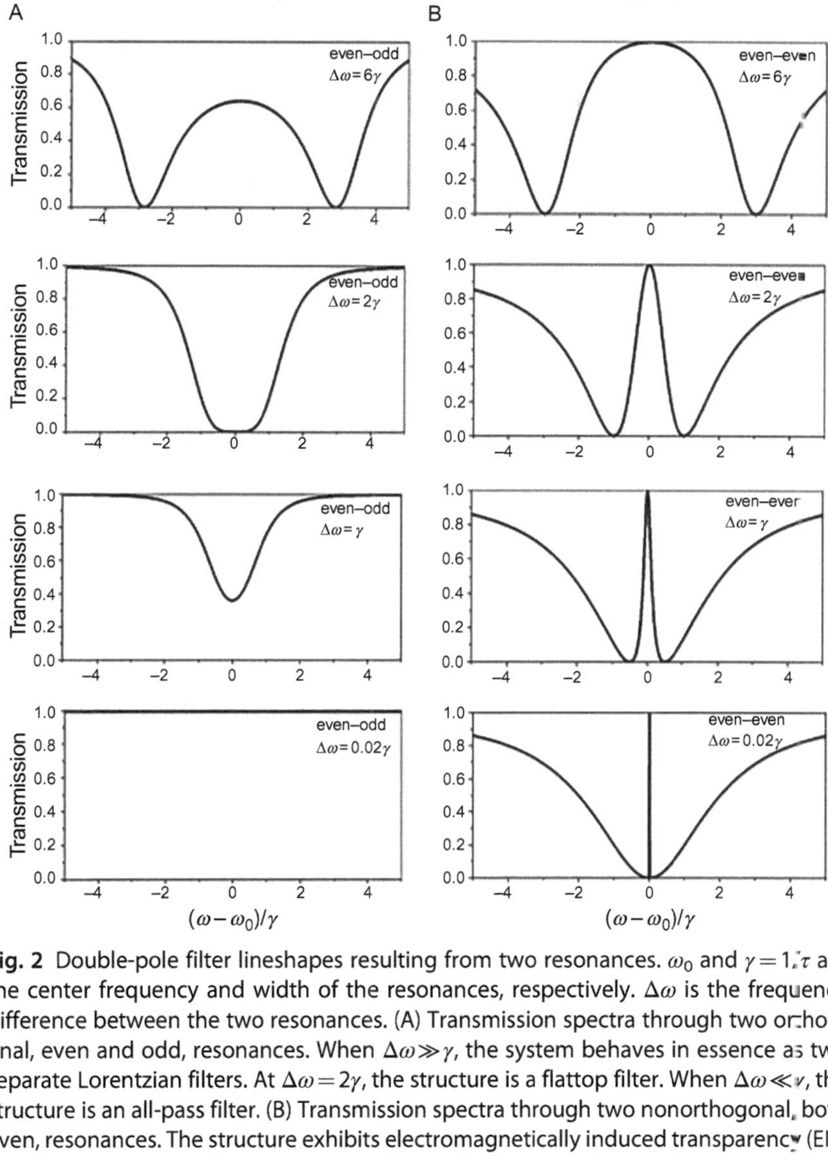

Fig. 2 Double-pole filter lineshapes resulting from two resonances. ω_0 and $\gamma=1/\tau$ are the center frequency and width of the resonances, respectively. $\Delta\omega$ is the frequency difference between the two resonances. (A) Transmission spectra through two orthogonal, even and odd, resonances. When $\Delta\omega \gg \gamma$, the system behaves in essence as two separate Lorentzian filters. At $\Delta\omega=2\gamma$, the structure is a flattop filter. When $\Delta\omega \ll \gamma$, the structure is an all-pass filter. (B) Transmission spectra through two nonorthogonal, both even, resonances. The structure exhibits electromagnetically induced transparency (EIT) lineshapes. *Reproduced with permission from Suh, W., Wang, Z., Fan, S., 2004. Temporal coupled-mode theory and the presence of non-orthogonal modes in lossless multimode cavities. IEEE J. Quantum Electron. 40 (10), 1511–1518.*

Because the resonances are orthogonal, similar to Eq. (1), the complex amplitude transmission coefficient can be written as

$$t=1-\frac{\gamma_{\text{even}}}{i\omega-i\omega_{\text{even}}+\gamma_{\text{even}}}-\frac{\gamma_{\text{odd}}}{i\omega-i\omega_{\text{odd}}+\gamma_{\text{odd}}} \tag{3}$$

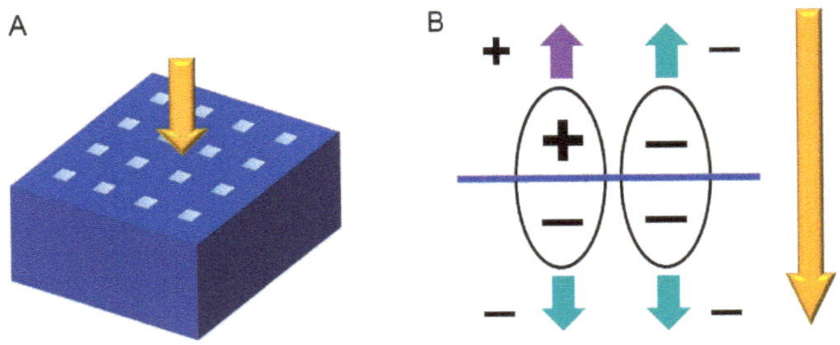

Fig. 3 (A) Photonic crystal slab structure consisting of a square lattice of air holes introduced into a high-index dielectric slab. The slab is 2.05*a*, thick and the radius of the holes is 0.12*a*, where *a* is the lattice constant. The *arrow* represents the direction of the incident light. (B) Schematic of a theoretical model for a resonator system that supports two resonant modes with opposite symmetry with respect to the mirror plane perpendicular to the incident light. *Reproduced with permission from Suh, W., Fan, S., 2004. All-pass transmission or flattop reflection filters using a single photonic crystal slab. Appl. Phys. Lett. 84, 4905.*

When $\gamma_{\text{even}} = \gamma_{\text{odd}} = \gamma$, $\omega_{\text{even}} = \omega_{\text{odd}} = \omega_0$, Eq. (3) can be simplified to:

$$t = \frac{i(\omega - \omega_0) - \gamma}{i(\omega - \omega_0) + \gamma} \tag{4}$$

and the photonic crystal slab becomes an all-pass filter. The intensity transmission is uniform for both on and off resonances. However, the phase delay varies from 0 to 2π in the vicinity of the resonance. This can be realized in the structure if we choose a dielectric constant of 10.07, as shown in Fig. 3. The all-pass transmission and the phase delay spectra are shown in Fig. 4. The theoretical derivation (Eq. 4) is verified by FDTD simulations (Suh and Fan, 2004).

When $\gamma_{\text{even}} = \gamma_{\text{odd}} = \gamma$, $|\omega_{\text{even}} - \omega_{\text{odd}}| = 2\gamma$, Eq. (3) becomes

$$T = |t|^2 = \frac{(\omega - \omega_0)^4}{(\omega - \omega_0)^4 + 4\gamma^4} \tag{5}$$

where $\omega_0 = (\omega_{\text{even}} + \omega_{\text{odd}})/2$ and flattop reflection is achieved in the vicinity of the resonance. This can also be realized in the structure shown in Fig. 3, if we choose a dielectric constant of 10.9. In Fig. 5, we plot the transmission spectrum. The theoretical derivation (Eq. 5) is again verified against FDTD simulations (Suh and Fan, 2004).

1.2 Coupled double-layer filters

All the filters we have shown so far are constructed on a single PCS. These filters can also be made with two coupled PCSs. This can be assumed because with additional layers, PCS filters offer higher quality factor and

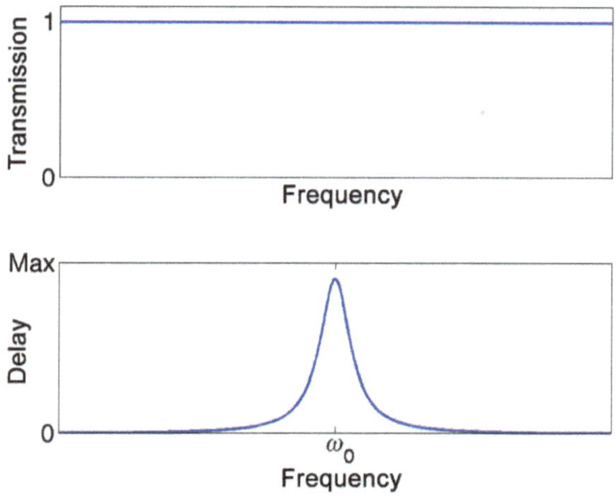

Fig. 4 Spectral response functions (intensity transmission and group delay) for the one-slab structure in Fig. 3 with a dialectic constant of 10.07. *Reproduced with permission from Zhou, W., Zhao, D., Shuai, Y.-C., Yang, H., Chuwongin, S., Chadha, A., Seo, J.-H., Wang, K. X., Liu, V. & Ma, Z., 2014. Progress in 2D photonic crystal Fano resonance photonics. Prog. Quantum Electron. 38 (1), 1–74.*

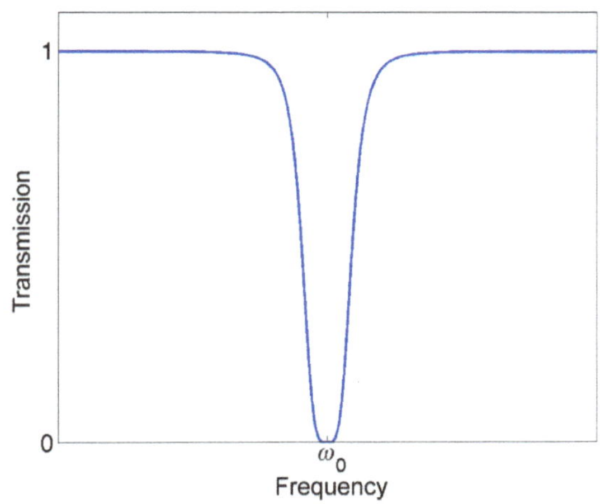

Fig. 5 Intensity transmission for the one-slab structure in Fig. 3 with a dialectic constant of 10.9. *Reproduced with permission from Zhou, W., Zhao, D., Shuai, Y.-C., Yang, H., Chuwongin, S., Chadha, A., Seo, J.-H., Wang, K. X., Liu, V. & Ma, Z., 2014. Progress in 2D photonic crystal Fano resonance photonics. Prog. Quantum Electron. 38 (1), 1–74.*

dispersion engineering capabilities (Shuai et al., 2013b). For example, by spacing two PCSs together closer than the wavelength, one can construct a structure for which the transmission is highly sensitive to the distance between the slabs, providing single-pole Fano and Lorentzian filters (Suh et al., 2005), as well as double-pole all-pass and flattop filters (Suh and Fan, 2003).

It is relatively straightforward to couple two resonances with the same symmetry in double-layer structures. Because the resonances are nonorthogonal, the transmission coefficient can no longer be intuitively written as a sum of the three pathways (two resonances plus the direct pathway) as in Eq. (3). Instead, one needs to solve the more general temporal coupled-mode equations (Suh et al., 2004). The amplitude transmission coefficient can be calculated as

$$t = t_d \mp \frac{(r_d \pm t_d)[\gamma_1(i\omega - i\omega_1) + \gamma_2(i\omega - i\omega_2)]}{(i\omega - i\omega_1 + \gamma_1)(i\omega - i\omega_2 + \gamma_2) - \gamma_1\gamma_2} \qquad (6)$$

where the top signs are used when both modes are even, and the bottom signs are used when both modes are odd. In Fig. 2B, we assume that both modes are even, and that $\omega_1 = \omega_0 - \Delta\omega/2$, $\omega_2 = \omega_0 + \Delta\omega/2$, and $\gamma_1 = \gamma_2 = \gamma$. The transmission behavior in Fig. 2B resembles that of electromagnetically induced transparency (EIT). At the resonant frequencies ω_1 and ω_2, the transmission is zero regardless of the spacing between the two resonant frequencies; while at ω_0, the transmission is at unity. This configuration was demonstrated using coupled PCSs (Suh and Fan, 2003).

In coupled PCSs, there exists extra opportunity to engineer dark resonances (Liu et al., 2009). A dark resonance is a guided resonance decoupled from particular free space radiation modes due to symmetry mismatch. We also refer to externally excited guided resonances as bright resonances. A dark resonance in coupled PCSs can arise from the coupling of dark resonances or bright resonances in individual slabs.

For two PCSs that each possess a dark resonance, bringing them close to each other allows their resonances to couple. If both slabs are aligned such that the proper symmetries are preserved, these coupled dark resonances remain dark. If we shift the relative position of these two slabs to break the symmetry, these coupled guided resonances are externally excited and become bright. This will result in two Fano resonances in close proximity, and the quality factors depend strongly on the lateral displacement. This type of dark resonance is illustrated in Fig. 6.

It is also possible to couple two bright resonances together to make them dark. For a pair of PCSs each with a bright resonance, we can decouple them from radiation modes by placing a Fabry-Perot resonance between the two slabs. At a frequency where the reflectivity of each individual slab is at unity, the bright resonances lose access to external radiation. This type of dark resonance is illustrated in Fig. 7.

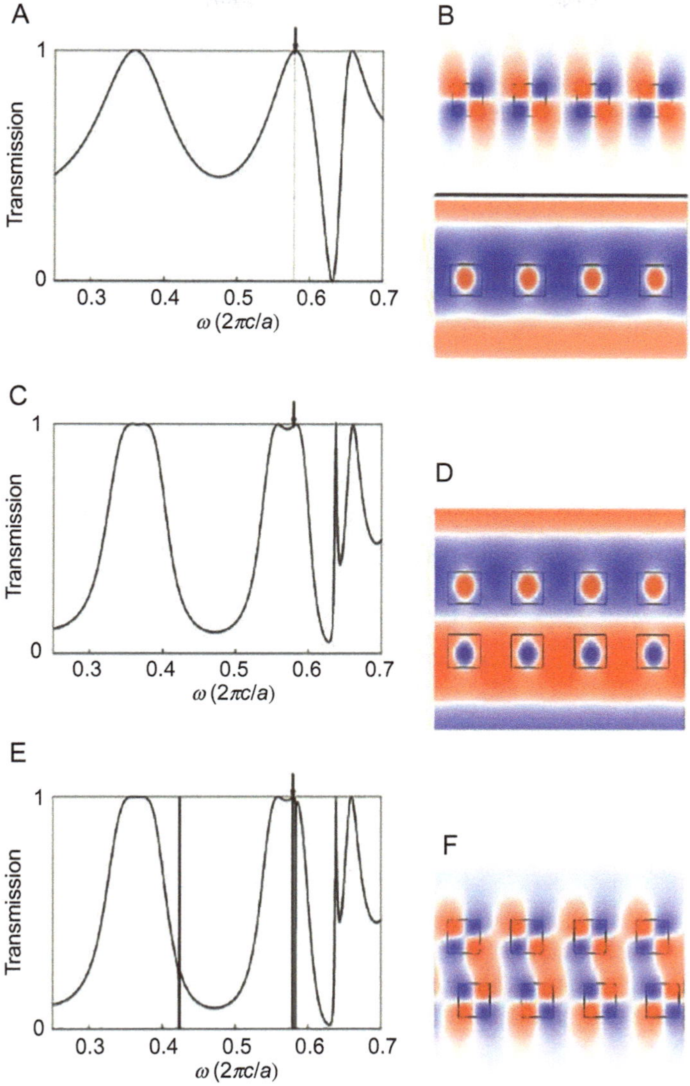

Fig. 6 Dark states in a two photonic crystal slab system that arise from coupled dark resonances from single slabs. (A) Single-slab transmission spectrum. (B) Mode profile at the dark state frequency (*top*) and field profile at the frequency indicted by the *arrow* in (A). (C) Double-slab transmission spectrum for $d = 0.5a$, where d is the spacing and a is the lattice constant. (D) Field profile at frequency indicated by the *arrow* in (C). (E) Double-slab transmission spectrum for $\Delta x = 0.15a$, where Δx is the lateral displacement. (F) Field profile at lowest frequency resonance from (E). *Reproduced with permission from Liu, V., Povinelli, M., Fan, S., 2009. Resonance-enhanced optical forces between coupled photonic crystal slabs. Opt. Express 17 (24), 21897–21909.*

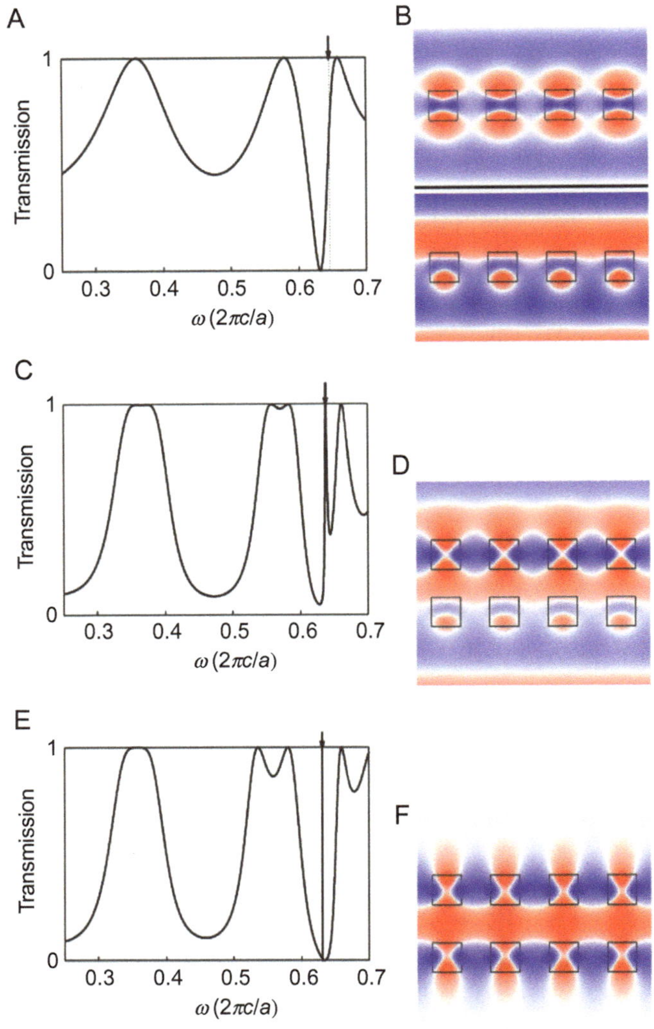

Fig. 7 Dark states in a two photonic crystal slab system that arise from coupled bright resonances from single slabs. (A) Single-slab transmission spectrum. (B) Mode profile at the dark state frequency (*top*) and field profile at the frequency indicted by the *arrow* in (A). (C) Double-slab transmission spectrum for $d=0.5a$, where d is the spacing and a is the lattice constant. (D) Field profile at frequency indicated by the *arrow* in (C). (E) Double slab transmission spectrum for $d=0.65a$. (F) Field profile at peak of resonance in (E). *Reproduced with permission from Liu, V., Povinelli, M., Fan, S., 2009. Resonance-enhanced optical forces between coupled photonic crystal slabs. Opt. Express 17 (24), 21897–21909.*

1.3 Fano resonance filter configurations

1.3.1 Single-layer Fano resonance filters

Following initial experimental demonstrations (Kanskar et al., 1997) and theoretical investigations (Fan and Joannopoulos, 2002; Peng and Morris, 1996; Suh and Fan, 2004), 2D PCS Fano resonance filters have been reported from a few other groups (Fan and Joannopoulos, 2002; Lin et al., 2005; McKerracher et al., 2013; Rosenberg et al., 2005).

Over the last few years, based on the PDMS transfer printing techniques, many different types of Fano resonance filters have been reported, based on single- or double-layer PCSs on Si, glass, and flexible substrates (Fig. 8). The key lattice parameters are the air hole radius (*r*), the lattice period (*a*),

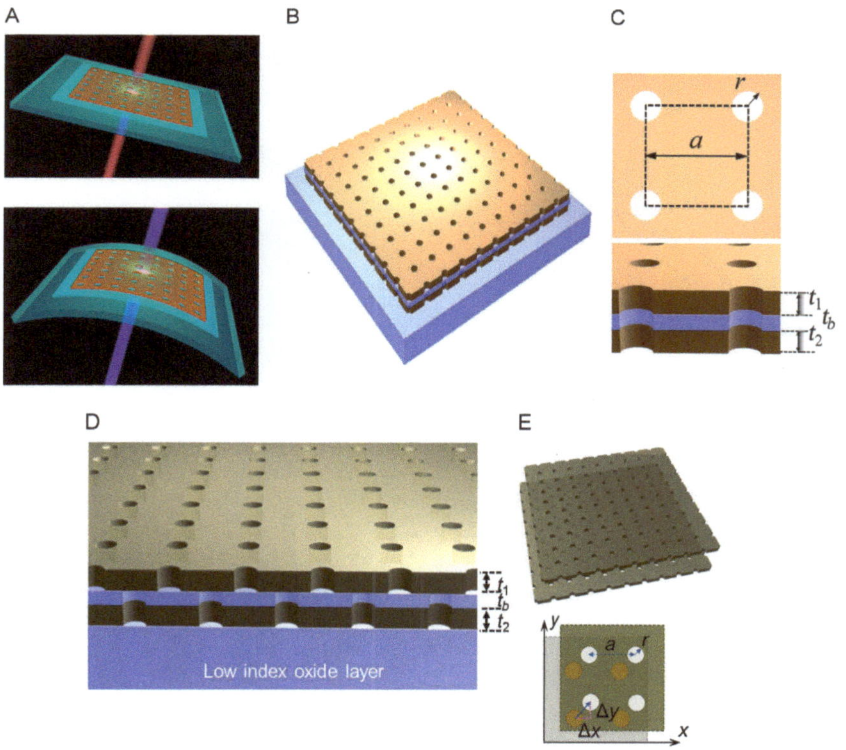

Fig. 8 Schematics of Fano resonance PCS filter configurations: (A) Single-layer PCS on glass (*top*) or flexible (*bottom*) substrates; (B) double-PCS Fano filter 3D sketch; (C) Key parameters defined for the square lattice double-layer PCS filter including lattice constant (*a*), air hole radius (*r*), thin oxide buffer layer thickness (t_b), and two single layer PCS layer thicknesses (t_1 and t_2); (D) 3D close-up view of double-PCS Fano filters with lattice displacement; and (E) Illustration and definition of key design parameters, including lattice displacement (Δx, Δy) between two layers.

thicknesses for the top (t_1) and bottom Si PhC layers (t_2), and the oxide buffer layer in between (t_b).

Yang et al. (Qiang et al., 2008) reported the first experimental surface-normal Fano filters based on transferred silicon nanomembranes on glass and on flexible PET substrates. Subsequently, with improved designs and processes, single-layer PCS Fano filters have been demonstrated with much improved performance. Shown in Fig. 9A and B are scanning electron microscope (SEM) images of the fabricated single-layer Fano filters on glass substrate. The measured (blue solid line) and simulated (red dashed line) transmission spectra are shown in Fig. 9C, with enlarged images shown in Fig. 9D. Two transmission dips were found at 1529.88 and 1564.62 nm. For the 1564.62 nm dip, the Q value was 1737, with a 26 dB extinction ratio.

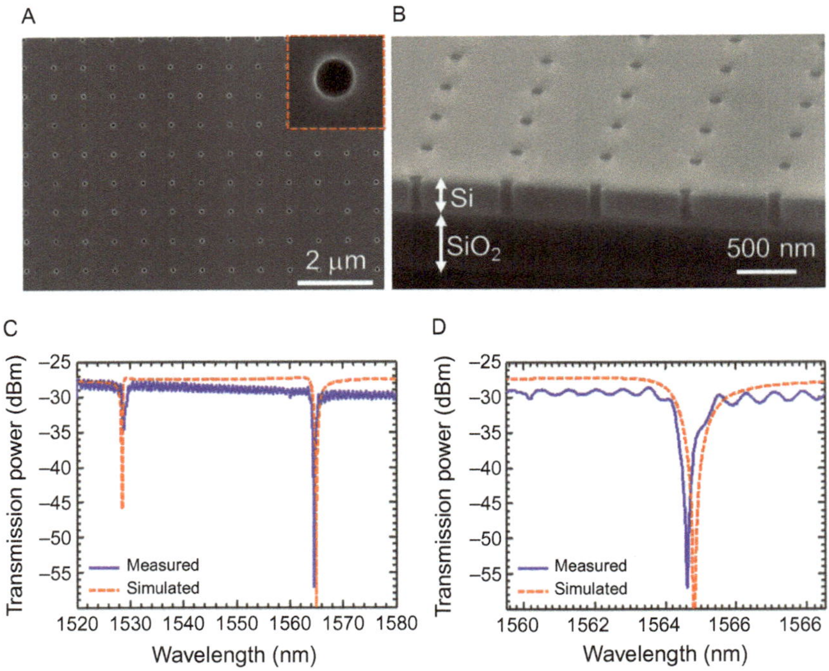

Fig. 9 Experimental results for Design S_1: (A) Top and (B) cross-section views of fabricated single-layer PhC Fano resonance filters on oxide buffer; (C) Measured (*blue solid line*) and simulated (*red dash line*) transmission spectra for the fabricated single-layer PhC Fano resonance filter transferred on glass substrates; and (D) Enlargement of (C) over the second dip ($\lambda = 1564.62$ nm) region. *Reproduced with permission from Shuai, Y., Zhao, D., Tian, Z., Seo, J., Plant, D. V., Ma, Z., Fan, S., Zhou, W., 2013b. Double-layer Fano resonance photonic crystal filters. Opt. Express 21 (21), 24582–24589.*

The angle and polarization properties of these filters have been investigated in detail by Qiang et al.(Qiang et al., 2008) and Li et al. (Chen et al., 2009), focusing on measurement and simulated dispersion properties. Detailed analysis of the angle-dependent Fano resonant mode dispersion properties is presented in Fig. 10. Following the conventional definition, the light incident angles are specified by two polar angles, the colatitude angle θ (an angle from the surface-normal direction) and the azimuth angle ϕ (an angle from the positive x-axis to the orthogonal projection of the incident beam in the x–y plane) (Lousse et al., 2004). The incident beam with the wave-vector k was kept within the Γ–X direction in the k-space, i.e., for the case shown in Fig. 10A with $\phi=0^\circ$, the momentum

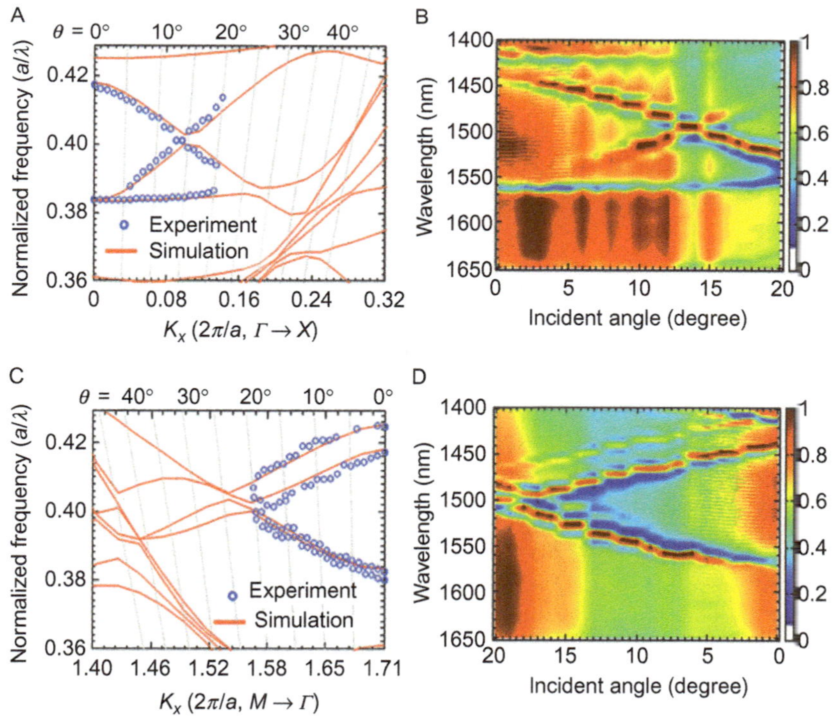

Fig. 10 (A, C) Simulated dispersion plot (*red lines*) zoomed-in along (Γ–X, Γ–M) directions, along with measured Fano resonances (*blue dots*) for different incident angles θ with (A) $\phi=0^\circ$ and $\Psi=0^\circ$, and (C) $\phi=45^\circ$ and $\Psi=45^\circ$. The *straight gray lines* represent the relationships between the normalized frequency and the normalized wave-vectors along (Γ–X, Γ–M) directions. (B, D) Measured transmission intensity contour plots for angle-dependent transmissions at different incident angles θ with (B) $\phi=0^\circ$ and $\Psi=0^\circ$, and (D) $\phi=45^\circ$ and $\Psi=45^\circ$. *Reproduced with permission from Qiang, Z., Yang, H., Chen, L., Pang, H., Ma, Z., Zhou, W., 2008. Fano filters based on transferred silicon nanomembranes on plastic substrates. Appl. Phys. Lett. 93, 061106.*

vector k_x relates to the normalized momentum $k_{\Gamma-X}$ along the Γ–X direction as $k_x = k_{\Gamma-x}\left(\frac{2\pi}{a}\right)\cos\phi$. Thus, the normalized frequency (a/λ) can be related to the normalized momentum $k_{\Gamma-X}$ in the dispersion plot as $k_{\Gamma-x} = \left(\frac{a}{\lambda}\right)\sin\theta$.

A set of straight lines can be superimposed on the normalized dispersion plot along the Γ–X direction, as shown in Fig. 10A. The measured Fano resonant modes were normalized and plotted in Fig. 10A, blue circles. Excellent agreement was achieved between the experimental and simulation results. The transmission intensity contour plots for the measured angle-dependent transmission spectra, with $\phi = 0^{\circ}$, are also shown in Fig. 10B, where the intensity dips/peaks are shown in blue and red. The intensity contour plot clearly resembles the spectra shifting behavior shown in Fig. 10A.

Following a similar procedure, the measured transmission results shown in Fig. 10B, with $\phi = 45^{\circ}$, were analyzed such that wave vector k lies in the Γ–M plane, with results shown in Fig. 10C and D. Again, the experimental and simulation results agree very well. It was confirmed that angle-independent filters are feasible based on the dispersion engineering of the Fano resonances in PCS. Details can be found in a short review by Zhou et al. (2009).

1.3.2 Double-layer Fano resonance filters

On the other hand, much higher Q's can be achieved in coupled double-layer Fano resonance PhC structures. Liu and Fan et al. (Liu et al., 2009) reported earlier that much higher Q's can arise from the coupled dark states in double-layer stacked PCS structures. Suspended double-layer PCS filters have also been experimentally demonstrated in GaAs (Stomeo et al., 2010), InP (Roh et al., 2010), and Si (Shuai et al., 2013b) materials. Both single- and double-layer Fano resonance Si PhC filters were designed and optimized for high Q's around 1550 nm on the spectral band. Shown in Table 1 are key design parameters for a few optimized design structures, where S_1–S_3 are for single-layer designs and D_1–D_3 are for double-layer designs. Notice that the Fano filter Q increases from 4500 for single-layer filter to over 98,000 for a double-layer filter. Shown in Fig. 11 are the simulated transmission spectra for Designs S_3 and D_3, with Q's of 4500 and 98,000, respectively (Shuai et al., 2013b).

Additionally, it was predicted that the double-layer PhC structure could produce an extremely high Q mode (infinite in theory), by varying the coupling condition between two PhC layers (Liu et al., 2009). Based on the Design D_2 parameters, transmission spectra were simulated by varying the buffer layer oxide thickness t_b.

Table 1 Key design parameters and *Q*'s for selected single (S1–S3)- and double (D1–D3)-layer filters

	a	*r/a*	t_1	t_2	t_b	*Q*
Unit	**nm**		**nm**	**nm**	**nm**	
S_1	765	0.08	260	–	–	4100
S_2	765	0.1	260	–	–	1900
S_3	780	0.08	260	–	–	4500
D_1	1000	0.2	230	230	160	1.2×10^4
D_2	1000	0.08	230	230	20	2.2×10^4
D_3	1000	0.05	230	230	20	9.8×10^4

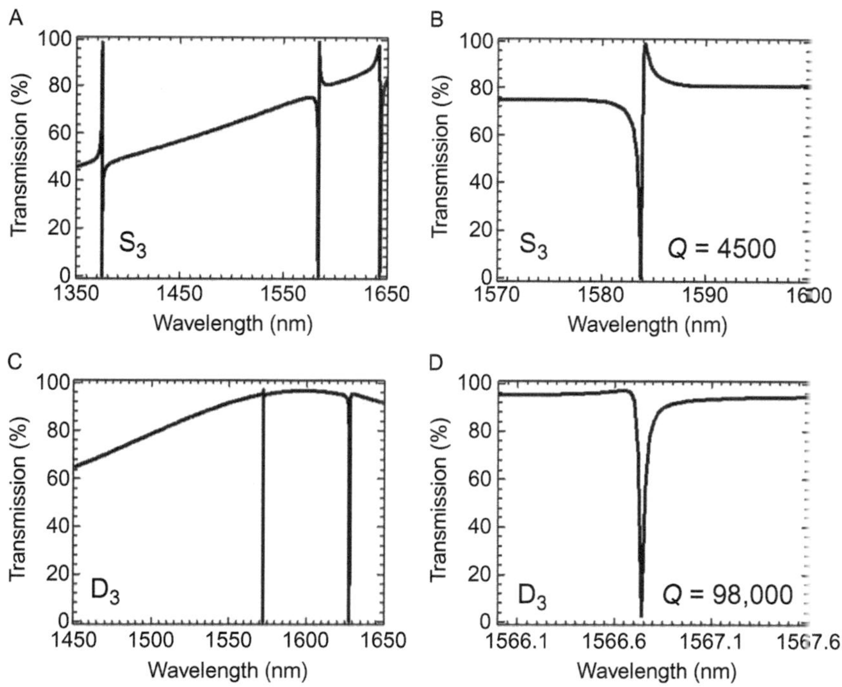

Fig. 11 Simulated transmission spectra for (A, B) single- and (C, D) double-layer Fano resonance PhC filters, where (B) and (D) are enlarged plots of (A) and (C), respectively. The design parameters are summarized in Table 1 for Case S_3 and D_3, respectively. *Reproduced with permission from Shuai, Y., Zhao, D., Tian, Z., Seo, J., Plant, D. V., Ma, Z., Fan, S., Zhou, W., 2013b. Double-layer Fano resonance photonic crystal filters. Opt. Express 21 (21), 24582–24589.*

For the double-layer structure, the simulated transmission spectra are plotted in Fig. 12A, with oxide buffers ranging from 0 to 160 nm thick. With the increased oxide buffer layer thicknesses, the high Q modes (shorter wavelength modes shown in Fig. 12A shift toward shorter wavelengths, with the filter Q value maximizing around 10,000,000 for a buffer layer of $t_b = 60$ nm, as shown in Fig. 12B. Shown in Fig. 12C is the magnified spectral plot for the transmission dip with a Q of 10,000,000. Simulated field distribution profiles for three cases close to the maximum Q are shown in Fig. 12D, where strong field confinement is evident for the high Q transmission dips at the optimal buffer layer thickness.

Two types of structures were prepared for double-layer PhC Fano filters on silicon and on quartz substrates. For the double-layer PhC Fano filters on

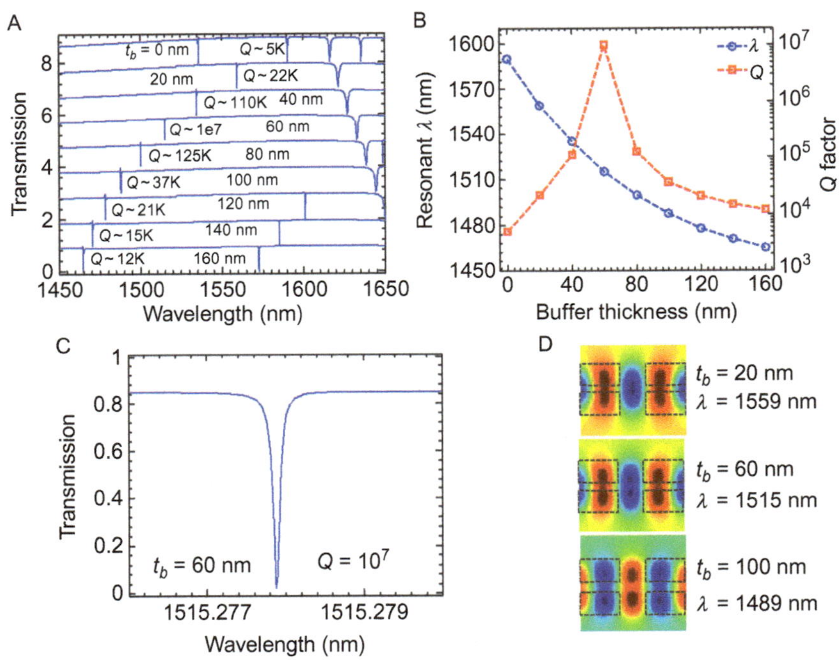

Fig. 12 Simulation results for Design D_2 with different buffer layer thicknesses t_b: (A) Transmission spectra with different t_b from 0 to 160 nm; (B) High *Q* resonant wavelengths and the corresponding *Q* values for different buffer thicknesses t_b; (C) Zoom-in spectrum for the buffer thickness $t_b = 60$ nm and filter *Q* of 10,000,000; and (D) Simulated E-field intensity profile at resonant wavelengths for three different t_b values, where $t_b = 60$ nm representing the highest *Q* condition for this design. *Reproduced with permission from Shuai, Y., Zhao, D., Tian, Z., Seo, J., Plant, D. V., Ma, Z., Fan, S., Zhou, W., 2013b. Double-layer Fano resonance photonic crystal filters. Opt. Express 21 (21), 24582–24589.*

SOI, a low index oxide buffer layer was first formed using thermal oxidation of a single-crystalline Si layer on the SOI substrate, followed by a low pressure chemical vapor deposition (LPCVD) poly-Si deposition process, to form a poly-Si/thermal SiO_2/crystalline-Si double-Si-layer structure. The other double-layer PhC Fano filters on quartz substrates were formed by two rounds of the LPCVD poly-Si deposition process, with a plasma-enhanced chemical vapor deposited (PECVD) SiO_2 layer sandwiched in between the two LPCVD poly-Si layers. Shown in Figs. 13A and 5C are cross-sectional SEM images for the double-layer poly-Si/SiO_2/c-Si filter structures, with different thermal oxide thicknesses. The same E-beam patterning and etching processes were utilized for the 2D-PhC patterning. An SEM image is shown in Fig. 13D.

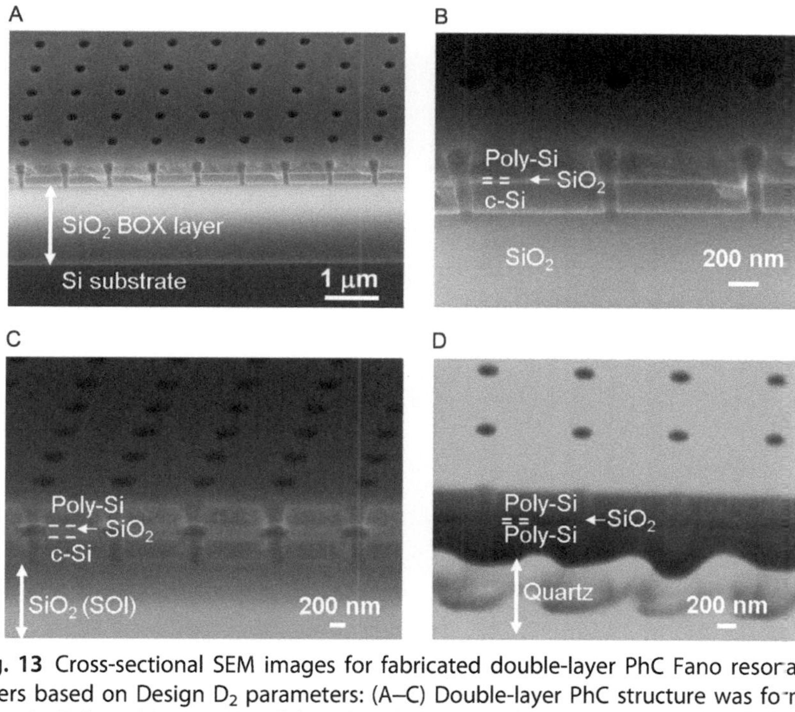

Fig. 13 Cross-sectional SEM images for fabricated double-layer PhC Fano resonance filters based on Design D_2 parameters: (A–C) Double-layer PhC structure was formed by poly-Si deposition on top of the SOI substrates; and (D) Double-layer PhC structure was formed by two steps of poly-Si deposition on quartz substrates. Notice the oxide buffer thicknesses are 20, 160, and 20 nm for cases (B), (C), and (D), respectively. *Reproduced with permission from Shuai, Y., Zhao, D., Tian, Z., Seo, J., Plant, D. V., Ma, Z., Fan, S., Zhou, W., 2013b. Double-layer Fano resonance photonic crystal filters. Opt. Express 21 (21), 24582–24589.*

The double-layer Fano filters were characterized by measuring transmission or reflection spectra for the two different configurations. For the transmission measurement on the double-layer filter quartz substrate, a transmission dip at 1545.2 nm was obtained, with an estimated Q of 5000 and an extinction ratio of 2.7 dB (Figs. 14A and 6B). For the reflection measurement on the double-layer filter on SOI substrate, a reflection peak was obtained at 1567 nm with a Q factor of 9734 and an extinction ratio of 8 dB (Figs. 14C and 6D). All of these measured spectra match well with the simulated ones at these resonance locations (with spectral dips or peaks). The measured Q values are lower than the simulated ones, which may be due to the imperfect process of etching the air holes, such as the conical shape and different hole sizes in the trilayer structure. We expect the filter performance to be improved with much higher Q factors when we optimize the fabrication process.

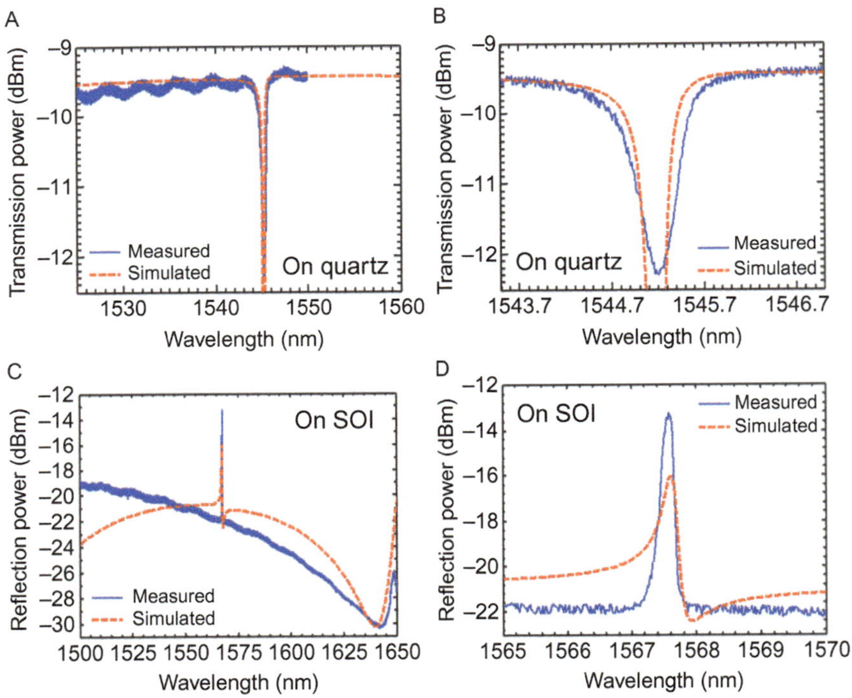

Fig. 14 (A, B) Measured (*blue solid line*) and simulated (*red dash line*) transmission spectra for the double-layer PhC Fano resonance filters on quartz; and (C, D) Measured (*blue solid line*) and simulated (*red dash line*) reflection spectra for the double-layer PhC Fano resonance filters on SOI. (B) and (D) are magnifications of (A), (C), respectively. *Reproduced with permission from Shuai, Y., Zhao, D., Tian, Z., Seo, J., Plant, D. V., Ma, Z., Fan, S., Zhou, W., 2013b. Double-layer Fano resonance photonic crystal filters. Opt. Express 21 (21), 24582–24589.*

1.3.3 Double-layer Fano resonance filters with controlled lattice displacement

It was theoretically understood that allowing lateral displacement of the lattices in the two layers with respect to each other enables an additional degree of freedom in controlling the filter characteristics of the two-layer structure (Liu et al., 2009; Suh et al., 2003). However, experimental demonstration of coupled double-layer PCS structures with controlled lattice displacement has never been reported, due to the challenges associated with the fabrication. But, by employing the PDMS (polydimethylsiloxane) nanomembrane transfer printing technique, it is possible to stack multiple layers of semiconductor and other dielectric materials (Rogers et al., 2011; Yang et al., 2012). Recently Shuai et al. (2013a) reported experimental demonstration of coupled double-layer PCS with precisely controlled lattice displacement, based on transfer printing and multilayer e-beam patterning alignment processes (Rogers et al., 2011; Zhang et al., 2012). Both simulation and measurement suggest that optical filter Q factors are very sensitive to the lattice displacement. For one structure with relatively large lattice displacement, they demonstrated a Q factor of 80,000 experimentally. Further design and process optimization leads them to expect much higher Q-factors, particularly due to a simulation Q factor of 2.11×10^8 for coupled double-PCS structures with optimal lattice displacement (Shuai et al., 2013a).

Shown in Fig. 15 are simulated transmission and reflection spectra for double-layer stacked Fano filters without (Fig. 15A and C) and with (Fig. 15B and D) lattice offset ($\Delta x = 0.2a$, $\Delta y = 0.5a$), with the insets showing the magnified plot for the dominant (highest Q) resonance in each case. Also shown in the insets are the schematics of the structures considered in the simulation. For the transmission spectra shown in Fig. 15A and B, a glass substrate was used. For the reflection spectra shown in Fig. 15C and D, an SOI substrate was employed. The design parameters are $a = 1000$ nm, $r/a = 0.08$, $t_1 = t_2 = 230$ nm, and $t_b = 20$ nm. Simulations were carried out to find the Q factors for the different lattice displacements. The results shown in Fig. 15B and D correspond with the highest Q value obtained for this set of design parameters. Notice that the Fano filter Q increases from 26,000 and 28,000 for perfectly aligned lattice structures (Fig. 15A and C) to 2.11×10^8 and 1.76×10^8 for lattice misaligned structures (Fig. 15B and D) with optimal lattice displacement being $\Delta x = 0.2a$ and $\Delta y = 0.5a$. It is also worth mentioning that the low index buffer layer thickness (e.g., SiO_2 used here) can also be varied for Q-factors approaching infinity (Liu et al., 2009; Shuai et al., 2013b).

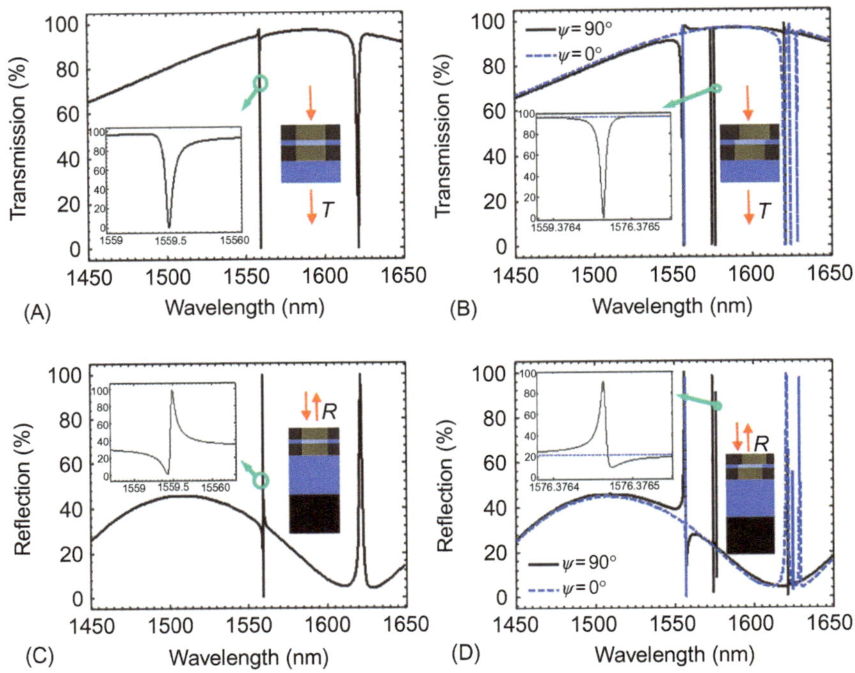

Fig. 15 Simulated transmission and reflection spectra for double-layer stacked Fano filters with (B, D) and without (A, C) lattice offset ($\Delta x = 0.2a$, $\Delta y = 0.5a$), with the *insets* showing enlarged plots for the dominant (high Qs) resonance for each cases. Also shown in the *insets* are the schematics of the structures considered in the simulation. For the transmission spectra shown in (A, B), glass substrate was considered. For the reflection spectra shown in (C, D), SOI substrate was considered. Ψ represents the incident beam polarization (E-vector) defined as the angle from the positive *x*-axis to the polarization direction. *Reproduced with permission from Shuai, Y., Zhao, D., Chadha, A. S., Seo, J.-H., Yang, H., Fan, S., Ma, Z., Zhou, W., 2013a. Coupled double-layer Fano resonance photonic crystal filters with lattice-displacement. Appl. Phys. Lett. 103 (24), 241106.*

Following a short thermal oxidation process for the formation of a thin SiO_2 buffer layer ($t_b = 20$ nm) on top of the silicon-on-insulator (SOI) substrate, single-layer Fano filters were fabricated with standard electron-beam lithography and dry etching processes. An unpatterned single crystalline Si nanomembrane was then stacked onto the patterned single-layer Fano filter using the PDMS stamp-assisted membrane transfer printing process. Filters with different displacements were fabricated. Shown in Fig. 16A–C are scanning electron microscope (SEM) images of fabricated double-layer Si-NM Fano filters with displaced lattices between the coupled layers. Shown in

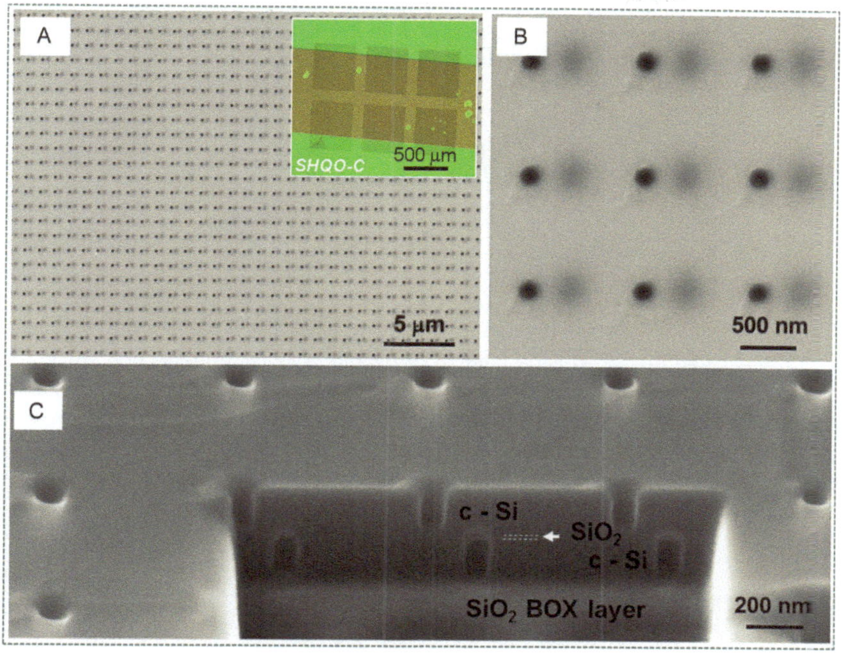

Fig. 16 Fabricated Fano filter scanning electron micrograph (SEM) images: (A, B) Top view and (C) cross-sectional view of double-layer Fano filter on SOI, with controlled lattice misalignment offset ($\Delta x = 0.2a$, $\Delta y = 0$). *Reproduced with permission from Shuai, Y., Zhao, D., Chadha, A. S., Seo, J.-H., Yang, H., Fan, S., Ma, Z., Zhou, W., 2013a. Coupled double-layer Fano resonance photonic crystal filters with lattice-displacement. Appl. Phys. Lett. 103 (24), 241106.*

Fig. 17 are the results obtained for the coupled double-layer PCSs, with large lattice displacement ($\Delta x = 0.495a$, $\Delta y = 0.45a$). A measured reflection spectrum is shown in Fig. 17A, along with a top SEM image shown in the inset. We obtained two resonant locations, around 1548.6 and 1567 nm, with a magnified plot around 1548.6 nm resonant shown in Fig. 17B. Again, simulation and Fano fit were carried out and the results match the experimental results well. In this case, we obtained a Q factor of 80,000. This value is the highest experimental Q factor reported so far, for all the different cases we fabricated. While the value is still far from the optimal Q factor of 2.11×10^8 the results demonstrate the great potential of coupled double-layer PCS structures with controlled lattice displacement. During testing, we also noticed that the performance of these filters is very sensitive to incident beam direction, collimation, and polarization orientation. We anticipate much higher Q-factors can be obtained with precisely controlled displacement and an optimized fabrication process.

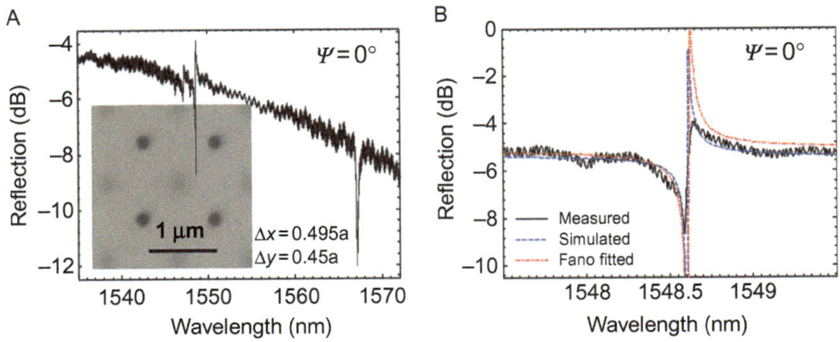

Fig. 17 Measured displaced Fano filter ($\Delta x = 0.495a$, $\Delta y = 0.45a$) reflection spectra for p-polarization state, with theoretical and experimental *Q*-factors of 115,000 and 72,000, respectively. Shown in the *inset* is the magnification of the measured reflection spectrum. *Reproduced with permission from Shuai, Y., Zhao, D., Chadha, A. S., Seo, J.-H., Yang, H., Fan, S., Ma, Z., Zhou, W., 2013a. Coupled double-layer Fano resonance photonic crystal filters with lattice-displacement. Appl. Phys. Lett. 103 (24), 241106.*

2. Fano resonance photonic crystal cavity modulators

The basic idea of the modulator discussed here is illustrated in Fig. 18 (Shuai et al., 2017). Two 230 nm thick silicon photonic crystal slabs, each patterned with an identical square lattice of circular air holes, are separated by a 20 nm layer of SiO_2. The silicon slabs are p- and n-doped, with the oxide layer providing electrical isolation between them. The lattice periodicity "*a*" is 1 μm and the hole radius "*r*" is 60 nm. These parameters are chosen so as to offer a high quality optical resonance in the vicinity of 1550 nm. The resonance can be excited by surface-normal incident light and the resulting field distribution can be calculated using rigorous coupled-wave analysis. Fig. 18B illustrates the electric field distribution of the high-Q resonance, calculated for our parameters using the free software package S4 (Liu and Fan, 2012).

Electrical coupling is enabled by the plasma dispersion effect (Chmielak et al., 2011; Jacobsen et al., 2006; Soref and Bennett, 1987), in which the concentration of the free carriers in silicon changes the real and imaginary parts of the refractive index. This mechanism was studied extensively in silicon (Soref and Bennett, 1987). The doping configuration employed in our device is that of a PIN structure, comprised of a p-doped (P) bottom layer of crystalline silicon, an electrical isolation (I) layer of SiO_2, and an n-doped (N) top layer of polycrystalline silicon (Fig. 18C).

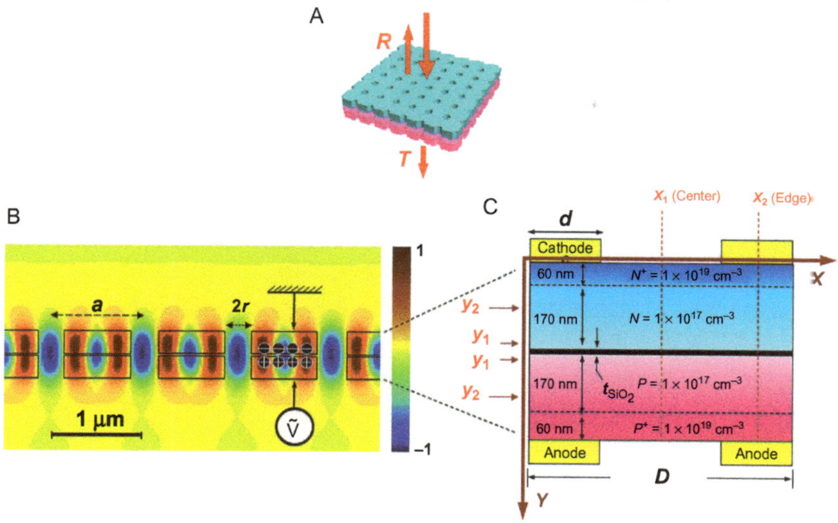

Fig. 18 A 3D sketch of stacked double-layer Photonic Crystal Slabs (PCSs) of two-dimensional square lattice holes' array was shown in (A). Light incidence from *top* was reflected or transmitted as the *arrows* in *red*. A high resonance mode excited by surface normal incidence was built in the stacked double-layer PCSs. The electrical field intensity was monitored in *color* as defined from "−1" to "1" as shown in (B). In simplified electrical simulation 2D model (C), electrical bias was applied to the capacitance-like stacked bilayer PCSs structure for electro-optical modulation. Device lateral size is "*D*" = 20 μm and the electrodes widths are set to be "*d*" = 5 μm. Total thickness of top or bottom Si layer is 60 nm + 170 nm = 230 nm as shown in (C). The "$tSiO_2$" is thickness of SiO_2 between two Si layers. In the *X–Y* plane, we set a few monitoring positions "x_1" center and "x_1" edge of the device; "y_1" near Si/SiO_2 interfaces and "y_2" at middle region of the Si layers. *Reproduced with permission from Shuai, Y.-C., Zhao, D., Liu, Y., Stambaugh, C., Lawall, J., Zhou, W., 2017. Coupled bilayer photonic crystal slab electro-optic spatial light modulators. IEEE Photonics J. 9 (2), 1–11.*

The SiO_2 layer provides electrical isolation between the doped layers, so that the device functions electrically as a parallel-plate capacitor. When a potential difference is imposed between the slabs, the accompanying charge accumulation changes their refractive indices, modifying the optical frequency of the high-Q mode. For a fixed optical frequency, the transmission and reflection are both affected, with slope efficiency proportional to the optical quality factor. We modeled the electrical response of our device in 2D (*X–Y*) using Taurus Medici (Synopsys) (Synopsys, 2016), a semiconductor device simulator. In the simulations, the doping level was taken to 10^{19} cm^{-3} at the 60 nm outer regions of the silicon slabs and 10^{17} cm^{-3} at the 170 nm inner regions (Fig. 18C). The device width was taken to be $D = 20\,\mu m$ limited by the allowed number of grid nodes, and the electrode width to be $d = 5\,\mu m$, as illustrated in Fig. 18C. The nonuniformity in the electric charge distribution owing to the etched holes was ignored.

Optimized device geometry and $Si/SiO_2/Si$ layers' thicknesses are determined by systematic investigation and analysis considering many factors in our device model. In the proposed $Si/SiO_2/Si$ PIN structure, different sandwiched SiO_2 thicknesses and Si doping levels are compared by combination of optical and electrical analysis. Sandwiched SiO_2 thickness of 20 nm was chosen to make a balance between optical and electrical effects. Multiple factors were considered such as: its desired optical Q value, stacked $Si/SiO_2/Si$ layers RC effect, free carrier absorption in Si, effective refractive index changes, sufficient electrical isolation between Si layers as well as process control preference.

In our optical study, we started from a transmission configuration as in Fig. 19A. Different SiO_2 thicknesses offer Q-factors varying from 5000 to 15,000 as shown in Fig. 19B. Two SiO_2 thickness cases are compared in Fig. 19C and D when applying forward bias, the 10 nm oxide case shows larger spectral shifting than 20 nm oxide case; however, strong free carrier absorption starts to occur with decreased oxide thickness, resulting in reduced Q factor and extinction ratio. To avoid unintended free carrier absorption and possible thermal effect, 20 nm oxide thickness is determined for later reflection configuration design. It was noticed that, in Fig. 19E, this interfaces accumulation (as "y_1" positions, in Fig. 18C) produces a larger blue shifting on spectra for the resonance mode when comparing to free carrier accumulation appearing in the middle of each Si layer (as "y_2" positions, in Fig. 18C). Owing to the semiconductor carriers equilibrium in these two stacked Si PCS, our proposed P/N doping structure (Fig. 18C) can be modulated by either forward or reverse bias for blue or red spectral shift, respectively, as shown in Fig. 19F. By applying the carrier refraction (Soref and Bennett, 1987) and semiconductor carriers equilibrium $np = ni^2$, one could imagine there would be a returning point of red-shifting with increased reverse bias voltage and this returning point was observed during device characterization.

A reflection configuration was further investigated by optical simulation. In Fig. 20A, a 2D sketch shows the device geometry design for two kinds of device sizes. It shows our electrical bias convention. We applied 1 μm buffer oxide layer and 380 nm PECVD oxide for effective reflection spectra as well as for electrical interconnection purposes. Simulation results show a dramatic spectral shifting in Fig. 20B when applying electrical bias. It reveals

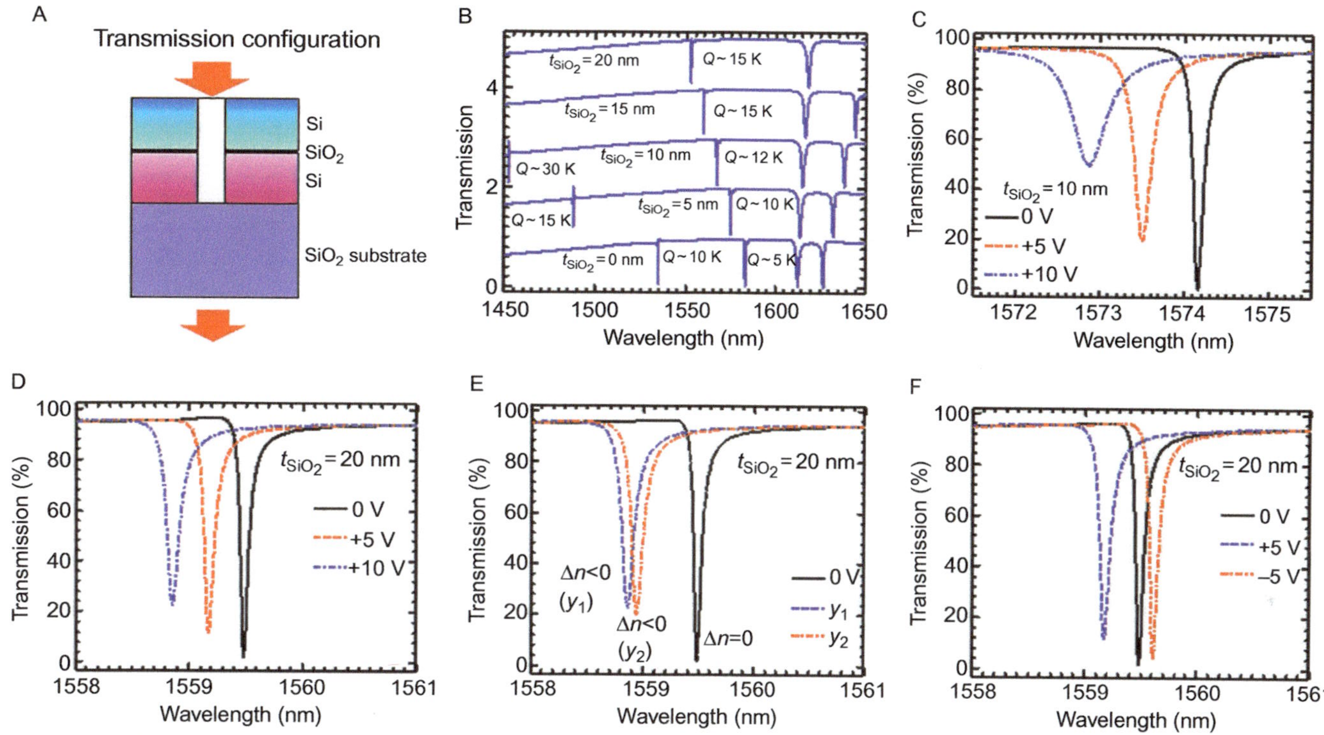

Fig. 19 Device electro-optical (EO) study in transmission configuration: a 2D sketch of the simulation lattice is shown in (A); for lattice constant "a" = 1 μm and "r/a" = 0.08, the transmission spectral centered at 1550 nm are shown in (B) with different sandwiched SiO_2 thicknesses varying from 0 to 20 nm. Two different sandwiched SiO_2 thicknesses, 10 nm in (C) and 20 nm in (D), are also compared for EO spectral shifting; Spectral shifting by free carrier accumulation occurring at either middle or interfaces of Si layers is compared in (E), by 5 V forward bias. Spectral shifting is also compared by 5 V forward/reverse bias, as shown in (F). *Reproduced with permission from Shuai, Y.-C., Zhao, D., Liu, Y., Stambaugh, C., Lawall, J., Zhou, W., 2017. Coupled bilayer photonic crystal slab electro-optic spatial light modulators. IEEE Photonics J. 9 (2), 1–11.*

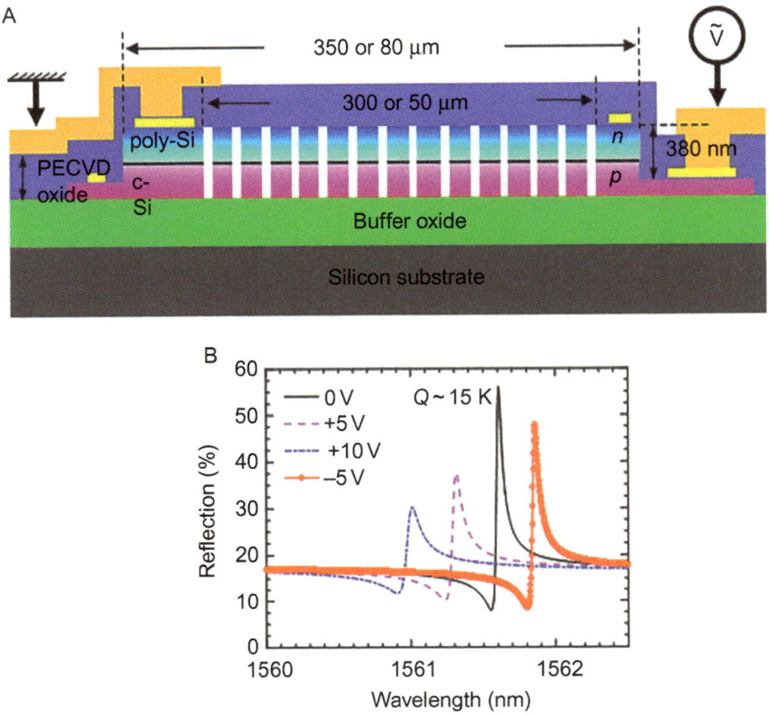

Fig. 20 (A) A sketch of device geometry cross-sectional view shows device design with two kinds of lateral dimension (350 μm mesa with 300 μm PC; 80 μm mesa with 50 μm PC) that are investigated in experiments. It has the same sandwiched SiO_2 thickness, "*r*," "*a*," Si layer thickness and doping profiles as those applied in transmission configuration (Fig. 19D and F). The simulated reflection spectra with different biases are studied in (B) with correspondence to transmission configuration. *Reproduced with permission from Shuai, Y.-C., Zhao, D., Liu, Y., Stambaugh, C., Lawall, J., Zhou, W., 2017. Coupled bilayer photonic crystal slab electro-optic spatial light modulators. IEEE Photonics J. 9 (2), 1–11.*

that the free carrier absorption-induced extinction ratio and Q-factor decreasing is more critical in reflection configuration.

For experimental demonstration, we fabricated devices with a somewhat larger lateral size than the simulated one (20 μm lateral size) for better surface-normal light coupling from free space and to facilitate optical alignment during characterization. In practice, we applied two lateral sizes for individual device. Photolithography and reactive ion etching defined a 350 μm × 350 μm (or 80 μm × 80 μm) mesa area inside of which e-beam lithography defined a 300 μm × 300 μm (or 50 μm × 50 μm) photonic crystal. Fabrication started with a SOI wafer with a 260 μm silicon thickness. A thermal oxide layer was grown on top of the Si. The growth process consumed a 30 nm layer of silicon from the surface, leaving 230 nm. The thermal grown SiO_2 layer was subsequently thinned down to 20 nm by means of a HF wet etching. A polycrystalline silicon layer of 230 nm thickness was deposited atop the thermal oxide layer by means of low-pressure chemical vapor deposition. The stacked bilayers were subsequently boron/phosphorous (p/n) doped by ion implantation into different depths, followed by rapid thermal activation. The array of holes for the photonic crystals was then patterned by electron-beam lithography on a 60 nm Cr layer that is used as a hard mask. The stacked poly-Si/SiO_2/Si layers were etched through to the bottom buffer oxide layer by reactive ion etching, after which the Cr mask was chemically removed. Finally, a 380 nm SiO_2 layer was deposited on top by PECVD for interconnection.

Fig. 21A shows a microscope image of a 6 × 6 array of devices fabricated with 350 μm × 350 μm mesa area. The optical resonance mode reflection measurement for an individual 350 μm × 350 μm mesa area device shows a good match to the simulation as shown in Fig. 21B. The measured optical power for each wavelength was collected after passing through two crossed polarizers to improve the extinction ratio (Nazirizadeh et al., 2008). Its measured reflection resonance peak wavelengths were recorded for applied electrical biases varying from −10 to +10 V. Spectral blue-shifting and red-shifting were found when applying forward and reverse bias, respectively. A maximum red-shifting returning point was observed at (-5 ± 1) V as shown in Fig. 21C. This spectral shifting is induced by overall free carrier concentrations variation in two Si layers when electrical charges were applied onto this capacitor like structure. Both spectral blue- and red-shift by forward and reverse bias were obtained in the p/n-type stacking silicon layers.

A 6 × 6 array of 80 μm × 80 μm mesa devices was also fabricated for high frequency EO modulation (Fig. 22). It was attached to a dielectric substrate with a coplanar waveguide (CPW) line that was designed for 10 GHz operation. A single 80 μm × 80 μm mesa area device was then wire-bonded. The top and bottom electrodes were connected to 150 μm ground/signal pads for wire-bonding. A scanning electron microscope (SEM) top view of a device region with 80 μm × 80 μm mesa area and a zoomed-in SEM image of the 50 μm × 50 μm e-beam patterned array of holes are shown in Fig. 22B and C.

While a modulator such as the one described here can in principle function in either transmission or reflection, we focus on the reflection mode of operation in this work. A schematic diagram of the setup used for tests is

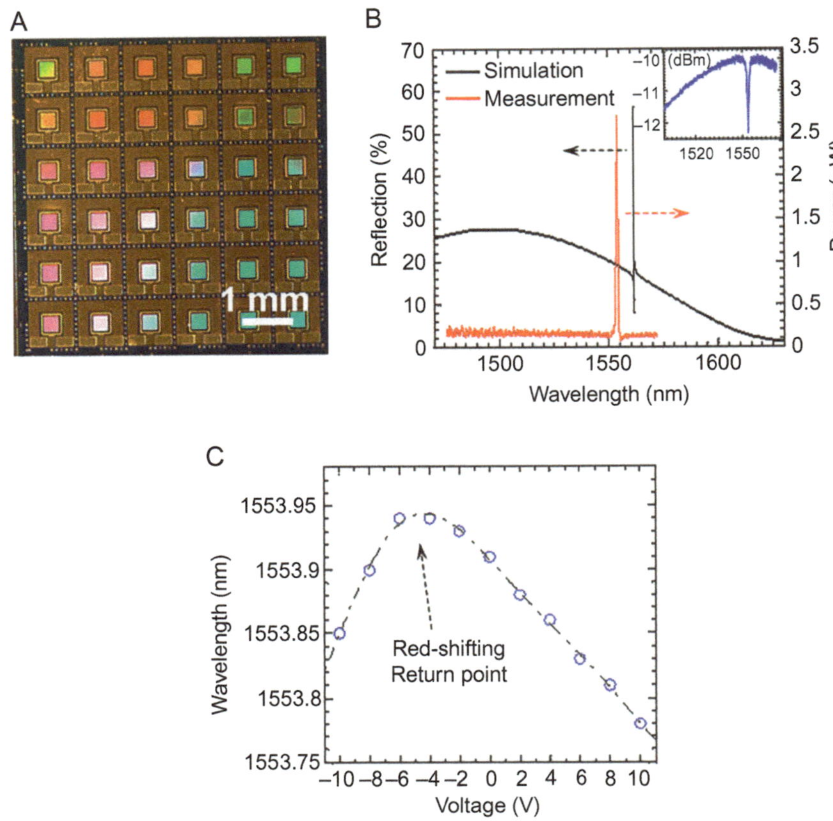

Fig. 21 A microscope top-view of a 6 × 6 array of 350 μm × 350 μm mesa area devices is shown in (A). Optical resonance mode reflection spectral for a single device is measured in (B) with crossed-polarizers method to improve the extinction ratio (Nazirizadeh et al., 2008). As shown in (C), continuous resonance mode spectral peak shifting was observed by applying electrical bias from −10 to +10 V. The measurement points in *blue circles* were connected by *line*, which shows a *red*-shifting return point between −4 and −6 V of reverse bias. *Reproduced with permission from Shuai, Y.-C., Zhao, D., Liu, Y., Stambaugh, C., Lawall, J., Zhou, W., 2017. Coupled bilayer photonic crystal slab electro-optic spatial light modulators. IEEE Photonics J. 9 (2), 1–11.*

shown in Fig. 23A. Light from a narrow-linewidth tunable laser is sent through a circulator, collimated, and passed through a polarizing beam splitter (PBS) to provide a well-defined linear polarization. The light is focused and directed at the Si-NM device such that the waist of the incident beam falls on the device at normal incidence. A quarter-wave plate can be set with its fast axis parallel to the polarization of the incident light, so that light reflected from the device retraces itself through the collimator and is

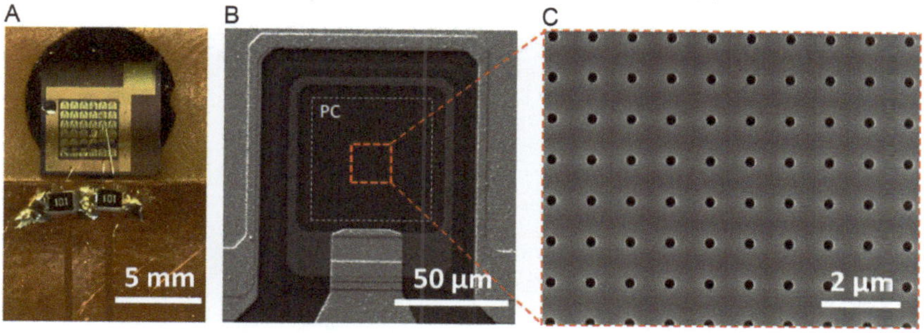

Fig. 22 A die of 6 × 6 array of 80 μm × 80 μm mesa devices is shown in (A), one selected device was wire-bonded to CPW line. In (B), a SEM top-view image for a single modulator device with 80 μm × 80 μm mesa area shows the details of interconnection and electrodes, with a 50 μm × 50 μm PCS area that is defined by the *white square dash line*. A zoomed-in SEM top-view of its PCS center region is shown on *right side* in (C). *Reproduced with permission from Shuai, Y.-C., Zhao, D., Liu, Y., Stambaugh, C., Lawall, J., Zhou, W., 2017. Coupled bilayer photonic crystal slab electro-optic spatial light modulators. IEEE Photonics J. 9 (2), 1–11.*

diverted by the circulator for detection, or with its fast axis at 45° to the incident polarization, so that light falling on the device is circularly polarized and the reflected light is deflected by the PBS and subsequently detected.

We supplemented the spectral shifting measurements with frequency-domain EO modulation measurements using a network analyzer. RF amplifier and a high-speed receiver were also used for device EO modulation frequency domain characterization. The predicted optical resonance was found near 1497.5 nm for an 80 μm × 80 μm mesa area device. For subsequent frequency domain EO modulation tests, the laser wavelength was fixed at 1499 nm where the slope of the reflection spectrum takes its highest value. Fig. 23B shows the measured frequency response of the particular 80 μm × 80 μm mesa device, where 0 dB corresponds to the static (low-frequency) response. As shown in the figure, the response falls off by 3 dB at a frequency of (200 ± 20) MHz.

In summary, we have experimentally demonstrated a novel electro-optic intensity modulator exploiting a Fano resonance in a bilayer photonic crystal architecture. Both blue and red spectral shifts in the resonance were observed, by applying opposite electrical biases. The 3 dB bandwidth of one of our devices, with a lateral size of 80 μm × 80 μm, was measured to be (200 ± 20) MHz. These spatial light modulators are polarization independent owing to highly symmetric 2D square lattice photonic crystal structure. Calculations suggest that bandwidths in the GHz range can be achievable if

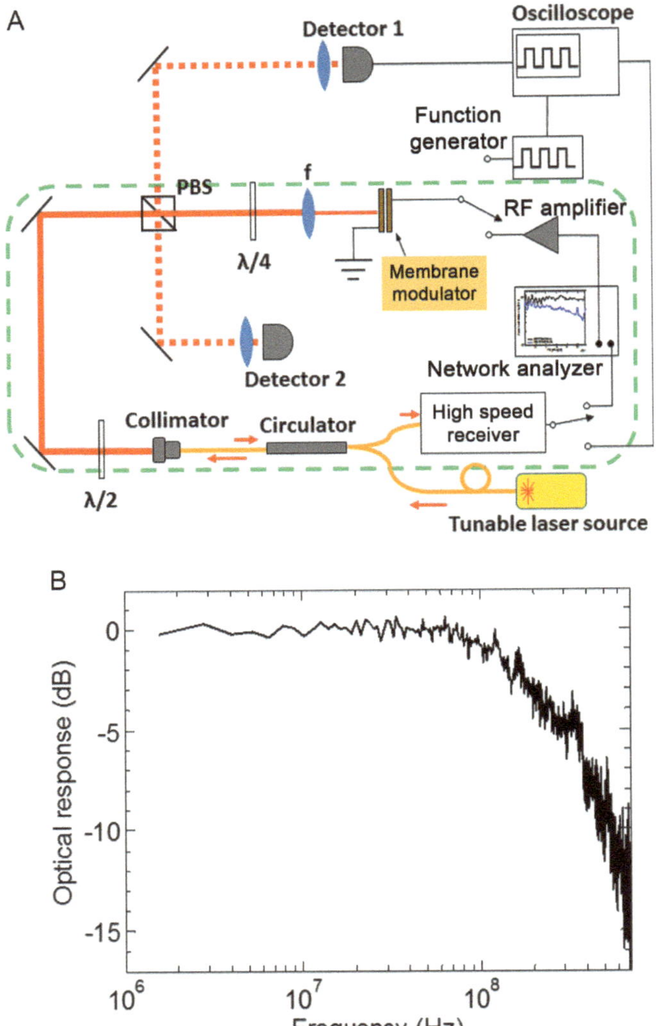

Fig. 23 A characterization setup shown in (A) was used for device optical reflection spectral collection and electro-optical (EO) modulation characterization. A network analyzer, RF amplifier, and a high-speed receiver were used for device EO modulation frequency domain characterization. As shown in (B), a 200 MHz 3 dB cutoff frequency, which is about 100 times of a 350 μm × 350 μm mesa area device's performance, was found for the device with 80 μm × 80 μm mesa area. *Reproduced with permission from Shuai, Y.-C., Zhao, D., Liu, Y., Stambaugh, C., Lawall, J., Zhou, W., 2017. Coupled bilayer photonic crystal slab electro-optic spatial light modulators. IEEE Photonics J. 9 (2), 1–11.*

the device size is reduced. High modulation depth should be possible as well, by using a higher optical Q factor PCS stacking structure (Shuai et al., 2013a, 2013b) with process-controlled PCS hole size. These bilayer coupled PCS-based modulators offer novel opportunities in vertical photonic integration, including free-space communications, 3D optical interconnect, and spatial control of light via extended horizontal arrays.

3. Fano resonance photonic crystal cavity optomechanics

3.1 Optical forces in nanoscale cavities

Maxwell, dealing mathematically with the stresses in an electromagnetic field, reached the conclusion that "in the medium in which waves are propagated there is a pressure normal to the waves and numerically equal to the energy in unit volume" (Kilic et al., 2007). Universal to physics from mesoscale and microscale down to nanoscale, optomechanical interactions are a rich subject of research and enable an especially rich set of phenomena and opportunities because of the presence of both attractive and repulsive forces (Nxcnons and Huu, 1901). Rapid developments have been made in this area in the past decade. Efforts were made to higher optomechanical coupling coefficients and several experimental demonstrations and approaches for enhancing the strength of the effect were achieved during 2007–09 (Nxcnons and Huu, 1901).

Optical forces resulting from interacting modes and cavities can scale to remarkably large values as the optical modes shrink to nanometer dimensions (Kippenberg and Vahala, 2008). It has been noted that the remarkable spatial confinement and large field enhancements afforded by nanometer-scale guided modes can result in significant forces when optical modes and cavities interact (Rakich et al., 2007; Van Thourhout and Roels, 2010). Scaled down to nanoscale dimensions, transverse optical forces in an integrated silicon photonic circuit through an embedded nanomechanical resonator were reported in 2008 (Povinelli et al., 2005a). In addition, intensive progresses were made in 2009. Both attractive and repulsive optical forces can be obtained between planarly coupled nanophotonic waveguides (Povinelli et al., 2005b); force is tuned from attractive to repulsive by controlling the relative phase of the optical fields injected into the waveguides (Li et al., 2008). A static mechanical deformation of up to 20 nm in a silicon nitride structure was experimentally demonstrated in the stacked double-ring resonator structure with resonance shifting spanning 80 times the intrinsic resonance linewidth with low optical power (Li et al., 2009). Micrometer-scale displacements of the waveguide are observed for milliwatt-level optical input powers in a structure with a high-Q optical microresonator coupled with a micrometer-scale waveguide (Roels et al., 2009). Seamless wavelength routing over a range of 3000 times the intrinsic channel width was achieved in a specially designed nanooptomechanical system (Wiederhecker et al., 2009). Eichenfield et al. proposed the idea of "optomechanical crystals," i.e., that in a photonic crystal, the periodicity of the host medium is used to manipulate the properties of light, whereas a phononic crystal uses periodicity to manipulate mechanical vibrations (Eichenfield et al., 2007). A picogram- and nanometer-scale in a photonic-crystal optomechanical cavity was measured in a pair of specially patterned nanoscale beams, resulting in a scale of the per-photon force and the mass of the structure (Eichenfield et al., 2009).

All of these results indicate that there is potential for radiation pressure actuated devices to be used in a variety of photonic applications, such as channel routing/switching, buffering, dispersion compensation, pulse trapping/release, widely tunable lasers, and other light signal manipulations on chip (Ma and Povinelli, 2012).

3.2 Optical forces in coupled Fano resonance PCSs

This concept was extended to 2D PCS's where the radiation force can be further enhanced due to the stored and strongly localized electromagnetic field inside the PCS's. Many systems have been investigated, including coupled waveguides, waveguides coupled to substrates, as well as a variety of resonator structures such as microrings, disks, or toroids that support whispering gallery modes, and point defects in photonic crystal slab structures that support standing-wave optical modes. Recently, the behavior of lateral and normal optical forces between coupled photonic crystal slabs attracts a lot of interest, because near the frequencies of guided resonances, we see significant enhancement of the optical forces (Liu et al., 2009; Notomi et al., 2006).

Optical forces in coupled photonic crystal slabs have received a great deal of attention recently (Notomi et al., 2006; Rodriguez et al., 2011; Roh et al., 2010; Woolf et al., 2013). It was shown theoretically that resonances arising from coupled photonic crystal slabs can generate optical forces that are proportional to the quality factor of the resonances (Liu et al., 2009). These forces can be attractive or repulsive and may also be engineered in

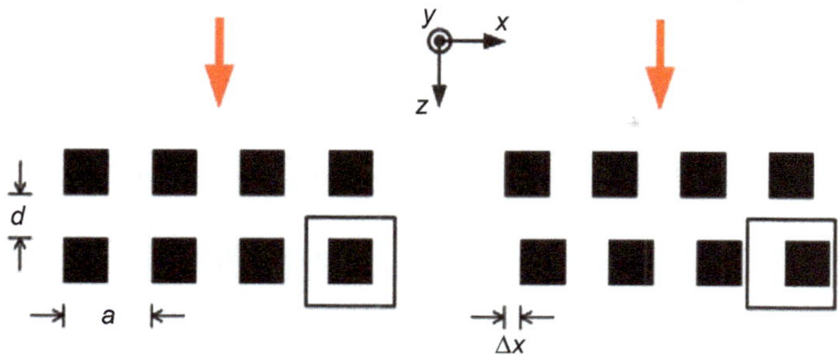

Fig. 24 Schematic of the double-slab system. The *red arrows* indicate the directions of the incident light. Each slab consists of an array of high index rods ($\varepsilon = 12$) surrounded by air ($\varepsilon = 1$). The *empty rectangle* indicates the surface over which the integration of the stress tensor is performed. In the *right panel*, the bottom slab is shifted relative to the first slab by a distance of Δx. *Reproduced with permission from Liu, V., Povinelli, M., Fan, S., 2009. Resonance-enhanced optical forces between coupled photonic crystal slabs. Opt. Express 17 (24), 21897–21909.*

asymmetric structures (Rodriguez et al., 2011). We will first review some of the basic principles for manipulating optical forces in coupled slab structures, as shown in Fig. 24. Due to their different origins, the two types of dark states in Section 1.2 exhibit qualitatively different force divergence behaviors. Three cases were considered (Liu et al., 2009).

In the first case shown in Fig. 24A, we consider the dark resonance from coupled bright resonances as in Fig. 7. As the slab separation is changed, the quality factor of the dark resonance becomes finite. The resonant properties are plotted in Fig. 25. Since the resonant decay rate must vanish at the separation, during which the resonance is decoupled from radiation, the lowest order term in the resonant decay rate is proportional to the square of the separation. Hence the optical force, proportional to the quality factor, diverges quadratically with respect to the separation (Fig. 26).

In the second case, shown in Fig. 24B, we consider a dark state from coupled dark resonances as in Fig. 6. With a relative shift, the dark state splits into two near-dark resonances, and the resonant properties of these two near-dark resonances are plotted in Fig. 27. The quality factor diverges as Δx^{-2}, which is the lowest order of dependence. The normal force, F_z, diverges in the same way as Q, but the lateral force, F_x, diverges as Δx^{-1} because the average lateral force per photon to the lowest order is proportional to Δx. The divergence behavior of the optical forces is plotted in Fig. 28.

In the third case, we consider a dark state from coupled bright resonances, similar to Case 1, except that the two slabs are shifted relative to each

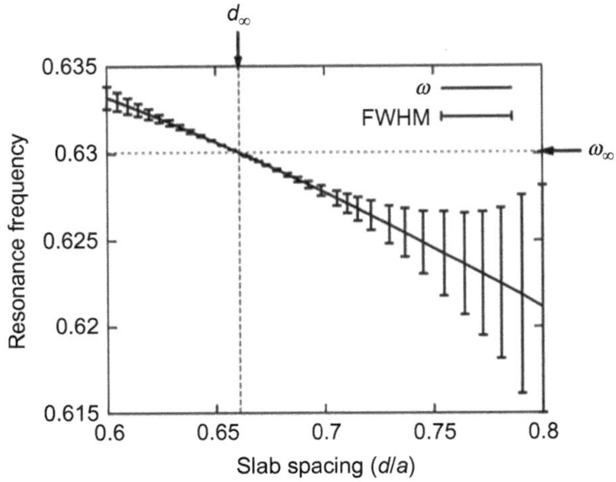

Fig. 25 Resonance peak and linewidth in the vicinity of the dark state arising from coupled bright resonances as in Fig. 7 (Case 1). *Reproduced with permission from Liu, V., Povinelli, M., Fan, S., 2009. Resonance-enhanced optical forces between coupled photonic crystal slabs. Opt. Express 17 (24), 21897–21909.*

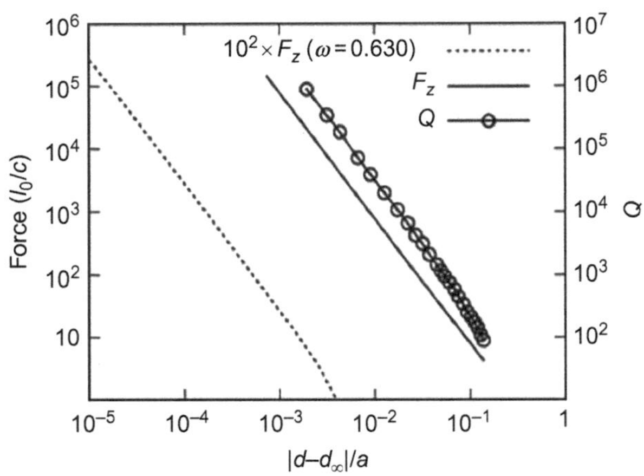

Fig. 26 Divergence behavior of normal force on resonance compared to the Q as the slab spacing is varied away from the dark state in Case 1. For reference, the force variation at a constant frequency near the dark state frequency is also plotted with a *dashed line*. *Reproduced with permission from Liu, V., Povinelli, M., Fan, S., 2009. Resonance-enhanced optical forces between coupled photonic crystal slabs. Opt. Express 17 (24), 21897–21909.*

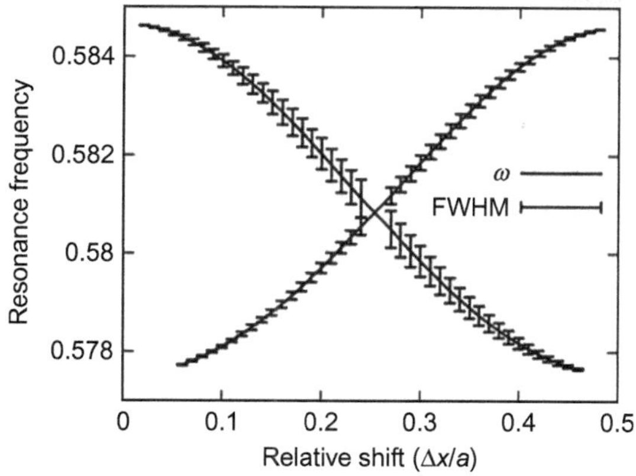

Fig. 27 Resonance peak and linewidth in the vicinity of the dark state arising from coupled dark resonances as in Fig. 6(Case 2). *Reproduced with permission from Liu, V., Povinelli, M., Fan, S., 2009. Resonance-enhanced optical forces between coupled photonic crystal slabs. Opt. Express 17 (24), 21897–21909.*

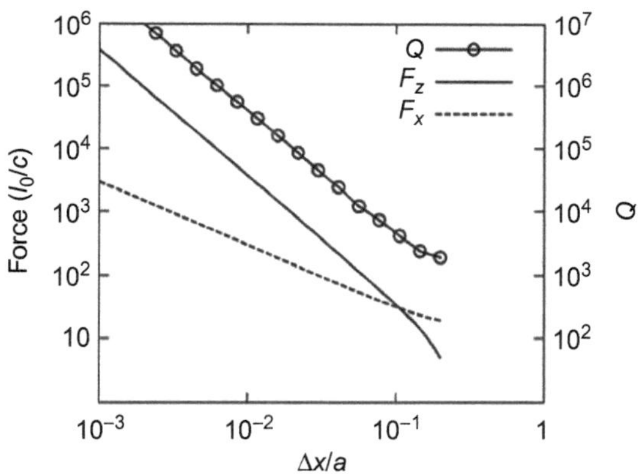

Fig. 28 Divergence behavior of normal and lateral forces on resonance compared to the *Q* for various shifts near the symmetry point $\Delta x = 0$ for Case 2. *Reproduced with permission from Liu, V., Povinelli, M., Fan, S., 2009. Resonance-enhanced optical forces between coupled photonic crystal slabs. Opt. Express 17 (24), 21897–21909.*

other by $0.2a$, where a is the lattice constant. A dark state still exists, as seen in Fig. 29. As the slab separation is changed, both the normal force, F_z, and the lateral force, F_x, vary in the same way as Q, inversely proportional to the square of the separation (Fig. 30).

In conclusion, in a coupled photonic crystal slab system, the optical forces are strongly influenced by the symmetries of the system as well as its resonance properties. It is worth noting that these proposed structures

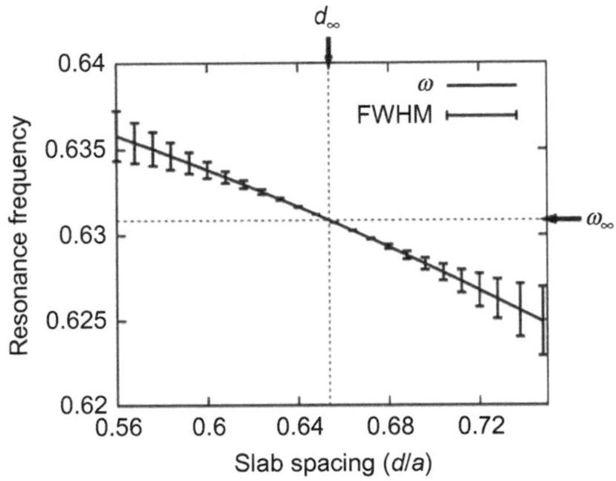

Fig. 29 Resonance peak and linewidth in the vicinity of the dark state for Case 3 at $\Delta x = 0.2a$. *Reproduced with permission from Liu, V., Povinelli, M., Fan, S., 2009. Resonance-enhanced optical forces between coupled photonic crystal slabs. Opt. Express 17 (24), 21897–21909.*

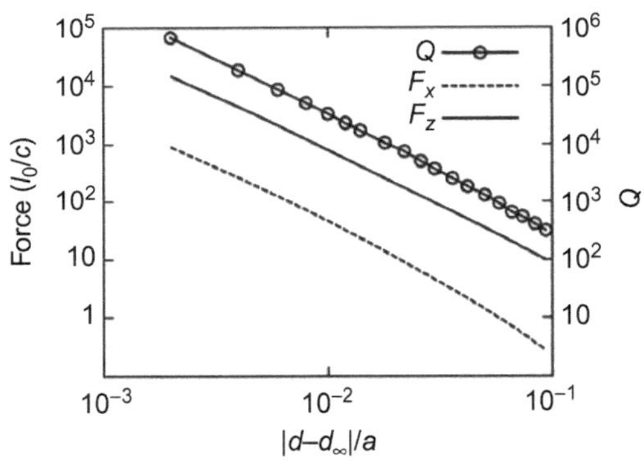

Fig. 30 Divergence behavior of normal and lateral forces on resonance compared to the Q for various slab spacings near the dark state in Fig. 6. *Reproduced with permission from Liu, V., Povinelli, M., Fan, S., 2009. Resonance-enhanced optical forces between coupled photonic crystal slabs. Opt. Express 17 (24), 21897–21909.*

can lead to enhanced light-induced pressure over larger areas, in a configuration that is directly accessible to the externally incident, free-space optical beams. The force scales with the Q of the resonances in the structure. By placing the system configuration in the vicinity of a dark state, a dramatic enhancement of the optical forces may be obtained due to the large resonant enhancement of the fields.

Followed by initial theoretical investigations of high-Q and strong optomechanical coupling in defect-mode double-layer PCS cavities

(Notomi et al., 2006), Notomi et al. (Roh et al., 2010) reported an experimental demonstration of strong optomechanical interactions in InP-based defect-free bilayer PCS cavities, based on bandedge or Fano resonance modes. Significant radiation force of the stored light (*F*) should appear as the electromagnetic energy stored in the cavity (*U*) changes with the change in spatial separations (*z*) between two coupled PCS slabs, as shown in the following equation:

$$F = -\frac{\mathrm{d}U}{\mathrm{d}z} = -\left(\frac{U}{\omega}\right)\left(\frac{\mathrm{d}\omega}{\mathrm{d}z}\right) = -\left(\frac{PQ}{\omega^2}\right)\left(\frac{\mathrm{d}\omega}{\mathrm{d}z}\right)$$

where z is the spatial coordinate of vertical direction, and the electromagnetic energy stored in the cavity is $U = N\hbar\omega_c = PQ/\omega$. P is the incident optical power. For the mode with $\frac{\mathrm{d}\omega}{\mathrm{d}z} > 0$, $F < 0$, attractive force appears between the two PCS's. Based on the modal dispersion properties (Notomi et al., 2006), the estimated radiation force per unit of electromagnetic energy in the cavity is ~0.23 μN/pJ, with the force estimated to be ~0.4 μN/pJ. For an incident power of 19.7 mW, a displacement of 3.6 nm was obtained for a cavity Q of 1600 at 1568 nm, as a result of an optical force of 10.8 nN (Fig. 31).

Theoretical analysis showed that the optical force is periodic, corresponding with displacement, resulting in stable and unstable equilibrium positions. Both attractive and repulsive force were induced depending

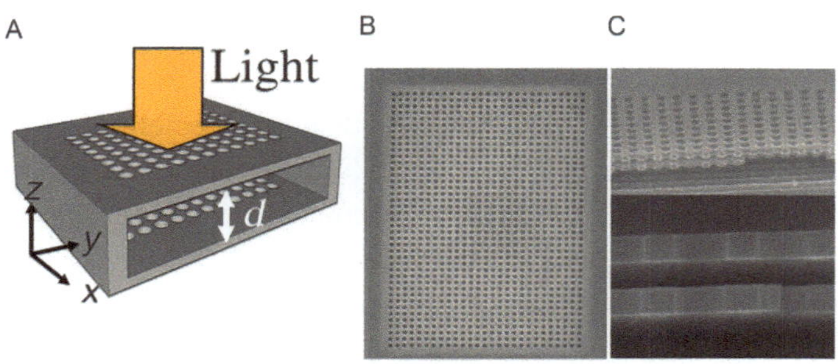

Fig. 31 (A) Schematics of the experiment. The structure is fully clamped (all-side clamped); (B) Top-view scanning electron microscope image; and (C) Bird's-eye view (*upper*) and cross section (*lower*) of a bilayer PCS. The period is 750 nm and the hole radius at the center is 270 nm. *Reproduced with permission from Roh, Y.-G., Tanabe, T., Shinya, A., Taniyama, H., Kuramochi, E., Matsuo, S., Sato, T., Notomi, M., 2010. Strong optomechanical interaction in a bilayer photonic crystal. Phys. Rev. B 81 (12), 121101.*

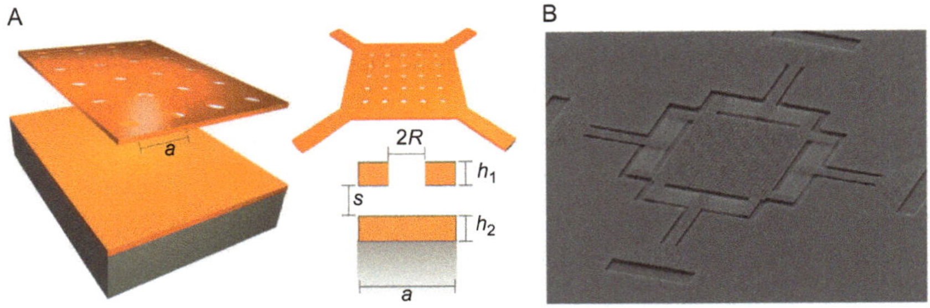

Fig. 32 (A, B) Schematic and (C) SEM image of single-membrane (asymmetric) structure: a photonic-crystal (holey) membrane consisting of a square-lattice of air holes on silicon is suspended (separation *s*) on *top* of an unpatterned (homogeneous) silicon slab (thickness $h_2 = 0.2a$) sitting on *top* of a semiinfinite silica substrate. Light is apparent on the membrane from the normal direction (*top*). Key parameters are: $a = 920$ nm, $R = 207$ nm, $h_1 = h_2 = 185$ nm, and $s = 265$ nm. *Reproduced with permissions from Rodriguez, A. W., McCauley, A. P., Hui, P.-C., Woolf, D., Iwase, E., Capasso, F., Loncar, M., Johnson, S. G., 2011. Bonding, antibonding and tunable optical forces in asymmetric membranes. Opt. Express 19 (3), 2225–2241; Hui, P.-C., Woolf, D., Iwase, E., Sohn, Y.-I., Ramos, D., Khan, M., Rodriguez, A. W., Johnson, S. G., Capasso, F., Loncar, M., 2013. Optical bistability with a repulsive optical force in coupled silicon photonic crystal membranes. Appl. Phys. Lett. 103 (2), 021102 (021102-5).*

on the spatial symmetry of the stored electromagnetic field profile in the double-layer PC slab cavities and air-slot cavities (Roh et al., 2010).

Similar to the symmetric double-layer PCS systems, tunable attractive and repulsive forces can also arise in highly asymmetric structures coupled with external radiation. One recent theoretical study by Rodriguez et al. (2011) shows this generality by demonstrating that tunable attractive and repulsive forces can also arise in highly asymmetric structures coupled with external radiation (Fig. 32A). Recently, Hui et al. (2013) reported an experimental observation of repulsive forces with a displacement of 1 nm/mW in such an Si-based hybrid PCS membrane/unpatterned Si membrane cavity (Fig. 32B). It is also predicted that such configurations can extend the displacement range to many tens of nanometers, providing a simple, nonintrusive solution to extending the actuation range of microelectromechanical systems (MEMS) devices.

3.3 Optomechanical control and MOMES applications

There is a variety of all-optical operations that can be created by this method, ranging from all optical switching and tuning (for telecom, sensing, and imaging applications) to adaptive dispersion and filter synthesis (for applications such as adaptive filtering of laser lines and optical clock recovery). MEMS's are the promising mechanisms for introducing variable functions into PC devices. Although several devices with variable structures of PC slabs combined with MEMS were considered, there have been very few reports on the experimental studies because of the difficulty in fabricating hybrid structures (Hui et al., 2013; Rodriguez et al., 2011). A few experimental results, as well as principle design, have been reported (Lu and Lee, 2009; Stomeo et al., 2010). In order to introduce variable function to the Fano resonances in PC slabs, Suh et al. (2003) studied two PC slabs facing each other analytically for a wavelength-tunable PC filter. Kanamori et al. (2007) experimentally demonstrated a wavelength-selective variable reflection filter controlled by changing the gap with a microelectromechanical actuator (Fig. 33).

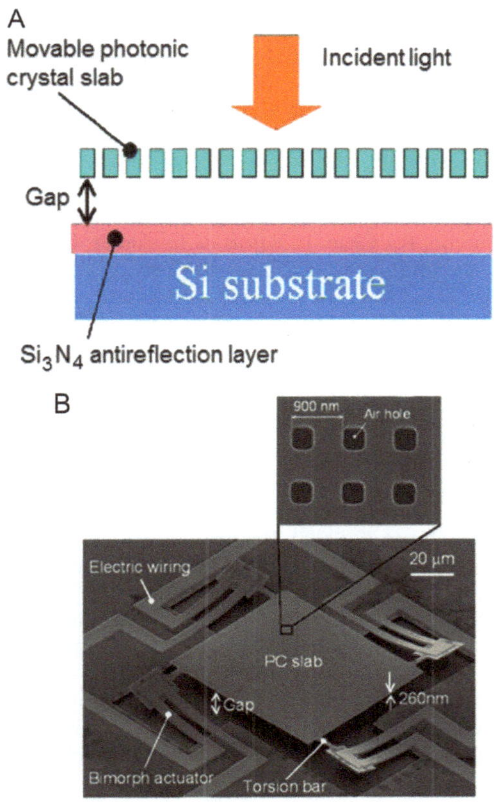

Fig. 33 (A) Cross-section diagram of the MEMS tunable Fano filter design and (B) SEM image of the fabricated Fano filter. The *inset* shows the magnified view of the fabricated PC slab. *Reproduced with permission from Kanamori, Y., Kitani, T., Hane, K., 2007. Control of guided resonance in a photonic crystal slab using microelectromechanical actuators. Appl. Phys. Lett. 90, 031911.*

Acknowledgments

The authors acknowledge contributions from their current and former students, as well as collaborators involved in the research work reviewed here. We also thank funding agencies and program directors for their instrumental support from US Air Force Office of Scientific Research (AFOSR, Dr. Gernot Pomrenke), US Army Research Office (ARO, Dr. Michael Gerhold), US National Science Foundation (NSF, Drs. John Zavada and Dominique Dagenais). We also thank the fabrication support from the University of Texas at Arlington Nanotechnology Research Center, and the University of Texas at Austin, part of NSF NNIN.

References

Chen, L., Qiang, Z., Yang, H., Pang, H., Ma, Z., Zhou, W.D., 2009. Polarization and angular dependent transmissions on transferred nanomembrane Fano filters. Opt. Express 17 (10), 8396–8406.

Chmielak, B., Waldow, M., Matheisen, C., Ripperda, C., Bolten, J., Wahlbrink, T., Nagel, M., Merget, F., Kurz, H., 2011. Pockels effect based fully integrated, strained silicon electro-optic modulator. Opt. Express 19 (18), 17212–17219.

Crozier, K.B., Lousse, V., Kilic, O., Kim, S., Fan, S., Solgaard, O., 2006. Air-bridged photonic crystal slabs at visible and near-infrared wavelengths. Phys. Rev. B 73 (11), 115126.

Eichenfield, M., Michael, C.P., Perahia, R., Painter, O., 2007. Actuation of micro-optomechanical systems via cavity-enhanced optical dipole forces. Nat. Photonics 1 (7), 416–422.

Eichenfield, M., Camacho, R., Chan, J., Vahala, K.J., Painter, O., 2009. A picogram- and nanometre-scale photonic-crystal optomechanical cavity. Nature 459 (7246), 550–555.

Fan, S., Joannopoulos, J.D., 2002. Analysis of guided resonances in photonic crystal slabs. Phys. Rev. B 65 (23), 235112.

Hui, P.-C., Woolf, D., Iwase, E., Sohn, Y.-I., Ramos, D., Khan, M., Rodriguez, A.W., Johnson, S.G., Capasso, F., Loncar, M., 2013. Optical bistability with a repulsive optical force in coupled silicon photonic crystal membranes. Appl. Phys. Lett. 103 (2), 021102, 021102-5.

Jacobsen, R.S., Andersen, K.N., Borel, P.I., Fage-Pedersen, J., Frandsen, L.H., Hansen, O., Kristensen, M., Lavrinenko, A.V., Moulin, G., Ou, H., 2006. Strained silicon as a new electro-optic material. Nature 441 (7090), 199–202.

Kanamori, Y., Kitani, T., Hane, K., 2007. Control of guided resonance in a photonic crystal slab using microelectromechanical actuators. Appl. Phys. Lett. 90, 031911.

Kanskar, M., Paddon, P., Pacradouni, V., Morin, R., Busch, A., Young, J.F., Johnson, S.R., MacKenzie, J., Tiedje, T., 1997. Observation of leaky slab modes in an air-bridged semiconductor waveguide with a two-dimensional photonic lattice. Appl. Phys. Lett. 70 (11), 1438–1440.

Kilic, O., Digonnet, M., Kino, G., Solgaard, O., 2007. External fibre Fabry-Perot acoustic sensor based on a photonic-crystal mirror. Meas. Sci. Technol. 18 (10), 3049–3054.

Kippenberg, T., Vahala, K., 2008. Cavity optomechanics: back-action at the mesoscale. Science 321 (5893), 1172–1176.

Li, M., Pernice, W., Xiong, C., Baehr-Jones, T., Hochberg, M., Tang, H., 2008. Harnessing optical forces in integrated photonic circuits. Nature 456 (7221), 480–484.

Li, M., Pernice, W., Tang, H., 2009. Tunable bipolar optical interactions between guided lightwaves. Nat. Photonics 3 (8), 464–468.

Lin, C., Lu, Z., Shi, S., Jin, G., Prather, D.W., 2005. Experimentally demonstrated filters based on guided resonance of photonic-crystal films. Appl. Phys. Lett. 87, 091102.

Liu, V., Fan, S., 2012. S 4: a free electromagnetic solver for layered periodic structures. Comput. Phys. Commun. 183 (10), 2233–2244.
Liu, V., Povinelli, M., Fan, S., 2009. Resonance-enhanced optical forces between coupled photonic crystal slabs. Opt. Express 17 (24), 21897–21909.
Lousse, V., Suh, W., Kilic, O., Kim, S., Solgaard, O., Fan, S.H., 2004. Angular and polarization properties of a photonic crystal slab mirror. Opt. Express 12 (8), 1575–1582.
Lu, T.-W., Lee, P.-T., 2009. Ultra-high sensitivity optical stress sensor based on double-layered photonic crystal microcavity. Opt. Express 17 (3), 1518–1526.
Ma, J., Povinelli, M.L., 2012. Applications of optomechanical effects for on-chip manipulation of light signals. Curr. Opin. Solid State Mater. Sci. 16 (2), 82–90.
McKerracher, I., Fu, L., Tan, H., Jagadish, C., 2013. Integration of bandpass guided-mode resonance filters with mid-wavelength infrared photodetectors. J. Phys. D Appl. Phys. 46 (9), 095104.
Nazirizadeh, Y., Müller, J.G., Geyer, U., Schelle, D., Kley, E.-B., Tünnermann, A., Lemmer, U., Gerken, M., 2008. Optical characterization of photonic crystal slabs using orthogonally oriented polarization filters. Opt. Express 16 (10), 7153–7160.
Notomi, M., Taniyama, H., Mitsugi, S., Kuramochi, E., 2006. Optomechanical wavelength and energy conversion in high-Q double-layer cavities of photonic crystal slabs. Phys. Rev. Lett. 97 (2), 023903.
Nxcnons, E., Huu, G., 1901. A preliminary communication on the pressure of heat and light radiation. Phys. Rev. 13, 307.
Peng, S., Morris, G., 1996. Resonant scattering from two-dimensional gratings. J. Opt. Soc. Am. A 13 (5), 993–1005.
Povinelli, M., Johnson, S., Lonèar, M., Ibanescu, M., Smythe, E., Capasso, F., Joannopoulos, J., 2005a. High-Q enhancement of attractive and repulsive optical forces between coupled whispering-gallery-mode resonators. Opt. Express 13 (20), 8286–8295.
Povinelli, M.L., Loncar, M., Ibanescu, M., Smythe, E.J., Johnson, S.G., Capasso, F., Joannopoulos, J.D., 2005b. Evanescent-wave bonding between optical waveguides. Opt. Lett. 30 (22), 3042–3044.
Qiang, Z., Yang, H., Chen, L., Pang, H., Ma, Z., Zhou, W., 2008. Fano filters based on transferred silicon nanomembranes on plastic substrates. Appl. Phys. Lett. 93, 061106.
Rakich, P.T., Popović, M.A., Soljačić, M., Ippen, E.P., 2007. Trapping, corralling and spectral bonding of optical resonances through optically induced potentials. Nat. Photonics 1 (11), 658–665.
Rodriguez, A.W., McCauley, A.P., Hui, P.-C., Woolf, D., Iwase, E., Capasso, F., Loncar, M., Johnson, S.G., 2011. Bonding, antibonding and tunable optical forces in asymmetric membranes. Opt. Express 19 (3), 2225–2241.
Roels, J., De Vlaminck, I., Lagae, L., Maes, B., Van Thourhout, D., Baets, R., 2009. Tunable optical forces between nanophotonic waveguides. Nat. Nanotechnol. 4 (8), 510–513.
Rogers, J.A., Lagally, M.G., Nuzzo, R.G., 2011. Synthesis, assembly and applications of semiconductor nanomembranes. Nature 477 (7362), 45–53.
Roh, Y.-G., Tanabe, T., Shinya, A., Taniyama, H., Kuramochi, E., Matsuo, S., Sato, T., Notomi, M., 2010. Strong optomechanical interaction in a bilayer photonic crystal. Phys. Rev. B 81 (12), 121101.
Rosenberg, A., Carter, M., Casey, J., Kim, M., Holm, R., Henry, R., Eddy, C., Shamamian, V., Bussmann, K., Shi, S., Prather, D.W., 2005. Guided resonances in asymmetrical GaN photonic crystal slabs observed in the visible spectrum. Opt. Express 13 (17), 6564–6571.
Shuai, Y., Zhao, D., Singh Chadha, A., Seo, J.-H., Yang, H., Fan, S., Ma, Z., Zhou, W., 2013a. Coupled double-layer Fano resonance photonic crystal filters with lattice-displacement. Appl. Phys. Lett. 103 (24), 241106.

Shuai, Y., Zhao, D., Tian, Z., Seo, J.-H., Plant, D.V., Ma, Z., Fan, S., Zhou, W., 2013b. Double-layer Fano resonance photonic crystal filters. Opt. Express 21 (21), 24582–24589.
Shuai, Y.-C., Zhao, D., Liu, Y., Stambaugh, C., Lawall, J., Zhou, W., 2017. Coupled bilayer photonic crystal slab electro-optic spatial light modulators. IEEE Photonics J. 9 (2), 1–11.
Soref, R., Bennett, B., 1987. Electrooptical effects in silicon. IEEE J. Quantum Electron. 23 (1), 123–129.
Stomeo, T., Grande, M., Rainò, G., Passaseo, A., D'Orazio, A., Cingolani, R., Locatelli, A., Modotto, D., De Angelis, C., De Vittorio, M., 2010. Optical filter based on two coupled PhC GaAs-membranes. Opt. Lett. 35 (3), 411–413.
Suh, W., Fan, S., 2003. Mechanically switchable photonic crystal filter with either all-pass transmission or flat-top reflection characteristics. Opt. Lett. 28 (19), 1763–1765.
Suh, W., Fan, S., 2004. All-pass transmission or flattop reflection filters using a single photonic crystal slab. Appl. Phys. Lett. 84, 4905.
Suh, W., Yanik, M.F., Solgaard, O., Fan, S., 2003. Displacement-sensitive photonic crystal structures based on guided resonance in photonic crystal slabs. Appl. Phys. Lett. 82 (13), 1999.
Suh, W., Wang, Z., Fan, S., 2004. Temporal coupled-mode theory and the presence of non-orthogonal modes in lossless multimode cavities. IEEE J. Quantum Electron. 40 (10), 1511–1518.
Suh, W., Solgaard, O., Fan, S., 2005. Displacement sensing using evanescent tunneling between guided resonances in photonic crystal slabs. J. Appl. Phys. 98, 033102.
Synopsys, 2016. Taurus Medici [Online]. Sysnopsys.
Thurman, S.T., Morris, G.M., 2003. Controlling the spectral response in guided-mode resonance filter design. Appl. Optics 42 (16), 3225–3233.
Van Thourhout, D., Roels, J., 2010. Optomechanical device actuation through the optical gradient force. Nat. Photonics 4 (4), 211–217.
Wiederhecker, G.S., Chen, L., Gondarenko, A., Lipson, M., 2009. Controlling photonic structures using optical forces. Nature 462 (7273), 633–636.
Woolf, D., Hui, P.-C., Iwase, E., Khan, M., Rodriguez, A.W., Deotare, P., Bulu, I., Johnson, S.G., Capasso, F., Loncar, M., 2013. Optomechanical and photothermal interactions in suspended photonic crystalmembranes. Opt. Express 21 (6), 7258–7275.
Yang, H., Zhao, D., Chuwongin, S., Seo, J.H., Yang, W., Shuai, Y., Berggren, J., Hammar, M., Ma, Z., Zhou, W., 2012. Transfer-printed stacked nanomembrane lasers on silicon. Nat. Photonics 6 (9), 615–620.
Zhang, K., Seo, J.H., Zhou, W., Ma, Z., 2012. Fast flexible electronics using transferrable silicon nanomembranes (topical review). J. Phys. D Appl. Phys. 45 (14), 143001.
Zhou, W., Ma, Z., Yang, H., Qiang, Z., Qin, G., Pang, H., Chen, L., Yang, W., Chuwongin, S., Zhao, D., 2009. Flexible photonic-crystal Fano filters based on transferred semiconductor nanomembranes. J. Phys. D Appl. Phys. 42, 234007.

CHAPTER SEVEN

On-chip photonic crystal surface-emitting lasers

Weidong Zhou*
Department of Electrical Engineering, University of Texas at Arlington, Arlington, TX, United States
*Corresponding author: e-mail address: wzhou@uta.edu

Contents

1. Introduction

Silicon-based light sources have witnessed significant advances over the last decade, with one of the most promising approaches being heterogeneous integration of direct-bandgap group III–V materials on various Si-based laser cavities on silicon-on-insulator (SOI) substrates (Bakir et al., 2006; Balakrishnan et al., 2006; Chen et al., 2011; Liang et al., 2010; Mi et al., 2005; Park et al., 2006; Roelkens et al., 2010; Stankovic et al., 2011;

Semiconductors and Semimetals, Volume 100
ISSN 0080-8784
https://doi.org/10.1016/bs.semsem.2019.04.002

Van Campenhout et al., 2007). Photonic crystals have been known to be an excellent candidate for ultracompact energy efficient lasers via spontaneous emission control and dispersion engineering. Based on defect-mode cavities and defect-free bandedge effects, lasers have been reported for low power single-photon sources as well as high power single-mode broad-area laser sources, mostly on III–V substrates (Altug et al., 2006; Hirose et al., 2014; Noda, 2010; Painter et al., 1999; Park et al., 2004; Yoshie et al., 2004; Zhou et al., 2001). Additionally, based on Fano resonance or guided resonance principles, broadband single-layer membrane reflectors (MRs) can also be designed based on photonic crystal cavities in constructing VCSELs (Fan and Joannopoulos, 2002; Yang et al., 2012; Zhou et al., 2014).

Here, we report design and analysis of photonic crystal surface-emitting membrane lasers for silicon photonics, based on printed IV–V QW heterostructures heterogeneously integrated on Si-based photonic crystal cavities. Based on the evanescent mode coupling, membrane lasers can be realized by vertical coupling of III–V QW membranes on defect-mode and bandedge photonic crystal cavities on Si. The effect of coupling strength and mode overlap on confinement factor and lasing threshold will be evaluated and optimized. On the other hand, based on photonic crystal MRs, MR-VCSEL cavities can be design with the proper selection of low index buffer layers between the photonic crystal MRs and III–V QW structures. Both photonic crystal guided resonance mode and Fabry–Perot (FP) lasing cavity mode need to be properly controlled.

Based on the membrane transfer printing technique for heterogeneous integration of III–V with Si materials (Yang et al., 2012), we demonstrated printed large-area single-mode photonic bandedge lasers (PBELs) as well as MR-VCSELs on SOI substrates. Optically pumped laser characteristics will be discussed, with the results compared with the gain threshold conditions for lasing model property analysis. Cavity scaling and electrical injection will also be discussed for practical and energy efficient source on-chip for integrated photonics.

2. Overview of photonic crystal lasers

System integration requires photonic devices to be *compact yet efficient, high performance yet low power consumption, highly uniform and simply controled, alignment free and scalable, low cost and reliable.* In search for the ideal 2D surface-emitting laser arrays, vertical-cavity surface-emitting laser (VCSEL) seems to be the most mature technology to date, with successful demonstrations of ultralow threshold, high single-mode power, and large size of arrays, and arrays on silicon substrates (Iga, 2000). VCSELs are routinely used for short distance data communication application, where typical array sizes are 6 or 12. It is expected that large array sizes (10s to 100s) are also needed for future optical interconnect and image processing systems. Size, uniformity, and power dissipation will be the dominant factors in realizing these large array sizes on surface-emitting lasers. In addition, current VCSEL technologies also face a number of challenges, such as transverse and polarization mode control, aperture size uniformity control, linewidth reduction, DBR resistance, limited thermal performances, and challenges on suitable DBRs for long wavelength VCSELs and complicated growth requirement (Iga, 2000).

Photonic crystals offer the most promising alternative to the current DBR-based VCSEL technology. Since the first proposal of spontaneous emission control with photonic crystals (Huang et al., 2007), a large variety of light-emitting devices have been demonstrated, mostly with 2D photonic crystal slab (PCS) air hole lattice structures. These structures are incorporated in LEDs for improved extraction efficiency and directionality, in edge-emitting lasers as cleave free end mirrors and/or PBG line defect defined waveguide, in DBR-based VCSELs for mode control, and in photonic band edge surface-emitting lasers (PBELs) as 2D DFB lasers. Most of these applications utilize the photonic bandgap (PBG) purely as mirrors or filters, where there is little impact on spontaneous emission rate. On the other hand, spontaneous emission control is enabled in photonic bandgap defect-mode lasers (PBDLs) by embedding the active region *inside* the photonic crystals, where 2D air columns are etched through the cavity, including active regions. A summary of different surface-emitting lasers is given in Table 1.

Broadly speaking, there are four types of PCS-based light sources:

(I) *Photonic crystal confined VCSELs*: By etching into the top portion of the DBRs in the otherwise conventional DBR-based VCSELs, transverse modal control is possible (Furukawa et al., 2004; Lundeberg et al., 2005; Song et al., 2002; Yokouchi et al., 2003). With the proper design of these photonic crystal structures, VCSEL beam properties can be engineered for improved performances.

(II) *PBDL*: By etching through the active region of the light-emitting structure, the spontaneous emission process can be altered in the defect (point and line defect, surface wave modes) cavity region, where enhanced/suppressed spontaneous emission is feasible when

Table 1 Photonic crystal surface-emitting lasers (PCSELs)

Configurations		PhC functions	Characteristics	References
DBR-Based VCSEL	**VCSEL** 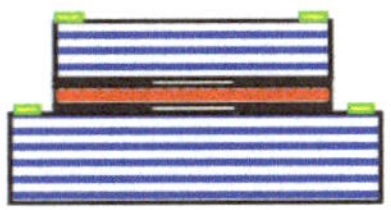	*Vertical-cavity surface-emitting lasers:* • DBR (1D PhC): HR mirror for high Q cavity formation	*Pros:* • Mature technology • Short vertical cavity • Relative low threshold	Iga (2000)
	PC_VCSEL 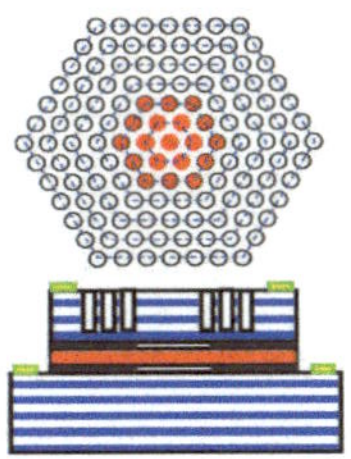	*Photonic crystal confined VCSELs:* • DBR (1D PhC): HR mirror for high Q cavity formation • 2D PhC air hole: Lateral mode confinement	• High output power • Proven reliability • Possible high power single mode in PC-VCSELs *Cons/Challenges:* • Wafer growth for DBRs • Long wavelengths • Uniform small aperture size control • Wavelength specific design	Song et al. (2002), Yokouchi et al. (2003), Furukawa et al. (2004), and Lundeberg et al. (2005)

Table 1 Photonic crystal surface-emitting lasers (PCSELs)—cont'd

Configurations		PhC functions	Characteristics	References
DBR-free PCSEL	**PBDL** 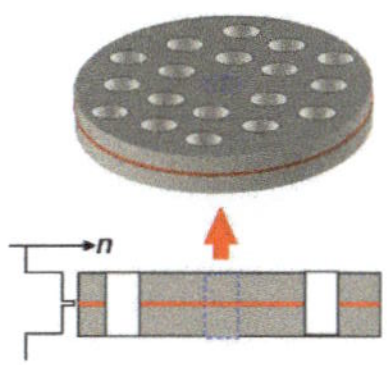 	*Photonic bandgap defect-mode lasers:* • 2D PhC air hole: Defect mode at the center of the PBG for enhanced spontaneous emission	*Pros*: • DBR free with simple wafer growth (LED like structure) • Spontaneous emission control with high spontaneous emission factor	Hwang et al. (2000), Painter et al. (1999), Park et al. (2004), and Zhou et al. (2000)
	PCDFB/PBEL 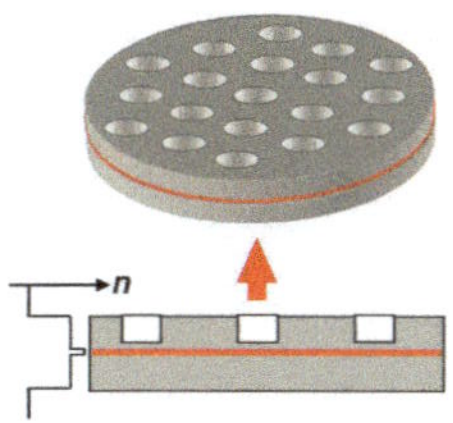 	*Photonic crystal DFBs or photonic bandedge lasers:* 2D PhC air hole: • Distributed feedback • Enhanced gain at bandedge due to flat dispersion (zero group velocity)	• Potentially zero threshold • Wavelength agile fabrication • Energy efficient • Scalable power output • Precise cavity control enable high uniformity *Cons/Challenges*: • Efficient electrical injection • Simplified process development	Dowling et al. (1994), Meier et al. (1999b), Noda et al. (2001), Susa (2001), Monat et al. (2003), and Kwon et al. (2004)

Continued

Table 1 Photonic crystal surface-emitting lasers (PCSELs)—cont'd

Configurations		PhC functions	Characteristics	References
DBR-free VCSEL	**MR-VCSEL** 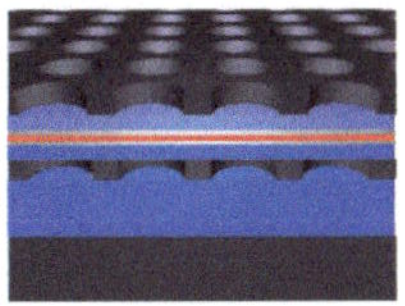	*Photonic crystal membrane reflector VCSELs:* • 1D high contrast grating or 2D photonic crystal-based single-layer reflectors • Either one or both MRs are used to replace DBRs	*Pros:* • DBR free with simple wafer growth (LED like structure) • Spontaneous emission control with high spontaneous emission factor • Ultracompact cavity • Simple cavity resonance tuning with patterning process *Cons/Challenges:* • Efficient electrical injection • Simplified process development	Sciancalepore et al. (2012) and Yang et al. (2012)

the active light emission region is placed inside the PBG defect cavity (Noda et al., 2007; Yablonovitch, 1987). With extremely high Q/V_{mode} (Q and V_{mode} being the corresponding quality factor and modal volume, respectively), strong spontaneous emission control is feasible for ultralow threshold single-mode lasers. Typically these defect modes are located below lightline. This configuration can lead to potentially thresholdless lasers (or single-mode LEDs) and single-photon sources (Bhattacharya et al., 2005; Colombelli et al., 2003; Ellis et al., 2011; Hwang et al., 2000; Painter, 1999; Park et al., 2004; Zhou et al., 2001). Another architecture reported recently (Wu et al., 2015) is the coupling of unpatterned monolayer tungsten diselenide monolayer gain medium with the defect-mode photonic crystal cavity. This configuration can lead to higher efficiency lasers as no-patterned or etching is done on the gain medium. Challenges may lie on the trade-offs in the desired high Q cavity inside the PBG and the high coupling efficiency between the high index gain medium and photonic crystal cavities.

(III) *PBELs*: The bandedge effect in defect-free photonic crystal structures, where the group velocity is close to zero, has advantages in achieving high performance semiconductor surface-emitting lasers, including low lasing threshold, single longitudinal and transverse mode over a large lasing area, narrow linewidth, high power output, small beam divergence angle, polarization control, and output beam pattern control (Chua et al., 2014; Hirose et al., 2014; Imada et al., 1999; Meier et al., 1999b; Miyai et al., 2006; Noda et al., 2001). Based on the bandedge effect, high performance lasers on GaAs have been demonstrated, including a very recent demonstration of high power, single mode, and large-area PC bandedge lasers with unpatterned QW active region using a regrowth process (Hirose et al., 2014). Different from PBDL discussed above, the modes considered here are typically located above lightline, thus these lasers are intrinsically surface emitting with excellent beam qualities.

(IV) *Membrane reflector VCSEL (MR-VCSELs):* Based on the Fano resonance principles, broadband reflection can be achieved in single-layer MRs based on 1D or 2D PCS. These ultracompact high performance MR structures have been proposed to replace the bulky DBR structures, to create ultracompact MR-VCSELs (Sciancalepore et al., 2012; Yang et al., 2012). Taking advantage of Fano resonance principles, ultracompact high performance VCSEL cavities can be formed on any substrate, such as III–V and Si, because the mirror cavity and the gain medium can be heterogeneously integrated with different material systems. This is one of the most promising ways to form lasers on Si, and to form high performance long wavelength VCSELs, in which DBR remains a great challenge.

Among four configurations discussed earlier, PBDL utilizes the localized or defect mode in the PBG. On the other hand, for the latter two configurations (PBEL and MR-VCSEL), it is not essential to design PCS's with PBG, as both utilize propagating modes in defect-free PCS structures. While PBEL can utilize the mode both below and above the light lines, MR-VCSEL's only use the modes above the light lines. Fano resonance modes can be designed in PBEL, where lasing is feasible due to distributed feedback (Chua et al., 2011). Most notably, the other type of Fano resonance-based laser is the recently demonstrated MR-VCSEL, where Fano resonance MRs are incorporated in the VCSEL cavity design to form ultracompact VCSEL's and lasers on Si. It is also worth mentioning that since PBG is not required in PBEL, the requirement on high index contrast can be relaxed. Lasing cavities can be built directly on patterned bulk semiconductor InP or Si substrates.

In this chapter, we will focus on detailed discussions related to on-chip photonic crystal lasers where the lasing cavity is formed directly on silicon substrates, the active gain medium is a conventional compound semiconductor quantum well (QW) structures.

3. Photonic crystal MR-VCSEL cavity design

3.1 Photonic crystal MRs

One of the key design targets for broadband reflectors is high reflectivity spectral bandwidth. The spectral bandwidth and reflection properties have been well investigated in multilayer DBR-based reflectors, where ultra-broadband reflectors can be achieved with the control of index contrast between pairs of dielectric layers. However, it remains a challenge to understand the correlations between spectral bandwidth and the design parameters in 2D PCS-based MR structures. On the other hand, details of theoretical investigations have been reported indicating the physical basis for wideband resonators in 1D grating structures. Magnusson et al. attributed the broadband reflection in 1D grating to the coupling of leaky modes and their excitation conditions (Magnusson and Shokooh-Saremi, 2008). Karagodsky et al. (2010) presented an analytic analysis of the ultrahigh reflectivity of subwavelength high contrast grating reflectors, and it was classified as a destructive interference effect between the two grating modes. For 2D PCS-based MRs, the designs are largely based on the trial-and-error approaches based on complete 3D computations, such as rigorous coupled-wave analysis and finite-difference time domain (FDTD) techniques. Based on the different designs for high Q filters and broadband MRs, researchers found that the Q factor (spectral linewidth) can be controlled by varying the radius, with very small r/a values (~0.1) favoring high Q filter designs and large r/a values (0.3–0.49) favoring broadband reflector designs. Additionally, the layer thickness is optimal at around 0.5λ, where λ is the target wavelength. Other design parameters include the lattice structure and the air hole shape, the buffer layer selection below and above the functional layer, effective indices, and so on (Zhou et al., 2014).

High reflectivity for surface-normal direction originates from the modal coupling between vertical radiation modes and discrete in-plane guide modes. As shown in Fig. 1A, multiple Fano resonance modes can be identified from the simulated transmission spectrum (log plot). At Fano resonances (e.g., 1500 nm shown in Fig. 1B), there is a coupling between in-plane guided mode (inside Si-MR) and vertical-cavity mode which results in a destructive interference in the direction of transmission.

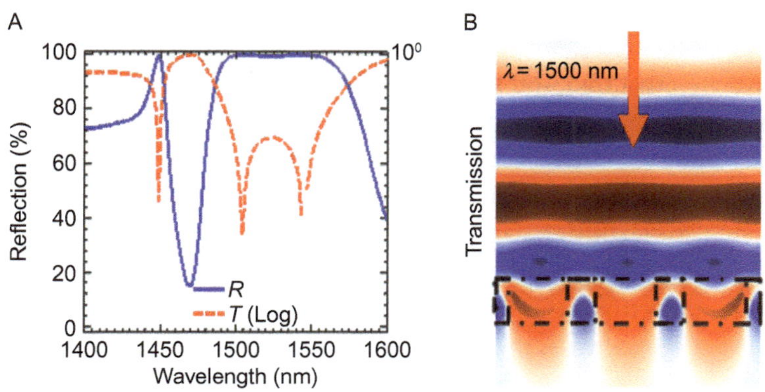

Fig. 1 Fano resonance membrane reflector (MR) design and modal coupling: (A) simulated reflection (*blue*) and transmission (*red*) spectra for the Fano MR and (B) simulated field distribution profiles for the design shown in (A) at one Fano resonance location ($\lambda = 1500$ nm).

Designs of MRs on SOI were reported by Qiang et al. (2010) who demonstrated control of reflection bands based on the control of vertical confinement with controlled refractive indices for the buffer layers below and above the MR layer. Detailed experimental results were subsequently reported by Yang et al. (2009) with a controlled index tuning range over 50 nm for both blue and red shifts, in which they controlled the buffer oxide index with controlled etching and the index above the slab with controlled oxide deposition.

Further design optimization leads to a much broader reflection band in Si-based 2D PCS MRs. In addition to the lattice parameters, the refractive index of the buffer layer is also very important for achieving high performance MRs. Though the requirement of high index contrast is relaxed here as PBG is not essential in these MR designs, it was found that sufficient index contrast below and above the single-layer MR is still critical in achieving high reflection with broad spectral band.

3.2 Energy and phase penetration properties in MRs

In Kim et al. (2007), the authors reported excellent work on the phase discontinuity Φ_r and the energy penetration depth L_e, which were estimated from the mode spacing in the FP cavity. In 2009, the in-plane optical property of 2D PCS was reported by Sauvan et al. (2009). With fully vectorial calculations, in heterostructure mirrors, the penetration length associated with the delay was much larger than the characteristic damping length of the energy distribution inside the mirror. This amazing property is evidence that these two characteristic lengths are not necessarily correlated. Consequently, the usual trade-off between short damping lengths and large penetration lengths that is encountered in classical DBRs can be overcome with carefully designed PhC structures. Later Zhao et al. (2010) investigated in more detail the phase and energy penetration depths, and field distributions of 1D subwavelength grating and 2D PCS reflectors.

In dielectric mirrors, the reflection is not an instantaneous process. It includes a reflection time delay (τ) and energy storage in the mirrors. The reflection delay increases the laser cavity round-trip time or the photon lifetime. The energy storage in mirrors decreases the modal volume and the confinement factor in the small cavity, such that the cavity round-trip time and the cavity volume have a comparable magnitude as the reflection delay and the mirror storage (Babic and Corzine, 1922; Sauvan et al., 2009). The reflection delay is directly related to the slope of the reflection phase shift Φ_r.

The relation can be expressed as $\tau = \partial \Phi_r / \partial \omega$. The phase penetration depth, L_p, is defined as the half-distance that light propagates in the incident medium during this delay time,

$$L_p = \frac{1}{2} v_g \tau = \frac{v_g}{2} \frac{\partial \Phi_r}{\partial \omega} \tag{1}$$

where v_g is the group velocity of the incident wave.

The energy storage is always associated with the parameter of energy penetration depth, L_e. It is the length that the field intensity decays to $1/e$ of its maximum from the edge of the cavity into the mirrors. However, this method is not suitable for calculating the L_e of the PC mirrors we discussed here, because the guided modes are excited inside the mirrors. But L_e can be estimated from the mirror transmission or reflection based on the following equation,

$$T = 1 - R = \exp\left(-\frac{h}{L_e}\right) \tag{2}$$

where T is the transmission and h is the mirror thickness (Sauvan et al., 2009; Zhao et al., 2010). While, for DBRs, L_e can be obtained with (Coldren and Corzine, 1995):

$$L_e = \frac{m_{\text{eff}}}{2}\left(\frac{\lambda}{4n_1} + \frac{\lambda}{4n_2}\right) \tag{3}$$

where $m_{\text{eff}} = \tanh(2mr)/(2r)$ is the effective period number seen by the incident light. $r = (n_1 - n_2)/(n_1 + n_2)$ and m are the actual period number in DBRs. n_1 and n_2 are the refractive index of the two materials in DBR. In the following we will numerically investigate the phase and energy penetration depth according to the above definitions.

Based on the simulation results shown in Fig. 2, compared to the DBR reflectors, these new types of single-layer ultracompact broadband reflectors can have larger and more complicated phase delays and smaller energy penetration properties, which can be engineered via dispersion engineering for large spectral-dependent phase delays, and ultrasmall energy penetration depths. The work reported was mostly based on one set of optimized design parameters for 1550 nm band reflectors. Following similar procedures, other design parameters can be used for reflectors with different reflection requirements, as well as different phase delays, energy penetrations, and field distributions. All the results and conclusions can be very helpful for the design of resonant cavities for a wide range of photonic devices.

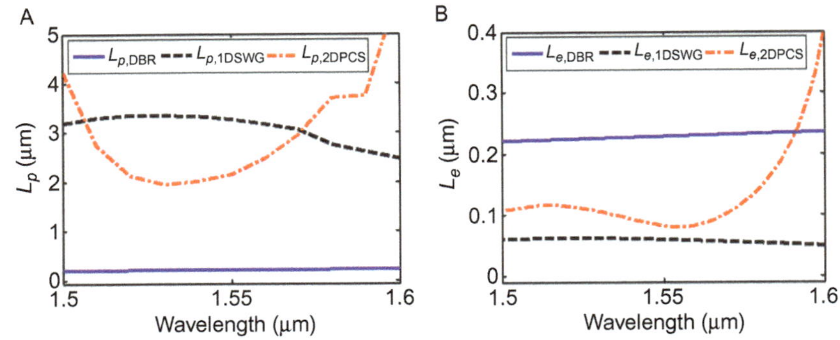

Fig. 2 (A) Calculated phase penetration depths for three types of reflectors and (B) the energy penetration depths for three types of reflectors. *Reproduced with permission from Zhao, D., Ma, Z., Zhou, W., 2010. Field penetrations in photonic crystal Fano reflectors. Opt. Express 18 (13), 14152–14158.*

3.3 MR-VCSEL

Once two good MRs with high reflectivity and broader overlapped spectral band are obtained, a multilayer Si-MR/QW/Si-MR (MR-VCSEL) cavity can be build up (Fig. 3A). Here, we need to decide the cavity resonant (longitude) mode by selecting a suitable cavity length, i.e., the thickness of the buffer layers between Si-MR and QW. Fig. 3B shows the reflection spectrum (blue line) of one designed MR-VCSEL cavity. To further confirm this cavity mode, the phase resonant condition (total phase change of one round-trip in cavity is equal to an integer times of 2π) is also calculated according the reflection phase change ϕ of the top and bottom MRs, which is plotted as red-dotted line in Fig. 3B. One can find the mode located at 1478 nm has a $m\cdot 2\pi$ phase change, which means 1478 nm is a real cavity mode.

By employing FDTD, a short temporal Gaussian pulse is used to excite the cavity modes and the quality factor of the cavity mode at 1478 nm is calculated to be 4300 according to $Q = \mathrm{Re}(\omega)/2\mathrm{Im}(\omega)$ (Oskooi et al., 2010). Then a longer temporal Gaussian pulse is used to excite the single cavity mode and the stable field is recorded. The E-field distribution is demonstrated in Fig. 3C with the red line, where the cavity index profile is also plotted with the blue line.

Another design concern that needs to be taken care of is the separation the lasing cavity modes from the QW layer as mentioned in Zhao et al. (2012). Despite the relatively large field concentration in Si-MR layers, the lasing mode field confinement factor is 5.6% for MR-VCSELs, which is comparable to DBR-based VCSELs. This critical feature enables low threshold lasing in the MR-VCSEL cavities.

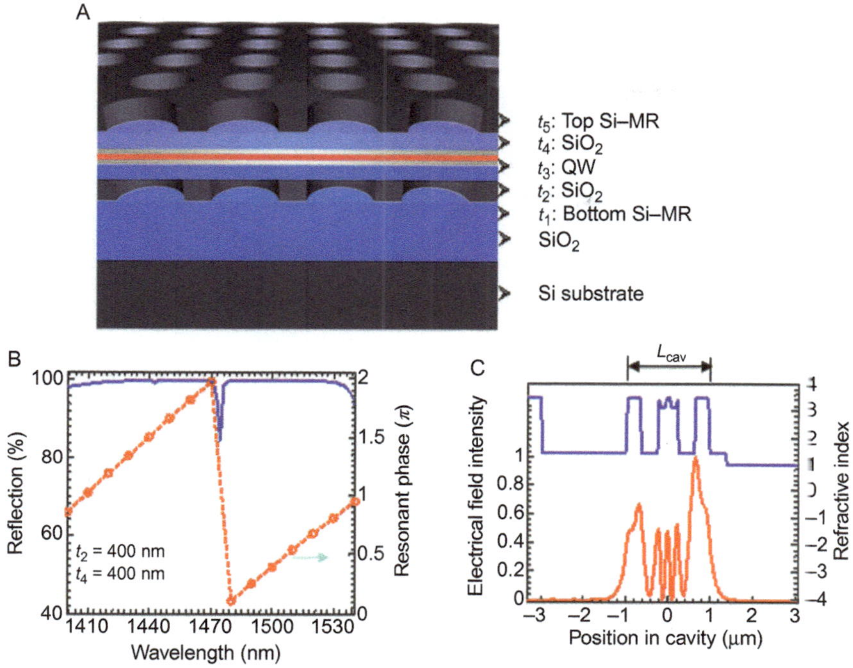

Fig. 3 Characteristics of designed MR-VCSEL cavity: (A) schematic of MR-VCSEL; (B) calculated cavity resonance mode based on cavity reflection and phase resonant condition; and (C) field distribution of cavity mode. *Reproduced with permission from Zhao, D., Yang, H., Chuwongin, S., Seo, J., Ma, Z., Zhou, W., 2012. Design of photonic crystal membrane-reflector-based VCSELs. IEEE Photonics J. 4 (6), 2169–2175.*

4. Photonic crystal bandedge surface-emitting laser cavity design

The principle of the lasing modes in the photonic crystal surface-emitting laser (PCSEL) cavities is based on the band edge modes of two-dimensional (2D) PC close to the second-order Γ-point (Imada et al., 2002; Meier et al., 1999a; Sakai et al., 2005), where the group velocities of these modes are close to zero. In such 2D PC cavities, the periodic patterned structure is designed close to the unpatented active region to provide strong in-plane distributed feedback (DFB) and the standing wave is formed in multiple directions due to the 2D DFB effect (Imada et al., 2002). Simultaneously, the in-plane propagating waves are diffracted toward the vertical direction by the PC structure itself due to the first-order Bragg diffraction. It is because of this multidirectional Bragg diffraction in these 2D PC

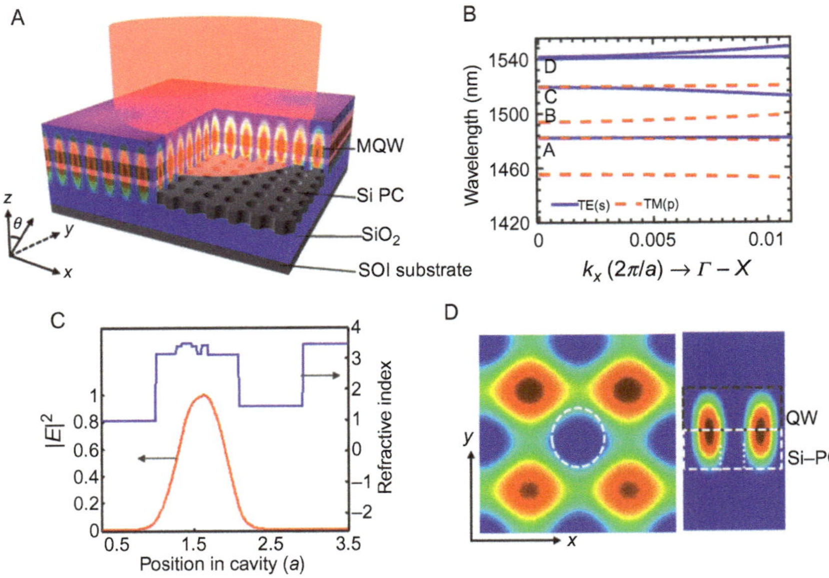

Fig. 4 (A) Schematic of PC bandedge membrane laser on SOI. (B) Photonic band diagram along *Γ–X* direction of Si-PC bandedge membrane laser cavity structure, with the transferred InGaAsP QW mesas on top. (C) Field distribution of cavity mode and (D) simulated field distributions. *Reproduced with permission from Liu, S.-C., Zhao, D., Liu, Y., Yang, H., Sun, Y., Ma, Z., Reuterskiöld-Hedlund, C., Hammar, M., Zhou, W., 2017. Photonic crystal bandedge membrane lasers on silicon. Appl. Optics 56 (31), H67–H73.*

structures, the cavity mode can lase over a large 2D area and gives a high power output (Chua et al., 2014; Hirose et al., 2014).

A schematic of the PC bandedge membrane laser on Si is shown in Fig. 4A. The laser cavity consists of an InGaAsP MQW membrane heterostructure, and a Si-PC slab patterned on SOI with the BOx layer thickness of 400 nm. The InGaAsP MQW layer design is based on the previously reported low temperature (LT) PC bandedge membrane laser, which consists of a bottom InGaAs contact layer, top and bottom InP cladding layers, and eight pairs of strain-compensated $In_{0.76}Ga_{0.24}As_{0.83}P_{0.17}$/$In_{0.485}Ga_{0.515}As_{0.83}P_{0.17}$ with the center emission wavelength of 1535 nm at room temperature (Zhao et al., 2016). In order to achieve room temperature PC bandedge membrane lasers in such a cavity, it is highly desirable to design a cavity with the following characteristics: (1) two dimension square lattices Si-PC structure is designed to have a lattice constant of 480 nm, based on the second-order Bragg diffraction condition (Chua et al., 2014; Hirose et al., 2014; Imada et al., 1999, 2002; Martınez et al., 2009;

Meier et al., 1999; Noda et al., 2001). (2) The cavity mode wavelength should match with MQW emission peak, with the hole radius and thickness of Si-PC slab designed to be 72 nm ($r = 0.15a$) and 190 nm, respectively. (3) The separation between the InGaAsP MQW layer and Si-PC slab is reduced to enhance the confinement factors and the in-plane coupling efficiency (Sakai et al., 2006; Sato et al., 2012).

4.1 Bandedge mode design

The approximate effective refractive index n_{eff} of the whole cavity (InGaAsP MQW active layer and Si-PC) is around 3.2 based on the average dielectric constant in the initial design. In order to achieve surface emission at the wavelength close to the communication wavelength range of 1550 nm, the lattice constant a of the cavity is calculated to be 480 nm according to second Bragg diffraction condition $a = \lambda / n_{\text{eff}}$. Second, the air hole radius is designed to be 72 nm ($r = 0.15a$) in order to reduce the vertical radiation loss to obtain high Q and low threshold (Zhao et al., 2016). Finally, to optimize the cavity mode to match the MQW emission peak, the thickness of the Si-PC slab is chosen to be around 190 nm (Kurosaka et al., 2009).

The band diagram for the Si-PC bandedge lasing cavity is shown in Fig. 4B. It is obtained from the Fano resonances peaks or dips in the simulated reflection or transmission spectra according to the relation between the in-plane wave vector $k_{//}$ and the incident angle θ, $k_{//} = 2\pi/\lambda \sin\theta$ (Lee et al., 2012), where λ is the wavelength and θ is the incident angle off the surface-normal direction. By changing the incident angle from the surface-normal direction (z-axis) to in-plane along the x-axis, the photonic bands in the Γ–X direction can be obtained. The Γ-point corresponds to the surface-normal incidence $\theta = 0$ degree. Here we mainly focus on a small value range of incident angle θ from 0 to 2 degrees, which corresponds to the $k_{//}$ range from 0 to ~0.01 ($2\pi/a$) in the vicinity of the band edge. The spectral simulation is done by the Fourier modal method with stanford stratified structure solver (S^4) software package (Liu and Fan, 2012). In Fig. 4B, the four bands in the normalized wavelength range (1520 nm, 1550 nm) are displayed with blue-solid lines for the transverse electric (TE, or s) polarization, and red-dashed lines for the transverse magnetic (TM, or p) polarization. One can see that all bands are very flat at the edges close to the Γ point, which indicate a very small group velocity and longer light-matter interaction time for these bandedge modes.

The field profiles of mode D at $\lambda = 1540\,\text{nm}$ is displayed in Fig. 4C. The intensity of E-field is plotted with a red line, and the cavity index profile is plotted with a blue line. The confinement factor can be calculated as follows:

$$\Gamma_{QW} = \frac{\int_{QW} \varepsilon |E|^2 \bullet dv}{\int_{cav} \varepsilon |E|^2 \bullet dv} \tag{4}$$

The quality factor and the field distribution of this cavity mode are also investigated. Their quality factors are obtained by Fano fitting the guided resonances in the reflection spectra (Luk'yanchuk et al., 2010). In our calculations, only the vertical radiation loss is considered as the in-plane loss and the internal loss are negligible (Hirose et al., 2014). Hence, only the radiation loss is considered for the infinite cavity using the following relationship:

$$\alpha = \frac{2\pi}{Q \bullet a} \tag{5}$$

Therefore, the gain threshold can be derived as follows:

$$g_{th} = \frac{\alpha}{\Gamma_{QW}} \tag{6}$$

The E-field along the center of cavity (left) and the cross-sectional views of the E-field intensity of modes (right) are shown in Fig. 4D. One can see the field is strongly confined inside the cavity. Based on the above fundamental equations, the confinement factor, quality factor, and radiation loss of mode D ($\lambda = 1540\,\text{nm}$) is calculated to be 16.83%, 1.9×10^5 and $0.68\,\text{cm}^{-1}$, respectively. Single-mode lasing can be predicted to occur at mode D which has the lowest gain threshold of $4.04\,\text{cm}^{-1}$.

4.2 Lateral cavity scaling

However, the scaling of on-chip lasers demands compact cavity sizes in order to achieve high energy efficiency and high speed operation. The reduction in lateral cavity size for PCSELs may result in a reduced in-plane and out-of-plane confinement and thus increased lasing threshold (Liang et al., 2012).

To investigate the lateral confinement effect, we considered two control parameters. (Liu et al., 2018) First, we adjust the lateral cavity size by controlling the pumping beam spot size, as shown in Fig. 5C. We also

A

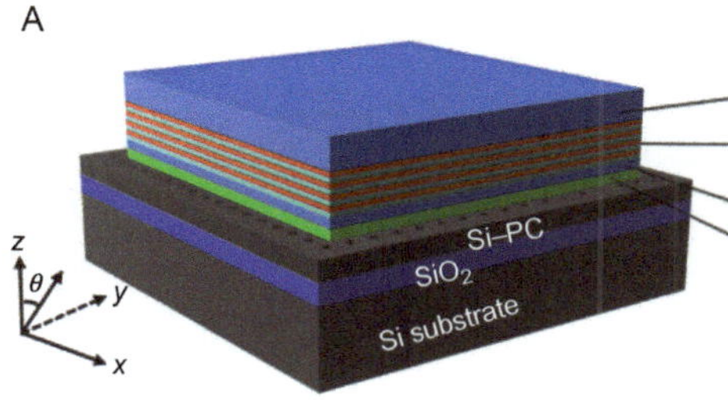

B

Layer	Thickness
p-InP	130 nm
InGaAsP MQW	127 nm
n-InP	30 nm
n-InGaAs	40 nm
Si–PC	190 nm, a = 480 nm, and r/a = 0.14
SiO_2	400 nm
Si substrate	525 μm

C

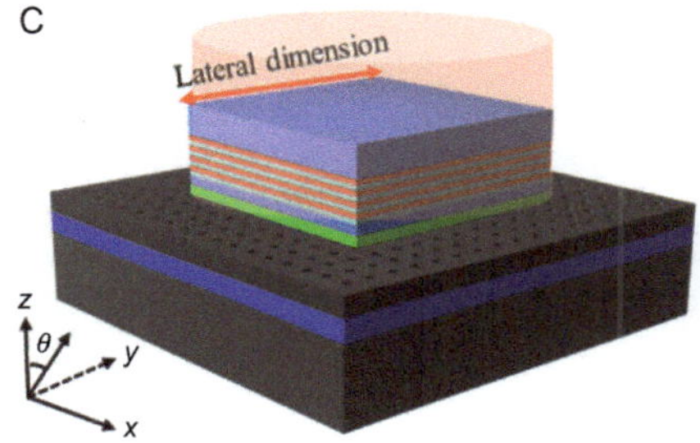

D

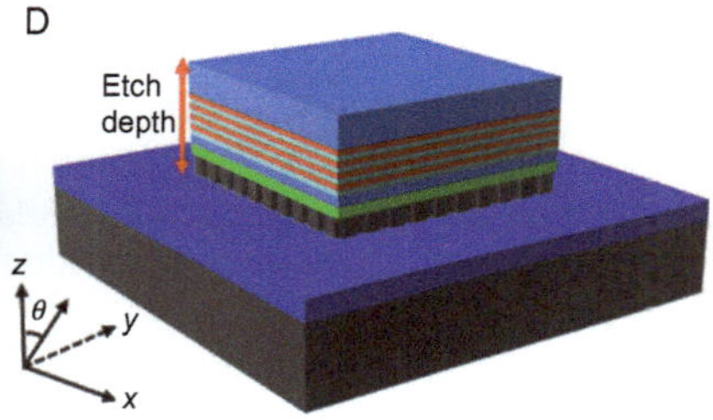

Fig. 5 (A) Schematic of PCSEL with 600 × 600 μm^2 size. (B) Cavity layer definitions. (C) Schematic of PCSEL with variation in lateral dimension (*L*). (D) Schematic of PCSEL with variation in vertical etch depth. *Reproduced with permission from Liu, S.-C., Zhao, D., Ge, X., Reuterskiöld-Hedlund, C., Hammar, M., Fan, S., Ma, Z., Zhou, W., 2018. Size scaling of photonic crystal surface emitting lasers on silicon substrates. IEEE Photonics J. 10 (3), 1–6.*

experimentally fabricated cavities with various lateral dimensions. Second, for a fixed lateral cavity size, we adjust the lateral confinement by controlling the etch depth defining the cavity region as shown in Fig. 5D. Theoretical investigations were first carried out to estimate the cavity Q factor and the associated gain thresholds for different lateral and vertical dimensions and then the corresponding threshold gains were estimated (Hirose et al., 2014; Liu et al., 2017). In our calculations, only the in-plane loss and vertical radiation loss are considered, and the internal loss is ignored (Hirose et al., 2014).

Shown in Fig. 6A are the calculated threshold gains and Q factors for different lateral cavity sizes. The Q factor drops rapidly while the threshold gain increases with decreasing device size from 80 to 20 μm. This indicates that the influence of in-plane loss has to be taken into account for finite device sizes. As a result, power escapes from the edges of the laser cavity due to weak lateral confinement.

To reduce the cavity loss due to the finite lateral dimensions, some lateral confinement can be introduced. Here we investigate a simple lateral

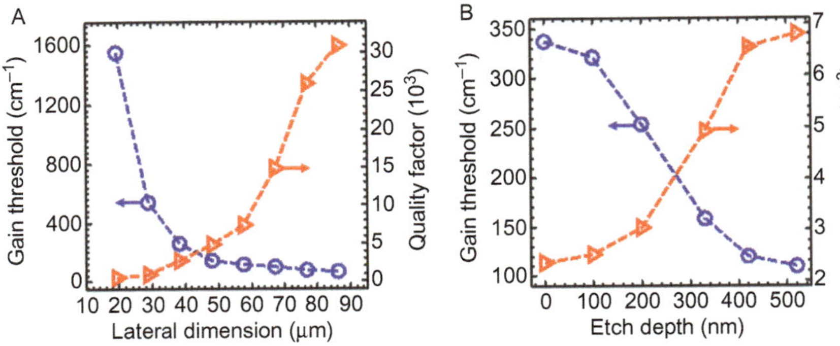

Fig. 6 (A) The simulated gain thresholds and *Q* factors for different device lateral dimensions and (B) the simulated gain thresholds and *Q* factors for a device with lateral dimension of 48 μm ($L=100\ a$) at different etching depths. *Reproduced with permission from Liu, S.-C., Zhao, D., Ge, X., Reuterskiöld-Hedlund, C., Hammar, M., Fan, S., Ma, Z., Zhou, W., 2018. Size scaling of photonic crystal surface emitting lasers on silicon substrates. IEEE Photonics J. 10 (3), 1–6.*

confinement scheme by controlling the etch depth of the lasing cavity (lateral index contrast). The result is shown in Fig. 6B for a PCSEL on Si with a lateral cavity dimension of 48 μm ($L=100\ a$). By increasing the etch depth from 0 to 500 nm, which corresponds to the complete removal of the InGaAsP QW heterostructure and Si-PC cavity (Fig. 5D), the cavity Q factor increases from 2300 to 6800. We attribute this to the increased lateral confinement due to the increased lateral index contrast.

5. Printed membrane laser cavities

The integration of InGaAsP QW heterostructure with the Si photonic crystal structure was carried out with a printing process. For Si-MR/InGaAsP QW/Si-MR MR-VCSEL cavity, two step printing processes are involved (Yang et al., 2012). For the desired vertical index contrast, low index SiO_2 buffer layers are prepared between Si-MR and InGaAsP QW layers. It is noted such a requirement on the low index buffer can limit device performance due to poor thermal conductivity of the presence of low index SiO_2 buffer layers.

Different from the MR-VCSEL (Yang et al., 2012), PCSEL structure only requires a one-step printing process for the heterogeneous integration of III–V gain structure on top of the patterned Si-PC cavity. It does not require a low index buffer layer to be placed in between the active region and the Si-PC layer. In fact, it is critical to minimize the separation between

QW layer and Si-PC cavity layer in order to obtain a high confinement factor which is critical to low threshold lasing.

An SOI wafer with 230 nm top Si layer and 400 nm BOx layer was used here. The top Si layer was first thinned down to 190 nm by thermal oxidation and selective wet etching processes. Next, the 2D square lattice PC pattern with a size of $1 \times 1\,mm^2$ was defined with ZEP-520A resist by using the standard electron beam lithography (EBL). The sample was then etched with the ZEP-520A resist pattern as the mask to create a high index contrast between Si and air by using the dry etching process. The scanning electron microscopic (SEM) top view of Si-PC on SOI substrate is shown in Fig 7A. The lattice constant *a* and air hole radius *r* of the fabricated PC are 480 and 72 nm, respectively.

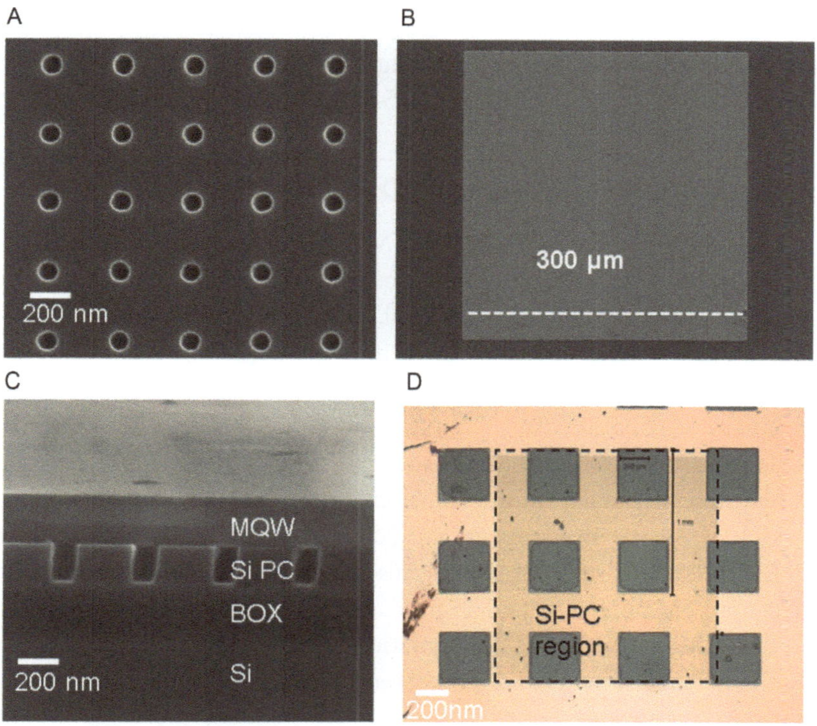

Fig. 7 (A) Zoomed-in SEM top view of the defect-free Si-PC cavity. (B) An SEM image of InGaAsP MQW mesa transferred onto a Si-PC. (C) Cross-sectional view of the cavity and (D) micrograph image of the transferred mesa array on a Si-PC. *Reproduced with permission from Liu, S.-C., Zhao, D., Liu, Y., Yang, H., Sun, Y., Ma, Z., Reuterskiöld-Hedlund, C., Hammar, M., Zhou, W., 2017. Photonic crystal bandedge membrane lasers on silicon. Appl. Optics 56 (31), H67–H73.*

InGaAsP MQW structure was grown on (001) *n*-InP substrate by metal organic chemical vapor deposition, with a 500 nm InGaAs sacrificial layer between the top QW structure and the bottom InP substrate. Substrate removal and wax protection processes were used here for the release of the large-area InGaAsP QW structure (Zhao et al., 2016). The transfer printing process was utilized to attach the InGaAsP MQW to the SOI substrate. The Si-PC on SOI substrate was first cleaned by solvent, followed by standard RCA cleaning to remove organic residues from the surface. The lifted InGaAsP MQW membrane is then carefully printed on a patterned Si-PC cavity. Subsequently, wax was removed with trichloroethylene (TCE) and isopropyl alcohol (IPA) solutions. Finally, a $300 \times 300\,\mu m^2$ active MQW square size is defined by photolithography and selective wet etching process. Fig. 7B and C shows the SEM top view and cross-sectional view of a transferred InGaAsP MQW membrane on a Si-PC cavity on a SOI substrate. Shown in Fig. 7D is a micrograph image of PC bandedge membrane lasers on SOI substrate. It is worth noting that this process could be applied at room temperature and is capable of a 5-μm alignment accuracy.

6. MR-VCSELs based on 1D PCS/grating mirrors

In addition to the 2D PCS-based MRs, 1D photonic crystal mirrors or high contrast grating structures have been proposed and demonstrated as well (Boutami et al., 2007; Huang et al., 2007). The most straight forward way of incorporating MR's is to grow the top MR layer monolithically on the laser heterostructure, to form hybrid VCSEL cavities with top MR and bottom DBR configurations. Huang et al. (2007) reported an electrically pumped 850 nm VCSEL on a GaAs substrate (Fig. 8). Excellent performance was reported with single-mode emission and a side-mode suppression ratio (SMSR) of 45 dB. Boutami et al. (2007) reported an optical 1550 nm VCSEL on an InP substrate, with pulsed operation at room temperature and a threshold power around 15 mW.

The ultimate goal for this configuration is to replace both top and bottom MR's with single-layer MR's. However, in this case, monolithic growth is not possible. One straight forward technique is to use the wafer bonding technique for the integration of a gain medium on top of the bottom MR's. Sciancalepore et al. (2012) reported an optically pumped 1550 nm VCSELs on Si. They used a double 1D PCM employing III–V/SiO_2 molecular bonding and amorphous-Si deposition processes. A 2-in. III–V wafer was bonded to a SiO_2 coated 200 mm SOI wafer, followed by an InP

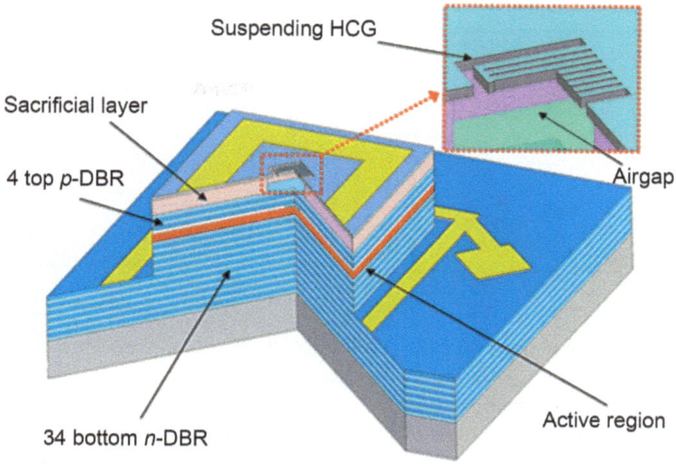

Fig. 8 Hybrid top MR/bottom DBR VCSEL configuration: suspended GaAs high contract grating (HCG) reflector monolithically grown on GaAs substrate. *Reproduced with permission from Huang, M.C.Y., Zhou, Y., Chang-Hasnain, C.J., 2007. A surface-emitting laser incorporating a high-index-contrast subwavelength grating. Nat. Photonics 1 (2), 119–122.*

substrate removal process. The polarization-dependent top 1-D PCM was formed on a deposited amorphous-Si film using deep-UV lithography and reactive ion etching.

7. MR-VCSEL characterization

Optically pumped devices are mounted inside a cryostat and characterized with a monochromator-based microphotoluminescence (μ-PL) set up using quasi-continuous wave (c.w.) 532 nm laser pumping (with 50% duty cycle).

Shown in Fig. 9A is the L–L plot (light output for different pump powers) and the corresponding spectral linewidths measured at $T = 50\,K$. The threshold pump power is ~8 mW, or 0.32 kW/cm^2. The measured spectral linewidths reduce from 30 nm below threshold to 0.6–0.8 nm above threshold. The measured spectral outputs are shown in Fig. 9B, for pump powers below, at, and above thresholds (points (i, ii, iii, iv) at L–L curve). The lasing spectral linewidth is ~8 Å, which is limited by the measurable resolution of monochromator. The relative peak location shift shown in Fig. 9B from bias levels (ii) and (iii, iv) is mostly related to mode hopping and temperature rise inside the active region at higher pump power levels.

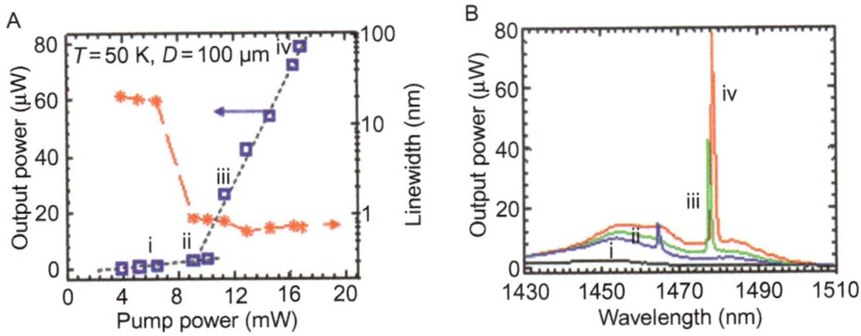

Fig. 9 (A) The L–L plot (light output for different pump powers) and the corresponding spectral linewidths at $T=50$ K and (B) the measured spectral outputs for pump powers below, at, and above thresholds. *Reproduced with permission from Yang, H., Zhao, D., Chuwongin, S., Seo, J.-H., Yang, W., Shuai, Y., Berggren, J., Hammar, M., Ma, Z., Zhou, W., 2012. Transfer-printed stacked nanomembrane lasers on silicon. Nat. Photonics 6 (9), 615–620.*

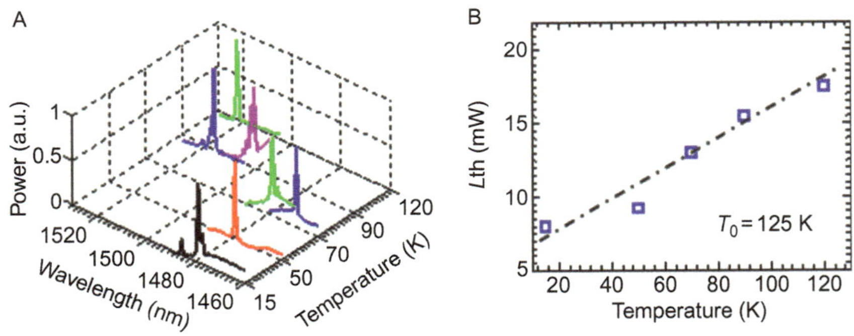

Fig. 10 (A) Measured lasing spectral at different temperatures under 532 nm laser pumping and (B) lasing threshold as a function of temperature. *Reproduced with permission from Yang, H., Zhao, D., Chuwongin, S., Seo, J.-H., Yang, W., Shuai, Y., Berggren, J., Hammar, M., Ma, Z., Zhou, W., 2012. Transfer-printed stacked nanomembrane lasers on silicon. Nat. Photonics 6 (9), 615–620.*

This MR-VCSEL LT device was also characterized at different temperatures up to 120 K. Fig. 10A shows the normalized lasing spectra (arbitrary units, a.u.) above the threshold at $T=15$, 50, 70, 90, and 120 K. The lasing peak red shifts as T increases and multimode lasing for most T cases at a rate ($d\lambda_c/dT$) close to the simulated 0.088 nm/K. There is a mode hopping occurred below and above the operation temperature of 80 K. As T rises higher than 125 K, there was no lasing any more, which is mainly limited by the MR reflector bandwidth. The LT design cavity has three lasing wavelengths of 1448 nm (at 10 K), 1478 nm (at 50 K), and 1520 nm (at 120 K) which match well with the cavity resonances. Fig. 10B shows the lasing threshold (L_{th}) as a function of T. L_{th} lineally increases with T, which matches the QW PL efficiency linearly decreases as T.

In order to estimate the lasing gain threshold in MR-VCSEL cavity, detailed investigation was also carried out to estimate the associated phase discontinuities in MR mirrors, as well the field penetration depth and effective cavity length in the optimized MR-VCSEL cavity. The results are summarized in Table 2. Based on lasing gain threshold conditions,

$$\Gamma g_{th} = \alpha_i + \alpha_{Si} + 1/_{L_{cav}} \ln(1/_{R_t R_b}) \quad (7)$$

Finally, lasing gain threshold can then be estimated, based on the cavity gain and loss characteristics reported earlier from similar QW structures (Rapp et al., 1999). Based on our design, MR-VCSEL can lase with a gain threshold of 840–1300 cm^{-1}, depending on the MR reflectivity.

The cavity design here can definitely lead to high performance MR-VCSELs with very low threshold and high efficiency, as shown in Table 2. For comparison, also shown in Table 2 are the typical parameters for long wavelength VCSELs based on conventional InGaAsP/InP DBRs. As can be seen, the MR-VCSEL has a very similar lasing gain threshold with conventional InGaAsP/InP DBR-based VCSEL, even though MR-VCSELs have smaller cavities with much shorter effective cavity lengths (2.4 μm vs 5–20 μm) (Karim et al., 2000; Rapp et al., 1997).

Table 2 MR-VCSEL lasing threshold estimates

Cavity types	**MR-VCSEL on Si**	**DBR VCSEL on InP**
Top reflector	Si NM MR	Five pairs GaAs/Al_xO_y DBR
Bottom reflector	Si NM MR	InGaAsP/InP DBR
Wavelength λ_{res}	1.513 μm	1.54 μm
Reflection R_t	98.2% (99.6%)	99.6%
Reflection R_b	98.9% (99.6%)	99.6%
Confinement Γ	5.6%	4.5%
Cavity length L_{cav}	1.21 μm	~1.0 μm
Effective cavity L_{eff}	3.783 μm	~13.1 μm
Cavity loss $\alpha_i + \alpha_{Si}$	40 cm^{-1}	43.1 cm^{-1}
Gain threshold g_{th}	1283 cm^{-1} (843 cm^{-1})	1224 cm^{-1}

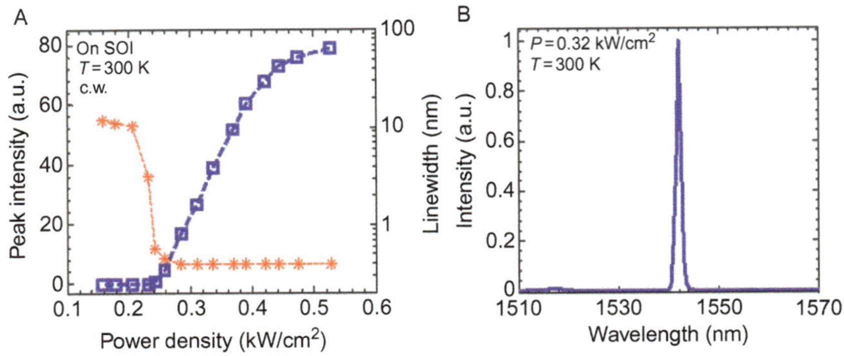

Fig. 11 (A) Lasing power and linewidth vs input power density at room temperature and (B) emission spectrum of the laser above threshold. *Reproduced with permission from Liu, S.-C., Zhao, D., Liu, Y., Yang, H., Sun, Y., Ma, Z., Reuterskiöld-Hedlund, C., Hammar, M., Zhou, W., 2017. Photonic crystal bandedge membrane lasers on silicon. Appl. Optics 56 (31), H67–H73.*

8. PCSEL characterization

PCSEL devices were optically pumped from the surface-normal direction via a long working distance objective lens (10 ×) by a c.w. green (532 nm) laser. The diameter of the excitation spot is around 110 μm. The optical characteristics of this PC bandedge membrane laser device were then measured by a monochromator-based μ-PL set-up. Fig. 11A shows the light output and linewidth as a function of the pump power measured on the surface of the focusing plane. A threshold power density of 0.25 kW/cm^2 was obtained. The measured spectral linewidth was reduced from ~14 nm (below threshold) to 0.54 nm (above threshold). Notice a thermal roll-over was observed at optical pumping intensities higher than 0.4 kW/cm^2. A spectrum of the laser above the threshold is shown in Fig. 11B where the lasing wavelength is 1542 nm, very close to the predicted value of mode *D* in the simulation. The lower threshold pump power density is a result of the high *Q* of mode *D*. All the other modes are possible to get lasing with sufficient pump energy.

Fig. 12 shows the lasing spectrum measured at a pump power density of 0.32 kW/cm^2. A clear single-mode operation was observed. A zoomed-in lasing spectrum plotted in semilog scale is shown in Fig. 12A. The peak wavelength, full width at half maximum (FWHM) of the spectrum, and SMSR are 1542 nm, 0.54 nm, and 31.8 dB, respectively. Fig. 12B is a measured far-field lasing profile with single-mode output at the focusing plane of the collimator.

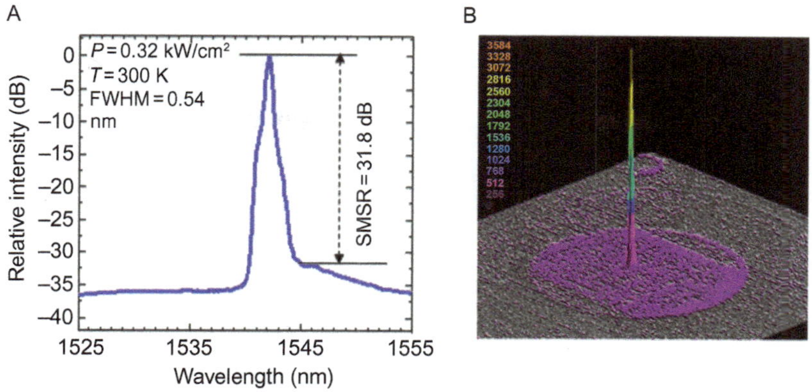

Fig. 12 (A) Lasing spectral output plotted in semilog scale above the pumping threshold and (B) 3D far-field laser profile. *Reproduced with permission from Liu, S.-C., Zhao, D., Liu, Y., Yang, H., Sun, Y., Ma, Z., Reuterskiöld-Hedlund, C., Hammar, M., Zhou, W., 2017. Photonic crystal bandedge membrane lasers on silicon. Appl. Optics 56 (31), H67–H73.*

9. PCSEL on bulk silicon substrate

As mentioned earlier, the low index buried oxide layer underneath the Si-PC layer limits the heat dissipation performance of the laser cavity due to relatively low thermal conductivity of oxide. Thus it is of great importance to introduce the PC bandedge membrane laser directly built on bulk Si substrate, where the Si-PC structure is used as the bottom cladding layer and the MQW membrane can be printed directly on bulk Si substrate without low index oxide buffer layer. The design parameters were carefully chosen for the bottom Si-PC as more restrict index constrain was involved as compared with laser on SOI substrate with low index buffer. A schematic of a successful design of this structure is shown in Fig. 13A, which consists of a transferred InGaAsP MQW membrane on a Si-PC cavity on a bulk Si substrate. A bulk Si wafer with 550 μm thickness was used here. The 2D square lattice PC pattern on bulk Si substrate was made following the same EBL patterning and dry etching processes. The SEM top view of Si-PC cavity on bulk Si substrate is shown in Fig. 13B. The lattice constant (*a*), air hole radius (*r*), and etched depth of the fabricated PC are 540, 243, and 400 nm, respectively. The approximate refractive index n_{eff} of the InGaAsP MQW active layer and Si-PC was around 3.2 and 2.2 based on the average dielectric constant in the design. Different from the PC bandedge membrane laser on SOI substrate, the PC structure here was considered as the low refractive index cladding layer of the cavity. The field profile of the lasing mode is displayed

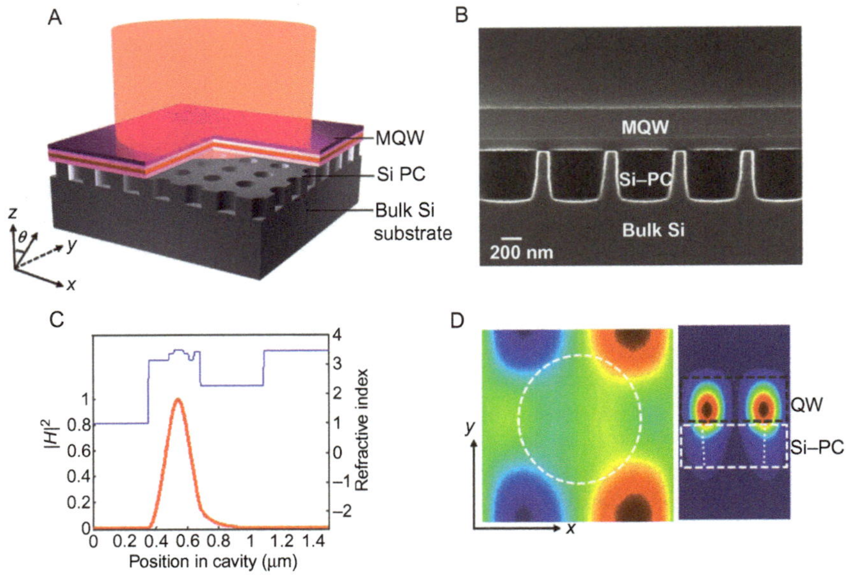

Fig. 13 (A) Schematic of PC bandedge membrane laser on bulk Si substrate. (B) Cross-sectional view of the Si-PC bandedge membrane laser cavity structure on bulk Si substrate, with the transferred InGaAsP QW mesas on top. (C) Field distribution of the cavity mode and (D) simulated in-plane and cross-sectional field distributions. *Reproduced with permission from Liu, S.-C., Zhao, D., Liu, Y., Yang, H., Sun, Y., Ma, Z., Reuterskiöld-Hedlund, C., Hammar, M., Zhou, W., 2017. Photonic crystal bandedge membrane lasers on silicon. Appl. Optics 56 (31), H67–H73.*

in Fig. 13C, where the integrated *H*-field intensity of the whole cavity along the *z*-axis direction is plotted together with the cavity index profile. The cross-sectional view of *H*-field intensity of lasing mode is shown in Fig. 13D. Note here that the large air hole radius was designed to ensure the high refractive index contrast. One can see the field was confined inside the thin cavity due to high index contrast above and below the laser cavity. Moreover, the thickness of the Si-PC slab was chosen to be 400 nm to obtain high Q value by reducing the vertical radiation loss. Finally, The Q value of the lasing mode can be theoretically calculated around 3000, which would be able to achieve the single-mode lasing in the 160 K operation temperature (Zhao et al., 2016).

To evaluate the laser performance experimentally, the PC bandedge membrane laser on bulk Si substrate was mounted on a temperature-controlled stage inside a cryostat and characterized by the same μ-PL setup system in our previous works (Yang et al., 2012; Zhao et al., 2016). Fig. 14A shows the light output and linewidth as a function of the pump

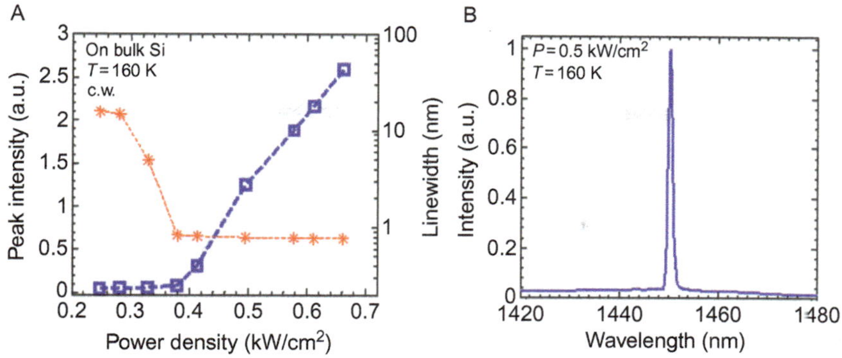

Fig. 14 (A) Lasing intensity and linewidth vs input power density at $T=160\,K$ for PCSEL on bulk Si substrate and (B) emission spectrum of the laser above threshold. *Reproduced with permission from Liu, S.-C., Zhao, D., Liu, Y., Yang, H., Sun, Y., Ma, Z., Reuterskiöld-Hedlund, C., Hammar, M., Zhou, W., 2017. Photonic crystal bandedge membrane lasers on silicon. Appl. Optics 56 (31), H67–H73.*

power measured on the surface of the PC bandedge membrane laser on bulk Si substrate. The laser threshold pump power density of $0.4\,kW/cm^2$ was obtained, where it was around a factor of two greater than PC bandedge membrane laser on SOI substrate at similar operation temperatures. We attribute this to the much lower optical confinement factor and less Q value within the cavity due to larger air hole radius. The calculated Q factor is 3.27×10^3, which is two orders lower than the Q for PCSEL on SOI. However, there are few strategies to improve the laser characteristics of present cavity design on bulk Si substrate, such as the introduction of vertical asymmetry in the air hole structure and effective utilization of tailoring the vertical etched profile. The Q value can be theoretically increased much higher above 3000, which would achieve lower threshold pump power density in the 160 K operation temperature. Fig. 14B presents the measured lasing spectrum above the threshold. The single-mode lasing peak wavelength and linewidth are 1452 and 0.87 nm in the 160 K operation temperature.

10. PCSEL thermal performance analysis

In order to compare the thermal performance, here we prepared two PC bandedge membrane lasers built onto bulk Si substrate and SOI substrate, respectively. The laser cavity for both devices is designed to operate at similar temperatures and emission wavelengths. The thicknesses of each layer of the InGaAsP MQW membrane and the parameter of Si-PC are listed in Table 3.

Table 3 Cavity design structure of laser on bulk Si and SOI

Layer materials	Laser on bulk Si	Laser on SOI
p-InP	130 nm	
InGaAsP MQW	127 nm	
n-InP	130 nm	
n-InGaAs	40 nm	
Si-PC	400 nm, $a = 540$ nm $r/a = 0.45$	190 nm, $a = 480$ nm $r/a = 0.15$
SiO_2	0	400 nm
Si substrate	525 μm	

The bulk Si substrate is expected to dissipate more heat along the vertical direction from the surface of the cavity to the bottom substrate and heat sink. A figure of merit to quantify the heat dissipation out of the cavity is the thermal resistance (Ferrara et al., 2015; Flick et al., 2011). By extracting the shift of emission peak wavelength with varying the temperature and pump power, the thermal resistance of the laser can be mathematically defined by the following ratio:

$$R_{\mathrm{th}} = \frac{\frac{\Delta\lambda}{\Delta P_{\mathrm{heat}}}|T = \mathrm{const}}{\frac{\Delta\lambda}{\Delta T}|P = \mathrm{const}} \sim \frac{\frac{\Delta\lambda}{\Delta P_p}|T = \mathrm{const}}{\frac{\Delta\lambda}{\Delta T}|P = \mathrm{const}} \qquad (8)$$

Here, the ΔP_{heat} is the heat generated from the cavity and ΔP_p is the total pump power. We assume heat generation is primarily in the QW section (Lindberg et al., 2005), and it is linearly proportional to the optical pump power for the operation window between threshold and roll-over. We first characterize the PC bandedge membrane laser on bulk Si substrate. The wavelength spectrum dependence of the pumping power from the PC bandedge membrane laser on bulk Si substrate was measured as shown in Fig. 15A. The peak of the emission wavelength slightly shifted toward longer wavelengths when the pump power increased from 25 to 50 mW at 160 K. In our measurement, the spectral shift vs power ratio presents the spectral shift vs power generated from the QW region. Second, the spectral dependence of the operation temperature varying from 120 to 200 K at a constant incident power was also measured and shown in Fig. 15B. Please note the cavity design of the PC bandedge membrane laser

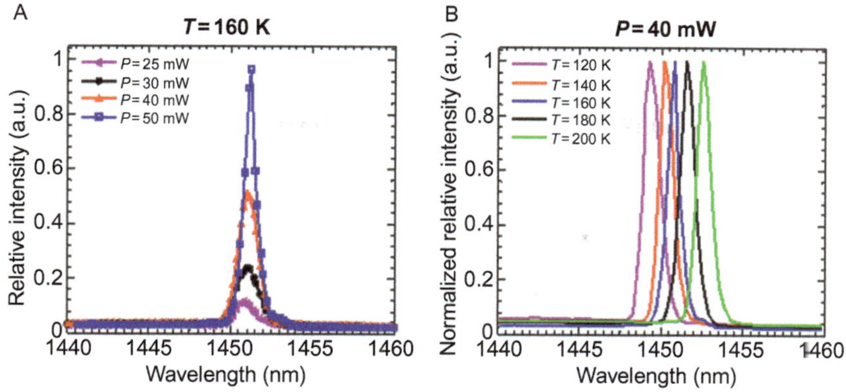

Fig. 15 (A) Lasing spectra at different pump power above threshold at constant 160 K. (B) Lasing spectra at different operation temperature with constant 40 mW pump power. *Reproduced with permission from Liu, S.-C., Zhao, D., Liu, Y., Yang, H., Sun, Y., Ma, Z., Reuterskiöld-Hedlund, C., Hammar, M., Zhou, W., 2017. Photonic crystal bandedge membrane lasers on silicon. Appl. Optics 56 (31), H67–H73.*

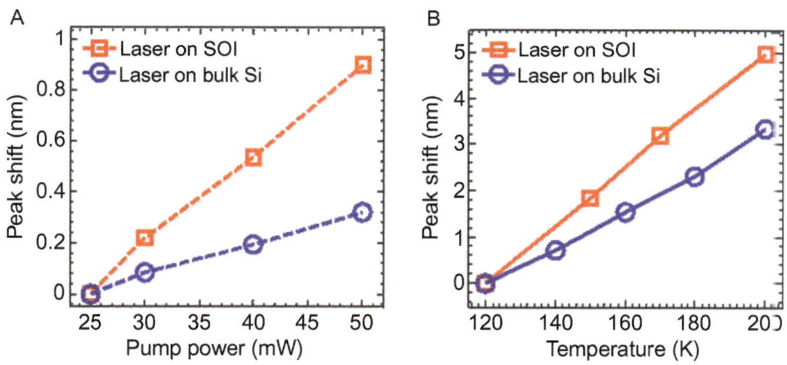

Fig. 16 (A) Emission peak wavelength shift vs pump powers for PC bandedge membrane laser on SOI and on bulk Si substrate. (B) Emission peak wavelength shift vs operation temperatures for PC bandedge membrane laser on SOI and on bulk Si substrate. *Reproduced with permission from Liu, S.-C., Zhao, D., Liu, Y., Yang, H., Sun, Y., Ma, Z., Reuterskiöld-Hedlund, C., Hammar, M., Zhou, W., 2017. Photonic crystal bandedge membrane lasers on silicon. Appl. Optics 56 (31), H67–H73.*

on bulk Si substrate was optimized for device operation around 160 K. Single-mode lasing operation was obtained at the 120–200 K range for the pump power at 40 mW. The wavelength shift vs pumping power is calculated to be 0.0125 nm/mW, and the shift vs temperature was 0.0412 nm/K.

Fig. 16A and B shows the measured emission wavelength peak shift at different pump powers and varying operation temperatures for both PC bandedge membrane lasers on bulk Si substrate and on SOI substrate operation at 160–180 K. For the PC bandedge membrane laser on bulk Si substrate, the thermal resistance was extracted from Fig. 16A and B and calculated to be 303.4 K/W. On the other hand, the thermal resistance of PC bandedge membrane laser on SOI was calculated to be 558.5 K/W. The thermal resistance of PC bandedge membrane laser on bulk Si substrate is therefore around a factor of two times lower than that of a structure with a BOx layer. We attributed the higher thermal resistance to the low thermal conductivity of the BOx layer. It is also worth noting that the thermal resistance values calculated here is a relative value, as the absolute value should be based on the exact heat source (ΔP_{heat}), which is not known here.

11. Electrically pumped PCSEL characterization

Finally, we present some recent results on electrically pumped PCSELs. A buried tunnel junction (BTJ) is incorporated here for efficient charge injection and lateral carrier confinement. As shown schematically in Fig. 17A, the PCSEL cavity consists of a transferred InGaAsP multiquantum well (MQW) heterostructure active region, printed on a single-layer Si-PC cavity. The thicknesses of each layer of the InGaAsP MQW membrane and the parameters of Si-PC are listed in Fig. 17B. BTJ n^{++}/p^{++} current injection structures with different sizes were formed using an etch and epitaxial regrowth process, including pass marks for continued processing on the Si substrate. The laser cavity structure parameters are optimized for low threshold lasing, following the similar procedures reported earlier (Liu et al., 2017; Zhao et al., 2016). Shown in Fig. 18A are the simulated

A

B

Description	Materials	BTJ-PCSEL
Contact layer: *n*	InGaAs	40 nm
Cladding layer: *n*	InP	80 nm
BTJ: n^{++}	InGaAs	20 nm
BTJ: p^{++}	InAlGaAs	10 nm
Cladding layer: *p*	InP	265 nm
MQW	InGaAsP	127 nm
Cladding layer: *n*	InP	31.5 nm
Contact layer: *n*	InGaAs	40 nm
Photonic crystal	Air hole Si–PC	125 nm, $r/a = 0.15$
BOX	SiO_2	400 nm
Substrate	Si substrate	525 μm

Fig. 17 Electrically pumped PCSELs on SOI substrate: (A) schematic and (B) detailed heterostructure and cavity design parameters.

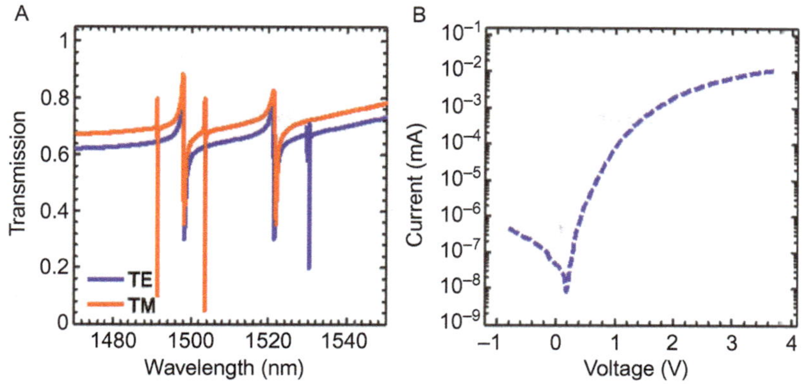

Fig. 18 Electrically pumped PCSELs on SOI substrate: (A) simulated transmission spectra for both TE and TM modes and (B) measured IV curve for the electrically pumped PCSEL.

transmission spectra of the cavity, with blue lines for the transverse electric (TE, or *s*) polarization, and red lines for the transverse magnetic (TM, or *p*) polarization. Substrate removal process is used in the device fabrication for the release of top large-area InGaAsP MQW membrane from the native InP substrate. The membrane is then transferred onto Si-PC cavity. Top *n*-type ring contact is formed by *e*-beam evaporation, with alloy contact consisting of Ni/Ge/Au/Ti/Au (10/20/30/10/200 nm). Mesa etching is then carried out down to the bottom *n*-contact layer, followed with another *e*-beam evaporation process for the bottom *n*-contact formation. A plasma-enhanced chemical vapor deposition SiO_2 layer is deposited for laser planarization, followed by interconnect layer formation. The complete device is then mounted on a chip carrier with wire bonding process for device characterization. The current–voltage (IV) characteristics were measured and evaluated for different BTJ sizes, with one curve plotted in semilog scale for electrically pumped PCSEL shown in Fig. 18B. Excellent IV characteristics were shown here, essential for efficient charge injection and lasing.

The device characterization was carried out with c.w. current injection at room temperature. The output was measured in the direction normal to the surface via a long working distance objective lens and sent into a monochromator. Fig. 19A shows the measured electroluminescence (EL) spectrum at various biased currents for a PCSEL with an $80 \times 80\,\mu m^2$ BTJ mesa region. A dominant single-mode lasing emission peak at 1504 nm appears above the threshold current. Fig. 19B shows the light–current–voltage (LIV) characteristics. The turn on voltage is about 0.8 V. The threshold current is about 2 mA.

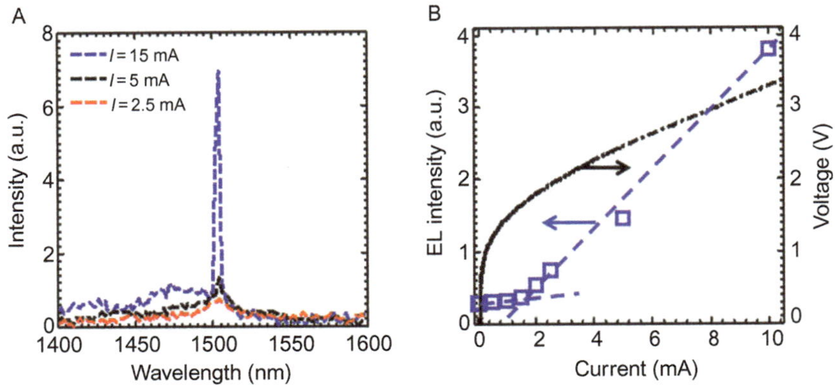

Fig. 19 (A) The emission spectra at different current injection levels measured at room temperature and (B) the light–current–voltage (LIV) characteristics for an $80 \times 80\,\mu m^2$ electrically pumped PCSEL on Si.

12. Summary

In summary, we review here recent progresses on photonic crystal surface-emitting membrane lasers for on-chip integration with CMOS electronics. Defect-free photonic crystal structures are incorporated into laser cavities as ultracompact MRs for DBR-free VCSELs and as bandedge cavities evanescently coupled with gain medium for PCSELs. Lateral cavity size effect was also investigated for cavity scaling. Electrically injected PCSELs with BTJ-based InGaAsP QW heterostructures were also demonstrated.

Further research is needed to address device efficiencies. While high efficiency VCSELs and PCSELs on native III–V substrates have been reported earlier with wall-plug efficiencies greater than 50%, the efficiencies in the heterogeneously integrated DBR-free MR-VCSELs and Si-based PCSELs are largely impacted by the cavity coupling and thin membrane surface and interface qualities. Device scaling in both vertical and lateral directions can also lead to improved device modulation speed and operation data transfer efficiency, with ultimate objective of subfemto joule per bit energy efficiency.

Thermal and reliability are another set of issues to be addressed. Since PBG is not essential in these photonic crystal structures reported here, relaxed index contrast is possible to have photonic crystal lasing cavities built on low index contract material systems. However certain index contract is still needed, which may result in the need of low index buffers in the lasing cavity. This issue needs to be further addressed to avoid the use of low index dielectric (e.g., SiO_2, Si_3N_4) due to the poor thermal conductivities and the strain-induced reliabilities issues.

On the other hand, new functional properties should also be expected from photonic crystal surface-emitting membrane lasers, as the modulation of photonic crystal cavity and the gain/loss modulation of the active media can result in lasers with unique spectral, spatial, and beam properties. Engineered beam shapes can be designed. Large-area single-mode lasers are also feasible. Laser beams can also be steered with certain angles. With the proper dispersion engineering, it is possible to have beam emission in-plane.

Acknowledgments

The authors acknowledge contributions from their current and former students, as well as collaborators involved in the research work reviewed here. Special thanks to Mattias Hammar from Royal Institute of Technology who provided QW materials for the work. We also thank the fabrication support from the University of Texas at Arlington Nanotechnology Research Center, and the University of Texas at Austin, part of NSF NNIN.

References

Altug, H., Englund, D., Vučković, J., 2006. Ultrafast photonic crystal nanocavity laser. Nat. Phys. 2 (7), 484–488.

Babic, D., Corzine, S., 1922. Analytic expressions for the reflection delay, penetration depth, and absorptance of quarter-wave dielectric mirrors. IEEE J. Quantum Electron. 28, 514–524.

Bakir, B.B., Seassal, C., Letartre, X., Viktorovitch, P., Zussy, M., Di Cioccio, L., Fedeli, J.M., 2006. Surface-emitting microlaser combining two-dimensional photonic crystal membrane and vertical Bragg mirror. Appl. Phys. Lett. 88, 081113.

Balakrishnan, G., Jallipalli, A., Rotella, P., Huang, S., Khoshakhlagh, A., Amtout, A., Krishna, S., Dawson, L.R., Huffaker, D.L., 2006. Room-temperature optically pumped (Al) GaSb vertical-cavity surface-emitting laser monolithically grown on an Si (100) substrate. IEEE J. Sel. Top. Quantum Electron. 12 (6), 1636–1641.

Bhattacharya, P., Sabarinathan, J., Topol'ancik, J., Chakravarty, S., Yu, P., Zhou, W., 2005. Quantum dot photonic crystal light sources. Proc. IEEE 93 (10), 1825–1838.

Boutami, S., Bakir, B., Regreny, P., Leclercq, J., Viktorovitch, P., 2007. Compact 1.55 μm room-temperature optically pumped VCSEL using photonic crystal mirror. Electron. Lett. 43, 282.

Chen, R., Tran, T.T.D., Ng, K.W., Ko, W.S., Chuang, L.C., Sedgwick, F.G., Chang-Hasnain, C., 2011. Nanolasers grown on silicon. Nat. Photonics 5 (3), 170–175.

Chua, S.L., Chong, Y., Stone, A.D., Soljacic, M., Bravo-Abad, J., 2011. Low-threshold lasing action in photonic crystal slabs enabled by Fano resonances. Opt. Express 19, 1539–1562.

Chua, S.L., Lu, L., Bravo-Abad, J., Joannopoulos, J.D., Soljacic, M., 2014. Larger-area single-mode photonic crystal surface-emitting lasers enabled by an accidental Dirac point. Opt. Lett. 39 (7), 2072–2075.

Coldren, L., Corzine, S., 1995. Diode Lasers and Photonic Integrated Circuits. Wiley, New York.
Colombelli, R., Srinivasan, K., Troccoli, M., Painter, O., Gmachl, C.F., Tennant, D.M., Sergent, A.M., Sivco, D.L., Cho, A.Y., Capasso, F., 2003. Quantum cascade surface-emitting photonic crystal laser. Science 302 (5649), 1374–1377.
Dowling, J.P., Scalora, M., Bloemer, M.J., Bowden, C.M., 1994. The photonic band edge laser: a new approach to gain enhancement. J. Appl. Phys. 75 (4), 1896–1899.
Ellis, B., Mayer, M.A., Shambat, G., Sarmiento, T., Harris, J., Haller, E.E., Vučković, J., 2011. Ultralow-threshold electrically pumped quantum-dot photonic-crystal nanocavity laser. Nat. Photonics 5 (5), 297–300.
Fan, S., Joannopoulos, J.D., 2002. Analysis of guided resonances in photonic crystal slabs. Phys. Rev. B 65 (23), 235112.
Ferrara, J., Yang, W., Zhu, L., Qiao, P., Chang-Hasnain, C.J., 2015. Heterogeneously integrated long-wavelength VCSEL using silicon high contrast grating on an SOI substrate. Opt. Express 23, 2512–2523.
Flick, T., Becks, K.H., Dopke, J., Mättig, P., Tepel, P., 2011. Measurement of the thermal resistance of VCSEL devices. J. Instrum. 6, C01021.
Furukawa, A., Sasaki, S., Hoshi, M., Matsuzono, A., Moritoh, K., Baba, T., 2004. High-power single-mode vertical-cavity surface-emitting lasers with triangular holey structure. Appl. Phys. Lett. 85, 5161–5163.
Hirose, K., Liang, Y., Kurosaka, Y., Watanabe, A., Sugiyama, T., Noda, S., 2014. Watt-class high-power, high-beam-quality photonic-crystal lasers. Nat. Photonics 8 (5), 406–411.
Huang, M.C.Y., Zhou, Y., Chang-Hasnain, C.J., 2007. A surface-emitting laser incorporating a high-index-contrast subwavelength grating. Nat. Photonics 1 (2), 119–122.
Hwang, J., Ryu, H., Song, D., Han, I., Park, H., Jang, D., Lee, Y., 2000. Continuous room-temperature operation of optically pumped two-dimensional photonic crystal lasers at 1.6 um. IEEE Photon. Technol. Lett. 12 (10), 1295–1297.
Iga, K., 2000. Surface-emitting laser-its birth and generation of new optoelectronics field. IEEE J. Sel. Top. Quantum Electron. 6 (6), 1201–1215.
Imada, M., Noda, S., Chutinan, A., Tokuda, T., Murata, M., Sasaki, G., 1999. Coherent two-dimensional lasing action in surface-emitting laser with triangular-lattice photonic crystal structure. Appl. Phys. Lett. 75, 316.
Imada, M., Chutinan, A., Noda, S., Mochizuki, M., 2002. Multidirectionally distributed feedback photonic crystal lasers. Phys. Rev. B 65 (19), 195306.
Karagodsky, V., Sedgwick, F.G., Chang-Hasnain, C.J., 2010. Theoretical analysis of subwavelength high contrast grating reflectors. Opt. Express 18 (16), 16973–16988.
Karim, A., Bjorlin, S., Piprek, J., Bowers, J., 2000. Long-wavelength vertical-cavity lasers and amplifiers. IEEE J. Sel. Top. Quantum Electron. 6 (6), 1244–1253.
Kim, J., Chrostowski, L., Bisaillon, E., Plant, D., 2007. DBR, sub-wavelength grating, and photonic crystal slab Fabry–Perot cavity design using phase analysis by FDTD. Opt. Express 15 (16), 10330–10339.
Kurosaka, Y., Iwahashi, S., Sakai, K., Miyai, E., Kunishi, W., Ohnishi, D., Noda, S., 2009. Band structure observation of 2D photonic crystal with various V-shaped air-hole arrangements. IEICE Electron. Exp. 6, 966–971.
Kwon, S.H., Kim, S.H., Kim, S.K., Lee, Y.H., 2004. Small, low-loss heterogeneous photonic bandedge laser. Opt. Express 12 (12), 5356–5361.
Lee, J., Zhen, B., Chua, S.-L., Qiu, W., Joannopoulos, J.D., Soljacic, M., Shapira, O., 2012. Observation and differentiation of unique high-Q optical resonances near zero wave vector in macroscopic photonic crystal slabs. Phys. Rev. Lett. 109, 067401.
Liang, D., Roelkens, G., Baets, R., Bowers, J., 2010. Hybrid integrated platforms for silicon photonics. Materials 3 (3), 1782–1802.

Liang, Y., Peng, C., Sakai, K., Iwahashi, S., Noda, S., 2012. Three-dimensional coupled-wave analysis for square-lattice photonic crystal surface emitting lasers with transverse-electric polarization: finite-size effects. Opt. Express 20 (14), 15945–15961.

Lindberg, H., Strassner, M., Gerster, E., Bengtsson, J., Larsson, A., 2005. Thermal management of optically pumped long-wavelength InP-based semiconductor disk lasers. IEEE J. Sel. Top. Quantum Electron. 11, 1126–1134.

Liu, V., Fan, S., 2012. S4: a free electromagnetic solver for layered periodic structures. Comput. Phys. Commun. 183 (10), 2233–2244.

Liu, S.-C., Zhao, D., Liu, Y., Yang, H., Sun, Y., Ma, Z., Reuterskiöld-Hedlund, C., Hammar, M., Zhou, W., 2017. Photonic crystal bandedge membrane lasers on silicon. Appl. Optics 56 (31), H67–H73.

Liu, S.-C., Zhao, D., Ge, X., Reuterskiöld-Hedlund, C., Hammar, M., Fan, S., Ma, Z., Zhou, W., 2018. Size scaling of photonic crystal surface emitting lasers on silicon substrates. IEEE Photonics J. 10 (3), 1–6.

Luk'yanchuk, B., Zheludev, N.I., Maier, S.A., Halas, N.J., Nordlander, P., Giessen, H., Chong, C.T., 2010. The Fano resonance in plasmonic nanostructures and metamaterials. Nat. Mater. 9, 707–715.

Lundeberg, L.D., Boiko, D.L., Kapon, E., 2005. Coupled islands of photonic crystal heterostructures implemented with vertical-cavity surface-emitting lasers. Appl. Phys. Lett. 87, 241120.

Magnusson, R., Shokooh-Saremi, M., 2008. Physical basis for wideband resonant reflectors. Opt. Express 16 (5), 3456–3462.

Martınez, L.J., Alen, B., Prieto, I., Galisteo-Lopez, J.F., Galli, M., Andreani, L.C., Seassal, C., Viktorovitch, P., Postigo, P.A., 2009. Two-dimensional surface emitting photonic crystal laser with hybrid triangular-graphite structure. Opt. Express 17, 15043–15051.

Meier, M., Mekis, A., Dodabalapur, A., Timko, A., Slusher, R.E., Joannopoulos, J.D., Nalamasu, O., 1999. Laser action from two-dimensional distributed feedback in photonic crystals. Appl. Phys. Lett. 74, 7–9.

Mi, Z., Bhattacharya, P., Yang, J., Pipe, K., 2005. Room-temperature self-organised In0.5Ga0.5As quantum dot laser on silicon. Electron. Lett. 41 (13), 742–744.

Miyai, E., Sakai, K., Okano, T., Kunishi, W., Ohnishi, D., Noda, S., 2006. Photonics: lasers producing tailored beams. Nature 441 (7096), 946.

Monat, C., Seassal, C., Letarte, X., Regreny, P., Gendry, M., Romeo, P.R., Viktorovitch, P., D'Yerville, M.L.V., Cassagne, D., Albert, J.P., Jalaguier, E., Pocas, S., Aspar, B., 2003. Two-dimensional hexagonal-shaped microcavities formed in a two-dimensional photonic crystal on an InP membrane. J. Appl. Phys. 93 (1), 23.

Noda, S., 2010. Photonic crystal lasers—ultimate nanolasers and broad-area coherent lasers [invited]. JOSA B 27 (11), B1–B8.

Noda, S., Yokoyama, M., Imada, M., Chutinan, A., Mochizuki, M., 2001. Polarization mode control of two-dimensional photonic crystal laser by unit cell structure design. Science 293 (5532), 1123–1125.

Noda, S., Fujita, M., Asano, T., 2007. Spontaneous-emission control by photonic crystals and nanocavities. Nat. Photonics 1 (8), 449–458.

Oskooi, A.F., Roundy, D., Ibanescu, M., Bermel, P., Joannopoulos, J., Johnson, S.G., 2010. MEEP: a flexible free-software package for electromagnetic simulations by the FDTD method. Comput. Phys. Commun. 181 (3), 687–702.

Painter, O., 1999. Two-dimensional photonic band-gap defect mode. Science 284 (5819), 1999.

Painter, O., Lee, R.K., Scherer, A., Yariv, A., O'Brien, J.D., Dapkus, P.D., Kim, I., 1999. Two-dimensional photonic band-gap defect mode laser. Science 284 (5421), 1819–1821.

Park, H.-G., Kim, S.-H., Kwon, S.-H., Ju, Y.-G., Yang, J.-K., Baek, J.-H., Kim, S.-B., Lee, Y.-H., 2004. Electrically driven single-cell photonic crystal laser. Science 305 (5689), 1444–1447.

Park, H., Fang, A.W., Cohen, O., Jones, R., Paniccia, M.J., Bowers, J.E., 2006. Design and fabrication of optically pumped hybrid silicon-AlGaInAs evanescent lasers. IEEE J. Sel. Top. Quantum Electron. 12 (6), 1657–1663.

Qiang, Z., Yang, H., Chuwongin, S., Zhao, D., Ma, Z., Zhou, W., 2010. Design of Fano broadband reflectors on SOI. IEEE Photon. Technol. Lett. 22 (15), 1108–1110.

Rapp, S., Piprek, J., Streubel, K., Andre, J., Wallin, J., 1997. Temperature sensitivity of 1.54-m vertical-cavity lasers with an InP-based Bragg reflector. IEEE J. Quantum Electron. 33 (10), 1839.

Rapp, S., Salomonsson, F., Streubel, K., Mogg, S., Wennekes, F., Bentell, J., Hammar, M., 1999. All-epitaxial single-fused 1.55μm vertical cavity laser based on an InP Bragg reflector. Jpn. J. Appl. Phys 38, 1261–1264.

Roelkens, G., Liu, L., Liang, D., Jones, R., Fang, A., Koch, B., Bowers, J., 2010. III–V/silicon photonics for on-chip and intra-chip optical interconnects. Laser Photonics Rev. 4 (6), 751–779.

Sakai, K., Miyai, E., Sakaguchi, T., Ohnishi, D., Okano, T., Noda, S., 2005. Lasing band-edge identification for a surface-emitting photonic crystal laser. IEEE J. Sel. Areas Commun. 23 (7), 1335–1340.

Sakai, K., Miyai, E., Noda, S., 2006. Coupled-wave model for square-lattice two-dimensional photonic crystal with transverse-electric-like mode. Appl. Phys. Lett. 89, 021101.

Sato, Y., Tanaka, Y., Upham, J., Takahashi, Y., Asano, T., Noda, S., 2012. Strong coupling between distant photonic nanocavities and its dynamic control. Nat. Photonics 6, 56–61.

Sauvan, C., Hugonin, J., Lalanne, P., 2009. Difference between penetration and damping lengths in photonic crystal mirrors. Appl. Phys. Lett. 95, 211101.

Sciancalepore, C., Bakir, B.B., Letartre, X., Harduin, J., Olivier, N., Seassal, C., Fedeli, J., Viktorovitch, P., 2012. CMOS-compatible ultra-compact 1.55-um emitting VCSELs using double photonic crystal mirrors. IEEE Photon. Technol. Lett. 24 (6), 455–457.

Song, D.S., Kim, S.H., Park, H.G., Kim, C.K., Lee, Y.H., 2002. Single-fundamental-mode photonic-crystal vertical-cavity surface-emitting lasers. Appl. Phys. Lett. 80 (21), 3901–3903.

Stankovic, S., Jones, R., Sysak, M.N., Heck, J.M., Roelkens, G., Van Thourhout, D., 2011. 1310-nm hybrid III–V/Si Fabry–Perot laser based on adhesive bonding. IEEE Photon. Technol. Lett. 23 (23), 1781–1783.

Susa, N., 2001. Threshold gain and gain-enhancement due to distributed-feedback in two-dimensional photonic-crystal lasers. J. Appl. Phys. 89 (2), 815–823.

Van Campenhout, J., Rojo Romeo, P., Regreny, P., Seassal, C., Van Thourhout, D., Verstuyft, S., Di Cioccio, L., Fedeli, J.M., Lagahe, C., Baets, R., 2007. Electrically pumped InP-based microdisk lasers integrated with a nanophotonic silicon-on-insulator waveguide circuit. Opt. Express 15 (11), 6744–6749.

Wu, S., Buckley, S., Schaibley, J.R., Feng, L., Yan, J., Mandrus, D.G., Hatami, F., Yao, W., Vuckovic, J., Majumdar, A., Xu, X., 2015. Monolayer semiconductor nanocavity lasers with ultralow thresholds. Nature 520, 69–72. https://doi.org/10.1038/nature14290(.

Yablonovitch, E., 1987. Inhibited spontaneous emission in solid-state physics and electronics. Phys. Rev. Lett. 58 (20), 2059–2062.

Yang, H., Chuwongin, S., Qiang, Z., Chen, L., Pang, H., Ma, Z., Zhou, W., 2009. Resonance control of membrane reflectors with effective index engineering. Appl. Phys. Lett. 95, 023110.

Yang, H., Zhao, D., Chuwongin, S., Seo, J.-H., Yang, W., Shuai, Y., Berggren, J., Hammar, M., Ma, Z., Zhou, W., 2012. Transfer-printed stacked nanomembrane lasers on silicon. Nat. Photonics 6 (9), 615–620.

Yokouchi, N., Danner, A., Choquette, K.D., 2003. Two-dimensional photonic crystal confined vertical-cavity surface-emitting lasers. IEEE J. Sel. Top. Quantum Electron. 9 (5), 1439–1445.
Yoshie, T., Scherer, A., Hendrickson, J., Khitrova, G., Gibbs, H., Rupper, G., Ell, C., Shchekin, O., Deppe, D., 2004. Vacuum Rabi splitting with a single quantum dot in a photonic crystal nanocavity. Nature 432 (7014), 200–203.
Zhao, D., Ma, Z., Zhou, W., 2010. Field penetrations in photonic crystal Fano reflectors. Opt. Express 18 (13), 14152–14158.
Zhao, D., Yang, H., Chuwongin, S., Seo, J., Ma, Z., Zhou, W., 2012. Design of photonic crystal membrane-reflector-based VCSELs. IEEE Photonics J. 4 (6), 2169–2175.
Zhao, D., Liu, S., Yang, H., Ma, Z., Reuterskiöld-Hedlund, C., Hammar, M., Zhou, W., 2016. Printed large-area single-mode photonic crystal bandedge surface-emitting lasers on silicon. Sci. Rep. 6.
Zhou, W.D., Sabarinathan, J., Kochman, B., Berg, E., Qasaimeh, O., Pang, S., Bhattacharya, P., 2000. Electrically injected single-defect photonic bandgap surface-emitting laser at room temperature. Electron. Lett. 36 (18), 1541–1542.
Zhou, W., Sabarinathan, J., Bhattacharya, P., Kochman, B., Berg, E.W., Yu, P.-C., Pang, S.W., 2001. Characteristics of a photonic bandgap single defect microcavity electroluminescent device. IEEE J. Quantum Electron. 37 (9), 1153–1160.
Zhou, W., Zhao, D., Shuai, Y.-C., Yang, H., Chuwongin, S., Chadha, A., Seo, J.-H., Wang, K.X., Liu, V., Ma, Z., 2014. Progress in 2D photonic crystal Fano resonance photonics. Prog. Quantum Electron. 38 (1), 1–74.

Index

Note: Page numbers followed by "*f*" indicate figures, and "*t*" indicate tables.

M

U

V

W

Y